TRANSACTIONS

OF THE

INTERNATIONAL

ASTRONOMICAL UNION

VOLUME XXIB - PROCEEDINGS

INTERNATIONAL ASTRONOMICAL UNION
UNION ASTRONOMIQUE INTERNATIONALE

TRANSACTIONS

OF THE

INTERNATIONAL ASTRONOMICAL UNION

VOLUME XXIB

PROCEEDINGS OF THE TWENTY-FIRST

GENERAL ASSEMBLY

BUENOS AIRES 1991

Edited by

JACQUELINE BERGERON
General Secretary of the Union

Springer-Science+Business Media, B.V.

Library of Congress Cataloging-in-Publication Data

International Astronomical Union. General Assembly (21st : 1991 :
  Buenos Aires, Argentina)
    Proceedings of the Twenty-first General Assembly, Buenos Aires, 1991
  / edited by Jacqueline Bergeron.
        p.   cm. -- (Transactions of the International Astronomical
  Union ; vol. 21B)
    English and Spanish.

    1. Astronomy--Congresses. 2. International Astronomical Union.
  General Assembly (21st : 1991 : Buenos Aires, Argentina)
  I. Bergeron, J. (Jacqueline) II. Title. III. Series: Transactions
  of the International Astronomical Union ; v. 21B.
  QB1.I627 1991
  520'.6'01--dc20   92-25559

ISBN 978-0-7923-1929-0   *ISBN 978-94-011-2820-9 (eBook)*
DOI 10.1007/978-94-011-2820-9

Printed on acid-free paper

PRÉSIDENT DE L'UNION ASTRONOMIQUE INTERNATIONALE

YOSHIHIDE KOZAI

PRESIDENT OF THE INTERNATIONAL ASTRONOMICAL UNION

1988–1991

CONTENTS

AIG/IAG	*Association internationale de géodésie*
	International Association of Geodesy
ASE/ESA	*Agence spatiale européenne*
	European Space Agency
BCT/TCB	Barycentric Coordinate Time
	Temps-coordonnée barycentrique
BDT/TDB	Barycentric Dynamical Time
	Temps dynamique barycentrique
BIPM/IBW	*Bureau international des poids & mesures*
	International Bureau of Weights & Measures
CASLEO	Complejo Astronomico El Leoncito (Argentina)
CCDS	Consultative Committee for the Definition of the Second
	Comité consultatif pour la définition de la seconde
CCIR	Consultative Committee on International Radiocommunications
	Comité consultatif international des radiocommunications
CIE	*Compagnie Internationale de l'Eclairage*
CNIE	Comision Nacional de Investigaciones Espaciales (Argentina)
CNRS	*Centre National de la Recherche Scientifique (France)*
CODATA	Committee on Data for Science & Technology (ICSU)
CONICET	National Council of Scientific & Technical Research (Argentina)
COSPAR	Committe on Space Research (ICSU)
	Comité pour la recherche spatiale (ICSU)
COSTED	Committee on Science & Technology in Developing Countries (ICSU)
CRAF	Comet Rendezvous Asteroid Flyby (NASA/ESA)
CTS	Committee on the Teaching of Science (ICSU)
EPS	European Physical Society
ESA/*ASE*	European Space Agency
	Agence spatiale européenne
FAGS	Federation of Astronomical & Geophysical Services
GCT/TCG	Geocentric Coordinate Time
	Temps-coordonnée géocentrique
IAF	International Astronautical Federation
IAFE	Instituto de Astronomia y Fisica del Espacio (Argentina)
IAGA	International Association of Geomagnetism & Aeronomy
IAMAP	International Association of Meteorology & Atmospheric Physics
IAR	Instituto Argentino de Radioastronomia (Argentina)
IAT/TAI	International Atomic Time
	Temps atomique international
IAU/UAI	International Astronomical Union
	Union astronomique internationale
ICSU	International Council of Scientific Unions
	Conseil international des unions scientifiques
IERS	International Earth Rotation Service
	Service international de la rotation de la terre
IGBP	International Geosphere-Biosphere Programme
IMU	International Mathematical Union
INSU	*Institut National des Sciences de l'Univers (CNRS)*
IS/SI	International Standards (units)
	Système International (units)
ISYA	International School for Young Astronomers
ISY	International Space Year
IUCAF	Interunion Commission on Frequency Allocations for Radio Astronomy & Space Sciences
IUGG/UGGI	International Union of Geodesy & Geophysics
	Union internationale de géodésie & géophysique
IUPAC/UICPA	International Union of Pure & Applied Chemistry
	Union internationlale de chimie pure & appliquée

IUPAP/*UIPPA*	International Union of Pure & Applied Physics
	Union internationale de physique pure & appliquée
ITU/*UIT*	International Telecommunication Union
	Union internationale des télécommunications
IUWDS	International Ursigram & World Day Service (IAU/*UAI*)
LLR	Lunar Laser Ranging
MERIT	Monitor Earth Rotation & Intercompare Techniques
NASA	National Aeronautics & Space Administration (USA)
OAC	Observatorio Astronomico de Cordoba (Argentina)
OAFA	Observatorio Astronomico Felix Aguilar (Argentina)
OALP	Observatorio Astronomico de La Plata (Argentina)
OMM/WMO	*Organisation météorologique mondiale*
	World Meteorological Organization
QBSA	Quarterly Bulletin on Solar Activity (IAU/*UAI*)
RAS	Radio Astronomy Service
	Service de radioastronomie
RDSS	Radio Determination Satellite Service
	Service de Détermination Radio par Satellite
SCAR	Scientific Committee on Antarctic Research
	Comité Scientifique pour la Recherche en Antarctique
SCOPE	Scientific Committee on Problems of the Environment (ICSU)
SCOSTEP	Scientific Committee on Solar-Terrestrial Physics
SI/IS	*Système International* (units)
	International Standards (units)
SNC	Special Nominating Committee (IAU/*UAI*)
TAI/IAT	*Temps atomique international*
	International Atomic Time
TDB/BDT	Barycentric Dynamical Time
	Temps dynamique barycentrique
TCB/BCT	Barycentric Coordinate time
	Temps-coordonnée barycentrique
TCG/GCT	*Temps-coordonnée géocentrique*
	Geocentric Coordinate Time
TDT	Terrestrial Dynamical Time
	Temps dynamique terrestre
TT	*Temps terrestre*
UAI/IAU	*Union astronomique internationale*
	International Astronomical Union
UGGI/IUGG	*Union internationale de géodésie & géophysique*
	International Union of Geodesy & Geophysics
UICPA/IUPAC	*Union internationlale de chimie pure & appliquée*
	Inernational Union of Pure & Applied Chemistry
UIPPA/IUPAP	*Union internationale de physique pure & appliquée*
	International Union of Pure & Applied Physics
UIT/ITU	*Union internationale des télécommunications*
	International Telecommunication Union
UNESCO	United Nations Educational, Scientific & Cultural Organization
URSI	Union radio-scientifique internationale
USNO	United States Naval Observatory
VLBI	Very Long Baseline Interferometry
VLP	Visiting Lecturer's Programme (IAU/UAI)
WARC	World Administrative Radio Conference (ITU/*UIT*)
	Conférence administrative mondiale des radiocommunications
WGPSN	Working Group on Planetary System Nomenclature (IAU/*UAI*)
WMO/*OMM*	World Meteorological Organization
	Organisation météorologique mondiale

FOREWORD

The XXIst General Assembly of the International Astronomical Union was held in Buenos Aires from July 23 to August 1, 1991. The activities of General Assemblies are designed to provide opportunities for scientific exchanges within scientific commission meetings or during informal exchanges and for establishing new personal links between astronomers from different countries and continents, and to review and conduct the administration of the Union.

At the Inaugural ceremony on July 23, the Union was honored by the presence of the President of the Republic of Argentina, Dr. Carlos S. Menem, the representatives of the Argentine government, of the city of Buenos Aires and of the National and Local Organizing committees.

The excellent scientific programme in Buenoes Aires was organized by the Presidents of the fourty IAU Commissions and coordinated by the IAU General Secretary (1988-1991), Prof. Derek McNally. The local arrangements were taken care of by Dr. Fernando R. Colomb, and his collaborators of the National Organizing Committee and by the Local Organizing Committee under the supervision of its chairman, Dr. Roberto H. Mendez.

The present volume, IAU Transactions XXIB, summarizes the work of the XXIst General Assembly. The discourses given during the Inaugural and Closing Ceremonies are reproduced in chapters I & III respectively. The proceedings of the two sessions of the General Assembly will be found in chapter II, which includes the Resolutions, the report of the Finance Committee and the Accounts and other aspects of the administration of the Union. Together with the report of the Executive Committee for this last triennium (chapter IV), they provide the permanent record for the Union in the period 1988-1991. This volume also contains the Commission reports from Buenos Aires compiles by the Presidents of Commissions (chapter V). The Statutes, By-Laws and a few working rules of the Union are published in chapter VI. Finally, chapter VII contains the list of countries adhering to the Union and the alphabetical, geographical and commission membership lists of over 7350 individual IAU members. The IAU still appears to be unique among the scientific Unions in maintaining this category of individual membership which contributes in a crucial way to the spirit and the aims of the Union. These lists will also be published separately.

The production of this volume would not have been possible without the special co-operation of the Presidents of Commissions who provided camera-ready manuscripts, the members of the Resolution Committee, to Prof. Jean-Claude Pecker who supervised the French translation of the Resolutions and to Dr. Eduardo Simonneau who supervised the reproduction of the different speeches in Spanish. I am also indebted to Prof. Jorge Sahade for his help in obtaining the official discourses. I am most grateful to Monique Léger-Orine who, with the assistance of Julie Crook worked on the edition of this document.

J. Bergeron
General Secretary, IAU
January 24, 1992

CHAPTER I

INAUGURAL CEREMONY 1991 JULY 23

The Inaugural Ceremony was held in the San Martin Cultural Centre, Buenos Aires, in the presence of the President of the Republic of Argentina, Dr. Carlos S. Menem, the National Secretary for Science and Technology, Dr. Raúl F. Matera, the Secretary for Education and Culture of the City of Buenos Aires, Lic. Osvaldo E. Devries, the President of the Argentinian Astronomical Association, Dr. Esteban Bajaja and the Chairman of the National Organising Committee, Dr. Fernando R. Columb.

The Chair was taken by Dr. Roberto H. Mendez, Chairman of the Local Organising Committee from the Institute of Astronomy and Physics of Space (IAFE), Buenos Aires.

A stylish musical interlude of works by Mozart, Piazzola, Belloso and Ocampo was provided by the La Plata University Brass Quintet.

Address by Dr. Roberto H. Mendez, Chairman, Local Organizing Committee

> "On behalf of the Local Organizing Committee it is a pleasure to welcome you all to Buenos Aires. Given the local economic situation, and the consequent restrictions in our budget, we have made a great effort to ensure the efficiency of scientific activities during this General Assembly. Your presence here is stimulating and will surely become a landmark for the younger generation of Argentine Astronomers. We thank you for coming and we wish all participants an invigorating astronomical experience. Thank you again, and good luck."

Address by Dr. Fernando R. Colomb, Chairman, National Organizing Committee

> "It is a great honour for me to welcome you to our country on behalf of the Argentine Astronomical Community.
>
> If meetings of this kind, which bring together men and women who search fervently to provide answers to the enigmas of the Universe, are always important, this one conference held in the year 1991 offers us the privilege to meet on the threshold of the commemoration of the fifth centennial of the discovery of America, the crucial step forward in the integration of humanity, in whose celebration the year 1992 has been declared the International Space Year.
>
> I believe that this particular circumstance, filled with special significance, constitutes a special motivation for the achievement of a more fruitful exchange of information, expectations and queries in our quest of the place man has been reserved in the cosmos.
>
> We are glad that the conference Agentina is hosting today is being attended by so many outstanding scientists from all over the world. Argentina has had an active tradition in astronomical research since 1871, when President Sarmiento inaugurated the Observatory in Cordoba, which allowed Benjamin Gould to level up the knowledge of

the southern hemisphere skies, information which was added to the data already existing for the northern hemisphere. At present there are six institutions devoted to astronomical research, over two hundred internationally renowned specialists working in various related branches, and the recently set up astronomical reserve of El Leoncito, in the province of San Juan, a space of over sixty thousand hectares, protected from interferences by a provincial law, offering the possibility of a pole for the development of observations in different wavelengths.

We are convinced that astronomical investigation and the answers it provides can contribute solutions for the welfare of humanity, at a time of explosive scientific and technological advance.

On behalf of the astronomers of Argentina, who feel very proud for having been chosen as the first Latin American country to organize the General Assembly of the International Astronomical Union, let me welcome you once more and hope that your stay in the city of Buenos Aires will be a truly memorable event for each one of you."

Address by Dr. Esteban Bajaja, President Argentina Astronomical Association

"Distinguished Guest, Ladies and Gentlemen,

It is a great honour and a great pleasure for me to be here now welcoming you to our country. A great honour and also a responsiblilty, because of the number and level of the scientists present here; because, I am speaking on behalf of the members of the Argentine Astronomical Association, a majority of them decided several years ago, when nobody imagined the critical times to come, to propose Argentina as host country for this General Assembly; and finally, because 120 years of professional astronomy in this country demand from us a corresponding competence in the organization of this meeting.

It is also a great pleasure because I am able this time to welcome to my country many of you whom I met in other parts of the world. I regret, however, that I miss the chance to welcome many other astronomers who are not coming just because this is a General Assembly, which, in their opinion, means too many people, too many subjects, too much time. In spite of this attitude, which will have to be considered when planning future General Assemblies, and other facts like the war in the Persian Gulf and the outbreak of cholera in some countries of South America, we are glad to see that the attendance exceeds our expectations.

To organize a General Assembly of the IAU is by itself a difficult job and implies a large responsibility, but when the host country has the economical situation that our country had, particularly during the last few years, such job is much more difficult to be carried out. The members of the LOC might write a book describing the all sort of problems they met in their long and painful way to reach this moment. Because of these difficulties we expect from you some indulgence when judging some details in the organization that might have gone or may go wrong.

The important thing is that you are here now and we shall try to make your stay as comfortable and fruitful as possible. For the latter you know what you expect from this General Assembly. Let me tell you what we expect from it, why we consider that it was worth the effort. As result of the 120 years of tradition in astronomy, during most of which Argentina had the leadership in South America, we have at present six observatories, two schools on astronomy and 180 members in our Association, more than two thirds of which are young astronomers and students, covering most of the fields in Astronomy. We have also a very good place for astronomical observations in El Leoncito, Prov. of San Juan, where, among others, we have a 2.15m optical telescope operating since 1986. There is even a project for a scientific satellite. On the basis of this picture we may conclude that we are in a rather good situation.

Being, however, in the southern end of South America and not being able to travel easily, means for us essentially to be isolated from the rest of the world. Scientific publications and different forms of communications make this isolation not absolute but they are not enough for the interaction needed for the scientific research. This isolation, together with a poor economical situation along many years, with low salaries and insufficient funds for equipment and maintenance, produced a slow but constant decrease in the intensity and, necessarily, in the level of the activity in astronomy. We are aware of this but it becomes painfully clear when we compare our activity with the activities in other countries.

In which way do we expect that this General Assembly will help us to correct this situation? The first result will be, certainly, that we shall be able to know each other, which will make, we hope, easier the communications and smaller the isolation. Our main problem, however has been the low financial support for our activity. The situation has improved recently but we need an increased and stable support and this will be only possible when the economical situation of the country improves. We hope that this will happen soon but then we still have to convince the government, and even ourselves, that our activity has to be fully supported.

The fact that the General Assembly is taking place in this country might help us to show to our people that there are many astronomers like we in the world, trying to discover the mysteries of the Universe. That there are large national efforts supporting this activity in many countries, in some of which the economical situation is even worse than in ours. That this science is for the mankind, the most important, not only for the knowledge of the Universe but also as a way to know itself. That this science contains all the other sciences.

Furthermore, in this General Assembly you are going to tell us about your magnificent observational and computational facilities, and the astronomical results obtained with them. We shall be very much impressed, I am sure, but I hope that we can overcome the depressing feeling that can invade us when comparing them with our facilities. Instead of that I hope that they will awake our enthusiasm, specially of the young people whose impulse has not been yet weakened by years of frustration and depression. I expect that they will see this display as a challenge to their own capacities to work with the modern technology and that they will try, and will get the help, to use those facilities.

I expect they will not want to be just witnesses of your progress, that they will want to participate, like the many Argentine astronomers around the world, some of them present here, who have chosen to live abroad to work and to live better, proving at the same time that our schools were good enough to enter in the world of science at international level. We expect, especially from them, that they will not forget their colleagues in this country working under inferior conditions.

In conclusion, I consider that this, more than a meeting to honour the 120 years of past Argentine astronomy, is a meeting for the young people, for those who are supposed to build the astronomy in this country for the next 50 years. With this in mind, I wish you and ourselves much success during this General Assembly and a happy stay."

Address by Lic. Osvaldo E. Devries, Secretario de Education y Cultura, Municipalidad de la Ciudad de Buenos Aires (Municipality)

"Ladies and Gentlemen, Members of the XXIst General Assembly of the International Astronomical Union:

In representation of the Mayor of Buenos Aires, Mr. Carlos Grosso, it is a pleasure to welcome you to this city.

It is a honour for us to receive these important representatives of the scientific field of Astronomy and Astrophysics of such different countries.

As a man of sciences and as Secretary of Education and Culture of the Town Hall of Buenos Aires (Municipality) I want to say we hope this city-site can be a good seat for the development and interchanges of knowledge in this congress and I am sure this meeting shall be a centre of up-to-date and advances in the sciences.

This gratitude is wider as we know that all serious and conscious development of the nature knowledges is always offering chances for the men in the difficult art of government. These knowledges -we know- are the channels and instruments of applying all the scientific performances in order to get a final destiny to benefit the peace, the growth and happiness of all the towns of the world.

To all of you I just have to say: I wish you the best success for this Congress and in the name of the citizens of Buenos Aires: Welcome! this city is yours."

Address by Dr. Raul F. Matera, Secretario de Ciencia y Tecnologia de la Presidencia de la Nacion

"As the Secretary of Science and Technology, I have the pleasure and honour to welcome to Argentina the International Astronomical Union members, their spouses, invited participants, representatives of other international organizations, young astronomers, all of whom have come to participate to this important event which will gather a significant part of the international astronomical community.

I also have the pleasure and honour to express the gratitude of our scientific community for having chosen the city of Buenos Aires as the site of the General Assembly, the first realized by the International Astronomical Union, along its seventy one year life, in an Ibero-American country. We feel obliged, very gratified and at the same time, very compromised for this significant attitude.

I know very well goals and reasons that give support to the existence of the International Astronomical Union, therefore I would like to remark, although it may not be necessary, that astronomical research and international co-operation in Astronomy have a long tradition among us. This tradition starts with the creation of the National Astronomical Observatory, today Astronomical Observatory of the Cordoba National University, inaugurated on the 24 th of October, 1871, and whose Director was the well-known and eminent astronomer from the USA, Dr. Benjamin Gould. And as a symbol of the spirit with which Argentine Astronomy is born, let me emphasize this: the official inauguration of our first observatory took place about a year after observational work was initiated.

After one hundred and twenty years, the development of Argentine astronomy has resulted in half a dozen of institutes, two Astronomy University Schools, the first one of them established in 1935, and an Astronomical Association which was born in 1958.

The Argentine government realizes the importance of science and technology, particularly in the present world, and of the unconditioned support that should be given to basic science, since if it is not carried on thoroughly, applied science cannot be developed, neither could be technology. The fact that the Secretariat under my charge depends directly on the Nation Presidency is a proof of that. This support which I mention is granted by means of the National Council of Scientific and Technical Research (CONICET), which depends on the Secretariat of Science and Technology, by means of subsidies, aids to pluri-annual research programmes, scholarships, the Research Career and other mechanisms, subject to the evaluations made by different Scientific Commissions.

In the particular case of Astronomy, let me mention the Astronomical Complex "El Leoncito", which was inaugurated in 1986, which is entirely financed by CONICET and the Secretariat under my charge together with CONICET and the National Universities of La Plata, which started the project, Cordoba and San Juan. All of these take part, by means of their representatives, in the Directive and Scientific Committee of this Astronomical Complex.

The economical difficulties that the Argentine Republic has been suffering and still undergoes have affected all national activities. In spite of this, I think that funds destined to science are important, the goal of my Secretariat being to obtain from the national budget a 1% of the National Gross Product for the Science and Technology Sector, that is to say, two and a half times more than what has been the norm up to now.

Together with this purpose of supporting scientific development I would like to mention the economical support that the Science and Technology Secretary and CONICET have compromised, with the intervention of the Economy Ministry, to materialize the construction of the first Argentine satellite of scienctific applications, denominated SAC-B, which will bring on board one Argentine experiment and three American experiments and which will be launched by NASA in 1994. I would like to emphasize that all experiments in its payload refer to important Astronomy projects.

Finally, I want to thank you again for your decision of making Buenos Aires, for ten days, the worldwide center of astronomical activity, and wish you a pleasant and fruitful stay."

Address by Professor Kozai

"President Menem, Ladies, Gentlemen and Distinguished Guests.

It is very fortunate for the IAU to come to Argentina to hold the XXIst General Assembly upon the invitation of the Association Argentina de Astronomia, Argentine Astronomical Society, with the support of the Government. We are particularly honored that the Opening Ceremony can be held with the presence of President Menem and Dr. Matera, Secretary for Science and Technology, as well as the Secretary for Education and Culture of the City of Buenos Aires, who kindly and warmly welcome us. The arrangements of the General Assembly were made by the National and Local Organising Committees with the advices by the IAU Executive Committee, particularly his General Secretary who had visited here several times for the past three years. We have appreciated very much their efforts towards the success of the General Assembly under difficult financial conditions.

Indeed it is the first time that the IAU has held the General Assembly in Latin America and the second time in the southern hemisphere.

We, astronomers, must observe stars and galaxies both in north and south to understand the universe and, therefore, we must have observatories in both hemispheres. Argentina is the oldest member of the IAU in the southern hemisphere, as Argentina adhered to the IAU already in 1927. And all of us know my predecessor Prof. Jorge Sahade who has played an important role in developing astronomical research in this country.

History of astronomical research in Argentina can be traced further
back.   Systematic astronomical research started in 1871 with
the foundation of the Astronomical Observatory of Cordoba.   In fact
in 1870, Argentinian President Sarmiento, after whom a nearby
street is named, invited a US astronomer, Benjamin A. Gould, who
founded the Astronomical Journal and intended to extend his
researches to the southern hemisphere, to come to Argentina.   It is
indeed exciting to know that astronomical researches were initiated
according to the desire of the President of Argentina.   In fact we
always ask Governments to support astronomical researches, and in
many cases, without great success.   Then, with the support of
the President   and   the Province   of   Cordoba,   he supervised
the construction of the Observatory of Cordoba.   His work,
Uranometria Argentina, deserved the immediate appreciation of
astronomers throughout the world.   John M. Thome, the second
Director, completed Gould's work and examined the positions and
brightnesses of 600,000 stars.

Astrophysical researches were introduced by the foundation of
Bosque Alegre Station, at an altitude of 1,250 meters with
a 60 inch telescope.   Several instruments were designed and
constructed by Argentinian astronomers including a worldwide famous
stellar spectrograph.   The Astronomical Observatory of La Plata was
founded in 1882.   Six to seven observatories and institutes were
created in this century.   They were connected to the university
system and for Cordoba and La Plata they turned into faculties or
colleges of the universities.

There are over 50 members of the IAU working in several fields of
astronomy.   Among the instruments available, there are optical
telescopes of 2.15 and 1.5 meter diameter as well as others of
smaller size and a radio telescope of 30 meter diameter at the
Argentine Radioastronomy Institute established in 1962.
The Institute of Astronomy and Space Physics was established in
1969 in the campus of the University of Buenos Aires.   During our
stay in Argentina I hope that many colleagues will have the chance
to visit some of these observatories and institutes.

Everybody knows that the IAU General Assembly is held every three
years.   In the old days it was the only opportunity for many
astronomers in all the fields to meet each other.   In fact, besides
the General Assembly, there were then very few international
astronomical meetings.   As the IAU membership has substantially
increased and, therefore, the number of the participants, it was
found that there was not enough time to discuss science during
the General Assembly because too many papers were presented and too
many meetings were held.   Then a new idea to hold symposia on
specified topics came out.   The IAU organised many symposia and
colloquia, which the Executive Committee selects among many
proposals.   Now many astronomers found that even at symposia and
colloquia there is not sufficient time for stimulating discussions
and workshop-style meetings on more specialised topics with limited
number of qualified participants have been organised without IAU
sponsorship.   And some of them have attracted many participants.

Therefore, it seems to me that the relative weight of the General
Assembly has decreased.   However, we still believe that the IAU
General Assembly should be the unique chance for many astronomers
in all fields and of many nationalities, from both developed and
developing countries, to get together to discuss and exchange

directly information on a wide range of astronomical problems. The Executive Committee has tried to improve the format of the General Assembly to attract more members, although we could not change the system substantially. Still I hope that such problems will be discussed here, again, and in the future.

Of course, to discuss scientific problems is more important and we can expect that new results obtained by scientific satellites, particularly those launched since the previous General Assembly, and major ground-based facilities with high technology as well as those derived from theoretical works will be presented. And we expect that new agreements on international co-operations will be made through discussions among many astronomers. In fact one of the main objectives of the IAU is to develop astronomy through international co-operation. Indeed, international co-operation is very important, particularly, in astronomy. Also we expect to hear about the present situations of several exciting projects, particularly, for large telescopes of the 8-10 meter class and development of sophisticated astronomical techniques. Therefore, there are many reasons to expect that the XXIst General Assembly will be very successful.

Now I would like to ask President Menem to declare the XXIst General Assembly of the International Astronomical Union open."

Address by Dr. Carlos S. Menem, Presidente de la Republica Argentina
(Palabras del Presidente Menem en el acto de inauguración de la XXI Asamblea General de la Unión Astronómica Internacional)

"Señor Presidente de la Asamblea; señores miembros de la Unión Astronómica Internacional; Señor secretario de Ciencia y Tecnología; cuerpo diplomático; señoras, señores, hermanas y hermanos del mundo: cuando comenzó su disertación el Presidente de la Asamblea, le preguntaba a la señora traductora, Ana Brown, por que no hacía la traducción del inglés al español, y me comento que este Congreso se iba a realizar en inglés.

Siempre que vienen algunos visitantes a la República Argentina que no saben el español, me preguntan si yo se inglés; les contesto permanentemente, que en mi país hablo el español. Cuando salgo de Argentina y me preguntan si se el inglés, digo lo mismo: fuera ded mi país sigo hablando el español.

La verdad es que no se hablar en inglés y lamento no poder comunicarme con ustedes en el idioma que se va convirtiendo en universal.

Les quiero expresar, como Presidente de todos los Argentinos la enorme satisfacción de que este Congreso se llieve a cabo en nuestra Patria, en nuestro territorio y en la Capital de la República, que algunos ya conocen y otros están conociendo.

Por supuesto que deseamos el mayor de los éxitos a esta Asamblea Internacional en el campo de esta fascinante ciencia como es la Astronomía.

Albert Einstein, uno de los grandes sabios que dió la humanidad al mundo, decía con una claridad mediana, total, absoluta, que Dios no jugaba a los dados.

La Astronomía es una ciencia que puede tener dos tipos de consideración: uno que deriva del aspecto teológico y otro del aspecto totalmente científico.

Los que creemos en Dios, decimos que nada nace de la casualidad, sino que todo es producto de la causalidad; que esta armonía, que este universo de los astros, que el girar permanente de los mismos, es producto de la presencia de Dios. Por otra parte, esta es una de las explicaciones que da Santo Tomas en cuanto a la existencia de Dios.

Además, está la explicación producto del conocimiento de los hombres, de la técnica y, fundamentalmente, de una ciencia como la Astronomía en su más pura esencia, que es lo que hacen permanentemente los hombres que se han dedicado a esta cuestión, a esta area tan importante de la vida de la humanidad. Es lo que ustedes -reitero- están haciendo en este momento y que nosotros avalamos en forma total y absoluta.

Esta ciencia pretende demostrar, a partir de esta armonía y de esta causalidad, el origen -diría- de los astros, de la humanidad y de la existencia misma en la Tierra desde el punto de vista científico.

Es por eso que Argentina como un país joven, nunca descuidó este aspecto que hace, evidentemente, a la vida de los pueblos y de las comunidades desde el punto de vista de lo que ocurre en el universo y como se proyecta éste hacia el planeta Tierra. También, está la búsqueda permanente desde la Tierra hacia las otras latitudes que hacen al universo, a esto que nos apasiona a nosotros como habitantes del planeta Tierra y a ustedes como científicos.

Cuando veo a científicos de todas partes del mundo reunidos en la República Argentina, cuando constato la presencia de científicos de Iberoamérica, tengo que remontarme inmediatamente a lo que en forma conexa hemos vivido hace pocas horas en Guadalajara, en la República de México, en un encuentro cumbre de todos los jefes de Estado y Presidentes de los países que integran este contexto iberoamericano.

Por supuesto que esta cuestión que hace a la Astronomía no estuvo ausente en las deliberaciones, al menos hubo una introducción por parte de algunos mandatarios respecto de este tema. Este Presidente lo ha hecho, fundamentalmente en lo que se refiere a este programa de satelización de las comunicaciones a nivel planetario, producto, por supuesto, de lo que aportan los científicos en el campo de la astronomía, sin ninguna duda.

Yo diría, como lo puede hacer el común de la gente que ha leído todo lo que hace a la armonía del Universo, que la Astronomía no tan sólo es la ciencia mas antigua del mundo, sino que también es la ciencia mas joven, porque se renueva en forma constante y permanente, como dije anteriormente. Esto nos ayuda a los hombres que nos hemos dedicado a esta verdadera pasión como es la política, a proyectar algunas respuestas y conclusiones de la Astronomía al campo de las relaciones entre las conmunidades nacionales e internacionales.

Todo esto es lo que hace al derecho internacional, cosa que ha sido motivo también de un tratamiento especial en esta cumbre iber americana que ha finalizado hace pocas horas en Guadalajara, México.

El señor secretario de Ciencia y Tecnología, asi como el Presidente, se han encargado de hacer referencia a los aspectos técnicos y a como evolucionó la Astronomía en la República Argentina, conceptos que, por supuesto, ratifico en estos momentos. Además, ratifico que estamos empeñados para que en el manor plazo posible estemos poniendo en orbita un satelite, el SAT 1, desde la República Argentina, para poder avanzar en este ámbito tan especial y tan deslumbrante -como dije antes- que es la Astronomía.

Queremos seguir iluminando desde la República Argentina algunas espectos que hacen a esta ciencia; marcar rumbos en Iberoamérica y creo que con el empeño que ponen hombres como el doctor Matera y sus colaboradores, lo estamos consiguiendo, al mismo tiempo que vamos a conseguir otros logros en forma rápida y decisiva en los años que se avecinan en nuestro país.

Los Argentinos aspiramos -si se me permite la expresión- a seguir brillando con luz propia en este terreno y a recibir y a receptar la cooperación de quienes mas han avanzado teniendo en cuenta la antiguedad de sus países. Reitero que el tratamiento de este tema es fundamental y trascendental para el futuro de las relaciones entre los hombres y de las relaciones interespaciales que evidentemente se van a dar en un futuro no muy lejano, a partir de lo que ya esta ocurriendo cuando el hombre ha llegado a la Luna y cuando se propone llegar a otros planetas de esta gran obra de Dios que es el Universo.

Finalmente, les quiero agradecer desde lo mas profundo de mi corazón el haberme invitado a participar de este acto y darme la posibilidad de dejar inaugurado este Congreso de Astronomía.

Se que en la Capital Federal, en Argentina, hay mil doscientos astrónomos, que vale tanto como decir que hay mil doscientos astros de la ciencia paseando por las tierras de la República Argentina. Hagan de cuenta que están en sus propias casas, hagan de cuenta que están en su propio firmamento, en su propio cielo y ven en cada uno de los Argentinos a un hermano de causa y de lucha por la justicia y la paz en la humanidad.

Muchísimas gracias, que Dios los bendiga y los abrazo sobre mi corazón."

Buenos Aires, 23 de julio de 1991.

TWENTY-FIRST GENERAL ASSEMBLY

First Session
held in the San Martin Cultural Centre
1991 July 23 16.30
Professor Y. Kozai, President, in the Chair

1.   Formal Opening by the President

"Ladies and Gentlemen,

After the inaugural ceremony we start the first session of the General Assembly. Two sessions of National Representatives of Adhering Countries to the IAU and of IAU Members are planned during the whole General Assembly. And this is the first one of the administrative sessions. According to the Statutes of the Union, the work of the Union is directed by the General Assembly. Therefore, we are supposed to report to the representatives of the adhering bodies and to the members what has happened in the Union since the previous General Assembly at Baltimore in 1988 and what actions have been taken by the Executive Committee to implement the decisions of the General Assembly and to direct the affairs of the Union in the interval between meetings of two consecutive General Assemblies.

Then we would like to nominate and/or identify the representatives of adhering countries and to the nominating and finance committees, and the members of the resolution committee who will work during the General Assembly for their objectives and who will report to you at the final session of the General Assembly. Here I would like to inform you that the number of the adhering bodies has been reduced by one by the unification of the two Germanies. Now a total of 56 countries adhere to the Union including 2 Associate Members. There are also more than 10 individual members in 9 countries which do not adhere to the Union, and it is expected that these numbers will be increased during the General Assembly.

In the past three years, at almost all the meetings of the Executive Committee, we spent much time to allocate funds to several scientific projects, such as symposia, colloquia, International Schools for Young Astronomers, central bureaux and data centers to sponsor projects proposed by the members. And we have tried to reduce the cost of administration. Now in this General Assembly we submit our proposal to improve the financial situation of the Union and I expect that we can find some solutions by the end of the Assembly.

I am afraid that shortage of funds is met not only in the IAU but also in almost all of the observatories and institutes in the world now. At several observatories, operations of some of their facilities had to be stopped for financial reasons, and several new projects cannot be started and/or are delayed due to similar reasons. Moreover, more severe problems exist, according to my view, in developing countries, where able astronomers are working without adequate funds and facilities and where even the one unit of contribution to adhere the Union, text-books on astronomy and tickets to go abroad to study and to attend meetings are too expensive to pay. IAU must try to help astronomers there. Such problems will be also discussed during the General Assembly. I believe that it is also one of the objectives of the IAU.

During the General Assembly, there are, of course, meetings of commissions, which are very important organisations in the IAU in any sense. Most of the participants will spend much of their time at commission meetings as well as Joint Discussions, Joint Commission Meetings and Invited Discourses, which were chosen by the Executive Committee out of many proposals by Commission Presidents. As much progress has been made and many new ideas have come out in astronomical researches in the past three years, I expect that participants will benefit very much by attending such meetings. As you may know from the final programme, poster sessions have been introduced in this General Assembly. I am sure that the poster sessions will stimulate more scientific discussions during the General Assembly. Still I believe that the structure of the scientific meetings can be improved by introducing new ideas which will also stimulate the creation of a new structure of the Union.

Besides scientific meetings there will be business meetings for the commissions for their own administrative business and for several items such as international co-operative projects. Among the commissions there are a few which are not purely scientific, like astronomical telegrams, protection of existing and potential observatory sites, exchange of astronomers and teaching of astronomy, all of which being of important concern for the IAU. It is also expected that several resolutions and recommendations will be submitted by the commissions and discussed in the final session of the General Assembly.

Indeed a very busy schedule will be in front of us during the General Assembly. I hope that all the meetings will be very successful and your participation will prove to be valuable to all of us and for you.

I am very happy to extend once again a warm welcome to all of you, Members of the Union, invited participants and guests.

I would like now to propose to send telegrams to the past Presidents and General Secretaries of the Union who are not able to be with us at the General Assembly, namely:

      V.A.   Ambartsumian
      A.A.   Blaauw
        C.   de Jager
        G.   Contopoulos
      J.H.   Oort
        R.   Hanbury Brown
        P.   Wayman
        E.   Müller.

I am happy to extend a warm welcome to members who have in the past served on the Executive Committee of the Union and who are here with us:

      J.C.   Pecker
        R.   West

and to the official representatives of the Adhering Organisations which support the Union:

Furthermore, I extend a hearty welcome to the official representatives of Sister Unions, ICSU Committees and other bodies, namely:

J.	Sahade	ICSU
C.	Segovia	IMU
P.	Pâquet	IUGG
R.D.	Eckers	URSI
W.C.	Martin	CODATA
E.	Tandberg-Hanssen	FAGS
B.	Robinson	IUCAF
S.T.	Wu	SCOSTEP
R.M.	Bonnet	ESA
M.C.	Huber	ESA
M.D.	Papagiannis	IAF
B.J.	Robinson	IUT/CCIR
D.	Huenemoerder	NASA
F.	Repetto	UNESCO

I would like to mention, particularly, the name of Prof. Jorge Sahade, who was the President of the Union in 1985-88 and an advisor to the Executive Committee in the past three years and also the official delegate of ICSU, the International Council of Scientific Unions, which is the umbrella for our Union also. I would like to ask him to give us a few words now."

Address by Prof. J. Sahade, ICSU Representative:

"Mr. President, Members of the Union, Ladies and Gentlemen,

It is my privilege and my pleasure to bring to you, at this first session of your XXIst General Assembly, the greetings of ICSU, the International Council of Scientific Unions, and wish you, on ICSU's behalf, a very fruitful and enjoyable gathering.

You have a very compact and very important programme before you, and, as a consequence, I will endeavour to be brief and to the point.

If you permit me, I would take the opportunity just to make you aware of the fact that ICSU is now quite conscious of and shares your deep concern in regard to the increasing dangers to research that arise from space debris and the different types of space pollution.

Your General Secretary, as your representative in the General Committee of ICSU, has been extremely convincing in the presentation of the case, and has also warned us about some weird proposals that would even be more detrimental - I should perhaps say more deadly- to astronomical research at large than it is generally appreciated.

As a result, the 26th Meeting of ICSU's General Committee, that took place in Lisbon on October 11-14, 1989, requested "ICSU to use all means at its disposal, notably through COSPAR, URSI, IUCAF, and ICSU's contacts in the Space Agencies, to seek effective measures to reduce or avoid present or possible future consequences of such space activities (those mentioned in the whereas section of the Resolution) on astronomy". This 1989 Resolution of the General Committee of ICSU led to a couple of Resolutions adopted at its 23rd General Assembly, held in Sofia, Bulgaria, on October 1-5, 1990:

Space debris

The General Assembly <u>noting</u> with satisfaction the studies already carried out on the hazard to space missions of impacts with space debris, <u>requests</u> COSPAR to continue this important study in the interests of astronomical and earth observations and in particular:

a. to refine the quantification of the hazards to current and future space missions from impact with space debris;

b. to contine strongly to urge the Spage Agencies to <u>adopt</u> more effective ways and means of reducing the hazard in the short term and to work towards its elimination in the longer term future.

The General Assembly also <u>requests</u> COSPAR to maintain an overview of any activities in space which might lead to pollution of nearby space bodies by matter of terrestrial origin.

Multi-wavelength interference with astronomical observation

<u>noting</u> the harmful increase in the level of electromagnetic pollution, ranging from light pollution to radio interference;

<u>recognizing</u> the impact of such pollution on Earth observation, astronomical observation and radio science;

The General Assembly

<u>strongly urges</u> all concerned administrations, agencies and regulatory bodies to recognize the importance of the continuing detection of faint electromagnetic signals at all frequencies in exploring the universe, in seeking the origins of life, in monitoring the natural resources of the Earth and the fragile balance of the Earth's ecosystem. Effective steps to reduce pollution are urgently required to restore the quality of observing conditions;

<u>further stresses</u> that the frequencies used in Earth exploration, radio astronomy and space research need urgent and specific protection from radio interference especially from telecommunications, navigation satellites and other air borne and space emissions;

<u>specially requests</u> to that end, that the World Administrative Radio Conference in 1992 ensure the rational use, conservation and protection of the scarce radio-frequency spectrum, taking into account the astrophysical signficance of the spectral lines listed in Resolution A7 of the 20th General Assembly of the IAU (1988, Baltimore) and <u>supports</u> the strenuous efforts of IUCAF to obtain such protection.

I would be happy to convey to ICSU any suggestion or relevant input that you might like to make whenever appropriate during the General Assembly.

Thank you very much."

After Professor Sahade's address, Professor Kozai continued the ceremony:

"I now ask those present to stand while the General Secretary reads the names of the members who have died since the 20th General Assembly".

The General Secretary read the list of members deceased since the XXth General Assembly on the invitation of the President (see Chapter III, Report of the Executive Committee, Section 9, pp. 111-112).

The Assembly observed a period of silence in their memory.

2.   Appointment of Official Interpreters

The General Secretary announced that J.-C. Pecker (English-French) and J. Rountree-Lesh (French-English) had agreed to serve as Official Interpreters.

3.   Report of the Executive Committee 1988-1991

The President invited the General Secretary to present the Report of the Executive Committee 1988-1991. The Report covers the period 1988 September - 1991 June 30.   (A summary of the Report of the Executive Committee 1988-1991 was published as Section 2-1 pp. 9-26 of Information Bulletin No.66 and appears in full in Chapter IV pp. 93-120).

The General Secretary summarised the Report highlighting the following points:

"The Executive Committee has devoted considerable attention during the triennium to reviewing the ongoing activities of the Union -format of General Assemblies, the Union's publications, fund raising and the Union's sources of income, the Working Group in the World Wide Development of Astronomy, adverse environmental impacts on astronomy. Many of the issues addressed have not yet reached finality but the decisions to be taken will have a profound effect on the future effectiveness of the Union. The format of General Assemblies is still being debated and will be discussed during this Assembly. The Transactions of the Union will also be discussed during this General Assembly. Members will have noted that changes have taken place in the Information Bulletin -two membership updates are published in the second and third January issue between General Assemblies, the Information Bulletin has been used to give in extenso information on the General Assembly including draft and final programmes. It is hoped these changes are advantageous. A new contract for IAU Publications has been signed with Kluwer Academic Publishers for the period 1992-1997, again providing royalty income for the Union. The Executive Committee has also considered the Union's

financial resources and two resolutions, as a result, will be
considered by this General Assembly. Default on subvention to
the Union has been growing but it is hoped that the situation
will improve as a result of action already taken. The Working
Group on the World Wide Development of Astronomy, established
at the XXth General Assembly, has used the triennium to
consider how it can best implement its terms of reference and
will be holding two meetings during this General Assembly to
discuss   for   future   action.   As   a result   of
the withdrawal of a resolution presented by the Swedish
National Committee for Astronomy to the XXth General Assembly,
the Executive Committee established a subcommittee to examine
the issues raised. That committee found reason for concern and
their conclusions are given in the full Report.   Steps have
been taken to implement the committee's suggested action and
an IAU/ICSU/UNESCO meeting will be held in the summer of 1992
in Paris to make governments and public aware of the threats
now posed to continued astronomical observation from the ground
and   from   space.   A presentation   to   the UN   Committee on
the Peaceful Uses of Outer Space is under consideration.
The IAU has been given every encouragement and support by ICSU
in its endeavours to publicise these threats.   The mandate of
Commission 50 is to be extended to include all threats to
astronomical observation.

The Secretariat is well settled into its new premises -which
are well suited to its purposes.   A new computer system
(replacing an IBM PC and a Rank Xerox word-processing system)
of three linked Compaq PCs has been installed. The new system
permits tape backup so improving the security of Union records.
Mrs. H. Gigan resigned   from   the service   of   the Union   in
January 1990 and was replaced by Mrs. J. Crook from July 1990.
Monique   Léger-Orine   continues   to   direct   the day-to-day
operation of the Secretariat with great efficiency.

At the XXth General Assembly, four countries were admitted as
Members of the Union and two as Associate Members. Because of
continued default on Union Subventions, the Executive Committee
has recommended that two countries resign from the Union.
However, both countries will be offered Associate Member Status
and may reapply for full membership once regular financial
arrangements can be resumed.   The status of the individual
members of the Union from the countries deemed to have resigned
from the Union is unaltered. At the close of the XXth General
Assembly   the Union   had   6711 individual   members   and   had
6624 individual members as of 1991 June 30.

The Report indicates that the administrative Commissions of
the Union   have   been   active   during   the Triennium.
The scientific programme of this General Assembly and Reports
on Astronomy (IAU Trans. XXIA) indicate that the remaining
Commissions have been active both singly and conjointly. It is
a particular pleasure to record that Commission 38 assisted
the travel of 25 astronomers mostly from developing countries.
Commission 46 has maintained its Visiting Lecturer Programme in
Peru and Paraguay and held three International Schools for
Young Astronomers -No. 16 in Cuba, No. 17 in Malaysia and
No. 18 in Morocco. This last was the first francophone school
but sadly the last to be organised on behalf of the Union by

J. Kleczek. Josip Kleczek pioneered and nurtured the ISYA and personally directed all 18 schools since their inception in 1966. He brought to the task a particular talent for organisation of an effective school in all parts of the world as well as a great warmth of personality. He kept a remarkable balance of practicality of execution. Many times an Executive Committee, even Commission 46, set down ideal objectives for ISYA. Josip translated ideals into the reality of the developing world. Perhaps the greatest test of the success of the schools is the fact that the co-organiser of the 17th School was, herself, a graduate of an earlier School. This Union owes Josip Kleczek a great debt of gratitude for his dedication to the realisation of a fine educational enterprise (Applause).

It is unfortunate that he cannot be here today to receive your ovation. We have two worthy successors in that Donat Wentzel will take over as Secretary to the ISYA and he will be ably assisted by Michèle Gerbaldi.

The Union held 12 Symposia, 22 Colloquia, 4 Regional Meetings and cosponsored 11 meetings with other ICSU Unions and Committees. The Union was represented at 21 other Unions, Committees and other important international bodies.

Finally, it will be seen that the Union has ended the triennium in a financial state which allows the maintenance of a balance which is about equal to the expected annual expenditure in the next triennium. Such a balance is necessary for the prudent operation of the Union but that balance is susceptible to erosion by currency inflation and by default on subvention. The proposed budget for 1992-1994 has been devised on the basis of ending a six year level funding period with moderate annual increases in support of the Union's scientific activities."

The President invited discussion of the Report of the Executive Committee 1988-1991. The Report has been given consideration by the Official Representatives of the Adhering Organisations. The financial part of the Report will be scrutinised by the Finance Committee whose report will be presented at the second session of the General Assembly. There being no points raised by Members of the Union from the Floor, the General Assembly unanimously approved the Report of the Executive Committee 1988-1991 subject to receiving the Report of the Finance Committee.

4.   Report of the work of the Special Nominating Committee

The President informed the Assembly that the Special Nominating Committee had selected the following IAU members for proposal as members of the Executive Committee from 1991 August 01.

President:	A.A. Boyarchuk	USSR
General Secretary:	J. Bergeron	France
Assistant General Secretary:	I. Appenzeller	Germany
President-Elect:	L. Woltjer	Netherlands

Vice-Presidents:	D.S.	Mathewson	Australia
	F.	Pacini	Italy
	V.	Radhakrishnan	India
	M.S.	Roberts	USA
	J.I.	Smak	Poland
	Shu Hua	Ye	China PR
Advisers:	Y.	Kozai	Japan
	D.	McNally	UK

5.   Announcement of the Official Representatives of Adhering Organisations and the Representatives to serve on the Nominating Committee

Country	National Committee Representatives		Nominating Committee Representatives	
Algeria	----	----------	----	----------
Argentina	A.	Feinstein	G.	Lopez Garcia
Australia	R.D.	Eckers	R.D.	Eckers
Austria	H.	Haupt	H.	Haupt
Belgium	P.	Smeyers	P.	Pâquet
Brazil	----	----------	M.T.	Pastoriza
Bulgaria	N.S.	Nikolov	M.K.	Tsvetkov
Canada	S.	van den Bergh	----	----------
Chile	A.	Gutierrez-Moreno	M.T.	Ruiz
China Nanjing	Li	Qi-bin	Li	Qi-bin
China Taipei	H.H.	Wu	Chow	C.K.
Colombia	E.	Brieva	E.	Brieva
Cuba	----	----------	----	----------
Czechoslovakia	----	----------	----	----------
Denmark	L.K.	Kristensen	L.K.	Kristensen
Egypt AR	A.Z.	Aiad	M.A.	Soliman
Finland	K.	Muiononen	K.	Muinonen
France	J.	Kovalevsky	B.	Morando
Germany	R.	Wielen	R.	Wielebinski
Greece	P.	Laskaridis	P.	Laskaradis
Hungary	B.	Szeidl	B.	Szeidl
Iceland	----	----------	----	----------
India	G.	Swarup	----	----------
Indonesia	----	----------	S.D.	Wiramihardja
Iran	----	----------	----	----------
Iraq	----	----------	----	----------
Ireland	M.	de Groot	M.	de Groot
Israel	----	----------	G.	Shaviv
Italy	V.	Castellani	L.	Padrielli
Japan	D.	Sugimoto	D.	Sugimoto
Korea DPR	----	----------	----	----------
Korea RP	H.S.	Yun	Chun	M.S.
Malaysia	----	----------	----	----------
Mexico	P.	Pismis	S.	Torres-Peimbert
Morocco	----	----------	----	----------
Netherlands	C.	Zwaan	C.	Zwaan
New Zealand	E.	Budding	E.	Budding
Nigeria	----	----------	----	----------
Norway	E.	Jensen	K.	Aksnes
Peru	----	----------	----	----------

Poland	J.	Smak	J.	Smak
Portugal	J.P.	Osorio	J.P.	Osorio
Rumania	----	----------	----	----------
Saudi Arabia	A.	Niazry	----	----------
South Africa	M.W.	Feast	E.E.	Baart
Spain	J.	Gomez-Gonzalez	J.	Gomez-Gonzales
Sweden	R.	Booth	A.	Hjalmarson
Switwerland	B.	Hauck	J.	Stenflo
Turkey	H.	Kirbiyik	H.	Kirbiyik
UK	F.	Graham-Smith	C.	Jordan
Uruguay	J.	Fernandez	J.	Fernandez
USA	K.I.	Kellermann	A.	Cox
USSR	N.V.	Steshenko	B.M.	Shuster
Vatican City State	C.	Corbally	C.	Corbally
Venezuela	G.	Bruzual	G.	Bruzual
Yugoslavia	A.	Cadez	A.	Cadez

6.   Acting Presidents of Commissions

Commission	Acting President	
4	P.K.	Seidelmann
5	G.A.	Wilkins
6	E.	Roemer
7	J.	Henrard
8	M.	Miyamoto
9	J.	Davis
10	E.R.	Priest
12	J.W.	Harvey
14	S.	Sahal-Brechot
15	J.	Rahe
16	A.	Brahic
19	M.	Feissel
20	R.M.	West
21	A.-C.	Levasseur-Regourd
22	C.S.	Keay
24	W.F.	van Altena
25	I.S.	McLean
26	H.A.	McAlister
27	M.	Breger
28	G.A.	Tammann
29	P.S.	Conti
30	D.W.	Latham
31	P.Eg.	Pâquet
33	M.	Mayor/L. Blitz
34	J.S.	Mathis
35	A.	Maeder
36	D.F.	Gray
37	C.	Pilachowski
38	F.G.	Smith
40	P.G.	Metzger
41	J.D.	North
42	R.H.	Koch
44	E.B.	Jenkins
45	M.	Golay
46	A.	Sandqvist
47	K.	Sato

48	J. Ostriker
49	B. Buti
50	D.L. Crawford
51	G. Marx
WGPSN	K. Aksnes
WGWDA	A. Batten

7.   Appointment of the Finance Committee

In accord with Statute 22(a), the General Assembly appointed
the following Finance Committee consisting of one representative
from each Adhering Organisation:

Country	Category	Units	Representative
Algeria	I	1	---- ----------
Argentina	III	4	Z. Lopez Garcia
Australia	III	4	L. Keay
Austria	I	1	H. Haupt
Belgium	IV	6	P. Smeyers
Brazil	II	2	---- ----------
Bulgaria	I	1	M.K. Tsvetkov
Canada	VI	14	D.C. Morton
Chile	I	1	B. Moreno
China Nanjing	V	10	Hong-Jun Su
China Taipei	I	1	H.H. Wu
Colombia	I	1	E. Brieva
Cuba	I	1	---- ----------
Czechoslovakia	III	4	---- ----------
Denmark	II	2	L. Helmer
Egypt AR	III	4	M. Soliman
Finland	I	1	K. Muiononen
France	VII	20	J.C. Pecker
Germany	VII	20	M. Grewing
Greece	II	2	P. Laskaridis
Hungary	II	2	I. Almar
Iceland	I	1	---- ----------
India	III	4	G. Swarup
Indonesia	I	1	---- ----------
Iran	I	1	---- ----------
Iraq	I	1	---- ----------
Ireland	I	1	M. de Groot
Israel	II	2	---- ----------
Italy	V	10	L. Padrielli
Japan	VII	20	D. Sugimoto
Korea DPR	I	1	---- ----------
Korea RP	I	1	M.S. Chun
Malaysia	I	1	---- ----------
Mexico	II	2	A. Serrano
Morocco	I	1	---- ----------
Netherlands	IV	6	C. Zwaan
New Zealand	I	1	E. Budding
Nigeria	I	1	L.I. Onuora
Norway	I	1	E. Leer
Peru	I	1	---- ----------
Poland	III	4	J. Smak
Portugal	II	2	J.J. Osorio
Rumania	II	2	---- ----------

Saudi Arabia	I	1	----	----------	
South Africa	III	4	G.D.	Nicolson
Spain	II	2	J.	Gomez-Gonzales
Sweden	III	4	A.	Ardeberg
Switwerland	III	4	G.	Burki
Turkey	I	1	H.	Kirbiyik
UK	VII	20	R.D.	Davies
Uruguay	I	1	J.	Fernandez
USA	VIII	30	P.	Boyce
USSR	V	10	B.M.	Shuster
Vatican City State	I	1	R.	Boyle
Venezuela	I	1	G.	Bruzual
Yugoslavia	II	2	A.	Cadez

8. **Appointment of the Resolutions Committee**

The President informed the Assembly that the Executive Committee proposed the establishment of a Resolutions Committee under the Chairmanship of Professor R.D. Davies, with Drs A.H. Batten (Executive Committee Representative), E. Budding, S. Isobe and J.-C. Pecker as members. The General Assembly unanimously agreed to this composition of the Resolutions Committee.

9. **Revision of Bye-Law 24**

The President asked the General Secretary to amplify the Executive Committee's proposal to extend the number of categories of adherence. The General Secretary explained the background and need for the proposed change formally set out as:

The Executive Committee of the International Astronomical Union

Recognising

a) the continued growth in the number of individual members of the Union and thereby, the increasing number of members of the Union from Adhering Country;

b) the need for additional financial support for the Union to meet the costs incurred by the Union arising from that growth in membership;

c) that the Union will continue to derive all but an insubstantial fraction of its income from its Adhering Organisations;

Recommends

that Bye-Law 24 of the Union be amended by the creation of 4 new categories of adherence IX, X, XI, XII paying 40, 55, 75, 100 units of contribution respectivily to allow Adhering Countries to assume a higher category of adherence than is currently possible.

Bye-Law 24 as revised would read

"Each Adhering Country pays annually to the Union a number of units of contribution corresponding to its category as follows:

Category as defined in Statute 8:

I   II   III   IV   V   VI   VII   VIII   IX   X   XI   XII

Number of units of contribution:

1   2   4   6   10   14   20   30   40   55   75   100"

10.   Resolutions submitted by Adhering Organisations

No Resolutions were proposed to the XXIst General Assembly by Adhering Organisations.

11.   Resolutions submitted by Commissions or Associated Inter-Union Commissions

No Resolutions were proposed to the XXIst General Assembly by Commissions or by Associated Inter-Union Commissions.

12.   A proposal by the Executive Committee that the Union should establish an IAU Trust Fund

The President asked the General Secretary to amplify the Executive Committee proposal to move towards the institution of an IAU Trust Fund. The General Secretary explained the background which led to the proposal finally set out as follows:

"The Executive Committee of the IAU

recognising

i) the severe financial restraints under which Adhering Organisations must operate

ii) the need for further financial support to develop new Union services and activities in support of astronomical research asks the General Assembly to signify their support/lack of support for the Executive Committee to proceed to a proposal for an International Astronomical Union Trust Fund, to be governed by a Board of independent Trustees answerable to the General Assembly, for consideration by the XXIInd General Assembly".

The Scientific Unions of the ICSU family derive their income directly or indirectly from the governments of the Adhering Member Countries usually through a body such as a National Academy. In recent years the worldwide finance available for science has come under considerable pressure and the IAU has had several periods of level funding for its science in recent years. This has meant a reduction in the support the Union has been able to give to the science of astronomy. Inflation inexorably erodes Union income. This means that the Union can only maintain existing commitments at a devaluing rate and cannot undertake new initiatives which are necessary in a rapidly changing world.

The concept of a Union Trust Fund was introduced in Information Bulletins 63 (p. 12) and 64 (p. 4). It has proved impossible to gauge Union opinion on this important matter and accordingly the Executive Committee has put forward a resolution to the XXIst General Assembly to test the reaction of the members attending to such a proposal.

The General Secretary stressed that members should only vote for the proposal at the second session of the General Assembly were they willing to contribute such a Fund.

The President then formally adjourned the meeting until 1991 August 01 at 10.00 and closed the meeting with a word of thanks to the participants.

TWENTY-FIRST GENERAL ASSEMBLY

Second Session
held in the San Martin Cultural Centre
1991 August 01 10.00
Professor Y. Kozai, in the Chair

The President asked the General Secretary to report on the fire[*]
which had occurred in the Cultural Centre the previous day
(1991 July 31). The General Secretary reported that prior to 08.00
on July 31, the final day of scientific sessions at the General
Assembly, a fire had ignited in some rubbish in one of the basement
car park levels of the San Martin Cultural Centre. Although
the fire was confined to the basement level, dense smoke had
permeated the remainder of the building and it was clear that some
days would elapse before the building was again usable.
The General Secretary summoned an emergency meeting of
the Executive Committee to consider the situation. The General
Secretary was empowered to take such action he deemed necessary to
retrieve as much of the scientific programme as was possible under
the circumstances and to arrange a venue for this second session of
the General Assembly.

It was fortunate that the General Assembly was divided between two
conference centres -eight parallel sessions were held in
the San Martin Cultural Centre and four were held in the La Plaza
development. La Plaza also had an open air amphitheatre where it
was possible to hold an information session for the participants
and all scientific activity was transferred to this site plus
a nearby cinema which was made available for the afternoon.
The LOC, their Conference Consultants Annajuan and the Management
of La Plaza made Herculean efforts to provide as much accommodation
as could be found and we are very much in the debt of all those in
and around La Plaza who went out of their way to assist us.
The efforts of Juan O'Farrell of Annajuan must be mentioned
especially (Professor A. Wolfendale gave the General Secretary
a large bottle for award to the most notable person of the General
Assembly -the General Secretary informed the Assembly he had
presented it to Juan O'Farrell): Juan seemed to conjure meeting
rooms out of nowhere and to generate notices as soon as a need was
identified. Without Juan, our best efforts would have been pale
indeed. The General Secretary expressed his appreciation of
the help he had received from those organising scientific meetings
-particularly from those who agreed to cancel their meetings; from
those, whose meetings were possible, for agreeing to hold them in
the lunch-hour and on into the evening; and from Bernard Hauck and
George Wilkins for volunteering the cancellation of Joint
Commission Meeting VIII Archiving Current Astronomical Data. This
Joint Commission Meeting was of great timeliness and papers
contributed to it will be published in Highlights Vol. 9.
Unfortunately, the discussion and visibility that the meeting would
have generated for this important topic has been lost. The General
Secretary hoped that he had managed to accommodate all
the Commissions who needed a final meeting and asked Presidents of
Commissions who kindly cancelled their meetings to report

the science to be presented in their Transactions XIIB reports. Regrettably the final day's poster session has been lost.

The General Secretary reported that the LOC and the IAU staff mounted a rescue operation in co-operation with the Fire Service and recovered the IAU records, papers, equipment and speakers' slides, not to mention airline tickets being reconfirmed. The IAU computer worked upon rescue and it seems the IAU data records are intact. The General Secretary asked for the Assembly's forebearance during this second session if some information needed correction -the Secretariat had planned to spend the previous day in peace and quietness preparing for the second session of the General Assembly. Nevertheless, the fact that a session was in progress, with only a faint smell of smoke to remind the Assembly of yesterday's traumatic events, is a tribute to the efforts of the LOC, the Secretariat and all those who gave unstintingly of their help to ensure the conclusion of business in some semblance of good order.

The General Secretary also reported that the two firemen, who were overcome by the smoke in their efforts to deal with the fire, were now recovering. Fortunately no-one was seriously injured in this fire and it was merciful that it occurred before the start of the working day when the building would have been full.

(*) The following information was received from Hugo Levato:

"The reason for the fire seems to be the following: At the fourth underground level, garbage was stored in the wrong place. This place has ventilation pipes which go to the street level. It seems that some pedestrian passed walking through and threw a cigarette or a match and it falled on the garbadge (mainly papers for discarding). It was a fire that during working hours would have been easily extinguised, but very early in the morning (5 a.m. approximately), no personnel was working there and the fire became more important."

Before passing to the Agenda, the General Secretary was asked by the President to review the voting procedure. The Rules for Voting may be summarised as follows:

a) only National Representatives may vote on financial and administrative matters (Statute 15 (a) (b));

b) all members of the Union present have one vote each on scientific matters (Statute 16);

c) the vote of the National Representatives is valid only if at least 2/3 of the Adhering Countries having the right to vote participate -in this instance 24 (Statute 15 (b);

d) the vote is determined by simple majority subject to (c) above. Countries in arrears with their subvention as of 1990 December 31 may not vote -such countries were advised prior to the session by the General Secretary (Statute 15 (a) (b));

e) on financial matters each Adhering Country has a number of votes equal to their category of adherence plus one (Statute 15 (b)). On administrative matters each Adhering Country has one vote (Statute 15 (a)).

A quorum having been established the General Assembly appointed H. Levato, E. Sadler and V. Trimble as Tellers.

13.   Report on the Finance Committee

The General Secretary announced that copies of the Report of the Finance Committee had been made available to the National Representatives and he invited P. Boyce, Chairman of the Finance Committee, to present the report.

Report of the IAU Finance Committee

"The Finance Committee appointed a subcommittee to examine the accounts and proposed budget of the Union and to make recommendations to the General Assembly.

The Committee believed that the summary of the accounts as printed closely reflects the result of the official audit of the Union's accounts.

The Finance Committee recommends that the accounting system be changed to reflect the revenues and expenses incurred in a given calender year, regardless of when the payments or receipts are made (known in US terms as "accrual accounting"). The Committee further requests that all triennium reports and budgets start with the year of the General Assembly. The budget will then have to include one additional year for the future General Assembly.

The Finance Committee recommends two cutting measures:

1. One particularly costly item for this General Assembly is the payment of all expenses for nearly all the incoming and outgoing Executive Committee for the duration of the General Assembly. There does not seem to be a uniform policy on this matter. The Finance Committee thus recommends that the Executive Committee consider whether the full payment of travel and per diem for such a long stay provides benefits commensurate with the high cost involved, and

2. that the Executive Committee should ask each National Committee to check the membership list annually and to eliminate individuals no longer interested in the Union's activities. The Committee further notes with approval that the Executive Committee has taken the steps of terminating the membership in the Union for two countries and strongly encourages the Executive Committee to strengthen the efforts to ensure prompt payment of dues.

The Finance Committee strongly urges that the reserve funds of the Union be maintained at a level equal to an average year's operations and points out that this requires a net gain equal to the inflation rate.

The Committee recommends that the General Assembly adopt the following rates as proposed by the General Secretary which are based upon assuming an average inflation rate of five percent per year in those countries where the Union conducts business:

1992:   2345 SwF,
1993:   2460 SwF and
1994:   2580 SwF

The Finance Committee recommends amending the proposed change to the levels of adherence as listed in IB 66 to the following pattern which reduces the absolute sizes of the steps for the higher categories:

Category

I	II	III	IV	V	VI	VII	VIII	VIII½	IX	X

Votes

2	3	4	5	6	7	8	9	9	10	11

Units

Present

1	2	4	6	10	14	20	30	-	-	-

Proposed

1	2	4	6	10	14	20	27	30	35	45

Since the United States currently adheres at a level of 30 units we suggest retaining that level temporarily as category VIII½. If the United States decides to change its level of adherence that categorry would be automatically dropped and it is intended that if additional categories are added, they should be in constant steps of 10 units.

The Committee recommends approval of the proposed budget with the proviso that the Executive Committee make every effort to obtain additional income through increased levels of adherence to the Union and better honoring of the dues and by maintaining tight control over costs.

Having done so at the past two General Assemblies, the Committee again recommends that, with a budget of the present size, prudent management practice would be to make a fiscally knowledgeable person always available to the Executive Committee and suggests that the Executive Committee consider establishing the office of Treasurer, but without expanding the size of the Executive Committee.

The Finance Committee encourages the Executive Committee to study the concept of an IAU Trust Fund structured along the lines as outlined in IB 63. The study should be done in consultation with the National Adhering Bodies and should include possible negative consequences upon the overall support of the Union."

The President thanked Dr. Boyce for his Chairmanship of the Finance Committee.

Vote on the Report of the Finance Committee:

The President conducted the vote on the Report of the Finance Committee as follows:

The audited Accounts for 1988, 1989 and 1990 were unanimously accepted together with the residual budget for 1991.

The increased Units of Contribution for 1992, 1993, 1994 were approved by 145 votes, there were two abstentions.

The proposed budget for 1992-1994 was unanimously accepted.

14. Vote on the Change of Bye-Law 24

An Amendment to the Executive Committee's proposed change of Bye-Law 24 was proposed by the National Representatives. They proposed that a redefined category VIII carrying 27 units of contribution should be inserted, that a special category VIII½ carrying 30 units of contribution should be created temporarily to accommodate the United States (currently paying 30 units of contribution) and that the step between units of contribution reduced from that proposed by the Executive Committee such that category IX carried 35 units of contribution, category X, carried 45 units of contribution and thereafter additional categories would carry units of contribution increasing in fixed steps of 10 units. Category VIII½ would be eliminated as and when the United States choose to alter its category of adherence. The proposed amendment led to the following form for Bye-Law 24. The associated number of votes for each category is appended.

The revised form of Bye-Law 24 proposed by the National Representatives as an amendment to the change proposed by the Executive Committee (pp. 21-22). "Each Adhering Country pays annually to the Union a number of units of contribution corresponding to its category as follows:

Category as defined in Statute 8:

I	II	III	IV	V	VI	VII	VIII	VIII½	IX	X

Number of units of contribution:

1	2	4	6	10	14	20	27	30	35	45.

If further Categories of Adherence are required in the future, the step in the number of units shall be 10 units/category."

This Amendment was carried by 142 votes: there were 5 abstentions.

The proposal to change Bye-Law 24 by the Executive Committee, therefore, was lost.

15. A proposal to institute an IAU Trust Fund

After some clarification from the General Secretary, the President explained that while strictly, the proposal to consider the formation of an IAU Trust Fund was financial, its sources of income lay outside the Adhering Countries -in particular from contributions by individual members of the Union -it was appropriate that opinion of all members of the Union present should be tested on this important issue. On the motion of the President, a substantial majority of the membership indicated support for the motion as set out in Section 12. Only one member wished to record an abstention. The question of an IAU Trust must now be considered by the Executive Committee.

16.   Resolutions submitted by the Executive Committee

The following resolutions were submitted by the Executive Committee:

Resolution A1: Sharing Hydroxyl Band with Land Mobile Satellite Services

The XXIst General Assembly of the International Astronomical Union,

considering

a) that the 1660-1660.5 MHz band is allocated to the Radio Astronomy Service on a shared, primary basis, and is used to observe hydroxyl lines, which are of the highest astrophysical importance, in many galaxies in the nearby Universe;

b) that the World Administrative Radio Conference for the Mobile Services (WARC MOB-87) has also allocated the 1660-1660.5 MHz band to the land mobile satellite service;

c) that WARC MOB-87 has added Footnote 730A to the Regulations, allowing administrations to authorize aircraft stations and ship stations to communicate with space stations in the land mobile satellite service in the 1660-1660.5 MHz band.

urges

that administrations adhering to the International Astronomical Union and to the International Telecommunication Union bear in mind at WARC 92 the importance of the primary allocation to the Radio Astronomy Service in the band 1660.0-1660.5 MHz;

and instructs the President

to support IUCAF strongly in its efforts to bring this resolution to the attention of delegations participating in WARC 92.

IUCAF   *Inter-Union Commission on Frequency Allocations for Radio Astronomy and Space Science*

WARC MOB   *World Administrative Radio Conference for Mobile Services*

Résolution A1: Partage de la bande de l'hydroxyle avec les services mobiles au sol

La XXIe Assemblée générale de l'Union Astronomique Internationale,

considérant

a) que la bande 1660-1660.5 MHz est attribuée au Service de la Radioastronomie sur une base de priorité et de partage et qu'elle est utilisée pour l'observation des raies de l'hydroxyle qui sont de la plus grande importance astrophysique dans de nombreuses galaxies appartenant à l'Univers proche ;

b) que la Conférence Administrative Mondiale des Radiocommunications pour les services mobiles (WARC MOB-87) a aussi attribué la bande 1660-1660.5 MHz aux services mobiles au sol associés aux satellites ;

c) que la WARC MOB-87 a ajouté la note 730A aux Réglementations Radio, permettant ainsi aux administrations d'autoriser les stations embarquées sur avion ou sur bateau de communiquer avec les stations spatiales par des services mobiles dans la bande 1660-1660.5 MHz ;

recommande de façon pressante

que les administrations adhérant à l'Union Astronomique Internationale et à l'Union Internationale des Télécommunications aient présente à l'esprit, lors de la Conférence Administrative Mondiale des Radiocommunications (WARC 92), l'importance d'attribuer en premier au Service de la Radioastronomie la bande 1660.0-1660.5 MHz ;

et demande au Président

d'appuyer fortement la Commission Inter-union pour les Allocations des Bandes de Fréquences à la Radioastronomie et à la Recherche Spatiale (IUCAF) dans ses efforts en vue de porter cette résolution à l'attention des délégations participant à la WARC 92.

IUCAF   Commission Inter-Union pour les Allocations des Bandes de Fréquences pour la Radioastronomie & la Recherche Spatiale

WARC MOB   Conference Administrative Mondiale des Radiocommunications pour les services mobiles

Resolution A2: Revision of Frequency Bands for Astrophysically Significant Lines

The XXIst General Assembly of the International Astronomical Union,

recalling

a) resolutions passed by the International Astronomical Union in 1979 and 1982 recommending the provision by national administrations of frequency bands for the astrophysically most important spectral lines;

b) the need expressed in those resolutions to protect these frequency bands from in-band, band-edge and harmonic emissions, especially from space-borne transmitters;

c) the documentation of Study Group 7 of the CCIR in Recommendation 314 and Reports 224 and 697 concerning harmful interference to the Radio Astronomy Service;

and considering

the careful reviews by the International Astronomical Union in
the period 1983-1991 of the astrophysically most important spectral
lines;

recommends

that the International Astronomical Union take note of the revision
of the frequencies of the astrophysically most important spectral
lines listed in Tables 1 and 2 below;

and instructs the President

to bring the resolution to the attention of the General Secretary
of the International Telecommunication Union, and to support IUCAF
strongly in its efforts to bring this resolution to the attention
of delegations participating in WARC 92.

CCIR   *Consultative Committee on International Radiommunications*

IUCAF   *Inter-Union Commission on Frequency Allocations for Radio Astronomy and
            Space Science*

WARC   *World Administrative Radio Conference*

Résolution A2: Révision des bandes de fréquences pour les raies d'intérêt astrophysique

La XXIe Assemblée générale de l'Union Astronomique Internationale,

rappelant

a) les résolutions de l'Union Astronomique Internationale de
   1979, 1982 and 1988 recommandant la mise à disposition, par les
   administrations nationales, de bandes de fréquences pour les
   raies spectrales de plus grande importance en Astrophysique ;

b) la nécessité exprimée dans ces résolutions de protéger ces
   bandes de fréquences des émissions dans la bande, en bordure de
   bande et des harmoniques, en particulier celles provenant de
   transmetteurs spatiaux ;

c) la documentation du Groupe d'étude 7 du Comité Consultatif
   Radio International (CCIR) dans sa Recommandation 314 et les
   rapports 224 et 697 traitant des interférences nuisibles au
   Service de la Radioastronomie ;

et considérant

les revues faites avec grand soin par l'Union Astronomique
Internationale, au cours de la période 1983-1991, des raies
spectrales de plus grande importance en Astrophysique ;

<u>recommande</u>

que l'Union Astronomique Internationale prenne note de la révision de la liste des raies spectrales les plus importantes en astrophysique, telles que répertoriées dans les Tableaux 1 et 2 annexés ;

<u>et demande au Président</u>

de porter cette résolution à l'attention du Secrétaire Général de l'Union Internationale des Télécommunications et d'appuyer fortement la Commission Inter-Union pour les Allocations des Bandes de Fréquences pour la Radioastronomie et la Recherche Spatiale (IUCAF) dans ses efforts pour porter cette résolution à l'attention des délégations participant à la WARC 92.

CCIR   *Comité Consultatif Radio International*

IUCAF   *Commission Inter-Union pour les Allocations des Bandes de Fréquences pour la Radioastronomie & la Recherche Spatiale*

WARC   *Conférence Administrative Mondiale des Radiocommunications*

TABLE I

Radio-frequency lines of the greatest importance to radio astronomy at frequencies below 275 GHz

(Raies de fréquence radio d'importance majeure pour la Radioastronomie aux fréquences inférieures à 275 GHz)

Substance *Corps Composé*	Rest frequency *Fréquences au repos*	Suggested minimum band *Bande minimum suggérée*	Notes (1)
Deuterium (DI) *Deuterium*	327.384 MHz	327.0 - 327.7 MHz	
Hydrogen (HI) *Hydrogène*	1420.406 MHz	1370.0 - 1427.0 MHz	(2),(3)
Hydroxyl radical (OH) *Radical oxhydrile*	1612.231 MHz	1606.8 - 1613.8 MHz	(3),(4)
Hydroxyl radical (OH) *Radical oxhydrile*	1665.402 MHz	1659.8 - 1667.1 MHz	(4)
Hydroxyl radical (OH) *Radical oxhydrile*	1667.359 MHz	1661.8 - 1669.0 MHz	(4)
Hydroxyl radical (OH) *Radical oxhydrile*	1720.530 MHz	1714.8 - 1722.2 MHz	(3),(4)
Methyladyne (CH) *Méthyladyne*	3263.794 MHz	3252.9 - 3267.1 MHz	(3),(4)

Methyladyne (CH) *Méthyladyne*	3335.481 MHz	3324.4 - 3338.8 MHz	(3),(4)
Methyladyne (CH) *Méthyladyne*	3349.193 MHz	3338.0 - 3352.5 MHz	(3),(4)
Formaldehyde (H_2CO) *Formaldehyde*	4829.660 MHz	4813.6 - 4834.5 MHz	(3),(4)
Methanol (CH_2OH) *Méthanol*	6668.518 MHz	6661.8 - 6675.2 MHz	(3),(6)
Ionized helium isotope (^3HeII) *Isotope ionisé de l'Hélium^3*	8665.650 MHz	8660.0 - 8670.0 MHz	
Methanol (CH_3OH) *Méthanol*	12.178 GHz	12.17 - 12.19 GHz	(3),(6)
Formaldehyde (H_2CO) *Formaldéhyde*	14.488 GHz	14.44 - 14.50 GHz	(3),(4)
Cyclopropenylidene(C_3H_2) *Cyclopropénylidène*	18.343 GHz	18.28 - 18.36 GHz	(3),(4),(6)
Water vapour (H_2O) *Vapeur d'eau*	22.235 GHz	22.16 - 22.26 GHz	(3),(4)
Ammonia (NH_3) *Ammoniac*	23.694 GHz	23.61 - 23.71 GHz	(4)
Ammonia (NH_3) *Ammoniac*	23.723 GHz	23.64 - 23.74 GHz	(4)
Ammonia (NH_3) *Ammoniac*	23.870 GHz	23.79 - 23.89 GHz	(4)
Silicon monoxide (SiO) *Monoxyde de silicium*	42.821 GHz	42.77 - 42.86 GHz	
Silicon monoxide (SiO) *Monoxyde de silicium*	43.122 GHz	43.07 - 43.17 GHz	
Carbon monosulphide (CS) *Monosulfure de carbone*	48.991 GHz	48.94 - 49.04 GHz	
Deuterated formylium (DCO^+) *Formylium deutéré*	72.039 GHz	71.96 - 72.11 GHz	(3)
Silicon monoxide (SiO) *Monoxyde de silicium*	86.243 GHz	86.16 - 86.33 GHz	
Formylium ($H^{13}CO^+$) *Formylium*	86.754 GHz	86.66 - 86.84 GHz	
Silicon monoxide (SiO) *Monoxyde de silicium*	86.847 GHz	86.76 - 86.93 GHz	
Ethynyl radical (C_2H) *Radical Ethynil*	87.300 GHz	87.21 - 87.39 GHz	(5)
Hydrogen cyanide (HCN) *Cyanure d'hydrogène*	88.632 GHz	88.34 - 88.72 GHz	(4)

Formylium (HCO$^+$) *Formylium*	89.189 GHz	88.89 - 89.28 GHz	(4)
Hydrogen isocyanide (HNC) *Isocyanure d'hydrogène*	90.664 GHz	90.57 - 90.76 GHz	
Diazenylium (N$_2$H$^+$) *Diazénylium*	93.174 GHz	93.07 - 93.27 GHz	
Carbon monosulphide (CS) *Monosulfure de Carbone*	97.981 GHz	97.65 - 98.08 GHz	(4)
Carbon monoxide (C^{18}O) *Monoxyde de carbone*	109.782 GHz	109.67 - 109.89 GHz	
Carbon monoxide (^{13}CO) *Monoxyde de carbone*	110.201 GHz	109.83 - 110.31 GHz	(4)
Carbon monoxide (C^{17}O) *Monoxyde de carbone*	112.359 GHz	112.25 - 112.47 GHz	(6)
Carbon monoxide (CO) *Monoxyde de carbone*	115.271 GHz	114.88 - 115.39 GHz	(4)
Formaldehyde (H$_2$^{13}CO) *Formaldéhyde*	137.450 GHz	137.31 - 137.59 GHz	(3),(6)
Formaldehyde (H$_2$CO) *Formaldéhyde*	140.840 GHz	140.69 - 140.98 GHz	
Carbon monosulphide (CS) *Monosulfure de carbone*	146.969 GHz	146.82 - 147.12 GHz	
Water vapour (H$_2$O) *Vapeur d'eau*	183.310 GHz	183.12 - 183.50 GHz	
Carbon monoxide (C^{18}O) *Monoxyde de carbone*	219.560 GHz	219.34 - 219.78 GHz	
Carbon monoxide (^{13}CO) *Monoxyde de carbone*	220.399 GHz	219.67 - 220.62 GHz	(4)
Carbon monoxide (CO) *Monoxyde de carbone*	230.538 GHz	229.77 - 230.77 GHz	(4)
Carbon monosulphide (CS) *Monoxyde de carbone*	244.953 GHz	244.72 - 245.20 GHz	(6)
Hydrogen cyanide (HCN) *Cyanure d'hydrogène*	265.886 GHz	265.62 - 266.15 GHz	
Formylium (HCO$^+$) *Formylium*	267.557 GHz	267.29 - 267.83 GHz	
Hydrogen isocyanide (HNC) *Isocyanure d'hydrogène*	271.981 GHz	271.71 - 272.25 GHz	

(1)   If Note (4) or Note (2) are not listed, the band limits are the Doppler-shifted frequencies corresponding to radial velocities of ± 300 km/s (consistent with line radiation occurring in our galaxy).

        (Les limites des bandes, pour toutes les raies spectrales figurant dans ce Tableau à l'exception de celles qui portent la note (4) ou la note (2), sont les fréquences décalées par l'effet Doppler correspondant à des vitesses radiales de ± 300 km/s (compatible avec l'émission spectrale se produisant dans notre Galaxie).

(2)   An extension to lower frequency of the allocation of 1400-1427 MHz is required to allow for the higher Doppler shifts for HI observed in distant galaxies.

        (Une extension vers les basses fréquences de l'attribution de la bande 1400-1427 MHz est nécessaire afin de tenir compte des effets Doppler importants pour la raie de HI observée dans les galaxies éloignées).

(3)   The current international allocation is not primary and/or does not meet bandwidth requirements. See the Radio Regulations for more detailed information.

        (L'attribution internationale actuelle n'est pas une attribution principale et/ou ne répond pas aux besoins pour la largeur de bande. On trouvera à ce sujet des précisions dans le Règlement des radiocommunications).

(4)   Because these line frequencies are also being used for observing other galaxies, the listed bandwidths include Doppler shifts corresponding to radial velocities of up to 1000 km s^{-1}. It should be noted that HI has been observed at frequencies redshifted to 500 MHz, while some lines of the most abundant molecules have been detected in galaxies with velocities up to 50000 km s^{-1}, corresponding to a frequency reduction of up to 17%.

        (Ces raies spectrales étant utilisées également pour l'observation d'autres galaxies, les largeurs des bandes mentionnées ci-dessus tiennent compte des effets Doppler correspondant à des vitesses radiales allant jusqu'à 1000 km s^{-1}. Il est à noter que HI a été observé à des fréquences décalées vers le rouge jusqu'à 500 MHz, et que quelques raies spectrales des molécules les plus abondantes ont été détectées dans des galaxies ayant des vitesses allant jusqu'à 50000 km s^{-1}, ce qui correspond à une diminution de fréquence pouvant atteindre 17%.

(5)   There are six closely spaced lines associated with this molecule at this frequency. The listed band is wide enough to permit observations of all six lines.

        (Six raies spectrales très proches les unes des autres à cette fréquence sont associées à cette molécule. La bande indiquée est suffisamment large pour permettre d'observer toutes ces raies).

(6)   This line frequency is not mentioned in Article 8 of the Radio Regulations.

        (Cette raie spectrale n'est pas mentionnée par le Règlement des radiocommunications. Article 8).

TABLE II

Radio-frequency lines of the greatest importance to radio astronomy
at frequencies between 275 and 811 GHz

(not allocated in the Radio Regulations)

(Raies de fréquence radio d'importance majeure
pour la Radioastronomie
aux fréquences comprises entre 275 et 811 GHz)

(Dans le Règlement des radiocommunications, il n'existe aucune attribution sur ces
fréquences)

Substance Corps Composé	Rest frequency Fréquences au repos	Suggested minimum band Bande minimum suggérée
Diazenylium (N_2H^+) Diazénylium	279.511 GHz	279.23 - 279.79 GHz
Carbon monoxide ($C^{18}O$) Monoxyde de carbone	329.330 GHz	329.00 - 329.66 GHz
Carbon monoxide (^{13}CO) Monoxyde de carbone	330.587 GHz	330.25 - 330.92 GHz
Carbon monosulphide (CS) Monosulphure de carbone	342.883 GHz	342.54 - 343.23 GHz
Carbon monoxide (CO) Monoxyde de carbone	345.796 GHz	345.45 - 346.14 GHz
Hydrogen cyanide (HCN) Cyanure d'hydrogène	354.484 GHz	354.13 - 354.84 GHz
Formylium (HCO^+) Formylium	356.734 GHz	356.37 - 357.09 GHz
Diazenylium (N_2H^+) Diazénylium	372.672 GHz	372.30 - 373.05 GHz
Water vapour (H_2O) Vapeur d'eau	380.197 GHz	379.81 - 380.58 GHz
Carbon monoxide ($C^{18}O$) Monoxyde de carbone	439.088 GHz	438.64 - 439.53 GHz
Carbon monoxide (^{13}CO) Monoxyde de carbone	440.765 GHz	440.32 - 441.21 GHz
Carbon monoxide (CO) Monoxyde de carbone	461.041 GHz	460.57 - 461.51 GHz
Heavy water (HDO) Eau lourde	464.925 GHz	464.46 - 465.39 GHz

Carbon (CI) *Carbone*	492.162 GHz	491.66 - 492.66 GHz
Water vapour ($H_2^{18}O$) *Vapeur d'eau*	547.676 GHz	547.13 - 548.22 GHz
Water vapour (H_2O) *Vapeur d'eau*	556.936 GHz	556.37 - 557.50 GHz
Ammonia ($^{15}NH_3$) *Ammoniac*	572.113 GHz	571.54 - 572.69 GHz
Ammonia (NH_3) *Ammoniac*	572.498 GHz	571.92 - 573.07 GHz
Carbon monoxide (CO) *Monoxyde de carbone*	691.473 GHz	690.78 - 692.17 GHz
Hydrogen cyanide (HCN) *Cyanure d'hydrogène*	797.433 GHz	796.64 - 798.23 GHz
Formylium (HCO^+) *Formylium*	802.653 GHz	801.85 - 803.85 GHz
Carbon monoxide (CO) *Monoxyde de carbone*	806.652 GHz	805.85 - 807.46 GHz
Carbon (CI) *Carbone*	809.350 GHz	808.54 - 810.16 GHz

Resolution A3: Preservation of Radio Frequencies for Radio Astronomy

The XXIst General Assembly of the International Astronomical Union,

noting

a. the long-standing concern of the International Astronomical Union for protecting radio astronomy from interference, particularly through resolutions passed at the General Assemblies in 1979, 1982, 1985 and 1988;

b. the increasing levels of harmful interference to radio astronomy, particularly from space and airborne transmitters, which diminish the advantages of locating observatories at remote sites;

c. the particularly high levels of harmful interference experienced consistently in the sub-band 1610.6-1613.8 MHz from navigation satellites which make observations of an astrophysically important hydroxyl line increasingly difficult;

d. that the 1612 MHz hydroxyl line has assumed greatly increased importance since the 1979 World Administrative Radio Conference due particularly to the discovery of numerous OH/IR stars which have been used for absolute distance determination in the Galaxy and for understanding stellar evolution;

e. that the World Administrative Radio Conference for the Mobile Services (WARC MOB-87) has also allocated the band 1610-1626.5 MHz to the Radio-Determination Satellite Service (RDSS), subject to footnote 733E of the Radio Regulations, which states that in Regions 1 and 3 harmful interference shall not be caused to the Radio Astronomy Service (RAS), and that in Region 2 several administrations have agreed to limited protection for the RAS;

f. that the WARC MOB-87 in Resolution PLEN/1 has invited the CCIR to continue its studies in order to obtain more precise results concerning the conditions of sharing in the band 1610-1625.5 MHz between the RDSS on the one hand and the RAS, among other services, on the other;

<u>urges</u>

1. that administrations adhering to the International Astronomical Union and the International Telecommunication Union strive for improved protection of the RAS in the 1610.6-1613.8 MHz band by upgrading the allocation status of the RAS to that of primary service in this sub-band at WARC 92;

2. that national administrations cooperate with IUCAF to examine means to prevent harmful interference to observations in the band 1610.6-1613.8 MHz from global navigation satellite systems, particularly in designing changes to existing systems and planning new systems;

3. that IUCAF, representing the IAU, respond rapidly to the invitation to continue studying in Study Group 7 of the CCIR the conditions for successfully sharing the band 1610-1626.5 MHz;

4. that administrations operating satellites or satellite systems in the aeronautical navigation satellite service at 1.5/1.6 GHz frequencies protect the RAS from harmful interference by appropriately filtering unwanted emissions;

<u>and instructs the President</u>

to bring this Resolution to the attention of the Secretary General of the International Telecommunication Union, and to support IUCAF strongly in its efforts to bring this resolution to the attention of delegations participating in WARC 92.

CCIR   Consultative Committee on International Radiocommunications

IUCAF   Inter-Union Commission on Frequency Allocations for Radio Astronomy and Space Science

RAS   Radio Astronomy Service

RDSS   Radio Determination Satellite Service

WARC   World Administrative Radio Conference

Résolution A3: Préservation des fréquences radio pour la Radioastronomie

La XXIe Assemblée générale de l'Union Astronomique Internationale,

notant

a. l'intérêt manifesté de longue date par l'Union Astronomique Internationale pour la protection de la Radioastronomie des interférences, en particulier par les résolutions adoptées lors des Assemblées Générales de 1979, 1982, 1985 et 1988 ;

b. le niveau croissant des interférences nuisibles affectant la Radioastronomie, et en particulier celles provenant de transmetteurs spatiaux ou aéroportés, réduisant de ce fait l'intérêt de situer les observatoires dans des sites éloignés ;

c. le niveau particulièrement élevé des interférences nuisibles régulièrement observées dans la sous-bande 1610.6-1613.8 MHz provenant de satellites de navigation qui rendent de plus en plus difficiles les observations d'une raie importante en astrophysique de l'hydroxyle ;

d. que la raie de l'hydroxyle à 1612 MHz a pris un intérêt croissant depuis la Conférence Administrative Mondiale des Radiocommunications (WARC) de 1979 en raison, en particulier, de la découverte de nombreuses étoiles OH/IR qui ont été utilisées pour la détermination des distances absolues dans la Galaxie et pour la compréhension de l'évolution stellaire ;

e. que la WARC pour les services mobiles (WARC MOB-87) a également attribué la fréquence 1610-1626.5 MHz au Service de Détermination Radio par Satellite (RDSS), en application de la note 733E du Règlement des radiocommunications, laquelle précise que dans les Régions 1 et 3 aucune interférence nuisible ne doit être causée aux services de Radioastronomie (RAS), et que, dans la Région 2, plusieurs administrations ont accepté une protection minimale du RAS ;

f. que la WARC MOB-87 dans sa Résolution PLEN/1 a invité le Comité Consultatif Radio International (CCIR) à poursuivre ses études en vue d'obtenir des résultats plus précis sur les conditions de partage de la bande 1610-1625.5 MHz entre le RDSS d'une part, et le RAS, entre autres services, d'autre part ;

demande expressément

1. que les organisations adhérant à l'Union Astronomique Internationale et à l'Union Internationale des Télécommunications s'efforcent d'obtenir une protection accrue du Service de Radioastronomie dans la bande de fréquence 1610.6-1613.8 MHz en élevant la classe de l'allocation au Service de Radioastronomie à celle d'un service principal pour cette sous-bande lors de la WARC 92 ;

2. que les administrations nationales coopèrent avec l'Inter-Union
    Commission pour les Allocations des Bandes de Fréquence pour
    la Radioastronomie et la Recherche Spatiale (IUCAF) afin
    d'étudier les moyens d'éviter les interférences nuisibles aux
    observations dans la bande 1610.6-1613.8 MHz dues aux systèmes
    de satellites de navigation globale, particulièrement par la
    planification de la modification des systèmes existants et de
    la conception de nouveaux systèmes ;

3. que l'IUCAF, représentant l'Union Astronomique Internationale,
    réponde rapidement à l'invitation à poursuivre, dans le cadre
    du Groupe d'étude n° 7 du CCIR, l'étude des conditions d'un
    partage réussi de la bande 1610-1626.5 MHz ;

4. que les administrations exploitant des satellites ou des
    systèmes de satellites dans le service de satellites de
    navigation aéronautiques à des fréquences comprises entre 1.5
    et 1.6 GHz protègent des interférences nuisibles le Service de
    Radioastronomie par un filtrage adéquat des émissions
    indésirables ;

et demande au Président

de porter cette Résolution à l'attention du Secrétaire Général de
l'Union Internationale des télécommunications, et d'appuyer
fortement l'IUCAF dans ses démarches pour porter cette résolution à
l'attention des délégations participant à la WARC 92.

CCIR   Comité Consultatif Radio International

IUCAF   Commission pour les Allocations des Bandes de Fréquence pour
            la Radioastronomie et la Recherche Spatiale

RAS   Service de Radioastronomie

RDSS   Service de Détermination Radio par Satellite

WARC   Conférence Administrative Mondiale des Radiocommunications

Resolution A4: Recommendations from the Working Group on Reference Systems

Recommendations I to IX

The XXIst General Assembly of the International Astronomical Union,

RECOMMENDATION I

considering,

that it is appropriate to define several systems of space-time
coordinates within the framework of the General Theory of
Relativity,

recommends,

that the four space-time coordinates $(x^0 = ct, x^1, x^2, x^3)$ be selected in such a way that in each coordinate system centred at the barycentre of any ensemble of masses, the squared interval ds^2 be expressed with the minimum degree of approximation in the form:

$$ds^2 = -c^2 d\tau^2$$

$$= - (1 - \frac{2U}{c^2}) (dx^0)^2 + (1 + \frac{2U}{c^2}) [(dx^1)^2 + (dx^2)^2 + (dx^3)^2],$$

where c is the velocity of light, τ is proper time, and U is the sum of the gravitational potentials of the above mentioned ensemble of masses, and of a tidal potential generated by bodies external to the ensemble, the latter potential vanishing at the barycentre.

Notes for Recommendation I

1. This recommendation explicitly introduces The General Theory of Relativity as the theoretical background for the definition of the celestial space-time reference frame.

2. This recommendation recognizes that space-time cannot be described by a single coordinate system because a good choice of coordinate system may significantly facilitate the treatment of the problem at hand, and elucidate the meaning of the relevant physical events. Far from the space origin, the potential of the ensemble of masses to which the coordinate system pertains becomes negligible, while the potential of external bodies manifests itself only by tidal terms which vanish at the space origin.

3. The ds^2 as proposed gives only those terms required at the present level of observational accuracy. Higher order terms may be added as deemed necessary by users. If the IAU should find it generally necessary, more terms will be added. Such terms may be added without changing the rest of the recommendation.

4. The algebraic sign of the potential in the formula giving ds^2 is to be taken as positive.

5. At the level of approximation given in this recommendation, the tidal potential consists of all terms at least quadratic in the local space coordinates in the expansion of the Newtonian potential generated by external bodies.

RECOMMENDATION II

considering.

a) the need to define a barycentric coordinate system with spatial
    origin   at the centre   of   mass   of   the solar   system   and
    a geocentric   coordinate   system   with   spatial   origin
    at the centre of mass of the Earth, and the desirability of
    defining analogous coordinate systems for other planets and for
    the Moon,

b) that the coordinate systems should be related to the best
    realization of reference systems in space and time, and,

c) that the same physical units should be used in all coordinate
    systems,

recommends that,

1. the space coordinate grids with origins at the solar system
    barycentre and at the centre of mass of the Earth show no
    global rotation with respect to a set of distant extragalactic
    objects,

2. the time coordinates be derived from a time scale realized
    by atomic clocks operating on the Earth,

3. the basic physical units of space-time in all coordinate
    systems be the second of the International System of Units (SI)
    for proper time, and the SI meter for proper length, connected
    to the SI second by the value of the velocity of light c
    $= 299792458 \text{ ms}^{-1}$.

Notes for Recommendation II

1. *This recommendation gives the actual physical structures and quantities that
    will be used to establish the reference frames and time scales based
    upon the ideal definition of the system given by Recommendation I.*

2. *The kinematic constraint for the rate of rotation of both the geocentric and
    barycentric reference systems cannot be perfectly realized. It is assumed that
    the average   rotation of   a large   number   of   extragalactic   objects
    can be considered to represent the rotation of the universe which is assumed to
    be zero.*

3. *If the barycentric reference system as defined by this recommendation is used
    for studies of dynamics within the solar system, the kinematic effects of
    the galactic geodesic precession may have to be taken into account.*

4. *In addition, the kinematic constraint for the state of rotation of
    the geocentric reference system as defined by this recommendation implies that
    when the system is used for dynamics (e.g., motions of the Moon and Earth
    satellites), the time dependent geodesic precession of the geocentric frame*

relative to the barycentric frame must be taken into account by introducing corresponding inertial terms into the equations of motion.

5. *Astronomical constants and quantities are expressed in SI units without conversion factors depending upon the coordinate systems in which they are measured.*

RECOMMENDATION III

considering,

the desirability of the standardisation of the units and origins of coordinate times used in astronomy,

recommends that,

1. the units of measurement of the coordinate times of all coordinate systems centred at the barycentres of ensembles of masses be chosen so that they are consistent with the proper unit of time, the SI second,

2. the reading of these coordinate times be 1977 January 1, $0^h \ 0^m \ 32.184^s$ exactly, on 1977 January 1, $0^h \ 0^m \ 0^s$ TAI exactly (JD = 2443144.5, TAI), at the geocentre,

3. coordinate times in coordinate systems having their spatial origins respectively at the centre of mass of the Earth and at the solar system barycentre, and established in conformity with the above sections (1) and (2), be designated as Geocentric Coordinate Time (TCG) and Barycentric Coordinate Time (TCB).

Notes for Recommendation III

1. *In the domain common to any two coordinate systems, the tensor transformation law applied to the metric tensor is valid without re-scaling the unit of time. Therefore, the various coordinate times under consideration exhibit secular differences. Recommendation 5 (1976) of IAU Commissions 4, 8 and 31, completed by Recommendation 5 (1979) of IAU Commissions 4, 19 and 31, stated that Terrestrial Dynamical Time (TDT) and Barycentric Dynamical Time (TDB) should differ only by periodic variations. Therefore, TDB and TCB differ in rate. The relationship between these time scales in seconds is given by:*

$$TCB - TDB = L_B \times (JD - 2443144.5) \times 86400.$$

The present estimate of the value of L_B is 1.550505×10^{-8} ($\pm 1 \times 10^{-14}$) (Fukushima et al., Celestial Mechanics, 38, 215, 1986).

2. *The relation TCB -TCG involves a full 4-dimensional transformation*

$$TCB - TCG = c^{-2} [\int_{t_o}^{t} (v_e^2/2 + U_{ext} (x_e))dt + v_e \cdot (x - x_e)],$$

x_e and v_e denoting the barycentric position and velocity of the Earth's centre of mass and x the barycentric position of the observer. The external potential U_{ext} is the Newtonian potential of all solar system bodies apart from the Earth. The external potential must be evaluated at the geocentre. In the integral, t = TCB and to is chosen to agree with the epoch of Note 3. As an approximation to TCB -TCG in seconds one might use:

$$TCB - TCG = L_C \times (JD -2443144.5) \times 86400 + c^{-2} v_e \cdot (x - x_e) + P.$$

The present estimate of the value of L_C is 1.480813 $\times 10^{-8}$ ($\pm 1 \times 10^{-14}$) (Fukushima et al., Celestial Mechanics, 38, 215, 1986). It may be written as [3GM/2c^{2}a] + $\in$ where G is the gravitational constant, M is the mass of the Sun, a is the mean heliocentric distance of the Earth, and $\in$ is a very small term (of order 2 $\times 10^{-12}$) arising from the average potential of the planets at the Earth. The quantity P represents the periodic terms which can be evaluated using the analytical formula by Hirayama et al., ("Analytical Expression of TDB-TDT$_o$", in Proceedings of the IAG Symposia, IUGG XIX General Assembly, Vancouver, August 10-22, 1987). For observers on the surface of the Earth, the terms depending upon their terrestrial coordinates are diurnal, with a maximum amplitude of 2.1 µs.

3. *The origins of coordinate times have been arbitrarily set so that these times all coincide with the Terrestrial Time (TT) of Recommendation IV at the geocentre on 1977 January 1, 0^h 0^m 0^s TAI. (See Note 3 of Recommendation IV.)*

4. *When realizations of TCB and TCG are needed, it is suggested that these realizations be designated by expressions such as TCB(xxx), where xxx indicates the source of the realized time scale (e.g., TAI) and the theory used for the transformation into TCB or TCG.*

RECOMMENDATION IV

considering,

a) that the time scales used for dating events observed from the surface of the Earth and for terrestrial metrology should have as the unit of measurement the SI second, as realized by terrestrial time standards,

b) the definition of the International Atomic Time, TAI, approved by the 14th Conférence Générale des Poids et Mesures (1971) and completed by a declaration of the 9th session of the Comité Consultatif pour la Définition de la Seconde (1980), recommends that,

1) the time reference for apparent geocentric ephemerides be Terrestrial Time, TT,

2) TT be a time scale differing from TCG of Recommendation III by a constant rate, the unit of measurement of TT being chosen so that it agrees with the SI second on the geoid,

3) at instant 1977 January 1, $0^h 0^m 0^s$ TAI exactly, TT have the reading 1977 January 1, $0^h 0^m 32.184^s$ exactly.

Notes for Recommendation IV

1. *The basis of the measurement of time on the Earth is International Atomic Time (TAI) which is made available by the dissemination of corrections to be added to the readings of national time scales and clocks.   The time scale TAI was defined by the 59th session of the Comité International des Poids et Mesures (1970) and approved by the 14th Conférence Générale des Poids et Mesures (1971) as a realized time scale.   As the errors in the realization of TAI are not always negligible, it has been found necessary to define an ideal form of TAI, apart from the 32.184^s offset, now designated Terrestrial Time, TT.*

2. *The time scale TAI is established and disseminated according to the principle of coordinate synchronization, in the geocentric coordinate sytem, as explained in CCDS, 9th Session (1980) and in Reports of the CCIR, 1990, annex to Volume VII (1990).*

3. *In order to define TT it is necessary to define the coordinate system precisely, by the metric form, to which it belongs.   To be consistent with the uncertainties of the frequency of the best standards, it is at present (1991) sufficient to use the relativistic metric given in Recommendation I.*

4. *For ensuring an approximate continuity with the previous time arguments of ephemerides, Ephemeris Time, ET, a time offset is introduced so that TT-TAI = 32.184^s exactly at 1977 January 1, 0^h TAI.   This date corresponds to the implementation of a steering process of the TAI frequency, introduced so that the TAI unit of measurement remains in close agreement with the best realizations of the SI second on the geoid.   TT can be considered as equivalent to TDT as defined by IAU Recommendation 5 (1976) of Commissions 4, 8 and 31, and Recommendation 5 (1979) of Commissions 4, 19 and 31.*

5. *The divergence between TAI and TT is a consequence of the physical defects of atomic time standards.   In the interval 1977-1990, in addition to the constant offset of 32.184^s, the deviation probably remained within the approximate limits of ± 10µs.   It is expected to increase more slowly in the future as a consequence of improvements in time standards.   In many cases, especially for the publication of ephemerides, this deviation is negligible.   In such cases, it can be stated that the argument of the ephemerides is TAI + 32.184^s.*

6. Terrestrial Time differs from TCG of Recommendation III by a scaling factor, in seconds:

$$TCG - TT = L_G \times (JD -2443144.5) \times 86400.$$

The present estimate of the value of L_G is 6.969291×10^{-10} ($\pm 3 \times 10^{-16}$). The numerical value is derived from the latest estimate of gravitational potential on the geoid, $W = 62636860$ (± 30) m^2/s^2 (Chovitz, Bulletin Géodesique, 62, 359, 1988). The two time scales are distinguished by different names to avoid scaling errors. The relationship between L_B and L_C of Recommendation III, notes 1 and 2, and L_G is, $L_B = L_C + L_G$.

7. The unit of measurement of TT is the SI second on the geoid. The usual multiples, such as the TT day of 86400 SI seconds on the geoid and the TT Julian century of 36525 TT days, can be used provided that the reference to TT be clearly indicated whenever ambiguity may arise. Corresponding time intervals of TAI are in agreement with the TT intervals within the uncertainties of the primary atomic standards (e.g., within $\pm 2 \times 10^{-14}$ in relative value during 1990).

8. Markers of the TT scale can follow any date system based upon the second, e.g., the usual calendar date or the Julian Date, provided that the reference to TT be clearly indicated whenever ambiguity may arise.

9. It is suggested that realizations of TT be designated by TT(xxx) where xxx is an identifier. In most cases a convenient approximation is:

$$TT(TAI) = TAI + 32.184^S.$$

However, in some applications it may be advantageous to use other realizations. The BIPM, for example, has issued time scales such as TT(BIPM90).

RECOMMENDATION V

considering,

that important work has already been performed using Barycentric Dynamical Time (TDB), defined by IAU Recommendation 5 (1976) of IAU Commissions 4, 8 and 31, and Recommendation 5 (1979) of IAU Commissions 4, 19 and 31,

recognizes,

that where discontinuity with previous work is deemed to be undesirable, TDB may be used.

Note to Recommendation V

Some astronomical constants and quantities have different numerical values depending upon the use of TDB or TCB. When giving these values, the time scale used must be specified.

CHAPTER II

RECOMMENDATION VI

considering,

the desirability of implementing a conventional celestial barycentric reference system based upon the observed positions of extragalactic objects, and,

noting,

the existence of tentative reference frames constructed by various institutions and combined by the International Earth Rotation Service (IERS) into a frame used for Earth rotation series,

recommends,

1. that intercomparisons of these frames be extensively made in order to assess their systematic differences and accuracy,

2. that an IAU Working Group consisting of members of Commissions 4, 8, 19, 24, 31 and 40, the IERS, and other pertinent experts, in consultation with all the institutions producing catalogues of extragalactic radio sources, establish a list of candidates for primary sources defining the new conventional reference frame, together with a list of secondary sources that may later be added to or replace some of the primary sources, and,

requests,

1. that such a list be presented to the XXIInd General Assembly (1994) as a part of the definition of a new conventional reference system,

2. that the objects in this list be systematically observed by all VLBI and other appropriate astrometric programmes.

Note for Recommendation VI

This recommendation essentially describes the first part of the work that must be done to prepare the realization of the reference system defined by Recommendations I and II. The choice of objects must be made in the first place by considering their observability by VLBI, but special care should be taken to include a large proportion of extragalactic radio sources with well identified optical counterparts.

RECOMMENDATION VII

considering,

a) that the new conventional celestial barycentric reference frame
    should be as close as possible to the existing FK5 equator and
    equinox and the dynamical equinox which are referred to
    J2000.0,

b) that it should be accessible to astrometry in visual as well
    as in radio wavelengths,

recommends,

1. that the principal plane of the new conventional celestial
    reference system be as near as possible to the mean equator
    at J2000.0 and that the origin in this principal plane
    be as near as possible to the dynamical equinox of J2000.0,

2. that the positions of the extragalactic objects selected in
    accordance with Recommendation VI and representing
    the reference frame be computed initially for the equator and
    equinox J2000.0 using the best available values of
    the celestial pole offset with respect to the IAU expressions
    for precession and nutation,

3. that a great effort be made to compare reference frames of all
    types, in particular the FK5, solar system and extragalactic
    reference frames,

4. that observing programmes be undertaken or continued in order
    to relate planetary positions to radio and optical objects, and
    to determine the relationship between catalogues of
    extragalactic source positions and the best catalogues of star
    positions, in particular the FK5 and Hipparcos catalogues.

Notes for Recommendation VII

1. This recommendation specifies the choice of the coordinate axes that will
    be adopted in the final reference frame and describes the work to be done
    before such a frame can be constructed. Although the considerations call for
    visual and radio wavelengths for the primary catalogue, other observable
    wavelengths are not excluded. Positions of objects observed in other
    wavelengths should also be referred to the same system.

2. The objective set by this recommendation is that there should be no
    discontinuity in the positions of stars when the present FK5 frame is replaced
    by the extragalactic reference frame. This means that the position of
    the extragalactic objects should be in the FK5 system for J2000.0. It is
    acknowledged that the best values of precession and nutation must be used in
    order to avoid introducing spurious proper motions into the positions of

extragalactic objects.   The final transfer to the preferred equinox and principal plane will be done by applying a rotation at J2000.0.

3. *The dynamical equinox in this recommendation is defined as the intersection of the mean equator and the ecliptic.   The latter is defined as the uniformly rotating plane of the orbit of the Earth-Moon barycentre averaged over the entire period for which the ephemerides are valid. Since it is ephemeris dependent, the choice of the equinoctial point will be made using the most accurate and generally available ephemerides of the solar system at the time.*

4. *The definition given to the reference system by Recommendations I and II implies the stability in time of the system of coordinates realized by the celestial reference frame. The directions of the coordinate axes should not be changed even if at some later date the realizations of the dynamical equinox or the celestrial ephemeris pole are improved.   Similarly, modifications to the set of extragalactic objects realizing the reference system should be made in such a way that the directions of the axes are not changed.   This means that once the coordinate axes have been specified, in the way described in the first part of the recommendation, the connection between the definition of the conventional reference system and the peculiarities of the Earth's kinematics will have been severed.*

5. *As long as the relationship between the optical and the extragalactic radio frames is not sufficiently accurately determined, the FK5 talogue shall be considered as a provisional realization of the celestial reference system in optical wavelengths.*

RECOMMENDATION VIII

<u>recognizing</u>,

a) the importance to astronomy of adopting conventional values of astronomical and physical constants,

b) that values of these constants should be unchanged unless they differ significantly from their latest estimates,

c) that estimates of these constants should be improved frequently to represent the current status of knowledge,

d) the necessity of providing standard procedures using these numerical values, and,

<u>noting</u>,

a) that the MERIT Standards and IERS Standards have contributed significantly to the progress of astronomy and geodesy,

b) that numerical values in these standards have served as a system of constants in analyzing observations of high quality, and

<u>considering</u>,

that procedures in these standards do not cover the whole of
fundamental astronomy,

<u>recommends</u>,

that a permanent working group be organized by Commissions 4, 5,
8, 19, 24 and 31, in consultation with the IAG and the IERS, in
order to update and improve the system of astronomical units and
constants, the list of estimates of fundamental astronomical
quantities and standard procedures; this group shall:

1. prepare a draft report on the system of astronomical units and
   constants at least six months before the XXIInd General
   Assembly (1994),

2. prepare a draft list of best estimates of astronomical
   quantities at least six months before each following General
   Assembly,

3. prepare, at least six months before each following General
   Assembly, a draft report on standard procedures needed in
   fundamental astronomy, which,

   a) should have a maximum degree of compatibility with the IERS
      Standards,

   b) should include the implementations of procedures in the form
      of tested software and/or test cases,

   c) should be available not only in written form, but also in
      machine-readable form,

4. prepare a draft report on possible electronic access to these
   units, constants, quantities and procedures at least six months
   before the XXIInd General Assembly (1994).

RECOMMENDATION IX

<u>recognizing</u>,

that a generally accepted non-rigid Earth theory of nutation,
including all known effects at the one tenth milliarcsecond level,
is not yet available,

<u>recommends</u>,

1. that those satisfied with accuracy of the nutation angles (ϵ or
   $\psi\ sin\epsilon_0$) numerically greater than $\pm$ 0.002" (one sigma rms) may
   continue to use the 1980 IAU Nutation Theory (P.K. Seidelmann,
   Celestial Mechanics, 27, 79, 1982),

2. that those requiring values of the nutation angles more
   accurate than ± 0.002" (one sigma rms) should make use of
   the Bulletins of the IERS which publish observations and
   predictions of the celestial pole offsets accurate to about
   ± 0.0006" (one sigma rms) for a period of up to six months in
   advance,

3. that the IUGG be encouraged to develop and adopt an appropriate
   Earth model to be used as the basis for a new IAU Theory of
   Nutation.

CCIR   *Consultative Committee on International Radiocommunications*

IAG   *International Association of Geodesy*

IUGG   *International Union of Geodesy & Geophysics*

IERS   *International Earth Rotation Service*

Résolution A4: Recommandations du Groupe de travail sur les Systèmes de référence

Recommandations I à IX

La XXIe Assemblée générale de l'Union Astronomique Internationale,

RECOMMANDATION I

considérant

qu'il convient de définir, dans le cadre de la théorie de
la Relativité générale, plusieurs systèmes de coordonnées spatio-
temporelles,

recommande

que les coordonnées spatio-temporelles (x^0 = ct, x^1, x^2, x^3) soient
choisies de telle façon que dans chaque système de coordonnées,
centré au barycentre de tout ensemble de masses, le carré ds^2 de
l'intervalle soit exprimé au plus faible niveau d'approximation
sous la forme :

$$ds^2 = -c^2 d\tau^2$$

$$= -(1 - \frac{2U}{c^2})(dx^0)^2 + (1 + \frac{2U}{c^2})[(dx^1)^2 + (dx^2)^2 + (dx^3)^2],$$

où c est la vitesse de la lumière, τ le temps propre et U la somme
des potentiels de gravitation de l'ensemble de masses considéré et
d'un potentiel de marée engendré par les corps extérieurs à cet

ensemble, ce potentiel étant écrit de façon à s'annuler au barycentre.

Notes pour la recommandation I

1. *Cette recommandation introduit explicitement la théorie de la Relativité générale comme base pour la définition du repère de référence spatio-temporel céleste.*

2. *Cette recommandation reconnaît que l'espace-temps ne peut pas être décrit par un système de coordonnées unique parce qu'un choix judicieux d'un système de coordonnées peut faciliter de façon significative la résolution des problèmes et clarifier la signification physique des phénomènes qui s'y rapportent. Lorsqu'on se trouve loin de l'origine spatiale, le potentiel de l'ensemble de masses auquel appartient le système de coordonnées devient négligeable alors que le potentiel des corps extérieurs se manifeste seulement par des termes de marées qui disparaissent à l'origine.*

3. *Le ds^2 proposé ne comprend que les termes nécessaires au niveau actuel de la précision des observations. Des termes d'ordre plus élevé peuvent être ajoutés si les utilisateurs le jugent utile. Si l'UAI l'estime nécessaire de façon générale, on ajoutera d'autres termes. Cette addition pourra être faite sans changer le reste de la recommandation.*

4. *Le potentiel dans la formule donnant le ds^2 doit être défini avec le signe plus.*

5. *Au niveau de l'approximation qu'implique cette recommandation, le potentiel de marée comprend tous les termes du deuxième degré au moins par rapport aux coordonnées spatiales locales dans le développement du potentiel newtonien créé par les corps extérieurs.*

RECOMMANDATION II

considérant

a) la nécessité de définir un système de coordonnées barycentriques ayant pour origine spatiale le centre de masse du système solaire et un système de coordonnées géocentriques ayant pour origine spatiale le centre de masse de la Terre, ainsi que l'avantage qu'il y aurait à définir les systèmes de coordonnées analogues pour d'autres planètes et pour la Lune,

b) que les systèmes de coordonnées devraient correspondre aux meilleures réalisations des systèmes de référence spatiaux et temporels,

c) que les mêmes unités physiques devraient être utilisées dans tous les systèmes de coordonnées,

<u>recommande que</u>

1. les réseaux de coordonnées spatiales ayant pour origine le barycentre du système solaire et le centre de masse de la Terre ne présentent pas de rotation globale par rapport à un ensemble d'objets extragalactiques éloignés,

2. les coordonnées temporelles dérivent d'une échelle de temps construite en utilisant des horloges atomiques en fonctionnement sur la Terre,

3. les unités physiques de base pour l'espace-temps dans tous les systèmes de coordonnées soient la seconde du Système international d'unités (SI) pour le temps propre et le mètre SI pour les longueurs propres, lié à la seconde SI par la valeur de la vitesse de la lumière $c = 299792458$ ms^{-1}.

Notes pour la recommandation II

1. *Cette recommandation indique les structures et les quantités physiques qui seront utilisées pour construire les repères de référence et les échelles de temps basés sur la définition idéale du système donnée par la recommandation I.*

2. *La contrainte cinématique relative à la rotation des systèmes de référence géocentrique et barycentrique ne peut être réalisée de façon parfaite. On fait l'hypothèse que la rotation moyenne d'un grand nombre d'objets extragalactiques peut être considérée comme représentant la rotation de l'Univers que l'on admet être nulle.*

3. *Si le système de référence barycentrique, tel qu'il est défini par cette recommandation, est utilisé dans des études de dynamique dans le système solaire, les effets cinématiques de la précession géodésique d'origine galactique peuvent devoir être pris en compte.*

4. *De plus, la contrainte cinématique relative à l'état de rotation du système de référence géocentrique, tel qu'il est défini par cette recommandation, implique que si on effectue des études de dynamique dans ce système (par exemple les mouvements de la Lune ou des satellites de la Terre), les effets variables avec le temps de la précession géodésique du repère géocentrique par rapport au repère barycentrique doivent être pris en compte en introduisant les termes d'inertie correspondants dans les équations du mouvement.*

5. *Les constantes et grandeurs astronomiques seront exprimées en unités du Système international (SI) sans facteurs de conversion qui dépendraient des systèmes de coordonnées dans lesquels elles sont mesurées.*

RECOMMANDATION III

considérant

qu'il est souhaitable de normaliser les unités et les origines des temps-coordonnées utilisés en astronomie,

recommande que

1. les unités d'échelle des temps-coordonnées de tous les systèmes de coordonnées centrés au barycentre d'ensembles de masses soient choisies de sorte qu'elles soient toutes compatibles avec l'unité de temps propre, la seconde du SI,

2. les lectures de ces temps-coordonnées soient 1977 janvier 1, $0^h \ 0^m \ 32,184^s$ exactement pour 1977 janvier 1, $0^h \ 0^m \ 0^s$ TAI exactement (JD = 2443144,5, TAI), au géocentre,

3. les temps-coordonnées dans les systèmes de coordonnées qui ont leur origine spatiale respectivement au centre de masse de la Terre et au barycentre du système solaire et qui sont établis conformément aux sections 1. et 2. ci-dessus soient désignés par Temps-coordonnée géocentrique (TCG) et Temps-coordonnée barycentrique (TCB).

Notes sur la recommandation III

1. *Dans le domaine commun à deux systèmes de coordonnées quelconques, la loi de transformation tensorielle appliquée au tenseur métrique est valable sans modification supplémentaire de l'unité de temps. En conséquence, la différence des temps-coordonnées de ces systèmes présente une variation séculaire. La Recommandation 5 (1976) des commissions de l'UAI 4, 8 et 31, complétée par la Recommandation 5 (1979) des commissions de l'UAI 4, 19 et 31, spécifie que le Temps dynamique terrestre (TDT) et le Temps dynamique barycentrique (TDB) ne doivent différer que par des variations périodiques. Il en résulte que TDB et TCB ont une différence de marche. La relation entre ces échelles de temps, en secondes, est donnée par :*

$$TCB - TDB = L_B \times (JD - 2443144,5) \times 86400.$$

La valeur actuellement estimée de L_B est $1,550505 \times 10^{-8}$ ($\pm 1 \times 10^{-14}$) (Fukushima et al., Celestial Mechanics, 38, 215, 1986).

2. *La relation TCB - TCG exige une transformation quadri-dimensionnelle complète :*

$$TCB - TCG = c^{-2} \ [\int_{t_o}^{t} (v_e^2/2 + U_{ext} \ (x_e))dt + v_e \cdot (x - x_e)],$$

x_e et v_e désignant la position et la vitesse barycentriques du centre de masse de la Terre et x la position barycentrique de l'observateur. Le potentiel extérieur U_{ext} est le potentiel newtonien de tous les corps du système solaire, sauf la Terre. Dans l'intégrale, t = TCB et t_o est choisi pour être en accord avec les origines spécifiées par la note 3. Comme approximation de TCB - TCG, exprimé en secondes, on peut utiliser :

$$TCB - TCG = L_C \times (JD - 2443144.5) \times 86400 + c^{-2}v_e \cdot (x - x_e) + P.$$

La valeur actuellement estimée de L_C est $1,480813 \times 10^{-8}$ ($\pm 1 \times 10^{-14}$) (Fukushima et al., Celestial Mechanics, 38, 215, 1986). L_C peut être exprimé par $[3GM/2c^2a] + \in$ où G est la constante de la gravitation, M est la masse du Soleil, a est la distance héliocentrique moyenne de la Terre, et où $\in$ est un très petit terme (de l'ordre de 2×10^{-12}) provenant du potentiel des planètes au niveau de la Terre. La quantité P représente les termes périodiques qui peuvent être évalués en utilisant la formule analytique de Hirayama et al. ("Analytical Expression of TDB-TDT$_0$", in Proceedings of the IAG Symposia, UGGI XIXe Assemblée générale, Vancouver, 10-22 août, 1987). Pour des observateurs sur la surface de la Terre, les termes dépendant de leurs coordonnées terrestres sont diurnes, avec une amplitude maximale de 2,1 µs.

3. Les origines des temps-coordonnées ont été arbitrairement fixées de sorte que ces temps coïncident tous avec le Temps terrestre (TT) de la Recommandation IV, au géocentre, pour 1977 janvier 1, 0^h 0^m 0^s (Voir note 4 de la Recommandation IV).

4. Quand des réalisations de TCB et TCG sont nécessaires, il est suggéré que ces réalisations soient désignées par des expressions telles que TCB(xxx), où xxx indique la source de l'échelle de temps réalisée (par exemple TAI) et la théorie utilisée pour la transformation en TCB ou en TCG.

RECOMMANDATION IV

considérant

a) que les échelles de temps utilisées pour dater les événements observés depuis la surface de la Terre ainsi que pour la métrologie terrestre doivent avoir comme unité d'échelle la seconde du SI, telle qu'elle est réalisée par des étalons terrestres de temps,

b) la définition du Temps atomique international (TAI), approuvée par la 14e Conférence générale des poids et mesures (1971) et complétée par une déclaration de la 9e session du Comité consultatif pour la définition de la seconde (1980),

recommande que

1) la référence temporelle pour les éphémérides apparentes géocentriques soit le Temps terrestre (TT),

2) TT soit une échelle de temps différant du TCG de la Recommandation III par une marche constante, l'unité d'échelle de TT étant choisie de sorte qu'elle s'accorde avec la seconde du SI sur le géoïde,

3) à l'instant 1977 janvier 1, 0^h 0^m 0^s TAI exactement, la lecture de TT soit 1977 janvier 1, 0^h 0^m $32,184^s$ exactement.

Notes sur la Recommandation IV

1. La base de la mesure du temps sur la Terre est le Temps atomique international
    (TAI) qui est mis à la disposition de ses utilisateurs par la publication de
    corrections à ajouter aux lectures des échelles de temps et horloges
    nationales. L'échelle de temps TAI a été définie par la 59e session du Comité
    international des poids et mesures (1970) et approuvée par la 14e Conférence
    générale des poids et mesures (1971) comme une échelle de temps réalisée. Comme
    les erreurs dans la réalisation du TAI ne sont pas toujours négligeables, on
    a jugé nécessaire de définir une forme idéale du TAI, mis à part le décalage de
    32,184s, qui est maintenant désignée par Temps terrestre, TT.

2. L'échelle de temps TAI est établie et disséminée suivant le principe de
    la synchronisation coordonnée, dans le système de coordonnées géocentrique,
    comme cela est expliqué dans les documents CCDS, 9e session (1980) et Rapports
    du CCIR, 1990, annexe au Volume VII (1990).

3. Afin de définir TT, il est nécessaire de définir précisément le système de
    coordonnées auquel il appartient en donnant sa métrique relativiste. Compte
    tenu des incertitudes en fréquence des meilleurs étalons, il suffit à présent
    (1991) d'employer la métrique donnée dans la Recommandation I.

4. Pour assurer une continuité approximative avec l'argument temporel précédemment
    utilisé pour les éphémérides, le Temps des éphémérides TE, un décalage de temps
    est introduit de sorte que TT - TAI = $32,184^S$ exactement pour 1977 janvier 1,
    0^h TAI. Cette date correspond à la mise en pratique d'un pilotage de
    la fréquence du TAI, introduit pour que l'unité d'échelle du TAI reste en
    accord étroit avec les meilleures réalisations de la seconde du SI sur le
    géoïde. On peut considérer que TT est équivalent au TDT défini par
    la Recommandation 5 (1976) des commissions de l'UAI 4, 8 et 31 et par
    la Recommandation 5 (1979) des commissions de l'UAI 4, 19 et 31.

5. La divergence entre TAI et TT est une conséquence des défauts physiques des
    étalons atomiques de temps. Dans l'intervalle 1977-1990, outre le décalage
    constant de $32,184^S$, l'écart entre TAI et TT est probablement resté entre les
    limites approximatives de ± 10µs. On espère que cet écart s'accroîtra plus
    lentement à l'avenir, par suite de l'amélioration des étalons de temps. Dans
    bien des cas, en particulier pour la publication d'éphémérides, cet écart est
    négligeable. Dans ces cas, on peut déclarer que l'argument des éphémérides est
    $TAI + 32,184^S$.

6. Le Temps terrestre TT diffère du TCG de la Recommandation III par un facteur
    d'échelle; on a, en secondes:

    $$TCG - TT = L_G \times (JD - 2443144,5) \times 86400.$$

    La valeur actuellement estimée de L_G est $6,969291 \times 10^{-10}$ (± 3×10^{-16}). Cette
    valeur numérique est déduite de la dernière estimation du potentiel
    gravitationnel sur le géoïde, $W = 62636860$ (± 30)m^2/s^2 (Chovitz, Bulletin
    Géodésique, 62, 359, 1988). Les deux échelles de temps sont distinguées par
    des noms différents afin d'éviter les erreurs de facteur d'échelle. La relation

entre les quantités L_B et L_C de la Recommandation III, notes 1 et 2, et L_G est
$L_B = L_C + L_G$.

7. *L'unité d'échelle de TT est la seconde du SI sur le géoïde. Les multiples usuels, tels que le jour de TT de 86400 secondes du SI sur le géoïde et le siècle julien de TT de 36525 jours de TT, peuvent être employés, pourvu que la référence au TT soit clairement indiquée chaque fois qu'il peut y avoir ambigüité. Les intervalles d'échelle correspondants de TAI et de TT ont des durées qui s'accordent dans la limite des incertitudes des étalons atomiques primaires (par exemple à moins de $\pm 2 \times 10^{-14}$ en valeur relative, en 1990).*

8. *Les repères de l'échelle TT peuvent suivre n'importe quel système de datation basé sur la seconde, par exemple la date du calendrier habituelle ou la Date julienne, pourvu que la référence au TT soit clairement indiquée chaque fois qu'il peut y avoir ambigüité.*

9. *Il est suggéré que les réalisations de TT soient désignées par TT(xxx) où xxx est un identificateur. Dans la plupart des cas une approximation convenable est:*

$$TT(TAI) = TAI + 32,184^S.$$

Cependant, dans certaines applications, il peut être avantageux d'utiliser d'autres réalisations. Le BIPM, par exemple, a produit des échelles de temps telles que TT(BIPM90).

RECOMMANDATION V

considérant

que des travaux importants ont déjà été réalisés en employant le Temps dynamique barycentrique (TDB), défini par la Recommandation 5 (1976) des commissions de l'UAI 4, 8 et 31 et la Recommandation 5 (1979) des commissions de l'UAI 4, 19 et 31,

reconnaît

que lorsqu'une discontinuité avec les travaux antérieurs est jugée indésirable, TDB peut être utilisé.

Note sur la Recommandation V

Certaines grandeurs et constantes astronomiques ont des valeurs numériques qui dépendent de l'usage de TDB ou de TCB. Quand on donne ces valeurs, l'échelle de temps employée doit être spécifiée.

RECOMMANDATION VI

considérant

qu'il est souhaitable de réaliser un système de référence céleste barycentrique conventionnel basé sur les positions observées d'objets extragalactiques et

notant

l'existence de repères de référence expérimentaux, construits par divers établissements et dont la combinaison établie par le Service international de la rotation terrestre (IERS) constitue un repère utilisé pour décrire la rotation de la Terre,

recommande

1. qu'on effectue de façon extensive des comparaisons entre ces repères afin d'établir leurs différences systématiques et leur exactitude,

2. qu'un groupe de travail de l'UAI, comprenant des membres des Commissions 4, 8, 19, 24, 31, 40 et de l'IERS ainsi que d'autres experts, en consultation avec tous les instituts produisant des catalogues de radio-sources extragalactiques, établisse une liste de sources primaires définissant le nouveau repère conventionnel de référence ainsi qu'une liste de sources secondaires qui pourraient ultérieurement être ajoutées, ou remplacer certaines sources primaires,

et demande

1. que cette liste soit présentée à la XXIIe Assemblée générale de l'UAI en 1994 à titre de contribution à la définition du nouveau système conventionnel de référence,

2. que les objets de cette liste soient systématiquement observés en radio-interférométrie à longue base et par d'autres programmes astrométriques appropriés.

Note pour la Recommandation VI

Cette recommandation décrit essentiellement la première partie du travail qui doit être effectué pour préparer la réalisation du système de référence défini par les recommandations I et II. Le choix des objets de référence doit être fait en tenant compte en premier lieu de leur observabilité par la radio-interférométrie à longue base, mais on prendra soin d'inclure une proportion importante de radio-sources extragalactiques ayant des contreparties optiques bien identifiées.

RECOMMANDATION VII

<u>considérant</u>

a) que le nouveau repère barycentrique céleste conventionnel de référence devrait se rapprocher autant que possible des équateur et équinoxe FK5 existants ainsi que de l'équinoxe dynamique, rapportés à l'époque J2000.0,

b) qu'il devrait être accessible aux instruments astrométriques aussi bien en lumière visible qu'en ondes radio,

<u>recommande</u>

1. que le plan principal du nouveau système céleste conventionnel de référence soit aussi proche que possible de l'équateur moyen de J2000.0 et que l'origine sur ce plan principal soit aussi proche que possible de l'équinoxe dynamique de J2000.0,

2. que les positions des objets extragalactiques sélectionnés en conformité avec la recommandation VI et représentant le repère de référence, soient initialement calculées pour l'équateur et l'équinoxe de J2000.0 en utilisant les meilleures corrections disponibles à la position du pôle céleste de J2000.0 donnée par les expressions adoptées par l'UAI pour la précession et la nutation,

3. qu'un grand effort soit fait pour comparer les repères de référence de tous types, en particulier le FK5, les repères dynamiques dans le système solaire et les repères extragalactiques,

4. que l'on entreprenne ou continue tous programmes d'observation destinés à rapporter les positions des planètes à des objets émettant en ondes radio ou en lumière visible et à déterminer les relations qui existent entre les catalogues de positions de sources extragalactiques et les meilleurs catalogues de positions d'étoiles, notamment les catalogues FK5 et HIPPARCOS.

Notes pour la Recommandation VII

1. *Cette recommandation spécifie les axes de coordonnées qui seront adoptés pour le repère de référence définitif et décrit le travail qu'il faut effectuer avant que l'on puisse construire un tel repère. Bien que les considérants se réfèrent, pour le catalogue primaire, aux longueurs d'onde visuelles et radio, d'autres longueurs d'onde observables ne sont pas exclues. Les positions des objets observés dans d'autres longueurs d'onde devront également être rapportées au même système.*

2. *L'objectif établi par cette recommandation est qu'il ne doit pas y avoir de discontinuité dans les positions des étoiles lorsque le repère FK5 actuel sera remplacé par le repère de référence extragalactique. Ceci signifie que les*

positions des objets extragalactiques devront être données dans le système FK5 pour J2000.0. On reconnaît que, pour ce faire, les meilleures valeurs de la précession et de la nutation doivent être utilisées afin d'éviter que des mouvements propres fictifs soient introduits dans la position des objets extragalactiques. La transformation finale pour se référer à l'équinoxe et au plan principal adoptés sera faite en appliquant une rotation à l'instant J2000.0.

3. *L'équinoxe dynamique, dans cette recommandation, est défini comme l'intersection de l'équateur moyen et de l'écliptique. Ce dernier est considéré comme le plan de l'orbite du barycentre Terre/Lune, en rotation uniforme moyennée sur l'intervalle de validité des éphémérides. Comme le choix de l'équinoxe dépend ainsi des éphémérides, on prendra les éphémérides du système solaire, généralement disponibles, qui seront les plus exactes en temps voulu.*

4. *La définition du système de référence donnée par les Recommandations I et II implique la stabilité dans le temps du système de coordonnées réalisé par le repère de référence céleste. Les directions des axes de coordonnées ne doivent pas être changées, même si à quelque date ultérieure les réalisations de l'équinoxe dynamique ou du Pôle céleste des éphémérides sont améliorées. De la même manière, des modifications de l'ensemble des objets extragalactiques qui réalisent le système de référence doivent être faites de telle sorte que les directions des axes ne soient pas changées. Cela signifie qu'une fois que les axes de coordonnées ont été spécifiés, comme cela est indiqué par la première partie de la recommandation, la connexion entre la définition du système conventionnel de référence et les particularités des mouvements de la Terre doit être abandonnée.*

5. *Tant que la relation entre le repère optique et le repère extragalactique radio n'est pas établie avec une exactitude suffisante, le catalogue FK5 sera considéré comme une réalisation provisoire du système de référence céleste, pour les longueurs d'onde optique.*

RECOMMANDATION VIII

reconnaissant

a) l'importance pour l'astronomie de l'adoption de valeurs conventionnelles de constantes astronomiques et physiques,

b) que les valeurs de ces constantes doivent demeurer inchangées à moins qu'elles ne diffèrent significativement de leur plus récente estimation,

c) que l'estimation de ces constantes doit être fréquemment améliorée afin de représenter l'état actuel des connaissances,

d) la nécessité de fournir des procédés de calcul normalisés pour utiliser ces valeurs numériques et

<u>notant</u>

a) que les "MERIT Standards" et les "IERS Standards" ont apporté
une contribution significative au progrès de l'astronomie et de
la géodésie,

b) que les valeurs numériques de ces normes (standards) ont servi
de système de constantes pour analyser des observations de
grande qualité et

<u>considérant</u>

que les procédés de calcul fournis par ces normes ne couvrent pas
la totalité de l'astronomie fondamentale,

<u>recommande</u>

qu'un groupe de travail permanent soit organisé par les commissions
4, 5, 8, 19, 24 et 31, en consultation avec l'AGI et l'IERS, afin
de mettre à jour et d'améliorer le système des unités et constantes
astronomiques, la liste des estimations de grandeurs astronomiques
fondamentales et des procédés de calcul normalisés; ce groupe devra

1. préparer un rapport provisoire sur le système des unités et
constantes astronomiques au plus tard six mois avant
la XXIIe Assemblée générale (1994),

2. préparer une liste provisoire des meilleures estimations de
grandeurs astronomiques au plus tard six mois avant chaque
Assemblée générale suivante,

3. préparer, au plus tard six mois avant chaque Assemblée générale
suivante, un rapport provisoire sur les procédés de calcul
normalisés dont on a besoin en astronomie fondamentale, lequel

    a) devrait avoir un maximum de compatibilité avec les "IERS
    Standards",

    b) devrait inclure des mises en pratique des procédés de calcul
    sous forme de programmes informatiques ou d'exemples
    éprouvés,

    c) devrait être disponible, non seulement sous forme écrite,
    mais aussi sous forme informatisée.

4. préparer un rapport provisoire sur un possible accès
électronique à ces unités, constantes, grandeurs et procédés de
calcul, au plus tard six mois avant la XXIIe Assemblée générale
(1994).

RECOMMANDATION IX

reconnaissant

qu'il n'existe pas, pour le moment, de théorie de la nutation d'une Terre non rigide incluant tous les effets au niveau de 10^{-4} seconde de degré qui fasse l'objet d'un consensus général,

recommande

1. que, pour des besoins ne nécessitant pas une exactitude sur les angles de nutation (ϵ or $\psi \sin\epsilon_o$) meilleure que ± 0,002" (à 1 o), on peut continuer à utiliser la théorie de la nutation UAI 1980 (P.K. Seidelmann, Celestial Mechanics, 27, 79, 1982),

2. que, lorsqu'on a besoin d'une exactitude meilleure que ± 0,002" (à 1 o), il faut utiliser les bulletins de l'IERS qui donnent les observations et les prédictions des écarts de position du pôle céleste avec une exactitude de l'ordre de ± 0,0006" (à 1 σ) pour une période allant jusqu'à 6 mois à l'avance,

3. que l'on encourage l'UGGI à établir et à adopter un modèle de Terre adéquat pour servir de base à une nouvelle théorie de la nutation de l'UAI.

CCIR   Comité Consultatif International des Radiocommunications

IAG/AIG   Association Internationale de Géodésie

IUGG   Union Internationale de Géodésie & Géophysique

IERS   Service International de la Rotation de l'Heure

Resolution A5: Encouraging International Development of Antarctic Astronomy

The XXIst General Assembly of the International Astronomical Union,

recognising

1. the potential for making some important classes of astronomical observations from Antarctica that are not possible from elsewhere on the Earth's surface,

2   the fact that the extremely dry, cold and tenuous atmosphere, above the Antarctic Plateau provides the best observing conditions on Earth in the infrared, sub-mm and mm wavelength range, and

3. the unique opportunities Antarctica offers for establishing truly international bases for scientific cooperation,

and noting that

1. technological advances are greatly widening the scope for exploiting the astronomical merits of Antarctica,

2. a Working Group of the ICSU Scientific Committee on Antarctic Research has formally recommended (*) serious international consideration be given to participation in designing, building and operating a new station in the highest part of the inland plateau,

3. there is widespread concern to ensure any development in Antarctica is compatible with preservation of the natural environment,

4. some astronomical instruments in Antarctica will be well suited to studies of global environmental problems,

5 astronomical activities and planning for new instrumentation in Antarctica have greatly increased over the last few years, and

6. international links should be increased to enhance scientific returns,

urges

National Committees for Astronomy and National Antarctic agencies
    to establish an international astronomical base on the high
    plateau

and resolves

to create a Working Group to encourage international cooperation in site testing and in designing and constructing new Antarctica astronomical facilities.

(*) *Recommendation 6 of the Atmospheric Sciences Working Group (now divided into two groups: Solar, Terrestrial and Astrophysical Research & and Physics and Chemistry of the Atmosphere) at the biennial meeting in 1990 of the Scientific Committee for Antarctic Research.*

Résolution A5: Encouragement pour le développement international de l'Astronomie en Antarctique

La XXIe Assemblée générale de l'Union Astronomique Internationale,

reconnaissant

1. la possibilité de faire depuis l'Antarctique plusieurs types d'observations astronomiques importantes qui ne peuvent être effectuées d'aucun autre endroit sur la surface terrestre ;

2. le fait que l'atmosphère extrêment sèche, froide et peu dense au-dessus du plateau antarctique fournit les meilleures conditions d'observation sur terre pour les domaines de longueurs d'onde infrarouge, submillimétrique et millimétrique ;

3. l'occasion unique qu'offre l'Antarctique d'établir des bases réellement internationales pour la coopération scientifique ;

et notant

1. que les progrès technologiques augmentent considérablement les possibilités d'exploiter les qualités astronomiques de l'Antarctique ;

2. qu'un Groupe de travail du Comité Scientifique pour la Recherche en Antarctique (SCAR, de l'ICSU) a formellement recommandé (*) que soit sérieusement envisagée une participation internationale à la conception, la construction et l'exploitation d'une nouvelle station sur la partie la plus élevée du plateau central ;

3. qu'il y a une volonté très large de s'assurer que tout développement en Antarctique est compatible avec la protection de l'environnement naturel ;

4. que certains instruments astronomiques en Antarctique seront bien adaptés à l'étude de problèmes d'environnement global ;

5. que les activités astronomiques et les projets de nouvelle instrumentation en Antarctique ont considérablement augmenté au cours des dernières années et

6. que les coopérations internationales pour ces projets doivent être développées pour améliorer les résultats scientifiques ;

demande expressément

que les instances nationales d'astronomie, ainsi que les Agences nationales pour l'Antarctique établissent une base internationale pour l'astronomie sur le haut plateau ;

et décide

de mettre en place un Groupe de travail afin d'encourager la coopération internationale pour le test des sites, la définition et la construction de nouvelles installations et équipements pour l'astronomie en Antarctique.

(*) 6e Recommandation du Groupe de travail en sciences atmosphériques (maintenant divisé en deux sous-groupes: Recherche astrophysique, Terre, Soleil et Physique et chimie de l'atmosphère) lors de la réunion biennale de 1990 du Comité Scientifique pour la Recherche en Antarctique (SCAR).

Resolution A6: Working Group on the Prevention of Interplanetary Pollution

The XXIst General Assembly of the International Astronomical Union,

recognising

that the pollution of the space environment in the close vicinity of the Earth is now of serious concern, and that pollution of the remainder of the solar system is only a matter of time,

recommends

that steps be taken immediately to ensure that interplanetary space throughout the solar system is protected as far as possible from all forms of pollution,

urges

the International Astronomical Union to establish an inter-Commission Working Group on the Prevention of Interplanetary Pollution and that the Working Group should consult widely with COSPAR, other relevant Unions, Space Agencies and the United Nations Committee for the Peaceful Uses of Outer Space.

COSPAR   Committee on Space Research

Résolution A6: Groupe de travail sur la prévention de la pollution interplanétaire

La XXIe Assemblée générale de l'Union Astronomique Internationale,

reconnaissant

que la pollution de l'espace dans un environnement proche de la Terre pose maintenant un sérieux problème et que la pollution du reste du système solaire n'est qu'une question de temps,

recommande

que des mesures soient prises immédiatement pour protéger de toutes formes de pollution l'ensemble de l'espace interplanétaire du système solaire aussi loin que possible.

et demande instamment

à l'Union Astronomique Internationale d'établir un groupe de travail inter-commission sur la prévention de la pollution interplanétaire et que ce dernier consulte largement le Comité de la Recherche Spatiale (COSPAR), les autres unions concernées, les agences spatiales et le Comité des Nations Unies pour assurer une utilisation pacifique de l'espace.

Resolution A7: Joint IUGG/IAU Working Group

The XXIst General Assembly of the International Astronomical Union,

recognizing

the importance of rapid determinations of Earth rotation recommended by the International Workshop "Interdisciplinary role of space geodesy" held in Erice (Italy) in 1988, and

considering

the proposal made to the International Association of Geodesy by its Special Study Group 5.98 on "Atmospheric excitation of the Earth's rotation" to set up a Working Group on "High time resolution measurements of Earth rotation",

requests

the Executive Committee of the International Astronomical Union to approach the International Association of Geodesy in order to consider the possibility of organizing a joint IUGG/IAU Working Group for such activity.

IUGG   International Union of Geodesy & Geophysics

Résolution A7: Groupe de travail conjoint IUGG/IAU

La XXIe Assemblée générale de l'Union Astronomique Internationale,

reconnaissant

l'importance des déterminations à haute résolution temporelle de la rotation de la Terre recommandées par le groupe de travail international "Rôle interdisciplinaire de la géodésie spatiale" tenu à Erice (Italie) en 1988, et considérant

la proposition faite à l'Association Internationale de Géodésie par
son groupe d'étude spécialisé 5.98 sur "l'excitation par
l'atmosphère d'irrégularités de la rotation de la Terre" de
constituer un groupe de travail sur les "Mesures à haute résolution
temporelle de la rotation de la Terre",

demande

au Comité Exécutif de l'Union Astronomique Internationale de
prendre contact avec l'Association Internationale de Géodésie et de
Géophysique pour envisager la possibilité d'organiser un groupe de
travail conjoint IUGG/UAI pour une telle étude.

IUGG   Union internationale de géodésie & de géophysique

Resolution A8: Catalogue Compilation

The XXIst General Assembly of the International Astronomical Union,

recognizing

the great value to astronomical research of comprehensive
catalogues of critically evaluated data on celestial objects of
particular types,

urges

that appropriate support be provided by institutions and funding
agencies to those experts who are willing to devote time to
the long-term task of compiling such catalogues.

Résolution A8: Compilation de Catalogues

La XXIe Assemblée générale de l'Union Astronomique Internationale,

reconnaissant

l'importance majeure des grands catalogues complets de données
critiquement sélectionnées concernant des objets célestes de
différents types,

demande instamment

qu'une aide appropriée soit apportée par les institutions et
agences de financement aux experts qui sont prêts à consacrer du
temps à ce travail à long terme de compilation de tels catalogues.

Resolution A9: Hazardous Near-Earth Objects

The XXIst General Assembly of the International Astronomical Union,

considering

that various studies have shown that the Earth is subject to occasional impacts by minor bodies in the solar system, possibly with serious results, and

noting

that there is a well-founded evidence that only a very small fraction of NEO's (Natural Near-Earth Objects: minor planets, comets and fragments thereof) has actually been discovered and has well-determined orbits,

affirms

the importance of expanding and sustaining scientific programmes for the discovery, continued surveillance and in-depth physical and theoretical study of potentially hazardous objects, and

resolves

to establish an ad-hoc Joint Working Group on NEO's, with participation of Commissions 4, 7, 9, 15, 16, 20, 21 and 22, to:

1. assess and quantify the potential threat, in close interaction with other specialists in these fields;

2. stimulate the pooling of all appropriate resources in support of relevant national and international programmes;

3. act as an international focal point and contribute to the scientific evaluation, and

4. report back to the XXIInd General Assembly of the IAU in 1994 for possible further action.

Résolution A9: Objects hasardeux proches de la Terre

La XXIe Assemblée générale de l'Union Astronomique Internationale,

reconnaissant

que diverses études ont montré que la Terre est soumise à des impacts occasionnels avec de petits corps du système solaire, impacts qui ont parfois des conséquences graves, et

<u>notant</u>

qu'il est clairement établi que seulement une très faible proportion des objets naturels proches de la Terre (planètes mineures, comètes et leurs fragments) ont de fait été découverts et ont des orbites déterminées avec précision,

<u>confirme</u>

l'importance qu'il y a d'accroître et de soutenir les programmes scientifiques de découverte, de surveillance continue, et d'étude de fond, physique et théorique, des objets potentiellement dangereux, et

<u>décide</u>

d'établir un groupe de travail ad hoc inter-commissions sur les objets naturels proches de la Terre avec la participation des Commissions 4, 7, 9, 15, 16, 20, 21 et 22, pour :

1. évaluer et quantifier le danger potentiel, en concertation avec les autres spécialistes de ce domaine ;

2. encourager la mise en commun de toutes les ressources nationales et internationales consacrées à ces programmes ;

3. servir de point centralisateur international et contribuer à l'évaluation scientifique des travaux, et

4. rendre compte à la XXIIe Assemblée générale de l'Union Astronomique Internationale en vue d' actions ultérieures possibles.

17.   Resolution proposed by the Resolutions Committee

Resolution B1: Endorsement of Commission Resolutions

The XXIth General Assembly of the International Astronomical Union;

<u>having</u>

full confidence in its Commissions,

<u>endorses</u>

the Resolutions submitted by them to the Resolutions Committee (pp. 71-77).

Resolution B1: Soutien des Résolutions des Commissions

La XXIe Assemblée générale de l'Union Astronomique Internationale,

accordant

son entière confiance à ses Commissions,

souscrit

aux résolutions qu'elles ont soumises au Comité des Résolutions (pp. 71-77).

18.   Resolutions proposed by the Commissions

Resolution C1: Directory of Astronomical Software

Commission 5

considers

that the establishment of an electronically accessible directory of astronomical software that is available and suitable for general use would be of great value to the astronomical community,

further considering

the need for an appropriate institution to providing facilities for the compilation and maintenance of such a directory, and

requests

that astronomers provide appropriate details for inclusion in the directory.

Resolution C2: Editorial Instructions

Commission 5

welcomes

the efforts being made by the editors of the major astronomical journals to standardise their instructions to authors and

urges

all editors to include the recommendations adopted by the XXth IAU General Assembly concerning the use of SI units, the designation of celestial objects, and the abbreviations for the titles of journals.

Resolution C3: Astronomical Telegrams

Commission 6

noting

the indispensable   character   of   the service   rendered   to
the international astronomical community by the Central Bureau for
Astronomical   Telegrams   by   rapid   communication   of   critical
information,

calls attention

to the importance of the token subvention as a demonstration of
the support of the IAU for this crucial activity,

strongly urges

the continuation of this subvention, and

further urges

the General Secretary to maintain an appropriate subvention in
the IAU budget and negotiate a payment schedule with the Director
of the Central Bureau for Astronomical Telegrams.

Resolution C4: Publication of Solar Eclipse Information

Commissions 10 & 12

considering

that the United States Naval Observatory (USNO) has for more
than forty years generously provided crucial information to assist
scientists who observe solar eclipses for scientific purposes (in
the form of the Central Solar Eclipse Circulars and other
specialized calculations) and,

recognising

that the USNO plans to cease publication of the Eclipse Circulars
due to programmatic changes and plans to continue to support
scientific observations by publishing eclipse circumstances in
the Astronomical Almanac, and by providing specialized eclipse
calculations to scientific researchers,

commend

on behalf of past and present eclipse researchers, the management
and staff of the USNO responsible for the preparation and
publication of the calculations and,

request

that the USNO continue to provide advance calculations for
a variety of sites in order to aid site selection and to publish
this information in Circulars or by other means and,

further commend

all official national organizations that prepare eclipse
calculations, and urge that they continue their efforts.

Resolution C5: Long-term Solar Observations

Commissions 10 & 12

considering

long-term observations are essential to understand the behaviour of
such quasi-periodic phenomena that characterize solar and stellar
activity and that link the Sun to our terrestrial environment;

recommend

1. strong support for continuing data-gathering programmes and
   observational facilities that are essential to long-term
   research;
2. the optimization of data-gathering enterprises in order to
   improve services to the research community.

Resolution C6: Comet Rendezvous Asteroid Flyby (CRAF) Mission

Commission 15

considering

current plans for the Comet Rendezvous Asteroid Flyby (CRAF)
mission, and

recognising

the need to study many alternative scenarios because of the lack of
basic data for possible target comets (P/Kopff, P/Tempel 2,
P/Tempel 1, P/Wild 2, P/d'Arrest, and P/Wirtanen), and

noting

that the CRAF Project Science Group is strongly encouraging
observations of these comets with ground-based, airborne, rocket-
borne, and Earth-orbit instruments; particularly to determine
the size, and spin-period of the nucleus, the orientation of
the spin-axis, and the development of outgassing and dust ejection
as the comets approach the Sun; and

that such observations are particularly desirable at the next
aphelion and perihelion passages through 1996; and

that such observations, must be made with good temporal coverage
and, especially near aphelion, require instruments with the highest
sensitivity,

strongly urges

all observing-time allocation committees to take account of these
considerations in their decisions about relevant programmes
submitted to them.

Resolution C7: Expansion of the Minor Planets Names Committee

Commission 20,

noting

the recent disagreement between it and the Working Group on
Planetary System Nomenclature (WGPSN) concerning the proposed names
for the recently discovered satellites of Neptune,

drawing attention to

its 1985 resolution to minimize the duplication between the names
of minor planets and natural satellites, and

considering

the vast potential for drawing on a number of different cultures
for the selection of names,

recommends

that its Minor Planet Names Committee, currently consisting of
the President, the Vice-President and the Director of the Minor
Planet Centre, be expanded to include more effective liaison with
the WGPSN, and

charges

the expanded Committee to take a more active rôle in both choosing
names and writing completed, concise citations.

WGPSN   Working Group for Planetary System Nomenclature

Resolution C8: Long-term Observation of Fifteen Minor Planets

Commission 20,

welcoming

the proposal of the Institute of Theoretical Astronomy (USSR) to prolong the observational programme for 15 selected minor planets (Nos. 1, 2, 3, 4, 6, 7, 11, 18, 35, 39, 40, 148, 382, 532 and 704) for the period 1991-2000, and

encouraging

all observatories which have astrographs of focal length 2 m to take part in this programme,

recommends

that the most precise reference catalogues, PPM (Positions and Proper Motions), Fokat (Fotograficheskij Katalog, Pulkovo) and ACRS (Astrographic Catalogue Reference System), are used for the determination of the spherical coordinates of the planets.

Resolution C9: Ephemerides of Minor Planets

Commission 20

supports

the activities of the Institute of Theoretical Astronomy (USSR) on the elaboration of PC software packages for the provision of ephemerides of minor planets, and

suggests

that systems like "STAMP" may be used together with the printed annual volumes "Ephemerides of Minor Planets".

Resolution C10: Data Centre at the Bureau des Longitudes

Commission 20

having heard

the report of the Chairman of the Working Group on Satellites, in which is proposed the creation of a Data Centre at the Bureau des Longitudes (France),

supports

this proposal, and

<u>recommends</u>

that this centre develops into an International Data Centre, as
defined in Internal Resolution of Commission 20, adopted on August
8, 1988, during the XXth IAU General Assembly.

Resolution C11: Variable Star Observations

Commissions 27 & 42

<u>considering</u>

that the systematic coverage of the long-term behaviour of
the population of variable stellar objects, such as the sixty years
of measurements made at Sonneberg Observatory, makes a major
contribution to Astronomy and Astrophysics,

<u>recommend</u>

that all efforts be undertaken to continue these important
measurements and to ensure the appropriate maintenance and
availability of the data archives.

Resolution C12: Space Schmidt Telescope

Commission 28

<u>recognizing</u>

the important scientific opportunities inherent in the ASCHOT
(80 cm Space Schmidt) project to be placed on board of the Soviet
space station in 1996, and

<u>noting</u>

that several of the available focal positions have not yet been
equipped with optimum detectors,

<u>urges</u>

astronomical institutions with detector capabilities to consider
participation in this project and further to enhance its scientific
return.

Resolution C13: Archiving Spectroscopic Results

Commission 29

considering

that a large amount of spectroscopic data has been collected on photographic plates,

that the widest use of electronic detectors has generated a rapid growth in the build-up of raw spectroscopic data files, and

that the information contained in such data could represent an important source for future studies,

recognizing

the importance of safeguarding such data, and the need to create an accessible archive of the observations,

recommends

that an IAU Working Group for spectroscopic Data Archives be set up in order to establish agreed means of archiving and distributing the spectroscopic data.

Resolution C14: Astronomical Archives

Commissions 41 and 5

recommend

that the Union supports the initiatives taken by them

1. to establish a register of the whereabouts of all extant astronomical archives of historical interest;

2. to impress on observatories and other institutions their responsibility for the preservation, conservation, and where possible, cataloguing of such archives;

3. to search for an institution that will allocate space and funds for maintaining such a register and publishing it.

Resolution C15: The use of Vacuum Wavelengths in Astronomy

Commission 44

recognizing

that with the increasing availability of spectroscopic observations in the middle and far ultraviolet provided by spectrometers in orbit, it has become desirable to provide a uniform wavelength scale by removing the traditional discontinuity in the expression of wavelengths from vacuum to air across 2000 Å, and

<u>whereas</u>

(1) there is a trend in both space and ground-based astronomy to replace air wavelengths with vacuum values,

(2) the IAU has already agreed upon and published a simple conversion formula between the two systems,

(3) neglect by authors to indicate which standard they are using in this time of flux is a source of confusion,

<u>recommends</u>

that the IAU favors a uniform expression of vacuum wavelengths across the entire spectrum in astronomy,

<u>urges</u>

since we are in a transition period, that all publications clearly indicate which convention is being used, and

<u>further urges</u>

that the IAU conversion algorithm (current reference: Oosterhoff, P.T. 1957 Trans IAU Vol. IX pp. 69, 202) be used and referenced in articles expressing wavelengths in air.

<u>**Resolution C16: Concerning Extraterrestrial Intelligence**</u>

Commission 51

<u>considering</u>

that searches for evidence of technologically developed life elsewhere in the universe have been conducted for more than 30 years by means of astronomical instruments, and

that much more extensive searches, using large radio telescopes around the world, are about to commence,

<u>recommends</u>

that the astronomical community follows the guidelines for verifying the nature of a candidate extraterrestrial intelligent signal and announcing its detection, as presented in the document from the International Academy of Astronautics and the International Institute of Space Law entitled "Declaration of Principles Concerning Activities Following the Detection of Extraterrestrial Intelligence".

19.   Appointment of the Special Nominating Committee 1991-1994 (SNC)

The President asked the General Secretary to announce the names of the members proposed for appointment by the General Assembly to the Special Nominating Committee 1991-1994. These persons will be convened by the President of the IAU for the purpose of proposing names to the XXIInd General Assembly (1994) for IAU Executive Committee membership (1994-1997). The four persons appointed are:

H. Abt   (USA)
K.C. Freeman   (Australia)
B. Hauck   (Switzerland)
D.C. Morton   (Canada)

The member of the SNC appointed by the Executive Committee is

A. Feinstein   (Argentina)

These appointments were unanimously confirmed by the General Assembly.

Note: The President and Past President are members of the SNC ex officio. The General Secretary and Assistant General Secretary are consultants to the SNC.

20.   Nomination of New Members of the Union

The General Secretary announced that the Executive Committee had, on the proposal of the Adhering Bodies and with the advice of the Nominating Committee, admitted 760 new members to the Union. The names of the new members had been displayed at the entrance to Room P in the La Plaza Centre during the course of the General Assembly. The names will be incorporated in the alphabetical list of IAU Members to appear in IAU Transactions XXIB.

21.   Applications for IAU Membership

No country had applied for IAU membership during the triennium.

22.   Changes in Commissions

The General Secretary read the proposals of the Executive Committee as regards the Presidents & Vice-Presidents of Commissions for the triennium 1991-1994:

Commissions/ Working Groups	President		Vice-president	
04	B.D.	Yallop	H.	Kinoshita
05	B.	Hauck	O.B.	Dluzhnevskaya
06	J.E.	Grindlay	R.M.	West
07	A.	Deprit	S.	Ferraz-Mello
08	L.V.	Morrison	C.A.	Smith
09	J.C.	Bhattacharyya	G.	Lelièvre
10	V.	Gaizauskas	O.	Engvold
12	J.O.	Stenflo	F.	Deubner
14	W.L.	Wiese	W.H.	Parkinson
15	A.	Harris	V.M.F.	A'Hearn
16	D.	Morrison	M.	Marov
			C.	de Bergh
19	B.	Kolaczek	J.	Vondrak
20	A.	Carusi	D.	Yeomans
21	M.	Hanner	M.	Hauser
22	J.	Stohl	I.	Williams
24	C.	de Vegt	C.	Turon

25	A.T.	Young	J.D.	Landstreet
26	H.	Abt	C.E.	Worley
27	J.R.	Percy	M.	Jerzykiewics
28	E.	Ye Khachikian	V.L.	Trimble
29	D.	Lambert	M.	Bessel
30	G.	Burki	C.D.	Scarfe
31	E.	Proverbio	H.	Fliegel
33	L.	Blitz	J.	Binney
34	H.	Habing	D.R.	Flower
35	P.	Demarque	C.	Chiosi
36	W.	Kalkofen	L.	Cram
37	J.C.	Mermilliod	A.	Feinstein
38	J.	Sahade	H.E.	Jorgensen
40	M.	Morimoto	J.B.	Whiteoak
41	S.	Debarbat	S.M.R.	Ansari
42	Y.	Kondo	M.	Rodono
44	J.	Trümper	G.	Fazio
45	D.	MacConnel	H.	Levato
46	L.	Gougenheim	J.	Percy
47	R.B.	Partridge	J.	Narlikar
48	J.	Ostriker	G.	Srinivasan
49	B.	Buti	H.	Ripken
50	P.	Murdin	S.	Isobe
51	R.D.	Brown	J.	Tarter
WGPSN	K.	Aksnes		
WGWDA	A.H.	Batten		
WGDAA	P.	Gillingham		

This proposal was received with acclamation by the General Assembly.

Organising Committees of Commissions

The General Secretary announced that, in the interest of economy of time, it had been decided not to present the lists of Members in the Organising Committees of Commissions, but that the lists were available at the IAU Secretariat for inspection. They will be printed in IAU Transactions XXIB in Chapter VI.

23.   Change of Name of Commission 12

The Executive Committee on the advice of Commissions 12 and 35 proposed that the name of Commission 12 currently "Radiation and Structure of the Solar Atmosphere" be changed with immediate effect to "Solar Radiation and Structure".

This change of name was unanimously agreed by the General Assembly.

24.   IAU representatives to other ICSU & International Institutions

Acronyms	Organisation	Representative
ICSU	International Council of Scientific Unions General Committee	J. Bergeron
BIPM/ CCDS	Bureau international des poids et mesures Working Group on the Temps Atomique International of the Consultative Committee for the Definition of the Second	G. Winkler
CCIR	Consultative Committee on International Radiocommunications	
	Study Group 2	J. Whiteoak/ A.R. Thompson
	Study Group 7	S. Leschiutta
CIE	Compagnie internationale de l'éclairage	D. Crawford
CODATA	Committee on Data for Science & Technology	A. Heck
COSPAR	Committe on Space Research	J. Bergeron
	COSPAR ISC B	J. Rahe
	COSPAR ISC D	S. Grzedzielski
	COSPAR ISC E	Y. Kondo/J.-P. Swings
	COSPAR Sub. Committee E1	R. Sunyaev
	COSPAR Sub. Committee E2	O. Engvold
COSTED	Committee on Science & Technology in Developing Countries	J. Bergeron
CTS	Committee on the Teaching of Science	J.M. Pasachoff
EPS	European Physical Society Conference Committee	J. Bergeron
FAGS	Federation of Astronomical & Geophysical Services	E.Tandberg-Hanssen
IAF	International Astronautical Federation	Y. Kondo
IERS	International Earth Rotation Service	Ya. Yatskiv
IGBP	International Geosphere-Ionosphere Programme	J. Eddy
ISY	International Space Year	L. Gouguenheim
IUCAF	Inter-Union Commission on Frequency Allocation for Radio Astronomy & Space Science	B.A. Doubinsky M. Ishiguro V.L. Pankonin G. Swarup
IUPAP	International Union of Pure & Applied Physics	V. Trimble
IUWDS	International Ursigram & World Day Service	H. Coffey
QBSA	Quarterly Bulletin on Solar Activity	S.T. Wu
SCOPE	Scientific Committee on Problems of Environment	R. Cayrel
SCOSTEP	Scientific Committee on Solar-Terrestrial Physics	S.T. Wu
URSI	Union Radio-Scientifique Internationale	J. Baldwin
WMO	World Meteorological Organization	G. Wallerstein

25.   Place and Date of the XXIInd General Assembly

The President called upon Professor H. Habing to present the invitation of the Netherlands to hold the XXIInd General Assembly in the Hague.   Professor Habing read the invitation to the Union from Prof. Dr. C. de Jager on behalf of the Board of the Royal Netherlands Academy of Arts and Sciences warmly inviting the Union to the Netherlands.   Professor Habing outlined some of the many attractions of the Netherlands and the Hague and promised a worthy scientific meeting.

The General Assembly accepted this invitation with acclamation and the President asked Professor Habing to convey the acceptance and gratitude of the Union to Prof. Dr. C. de Jager and the Royal Netherlands Academy of Arts and Sciences.

The proposed dates for the XXIInd General Assembly are 1994 August 14-27.

26.   Election to the Union of a President, a President-elect, six Vice-Presidents, a General Secretary and an Assistant General Secretary.

The General Assembly approved by acclamation the proposal of the President that Academician A.A. Boyarchuk be elected the new President of the Union, for the term 1991-1994.

The General Assembly approved by acclamation the proposal of the President that Professor L. Woltjer be elected the new President-elect of the Union, for the term 1991-1994.

The President announced that Professor V. Radhakrishnan, Dr. M.S. Roberts, Dr. Ye Shu Hua would continue as Vice-Presidents of the Union for the term 1991-1994 and the General Assembly approved by acclamation the proposal of the President that Professors D. Mathewson, F. Pacini and J. Smak be elected Vice-Presidents of the Union for the term 1991-1994.

The President finally proposed that Dr. J. Bergeron be elected General Secretary of the Union and Professor I. Appenzeller be elected Assistant General Secretary of the Union for the term 1991-1994.   This proposal was received with acclamation by the General Assembly.

The President then invited Professors L. Woltjer, D. Mathewson, F. Pacini, J. Smak and I. Appenzeller to join the Executive Committee on the platform.

Following these elections, the IAU Executive Committee for the period 1991-1994 will be:

President:	A.A. Boyarchuk	USSR
General Secretary:	J. Bergeron	France
Assistant General Secretary:	I. Appenzeller	Germany
President-Elect:	L. Woltjer	Netherlands

Vice-Presidents:	D.S.	Mathewson	Australia
	F.	Pacini	Italy
	V.	Radhakrishnan	India
	M.S.	Roberts	USA
	J.I.	Smak	Poland
Shu Hua		Ye	China PR
Advisers:	Y.	Kozai	Japan
	D.	McNally	UK

CHAPTER III

CLOSING CEREMONY

The President warmly thanked the Local Organising Committee and all those who had assisted them in their task, especially to all those who had worked so very hard the previous day assisting the General Secretary retrieve a residual scientific programme following the fire in the San Martin Cultural Centre.

Address by the President 1988-1991

"Dear Colleagues and Friends,

Now it is the time for the new President, Dr. Alexander Boyarchuk, to make a speech. However, before that I would like to ask your permission to spend a few minutes to talk.

Now I am relieved from my heavy duty as the President of the International Astronomical Union. It was indeed a heavy and difficult job for me while serving as the Director of the National Astronomical Observatory, Japan, which was founded three years ago and is composed from three institutes including Tokyo Astronomical Observatory and has tried to obtain funds to build an 8 meter infrared-optical telescope as well as other new instruments. Still I feel that I have spent an exciting time in the past three years, and that I could not have completed my job without the strong support and assistance provided by Drs. Derek McNally and Jacqueline Bergeron. Indeed the jobs of the General Secretary and the Assistant General Secretary are very heavy, much heavier than mine. Still they have completed their jobs for the Union with helps by their own secretaries who came here to work with them and for the participants.

Also the President-Elect and the six Vice-Presidents have been wonderful and friendly members of the Executive Committee and gave me much fruitful advice. I would like to express my hearty thanks to all of them, particularly the three Vice-Presidents, Professors A.H. Batten, R. Kippenhahn and P.O. Lindblad who will complete their terms today. Our advisors, Professor Sahade and Dr. J.-P. Swings, have contributed very much to the activities of the Executive Committee.

Of course, our secretaries in Paris secretariat, Monique Léger-Orine and Julie Crook, have worked very hard and handled many administrative jobs for us. I believe that you also appreciated their efforts during this General Assembly. I should like to mention, particularly, the name of Monique Léger-Orine who worked here day and night, even after mid-night and on Saturday and Sunday for us as the Administrative Assistant of the Union. We are grateful to the efforts made by the Local Organising Committee, particularly, Dr. Roberto Mendez, to make the General Assembly successful: we will all not forget how quickly the General Assembly resumed to its usual condition after the fire yesterday prevented us to use the pre-assigned meetings rooms. At my home institute, National Astronomical Observatory, I was assisted by Mrs. Hatsue Toya Suzuki who has handled IAU jobs as her unpaid extra works, which I have also appreciated very much.

As I told you three years ago, I was educated and trained as
an astronomer in a remote country in the far east in the period
when no major facilities existed and we could find very few chances
to attend international meetings like the General Assembly.   Then
I always dreamed that in future we can do much better work and
always hoped and still hope that tomorrow will be better than today
and next year will be better than this year in not only
astronomical research but also everyday's life.   And it has been
realised in some sense for us.   Still I worry that the speed of
development is very slow and sometimes we meet very difficult time.
I am afraid that now is not an easy time for the IAU and many
astronomers, particularly for financial reason.   Still I believe
that important astronomical researches have been achieved as you
have seen during the General Assembly and will further evolve, both
in developed and developing countries, as astronomy is supported by
many people and can give them dreams to imagine how the universe
looks and was created.   I believe that IAU must play an important
role promoting astronomical research activities.   Therefore, at
the end of my speech I would like to ask all the members,
particularly, Presidents of the Commisions to play a positive role
in activities of IAU in all the aspects so that we can expect to
see a better IAU at the next General Assembly in the Netherlands,
where I desire to see many more members of the IAU and their
guests.

I would like now to ask the new President, Dr. Alexander Boyarchuk,
to come here to speak."

Address by the President 1991-1994

"Dear Professor Kozai, Members of the Union, Ladies and Gentlemen.

It is indeed a great honour for me to be nominated and to serve as
the President of the International Astronomical Union.   I regard
this nomination as the recognition of the contribution made by
astronomers of my country to the development of astronomy.   And
though for some economic and political reasons this contribution
has somewhat decreased last years, I believe that the situation
will improve in the future.

At the moment astronomy is at a very interesting phase of its
development.   Due to the progresses made in science and technolgy,
observational astronomy has reached a qualitatively new level.   The
results of observations with Hipparcos will allow the improvement
of the stellar coordinates accuracy by almost a 100 times.   The
Hubble telescope observations have improved space resolution as
many as five times.   Much progress has been made in this field with
ground-based telescope observations using adaptive optics.

The 10-meter Keck-telescope now opens a new era of supergiant
ground-based telescopes.   I believe that in the year 2000 we shall
see 10 telescopes of 8 and more meters in diameter.   In the coming
five years ground-based radio interferometers with antennas of 25-
32 meters in diameter, ten of thousands kilometers apart, will come
into operation in the USSR and the USA.   Many remarkable space
experiments are planned to be made in the future.   Astronomy
becomes more and more an international science.   Owing to the

Our meetings have run smoothly thanks to the efforts of our ever present Secretary General, Dr. Derek McNally, his staff, and an impressive cadre of projectionists and assistants. We all witnessed that operation in action as they and the LOC reorganized most of yesterday's activities following the fire in the San Martin Cultural Center. Some of you may have heard the rumor that the fire was just one more attempt to sabotage the Hubble Space Telescope Project. I understand that the HST Project is considering hiring an Exorcist to help improve their luck in the future!

We have many positive memories of Buenos Aires and Argentina to carry home with us. We were privileged to hear addresses by the President of Argentina and the Minister of Education demonstrating their support for our Science. We were able to attend a concert at the magnificent Colon Theater; take part in a classic Asado Criollo; witness the intricacies of the footwork in the Tango; and participate in the Closing Dinner, where we were treated to a wide variety of delicacies.

Finally, a General Assembly is an expensive operation and it is impossible to run one without the help of the local government. In this context, we would like to express our gratitude to the Secretary of Science and Technology for financial assistance through the CONICET. In addition, we are grateful to the Director of the La Plaza Complex for providing us, without cost, rooms to continue our meetings following the fire in the San Martin Cultural Center.

We leave Buenos Aires and Argentina with many fond memories and enthusiastically thank our local hosts for their many efforts on our behalf.

Quisicra ahora repetir mi agradecimiento, en mi limitado espanol, para todos aquellos, que han hecho posible, que esta Asamblea General, se realicea en Argentina. Muchas Gracias."

The President then invited Miss A. Kozai to address the General Assembly on behalf of the registered guests.

Address by Miss A. Kozai

"On behalf of the Registered Guests I would like to express our appreciation to everyone from the National and Local Organising Committees for making our stay in Buenos Aires unforgettable. You were always friendly and helpful.

Visits to museums, the National Congress and many places of cultural interest helped us to satisfy our curiosity about this charming city. At the Colón Theater we listened to concertos composed and played by Argentinian musicians. On Sunday, finally, we could meet famous Gauchos. We enjoyed every single moment so much that I couldn't believe we had to say good-bye to old and new friends at the Closing Banquet last night. What fun it is to be families and friends of astronomers!

Our time here has also been greatly enriched by the hospitality, patience and efforts of the staff and volunteers, as well as the editorial staff of Cruz del Sur, which contains very useful information for us all.

Les agradezco su hospitalidad. Me gustaría volver a verlos de nuevo. Muchísimas gracias por todo.

Adios".

Address by the retiring General Secretary, Dr. D. McNally

"Monsieur le Président, Membres du Comité Executif, Membres de l'Union, Mesdames, Mesdemoiselles et Messieurs,

Je retombe maintenant dans le rang des astronomes de base.

Avant que je n'exprime mes remerciements à tous ceux qui ont oeuvré au succès de cette Assemblée Générale, je tiens à remercier tout particulièrement la communauté astronomique française pour l'aide que le Centre National de la Recherche Scientifique a apportée à notre Union en accueillant dans ses locaux de l'INSU de Paris le Secrétariat permanent de l'Union et en autorisant la mise à disposition de notre Union de l'un de ses personnels.

J'ai, durant ces trois années, pu apprécier mes séjours à Paris et, plus particulièrement, son atmosphère. Bien que je n'aie pu pleinemènt profiter de Paris en tant que touriste autant que je l'aurais souhaité, j'ai été amené à expérimenter certains de ses aspects, probablement ignorés de la plupart des touristes...

Enfin, j'ai pu constater l'intérêt suscité et l'aide accordée à l'enseignement et à la recherche scientifique par les autorités de tutelle de ce pays. Je ne peux que formuler le souhait qu'il en soit rapidement de même outre Manche.

J'adresse, à toutes et à tous, mes plus vifs remerciements pour l'hospitalité que vous avez bien voulu me manifester durant ces trois dernières années.

Mr. President, Members of the Executive Committee, Members of the Union, Ladies and Gentlemen,

I think it can be said, without fear of contradiction, that this has been an exciting General Assembly. I am not just referring to the events of yesterday! There has been a great deal of exiting science in your meetings.

Following today's election, I am now free to do Astronomy and to teach. It is now my very pleasant task to thank all those very many people whose efforts have so materially contributed to the organisation of this General Assembly.

I would like at the outset to pay a special tribute to Monique Léger-Orine.   On her falls the responsibility for the smooth operation of the Secretariat. Every month I made a week long visit to Paris, created chaos in the Secretariat and by my next visit all was ordered once again. You have all noticed her work load here which involved a certain burning of the midnight oil.   Even after yesterday morning's fire, she had reorganised the Secretariat so that by the afternoon, work was calmly proceeding. Monique, we all owe you a great deal.

Monique is ably assisted in Paris by Julie Crook. I can only guess at what Julie must now be thinking of the ordered progress of a General Assembly. Monique and Julie were assisted here in Buenos Aires by my secretary -Valerie Peerless and by Jacqueline Bergeron's secretary -Valerie Demailly. I think you will agree that Julie and the two Valeries have done a fine job.

It has been my very great pleasure to serve two successive but effective Executive Committees. They had different approaches to science -both were very concerned to ensure that the IAU selected the best science for their Symposia, Colloquia and Regional Meetings. Both were noteworthy for their companionable good fellowship.

I received an effective training in General Secretaryship from Jean-Pierre Swings. He is a believer in that philosophy of teaching swimming expeditiously which involves pushing the pupil in at the deep end and shouting encouragement to keep one's head above water. He was always ready to answer questions and draw on his experience when I sought advice -your telephone will be quieter from now on Jean-Pierre.

It was a pleasure to work with Jacqueline Bergeron who undertook, with vigour, the administration of the Union's Scientific Meetings. As you may have noticed, it is likely she will display the same vigour in administering the affairs of the Union. However, I hope her final day as General Secretary will be a good deal less fraught than yesterday. Jacqueline, every success in the exacting task of General Secretary.

It has been a great pleasure to work with our past President, Jorge Sahade. He told me, I should think adventurously when he proposed holding a General Assembly in Argentina. Were it not for his adventurous thinking, I would not have fulfilled a childhood dream of seeing the Andes. I have also been able to visit the Argentinian observatories of Cordoba, La Plata, the Argentinian Institute of Radio Astronomy in Villa Elisa and the magnificently sited Leoncito. Thank you, Jorge.

It has been a pleasure to have worked with Yoshihide Kozai. He has carefully steered the Executive Committee through those long agendas that they must adjudicate. It was always a matter of some surprise that after long, intense debate we were usually at that place in the agenda at the time he and I had thought would be about right. Yet he never seemed to urge us on or hold us back -just skilful steering.

Now I must turn to the Local Organising Committee. Roberto Mendez and his team worked extremely hard. If a visiting General Secretary reinvented the wheel, they were too polite to say so. They worked hard to arrange the facilities for a good scientific meeting. They had to work during a period of high inflationary pressure. It is hard to realise that the current stability of exchange rate was only achieved from April 01 this year -it is hard for those of us visiting Argentina now to realise what it is like to have to live with 100% inflation in a week as was the case during one of my visits to Buenos Aires. Roberto quietly steered me from CONICET to Science and Technology and now that it has all happened, he must have got me to say the right things or perhaps diplomatically translated into Spanish what he thought I should have said. I would like to express my thanks to all of the Local

Organising Committee, to the National Organising Committee who supported them and to the Argentinian Astronomical Society who are our hosts on this occasion.

I would also like to say once again a special thank you to Juan O'Farrell who did so much to help us find meeting rooms after the Conference Centre fire yesterday. He performed miracles.

Let me not forget to express my appreciation to Patrick Moore and John Mason for their sterling work on Cruz del Sur. Despite the fire, you got a full 10 issues. The editing of the General Assembly newspaper is hard work and involves the keeping of "astronomical" hours. Well done.

I must also express my sincere thanks to Professor H.S. Ghielmetti, Director of the Instituto di Astronomia y Fisical del Espacio. He kindly extended the facilities of IAFE to the IAU, making his conference room available for my meetings with the LOC and accommodating the Secretariat before the General Assembly. His hospitality and his philosophy were much appreciated.

Finally may I say a special thank you to less visible people who have also done much to help me over the last six years. To Professors Sir Robert Wilson and Franz Heymann -the Heads of Department who agreed I could have time to undertake IAU duties, to Sir James Lighthill, Provost of University College London, for agreeing enthusiastically that I should become half time on full pay. To my secretary, Valerie Peerless, for undertaking a great deal of extra load and travel arranging with great cheerfulness -she can look forward to less frenetic times.

Hanbury Brown got me into this game. He said it would be fun and rewarding. It was.

I owe a great deal to my wife, more than perhaps I now realise, who has put up with many long absences from home. I hope she still recognises me. The IAU would not be the IAU without you, the Members. Thank you for coming and thank you for participating".

Address by the incoming General Secretary, Dr. J. Bergeron

"Dear Colleagues,

Being the last one to speak today, most of what I wanted to say has already been said, and I will thus not hold you very long.

It is a great honor and pleasure to have been chosen as the new General Secretary by the Special Nominating Committee and by you.

First, I wish to thank again the Local Oraganising Committee in particular for the speed at which they have reacted yesterday, after the fire broke out in the San Martin Cultural Center, and have been able to find another large amphitheatre to hold the General Assembly. I also wish to acknowledge the solidarity of the people of the Plaza and San Martin Cultural Centers and of Buenos Aires who have helped the Local Organising Committee and us so kindly.

During the last three years, I have greatly appreciated and relied upon the experience and advice of my predecessors Derek McNally, Jean-Pierre Swings, Richard West and Jean-Claude Pecker. Je compte beaucoup sur leur aide et leurs conseils dans les trois années à venir et j'espère qu'ils continueront à m'accorder leur soutien.

Let me now come to the future. I wish to briefly expose to you my personal ideas concerning the format of the next General Assembly that I will propose to the Executive Committee for consideration. I would like that longer scientific sessions be held and this could be done in three to four day symposia which could run in parallel with Commission Meetings. In 1994, about half of the symposia and colloquia could be part of the General Assembly in the Hague, with two to three symposia in parallel at a given time. This does not at all imply dropping out Joint Commission Meetings. Indeed one day meetings are important and should complement the symposia. I would greatly value to have your comments on this suggestion.

Finally, I hope that during the forthcoming triennium additional countries will seek to adhere to the Union.

Thank you again for your confidence and your patience."

CHAPTER IV

REPORT OF THE EXECUTIVE COMMITTEE

1988-1991

1. Composition of the Executive Committee

President:	Y.	Kozai	Japan
General Secretary:	D.	McNally	UK
Assistant General Secretary:	J.	Bergeron	France
President-Elect:	A.A.	Boyarchuk	USSR
Vice-Presidents:	A.H.	Batten	Canada
	R.	Kippenhahn	Germany
	P.O.	Lindblad	Sweden
	V.	Radhakrishnan	India
	M.S.	Roberts	USA
	Shu Hua	Ye	China PR
Advisers:	J.	Sahade	Argentina
	J.-P.	Swings	Belgium

2. Administration

Executive Committee Meetings

Four Meetings of the Executive Committee were held:

57th Meeting:	31.07-01.08.88	Baltimore, USA
		J. Sahade in the Chair
58th Meeting:	11.08.88	Baltimore, USA
		Y. Kozai in the Chair
59th Meeting:	01-04.09.90	Stockholm, Sweden
		Y. Kozai in the Chair
60th Meting:	07-10.09.90	Munich, Germany
		Y. Kozai in the Chair

The activities of the Executive Committee are recorded in the Minutes of the Executive Committee and summaries of these Minutes are published in the Information Bulletins (EC 57, 58 are summarised in IB 61 (pp. 10 & 11), 59 in IB 63 (pp. 10 & 11), 60 in IB 65 (pp. 6 & 7). The activities directed towards improving the Union's service to astronomy include the following:

i. A proposal to extend the Executive Committee, by the addition of a President-Elect, was put to the Extraordinary General Assembly which preceded the XXth General Assembly and carried. The object of this extension is to give greater continuity in the affairs of the Union for the holder of the office of President. The holder of the Office serves the Union for nine years -three as President-Elect, three as President and three as Advisor to the President.

ii. A proposal to institute a new class of National Member
-Associate Member- was put to the Extraordinary General
Assembly preceding the XXth General Assembly and carried.
The object of this proposal was to allow closer integration
to the Union of countries currently embarking on
the development of a commitment to astronomical science. Two
countries -Malaysia, Peru- were admitted to the Union as
Associate Members by the XXth General Assembly of the Union.

iii. The 57th Executive Committee established the Working Group
for the Worldwide Development of Astronomy under
the Chairmanship of Vice-President A.H. Batten with a wide-
ranging mandate to review, consider and encourage ways and
means whereby the Union could further enhance and extend
participation in astronomical science.

iv. The 58th Executive Committee established a subcommittee under
the Chairmanship of J. Sahade (with members -F. Drake,
A. Massevich, M. McCarthy & F.G. Smith) to discuss
the important issues raised by a Resolution submitted to
the XXth General Assembly by the Swedish National Committee
for Astronomy and subsequently withdrawn. The subcommittee
after consultation by post and a meeting at the Union
Secretariat reported:

 . that the IAU should approach ICSU, COSPAR and other
    scientific unions concerning the on-going and planned
    civilian and military activites in space that would
    pollute the space environment and would be harmful to
    the free pursuit of scientific research and convene, at
    the earliest possible time, an IAU/ICSU conference to
    discuss these problems and suggest appropriate
    activities;

 . that the IAU set up a Standing Advisory Committee to
    examine activities proposed on the ground and in space
    which might damage Astronomy;

 . that the IAU take a strong stand, with ICSU and before
    the relevant bodies calling for free exchanges of
    information and urging that all astronomically useful
    technological developments be made available for
    application in the progress of astronomy;

 . that the IAU urge the National Committees for Astronomy
    to take the necessary steps that their own countries
    would act individually and internationally in support of
    the aims of the subcommittee;

 . that the Executive Committee report on the result of
    their endeavours in respect of the conclusions of
    the subcommittee at the next IAU General Assembly.

The EC accepted the report of the subcommittee at their 60th
Meeting (Munich, 1990). The Executive Committee have

 . begun the organisation of an Expository Meeting with ICSU
    and UNESCO support to be held 1992 June 30, July 01, 02
    at UNESCO Headquarters, Paris;

.   had IAU resolutions on Adverse Environmental Impacts on
    Astronomy (in the widest terms) accepted by the General
    Committee of ICSU (1989) and the General Assembly of ICSU
    (1990) (in association with URSI);

.   with IUPAC and several other ICSU Unions asked ICSU to
    examine the impact of all weapons of mass destruction;

.   begun to consider the possibility of making a submission
    to the UN Committee on the Peaceful Uses of Outer Space;

.   proposed that Commission 50 considers the extension of
    its mandate to include all forms of Adverse Environmental
    Impact on Astronomy as a more appropriate alternative to
    a Standing Committee;

v. A continuing concern for the Executive Committee has been
    the financing of the Union. As national financial
    constraints bite deeper, an increasing number of countries
    are finding difficulty in maintaining their subventions to
    the Union. Currently the arrears of subvention are at
    a level of 75% of the annual expenditure of the Union on
    Symposia and Colloquia. The EC is aware that the financial
    contributions it can make to the scientific work of the Union
    can only be a small (and decreasing) fraction of their total
    cost. There is no money to support new and much needed
    ventures since the Union cannot increase its subvention
    unrealistically at a time when Adhering Bodies are under
    severe pressure to reduce their budgets.

vi. The Union has been considering fund-raising activities over
    a number of triennia and the EC will be putting a proposal to
    the XXIst General Assembly designed to measure support for
    the establishment of an IAU Trust Fund based on individual
    subscription by members, donations and bequests. The primary
    object of such a Trust Fund would be the accumulation of
    capital and income to finance new Union projects.

vii. The EC has reviewed the publication policy of the Union. It
     has put out the contract for the Union publications to tender
     and the contract has been retained by the Union's present
     publisher, Kluwer Academic Publishers for a further period of
     six years 1992-1997 against a tightly contested field.
     The Union will continue to benefit from the royalties
     generated by its publications. The EC has also looked at
     the continued publications of Transactions. Transactions B
     will be continued as the public record of the Union and
     a review of Transactions A is in progress. During
     the triennium it was decided to publish General Assembly
     information in the Information Bulletin in order to cut down
     on proliferation of documents and to keep a single source of
     information on Union activities.

viii. Following concern expressed during the XXth General Assembly
(in particular the levelling of the participation rate at
General Assemblies at about 2000 individuals despite
a doubling of Union membership every 14 years) the Executive
Committee has devoted considerable attention to this problem.
EC 58 set up a subcommittee to examine the matter under
the Chairmanship of the President (members: J. Lequeux,
D. McNally, E. Roemer, J.-P. Swings). A proposal, from
the subcommittee to shorten the General Assembly has been
made and the EC has referred the debate both to
the Presidents of Commission and the membership to seek
a consensus. Some recommendations from the subcommittee for
improving the presentation of science at the General Assembly
will be implemented with effect from the XXIst General
Assembly viz:

   .   Vice-Presidents and President-Elect serving ex officio on
       the Scientific Organising Committees of Joint
       Discussions.

   .   reduction in the number of Symposia/Colloquia associated
       with the General Assembly

   .   encouraging the greater use of Joint commission meetings

   .   the introduction of a limited display of posters.

Officers Meetings

The Officers -President, General Secretary, Assistant General
Secretary- met 7 times:

30.07.88	Baltimore, USA
10.08.88	Baltimore, USA
23-24.02.89	Paris, France
31.08.89	Stockholm, Sweden
13-14.02.90	Paris, France
06.09.90	Munich, Germany
07-02.91	Paris, France

IAU Secretariat

The Secretariat is now established in its new premises and
hosted by INSU. The postal address, telephone, fax, telex and
e-mail for the Secretariat are as follows:

Address:	Room 318/319, 98bis, Boulevard Arago, 75014 Paris, France
Telephone:	(33) 1-43-25-83-58
Telefax:	(33) 1-40-51-21-00
Telex:	205671 IAU F
e-mail:	IAPOBS::IAU or IAU@FRIAP51

Following the failure of the Union's IBM PC in the summer of
1989, a new computer system was installed to replace both
the failed PC and the Xerox Word Processing System. The new
system comprises 3 x Compaq 386s PCs supporting 2 dot matrix
printers, a laser printer and a tape streamer back up drive.
The new system with Word V, Lotus 1,2,3 and DBase III+ provides
an excellent service. The computers are linked using
Mainlan software.

In the spring of 1991 the Secretariat's computers were linked to the central computer of the Institut d'Astrophysique to provide an e-mail service.

IAU Staff

Mrs. H. Gigan resigned as part-time bilingual secretary with effect from 1990 January 31 and was replaced, with effect from 1990 July 09, by Mrs. J. Crook as full-time secretary/typist.

Following changes in French law, the position of Monique Léger-Orine, Administrative Assistant was regularised by CNRS. Monique Léger-Orine remains an employee of CNRS while seconded to the IAU.

Adhering Countries

Algeria, Iceland, Morocco & Saudi Arabia were admitted as Members of the Union at the XXth General Assembly. Malaysia & Peru were admitted as the first Associate Members of the Union at the XXth General Assembly.

Individual Membership

792 new individual Members were admitted to the Union at the XXth General Assembly, bringing the total membership to 6711 as of 1988 November 30. As of 1991 March 31 the membership of the Union was 6624.

Consultants

The 40 Commissions of the Union and the associated Working Groups were assisted by 223 Consultants during the triennium.

3.   Commissions of the IAU

Commission Reports for the period 1987 July 01-1990 June 30 have been published as IAU Transactions XXIA -Reports on Astronomy (619 pages). The triennium has been one of characteristic activity for the Commissions of the Union. Several Commissions have commented that the striking feature of the triennium was the launch of several major "observatories" into space -HIPPARCOS, the Hubble Space Telescope, ROSAT- which will do much to change the perspective of astronomical science in the current decade and beyond. Commissions offering specific services are listed below.

Commission 5: Documentation & Astronomical Data

The IAU Style Manual prepared by G.A. Wilkins was published in IAU Transactions XXB -Proceedings of the XXth General Assembly and was dedicated to the memory of D.H. Sadler, General Secretary of the Union 1958-1964. The Style Manual is available on request from the IAU Secretariat. A summary of the Recommendations contained in the Manual was published in IB 65 (pp. 8-13).

Commission 6: Astronomical Telegrams

The IAU Central Bureau for Astronomical Telegrams (Director: B.G. Marsden) issued the following number of *Circulars & Telegram Books* in 1988, 1989, 1990 respectively: *175, 35; 240, 59; 216, 43.* 1989 was the first full year of e-mail dissemination of the Circulars -by the end of 1990 there were 207 subscribers to this service, together with over 700 subscribers to the printed Circulars.

Commission 19: Rotation of the Earth

The period has been marked by the start of the new International Earth Rotation Service (IERS), which benefits from a tight co-operation between astronomers, geodesists and specialists in satellite geodesy, as well as meteorologists. The scope of the IERS covers not only the Earth's rotation *per se*, but also the conventional terrestrial reference frame, of direct interest to the International Association of Geodesy, and a high accuracy (0.001") celestial reference frame based on extragalactic compact sources observed in Very Long Baseline Interferometry. The IERS conventional celestial reference frame is consistent with the FK5 within the uncertainties of the latter (0.04"). The IERS Standards (1989) which contain the current best estimates of astronomical models and constants are used in many fields of astronomy and geodesy.

Shortly after the adoption of the IAU 1980 Theory of Nutation, VLBI and LLR observations started to show evidence of inaccuracies in some of its components, at the level of 0.001" to 0.01"; in addition, these observations show that the IAU 1976 precession constant requires a correction of about -0.27"/cy. To avoid systematic errors that would result from the use of the conventional models, IERS has complemented the usual three orientation parameters (x, y which describe the motion of the rotation axis in the Earth, and universal time) by two additional parameters, the celestial pole offsets in longitude and in obliquity. These two parameters can be directly interpreted in terms of nutations in space of the rotation axis.

In parallel to the development of the observing techniques and to their efficient coordination in IERS, the high accuracy (better than 0.001") of today's Earth orientation determinations has encouraged deeper research in the contribution of the various parts of the planet to its rotation variations. Independent analyses converge to ascribe the irregular oscillations in polar motion and universal time to the atmospheric influence, for periods from a few days to several years. The corresponding atmospheric mechanisms are progressively unveiled for the rapid oscillations, but much remains to be understood for the lower frequencies (Chandler term, Southern oscillation) and for the suspected relationships with the solar emissions. The influence of the oceans and groundwater is being clarified. The study of solid Earth effects is also progressing, thanks to the parallel developments of theory and geophysical measurements.

Commission 20: Positions & Motions of Minor Planets,
       Comets & Satellites

Minor Planet Circulars were issued, under the direction of B.G.
Marsden, to a total of 1448, 1528 & 1872 pages in 1981, 1989,
1990 respectively. In 1988, 1989 & 1990, 236, 3339 & 384 minor
planets were numbered.

Commission 38: Exchange of Astronomers

Since September 1988, Commission 38 assisted the travel of
5, 12 and 8 astronomers in 1988, 1989 and 1990 respectively,
disbursing a total of 56,000 Sw.F. The bulk of these grants
was to assist astronomers from the developing world spend
a period of three months or longer at a host institution to
further their astronomical research.

The list of the grantees (September 1988-December 1990) is
given below:

Name	Home Institution *Host Institution*	Country
Bhattacharya D.	Raman Institute *University of Amsterdam*	India *Netherlands*
Beauge C.	Observatorio Astronomico *Institute of Astronomy* *Cambridge*	Argentina *UK*
Clocchiatti A.	Observatorio Astronomico *University of Texas* *McDonald Observatory*	Argentina *USA*
Das Gupta	IUC Astronomy & Astrophysics *University of Wales Cardiff*	India *UK*
Dwivadi B.N.	Banaras Hindu University *Glasgow University*	India *UK*
Ferraz-Melo S.	Sao Paulo University *Observatoire de Paris-Meudon*	Brazil *France*
Ghosh P.	Tata Institute *University of Illinois*	India *USA*
Gomez M.	Observ. Astronomico Cordoba *Smithsonian Astrophys. Obs.*	Argentina *USA*
Ibanez M.H.	University de Los Andes *St Andrews*	Venezuela *UK*
Jerzykiewicz M.	Wroclaw Univ. Observatory *SAAO*	Poland *South Africa*
Jorissen A.	Institut Astronomie Brussels *University of Texas Austin*	Belgium *USA*
Kim Ho-Il	Yonsei Univ. Observatory, Seoul *Univ. Nebraska, Lincoln*	Korea R *USA*
Kubacek D.	Astronomical Institute *Harvard-Smithsoninan*	Czechoslovakia *UK*

Li Zhongyuan	University of Science & Technology *MSSL*	China PR *UK*
Mauas R.	IAFE *Harvard Smithsonian*	Argentina *USA*
Nagendra K.N.	Indian Inst. of Astrophysics *Rensselaer Polyt. Inst. NY*	India *USA*
Ryan S.	Mount Stromlo Observatories *Yale University/* *Institute of Astronomy* *Cambridge*	Australia *USA* *UK*
Sagar R.	Indian Inst. of Astrophysics *AAO*	India *Australia*
Saha S.K.	Indian Inst. Astrophysics Bangalore *CERGA-OCA, Grasse*	India *France*
Sattyaprakash B.S.	IUC Astronomy & Astrophysics *University of Wales Cardiff*	India *UK*
Seimis J.	University of Athens *Inst. Theoretical Astronomy* *Leningrad/* *N. Copernicus Astron. Center*	Greece *USSR* *Poland*
Singal A.K.	TIFR *Inst. für Astrophysik Bonn*	India *Germany*
Stefl S.	Ondrejov Observatory *University of Tucson Arizona* *University of Toronto*	Czechoslovakia *USA* *Canada*
Tancredi G.	University Rep. Montevideo *Uppsala University*	Uruguay *Sweden*
Wang Ruyou	Shangai Observatory *NOAO*	China PR *USA*

The rules adopted by Commission 38 for its travel grants are published annually in the January issue of the Information Bulletin.

Commission 46: *Teaching of Astronomy*

- Visiting Lecturer Programme

Two VLP programmes -Paraguay & Peru- have been supported during the triennium.

The VLP in Peru is now in its sixth year of operation and, in principle, is due to terminate (but see later).

The VLP in Paraguay will not end until 1992 and can be extended.

These VLPs have stimulated interest in astronomical studies in the host countries. However, that very success has identified further problems not anticipated initially. Although the VLP in Peru should end in 1991, the IAU has only been able to supply four out of six lecturers specified in the contract. It is proving very difficult in the present financial climate for lecturers to obtain leave of absence with pay for the duration of a lecture course.

- International Schools for Young Astronomers (ISYA)

Three International Schools for Young Astronomers have been
held during the triennium:

No. 16   Cuba (Havana),   1989 July 31-August 20,
No. 17   Malaysia (Malacca),   1990 May 28-June 15,
No. 18   Morocco (Marakesh),   1990 September 10-28,
              this last one with support from UNESCO.

All 18 ISYA have been organised for the IAU by J. Kleczek.
Following his retirement, he has relinquished the direction
of the ISYA. The IAU is very grateful to Dr. Kleczek for his
devotion to the ISYA, for his patience and hard work in
organising them and for his active participation in each
School.   He has established a fine institution and
the success of the ISYA is no more clearly demonstrated
than by the fact that Dr. Kleczek's co-organiser of the 17th
ISYA was   herself   a graduate   of   an earlier   School.
Dr. Kleczek's successor will be Prof. D. Wentzel, assisted by
Dr. M. Gerbaldi.

4.   Scientific meetings held

Symposia (14)

    134   *Active Galactic Nuclei*
           15-18.08.88, Santa Cruz, USA

    137   *Flare Stars in Star Clusters, Associations
           & the Solar Vicinity*
           23-27.10.89, Byurakan, USSR

    138   *Solar Photosphere: Structure, Convection
           & Magnetic Fields*
           15-20.05.89, Kiev, USSR

    139   *Galactic & Extragalactic Background Radiation
           -Optical, Ultraviolet & Infrared Components*
           12-16.06.89, Heidelberg, Germany

    140   *Galactic & Intergalactic Magnetic Fields*
           19-25.06.89, Heidelberg, Germany

    141   *Inertial Coordinate System on the Sky*
           07-21.10.89, Leningrad, USSR

    142   *Basic Plasma Processes on the Sun*
           01-05.12.89, Bangalore, India

    143   *Wolf-Rayet Stars & Interrelations with
           other Massive Stars in Galaxies*
           18-22.06.90, Denpasar, Indonesia

    144   *The Interstellar Disk-Halo Connection in Galaxies*
           18-22.06.90, Leiden, Netherlands

    145   *Evolution of Stars:
           The Photospheric Abundance Connection*
           27-31.08.90, Bruzba, Bulgaria

    146   *Dynamics of Galaxies
           & their Molecular Cloud Distributions*
           05-09.06.90, Paris, France

147   *Fragmentation of Molecular Clouds & Star Formation*
       11-15.06.90, Grenoble, France

148   *The Magellanic Clouds & their Dynamical*
       Interaction with the Milky Way
       09-13.07.90, Sydney, Australia

152   *Chaos, Resonance & Collective Dynamical*
       Phenomena in the Solar System
       July 15-19, 1991, Angra dos Reis, Brazil

Colloquia (22)

104   *Solar & Stellar Flares*
       15-19.08.88, Stanford, USA

107   *Algols*
       15-19.08.88, Victoria, Canada

111   *The Use of Pulsating Stars in Fundamental Problems*
       of Astronomy
       15-17.08.88, Lincoln, USA

112   *Light Pollution, Radio Interference & Space Debris*
       15-17.08.88, Washington, USA

113   *Physics of Luminous Blue Variables*
       15-18.08.88, Quebec, Canada

114   *White Dwarfs*
       14-19.08.88, Hanover, USA

115   *High Resolution X-Ray Spectroscopy of Cosmic Plasmas*
       22-25.08.88, Cambridge, USA

116   *Comets in the post-Halley Era*
       24-29.04.89, Bamberg, Germany

117   *Dynamics of Solar Prominences*
       25-29.09.89, Hvar, Yugoslavia

120   *Structure & Dynamics of the Interstellar Medium*
       16-19.04.89, Granada, Spain

121   *Inside the Sun*
       22-26.05.89, Versailles, France

122   *Physics of Classical Novae*
       27-30.06.89, Madrid, Spain

123   *Observations in Earth Orbit & Beyond*
       24-27.04.90, Greenbelt, USA

124   *Paired & Interacting Galaxies*
       04-07.12.89, Tuscaloosa, USA

125   *Radio Recombination Lines: 25 years*
       of Investigations
       11-16.09.89, Pushino, USSR

126   *Origin & Evolution of Interplanetary Dust*
       27-30.08.90, Kyoto, Japan

127   *Reference Systems*
       14-20.10.90, Virginia Beach, USA

128 *Magnetospheric Structure*
     & Emission Mechanisms of Radio Pulsars
     17-23.06.90, Lagow, Poland

129 *Structure & Emission Properties*
     of Accretion Disks
     02-06.07.90, Paris, France

130 *The Sun & Cool Stars: Activity, Magnetism, Dynamics*
     17-21.06.90, Helsinki, Finland

131 *Radio Interferometry*
     -Theory, Techniques & Applications
     8-12.10.90, Socorro, USA

132 *Instability, Chaos & Predictability*
     in Celestical Mechanics & Stellar Dynamics
     10-13.10.90, Delhi, India

Regional Meetings (4)

11th European Regional Astronomy Meeting
03-07.07.89, Tenerife, Spain

6th Latin-American Regional Meeting
16-20.10.89, Gramado, Brazil

5th Asian-Pacific Regional Meeting
16-20.07.90, Sydney, Australia

12th European Regional Meeting
European Astronomers look to the Future
8-11.10.90,Davos, Switzerland

Co-Sponsored Meetings (11)

Radio Astronomy Seeing -Tropospheric
& Ionospheric Effects
15-19.05.89, Beijing, China PR, with URSI

Physics of the Outer Heliosphere
19-22.09.89, Warsaw, Poland
with COSPAR & the Polish Academy of Sciences

7th Quadriennial *Solar Terrestrial Physics* Symposium
25-30.06.90, the Hague, Netherlands
Co-sponsored by SCOSTEP, COSPAR, IAGA, IAMAP, URSI & IUPAP.

COSPAR XXVIII: the Hague, Netherlands:
     Plenary Meeting, 25.06-06.07.90

     Magnetospheres of the Outer Planets, 25-27.06.90

     The Infrared Submillimeter Universe at high Redshifts
     25-29.06.90

     Astrometry from Space, 27-28.06.90

     Space Observations of the Solar Corona
     & the Origin of the Solar Wind
     27-29.06.90

     Latest Results on Mars & Phobos Studies
     28-30.06.90

     Neptune after Voyager, 02-04.07.90

General Assembly

The XXth General Assembly was held from 02 through 11.08.88 in
Baltimore, USA. IAU Transactions XXB (1990) contain the Report
of the Proceedings of the Extraordinary General Assembly,
the General Assembly of the Union, the Resolutions adopted by
the General Assembly and the report of the business and
scientific sessions of the Union's 40 Commissions during
the General Assembly.

5.   Publications

Publisher

The tender for the IAU Publishing Contract was issued in late
1989 and a choice was made between four highly competitive
tenders.   Kluwer   Academic   Publishers   -the current   IAU
publisher- offered the most advantageous terms and a new
contract was signed in 1991 March.   The new contract will run
for six years from 1992 January 01-1997 December 31.

Sales

Number of copies sold (**hard,** soft) by Kluwer Academic
Publishers in the period 1988-1990:

Transactions

XIIIA (1), XIVA (1), XVA (1), XVIA Part 2 (5),
XVIIA Part 2 (3), XVIIB (4), XVIIIA (6), XIXA (11),
XIXB (29), XXA (326), XXB (244).

Highlights

1 (2), 2 (2), 3 (4), 4 Part 1 (2,1), 4 Part 2 (11,2), 5 (4),
6 (10), 7 (35,25), 8 (210,157).

Symposia Proceedings

32 (7), 35 (4), 37 (2), 38 (3), 39 (5), 40 (3), 43 (4),
44 (7), 45 (5), 46 (3), 47 (8), 48 (6), 49 (1), 50 (7),
51 (2,1), 52 (5), 53 (2), 54 (2,1), 55 (3,1), 56 (3),
57 (4,1), 58 (3), 59 (2,1), 60 (2), 61 (7,1), 62 (3,1),
63 (1), 64 (3,3), 65 (4), 66 (3), 67 (5,4), 68 (3,2),
69 (4,3), 70 (3,2), 71 (4,3), 72 (3,2), 73 (6,2), 74 (6,3),
75 (3,5), 76 (4,2), 77 (2,1), 78 (4,1), 79 (7,3), 80 (3,2),
81 (5,3), 82 (18,4), 83 (2,1), 84 (4), 85 (21), 86 (7,1),
87 (6), 88 (6,3), 89 (6,1), 90 (4,4), 91 (5,3), 92 (4),
93 (4,2), 94 (2,1), 95 (10), 96 (6), 97 (8), 98 (2,9),
99 (2), 100 (3,4), 101 (5,1), 102 (9,4), 103 (8,4),
104 (9,10), 105 (11,7), 106 (11,13), 107 (7,11), 108 (6,8),
109 (11,10), 110 (11,14), 111 (11,14), 112 (28,34),
113 (18,19), 114 (12,44), 115 (17,39), 116 (5,17),
117 (34,76), 118 (33,37), 119 (46), 120 (12,34), 121 (30,64),
122 (49,112), 123 (47,81), 124 (79,135), 125 (119,126),
126 (294,176), 127 (54,177), 128 (345,132), 129 (326,176),
130 (318,206), 131 (280,182), 132 (286,151), 133 (309,157),
134 (297,267), 135 (289,182), 136 (297,222), 137 (236,115),
138 (297,150), 139 (263,109), 140 (309,165), 141 (261,144),
142 (258,110).

Information Bulletin

Six issues of the Information Bulletin (Nos. 61-66) were published during the triennium. Members of the Union, its Consultants, Adhering Organisations, sister Unions and selected Scientific Institutions receive the Information Bulletin free of charge. The print run is now 7700 copies.

6.   Relations with other organisations

International Council of Scientific Unions (ICSU)

The IAU was represented by the General Secretary at the following meetings of (ICSU):

General Assemblies:
XXIInd (Beijing, 1989) & XXIIIrd (Sofia, 1990)

Meetings of the General Committee:
24th, 25th (Beijing, 1988), 26th (Lisbon, 1989), 27th & 28th (Sofia, 1990)

ICSU, with immediate effect, moves to a three year cycle of General Assemblies, the 24th will be held in 1993, though the General Committee will continue to meet annually between General Assemblies.

ICSU's flagship project -The International Geosphere Biosphere Programme- moves from its planning to implementation phase in 1991; the Committee on Ethical Problems of Science has been subsumed as a function of ICSU's Executive Board; the Executive Board is to carry out a study into the Freedom of Access to Information.
ICSU has given strong support to the IAU, URSI, IUCAF in their endeavours to minimise Adverse Environmental Impacts on Astronomy and is cosponsoring, together with UNESCO, an Expository IAU Meeting on this topic in mid-1992.

The IAU was represented at the XXVIIth Plenary Meeting (The Hague, 1990) of COSPAR (The ICSU inter-Union Committee on Space Research) by the General Secretary & L. Perek.

IAU representatives to other ICSU & International Institutions

The following representatives were active during the period 1988-1991:

Acronyms	Organisation	Representative
ICSU	International Council of Scientific Unions General Committee	D. McNally
BIPM/ CCDS	Bureau international des poids et mesures Working Group on the Temps Atomique International of the Consultative Committee for the Definition of the Second	G. Winkler
CCIR	International Radio Consultative Committee Study Group 2	J. Whiteoak/ A.R. Thompson
	Study Group 7	S. Leschiutta
CIE	Compagnie internationale de l'éclairage	D. Crawford

CODATA	Committee on Data for Science & Technology	G. Westerhout
COSPAR	Committe on Space Research	D. McNally
	COSPAR ISC B	J. Rahe
	COSPAR ISC D	S. Grzedzielski
	COSPAR ISC E	Y. Kondo
	COSPAR Sub. Committee E1	R. Sunyaev
	COSPAR Sub. Committee E2	M. Pick
COSTED	Committee on Science & Technology in Developing Countries	D. McNally
CTS	Committee on the Teaching of Science	L. Gouguenheim
EPS	European Physical Society Conference Committee	J. Bergeron
FAGS	Federation of Astronomical & Geophysical Services	E.Tandberg-Hanssen
IAF	International Astronautical Federation	Y. Kondo
IERS	International Earth Rotation Service	Ya. Yatskiv
IGBP	International Geosphere-Ionospehre Programme	J. Eddy
ISY	International Space Year	L. Gouguenheim
IUCAF	Inter-Union Commission on Frequency Allocation for Radio Astronomy & Space Science	B.A. Doubinsky, N. Kaifu, V.L. Pankonin, G. Swarup
IUPAP	International Union of Pure & Applied Physics	V. Trimble
IUWDS	International Ursigram & World Day Service	H. Coffey
QBSA	Quarterly Bulletin on Solar Activity	E. Hiei
SCOPE	Scientific Committee on Problems of Environment	R. Cayrel
SCOSTEP	Scientific Committee on Solar-Terrestrial Physics	S.T. Wu
URSI	Union Radio-Scientifique Internationale	J. Baldwin

7.   Financial matters

The triennial accounts for 1988, 1989 & 1990 are summarised below. The information is extracted from the "Vérification des Comptes" for 1988, 1989 & 1990 as certified by the IAU Auditors, P. Meyssonnier EC (1988) & J.-M. Malaize (1989 & 1990).

The accounts are presented in terms of Sw.F.

The ICSU mean annual conversion rates for Swiss francs for 1988-1990 are:

	US $	DFl.	FF	£
1988	1.4492	0.7397	0.2463	2.5883
1989	1.6267	0.7679	0.2560	2.6773
1990	1.4008	0.7630	0.2562	2.4618

The audited accounts are given to two decimal places. The summary accounts are rounded to the nearest Sw.F.

The unit of subscription for 1988, 1989 & 1990 was 1975, 2145 & 2220 Sw.F respectively.

SUMMARY ACCOUNTS 1988-1990

INCOME

		BUDGET	1988	1989	1990	TOTAL
1.	CONTRIBUTIONS					
1.1.	Adh. Organisations	1457490	461105	476858	493041	1431004
1.2.	ICSU Allocation	87000	25122	25089	28333	78544
	TOTAL 1.	1544490	486227	501947	521374	1509548
2.	PUBLICATIONS					
2.1	Royalties)	165000	55478	57593	60482	173553
2.2.	Others)	-	73	0	3268	3341
	TOTAL 2	165000	55551	57593	63750	176894
3.	INTEREST AND OTHER RECEIPTS					
3.1.	Current accounts)		6394	457	797	7648
3.2.	Saving accounts)	85000	15621	42794	53694	112109
3.3.	Gains on Exchange)		45	0	634	679
	TOTAL 3	85000	22060	43251	55125	120436
4.	OTHER RECEIPTS					
4.1.	Sp. contributions)		4926	0	0	4926
4.2.	Grants)		0	0	0	0
4.3.	Refunds)	-	14444	3070	2604	20118
4.4.	Other)		0	2859	0	2859
	TOTAL 4.	0	19370	5929	2604	27903
	TOTAL INCOME	1794490	583208	608720	642853	1834781

Notes on Income:

1. Shortfall on contributions from Adhering Organisations and from ICSU. With increasing levels of default by Adhering Organisations and despite an expected rise from ICSU in the early part of the next triennium, the increase ICSU subvention due in 1993 will lead to further erosion in the value of the ICSU contribution.

2. Royalty income was moderately in increase of budget.

3. The apparent good result in interest obtained must be offset against losses on exchange.

4. Income from refunds etc was in excess of expectation.

    Total income was in excess of budget expectation though losses on exchange made an income in line with the budget expection.

EXPENDITURE

		BUDGET	1988	1989	1990	TOTAL
1.	EXECUTIVE COMMITTEE					
1.1.	EC mtgs	80000	623	39431	31116	71170
1.2.	Officer's mtgs	24000	2549	2346	4689	9584
1.3.	SNC expenses	5000		0		0
1.4.	Others	0		0		0
	TOTAL 1	109000	3172	41777	35805	80754
2.	PUBLICATIONS					
2.1.	Information Bulletin	56000	15662	21246	13144	50052
2.2.	Other Publications	0	1035	0	0	1035
2.3.	Devel. Countries)		1667	10159	3974	15800
2.4.	Exec. Committee)	32000	0	8217	5027	13244
2.5.	Post/Handling)		0	2056	947	3003
	TOTAL 2	88000	18364	41678	23092	83134
3.	SCIENTIFIC ACTIVITIES					
3.1.	MEETINGS					
3.1.1.	General Assembly	240000	337500	13337	104	350941
3.1.2.	Symposia & Colloquia	365000	105228	121824	109458	336510
3.1.3.	Reg. Astr. Meetings	65000	0	27000	19252	46252
3.1.4.	ISYA	50000	0	11669	19373	31042
3.1.5.	VLP	72000	2894	4261	11182	18337
3.1.6.	Others	0	0	6020	3000	9020
	Sub-total 3.1.	792000	445622	184111	162369	792102
3.2.	COMMISSION ACTIVITIES					
3.2.1.	Commission Expenses	10000	2702	2000	2009	6711
3.2.2.	Commission Projects					
	Exch. Astronomers	70000	15078	30665	15902	61645
	Telegram Bureau)			3253		3253
	Minor Planet Ctr)			3253		3253
	Variable Star Cat)	51500	2595	9235		11830
	Meteor Data Ctr)		0	0	1000	1000
	Others		0	0	0	0
	Sub-total 3.2.	131500	20375	48406	18911	87692
3.3.	RELATIONS WITH OTHER ORGANISATIONS					
3.3.1.	Subvention to ICSU	36000	6521	651	0	7172
3.3.2.	Inter-Union Commissions	21000	6159	11214	769	18142
3.3.3.	Repr. to other Organis.	24000	4028	2432	3405	9865
	Sub-total 3.3.	81000	16708	14297	4174	35179

3.4.	**OTHER ACTIVITIES**				
3.4.1.	EC projects	15000	0	3452	601	4053
3.4.2.	Other projects	0	0	0	0	0
	Sub-total 3.4.	15000	0	3452	601	4053
	TOTAL 3.	1019500	482705	250266	186055	919026

4.	**ADMINISTRATION**				
4.1.	**SECRETARIAT**				
4.1.1.	Salaries/Charges)		105689	115722	102473	323884
4.1.2.	Office Expenses)		45811	114756	44570	205137
4.1.3.	Gl Secr. Expenses)	592500	16181	16012	17674	49867
4.1.4.	Pdt's Expenses)		7117	0	0	7117
4.1.5.	AG Secr. Expenses)		0	0	649	649
	Sub-total 4.1.	592500	174798	246490	165366	586654
4.2.	**OTHERS**				
4.2.1.	Bank charges)		2438	3581	4464	10483
4.2.2.	Audit fees)	6000	2921	1586	2431	6938
4.2.3.	Loss on exchange)		10652	41004	4349	56005
4.2.4.	Miscellaneous)		0	2019	793	2812
	Sub-total 4.2.	6000	16011	48190	12037	76238
	TOTAL 4.	598500	190809	294680	177403	662892
	Additional Expenditure Symposium 130		740			740
	TOTAL EXPENDITURE	1815000	695790	628401	422355	1746546

Notes on Expenditure

1. Executive Committee Meetings within the budget (one meeting only just). Officers Meetings within budget -one Officer on site and another carefully scheduled other travels.

2. Information Bulletin appears within budget but cost of distribution of one issue is not included in 1990. Cost of Distribution to Developing countries and EC is variable depending on what is published in the year.

3. There was considerable overspend on the XXth General Assembly -however the figure does improperly include subventions to both Telegrammes & Minor Planet Centers.

    The VLP spend has been very much less than anticipated.

    Telegrammes & Minor Planets Circulars were fully paid in early 1991.

    Subvention to ICSU was brought up to date in early 1991 after changes in payment structure. The basis of payment to ICSU will increase in 1993 from 2.5% to 3.5% of receipts from Adhering Organisations.

4.   Salaries remain within budget as do Office Expenses. The increase in
            Office Expenses in 1989 reflected the move of the Secretariat and
            the acquisition of new Computer facilities.

            Bank charges and commission is increasing.

            Audit charges dropped over the triennium.

            There was a 5 month saving on salary for one member of the staff.

BALANCE 1988-1990

The Balance for the years 1988-1990 is set out below:

	1988	1989	1990
Balance at 31 December of preceding year	+781963	+661997	+658873
Revaluation of balance {+ = gain} {- = loss}	-7388	+16558	-19531
Income	+583208	+608719	+642853
Expenditure	-695791	-628401	-422355
Balance at end of year	+661997	+658873	+859840
Excess of Income over Expenses	-112582	-19682	+220498
Budget excess of Income over Expenditure	-106170	+46350	+44710

Notes on Balance 1988-199

1.   The balance at the end of 1990 is large but should be diminished by
            about 50,000 Sw.F. due in 1990 but unpaid until early 1991. The balance
            remaining is satisfactory.

2.   The average cost of running the Union in the triennium is budgeted to
            rise to an average of 730,000 Sw.F. annually.   The Union Balance is
            therefore only 80,000 in excess of 1 year of operation in the next
            triennium.

PRELIMINARY ACCOUNTS 1991 01 - JUNE 30

The preliminary accounts are set out below and follow the format of the Summary Accounts 1988-1990.   The Executive Committee Budget for 1991 is given, together with the sums remaining to be collected or disbursed.   The figures represent the provisional state of the Union's Income and Expenditure as of 1991 June 20; the figures are unaudited and do not include gain/loss on exchange.

The unit of contribution for 1991 was 2220 Sw.F.

INCOME

		INCOME	EC BUDGET	DIFFERENCE
1.	CONTRIBUTIONS * * Expected ICSU grant not yet received	451783	554360	102577
2.	PUBLICATIONS	0	55000	55000
3.	INTEREST	26206	25000	1206
4.	OTHER RECEIPTS	1588	0	1588
	TOTAL INCOME	479577	634360	154783

EXPENDITURE

		EXPENDITURE	EC BUDGET	DIFFERENCE
1.	EXECUTIVE COMMITTEE	7201	8000	799
2.	PUBLICATIONS			
2.1.	Information Bulletin	18622	22000	3378
2.2.	Publications Devel. Countries Exec. Committee	11473	10000	-1473
	TOTAL 2	30095	32000	1905
3.	SCIENTIFIC ACTIVITIES			
3.1.	MEETINGS			
3.1.1.	General Assembly	67083	240000	1972917
3.1.2/3	Symposia & Colloquia	6454	125000	118546
3.1.4.	Reg. Astr. Meetings	0	0	0
3.1.5.	ISYA 457	0	-457	
3.1.6.	VLP 0	24000	24000	
	Sub-total 3.1.	73997	389000	315006
3.2.	COMMISSION ACTIVITIES			
3.2.1.	Commission Expenses	0	5000	5000
3.2.2.	Exch. Astronomers	1786	24000	22214
	Other Projects	2375	13500	11125
	Sub-total 3.2.	4161	42500	38339

3.3.   RELATIONS WITH OTHER ORGANISATIONS

3.3.1.   Dues to ICSU *   31671   13200   -18471
3.3.2.   Other dues   18879   14000   -4879
          * 2 years of subvention to ICSU

          Sub-total 3.3.   50550   27200   -23350

3.4.   OTHER ACTIVITIES   3710   5000   1290

          TOTAL 3.   132415   463700   331285

4.   ADMINISTRATION

4.1.   Salaries/Training   39797   130000   90203
4.2.   Office Expenses   20532   60000   39468
4.3.   Officers' Expenses   10697   31000   20303
4.4.   Miscellaneous   5833   2500   -3333

          TOTAL 4.   76859   223500   331285

          TOTAL EXPENDITURE   246570   727200   480630

          Excess of Income over Expenditure

PROPOSED IAU BUDGET 1992-1994

This budget, expressed in Sw.F., is planned on the basis of 248 units of contribution and an allowance of 5%/annum has been made for inflation (i.e. about 20% over 3 years) + an additional 10 Sw.F./annum/unit of contribution to compensate the loss of the DDR subvention.

The unit of contribution will be:

$$1992: 2335 + 10 = 2345$$
$$1993: 2450 + 10 = 2460$$
$$1994: 2570 + 10 = 2580$$

INCOME

		1992-94	1992	1993	1994	1989-91
1.	CONTRIBUTIONS					
1.1.	Adh. Organisations	1831480	581560	610080	639840	1532720
1.2.	ICSU Allocation	93200	34000	28900	30300	78000
1.3.	Grants UNESCO	13000	13000	-	-	-
	TOTAL 1.	1937680	628560	638980	670140	1610720
2.	PUBLICATIONS					
	Royalties	186000	62000	62000	62000	165000
	TOTAL 2	186000	62000	62000	62000	165000
3.	INTEREST AND OTHER RECEIPTS					
3.1.	Bank interest	75000	25000	25000	25000	75000
3.2.	Other income *	-	-	-	-	-

* May be some additional income in 1994 on sale of IAU History

		1992-94	1992	1993	1994	1989-91
	TOTAL 3.	75000	25000	25000	25000	75000
Up 19%	TOTAL INCOME	2198680	715560	725980	757140	18507

EXPENDITURE

An effort has been made to increase scientific expenditure

		1992-1994	1992	1993	1994	1989-1991
1.	EXECUTIVE COMMITTE					
1.1	EC mtgs	95000	45000	50000 -		80000
1.2	Officer's mtgs	24000	8000	8000	8000	24000
1.3	SNC expenses	5000	-	5000	-	5000
Up 14%	TOTAL 1	124000	53000	63000	8000	109000

2.	PUBLICATIONS					
2.1	Information Bulletin	66000	20000	22000	24000	63000
2.2.	Other Publications *	25000	-	-	25000	-
2.3	Developing Countries	36000	12000	12000	12000	30000
2.4.	Executive Committee	15000	5000	5000	5000	-

* Allowance for IAU History in 1994

Up 53%	TOTAL 2	142000	37000	39000	66000	93000
3.	SCIENTIFIC ACTIVITIES					
3.1	MEETINGS					
3.1.1.	General Assembly	275000	-	-	275000	240000
3.1.2.	Symposia/Colloquia	414000	133000	138000	143000	365000
3.1.3	Regional Astr. Mtgs *	45000	26000	19000	-	65000
3.1.4.	ISYA	60000	30000	30000	-	50000
3.1.5.	VLP	84000	28000	28000	28000	72000
3.1.6.	AEI Meeting	44500	44500	-	-	-

* Only Latin America/Asia Pacific to fund

Up 14%	Sub-total 3.1.	922500	261000	215000	446000	792000
3.2.	COMMISSION ACTIVITIES					
3.2.1.	Commission expenses	15000	5000	5000	5000	15000
3.2.2.	Commission projects					
	Exch. of Astronomers	76000	25000	25000	26000	70000
	Telegram Bureau	11000	3650	3650	3700	10500
	Minor Planet Ctr	11000	3650	3650	3700	10500
	Variable Star Cat	11000	3650	3650	3700	10500
	Meteor Data Ctr	3200	1050	1050	1100	3000
	Others	6000	2000	2000	2000	6000
Up 6%	Sub-total 3.2.	133200	44000	44000	45200	125500
3.3.	RELATIONS WITH OTHER ORGANISATIONS					
3.3.1.	Subvention to ICSU	58400	14600	21400	22400	38500
3.3.2.	Inter-Union Commiss.					
	FAGS	19500	6000	6500	7000	13000
	IUCAF	21500	8000	6000	7500	-
	CTS	1800	600	600	600	-
	GLOBMET	2800	500	800	1500	-
3.3.3.	Repr. to other Organisations	25500	8000	8500	9000	24000
Up 71%	Sub-total 3.3.	129500	37700	43800	48000	75500
3.4.	OTHER ACTIVITIES					
3.4.1.	EC projects	15000	5000	5000	5000	15000
3.4.2.	Other projects	6250	2000	2750	1500	-
	Sub-total 3.4.	21250	7000	7750	6500	15000
Up 10%	TOTAL 3	1206450	350200	310550	545700	1008000

4.   ADMINISTRATION

4.1.   SECRETARIAT

4.1.1.	Salaries/Charges	426000	135000	142000	149000	375000
4.1.2.	Training courses	9000	3000	3000	3000	-
4.1.3.	Office supplies	191000	61000	64000	66000	165000
4.1.4.	Gl Secr. Expenses	39000	12500	13000	13500	57000
4.1.5.	Pdt's expenses	21000	7000	7000	7000	18000
4.1.6.	AG Secr. Expenses	12750	4250	4250	4250	12000
4.1.7	Local help to AG Secretary	3750	1250	1250	1250	3000

Up 13%   Sub-total 4.1.   706500   224000   234500   244000   630000

4.2.   OTHER EXPENSES

4.2.1.	Bank charges	4050	1250	1350	1450	3000
4.2.2.	Audit fees	9750	3000	3250	3500	4500
4.2.3.	Loss on exchange	-	-	-	-	-
4.2.4.	Miscellaneous	7500	2500	2500	2500	-

Up 182% Sub-total 4.2.   21300   6750   7100   7450   7500

Up 18%   TOTAL 4.   723800   230750   244600   251450   637500

Up 18%   TOTAL EXPENDITURES   2196250   670950   654150   871150   1852500

Up 18%   TOTAL INCOME   2198680   715560   725980   757140   1850720

         SURPLUS/LOSS   2430   47110   67830   -114010   -1780

Surplus is less than 1% of income and is too small a balance and is easily worked
out by currency fluctuations and default in payment of subventions.

8.   List of Adhering Organisations

Country	Category	Unit(s)
Algeria	I	1
Argentina	III	4
Australia	III	4
Austria	I	1
Belgium	IV	6
Brazil	II	2
Bulgaria	I	1
Canada	VI	14
Chile	I	1
China Pr	V	10
China R	I	1
Colombia	I	1
Cuba	I	1
Czechoslovakia	III	4
Denmark	II	2
Egypt Ar	III	4
Finland	I	1
France	VII	20
Germany	VII	20
Greece	II	2
Hungary	II	2
Iceland	I	1
India	III	4

Indonesia		I	1
Iran		I	1
Iraq		I	1
Ireland		I	1
Israel		II	2
Italy		V	10
Japan		VII	20
Korea Dpr		I	1
Korea Rp		I	1
Malaysia	(Associate Member)	I	1
Mexico		II	2
Morocco		I	1
Netherlands		IV	6
New Zealand		I	1
Nigeria		I	1
Norway		I	1
Peru	(Associate Member)	I	1
Poland		III	4
Portugal		II	2
Rumania		II	2
Saudi Arabia		I	1
South Africa		III	4
Spain		II	2
Sweden		III	4
Switwerland		III	4
Turkey		I	1
UK		VII	20
Uruguay		I	1
USA		VIII	30
USSR		V	10
Vatican City State		I	1
Venezuela		I	1
Yugoslavia		II	2

9.   List of Members deceased

Alania I. F.	USSR	17/05/91
Ahnert Paul	Germany	27/02/89
Alvarez Luis	USA	
Atkinson Robert D'E.	USA	
Barker Bruce M.	USA	91
Barkhatova Klaudia	USSR	09/01/90
Bates Richard Heaton T.	New Zealand	03/11/90
Beardsley Wallace R.	USA	16/03/91
Bennett John Caister	South Africa	11/04/90
Bielicki Maciej	Poland	27/11/88
Boyer Charles	France	09/89
Charvin Pierre	France	24/01/90
Christophe-Glaume J.	France	08902/01
Clauzet L. B. Ferreira	Brazil	23/12/90
Collinson Edward H.	UK	26/09/90
Costain Carman H.	Canada	21/12/89
Cowling Thomas G.	UK	16/06/90
Cunningham Leland	USA	31/05/89
Davies John G.	UK	
De Brito E Abreu J. C.	Portugal	20/01/89
Dobrovolsky Oleg V.	Ussr	12/12/89
Douglas A. V.	Canada	27/07/88
Engelhard E. J. G.	Germany	10/11/84

Fireman Edward L.	USA	29/03/90
Freiesleben H. C.	Germany	14/11/87
Fuchs Josef	Austria	09/04/89
Giese Richard H.	Germany	88
Gledhill J. A.	South Africa	19/06/88
Hardie R.	USA	19/12/89
Hardorp Johanes	USA	11/12/88
Haupt Ralph F.	USA	
Huruhata Masaaki	Japan	88
Jackson D.	South Africa	19/02/88
Järnefelt Gustaf J.	Finland	09/08/89
Javet P.	Switzerland	21/07/89
Jeffreys Harold Sir	UK	18/03/89
Ledoux Paul	Belgium	06/10/88
Levin Boris	USSR	10/04/89
Luud Lauri	USSR	12/05/89
Martel Marie-Therese	France	
Martynov D. Ya	USSR	22/10/89
Masursky Harold	USA	24/08/90
Millman Peter M.	Canada	10/12/87
Müller Helmut O.	Switzerland	17/05/90
Oki Tosio	Japan	28/07/90
Oskanyan V.	USSR	10/01/89
Pearce J. A.	Canada	07/09/89
Pearse R. W. B.	UK	13/04/89
Penston Michael V.	UK	24/12/90
Plakidis Stavros	Greece	30/01/91
Ramberg Jöran M.	Sweden	13/02/91
Roberts Walter Orr	USA	03/90
Rybka Eugeniuze	Poland	08/12/88
Sagitov Marat	USSR	14/11/88
Salomonovich A. E.	USSR	09/03/89
Sanduleak Nicholas	USA	07/05/90
Stodolkiewicz Jerzy S.	Poland	26/07/88
Tanaka Katsuo	Japan	02/01/90
Terrazas L. Rivera	Mexico	04/04/89
Thoren Victor E.	USA	09/03/91
Tuominen J.	Finland	31/01/89
Van Hoof Armand	Belgium	07/02/89
Vasilevskis Stanislas	USA	01/07/88
Vinti John P.	USA	09/90
Von Soher Herman	Austria	
Wagner C. U.	Germany	24/06/89
Warzee J.	Belgium	25/07/86
Webster B. Louise	Australia	
Wilson Raymond H.	USA	21/12/89
Zhou Xing-Hai	China PR	

10.   List of IAU publications

Transactions and Highlights

    Highlights of Astronomy
    Vol. 8
    Ed. D. McNally

    Transactions of the International Astronomical Union
    Vol. XXB, Proceedings of the General Assembly,
    Baltimore 1988

Transactions of the International Astronomical Union
Vol. XXIA, Reports on Astronomy
Ed. D. McNally

IAU Symposia Volumes (Kluwer Academic Publishers)

132   *The Impact on Very High S/N Spectroscopy on Physics*
      Eds. G. Cayrel de Strobel & M. Spite

133   *Mapping the Sky -Past Heritage & Future Directions*
      Eds. S. Debarbat, J.A. Eddy, H.K. Eichhorn & A.R. Upgren

134   *Active Galacic Nuclei*
      Eds. D.E. Osterbrock & J.S. Miller

135   *Interstellar Dust*
      Eds. L.J. Allamandola & A.G.G.M. Tielens

136   *The Center of the Galaxy*
      Ec. M. Morris

137   *Flare Stars in Star Clusters, Associations*
      and the Solar Vicinity
      Eds. B.R. Pettersen, L.V. Mirzoyan & M.K. Tsvetkov

138   *Solar PhotosphereStructure, Convection*
      and Magnetic Fields
      Ed. J.O. Stenflo

139   *Galactic and Extragalactic Background Radiation*
      -Optical, Ultraviolet and Infrared Components
      Eds. S.A. Bowyer & Ch. Leinert

140   *Galactic and Extragalactic Magnetic Fields*
      Eds. R. Beck, P.P. Kronberg & R. Wielebnski

141   *Inertial Coordinate System on the Sky*
      Eds. J.H. Lieske & V.K. Alabakin

142   *Basic Plasma Processes on The Sun*
      Eds. E.R. Priest & V. Krishan

143   *Wolf-Rayet Stars & Interrelations with*
      other Massive Stars in Galaxies
      Eds. K. van der Hucht & B. Hidayat

145   *Evolution of Stars: the Photospheric Abundance Connection*
      Eds. G. Michaud & A. Tutukov

146   *Dynamics of Galaxies &*
      their Molecular Cloud Distributions
      Eds. F. Combes & F. Casoli

148   *The Magellanic Clouds*
      Eds. R. Haynes & D. Milne

IAU Colloquia

104   *Solar & Stellar Flares*
      Eds. B.M. Haisch & M. Rodono
      Kluwer Academic Publishers

104   *Solar & Stellar Flares Poster Papers*
      Eds. B.M. Haisch & M. Rodono
      Special publication-1989, Città Universitaria,
      95125 Catania, Italy
      Catania Astrophysical Observatory Series

105   *The Teaching of Astronomy*
      Eds. J.M. Pasachoff & J. Percy
      Cambridge University Press

106   *Evolution of Peculiar Red Giant Stars*
      Eds. H.R. Johnson & B. Zuckerman

107   *Algols*
      Ed. A.H. Batten
      Kluwer Academic Publishers

110   *Library & Information Services in Astronomy*
      Eds. G.A. Wilkins & S. Stevens-Rayburn
      US Naval Observatory

111   *The Use of Pulsating Stars in* Fundamental Problems
      of Astronomy
      Ed. E.G. Smith
      Cambridge University Press

113   *Physics of Luminous Blue Variables*
      Eds. K. Davidson, A.F.J. Moffat & H.J.G.L.M. Lamers
      Kluwer Academic Publishers

114   *White Dwarfs*
      Ed. G. Wegner
      Springer-Verlag, Lecture Notes in Physics

115   *High Resolution X-ray Spectroscopy of Cosmic Plasmas*
      Eds. P. Gorenstein & M. Zombeck
      Cambridge University Press

118)
) Abandoned
119)

120   *Structure and Dynamis of the Interstellar Medium*
      Eds. G. Tenorio-Table, M. Moles & J. Melnick
      Springer-Verlag, Lecture Notes in Physics, 350

121   *Inside the Sun*
      Eds. G. Berthomieu & M. Cribier
      Kluwer Academic Publishers

Other Publications

      The Abundance   Spread   within   Globuler   Clusters:
      Spectrosocopy of Individual Stars
      Eds. G. Cayrel de Strobel, M. Spite & T. Lloyd Evans
      Report of JCM5 & CM 37/3, 20th General Assembly, IAU
      Observatoire de Paris, Publ.

Commission Publications list

Commission 05: *Documentation & Astronomical Data*

      IAU Style Manual

Commission 10: *Solar Activity*

      Quarterly Bulletin on Solar Activity.
      Published at the Tokyo Astronomical Observatory,
      F. Moriyama, Mitaka, Tokyo 181, Japan

Commission 20:   *Positions & Motions of Minor Planets, Comets &*
         Satellites

         Minor Planet Circulars.
         Issued by the Minor Planet Center, B.G. Marsden
         Centr for Astrophysics, 60 Garden Street,
         Cambridge MA 02138, USA

Commission 27: *Variable Stars*

         Information Bulletin on Variable Stars
         Prepared and distributed by the Konkoly Observatory
         of the Hungarian Academy of Sciences
         1525 Budapest XII, Box 67, Hungary

         Catalogue of Variable Stars
         Catalogue of Suspected Bariable Stars
         Name-lists of Variable Stars
         Distributed by Publishing House Nauka, Moscow, USSR

Commission 46: *Teaching of Astronomy*

         Newsletter on the Teaching of Astronomy

         Astronomy Educational Material

CHAPTER V

REPORTS OF MEETINGS OF COMMISSIONS

Commission 4: Ephemerides

President: P. K. Seidelmann   Secretary: C. A. Williams

Commission 4 was convened in 9 sessions by President, P. K. Seidelmann.

Business Meeting, Session 1, July 24: A tribute was made to the commission's deceased member R. F. Haupt.

C. A. Williams was appointed by the President as secretary.

The President reported that the executive committee of the IAU had begun discussions of possible reorganization of some commissions, including Commission 4. It has also been suggested that the General Assembly contain more science. In keeping with this spirit, the sessions of Commission 4 at this assembly will emphasize scientific papers.

The elected officers of Commission 4 for 1991–1994 are President, B. D. Yallop and Vice President, H. Kinoshita.

The following membership of the organizing committee was approved:

V. K. Abalakin   Y. Kubo   H. Schwan   E. M. Standish
J. Chapront   B. Morando   P. K. Seidelmann Fu Tong
L. E. Doggett

The following new members were elected:

T. Fukushima   G. H. Kaplan   Mitsuru Soma   C. A. Williams
Miaofu He   Ciyuan Liu   Nguyen Mau Tung

No consulting members were announced.

A special report was presented by G. I. Eroshkin on the topic "Numerical Modeling of the Planetary Orbital Motion and Lunar Orbital and Rotational Motions for the AE91 Ephemeris". Eroshkin also presented a report entitled "On the Scientific Programs of ITA in the Field of Ephemeris Astronomy," authored by N. I. Glebova, G. I. Eroshkin, M. A. Fursenko, V. N. Lvov, M. L. Sveshnikov, V. I. Skripnichenko, R. I. Smekhacheva, V. I. Valyaev, and A. A. Shiryaev. Eroshkin offers the following summary of these reports:

" 'Integral program systems' or program complexes with features of informative and inquiry systems as well as of data base control systems with graphic means and advanced interface are under active development in the Institute of Theoretical Astronomy (ITA, St. Petersburg). Among these systems are the following:

1. ephemerides of major bodies of the Solar System,
2. ephemeris maintainance of the observations of major bodies of the Solar System as well as predictions of star occultations by these bodies,
3. acquisition, storing and fitting of astronomical observations.

Experimental versions of AE91 ephemeris of the Sun, major planets, and the Moon, based on the high-precision numerical theory are also obtained in ITA."

Ephemerides: Present and Future, Issues and Uses, Scientific Sessions 2 and 3, July 24: The following 13 scientific papers were given.

1. R. West, *User Requirements*   Very accurate ($\pm 0''\!.1$) source positions (for pointing) and source motions (for tracking) are needed when objects cannot be seen because they are

faint or the observation is not being made in the visible part of the spectrum, when the field of view of the telescope is small, especially true with adaptive optics and, when telescope time is at a premium, for reasons of efficiency. Data in machine readable form is essential for many reasons: it can be easily read into the small computers that control telescopes but cannot be stored on them, circumstances may change during observation when other data will be required that can be retrieved quickly; and it minimizes errors and increases the performance efficiency of observers who use pc's containing the data and algorithms for transforming it to formats needed for specific observations.

2. B. Morando, *Publications: Contents and Future Requirements*   Ephemerides are published in France primarily by the Bureau des Longitudes and by some private groups. BDL offers some tabulated ephemerides and software for computation of ephemerides in machine readable form. For example, ephemerides of satellites are available for a pc or a Macintosh. An astronomical ephemeris was also prepared for the "MINITEL" service on the french telephone system. From telephones, equipped with keyboard and monitor, one simply dials 36 16 BDL to get information on the calendar, rise and set times, eclipses, phases of the Moon, and other information of public interest.

3. C. Smith, *Star Catalogs*   The work reported here was done in collaboration with Schwan. Catalogs currently available are listed in order of decreasing accuracy, together with the number of stars they contain, the mean epoch, mean error in position, and mean error in proper motion.

Catalog	Stars	Epoch	m e ($''$)	m e ($''/cy$)
FK5 basic	1535	1949	0.019	0.071
FK5 extension	3117	1944	55	255
IRS, Part 1	29163	1949.5	80	430
IRS, Part 1,2 ($\delta > 0$)	17,431	1952	*	*
IRS, Part 1,2 ($\delta < 0$)	18,596	1945	*	*
ACRS Part 1	250,052	1950	120	470
ACRS Part 2	70,159	($\delta > 0$)1945	*	*
		($\delta < 0$)1955	*	*
PPM South	144,787	1950	140	550
PPM North	181,731	1931	100	430

* Discordances and poor observational history make it difficult to give meaningful statistics on mean errors.

4. B. D. Yallop, *Compact Data for Navigation*   The level of precision required for navigation is $\pm0\overset{\prime\prime}{.}1$ for ephemeris data, with approximately 3 coefficients required to represent motion of a body for 15^0. Presentation of the data in book form is currently required by law for some vessels. Presentation by computer requires decisions regarding computer type, size and disk densities, languages and specific user requirements. Presentation with optical disk could provide sound, pictures, moving pictures, as well as explanations and calculations.

5. B. Morando (for J. Chapront), *Connaissance des Temps*   Data for the Sun, Moon, planets, four minor planets and the Galilean satellites are presented in the *Connaissance des Temps*. The satellite data are given as differential coordinates with respect to Jupiter. Data are available as mean or apparent place with respect to the epoch J2000. The planetary theory of Bretagnon, VSOP87, is available in Fortran subroutines for the planets Mercury to Neptune, in several sets of variables including heliocentric, geocentric, rectangular or

spherical.

6. L. E. Doggett, *Almanacs on Disks*   MICA stands for Multi-Year Interactive Computer Almanac. It is a menu driven, compiled program available for the Macintosh and MS–DOS systems to provide almanac data with a precision of $\pm 1''$. The computation kernel uses DE/LE 200 and provides data for 10 years in a variety of coordinate systems (barycentric, helicentric, geocentric, rectangular, spherical, equatorial, ecliptic) and several different epochs. The program is in β testing at the present time. Questions still being considered concern maintainance, production of higher versions, and a contact office for users to get help with problems.

7. T. Fukushima, *Standardized Software*   Standard software refers to a library of subroutines, similar to the IMSL library, providing numerical recipes for astronomical phenomena such as precession, nutation, time transformations, prediction of phenomena, planet/lunar ephemerides. Open questions concern not only construction but maintainance. A possible organizational structure would be to have a central bureau serving local computer centers and private users, while maintaining a research group and receiving outside contributions which they would have reviewed by an independent editorial board.

8. G. H. Kaplan, *Bulletin Board Almanac*   Bulletin Boards have the advantage of providing data with frequent updates, being easily accessed and giving faster response times. Their disadvantages are that the computer connections are sensitive to the quality of the telephone lines and they present security risks. Another consideration is that professionals may have difficulty accessing them when there is a phenomena occurring that has wide popular appeal, such as an eclipse.

9. V. K. Abalakin, *Minor Planet Data*   At ITA, pc's and main frames are now available. In addition to ephemerides and integration software for distribution, about 30,000 observations of stellar occultations by minor planets are available with an rms error of about $\pm 0''.5$. These observations may be used for detecting systematic errors in catalogs as well as determining planetary masses, predicting future occultations, and helping to connect the optical – radio reference frames. The new software for ephemeris production is called STAMP and for data base management, CERES.

10. J. E. Arlot, *Satellite Ephemerides*   Expressions using Chebyshev polynomials or mixed functions (secular and periodic terms) are used for satellite ephemerides. Mixed functions are used primarily for telescope guiding. Differential positons typically require 16 to 20 coefficients for sufficient accuracy. A permanent effort in observing is needed to provide consistent ephemerides.

11. D. Pascu, *Stellar Data – Calibration Sources*   For the past decade the redesigned *Astronomical Almanac* has carried a considerably expanded list of stars and other astrophysical data. The motivation for this effort was to make the Almanac useful to a wider group of scientists as well as to aid the IAU in the standardization of data. Lists of calibrational objects and finding lists for the most commonly observed objects comprise the contents of Section H of the Almanac. A system of experts from the international astronomical community is used to provide data and technical information.

12. E. M. Standish, *Principal Planets and Moon*   The quality of an ephemeris depends on the numerical integration algorithm, the equations of motion, the initial conditions and the values of the constants. At the present time, numerical integration methods and the equations of motion are equal to the level of accuracy of the observations. The quality, then,

depends on establishing accurate values of the initial conditions and constants. There is a need to archive data and to keep track of versions of software used in data reductions. The ability to trace the values of constants used in older solutions and to access the original data is a valuable asset.

13. B. D. Yallop, *Astronomical Phenomena*   Many phenomena, such as the time of conjunctions or of maximum elongations, can be calculated as the root of an equation of the type $f(t) = 0$ or $f'(t) = 0$. Ambiguities in definitions or in the connection of a definition to a particular coordinate system may lead to inconsistencies in predicting phenomena. This suggests a need to make definitions independent of the choice of the coordinate system and to adopt definitions with simpler algorithms.

Business Meeting, Session 4, July 27: After much discussion, the resolution on Earth-crossing asteroids was approved for submission to the General Assembly. It was also approved by Commissions 7, 9, 15, 16, 20, 21, and 22.

IAU/IAG/COSPAR Working Group on Cartographic Coordinates and Rotational Elements of Planets and Satellites, Session 5, July 26: The report of the working group was given by M. E. Davies, with special attention given to the data that have been revised since the last report. The session was held jointly with Commission 16. The report was approved.

Ephemerides Research, Scientific Sessions 6 and 7, July 26: These sessions were held jointly with Commission 7. The following papers were contributed:

J. Laskar	*Long-Term Planetary Motions*
E. M. Standish	*Ephemerides Accuracy Limitations*
C. A. Veillet	*Present and Future Observational Accuracies*
V. A. Brumberg	*Relativistic Effects - 2nd Order*
J. L. Hilton	*Force Model Improvements*
E. M. Standish	*Ephemerides Solution Parameters*
A. Deprit	*Preparing a Problem for Mathematica*

Working Group on Reference Frames, Sessions 8 and 9, July 29: A joint discussion was held with Commissions 7, 8, 19, 20, 24, 31, 33, 40 to discuss the report of this working group. Nine resolutions were accepted, with only minor changes to the versions printed in the IAU Final Program, for presentation to the General Assembly for adoption.

COMMISSION 5: DOCUMENTATION AND ASTRONOMICAL DATA
DOCUMENTATIONS ET DONNEES ASTRONOMIQUES

PRESIDENT: G. A. Wilkins

Report of Meetings at Buenos Aires: July 1991

Session 1.	26 July 1991	09.00 - 10.30
BUSINESS MEETING		Chairman: G. A. Wilkins

1.   **Introduction.**   The President opened the meeting by reviewing the agenda and the schedule of the meetings of the Commission. He regretted that it would be not be possible to hold a second business meeting towards the end of the Assembly to allow for a review of the future organisation and programme of the Commission after the discussions of the coming week. [This report includes some details that were actually decided at later meetings.] He noted that the chairmen of three of the working groups were not present in Buenos Aires, and so he would chair the meetings concerned. He also drew attention the Joint Commission Meeting on the archiving of current observational data; this meeting had been initiated by the Commission, but it involved many other Commissions. [Unfortunately, owing to the fire in the basement of the San Martin Centre on 31 July, this meeting had to be cancelled, although the papers will be published in *Highlights of Astronomy*.] The total attendance at the meeting was 23, including 11 members, 1 consultant, 9 nominees for membership and 2 others.

2.   **Report of the President for 1988-1991.**   The President noted that his report for the period 1 November 1987 to 30 June 1990 had been published in *Trans. IAU* 21A, 7-12, and that an updated version had been distributed widely in the Commission's *Newsletter No.5* earlier in the month. He expressed his gratitude to the Vice-President, B. Hauck, who had agreed to distribute the Newsletter at short notice when it was unexpectedly found that the Royal Greenwich Observatory would not distribute it. In general, the levels of activity of the working groups had been fairly low but, as the following reports on the specialised meetings show, useful progress has been made on many of the projects. The President expressed his thanks to all those who had contributed to the work of the Commission during the past 6 years, and especially to the Vice-President, the chairmen of working and task groups, and the other members of the Organising Committee.

3.   **Working Groups for 1991-1994.**   It was agreed that the following working groups and task groups should be continued, with a change of chairman in some cases.

W.G. Astronomical data	A. Heck
Computer communications and	
astronomical software	K. Turner
Designations	P. Dubois
Editorial Policy	G. A. Wilkins
FITS standards	P. J. Grosbol
Information retrieval	L. D. Schmadel
T.G. IAU Thesaurus	R. M. Shobbrook
UDC 52	G. A. Wilkins
J.W.G. Astronomical libraries	W. H. Warren
with co-chairman	B. G. Corbin
(nominated by the Special Libraries Association)	

In addition it was agreed that, in accordance with the wishes of the IAU Executive Committee, there should be a new Working Group that would be made up of representatives of the principal international data centres. It proved to be impracticable to set up this working group at Buenos Aires, and so C. Jaschek will act as convenor of a meeting that will adopt a constitution and elect the chairman of the Group.

The Commission was invited to nominate a representative on the new IAU Working Group on standards for astronomical reference systems; it was agreed that this would be appropriate and that G. Kaplan would be the Commission's nominee. It was also agreed that the Commission should participate in the Commission 41 Working Group on the preservation of astronomical documents; B. Hauck will be a member of the Group, one of whose objectives will be to set up an international registry of astronomical archives of historical interest that are held at astronomical institutions.

4.   Officers and Organising Committee for 1991-1994. It was agreed that the officers for the coming triennium should be:

President: B. Hauck   Vice-President: O. B. Dluzhnevskaya

It is customary for the President of Commission 5, which is regarded as a sub-committee of the IAU Executive Committee, to serve for two terms.

It was also agreed that the chairmen of the working groups should be members of the Organising Committee; the complete list is:

M. Creze, P. Dubois, P. J. Grosbol, A. Heck, Li Qibin, J. M. Mead,
L. D. Schmadel, K. Turner, W. H. Warren, G. Westerhout, G. A. Wilkins.

It may be noted that Gart Westerhout was added to the list in the recognition of his past major contributions to the work of the Commission and in the expectation that he will continue to be one of its most active members.

5.   New members and consultants. The following persons will (subject to confirmation by the IAU Secretariat) become members of the Commission:

H. Andernach, M. A. Albrecht, P. Alvarez, Y. Chu, B. Cogan, M. Creze,
G. Helou, A. S. Kharin, Li Qibin, P. Linde, L. S. Lyubimbov, S. M. Matz,
G. Paturel, G. Riegler, S. Roessiger, V. I. Sczipnichenko, A. P. G. Serrano,
A. G. Sokolsky, K. Turner, M. Tsvetkov, R. Wielen, Wu Zhiren.

The following persons will be invited to be consultants for the coming triennium:

S. Borde, E. Bouton, B. Corbin, H. Knudsen, M. Kurtz, S. Laloe,
D. Lubowich, W. Luck, A. Ratnakar, R. M. Shobbrook, M. Vargha.

6.   Resolutions. The Commission adopted three resolutions, which are briefly summarised as follows; the full texts are given with the report of the General Assembly.

1. That support be provided to experts for the compilation of catalogues of critically evaluated data.

2. That a directory of astronomical software for general use should be compiled and maintained.

3. That editors of astronomical journals should include in their instructions to authors the IAU recommendations on the use of SI units, the designation of celestial objects, and the abbreviations for the titles of journals.

In addition, the Commission agreed to support the proposal to Commission 41 on the establishment of a register of astronomical archives of historical interest.

7. Terms of reference. The President pointed out that it is the policy of the Union that each commission should have an agreed set of terms of reference and rules of procedure. As a step towards this he put forward the following statement of the role of the Commission, which he had prepared for presentation at the meeting of the IAU Working Group on the Worldwide Development of Astronomy.

"The principal objective of Commission 5 is to improve and extend the flow of information to and within the astronomical community. The Commission provides advice to the IAU Executive Committee and carries out the following principal activities in support of this objective:

1. The development of standards and techniques for the dissemination, archiving and retrieval of information.

2. The organisation of cooperative activities relating to information that is published in printed form or made available in electronic, magnetic, or optical-disc formats."

It was agreed that this statement should be used pending further consideration.

8. Future meetings. The President put forward the view that working groups were more productive, and cooperative activities were more successful, when those taking part were able to meet between assemblies. The members and consultants should look out for, and draw attention to, conferences where it might be appropriate to hold 'splinter' meetings relating to the Commission's activities before the next General Assembly; this is particularly desirable as many members had been unable to attend this meeting.

9. Other business. A. Serrano suggested that the editor of the daily newspaper of the Assembly be asked to include a note to encourage participants to make known their e-mail addresses to the IAU Office and to the Chairman of the Working Group on Computer communications. [This was not done, but such a note could be included in the IAU Information Bulletin.] There being no other business, the President closed the meeting after thanking those present for their participation.

Sessions 2 & 3.   25 July 1991   09.00-12.30

ASTRONOMICAL DATA   Chairman: G. Westerhout

1. Introduction. The Chairman of the Working Group on Astronomical Data, Gart Westerhout, opened the meeting by reviewing the agenda. At least 30 members and other participants were present. Much detail has had to be omitted from the following report, but an extended report is given in the Newsletter No. 3, dated 20 August 1991, that has been prepared and distributed by the Chairman.

2. Officers and membership. After a short discussion the officers and organising committee for the coming triennium were appointed:

Chairman: Andre Heck; Vice-Chairman: Jaylee Mead;
Organising Committee: L. Benacchio, M. Creze, O. B. Dluzhnevskaya,
P. J. Grosbol, B. Hauck, G. Lynga, S. Nishimura, W. H. Warren.

Several new members were approved, bringing the total membership to 52.

3. Reports from data centres. The Chairman read the report by O. Dluzhnevskaya on the data centre in Moscow. M. Creze presented his report on the CDS in Strasbourg, and drew particular attention to the improved facilities and increased usage of SIMBAD; he also paid tribute to the work of his predecessor, Carlos Jaschek, who had initiated the CDS and encouraged the development of regional data centres in other countries. W. H. Warren reported on the NASA/GSFC centre. A CD-ROM disc containing 31 catalogues had been prepared and distributed successfully; a two-disc set containing 114 catalogues has since been prepared with IAU support. The on-line system now allows direct file transfers. Li Qibin reported on the Chinese astronomical data centre in Beijing; he plans to organise a workshop on data centres in 1992. The Chairman distributed a report provided by B. Madore on the services available from the NASA/JPL Extra-galactic Database (NED). T Spoelstra reported that all the radio data from the Westerbork Synthesis Telescope (since 1970) are now available in computer-readable form.

During the discussion, concern was expressed that the costs of access to some data centres, including SIMBAD, made it impossible for astronomers in some countries (for example, those with hard-currency restrictions) to make use of the facilities. Creze pointed out that SIMBAD had been developed by a small research institute; he hoped to obtain more support from the major user countries so that access charges could be waived for small users, who would still have to pay their own telecommunication charges. It was agreed that the IAU should request data centres to provide free access for astronomers in developing countries; it was subsequently decided that the chairman should write on behalf of the Working Group, and he has done this.

4. New Working Group on Data Centres.   C. Jaschek stated that he had proposed the setting up of a Federation of Astronomical Data Centres and Data Services, and he outlined the aims of such a federation. The IAU Executive Committee had, however, suggested that it would be more appropriate to set up within Commission 5 a new working group whose members would be representatives of the major organisations that provide data services to the international community. It proved to be impossible to hold a meeting of representatives during the General Assembly, and so Jaschek will convene a meeting at which his proposal may be discussed and, if it is adopted, the objectives and structure may be agreed and the officers of the Group may be elected. The matter will then be considered by the President of the Commission.

5. FITS standards.   P. Grosbol, the chairman of the FITS Working Group, reported that a comprehensive document on the FITS standards (for data on images) is being prepared by the NASA Standards Office. The draft is under study, and the final version will be subject to approval by the IAU Working Group.

6. Data on radio sources.   H. Andernach drew attention to the lack of a proper archive of radio-source data. He has started to systematically collect such data, and Commission 40 has set up a working group on 'archiving and databases for radio astronomy'. During the discussion it was pointed out that existing data centres could assist in the distribution of the final database, but they did not have the expertise for its compilation.

7. Long-term cataloguing projects.   M. Barbier made an impassioned plea for the support of long-term cataloguing projects, and gave as an example her own work on the compilation of a catalogue of radial velocities. Such 'routine' projects require considerable expertise, but it is difficult to get adequate support as reseach grants are usually only given when results can be expected immediately. After discussion, a resolution urging that support be given to such projects was adopted; this resolution was subsequently endorsed by Commission 5 and by the General Assembly.

   8. Astronomical software.   S. West, on behalf of T. Banks, announced the availability of an archive of computer software for the analysis of data on variable stars at the Victoria University of Wellington, New Zealand. This led to a discussion on the problems of such archives: for example, who is responsible for the quality and maintenance of the code and of its documentation? Eventually, it was generally agreed that a directory of software is more useful than an archive, so that users can contact directly the person(s), or the organisation, responsible for the original software or for its subsequent development and distribution. C. R. Benn had started to compile such a directory as part of the activities of the Working Group on Computer Communications and Software (see report on the joint session with Commission 6), and the matter is being considered by a working group of the American Astronomical Society; it was suggested that NASA's Astrophysics Data System would be ideally suited to house such a directory. It was agreed to submit a resolution endorsing the establishment and maintenance of a software directory service.

   9. Electronic mail.   There was a short discussion on the usefulness of directories of the e-mail addresses, and on the problems associated with their compilation, maintenance and unwanted use for "junk-mail". This topic was discussed again at a joint session with Commission 6, and K. Turner was appointed chairman of the Working Group on Computer communications and astronomical software.

10. Other matters. The Chairman drew attention to the relevance of CODATA activities to astronomy. R Hanisch mentioned that the Space Telescope Science Institute is plannng to make Palomar–plate scans available on CD–ROMs. The Chairman was thanked for his work on behalf of the Group during the past six years.

Session 4.   27 July 1991   09.00–10.30

DESIGNATION OF ASTRONOMICAL OBJECTS   Chairman: F. Spite

1. Preliminaries. The Chairman considered that it was time for him to retire from chairmanship of the Working Group on Designations, and it was agreed that P. Dubois should be the Chairman for the next triennium. The membership of the Group was reviewed and is now: H. E. Bond, K. S. de Boer, T. E. Corbin, H. R. Dickel, O. B. Dluzhnevskaya, G. Helou, P. W. Hodge, C. Jaschek, M.–C. Lortet, J. M. Mead, J.–C. Mermilliod, F. Ochsenbein, G. Paturel, F. Spite and W. H. Warren. Any of these members may be consulted if there is any doubt about the designation of newly discovered sources outside the Solar System.

2. Report on progress. The Chairman reported that the summary of the 'main guidelines' had been published recently by several journals with their instructions to authors; it has the title *Specifications concerning names, designations and nomenclature for astronomical radiation sources outside the Solar System*. It was also sent to the presidents of the commissions of the IAU for comment. In addition, the list of current abbreviations and acronyms for catalogues and lists has been made available on–line to users of the SIMBAD database, and will be accessible early in 1992 at the Paris Observatory by e–mail (contact S. Borde). Two papers by S. Borde and M.–C. Lortet to be published in *Inf. Bull. CDS* give the current status of the list. H. Dickel has reported to Commmission 34 on work on the designation of complicated sources.

3. New actions. The text of the 'Specifications' summary was modified in the light of comments received from commission presidents and others. It was agreed that a very short summary of this summary should be prepared; it would contain references to the publication of the full summary so that editors could publish it regularly to remind authors and referees of the importance of the use of unambiguous designations. Further efforts will be made to encourage authors (and data centres) to ensure that all new lists of objects satisfy the guidelines and observatories to register carefully the designations of all objects that are observed; a short document on this matter is to be prepared and distributed to all observatories.

Further thought must be given to finer details, such as: precise formats for machine–readable lists, including the coding of exponents and indices, the use of separators and concatenators, and the handling of insertions; the designation of sub–components, of stars in clusters, of complicated sources, and of overlapping sources; and the identification of sources seen in different wavelengths. The Working Group would welcome suggestions on these matters so that the guidelines may be extended appropriately.

Session 5.   29 July 1991   14.00 – 15.30

EDITORIAL POLICY   Chairman: G. A. Wilkins

1. Introduction. The Chairman apologised for the absence of the Chairman of the Working Group on Editorial Policy, P. A. Wayman, and for the lack of a report on the activities of the Group. He expressed his disappointment at the lack of clear progress in the adoption by the astronomical community of the resolution on publications that was adopted at the last IAU General Assembly. He drew attention to the availability (without charge) of the *IAU Style Manual* from the IAU Office in Paris and to the summary that was reprinted in *IAU Information Bulletin 65* (January 1991). He stated that he would be the next chairman of the working group, and that he would be glad to hear from persons who would would be willing to participate in its activities. He saw the main tasks as being the improvement of the Style Manual and the continuation of the attempts to get its recommendations adopted more widely. The meeting was attended by (at least) 8 members of the Commisssion and 9 other participants.

2. The Style Manual. The main concerns that were expressed during the review of the Style Manual related to the reluctance of many (senior) astronomers to use SI units, even though they are very widely taught in schools and colleges and are obligatory in many fields of science and technology, and to the difficulties in making the recommendations on designations better known and more widely used. (The latter topic was discussed in greater depth at session 4.) It was suggested that efforts should be made to persuade the producers of major databases to use SI units. W. H. Warren drew attention to the need for standard abbreviations for special characters, such as Greek letters, for use in electronic communications of data and texts.

J.–C. Pecker considered that a French translation of the Style Manual should be made since French is an official language of the Union. He also felt that translations into other languages would be useful as the advice and recommendations are largely independent of the language of publication.

3. Use of TEX and related systems. R. Hanisch reviewed recent developments in the growing use of TEX and LATEX in conjunction with style files, such as AASTEX and SPACEKAP, for the submission of papers for publication. This technique also allows papers to be distributed by electronic–mail and to be archived in computer databases. It was agreed that authors should not be expected to deal with typesetting details, such as the choice of fonts. The IAU Style Manual should include guidance on the use of such systems.

Session 6.   29 July 1991   16.00 – 17.30

INFORMATION RETRIEVAL   Chairman: G. A. Wilkins

1. Preliminaries. The Chairman gave the apologies of the Chairman of the Working Group on Information Retrieval, L. D. Schmadel, for his absence and stated that he had received a report from him. Schmadel will continue as Chairman of the Group, but he would appreciate more help from members of the Commission; he hopes to work more closely with the working groups on designations and astronomical data. The total attendance was 22, of whom at least 9 were members of Commission 5.

2. Abstracting services. *Astronomy and Astrophysics Abstracts* (AAA) has continued to appear regularly; much of the information given in the printed volumes will be incorporated into the on-line database PHYS of the Fachinformationszentrum at Karlsruhe. A new edition of a booklet with the title "User Aids: Astronomy and Astrophysics" has been published as no. 3–9 in the *Phys Database Reference Series* (ISSN 0721-5274). The IAU Vocabulary of astronomical terms will not after all be published in the information bulletin of the Strasbourg data centre since it may be superseded by the IAU Thesaurus. It was noted that the editors of several of the leading astronomical journals have issued a common list of terms to be used for subject headings in their indexes.

3. IAU Thesaurus. B. Corbin presented the report of the Chairman of the Task Group on the IAU Thesaurus on behalf of R. M. Shobbrook, who was unable to attend. Version 3 of the thesaurus has been compiled, but further detailed checking will be carried out before it is issued. The chairman also presented a brief report by W. Lück on the checks that he and Schmadel had carried out on the earlier version. It was generally agreed that the thesaurus should be made available on computer disc, either instead of or as well as in printed form. It is hoped that journal editors will then use the thesaurus to select the terms to be used in their indexes. The thesaurus is being translated into French, Italian and Spanish by interested librarians. A German translation may also be be made. The Commission is greatly indebted to Robyn Shobbrook for the very considerable effort that she has devoted to this task; the support of the Anglo–Australian Observatory is also gratefully acknowledged.

4. Classification systems. G. Wilkins expressed his regrets that the Task Group on the Revision of UDC 52 had as yet been not made any contribution to the revision of the classification schedule for astronomy; he hoped that proposals for the revision would be made during the coming year. A table of the relationship between the PHYS (= PACS) and AAA classification systems has been completed and is given in AAA 1990 (52A).

5. Other matters. The object designation list used for AAA and PHYS has been enlarged, but it has not yet been reviewed by the Working Group on designations. P. Dubois pointed out that there is a longer list of catalogue designations on-line at the Strasbourg CDS.

The first volume of the multilingual *Space Sciences Dictionary* that is being prepared by J. Kleczek has been published by Elsevier; it deals with radiation and matter.

J. M. Watson has prepared a survey of 'Astronomical bibliography from commercial databases'; it has been published with many other relevant papers in *Databases & on-line data in astronomy*, edited by M. A. Albrecht & Daniel Egret, and published by Kluwer (1991).

The meeting concluded with review by M. Kurtz of the status of NASA's Astrophysics Data System (ADS), which is primarily intended to allow remote use of many numerical databases with a common user interface, and of an associated project (now called STELAR) to explore the use of electronic means for improving access to the literature of astronomy. M. Albrecht pointed out ESA's European Space Information System (ESIS) will be similar to ADS.

Session 7.   30 July 1991   9.00–10.20

ASTRONOMICAL LIBRARIES   Chairman: W. H. Warren

1. Preliminaries. The Chairman of the Working Group on Astronomical Libraries, W. H. Warren, stated that the Group had been set up as result of the discussions at Washington and Baltimore in 1988.

The President of Commission 5, G. A. Wilkins, was invited to comment on how the Group could influence the IAU. He first reviewed the background and pointed out that there was already a great deal of cooperation betwen the librarians themselves, especially through the PAM–Group of the Special Libraries Association (SLA) and the European Group of Astronomy Librarians (EGAL). The main objective of the Group should be to foster greater cooperation between librarians and astronomers at both local and international levels. [In particular, there is now often only a poor exchange of information between them about changes in astronomical practices and requirements and about new developments in information processing techniques. As a consequence, the expertise of the librarians and the resources of the libraries are often under-utilised.] The Group could publicise its activities through reports in the *IAU Information Bulletin*, and it could write on particular issues to the IAU Executive Committee or to individual directors. Very few librarians are able to attend IAU General Assemblies and so the Group could also consider arranging another colloquium with the aim of involving a good mix of astronomers, database experts and librarians in the discussions. The attendance at the session (over 30, from 17 countries, but of whom only about 5 were members of Commission 5 and only 3 were librarians) showed the breadth of the interest in these matters.

2. Reports and comments on activities. Brenda Corbin and others reported on some current activities, some of which are mentioned in the reports of other sessions.

A second edition of the Union List of Astronomy Serials is in preparation by J. Bausch; it will contain information from at least 41 libraries in 7 countries, but more contributions, especially from Europe, would be welcomed.

An e-mail service, known as ASTROLIB and giving information and requests, is provided by Ellen Bouton from NRAO. Items for distribution should be sent to her (library@nrao.edu). The astronomers present were encouraged to send her information about new data products that might be of general interest. They were also requested to draw the attention of their librarians to the 'twinning programme' organised by M. Cummins (University of Toronto).

There were also complaints that the prices of IAU publications are very high with that the result that some libraries cannot afford to buy them. It was pointed out that the IAU received high royalty payments, and that in recent years the US dollar had depreciated considerably against the Dutch guilder.

Joint Meeting of Commissions 5 & 6

ELECTRONIC MAIL & COMPUTER SOFTWARE   Co-chairmen: E. Roemer & G. A. Wilkins

1. Preliminaries. The meeting first dealt with some items of business of Commission 6. It was also reported that the new chairman of the Commission 5 Working Group on Computer Communications and Astronomical Software is K. Turner. The total attendance was over 40, of whom only about one-third were members of Commissions 5 or 6.

2. Electronic mail. B. G. Marsden, Director of the IAU Central Bureau for Astronomical Telegrams, opened the discussion by referring to the increasing use of e-mail for the distribution of IAU 'telegrams' and to the problems that were encountered, especially those resulting from the current lack of automatic transfers from one network to another and from the lack of standardisation in the forms of addresses. One specific recommendation that received general approval is that all astronomical institutions should have a username POSTMASTER, and should arrange to forward mail promptly to the person named in the message. There was also general agreement that the international e-mail directory produced by C. R. Benn and R. Martin at the Royal Greenwich Observatory is extremely useful, and that every effort should be made to ensure its continued maintenance, extension (by, for example, the inclusion of the addresses of AAS members) and distribution. Information on addresses, changes in networks, etc should be sent to Benn or Turner. It was suggested that the directory should also give FAX addresses, since although this system is often expensive it is useful for some purposes, such as sending graphical information.

3. Astronomical software. Turner stated that he will endeavour to update and extend the directory of astronomical software that had been started by Benn, and he requested that others help him by drawing his attention to the availability of suitable software. P. Linde drew attention to the world-wide non-commercial electronic bulletin board known as USENET. It contains many specialist newsgroups, of which SCI.ASTRO and SCI.SPACE are of interest to professional astronomers even though much of the material is contributed by amateurs. It would be feasible to create a special newsgroup for astronomical software, and D. Wells has recently created a new group, ALT.SCI.ASTRO.FITS to discuss issues related to the FITS standards. Attention was drawn to the problems of distributing software by e-mail, but it was generally felt that the main effort should be devoted to the development of the directory, and that for the time being the programs and documentation could be distributed on tape or disc by ordinary post.

COMMISSION 6: ASTRONOMICAL TELEGRAMS (TELEGRAMMES ASTRONOMIQUES)

Report of Meeting, 24 July 1991

President: E. Roemer   Secretary: B. G. Marsden

On opening the meeting, President Roemer requested participants to stand for a few moments in memory of the three members who had died since the previous meeting: Leland E. Cunningham, Edgar Everhart and Dmitrij Ya. Martynov. She noted that several non-members—but only four members—were present and that extra copies of the Commission's report in IAU *Trans.* XXIA were available. Jinming Liu and D. J. Tholen were approved as new members of the Commission. J. E. Grindlay and R. M. West were confirmed as incoming President and Vice President for 1991-1994, with S. Isobe, B. G. Marsden and E. Roemer as the members of the Organizing Committee. D. W. E. Green, Associate Director of the Central Telegram Bureau, was made a Consultant.

Director Marsden reported that the Central Telegram Bureau had been functioning smoothly and that apart from the elimination of the old mechanical TWX machines in Sept. 1988 there had been no essential changes in operation since the Commission's previous meeting. With the appearance of SN 1987A and the availability of the IAU *Circulars* by "setting host" over SPAN, followed in mid-1988 by the regular distribution by electronic mail, the rate of publication increased to what is now a rather steady 250 issues per year. However, subscriptions were converted from a per-issue to a per-month charge in 1986, when the rate at which *Circulars* were being issued was only about half what it is now. The printed version of the *Circulars* still commands the vast majority of the subscribers, and the bill for photo-offset printing has therefore more than doubled since the subscription rates were set; furthermore, although almost all the material received for the *Circulars* arrives in computer form, the preparation of them in a timely manner at all hours of the day and many of the night, during the week and at weekends, often involves considerable effort. A fairly substantial increase, $33\frac{1}{3}$ percent, in the subscription rate for the printed edition was therefore made in Sept. 1990. This increase was made reluctantly, for the deficit could have been eliminated if all those who see the electronic version of the *Circulars* were actually to take out subscriptions. It is unfortunate that extensive pirating occurs, with copies of the electronic versions of the *Circulars* being widely distributed to astronomers (principally professional) over much of the world, and the concomitant increase in the electronic subscription rates by $66\frac{2}{3}$ percent is an added burden only for the 200 or so honest electronic users. The number of electronic subscribers does continue to climb, some of the recent increase having come from amateur astronomers who receive the electronic *Circulars* at CompuServe addresses.

It is not easy to state a precise figure for the Bureau's annual income, for the subscriptions are combined with those for the *Minor Planet Circulars*, and in any case, the electronic component is conducted as a joint venture with the Minor Planet Center. A working figure might be $100 000, some $20 000 of which is then spent on printing and another $20 000 of which concerns subscriptions to the Bureau's telegram service, essentially at cost. A third $20 000 goes to the Smithsonian Institution in overhead charges, but this does not come close to meeting the Smithsonian's contribution through postage, computer equipment, office rent and the salary of the Director. The remaining $40 000 goes toward salary and benefits for the Associate Director, as well as for secretarial and occasional student assistance and for supplies, some of which are also shared with the Minor Planet Center.

In the ensuing discussion, Isobe asked whether service to subscribers known to be making illicit redisseminations of the *Circulars* could be refused, and Tholen enquired about the copyright of the material. To copyright the material would not be practical and a disservice to astronomers, and summarily to cut off a subscriber who appeared to have paid in good faith should only be done as a very last resort. After all, the purpose of the *Circulars* is to inform all astronomers of new discoveries and other urgent information as rapidly as possible. That does not imply that users with means should not bear some responsibility for helping support the service, however, and there was a consensus that the best solution would be to explore the possibility of charging a considerably higher rate to those who make extensive redistributions. D. Steel wondered whether some kind of corporate sponsorship might be sought.

West enquired whether the printed *Circulars*, which are now sometimes held for as much as a week before being printed, and then take at least as long for delivery, could be discontinued. Marsden responded that there were still 500 people who subscribe only to the printed version and that the revenue from them is greater than that from the 200 who also receive the electronic version. W. Liller suggested that postcard reminders should be sent to subscribers when it was time to renew. Marsden responded that the special low subscription rate was established specifically to reduce this kind of administrative burden; invoices can certainly be sent to anyone who needs a reminder, but the regular subscription rate, which is $66\frac{2}{3}$ percent higher, is then charged. The regular rate can perhaps be viewed as a "library" rate, but there is no prohibition on librarians who wish to avail themselves of the lower rate.

The matter of line charges was brought up and whether an inability to pay them played a role in deciding whether items should be published. Stressing that the line charges were an important part of the Bureau's income, Marsden responded that the number of rejections for this reason was almost negligible, and that if there were any, they were due, not to inability to pay, but to reluctance to pay. Items from amateurs are exempted from line charges, but great selectivity is then exercised over what is published—mainly magnitude estimates of comets and novae used as fillers.

Noting that the IAU Secretariat had been slow to forward the subventions to the Central Bureau in recent years, it was agreed to request the General Secretary automatically to include an appropriate subvention in the IAU budget and to negotiate a payment schedule with the Bureau's Director.

The following resolution was then adopted:

Commission 6,

noting the indispensable character of the service rendered to the international astronomical community by the Central Bureau for Astronomical Telegrams by rapid communication of critical information,

calls attention to the importance of the token subvention as a demonstration of the support of the IAU for this crucial activity and strongly urges its continuation.

Report of Meeting, 30 July 1991

Although originally planned as a separate meeting on the services—particularly the electronic services—of the Central Bureau, this meeting became a joint one with Commission 5 on electronic communication generally. In the portion of particular interest to Commission 6, Marsden noted that there had been since 1988 a considerable increase in the number of countries to which the IAU *Circulars* are sent by e-mail, the only notable holdouts now being Bulgaria, China, Czechoslovakia and Romania. The movement toward the INTERNET system, in which each address ends in a two-letter country code (apart from the U.S.), was generally welcomed, but in several cases subscribers sent in addresses that were not operative, and in others addresses simply would not function with the mailing system (involving UUNET) being used for the e-mail distribution of the *Circulars*. A particular problem involving both INTERNET and BITNET addresses in Germany (.DE) was confirmed by others at the meeting.

Some subscribers, notably in South Africa, remarked that electronic delivery of the *Circulars* sometimes took as long as a week. It was agreed that the electronic mail systems used by astronomers were still far from perfect, G. Winkler remarking that one got what one paid for, and astronomers make no financial contribution to the upkeep of INTERNET. Marsden agreed that some of the delay was in computers used by the Central Bureau and that he hoped this situation could be improved during the course of the next few months.

COMMISSION 7 : CELESTIAL MECHANICS (MÉCANIQUE CELESTE)

Report of Meetings, 24, 27 and 29 July 1991

PRESIDENT : J. Henrard
VICE-PRESIDENT : A. Deprit
SECRETARY : S. Ferraz-Mello

Business Meeting

The business session was held on July 24th. A second brief meeting was held on July 29th. Items addressed were:

1. Election of Organizing Committee

The commission elected the following officers and members of the Organizing Committee for the term 1988 to 1991:

President:	A.Deprit
Vice President:	S.Ferraz-Mello
Members:	K.B.Bhatnagar
	J.Chapront
	Cl.Froeschlé
	J.Henrard (Past President)
	K.Kholshevnikov
	H.Kinoshita
	J.H.Lieske
	He Miaofu
	A.Milani
	S.Peale
	A.E.Roy
	A.G.Sokolsky

2. Election of New Members of the Commission

The following IAU members were proposed as new members of Commission 7 and were approved after a short presentation: R.Branham (Argentina), C.Edelman (France), L. El Bakali (Morocco), Z.Knežević (Yugoslavia), Xu Jihong (China), R.Orellana (Argentina), M.E.Soffel (Germany), A.G.Sokolsky (URSS), G.B.Valsecchi (Italy), R. Vieira Martins (Brazil) and Liao Xinhao (China). The proposals were done by the National Committees or by members of the Commission present at the meetings.

3. Consultants

The Commission elected the following consultants for the term 1988-1991: V.I.Arnold (USSR), A.Chenciner (France), K.Meyer (USA), J.Moser (Switzerland), D.Saari (USA) and C.Simó (Spain).

4. Deceased Members

The commission paid tribute to the memory of the distinguished scientists devoted to Celestial Mechanics Drs. Leland E. Cunningham, Edgard Everhart and John P. Vinti, members of the commission, and John V. Breakwell, deceased since the past General Assembly.

5. Working Group on Near Earth Objects

In the first session, Dr. B.Marsden reported to the commission the contents of the proposals concerning the support to programs designated to study near-earth objects. In the following session the Commission discussed and approved the creation of an ad-hoc Working Group on natural near-Earth objects (NEO) and indicated Drs. P.K.Seidelman and A.G.Sokolsky to represent the commission in this Working Group.

6. Recommendations on Reference Systems

The commission participated in the joint discussion held on July 25th.

7. Past and future Colloquia and Symposia

Reports were presented concerning IAU Colloquium No. 132 held in Delhi (India) in October, 1990 and IAU Symposium No. 152 held in Angra dos Reis (Brazil), in July 1991. The president of the Commission informed that IAU Colloquium No. 118, scheduled to be realized in June 1989, in China, has been postponed because of the internal political situation of China at that time and then cancelled because of the low number of people that confirmed their participation in the meeting at the new scheduled date (mid 1990). Two possible proposals of Colloquia for 1994, one in URSS and the other in Belgium were announced. Informations were also given on meetings interesting this commission to be held in 1992 in France and in Czechoslovakia.

Scientific Sessions

Five invited reports were presented in the sessions held on July 27th and 29th:

A. Present status of rigid Earth rotation theory (H.Kinoshita)
B. The stellar problem of three bodies (S.Ferrer)
C. Chaotic motion in the Solar system. A review. (J.Laskar)
D. Integration of elliptic functions by computer (A.Deprit)
E. The scientific projects of ITA on the computer methods of Celestial Mechanics (A.G.Sokolsky)

The following abstracts were provided by the lecturers.

Present Status of Rigid Earth Rotation Theory (H.Kinoshita)

The present IAU Nutation Series, which is based on Kinoshita's (1976) rigid Earth theo-ry modified by Wahr's (1979) linear response theory for the non-rigidity of the Earth, is now not compatible with the recent precise observations such as VLBI and LLR. In these circum-stances, Kinoshita and Souchay (1990) have reconstructed a rigid Earth rotation theory by taking account of

1) more precise orbital theories of the Moon and the Sun,

2) both direct and indirect planetary perturbations and

3) the coupling effect between the rotational motion and the orbital motion.

The truncation level of the new trigonometric series is 0.005 millisecond of arc.

Then Souchay and Kinoshita (1991) compared the new nutation series with numerical inte-gration and showed that the internal accuracy of the new rigid rotation theory is 0.1 millisecond of arc, which is much better than the precision of the present observations. They found, how-ever, that the residuals in the comparison still have systematic components which correspond to the term with argument Ω, the semi-monthly term, and the semi- annual term.

The Stellar Problem of Three Bodies (S.Ferrer)

Presentation of a research did jointly with Dr. Osácar to clarify and sharpen the qualitative analysis of this problem made by Lidov and Ziglin. The authors explained how an extensive symbolic reduction carried by computer led them to realize that the reduced phase space is a two-dimensional sphere. The diagrams corresponding to each type of phase flow according to a complete discussion of all possible bifurcations determined by the four integrals of the system were presented.

Chaotic Motion in the Solar System: a Review (J.Laskar)

The number of publications on chaotic behaviour in the solar system has increased very much in the previous years. It reflects that we begin now to really take into account the result of Poincaré that the equations of celestial mechanics are not integrable and this consideration necessitate for us the use of more sophisticated mathematics. One of the first evidence of chaotic motion in the solar system was the chaotic tumbling of Hyperion. In case of much smaller perturbations, it is possible to trap the chaotic region between KAM invariant curves, in order to obtain stability for all time. Even in a chaotic region, like the vicinity of an unstable Lagrangian point, the computation of invariant manifolds allows to make precise orbit determination and satellites station keeping. In the vicinity of a stable Lagrangian point, improved Nekhoroshev's theorems give regions of practical stability of significant size. Secondary resonances are found to play an important role in the tidal evolution of the Uranian satellites in order to allow Miranda to increase its inclination and then escape from resonance. The solar system and in particular the inner solar system was found to be chaotic because of secular resonances, with a Lyapunov exponent of 1/5 million years. This has been confirmed by direct numerical integration. Ref: Chaotic Behaviour of the solar system (J. Laskar) Transactions Vol. 21 A (Commission 7 report).

Integration of Elliptic Functions by Computer (A.Deprit)

In collaboration with Prof. Alberto Abad, Prof. Vincent Coppola and Dr. Bruce Miller, a code in Mathematica to calculate in symbolic form the integral of a product of powers, positive or negative, of Jacobian elliptic functions has been developed. The author pointed to some of the problems in celestial mechanics where an automated processor of elliptic functions and elliptic integrals is likely to open new vistas.

The Scientific Projects of I.T.A. on the Computer Methods of Celestial Mechanics (A.G.Sokolsky)

Some important problems in Celestial Mechanics are described: algorithmization and revision of local methods; quasiglobal methods; global (evolutionary) models; economization of algorithms and the theory of potential; connection with ephemeris astronomy. All such problems lead us to the idea that we need new approaches in computer realization of modern methods of astronomy. The main tools of this *informational technology* are:

1. In the field of computational algorithms the basic efforts must be concentrated on the construction of new fast integrators which can be adapted to the concrete form of integrated equations; the construction of summators for long analytical expressions; the application of modern observational data processing and filters well suited for astronomy; the accomplishment of parallel computations.

2. In the field of analytic calculations, there are two important problems - how to find the equilibrium point between universal CAS (computer algebra systems) and specialized CAS which are constructed for specific narrow classes of tasks; how to organize an interface between numeric and analytical components of algorithms.

3. In the field of representation of voluminous arrays of observational data in the computer form, one should think in terms of data bases. Modern astronomical data bases must include: banks of observational data, banks of theories, banks of requests.

4. At last, the modern program product must have a developed user interface which includes the shell (help-files, the menu subsystem, etc.) and gives the opportunity for the user to educate himself when using this program product.

As examples of realization of such an approach some computer products of ITA are described:

"STAMP" – electronic book for representation of ephemerides of minor planets for the current year. This program product is a computer version of the known year-book "Ephemerides of Minor Planets", published by ITA in the last 45 years.

"CERES" – integrated program system for studies of minor planets of the Solar system. It allows to work with data bases of numbered minor planets, to accomplish ephemeris computations, attachment to the observations, loaning from the fundamental stellar catalog, visualization of results, use of an informational help-subsystem.

"CELESTA" – integrated package for application of local methods of nonlinear analysis to the problems of classical and celestial mechanics. It contains special procedures of the REDUCE CAS and a shell implementing the user-interface.

COMMISSION 8 : POSITIONAL ASTRONOMY (ASTRONOMIE DE POSITION)

PRESIDENT : M. Miyamoto
VICE-PRESIDENT : L. V. Morrison
ORGANIZING COMMITTEE : P. Benevides-Soares, D. P. Duma, L. Helmer, J. A. Hughes, J. Kovalevsky, L. Lindegren, Hu Ning-sheng, F. Noël, G. I. Pinigin, Y. Réquième, H. Schwan, C. A. Smith, M. Yoshizawa

Introduction

Considering the increase of common interests and cooperative activities among related commissions, the commission held four scientific sessions jointly with the Commissions 24, 26, and 40. These sessions provided worthwhile forums for discussing the common issues such as the global astrometry over the northern and southern hemispheres, and the linking the optical and radio reference frames.

The reporter (now Ex-President) wishes here to thank the commission presidents involved who made these meetings very worthwhile, and to especially acknowledge Prof. W. F. van Altena, the president of Comm. 24, whose contribution was noteworthy. The Joint Discussions of interest to the commission on **"Reference Systems : What are they & What's the Problem ?"** (co-sponsored by Comms. 4, 7, 8, 19, 20, 24, 31, 33, and 40) and on **"HIPPARCOS — an assessment"** (co-sponsored by Comms. 4, 8, 19, 24, 31, 33, 40, and 44) are reported elsewhere.

Business Meeting
Chairman : M. Miyamoto

The first Business Meeting was held on 24 July, 1991. The President welcomed the members of the commission to Buenos Aires and announced the appointement of E. F. Arias as the commission secretary during the XXIst General Assembly. Then, he asked for a moment of silence to pay tribute to the memory of Dr. L. B. F. Clauzet, who had been a very active member of the commission for many years.

1. Organization and Membership

The President announced the results of the election of the commission officers for the coming triennium, and the commission approved unanimously the following officers :

President : L. V. Morrison
Vice-President : C. A. Smith

Organizing Committee : P. Benevides-Soares, T. E. Corbin, L. Helmer, J. Kovalevsky, L. Lindegren, J. A. López, M. Miyamoto(ex-officio), Hu Ning-sheng, F. Noël, D. D. Polojentsev, Y. Réquième, H. Schwan, M. Yoshizawa.

The followings were proposed and approved as new members of the commission : W. F. van Altena, Robert W. Argyle, Mireille L. Bougeard, Beatrince Bucciarelli, Ole Einicke, M. Froeschlè, M. Fujishita, Gao Buxi, G. H. Kaplan, K. Kurzynska, Maria Sarasso, Qi Guan-Rong, J. Reynolds, and Laurence G. Taff. Since the XXth General Assembly, four members have resigned from the commission. The total membership is now 160. The President announced that the commission will cosponsor the following two IAU Symposia :

Developements in Astrometry and their Impact on Astrophysics and Geophysics
(proposers : I. I. Mueller and Ye Shuhua)
Shanghai, China; September 1992.
Sub-arcsecond Radio Astronomy
(proposers : R. D. Davies and R. S. Booth)
Univ. of Manchester, UK; July 1992.

The second short Business Meeting was held on 31 July immediately before the last scientific session chaired by L. V. Morrison. Considering that the commission had not received any report of the activities of the **Working Group on Astronomical Refraction** and the **Working Group on Horizontal Meridian Circle**, the continuation of the two working groups was considered. Recognizing the importance of astronomical refraction in ground-based optical and radio astrometry and the need for increased accuracy, the commission decided to set up the following new Working Group on Astronomical Refraction :

J. A. Hughes (chairman), D. G. Currie, A. I. Nefedeva, T. Rafferty, T. A. Th. Spoelstra, M. Yoshizawa.

The commission postponed a final decision on the discontinuation of the WG on H. M. C. till the next General Assembly.

2. Working Group on Star Lists

T. E. Corbin gave only a brief report on the present status of the Working Group, because details of the activities were to be presented later at the scientific meeting jointly held with Comm. 24. In view of the universal need for a unified list of fundamental stars fainter than the FK5 Extension, Commissions 8 and 24 formed at the XIXth G. A., the following Working Group on Star Lists :
T. E. Corbin (Chairman), G. Carrasco, L. Helmer, Luo Ding-Jiang, M. Miyamoto, L. V. Morrison,
D. D. Polojentsev, Y. Réquième, and S. Röser.

T. E. Corbin announced that the Working Group would propose the following recommendations at the scientific session on 31 July :
a) that the list of about 3000 stars being prepared by the Working Group be designated as future fundamental stars in the apparent magnitude range 9.5 to 13.0,
b) that this list be generally distributed — especially to those who are actively involved in programs to determine star positions,
c) that these stars be observed to relate their positions to the extragalactic system and the dynamical system, in accordance with the recommendations of the Working Group on Reference Systems.

These recommendations were approved at the scientific meeting "Optical Reference Frame and Catalogues — Fainter Extension of the Optical Frame" on 31 July.

3. Working Group on Astrolabes

The report on activities of the WG was presented by F. Chollet. After the MERIT Campaign, the astrolabe objectives were shifted to star catalog observations. In order to cope with these activities, the WG was reorganized and the following membership confirmed : F. Chollet (Chairman), V. Gubanov, Li Dong Ming, and F. Noël. In order to increase the quality and scope of the astrolabe, this triennium was devoted to instrumental modification at Santiago (Chile), Sao Paulo (Brazil), San Fernando (Spain), CERGA (France), Paris (France). Some projects to install modern astrolabes are in progress in Turqia and Marocco. The following results obtained in this triennium should be noted : The photoelectric astrolabe of CERGA has been devoted to the star catalog observation, obtaining very good precision of $0\rlap{.}''17$, and the solar astrolabe has continued to observe the solar diameter with a precision not far from $0\rlap{.}''2$, by using a CCD micrometer.

D. J. Hutter and J. A. Hughes reported on the current status of the design and construction of the U. S. Naval Observatory Astrometric Optical Interferometer (AOI), a dedicated astrometric interferometer observatory which will be in operation at a site in the western U. S. in 1993.

The AOI will be built on the experience gained from the Mark III Optical Interferometer on Mt. Wilson, which was a joint project involving NRL, SAO, USNO, and MIT. The Mark III has already demonstrated wide-angle astrometry with uncertainties of 5 to 10 mas from observations repeated over several nights and systematic agreement with FK5 catalog positions within the larger quoted uncertainties. The new AOI will consist of 4 elements in a fixed array, with baselines arranged for instantaneous position measurement in two dimensions. State-of-the-art delay lines as well as extensive laser metrology will be incorporated. Several beam combination techniques will be used to increase the precision of astrometric measurements. The AOI will achieve unprecedented precision in the optical regime, with statistical and systematic errors below 5 mas and 1 mas for wide angle and narrow angle measurements. This array will establish a catalog of positions of over 1000 stars.

R. C. Stone and D. G. Monet reported on the CCD Transit Telescope (its detector, environmental and laser metrology systems, etc.) under development at the U. S. Naval Observatory Flagstaff Station. The telescope (20 cm ϕ) uses a CRAF 1024 × 1024 (12 micron pixel) cooled chip for scanning regions of the sky and can observe objects with magnitudes between $V = 4.0$ to 17.5. Various scanning techniques are used that enable stars to be scanned from the equator to the pole with little distortion of the images. The telescope can observe about 9000 stars an hour in long strip scans and accurate positions ($\sigma \sim \pm 0\farcs1$) can be determined for them that are closely tied into the extragalactic VLBI reference frame. Further improvements in accuracy are hoped for by numerical modeling of refraction and the use of laser interferometry to monitor instrumental motions in real time.

L. Helmer reported on the recent implementation of CCD technique on the Carlsberg Automatic Meridian Circle. A new CCD camera, intended for reading the divided circle, has been developed at Copenhagen University Observatory. The camera has been tested in the laboratory on a graduated circle from Heidenhein, and it has been shown that the standard deviation of the reading of one division line is as low as $0\farcs003$, with 1 second integration and evaluation time. This system together with the new circle will be mounted on La Palma in 1992. Furthermore, a more sophisticated CCD camera for detecting star transits is under development, and will be tested in both fixed frame and drift-scan modes on old Cooke Transit Circle now at Copenhagen University Observatory.

M. Yoshizawa presented a progress report on development of a Drift-Scanning CCD Micrometer. Tokyo PMC group has constructed a prototype model of the micrometer "DISC" (DIgital strip Scanning Ccd micrometer), in which the electric charges stored in CCD pixel are successively transferred to the next pixel in order to follow the diurnal motion of celestial objects without any mechanical motion of the micrometer. The prototype DISC consists of a single-field CCD image sensor TH7883 (Thomson-CSF) cooled by liquid nitrogen down to about 170°K, CCD drive board, 16 bit A/D converter, and EWS to control the whole system. The prototype DISC was attached to the Gautier Meridian Circle at Mitaka and tested in the last winter, observing stars and open star clusters with good photometric standards. It was shown that DISC can observe stars down to $V = 14.5$ in the rather bright night sky of Mitaka. The construction of a next generation DISC with a larger CCD chip, EEV 88331, has been started as a result of this encouraging experience.

F. Laclare gave a presentation of a newly developed CCD camera to be fitted to the Solar Astrolabe of CERGA, so as to make measurement of the solar diameter more objective.

Variations of the solar diameter have been detected in a long series of observations made with the solar astrolabe. But, there remains still a suspicion that the cause of the variations may be an observational or atmospheric effect. The results obtained during the two experimental observations with the CCD camera and the Danjon micrometer show very good agreement between the impersonal and visual observations. The dispersion of the daily measurements of the solar diameter with the camera is about $0\rlap{.}''1$, which is a significant improvement in quality over visual observations ($0\rlap{.}''18$). Moreover, the two dimensional analysis of the digitized solar image should allow the evaluation of a possible atmospheric contribution to the observed variations to be made.

Progress Report on Observing Programs, 26 July 1991
Chairman : L. Helmer

<u>T. E. Corbin, E. Holdenried, G. Wycoff, and T. Rafferty</u> reported on the current status of the USNO Transit Circles. The 6″ Transit Circle in Washington and the 7″ Transit Circle at Black Birch (New Zealand) are both currently in the second circle of the Pole-to-Pole Fundamental Program. The 6″ TC started its second circle in Feb. 1990 and to date 53457 observations have been made. Combined with the first circle results this makes a total of 175511 observations on the 6″ TC thus far in the program. The total for the first circle of the 7″ TC is 156472 observations, and the second circle was begun in May 1991. The 7″ TC has had particular success with its observations of day stars with 18330 made during the first circle. It appears possible to reduce and model the 7″ TC day observations separately from the night, thereby giving a much stronger solution for the observations of the Sun, Mercury and Venus.

Comparisons of the absolute observations with the FK5 generally show the same features as found by the CAMC, Bordeaux AMC, and Tokyo PMC programs north of $-20°$. South of that declination there is weak oscillating feature in right ascension going to about $-50°$ and a strong positive trend, in the sense of Black Birch − FK5, south of there. The declination differences mainly show a strong positive feature from $-30°$ to $-65°$.

<u>L. V. Morrison</u> reported on CAMC observations of solar system objects. The CAMC spends about 2% of its time on observing solar system objects. Since the beginning of observing on La Palma the CAMC has made a total of about 13000 observations of Mars, Callisto, Saturn, Titan, Iapetus, Uranus, Neptune, Pluto and 60 minor planets. The accuracy of an observation ranges between $0\rlap{.}''14$ and $0\rlap{.}''25$ depending on the zenith distance and magnitude, as described in the Introductions to the Carlsberg Meridian Catalogs.

A detailed analysis of the Titan and Iapetus observations combined with those made at Bordeaux produce very consistent opposition corrections to JPL DE202 for Saturn for the years 1987–89 (D. B. Taylor et al., 1991). These corrections are much more reliable than those derived from direct observations of Saturn's disk. An analysis of the CAMC observations of Pluto (the first ever made by a meridian circle) gives corrections of $+0\rlap{.}''4$ and $+0\rlap{.}''2$ to R.A. and Dec. of JPL DE202, respectvely. These are consistent with Klemola's astrographic results of 1988 and Mink's occultation result of 1988.

<u>M. Yoshizawa and M. Miyamoto</u> gave a progress report on the First Tokyo PMC Program which started in 1985 and is to end in 1992. The Program comprises about 33000 stars and the solar system objects (the Sun, 6 major planets, Ceres, Pallas, Juno, Vesta, Hebe, Iris, Flora, Metis, and Eunomia). The program stars are composed of FK5 stars, NPZT stars, AGK3RN stars, Zodiacal stars, OB stars, H_2O masers, and faint reference stars around QSO's. The three annual catalogs, Tokyo PMC 85 (1007 stars), 86 (3974 stars), and 87 (5748 stars) have been published. The fourth catalog, Tokyo PMC 88 (3845 stars) will be published soon. A local systematic error of about $0\rlap{.}''1$ in the FK5 system is evident in Tokyo PMC 86 and 87, suggesting an erroneous proper motion system of the FK5 in some parts of the sky. The final goal of the First Tokyo PMC Program is to provide absolute positions of the program stars, consistent with the dynamical theory of the planetary system. The compilation of the absolute Tokyo PMC catalog will start in 1992.

S. Sadžakov, M. Dačić, and Z. Cvetković reviewed the recent observations of solar system objects (the Sun, Mercury, Venus, and Mars) performed with the Large Meridian Circle of Belgrade Astronomical Observatory, and gave an error estimate of these observations. They stressed that the quality of the Belgrade observations with regard to both systematic and random errors is comparable to other modern observations.

A paper by D. D. Polojentsev and R. Zalles was read by V. Abalakin. The paper reviewed the current astrometric activity with the astrograph ($D = 23$ cm, $F = 230$ cm, field $4° \times 4°$, scale $90''$/mm) in the Soviet-Bolivian Astronomical Observatory at Tarija. The observing programs being carried out are : FOKAT-S Catalog to provide precise positions of about 200,000 southern stars up to 11th magnitude, Bright Star Catalog to supplement bright southern stars not included in the FK4 catalog, and Equatorial Star Catalog to provide reference stars with density of 40 stars per square degree.

Another paper by A. S. Kharin was also read. The paper presented a list of 836 FK5 stars observed in the infrared range. These stars will be used as reference stars in IR range with positions and proper motions given in the FK5 system. The diskette of the list will be provided on request.

Astrometry in the Southern Hemisphere
(joint with Comms. 24 and 26), 27 July 1991
Chairman : J. A. López

The activities of this session are described by W. F. van Altena in the Report of Commission 24.

Linking Optical and Radio Reference Frames
(joint with Comms. 24 and 40), 29 July 1991
Chairman : T. E. Corbin

K. Johnston reviewed the present status of linking the optical and radio reference frames. The radio reference frame appears to be well defined at the mas level in the northern hemisphere and down to a declination of $-30°$. Observations initiated over the past two years for sources south of $-30°$ have resulted in accuracies of order ten mas level (Russell). The initiation of dual frequency S/X VLBI observation in the southern hemisphere will result in positional accuracies at the mas level for a large number of sources by 1993. Observations for a global radio reference frame of 400 sources are being carried out (Russell) to extend the present number of reference sources obtained through geodetic programs. The optical reference frame now in use is the FK5 which is believed to be accurate to 50 mas level with zonal errors in the southern hemisphere sometimes approaching a hundred mas.

There are three methods being followed to relate these two frames. The first method is to determine the optical positions of the extragalactic radio sources making up the radio frame (Johnston, Russell, de Vegt). The optical positions of the radio sources can be measured relative to nearby reference stars with a precision at the 10 mas level on prime focus plates obtained with a 4m class telescope. These reference stars can be related to FK5 stars by wide field astrograph plates. The accuracy of the position of the radio sources on the FK5 frame is estimated to be at the 50 mas level (de Vegt). The second method is to determine the positions of bright stars which also display radio emission (Lestrade, de Vegt). Many modern transit circle programs are being undertaken to refine the positions of these stars on the FK5 system. Meanwhile, VLA and VLBI observations of about fifty stars which emit continuum radiation have been made (Johnston). For stars emitting continuum radiation, it is probable that the optical and radio centers coincide at a few mas level. Preliminary comparison of the optical and radio observations shows that the reference frames in the northern hemisphere coincide at the 100 mas level when appropriate zero point alignments and rotations are made. In the third method, bright stars which display maser emission are used to relate the reference frames. Sufficient observations have to be made to relate the photocenter to the circumstellar shell emitting maser radiation. The masers which are located closest to the stellar photosphere, such as the SiO masers, may be capable of relative accuracies at the 10 mas level; but this has still to be demonstrated.

Although the present accuracies of the frame-tie measurements are limited to the 50 mas level (cf. FK5) the success of the HIPPARCOS mission will improve the accuracy of the optical frame to a few mas level. With the advent of this improved accuracy, future frame-tie observations must be made with mas accuracy to relate the both frames.

R. C. Stone gave a description of the use of a CCD Transit Telescope for linking reference frames and presented preliminary observational results obtained with it. The U. S. Naval Observatory at Flagstaff has developed a CCD transit telescope capable of observing both FK5 stars as well as extragalactic sources defining the radio reference frame. Objects as faint as $V = 17.5$ can be observed. With a screen placed in front of the objective, the bright limit is extended to $V = 3.5$. The accuracy of the position in both coordinates is about $0''.1$. Thus, observations of FK5 stars and VLBI extragalactic sources can be observed directly with this instrument, and reductions can be made on a nightly basis. The transit circle thus developed can be used for determining the offsets between the optical and radio positions of extragalactic objects, and can determine accurate positions for a large numbers of faint stars that are tied into the extragalactic radio reference frame.

An analysis of the positions of VLBI extragalactic sources obtained with the instrument indicates that some sources have large differences between their radio and optical positions. Observations of radio stars and FK5 stars indicate that a difference between the VLBI and FK5 reference frames can exceed $0''.1$, but a better agreement of both frames is found for the radio stars. These preliminary results will be improved by further observation.

M. Yoshizawa and S. Suzuki presented the optical position of 3C273B determined by a CCD meridian circle. A prototype model of the CCD meridian circle developed at National Astronomical Observatory of Japan was used to determine directly the optical position of 3C273B on the FK5 system. The right ascensions of 3C273B and faint stars around it were determined with "DISC" (DIgital strip Scanning Ccd micrometer) fitted to the Gautier meridian circle at Mitaka. The observed position of 3C273B was found to coincide with its best available radio position (R.A.) to within $0''.07$.

L. V. Morrison reported on the photographic astrometry of the four radio stars, Z Her, 9 Sgr, HD193793, and SV Cam (R. W. Argyle et al., 1991), carried out in a test program using the Wide Field Camera (f/8) of the Jacobus Kapteyn Telescope in conjunction with the automatic plate measuring machine at Cambridge. In all cases the positions of the reference stars used were provided by the Carlsberg Automatic Meridian Circle catalogs. From this test program an accuracy of about $0''.04$ was obtained for the optical positions of radio stars.

It was found that there is no detectable magnitude equation even though target and reference stars differ by up to 5 mag. Thus, by observing the reference stars directly with the CAMC there is no need to construct a secondary frame of reference to connect the radio stars to the FK5 system. This encouraging result has led to the expansion of the test program to include all radio stars with positions measured by the VLA, directly by VLBI and in progress with MERLIN. The aim of the expanded program is two-fold : to link the optical reference frame defined by the FK5 to the VLBI radio frame and to link the HIPPARCOS frame similarly.

C. Fabricius gave a progress report on the CAMC program for linking the optical and radio reference frames. Nine QSO's with VLBI radio positions are being observed. Furthermore, 9300 reference stars of $V \sim 11$ in 360 fields have already been observed. A new observing list of fainter stars of $12 \leq V \leq 14$ within a field of $30' \times 30'$ around the 400 VLBI sources is being prepared. This list will be suitable for a wide field CCD micrometer.

Y. Réquième reported on the Bordeaux Observations of radio stars. Optical positions of 221 radio stars have been obtained with the Bordeaux Automatic Meridian Circle in the period of 1984 to 1990 (Réquième and Mazurier, 1991). The observations were strictly differential, with 6 to 8 FK5 stars per hour in the range of $-25° < \delta < +75°$ observed in order to determine the 8 instrumental parameters. The mean positions given at J2000 were corrected for chromatic refraction and also for photocenter effect in the case of duplicity. The expected accuracy is about $0\overset{''}{.}05$ at least for the northern stars which were observed 32 times on the average. The mean position of 3C273B was also obtained from 35 observations under the same conditions.

P. Hemenway and N. Argue reported on the HIPPARCOS Link Plans. Assuming that the VLBI extragalactic reference frame represents a close approximation to an inertial frame, the HIPPARCOS Project has been pursuing observations to link the HIPPARCOS frame to the VLBI frame in both position and rotation. Since HIPPARCOS cannot observe extragalactic objects directly, two indirect approaches are being taken. First, the positions and rotations of about 40 radio stars included in HIPPARCOS are being observed with the VLA and VLBI. Second, optical counterparts of VLBI radio sources are being observed optically with respect to the HIPPARCOS frame. The highest accuracy observations will be made of a few tens of sources with the Hubble Space Telescope and ground-based CCD devices. Furthermore, several hundred sources will be observed photographically, sometimes combined with CCD observations. The final link accuracy is expected to be between $0\overset{''}{.}002$ and $0\overset{''}{.}0001$ in positional offset, and $0\overset{''}{.}002$ /yr and $0\overset{''}{.}0001$ /yr in rotation.

R. J. Davis stressed the importance of observations of radio stars with "MERLIN" at an intermediate resolution for linking the optical and radio reference frames. The radio stars are not always point sources at radio frequencies and the stellar radio centroid depends on the resolution of the interferometer system used. It is known that thermal and non-thermal components exist in these stars and these components may be spatially separated and related to material in different orbits around the central star. The VLBI is sensitive only to the non-thermal component, while the MERLIN with its resolution intermediate between the VLA and VLBI retains surface brightness sensitivity to detect the thermal as well as the non-thermal emission. The intermediate resolution will therefore play an important part in establishing the location of the emission as a function of orbital phase and during any radio/optical outbursts. Thus, the MERLIN link observations should give the global rotation matrix needed to bring the optical frame into alignment with the VLBI frame.

Progress Reports on Catalogs, 29 July 1991
Chairman : C. Smith

M. Yoshizawa and S. Suzuki presented the third and fourth annual catalogs of the Tokyo PMC 87 and 88 of 5748 and 3845 stars, respectively. The mean systematic trends of the observed positions minus those of FK5 basic stars agree quite well with each other for all the Tokyo PMC catalogs. It is confirmed that the FK5 system has a systematic error $\Delta\delta_\delta$ of about $0\overset{''}{.}1$ in the declination zone $40° < \delta < 60°$. The systematic error must be explained by an imperfection in the FK5 proper motion system.

C. Fabricius reported on the Carlsberg Meridian Catalogs No.5–No.7, which are based on the observations with the new moving slit micrometer fitted in May 1988. Catalogs Nos. 5 and 6 have yielded the positions of about 35000 stars with a zenith mean error of $0\overset{''}{.}08$ in R. A. and Dec., and $0\overset{m}{.}04$ in magnitude. The proper motions of these stars have been derived by combining the positions with these in catalogs at earlier epochs. The mean errors of the proper motions are between $0\overset{''}{.}002$ /year and $0\overset{''}{.}003$ /year. The Carlsberg Meridian Catalogs (Nos. 1–7) constitute a reference net with a density of 2.5 stars per square degree.

R. A. Carestia and C. C. Mallamaci proposed an alternative format for presenting positions and proper motions of the FK5 stars in equatorial rectangular coordinates.

G. Carrascco and P. Loyola reported on the results of differential observations of the FK5 stars, carried out with Repsold Meridian Circle at Cerro Calán National Observatory. A total of 10153 observations were obtained for the declination zone $+40°$ to $-90°$. The mean observational epoch is 1988.3 and the mean internal error of the positions is $\pm 0''.08$.

M. Yoshizawa and M. Miyamoto reported on the optical positions of 51 H_2O maser sources (mainly Mira-type variables), observed in 1987 with Tokyo PMC. The observed sources are distributed on the sky rather uniformly, which is benefited for the global link of the optical and radio reference frames. The positions of a set of about 60 H_2O maser sources were also observed with VLA in 1986. Fifteen of those sources are common to these observed to the Tokyo PMC. The optical and radio positions of these common sources are being compared.

E. F. Arias reviewed on the IERS Celestial Reference Frame (IERS CRF). The IERS is responsible for defining and maintaining a celestial reference frame based on the positions of extragalactic radio sources with VLBI.

The first realization of the IERS CRF (Arias et al., 1988) fixed the initial definition of axes of the frame. Each year, as new individual realizations are submitted to IERS, a new version is computed. The IERS Ox axis is in agreement with the FK5 origin of R.A. The Oz axis points in the direction of the mean pole at J2000.0 as defined by the conventional models for precession and nutation. The successive realizations of IERS CRF elaborated up to now have maintained the initial definition of the axes within $0''.00001$. The last realization of the IERS CRF, RSC (IERS) 91 C 01, is based on the observations of a total number of 396 compact extragalactic radio sources uniformly distributed on the sky in the range $-83° < \delta < +85°$. These VLBI data have been compiled on the basis of two global solutions independently obtained by GSFC and JPL. The axes of RSC (IERS) 91 C 01 are fixed by 57 primary sources selected by statistical tests.

Optical Reference Frame and Catalogues — Fainter Extension of the Optical Frame — (joint with Comm. 24), 31 July 1991
Chairman : K. Johnston

T. E. Corbin and H. Schwan reviewed the extension to the FK5 catalog. Work on the "Basic FK5" giving the classical fundamental stars has been completed and the printed version published. In the second part of the FK5 (the "FK5 Extension"), improved positions and proper motions for 3117 new fundamental stars will be given. There are 992 stars selected from the FK4 Sup catalog and 2125 stars from the IRS list, extending the fundamental system to about $V = 9.5$. The determination of the mean positions and proper motions for the new fundamental stars has been completed and the compilation of various auxiliary data on a uniform basis, such as parallaxes or apparent magnitudes, has been finished.
The following data represent the characteristics of the FK5 Extension :
  Average mean epoch : 1944
  Average mean error of mean positions at mean epoch : $0''.055$
  Average mean error of proper motions : $0''.255$/cy
The final catalog is given in the system of the Basic FK5 and based on the IAU (1976) System of Astronomical Constants. The format is, as far as possible, the same as for the Basic FK5. A tape version of the catalog was sent to the astronomical data centers in June 1991.

T. E. Corbin and L. V. Morrison gave a progress report from the Working Group on Star Lists, concerning the Intermediate Fundamentals (IF) — a unified list of fundamental stars fainter than the FK5 Extension. The Working Group has discussed and approved that a list of about 3000 stars in the magnitude range $9^m.5$ to $13^m.0$ with roughly equal numbers of stars from $9^m.5$ to $11^m.0$ and $11^m.5$ to $13^m.0$ should be selected. It was further agreed that first consideration should be given to stars in the HIPPARCOS Input Catalog and to those included in current observing programs such as the CAMC reference stars near extragalactic benchmark radio sources.

The IF1 list (9.^{m}5 to 11.^{m}0 list) was compiled by L. V. Morrison, resulting in 1672 stars showing an even distribution over the sky, while the IF2 list (11.^{m}0 to 13.^{m}0 list) was selected by T. E. Corbin. The total number of the IF2 stars amounts to 1143, composed of 10 HIPPARCOS High Priority Radio Stars, 11 HST–HIPPARCOS Link Stars, 1032 HIPPARCOS Input Stars, and 90 CAMC Reference Stars around benchmark radio sources. Known double stars, variable stars, and high proper motion stars were eliminated from the above lists.

There remain, at this time, several gaps in the distribution of the stars. These gaps cannot be filled with stars from the above-mentioned sources. The Working Group will fill the gaps in the southern hemisphere with stars from CPC2 / FOKAT-S, and those in the northern hemisphere with stars from USNO Astrograph plates, and add other recommended stars such as O·B stars proposed by M. Miyamoto.

After Corbin's presentation and some discussions, the recommendations announced at Business Meeting on 24 July were approved by Commissions 8 and 24.

S. Röser and U. Bastian presented a reference star catalog of positions and proper motions (PPM) of about 360000 stars covering the whole sky. The PPM is produced at Astronomisches Rechen-Institut, the southern part in cooperation with 3 institutes in the USSR : Pulkovo observatory, Kiev observatory and Sternberg institute. At present the status of the PPM is the following. The northern hemisphere portion was completed in 1989; a tape version is available through international astronomical data centers and a printed version appeared in April 1991. A preliminary version of the southern hemisphere part was completed in mid-1990 and its tape version is also available through the data centers. The final version will be completed at the end of 1991. The main characteristics of the PPM at present are the following :

	Stars	Mean error of position		Mean error of p.m.	
		R.A.	Dec.	R.A.	Dec.
		at epoch 1990 (arcsec)		(arcsec /cy)	
PPM North	181731	0.27	0.27	0.43	0.42
PPM South (Pr.)	144787	0.26	0.26	0.55	0.55
PPM South (4.91)	144787	0.12	0.12	0.40	0.40
Expected properties of the final PPM South :					
PPM South	180000	0.12	0.12	0.30	0.30

Since the summer of 1990 a new reference catalogue for the whole sky has been available which surpasses the SAO catalogue's accuracy by a factor of 3 to 6 in positions and proper motions. For the time being, a version of PPM on the old system of B1950 FK4 is also supplied for observers to take advantage of its higher precision compared to SAOC. The final version of the southern part will contain some 40 000 stars more than the preliminary one. This is especially important in the regions $0° < \delta < -20°$, and south of $-64°$. There the density of stars per sq. deg. will be increased from about 5 and 3, respectively, in SAOC to about 9 in the final PPM South.

T. E. Corbin and S. Urban reported on the compilation of the catalogs of the International Reference Stars (IRS) and the Astrographic Catalog Reference Stars (ACRS), both of which are now available at the international astronomical data centers.

The IRS is based on the AGK3R and SRS lists. The AGK3R portion has been completed for some time, and the recent completion of the SRS has made it possible to complete the IRS. The southern IRS was compiled from 87137 star positions in 94 catalogs, resulting in 19827 mean positions and proper motions. The combination of the southern and northern portions has produced the IRS of 36027 mean positions and proper motions with an average epoch of 1949.5. The IRS is divided into two parts : Part I contains the stars with good observational histories and has 29163 stars with average proper motion mean errors of 0."43 /cy in R.A. and 0."42 /cy in Dec. The catalog is available on B1950 FK4 and J2000 FK5.

The ACRS is the result of combining data from photographic and meridian circle catalogs in order to give the best proper motions at the highest density possible. The basic lists are the AGK3 in the north and the CPC2 in the south. A total of 1643783 catalog positions from 167 catalogs were reduced with the combined FK4, improved FK4 Sup and IRS to produce 320211 mean positions and proper motions with an average epoch of 1949.5. The ACRS is also divided into two parts : Part I contains 250052 stars with average proper motion mean errors of 0."47 /cy in R.A. and 0."46 /cy in Dec. Part II, like that of the IRS, is mostly stars whose proper motions are derived from only two catalog positions. The positions in the catalog are given for both B1950 FK4 and J2000 FK5.

H. Jahreiss gave a report on the successive compilations of extensive astrometric data entitled "the CDA" (Catalogue de Donnees Astrometrique), in which the "best" astrometric parameters and cross identifications are given for every star. The first version of the CDA was created in 1984 out of the need to get a quick overview of existing astrometric catalogs fulfilling the required positional accuracy of 1."5 at the epoch of 1990 for the HIPPARCOS satellite. The second version of the CDA compiled in 1988 took into account the individual errors of positions and proper motions of the stars. This version contains positions for 463615 stars and proper motions for 327477 stars. The third version of the CDA has recently been started with the intention to compile astrometric data for as many stars as possible. In particular, stars of astrophysical interest (nearby stars, high proper motion stars, etc.) will be added. At present the CDA contains 466072 positions and 337478 proper motions, providing a mean positional accuracy of 1" at the epoch of 2000.0 (except 555 stars with large errors). The content of the CDA will soon increase to about 550000 stars when the HIPPARCOS Input Catalog is released.

W. R. Dick reported on the optical work at Bonn, Potsdam and Tautenburg, concerning the linkage of the HIPPARCOS system to the extragalactic objects. Besides the link of the HIPPARCOS reference frame to extragalactic objects via radio stars or by the HST, photographic astrometry is also able to calibrate the HIPPARCOS proper motions with respect to an inertial system. Numerical simulations have shown that even with a very small number of well-distributed link fields, the photographic method is competitive with other techniques (P. Brosche et al., 1991).

Therefore, the Bonn and Potsdam astrometric groups are engaged in programmes for the HIPPARCOS extragalactic link with ground-based observations. The Bonn programme is based on plates from different telescopes with an epoch difference of up to 90 years. Proper motions have been derived for 45 HIPPARCOS stars around 4 bright QSO's, together with "fictitious" proper motions of the QSO's. The Potsdam / Tautenburg programme uses exclusively plates from the Tautenburg Schmidt Telescope. Measurements of 19 proper motion fields are now complete. On the basis of these recent results, an accuracy of up to 0."1/cy has been estimated for both programmes. Additionally, the data from proper motion studies of globular clusters can be used for the HIPPARCOS extragalactic link. Both at Bonn and Potsdam optical positions of extragalactic radio sources have been derived for the radio / optical link. The external random error is about 0."15 to 0."20. Improvements to reduction techniques have been also proposed.

Optical Reference Frame and Catalogues — Fainter Extension of the Optical Frame — (joint with Comm. 24), 31 July 1991
Chairman : L. V. Morrison

The activities of this session are described by W. F. van Altena in the Report of Commission 24.

COMMISSION 9: INSTRUMENTS AND TECHNIQUES (INSTRUMENTS ET TECHNIQUES)

PRESIDENT: J. Davis   VICE-PRESIDENT: J.C. Bhattacharyya

Summaries of the business and scientific sessions held by Commission 9 during the 21st IAU General Assembly in Buenos Aires are given in this report. As part of the General Assembly program Commission 9 also co-sponsored two Joint Commission Meetings and joined with Commission 25 in a joint meeting on "Contributions of Polarimetry to Stellar Astrophysics". JCM II, on "Automated Telescopes for Photometry and Imaging" and "Performance and Results with IR Arrays", was co-sponsored with Commission 25 and the report can be found in Highlights of Astronomy (Volume 9). JCM IV, on "The Development of Antarctic Astronomy", was co-sponsored with Commissions 40 and 50 and its report will also appear in Highlights of Astronomy. The report on "Contributions of Polarimetry to Stellar Astrophysics" appears in the contribution from Commission 25 to this volume.

BUSINESS

Business Meeting: 30 July 1991

Chair: J. Davis

The business session opened with brief reports from the Commission's three Working Groups. The future of the Working Groups was discussed and it was agreed that the Working Groups on "Detectors" and on "High Angular Resolution Interferometry" should both continue unchanged. Both represented active and developing areas of great interest to the Commission and were playing significant roles in the Commission's program. The Working Group on "Photography" was the subject of some concern. Although it had been active between General Assemblies, with a meeting held in Garching, Germany, in October 1990, it had no organised program in Buenos Aires due to illness of the Chair of the Working Group. G. Westerhout proposed that an informal group of astronomers interested in large area plate measuring instruments merge with the Working Group on Photography. After discussion it was suggested and accepted that the Working Group should widen its scope to include large area plate measuring instruments and large area electronic detectors as well as photography. The Commission agreed to change the name of the Working Group from "Photography" to "Wide Field Imaging". R.M. West was elected by the Commission to chair the Working Group for the period 1991-94. Further details are given in the Working Group report.

The Commission discussed two resolutions and agreed to support both. The first sought the establishment of an ad-hoc Joint Working Group on Natural Near-Earth Objects with the participation of Commissions 4, 7, 9, 15, 16, 21 and 22. The Commission nominated G. Lelièvre to be its representative on the Working Group. The second resolution "Encouraging International Development of Antarctic Astronomy" also sought the establishment of a Working Group. Its role would be to "encourage international cooperation in site testing and in designing and constructing new Antarctica astronomical facilities". This resolution was supported by Commission's 9, 40, 44 and 50. P. Gillingham was nominated as Commission 9's representative for this Working Group.

The President reported on the IAU Colloquia and Symposia that the Commission had proposed or co-sponsored in the period 1988-91 and on meetings currently being planned or proposed. Future meetings with which Commission 9 is associated are:

Colloquium 136: "Stellar Photometry - Current Techniques and Future Developments" (jointly with Commission 25), Dublin, Ireland, 4-7 August, 1992.

Meetings currently proposed by Commission 9 and under consideration by the IAU Executive are:

A Symposium on "Very High Angular Resolution Imaging" to be held in Sydney, Australia in
            January 1993 (jointly with Commission 40) .
A Colloquium on "Schmidt Telescopes" to be held in Indonesia in 1993.

The President reminded the members of the need to support the incoming President and his Organising
Committee. In particular, it is important to respond to requests for information, suggestions for
meetings, or for views on Commission and IAU matters in general. Often it is has proved necessary to
respond to the General Secretary or Executive without the benefit of the views of the membership, or
even the Organising Committee, because so few people respond to requests.

The following were nominated and elected as officers of the Commission for the period 1991-94:
President: J.C. Bhattacharyya; Vice-President: G. Lelièvre; Organising Committee: M. Cullum,
J. Davis, C.M. Humphries, I.S. McLean, F. Merkle, W.J. Tango, R.M. West.

The following were welcomed as new members of the Commission: P.Alvarez, C. Baffia, A. Barcia,
D.M. Gibson, P. Grosbøl, Li, Z-g., P.J. McGregor, Qiu P., M.H. Slovak, Wang, Z-m., and Zhang,
X-z.

The Commission also noted the resignations of: B. Campbell, P. Fellgett, J. Jelley and G. Zambon.
The members present at the meeting stood in silence in memory of past member P. Charvin.

It was agreed that the Commission Newsletter, commenced in 1990, should be continued. The
President urged members to assist by submitting items of news and interest to the incoming President
for inclusion in forthcoming issues.

The incoming President, J.C. Bhattacharyya, addressed the meeting, thanked the members for their
support and, on behalf of the Commission, thanked the outgoing President for his work for the
Commission.

SCIENTIFIC SESSIONS AND WORKING GROUP MEETINGS

New Instruments and Techniques: 24 July 1991

Chair: J. Davis

In this double session invited reports were presented on a number of major new astronomical
instruments, on adaptive optics, and on the Steward Observatory Mirror Laboratory program. Reports
were also presented on areas covered by two of the Commission's Working Groups, namely Detectors
and High Angular Resolution Interferometry. The following summaries follow the order in which the
reports were presented except for the report on "Adaptive Optics" which was given on 30 July prior to
the Commission's business meeting.

Detectors - G. Lelièvre (Chair, Working Group on Detectors)
The new generation of large format Charge Coupled Devices (CCDs) now appearing on the
commercial market is starting a general revolution for observations at optical wavelengths in all
observatories. The evolution of CCDs and their applications in astronomy can be traced through the
proceedings of various dedicated conferences (for a list see Commion 9's report for the period 1987-
90 in Transactions of the IAU, XXIA, pp.48-9). The notable features are the increased range of CCDs
available, their increased use in observatories, and the improved performance through lower readout

noise, improved quantum efficiency, UV coatings, thinning, etc. CCDs are now available in a form that allows them to be butted together (edge-buttable) opening up the possibility of achieving very large CCD arrays. These developments have had an impact on many fields of astronomy. For example, CCDs are now used for high precision photometry and spectrophotometry. They are used for deep imaging and high resolution imaging and have found applications using small to medium sized telescopes. In spectroscopy, large two-dimensional detectors with low readout noise have enabled high spectral resolution combined with high signal to noise ratio to be obtained on echelle spectrographs and with image slicers. CCD detectors are also being used in Fabry-Perot and Fourier Transform Spectrometers. Fundamental astrometry is also being enhanced by the use of CCDs, notably at the United States Naval Observatory's Flagstaff Station. In addition, new CCD controllers are being considered in most observatories with several under development.

Recent developments in photoelectric detectors include the use of a GaAs cathode in a Ranicon detector (Space Telescope Science Institute), the development of an improved CP40 photon counting detector (Observatoire de Paris and Observatoire de la Côte d'Azur), the development of a second generation Image Photon Counting System (IPCS) (European Space Agency), improved coding and arrays in MAMA detectors for Space missions (Stanford University), and the successful use of an electron bombarded CCD (EBCCD) as a wavefront sensor detector in the COME-ON adaptive optics experiment (see the contribution on adaptive optics in this report).

In conclusion, there is still a need for fast, high quantum efficiency detectors for applications in adaptive optics, wavefront sensors, and interferometry. At present CP40, PAPA and EBCCD detectors are used for this purpose but there is room for significant improvement.

High Angular Resolution Interferometry - W.J. Tango (Chair, Working Group on High Angular Resolution Interferometry)
In high angular resolution stellar interferometry the light from two or more widely separated apertures is brought together and coherently combined. The angular resolution of an interferometer depends on the separation of the apertures and the operating wavelength. Typically it will be in the range 10^{-2} to 10^{-5} arcseconds. The technique enables fundamental properties of stars such as surface fluxes, effective temperature, radius, and mass to be determined, and it opens up a wide range of stellar studies many of which cannot be tackled in any other way.

Stellar interferometry poses major instrumental and technical challenges, as the tolerances on the optical paths within the instrument are of the order of a wavelength of light and the effects of atmospheric "seeing" are more deleterious than in conventional astronomical imaging. Significant developments in the key areas of active/adaptive optics, detector technology, thin film coatings, metrology and computing/control have had a major impact and there are now a number of interferometers either operating, or in an advanced planning stage, which incorporate these new technologies.

Some of these projects and their current status will be highlighted very briefly; details of individual programs will be given at the Working Group session on 31 August:

The Center for High Angular Resolution (CHARA) Project (Georgia State University): The CHARA project is to build a multiple telescope array capable of submilliarcsecond resolution at optical wavelengths. The project has reached the detailed planning stage and a funding application is under consideration.
The Cambridge Optical Aperture Synthesis Telescope (COAST) (University of Cambridge): COAST is currently operating as a two aperture interferometer, but it is planned to expand the instrument into a four aperture array with baselines of the order of 100 m.

The Infrared Spatial Interferometer (ISI) (University of California, Berkeley): ISI is a two aperture infared heterodyne interferometer located on Mt. Wilson. It is being used to observe the dust envelopes around cool stars. There is an ongoing program to upgrade and extend the capabilities of the instrument.

The Infrared-Optical Telescope Array(IOTA): IOTA is a collaborative project involving a number of US institutions and the instrument is currently under construction at the Whipple Observatory (Mt. Hopkins, Arizona). It will have two 0.45 m telescopes initially and baselines up to the order of 100 m.

CHARON and GI2T (Observatoire de la Côte d'Azur): CHARON is a development of the Interféromètre à deux Télescopes (I2T) and it is planned to become a multiple aperture, imaging instrument. The Grand Interféromètre à deux Télescopes (GI2T) is in operation.

The Sydney University Stellar Interferometer (SUSI): SUSI is a two aperture interferometer with baselines up to 640 m. It is undergoing commissioning tests and is expected to commence a regular observing program in 1992.

The Mark III Astrometric Interferometer (US Naval Research Laboratory, Smithsonian Astrophysical Observatory, US Naval Observatory, Massachusetts Institute of Technology): The Mark III instrument is engaged in an extensive observational program. The USNO is developing plans for a dedicated astrometric instrument while the USNRL is developing plans for the Big Optical Array (BOA).

Optical Fiber Spectrograph Feeds - P. Gray (given by P. Gillingham) (Anglo-Australian Observatory)
The use of optical fibres in astronomical instrumentation continues to grow. The technology to efficiently use them as light guides, and successfully employ them in specific applications, has been developing steadily over the past 10 years. In addition, new applications are being found for their use. The use of fibres in astronomy can be broadly divided into two main areas.

Firstly, they are being employed in applications which enhance the capability and performance of existing instruments. These applications are usually relatively small budget projects and demonstrate that fibre feeds are a cost effective way of utilising existing instruments in new ways. Some examples of these systems include: simple plug-plate multi-object fibre spectroscopy systems feeding existing Cassegrain spectrographs; long fibre feeds to Coudé instruments; and two-dimensional area spectroscopy using short fibre arrays.

Secondly, several ambitious new projects are underway, or are being proposed, in which a major new telescope facility is being based solely on the use of optical fibres. The majority of these projects involve multi-object spectroscopy of large numbers of objects (400-600). Some involve a major upgrade to an existing telescope. For example, the two degree field (2dF) project at the Anglo-Australian Observatory which will involve a new prime focus with a large field corrector, multi-fibre robotic positioner and dedicated fibre spectrographs. Other projects have been proposed which involve the construction of a dedicated 2-3m class telescope built specifically for multi-fibre spectroscopy. An example is the 2.5m digital survey telescope (DSS) which has been proposed by the ARC consortium in the USA.

Other areas of development of fibre feed instrumentation include the use of mono-mode fibres for interferometric links between telescopes and the use of infrared transmitting fibres to feed infrared instrumentation.

The ESO New Technology Telescope (NTT) and Very Large Telescope (VLT) - M. Tarenghi (European Southern Observatory)
The principal new technology features of the NTT are the active optical control system and the telescope building design. The NTT is an alt-az mounted telescope with a 3.58 m diameter primary

and actively controlled mirrors. Its housing rotates with the telescope but is open at both ends to allow maximum natural ventilation. The first direct images of astronomical objects with the optics of the NTT in an optimised state were obtained in March 1989. Images of 0.33 arcsecond FWHM have been obtained and the contribution of the telescope enclosure to the seeing has been shown to be negligible. The first regular astronomical observations commenced in January 1990 and the inauguration of the telescope took place on 6 February 1990.

The VLT will have four 8 m aperture telescopes giving it the equivalent collecting area of a 16 m telescope. It is to be located on Cerro Paranal, an isolated, 2664 m high mountain in the central part of Chile's Atacama desert. The telescopes may be used alone, in combined mode, or in a combined interferometric mode. The four telescopes will be positioned in a two-dimensional array which gives good (u,v) plane cover for the interferometric mode and also preserves the best seeing for each of the individual telescopes. Some details of the interferometric mode are given in the report to the Working Group on High Angular Resolution Interferometry and tests of the COME-ON prototype adaptive optics system for the VLT are given in the following report. A range of instruments is planned for the VLT and a review article on them has been published (Journal of Optics, 1991, **22**, p.85).

A contract has been signed for the levelling of the top of the Paranal mountain so that it can accommodate the entire array of 8 m and auxilliary telescopes, as well as associated buildings that together make up the VLT Observatory. This work should be completed by March 1992. A contract has also been signed for an in-depth engineering design study of all the structures and buildings to be erected at Paranal as well as the optimal layout of the access roads. Actual construction is expected to commence on site in the second half of 1992.

Details of the VLT and progress reports appear regularly in the ESO publication "The Messenger".

Adaptive Optics - Fritz Merkle (European Southern Observatory)
Since the IAU General Assembly in 1988 significant progress has been achieved in the area of adaptive optics. At that time several projects were under way but a demonstration of the feasibility and gain which could be obtained for high resolution imaging in astronomy was still missing.

The adaptive optics program of the European Southern Observatory will be summarised after a brief review of worldwide activity in the field.

The adaptive optics project of the University of Hawaii, USA, is now working in the laboratory and will go to the telescope in the fourth quarter of 1991. This system, which is based on wavefront curvature sensing and a bimorph mirror, will be operated at visible wavelengths.

Another approach has been used by the Steward Observatory in collaboration with the Thermo Electron Corporation, USA. A neural network is used to determine directly the correction values for the mirror from image information. The first tests to phase the Multiple Mirror Telescope mirrors demonstrated the feasibility of this method.

The Martini system of the University of Durham, UK, which allows the coalignment of six masked subapertures for the 4 m William Herschel Telescope, La Palma, using tip/tilt mirrors, now regularly produces sharpened images.

A test of an adaptive optics system has been carried out on an astronomical telescope at Yunnan Observatory, People's Republic of China. The system was developed at the Institute of Optics and Electronics, Chengdu, but it was not designed specifically for astronomical applications and it suffers from severe sensitivity problems. Some results for very bright object can be expected.

There are adaptive optics projects in progress at the Johns Hopkins University and the University of Illinois in the USA.

A quite significant input to adaptive optics came in May 1991 from American groups at the MIT Lincoln Laboratory and Philips Laboratory who are working on defence oriented applications. Both teams made public experiments which successfully demonstrated that the principle of artificial reference stars for wavefront sensing, created by scattering of a laser beam, works and will be applicable to full sky coverage from infrared to visible wavelengths. Some of these tests date back to 1983, two years before it was proposed by Foy and Labeyrie in France. These results will give impetus to further developments and it is already obvious that very large telescope projects, like Gemini, the ESO-VLT and others, will include this technique in their adaptive optics programs.

In fact, adaptive optics is already one of the main features of the ESO - Very Large Telescope (VLT) program. First results with the so-called COME-ON prototype system developed for the VLT demonstrated the feasibility and the significant gain of this technology for astronomical imaging. This system gave the first diffraction limited images with adaptive optics in the near infrared for wavelengths between 2.2 and 4.8 microns at the Observatoire de Haute-Provence 1.52 meter telescope in October 1989. Between April 1990 and May 1991 five successful observing runs took place with the ESO 3.6 meter telescope at La Silla, two of these exclusively for scientific programs. For objects like Eta Carinae, Ceres, NGC 1068, Titan, and others, diffraction limited near infrared images have been obtained and the reduction and interpretation are under way.

The COME-ON prototype system, developed as a collaboration between the European Southern Observatory, the Observatoire de Paris-Meudon, ONERA, and the French company Laserdot, is based on a 19 actuator deformable mirror with discrete piezoelectric actuators. The stroke of each actuator is ± 7.5 micrometers. Global wavefront tilt correction is separated from the higher orders of aberration by using an additional tip/tilt mirror. The wavefront sensor is of the Shack-Hartmann type with 5 by 5 subapertures. An electron bombarded CCD makes it photon noise limited. The computing power for the control system comes from a dedicated hardware device with a 68020 processor based host computer. A bandwidth of 25 Hz has been reached. The system is currently equipped with a 32 x 32 IR-array camera.

Presently the system is being upgraded with a deformable mirror with 52 piezoelectric actuators and an improved control system to achieve 40 Hz bandwidth. In a second phase it is planned to transform the system into a standard user instrument for the La Silla observatory. For this purpose it is anticipated that it will be equipped with an expert system to assist the astronomer in finding the best operational parameters based on seeing, brightness and type of object and/or reference source, wavelength, etc. The infrared camera will be equipped with a 128 x 128 and/or 256 x 256 sensor array. The first tests with the upgraded system are expected for mid 1992.

The final definition of the adaptive systems for the 8 meter telescopes has started, including the possible implementation of artificial guide stars.

In conclusion, it can be stated that during the last three years it has been demonstrated that adaptive optics is feasible, that it is important for astronomy, and that new technologies will make it even more powerful and applicable to nearly all high resolution imaging problems from the ground and, in particular, to long baseline interferometry.

The LEST Project - O. Engvold (LEST Foundation)
The Large Earth-based Solar Telescope (LEST), which is run jointly by Germany, Israel, Italy, Norway, Spain, Switzerland and the USA, will be a powerful, next-generation solar telescope with

unprecedented angular resolution and high accuracy polarimetry. The very promising development of techniques for controlling telescope aberrations ("live optics") and seeing ("adaptive optics") and the access to very good sites, means that spatial resolution close to 0.1 arcseconds can in the near future be obtained with ground-based solar observations in the visible and near-infrared.

LEST will be sited on La Palma, Canary Islands, near the Caldera rim on the Roque de los Muchachos observatory. This site offers superb seeing conditions which will enable LEST to reach its ambitious scientific goals.

The conceptual LEST design was completed in 1990. The telescope is a 2.4 m aperture, "polarization-free" concept based on a modified Gregorian optical system. A fast polarization modulator will be located close to the secondary focus of the system. An actively controlled NTT-type main mirror, a high precision pointing and tracking system, a helium-filled light path and thin entrance window, together with an integrated adaptive optics system, will provide near diffraction-limited performance of the telescope.

The construction of LEST will begin in early 1993, and the telescope will be ready for "first light" in 1996.

The Steward Observatory Mirror Lab Program - R. Thompson (Steward Observatory)
The Steward Observatory Mirror Laboratory has the fabrication of mirrors for 8 m class telescopes, including three in which the University of Arizona is a partner, as its primary goal. The mirror design is a honeycomb sandwich which has the properties of being stiff, light and thermally responsive combined with short focal length. The technical innovations that have made the fabrication of such mirrors possible include spin-casting of the monolithic honeycomb sandwiches to give structural efficiency, and rapid tracking of ambient temperature. This process results in a saving of some 20 tons of glass and 9 months of annealing time for an 8 m mirror. The technique of polishing with actively stressed laps has also been developed to allow the production of high quality aspheric surfaces as fast as F/1.

The Mirror Lab successfully cast three 3.5 m mirrors in 1988 and 1989. Preparations are under way to cast a 6.5 m mirror to convert the Multiple Mirror Telescope to a single 6.5 m aperture, doubling its aperture. This casting is scheduled for January 1992 and will be followed by a series of mirrors of 8 to 8.4 m diameter.

The stressed-lap polishing process has been developed to finish fast mirrors to an accuracy consistent with the very best telescope sites. Projects currently under way are the polishing of a 1.8 m F/1 primary mirror for the Vatican Advanced Technology Telescope, and a 3.5 m F/1.5 primary mirror for a US Air Force telescope. Polishing of the Vatican mirror is nearly complete. The specification is to produce images better than 1/8 arcsecond FWHM.

A polishing facility large enough for simultaneous polishing and testing of two 8 m mirrors is in place together with a 24 m vibration isolated test tower. Stressed laps up to 1.2 m diameter have been built and a 2.5 m diameter lap is being designed for use on the large mirrors.

The Keck Telescope - S.G. Djorgovski (California Institute of Technology)
The Keck 10-meter telescope, a joint venture of the University of California, the California Institute of Technology and the University of Hawaii, is now nearing completion at the summit of Mauna Kea. The telescope is still undergoing engineering tests, and the first scientific observations are anticipated in late 1991 or early 1992. The construction of the second 10-meter telescope, Keck-2, is expected to start shortly, and to be complete by 1996. The two telescopes are expected to be used for optical/IR

interferometry for at least a fraction of the time. Plans are also being developed for a subsidiary array of 1.5 m telescopes (Side-Kecks!) to be used as an independent interferometric array but also with the capability of working with the two 10 m telescopes in a combined interferometric mode.

Working Group on Photography: 26 July 1991

Due to the illness of the Chair of the Working Group, J-L Heudier, no scientific program was held. The President of the Commission chaired the session in which the future of the Working Group was discussed. It was clear that there was a strong feeling that the WG should continue but that its area of interest should be broadened and that its name should possibly be changed to reflect this. G. Westerhout proposed that a group of astronomers, which had held informal meetings to discuss large area measuring instruments, should merge with the Working Group. It was agreed that further discussions should take place prior to the Commission's business session on 30 July and that proposals for the future of the WG should be put to the Commission for consideration. This was done and the outcome is reported briefly in the report of the business session given earlier. Details of the plans for the WG's immediate future follow.

The Working Group will have the title "Wide Field Imaging" and its main areas of concern will be : 1. Sky Surveys and Patrols; 2. Photographic Techniques; 3. Digitization Techniques; and 4. Archiving and Retrieval of Wide-Field Data.   Membership will be established by merging members of the former Working Group on "Photography" and the informal digitization groups. For the period 1991-4 the Chair will be R.M. West and the Organising Committee will include (to be confirmed): J. Guibert, R. Humphreys, K. Ishida, B. Lasker, H. Lorenz, H. McGillivray, D. Malin, N. Reid, and M. Tsvetkov. A meeting of the re-formed WG is being considered for autumn 1992 and a newsletter is also under consideration.

Working Group on Detectors: 30 July 1991

Chair: G. Lelièvre

CCDs at the Flagstaff Station - M.H. Ables (US Naval Observatory, Flagstaff):
Astrometric results obtained with CCD detectors have given accuracies equal to or better than 0.0010 arcseconds. Developments for astrometry include the advanced astrometric mosaic array with 6 CCDs and a large format (5 cm x 5 cm) Tektronic 2048 x 2048 CCD.

A transputer-based CCD controller - I.S. Glass(South African Astronomical Observatory):
A compact CCD controller has been developed by groups at the Royal Greenwich Observatory and the South African Astronomical Observatory. It occupies a small Eurocard crate attached to the side of a dewar containing the detector. It offers the possibility of extension to multiple readout devices, control of filter wheels, etc. and fiber-optic communication with the control computer. Ten examples are being constructed for the control of infrared and optical CCDs as well as for CCD-based acquisition cameras.

Astronomical detectors at ESO - M. Cullum (given by F. Merkle) (European Southern Observatory):
Since the last IAU General Assembly in Baltimore, several new large format optical CCDs have been commissioned at ESO's La Silla Observatory. These include several 2048 x 2048 chips from Loral, and 1024 x 1024 chips from both Thomson and Tektronix. The Loral and Thomson CCDs are front illuminated and most have been coated to enhance their UV response. The Tektronix chips are thinned and exhibit a RQE of between 10-12% at 350 nm. Readout noise figures have also evolved

with the current record being held by Thompson CCDs with values down to 2.8 e⁻ rms measured on the telescope.

The development of infrared array detectors has been rapid. Current arrays in service include Philips Components 64 x 64 *MCT* and a Santa Barbara 62 x 58 *InSb* detector. A Cincinatti 64 x 64 *InSb* detector is currently being evaluated in ESO's Garching laboratories and looks extremely promising in terms of large full-well capacity and low dark current. ESO also has a Rockwell 256 x 256 *MCT* detector under test which should be ready for installation at La Silla during the first half of 1992.

Electronic developments related to CCDs have centred on a new VME-based controller. ESO currently has about seven VME controllers of an earlier design in service at La Silla. The new design is necessary to meet the rather special requirements of the VLT project. These include a reduction in the unit production costs and make the system more amenable to larger-scale production methods, the possibility of setting all analogue clock and bias voltages remotely, and a reduction in the size and heat dissipation of the CCD head.

ESO's MAMA photon-counting detector has only been used for UV spectroscopy to date. Due to a lack of a recording system for time-tagged events, it has not been possible so far to make use of the system for time-resolved applications. With the help of outside collaboration, it is hoped that some initial experiments in this direction can be started within the next 6 months.

Review of work at the Observatoire de Paris - G. Lelièvre (Observatoire de Paris)
Work relating to wavefront sensors, adaptive optics, and interferometry by P. Lena, J-M. Mariotti, R. Foy et al. was reviewed. It was noted that fast and efficient detectors are still required in order to obtain the high temporal resolution of a few milliseconds necessary to avoid the effects of phase smearing due to atmospheric turbulence during an observational sample time.

An electron bombarded CCD (EBCCD) has been provided by Laboratoire d'Electronique Philips (LEP) and successfully tested as a detector for the wavefront sensor in the COME-ON experiment at the Observatoire de Haute-Provence and the European Southern Observatory. This device allows discrimination against multi-electron impacts, an advantage over classical photon counting cameras.

An improved CP40 camera (a photon counting camera with 3200 x 2400 pixels) has been completed at Paris Observatory.

Speckle and direct imaging using deconvolution techniques based on reconstruction of the wavefront is under development for the visible domain (R. Foy and G. Lelièvre).

In the ATLAS Experiment a laser made artificial star has been tested (M. Tallon and ONERA) at the Lunar Laser Station (CERGA) and has been shown to provide a 6 arcsecond artificial star image at an altitude of 15 km.

G. Lelièvre summarised reports received from groups unable to attend the General Assembly:

From P.R. Jorden (Royal Greenwich Observatory): The Royal Greenwich Observatory has been particularly active in recent years since the William Herschel 4.2 m Telescope saw "first light" in 1987. Detectors include integrated TV for field-viewing, CCD-based autoguiders, Image Photon Counting System (IPCS), cryogenic CCD cameras, an embedded CCD for a faint object spectrograph, and the main grating spectrograph (ISIS) uses separate CCDs in its red and blue arms. EEV CCD-05-30 chips are used. They provide excellent performance including 3e⁻ readout noise, 50% peak quantum efficiency and no defective columns in a super-grade 1280 x 1180 format array. A new integrated CCD controller was designed to give high performance, compact, multi-chip CCD operation. A development program for thinned CCD by EEV and financed by the AAT is also complete. Development plans include multi-chip mosaics and lower readout noise devices.

From G.J. Monnet (Canada-France-Hawaii Telescope (CFHT)):   A Ford 2048 x 2048 CCD is in general use by observers. A new generation controller with improved readout time is being designed for various chips and for an infrared array (256 x 256 from Rockwell).

The archiving of CFHT data will be handled by the Canadian Astronomical Data center at the Dominion Astrophysical Observatory.

An adaptive optics bonette is under construction to correct static effects on a 4 arcminute field and atmospheric turbulence on a small field (resolution better than 0.2 arcseconds in the 0.5 to 2.5 mm wavelength range). Future planned developments include the study of large CCD mosaics.

From Bartoletto (Padova):   A CCD working group was formed to equip the new Galileo Telescope. A collaboration with the Cerro Tololo Inter-American Observatory (CTIO) was arranged for the development of a multiple CCD controller based on the CTIO transputer sequences and having the capability of reading various chips and mosaics (4 output CCD controller). The detectors are required for focal plane instrumentation, autoguiding and active optics systems.

From J.G. Timothy (Stanford University):   Several MAMA detectors are under test for space experiments. For the latest, an array of 726 x 8096 pixels is proposed. Improvements have been made in the x-y coding of MAMA detectors.

The Organising Committee for the Working Group for 1991-4 is: M. Cullum (Chair), B. Fort, J. Geary, P.R. Jorden, G. Lelièvre and J.G. Timothy.

Working Group on High Angular Resolution Interferometry: 31 August 1991

Chair: W. J. Tango

Nine papers were submitted for the scientific program of the Working Group:

COAST - The Cambridge Optical Aperture Synthesis Telescope - D. Green (Cambridge University):
COAST is planned to be an array of four 40 cm telescopes for imaging in the red and infrared, using baselines up to approximately 100 m.   Two telescopes are currently operational, together with equipment for acquisition, guiding, and variable optical path delay compensation, all housed in a passive thermally stable building. Fringes are, weather permitting, currently being obtained over a wide range of declinations and hour angles.

The Infrared-Optical Telescope Array (IOTA) - N. Carleton, P. Horowitz, M. Lacasse, P. Nisenson, C. Papaliolios, M. Pearman, R. Reasenberg, W. Traub (Center for Astrophysics), R. Predmore, P. Schloerb, S. Strom (University of Massachusetts), D. Gibson (MIT Lincoln Lab), J. Benson, M. Dyck (University of Wyoming):
IOTA is planned to operate initially with two 0.45 m telescopes and baselines of the order of 100 m at visible and infrared wavelengths. The instrument will be sited at Mt. Hopkins and is scheduled to begin operating in the first half of 1992. Flat siderostat mirrors will feed light into fixed beam compression telescopes which will reduce the beam diameter by a factor of ten. The light is then

transferred via relay optics to the delay lines and the beam combining optics. The relay optics and delay lines are all in vacuum. The visible light fringes will be detected using a grism spectrometer as described by Traub (JOSA A, 1990, **7**, p.1779), while the infrared detection will be with two single element detectors operating in the spatio-spectral interferometric mode.

The Big Optical Array - K. Johnston (U. S. Naval Research Laboratory):
The U.S. Naval Research Laboratory is developing the design of a large optical interferometer known as the Big Optical Array (BOA). It will be a phase closure, imaging instrument with six 0.5 m siderostats moveable to various stations on a Y shaped array with available baselines from 4 m to 470 m. It will be a single r_0 instrument with a magnitude limit of the order of 9.5 for stellar angular diameter measurements and 4.5 for full image mapping. It is planned to site BOA on Anderson Mesa, near Flagstaff, Arizona.

Laser Reference Stars: first results of the ATLAS experiment - M. Tallon (Observatoire de Paris):
A bright reference star lying in the isoplanatic patch surrounding the observed object is needed to operate adaptive optics devices. The probability of finding a reference star in a given direction in the sky is low, particularly for fully correcting wavefronts at visible wavelengths. To operate adaptive optics in any direction in the sky, even in the visible, it is possible to create an "artificial reference star" by using the backscattered light of a laser beam, focused in the atmosphere. The ATLAS experiment is being carried out by ONERA, Observatoire de Paris, and CERGA. The first aim of the experiment is to create a laser reference star and to compare the wavefronts restored from simultaneous observations of a real bright star and the laser spot. An image of a reference star has been obtained.

Optimum Exposure Times for High Resolution Imaging - B. Lopez (European Southern Observatory):
It is important to be able to estimate the stability time for interference fringes since this affects the signal level that can be detected. A theory has been developed which relates this time to the standard deviation of the wind speed of the turbulent layers and Fried's r_0 parameter. This was tested at La Silla by monitoring the wavefront tilts at two separate positions on the wavefront. Values of the "speckle lifetime" were observed in the range 4 to 16 ms over seven nights.

The Sydney University Stellar Interferometer (SUSI) - J. Davis (University of Sydney):
The Sydney University Stellar Interferometer (SUSI) is sited at the Paul Wild Observatory in northern New South Wales alongside the Australia Telescope. Construction commenced in October 1987 and all civil and building works were completed by early 1990. Since then an extensive installation, alignment, and commissioning program has been pursued. SUSI has 12 siderostat stations distributed along a North-South line to give baselines in the range 5 m to 640 m. Currently the inner baselines are being commissioned. Four stations are equipped with siderostats and the remaining siderostats are to be progressively installed over the next 12 months. Two stations have been fully commissioned along with an intensified CCD based acquisition system and wavefront tilt correcting servos. The beam-combining and visibility detection systems from the SUSI prototype have been installed. The optical path length compensation system, which will have two compensation carriages running on parallel tracks, will give a total compensation range of ±420 m. The fine control carriage has been installed and partialy commissioned along with the laser metrology system. The SUSI project is jointly funded by the Australian Research Council and the University of Sydney. At the time that this report was given, attempts to detect fringes had not been made. However, shortly after the General Assembly the first fringe detection with starlight was reported.

The CHARA Array Project - H. McAlister (Georgia State):
Astronomers from Georgia State University's Center for High Angular Resolution Astronomy (CHARA), in a collaboration with scientists and engineers from the Georgia Institute of Technology, have proposed the construction of a multiple telescope array capable of submilliarcsecond resolution at optical wavelengths. The CHARA Array will consist of seven 1 m aperture telescopes in a Y-shaped array contained within a 400 m diameter circle. The facility will be located at a site in the southwestern USA. Each collecting telescope will have an independent delay line, and thus, while the initial observations will be made by pairwise interferometry, the instrument will be developed into a fully imaging array using all apertures together. It is expected that start-up funds in the second half of 1991 will permit the initiation of detailed analysis and design of critical subsystems of the Array. This work was supported during a feasibility study phase by the US National Science Foundation.

The VLT Interferometer - F. Merkle (European Southern Observatory):
The VLT Interferometer (VLTI) consists of an array of four adaptive optics aided 8 m telescopes as well as two or more 1.8 m Auxiliary Telescopes. The array of smaller telescopes, also referred to as the VISA (the VLT Interferometric Sub-Array), will be 100% dedicated to interferometry, and will be used on its own when the sensitivity of the large 8 m telescopes is not needed. The VISA telescopes will be moveable, and it will be possible to combine them with the 8 m telescopes when so desired since all telescopes will share the same interferometric infrastructure. The maximum baseline between the VISA telescopes will be 190 m and, between the 8 m telescopes, 128 m. The VLTI design specifies a large co-phased/coherenced interferometric field of view (8 arcsec) and an ability to "coherence guide" on objects 15 arcmin away.

The USNO Astrometric Optical Interferometer - D. Hutter and J. Hughes (US Naval Observatory):
The US Naval Observatory Astrometric Optical Interferometer (AOI) will be a dedicated astrometric interferometer. It is currently being designed and is planned to be in operation at a site in the western USA in 1993. The AOI will be built upon the experience gained from the Mark III Optical Interferometer on Mt. Wilson, California, which has been a joint project involving the Naval Research Laboratory, the Smithsonian Astrophysical Observatory, the US Naval Observatory and the massachusetts Institute of Technology. The Mark III has demonstrated wide-angle astrometry with uncertainties of 5 to 10 milliarcseconds from observations repeated over several nights and systematic agreement with FK5 catalog positions with the (larger) quoted uncertainties of the catalog. The new AOI will consist of four elements in a fixed array, with baselines arranged for instantaneous position measurement in two dimensions. State-of-the-art delay lines as well as extensive laser metrology will be incorporated. Several beam combination techniques will be used to increase the precision of astrometric measurements. The USNO AOI will achieve unprecedented precision in the optical regime (450 to 900 nm wavelength), with statistical and systematic errors below 5 milliarcseconds for wide angle measurements and 1 milliarcsecond for narrow angle measurements (object separations less than one degree). This array will establish a catalog of the positions of over 1000 stars.

Working Group business: Due to the disruption of the day's program caused by the fire in the Centro San Martin, it was not possible to hold a formal business session. After informal discussion with the WG members who were present and the Commission President, the organising committee will have the following membership for the next three years: J.E. Baldwin (UK), P. Lena (France), H.A. McAlister (USA), F. Merkle (European Southern Observatory), S.T. Ridgway (USA), W. J. Tango (Australia) (Chair) and C. H. Townes (USA).

COMMISSION 10: SOLAR ACTIVITY (ACTIVITE SOLAIRE)

COMMISSION 12: SOLAR RADIATION AND STRUCTURE
(RADIATION ET STRUCTURE SOLAIRE)

Report of Meetings, 24, 25, 26, 29 and 30 July 1991

PRESIDENT (10): E. Priest   SECRETARY (10): B. Schmieder
PRESIDENT (12): J. Harvey   SECRETARY (12): J. Stenflo

This report was prepared jointly by Commissions 10 and 12 and reflects the combined nature of the activities of these commissions at the 21st General Assembly. In addition to the report of activities given here, readers should consult reports of Joint Commission Meetings co-sponsored by Commissions 10 and 12 (JCM 1, 3, 6) found elsewhere in this volume or in Highlights of Astronomy (Volume 9). Also, a meeting entitled "Nonlinear and Turbulent Processes in the Solar Wind and Astrophysical Plasmas" was held jointly under the sponsorship of Commission 49.

1. Joint Business Meeting (25 July 14:00 - 15:30)

The meeting was opened by E. Priest with greetings to the members of both commissions. J. Harvey then read a list of deceased members of both commissions. The members rose and observed a moment of silence. Commission 10 held an election for new officers with the results for the period 1991-4: President: V. Gaizauskas. Vice-President: O. Engvold. Organizing Committee: Ai G., E. Antonucci, A. Benz, T. Forbes, A. Galal, E. Hildner, J. Jakimiec, M. Karlický, I. Kim, M. Machado, E. Priest, T. Sakurai, B. Schmieder, M. Schüssler, V. Stepanov, and E. Tandberg-Hanssen. Commission 10 then elected 69 new members. Next, Commission 12 held its election for new officers for the 1991-4 period with these results: President: J. Stenflo. Vice-President: F. Deubner. Organizing Committee: T. Ayres, J. Harvey, V. Karpinsky, E. Landi Degl'Innocenti, Å. Nordlund, J. Pasachoff (Chairman, Working Group on Solar Eclipses), K. Sivaraman, J. Staude, Wang J.-X., P. Wilson and H. Yoshimura. Commission 12 then elected 37 new members.

E. Priest reported on the state of Commission 10. He noted the vigor of solar research in many countries as evidenced by new projects both on the ground and in space. The aims of Commission 10 are to sponsor meetings and encourage international cooperation in the study of solar activity and to facilitate cooperation between observers and theorists. In order to inform members he had sent out a regular newsletter during the previous three years. Four goals for the future were suggested: to maintain high standards of scientific research, to inspire young scientists to study the Sun, to develop links with other astronomers, and to help break down barriers of prejudice and deepen warmth and collaboration in our international community.

J. Harvey reported on activities of Commission 12. He noted the healthy state of solar research by reference to the Commission 12 contribution to Reports on Astronomy (Vol. XXIA). Disappointment was expressed about failure to obtain IAU sponsorship of all proposed solar meetings. On the positive side, some of these meetings were held without IAU sponsorship with very successful results. The issue of merging Commissions 10 and 12 was raised during the 20th General Assembly. The memberships of both commissions were polled with a clear result to not merge the commissions. At the same time, the idea of name changes was also tested by poll. There was no strong sentiment for a change of the name of Commission 10 but a small majority favored a change of the name of Commission 12. After discussion with Commission 35, the name Solar Radiation and Structure was submitted to the IAU Executive Committee and approved.

1.1. WORKING GROUP AND REPRESENTATIVE REPORTS

1.1.1. *Working Group on Solar Eclipses* (E. Hiei). The Working Group was quite active during the previous three years in preparation for the total eclipses of 22 July 1990 and the extraordinary one of 11 July 1991. National coordinators and committees were identified for both eclipses in a number of countries. During the business meeting held at the General Assembly, the total eclipse of 30 June 1992 was discussed. A resolution concerning publication of eclipse information was discussed and adopted. This resolution was also discussed at the general business meeting and was accepted as given below. J. Pasachoff was elected the new Chairman of the Working Group.

1.1.2. *Quarterly Bulletin on Solar Activity* (E. Hiei). With IAU sponsorship, the QBSA is published by the National Astronomical Observatory (Japan) in five sections. More than 135 observatories are contributing data to this valuable compilation which now covers five solar activity cycles.

1.1.3. *International Ursigram and World Days Service* (E. Tandberg-Hanssen for H. Coffey). The IUWDS is a permanent service of URSI, IAU and IUGG which alerts the world scientific community about transient events and helps coordinate scientific observations that cannot be carried out continuously. The backbone of the IUWDS network is a group of Regional Warning Centers and World Data Centers. A move toward digital recording and dissemination of data is a strong trend of the past three years. An important issue that has not resolved is how to distribute such data to researchers not yet equipped with suitable digital facilities.

1.1.4. *Sunspot Index Data Center* (based on a written report by A. Koeckelenbergh and P. Cugnon). The SIDC is responsible for preparation of the International Sunspot Number Ri. The forecast for sunspot maximum made three years ago at the Baltimore General Assembly proved to be quite accurate. More than 115 stations provide data to the SIDC of which about 80 are regularly used for the preparation of the index. Compilations are mailed to about 450 sites and telefax and telex messages are sent to additional subscribers. Electronic mail is becoming an important new way for distributing data products.

1.1.5. *Federation of Astronomical and Geophysical Data Analysis Services* (E. Tandberg-Hanssen). The FAGS is an umbrella organization established by ICSU to oversee and help fund the activities of 10 services including the QBSA, IUWDS and SIDC of particular interest to Commission 10. The IAU Commission 10 representative is E. Tandberg-Hanssen, who is also president of FAGS. At the Grenoble General Assembly, Commission 10 asked the Debrecen Heliophysical Observatory to continue the work of the Greenwich Photoheliograph program. One year of data was published but a backlog has built up. To continue the service requested by Commission 10, Debrecen has applied to FAGS for funding assistance. The IAU has agreed to grant Debrecen $500 per year for each year of data that is reduced and published for ten years of data. To provide more flexibility in the future, it was agreed to concentrate all of the solar services supported by FAGS in a single panel named International Services on Solar Activity. The panel membership would be the President or Vice-President of Commission 10, representatives of QBSA, SIDC, IUWDS, Debrecen, IAU representatives to FAGS, and archiving experts. The terms of reference are: (i) to ensure the services are responsive to the needs of the community, (ii) to make a recommendation to the Commission 10 business meeting at each IAU General Assembly on the apportionment of funds provided by FAGS, (iii) to request reports as necessary, (iv) to consider modern formats for archiving data. A meeting is planned prior to April 1992.

1.1.6. *SCOSTEP* (S.T. Wu). The Solar-Terrestrial Energy Program (STEP) is a major program of SCOSTEP that started in 1990 and will continue through 1995. This program has focussed work on process that couple energy from the Sun to Earth. SCOSTEP sponsored a Solar Terrestrial Symposium in 1990 and another major symposium is scheduled during 1992.

1.1.7. *Flares 22* (V. Gaizauskas for M. Machado). This program is described in section 2.2.4.

1.1.8. *Commission 10 Representatives to Organizations.* The following representatives to organizations were elected: E. Tandberg-Hanssen (FAGS), H. Coffey (IUWDS), T. Hirayama (QBSA), S.T. Wu and E. Priest (SCOSTEP), and O. Engvold (COSPAR).

1.2. RESOLUTIONS

Resolutions were proposed by V. Gaizauskas and J. Harvey during the business meeting. After extended discussions, amendments and changes, both were adopted by Commissions 10 and 12. They were submitted to the IAU Executive Committee and the final General Assembly and were adopted. The text of these resolutions follows.

Commissions 10 and 12,

considering long-term observations are essential to understand the behaviour of such quasi-periodic phenomena which characterize solar and stellar activity and which link the Sun to our terrestrial environment;

The IAU, meeting in General Assembly,

recommends (1) strong support for the continuation of data-gathering programmes and observational facilities that are essential to long-term research; (2) optimization of data-gathering enterprises in order to improve services to the research community.

Commissions 10 and 12,

considering that the United States Naval Observatory has for more than forty years generously provided crucial information to assist scientists who observe solar eclipses for scientific purposes (in the form of the Central Solar Eclipse Circulars and other specialized calculations) and,

recognising that the USNO plans to cease publication of the Eclipse Circulars due to programmatic changes and plans to continue to support scientific observations by publishing eclipse circumstances in the Astronomical Almanac, and by providing specialized eclipse calculations to scientific researchers,

commend on behalf of past and present eclipse researchers, the management and staff of the USNO responsible for the preparation and publication of the calculations and,

request that the USNO continue to provide advance calculations for a variety of sites in order to aid site selection and to publish this information in Circulars or by other means and,

further commend all national organizations that prepare eclipse calculations and urge that they continue their efforts.

2. Scientific Meetings

In the following reports, authors and coauthors of presentations are listed with the actual presenter underlined.

2.1. RESULTS OF 1990 AND 1991 ECLIPSES (24 July 09:00 - 12:30; E. Hiei)

Approximately 60 people attended presentations of results from the 1990 and 1991 total solar eclipses. I. Kim reported that an effort to record coronal fine structure using 3 cameras widely spaced along the 1990 path of totality in the USSR was frustrated by poor weather at two of the sites. At the same eclipse P. Kaufmann used the 13.7m Itapetinga radio telescope to detect Fresnel fringes at 22 GHz. He concluded that the fringes were caused by hot, quiescent spots in the corona.

The 1991 eclipse occurred just days before the meeting, so only very preliminary reports were available. This eclipse will long be remembered for its passage across the large telescopes on Mauna Kea (Hawaii) and for the outstanding beauty of the corona for those lucky enough to see it. It was also the first eclipse studied intensively with CCD detectors both in the visible and infrared.

J. Pasachoff presented an extensive list of observations planned to be done in Hawaii. He reported that weather conditions were unsettled at the time of the eclipse and that Mauna Kea was affected by cirrus clouds and dust from the eruption of Mount Pinatubo. J. Vial reported on successful recording of coronal fine structure and rapid changes using cameras at the focal plane of the 3.6m CFHT. J. Harvey read reports from several experiment groups on Mauna Kea: Sub-millimeter and millimeter observations of the chromosphere and prominences by two groups using different facilities were successful. The chromosphere was observed to be much higher than predicted by models. IRTF observations of the 12μm Mg I emission line indicated that the line is formed in the upper photosphere. Video and CCD recordings using the 2.2m Hawaii telescope showed a wealth of fine structure in various emission lines. Three experiments were attempted to observe thermal emission from dust around the Sun. All operated successfully but no reports of dust were made.

S. Isobe reported on four Japanese experiments done from Mexico. These were a timing measurement by the Hydrographic Office, a Kyoto University measurement of the thermal balance of coronal loops, an NAO experiment to detect cool material in the corona and an heroic infrared measurement of coronal structure by Isobe and collaborators at very high altitude. J.

Stohl reported on successful observations of the polarization of the coronal green line and high resolution images made by Skalnate Pleso observers at La Paz (Mexico). T. Jurriens and J. Fierro described popular level observations made by 15 international groups at La Paz. I. Kim reported that the USSR expedition to Brazil was successful in obtaining high resolution images of the corona as part of an international set of cameras stationed along the eclipse path. Sites in Mexico and Hawaii were also successful. A highlight of the meeting was E. Hiei's showing of video recordings of the eclipse made on Mauna Kea using the latest high-resolution video equipment.

2.2. COLLABORATIVE PROGRAMS IN SOLAR ACTIVITY STUDIES (25 July 16:00 - 17:30; V. Gaizauskas)

2.2.1. *Radiative Inputs of the Sun to Earth - The RISE Project* (J. Pasachoff for P. V. Foukal). The aim of the RISE project is to measure and understand the variable radiative outputs of the Sun at the present time. Its ultimate goal is to respond to mounting concern over changes in Earth's climate and atmospheric chemistry by supplying precise information on changes due exclusively to the Sun's variability. Trends over astronomical time scales will be clarified by extending photometric studies to nearby sun-like stars.

Although it has been conceived in the USA as a national effort, RISE depends on international collaboration for 24-hr coverage of key parameters. Workshops held in 1987 and 1989 led to the formation of 6 working groups concerned with: multi-wavelength measurements, data analysis, theory, and space observations of the total irradiance as well as the EUV and UV spectral irradiances. Design for a precision solar photometric telescope has been completed. One instrument will be located in Tucson, Arizona; an identical twin will be operated at a site suitably displaced about 12 hr in longitude. Funds totalling $11.1 million (US) for a 5-yr programme have yet to be approved by the National Science Foundation.

2.2.2. *Collaborative Solar Radio Programmes* (M.R. Kundu). The application of supersynthesis arrays to arc-sec imagery of solar structures at microwave frequencies has promoted strong collaborations between radio and optical astronomers for over a decade. Now that pulse-like emissions have been detected from some flares at both microwave and hard X-ray wavelengths, links have been forged between groups of astronomers with instruments responding to an extremely broad energy band. A major collaborative effort is underway to identify the accelerative process leading to this 'fragmentation' of flare emission. One of the more unusual collaborations made possible by technological advances involves detection of flare emission from very high energy particles at opposite ends of the electromagnetic spectrum - gamma-rays with the Gamma Ray Observatory (GRO) and mm-waves with the large Berkely-Illinois-Maryland (BIMA) interferometer.

The loss in recent years of the Culgoora and Clark Lake radioheliographs has been a serious blow to investigations of transient coronal emission. Fortunately the gap will soon be filled by the construction, now underway at Pune (near Bombay, India) of the Giant Meter-Wave Radio Telescope, a radioheliograph operating between 28 and 1410 MHz.

2.2.3. *Collaborative Magnetographic Studies: China - USA* (Ai Guoxiang and S.F. Martin). The existence of similar videomagnetographs at Big Bear (Caltech) and Huairou (Beijing) Solar Observatories makes it possible to conduct collaborative studies of solar magnetic structures on a round-the-clock basis. The evolution of active regions and of the photospheric network can now be tracked for days in succession with few interruptions. Video movies were presented which show the immense possibilities now opening for studying flare build-up in active regions. Most magnetic structures change slowly over a period of days, but some important changes occur on shorter time scales and would be missed without the extended coverage.

2.2.4. *FLARES 22* (V. Gaizauskas for M.E. Machado). FLARES 22 is a project sponsored by SCOSTEP's Solar Terrestrial Panel (STEP), endorsed by COSPAR and by the IAU. It is also designated an official International Space Year (ISY) project by the Space Agency Forum for ISY. The scientific aims of FLARES 22 are: (a) to identify and prioritize problems for individual and cooperative research on solar flares during the maximum phase of Solar Cycle 22; (b) to assign active-region targets for collaborative observing campaigns by participating observatories in co-operation with MAX '91 (a national program in the USA supported by NASA); (c) to conduct workshops in which theory is confronted with observations of flare-relevant phenomena.

The first FLARES 22 campaign (Aug. 10 - 25, 1990) was initiated by solar scientists in the USSR. A joint FLARES 22/MAX '91 campaign ("Energetic Solar Phenomena") was held 07 Dec 1990 - 26 Jan 1991 in conjunction with a long-duration circumpolar (Antarctic) flight of a high altitude balloon for observing high-energy solar X-rays. The VLA was dedicated for solar observations during the same period. Another joint FLARES 22/MAX '91 campaign ("Gamma-Ray, Hard X-ray and Neutron Studies of Solar Flares") is scheduled for 03 - 17 October, 1991 in conjunction with NASA's Gamma-Ray Observatory satellite.

The first FLARES 22 workshop, attended by 83 participants in Chantilly, France, from 16 to 19 October 1990, had its proceedings published in February 1991 (editors: E.R. Priest and B. Schmieder). The proceedings contain the results gathered by 5 working groups assigned to each of the following topics: energy storage; rapid fluctuations in flare emissions; microflares; flare loops and giant arches; material ejections. The main priorities for future campaigns and workshops include:
- Magnetic Field Configurations (e.g. monitoring the vector magnetic field; measuring electric fields in active regions).
- Rapid fluctuations (e.g. determining whether fragmentation of the energy process or modulation of the emission process is the cause).
- Material Ejections (e.g. investigating the relationship of large-scale coronal structures to flares and coronal mass ejections).

A second FLARES 22 Workshop is planned for October 1992 in the Crimea (USSR).

2.3. DYNAMICS AND STRUCTURE OF PROMINENCES (26 July 09:00 - 12:30; E. Tandberg-Hanssen)

This meeting was organized in two parts with a break between them. Four invited talks were given according to the following program:

E. Tandberg-Hanssen: Introduction and definition of the subject.
B. Schmieder: Structural Elements of Filaments.
O. Engvold: Dynamic Nature of Filaments.
J. C. Vial: Structural Characteristics of Eruptive Prominences.
T. Forbes: The Eruption Mechanisms.

The four invited speakers gave excellent and up-to-date reviews of our understanding of the nature of quiescent prominences, both in their quiescent stage and during disparition brusque (eruptions). The format of the session allowed considerable and valuable discussion on most topics raised. In addition to the invited talks a number of interesting poster papers were presented on the prominence topics, viz.

Mein, P. and Mein, N. "Fine Structure of a Quiescent Prominence"
Moore, R.L., Hagyard, M.J., Davis, J. M., and Porter, J.G. "The MSFC Vector Magnetograph, Eruptive Flares, and the Solar-A X-Ray Images"
Oliver, R. and Ballester, J.L. "The Effect of Coronal Magnetic Arcades on Prominence Properties"
Pande, M.C. and Bondal, K.R. "On the Rotational Motion in Quiescent Prominences"
Schmieder, B., Dere, K.P. and Wiik, J.E. "Dynamics in the Prominence-Corona Transition"
Wiik, J.E., Schmieder, B., Noens, J.C., and Heinzel, P. "Fine Structure Analysis of Prominences in Hα and Coronal Lines"
Gopalswamy, N., White, S.M., and Kundu, M.R. "A Study of Solar Radio Filaments at 20 cm"
Zhang, H., Ai, G., and Shi, Z. "The Observations of Magnetic and Velocity Fields with Video Vector Magnetograph in the Filaments of the Active Regions"
Martin, S.F. "The Essential Role of Rotational Discontinuities in the Formation of Filaments"

2.4. HIGH-RESOLUTION OBSERVATIONS AND THEORY OF SOLAR MAGNETISM AND CONVECTION (29 July 09:00 - 12:30; J.O. Stenflo)

2.4.1. *High Resolution Observations of Solar Magnetic Fields* (C.U. Keller).

Recent improvements of instruments and data reduction techniques make it possible to reach 0.3 sec of arc resolution in spectrograms and magnetograms of small-scale solar magnetic structures. Some of the most puzzling observations of the last few years are reported in the present review. While variations of the longitudinal velocity and the magnetic field strength of penumbral filaments are not related to brightness variations, there exist indications that the brightest filaments are associated with more vertical fields. Magnetic elements, which can be detected down to the diffraction limit, may appear bright as well as dark in the continuum. Some examples of network bright points do not show any measurable polarization. Oscillations in magnetic elements as determined by the shifts of the Stokes V zero-crossing correspond to the ones seen in the quiet Sun in Stokes I but with a reduced amplitude. There are no indications from spatially resolved observations for the large amplitude oscillations required by high spectral resolution observations to explain the width of Stokes V profiles.

2.4.2. *Results from the La Palma Observations by the Lockheed Group* (P.C.H. Martens , A.M. Title, Z.A. Frank, R.A. Shine, T.D. Tarbell).

High resolution observations of photospheric velocity and magnetic fields obtained with the Swedish optical telescope and Lockheed tunable optical filter at La Palma have been Fourier analysed. The 2-D Fourier spectra of the horizontal velocity field and the line-of-sight magnetic field are found to be isotropic, forming power laws as a function of the absolute wave vector. The relation between the power law indices of the magnetic and velocity field is consistent with the predictions for passively advected scalars in Kolmogorov type turbulence.

In addition it is found that the power-law index of the Poynting flux that enters the corona and presumably causes coronal heating is consistent with the power law dependence of the Fourier transformed X-ray emission observed with the rocket-borne NIXT instrument. These results suggest that MHD turbulence theory is the correct theoretical framework for the analysis of photospheric magnetic motions and coronal heating.

2.4.3. *Stokes V Diagnostics in the Infrared* (I. Rüedi, S. K. Solanki, W. C. Livingston, and J. O. Stenflo). Information on the intrinsic properties of solar magnetic elements beyond the attainable spatial resolution can be obtained by indirect methods, e.g. using the line-ratio method or observing the complete Zeeman splitting in the infrared. The most powerful method is to apply a line-ratio method in the infrared, which has been done in the present work by analysing the Stokes V line profiles of two Fe I lines near 1.56 μm, having greatly different Landé factors (3.0 and 1.5). This allows the Zeeman broadening (due to a distribution of field strengths) to be separated from the Doppler broadening. The analysed spectra have been obtained with the McMath spectrograph and a 3×3 arcsec^2 sampling aperture in a large number of locations near the center of the solar disk.

The analysis shows that the Zeeman broadening can be fully accounted for by the field-strength variation along the line of sight due to the height divergence of self-consistently modelled magnetohydrostatic fluxtubes. There is no need for any horizontal field-strength distribution, except for the need of using two distinctly different fluxtube components instead of one within the sampling aperture. The traditional approach has been to use a two-component model, with one magnetic and one non-magnetic component, and this is sufficient for reproducing many of the infrared Stokes V spectra. For a considerable number of cases, however, a three-component model is necessary (two magnetic and one non-magnetic component), but no case has been found for which more than three components are needed (note that without the height divergence of the field, many more components would be required to obtain a good fit).

Most of the magnetic elements have field strengths of 1.5-1.6 kG, but many components are found to be intrinsically weak (well below kG), although about 90% of the flux in the present data set is in kG form. The plasma β is typically 0.3 in the magnetic elements, but for the intrinsically weaker components the values are higher.

2.4.4. *Size Distribution of Solar Magnetic Flux* (T. Bogdan). Statistical analysis of the sizes of sunspots, sunspot groups, and active regions have led to the following conclusions:
(1) Sunspot umbral area measurements seem to be the most unambiguous selection-independent data. Their areas are distributed lognormally over a range of areas from 2 to 141 millionths of a solar hemisphere. Little (if any) variation in the shape of this distribution with solar cycle is evident. It is suggested that their origin is connected with the (complete) fragmentation of large regions of magnetic flux. Whether this distribution obtains for smaller structures like pores, knots, and elements is an open question due to observational impediments, chiefly lack of good spatial resolution and contrast.
(2) Sunspot groups *seem* to have a distribution of sizes similar to the sunspots from which they are constructed. Their decay rates are distributed lognormally, and there is some preference for them to decay at a rate proportional to their circumference.
(3) Active regions area distributions show mixed results suggesting that they are not free from measurement biases. While Tang et al. have found an exponential distribution at solar maximum but a somewhat different distribution at solar minimum, the results of K.L. Harvey indicate no apparent variation with solar cycle, and the log dN/dA vs. log A plot requires terms up to a cubic in log A for a reasonable fit.

2.4.5. *Selfsimilar Magnetic Fields* (K. Galsgaard and Å. Nordlund). Recent numerical simulations of the solar convection zone have indicated that the convection zone may have a selfsimilar structure, with similar patterns repeated over a range of scales. By inference, such a selfsimilar structure would carry over to the magnetic field, which is convectively controlled in the subsurface layers of the convection zone. This is supported by observations of the horizontal distribution of the magnetic fields by Tarbell et al., who have shown that the surface distribution of solar magnetic fields is selfsimilar, with a fractal dimension of approximately 1.6 over several decades in size.

A fully dynamic treatment of a magnetic field which is selfsimilar over a large range of scales is beyond the capabilities of even the fastest supercomputers of today. Therefore in the present work an approach has been chosen where semianalytical models of stressed magnetic fields are built. Fourier series with complex coefficients are used to generate scalar and vector potentials representing spatially complicated magnetic fields, which are selfsimilar over a limited range of scales. The external stress is simulated by applying analytical deformation-transformations to the fields, and the potential for dynamic and catastrophic events in these toy-fields is explored by numerical and graphical means.

2.4.6. Correspondence between X-Ray Bright Points and Evolving Magnetic Features in the Quiet Sun (D.F. Webb , S.F. Martin, D. Moses, J. Harvey). Coronal bright points, first identified as X-ray Bright Points (XBPs), are compact (about 20-30 sec of arc), short lived (about 8 hr), and associated with small-scale bipolar magnetic flux features in the quiet Sun. Several studies have yielded contradictory results suggesting that coronal bright points are either primarily a signature of emerging flux in the quiet Sun, or of the disappearance of pre-existing, opposite-polarity flux. Results are presented, based on the use of coordinated data obtained during X-ray sounding rocket flights on 15 August and 11 December 1987 to determine the correspondence of XBPs with time-lapse ground-based observations of evolving bipolar magnetic structures. These results indicate that, at least during this phase of the solar cycle, XBPs are much more frequently associated with random encounters of pre-existing magnetic flux of opposite polarity which are cancelling than with emerging flux regions. It is suggested that this type of XBP is prevalent around solar minimum because of the dominance of mixed polarities leading to random flux cancellations, and that these XBPs are the result of reconnection involved in the cancellations.

2.4.7. A Quiescent Filament and Associated Supergranulation Network (Zhang Yi, O. Engvold). A quiescent filament and its association with the adjacent photospheric regions has been studied from high-resolution filtergram observations obtained with the Swedish solar telescope at La Palma using the Lockheed tunable, narrow-band filter. The observations include time series at 7 positions in Hα, for studies of the fine structure of the filament and chromosphere, 3 positions in the line Fe I 5576 Å for determination of the photospheric Doppler velocities, and two orientations of circular polarization in the line wing of Fe I 6302 Å , for determination of the photospheric magnetic field. Continuum images of the photospheric granulation are used to determine photospheric flows. The relations between the chromospheric and photospheric network and the presumed foot points of the filament are demonstrated and discussed.

2.5. STATUS OF NEW GROUND AND SPACE SOLAR PROJECTS (29 July 14:00 - 15:30; E. Antonucci)

The program of presentations was as follows. Solar-A was successfully launched after the meeting and renamed Yohkoh. Both it and Ulysses are returning excellent data. The other projects remain further in the future.

2.5.1. Status of the LEST Project (O. Engvold).
2.5.2. Physics of the Corona with Ulysses (G. Noci).

2.5.3. *Physics of the Active Corona with the Solar-A Mission* (E. Hiei).
2.5.4. *Physics of the Solar Corona with SOHO* (M. Huber).
2.5.5. *The Orbiting Solar Laboratory* (J. Harvey for D. Spicer).

2.6. PROGRESS IN HELIOSEISMOLOGY (29 July 16:00 - 17:30; J. Harvey)

2.6.1. *Solar Oscillations, Solar Opacities, and the Solar Convection Zone Helium Content* (J. Guzik and A. Cox). A model of the Sun with improved opacities and equation of state was constructed. The model predicts frequencies in good agreement with observations at degrees up to 200; above this value discrepancies are larger, probably due to larger observational errors. Modes with degrees 300-600 are sensitive to the helium abundance (Y) of the convection zone to about 0.01. Comparison with observation favors a Y of 0.24, 0.03 less than the primordial value, and consistent with that expected due to element diffusion and settling.

2.6.2. *Comparison of the Acoustic Spectrum of the Sun and the Standard Solar Model* (D. Guenther). A set of solar models has been constructed, each based on a single modification to the physics of a reference model. Additionally, a 'best' solar model was produced that incorporates improved nuclear reaction rates, equation of state, opacities, and treatment of the atmosphere. A comparison of the new solar model with observed p-mode frequencies at low degrees shows agreement within errors associated with uncertainties of the model physics (primarily opacities).

2.6.3. *Solar G-Mode Signatures in P-Mode Signals* (J.R. Kennedy, S.M. Jefferies and F.Hill). An internal g mode should induce oscillatory perturbations in the thermodynamic parameters of p-mode cavities. This will cause frequency modulation of the associated p modes and will result in a pair of sidelobes symmetrically placed about the p-mode frequency by an amount equal to the g-mode frequency. Their amplitudes will be proportional to the ratio of the p-mode frequency deviation to the g-mode frequency. Thus, a replica of the integrated light g-mode spectrum should appear about each p-mode signal. The amplitude of a single g-mode sideband is about 10% of that expected for a direct observation of the g mode, however, averaging many sidebands may improve the sensitivity of g-mode observation attempts over direct methods.

2.6.4. *G-Mode Research at SCLERA* (H. Hill). A brief summary of g-mode detection using SCLERA data was presented. The benefit of g-mode observations is the strong diagnostic potential of the deep interior. The intriguing possibility of detecting solar g-mode oscillations by means of gravitational wave detection systems under consideration was discussed.

2.6.5. *BISON and Some Recent Results* (G. Isaak). The Birmingham Solar Oscillation Network (BISON) has been operating for more than one sunspot cycle at several sites around the world. Measurements of unimaged sunlight provide data about low degree oscillations. Particular emphasis has been given to studies of the cycle variation of frequencies of p-modes and to comparison of low degree frequencies with models that posit mixing of the solar core during the evolution of the Sun. Cycle variation of p-mode frequencies is small but is well established. The issue of possible mixing of the solar core is still not resolved.

COMMISSION 14

Atomic and Molecular data (Données atomiques et moléculaires)

Report of meetings on 24, 25 and 31 July 1991

President   : Sylvie Sahal-Bréchot
Vice President   : Wolfgang L. Wiese
Secretary   : Peter L. Smith

Activities of Commission 14 at the XXIst General Assembly consisted in a business meeting session (August 24) and two scientific meetings: one open meeting of one session (August 31) and one specific interdisciplinary meeting *"Astrophysical Opacities"*, of 4 consecutive sessions (August 25) complementary to that organized by Commission 35. In addition a Joint Commission meeting *"Atomic and Molecular data for Space Astrophysics, needs and availability"*, was organized together with Commission 44, and sponsored by Commission 10, 12, 15, 16, 29, 34, 35, and 36; it held on August 26, and is reported elsewhere.

1. Business meeting (July 24, session 2)

The President welcomed the members of the Commission and Peter L. Smith was appointed as Secretary.

1.1. REVIEW OF MEMBERSHIP OF THE COMMISSION:

Decease:
> S. L. Mandelstam, Organizing Committee member.

Resignations:
> Dr. F.Lovas resigned from the Organizing Committee.
> Dr. A. Monfils resigned from Commission 14.

New members for Commission 14 (proposed by the President):

New IAU elect members:
> Dr. Nicole F.Allard, Dr. Michèle Eidelsberg, Dr. Jean W. Gallagher, Dr. Sydney Leach, Dr Jacques Le Bourlot, Dr Jean-Louis Lemaire, Prof. Michael E. Mickelson, Dr. Annie Spielfiedel, Dr. Lydia Tchang-Brillet.

IAU members:
> Dr Milan Dimitrijević, Dr. Hoang Binh Dy, Dr. Louis d'Hendecourt, Dr. David L. Lambert, Dr. François. Rostas, Dr. K. Sinha, Dr. Tran Minh Nguyet.

2.2. PRESIDENT, VICE-PRESIDENT AND OFFICERS FOR THE TRIENNUM 1992-1995:

After discussion, the new Organizing Committee for the triennum 1992-1995 was approved:

President: Dr. Wolfgang L. Wiese
Vice-President: Dr. William H. Parkinson
Members: Dr. Saül Adelman, Dr. Jacques Dubau, Dr. Alan H. Gabriel, Dr. Kato Takako, Prof. Ralph W. Nicholls, Prof. Harry Nussbaumer, Dr. Zenonas B. Rudzikas, Dr. Sylvie Sahal-Bréchot, Dr. Peter Smith.

2.3. SCOPE OF THE COMMISSION AND WORKING GROUPS:

Dr. Jean Gallagher, who could only be appointed as a Consultant to Working Group 3 in the 1989-1991 triennum, since she was not a IAU member, was officially appointed as Chairperson of that Working Group.

The subject matter and the objectives of the Commission were rewiewed and discussed: the fertilization of the cross-discipline interaction between atomic and molecular physics is the major scope of the Commission. The increase of the need for a variety of atomic and molecular data in gaseous phase for astrophysics was noted, and thus the five preceding Working Groups were confirmed. In addition, owing to the increasing interest of solid phase chemistry (interaction gaz-surface), such as chemical reactions on grains, heterogeneous catalyse, photolyse, adsorption and desorption processes, it was decided to create a new Working Group on that topic; Sydney Leach was proposed to chair it; (contacted by the President after the General Assembly, he accepted the proposition).

Therefore the Working groups and their Chairmen for 1992-1995 were approved as follows:

1. *Atomic spectra and wavelengths standards* (W.C. Martin)
2. *Atomic transition probabilities* (W.L. Wiese)
3. *Collision processes* (J.W. Gallagher)
4. *Line broadening* (N. Feautrier)
5. *Molecular structure and transition data* (W.H. Parkinson)
6. *Molecular reactions on solid surfaces* (S. Leach)

* *

2. Astrophysical Opacities (July 25, sessions 1-4)

Scientific Organizing Committee: A.E. Lynas-Gray, A.K. Pradhan, M.J. Seaton, C.J. Zeippen.

INTRODUCTION - *Chair: W.L. Wiese*

STELLAR ENVELOPE OPACITIES
F. J. ROGERS
Lawrence Livermore National Laboratory, P.O. Box 808, Livermore, California 94550, USA

Until recently most studies of stellar evolution and pulsation have relied on opacities calculated at Los Alamos National Laboratory in the 60's and 70's; prior to the advent of supercomputers. Consequently, a number of approximations were made to facilitate calculational tractibility. For example, they used of hydrogenic oscillator strengths and neglected configurational term splitting in bound-bound transitions. As observational capabilities increased, a number of substantial disagreements with theoretical prediction were noted. Several studies found that increasing the opacity in specific temperature regions could reduce or eliminate these disagreements (Stellingwerf, 1978; Simon, 1982). This inspired Norman Simon to call for new independent calculations of opacity. Several groups have responded to this plea and a in depth discussion of these efforts will be presented in this session. In addition, results from some recent, very encouraging, studies that have used the new opacities will be presented.

Stellingwerf, R.F, 1978, *Astron.J.*, **83**, 1184
Simon, N. R., 1982, *Ap. J. (Letters)*, **260**, L87

MOLECULAR OPACITIES

The introductory remarks on molecules are reported along with the following report on the talk about molecular data presented by U. G. Jørgensen.

2.1. SESSION 1. ATOMIC AND MOLECULAR DATA - *Chair: W.L. Wiese*

R-MATRIX METHODS FOR CALCULATION OF ATOMIC RADIATIVE DATA

M.J SEATON[*] , A.K. PRADHAN[+] and J.A. TULLY[†]

* : *University College London, Gower Street, London WCIE 6BT, U.K.*
+ : *Department of Astronomy, Ohio State University, Columbus, Ohio 43210,USA*
† : *Observatoire de la Côte d'Azur, URA CNRS 1362, B.P. 139, O6003 Nice CEDEX, France*

1. General Description (Seaton)

The **Opacity Project (OP)** is an international collaboration (see Seaton *et al.*, 1991) concerned with the calculation of opacites and of the required atomic data. I will give a general description of the methods used in the atomic-physics work and two of my colleagues will give some examples of results obtained. Let me first recapitulate some basic ideas of atomic-structure theory.

- **The Central-Field (CF) model.** An electron in a central field has a wave function $\phi_{nlm} = Y_{lm}(\hat{r})(1/r)P_{nl}(r)$.

- **Configurations.** A configuration is defined by a set of nl quantum numbers and has a wave function $\Phi(x_1, ...,x_N)$ constructed using functions ϕ_{nlm} and anti-symmetrised in space- and-spin coordinates $x_i = (\hat{r}_i, \sigma_i)$.

- **Configuration Interaction (CI).** The wave functions are $\Psi_k = \sum_j \Phi_j c_{jk}$ with the coefficients c_{jk} such as to diagonalise $(\Psi_k \mid H \mid \Psi_{k'})$.

- **Close-Coupling (CC).** The wave functions are $\Psi = \mathcal{A} \sum_i \psi_i(x_i,..., x_N)\, \theta_i(x_{N+1})$ where ψ_i is a CI function for an N-electron "target" state, θ_i a function for an added electron and $\mathcal{A}$ an anti-symmetrisation operator.

In the OP work the CI and CC approaches are combined, the wave functions being $\Psi = \mathcal{A} \sum_i \psi_i(\theta_i) + \sum_j \Theta_j c_j$. The functions θ_i and coefficients c_j are optimised using R-matrix methods (Berrington *et al.*, 1987). Calculations of energy-levels, f-values and photoionisation cross sections have been made for all cosmically abundant elements in all ionisation stages, and for all excited states up to n = 10. Availability of the atomic data is discussed by Mendoza (1991).

For systems containing up to 12 electrons the calculated f-values and photoionisation cross sections should be accurate to within about 10 %(except for transitions involving delicate cancellations). For the more complex systems the data may be less accurate (due to truncations in the expansions) and less complete (some supplementary data have been calculated using more approximate methods). Methods used for the calculation of line profiles are described by Seaton (1987a, 1987b, 1988, 1989, 1990).

2. Photoionisation of Fe II (Pradhan).

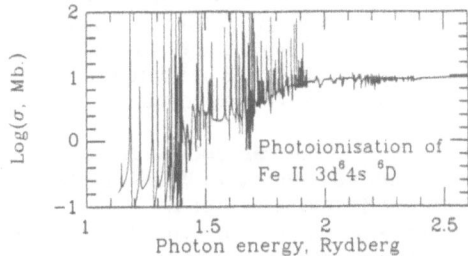

Figure 1

Some of the most OP extensive calculations have been carried out for the ions Fe I - VI. As an example consider the (electron + Fe III) system to obtain oscillator strengths and photoionization cross sections for Fe II. In Fig. 1 we present the photoionization cross section for the ground state $3d^6$ (^{5}D) $4s$ ^{6}D, including 21 Fe II target states (Le Dourneuf *et al.*, 1991).

3. Results for Be-like ions (Tully)

We have computed energy levels, oscillator strengths and photoionisation cross sections for 15 members of the Be sequence. Jannitti et al. (1986) have measured photoionisation of B^+ $2s^2\,^1S$ and $2s2p\,^3P°$. Their threshold cross-sections are in quantitative agreement with our data and their resonances in the $^1P°$ continuum resemble those we predict. Lang et al. (1987) have performed lifetime measurements in N IV, O V and Ne VII. Six of their 15 A values agree with ours to better than 10 per cent. But there are big differences for 7 transitions with fairly small A values. The OP results agree however with those from independent CI calculations (see Tully et al. 1990). Fawcett (1984) has calculated gf values for many ions in the sequence, taking into account spin-orbit coupling and CI. For some transitions our results are quite close to his while for others they disagree by more than 50 per cent. (see Tully et al. 1991 and Seaton et al. 1991). One possible explanation is the breakdown in LS coupling.

References

K.A. Berrington, P.G. Burke, K. Butler, M.J. Seaton, P.J. Storey, K.T. Taylor and Yu Yan. *J. Phys. B.*, **20**, 6379, 1987.

M. Le Dourneuf, S.N. Nahar and A.K. Pradhan, *J.Phys.B* (to be published in *J.Phys.B*).

B.C. Fawcett, *At. Data Nucl. Data Tables*, **30**, 1, 1984; erratum **33**, 479, 1985.

E. Jannitti, F. Pinzhong and G. Tondello, *Phys. Scr.* **33**, 434, 1986.

J. Lang, R.A. Hardcastle, R.W.P. McWhirter and P.H. Spurrett. *J. Phys. B.*, **20**, 43, 1987.

C. Mendoza. IAU 21st General Assembly, Joint Commission Meeting on "Atomic and Molecular Data for Space Astronomy"

M.J. Seaton, *J. Phys. B*, **20**, 6363, 1987a; **20**, 6431, 1987b; **21**, 3033, 1988; **22**, 3603, 1989; **23**, 90, 1990.

M.J. Seaton, C.J. Zeippen, J.A. Tully, A.K. Pradhan, C. Mendoza, A. Hibbert and K.A. Butler. Proceedings of Workshop on Astrophysical Opacities (Caracas, July 1991), to be published in *Revista Mexicana de Astronomia y Astrofisica*.

J.A. Tully, M.J. Seaton and K.A. Berrington. *J. Phys. B.*, **23**, 3811, 1990.

J.A. Tully, M.J. Seaton and K.A. Berrington. *J. Phys. (Paris) IV, C1, Suppl. J. Phys. II*, **1**, 169, 1991.

MOLECULAR DATA
Uffe Gråe JØRGENSEN

Niels Bohr Institute, Blegdamsvej 17, DK-2100 Copenhagen, Denmark

The most common uses of molecular data for astrophysical purposes are probably identification of spectral features, determination of isotopic ratios, abundance determination, and computation of stellar model atmospheres. For the two first purposes information about only the lines of interest may sometimes be sufficient, and of the order half a million lines are known from laboratory measurements today. This is, nevertheless, only a tiny fraction of the line-data needed for the two latter purposes, but it would of course be meaningless to try to measure all the lines, and good methods that can combine molecular constants from laboratory experiments with quantum mechanically computed frequencies and intensities have also been developed. The advantage of the ab initio method is that it can handle a complete computation of all lines up to the dissociation energy of the molecule. The disadvantage is that it generally does not allow for an internal uncertainty estimate. Tight interaction between theoretical predictions and laboratory experiments is therefore crucial. Preferably, data for all molecules of interest should fulfill the following four criteria.

1) The strongest bands to a certain limit should be measured with good accuracy in the laboratory to assure that the absorption coefficient integrated over the whole spectrum is correct, and that the lines easily visible in stellar spectra can be traced accurately.

2) The complete set of lines up to the dissociation energy should be calculated to assure that the right effect on the model atmospheric structure and low resolution spectra (including the continuum drawing) can be predicted.

3) A sample of very weak bands and lines that include high energy upper or lower states should be measured in the laboratory and compared with the computations to test the quality of the most uncertain part of thecalculated potential and dipole moment functions.

4) The temperature dependence of the absorption coefficient in low resolution should be measured to ensure that the computed integrated absorption coefficient has the right temperature dependence.

These criteria are not fulfilled for any of the molecules we are interested in, but hopefully a greater collaborative effort will allow this. Point 4) above is only fulfilled for water, point 3) only for HCN, and point 2) only described for CN, whereas point 1) is fulfilled for most astrophysically important molecules except some of the radicals. The expected wealth of data from the coming satellites with high resolution spectroscopic capacity in the infrared will be hard to use satisfactorily without a considerably better astrophysical molecular data base.

The most extensive compilations of empirical data are the HITRAN (L.S. Rothman *et al.* 1987; *Applied Optics*, **26**, 4058) and the GISA (L. Husson *et al.* 1986, *Ann. Geophys.*, **4A**, 185) data bases. The latest released edition of the HITRAN data base contained strength, frequency and excitation energy for 348,000 lines from 28 molecules (70 isotopic varieties), and the coming release is expected to contain more than 600,000 lines. The GISA data base contain basically the same data as the HITRAN, plus a few extra molecules of mainly environmental interest, and the next release is expected to contain slightly more than 700,000 lines in total. Laboratory data on lines from a smaller number of molecules (including TiO, ZrO, YO, CN, FeH and LaO) of particular astrophysical interest, are available from J.G. Phillips and S.P. Davis (1987, *Publ. Astron. Soc. Pac.*, **99**, 1105; and preceeding research).

Line strength, line frequency andexcitation energy are listed for a total of 22 million lines of 20 diatomic molecules and 35 million lines of 4 polyatomic moleculesin the astrophysical data bases. Only little is known about the quality of the data for the weak lines and bands, and no comparison of data computed by different groups exists.

R.L. Kurucz (1991; in *Stellar Atmospheres: Beyond Classical Models*, eds. Crivellari *et al.*, NATO ASI, p.421) has for many years offered a compilation of molecular lines to the community, which now contain 15.7 million lines from the molecules H_2, CH, NH, OH, MgH, SiH, CO, SiO, CN, C_2, TiO, in various isotopic varieties.

U.G. Jørgensen and M. Larsson (1990, *Astron. Astrophys.*, **238**, 424) computed a line list of 3.3 million lines from the "red" (A-X) system of CN. The list is the first to include a completeness study, in the sense that the Boltzman factor of the energy levels contribute more than 99% of the theoretical partition function for all temperatures of astrophysical interest. The haze of weak lines was found to influence the low-resolution synthetic spectra substantially. The line list is available on request together with similar lists for CH (in preparation), H_2O (in preparation) and HCN (up to 10 million lines; U.G. Jørgensen *et al.* 1985, *J. Chem. Phys.*, **83**, 3034).

D.R. Alexander *et al.* (1989, *Astrophys. J.*, **345**, 1014) computed, and offers, a list of 4.5 million lines from the electronic ground state of the water molecule. The method developed by the authors use observed lines to define a statistical distribution function of line strengths and lower excitation levels, and assume pre-specified values of line spacing and number of lines to distribute observed low resolution absorption data into individual lines. The observational low resolution data is available from C.B. Ludwig *et al.* (1973, *Handbook Infrared Rad. Combustion Gases*, NASA SP-3080).

J.M. Brett (1990, *Astron. Astrophys.*, **231**, 440) estimated the f-values of the A-X and B-X system of VO and the f-value of the TiOε system, as the ratio between the intensity of these bands and the band from TiOδ in observed stellar spectra, and offers a tape with 2 million lines from the two VO systems.

F. Querci *et al.* (1974, *Astron. Astrophys.*, **31**, 265) compiled, and offers, a tape with gf-values, frequencies and excitation energies (but only identification of which molecule and isotope each line belongs to) of one million lines from CN, C_2 and CO.

J.E. Littleton and S.P. Davis (1985, *Astrophys. J.*, **296**, 152) measured f-values of bands from ZrO and YO, and calculated from this (in yet unpublished work) the frequency, strength and excitation energy for 332,000 ZrO lines and 1,200 YO lines. The data are available from J.E. Littleton on request.

A.L. Piñeiro *et al.* (1987, *J. Mol. Spec.*, **125**, 184; and references therein) offers a compilation of 14,000 lines of SiO, 35,000 lines of CO, 35,000 lines of CS, and 600 lines of SiS. Their method has been to use dipole moment computations from the literature and empirical data to fit a Padé approximant to the dipole moment function, and from this calculate all transitions within what they consider the well determined part of the dipole moment function.

2.2 SESSION 2: EQUATIONS OF STATE - *Chair: A.K. Pradhan*

EQUATIONS OF STATE: COMPARISONS OF RESULTS FROM TWO COMPETING FORMALISMS.
Werner DÄPPEN

Departments of Physics and Astronomy , University of Southern California, Los Angeles , USA

Currently there are two basic approaches to the equation of state for stellar envelopes and interiors. Both want to realize the same goal, equilibrium statistical mechanics of reacting plasmas, but their methods differ. One (the older) method chooses the so-called "chemical picture", in which the notion of atoms is maintained despite the plasma environment. A mixture of atoms, molecules, ions, electrons and nuclei is considered, and the occurring ionization and dissociation reactions (thus the name chemical picture) are treated according to the entropy-maximum (or free-energy minimum) principle, which is, in the language of the astrophysicists, the Saha equation. Interactions of the plasma with atoms and ions are introduced separately in a heuristic way. The other (more recent) method is based on the so-called "physical picture", where only fundamental particles (electrons, nuclei) explicity enter. Furthermore, through the means of activity expansions, the problems of plasma physics and statistical mechanics are treated on the same footing.

The international "Opacity Project" (OP) uses an equation of state realized in the chemical picture (Mihalas, Hummer, Däppen , MHD), and the opacity effort pursued at Livermore by Iglesias and Rogers (OPAL) one in the physical picture (hereinafter Livermore equation of state). Since the equation of state plays an important role in any opacity calculation, because it predicts both level populations and ionization degrees, and since at the desired level of accuracy no laboratory experiments exist, comparisons of theoretical formalisms are so far the only means to estimate the source of opacity uncertainty due to the equation of state.

First comparisons concentrated on the case of conditions as found in the H-He ionization zones of the Sun. For both equations of state, pressure and thermodynamic quantities were compared with those of simple mixtures of ideal gases, whose ionization equilibrium is determined by a Saha equation that contains ground states only. It was found that in these regions of the Sun, the predicitons of the MHD and Livermore equations of state are virtually the same, despite the fact that they differ from the simple Saha results by a few percent [1,2].

This agreement was by no means expected, and it turned out that it was due to the Coulomb interaction, present in the MHD and Livermore equation of state, which is the dominating non-ideal effect under these conditions. And yet, despite the common Coulomb treatment, one would intuitively expect a greater difference, essentially because of the treatment of the excited states of atoms and ions in the MHD equation of state. That this is not the case could indicate a perhaps accidental cancellation of contributions within the MHD formalism, and the question is currently being examined.

Very recently, Forrest Rogers and I again compared results, this time for higher temperatures and densities ($\rho = 10^{-2} \dots 10$ g/cm^3) [3]. We found a good agreement for solar conditions, that is for the temperatures as they are in the Sun at these densities. However, the Sun just marginally passes: for slightly less massive stars, the discrepancy soon becomes very important indeed. Finally, we also considered the case of heavier elements. Here, for the first time we succeeded in establishing a clear case of disagreement between the MHD and Livermore results. For a solar composition, at a temperature of about 2 10^5 K and a density of 5 10^{-3} g/cm^3, the predicted ionization degrees of C, N and O are quite drastically different, with the Livermore values lying closer to the simple Saha results. Despite the fact that of course only some 2% of the matter in the Sun consists of elements heavier that H and He, the present state of helioseismology carries the potential of direct thermodynamic tests (via sound speed) of the equation of state under these conditions [4]. Furthermore, since there are (though quite localized) zones with distinctly different ionization degrees (by up to a factor of 2), opacity will undoubtedly also be influenced by this equation of state issue, and stellar cases might be found that could test the equation of state via the opacity.

[1] W. Däppen , Y. Lebreton, F. Rogers, 1990: *Solar Physics* , **128**, 35-47.
[2] W. Däppen , 1990: in *"Progress of seismology of the sun and stars"* (eds. Y. Osaki & H. Shibahashi, Springer)
[3] W. Däppen , to appear in the proceedings of the workshop "Astrophysical Opacities" (Caracas, July 1991)
[4] J. Christensen-Dalsgaard, W. Däppen : 1992, *Astronomy and Astrophysics Review*.

MOLECULAR EQUILIBRIA WITH CONDENSATION IN ASTROPHYSICAL GASES
C.M. SHARP

Service P.T.N., Centre d'Etudes de Bruyères-le-Châtel, BP 12,   F-91680 Bruyères-le-Châtel,   France

1. Introduction

In gases at temperatures below about 6000 K, molecules play an important rôle in the opacity, and below about 2000 K grain scattering can increase the opacity very much further. These conditions are very typical of stars and planets during their formation processes, together with brown dwarfs and circumstellar shells. If thermodynamic equilibrium is assumed, the abundances depend on the element composition and the temperature and pressure. However, most previous work on molecular equilibria examined the effects of temperature and pressure, with at best only a few compositions.

As is well known, the abundances of the molecular species can depend in a very sensitive and non-linear way on the element composition, particularly the C/O ratio due to CO being the most tightly bound of all diatomic molecules, together with the presence of any condensed species. In order to reduce the parameter space to two dimensions, we have calculated molecular abundances for a large number of temperatures and C/O ratios for a fixed total pressure, with condensates included when appropriate. Only the gas phase and how it is modified by condensed phases is discussed here.

2. Discussion of Calculations

73 mixtures with compositions of log(C/O) = - 0.37 to 0.35 in steps of 0.01 were computed in one degree intervals from 4000 to 1000 at a constant total pressure of 500 dyne/cm^2, (4.935x10^{-4} atmospheres). Each mixture had the 20 most abundant elements with, excepting carbon, the Anders and Grevesse (1989) solar mixture with suggested modifications by Meyer (1991). For the different mixtures only carbon was changed, with the ratio of all other elements with respect to hydrogen being kept at the solar values. For the solar mixture log(C/O) = - 0.33.

The gas phases consisted of the 20 neutral atoms, and 68 diatomic and 74 polyatomic molecules. Additionally above 2000 K, 39 charged species consisting of free electrons, and atomic and molecular ions, were included, and below 2000 K, 54 condensed species were specified, of which a proportion, depending on temperature and composition, would appear in the condensed phases. The most important condensates were the oxides, silicates, carbides and graphite, and included in the calculations was a solid solution of melilite ($Ca_2Al_2SiO_7$ and $Ca_2MgSi_2O_7$).

Most of the thermodynamic data were obtained from the JANAF tables (Chase, 1982), which were put in a form suitable for computation. For each temperature and composition, the abundances of each species in the gas phase, together with the condensates below 2000 K, were obtained by minimising the total Gibb's free energy of the system, together with the constraints of mass balance and total charge neutrality, using the computer code SOLGASMIX (Besmann, 1977), which we have substantially improved (Sharp and Huebner, 1990).

3. Results

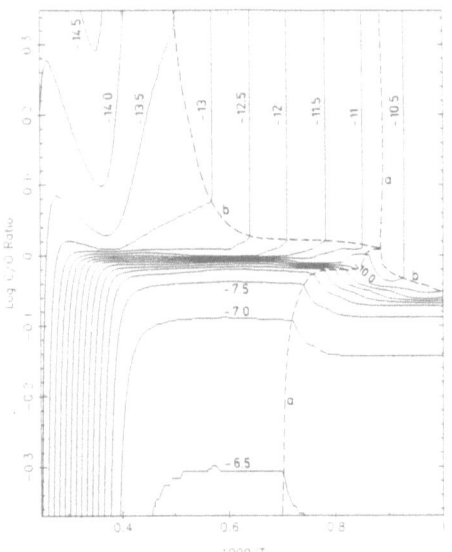

Fig.1 Contours of log partial pressures in atmospheres of H_2O for different temperatures and compositions at a constant total pressure of 4.935×10^{-4} atmospheres. The dashed curves "a" and "b" mark the high temperature limits of the condensates Mg_2SiO_4 and graphite respectively.

Figure 1 is a contour plot of the log partial pressure of H_2O, which can be a particularly important opacity source at low temperatures, in atmospheres as a function of log(C/O) and 1000/T. It is seen that for small changes of the C/O ratio in the region where that ratio is close to unity, the abundance of H_2O changes very rapidly, and is due to CO controlling the chemistry of the gas phase At higher temperatures H_2O dissociates, and when the temperature is decreased for the oxygen rich case the abundance also falls due to oxygen being taken out of the gas phase by the condensation of oxides and silicates. The most important sink of oxygen is the condensate Mg_2SiO_4, which appears to the right of the dotted curve "a".

In the carbon rich case the increase in the H_2O abundance with decreasing temperature is due to carbon being taken out of the gas phase mostly by graphite, this reduces the abundance of CO which frees up some oxygen that can form H_2O. The vertical contours are equally spaced and reflect the vapour pressure curve of graphite, which appears to the right of the dotted curve "b". Note that in this region curve "a" no longer plays an important rôle.

Plots of many molecules containing carbon, such as HCN, are almost complementary to H_2O with the maxima and minima interchanged. The effects of condensation on HCN abundances for different compositions were discussed in Sharp (1990).

4. Conclusions

We have shown here that the abundance of H_2O is very sensitive to the C/O ratio when that ratio is close to unity in the temperature range considered, and for certain conditions the abundance can also be sensitive to the formation of condensed species. This region separates areas where the abundance is relatively insensitive to changes of temperature and composition, and is seen for many other molecules. This will be the subject of a more detailed future paper.

References

Anders, E. and Grevesse, N.: 1989, *Geochim. Cosmochim. Acta*, **53**, 197.

Besmann, T.M.: 1977, "SOLGASMIX-PV, *A computer Program to calculate Equilibrium Relationships in Complex Chemical Systems*"(Oak Ridge National Laboratory Report TM-5775).

Chase, M.W.: 1982, JANAF Thermodynamic Tables, Magnetic Tape Version (Midland, MI: Dow Chemical Co.).

Meyer, J.-P.: 1991, Private communication.

Sharp, C.M.: 1990, *Astrophysics and Space Science*, **171**, 185.

Sharp, C.M. and Huebner, W.F.: 1990, *Ap. J. Suppl. Ser.*, **72**,417.

2.3. SESSION 3. OPACITIES - *Chair: T.R. Carson*

RADIATIVE ATOMIC ROSSELAND MEAN OPACITY TABLES
Carlos A. IGLESIAS and Forrest J. ROGERS

Lawrence Livermore National Laboratory, P.O. Box 808, Livermore, California 94550, USA

We present radiative Rosseland mean opacity tables calculated with the OPAL code (Iglesias and Rogers 1991) which removes several approximations present in previous calculations. In particular, improvements in the atomic physics have been incorporated which yield reasonably accurate photon absorption data (accuracy comparable to single-configuration, self-consistent field calculations with relativistic corrections) and include the configuration term splitting in the LS-coupling scheme. In some cases, the latter can increase the number of spectral lines by orders of magnitude compared to previous opacity codes. OPAL also employs an equation of state method which avoids the usual ad hoc cut offs introduced in free energy minimizaton schemes. The opacity tables are given in terms of temperature and R, where $R = \rho/T^3$ with ρ the density and T the temperature. Extensive results are given for the Anders-Grevesse (1989) metal abundances which allow accurate interpolation in temperature, density, hydrogen mass fraction, and metal mass fraction. The range of T and R cover typical stellar conditions from the interior to the hotter atmospheres. Cool atmospheres are not considered since photoabsorption by molecules is neglected. Only radiative processes are taken into account so that electron conduction is not included.

Comparison to the Los Alamos Opacities (Huebner 1986; Cox and Tabor 1976) show factors of 2-3 opacity enhancements at temperatures around 300,000K, which are mostly due to the improved atomic physics of partially filled M-shell ions. As a result, the OPAL opacities show considerably more sensitivity to the metallicity than earlier calculations. Uncertainties in this new opacity bump near $T = 300,000$ K have been investigated. It is found that several issues such as line broadening, the angular momentum coupling in the atomic data generation, and metal composition can significantly affect the results. More modest opacity enhancements are reported (~20%) at 3 million degrees which are due to the improved atomic data at lower densities and to the equation of state at the higher densities. Not all differences result in increased opacities. Reductions of ~20% are found in the region 10,000 K< T <100,000 K. These opacity decreases have been traced to the treatment of the hydrogen Lyman-α line wing. There are also opacity decreases at the highest temperatures and densities (~15%) where differences in the treatment of degeneracy affect the photon scattering from free electrons.

References

Anders, E. and Grevesse, N. 1989, *Geochim. Cosmochim.Acta.* , **53**, 197

Cox, A. N. and Tabor, J. E. 1976, *Ap. J. Suppl.* , **31**, 271

Iglesias, C. A. and Rogers, F. J.1991, *Ap. J. (Letters)*, **371**, L73
Huebner, W. F. 1986, in *Physics of the Sun*, ed. P. A. Sturrock, T. E. Holzer, D. M. Mihalas, and
R. K. Ulrich (Dordrecht: Reidel), **1**, 33

THE OPACITY PROJECT: OPACITIES
YU YAN
Department of Astronomy, University of Illinois, 1002 West Green Street, Urbana, Illinois 61801, USA.

1. Introduction

The current interests in revisions of stellar opacities stem largely from a study by Simon (1982), who suggested that the heavy element contribution to the opacity might have been under-estimated by a factor of 2--3 in previous opacity calculations (for example, Cox and Tabor 1976). In 1983 a number of researchers in atomic physics and astrophysics came together to discuss a joint effort to recalculate stellar opacities; parallel works on the equation of state, line broadening theories and streamlined production of accurate atomic data were pursued at various research centres in five countries. This was later known as the Opacity Project (OP; see, for example, Seaton 1987). Another major effort at reexamining heavy element opacities is the OPAL project (Iglesias, Rogers and Wilson 1987, 1990). Here we discuss briefly the methods used in the OP calculation of stellar opacities, and present selected results in a format for easy comparison with those of OPAL.

2. Methods and selected results

We use the MHD equation of state (Hummer and Mihalas 1988, Mihalas, Däppen and Hummer 1988). It is based on the free energy minimisation method, and takes into account non-ideal effects due to various plasma interactions. The chemical picture assumed by this equation of state is especially suited to opacity calculations, as the relevant spectroscopic notions find a direct equivalence in the formalism.

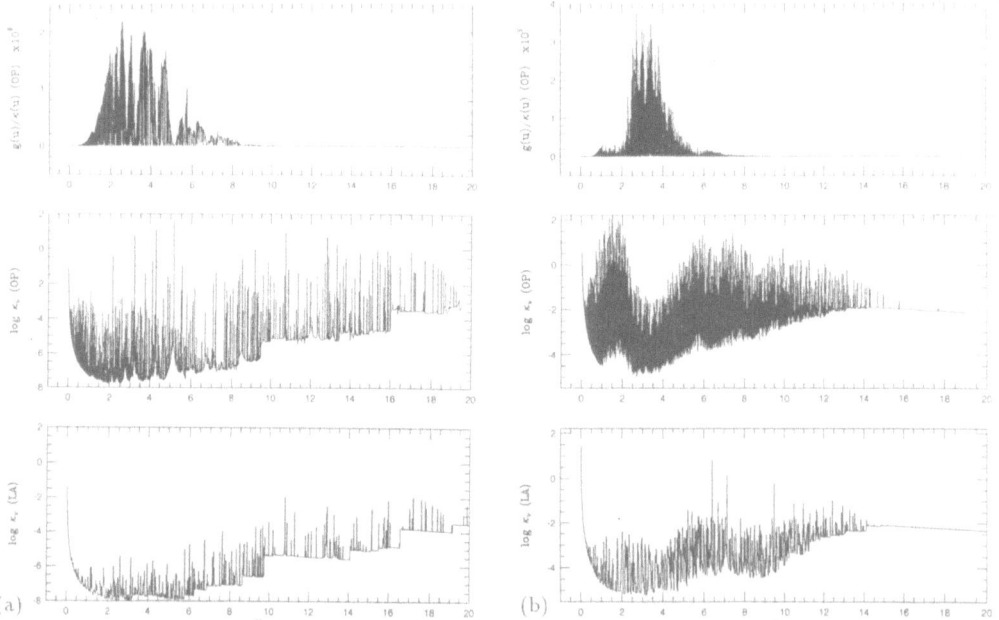

Fig. 1.— Monochromatic opacities from LAAOL and OP (lower, middle plots respectively) and the corresponding OP Rosseland integrand (upper plot) for (a) pure carbon at log T = 4.4626, log ρ = 8.033 and (b) pure iron at log T = 5.5417, log ρ = -5.241.

We use close-coupling type atomic data, supplemented at high target energies by a "top-up" procedure (see the paper by M.J. Seaton in this volume). Figure 1(a) compares pure carbon monochromatic opacities at one (T, ρ) tabulated in the Los Alamos Astrophysical Opacity Library LAAOL; see Hüebner *et al.* 1977), lower plot, to those obtained using the OP method, middle plot. The effect of including detailed auto-ionisation features on the mean opacity is most clearly seen in the Rosseland integrand of the OP spectrum, upper plot. The corresponding Rosseland means are 3.16 (LAAOL) and 6.76 (OP) cm^2/g. Figure 1(b) compares the pure iron monochromatic opacities at another (T, ρ). The higher opacities obtained by the OP can mostly be attributed to the much improved atomic model.

The OP atomic data and the opacity codes have undergone extensive checks; several full tables of mean opacities were produced in the process. Figures 2(a,b) compare the OP and OPAL Rosseland means for $X = 0.7$, $Z = 0.02$ and "AG abundances" (Anders and Grevesse 1989) at a range of temperatures and two values of the density parameter $R = \rho / T_6^3$, where T_6 is temperature in 10^6 K. The current agreement between the two calculations is in general quite good; some rather significant discrepancies develop at low densities and temperatures above 10^5 K (the so-called "Z-bump"), the source of which is yet to be determined. We note, however, that the OP results may undergo *upward* revisions in the near future, due mainly to inclusion of the following effects: (i) fine structure and intermediate coupling, (ii) transitions involving highly excited target states, (iii) ions of nickel. For a fuller discussion of future revisions in stellar opacity calculations see Yu Yan (1991).

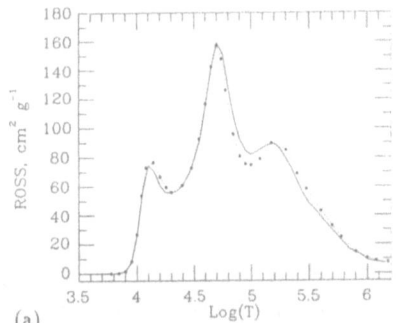 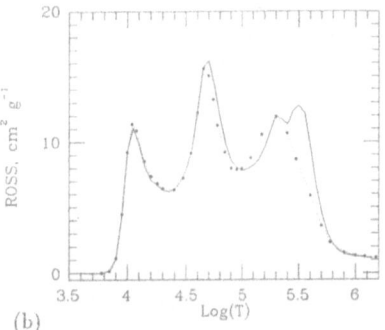

(a)   (b)

Fig. 2.—Rosseland mean opacities for $X = 0.7$, $Z = 0.02$ (AG abundances) and (a) log $R = -2.5$, (b) log $R = -3.5$. Curves are: ———— OP   · ·· ····· OPAL

3. The Opacity Library

The many man-years of research effort committed to this work will culminate in an Opacity Project User Service (OPUS), which will provide a user-oriented, publicly accessible opacity library for the astrophysical community, capable of computing mean and multi-group mean opacities, opacity distribution functions and monochromatic opacities of arbitrary composition. The design of the opacity library has been made simple by the approximation, which was verified to be sufficiently good, that the statistical-mechanical properties of a chemical element is independent of the plasma composition at a given *electron density* N_e. In practice we have computed and archived *single element* monochromatic opacities using occupation numbers from the solar mixture on the grid $3.5 \leq \log T \leq 7.2$, $\Delta\log T = 0.025$, $\Delta\log N_e = 0.25$ and lower, upper bounds for log N_e varying with log T. For this grid and 14 chemical elements the amount of monochromatic opacity data after packing is about 12 gigabytes. Mixing codes have been developed to construct mixture opacities from the archive of single element monochromatic opacities. The archive and the mixing codes form the Opacity Library.

The production of mixture opacities from the Opacity Library is computationally straightforward. This simplified mixing scheme is the basis for OPUS, which will evolve in several stages towards increasing autonomy and ease of use. Given the rapid advances in workstation technology, it is hopeful that a stand-alone, locally maintained facility may soon provide the entire resources for OPUS, allowing all manners of network access.

Acknowledgment

Other members of the OP involved in the opacity calculations are D.M. Mihalas, A.K. Pradhan and M.J. Seaton.
We thank F.J. Rogers and C.A. Iglesias for providing data for comparison in advance of publication.
Work supported by National Science Foundation grants AST85-19209 and AST89-14143, and by unrestricted research funds from the University of Illinois.

References

Anders, E. and Grevesse, N. 1989, *Geochim. Cosmochim Acta.*, **53**, 197.
Cox, A.N. and Tabor, J.E. 1976, *AP. J. Supp.*, **31**, 271.
Hüebner, W.F., Merts, A.L., Magee, N.H.Jr. and Argo, M.F. 1977, *Astrophysical Opacity Library*, Los Alamos Scientific Laboratory Report LA-6760-M.
Hummer, D.G. and Mihalas, D.M. 1988, *Ap. J.*, **331**, 794.
Iglesias, C.A., Rogers, F.J. and Wilson, B.G. 1987, *Ap. J.(letters)*, **322**, L45.
—— 1990, *Ap. J.*, **360**, 221.
Mihalas, D.M., Däppen, W. and Hummer, D.G. 1988, *Ap. J.*, **331**, 815.
Seaton, M.J. 1987, *J. Phys. B: At. Mol. Phys.*, **20**, 6363.
Simon, N.R. 1982, *Ap. J.*, **260**, L87.
Yu Yan 1991, *Rev. Mexicana Astron. Astrof.*, in print.

2.4. SESSION 4. APPLICATIONS - *Chair: S. Sahal-Bréchot*

NEW OPACITIES
Robert L. KURUCZ

Harvard-Smithsonian Center for Astrophysics, 60 Garden Street, Cambridge, MA 02138, U.S.A.

I have described my atomic and molecular line data calculations at the IAU general Assembly in Baltimore (Kurucz 1988) and at a NATO workshop in Trieste (Kurucz 1991a). I now have data for 58 million atomic and diatomic molecular lines.

The opacities are calculated with a version of my spectrum synthesis computer program that has been under development since 1965 and has been described by Kurucz and Furenlid (1981) and by Kurucz and Avrett (1981). The algorithms for computing the total line opacity are extremely fast because maximum use is made of temperature and wavelength factorization and pretabulation. There is no limit to the number of spectrum lines that can be treated. The equation of state and continuum opacities are computed with a version of the model atmosphere program ATLAS (Kurucz 1970). Photoionization continua are put in at their exact positions, each with its own cross-section and with the series of lines that merge into each continuum included so that there are no discontinuities in the spectrum. Hydrogen line profiles are computed using a routine from Peterson (1979). Autoionization lines have Shore-parameter Fano profiles. Other lines have Voigt profiles that are computed accurately for any value of the damping parameter a which includes radiative, Stark, and van der Waals broadening.

In late 1988 I used the line data to compute new solar abundance opacity tables for use in my stellar atmosphere modelling. The calculations included all 58,000,000 lines for 56 temperatures from 2000K to 200000K, 21 log pressures from -2 to +8, and 5 microturbulent velocities 0, 1, 2, 4, 8 km/s, and took a large amount of Cray time at the San Diego Supercomputer Center. The spectral resolution was 500000 for 3,500,000 wavelength points in the wavelength range is 8.97666 nm to 10000 nm. The opacity is tabulated both as 12-step distribution functions for intervals on the order of 1 to 10 nm, and as opacity sampling where, simply, every hundredth wavelength point in the calculation was saved. There are actually two sets of distribution functions: a higher resolution version with 1212 "little" intervals, and a lower resolution version with 328 "big" intervals. The "little" wavelength intervals are nominally 1 nm in the ultraviolet and 2 nm in the visible.

I have rewritten my model atmosphere program to use the new line opacities, additional continuous opacities, and an approximate treatment of convective overshooting. The opacity calculation was checked by computing a new solar model (Kurucz 1991b) that matches the observed irradiance (Neckel and Labs 1984; Labs et al. 1987). I am confident that I have solved the missing opacity problem.

Since the beginning of 1990 I have been able to take tremendous advantage of the new Cray YMP at the San Diego Supercomputer Center. In a few months I ran more than I had expected to do in two years. I computed opacities ranging from 0.00001 solar to 10 times solar, enough to compute model atmospheres ranging from the oldest Population II stars to high abundance Am (metallic line) and Ap (peculiar) stars. The hardest part was transmitting the results (200 tapes) back to Cambridge over Internet, but even that usually worked quite well. The exact abundances are [+1.0], [+0.5], [+0.3], [+0.2], [+0.1], [+0.0], [-0.1], [-0.2], [-0.3], [-0.5], [-1.0],[-1.5], [-2.0], [-2.5], [-3.0], [-3.5], [-4.0], [-4.5], [-5.0], [+0.0, no He]. The final files for each abundance require two 6250 bpi VAX backup tapes. I have begun to distribute copies of the tapes. I plan to produce 600 megabyte CD-ROMs of these opacities that can be read on any workstation with a CD reader.

Rosseland mean opacities have been computed for each abundance for temperatures up to 200000K for use in interiors and envelope calculations. The mass density and electron number were also saved to allow changing variables. These Rosseland opacities should be much better than the Los Alamos opacities in the same temperature region because they treat all the lines explicitly and with detailed profiles, and because the equation of state and the opacity spectra are computed for the actual composition, not as a post facto mixture. The numerical accuracy of any sort of mean opacity must be more reliable in my calculation.

I am open to suggestions for computing opacities for other abundance mixes. At the present time I am planning to compute C/O variations, Population II opacities with enhanced light even elements, and some Am and Ap mixes.

Acknowledgements

This work is supported in part by NASA grants NSG-7054, NAG5-824, and NAGW-1486, and has been supported in part by NSF grant AST85-18900. The most important contribution to this work is a large grant of Cray computer time at the San Diego Supercomputer Center.

References

Kurucz, R.L., 1970: *SAO Spec. Rep.* No. 309, 291 pp.
Kurucz, R.L., 1991a: pp.440-448 in Stellar Atmospheres: *Beyond Classical Models* (ed. by L. Crivellari, I. Hubeny, and D.G. Hummer), NATO ASI Series, Kluwer, Dordrecht.
Kurucz, R.L., 1991b: pp. 27-44 in Precision Photometry: *Astrophysics of the Galaxy* (ed. A.G.Davis Philip, A.R. Upgren, and K.A. Janes) L. Davis Press, Schenectady.
Kurucz, R.L. and Avrett, E.H., 1981: *SAO Spec. Rep.* No. 391, 145 pp.
Kurucz, R.L. and Furenlid, I.,1981: *SAO Spec. Rep.* No. 387, 142 pp.
Labs, D., Neckel, H., Simon, P.C., Thuillier, G. 1987: *Solar Phys* ics, **107**, 203-209.
Neckel, H. and Labs, D., 1984: *Solar Physics,* **90**, 205-258.
Peterson, D.M., 1979 : Personal communication.

OPACITIES AND CEPHEID MODELS

Shashi KANBUR

Department of Physics and Astronomy, 260 Behlen, University of Nebraska, Lincoln, Nebraska, NE 68588, USA.

Beat Cepheids are classical Cepheids which pulsate simultaneously in the fundamental and first overtone. The two periods can be accurately disentangled and extracted from observations. In attempting to analyze such stars, use is made of the Petersen diagram where the ratio of the first overtone to fundamental period, P1/P0,is plotted against the fundamental period P0. One approach in modelling such stars is to use linear non-adiabatic theory, with masses and luminosities mandated by stellar evolutionary calculations, to compute the fundamental and first overtone modes.

The theoretical and observed period ratios P1/P0 can then be plotted on the same Petersen diagram. Figure 1 of Moskalik, Buchler and Marom (1991) shows such a diagram. The dashed curves denote models calculated with the existing Los Alamos opacities, whilst the dots denote observations.The figure shows a clear disagreement between observations and theory.

These and other discrepancies between pulsation and evolutionary theory led to the suggestion that the opacities may be in error (Simon 1982). Two groups (Iglesias and Rogers 1987 and Seaton 1987) have now recalculated opacities. Moskalik, Buchler and Marom (1991) have used the OPAL (Iglesias and Rogers 1987). Figure 1 of that paper shows that now the theoretical period ratios lie in the loci of observations. Figure 1 of Simon and Kanbur shows calculations made with the Opacity project opacities (Seaton 1987). Again is good agreement between observations and theory.

Since the OPAL and Opacity Project opacities have been calculated using different techniques, but yield similar (though not exactly the same) results, it seems that the conjecture of Simon (1982) was correct.

Simon and Kanbur (1991) give a fuller description of this another problems in pulsation theory which could be alleviated by the use of both the OPAL and Opacity Project opacities.

References

Iglesias, C.A. and Rogers, F.J., 1991, *Ap.J*, **371**, 408.
Moskalik, Buchler and Marom, 1991, *Ap.J.*, submitted.
Seaton, M.J., 1987, *J.Phys. B.*, **20**, 6363.
Simon, N.R., 1982, *Ap.J*,. **260**, L87.
Simon, N.R. and Kanbur, S.M. (1991) *Revista Mexicana de Astronomia y Astrofisica*.

✳ ✳

3. Open scientific meeting *(July 31, session 3)*

3.1. HIGHLIGHTS AND UPDATE OF THE TRIENNUM REPORT

3.1.1. Working Group 1 : Atomic Spectra and Wavelength Standards (Report Update)
W. C. MARTIN

Critical compilations of wavelengths and their energy level classifications have been published for all spectra of Mg [1] and Al [2], and for Cu X-XXIX [3].

O'Brian *et al.* [4] give accurate new values for 442 Fe I levels; many improved wavelengths can be obtained from these levels. Johansson and Learner's new measurements of Fe I infrared lines are available [5], and Nace *et al.* [6] give new high-accuracy wavelengths for 539 Fe I, II ultraviolet lines. Persson *et al.* [7] have completed a large extension of the Ne III analysis.

References:
[1] Kaufman, V., Martin W. C. : 1991, *J. Chem. Ref. Data*, **20**, 83-152.and [2], *ibidem*, 775-858.
[3] Shirai, T., *et al.* : <u>ibid</u>., 1-81.

[4] O'Brian, T.R., et al.: 1991, J. Opt. Soc. Am. B 8, 1185-1201.

[5] Johansson, S., Learner, R.C.M.: 1990, Astrophys. J. 354, 755-762.

[6] Nave, G., et al.: 1991, J. Opt. Soc. Am. B 8, 2028-2041.

[7] Persson, W., et al.: 1991, Phys.Rev. A 43, 4791-4823.

3.1.2. Supplemental Report of Working Group 2 : Atomic Transition Probabilities
W.L. WIESE

Since the completion of the working group report in the summer of 1990, several significant new papers on atomic transition probability data of astrophysical interest were published and some other developments have occurred that are worth noting.

A comprehensive paper containing transition probabilities for 1800 lines of Fe I, obtained by combinaison of atomic lifetime and branching ratio measurements, was published by O'Brian et al. (1991). Typically, the data are estimated to be accurate in the range from 5% to 15% and they agree very closely with the data in the critical compilation by Fuhr et al. (1988). For lines from highly excited levels the new data set should be more accurate and is therefore recommended.

Bell et al. (1990, 1991), Hibbert et al. (1991) and Biémont et al. (1991) have done a series of sophisticated atomic structure calculations, mainly with the CIV3 code, for numerous multiplets of neutral nitrogen and neutral and singly ionized oxygen. Comparisons with a number of recent experimental data, especially from lifetime measurements indicate that these data should be typically in the 5-10 % uncertainty range.

An additional paper in the series "Atomic Data for Opacity Calculations" has been published by Tully et al. (1990), which contains numerous oscillator strength data for the beryllium sequence. An extensive comparison of these "Opacity Project" data with experimental results has been carried out by Allard et al. (1990) for the three Be-like ions C III, N IV, and O V. The experimental results and the opacity data typically agree within ± 10%. A similar, even more extensive comparison study is underway in the Data Center on Atomic Transition Probalities at NIST. From hundreds of comparisons involving the carbon, nitrogen, and oxygen atoms as well as various ions of these elements, one can draw the - still preliminary - conclusion that the opacity data are especially good for ions where they agree within 5-10 % with the best lifetime results (the latter are also estimated to be accurate within the same uncertainty range). However, for neutral atoms, especially for N I, differences up to 40 % occur. Thus, for the neutral atoms the above mentioned calculations by Bell, Hibbert, and colleagues - which agree much closer with the best experimental data - are recommended in preference to the opacity results.

Fuhr and Wiese (1990) recently revised and expanded their listings of atomic transition probabilities for the 1990-1991 edition of the CRC Handbook of Chemistry and Physics. Several significant additions and revisions were made, especially for Ba I, Hg I, Mo I and the element of the Fe-group.

Finally, we would like to draw attention to the proceedings of a 1990 International Colloquium, Edited by J. E. Hansen, which has been largely on the subject of atomic oscillator strengths. Many recent results are given or referenced, and the principal experimental and theoretical approaches are reviewed.

References :

1. O'Brian, T. R., M. E. Lawler, J. E. Whaling, W., Brault, J. W.: J. Opt. Soc. Am. B. 8, 1185 (1991).

2. Bell, K. L., Hibbert, A.: J. Phys. B. 23, 2673 (1990).

3. Bell, K. L., Hibbert, A., McLaughlin, B. M., Higgins, K.: J. Phys. B. 24, 2665 (1991).

4. Hibbert, A., Biémont, E., Godefroid, M., Vaeck, N.: Astron. Astrophys., Suppl. Ser. 88, 505 (1991).

5. Biémont, E., Hibbert, A., Godefroid, M., Vaeck, N., Fawcett, B. C.: Astrophys. J. 375, 818 (1991).

6. Tully, J. A., Seaton, M. J., Berrington, K. A.: J. Phys. B. 23, 3811 (1990).

7. Allard, N., Artru, M.-C., Lanz, T., Le Dourneuf, M.: Astron. Astrophys., Suppl. Ser. 84, 563 (1990).

8. Fuhr, J. R., Wiese, W. L.: "Atomic Transition Probalities" CRC Handbook of Chemistry and Physics, (1990-1991 Edition, pages 10-128-179, D. L. Lide, Ed., CRC Press, Boca Raton (1990)).

9. Hansen, J. E., Ed., "Atomic spectra and Oscillator Strengths for Astrophysics and Fusion Research", North-Holland, Amsterdam, (1990).

3.1.3. Highlights Working Group 5: Molecular Structure and Transitions Data
W. H. PARKINSON

The highlights of significant achievement occured over the full spectral range and involved both laboratory study and astronomical detection.

The increased activity in spectroscopy of silicon-bearing carbon molecules and of hydrocarbon radicals is indicative of the achievements and discoveries at longer wavelengths. For example, the laboratory measurements by the groups at NIST (Lovas *et al.* 1989[1]) and at Harvard-Smithsonian (Gottlieb *et al.* 1989[2]) have led to basic spectroscopic parameters and rotational analysis of the silicon dicarbide molecule (Si CC).

Frank Lovas has emphasized (Lovas and Suenram 1989[3], Lovas 1990[4]) the increase in the number of hydrocarbon radicals observed in the microwave region. Spectacular examples of these are the new laboratory spectroscopic studies of carbon molecules. The first was the work on the carbene ring C_3H_2 (Vrtilek *et al.* 1987[5]), followed, recently, by the laboratory detection of two molecules in the sequence of linear carbenes, H_2CCC (Vrtilek *et al.* 1990[6]), and H_2CCCC (Killian *et al.* 1990[7]). Astronomical detection of both of these molecules followed the laboratory observation (Cernicharo *et al.* 1991[8], and Cernicharo *et al.* 1991[9]).

Infrared observations of auroral activity on Jupiter and of laboratory light sources have provided significant spectroscopic informations of the fundamental astrophysical species, H_3^+ (eg. Maillard *et al.* 1990[10] and Oka and Geballe 1990[11]).

There has been interesting new work in the area of molecular structure and spectroscopy involving studies of high-spin multiplicity systems of N_2, never before characterized (Partridge *et al.* 1988[12] and Huber and Vervloet 1988[13]).

Considerable new theoretical and experimental effort has been directed toward obtaining quantitative spectroscopic data for CO in the ultraviolet (Kirby and Cooper 1989[14], Eidelsberg and Rostas 1990[15], and Stark *et al.* 1991[16]).

The predissociation line widths of the Schumann-Runge bands of $^{16}O_2$, $^{18}O_3$, and $^{16}O^{18}O$ have been derived from their measured absolute cross sections (Yoshino *et al.* 1989[17], 1990[18], 1990[19]).

R.W. Nicholls and colleagues have worked to enhance significantly their spectral synthetis codes which they have recently applied to O_2 Herzberg system (1991[20]; 1991[21]).

Phillips and Davis and colleagues have made significant extension to the analysis and understanding of CaH and FeH (1988[22]).

References

(1) Suenram, R.D., Lovas, F.J. and Matsumara, K., 1989, *Astrophys. J. Lett.* **342**, L103.
(2) Gottlieb, C.A., Vrtilek, J.M., and Thaddeus, P., 1989, *Astrophys. J. Lett.* **349**, L29.
(3) Lovas, F.J., and Suenram, R.D., 1989, *J. Chem. Ref. Data* **18**, 1245.
(4) Lovas, F.J., 1990, Private communication.
(5) Vrtilek, J.M., Gottlieb, C.A., and Thaddeus, P., 1987, *Astrophys. J.* **314**, 716.
(6) Vrtilek, J.M., Gottlieb, C.A., Gottlieb, E.W., Killian, T.C., and Thaddeus, P., 1990, *Astrophys. J. Lett.* **354**, L53.
(7) Killian, T.C., Vrtilek, J.M., Gottlieb, C.A., Gottlieb, E.W., and Thaddeus, P., 1990, *Astrophys. J. Lett.* **365**, L89.
(8) Cernicharo, J., Gottlieb, C.A., Guélin, M., Killian, T.C., Paubert, G., Thaddeus, P., and Vrtilek, J.M., 1991, *Astrophys. J. Lett.* **368**, L39.
(9) Cernicharo, J., Gottlieb, C.A., Guélin, M., Killian, T.C., Thaddeus, P., and Vrtilek, J.M., 1991, *Astrophys. J. Lett.* **368**, L43.
(10) Maillard, J.P., Drossart, P., Watson, J.K.G., Kim, S.J., and Caldwell, J., 1990, *Astrophys. J.* **363**, L37.
(11) Oka, T. and Geballe, T.R., 1990, *Astrophys. J.* **361**, L53.
(12) Partridge, H., Langhoff, S.R., Bauschlicher, C.W., and Schwenke, O.W., 1988, *J. Chem. Phys.* **88**, 3174.
(13) Huber, K.P. and Vervloet, M., 1988, *J. Chem. Phys.* **89**, 5957.
(14) Kirby, K. and Cooper, D.L., 1989, *J. Chem. Phys.* **90**, 4895.
(15) Eidelsberg, M. and Rostas, F., 1990, *Astron. Astrophys.* **235**, 472.
(16) Stark, G., Yoshino, K., Smith, P.L., Ito, K., and Parkinson, W.H., 1991, *Astrophys. J.* **369**, 574.
(17) Freeman, D.E., Cheung, A.S-C., Yoshino, K., and Parkinson, W.H., 1989, *J. Chem. Phys.* **91**, 6538.

(18) Cheung, A.S.-C., Yoshino, K., Esmond, J.R., Chiu, S.S-L., Freeman, D.E., and Parkinson, W.H., 1990, *J. Chem. Phys.* **92**, 842.

(19) Chiu, S.S-L., Cheung, A.S-C., Yoshino, K., Esmond, J.R., Freeman, D.E., and Parkinson, W.H., 1990, *J. Chem. Phys.* **93**, 5539.

(20) Cann, M.W.P., and Nicholls, R.W., 1991, *Can. J. Phys.* (in press).

(21) Cann, M.W.P., Lui, C.W., and Nicholls, R.W., 1991, *JQRST* (in press).

(22) Phillips, J.G., and Davis, S.P., 1988, *Astrophys. J. Suppl. Ser.* **66**, 227.

3.2. CONTRIBUTED PAPERS

3.2.1. *Accurate oscillator accurate oscillator strengths of astrophysical interest for neutral carbon, nitrogen and oxygen.*

E. BIÉMONT[1], A. HIBBERT[2], M. GODEFROID[3] AND N. VAECK[3]

[1] *Université de Liège, Belgique,* [2] *The Queen's University of Belfast, UK ,* [3] *Université Libre de Bruxelles, Belgique*

C, N, O are major volatiles which are incompletely condensed in meteorites. Consequently, the sun remains the basic source of information for the solar system abundances. A detailed consideration of correlation effects and of departures from LS-coupling is necessary for solar abundance studies. For these reasons, new transition probabilities based on configuration interaction wavefunctions have been calculated in the dipole length and dipole velocity forms of the matrix elements for allowed and spin-forbidden transitions of N I and O I [1,2]. For C I the calculations are in progress and will be published soon. In the framework of the Breit-Pauli approximation, configuration interaction was included in the calculations performed with the CIV3 code of Hibbert for all the transitions connecting the $n = 3$ and $n = 4$ energy levels. In order to improve upon the agreement vetween eigenvalue and experimental energy differences, empirical corrections have been introduced in CIV3.

The accuracy of the transition probalities has been tested from comparisons with the most accurate theoretical or experimental results presently available and can be estimated to be of a few percent.

The results have been used, combined with the best solar data available, to improve the photospheric abundance determination of these elements [3,4]. The final results, considered as the most accurate presently available, are $A_N = 7.99 \pm 0.04$ and $A_O = 8.86 \pm 0.04$, respectively, in the usual logarithmic scale.

References
[1] Hibbert, A., Biémont, E., Godefroid, M. and Vaeck, N., 1991, *Astron. Astrophys. Suppl.* **88**, 505

[2] Hibbert, A., Biémont, E., Godefroid, M. and Vaeck, N., 1991, *J. Phys. B* (in press)

[3] Biémont, E., Froese Fischer, C., Godefroid, M., Vaeck, N., Hibbert, A., 1990, Proceedings of the meeting *Atomic Data and Oscillator Strengths for Astrophysics and Fusion Research*, Ed. J.E. Hansen, Amsterdam, p.59.

[4] Biémont, E., Hibbert, A., Godefroid, M., Vaeck, N., Fawcett, B.C., 1991, *Astrophys. J.* **375**, 818.

3.2.2. *Spectral study of highly ionized noble gases in high temperature plasmas*

M. GALLARDO +, J. REYNA ALMANDOS *, F. BREDICE * AND M. RAINERI **

Atomic Spectroscopy Group, Optical Research Center (CIOp), C C 124, (1900) La Plata, Argentina

The spectra of highly charged ions of NEON, ARGON, KRYPTON and XENON with few valence electrons have been studied in the 300-2100 A range using a pulsed discharge tube. For Ne spectra we found lines up to the sixth degree of ionization, and we found lines corresponding to Ar VIII, Kr VIII and Xe IX.

In the spectral analysis we used isoelectronic comparisons and atomic calculations. We found new energy levels and classified lines and, in the existing levels values, the uncertainty has been considerably decreased.

+ *Research of the CONICET of Argentina.,* * *Research of the C.I.C. of Argentina.,* **Fellow of the C.I.C. of Argentina.*

3.2.3. A new technique for storing and interpolating rate coefficients

ALAN BURGESS* AND JOHN A. TULLY†
* : D.A.M.T.P., Silver Street, Cambridge CB3 9EW, England.
† : Observatoire de la Côte d'Azur, URA 1362 du C.N.R.S., B.P. 139, 06003 Nice Cedex, France.

We have developed a powerful technique for compacting and assessing atomic data for electron impact excitation of positive ions. Rate coefficients for such processes are derived from thermally average collision strengths $\Upsilon(T)$ which usually vary smoothly with temperature. The present method begins by scaling Υ in order to produce a reduced form Υ_r. This is then plotted as a function of the scaled temperature T_r which maps the entire range of T onto the interval (0,1). Υ_r can be accurately fitted by a 5-point cubic spline, so economising on storage while at the same time providing a convenient means for interpolating and extrapolating data. Straightforward tabulation normally requires more data points than this to cover a limited temperature range. With increasing amounts of atomic data becoming available, the problem of storage needs careful consideration. The present method is a possible solution. It forms the basis of an interactive computer program called OMEUPS which uses graphical display and is designed to be convenient for use by astrophysicists as well of those working in atomic collision theory. Some graphic examples will be presented to illustrate the method.

3.2.4. New data for the ground electronic state of molecular hydrogen

M.E. MICKELSON*, L.E. LARSON*, D.W. FERGUSON† AND K. NARAHARI RAO†
* : (Denison University, Granville Ohio 43023, USA)
† : (The Ohio State University, Columbus, Ohio 43210, USA)

The precision and accuracy of modern high resolution spectroscopic observations of cosmic bodies continues to drive the need for improved laboratory data and theories of atoms and molecules. Inparticular, recent measurements of asymmetric molecular hydrogen quadrupole lines in the absorption spectrum of Uranus and observations of high J rotational transitions in the infrared emission spectrum of molecular hydrogen in the Orion nebulae point to the need for high resolution and high accuracy data. We have recently completed the measurement of pressure shifts, line strengths and shapes, broadening coefficients and line positions for the S(0) and S(1) lines of molecular hydrogen in the 4-0 vibration-rotation band. We have also detected for the first time in the laboratory the 5-0 S(1) line and confirm its detection in Uranus. With these and other high quality measurements of the overtone spectrum of molecular hydrogen in the infrared and visible, a consistent set of parameters has emerged which should be useful to astrophysicists and may challenge theorists to improve their ab-initio calculations.

3.2.5. An investigation of Brueckner's theory of line broadening with application to the sodium D-lines

S.D. ANSTEE AND B.J. O'MARA
Department of Physics, the University of Queensland, St Lucia, Queensland 4072, Australia

Approximations in Brueckner's theory of spectral broadening by collisions with neutral hydrogen atoms relevant to a solar-type atmosphere have been discussed, and a modified theory for s-p transitions has been presented. The theory utilises explicit expressions for the interatomic interaction energy between a hydrogen atom in its ground state and general $m = 0, \pm 1$ p-states, derived from second-order perturbation theory without exchange, allowing for removal of the Lindholm-Foley average over m-states in the original Brueckner model. Approximate upper and lower bounds for the linewidth of the sodium D-lines are derived, and these values are contrasted with available theoretical, experimental and solar empirical results. The removal of the Lindholm-Foley average is shown to reduce the D-state linewidths by about 30%, and an analysis of the interatomic separations important in the line-broadening cross-section for the D-lines has shown that there is little atomic overlap at the separations that are important.

COMMISSION 15: PHYSICAL STUDIES OF COMETS, MINOR PLANETS AND METEORITES
(L'ETUDE PHYSIQUE DES COMETES, DES PETITES PLANETES ET DES METEORITES)

Report of Meetings on 25, 26, 29 and 31 July 1991

PRESIDENT: J. Rahe.   VICE-PRESIDENT: A.W. Harris.   SECRETARY: D.I. Steel.

BUSINESS MEETING OF COMMISSION 15 : 26 July 1991, Session 1

The meeting was called to order by the President, J. Rahe, with approximately 25 members of Commission 15 being present. Those assembled stood in silence in remembrence of the following of our co-workers who had died in the last triennium: O.V. Dobrovolsky, E. Everhart, P.M. Millman, C.B. Opal and Zhou Z.-H.

Officers and Organizing Committee

A.W. Harris and M.F. A'Hearn were proposed by the OC for President and Vice-President, respectively, of the Commission, and were confirmed by the Executive Council of the Union, for the triennium 1991-1994. D.I. Steel was appointed Secretary. Following long tradition, members of the Organizing Committee serve for two three-year terms. Those retiring from the Committee include: C. Arpigny, J. Wasson, O.V. Dobrovolsky (deceased), H. Haupt, and S. Wyckoff, and are thanked for their service. Newly appointed members are M.A. Barucci, W.F. Huebner, H.U. Keller, N.N. Kiselev and Z. Knezevic. Continuing members are M. Belton, J. Brandt, E. Grün, D.W. Hughes, D. Lupishko, H. Rickman, and E. Tedesco. Following two terms of service, C. Arpigny stepped down as Chairman of the Working Group on Comets, and was replaced by H.U. Keller. V. Zappalá will continue as Chairman of the Working Group on Minor Planets and Meteorites through 1994. After a brief discussion of the relationship of the meteoritical community within the IAU, it was agreed to combine the Working Group on Meteorites with the Working Group on Minor Planets.

Membership

The following were admitted as new members of Commission 15: M. Barucci, R. Binzel, F. Capaccioni, M.T. Capria, G. Carruthers, U. Carsenty, A. de Almeida, G. Forti, P. Gamelgaard, W.T. Huntress, H. Karttunen, C.S.L. Keay, N.N. Kiselev, A.L. Lane, K. Meech, T. Michalowski, K.S. Noll, C.B. Opal (deceased), C.T. Pillinger, T. Sato, M.V. Sykes, I. Toth, T. Van Flandern, A.S. Walker, J.-I. Watanabe. Five persons were appointed Consultants for the next triennium: D. Boice, R. Gil-Hutton, S. Hoban, G. Tancredi, and T. Yanai.

Communications

Future meetings of interest to members of the commission were discussed, amongst them that at Liège in June 1992, and that on 'Meteoroids and their Parent Bodies' to be held in Bratislava in early September 1992; the incoming President of Commission 22, J. Stōhl, warmly extended an invitation to all present to attend that meeting.

A communication from the Giotto Extended Mission Office at ESOC was received, requesting astrometric observations of P/Grigg-Skjellerup (the next target for the Giotto spacecraft) prior to the July 1992 encounter.

Resolutions

The various resolutions before the GA and Commission 15 were then discussed. The GA resolution on the detection and study of Near-Earth Objects was adopted unopposed. The GA resolution on Interplanetary Pollution was discussed with some dissension on the grounds that it might limit future spacecraft missions and hence the activities of the commission; the resolution was adopted by a majority vote. A Commission 15 resolution advocating the encouragement of investigations of likely target comets for the CRAF spacecraft was adopted unopposed. A proposal to set up a Working Group on disruptive impacts between minor planets was introduced and discussed, but dropped. Finally, a communication from Wamsteker asking the support of Commission 15 for world astronomy days and coordinated observation programs was tabled; the fact that 1992 is International Space Year was noted.

SCIENTIFIC PRESENTATIONS ON COMETS : 25 July 1991, Session 1

A session dedicated to comets was organized and chaired by J.C. Brandt.

A.-C. Levasseur-Regourd (University of Paris VI), with J.B. Renard and E. Hadamcik, described polarimetric mapping of the coma of Comet Levy (1990c), and showed that in the inner coma the slope of the polarization versus phase angle curve is higher than at distances above 2000 km. This may indicate that the dust grains are smaller than typical, with dust jets like those in P/Halley. Future observations should try to resolve the inner and outer coma since differing properties are found.

H.J. Reitsema with co-workers M. Descour and W.A. Delamere (Ball Aerospace, Boulder, CO, USA) reviewed the Giotto observations of the coma dust distribution within 2000 km of the nucleus of P/Halley and presented new findings on the changes observed during the 3-hour period of camera coverage.

M. Wallis (University of Wales, Cardiff), with N. Meredith (U.C. London), discussed imaging of P/Grigg-Skjellerup at the 630 nm wavelength of [OI] with a new photon detector which alleviates contamination by NH_2. They find that the H_2O parent, which in turn produces the [OI], has a scale height of $\sim$2000 km at 1 AU, several times the [OI] collisional quenching scale.

Finally J.C. Brandt (University of Colorado) described the status of our knowledge of the cause of cometary disconnection events. He reviewed the various theories and the problems in investigating these DE's, concluding that progress is being made, and that the latest results continue to favour the mechanism of magnetic reconnection at interplanetary sector boundaries as the origin of the observed phenomenon.

SCIENTIFIC PRESENTATIONS ON MINOR PLANETS : 26 July 1991, Session 2

D. Morrison (NASA-Ames) reviewed the recent meeting on 'Near-Earth Asteroids' (NEA's) held in June-July at San Juan Capistrano, California. In particular he discussed the plans now being developed for an international program to discover essentially all near-Earth objects down to 0.5 km ($\sim$10,000 in total) within the next 25 years.

B.G. Marsden (Harvard-Minor Planet Center) gave an overview of the discussions of families of asteroids which occurred at the recent meeting in Flagstaff, Arizona. From the times when J. Williams suggested many families to those of A. Carusi and G. Valsecchi who suggested only three families existed, this subject has come a long way. There are now suggestions of not only distinct families but also, from V. Zappalá, the idea of clumps, clans and clusters.

K. Lumme (Helsinki) discussed how to derive some global properties (such as the shape, albedo and pole orientation) of convex minor planets using spherical harmonic expansions of their light-curves.

D.K. Yeomans presented a paper on behalf of S. Ostro (Jet Propulsion Laboratory) in which radar observations of near-Earth asteroids between 1980 and 1991 from Arecibo and Goldstone were reviewed. Of particular interest were the results which showed that 1986 DA appears to be metallic in nature, whilst 1989 PB is a dumbell shape, possibly a contact binary. Radar data are also invaluable in refining the orbit of a NEA.

E. Tedesco (Jet Propulsion Laboratory) discussed the status of the second version of the IRAS asteroid catalog scheduled for release in April 1992. He described how the flux over-estimation and systematic difference between the albedos and diameters obtained at 12 and 25 μm were dealt with, and showed that the accuracy of the standard thermal model used to derive diameters from the IR flux is of order 5%.

T. Van Flandern (Meta Research, Washington, DC) discussed the evidence that the standard model of the origin of the minor planets is incorrect, and predicted that the Galileo spacecraft will detect many small satellites around Gaspra when it encounters that asteroid in late October 1991. This would support his hypothesis that the minor planets originated in the explosion of a larger body in the astronomically-recent past.

DISCUSSION OF COMETARY DATABASES : 29 July 1991, Session 1

A session dedicated to cometary databases was chaired by M.F. A'Hearn, advance organization having been carried out by C. Arpigny. The discussion commenced with a general overview by M.F. A'Hearn of the items to be possibly included in the database.

B.G. Marsden then reviewed the discussions on this topic which occurred at the recent meeting in Flagstaff. He pointed out the problems that existed for comets due to their diversity (compared to minor planet data), and also that the volume of data was very large, with little uniformity.

R. West suggested that data should be stored in 'as original as possible' a state, but simply intelligible. Due to the volume of data he suggested that as a first step the planned database should perhaps be just an index to the raw data sets which would be held by their originators. This met with general approval. Further suggestions were made that the most important thing required at present was a bibliography of publications on each comet, and a listing of people who hold data which are available to others.

Mechanisms for making the cometary database available to others were then discussed, with the idea that the data should be accessible through an electronic mail network such as INTERNET being favoured. The possibility that a bulletin board might be set up for comets and minor planets was mooted, although this would require a large amount of time for one person to administer. The consensus was that direct contacts between people who are active in the field was the most likely method to succeed: an informal but effective system.

JOINT MEETING WITH COMMISSION 20 ON COMET-MINOR PLANET INTERRELATIONS : 31 July 1991, Session 4

This joint meeting, organized by A.W. Harris and R. West, was originally scheduled to occupy two sessions. Because of the fire in the San Martin Conference Building, it was necessary to compress it to a single session, chaired by A.W. Harris.

D. Steel (Anglo-Australian Observatory) discussed the evidence for meteoroid streams associated with Apollo-type asteroids; such streams might indicate the asteroids to be extinct or dormant comets. The available database on meteor orbits, and its analysis, were described. The calculation of theoretical meteor radiants was also mentioned, and it was shown that methods used to date often give misleading results: in order to get a realistic radiant it is necessary to integrate orbits of comets or minor planets to the points at which they have a node at 1 AU, and then calculate the radiant.

R. West (European Southern Observatory) described methods for detecting low-level activity in distant comets so as to determine whether they should be classified as comets or minor planets (*e.g.* Chiron). Three types of detections are possible which may indicate comet-like activity: (a) a tail (to 6–7 AU); (b) a coma (to ~17 AU); or (c) increased brightness (to 17 AU+). Observations have been made of various distant minor planets, Trojans, and 1991 DA, and now data collection for 6 comets beyond 15 AU is planned.

B. Gustafson (University of Florida) showed his backward orbital integrations of meteoroids from the Geminid shower, compared these to 3200 Phaethon, and found that the meteoroids could have been released from that minor planet during a cometary phase of its evolution within the past few millenia. Gustafson then gave a paper, jointly with S.F. Dermott and colleagues from Florida and the Brazilian National Observatory, on IRAS data fitting to models of the zodiacal dust cloud. They find that the cloud is not rotationally symmetric, and believe that particles released in an asteroid break-up are largely responsible for the origin of the cloud; in particular they implicate the asteroids in the Hirayama family.

J. Stōhl, with V. Porubčan (Slovak Academy of Sciences, Bratislava), detailed an investigation of the association of meteoroid streams with near-Earth asteroids, finding a strong association for 10 Apollos and 5 Amors which come within 0.1 AU of the Earth.

H. Rickman (Uppsala Observatory) presented evidence for rapid dust mantling on short-period comets, dependence being found upon the orbital types and physical parameters of the nuclei. Marginally stable mantles are often formed within ~10 revolutions about the Sun, but these are only just stable and so may be disrupted by a change in the orbit (*e.g.* $\Delta q > -0.5$ AU). A new parameter to measure the 'freshness' of the nucleus was introduced, and its values for different comets discussed.

I.P. Williams (Queen Mary and Westfield Colleges, London), with C.-I. Lagerkvist (Uppsala), A. Fitzsimmons (Queen's University, Belfast) and P. Magnusson (QMWC, London), described 400–800 nm spectra from minor planets suspected to be 'dead comets': mostly Trojans and distant main-belt asteroids. For 20 such objects plus 3 'real' comets they found emission features only for the comets plus Chiron. A peculiar time-dependent absorption feature was seen in the spectrum of Thule near 520 nm which requires an explanation; a similar feature was detected in Rolandia.

To close this session, and the Commission 15 activities at the GA, B.G. Marsden (Harvard-Minor Planet Center) was delegated to lead a discussion on whether 'active' minor planets should be re-designated as comets, and whether the contrary should be the case for defunct comets. This subject has become topical due to the activity of 2060 Chiron which has been identified in the past few years. Historically there are 28 cases of objects being given preliminary designations as minor planets which have later proven to be comets, and 9 cases of 'comets' which were later recognized to be minor planets. Some questions in this area have yet to be resolved. The consensus was that objects should not be re-designated (*i.e.* 2060 Chiron should continue to be described that way), and that if it is not possible to establish the cometary character of an object soon after its discovery then it should be designated as a minor planet. Re-designation as a comet prior to numbering as a minor planet is straightforward.

PHYSICAL STUDIES OF PLANETS AND SATELLITES

ETUDE PHYSIQUE DES PLANETES ET DES SATELLITES

Report of Meetings

PRESIDENT: André Brahic

Business Meeting:

At the 21th General Assembly of the I.A.U., Commission 16 held its business meeting, which opened at 2:00 p.m. on 30 July 1991, under the chairmanship of the President, Pr. André Brahic (France).

It is with great sadness that we have to report the death of Harold Masursky on August 24, 1990 at the age of 66. Harold Masursky was the President of the Working Group for Planetary Nomenclature and a member of the Organizing Committee of Commission 16. He was well known for his active participation to space exploration of the planets. He was one of the pioneers of planetary geology. He wrote many important contributions on planetary surfaces.

The minutes of Commission 16's last business meeting of 10 August 1988 were distributed, discussed and accepted by the Commission. Pr. Brahic made a brief report of his activities on behalf of the Commission.

Planetary nomenclature:

Dr. Kaare Aksnes (Norway), President of the Working Group for Planetary System Nomenclature (W.G.P.S.N.), presented a report on this group's activities during the last triennum. W.G.P.S.N. held its 17th meeting in Paris and Fréhel, France (19-22 October 1989), its 18th meeting in Den Haag, Holland (2 July 1990), its 19th meeting in Florence, Italy (29-30 April 1991) and the 20th meeting in Buenos Aires, Argentina (30 July 1991) during the I.A.U. General Assembly. Largely due to the discoveries with the Voyager and Magellan spacecrafts, there has been an increasing need for new nomenclature. The following nomenclature has been recommended to the I.A.U. by W.G.P.S.N.(they have the approval of the I.A.U. Executive Committee since August 1991):

1. Surface features on Moon, Venus, Mars, and Triton. The corresponding names are listed in Reports on Astronomy, vol. XXIA, 613-619 (1991) and in vol. XXIB in the Transactions from the I.A.U. General Assembly in Buenos Aires. In particular, three new feature categories are introduced for Venus:

RETICULUM, RETICULA	reticular form (netlike) pattern
FARRUM, FARRA	row of pancake-like structures
VALLIS, VALLES	sinuous valley

2. Satellites of Saturn and Neptune.

Temporary Designation	Permanent designation	Name	Distance from Planet's Center
1981S13	Saturn XVIII	Pan	133 600 km
1989N6	Neptune III	Naiad	48 230 km
1989N5	Neptune IV	Thalassa	50 070 km
1989N3	Neptune V	Despina	52 530 km
1989N4	Neptune VI	Galatea	61 950 km
1989N2	Neptune VII	Larissa	73 550 km
1989N1	Neptune VIII	Proteus	117 640 km

3. Neptune Rings and Ring Arcs.

Temporary Designation	Name	Distance from Planet's Center
1989N3R	Galle (N42)	41 900 km
1989N2R	Leverrier (N53)	53 200 km
1989N1R	Adams (N63)	62 900 km
"Leading" ring arc	Liberté	62 900 km
"Equidistant" ring arc	Egalité	62 900 km
"Following" ring arc	Fraternité	62 900 km

For the next triennum, W.G.P.S.N. will have the following membership:

President:	K. Aksnes	Norway
Members:	R. Batson	U.S.A.
	A. Brahic	France
	M. Ya Marov	U.S.S.R.
	D. Morrisson	U.S.A.
	T.C. Owen	U.S.A.
	V.V. Shevchenko	U.S.S.R.
	B.A. Smith	U.S.A.
	L.A. Soderblum	U.S.A.
Consultants:	J.M. Boyce	U.S.A.
	P. Masson	France

Changes in the membership in the Task Groups under W.G.P.S.N. will be decided on later.

Election of officers:

The Organizing Committee of Commission 16 was then discussed. The following slate was presented and elected unanimously by the members present:

President: David Morrison (U.S.A.)
Vice Presidents: Catherine de Bergh (France)
                 Michael Ya. Marov (U.S.S.R.)
Members: Alexandr Basilevsky (U.S.S.R.)
                 Michael Belton (U.S.A.)
                 André Brahic (France)
                 Dale P. Cruikshank (U.S.A.)
                 Merton Davies (U.S.A.)
                 Thérèse Encrenaz (France)
                 Daniel Gautier (France)
                 Vasily I. Moroz (U.S.S.R.)
                 Tobias Owen (U.S.A.)
                 Jurgen Rahe (U.S.A.)
                 Bruno Sicardy (France)
                 Bradford Smith (U.S.A.)
                 David Tholen (U.S.A.)

New Members of the Commission:

The following list of new members of the Commission was unanimously endorsed by the members present:

Richard P. Benzel	(U.S.A.)	Commissions 15, 16, 20
Maria Teresa Capria	(Italy)	Commissions 15, 16
Maria M. Carsmaru	(Rumania)	Commission 16
Jean Claude M. C. Gérard	(Belgium)	Commission 16
Alan Harris	(U.S.A.)	Commissions 15, 16
M.A. Lopez-Valverde	(Spain)	Commission 16
Takafumi Matsui	(Japan)	Commission 16
Michael E. Mickelson	(U.S.A.)	Commissions 14, 16, 41
Keith Stephen Noll	(U.S.A.)	Commissions 15, 16
Jean Marc Petit	(France)	Commissions 7, 16, 28
Colin Pillinger	(G.B.)	Commissions 15, 16, 22
Zdenek Pokorny	(Czechoslovakia)	Commissions 16, 46
Carolyn Porco	(U.S.A.)	Commission 16
Irina Predeanu	(Rumania)	Commission 16
Jurgan Rahe	(U.S.A.)	Commissions 15, 16
Françoise Roques	(France)	Commission 16
Agustin Sanchez-Lavega	(Spain)	Commission 16
Bruno Sicardy	(France)	Commission 16
Alta Sharon Walker	(U.S.A.)	Commissions 15, 16, 51

The Commission Membership is the following: U.S.A. (129), U.S.S.R. (21), France (19), U.K. (12), Spain (7), Germany (7), Japan (4), ...

Future activities:

The future activities of Commission 16 were discussed with Pr. Brahic leading the discussion. Several items were discussed, among them:
- We had two good scientific sessions, with respectively 150 to 200 participants and 30 to 50 participants, but most of the attendees were interested astronomers from other areas rather than members of the planetary community.
- Concern was expressed about the declining number of planetary scientists attending the I.A.U. General Assembly, and especially the decreasing number of planetary scientists active in Commission 16. In order to assure continued interest in planetary science, especially by young scientists at the beginning of their career, the establishment of summer schools for graduate and undergraduate planetary science students was discussed as well as issuing a Commission 16 Newsletter on a regular basis.
- The need for descriptions of current missions and results derived from them, was particularly felt at this General Assembly, and a consensus was reached that a special effort should be made to present highlights of these missions during the next General Assembly in teh Netherlands.
- Already during the Baltimore General Assembly it was suggested to broaden the scope of Commission 16 by including the search for and characterization of other planetary systems without changing the Commission's name. This recommandation was reiterated in Buenos Aires but a decision was not made.

Future meetings:

Future meetings of interest to Commission Members were discussed:
- The Conference on Solar System Astronomy, which had scheduled to take place in China and which was discussed already in the 1988 meeting, has been canceled. Dr. Morrrison was asked to pursue this idea further, and the Commission agreed to endorse it should it be proposed.
- Dr. D. Cruikshank's and Dr. T. Gehrels' plan was discussed to organize a meeting on Neptune in 1992, i.e. one year before A. Brahic plans to hold another meeting on the same planet in France to commemorate the birthday of its discovery. The Commission recommended that the organizers of the two meetings coordinate their plans.
- Finally it was mentionned that Dr. D. Tholen is contemplating to organize a conference on Pluto in 1992 or 1993. Again, no formal proposal was submitted.

When the meeting adjourned at about 15:45 p.m., approximately 15 to 20 people were in attendance.

Working Group on Cartographic Coordinates:

A meeting of the I.A.U./I.A.G./C.O.S.P.A.R. Working Group on Cartographic Coordinates and Rotational Elements of the Planets and the Satellites was held with members of both Commissions 4 and 16 in attendance under the chairmanship od Dr. Merton E. Davies (U.S.A.). The report of the Working Group is included in the I.A.U. Transactions.

Scientific Meetings:

Good scientific sessions with a large attendance had been organized. Most of the attendees were interested astronomers from other areas rather than members of the planetary community. Many planetary astronomers did not come to Buenos Aires. As a comparison, it has been observed that the attendance to the Commission 16 meetings was specially poor during the 1985 General Assembly in New Delhi, and was rather large during the 1988 General Assembly in Baltimore. Several participants to the Scientific Meetings expressed their satisfaction concerning the inter-disciplinary exchanges during these meetings. The program of the Joint Discussion Meeting and of the Commission Meeting is given below:

1. Joint Discussion with Commissions 7, 15, 20, 22, 28, 33, 35, 37, 45, 48, and 51
   Organizing Committee: A.H. Batten, A. Brahic (Chairman), M. Golay, G.L. Harris, A. Henrard, C.S.L. Keay, A. Maeder, G. Marx, M. Mayor, R.A. Sunyaev, G.A. Tamman, R. West.

- Planetary Atmospheres and Chemical Constraints	T. Owen
- Comets and Solar System Formation	A.C. Levasseur-Regourd
- Planetary Rings and Dynamical Constraints	A. Brahic
- From Planetoids to Planets	P. Barge and R. Pellat
- Search for Extra Solar Planets from Space	R. Terrile
- The ß Pictoris Disk	B. Smith
- Circumstellar Disks and Star Formation	L. Hartmann and M. Gomez
- Circumstellar Environnement of Young Stellar Objects	T. Montmerle
- Magnetic Activity around T Tauri Stars	T. Montmerle
- Effects of Interactions Involving Protostellar Disks	R. Larson

2. Commission 16 Scientific Meeting
- Study of Rings from Voyager to Cassini   C. Porco
- International Jupiter Watch   D. Morrison
- The Pluto - Charon System   D. Tholen
- Possible Detection of the Surface of Titan   K. Noll
- Gallery of Magellan Venus Photos   M. Davies and D. Morrison
- Origin of Inner Planet Atmospheres   T. Owen
- Possible Lunar Transient Event   J. Seiradakis
- Possible Venus Transient Event   H. Varvoglis
3. Posters
- Vertical Structure and Granulometry of High Martian Clouds
from Solar Occultation   E. Chassefière and J.E. Blamont
- Mutual Phenomena of Jupiter Satelittes observed with
a CCD camera   J. Colin, J.F. Le Campion, M. Rapaport, G. Montignac,
                                                             J.M. Desbats, F. Chauvet and G. Dourneau
- The 5:3 Jovian Resonance   S. Fernandez and M. Mosconti
- Solid Bodies around ß Pictoris?   A.M. Lagrange-Henri, H. Beust, A. Vidal-Madjar and
                                                             E. Chassefière

- Hydrocarbons Abundances in the Titan's Atmosphere. Comparison
between a Photochemical Model and Observations   L.M. Lara, A. Coustenis, R. Rodrigo, P. Romani,
                                                             J.J. Lopez-Moreno and E. Chassefière

COMMISSION 19 : ROTATION OF THE EARTH (ROTATION DE LA TERRE)

President :   M. Feissel
Vice-President: B. Kolaczek

Organizing Committee:
P. Brosche	W.E. Carter	J.O. Dickey
D.M. Djurovic	Jin W.-J.	N. Mironov
D.D. McCarthy	M.G. Rochester	T. Sasao
B.E. Schutz	J. Vondrak	G.A. Wilkins

INTRODUCTION

The main topics covered during this General Assembly were the nine resolutions proposed by the Working Group on Reference Systems (WGRS), of which Commission 19 was one of the sponsoring commissions, and the achievements of the newly created International Earth Rotation Service (IERS), concerning not only the precise determination of the Earth's rotation irregularities, but also the astronomical constants and models, and the extragalactic celestial reference frame.

The observing techniques used for the Earth rotation studies are radioastrometry (VLBI), Lunar Laser Ranging and satellite geodesy (LAGEOS, Global Positioning System), which insure a sub-milliarcsecond accuracy. However, the less precise method of optical astrometry has unique capabilities in specific fields, where Commission 19 made steps to help their efficient use: the commission supports a Working Group to reanalyze the 1900-1990 observations in reference frames provided by IERS and HIPPARCOS, and it encourages to seek cooperation with the International Association of Geodesy (IAG) to extract the geophysical signal from these observations (see the two Commission Resolutions listed at the end of this report).

Research on the Earth's rotation is an interdisciplinary field between astronomy and geophysics. Commission 19 has always had common members with the IUGG and/or IAG Working groups on this subject. The IAU Resolution listed at the end of this report re-affirms the intent of the commission to continue joint activities in this field.

BUSINESS SESSION 1.

The President presented a summary of activities of Commission 19 in the term 1988-1991. Following a resolution passed at the 20th General Assembly, the Presidents of Commissions 19 and 31 (Time) presented to the IAU Executive Committee a report to justify the request of an increase in the IAU support to the Federation of Astronomical and Geophysical data analysis Services (FAGS). This request was also supported by 10 commissions (5, 8, 10, 24, 25, 26, 33, 34, 40, 45) on the basis of their need of 5 of the 11 FAGS services (IERS, QBSA, IUWDS, SIDC, CDS). As a result of this action, the IAU Budget for 1991-1994 includes a raise of the IAU grant to FAGS.

The General Secretary proposed in 1990 to amalgamate commissions 19 and 31. The President of Commission 19, after consulting the Organizing Committee, refused the amalgamation, as did Commission 31. On this occasion, the interests of Commission 19 were defined as follows.

- Understanding of the irregularities of the Earth's rotation (polar motion, duration of the day, precession-nutation).

- Provision of the Earth's orientation to the astronomical community (operational and scientific solutions).

- Astrometric modelling.

- Extragalactic reference frames.

- Relationship among reference frames linked to quasars, to stars, to objects of the solar system, and to artificial satellites of the Earth.

- Terrestrial reference frames.

Commission 19 is one of the sponsoring commissions of the Working Group on Reference Systems (WGRS, Chairman : J. Hughes) ; 17 commission members formally belong to the WGRS, and others made contributions to its work. The nine recommendations of the WGRS, later adopted as IAU Resolutions, establish new basic concepts for the space-time reference systems to be used in astronomy, defined homogeneously in the framework of the theory of General Relativity. Detailed analysis of the concepts and of the consequences of their use can be found in the proceedings of IAU Colloquium 127 and of the Joint Discussion organized during this General Assembly by the WGRS. Of special importance to Commission 19 is the adoption of an extragalactic celestial system as primary system, as is already the case in the study of the Earth's rotation.

Commission 19 has another Working Group, on "Earth rotation in the HIPPARCOS reference frame", chaired by J. Vondrak. Refer to the report on scientific session 2 for detailed information. Commission 19 has high interest in the work of the International Earth Rotation Service (IERS), a service of FAGS sponsored by IAU and IUGG. The scientific meeting 3 is devoted to IERS.

Two Scientific meetings co-sponsored by Commission 19 were held since the IAU 20th General Assembly:
- IAU Symposium 141, 17-21 October 1989, Leningrad (USSR) "Inertial coordinate system on the sky". Proceedings published by Kluwer (1990), Lieske and Abalakin eds.
- IAU Colloquium 127, 14-20 October 1990, Virginia Beach (USA) "Reference systems". Proceedings published by USNO (1991), J.A. Hughes, C.A. Smith, G.H. Kaplan eds.

Three joint meetings co-sponsored by Commission 19 are to be held during the General Assembly, with Proceedings to appear in Highlight of Astronomy, vol. 7 :
- Rotation of the solar system bodies, with Commissions 15, 16, 20;
- Reference systems : what are they and what's the problem ?, with Commissions 4,7,8, 20,24,31,33,40;
- HIPPARCOS - an assessment, with Commissions 4, 8, 19, 24, 31, 40, 44;

Two proposed symposia are supported by Commission 19 :
- Developments in astrometry and their impacts on astrophysics and geodynamics, 15-19 September 1992, Shanghai (China),
- VLBI Technology, progress and future observational possibilities, 16-20 August 1993, Kyoto (Japan).

REFERENCE FRAMES : REALIZATION AND CONNECTION.
(Chair : D. McCarthy ; Secretariat : J. Luck).

      O. Sovers (JPL) reported on a global VLBI catalog solution made using a number of observations from the Crustal Dynamics Project and IRIS. The largest errors are due to the treatment of possible source structure, source variability, and the overall connection to existing reference frames.

      F. Arias (La Plata Observatory) reported on the connection of a radio source catalog to optical catalogs using RS CVn stars. Using simulated HIPPARCOS results she expects errors in the resulting link of about 0.001" in positions and 0.001"/yr in the angular velocity.

      M. Feissel gave information on current discussions with the IAU General Secretary about an overall re-organization of commissions dealing with astrometry.

OPTICAL ASTROMETRY FOR EARTH ROTATION
(Chair : J. Vondrak ; Secretariat : F. Arias).

Jan Vondrak presented the report of the Working Group on Earth rotation in the HIPPARCOS reference frame, created by Commission 19 in Baltimore, 1988. The WG has 14 members and worked mainly by correspondence. It met only twice : October 1989 in Leningrad (IAU Symp. 141) and June 1991 in Paris (4 èmes Journées Systèmes de Référence Spatio-temporels). The WG prepared a list of observatories with best results to participate in the project of the prepared new reduction of optical astrometry observations (latitude since 1990, clock corrections since 1955) based on star-by-star data; 53 instruments were selected, out of which 41 already answered positively. The problems connected with homogeneisation of the observed data were shown and algorithms to be used were outlined (in order to bring all the observations into the unique celestial reference frame, using the same model for precession, nutation, aberration, gravitational deflection of light, refraction etc). Observation equations for all types of instruments have been derived and the necessary constraints to keep the terrestrial reference frame consistent with that of IERS formulated. Then the form of the normal equations of the final adjustment was presented and the problems connected with the identification and exclusion of outliers and selection of parameters to be adjusted were shown. The following discussion revealed the possibility of correlations between some of the adjusted parameters, e.g., UT1-UTC and motion of the celestial pole, expected due to nonregular distribution of observations in time.

      Ye Shu-hua presented the report of the Analysis Center of Optical Astrometry in Shanghai, endorsed by Commission 19. It collected and analysed the data since 1988.0. The number of observations progressively decreased during the past three years, and the precision of the ERP determinations accordingly degraded. Shanghai Observatory published and distributed 12 quarterly reports; an astrometric database containing the observations from 1962.0 has been created. At present, there are 40 optical astrometric instruments active. It would be advisable to change the scientific goals of the Center, and to use these results to monitor the variations of the local verticals, using the ERP as defined by IERS and better star positions as given by HIPPARCOS as standards. Thus it will be possible to separate the local terms from the observational errors more efficiently. Shanghai Observatory is ready to undertake such a task. She also stressed that a great progress had been made lately in developing automatic photoelectric astrolabe in China, able to observe stars up to 11th magnitude. These observations can be used for prediction id earthquakes, as indicated in the papers by Hu Hui et al. and Zhang Guodang, whose abstracts were presented also by Ye Shu-hua.

K. Yokoyama shortly informed the audience about the works of the National Astronomical Observatory at Mizusawa related to this WG. The compilation of a unified catalog of Paris and Mizusawa astrolabe stars is on the way. The method used is a star-by-star comparison which is more informative and more useful for any research than using the averaged data. They can estimate ERP as well if they collect data from many stations.

B. Kolaczek presented the paper of M. Barlik and J. Rogowski: Variations of the plumb-line direction obtained from astronomical and gravimetric observations. During the past 15 years, these variations were systematically measured at the latitude station Jozefoslaw by the Warsaw University of Technology. There is a correlation between the secular drifts of both astrometrically and gravimetrically determined variations; the variations of the groundwater level show similar trend, but they cannot explain the variation of the meridien component.

M. Feissel then presented the draft versions of two resolutions, concerning 1) the future work of the WG on Earth rotation in the HIPPARCOS reference frame and 2) the Shanghai center, to be voted in the last bussiness meeting of Commission 19. The text of the first resolution was accepted by the attendants, while there were some objections to the wording of the second one; the participants thought that scientific considerations justifying the change should be more specific, even if they are out of direct interest of Commission 19. It was also stressed that the only long series of data existing are the ones with optical astrometry and it would be good not to interrupt them. As a consequence, a committee (J. Kovalevsky, D. McCarthy and B. Kolaczek) was designated to improve the formulation of the second resolution before the last bussiness meeting.

INTERNATIONAL EARTH ROTATION SERVICE (IERS)
(with Commission 31 ; Chair : K. Yokoyama, Chairman of the IERS Directing Board)

This session was organized to report to the IAU science community about the scientific and organizational achievements made by the various components of IERS during three and a half years since the beginning of the service in 1988. In the context of IAU interest, IERS has contributed considerably to constructing IAU reference systems, by compiling the IERS extra-galactic radio reference frame, as well as the IERS Standards as its fundamental basis. In fact during this General Assembly, IERS achievements have been reported at the Joint Discussion on Reference Systems, and the importance of the IERS work for the IAU reference system issues was clearly stated in the passed resolutions.

The meeting was attended by about 100 scientists. The session was started by an overview of the IERS activities by K. Yokoyama and "Summary Report of the IERS for the Period from August 1988 through July 1991" was distributed to the attendants. Then reports by the Central Bureau (CB, M. Feissel), Sub-Bureau for Rapid Service and Predictions (SBRSP, D.D. McCarthy), Sub-Bureau for Atmospheric Angular Momentum (SBAAM, D. Salstein, presented by M. Feissel), four technique Coordinating Centers (CC, VLBI, W.E. Carter; SLR, B.E. Schutz; GPS, W.G. Melbourne; LLR, Ch. Veillet; former three were presented by K. Yokoyama), and the advisor of the IERS Standards (D.D. McCarthy) followed.

K. Yokoyama summarized the report on the meetings, publications, dissemination of data and information, relations to the international bodies, etc., as well as the following special projects and events.

1) The Sub-Bureau for the Atmospheric Angular Momentum was set up on October 1, 1989 at the National Meteorological Agency (USA) led by J.A. Miller.

2) "IERS Standards (1989)" was published in November 1989 as the IERS Technical Note No 3 (ed. D.D. McCarthy). Improvement is being explored to cope with the millimeter accuracy achieved by the improvement of the observing techniques.

3) GPS is now one of the four observing techniques of IERS, with the Jet Propulsion Laboratory (USA) being its Coordinating Center. GPS CC conducted the first IERS global GPS campaign called GIG'91 during January to February 1991, with over 130 participating stations and 20 Analysis Centers. The first results of the polar motion analyzed by an MIT group was reported. Agreement with the VLBI results is excellent and this insures the usefulness of GPS in the future of the Earth rotation study.

4) VLBI technical development centers have been set up at the Haystack Observatory (USA) and the Communications Research Laboratory (Japan).

5) Intensive campaigns to detect short periodic variations in the Earth's rotation and relate them to atmospheric excitation have been decided to be conducted during 1991 to 1992 putting emphasis on synchronized observation with multiple observing techniques.

M. Feissel reported the activities of CB. The IERS Celestial Reference Frame (ICRF) has been implemented at the celestial frame section of the IERS/CB based on the individual radio reference frames provided by the VLBI Analysis Centers. It includes 396 objects between declinations -84° and +85° with 109 objects having uncertainties smaller than 0.001". The consistency among the individual frames after removing biases is about 0.0005", and the origins on the equator agree within ±0.0007". Uncertainties of the angles between the frames are at the level of 0.0001"~0.0003". The current efforts are to extend the Southern hemisphere coverage by VLBI observations.

The IERS Terrestrial Reference Frame (ITRF) has been implemented by the terrestrial frame section of the IERS/CB based on sets of station coordinates by various techniques provided by Analysis Centers. The scale and origin of the ITRF are defined by LAGEOS based coordinates computed by the Center for Space Research, University of Texas. In the latest realization, 134 sites are included, with 51 primary sites and 28 colocation sites. The uncertainty of the coordinates of most stations are smaller than a few cm.

Earth orientation parameters (EOP) have been computed at the Earth orientation section of the IERS/CB keeping consistency with the ICRF and ITRF. In the Annual Report for the Year 1990, three VLBI series, GSFC, NGS and USNO, and three SLR series, CSR, DGFII and GSFC were adopted to produce the 5-day spacing polar motion; the systematic corrections model by constant bias and linear trend. On the other hand, a daily UT1 series was produced from two VLBI series, NGS and GSFC, as well as the NGS daily UT1 series and the CSR SLR series. The length of day values are made by differenciating the daily UT1 series. Uncertainties of the 5-day raw normal values of polar motion are usually better than 0.0003" and those of 1-day UT1 are better than 0.1ms. Series of the celestial pole offsets were also produced on the basis of the VLBI series, with a precision better than 0.0002".

D.D. McCarthy's report of SBRSP was as follows. The major contributors to the SBRSP maintained by the National Earth Orientation Service (NEOS, USA) are: LAGEOS day (SLR), Delft 5-day (SLR), IRIS 5-day (VLBI), IRIS 1-day, Texas LLR. The series given in the Bulletin A for rapid service (NEOS rapid) has some systematic biases compared to the refined series of the NEOS and Central Bureau final values, but smaller than 0.001".

D. Salstein reported that SBAAM serves as the focal point for the collection of atmospheric data from the four centers, U.S. National Meteorological Center, European Centre for Medium Weather Forecast, United Kingdom Meteorological Office, and Japan Meteorological Agency. The primary quantities are analysis and forecast values of the effective atmospheric angular momentum functions, at 00 and 12 UT. Recently, three centers except JMA are preparing for providing 6-hour resolution data. Some data are available on a daily basis via a dial-up system.

W.E. Carter reported various achievements and plans of the VLBI community. IRIS-A, NAVNET, IRIS-P, IRIS-S and DSN networks are running regularly, and especially IRIS-A and NAVNET are producing 3.5-day spacing EOP's. Southern extension of the VLBI networks was emphasized. Important projects, such as VLBA, USSR QUASAR and Japanese VSOP were introduced. It was particularly important that the doubling of the bandwidths (both in s and x bands) of the US Mark-3 system realized mm accuracy repeatability in station coordinates determination.

B.E. Schutz reported that the operation of LAGEOS tracking by SLR has been done regularly. Following the launching of the USSR ETALON satellites (ETALON-1 in January 1989, ETALON-2 in May, 1989), a special campaign to evaluate the ETALON satellites for Earth rotation applications took place during September 1 to December 1, 1990.

Ch. Veillet reported that the operation of LLR has been conducted at Haleakala, CERGA and McDonald. In addition to the regular three stations, Wettzell and Orroral are attempting lunar laser ranging. Recently, Haleakala has met a serious budgetary cut from NASA for ranging to the Moon.

D.D. McCarthy reported that the revision of "IERS Standards (1989)" is being explored and a new version will be finalized by the end of 1991, by incorporating geophysical models hitherto unmodeled, as well as improved versions of the models in the current standards.

In order to advertise the IERS activities in the IAU, an article titled "The International Earth Rotation Service" prepared by F. Arias and K. Yokoyama appeared on the IAU newspaper, "Cruz del Sur", No. 10 issued on August 1.

At the end of the session, J. Kovalevsky made the following acknowledgement for the activities of IERS.

"We just heard the reports on the remarkable results obtained by IERS. This complex, and increasingly complex organization is working amazingly smoothly, and this by itself is an achievement. The precision of results is outstanding. So I believe that I can speak not only for myself but also on behalf of the whole scientific community here present to warmly congratulate all the participants of IERS from the Central Bureau, Coordinating Centers, Analysis Centers and observing stations."

RELATIVITY AS ITS AFFECTS REFERENCE SYSTEMS
(with Commission 31; Chair: P.Pâquet)

See the report of Commission 31, this volume.

VARIATIONS IN THE EARTH'S ROTATION.
(Chair : B. Kolaczek ; Secretariat : J. Popelar)

The session was chaired by Dr. B. Kolaczek who welcomed all and briefly introduced the program of the session and speakers.

V. Dehant reported on "The Effects of the Free Inner Core Nutation (FICN) on the Earth's Nutation". Computations of the FICN have been done analytically and corrections to the IAU 1980 nutation series have been obtained for the annual, the semi-annual and the 13.6 day terms. Comparisons with observationnal results show the best agreement for the semi-annual at 0.1 mas whereas unexplained residuals still remain for the 13.6 day and the annual terms. Further improvement for the annual term is expected from introduction of atmospheric pressure effects to be considered in the future work. Importance of ocean and gravity effects was pointed out by Yokoyama in the discussion and Matsakis mentioned that instabilities in some numerical solutions may account for differences between the analytical and numerical results.

D. Djurovic presented a paper co-authored by P. Pâquet "A Quasi Biennial Oscillation in Earth Rotation Fluctuations and Activity of the Solar Corona". He described the data sources and processing techniques which have been used to detect a significant correlation between the Sun corona indices and the UT1-TAI time series between 1964 and 1990. Solar activity effects on the atmosphere and the Earth's magnetic field have been identified to facilitate such an interaction with the solid Earth. Questions about the length of the series used for the analysis and other possible periods have been addressed in the discussion which also emplasizes the need for greather collaboration with solar physicists.

D. McCarthy presented the paper by J.O. Dickey on "High Time Resolution Measurements of Earth Rotation". The work started in 1988 when Earth rotation measurements by several techniques were analysed with objectives to refine geophysical models and improve predictions of changes in the Earth rotation. A significant coherence between LOD and AAM down to about 8 days has been shown and a mismatch at 13.6 day period has been removed by the introduction of the Brosche ocean tide model. The AAM data show increased noise levels at lower frequencies. Future campaigns are expected to improve accuracy and increase frequency of geodetic measurements to be coordinated by the IERS. In the discussion Kolaczek pointed out the great importance of this work for the IAU Commission 19 and supporting resolution should be strongly endorsed.

K. Yokoyama reviewed "Results of the IRIS-P Burst Earth Rotation Observations in February 1990". Several 24-hour VLBI observing sessions by different networks (IRIS-P, IRIS-A, NAVNET) have been combined to obtain 2-hour EOP solutions over a pariod of one week. Using fixed station coordinates obtained from a global solution, short term variations in EOP have been determined but comparison with other independent techniques is needed for verification. Also, data reduction procedures will have to be studied to eliminate any possible systematic effects on the 2-hour EOP solutions due to nutation, atmospheric and other modelling. The design of future new VLBI systems should facilitate detection of short term EOP variations. The discussion stressed the preliminary character of the present results which need further studies and analyses.

M. Feissel summarized "Comments on the Precision of Polar Motion Determination" based on the analyses of six series of VLBI and SLR observations between 1984 and 1990. As a probable effect of unequal longitude distribution of stations in the VLBI and SLR networks used in the determination of polar motion, the uncertainty ellipses of the 5-day polar coordinates have a preferred direction of the major axis at about 45° E longitude. The direction obtained from the Allan variance analysis of the series of results match those determined from the variance-covariance matrices of each series. As a result, the IERS pole positions at 5-day intervals have an error ellipse with semi-major axis in the 45° E direction with an amplitude of about 0.4 mas. During the interval 1984-1990 the instability of polar motion for periods under 100 days attributable to geophysical causes have a noise ellipse with the semi-major axis in the direction 15-30° and the ratio between the semi-minor and the semi-major axes of 0.5. For higher frequencies the noise ellipse of the geophysical effects matches the error ellipse of the measurements for periods of about 10 days. In the discussion it was agreed that a particular attention must be given to a complete removal of any systematic differences between the series.

B. Kolaczek presented the last paper on "Analyses of Variations of Chandler Nutation and Seasonal Variations of Polar Motion between 1846 and 1988" based on the pole coordinates series compiled by the Kiev Main Astronomical Observatory. The FFT spectrum shows two peaks for the Chandler range and a single peak for the annual variations. The Chandler and annual nutations of pole coordinate variations were computed by the FFT filter as differences between input data and data obtained by inverse FFT with a selected frequency band removed. The choice of the removed frequency band has a considerable effect on the amplitude modulations of the Chandler and annual oscillations. The most pronounced oscillations of amplitude variations for the Chandler and annual oscillations have the period of about 40 years which may suggest a common source of their perturbations. The computed models of the amplitude modulation of the Chandler nutation consistent with the past oscillations predict a deep minimum for the Chandler wobble amplitude between 2010 and 2020.

A general discussion on a proposed resolution supporting greater cooperation with meteorologists and geophysicists in studies of the Earth rotation followed. It was decided that the resolution will be finalized and presented for a vote at the meeting of the commission on July 31, 1991. B. Kolaczek thanked all speakers and participants and adjourned the session.

THE WGRS RECOMMENDATIONS.
(With Commissions 4,7,8,20,24,31,33,40; Chair: J. Hughes)

The scientific aspects of the WGRS recommendations had been covered during the Joint Discussion earlier in the General Assembly. The discussion focused on their implementation. It was agreed that the recommendations are meant to *propose* a high precision consistent framework, and that actual implementations would be adapted to each particular case.

BUSINESS SESSION 2.
(Chair : M. Feissel ; Secretariat : J. Vondrak)

The General Secretary had made a proposal to merge some commissions, among which 19 (Rotation of the Earth) and 31 (Time) ; after receiving responses from the commissions concerned which emphasized the growing independence of their matters of concern, he withdrew his proposal. However, it is understood that the Executive Committee of IAU (EC) still wishes to reduce the number of commissions. Due to lack of information on the EC precise intents, the question could not be examined by the commission members. This situation gave rise to a general discussion on the functionning the IAU ; the main ideas expressed in the discussion are as follows. There is a lack of communication between the EC and the Commission Presidents which precludes mutual understanding ; the quick turnaround of functions (3 years) makes it impossible to pursue large scale actions ; the General Secretary should stay in function for a longer time, e.g. three successive terms ; if a reorganisation of the IAU is to be considered, proposals should be made by the EC to the National Committees for consideration, rather than to the Commission Presidents.

Information was given on the format proposed by the incoming General Secretary for the next General Assembly, to be held about 14-27 August 1994 in Den Haag, The Netherlands. Some of the symposia and colloquia sponsored by IAU in conjunction with the General Assembly would be held during the General Assembly itself, in parallel with commission meetings. The detailed setting will be worked out by the Executive Committee.

The establishment and membership of the two Working Groups recommended by the WGRS was discussed.

The Working Group on the Reference Frame is formed by IAU Commissions 4, 8, 19, 24, 31, 40, and IERS. A draft list of members was set up by the chairman of the WGRS, without consultation of the commissions involved. The incoming President of Commission 19, B. Kolaczek, is asked by the commission members to establish formal relationships with the chairman of the Working Group, C. De Vegt, and to nominate representatives of the commission.

The Permanent Working Group on Astronomical Standards is organized by Commissions 4, 8 , 19 , 24 and 31, in consultation with the IAG and the IERS ; its chairman is T. Fukushima. Commission 19 nominated N. Capitaine, V. Dehant and D. McCarthy as its representatives in this Working Group.

The three resolutions presented during the previous meetings were further discussed and adopted. In the discussion of the first one, it was recognized that the schedule set by the IAU for the Business meetings was inconsistent with the deadline for submitting resolutions to the Resolution Committee ; as a consequence, the draft resolution had to be submitted to the latter before the vote could take place in the commission. The final text of the resolutions are listed at the end of the Commission report.

The commission officers for 1991-1994 listed hereafter were elected.

President : B. Kolaczek
Vice President : J. Vondrak

Organizing Committee

N. Capitaine (1)	N. Mironov (2)
J.O. Dickey (2)	W.G. Melbourne (1)
S. Dickman (1)	L. Morrison (1)
M. Feissel (1)	D. Robertson (1)
Jin W.J. (2)	T. Sasao (2)
D.D. McCarthy (2)	P. Wilson (1)

(1) First term ; (2) Second term

The following list of new commission members and consultants was adopted.

New members		Consultants	
A. Banni	Italy	K.D. Aldridge	Canada
G. Beutler	Switzerland	B.F. Chao	USA
M. Bougeard	France	R.Eanes	USA
J. Boytel	Cuba	R.S. Gross	USA
A. Brzezinski	Poland	T. Herring	USA
V. Dehant	Belgium	H. Jochmann	Germany
D. Gambis	France	Z.S. Li	China
J. Hefty	Czechoslov	C. Ma	China
O. Kameya	Japan	K. Nurutdinov	USSR
Z.A. Li	China	N. Pejovic	Yugoslavia
C.Y. Liu	China	B. Richter	Germany
G. Petit	France	R. Sabadini	Italy
M. Soffel	Germany	H. Schuh	Germany
K. Steinert	Germany	G. Soltau	Germany
J.Y. Xu	China	M. Stavinschi	Romania
Y.Z. Zhu	China	M. Tsesis	USSR
		C. Wilson	USA
		V.E. Zharov	USSR
		Y.Z. Zhu	China
		J.Y. Xu	China

Three members died in the 1988-1991 term : J.G. Davies (UK), H. Jeffreys (UK), and A. Stoyko (France) ; three members resigned : K. Lambeck (Australia), N.P.J.

O'Hora (UK), and A Orte (Spain) ; 48 other individuals who did not confirm their intent to remain members of Commission 19 had their names deleted from the membership list.

RESOLUTIONS

HIGH TIME RESOLUTION MEASUREMENTS OF THE EARTH'S ROTATION

The 21st General Assembly of the International Astronomical Union

recognizing
>the importance of rapid determinations of Earth rotation recommended by the International Workshop "Interdisciplinary role of space geodesy" held in Erice (Italy) in 1988, and

considering
>the proposal made to the International Association of Geodesy by its Special Study Group 5.98 on "Atmospheric excitation of the Earth's rotation" to set up a Working Group on "High time resolution measurements of Earth rotation",

requests
>the Executive Committee of International Astronomical Union to approach the International Association of Geodesy in order to consider the possibility of organizing a joint IUGG/IAU Working Group for such activity.

THE EARTH'S ROTATION IN THE HIPPARCOS REFERENCE FRAME

Commission 19, Rotation of the Earth

Considering
- that the HIPPARCOS program is expected to provide star coordinates, yearly proper motions and parallaxes at the 0.002" level of accuracy,

- the conclusion of the Working Group on Earth Rotation in the HIPPARCOS reference frame stating that the re-reduction of past optical observations is feasible,

- that the new data collection and the test of the new reduction scheme proposed by the Working Group will require time for preparation,

thanks
- the observatories which have accepted to contribute their observations to the new analysis,

- the Working Group on Earth Rotation in the HIPPARCOS reference frame for its achievements under the efficient chairmanship of J. Vondrak,

recommends
- that the Working Group on Earth Rotation in the HIPPARCOS reference frame extends its activity to the actual implementation of the analysis scheme proposed, under the present chairmanship
- that the observatories provide the past observations to the Working Group,
- that the Astronomical Institute of the Czechoslovak Academy of Sciences collects the observations and performs their global analysis,
- that the contributing observatories be given the same priority of access to the results of the analysis as the Working Group members.

APPLICATIONS OF OPTICAL ASTROMETRY TIME AND LATITUDE PROGRAMS

Commission 19, Rotation of the Earth

Considering
- that modern astrometric observations provide a unique set of data sensitive to variations in the deflection of the vertical,
- that optical astrometric data previously used to measure the rotation of the Earth have been shown to measure the variations in the deflection of the vertical,
- that the collected astrometric data contain valuable information on star positions including radio stars,
- that Recommendation 7 of the Working Group on Reference Systems calls for new comparisons between reference frames,

thanks
the Shanghai Observatory for establishing and operating an analysis centre for optical Earth rotation data, and

recommends
that optical astrometric data continue to be collected by the Shanghai Observatory in order
    1) to investigate the possibility of deriving long-term variations in the deflection of the vertical within the reference frame provided by HIPPARCOS, and that the International Association of Geodesy be invited to consider undertaking this project,
    2) to provide data for the connection of celestial reference frames.

COMMISSION 20: POSITIONS AND MOTIONS OF MINOR PLANETS, COMETS AND SATELLITES (POSITIONS ET MOUVEMENTS DES PETITES PLANETES, DES COMETES ET DES SATELLITES)

Report of meetings on 24th, 25th, 26th, 30th and 31st of July, 1991

PRESIDENT: R.M. West   VICE-PRESIDENT: A. Carusi   SECRETARY: H. Rickman

1. ADMINISTRATION AND SCIENTIFIC SESSION ON OCCULTATIONS AND MINOR PLANETS (July 24)

The President welcomed members of the Commission to Buenos Aires and opened the administrative session by reporting on the Commission activities during the past triennium and on some problems to be dealt with in the future. The progress was highlighted by the recent recoveries at the Minor Planet Center of (724) Hapag by S. Nakano and (878) Mildred by G.V. Williams; only (719) Albert is still lost. The work on the transition from the FK4/B1950.0 to the FK5/J2000.0 coordinate system has been completed, and the J2000.0 system will be used in the Ephemerides of Minor Planets (EMP) from the 1992 edition. The recommended procedures are described in detail in the report of the System Transition Committee delivered by its chairman, D.K. Yeomans, in August 1990.

The recent problem of duplication of minor planet names for some of the newly discovered Neptunian satellites was meantioned as the topic of a proposed commission resolution (see below). Of some concern is the question about which planets should be included into orbital calculations involving planetary perturbations. A standard is desirable, but the current one of including Pluto seems inappropriate. Furthermore, recent years have seen an increasing number of discoveries indicating a transitional nature of many objects with respect to the asteroid/comet classification. Thus problems arise regarding the naming of objects, and the question whether any change of current practice is warranted was foreseen as a topic of debate during the joint session with Comm. 15 on Comet-Minor Planet Interrelations. Finally, the President called attention to the lack of coordination between the many meetings of interest to Comm. 20 members that have recently been held in different countries and continents.

All present stood in silence for a short while in remembrance of the Commission members who had died during the past triennium: Willi Strobel, Maciej Bielicki, Leland E. Cunningham, Edgar Everhart III, Jack Bennett, Minoru Honda and Peter Millman.

The current membership of the Commission and its committees and working groups was then reviewed, and suggestions for new members and consultants as well as regarding the composition of the committees and working groups were invited. The nomination by the OC of A. Carusi as President and D.K. Yeomans as Vice-President of Commission 20 for the coming triennium had in the meantime been agreed to by the IAU Executive Committee and was confirmed by acclamation. Likewise, H. Rickman was unanimously appointed as Secretary of Commission 20 (1991 - 1994). Yeomans' place in the OC, was left vacant until further notice. In the WG on Comets, R.M. West stepped back as a member, and it was noted that N.A. Belyaev's membership is no longer appropriate since he does not work in astronomy any more. The future replacement by another member of the ITA staff was considered desirable. L.K. Kristensen wanted to step back as a member of the WG on Occultations. D.K. Yeomans was proposed to replace R.M. West in the Satellite Nomenclature Liaison Committee and the Minor Planet Names Committee, while in the Comet Nomenclature Committee as well as in the

Standing Committee to Oversee the Publication of Photometric Data for Minor Planets, West will be automatically replaced by the new President, A. Carusi. The Chairman of the Satellite Nomenclature Liaison Committee, K. Aksnes, raised the question of the continued usefulness of this committee, but no action was proposed. L.D. Schmadel, having replaced E. Bowell, was confirmed as the Chairman of the Study Group on Minor Planet Names. L.K. Kristensen wished to withdraw from this study group.

A number of resolutions were presented for possible endorsement by the Commission. It was proposed to submit to the General Assembly a joint commission resolution concerned with the assessment of the perceived need for better knowledge and surveillance of the population of Near-Earth Objects (NEO's) and better understanding of the hazard they present to mankind. The attitude to this proposal was generally positive and since the wording was recognized as a matter of importance, it was decided to call a common meeting among all concerned commissions on 25 July (see below). The opinion of Comm. 20 on two further resolutions before the GA was asked, one of them prepared by the IAU WG on Reference Systems and the other recommending the avoidance of interplanetary pollution. Several internal Commission resolutions were also proposed. Two of them concerned the activities of the ITA in S:t Petersburg, dealing with the observational programme for selected minor planets and special software packages to be used with the Ephemerides of Minor Planets. One resolution concerned the establishment of a Data Centre on Satellites at the Bureau des Longitudes in Paris. Two resolutions dealt with nomenclature business. One of them suggested the future amalgamation of the Commission's committees (currently at least four) dealing with the naming of solar system objects. The other recommended an expansion of the Minor Planet Names Committee to deal more actively and efficiently with selection of names and citations and to secure a closer liaison with the IAU WGPSN, thus hopefully avoiding further undue duplication of names.

A plea from the European Space Agency to obtain astrometric observations of comet P/Grigg-Skjellerup in preparation for the July 1992 Giotto encounter was noted. Another matter for discussion at the following administrative session was brought up by the President. In 1969 L. Kohoutek had discovered a minor planet which he had wished to name "Palach" in memory of the Czech student who died early that year in protest against the occupation of Czechoslovakia in 1968. For personal reasons, he had at that time hesitated to make a formal claim for the name Palach, but his wish was known at the Minor Planet Centre. Since the resolution of Comm. 20 not to name minor planets after political persons unless they have been dead for 100 years was passed only in 1985, the Commission was asked on Kohoutek's behalf whether the asteroid (1834) could now receive the name Palach since it had been conceived long before the resolution entered into force.

The President read the report by L. Wasserman on results obtained from occultations by minor planets and satellites during the past triennium. Special attention was given to the 3 July 1989 occultation of 28 Sgr by Titan. In general, few events of good quality had been predictable – among them occultations by (529) Brixia, (9) Metis and (4) Vesta. The preliminary report by D.W. Dunham *et al.* on additional occultations by (216) Kleopatra and (381) Myrrha at the ACM-91 meeting in Flagstaff was also highlighted. In view of L.K. Kristensen's wish to withdraw from the WG on occultations, the need for a replacing person from Northern Europe was stressed and suggestions were solicited.

B.G. Marsden then took the chair for the first scientific session on minor planets. He himself gave the first report about the plans of the Minor Planet Center for the J2000.0 transition. A time-table including actions both taken and foreseen was shown. Due to the mass conversion of observations, orbits and computer programmes to be effectuated during the period Nov. 1991 to

Jan. 1992, possible closure of the MPC for part of that period was foreseen, although genuine emergencies would be handled by the Central Bureau for Astronomical Telegrams. As future standard for planetary perturbations, Mercury-Neptune plus Ceres was proposed. A review of mass determinations for Ceres was given, leading to a recommended value of $5.0 \cdot 10^{-10}$ solar masses. Special attention was also given to the choice of the best system of J2000.0 solar coordinates, as well as the availability of J2000.0 star catalogues.

A.G. Sokolsky then presented ITA proposals on the programme of Selected Minor Planets (SMP) and related issues. About 30,000 observations of SMP made since 1949 at 35 observatories around the world have been collected and checked at ITA. Sokolsky surveyed the purposes of the programme both for checking and comparing astrometric catalogues and for doing various important asteroid research. To improve on the programme, several recommendations were made. Only the brightest 15 objects, of mean opposition magnitude < 13, should be considered, allowing to extend the observational arcs to near quadrature. Modern reference catalogues, and telescopes with focal lengths $\gtrsim 2 - 3$ m and image scales $\gtrsim 70 - 100$ arcsec/mm, should be used to reach high precision. Observing minor planets near the crossing-points of their sky tracks serves the same purpose. The software packages STAMP and CERES developed in the USSR for minor planet work on IBM-PC compatible machines in MS DOS environment were introduced and their benefits described. The EMP are already available on diskettes for easy use with STAMP and CERES, and this is the preferred means of distribution in the future. B.G. Marsden asked the audience how many would still like to have printed ephemerides, and the answer came out as roughly one person out of four.

R.M. West read a report, on behalf of L.D. Schmadel, on the progress of the Study Group on Minor Planet Names. Statistics of the current situation were summarized. As of 27 June 1991, there were 3858 named asteroids. 65% of these can be regarded as secure in the sense of complete knowledge of the meaning of the name, and a further 28% are secure but incomplete. The rest ($\sim 7\%$) are uncertain, or have questionable or unknown meanings. It is hoped that this number can still be halved. A Dictionary of Minor Planet Names is to be issued by Springer Verlag in the spring of 1992. Finally, the counting of votes given for the naming of minor planet nr. 5000 (foreseen in late 1991) was performed by R.M. West and O. Hainaut. Out of ten suggestions, the winning one was *IAU* with 6 votes out of a total of 20.

2. JOINT COMMISSION MEETING ON THE PROPOSED SEARCH FOR NEAR-EARTH OBJECTS (July 25)

At short notice a Joint Commission Meeting (involving members from Commissions 4, 7, 9, 15, 16, 20, 21 and 22) was convened in order to discuss the proposed resolution to the IAU GA on the hazard posed to the Earth's biosphere by Near-Earth Objects (NEO's: minor planets, comets, and fragments thereof). A wide-ranging discussion ensued in which the scope and wording of the resolution was debated. Some of those present wanted the resolution expanded and to include various other possible aims of the search beyond the hazard aspect: especially follow-up and physical observations for purely scientific goals, but also to identify future spacecraft targets, and assess mineral resources deriveable from NEO's. The fact that an international effort is essential was clear, which makes the adoption of the resolution by the IAU of paramount importance.

The views of the Presidents of each of the commissions involved were canvassed. They were all principally in favour. Plans for the membership of the Working Group to be set up should the resolution be carried by the GA (as actually occurred on 1st August) were discussed. The final wording of the draft resolution to be submitted to the IAU Executive Committee was as follows:

The XXIst General Assembly of the International Astronomical Union,
Considering that various studies have shown that the Earth is subject to occasional impacts by minor bodies in the solar system, sometimes with catastrophic results, and
Noting that there is well-founded evidence that only a very small fraction of NEO's (natural Near-Earth Objects: minor planets, comets and fragments thereof) has actually been discovered and have well-determined orbits,
Affirms the importance of expanding and sustaining scientific programmes for the discovery, continued surveillance and in-depth physical and theoretical study of potentially hazardous objects, and
Resolves to establish an *ad hoc* Joint Working Group on NEO's, with the participation of Commissions 4, 7, 9, 15, 16, 20, 21 and 22, to:
1. Assess and quantify the potential threat, in close interaction with other specialists in these fields;
2. Stimulate the pooling of all appropriate resources in support of relevant national and international programmes;
3. Act as an international focal point and contribute to the scientific evaluation; and
4. Report back to the XXIInd General Assembly of the IAU in 1994 for possible further action.

Upon decision that each of the commissions involved select two members of the Working Group, the following were eventually proposed:

A. Carusi (Italy; Convenor), A.T. Basilevsky (USSR), B.Å.S. Gustafson (Sweden/USA), A.W. Harris (USA), Y. Kozai (Japan), G. Lelièvre (France), A.C. Levasseur-Regourd (France), B.G. Marsden (USA), A. Milani (Italy), D. Morrison (USA), K. Seidelman (USA), E.M. Shoemaker (USA), A.G. Sokolsky (USSR), D. Steel (Australia/UK), J. Štohl (Czechoslovakia), Tong Fu (P.R. China).

3. SCIENTIFIC SESSION ON MINOR PLANETS (July 26)

B.G. Marsden chaired the second session on minor planets spanning a wide range of topics with five scientific presentations and one short communication.

D.K. Yeomans started the session by reviewing the situation regarding acquisition of radar astrometric data (line-of-sight velocity and range) for minor planets and comets, and the benefits of their incorporation into orbit determinations. Such data currently exist for four comets, more than 30 NEO's and more than 20 main-belt asteroids. Moreover, after upgrading of the Arecibo and Goldstone antennas, the detection limits will soon be further improved. Radar data are complementary to, and more precise than, optical observations. In combination with the latter, they can dramatically improve the recovery ephemerides of NEO's after their first passage near the Earth. This was exemplified by the case of the Earth-approacher 1989 PB.

E. Helin gave a summary of the international conference held in San Juan Capistrano 30 June – 3 July 1991 on Near-Earth Asteroids. The meeting had an attendance of about 150 people: scientists, engineers and general public. Many of the contributions were briefly reviewed, in particular those dealing with search programmes and follow-up observations for securing reliable orbits, and related papers dealing with statistical and evolutionary aspects of the NEO population.

K. Muinonen then presented a paper, co-authored by E. Bowell, on orbital error analysis as a criterion for numbering asteroids in an objective and automated way. The principle should be to compute a reference orbit omitting bad observations. If necessary, single-night apparitions and single-night observations with a large individual influence on the ephemeris uncertainty should be excluded. The maximum ephemeris uncertainty arising from the exclusion of single observations from the material over, e.g., 10 years should then be less than, say, 10 arcsec in order for the asteroid to be numbered. B.G. Marsden pointed out a problem in dealing with systematic trends in the residuals, to which the proposed method is not sensitive.

The current situation regarding absolute magnitudes of asteroids was reviewed by E.F. Tedesco. Reasons for modifying the H,G system used in the Sep. 1986 listing were given, in particular concerning the non-uniqueness problems of selecting a default value of G among several alternatives. The new procedure is to take $G = 0.15$ as an average over all albedo groups unless G can be derived from least-squares fits to phase data. The current data base contains 113 asteroids with H,G values thus derived – less than before since solutions are no longer attempted when the material spans a wide range of aspects. L.K. Kristensen proposed G derived as a function of albedo when infrared magnitudes are available. For other cases and new objects there is a selection effect in favour of high albedos, so the default value should be larger than the mean for low-numbered planets. In response to this, Tedesco mentioned that the mean value of G is based on the observed distribution for well-observed asteroids – a sample that is highly biased toward close, large and high-albedo objects. While the mean for this sample is 0.17-0.18, corresponding to a high-albedo and a low-albedo half with values near 0.25 and 0.1 respectively, correction for the fact that the total observed sample has about 3/4 low-albedo objects leads to a mean value near 0.15.

The status of the PPM catalogue was summarized by S. Röser. Its northern part (authors: S. Röser and U. Bastian) is available since 1991 in printed form (two volumes; Spektrum Akad. Verlag). The southern part already has a preliminary version published in *Astron. Astrophys.* **187**, 159 (1991) by Bastian *et al.* and will be available in final form in Dec. 1991 aiming at a book to appear in early 1992. Both parts are also available on magnetic tape (from CDS at Strasbourg and NSSDC at NASA-Goddard SFC) and on 3.5-inch diskettes (15 per hemisphere). PPM North has 181731 stars with a mean error of 0.27 arcsec in R.A. and Decl. at epoch 1990 and a mean proper motion error of somewhat more than 0.4 arcsec/century in both coordinates. In PPM South, about 180000 stars are expected with positional mean errors of 0.12 arcsec and proper motion errors of 0.3 arcsec/century.

Some problems of naming minor planets were finally discussed by S. Isobe. Since the number of Japanese discoveries is already rising rapidly and the number of efficient telescopes (> 60 cm aperture) in Japan will soon be doubled, the expected discovery rate is ~ 1000 per decade. Isobe suggested that measures might be taken to facilitate the work spent on naming so many objects, like stopping the naming of minor planets after nr. 5000, or having an IAU naming committee instead of soliciting proposals from the discoverers.

4. ADMINISTRATIVE SESSION (July 26)

The President chaired this session and started by asking the Commission's approval of seven resolutions. Following a brief, clarifying discussion, all of them were unanimously agreed upon. The first resolution, on Reference Frames, was still being discussed within the Working Group on Reference Frames, but the members present expressed their confidence that the final text would be acceptable for Comm. 20 and its representative to this WG, D.K. Yeomans. The second resolution concerned the search for Near-Earth Objects (see above). The following five Commission resolutions were then passed:

Commission 20,
Noting the recent disagreement between it and the Working Group on Planetary System Nomenclature (WGPSN) concerning the proposed names for the recently discovered satellites of Neptune,
Drawing attention to its 1985 resolution to minimize the duplication between the names of minor planets and natural satellites, and
Considering the vast potential for drawing on a number of different cultures for the selection of names,
Recommends that its Minor Planet Names Committee, currently consisting of the President, the Vice-President and the Director of the Minor Planet Center, be expanded to include more effective liaison with the WGPSN, and
Charges the expanded Committee to take a more active role in both choosing names and writing completed, concise citations.

Commission 20,
Considering that it currently has at least four committees that deal with some aspect of the naming of solar system objects,
Suggests that they be at some time amalgamated into a single Nomenclature Committee.

Commission 20,
Welcoming the proposal of the Institute of Theoretical Astronomy to prolong the observational programme for 15 selected minor planets (nos. 1, 2, 3, 4, 6, 7, 11, 18, 25, 39, 40, 148, 389, 532 and 704) for the period 1991-2000,
Encourages all observatories which have astrographs of focal length > 2 m to take part in this programme, and
Recommends that the most precise reference catalogues, PPM (Positions and Proper Motions), Fokat (Fotograficheskij Katalog, Pulkovo) and ACRS (Astrographic Catalog Reference System), are used for the determination of the spherical coordinates of the planets.

Commission 20
Supports the activities of the Institute of Theoretical Astronomy (ITA) on the elaboration of PC software packages for the provision of ephemerides of minor planets, and
Suggests that systems like "STAMP" may be used together with the printed, annual volumes "Ephemerides of Minor Planets".

Commission 20,
Having heard the report of the Chairman of the Working Group on Satellites in which is proposed the creation of a Data Centre at the Bureau des Longitudes (France),
Supports this proposal, and
Recommends that this Centre develops into an International Data Centre, as defined in Internal Resolution of Commission 20, adopted on August 8, 1988, during the XX IAU General Assembly.

The president projected the lists of the Commission's proposed new members and consultants and of the members and chairmen of the various Working Groups and Committees. The names were reviewed and then approved by the Commission members, as listed below:

New members: C. Blanco (Italy), C. Edelman (France), E. Elst (Belgium), G. Hahn (UK/ Sweden), S. Isobe (Japan), R. Jacobson (USA), H. Kosai (Japan), A. Manara (Italy), R. Mc-Naught (Australia), A.K. Pandey (India), V. Protitch-Benishek (Yugoslavia), R. Rajamohan

(India), S. Röser (Germany), K. Russell (Australia), M. Sato (Brazil), A.G. Sokolsky (USSR), D. Steel (Australia), J.B. Tatum (Canada), M. Tsuchida (Brazil).

Consultants for 1991-94: C.M. Bardwell (USA), S.J. Bus (USA), K.I. Churyumov (USSR), R.W. Farquhar (USA), E. Kazimirchak-Polonskaya (USSR), S.M. Milbourn (UK), K. Muinonen (USA), Z.M. Pereyra (Argentina), H. Oishi (Japan), J.G. Sanguin (Argentina), N. Samojlova-Yakhontova (USSR), T. Seki (Japan), C.S. Shoemaker (USA), G. Tancredi (Uruguay), G.V. Williams (USA).

Organizing Committee: K. Aksnes, J.-E. Arlot, L. Kresák, B.G. Marsden, T. Nakamura, H. Rickman, V. Shor, L. Wasserman, R.M. West, J.X. Zhang.

Working Group on Comets: M.E. Bailey, M.P. Candy, A. Carusi, A. Gilmore, L. Kresák, B.G. Marsden, S. Nakano, H. Rickman (ch.), E. Roemer, G. Sitarski, P. Wild, D.K. Yeomans.

Working Group on Satellites: K. Aksnes, J.-E. Arlot (ch.), S. Ferraz-Mello, P.A. Ianna, R.A. Jacobson, J.H. Lieske, B. Morando, J.D. Mulholland, T. Nakamura, D. Pascu, M. Rapaport, P.K. Seidelman, V. Shor, D.B. Taylor, R. Vieira-Martins.

Working Group on Occultations: J.C. Bhattacharyya, C. Blanco, G.L. Blow, D.W. Dunham, M.-F. He, A.R. Klemola, R.L. Millis, M.D. Overbeek, V. Shor, M. Soma, G.E. Taylor, L. Wasserman (ch.).

Satellite Nomenclature Liaison Committee: K. Aksnes (ch. & delegate to WGPSN), J.-E. Arlot, A. Carusi, P.K. Seidelman (vice-ch. & alternate delegate to WGPSN), D.K. Yeomans.

Standing Committee to Oversee Publication of Photometric Data for Minor Planets: E. Bowell, A. Carusi, A.W. Harris, B.G. Marsden.

Minor Planet Names Committee: K. Aksnes, A. Carusi, Y. Kozai, B.G. Marsden, L.D. Schmadel, V.A. Shor, D.K. Yeomans.

Comet Nomenclature Committee: A. Carusi, B.G. Marsden, H. Rickman.

Study Group on the Origins of Minor Planet Names: V.K. Abalakin, E. Bowell, F. Edmondson, H. Haupt, B.G. Marsden, J.D. Mulholland, E. Roemer, L.D. Schmadel (ch.), K. Tomita, I. Van Houten-Groeneveld, J.X. Zhang.

The Commission finally discussed whether minor planet (1834) should be named "Palach", as suggested by B.G. Marsden on L. Kohoutek's behalf. Different opinions were voiced and a secret vote was called for. A majority was in favour of accepting the name (1834) Palach (16 votes against 6) and the naming has since been made official on MPC 18643.

SCIENTIFIC SESSION ON COMETS (July 30)

H. Rickman chaired this session, mainly devoted to recent advances in cometary dynamics but with other important progress being covered as well.

M.E. Bailey reviewed the problem of the capture of short-period (SP) comets. The flux of comets in near-parabolic orbits and their average capture probability was discussed, and it was concluded that there is a shortage in the steady-state number of SP comets thus predicted. Additional sources hence appear required, and a trans-Neptunian belt and an inner core of the Oort cloud were discussed from the point of view of comet formation scenarios as well as of their ability to explain the characteristics of the Jupiter family. The approximate dynamical techniques hitherto used for elucidating the latter aspect were reviewed and criticized. Several additional problems that merit further attention were mentioned, and as a general conclusion the source of SP comets was deemed still unresolved.

The problems of modelling cometary dynamics and the essential properties of cometary chaos were then further highlighted by C. Froeschlé. As a word of caution, he remarked that these are still largely open issues: if he knew how to model cometary chaos, he would have done it! He then went on to review the various mapping techniques, either deterministic (Keplerian map for near-parabolic comets; planetary conjunction "kicks" for near-circular orbits in the giant planet region; synthetic map for interpolation on a phase-space grid) or stochastic (Monte Carlo simulations; Markov chain modelling). Particular attention was paid to the problem of injections from the trans-Neptunian "Kuiper belt" into Neptune-crossing orbits.

A. Carusi, in collaboration with L. Kresák and G.B. Valsecchi, talked about the complementary aspect of accurately representing the motions of real SP comets. He thus focussed on results obtained when preparing the 2nd edition of the atlas of long-term evolutions of short-period comets, for which the integrations are now finished. The new atlas will contain 155 comets, of which 93 have been observed at several apparitions. Statistics on the number of planetary encounters and temporary librations were given, and examples of remarkable behaviour were shown (e.g., the temporary Jovian satellite capture of P/Helin-Roman-Crockett around the year 2075).

The next talk was given by A. Smette and R.M. West, with O. Hainaut as co-author, on observations of distant comets – P/Halley and others. The emphasis was naturally placed on the recent outburst of P/Halley, first observed on 12 Feb. 1991 using the Danish telescope at ESO, La Silla, but apparently dating back to late December 1990. Images were presented showing the time-evolution of this event. Such observations strengthen the case for obtaining images of other distant comets – for accurate astrometry and also to learn whether outbursts are a common feature beyond Saturn's orbit. Recommendations for observing techniques were made, and problems such as the large number of background galaxies at $m > 25$ were discussed. P/Halley will probably be followed all the way to aphelion.

Finally G.B. Valsecchi presented the recent calculations, performed together with A. Carusi, L. Kresák and M. Kresáková, on the orbit of comet P/d'Arrest linking it to La Hire's comet of 1678. This one actually topped the authors' preliminary list of ancient comets that might prove identical to known SP comets discovered at later dates, although they were not aware of Valz' (1851) suggestion. The linkage is remarkable in one particular regard, i.e., that four encounters to within 0.5 AU of Jupiter occurred between 1678 and 1851. Several integrations using different dynamical models were shown, and it was demonstrated that a nongravitational effect similar to the present one must exist over the entire interval in order to fit the observations satisfactorily. In the discussion, D.K. Yeomans drew attention to the fact that the nongravitational effect of P/d'Arrest correlates very well with the comet's strongly asymmetric light curve, indicating a constancy of its outgassing behaviour.

5. SCIENTIFIC SESSION ON SATELLITES (July 30)

This session was chaired by J.-E. Arlot who also gave the first paper on the activities and goals of the Working Group on Satellites. In particular, he gave an account of the plans to set up a Data Centre for astrometric data on natural satellites that had been initiated at the previous IAU GA in Baltimore 1988. Among the tasks of the Data Centre, he listed the classification and documentation of both raw and reduced data, the publication of catalogues with evaluation of the accurary, and the availability of these data by circulars or by log-in through electronic networks. In addition, the search for old observations and conversion of these into a standard reference frame, the collection of bibliographic information on both theoretical dynamics and physical studies, and the provision of ephemerides available through electronic networks, were mentioned as important goals.

D. Pascu then reviewed CCD observations of satellites, as performed ground-based by seven groups around the world. These had been preceded by observations with electronographic cameras, for instance during the 1980 Saturn ring plane crossing leading to several discoveries of new satellites. The obvious strengths of CCD's are their speed and their linear response, facilitating background subtraction. Their small fields of view so far make them best suited for work on faint inner satellites. Problems arising in doing CCD astrometry of satellites were then described.

L. Duriez reviewed theoretical work on Saturnian satellites (except rings and shepherds) and concluded that first, a great deal of work has been done to improve both observations and theories of motion. Thus for the major satellites there is hope of new theories soon representing even the best observations to the level of their accuracy (< 350 km). However, in spite of this progress, large improvements are still needed in order to be ready for the CASSINI mission with its extraordinary requirements (precision $<$ a few km). Duriez nonetheless expressed the feeling that recent works on major satellites are already on the right tracks to succeed in this goal.

Finally, J.-E. Arlot presented a paper on the influence of surface effects on the astrometry of Galilean satellites on behalf of W. Thuillot and P. Descamps. These involve various limb darkening effects and asymmetries of the albedo distributions, and the astrometry is mainly affected by a displacement of the photocenter relative to the true disk center amounting to a maximum of 0.10 arcsec for Ganymede. In the observations of mutual events, which reach accuracies of 0.02 arcsec, surface effects become noticeable and have been known for some time to cause the systematic residuals occurring in such observations. Progress in the photometric modelling of the satellites and in the relevant astrometric observations was reviewed.

SPECIAL SCIENTIFIC SESSION ON NONGRAVITATIONAL EFFECTS

This session was affected by the fire in the San Martin Centre so that all presentations had to be drastically shortened. H. Rickman was the chairman, but D.K. Yeomans agreed to act as substitute chairman for most of the talks given by Rickman on behalf of himself or absent friends.

The first speaker was D.K. Yeomans who gave a review of recent advances in nongravitational force modelling. Different lines of approach were surveyed, such as Sitarski's long-term extrapolation of P/Halley's motion fitting the observed times of perihelion passage to a model variation of the semimajor axis, Rickman's mass estimates for P/Halley and, in general, attempts of modelling a force that is asymmetric w.r.t. perihelion. The standard, symmetric force model was deemed to be fine for ephemeris predictions but inappropriate for physical studies. Asymmetric models should fit the water production curves but remain simple as regards extra parameters to solve for. Examples of asymmetric models were given with emphasis on P/d'Arrest and P/Halley. Last but not least, improved modelling must depend upon improved astrometry, thus highlighting the need to work with improved star catalogues, long-focus telescopes and the best available detectors and measuring equipment. The inclusion of radar data was shown to be important in some cases, exemplified by the detection of slight nongravitational effects in the motion of (1566) Icarus.

H. Rickman then reviewed his own work, mostly in collaboration with M.C. Festou and L.

Kamél, on prediction and physical interpretation of nongravitational effects. After a background to the idea that the easily measurable perihelion delays ΔP of SP comets should correlate with the light curve asymmetries both in sign and modulus, the introduction of an asymmetry parameter E was described and the correlation between E and ΔP was demonstrated. It was shown how this statistical regression had been used to predict P/Brorsen-Metcalf's ΔP just before it was recovered in 1989. B.G. Marsden pointed out that P/Swift-Tuttle may possibly be expected back at perihelion in late 1992 and suggested that light curve data from the 1862 apparition could help constraining the time of perihelion passage.

On behalf of G. Sitarski, H. Rickman presented recent progress by Sitarski and co-workers in solving directly for physical properties of the nucleus in multi-apparition linkages involving nongravitational forces. Results concerning comets P/Honda-Mrkos-Pajdušáková, P/Kearns-Kwee, P/Grigg-Skjellerup and P/Kopff were shown. Usually the parameters solved for in addition to the six osculating elements are A (the modulus of the jet acceleration at heliocentric distance 1 AU), η (the lag angle = deviation of the jet acceleration vector from the antisolar direction), I (the equatorial obliquity) and Φ (the cometocentric solar longitude at perihelion), and the two time derivatives dI/dt and $d\Phi/dt$ modelled as constants. The linkage intervals thus obtained range from 4 to 16 apparitions, and the mean residuals are around 2 arcsec or less for $\gtrsim 100$ observations.

H. Rickman then gave a report on recent developments in nongravitational force modelling at ITA on behalf of Yu.V. Batrakov, Yu.D. Medvedev and Yu.A. Chernetenko. One investigation, performed by Chernetenko, concerns P/Encke, and observations from 1901-1987 are used. The physical model to link all these apparitions is similar to that of Whipple and Sekanina (1979: *Astron. J.* 84, 1894), but light shifts (offsets of the photocenter from the nucleus) are introduced for the different apparitions individually. The other project, led by Medvedev, involves a detailed representation of P/Halley's motion during the 1985-86 apparition. A thermal model of the nuclear surface layers is used in order to compute the jet force and torque as functions of the parameters used. These are the spin parameters and up to 13 coefficients describing the shape of the nucleus in terms of 3rd-order spherical harmonics. Reduced solutions limiting the number of unknowns have yielded shape coefficients consistent with other knowledge based on close-up imaging.

P. Colom was the last speaker, reporting on the work of the Meudon group on radio-astronomical measurements of the outgassing asymmetries and jet forces affecting the cometary nuclei. Analysis of the OH emission lines yields the geocentric radial velocity of the centroid of the OH cloud and thus also of the parent H_2O cloud. The difference w.r.t. the geocentric ephemeris velocity of the nucleus then yields the line-of-sight component of the mean outflow velocity with good accuracy. Results for P/Halley (already published) and a few more recent comets were presented and compared.

The final joint session on Comet-Minor Planet Interrelations is dealt with in the report of Comm. 15. At the end of that session, the Comm. 20 members present thanked the President, R.M. West, for his contributions to the running of the Commission during the past triennium.

COMMISSION 21 : THE LIGHT OF THE NIGHT SKY
LUMIERE DU CIEL NOCTURNE

Report of the meetings 25, 26, 27 and 31 July 1991

President A.C. LEVASSEUR-REGOURD   Vice-President M.S. HANNER

26 July 1991
Business session

The President opened the meeting at 9:00 by welcoming the participants ; twelve Commission members were present.

I. Activities during the past triennium

Commission 21 has sponsored two major scientific meetings :

IAU Symposium 139, The galactic and extragalactic background radiation, Heidelberg, June 1989, S. BOWYER and C. LEINERT eds., Kluwer Academic Publishers, 1990.

IAU Colloquium 126, Origin and evolution of interplanetary dust, Kyoto, August 1990, A.C. LEVASSEUR-REGOURD and H. HASEGAWA eds., Kluwer Academic Publishers, 1991.

The triennial report for the IAU transactions has been prepared ; reprints were distributed to the Commission members. Four scientific sessions have been organized for the IAU General Assembly, with a total of fifteen presentations. Extended abstracts were distributed to the participants and mailed to the Commission members.

A letter poll on the emphasis for Commission 21 was analyzed by S. BOWYER. The results are that the Commission emphasis should by equally divided around the studies of zodiacal light, diffuse galactic light and extragalactic light, with airglow as an additional significant topic. These foreground and background sources of radiation are necessarily observed together ; all of these sources need to be understood in order to separate the contribution from anyone component accurately. In recent years, space measurements from rockets, satellites, and space probes have been emphasized and the wavelength coverage has been extended to include both the ultraviolet and the infrared.

II. Officers and Membership

The following officers, proposed by the Organizing Committee, were elected :

President   Martha S. HANNER (USA)
Vice-President   Michael G. HAUSER (USA)
Organizing Committee   S. BOWYER (USA), R. DUMONT (France), Yu. I. GALPERIN (Russia), S.S. HONG (S. Korea), J. HOUCK (USA), Ph. LAMY (France), Ch. LEINERT (Germany), A.Ch. LEVASSEUR-REGOURD (France), T. MUKAI (Japan).

The roster of Commission 21 members was reviewed. After some discussion, it was agreed that the President will send a letter to all the Commission members, requesting an indication of continued interest with, three months thereafter, a follow-up letter to those who have not replied. Those not responding either to this second letter will be dropped from the Commission membership. A new consultants list was also approved.

III. IAU Resolutions and Working Groups

The following resolutions have been approved.

The resolution (with commissions 4, 7, 9, 15, 16, 20, 22) related to the importance of expanding and sustaining scientific programmes for the discovery, continued surveillance and in-depth physical and theoretical study of potentially hazardous Natural Near-Earth Objects was approved. B.A.S. GUSTAFSON and A.C. LEVASSEUR-REGOURD were appointed to represent Commission 21 in the Joint Working Group on NEO's.

The resolution (with Commission 15, 20, 22 and 51) related to the prevention of pollution of the interplanetary space environment was approved, after some discussion and the proposal of a revised version.

M.S. HANNER was appointed to represent Commission 21 in the light pollution working group (with Commission 50).

A resolution regarding surface brightness units, in agreement with IAU recommendation, was approved (see last paragraph).

25 July 1991

Scientific session 1 on the extragalactic and galactic components of the light of the night sky

M.G. HAUSER, Chair

The two comprehensive talks in this session provided an up-to-date view of current knowledge about diffuse cosmic radiations in the infrared (R. SILVERBERG) and the far and extreme ultraviolet (S. BOWYER) spectral domains. Some similarities in the status of the observations at these wavelengths were apparent : the observed sky brightness is dominated by sources within the Galaxy and, though many mechanisms for generating extragalactic backgrounds have been suggested, unambiguous detection of such backgrounds has yet to occur. New data in both fields promise a much clearer picture by the time of the next General Assembly.

26 July 1991

Scientific session 2 on the atmospheric component of the light of the night sky and adverse impact on observations

A.C. LEVASSEUR-REGOURD, Chair

In order to understand the observed brightness distribution of the atmospheric diffuse light, a solution to the problem of the radiative transfer in an anisotropically scattering spherical atmosphere was proposed by S.M. KWON, S.S. HONG and Y.S. PARK, using a ray tracing technique and different values of the asymmetry factor in the Henyey-Greenstein function.

The importance of irregular space and time density variations, mainly due to atmospheric turbulence in the nightglow, was stressed by H. TEITELBAUM. A review of the observed variations was presented, with emphasis on the green and red oxygen lines and the hydroxyl Meinel bands.

A report on the airglow regular observation programme at Tokyo national astronomical observatory was presented by H. TANABE. The observations have been carried out on a regular basis from 1957 to 1990.

An interpretation of the different O_2 emissions in the airglow was presented by M.J. LOPEZ-GONZALEZ, J.J. LOPEZ-MORENO, and R. RODRIGO. The photochemical scheme for the OH emissions was also presented, and information on the density of atomic hydrogen and ozone was inferred from simultaneous measurements of these emissions.

The session concluded with a presentation by D.L. CRAWFORD on the increasing adverse environmental impacts on astronomy and on some of the steps being taken to solve this growing problem.

26 July 1991

Scientific session 3 on the interplanetary component of the light of the night sky (zodiacal light)

M.S. HANNER, Chair

The IRAS sky background survey was described by J.M. VRTILEK. The zodiacal emission, for $60° < \varepsilon < 120°$ in four bandpasses as a function of time is contained in the zodiacal observation history file, available on magnetic tape. The main features of these observations were described.

The recent sky survey by the DIRBE instrument on board COBE was described by R. SILVERBERG. The absolutely calibrated fluxes, from 1 to 200 μm, and for $64° < \varepsilon < 124°$ will provide a very reliable data base for the zodiacal emission.

A model of the zodiacal emission based on a dynamical model for the interplanetary dust cloud, was presented by B.A.S. GUSTAFSON, R.H. ZERULL, E. CORBACH and K. SCHULTZ, as well as scattering measurements performed at the Ruhr University microwave scattering laboratory, shortly before it was closed down.

The session concluded with a discussion by R.K. SOBERMAN of jetting cosmoids as a possible explanation for the Pioneer 10 experimental data and as a major source of the zodiacal light. These are fluffy aggregates of volatile material which are impulsively disrupted by solar heating.

30 July 1991

Scientific session 4 on the interplanetary component of the light of the night sky, continued

H. TANABE, Chair

This last session started with a presentation by A.C. LEVASSEUR-REGOURD, J.B. RENARD and R. DUMONT of the results obtained on local optical properties of the zodiacal dust, with emphasis on the variations of the average dust properties (albedo, polarization) with location, both with radial distance in the symmetry plane and with elevation perpendicular to it. The results obtained out of the symmetry plane suggest that the cloud is a complex mixture of grains of different origins.

The status of zodiacal light and zodiacal emission studies was reviewed by M.S. HANNER. Some of the still unanswered questions were pointed out, e.g. infrared emission at small elongations or above 50 µm, physical changes that occur with heliocentric distance or with distance to the symmetry plane, near the Sun dust distribution, and sublimation processes.

Studies about exact electromagnetic scattering by aggregates of spherical particles were presented by K. LUMME and K. MUINONEN, with emphasis on the coherent backscattering mechanism and the reversal of polarization at small phase angles, that play an important role in the opposition effect and negative polarization of atmosphereless Solar System bodies and the zodiacal cloud.

The session concluded with a presentation by H. TANABE of some images of the July 1991 solar eclipse and with a presentation by C. CESARSKY of the Infrared Space Observatory (ISO) to be launched in 1993, with special emphasis on the zodiacal emission observational possibilities.

27   July   1991

Workshop   on   standard   tables   and   units

S.   BOWYER,   Chair

The session was devoted to a discussion of units to be used for diffuse radiation. A wide variety of units were discussed at length. A strong endorsement was made by Commission 21 that the IAU recommendation on SI units is followed.

The advantages of alternative units for subsections of the electromagnetic spectrum were acknowledged and it was agreed that if alternate units were employed for specialized papers, the authors should provide a conversion to SI units in these papers, taking into account the particular difficulties of converting S_{10} and other magnitude based units to SI units.

The following resolution of Commission 21 regarding surface brightness units was formulated and approved.

In accord with IAU recommendation, Commission 21 recommends the use of SI units for surface brightness, either $W\ m^{-2}\ sr^{-1}\ Hz^{-1}$ or $W\ m^{-2}\ sr^{-1}\ m^{-1}$. We recommend against the future use of S_{10} units and other magnitude based fundamental units. We strongly recommend that, if other units are used, the conversion to SI units be clearly stated.

[NOTE: Due to a communication problem the following report was not received in time for inclusion in the previous volume of Transactions.]

COMMISSION 22 - METEORS AND INTERPLANETARY DUST
(METEORS ET LA POISSIERE INTERPLANETAIRE)

Report of Meetings, 3 and 9 August 1988

PRESIDENT: P. B. Babadzhanov   SECRETARY: D. Olsson-Steel

3 August 1988

I.   WELCOME AND TRIENNIAL REPORT

The President welcomed 22 members and participants to the meeting and expressed pleasure that P. M. Millman a long-standing member of the Commission, was able to be present.

The President called on members to stand in silence to honour the memory of deceased members of the Commission: J. Hoppe and R. H. Giese.   J Weinberg noted that the death of Professor Giese was also a great loss to Commission 21.

The President presented the report of the Commission for the triennium and thanked members for their contributions.   He noted many achievements, including a second GLOBMET meeting (Kazan, USSR, 1988), and a symposium on Asteroids, Comets and Meteors II in Uppsala, Sweden in 1985.   Much had been done in the investigation of the evolution of meteoroid streams, the physics of meteors, the origin of tektites and the distribution of interplanetary dust.

II.   APPOINTMENT OF NEW MEMBERS AND CONSULTANTS.

Nine new members were confirmed: S. V. Clube, S. Djorgovski, R. Hawkes, V. Kruchinenko, W. Napier, D. Olsson-Steel, I. Shestaka, J. Swestka, Wang De-Chang.

Seven consultants were confirmed: G. V. Andreev, M. Dubin, G. Eichhorn, M. Koseki, J. Mason, Yu. V. Obrubov, E. T. Rusk, W-X Wang.

III.   PROPOSALS FOR INCOMING OFFICERS.

The President proposed the names of C. S. L. Keay as incoming President and J. Stohl as Vice-President which were endorsed unanimously.

The Organising Committee for the next triennium was proposed and endorsed as follows: P. B. Babadzhanov, W. J. Baggaley, O. Belkovich, Z. Ceplecha, E. Grün, I. Hasegawa, J. Jones, C. Koeberl, D. Revelle, I. Williams.

IV.   RESOLUTION

The following Resolution was unanimously approved by the meeting:

That Commission 22 commends the Lund Meteor Data Centre for the quality of the archived meteor data and strongly recommends continued financial support by the IAU.

(Financial support has been included in the current IAU budget.)

In supporting the resolution, J. A. O'Keefe asked whether the meteor orbit data had been searched for evidence of lunar ejecta. The associated problems were discussed and none present knew of any such search.

V.   GENERAL BUSINESS

J. Stohl proposed a Joint Symposium should be held in Czechoslovakia in 1992, the title to be "Meteoroids - Their Dynamics, Physics, Chemistry and Sources". C. S. L. Keay agreed to seek the support of Commission 15, 20 and 21 and find an acceptable date for the Symposium.

C. S. L. Keay, as nominated President for 1988-91, sought the guidance of the members present on matters to be pursued during the triennium. P. M. Millman and I. Williams considered that the existing Commission structure was satisfactory and that there was no case for change as far as Commission 22 was concerned. J. A. Nuth pointed out that there was some difficulty for new IAU members in knowing which Commission to apply to join, as such information was not widely available. J. Stohl relayed a comment in absentia from L. Kresak (President of Commission 15) who had received the suggestion from H. Fechtig that small overlapping Commissions such as 15, 21, and 22 should be amalgamated in view of the small attendances at some General Assemblies. Kresak did not agree with this point of view but considered that there should be closer cooperation between the three Commissions.

C. S. L. Keay also drew to the attention of the Commission the widespread concern at the level of pollution of all kinds in near-Earth space, and urged the Commission to give thought to the prospect of increasing pollution in interplanetary space, since interplanetary dust was a major subject of the Commission. Keay argued that this topic should be addressed while there was a chance to do so before it became serious, unlike the post-hoc concern over Earth-orbit pollution. He suggested that a working group should be set up and, through correspondence, draw up recommendations for a joint IAU/COSPAR Panel on Interplanetary Pollution. G. Eichhorn, P. M. Millman, J. Weinberg, J. O'Keefe, D. Olsson-Steel, R. Soberman and N. Misconi spoke strongly in favour of the proposal. It was agreed that I. Williams should act as convener of a Working Group comprising R. Soberman, J. Weinberg, P. B. Babadzhanov, D. Olsson-Steel and the Presidents and nominees from interested IAU Commissions and COSPAR. D. Kessler, S. Fred Singer, V. N. Lebedinets, H. Fechtig, G. Schwehm and I. Kapisinsky are to be invited to participate, and the Convener is to have power to co-opt. All nations with launch-vehicle capability should be represented.

There was some discussion of the sphere of interest for the Working Group. It was generally felt that the Group should be concerned with space operations in cis-lunar space and beyond, as the polluted state of near-Earth space was the subject of the IAU Colloquium 112 in Washington from August 13-16, 1988, on Light Pollution, Radio Interference and Space Debris - The Increasing Environmental Impacts on Astronomy. C. S. L. Keay suggested that, inter alia, topics of concern should include propellants, explosives and discarded materials. J. A. Nuth added that mass-transfer schemes should also be of concern.

9 August 1988

VI.   SCIENTIFIC PRESENTATIONS

P. B. Babadzhanov, N. Mahmudov and Yu. V. Obrubov:
                    Meteoroidal Complex of the Comet Encke.

     P. B. Babadzhanov discussed the meteoroidal complex produced by comet P/Encke on the basis of differing ejection velocities, following the effects of planetary perturbations over 5000 years, and showed that the various showers in this complex are explicable in this way.

I. P. Williams: Origin and Evolution of the Geminid Stream.

     I. P. Williams considered the formation of the Geminid stream and showed that the mechanism whereby 3200 Phaethon liberates meteoroids may be a radiative-equivalent of the Roche lobe: This would apply only to parent objects like Phaethon which have very small perihelion distances.

B. McIntosh and M. Simek: The Structure of the Geminids.

     B. A. McIntosh reported on observations of the Geminid shower with radars at Ottawa and Ondrejov which show a skewed profile and a notable lack of large meteoroids.

     B. McIntosh (with J. Jones) also discussed numerical integrations of the orbits of Halleyid meteoroids, and showed that the Eta Aquarid shower mainly consists of meteoroids released about 500 years ago, with The Orionids being rather older (particles released about 4000 years ago).

P. B. Babadzhanov, A. Hajduk, Yu. V. Obrubov and A. N. Pushkarev:
                    Orbital evolution of particles
                    released from comet Halley.

     P. B. Babadzhanov additionally reported upon integrations of Halleyid meteoroids over eleven millenia for differing ejection velocities, and found that very little dispersion occurs in the elements: The orbits of the meteoroids and the comet are very stable.

I. Kapisinsky and J. Stohl: On Lifetimes of the Taurid
                    Meteoroids.

     J. Stohl presented some ideas upon the erosion of meteoroids, particularly those in the Taurid complex, and showed that the Poynting-Robertson lifetimes may be reduced by a factor of ten in this way.

M. Dubin and R. K. Soberman: Particles Jetting as Explanation for
                    Anomalous Optical Measurements of
                    Interplanetary   Particulates   from
                    Pioneer 10/11

     R. K. Soberman reported upon a recent re-analysis of the data from the meteoroid optical detectors on Pioneer 10 and 11. Their new interpretation of the 283 events observed indicates the existance of jetting of gas and fine dust from a large population of small (size about 5 cm) low-albedo comet-like objects which they term 'cosmoids'.

Ch. Koeberl: Recent Developments in Impact Cratering and Textite Research.

    Ch. Koeberl reviewed recent developments in the study of Tektites, in particular the Muong-Nong tektites from Laos.   These are part of the Australasian strewn field, and may be from an oceanic impact site in Indochina.   A terrestrial origin for tektites on the basis of their chemistry was emphasized.

J. A. O'Keefe: Textites and Biological Extinctions.

    J. A. O'Keefe argued for a lunar origin of tektites, and discussed the implications for life on Earth if a debris ring was formed in this way.

D. Olsson-Steel: On the Interrelationship of Comets, Asteroids and Meteors

    D. Olsson-Steel reported on a re-analysis of radar meteor orbit data gathered as part of the Adelaide, Harvard, Obninsk and Kharkov surveys. Strong evidence exists for meteoroid streams associated with several Apollo asteroids including 2201 Oljato, 1982 TA, 1984 KB, 1566 Icarus, and 5025 P-L.

D. Olsson-Steel: On Recent Meteor Observations by 2, 6 and 54 MHz radars.

    Olsson-Steel also presented the results of recent radar observations in the HF band which show that VHF meteor radars previously used have only been able to detect a few percent of the terrestrial influx of microgram-milligram particles: This has implications for the supply of such particles from comets and asteroids.

V. Porubcan, J. Stohl: The Taurid Meteor Complex and its Cometary and Asteroidal Associations

    J. Stohl discussed the various meteor showers in the Taurid complex and their relationship to comet P/Encke and various Apollo asteroids.   A search for Taurid fireballs is currently underway, and the orbit/radiant database is also being examined to more fully understand this long-duration shower complex.

K. Ohtsuka: Possible Association of December Monocerotids with Periodic Comet Mellish (1917 I).

    K. Ohtsuka presented recent Japanese photographic observations of the December Monocerotid meteors which has enabled this stream to be definitely linked to comet P/Mellish.

I. Hasegawa: Meteors Associated with Halley's Comet observed by Members of the Nippon Meteor Society.

    I. Hasegawa reviewed the activities of the Nippon Meteor Society, including visual, telescopic, photographic, FM radio and television observations.   Rate plots for the Eta Aquarid shower from 1982-87 show no increase coinciding with the return of the parent comet, as expected.

Y. Taguchi: Four-colour Observations of Persistent Meteor Trains.

    Y. Taguchi reported four-colour photographic observations of a persistent meteor train.   Strong red emission detected was not due to the sodium D lines, but might be due to molecular oxygen or nitrogen.

COMMISSION 22 - METEORS AND INTERPLANETARY DUST
(METEORS ET LA POISSIERE INTERPLANETAIRE)

Report of Meetings, 1991 July 26, 27 and 29

PRESIDENT: C.S.L. Keay   SECRETARY: D.I. Steel

<u>1991 July 26</u>

I. WELCOME

The President welcomed 15 members and several consultants and members of other commissions to the meeting.

Those present were requested to stand in silence to honour the memory of deceased members of the Commission: E.I. Fireman, B.J. Levin and P.M. Millman.

II. INCOMING OFFICE BEARERS.

The President proposed the names of J. Stohl as incoming President and I.P. Williams as Vice-President which were endorsed unanimously.

III. PRESIDENT'S REPORT.

The President presented the scientific report of the Commission for the triennium and thanked members for their contributions. He noted with regret the resignation of E. Anders from the commission. The commission membership had fallen to 98. During the last triennium the commission had been requested to consider amalgamation with other commissions. The response from members had been overwhelmingly against any such consolidation on the grounds that the major activities of Commission 22 were unique to the commission.

Support for the Lund Meteor Data Center was being maintained and it was noted that a small but significant allocation for the GLOBMET program had been proposed for the 1992-1994 triennium.

The President referred to initiatives taken at the Baltimore meeting and expressed his gratitude to J. Stohl for his support and for his efforts in preparing for a conference on Meteoroids and their Parent Bodies to be held near Bratislava in early July 1992; D. Meisel for coordinating and producing a report on Professional-Amateur Cooperation in Meteor Science; and I. Williams for assembling information and drafting a Resolution on Interplanetary Pollution Prevention.

IV. METEOR DATA CENTER REPORT (B.A. Lindblad).

The Meteor Data Center in Lund continues to archive, document and disseminate information on meteoroid orbits. Orbital data has been gathered from a large number of sources such as published catalogues, internal observatory reports, data tapes, etc. At present some 5,300 photographic, two-station orbits and more than 60,000 radio orbits are archived. The photographic meteor records in most cases include Earth encounter data, such as velocities, heights and meteor magnitudes.

V. RESOLUTIONS

1. <u>Professional-Amateur Cooperation</u>

The report of the initial Working Group on Professional-Amateur Cooperation (Convener: D. Meisel) was received. It was resolved to set up a continuing Working Group consisting of 6 paired professionals and amateurs to achieve as broad geographical coverage as possible, the Convenor to be an IAU member and have a casting vote. Of the 6 amateurs, it was decided that 3 (at least) should be drawn from the membership of the International Meteor Organization. Appointment of the membership was deferred to the next Business Session to allow informal discussions to proceed.

2. Interplanetary Pollution Prevention

The report of the Working Group on the Prevention of Interplanetary Pollution (Convener: I. Williams), initiated by Commission 22 at the previous General Assembly, was received and the recommended resolution debated. It was passed unopposed after re-wording to make it more widely acceptable.

3. Search for Near-Earth Objects

A resolution originating from Commission 20 to set up a Working Group to facilitate the search for near-Earth objects was discussed at length and approved. Two members of Commission 22 were nominated to serve on the Working Group following adoption of the resolution by the General Assembly. The nominees are D.I. Steel and J. Stohl.

VI. SCIENTIFIC PRESENTATIONS

This session focussed on the subject of the Taurid Complex of Earth-crossing objects and related science. Apart from four well-known meteor showers this complex also includes P/Comet Encke plus several Apollo-type asteroids. The final paper dealt with the solution of a long-standing meteor fireball mystery.

D.I. Steel, D. Asher and V. Clube: **The Taurid Complex.**

The origin and evolution of the Taurid Complex was discussed in terms of the breakup of a giant comet 10-20,000 years ago. Using meteor orbits from the IAU Meteor Data Center they identified the gross characteristics of the four meteor showers in terms of orbital parameter dependence upon solar longitude, and found that they can match the trend with a model involving the gradual disintegration of a large parent comet, perturbations due mainly to Jupiter then causing differential dispersion of the meteoroids depending upon their initial orbital elements. Quite large ejection velocities (up to a few km/sec) are required by this model, which may argue for catastrophic disruption of the giant comet, possibly in an asteroid belt collision.

J. Stohl and V. Porubcan: **The Taurid Complex.**

A detailed investigation of the full extent of the Taurid Complex has revealed that it is rather more extensive than previously believed. The nighttime Taurid showers appear to continue from October, or earlier, through to at least February. This also argues for the dispersal of a very significant initial source, with the flux and spatial density of particles in previous millenia being much higher than in this epoch.

J. Mason: **Ten-year Analysis of the Taurid Complex**

With a low count rate but prolonged activity, the Taurids have proven to be a rather more difficult pair of showers to study than, say, the Geminids with high activity lasting only for a few days. Data gathered by the Meteor Section of the BAA shows evidence of structure within the complex which has not previously been recognized.

D.I. Steel and V. Clube: **Radar Meteors and the Small Comet Hypothesis.**

L. Frank and his co-workers claim that small comets strike the Earth at a rate of 20 per minute. They showed an apparent correlation with radar forward scatter meteor rate data. Their selection and interpretation of old forward-scatter data is open to severe criticism. The apparent agreement between the "atmospheric hole" rates of Frank et al. and the meteor rates dissappears if the more extensive, well-controlled, back-scatter data collected later from Ottawa are used. This weakens the case of Frank et al., but it was argued that the small comet phenomenon, if real, might in fact be related to the Taurid Complex with the forward-scatter data preferentially picking up the low-density meteoroids from that stream, the atmospheric hole data having been collected at the Taurid time of year (November-December).

R. Sobermann and M. Dubin: **Cosmoid Meteor Signature.**

The anomalous (in terms of standard models) Pioneer 10 and 11 dust data has been reinterpreted in terms of the disruption of largely-volatile meteoroids of mass near 1 kg and near-parabolic or hyperbolic orbits, which are termed "cosmoids". It was suggested that the atmospheric hole phenomenon of Frank et al. and various other enigmatic observations can also be explained in terms of the cosmoid model without contradicting the many counter arguments put up against the Frank model of 100 tonne small comets, such as the lack of a large seismically-detected lunar impact rate.

C.S.L. Keay: **Electrophonic Meteor Fireballs.**

The long-term problem of electrophonic sounds associated with bright fireballs were described. Such sounds have been reported for many years but have usually been discounted as being purely psychological in origin. After analysing many reports of such sounds occurring when a fireball was observed crossing south-eastern Australia in the late 1970's, a model was developed for the propagation and detection of this phenomenon by ELF radio waves generated in the meteor wake. This has recently been verified by two Japanese groups who recorded a burst of ELF radio emission from a bright Perseid fireball which was simultaneously photographed and also heard electrophonically by an observer.

1991 JULY 27

I.   SCIENTIFIC PRESENTATIONS.

The unifying theme of this session was meteor streams and surveys.

C.S.L. Keay and L. Rogers: **Newcastle Meteor Radar - First and Last Results.**

The Newcastle radar was one of the first to have microprocessor data analysis on-line. Lack of manpower and funding support frustrated its completion and has now forced it to close. Selected data including activity profiles of some southern meteor showers was presented.

D.I. Steel, W.J. Baggaley and A.D. Taylor: **New Zealand Meteor Orbit Radar.**

First results from a new meteor orbit radar were presented. Detecting echoes from particles down to 100 microns in size, 133,000 individual orbits have been measured so far, about twice the total number available from all other surveys to date. Certain specific meteor showers associated with known comets have been targeted, and observations made at the times at which showers from Apollo asteroids might occur.

V. Porubcan and J. Stohl: **Burst of Lyrids in 1982.**

Radar observations revealed increased Lyrid activity in April 1982, being high above the normal annual counts. This could be interpreted in terms of the Earth passing through a high spatial density grouping within the stream. Since the comet known to be associated with the Lyrid stream (Comet Thatcher) is well away from that position it is suggested that there may be at least one other large cometary fragment in the stream, and that this has released meteoroids during the last few millenia.

J. Mason: **Geminid Results for 1990 and 1980 Compared.**

Observations of the Geminid shower in 1980 and 1990 showed how its activity profile has changed due to secular precession. This shower has only been known for a century and a half, and will disappear at some time within another century or two.

I.P. Williams, Z. Wu, P.B. Babadzhanov and D.W.Hughes: **Leonids and Bielids.**

The past orbit of P/Comet Biela showed that in 1833 it passed the Leonid meteoroid stream at a distance indistinguishable from zero.   The break-up of that comet (the parent of the Bielid/Andromedid shower) in the 1840's may have been caused by high-velocity impacts by the Leonid meteoroids upon the cometary nucleus.

I. Hasegawa: **Meteors Associated with Comet Levy (1991 Q).**

Calculations of the theoretical meteor radiants to be expected from meteoroids from P/Comet Levy (1991q), as derived by K.Ohtsuka, were presented. If this comet has released a substantial number of meteoroids, then enhanced activity from a deep southern radiant (RA = 321 deg, Dec = -62 deg) would be expected around the end of August.

1991 July 29

I.   BUSINESS SESSION.

The Organising Committee for the next triennium was proposed and endorsed as follows: P.B. Babadzhanov, Z. Ceplecha, I. Hasegawa, J. Jones, C.S.L. Keay, C. Koeberl, R. Soberman, D. Steel, E. Tedesco.

Nine new members were elected, including five new members of the IAU: M. Dubin (USA), B. Gustafson (Sweden/USA), E. Helin (USA), J. Mason (UK), C. Murray (UK), C. Pillinger (UK), Xu Pinxin (China P.R.), V. Quesada (Italy), and M. Sykes (USA).

Nine consultants were appointed: G.V. Andreev (USSR), W. Jones (UK), D.J. Kessler (USA), K. Nagasawa (Japan), H. Nakamura (Japan), Yu. V. Obrubov (USSR), A.A. Voloshchuk (USSR) and H.A. Zook (USA).

The required twelve members of the Working Group on Professional-Amateur Cooperation were appointed:

Professionals (IAU members)	Amateurs
V. Porubcan (Convener)(Czecho-Slovakia)	D. Ocenas (Czecho-Slovakia)
I. Hasegawa (Japan)	K. Ohtsuka (Japan)
R. Hawkes (Canada)	P. Brown (Canada)
J. Mason (United Kingdom)	G. Spalding (United Kingdom)
D.I. Steel (Australia)	J. Wood (Australia)
V. Terentjeva (Russia)	J. Rendtel (Germany).

The 6 amateurs were formally approved as additional Consultants to the Commission.

In response to an official request, the membership of Commission 22 was reviewed, revealing that of its current membership 7 are members of more than the limit of three Commissions. It was considered that this is due largely to the difficulties experienced in resigning from other Commissions. All 7 are active in the field covered by Commission 22 and should remain among its members.

24. PHOTOGRAPHIC ASTROMETRY (ASTROMETRIE PHOTOGRAPHIQUE)

PRESIDENT: William F. van Altena.
VICE PRESIDENT: Christian de Vegt.
ORGANIZING COMMITTEE: A. N. Argue, P. Brosche, R. B. Hanson, P. A. Ianna,
                                    D. D. Polojentsev, C. Turon, A. R. Upgren.

1.0 Introduction

Commission 24 had seven sessions during the XXIst General Assembly of the IAU in Buenos Aires, one to discuss the business of the Commission and six to report and plan the scientific activities of the Commission members. In addition, numerous members were involved in meetings of the Working Group on Reference Frames (WGRS) and the Joint Discussions on Hipparcos and the Hubble Space Telescope. The highlights of the meeting for many of our members were the reports of the continuing successful observations with Hipparcos. Preliminary results were shown that demonstrated that the accuracy of the positions, parallaxes and photometry were in accord with the predictions for the revised mission. Comparisons were made with ground based observations of the same objects which showed no systematic errors other than those that could be attributed to the ground based positions, parallaxes and photometry. In addition, reports were given on the first astrometric results obtained with the HST, which indicate that high accuracy binary star observations can be made and that the positional accuracy of the Fine Guidance Sensors may be able to reach their design goal, in spite of the optical problems. Further information on both Hipparcos and the HST is contained later in this report and in the Transactions of the two Joint Discussions. Similarly, a full report of the Joint Discussion on Reference Systems can be found in those Transactions.

2.0 Business Meeting

The Business Meeting was held during the second afternoon session of July 24; 23 members were present. The first order of business, following the welcoming remarks by the President was the certification of the election results for the officers for 1991 through 1994. President: Chr. de Vegt (Germany); Vice President: C. Turon (France); Scientific Organizing Committee: A. N. Argue (UK), P. D. Hemenway (USA), H. Jahreiss (Germany), D. D. Polojentsev (USSR), J. Stock (Venezuela), W. F. van Altena (USA), J.-J. Wang (PRC). The rules of our Commission state that the Vice President normally assumes the Presidency in the following triennium, and that the Scientific Organizing Committee consists of six members and the outgoing president. The above results of the election for the period 1991 - 1994 were accepted by the members of the Commission present.

The Members observed a minute of silence in respect for two of our members who died during the past triennium: Past President S. Vasilevskis (USA) and member X.-H. Zhou (PRC).

Thirteen new members of the Commission were approved: R. Bouigue (France); W. Dick (Germany); M. Hemenway (USA); I. Kumkova (USSR); D. J. MacConnell (USA); M. Miyamoto (Japan); L. Morrison (UK); W. Osborn (USA); G. Pizzichini (Italy); I. Platais (USSR); P. J. Shelus (USA); T.-g. Yang (PRC); N. Zacharias (Germany). T. Girard (USA) was approved as a Consultant to the Commission and one resignation was received from L. Stange (Germany) who retired recently. This brings our membership in the Commission to 140.

A report from the Working Group on Parallax Standards was given by A. R. Upgren, who reported that the standards are being observed at the parallax observatories that are still active. It was agreed that an inventory of standard fields and plates taken at each observatory should be made and an effort made to begin measurement of the plates. The Working Group will report on the results of the survey and reductions at the next General Assembly.

A brief report from the Working Group on Star Lists was given by T. E. Corbin, who indicated that the bright portion of the list from 9th to 11th mag. had been prepared at the Royal Greenwich

Observatory by L. Morrison and the fainter part from 11th to 13th mag. had been prepared at the US Naval Observatory. While the bright part was relatively easy to prepare based on the INCA database, considerable difficulties had been encountered in preparing the fainter portion and some gaps in the spatial distribution still exist. More details of this report can be found in the Commission 8 report.

Many problems were encountered by our members in dealing with and obtaining service from the IAU Travel Agent, Marsans. A resolution was therefore passed directing the President to ask the Executive Committee to instruct Marsans to drastically improve its service to the IAU members during the meetings. The President was informed by the Executive Committee that due to the large number of complaints that it had received, such action was being taken.

3.0 Triennial Report

W. J. Luyten writes that he is continuing his search for very faint large proper motions stars and low-luminosity objects in the far southern hemisphere on plates taken with the ESO telescopes. His progress has been slow however since he is doing the work with a blink microscope.

4.0 Poster Papers

A new feature of this General Assembly was the institution of "Poster Papers". Three poster papers dealing with the scientific activities of our Commission were presented.

C. E. López displayed some results of his ongoing work on the "Astrometry of Southern Variable and Suspected Variable Stars and Suggestions for a New Format of the Left-Hand Page of the General Catalogue of Variable Stars". As a part of the Yale-San Juan Southern Proper Motion program he has determined improved positions of over 2000 confirmed variable and 500 suspected variable stars from measurements of the first-epoch plates of the SPM . The new positions are referred to the system of the SRS and have an average single coordinate standard error in their positions of 0.7". The region south of declination -67° has been completed while about 1300 objects located in Ara, CrA, and northern Pav are being processed; work is now continuing in the region of the South Galactic Pole. He also proposed that the GCVS include a position quality code, and that additional information on improved positions and finding charts in the literature be included.

G. Pizzichini and M. R. Cristallo reported on a search for x-ray emission that they have initiated among several hundred Lowell Observatory proper motion stars that lie within the fields observed with the Einstein IPC detector. They hope to add to the number of nearby stars observed for x-ray emission given that a selection of stars based on proper motion will be dominated by nearby stars. In addition they plan to study the number of high velocity stars in the sample that are found to have x-ray emission.

S. Roeser presented a paper on the PPM Catalogue which detailed the characteristics of the 380,000 stars included in the Catalogue. More details about this paper and a companion oral paper on the same subject can be found in the report of Commission 8 for 31 July entitled "Optical Reference Frames and Catalogues - Fainter Extension of the Optical Frame, I"

5.0 Scientific Sessions

At the urging of the General Secretary, Dr. Derek McNally, the commissions were encouraged to hold joint meetings between those commissions with scientific issues of common interest. It was in that spirit that the Presidents of Commissions 8 and 24 decided to hold four of their sessions jointly. In addition, one of the four included members of Commission 26 while another included members of Commission 40. The organization of so many joint commission meetings was complicated and very time consuming and I would like to express my gratitude to the President of Commission 8, Dr. M. Miyamoto for the enormous effort that he put in to making the meetings a success. Thanks are also due to my colleague Dr. I. Platais, who took over my responsibilities during my absence from New Haven prior to the General Assembly. In order to avoid duplication, the two joint sessions 27 July "Astrometry in the Southern Hemisphere" and 31 July "Optical Reference Frames and Catalogues - Fainter Extension of the Optical Frame, II" are described here while the other two sessions of 29 July "Linking the Optical and Radio Reference Frames" and 31 July "Optical Reference Frames and Catalogues - Fainter Extension of the Optical Frame, I" are described in the Transactions of Commission 8.

5.1 24 July "Progress Reports on Observing Programs and Catalogues" chaired by Chr. de Vegt.

H. Jahreiss and W. Gliese reported on the status of "The Third Catalogue of Nearby Stars", which is nearing completion at the Astronomisches Rechen Institut. A preliminary version of the CNS3 is now available on a new CD-ROM that has just been released by the NASA Astronomical Data Center. It contains all known stars within 25 parsecs of the Sun, as determined by the 1989 preliminary version of the General Catalogue of Trigonometric Parallaxes now being completed at Yale and a combination of photometric and spectroscopic parallaxes. The CNS3 contains 3804 objects, among which are 552 systems with 1091 components. In addition there are 195 spectroscopic binaries, 45 suspected SB's and 43 objects with variable radial velocity. Every star with a trigonometric parallax greater than 0.039" has been included, even though other evidence indicates that 505 stars have larger distances. Only photometric parallaxes exist for 1165 stars and for 734 stars the trigonometric parallax was replaced by the photometric parallax. Contrary to the practice of the CNS2 (Gliese 1969), the different types of parallaxes were not combined. Almost half of the stars have parallaxes with a precision better than 15%, while the errors in the absolute magnitudes for most of the remaining stars are on the order of 0.4 mag. A comparison of the CNS3 with the CNS2 reveals no significant systematic differences in the parallaxes.

W. van Altena reported on the status of the new edition of "The General Catalogue of Trigonometric Parallaxes" (YPC) which is being prepared at the Yale University Observatory in collaboration with J. T. Lee and E. D. Hoffleit. As with the CNS3, a preliminary version of the YPC has been released on the new CD-ROM issued by the Astronomical Data Center. A new weighting system has been used for the YPC which is based on the root-square-sum of the scaled published error and a constant error for the observatory, or series in question. Of the 15,346 parallaxes for 7,879 stars included in this version, 688 are negative and approximately 1500 are usable for the calibration of stellar masses, distances and absolute magnitudes.

F. Crifo described the Hipparcos Input Catalogue in a paper co-authored by C. Turon. The Input Catalogue was prepared as the observing programme for ESA's highly successful Hipparcos astrometric satellite. It includes 118,000 stars drawn from the lists of many proposers according to their scientific priority and a "survey" (basic list of bright stars), and the operational requirements of the satellite. The INCA consortium has either selected the best data from the published literature, or in the many cases where the data were of low accuracy, set up new ground based observational programs to determine the needed stellar parameters. The printed version of the Input Catalogue is now in press. It includes a brief summary of the selection criteria, and a detailed description of the the stellar content as a function of magnitude, spectral type, duplicity, variability and location on the sky. Identification charts are also given for some 10,000 faint stars, in addition to charts for stars in open galactic clusters and in the Magellanic Clouds. Details on each component of double/multiple systems are also available. A great deal of additional information on the status of the Hipparcos satellite can be found in this volume in the Joint Discussion on Hipparcos.

W. van Altena described a new HST Fixed Head Star Tracker (FHST) catalogue that T.-g. Yang is preparing at the Yale University Observatory. The new FHST catalogue will be used in the operations of the HST to preview future updates to the pointing of the HST. False updates caused by missing stars in the current catalogue have led to spurious corrections to the pointing of the HST and a subsequent loss of observing time. The INCA consortium supplied a base catalogue of 92,560 stars extracted from the INCA databases and the TIC with $V \leq 8.5$, or $B \leq 9.2$, when the visual magnitude was not available. Yang has matched those stars with numerous other catalogues and created an appendix file including stars within the magnitude limit but not in the base catalogue. A total of 28,700 stars has been added to the new FHST Catalogue to $V \leq 8.5$, or $B \leq 9.2$. Yang is also preparing an additional appendix that includes data on the duplicity and variability of each star in the catalogue.

In the absence of A. R. Klemola, R. B. Hanson and B. F. Jones, van Altena gave a report on the status of the Lick Northern Proper Motion Program. The NPM program consists of the determination of the absolute proper motions with an rms accuracy of 0.005"/yr (5 mas/yr) for approximately 300,000 stars with respect to 70,000 faint galaxies north of declination -23°. In addition, accurate positions and B, V photometry are being determined for those stars. Of the 300,000 stars, 116,000 have been selected by Klemola from over 600 references to form an Input Catalogue of Special Stars. The Yale-San Juan extension of this program to the southern hemisphere is described in the Transactions of the Commission

24 session on the Southern Hemisphere. To date the plate surveys and measurements have been completed for 750 fields in all declinations zones outside the Milky Way from -5° to +90°. The -10° zone of 50 fields is nearly complete and approximately 100 fields remain in the -15° and -20° zones. The astrometric reductions are complete for the northern sky and are in progress for the southern part of the NPM program. Tests show that the precision of a single faint NPM proper motion is 5 mas/yr while the accuracy is 2 mas/yr for the absolute zero point of an individual NPM field; the overall systematic zero-point error is 0.6 mas/yr. The photometric reductions are complete for the sky north of the equator and south of +65°, while they are in progress for the balance of the northern sky. Investigations related to the structure of the Galaxy utilizing the new absolute proper motions are continuing, and a comparison with other catalogues is planned. The first part of the Lick NPM Catalogue is planned to be available on magnetic tape through the data centers in mid-1992.

An abstract entitled "Progress Reports on the Kiev Observing Programs and Catalogues" was submitted by Ya. S. Yatskiv, in collaboration with N. V. Kharchenko, V. S. Kislyuk, L. K. Pakulyak, and S. P. Rybka. Yatskiv planned to describe the current status of the Photographic Four-Fold Coverage of the Northern Hemisphere (FON) as well as their study of the Main Meridional Section of the Galaxy (MEGA). The observational part of the FON project, which is carried out with six wide-angle astrographs, is nearly 90% complete. The proper motions are determined using the Astrographic Catalogue for first epoch positions. As a part of the MEGA project, a general catalogue of astrometric data of 26,500 stars has been compiled and corrections to the precession constants and stellar secular parallaxes have been determined. In addition, as a part of the KSZ project, a general catalogue of absolute proper motions of 21,817 stars in 75 regions of the sky with galaxies was compiled.

N. G. Rizvanov and I. E. Tselishchev also submitted an abstract on the "Rereduction of Carte du Ciel Plates in the PPM Reference Frame in the Praesepe Region". They have rereduced 35 Paris zone plates and 10 Bordeaux zone plates of the CdC into the PPM system as a part of the establishment of a wide-angle astrometric standard region in Praesepe. The resulting catalogue, which they have compared with Fresneau's 1983 reductions, has a positional accuracy of 0.17".

A. R. Upgren presented the results of an investigation of the "Parallaxes of 30 Subluminous Stars and the Calibration of the Old Main Sequence", which was prepared in collaboration with H. Zhao of the Van Vleck Observatory and J. T. Lee of Yale. Using 15 newly determined parallaxes based on plates taken with the Van Vleck refractor and 15 already published, they have improved the data on the position of the old main sequence for the very metal poor stars.

5.2 24 July "Developments in Space- and Ground-Based Instrumentation" chaired by J A. Hughes.

K. Johnston reviewed the status of the many existing and planned "Interferometers in Space and on the Ground". At the present time the Mount Wilson Interferometer is the only long baseline instrument providing astrometric observations. Precise positions for a small number of bright stars have been determined with a precision on the order of 10 mas and the orbits of a few binary stars have also been determined. Plans are underway for extended ground based arrays by the U. S. Naval Observatory for basic astrometric measurements over both wide and narrow angles, and the CHARA consortium led by Georgia State Univ. to measure binary star orbits. Many instruments are proposed for measuring the diameters of stars and imaging applications. The Mount Wilson interferometer has now measured the diameters of about fifty bright stars to one percent accuracy. The Sydney Univ. Stellar Interferometer is beginning observations of stellar diameters. The Univ. of Wyoming's IR Multiaperture Array has achieved first light. Other interferometer arrays such as CERGA's I2T and GI2T, SAO's Imaging Optical Telescope Array, and the adjunct to the VLT have achieved preliminary results or are in the planning stage. In space the concepts are as follows. Two low accuracy (1-10 mas) satellites, Lomonosov and Regatta-Astro (Space Research Institute, Moscow), have been proposed for measuring the positions of a large number of stars. Two highly accurate astrometric missions have also been proposed: POINTS (Smithsonian Astrophysical Observatory, USA) and OSI (Jet Propulsion Laboratory, USA). It is estimated that both instruments will be capable of astrometric measurements at the ten microsecond level while OSI will also be capable of imaging celestial objects.

V. K. Abalakin provided a short paper which described the Soviet AIST project (described in more detail at IAU Symposium 141), which proposes to provide a second epoch Hipparcos-like spacecraft. The main purpose of the project is to construct a fundamental system of coordinates on the celestial sphere with a mean accuracy of 1 mas. The goal is to obtain a density of about 10 stars per square degree, or about 400,000 stars in total, among which are included the 120,000 stars of the Hipparcos list. In addition, it is planned to fly the satellite with an epoch difference of 10 years from Hipparcos, which would yield proper motions with an accuracy of 1 mas/yr. Finally, a Tycho-like star mapper would be included to construct a catalogue of at least one million stars with an accuracy of 20 - 30 mas.

H. McAlister discussed "The Status of the CHARA Array", which is a multi-telescope array for very high resolution astronomy at optical wavelengths. The project is a joint effort involving astronomers from GSU/CHARA and engineers from the Georgia Tech Research Institute (GTRI) of the Georgia Institute of Technology. The Array would consist of seven 1-meter aperture telescopes configured in a non-redundant Y-shaped array contained within a circle of 400 meters diameter. This Array would provide a limiting resolution of 0.2 mas and a nominal limiting magnitude of V = 9 to 11 depending largely upon the seeing conditions. The initial scientific goals would be the determination of fundamental astrophysical parameters of stars including masses, radii, luminosities, distances, etc. through the resolution of spectroscopic binaries and single star photospheres. The Array will also be configured so as to carry out imaging experiments for classes of objects with complex extended structures, such objects being particularly prevalent in the infrared. The feasibility of incorporating or retrofitting adaptive optics will be explored in order to make each collecting aperture fully coherent. They anticipate some funding beginning during the late fall of 1991 that will permit the further exploration of certain technical aspects such as delay lines, adaptive optics, imaging, and detectors as well as the final selection of a site for the Array. The full embarkation upon the project could then begin in late 1992 or early 1993.

Chr. de Vegt described his proposed "Plans for New Large Astrometric Telescopes" that consist of a modified Schmidt-Cassegrain layout. The design attaches the secondary directly to the achromatic corrector plate to avoid any noncircular diffraction effects. The design study is based on a 1.5 meter, F/5 layout and provides a 2.8 degree diameter, perfectly flat field, with a minimal regular third order distortion. Using standard photographic emulsions, he indicates that astrometrically measurable images of stars to magnitude 18.5 should be obtained with an exposure time of 10 minutes.

H. D. Ables presented a review of "New Developments in CCD Astrometry" at the U. S. Naval Observatory Flagstaff Station, in collaboration with C. C. Dahn and D. G. Monet. Results from the first list of 72 stars observed as a part of the USNOFS CCD parallax program were shown which demonstrated that the accuracy of the relative parallaxes ranged from 0.5 to 2.7 mas (s.e.), with a median accuracy of 1.0 mas for stars in the magnitude range $15 < V < 20$. In addition, a new array of six CCD's with a large format (5 cm x 5 cm) was shown that is currently being tested for use in the parallax program. One of the CCD's will be read out at a relatively rapid rate to avoid saturation while observing a bright parallax star, while the other five will integrate for a time sufficient to obtain an adequate signal-to-noise on the fainter reference stars. The 8-inch transit telescope is currently outfitted with a 1024 x 1024 CCD. This system is used to obtain positions with an accuracy of 10 mas by strip scanning over lengths of many degrees. Future work is anticipated at higher accuracies with a larger aperture transit telescope using custom designed CCD arrays. Finally, a new generation precision plate measuring machine is now undergoing acceptance tests at the USNOFS which uses two cosmetically perfect 1320 x 1035 CCD cameras. It is planned to begin measuring the first and second epoch Palomar Sky Survey plates with an expected astrometric accuracy of 0.1".

5.3 27 July "Astrometry in the Southern Hemisphere" chaired by J. A. López.

To celebrate this first General Assembly to be held in Latin America, C. E. López presented a review of "Argentinian Contributions to Astrometry" in a session that was held jointly with Commissions 8 and 26. He outlined the principal contributions to optical astrometry from the beginning, with the dedication of the Cordoba Observatory on October 24, 1871. The first catalogue, Uranometria Argentina, was observed from September 1870 to October 1871 and presented on the dedication day! Altogether, during the last 120 years, Argentinian astronomers have contributed to the improvement of the astrometric data for more than 1.3 million stars using meridian circles, astrolabes and astrographs.

Chr. de Vegt described the Second Cape Photographic Catalogue, CPC2, which contains the positions and visual magnitudes of 276,131 stars in the approximate magnitude range 6.5 < V < 10.5. 5820 plates, with two 3-minute exposures, were taken in a four-fold overlap of the southern hemisphere during the period 1962 - 1972, using a newly designed 4-element lens (F/10, f = 2000 mm, 4.1 x 4.1° field, scale = 100"/mm). The plates were measured on the Royal Greenwich Observatory GALAXY machine and have been reduced using a classical plate constant method into the SRS system both in B1950 and J2000. A small magnitude equation in declination amounting to 75 mas from 7 < V < 10 has been found, but there are no apparent color effects. Quality classes in the catalogue range from 195,000 stars with positional errors of 54 mas (s.e.), 53,300 with 80 mas, and some 27,000 for identification purposes only with 200 mas errors. Preliminary proper motions have been derived for the 120,000 SAO stars in the catalogue. Final proper motions will be obtained when the recently finished FOKAT-S catalogue is available.

C. E. López and I. Platais described progress on the Yale-San Juan Southern Proper Motion (SPM) program in a report done in collaboration with T. Girard and W. van Altena. The SPM is the extension of the Lick NPM into the southern hemisphere which was described earlier. To date, of the 958 fields in the program, more than 150 second epoch fields have been repeated. The procedures for the plate measurement and reductions have been developed and catalogues of objects to be measured in an extended region around the south galactic pole are being collected. Extensive use is currently being made of the new ROE catalogue of stars and galaxies measured on the SRC Schmidt survey at Edinburgh. These catalogues are used to select stars and galaxies for measurement, in combination with PDS scans of the SPM plates which are displayed on a workstation. A continuing problem is the location of stars with poorly determined coordinates, such as variables, proper motion stars, PHL blue objects, etc. The local amateur Astronomical Society of New Haven is assisting in that part of the project. A new image centering algorithm using cubic splines has been developed and tested. It has improved convergence properties for faint stars and galaxies, but about 10 - 15% lower precision for the derived positions than the two-dimensional gaussian fitting method. It has been found that when scanning with certain combinations of speed and step size, the PDS microdensitometer oscillates in the direction perpendicular to the scan direction with an amplitude of several microns. This resonance can potentially lead to much lower accuracy in the derived positions when they are derived from the marginal distributions.

M. Catalan discussed the "Possibilities of Relocating the San Fernando Automatic Meridian Circle in the Southern Hemisphere". The plans are to modernize and automate the Meridian Circle so that it is similar in capabilities to the Carlsberg Automatic Meridian Circle at La Palma. If the automated Meridian Circle were then moved to the Southern Hemisphere, it would, in collaboration with the CAMC, provide full sky coverage with high accuracy instruments. The automation of the San Fernando instrument is planned to be completed in 1991 and testing will commence in 1992. Informal discussions are currently in progress with individuals at the potential sites.

J. Russell described progress that is being made in "Radio Astrometry in the Southern Hemisphere". The program for the establishment of the radio/optical reference frame link has made its greatest advances in the southern hemisphere in an international cooperative effort of astronomers from many countries, most notably Australia, South Africa, USA, Japan and Germany. They have continued to observe new sources to fill in the gaps in the distribution of sources in the southern hemisphere and to reobserve the previously catalogued sources to improve the radio positions. In the zone between -25° and -40°, where there were very few sources before, there are now positions for an additional 34 with an error of roughly 1 mas. Sources south of -40° have been reobserved in both X- and S-bands to allow for ionospheric corrections. The new observations improved the accuracy to a few mas and included 33 of the 41 observed previously; other sources will be reobserved soon. Recently, an additional 51 new sources have been observed with an accuracy of 2-3 mas between the Australian antennas at Hobart and Tidbinbilla. This brings the total number of sources to 192, close to the goal for the radio/optical link of 200 sources per hemisphere. The sources are well distributed across the sky, with the exception of the galactic plane, and especially the galactic center. The task in the southern hemisphere is made more difficult than in the north by the lack of antennas and the placement of the southern landmasses -- there are no intermediate length baselines between 1000 and 9000 km.

C. Anguita and M.-T. Ruiz described the status of "Parallax Observations in the Southern Hemisphere", which consists of their work at CTIO and that of P. A. Ianna at Mt. Stromlo and Siding Springs Observatories. Both of the southern parallax programs are using the methods developed by Monet and Dahn for the determination of trigonometric parallaxes with CCD detectors. Ianna at the Siding Springs Observatory has initiated a CCD parallax program with the one-meter reflector that concentrates on astrophysically interesting stars and faint high proper motion stars. In addition, he is also continuing with the more traditional photgraphic parallax program with the old Yale refractor at Mt. Stromlo, where the concentration is on high proper motion stars and members of the Hyades cluster. Anguita and Ruiz are using the CTIO 1.5-m telescope with the f/13 Cassegrain focus and a CCD detector. The program stars are selected from the LHS Catalogue and from a survey by Ruiz and her collaborators. Most of the stars have turned out to be sub-dwarfs, cold white dwarfs and red dwarfs. The precision of the x,y image positions, which were obtained using the algorithms of DAOPHOT software package, is on the average 6 mas (0.02 pixel) in the range 2,000 to 30,000 electron counts, with a sky background of about 800 electron counts. The precision deteriorates exponentially below 2000 electron counts. A comparison of the CTIO parallax of vB10 with the U.S. Naval Observatory's CCD parallax for the same star shows excellent agreement. The current program is continuing with a sample of about 50 southern faint high proper motion stars.

H. McAlister described the slowly improving situation of "Double Star Observations in the Southern Hemisphere". Observers of southern double stars have included some of the greatest names in the field (Herschel, Rossiter, van den Bos, Finsen), but their programs of discovery and measurement have been followed by long periods of inactivity. In recent years, southern double star astrometry has largely been conducted by micrometer observers based in the north who make occasional trips to southern observatories. Many systems discovered in the south have not been adequately observed over the years and hence the elucidation of their orbital motions has been delayed. The GSU/CHARA program of binary star speckle interferometry has been conducted from Kitt Peak since 1975 with observing runs scheduled two to three times per year. This program was extended to full sky coverage in 1988 with twice yearly trips to CTIO where the GSU/CHARA speckle camera system is used on the 4-m telescope. Several thousand measures have been reduced to date, and the first series included the discoveries of five new southern binaries including an additional close component to HR 6027. It is their intention to maintain this southern hemisphere speckle program as long as resources and access to telescope time permit.

5.4 29 July "Linking the Optical and Radio Reference Frames" chaired by T. E. Corbin.

This session was held jointly with Commissions 8 and 40. The reports can be found in the Transactions of Commission 8.

5.5 31 July "Optical Reference Frames and Catalogues: Fainter Extension of the Optical Frame" chaired by L. Morrison.

This session is a continuation of a Joint Meeting with Commission 8 that bears the same title, but was chaired by K. Johnston. The reports from that first session can be found in the Transactions of Commission 8.

Chr. de Vegt reported on joint work with J. A. Hughes concerning "The USNO/Hamburg Global Astrometric Faint Star Catalogue Project". They propose that a high precision astrometric catalogue using a modern astrograph be established with a limiting magnitude to at least V = 14 in both hemispheres. The standard four-fold plate overlap procedure combined with the Hipparcos reference frame and a global block adjustment should yield a positional accuracy of 0.05" and a global system accuracy of 0.01". The observing program should take less than three years, and with parallel plate measuring, the whole project could be completed in five years. This proposed catalogue would provide for the first time a homogeneous, global new epoch for the final reduction of the Astrographic Catalogue and a rigorous basis for the astrometric calibration of all large Schmidt sky surveys.

L. Helmer described the "Candidate Catalogue for a Faint Extension of the Fundamental Reference Frame". This Catalogue is intended to create a net of faint reference stars in the magnitude range 11.5 < V < 12.0, which would be useful for photographic work. At the Copenhagen University Observatory, stars in the northern sky have been selected at a density of about one per square degree, mainly from the Astrographic Catalogue. The positions have been transformed to the FK5 system and proper motions derived where possible. The candidate stars are proposed for reobserving on automatic meridian circles and have been included in the program for the CAMC on La Palma. The first results from the CAMC have been compared with observations made with the Bordeaux automatic meridian circle and systematic errors do not exceed 0.02", with a success rate close to 98.5%. The selection is now complete from declinations +90° through -52°, while the observations are 31% complete.

I. Platais reported on "The Yale-San Juan Plans for a Faint Secondary Reference Frame" in collaboration with T. Girard, C. E. López and W. van Altena. In previous papers, proposals have been presented to extend the faint reference frame to V < 14, however there is still a need for fainter reference stars for the large reflectors and deep Schmidt photographs. These instruments are used to determine positions of stars, galaxies, QSO's and other objects for the linkage of the radio and optical frames and for multi-object spectrographs and other instruments. Using plates from the Southern Proper Motion program, a network of faint secondary standards south of declination - 17° is being determined from approximately the 12th to the 18th magnitudes with a density of 10 to 15 stars per square degree at the faint end of the sequence. The positional accuracy (at the mean epoch of 1980) of the equatorial coordinates for the faint standards is hoped to be about 0.10", while the accuracy of the absolute proper motions should be about 4 to 5 mas/yr.

C. Turon highlighted the "Hipparcos Contributions" to the establishment of a faint reference frame and noted that many of the details could be found in the proceedings of the Joint Discussion on Hipparcos in this volume. The faint part of the program consists of 2400 stars fainter than V = 11 and about 30,000 stars in the range 9 < V < 11. T. Corbin has selected stars for the extension of the fundamental system from the faintest Hipparcos stars, while L. Morrison has been responsible for the next brighter group. Approximately 1000 IRS stars are not being observed by Hipparcos and C. Smith is working on the addition,or replacement, of those IRS stars by candidate Hipparcos stars. In addition, the list of radio-stars and "link" stars observed by Hipparcos, that are close to QSO's, have been distributed to the relevant individuals in the HST project and to the ground-based observers.

P. Hemenway outlined the "Status of the Hubble Space Telescope for Making Observations Towards Linking the Hipparcos and Extragalactic Reference Frames" on behalf of the Space Telescope Astrometry Team. The status for astrometric measurement of the Fine Guidance Sensors (FGS) of the Hubble Space Telescope, including a description of their operation, was given at the IAU Colloquium 127. Since that report, measurements of about 25 stars at each of 10 pointings have been reduced for a preliminary Optical Field Angle Distortion (OFAD) calibration. The observations were made with the guiding system in Coarse Track and the astrometer FGS operating in Fine Lock. The results show residuals from the overlapping plate solution of 3-4 mas. Because the spacecraft stability was poorer at that time, the accuracy in the future may be as good as originally expected, i.e., 2 mas rms per observation. Once the final collimation position for the secondary mirror is determined, probably in early November 1991, it should then be possible to start the astrometric calibration and science observations. Due to the reduced amplitude of the FGSs transfer functions, it may be necessary to restrict observations to objects brighter than 15th mag. In that case, it will be necessary to extend the observations for the Hipparcos Link to bright QSOs and BL Lacs, which may be radio-quiet. As a result, the emphasis would be on determining the rotation of the Hipparcos instrumental system with respect to the extragalactic frame, which with 15 objects will be at the 2 mas/yr level. If it is possible to obtain the proper motions for more objects, and over a longer time base, the expected accuracy may be as good as 0.5 mas/yr to the system rotation.

F. Chollet and N. Capitaine described a new procedure for computing the apparent places of the stars referred to the intermediate frame. This frame is linked to the nonrotating origin for estimating the Earth's rotation for astrometric observations. The procedure is based on the matrix transformation of vector components between the geocentric celestial and terrestrial reference systems. It uses the nonrotating origin for reckoning the Earth's angle of rotation and the celestial coordinates of the Celestial ephemeris Pole to account for the effects of precession and nutation. The procedure was derived from the one used to compute the apparent positions of stars for astrolabe reductions using the FK5 and can be applied to the Hipparcos frame.

COMMISSION 25: STELLAR PHOTOMETRY AND POLARIMETRY
Report of Meetings, 26, 27, 29 July 1991

PRESIDENT: Ian S. Mclean   **SECRETARY**: E. Milone

26 July 1991

The Business Meeting of Commission 25 at the XXI IAU General Assembly was held during Session 2 on July 26, 1991 in Room G, San Martin Cultural Center. The President Ian McLean called the meeting to order, welcomed members and attendees, and requested approval of the Agenda for the meeting. With the Agenda approved by the members present, Prof. McLean reported on changes in membership. There had been 2 resignations and 20 new proposed Comm 25 members (14 at the GA), giving a total new membership of 200. There were no objections to any of the proposed new members. The President reported that 2 IAU Symposia (Nos. 143 and 148) had been supported and 3 Colloquia were being sponsored by Comm. 25. Also, a half-yearly Newsreport had been initiated. He reviewed the triennial report and distributed copies.

Elections

Voting on new officers was carried out. New Organizing Committee members are: S. Adelman, R. Genet, T. Moffett, A. Landolt and C. Sterken. Continuing to serve are: M. Breger, I.K. Knude; J.D. Landstreet, J. Lub, J. Menzies, V. Straizys, F.J. Vrba and the retiring president I.S. McLean. The President placed the name of John Landstreet (not present) in the nomination for Vice-President. Landstreet was elected unanimously. As is the custom, Dr. A.T. Young our current Vice-president was nominated for President, Dr. M. Breger seconding the nomination. Andy was was elected by a unanimous vote.

Discussion on the future of the Commission

The President invited discussion on the future of the commission, stating that the IAU had requested that commissions reappraise their role, so that consolidation of some commissions could be carried out. A spirited discussion among the members ensued with the conclusion being that we desire to continue as a commission and have the arguments to defend our position.

New Business

Working groups: Landstreet is a representative on a W.G. on Ap stars. Another multi-commission WG is the one on synthetic photometry; Comm 25 is represented by Buser. A small W.G. on the infrared was formed at the last GA by Milone; input to this working group would be welcome. Garrison is responsible for standard stars and Chris Corbally is the new newsletter chairman in this WG. Sterken: Important questions which we need to consider are how many and how faint. Garrison: Standardization of filter systems could be included in the newsletter. Crawford: Said dialogue on these matters is important and MOVED that a small W.G. be formed to report back in three years. A.T. Young will coordinate.

27, July 1991

PHOTOMETRY SESSION: Chair - I.S. McLean

In the absence of any of the authors of the report on the Vilnius system, A.T.Young read the short note "A Possibility of Photometric Three-dimensional Classification of Stars at V=20" by V. Straizys, F. Smriglio, and R. P. Boyle. The 7-band Vilnius system allows photometric determination of spectral class, absolute magnitude, and metallicity, regardless of interstellar reddening, and permits recognition of many types of stellar peculiarities (Straizys, "Multicolor Stellar Photometry"; Pachart, 1990).

C. Sterken discussed "The future of existing photometric systems." Every new detector brings more systems, incompatible with the old ones. Standardization is very difficult. Problems connected with colors remain unsolved since the 1920's. It is difficult to escape from these problems for many reasons. We must return to teaching fundamentals of photometry to all observers, develop standard reduction programs and we need a new standard system that can be reproduced accurately; it must be well-calibrated, and "open". An approach is to buy many identical filter sets and install them in a world-wide network of APTs.

A. Landolt reported on his UBVRI (Cousins RI) standards to V=16 and 21. Extinction was measured every night at CTIO; though he reported mean values, the night-to-night variations are clear and significant. He has observed 20 to 30 standard stars per night. An interesting finding is that the RMS error changes considerably with time during the night. Of 526 candidate stars, 298 survived as standards, observed about 29 times each. The transformations are nonlinear. Typical mean errors of the final means are 3 millimags. The response functions used will be published, both for filters and for the PMTs used.

A. T. Young then discussed "Transformations and accuracy: problems and possibilities." Menzies et al. (MN 248, 642, 1991) find systematic differences of several hundredths of a magnitude from Landolt, in measuring the Landolt equatorial standards in the Cousins version of UBV. These problems are basically aliases, because the bands under-sample the spectrum. The problems lie in the fundamentals of photometry. The Stromgren-King theory uses a Taylor-series expansion to obtain terms that are products of moments and derivatives (Stromgren 1937, in "Handbuch der Experimentalphysik"; King 1952 in AJ 57, 253). It turns out that the derivatives increase as fast as the moments decrease when passband width changes; the terms maintain their relative size, and only the factorials in the denominators allow convergence. Terms of high order MUST be retained. Accurate transformations should be possible if bands that sample the spectrum correctly are used.

E. F. Milone reported status of infrared photometry: There have been a few developments in IR systems since the meeting on Infrared Extinction and Standardisation [A volume was published by Springer Verlag in the series 'Lecture Notes in Physics', No. 341, 1989.] (hereafter IRES) at the Baltimore General Assembly. In system definition and transformation: Bessell and Brett (summary in IRES, pp. 61, 1989; details in PASP, 100, 1134, 1988) review previous work and compare several JHKLL'M systems in use at that time (SAAO, ESO, CIT/CITO, MSO, AAO, and Arizona), and derive linear transformation coefficients and zero points among them. They also proposed a "homogenized" system, the MSO JHKL system. Establishment of a standard system involving fainter stars is an increasingly important need (McLean, IRES, p. 66, 1989). A working group led by Milone is looking at defining a new system ; all interested parties are urged to contact and perhaps join the working group.

29 July, 1991

CONTRIBUTIONS OF POLARIMETRY TO STELLAR ASTROPHYSICS
(A joint meeting with Commissions 9, 27, and 42)
Chair: Bob Koch (morning), Ian McLean (afternoon)
The following papers were presented:

A review of physical processes - K. Nordsieck
From periodic and non-periodic variability of Q and U one can ascertain information such as inclination, axis orientation, sense of rotation, location of scatterer. Spectropolarimetry can separate different kinds of scattering by wavelength dependence e.g. electron, Rayleigh, and Mie scattering. For example, nebular polarization is strongly depolarized by light from the central star: ie. PMS stars seen only in scattered nebular light have big polarizations, those seen directly have much smaller polarizations. When $\tau \ll 1$ the problem is that the dust envelope is poorly constrained whereas for $\tau \gg 1$ the problem is that radiative transfer is not trivial.

Array detectors in astronomical polarimetry - J. Tinbergen and I. McLean
After first generation of polarimeters, area detectors have given a major advance - CCDs being the best of these. But the problem is how to do rapid modulation when the CCD readout is normally slow. One solution is to split E and O rays, image them on different pixels, and find Q and U from the ratio. Current CCD and IR polarimeters were reviewed and details were given of the ISIS spectropolarimeter for the 4.2-m W.H.T.

Interstellar polarization in the environments of star forming regions - F.J. Vrba, H.G. Marraco, S.E. Strom, L. Hillenbrand. For Herbig Ae/Be stars we have obtained new optical and infared photometry, new polarimetry of the stars and surrounding local ISM, and determined jet/axial rotation directions from optical and molecular mapping and disk models from the literature. The resultant Spectral Energy Distributions group into two: those that follow a wavelength to the -4/3 power (indicative of optically thick reprocessing or accretion disks) or those with flat or rising SEDs (indicative of the presence of material outside of the accretion disk). The results support the view that YSO polarizations are due to scattered light rather than aligned grains in a disk.

Problems with polarization standards - J. Dolan
For the high polarization standards used for position angle (PA) calibration, it is difficult to do better than about $\sigma = 1°$. To get $\sigma < 1$ degree, we need that the PA is NOT a function of time or wavelength, although the latter is OK if bandpasses are well-defined. The problem with high polarization stars is that they are typically distant giants with a component of intrinsic polarization due to scattering from extended envelopes. What to do about this?
a) observe asteroids since scattering plane between Earth-asteroid-Sun is well-determined
b) mechanically align Glan-Thompson prism to N-S, can be done to 0.1 degree if careful.

A new polarization study of the magnetic Babcock's star - M. Breger and N.S. Polosukhina
Data in literature appear to show variable polarization for this star, but not when you take into account error bars. New UBVRI polarimetric observations show no variability and that observed

polarization is completely consistent with ISM polarization.

Spectropolarimetry of massive hot stars - R. Schulte-Ladbeck
Summary of results: EZ CMa - P and theta changes correlated with emission line flux. HD 191765 - P, but not theta correlated eith emission line flux. If P(line) not same as P(cont) this implies intrinsic polarization and the quantities P(line)/P(cont), deltaQ, and deltaU reveal information about atmospheric structure and winds. Polarized lines are slightly red-shifted which implies outflowing winds.

WUPPE results: hot unevolved and evolved objects - K. Nordsieck
Zeta Tau - Be star w/ constant polarization - line blanketing w/ Balmer jump in polarization. Pi Aqr - Be star w/ variable polarization - line blanketing but w/o Balmer jump in polarization. Kappa Cas - hot supergiant - only interstellar polarization. Alpha Cam - hot supergiant - only interstellar polarization. P Cyg - hot supergiant - high states of polarization during WUPPE obs Alpha Ori - cool supergiant - see polarization features coincident with TiO bands - actually 2 sets at 45 degree different theta - (a) narrow TiO theta, (b) at blue upturn theta. Also see Mg II 2800 chromospheric lines in polarization at above thetas. Far UV results imply process in atmosphere not in circumstellar shell.

Environments of luminous late-type variables - A.M. Magalhaes
L2 Pup - non-uniform stellar disk, grain size changes during course of years, stable symmetry plane from IR observations: V CVn - P anti-correlated with brightness: o Ceti - optical p(lambda) closely related to shockwave, P increases until shock dies out.

Miras imaged by speckle and their polarization signatures - R. Boyle
Mira A and B studied with polarimetric coverage during 4 different epochs between Nov 1983 and Dec 1988. Before phase 0.8 (eruption point) P slowly decreases, after phase 0.8 P suddenly grows with sudden increase of grains in shell. Dust located at about 3 stellar radii by 11 micron spatial interferometry at T = 1200 K. The radius depends on wavelength (0.030-0.100"); outside of TiO bands it is typically 0.04".

Polarization of young stellar objects: general properties and a few examples - P. Bastien
Unresolved observations of T Tauri, Herbig Ae/Be stars yield no standard polarization lambda dependence, P can be large: e.g. HL Tau 12%, V376 Tau 21%. Stars with IR excesses have larger P and observed circular polarization implies dust not electron-scattering. Resolved observations reveal extended, centro-symmetric polarization patterns so you can find the illuminating source. Reported 56 galactic sources with polarization maps (mostly YSOs): 20% centro-symmetric, 60% aligned vectors and 11% peculiar. Multiple scattering is the cause - not aligned grains.

Scattering envelopes in Wolf-Rayet stars - A. Moffat
16 early-type binaries with polarization observed - of these 11 are Wolf-Rayet stars.
There are two types of polarization: (a) No obvious periods in polarization - great deal of variability in many cases, in other cases no variability or even any polarization - implies winds are symmetric (b) Periodic polarization variability - due to binary modulation - sometimes modes a) and b) are seen together. Interpretation - Polarization is produced by electron-scattering with sufficient asymmetry to produce polarization. From these observations one can get orbits. Several circular and eliptical orbits have now been fit.

COMMISSION 26: DOUBLE AND MULTIPLE STARS
(ETOILES DOUBLES ET MULTIPLES)

Report of Meetings Held at the 21st General Assembly

PRESIDENT: Harold A. McAlister   VICE-PRESIDENT: Helmut A. Abt

INTRODUCTION

The program for Commission 26 in Buenos Aires consisted of opening and closing business sessions on 24 and 27 July 1991. During two scientific sessions on 26 July, the following papers were presented:

A Catalogue of Variable Components in Visual Double & Multiple Stars
Patricia Lampens, *Observatoire Royal de Belgique*

Status of the Washington Double Star Catalog
Charles Worley, *U.S. Naval Observatory*

The CHARA Speckle Programs
Harold McAlister, *Georgia State University*

The U.S.N.O. Speckle Interferometry Program
Charles Worley, *U.S. Naval Observatory*

Lunar Occultations and the Hyades
Deane Peterson, *S.U.N.Y. Stony Brook*

CCD Photometry of Possible Trapezium Systems
Helmut Abt, *Kitt Peak National Observatory*

Near-IR Studies of Young Binary Stars: Results from Direct Imaging,
Speckle Interferometry, and Lunar Occultations
Hans Zinnecker, *Universität Würzburg*

Double Star Astronomy with the Hubble Space Telescope
Otto Franz, *Lowell Observatory*

The Commission also participated with Commissions 8 and 24 in the Joint Commission Meeting on *Astrometry in the Southern Hemisphere*.

SCIENTIFIC SESSIONS

Approximately 50 people were present during the Commission's scientifc sessions. Authors of the papers presented to the Commission have provided the following summaries of their works.

A Catalogue of Variable Components in Visual Double & Multiple Stars – Patricia Lampens, Observatoire Royal de Belgique — A machine readable catalogue of variable components in visual double and multiple stars has been compiled by intersecting two recent catalogues, the "General Catalogue of Variable Stars" by P.N. Kholopov *et al.* (Vols I-III, 4th ed., Moscow, Nauka, 1987) for the variabliity information and the "Catalogue of Components of Double and Multiple Stars" by J. Dommanget (Cont. Van Vleck Obs., vol. 8, pp. 77-82, 1989) for the astrometric information. 468 variable components have been identified. Accurate positions ($\sigma_\alpha = \pm 0\overset{s}{.}01$; $\sigma_\delta = \pm 0\overset{''}{.}1$), component designations and additional identifiers such as HD and DM numbers are provided. Relative positions are also given since they are helpful in correctly recognizing the variable component at the telescope. The study of these components in physical double stars is of particular relevance in the fields of stellar structure and evolution.

Status of the Washington Double Star Catalog – Charles Worley, U.S. Naval Observatory — The WDS consists of two compilations of data, plus various ancillary tables and inverse files. The Observation Catalog now contains approximately 433,000 means. It is complete for data published since 1927. For many years, work has proceeded to access the earlier observations, of which 153,000 are now present (estimated completion of 85-90%). The Index Catalog now contains over 77,000 pairs and has been reformatted to better accomodate modern needs. A new tape version should be available in a few years.

The CHARA Speckle Programs – Harold A. McAlister, Georgia State University — The GSU/CHARA speckle program was initiated in 1977 and since 1980 has been carried out at the Kitt Peak 4-m telescope using a remotely controlled speckle camera system employing an intensified CCD array. Participants in this program include W.I. Hartkopf, W.G. Bagnuolo, and the author (all from GSU) and O.G. Franz (Lowell Observatory). The emphasis has been the continuing collection of precise, high resolution measurements of visual, spectroscopic, occultation, and composite spectrum binaries. The CHARA program has now published some 13,000 measures and discovered ~250 new binary stars. Methods have been developed for real-time data processing to obtain astrometric information, including the elimination of the 180° quadrant ambiguity inherent in classical speckle interferometry. An algorithm for extracting differential photometric information from speckle data is also near maturation. Since 1988, the CHARA speckle program has been expanded to all-sky coverage with twice yearly trips to the CTIO 4-m telescope.

The U.S.N.O. Speckle Interferometry Program – Charles Worley, U.S. Naval Observatory — In October, 1990, the U.S. Naval Observatory received a new speckle camera system based on an improved version of the successful Georgia State University instrument. With efficient software supplied by the GSU/CHARA group, this system has already made well over a thousand measures at the Washington 26-inch refractor. We would like to emphasize that this instrument arrived ahead-of-schedule, below budget, and worked well at the first instance of its use.

Lunar Occultations and the Hyades – Deane Peterson, S.U.N.Y. Stony Brook — Obtaining and using lunar occultation data is at once a powerful and very limited technique. Only a fraction of the sky is covered by the Moon's disk during its

18-yr cycle, but in this region objects can be usefully observed down to about 10th magnitude at good sites. For binary systems, differential photometry can be obtained (to ±0.1 mag) along with projected separations (to ±1 mas) and angular diameters (to ±1 mas for $d \geq 2$ mas). We discuss four Hyades objects for which we have observed, or have tried to observe, occultations, and the impact those data have had on our understanding of the systems and the cluster.

The first of these was Finsen 342, a 13-yr system noticable as a large residual in McClure's solution for the cluster distance. (McClure's technique involves forcing the eclipsing binary, vB22, onto the mass–luminosity relation defined by the visual binaries thereby providing a distance estimate). The occultation trace showed the secondary on the wrong side, leading to a proposal (see Peterson and Solensky 1987 for details) that a 6-yr period was appropriate. This has since received strong support from speckle observations.

The second system, 51 Tau, was shown by Peterson and Solensky (1988) to serve as a replacement for vB22 in establishing the distance scale. More importantly, it could be added to vB22 and the visual binaries in establishing an even higher weight solution for the distance, the Δm being the major remaining uncertainty. The system escaped observation in the late 1980's, and will do so again in the early 1990's unless observations can be made from the southern hemisphere.

The third system, θ^2 Tauri, has been known to be a spectroscopic binary (SB1) for over 80 years, but it was not until the Moon's passages over the cluster center around 1980 that it was discovered that the secondary is only about 1.1 mag fainter than the primary. This lead us to search for the secondary spectrum using high dispersion plates and cross-correlation techniques. The detection of the secondary spectrum was reported by us last year (1990 ASP Conf. on Clusters). Since then the Mark III interferometer has resolved the system and, with the mass ratio, has found the mass and distance to the system. Beyond that, the pair is important in that it consists of a main sequence star and a turnoff star of the same color and a known magnitude difference, which provides a much more stable way of using theoretical isochrones to evaluate the age of the cluster.

The last system, θ^1 Tauri, has a potentially significant contribution. The primary is a K giant and the secondary, ~3.5 magnitudes fainter and a late F dwarf, was detected during a series of occultations in the early 1980's from which an angular diameter was also determined for the primary. The system has since been resolved by speckle techniques, and the motion of the primary is being followed by the radial velocity observers as well. By using the measured Δm and colors, placing the secondary on the cluster main sequence will establish the distance. Both the orbital solutions and our knowledge of the mass of the cluster turnoff will provide the mass of the primary. The angular diameter will give the radius, and the characteristics of the cluster provide the age and the initial composition. The only other convective star whose parameters are known to this extent is the Sun. This will provide an important check on the assumption that the mixing length ratio used in interior calculations is the same for all phases of evolution.

CCD Observations of Possible Trapezium Systems – Helmut Abt, Kitt Peak National Observatory, and Christopher Corbally, S.J., Vatican Observatory Research Group — Trapezium systems are dynamically unstable and should therefore be young. Ambartsumian and Sharpless found them in HII regions, but what are their lifetimes? Allen, Tapia, and Parrao (Rev. Mex. Astr. Ap., vol. 3, p. 119, 1977) scanned 20,000 multiple-star sytems in the IDS and listed 968 apparent Trapezium systems that seemed to be free of optical companions. But they found that the most frequent primary type is F, which is not that of young stars.

We obtained CCD observations for 268 of those systems with a 1-m Kitt Peak telescope. They yielded an accuracy in Johnson UBV photometry of 0.06 mag or better. Surprisingly, the mean astrometric accuracy is $0°.06$ in position angle and $0''.07$ in angular separation. But we also needed one MK spectral type per system to determine the distance and reddening. Types have been obtained for 62 systems. Of those, 57 have enough optical components that they are not Trapezium systems. The five remaining Trapezium systems have primary types of O5–B1, so the maximum age of a Trapezium system is about 10^7 years.

Near-IR Studies of Young Binary Stars: Results from Direct Imaging, Speckle Interferometry, and Lunar Occultations – Hans Zinnecker, Universität Würzburg — A selection of recent near-infrared (NIR) results on young, low-mass, spatially resolved binary stars with separations in the range of $0''.1$ to $10''$ is reported. These results are based on collaborative work with Leinert and Haas (MPIA/ Heidelberg), Christou and Ridgway (NOAO/Tucson), Richichi (Arcetri), and Moneti (ESO/Chile). NIR studies, which are differential due to the coeval nature of the components of young binary systems, bear on binary star formation and on pre-main sequence stellar and circumstellar evolution. From direct NIR array imaging of known T Tauri binaries, we found (A&A, vol. 242, p. 428) that the stronger NIR color excess tends to reside with the more luminous component at $2\,\mu$, possibly implying active accretion disks to be present around each individual stellar component or preferentially around the brighter component. With slit scanning (1D) speckle interferometry we detected XZ Tau to be a binary with projected separation of $0''.3$ (A&A vol. 230, p. L1). The infrared companion of XZ Tau brings the roster of T Tauri stars with infrared companions to 5 (the others being T Tau, Haro 6-10, Glass I, and Elias 22). By the same means we also identified GG Tau as a hierarchical quadruple system. Thus, speckle data have shown that both GG Tau and GG Tau/c ($10''$ south) are themselves binaries ($0''.3$ and $1''.4$ apart, respectively). Using a NIR speckle camera, we investigated the known binaries S CrA, AS 205, and T Tau in 2D for the first time, and discovered the duplicity of Elias 49 (MWC 863), DD Tau, FX Tau, and LkCa 3. We confirmed the FU Ori object Z CMa to be a binary with a projected separation of $0''.10$ (following private communication by Beckwith) and suggested that tidal interaction might trigger the FU Ori outburst in the circumstellar disk of either the primary or the secondary. Finally, lunar occultations at $2\,\mu$ (similar to those pioneered by Simon *et al.*) confirmed the binary nature of FV and FW Tau. However, no spatially extended NIR emission around the individual components in these systems could be detected, contrary to the single T Tauri star DG Tau where we have seen a halo of size 45 mas or about 6 AU, the smallest structure resolved up to now in any young star.

Double Star Astronomy with the Hubble Space Telescope – Otto Franz, Lowell Observatory — An orbital solution from visual micrometry measures necessarily made mostly near apastron for the binary star ADS 11300 = Hu 581 = WDS 18229+1458 predicted periastron passage in 1984 at a separation not resolvable by visual observers. Realizing that a single high resolution observation at the present time would provide crucial information to define the orbit, ADS 11300 was observed with the Hubble Space Telescope Fine Guidance Sensors on 1 November 1990. The data yielded an angular separation of 0."066. This new result permitted a revised prediction of periastron passage during 1992 at a separation below 0."01, making ADS 11300 a challenging target for continued HST-FGS astrometric measurement.

COMMISSION BUSINESS

During the first business session, the President welcomed those in attendance and asked for a moment of silence in memory of two Commission members who passed away since the last General Assembly, Wallace R. Beardsley and Raymond H. Wilson.

A brief discussion was conducted concerning the initiative from the General Secretary during the triennium to amalgamate certain commissions which appeared to have common scientific ties. It was the strong consensus of all commissions concerned not to amalgamate. It was recognized that a reassessment of commission structure was appropriate from time to time, but that such reassessments should be based solely upon scientific grounds and be carried out with extensive input from the membership.

The results of the election of new officers were announced – those results being:

President: H. Abt (USA)   *Vice President*: C. Worley (USA)

Organizing Committee – New Members: Y. Balega (USSR), F. Fekel (USA), C. Scarfe (Canada), and H. Zinnecker (Germany); *Continuing Members*: P. Bernacca (Italy), H. McAlister (USA), and E. Van Dessel (Belgium).

On behalf of a letter written by J. Dommanget (Belgium) to the Commission President and distributed to other astronomers as well, K.Aa. Strand (USA) brought up the question of accessibility of the Washington Double Star Catalogue (WDS) to the scientific community. The President responded that this was not appropriately Commission business because, whereas the Commission has always strongly supported the efforts involved in maintaining the WDS and its predecessor catalogs, the WDS was a scientific effort carried out and funded by the U.S. Naval Observatory. The USNO Scientific Director G. Westerhout reiterated this point and emphasized that the data within the WDS are routinely and promptly provided to any requesting astronomer without charge as a service of the USNO.

C. Worley explained the impracticality of maintaining the WDS data base, which is essentially not a complete collection and hence is being continuously revised, at the various data centers. This argument was strongly seconded by W. Warren of the Astronomical Data Center at Goddard SFC.

Upon the recommendation of R.S. Harrington (U.S. Naval Observatory), we remind the orbit computers of the need for more accurate ephemeris calculations to support the new observing techniques becoming available. There will soon be the need to allow for variations of elements in many systems, and in one or two cases even now. These variations may be due to change in perspective due to intrinsic motion or to dynamical variations in multiple systems. To this end, parameters that must be included are the epoch of orientation of the node and the epoch of osculation of the orbit. If the epoch of osculation is not available, give at least the weighted mean epoch if appropriate. We recommend starting these practices now so that they will be standard when required.

Vice President Abt chaired the final business session. New members of the IAU who selected Commission 26 for their affiliation were announced by H. McAlister to be: Yuri Balega (USSR), Silva Gaudenzi (Italy), Jon E. Hakkila (USA), Patricia Lampens (Belgium), Roberto Morbidelli (Italy), Corinne Rossi (Italy), Dimitrios Sinachopoulos (Belgium), and James R. Sowell (USA). Astronomers already holding IAU membership who newly wished to join the Commission were: John Davis (Australia), Hartmut Jahreiss (Germany), and Deane Peterson (USA). These additions to the Commission bring the total membership to 109 astronomers.

As at the last General Assembly, the future of the Commission's *Circulaire d'Information* was discussed. The *Circulaire* has been published at their expense by Drs. Muller and Couteau at the Observatoire de Nice, an activity much appreciated by the Commission and by the entire double star community. It is hoped that this publication can continue and its use be expanded.

H. Abt conducted a thought provoking discussion of the status of various subfields of double star astronomy including astrometry (positions and parallaxes), visual micrometry, interferometry, occultation studies, photometry, radial velocities, spectroscopy (MK types and abundances), evolutionary studies, and dynamical studies. The most mature of these areas was considered to be the several aspects of theoretical dynamics while virtually every other area was far from saturated by both results and potential contributers.

The question of "standard stars", or more precisely systems commonly observed by groups employing different observational techniques for measuring (θ, ρ), was raised by C. Worley. This is particularly important for speckle and other types of interferometry where spatial calibration may not be straightforward and for space observations from HIPPARCOS and HST. It was agreed that a committee consisting of O.G. Franz, K. Johnston, H. McAlister, and D. Peterson would work toward the goal of suggesting a list of "standards" along the lines of a previous attempt by McAlister and W.I. Hartkopf (P.A.S.P., vol. 95, p. 778, 1983).

COMMISSION No. 27

VARIABLE STARS (ETOILES VARIABLES)

Report of Meetings: July 26, 31, 1991

PRESIDENT: Michel Breger   SECRETARY: Thomas Barnes

1. BUSINESS MEETING (July 26, 1991)

1.1. M. Breger called the business meeting to order at 1400 hours and welcomed the members of the Commission. He presented an introduction to the meetings of the Commission during the General Assembly. Approximately 50 members of the Commission were in attendance.

1.2. There followed a brief discussion of the availability of the General Catalogue of Variable Stars. H. Bond enquired as to how members in the USA may obtain a personal copy. J.A. Mattei responded that the GCVS is distributed in the USA by the American Association of Variable Star Observers through an agreement with the Sternberg State Astronomical Institute. She noted that the first four volumes are currently available for $50 plus $5 for shipping per volume. Requests should be sent to the AAVSO, 25 Birch Street, Cambridge, Massachusetts 02138, USA. A fifth volume on extragalactic variables is in preparation in the USSR.

1.3. B. Szeidl, editor of the Information Bulletin on Variable Stars, reported on the status of the Bulletin. He noted that the Bulletin is now in its thirtieth year of publication, and he expressed his hope that it will continue for another thirty years. However, after twenty-five years as editor, he decided that it was time to retire from that position. L. Szabados and K. Olah, both of Konkoly Observatory, accepted to be editor and co-editor effective immediately. J. Smak rose to offer very great appreciation from the variable star community to B. Szeidl for his many years of outstanding service as editor of the Bulletin. These remarks were followed by great applause.

   B. Szeidl also drew the members' attention to serious economic problems in publication of the Bulletin. At present the Bulletin is mailed to approximately 600 addresses, which includes 350 institutions. The financial resources of the Konkoly Observatory no longer permit this number of mailings, and B. Szeidl suggested that the Bulletin be distributed only to institutions. There followed an extensive discussion of the finances of the Bulletin. M. Breger said that Commissions 27 and 42 (Close Binary Stars) have asked the IAU Secretariat for $500 over three years to assist with the distribution of the Bulletin. B. Koch (President of Commission 42) remarked that Commission 42 strongly supports financial assistance to the Bulletin as about 50% of the activity of that Commission involves variable stars. The Secretariat has not yet indicated its willingness to provide this financial assistance. E. Malone moved that Commission 27 request all possible financial support for the Information Bulletin on Variable Stars from the Secretariat. This motion was seconded and accepted by the members. E. Malone also suggested that the Commission may wish to appoint an editorial board to assist with publication of the Bulletin. M. Breger reported that the new editors of the Bulletin have proposed an editorial board of L. Balona, M. Breger, M. deGroot, D.S. Hall, R. Koch, J.M. LeContel, J. Percy, M. Rodono, J. Smak, and C. Sterken. A. Landolt moved the acceptance of this editorial board by the Commission. The motion was seconded and accepted by the members. Concerning the suggestion that personal copies of the Bulletin be discontinued, numerous members said that they have access to the Bulletin only through their personal subscriptions and they requested that these be continued.

H. Bond enquired whether submissions to the Bulletin are reviewed. B. Szeidl responded that the editors act as referees in the interest of rapid publication, and that 15–20% of the submissions are rejected. In questionable cases, an informal referee system is used.

1.4.   M. Breger reported on the IAU Archives for Unpublished Photoelectric Observations of Variable Stars. The number of files in the archive has increased from 174 in 1988 to 221 (not including 14 incomplete files) at present. Copies of the archives now exist in France, the USSR, and the United Kingdom. Researchers may submit and retrieve files electronically, with details of the procedure given in the current Transactions of the IAU. Approximately 1–2 requests come each month to each center for access to the files.

1.5.   M. Breger also reported on the Delta Scuti Newsletter, which has now grown to 22 pages per issue. There have been three issues of the Newsletter and the next issue is anticipated for distribution in November. Copies are sent to interested individuals active in the field only. M. Breger expressed his appreciation to all contributors to the Newsletter.

1.6.   The following slate of officers was proposed for Commission 27: President: J. Percy; Vice-President: M. Jerzykiewicz; Organizing Committee: T. Barnes, M. Breger, J. Christensen-Dalsgaard, R. Gershberg, D. Kurtz, J. Mattei, M. Rodono, N. Samus, M. Smith, B. Szeidl, and M. Takeuti. The members approved these officers by unanimous vote. Approximately 50 new members were proposed for the Commission and were approved by vote of the members. This brings the membership of the Commission to about 400 members.

1.7.   M. Breger reported on the situation at Sonneberg Observatory which is proposed to be closed by the German government. The Sonneberg Observatory has conducted important research for 66 years in the field of variable stars and has a valuable, archival plate collection. M. Breger submitted to the members for discussion a joint resolution with Commission 42 to the General Assembly in support of the research of the Sonneberg Observatory. The (final) text of the resolution is as follows:

"IAU Commission 27 considering that the systematic coverage of the long-term behavior of the population of variable stellar objects, such as the sixty years of measurements made at Sonneberg Observatory, makes a major contribution to Astronomy and Astrophysics, recommends that all efforts be undertaken to continue these important measurements and to ensure the appropriate maintenance and availability of the data."

Considerable discussion on how this important work could best be supported followed. In particular, J. Smak noted that the context of this resolution is to assist the Sonneberg astronomers in finding support for their work and their archive and not to interfere in any decisions being taken by national governments. The members then voted on the resolution, 43 votes for the resolution and no votes opposed. The members of the Commission also authorized M. Breger to make minor editorial changes to the resolution (if necessary) and to present this resolution to the IAU General Assembly.

1.8.   M. Breger brought forward the suggestion of the Secretariat that each Commission examine its reason for being, and he proposed for discussion a merger with Commission 26 (Double & Multiple Stars) or 42 (Close Binary Stars). The members present opposed this possibility and strongly preferred to retain the present format of the Commission. Several members encouraged more interaction among the various astronomical fields allied with variable stars. J. Percy recommended that the Organizing Committee take an active role in proposing meetings involving more than one commission.

1.9.   J. Percy raised the question of whether the members find the current Reports on Astronomy useful to them. Several of the Commission members spoke very positively about the value of the Reports. J. Matthews and O. Osborn suggested that the Reports are so useful that they should be published more widely than just in the Transactions of the IAU.

1.10.   W. Wamstecker proposed a resolution to the Commission in support of World Astronomy Days. The text of the resolution was "Recognizing that the detailed planning of the World Astronomy Days in the context of the International Space Year is in full concordance with the

IAU Resolution A4 (XXth General Assembly), Commission 27 supports the activities and hopes that ALL OBSERVATORIES will be able to cooperate to make these World Astronomy Days a successful activity involving astronomers from all IAU member countries." W. Wamstecker explained that the resolution was intended to encourage all observatories to cooperate in multi-site observations during World Astronomy Days. He reported that Commissions 28, 42, and 44 had approved the resolution and that other commissions would be considering it in their business meetings. Various opinions on the import of the resolution were offered by the members. C. Sterken noted that multi-site observations are of considerable value for standard star observations because good estimates of the external errors can be obtained. D. Kurtz expressed doubt that major observatories will be able to devote their larger telescopes to such a program. E. Malone suggested that the scientific value of the observations should be the driving force behind such collaborations. The resolution was approved by the members by a vote of 27 in favor to none opposed.

2. FIRST SCIENTIFIC SESSION (JULY 26, 1991)

New Developments in Variable-Star Research

CHAIRMEN: M. BREGER AND J. PERCY

The first scientific session of Commission 27 began immediately following the close of the business meeting. Seven papers were given, each followed by lively discussion. Approximately 50 people attended the session.

P.A. Whitelock reported on the long-term variability of luminous supergiant stars from infrared JHKL observations. She showed a quasi-periodic variability of Eta Car on a time scale of 2000 days, as well as a secular increase for the same star over the period 1974–1991. VY CMa showed similar behavior: a secular decrease with a quasi-period near 1500 days. For the star AG Car, she noted brightening in an emission shell phase in 1982 and a rapid increase in brightness in 1990. For the latter star, additional observations are urgently needed.

H. Bond discussed pulsations of central stars of planetary nebulae. Six such stars are now known. They are non-radial pulsators with typical periods of 16–31 minutes. All are extremely hot, carbon and nitrogen rich objects, supporting a pulsation mechanism involving cyclical ionization of C and O. Analysis of the pulsations may lead to determinations of the stellar mass, interior structure and evolutionary contraction rate.

The Livermore OPAL opacities have been used by A. Cox to calculate models of double-mode RR Lyrae variables (RRd class). The opacity decreases in the region below 10^5 K, changing the models and making all radial mode periods larger. This results in lower period ratios and higher RRd masses than other opacity codes. These masses are now possibly even larger than evolution masses. If so, new horizontal branch evolution tracks should be calculated.

T. Barnes presented recent work done in collaboration with T. Wilson, S. Hawley, and W. Jefferys on the Cepheid distance scale. A maximum-likelihood statistical parallax analysis of classical Cepheids was performed to determine kinematical and absolute magnitude parameters. The proper motion data used were taken from the extensive compilation by D. K. Karimova and E.D. Pavlovskaya. The best estimate for the mean absolute magnitude at log (period) = 0.8 mag. is –3.46 ± 0.33 mag. He emphasized that any improvement in this statistical parallax result will require marked improvement in the proper motions.

M. Breger presented new results on Tau Peg which impact our understanding of the singly-periodic Delta Scuti stars. Although Tau Peg shows a dominant pulsation frequency of 18.4052 cycles per day, it is not a radial pulsator. He identified the frequency as a non-radial p_3 or p_4 mode with $l = 2$ by examining the amplitude ratios and phase differences between observed light and color variations as well as from the derived Q value of 0.016 ± 0.003 days. Thus for small amplitude stars, the existence of a single frequency of oscillation cannot be taken as evidence of radial pulsation.

C. Sterken communicated a study of period changes in Beta Cephei by A. Pigulski and D. Boratyn. Beta Cep has a pulsation period of 0.1905 days and is also a member of a multiple system. One visual companion is located at 13.4 arcsec and another has separation decreasing from 0.25 to 0.07 arcsec in less than 20 years. The pulsation period of Beta Cep underwent a sudden decrease of about 10^{-5} days in 1920. Period changes are common in Beta Cep stars, but have never been solidly explained by any mechanism. This work analyzed 70 years of photometric and spectroscopic data and showed that the period variations of Beta Cep could be completely explained by the light-time effect in the binary system. From this, they also obtained a preliminary solution for the orbit of the system, which has a period of 92 years.

The observed period ratio of double mode Cepheids, approximately 5/7, is larger than the theoretical one obtained from models having standard evolutionary masses. This disagreement is not yet resolved to everyone's satisfaction. M. Takeuti discussed the discrepancy in terms of the coupling between two different modes of pulsation. The coupling can decrease the period ratio until a synchronized state is reached with ratio 1/2. Using LNA wave functions, the strength of non-adiabatic coupling was calculated. He found that the non-adiabatic coupling is stronger than the adiabatic one for radiative model Cepheids which suggests a possible mechanism for the observed period ratio.

3. SECOND SCIENTIFIC SESSION (JULY 31, 1991)

The Role of Rotation in Stellar Variability

CHAIRMEN: J. PERCY AND M. BREGER

The final scientific session of Commission 27 was re-scheduled to the morning of July 31 because of the fire in the conference hall. As not all members and speakers could be notified of the new time and place of the session, only about 25 people attended the session.

The roAp star HR 3831 has a pulsational period of 11.677 minutes and a variable amplitude of 0 to 5 mmag. D.W. Kurtz presented results by himself and his collaborators A. Kanaan, P. Martinez, and P. Tripe based on 516 hours of new high-speed photometry. They showed that the period of pulsation amplitude modulation is equal to the period of mean light variation. They also demonstrated that the times of pulsation maximum and magnetic extremum coincide and are different from the time of mean light extremum. From these circumstances, they concluded that HR 3831 is an oblique rotator with the magnetic axis and pulsational axis aligned, rather than a spotted pulsator with amplitude modulation caused by surface inhomogeneities.

L. Mestel presented a discussion of the rotational evolution of solar-type stars with different degrees of core-envelope coupling. The results suggested that weak coupling can give a reasonable agreement to observations of rotational properties in the Alpha Per, Pleiades and Hyades clusters. In weak coupling the supply of angular momentum from the core does not affect the surface rotation up to the Pleiades age, but does have some influence at the Hyades age. This model also predicts that the Sun began main-sequence life with rotation at 40 km/s.

The interaction of rotation and pulsation in early-type stars was discussed by D. Baade. Among OB stars, variability is ubiquitous; only stars with strong magnetic fields and/or surface abundance anomalies may be exceptions. In un-evolved and moderately evolved stars, amplitudes drop steeply beyond B7. Although numerous subclasses of variables have been suggested, most lack physical justification. Only the Beta Cephei stars, with short periods and (usually) radial modes, may be a distinct class. In young stars, rotation is important, and the distribution of v sin i values shows a lack of narrow-lined O stars, whereas in B stars there is a slowly rotating population (largely Cp stars) and a rapidly rotating population (largely Be stars). Non-radial pulsation (NRP) can change a star's rotational profile, including its apparent surface rotation rate. Conversely, homogeneous evolution in the presence of rapid rotation would modify the opacity profile and so give rise to different pulsation properties. This may explain the general weakness of low-order NRP modes in Bn stars, whereas they are predominant in Be stars; indeed, they may contribute to the episodic mass loss.

GALAXIES

Report of meetings of July 24, 25, 26, 27, 29, 30 and 31, 1991

PRESIDENT: G.A. Tammann   SECRETARY: E.Ye. Khachikian

Commission 28 was involved in an all-day Joint Commission Meeting, of which the proceedings will be published in the report of Commission 29:
30 July: **Early Nucleosynthesis in Galaxies** (Commissions 28, 29, 35).
One business session and one scientific session on "New Results" were held, while four Working Groups each organized one or two half-day session(s). Short abstracts, where available, of the papers presented are given in the following.

25 July: Business

CHAIR: G.A. Tammann

The following items were covered in the business section:
Draft report. The Commission adopted the report which was distributed to the members of the Commission and which has already appeared in *Report on Astronomy* vol. XXI A. The desirability of the reports is discussed. Of 40 members present, only 28 have read the report, but 38 members voted for a continuation of the reports.
Officers. The Commission unanimously elected E.Ye. Khachikian (USSR) as President for the following three years and V.L. Trimble (USA) as Vice-President. Outgoing members of the Organizing Committee are J. Lequeux (France), Li Qi-bin (China), H. Quintana (Chile), and P. van der Kruit (Netherlands). They are thanked for their in part very important services for the Commission. Continuing members are F. Bertola (Italy), R.S. Ellis (UK), K.C. Freeman (Australia), J.S. Gallagher (USA), S. Okamura (Japan), and G.A. Tammann (Switzerland). The Commission elected as new members G. Bruzual (Venezuela), F. Israel (Netherlands), Jian-sheng Chen (China), and M.H. Ulrich (ESO). The Commission took note of a list of 95 new members who the President has accepted during his three-year term.
Resolutions. The Commission approved the resolution on *World Astronomy Days* as introduced by W. Wamsteker and the resolution on ASCHOT (80 cm space Schmidt) as introduced by H.C. Arp.
Working Groups. C.J. Peterson informed that the WG on "Internal Motions in Galaxies" and "Galaxy Photometry and Spectrophotometry" will be joint. W. Liller announced his wish to pass on his chair of the WG on "Supernovae". The WG on "Redshifts of Galaxies" under J. Huchra and on "Nomenclature" under K.S. de Boer had no sessions during the General Assembly.
Other business. The Commission was informed of forthcoming conferences on galaxies.

24 July: New Results

CHAIR: G.A. Tammann

Abstracts of the following papers were received:

S. Djorgovski, Pasadena: **Systematics of Galaxy Properties: Hints about their Formation.**

Observed ranges of galaxian global properties and the scaling laws and correlations between them reflect processes of galaxy formation and evolution, and thus contain interesting cosmological information.

The basic correlations which tie most of the global properties of galaxies are intrinsically bivariate, both for ellipticals and spirals. The origins of these correlations are not yet fully understood, but some hints about galaxy formation can be deduced from them. Slight environmental variations, possibly caused by differences in the formation histories, may exist in correlations used as distance indicators, and appear as spurious peculiar velocities. Galaxies form two-dimensional sequences in the parameter space whose axes reflect the size (mass, luminosity, or radius), density or surface brightness, and kinetic temperature (velocity dispersion, circular speed for disks). This galaxy parameter space promises to become an equivalent of the H-R diagram for galaxies, a new organizing framework for extragalactic astronomy and cosmology.

A.P. Fairall, Cape Town: **A Southern Redshift Catalogue.**

An updated version of "Southern Redshifts - Catalogue and plots" (with A. Jones) will be available from October. It provides a "best-estimate" of the heliocentric radial velocities of almost 13000 galaxies south of Declination $0°$. It is based on towards 17000 redshift measurements, from over 200 sources, either published or otherwise made public. Each entry is identified as optical, radio (or both) and carries an indication of optical emission lines where present.

Where two or more redshift measurements are available for the same galaxy, an assessment of redshift errors is possible. As a general rule, external errors are twice those claimed by authors, while approximately 1% of redshifts from each reference are erroneous. A significant number of discrepant velocities occur - many from recent fibre-optic spectroscopy.

Redshift plots - "slices" in R.A. and Declination can be generated from the catalogue; if required, various emission-line categories can be identified. These map structures in the southern skies, including superclusters, great and lesser walls, and numerous voids.

V.G. Gurzadyan, Yerevan: **The Dynamics of Galaxies and Clusters: The Concept of Ergodic Theory.**

The metrical theory of Dynamical Systems, Ergodic Theory enables one to investigate most essential topics of the dynamics of galaxies and clusters. Among the important ones is the result that relaxation of real stellar systems, globular clusters and elliptical galaxies occurs not due to binary encounters, as was considered for decades, but by means of collective interaction of all stars. The corresponding collective relaxation time is smaller than the binary one. This fact has numerous fundamental consequences, including the solution of Zwicky's paradox for elliptical galaxies. An interesting observational indication of this fact was obtained recently by Enrico Vesperini of the Scuola Normale Superiore of Pisa while investigating the core collapse phenomena for 127 globular clusters: the binary relaxation time scale does not fit the observational data while the formula for collective one does it quite well.

Disk stellar systems investigated by means of the properties of groups of diffeomorphisms on right-invariant Riemannian metrical spaces demonstrate different features: the motion is exponentially unstable, while the velocity field remains stable. The corresponding mixing time (not relaxation) for the Galaxy and Solar neighbourhood yields 100 million years.

The methods are very useful while investigating many non-trivial properties of the dynamics of other types of stellar systems (including those derived by means of numerical experiments).

Slobodan Ninkovic, Zlatko Catovic: **On the Potential in the Local Group of Galaxies.**

Assuming a rectilinear approaching of the two most massive Local Group galaxies, M31 and our Galaxy, the motion of a test particle in their field is examined. For the potential of an individual galaxy we assume the form $GM/r^2 + b^2)^{1/2}$; G - gravitation constant, M - mass of the galaxy, r - distance to its center, b - constance. The examined cases indicate that even if masses of the Milky Way and of the Andromeda Galaxy are not too high (say about $5 \times 10^{11} M_o$) and the masses are nearly equal, the apogalactic distances at a few hundreds kpc seem not to result in stable orbits.

Slobodan Ninkovic, Zlatko Catovic: **An Investigation of the Andromeda Galaxy.**

Assuming the spherical symmetry with a constant circular velocity the present author studies the Andromeda Galaxy using a sample of 158 globular clusters. The conclusions are that the eccentricities of the clusters seem to be high and that their line-of-sight velocities suggests a value of about 175 km s^{-1} as a lower limit to the circular velocity.

I.A. Issa, Cairo: **Archimedes Spiral and the Surface Distribution of Dark Clouds in M31.**

The surface distribution of dark clouds along the major and minor axes were combined to give a general formula expressing the surface distribution along both axes simultaneously. Absorption values were determined, assuming the parameters of the standard cloud model as a function of the distance along both axes. Isocloud line numbers were drawn as a function of the distance along both axes. Two knots appeared in the new distribution function. This may announce the beginning of spiral arms. The knots may indicate the least possible numbers of dark clouds per kiloparsec for a spiral arm to appear.

M. Kalinkov and I. Kuneva, Sofia: **Superclusters, Supervoids, and Minivoids among Rich Clusters of Galaxies.**

Multivariate redshift estimates for AC0 clusters of galaxies with unknown redshift are made. A catalog of 196 superclusters is presented (density contrast 40 and percolation radius of 100 Mpc). We found 96 doublets of clusters, 10 supervoids and 39 minivoids among rich clusters of galaxies.

I. Kuneva and M. Kalinkov, Sofia: **Reference Catalog of AC0 Clusters of Galaxies.**

The Reference Catalog of all AC0 (Abell, Corwin, and Olowin, 1989) clusters of galaxies is completed. The Catalog contains optical, radio, and X-ray data, gathered from the literature. In fact many catalogs (Abell, AC0, Leir and van den Bergh, Owen et al., Struble and Rood, ...) are merged and some of the characteristics were reduced into a homogeneous system ($H_o = 100$ km s^{-1}Mpc^{-1} and $q_o = 1/2$). The Catalog contains references for observations or theoretical investigations of the clusters. The references are given with their titles and key words.

25 July: WG on Space Schmidt Survey

CHAIR: H. Arp

The necessity for a wide-angle UV space telescope for astrophysical investigations was discussed. Some of the main scientific goals such as full sky survey, study of UV emission of extragalactic and galactic objects as well as faint-surface brightness research both in UV and near IR which may be achieved by the use of an orbital Schmidt-type telescope were presented.

As an introduction to the discussion the following paper was given: H. Lorenz (Germany): The Astrophysical Orbital Telescope ASCHOT. The project has been developed in the Bjurakan Astrophysical Observatory (Armenia) and the Central Institute for Astrophysics Potsdam (Germany). The optical system of the telescope is of fully reflecting Schmidt type. The diameters of the main spherical mirror and the correcting mirror are 1200 mm and 800 mm, respectively. The focal length is 2300 mm and the field of view 5°. The displacement of the focal surface out of the telescope permits to use several detectors for different wavelength regions. The sensitivity of the telescope is estimated to be about 24^m for stellar objects at 1500 A.

The telescope will be launched in 1996.

26 and 29 July: WG on the Magellanic Clouds

CHAIR: M. Feast

The following Organizing Committee was elected: Walborn (USA; Chairman), Wood (Australia), De Boer (Germany), Suntzeff (USA, Chile) and Feast (South Africa).

The following papers were read:

(A)   HST Observations of the Magellanic Clouds.
N. Panagia: The SN 1987A Ring and the Distance of the LMC.
B. Campbell: Wide Field/Planetary Camera Images of the 30 Doradus Central Cluster.
S. Shore: Goddard High Resolution Spectrograph Observations of Melnick 42 and R136 in 30 Doradus.
N. Walborn: Spectral Morphology of Melnick 42 and Faint Object Camera Images of R136 (for G. Weigelt).
C. Blades: Faint Object Camera Images of Planetary Nebulae in the Magellanic Clouds.

(B)   General
M. Bessell: Chemical Abundances of K Giants in the Magellanic Clouds.
D. Hatzidimitriou, R. Cannon and M. Hawkings: The Kinematics of intermediate-age and old stars in the outer parts of the SMC.
H. Dottori and E. Bica: Magellanic Cloud Clusters.
R. Wielebinski: The Status of SMC/LMC Radio Continuum Surveys.
A. Feinstein, R. Vazques and W. Seggewiss: The Young Open Clusters NGC 1962-65-66-70 in the LMC.
W. Seggewiss, K.S. de Boer, T. Richtler et al.: Comparison of Cluster and Field Star Populations in the Magellanic Clouds.
S. Shore: Studies of Novae in the Magellanic Clouds.
N. Martin, H. Lindgren and M.C. Lortet: Star Formation History in the 30 Doradus and Shapley II Constellation Areas.
R. Wing: Statistics of K supergiants in the LMC.

27 July: WG on Internal Motions in Galaxies and Galaxy Photometry and Spectrophotometry

CHAIR: C.J. Peterson

Abstracts of the following papers have been received.

P. Pismis, Mexico: **Annular Features in Spiral Galaxies Rather Spirals than Rings.**

Brightness enhanced annular structures within early type spiral galaxies particularly of the barred type are known to exist. These structures are referred to as "rings". In no case are they genuine rings but rather spirals. Bi-symmetrical details shown in well observed high resolution images attest clearly that the annular features consist of section of symmetrical spirals and are not torus-like structures.

To insist that the annular structures should be *termed tight spirals and not rings is not a matter of semantics*. To call such features "rings" disregards their bi-symmetrical nature, concealing thus the very property which is crucial in deciphering the origin of the so-called rings.

A classification of "rings" is currently accepted; it contains: rings galaxies, polar rings, ringed galaxies in which are defined outer, inner and nuclear "rings". In all cases they are not genuine rings as hinted above. We call attention in particular to nuclear annuli of early type, mostly barred galaxies. Nuclear "rings" are unmistakably tightly wrapped spirals. These show rotation around the nucleus as well as radial motions; the latter may be outward or inward.

In a series of papers (Pismis and Moreno 1984 and on) we have developed a model which explains the formation of tight nuclear spirals. The model: a) Matter (plasma) is ejected from localized areas on a galactic nucleus (ejection is thus neither isotropic nor toroidal). b) These areas are located in a bi-polar fashion on the equator of a rotating nucleus. c) Collimation of the outflow is ensured by magnetic lines of force which entrain the plasma.

The equation of the locus of the ejecta t years after the onset of ejection is given by Pismis and Moreno (1987, 1990). We have applied our model to the central spiral of NGC 4736 adopting its physical parameters from van der Kruit (1976). The computed spiral agrees remarkably well with the observed kinematics as well as with the morphology. Also our model predicts aside from outward motions, which decrease along the opening of the spiral, also inward motions detected in some galaxies. These circumstances constitute a crucial test to our model.

As to energy requirements, to produce an expanding ring will require energy at least three orders of magnitude higher than the energy to produce an expanding tight spiral of similar dimensions.

We have undertaken multi-color photometric as well as spectrophotometric studies of the bulges of a selected group of early type spirals with the aim to determine the detailed morphology and the velocity structure of their nuclear spirals and thus be able to check further our model for the origin and development of the tightly wrapped nuclear spirals.

T.K. Chatterjee, Mexico: **The Frequency of Merging Galaxies and their Dynamical Implications.**

A study of the expected frequency of merging galaxies (not considering the galaxies to be embedded in massive halos), on the basis of the collisional theory of their formation, is conducted by studying collisions with different parameters and with different progenitor pairs. Assuming that ~ 10% of all galaxies are ellipticals, our results indicate that ~ 0.1% of all ellipticals are expected to be merger products. In the light of the current observational evidence, especially of the existence of a "fundamental parameter plane" where the global properties of ellipticals lie, a strong regularity in elliptical galaxy formation is indicated. The results favour the formation of ellipticals by dissipative collapse, followed by secular evolution by mergers.

29 July: WG on Supernovae

CHAIR: V. Trimble

Nine papers were scheduled and delivered in this session. Abstracts of three of them, provided by the authors, conclude this section. Highlights of the others include:

Radio light curves (N. Panagia). Five Type II events, and no Ia's, have now been seen. Radio light curves can be interpreted as ejecta shocking red giant winds of progenitors. Based on this interpretation, the progenitor of 1979C was at least 13 M_o and shed a wind of about 10^{-4} M_o/yr, while 1986J was at least 20 M_o and had a somewhat stronger wind. Recent VLBI data show the growing remnant to be somewhat elongated.

Late-time spectroscopy of 1988Z, a peculiar SN II (E. Sadler). These results have now been published by R.A. Stathakis and E.M. Sadler (1991, MNRAS 250, 786).

Supernova remnants observed by ROSAT (B. Aschenbach). The ROSAT survey has resulted in the first SNRs to be discovered first as X-ray sources; the complete sky is expected to contain about 60 or these. 1987A has not yet been detected, down to a level of 2.5 x 10^{34} erg/s. Magnificent images have been produced for RCW 103, SNR 18.9-1.1, and other remnants.

Supernova remnants observed by BBXRT (R. Petre). A scan of the LMC region has not detected SN 1987A, somewhat below the ROSAT limit and considerably below the flux reported by Ginga in 1988. The spectrum of N132D reveals lines of O, Mg, Si, S, and Fe, in relatively high ionization states. An abundance analysis shows that this thousand-year-old remnant must consist largely of swept up interstellar matter, since the metallicity is about 0.2 that of the sun, the same as the general ISM around it. [Fe/O] is negative, as expected for metal-deficient material.

Supernova remnants observed by ASTRO (T. Stecher). Five SNRs were scanned with this ultraviolet experiment, yielding a two-sigma detection of 1987A and images of the Crab Nebula that resemble those in the higher-excitation optical lines.

The Cygnus superbubble (M. Shull). IRAS maps of this recent star forming region show a superbubble in the process of formation, with dust at 30-35K and energy input from the O stars of Cyg OB1 as well as one or two Wolf Rayet stars and (probably) 5-10 supernovae. The stars of Cyg OB2 have also contributed to the superbubble of the Cygnus X region revealed in the ROSAT images.

M. Hamuy: **The Optical Light Curves of Type Ia Supernovae 1980N and 1981D in Fornax A.**

We present optical photometry of the two supernovae, 1980N and 1981D, which appeared in the peculiar D-type galaxy NGC 1316 (Fornax A). These data are combined with published observations to produce definitive optical light curves. We find that the maximum-light magnitudes of both supernovae were the same to within ±0.1 mag, in agreement with infrared light curve observations. The shapes of the UBV light curves of the best observed of the two supernovae, 1980N, closely resembled those of the type Ia prototype SN 1981B. We also show that an optical spectrum of SN 1980N taken 30 days after B maximum was virtually identical to a spectrum of SN 1981B obtained at the same point in its evolution. These findings lend support to claims that the majority of type Ia supernovae form a highly homogeneous class of objects. Nevertheless, the B-V colors at B maximum of the NGC 1316 supernovae were 0.3-0.5 mag redder than previous estimates of the intrinsic B-V color of type Ia supernovae at this phase. Although dust extinction within NGC 1316 could explain this difference, there is little evidence to support such a large reddening. By comparing the photometric data of SN 1980N with the light curves of SN 1984A in NGC 4419, and by assuming that the absolute magnitudes of the majority of type Ia supernovae are indeed very similar, NGC 1316 would appear to be at essentially the same distance as the core of the Virgo Cluster.

Cecilia Kozma and Claes Fransson: **Radioactive Excitation in SN 1987A and other Supernovae.**

We have calculated the gamma-ray deposition in different composition zones in core collapse supernovae using the Spencer-Fano equation. For these results we have given analytical expressions. The same calculations have been done for pure helium, oxygen, and iron plasmas. We have also calculated the fraction of the absorbed gamma-ray energy which is re-emitted as photons with energy about 3.4 eV, the n = 2 threshold for hydrogen. We have applied these results to calculate the H-alpha luminosity from SN 1987A, and we find good agreement with observations.

P. Lundqvist: **Narrow Lines from SN 1987A.**

New models, based on a ring geometry, have been used to model the narrow line emission from SN 1987A. With a tilt angle of 45 degrees for the ring, the ring radius is ~ 6 x 10E17 cm. Gas density, elemental composition and properties of the ionizing source are close to what has previously been found for a shell geometry (1). The most important improvement compared to the old models, is that ring models may be optically thick to the ionizing burst. This results in better agreement with observations for low-ionization lines like N III] 1750 and N IV] 1486. An improved treatment for the diffuse emission, as well as escape probabilities gives more [O III] emission at late times, also in agreement with observations. Finally, it is noted that several lines should be observable, especially lines of Fe II and Fe III. It is likely that a recently observed line at 5159 A, and which was identified as a Fe VII line (2), is instead a Fe II line. From calculated line intensities [Fe II] 5159 is expected to be as strong as observed Fe II lines at 7155A (2, 3) and 8617A (2).
(1) Lundqvist, P., and Fransson, C., 1991, ApJ, in press. (2) Wang, L., 1991, preprint. (3) Meikle, P., and Cumming R., 1991, priv.comm.

31 July: WG on Supernovae

CHAIR: W. Liller

I. SN 1987A Update (total time = 30 min.)   Suntzeff, N.B., CTIO. Bolometric light curves.
                                                                        Bouchet, P., ESO/La Silla. ESO observations.
                                                                        Meikle, W.P., Imperial College London. IR spectroscopy.

II. Current Status of our Understanding of SNe (60 min.)

Phillips, M.M., CTIO. Optical spectra.
Dubner, G., IAFE. VLA observations of SNe.
Maza, J., U. of Chile. SN Searches and results.
Van den Bergh, S., DAO. Rates, distribution.
Wheeler, J.C., U. Texas. Theoretical spectra.
Chugai, N.N., U.Moscow. Physics of Type II SNe.

Nicholas B. Suntzeff, M.M. Phillips, D.L. Depoy, J.H. Elias, and A.R. Walker: **The Late-Time Bolometric Evolution of SN 1987A.**

We review our results on the optical, infrared, and bolometric luminosity evolution of SN 1987A through day 1444 since the explosion of the supernova. We find that both the optical and mid-infrared magnitudes are slowly leveling out after their rapid decline in brightness during the epoch of dust formation. The optical magnitudes are declining with an e-folding time of near 400 days, which is quite close to the rate expected from the decay of ^{57}Co. The near infrared colors initially reddened after dust formation around day 500, but by day 1000, they began to evolve to the blue, perhaps due to the thinning of the dust. Optical spectra obtained through day 1500 show no evidence for a sudden drop in the optical emission-line luminosities. This suggests that the region of optical emission has not undergone the so-called "infrared catastrophe", when the cooling of the ejecta becomes dominated by the infrared fine-structure lines, at least in the case described by the simple models of Fransson and Chevalier (1987).

The infrared data have been combined with optical photoelectric and CCD UBVRI photometry obtained at CTIO to study the temporal evolution of the bolometric luminosity of SN 1987A. By day 1000, another source of energy beside the radioactive decay of ^{56}Co is needed to explain the slow leveling off of the bolometric luminosity decline. This extra source of energy must be declining in time, and a *constant* energy source such as ^{44}Ti can be ruled out. The best fit to the data is for 5 ± 1 times the predicted ("solar") amount of ^{57}Co/^{56}Co. If a pulsar or other compact energy source is present, the energy input must have declined from ~ 37.5 dex ($\log_{10}(\mathrm{erg\ s}^{-1})$) on day 1000 to ~ 36.8 dex on day 1444. The enhanced ^{57}Co which provides the best fit to the data is in conflict with published hard x-ray and infrared spectral data. However, a careful reanalysis of the uncertainties in the observed hard X-ray and uvoir fluxes, and the model predictions, suggests that *all* the data can be fit by the energy deposition from $0.075 M_O$ of ^{56}Ni, a ^{57}Co/^{56}Co between 2.5 and 4 times "solar", and "solar" ^{44}Ti.

P. Bouchet, I.J. Danziger, C. Gouiffes, L.B. Lucy and M. Della Valle - SN 1987A: **Dust and line luminosities: Observations of the later phases at ESO.**

We review the evidence for dust formation and its spectral characteristics. We show that dust is still present and is absorbing more strongly than ever at least up till day 1400. We present a simplified qualitative model of the expanding envelope with 2 zones: a higher velocity hotter region, and a lower velocity cooler region where the IR catastrophe could have started at day ~ 760. Our observations exclude a value of the ^{56}Co/^{57}Co ratio as high as 4 x solar. Then, the possible source of energy required to give account to the observed flattening of the bolometric light curve is probably a compact central object, radiating as a pulsar, or surrounded by an accretion disk depositing matter either continuously or at varying intervals onto the collapsed object.

The presence of such a source of energy is also required to keep the gas in the circumstellar ring surrounding SN 1987A excited, as we show that the excitation of this material cannot be due solely to the initial UV outburst.

Peter Meible: **IR Spectroscopy of SN 1987A - an update.**

The introduction of more sensitive near infrared detectors has meant that we can still observe SN 1987A spectroscopically. Using the new infrared facility, IRIS, at the AAT, we have extended our spectroscopic coverage to around day 1500 (after the explosion). The spectra cover 0.9 to 2-4 μm at a resolution of 400 and aperture of 5" x 12". Circumstellar lines are increasingly prominent, especially H α I 1.083 μm. Resolved features arising from the ejecta can still be seen. Ejecta [Fe II] lines have faded significantly relative to other features, over the last ~ 500 days. This may be due to recombination of Fe^+ to Fe^0. [Fe I] lines are still prominent.

Since about day 700, lines of HI, Na I, Si I have *matched* the radioactive luminosity decay. This suggests that there has been no increase in the transparency of the emission regions to the χ-rays over a doubling of the supernova age. Exploration of this may require the persistence of clumps which are optically thick to the χ-rays.

Preliminary modeling of the IR spectra (based on model 10 HM abundances) confirms more line identifications.

M.M. Phillips: Optical Spectroscopy of Type I Supernovae.

Optical spectroscopic observations of several recent type I supernovae are presented. The type Ia event 1991 T is of particular interest since the first spectra, obtained approximately one week before maximum, did not show any trace of the Si II $\lambda6355$ and Ca II H k K absorption which normally characterizes these early phases. These lines finally did appear around maximum light, and by ten days after maximum the spectrum of SN 1991 T had evolved to closely resemble the spectra of other well-observed type Ia supernovae at comparable epochs. These data provide dramatic evidence that type Ia supernovae are not a perfectly homogeneous class of objects. The spectral differences observed for SN 1991 T suggest that the distribution of intermediate mass elements in the outer layers of the ejecta of these objects can differ significantly from one event to another.

Spectra of several recent type Ib (helium-rich) and Ic (helium-poor) supernovae are also presented. These objects show a fairly large dispersion in relative line strengths and widths. Indeed, evidence is found for a continuous range of helium line strengths, which suggests that the Ib and Ic designations are actually the extremes of a single class of supernovae.

G. Dubner[1], M. Goss[2], F. Winkler[3] and D. Moffett[2]: VLA Observations of Supernova Remnants.

[1] IAFE, Buenos Aires, Argentina; [2] NRAO, Socorro, USA; [3] Middlebury College, Vermont, USA

We report radio continuum observations of 14 catalogued galactic SNRs located very close to the galactic plane, between $\iota = -10°$ and $\iota = +47°$. The observations were done in 1990 and 1991 with the VLA in the D and C/D configurations, providing at 1465 MHz an angular resolution of about 50" and 35" respectively, with an average noise of about 6 mJy/beam. Two out of 14 sources were found to show a "composite-type" morphology. Other six remnants appear as almost complete circular shells; the rest are incomplete shells with the visible portion aligned with the galactic plane.

José Maza: Supernova Searches and Results.

A total of 766 SNe have been detected since S Andromedae in 1885. Of those 714 have been discovered since 1950. The number of SNe per year have increased from an average of 10 in 1950 up to 20 in 1985. In 1991 SN searches have produced a record of 37 in half a year! From 1979 to 1984 a total of 46 SNe were discovered at Cerro El Roble (University of Chile), 43% of the total number discovered in the world during that period. Now a new program of SN search has been started in 1990, involving Cerro Tololo Interamerican Observatory and the University of Chile. We have detected 9 new SNe in 10 months.

S. van den Bergh: The Galactic Supernova Rate.

Historical observations yield a "local" supernova rate of 5 per millennium within 3 kpc of +4 sun. This value is *much* greater than that calculated from various indirect techniques (see table).

Method	SNe/1000 yr
Historical (R < 3 kpc)	5
Stellar LF new Sun	~ 0.3
Comparison with LMC WR Stars	0.2 ± 0.1

Gas depletion timescale 5 Gyr.   0.2
Shapley-Ames galaxies (2.2 ± 1.0) h^2

The "historical" SN rate would correspond to a *total* Galactic supernova rate of 15-20 *per century*. Or put in another way it would imply that the SN rate in all of M33 is lower than that in the region within 3 kpc of the Sun.

No plausible explanation for this discrepancy has been found.

J. Craig Wheeler: **Theoretical Spectra of Supernovae.**

Atmosphere models of Type Ia supernovae near maximum light show that Nomoto's carbon defla-gration model provides an especially good agreement with observations compared to other similar mod-els. The same model does not agree as well with early spectra of SN 1990N, but suggests that the absorp-tion feature at 3200A is not radioactive cobalt, but a blend of titanium lines.

Observations of the nebular spectra of SN 1987A at 2200 days are consistent with a significant con-tribution to the emission lines coming from the solar abundance of heavy elements in the outer envelope. The helium in the envelope is sufficient to give a strong line of He I λ 10830. Until the core and envelope contributions are adequately deconvolved, it will be difficult to determine the quantity of freshly synthe-sized oxygen.

Models for Type Ib/c supernovae have been computed based on helium cores of moderately massive stars. Such models are too cool near maximum and overly contaminated by Fe II. Models of the nebular phase suggest that the optical spectrum is rather insensitive to significant changes in the composition but that He I λ 10830 is a key diagnostic of the helium abundance.

N.N. Chugai: **SNe II Powered by Ejecta-wind Interactions.**

The interaction of the expanding ejecta of the type II supernova with the pre-SN wind may con-tribute noticeably to the optical output of SN II, particularly at late times. The strong excess of the Hα flux over the model of the radioactive source is revealed in the late-time behavior of SNe II 1979C, 1980K, and 1987F. The bulk of observational properties strongly suggests that this excess is energized by the ejecta-wind interaction. Furthermore, the extremely high luminosity of SN 1987F (as well as its coun-terpart, SN 1988Z) is probably also powered by the interaction of the ejecta with the very dense wind $(M/u_w \sim 10^{17}$ g cm$^{-1})$.

COMMISSION 29 - STELLAR SPECTRA

PETER S. CONTI - PRESIDENT
Joint Institute for Laboratory Astrophysics
University of Colorado
Boulder, Colorado 80309-0440 USA

The triennial Business Meeting of the Commission was held on 26 July with about 50 members in attendance. The IAU Executive Committee formally approved the new Commission President to be D. Lambert, the new Vice-President, M. Bessell. New Scientific Organizing Committee (SOC) members elected from among 14 nominations were D. Baade, A. Cassatella, C. Pilachowski, K. Sadakane, B. Barbuy and B. Gustafson; P. Conti continues as Past President. The President thanked retiring SOC members A. Boesgaard, G. Cayrel de Strobel, M. Spite and J. Smolinski for their service to the Commission.

There have been three newsletter mailings to Commission members over the past three years, and 250 astronomers responded positively in their desire for continued membership; 10 resigned, or were deceased. About 40 others gave no indication of further interest in Commission 29 membership and, following guidelines laid out by the SOC in 1988, were recommended to the IAU Secretariat to be purged from our rolls. A list of another 50 or so names who had indicated, from various sources, they wished to become members of Commission 29, was transmitted to the Secretariat for addition to our membership.

Commission 29 has sponsored, or co-sponsored, the following IAU Symposia that have been held since the last General Assembly: #143 Wolf-Rayet Stars, Denpasar, Indonesia June 1990; #145 Photospheric Abundances, Druzba, Bulgaria August 1990; #148, Magellanic Clouds, Sydney, Australia July 1990. We were also involved with IAU Symposia #149 on Stellar Populations in Galaxies, Angra dos Reis, Brazil and #151 Evolutionary Processes in Binary Stars, Cordoba, Argentina, held right after the General Assembly. Commission 29 is also sponsor or co-sponsor for proposed IAU Symposia or Colloquia as follows: Peculiar Versus Normal Phenomena in A-Type and Related Stars, Trieste, Italy, July 1992; Inside the Stars; Stars at the Main Sequence or Close to It, Vienna, Austria, April 1992; The Sun as a Variable Star, Boulder, Colorado, 1993. The first two of these have been approved by the IAU.

A full day Scientific Workshop on "Early Nucleosynthesis in Galaxies", sponsored jointly by Commissions 28 and 35, was organized by P. Conti and A. Maeder. The aim was to bring together new observations and theoretical advances on the abundances, nucleosynthesis and stellar evolution in the early stages of galaxy development. About 80 astronomers listened to 10 review talks and participated in stimulating discussions. An introductory talk by P. Conti posed questions concerning the chemical gradients over SPACE and TIME in our galaxy. How well established are they? He asked workers to be careful when discussing composition and metal abundance. For example, using the most recent data, halo stars have [O/Fe] > 1 but in the Magellanic Clouds [O/Fe] < 1. Clearly the halo stars and the Clouds have followed different star formation histories. The most recent discussions of abundance gradients with galacto-centric distance, using B stars and HII regions, do NOT show a uniform decrement with distance, rather merely a tendency toward higher abundances inward from the sun, and lower outward, with considerable scatter. D. Lambert gave a detailed review of abundances in Pop II stars and C. Pilachowski discussed her work with T. Armandroff on CNO abundances in Globular Clusters. They found that 1) [C+N+O/Fe] is within the range 0.0 to +0.5 in globular clusters over a wide range in [Fe/H]; 2) [C+N+O/Fe] is constant within a cluster with the possible exceptions of Omega Centauri and M22; 3) Log (O) vs Log (C+N) diagrams for the clusters M92, M13, M4, and 47 Tuc are consistent with the mixing of CNO processed material either in the star itself or in an earlier star, and 4) the constancy of [C+N+O / Fe] within clusters argues strongly that CN variations found in unevolved stars in clusters are not due to contamination by products of triple-alpha burning. M. Bessell gave a comprehensive review of the composition of stars within the Magellanic Clouds.

J. Bergeron followed with a discussion of abundances seen in the absorption line systems toward QSOs. These are currently examples of the least processed material yet studied and give us unique information on the early history of the Universe. B. Pagel presented his recent work about the helium abundances in metal deficient HII regions and HII galaxies. Previous estimates of helium abundances from recombination lines have been improved by selecting objects with negligible neutral helium and careful segregation of objects with W-R spectra (W-R galaxies) which had biased earlier determinations of the primordial helium abundance Yp and dY/dZ. Using the 20 remaining objects, one obtains the maximum likelihood regression Y = 0.227 + 120 (O/H) and dY/dZ = 6, considerably larger than predicted recently by Maeder. A. Maeder followed with a review of his work on helium and metal synthesis in massive stars with low Z. This is particularly applicable to understanding the chemical evolution of the Magellanic Clouds.

The contribution of type I and II supernovae to Galactic nucleosynthesis was the topic offered by F.-K. Thielemann. These SNeI and SNeII represent the major sources of heavy element production in galaxies. Stellar winds of intermediate mass stars only contribute significantly to C and N isotopes and are the only source for the STRONG s-process component. Thus, all nuclei from O through the Fe-group, and for heavier nuclei, the WEAK s-process component and all r-process nuclei originate from supernovae. The time variation in the rates for both types of events is different. SNeII, due to their origin from fast evolving massive stars, dominate the heavy element production in the early phase of galactic evolution. SNeIa, originating from mass transfer in binary white dwarf systems, leading to a growth of the more massive star beyond the Chandrasekhar mass, occur only with much longer time scales. The solar system composition is only a snapshot in time at a specific location in the galactic disk. By now, the major solar system abundance features can be understood in this framework, due to the fact that models of SNIa and SNII events (over the mass range of progenitor stars) have become available. M. Grenon discussed the sites of nucleosynthesis for the halo and bulge stars and C. Chiosi finished the day's Workshop with a talk on the chemical evolution in the early Galaxy.

A number of members of Commission 29 met on 25 July to discuss the problem of archiving and distribution of spectroscopic data. This meeting was organized and chaired by R. Viotti. The 35 participants considered that (i) a large amount of unique information has been collected on spectroscopic plates, and (ii) the widespread application of electronic detectors has generated an extremely rapid growth of raw spectral files. The importance of safeguarding these data, and building an accessible catalogue, was recognized. It was suggested that an IAU Working Group be created in order to establish agreed means of archiving and distributing spectroscopic data. The participants included representatives of many important astronomical observatories, and of the two main astronomical data centers (CDS and NASA ADC). Professor Alexander Boyarchuck, IAU President-elect, gave the introductory talk, followed by contributions of Roberto Viotti (the requests of the Astronomical Community), Elena Terlevich (the La Palma Data Archiving System), Miguel Albrecht (the ESO Archive Facility), Robert Garrison (data archives), Willem Wamsteker (the IUE Archive), Carlos Jaschek (CDS and data documentation), and Richard West (spectroscopic plates). There were lively discussions followed by the approval of a specific recommendation to IAU Executive Committee. The Proceedings of the Meeting will be published in 1991 as an Internal Report of the Istituto di Astrofisica Spaziale edited by R. Viotti and G.B. Baratta.

The Working Group on Peculiar Red Giant Stars (joint with Commission 45) held its business meeting on 29 July under the leadership of H.R. Johnson. A twice yearly Newsletter is currently being sent to over 150 astronomers; S. Yorka is continuing as editor. R.F. Wing was elected the new Chairman of the WG and H.R. Johnson, U.G. Jorgensen, J. Magalhaes, M. Querci, R. Stencel, T. Tsuji and S. Yorka were selected as members of the SOC. The directions of future interest of the WG were broadly discussed. Some felt the WG should be more narrowly focussed on chemically peculiar red giants; others suggested yellow giants should be included. The new Chair thought the WG should focus on S and carbon stars. While consensus on this issue was not reached, it was agreed that a new IAU Meeting on red giant stars should be promoted within the next 3 years.

The Working Group on Be stars held its business meeting during two sessions on 31 July, chaired by D. Baade. A new SOC was elected with L. Balona (chair), A. Dachs, D. Gies, P. Harmanec, J. Percy, G. Peters, M. Smith and R. Waters. A symposium on Be and related stars planned for October 1993 at the Cote d'Azur was discussed. Peters will continue to be editor of the Be star Newsletter with financial support from ESO. The first half of the scientific session, chaired by Baade, consisted of a review of the differences between Be and non-Be stars by Balona and a talk by Hearn on what theorists would like observers to do. Ringuelet and Doazan presented short contributions. The second half consisted of a lively discussion of new ideas on the Be phenomenon, chaired by Hearn. These included pulsation versus rotational modulation, the bi-stability mechanisms, spectral transients, the effect of increased iron opacities, etc.

The Working Group on Chemically Peculiar stars held its business meeting on 25 July under the leadership of K. Sadakane, with some 25 members in attendance. The new editor of the A Peculiar Newsletter is S. Ansari. New members of the SOC were elected: M. Gerbaldi (Chair), V. Khoklova, D. Leckrone, G. Mathys, G. Michaud, K. Sadakane, and S. Ansari. Short scientific presentations were made by G. Mathys ("Geometry of Ap Stars Magnetic Field; New Constraints from Spectro-polarimetric Observations"), J. Mathews ("Measuring the Limb

Darkening of Pulsating Ap Stars"), and D. Leckrone ("Ultraviolet Spectroscopy of Chi Lupi with the HST GHRS").

The Joint Working Group on Standard Stars (with Commissions 30 and 45) met on 25 July under the Chairmanship of A. Batten. After spirited discussion about the future of this WG, it was decided to continue its existence, perhaps with new emphasis. R. Garrison was elected the new Chairman, and C. Corbally and M. Mccarthy graciously volunteered to serve as editors of the Newsletter which would be printed and distributed by the Vatican Observatory. S. Adelman, A. Batten, P. Egret, I. Glushneva, D. Gray, M. Mccarthy, E. Olsen and L. Pasinetti were elected to the SOC. Brief scientific communications were presented by A. Batten, J. Rountree-Lesh, R. Viotti, R. Garrison and M. Berger.

There was an informal gathering on 27 July, organized by R. Viotti, of about 25 astronomers engaged in work on Symbiotic Stars. There is an apparent world-wide interest in active exchange of information concerning these enigmatic stars. It was suggested that interested astronomers could gather information through e-mail contacts. R. Viotti (40058::viotti), S. Shore (iue::shore), P. Whitelock (paw@saao.ac.za) and J. Ratter (cfa::cfa8::aavso) volunteered to act as nodes for general, UV, IR, and optical correspondence, respectively. It was agreed to consider organizing a IAU Colloquium on Symbiotic Stars and Related Objects in the near future. It was also thought that there was no need, at the present time, to organize a formal working group within the structure of Commission 29.

COMMISSION 30: RADIAL VELOCITIES (VITESSES RADIALES)

Report of Meetings 24 July - 1 August 1991

PRESIDENT: D. W. Latham   SECRETARY: R. P. Stefanik

I. *Business Meeting.*

The business meeting of Commission 30 was held on Friday 26 July 1991 (session 4). The following items were on the agenda:

I.1. PRESIDENT'S REPORT.

Measured in terms of the number of new high-quality radial velocities obtained for stars and galaxies, the last three years must be the most productive in the history of the Commission. On the stellar side, the CORAVEL and CfA groups continued to mass produce thousands of velocities for a variety of ambitious survey projects, usually of quite faint objects and with precisions often better than 500 m s^{-1}, while Roger Griffin has continued to observe in Cambridge, publishing spectroscopic orbits like clockwork and rapidly closing in on photoelectric orbit number 100. Other traditional centers for stellar velocity work in Canada, Argentina, Australia, and the United States continued to be active, while several new efforts in New Zealand, the Soviet Union, and the United States started producing results, soon to be followed by an effort in Beijing. Together these efforts have produced fundamentally important new insights in the overall characteristics of the binaries in various stellar populations. Theories of star formation will now have to confront these new observational results.

The long-standing Canadian effort to monitor a small sample of bright stars with very high precision was joined by sustained efforts in Texas, Arizona, and California. Altogether, more than 100 stars are now being monitored at the 5 to 20 m s^{-1} level, and we may expect this activity to expand substantially in the future, as interest in the search for extrasolar planets grows, both in support of NASA's ambitious SETI Microwave Observing Project Targeted Search, which is fully funded in 1991 and expects to run through the end of the millennium, and as part of NASA's proposed new initiative, Towards Other Planetary Systems, which is not yet funded. The one sour note here is that Bruce Campbell has decided to leave astronomy and has resigned from Commission 30 and the IAU.

On the galaxy side, the interest in the large-scale structure and velocity flows in the universe has been intense, and enormous progress has been made with redshift surveys. Originally we had planned a scientific session in Commission 30 to discuss the latest results in this field, but there was so much duplication with sessions scheduled in the Cosmology and Galaxy Commissions, that we chose to avoid heavy overlap and changed the schedule. There have been at least two significant developments in the technology of measuring galaxy redshifts. New CCD spectrographs are coming into use, often with multiple fiber feeds. Several groups are now able to obtain more than 100 redshifts per night of telescope time, and this number may eventually approach 1000. There is even an effort to build a special telescope facility specifically for the project to measure a million redshifts.

Thus there should be little wonder that the task of keeping track of all the new observational results is rapidly becoming very daunting. Through the selfless efforts of M. Barbier, and D. S. Evans before her, the stellar radial velocity catalogs have almost been kept up to date. However, M. Barbier is scheduled to retire in 1994, and I have invited her to review her efforts so that we may discuss how this work might be continued in the future. On the galaxy side there are at least three major efforts to keep track of the redshifts that have been observed world wide, but here I suspect the situation will rapidly get out of hand, and not just because the numbers of observations are increasing so rapidly. There is the additional problem that most of the objects are so faint that they do not appear in any catalogs. Indeed, the first step for most redshift surveys is to make the photometric and astrometric observations necessary to create an observing list! The galaxy observers have one advantage anyway, namely that galaxies don't have variable velocities due to orbital motion or pulsation, and multiple observations are not needed to test for variable velocity.

With stars, one of the main interests is to detect and interpret velocity variations, especially at low amplitudes. To detect long-period low-amplitude variations there must be good stability in the instrumental zero points, and to combine

269

data from one telescope to the next, the relative zero points must be well determined. These are two of the reasons why there has been so much effort to establish a new set of Radial Velocity Standard Stars. Among the galaxy observers, the attitude towards standards and comparing velocity zero points is much more casual. This is a problem which should be faced. Perhaps this Commission can force some progress by proposing a list of radial velocity standard galaxies, both emission and absorption line objects.

As an indication of the scientific vitality of the Commission, I mention three of the meetings and conferences that we have cosponsored and helped plan:

-Workshop on Large-Scale Structures and Peculiar Motions in the Universe. Rio de Janeiro, Brazil, 23-27 May 1989.
-Evolutionary Processes in Interacting Binary Stars. IAU Symposium No. 151. Córdoba, Argentina, 5-8 August 1991.
-New Frontiers in Double and Multiple Star Research. IAU Colloquium No. 135, Atlanta, Georgia, 5-10 April 1992.

Inside the IAU bureaucracy there are two ongoing debates, one about the Transactions A, the Commission Reports on Astronomy, and the other about the format of the General Assembly. I have supported the view that the time has come for the Reports on Astronomy to be discontinued. In my experience the Transactions A serve very little use, but their preparation extracts a heavy penalty of forced labor from the Commission Presidents. In support of this view I refused to prepare a report for Commission 30, so none appears in Transactions XXIA. (At this point a majority of the Commission members present indicated support for continuing the Reports on Astronomy. One member, an ex-president of another Commission, indicated strongly that they should be discontinued.)

For some purposes the Commission structure of the IAU is reasonably effective. It provides a well-defined population of senior scientists that can be called on for advice and services, for example to referee proposals for Symposia and Colloquia, to prepare Reports on Astronomy, and to organize the scientific sessions at the General Assemblies. In principle the Commission officers have all indicated their willingness to accept these responsibilities, although in practice many officers are very busy people who do not have lots of time to spare. When it comes to organizing the scientific sessions at the General Assemblies, the performance of the various Commissions can be quite spotty. Another problem is that the Commission officers tend to be the "old boys," and this is one of several reasons why new work by young astronomers often fails to get represented properly.

How can the scientific sessions at the General Assemblies be made more consistently excellent? Inevitably this will only happen if the session organization is taken out of the hands of the individual Commission officers. As documented in Information Bulletin 66, Derek McNally has suggested that scientific sessions be organized by Program Committees for the eight sectors: solar system, sun, stars, ISM, the Galaxy, galaxies, cosmology, and instrumentation. This might indeed be a good way to improve the quality and uniformity of the scientific sessions. Of course, some activities are best left with the Commissions and Working Groups. I have in mind such things as efforts to set up new systems of standards.

I have run into a serious problem in my efforts to organize the Commission 30 scientific sessions for General Assembly XXI. Several of the speakers who I invited to give key papers decided not to attend the General Assembly, but instead to attend one of the associated Symposia and Colloquia. They explained that they could not afford to be away for the extended time needed to attend both the General Assembly and the associated Symposium or Colloquium. In several cases I was told that the associated Symposium or Colloquium would provide a more stimulating use of the limited time and travel resources available. Thus, it appears that the associated Symposia and Colloquia are diluting the scientific content of the General Assembly. Perhaps the prohibition of associated Symposia and Colloquia would solve this problem. Another far out idea might be to run the associated Symposia and Colloquia exclusively at the General Assembly, thus providing the lion's share of the science. If you can't lick them, join them. The bottom line is that I agree the time has come to make some adjustments to the format of the General Assembly, and I like many of the suggestions that D. McNally published in Information Bulletin 66. The next General Secretary, J. Bergeron, also wants to make some changes in the way scientific sessions are organized, so look for some evolution in the format of the General Assembly starting in 1994.

I.2. MEMBERSHIP.

The Commission voted to welcome the following new members and consultants of Commission 30;

T. R. Beers (USA)	N. Samus (USSR)
R. J. Davis (USA)	A. Tokovinin (USSR)
A. W. Irwin (Canada)	J. L. Tonry (USA)
J.-C. Mermilliod (Switzerland)	G. Solivella (Argentina)
N. I. Morrell (Argentina)	G. A. H. Walker (Canada)
P. S. Pellegrini (Brazil)	A. Duquennoy (consultant, Switzerland)

I.3. COMMISSION OFFICERS.
The Commission voted to approve the following slate of officers for 1991-1994:

President: G. Burki (Switzerland), *Vice-President*: C. Scarfe (Canada)
Organizing Committee: J. Andersen (Denmark), L. N. da Costa (Brazil), A. P. Fairall (South Africa), K. Freeman (Australia), J. B. Hearnshaw (New Zealand), and D. W. Latham (USA).

I.4. WORKING GROUP ON RADIAL VELOCITY STANDARD STARS.
A scientific session was devoted to reports on the progress towards a new set of IAU Radial Velocity Standard Stars (see report below). It was clear that a great deal of work remains to be done in this area, and the Commission voted to continue the Working Group for the period 1991-1994, with the following members: R. P. Stefanik, (chairman, USA), J. Andersen (Denmark), A. Duquennoy (Switzerland), J. B. Hearnshaw (New Zealand), R. D. Mathieu (USA), and C. Scarfe (Canada). The Working Group itself met for two hours on Tuesday 30 July 1991 (see report below).

I.5. CATALOGS.
M. Barbier reported on her work with radial velocity catalogs in Marseille. The Bibliographic Catalogue, published by M. Barbier and M. Petit, (A&AS 85, 885, 1990), takes into account references from 1970 to 1985. The first Radial Velocities Catalogue was published (A&AS 80, 77, 1989) and gives new mean radial velocities till 1980. The next Catalogue is expected to be published in 1992, and will give new velocities up till 1985. Unfortunately, M. Barbier plans to retire in 1994, and she asked for help in identifying a way to continue these cataloging efforts after her retirement. The Commission asked the Working Group on Radial Velocity Standard Stars to review the catalog situation and to make recommendations. The Commission considered reestablishing a Working Group on Mean Radial Velocities, but recommended instead that issues related to combining data from different observatories should be handled by the Working Group on Radial Velocity Standard Stars.

II. *Scientific Meetings*.

Commission 30 held a total of five formal scientific sessions, one of them jointly with Commission 33. In addition, Commission 30 members presented several papers in the dual session "Binary stars, radial velocities, and open cluster memberships," organized by Commission 37, the dual session "Distribution of gas and stars in the Milky Way," organized by Commission 33, and in Joint Commission Meeting VII "High-redshift galaxies and large-scale structure."

II.1. BINARIES IN STELLAR POPULATIONS.
This meeting was organized and chaired by D. W. Latham. It took place on Wednesday morning 24 July 1991 (sessions 1 and 2). The purpose was to report on the great strides which are being made in our understanding of the characteristics of binaries in several different stellar populations, for example in the disk and halo of our Galaxy, and in open clusters. Much of the progress is coming with spectroscopic binaries, where most of the orbital periods are less than 1,000 days. To a large extent this is a fruition of several large efforts begun 10 and even 20 years ago. Of special importance are the large surveys of carefully selected samples that have been monitored for several years. These samples are just beginning to get large enough so that the distributions of the orbital characteristics for the binaries have some statistical reliability. As a result, we are gaining fundamentally important results about the statistical properties of binaries, which will now confront the theorists who work on the formation and evolution of binary and multiple systems.

As the time duration of some of the longer-lasting surveys is extended, we find that quite a few spectroscopic orbits with periods in the range 1,000 to 10,000 days are being derived. Thus the gap between spectroscopic orbits and visual orbits is being narrowed. In recognition of this growing overlap between two techniques which have historically been independent, there will be an IAU Colloquium, No. 131, on New Frontiers in Double and Multiple Star Research, to be held near Atlanta, 5-10 April 1992.

A session on Binaries in Stellar Populations should begin with a presentation on pre-main-sequence binaries. Ten years ago this would have been a very easy paper to present, because only one or two pre-main-sequence binaries were known. Actually, this was pretty embarrassing, because we knew that there were lots of main-sequence binaries. Was it really possible that stars decided to get together to become binaries only after they arrived on the main sequence? But radial-velocity studies of large samples in star-forming regions such as Taurus-Auriga, Orion, NGC 2264, and Ophiucus have turned up quite a number of spectroscopic binaries, and now there are more than ten published orbital solutions. An especially interesting case is the classical T Tauri star GW Orionis, with a circular orbit despite the relative long period of 242 days. The spectrum for the infrared excess from this system has a dip near 10 μ, indicating a gap in its accretion disk at the distance where the companion is circling.

One of the interesting issues in the study of the orbital characteristics of binaries in a coeval sample is the transition between circular orbits at shorter periods and a distribution of eccentric orbits at the longer periods. There seems little doubt that the short period binaries originally had also a distribution of orbital eccentricities, but tidal mechanisms have circularized all the orbits with close separations, and therefore short periods. Some experts argue that orbital circularization happens mostly during the pre-main-sequence stage while the stars are still swollen and collapsing. Other experts argue that circularization happens mostly during the long main-sequence lifetime of the binaries. This is an issue where observations should be able to make a contribution, but the results are still ambiguous. Perhaps both mechanisms are important.

The new results show that the binary populations in the disk and halo must have started out with essentially the same characteristics, although both stellar evolution and orbital evolution have effected the characteristics observed now for the halo. In particular, the frequency of spectroscopic binaries in the halo appears to be about 20%, indistinguishable from the disk and open clusters.

The following scientific contributions were presented:
David W. Latham: An Overview of Binaries in Stellar Populations.
Virginia Trimble: The Origin of the Halo Binary Myth.
Gilbert Burki: Binaries amongst the Supergiants and Cepheid Variables.
Michel Mayor: The Characteristics of Binaries amongst Solar-Type Dwarfs in the Disk.
David W. Latham: The Characteristics of Binaries amongst Halo Dwarfs.
Michel Mayor: The Characteristics of Binaries in Open Clusters.

II.2. THE IMPACT OF RADIAL VELOCITY OBSERVATIONS ON GALACTIC STRUCTURE AND EVOLUTION.

This joint meeting with Commission 33 was organized by D. W. Latham and chaired by M. Mayor. It took place on Saturday morning 27 July 1991 (session 1). The purpose was to report on the progress which has been made towards understanding the structure, kinematics, chemistry, and evolution of the disk and halo of the Milky Way, from radial-velocity observations of both stars and gas.

The following scientific contributions were presented:
Leo Blitz: The Shape of the Disk.
David W. Latham: The Structure, Kinematics, Chemistry, and Evolution of the Thick Disk and Halo.
Michel Grenon: Connections between the Disk, Halo and Bulge.
Dante Minniti: The Kinematics and Chemistry of the Bulge.

II.3. RADIAL VELOCITY STANDARD STARS.

This meeting was organized and chaired by D. W. Latham. It took place on Saturday morning 27 July 1991 (session 2). The purpose was to report on the progress made towards the goal of establishing a new set of late-type IAU Radial Velocity Standard Stars with individual mean velocities and absolute zero point of the system good to 100 m s^{-1}. The agreement on the zero point for solar-type dwarfs, based on observations of minor planets at three observatories, appears to be close to the 100 m s^{-1} goal. However, there is a significant color effect, resulting in a disagreement of 1 km s^{-1} or more for the reddest stars. At the blue end of the main sequence, new techniques have been applied and progress is being made in the monitoring of potential candidates for standard stars. As more observations accumulate, the evidence grows that most of the late-type giants have low-amplitude velocity variations, as do some of the late-type dwarfs. The session ended with a lively discussion of what work should be done next.

The following scientific contributions were presented:
David W. Latham: An Overview of the Efforts towards New Radial Velocity Standard Stars.
Alan H. Batten: Standard Star Observations at Victoria.
Robert P. Stefanik: Standard Star Observations at the Center for Astrophysics.
Michel Mayor: Standard Star Observations with CORAVEL.
Helmut A. Abt: Fekel's Observations of Early-Type Candidate Standards at Kitt Peak.

II.4. NEW TECHNIQUES.

This meeting was organized and chaired by D. W. Latham. It took place on Tuesday afternoon 30 July 1991 (session 3). The original purpose was to review the efforts at several observatories to develop techniques for very high precision in the range 5 to 50 m s^{-1}. Unfortunately, only two of the nine groups doing this work were represented at the General Assembly.

The following scientific contributions were presented:
David W. Latham: Towards Higher Precision; An Overview of Efforts Around the World.
Thomas G. Barnes: The Texas Program.
R. de la Reza: A Search for T Tauri Binaries among IRAS Stars.

III. *Working Group on Radial-Velocity Standard Stars.*

A meeting of the Working Group on Radial Velocity Standard Stars took place on Tuesday morning 30 July 1991, attended by those members present at the General Assembly (R. P. Stefanik, M. Mayor, G. Burki, and D. W. Latham). The Working Group discussed recommendations in the following areas:

-Procedures for reporting to the astronomical community the present status of the Radial Velocity Standard Stars.
-Work on Radial Velocity Standard Stars to be done over the next three years.
-Priorities and guidelines for cataloging efforts in the future.
-Publication procedures and formats for all results.

Despite the hopes expressed at General Assembly XX in Baltimore, we are still not in a position to publish a definitive list of new IAU Radial Velocity Standard Stars, because of two problems. First, we have the vexing question of the color dependence of the zero point comparison between Victoria, CORAVEL, and CfA. The fact that the Victoria and CfA zero points agree quite well should not be interpreted as conclusive evidence that the problem lies with CORAVEL, because there are several types of systematic problems which could still plague the CfA and Victoria systems. Further work should be done to clarify the source of the color problem. Second, quite a number of the present set of IAU Radial Velocity Standard Stars are clearly variable, and several now have reliable orbital solutions. Even among the "future primary standard star candidates" published in the Commission 30 report in Transactions XXB, where an attempt was made to eliminate stars with significantly variable velocities, several additional candidates are now suspected to be variable. The problem is that there will always be suspected variables as more data with better precision are accumulated. Where should the line be drawn?

III.1. REPORTING THE STATUS OF THE RADIAL VELOCITY STANDARD STARS.

An attempt should be made to communicate to the astronomical community the identity of those IAU Radial Velocity Standard Stars which are variable at a semi-amplitude larger than 1 km s^{-1} or whose IAU velocity appears to be in error by more than 1 km s^{-1}. The Working Group agreed that a revised list of IAU Radial Velocity Standard Stars should be provided to the *Astronomical Almanac* in which such stars have been removed from the list of approved standards. To avoid possible confusion that might result from the publication of interim velocities, only the present IAU velocities should be listed. To inform astronomers which stars were removed from the primary list, a second table should be provided in the *Astronomical Almanac*, listing those stars and giving a short reason why they were demoted. This way astronomers may have some hint about the reasons for the fall from grace of their favorite standard stars.

To stimulate progress towards the establishment of a new set of primary IAU Radial Velocity Standard Stars, the Working Group recommended that the teams who are working on this project at individual observatories should each publish a status report describing their efforts and listing their individual velocity measurements. This will make it much easier for the Working Group to compare the results from the different teams, which can not be done at one or two short meetings during the crowded schedule of a General Assembly.

The chairman of the Working Group should assemble and maintain a catalog of the IAU Radial Velocity Standard Stars plus the candidates for the new list of standards, listing the best value presently available from all the teams measuring these stars as well as the IAU velocity, and containing notes on any variability either confirmed or suspected. This catalog should be made available in electronic and/or paper form upon request to the chairman, R. P. Stefanik at the Center for Astrophysics: stefanik@cfa.harvard.edu (internet) or stefanik@cfa (bitnet).

III.2. WORK FOR THE NEXT THREE YEARS.

In consideration of the good agreement between CORAVEL, CfA, and Victoria for the G dwarfs, where the zero point has been established by observations of minor planets, a modest set of G dwarf stars should be selected as candidates for a subset of primary standards which might be ready for adoption by the next General Assembly. The extensive monitoring of nearby solar dwarfs carried out by the CORAVEL team over the past decade should be drawn upon to define a list of one or two dozen candidate stars which are unlikely to have large velocity variations. The candidates should be distributed around the equatorial region of the sky to allow access from both hemispheres. These stars should then be observed by other teams with the highest precision possible, both to allow comparisons of the zero points and to further check for variable velocities.

To investigate possible sources of the color term in the zero-point comparison, observations of the main sequence of nearby open clusters should be used, to as faint and red limits as possible. The Hyades cluster is a good candidate, but care should be taken to adopt exactly the same convergent point when comparing the results for specific stars in this cluster as measured by various instruments. Attempts should also be made to develop calculated spectra for use as templates with instruments which record spectra, where the correlations are computed digitally. Unfortunately, the models and molecular line lists are not yet good enough to extend this approach to very cool stars.

For those stars which appear to have velocity variations which are strictly periodic, attempts should be made by the various teams to see if they get the same orbital solutions from their independent data sets. Here the goal is to extend our knowledge of low-mass companions of stars.

Establishing a satisfactory system of standard stars of early spectral types (O-B-A) will require several years of additional effort. Continued monitoring of Fekel's list of candidates is strongly encouraged.

III.3. CATALOG PRIORITIES.
The Working Group discussed the various catalogs which are needed, and recommended the following priorities:

1. Maintain a bibliographic catalog. One of the main goals of this effort is to make sure that the SIMBAD data base is as accurate and up-to-date as possible. The team in Strasbourg does not have the resources or expertise to evaluate the literature critically, so this must be done by someone with considerable experience in the radial velocity field. Ideally this catalog should be updated on an annual basis. An astronomer must be found who can continue this work after M. Barbier retires in 1994.

2. Maintain a spectroscopic orbit catalog. Alan Batten has already retired and will no longer be able to work towards a ninth edition of his catalog of spectroscopic orbits. Here again it is important to have an expert who can make a critical evaluation of the published orbits, following the tradition established by Batten. However, the Working Group recommended that all published orbital solutions for each star should be included in a concise format, to allow the user to make an independent evaluation of the status of the solutions. David Latham volunteered that this work could move to the Center for Astrophysics, where it could be included as part of the effort to document the characteristics of the target star candidates for NASA's SETI Microwave Observing Project.

3. Maintain a mean velocity catalog. Hopefully an astronomer can be found to continue this work after M. Barbier retires in 1994. Such a catalog can be used for statistical studies and research on Galactic structure, so it would be useful if it included photometry where available. With the increase in the number of different instruments being used for velocity measurements, a new weighting scheme should be devised, based on the actual precision achieved by each system.

For all these catalogs epoch 2000 positions should be used, as officially recommended by the IAU, and the names of stars should follow IAU recommendations as closely as possible (IAU Transactions XXB, p. S36, 1988).

III.4. PUBLICATION OF RESULTS.
The Working Group supports strongly the long-standing IAU recommendation that all individual velocity measurements should be published, not just for variable stars, but also for stars that appear to be constant. With each velocity (expressed in km s^{-1}) an estimate of the error should be given. The heliocentric Julian Day for the middle of the exposure should be used, not the modified Julian Day. The values for the velocities and times should be given to one more place than is significant. In the header for each star, not only should the star name be given, following the IAU recommendations documented in the preceding section, but also the epoch 2000 coordinates should be listed. This redundancy is very valuable for identifying errors, and for assisting in the preparation of observing lists.

To assist in the preparation of merged catalogs of individual velocities from various observatories and in the preparation of a mean velocity catalog, the Working Group recommends that an electronic and/or magnetic version of the results in ASCII format be prepared when each publication is accepted, and should be sent to a central repository, presumably the astronomer who takes over the responsibility for the mean velocity catalog. The recommended format for every velocity includes the header information, to facilitate chronological merging:

Star name (16 characters); position "hh:mm:ss.s sdd:mm:ss" (epoch 2000); instrument code (4 characters); velocity "svvv.vvv" and error "vv.vvv"; heliocentric Julian Day minus 2,400,000 "ddddd.dddd"; and comments, such as alternative names. It is envisioned that the person in charge of the central repository would assign a numerical code to identify the instrument and publication, and this would also be included with each entry. Each field would be separated by a blank space, to enhance readability and ease of proofreading.

III.5. CANDIDATE RADIAL VELOCITY STANDARD STARS

For more than a decade, Francis K. Fekel has been monitoring the velocities of bright early type stars which might be suitable candidates for eventual service as radial velocity standard stars, using a variety of instruments: Kitt Peak photographic, McDonald Observatory photographic, McDonald Observatory Reticon blue (4500A) and red (6375A), and Kitt Peak CCD blue (4500A) and red (6430A). His three primary early type standards are 68 Tau, θ Leo, and o Peg, with adopted velocities of +38.9, +7.7, and +8.5 km s^{-1}. In Table 1 his mean velocities are compared with published values from Lick (Publ. Lick. Obs. vol. 16), Morse, Mathieu, and Levine (1991 AJ, 101, 1495 and private communication)

Table 1. Candidates for radial velocity standard stars with early spectral types

HD	HR	Name		α (2000) δ		V	Sp	v_{rot}	Fekel	Lick	Morse	Liu
1438	70	26	And	00 18 42	+43 47 28	6.11	B8 V	20			+3.3	
3360	153	17 ζ	Cas	00 36 58.2	+53 53 49	3.66	B2.5 IV	18	+0.5	+2.0	−0.3	+2.1
14252	675	10	Tri	02 18 57.0	+28 38 33	5.03	A2 V	18	−0.1	−0.2	−0.6	+0.6
23408	1149	20	Tau	03 45 49.5	+24 22 04	3.87	B8 III	35			+6.5	+7.6
23607				03 47 19.2	+24 08 23	8.25	A7 V	12			+5.1	+4.4
23873				03 49 21.6	+24 22 53	6.60	B9.5 V	85			+6.5	+6.3
26912	1320	49 μ	Tau	04 15 32.0	+08 53 32	4.29	B3 IV	89			+14.9	
27638	1369	59 χ	Tau	04 22 34.9	+25 37 45	5.37	B9 V	250			+14.4	
27962	1389	68 δ^3	Tau	04 25 29.3	+17 55 41	4.29	A2 IV-V	8	+38.9	+36.3	+38.7	+38.4
28114	1397			04 26 21.0	+08 35 24	6.06	B6 IV		+13.4			
34078	1712	AE	Aur	05 16 18.2	+34 18 43	5.96	O9.5 V	5			+52.8	
35708	1810	114	Tau	05 27 38.0	+21 56 13	4.88	B2.5 IV	20	+18.4	+15.8	+17.0	+20.4
38899	2010	134	Tau	05 49 32.9	+12 39 04	4.91	B9.5 V	21	+20.9	+22.0	+21.6	+24.4
43112	2222			06 15 08.5	+13 51 04	5.91	B1 V	25			+37.3	
58142	2818	21	Lyn	07 26 42.8	+49 12 42	4.64	A0mA1 IV	14	+26.9	+25.9	+26.0	
61555	2948			07 38 49.3	−26 48 07	4.5	B3 V	60			+24.1	
61556	2949			07 38 49.7	−26 48 13	4.8	B7 V	80			+25.5	
65900	3136			08 01 13.8	+04 52 47	5.65	A1 V	29	+44.2		+43.9	
72688	3383			08 34 01.5	−02 09 05	5.81	A1 V	4	+3.7			
77484				09 02 50.6	+00 24 30	7.8	B9.5 V	100			+27.2	
92728	4187	39	UMa	10 43 43.3	+57 11 57	5.80	A0 V	14	−14.9		−16.9	
95418	4295	48 β	UMa	11 01 50.4	+56 22 56	2.37	A0mA1 IV	40	−13.4		−14.4	
97633	4359	70 θ	Leo	11 14 14.3	+15 25 46	3.34	A2 IV	18	+7.7		+5.3	
107966	4717	13	Com	12 24 18.4	+26 05 55	5.18	A3 IV	44	+0.3		−2.5	
112413	4915	12 α^2	CVn	12 56 01.6	+38 19 06	2.90	A0 IIIp	20			+0.4	
128167	5447	28 σ	Boo	14 34 40.7	+29 44 42	4.46	F2 V	8	+0.0			
145570	6031	15 ψ	Sco	16 11 59.9	−10 03 51	4.94	A3 IV	34	−7.6			
145647	6035			16 11 28.7	+16 39 56	6.08	A0 V	40	−13.5			
147394	6092	22 τ	Her	16 19 44.3	+46 18 48	3.89	B5 IV	37	−17.8	−13.8		
155763	6396	22 ζ	Dra	17 08 47.1	+65 42 53	3.17	B6 III	37	−16.2			
166182	6787	102	Her	18 08 45.4	+20 48 52	4.36	B2 IV	37	−14.3			−14.1
179761	7287	21	Aql	19 13 42.6	+02 17 38	5.15	B8 II-III	12	−4.9	−4.8	−5.9	−3.7
186568	7512			19 43 51.3	+34 09 45	6.05	B8 III	14	−8.4		−8.8	−6.5
193432	7773	8 ν	Cap	20 20 39.7	−12 45 33	4.76	B9.5 Va+	18	−2.2	−2.8	+0.5	
196426	7878			20 37 18.2	+00 05 50	6.22	B8 IIIp	6	−18.6			
196821	7903			20 39 10.5	+21 49 03	6.08	A0 III	18	−32.8			−29.2
209459	8404	21	Peg	22 03 18.9	+11 23 11	5.80	B9.5 V	4	+0.4		+0.6	+4.5
214994	8641	43 o	Peg	22 41 45.3	+29 18 27	4.79	A1 IV	4	+8.5	+7.9	+9.1	+10.6
217811	8768			23 02 45.1	+44 03 32	6.39	B2 V	10	−10.2		−11.2	

and Liu, Janes, and Bania (1991 ApJ, 377, 141). Other published velocities for some of these stars may be found in Wolf (1978 ApJ, 222, 556), and Abt and Levy (1978 ApJS, 36, 241). In addition to the Fekel list, Table 1 also includes three candidates from Morse *et al.* with very early spectral types, two candidates in the Pleiades proposed as possible standards by Liu *et al.*, and six candidates from the Lindroos catalog of wide pairs suggested as possible standards by Mathieu, Latham and Morse (the velocities for these six are listed in the column under Morse).

For more than a decade there have been major efforts at the Center for Astrophysics, Geneva Observatory, and Dominion Astrophysical Observatory to monitor the IAU radial velocity standard stars with late spectral types. The status of these efforts is summarized in table 2, where stars have been removed if their velocity varied by more than 1 km s^{-1}.

Table 2. Candidates for radial velocity standard stars with late spectral types

HD	HR	Name		α (2000) δ	V	Sp	IAU	CfA	COR	DAO
693	33	6	Cet	00 11 15.8 −15 28 05	4.89	F6 V	+14.7	+14.69	+14.92	+15.09
3712	168	18 α	Cas	00 40 30.4 +56 32 15	2.23	K0-IIIa	−3.9			−4.26
3765				00 40 49.1 +40 11 14	7.36	dK5	−63.0	−62.80		
4128	188	16 β	Cet	00 43 35.3 −17 59 12	2.04	G9.5	+13.1	+13.49	+12.71	+13.16
4388				00 46 27.0 +30 57 06	7.51	K3 III	−28.3	−27.63		
8779	416			01 26 27.2 −00 23 55	6.41	gK0	−5.0	−3.83	−4.72	
9138	434	98 μ	Psc	01 30 11.1 +06 08 38	4.84	K4 III	+35.4	+34.25	+33.59	
12029				01 58 41.8 +29 22 47	7.80	K2 III	+38.6			
12929	617	13 α	Ari	02 07 10.3 +23 27 45	2.00	K2-IIIab	−14.3		−15.02	−14.51
18884	911	92 α	Cet	03 02 16.7 +04 05 23	2.53	M1.5 IIIa	−25.8	−25.51	−26.79	−25.30
22484	1101	10	Tau	03 36 52.3 +00 24 01	4.28	F9 IV-V	+27.9	+28.06	+27.85	+27.95
23169				03 43 52.8 +25 43 36	8.75	G2 V	+13.3			
26162	1283	43	Tau	04 09 09.9 +19 36 33	5.50	K1 III	+23.9	+25.02	+24.26	+24.79
29139	1457	87 α	Tau	04 35 55.2 +16 30 33	0.85	K5+ III	+54.1	+54.51	+53.55	+54.25
29587				04 41 36.3 +42 07 06	7.29	dG2	+112.4	+111.81		
32963				05 07 55.7 +26 19 42	7.72	G2 V	−63.1			
36079	1829	9 β	Leo	05 28 14.7 −20 45 35	2.84	G5 II	−13.5	−13.81	−14.37	
42397				06 11 34.7 +25 00 35	8.03	G0 IV	+37.4			
51250	2593	18 μ	CMa	06 56 06.6 −14 02 37	5.00	K2 III + B9 V	+19.6	+18.46	+17.53	
62509	2990	78 β	Gem	07 45 18.9 +28 01 34	1.14	K0 IIIb	+3.3	+2.83	+3.23	
65583				08 00 32.2 +29 12 44	7.00	dG7	+12.5	+14.68		
65934				08 02 11.1 +26 38 18	7.94	G8 III	+35.0			
66141	3145			08 02 15.8 +02 20 04	4.39	K2 III	+70.9	+72.07	+71.14	+71.45
75935				08 53 49.9 +26 54 48	8.63	G8 V	−18.9			
80170	3694			09 16 57.1 −39 24 05	5.33	K5 III-IV	+0.0		+0.00	
81797	3748	30 α	Hya	09 27 35.2 −08 39 31	1.98	K3 IIIa	−4.4	−4.18	−5.31	−4.35
84441	3873	17 ε	Leo	09 45 51.0 +23 46 27	2.98	G1 IIab	+4.8		+4.38	+4.40
86801				10 01 34.3 +28 33 59	8.88	G0 V	−14.5			
89449	4054	40	Leo	10 19 44.1 +19 28 15	4.79	F6 IV	+6.5	+6.41	+6.00	+6.43
90861				10 29 53.6 +28 34 52	7.20	K2 III	+36.3	37.83		
92588	4182	33	Sex	10 41 24.1 −01 44 29	6.26	sgK1	+42.8	+42.88	+42.24	
102494				11 47 56.4 +27 20 25	7.44	G8 IV	−22.9	−21.94		
102870	4540	5 β	Vir	11 50 41.6 +01 45 53	3.61	F9 V	+5.0	+4.48	+4.31	+4.38
103095	4550			11 52 58.7 +37 43 07	6.45	G8 Vp	−99.1	−98.51		−98.58
107328	4695	16	Vir	12 20 20.9 +03 18 45	4.96	K0.5 IIIb	+35.7	+36.67	+35.91	+36.48
108903	4763	γ	Cru	12 31 16.7 −57 04 51	1.63	M4 III	+21.3		+20.25	
109379	4786	9 β	Crv	12 34 23.2 −23 23 48	2.65	G5 IIb	−7.0	−7.86		
112299				12 55 28.1 +25 44 23	8.66	F8 V	+3.4	+4.53		
114762				13 12 19.7 +17 31 00	7.31	dF7	+49.9	+49.32	+49.03	
122693				14 02 52.1 +24 33 48	8.21	F8 V	−6.3	−5.34		

Table 2. Continued

HD	HR	Name	α	(2000)	δ	V	Sp	IAU	CfA	COR	DAO
123782	5300	13 Boo	14 08 17.2	+49 27 29		5.25	M2 IIIab	-13.4	-14.25		
124897	5340	16 α Boo	14 15 39.6	+19 10 57		-0.04	K1.5 III	-5.3	-5.01	-5.98	-5.30
126053	5384		14 23 15.2	+01 14 30		6.27	G1 V	-18.5	-19.40	-19.47	
132737			14 59 52.5	+27 09 35		8.02	K0 III	-24.1	-22.57		
136202	5694	5 Ser	15 19 18.7	+01 45 55		5.06	F8 IV-V	+53.5	+54.66	+54.30	+54.45
140913			15 45 07.7	+28 28 10		8.21	G0 V	-20.8	-19.85		
144579			16 04 56.7	+39 09 23		6.66	dG8	-60.0	-59.50		
145001	6008	7 κ Her	16 08 04.4	+17 02 49		5.00	G5 III	-9.5	-10.36	-10.77	-10.33
146051	6056	1 δ Oph	16 14 20.6	-03 41 39		2.74	M0.5 III	-19.8	-19.00	-20.24	-19.14
149803			16 35 54.2	+29 44 44		8.40	F7 V	-7.6	-6.54		
150798	6217	α TrA	16 48 39.8	-69 01 39		1.92	K2 IIb-IIIa	-3.7	-3.50		
154417	6349		17 05 16.7	+00 42 09		6.01	G0 V	-17.4	-16.87	-16.81	-16.92
157457	6468	κ Ara	17 25 59.8	-50 38 01		5.23	G8 III	+17.4		+17.22	
161096	6603	60 β Oph	17 43 28.2	+04 34 02		2.77	K2 III	-12.0	-12.33	-12.86	-12.18
168454	6859	19 δ Sgr	18 20 59.5	-29 49 42		2.70	K2.5 IIIa	-20.0		-20.84	
171232			18 32 35.9	+25 29 21		7.73	G8 III	-35.9	-36.00		
171391	6970		18 35 02.2	-10 58 38		5.14	G8 III	+6.9	+7.39	+7.10	
182572	7373	31 Aql	19 24 58.0	+11 56 39		5.16	G7 IV	-100.5	-100.26	-100.33	-100.27
		+28 3402	19 35 00.2	+29 05 13		9.05	F7 V	-36.6	-36.67		
186791	7525	50 γ Aql	19 46 15.4	+10 36 48		2.72	K3 II	-2.1		-3.38	-1.97
187691	7560	54 o Aql	19 51 01.5	+10 24 56		5.11	F8 V	+0.1	+0.02	-0.19	-0.05
194071			20 22 37.4	+28 14 47		8.13	G8 III	-9.8	-9.43		
203638	8183	33 Cap	21 24 09.5	-20 51 08		5.77	K0 III	+21.9	+22.27	+21.65	
204867	8232	22 β Aqr	21 31 33.3	-05 34 16		2.91	G0 Ib	+6.7	+6.68	+6.02	+6.71
206778	8308	8 ε Peg	21 44 11.0	+09 52 30		2.39	K2 Ib-II	+5.2		+2.63	
212943	8551	35 Peg	22 27 51.4	+04 41 44		4.79	K0 III-IV	+54.3	+54.39	+53.85	+54.26
213014			22 28 11.3	+17 15 48		7.70	dG8	-39.7	-39.57	-40.27	
213947			22 34 36.4	+26 35 52		7.53	K4 III	+16.7	+15.91		
222368	8969	17 ι Psc	23 39 56.9	+05 37 35		4.13	F7 V	+5.3	+5.52	+5.58	+5.60
223094			23 46 25.3	+28 42 13		7.45	K5 III	+19.6	+20.22		
223311	9014		23 48 32.3	-06 22 50		6.07	gK4	-20.4	-20.22	-20.76	
223647	9032	1 γ Oct	23 52 06.7	-82 01 08		5.11	G5 III	+13.8		+14.50	

References:

CfA: R. P. Stefanik and D. W. Latham, CfA Digital Speedometers, Cassegrain echelle spectrographs/photon-counting Reticon detectors at 1.5-m Wyeth Reflector, Oak Ridge Observatory, and 1.5-m Tillinghast reflector, Whipple Observatory, as reported at IAU General Assembly XXI, Buenos Aires, 1991; V(minor planet system) = V(table) + 0.00.

COR: M. Mayor, G. Burki and A. Duquennoy, CORAVEL Spectrometers on the 1.5-m Danish Telescope, European Southern Observatory, La Silla, as reported at IAU General Assembly XX, Baltimore, 1988. V(minor planet system) = V(table) + 0.40.

DAO: C. D. Scarfe, A. H. Batten and J. M. Fletcher, photographic spectrometer, coude spectrograph, 1.2-m reflector, Dominion Astrophysical Observatory, Publ. Dom. Astrophys. Obs., 28, 21, 1990.

For more than a decade, the CORAVEL team has been monitoring a sample of nearby solar-type stars drawn from the Gliese Catalogue. In Table 3 we list their solar-type stars which have no obvious velocity variations. The final five columns give the mean CORAVEL radial velocity, the velocity *rms*, the number of observations, the time span of the observations in days, and the projected equatorial velocity. These stars should be useful for comparing the velocity zero points of different systems with an absolute zero point based on observations of minor planets. In particular, the use of these stars should help eliminate possible systematic errors as a function of color or luminosity.

Table 3. CORAVEL candidates for radial velocity standard stars with solar-type spectra

HD	HR	Name			α	(2000)	δ	V	Sp	CORAVEL		N_{obs}	Days	$v \sin i$
1461	72				00 18 40.4	−08 03 04		6.46	G0 V	−10.45	±0.26	12	3658	2.4
1835	88	9		Cet	00 22 50.3	−12 12 37		6.38	G3 V	−2.58	±0.28	20	2926	6.8
16895	799	13	θ	Per	02 44 10.2	+49 13 47		4.12	F7 V	+24.38	±0.27	11	4456	8.9
22879					03 40 19.7	−03 12 51		6.74	F9 V	+120.32	±0.28	59	4841	3.5
31966					05 00 33.6	+14 22 59		6.72	G5 V	−18.38	±0.24	13	2624	2.2
32923	1656	104	μ	Tau	05 07 25.1	+18 38 41		5.01	G4 V	+20.33	±0.31	22	4788	1.5
39881	2067				05 56 01.6	+13 56 04		6.67	G5 IV	+0.09	±0.25	16	2618	1.5
65583					08 00 32.8	+29 13 43		6.94	G8 V	+14.54	±0.25	81	4385	2.3
111395	4864				12 48 48.2	+24 50 29		6.31	G7 V	−9.20	±0.24	16	3320	3.0
118576A					13 37 13.0	+30 05 04		9.32	G8 V	+3.40	±0.40	14	4431	2.8
118576B					13 37 14.3	+30 05 12		10.53	G0	+3.59	±0.42	13	4431	1.4
126053	5384				14 23 14.5	+01 14 54		6.30	G1 V	−19.49	±0.26	198	4215	2.3
130948	5534				14 50 15.2	+23 54 41		5.85	G1 V	−2.64	±0.35	19	4871	6.7
131156	5544	37	ξ	Boo	14 51 22.7	+19 06 09		4.55	G8 V	+0.86	±0.22	15	4854	5.1
140538	5853	23	ψ	Ser	15 44 01.8	+02 31 02		5.88	G2.5 V	+18.69	±0.30	15	4867	2.1
146233	6060	18		Sco	16 15 36.4	−08 21 45		5.50	G2 Va	+11.56	±0.16	13	3812	3.1
154417	6349				17 05 16.8	+00 42 26		6.01	F8.5 IV-V	−16.95	±0.29	230	4797	6.0
161797	6623	86	μ	Her	17 46 28.6	+27 43 52		3.41	G5 IV	−16.71	±0.30	20	3743	1.8
187691	7560	54	o	Aql	19 51 00.7	+10 25 03		5.10	F8 V	−0.16	±0.27	229	4826	4.0
190360	7670				20 03 34.7	+29 54 15		5.71	G6 IV+	−45.90	±0.25	20	4787	1.6
197076	7914				20 40 44.7	+19 55 52		6.45	G5 V	−35.86	±0.24	15	4312	2.2
217014	8729	51		Peg	22 57 27.1	+20 46 05		5.49	G2.5 IVa	−33.73	±0.23	22	4423	1.9

For more than a decade, a Canadian team has been monitoring about a dozen bright solar-type stars (Campbell, Walker, and Yang 1988, ApJ, 331, 902; Walker, Yang, Campbell, and Irwin 1989, ApJ, 343, L21) using a hydrogen-fluoride technique on the Canada France Hawaii Telescope. In Table 4 we list the nine solar-type stars in their program which show a total velocity range of less than 200 m s^{-1}. These stars should also be useful for comparing the velocity zero points of different systems.

Table 4. CFHT candidates for radial velocity standard stars with solar-type spectra

HD	HR	Name		α	(2000)	δ	V	Sp	Range
10700	509	τ	Cet	01 44 10.0	−15 56 58		3.50	G8V	80
19373	937	ι	Per	03 08 57.5	+49 36 53		4.05	G0 V	70
20630	996	κ^1	Cet	03 19 20.7	+03 22 08		4.83	G5 V	100
90839	4112	36	UMa	10 30 38.6	+55 58 52		4.80	F8 V	120
102870	4540	β	Vir	11 50 39.2	+01 46 07		3.61	F8 V	90
114710	4983	β	Com	13 11 55.3	+27 51 57		4.26	G0 V	90
115617	5019	61	Vir	13 18 28.0	−18 17 47		4.74	G6 V	90
131156	5544	ξ	Boo A	14 51 22.7	+19 06 09		4.55	G8 V	180
188512	7602	β	Aql	19 55 18.5	+06 24 48		3.71	G8 IV	80

Commission 31 : Time (Heure)

Report of Commission Meetings

During this General Assembly, Commission 31 held 1 business and 5 scientific sessions among which 2 were organized jointly with Commission 19.

Moreover Commission 31 participated to the Joint Discussions on "Reference Frames" and "Hipparcos" which involved Commissions 4, 7, 8, 19, 20, 24, 31, 33, 40 and 4, 8, 19, 24, 31, 33, 40, 44 respectively. The Joint Discussion on Reference Frames led to the adoption by the IAU General Assembly of a set of 9 resolutions which introduce relativistic theory in the definition and realization of time scales and reference frames. Of particular interest for Commission 31, they introduced also more precise definitions of existing time scales and new designations: TT = Terrestrial Time,TCG = Geocentric Coordinate Time, TCB = Barycentric Coordinate Time. The definitions are given in the set of resolutions adopted by IAU; the attached figure will help to clarify the relations between those time scales. In Celestial Mechanics 1991, a paper of V.Brumberg and M.Soffel could also highlights the resolutions mentioned above.

The first session (24 July) was dedicated to the presentation of scientific papers and to the reports of members of Commission 31 who are representatives in different agencies or commissions.

Dr. Winkler gave a full report concerning the 11th session (19-20 April, 1989) of the Comité Consultatif pour la Définition de la Seconde (CCDS) at the BIPM. The meeting discussed progress in time links (GPS, TV, LORAN-C, VLBI) and in frequency standards technology (mercury ion trap frequency standard). A special working group chaired by David Allan is preparing recommendations to be used to improve the uniformity of GPS measurements made in laboratories worldwide. Two declarations have been produced concerning the opportunity for BIPM to establish other working groups to define precise conditions on the operations of various time link systems. Five recommendations are also taken. The first recommendation suggests that the offset of 1 microsecond between UTC(k) and UTC, requested by CCIR, be changed to the "region of a few microseconds". Recommendation (2) aims a general improvement of the performance of the clocks contributed to TAI. Recommendation (3) deals with a possible improvement of the reference coordinates of antennas used in one-way time intercomparison, while Recommendation (4) recommends BIPM to acquire clock and time comparison hardware to provide a more intimate connection of BIPM with the problems of high precision time intercomparison. Recommendation (5) requests the BIPM to study and organize an optimum network of time links among the contributors to TAI in order better to utilize the major time intercomparison methods (GPS, GLONASS, two-way satellite time transfers).

Report of TAI determination carried out at BIPM is given by G. Petit. Since the 1st of January 1988, BIPM is officially in charge of the establishment and dissemination of TAI and UTC. The report deals with the evolutions that have occurred since that time and on the associated researches that have been carried out at BIPM. GPS is now by far the main technique used for time transfer. The fraction of laboratories equipped has grown from 60% at the end of 1987 to 85% at the end of 1990. Conversely the use of LORAN-C and TV is declining. GLONASS has appeared in 1990 to link UTC (SU). Steering was resumed in 1989, after a period of more than 5 years when it was not found necessary. It has been performed 5 times, in steps of 5.10^{-15}, the general rule being that the frequency adjustments should be below the intrinsic instability of the scale. The general trend is still that the frequency of EAL tends to decrease with respect to the frequency standards.

Different studies are being conducted by the BIPM concerning use of Kalman Filter for time scales, and correlations between contributing clocks. It has been shown that the main cause is change in environment, especially humidity, but apparent correlations due to the time links or to TAI may appear.

The coordinates of the GPS antennas of 23 laboratories have been determined in the global reference frame ITRF88, either by compilation of existing data or by processing of the time data themselves.

A comparative study of two codeless ionospheric calibrators has been carried out. It has been shown that the use of such ionospheric measurements, associated with post processed precise ephemerides, can greatly improve the precision and accuracy of the GPS time transfer. The CCDS has the task to ensure the standardization of hardware, software and procedures, which is essential in case of Selective Availability.

Finally BIPM participated in the coordination of two-way time transfer and started to study GLONASS.

Dr. J. Luck gave a report on the status on Time development in Australia. The National Standard Commission (NCS) which administers the National Measurement Act 1960, has formed a National Time Committee, and has initiated Commonwealth legislation to have UTC (AUS) legally adopted as the national time scale, as close as possible to UTC (BIPM). The legislation will define time zones and daylight saving based on UTC (AUS); they were formerly the responsibility of individual states and territories and based on nineteenth-century concepts adopted in 1983. The standard frequency and time signal service VNG, discontinued by TELECOM Australia in 1987, was re-activated experimentally in 1988. A permanent service was inaugurated on 3 July 1991, broadcasting on 5.0, 8.638, 12.984 and 16.0 MHz.

The National Time Committee is to study the introduction of low frequency time signals for Australia, similar to DCF 77 in Europe. Low cost medium-resolution GPS timing receivers are also being studied. Since 1988, UTC (AUS) has been defined as equal to UTC (USNOMC), the realization being the 1 pps output from a GPS Time Transfer Unit at CSIRO National Measurement Laboratory corrected by GPS Time variations published in USNO Series 4. The free-running time scale, formerly known as UTC (AUS), is now designated TA (AUS) and is based on selections from 20 caesium standards and 2 hydrogen masers using an algorithm unchanged since 1980. Clock comparisons are made by GPS, local TV and AUSSAT TV. An annual term of amplitude 150-200 ns in TA(AUS)-UTC(BIPM) is evident from the connection by GPS during 1983-1991.

On the problem of the introduction of relativistic corrections for new time scales and Reference Frames, Prof. R.O. Vicente expressed some critical considerations, arguing for a more conservative approach to avoid changes which do not appear as fundamental for practical users.

S. Debarbat summarized contributions presented in 1989 at the meeting "Journées Systèmes de référence spatiotemporels", organized yearly at the Paris observatory; they concern the relativistic effects in astrometry and are related with the set of recommendations to be analyzed by the Joint Discussion on "Reference Frame". S. Debarbat mentioned several relevant papers (Th. Damour, M. Soffel, C. Xu,C.Boucher) published in the Proceedings of the meeting above (available on request, at the Observatoire de Paris, Ed. N. Capitaine). They took steps that conducted to an evolution of concepts and considerations, allowing one to reach some consensus between the authors above; in a common paper they studied equations of motion with the integrated ephemeris from JPL.

P. Pâquet presented the procedure used to realize the reference time scale UTC(ORB) in use at the Royal Observatory of Belgium since early 1991. A mathematical clock is deduced from 3 atomic clocks whose long term drift is measured with respect to an external frequency (GPS or TAI); this mathematical clock is realized by frequency adjustment of a Rubidium.

The precision is of the order of 2×10^{-13}.

The second scientific session (24 July) was dedicated to the activities of the Working Group on Time Transfer, chaired by H. Fliegel.

Since D. Allan was not able to attend the General Assembly, H. Fliegel presented the report of the Working Group on Pulsars; he summarized interesting results obtained in Berkeley, Princeton, Arecibo, Meudon. D. Allan is willing to provide copy of the report.

G. Petit, addressed the assessment of GPS time transfer accuracy by the condition of closure around the world. The precision of time transfer by the GPS common view method is routinely at the level of 10 to 20 ns, even over intercontinental distances. When special care is taken to correct the main sources of error (satellite ephemerides, ionospheric delay, station coordinates), 4-5 ns can be achieved.

Using three intercontinental links around the world, it is possible to establish a closure condition by adding the three independent time transfer values, which should add to zero. This provides a direct estimation of the accuracy of the time transfer technique, except for the cancellation of constant or slowly varying receiver biases.

W. Lewandowski, G. Petit, C. Thomas computed such a closure condition for more than six months using data recorded at the Paris Observatory in Paris (France), at the Communication Research Laboratory in Tokyo (Japan) and at the National Institute of Standards and Technology in Boulder,Co (USA). They used GPS ionospheric measurements at the three sites and precise ephemerides provided by the US Defense Mapping Agency to correct the GPS time measurements. The closure condition is verified within a few nanoseconds. The remaining bias is significant, and an attempt is made to explain its origin in several remaining sources of error, and to estimate their respective effect.

Mrs. Ye presented a paper on Time Synchronization in China. At the end of the 70s' a special LF station transmitting BPL signals, which is similar to Loran-C, was established at the centre of China for domestic precise time comparison. The transmitting accuracy of BPL is better than 100 ns; meanwhile, the methods of passive and active television via satellites, and portable clock, were used for time comparison at Shaanxi Observatory, Shanghai Observatory, Beijing Observatory, Institute of Geodesy and Geophysics, Institute of Beijing Radio Metrology and Institute of National Metrology. Significant results were obtained in China in the 1980s'. Starting from 1990, most of laboratories in China, have GPS receivers made in China and GPS is the main method of synchronization with BIPM and Foreign time and Frequency laboratories. In recent years the precision obtained by different methods is as follows : LF ground wave (500 ns), Laser pulses (few ns), VLBI (few ns), GPS (10 to 20 ns), satellite SIRIO (better than 100 ns), portable clocks (better than 100 ns).

J.Zhang and Wu Bijun estimated the influence of ionosphere in Wuhan region on GPS signal time delay.

Using the TEC and sunspot data, the authors analyzed the GPS signal propagation time delay variation caused by the ionosphere. By using TEC monthly middle data in the Wuhan region to correct additional time delay caused by the ionosphere, they can make GPS single frequency altimetry accuracy and timing measurements match the dual frequency altimetry accuracy and common view time synchronous accuracy. There is a potential for further improving this accuracy.

Two papers were dedicated to time transfer by geostationary satellites. J. Luck indicated that TV signals from the geostationary communications satellite AUSSAT are used routinely for clock comparisons in the South East quadrant of the continent. They are adequate to 100 ns with orbital corrections from radar ranging, and are used, for example, to transfer UTC (AUS) to the NGS VLBI/CIGNET station at Hobart, Tasmania. Portable GPS-TTU visits have confirmed the stability of AUSSAT TV biasses to 50 ns, but reveal microsecond-level anomalies in 1 pps outputs from GPS geodetic receivers. Experiments are under way within AUSLIG to extend coverage over the whole country by receiving PAL transmissions from national beam.

The AUSSAT orbit can be refined by converting the pseudoranges measured by TV to range differences at pairs of stations whose clock differences are known from GPS. Experiments to date, solving for eccentricity, inclination and station biasses, have reduced time transfer errors below 10 ns.

The next generation of AUSSAT spacecraft, to be launched during 1992, will carry an array of 14 retroreflectors, 38 mm diameter each. Auroral Laser Ranging Observatory has been upgraded to provide the laser power, efficiency and reliability to support this operation with 1 cm precision in range and 3 arcsecond precision in pointing. 1 ns time transfer accuracy over much of Australia, independent of GPS, will result from the better orbit determination.

P. Pâquet reviewed the results obtained in 1989-1990 by a European group who performed time comparison by geostationary satellites; it was not possible to reach a better precision than 100 to 200 ns because the error of the satellite position was of the order of one to two kilometers. On the other hand, a European Working Group "COGEOS" aims to locate the geostationary satellites for the study of zonal harmonics of the Earth gravity field. Since photographic results have been disappointing, an experiment to locate the satellites using time comparison technique is proposed. A first experiment is proposed to be conducted by the end of 1991. The analysis will combine both orbit and time differences determination, or orbit only if GPS is used for time comparison.

C. Veillet reports on the LASSO experiment; which is made mainly of a laser pulse detector associated to an event-timer and a stable oscillator. It is flying on a geosynchronous satellite Meteosat 3/P2 since

August 1988. Ground based laser ranging stations at TUG (Graz, Austria) and OCA/CERGA (Grasse, France) succeeded in November 1991 in getting echoes from the satellite retroflectors and arrival times of the laser pulses at the satellite on the on board time scale. The analysis of these data demonstrated that LASSO is capable of linking time scales at two remote sites with a subnanosecond accuracy.

In September 1991 should begin a new phase of LASSO involving OCA/CERGA and two laser stations in the USA (MLRS, Texas, and 48" Goddard, Maryland), as Meteosat 3/P2 is located at 50 deg. W in the geosynchronous belt since the end of July 1991. NASA, USNO and NIST will cooperate with OCA/CERGA for the success of this unique experiment whose results will be compared to other time transfer techniques (GPS and Mitrex two-way).

In conjunction with LASSO, time transfer comparisons have been performed between GPS and Mitrex two-way through a nine months campaign at TUG and Grasse. A preliminary analysis shows an internal precision of 0.7 ns for a typical two-way session. A rough comparison between GPS and two-way for the first three months shows a 2.7 ns rms of the differences. This campaign will be extended by adding sites in the USA in the frame of LASSO.

Together with the members of Commission 19, the second part of the session (2) was devoted to pay homage to Dr.B.Guinot who has recently retired and was past president of the Commission 31. Dr. Pâquet and Dr. Winkler gave to the members of Commissions 19 and 31 a brief but hearty summary of the scientific and organizational activity carried out by Dr. B. Guinot during about 35 years as director of the BIH and after 1985 at the BIPM.

One session has been entirely dedicated to a seminar given by Prof. V. Brumberg who states several points concerning relativistic aspects for the realization of Time scales and references systems.

It is not possible to summarize the Brumberg's bright presentation but the figures 1 to 4 are extremely useful to return to papers which addressed this problem since few years; they concern the comparisons of observations with respect to computed quantities (fig. 1), the dynamical and non-rotating reference systems (fig. 2), the hierarchy of relativistic reference systems (fig. 3), possible evolution of relativistic Time scales (fig. 4).

A session has been dedicated to a general discussion concerning proposals issued from the IAU Executive Committee, and more particularly, concerning the amalgamation of Commissions and format of future General assemblies. The proposal to reduce the number of Commissions is related to the decision to re-organize the format of the General Assemblies. Participants recognize that it is mandatory to change the format of the assemblies, but nevertheless they confirm what has already been expressed two years ago by exchange of correspondence, that Commission 31 has its own specificity and does not require any amalgamation. Moreover since the Field of timing and reference systems is rapidly changing, no reorganization could be valid for long. The existence of Commission 31 must be preserved to prevent miscommunication between physicists and astronomers, and contact must also be maintained with advanced engineering users.

For the format of the General Assemblies, the participants were not in favour of organizing Symposia before or after an Assembly. It is porposed to restrict the scientific sessions to workshop or symposia organized in common by several commissions; one or two business meetings would be held by Commission.

The participants feel that in its preparation for the next General Assembly, the Organizing Committee must address this problem rapidly.

A new IAU working group on "Astronomical Constants" has been initiated by Dr. Fukushima; the Commission will be represented by V. Brumberg and G. Petit.

The two existing working groups continue under the chairmanship of D. Allan "The use of Millisecond Pulsars and Timing of Pulsars" and J. Luck "Modern Techniques of Time Transfer".

The new organizing Committee of the Commission is as follows :
 President : . E. Proverbio
 Vice-President : H. Fliegel
 Members :   D. Allan, V.A. Brumberg, F. Fujimoto M. Granveaud, B. Guinot,
        W. Klepezynski, J. Kovalevsky, J. Luck, I. Mueller, P. Pâquet, Ye Shu Hua.
 The representatives to international organization are :
- to CCDS and BIPM : W. Winkler
- to FAGS : J. Kovalevsky
- to CCIR : S. Leschiutta

Relations between Time scales

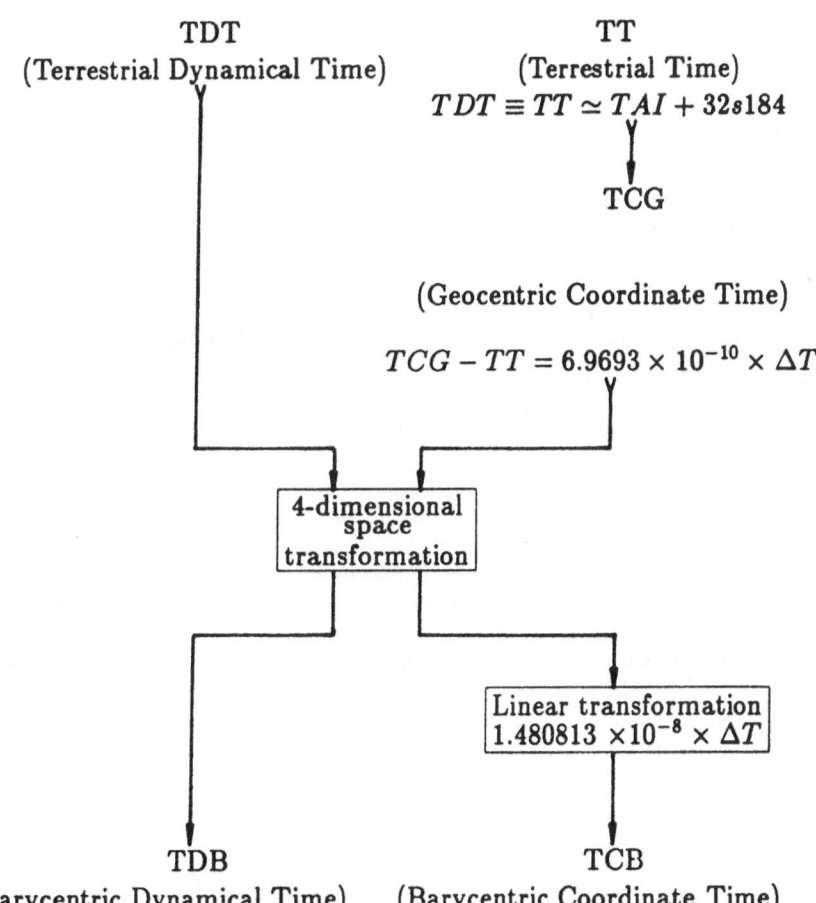

Recommendation 1976   Recommendation 1991

TDT   TT
(Terrestrial Dynamical Time)   (Terrestrial Time)

$$TDT \equiv TT \simeq TAI + 32s184$$

TCG

(Geocentric Coordinate Time)

$$TCG - TT = 6.9693 \times 10^{-10} \times \Delta T$$

4-dimensional
space
transformation

Linear transformation
$1.480813 \times 10^{-8} \times \Delta T$

TDB   TCB
(Barycentric Dynamical Time) (Barycentric Coordinate Time)

$$TCB = TDB + 1,550505 \times 10^{-8} \times \Delta T$$
$$\Delta T = (\text{date in days - 1977 January 1, } 0^h)_{TAI} \times 86400 \text{ sec}$$

Fig.1 O-C, comparison in relativistic astronomy

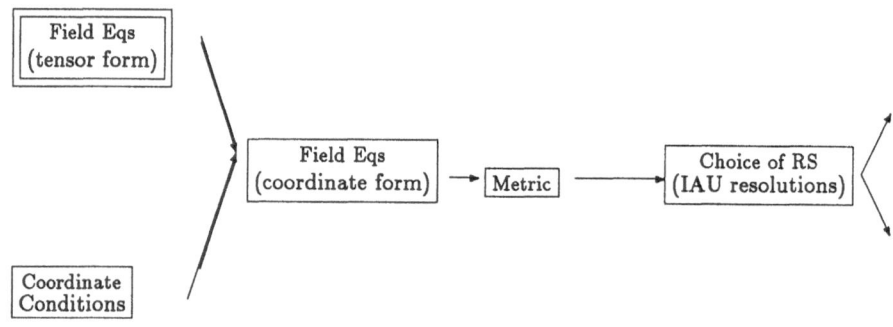

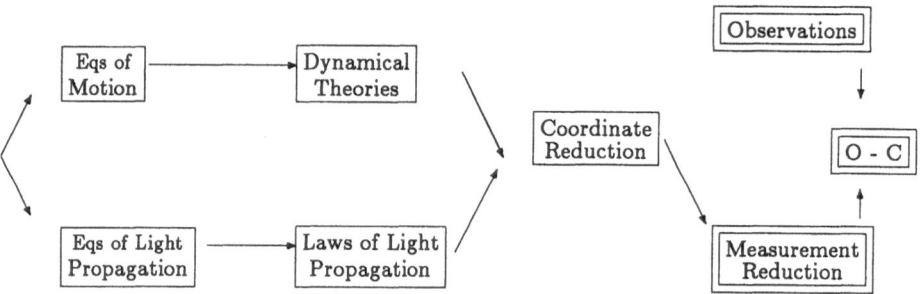

In contrast to the inertial coordinates of Newtonian astronomy there exist no global co-ordinates in relativistic astronomy which may be treated as physically meaningful measurable quantities. Therefore, to perform the physically meaningful O-C analysis one has to convert dynamical effects of motion of the bodies and kinematical effects of light propagation, to be considered in one and the same Reference System (RS), into the measurable quantities (coor-dinate reduction). The various steps of this process are shown in this figure. The coordinate independent steps are marked by double lines.

Fig.2. Dynamical (DRS) and Kinematical (KRS) Non-Rotating Reference Systems.

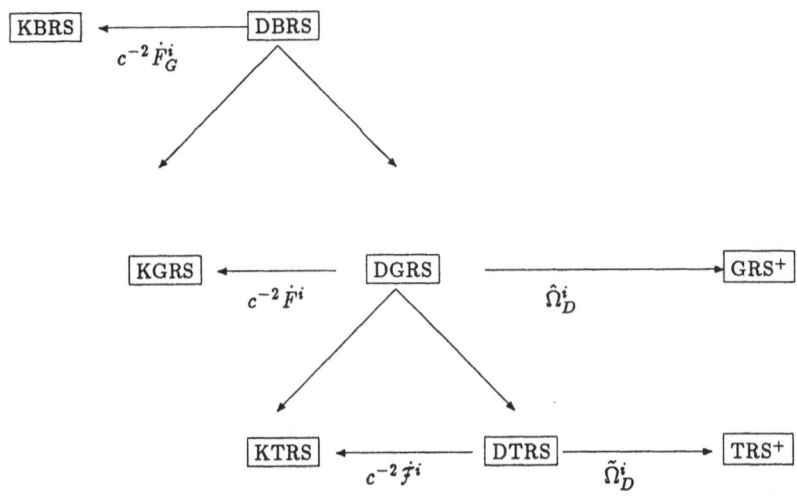

In contrast to Newtonian astronomy one should distinguish in relativistic astronomy between DRS and KRS. Absence of kinematical rotation of any RS means that the transformation between the spatial axes of this RS and generating DRS of more high level contains no rotation terms. With respect to the DRS of the same level such a KRS rotates with the angular velocity of the relativistic order of smallness. KBRS, KGRS and KTRS differ from DBRS, DGRS and DTRS by the amount of galactic $(c^{-2}\dot{F}_G^i)$, geodesic $(c^{-2}\dot{F}^i)$ and topocentric $(c^{-2}\dot{f}^i)$ precession, respectively. For astrometric purposes one introduces the systems GRS^+ and TRS^+ rotating (at the average) with the Earth with the angular velocities $\hat{\Omega}_D^i$ and $\tilde{\Omega}_D^i$ with respect to DGRS and DTRS, respectively. GRS^+ rotates with respect to KGRS with the angular velocity $\hat{\Omega}_K^i = \hat{\Omega}_D^i - c^{-2}P_{ij}\dot{F}^j$ with P_{ij} being the (Newtonian) orthogonal matrix of spatial transformation from GRS to GRS^+. Similar relation holds for $\tilde{\Omega}_D^i$ and $\tilde{\Omega}_K^i$.

Fig.3. Hierarchy of Relativistic Reference Systems for Solar System Astronomy

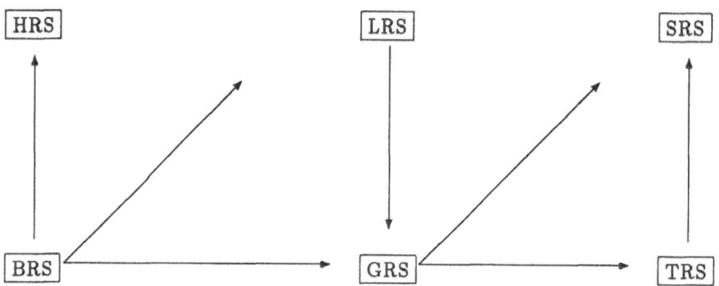

This hierarchy includes solar system barycentric RS (BRS), heliocentric RS(HRS), Earth-Moon local RS (LRS), geocentric RS (GRS), topocentric RS (TRS) and satellite RS(SRS). The first four systems are used for the representation of motion of the solar system bodies. The last two systems are suitable for the description of observations made by a ground observer (TRS) or on board of an Earth satellite (SRS). Construction of all systems is based on three main principles: (1) using of one and the same coordinate conditions, (2) compatibility with the principle of equivalence involving the representation of the external mass influence only in the form of tidal terms, and (3) absence of dynamical rotation, i.e. non-existence of Coriolis and inertial terms in the metric of RS

Fig.4. Relativistic Time Scales

	1976	1991	?
Time	TDT	TCB	TCB
	TDT	TCG	TCG
Scales	TAI	TAI	TAIM
Attributes	L_c	-	-
	L_G	L_G	-
	-	TT	-
	geoid	geoid	-

1976 time scales involve two ill-defined constants L_C and L_G and vague defined notion of the geoid. 1991 time scales remove L_C but still retain L_G, the auxiliary time scale $TT(= TDT)$ and the notion of the geoid. It may be possible in future to remove all these attributes introducing TAIM (TAI modified) as the physical realization of TCG (just as TAI represents now the physical realization of TT)

33. STRUCTURE AND DYNAMICS OF THE GALACTIC SYSTEM

PRESIDENT: M. Mayor (Switzerland)
VICE–PRESIDENT: Leo Blitz (USA)

1 BUSINESS MEETING

I. IAU report on astronomy, symposia etc.

For several triennia commission 33 had maintained the tradition to publish a "long report" in addition to the IAU report published in Transactions A. In view of the now existing excellent publication of Astronomy and Astrophysics Abstracts it appears that the 35 pages or more of the IAU report are sufficient to reasonably cover the most important progresses done in the field of Galactic Structure. Therefore the "long report" of commission 33 has been suppressed.

Commission 33 has cosponsored the following IAU symposia or colloquia:

IAU symposium 139 *"Galactic and extragalactic background radiation"* held in Heidelberg (June 1989); IAU symposium 144 *"The interstellar disk–halo connection in galaxy"* held in Leiden (June 1990); IAU symposium 148 *"The Magellanic Clouds and their dynamical interaction with the Milky Way"* held in Sydney (July 1990); IAU symposium 149 *"The stellar populations of galaxies"* held in Angra dos Reis (August 1991); IAU symposium 150 *"The astrochemistry of cosmic phenomena"* held in Campos de Jordao (August 1991); IAU symposium 153 *"Galactic bulges"* to be held in Ghent in August 1992; IAU symposium 156 *"Developments of astrometry and their impacts on astrophysics and geodynamics"* to be held in Shanghai in September 1992; IAU symposium 155 *"Planetary nebulae"* to be held in Innsbruck in July 1992; IAU colloquium 127 *"Reference systems"* held in Virginia Beach (Oct. 1990); and IAU colloquium 132 *"Instability, chaos, predictability in celestial mechanics and stellar dynamics"* held in Delhi (Oct. 1990).

II. Membership

Commission 33 approved and welcomed new members proposed by the various national committees as well as several IAU members who requested to join the commission. These included:

Aguilar Chin Luis, USA; Bienaymé Olivier, France; Chapman Jessica, Australia; Cubarsi Rafael, Spain; Fujiwara Takao, Japan; Gemmo Alessandra, Italy; Gupta Sunil, India; Hakkila Jon Eric, USA; Hanami Hitoshi, Japan; Lee Hyung Mok, Korea; Leisawitz David, USA; Polyachenko Valerij, USSR; Polymilis Chronis, Greece; Raharto Moedj, Indonesia; Robin Annie, France; Ruelas–Mayorga R, Mexico; Sanz I Subirana Joanne, Spain; Schechter Paul, USA; Seimenis John, Greece; Spergel David, USA; Sygnet Jean–François, France; Wyse Rosemary, USA; Zachilas Loukas, Greece;

as well as:

Aizu Ko, Japan; Andersen Johannes, Denmark; Banhatti D., India; Dejonghe Herwig, Belgium; Fridmann Aleksey, USSR; Gottesman Stephen, USA; Grenon Michel, Switzerland; Latham David, USA; Liebert James, USA; Lu Philip, USA; Matteucci Francesca, Italy; Murray Andrew, UK; Nordstroem Birgitta, Denmark; Norman Colin, USA; Oblak Edouard, France; Pandey A.K., India; Pavlovskaya E., USSR; Rebeirot Edith, France; Sargent Anneila, USA; Sobouti Yousef, Iran; Valtonen Mauri, Finland; Whitelock Patricia, Australia; Xiang Delin, China.

We regret the death of our colleagues Prof. Jöran Ramberg from Uppsala and Kyrill Ogorodnikov from Leningrad.

III. Officers

The new President of commission 33 is Leo Blitz of USA. James Binney from U.K. is the new Vice-President. The members of the Organizing Committee for the upcoming triennium 1991–1994 are Michel Mayor, Switzerland (past President), Catherine Cesarsky, France, Gerard Gilmore, UK, and the new members: Francesca Matteucci, Germany, Yuzuru Yoshii, Japan, Aleksey Fridmann, USSR, Agris Kalnajs, Australia, Johannes Bloemen, Netherlands and Mark Morris, USA.

IV. Working Group on "Galactic Constants"
F.J. Kerr: "Galactic Constants"

J.F. Kerr presented a short report on recent determinations of the Galactic constants R_O and Θ_O, on behalf of the Commission's Working Group on Galactic Constants. This covered a period of six years since the last major report on this subject. The unweighted mean of ten determinations of R_O came to 7.8 ± 0.4 kpc, and the mean of four determinations of the circular velocity was 216 kms^{-1}. Neither of these is sufficiently different from the currently–accepted values to lead to any recommendation for change.

The Working Group was terminated at this point, with thanks. Kerr was asked to continue to monitor the situation for future developments, including also the constants A and B. Any new information on any of the four constants should be sent to him.

In the very near future the space astrometric mission HIPPARCOS with its large number of high precision stellar proper motions will probably allow to fully reconsider the domain of galactic constants. Michel Crézé is invited to establish the contact between our commission and the HIPPARCOS team and, if necessary, to report at the next General Assembly.

2 SCIENTIFIC MEETINGS

At the General Assembly Commission 33 was among the cosponsoring commissions of three Joint Discussions: JDII *"Reference Systems"* chaired by J.A. Hughes; JDV *"Origin of Stars and Planetary Systems"* chaired by André Brahic and JDVI *"HIPPARCOS - An Assessment"* chaired by Catherine Turon.

"Impact of radial velocity observations on Galactic Structure and Evolution"
Chaired by David Latham and Michel Mayor

The speakers were Leo Blitz (Maryland) reviewing works on the shape of the galactic disk; David Latham (CfA) described the kinematics of the thick disk; Michel Grenon (Geneva Observatory) reported on connections between disk, halo and bulge and finally Dante Minniti (Arizona) presented bulge kinematics.

"Large Scale Distribution of Gas and Stars in the Milky Way"
Chaired by Leo Blitz

It was a double session lasting from 9.00 to 12.30. The speakers were David Spergel (Princeton), Michael Rich (Columbia), Patricia Whitelock (SAAO), David Latham (CfA), Harvey Liszt (NRAO), Leonardo Bronfman (Chile) and Tetsuo Hasegawa (Tokyo). The session was divided into two parts. The first dealt with the bulge of the Galaxy, and the second with the disk and halo. The session on the bulge featured direct new evidence that the bulge of the Galaxy is actually a bar, and an independent confirmation of this picture from IRAS observations of Mira variables. An important study of the metallicity of the bulge shows that the stars have evolved as if they were in a closed box without significant mixing of the inner disk population. The work on the disk and the halo included a talk on the evidence for metallicity differences between the thin and thick disk populations. Papers on the gas discussed an overview of the distribution of molecular gas as well as new evidence that Giant Molecular Clouds become less dense with increasing distance to the galactic center. Work was also presented describing outstanding anomalies in our understanding of the distribution of atomic gas. The papers will be combined with two papers from the session on the galactic center organized by D.Y. Gezari (GSFC), and published by Kluwer.

"The Galactic Nucleus"

Chaired by Dan Gezari

This session was held jointly with and immediately following the Commission 40 session "The Galactic Center Region and Pulsars" (Chairman: Peter Mezger). D. Gezari presented extensive infrared eight–color (5–18 μm) imaging array observations of the Sgr A West complex with his Goddard 58x62 pixel Array Camera. Detailed modelling of the dust density and temperature structure of the region suggest that the brightest compact IRS sources are internally heated. This result as well as a correlation found between temperature enhancements on the extended ridge and newly discovered HeI emission line stars imaged by Genzel et al. indicate that imbedded sources distributed through the ridge could provide the luminosity observed in the Galactic Center, and that an exotic central engine or black hole at Sgr A* need not be invoked. D. Aitken described the first polarimetry with an array detector, using the Goddard infrared camera at 12.4 μm, of the central parsec of the Sgr A West complex. Diffraction–limited sampling revealed smooth magnetic field structure through the emission ridge made up of the northern arm and east–west bar, unperturbed by the cluster of sources at IRS16. This implies that the IRS16 sources must be out of the plane of the ridge system. This field structure is modeled by an elliptical flow pattern of material with one focus at Sgr A*. P. Mezger described new constraints on the spectrum of Sgr A* over a wide range of radio, sub–millimeter and infrared wavelengths. The revised spectrum was modeled as emission from a 500 $M_\odot$ dust cloud which becomes optically thick at 1–mm. H. Zinnecker reported new 1μm CCD images of the Galactic Center showing astrometry of several new near–infrared point sources near Sgr A*, including candidates for stellar objects at the Galactic Center, but no definitive identification with Sgr A* could be made. J. Pipher discussed extensive infrared Brackett line imaging observations and summarized detection of point–like sources at the Galactic Center which have no radio counterparts, possibly Wolf–Rayet stars, and measured velocities indicative of stellar winds. Lunar occultations also revealed double sources at IRS1 and IRS13. F. Yusef–Zadeh described high resolution VLA observations at 2–cm with 0.3 arcsec resolution which showed bow–shock structure associated with the mass–loosing envelope of the supergiant star IRS7 lying 1 light year in projection from the Galactic Center (IRS16/Sgr A*). This bow shock structure was modelled in terms of two colliding winds, one from the IRS16 region and the other from IRS7. In addition, a tail of ionized gas on IRS7 was detected trailing directly away from the Galactic Center, seen as further evidence of this complex interaction.

Commission 34
Interstellar Matter (Matiere Interstellaire)

President: John S.Mathis (USA)
Vice-President: Harm Habing (Neth)
Scientific Organizing Committee: K. S. de Boer (Ger), M. Dopita (Australia), A. Fabian (UK), David Flower (UK), R. Genzel (Ger), N. Kaifu (Japan), J. Lequeux (France), B. Shustov (USSR), C. Lada (USA), L. Rodríguez (Mex), D. York (USA). Chairs of Working Groups: Y. Terzian (Planetary Nebulae); Helene R. Dickel (Nomenclature); A. Dalgarno (Astrochemistry).

During the XXI General Assembly, the Commission was the prime sponsor of Joint Discussion I ("An Overview of the Interstellar Medium"), Joint Commission Meeting V (Late Evolution of Low-Mass Stars), and was a co-sponsor of JCM III (Atomic and Molecular Data for Space Astronomy). It also held a joint commission meeting with Comm. 35 (Stellar Structure), and held several commission meetings on specialized topics regarding the interstellar medium (ISM). Sessions at the GA included only invited reviews, but poster papers describing new results were also presented. This summary provides an overview of the review talks and of the business meeting of the commission by the president; any errors are probably his and not the speakers'.

The program began with the Joint Discussion I (with Comm 28, 40, 44, and 48) of "An Overview of the Interstellar Medium", to be published in *Highlights of Astronomy*.

D. Lambert opened with a review of the chemical and physical properties of clouds. Two general approaches to understanding the conditions within clouds have been taken: (a) Correlations among observable characteristics (molecular column densities, line widths, etc.) from many lines of sight in several clouds, and (b) detailed modelling of individual clouds. Many observations (fine-structure lines of [C I], [C II], and [O I]; CO and H_2 arising from various levels; many molecular lines, etc.) must be fitted to produce a self-consistent model with gas-phase chemistry (excepting the formation of H_2 on grain surfaces). However, there are also many adjustable parameters: the temperatures and densities within the clouds, many cross sections, the properties of grains, and the rate of ionization and heating by cosmic rays. Generally the agreement between models and observations is quite good, but CH^+ and NH are observed stronger than predicted by 2 - 3 orders of magnitude. Shocks may play vital roles in many cases.

H. Lizst discussed the intercloud medium, and the organization (spatial characteristics) of various components. He referred to the conventional dense regions as "clowds" to emphasize that there is a continuous variation of physical conditions, but the convenience of referring to typical dense regions by a name is irresistible. The "interclowd" medium has at least three components itself: the neutral H, that may be cool ($\approx$ 100 K) or warm ($\approx$ 8000 K), the diffuse, warm, ionized component ($\approx$ 8000 K), and hot ($\approx 10^6$ K). The general properties of the components were mentioned. Detailed studies show that there must be a *gradual* change in the neutral gas from low-T, high-density conditions near the centers of clouds to low-T, high n material in the outer regions.

J.-L. Puget reviewed interstellar dust. Only one parameter, commonly taken to be R [$\equiv$ A(V)/E(B-V)], characterizes the wavelength variation of the extinction cross section among various lines of sight over the range 1 μm - 0.1 μm. The depletion of refractory elements onto dust grains also varies among sight lines. Another very important aspect of interstellar dust is the emission in the NIR bands that dominate the *IRAS* 12 μm and 25 μm filters for the ISM and contribute heavily to the 60 μm window. A very promising explanation: a continuum of grain sizes extending from large molecules (2.5 Å - 12 Å) through "Very Small Grains" (size ≈ 12 Å - 50 Å) to the conventional grains, ranging up to about 250 Å.

R. Wielebinski reviewed the information that low-frequency radio observations taken at the 100-meter telescope at Effelsberg can provide regarding the magnetic field in spirals, with some comments about our own galaxy. The direction of the field can be partially inferred from the polarization and its frequency dependence, as well as the Faraday rotation of supernova remnants. The field geometry seems complex and varies among objects.

The high-energy component of the ISM was discussed by C. Cesarsky. The smooth power-law spectrum for cosmic-ray (CR) proton energies in the 10 - 10^6 GeV range, n $\propto$ $E^{-2.73 \pm 0.09}$, suggests a Fermi-type acceleration mechanism. The isotropy indicates confinement and diffusion within the Galaxy for the lower-energies, and the large abundances of the very light elements (Li, Be, and B) produced through spallation provide estimates of the amount of material through which the CRs have passed. The higher-energy CRs escape from the Galaxy, and the injection spectrum is about $E^{-2.1}$. Shocks from supernovae should provide approximately that type of injection spectrum, at least up to energies of ≈ 10^{14} eV. However, it is difficult to see how the maximum energies of CRs (>10^{20} eV) can be produced by supernova shocks; perhaps they arise from shocks produced by collections of supernovae. The amount of secondary CRs (produced from the original CRs by collisions with the interstellar gas) is inconsistent with the low supernova rate in the ISM. The decrease of the ratio of primary/secondary CRs with energy shows that there is more than simple Fermi acceleration involved. The ratio of positrons to electrons, and its energy dependence, is also a mystery.

The center of the Galaxy was reviewed by F. Yusef-Zadeh, starting with the overall structure for the inner 500 pc and narrowing to the inner cluster presumably at the center. The stellar velocity dispersion decreases with distance from the center, and the center of the stellar cluster may contain a compact object of about few x 10^6 $M_\odot$. If there is no black hole, one needs n(stars) $\propto$ $r^{-2.8}$ for the kinematics, inconsistent with the light if (M/L) = constant for the stars. A black hole at the center will show an Einstein ring from the background at about 30 milliarcsec.

T. van der Hulst discussed the morphology of the ISM in various galaxies; M51, M31 have CO and H I coincident, and H I in both cold and warm clouds. M33 has holes in H I from a few x 10 pc in size up to kpc with the ages of a few x 10^6 yr. Many other types of galaxies were mentioned, especially NGC 891 (an edge-on spiral, in which the ISM at large z can be studied) and M 101, a grand design Sc with empty regions along its arms. The pressure required to produce the holes requires hundreds of supernovae, but this might be plausible. There might be gas infalling into the holes.

The theory of the ISM was summarized by J. M. Shull. What supports the ISM at large z from the Galactic plane: is pressure supplied by thermal motions, kinetic energy of clouds, magnetic pressure, or CRs? Possibly all of the above! Gas pressure alone is not enough to support the total weight of all of the known gas. A magnetic field of ≈ 5 µG would help greatly, and possibly heating by MHD waves increases the gas pressure at large z. Another mystery is the source of ionization of the Hα observed by Reynolds (see below). Possibly the H$^+$ is produced in a layer of turbulent mixing between hot and cold regions in the ISM, a process that should produce about the right spectrum. Shull also discussed galaxy mergers as suggested by N-body calculations, and the recent observations of the "Lyman-α forest" lines in the spectrum of 3C273 by the High Resolution Spectrograph of the Hubble Space Telescope.

The commission also held a joint meeting with Commission 35 (Stellar Structure). P. Myers reviewed molecular cloud cores and star formation, explaining the variation in physical conditions in clouds from isolated diffuse clouds through giant molecular clouds, with special attention to their cores. He reviewed the determination of cloud conditions (T, mean density <n>, and optical depth) via the NH_3 hyperfine lines. Some clouds contain embedded OB stars while others with very similar line widths, shapes (elongated, with about 2:1 aspect ratios), and sizes do not. Most clouds seem prolate (cigar-shaped). The observed relation between velocity dispersion and column density is that predicted by virial equilibrium, with nonthermal motions unimportant for the smallest clouds but dominant for giant clouds. Rotation is not significant in providing support for the clouds. Outflows and magnetic heating are sufficient for some clouds, since magnetic heating should be comparable to CR heating for fields of $10^{-4} - 10^{-3}$ G observed in clouds.

F. Adams reviewed theories of star formation. The standard scenario is that the material with the lowest angular momentum falls into the center of the cloud and produces a central protostar. Quickly a disk forms surrounded by the spheroidal collapsing gas/dust cloud in which rotation is not dominant. The collapse is far from homologous. The source of energy is the shock heating of the disk by infalling material as well as from the compression of the central star. A polar outflow develops along with the equatorial infall through the cloud and disk. The last stage before a single star is a naked star plus disk. Outflows are observed from molecular microwave lines in clouds; disks are seen in T Tauri stars. Both the pure inflow and the outflow-plus-inflow stages have about the same luminosity but a spectrum much broader than any single Planck function. The disks can be "passive", producing no energy of their own but simply reradiating stellar energy, or "active", emitting their own energy from viscous dissipation. Active disks are massive $(0.1 - 1 \, M_\odot)$ and luminous $(\approx L_*)$, with sizes of ≈ 100 AU. The actual disk physics is still obscure, but nonaxisymmetric modes (m = 1) are unstable and the most global. The mass contained within the instability is sufficient to displace the star from the center of gravity of the system, perturbing the disk and increasing the instability. However, the formation of binary system and prediction of the mass spectrum of the forming stars from first principles are still very obscure.

The effects of star formation on the surrounding ISM was discussed by L. Rodríguez. The 150 km s^{-1} outflows produce cavities in the clouds and also shells of H_2 (as seen in ^{18}CO); these flows make the whole cloud core more turbulent by 1 km s^{-1}, which is supersonic. The timescale of dissipation of the turbulence, (cloud size)/(line width), is 10^7 yr, shorter than the timescale of consumption of the gas in the cloud cores (10^8 yr). Maybe the clouds have long dissipation times because the magnetic field controls the motions.

There are about 10 times more ultracompact H II regions (detected from their almost unique set of IRAS colors, confirmed by VLA observations of the free-free radiation) than expected if the freshly ionized gas expands freely from the central star and star formation rate is steady. Probably the gas is confined by the ram pressure of the star's motion through the surrounding cloud, as in a cometary nebula.

Infrared observations of star formation in other galaxies were reviewed by C. Telesco. At 2 μm one observes the old stars directly; at 10 μm, the reradiated radiation from warm dust. Interacting galaxies produce a high 60/100 μm color temperature and a large blue luminosity relative to FIR. It is clear that interaction triggers star formation, as is also shown in barred spirals. Perhaps even M 82, the classical interacting galaxy (with M 81), has a bar, since it is bilobal in 10 μm radiation. There are knots of ≈ 1 kpc size (= 1 arcsec) with $L = 10^8 \, L_\odot$. The mass associated with a given luminosity depends strongly upon the low-mass cutoff in the IMF; if this cutoff is 0.1 $M_\odot$ in the center of M 82, *all* of the mass would have to be in very young stars (a very doubtful assumption). The low-mass end of the IMF is presumably greatly reduced there.

P. Cox reviewed FIR and sub-mm radiation as diagnostics of star formation in galaxies. One needs to know dust properties and the geometrical association of stars and dust to make accurate predictions. Most of the FIR production (70%) arises from atomic gas (the "cirrus" discovered by *IRAS*), ionized gas most of the rest, and molecular clouds only 7%. About 70% of the FIR arises from OB stars. But the Rosette Nebula region emits most of its FIR in the molecular clouds, not near the OB stars in the cluster; about 30% of the luminosity of the OB stars is reradiated as FIR. In general, the FIR production of the OB stars is spread through the ISM by their contribution to the general interstellar radiation field, not absorbed locally in their natal molecular clouds. Within the 4 kpc ring in the Galaxy, recently formed stars (types OBA) contribute about 70% of the dust heating, while the old disk stars the remainder. The overall conclusion is that the FIR luminosity of galaxies indicates mainly the star formation rate within the last Gyr or so, except for starburst galaxies where the rate is very nonsteady.

The fundamentals of using molecules as diagnostics of the physical conditions in molecular clouds were reviewed by A. Hjalmarson. The temperature must be derived from an optically thick molecular line (e.g., CO), while clumping within the clouds, on a scale different for each molecule, is a major problem in comparing species. The transfer of line radiation in various situations of level populations was discussed. Turbulence occurs on all scales, from winds from bipolar nebulae, MHD waves, and cascading turbulence from galactic rotation.

Shocks are an important ingredient of the ISM, leading to heating, compression, excitation and dissociation of molecules. D. Neufeld gave an overview of their classification and physical effects on the gas chemistry. When the relative speed of the shocked and unshocked material exceeds the speed of sound there is a jump in the physical conditions, producing a J-shock. However in the presence of a magnetic field the ISM behaves as two fluids: one is the charged particles (ions, electrons, small grains), frozen to the field, and the other the bulk of the neutral gas. If the difference in the upstream and downstream speeds is supersonic but less than the Alfven speed in the charged fluid alone (recall that the Alfven speed contains the fluid density), and the cooling time is less than the time of relaxation of the ions and neutrals, there is not a jump in the physical parameters (T, n, etc). Rather, there is a continuous change in n and T, and the shock is called a C (= "continuous") shock. The chemistry of the two types is markedly different. Behind slow J shocks in molecular clouds ("slow" means H_2 is not dissociated, v < 50 km s^{-1}) the temperature rises enough so that some endothermic reactions (esp. forming CH^+) occur and H_2 is excited to high levels. Most of the predictions look good. The Orion KL region is a good laboratory for shock diagnostics, with observations of the infrared transition in H_2 seen with broad line wings to ± 100 km s^{-1}. Temperature-sensitive ratios are surprisingly constant. A possible solution is a bow-shock model that has a scale-invariant geometry. Water masers might also be produced in fast shocks.

Since H_2 produces no observable lines unless it is relatively hot, while CO is readily observable, one of the most important numbers in astronomy is $X \equiv N(H_2)/[\int T(CO)\, dv]$, expressed in units of (10^{20} molecules K^{-1} km^{-1} s). T. Dame explained that this number can be found indirectly through correlating dust extinction with both $N(H_2)$ and $\int T(^{13}CO)$ dv in molecular clouds, and using the $T(^{13}CO)/T(^{12}CO)$ found in clouds. A second approach is to use the absorption at 2.4 μm of clouds seen against the background galactic bulge, and obtaining the mass of each cloud through the virial theorem. Both methods relate extinction to H_2 mass by the value in the *diffuse* ISM, and its variation with metallicity throughout the Galaxy. Clouds seem to be in virial equilibrium, and determinations of X in the outer galaxy (same metallicity as local) agree with those from the inner regions. Another independent method of finding X is through γ-rays produced by interaction of CRs with nuclei in the ISM. One can find X by considering molecular clouds and subtracting the estimated atomic H contribution, with no assumptions regarding the extinction/H_2 ratio. However, one must estimate the galactic density of CRs and their penetration into clouds.

A. Wolfendale gave a short presentation from the perspective of a γ-ray astronomer. He basically agreed with Dame's interpretation of the uncertainties in the methods. The "final" value of X preferred by Dame was between 3 and 1.8, with 2.2 preferred. Wolfendale assessed it to be 1 - 1.5. Older values in the literature are up to 3.6. There are exceptional regions: in the Galactic center, $X \approx 0.2$, and in the outer Galaxy, $X \approx 4 - 9$.

The session on diffuse emission started with an excellent talk by R. Reynolds on the ionized diffuse emission. Reynolds's basic observation is of very faint $H\alpha$ seen in every direction, with a required ionizing luminosity equivalent to 100% of all supernovae, or 14% of O stars. The local n_e is 0.1 cm^{-3}; T from the lines strengths is $\approx$ 8000 K; the volume filling factor is about 0.2. The distribution of n_e with z contains 2 components; one with a scale height of 170 pc, the other very smooth and exponential with a scale height of 900 pc. The mass is in the extended component. Most of the ISM is ionized above 700 pc. The spectrum is strong [S II], weak [O III], [O I], and [N II]. Clouds are resolved in velocity; sometimes they are correlated with H I, sometimes anticorrelated. The source of ionization could well be O stars, but the problem is the penetration of the ionizing radiation from the plane to high above the plane. Perhaps the ISM is in strings or other highly clumped structures.

The ISM far from the Galactic plane, except for the H$^+$ layer, was discussed by C. Blades. The observational data are the absorption line components of various ions as seen against the spectra of background stars. The material is in very patchy clouds (i.e., discrete absorption components). The recent HST spectrum of 3C273 is most interesting, with absorption clouds at -650, -325, and -50 pc if the Galactic rotation law is assumed. Low stages of ionization (C II, Mg II, etc.) are ubiquitous, and high stages (C IV, Si IV, N V, etc) appear at 1 - 3 kpc. C IV is very patchy out to 1 kpc. Ti II, found with H I, is very smooth up to 1 kpc. There are strong depletions in the abundances of most elements, and the higher velocity clouds have lower depletions than do the lower. The High Resolution Spectrograph on HST, with its vastly better resolution and signal/noise ratio, will make a vastly better analysis of the clouds possible.

The crucial role of supernova remnants (SNR) in the ISM was reviewed by Joel Bregman. Massive stars stir, ionize, and heat the ISM by their radiation and winds, but most of the kinetic energy is deposited by supernovae. The phases of SNR are well understood: a free expansion at first, followed by a blast wave solution of expansion, then a radiative shock, and finally a pressure-driven snowplow phase, during which the interior of the SNR cools.

About a half of SNe ought to occur in OB associations of massive stars of similar ages, with the other half in isolated stars. In the OB association, if a second and other later SNe (perhaps up to 30) occur in the remnant during the snowplow phase, a superbubble with a shell of accelerating H I is formed. The acceleration produces Rayleigh-Taylor instabilities that form clumps. "Chimneys" occur when the thickness of the superbubble is comparable to the scale height of the gas. The magnetic field has a considerable effect upon the evolution if its strength is 5 - 7 μG, but not if |B| $\leq$ 3 μG. Conduction is inhibited if the strong field is tangled but can take place along the field if it is aligned. The O VI absorption lines observed by the *Copernicus* satellite might arise from old SNRs cooling by radiation rather than by conduction. If the field is strong, the energy is confined to the disk of the Galaxy, but an aligned field can allow bubbles.

An important question: is the H I in clouds or sheets? An answer is provided by the number of Ly-α components relative to the 21-cm intensity along a line of sight. Almost any sheet would be thick enough to introduce a Ly-α component, while the 21-cm arises from the volume emissivity. Models fitted to the observational data suggest that clouds are too clumpy (or mottled) in appearance, or that there should be more dispersion in the strengths of the Ly-α absorptions. Also the *IRAS* cirrus suggests a sheetlike structure. The ISM seems to be dominated by a continuous medium pushed around by SNRs. Numerical simulations suggest a similar picture.

A major activity of the Commission was the Joint Commission Meeting V, "Late Evolution of Low-Mass Stars", chaired by Y. Terzian. The talks reviewed observations of red giants and Mira variables, theoretical evolution of AGB stars, the planetary nebula phase (of primary concern to Commission 34), and white dwarfs. M. Peimbert discussed the types and abundances of planetaries, while S. Kwok covered planetary nebula evolution and the formation of multiple shells. The talks were in general excellent and will be recorded in *Highlights of Astronomy*, so they will not be summarized here for reasons of economy of space.

The final session of the Commission was to have taken place on 31 July. Unfortunately a fire prevented any meetings from taking place in the San Martín Cultural Center. The commission agreed to relinquish its room in the Plaza center, and the final reviews, by A. G. G. M. Tielens ("IR emission bands from the ISM") and F. Boulanger ("IR Continuum Emission from the ISM") could not be heard. Although disappointed at this turn of events, we realized that many people were interested in the early results from the Hubble Space Telescope.

During the business meeting, the deaths of 3 members of the commission during the preceding triennium were noted with deep regret: Drs. A. H. Barrett, M. V. Penston, and B. L. Webster. They have all made their marks on the understanding of interstellar matter and upon the world around them as well.

The Working Group on Astronomical Nomenclature, chaired by Helene R. Dickel, was disbanded at her request. The members of this WG have worked for over ten years and have, in collaboration with WGs from other commissions, adopted a set of guidelines for the naming and designation of astronomical objects. These guidelines have been published in *Publ. Astr. Soc. Pacific*, **102**, 1231, 1990 and also in A&A, Supplementary issue, May 1991, pp. A11 - A13 and partially in the July 1991 issue of A. J., p iv. Other major astronomical journals have been asked to publish the guidelines in due time. The Commission strongly commended its members who have worked on this project over the years. In addition to Dr. Dickel, they are: T. Chester, K. S. de Boer, J. Dickey, M. Felli, L. Higgs, L. Kohoutek, M. Kutner, M.-C. Lortet, R. Manchester, J. M. Meade, J. Moran, N. Panagia, and R. Schwartz.

The commission elected the following officers to serve during the 1991-1994 triennium: President, H. Habing (Neth); Vice-President, D. Flower (UK); new members of Scientific Organizing Committee: F. Bruhweiler (USA), E. Falgarone (France), T. Lozinskaya (USSR), P. Martin (Canada), P. Myers (USA), S. Pottasch (Neth), and M. Rosa (Ger). The president thanked the departing members (Drs. de Boer, Lada, Lequeux, Shustov, and York), as well as the continuing members, for their advice and assistance. Dr. Habing thanked the departing president for his efforts on the commission's behalf during the past period.

The commission discussed whether or not the Reports on Astronomy, the triennial review of the field with extensive bibliographic references, was worth the considerable effort that goes into its production. A few of the attendees had used the Reports of the past (fewer than a similar poll of commission presidents and vice-presidents suggested later in the Assembly). The general consensus was that a reduced version of the Report is appropriate for Commission 34.

The Commission will sponsor an IAU Symposium on "Planetary Nebulae" in Innsbruck, Austria, 13 - 17 July 1992.

35. STELLAR CONSTITUTION (CONSTITUTION DES ÉTOILES)

PRESIDENT: A. Maeder (Switzerland)
VICE–PRESIDENT: P. Demarque (USA)
SECRETARY: Y. Lebreton (France)

24 July 1991

1 BUSINESS MEETING

The meeting starts by a moment of silence with the members standing in respect for three members of the Commission who passed away in the last three years. They were T.G. Cowling, President of the Commission from 1952 to 1958, P. Ledoux, President of the Commission from 1964 to 1967, and J. Tuominen.

Commission 35 has democratically chosen the new Officers according to a procedure initiated after the Montreal General Assembly. Firstly, a call for candidatures was made in January 1990 and then a ballot was performed in March 1990 to designate a new Vice President, who in principle at the next term becomes President, and new members of the Organising Committee, to replace the former members having served for 6 years. The poll was satisfactory, 35% of the Commission members participated in the vote. For the term 1991-1994, the elected President is Pierre Demarque (USA) and the Vice President Cesare Chiosi (Italy). The members of the Organising Committee are: R. Canal, I. Iben, A. Maeder, G. Michaud, K. Nomoto, A. Renzini, A.V. Tutukov, D. VandenBerg and G. Vauclair. The President, Vice–President and the members of the Organising Committee were endorsed by the Executive Committee at the General Assembly.

The Commission Report with contributions of F. D'Antona, P. Demarque, Y. Totsuka, E.L. Fitzpatrick, S. Kwok, J.C. Wheeler, E.P.J. van den Heuvel and G. Meynet was sent to all Commission members. In answer to questions the General Secretary addressed to Commission Presidents, clear statements were expressed at the business meeting in favour of the continuation of this form of Commission Report.

During the last triennium, 3 circular letters were sent to all Commission members and 14 to the Committee members. The President mentions that there had been a great increase in the number of requests for sponsorship of scientific meetings. Each member of the Organising Committee has been consulted. The general philosophy is that we are not to approve anything, but to help and support valuable meetings, since the Executive Committee needs pertinent advices to make a choice among numerous proposals. In this way, our Commission has given sponsorship to IAU colloquia or symposia. The titles are 1) Nonlinear phenomena in stellar variability, 2) Planetary nebulae, 3) The cosmic dynamo, 4) The bulges of galaxies, 5) Inside the stars, 6) New perspectives on stellar pulsation and pulsating variable stars.

The President closes the business meeting by some general considerations on the present status of the field of Stellar Constitution set into the general astronomical context. He notes that this domain remains, and is going to remain, at the crossroads of most research lines in astrophysics. The observational input keeps going to be very large; just to mention SN 1987A, the X and gamma bursts, the solar and stellar oscillations, the data from high resolution spectroscopy and the observations of stars in external galaxies. All these areas need large developments in theoretical models and important investments on physical ingredients. This is the essence of the field of Stellar Constitution, which presently experiences such an impressive expansion, that the domain can no longer be mastered by a single researcher. In that respect, Stellar Constitution is meeting just what Science experienced a few centuries back and what Astronomy encountered a few decades ago.

2   SCIENTIFIC MEETINGS

Commission 35 held a total of 10 scientific sessions, in addition to the business meeting. Two sessions were devoted to "Evolution of Stars and Star Clusters in the Magellanic Clouds" and are summarized below. One session was a "Progress Report on Opacities, Equation of State and their Effects on Solar and Stellar Models" and is also summarized below. Two sessions were held with Commission 34 on "Star Formation and the Interstellar Medium", see Report of Commission 34; one session on "Progress in Helioseismology" was held together with Commissions 10 and 12, see Report of Commission 10 and 12. Four sessions were organized on "Early Nucleosynthesis in Galaxies" with Commissions 28 and 29; due to space requirements, only the titles of the contributions are given below. A session on the "Progenitor Evolution of SN 1987A" had unfortunately to be cancelled due to the fire in the Conference Center.

Commission 35 also co–sponsored a Joint Discussion on Origin of Stars and Planetary Systems and two Joint Commission Meetings on Atomic and Molecular Data for Space Astronomy and on Late Evolution of Low Mass Stars. Proceedings will be found in Highlights of Astronomy.

24 July 1991

Evolution of stars and star clusters in the Magellanic Clouds

Chaired by C. Chiosi and D. Sugimoto

M. Mateo	Constraints on the IMF from Magellanic Cloud star clusters
C. Chiosi	Confrontation of stellar evolution theory with observation of Magellanic Cloud clusters (summary not received)
A. Maeder	WR and O stars in the Magellanic Clouds
S.Van den Bergh	Age distribution of clusters in the LMC and SMC
G. Meylan	Observational constraints on the dynamics of Magellanic Clouds globular clusters
D. Sugimoto	Dynamical evolution of single and binary globular clusters
N. R. Walborn	Two-stage starbursts in the LMC
A. Ray,	
N. Rathnasree	Evolution of massive binary stars in the LMC and its implication for pulsar population
M. Kigushi, S. Narita and C. Hayashi	Stability of a nearly Keplerian disk embedded in corona

Mateo reviewed photometric determinations of the IMF for stars between 1-10 $M_\odot$ in Magellanic Cloud star clusters. The earliest studies by Mateo, Elson et al. yielded very different results, although there were at that time no clusters in common to both samples. Recent CCD studies suggest that crucial differences in methodology can account for the earlier discrepant results. Specifically, the Elson et al. photographic results appear to be erroneous: recent CCD studies of some of the clusters in their sample consistently yield much steeper slopes. Differences in the results derived from various CCD studies are systematically correlated with the manner in which completeness corrections are determined and applied. Current studies show that the observed slopes of the IMF of Magellanic Cloud clusters range from 1.0 to 2.5 (the Salpeter value being 1.35), with a mean value near 1.8. Because of the systematic differences among various studies, the actual range in IMF slopes is surely smaller than this; a single, universal IMF slope near 1.8 may be applicable to most clusters. The present results provide no evidence for significant variations in the IMF slope correlated with age or metallicity.

Maeder discussed WR and O stars distributions in the Magellanic Clouds. Among the stellar properties showing very large differences from galaxy to galaxy, the number frequencies of WR stars and the number ratios of WN/WC stars are extreme cases. In the solar neighbourhood, the LMC and SMC the number ratios of WR/O stars are respectively 0.12, 0.04, 0.015; for the WC/WN ratios, the values are 1.9, 0.26, 0.14. The studies by Maeder show that the origin of these differences mainly lies in the stellar evolution. For massive stars, differences in metallicity Z mainly act through stellar winds, which according to recent wind models are higher at larger metallicities. Thus, a larger Z implies more peeling off and more WR stars formed, with higher WC/WN ratios. Quantitative comparisons between the Maeder stellar models and observations in the Galaxy, the LMC, the SMC and other galaxies very well support this view. The link Z-mass-loss-evolution has also a great impact on the nucleosynthesis and chemical yields.

Van den Bergh reviewed evidence on the rates of star and cluster formation in the Magellanic Clouds. In the SMC star formation seems to have turned on gradually with the oldest cluster (NGC121) having an age of about 12 Gyr. In the LMC an initial burst of activity about 15 Gyr ago formed a number of massive globular clusters and numerous field RR Lyrae stars. After this the large cloud remained quiescent for about 10 Gyr. A second burst of star and cluster formation started 3-4 Gyr ago and continues to the present day. The fact that star and cluster formation rates in the LMC and SMC are not correlated proves that such bursts of star formation did not result from gravitational interactions between the clouds. The fact that clusters with ages of 5-10 Gyr are seen in the SMC shows that their absence in the LMC is NOT due to selective effects.

Meylan reviewed the problems related to the dynamics of globular clusters. Because of the large spread in age among clusters, the Magellanic Clouds provide a unique opportunity to learn about cluster formation and evolution. Dynamical models (e.g. King-Michie) need three observational constraints: 1) the surface brightness profile gives an indication on the concentration. From a survey of Magellanic clusters, it appears that, among old clusters, the fraction of collapsed to King-Michie type clusters is the same in the Galaxy and in the LMC. 2) the color-magnitude diagram gives a precise determination of the cluster age when compared with stellar models, for NGC 2070, comparison with isochrones from Maeder and Meynet (1991) gives an age smaller than 3×10^6 yr. Contrary to Melnick (1985), no turnoff is observed in the Meylan et al. data which concern only the Melnick's field. This could be an indication of delayed star formation in the central parts of the cluster. The luminosity function is derived from the color-magnitude diagram stars; for NGC 2070 an IMF exponant $x = 0.7 \pm 0.4$ is found. 3) the velocity dispersion is obtained from integrated light spectra (Dubath et al. 1990). Combined with the surface brightness profile, the velocity dispersion constraints a King-Michie dynamical model and allows determination of mass and mass-luminosity ratio. Core velocity dispersions have been obtained for 23 galactic clusters, 28 LMC clusters, and 3 clusters of the Fornax dwarf spheroidal. All the old globular clusters studies in the Galaxy, the LMC and Fornax display strong similitudes in the ranges for velocity dispersion, mass, and M/Lv ratio, indicating a possible universality of their characteristics and formation mode.

Main objectives of Sugimoto's talk is to extend cooperations between the two fields, stellar evolution and stellar dynamics. The gravothermal catastrophe and succeeding post-collapse evolution with binary hardening of globular cluster are the same physical processes as the gravitational contraction and the nuclear burning of a single star, respectively. Evolution and merging of binary globular clusters, which are observed in LMC, are quite similar to Roche lobe overflow and coalescence of close binary stars. Formation of binary stars and millisecond pulsars in the core of globular clusters are one of the common subjects. Differences in chemical evolution among LMC, SMC and the Galaxy as discussed in the preceeding talks should also be interpreted under both lights of stellar evolution and the dynamical history of LMC-SMC-Galaxy system. Special purpose computer being developed for N-body problem is also applicable to smoothed hydrodynamics of stars.

Walborn reviewed two-stage starbursts in the LMC. A study of the stellar content of the LMC giant shell H II region N11 by Parker, Garmany, Massey, and Walborn has revealed significant differences between the concentrated association occupying the central, evacuated cavity and those ionizing the surrounding nebulosities. The former contains an earliest spectral type of O6, OB supergiants, and a WC + late-O system, implying an age of $5 \ 10^6$ yr, while the latter contains (newly discovered) O3 stars, candidate ZAMS O stars, and have a flatter IMF, consistent with an age of $3 \ 10^6$ yr. This morphology is remarkably reminiscent of that in 30 Doradus, where IR protostars (age less than 10^6 yr) have recently been found surrounding the central 3-million-year-old cluster. Thus N11 appears as an aged 30 Dor; its central WC + O system may even be an evolved R136. Furthermore, N11, which is the second-ranked H II region in the LMC, is located diametrically opposite 30 Dor, off the other end of the Bar. A hypothesis emerges whereby the initial, centrally concentrated starbursts are caused by events on a galactic scale, and winds/supernovae from its massive stars trigger a secondary burst around the periphery about $2 \ 10^6$ yr later. The two mechanisms may produce different IMFs.

The LMC Hertzsprung-Russel diagram compiled by Fitzpatrick and Garmany shows supergiants immediatly redwards of the main sequence while theoretical models of massive stars with normal hydrogen abundance predict that this region should be unpopulated. Supergiants which have accreted helium from a binary companion, further evolve in a way so that the models and observed data are consistent (Tuchman and Wheeler). Ray and Rathnasree compare the optical data on OB supergiants with models of massive stars with a LMC metallicity with and without He-enriched envelopes and conclude that about 60% of supergiants may occur in binaries. These binaries will evolve into massive X-ray binaries, the observed number and orbital period distribution of which constrain the scenarios of supergiant binaries evolution. The distribution of post main sequence binaries and related systems like WR + O stars so obtained is bimodal with close and wide binaries in which the latter type is dominating. The expected space velocity distribution of runaway O-stars arising from the first explosion in the binary system are calculated. The second Supernova disrupts the binary giving rise to two runaway neutron stars. The steady-state spin and radio luminosity distributions of single pulsars born from the massive stars is determined under simple assumptions. A small but significant number of observable single radio pulsars arising out of the disrupted massive binaries may appear in the few millisecond spin period range. Most pulsars coming from long period binaries will have a low velocity of ejection and might cluster around the OB associations in the LMC.

Kigushi, Narita and Hayashi have carried out a numerical simulation for the time evolution of a proto-solar nebula enclosed by a corona. The initial condition for the calculation is a equilibrium configuration of the type of $f(r) \cdot g(\theta)$. For the configuration used the density is 300 times greater in the equator than the pole, the temperature is 30 times greater in the pole than the equator, and the thickness of the disk is $\delta\theta$ about 0.15. This configuration is stable when the adiabatic condition is imposed. However if temperature is assumed to be a function of position, turbulence develops just above the disk and penetrates into the disk. The resulting disk accretion parameter α is greater than about 0.2 in the disk, which is a marginal value below which Roche instability develops. This shows the importance of the radiative transfer for the estimate of the angular momentum transfer in the disk.

27 July 1991

Progress report on opacities, equation of state and their effects on solar and stellar models

Chaired by A. Maeder

W. Däppen	Present status of the opacity project
A.N. Cox	Opacity calculations, Cepheids and OB star instabilities
D. Guenther	Input physics and effects on solar models and oscillations

Däppen gave information on the present status of new opacity calculations. Currently, there are two large projects to compute solar and stellar opacities, one being the international so-called "Opacity Project (OP)" led by Seaton and Mihalas, the other an effort pursued at Livermore by Rogers and Iglesias, called "OPAL". While OP is based on very detailed ab-initio atomic physics, it relies on a more heuristic description of the plasma. OPAL is based on a systematic treatment of the quantum statistical mechanics of the plasma, but it uses more simplified atomic physics (parametric potentials). At a recent meeting held in Caracas, these two groups compared results, and it emerged that both achieved significant progress by yielding a factor of 2-3 enhancement of the heavy-element opacity (compared with the older Los Alamos opacities), while agreeing with each other to 10-20%. Preliminary calculations have shown that these new opacities can indeed solve the problem with the mass of beat and bump Cepheids. It looks therefore that after some 25 years of great successes with Los Alamos astrophysical opacities a new era has begun, and promising applications of both OP and OPAL data are on the way.

Cox shows that recent stellar (OPAL) opacities from the Lawrence Livermore National Library have been very successful in explaining problems in stellar structure and stability. OPAL include numerous improvements in all the basic photon absorption processes, but their most dramatic effect is the inclusion of the millions of same-shell formally forbidden iron lines. These M shell transitions give the most lines and the most increase at a temperature centered around 250,000 K at a density of near 10^{-5} g cm^{-3}. These iron lines increase the opacity for yellow giants envelopes by a factor of three typically. Such opacity increases have been shown to change the yellow giant pulsator envelope structures to decrease their apparent density concentration for the pulsation modes so that the period ratio between the first radial overtone and the fundamental mode decreases enough to agree well with that observed using traditional mass models. These larger opacities also can deepen the convection zone during pre-main sequence lithium burning for stars now seen as Li deficient G giants in the Hyades. The most impressive effect of the new larger and more rapidly rising opacities with temperature is the large enough kappa effect in OB stars envelopes so that the beta Cephei pulsations can be predicted. This prediction of pulsational instability now depends on the iron abundance, and stable stars can exist if there is not enough iron. To explain variability in all kinds of OB stars (including Be stars, luminous blue variables, and even very hot helium deficient stars) and their episodes of constancy and variability, apparently levitation of iron in the pulsation driving layers by photon momentum absorption is required. A small but sufficient iron concentration increase can initiate pulsations, but they then might cause enough shear to remix the envelope to a stable condition again. The OB star pulsation problems have now been reduced to composition problems in the layers about 10^{-6} of the mass into the star.

Guenther presented several solar models each of them differing in the input physics adopted (table 1) and compared them to observations. In conclusion, with respect to improving the p-modes, the most important new physics are: low-T opacities, T-tau relation for atmosphere, equation of state's effect on thermodynamic variables. None of these affect the interior structure. With respect to changing the solar interior, the most important new physics are: high-T opacities and P, rho, T changes via Coulomb corrections in equation of state. All solar models have an initial helium abundance Y of about 0.28. The OPAL opacities deepen the solar convection zone to a depth consistent with p-mode inversions. The neutrino flux is not significantly changed.

Table 1: Model characteristics

Model	Y	α	log T_c	log ρ_c	X conv	M conv/$M_\odot$	log T conv
Standard model	0.2875	1.195	7.186	2.166	0.745	0.0153	6.284
MHD equ. of state	0.2736	1.256	7.188	2.159	0.734	0.0190	6.308
D.-H. correction	0.2771	1.223	7.185	2.160	0.740	0.0167	6.294
Cox low T opacities	0.2876	2.066	7.186	2.166	0.744	0.0153	6.285
Kurucz low T opac.	0.2875	1.559	7.186	2.166	0.745	0.0152	6.283
OPAL opacities	0.2918	1.248	7.193	2.168	0.724	0.0211	6.331
Ross-Aller mixture	0.2843	1.242	7.190	2.166	0.738	0.0177	6.299
Bahcall nuclear rates	0.2843	1.194	7.189	2.181	0.744	0.0151	6.284
K. S. atmosphere	0.2875	1.423	7.186	2.166	0.745	0.0155	6.284

COMMISSION 35

30 July 1991

Early nucleosynthesis in galaxies

Chaired by P. Conti and A. Maeder

P. Conti	Early nucleosynthesis – Nature of the problem
D. Lambert	Pop II stars: CNO abundances and light metals
C. Pilaichowski	Abundances in old clusters
M. Bessel	Stellar composition in the Magellanic clouds
H. Lamers	Abundances of low-mass post AGB stars
J. Bergeron	Abundances in absorption line systems towards QSOs
D. Hollowell	AGB star model at low Z
A. Maeder	Helium and metal synthesis in massive stars at low Z
B. Pagel	Helium and metal abundances in HII regions
F. Thielemann	SN and chemical synthesis for metal poor stars
M. Grenon	The sites of nucleosynthesis: halo and bulge
C. Chiosi	Chemical evolution in the early Galaxy

The above contributions covering the major aspects of early nucleosynthesis in galaxies were delivered in a full day meeting.

THEORY OF STELLAR ATMOSPHERES
(THEORIE DES ATMOSPHERES STELLAIRES)

President: David F. Gray   Vice President: Wolfgang Kalkofen

New Officers

It has become the practice of the commission for the president, with the assistance of his organizing committee, to assemble names for the new organizing committee from suggestions of the commission members. It is the president's responsibility to see that a reasonable balance is maintained in geographical distribution and areas of research expertise. Normally the vice president ascends to the presidency. Continuous membership on the organizing committee is limited to two consecutive terms except for the president and vice president. To make this procedure function, We depend on suggestions from the membership for new members of the organizing committee.

The new organizing committee for 1991-94 consists of W. Kalkofen, USA (president); L. Cram, Australia (vice president); Y. Cuny, France; D. Dravins, Sweden; J.L. Linsky, USA; R. Pallavicini, Italy; A. Peraiah, India; A. Sapar, USSR; T. Tsuji, Japan; R. Wehrse, Germany; L.A. Willson, USA (new); and C. Zwaan, Netherlands (new).

Thirty new members joined our commission at this General Assembly: Baade, D., Baliunas, S., Balona, L.A., Basri, G., Bopp, B.W., Catalano, F., Catalano, S., Cuntz, M., Doazan, V., Fontaine, G., Giampapa, M.S., Gigas, D., Glebocki, R., Hall, D.S., Judge, P., Kondo, Y., Lamers, H.J.G.L.M., Luck, R.E., Luttermoser, D.G., Mathys, G., Piskunov, N.E., Rodono, M., Rucinski, S., Ryabchikova, T.A., Schrijver, C.J., Tuominen, I., Walter, F., Vaughan, A.H., Vilhu, O., Wolff, S.C. The commission membership now stands at 266.

Working Groups

Commission 36 has continued to sponsor the following working groups: Ap/CP Stars, Astrochemistry, Atomic and Molecular Data, Be Stars, Peculiar Red Giants, and Synthetic Photometry.

Meetings Associated with Commission 36

The following meetings were sponsored or otherwise associated with our commission. Not all were successful in gaining IAU status. Some are yet to be held.

> 1992 July, "New Perspectives on Stellar Pulsation and Pulsating Variable Stars", Victoria, D. Welch, J. Nemec.
> 1992 July, "Planetary Nebulae", Innsbruck, A. Acker.
> 1992 July, "Peculiar versus Normal Phenomena in A-Type & Related Stars", Trieste, M. Hack.
> 1992 March, "Stellar Chromospheres", Cambridge, U.K., C.S. Jeffery.
> 1992 February, "Infrared Solar Physics", Tucson D. Deming, D. Rabin.
> 1992 February, The Tenth Colloquium on "Ultraviolet and X-Ray Spectroscopy of Astrophysical and Laboratory Plasmas", Berkeley, S.M. Kahn.
> 1992 January, "Non-Linear Phenomena in Stellar Variability", Mito, M. Takeuti.

1991 August, "The Astrochemistry of Cosmic Phenomena", Campos de Jordao, A. Dalgarno.

1991 August, "Evolutionary Processes in Interacting Binary Stars", Cordoba, Y. Kondo.

1991 July-August, "Eruptive Solar Flares", Buenos Aires, B.V. Jackson.

1991 July-August, "Solar and Stellar Coronae", Buenos Aires, R. Pallavicini.

1991 July-August, "Chromospheres and Circumstellar Envelopes of Red Giant Stars", Buenos Aires, H.R. Johnson.

1991 July-August, "Astrophysical Opacities", Buenos Aires, S. Sahal-Brechot.

1990 September, "Stellar Atmospheres: Beyond Classical Models", Trieste, "L. Crivellari.

1990 September, "Rotation and Angular Momentum of Low-Mass Stars", Noto, S. Catalano.

1990 August, "Evolution of Stars: the Photospheric Abundance Connection", Druzbe, G. Michaud.

1990 July, "The Sun and Cool Stars: Activity, Magnetism, Dynamos", Helsinki, I. Tuominen.

1990 July," Surface Inhomogeneities in Late-Type Stars", Armagh, J.G. Doyle.

1990 June, "Mechanisms of Chromospheric and Coronal Heating", Heidelberg, P. Ulmschneider.

1990 June, "Wolf-Rayet Stars and Interrelations with other Massive Stars in Galaxies", Denpasar, K.A. van der Hucht.

1990 May, "Confrontation Between Stellar Pulsation and Evolution", Bologna, C. Cacciari.

1989 December, "Basic Plasma Processes in the Sun", Bangalore, V. Krishan.

1989 October, "Angular Momentum and Mass Loss for Hot stars", Ames, L.A. Willson.

Meetings held at the General Assembly

Two scientific sessions were held by Commission 36 during the General Assembly in Bueons Aires. The papers presented are as follows.

Session 1 Extended Atmospheres (Chairman, D.F. Gray)

S.V. Mallik, "The H_α Line as a Diagnostic of Cool Supergiant Chromospheres."
W.H. Wehlau, "Surface Features on Ap Stars."
J.C. Pecker, "Comments on Non-Sphericity."
A. Peraiah, "Departures from Sphericity in Stellar Atmospheres."
R. Wehrse, "The Solution of the 3D Radiative Transfer Equation."

Session 2 Red Giant Stars (Chairman, H.R. Johnson)

K.G. Carpenter, "Ultraviolet and Visual Observations of Chromospheres."
P. Ulmschneider, "Mechanisms for Heating the Chromosphere."
W. Van der Veen, "Observations of CS Gas and Dust."
A.G. Hearn, "Mechanisms of Mass Loss."

Commission 36 also co-sponsored the Joint Discussion on Solar & Stellar Coronae (with Commissions 10, 12, and 44).

STAR CLUSTERS AND ASSOCIATIONS (AMAS STELLAIRES ET ASSOCIATIONS)

PRESIDENT: G.L.H.HARRIS   SECRETARY: C. PILACHOWSKI

BUSINESS SESSION

I.   IAU REPORT ON ASTRONOMY, COLLOQUIA ETC.

The Business Meeting of Commission 37 was called to order by Acting President and Vice President C. Pilachowski. She conveyed the regrets of President G. Harris at being unable to attend the General Assembly. Other members of the Organizing Committee present included J. Hesser and J. Zhao.

II. MEMBERSHIP:

Commission 37 approved and welcomed new members proposed by the various national committees, as well as several IAU members who requested to join the commission. These included Eugene Milone of Canada, David Latham of the US, M. Tsvetkov of Bulgaria, Georges Meylan of the US, and Mariano Mendez of Argentina.

III. OFFICERS:

The new President of Commission 37 is Jean–Claude Mermilliod of Switzerland. Alejandro Feinstein of Argentina is the new Vice President. The members of the Organizing Committee for the triennium 1991–4 are Past President G. Harris of Canada, K. Janes of the U.S., D. Vandenberg of Canada, J. Claria of Argentina, J. Zhao of the P.R.C., R. Buonanno of Italy, G. Da Costa of Australia, and G. Meylan of the U.S.

SCIENTIFIC SESSIONS

The first scientific session of Commission 37 was held on Monday, 29 July 1991. on the subject of Binary Stars, Radial Velocities, and Open Cluster Memberships. The double session was chaired by C. Pilachowski and J. Hesser, and the following 7 invited talks were presented:

1   "ECLIPSING BINARY STARS AS A DIAGNOSTIC TOOL
FOR CLUSTER EVOLUTION STUDIES"
E.F. Milone (RAO, U. of Calgary)

The improvement of synthetic eclipsing light curve modeling codes over the past decade now permits simultaneous modeling in multiple passbands, radial velocities, and, to handle asymmetries, star spots. We have modified the Wilson–Devinney code by replacing its Carbon–Gingerich atmospheres option with one based on Kurucz's atmospheres. With empirical corrections of Buser and Kurucz, unprecedented accuracy in modeling UV light curves can now be carried out, and a diagnostic tool to probe chemical composition of systems in clusters is at hand. A simplex-enfolding version of the WD code can now probe parameter space and thus check the uniqueness of the WD least-squares solution. The contact system H235 in NGC 752 provides a timely example.

2   MEMBERSHIPS AND THE BINARY POPULATION IN THE HYADES
Robert P. Stefanik and David W. Latham, Harvard–Smithsonian Center for Astrophysics

We have been monitoring the radial velocities of nearly all the proper-motion and photometric candidate members of the Hyades for almost a decade, from the F stars down to a V magnitude limit of almost 15. About 25% of the proper-motion candidates are not members, and therefore should not be included in any convergent- point solutions for the distance to the Hyades. We confirm that the transition period between circular and eccentric orbits for the main-sequence spectroscopic binaries is about 5.5 days. The frequency of spectroscopic binaries is consistent with the value of about 20% recently derived by the CORAVEL and CfA teams for various populations of field stars.

3   THE BINARY POPULATION IN M67

David W. Latham, Harvard–Smithsonian Center for Astrophysics; Robert D. Mathieu, University of Wisconsin; Alejandra A. E. Milone, Harvard–Smithsonian Center for Astrophysics and University of Cordoba; Robert J. Davis, Harvard–Smithsonian Center for Astrophysics

Precise radial velocity observations spanning almost 20 years have been used to derive spectroscopic orbits for 22 binaries brighter than $V = 12.7$ in the field of the old open cluster M67 (Mathieu, Latham, and Griffin). All the binaries near the main sequence and with periods shorter than 11 days have circular orbits. This is consistent with tidal circularization being effective on the main sequence over the 5 Gyr age of the cluster.

Our original survey did not extend much below the cluster turn off, and evolutionary effects among the subgiants and giants complicate the tidal circularization arguments. To alleviate this ambiguity we are now extending our binary survey in M67 to a magnitude limit of nearly $V = 16$, well down the main sequence. We have identified about 45 new main-sequence binary candidates and have already derived preliminary orbits for 15 of these. The value we determine for the transition period remains unchanged.

If there were no primordial binaries in M67, the core would have undergone gravothermal collapse in a time very short compared to the age of the cluster. Thus a knowledge of the population of main-sequence binaries is essential for N-body models of the dynamical evolution of the cluster.

4   MEMBERSHIP AND BINARY FREQUENCY FOR BLUE STRAGGLERS

Alejandra A. E. Milone, Harvard Smithsonian Center for Astrophysics and University of Cordoba David W. Latham, Harvard–Smithsonian Center for Astrophysics

We have attempted to measure the radial velocities of 62 candidate blue stragglers in the five open clusters NGC 752, 2360, 2420, 2682, and 7789 using the CfA systems on Mt. Hopkins; 42 of these candidates have yielded reliable velocities.

The unusually large population of blue straggler candidates in NGC 7789 has proven to be a result of serious contamination by non members.

Based on multiple velocity measurements over a few hundred days, we estimate that at least 40% of the blue stragglers that yield CfA velocities are binaries.

5   BINARY FREQUENCIES IN TWO OPEN CLUSTERS

N. Morrell, Universidad Nacional de La Plata, and H. Abt, National Optical Astronomy Observatories

A search for binaries was carried out among the brightest main sequence members of IC 4665 and the Alpha Persei cluster, through coudé–CCD spectroscopy (15 Å/mm) in order to compare binary frequencies in both clusters. For IC 4665, among 15 stars observed, we found 4 spectroscopic binaries (two double-lined) with periods ranging from 6 to 11 days. We observed 26 probable members of the Alpha Persei cluster, 3 of them are binaries (1 double-lined) and 4 are possible binaries (preliminary orbital solutions were found for 3 of them). All the periods are longer than 20 days, and the half-amplitudes lower than 13 km/s, which suggests relatively low-mass secondary components.

6   EVOLUTIONARY CHANGES IN BINARIES IN OPEN CLUSTERS

H. Abt, National Optical Astronomy Observatories, and N. I. Morrell, Universidad Nacional de La Plata

Originally we thought that binaries with $P > 10$ years were formed by fission (because the secondary mass function was flat) and the longer ones by capture (because the secondaries fit the van Rhijn function). But fission does not work for compressible gases. We found that all characteristics of cluster and field binaries can be explained by capture alone. Simulations by Aarseth and Hills indicate that initially high-mass stars will form long-period binaries with low-mass stars because they are so plentiful, but after many captures and disruptions the high-mass stars tend to form short-period binaries with high-mass stars. That is consistent with the observations that in young clusters (Orion Neb., Alpha Persei) the mass ratios are all large and the periods are long, but after 10–100 free-fall times the mass ratios are mostly near 1 and the periods are shorter (as in IC 4665).

7 BINARITY AND MEMBERSHIP FROM CORAVEL RADIAL VELOCITY OBSERVATIONS IN OPEN CLUSTERS

J.C. Mermilliod, Institut d'Astronomie de l'University de Lausanne

Results of the systematic observing program in open clusters undertaken with the Coravel radial-velocity scanners at the Haute–Provence and La Silla (ESO) Observatories have been presented. It includes about Berenices, NGC 752, 6475, 7092, IC 2391 and 2602 and Blanco 1) and about 1000 red giants in 175 open clusters. The main motivations are the determi- nation of membership, the determination of orbital elements and the study of their distribution, and the measurement of the rotation (dwarfs). Numerous binaries have been discovered and observed frequently. 50 orbits have been obtained so far for the dwarfs and 80 for the red giants. The overall frequency of spectroscopic binaries among the red giants is 23% (150 SB). The binary frequency in Praesepe seems to be twice that of the Pleiades in the interval F5–K0.

On July 30, Commission 37 organized a session on the subject of globular cluster ages. The Session was chaired by C. Pilachowski. Four papers were presented, as follows:

8 OLD CLUSTER AGES

P. Demarque, Yale University

The ages of the old star clusters provide basic information on the formation and evolution of the Galaxy. Recent advances indicate that an age spread of up to 4–5 Gyr exists among globular clusters, suggesting a long time scale of formation for the galactic halo. Absolute ages are more uncertain: current research focusses on the effects of helium diffusion near the main sequence turnoff, the dependence of the RR Lyrae luminosity on metallicity, and the [O/Fe] ratio in metal poor stars. For M92, the age is 16 ± 3 Gyr. An age below 13 Gyr would cause inconsistencies with several other pieces of astronomical data.

Concerning the age of the old disk, it was pointed out that a solar calibration for the helium content and mixing-length of the turnoff stars in NGC 188 yields an age of 6-8 Gyr (depending on reddening) for the cluster. NGC 6791 is 1 Gyr older if of solar metallicity. If more metal rich than NGC 188, its age is reduced.

Finally, it was noted that recent data from Walker et al. on RR Lyrae variables in the galactic bulge, when compared to synthetic HB models (by Y.-W. Lee), suggest that the field RR Lyrae population in the bulge is 1-1.5 Gyr older than M92. The globular clusters may not contain the oldest stars in the Galaxy !

9 ALPHA ELEMENT ABUNDANCES IN THE DRACO DWARF SPHEROIDAL GALAXY

J.E. Hesser, DAO/HIA/NRC

In a program conducted with M.D. Lehnert, R.A. Bell and J.B. Oke, abundances have been measured for 14 giant stars with $+0.2 < Mv < -2.5$ ($19.6 < V < 16.9$) using Oke's double spectrograph equipped with CCD detectors on the 5-m Hale Telescope. A synthetic spectral analysis was performed of the calcium infrared triplet and the strongest line (5183 Å) of the magnesium 'b' triplet. Abundances inferred from the two elements correlate closely and range from -1.4 to -2.4, with an average of -1.9 ± 0.4. The abundances seem to fall into two groups, one with a $<[Fe/H]> -1.6 \pm 0.2$ and the other with $<[Fe/H]> -2.3 \pm 0.2$. These results confirm and extend in several ways other claims for an abundance spread among the stars of the Draco Dsph; in particular they reach to much less evolved giants. It seems likely that the dominant factor in the color spread among giants in the color-magnitude diagram is the wide range of [Fe/H] at a given luminosity.

10 OXYGEN ABUNDANCES IN HALO STARS

M. Bessell, Mt. Stromlo and Siding Spring Observatories

The oxygen abundance has been determined for a sample of metal-poor G dwarfs ($-1.2 < [Fe/H] < -2.7$) in two independent ways; by analysis of OH lines between 3080 - 3200Å and the permitted high excitation far-red OI triplet (7771 - 7775Å, c= 9.1eV). The oxygen abundances determined from the low excitation OH lines are up to 0.55 dex lower than those measured from the high excitation OI lines. The abundances for the far-red OI triplet lines agree with those re-derived from Abia and Rebolo and the abundances from the OH lines in dwarfs and giants are in agreement with the re-derived O abundances of Barbuy and others from the forbidden resonance OI line (6300Å, c = 0.0eV). Because the c = $0 - 1.7$eV OH lines are formed in the same layers as the majority of Fe, Ti and other neutral metal lines used for abundance analyses we believe that the OH lines and the forbidden OI line yield the true oxygen abundances relative to the metals. The most likely explanation for the systematic overabundances derived from high excitation lines is that metal-poor stars are much hotter in the deep layers where these lines are formed than are the model atmospheres used in the analyses. Discounting all far-red OI triplet abundances the behaviour of O/Fe with decreasing metallicity is very similar to that of the other alpha-elements, especially Ca, and shows a linear increase in [O/Fe] from 0 at [Fe/H]= 0 to +0.5 at around [Fe/H]= -2. [O/Fe] remains at about 0.5 dex for [Fe/H] between -2 and -4.

11   CNO ABUNDANCES IN GLOBULAR CLUSTERS
C. A. Pilachowski and T. Armandroff, National Optical Astronomy Observatories

New observations as well as published observations of carbon, nitrogen, and oxygen abundances in globular cluster stars are reviewed in order to pursue the questions of the original C+N+O abundance in unevolved stars in globular clusters, the variation of C+N+O from star to star within a cluster, and the role of O>N cycle processing in giants. We find that: 1) [C+N+O/Fe] is within the range 0.0 to +0.5 in globular clusters over a wide range in [Fe/H]; 2) [C+N+O/Fe] is constant within a cluster with the possible exceptions of omega Centauri and M22; 3) Log (O) vs Log (C+N) diagrams for the clusters M92, M13, M4, and 47 Tuc are consistent with the mixing of CNO processed material either in the star itself or in an earlier star, and 4) the constancy of [C+N+O/Fe] within clusters argues strongly that CN variations found in unevolved stars in clusters are not due to contamination by products of triple-alpha burning.

Commission 37 concluded its scientific program on July 31 with a variety of papers contributed on subjects related to globular and open clusters. The Session was compressed from 3 hours to only 90 minutes due to the fire in the San Martin Center that morning. The session was chaired by C. Pilachowski. The following individuals presented their work:

G. Meylan	"Two High Velocity Stars Ejected Out of 47 Tuc"
F. Graham–Smith	"10 Milli–Second Pulsar in 47 Tuc"
J. Colin	"Interaction between a Galactic Disc and a Globular Cluster"
D. Minniti	"Linear Polarization of Stars in 7 Metal Poor Globular Clusters"
J. C. Forte	"Dust in Globular Clusters"
I. Platais	"Luminosity Functions of Poorly Populated Open Clusters"
N. Nikolav	"Luminous Stars in Spiral Arms of M31"
R. Costero	"Origins of Runaway and Field O Stars"
A. J. Delgado	"Photometry of Young Open Clusters - NGC 1502 and NGC 2169"
E. Alfaro	"CCD BVR Photometry of the Open Cluster IC 1311"
M. Rabolli	"Globular Clusters in Fornax Ellipticals"
M. Tsvetkov	"Muenster–ESO Flare Star Search Project"

At the conclusion of this session Jean–Claude Mermilliod summarized recent progress with the Open Cluster Database.

COMMISSION 38: EXCHANGE OF ASTRONOMERS

COMMISSION 38: ECHANGE D'ASTRONOMES

President: F. Graham Smith   Vice-President: J. Sahade

Business: The following Members are appointed for the triennium 1991-1994 in the Organizing Committee: D.M. Chitre, J.R. Ducati, H.E. Jorgensen (Vice-President),   G. Krishna,   K.C. Leung,   M. Morimoto,   M.S. Robert, J. Sahade (President) & F.G. Smith (ex officio).

Since September 1988, Commission 38 assisted the travel of 5, 12 and 8 astronomers in 1988, 1989 and 1990 respectively, disbursing a total of 56,000 Sw.F.   The bulk of these grants was to assist astronomers from the developing world spend a period of three months or longer at a host institution to further their astronomical research.

The list of the grantees (September 1988-December 1990) is given below:

Name	Home Institution *Host Institution*	Country
Bhattacharya D.	Raman Institute *University of Amsterdam*	India *Netherlands*
Beauge C.	Observatorio Astronomico *Institute of Astronomy Cambridge*	Argentina *UK*
Clocchiatti A.	Observatorio Astronomico *University of Texas McDonald Observatory*	Argentina *USA*
Das Gupta	IUC Astronomy & Astrophysics *University of Wales Cardiff*	India *UK*
Dwivadi B.N.	Banaras Hindu University *Glasgow University*	India *UK*
Ferraz-Melo S.	Sao Paulo University *Observatoire de Paris-Meudon*	Brazil *France*
Ghosh P.	Tata Institute *University of Illinois*	India *USA*
Gomez M.	Observ. Astronomico Cordoba *Smithsonian Astrophys. Obs.*	Argentina *USA*
Ibanez M.H.	University de Los Andes *St Andrews*	Venezuela *UK*

Jerzykiewicz M.	Wroclaw Univ. Observatory	Poland
	SAAO	*South Africa*
Jorissen A.	Institut Astronomie Brussels	Belgium
	University of Texas Austin	*USA*
Kim Ho-Il	Yonsei Univ. Observatory, Seoul	Korea R
	Univ. Nebraska, Lincoln	*USA*
Kubacek D.	Astronomical Institute	Czechoslovakia
	Harvard-Smithsoninan	*UK*
Li Zhongyuan	University of Science	China PR
	& Technology	
	MSSL	*UK*
Mauas R.	IAFE	Argentina
	Harvard Smithsonian	*USA*
Nagendra K.N.	Indian Inst. of Astrophysics	India
	Rensselaer Polyt. Inst. NY	*USA*
Ryan S.	Mount Stromlo Observatories	Australia
	Yale University/	*USA*
	Institute of Astronomy	*UK*
	Cambridge	
Sagar R.	Indian Inst. of Astrophysics	India
	AAO	*Australia*
Saha S.K.	Indian Inst. Astrophysics Bangalore	India
	CERGA-OCA, Grasse	*France*
Sattyaprakash B.S.	IUC Astronomy & Astrophysics	India
	University of Wales Cardiff	*UK*
Seimis J.	University of Athens	Greece
	Inst. Theoretical Astronomy	*USSR*
	Leningrad/	
	N. Copernicus Astron. Center	*Poland*
Singal A.K.	TIFR	India
	Inst. für Astrophysik Bonn	*Germany*
Stefl S.	Ondrejov Observatory	Czechoslovakia
	University of Tucson Arizona	*USA*
	University of Toronto	*Canada*
Tancredi G.	University Rep. Montevideo	Uruguay
	Uppsala University	*Sweden*
Wang Ruyou	Shangai Observatory	China PR
	NOAO	*USA*

The rules adopted by Commission 38 for its travel grants are published annually in the January issue of the Information Bulletin.

COMMISSION No. 40

RADIO ASTRONOMY (RADIOASTRONOMIE)

Report of Meetings, 24, 25, 26, 29, 30, 31 July, 1991

PRESIDENT: P.G. Mezger   VICE-PRESIDENT: M. Morimoto

Business Meetings   SECRETARY: T.L. Wilson

I   NEW OFFICERS

M. Morimoto (Japan) was elected President and J.B. Whiteoak (Australia) was elected Vice-President for the period 1991-1994. The term of members of the Scientific Organizing Committee (SOC) is 6 years. Continuing members of the SOC are D.C. Backer (USA), R. Fanti (Italy), R. Guesten (Germany), J. Moran (USA) and J.M. van der Hulst (Netherlands). The following were elected members of the SOC: R. Davis (UK), J. Gomez-Gonzales (Spain), F.R. Colomb (Argentina), Shuhua Ye (China), P.G. Mezger (Germany), E. Gerard (France), T. Velusamy (India), L. Baath (Sweden), E.E. Baart (S.Africa) and I. Gossachinskiy (USSR). Those leaving the SOC are: J.E. Baldwin (UK), A. Baudry (F), R.S. Booth (Sweden), D.L. Jauncey (Australia), N. Kaifu (Japan), V.K. Kapahi (India), L. Matveyenko (USSR), G.D. Nicolson (South Africa), E. Seaquist (Canada) and Q.F. Yin (China).

II   NEW MEMBERS

Eighty-two new members were admitted to Commission 40, bringing the total member of members to 834. The Commission continued its policy of not restricting further growth in its numbers. The advantages of a very large Commission with common interests in technique aplied to a wide range of astronomical interests were thought to outweigh the difficulties of communication with the members. The full programme of Commission meetings together with its involvement in many Joint Discussions and Joint Commission Meetings supported this view.

III   IUCAF

The purpose and workings of IUCAF were explained by J. Findlay at the 1988 General Assembly. His talk is given in the Proceedings of the Twentieth General Assembly, 1990, Kluwer, Dordrecht, pp 305-312.

In view of the upcoming World Administration Radio Conference (WARC), all IUCAF and CCIR representatives will be retained, as far as possible. These are: IUCAF (B.J. Robinson (Chairman) B.A. Doubinsky, M. Ishiguro, A.R. Thompson, and G. Swarup).

The CCIR representatives are: Study Group 7 (J.B. Whiteoak and A.R. Thompson, S. Leschiutta) as requested at the last IAU General Assembly, a short list of CCIR report titles, provided by A.R. Thompson, were sent to all members of Comm.40.

IV   WORKING GROUP ON FREQUENCY PROTECTION

B.J. Robinson is the Chairman. Due to illness he was unable to attend the meeting in Buenos Aires, and was replaced by J.G. Whiteoak. The meeting was held on 29.7. Resolution A1 (Sharing the Hydroxyl Band with Land Mobile Satellite Services) and A2 (Revision of Frequency Bands for Astrophysically Significant Lines) were drafted and submitted for IAU approval.

V   WORKING GROUP ON DATA BASES
      Discussions were begun to determine the best methods of exchanging radio
astronomical data, including maps. A Discussion Group was formed, under the
chairmanship of H. Andernach (Inst. de Astrofisica de Canarias).

Scientific Sessions of Comm.40

Instrumentation for High Angular Resolution

July 24, 1991 (Chairman: M. Morimoto)

R.J. Davis	– MERLIN
G. Swarup	– Giant Meterwave Radiotelescope (GMRT)
C. Walker	– The Very Long Baseline Array (VLBA)
J.M. Moran	– The Submillimeter Array (SMA)
G. Sielstad	– The Green Bank Telescope (GBT)
N. Kardashev	– The Radioastron Project
N. Schuch	– A Large Southern Hemisphere Telescope
W.M. Goss	– The Very Large Array (VLA)
G. Nicholson	– The Hartebeesthoek Observatory
B. McAdam	– The Molonglo Synthesis Telescope
T.L. Wilson	– Millimeter VLBI in the framework of the MPIfR/IRAM collaboration
M. Morimoto	– VSOP

Measurements of Radio Galaxies and QSO's

July 25, 1991 (Chairman: R.S. Booth)

R. Davis	– Observations of Superluminal Motions at Long Wavelengths: Global Observations of 3C273
H. Andernach	– Radio Source Catalogs and Data Bases
R.S. Booth,	
L. Baath	– Millimeter Very Long Baseline Observations
R. Wielebinski	– CO in Active Galaxies
J. Wrobel	– CO Line Observations of the Lenticular Galaxy NGC4710

High Resolution Maps of External Galaxies

July 26, 1991 (Chairman: J.M. van der Hulst)

W.M. Goss	– Imaging of NGC253 in H92α with the VLA
R.D. Davies	– Radio AGN's at high resolution
R.D. Ekers	– New Results from the AT
R. Braun	– Resolved HI in 11 nearby galaxies
D. Puche	– Holmberg II: A Laboratory for studying the violent ISM

The Galactic Center Region

July 30, 1991 (Chairman: P.G. Mezger)

R. Zylka   – The Inner 50pc Region
M. Lindquist   – The 5–100pc Region
D. Gezari   – A Summary of Infrared Measurements
W. Miller Goss   – Sgr A in the H92α line with the VLA
R.D. Ekers   – Observations of Sgr A*
J. Pipher   – Infrared Observations of the Sgr A Region

Fine Scale Structure of Galactic Molecular Clouds, SNR's and HII Regions

July 31, 1991 (Chairman: T.L. Wilson)

T.L. Wilson   – Thermal, Quasi-thermal Emission and Absorption
J.M. Moran   – Maser Line Results
E.E. Baart   – The Rhodes 2.3GHz Survey
Lewis Ball   – Radio Emission from SN 1987A

Pulsars

July 31, 1991 (Chairman: T.L. Wilson)

F. Graham Smith   – Crab Pulsar Glitches
G. Nicholson   – Vela Pulsar Glitch 9
W.M. Goss   – SN remnants & Pulsars
Alok Ray   – Origin of the Planet Companion to PSR 1829

Joint Discussions were held on the following subjects

An Overview of the Interstellar Medium,
The Development of Antarctic Astronomy,
Reference Systems, What are they and what's the Problem?
The Cosmic Background,
Hipparcos – an Assessment.

Scientific sessions were held jointly with

Comm. 8: *Linking Optical and radio Reference Frames,*
and
Comm. 49: *Non-Linear and Turbulent Processes in the Solar Wind and Astrophysical Plasmas*

The results from these meetings are described elsewhere in this volume.

The following is a list of future IAU Symposia and Colloquia to be held with the support of Comm.40 (with the contact person in parentheses):

- *Astronomy with Millimeter and Submillimeter Wace Interferometry*, Hakone, Japan (I. Ishiguro)
- *Symposium on VLBI*, Kyoto, Japan (M. Morimoto)
Particle Acceleration Phenomena in Astrophysical Plasmas, College Park. Md, USA (E.C. Chupp, A.O. Benz)
- *Very High Angular Resolution Imaging*, Sydney, Australia (J. Davis)
- *Physics of Neutron Stars*, Leningrad, USSR (D.A. Varshalovich)
- *Active Galactic Nuclei across the Electromagnetic Spectrum*, Geneva, Switzerland (T. Courvoisier),
- *Developments of Astrometry and its Impacts on Astrophysics and Geodynamics*, Shanghai, China (Shu Hua Ye)

COMMISSION 41: HISTORY OF ASTRONOMY
(HISTOIRE DE L'ASTRONOMIE)
(a joint IAU-IUHPS Commission)

J. D. NORTH
Filosofisch Instituut
Rijksuniversiteit Groningen
A-weg 30, 9718 CW Groningen
The Netherlands

The scientific programme at the 21st General Assembly of the I. A. U., held at Buenos Aires, Argentina, between 23 July and 3 August 1991, occupied four sessions. The general theme of the historical meetings was 'The Preservation and Conservation of Astronomical Archives and Instruments', and thirteen papers were presented by members, or on their behalf. These were: *The demise of the Bush Barrow lozenge* (J. D. North, Presidential Address); *La sauvegarde du patrimoine - source de documentation historique* (S. Débarbat, President Elect); *Masonry instruments at Delhi - Jantar Mantar and the programme for their restoration* (G. S. D. Babu); *Initiatives taken by the Italian Astronomical Society for the cataloguing and preservation of archives and instruments of historical interest* (E. Proverbio); *Safeguarding the history of a national observatory: the U. S. National Observatory* (S. J. Dick); *The preservation of Peruvian archaeoastronomical sites* (M. L. Aguilar); *On the conservation of the astronomical instruments in Romania* (E. Botez and T. Oproiu); *The archives of a French astronomer: Henri Chrétien* (F. Le Guet Tully and others); *The probable ancient determination of the planetary periodic time of reference* (Y. Maeyama); *An overview of extant astronomical records from the pre-telescopic period* (K. Yau); *Three centuries of astronomy in Marseilles* (R. Augarde, Y. P. Georgelin, M.-J. Meynent, M.-L. Prévot); *Joannes Exarch and the distribution of ancient Greek astronomy among Slav peoples* (N. S. Nikolov).

A video film prepared by Peter Parodi and Horacio Tignanelli, both of Argentina, 'Recent research into the astronomy of Tiwanacu', was shown during one session.

Officers for the session 1988-1991 were as follows: J. North, President; S. Debarbat, Vice-President; J. Eddy (former President), A. A. Gorstein, D. DeVorkin, Xi Ze-Zong. At the business meeting of the Commission held on 26 July, the following officers were duly elected, taking into account the earlier postal ballot among members:

President:	S. Débarbat
Vice-President:	S. M. R. Ansari
Organising Commission:	S. J. Dick
	M. G. Firneis
	K. P. Moesgaard
	J. D. North

At the business meeting a review was made of various assemblies held in the period 1988-1991, at which the Commission has participated, notably the 18th General Assembly of the I.U.H.P.S. at Hamburg and Munich (August 1989) and the conference held in Vienna on the theme 'The Interaction of European and Asian Astronomy' in September 1990. This last was organised locally by

M. G. Firneis and her colleagues at the Astronomical Institute and the Austrian Academy of Sciences. It had been hoped to follow the Buenos Aires meetings with an archaeoastronomy meeting in Peru, that is, in August 1991, but troubles that hit that country unexpectedly have meant that it will be held instead in 1992. The organiser will be M. L. Aguilar, U. S. M., San Marcos University, Ap. postal no. 11, 0481 Lima.) Various other meetings are under consideration by the Commission, among them one planned for 1993 relating to the Struve family (organiser H. Eelsalu, Tartu Astrophysical Observatory, Estonia), and another to commemorate the bicentenary of the birth of John Herschel (organiser B. Warner & others, Royal Society of South Africa, for 1992). A meeting in which the Commission will participate is due to take place at the B. M. Birla Science Centre in Hyderabad and at Jaipur between 12 and 16 December 1991 ('Indian and other Asiatic Astronomies', B. G. Sidharth or S. M. R. Ansari).

Further discussion took place around the state of the *General History of Astronomy*, supported by the Commission since its inception (see below). In view of the main theme of the Commission's meetings, it was decided on a unanimous vote of those members present to make a formal resolution to be presented to the General Assembly in its closing session. This was done jointly by Commission 41 and Commission 5, and was

> that the Union supports an initiative taken by Commissions 41 and 5 (1) to establish a register of the whereabouts of all extant astronomical archives of historical interest; (2) to impress on observatories and other institutions their responsibility for the preservation, conservation, and where possible cataloguing of such archives; (3) to search for an institution that will allocate space and funds for maintaining such a register and publishing it.

It was not felt that the proposal should take instruments into account at this stage, since the problems presented there are of a very different character; but if the one project makes progress, the other should certainly follow in its train. The first objective is to stop the wholesale destruction of materials of historical value; the second is to make the whereabouts of material better known to scholars. Many instances were reported at this meeting of situations where a series of directors of institutes, librarians, and others invested great care in the preservation of materials over a long period, only to be followed by others entirely without a feeling for the past, prepared to sell or destroy their inheritance. The General Assembly endorsed the resolution. The following Working Group was established to look into the possibilities mentioned in the resolution: S. Débarbat (ex officio, President), S. J. Dick, E. Proverbio, B. Hauck (comm. 5), D. Dewhirst (comm. 5).

During the Commission meetings, three members who had died since the last General Assembly were remembered: K. A. F. Fischer, H. C. Freiesleben, and V. Thoren. Victor Thoren was a former colleague of Frank K. Edmondson at Indiana University, who, having the right to propose a name for minor planet (3717)=1964 CG, discovered during the minor planet program on 15 February 1964, wishes to call it (3717) THOREN. The Commission heartily endorsed this idea. Thoren's book on Tycho Brahe had been presented to him only a few days before his untimely and unexpected death, serving perhaps as a reminder of an earlier astronomer than Tycho.

Publications

The General History of Astronomy was a project that originated within the Commission. Of its four volumes in seven parts, two had appeared at the time of the meeting: vol. ii, *Planetary Astronomy from the Renaissance to the Rise of Astrophysics*, Part A; and vol. iv, *Astrophysics and Twentieth-Century Astronomy to 1950*, part A. Of the remaining five parts, four were well advanced in preparation.

COMMISSION NO. 42

CLOSE BINARY STARS (ÉTOILES DOUBLES SERRÉES)

PRESIDENT: R. H. Koch   SECRETARY: E. Lapasset

I. BUSINESS SESSION

The Business Session of the Commission was called to order at 11:00 in Room F of the San Martín Cultural Center on July 25, 1991. Thirty seven members and guests attended.

Over the past triennium, the Commission suffered the resignation of one member and the death of D. Ya. Martynov. The attending members observed a moment of silence in memory of our deceased colleague. A total of 38 Union members (3 established members and 35 new ones) had solicited Commission membership and their names were approved unanimously by the current OC and attending members. The new members represent 19 different adhering countries and bring the total Commission membership to 348. This number is certain to grow significantly before the end of the century and communication among the officers and general membership will become a matter of concern.

From the OC, six members (A. Cherepashchuk, E. Budding, K.–C. Leung, J. Rahe, G. Shaviv, and J. Smak) retired. The President took particular note of the administrative and financial support rendered by Dr. Rahe and the personal and scientific support of Profs. Leung and Smak. To replace the retiring OC members, a mechanism of at–large nomination and OC approval voting resulted in new members: P. Eggleton, E. F. Guinan, G. Hill, P. Szkody, A. V. Tutukov, and D.–S. Zhai. The composition of the new OC was ratified by a show of hands. The OC had previously recommended to the Union that Y. Kondo and M. Rodonò succeed as President and Vice–President, respectively, and these choices were unanimously endorsed by the attending members. During thte current triennium, R. S. Polidan will serve as Commission Secretary.

The President commented on the current *Draft Report* and its availability and repeated his appreciation of Dr. Rahe's assistance in its copying and distribution.

Two items of Old Business were aired. The first concerned the conceptual and practical future of critical catalogues of the orbits of spectroscopic binaries such as that in 1989, *Publ. DAO Victoria* 17, 1. The whole issue is a multi–sided one engrossing experienced and scholarly editors, a possibly–enduring need for a paper–medium catalogue as well as versions on magnetic or optical disks, the very significant costs of publication and distribution, the phenomenal productivity of several observing groups employing photoelectric detection, global diminution of funds for library acquisitions, the utility of catalogue supplements such as those from Toulouse, and possible roles for NASA and ESA and the Strasbourg CDS in disseminating catalogued information. The discussion among A. H. Batten, G. F. Peters, Y. Kondo, W. Wamsteker, M. Rodonò, T. J. Herczeg, J. Smak, A. U. Landolt, and P. B. Etzel led to no comprehensive agreement. The sense of the meeting, however, was that practical constraints and an enduring need will enforce a resolution of the matter in the relatively near future and that the enduring problem will remain the assumption of editorial responsibility.

The second item of Old Business noted that there is no expectation of a new edition of *A Finding List for Observers of Interacting Binaries*. The President remarked on the magnitude of such an editorial effort and also on the formidable chore of making a machine–readable version of the U. Florida *Card Catalogue of Eclipsing Binaries*. F. B. Wood, (whose personal papers include this *Catalogue*) noted that these documents serve as guides to observing programs and local responsibility for the *Card Catalogue* has been assumed largely by K.-Y. Chen. For the foreseeable future, Prof. Chen will continue to respond to requests for information from the *Catalogue*.

At the invitation of the President, New Business was opened by M. Breger, President of Commission 27. Prof. Breger enlarged on the attempts by Konkoly Observatory to sustain institutional and individual distribution of the *IBVS* in the face of financial stringencies. An approach to a solution, very much in the interests of Commission 42 people, appears at hand with a measure of financial assistance from the Union and structural changes in the editorial handling of the *IBVS* content itself. The attending members signified full support for these changes in the directions of stability and continued publication and distribution. The interested reader is referred to the Report of Commission 27 in this volume for the culmination of this matter.

Prof. Breger then described the likely closing of the Sonneberg Observatory which has made effective and abundant contributions to photometry of many northern and southern close binaries as well as of intrinsic variables. This event appears to be just one of the numerous unhappy episodes befalling German scientific enterprises and is a German domestic matter to be decided in the best national scientific interests. Commission 42 needed no urging to recognize that the very large Sonneberg plate collection should be preserved not only as archival material but also as a working resource which has not come close to exhaustion. A very aninmted discussion on the niceties of how to move in this direction resulted among R. E. Wilson, and Profs. Smak, Herczeg, Rodonò, Leung, and Breger. The attending members then gave assent to the President's engaging this topic with the intention of generating a Resolution jointly from Commissions 27 and 42, such Resolution to speak to the matters under discussion and to be presented to the Second General Assembly for its consideration. As may be noted elsewhere in this volume, the Resolution was passed.

The triennial administrative and financial report concerning the *Bibliography and Program Notes on Close Binaries* had been prepared by the Editor–in–Chief, A. Yamasaki, and was read to the attending members. Some concern was voiced that Prof. Yamasaki seemed to commit himself to only a short–term continuaton of his role. It was left to the incoming President to resolve the matter.

The President then called upon Prof. Wamsteker to explain the realization of several World Astronomy Days (WADs) in the International Space Year, 1992. The novelty and utility and the constraints upon such an enterprise were developed by F. Giovanelli and Profs. Rodonò, Herczeg, Smak, and Peters. The assembled members signified their expectation that the Commission interest itself in initial planning stages so as to realize this concept and associate itself with a recommendation for the WADs. At a meeting on July 29, 1991 attended by Prof. Kondo (then the Vice–President) CVs and X–ray close binaries were singled out as two of the few possible types of targets which the WADs could exploit. The winds from hot stars were also noted as likely phenomena which could profit from WAD attention and, as is well- known, very many such stars are really interacting binaries and that clashing winds from the binary components will be an important ingredient in the eventual comprehensive understanding of winds. Commissions 15, 27, 28, 42, and 44 were all associated with a Resolution to the Second General Assembly endorsing the WADs and their realization. The Resolution was accepted.

Having already signified his assent to the current thinking of the Working Group *Designations* of Commission 5, the President did not attend the July 27, 1991 meeting of the WG. The concern of the WG is with adequate identification of a celestial target when the research concerning it is eventually published. Problems arise particularly for newly–discovered objects for which only "non–standard" names are given in the publication making it difficult for other investigators to attempt verification or subsequent observation.

Prof. Lapasset remarked on the upcoming IAU Symposium No. 151 to convene in Cordoba when the General Assembly ended. The President then summarized some news regarding an aborted neutron– stars meeting and a projected IAU Sympoium for which Commission 42 co–sponsorship would not have been appropriate. Finally, Prof. Budding summarized the state of tentative preparation for a meeting in the mid–future of interest to the Commission and to be held in New Zealand.

The meeting adjourned at 12:30.

II. SCIENTIFIC SESSIONS 2 AND 3

As planning started for the Commission's scientific program, there arose the possibility for time dedicated to the methodology of light curve analysis. This concept, central of course to the Commission, had been advanced by E. F. Milone in the recent past so it seemed opportune to create it for the current General Assembly under the chairmanship of Prof. Milone. An SOC was composed in good time and a program developed and accepted by the Union Executive.

Two 90–minute sessions were scheduled for July 31, 1991 but the fire that morning in the Cultural Center annihilated this schedule. Through the good offices of the General Secretary, it was possible to meet for 90 minutes in a bistro near the Cultural Center. For this situation Prof. Milone re–worked the schedule so as to give time only to speakers who were not attending the Cordoba Symposium. Even this arrangement meant that some speakers had to abridge their presentations and none of the scheduled posters could be displayed. Prof. D. P. Hube acted as Secretary for this session, which was attended by about 45 members. Seven oral papers were presented. The first, by R. E. Wilson, elaborated on the opportunities within the WD code to extend the modelling parameters beyond the conventional light and velocity ones in the sense, for example, of interpreting color and line index and polarization information. On behalf of Prof. Wilson and himself, D. C. Terrell then described attempts to calculate assorted binary gas flows and the observational consequences of these flows. J. Kallrath's paper concerned the *SIMPLEX* algorithm and the power of its application at the appropriate stage of modelling. R. L. Kurucz next presented a brief summary of his modernized stellar atmospheres built around more realistic continuum opacities and line inventories than had been available hitherto. W. van Hamme then desribed the new treatment of the "reflection" effect now developed for the WD code, its application to the binary BF Aur, and the change in interpretation fo this binary that now eventuates. Prof. Wilson then summarized briefly the attempts by J. Mukherjee, Dr. Peters, and himself to determine rotational velocities from line profiles. Finally, Prof. Leung (on behalf of his co-author D.–Q. Zhou) elaborated on the possibility that light–curve asymmetries arise in part from stellar atmospheric circulation patterns. Despite the curious venue and crowded schedule, vigorous questioning and discussion followed the papers.

Through the procedural and financial assistance of the LOC for the Cordoba Symposium, it was possible to schedule a 60- minute session in a meeting room of the Gran Hotel Dorá in Cordoba on August 7, 1991. About 25 delegates attended this rump session, Prof. Milone again in the Chair. Prof. Budding first offered his appreciation of the posters deferred from presentation in Buenos Aries. Only one of these, that by I.–S. Nha and H.–I. Kim, concerned application of modelling procedures to real light curves, in this case the intrinsically variable curves of RX Cas. A poster by J. Díaz–Cordoves, A. Claret, and A. Giménez concentrated on the subtle effects between linearized and non–linearized darkening laws while another paper by Profs. Giménez and Díaz–Cordoves remarked on two particular prospective improvements to the *EBOP* code. An *ad hoc* summary of a poster by T. Banks represented the state of the software package ILOT.

Four oral papers were then given. Prof. Etzel developed at some length his expectations for improvements in *EBOP* beyond those already mentioned and laid emphasis on the limitations of applicability of the code. On behalf of D. H. Bradstreet, E. F. Guinan gave a short description of a pedagogical package *Binary Maker* which is basically a graphics tool for displaying light and velocity curves and aspects of close binaries at chosen orbital phases. Prof. Guinan then rehearsed his model for Eps Aur wherein a hollow, tilted disk can account for a major fraction of the eclipse detail. Finally, and on behalf of C. R. Stagg, Prof. Milone briskly summarized prospective changes and improvements to the WD code that are to be implemented at Calgary. A short but spirited and rather far–ranging discussion closed this session.

The proceedings of the sessions are to be refereed and edited by Prof. Milone for a Kluwer volume.

III. SCIENTIFIC SESSIONS 4 THROUGH 7

For almost a year planning had progressed for a joint meeting among Commissions 9, 25, 27, and 42 concerning modern stellar polarization. The Union Executive eventually accepted this as a joint–meeting topic with Commission 25 acting in the lead role. Four 90–minute sessions were scheduled to occupy the entire working day of July 29, 1991. Between 55 and 60 people were in attendance.

The morning sessions, chaired by R. H. Koch, concerned itself primarily with the physical processes generating polarizated radiation from stars, modern instrumentation for medium and large ground–based instruments, the character and magnitude of polarized signals from star–forming regions, precision currently attributable to polarization standards, and the apparent constancy of the visible–band polarized radiation from Plaskett's Star.

The afternoon sessions were chaired by I. S. McLean and spoke more closely to interests of Commission 42 members. Three contributions from the Wisconsin group emphasized that spectropolarimetry of moderately–bright stars can now be accomplished both from ground and Earth–orbit with even modest– aperture telescopes. R. Schulte–Ladbeck and K. Nordsieck elaborated on the Pine Bluff and WUPPE observations of hot, massive stars whereas M. Magalhães presented his and others' work on luminous late–type variables. R. Boyle's paper was concerned with the polarization of Miras and P. Bastien's with that from young stars. A second paper from the Montreal group was given by A. Moffat and showed impressively that the polarization parameters from at least selected Wolf–Rayet stars can remain stable and very well–defined over at least brief time intervals. This, leading to evaluation of the orbital plane inclinations for W–R close binaries, permits accurate determinations of masses and mass functions so that it is possible to make an orderly evolutionary sorting between subtypes of the WC and WN classes.

More detailed presentation of the results of these sessions may be found in the Commission 25 Report.

COMMISSION No. 44

ASTRONOMY FROM SPACE
(L'ASTRONOMIE À PARTIR DE L'ESPACE)

Report of Business Meeting, August 24, 1991

PRESIDENT: E. B. Jenkins   SECRETARY: K. G. Carpenter

1. New Officers of the Commission

A motion was adopted that approved the usual transition of leadership, where the current vice-president will assume the office of president for the forthcoming triennium. Thus, J. Trümper will replace E. B. Jenkins, and Dr. Jenkins will remain on the Scientific Organizing Committee (SOC) as a consultant to the new president. G. G. Fazio was selected to serve as the new vice-president; this choice was approved earlier by the Executive Committee of the IAU.

2. Scientific Organizing Committee

The current membership of the SOC is as follows: B. F. Burke, G. W. Clark, G. G. Fazio, J. B. Hutchings, S. D. Jordan, Y. Kondo (Past President), K. A. Pounds, J. Rahe, B. D. Savage, G. B. Sholomitsky, R. A. Sunyaev, Y. Tanaka, J. Trümper (Vice President) and W. Wamsteker. There was an open discussion about the rotation of some members of the SOC, and various people made recommendations for replacements representing certain geographical areas or scientific disciplines. Drs. Clark, Kondo, Pounds, Rahe, Sholomitsky and Tanaka have already served for two terms, and thus are due for rotation off the committee. The final composition of the new SOC will be dependent on specific individuals agreeing to serve.

2. Status of Working Groups

At the previous business meeting of Commission 44 in Baltimore, two working groups were established. Their status and accomplishments are reviewed below.

2.1 MULTI-WAVELENGTH ASTROPHYSICS

Over the past three years, there have been several multi-wavelength observing campaigns organized under independent initiatives, but the working group itself has not undertaken any organized activity. Also, an interactive data base for the coordination of multi-wavelength observations, called MultiWaveLink, has been established for the general astronomical community at Pennsylvania State University. After a discussion that questioned the effectiveness of the working group and its lack of a clear mandate, a motion was offered and carried to terminate Commission 44's support of the working group.

2.2 LUNAR-BASED ASTRONOMY

There has been considerable interest in planning for future installations on the moon that could provide meaningful benefits to astronomy. NASA has sponsored studies that highlighted some observing concepts, and ESA has its own working group to study Lunar Science, including astronomy. Japan has plans for advancing to the moon, starting with a penetrator mission in 1996, and then progressing to more ambitious, manned missions thereafter. A chapter entitled *Astronomy from the Moon* appears in the report of the Astronomy and Astrophysics Survey Committee, chaired by J. Bahcall and sponsored by the US National Research Council.

The working group on Lunar Astronomy was proposed three years ago by H. Smith, and he agreed to serve as its chair. Unfortunately, a serious health problem prevented him pursuing his original intention of providing the needed leadership. There was a general consensus that there is still a good rationale for continued support of the Lunar Astronomy working group by Commission 44. A motion was proposed and adopted that a new attempt should be made to activate such a working group, and that it should include representatives from the US, the European community, and Japan.

2.3 NEW PROPOSALS

A proposal was made that Commission 44 form a working group on pollution in high earth orbit, but the motion failed to carry. Questions were raised about how the work of such a group would fit in with the efforts already underway elsewhere in the IAU (such as those discussed in the General Assembly). It was agreed that W. Wamsteker would pursue space pollution issues with other commissions, and Y. Kondo would also act as a liaison on this problem between our commission and COSPAR.

3. Program for the XXIst General Assembly

Commission 44 agreed to organize or sponsor the following special scientific sessions.

3.1 JOINT DISCUSSIONS

   a. An Overview of the Interstellar Medium (*SOC Chair:* B. G. Elmegreen; *other commissions involved:* 28, 34, 40, and 48)
   b. Hipparcos - an Assessment (*SOC Chair:* C. Turon; *other commissions involved:* 4, 19, 24, 31, and 40)
   c. Cosmic Background (*SOC Chair:* R. B. Partridge; *other commissions involved:* 40, 47 and 48)
   d. Results from ROSAT, GRO and other recent High Energy Astrophysical Missions (*SOC Chair:* J. Trümper; *one other commission involved:* 48)
   e. First Results from the Hubble Space Telescope (*SOC Chair:* C. A. Norman; *other commissions involved:* 16, 24, 28, 29, 33, 34, 47, and 48)

3.2 JOINT COMMISSION MEETINGS

a. High Redshift Galaxies (*SOC Chair:* K. Sato; *other commissions involved:* 28, 47, and 48)

b. Solar & Stellar Coronae (*SOC Chair:* R. Pallavacini; *other commissions involved:* 10, 12, 36, and 49)

c. Atomic and Molecular Data for Space Astronomy: Needs & Availability (*SOC Chair:* P. Smith ; *other commissions involved:* 10, 12, 14, 15, 16, 29, 34, 35, 36)

d. A Proposal for an International Antarctic Observatory (*SOC Chair:* P. Gillingham; *other commissions involved:* 9, 40, 50)

3.3 MEETING SPONSORED BY COMMISSION 44 ONLY

a. Half day scientific session on "Results from Ultraviolet Instruments on Astro" (*SOC Chair:* T. Gull)

4. Future Symposia and Colloquia

The following events co-sponsored by Commission 44 have been approved by the Executive Committee of the IAU:

1. Symposium No. 150: The Astrochemistry of Cosmic Phenomena (*SOC Chair:* A. Dalgarno; *other commissions involved:* 22, 33, 34, and 36)

2. Symposium No. 151: Evolutionary Processes in Interacting Binary Stars (*SOC Chair:* Y. Kondo; *other commissions involved:* 42 + others?)

3. Colloquium No. 129: Structure and Emission Properties of Accretion Disks (*SOC Chairs:* C. Bertout and S. Collin-Souffrin; *other commissions involved:* 27, 36, 42 and 48)

5. New Proposals for Meetings

Commission 44 has agreed to support proposals to the Executive Committee for the following future meetings:

1. A colloquium entitled, "Particle Acceleration Phenomena in Astrophysical Plasmas" to be held in January 1993 at College Park, Maryland.

2. A colloquium entitled, "The Sun as a Variable Star" to be held in August 1993 in Boulder, Colorado.

3. A colloquium entitled, "Inside the Stars" to be held in April 1992 in Vienna, Austria.

4. A colloquium entitled, "Solar Coronal Structures" to be held in September 1993 in Tatranská Lomnica, Czechoslovakia.

6. Resolution on the Adoption of Vacuum Wavelengths for Space Astronomy Data

The large accumulation of spectroscopic data in the far and middle ultraviolet by IUE and HST now accentuates the transformation in the traditional way that wavelengths are expressed on either side of $\lambda = 2000$Å. The current convention is that transitions above this boundary (up to the infrared) are expressed using wavelengths in air at standard conditions, while those below are expressed with vacuum wavelengths. So that space and ground based astronomy enterprises could operate with a uniform system, Commission 44 was asked to consider the merits of recommending an elimination this difference — one that otherwise seemed so natural in laboratory spectroscopic contexts. An important incentive for working with vacuum wavelengths above 2000Å is the eradication of an awkward discontinuity in the wavelength scales of spectra that cross the boundary. The Commission also recognizes that a vacuum scale in the visible simplifies the derivation of accurate transition wavelengths from energy level differences and also makes the identification of redshifts for far-uv lines shifted into the visible more straightforward. Offsetting these advantages is the confusion created by any change in convention, that is, when the substitution of new wavelengths for certain lines clash with the more familiar identifications and older tabulations. This problem should be ameliorated by the current trend toward computer databases that permit one to convert between the two systems, and, for data appearing in publications, the growing practice of having both vacuum and air wavelengths appear in the tabulations.

On the basis of the above arguments and some informal consultations with members of Commission 14, Commission 44 submitted a resolution that endorses the adoption of vacuum wavelengths across the entire spectrum and also urges publications to identify clearly which convention is being used during the transition period. The text of the resolution appears elsewhere in this volume.

7. World Astronomy Days

A motion was proposed and passed that Commission 44 will support the activities associated with World Astronomy Days.

STELLAR CLASSIFICATION (CLASSIFICATION STELLAIRE)

PRESIDENT: M. GOLAY   SECRETARY: N.CRAMER

BUSINESS SESSION, July 25, 1991

Report of the President:

I.   IAU REPORT ON ASTRONOMY, COLLOQUIA ETC.

We have kept the same headings as in the previous report reflecting again the increase in activity in the field of automatic classification and classification of extra-atmospheric spectra. We have supported the proposal for a meeting in Trieste "Peculiar versus normal phenomena in A-type and related stars"

II.   MEMBERSHIP:

New Members:	E. M. Sion (USA), M. Grenon (Switzerland), E. Fitzpatrick (USA), C. Garmany (USA), R. O. Gray (USA), S. Shore (USA), A. Baglin (France), A. A. Ferro (Mexico).
Consultant members:	N. Cramer (Switzerland), M. Kurtz (USA).
Deceased Members:	Nicholas Sanduleak (USA) and Nina Morguleff (France).

III.   OFFICERS:

A	President:	D. J. MacConnell (USA)
	Vice-President:	O. H. Levato (Argentina)
B	Scientific Organizing Committee:	
	1. Remaining:	M. Golay (Switzerland) as past-President, T. Lloyd-Evans (South Africa), C. Corbally (Vatican), N. R. Walborn (USA), R. Wing (USA), K. Zdanavicius (Lithuania).
	2. Proposed and adopted:	M. Grenon (Switzerland), R. A. Bartaya (USSR), D. Egret (France), H. Maehara (Japan).
	3. Retiring :	N. Houk (USA), R. F. Garrison (Canada), E. H. Olsen (Denmark)

IV.   REPORT OF THE WORKING GROUPS :

a)   Photometric and spectroscopic data

A. G. Davis Philip (Chairman) Members have reported to me on the creation of new catalogs and I forward the information to the Stellar Data Center in Strasbourg. For the next triennium, I will finish my work with D. Egret on the creation of a catalog of standards, to be published by the CDS. The committee has agreed to act as advisors on the project. When I have a nearly final version of the catalog, it will be circulated to the committee members for their suggestions and corrections. We will continue the work on compiling data on new catalogs.

b)   Peculiar red giants

H.R. Johnson (Indiana, USA), (past Chairman) has reported on the activities of the past three years. See report Commission 29 in the proceedings of the General Assembly. R.F. Wing (Ohio, USA) will be Chairman for the next triennium.

SCIENTIFIC SESSIONS

Scientific Session held on July 26

1   Recent Developments in the Spectral Classification of OB Stars in the Magellanic Clouds
N. R. Walborn

Several substantial results of systematic OB classification programs in the MCs during the past three years were reviewed. They include applications of advanced digital, multi–object, and UV techniques. Fitzpatrick has obtained high–quality optical data for a large sample of B supergiants in the LMC, discovering substantial variations in the relative CNO line strengths at a given spectral type, which have significant implications for understanding the evolution of massive stars in general and the SN 1987A progenitor in particular. Schild and Testor at ESO and Garmany, Massey, and Parker at CTIO have been conducting extensive programs in the O associations, which have revealed many previously uncatalogued extremely early O objects, with clear implications for our knowledge of the MC massive stellar content, the IMF, and sequential star formation. Finally, I presented new optical and HST/IUE spectrograms of Melnick 42, 03 If*/WN6–A in 30 Doradus, together with appropriate comparison objects; the former is a good candidate for one of the most massive stars known.

2   Spectra of Wolf–Rayet stars in the Milky Way and Magellanic Clouds
P. S. Conti

Far–UV, optical, and near–IR (to 1 micron) spectra of WR stars in the Milky Way and MCs were compared and contrasted. Among WN types, the Galaxy and LMC spectra are similar; the SMC stars seem to have weaker lines, but the issue is clouded by their binary status. Among WC types, the Galactic and LMC stars have different mean spectral subtype properties; the one SMC example is an unusual type. These differences may be related to distinctive initial metallicity environments which affect massive star evolution.

Scientific Session held on July 29

3   Comparison of the Reliability of Spectroscopic Data from Naked CCD's versus Intensified CCD's
R.F. Garrison, B. Beattie, K. Kamper, M. Pedreros and J. Thomson

It is important, when doing MK classification work, not to lose the level of discrimination that has been achieved with the best photographic work. Having used a variety of different types of digital detections, I have learned something about the problems that can be encountered. During the past fifteen years, I have used Wampler's Image Dissector Scanner, Schectographs, 2D–Fruttis and naked Reticon and CCD's. There are many pitfalls in digital work, but because the work is quantitative, it is sometimes more difficult to be aware of them.

1. As pointed out by Walborn in 1990. noise in tracings can be more easily confused with signal than in photographic work.
2. Signal-to-noise ratios of 200–300 is the optimum; with less, the reliability is too low. More is always nice (e.g. 1000) but is not essential for good classification discrimination.
3. The more bias frames and flat fields used, the better.
4. It is best to expand the tracings sufficiently that the eye is attracted to the line strength, not to the line depths (as when the lines appear in the form of spikes).

With care, digitally determined types can be obtained with equal to greater discrimination to those of the best photographic work.

We have compared tracings from an intensified CCD with those of a naked CCD and have found that the lines from the former are systematically shallower than those from the latter. Obviously, the intensified CCD must be used for faint work. With proper attention to taking standards with exactly the same setup, it should be possible to achieve reasonable classifications. However, it is not recommended to use the intensified CCD's for fundamental work on MK standards.

4   Achieving Accurate MK Classification with Digital Spectra
C.J. Corbally

The correct techniques for getting accurate MK classifications from photographic spectra have been well–established since the days of the original MKK atlas. This session considered what techniques and parameters are needed to achieve the same kind of accuracy via digital detectors.

Clearly, the higher the S/N of spectra, the greater becomes their reliability for MK classification or any other purpose. At the telescope, though, some decision has to be made as to what S/N is adequate for the programme in hand.

Experience with digital spectra indicates that, for a star with a normal spectrum, a S/N of 35 to 50 is adequate to type it within a spectral subclass or luminosity interval. (The normal stars would include "normal" weak–line stars too). However, one may not always be certain that the spectrum is completely normal. For stars with subtle peculiarities, such as the λ Boötis–type stars, a greater S/N of 150 will bring greatly increased confidence in classification that is essential to the study of such stars. So, for really reliable work, spectra with a S/N of 150, and even as high as 300, together with repeated spectra of the same star, are needed; for more statistical work on normal stars, a lower S/N is workable.

In discussion, it was emphasized that, as with photographic work, there is no substitute for an adequate grid of primary MK standards. There was no firm conclusion on how exchangeable were standard spectra between similar digital detectors. This probably depends on the accuracy required. Further experimentation, such as the effect of widening spectra on 2–D arrays is to be encouraged.

5   Spectral Classification and Projected Rotational Velocities for A–type Stars
N. Morell and H. Abt

We presented a progress report on a program devoted to obtaining MK classification and *vsini* for 200 A–type stars (all stars with types between B9 and F0 in the 3[rd]edition of the BSC with $\delta > -20°$).

The aim of the project is to re–discuss the distribution of *vsini* for this spectral range, which was claimed to be bimodal by van den Heuvel (BAN **19**, 309, 1968).

The classification is carried out on 39 Å/mm photographic spectra. The *vsinis* (in Slettebak's system, Slettebak et al. ApJS **29** 137, 1975) are derived from Coudé–CCD observations centred at λ 4481 Å.

At present, we have analyzed all the Coudé data and 1/3 of the classification work is done.

6 A–type Stars with Spectral Peculiarities
C. and M. Jaschek

During the past years, together with Y. Andrillat, we have continued our work on Ae and A shell–type stars in the blue and near–IR. In collaboration with Malaroda and Levato, a radial velocity study of the shell lines has shown that shells move with velocities of a few km/s with respect to the photospheres.

It was also found that many shell stars display strong IR excesses in the IRAS bands.

In collaboration with Pedoussant, Ginestet, and Carquillat, a search for the spectra of the companions of Am stars was carried out in the near-IR. Preliminary results show that often no trace of the companion is found.

We hope to publish in the near future an Atlas of O–F5 stars in the Paschen 20–P12 region, in collaboration with Y. Andrillat.

7 Effective Temperatures and Gravities of A–type Stars
R. Faraggiana, A. Garcia, M. Gerbaldi, J. Zorec

The Teffs, log gs, and angular diameters of over 600 normal and chemically peculiar A-type stars were calculated using a modified Blackwell and Shally (1977) method. The Teffs obtained are systematically lower by 2% than those estimated by Code et al. (1976), and the angular diameters are 6% larger. These differences are due to a small difference in the calibration of visible fluxes used in this work (Johnson and Mitchell, 1975) compared to those adopted by Code et al. (1976) (Oke and Schild, 1970).

Log g was estimated using the Lester et al. (1986) calibration of the $H\beta$ photometric index. For normal A stars, these gravities coincide with those obtained from stellar evolutionary models (Maeder and Meynet, 1988) while for chemically peculiar A stars, the gravities estimated from the $H\beta$ index are systematically lower than those calculated from evolutionary tracks.

8 An Ultra–high S/N Atlas of Vega
A.F. Gulliver, S.J. Adelman, K. Sadakane, M. Takada-Hidac, G.C.L. Aikman

High dispersion, 2.4 and 4.8 Åmm−1, ultra-high S/N Reticon spectra have been obtained with the Coudé spectrographs of the 1.2 m telescope of the Dominion Astrophysical Observatory. A mean S/N of 3000 over the spectral region $\lambda\lambda$ 3800-8800 is being achieved. The data will be presented as a spectral atlas, available on magnetic tape or paper copy. Modelling of Vega will be accomplished using ATLAS69 and SYNTHE.

9 Ap Stars with Resolved, Magnetically Split Lines
G. Mathys

High–resolution ($\lambda/\Delta\lambda = 6\ 10^4$ to 10^5) spectroscopic observations of Ap stars with resolved, magnetically split lines were reported. Nine of these stars have been recently discovered which raises the total number of known Ap stars with resolved, magnetically split lines to 21. The differential magnetic intensification of the two Fe II lines λ 6147.7 and λ 6149.2, which undergo partial Paschen-Back effect, was demonstrated. The previously advocated existence of a correlation between the index Z of Geneva photometry and the mean stellar magnetic field modulus cannot be confirmed.

Discussion: N. Cramer stressed that, even though the Z-magnetic field relation may have been purely coincidental, Z remains a good peculiarity indicator. S. Adelman recalled that Preston had observed a partial Paschen-Back effect in HD 215441.

10 Photometric and Spectroscopic Observations of the λ Boo Stars in the Infrared and the Ultraviolet M. Gerbaldi

The true λ Boo stars show a deficiency of Fe group elements by a factor of about 3, a higher deficiency for Mg, Ca and Al, and normal or slightly overabundant C, N, O, and S.

The peculiarities of these stars imply that the behaviour of the photometric indices must be checked in order to determine if their calibration can be used to determine Teff. The relationship between several photometric indices over a large spectral range has been established for a set of metal-weak candidates and normal stars. Observations in the J, H, K system have been done at ESO (La Silla).

The relationships between (b-y), (B–V), and (V–K), (J–K) have been established after dereddening.

It is clear that true λ Boo stars behave like normal stars from the blue to the near IR. So, calibration of photometric indices can be safely used for determination of Teff.

11 The Photometric Multicolour Boxes as a Simple Tool for Automatic Photometric Classification of Stars M. Golay and B. Nicolet

A star belongs to the same box as that of a central star if none of its colours deviate by more than a given quantity from the corresponding colours of the central star. The same definition is adopted for boxes made with reddening–free photometric parameters (d, δg for the U, B_1, $B_2 V_1 G$ photometric system).

The concept of photometric boxes can be applied only to multicolour photometric systems having a large number of measured stars and a very great stability of the passbands (Geneva photometric system, 40'000 stars measured in seven colours).

In a given box, stars have (within some limits) the same spectral classifications, same absolute visual magnitudes, same metal abundances and metallic characteristics, peculiarities (difficult, however, to define the type of peculiarity), and the same reddenings. To be able to efficiently apply the photometric box method, we must have a very large number of stars also very well classified spectroscopically (for example, a classification quality of at least 1 by Houk), a large number of various types with good metallicity criteria (for example, Gray's catalogue, but with more stars). The boxes can often be very well populated, and the security of the classification is increased if we can find at least one or two stars with well-defined spectroscopic characteristics in each box.

12 The Precision Photometry and the Selection of Stars for Spectroscopic Study
       A.G. Davis Philip

CCD photometry of faint stars has shown that they can be measured with high precision. In many photometric systems, the increased accuracy allows one to determine astrophysical parameters for stars which could not be reached successfully by single–channel photometry. In my program, the Strømgren four–colour system has been used to study the characteristics of blue HB stars in globular clusters and in the field. It is possible to identify stars on globular cluster HBs which are evolving to the blue and to separate them from more highly evolve d stars whose evolutionary tracks have turned and are now evolving towards the AGB. Such information can be used to select stars for further spectroscopic investigation in studies of stellar evolution. This is one example of how precise photometric observations can be used to select interesting stars; many other examples exist in the four–color and other photometric systems.

13 The Third Dimension in Spectral Classification
       C. Jaschek

One of the consequences of classifying fainter stars, farther from the sun than was possible previously, is that the third parameter, weakening or stengthening, of metallic lines will become more important than ever.

The problem of line weakening dates back to Roman (1950) and Keenan, but came back in force in recent years through the work of Abt (1986), Corbally (1987), Gray (1989) and Jaschek et al. (1989).

Three basic facts emerge from this work:

a) a large fraction of weak–lined stars exists in F and G type stars, although extreme cases are rare.

b) classification of line weakening agrees well with results from both Strömgren and Geneva photometry

c) it is necessary to unify the existing scales of line weakening (Abt, Keenan, Jaschek et al., Gray) and bring forth a consistent scheme with appropriate standards.

The notation used for MK classification should be simplified so that classifications referring to the line strength of isolated elements be put into notes rather than in the classification itself.

Scientific Session held on July 31

14 The Current Status of Models for Intermediate Temperatures
       R. L.Kurucz

High–volume classification is an empirical problem and a computer problem. The state of model atmosphere prediction is still too primitive to approach this problem theoretically. Once the classifications have been made, theory will be much better, but for now I will list a few problems.

There is not enough information in a low–resolution spectrum or in any color system, even at infinite signal-to-noise, to determine the properties of a star. That information has been integrated away. That is also true of the model atmospheres which have been integrated with statistical opacities. With low-resolution data, one can only determine a classification box. One has to find a representative star in that same box and study it at high resolution and signal-to-noise in order to determine the properties for that box. However, it is still beyond the state-of-the-art to compute the spectrum of any star theoretically because we do not know all the line and opacity data that affect it. With high–quality spectra, we can at least tell which lines match and which do not, or are missing and try to fix them.

Metal abundance parameters such as [Fe/H] are not good descriptors. In Pop II stars, the light alpha process elements are enhanced. Abundances are not well known even in the sun. There is a difference of 0.17 in the log between abundances determined from Fe I and Fe II. (Any time an abundance is quoted relative to the Sun, the solar abundance used must also be given so the abundance can be corrected when the solar abundance changes.) The He abundance is not known and, especially in early–type stars, can be highly variable because of diffusion. I expect that there is a range of models with different temperatures, gravities, and He abundances that will produce the same fluxes, colors and Balmer–line profiles.

The line blocking and structure of an atmosphere depend strongly on the microturbulent velocity; it is in my new models.

People tend to forget that lines that are washed out in rotating stars and are not visible at low resolution or low signal-to-noise are still there as opacity and in blends and must be considered in an analysis.

The state of absolute spectrophotometry is dreadful. In order to do a reliable analysis, there should also be proper spectrophotometry covering the UV, visible, and IR for a reference star from each box.

15   Maintaining and Refining the MK System in an Era of Automated Classification
          R. F. Garrison

While surveys are an important function of classification, there will always be a need to refine the system in the light of new discoveries and to look at peculiar objects from different perspectives. With the data rates anticipated from the new generation of multi-object spectrographs, new categories of peculiar or variable stars inevitably will be discovered, and there will be a need to integrate them into the system with care.

When the computer isolates interesting stars that do not fit into the system, an experienced classifier will want to examine them, because the MK process gives a unique perspective for the understanding of complex phenomena. This perspective is complemented by the automation techniques, as well as by photometric techniques and by detailed high–dispersion studies.

New wavelength regions will be surveyed, new instrumentation used, and new techniques devised. Using the MK Process, new standards will need to be set up, or the old standards will need to be tested under the new conditions.

This requires the interaction of experienced classifiers. More, not fewer, astronomers experienced in spectral classification techniques will be needed to interpret intelligently the results of the automation, mainly due to the large numbers of stars surveyed.

16   The Need for Automation in the Reduction and Analysis of Stellar Spectra
          M. J. Kurtz

Advances in instrumentation and optics have brought us to the edge of a new era in spectroscopy. Beginning in 1995, it will be possible for a single instrument, the upgraded MMT, using a 300–fiber spectrograph, to obtain 1000 classification quality spectra per hour for stars with V < 14. Several other similar projects, both larger and smaller, are planned.

The ability of individual researchers to obtain 100,000 spectra per year will cause substantial changes in the way they are used. For results to be understood, a standard nomenclature describing normal spectra must be used, and it must be possible to obtain these descriptors automatically. The MK system can and should serve as the basis for this system of description; an automated procedure for obtaining MK types must be developed.

17   The Revised White Dwarf Spectral Classification System
          E. M. Sion

The lack of temperature distinctions among the H- and He-dominated spectral sequences of WDs, the discovery of hybrid composition WDs, and the many composition subclasses further complicated by variable polarization, and other indicators of magnetism all spoke to the need for a new system of WD classification. The new scheme (described in Sion et al. 1983, ApJ **269**, 253) retains a link with the old system but introduces a better description of what the spectrum actually exhibits and provides quantitative temperature information. These changes are consistent with the expressed needs of investigators in the field. The combination of symbols consists of (1) an uppercase D for degenerate; (2) an uppercase letter for primary spectroscopic type in the optical spectrum; (3) an uppercase letter for secondary spectroscopic features, if present; and (4) a temperature index from 0 to 9 defined by $10 \times \theta_{eff}$ (=50400/T). Examples of classified spectra were presented. Classification anomalies on this system, among both very cool and very hot WDs with exotic surface compositions, were also discussed.

18   Spectroscopic Classification of White Dwarfs: Hydrogen or Helium dominated Atmospheres
          G. L. Hammond

In contrast to the majority of WDs, the spectral classification scheme for the degenerates with Teffs less than approximately 11,000 K indicates only the principal chemical **contaminant** in the line–forming layers. Although it is difficult to derive H/He abundance ratios because of many uncertainties in the line–broadening processes in these cool, dense atmospheres, it is essential to obtain good estimates of these abundances to assess the roles of convective mixing, gravitational diffusion, and accretion from the interstellar medium on the chemical evolution of their envelopes.

We reviewed the principal H/He ratio diagnostic schemes: variations in the Balmer decrement, large Balmer–line profile variations due to quenching, and wavelength shifts in the Ca II H and K lines due to neutral H and He collisions. We presented a new calibration of the Balmer decrement vs. H/He ratio based on the best available data and theory of neutral, non-resonant broadening of Balmer lines, and we presented the results of all these diagnostics in a plot of H/He ratios vs. Teff for a large sample of cool WDs. We concluded that an H-rich (DA) WD with Teff less than 8000 K is an **extremely** rare object.

19   Recent Models of Red-Giant Stars
          H. R. Johnson

All available atmospheric models for red-giant stars are based on the principles of hydrostatic equilibrium and local thermodynamic equilibrium, but they differ in the use of plane-parallel geometry (PPG) or spherical geometry (SG) and especially by the increasingly realistic molecular opacities and improved opacity treatments, including the ODF and OS. We reviewed and compared several recent sets of photospheric models: (1) models for K, M giants with OS opacities and PPG (Brown et al., 1989); (2) Mira models with mean opacities and SG (Bessell et al., 1989); models for K, M giants with OS opacities and SG (Plez, 1991); (4) models for C stars with OS opacities and SG (Jorgensen et al., 1991); and (5) pulsating models for Miras with mean opacities and SG (Bowen, 1988).

20   The Current Status of Models for Intermediate Temperatures
        R. L. Kurucz

I have used my newly calculated Fe group line list together with my earlier atomic and molecular line data, 58,000,000 lines total, to compute new opacities for the temperature range 2000 K to 200,000 K. The new models have been presented at the workshop "Precision Photometry: Astrophysics and Galaxy", to appear in L. Davis Press, Schenectady 1991.

The models, fluxes, and colors are available on magnetic tape and will also be distributed on CD-ROMs.

21   Comments on Models for Hot Stars
        R. L. Kurucz

Kudritzki and Hummer have written a review paper "Quantitative Spectroscopy of Hot Stars" in Annual Reviews, **28**, 303-45, 1990. More work is required in the following areas:

1. Current models do not include enough metal–line opacity. I found that increasing the number of atomic lines from one to 42 million decreases the log g by a factor of 3 for stars near the radiation pressure limit. Hot stars may have higher gravities than the hot models have indicated. The additional opacity also helps to thermalize the radiation field so the non-LTE effects may be overestimated, although still strong.

2. Most hot stars are rapidly rotating. Collins and co–workers have shown dramatic effects of gravity darkening with the poles as much as 5 to 10% hotter than the equator. H II regions should be prolate in ions sensitive to the far UV flux.

3. The existing observations, especially those from IUE, are not of high quality; S/N is low. Line profiles and equivalent widths can have large systematic errors.

Humphreys, Kudritzki, and Groth (A&A **245**, 593-603, 1991) showed an interesting result producing strong H lines by adopting a high He abundance in an A supergiant. That suggests to me that model atmospheres for a range of T_{eff}, log g, and He may predict the same fluxes, colors, and Balmer–line profiles. Because of diffusion, it may be that the He abundance in all early–type stars must be considered a free parameter.

22   Stellar Parameters from Stellar Wind Models for Early-types Stars
        R. Blomme

Early-type stars lose a large amount of material due to their stellar wind. It is driven by radiative acceleration in the spectral lines. Once the stellar parameters are known, the hydrodynamical equations can be solved to give the mass–loss rate, and the parameter can be adjusted it until the calculated terminal velocity equals the observed one. For a sample of early–type stars, we found that the resulting masses are only half the evolutionary masses. However, trial calculations including the effect of rotation, limb-darkening, and turbulence show that the observed terminal velocity can also be obtained using the evolutionary masses. Therefore, the derivation of the mass from a stellar wind code does not appear to be possible at this time.

23   Physical Processes in the Spectral Evolution of White Dwarfs
        G. Fontaine and F. Wesemael

We briefly reviewed our current understanding of the theory of the spectral evolution of WD stars. These stars have atmospheres dominated by either H or He but often show a wide and puzzling variety of trace element constituents. Such unusual abundances are believed to be due to the simultaneous operation, in the outer layers of these stars, of a number of physical processes which also erase the abundances in the photosphere at the onset of cooling. We described these mechanisms and their interactions.

24   Spectral Classification of Hot Subdwarfs, Planetary Nebula Nuclei and other Evolved
        and Faint Blue Stars   *J. Liebert and R.A. Saffer*

Stars in the left part of the H-R diagram exhibit a wide range of (high) Teff, log g, and photospheric abundance. Faint blue stars found especially at high galactic latitude are mainly in one of several late phases of stellar evolution but also include stars which appear to be on the upper main sequence. Historically, the schemes used for classifying their spectra are incomplete, inconsistent, and confusing. Yet, in order to maintain continuity with existing literature, we attempt to combine or adapt the spectral classification systems used for hot subdwarfs, PN nuclei, WDs, main sequence stars and HB stars to form a self-consistent, unified system.

COMMISSION 46: TEACHING OF ASTRONOMY
(L'ENSEIGNEMENT DE L'ASTRONOMIE)

Report of meetings, 22, 24, 26, and 30 July 1991

PRESIDENT: Aa. Sandqvist   SECRETARY: J. Pasachoff

First Business Meeting, 26 July 1991

I. MEMBERSHIP
The commission membership, as well as the vice-presidency and organizing committee, for 1991-1994 were discussed, which resulted in the following configuration:

President: L. Gouguenheim;
Vice-president: J. Percy (Newsletter, Travelling Telescope)
Organizing Committee: J. Fierro, M. Gerbaldi (International School for Young Astronomers), L. Houziaux (Newsletter), S. Isobe, C. Iwaniszewska, J. Pasachoff (ICSU-Committee on Teaching of Science), R. Robbins (Astronomy Education Material), Aa. Sandqvist (past president), D. Wentzel (International School for Young Astronomers, Visiting Lecturers Program).

There are three types of membership in Commission 46: National Representatives with duties including national reports and liaison with home countries, regular members and consulting members.
The National Representatives for 1991-1994 are:
Argentina–A. Feinstein; Australia–A. Rodgers; Austria–A. Hanslmeier; Belgium–A. Noels; Brazil–W. Maciel; Bulgaria–N. S. Nikolov; Canada–R. Bochonko; Chile–J. Maza; China Nanjing–K.-J. Feng; China Taipei–C.-S. Shen; Colombia–E. Brieva; Czechoslovakia–J. Siroki; Denmark–H. J. Fogh Olsen; Egypt–A. Aiad; Finland–H. Oja; France–L. Gouguenheim; Germany–W. Schlosser; Greece–L. N. Mavridis; Hungary–G. Szecsenyi-Nagy; India–S. Ramadurai; Indonesia–B. Hidajat; Ireland–J. Haywood*, Italy–E. Proverbio; Japan–S. Isobe; Korea Rep.–J.-O. Woo; Malaysia–M. Othman; Mexico–S. Torres-Peimbert; Netherlands–L. L. E. Braes; New Zealand–E. Budding; Nigeria–S. Okoye; Norway– J.-E. Solheim; Paraquay–A. E. Troche-Boggino; Poland–C. Iwaniszewska; Portugal–J. Osorio; Romania–E. Botez; South Africa–A. Fairall; Spain–M. Catala-Poch; Sweden–Aa. Sandqvist; Switzerland–L. Martinet; United Kingdom–D. Clarke; Uruguay–J. Fernadez; USA–J. Pasachoff; USSR–E. Kononovich; Vatican City State–M. F. McCarthy; Venezuela–N. Calvet; Yugoslavia–J. Milogradov-Turin.

Lucienne Gouguenheim, executing one of her vice-presidential responsibilities, had written all the members to survey them. Out of 121 members she received 79 answers. She will follow up with individual letters to redefine the active list; members not answering the follow-up letter will be terminated at the end of the triennium 1991-1994. John Percy suggested that National Adhering Organizations be notified of the National Representative on the Commission list and asked to confirm their representative.
The following previous National Representatives become regular members: S. Ferraz-Mello, H. F. Haupt and H. Zimmerman; the following are new regular members: M. T. Brück,

C. R. Fleck, M. K. Hemenway, S. Lai, Q. Li, J. Narlikar, I. Nha, L. I. Onuora, U. S. Pandey, Z. Pokorny, N. S. R. Prabhakaran, J. Sahade, M. A. Seeds, D. Sukartadiredja and S. Wang; the following members have retired from the Commission: J. Kleczek, M. Rigutti and B. Sevarlic; the following are consulting members for 1991-1994: P. S. Bretones, V. Gonzalo, B. W. Jones, J. Nevatia Chordia, R. Szostak, H. L. Tignanelli and P. Viet Trinh.

II. INTERNATIONAL SCHOOL FOR YOUNG ASTRONOMERS (ISYA)

Eighteen schools have been held on four continents since ISYA was inaugarated in 1966: Africa, Asia, Europe and South America. They were in Argentina (twice), Brazil, China, Cuba (in 1989), Egypt, England, Greece, India, Indonesia (twice), Italy, Malaysia (in 1990), Morocco (in 1990), Nigeria, Portugal, Spain and Yugoslavia. Their purposes include advancing astronomy in astronomically developing countries by broadening the perspective of students and making local astronomy prominent. The secretary for all 18 schools has been Josip Kleczek, who is now retiring. *The Commission voted to express its appreciation to Josip Kleczek for the tremendous amount of work he has done over the years on behalf of the Commission and the IAU, and to send him a letter on this matter.* Donat Wentzel was then appointed as the new secretary for ISYA and Michelle Gerbaldi as assistant secretary. Hosts for future ISYA are China (summer 1992), India (tentatively in December 1992) and Egypt (under discussion).

III. VISITING LECTURERS PROGRAM (VLP)

Don Wentzel, the coordinator of the VLP, presented the status of the Peru and Paraguay VLPs. The IAU pays only travel support for the VLP. For VLP lecturers, the host organization often pays supplements, including housing. Such lectureships are for 3-month periods, and it is often difficult to get suitable lecturers. There is some concern about how to assess the long-range effects of the programs. The chairman, W. N. Cristiansen, of the organizing committee of a possible future China VLP has retired and will most likely be replaced by J. Wampler.

IV. TRAVELLING TELESCOPE (TT)

John Percy, who is the project coordinator, reported on the Travelling Telescope. The TT was funded in funded in 1986 by the Canadian Commission for UNESCO and exhibited at the IAU Colloquium #105, Williamstown, and the IAU General Assembly, Baltimore, in 1988. Packed in 4 crates weighing 300 kg, it travelled to ISYA in Malaysia in 1990. The transport for the telescope and an accompanying technician was provided by Air Canada; limited success was achieved due to some technical problems with the telescope. In 1991, repair and refurbishment of the TT were needed. Some new instrumentation was required and well over $ 1000 were raised from private sources and the IAU. The TT will travel to the Paraguay VLP this coming autumn and winter. Some funding is needed, possible sources include American Airlines. In 1992, the TT may travel to ISYAs in China and/or India; funding is sought from national airlines in the form of free transport for the telescope and technician.

Some problems were discussed: • Funding; national airlines may be the best solution. • Liaison; a good contact person is needed in each host country. Attempts to get the TT to the ISYA in Cuba failed, and there is still no satisfactory contact person in Cuba. • Lack of a technician. • Limited time and energy of the present coordinator. Wayne Osborn has volunteered to work with John Percy on the TT project and may take over after the triennium.

V. MISCELLANEOUS

Syuzo Isobe reported on the October 1987 Asian-Pacific meeting in Beijing. He is publishing an Asian-Pacific Teaching Bulletin, three issues have been published so far

(February 1990, September 1990, and March 1991), with #4 due in October 1991. A meeting on Teaching of Astronomy in the Asian-Pacific Region was held on 18 July 1990 in Australia.

Reports were presented on the meeting for the Teaching of Astronomy in Barcelona in September 1990 and on planetaria, including the new Tycho Brahe Planetarium in Denmark.

The comprehensive course on astronomy, compulsory for the majority of all Grade 10 pupils in the German Democratic Republic, has ended with the demise of that country.

Second Business Meeting, 30 July 1991

I. NEWSLETTER

John Percy, the editor, reported on the Commission 46 Newsletter: The Teaching of Astronomy. Issues appeared as follows - # 26: November 1988 (AEM); # 27: January 1989; # 28: August 1989; # 29: February 1990; # 30: November 1990; # 31: April 1991 (National Reports); # 32: April 1991 (AEM). The following issues are scheduled: # 33: August 1991 (General Assembly issue); # 34: December 1991. The desired rate of publication is twice a year. The upgrading of the Newsletter to a journal was discussed. The usual problem exists: a lack of material to publish. Possible solutions are (1) ask the National Representatives to supply at least one short item (in addition to the National Reports) each three years; (2) encourage members of the Organizing Committee to submit articles or reports on their areas of responsibility; (3) appoint regional correspondents to provide news from their areas; (4) publish abstracts of education articles from other sources. Jay Pasachoff reported that *Astronomy* magazine in the United States tried an alternate version with some pages on education for a few months, but that version did not attract enough subscribers for them to continue it. Other suggestions were that the Newsletter should just include a bibliography of all teaching of astronomy articles; National Representatives should get a mailing list within a country. Everybody should send abstracts of published articles to the editor.

The President presented a report on the printing and distribution of the Newsletter. Leo Houziaux has been in charge of printing and distribution for the last 6 years. This triennium, about 150 pages were published at a cost of $5344 for a mailing list with about 500 addresses. A budget of about $7500 is needed for the next triennium. Houziaux is ready to pass on the job to anyone interested. After discussion, it was decided that the regular issues will be continued to be edited by John Percy, who will also be responsible for their printing and distribution for the next three years. Leo Houziaux, in collaboration with the new president, Lucienne Gouguenheim, will be responsible for publishing and distributing the National Reports and the Astronomy Educational Material. Julieta Fierro has volunteered to help working on the Newsletter publication with the possibility of taking over after the next triennium.

II. ASTRONOMY EDUCATION MATERIAL (AEM)

The Astronomy Education Material (AEM) consists of large lists of books, atlases, magazines, lecture notes, films, slides etc. together with addresses of distributors and other relevant information. At the 1988 Baltimore meeting, the AEM was revised to include only four languages: Part A English (R. R. Robbins), Part B Russian (C. Iwaniszewska), Part C French (M. Gerbaldi) and Part D Spanish (J. Fierro). Parts C and D have been published in the Newsletter #32 (April 1991). Part B has not materialized due to the lack of communication with the Soviet Union. Part A will be published in the fall of 1991.

In a letter to the President, R. Robbins has suggested that Part A be replaced this year by "Astronomy Education and Instructional Aids" prepared by H. J. Augensen, Widener University, Chester, Pennsylvania, USA for Springer-Verlag's future compendium *Astronomy: A Handbook*. Robbins suggested that the overlap with Augensen's work was too great this time to make it effective to publish a separate study. Also, he enclosed a set of booklets from the Astronomical Society of the Pacific. Though they are not complete in coverage, they select important materials and so may be of special use in the developing world. Robbins furthermore suggested assembling a volume of reprints of book reviews (or, if there is a problem with gaining permissions, abstracts of book reviews). The President presented the results of his negotiations with Augensen and Springer and recommended to the Commision to adopt Augensen's chapter as a replacement for this triennium's AEM Part A. This recommendation was approved and Jay Pasachoff then moved a vote of thanks to H. J. Augensen, which was carried. The cost of distributing the Augensen report, which is 76 pages long, through the Newsletter will be about $2000; the IAU is contributing $1000, so another $1000 must be found from a Foundation or elsewhere. Leslie Onuora suggested that, for financial reasons, the booklet of Augensen's materials be distributed only on request. John Percy reported that there is a questionnaire in the next Newsletter about continuing an individual's Newsletter subscription, and that he can add a question about receiving the report. David Crawford stressed the Astronomical Society of the Pacific's educational function. Jay Pasachoff mentioned their Newsletter available to teachers on request, with a circulation reported by Mary Kay Hemenway to be 13,000 per issue.

The President then raised the question whether Part B (Russian) should be stopped. Jay Pasachoff suggested suspension since there might exist some future opportunity to continue this project. A vote for suspension was carried.

For Parts C and D, Julieta Fierro suggested a minimum library of books for a Spanish-speaking library rather than a complete set of references. Michelle Gerbaldi reported on French materials, and said that there has been little published recently. She reported that the French version is already done.

III. EUROPEAN ASTROPHYSICS DOCTORAL NETWORK (EADN)

The President reported on the European Astrophysics Doctoral Network (EADN) which started in 1988. The aim of the EADN is to stimulate the mobility of graduate students who are preparing their doctoral theses within Europe. Students are encouraged to spend from three to twelve months at a foreign EADN university and receive about a 20% subsidy for extra expenses caused by their foreign studies. The EADN also organizes summer schools for graduate students at the beginning of their doctoral research in astrophysics. Four summer schools have been held so far in Les Houches - France, Ponte de Lima - Portugal, Dublin - Ireland, and Graz - Austria. The 1992 EADN Predoctoral Astrophysics School will probably take place in Berlin - Germany, and the 1993 School possibly in Greece. The General Coordinator of the EADN is Jean Heyvaerts, Observatoire de Meudon - France.

IV. ICSU–CTS AND THE INTERNATIONAL SPACE YEAR (ISY)

Lucienne Gouguenheim reported on the International Council of Scientific Unions – Committee on the Teaching of Science (ICSU–CTS). This committee met in April 1989 and May 1991 in Paris. In January 1991, a group met in Zimbabwe to prepare the "Innovational Teaching Materials for Science Education" for African schools. Their new project concerns "Strengthening Science Training and Research in the Third World". Gouguenheim has been the IAU representative on ICSU–CTS for 9 years and is now retiring from this committee. The President announced that she will be replaced by Jay Pasachoff. A vote of thanks was moved and passed for Gouguenheim.

Gougenheim also reported on the International Space Year (ISY), which will be 1992. There will be a Space Science Project for Teachers of Science. An inventory of available and locally produced material is being issued by J. P. Stoltman, Department of Geography, Kalamazoo, Michigan 49008-5053, USA, FAX (618) 387 0958.

V. MISCELLANEOUS

(1) John Percy reported on six free subscriptions to *Sky & Telescope* that are distributed to places in the developing world. Six additional free subscriptions are being added, for distribution in Eastern Europe. The Smithsonian Astrophysical Observatory, Cambridge, Mass., through James Cornell, is making available 25 sets of the new set of 125 slides of astronomical objects across the spectrum. The Commision voted to send letters of thanks.

(2) The President reported that the Strasbourg catalogues are now available on CD-ROM. The problem was raised that they may not be free, and that the developing countries are less likely to have CD-ROM players available.

(3) The Commission discussed a future Colloquium on the Teaching of Astronomy. John Percy suggested a Joint Discussion at the next General Assembly, an idea that met with overwhelming approval. Most of the members of the IAU are involved with education in some way and a Joint Discussion would bring more prominence to the teaching function. Perhaps another IAU Colloquium could follow that Joint Discussion.

(4) Gonzalo Vicino reported on the Uruguay meeting, La Cultura Astronomical en la Sociedad Moderna, held in Montevideo 16-20 July 1991.

(5) Julieta Fierro reported on plans for a new science museum in Mexico. Two exhibits on astronomy at subway stations have been very well received.

(6) David Crawford requested a liaison with Commission 50 on the preservation of observatory sites and stressed the importance of education. He suggested Syuzo Isobe, a member of both Commissions, as liaison, which was approved. He also commented that the International Dark-Sky Association (IDA) will be translating certain information sheets.

(7) The President discussed a link with the International Planetarium Society (IPS). Jay Pasachoff is an IPS member and will serve as liaison. Julieta Fierro said that such a link is important particularly in Mexico where most planetariums are chaired by non-professionals.

Scientific Meetings, 24 July 1991

The topic of a two-session scientific meeting on the afternoon of 24 July 1991 was Introducing Modern Astronomy and Astrophysics into Classroom Exercises - At High School, University Undergraduate and Graduate Levels. The first session was chaired by the President, the second session by the Vice-President. Abstracts of the papers presented are printed below:

I. UNDERGRADUATE EXERCISES IN MODERN ASTROPHYSICS (L. BOTTINELLI)

(1) The distance of the cepheid star RS Puppis is determined through a geometrical method using the difference of phase between the light curve of the star and those of two nebulae surrounding the star and reflecting the star light (data from *A. & A.* $\underline{16}$, 252). (2) Discussion of the detectability of cepheid stars according to their periods, through the period-luminosity relationship, by using the Hubble Space Telescope, in relation with the extragalactic distance scale. (3) Given the location of (i) white dwarf and (ii) supergiant stars, in a mass-radius diagram, discuss the location of other classes of stars (main sequence, neutron stars and black holes) and comment on the mean stellar densities. (4) Jean's criterion is established in terms of mean density as a function of temperature, total mass and chemical composition; the contraction of interstellar clouds, leading to star formation, is then discussed in a mass-mean density diagram according to various physical conditions relevant to the interstellar medium. (5) From the observed radial velocity curve of the visible component of the variable double star V616 Monocerotis, and that of the accretion disk of the unobserved

compact component, the mass derived for this last component is discussed and it is concluded that this component is most probably a black hole (data from *Ap.J.* 345, 492 and 359, L47).

II. UNDERGRADUATE PROJECTS USING ASTROPHYSICS AND ASTRONOMICAL DATA BASES (M. GERBALDI)

We shall present activities that my colleagues and myself have developed these last years for science students at the Paris XI University, at two different levels: (1) second year students (2) fourth year students. (1) The students develop a project using a large computer where an Astronomical Data Base has been implemented. They first use a software package to select observational data from this Base for a selected sample of stars that they have previously defined according to their project. Then they analyse the data with a Fortran program written by themselves. All of them have learnt a computer language the year before. The assessment of their projects gives the students credit for adequate work done in programming and for ability to draw astrophysical conclusions from their analysis. (2) The students develop a project using Macintosh microcomputers. These projects are developed using the Astronomical Data Bank "SIMBAD" in order to get the student familiar with a real and large Data Base. As in (1) the student selects specific data which are then analysed in terms of astrophysical parameters. Such analysis tries to exploit fully the large capacity of this PC in terms of software. The project done in this frame is considered as an introduction to a research project.

III. UNDERGRADUATE ASTRONOMY LABORATORY EXERCISES (M. K. HEMENWAY)

The University of Texas offers laboratory experiences to a wide range of students. Non-science majors meet two hours per week for 15 weeks (one credit hour) to perform both indoor and outdoor laboratories. A six hour per week course (three credit hours) is offered in two formats: for the future science teacher and for experienced teachers. These courses are offered using a modified Keller-method of self-paced instruction. This allows both intensive student-teacher interaction and reduces the demand for equipment. Astronomy majors in their third year of studies use a 41-cm telescope with modern equipment for their one credit hour course. They can perform observations with eyepieces, a filar micrometer, a camera, a spectrograph, a photometer, and a CCD-camera. The benefits and problems associated with teaching these different classes will be discussed. This work is partially supported by the NSF under grant TPE 9050289. Reference: Hemenway, M.K. and Robbins, R.R. *Modern Astronomy: An Activities Approach*, second edition, University of Texas Press, 1991.

IV. DATA DISTRIBUTION THROUGH TELEPHONE-LINKED PERSONAL COMPUTER (S. ISOBE)

Our section of amateur astronomer education is promoting some programs: (1) Exchanging of astronomical data with the Central Bureau for Astronomical Telegrams of the IAU. (2) Confirmation of newly discovered celestial objects. (3) Spring school for leading amateur astronomers. (4) Arrangement of open house of our observatory. (5) Collection of materials for future exhibition center. (6) Response to questions from the general public through telephone calls and letters. Since number of telephone calls increases up to 30 per day, on some busy days our staff has to answer continuously. Moreover, needs for our services become very wide, and some amateur observers claim data just relating to their observations. Therefore, we started a new service from December, 1990, which is data distribution program through telephone-linked personal computers. In Japan, we easily transfer data through the telephone lines. Our system is composed of a PC personal computer with 40 Mbyte data disk connected with telephone line by a 2400-bps modem. One who has a similar system can connect by dialing our telephone number. At the moment, all the data can be accessed without any restriction. However, we have a plan to give user's identification number to registered users who can see some special data. The number of connected users is increasing and was 20 connections per day in May and June. Many users enjoy seeing the IAU Circular which should be the case for future negotiation in charging for the data. We are intending to develop our services to the general public in different ways.

V. THE AAVSO AND VARIABLE STAR ACTIVITIES AND PROJECTS FOR SCHOOLS AND UNIVERSITIES (J. A. MATTEI)

The American Association of Variable Star Observers (AAVSO) is the largest organisation of variable star observers in the world with members in 42 countries. The purpose of the AAVSO is to coordinate variable star observing, mostly by amateur astronomers, evaluate the accuracy of these observations, compile, process, and publish them, and make them available to astronomers, researchers, and educators. Over 6.5 million observations have been compiled since the AAVSO was founded in 1911. About 250,000 observations are submitted to and archived by the AAVSO each year. Over 200 requests are made by astronomers and educators for AAVSO data and services each year. The AAVSO has been helping teachers and students on an ad hoc basis in setting up observing programs and science projects on variable stars. Now the AAVSO has embarked on a new education initiative of developing a flexible set of hands-on educational activities and projects based on its unique electronic database of variable star measurements. Students will be able to experience the excitement of doing real science with real data - making new measurements and new discoveries. By carrying out all aspects of the research process, they can develop and integrate skills in science, mathematics, physics, and computer science. If the project is funded, together with John R. Percy at University of Toronto, Canada, we will prepare a package of information, consisting of computer-readable AAVSO data, computer programs for data analysis, Students and Teachers' Manuals, instructional video tape, set of slides, and finder charts, and test them in Teachers' workshops and selected classrooms before it is ready for distribution.

VI. THE INCLUSION OF CONTEMPORARY ASTRONOMY INTO THE STUDY PLANS OF ASTRONOMY (J. M. PASACHOFF)

We describe methods of including the exciting results of contemporary astronomical research into courses on all levels, including elementary school, junior-high and high school, and university. We have included such results in our textbooks on all these levels, including the elementary-school set *Discover Science*, which exists in a Spanish-language translation (*Descubre las sciencias*). We also describe the current efforts in the United States to place more emphasis on teaching astronomy, including concrete steps recommended in the decennial report on astronomy and astrophysics released in 1991 and the establishment of a prize for education and of a new Working Group on Astronomy by the American Astronomical Society. Finally, we discuss methods of obtaining illustrative materials in the form of slides, videotapes, or overhead transparencies to show recent discoveries.

VII. ACTIVITIES FOR AN ASTRONOMY UNIT IN A SENIOR HIGH SCHOOL PHYSICS COURSE (J. R. PERCY)

Within the grade 12 physics course in the Ontario school science curriculum, there is a 10-hour astronomy unit which includes such "physical" topics as: the Sun, telescopes, spectroscopy, parallax, the inverse-square law of brightness, physical properties and classification of stars, their structure and energy sources, nucleosynthesis and supernovae. We have developed and tested a set of a dozen activities which can be used in this astronomy unit, or in other senior high school physical science courses. Copies of the activities are available on request.

VIII. THE SUN AND SOLAR ACTIVITY - A LABORATORY EXERCISE ON SOLAR PHOTOS (D. G. WENTZEL)

Using five sheets of photographs (taken on or within a few days of the total eclipse of 7 March 1970) - a corona picture taken during the eclipse, an X-ray photograph, sheets of spectroheliograms, and a magnetogram - the student is expected to

(1) Observe and recognize various solar features on several types of photographs.

(2) Relate the appearance of features seen against the dark sky to the appearance of the same features in front of the solar disk.

(3) Relate the various features to each other by tracing the influence of sunspots through several layers of the solar atmosphere.

IX. UNDERGRADUATE LABORATORY EXERCISES WITH A CCD DETECTOR (D. MCNALLY)

We have installed a CCD detector on our 24" reflector and we find that it allows us to image extended nebulous objects - we have not been able to do this photographically for 15 years in the bright skies of Mill Hill. Exposure times have dropped from hours to minutes and real observing projects are once again possible. There are optimists who talk about research........ We allow the undergraduates to use STARLINK reduction packages on a microVax. The reaction has been encouragingly positive from the undergraduates. We mainly carry out spectroscopic studies rather than direct imaging.

The meeting was closed with brief presentations by N. S. Nikolov on a new Bulgarian text book for high school astronomy, and by B. Monsignori-Fossi on the Italian Astronomical Society.

Astronomer–Schoolteacher Meeting, 22 July 1991

The traditional astronomer - local schoolteacher meeting was held at the Buenos Aires Planetarium in the afternoon of July 22, 1991. The subject of the meeting was *Methods and Means of Teaching Modern Astronomy* and there were about one hundred participants. They included astronomy teachers from both elementary and high schools, staff members from the Buenos Aires and Rosario (Argentina) and Montevideo (Uruguay) Planetariums, astronomers directly interested in educational aspects, and also astronomy students from La Plata Observatory and amateurs. The speakers were: J. L. Sersic (Argentina), R. M. West (ESO), J. and N. Pasachoff (USA), S. Torres-Peimbert (Mexico), R. Garrison (Canada), M. Gerbaldi (France) and J. Fierro (Mexico). The audience participated actively, and there were very interesting discussions following every talk. Books and educational material were exhibited by Pasachoff and Fierro and given in donation to Argentine institutions. J. C. Muzzio (Argentina) organized the meeting, in close collaboration with the Presidents of IAU Commission 46, Aa. Sandqvist (Sweden), and of the Local Organizing Committee of the IAU General Assembly, R. Mendez (Argentina).

COMMISSION No. 47

COSMOLOGY(COSMOLOGIE)

Report of Meetings on 25, 26 and 27 July 1991

PRESIDENT: Katsuhiko Sato   VICE PRESIDENT: R. B. Partridge

I. BUSINESS MEETING

At the beginning of the meeting, the president asked for a short silence to commemorate the recent death of an eminent member of the Commission, Dr. Hidekazu Nariai, Professor of emeritus, Hiroshima University, whose contribution to the advancement of cosmology has been outstanding. He served as a member of the organizing committee in term of 1970- 1976.

The president informed that all the commission members who answered the questionnaire circulated on April 1990 expressed the opinion that the vice president R. B. Partridge takes over the office of the president. Accordingly, he has proposed to IAU Executive Committee the nomination of R. B. Partridge (U.S.A.) as the president commission 47 for the period of 1991 - 1994. He also informed members that he has proposed J. V. Narlikar (India) as vice president following the result of questionnaire. These nominations were endorsed by the participating members of the commission.

Following new members of the organizing committee were elected in this meeting: J. Bond (Canada), M. Geller (U.S.A.), A. Szalay (Hungary) and D. Jauncey (Australia). Including members who remain in office, A. Dressler, M. Rees, H. Reeves, P. Shaver and S. Shandarin and K. Sato as the outgoing president, ten people serve as the members of organizing committee.

The president asked participating members to discuss the rule of nominating new vice president and new organizing committee. He reported rules of other commissions and also the opinion of the president of commission 35 in the commission president meeting held on 23 July that they should be elected by ballot. It was agreed that 1)at present stage commission 47 does not adopt the direct ballot, but 2) according to the organizing committee, next president, R. B. Partridge investigates the rule including the possibility that the new vice president is nominated by ballot in the organizing committee.

II. SCIENTIFIC SESSIONS

Four sesssions were planned originally, but the last sesssion was cancelled by fire accident. The first two sessions (25 July), Early Universe, which were organized K. Sato and chaired by him and J. V. Narlikar. The last two sessssion, Cosmological Parameters, were organized and chaired by V. Trimble.

EARLY UNIVERSE

Primordial Nucleosynthesis in Inhomogeneous Universe

Katsuhiko Sato (The University of Tokyo) and N. Terasawa

From recent studies of nucleosynthesis in inhomogeneous universe, It has been pointed out that 1) $\Omega_B = 1$ model becomes consistent with light element observations, 2) heavy elements including r-process elements are abundantly synthesized, and 3) ^{9}Be plays a role of the indicator of inhomogeneity since ^{9}Be is also abundantly synthesized. These suggestions were, however, based on a simple two-zone model in which back diffusion of neutrons are neglected during the nucleosynthesis. In this short review, It is shown that 1)the upper limit of Ω_B which is consistent with light element observations is almost the same as that of homogeneous case, 2) the ^{9}Be abundance in the simple two-zone model is overestimated at least two orders of magnitude in plausible parameter space and 3) the abundances of CNO elements are very small in the proper range of baryon/photon ratio consistent with the abundances of light elements.

Baryon Isocurvature Scenario in Inflationary Cosmology

J. Yokoyama (Fermilab/U Tokyo), M. Sasaki and Y. Suto

We derive a proper expression of the power spectrum of baryon- number fluctuations arising from models of baryogenesis with both intrinsic and spontaneous CP violation. It is explicitly calculated in a power-Law inflation model and found to be almost scale-invariant on small scales and white-noise on large scales. Under appropriate choice of particle physics model, it is possible to account for the necessary initial condition for the baryon isocurvature scenario proposed by Peebles.

Natural Inflation and Large-Scale Structure

Joshua Frieman (NASA/Fermilab Astrophysics Center)

For ten years now, the inflationary scenario has been in a state of theoretical limbo: it is a beautiful idea in search of a compelling model. The problem arises from the required tiny self-coupling of the inflation scalar field, $\lambda_\phi \gtrsim 10^{-14}$. In most models, this small number is either unnatural or unexplained. We show that a pseudo-Nambu-Goldstone boson, with a potential of the form $V(\phi) = \Lambda^4[1 \pm \cos(\phi/f)]$, can naturally give rise to inflation if $f \sim m_{pl}$ and $\Lambda \sim M_{GUT}$. In addition, the primordial power spectrum $|\delta_k|^2 \sim k^n$, with spectral index $n = 1 - m_{pl}^2/8\pi f^2$, can have substantial extra power or large scales (compared to the scale-invariant, Harison-Zeldovich $n = 1$ spectrum). In order to fit the galaxy angular correlation function $w_{gg}(Q)$ measured by the APM survey, we find that a spectrum with $n \simeq 0 - 0.4$ is required, within the context of inflation. However, microwave quadrupole anisotropy limits from RELICT and COBE require $n \geq 0.3$ (for a bias parameter $bleq2$). As a result, the current anisotropy bonds must be nearly saturated if the extra power seen an $30 - 100h^{-1}$Mpc scales is to be explained by inflation and cold dark matter, without additional ingredients.

Galaxy Clustering in a Bubbly Universe

Angela V. Olinto (Univ. of Chicago)

Recent redshift surveys suggest that most galaxies are distributed on the surface of bubbles surrounding large voids. To investigate the quantitative consistency of this qualitative picture of large-scale structure, we study analytically the clustering properties of galaxies in a Universe filled with spherical shells. This phenomenological model comprises three galactic populations: shell galaxies, placed at random on spherical shells distributed randomly in space; cluster galaxies, located at the points where three shells intersect; and a random, unclustered component of background galaxies. We calculate the two-point galaxy correlation function, the three point galaxy correlation function, the galaxy-cluster cross-correlation function, and the void probability function for models with an arbitrary distribution of shell sizes. With 20% of galaxies in clusters, the observed $\xi_{gg}(r)$ can be reproduced with the same power law distribution of shell sizes ($\sim R^{-\alpha}$, $\alpha \simeq 4$) which fits the cluster-cluster two-point correlation function, $\xi_{cc}(r)$.

Large Scale Anisotropies of the Cosmic Microwave Background

N. Sugiyama (The Univ. of Tokyo), N. Gouda and M. Sasaki

We formulate a general method to calculate systematically any multipole moment of the cosmic microwave background anisotropy in an open universe. Using this method, we evaluate the effect of negative curvature to the quadrupole moment in the three representative cosmological models, pure baryonic, cold dark matter and hot dark matter models, for both adiabatic and isocurvature initial density fluctuations with power law. The anisotropy can be divided into three parts, namely, the intrinsic photon fluctuation at decoupling time, the Sachs-Wolfe term which can be written as the difference of the gravitational potential between present and the decoupling time and the effect caused by a time derivative of the potential. Sometimes the last two terms are called the generalized Sachs-Wolfe term. From numerical calculations, we find that the generalized Sachs-Wolfe effect dominates the quadrupole moment in almost all of the low density model. On the other hand, the intrinsic term plays an important role for the isocurvature models and dominates on low density. Though constraints obtained by comparing our numerical results with recent observations for quadrupole moment are still compatible with those of small angle observations, our constraints are severer since the quadrupole anisotropy would hardly be affected by phenomena which might have occurred after decoupling, such as reionization or gravitational lensing.

On the Interpretation of Observations of Cosmic Background Radiation

V. G. Gurzadyan (Yerevan Physics Institute)

The high level of isotropy of Cosmic Microwave Background Radiation is one of the basic facts of observational cosmology. According to the conventional viewpoint the perturbations of MwB should be the same since the epoch of last scattering. However, in our recent paper (Gurzadyan V.G., Kocharyan A.A. *Preprint ICTP*, 90/318, 1990, Trieste, *Astr. Ap.*, in press) an effect leading to decrease of perturbations of MwB at postscattering epoch is investigated. The consideration of the behavior of correlation functions for a flow of null geodesics by means of the methods of ergodic theory enables one to evaluate this effect based purely on geometrical and topological properties of the Universe. The anisotropy of MwB is shown to decrease in open Friedmannian Universe to a level far below present experimental accuracy; dependence of angular scale and the value of Ω ($\Omega \leq 1 - \epsilon; \epsilon \ll 1$) is weak. This effect can have great consequences for the interpretation of the observations. In particular, anisotropy at a level $10^{-5} \div 10^{-7}$, if discovered in near future, will probably indicate that our Universe is closed, $\Omega > 1$, and thus at least by 90% is filled with dark matter. The absence of such anisotropy should not have as simple influence on the theories of formation of galaxies as it is considered at present.

Correlated Fluctuating Signals in the Analysis of the Large Scale Struvture of the Universe

M. D. Suran (Astronomical Institute of Romanian Academy)

Recent observations of the large scale of the Universe suggest inhomogeneities on scale larger than previously though, for example: filaments, Great Walls and some other "monsters". We are interested in the studies of the signature of such types of inhomogeneities. As an observational method, we purpose to study the fluctuations in the long tail of the correlation function, as sign of inhomogeneities. As observational material we have used three different 3D catalogs Las Campanas Deep Redshift Survey in polar caps, for galaxies, The Northern Cone of Metagalaxy, for clusters and The Northern Cone of Superclusters, for superclusters distribution. For these catalogs we have compared the two point correlation function in the long tail. In this case the signal processing is in the form: $\xi(r) = \xi_N + \xi_S, r\; r_i, \; r_i = r_g, \; r_c, \; r_{sc}$ and where the noise signal: $\xi_N(r) = \xi_{cat} + \xi_{gal}$ is a sum of the catalog and the galaxies signals. We demonstrate that the signal, compared between the three 3D catalogs, shows: *periodical* and *phase correlated fluctuations, in the long tail of correlation function.* For revealing such effects we have used two different methods: Fourier analysis and the statistical Whittaker-Robinson method. We show that in the long tail of the correlation function we can decelate two types of periodicities: one on scale $\sim$ 30Mpc(the same as ξ_{gal}) and the second on scale 130Mpc(possible related to supercluster objects). The comparison between the 3D catalogs with the 1D coherent signal (BEKS 1991) demonstrate also that the fluctuations are correlated in phase. The comparison with the theoretical model of 3D lattice topology (Peebles 1989) show two adiacent differences: a possible phase retard of Peebles model and a possible 3D unamortized signal.

Abundances of Li and Be in Helo Stars

P. Demarque

The physical processes responsible for the surface dilution of Li in halo dwarfs during pre-main sequence and post-main sequence evolution are discussed. In addition to standard stellar evolution theory, the effects of diffusion of Li and He (primarily gravitational settling and thermal diffusion), and also of rotation are considered. The initial Li_p abundance derived from the standard models is $2.17^{+0.04}_{-0.13}$ (2σ)(Deliyannis, Demarque and Kawala 1990 *Ap.J.S.* **73**, 21); for diffusive models $Li_p \leq 2.35$ (Chaboya et al. 1991, *BAAS*, **22**, 1205); for rotating models $Li_p < 3.1$ (Pinsonneauet, Deliyannis and Demarque 1991 *Ap.J.* in press). Constraints on the combined effects of diffusion and rotation can be placed by the flatness of the observed Spite plateau in the ($Livs T_{eff}$)-plane- consequences for cosmology are considered. In the case of Be, there is no depletion during stellar evolution (Deliyannis and Pinsonneauet 1990 *Ap.J.L.* **326**, L23). There is also evidence for galactic enrichment in Be. Recent work by Ryan et al. (1991) shows possible evidence for a plateau at very low metalicity, corresponding to a Be abundance several orders of magnitude larger than predicted by standard BBN.

The Spectrum Distortion of Relic Radiation in Moment of Universe Recombination

V. V. Burdyuzha (Lebedev Physical Inst.), A. N. Chekmezov (Inst. of General Phys.) V. N. Lukash (Lebedev)

The distortion of the CMB from the Planck spectrum may necessarily arise because of the production of energetic quanta in the process of recombination, which leads to its retarding and the distortions in the Wein and R-J regions. We reanalyzed the recombinational dynamics from point of view of the nonequilibrium kinetics after the papers Peebles (1968), Zeldovich, Kurt and Sunyev (1969), Jones and Wyse (1985) and Krolik (1989, 1990). We argue that the CMB distortion in L_α line takes place in the region $\lambda \sim 100 - 150\mu$. This is caused by the deviation of a population of the ground state of H from the equilibrium population which can be calculated due to the Saha formula. We present the dependence of a distortion from baryonic densities.

COSMOLOGICAL PARAMETERS (reported by V. Trimble)

Twelve papers were scheduled for these sessions, focusing primarily on distance indicators, age indicators, and handles on dark matter, including baryonic dark matter from nucleosynthesis constraints. Six of the papers were delivered and six pre-empted by the fire. Abstracts of one of the papers presented and five of the six not presented (provided by the authors) come at the end of this section. Highlights of the other papers presented include:

The Hubble parameter (R. Terlevich). Astronomers have considered a very large number of distance indicators in the effort to calibrate H_0. All of these are subject to both random and systematic errors, including the speaker's own work on H-beta flux vs. line width in HII regions. Two controversial suggestions were made: first, that (despite the large number of careful investigations that have found largish values of H_0) the correct value must be at the small end of the possible range; and that additional standard candles in field galaxies are unlikely to be useful, and investigators should focus on clusters.

D/H in the interstellar medium (J. Pasachoff). The D/H ration is a strong constraint on the cosmic density of baryonic material. The Lyman alpha line has been seen, but is difficult to separate from the stronger H^1 line; better data will probably come only with FUSE. A tentative detection of the 92 cm analog of the 21 cm line has recently been reported, in absorption against Cas A. It shares the local H_2, not neutral atomic gas, velocity, which may be grounds for concern. Initial analysis indicates 11% of closure density in baryons, for $H = 50$.

Helium in HII galaxies (E. Terlevich). A recent reanalysis of the galaxy SBS 0335-052 (claimed by others as more metal deficient than I Zw 18) shows that they are actually about the same. The new galaxy, however, yields a somewhat lower helium abundance associated with $O/H = 0.022$ solar, and so implies a very low primordial value for $Y = 0.21 \pm .005$.

Primordial(?) boron and inhomogeneous nucleosynthesis (D.L. Lambert). The presence of boron has recently been established in stars with Z only about 1% that of the sun. There is currently no evidence for a minimum boron abundance, uncorrelated with Fe/H, that might be evidence for a primordial (inhomogeneous) contribution but the B/Be ratio in metal-deficient stars differs slightly from the best predictions of the abundances due to cosmic ray spallation, leaving open the possibility of a primordial contribution to one or both.

Do we need baryonic dark matter? (J.E. Felten). The "best- fit" value of the total mass density of the universe based on dynamical considerations appears to exceed significantly the best estimate of the density in luminous matter, based on a luminosity density and the mass-to light ratio appropriate for a normal stellar population. The speaker believes, however, that the error bars are large enough that the case for non-luminous baryonic material has not yet been firmly made. The issue is somewhat confused by disagreement about whether X-ray emitting gas in rich clusters of galaxies should or should not be counted as "dark".

Are SNIa good Standard Candles?

S. van den Bergh

All published UBV photometry of $SNIa$ has been used to form a list of 31 objects for which the error of $m(\max)$ is less than about $0.5mag$. A Hubble diagram for these objects shows large scatter, partly due to dust absorption and, probably, Malmquist bias. For the five objects in E/So galaxies (SN 1986G excluded), the dispersion at maximum light is $0.9mag$. Allowing for observational errors and deviations from smooth Hubble flow probably reduces the real dispersion in $m(\max)$ for SNIa to below $0.9mag$. For 6 SNIa in the Virgo cluster, $11.2 \leq m(\max) \leq 14.05$, or $11.2 \leq m(\max) \leq 1.27$ if SN 1971G in the edge-on galaxy NGC 4165 is excluded. It is noted that the apparently unreddened SN 1984A (described by Hamuy in another session) has $B(\max) = 12.7$, which is $1.5mag$ fainter than SN 1961H, which occurred in the Virgo E galaxy NGC 4564. It is concluded that neither the Hubble diagram nor the SNe in Virgo yet give strong support for the hypothesis that SNIa are good standard candles. Clearly, more photometric observations are needed to establish whether SNIa are standard candles with a small luminosity dispersion at maximum light.

A Systematic Effect in the Use of Planetary Nebula as Standard Candles

L. Bottinelli(Observatoire de Paris and Universite Paris XI) , L. Gougurnheim, G. Paturel AND P. Teerikorpi

The distances determined by Jacoby et al.(1990a) for 6 early type galaxies members of the Virgo cluster of galaxies, using the method of the planetary nebulae luminosity function (PNLF) are correlated with the magnitude of the parent galaxy. A similar trend is found within the Leo I group of galaxies and is confirmed by a comparison between the PNLF distances and distance estimates from the surface brightness fluctuation method. This effect works in the sense that more luminous galaxies are derived smaller distances. Another strange systematic feature is the increase of the specific PN density from luminous to faint galaxies, clearly seen in Virgo and Leo I data. We show that both these systematic trends can be due to an exponential part in the bright end of the PNLF. If this is the case, the short value derived for the Virgo distance by the PNLF method should be regarded uncertain. We analyze shortly the use of an exponential Lf as a distance indicator and note that if it is applied to the planetary nebulae, one should select the calibrators to match at least the Hubble type of the measured galaxies. As a preliminary example the relative distance $\mu_{Vir} = \mu_{Leo}$ is derived by the exponential LF method to be 1.15 ± 0.04 (with a fixed value for the slope of the LF). We emphasize the need of a detailed study of the PNLF in galaxies within Virgo and closer with a large luminosity and type range.

Tully-Fisher Distances

L. Gouguenheim (Observatoire de Paris and Univ. Paris), P. Teerikorpi, L. Bottinelli and G. Paturel

We have investigated the velocity-distance relation close to the direction of the Virgo cluster, with distances from the Tully- Fisher relation, and with attention to the Malmquist bias. For $\Theta > 8deg$ a behaviour is releaved which is as expected from the Tolman-Bondi solutions for an expanding spherical mass distribution, previously discussed by Tully and Shaya (1984;TulSha) using a smaller sample of galaxies. 1) Various density distributions, constrained by the mass inside the Local Group distance (required to produce V_{vir}), agree with the observations, but only if the mass within the Virgo 6 degree region is close to or larger than the standard Virgo virial mass values. This is so independently of the value of q_0, of the slope of the density distribution outside of Virgo, and of the values adopted for Virgo distance and velocity. The v vs. d relation (velocity and distance from the Virgo center) shows directly that the "zero velocity surface" lied at $d \simeq 0.45$. 2)Generally, the observations imply that the central 6 deg mass has the standard M(virial) as a rough lower limit. From this follows, together with the light ratios within the local supercluster and the light enhancement relative to the general field, that light does no trace mass. Analysis of the data below $\Theta = 8$deg and inside the standard 6 deg circle led to the following conclusions; 3)The Tolman-Bondi behaviour may be discerned close to the Virgo center, producing high velocities $V_0 > 2000$km/s for infalling galaxies with distances somewhat smaller than the Virgo distance. 4)An expanding component, as proposed by de Vaucouleurs (1982), causes the negative velocities for several galaxies nearer than r_{Vir}. Hydrogen deficient galaxies prefer this expanding component, and there is evidence that HI deficiency causes underestimates of Tully-Fisher distances, e.g. transferring negative-velocity galaxies too much to the foreground. 5)Background contamination produces an asymmetrical distribution of velocities behind r_{Vir} (behind the high-velocity expanding component). Hence the large spiral velocities in the Virgo cluster, $V_0 > 1800$km/s, can be ascribed to three origins: TB-infall, expanding component, Malmquist-biased background. Small velocities are mostly due to the expanding component. 6)The dynamical components of the Virgo region are differently distributed in the sky: The expanding component has a flattened distribution along the line M87/M84-M59. The infalling high-velocity galaxies show a distribution which is a continuation into the 6 deg circle of the Southern Extension infall previously suggested by TulSha. The other spirals do not show a concentration in the 6 deg circle. 7)The tight angular concentration of negative-velocity galaxies can be understood by a combined effect of projection and quick deceleration of the initial high expansion velocities (when the central mass is high enough, close to that required by the TB inflow).

Globular Cluster Ages

P. Demarque (Yale University)

The ages of the oldest globular clusters provide a lower limit to the age of the Universe. Recent advances indicate that an age spread of up to $4 - 5gyr$ exists among globular clusters, suggesting a long time scale of formation for the galactic halo. Absolute ages are more uncertain; current research focuses on the effects of helium diffusion near the main sequence turnoff, the dependence of the RR Lyrae luminosity on metallicity, and the O/Fe ratio in the oldest stars. For the old cluster M92, the age is $16 \pm 3gyr$. An age below $13gyr$ for M92 would cause inconsistencies with several other pieces of astronomical data.

It was noted in conclusion that there is recent evidence, based on a comparison of the RR Lyrae distribution in the galactic bulge with synthetic HB models by Y.-W. Lee, that most the RR Lyrae variables in the bulge are older than M92 by about $1 - 1.5gyr$. Thus globular clusters may not contain the oldest stars in the Galaxy! These ages put a strong constraint on cosmological models.

Stellar Th/Eu Ratios and the Age of the Galaxy

B.E.J. Pagel

While the best estimate of the age of the universe comes from globular cluster HR diagrams, it would be desirable to have an independent check, e.g., from radioactive cosmochronology. However, solar system actinide ratios are of no use for this purpose, because one can get any age upwards of $6.5 Gyr$ according to ones favourite model of Galactic chemical evolution. Butcher (1987) invented the Th/Nd method, which holds promise for age-dating individual stars, provided certain improvements are added: replaced Nd with Eu (r process) and correct for the line blend with CoI using new oscillator strengths after Lawler et al. (Nature 1990). Having done this, one can deduce ages in a way that is comparatively insensitive to galactic chemical evolution models, and Eu/Nd ratios measured by da Silva et al. plus Lawler et al.'s Th/Nd ratios give Th/Eu ratios completely consistent with ages of up to $19 \pm 4 Gyr$, deduced from positions of eild stars of known parallax in the HR diagram, contrary to the result originally claimed by Butcher. Very recently, P. Francqis has measured Th/Eu ratios in halo stars which one would expect to have similar age to Butcher's oldest star HY 3018, with $[Fe/H] = -1.0$. Francqis's Th/Eu ratios agree, as expected, with HR 3018 down to $[Fe/H] = -2.5$ or so, but at stell lower metallicities (where Eu/Fe goes down), Th/Eu goes up, suggesting a change in r-process production ratios at the very lowest metallicities.

Nearby Galaxy Flows Modeled by the Light Distribution
                                      E.J. Shaya, R.B. Tully, and M.J. Pierce

The observed distribution of light in the Local Supercluster has been used to determine the expected velocities of galaxies if the assumptions are valid that non-expansion motions are generated by gravitaional perturbations and that mass is distributed like the light. Since detailed knowledge of the light distribution extends to only 3000 km/s, three extra sources are added at large distances: one associated with the Great Attractor, one loosely with the Perseus-Pisces Supercluster region, and one with the Shapley Concentration at a distance corresponding to 13,800 km/s. The nearer two sources are motivated by improvements they offer to χ^2 fits while the distant source is required to get agreement with the cosmic microwave background dipole. Comparison is made with observed velocity field maps based on 301 high quality distance estimates in 142 groups and 53 individual galaxies. The assumption that gravitaional perturbations must dominate the generation of peculiar velocities is substantiated and $M/L \simeq 144 h \Omega_0^{0.4}$ is found. A surprisingly strong conclusion can be drawn that the clumped mass is clustered on scales $< 1\mathrm{Mpc}$ with $M/L \sim 100$ and that an *insignificant* amount of additional mass is clustered on scales between 1 Mpc and $\sim 20\mathrm{Mpc}$. A value of $\Omega_{gal} \sim 0.08$ is associated with clumped mass. There are hints of velocity streaming with coherence over 20,000 km/s and mass fluctuations on $10^{17} M_\odot$ scales. The model provides a natural description of the 'local velocity anomaly'. The χ^2 fit for the preferred model results in an equivalent *rms* uncertainty in the difference between observed and model velocities of 18% of the observed distance.

COMMISSION 49:THE INTERPLANETARY PLASMA & THE HELIOSPHERE

Report of Meetings, July 25 and 31, 1991

PRESIDENT : L.F. BURLAGA   ACTING PRESIDENT : B. BUTI

July 25, 1991

BUSINESS MEETING
        President, Dr. Burlaga could not attend the General Assembly;
B. Buti, Vice President Commission 49, convened the meeting on July
25 morning. The attendance at the meeting was poor because of the
clash with the business meeting of Commission 10, which has a large
number of members common with Commission 49. The members were apprised
of the activities of the Working Group on Plasma Astrophysics during
the last three years. The future scientific activities of the Commission were also discussed.

July 31, 1991

SCIENTIFIC MEETING
        One day meeting on 'NONLINEAR AND TURBULENT PROCESSES IN SOLAR
AND ASTROPHYSICAL PLASMAS' was held jointly with Commissions 10/12,
40 and 48. The meeting was organized and chaired by B. Buti. Eight
invited review talks were given.

HIGHLIGHTS OF INVITED REVIEWS

Radio Emission Processes in Astrophysics (D.B. Melrose):
        Coherent radio emission processes, which can account for in-
tense observed radiations, were briefly reviewed. The electron cyclotron maser emission (ECME) process was discussed in details; positive
and negative points of ECME were clearly pointed out. For pulsar radio
emission, curvature emission by bunches was presented.

Reconnection Processes in Astrophysical Plasmas (T.G. Forbes):
        The conversion of magnetic field energy into heat and particle
acceleration is possible through magnetic reconnection. Collisional
models based on MHD were discussed in details. However, it was clearly
pointed out that collisionless models of reconnection are needed to
determine the small scale length of the field, which is an essential
ingredient for the rapid reconnection. Only the latter can explain
explosive like phenomena e.g., flare stars.

Chaos in Magnetoplasmas (B. Buti):
        Starting from the basic equations governing the dynamics of
magnetoplasmas, the evolution of nonlinear Alfven waves, which are
observed in many astrophysical plasma, was demonstrated through Poincare map analysis. Interesting phenomena of chaos, strange attractors
and inertial stabilization due to heavy ion in multispecies plasmas
were elaborated.

Strong Langmuir Turbulence in Lower Solar Corona
            and other Magnetized Plasmas (M.V. Goldman):
     Lower solar corona and other plasmas with moderate magnetic
fields provide an environment for radio emissions associated with
strong Langmuir fields observed in a variety of space plasmas. Theoretical as well as numerical results for the evolution of nonlinear
Langmuir waves were presented. Potential problems, still to be studied,
were pointed out.

Gravitational Clustering of Galaxies (S. Inagaki):
     The process of gravitational clustering of galaxies was com-
pared with electrostatic plasma turbulence by considering both the
static and the exanding universe. The development of correlations,
in linear instability phase of galaxy clustering, was reviewed. Beauti-
ful results of simulation of galaxy clustering were shown on video.

Stochastic Acceleration and Diffusion in a Turbulent Plasma (J. Cary):
     Beam-driven instabilities, in stellar winds, can produce turbu-
lent fields. By means of nonlinear dynamics techniques, it was shown
that such turbulent electrostatic fields can lead   to   stochastic
acceleration and velocity diffusion. Self-consistent simulation results
were presented; these results were in agreement with the author's
mapping theory but did not agree with the turbulent trapping theory.

Interstellar Plasma Turbulence: Observation and Theory (S.R. Spangler):
     The direct data on interstellar turbulence is difficult to
obtain because of small spatial scales on which this turbulence occurs.
Different mechanisms for generation of this turbulence e.g., the ones
based on solar system analogs; turbulence from shock reflection at
interstellar clouds etc. were reviewed. The role of scintillation
observations, to constrain the above mentioned theoretical models,
was discussed. In the end, anticipated observational techniques, for
measuring density and magnetic field fluctuations in interstellar
turbulence, were pointed out.

Cometary Plasma Waves and Turbulence: Observations (B.T. Tsurutani):
     The in situ measurements of cometary plasma waves, from 1985-
1986 satellite encounters with comets Giacobin-Zinner and Halley,
were reviewed. Temporal and spatial evolution of steepened magnetosonic
waves and their precursor large amplitude whistler waves was illus-
trated and discussed at length.

POSTER PAPERS

     There were about 15 posters related to topics on plasma astro-
physics; these were displayed on different days.

COMMISSION 51: BIOASTRONOMY (BIOASTRONOMIE)
Report of Meetings, on 25,26,30 July 1991

PRESIDENT: George Marx   VICE-PRESIDENT: Ron Brown
SECRETARY: Jean Heidman

26. July 1991: Business meeting

Commission 51 deals with scientific problems related to Bioastronomy, including discussions concerning the emergence of life and intelligence in the Universe, designation of the promising sites, empirical search for life and technology, finally the possible connection of the terrestrial life to astronomical factors. This Commission was formed at the iniciative of Michael Papagiannis (Boston University) in 1982 and it worked under the chairmanship of Michael Papagiannis, Francis Drake and George Marx since then. The Commission counts well above 300 members from astronomy and interdisciplinary fields. Members of the Commission were reached by the Bioastronomy News, edited by Michael Papagiannis. The president of the Commission 51 reported about the activities as follows:

Meetings on Bioastronomy at the IAU General Assemblies were well attended by the members of IAU (beyond the membership of the Commission 51), and by invited experts from other sciences since 1979. This proves that the interdisciplinary approach of our Commission corresponds to a definite need. The advances of our scientific understanding of planet and comet formation, the exploration of the Solar System by space probes, the empirical and theoretical investigation of the planetary atmospheres, the clarification of the early history of our own planet, the confrontation of astronomical and geological evidences with the laboratory experiments concerning chemical evolution, the grand theories of the origins of life based upon our most recent scientific knowledge, finally the spectacular advances of radio astronomy due to the fast development of computer technique made bioastronomy to a solid and respected branch of research in the 1980-es. This explains the increased interest of the scientific community and funding agencies.

Members of the outgoing Organizing Committee of IAU Commission 51 (1988-1991) were George Marx, president (Hungary), Ronald D. Brown, vice-president (Australia), Michael Papagiannis, founding president and present secretary (U.S.A.), furthermore Frank Drake, past president (U.S.A.), Bruce Campbell (Canada), Samuel Gulkis (U.S.A.), Jun Jugaku (Japan), Jean Heidman (France), Nikolai Kardashev (Soviet Union), V.I. Slysh (Soviet Union), Jill C. Tarter (U.S.A.).

The members of the incoming Organizing Committee (1992-1994) are Ronald D. Brown, president (Australia), Jill C. Tarter, vice-president (U.S.A.), Jean Heidman, secretary (France), Michael Papagiannis, past president and editor for Bioastronomy News (U.S.A.), Frank Drake, past president (U.S.A.), George Marx, past president (Hungary), furthermore Bruce Cambell (Canada), F. Colomb (Argentina), H. Hirabayashi (Japan), Mike Klein (U.S.A.), S. Slysh (Soviet Union).

The highlights of the first decade (1982-1991) of IAU Commission 51 were the Bioastronomy Symposia in Boston (U.S.A., 1984), at Balaton (Hungary, 1987) and in the Alps (France, 1990). These were well attended by about 100 experts from the fields of planetary astronomy, radio astronomy, geology, climatology, organic chemistry, biochemistry, basic biology and psychology. Prominent astronomers joined the symposia of the Commission who considered the place and role of life a question worth of special interest.

The Third Bioastronomy Symposium was organized by Jean Heidman (radio-astronomer, Paris), in Val Cenis, at the heart of the French Alps in July 1990. The programme of the conference was designed by Michael Klein (Jet Propulsion Laboratory, U.S.A.). The Proceedings of the Third Bioastronomy symposium is edited by Jean Heidmann and is printed by the Springer Verlag. – The discovery of the infrared excess related to main sequence stars, the sighting of circumstellar dust disks and the study of periodic variation in the Doppler shift of stellar spectra strengthened our conviction in the abundance of planetary systems. The study of interstellar molecular clouds, the missions to comets, the observation of bombardment of the terrestrial atmosphere by ice meteorites contributed to the empirical knowledge of chemical evolution in space. Comparative studies of the atmospheric chemistry of Venus, Mars, Titan and the reconstruction of the early terrestrial atmosphere indicated the initial conditions for the emergence of life. Finally, the spectacular advances of electronics made the scanning of sky for signals indicating advanced extraterrestrial technology more economic and promising than ever since the first venture of Frank Drake. – At the 30th anniversary of Project Ozma friends and followers greeted Prof. Frank Drake in Santa Cruz in November 1990.

The Commission 51 discussed and unanimously supported the Declaration of Principles concerning activities following the detection of extraterrestrial intelligence, formulated by the International Academy of Astronautics, and proposed its acceptance to the International Astronomical Union. – The proposal for the creation of global information network for the Search of Extra terrestrial Intelligence (SETI) was accepted, and Jean Heidman (Observatory of Paris) was charged with its organization.

The Symposia of IAU Commission 51 were not the only activities in the field of bioastronomy. The Planetary Society organized an international SETI conference in Toronto (1988). The International Academy of Astronautics has a Committee on Search for Extraterrestrial Intelligence, headed by John Billingham, Ivan Almar, Jill Tarter, and it keeps close contact to Commission 51 of IAU. They organized SETI sessions in Bangalore (1988), Torremolinos (1989), Dresden (1990) and they cosponsored the Third Symposium on Biosatronomy (1990).

The activity in search for extraterrestrial intelligence will speed up in the 1990es. Argentina, the host country of this General Assembly will establish the first SETI dedicated facility for the Southern Sky under the direction of F. Colomb, in cooperation with the Planetary Society. NASA launches two major SETI programs: a targeted search (directed by NASA Ames) and an All-Sky-Survey (directed by the Jet Propulsion Laboratory). The intended starting day is 12 October 1992, the 500th/a Anniversary of landing of Columbus in the New World. The planned location of the 4th Bioastronomy Symposium is understandably: California, 1993.

25. July 1991: Astronomical Impacts on Life and Extinction

George Marx (Budapest, Hungary): Astrophysics, Climate, Technology. The inner planets of the Solar System (Venus, Earth, Mars) started evolving under very similar conditions but only the Earth has been able to develop an atmosphere sustaining liquid water through billions of years, what is a prerequisite for biological evolution. This raises the question about the nature of the terrestrial thermostat and its stability against external and internal impacts. Understanding this problem is of vital importance for the anticipation of future variations in the terrestrial climate, what is nowadays under attack by human industrial releases.

David Schwartzmann (Washington D.C., USA): The Relative Stability of the Biosphere Geophysiology. A recognition of the role of biota in the enhancement of chemical weathering on the Earth's continents leads to important consequences:

1. The habitability of Earth is a result not only of geochemistry but geophysiology. Were it not for microbial enhancement of weathering, surface temperature on Earth would have remained fit only for thermophilic bacteria ($>50°C$).

2. The transition from an abiotic to biotic surface was followed soon after by colonization of land by thermophilic bacteria, leading to sharp reduction in CO_2 and temperature, opening up the possibility of the evolution of low-temperature life.

3. The global surface temperature hovered near 20°C as a result of microbal and higher plant amplification. Leaving aside human intervention, temperatures will climb above 50°C in 3 billion years, appropriate only for thermophilic bacteria.

4. The geophysiological climatic regulator is a necessary condition for the evolution of complex life including intelligence around other sun-like stars.

Mikio Shimizu (Kanagawa): Water planet Earth should selects Nucleic Acid. Nucleic acid bases were selected for the terrestial life by water, which characterized the planet Earth in the Solar System. In water, two (hydrophobic) nucleotides stack in order to decrease the area in contact to water. (In contrast, two nucleotides make strong Watson-Crick type hydrogen bonds in a hydrophobic solution such as chloroform or benzene.) Under this conformation, the positions of the hydrogen bond aceptors and donors are uniquely determined to be able to recognize some biomolecules, other bases or amino acids.

L.R. Doyle (SETI Institute, USA): Astrophysical and Planetary Constraints on Exobiological Habitats. Liquid water seems to be essential for the origin and evolution of biology around a star. The ecospheric zones in the Solar System are presentaEarth, early Mars, possibly the Jovian moon Europa, and speculatively early large cometary nuclei. The heating source for the Earth and early Mars is solar flux, complimented by atmospheric green house warming. Tidal heating of Europa, caused by the pumping of its orbital eccentricity by the other outer satellites could possibly maintain a liquid water environment a few kilometers below the ice surface. In addition, radiogenic heating of cometary interiours by [26]Al may have also maintained a liquid water environment for an exobiologically significant time period. – The cratered terrain runoff channels on early Mars indicate large amounts of liquid water existing there, yet recent attempts to create such an environment have run into difficulties with CO_2 condensation. A reexamination of the standard early

solar model may be necessary requiring early solar mass loss (one of the explanations for
the solar lithium depletion.)

C.N. Matthews (Chicago): Life and Death from Comets. In the presence of a base
such as ammonia, liquid HCN (bp 25°C) polymerizes spontaneously at room temperature
to a black solid. The polymers are stable ladder structures with conjugated =C=N– bonds,
and polyamidines are readily converted by water to polypeptides. These macromolecules,
so easily formed in a reducing environment, could be major components of the dark matter
observed on many bodies in the outer Solar System. The nonvolatile black crust of comet
Halley may consist largely of such polymers. HCN polymers are the dark bearing solids
identified spectroscopically in the dust of new comets, on the surfaces of several asteroids of
spectral class D, within the rings of Uranus, and covering the dark hemisphere of Saturn's
satellite Iapetus. HCN polymerization could account also for the yellow-orange-brown
coloration of Jupiter and Saturn, as well as for the orange haze high in Titan's atmosphere.
– Direct evidence for the extraterrestrial presence of HCN polymers may be in hand if
mass extinctions of life were brought about by bolide impacts, particularly during the
Cretaceous/Tertiary period 65 million years ago. The alpha-amino acids detected in such
boundary regions, presumably imported by comets or meteorites, may well be hydrolysis
products of cyanide polymers.

I. Almár (Budapest): Supernovae, Co-rotation, Life. Since both supernovae and giant
molecular clouds are concentrated in spiral arms, the motion of the Solar System with
respect to the spiral arms of the Galaxy is crucial. According to Balázs, Marochnik and
Mukhin, Sun is near the co-rotation circle, there would be no crossing of any spiral arm
during the lifetime of Earth.

C.A. Olano (Buenos Aires): Encounters with Giant Interstellar Clouds. These might
be the prime motors in the catastrophic extinctions of species. Such encounters can per-
turbe the Oort cloud to induce comet showers. The past galactic paths of the Sun and
molecular clouds have been traced, taking into account the gravitational perturbations of
the Sun's orbit by giant clouds. The Sun passed near Ori OB1 around $1.5 \cdot 10^7$ years ago
and near Mon OB1 around $3.4 \cdot 10^7$ years ago, in approximate correspondence with the
dates of the Miocene and the major Eocene mass extinctions.

*M.D. Papagiannis (Boston): Impacts, Mass Extinctions and the Nemesis Hypothe-
sis.* Life may help to avoid a run-away glaciation or run-away greenhouse-effect, which
respectively have made Mars and Venus uninhabitable. Impacts of meteors (asteroids,
comets) may produce dust clouds, resulting in stopping of photosynthesis and global drop
of temperature, what might result in mass extinction. Evidence for such K/T impact was
found recently on the Yucatan peninsula. The highly excentric orbit of the hypothetical
dwarf star Nemesis may explain periodic extinctions: by distrubing the cometary belt, it
can cause cometary showers on Earth, supplying organics and causing "cosmic winters".

F. Graham-Smith (Jodrell Bank, UK): The Discovery of a New Planet. The first new
planet outside the Solar System has been found by the radioastronomers Andrew Lyne,
Matthew Bailes and Setnam Shemar. The survey of the galactic plane by the 76-metre
Lovell radio telescope revealed 40 new pulsars. They fitted previous models of solitary or
binary systems but PSR 1829-10, with a period of 0.330s. Observations extending over

three years show that the arrival times of the pulses vary by 0.008s sinusoidally with a period of 184 days. This corresponds to the motion of a neutron star with a mass of 1.4 solar masses under the gravitational pull of a 10 Earth-mass planet at a distance of 0.7 A.U. – The measured period derivative of the pulsar gives the characteristic age of 1.25 million years. The origin of the planet is the subject of speculation. The standard scenario of red giant followed by Type II supernova would disrupt any planetary system. An accreting white dwarf may have collapsed to a neutron star, the collapse being slowed down by the same rapid rotation that formed a disk producing the planet. The excentricity of the planetary orbit is less than 0.05, what can be explained by interaction with the disk.

30 June 1991: Search for Extraterrestrial Technology

Steven J. Dick (U.S. Naval Observatory): Bioastronomy–Birth of a New Scientific Discipline. Since the dawn of the Space Age four research specialties have converged to form the new discipline of Bioastronomy: planetary science, planetary systems science, origin of life studies and the search fore extraterrestrial intelligence (SETI). Bioastronomy is being institutionalized by meetings, recognition in international societies and funding from several sources, but a journal unique to the discipline is still lacking.

M.D. Papagiannis (Boston): Tritium Line May Reveal Alien Observing Stations. The nearest civilization may have detected the presence of oxygen in our atmosphere, which is indication of life, and may have dispatched an automated probe to our Solar System to keep them informed about developments on our planet. The space station may get energy, nuclear fusion of deuterium into helium, a by-product of which is tritium. Tritium has a short half-life, therefore its existence in significant amounts would indicate the presence of a nuclear fusion plant in our solar system. Valdez and Freitas searched for tritium in 53 nearby stars. They did not search, however, for tritium in our Solar System, what we hope to do with the   radio telescope at Green Bank.

Ivan Almar (Hungary): Selecting targets of both astrophysical and dynamical interest. Several authors have suggested that astrophysically improbable objects like blue stragglers, Ap and carbon stars are modified by advanced civilizations. According to Balázs, Marochnik and Mukhin only stars in a narrow zone around the co-rotation circle had time long enough to develop ETI. Therefore all stars modified by astroengineering activity must have a strong concentration near $l = 90°$ and $270°$. Although there is some deficiency of N and S stars near the direction of the galactic center, the real distribution function is different from what has been derived from the "belt of life" hypothesis.

Fernando R. Colomb (Buenos Aires): SETI Activities in Argentina. Since October 12, 1990, META II is observing the southern sky. It uses a 30 m radiotelescope of the Instituto Argentino de Radioastronomia (IAR) during 12 hs. a day, to cover all the sky south of -10° at the frequency of the 21-cm HI line. Successive spectra of 400 KHz at 0.05 Hz resolution are searched for features characteristics of a narrowband beacon transmission. META II was built at Harvard University by J.C. Olalde and E. Hurrell under the direction of P.Horowitz and with funds provided by The Planetary Society.

Stuart Bowyer, Dan Wertheimer, Chuck Donnelly, Michael Lampton(Berkeley): Recent Results and Progress on SERENDIP III. The SERENDIP project is an ongoing program of monitoring and processing broadband radio signals acquired by existing radio astronomy observatories. SERENDIP operates in a piggyback mode: it makes use of whatever observing plan is under way at its host observatory. The data acquisition system operates autonomously. This approach makes is possible to obtain large amounts of high quality observing time in a manner that is economical. The SERENDIP II system monitored some 64 000 channels simultaneously and was installed at the NRAO 3300-foot telescope at Green Bank. It was operated there for two years until the collapse of that telescope. SERENDIP III will monitor more than 4 000 000 channels simultaneously.

John Whiteoak (Australia): SETI Activities in Australia. Observations with the Parkes Radiotelescope. A group of scientists from the University of Western Australia, Division of Radiophysics CSIRO, and Australia Telescope National Facility CSIRO, have used the 64-m radiotelescope at a frequency of 4.46GHz to search for narrow spectral-line emission from selection of stars and globular clusters. Null results are used to estimate an upper period of 100 million years for extraterrestrial civilizations trying to make contact.

Kent Cullers (NASA Ames): Targated Search Signal Detection from Theory to Practice. NASA's Targeted Search for Extraterrestrial Intelligence examines narrowband data for evidence of drifting continuous and pulsed sinusoids. The pulsed sinusoidal signals are sought from approximately 1Hz to 30Hz in octave steps. The Targeted Search Signal Detectors process the multiple resolution output of the MCSA, a 10 MHz spectrometer with about 14 million channels at each of six resolutions. Development of efficient pulse detection and CW path accumulation algorithms allowed computation rates to drop into the range of commercially available hardware. Current plans are expected to result in the first full signal detector implementation late this year. The pulse detector is being developed at NASA-Ames. It uses a parallel architecture. Its heart is the 1860 processor.

S. Gulkis, D.J.Burns, C.Foster, M.F.Garyantes, M.J.Klein, S.Levin, E.T.Olsen, H.C. Wilck, and G.Zimmerman (JPL): Status of the NASA SETI Sky Survey. The primary objective is to search the entire sky over the frequency range 1GHz to 10GHz for evidence of narrow band signals of extraterrestrial, ingelligent origin. A frequency resolution near 20Hz will be used across the entire band. A spectrum analyzer with upwards of the million channels will be used to keep the survey time to approximately six years. Data rates in excess of 10 gigabits per second will be generated in the data taking process. Sophisticated data processing techniques will be required to determine the ever changing receiver baselines, and to detect and archive potential SETI signals. Existing radio telescopes, including several of NASA's Deep Space Network (DSN) 34 meter antennas located at Goldstone, CA and Tidbinbilla, Australia, will be used for the observations. The Jet Propulsion Laboratory in Pasadena, California, has the primary responsibility to develop and carry out the Sky Survey. – In order to lay the foundation for the full scale SETI Sky Survey, a prototype system is being developed at the Jet Propulsion Laboratory. This system will be used to provide a proof of concept model for the Sky Survey observing project. It will used to test and refine real time signal detection algorithms, to test scan strategies and observatory control functions, and to test RFI rejection schemes.

WORKING GROUP FOR PLANETARY SYSTEM NOMENCLATURE (WGPSN)

(GROUP DE TRAVAIL POUR LA NOMENCLATURE DU SYSTEM PLANETAIRE)
(Committee of the Executive Committee)

PRESIDENT: K. Aksnes
MEMBERS: R. Batson, A. Brahic, M. Fulchignoni, M.Ya. Marov
                     D. Morrison, T.C. Owen, V.V. Shevchenko, B.A. Smith
                     L.A. Soderblom
CONSULTANTS: J.M. Boyce, P. Masson

We are sad to report that WGPSN's first President, Dr. Peter Millman, passed away on December 11, 1990 at the age of 84. He was a world authority on meteors, and a leading figure in Canadian astronomy for many years. Dr. Millman was a very dedicated leader of WGPSN from its inception in 1973 until 1982.

WGPSN held its 19th meeting in Florence, Italy 29–30 April 1991 and the 20th meeting in Buenos Aires, Argentina 30 July 1991 during the IAU General Assembly. Much work is also being done by correspondance, increasingly by electronic mail.

Names were recommended for one satellite of Saturn and for six satellites, three rings, and three ring arcs of Neptune. These objects had all been discovered in Voyager mission photographs. The success of the Magellan mission in mapping Venus' surface in great detail has created a need for many new Venus names of which 121 are introduced below. Also 79 new Mars features have been named; they will appear on a new Mars map needed for the Mars Observer mission in a few years time.

The Working Group is asking for IAU approval of the following new resolutions:

1. In proposing names for newly discovered satellites of the Solar System, it is recommended that first consideration be given to procedures already established within the IAU, and that confusing duplication with asteroids be avoided. (Amendment of Res. 1 in IAU Trans. XXB, 1988).

2. Three new feature categories are introduced for Venus:

   RETICULUM, RETICULA   reticular (netlike) pattern
   FARRUM, FARRA   row of pancake-like structures
   VALLIS, VALLES   sinuous valley

———

The following names, and those listed in Reports on Astronomy, Vol. XXIA, 613–619,1991, have the approval of the IAU Executive Committee (since August 1991):

MARS

Name	Lat	Long	Attribute
CRATER			
Ayacucho	38.5N	92.0W	Town in Bolivia
Bentham	56.0S	40.3W	Town in England
Cangwu	42.1N	89.5W	Town in China
Charlieu	38.3N	84.1W	Town in France
Chatturat	35.7N	94.8W	Town in Thailand
Dunhuang	80.9S	48.7W	Town in China
Dzeng	80.5S	70.5W	Town in Cameroon
Ellsley	36.5N	83.0W	Town in England
Escorial	77.0N	54.3W	Town in Spain
Fenagh	34.6N	215.7W	Town in Ireland
Gandzani	34.5N	90.8W	Town in Georgian SSR
Handlova	37.9N	88.4W	Town in Czechoslovakia
Irharen	34.8N	219.2W	Town in Algeria
Jampur	38.8N	81.0W	Town in Pakistan
Kamativi	20.7S	260.0W	Town in Zimbabwe
Kamloops	53.9S	32.1W	Town in Canada
Kisambo	34.3N	89.0W	Town in Zaire
Krasnoye	36.1N	216.2W	Town in Russia
Leleque	36.7N	221.9W	Town in Argentina
Lodwar	55.4S	43.0W	Town in Kenya
Lomela	81.7S	56.0W	Town in Zaire
Lowbury	42.8N	93.0W	Town in New Zealand
Mari	52.4S	45.7W	Ruined city in Syria
Mendota	36.1N	221.8W	Town in USA
Milford	52.6S	45.1W	Town in Utah, USA
Nipigon	34.0N	81.9W	Town in Canada
Ocampo	32.9N	221.7W	Town in Mexico
Oodnadatta	52.7S	34.8W	Town in Australia
Rayadurg	18.6S	257.6W	Town in India
Reykholt	40.8N	85.8W	Town in Iceland
Sarn	77.5S	54.5W	Town in Wales
Suata	19.2S	253.3W	Town in Venezuela
Woolgar	34.8N	85.5W	Town in Australia
CATENA			
Apodis Catena	27.2S	256.8W	Classical albedo name
Baphyras Catena	38.8N	84.0W	Classical river
CAVUS			
Argyre Cavi	49.1S	40.2W	Albedo name
Octantis Cavi	52.7S	45.6W	Albedo name
CHAOS			
Aurorae Chaos	80.6S	34.5W	Albedo name

MARS(Cont.)

Name	Lat	Long	Attribute
CHASMA			
Arsia Chasmata	7.9S	119.4W	Albedo name
Ascraeus Chasmata	8.7N	105.5W	Classical albedo name
Pavonis Chasma	3.8N	111.1W	Albedo name
COLLES			
Tempe Colles	33.9N	82.7W	Classical albedo name
DORSUM			
Atrax Dorsum	38.3N	89.1W	Classical town
Auxo Dorsum	56.1S	41.9W	One of the Graces
Charis Dorsum	55.8S	41.2W	One of the Graces
Cleia Dorsum	55.3S	46.3W	One of the Graces
Hegemone Dorsum	55.3S	44.9W	One of the Graces
Pasithea Dorsum	55.7S	41.8W	One of the Graces
Phaenna Dorsum	54.3S	43.2W	One of the Graces
FOSSA			
Chalce Fossa	51.9S	40.0W	Albedo name
Mangala Fossa	16.5S	148.8W	Named for nearby vallis
Pavonis Fossae	4.2N	111.2W	Albedo name
Tanais Fossae	38.6N	85.3W	Classical albedo name
Tyrrhena Fossae	22.1S	254.5W	Classical albedo name
MENSA			
Ascraeus Mensa	11.7N	107.8W	Classical albedo name
MONS			
Chalce Montes	53.8S	37.0W	Albedo name
Galaxius Mons	35.0N	217.7W	Classical albedo name
Gonnus Mons	41.6N	90.8W	Classical town
Horarum Mons	51.3S	36.4W	Albedo name
Octantis Mons	55.5S	42.5W	Albedo name
Pindus Mons	39.7N	88.9W	Mountains near Vale of Tempe
Tanaica Montes	39.7N	90.8W	Classical albedo name
Tyrrhena Mons	24.5S	258.7W	Classical albedo name
PATERA			
Angusta Patera	80.7S	79.5W	Albedo name
Australis Patera	80.2S	51.5W	Albedo name
Issedon Paterae	38.8N	89.9W	Classical albedo name
PLANUM			
Ascuris Planum	39.5N	84.0W	Classical albedo name

MARS(Cont.)

Name	Lat	Long	Attribute
SULCUS			
Arsia Sulci	6.4S	129.7W	Albedo name
Ascraeus Sulci	11.9N	108.7W	Classical albedo name
Pavonis Sulci	3.9N	117.6W	Albedo name
THOLUS			
Issedon Tholus	36.3N	94.6W	Classical albedo name
E. Mareotis Tholus	36.1N	85.0W	Classical albedo name
N. Mareotis Tholus	36.8N	86.0W	Classical albedo name
W. Mareotis Tholus	35.8N	87.5W	Classical albedo name
VALLES			
Dzigai Vallis	59.7S	31.3W	Valley in Navajo
Enipeus Vallis	37.5N	93.1W	Classical river
Nia Vallis	54.3S	33.0W	Lowell canal name
Pallacopas Vallis	54.5S	21.2W	Lowell canal name
Surius Vallis	60.3S	51.0W	Lowell canal name

VENUS

Name	Lat	Long	Diam (km)	Attribute
CRATER				
Adivar	8.9N	75.9E	30	Turkish educator
Aglaonice	26.5S	339.9E	66	Ancient astronomer
Agnesi	39.5S	37.8E	40	Italian mathematician
Agrippina	33.2S	65.2E	37	Roman empress
Alcott	59.5S	354.5E	71	American author
Al-Taymuriyya	32.9N	336.2E	22	Egyptian author
Amalasthuna	11.5S	342.4E	18	Ostrogoth queen
Amaya	11.3N	89.1E	32	Spanish Gypsy dancer
Amenardes	15.0N	54.1E	25	Egyptian princess
Andami	17.5S	26.3E	28	Iranian doctor
Anicia	26.4S	31.1E	30	Greek physician and poet
Annia Faustina	22.1N	4.6E	20	Roman empress
Astrid	21.4S	335.5E	12	Scandinavian first name
Aurelia	20.3N	331.8E	31	Julius Caesar's mother
Avviyar	18.0S	353.6E	21	Tamil poet
Badarzewska	22.6S	137.0E	28	Polish composer
Ban Zhao	17.2N	146.9E	38	Chinese historian
Barrera	16.6N	109.3E	25	16th Cen. medical writer
Barton	27.4N	337.5E	50	American Red Cross founder
Bassi	19.0S	64.6E	35	Italian physicist
Behn	32.5S	141.8E	25	English writer
Bergolts	28.1S	80.8E	30	Russian poet

VENUS(Cont.)

Name	Lat	Long	Diam (km)	Attribute
CRATER				
Blixen	59.9S	145.6E	22	Danish writer
Bonnevie	36.1S	126.8E	85	Norwegian biologist
Boulanger	26.5S	99.3E	57	French composer
Bourke-White	21.2N	147.8E	31	American photo-journalist
Bridgit	45.3S	348.9E	11	First name from Ireland
Buck	5.7S	349.6E	22	American writer
Budevska	0.5N	143.0E	20	Bulgarian princess
Callas	2.4N	26.9E	30	Greek opera singer
Callirhoe	21.3N	140.6E	32	Greek sculptor
Carreno	3.9S	16.1E	57	Venezuelan musician
Carson	24.2S	344.2E	41	American biologist,author
Chapelle	6.4N	103.8E	23	American photo-journalist
Chiyojo	47.8S	95.2E	35	Japanese (Edo) poetess
Cori	25.4N	72.7E	50	Czech biochemist
Cunitz	14.5N	350.9E	48	Polish astronomer
Cynthia	16.7S	347.5E	19	Greek first name
Danilova	26.4S	337.3E	50	Russian ballet dancer
De Beauvoir	2.0N	96.0E	40	French writer
De Lalande	20.3N	354.9E	20	French astronomer
Deloria	32.0S	97.0E	38	American anthropologist
Devorguilla	15.3N	3.8E	22	Irish heroine
Erxleben	50.9S	39.3E	28	German scholar
Ferber	26.4N	13.0E	23	American author
Ferrier	15.8N	111.1E	30	English singer
Festa	11.5N	27.2E	25	Italian painter
Flagstad	54.3S	18.9E	48	Norwegian opera singer
Frank	13.2S	12.9E	20	Dutch author
Fredegonde	50.7S	92.9E	26	Frankish queen
Germain	38.0S	63.5E	33	French mathematician
Goppert-Mayer	59.8N	26.5E	35	U.S. physicist(Polish-born)
Greenaway	22.9N	145.1E	85	English author,artist
Guilbert	57.9S	13.3E	30	French singer
Halle	19.8S	145.4E	23	Austrian violonist
Hellman	4.8N	356.2E	24	American playwright
Henie	52.0S	145.8E	70	Norwegian figure scater
Hepworth	5.2N	94.7E	54	English sculptoress
Himiko	19.0N	124.2E	35	Japanese queen
Holiday	46.7S	12.7E	24	American singer
Horner	23.4N	97.5E	28	19th Cen. naturalist
Hua Mulan	86.8N	337.7E	23	Chinese warrior
Huang Daopo	54.2S	165.1E	27	Chinese engineer
Hwangcini	6.3N	141.7E	30	16th Cen. Korean poet
Joliot-Curie	1.6S	62.1E	80	French physicist
Kartini	57.8N	333.0E	24	Javnese educator
Kollwitz	25.2N	133.6E	30	German artist

VENUS(Cont.)

Name	Lat	Long	Diam	Attribute
CRATER			(km)	
Kushinada	5.7S	76.6E	23	Japanese (Izumo) poetess
Lachappelle	26.7N	336.5E	35	French researcher
Landovska	84.5N	74.2E	33	Polish instrumentalist
Lehmann	44.1S	38.7E	20	Danish geophysicist
Li Qingzhao	23.7N	94.3E	21	Chinese poetess
Lullin	23.1N	81.0E	24	Swiss entomologist
Manzolini	25.7N	91.1E	42	Italian anatomist,teacher
Maria Celeste	23.5N	140.5E	90	Galileo's daughter(Ital.)
Marsh	63.7S	46.7E	35	New Zealand writer
Mead	12.5N	57.4E	280	American anthropologist
Merian	34.5N	76.2E	20	Dutch entomologist
Millay	24.4N	110.9E	45	American poet
Monna Lisa	25.6N	25.3E	80	Ital. model for L. da Vinci
Mu Guiying	41.2N	80.7E	25	Chinese warrior
Nemcova	5.9N	125.0E	24	Czech novelist,poet
Nijinskaya	25.9N	122.3E	30	Russian dancer
O'Connor	26.0S	143.8E	27	American novelist
Parra	20.5N	78.1E	50	Chilean writer
Piaf	0.8N	5.1E	30	French singer
Recamier	12.5S	57.9E	25	French patron of letters
Riley	14.0N	72.2E	25	English botanist
Roxanna	26.5N	334.6E	9	Persian first name
Samintang	39.0S	80.6E	24	Korean poetess
Saskia	28.6S	337.2E	40	Rembrandt's wife(Dutch)
Scarpellini	23.4S	34.4E	25	Italian astronomer
Simonenko	26.9S	97.3E	35	Soviet astronomer
Stein	30.0S	345.5E	24	American writer
Stuart	30.8S	20.2E	67	Mary,Quen of Scots
Vigier Lebrun	17.3N	141.3E	53	French painter
Von Siebold	52.0S	36.7E	36	German physician
Wilder	17.4N	122.4E	35	American author
Woolf	37.7S	27.1E	25	British writer
Xantippe	10.9S	11.7E	41	Greek wife of Socrates
Xiao Hong	43.6S	101.5E	37	Chines novelist
Zhu Shuzhen	26.5S	356.6E	32	Chinese poetess
CORONA				
Beyla Corona	25.0N	15.5E	400	Norse Earth goddess
Eithinoha Corona	57.5S	7.5E	500	Iroquois Earth goddess
Gertjon Corona	30.0S	276.0E	250	Teutonic fertility goddess
Hervor Corona	25.5S	269.0E	250	Norse fertility goddess
Kamui-Huci Corona	63.5S	322.5E	300	Jap.(Ainu) Earth goddess
Otygen Corona	57.0S	30.5E	400	Mongolian Earth mother
Quetzalpetlatl Corona	68.0S	354.5E	400	Aztec fertility goddess
Rigatona Corona	33.5S	278.5E	300	Celtic fertility goddess

VENUS(Cont.)

Name	Lat	Long	Diam (km)	Attribute
FLUCTUS				
Eriu Fluctus	38.0S	1.0E	1200	Irish Earth mother
Kaiwan Fluctus	48.0S	1.5E	1200	Ethiopian Earth mother
Mylitta Fluctus	56.0S	353.5E	1250	Semitic mother goddess
FOSSA				
Enyo Fossae	63.0S	347.0E	900	Greek war goddess
Nike Fossae	59.5S	339.5E	850	Greek goddess of victory
LINEA				
Kalaipahoa Linea	60.5S	336.5E	2400	Hawaiian war goddess
Morrigan Linea	54.5S	311.0E	3200	Celtic war goddess
Penardun Linea	54.0S	344.0E	975	Celtic sky goddess
MONS				
Nepthys Mons	33.0S	317.5E	350	Egyptian land goddess
REGIO				
Dione Regio	31.5S	323.0E	2300	Greek Titaness
Hyndla Regio	24.0N	294.5E	2300	Norse wood Giantess

SATELLITES OF SATURN AND NEPTUNE

Temporary designation	Permanent designation	Name	Distance from planet's center (km)
1981S13	Saturn XVIII	Pan	133,600
1989N6	Neptune III	Naiad	48,230
1989N5	Neptune IV	Thalassa	50,070
1989N3	Neptune V	Despina	52,530
1989N4	Neptune VI	Galatea	61,950
1989N2	Neptune VII	Larissa	73,550
1989N1	Neptune VIII	Proteus	117,640

NEPTUNE RINGS AND RING ARCS

1989N3R	Galle(N42)	41,900
1989N2R	Leverrier(N53)	53,200
1989N1R	Adams(N63)	62,900
"Leading" ring arc	Liberte	62,900
"Equidistant" ring arc	Egalite	62,900
"Following" ring arc	Fraternite	62,900

WORKING GROUP FOR THE WORLDWIDE DEVELOPMENT OF ASTRONOMY

(GROUPE DE TRAVAIL POUR LE DEVELOPPEMENT MONDIAL DE L'ASTRONOMIE)

Report of Meeting held on Friday, 26th July, 1991

Chairman: A.H. Batten

About 30 IAU members, representing some 15 countries,
attended the meeting. The Chairman pointed out that most of the
countries joining the Union since the end of the Second World War
were newly independent with little or no tradition of modern
scientific research. The astronomical communities in these
countries are often small, isolated, and without the facilities
and funds that many of us take for granted, even in times of
restricted funding. These considerations led the Executive
Committee to form the Working Group, whose present membership -
in addition to the Chairman - comprises Kippenhahn, Kozai, Léna
and Sahade. This is the first General Assembly at which the
Group could hold an open meeting. The Chairman hoped that the
meeting would make members of the Union aware of the Group, and
provide an opportunity for those we wish to help to express their
needs. The first part of the meeting would provide a survey of
what the IAU is already doing, and the second would be devoted to
local initiatives already being taken, independently of the IAU.
      F.G. Smith spoke of the travel grants provided by Commission
38 (Exchange of Astronomers) to established astronomers making
temporary working visits to countries other than their own.
Clear, simple guidelines enable individual decisions to be made
quickly. Grants to graduate students (e.g. to make observing
trips) are not permitted by the guidelines. It might be better to
try to find funds for such purposes by pointing out the need to
various national academies, rather than by attempting to expand
the IAU's own programme to include other kinds of visits than
those for which it was instituted.
      A. Sandquist summarized several programmes of Commission 46
(Teaching of Astronomy), particularly the Visiting Lecturer's
Programmes (VLP), the International Schools for Young Astronomers
(ISYA), the travelling telescope, the Newsletter (listing
educational materials) and the IAU Colloquium of three years ago:
The Teaching of Astronomy. The value of both the travelling
telescope and VLPs was questioned in discussion, but there were
also strong advocates of each of them. The value of experience
with even small telescopes to students in countries that have

none at all was stressed. The travelling telescope could provide
such experience. It was suggested that the IAU might compile a
central register of small instruments, no longer used, that
established observatories might be prepared to give away or at
least to make available on long-term loan. Discussion of VLPs
centred round the difficulty of getting lecturers for three-month
terms. Some participants thought that shorter more intensive
courses might overcome this difficulty, but most felt that the
longer interaction between teacher and students, permitted by the
three-month courses, was valuable. D.L. Wentzel spoke of his
concerns that both VLPs and ISYAs were often too dependent on one
individual in the host institution, and that a mechanism for
assessing the lasting effects of these operations is needed.

        G.A. Wilkins spoke of the concern of Commission 5
(Documentation and Astronomical Data) for all member countries of
the Union. He discussed such initiatives as J. Mead's Selected
Astronomical Catalogs on CD-ROM. It was emphasized that the IAU
has made a grant to defray the costs of distributing this to
small institutions in developing countries, and that the
equipment needed to read the disks is relatively inexpensive.
The difficulty of communicating with these small institutions was
discussed. Wilkins thought that Commission 5 could advise them
on how to link themselves to e-mail networks. Onuora said that
in Nigeria communication had been improved by association with a
specific institution.

        The importance of introducing astronomy teaching into
secondary schools, where it could serve as a paradigm for all
science teaching, was stressed. However, it is clearly
impracticable for the IAU to reach all the world's secondary-
school teachers. One IAU lecture in a developing country, even
if given to school teachers, is of limited value. Moreover, the
IAU's prime function is to encourage research in astronomy.
Nevertheless, there was a strong feeling that we should try to
influence education at secondary-school level, and even at the
popular level. One proposal was for an exhibition of all the
world's astronomy text-books at the XXII GA.

        The local initiatives and problems discussed in the second
part of the meeting included the Inter-University Centre for
Astronomy and Astrophysics at Pune, India (Narlikar), teaching
astronomy in the Asian-Pacific region (Isobe), the situation of
astronomers in China (Li Qi Bin) and the Soviet Astronomical
Society (Bochkarev). Wamsteker also discussed the proposed World
Astronomy Days that would provide access to modern instrumentation
to astronomers that do not normally have it. The audience present
expressed its support for the concept.

                                        - end of report -

CHAPTER VI

FUNCTIONS AND STATUTES OF THE UNION

1.   Useful addresses

The addresses of the Secretariat is given and members of the Executive Committee are noted, pp. 17-18.

The addresses of the presidents and vice-presidents of the Commissions have been published in IB 67 pp. 22-33.

2.   History of the Union

A brief guide to the former Presidents, General Secretaries, etc... of the Union will be published in IB 68.

3.   IAU Representatives to other Organisations

The list of the triennium 1991-1994 is given in IB 67 pp. 44-45.

4.   Services of the Union

A summary of the activities of the Union Services may be found in Transactions XVIIB, pp. 387-393.

5.   Statutes, By-Laws & Working Rules

See hereafter.

STATUTS

I. **Dénomination, Buts et Domicile**

I.1. L'Union Astronomique Internationale (ci-après dénommée l'Union) est une organisation non-gouvernementale, qui a pour buts de :

a) développer l'astronomie par la coopération internationale,

b) encourager l'étude et le développement de l'astronomie sous tous ses aspects,

c) servir et sauvegarder les intérêts de l'astronomie.

I.2. L'Union a son siège légal à Bruxelles.

II. **Affiliation de l'Union**

II.3. L'Union adhère au Conseil International des Unions Scientifiques.

III. **Membres de l'Union**

III.4. L'Union a pour membres :

a) des personnes morales (Pays adhérents) ;

b) des personnes morales associées (Pays associés) ;

c) des membres individuels (Membres).

IV. **Organisations affiliées**

IV.5. L'Union peut accepter l'affiliation d'organisations internationales non-gouvernementales qui contribuent au développement de l'astronomie.

V. **Pays adhérents**

V.6. Les pays adhèrent à l'Union soit :

a) par l'intermédiaire de l'organisation par laquelle ils adhèrent au Conseil International des Unions Scientifiques, ou par l'intermédiaire d'un Comité National d'Astronomie approuvé par cette organisation,

soit :

b) s'ils n'adhèrent pas au Conseil International des Unions Scientifiques, par l'intermédiaire d'un Comité National d'Astronomie reconnu par le Comité exécutif de l'Union.

c) Les Organisations ou Comités mentionnés à l'article 6(a) et les Comités Nationaux d'Astronomie mentionnés à l'article 6(b) sont dénommés ci-après organismes adhérents.

V.7. L'adhésion d'un pays à l'Union est proposée par le Comité exécutif et approuvée par l'Assemblée générale : elle prend fin si le pays se retire de l'Union ou si le pays n'a pas payé sa contribution durant cinq ans.

V.8. Les Pays adhérents sont répartis en catégories. Le nombre des catégories est fixé par le Règlement. Un pays qui sollicite son adhésion indique la catégorie dans laquelle il désire être classé. La proposition peut être refusée par le Comité exécutif si la catégorie est manifestement inadéquate.

STATUTES

I.   **Denomination, Objects and Domicile**

I. 1.   The International Astronomical Union (referred to as the Union) is a non-governmental organization, whose objects are:

        a) to develop astronomy through international co-operation,

        b) to promote the study and development of astronomy in all aspects,

        c) to further and safeguard the interests of astronomy.

I. 2.   The legal domicile of the Union is Brussels.

II.   **Adherence to the Union**

II. 3.   The Union adheres to the Internation Council of Scientific Unions.

III.   **Composition of the Union**

III. 4.   The Union is composed of:

        a) full members (Adhering Countries);

        b) associate members (Associate Countries);

        c) individual members (Members).

IV.   **Affiliated Organizations**

IV. 5.   The Union may admit the affilation of international non-governmental organizations which contribute to the development of astronomy.

V.   **Adhering Countries**

V. 6.   Countries adhere to the Union either:

        a) through the organization by which they adhere to the International Council of Scientific Unions, or through a National Committe of Astronomy approved by that organization.

        or:

        b) if they do not adhere to the International Council of Scientific Unions, through a National Committee of Astronomy recognized by the Executive Committee of the Union.

        c) The Adhering Organization and National Committee of Astronomy are referred to as adhering bodies.

V. 7.   Adherence of a country to the Union is approved, on the proposal of the Executive Committee, by the General Assembly; it terminates if the Country withdraws from the Union or if the country has not paid its dues for five years.

V. 8.   Adhering Countries are classified in categories. The number of categories shall be specified in the By-laws. A country requesting adherence shall specify the category in which it desires to be classed. The specification may be declined by the Executive Commitee if the category proposed is manifestly inadequate.

VI.	**Pays associés**
VI.9.	Les pays souhaitant faire partie de l'Union tout en développant l'astronomie dans leur territoire peuvent le faire à titre de Membres associés.
VI.10.	L'organisme adhérent d'un pays associé peut être soit l'organisation par l'intermédiaire de laquelle le pays adhère au Conseil International des Unions Scientifiques, soit une institution d'éducation supérieure soit un conseil scientifique national.
VI.11.	Les pays sont acceptés en qualité de Membres associés par l'Assemblée générale, sur proposition du Comité exécutif, pour une période maximale de neuf ans au terme de laquelle ils deviennent membres à part entière, ou se retirent de l'Union.
VI.12.	Durant la période probatoire, l'Union peut accepter, à la requête de l'organisation adhérente, d'aider au développement de l'astronomie dans ce pays via le Programme de Professeurs Visiteurs et/ou de tout programme adéquat.
VII.	**Membres**
VII.13.	Les Membres sont admis dans l'Union par le Comité exécutif, sur proposition de l'un des organismes adhérents mentionnés à l'article 6, en considération de leur activité dans une branche de l'astronomie.
VIII.	**Assemblée générale**
VIII.14.	

a) L'activité de l'Union est dirigée par l'Assemblée générale des représentants des Pays adhérents et des Membres. Chaque Pays adhérent nomme un représentant autorisé à voter en son nom.

b) L'Assemblée générale rédige un Règlement qui précise les modalités d'application des Statuts.

c) Elle nomme un Comité exécutif chargé d'exécuter les décisions de l'Assemblée générale, et d'administrer l'Union pendant la période séparant les réunions de deux Assemblées générales ordinaires successives. Le Comité exécutif rend compte de sa gestion à l'Assemblée générale. L'Assemblée générale, en acceptant le rapport du Comité exécutif, le décharge de sa responsabilité.

VIII.15.

a) Sur les questions concernant l'administration de l'Union, sans implication budgétaire, le vote à l'Assemblée générale a lieu par Pays adhérent, chaque pays disposant d'une voix. Les Pays adhérents qui ne sont pas à jour de leurs cotisations annuelles au 31 décembre de l'année précédant l'Assemblée générale ne peuvent pas participer aux votes.

b) Sur les questions engageant le budget de l'Union, le vote a lieu de même par Pays adhérent, dans les conditions et avec les réserves prévues à l'article 15(a), le nombre de voix de chaque Pays adhérent étant égal à l'indice de sa catégorie, définie conformément à l'article 8, augmenté d'une unité.

VI.   **Associate Countries**

VI. 9.   Countries that would like to join the Union while developing
           Astronomy in their territory may do so as Associate Members.

VI.10.   The Adhering Body for Associate Members may be the organization by
           which the country adheres to the International Council of Scientific
           Unions or through an institution of higher learning or a National
           Research Council.

VI.11.   Countries are accepted as Associate Members by the General Assembly,
           on the proposal of the Executive Committee, for a maximum interval
           of nine years, at the end of which they either become a full Member,
           or they resign from the Union.

VI.12.   During the probationary period, the Union, if asked by the Adhering
           Organization, may agree to help in the development of Astronomy in
           that country through the Visiting Lecturers' Programme and/or any
           other appropriate programme.

VII.   **Members**

VII.13.   Members are admitted to the Union by the Executive Committee, on
           the proposal of an adhering body referred to in article 6, with
           regard to their achievements in some branch of astronomy.

VIII.   **General Assembly**

VIII.14.

   a) The work of the Union is directed by the General Assembly of
      representatives of Adhering Countries and of Members.   Each
      Adhering Country appoints a representative authorised to vote in
      its name.

   b) The General Assembly draws up By-laws governing the application
      of the Statutes.

   c) It appoints an Executive Committee to implement the decisions of
      the General Assembly, and to direct the affairs of the Union in
      the interval between meetings of two successive ordinary General
      Assemblies.   The Executive Committee reports to the General
      Assembly.   The General Assembly, in accepting the report of
      the Executive Committee, discharges it of liability.

VIII.15.

   a) On questions concerning the administration of the Union, not
      involving its budget, voting at the General Assembly is by
      Adhering Country, each country having one vote.   Adhering
      Countries which have not paid their annual contributions up to
      31 December of the year preceding the General Assembly may not
      participate in the voting.

   b) On questions involving the budget of the Union, voting is
      similarly by Adhering Country, under the same conditions and with
      the same reservations as in article 15(a), the number of votes
      for each Adhering Country being one greater than the number of
      its category, as defined in atricle 8.

c) Les Pays adhérents peuvent voter par correspondance sur les questions figurant à l'ordre du jour de l'Assemblée générale.

d) Un scrutin n'est valable que si au moins deux tiers des Pays adhérents disposant du droit de vote en vertu de l'article 15(a) y prennent part.

e) Les Pays associés ne peuvent voter que sur des questions concernant les Membres associés.

VIII.16. Sur les questions scientifiques n'engageant pas le budget de l'Union, les Membres de l'Union disposent chacun d'une voix.

VIII.17. Sur toutes les questions prévues aux articles 15 et 16, les décisions sont prises à la majorité absolue des suffrages. Cependant, une décision de modification des Statuts n'est valable que si elle a été prise à la majorité des deux tiers des voix des Pays adhérents qui disposent du droit de vote en vertu de l'article 15a).

VIII.18. Une proposition de modification des Statuts ne peut être discutée que si elle figure, en tant que telle, à l'ordre du jour de l'Assemblée générale.

IX. **Comité exécutif**

IX.19. Le Comité exécutif se compose du Président de l'Union, du "Président-elect", de six Vice-Présidents, du Secrétaire général et du Secrétaire général adjoint, élus par l'Assemblée générale sur la proposition du Comité Spécial des Nominations. Le "Président-elect" deviendra normalement le Président du prochain Comité exécutif.

X. **Commissions de l'Union**

X.20. L'Assemblée générale crée des Commissions en vue d'assurer la réalisation des buts qu'elle se propose.

XI. **Représentation légale de l'Union**

XI.21. Le Secrétaire général est le représentant légal de l'Union.

XII. **Budget et Cotisations**

XII.22.

a) Pour chaque Assemblée générale ordinaire, le Comité exécutif prépare un projet de budget pour la période à courir jusqu'à l'Assemblée générale ordinaire suivante, ainsi que les comptes de l'Union pour la période précédente. Il les soumet au Comité des Finances pour examen ; ce Comité des Finances est composé de membres nommés par les organismes adhérents, à raison d'un membre par organisme, et il est approuvé par l'Assemblée générale. Lors de sa première séance pendant l'Assemblée générale, le Comité des Finances élit un Président parmi ses membres.

b) On questions involving the budget of the Union, voting is similarly by Adhering Country, under the same conditions and with the same reservations as in article 15(a), the number of votes for each Adhering Country being one greater than the number of its category, as defined in atricle 8.

c) Adhering Countries may vote by correspondence on questions on the agenda for the General Assembly.

d) A vote is valid only if at least two thirds of the Adhering Countries having the right to vote by virtue of article 15(a) participate in it.

e) Associate Countries have the right to vote only on questions concerning associate membership.

VIII.16. On scientific questions not involving the budget of the Union the Members of the Union each have one vote.

VIII.17. On all questions in article 15 and 16, decisions are taken by an absolute majority of the votes cast. However, a decision to change the Statutes is only valid if taken with the approval of at least two thirds of the votes of the Adhering Countries having the right to vote by virtue of article 15(a).

VIII.18. A motion to change the Statutes can only be discussed if it appears, in specific terms, on the agenda for the General Assembly.

IX. **Executive Committee**

IX.19. The Executive Committee consists of the President of the Union, the President-elect, six Vice-Presidents, the General Secretary and the Assistant General Secretary elected by the General Assembly on the proposal of a Special Nominating Committee. The President-elect will normally become President of the succeeding Executive Committee.

X. **Commissions of the Union**

X.20. The General assembly forms Commissions for such purposes as it may decide.

XI. **Legal Representation of the Union**

XI.21. The General Secretary is the legal representative of the Union.

XII. **Budget and Dues**

XII.22.

a) For each ordinary General Assembly the Executive Committee prepares a budget proposal covering the period to the next ordinary General Assembly, together with the accounts of the Union for the preceding period. It submits these to the Finance Committee for consideration; this Finance Committee consists of one member nominated by each adhering body and approved by the General Assembly. At its first meeting during the General Assembly, the Finance Committee elects a chairman from among its members.

b) Le Comité des Finances examine les comptes de l'Union pour voir si les dépenses engagées ont été conformes aux voeux émis lors de la précédente réunion de l'Assemblée générale et il s'assure que le budget proposé vise à la poursuite de la politique de l'Assemblée générale, telle qu'elle est interprétée par le Comité exécutif. Il présente des rapports sur ces questions qu'il soumet à l'Assemblée générale pour approbation des comptes, et pour décision sur le budget.

c) Chaque Pays adhérent verse annuellement à l'Union un nombre d'unités de cotisation qui est fonction de sa catégorie. Le nombre d'unités de cotisation pour chaque catégorie est fixé par le Règlement.

d) La cotisation annuelle des Pays associés s'élève à une unité de contribution.

e) Le montant de l'unité de cotisation est fixé par l'Assemblée générale, sur la proposition du Comité exécutif et avec l'avis du Comité des Finances.

f) Le paiement des cotisations est à la charge des organismes adhérents. La responsabilité de chaque Pays adhérent envers l'Union est limitée au montant des cotisations dues par ce pays à l'Union.

g) Un Pays adhérent qui cesse d'adhérer à l'Union renonce de ce fait à ses droits sur l'actif de l'Union.

XIII.   **Dissolution de l'Union**

XIII.23.   La décision de dissoudre l'Union n'est valable que si elle est prise à la majorité des trois quarts des voix des Pays adhérents qui disposent du droit de vote en vertu de l'article 15(a).

XIV.   **Dévolution de l'Autorité en Cas de Force majeure**

XIV.24.   Si, par suite d'événements indépendants de la volonté de l'Union, des circonstances apparaissent qui rendent impossible le respect des clauses de ces Statuts et du Règlement établi par l'Assemblée générale, les organes et membres du Comité exécutif de l'Union, dans l'ordre fixé ci-dessous, prendront toutes dispositions qu'ils jugeront nécessaires pour la continuation du fonctionnement de l'Union. Ces dispositions devront être soumises à une autorité supérieure dès que cela deviendra possible, jusqu'à ce qu'une Assemblée générale extraordinaire puisse être réunie. L'autorité est dévolue dans l'ordre ci-dessous : l'Assemblée générale ; une Assemblée générale extraordinaire ; le Comité exécutif, réuni ou par correspondance ; le Président de l'Union ; le Secrétaire général ou, à défaut de la possibilité de recourir à l'une de ces autorités ou de leur disponibilité, un des Vice-Présidents.

XV.   **Clauses finales**

XV.25.   Ces Statuts entrent en vigueur le 1er août 1991.

XV.26.   Les présents Statuts sont publiés en versions française et anglaise. En cas d'incertitude, la version française fait seule autorité.

b) The Finance Committee examines the accounts of the Union from the point of view of responsible expenditure within the intent of the previous General Assembly, and it considers whether the proposed budget is adequate to implement the policy of the General Assembly, as interpreted by the Executive Committee. It submits reports on these matters to the General Assembly for the approval of the account and decision on the budget.

c) Each Adhering Country pays annually to the Union a number of units of contribution according to its category. The number of units of contribution for each category shall be specified in the By-laws.

d) Associate Countries pay annually one unit of contribution.

e) The amount of the unit of contribution is determined by the General Assembly, on the proposal of the Executive Committee and with the advice of the Finance Committee.

f) The payment of contributions is the responsability of the adhering bodies. The liability of each Adhering Country in respect of the Union is limited to the amount of that country's dues to the Union.

g) An Adhering Country that ceases to adhere to the Union resigns at the same time its rights to a share in the assets of the Union.

XIII.   **Dissolution of the Union**

XIII.23. The decision to dissolve the Union is only valid if taken with the approval of three quarters of the votes of the Adhering Countries having the right to vote by virtue of article 15(a).

XIV.   **Emergency Powers**

XIV.24.   If, through events outside the control of the Union, circumstances arise in which it is impracticable to comply with the provisions of these Statutes and of the By-laws drawn up by the General Assembly, the organs and officers of the Union, in the order specified below, shall take such actions as they deem necessary for the continued operation of the Union. Such action shall be reported to a higher authority immediately this becomes practicable until such time as an extraordinary General Assembly can be convened. The following is the order of authority:

The General Assembly; an extraordinary General Assembly; the Executive Committee in meeting or by correspondence; the President of the Union; the General Secretary; or failing the practicability or availability of any of the above, one of the Vice-Presidents.

XV.   **Final Clauses**

XV.25.   These Statutes enter into force on 1 August 1991.

XV.26.   The present Statutes are being published in French and English version. In case of doubt, the French version is the only authority.

REGLEMENT

I.   **Les Membres de l'Union**

I. 1.   Les demandes d'adhésion des pays à l'Union Astronomique Internationale (ci-après dénommée l'Union) sont examinées par le Comité exécutif et soumises à l'approbation de l'Assemblée générale.

I. 2.   Les propositions de modifications de la liste des Membres sont, après examen attentif des suggestions des Présidents de Commissions, soumises pour avis au Comité des Nominations, composé d'un représentant de chaque Pays adhérent désigné par l'organisme adhérent habilité, avant la décision du Comité exécutif.

I. 3.   Les Commissions peuvent, avec l'approbation du Comité exécutif, coopter des consultants qu'elles jugent en mesure d'apporter une contribution utile à leur travail. L'adhésion des consultants a pour terme le dernier jour de la première Assemblée générale ordinaire qui suit leur admission, à moins qu'elle ne soit renouvelée.

I. 4.   Une organisation affiliée peut participer au travail de l'Union dans les conditions fixées par accord entre l'organisation et le Comité exécutif.

II.   **L'Assemblée générale**

II. 5.   L'Union se réunit en Assemblée générale ordinaire régulièrement une fois tous les trois ans. Si le lieu et la date de l'Assemblée générale ordinaire n'ont pas été décidés lors de la précédente Assemblée générale, ils sont fixés par le Comité exécutif et communiqués aux organismes adhérents au moins six mois à l'avance.

II. 6.   Le Président peut convoquer, avec l'accord du Comité exécutif, une Assemblée générale extraordinaire. Il est tenu de le faire à la demande du tiers des Pays adhérents.

II. 7.   L'Ordre du Jour de chaque Assemblée générale ordinaire est arrêté par le Comité exécutif et communiqué aux Organismes adhérents au moins quatre mois avant le premier jour de la réunion. Il devra inclure la proposition du Comité exécutif concernant le montant de l'unité de cotisation qui permet l'application de l'article 24.

II. 8.

a) L'Ordre du Jour doit inclure toute motion ou proposition reçue par le Secrétaire général au moins cinq mois avant le premier jour d'une Assemblée générale ordinaire, qu'elle émane d'un organisme adhérent, d'une Commission de l'Union ou d'une Commission mixte dans laquelle l'Union est représentée.

BY - LAWS

I.   **Membership**

I. 1.   Applications of countries for adherence to the International
           Astronomical Union (referred to as the Union) are examined by
           the Executive Committee and submitted to the General Assembly for
           approval.

I. 2.   Proposed changes in the list of Members are, with due regard to
           the suggestions of the Presidents of Commissions, submitted for
           advice to the Nominating Committee, consisting of one representative
           of each Adhering Country designated by the appropriate adhering
           body, before decision by the Executive Committee.

I. 3.   Commissions may, with the approval of the Executive Committee, co-
           opt consultants whom they consider may contribute to their work.
           The adherence of consultants expires on the last day of the ordinary
           General Assembly next following their admission, unless renewed.

I. 4.   An affiliated organization may participate in the work of the Union
           as mutually agreed between the organization and the Executive
           Committee.

II.   **General Assembly**

II. 5.   The Union meets in ordinary General Assembly, as a rule, once every
           three years. The place and date of the ordinary General Assembly
           unless determined by the General Assembly at its previous meeting,
           shall be fixed by the Executive Committee and communicated to
           the adhering bodies at least six months beforehand.

II. 6.   The President, with the consent of the Executive Committee, may
           summon an Extraordinary General Assembly. He must do so at
           the request of one third of the Adhering Countries.

II. 7.   The agenda of business for each ordinary General Assembly is
           determined by the Executive Committee and is communicated to
           the adhering bodies at least four months before the first day of
           the meeting. It shall include the proposal of the Executive
           Committee in regard to the unit of contribution as called for in
           article 24.

II. 8.

           a) Any motion or proposal received by the General Secretary at least
              five months before the first day of an ordinary General Assembly,
              whether from an adhering body, from a Commission of the Union,
              from an Inter-Union Commission on which the Union is represented,
              must be placed on the agenda.

b) Une motion ou proposition concernant l'administration ou le budget de l'Union qui ne figure pas à l'Ordre du Jour, préparé par le Comité exécutif, ou tout amendement à une motion qui figure à l'Ordre du Jour, ne peut être discuté qu'avec l'accord préalable des deux tiers au moins des voix des Pays adhérents représentés à l'Assemblée générale et disposant du droit de vote en vertu de l'article 15(a) des Statuts.

II. 9.   S'il y a doute sur le caractère administratif ou scientifique d'une question donnant lieu à un vote, l'avis du Président est prépondérant.

II.10.   En cas de partage égal des voix, le Président a voix prépondérante.

II.11.   Le Président peut inviter des représentants d'autres organisations, des scientifiques et de jeunes astronomes à participer à l'Assemblée générale. Avec l'accord du Comité exécutif, il peut déléguer ce privilège au Secrétaire général en ce qui concerne les représentants d'autres organisations, aux organismes adhérents en ce qui concerne les scientifiques et les jeunes astronomes.

III.   **Le Comité spécial des Nominations**

III.12.

a) Les propositions pour les élections du Président de l'Union, du Président-elect, des six Vice-Présidents, du Secrétaire général et du Secrétaire général adjoint sont soumises à l'Assemblée générale par le Comité spécial des Nominations. Ce Comité se compose du Président en fonction et du Président sortant, d'un membre proposé par le Comité exécutif sortant et n'appartenant ni au Comité exécutif actuel ni au Comité exécutif précédent, et de quatre membres élus par le Comité des Nominations parmi douze membres proposés par les Présidents de Commissions. A l'exception du Président en fonction et du Président sortant, les membres actuels et les anciens membres du Comité exécutif ne doivent pas faire partie du Comité spécial des Nominations. Les membres du Comité spécial des Nominations doivent tous appartenir à des pays différents.

b) Le Secrétaire général et le Secrétaire général adjoint participent au travail du Comité spécial des Nominations à titre consultatif.

c) Le Comité spécial des Nominations est nommé par l'Assemblée générale et est responsable directement devant elle. Il reste en fonction jusqu'à la fin de l'Assemblée générale ordinaire qui suit immédiatement sa nomination, et il peut combler toute vacance survenant parmi ses membres.

b) A motion or proposal concerning the admininstration of budget of the Union which does not appear on the agenda prepared by the Executive Committee, or any amendment to a motion that appears on the agenda, shall only be discussed with the prior approval of at least two thirds of the votes of Adhering Countries represented at the General Assembly and having the right to vote by virtue of Statute 15(a).

II. 9. If there is doubt as to the administrative or scientific character of a question giving rise to a vote, the President determines the issue.

II.10. Where there is an equal division of votes, the President determines the issue.

II.11. The President may invite representatives of other organizations, scientists and young astronomers to participate in the General Assembly. Subject to the agreement of the Executive Committee he may delegate this privilege concerning representatives of other organizations to the General Secretary, and concerning scientists and young astronomers to the adhering bodies.

III. **Special Nominating Committee**

III.12.

a) Proposals for elections to the President of the Union, a President-elect, six Vice-Presidents, the General Secretary and the Assistant General Secretary are submitted to the the General Assembly by the Special Nominating Committee. This consists of the President and past President of the Union, a member proposed by the retiring Executive Committee, and four members elected by the Nominating Committee from among twelve Members proposed by Presidents of Commissions. Other than the President and immediate past President, present and former members of the Executive Committee shall not serve on the Special Nominating Committee. No two members of the Special Nominating Committee shall belong to the same country.

b) The General Secretary and the Assistant General Secretary participate in the work of the Special Nominating Committee in an advisory capacity.

c) The Special Nominating Committee is appointed by the General Assembly to which it reports direct. It remains in office until the end to the ordinary General Assembly next following that of its appointment, and it may fill any vacancy occurring among its members.

IV. **Le Comité exécutif et ses Membres**

IV.13.

    a) Le Président de l'Union reste en fonction jusqu'à la fin de l'Assemblée générale ordinaire qui suit immédiatement celle de son élection ; les Vice-Présidents restent en fonction jusqu'à la fin de la deuxième Assemblée générale ordinaire qui suit celle de leur élection. Ils ne sont pas rééligibles immédiatement pour les mêmes fonctions.

    b) Le Secrétaire général et le Secrétaire général adjoint restent en fonction jusqu'à la fin de l'Assemblée générale ordinaire qui suit immédiatement celle de leur élection. Normalement, le Secrétaire général adjoint succède au Secrétaire général, mais l'un et l'autre peuvent être réélus aux mêmes fonctions pour une seconde période consécutive.

    c) Les élections ont lieu au cours de la dernière réunion de l'Assemblée générale, les noms des candidats proposés ayant été annoncés au cours d'une réunion antérieure.

IV.14. Le Président sortant et le Secrétaire général sortant deviennent conseillers du Comité exécutif jusqu'à la fin de l'Assemblée générale ordinaire qui suit immédiatement celle de la fin de leur mandat. Ils participent au travail du Comité exécutif et assistent à ses réunions sans droit de vote.

IV.15. Le Comité exécutif peut combler toute vacance survenant en son sein. Toute personne ainsi nommée reste en fonction jusqu'à l'Assemblée générale ordinaire suivante.

IV.16. Le Comité exécutif peut rédiger et publier des Directives pour expliciter les Statuts et le Règlement.

IV.17. Le Comité exécutif nomme le représentant de l'Union qui doit siéger au sein du Conseil International des Unions Scientifiques ; si ce représentant n'est pas déjà un membre élu du Comité exécutif, il devient conseiller.

IV.18.

    a) Le Secrétaire général est responsable auprès du Comité exécutif des dépenses qu'il engage, qui ne doivent pas dépasser le montant des fonds mis à sa disposition.

    b) Un bureau administratif, sous la direction du Secrétaire général, est chargé de la correspondance, de la gestion des fonds de l'Union, et de la conservation des archives.

V. **Commissions**

V.19.

    a) Les Commissions de l'Union poursuivent les buts scientifiques de l'Union par des moyens tels que l'étude de domaines particuliers de l'Astronomie, l'encouragement de recherches collectives et la discussion de questions relatives aux accords internationaux et à la standardisation.

    b) Les Commissions de l'Union établissent des rapports sur les sujets qui leur ont été confiés.

IV.   Officers and Executive Committee

IV.13.

a) The President of the Union remains in office until the end of the ordinary General Assembly next following that of his election; the Vice-Presidents remain in office until the end of the second ordinary General Assembly following that of their election. They may not be re-elected immediately to the same offices.

b) The General Secretary and the Assistant General Secretary remain in office until the end of the ordinary General Assembly next following that of their election. Normally the Assistant General Secretary succeeds the General Secretary though both officers may be re-elected for another term.

c) The election takes place at the last session of the General Assembly, the names of the candidates proposed having been announced at a previous session.

IV.14.   The retiring President and the retiring General Secretary become advisers to the Executive Committee until the end of the ordinary General Assembly next following that of their retirement. They participate in the work of the Executive Committee and attend its meetings without voting right.

IV.15.   The Executive Committee may fill any vacancy occurring among its members. Any person so appointed remains in office until the next ordinary General Assembly.

IV.16.   The Executive Committee may draw up and publish Working Rules to implement the Statutes and By-laws.

IV.17.   The Executive Committee appoints the Union's representative to the International Council of Scientific Unions; if not already an elected member of the Executive Committee, this representative will become its adviser.

IV.18.

a) The General Secretary is responsible to the Executive Committee for not incurring expenditure in excess of the funds at his disposal.

b) An Administrative office, under the direction of the general Secretary, conducts the correspondence, administers the funds, and preserves the archives of the Union.

V.   Commissions

V.19.

a) The Commissions of the Union shall pursue the scientific objects of the Union by activities such as the study of special branches of astronomy, the encouragement of collective investigations, and the discussion of questions relating to international agreements or to standardization.

b) The Commissions of the Union shall prepare reports on the work with which they are concerned.

F18

V.20.   Chaque Commission se compose :

a) d'un Président et au moins un Vice-Président élus par l'Assemblée générale sur la proposition du Comité exécutif. Ils demeurent en fonction jusqu'à la fin de l'Assemblée générale ordinaire qui suit immédiatement celle de leur élection. Ils ne sont pas normalement rééligibles,

b) d'un Comité d'Organisation, dont les membres sont désignés par la Commission sous réserve de l'approbation du Comité exécutif. Le Comité d'Organisation assiste le Président et le(s) Vice-Président(s) dans leur tâche. Une Commission peut décider qu'elle n'a pas besoin de Comité d'Organisation,

c) des membres de l'Union, nommés par les Présidents, Vice-Président(s) et Comité d'Organisation, en considération de leurs spécialités ; leur désignation est soumise à confirmation par le Comité exécutif.

V.21.   Entre deux Assemblées ordinaires, les Présidents de Commissions peuvent coopter, parmi les Membres de l'Union, de nouveaux membres des Comités d'Organisation et des Commissions elles-mêmes.

V.22.   Les Commissions rédigent leur propre règlement. Les décisions sont prises, à l'intérieur des Commissions, par un vote de leurs membres et elles deviennent d'application après approbation par le Comité exécutif.

VI.   **Organismes adhérents**

VI.23.   Le rôle des Organismes adhérents est d'encourager et de coordonner, sur leurs territoires respectifs, l'étude des diverses branches de l'astronomie, particulièrement en ce qui concerne leurs besoins sur le plan international. Ils ont le droit de soumettre au Comité exécutif des propositions pour discussion par l'Assemblée générale.

VII.   **Finances**

VII.24.   Chaque Pays adhérent verse à l'Union une cotisation annuelle, qui est multiple de l'unité de cotisation en fonction de sa catégorie, comme suit :

Catégories définies conformément à l'Article 8 des Statuts :

I   II   III   IV   V   VI   VII   VIII   VIII½   IX   X

Nombre respectif d'unités de contribution :

1   2   4   6   10   14   20   27   30   35   45

Si des catégories d'adhésion doivent être ajoutées ultérieurement, le pas du nombre d'unités sera de 10 unités par catégorie.

V.20.   Each Commission consists of:

a) a President and at least one Vice-President elected by the General Assembly on the proposal of the Executive Committee. They remain in office until the end of the ordinary General Assembly next following that of their election. They are not normally re-eligible,

b) an Organizing Committee, whose members are appointed by the Commission subject to the approval by the Executive Committee. The Organizing Committee assists the President and Vice-President(s) in their duties. A Commission may decide that it needs no Organizing Committee,

c) Members of the Union, appointed by the President, Vice-President(s) and the Organizing Committee, in consideration of their special interests; their appointment is subject to the confirmation by the Executive Committee.

V.21.   Between two ordinary General Assemblies, Presidents of Commissions may co-opt, from among Members of the Union, new members to the Organizing Committees and to the Commissions themselves.

V.22.   Commissions draw up their own rules. Decisions within Commissions are taken according to the vote of their members, and they become effective once they are approved by the Executive Committee.

VI.   **Adhering Bodies**

VI.23.   The functions of the Adhering Bodies are to promote and co-ordinate, in their repective territories, the study of the various branches of astronomy, more especially in relation to their international requirements. They are enlitled to submit to the Executive Committee motions for discussions by the General Assembly.

VII.   **Finances**

VII.24.   Each Adhering Country pays annually to the Union a number of units of contribution corresponding to its category as follows:

Category as defined in Statute 8:

I	II	III	IV	V	VI	VII	VIII	VIII½	IX	X

Number of units of contribution:

1	2	4	6	10	14	20	27	30	35	45

If further Categories of Adherence are required in the future, the step in the number of units shall be 10 units/category.

VII.25.   Les ressources de l'Union sont consacrées à la poursuite de ses buts, y compris :

a) les frais de publication et les dépenses administratives ;

b) l'encouragement des activités astronomiques qui nécessitent la coopération internationale ;

c) la cotisation due par l'Union au Conseil International des Unions Scientifiques.

VII.26.   Les ressources provenant de dons sont utilisées par l'Union en tenant compte des voeux exprimés par les donateurs.

VIII.   **Publications**

VIII.27.   L'Union a la propriété littéraire de tous les textes imprimés dans ses publications, sauf accord différent.

VIII.28.   Les Membres de l'Union ont le droit de recevoir les publications de l'Union gratuitement ou à prix réduit, à la discrétion du Comité exécutif qui décide en fonction de la situation financière de l'Union.

IX.   **Clauses finales**

IX.29.   Ce règlement entre en vigueur le 1er août 1991. Il peut être modifié avec l'approbation de la majorité absolue des voix des Pays adhérents qui disposent du droit de vote en vertu de l'article 15(a) des Statuts.

IX.30.   Le présent règlement est publié en versions française et anglaise. En cas d'incertitude, la version française fait seule autorité.

VII.25.   The income of the Union is to be devoted to its objects, including

   a) costs of publication and expenses of administration;

   b) the promotion of astronomical enterprises requiring international co-operation;

   c) the contribution due from the Union to the International Council of Scientific Unions.

VII.26.   Funds derived from donations are used by the Union in accordance with the wishes expressed by the donors.

VIII.   **Publications**

VII.27.   The Union has copyright to all materials printed in its publications, unless otherwise arranged.

VII.28.   Members of the Union are entitled to receive the publications of the Union free of charge or at reduced prices at the discretion of the Executive Committee taking due regard of the financial situation of the Union.

IX.   **Final Clauses**

IX.29.   These By-laws enter into force on 1 August 1991. They çan be changed with the approval of an absolute majority of the votes of the Adhering Countries having the right to vote by virtue of Statute 15(a).

IX.30.   The present By-laws are being published in French and English versions. In case of doubt, the French version is the only authority.

DIRECTIVES

I.   **Publications**

I. 1.   Les publications de l'Union Astronomique International,
       approuvées dans le budget par l'Assemblée générale, sont
       préparées par le Bureau administratif l'Union.

I. 2.   Les Commissions de l'Union peuvent, avec l'approbation du
       Comité exécutif, avoir leurs propres publications.

I. 3.   Le Comité exécutif décide, sur la proposition du Secrétaire
       général, des modalités de distribution des publications de
       l'Union.

I. 4.   Les Membres de l'Union peuvent acquérir les publications de
       l'Union à un prix réduit.

II.   **Appartenance à l'Union**

       A. **Pays adhérents**

II. 5. Les demandes d'adhésion à l'Union formulées par les pays sont
       examinées par le Comité exécutif compte tenu des points
       suivants :

       a) justesse du choix de la catégorie dans laquelle le pays
          souhaite être classé ;

       b) situation actuelle de l'Astronomie dans le pays formulant
          la demande, et ses possibilités de développement ;

       (c) mesure dans laquelle le futur organisme adhérent est
          représentatif de l'activité astronomique de son pays.

II. 6. Les demandes proposant une contribution annuelle appropriée
       seront soumises pour décision à l'Assemblée générale, avec
       la recommandation du Comité exécutif.

       B. **Membres**

II. 7. Les personnes proposées pour devenir Membres de l'Union doivent
       en principe être choisies parmi des astronomes et
       des chercheurs dont les activités sont liées à l'astronomie,
       compte tenu de :

       a) la qualité de leur oeuvre scientifique ;

       b) la mesure dans laquelle leur activité scientifique implique
          des recherches astronomiques ;

       c) leur désir de contribuer à la poursuite des buts de l'Union.

II. 8. Les jeunes astronomes doivent être considérés comme pouvant
       devenir Membres de l'Union dès qu'ils ont fait la preuve de
       leur capacité (en principe par une thèse de doctorat ou son
       équivalent) et de leur aptitude (quelques années d'activité
       fructueuse) à mener une recherche personnelle.

II. 9. Pour les astronomes professionnels, leur contribution à
       l'astronomie peut consister soit en des recherches
       personnelles, soit en une collaboration assidue à
       des programmes importants d'observations.

WORKING RULES

I.   **Publications**

I.1.   The publications of the International Astronomical Union, approved in the budget by the General Assembly, are prepared by the Administrative Office of the Union.

I.2.   Commissions of the Union may, with the approval of the Executive Committee, issue their publications independently.

I.3.   The distribution of publications of the Union is decided, on the proposal of the General Secretary, by the Executive Committee.

I.4.   Members may purchase the publications of the Union at reduced prices.

II.   **Membership**

A. Adhering Countries

II.5.   Applications of countries for adherence to the Union are examined by the Executive Committee for:

a) the adequacy of the category in which the country wishes to be classified;

b) the present state and expected development of astronomy in the applying country;

c) the degree to which the prospective adhering body is representative of its country's astronomical activity.

II.6.   Applications proposing an adequate annual contribution to the Union shall, with the recommendation of the Executive Committee, be submitted to the General Assembly for decision.

B. Members

II.7.   Individuals proposed for Union Membership should, as a rule, be chosen from among astronomers and scientists, whose activity is closely linked with astronomy taking into account:

a) the standard of their scientific achievement;

b) the extent to which their scientific activity involves research in astronomy;

c) their desire to assist in the fulfilment of the aims of the Union.

II.8.   Young astronomers should be considered eligible for membership after they have shown their capability (as a rule Ph.D. or equivalent) of and experience (some years of successful activity) in conducting original research.

II.9.   For full time professional astronomers the achievement in astronomy may consist either of original research or of substantial contributions to major observational programs.

II.10. Les autres personnes ne peuvent devenir Membres de l'Union que si certains de leurs travaux originaux concernent étroitement la recherche astronomique.

II.11. Huit mois avant une Assemblée générale ordinaire, il sera demandé aux organismes adhérents de proposer de nouveaux Membres. Les propositions devront parvenir au Secrétaire général au moins cinq mois avant la première session de l'Assemblée générale. Les propositions reçues après cette date limite ne seront prises en considération que si des circonstances exceptionnelles justifient le retard.

II.12. Chaque proposition de nouveau Membre doit être présentée séparément et indiquer le nom, les prénoms et l'adresse postale du candidat (de préférence celle de son Institut ou Observatoire), ses date et lieu de naissance, l'Université devant laquelle il a soutenu sa thèse ou le diplôme équivalent, la date de soutenance, la situation actuelle du candidat, les titres et renseignements bibliographiques de deux ou trois de ses articles ou publications les plus significatifs et, s'il y a lieu, tous les renseignements susceptibles d'être pris en considération par le Comité des Nominations.

II.13.

a) Les Présidents de Commissions qui désirent suggérer de nouveaux membres doivent adresser leurs suggestions au Secrétaire général au moins cinq mois avant la première session d'une Assemblée générale ordinaire. Les propositions devront fournir les mêmes renseignements que ceux mentionnés à l'article 12.

b) Le Secrétaire général fait part de ces suggestions aux organismes adhérents intéressés.

II.14. Le Secrétaire général préparera deux listes pour le Comité des Nominations

a) l'une contenant les noms des candidats proposés par les organismes adhérents,

b) l'autre contenant les noms des candidats proposés par les Présidents de Commissions, mais qui ne sont pas déjà inclus dans les propositions des organismes adhérents.

II.15. A partir des deux listes mentionnées à l'article 14, le Comité des Nominations prépare les propositions définitives de nouveaux membres de l'Union.

II.16. Les organismes adhérents peuvent proposer la radiation de Membres ayant abandonné le domaine de l'astronomie pour d'autres activités, à moins qu'ils ne continuent à apporter une contribution à l'astronomie. Ces propositions doivent être portées à la connaissance du Secrétaire général et du Membre concerné.

II.17. Le Secrétaire général publiera la liste alphabétique des Membres de l'Union dans les Transactions de chaque Assemblée générale ordinaire.

II.10.   Others are eligible for membership only if they are making original contributions closely linked with astronomical research.

II.11.   Eight months before an ordinary General Assembly, adhering bodies will be asked to propose new Members. The proposals should reach the General Secretary not later than five months before the first session of the General Assembly. Proposals received after the closing date will only be taken into consideration if the delay is justified by exceptional circumstances.

II.12.   Each proposal shall be written seprarately. It should include the name, first names and postal address of the candidate, preferably that of his/her Institute or Observatory, his/her place and date of birth, the University and the year of his/her Ph.D. or equivalent title, his/her present occupation, titles and bibliographic data of two or three of his/her more important papers or publications, and details, if any, worthy to be considered by the Nominating Committee.

II.13.
          a) Presidents of Union Commissions wishing to suggest new Members for admittance should address their suggestions to the General Secretary five months before the first session of an ordinary General Assembly. The proposals should contain particulars as in article 12.
          b) The General Secretary notifies the adhering bodies in questions of such suggestions.

II.14.   The General Secretary shall prepare two lists for the Nominating Committee:
          a) one containing the candidates proposed by the adhering bodies;
          b) the other containing those suggested by President of Commissions, but not included among the proposals of the adhering bodies.

II.15.   The Nominating Committee prepares the final proposals for Union membership from the two lists as mentioned in article 15.

II.16.   Adhering Bodies should propose cancellation of Members who have left the field of astronomy for other interests, unless they continue to contribute to astronomy. Such proposals should be announced to the Member concerned and to the General Secretary.

II.17.   The alphabetical list of Union Members will be published by the General Secretary in the Transactions of each ordinary General Assembly.

III.   **Membres des Commissions**

III.18.   Les membres des Commissions de l'Union sont cooptés par les Commissions ; Cette procédure est régie par des règles établies par les Commissions elles-mêmes.

III.19.   Les Commissions devraient choisir, ou approuver, la liste des membres de leurs commissions compte tenu de la spécialité de ces personnes, en particulier de leur activité scientifique dans le domaine de recherche de la Commission, et leur contribution au travail de la Commission. Elles peuvent

a) inviter les Membres de l'Union à devenir membres de la Commission,

b) radier les membres de la Commission qui n'ont pas contribué à son activité,

c) accepter ou refuser les demandes présentées par des Membres de l'Union, ou par des personnes proposées comme tels, en vue d'appartenir à la Commission,

d) suggérer l'élection comme Membres de l'Union de personnes n'y appartenant pas, ce qui leur permettrait alors de devenir membres de la Commission.

III.20.   Les Membres de l'Union ne peuvent pas, en règle générale, appartenir à plus de trois Commissions.

III.21.   Les Membres de l'Union peuvent demander à être admis dans une Commission en écrivant au Président de cette Commission. Ils ne devraient faire cette demande que si leur propre activité rentre dans le cadre des recherches de la Commission et s'ils sont décidés à contribuer au travail de la Commission.

III.22.   Les membres des Commissions peuvent se retirer d'une Commission en écrivant à son Président.

III.23.   En envoyant leur propositions de nouveaux Membres, les organismes adhérents peuvent également suggérer le choix d'une Commission pour chaque candidat.

III.24.   Le Secrétaire général enregistrera et analysera la liste des membres des Commissions ; si cela est nécessaire, il tentera de trouver une solution aux anomalies évidentes.

III.25.   Le Secrétaire général publiera la liste des membres des Commissions dans les Transactions de chaque Assemblée générale ordinaire.

IV.   **Consultants**

IV.26.   Peuvent être élus Consultants des personnes qui ne sont pas astronomes, mais qui sont susceptibles de servir les intérêts de l'astronomie.

IV.27.   Les Commissions doivent en principe envoyer, pour approbation, leurs propositions de consultants au Secrétaire général au moins cinq mois avant la première session d'une Assemblée générale ordinaire.

III.   Commission Membership

III.18.   Members of Union Commissions are co-opted by Commissions. The rules governing the procedure of such co-option are drawn up by the Commissions themselves.

III.19.   Commissions should choose, or approve of, Commission members taking into account their special interests, in particular their scientific activity in the appropiate fields of research and their contribution to the work of the Commission. They may:

    a) invite Members to become members of their Commission;

    b) remove members who have not contributed to the work of the Commission;

    c) accept or reject applications for membership from existing or proposed Members;

    d) suggest non-members for election as Members, thus enabling them to become members of the Commission

III.20.   Members may not, as a rule, be members of more than three Commissions.

III.21.   Members may apply for Commission membership by writing to the President of the Commission concerned. Such applications should only be made if the Member is actively engaged in the appropriate field of research and is prepared to contribute to the work of the Commission.

III.22.   Members of Commissions may resign from a Commission by writing to its President.

III.23.   Adhering Bodies, in sending in their proposals for new Members, may also suggest one Commission for each candidate.

III.24.   The General Secretary will record and analyse the lists of members of Commissions, if necessary he will try to resolve any outstanding anomalies.

III.25.   The list of Commission members will be published by the General Secretary in the Transactions of each ordinary General Assembly.

IV.   Consultants

IV.26.   Eligible as Consultants are non-astronomers in a position to further the interest in astronomy.

IV.28.   Le Secrétaire général préparera une liste des personnes proposées comme consultants et la soumettra pour approbation au Comité exécutif.

IV.29.   Le Bureau administratif établira une liste alphabétique des consultants.

IV.30.   Les consultants peuvent participer aux réunions de l'Union. Ils peuvent avoir droit de vote dans leurs Commissions respectives. Ils reçoivent gratuitement le Bulletin d'Information de l'Union.

V.   Réunions Scientifiques

V.31.   Le Secrétaire général publiera un règlement pour les réunions scientifiques organisées ou parrainées par l'Union.

VI.   Contacts Extérieurs

VI.32.   Aucune relation avec des tiers, imputable à l'Union, ne sera entreprise par quiconque membre de l'Union, si ce n'est sous l'autorité du Secrétaire général.

VI.33.   Les représentants de l'Union dans d'autres organisations, en particulier les Comités de l'ICSU et les Commissions Inter-Unions, seront désignés par le Comité exécutif. Les noms sont proposés par les Présidents des Commissions concernées.

VI.34.   Les dépenses encourues par les représentants de l'Union dans d'autres organisations seront remboursées à la discrétion du Secrétaire général, dans les limites du Budget adopté par l'Assemblée générale. Les représentants sont priés d'obtenir l'accord préalable du Secrétaire général avant d'engager ces dépenses.

VII.   Assemblées générales

VII.35.   Huit mois avant l'Assemblée générale, le Secrétaire général envoie aux Comités Nationaux d'Astronomie et aux Organisations adhérentes le budget préparé par le Comité exécutif, pour commentaires.

IV.27. Proposals of Commissions for the approval of consultants should, as a rule, reach the General Secretary not later than five months before the first session of an ordinary General Assembly.

IV.28. The General Secretary shall prepare a list of those proposed for admission as consultants and submit it to the Executive Committee for approval.

IV.29. The Administrative Office will maintain an alphabetical list of consultants.

IV.30. Consultants may participate in the meetings of the Union. They may have voting right in the respective Commission. They receive, free of charge, the Information Bulletin of the Union.

V. **Scientific Meetings**

V.31. The General Secretary shall publish rules for scientific meetings organized or sponsored by the Union.

VI. **External Contacts**

VI.32. No dealings with third parties, attributable to the Union, shall be undertaken by any Member of the Union except on the authority of the General Secretary.

VI.33. Representatives of the Union in other bodies, especially ICSU Committees and ICSU Inter-Union Committees, shall be appointed by the Executive Committee. Nominations are sought from Presidents of appropriate Commissions.

VI.34. Expenses incurred by Representatives of the Union in other bodies will be reimbursed at the discretion of the General Secretary, within the provisions of the Budget Estimate adopted by the General Assembly. Representatives are required to obtain prior approval of the General Secretary before incurring such expenses.

VII. **General Assemblies**

VII.35. The General Secretary distributes the budget prepared by the Executive Committee to National Committees of Astronomy and/or Adhering Organizations for comments eight months before the General Assembly.

CHAPTER VII

MEMBERSHIP

EXPRESSION OF INTENT

In the application of Statute 5, it is the current intention of the Executive Committee that, Individual Members of the Union in countries that cease participation retain full individual membership.

This chapter is composed as follows:

1.   List of adhering countries[*]

2.   Membership of commissions

3.   Geographical repartition[**] of members
          . within adhering countries
          . within non adhering countries

4.   Alphabetical list of members

[*]   The situation of countries which is published corresponds to the situation as of December 31, 1991.

[**]   This repartition is done on the basis of the country in which IAU members exercise their work and not their citizenship.

1.   List of Adhering Organizations

The year of adherence and approximate number of IAU members
residing in the different Adhering Countries are indicated (as of
December 1991).

(*) Associate National Members as opposed to Full National
    Members (Article 4 of the Statutes).

USSR (which joined the Union in 1935) is replaced by the
following states in which individual IAU members are working:

	Members
Armenia	18
Azerbaidzhan	8
Estonia	11
Georgia	18
Kazakhstan	9
Latvia	9
Lithuania	3
Russia	264
Tajikistan	4
Ukraine	84
Uzbekistan	5

	Year of Adherence	Members
Algeria Commissariat à la Recherche (CRAAG) BP 63 Bouzareah Alger	1988	1
Argentina Presidencia de la Nacion Secretaria de Ciencia/Tecnologia Consejo Ncl de Investigaciones Cientificas y Técnicas (CONICET) Rivadavia 1917 1033 Buenos Aires	1927	65
Australia Australian Academy of Sciences Attn Executive Secretary Box 783 Canberra City. ACT 2601	1939	173

Austria	Bundestministerium für Wissenschaft und Forschung Minoritenplatz 5 1010 Wien	1955	29
Belgium	Administration Affaires Communes ETS Scientifiques de L'Etat Bd Pacheco 6e Etage B 1000 Brussels	1920	81
Brazil	Conselho Ncl Desenvolvimento Científico/Tecnológico -CNPq Av W3 Norte Quadra 507 B Caixa Postal 11-1142 70740 Brasilia DF	1961	87
Bulgaria	Bulgarian Sciences Academy 7 November Street 1 1000 Sofia	1957	48
Canada	National Research Council International Affairs Montreal Road Ottawa Ontario K1A 0R6	1957	206
Chile	Dr. Claudio Anguita, Chairman Chilean Ntl Astronomy Cttee Univ. Chile, Faculd. Ciencias Casilla 36D Santiago de Chile	1947	43
China Nanjing	Chinese Astronomical Society Purple Mountain Observatory Academia Sinica Nanjing	1935	292
China Taipei	Vice President Academia Sinica Taiwan Taipei 11529	1959	19
Colombia	Mrs. M. A. Velasco Facultad de Sciencias Universidad Ncl de Colombia Apartado Aereo 5997 Observatorio Astronomico Ncl Bogota D E	1967	2

Cuba		1970	6
	Academia Ciencias de Cuba		
	Capitolio Ncl		
	La Habana		
Czechoslovakia		1922	94
	Czech & Slovak National IAU Committee		
	Observatory Ondrejov		
	251 65 Ondrejov		
Denmark		1922	50
	Kge Danske Vidensk Selskab		
	H C Andersen Boulevard 35		
	DK 1553 København V		
Egypt AR		1925	38
	Academy Scientific Research		
	Scientific & Cultural Relations		
	101 Kasr El-Einy Street		
	Cairo		
Finland		1948	29
	Delegation of the Finnish Academy		
	of Sciences & Letters		
	Mariankatu 5		
	SF 00170 Helsinki		
France		1920	561
	Académie des Sciences		
	COFUSI		
	23, quai Conti		
	F 75006 Paris		
Germany		1951	429
	President, German Council		
	of German Observatories		
	Universitäts Sternwarte		
	Geismarlandstrasse 11		
	D 3400 Göttingen		
Greece		1920	89
	Academy of Athens		
	28 Panepistimiou Street		
	GR 106 79 Athens		
Hungary		1947	36
	Hungarian Academy of Sciences		
	Box 6		
	Roosevelt Ter 9		
	H 1361 Budapest V		
Iceland		1988	3
	Ministry of Education		
	Science Institute		
	Universtiy of Iceland		
	Hverfisgötu 4-6		
	IS 101 Reykjavik		

India	1964	208

Indian Ntl Science Academy
Bahadur Shah Zafar Marg
New Delhi 110002

Indonesia	1979	9

Scientific and Technologic Cooperation
Indonesian Institute of Sciences
Lembaga Ilmu Pengetahuan Indones
Jl. Jend. Gatot Subroto No. 10
Jakarta Selatan

Iran	1969	9

University of Tehran
Office of International Relation
Tehran

Iraq	1976	8

Council for Sc Research
Astronomy & Space Res Center
Box 255
Baghdad

Ireland	1947	26

The Royal Irish Academy
19 Dawson Street
Dublin 2

Israel	1954	44

Israel Sciences Academy and Humanities
Albert Einstein Square
Talbieh
Jerusalem 91040

Italy	1920	373

Consiglio Nzle Ricerche
Servizio Relazioni Internl
Piazzale Aldo Moro 7
I 00100 Roma

Japan	1920	350

Science Council of Japan
22-34 Roppongi 7 chome
Minato-ku
Tokyo 106

Korea DPR	1961	21

Academy of Sciences of DPRK
Pyongyang

Korea RP	1973	32

Korean Astronomical Society
Dpt of Astronomy
College of Natural Sciences
Seoul National University
Seoul 151

Malaysia (*)		1988	5
National Space Science Education Center			
Prime Minister's Department			
c/o Islamic Center			
Jalan Perdana			
50519 Kuala Lumpur			
Mexico		1921	55
Instituto de Astronomia UNAM			
Apartado Postal 70 264			
Cd. Universitaria			
Mexico 04510 DF			
Morocco		1988	1
CNCPRST			
52 Charia Omar Ibn Khattab			
B.P. 1346 RP			
Agdal			
Netherlands		1922	162
Koninklijke Nederlandse Akademie			
van Wetenschappen			
Kloveniersburgwal-29			
NL 1011 JV Amsterdam			
New Zealand		1964	22
Mrs. S M Usher			
Assistant Executive Officer			
Royal Society of New Zealand			
Box 598			
Wellington			
Norway		1922	20
Norske Videnskaps-Akademi			
Drammensveien 78			
N 0271 Oslo 2			
Peru (*)		1988	1
Ing Carlos del Rio Cabrera			
President			
Camilo Carrillo 118 9 Piso			
Lima 11			
Poland		1922	94
Polskiej Akademii Nauk			
Palac Kultury I Nauki			
Skrytka pocztowa 24			
00 901 Warsaw			
Portugal.		1924	16
Secçao Portuguesa Unioes Intern.			
Astronomica/Geodesica/Geofisica			
SPUIAGG			
Praça de Estrela			
P 1200 Lisboa			

Saudi Arabia	1988	6
King Abdulaziz City		
for Science & Technology		
Directorate of Technology Transfer		
and International Cooperation		
PO Box 6086		
Riyadh 11442		
South Africa	1938	35
South African ICSU Secretariat		
Foundation for Research Dvlpt		
Box 2600		
Pretoria 0001		
Spain	1922	161
Comisión Ncl Astronomia		
Instituto Géografico/Cadastral		
General Ibañez 3		
E Madrid 3		
Sweden	1925	84
The Foreign Secretary		
The Royal Swedish Academy of Sciences		
Box 50005		
S 104 05 Stockholm		
Switzerland	1923	55
Schweizerische Akademie		
der Naturwissenschatfe		
Bärenplatz 2		
Postfach 8120		
CH 3001 Bern		
Turkey	1961	47
Türk Astronomi Dernegi Baskani		
The Dean of the Faculty		
of Arts & Sciences		
Middle East Technical University		
06531 Ankara		
UK	1920	481
The Royal Society		
ICSU Information Officer		
6 Carlton House Terrace		
London SW1Y 5AG		
Uruguay	1970	1
Ministerio Relaciones Exteriores		
Avenida 18 de Julio 1205		
Montevideo		
USA	1920	2069
Office of International Affairs		
National Academy of Sciences		
FO 2020		
2101 Constitution Avenue NW		
Washington DC 20418		

Vatican City State   1932   5
        Governatorato Citta Vaticano
        Castel Gandolfo 3
        V 00120 Citta del Vaticano

Venezuela   1953   7
        Centro de Investigaciones
        de Astronomia CIDA
        Apartado Postal 264
        Merida 5101 A

Yugoslavia   1935   39
        Savez Drustava Matematicara,
        Fisica i Astronoma Jugoslavije
        Institut za Matermatiku i Fisiku
        Cetinjski put bb
        8100 Titograd

        Total members in Adhering Countries   7260

N.B.   Nigeria and Rumania have been proposed the statute of Associate
        Member.

        Total members in non Adhering Countries   41

        Total membership of the IAU   7301

2. Membership of commissions

Number of Members

COMPOSITION OF COMMISSION 04 1991-1994

President   : YALLOP BERNARD D DR

Vice-President(s) : KINOSHITA HIROSHI DR

Organizing Committee: ABALAKIN VICTOR K DR
                       CHAPRONT JEAN DR
                       DOGGETT LEROY E DR
                       KUBO YOSHIO
                       MORANDO BRUNO L DR
                       SCHWAN HEINER DR
                       SEIDELMANN P KENNETH DR
                       STANDISH E MYLES DR
                       TONG FU

Members:

AOKI SHINKO PROF	ARIAS DE GREIFF J PROF	ARLOT JEAN-EUDES
BANDYOPADHYAY A PROF	BEC-BORSENBERGER ANNICK	BHATNAGAR ASHOK KUMAR
BRETAGNON PIERRE DR	BRUMBERG VICTOR A DR	CAPITAINE NICOLE
CATALAN MANUEL DR	CHAPRONT-TOUZE MICHELLE	CHOLLET FERNAND DR
COMA JUAN CARLOS	DAVIES MERTON E MR	DE CASTRO ANGEL DR
DEPRIT ANDRE PROF	DI XIAO-HUA	DICKEY JEAN O'BRIEN
DUNCOMBE RAYNOR L DR	DUNHAM DAVID W	FIALA ALAN D DR
FOMINOV ALEXANDR M DR	FUKUSHIMA TOSHIO DR	FURSENKO M A DR
GLEBOVA NINA I DR	GONDOLATSCH FRIEDRICH PRF	HE MIAO-FU
HENRARD JACQUES PROF	ILYAS MOHAMMAD DR	JANICZEK PAUL M DR
JOHNSTON KENNETH J	KAPLAN GEORGE H DR	KING ROBERT WILSON JR DR
KLEPCZYNSKI WILLIAM J DR	KOLACZEK BARBARA DR	KRASINSKY GEORGE A DR
LAHIRI N C	LASKAR JACQUES DR	LEDERLE TRUDPERT DR
LEHMANN MAREK DR	LI GI MAN	LI HYOK HO
LI NENG-YAO	LIESKE JAY H DR	LIU BAO-LIN
MAJID ABDUL BIN A H DR	MORRISON LESLIE V DR	MUELLER IVAN I PROF
NEWHALL X X DR	OESTERWINTER CLAUS	O'HANDLEY DOUGLAS A DR
REASENBERG ROBERT D DR	ROMERO PEREZ M PILAR	ROSSELLO GASPAR
SALAZAR ANTONIO DR	SHAPIRO IRWIN I PROF	SHIRYAEV ALEXANDER A DR
SIMON JEAN-LOUIS MR	SINZI AKIRA M DR	SOCHILINA ALLA S DR
SOMA MITSURU DR	TING YEOU-TSWEN	TUNG NGUYEN MAU DR
VAN FLANDERN THOMAS DR	WACKERNAGEL H BEAT DR	WIELEN ROLAND PROF DR
WILKINS GEORGE A DR	WILLIAMS CAROL A	WILLIAMS JAMES G DR
WINKLER GERNOT M R DR	XIAN DING-ZHANG	YAMAZAKI AKIRA DR
ZAMBRANO ALEJANDRO DR		

COMPOSITION OF COMMISSION 05 1991-1994

President   : HAUCK BERNARD PROF

Vice-President(s) : DLUZHNEVSKAYA O B DR

Organizing Committee: CREZE MICHEL DR
                       DUBOIS PASCAL DR
                       GROSBOL PREBEN JOHNSON DR
                       HECK ANDRE DR
                       LI QI-BIN
                       MEAD JAYLEE MONTAGUE DR
                       SCHMADEL LUTZ D DR
                       TURNER KENNETH C DR
                       WARREN WAYNE H JR DR
                       WESTERHOUT GART DR
                       WILKINS GEORGE A DR

Members:

ABALAKIN VICTOR K DR	ABT HELMUT A DR	ALBRECHT MIGUEL A DR
ALVAREZ PEDRO DR	ANDERNACH HEINZ DR	A'HEARN MICHAEL F DR
BAKER NORMAN H PROF	BENACCHIO LEOPOLDO	BENN CHRIS R DR
BESSELL MICHAEL S DR	BIDELMAN WILLIAM P PROF	BOUSKA JIRI DR
CHU YAOQUAN	COGAN BRUCE C DR	COLUZZI REGINA DR
DAVIS MORRIS S PROF	DAVIS ROBERT J DR	DE BOER KLAAS SJOERDS DR
DEWHIRST DAVID W DR	DICKEL HELENE R DR	DIXON ROBERT S DR
DUCATI JORGE RICARDO DR	DUNCOMBE RAYNOR L DR	EGRET DANIEL DR
GARSTANG ROY H PROF	GRIFFIN ROGER F DR	GUIBERT JEAN DR
HANISCH ROBERT J DR	HARVEL CHRISTOPHER ALVIN	HEFELE HERBERT PH D
HEINRICH INGE	HEINTZ WULFF D DR	HELOU GEORGE DR
HUANG BI-KUN	JASCHEK CARLOS O R PROF	JENKNER HELMUT DR
KADLA ZDENKA I DR	KALBERLA PETER	KAPLAN GEORGE H DR
KHARIN A S DR	KLECZEK JOSIP DR	KRISTENSEN LEIF KAHL
LANTOS PIERRE DR	LEDERLE TRUDPERT DR	LEQUEUX JAMES DR
LINDE PETER DR	LIU JINMING	LONSDALE CAROL J DR
LORTET MARIE CLAIRE	LYNGA GOSTA DR	LYUBIMKOV LEONID S DR
MATZ STEVEN MICHEAL DR	MCLEAN BRIAN JOHN	MCNALLY DEREK DR
MCNAMARA DELBERT H DR	MEADOWS A JACK PROF	MEIN PIERRE
MERMILLIOD JEAN-CLAUDE DR	MITTON SIMON DR	NISHIMURA SHIRO DR
OCHSENBEIN FRANCOIS DR	PAMYATNIKH A A DR	PASINETTI LAURA E PROF
PATUREL GEORGES	PECKER JEAN-CLAUDE PROF	PHILIP A G DAVIS
PIZZICHINI GRAZIELLA	POLECHOVA PAVLA DR	PUCILLO MAURO DR
QUINTANA HERNAN DR	RAIMOND ERNST DR	RATNATUNGA KAVAN U
REMY BATTIAU LILIANE G A	RENSON P F M DR	RIEGLER GUENTER R DR
ROESSIGER SIEGFRIED DR	ROMAN NANCY G DR	RUSSO GUIDO DR
SARASSO MARIA DR	SCHILBACH ELENA DR	SCHLUETER A PROF DR
SCHMIDT K H DR	SEDMAK GIORGIO PROF	SERRANO ALFONSO DR
SHAKESHAFT JOHN R DR	SHCHERBINA-SAMOJLOVA I DR	SOKOLSKY ANDREJ G DR
SPITE FRANCOIS M DR	TERASHITA YOICHI PROF	TRITTON SUSAN BARBARA
TSVETKOV MILCHO K DR	UESUGI AKIRA DR	WALLACE PATRICK T MR
WAYMAN PATRICK A PROF	WEIDEMANN VOLKER PROF	WELLS DONALD C III DR
WENGER MARC	WIELEN ROLAND PROF DR	WORLEY CHARLES E DR
WRIGHT ALAN E DR	WU ZHIREN DR	

COMPOSITION OF COMMISSION 06 1991-1994

President   : GRINDLAY JONATHAN E DR

Vice-President(s) : WEST RICHARD M DR

Organizing Committee: ISOBE SYUZO DR
                       MARSDEN BRIAN G DR
                       ROEMER ELIZABETH PROF

Members:

AKSNES KAARE DR	BIRAUD FRANCOIS DR	CANDY MICHAEL P MR
FILIPPENKO ALEXEI V DR	GILMORE ALAN C MR	HERS JAN MR
KOZAI YOSHIHIDE PROF	LIU JINMING	MRKOS ANTONIN DR
NAKANO SYUICHI	POUNDS KENNETH A PROF	ROSINO LEONIDA PROF
SHAROV A S DR	THOLEN DAVID J DR	

COMPOSITION OF COMMISSION 07 1991-1994

President   : DEPRIT ANDRE PROF

Vice-President(s) : FERRAZ-MELLO S PROF DR

Organizing Committee: BHATNAGAR K B DR

                       CHAPRONT JEAN DR

                       FROESCHLE CLAUDE DR

                       HE MIAO-FU

                       HENRARD JACQUES PROF

                       KHOLSHEVNIKOV K V DR

                       KINOSHITA HIROSHI DR

                       LIESKE JAY H DR

                       MILANI ANDREA

                       PEALE STANTON J PROF

                       ROY ARCHIE E PROF

                       SOKOLSKY ANDREJ G DR

Members:

ABAD ALBERTO J DR	ABALAKIN VICTOR K DR	AHMED MOSTAFA DR
AKIM EFRAIM L DR	AKSENOV E P PROF DR	AKSNES KAARE DR
ALEXANDER MURRAY E DR	ALTAVISTA CARLOS A DR	ANTONACOPOULOS GREG PROF
AOKI SHINKO PROF	BAGHOS BALEGH B DR	BALMINO GEORGES G DR
BARBERIS BRUNO	BATRAKOV YU V DR	BEC-BORSENBERGER ANNICK
BENEST DANIEL DR	BETTIS DALE G PROF	BOIGEY FRANCOISE
BORDERIES NICOLE	BOZIS GEORGE PROF	BRANHAM RICHARD L JR
BRETAGNON PIERRE DR	BRIEVA EDUARDO PROF	BROOKES CLIVE J DR
BROUCKE ROGER DR	BRUMBERG VICTOR A DR	BRUNINI ADRIAN DR
CALAME ODILE DR	CANDY MICHAEL P MR	CARANICOLAS NICHOLAS DR.
CARPINO MARIO DR	CEFOLA PAUL J DR	CHAPRONT-TOUZE MICHELLE
CHEN ZHEN	CHOI KYU-HONG	CID PALACIOS RAFAEL PROF
CONTOPOULOS GEORGE PROF	COOK ALAN H PROF	COUNSELMAN CHARLES C PROF
CUI DOU-XING	DANBY J M ANTHONY DR	DAVIS MORRIS S PROF
DEMIN V G PROF DR	DIKOVA SMILIANA D	DOGGETT LEROY E DR
DORMAND JOHN RICHARD DR	DOURNEAU GERARD DR	DROZYNER ANDRZEJ
DUNCOMBE RAYNOR L DR	DURIEZ LUC DR	DVORAK RUDOLF DR
EDELMAN COLETTE DR	EICHHORN HEINRICH K DR	EL BAKKALI LARBI DR
ELIPE SANCHEZ ANTONIO	EMELIANOV NIKOLAJ V DR	ERDI B DR
FARINELLA PAOLO DR	FERNANDEZ SILVIA M DR	FERRER MARTINEZ SEBASTIAN
FIALA ALAN D DR	FONG CHU-GANG	GALIBINA I V DR
GALLETTO DIONIGI	GAPOSCHKIN EDWARD M DR	GARFINKEL BORIS DR
GASKA STANISLAW DR	GIACAGLIA GIORGIO E PROF	GOLDREICH P DR
GOMES RODNEY D S DR	GONZALEZ CAMACHO ANTONIO	GOUDAS CONSTANTINE L PROF
GREBENIKOV E A PROF DR	GREENBERG RICHARD DR	GROUSHINSKY N P PROF DR
HADJIDEMETRIOU JOHN D	HAMID S EL DIN PROF	HANSLMEIER ARNOLD
HEGGIE DOUGLAS C DR	HELALI YHYA E DR	HENON MICHEL C DR
HORI GENICHIRO PROF	HUANG CHENG DR	HUANG TIANYI
IVANOVA VIOLETA DR	IZVEKOV V A DR	JANICZEK PAUL M DR

JEFFERYS WILLIAM H DR   JOURNET ALAIN   JOVANOVIC BOZIDAR
JUPP ALAN H DR   KAMMEYER PETER C DR   KATSIS DEMETRIUS DR
KAULA WILLIAM M PROF   KING-HELE DESMOND G DR   KLOKOCNIK JAROSLAV DR
KNEZEVIC ZORAN   KOVALEVSKY JEAN DR   KOZAI YOSHIHIDE PROF
KRASINSKY GEORGE A DR   KUSTAANHEIMO PAUL E PROF   LALA PETR DR
LASKAR JACQUES DR   LAZOVIC JOVAN P PROF   LEMAITRE ANNE DR
LIAO XINHAO DR   LISSAUER JACK J DR   LU BEN-KUI
LUNDQUIST CHARLES A DR   MACIEJEWSKI ANDRZEJ J DR   MAGNARADZE NINA G DR
MARCHAL CHRISTIAN DR   MARKELLOS VASSILIS V DR   MARSDEN BRIAN G DR
MATAS VLADIMIR R DR   MAVRAGANIS A G PROF   MEIRE RAPHAEL
MELBOURNE WILLIAM G DR   MERMAN G A DR   MESSAGE PHILIP J DR
MIGNARD FRANCOIS DR   MIKKOLA SEPPO DR   MIOC VASILE DR
MOONS MICHELE B M M   MORANDO BRUNO L DR   MULHOLLAND J DERRAL DR
MUSEN PETER DR   MYACHIN VLADIMIR F DR   NACOZY PAUL E DR
NAHON FERNAND PROF   NOBILI ANNA M   NOSKOV BORIS N DR
NOVOSELOV V S PROF DR   OESTERWINTER CLAUS   OMAROV TUKEN B PROF
ORELLANA ROSA BEATRIZ DR   ORUS JUAN J PROF   OSORIO JOSE J S P PROF
O'HANDLEY DOUGLAS A DR   PAL ARPAD PROF DR   PAUWELS T DR
PETIT JEAN-MARC DR   PETROVSKAYA M S DR   PIERCE A KEITH DR
POPOVIC BOZIDAR PROF DR   ROBINSON WILLIAM J DR   RODRIGUEZ-VILLAMIL R DR
RYABOV YU A PROF DR   SAGNIER JEAN-LOUIS DR   SANSATURIO MARIA E DR
SCHOLL HANS DR   SCHUBART JOACHIM DR   SCONZO PASQUALE DR
SEGAN STEVO   SEHNAL LADISLAV DR   SEIDELMANN P KENNETH DR
SEIN-ECHALUCE M LUISA DR   SESSIN WAGNER DR   SHAPIRO IRWIN I PROF
SHARAF SH G DR   SIDLICHOVSKY MILOS DR   SIMA ZDISLAV DR
SIMON JEAN-LOUIS MR   SINCLAIR ANDREW T DR   SIRY JOSEPH W
SKRIPNICHENKO VLADIMIR DR SOFFEL MICHAEL DR   STANDISH E MYLES DR
STELLMACHER IRENE DR   SULTANOV G F ACAD   SUN YI-SUI
SZEBEHELY VICTOR G PROF   TABORDA JOSE ROSA DR   TATEVYAN S K DR
TAWADROS MAHER JACOUB DR   TAYLOR DONALD BOGGIA DR   THIRY YVES R PROF
TONG FU   VALSECCHI GIOVANNI B DR   VALTONEN MAURI J PROF
VARVOGLIS H DR   VASHKOV'YAK SOF'YA N DR   VEILLET CHRISTIAN
VIEIRA MARTINS ROBERTO DR VILHENA DE MORAES R DR   WALCH JEAN-JACQUES
WALKER IAN WALTER   WHIPPLE ARTHUR L DR   WILLIAMS CAROL A
WNUK EDWIN   WU LIAN-DA   XU JI-HONG DR
XU PINXIN   YAROV-YAROVOJ M S DR   YI ZHAO-HUA
YOKOYAMA TADASHI DR   YOSHIDA HARUO   YOSHIDA JUNZO PROF
YUASA MANABU DR   ZAFIROPOULOS BASIL DR   ZARE KHALIL DR
ZHANG SHENG-PAN   ZHENG JIA-QING   ZHENG XUE-TANG
ZHOU HONG-NAN   ZHU WEN-YAO

COMPOSITION OF COMMISSION 08 1991-1994

President   : MORRISON LESLIE V DR

Vice-President(s) : SMITH CLAYTON A JR DR

Organizing Committee: BENEVIDES SOARES P DR
                       CORBIN THOMAS ELBERT DR
                       HELMER LEIF
                       HU NING-SHENG
                       KOVALEVSKY JEAN DR
                       LINDEGREN LENNART DR
                       LOPEZ JOSE A ING
                       MIYAMOTO MASANORI DR
                       NOEL FERNANDO
                       POLOZHENTSEV DIMITRIJ DR
                       REQUIEME YVES DR
                       SCHWAN HEINER DR
                       YOSHIZAWA MASANORI DR

Members:

ANGUITA CLAUDIO A DR	ARGYLE ROBERT WILLIAM MR	ARGYRAKOS JEAN PROF DR
BACCHUS PIERRE PROF	BACKER DONALD CH DR	BAGILDINSKIJ BRONISLAV K
BEM JERZY DR	BIEN REINHOLD DR	BILLAUD GERARD J
BOUGEARD MIREILLE L DR	BROUW W N DR	BUCCIARELLI BEATRICE DR
BYKOV MIKLE F DR	CARESTIA REINALDO A DR	CARRASCO GUILLERMO DR
CATALAN MANUEL DR	CHA DU JIN	CHAMBERLAIN JOSEPH M DR
CHERNEGA N A A DR	CHIUMIENTO GIUSEPPE	CHLISTOVSKY FRANCA DR
CHOLLET FERNAND DR	COSTA EDGARDO DR	COUNSELMAN CHARLES C PROF
CRIFO FRANCOISE DR	DE VEGT CH PROF DR	DEBARBAT SUZANNE V DR
DEJAIFFE RENE J DR	DICK STEVEN J	DJUROVIC DRAGUTIN M DR
DRAVSKIKH A F DR	DUMA DMITRIJ P DR	DUNCOMBE RAYNOR L DR
EICHHORN HEINRICH K DR	EINICKE OLE H DR	FABRICIUS CLAUS V
FEDOROVA RIMMA T DR	FEISSEL MARTINE DR	FOMIN VALERY A DR
FROESCHLE MICHEL DR	FUJISHITA MITSUMI DR	GAO BUXI
GAUSS F STEPHEN	GRUDLER PIERRE	GUBANOV VADIM S DR
GULYAEV A P DR	HEINTZ WULFF D DR	HEMENWAY PAUL D DR
HOEG ERIK DR	HUA YING-MIN	JACKSON PAUL DR
JIANG CHONG-GUO	JOHNSTON KENNETH J	JOURNET ALAIN
KAPLAN GEORGE H DR	KHARIN A S DR	KLOCK B L DR
KOKURIN YURIJ L DR	KONIN V V DR	KOSIN GENNADIJ S DR
KURZYNSKA KRYSTYNA DR	LACLARE F MR	LACROUTE PIERRE A PROF
LATTANZI MARIO G	LEDERLE TRUDPERT DR	LEHMANN MAREK DR
LI DONG-MING	LI NENG-YAO	LI ZHIGANG
LI ZHI-FANG	LOYOLA PATRICIO DR	LU CHUN-LIN
LUO DING-JIANG	MANRIQUE WALTER T PROF	MAO WEI
MAVRIDIS L N PROF	MELCHIOR PAUL J PROF DIR	MITIC LJUBISA A DR
MUINOS JOSE L DR	MURRAY C ANDREW	NAKAJIMA KOICHI DR
NEFEDEVA ANTONINA I PROF	NEMIRO ANDREJ A DR PROF	NIKOLOFF IVAN DR

OLSEN FOGH H J .... OSORIO JOSE J S P PROF .... PAKVOR IVAN

PERRYMAN MICHAEL A C DR .... PETROV G M DR .... PHAM-VAN JACQUELINE MME

PILOWSKI K PROF DR .... PINIGIN GENNADIJ I DR .... POLNITZKY GERHARD DR

POMA ANGELO DR .... PROVERBIO EDOARDO PROF .... PUGLIANO ANTONIO PROF

QI GUAN RONG .... QIAN ZHI-HAN DR .... QUIJANO LUIS

RAIMOND ERNST DR .... REIZ ANDERS PROF .... REYNOLDS JOHN DR

ROESER SIEGFRIED DR .... ROUSSEAU JEAN-MICHEL MR .... RUSSELL JANE L DR

RUSU I DR .... SADZAKOV SOFIJA DR .... SALETIC DUSAN

SANCHEZ MANUEL .... SARASSO MARIA DR .... SATO KOICHI DR

SCHMEIDLER F PROF DR .... SEVARLIC BRANISLAV M PROF .... SHEN KAIXIAN

SHI GUANG-CHEN .... SIMS KENNETH P DR .... SOEDERHJELM STAFFAN DR

SOLARIC NIKOLA .... SOMA MITSURU DR .... SPOELSTRA T A TH DR

STANGE LOTHAR .... STONE RONALD CECIL .... TAFF LAURENCE G DR

THOBURN CHRISTINE .... THOMAS DAVID V DR .... TURON C DR

VAN ALTENA WILLIAM F PROF .... VAN LEEUWEN FLOOR DR .... VON DER HEIDE JOHANN DR

WALLACE PATRICK T MR .... WALTER HANS G DR .... WESTERHOUT GART DR

WIELEN ROLAND PROF DR .... XIA YI-FEI .... XIE LIANGYUN

XU BANG-XIN .... XU TONG-QI .... YAMAZAKI AKIRA DR

YASUDA HARUO PROF DR .... YATSKIV YA S DR .... YE SHU-HUA

YU KYUNG-LOH PROF .... ZHANG HUI

COMPOSITION OF COMMISSION 09 1991-1994

President   : BHATTACHARYYA J C PROF

Vice-President(s) : LELIEVRE GERARD DR

Organizing Committee: CULLUM MARTIN DR
                      DAVIS JOHN PROF
                      HUMPHRIES COLIN M DR
                      MCLEAN IAN S DR
                      MERKLE FRITZ DR
                      TANGO WILLIAM J DR
                      WEST RICHARD M DR

Members:

ABLES HAROLD D DR	AI GUOXIANG	AIME C DR
ALBRECHT RUDOLF DR	ALVAREZ PEDRO DR	APARICI JUAN DR
ARNAUD JEAN PAUL	ASHOK N M DR	ASSUS PIERRE DR
ATHERTON PAUL DAVID	BABA NAOSHI DR	BABCOCK HORACE W DR
BAFFA CARLO DR	BAO KEREN	BARANNE A DR
BARCIA ALBERTO DR	BARROSO JR JAIR	BARUCH JOHN DR
BARWIG HEINZ	BAUM WILLIAM A DR	BECKLIN ERIC E DR
BEER REINHARD DR	BENSAMMAR SLIMANE DR	BINGHAM RICHARD G DR
BLITZSTEIN WILLIAM DR	BONANNO GIOVANNI DR	BONNEAU DANIEL
BORGNINO JULIEN DR	BOYCE PETER B DR	BRAULT JAMES W DR
BRECKINRIDGE JAMES B DR	BREJDO IZABELLA I DR	BRIDGELAND MICHAEL DR
BURTON W BUTLER DR	CAO CHANGXIN	CHELLI ALAIN
CHRISTY JAMES WALTER DR	CLARKE DAVID DR	COHEN RICHARD S
COOKE JOHN ALAN	CORNEJO ALEJANDRO A DR	CRAWFORD DAVID L DR
CURRIE DOUGLAS G DR	DALL'OGLIO GIORGIO DR	DAN XHI-XIANG
DESAI JYOTINDRA N	DIEGO FRANCISCO DR	DOBRONRAVIN PETER DR
DOKUCHAEVA OLGA D DR	DOUGLAS NIGEL DR	DRAVINS DAINIS PROF
DREHER JOHN W	DUCHESNE MAURICE DR	DUNKELMAN LAWRENCE
EDWIN ROGER P	ENGELS DIETER DR	ENGVOLD ODDBJOERN DR
FABRICANT DANIEL G	FEHRENBACH CHARLES PROF	FELENBOK PAUL DR
FLETCHER J MURRAY	FOMENKO ALEXANDR F DR	FORD W KENT JR DR
FORT BERNARD P DR	FOY RENAUD DR	FU DELIAN
GALAN MAXIMINO J	GAO BILIE	GAUSS F STEPHEN
GAY JEAN DR	GIBSON DAVID MICHAEL DR	GILLINGHAM PETER MR
GLASS IAN STEWART DR	GONG SHOU-SHEN	GRAY PETER MURRAY
GRIFFITHS RICHARD E DR	GRIGORJEV VICTOR M DR	GROSBOL PREBEN JOHNSON DR
GRUNDMANN WALTER	GUIBERT JEAN DR	GUTCKE DIETRICH
HADLEY BRIAN W	HALLAM KENNETH L DR	HAMMERSCHLAG ROBERT H DR
HANISCH ROBERT J DR	HAO YUN-XIANG	HARMER CHARLES F W MR
HARMER DIANNE L MRS	HECKATHORN HARRY M	HEUDIER JEAN-LOUIS DR
HEWITT ANTHONY V DR	HILLIARD R DR	HONEYCUTT R KENT PROF
HOOGHOUDT B G IR	HOUGH JAMES DR	HU JING-YAO
HU NING-SHENG	HUANG TIE-QIN	HYSOM EDMUND J
ILYAS MOHAMMAD DR	JAYARAJAN A P MR	JEFFERS STANLEY DR

JENKNER HELMUT DR / JIANG SHI-YANG / JONES BARBARA
JOSEPH CHARLES LYNN DR / KARACHENTSEV I D DR / KARPINSKIJ VADIM N DR
KIPPER TONU DR / KISSELL KENNETH E DR / KLOCK B L DR
KLOCOK LUBOMIR DR / KOEHLER H PROF DR / KOEHLER PETER
KOPYLOV I M DR / KOROVYAKOVSKIJ YURIJ P DR / KOVACHEV B J DR
KREIDL TOBIAS J N / KUEHNE CHRISTOPH F / KULKARNI PRABHAKAR V PROF
LABEYRIE ANTOINE DR / LAQUES PIERRE DR / LASKER BARRY M DR
LEMAITRE GERARD R DR / LI DEPEI / LI TING
LI ZHIGANG / LIVINGSTON WILLIAM C / LOCHMAN JAN
LU RUWEI DR / LYNCH DAVID K / MACK PETER DR
MAHRA H S DR / MAILLARD JEAN-PIERRE DR / MAJOR JOHN DR
MALIN DAVID F MR / MALKAMAEKI LAURI J DR / MARTINS DONALD HENRY DR
MATZ STEVEN MICHEAL DR / MCGREGOR PETER JOHN DR / MCMULLAN DENNIS DR
MEGEVAND DENIS DR / MEINEL ADEN B PROF / MENG XINMIN
MERTZ LAWRENCE N DR / MIKHELSON NIKOLAJ N DR / MILLIKAN ALLAN G MR
MINAROVJECH MILAN / MORGAN BRIAN LEALAN / MORRIS MICHAEL C
MORTON DONALD C DR / MURRAY STEPHEN S DR / NAKAI YOSHIHIRO
NELSON JERRY E DR / NEWTON GAVIN DR / NIEMI AIMO
NISHIMURA SHIRO DR / NUNES ROGERIO S DE SOUSA / ODGERS GRAHAM J DR
OHTSUBO JUNJI DR / O'DELL CHARLES R DR / PASIAN FABIO
PENNY ALAN JOHN DR / PERRYMAN MICHAEL A C DR / PETFORD A DAVID DR
PETROV PETER P DR / PICAT JEAN-PIERRE DR / PRITCHET CHRISTOPHER J DR
PROKOF'EVA VALENTINA V DR / PUCILLO MAURO DR / QIU PUZHANG ASS PROF
RACINE RENE DR / RAKOS KARL D PROF / RAMSEY LAWRENCE W DR
REAY NEWRICK K DR / REDFERN MICHAEL R DR / RICHARDSON E HARVEY DR
RING JAMES PROF / ROBERTSON NORNA DR / ROBINSON LLOYD B DR
RODDIER CLAUDE DR / RODDIER FRANCOIS PROF / ROSCH JEAN PROF
ROUNTREE JANET DR / RUDER HANNS / RUPPRECHT GERO DR
RUSCONI LUIGIA DR / RYLOV VALERIJ S DR / SAULT ROBERT DR
SAXENA A K DR / SCHROEDER DANIEL J PROF / SCHULTZ ALFRED BERNARD DR
SCHULTZ G V DR / SCHUMANN JOERG DIETER DR / SEDMAK GIORGIO PROF
SERVAN BERNARD / SHAKHBAZYAN YURIJ L DR / SHCHEGLOV P V DR
SHEN CHANGJUN / SHEN PARN-AN / SHIVANANDAN KANDIAH DR
SIM MARY E MISS / SLOVAK MARK HAINES DR / SMITH CHARLES DITTO
SMYTH MICHAEL J DR / SNEZHKO LEONID I / STESHENKO N V DR
STESHENKO N V DR / STOREY JOHN W V DR / SU DING-QIANG
SWINGS JEAN-PIERRE DR / TRAUB WESLEY ARTHUR / TUEG HELMUT DR
TULL ROBERT G / ULICH BOBBY LEE / VAKILI FARROKH DR
VALNICEK BORIS DR / VALTONEN MAURI J PROF / VAN CITTERS GORDON W DR
VELKOV KIRIL / VLADIMIROV SIMEON / VRBA FREDERICK J DR
WALKER ALISTAIR ROBIN DR / WALKER DAVID DOUGLAS DR / WALKER GORDON A H PROF
WALKER MERLE F PROF / WALLACE PATRICK T MR / WAMPLER E JOSEPH PROF
WANG LAN-JUAN / WANG YANAN / WANG YIMING
WANG ZHENG MING / WARD HENRY DR / WATSON FREDERICK GARNETT
WEISS WERNER W DR / WESTPHAL JAMES A PROF / WILCOCK WILLIAM L PROF
WINDHORST ROGIER A DR / WLERICK GERARD DR / WOEHL HUBERTUS DR
WORDEN SIMON P DR / WORSWICK SUSAN / WU LIN-XIANG
WYLLER ARNE A PROF / WYNNE CHARLES G PROF / YANG SHI JIE
YAO ZHENG-QIU / YE BINXUN / ZACHAROV IGOR DR
ZEALEY WILLIAM J DR / ZHANG XIU ZHONG / ZHANG YOUYI
ZHOU BIFANG DR / ZHU NENGHONG

COMPOSITION OF COMMISSION 10 1991-1994

President   : GAIZAUSKAS VICTOR DR

Vice-President(s) : ENGVOLD ODDBJOERN DR

Organizing Committee: AI GUOXIANG

                       ANTONUCCI ESTER DR

                       BENZ ARNOLD DR

                       DULDIG MARCUS LESLIE DR

                       FORBES TERRY G DR

                       GALAL A A PROF

                       HILDNER ERNEST DR

                       JAKIMIEC JERZY PROF

                       KARLICKY MARIAN

                       KIM IRADIA S

                       MACHADO MARCOS

                       PRIEST ERIC R PROF

                       SAKURAI TAKASHI DR

                       SCHMIEDER BRIGITTE DR

                       SCHUESSLER MANFRED DR

                       STEPANOV ALEXANDER V DR

                       TANDBERG-HANSSEN EINAR A

Members:

ABBASOV ALIK R DR	ABOUDARHAM JEAN DR	ABRAMI ALBERTO PROF
AHLUWALIA HARJIT SINGH DR	ALISSANDRAKIS C PH D	ALTROCK RICHARD C DR
ALTSCHULER MARTIN D PROF	ALY JEAN JACQUES DR	ALY M KHAIRY PROF
AMARI TAHAR DR	AMBASTHA A K DR	AMBROZ PAVEL DR
ANDERSEN BO NYBORG DR	ANDERSON KINSEY A PROF	ANTALOVA ANNA
ANTIOCHOS SPIRO KOSTA	ANZER ULRICH DR	ASCHWANDEN MARKUS DR
ATAC TAMER	ATHAY R GRANT DR	AURASS HENRY DR
AVIGNON YVETTE DR	BABIN ARTHUR DR	BABIN V G DR
BAGARE S P DR	BALLESTER JOSE LUIS DR	BALLI EDIBE PROF
BARROW COLIN H DR	BATCHELOR DAVID ALLEN DR	BECKERS JACQUES M DR
BEEBE HERBERT A	BELL BARBARA DR	BELVEDERE GAETANO DR
BERGER MITCHELL DR	BHATNAGAR ARVIND DR	BOCCHIA ROMEO DR
BOHN HORST-ULRICH	BOMMIER VERONIQUE DR	BONDAL KRISHNA RAJ DR
BORNMANN PATRICIA L DR	BOUGERET J L DR	BOYER RENE
BRANDENBURG AXEL DR	BRANDT PETER N	BRAY ROBERT J DR
BROWN JOHN C PROF	BROWNING PHILIPPA DR	BRUECKNER GUENTER E DR
BRUNER MARILYN E DR	BRUZEK ANTON DR	BUECHER ALAIN DR
BUECHNER JORG DR	BUMBA VACLAV DR	BUYUKLIEV GEORGI DR
CADEZ VLADIMIR	CALLY PAUL S DR	CANE HILARY VIVIEN
CARGILL PETER J DR	CARLQVIST PER A DR	CHAMBE GILBERT
CHANDRA SURESH DR	CHAPMAN GARY A DR	CHEN BIAO
CHEN ZHENCHENG	CHENG CHUNG-CHIEH DR	CHERNOV GENNADIJ DR
CHERTOPRUD V E DR	CHIUDERI-DRAGO FRANCA PR	CHIUEH TZIHONG DR
CHUPP EDWARD L DR	CLIVER EDWARD W	COFFEY HELEN E MS

COLLADOS MANUEL DR	COOK JOHN W	CORREIA EMILIA DR
COSTA JOAQUIM EDUARDO DR	COVINGTON ARTHUR E	CRAMER NEIL DR
CRANNELL CAROL JO DR	CSADA IMRE K DR	CULHANE LEONARD PROF
DATLOWE DAYTON DR	DAVILA JOSEPH DR	DEL TORO INIESTA JOSE DR
DENNIS BRIAN ROY DR	DERE KENNETH PAUL	DERMENDJIEV VLADIMIR DR
DEUBNER FRANZ-LUDWIG DR	DEZSO LORANT PROF	DIALETIS DIMITRIS DR
DING YOU-JI	DIZER MUAMMER PROF	DOLLFUS AUDOUIN PROF
DRYER MURRAY DR	DUBAU JACQUES DR	DUBOIS MARC A
DULK GEORGE A PROF	DUMITRACHE CHRISTANA	DUNCAN ROBERT A PROF
DUNN RICHARD B DR	DWIVEDI BHOLA NATH DR	EDDY JOHN A DR
ELSTE GUNTHER H DR	ELWERT GERHARD PROF	EMSLIE A GORDON
ENOME SHINZO PROF	FALCHI AMBRETTA	FALCIANI ROBERTO DR
FANG CHENG	FARNIK FRANTISEK	FEIBELMAN WALTER A DR
FENG KE-JIA	FISHER GEORGE HEWITT DR	FOING BERNARD H DR
FORTINI TERESA DR	FOSSAT ERIC DR	FRIEDMAN HERBERT DR
FU QI JUN	GABRIEL ALAN H	GALLOWAY DAVID DR
GARCIA DE LA ROSA JOSE I	GARY GILMER ALLEN DR	GELFREIKH GEORGIJ B DR
GERGELY TOMAS ESTEBAN DR	GIBSON DAVID MICHAEL DR	GILLILAND RONALD LYNN
GILMAN PETER A DR	GLATZMAIER GARY A	GLEISSBERG WOLFGANG PROF
GNEVYSHEVA RAISA S DR	GODOLI GIOVANNI PROF	GOEDBLOED JOHAN P PROF DR
GOKHALE MORESHWAR HARI PR	GOOSSENS MARCEL DR	GOPASYUK S I DR
GRANDPIERRE ATTILA DR	GU XIAO-MA	GURMAN JOSEPH B DR
GURTOVENKO E A DR	HAGEN JOHN P	HAGYARD MONA JUNE
HAMMER REINER	HANAOKA YOICHIRO DR	HANASZ JAN DR
HANSEN RICHARD T MR	HANSLMEIER ARNOLD	HARVEY JOHN W DR
HASAN SAIYID STRAJUL	HATHAWAY DAVID H DR	HAUG EBERHARD DR
HAYWARD JOHN	HEINZEL PETR DR	HENOUX JEAN-CLAUDE DR
HERMANS DIRK DR	HIEI EIJIRO DR	HIRAYAMA TADASHI PROF
HOLLWEG JOSEPH V	HOLMAN GORDON D	HOLZER THOMAS EDWARD DR
HONG HYON IK	HOOD ALAN	HOWARD ROBERT F DR
HOYNG PETER DR	HUANG YOU-RAN	HUDSON HUGH S DR
HURFORD GORDON JAMES	HYDER C L DR	IOSHPA B A DR
IVANCHUK VICTOR I DR	JAIN RAJMAL DR	JARDINE MOIRA MARY DR
JENSEN EBERHART PROF	JIANG YAO-TIAO	JOCKERS KLAUS DR
JONES HARRISON PRICE DR	JORDAN STUART D DR	JOSELYN JO ANN C DR
JOVANOVIC BOZIDAR	KAHLER STEPHEN W DR	KAI KEIZO DR
KALMAN BELA DR	KANE SHARAD R DR	KANG JIN SOK
KARPEN JUDITH T	KAUFMANN PIERRE PROF	KIM KAP-SUNG DR
KIPLINGER ALAN L DR	KJELDSETH-MOE OLAV DR	KLECZEK JOSIP DR
KLEIN KARL LUDWIG DR	KLIMCHUK JAMES A DR	KLVANA MIROSLAV
KNOSKA STEFAN	KOECKELENBERGH ANDRE DR	KOPECKY MILOSLAV DR
KOSTIK ROMAN I	KOSUGI TAKEO	KOTRC PAVEL
KOUTCHMY SERGE DR	KOVACS AGNES DR	KRAUSE F DR
KRIVSKY LADISLAV DR	KRUEGER ALBRECHT DR	KUBOTA JUN DR
KUENZEL HORST	KUKLIN G V DR	KUNDU MUKUL R DR
KUPERUS MAX PROF DR	KUROCHKA L N DR	KUROKAWA HIROKI DR
LANDMAN DONALD ALAN	LANG KENNETH R ASST PROF	LANTOS PIERRE DR
LEIBACHER JOHN DR	LEROY BERNARD DR	LEROY JEAN-LOUIS
LI CHUN-SHENG	LI SON JAE	LI WEI BAO
LIN YUANZHANG	LIRITZIS IOANNIS DR	LIU XINPING PROF
LIVSHITS M A DR	LOUGHHEAD RALPH E DR	LOW BOON CHYE
LUNDSTEDT HENRIK DR	LUO BAO-RONG	LUO XIANHAN
LUSTIG GUENTER DR	MACKINNON ALEXANDER L	MACQUEEN ROBERT M DR
MACRIS CONSTANTIN J PROF	MAKAROV VALENTINE I	MAKITA MITSUGU DR
MALHERBE JEAN MARIE DR	MALITSON HARRIET H MS	MALTBY PER PROF

MALVILLE J MCKIM PROF	MANN GOTTFRIED DR	MARIS GEORGETA DR
MARISKA JOHN THOMAS	MARTENS PETRUS C DR	MARTRES MARIE-JOSEPHE
MASON GLENN M	MATSUURA OSCAR T DR	MATTIG W PROF DR
MAXWELL ALAN DR	MCCABE MARIE K MS	MCINTOSH PATRICK S
MCKENNA LAWLOR SUSAN	MCLEAN DONALD J DR	MEERSON BARUCH DR
MEIN PIERRE	MELROSE DONALD B PROF	MENDES DA COSTA ARACY DR
MERGENTALER JAN PROF	MESSEROTTI MAURO DR	MICHALITSIANOS ANDREW
MOGILEVSKIJ EH I DR	MOISEEV I G DR	MONSIGNORI FOSSI BRUNA DR
MORENO-INSERTIS FERNANDO	MORETON G E	MORIYAMA FUMIO PROF
MOROZHENKO N N DR	MOTTA SANTO DR	MULLER RICHARD DR
MUSIELAK ZDZISLAW E DR	NAGASAWA SHINGO PROF	NAKAGAWA YOSHINARI DR
NAKAJIMA HIROSHI	NAMBA OSAMU DR	NEIDIG DONALD F DR
NELSON GRAHAM JOHN DR	NEUPERT WERNER M DR	NISHI KEIZO DR
NOCERA LUIGI DR	NOENS JACQUES-CLAIR DR	NOYES ROBERT W PROF
NUSSBAUMER HARRY PROF	OBRIDKO VLADIMIR N DR	OEKTEN ADNAN DR
OGIR MAYA DR	OHKI KENICHIRO DR	ORRALL FRANK Q PROF
OZGUC ATILA	PALLAVICINI ROBERTO DR	PALLE PERE-LLUIS DR
PALUS PAVEL DR	PAN LIANDE	PAP JUDIT
PARKINSON JOHN H DR	PARKINSON WILLIAM H DR	PASACHOFF JAY M PROF
PATERNO LUCIO PROF	PEDERSEN BENT M DR	PERES GIOVANNI DR
PETROSIAN VAHE PROF	PFLUG KLAUS DR	PHILLIPS KENNETH J H
PICK MONIQUE DR	PIDDINGTON JACK H RES FEL	PNEUMAN GERALD W
POLAND ARTHUR I DR	POLETTO GIANNINA PROF	POLUPAN P N DR
POQUERUSSE MICHEL	PORTER JASON G DR	PREKA-PAPADEMA P DR
PROKAKIS THEODORE J DR	RAADU MICHAEL A DR	RABIN DOUGLAS MARK
RAO A PRAMESH DR	RAOULT ANTOINETTE DR	RAY CHOUDHURI ARNAB DR
RAYROLE JEAN R DR	REES DAVID ELWYN DR	REEVES EDMOND M DR
REEVES HUBERT PROF	REGULO CLARA DR	REZENDE COSTA JOAQUIM DR
RIEGER ERICH DR	RIJNBEEK RICHARD DR	ROBINSON JR RICHARD D DR
ROCA CORTES TEODORO	ROEMER MAX PROF	ROMANCHUK PAVEL R DR
ROMPOLT BOGDAN DR	ROSCH JEAN PROF	ROUDIER THIERRY DR
ROXBURGH IAN W PROF	ROZELOT JEAN P	RUBASHEV BORIS M DR
RUEDIGER GNTHER DR	RUSIN VOJTECH	RUST DAVID M DR
RUZDJAK VLADIMIR DR	RUZICKOVA-TOPOLOVA B DR	RYBANSKY MILAN
RYUTOVA MARGARITA P DR	SAEMUNDSON THORSTEINN	SAHAL-BRECHOT SYLVIE DR
SAITO KUNIJI PROF	SAKURAI KUNITOMO PROF	SANCHEZ ALMEIDA JORGE DR
SAWYER CONSTANCE B DR	SCHATTEN KENNETH H DR	SCHINDLER KARL PROF DR
SCHLUETER A PROF DR	SCHMAHL EDWARD J DR	SCHMELZ JOAN T DR
SCHMIDT H U DR	SCHOBER HANS J DR	SCHRIJVER C J DR
SCHROETER EGON H PROF	SEMEL MEIR DR	SHAPLEY ALAN H
SHEA MARGARET A DR	SHEELEY NEIL R DR	SHI ZHONG-XIAN
SHIBASAKI KIYOTO	SHINE RICHARD A DR	SILBERBERG REIN DR
SIMNETT GEORGE M	SIMON GUY	SINHA K DR
SITNIK G F PROF	SLONIM E M DR	SMALDONE LUIGI ANTONIO
SMITH DEAN F DR	SMOL'KOV GENNADIJ YA DR	SOBOTKA MICHAL DR
SOLANKI SAMI K DR	SOLIMAN MOHAMED AHMED DR	SOMOV BORIS V DR
SOTIROVSKI PASCAL DR	SPADARO DANIELE DR	SPICER DANIEL SHIELDS DR
SPRUIT HENK C DR	STELLMACHER GOETZ	STENFLO JAN O DR
STEPANIAN N N DR	STESHENKO N V DR	STESHENKO N V DR
STEWART RONALD T MR	STIX MICHAEL DR	STOKER PIETER H
STRONG KEITH T DR	STURROCK PETER A PROF	SU QING-RUI DR
SUKARTADIREDJA DARSA	SUN KAI	SVESTKA ZDENEK DR
SYKORA JULIUS DR	SYLWESTER BARBARA DR	SYLWESTER JANUSZ
TAKAKURA TATSUO PROF EMER	TAKANO TOSHIAKI DR	TALON RAOUL DR
TAMENAGA TATSUO DR	TANDON JAGDISH NARAIN DR	TANG YU-HUA

TAPPING KENNETH F   TERNULLO MAURIZIO DR   TESKE RICHARD G PROF
THOMAS JOHN H PROF   THOMAS ROGER J DR   THOMPSON MICHAEL J DR
TIFREA EMILIA DR   TLAMICHA ANTONIN DR   TRELLIS MICHEL DR
TREUMANN RUDOLF A DR   TRITAKIS BASIL P DR   TROTTET GERARD DR
TSINGANOS KANARIS DR   TSUBAKI TOKIO PROF   TSUNETA SAKU DR
TUOMINEN ILKKA V DR   UNDERWOOD JAMES H DR   URPO SEPPO I
VALNICEK BORIS DR   VAN ALLEN JAMES A PROF   VAN DEN OORD BERT H J DR
VAN DRIEL-GESZTELYI L DR   VAN HOVEN GERARD DR   VAN'T VEER FRANS DR
VAUGHAN ARTHUR H DR   VECK NICHOLAS   VEKSTEIN GREGORY DR
VELKOV KIRIL   VELLI MARCO DR   VENKATESAN DORASWAMY DR
VERGEZ MADELEINE DR   VERHEEST FRANK PROF   VERMA V K DR
VIAL JEAN-CLAUDE   VILMER NICOLE DR   VINLUAN RENATO
VINOD S KRISHAN MRS DR   VITINSKIJ YURIJ I DR   VRSNAK BOJAN DR
VYALSHIN GENNADIJ F DR   WALDMEIER MAX PROF DR   WANG HAIMIN DR
WANG JIA-LONG   WANG JING-XIU   WANG YI-MING DR
WEBB DAVID F   WENTZEL DONAT G DR   WIEHR EBERHARD DR
WILD JOHN PAUL DR   WILSON PETER R PROF   WITTMANN AXEL D DR
WOEHL HUBERTUS DR   WOLFSON RICHARD DR   WOLTJER LODEWIJK PROF
WU LIN-XIANG   WU MING-CHAN   WU SHI TSAN DR
XANTHAKIS JOHN N PROF   XU AO-AO   XU ZHENTAO
YAKOVKIN N A DR   YAO JIN-XING   YE SHI-HUI
YEH TYAN DR   YOSHIMURA HIROKAZU DR   YOU JIAN-QI
YUN HONG-SIK PROF   ZACHARIADIS THEODOSIOS DR ZAPPALA ROSARIO ALDO DR
ZELENKA ANTOINE DR   ZHANG BAI-RONG   ZHANG HE-QI
ZHANG ZHEN-DA   ZHAO REN-YANG   ZHARKOVA VATENINA DR
ZHELYAZKOV IVAN DR   ZHOU DAOQI   ZHUGZHDA YUZEF D DR
ZIRIN HAROLD DR   ZLOBEC PAOLO DR   ZOU YI-XIN
ZWAAN CORNELIS PROF DR

COMPOSITION OF COMMISSION 12 1991-1994

President   : STENFLO JAN O DR

Vice-President(s) : DEUBNER FRANZ-LUDWIG DR

Organizing Committee: AYRES THOMAS R

                       HARVEY JOHN W DR

                       KARPINSKIJ VADIM N DR

                       LANDI DEGL'INNOCENTI E PR

                       NORDLUND AKE DR

                       PASACHOFF JAY M PROF

                       SIVARAMAN K R DR

                       STAUDE JUERGEN DR

                       WANG JING-XIU

                       WILSON PETER R PROF

                       YOSHIMURA HIROKAZU DR

Members:

ABOUDARHAM JEAN DR	ACTON LOREN W DR	ADAM MADGE G DR
AI GUOXIANG	AIME C DR	ALISSANDRAKIS C PH D
ALTROCK RICHARD C DR	ALTSCHULER MARTIN D PROF	ANDERSEN BO NYBORG DR
ANDO HIROYASU DR	ANSARI S M RAZAULLAH PROF	ANTIA H M DR
ARNAUD JEAN PAUL	ATHAY R GRANT DR	BALIUNAS SALLIE L
BALTHASAR HORST DR	BEARD DAVID B DR	BECKERS JACQUES M DR
BECKMAN JOHN E PROF	BEEBE HERBERT A	BEL NICOLE J DR
BENFORD GREGORY DR	BHATNAGAR ARVIND DR	BHATTACHARYYA J C PROF
BILLINGS DONALD E PROF	BLACKWELL DONALD E PROF	BLAMONT JACQUES E PROF
BOCCHIA ROMEO DR	BOEHM KARL-HEINZ PROF	BOEHM-VITENSE ERIKA PROF
BOHN HORST-ULRICH	BOMMIER VERONIQUE DR	BONNET ROGER M DR
BOOK DAVID L	BORNMANN PATRICIA L DR	BOUGERET J L DR
BRANDT PETER N	BRAULT JAMES W DR	BRAY ROBERT J DR
BRECKINRIDGE JAMES B DR	BRUECKNER GUENTER E DR	BRUNER MARILYN E DR
BRUNING DAVID H DR	BRUZEK ANTON DR	BUMBA VACLAV DR
CADEZ VLADIMIR	CAVALLINI FABIO	CEPPATELLI GUIDO DR
CHAMBE GILBERT	CHAN KWING LAM	CHAPMAN GARY A DR
CHEN BIAO	CHENG CHUNG-CHIEH DR	CHISTYAKOV VLADIMIR E DR
CHRISTENSEN-DALSGAARD J	CHVOJKOVA WOYK E DR	CLARK THOMAS ALAN DR
COLLADOS MANUEL DR	COOK JOHN W	COX ARTHUR N DR
CRAIG IAN JONATHAN D DR	CRAM LAWRENCE EDWARD PROF	CRAMER NEIL DR
DARA HELEN DR	DE JAGER CORNELIS PROF	DEL TORO INIESTA JOSE DR
DELACHE PHILIPPE J DR	DELBOUILLE LUC PROF	DELIYANNIS JEAN DR
DEMARQUE P PROF	DEMING LEO DRAKE DR	DEZSO LORANT PROF
DOGAN NADIR PROF	DRAVINS DAINIS PROF	DUMONT SIMONE DR
DUNKELMAN LAWRENCE	DUNN RICHARD B DR	DUVALL THOMAS L JR
EINAUDI GIORGIO	ELLIOTT IAN DR	ELSTE GUNTHER H DR
EPSTEIN GABRIEL LEO DR	ESSER RUTH DR	EVANS J V DR
FALCIANI ROBERTO DR	FANG CHENG	FELDMAN URI
FIALA ALAN D DR	FISHER GEORGE HEWITT DR	FOFI MASSIMO DR

FOMICHEV VALERI V DR | FONTENLA JUAN MANUEL DR | FOSSAT ERIC DR
FOUKAL PETER V DR | FRAZIER EDWARD N DR | FRIEDMAN HERBERT DR
FROEHLICH CLAUS | GABRIEL ALAN H | GAIZAUSKAS VICTOR DR
GARCIA-BERRO ENRIQUE DR | GAUR V P | GLATZMAIER GARY A
GODOLI GIOVANNI PROF | GOKDOGAN NUZHET PROF | GOLDMAN MARTIN V
GOMEZ MARIA THERESA DR | GOPALSWAMY N DR | GOPASYUK S I DR
GORDON CHARLOTTE PROF | GREVESSE N DR | GU XIAO-MA
GURTOVENKO E A DR | HAGYARD MONA JUNE | HAMMER REINER
HEINZEL PETR DR | HEJNA LADISLAV DR | HIEI EIJIRO DR
HILDNER ERNEST DR | HILL FRANK DR | HIRAYAMA TADASHI PROF
HOANG BINH DY DR | HOLWEGER HARTMUT PROF | HORTON BRIAN H DR
HOTINLI METIN DR | HOUSE LEWIS L DR | HOWARD ROBERT F DR
HOYNG PETER DR | ILLING RAINER M E | JABBAR SABEH RHAMAN
JACKSON BERNARD V DR | JEFFERIES JOHN T DR | JEFFERIES STUART DR
JONES HARRISON PRICE DR | JORDAN CAROLE DR | JORDAN STUART D DR
JOSHI G C DR | KALKOFEN WOLFGANG DR | KALMAN BELA DR
KARLICKY MARIAN | KARPEN JUDITH T | KATO SHOJI PROF
KAUFMANN PIERRE PROF | KAWAGUCHI ICHIRO PROF | KEIL STEPHEN L
KHETSURIANI TSIALA S DR | KLEIN KARL LUDWIG DR | KNEER FRANZ DR
KNOLKER MICHAEL DR | KONONOVICH EDWARD V DR | KOPECKY MILOSLAV DR
KOSTIK ROMAN I | KOTOV VALERY DR | KOTRC PAVEL
KOUTCHMY SERGE DR | KOYAMA SHIN PROF DR | KRAEMER GERHARD DR
KRUEGER ALBRECHT DR | KUBICELA ALEKSANDAR DR | KUKLIN G V DR
KULCAR LADISLAV DR | KUNDU MUKUL R DR | KUPERUS MAX PROF DR
KUROCHKA L N DR | LA BONTE BARRY JAMES | LABS DIETRICH PROF
LANDI DEGL'INNOCENTI M | LANDMAN DONALD ALAN | LANDOLFI MARCO
LANTOS PIERRE DR | LEIBACHER JOHN DR | LEIGHTON R B PROF
LEROY JEAN-LOUIS | LIN YUANZHANG | LINSKY JEFFREY L DR
LIVINGSTON WILLIAM C | LOCKE JACK L DR | LOPEZ-ARROYO M
LOUGHHEAD RALPH E DR | LUEST REIMAR PROF | LUSTIG GUENTER DR
MAKAROV VALENTINE I | MAKAROVA ELENA A DR | MAKITA MITSUGU DR
MARIK MIKLOS DR | MARILLI ETTORE DR | MARISKA JOHN THOMAS
MARMOLINO CIRO | MATSUSHIMA SATOSHI DR | MATTIG W PROF DR
MCKENNA LAWLOR SUSAN | MEIN PIERRE | MELROSE DONALD B PROF
MERGENTALER JAN PROF | MEWE R DR | MEYER FRIEDRICH DR
MICHARD RAYMOND DR | MIHALAS DIMITRI DR | MILKEY ROBERT W DR
MONSIGNORI FOSSI BRUNA DR | MOORE RONALD L DR | MORENO-INSERTIS FERNANDO
MORIYAMA FUMIO PROF | MOURADIAN ZADIG M DR | MUELLER EDITH A PROF
MULLER RICHARD DR | MUNRO RICHARD H DR | NAMBA OSAMU DR
NECKEL HEINZ DR | NESIS ANASTASIOS DR | NICOLAS KENNETH ROBERT
NICOLET MARCEL PROF | NISHI KEIZO DR | NOYES ROBERT W PROF
ORRALL FRANK Q PROF | OSTER LUDWIG F PROF DR | OWOCKI STANLEY PETER DR
PALLE PERE-LLUIS DR | PALUS PAVEL DR | PANDE MAHESH CHANDRA DR
PAPATHANASOGLOU D DR | PARKINSON WILLIAM H DR | PECKER JEAN-CLAUDE PROF
PEYTURAUX ROGER H PROF | PFLUG KLAUS DR | PHILLIPS KENNETH J H
PIERCE A KEITH DR | POQUERUSSE MICHEL | PRIEST ERIC R PROF
PROKAKIS THEODORE J DR | RABIN DOUGLAS MARK | RADICK RICHARD R DR
RAOULT ANTOINETTE DR | REES DAVID ELWYN DR | REEVES EDMOND M DR
REGULO CLARA DR | RIGHINI-COHEN GIOVANNA DR | RIGUTTI MARIO PROF
ROBERTI GIUSEPPE DR | ROCA CORTES TEODORO | RODDIER FRANCOIS PROF
ROLAND GINETTE DR | ROUDIER THIERRY DR | ROVIRA MARTA GRACIELA
RUSIN VOJTECH | RUTTEN ROBERT J DR | RYBANSKY MILAN
SAKAI JUNICHI | SAKURAI TAKASHI DR | SAMAIN DENYS DR
SANCHEZ ALMEIDA JORGE DR | SAUVAL A JACQUES DR | SCHMAHL EDWARD J DR
SCHMIDT WOLFGANG DR | SCHMITT DIETER DR | SCHOBER HANS J DR

SCHUESSLER MANFRED DR	SCHWARTZ STEVEN JAY	SEATON MICHAEL J PROF
SEMEL MEIR DR	SEVERINO GIUSEPPE	SHALLIS MICHAEL J DR
SHEELEY NEIL R DR	SHEN LONG-XIANG	SHINE RICHARD A DR
SIMON GEORGE W DR	SIMON GUY	SINGH JAGDEV DR
SINHA K DR	SITNIK G F PROF	SITTERLY CHARLOTTE M DR
SKUMANICH ANDRE PROF	SMITH PETER L DR	SOBOLEV VLADISLAV M DR
SOLANKI SAMI K DR	SONG MU-TAO	SOTIROVSKI PASCAL DR
SOUFFRIN PIERRE B DR	SPICER DANIEL SHIELDS DR	STATHOPOULOU MARIA DR
STEBBINS ROBIN	STEFFEN MATTHIAS DR	STIX MICHAEL DR
SUEMATSU YOSHINORI DR	SVESTKA ZDENEK DR	SWENSSON JOHN W DR
TANDBERG-HANSSEN EINAR A	TEPLITSKAYA R B DR	THOMAS JOHN H PROF
THOMAS RICHARD N DR	TORELLI M DR	TOUSEY RICHARD DR
TRIPATHI B M DR	TRUJILLO BUENO JAVIER DR	TSAP T T DR
TSIROPOULA GEORGIA DR	TSUBAKI TOKIO PROF	UCHIDA YUTAKA PROF
UNNO WASABURO PROF	UUS UNDO DR	VAN HOVEN GERARD DR
VASILEVA GALINA J DR	VAUGHAN ARTHUR H DR	VENKATAKRISHNAN P DR
VIAL JEAN-CLAUDE	VILMER NICOLE DR	VITINSKIJ YURIJ I DR
VOLONTE SERGE DR	VUKICEVIC K M PROF DR	WALDMEIER MAX PROF DR
WANG ZHEN-YI	WARWICK JAMES W DR	WEISS NIGEL O DR
WENTZEL DONAT G DR	WITTMANN AXEL D DR	WOEHL HUBERTUS DR
WORDEN SIMON P DR	WU HSIN-HENG DR	WU LIN-XIANG
WYLLER ARNE A PROF	YOU JIAN-QI	YOUSSEF NAHED H PROF
YUN HONG-SIK PROF	ZARRO DOMINIC M DR	ZELENKA ANTOINE DR
ZHOU DAOQI	ZHUGZHDA YUZEF D DR	ZIRIN HAROLD DR
ZIRKER JACK B DR	ZWAAN CORNELIS PROF DR	

COMPOSITION OF COMMISSION 14 1991-1994

President   : WIESE WOLFGANG L DR

Vice-President(s) : PARKINSON WILLIAM H DR

Organizing Committee: ADELMAN SAUL J DR
                       DUBAU JACQUES DR
                       GABRIEL ALAN H
                       KATO TAKAKO DR
                       NICHOLLS RALPH W PROF
                       NUSSBAUMER HARRY PROF
                       RUDZIKAS ZENONAS B
                       SAHAL-BRECHOT SYLVIE DR
                       SMITH PETER L DR

Members:

ALLARD NICOLE DR	ANDREW KENNETH L PROF	ARDUINI-MALINOVSKY M DR
ARTRU MARIE-CHRISTINE DR	BARNARD HANNES A J DR	BARROW RICHARD F DR
BATES DAVID R PROF	BELY-DUBAU FRANCOISE	BERRINGTON KEITH ADRIAN
BIEMONT EMILE DR	BLACK JOHN HARRY DR	BLAHA MILAN DR
BOMMIER VERONIQUE DR	BRANSCOMB L M DR	BRAULT JAMES W DR
BROMAGE GORDON E DR	BURGESS ALAN DR	CARBON DUANE F DR
CARROLL P KEVIN PROF	CARVER JOHN H PROF	COOK ALAN H PROF
CORLISS C H DR	CORNILLE MARGUERITE DR	CZYZAK STANLEY J DR
DALGARNO ALEXANDER PROF	DAVIS SUMNER P DR	DE FREES DOUGLAS J DR
DELSEMME ARMAND H PROF DR	DESESQUELLES JEAN DR	DIERCKSEN GEERD H F PH D
DIMITRIJEVIC MILAN	DRESSLER KURT PROF	DUFAY MAURICE PROF
EDLEN BENGT PROF	EIDELSBERG MICHELE DR	EPSTEIN GABRIEL LEO DR
FAUCHER PAUL DR	FEAUTRIER NICOLE DR	FEDERICI LUCIANA
FEDERMAN STEVEN ROBERT	FINK UWE DR	FLOWER DAVID R DR
FUHR JEFFREY ROBERT DR	GALLAGHER JEAN W DR	GARGAUD MURIEL DR
GARSTANG ROY H PROF	GARTON W R S PROF	GLAGOLEVSKIJ JU V DR
GOLDBACH CLAUDINE MME	GRANT IAN P DR	GREEN LOUIS C PROF
HEDDLE DOUGLAS W O PROF	HEFFERLIN RAY A PROF	HENDECOURT D' LOUIS DR
HEROLD HEINZ	HERZBERG GERHARD DR	HESSER JAMES E DR
HOANG BINH DY DR	HOUSE LEWIS L DR	HUBER MARTIN C E DR
HUEBNER WALTER F DR	ILIEV ILIAN	IRWIN ALAN W DR
JACQUINOT PIERRE DR	JOHANSSON SVENERIC DR	JOHNSON DONALD R DR
JOHNSON FRED M PROF DR	JOLY FRANCOIS DR	JORDAN CAROLE DR
JORDAN H L DR DIREKTOR	KENNEDY EUGENE T	KESSLER KARL G DR
KIELKOPF JOHN F DR	KIM ZONG DOK	KING R B DR
KINGSTON ARTHUR E PROF	KIPPER TONU DR	KIRBY KATE P DR
KOHL JOHN L DR	KROTO HAROLD PROF	LAGERQVIST ALBIN PROF
LAMBERT DAVID L PROF	LANDMAN DONALD ALAN	LANG JAMES DR
LAUNAY JEAN-MICHEL DR	LAWRENCE G M DR	LAYZER DAVID PROF
LE BOURLOT JACQUES DR	LE DOURNEUF MARYVONNE	LEACH SYDNEY DR
LEGER ALAIN DR	LEMAIRE JEAN-LOUIS DR	LESAGE ALAIN DR
LOULERGUE MICHELLE DR	LOVAS FRANCIS JOHN DR	LUTZ BARRY L DR

MAILLARD JEAN-PIERRE DR   MARTIN WILLIAM C DR   MASON HELEN E DR
MCWHIRTER R W PETER DR   MEWE R DR   MICKELSON MICHAEL E DR
MUMMA MICHAEL JON   NEWSOM GERALD H PROF   NOLLEZ GERARD DR
OBI SHINYA PROF   OETKEN L DR   OKA TAKESHI DR
OMONT ALAIN PROF   ORTON GLENN S DR   PEACH GILLIAN DR
PETRINI DANIEL DR   PETROPOULOS BASIL CH DR   PETTINI MARCO
PFENNIG HANS H DR   PHILLIPS JOHN G PROF   PIACENTINI RUBEN DR
PROKOF'EV VLADIMIR K PROF QUERCI FRANCOIS R DR   RAO K NARAHARI
RICHTER JOHANNES PROF   ROSS JOHN E R DR   ROSTAS FRANCOIS DR
ROUEFF EVELYNE M A DR   RUDER HANNS   SCHADEE AERT DR
SCHRIJVER JOHANNES DR   SEATON MICHAEL J PROF   SHARP CHRISTOPHER DR
SHORE BRUCE W   SINHA K DR   SITTERLY CHARLOTTE M DR
SMITH GEOFFREY DR   SMITH WM HAYDEN PROF   SOMERVILLE WILLIAM B DR
SORENSEN GUNNAR DR   SPIELFIEDEL ANNIE DR   STARK GLEN DR
STEENMAN-CLARK LOIS DR   STEHLE CHANTAL DR   STREL'NITSKIJ VLADIMIR DR
SUMMERS HUGH P DR   SWINGS JEAN-PIERRE DR   TAKAYANAGI KAZUO PROF
TATUM JEREMY B DR   TCHANG-BRILLET LYDIA DR   TOUSEY RICHARD DR
TOZZI GIAN PAOLO   TRAN MINH NGUYET DR   TREFFTZ ELEONORE E DR
VAN REGEMORTER HENRI DR   VAN RENSBERGEN WALTER DR VARSHALOVICH DIMITRIJ PR
VOELK HEINRICH J PROF   VOLONTE SERGE DR   VUJNOVIC VLADIS DR
WARES GORDON W DR   WENIGER SCHAME DR   WILSON ROBERT PROF SIR
WINNEWISSER GISBERT DR   WUNNER GUENTER   YOUNG LOUISE GRAY DR
YU YAN DR   ZEIPPEN CLAUDE DR   ZENG QIN DR
ZIRIN HAROLD DR

COMPOSITION OF COMMISSION 15 1991-1994

President   : HARRIS ALAN WILLIAM DR

Vice-President(s) : A'HEARN MICHAEL F DR

Organizing Committee: BARUCCI MARIA A DR
                       BELTON MICHAEL J S DR
                       BRANDT JOHN C DR
                       GRUEN EBERHARD DR
                       HUEBNER WALTER F DR
                       HUGHES DAVID W DR
                       KELLER HORST UWE DR
                       KISELEV NIKOLAI N DR
                       KNEZEVIC ZORAN
                       KRESAK LUBOR DR
                       LUPISHKO DMITRIJ F
                       RAHE JURGEN PROF
                       RICKMAN HANS DR
                       STEEL DUNCAN I DR
                       TEDESCO EDWARD F
                       ZAPPALA VINCENZO PROF

Members:

ALLEGRE CLAUDE PROF	ANDRIENKO DMITRY A DR	ARNOLD JAMES R DR
ARPIGNY CLAUDE PROF	AXFORD W IAN PROF	BABADZHANOV PULAT B DR
BAILEY MARK EDWARD	BARKER EDWIN S DR	BEARD DAVID B DR
BINZEL RICHARD P DR	BIRCH PETER MR	BLAMONT JACQUES E PROF
BOCKELEE-MORVAN DOMINIQUE	BOEHNHARDT HERMANN DR	BOUSKA JIRI DR
BOWELL EDWARD L G DR	BRECHER AVIVA DR PROF	BROWN ROBERT HAMILTON
BROWNLEE DONALD E PROF	BRUNK WILLIAM E DR	BURATTI BONNIE J DR
BURLAGA LEONARD F DR	BURNS JOSEPH A PROF	CAMPINS HUMBERTO DR
CANDY MICHAEL P MR	CAPACCIONI FABRIZIO DR	CAPRIA MARIA TERESA DR
CARRUTHERS GEORGE R DR	CARSENTY URI DR	CARUSI ANDREA
CELLINO ALBERTO DR	CEPLECHA ZDENEK DR	CERRONI PRISCILLA DR
CHANDRASEKHAR T DR	CHAPMAN CLARK R DR	CHAPMAN ROBERT D DR
CHEN DAO-HAN	CHEREDNICHENKO V I DR	CHERNYKH N S DR
CLAIREMIDI JACQUES DR	CLAYTON GEOFFREY C DR	CLAYTON ROBERT N DR
CLUBE S V M DR	COCHRAN ANITA L DR	COCHRAN WILLIAM DAVID DR
COMBI MICHAEL R DR	COSMOVICI BATALLI C DR	CRISTESCU CORNELIA G DR
CROVISIER JACQUES	CRUIKSHANK DALE P DR	CUYPERS JAN DR
DANKS ANTHONY C DR	DE ALMEIDA AMAURY A DR	DE PATER IMKE
DE SANCTIS GIOVANNI	DEBEHOGNE HENRI DR SC	DEGEWIJ JOHAN DR
DELSEMME ARMAND H PROF DR	DERMOTT STANLEY F	DEUTSCHMAN WILLIAM A DR
DI MARTINO MARIO	DONN BERTRAM D	DOSSIN F DR
DRYER MURRAY DR	DZHAPIASHVILI VICTOR P DR	ENCRENAZ THERESE DR
ERSHKOVICH ALEXANDER PROF	EVIATAR AHARON PROF	FARINELLA PAOLO DR
FECHTIG HUGO DR	FELDMAN PAUL DONALD DR	FERNANDEZ JEAN-CLAUDE DR
FERNANDEZ JULIO A DR	FERRIN IGNACIO	FESTOU MICHEL C DR

FORTI GIUSEPPE DR / FROESCHLE CHRISTIANE D DR / FUJIWARA AKIRA DR
FULCHIGNONI MARCELLO PROF / GAMMELGAARD PETER MAG SCI / GEHRELS TOM PROF
GEISS JOHANNES PROF / GERARD ERIC DR / GIBSON JAMES
GIOVANE FRANK / GRADIE JONATHAN CAREY / GREEN SIMON F
GREENBERG J MAYO DR / GREENBERG RICHARD DR / GROSSMAN LAWRENCE PROF
GRUDZINSKA STEFANIA DR / GUSTAFSON BO A S / HAJDUK ANTON DR
HALLIDAY IAN DR / HANNER MARTHA S DR / HAPKE BRUCE W DR
HARTMANN WILLIAM K / HARWIT MARTIN PROF / HASEGAWA ICHIRO DR
HASER LEO N K DR / HAUPT HERMANN F PROF / HELIN ELEANOR FRANCIS
HERZBERG GERHARD DR / HU ZHONG-WEI / HUNTRESS WESLEY T DR
IBADINOV KHURSANDKUL DR / IP WING-HUEN / IRVINE WILLIAM M PROF
ISOBE SYUZO DR / IVANOVA VIOLETA DR / JACKSON WILLIAM M DR
JOCKERS KLAUS DR / JOHNSON TORRENCE V DR / KARTTUNEN HANNU DR
KEAY COLIN S L PROF / KEIL KLAUS DR / KNACKE ROGER F DR
KOEBERL CHRISTIAN DR / KOHOUTEK LUBOS DR / KONOPLEVA VARVARA P DR
KOWAL CHARLES THOMAS / KRESAKOVA MARGITA DR / KRISHNA SWAMY K S DR
KRISTENSEN LEIF KAHL / LAGERKVIST CLAES-INGVAR / LAMY PHILIPPE DR
LANCASTER BROWN PETER / LANE ARTHUR LONNE DR / LARSON HAROLD P DR
LARSON STEPHEN M / LEBOFSKY LARRY ALLEN / LEE THYPHOON
LEVASSEUR-REGOURD A C PR / LILLER WILLIAM DR / LILLIE CHARLES F DR
LINDSEY CHARLES ALLAN / LIPSCHUTZ MICHAEL E DR / LISSAUER JACK J DR
LIU LIN-ZHONG / LIU ZONGLI / LOPES-GAUTIER ROSALY DR
LUEST RHEA DR / LUMME KARI A DR / LUTZ BARRY L DR
LYTTLETON RAYMOND A PROF / MALAISE DANIEL J DR / MARAN STEPHEN P DR
MARSDEN BRIAN G DR / MATSON DENNIS L DR / MATSUURA OSCAR T DR
MCCORD THOMAS B DR / MCCROSKY RICHARD E DR / MCDONNELL J A M PROF
MCFADDEN LUCY ANN DR / MCKENNA LAWLOR SUSAN / MEECH KAREN DR
MEISEL DAVID D DR / MENDIS DEVAMITTA ASOKA DR / MICHALOWSKI TADEUSZ DR
MILANI ANDREA / MILET BERNARD L DR / MILLER FREEMAN D PROF
MILLIS ROBERT L DR / MOEHLMANN DIEDRICH / MOORE ELLIOTT P PROF
MOROZ V I PROF DR / MORRISON DAVID PROF / MRKOS ANTONIN DR
MUKAI TADASHI DR / MUMMA MICHAEL JON / NAKAMURA TSUKO DR
NAPIER WILLIAM M DR / NEFF JOHN S / NEUKUM G DR
NEWBURN RAY L JR / NIEDNER MALCOLM B DR / NOLL KEITH STEPHEN DR
O'DELL CHARLES R DR / O'KEEFE JOHN A DR / PAOLICCHI PAOLO DR
PARISOT JEAN-PAUL / PELLAS PAUL DR / PEREZ-DE-TEJADA H A DR
PILCHER CARL BERNARD DR / PILLINGER COLIN DR / PITTICH EDUARD M DR
PRIALNIK-KOVETZ DINA DR / PROISY PAUL E DR / REITSEMA HAROLD J
REMY BATTIAU LILIANE G A / REVELLE DOUGLAS ORSON DR / ROEMER ELIZABETH PROF
SAGDEEV ROALD Z DR / SAITO TAKAO PROF / SCALTRITI FRANCO DR
SCHLEICHER DAVID G DR / SCHLOERB F PETER / SCHMIDT H U DR
SCHMIDT MAARTEN PROF / SCHOBER HANS J DR / SCHOLL HANS DR
SCHUBART JOACHIM DR / SEKANINA ZDENEK DR / SHARP CHRISTOPHER DR
SHIMIZU MIKIO PROF / SHKODROV V G DR / SHOEMAKER EUGENE M
SHOR VIKTOR A DR / SHUL'MAN L M DR / SIVARAMAN K R DR
SMITH BRADFORD A PROF / SMOLUCHOWSKI ROMAN PROF / SNYDER LEWIS E
SOLC MARTIN / SPINRAD HYRON PROF / STOHL JAN DR
SURDEJ JEAN M G / SVOREN JAN / SYKES MARK VINCENT DR
SZEGO KAROLY DR / TAKEDA HIDENORI DR / TANABE HIROYOSHI DR
TATUM JEREMY B DR / TERENTJEVA ALEXANDRA K DR / THOLEN DAVID J DR
TOMITA KOICHIRO MR / TOTH IMRE DR / VALSECCHI GIOVANNI B DR
VAN FLANDERN THOMAS DR / VANYSEK VLADIMIR PROF / VEEDER GLENN J DR
VEVERKA JOSEPH DR / VILAS FAITH DR / WALKER ALISTAIR ROBIN DR
WALLIS MAX K DR / WANG SI-CHAO / WASSON JOHN T
WATANABE JUN-ICHI DR / WDOWIAK THOMAS J DR / WEAVER HAROLD F PROF

WEHINGER PETER A DR   WEIDENSCHILLING S J DR   WEISSMAN PAUL ROBERT
WETHERILL GEORGE W   WHIPPLE FRED L DR   WILKENING LAUREL L DR
WILLIAMS IWAN P PROF   WISNIEWSKI WIESLAW Z   WOOD JOHN A DR
WOOLFSON MICHAEL M PROF   WOSZCZYK ANDRZEJ PROF   WYCKOFF SUSAN DR
YABUSHITA SHIN A PROF   YAVNEL ALEXANDER A DR   YEOMANS DONALD K DR
ZARNECKI JAN CHARLES DR   ZELLNER BENJAMIN H DR

COMPOSITION OF COMMISSION 16 1991-1994

President   : MORRISON DAVID PROF

Vice-President(s) : DE BERGH CATHERINE DR

                      MAROV MIKHAIL YA PROF

Organizing Committee: BAZILEVSKY ALEXANDR T

                       BELTON MICHAEL J S DR

                       BRAHIC ANDRE DR

                       CRUIKSHANK DALE P DR

                       DAVIES MERTON E MR

                       ENCRENAZ THERESE DR

                       GAUTIER DANIEL

                       MOROZ V I PROF DR

                       OWEN TOBIAS C PROF

                       RAHE JURGEN PROF

                       SMITH BRADFORD A PROF

                       THOLEN DAVID J DR

Members:

AKABANE TOKUHIDE DR	APPLEBY JOHN F	ARTHUR DAVID W G
ATREYA SUSHIL K	BARROW COLIN H DR	BATSON RAYMOND MILNER DR
BATTANER EDUARDO DR	BAUM WILLIAM A DR	BEEBE RETA FAYE DR
BEER REINHARD DR	BENDER PETER L DR	BERGE GLENN L DR
BERGSTRALH JAY T DR	BERTAUX J L DR	BHATIA R K DR
BINZEL RICHARD P DR	BLAMONT JACQUES E PROF	BOBROV M S DR
BONDARENKO L N DR	BOSMA PIETER B DR	BOSS ALAN P DR
BOYCE PETER B DR	BRECHER AVIVA DR PROF	BROADFOOT A LYLE DR
BROWN ROBERT HAMILTON	BRUNK WILLIAM E DR	BURATTI BONNIE J DR
BURNS JOSEPH A PROF	CALAME ODILE DR	CALDWELL JOHN JAMES
CAMERON WINIFRED S MRS	CAMICHEL HENRI DR	CAMPBELL DONALD B
CAPRIA MARIA TERESA DR	CARSMARU MARIA M DR	CATALANO SANTO DR
CHAMBERLAIN JOSEPH W PROF	CHAPMAN CLARK R DR	CHEN DAO-HAN
CLAIREMIDI JACQUES DR	COCHRAN ANITA L DR	COLOMBO G PROF DR
COMBI MICHAEL R DR	CONNES JANINE DR	COUNSELMAN CHARLES C PROF
DE PATER IMKE	DEGEWIJ JOHAN DR	DERMOTT STANLEY F
DICKEL JOHN R	DICKEY JEAN O'BRIEN	DOLLFUS AUDOUIN PROF
DRAKE FRANK D PROF	DROSSART PIERRE DR	DURRANCE SAMUEL T DR
DZHAPIASHVILI VICTOR P DR	ELLIOT JAMES L DR	ELSTON WOLFGANG E PROF
EL-BAZ FAROUK DR	ESHLEMAN VON R PROF	ESPOSITO LARRY W
FARINELLA PAOLO DR	FIELDER GILBERT DR	FINK UWE DR
FOX KENNETH DR	FOX W E MR	FUJIWARA AKIRA DR
GEAKE JOHN E DR	GEHRELS TOM PROF	GEISS JOHANNES PROF
GERARD JEAN-CLAUDE M C DR	GICLAS HENRY L MR	GIERASCH PETER J DR
GOLD THOMAS PROF	GOLDREICH P DR	GOLDSTEIN RICHARD M DR
GOODY R M	GORENSTEIN PAUL DR	GOUDAS CONSTANTINE L PROF
GREEN JACK PROF	GROSSMAN LAWRENCE PROF	GUERIN PIERRE DR
GUEST JOHN E DR	GULKIS SAMUEL DR	GURSHTEIN ALEXANDER A DR

HABIBULLIN SH T PROF DR	HAGFORS T DR	HALLIDAY IAN DR
HARRIS ALAN WILLIAM	HERZBERG GERHARD DR	HIDE RAYMOND PROF
HOLBERG JAY B	HOREDT GEORG PAUL DR	HOVENIER J W DR
HU ZHONG-WEI	HUBBARD WILLIAM B PROF	HUNT G E DR
HUNTEN DONALD M PROF	IRVINE WILLIAM M PROF	IWASAKI KYOSUKE DR
JOHNSON TORRENCE V DR	JURGENS RAYMOND F	KARANDIKAR R V PROF
KAULA WILLIAM M PROF	KILADZE R I DR	KISLYUK VITALIJ S DR
KOPAL ZDENEK PROF	KOWAL CHARLES THOMAS	KSANFOMALITI L V DR
KUMAR SHIV S PROF	KURT V G DR	KUZMIN ARKADII D PROF DR
LANE ARTHUR LONNE DR	LARSON HAROLD P DR	LARSON STEPHEN M
LEIKIN G A DR	LEWIS J S	LISSAUER JACK J DR
LOCKWOOD G WESLEY DR	LOPES-GAUTIER ROSALY DR	LOPEZ-MORENO JOSE JUAN
LOPEZ-PUERTAS MANUEL	LOPEZ-VALVERDE M A DR	LUMME KARI A DR
LUTZ BARRY L DR	MAHRA H S DR	MATSON DENNIS L DR
MATSUI TAKAFUMI DR	MAYER CORNELL H	MCCORD THOMAS B DR
MCELROY M B DR	MEADOWS A JACK PROF	MICKELSON MICHAEL E DR
MIDDLEHURST BARBARA M MS	MIKHAL JOSEPH SIDKY PROF	MILLIS ROBERT L DR
MIYAMOTO SIGENORI PROF	MOEHLMANN DIEDRICH	MOLINA ANTONIO
MOORE PATRICK DR	MORENO FERNANDO DR	MOROZHENKO A V DR
MOUTSOULAS MICHAEL PROF	MULHOLLAND J DERRAL DR	MUMMA MICHAEL JON
MURPHY ROBERT E DR	NAKAGAWA YOSHITSUGU DR	NESS NORMAN F DR
NEUKUM G DR	NOLL KEITH STEPHEN DR	OTTELET I J DR
O'KEEFE JOHN A DR	PANG KEVIN	PAOLICCHI PAOLO DR
PETIT JEAN-MARC DR	PETROPOULOS BASIL CH DR	PETTENGILL GORDON H PROF
PILLINGER COLIN DR	POKORNY ZDENEK DR	POLLACK JAMES B DR
POTTER ANDREW E DR	PREDEANU IRINA DR	RAO M N DR
RODRIGO RAFAEL	ROQUES FRANCOISE DR	ROSCH JEAN PROF
RUNCORN S K PROF	RUSKOL EUGENIA L DR	SAFRONOV VICTOR S DR
SAGAN CARL DR	SAISSAC JOSEPH DR	SANCHEZ-LAVEGA AGUSTIN DR
SCHLEICHER DAVID G DR	SCHLOERB F PETER	SHAPIRO IRWIN I PROF
SHEVCHENKO VLADISLAV V DR	SHIMIZU MIKIO PROF	SHIMIZU TSUTOMU PROF EMER
SHOEMAKER EUGENE M	SICARDY BRUNO DR	SINTON WILLIAM M
SJOGREN WILLIAM L MR	SMOLUCHOWSKI ROMAN PROF	SODERBLOM LARRY DR
SONETT CHARLES P PROF	STOEV ALEXEI	STONE EDWARD C DR
STROBEL DARRELL F	STROM ROBERT G PROF	STRONG JOHN D PROF
SYNNOTT STEPHEN P	TEJFEL VIKTOR G DR	TERRILE RICHARD JOHN
TOMBAUGH CLYDE W PROF	TRAFTON LAURENCE M DR	TRAN-MINH FRANCOISE DR
TROITSKY V S PROF DR	TYLER JR G LEONARD DR	VAN ALLEN JAMES A PROF
VAN FLANDERN THOMAS DR	VEVERKA JOSEPH DR	WALKER ALTA SHARON DR
WALKER ROBERT M A PROF	WALLACE LLOYD V DR	WASSERMAN LAWRENCE H DR
WASSON JOHN T	WEIDENSCHILLING S J DR	WEIMER THEOPHILE P F DR
WEST ROBERT ALAN	WETHERILL GEORGE W	WHITAKER EWEN A
WILDEY ROBERT L PROF DR	WILLIAMS IWAN P PROF	WILLIAMS JAMES G DR
WOOD JOHN A DR	WOOLFSON MICHAEL M PROF	WOSZCZYK ANDRZEJ PROF
YODER CHARLES F	YOUNG ANDREW T DR	YOUNG LOUISE GRAY DR
ZHARKOV VLADIMIR N DR		

COMPOSITION OF COMMISSION 19 1991-1994

President   : KOLACZEK BARBARA DR

Vice-President(s) : VONDRAK JAN DR

Organizing Committee: CAPITAINE NICOLE
                      DICKEY JEAN O'BRIEN
                      DICKMAN STEVEN R
                      FEISSEL MARTINE DR
                      JIN WEN-JING
                      MCCARTHY DENNIS D DR
                      MELBOURNE WILLIAM G DR
                      MIRONOV NIKOLAY T
                      MORRISON LESLIE V DR
                      ROBERTSON DOUGLAS S
                      SASAO TETSUO DR
                      WILSON P DR

Members:

ARABELOS DIMITRIOS DR	ARIAS ELISA FELICITAS	BANG YONG GOL
BANNI ALDO DR	BARLIER FRANCOIS E DR	BENDER PETER L DR
BEUTLER GERHARD PROF	BLINOV N S DR	BOUCHER CLAUDE DR
BOYTEL JORGE DEL PINO DR	BROSCHE PETER PROF	BRZEZINSKI ALEKSANDER DR
CARTER WILLIAM EUGENE	CHIUMIENTO GIUSEPPE	DEBARBAT SUZANNE V DR
DEHANT VERONIQUE DR	DEJAIFFE RENE J DR	DJUROVIC DRAGUTIN M DR
ELSMORE BRUCE DR	EL-SHAHAWY MOHAMAD PROF	ENSLIN HEINZ DR
FLIEGEL HENRY F	FONG CHU-GANG	FUJISHITA MITSUMI DR
GAMBIS DANIEL DR	GAO BUXI	GAPOSCHKIN EDWARD M DR
GROTEN ERWIN PROF	GUINOT BERNARD R PROF	HALL R GLENN DR
HAN TIANQI	HEFTY JAN DR	HEMMLEB GERHARD DR
HIDE RAYMOND PROF	HUA YING-MIN	IIJIMA SHIGETAKA PROF
JAKS WALDEMAR DR	KAKUTA CHUICHI DR	KAMEYA OSAMU DR
KLEPCZYNSKI WILLIAM J DR	KNOWLES STEPHEN H DR	LEHMANN MAREK DR
LI ZHENG-XIN DR	LI ZHIAN DR	LIESKE JAY H DR
LIU CI-YUAN DR	LUO DING-JIANG	LUO SHI-FANG
MANABE SEIJI DR	MARKOWITZ WILLIAM DR	MATSAKIS DEMETRIOS N
MEINIG MANFRED DR	MELCHIOR PAUL J PROF DIR	MERRIAM JAMES B
MOCZKO JANUSZ DR	MONET ALICE K B DR	MORGAN PETER DR
MUELLER IVAN I PROF	NAUMOV VITALIJ A DR	NEWHALL X X DR
NIEMI AIMO	OOE MASATSUGU DR	PAQUET PAUL EG DR
PICCA DOMENICO DR	PILKINGTON JOHN D H DR	POMA ANGELO DR
POPELAR JOSEF DR	PROVERBIO EDOARDO PROF	RANDIC LEO PROF DR
ROCHESTER MICHAEL G PROF	RUDER HANNS	RUNCORN S K PROF
RUSU I DR	RYKHLOVA LIDIJA V DR	SADZAKOV SOFIJA DR
SANCHEZ MANUEL	SATO KOICHI DR	SCHUTZ BOB EWALD
SEKIGUCHI NAOSUKE PROF	SEVARLIC BRANISLAV M PROF	SEVILLA MIGUEL J DR
SHAPIRO IRWIN I PROF	SIDORENKOV NIKOLAY S	SMITH HUMPHRY M
SOFFEL MICHAEL DR	STANILA GEORGE DR	STEPHENSON F RICHARD DR

SUGAWA CHIKARA DR	SUN YONGXIANG	TAPLEY BYRON D DR
TARADY VLADIMIR K DR	TORAO MASAHISA	TSAO MO PROF
TSUBOKAWA IETSUNE DR	VEILLET CHRISTIAN	VICENTE RAIMUNDO O PROF
WANG ZHENG MING	WILKINS GEORGE A DR	WILLIAMS JAMES G DR
WINKLER GERNOT M R DR	WU SHOU-XIAN	XIA JIONGYU
XIAO NAI-YUAN	XU JIA-YAN	XU TONG-QI
YANG FUMIN	YATSKIV YA S DR	YE SHU-HUA
YOKOYAMA KOICHI DR	YUMI SHIGERU PROF DR	ZHANG GUO-DONG
ZHAO MING	ZHENG DA-WEI	ZHU YAOZHONG DR
ZHU YONG-HE		

COMPOSITION OF COMMISSION 20 1991-1994

President   : CARUSI ANDREA

Vice-President(s) : YEOMANS DONALD K DR

Organizing Committee: AKSNES KAARE DR
                      ARLOT JEAN-EUDES
                      KRESAK LUBOR DR
                      MARSDEN BRIAN G DR
                      RICKMAN HANS DR
                      SHOR VIKTOR A DR
                      WASSERMAN LAWRENCE H DR
                      WEST RICHARD M DR
                      ZHANG JIA-XIANG

Members:

ABALAKIN VICTOR K DR	A'HEARN MICHAEL F DR	BABADZHANOV PULAT B DR
BAILEY MARK EDWARD	BATRAKOV YU V DR	BEC-BORSENBERGER ANNICK
BENEST DANIEL DR	BIEN REINHOLD DR	BLANCO CARLO DR
BLOW GRAHAM L	BOERNGEN FREIMUT DR PH	BOWELL EDWARD L G DR
BRANHAM RICHARD L JR	BURNS JOSEPH A PROF	CALAME ODILE DR
CANDY MICHAEL P MR	CHAPRONT-TOUZE MICHELLE	CHERNYKH N S DR
CHIO CHOL ZONG	CHURMS JOSEPH	CRISTESCU CORNELIA G DR
DE PASCUAL MARTINEZ M DR	DE SANCTIS GIOVANNI	DEBEHOGNE HENRI DR SC
DELSEMME ARMAND H PROF DR	DIRIKIS M A DR	DOLLFUS AUDOUIN PROF
DONNISON JOHN RICHARD DR	DOURNEAU GERARD DR	DOVAL JORGE PEREZ DR
DUNHAM DAVID W	DVORAK RUDOLF DR	EDELMAN COLETTE DR
EDMONDSON FRANK K PROF	ELLIOT JAMES L DR	ELST ERIC WALTER DR
EVDOKIMOV YU V DR	FERNANDEZ JULIO A DR	FERRAZ-MELLO S PROF DR
FERRERI WALTER	FORTI GIUSEPPE DR	FRANKLIN FRED A DR
FREITAS MOURAO R R DR	FROESCHLE CLAUDE DR	FURUKAWA KIICHIRO DR
GARFINKEL BORIS DR	GEHRELS TOM PROF	GIBSON JAMES
GICLAS HENRY L MR	GILMORE ALAN C MR	GREENBERG RICHARD DR
HAHN GERHARD J DR	HARPER DAVID DR	HARRINGTON ROBERT S DR
HARRIS ALAN WILLIAM DR	HASEGAWA ICHIRO DR	HAUPT HERMANN F PROF
HE MIAO-FU	HELIN ELEANOR FRANCIS	HEMENWAY PAUL D DR
HENRARD JACQUES PROF	HERS JAN MR	HEUDIER JEAN-LOUIS DR
HURNIK HIERONIM PROF	IANNA PHILIP A	ISOBE SYUZO DR
IVANOVA VIOLETA DR	IZVEKOV V A DR	KHATISASHVILI ALFEZ SH DR
KIANG TAO PROF	KINOSHITA HIROSHI DR	KISSELEVA TAMARA P
KLEMOLA ARNOLD R DR	KNEZEVIC ZORAN	KOHOUTEK LUBOS DR
KOSAI HIROKI	KOWAL CHARLES THOMAS	KOZAI YOSHIHIDE PROF
KRISTENSEN LEIF KAHL	LAGERKVIST CLAES-INGVAR	LAZZARO DANIELA DR
LIESKE JAY H DR	LINDBLAD BERTIL A DR	LOMB NICHOLAS RALPH DR
LOVAS MIKLOS	MACHADO LUIZ E DA SILVA	MAHRA H S DR
MANARA ALESSANDRO A DR	MCCROSKY RICHARD E DR	MCNAUGHT ROBERT H
MESSAGE PHILIP J DR	MILANI ANDREA	MILET BERNARD L DR
MILLIS ROBERT L DR	MINTZ BLANCO BETTY MRS	MORANDO BRUNO L DR

MRKOS ANTONIN DR   MULHOLLAND J DERRAL DR   MURRAY CARL D DR
NACOZY PAUL E DR   NAKAMURA TSUKO DR   NAKANO SYUICHI
NOBILI ANNA M   OTERMA LIISI PROF   OVERBEEK MICHIEL DANIEL
PANDEY A K   PASCU DAN DR   PAUWELS T DR
PIERCE DAVID ALLEN DR   PITTICH EDUARD M DR   POPOVIC BOZIDAR PROF DR
PROTICH MILORAD B   QUIJANO LUIS   RAJAMOHAN R DR
RAMSAY DONALD A DR   RAPAPORT MICHEL DR   REITSEMA HAROLD J
ROEMER ELIZABETH PROF   ROESER SIEGFRIED DR   RUSSELL KENNETH S DR
SAGNIER JEAN-LOUIS DR   SATO MASSAE DR   SCHMADEL LUTZ D DR
SCHOBER HANS J DR   SCHOLL HANS DR   SCHRUTKA-RECHTENSTAMM PR
SCHUBART JOACHIM DR   SCHUSTER WILLIAM JOHN DR   SEIDELMANN P KENNETH DR
SEKANINA ZDENEK DR   SHELUS PETER J DR   SHKODROV V G DR
SHOEMAKER EUGENE M   SINCLAIR ANDREW T DR   SITARSKI GRZEGORZ PROF
SOKOLSKY ANDREJ G DR   SOMA MITSURU DR   STANDISH E MYLES DR
STEEL DUNCAN I DR   STELLMACHER IRENE DR   SULTANOV G F ACAD
SVOREN JAN   SYNNOTT STEPHEN P   TATUM JEREMY B DR
TAYLOR DONALD BOGGIA DR   THOLEN DAVID J DR   TOMITA KOICHIRO MR
TORRES CARLOS DR   TSUCHIDA MASAYOSHI DR   VAGHI SERGIO DR
VALSECCHI GIOVANNI B DR   VAN FLANDERN THOMAS DR   VAN HOUTEN C J DR
VAN HOUTEN-GROENEVELD I   VAVROVA ZDENKA DR   VEILLET CHRISTIAN
VIEIRA MARTINS ROBERTO DR VU DUONG TUYEN DR   WEISSMAN PAUL ROBERT
WHIPPLE ARTHUR L DR   WHIPPLE FRED L DR   WILD PAUL PROF
WILLIAMS IWAN P PROF   WILLIAMS JAMES G DR   WROBLEWSKI HERBERT DR
YABUSHITA SHIN A PROF   YUASA MANABU DR   ZADUNAISKY PEDRO E PROF
ZAPPALA VINCENZO PROF   ZIOLKOWSKI KRZYSZTOF DR

COMPOSITION OF COMMISSION 21 1991-1994

President   : HANNER MARTHA S DR

Vice-President(s) : HAUSER MICHAEL G DR

Organizing Committee: BOWYER C STUART PROF
                       DUMONT RENE DR
                       GALPERIN YU I PROF
                       HONG SEUNG SOO DR
                       HOUCK JAMES R
                       LAMY PHILIPPE DR
                       LEINERT CHRISTOPH DR
                       LEVASSEUR-REGOURD A C PR
                       MUKAI TADASHI DR

Members:

ANDERSON KINSEY A PROF	ANGIONE RONALD J DR	BAGGALEY WILLIAM J PROF
BANOS COSMAS J DR	BATES DAVID R PROF	BELKOVICH O I DR
BLACKWELL DONALD E PROF	BLAMONT JACQUES E PROF	BROADFOOT A LYLE DR
CHAMBERLAIN JOSEPH W PROF	CLAIREMIDI JACQUES DR	DACHS JOACHIM PROF DR
DERMOTT STANLEY F	DIVARI N B DR	DUBIN MAURICE DR
DUFAY MAURICE PROF	DUNKELMAN LAWRENCE	ELSAESSER HANS PROF
FECHTIG HUGO DR	FELDMAN PAUL DONALD DR	FISHKOVA LUISA M PROF
FUJIWARA AKIRA DR	GADSDEN MICHAEL DR	GIOVANE FRANK
GREENBERG J MAYO DR	GRUEN EBERHARD DR	GUSTAFSON BO A S
HALLIDAY IAN DR	HARWIT MARTIN PROF	HAUG ULRICH PROF
HENDECOURT D' LOUIS DR	HENRY RICHARD C PROF	HOFMANN WILFRIED DR
IVANOV-KHOLODNY G S DR	JACKSON BERNARD V DR	JAMES JOHN F MR
JARRETT ALAN H PROF	JOUBERT MARTINE	KAPLAN J DR
KARANDIKAR R V PROF	KARYGINA ZOYA V DR	KOUTCHMY SERGE DR
KULKARNI PRABHAKAR V PROF	LEGER ALAIN DR	LILLIE CHARLES F DR
LOPEZ-GONZALEZ MARIA J DR	LOPEZ-MORENO JOSE JUAN	LOPEZ-PUERTAS MANUEL
LUMME KARI A DR	MAIHARA TOSHINORI DR	MATHER JOHN CROMWELL
MATSUMOTO TOSHIO DR	MATTILA KALEVI DR	MAUCHERAT J DR
MCDONNELL J A M PROF	MISCONI NEBIL YOUSIF DR	MOLINA ANTONIO
MORGAN DAVID H DR	MUINONEN KARRI DR	MUKAI SONOYO DR
NAWAR SAMIR DR	NEUZIL LUDEK DR	NEY EDWARD P PROF
NICOLET MARCEL PROF	NISHIMURA TETSUO DR	PARESCE FRANCESCO DR
PERRIN JEAN MARIE DR	PFLEIDERER JORG PROF	PITZ ECKHART DR
RADOSKI HENRY R DR	RAPAPORT MICHEL DR	RIPKEN HARTMUT W DR
ROACH FRANKLIN E	ROBLEY R DR	RODRIGO RAFAEL
ROOSEN ROBERT G DR	ROOSEN ROBERT G DR	ROZHKOVSKIJ DIMITRIJ A
SANCHEZ FRANCISCO PROF	SANCHEZ-SAAVEDRA M LUISA	SAXENA P P DR
SCHWEHM GERHARD DR	SHAROV A S DR	SHEFOV NICOLAI N
SOBERMAN ROBERT K DR	SPARROW JAMES G DR	STAUDE HANS JAKOB PH D
TANABE HIROYOSHI DR	TOLLER GARY N DR	TOROSHLIDZE TEIMURAZ I DR
TRUTSE YU L DR	TYSON JOHN A DR	VAN ALLEN JAMES A PROF
VAN DE HULST H C PROF DR	VRTILEK JAN M DR	WALLACE LLOYD V DR

WALLIS MAX K DR   WEILL GILBERT M DR   WEINBERG J L DR
WENIGER SCHAME DR   WESSON PAUL S DR   WILSON P DR
WITT ADOLF N DR   WOLSTENCROFT RAMON D DR WOOLFSON MICHAEL M PROF
YAMAKOSHI KAZUO   YAMAMOTO TETSUO DR   YAMASHITA KOJUN DR
ZERULL REINER H DR

COMPOSITION OF COMMISSION 22 1991-1994

President   : STOHL JAN DR

Vice-President(s) : WILLIAMS IWAN P PROF

Organizing Committee: BABADZHANOV PULAT B DR
                       CEPLECHA ZDENEK DR
                       HASEGAWA ICHIRO DR
                       JONES JAMES DR
                       KEAY COLIN S L PROF
                       KOEBERL CHRISTIAN DR
                       SOBERMAN ROBERT K DR
                       STEEL DUNCAN I DR
                       TEDESCO EDWARD F

Members:

ABBOTT WILLIAM N DR	BAGGALEY WILLIAM J PROF	BEARD DAVID B DR
BELKOVICH O I DR	BHANDARI N DR	BIBARSOV RAVIL'SH DR
BLACKWELL ALAN TREVOR	BROWNLEE DONALD E PROF	CARUSI ANDREA
CEVOLANI GIORDANO	CLIFTON KENNETH ST	CLUBE S V M DR .
DJORGOVSKI STANISLAV DR	DUBIN MAURICE DR	ELFORD WILLIAM GRAHAM DR
FECHTIG HUGO DR	FORTI GIUSEPPE DR	GLASS BILLY PRICE DR
GOSWAMI J N DR	GRUEN EBERHARD DR	GUSTAFSON BO A S
HAJDUK ANTON DR	HAJDUKOVA MARIA	HALLIDAY IAN DR
HANNER MARTHA S DR	HARVEY GALE A DR	HAWKES ROBERT LEWIS DR
HAWKINS GERALD S DR	HELIN ELEANOR FRANCIS	HEY JAMES STANLEY DR
HODGE PAUL W PROF	HONG SEUNG SOO DR	HUGHES DAVID W DR
JACCHIA LUIGI G DR	JENNISON ROGER C PROF	KAISER THOMAS R PROF
KAPISINSKY IGOR	KASHSCHEEV B L PROF DR	KOSTYLEV K V DR
KRAMER KH N DR	KRESAK LUBOR DR	KRESAKOVA MARGITA DR
KRUCHINENKO VITALIY G	KVIZ ZDENEK DR	LAMY PHILIPPE DR
LEBEDINETS VLADIMIR N DR	LEVASSEUR-REGOURD A C PR	LINDBLAD BERTIL A DR
LOVELL SIR BERNARD PROF	MARVIN URSULA B DR	MASON JOHN WILLIAM DR
MCCROSKY RICHARD E DR	MCDONNELL J A M PROF	MCINTOSH BRUCE A DR
MEISEL DAVID D DR	MILES HOWARD G MR	MISCONI NEBIL YOUSIF DR
MURRAY C ANDREW	NAKAZAWA KIYOSHI DR	NAPIER WILLIAM M DR
NEWBURN RAY L JR	NUTH JOSEPH A III	O'KEEFE JOHN A DR
PADEVET VLADIMIR DR	PECINA PETR	PILLINGER COLIN DR
PLAVEC ZDENKA DR	POLNITZKY GERHARD DR	PORUBCAN VLADIMIR DR
QUESADA VINICIO DR	RAJCHL JAROSLAV DR	REVELLE DOUGLAS ORSON DR
RIPKEN HARTMUT W DR	ROOSEN ROBERT G DR	RUSSELL JOHN A PROF
SEKANINA ZDENEK DR	SHAO CHENG-YUAN	SHESTAKA IVAN S DR
SIMEK MILOS DR	SVESTKA JIRI DR	TERENTJEVA ALEXANDRA K DR
TOMITA KOICHIRO MR	VERNIANI FRANCO PROF	WANG DE-CHANG
WEINBERG J L DR	WETHERILL GEORGE W	WHIPPLE FRED L DR
WOOD JOHN A DR	WOOLFSON MICHAEL M PROF	XU PINXIN
YAMAKOSHI KAZUO	YAVNEL ALEXANDER A DR	YEOMANS DONALD K DR
ZVOLANKOVA JUDITA		

COMPOSITION OF COMMISSION 24 1991-1994

President   : DE VEGT CH PROF DR

Vice-President(s) : TURON C DR

Organizing Committee: ARGUE A NOEL MR
                       HEMENWAY PAUL D DR
                       JAHREISS HARTMUT DR
                       POLOZHENTSEV DIMITRIJ DR
                       STOCK JURGEN D
                       VAN ALTENA WILLIAM F PROF
                       WANG JIA-JI

Members:

ABHYANKAR KRISHNA D PROF	BALLABH G M DR	BASTIAN ULRICH
BENEDICT GEORGE F DR	BLAAUW ADRIAAN PROF DR	BOUIGUE R
BRANHAM RICHARD L JR	BREAKIRON LEE ALLEN DR	BRONNIKOVA NINA M
BROSCHE PETER PROF	BUNCLARK PETER STEPHEN DR	CHIU LIANG-TAI GEORGE
CHRISTY JAMES WALTER DR	CHURMS JOSEPH	CLUBE S V M DR
CONNES PIERRE DR	CORBIN THOMAS ELBERT DR	CREZE MICHEL DR
CRIFO FRANCOISE DR	CUDWORTH KYLE MCCABE DR	DAHN CONARD CURTIS DR
DELHAYE JEAN PROF	DICK WOLFGANG DR	DOMMANGET J DR
DOUGLASS GEOFFREY G	EICHHORN HEINRICH K DR	ELSMORE BRUCE DR
FALLON FREDERICK W DR	FANSELOW JOHN LYMAN	FIRNEIS FRIEDRICH J DR
FIRNEIS MARIA G DR	FRACASTORO MARIO G PROF	FRANZ OTTO G DR
FREDRICK LAURENCE W PROF	FRESNEAU ALAIN DR	GALLOUET LOUIS DR
GATEWOOD GEORGE DIRECTOR	GICLAS HENRY L MR	GLIESE WIHELM PROF
GOYAL A N DR	GUIBERT JEAN DR	HANSON ROBERT B DR
HARRINGTON ROBERT S DR	HARTKOPF WILLIAM I DR	HARWOOD DENNIS MR
HEINTZ WULFF D DR	HERSHEY JOHN L DR	HEUDIER JEAN-LOUIS DR
HILL GRAHAM DR	HOFFLEIT E DORRIT DR	IANNA PHILIP A
IRWIN MICHAEL JOHN DR	JEFFERYS WILLIAM H DR	JOHNSTON KENNETH J
JONES BURTON DR	JONES DEREK H P DR	KANAEV IVAN I DR
KISLYUK VITALIJ S DR	KLEMOLA ARNOLD R DR	KLOCK B L DR
KOLCHINSKIJ I G DR	KOVALEVSKY JEAN DR	KUMKOVA IRINA I DR
LACROUTE PIERRE A PROF	LAPUSHKA K K DR	LATYPOV A A DR
LE POOLE RUDOLF S DR	LIPPINCOTT SARAH LEE DR	LOPEZ CARLOS LIC
LOZINSKIJ A M DR	LU PHILLIP K DR	LUTZ THOMAS E DR
LUYTEN WILLEM J PROF	MACCONNELL DARRELL J DR	MACHADO LUIZ E DA SILVA
MARSCHALL LAURENCE A	MCALISTER HAROLD A DR	MCLEAN BRIAN JOHN
MEINEL ADEN B PROF	MENNESSIER MARIE-ODILE DR	MIYAMOTO MASANORI DR
MONET DAVID G	MORRISON LESLIE V DR	MURRAY C ANDREW
NICHOLSON WILLIAM	NUNEZ JORGE DR	OJA TARMO PROF
ONEGINA A B DR	OSBORN WAYNE DR	PAN RONG-SHI
PASCU DAN DR	PERRYMAN MICHAEL A C DR	PIZZICHINI GRAZIELLA
PLATAIS IMANT K DR	POTTER HEINO I DR	PROCHAZKA FRANZ V DR
QIN DAO	QUIJANO LUIS	REQUIEME YVES DR
RIZVANOV NAUFAL G DR	ROEMER ELIZABETH PROF	ROESER SIEGFRIED DR

RUDER HANNS

RUSSELL JANE L DR

SANDERS W L PROF

SCHILBACH ELENA DR

SHELUS PETER J DR

SHI GUANG-CHEN

SIMS KENNETH P DR

SMITH CLAYTON A JR DR

STEIN JOHN WILLIAM

STEINERT KLAUS GUENTER DR

STONE RONALD CECIL

STRAND KAJ AA DR

THOMAS DAVID V DR

UPGREN ARTHUR R DR

VALBOUSQUET ARMAND DR

VAN DE KAMP PETER

VILKKI ERKKI U

WALTER HANS G DR

WAN LAI

WASSERMAN LAWRENCE H DR

WESSELINK ADRIAAN J DR

WESTERHOUT GART DR

WHITE GRAEME LINDSAY DR

WILLIAMS CAROL A

WORLEY CHARLES E DR

WROBLEWSKI HERBERT DR

YANG TING-GAO

YOUNIS SAAD M

ZACHARIAS NORBERT DR

COMPOSITION OF COMMISSION 25 1991-1994

President   : YOUNG ANDREW T DR

Vice-President(s) : LANDSTREET JOHN D PROF

Organizing Committee: ADELMAN SAUL J DR
                      BREGER MICHEL PROF DR
                      GENET R M DR
                      KNUDE JENS KIRKESKOV DR
                      LANDOLT ARLO U PROF
                      LUB JAN DR
                      MCLEAN IAN S DR
                      MENZIES JOHN W DR
                      MOFFETT THOMAS J PROF
                      STERKEN CHRISTIAAN LEO DR
                      STRAIZYS V PROF DR
                      VRBA FREDERICK J DR

Members:

ABLES HAROLD D DR	ALBRECHT RUDOLF DR	ANGEL J ROGER P PROF
ANGIONE RONALD J DR	ANTHONY-TWAROG BARBARA J	ARGUE A NOEL MR
ARNAUD JEAN PAUL	ARSENIJEVIC JELISAVETA	ASHOK N M DR
ASPIN COLIN DR	AXON DAVID	BAHNG JOHN D R PROF
BALDINELLI LUIGI DR	BALONA LUIS ANTERO DR	BARNES III THOMAS G DR
BARRETT PAUL EVERETT DR	BECK RAINER	BECKER WILHELM PROF
BEHR ALFRED PROF EMERITUS	BESSELL MICHAEL S DR	BLANCO VICTOR M DR
BLECHA ANDRE BORIS G DR	BOOKMYER BEVERLY B DR	BORGMAN JAN DR PROF
BORRA ERMANNO F DR	BROWN DOUGLAS NASON	BRUCK HERMANN A PROF
BUSER ROLAND DR	CARNEY BRUCE WILLIAM	CASTELAZ MICHEAL W DR
CELIS LEOPOLDO DR	CHUGAJNOV P F DR	CONNOLLY LEO PAUL
COUSINS A W J DR	COYNE GEORGE V DR	CRAWFORD DAVID L DR
CUYPERS JAN DR	DACHS JOACHIM PROF DR	DAHN CONARD CURTIS DR
DANFORD STEPHEN C DR	DENOYELLE JOZEF KIC	DESHPANDE M R DR
DOLAN JOSEPH F DR	DUBOUT RENEE	DUCATI JORGE RICARDO DR
EDWARDS PAUL J DR	EELSALU HEINO DR	FABREGAT JUAN DR
FEINSTEIN ALEJANDRO DR	FERNIE J DONALD PROF	FORTE JUAN CARLOS DR
GALLOUET LOUIS DR	GEHRELS TOM PROF	GEHRZ ROBERT DOUGLAS DR
GERBALDI MICHELE DR	GHOSH S K DR	GLASS IAN STEWART DR
GOLAY MARCEL PROF	GOY GERALD PROF	GRAHAM JOHN A DR
GRAUER ALBERT D	GRENON MICHEL DR	GREWING MICHAEL PROF
GUETTER HARRY HENDRIK	GUTIERREZ-MORENO A DR MRS	HALL DOUGLAS S DR
HARWOOD DENNIS MR	HAUCK BERNARD PROF	HAYES DONALD S DR
HECK ANDRE DR	HENSBERGE HERMAN	HILDITCH RONALD W DR
HILL PHILIP W DR	HOLMBERG ERIK B PROF	HU JING-YAO
HUANG LIN	HUOVELIN JUHANI DR	HYLAND A R HARRY DR
IRWIN ALAN W DR	IYENGAR K V K PROF	JERZYKIEWICZ MIKOLAJ DR
JORDI NEBOT CARME DR	JOSHI SURESH CHANDRA DR	JOSHI U C DR
KAWARA KIMIAKI	KEPLER S O	KILKENNY DAVID DR

KING IVAN R PROF	KOCH ROBERT H DR	KULKARNI PRABHAKAR V PROF
KUNKEL WILLIAM E DR	KURTZ DONALD WAYNE DR	KVIZ ZDENEK DR
LABHARDT LUKAS	LASKARIDES PAUL G ASSPROF	LASKER BARRY M DR
LENZEN RAINER DR	LI SIN HYONG	LINDE PETER DR
LOCKWOOD G WESLEY DR	LUNA HOMERO G DR	MAITZEN HANS M DR
MANFROID JEAN DR	MARKKANEN TAPIO DR	MARRACO HUGO G DR
MASANI A PROF	MAYER PAVEL DR	MCCARTHY MARTIN F DR
MENDOZA V EUGENIO E DR	MIANES PIERRE DR	MILLER JOSEPH S PROF
MILONE EUGENE F PROF	MINTZ BLANCO BETTY MRS	MORENO HUGO PROF
MORRIS STEPHEN C DR	MULLER A B DR	MUMFORD GEORGE S PROF
NICOLET BERNARD	NOTNI P DR	OBLAK EDOUARD
OESTREICHER ROLAND	ORSATTI ANA M DR	PAGE ARTHUR MR
PEDREROS MARIO DR	PEL JAN WILLEM DR	PENNY ALAN JOHN DR
PERRY CHARLES L DR	PFAU WERNER	PFEIFFER RAYMOND J
PHILIP A G DAVIS	PIIROLA VILPPU E DR	RAO P VIVEKANANDA DR
REGLERO-VELASCO VICTOR DR	ROBINSON EDWARD LEWIS DR	ROSLUND CURT DR
RUFENER FREDY G PROF	RYDGREN ALFRED ERIC JR DR	SARMA M B K PROF
SCHMIDT EDWARD G	SCHOENEICH W DR	SCHUSTER WILLIAM JOHN DR
SHAKHOVSKOJ NIKOLAY M DR	SHAWL STEPHEN J DR	SMITH CHARLES DITTO
SMYTH MICHAEL J DR	STEINLIN ULI PROF	STOCK JURGEN D
STOCKMAN HERVEY S JR DR	STONE REMINGTON P S DR	STROHMEIER WOLFGANG PROF
SULLIVAN DENIS JOHN DR	SZKODY PAULA DR	TANDON S N PROF
TAPIA-PEREZ SANTIAGO	TINBERGEN JAAP DR	TODORAN IOAN DR
TOKUNAGA ALAN TAKASHI DR	TOLBERT CHARLES R DR	TRODAHL HARRY JOSEPH DR
ULRICH BRUCE T PROF	URECHE VASILE DR	VARDANIAN R A DR
VAUGHAN ARTHUR H DR	VERMA R P DR	VISVANATHAN NATARAJAN DR
WALKER ALISTAIR ROBIN DR	WALLENQUIST AAKE A E PROF	WALRAVEN TH DR
WANG CHUAN-JIN	WARREN WAYNE H JR DR	WEISS WERNER W DR
WEISTROP DONNA DR	WESSELINK ADRIAAN J DR	WESSELIUS PAUL R DR
WHITE NATHANIEL M DR	WIELEBINSKI RICHARD PROF	WILLSTROP RODERICK V DR
WINIARSKI MACIEJ	WISNIEWSKI WIESLAW Z	WOO JONG OK
WRAMDEMARK STIG S O DR	YAMASHITA YASUMASA PROF	YIN JI-SHENG
ZIZNOVSKY JOZEF DR		

COMPOSITION OF COMMISSION 26   1991-1994

President   : ABT HELMUT A DR

Vice-President(s) : WORLEY CHARLES E DR

Organizing Committee: BALEGA YURI YU DR
                       BERNACCA P L PROF
                       FEKEL FRANCIS C
                       MCALISTER HAROLD A DR
                       SCARFE COLIN D DR
                       VAN DESSEL EDWIN LUDO DR
                       ZINNECKER HANS

Members:

ALLEN CHRISTINE	ARGUE A NOEL MR	ARMSTRONG JOHN THOMAS DR
BACCHUS PIERRE PROF	BAGNUOLO WILLIAM G JR DR	BAIZE PAUL DR
BATTEN ALAN H DR	BEAVERS WILLET I DR	BONNEAU DANIEL
BROSCHE PETER PROF	CABRITA EZEQUIEL DR	CAMPBELL ALISON DR
CESTER BRUNO PROF	CHEN ZHEN	COUTEAU PAUL PROF
CULVER ROGER BRUCE DR	DADAEV ALEKSANDR N DR	DOCOBO DURANTEZ JOSE A
DOMMANGET J DR	DOUGLASS GEOFFREY G	DUNHAM DAVID W
EICHHORN HEINRICH K DR	FERRER OSVALDO EDUARDO DR	FLETCHER J MURRAY
FRACASTORO MARIO G PROF	FRANZ OTTO G DR	FREDRICK LAURENCE W PROF
FREITAS MOURAO R R DR	FURENLID INGEMAR K DR	GATEWOOD GEORGE DIRECTOR
GAUDENZI SILVIA DR	GEYER EDWARD H PROF DR	GLIESE WIHELM PROF
HALBWACHS JEAN LOUIS DR	HARRINGTON ROBERT S DR	HARTKOPF WILLIAM I DR
HERSHEY JOHN L DR	HIDAJAT BAMBANG PROF DR	HILL GRAHAM DR
HOLDEN FRANK	IANNA PHILIP A	ISHIDA GORO DR
JASCHEK CARLOS O R PROF	KISELYOV ALEXEJ A DR	KOPAL ZDENEK PROF
KUMSISHVILI J I DR	LAMPENS PATRICIA DR	LATHAM DAVID W DR
LATTANZI MARIO G	LEINERT CHRISTOPH DR	LING J DR
LIPPINCOTT SARAH LEE DR	LODEN KERSTIN R DR	LODEN LARS OLOF PROF
LUYTEN WILLEM J PROF	MEYER CLAUDE DR	MIKKOLA SEPPO DR
MOHAN CHANDER DR	MORBEY CHRISTOPHER L	MORBIDELLI ROBERTO DR
MOREL PIERRE JACQUES DR	MULLER PAUL	OBLAK EDOUARD
OSWALT TERRY D DR	PANNUNZIO RENATO	PETERSON DEANE M DR
POPOVIC GEORGIJE DR	POVEDA ARCADIO DR	RAKOS KARL D PROF
RUSSELL JANE L DR	SALUKVADZE G N DR	SCARDIA MARCO
SCHMIDTKE PAUL C DR	SHUL'BERG A M DR	SINACHOPOULOS D DR
SMAK JOSEPH I PROF	SOWELL JAMES ROBERT DR	STEIN JOHN WILLIAM
STRAND KAJ AA DR	SZABADOS LASZLO PH D	TOKOVININ ANDREJ A DR
TRIMBLE VIRGINIA L DR	TSAY WEAN-SHUN DR	UPGREN ARTHUR R DR
VALBOUSQUET ARMAND DR	VALTONEN MAURI J PROF	VAN ALTENA WILLIAM F PROF
VAN DE KAMP PETER	VAN DER HUCHT KAREL A DR	WALKER RICHARD L
WEIS EDWARD W DR	WIETH-KNUDSEN NIELS P DR	YAN LIN-SHAN

COMPOSITION OF COMMISSION 27 1991-1994

President   : PERCY JOHN R PROF

Vice-President(s) : JERZYKIEWICZ MIKOLAJ DR

Organizing Committee: BARNES III THOMAS G DR
                       BREGER MICHEL PROF DR
                       CHRISTENSEN-DALSGAARD J
                       GERSHBERG R E DR
                       KURTZ DONALD WAYNE DR
                       MATTEI JANET AKYUZ DR
                       RODONO MARCELLO DR
                       SAMUS NIKOLAI N DR
                       SMITH MYRON A ASST PROF
                       SZEIDL BELA DR
                       TAKEUTI MINE DR

Members:

AIZENMAN MORRIS L DR	ALBINSON JAMES DR	ALFARO EMILIO JAVIER
ALPAR ALI DR	ANDO HIROYASU DR	ANTIPOVA LYUDMILA DR
ANTONELLO ELIO	ARELLANO FERRO ARMANDO	ARKHIPOVA V P DR
ARSENIJEVIC JELISAVETA	ASTERIADIS GEORGIOS DR	AVGOLOUPIS STAVROS DR
BAADE DIETRICH DR	BAGLIN ANNIE DR	BAKER NORMAN H PROF
BALONA LUIS ANTERO DR	BARTOLINI CORRADO	BARWIG HEINZ
BASTIEN PIERRE DR	BATESON FRANK M OBE DR	BATH GEOFFREY T DR
BAUER WENDY HAGEN	BEDOGNI ROBERTO	BELMONTE AVILES J A DR
BELSERENE EMILIA P	BELVEDERE GAETANO DR	BENSON PRISCILLA J DR
BERTHOMIEU GABRIELLE DR	BESSELL MICHAEL S DR	BIANCHINI ANTONIO DR
BOCHONKO D RICHARD DR	BOLTON C THOMAS PROF	BOND HOWARD E DR
BOPP BERNARD W DR	BOULON JACQUES J DR	BOWEN GEORGE H DR
BOYARCHUK A A DR	BOYARCHUK MARGARITA E DR	BROWN DOUGLAS NASON
BURKI GILBERT PROF	BUSKO IVO C DR	BUTLER C JOHN DR
BUTLER DENNIS DR	BYRNE PATRICK B DR	CAMERON ANDREW COLLIER DR
CATCHPOLE ROBIN M DR	CHAVIRA ENRIQUE SR	CHEREPASHCHUK A M PROF
CHRISTY ROBERT F DR	CHUGAJNOV P F DR	COHEN MARTIN DR
CONNOLLY LEO PAUL	CONTADAKIS MICHAEL E DR	COULSON IAIN M DR
COUTTS-CLEMENT CHRISTINE	COX ARTHUR N DR	CUTISPOTO GIUSEPPE DR
CUYPERS JAN DR	DANFORD STEPHEN C DR	DE GROOT MART DR
DELGADO ANTONIO JESUS	DEMERS SERGE DR	DEUPREE ROBERT G DR
DICKENS ROBERT J DR	DJORGOVSKI STANISLAV DR	DOWNES RONALD A DR
DUNLOP STORM	DUPUY DAVID L DR	DZIEMBOWSKI WOJCIECH PROF
EDWARDS PAUL J DR	EFREMOV YURY N DR	EL-BASSUNY ALAWY A A DR
ESKIOGLU A NIHAT	EVANS ANEURIN	EVANS NANCY REMAGE DR
EVREN SERDAR DR	FADEYEV YURI A	FEAST MICHAEL W PROF
FEIBELMAN WALTER A DR	FERLAND GARY JOSEPH	FERNIE J DONALD PROF
FITCH WALTER S DR	FRIEDJUNG MICHAEL DR	FROLOV M S DR
GAHM GOESTA F DR	GALLAGHER III JOHN S DR	GARRIDO RAFAEL
GASCOIGNE S C B DR	GENET R M DR	GEYER EDWARD H PROF DR

SINVHAL SHAMBHU DAYAL DR	SMAK JOSEPH I PROF	SMEYERS PAUL PROF
SOLIMAN MOHAMED AHMED DR	SRIVASTAVA RAM KUMAR DR	STARRFIELD SUMNER PROF
STELLINGWERF ROBERT F DR	STEPIEN KAZIMIERZ DR	STERKEN CHRISTIAAN LEO DR
STOBIE ROBERT S DR	STRASSMEIER KLAUS G DR	STROHMEIER WOLFGANG PROF
STROM KAREN M	STROM STEPHEN E	SZABADOS LASZLO PH D
SZATMARY KAROLY DR	SZECSENYI-NAGY GABOR DR	SZKODY PAULA DR
TAMMANN G ANDREAS PROF DR	TEMPESTI PIERO PROF	TERZAN AGOP DR
THOMPSON KEITH DR	TJIN-A-DJIE HERMAN R E DR	TORRES CARLOS ALBERTO DR
TREMKO JOZEF DR	TSIOUMIS ALEXANDROS DR	TSVETKOV MILCHO K DR
TURNER DAVID G DR	TUTUKOV A V DR	TYLENDA ROMUALD DR
USHER PETER D DR	VALTIER JEAN-CLAUDE DR	VAN AGT S L TH J DR
VAN GENDEREN A M DR	VERHEEST FRANK PROF	VIOTTI ROBERTO DR
VOGT NIKOLAUS DR	WAELKENS CHRISTOFFEL	WALKER EDWARD N MR
WALKER MERLE F PROF	WALKER WILLIAM S G	WALLERSTEIN GEORGE PROF
WALRAVEN TH DR	WARNER BRIAN PROF	WATSON ROBERT DR
WEBBINK RONALD F DR	WEHLAU AMELIA DR	WEHLAU WILLIAM H PROF
WEISS WERNER W DR	WELCH DOUGLAS L DR	WENZEL W DR
WESSELINK ADRIAAN J DR	WHITELOCK PATRICIA ANN DR	WILLIAMON RICHARD M
WILLSON LEE ANNE DR	WILSON LIONEL DR	WING ROBERT F PROF
WISNIEWSKI WIESLAW Z	WOOD PETER R DR	WRIGHT FRANCES W DR
XIONG DA-RUN	YAO BAO-AN	ZSOLDOS ENDRE DR
ZUCKERMAN BEN M DR		

COMPOSITION OF COMMISSION 28 1991-1994

President   : KHACHIKIAN E YE PROF

Vice-President(s) : TRIMBLE VIRGINIA L DR

Organizing Committee: BERTOLA FRANCESCO PROF
                       BRUZUAL GUSTAVO
                       CHEN JIAN-SHENG
                       ELLIS RICHARD S
                       FREEMAN KENNETH C PROF
                       GALLAGHER III JOHN S DR
                       ISRAEL FRANK P DR
                       OKAMURA SADANORI DR
                       TAMMANN G ANDREAS PROF DR
                       ULRICH MARIE-HELENE D DR

Members:

ABLES HAROLD D DR	AFANAS'EV VIKTOR L DR	AGUERO ESTELA L DR
AGUILAR CHIU LUIS A DR	AHMAD FAROOQ DR	ALCAINO GONZALO DR
ALLADIN SALEH MOHAMED DR	ALLEN RONALD J DR	ALLOIN DANIELLE DR
AMBARTSUMIAN V A PROF DR	ANDRILLAT YVETTE DR	ANN HONG BAE DR
ARDEBERG ARNE L PROF	ARKHIPOVA V P DR	ARP HALTON DR
ATHANASSOULA EVANGELIE DR	AZZOPARDI MARC DR	BAHCALL JOHN N PROF
BAILEY MARK EDWARD	BAJAJA E DR	BALDWIN JACK A DR
BALKOWSKI-MAUGER CH DR	BALLABH G M DR	BANHATTI DILIP GOPAL DR
BARBON ROBERTO PROF	BARCONS XAVIER DR	BARTHEL PETER DR
BASSINO LILIA P DR	BASU BAIDYANATH PROF	BATTANER EDUARDO DR
BATTINELLI PAOLO DR	BAUM WILLIAM A DR	BECK RAINER
BEGEMAN KOR G DR	BENDER RALF DR	BENDINELLI ORAZIO
BENEDICT GEORGE F DR	BERGERON JACQUELINE A DR	BERGVALL NILS AKE SIGVARD
BERKHUIJSEN ELLY M DR	BETTONI DANIELA DR	BHATTACHARYYA TARA DR
BICA EDUARDO L D DR	BIERMANN PETER L DR	BIJAOUI ALBERT DR
BINETTE LUC	BINGGELI BRUNO	BINNEY JAMES J DR
BIRETTA JOHN ANTHONY DR	BIRKINSHAW MARK	BLITZ LEO DR
BLOCK DAVID LAZAR PROF	BLUMENTHAL GEORGE R DR	BOERNGEN FREIMUT DR PH
BOESHAAR GREGORY ORTH DR	BOISSON CATHERINE DR	BOKSENBERG ALEC PROF
BONTEKOE ROMKE DR	BORCHKHADZE TENGIZ M DR	BOSMA ALBERT DR
BOTTINELLI LUCETTE DR	BOUCHET PATRICE DR	BRACCESI ALESSANDRO PROF
BRAUN ROBERT DR	BRECHER KENNETH PROF	BRINKMANN WOLFGANG
BRINKS ELIAS DR	BRODIE JEAN P	BROSCH NOAH DR
BROSCHE PETER PROF	BURBIDGE E MARGARET PROF	BURBIDGE GEOFFREY R PROF
BURNS JACK O'NEAL JR	BURSTEIN DAVID	BUTA RONALD J DR
BUTCHER HARVEY R PROF DR	BYRD GENE G DR	CAMERON LUZIUS MARTIN
CAMPUSANO LUIS E	CANNON RUSSELL D DR	CAPACCIOLI MASSIMO DR
CARIGNAN CLAUDE DR	CARRANZA GUSTAVO J DR	CARSWELL ROBERT F DR
CARTER DAVID DR	CASERTANO STEFANO DR	CEPA JORDI DR
CHALABAEV ALMAS DR	CHAMARAUX PIERRE DR	CHATTERJEE TAPAN K DR
CHEN ZHENCHENG	CHINCARINI GUIDO L DR	CHOU CHIH-KANG DR

CHU YAOQUAN	CHUGAI NIKOLAI N DR	CHUVAEV K K DR
CLAVEL JEAN	COHEN ROSS D DR	COLIN/JACQUES DR
COMTE GEORGES DR	CONTOPOULOS GEORGE PROF	COOK KEM HOLLAND DR
CORWIN HAROLD G JR	COUCH WARRICK DR	COURTES G PROF
COURVOISIER THIERRY J-L	COWSIK RAMANATH	CRANE PHILIPPE
DA COSTA NICOLAI L-A	DANKS ANTHONY C DR	DAVIDGE TIMOTHY J DR
DAVIDSEN ARTHUR FALNES DR	DAVIES RODNEY D PROF	DAVIS MARC DR
DE BOER KLAAS SJOERDS DR	DE BRUYN A GER DR	DE CARVALHO REINALDO DR
DE LA NOE JEROME DR	DE ROBERTIS M M DR	DE SILVA L N K DR
DE VAUCOULEURS GERARD PR	DE ZEEUW PIETER T DR	DEJONGHE HERWIG BERT DR
DEKEL AVISHAI	DEL OLMO OROZCO A DR	DEMERS SERGE DR
DETTMAR RALF-JRGEN DR	DI FAZIO ALBERTO	DI SEREGO ALIGHIERI S DR
DIAZ ANGELES ISABEL DR	DICKENS ROBERT J DR	DICKEY JOHN M
DONAS JOSE DR	DONNER KARL JOHAN	DOTTORI HORACIO A DR
DRESSEL LINDA L	DRESSLER ALAN	DUBOIS PASCAL DR
DUFOUR REGINALD JAMES	DULTZIN-HACYAN D DR	DURRET FLORENCE DR
DUVAL MARIE-FRANCE	D'ODORICO SANDRO DR	EBISUZAKI TOSHIKAZU DR
EDELSON RICK DR	EDMUNDS MICHAEL GEOFFREY	EFSTATHIOU GEORGE
EINASTO JAAN DR	EKERS RONALD D DR	ELMEGREEN DEBRA MELOY
ELVIS MARTIN S DR	ELVIUS AINA M PROF	EMERSON DAVID
EVANS ROBERT REV	EVANS ROGER G DR	FABBIANO GIUSEPPINA
FABER SANDRA M PROF	FABRICANT DANIEL G	FAIRALL ANTHONY P PROF
FALCO-ACOSTA EMILIO E DR	FALL S MICHAEL DR	FEAST MICHAEL W PROF
FEITZINGER JOHANNES PROF	FERLAND GARY JOSEPH	FERRINI FEDERICO
FIELD GEORGE B PROF	FILIPPENKO ALEXEI V DR	FLIN PIOTR
FLORSCH ALPHONSE DR	FOLTZ CRAIG B	FORD HOLLAND C RES PROF
FORD W KENT JR DR	FORTE JUAN CARLOS DR	FOUQUE PASCAL DR
FREEDMAN WENDY L DR	FRICKE KLAUS DR	FRIED JOSEF WILHELM DR
FROGEL JAY ALBERT DR	FTACLAS CHRIST	FUCHS BURKHARD DR
FUJIMOTO MASAYUKI DR	FUKUGITA MASATAKA DR	GALLETTA GIUSEPPE PROF
GAMALELDIN ABDULLA I DR	GARCIA LAMBAS DIEGO DR	GASCOIGNE S C B DR
GAVAZZI GIUSEPPE DR	GELLER MARGARET JOAN	GEORGIEV TSVETAN DR
GERHARD ORTWIN	GHIGO FRANCIS D DR	GHOSH P DR
GIACANI ELSA BEATRIZ DR	GIOVANARDI CARLO	GIOVANELLI RICCARDO DR
GLASS IAN STEWART DR	GONZALEZ SERRANO J I DR	GOODRICH ROBERT W DR
GORGAS GARCIA JAVIER DR	GOSS W MILLER PROF	GOTTESMAN STEPHEN T DR
GOUGUENHEIM LUCIENNE	GRAHAM JOHN A DR	GRANDI STEVEN ALDRIDGE DR
GRASDALEN GARY L DR	GREGG MICHAEL DAVID DR	GRIFFITHS RICHARD E DR
GUNN JAMES E PROF	GURZADIAN G A PROF DR	HAGEN-THORN VLADIMIR A DR
HAGIO FUMIHIKO DR	HAMABE MASARU DR	HAMMER FRANCOIS DR
HANAMI HITOSHI DR	HANDA TOSHIHIRO DR	HARA TETSUYA DR
HARDY EDUARDO	HARMS RICHARD JAMES DR	HARNETT JULIEINE DR
HASAN HASHIMA DR	HE XIANG-TAO	HECKMAN TIMOTHY M
HEESCHEN DAVID S DR	HEIDMANN JEAN DR	HELOU GEORGE DR
HENIZE KARL G ASTRONAUT	HENRY RICHARD B C DR	HERNQUIST LARS ERIC DR
HEWITT ADELAIDE	HEWITT ANTHONY V DR	HICKSON PAUL DR
HINTZEN PAUL MICHAEL N DR	HIOTELIS NICOLAOS DR	HJALMARSON AKE G DR
HODGE PAUL W PROF	HOLMBERG ERIK B PROF	HOPP ULRICH DR
HOYLE FRED SIR	HU FU-XING	HUA CHON TRUNG DR
HUANG JIE-HAO	HUANG KE-LIANG	HUANG SONG-NIAN DR
HUANG YONGWEI	HUCHRA JOHN PETER DR	HUCHTMEIER WALTER K DR
HUMMEL EDSHO	HUMPHREYS ROBERTA M PROF	HUNSTEAD RICHARD W DR
HUNTER CHRISTOPHER PROF	HUNTER JAMES H PROF	ICHIKAWA SHIN-ICHI DR
ICHIKAWA TAKASHI	ILLINGWORTH GARTH D DR	IMPEY CHRISTOPHER D DR
IRWIN JUDITH DR	ISSA ALI DR	IZOTOV YURI DR

JAFFE WALTER JOSEPH DR　　JOG CHANDA J DR　　JONES THOMAS WALTER DR
JOSHI MOHAN N PROF　　JOSHI U C DR　　JOY MARSHALL J DR
JUGAKU JUN DR　　JUNKKARINEN VESA T DR　　KALAFI MANOUCHER
KALINKOV MARIN P DR　　KALLOGLIAN ARSEN T DR　　KANDRUP HENRY EMIL DR
KANEKO NOBORU DR　　KAPAHI VIJAY K DR　　KARACHENTSEV I D DR
KAROJI HIROSHI DR　　KATGERT PETER DR　　KAUFMAN MICHELE DR
KEEL WILLIAM C　　KELLERMANN KENNETH I DR　　KENNICUTT ROBERT C JR
KERR FRANK J DR　　KING IVAN R PROF　　KINMAN THOMAS D DR
KIRSHNER ROBERT PAUL DR　　KLEIN ULRICH　　KNAPP GILLIAN R DR
KOCHHAR R K DR　　KODAIRA KEIICHI PROF　　KOGOSHVILI NATELA G
KOJOIAN GABRIEL DR　　KOLLATSCHNY WOLFRAM DR　　KONTIZAS EVANGELOS DR
KOO DAVID C-Y DR　　KORMENDY JOHN DR　　KRAAN-KORTEWEG RENEE C DR
KRISHNA GOPAL　　KRON RICHARD G　　KRUMM NATHAN ALLYN
KUNCHEV PETER DR　　KUNTH DANIEL　　KUSTAANHEIMO PAUL E PROF
LAFON JEAN-PIERRE J DR　　LARSON RICHARD B PROF　　LASKER BARRY M DR
LAUBERTS ANDRIS DR　　LAUSBERG ANDRE DR　　LAWRENCE ANDREW DR
LAYZER DAVID PROF　　LE FEVRE OLIVIER DR　　LEACOCK ROBERT JAY
LELIEVRE GERARD DR　　LEQUEUX JAMES DR　　LI JING
LI QI-BIN　　LI XIAO-QING　　LILLER WILLIAM DR
LILLY SIMON J DR　　LIN CHIA C PROF　　LINDBLAD PER OLOF PROF
LIPOVETSKY V A　　LIU RU-LIANG　　LIU YONG-ZHEN
LO KWOK-YUNG DR　　LONSDALE CAROL J DR　　LOOSE HANS-HERMANN DR
LOPEZ ROSARIO DR　　LORD STEVEN DONALD DR　　LORENZ HILMAR
LORTET MARIE CLAIRE　　LOW FRANK J DR　　LUCEY JOHN DR
LUGGER PHYLLIS M　　LUMINET JEAN-PIERRE　　LYNDEN-BELL DONALD PROF
LYNDS BEVERLY T DR　　LYNDS ROGER C DR　　MA ER
MACALPINE GORDON M　　MACCHETTO FERDINANDO DR　　MADORE BARRY FRANCIS DR
MARCELIN MICHEL　　MARQUES DOS SANTOS P PROF　　MARTIN MARIA CRISTINA DR
MARTINET LOUIS PROF　　MASEGOSA GALLEGO J DR　　MATHEWSON DONALD S PROF
MAURICE ERIC N　　MAVRIDES STAMATIA DR　　MCBREEN BRIAN PHILIP DR
MCCREA WILLIAM SIR　　MEDIAVILLA EVENCIO DR　　MEIER DAVID L
MEIKLE WILLIAM P S　　MEISENHEIMER KLAUS DR　　MENON T K PROF
MILET BERNARD L DR　　MILEY G K DR　　MILLER HUGH R PROF
MILLER JOSEPH S PROF　　MILLER RICHARD H DR　　MILLS BERNARD Y PROF
MINEVA VENETA DR　　MIRABEL IGOR FELIX DR　　MOLES MARIANO J DR
MOORWOOD ALAN F M　　MORENO EDMUNDO DR　　MORGAN WILLIAM W PROF
MOSS CHRISTOPHER DR　　MOULD JEREMY R　　MUNOZ-TUNON CASIANA
MURRAY STEPHEN S DR　　MUZZIO JUAN C PROF　　NAKAI NAOMASA DR
NARLIKAR JAYANT V PROF　　NAVARRO JULIO FERNANDO DR　　NITYANANDA R DR
NOGUCHI MASAFUMI DR　　NOONAN THOMAS W PROF　　NOREAU LOUIS DR
NORMAN COLIN A PROF　　NULSEN PAUL DR　　OEMLER AUGUSTUS JR DR
OKE J BEVERLEY PROF　　OLEAK H DR　　OORT JAN H PROF
OSMAN ANAS MOHAMED PROF　　OSTERBROCK DONALD E PROF　　OWEN FRAZER NELSON DR
O'CONNELL ROBERT WEST DR　　O'DEA CHRISTOPHER P DR　　PACHOLCZYK ANDRZEJ G PROF
PAGE THORNTON L DR　　PALMER PHILIP DR　　PALUMBO GIORGIO G C DR
PAN RONG-SHI　　PAPAYANNOPOULOS TH DR　　PARKER QUENTIN DR
PASTORIZA MIRIANI G DR　　PATUREL GEORGES　　PEIMBERT MANUEL DR
PENG QIU-HE　　PEREA-DUARTE JAIME D DR　　PEREZ ENRIQUE DR
PEREZ FOURNON ISMAEL DR　　PETERS WILLIAM L III DR　　PETERSON BRADLEY MICHAEL
PETERSON CHARLES JOHN DR　　PETROSIAN ARTASHES R DR　　PETROV GEORGY TRENDAFILOV
PFENNIGER DANIEL DR　　PHILLIPS MARK M DR　　PICKLES ANDREW JOHN DR
PISMIS DE RECILLAS PARIS　　POPOV VASIL NIKOLOV　　POVEDA ARCADIO DR
PRABHU TUSHAR P　　PRENDERGAST KEVIN H PROF　　PRESS WILLIAM H DR
PREVOT-BURNICHON M L DR　　PRIEUR JEAN-LOUIS DR　　PRITCHET CHRISTOPHER J DR
PRONIK I I DR　　PRONIK V I DR　　PROUST DOMINIQUE

QUINN PETER DR	QUINTANA HERNAN DR	RAFANELLI PIERO DR
RAMPAZZO ROBERTO DR	RANA NARAYAN CHANDRA DR	REAVES GIBSON PROF
REBEIROT EDITH DR	REICHERT GAIL ANNE DR	REPHAELI YOEL DR
RICHER HARVEY B DR	RICHSTONE DOUGLAS O DR	ROBERTS MORTON S DR
ROBERTS WILLIAM W JR PROF	ROBERTSON JAMES GORDON DR	RODRIGUEZ-ESPINOSA JOSE
ROESER HERMANN-JOSEF DR	ROOD HERBERT J	ROOS NICOLAAS DR
ROSA MICHAEL RICHARD DR	ROSADO MARGARITA DR	ROSE JAMES ANTHONY
ROTS ARNOLD H DR	RUBIN VERA C DR	RUDNICKI KONRAD PROF
RYDBECK GUSTAF H B DR	SADLER ELAINE MARGARET	SADUN ALBERTO CARLO DR
SALA FERRAN DR	SALVADOR-SOLE EDUARDO	SANAHUJA BLAS
SANCISI RENZO DR	SANDERS DAVID B DR	SANDERS ROBERT DR
SANROMA MANUEL DR	SAPRE A K DR	SARAZIN CRAIG L DR
SARGENT WALLACE L W DR	SASAKI TOSHIYUKI DR	SASLAW WILLIAM C PROF
SASTRY SHANKARA K	SAVAGE ANN DR	SCHECHTER PAUL L DR
SCHMIDT K H DR	SCHMIDT MAARTEN PROF	SCHUECKING E L DR
SCHULTZ G V DR	SCHULZ HARTMUT DR	SCHWARZ ULRICH J DR
SCHWEIZER FRANCOIS DR	SCIAMA DENNIS W DR	SCOVILLE NICHOLAS Z
SEARLE LEONARD DR	SEIDEN PHILIP E	SELLWOOD JERRY A
SERSIC J L DR	SETTI GIANCARLO PROF	SHAHBAZIAN ROMELIA K DR
SHAKESHAFT JOHN R DR	SHARPLES RAY DR	SHAVER PETER A DR
SHAYA EDWARD J DR	SHERWOOD WILLIAM A DR	SHIELDS GREGORY A DR
SHORE STEVEN N	SHOSTAK G SETH DR	SILLANPAA AIMO KALEVI DR
SIMIEN FRANCOIS DR	SIMKIN SUSAN M DR	SITKO MICHAEL L
SKILLMAN EVAN D DR	SLEZAK ERIC DR	SMITH BRUCE F DR
SMITH ERIC PHILIP DR	SMITH HARDING E JR DR	SMITH HAYWOOD C DR
SMITH MALCOLM G DR	SOBOUTI YOUSEF PROF	SOLTAN ANDRZEJ MARIA DR
SONG GUO-XUAN	SPARKS WILLIAM BRIAN	SPINRAD HYRON PROF
STAVELEY-SMITH LISTER DR	STEPANIAN A A DR	STEPANIAN JIVAN A DR
STIAVELLI MASSIMO DR	STIRPE GIOVANNA M DR	STONE REMINGTON P S DR
STORCHI-BERGMAN THAISA DR	STROM RICHARD G DR	STROM ROBERT G PROF
SU HONG-JUN	SUBRAHMANYAM P V DR	SULENTIC JACK W DR
SULLIVAN WOODRUFF T III	SUNDELIUS BJOERN DR	TACCONI LINDA J DR
TAKALO LEO O DR	TAKAMI HIDEKI DR	TAKASE BUNSHIRO PROF
TALBOT RAYMOND J JR DR	TANAKA YUTAKA D DR	TANIGUCHI YOSHIAKI DR
TELESCO CHARLES M DR	TERLEVICH ROBERTO JUAN	TERZIAN YERVANT PROF
THAKUR RATNA KUMAR DR	THOMAS PETER A DR	THOMPSON LAIRD A DR
THONNARD NORBERT DR	THUAN TRINH XUAN DR	TIFFT WILLIAM G PROF
TONG YI	TONRY JOHN DR	TOOMRE ALAR DR
TOVMASSIAN H M DR	TOYAMA KIYOTAKA	TREDER H J PROF DR
TREMAINE SCOTT DUNCAN	TRINCHIERI GINEVRA	TULLY RICHARD BRENT DR
TURNER EDWIN L DR	TYSON JOHN A DR	UNGER STEPHEN DR
URBANIK MAREK DR	VALENTIJN EDWIN A DR	VALTONEN MAURI J PROF
VAN ALBADA TJEERD S DR	VAN DEN BERGH SIDNEY PROF	VAN DER HULST JAN M DR
VAN DER KRUIT PIETER C DR	VAN DER LAAN H PROF DR	VAN DRIEL WILLEM DR
VAN GENDEREN A M DR	VAN GORKOM JACQUELINE H	VAN MOORSEL GUSTAAF DR
VAN WOERDEN HUGO PROF DR	VARMA RAM KUMAR PROF	VAUGLIN ISABELLE DR
VERON MARIE-PAULE DR	VERON PHILIPPE DR	VISVANATHAN NATARAJAN DR
VOGLIS NIKOS DR	VORONTSOV-VEL'YAMINOV B A	VRTILEK JAN M DR
WAKAMATSU KEN-ICHI DR	WALTERBOS RENE A M DR	WARD MARTIN JOHN
WARNER JOHN W DR	WEEDMAN DANIEL W PROF	WEHINGER PETER A DR
WEILER KURT W DR	WELCH GARY A DR	WESTERLUND BENGT E PROF
WHITE SIMON DAVID MANION	WHITFORD ALBERT E PROF	WHITMORE BRADLEY C
WIELEBINSKI RICHARD PROF	WIELEN ROLAND PROF DR	WIITA PAUL JOSEPH
WILD PAUL PROF	WILKINSON ALTHEA	WILLIAMS BARBARA A
WILLIAMS ROBERT E DR	WILLIAMS THEODORE B DR	WILLS BEVERLEY J DR

WILLS DEREK DR   WILSON ALBERT G DR   WINDHORST ROGIER A DR
WLERICK GERARD DR   WOOSLEY S E PROF   WORRALL DIANA MARY
WROBEL JOAN MARIE DR   WYNN-WILLIAMS C G DR   XIA XIAOYANG DR
YAMAGATA TOMOHIKO DR   YOUNG JUDITH SHARN   ZAMORANO JAIME DR
ZASOV ANATOLE V DR   ZAVATTI FRANCO   ZEILINGER WERNER W DR
ZHOU YOU-YUAN   ZINN ROBERT J DR

COMPOSITION OF COMMISSION 29 1991-1994

President   : LAMBERT DAVID L PROF

Vice-President(s) : BESSELL MICHAEL S DR

Organizing Committee: BAADE DIETRICH DR
                       BARBUY BEATRIZ DR
                       BOESGAARD ANN M PROF
                       CASSATELLA ANGELO DR
                       CONTI PETER S DR
                       GERBALDI MICHELE DR
                       GUSTAFSSON BENGT DR
                       NUGIS TIIT
                       PILACHOWSKI CATHERINE DR
                       SADAKANE KOZO DR

Members:

ABHYANKAR KRISHNA D PROF	ABT HELMUT A DR	ADELMAN SAUL J DR
AIKMAN G CHRIS L	ALECIAN GEORGES DR	ALLER LAWRENCE HUGH
ANDRILLAT HENRI L PROF	ANDRILLAT YVETTE DR	APPENZELLER IMMO PROF
ARAUJO FRANCISCO X DE DR	ARTRU MARIE-CHRISTINE DR	ATAC TAMER
BALIUNAS SALLIE L	BALLEREAU DOMINIQUE DR	BARATTA GIOVANNI BATTISTA
BARRY DON C DR	BASRI GIBOR B	BATALHA CELSO CORREA DR
BAUER WENDY HAGEN	BECKMAN JOHN E PROF	BELLAS-VELIDIS IOANNIS DR
BERGER JACQUES G DR	BOGGESS ALBERT DR	BOND HOWARD E DR
BONSACK WALTER K PROF	BOPP BERNARD W DR	BOUVIER JEROME
BOUVIER PIERRE PROF	BOYARCHUK A A DR	BRANDI ELISANDE ESTELA DR
BREYSACHER JACQUES	BROWN DOUGLAS NASON	BRUHWEILER FRED C JR
BRUNING DAVID H DR	BUES IRMELA D DR	BURKHART CLAUDE DR
BUTLER KEITH DR	CAMPBELL BRUCE DR	CARNEY BRUCE WILLIAM
CARPENTER KENNETH G DR	CASTELLI FIORELLA DR	CATALA CLAUDE DR
CATALANO SANTO DR	CATCHPOLE ROBIN M DR	CAYREL DE STROBEL GIUSA
CAYREL ROGER DR	CLIMENHAGA JOHN L PROF	COLUZZI REGINA DR
CORBALLY CHRISTOPHER	CORNIDE MANUEL	COTTRELL PETER LEDSAM
COWLEY ANNE P DR	COWLEY CHARLES R PROF	DAWANAS DJONI N DR
DE CASTRO ELISA	DE GROOT MART DR	DIVAN LUCIENNE DR
DOAZAN VERA DR	DOBRONRAVIN PETER DR	DOLIDZE MADONA V DR
DRAVINS DAINIS PROF	DUNCAN DOUGLAS KEVIN DR	DWORETSKY MICHAEL M DR
FARAGGIANA ROSANNA PROF	FEAST MICHAEL W PROF	FELENBOK PAUL DR
FERNANDEZ-FIGUEROA M J DR	FITZPATRICK EDWARD L DR	FLOQUET MICHELE DR
FOING BERNARD H DR	FOY RENAUD DR	FRANCHINI MARIAGRAZIA DR
FRANCOIS PATRICK DR	FRANDSEN SOEREN PROF	FREIRE FERRERO RUBENS G
FRIEDJUNG MICHAEL DR	FRIEL EILEEN D DR	FUJITA YOSHIO PROF
FURENLID INGEMAR K DR	GARMANY CATHERINE D DR	GARRISON ROBERT F PROF
GEHREN THOMAS PH D	GERSHBERG R E DR	GHOSH KAJAL KUMAR DR
GIAMPAPA MARK S	GILRA DAYA P DR	GIOVANNELLI FRANCO DR
GLAGOLEVSKIJ JU V DR	GLUSHNEVA I N DR	GOEBEL JOHN H DR
GRADY CAROL ANNE DR	GRATTON R G DR	GRAY DAVID F PROF

GREENSTEIN J L PROF	GRIFFIN RITA E M DR	GRIFFIN ROGER F DR
GROTH HANS G PROF DR	GUTHRIE BRUCE N G DR	HACK MARGHERITA PROF
HANUSCHIK REINHARD DR	HARMER CHARLES F W MR	HARMER DIANNE L MRS
HARTMANN LEE WILLIAM	HEARNSHAW JOHN B DR	HEBER ULRICH
HEINTZE J R W DR	HENIZE KARL G ASTRONAUT	HENRICHS HUBERTUS F DR
HERBIG GEORGE H DR	HESKE ASTRID DR	HINKLE KENNETH H
HIRAI MASANORI DR	HIRATA RYUKO	HOEFLICH PETER DR
HOUK NANCY DR	HOUZIAUX L PROF	HRON JOSEF DR
HUANG CHANG-CHUN	HUBENY IVAN	HUBERT HENRI DR
HUBERT-DELPLACE A-M DR	HUENEMOERDER DAVID P DR	HUNGER KURT PROF
HYLAND A R HARRY DR	JIANG SHI-YANG	JOHNSON HOLLIS R PROF
JORDAN CAROLE DR	JOSHI SURESH CHANDRA DR	JUGAKU JUN DR
KEENAN PHILIP C PROF EMER	KHOKHLOVA V L DR	KING R B DR
KIPPER TONU DR	KITCHIN CHRISTOPHER R DR	KODAIRA KEIICHI PROF
KOGURE TOMOKAZU DR	KOUBSKY PAVEL	KOVACHEV B J DR
KRAFT ROBERT P PROF	KREMPEC-KRYGIER JANINA DR	KWOK SUN DR
LAGO MARIA TERESA V T PR	LAIRD JOHN B DR	LAMERS HENNY J G L M DR
LAMONTAGNE ROBERT DR	LANDSTREET JOHN D PROF	LANGER GEORGE EDWARD DR
LANZ THIERRY DR	LE CONTEL JEAN-MICHEL	LECKRONE DAVID S DR
LESTER JOHN B DR	LEVATO ORLANDO HUGO DR	LIEBERT JAMES W DR
LITTLE-MARENIN IRENE R DR	LOCANTHI DOROTHY DAVIS DR	LOPES DALTON DE FARIA DR -
LUCK R EARLE DR	LUNDSTROM INGEMAR DR	MAGAIN PIERRE DR
MAGAZZU ANTONIO DR	MAILLARD JEAN-PIERRE DR	MAITZEN HANS M DR
MALARODA STELLA M DR	MARILLI ETTORE DR	MASSEY PHILIP L
MATHYS GAUTIER DR	MCGREGOR PETER JOHN DR	MCNAMARA DELBERT H DR
MEGESSIER CLAUDE DR	MIKULASEK ZDENEK DR	MOFFAT ANTHONY F J DR
MOLARO PAOLO DR	MOOS HENRY WARREN DR	MOROSSI CARLO
MORRISON NANCY DUNLAP DR	NECKEL HEINZ DR	NICHOLLS RALPH W PROF
NIEMELA VIRPI S DR	NISHIMURA SHIRO DR	NISSEN POUL E PROF
NORRIS JOHN DR	OKAZAKI ATSUO T DR	ORLOV MIKHAIL DR
OWOCKI STANLEY PETER DR	PAGEL BERNARD E J PROF	PARSONS SIDNEY B DR
PARTHASARATHY M DR	PASINETTI LAURA E PROF	PATERSON-BEECKMANS F
PEDOUSSAUT ANDRE	PEERY BENJAMIN F PROF	PERRIN MARIE-NOEL DR
PETERS GERALDINE JOAN DR	PETERSON RUTH CAROL DR	PLAVEC MIREK J PROF
POLCARO V F	POLIDAN RONALD S	PRADERIE FRANCOISE DR
PRINJA RAMAN DR	QUERCI FRANCOIS R DR	QUERCI MONIQUE DR
RAO N KAMESWARA	RAUTELA B S DR	REBOLO RAFAEL DR
REGO FERNANDEZ M DR	REIMERS DIETER PROF	RINGUELET ADELA E DR
RODGERS ALEX W DR	ROSE JAMES ANTHONY	ROSSI CORINNE DR
ROSSI LUCIO	RUTTEN ROBERT J DR	SAHADE JORGE PROF
SANCHEZ ALMEIDA JORGE DR	SANWAL BASANT BALLABH DR	SAREYAN JEAN-PIERRE DR
SCHILD RUDOLPH E DR	SCHOLZ GERHARD DR	SCHROEDER KLAUS PETER DR
SEGGEWISS WILHELM PROF	SHORE STEVEN N	SIMON THEODORE
SINNERSTAD ULF E PROF	SLETTEBAK ARNE PROF	SMITH GRAEME H DR
SMITH MYRON A ASST PROF	SMITH VERNE V DR	SMOLINSKI JAN DR
SNEDEN CHRISTOPHER A	SNOW THEODORE P PROF	SODERBLOM DAVID R
SONNEBORN GEORGE DR	SPITE FRANCOIS M DR	SPITE MONIQUE DR
STALIO ROBERTO DR	STAWIKOWSKI ANTONI DR	STECHER THEODORE P
STEFFEN MATTHIAS DR	STEFL STANISLAV DR	STENCEL ROBERT EDWARD
SUNTZEFF NICHOLAS B	SVOLOPOULOS SOTIRIOS PROF	SWENSSON JOHN W DR
SWINGS JEAN-PIERRE DR	TAKADA-HIDAI MASAHIDE DR	TALAVERA A DR
THEVENIN FREDERIC DR	TOMOV TOMA V DR	TUOMINEN ILKKA V DR
UNDERHILL ANNE B DR	UTSUMI KAZUHIKO DR	VALTIER JEAN-CLAUDE DR
VAN DER HUCHT KAREL A DR	VAN'T VEER-MENNERET CL DR	VASU-MALLIK SUSHMA DR
VILHU OSMI DR	VIOTTI ROBERTO DR	VLADILO GIOVANNI DR

VOGT NIKOLAUS DR   VOGT STEVEN SCOTT   VREUX JEAN MARIE DR
WALLERSTEIN GEORGE PROF   WATERWORTH MICHAEL DR   WEGNER GARY ALAN
WEHINGER PETER A DR   WEHLAU WILLIAM H PROF   WEISS WERNER W DR
WENIGER SCHAME DR   WILLIAMS PEREDUR M DR   WING ROBERT F PROF
WOLF BERNHARD PH D   WOLFF SIDNEY C DR   WOOD III H J DR
WYCKOFF SUSAN DR   YAMASHITA YASUMASA PROF   ZOREC JEAN DR
ZVERKO JURAJ DR

COMPOSITION OF COMMISSION 30 1991-1994

President   : BURKI GILBERT PROF

Vice-President(s) : SCARFE COLIN D DR

Organizing Committee: ANDERSEN JOHANNES
                       DA COSTA NICOLAI L-A
                       FAIRALL ANTHONY P PROF
                       FREEMAN KENNETH C PROF
                       HEARNSHAW JOHN B DR
                       LATHAM DAVID W DR

Members:

ABT HELMUT A DR	AZZOPARDI MARC DR	BALONA LUIS ANTERO DR
BARBIER-BROSSAT M DR	BATTEN ALAN H DR	BEAVERS WILLET I DR
BEERS TIMOTHY C DR	BERTIAU FLOR C PROF	BOULON JACQUES J DR
BREGER MICHEL PROF DR	BURNAGE ROBERT	CAMPBELL BRUCE DR
CARNEY BRUCE WILLIAM	CARQUILLAT JEAN-MICHEL	COCHRAN WILLIAM DAVID DR
CRAMPTON DAVID DR	DAVIS MARC DR	DAVIS ROBERT J DR
DE JONGE J K DR	DE VAUCOULEURS GERARD PR	DUFLOT MARCELLE DR
EDMONDSON FRANK K PROF	EELSALU HEINO DR	FEHRENBACH CHARLES PROF
FEKEL FRANCIS C	FLETCHER J MURRAY	FLORSCH ALPHONSE DR
FOLTZ CRAIG B	GEORGELIN YVON P DR	GIESEKING FRANK DR
GILMORE GERARD FRANCIS	GIOVANELLI RICCARDO DR	GOUGUENHEIM LUCIENNE
GRIFFIN ROGER F DR	HALBWACHS JEAN LOUIS DR	HEINTZE J R W DR
HEWETT PAUL	HILDITCH RONALD W DR	HILL GRAHAM DR
HRIVNAK BRUCE J	HUANG CHANG-CHUN	HUBE DOUGLAS P DR
HUCHRA JOHN PETER DR	IMBERT MAURICE DR	IRWIN ALAN W DR
KADOURI TALIB HADI	KARACHENTSEV I D DR	KRAFT ROBERT P PROF
LEVATO ORLANDO HUGO DR	LEWIS BRIAN MURRAY DR	LINDGREN HARRI
MARSCHALL LAURENCE A	MARTIN NICOLE DR	MATHIEU ROBERT D DR
MAURICE ERIC N	MAYOR MICHEL PROF	MAZEH TSEVI DR
MCCLURE ROBERT D PROF	MCMILLAN ROBERT S DR	MELNICK GARY J
MERMILLIOD JEAN-CLAUDE DR	MEYLAN GEORGES DR	MORBEY CHRISTOPHER L
MORRELL NIDIA DR	NORDSTROEM BIRGITTA DR	OETKEN L DR
PEDOUSSAUT ANDRE	PELLEGRINI PAULO S S DR	PERRY CHARLES L DR
PETERSON RUTH CAROL DR	PHILIP A G DAVIS	POPOV VICTOR S DR
PRESTON GEORGE W DR	PREVOT LOUIS DR	QUINTANA HERNAN DR
RATNATUNGA KAVAN U	REBEIROT EDITH DR	ROMANOV YURI S DR
RUBIN VERA C DR	SAMUS NIKOLAI N DR	SANWAL N B DR
SMITH MYRON A ASST PROF	SOLIVELLA GLADYS R LIC	STEFANIK ROBERT DR
STOCK JURGEN D	TOKOVININ ANDREJ A DR	TONRY JOHN DR
VAN DESSEL EDWIN LUDO DR	WALKER GORDON A H PROF	WEGNER GARY ALAN
WILLSTROP RODERICK V DR	YOSS KENNETH M DR	

COMPOSITION OF COMMISSION 31 1991-1994

President   : PROVERBIO EDOARDO PROF

Vice-President(s) : FLIEGEL HENRY F

Organizing Committee: ALLAN DAVID W MR
                      BRUMBERG VICTOR A DR
                      FUJIMOTO MASA-KATSU DR
                      GRANVEAUD MICHEL
                      GUINOT BERNARD R PROF
                      KLEPCZYNSKI WILLIAM J DR
                      KOVALEVSKY JEAN DR
                      LUCK J DR
                      MUELLER IVAN I PROF
                      PAQUET PAUL EG DR
                      YE SHU-HUA

Members:

ABELE MARIS K DR	AFANASJEVA PRASKOVYA M DR	ALLEY CARROL O DR
AOKI SHINKO PROF	BENAVENTE JOSE	BENDER PETER L DR
BLINOV N S DR	BOLOIX RAFAEL DR	BONANOMI JACQUES DR
CAPRIOLI GIUSEPPE PROF	CARTER WILLIAM EUGENE	CATALAN MANUEL DR
CHAMBERLAIN JOSEPH M DR	DICKEY JEAN O'BRIEN	DOMINSKI IRENEUSZ DR
DOUGLAS R J MR	ENSLIN HEINZ DR	FALLON FREDERICK W DR
FEISSEL MARTINE DR	GAIGNEBET JEAN DR	GALLIANO PIER GIORGIO
GALLIANO PIER GIORGIO	GAMBIS DANIEL DR	GRUDLER PIERRE
HALL R GLENN DR	HAN TIANQI	HARA KEN NOSUKE DR
HELLWIG HELMUT WILHELM DR	HEMMLEB GERHARD DR	HERS JAN MR
IIJIMA SHIGETAKA PROF	JIN WEN-JING	KAKUTA CHUICHI DR
KESSLER KARL G DR	KOBAYASHI YUKISAYU	KOLACZEK BARBARA DR
LESCHIUTTA S PROF	LIANG ZHONG-HUAN	LIU JINMING
LU BEN-KUI	LUO DINGCHANG	LUO SHI-FANG
MARKOWITZ WILLIAM DR	MATHUR B S DR	MATSAKIS DEMETRIOS N
MCCARTHY DENNIS D DR	MEINIG MANFRED DR	MELBOURNE WILLIAM G DR
MELCHIOR PAUL J PROF DIR	MIAO YONG-RUI	MORGAN PETER DR
NAUMOV VITALIJ A DR	NEWHALL X X DR	NIIMI YUKIO
NOEL FERNANDO	PETIT GERARD DR	PILKINGTON JOHN D H DR
POPELAR JOSEF DR	PUSHKIN SERGEY B DR	QI GUAN RONG
RANDIC LEO PROF DR	ROBERTSON DOUGLAS S	SCHULER WALTER DR
SMITH HUMPHRY M	SMYLIE DOUGLAS E DR	SONG JIN-AN
STANILA GEORGE DR	THOMAS CLAUDINE DR	TSUCHIYA ATSUSHI DR PROF
VEILLET CHRISTIAN	VICENTE RAIMUNDO O PROF	WACKERNAGEL H BEAT DR
WEBROVA LUDMILA DR	WIETH-KNUDSEN NIELS P DR	WILKINS GEORGE A DR
WINKLER GERNOT M R DR	WU SHOU-XIAN	XU BANG-XIN
YANG KE-JUN	ZHAI ZAOCHENG	ZHANG JINTONG
ZHAO GANG	ZHENG YING	ZHUANG QIXIANG

COMPOSITION OF COMMISSION 33 1991-1994

President   : BLITZ LEO DR

Vice-President(s) : BINNEY JAMES J DR

Organizing Committee: BLOEMEN JOHANNES B G M DR
                       CESARSKY CATHERINE J DR
                       FRIDMAN ALEKSEY M DR
                       GILMORE GERARD FRANCIS
                       KALNAJS AGRIS J DR
                       MATTEUCCI FRANCESCA DR
                       MAYOR MICHEL PROF
                       MORRIS MARK ROOT DR
                       YOSHII YUZURU DR

Members:

AARSETH SVERRE J DR	ADAMSON ANDREW DR	AFANAS'EV VIKTOR L DR
AGEKJAN TATEOS A PROF	AGUILAR CHIU LUIS A DR	AIZU KO PROF
ALTENHOFF WILHELM J DR	AMBARTSUMIAN V A PROF DR	AMBASTHA A K DR
ANDERSEN JOHANNES	ANTONOV VADIM A DR	AOKI SHINKO PROF
ARDEBERG ARNE L PROF	ASTERIADIS GEORGIOS DR	ATHANASSOULA EVANGELIE DR
BAHCALL JOHN N PROF	BALAZS LAJOS G DR	BALBUS STEVEN A DR
BALDWIN JOHN E DR	BANHATTI DILIP GOPAL DR	BARBANIS BASIL PROF
BARBERIS BRUNO	BASH FRANK N PROF	BASU BAIDYANATH PROF
BAUD BOUDEWIJN DR	BECKER WILHELM PROF	BERKHUIJSEN ELLY M DR
BIENAYME OLIVIER DR	BLAAUW ADRIAAN PROF DR	BLANCO VICTOR M DR
BOULON JACQUES J DR	BRONFMAN LEONARDO DR	BURKE BERNARD F DR
BURTON W BUTLER DR	CALDWELL JOHN A R	CANE HILARY VIVIEN
CARRASCO LUIS DR	CASWELL JAMES L DR	CESARSKY DIEGO A DR
CHAPMAN JESSICA DR	CHEN ZHEN	CHRISTODOULOU DMITRIS DR
CHURCHWELL EDWARD B DR	CIURLA TADEUSZ	CLEMENS DAN P DR
CLUBE S V M DR	COHEN RICHARD S	COLIN JACQUES DR
COMINS NEIL FRANCIS	CONTOPOULOS GEORGE PROF	COSTA EDGARDO DR
COURTES G PROF	CRAMPTON DAVID DR	CRAWFORD DAVID L DR
CREZE MICHEL DR	CUBARSI RAFAEL DR	CUDWORTH KYLE MCCABE DR
CUPERMAN SAMI PROF	DAVIES RODNEY D PROF	DE JONG TEIJE DR
DEJONGHE HERWIG BERT DR	DEKEL AVISHAI	DELHAYE JEAN PROF
DENOYELLE JOZEF KIC	DICKEL HELENE R DR	DICKEL JOHN R
DICKMAN ROBERT L DR	DIETER NANNIELOU H DR	DOWNES DENNIS DR
DRILLING JOHN S	DUCATI JORGE RICARDO DR	DZIGVASHVILI R M DR
EDMONDSON FRANK K PROF	EFREMOV YURY N DR	EGRET DANIEL DR
EINASTO JAAN DR	ELMEGREEN DEBRA MELOY	ELSAESSER HANS PROF
ELVIUS TORD PROF EMERITUS	EVANGELIDIS E DR	FABER SANDRA M PROF
FALL S MICHAEL DR	FEAST MICHAEL W PROF	FEHRENBACH CHARLES PROF
FEITZINGER JOHANNES PROF	FENKART ROLF P PROF DR	FIGUERAS FRANCESCA DR
FITZGERALD M PIM PROF	FREEMAN KENNETH C PROF	FUCHS BURKHARD DR
FUJIMOTO MASA-KATSU DR	FUJIWARA TAKAO DR	FUKUNAGA MASATAKA DR
GALLETTO DIONIGI	GARZON FRANCISCO DR	GEMMO ALESSANDRA DR

GENKIN IGOR L PROF DR	GEORGELIN YVON P DR	GEORGELIN YVONNE M DR
GLIESE WIHELM PROF	GOLDREICH P DR	GORDON MARK A DR
GOTTESMAN STEPHEN T DR	GRAYZECK EDWIN J DR	GRENON MICHEL DR
GUPTA SUNIL K DR	GYLDENKERNE KJELD DR	HABE ASAO
HABING H J DR	HAKKILA JON ERIC DR	HAMAJIMA KIYOTOSHI DR
HANAMI HITOSHI DR	HARTKOPF WILLIAM I DR	HAUG ULRICH PROF
HAWKINS MICHAEL R S	HAYLI AVRAM PROF	HEILES CARL PROF
HENON MICHEL C DR	HERBST WILLIAM DR	HERMAN JACOBUS DR
HOBBS ROBERT W DR	HORI GENICHIRO PROF	HRON JOSEF DR
HUANG SONG-NIAN DR	HUGHES VICTOR A PROF	HULSBOSCH A N M DR
HUMPHREYS ROBERTA M PROF	HUNTER CHRISTOPHER PROF	IKEUCHI SATORU DR
INAGAKI SHOGO DR	INNANEN KIMMO A PROF	IRWIN JOHN B PROF
ISOBE SYUZO DR	ISRAEL FRANK P DR	IWANISZEWSKA CECYLIA DR
IWANOWSKA WILHELMINA PROF	IYE MASANORI DR	JACKSON PETER DOUGLAS DR
JAHREISS HARTMUT DR	JASCHEK CARLOS O R PROF	JASNIEWICZ GERARD DR
JIANG DONG-RONG	JOG CHANDA J DR	JOHNSON HUGH M DR
JONAS JUSTIN LEONARD	JONES DEREK H P DR	KABURAKI MASAKI PROF
KALANDADZE N B DR	KASUMOV FIKRET K O DR	KATO SHOJI PROF
KERR FRANK J DR	KHARADZE E K PROF	KING IVAN R PROF
KINMAN THOMAS D DR	KLARE GERHARD DR	KNAPP GILLIAN R DR
KOLESNIK IGOR G DR	KOLESNIK L N DR	KORMENDY JOHN DR
KULSRUD RUSSELL M DR	KUTUZOV S A DR	LAFON JEAN-PIERRE J DR
LARSON RICHARD B PROF	LATHAM DAVID W DR	LECAR MYRON DR
LEE HYUNG MOK DR	LEE SANG GAK	LEISAWITZ DAVID DR
LI JING	LIEBERT JAMES W DR	LIN CHIA C PROF
LINDBLAD PER OLOF PROF	LOCKMAN FELIX J	LODEN KERSTIN R DR
LODEN LARS OLOF PROF	LU PHILLIP K DR	LUNEL MADELEINE DR
LUYTEN WILLEM J PROF	LYNDEN-BELL DONALD PROF	LYNGA GOSTA DR
MACCONNELL DARRELL J DR	MACRAE DONALD A PROF	MANCHESTER RICHARD N DR
MARK JAMES WAI-KEE DR	MAROCHNIK L S PROF DR	MARTINET LOUIS PROF
MATHEWSON DONALD S PROF	MAVRIDIS L N PROF	MCCARTHY MARTIN F DR
MCGREGOR PETER JOHN DR	MEATHERINGHAM STEPHEN DR	MENNESSIER MARIE-ODILE DR
MEZGER PETER G PROF	MIKKOLA SEPPO DR	MILLER RICHARD H DR
MIRABEL IGOR FELIX DR	MIRZOYAN L V DR PROF	MIYAMOTO MASANORI DR
MOFFAT ANTHONY F J DR	MONET DAVID G	MONNET GUY J DR
MUENCH GUIDO PROF	MURRAY C ANDREW	MUZZIO JUAN C PROF
NAHON FERNAND PROF	NECKEL TH DR	NELSON ALISTAIR H DR
NINKOVIC SLOBODAN	NISHIDA MINORU PROF	NISHIDA MITSUGU
NORDSTROEM BIRGITTA DR	NORMAN COLIN A PROF	OBLAK EDOUARD
OJA TARMO PROF	OKUDA HARUYUKI DR PROF	OLANO CARLOS ALBERTO DR
OLLONGREN A PROF DR	OORT JAN H PROF	OSTRIKER JEREMIAH P PROF
PALMER PATRICK E PROF	PALOUS JAN DR	PANDEY A K
PAPAYANNOPOULOS TH DR	PAULS THOMAS ALBERT DR	PAVLOVSKAYA E D DR
PEIMBERT MANUEL DR	PEREK LUBOS DR	PERRY CHARLES L DR
PESCH PETER DR	PHILIP A G DAVIS	PIER JEFFREY R DR
PILOWSKI K PROF DR	PISMIS DE RECILLAS PARIS	POLYACHENKO VALERIJ L DR
POLYMILIS CHRONIS DR	PRICE R MARCUS DR	PRIESTER WOLFGANG PROF
QIAN ZHONG-YU	RABOLLI MONICA DR	RAHARTO MOEDJI
RATNATUNGA KAVAN U	REBEIROT EDITH DR	REID NEILL
REIF KLAUS DR	RIEGEL KURT W DR	ROBERTS MORTON S DR
ROBERTS WILLIAM W JR PROF	ROBIN ANNIE C DR	ROBINSON BRIAN J DR
ROHLFS K PROF DR	RONG JIAN-XIANG	RUBIN VERA C DR
RUELAS-MAYORGA R A DR	RUIZ MARIA TERESA DR	RYBICKI GEORGE B DR
SAAR ENN DR	SALA FERRAN DR	SANCHEZ-SAAVEDRA M LUISA
SANDQVIST AAGE DR	SANZ I SUBIRANA JAUME DR	SARGENT ANNEILA I

SCHECHTER PAUL L DR | SCHMIDT HANS PROF | SCHMIDT K H DR
SCHMIDT MAARTEN PROF | SCHMIDT-KALER TH PROF | SEGGEWISS WILHELM PROF
SEIMENIS JOHN DR | SELLWOOD JERRY A | SHANE WILLIAM W DR
SHAROV A S DR | SHER DAVID DR | SHIMIZU TSUTOMU PROF EMER
SHU FRANK H PROF | SHUTER WILLIAM L H DR | SIMONSON S CHRISTIAN DR
SLETTEBAK ARNE PROF | SOBOUTI YOUSEF PROF | SOLOMON PHILIP M DR
SONG GUO-XUAN | SPARKE LINDA | SPERGEL DAVID N DR
SPIEGEL E DR | STECKER FLOYD W DR | STEFANOVITCH-GOMEZ A E DR
STEINLIN ULI PROF | STEPHENSON C BRUCE PROF | STIBBS DOUGLAS W N PROF
STROBEL ANDRZEJ DR | STURCH CONRAD R DR | SVOLOPOULOS SOTIRIOS PROF
SYGNET JEAN FRANCOIS DR | SZEBEHELY VICTOR G PROF | TAMMANN G ANDREAS PROF DR
TERZIDES CHARALAMBOS DR | THE PIK-SIN PROF | THIELHEIM KLAUS O DR
TOBIN WILLIAM | TOMISAKA KOHJI DR | TONG YI
TOOMRE ALAR DR | TOOMRE JURI | TORRA JORDI DR
TOSA MAKOTO DR | TREFZGER CHARLES F DR | TSIOUMIS ALEXANDROS DR
TURON C DR | UPGREN ARTHUR R DR | VALTONEN MAURI J PROF
VAN DER KRUIT PIETER C DR | VAN WOERDEN HUGO PROF DR | VANDERVOORT PETER O DR
VEGA E IRENE DR | VENUGOPAL V R DR | VERSCHUUR GERRIT L PROF
VETESNIK MIROSLAV DR | VOROSHILOV V I DR | WAYMAN PATRICK A PROF
WEAVER HAROLD F PROF | WEISTROP DONNA DR | WESTERHOUT GART DR
WESTERLUND BENGT E PROF | WHITE RAYMOND E DR | WHITELOCK PATRICIA ANN DR
WHITEOAK J B DR | WHITTET DOUGLAS C B DR | WIELEBINSKI RICHARD PROF
WIELEN ROLAND PROF DR | WOLTJER LODEWIJK PROF | WOODWARD PAUL R DR
WRAMDEMARK STIG S O DR | WYSE ROSEMARY F DR | XIANG DELIN
YOUNIS SAAD M | YUAN CHI PROF | ZACHILAS LOUKAS DR
ZHANG BIN | ZHAO JUN-LIANG |

COMPOSITION OF COMMISSION 34 1991-1994

President   : HABING H J DR

Vice-President(s) : FLOWER DAVID R DR

Organizing Committee: BRUHWEILER FRED C JR
                       DOPITA MICHAEL ANDREW DR
                       FALGARONE EDITH
                       GENZEL REINHARD DR
                       KAIFU NORIO DR
                       LOZINSKAYA TAT'YANA A DR
                       MARTIN PETER G PROF
                       MATHIS JOHN S PROF
                       MYERS PHILIP C
                       POTTASCH STUART R PROF
                       RODRIGUEZ LUIS F
                       ROSA MICHAEL RICHARD DR

Members:

AANNESTAD PER ARNE DR	ACKER AGNES PROF DR	ADAMS FRED DR
AIAD A PROF	AITKEN DAVID K DR	AKABANE KENJI A PROF
ALCOLEA JAVIER DR	ALDROVANDI S M VIEGAS DR	ALLER LAWRENCE HUGH
ALTENHOFF WILHELM J DR	ANANTHARAMAIAH K R DR	ANDREW BRYAN H DR
ANDRIESSE CORNELIS D DR	ANDRILLAT HENRI L PROF	ANDRILLAT YVETTE DR
ANGLADA GUILLEM DR	ARKHIPOVA V P DR	ARNY THOMAS T DR
AVERY LORNE W DR	AXFORD W IAN PROF	AZCARATE ISMAEL N DR
BAARS JACOB W M DR	BAART EDWARD E PROF	BACHILLER RAFAEL DR
BALDWIN JOHN E DR	BALUTEAU JEAN-PAUL DR	BANIA THOMAS MICHAEL
BARLOW MICHAEL J DR	BARNES AARON DR	BARRETT ALAN H PROF
BASH FRANK N PROF	BAUDRY ALAIN DR	BECKLIN ERIC E DR
BECKMAN JOHN E PROF	BECKWITH STEVEN V W	BEDOGNI ROBERTO
BEL NICOLE J DR	BENAYDOUN JEAN-JACQUES DR	BERGERON JACQUELINE A DR
BERKHUIJSEN ELLY M DR	BERNAT ANDREW PLOUS DR	BERTOUT CLAUDE
BHATT H C DR	BIANCHI LUCIANA	BIEGING JOHN HAROLD DR
BIGNELL R CARL DR	BINETTE LUC	BIRKLE KURT PH D
BLACK JOHN HARRY DR	BLADES JOHN CHRIS DR	BLAIR GUY NORMAN DR
BLESS ROBERT C PROF	BLITZ LEO DR	BOCHKAREV NIKOLAY G DR
BODE MICHAEL F	BODENHEIMER PETER PROF	BOESHAAR GREGORY ORTH DR
BOGGESS ALBERT DR	BOHLIN RALPH C DR	BOISSE PATRICK DR
BOLAND WILFRIED	BORGMAN JAN DR PROF	BOULANGER FRANCOIS
BOUVIER JEROME	BRAND PETER W J L DR	BRAUNSFURTH EDWARD PH D
BREITSCHWERDT DIETER DR	BRINKMANN WOLFGANG	BROMAGE GORDON E DR
BROWN RONALD D PROF	BUJARRABAL VALENTIN	BURGESS ALAN DR
BURKE BERNARD F DR	BURTON MICHAEL G DR	BURTON W BUTLER DR
BYSTROVA NATALIJA V DR	CANTO JORGE DR	CAPLAN JAMES
CAPPA DE NICOLAU CRISTINA	CAPRIOTTI EUGENE R DR	CAPUZZO DOLCETTA ROBERTO
CARDELLI JASON A DR	CARRUTHERS GEORGE R DR	CASTANEDA HECTOR
CASWELL JAMES L DR	CERNICHARO JOSE DR	CERRUTI-SOLA MONICA

CERSOSIMO JUAN CARLOS DR	CESARSKY CATHERINE J DR	CESARSKY DIEGO A DR
CHANDRA SURESH DR	CHEVALIER ROGER A DR	CHINI ROLF
CHOPINET MARGUERITE DR	CHU YOU-HUA	CHURCHWELL EDWARD B DR
CLARK FRANK OLIVER DR	CLEGG ROBIN E S DR	CODE ARTHUR D
COHEN MARSHALL H PROF	COLLIN-SOUFFRIN SUZY DR	COLOMB FERNANDO R DR
CORBELLI EDVIGE DR	COSTERO RAFAEL	COURTES G PROF
COWIE LENNOX LAUCHLAN DR	COX DONALD P PROF	COX PIERRE DR
COYNE GEORGE V DR	CRANE PHILIPPE	CRAWFORD IAN ANDREW DR
CROVISIER JACQUES	CRUVELLIER PAUL E DR	CUDABACK DAVID D DR
CUGNON PIERRE DR	CZYZAK STANLEY J DR	DAHN CONARD CURTIS DR
DALGARNO ALEXANDER PROF	DANKS ANTHONY C DR	DANLY LAURA DR
DAVIES RODNEY D PROF	DE BOER KLAAS SJOERDS DR	DE JONG TEIJE DR
DE LA NOE JEROME DR	DEGUCHI SHUJI DR	DEHARVENG LISE DR
DEWDNEY PETER E F DR	DI FAZIO ALBERTO	DICKEL HELENE R DR
DICKEL JOHN R	DICKEY JOHN M	DINERSTEIN HARRIET L
DISNEY MICHAEL J PROF	DOKUCHAEV VYACHESLAV DR	DOKUCHAEVA OLGA D DR
DONN BERTRAM D	DORSCHNER JOHANN DR	DOTTORI HORACIO A DR
DOWNES DENNIS DR	DRAINE BRUCE T	DRAPATZ SIEGFRIED W DR
DREHER JOHN W	DUBNER GLORIA DR	DUBOUT RENEE
DUFOUR REGINALD JAMES	DULEY WALTER W PROF	DUPREE ANDREA K DR
DWEK ELI	DYSON JOHN E DR	D'ODORICO SANDRO DR
ELITZUR MOSHE	ELLIOTT KENNETH H DR	ELMEGREEN BRUCE GORDON DR
ELMEGREEN DEBRA MELOY	ELVIUS AINA M PROF	EMERSON JAMES P
ENCRENAZ PIERRE J DR	ESCALANTE VLADIMIR DR	ESIPOV VALENTIN F DR
EVANS ANEURIN	EVANS NEAL J II ASS PROF	FALK SYDNEY W JR DR
FALLE SAMUEL A DR	FAN YING	FAULKNER DONALD J DR
FEDERMAN STEVEN ROBERT	FEIBELMAN WALTER A DR	FEITZINGER JOHANNES PROF
FELLI MARCELLO DR	FELTEN JAMES E DR	FERLET ROGER DR
FERRINI FEDERICO	FIELD DAVID	FIELD GEORGE B PROF
FIERRO JULIETA	FISCHER JACQUELINE	FLANNERY BRIAN PAUL DR
FLECK ROBERT CHARLES DR	FORD HOLLAND C RES PROF	FORSTER JAMES RICHARD DR
FRIDLUND MALCOLM DR	FRIEDEMANN CHRISTIAN DR	FRISCH PRISCILLA
FUKUI YASUO DR	FURNISS IAN	GARDNER FRANCIS F DR
GAUSTAD JOHN E PROF	GAY JEAN DR	GEBALLE THOMAS R DR
GEHRELS TOM PROF	GEORGELIN YVON P DR	GERARD ERIC DR
GEROLA HUMBERTO DR	GEZARI DANIEL YSA DR	GIACANI ELSA BEATRIZ DR
GILRA DAYA P DR	GIOVANELLI RICCARDO DR	GODFREY PETER DOUGLAS DR
GOEBEL JOHN H DR	GOLDREICH P DR	GOLDSMITH DONALD W DR
GOLDSMITH PAUL F DR	GOLDSTEIN SAMUEL J PROF	GOLDSWORTHY FREDERICK A
GOMEZ GONZALEZ JESUS DR	GORDON COURTNEY P PROF ,	GORDON MARK A DR
GOSACHINSKIJ I V DR	GOSS W MILLER PROF	GRAHAM DAVID A
GRASDALEN GARY L DR	GREENBERG J MAYO DR	GREWING MICHAEL PROF
GUELIN MICHEL DR	GUERTLER JOACHIN DR	GUESTEN ROLF
GULL THEODORE R DR	GURZADIAN G A PROF DR	GUSEINOV O H PROF
HACKWELL JOHN A DR	HANAMI HITOSHI DR	HARDEBECK ELLEN G DR
HARRINGTON J PATRICK DR	HARRIS ALAN WILLIAM	HARRIS STELLA
HARTEN RONALD H DR	HARTL HERBERT DR	HARTQUIST THOMAS WILBUR
HARVEY PAUL MICHAEL DR	HAYASHI SAEKO S DR	HAYNES RAYMOND F PROF
HECHT JAMES H DR	HEILES CARL PROF	HELFER H LAWRENCE PROF
HELOU GEORGE DR	HENDECOURT D' LOUIS DR	HENIZE KARL G ASTRONAUT
HENKEL CHRISTIAN	HENNING THOMAS DR	HERZBERG GERHARD DR
HIDAJAT BAMBANG PROF DR	HIGGS LLOYD A DR	HILDEBRAND ROGER H
HIPPELEIN HANS H DR	HIROMOTO NORIHISA DR	HJALMARSON AKE G DR
HJELLMING ROBERT M DR	HOBBS LEWIS M DR	HOEGLUND BERTIL PROF
HOLLENBACH DAVID JOHN DR	HOLLIS JAN MICHAEL DR	HONG SEUNG SOO DR

HOUZIAUX L PROF	HUA CHON TRUNG DR	HUGHES VICTOR A PROF
HULSBOSCH A N M DR	HUMMER DAVID G DR	HUTCHINGS JOHN B DR
HUTSEMEKERS DAMIEN DR	IRVINE WILLIAM M PROF	ISOBE SYUZO DR
ISRAEL FRANK P DR	ISSA ALI DR	ITOH HIROSHI DR
IYENGAR K V K PROF	JABIR NIAMA LAFTA	JACOBY GEORGE H
JACQ THIERRY	JAFFE DANIEL T	JENKINS EDWARD B DR
JENKINS L F MS	JENNINGS R E PROF	JOHNSON FRED M PROF DR
JOHNSON HUGH M DR	JOHNSTON KENNETH J	JONES FRANK CULVER DR
JOURDAIN DE MUIZON M DR	JURA MICHAEL DR	JUST ANDREAS DR
KAFATOS MINAS DR	KAFTAN MAY A DR	KAHN FRANZ D PROF
KALER JAMES B PROF	KAMIJO FUMIO PROF DR	KAZES ILYA DR
KEGEL WILHELM H PROF	KENNICUTT ROBERT C JR	KERR FRANK J DR
KHARADZE E K PROF	KHROMOV G S DR	KIMURA HIROSHI DR
KIRKPATRICK RONALD C DR	KIRSHNER ROBERT PAUL DR	KNACKE ROGER F DR
KNAPP GILLIAN R DR	KNUDE JENS KIRKESKOV DR	KOEPPEN JOACHIM DR
KOHOUTEK LUBOS DR	KOLESNIK IGOR G DR	KONDO YOJI DR
KOORNNEEF JAN DR	KOSTYAKOVA ELENA B DR	KOZASA TAKASHI
KRAUTTER JOACHIM DR	KREYSA ERNST	KRISHNA SWAMY K S DR
KUIPER THOMAS B H DR	KUMAR C KRISHNA DR	KUNDU MUKUL R DR
KUNTH DANIEL	KUNZE RUEDIGER DR	KUTNER MARC LESLIE DR
KWITTER KAREN BETH DR	KWOK SUN DR	KYLAFIS NIKOLAOS D DR
LADA CHARLES JOSEPH DR	LAFON JEAN-PIERRE J DR	LANGER WILLIAM DAVID DR
LASKER BARRY M DR	LAURENT CLAUDINE DR	LE SQUEREN ANNE-MARIE DR
LEE TERENCE J DR	LEGER ALAIN DR	LEISAWITZ DAVID DR
LEPINE JACQUES R D DR	LEQUEUX JAMES DR	LEUNG CHUN MING DR
LILLER WILLIAM DR	LIN CHIA C PROF	LINKE RICHARD ALAN DR
LISZT HARVEY STEVEN	LIZANO-SOBERON SUSANA DR	LO KWOK-YUNG DR
LOCKMAN FELIX J	LOREN ROBERT BRUCE DR	LORTET MARIE CLAIRE
LOUISE RAYMOND PROF	LOVAS FRANCIS JOHN DR	LOW FRANK J DR
LUCAS ROBERT DR	LYNDS BEVERLY T DR	MACIEL WALTER J DR
MACLEOD JOHN M DR	MAIHARA TOSHINORI DR	MALLIK D C V DR
MANCHADO ARTURO DR	MANCHESTER RICHARD N DR	MANFROID JEAN DR
MARSTON ANTHONY PHILIP DR	MARTIN ROBERT N DR	MARTIN-PINTADO JESUS
MASSON COLIN R	MATHER JOHN CROMWELL	MATHEWS WILLIAM G PROF
MATHEWSON DONALD S PROF	MATSUMURA MASAFUMI DR	MATTILA KALEVI DR
MCCALL MARSHALL LESTER DR	MCCRAY RICHARD DR	MCCREA J DERMOTT
MCGEE RICHARD X DR	MCGREGOR PETER JOHN DR	MCKEE CHRISTOPHER F PROF
MCKEITH CONAL D DR	MCNALLY DEREK DR	MEABURN J DR
MEBOLD ULRICH DR PROF	MEIER ROBERT R	MELNICK GARY J
MENDEZ ROBERTO H DR	MENON T K PROF	MENZIES JOHN W DR
MESZAROS PETER DR	MEZGER PETER G PROF	MICHALITSIANOS ANDREW
MILLAR THOMAS J DR	MILLER JOSEPH S PROF	MILNE DOUGLAS K DR
MININ I N PROF	MINN YOUNG KEY DR	MITCHELL GEORGE F DR
MIYAMA SYOKEN	MIZUNO SHUN	MO JING-ER
MONIN JEAN-LOUIS DR	MORENO CORRAL MARCO A DR	MORGAN DAVID H DR
MORIMOTO MASAKI DR	MORRIS MARK ROOT DR	MORTON DONALD C DR
MOUSCHOVIAS TELEMACHOS CH	MUENCH GUIDO PROF	MUFSON STUART LEE DR
NAKADA YOSHIKAZU DR	NAKAGAWA TAKAO DR	NANDY KASHINATH DR
NEUGEBAUER GERRY DR	NGUYEN-QUANG RIEU DR	NORDH H LENNART DR
NORMAN COLIN A PROF	NULSEN PAUL DR	NUSSBAUMER HARRY PROF
NUTH JOSEPH A III	OHTANI HIROSHI DR	OKUDA HARUYUKI DR PROF
OLOFSSON HANS	OMONT ALAIN PROF	ONAKA TAKASHI
OSAKI TORU DR	OSBORNE JOHN L DR	OSTERBROCK DONALD E PROF
OZERNOY LEONID M PROF	O'DELL CHARLES R DR	O'DELL STEPHEN L
PAGEL BERNARD E J PROF	PALLA FRANCESCO	PALMER PATRICK E PROF

PANAGIA NINO DR

PANKONIN VERNON LEE DR

PARKER EUGENE N

PAULS THOMAS ALBERT DR

PECKER JEAN-CLAUDE PROF

PEIMBERT MANUEL DR

PENA MIRIAM DR

PENZIAS ARNO A DR

PEQUIGNOT DANIEL

PERAULT MICHEL

PERINOTTO MARIO PROF

PERSI PAOLO

PETERS WILLIAM L III DR

PETROSIAN VAHE PROF

PHILLIPS JOHN PETER

PHILLIPS THOMAS GOULD DR

PINEAU DES FORETS G DR

PISMIS DE RECILLAS PARIS

POEPPEL WOLFGANG G L DR

PORCEDDU IGNAZIO E P DR

PRASAD SHEO S

PREITE-MARTINEZ ANDREA DR

PRICE R MARCUS DR

PRONIK I I DR

PSKOVSKIJ JU P DR

PUGET JEAN-LOUP DR

QIN ZHI-HAI

RADHAKRISHNAN V PROF

RAIMOND ERNST DR

RAWLINGS JONATHAN DR

RAYMOND JOHN CHARLES

REIPURTH BO

RENGARAJAN T N DR

REYNOLDS RONALD J DR

RICKARD LEE J DR

RIGHINI-COHEN GIOVANNA DR

ROBBINS R ROBERT PROF

ROBERGE WAYNE G DR

ROBERTS WILLIAM W JR PROF

ROBINSON BRIAN J DR

ROBINSON GARRY DR

ROCHE PATRICK F DR

ROELFSEMA PETER DR

ROESER HANS-PETER

ROGER ROBERT S DR

ROGERS ALAN E E DR

ROHLFS K PROF DR

ROSADO MARGARITA DR

ROSE WILLIAM K DR

ROSINO LEONIDA PROF

ROUAN DANIEL DR

ROXBURGH IAN W PROF

ROZHKOVSKIJ DIMITRIJ A

ROZYCZKA MICHAL

RUSSELL STEPHEN DR

SABANO YUTAKA DR

SABBADIN FRANCO DR

SAHU KAILASH C DR

SALINARI PIERO

SALPETER EDWIN E PROF

SANCHEZ-SAAVEDRA M LUISA

SANCISI RENZO DR

SANDELL GORAN HANS L DR

SANDQVIST AAGE DR

SARAZIN CRAIG L DR

SARGENT ANNEILA I

SARMA N V G PROF

SATO FUMIO DR

SATO SHUJI DR

SAVAGE BLAIR D DR

SAVEDOFF MALCOLM P PROF

SCALO JOHN MICHAEL

SCARROTT STANLEY M DR

SCHALEN CARL PROF

SCHATZMAN EVRY PROF

SCHERB FRANK PROF

SCHEUER PETER A G DR

SCHMIDT THOMAS DR

SCHMIDT-KALER TH PROF

SCHMID-BURGK J DR PROF

SCHULTZ G V DR

SCHULZ ROLF ANDREAS

SCHWARTZ PHILIP R DR

SCHWARTZ RICHARD D

SCHWARZ ULRICH J DR

SCOTT EUGENE HOWARD

SCOVILLE NICHOLAS Z

SEATON MICHAEL J PROF

SEKI MUNEZO DR

SHAH GHANSHYAM A DR

SHANE WILLIAM W DR

SHAO CHENG-YUAN

SHAPIRO STUART L

SHARPLESS STEWART PROF

SHAVER PETER A DR

SHAWL STEPHEN J DR

SHCHEGLOV P V DR

SHERWOOD WILLIAM A DR

SHIELDS GREGORY A DR

SHU FRANK H PROF

SHULL JOHN MICHAEL

SHULL PETER OTTO DR

SHUSTOV BORIS M DR

SHUTER WILLIAM L H DR

SILBERBERG REIN DR

SILK JOSEPH I PROF

SILVESTRO GIOVANNI

SIMONS STUART DR

SINGH PATAN DEEN DR

SITKO MICHAEL L

SIVAN JEAN-PIERRE DR

SKILLING JOHN DR

SMITH BARHAM W DR

SMITH CRAIG H DR

SMITH HOWARD ALAN

SMITH PETER L DR

SMITH ROBERT G DR

SNELL RONALD L

SNOW THEODORE P PROF

SOBOLEV V V DR

SOFIA SABATINO PROF

SOFUE YOSHIAKI PROF

SOLC MARTIN

SOLOMON PHILIP M DR

SOMERVILLE WILLIAM B DR

SPITZER LYMAN JR DR

STAHLER SETVEN W DR

STANGA RUGGERO

STECHER THEODORE P

STENHOLM BJOERN DR

STROM RICHARD G DR

SU BUMEI

SUH KYUNG-WON DR

SUN JIN

SUZUKI YOSHIMASA PROF

TAKAKUBO KEIYA PROF

TAKANO TOSHIAKI DR

TAMURA SHINICHI DR

TANAKA MASUO DR

TARAFDAR SHANKAR P DR

TAYLOR KENNETH N R PROF

TENORIO-TAGLE G DR

TERZIAN YERVANT PROF

THADDEUS PATRICK PROF

THE PIK-SIN PROF

THOMPSON A RICHARD DR

THONNARD NORBERT DR

THRONSON HARLEY ANDREW JR

TOKAREV YURIJ V DR

TORRELLES JOSE M DR

TORRES-PEIMBERT SILVIA DR

TOSI MONICA

TOWNES CHARLES HARD DR

TREFFERS RICHARD R

TURNER BARRY E DR

TURNER KENNETH C DR

ULRICH MARIE-HELENE D DR

URASIN LIRIK A DR

VAN DE HULST H C PROF DR

VAN DER LAAN H PROF DR

VAN DISHOECK EWINE F DR

VAN GORKOM JACQUELINE H

VAN WOERDEN HUGO PROF DR

VANDEN BOUT PAUL A

VANYSEK VLADIMIR PROF

VARSHALOVICH DIMITRIJ PR   VERSCHUUR GERRIT L PROF   VIALA YVES
VIALLEFOND FRANCOIS   VIDAL JEAN-LOUIS DR   VIDAL-MADJAR ALFRED DR
VINER MELVYN R DR   VISVANATHAN NATARAJAN DR   VOGEL MANFRED DR
VOLK KEVIN DR   VORONTSOV-VEL'YAMINOV B A VOSHCHINNIKOV NICOLAI DR
VRBA FREDERICK J DR   WALKER GORDON A H PROF   WALMSLEY C MALCOLM DR
WALTON NICHOLAS A DR   WANNIER PETER GREGORY DR   WATT GRAEME DAVID
WEAVER HAROLD F PROF   WEILER KURT W DR   WEISHEIT JON C DR
WENDKER HEINRICH J PROF   WESSELIUS PAUL R DR   WEYMANN RAY J PROF
WHITE GLENN J   WHITE RICHARD L   WHITELOCK PATRICIA ANN DR
WHITEOAK J B DR   WHITTET DOUGLAS C B DR   WHITWORTH ANTHONY PETER
WICKRAMASINGHE N C PROF   WILLIAMS DAVID A PROF   WILLIAMS ROBERT E DR
WILLIS ALLAN J DR   WILLNER STEVEN PAUL DR   WILSON ROBERT W DR
WILSON THOMAS L DR   WINNBERG ANDERS DR   WINNEWISSER GISBERT DR
WITT ADOLF N DR   WOLSTENCROFT RAMON D DR   WOLSZCZAN ALEXANDER DR
WOLTJER LODEWIJK PROF   WOODWARD PAUL R DR   WOOLF NEVILLE J
WOOTTEN HENRY ALWYN   WRIGHT EDWARD L DR   WU CHI CHAO DR
WYNN-WILLIAMS C G DR   XIANG DELIN   YABUSHITA SHIN A PROF
YORK DONALD G DR   YORKE HAROLD W DR   YOUNIS SAAD M
ZEALEY WILLIAM J DR   ZEILIK MICHAEL II DR   ZENG QIN DR
ZHANG CHENG-YUE   ZHOU ZHEN-PU   ZIMMERMANN HELMUT DR
ZUCKERMAN BEN M DR

COMPOSITION OF COMMISSION 35   1991-1994

President   : DEMARQUE P PROF

Vice-President(s)   : CHIOSI CESARE S DR

Organizing Committee: CANAL RAMON M DR
                       IBEN ICKO JR PROF
                       MAEDER ANDRE PROF
                       MICHAUD GEORGES J DR
                       NOMOTO KEN'ICHI DR
                       RENZINI ALVIO PROF
                       TUTUKOV A V DR
                       VANDENBERG DON DR
                       VAUCLAIR GERARD P DR

Members:

AIAD A PROF	AIZENMAN MORRIS L DR	ANAND S P S DR
ANGELOV TRAJKO	ANTIA H M DR	APPENZELLER IMMO PROF
ARAI KENZO DR	ARIMOTO NOBUO DR	ARNETT W DAVID PROF
ARNOULD MARCEL L DR	AUDOUZE JEAN PROF	BAGLIN ANNIE DR
BAKER NORMAN H PROF	BAYM GORDON ALAN DR	BEAUDET GILLES DR
BECKER STEPHEN A	BELMONTE AVILES J A DR	BENZ WILLY
BERTHOMIEU GABRIELLE DR	BISNOVATYI-KOGAN G S DR	BLUDMAN SIDNEY A PROF
BOCCHIA ROMEO DR	BODENHEIMER PETER PROF	BOEHM KARL-HEINZ PROF
BONDI HERMANN PROF SIR	BONO GIUSEPPE DR	BOSS ALAN P DR
BROWNLEE ROBERT R DR	BUCHLER J ROBERT PROF	BURBIDGE GEOFFREY R PROF
CALLEBAUT DIRK K DR	CALOI VITTORIA DR	CAMERON ALASTAIR G W PROF
CAPUTO FILIPPINA DR	CARSON T R DR	CASTELLANI VITTORIO PROF
CASTOR JOHN I DR	CAUGHLAN GEORGEANNE R	CHAN KWING LAM
CHAN ROBERTO DR	CHANDRASEKHAR S PROF	CHECHETKIN VALERIJ M DR
CHEVALIER CLAUDE DR	CHITRE SHASHIKUMAR M DR	CHIU HONG-YEE DR
CHKHIKVADZE IAKOB N	CHRISTENSEN-DALSGAARD J	CHRISTY ROBERT F DR
COHEN JEFFREY M DR	CONNOLLY LEO PAUL	COWAN JOHN J DR
DAS MRINAL KANTI	DAVIS CECIL G JR	DE GREVE JEAN-PIERRE DR
DE JAGER CORNELIS PROF	DE LOORE CAMIEL PROF	DE MEDEIROS JOSE RENAN DR
DEARBORN DAVID PAUL S DR	DEINZER W PROF DR	DESPAIN KEITH HOWARD DR
DEUPREE ROBERT G DR	DINGENS P PROF DR	DLUZHNEVSKAYA O B DR
DURISEN RICHARD H DR	DZIEMBOWSKI WOJCIECH PROF	D'ANTONA FRANCESCA DR
EDWARDS ALAN CH DR	EDWARDS TERRY W	EGGLETON PETER P DR
EMINZADE T A DR	ENDAL ANDREW S DR	EPSTEIN ISADORE PROF
ERGMA E V DR	ERIGUCHI YOSHIHARU DR	EZER-ERYURT DILHAN PROF
FADEYEV YURI A	FAULKNER DONALD J DR	FAULKNER JOHN PROF
FERNANZ MARGARITA DR	FLANNERY BRIAN PAUL DR	FONTAINE GILLES DR
FORBES J E DR	FOSSAT ERIC DR	FOWLER WILLIAM A PROF
FRANTSMAN YU L DR	FUJIMOTO MASAYUKI DR	GABRIEL MAURICE R DR
GALLINO ROBERTO	GEROYANNIS VASSILIS S DR	GHEIDARI S NASSIRI DR
GIANNONE PIETRO PROF	GIMENEZ ALVARO	GINGOLD ROBERT ARTHUR DR
GIRIDHAR SUNETRA DR	GLATZMAIER GARY A	GONG SHU-MO

GOUGH DOUGLAS O DR	GOUPIL MARIE JOSE	GRAHAM ERIC DR
GREGGIO LAURA DR	GURM HARDEV S PROF	HACHISU IZUMI DR
HAMMOND GORDON L DR	HASHIMOTO MASA-AKI DR	HAYASHI CHUSHIRO PROF
HENRY RICHARD B C DR	HILF EBERHARD R H PH D	HITOTSUYANAGI JUICHI PROF
HOLLOWELL DAVID EARL DR	HOSHI REIUN DR	HOYLE FRED SIR
HUANG RUN-QIAN	HUMPHREYS ROBERTA M PROF	ILIEV ILIAN
IMSHENNIK V S DR	ISAAK GEORGE R PROF	ISERN JORGE DR
ISHIZUKA TOSHIHISA DR	ITOH NAOKI DR	JAMES RICHARD A DR
KAEHLER HELMUTH DR	KAMINISHI KEISUKE PROF	KATO MARIKO
KHOZOV GENNADIJ V	KIGUCHI MASAYOSHI DR	KING DAVID S PROF
KIPPENHAHN RUDOLF PROF	KIZILOGLU NILGUEN DR	KNOLKER MICHAEL DR
KOCHHAR R K DR	KOESTER DETLEV DR	KOSOVICHEV ALEXANDER
KOTHARI D S DR	KOVETZ ATTAY PROF	KOYAMA KO-U-ICHI DR
KOZLOWSKI MACIEJ DR	KUMAR SHIV S PROF	KUSHWAHA R S PROF
KWOK SUN DR	LABAY JAVIER	LAMB DONALD QUINCY JR DR
LAMB SUSAN ANN DR	LANGER NORBERT DR	LARSON RICHARD B PROF
LASKARIDES PAUL G ASSPROF	LASOTA JEAN-PIERRE DR	LATOUR JEAN J
LEBOVITZ NORMAN R PROF	LEBRETON YVELINE DR	LEE THYPHOON
LEPINE JACQUES R D DR	LI ZONG-WEI	LIEBERT JAMES W DR
LINNELL ALBERT P PROF	LITTLETON JOHN E	LIVIO MARIO PROF
MAHESWARAN MURUGESAPILLAI	MALLIK D C V DR	MARX GYORGY PROF
MASANI A PROF	MASSEVICH ALLA G DR	MATTEUCCI FRANCESCA DR
MAZUREK THADDEUS JOHN DR	MAZZITELLI ITALO DR	MCCREA J DERMOTT
MELIK-ALAVERDIAN YU DR	MESTEL LEON PROF	MEYER-HOFMEISTER E DR
MEYNET GEORGES DR	MITALAS ROMAS ASSOC PROF	MIYAJI SHIGEKI DR
MOELLENHOFF CLAUS DR	MOHAN CHANDER DR	MONAGHAN JOSEPH J DR
MOORE DANIEL R DR	MORGAN JOHN ADRIAN	MORRIS STEPHEN C DR
MOSKALIK PAWEL DR	MOSS DAVID L DR	MUELLER EWALD
NADYOZHIN D K DR	NAKAMURA TAKASHI DR	NAKANO TAKENORI DR
NAKAZAWA KIYOSHI DR	NARASIMHA DELAMPADY DR	NARITA SHINJI DR
NEWMAN MICHAEL JOHN DR	NISHIDA MINORU PROF	NOELS ARLETTE DR
ODELL ANDREW P	OHYAMA NOBORU PROF	OKAMOTO ISAO DR
OSAKI YOJI DR	OSTRIKER JEREMIAH P PROF	OSWALT TERRY D DR
PACZYNSKI BOHDAN PROF	PAMYATNIKH A A DR	PANDE GIRISH CHANDRA PROF
PAPALOIZOU JOHN C B DR	PEARCE GILLIAN DR	PHILLIPS MARK M DR
PINES DAVID PROF	PINOTSIS ANTONIS D DR	PLAVEC MIREK J PROF
PORFIR'EV V V DR	POVEDA ARCADIO DR	PRENTICE ANDREW J R DR
PRIALNIK-KOVETZ DINA DR	PROFFITT CHARLES R DR	PROVOST JANINE DR
QU QIN-YUE	RAEDLER K H DR	RAMADURAI SOURIRAJA DR
RAY ALAK DR	REEVES HUBERT PROF	REIZ ANDERS PROF
RITTER HANS DR	ROOD ROBERT T DR	ROUSE CARL A DR
ROXBURGH IAN W PROF	RUBEN G PROF DR	SACKMANN I JULIANA DR
SAIO HIDEYUKI DR	SAKASHITA SHIRO PROF	SALPETER EDWIN E PROF
SANTOS FILIPE D DR	SATO KATSUHIKO PROF	SAVEDOFF MALCOLM P PROF
SAVONIJE GERRIT JAN DR	SCALO JOHN MICHAEL	SCHATTEN KENNETH H DR
SCHATZMAN EVRY PROF	SCHILD HANSRUEDI	SCHOENBERNER DETLEF PROF
SCHRAMM DAVID N PROF	SCHUTZ BERNARD F PROF	SCHWARZSCHILD MARTIN PROF
SCUFLAIRE RICHARD DR	SEARS RICHARD LANGLEY DR	SEIDOV ZAKIR F DR
SENGBUSCH KURT V DR	SHAVIV GIORA PROF	SHIBAHASHI HIROMOTO DR
SHIBATA YUKIO DR	SHUSTOV BORIS M DR	SIENKIEWICZ RYSZARD DR
SIGNORE MONIQUE DR	SILVESTRO GIOVANNI	SION EDWARD MICHAEL
SMEYERS PAUL PROF	SMITH ROBERT CONNON DR	SOBOUTI YOUSEF PROF
SOFIA SABATINO PROF	SOUFFRIN PIERRE B DR	SPARKS WARREN M DR
SPIEGEL E DR	SREENIVASAN S RANGA PROF	STARRFIELD SUMNER PROF
STELLINGWERF ROBERT F DR	STIBBS DOUGLAS W N PROF	STRITTMATTER PETER A PROF

SUDA KAZUO PROF
SUGIMOTO DAIICHIRO PROF
SWEET PETER A PROF
SWEIGART ALLEN V DR
TAAM RONALD EVERETT DR
TAKAHARA MARIKO
TASSOUL MONIQUE DR
TAYLER ROGER J PROF
THIELEMANN FRIEDRICH-KARL
THOMAS HANS-CHRISTOPH DR
THOMPSON MICHAEL J DR
TJIN-A-DJIE HERMAN R E DR
TOHLINE JOEL EDWARD
TOOMRE JURI
TRIMBLE VIRGINIA L DR
TRURAN JAMES W JR
TSCHARNUTER WERNER M DR
TUOMINEN ILKKA V DR
UCHIDA JUICHI DR
ULRICH ROGER K PROF
UNNO WASABURO PROF
UUS UNDO DR
VAN DEN HEUVEL EDWARD P J
VAN DER BORGHT RENE PROF
VAN DER RAAY HERMAN B
VAN HORN HUGH M PROF
VAN RIPER KENNETH A DR
VARDYA M S DR
VILA SAMUEL C PROF
VILHU OSMI DR
VILKOVISKIJ.EMMANUIL Y DR
WARD RICHARD A DR
WEAVER THOMAS A DR
WEBBINK RONALD F DR
WEIGERT ALFRED PROF
WEISS ACHIM DR
WEISS NIGEL O DR
WHEELER J CRAIG PROF
WILLSON LEE ANNE DR
WILSON ROBERT E PROF
WINKLER KARL-HEINZ A DR
WOOD PETER R DR
WOOSLEY S E PROF
XIONG DA-RUN
YORKE HAROLD W DR
YUNGELSON LEV R
ZAHN JEAN-PAUL DR
ZHEVAKIN S A PROF DR
ZIOLKOWSKI JANUSZ DR

COMPOSITION OF COMMISSION 36 1991-1994

President   : KALKOFEN WOLFGANG DR

Vice-President(s) : CRAM LAWRENCE EDWARD PROF

Organizing Committee: CUNY YVETTE J DR
                       DRAVINS DAINIS PROF
                       GRAY DAVID F PROF
                       LINSKY JEFFREY L DR
                       PALLAVICINI ROBERTO DR
                       PERAIAH ANNAMANENI DR
                       SAPAR ARVED DR
                       SEATON MICHAEL J PROF
                       TSUJI TAKASHI
                       WEHRSE RAINER DR
                       WILLSON LEE ANNE DR
                       ZWAAN CORNELIS PROF DR

Members:

ABBOTT DAVID C DR	ABHYANKAR KRISHNA D PROF	ALLER LAWRENCE HUGH
ALTROCK RICHARD C DR	ARPIGNY CLAUDE PROF	ATHAY R GRANT DR
AUER LAWRENCE H DR	AUMAN JASON R PROF	AVRETT EUGENE H DR
BAADE DIETRICH DR	BAIRD SCOTT R	BALIUNAS SALLIE L
BALONA LUIS ANTERO DR	BASCHEK BODO PROF	BASRI GIBOR B
BELL ROGER A DR	BERNAT ANDREW PLOUS DR	BERTOUT CLAUDE
BLANCO CARLO DR	BLESS ROBERT C PROF	BLOMME RONNY DR
BODO GIANLUIGI DR	BOEHM KARL-HEINZ PROF	BOEHME SIEGFRIED DR
BOEHM-VITENSE ERIKA PROF	BOESGAARD ANN M PROF	BOPP BERNARD W DR
BOWEN GEORGE H DR	BROWN ALEXANDER	BROWN DOUGLAS NASON
BUES IRMELA D DR	CARBON DUANE F DR	CARLSSON MATS DR
CARSON T R DR	CASSINELLI JOSEPH P DR	CASTOR JOHN I DR
CATALA CLAUDE DR	CATALANO FRANCESCO A DR	CATALANO SANTO DR
CAYREL DE STROBEL GIUSA	CAYREL ROGER DR	CHAN KWING LAM
CHEN PEISHENG	CHUGAI NIKOLAI N DR	CONTI PETER S DR
COWLEY CHARLES R PROF	CRIVELLARI LUCIO	CUGIER HENRYK DR
CUNTZ MANFRED DR	DAVIS CECIL G JR	DELACHE PHILIPPE J DR
DIMITRIJEVIC MILAN	DOAZAN VERA DR	DOMKE HELMUT PH D
DOYLE JOHN GERARD	DRAKE STEPHEN A	DUFTON PHILIP L DR
DUPREE ANDREA K DR	EDVARDSSON BENGT DR	ELSTE GUNTHER H DR
ERIKSSON KJELL DR	EVANGELIDIS E DR	FÅRAGGIANA ROSANNA PROF
FAUROBERT-SCHOLL M DR	FINN G D DR	FITZPATRICK EDWARD L DR
FONTAINE GILLES DR	FONTENLA JUAN MANUEL DR	FOY RENAUD DR
FREIRE FERRERO RUBENS G	FRIEND DAVID B DR	FRISCH HELENE DR
FRISCH URIEL DR	FROESCHLE CHRISTIANE D DR	GAIL HANS-PETER DR
GALLINO ROBERTO	GEBBIE KATHARINE B DR	GIAMPAPA MARK S
GIGAS DETLEF DR	GLEBOCKI ROBERT PROF	GOKDOGAN NUZHET PROF
GORDON CHARLOTTE PROF	GOUGH DOUGLAS O DR	GRANT IAN P DR
GREENSTEIN J L PROF	GREVESSE N DR	GRININ VLADIMIR P DR

GROTH HANS G PROF DR	GUSSMANN E A DR	GUSTAFSSON BENGT DR
HACK MARGHERITA PROF	HAISCH BERNHARD MICHAEL	HALL DOUGLAS S DR
HAMMAN WOLF-RAINER	HARTMANN LEE WILLIAM	HARUTYUNIAN HAIK A DR
HEARN ANTHONY G DR	HEASLEY JAMES NORTON	HEBER ULRICH
HEKELA JAN DR	HEROLD HEINZ	HITOTSUYANAGI JUICHI PROF
HOARE MELVIN DR	HOEFLICH PETER DR	HOLWEGER HARTMUT PROF
HOLZER THOMAS EDWARD DR	HOTINLI METIN DR	HOUSE LEWIS L DR
HUBENY IVAN	HUMMER DAVID G DR	HUNGER KURT PROF
HUSFELD DIRK DR	HUTCHINGS JOHN B DR	IVANOV VSEVOLOD V DR PROF
JEFFERIES JOHN T DR	JOHNSON HOLLIS R PROF	JUDGE PHILIP DR
KADOURI TALIB HADI	KAMP LUCAS WILLEM DR	KANDEL ROBERT S DR
KARP ALAN HERSH DR	KHOKHLOVA V L DR	KLEIN RICHARD I DR
KODAIRA KEIICHI PROF	KOESTER DETLEV DR	KOLESOV A K DR
KONDO YOJI DR	KONTIZAS EVANGELOS DR	KRIKORIAN RALPH DR
KRISHNA SWAMY K S DR	KUDRITZKI ROLF-PETER PH D	KUHI LEONARD V PROF
KUMAR SHIV S PROF	KURUCZ ROBERT L DR	KUSHWAHA R S PROF
LAMBERT DAVID L PROF	LAMERS HENNY J G L M DR	LANDSTREET JOHN D PROF
LEIBACHER JOHN DR	LIEBERT JAMES W DR	LINNELL ALBERT P PROF
LIU CAIPIN	LUCK R EARLE DR	LUTTERMOSER DONALD DR
LYUBIMKOV LEONID S DR	MADEJ JERZY	MARLBOROUGH J M PROF
MASSAGLIA SILVANO	MATHYS GAUTIER DR	MATSUMOTO MASAMICHI PROF
MATSUSHIMA SATOSHI DR	MICHAUD GEORGES J DR	MIHALAS DIMITRI DR
MIYAMOTO SIGENORI PROF	MNATSAKANIAN MAMIKON A DR	MUENCH GUIDO PROF
MUKAI SONOYO DR	MUSIELAK ZDZISLAW E DR	MUTSCHLECNER J PAUL DR
NAGIRNER DMITRIJ I DR	NARASIMHA DELAMPADY DR	NARIAI KYOJI PROF
NEFF JOHN S	NIKOGHOSSIAN ARTHUR G DR	NORDLUND AKE DR
ORRALL FRANK Q PROF	OWOCKI STANLEY PETER DR	OXENIUS JOACHIM DR
O'MARA BERNARD J PROF	PAGEL BERNARD E J PROF	PANEK ROBERT J DR
PASINETTI LAURA E PROF	PECKER JEAN-CLAUDE PROF	PETERS GERALDINE JOAN DR
PHILLIPS JOHN G PROF	PINTO PHILIP ALFRED DR	PISKUNOV NIKOLAI E DR
PORNCHAI P.-TANAKUN	POTTASCH STUART R PROF	PRADERIE FRANCOISE DR
PULS JOAHIM DR	QUERCI FRANCOIS R DR	QUERCI MONIQUE DR
RACHKOVSKY D N DR	RAMSEY LAWRENCE W DR	REIMERS DIETER PROF
RODONO MARCELLO DR	ROSS JOHN E R DR	ROVIRA MARTA GRACIELA
RUCINSKI SLAWOMIR M DR	RUTTEN ROBERT J DR	RYABCHIKOVA TANYA DR
RYBICKI GEORGE B DR	SAITO KUNIJI PROF	SAKHIBULLIN NAIL A DR
SCHARMER GOERAN BJARNE	SCHMALBERGER DONALD C DR	SCHMID-BURGK J DR PROF
SCHMUTZ WERNER	SCHOENBERNER DETLEF PROF	SCHOLZ M PROF
SCHRIJVER C J DR	SEDLMAYER ERWIN DR	SHINE RICHARD A DR
SHIPMAN HENRY L DR	SIMON KLAUS PETER	SIMON THEODORE
SIMONNEAU EDUARDO DR	SITNIK G F PROF	SKUMANICH ANDRE PROF
SNEZHKO LEONID I	SNIJDERS MATTHEUS A J DR	SOBOLEV V V DR
SOUFFRIN PIERRE B DR	SPIEGEL E DR	SPITE FRANCOIS M DR
SPITE MONIQUE DR	SPRUIT HENK C DR	STALIO ROBERTO DR
STEFFEN MATTHIAS DR	STEIN ROBERT F ASSOC PROF	STEPIEN KAZIMIERZ DR
STIBBS DOUGLAS W N PROF	STROM STEPHEN E	SWIHART THOMAS L DR
SZECSENYI-NAGY GABOR DR	TAKEDA YOICHI DR	TARAFDAR SHANKAR P DR
THEJLL PETER ANDREAS DR	THOMAS RICHARD N DR	TOOMRE JURI
TRAVING GERHARD PROF	TUOMINEN ILKKA V DR	UENO SUEO PROF
UESUGI AKIRA DR	ULMSCHNEIDER PETER PROF	UNDERHILL ANNE B DR
UNNO WASABURO PROF	VAKILI FARROKH DR	VAN REGEMORTER HENRI DR
VAN'T VEER FRANS DR	VAN'T VEER-MENNERET CL DR	VARDAVAS ILIAS MIHAIL
VARDYA M S DR	VASU-MALLIK SUSHMA DR	VAUGHAN ARTHUR H DR
VIIK TONU DR	VILHU OSMI DR	WALTER FREDERICK M
WATANABE TETSUYA	WATERS LAURENS B F M DR	WEBER STEPHEN VANCE

WEIDEMANN VOLKER PROF   WELLMANN PETER PROF DR   WHITE RICHARD L
WICKRAMASINGHE N C PROF   WILSON PETER R PROF   WILSON S J
WOEHL HUBERTUS DR   WOLFF SIDNEY C DR   WRIGHT KENNETH O DR
WYLLER ARNE A PROF   YANOVITSKIJ EDGARD G DR   YORKE HAROLD W DR
ZAHN JEAN-PAUL DR

COMPOSITION OF COMMISSION 37 1991-1994

President   : MERMILLIOD JEAN-CLAUDE DR

Vice-President(s) : FEINSTEIN ALEJANDRO DR

Organizing Committee: CLARIA JUAN DR
                      DA COSTA GARY STEWART DR
                      HARRIS GRETCHEN L H DR
                      JANES KENNETH A DR
                      MEYLAN GEORGES DR
                      VANDENBERG DON DR
                      ZHAO JUN-LIANG

Members:

AARSETH SVERRE J DR	ABOU-EL-ELLA MOHAMED S DR	AGEKJAN TATEOS A PROF
AIAD A PROF	ALCAINO GONZALO DR	ALFARO EMILIO JAVIER
ALKSNIS ANDREJS DR	ALLEN CHRISTINE	APARICIO ANTONIO DR
ARMANDROFF TAFT E DR	AURIERE MICHEL	BALAZS BELA A PROF DR
BECKER WILHELM PROF	BELL ROGER A DR	BIJAOUI ALBERT DR
BLAAUW ADRIAAN PROF DR	BOUVIER PIERRE PROF	BUONANNO ROBERTO
BURKHEAD MARTIN S	BUTLER DENNIS DR	BYRD GENE G DR
CALLEBAUT DIRK K DR	CALOI VITTORIA DR	CANNON RUSSELL D DR
CAPUTO FILIPPINA DR	CAPUZZO DOLCETTA ROBERTO	CARNEY BRUCE WILLIAM
CASTELLANI VITTORIO PROF	CHAVARRIA-K CARLOS	CHIOSI CESARE S DR
CHRISTIAN CAROL ANN	CHUN MUN-SUK DR	COLIN JACQUES DR
CUDWORTH KYLE MCCABE DR	CUFFEY J MR	DANFORD STEPHEN C DR
DAPERGOLAS A DR	DAUBE-KURZEMNIECE I A DR	DEJONGHE HERWIG BERT DR
DEMARQUE P PROF	DEMERS SERGE DR	DI FAZIO ALBERTO
DICKENS ROBERT J DR	DLUZHNEVSKAYA O B DR	D'ANTONA FRANCESCA DR
EFREMOV YURY N DR	EINASTO JAAN DR	ELMEGREEN BRUCE GORDON DR
EL-BASSUNY ALAWY A A DR	FALL S MICHAEL DR	FEAST MICHAEL W PROF
FITZGERALD M PIM PROF	FORBES DOUGLAS DR	FORTE JUAN CARLOS DR
FREEMAN KENNETH C PROF	FUSI PECCI FLAVIO	GASCOIGNE S C B DR
GEISLER DOUGLAS P DR	GOLAY MARCEL PROF	GRATTON R G DR
GREEN ELIZABETH M DR	GRIFFITHS WILLIAM K	GRINDLAY JONATHAN E DR
GRUBISSICH C PROF DR	GUETTER HARRY HENDRIK	HANES DAVID A DR
HARRIS HUGH C	HARRIS WILLIAM E DR	HARVEL CHRISTOPHER ALVIN
HASSAN S M PROF	HATZIDIMITRIOU DESPINA DR	HAWARDEN TIMOTHY G DR
HAZEN MARTHA L DR	HEGGIE DOUGLAS C DR	HENON MICHEL C DR
HERBST WILLIAM DR	HESSER JAMES E DR	HEUDIER JEAN-LOUIS DR
HILLS JACK G DR	HUT PIET	IBEN ICKO JR PROF
ILLINGWORTH GARTH D DR	INAGAKI SHOGO DR	ISHIDA KEIICHI PROF
JONES DEREK H P DR	JOSHI U C DR	KADLA ZDENKA I DR
KAMP LUCAS WILLEM DR	KANDRUP HENRY EMIL DR	KILAMBI G C DR
KING IVAN R PROF	KONTIZAS EVANGELOS DR	KONTIZAS MARY DR
KRAFT ROBERT P PROF	KRON GERALD E DR	KUN MARIA DR
LADA CHARLES JOSEPH DR	LANDOLT ARLO U PROF	LAPASSET EMILIO DR
LARSSON-LEANDER G PROF	LATHAM DAVID W DR	LAVAL ANNIE DR

LEISAWITZ DAVID DR
LLOYD EVANS THOMAS DR
LODEN LARS OLOF PROF

LU PHILLIP K DR
LYNDEN-BELL DONALD PROF
LYNGA GOSTA DR

MAEDER ANDRE PROF
MARKKANEN TAPIO DR
MARRACO HUGO G DR

MARSHALL KEVIN P
MARTINS DONALD HENRY DR
MATTEUCCI FRANCESCA DR

MAYOR MICHEL PROF
MENDEZ MARIANO DR
MENON T K PROF

MENZIES JOHN W DR
MIKKOLA SEPPO DR
MILONE EUGENE F PROF

MOFFAT ANTHONY F J DR
MOHAN VIJAY DR
MOULD JEREMY R

MURRAY C ANDREW
MUZZIO JUAN C PROF
NEMEC JAMES

NESCI ROBERTO
NEWELL EDWARD B DR
NISSEN POUL E PROF

OGURA KATSUO DR
ORTOLANI SERGIO
OSMAN ANAS MOHAMED PROF

PANDEY A K
PARSAMYAN ELMA S DR
PEDREROS MARIO DR

PENNY ALAN JOHN DR
PETERSON CHARLES JOHN DR
PETROVSKAYA M S DR

PHILIP A G DAVIS
PILACHOWSKI CATHERINE DR
PISKUNOV ANATOLY E

PLATAIS IMANT K DR
POPOVA MALINA D PROF DR
POVEDA ARCADIO DR

PRITCHET CHRISTOPHER J DR
QIAN BO-CHEN
RAM SAGAR DR

RENZINI ALVIO PROF
RICHER HARVEY B DR
RICHTLER TOM DR

ROSINO LEONIDA PROF
ROTH-HOPPNER MARIA LUISE
ROUNTREE JANET DR

RUPRECHT JAROSLAV DR
RUSSEVA TATJANA
SALUKVADZE G N DR

SAMUS NIKOLAI N DR
SANDERS W L PROF
SAWYER-HOGG HELEN B DR

SCARIA K K DR
SCHILD HANSRUEDI
SHAROV A S DR

SHAWL STEPHEN J DR
SHER DAVID DR
SHOBBROOK ROBERT R DR

SIMODA MAHIRO PROF
SMITH GRAEME H DR
SPURZEM RAINER DR

STETSON PETER B. DR
SUGIMOTO DAIICHIRO PROF
SZECSENYI-NAGY GABOR DR

TERZAN AGOP DR
THE PIK-SIN PROF
TORNAMBE AMEDEO

TRIPICCO MICHAEL J DR
TSVETKOV MILCHO K DR
TSVETKOVA KATIA

TURNER DAVID G DR
TWAROG BRUCE A
UPGREN ARTHUR R DR

VAN ALTENA WILLIAM F PROF
VAN DEN BERGH SIDNEY PROF
VAZQUEZ RUBEN ANGEL DR

WALKER GORDON A H PROF
WALKER MERLE F PROF
WALLENQUIST AAKE A E PROF

WAN LAI
WARREN WAYNE H JR DR
WEAVER HAROLD F PROF

WEHLAU AMELIA DR
WHITE RAYMOND E DR
WIELEN ROLAND PROF DR

WRAMDEMARK STIG S O DR
WU HSIN-HENG DR
XIRADAKI EVANGELIA DR

ZINN ROBERT J DR

COMPOSITION OF COMMISSION 38 1991-1994

President   : SAHADE JORGE PROF

Vice-President(s) : JORGENSEN HENNING E PROF

Organizing Committee: CHITRE DATTAKUMAR M DR
                       DUCATI JORGE RICARDO DR
                       KRISHNA GOPAL
                       LEUNG KAM CHING PROF
                       MORIMOTO MASAKI DR
                       ROBERTS MORTON S DR
                       SMITH F GRAHAM PROF

Members:

ALY M KHAIRY PROF	AL-SABTI ABDUL ADIM DR	BOYARCHUK A A DR
CACCIN BRUNO	FLORSCH ALPHONSE DR	HAUPT HERMANN F PROF
KOZAI YOSHIHIDE PROF	MACRAE DONALD A PROF	MARIK MIKLOS DR
MUELLER EDITH A PROF	NHA IL-SEONG DR	NINKOVIC SLOBODAN
OKOYE SAMUEL E PROF	ROUTLY PAUL M DR	SWARUP GOVIND PROF
TOLBERT CHARLES R DR	VAN DEN HEUVEL EDWARD P J	WANG SHOU-GUAN
WOOD F BRADSHAW PROF	YE SHU-HUA	

COMPOSITION OF COMMISSION 40 1991-1994

President   : MORIMOTO MASAKI DR

Vice-President(s) : WHITEOAK J B DR

Organizing Committee: BAART EDWARD E PROF
                       BAATH LARS B DR
                       BACKER DONALD CH DR
                       COLOMB FERNANDO R DR
                       DAVIS RICHARD J DR
                       FANTI ROBERTO
                       GERARD ERIC DR
                       GOMEZ GONZALEZ JESUS DR
                       GOSACHINSKIJ I V DR
                       GUESTEN ROLF
                       MEZGER PETER G PROF
                       MORAN JAMES M DR
                       VAN DER HULST JAN M DR
                       VELUSAMY T DR
                       YE SHU-HUA

Members:

ABDULLA SHAKER ABDUL AZIZ	ABLES JOHN G DR	ABRAMI ALBERTO PROF
ADE PETER A R DR	AIZU KO PROF	AKABANE KENJI A PROF
AKUJOR CHIDI E	ALEXANDER JOSEPH K	ALEXANDER PAUL DR
ALLEN RONALD J DR	ALLER HUGH D DR	ALLER MARGO F DR
ALTENHOFF WILHELM J DR	ANANTHARAMAIAH K R DR	ANDERNACH HEINZ DR
ANDREW BRYAN H DR	APARICI JUAN DR	ARNAL MARCELO EDMUNDO DR
ASCHWANDEN MARKUS DR	ASSOUSA GEORGE ELIAS DR	AURASS HENRY DR
AVERY LORNE W DR	AVIGNON YVETTE DR	AXON DAVID
BAARS JACOB W M DR	BACHILLER RAFAEL DR	BAGRI DURGADAS S
BALASUBRAMANIAN V DR	BALDWIN JOHN E DR	BALKLAVS A E DR
BALL LEWIS DR	BALLY JOHN DR	BALONEK THOMAS J DR
BANHATTI DILIP GOPAL DR	BARRETT ALAN H PROF	BARROW COLIN H DR
BARTEL NORBERT HARALD DR	BARTHEL PETER DR	BARVAINIS RICHARD DR
BASH FRANK N PROF	BASU DIPAK DR	BATTY MICHAEL DR
BAUDRY ALAIN DR	BECK RAINER	BENN CHRIS R DR
BENNETT CHARLES L DR	BENSON PRISCILLA J DR	BENZ ARNOLD DR
BERGE GLENN L DR	BERKHUIJSEN ELLY M DR	BHANDARI RAJENDRA DR
BHONSLE RAJARAM V PROF	BIEGING JOHN HAROLD DR	BIERMANN PETER L DR
BIGNELL R CARL DR	BIRAUD FRANCOIS DR	BIRETTA JOHN ANTHONY DR
BIRKINSHAW MARK	BLAIR DAVID GERALD	BLANDFORD ROGER DAVID DR
BLOEMHOF ERIC E DR	BOCKELEE-MORVAN DOMINIQUE	BOISCHOT ANDRE DR
BOLTON JOHN G	BOOTH ROY S PROF	BORIAKOFF VALENTIN
BOS ALBERT DR	BOTTINELLI LUCETTE DR	BOWERS PHILLIP F
BRACEWELL RONALD N PROF	BRAUDE SEMION YA PROF AG	BREGMAN JACOB D IR
BRIDLE ALAN H PROF	BRINKS ELIAS DR	BRODERICK JOHN DR
BRONFMAN LEONARDO DR	BROTEN NORMAN W	BROUW W N DR

BROWNE IAN W A DR	BUJARRABAL VALENTIN	BURBIDGE GEOFFREY R PROF
BURKE BERNARD F DR	CAROUBALOS C A PROF	CARR THOMAS D PROF
CASTETS ALAIN. DR	CASWELL JAMES L DR	CATARZI MARCO DR
CERNICHARO JOSE DR	CHAN KWING LAM	CHEN HONGSHENG
CHIKADA YOSHIHIRO DR	CHINI ROLF	CHO SE HYUNG DR
CHRISTIANSEN WAYNE A	CHRISTIANSEN WILBUR PROF	CHU HAN-SHU PROF
CLARK BARRY G DR	CLARK DAVID H DR	CLARK FRANK OLIVER DR
CLEMENS DAN P DR	COHEN MARSHALL H PROF	COHEN RAYMOND J DR
COHEN RICHARD S	COLE TREVOR WILLIAM PROF	COLEMAN PAUL HENRY DR
CONDON JAMES J DR	CONKLIN EDWARD K	CONWAY ROBIN G DR
CORDES JAMES M	COTTON WILLIAM D Jr	COUTREZ RAYMOND A J PROF
COVINGTON ARTHUR E	CRANE PATRICK C	CROOM DAVID L DR
CROVISIER JACQUES	CUDABACK DAVID D DR	DAGKESAMANSKY RUSTAM D DR
DAINTREE EDWARD J DR	DAISHIDO TSUNEAKI PROF	DAVIES RODNEY D PROF
DAVIS MICHAEL M DR	DAVIS ROBERT J DR	DE GROOT T DR
DE JAGER CORNELIS PROF	DE LA NOE JEROME DR	DE RUITER HANS RUDOLF
DE YOUNG DAVID S DR	DEGAONKAR S S DR	DELANNOY JEAN DR
DENISSE JEAN-FRANCOIS DR	DENT WILLIAM A PROF	DESHPANDE AVINASH
DESPOIS DIDIER DR	DEWDNEY PETER E F DR	DHAWAN VIVEK DR
DICKEL HELENE R DR	DICKEL JOHN R	DICKEY JOHN M
DICKMAN ROBERT L DR	DIETER NANNIELOU H DR	DIXON ROBERT S DR
DOUBINSKIJ B A DR	DOUGLAS JAMES N PROF	DOWNES DENNIS DR
DOWNS GEORGE S DR	DRAKE FRANK D PROF	DRAKE STEPHEN A
DRAVSKIKH A F DR	DREHER JOHN W	DUFFETT-SMITH PETER JAMES
DULK GEORGE A PROF	DWARAKANATH K S	DYSON F J DR
EDELSON RICK DR	EKERS RONALD D DR	ELGAROY OYSTEIN PROF
ELLIS G R A PROF	ELSMORE BRUCE DR	ELWERT GERHARD PROF
EMERSON DAVID	ENOME SHINZO PROF	EPSTEIN EUGENE E DR
ERICKSON WILLIAM C DR	ERIKSEN GUNNAR PROF	ESHLEMAN VON R PROF
EVANS KENTON DOWER DR	EWING MARTIN S	FACONDI SILVIA ROSA DR
FEIGELSON ERIC D DR	FEIX GERHARD DR	FELDMAN PAUL A DR
FELLI MARCELLO DR	FELTEN JAMES E DR	FERETTI LUIGINA
FERRARI ATTILIO DR	FIELD GEORGE B PROF	FINDLAY JOHN W DR
FINKELSTEIN ANDREJ M DR	FLEISCHER ROBERT DR	FLETT ALISTAIR M
FLORKOWSKI DAVID R DR	FOLEY ANTHONY DR	FOMALONT EDWARD B DR
FORT DAVID NORMAN DR	FORVEILLE THIERRY DR	FOUQUE PASCAL DR
FRATER ROBERT H DR	FRIBERG PER	FRIEDMAN HERBERT DR
FRISK URBAN DR	FUERST ERNST DR	FUKUI YASUO DR
GALT JOHN A DR	GARAY GUIDO DR	GARDNER FRANCIS F DR
GARRINGTON SIMON DR	GAYLARD MICHAEL JOHN	GEBLER KARL-HEINZ DR
GELDZAHLER BERNARD J	GELFREIKH GEORGIJ B DR	GENT HUBERT MR
GENZEL REINHARD DR	GERGELY TOMAS ESTEBAN DR	GHIGO FRANCIS D DR
GIL JANUSZ A PROF	GINZBURG VITALY L PROF	GIOIA ISABELLA M DR
GIOVANNINI GABRIELE	GOLD THOMAS PROF	GOLDSMITH PAUL F DR
GOLDSTEIN SAMUEL J PROF	GOLDWIRE HENRY C JR	GONZE ROGER F J IR
GOPALSWAMY N DR	GORDON MARK A DR	GORGOLEWSKI STANISLAW PR
GOSS W MILLER PROF	GOTTESMAN STEPHEN T DR	GOWER J F R DR
GRAHAM DAVID A	GREEN DAVID DR	GREGORINI LORETTA
GREGORY PHILIP C DR	GREWING MICHAEL PROF	GUELIN MICHEL DR
GUIDICE DONALD A DR	GULKIS SAMUEL DR	GULL STEPHEN F DR
GWINN CARL R DR	HADDOCK FRED T DR	HAGEN JOHN P
HALL PETER J DR	HAMILTON P A DR	HAN FU
HAN WENJUN	HANASZ JAN DR	HANBURY BROWN ROBERT PROF
HANDA TOSHIHIRO DR	HANISCH ROBERT J DR	HANKINS TIMOTHY HAMILTON
HARDEE PHILIP	HARNETT JULIEINE DR	HARRIS DANIEL E DR

MATVEYENKO L I DR	MAUERSBERGER RAINER DR	MAXWELL ALAN DR
MAY J	MAYER CORNELL H	MCADAM W BRUCE DR
MCCONNELL DAVID DR	MCCULLOCH PETER M DR	MCKENNA LAWLOR SUSAN
MCLEAN DONALD J DR	MEBOLD ULRICH DR PROF	MEEKS M LITTLETON DR
MEIER DAVID L	MENON T K PROF	MICHALEC ADAM
MILEY G K DR	MILLS BERNARD Y PROF	MILNE DOUGLAS K DR
MILOGRADOV-TURIN JELENA.	MIRABEL IGOR FELIX DR	MITCHELL KENNETH J DR
MOISEEV I G DR	MOLCHANOV A P PROF	MORISON IAN MR
MORITA KAZUHIKO	MORIYAMA FUMIO PROF	MORRAS RICARDO DR
MORRIS DAVID DR	MORRIS MARK ROOT DR	MUNDY LEE G DR
MURDOCH HUGH S DR	MUTEL ROBERT LUCIEN	MUXLOW THOMAS
MYERS PHILIP C	NADEAU DANIEL DR	NAGNIBEDA VALERY G DR
NAN REN-DONG	NEFF JOHN S	NGUYEN-QUANG RIEU DR
NICOLSON GEORGE D DR	NISHIO MASANORI DR	OEZEL MEHMET EMIN DR
OHISHI MASATOSHI DR	OKOYE SAMUEL E PROF	OKUMURA SACHIKO DR
OLBERG MICHAEL DR	ONUORA LESLEY IRENE DR	OORT JAN H PROF
OSTERBROCK DONALD E PROF	OWEN FRAZER NELSON DR	O'DEA CHRISTOPHER P DR
O'SULLIVAN JOHN DAVID DR	PACHOLCZYK ANDRZEJ G PROF	PADMAN RACHAEL
PADRIELLI LUCIA	PALMER PATRICK E PROF	PANKONIN VERNON LEE DR
PAPAGIANNIS MICHAEL D PRO	PAREDES JOSE MARIA DR	PARIJSKIJ YU N DR
PARKER EDWARD A DR	PARMA PAOLA	PARRISH ALLAN DR
PASACHOFF JAY M PROF	PAULINY TOTH IVAN K K DR	PAULS THOMAS ALBERT DR
PAYNE DAVID G	PEARSON TIMOTHY J	PEDLAR ALAN DR
PENG YUN-LOU	PENZIAS ARNO A DR	PEREZ FOURNON ISMAEL DR
PERLEY RICHARD ALAN	PETERS WILLIAM L III DR	PETTENGILL GORDON H PROF
PHILLIPS THOMAS GOULD DR	PICK MONIQUE DR	PLANESAS PERE
PONSONBY JOHN E B DR	POOLEY GUY DR	PORCAS RICHARD DR
PRESTON ROBERT ARTHUR	PREUSS EUGEN DR	PRICE R MARCUS DR
PRIESTER WOLFGANG PROF	PUSCHELL JEFFERY JOHN	QIAN SHAN-JIE
QIU YU-HAI	RADHAKRISHNAN V PROF	RAIMOND ERNST DR
RAMATY REUVEN DR	RAO A PRAMESH DR	RAOULT ANTOINETTE DR
RAY THOMAS P	RAZIN V A DR	READHEAD ANTHONY C S DR
REBER GROTE DR	REICH WOLFGANG	REID MARK JONATHAN DR
REIF KLAUS DR	REYES FRANCISCO DR	REYNOLDS JOHN DR
RIBES JEAN-CLAUDE DR	RICKARD LEE J DR	RICKETT BARNABY JAMES DR
RIIHIMAA JORMA J DR	RILEY JULIA M DR	ROBERTS DAVID HALL DR
ROBERTS MORTON S DR	ROBERTSON DOUGLAS S	ROBERTSON JAMES GORDON DR
ROBINSON BRIAN J DR	ROBINSON JR RICHARD D DR	RODRIGUEZ LUIS F
ROEDER ROBERT C PROF	ROELFSEMA PETER DR	ROENNAENG BERNT O DR
ROESER HANS-PETER	ROGER ROBERT S DR	ROGERS ALAN E E DR
ROGSTAD DAVID H DR	ROHLFS K PROF DR	ROMNEY JONATHAN D DR
ROWSON BARRIE DR	RUBIN ROBERT HOWARD	RUBIO MONICA DR
RUDNICK LAWRENCE DR	RUSSELL JANE L DR	RYDBECK GUSTAF H B DR
RYDBECK OLOF E H PROF	RYZHKOV NIKOLAI F DR	SAIKIA DHRUBA JYOTI DR
SALPETER EDWIN E PROF	SANAMIAN V A DR	SANDELL GORAN HANS L DR
SANDERS DAVID B DR	SARGENT ANNEILA I	SARMA N V G PROF
SASTRY CH V	SATO FUMIO DR	SAUNDERS RICHARD D E
SAVAGE ANN DR	SAWANT HANUMANT S DR	SCALISE JR EUGENIO DR
SCHAAL RICARDO E DR	SCHEUER PETER A G DR	SCHILIZZI RICHARD T DR
SCHLICKEISER REINHARD DR	SCHMIDT MAARTEN PROF	SCHUCH NELSON JORGE
SCHULTZ G V DR	SCHULZ ROLF ANDREAS	SCHWARTZ PHILIP R DR
SCHWARZ ULRICH J DR	SCOTT JOHN S DR	SCOTT PAUL F DR
SEAQUIST ERNEST R PROF	SEIELSTAD GEORGE A	SEIRADAKIS JOHN HUGH DR
SETTI GIANCARLO PROF	SHAFFER DAVID B DR	SHAKESHAFT JOHN R DR
SHAVER PETER A DR	SHERIDAN K V DR	SHEVGAONKAR R K DR

SHIMMINS ALBERT JOHN	SHOLOMITSKY G B DR	SHUTER WILLIAM L H DR
SIEBER WOLFGANG PH D	SIMON PAUL A DR	SINHA RAMESHWAR P
SKILLMAN EVAN D DR	SLADE MARTIN A III DR	SLEE O B DR
SLYSH VJACHOSLAV I DR	SMITH ALEX G PROF	SMITH DEAN F DR
SMITH F GRAHAM PROF	SMOL'KOV GENNADIJ YA DR	SOBOLEVA N S DR
SOFUE YOSHIAKI PROF	SOROCHENKO R L DR	SPENCER JOHN HOWARD
SPENCER RALPH E DR	SPOELSTRA T A TH DR	SRAMEK RICHARD A DR
STAHR-CARPENTER M DR	STANLEY G J	STANNARD DAVID DR
STEFFEN MATTHIAS DR	STEINBERG JEAN-LOUIS DR	STEWART PAUL DR
STEWART RONALD T MR	STONE R G DR	STOREY MICHELLE DR
STROM RICHARD G DR	STRUKOV IGOR A DR	SUGITANI KOJI DR
SUKUMAR SUNDARAJAN DR	SULLIVAN WOODRUFF T III	SWARUP GOVIND PROF
SWENSON GEORGE W JR PROF	TABARA HIROTO DR	TAKABA HIROSHI DR
TAKAGI KOJIRO PROF	TAKAKUBO KEIYA PROF	TAKAKURA TATSUO PROF EMER
TAKANO TOSHIAKI DR	TANAKA RIICHIRO PROF	TARTER JILL C DR
TATEMATSU KENICHI DR	TAYLOR A R DR	TERZIAN YERVANT PROF
THOMASSON PETER DR	THOMPSON A RICHARD DR	THUM CLEMENS DR
TLAMICHA ANTONIN DR	TOFANI GIANNI PROF	TOLBERT CHARLES R DR
TOMASI PAOLO DR	TOVMASSIAN H M DR	TOWNES CHARLES HARD DR
TRITTON KEITH P DR	TROITSKY V S PROF DR	TROLAND THOMAS HUGH
TRUONG-BACH	TSUBOI MASATO DR	TURLO ZYGMUNT DR
TURNER BARRY E DR	TURNER KENNETH C DR	TURTLE A J DR
TZIOUMIS ANASTASIOS DR	UDAL'TSOV V A DR	UDAYA SHANKAR N DR
UKITA NOBUHARU	ULRICH BRUCE T PROF	ULRICH MARIE-HELENE D DR
UNGER STEPHEN DR	UNWIN STEPHEN C	URPO SEPPO I
USON JUAN M DR	VALLEE JACQUES P DR	VALTAOJA ESKO
VALTONEN MAURI J PROF	VAN DE HULST H C PROF DR	VAN DER KRUIT PIETER C DR
VAN DER LAAN H PROF DR	VAN GORKOM JACQUELINE H	VAN NIEUWKOOP J DR IR
VAN WOERDEN HUGO PROF DR	VANDEN BOUT PAUL A	VATS HARI OM DR
VAUGHAN ALAN DR	VENUGOPAL V R DR	VERON PHILIPPE DR
VERSCHUUR GERRIT L PROF	VERTER FRANCES DR	VILAS FAITH DR
VILAS-BOAS JOSE W DR	VINER MELVYN R DR	VIVEKANAND M DR
VOGEL STUART NEWCOMBE DR	WALKER ROBERT C DR	WALL JASPER V DR
WALMSLEY C MALCOLM DR	WALSH DENNIS DR	WAN TONG-SHAN
WANG JING-SHENG	WANG SHOU-GUAN	WANNIER PETER GREGORY DR
WARDLE JOHN F C PROF	WARMELS REIN HERM DR	WARNER PETER J DR
WARWICK JAMES W DR	WEHRLE ANN ELIZABETH DR	WEI MINGZHI
WEIGELT GERD DR	WEILER EDWARD J DR	WEILER KURT W DR
WELCH WILLIAM J PROF	WELLINGTON KELVIN DR	WENDKER HEINRICH J PROF
WESTERHOUT GART DR	WESTFOLD KEVIN C PROF	WICKRAMASINGHE N C PROF
WIELEBINSKI RICHARD PROF	WILD JOHN PAUL DR	WILKINSON PETER N DR
WILLIS ANTHONY GORDON DR	WILLS BEVERLEY J DR	WILLS DEREK DR
WILLSON ROBERT FREDERICK	WILSON ANDREW S DR	WILSON ROBERT W DR
WILSON THOMAS L DR	WILSON WILLIAM J DR	WINDHORST ROGIER A DR
WINK JOERN ERHARD DR	WINNBERG ANDERS DR	WINNEWISSER GISBERT DR
WITZEL ARNO DR	WOLSZCZAN ALEXANDER DR	WOLTJER LODEWIJK PROF
WOODSWORTH ANDREW W.DR	WOOTTEN HENRY ALWYN	WRIGHT ALAN E DR
WROBEL JOAN MARIE DR	WU HUAI-WEI	WU NAILONG DR
WU SHENGYIN	WU XINJI	XIA ZHIGUO DR
XU PEI-YUAN	XU ZHI-CAI	YANG JIAN
YIN QI-FENG	YOUNIS SAAD M	ZAITSEV VALERII V DR
ZENSUS J-ANTON DR	ZHANG FU JUN	ZHELEZNIAKOV VLADIMIR V
ZHENG YI-JIA	ZHOU TI-JIAN	ZIEBA STANISLAW DR
ZLOBEC PAOLO DR	ZLOTNIK ELENA YA DR	ZUCKERMAN BEN M DR

COMPOSITION OF COMMISSION 41 1991-1994

President   : DEBARBAT SUZANNE V DR

Vice-President(s) : ANSARI S M RAZAULLAH PROF

Organizing Committee: DICK STEVEN J
                      FIRNEIS MARIA G DR
                      MOESGAARD KRISTIAN P
                      NORTH JOHN DAVID PROF

Members:

ARGYRAKOS JEAN PROF DR	BADOLATI ENNIO	BANDYOPADHYAY A PROF
BENSON PRISCILLA J DR	BERENDZEN RICHARD DR	BISHOP ROY L DR
BO SHU-REN	BROOKS RANDALL C DR	BRUNET JEAN-PIERRE DR
CARLSON JOHN B	CHEN ZUN-GUI	CORNEJO ALEJANDRO A DR
CUI ZHEN-HUA	DADIC ZARKO DR	DARIUS JON DR
DEEMING TERENCE J DR	DEKKER E DR	DEVORKIN DAVID H
DEWHIRST DAVID W DR	DOBRZYCKI JERZY PROF	DUMONT SIMONE DR
EDDY JOHN A DR	EDMONDSON FRANK K PROF	EELSALU HEINO DR
ERPYLEV N P DR	FERNIE J DONALD PROF	FERRARI D'OCCHIEPPO K DR
FLORIDES PETROS S PROF	FODERA SERIIO GIORGIA DR	FREITAS MOURAO R R DR
GINGERICH OWEN PROF	GURSHTEIN ALEXANDER A DR	HAWKINS GERALD S DR
HAYLI AVRAM PROF	HERRMANN DIETER PROF DR	HOSKIN MICHAEL A DR
HOWSE H DEREK	HYSOM EDMUND J	IDLIS G M DR
JACKISCH GERHARD DR	JIANG XIAO-YUAN DR	KENNEDY JOHN E PROF
KHROMOV G S DR	KIANG TAO PROF	KING DAVID S PROF
KING HENRY C DR	KRUPP EDWIN C DR	KUNITZSCH PAUL PROF
LANG KENNETH R ASST PROF	LEVY EUGENE H DR	LEVY JACQUES R DR
LI ZHI-SEN	LIU CI-YUAN DR	LOPES-GAUTIER ROSALY DR
MALIN STUART	MCKENNA LAWLOR SUSAN	MERLEAU-PONTY J PROF
NADAL ROBERT	NAKAYAMA SHIGERU DR	NICOLAIDIS EFTHYMIOS DR
ORCHISTON WAYNE DR	OSTERBROCK DONALD E PROF	PEDERSEN OLAF PROF
PETERSON CHARLES JOHN DR	PETRI WINFRIED PROF DR	PIGATTO LUISA DR
PINGREE DAVID PROF	POGO ALEXANDER DR	PORTER NEIL A PROF
POULLE EMMANUEL PROF	PROKAKIS THEODORE J DR	PROVERBIO EDOARDO PROF
QUAN HEJUN	RONAN COLIN A	RYBKA PRZEMYSLAW DR
SATO NAONOBU PROF	SBIRKOVA-NATCHEVA T	SCHAEFER BRADLEY E DR
SHUKLA K	SIGNORE MONIQUE DR	SOLC MARTIN
STEPHENSON F RICHARD DR	STOEV ALEXEI	SULLIVAN WOODRUFF T III
SUNDMAN ANITA DR	SVOLOPOULOS SOTIRIOS PROF	SWERDLOW NOEL PROF
TATON RENE PROF	VERDET JEAN-PIERRE DR	WANG DE-CHANG
WATTENBERG D PROF	WHITAKER EWEN A	WHITE GRAEME LINDSAY DR
WHITROW GERALD JAMES PROF	WILSON CURTIS A	WRIGHT HELEN GREUTER
XI ZE-ZONG	XU ZHENTAO	YABUUTI KIYOSHI PROF
YAU KEVIN K C DR	YEOMANS DONALD K DR	ZHANF SHOUZHONG DR
ZHANG PEIYU	ZHUANG WEIFENG	ZOSIMOVICH IRINA D

COMPOSITION OF COMMISSION 42 1991-1994

President   : KONDO YOJI DR

Vice-President(s) : RODONO MARCELLO DR

Organizing Committee: EGGLETON PETER P DR
                       GIMENEZ ALVARO
                       GUINAN EDWARD FRANCIS DR
                       HILDITCH RONALD W DR
                       HILL GRAHAM DR
                       KOCH ROBERT H DR
                       MUTEL ROBERT LUCIEN
                       POLIDAN RONALD S
                       SZKODY PAULA DR
                       TUTUKOV A V DR
                       WEBBINK RONALD F DR
                       YAMASAKI ATSUMA DR
                       ZHAI DI-SHENG

Members:

ABHYANKAR KRISHNA D PROF	AL-NAIMY HAMID M K DR	ANDERSEN JOHANNES
ANTIPOVA LYUDMILA DR	ANTONOPOULOU E DR	AQUILANO ROBERTO OSCAR DR
AWADALLA NABIL SHOUKRY DR	BARONE FABRIZIO DR	BARTOLINI CORRADO
BATESON FRANK M OBE DR	BATH GEOFFREY T DR	BATTEN ALAN H DR
BELL STEVEN DR	BLAIR WILLIAM P DR	BLITZSTEIN WILLIAM DR
BOLTON C THOMAS PROF	BONAZZOLA SILVANO DR	BOOKMYER BEVERLY B DR
BOPP BERNARD W DR	BRADSTREET DAVID H DR	BRANDI ELISANDE ESTELA DR
BREINHORST ROBERT A DR	BROGLIA PIETRO DR	BROWNLEE ROBERT R DR
BRUHWEILER FRED C JR	BUDDING EDWIN DR	BUNNER ALAN N DR
BUSSO MAURIZIO	CALLANAN PAUL DR	CATALANO SANTO DR
CESTER BRUNO PROF	CHAMBLISS CARLSON R DR	CHANMUGAM GANESAR PROF
CHAPMAN ROBERT D DR	CHAUBEY UMA SHANKAR DR	CHEN KWAN-YU PROF
CHEREPASHCHUK A M PROF	CHOCHOL DRAHOMIR	CHOI KYU-HONG
CILLIE G G PROF	CLARIA JUAN DR	CLAUSEN JENS VIGGO LEKTOR
COLLINS GEORGE W II PROF	COWLEY ANNE P DR	CRISTALDI SALVATORE DR
DADAEV ALEKSANDR N DR	DE GREVE JEAN-PIERRE DR	DE GROOT MART DR
DE KORT JULES J DR	DE LOORE CAMIEL PROF	DELGADO ANTONIO JESUS
DEMIRCAN OSMAN DR	DORFI ERNST ANTON DR	DOUGHTY NOEL A DR
DRECHSEL HORST DR	DUERBECK HILMAR W DR	DURISEN RICHARD H DR
DUSCHL WOLFGANG J DR	D'ANTONA FRANCESCA DR	EATON JOEL A DR
ECHEVARRIA JUAN DR	ETZEL PAUL B DR	FAULKNER JOHN PROF
FEKEL FRANCIS C	FERLUGA STENO DR	FERRARI D'OCCHIEPPO K DR
FERRER OSVALDO EDUARDO DR	FIRMANI CLAUDIO A PROF	FLANNERY BRIAN PAUL DR
FRACASTORO MARIO G PROF	FRANK JUHAN	FRANTSMAN YU L DR
FREDRICK LAURENCE W PROF	FRIEDJUNG MICHAEL DR	GARCIA DE MARIA J M DR
GARMANY CATHERINE D DR	GEYER EDWARD H PROF DR	GIANNONE PIETRO PROF
GIOVANNELLI FRANCO DR	GIURICIN GIULIANO	GOLDMAN ITZHAK DR
GRYGAR JIRI DR	GULLIVER AUSTIN FRASER DR	GURSKY HERBERT DR

GUSEINOV O H PROF

GYLDENKERNE KJELD DR

HADRAVA PETR

HALL DOUGLAS S DR

HAMMERSCHLAG-HENSBERGE G

HANAWA TOMOYUKI DR

HANTZIOS PANAYIOTIS DR

HARMANEC PETR DR

HASSALL BARBARA J M DR

HAZLEHURST JOHN DR

HEINTZ WULFF D DR

HELT BODIL E

HENSLER GERHARD PROF

HERCZEG TIBOR J PROF DR

HILLS JACK G DR

HJELLMING ROBERT M DR

HOFFMANN MARTIN DR

HOLT STEPHEN S

HONEYCUTT R KENT PROF

HORAK TOMAS B DR

HORIUCHI RITOKU DR

HRIC LADISLAV DR

HRIVNAK BRUCE J

HUANG RUN-QIAN

HUBE DOUGLAS P DR

HUTCHINGS JOHN B DR

IBANOGLU C DR

IMAMURA JAMES DR

IMBERT MAURICE DR

IRWIN JOHN B PROF

JABBAR SABEH RHAMAN

JASCHEK CARLOS O R PROF

JASNIEWICZ GERARD DR

JOSS PAUL CHRISTOPHER DR

JURKEVICH IGOR DR

KADOURI TALIB HADI

KAITCHUCK RONALD H

KALUZNY JANUSZ DR

KANDPAL CHANDRA D

KANG YOUNG WOON DR

KARETNIKOV VALENTIN G R

KAWABATA SHUSAKU PROF

KENYON SCOTT J DR

KHALESSEH BAHRAM DR

KITAMURA M PROF

KJURKCHIEVA DIANA DR

KOPAL ZDENEK PROF

KOUBSKY PAVEL

KRAFT ROBERT P PROF

KRAICHEVA ZDRAVSKA DR

KRAUTTER JOACHIM DR

KREINER JERZY MAREK DR

KRIZ SVATOPLUK DR

KRON KATHERINE GORDON

KRUCHINENKO VITALIY G

KRUSZEWSKI ANDRZEJ PROF

KRZEMINSKI WOJCIECH DR

KUMSIASHVILY MZIA I DR

KURPINSKA-WINIARSKA M DR

KVIZ ZDENEK DR

KWEE K K DR

LA DOUS CONSTANZE A DR

LACY CLAUD H DR

LAMB DONALD QUINCY JR DR

LANDOLT ARLO U PROF

LAPASSET EMILIO DR

LARSSON STEFAN DR

LARSSON-LEANDER G PROF

LAVROV M I PROF

LEE WOO-BAIK DR

LEE YONG-SAM DR

LEUNG KAM CHING PROF

LI ZHONGYUAN

LINNELL ALBERT P PROF

LIU QINGYAO DR

LIU XUEFU

LUCY LEON B PROF

LYUTY VICTOR M DR

MACDONALD JAMES

MACERONI CARLA

MAMMANO AUGUSTO DR

MARDIROSSIAN FABIO

MARILLI ETTORE DR

MARINO BRIAN F ENG

MARSH THOMAS DR

MATHIEU ROBERT D DR

MATTEI JANET AKYUZ DR

MAUDER HORST PROF DR

MAYER PAVEL DR

MAZEH TSEVI DR

MCCLUSKEY GEORGE E JR DR

MELIA FULVIO DR

MEYER-HOFMEISTER E DR

MEZZETTI MARINO

MIKOLAJEWSKA JOANNA DR

MIKULASEK ZDENEK DR

MILANO LEOPOLDO DR

MILONE EUGENE F PROF

MINESHIGE SHIN DR

MIYAJI SHIGEKI DR

MOCHNACKI STEPHAN W DR

MORGAN THOMAS H DR

MOUCHET MARTINE DR

MUMFORD GEORGE S PROF

NAKAMURA YASUHISA

NARIAI KYOJI PROF

NATHER R EDWARD

NELSON BURT DR

NEWSOM GERALD H PROF

NHA IL-SEONG DR

NIARCHOS PANAYIOTIS PH D

NIEMELA VIRPI S DR

NORDSTROEM BIRGITTA DR

OEZKAN MUSTAFA TUERKER DR

OH KYU DONG DR

OKAZAKI AKIRA DR

OLAH KATALIN DR

OLIVER JOHN PARKER DR

OLSON EDWARD C PROF

OSAKI YOJI DR

PACZYNSKI BOHDAN PROF

PADALIA T D DR

PANDEY UMA SHANKAR DR

PARK HONG SUH DR

PARTHASARATHY M DR

PATKOS LASZLO DR

PAVLOVSKI KRESIMIR

PETERS GERALDINE JOAN DR

PICCIONI ADALBERTO

PIIROLA VILPPU E DR

PLAVEC MIREK J PROF

POPPER DANIEL M PROF

PRINGLE JAMES E DR

PUSTYL'NIK IZOLD B DR

QIAO GUOJUN

RAFERT JAMES BRUCE

RAHE JURGEN PROF

RAHUNEN TIMO

RAKOS KARL D PROF

REFSDAL S PROF DR

REGLERO-VELASCO VICTOR DR

RICHARDS MERCEDES T DR

RITTER HANS DR

ROBB RUSSEL M

ROBERTSON JOHN ALISTAIR

ROBINSON EDWARD LEWIS DR

ROVITHIS PETER DR

ROVITHIS-LIVANIOU HELEN

ROXBURGH IAN W PROF

RUCINSKI SLAWOMIR M DR

RUSSO GUIDO DR

SADIK AZIZ R DR

SAHADE JORGE PROF

SAIJO KEIICHI

SAMEC RONALD G DR

SANWAL N B DR

SANYAL ASHIT DR

SAVONIJE GERRIT JAN DR

SCALTRITI FRANCO DR

SCARFE COLIN D DR

SCHILLER STEPHEN

SCHMID HANS MARTIN DR

SCHMIDT HANS PROF

SCHMIDTKE PAUL C DR

SCHOBER HANS J DR

SCHOEFFEL EBERHARD F DR	SEGGEWISS WILHELM PROF	SEMENIUK IRENA DR
SHAFTER ALLEN W DR	SHAKURA NICHOLAJ I DR	SHAVIV GIORA PROF
SHEN LIANG-ZHAO	SHU FRANK H PROF	SHUL'BERG A M DR
SIMA ZDISLAV DR	SIMMONS JOHN FRANCIS L	SINVHAL SHAMBHU DAYAL DR
SION EDWARD MICHAEL	SISTERO ROBERTO F DR	SLOVAK MARK HAINES DR
SMAK JOSEPH I PROF	SMITH ROBERT CONNON DR	SOBIESKI STANLEY DR
SOEDERHJELM STAFFAN DR	SOLHEIM JAN ERIK	SPARKS WARREN M DR
SRIVASTAVA J B DR	SRIVASTAVA RAM KUMAR DR	STAGG CHRISTOPHER DR
STARRFIELD SUMNER PROF	STEEL DUNCAN I DR	STEIMAN-CAMERON THOMAS DR
STEINER JOAO E DR	STENCEL ROBERT EDWARD	STROHMEIER WOLFGANG PROF
SUGIMOTO DAIICHIRO PROF	SUNDMAN ANITA DR	SVECHNIKOVA MARIA A DR
SZABADOS LASZLO PH D	SZAFRANIEC ROZALIA DR	TAAM RONALD EVERETT DR
TAN HUISONG	TEAYS TERRY J DR	THOMPSON KEITH DR
TODORAN IOAN DR	TOUT CHRISTOPHER DR	TREMKO JOZEF DR
TRIMBLE VIRGINIA L DR	URECHE VASILE DR	VAN DEN HEUVEL EDWARD P J
VAN HAMME WALTER	VAN PARADIJS JOHANNES DR	VAN'T VEER FRANS DR
VAZ LUIZ PAULO RIBEIRO	VETESNIK MIROSLAV DR	WADE RICHARD ALAN DR
WALKER RICHARD L	WALKER WILLIAM S G	WALTER KURT PROF DR
WARD MARTIN JOHN	WARGAU WALTER F DR	WARNER BRIAN PROF
WEHLAU WILLIAM H PROF	WEIGERT ALFRED PROF	WEILER EDWARD J DR
WELLMANN PETER PROF DR	WESSELINK ADRIAAN J DR	WEST DONALD K DR
WHEELER J CRAIG PROF	WILLIAMON RICHARD M	WILLIAMS GLEN A DR
WILLIAMS ROBERT E DR	WILSON ROBERT E PROF	WOOD DAVID B DR
WOOD F BRADSHAW PROF	WOOD JANET H DR	WRIGHT KENNETH O DR
ZEILIK MICHAEL II DR	ZHANG ER-HO DR	ZHANG JINTONG
ZHOU DAOQI	ZHOU HONG-NAN	ZHU CI-SHENG
ZIOLKOWSKI JANUSZ DR	ZUIDERWIJK EDWARDUS J	ZWITTER TOMAZ

COMPOSITION OF COMMISSION 44 1991-1994

President   : TRUEMPER JOACHIM PROF

Vice-President(s) : FAZIO GIOVANNI G DR

Organizing Committee: BURKE BERNARD F DR
                       HUTCHINGS JOHN B DR
                       JENKINS EDWARD B DR
                       JORDAN STUART D DR
                       SAVAGE BLAIR D DR
                       WAMSTEKER WILLEM DR

Members:

ACTON LOREN W DR	AGRAWAL P C DR	AHMAD IMAD ALDEAN DR
ALEXANDER JOSEPH K	ANDERSEN BO NYBORG DR	ASCHENBACH BERND PH D
AYRES THOMAS R	BADIALI MASSIMO	BALIUNAS SALLIE L
BARSTOW MARTIN ADRIAN DR	BENEDICT GEORGE F DR	BENNETT CHARLES L DR
BENNETT KEVIN DR	BERGERON JACQUELINE A DR	BERNACCA P L PROF
BIANCHI LUCIANA	BLAMONT JACQUES E PROF	BLEEKER JOHAN A M DR IR
BLESS ROBERT C PROF	BOGGESS ALBERT DR	BOGGESS NANCY W DR
BOHLIN RALPH C DR	BOKSENBERG ALEC PROF	BONNET ROGER M DR
BOUGERET J L DR	BOWYER C STUART PROF	BOYARCHUK A A DR
BOYD ROBERT L F PROF SIR	BRANDT JOHN C DR	BRINKMAN BERT C DR
BROWN ALEXANDER	BRUECKNER GUENTER E DR	BRUHWEILER FRED C JR
BRUNER MARILYN E DR	BUMBA VACLAV DR	BUNNER ALAN N DR
BURGER MARIJKE DR	BURTON WILLIAM M	BUTLER C JOHN DR
BUTTERWORTH PAUL	CAMPBELL MURRAY F	CARPENTER KENNETH G DR
CARROLL P KEVIN PROF	CARVER JOHN H PROF	CATURA RICHARD C DR
CHAPMAN ROBERT D DR	CHARLES PHILIP ALLAN	CHOCHOL DRAHOMIR
CHUBB TALBOT A DR	CLARK GEORGE W PROF	CLARK THOMAS ALAN DR
CODE ARTHUR D	CORDOVA FRANCE A D	COURTES G PROF
COURVOISIER THIERRY J-L	CRANNELL CAROL JO DR	CULHANE LEONARD PROF
DAVIDSEN ARTHUR FALNES DR	DAVIS ROBERT J DR	DE JAGER CORNELIS PROF
DENNIS BRIAN ROY DR	DI COCCO GUIDO	DOLAN JOSEPH F DR
DOMINGO VICENTE DR	DUNKELMAN LAWRENCE	DUPREE ANDREA K DR
ELVIS MARTIN S DR	EL-RAEY MOHAMED E PROF	FABRICANT DANIEL G
FARAGGIANA ROSANNA PROF	FELDMAN PAUL DONALD DR	FERRARI TONIOLO MARCO
FICHTEL CARL E DR	FISHER PHILIP C	FISHMAN GERALD J
FITTON BRIAN DR	FOING BERNARD H DR	FREDGA KERSTIN PROF
FRIEDMAN HERBERT DR	FRISK URBAN DR	FU CHENG-QI
FURNISS IAN	GABRIEL ALAN H	GEZARI DANIEL YSA DR
GIACCONI RICCARDO PROF	GILRA DAYA P DR	GLASER HAROLD DR
GOLD THOMAS PROF	GONDHALEKAR PRABHAKAR DR	GREWING MICHAEL PROF
GREYBER HOWARD D DR	GRIFFITHS RICHARD E DR	GULL THEODORE R DR
GURSKY HERBERT DR	GUSEINOV O H PROF	HACK MARGHERITA PROF
HADDOCK FRED T DR	HALLAM KENNETH L DR	HAN ZHENG-ZHONG
HANG HENG-RONG	HARMS RICHARD JAMES DR	HARTZ THEODORE R DR
HARVEY CHRISTOPHER C DR	HARVEY PAUL MICHAEL DR	HAUSER MICHAEL G DR

HAWKINS ISABEL DR	HEARN ANTHONY G DR	HECKATHORN HARRY M
HEISE JOHN DR	HELMKEN HENRY F DR	HELOU GEORGE DR
HENIZE KARL G ASTRONAUT	HENOUX JEAN-CLAUDE DR	HENSBERGE HERMAN
HESKE ASTRID DR	HINTEREGGER HANS E DR	HOFFMAN JEFFREY ALAN DR
HOLBERG JAY B	HOLT STEPHEN S	HOUZIAUX L PROF
HOWARTH IAN DONALD	HOYNG PETER DR	HU WEN-RUI
HUBER MARTIN C E DR	IMHOFF CATHERINE L DR	INOUE HAJIME DR
IYENGAR K V K PROF	JAMAR CLAUDE A J DR	JORDAN CAROLE DR
KAFATOS MINAS DR	KARPINSKIJ VADIM N DR	KASTURIRANGAN K DR
KESSLER MARTIN F DR	KIMBLE RANDY A DR	KOCH-MIRAMOND LYDIE DR
KONDO YOJI DR	KRAEMER GERHARD DR	KRAUSHAAR WILLIAM L PROF
KURT V G DR	LAMERS HENNY J G L M DR	LECKRONE DAVID S DR
LEMAIRE PHILIPPE DR	LEWIN WALTER H G PROF	LI TIPEI
LI ZHONGYUAN	LINDBLAD BERTIL A DR	LINSKY JEFFREY L DR
LINSLEY JOHN	LONG KNOX S DR	LOVELL SIR BERNARD PROF
LUEST REIMAR PROF	MA YU-QIAN	MACCHETTO FERDINANDO DR
MALAISE DANIEL J DR	MALITSON HARRIET H MS	MALKAN MATTHEW ARNOLD DR
MANARA ALESSANDRO A DR	MANDOLESI NAZZARENO	MARAN STEPHEN P DR
MARAR T M K	MAROV MIKHAIL YA PROF	MATHER JOHN CROMWELL
MATSUOKA MASARU DR	MCCLUSKEY GEORGE E JR DR	MCCRACKEN KENNETH G DR
MCWHIRTER R W PETER DR	MEAD JAYLEE MONTAGUE DR	MELIA FULVIO DR
MELNICK GARY J	MEWE R DR	MICHALITSIANOS ANDREW
MIYAMOTO SIGENORI PROF	MODISETTE JERRY L PROF	MONET DAVID G
MONFILS ANDRE G PROF	MOOS HENRY WARREN DR	MORGAN THOMAS H DR
MORTON DONALD C DR	MUELLER EDITH A PROF	MURDOCK THOMAS LEE
NEFF SUSAN GALE DR	NESS NORMAN F DR	NEUPERT WERNER M DR
NICHOLS-BOHLIN JOY DR	NORDH H LENNART DR	NORMAN COLIN A PROF
NOVICK ROBERT	NOVOTNY VACLAV	NOYES ROBERT W PROF
ODA MINORU PROF	ODA NAOKI	OERTEL GOETZ K DR
OGAWARA YOSHIAKI	OKUDA TORU	OLTHOF HINDERICUS DR
OWEN TOBIAS C PROF	O'BRIEN PAUL THOMAS DR	O'MONGAIN EON
PACIESAS WILLIAM S DR	PACINI FRANCO PROF	PAPAGIANNIS MICHAEL D PRO
PARKINSON JOHN H DR	PARKINSON WILLIAM H DR	PEACOCK ANTHONY DR
PERRY PETER M DR	PETERS GERALDINE JOAN DR	PETERSON LAURENCE E PROF
PHILLIPS KENNETH J H	PINKAU K PROF	PIPHER JUDITH L
POLIDAN RONALD S	POUNDS KENNETH A PROF	PRICE STEPHAN DONALD
PROKOF'EV VLADIMIR K PROF	PROSZYNSKI MIECZYSLAW	RAHE JURGEN PROF
RAO RAMACHANDRA V PROF	REES MARTIN J PROF	REEVES EDMOND M DR
RENSE WILLIAM A DR	RIEGLER GUENTER R DR	RIGHINI-COHEN GIOVANNA DR
ROMAN NANCY G DR	ROSENDHAL JEFFREY D DR	RUBEN G PROF DR
RUDER HANNS	SAGDEEV ROALD Z DR	SAHADE JORGE PROF
SANDERS WILTON TURNER III	SATO KATSUHIKO PROF	SCHMIDT K H DR
SCHOENEICH W DR	SCHULTZ G V DR	SCHWARTZ DANIEL A DR
SCHWARTZ STEVEN JAY	SCHWEHM GERHARD DR	SELVELLI PIERLUIGI DR
SEQUEIROS JUAN DR	SHEFFIELD CHARLES DR	SHIVANANDAN KANDIAH DR
SHOLOMITSKY G B DR	SILVESTRO GIOVANNI	SIMON PAUL A DR
SIMON PAUL C DR	SMITH BRADFORD A PROF	SMITH HOWARD ALAN
SMITH LINDA J	SMITH PETER L DR	SNOW THEODORE P PROF
SOFIA SABATINO PROF	SONNEBORN GEORGE DR	SPADA GIANFRANCO DR
SPEER R J DR	SPITZER LYMAN JR DR	STACHNIK ROBERT V
STAUBERT RUDIGER PROF DR	STECHER THEODORE P	STEINBERG JEAN-LOUIS DR
STEINER JOAO E DR	STENCEL ROBERT EDWARD	STERN ROBERT ALLAN
STIER MARK T	STOCKMAN HERVEY S JR DR	STONE R G DR
SU WAN-ZHEN	SUN WEI-HSIN DR	SUNYAEV RASHID A DR
TAKAKURA TATSUO PROF EMER	TANAKA YASUO PROF	THOMAS ROGER J DR

TOVMASSIAN H M DR	TRAUB WESLEY ARTHUR	TSUNEMI HIROSHI DR
UNDERHILL ANNE B DR	UNDERWOOD JAMES H DR	UPSON WALTER L II DR
VALNICEK BORIS DR	VALTONEN MAURI J PROF	VAN BEEK FRANK PROF DR
VAN DE HULST H C PROF DR	VAN DER HUCHT KAREL A DR	VAN DUINEN R J DR
VAN SPEYBROECK LEON P DR	VIAL JEAN-CLAUDE	VIDAL-MADJAR ALFRED DR
VILHU OSMI DR	VIOTTI ROBERTO DR	VRTILEK SAEQA DIL DR
WALSH DENNIS DR	WANG SHUI	WARNER JOHN W DR
WEHRLE ANN ELIZABETH DR	WEILER EDWARD J DR	WEILER KURT W DR
WEINBERG J L DR	WESSELIUS PAUL R DR	WESTPHAL JAMES A PROF
WILLIS ALLAN J DR	WILLNER STEVEN PAUL DR	WILSON ROBERT PROF SIR
WINKLER CHRISTOPH DR	WRAY JAMES D DR	WU CHI CHAO DR
WUNNER GUENTER	YAMASHITA KOJUN DR	ZARNECKI JAN CHARLES DR
ZOMBECK MARTIN V DR	ZOU HUI-CHENG	

COMPOSITION OF COMMISSION 45 1991-1994

President   : MACCONNELL DARRELL J DR

Vice-President(s) : LEVATO ORLANDO HUGO DR

Organizing Committee: BARTAYA R A DR
                       CORBALLY CHRISTOPHER
                       EGRET DANIEL DR
                       GOLAY MARCEL PROF
                       GRENON MICHEL DR
                       LLOYD EVANS THOMAS DR
                       MAEHARA HIDEO DR
                       WALBORN NOLAN R DR
                       WING ROBERT F PROF
                       ZDANAVICIUS KAZIMERAS DR

Members:

ALBERS HENRY PROF	ARDEBERG ARNE L PROF	ARELLANO FERRO ARMANDO
BABU G S D	BAGLIN ANNIE DR	BAHNG JOHN D R PROF
BARBIER-BROSSAT M DR	BARRY DON C DR	BELL ROGER A DR
BIDELMAN WILLIAM P PROF	BLANCO VICTOR M DR	BUSCOMBE WILLIAM PROF
BUSER ROLAND DR	CELIS LEOPOLDO DR	CESTER BRUNO PROF
CHEREPASHCHUK A M PROF	CHRISTY JAMES WALTER DR	CLARIA JUAN DR
COLUZZI REGINA DR	COWLEY ANNE P DR	CRAMPTON DAVID DR
CRAWFORD DAVID L DR	DIVAN LUCIENNE DR	DUFLOT MARCELLE DR
ELVIUS TORD PROF EMERITUS	FARAGGIANA ROSANNA PROF	FEAST MICHAEL W PROF
FEHRENBACH CHARLES PROF	FITZPATRICK EDWARD L DR	FUKUDA ICHIRO
GARMANY CATHERINE D DR	GARRISON ROBERT F PROF	GERBALDI MICHELE DR
GEYER EDWARD H PROF DR	GLAGOLEVSKIJ JU V DR	GUETTER HARRY HENDRIK
GURZADIAN G A PROF DR	HACK MARGHERITA PROF	HALLAM KENNETH L DR
HAUCK BERNARD PROF	HAYES DONALD S DR	HENIZE KARL G ASTRONAUT
HOUK NANCY DR	HUANG LIN	HUMPHREYS ROBERTA M PROF
JASCHEK CARLOS O R PROF	JASCHEK MERCEDES DR	KEENAN PHILIP C PROF EMER
KHARADZE E K PROF	KRON GERALD E DR	KURTZ DONALD WAYNE DR
LABHARDT LUKAS	LASALA GERALD J DR	LEE SANG GAK
LODEN KERSTIN R DR	LOW FRANK J DR	LUTZ JULIE H DR
LYNGA GOSTA DR	MALARODA STELLA M DR	MCCARTHY MARTIN F DR
MCCLURE ROBERT D PROF	MCNAMARA DELBERT H DR	MEAD JAYLEE MONTAGUE DR
MENDOZA V EUGENIO E DR	MORGAN WILLIAM W PROF	MOROSSI CARLO
NANDY KASHINATH DR	NICOLET BERNARD	NORTH PIERRE
NOTNI P DR	OJA TARMO PROF	OLSEN ERIK H
OSBORN WAYNE DR	PARSONS SIDNEY B DR	PASINETTI LAURA E PROF
PERRY CHARLES L DR	PHILIP A G DAVIS	PIZZICHINI GRAZIELLA
PRESTON GEORGE W DR	RAUTELA B S DR	ROMAN NANCY G DR
ROUNTREE JANET DR	RUDKJOBING MOGENS PROF	SANWAL N B DR
SCHILD RUDOLPH E DR	SCHMIDT-KALER TH PROF	SEITTER WALTRAUT C PROF
SHARPLESS STEWART PROF	SHORE STEVEN N	SINNERSTAD ULF E PROF
SION EDWARD MICHAEL	SLETTEBAK ARNE PROF	STEINLIN ULI PROF

STEPHENSON C BRUCE PROF   STOCK JURGEN D   STRAIZYS V PROF DR
STROBEL ANDRZEJ DR   UPGREN ARTHUR R DR   WALKER GORDON A H PROF
WARREN WAYNE H JR DR   WESSELIUS PAUL R DR   WESTERLUND BENGT E PROF
WILLIAMS JOHN A DR   WU HSIN-HENG DR   WYCKOFF SUSAN DR
YAMASHITA YASUMASA PROF   YOSS KENNETH M DR

COMPOSITION OF COMMISSION 46 1991-1994

President   : GOUGUENHEIM LUCIENNE

Vice-President(s) : PERCY JOHN R PROF

Organizing Committee: FIERRO JULIETA
                       GERBALDI MICHELE DR
                       HOUZIAUX L PROF
                       ISOBE SYUZO DR
                       IWANISZEWSKA CECYLIA DR
                       PASACHOFF JAY M PROF
                       ROBBINS R ROBERT PROF
                       SANDQVIST AAGE DR
                       WENTZEL DONAT G DR

Members:

ACKER AGNES PROF DR	AIAD A PROF	ANDRILLAT HENRI L PROF
ANSARI S M RAZAULLAH PROF	BACALOV MIHAIL	BENSON PRISCILLA J DR
BOCHONKO D RICHARD DR	BOTEZ ELVIRA DR	BOTTINELLI LUCETTE DR
BRAES L L E DR	BRIEVA EDUARDO PROF	BROSCH NOAH DR
BRUCK MARY T DR	BUDDING EDWIN DR	BUSCOMBE WILLIAM PROF
CALVET NURIA DR	CATALA POCH M A	CHAMBERLAIN JOSEPH M DR
CODINA LANDABERRY SAYD J	COUPER HEATHER MISS	CUI ZHEN-HUA
DARIUS JON DR	DUPUY DAVID L DR	DUVAL MARIE-FRANCE
EMERSON DAVID	FAIRALL ANTHONY P PROF	FENG KE-JIA
FERNANDEZ JULIO A DR	FERNANDEZ-FIGUEROA M J DR	FERRAZ-MELLO S PROF DR
FIENBERG RICHARD T DR	FLECK ROBERT CHARLES DR	GALLINO ROBERTO
GINGERICH OWEN PROF	GURM HARDEV S PROF	HAUPT HERMANN F PROF
HAYWOOD J	HEMENWAY MARY KAY M DR	HEUDIER JEAN-LOUIS DR
HIDAJAT BAMBANG PROF DR	HOFF DARREL BARTON	ILYAS MOHAMMAD DR
IMPEY CHRISTOPHER D DR	JARRETT ALAN H PROF	JONES BARRIE W DR
KELLER HUWE ULRICH DR	KENNEDY JOHN E PROF	KITCHIN CHRISTOPHER R DR
KONONOVICH EDWARD V DR	KOURGANOFF VLADIMIR PROF	KREINER JERZY MAREK DR
KRUPP EDWIN C DR	LAGO MARIA TERESA V T PR	LAI SEBASTIANA
LI QI-BIN	LITTLE-MARENIN IRENE R DR	LOMB NICHOLAS RALPH DR
MA XING-YUAN	MACIEL WALTER J DR	MADDISON RONALD CH DR
MARSH JULIAN C D	MARTINET LOUIS PROF	MAVRIDIS L N PROF
MAZA JOSE	MCCARTHY MARTIN F DR	MCNALLY DEREK DR
MEIDAV MEIR DR	MOMCHEV GOSPODIN	MOREELS GUY DR
MUZZIO JUAN C PROF	NARLIKAR JAYANT V PROF	NHA IL-SEONG DR
NICOLOV NIKOLAI S DR	NOELS ARLETTE DR	OJA HEIKKI DR
OKOYE SAMUEL E PROF	OLSEN FOGH H J	ONUORA LESLEY IRENE DR
OSBORN WAYNE DR	OSORIO JOSE J S P PROF	OTHMAN MAZLAN
OWAKI NAOAKI DR	PANDEY UMA SHANKAR DR	PARISOT JEAN-PAUL
POKORNY ZDENEK DR	PRABHAKARAN NAYAR S R DR	PROVERBIO EDOARDO PROF
RAMADURAI SOURIRAJA DR	ROBINSON LEIF J	RODGERS ALEX W DR
ROSLUND CURT DR	ROY ARCHIE E PROF	SAFKO JOHN L
SANAHUJA BLAS	SAXENA P P DR	SBIRKOVA-NATCHEVA T

SCHLEICHER DAVID G DR   SCHLOSSER WOLFHARD PROF   SCHMIDT THOMAS DR
SCHMITTER EDWARD F DR   SCHROEDER DANIEL J PROF   SEEDS MICHAEL AUGUST DR
SHEN CHUN-SHAN   SHIPMAN HENRY L DR   SIROKY JAROMIR DR
SOLHEIM JAN ERIK   STEFL VLADIMIR   STENHOLM BJOERN DR
STOEV ALEXEI   SUKARTADIREDJA DARSA   SVESTKA JIRI DR
SZECSENYI-NAGY GABOR DR   SZOSTAK ROLAND DR   TABORDA JOSE ROSA DR
TORRES-PEIMBERT SILVIA DR TROCHE-BOGGINO A E DR   VAUCLAIR SYLVIE D DR
VLADIMIROV SIMEON   VUJNOVIC VLADIS DR   WANG SHUNDE DR
WEST RICHARD M DR   WILLIAMON RICHARD M   WOO JONG OK
ZEALEY WILLIAM J DR   ZEILIK MICHAEL II DR   ZIMMERMANN HELMUT DR

COMMISSION 46 NATIONAL REPRESENTATIVES

ARGENTINA	FEINSTEIN ALEJANDRO DR
AUSTRALIA	RODGERS ALEX W DR
AUSTRIA	HANSLMEIER ARNOLD
BELGIUM	NOELS ARLETTE DR
BRAZIL	MACIEL WALTER J DR
BULGARIA	NICOLOV NIKOLAI S DR
CANADA	BOCHONKO D RICHARD DR
CHILE	MAZA JOSE
CHINA PR	FENG KE-JIA
CHINA R	SHEN CHUN-SHAN
COLOMBIA	BRIEVA EDUARDO PROF
CZECHOSLOVAKIA	SIROKY JAROMIR DR
DENMARK	OLSEN FOGH H J
EGYPT	AIAD A PROF
FINLAND	OJA HEIKKI DR
FRANCE	GOUGUENHEIM LUCIENNE
GERMANY	SCHLOSSER WOLFHARD PROF
GREECE	MAVRIDIS L N PROF
HUNGARY	SZECSENYI-NAGY GABOR DR
INDIA	RAMADURAI SOURIRAJA DR
INDONESIA	HIDAJAT BAMBANG PROF DR
IRELAND	HAYWOOD J
ITALY	PROVERBIO EDOARDO PROF
JAPAN	ISOBE SYUZO DR
KOREA R	WOO JONG OK
MALAYSIA	OTHMAN MAZLAN
MEXICO	TORRES-PEIMBERT SILVIA DR
NETHERLANDS	BRAES L L E DR
NEW ZEALAND	BUDDING EDWIN DR
NIGERIA	OKOYE SAMUEL E PROF
NORWAY	SOLHEIM JAN ERIK
PARAGUAY	TROCHE-BOGGINO A E DR
POLAND	IWANISZEWSKA CECYLIA DR
PORTUGAL	OSORIO JOSE J S P PROF
RUMANIA	BOTEZ ELVIRA DR
SOUTH AFRICA	FAIRALL ANTHONY P PROF
SPAIN	CATALA POCH M A
SWEDEN	SANDQVIST AAGE DR
SWITZERLAND	MARTINET LOUIS PROF
UK	CLARKE DAVID DR
URUGUAY	FERNANDEZ JULIO A DR
USA	PASACHOFF JAY M PROF
VATICAN CITY STATE	MCCARTHY MARTIN F DR
VENEZUELA	CALVET NURIA DR
YUGOSLAVIA	MILOGRADOV-TURIN JELENA
(EX USSR)	KONONOVICH EDWARD V DR

COMPOSITION OF COMMISSION 47 1991-1994

President   : PARTRIDGE ROBERT B PROF

Vice-President(s) : NARLIKAR JAYANT V PROF

Organizing Committee: BOND JOHN RICHARD
                       DRESSLER ALAN
                       GELLER MARGARET JOAN
                       JAUNCEY DAVID L DR
                       REES MARTIN J PROF
                       REEVES HUBERT PROF
                       SATO KATSUHIKO PROF
                       SHANDARIN SERGEI F DR
                       SHAVER PETER A DR
                       SZALAY ALEX DR

Members:

AIZU KO PROF	ALFVEN HANNES PROF	ALLAN PETER M
ANDREANI PAOLA MICHELA DR	ANDRILLAT HENRI L PROF	AUDOUZE JEAN PROF
AULUCK FAQIR CHAND PROF	AZUMA TAKAHIRO DR	BALDWIN JOHN E DR
BANERJI SRIRANJAN DR	BANHATTI DILIP GOPAL DR	BARBERIS BRUNO
BARDEEN JAMES M PROF	BARNOTHY JENO DR PROF	BARROW JOHN DAVID
BARTHEL PETER DR	BASU DIPAK DR	BECHTOLD JILL DR
BECKMAN JOHN E PROF	BEL NICOLE J DR	BELINSKY VLADIMIR DR
BENNETT CHARLES L DR	BERGERON JACQUELINE A DR	BERMAN MARCELO S DR
BERTOLA FRANCESCO PROF	BERTSCHINGER EDMUND DR	BETANCORT-RIJO JUAN DR
BHAVSAR SUKETU P	BICKNELL GEOFFREY V DR	BIGNAMI GIOVANNI F
BIRKINSHAW MARK	BLUDMAN SIDNEY A PROF	BOKSENBERG ALEC PROF
BONDI HERMANN PROF SIR	BONNOR W B PROF	BORGEEST ULF DR
BOYLE BRIAN DR	BRECHER KENNETH PROF	BURBIDGE GEOFFREY R PROF
CALVANI MASSIMO DR	CARR BERNARD JOHN	CASTAGNINO MARIO DR
CAVALIERE ALFONSO G PROF	CHANG KYONGAE DR	CHEN JIAN-SHENG
CHENG FU-HUA	CHENG FU-ZHEN	CHINCARINI GUIDO L DR
CHITRE DATTAKUMAR M DR	CHU YAOQUAN	CLARIA JUAN DR
COCKE WILLIAM JOHN PROF	COHEN JEFFREY M DR	COHEN ROSS D DR
COLES PETER DR	CONDON JAMES J DR	CRANE PATRICK C
CRANE PHILIPPE	CRISTIANI STEFANO DR	DADHICH NARESH DR
DANESE LUIGI DR	DAS P K DR	DATTA BHASKAR DR
DAVIDSON WILLIAM PROF	DAVIES PAUL CHARLES W	DAVIES ROGER L DR
DAVIS MARC DR	DAVIS MICHAEL M DR	DE RUITER HANS RUDOLF
DE VAUCOULEURS GERARD PR	DE ZOTTI GIANFRANCO DR	DEKEL AVISHAI
DEMARET JACQUES DR	DICKE ROBERT H PROF	DIONYSIOU DEMETRIOS PROF
DOROSHKEVICH ANDREI G DR	DULTZIN-HACYAN D DR	DYER CHARLES CHESTER DR
EFSTATHIOU GEORGE	EHLERS JURGEN PROF	EINASTO JAAN DR
ELLIS GEORGE F R PROF	ELLIS RICHARD S	ELVIS MARTIN S DR
ENGINOL TURAN B DR	FABER SANDRA M PROF	FALK SYDNEY W JR DR
FALL S MICHAEL DR	FANG LI-ZHI	FELTEN JAMES E DR
FIELD GEORGE B PROF	FILIPPENKO ALEXEI V DR	FLORIDES PETROS S PROF

FOCARDI PAOLA DR	FONG RICHARD	FORD HOLLAND C RES PROF
FORMAN WILLIAM RICHARD DR	FOUQUE PASCAL DR	FRANCESCHINI ALBERTO
FRENK CARLOS S	FUJIMOTO MITSUAKI DR	FUKUGITA MASATAKA DR
FUKUI TAKAO DR	GALLETTO DIONIGI	GARRISON ROBERT F PROF
GIALLONGO EMANUELE DR	GIURICIN GIULIANO	GODART ODON PROF
GOLD THOMAS PROF	GOLDSMITH DONALD W DR	GONG SHU-MO
GORET PHILIPPE DR	GOSSET ERIC DR	GOUDA NAOTERU DR
GREGORY STEPHEN ALBERT DR	GREYBER HOWARD D DR	GRISHCHUK L P DR
GUDMUNDSSON EINAR H	GUNN JAMES E PROF	HACYAN SHAHEN DR
HAGEN HANS-JUERGEN DR	HARA KEN NOSUKE DR	HARDY EDUARDO
HARMS RICHARD JAMES DR	HARRISON EDWARD R PROF	HAWKING STEPHEN W PROF
HAYASHI CHUSHIRO PROF	HE XIANG-TAO	HEAVENS ALAN DR
HEIDMANN JEAN DR	HELLER MICHAEL PROF	HEWETT PAUL
HEWITT ADELAIDE	HOYLE FRED SIR	HU ESTHER M DR
HUCHRA JOHN PETER DR	ICKE VINCENT DR	IKEUCHI SATORU DR
IMPEY CHRISTOPHER D DR	ISHIHARA HIDEKI DR	IYER B R DR
JAROSZYNSKI MICHAL	JIANG SHUDING	JONES BERNARD J T DR
JOSHI MOHAN N PROF	JUNKKARINEN VESA T DR	JUSZKIEWICZ ROMAN
KANDRUP HENRY EMIL DR	KAPOOR RAMESH CHANDER	KARACHENTSEV I D DR
KASPER U DR	KATO SHOJI PROF	KAWABATA KINAKI PROF
KAYSER RAINER DR	KELLERMANN KENNETH I DR	KEMBHAVI AJIT K
KIM JIK SU	KODAMA HIDEO	KOKKOTAS KONSTANTINOS DR
KOLB EDWARD W DR	KOO DAVID C-Y DR	KORMENDY JOHN DR
KOVETZ ATTAY PROF	KOZLOVSKY B Z DR	KRASINSKI ANDRZEJ PROF
KRISS GERARD A DR	KUNTH DANIEL	KUSTAANHEIMO PAUL E PROF
LACEY CEDRIC DR	LACHIEZE-REY MARC	LAHAV OFER DR
LAKE KAYLL WILLIAM DR	LASOTA JEAN-PIERRE DR	LAUSBERG ANDRE DR
LAYZER DAVID PROF	LEQUEUX JAMES DR	LI ZHI-FANG
LIEBSCHER DIERCK-E DR	LILJE PER VIDAR BARTH DR	LILLY SIMON J DR
LIU LIAO	LIU YONG-ZHEN	LONGAIR M S PROF
LONSDALE CAROL J DR	LU TAN	LUCCHIN FRANCESCO
LUMINET JEAN-PIERRE	LYNDEN-BELL DONALD PROF	MACCALLUM MALCOLM A H
MADDOX STEPHEN DR	MAEDA KEI-ICHI DR	MANDOLESI NAZZARENO
MANDZHOS ANDREJ V DR	MARANO BRUNO	MARDIROSSIAN FABIO
MAREK JOHN	MARTINEZ-GONZALEZ E DR	MATERNE JUERGEN DR
MATHER JOHN CROMWELL	MATSUMOTO TOSHIO DR	MATZNER RICHARD A PROF
MAVRIDES STAMATIA DR	MCCREA J DERMOTT	MCCREA WILLIAM SIR
MELOTT ADRIAN L PROF	MERAT PARVIZ	MERIGHI ROBERTO DR
MESZAROS ATTILA DR	MESZAROS PETER DR	MEYER DAVID M DR
MEYLAN GEORGES DR	MEZZETTI MARINO	MISNER CHARLES W PROF
MORRISON PHILIP PROF	MULLER RICHARD A	NAMBU YASUSADA DR
NARASIMHA DELAMPADY DR	NEEMAN YUVAL PROF	NISHIDA MINORU PROF
NOERDLINGER PETER D PROF	NOONAN THOMAS W PROF	NORMAN COLIN A PROF
NOTTALE LAURENT	NOVELLO MARIO DR	NOVIKOV I D DR
NOVOTNY JAN DR	OEMLER AUGUSTUS JR DR	OKOYE SAMUEL E PROF
OLOWIN RONALD PAUL DR	OMNES ROLAND PROF	ONUORA LESLEY IRENE DR
OORT JAN H PROF	OZERNOY LEONID M PROF	OZSVATH I PROF
O'CONNELL ROBERT WEST DR	PACHNER JAROSLAV PROF	PADMANABHAN T DR
PADRIELLI LUCIA	PAGE DON NELSON	PAN RONG-SHI
PEACOCK JOHN ANDREW	PECKER JEAN-CLAUDE PROF	PEEBLES P JAMES E
PENZIAS ARNO A DR	PERRYMAN MICHAEL A C DR	PERSIDES SOTIRIOS C
PETERSON BRUCE A DR	PETROSIAN VAHE PROF	PRESS WILLIAM H DR
PUGET JEAN-LOUP DR	QU QIN-YUE	RAMELLA MASSIMO
RAWLINGS STEVEN DR	RAYCHAUDHURI AMALKUMAR DR	REFSDAL S PROF DR
RINDLER WOLFGANG PROF	RIVOLO ARTHUR REX	ROBERTS DAVID HALL DR

ROBINSON I PROF   ROEDER ROBERT C PROF   ROSQUIST KJELL
ROWAN-ROBINSON MICHAEL DR ROXBURGH IAN W PROF   RUBIN VERA C DR
RUDDY VINCENT P DR   RUDNICK LAWRENCE DR   RUFFINI REMO
SAAR ENN DR   SADAT RACHIDA DR   SALVADOR-SOLE EDUARDO
SALZER JOHN JOSEPH DR   SANZ JOSE L DR   SAPAR ARVED DR
SARGENT WALLACE L W DR   SASAKI MISAO   SATO HUMITAKA PROF
SAVAGE ANN DR   SAZHIN MICHAIL DR   SCHATZMAN EVRY PROF
SCHECHTER PAUL L DR   SCHEUER PETER A G DR   SCHMIDT MAARTEN PROF
SCHNEIDER JEAN   SCHNEIDER PETER DR   SCHRAMM DAVID N PROF
SCHUCH NELSON JORGE   SCHUECKING E L DR   SCHULTZ G V DR
SCIAMA DENNIS W DR   SCOTT ELIZABETH L PROF   SEGAL IRVING E DR
SEIDEN PHILIP E   SEIELSTAD GEORGE A   SERSIC J L DR
SETTI GIANCARLO PROF   SHAVIV GIORA PROF   SHAYA EDWARD J DR
SHIVANANDAN KANDIAH DR   SIGNORE MONIQUE DR   SILK JOSEPH I PROF
SIMON RENE L E PROF   SISTERO ROBERTO F DR   SMITH HARDING E JR DR
SMITH RODNEY M DR   SMOOT III GEORGE F.   SOKOLOWSKI LECH
SONG DOO JONG DR   SPYROU NICOLAOS PROF   STECKER FLOYD W DR
STEIGMAN GARY PROF   STEWART JOHN MALCOLM DR   STOEGER WILLIAM R DR
STRUBLE MITCHELL F   STRUKOV IGOR A DR   SUBRAHMANYA C R
SUGIYAMA NAOSHI DR   SUNYAEV RASHID A DR   SURDEJ JEAN M G
SUTHERLAND WILLIAM DR   SUTO YASUSHI DR   TAKAHARA FUMIO DR
TAMMANN G ANDREAS PROF DR TANABE KENJI DR   TARTER JILL C DR
TAYLER ROGER J PROF   THOMPSON LAIRD A DR   THUAN TRINH XUAN DR
TIFFT WILLIAM G PROF   TIPLER FRANK JENNINGS DR TOMIMATSU AKIRA DR
TOMITA KENJI PROF   TONRY JOHN DR   TREDER H J PROF DR
TREMAINE SCOTT DUNCAN   TREVESE DARIO   TRIMBLE VIRGINIA L DR
TSAMPARLIS MICHAEL DR   TULLY RICHARD BRENT DR   TURNER EDWIN L DR
TURNER MICHAEL S   TYSON JOHN A DR   TYTLER DAVID DR
UMEMURA MASAYUKI DR   USON JUAN M DR   VAGNETTI FAUSTO
VAIDYA P C PROF   VAN DER LAAN H PROF DR   VANYSEK VLADIMIR PROF
VETTOLANI GIAMPAOLO   VISHNIAC ETHAN T   VISHVESHWARA C V PROF
VOGLIS NIKOS DR   VON BORZESZKOWSKI H H DR WAGONER ROBERT V PROF
WAINWRIGHT JOHN DR   WANG RENCHUAN   WEBSTER ADRIAN S DR
WEBSTER RACHEL   WEINBERG STEVEN DR   WESSON PAUL S DR
WHEELER JOHN A DR   WHITE SIMON DAVID MANION WHITROW GERALD JAMES PROF
WILKINSON DAVID T   WILL CLIFFORD M DR   WILSON ALBERT G DR
WINDHORST ROGIER A DR   WOLTJER LODEWIJK PROF   WRIGHT EDWARD L DR
XANTHOPOULOS B C DR   XIANG SHOUPING   YANG LAN-TIAN
YOSHII YUZURU DR   YOSHIMURA MOTOHIKO DR   ZAMORANI GIOVANNI
ZEL'MANOV A L DR   ZHANG JIA-LU   ZHANG ZHEN-JIU
ZHOU YOU-YUAN   ZHU SHI-CHANG   ZHU XINGFENG
ZIEBA STANISLAW DR   ZOU ZHEN-LONG   ZUIDERWIJK EDWARDUS J

COMPOSITION OF COMMISSION 48 1991-1994

President   : OSTRIKER JEREMIAH P PROF

Vice-President(s) : SRINIVASAN G

Organizing Committee: CESARSKY CATHERINE J DR
                      CLARK GEORGE W PROF
                      GIACCONI RICCARDO PROF
                      PACINI FRANCO PROF
                      QU QIN-YUE
                      SALPETER EDWIN E PROF
                      SCHEUER PETER A G DR
                      SCHRAMM DAVID N PROF
                      SUNYAEV RASHID A DR
                      TRIMBLE VIRGINIA L DR
                      WOLFENDALE ARNOLD W PROF
                      WOLTJER LODEWIJK PROF

Members:

ABRAMOWICZ MAREK DR	ACHARYA BANNANJE S DR	ADAMS DAVID J DR
AGRAWAL P C DR	AGUIAR ODYLIO DENYS DR	AHLUWALIA HARJIT SINGH DR
AIZU KO PROF	ALFVEN HANNES PROF	APPARAO K M V DR
ARAFUNE JIRO DR	ARNAUD MONIQUE	ARNOULD MARCEL L DR
ARONS JONATHAN	ASCHENBACH BERND PH D	ASSEO ESTELLE DR
AUDOUZE JEAN PROF	AXFORD W IAN PROF	BAAN WILLEM A
BARNOTHY JENO DR PROF	BASU DIPAK DR	BAYM GORDON ALAN DR
BECKER ROBERT HOWARD	BEGELMAN MITCHELL CRAIG	BENFORD GREGORY DR
BENVENUTO OMAR DR	BERGERON JACQUELINE A DR	BICKNELL GEOFFREY V DR
BIERMANN PETER L DR	BIGNAMI GIOVANNI F	BISWAS SUKUMAR DR
BLANDFORD ROGER DAVID DR	BLEEKER JOHAN A M DR IR	BLONDIN JOHN M DR
BLUDMAN SIDNEY A PROF	BONAZZOLA SILVANO DR	BONNET-BIDAUD J M DR
BONOMETTO SILVIO A DR	BOYD ROBERT L F PROF SIR	BRAGA JOAO DR
BRECHER KENNETH PROF	BUNNER ALAN N DR	BURBIDGE GEOFFREY R PROF
BURROWS ADAM SETH	CAMENZIND MAX DR	CAMERON ALASTAIR G W PROF
CARDINI DANIELA DR	CASH WEBSTER C JR	CASSE MICHEL DR
CATURA RICHARD C DR	CAUGHLAN GEORGEANNE R	CAVALIERE ALFONSO G PROF
CHAKRABARTI SANDIP K DR	CHAKRABORTY DEO K DR	CHANDRASEKHAR S PROF
CHECHETKIN VALERIJ M DR	CHIAN ABRAHAM CHIAN-LONG	CHITRE SHASHIKUMAR M DR
CHUBB TALBOT A DR	CHUPP EDWARD L DR	COHEN JEFFREY M DR
COLLIN-SOUFFRIN SUZY DR	CONDON JAMES J DR	COWIE LENNOX LAUCHLAN DR
COWSIK RAMANATH	CRUISE ADRIAN MICHAEL DR	CULHANE LEONARD PROF
CURIR ANNA	DA COSTA ANTONIO A DR	DA COSTA JOSE MARQUES DR
DADHICH NARESH DR	DAMLE S V DR	DAUTCOURT G DR
DAVIDSEN ARTHUR FALNES DR	DAVIDSON WILLIAM PROF	DAVIS LEVERETT JR PROF
DAVIS MICHAEL M DR	DE FELICE FERNANDO DR	DE GRAAF T DR
DE YOUNG DAVID S DR	DEBRUNNER HERMANN DR	DENNIS BRIAN ROY DR
DEWITT BRYCE S DR	DICKE ROBERT H PROF	DISNEY MICHAEL J PROF
DOLAN JOSEPH F DR	DOTANI TADAYASU DR	DRAKE FRANK D PROF

DRURY LUKE O'CONNOR DR   DUORAH HIRA LAL DR   DUROUCHOUX PHILIPPE
DUTHIE JOSEPH G PROF   EDELSON RICK DR   EDWARDS PAUL J DR
EICHLER DAVID DR   EILEK JEAN   ELVIS MARTIN S DR
EMANUELE ALESSANDRO DR   EVANS W DOYLE   FABIAN ANDREW C DR
FANG LI-ZHI   FAZIO GIOVANNI G DR   FELTEN JAMES E DR
FENTON K B DR   FERRARI ATTILIO DR   FICHTEL CARL E DR
FIELD GEORGE B PROF   FISHER PHILIP C   FORMAN WILLIAM RICHARD DR
FOWLER WILLIAM A PROF   FRANCESCHINI ALBERTO   FRANDSEN SOEREN PROF
FRANK JUHAN   FRANSSON CLAES   FRIEDMAN HERBERT DR
GAISSER THOMAS K   GALEOTTI PIERO PROF   GARMIRE GORDON P PROF
GINZBURG VITALY L PROF   GOLD THOMAS PROF   GOLDSMITH DONALD W DR
GONZALES-A WALTER D DR   GREENHILL JOHN DR   GREISEN KENNETH I PROF
GRENIER ISABELLE DR   GREWING MICHAEL PROF   GREYBER HOWARD D DR
GRIFFITHS RICHARD E DR   GRINDLAY JONATHAN E DR   GUNN JAMES E PROF
GURSKY HERBERT DR   GUSEINOV O H PROF   HALL ANDREW NORMAN
HANG HENG-RONG   HARWIT MARTIN PROF   HAUBOLD HANS JOACHIM
HAWKING STEPHEN W PROF   HAYMES ROBERT C PROF   HEISE JOHN DR
HELFAND DAVID JOHN   HENRIKSEN RICHARD N DR   HENRY RICHARD C PROF
HOFFMAN JEFFREY ALAN DR   HOLLOWAY NIGEL J DR   HOYLE FRED SIR
HUANG KE-LIANG   HUNT LESLIE DR   ICHIMARU SETSUO DR
INOUE HAJIME DR   IPSER JAMES R PROF   ISRAEL WERNER PROF
ITO KENSAI A PROF   ITOH MASAYUKI DR   JACKSON JOHN CHARLES DR
JAFFE WALTER JOSEPH DR   JOKIPII J R PROF   JONES FRANK CULVER DR
JONES THOMAS WALTER DR   JOSS PAUL CHRISTOPHER DR   JULIUSSON EINAR DR
KAFKA PETER   KAHN FRANZ D PROF   KAPOOR RAMESH CHANDER
KATZ JONATHAN I   KELLERMANN KENNETH I DR   KELLOGG EDWIN M DR
KEMBHAVI AJIT K   KII TSUNEO DR   KILLEEN NEIL DR
KIRK JOHN DR   KLINKHAMER FRANS DR   KOCHAROV GRANT E PROF
KOCH-MIRAMOND LYDIE DR   KOLB EDWARD W DR   KONDO MASAAKI DR
KOSHIBA MASA-TOSHI DR   KOUPELIS THEODOROS DR   KOYAMA KATSUJI
KOZLOWSKI MACIEJ DR   KREISEL E PROF   KRISTIANSSON KRISTER PROF
KULSRUD RUSSELL M DR   KUNDT WOLFGANG PROF DR   KURT V G DR
KUSUNOSE MASAAKI DR   LAMB DONALD QUINCY JR DR   LAMB FREDERICK K PROF
LAMB SUSAN ANN DR   LAMPTON MICHAEL   LASHER GORDON JEWETT DR
LATTIMER JAMES M DR   LEA SUSAN MAUREEN DR   LI QI-BIN
LI TIPEI   LI YUAN-JIE   LI ZONG-WEI
LIANG EDISON P DR   LINSLEY JOHN   LIU RU-LIANG
LONGAIR M S PROF   LOVELACE RICHARD V E DR   LU JU FU DR
LU TAN   LUEST REIMAR PROF   LUMINET JEAN-PIERRE
LYNDEN-BELL DONALD PROF   MA YU-QIAN   MACCACARO TOMMASO DR
MACCAGNI DARIO   MACCHETTO FERDINANDO DR   MAGGIO ANTONIO DR
MARTIN INACIO MALMONGE DR   MASON GLENN M   MASON KEITH OWEN
MATSUMOTO RYOJI DR   MATSUOKA MASARU DR   MATZ STEVEN MICHEAL DR
MAZUREK THADDEUS JOHN DR   MCBREEN BRIAN PHILIP DR   MCCRAY RICHARD DR
MEDINA JOSE DR   MEIER DAVID L   MEIKSIN AVERY ABRAHAM DR
MELROSE DONALD B PROF   MESTEL LEON PROF   MESZAROS PETER DR
MEYER FRIEDRICH DR   MEYER JEAN-PAUL DR   MICELA GIUSEPPINA DR
MICHEL F CURTIS PROF   MILLER JOHN C DR   MIYAJI SHIGEKI DR
MIYAMOTO SIGENORI PROF   MOON SHIN HAENG DR   MORRISON PHILIP PROF
MURAKAMI TOSHIO   NAIDENOV VICTOR O   NEEMAN YUVAL PROF
NITYANANDA R DR   NOMOTO KEN'ICHI DR   NORMAN COLIN A PROF
NOVICK ROBERT   NULSEN PAUL DR   ODA MINORU PROF
OEZEL MEHMET EMIN DR   OGAWARA YOSHIAKI   OKEKE PIUS N DR
OKOYE SAMUEL E PROF   OZERNOY LEONID M PROF   O'CONNELL ROBERT F PROF
O'SULLIVAN DENIS F   PACHOLCZYK ANDRZEJ G PROF PAGE CLIVE G DR

PALUMBO GIORGIO G C DR   PANDEY UMA SHANKAR DR   PARK MYEONG-GU DR
PARKER EUGENE N   PARKINSON JOHN H DR   PAULINY TOTH IVAN K K DR
PENG QIU-HE   PEROLA GIUSEPPE C DR   PETERSON BRUCE A DR
PETERSON LAURENCE E PROF   PETHICK CHRISTOPHER J DR   PETROSIAN VAHE PROF
PIDDINGTON JACK H RES FEL   PINKAU K PROF   PINTO PHILIP ALFRED DR
PIRO LUIGI DR   PORTER NEIL A PROF   POUNDS KENNETH A PROF
PRASANNA A R DR   PREUSS EUGEN DR   PROTHEROE RAYMOND J DR
QUINTANA HERNAN DR   RADHAKRISHNAN V PROF   RAMADURAI SOURIRAJA DR
RAUBENHEIMER BAREND C PR   RAZDAN HIRALAL   REES MARTIN J PROF
REEVES HUBERT PROF   REICHERT GAIL ANNE DR   RENGARAJAN T N DR
ROSNER ROBERT   ROSSI BRUNO B PROF   RUFFINI REMO
SALVATI MARCO   SANDERS WILTON TURNER III SANTOS NILTON OSCAR DR
SARTORI LEO PROF   SASLAW WILLIAM C PROF   SAVEDOFF MALCOLM P PROF
SCARGLE JEFFREY D DR   SCHAEFER GERHARD DR   SCHATTEN KENNETH H DR
SCHATZMAN EVRY PROF   SCHILIZZI RICHARD T DR   SCHNOPPER HERBERT W DR
SCHREIER ETHAN J DR   SCHWARTZ DANIEL A DR   SCIAMA DENNIS W DR
SCIORTINO SALVATORE DR   SCOTT JOHN S DR   SEIELSTAD GEORGE A
SETTI GIANCARLO PROF   SEWARD FREDERICK D   SHAHAM JACOB PROF
SHAKURA NICHOLAJ I DR   SHAPIRO MAURICE M PROF   SHAVER PETER A DR
SHAVIV GIORA PROF   SHIELDS GREGORY A DR   SHUKRE C S DR
SIGNORE MONIQUE DR   SIKORA MAREK   SILBERBERG REIN DR
SKILLING JOHN DR   SMITH BARHAM W DR   SOFIA SABATINO PROF
SPADA GIANFRANCO DR   STAUBERT RUDIGER PROF DR   STECKER FLOYD W DR
STEIGMAN GARY PROF   STEINER JOAO E DR   STEPANIAN A A DR
STEPHENS S A DR   STOCKMAN HERVEY S JR DR   STRONG IAN B DR
STURROCK PETER A PROF   SVENSSON ROLAND   SWANK JEAN HEBB
TAKAHARA FUMIO DR   TANAKA YASUO DR   TAYLER ROGER J PROF
TERRELL NELSON JAMES JR   THORNE KIP S PROF   TOMIMATSU AKIRA DR
TOTSUKA YOJI DR   TRUEMPER JOACHIM PROF   TRURAN JAMES W JR
TRUSSONI EDOARDO   TSURUTA SACHIKO DR   TSYGAN ANATOLII I PROF
TYLKA ALLAN J DR   VALTONEN MAURI J PROF   VAN DEN HEUVEL EDWARD P J
VAN DER WALT D J DR   VAN RIPER KENNETH A DR   VIDAL NISSIM V DR
VOELK HEINRICH J PROF   WANAS M I PROF   WANG DEYU
WANG SHOU-GUAN   WANG YI-MING DR   WANG ZHEN-RU
WEAVER THOMAS A DR   WEBSTER ADRIAN S DR   WEHRLE ANN ELIZABETH DR
WEISHEIT JON C DR   WEISSKOPF MARTIN CH DR   WENTZEL DONAT G DR
WESTFOLD KEVIN C PROF   WHEELER J CRAIG PROF   WHEELER JOHN A DR
WILL CLIFFORD M DR   WILSON JAMES R DR   WOLSTENCROFT RAMON D DR
WOLTER ANNA DR   WORRALL DIANA MARY   YANG LAN-TIAN
YOU JUNHAN   ZAMORANI GIOVANNI   ZHANG HE-QI
ZHANG JIA-LU   ZHANG ZHEN-JIU   ZOMBECK MARTIN V DR

COMPOSITION OF COMMISSION 49 1991-1994

President   : BUTI BIMLA PROF

Vice-President(s) : RIPKEN HARTMUT W DR

Organizing Committee: BURLAGA LEONARD F DR
                      KELLER HORST UWE DR
                      NAKAGAWA YOSHINARI DR
                      PERKINS FRANCIS W DR
                      RAADU MICHAEL A DR
                      SAGDEEV ROALD Z DR
                      SONETT CHARLES P PROF
                      SUESS STEVEN T DR
                      VINOD S KRISHAN MRS DR

Members:

AHLUWALIA HARJIT SINGH DR	ANANTHAKRISHNAN S	ANDERSON KINSEY A PROF
BARNES AARON DR	BARROW COLIN H DR	BARTH CHARLES A PROF
BERTAUX J L DR	BLACKWELL DONALD E PROF	BLANDFORD ROGER DAVID DR
BLUM PETER PROF	BOCHSLER PETER	BONNET ROGER M DR
BRANDT JOHN C DR	BUECHNER JORG DR	CHAMBERLAIN JOSEPH W PROF
CHASSEFIERE ERIC	CHEN BIAO	CHITRE DATTAKUMAR M DR
CHITRE SHASHIKUMAR M DR	CHOU CHIH-KANG DR	COUTURIER PIERRE
CRAMER NEIL DR	CUPERMAN SAMI PROF	DE JAGER CORNELIS PROF
DELACHE PHILIPPE J DR	DOLGINOV ARKADY Z PROF DR	DRYER MURRAY DR
DULDIG MARCUS LESLIE DR	DURNEY BERNARD DR	DYSON JOHN E DR
ERGMA E V DR	ESHLEMAN VON R PROF	EVIATAR AHARON PROF
FAHR HANS JOERG PROF DR	FEYNMAN JOAN DR	FIELD GEORGE B PROF
GOLDMAN MARTIN V	GOSLING JOHN T DR	GRZEDZIELSKI STANISLAW PR
HABBAL SHADIA RIFAI	HARVEY CHRISTOPHER C DR	HERAS ANA M DR
HEYVAERTS JEAN DR	HOLLWEG JOSEPH V	HOLZER THOMAS EDWARD DR
HUMBLE JOHN EDMUND DR	INAGAKI SHOGO DR	JOKIPII J R PROF
JOSELYN JO ANN C DR	KAKINUMA TAKAKIYO T PROF	LAFON JEAN-PIERRE J DR
LAI SEBASTIANA	LEVY EUGENE H DR	LI XIAO-QING
LOTOVA N A DR	LUEST REIMAR PROF	LUNDSTEDT HENRIK DR
MACQUEEN ROBERT M DR	MANGENEY ANDRE DR	MASON GLENN M
MATSUURA OSCAR T DR	MAVROMICHALAKI HELEN DR	MELROSE DONALD B PROF
MENDIS DEVAMITTA ASOKA DR	MESTEL LEON PROF	MICHEL F CURTIS PROF
MOUSSAS XENOPHON PH D	PARESCE FRANCESCO DR	PARKER EUGENE N
PFLUG KLAUS DR	PNEUMAN GERALD W	READHEAD ANTHONY C S DR
REAY NEWRICK K DR	RICKETT BARNABY JAMES DR	RIDDLE ANTHONY C DR
ROACH FRANKLIN E	ROSNER ROBERT	ROXBURGH IAN W PROF
RUSSELL CHRISTOPHER T	SARRIS EMMANUEL T PH D	SASTRI HANUMATH J DR
SAWYER CONSTANCE B DR	SCHATZMAN EVRY PROF	SCHERB FRANK PROF
SCHINDLER KARL PROF DR	SCHMIDT H U DR	SCHREIBER ROMAN
SCHWARTZ STEVEN JAY	SETTI GIANCARLO PROF	SHAWHAN STANLEY D DR
SHEA MARGARET A DR	SMITH DEAN F DR	STONE R G DR
STURROCK PETER A PROF	TRITAKIS BASIL P DR	VAINSTEIN L A DR

VAN ALLEN JAMES A PROF	VERHEEST FRANK PROF	VUCETICH HECTOR DR
WANG SHUNDE DR	WANG YI-MING DR	WATANABE TAKASHI DR
WELLER CHARLES S DR	WILD JOHN PAUL DR	WU SHI TSAN DR
YEH TYAN DR		

COMPOSITION OF COMMISSION 49 1991-1994

President   : BUTI BIMLA PROF

Vice-President(s) : RIPKEN HARTMUT W DR

Organizing Committee: BURLAGA LEONARD F DR
                      KELLER HORST UWE DR
                      NAKAGAWA YOSHINARI DR
                      PERKINS FRANCIS W DR
                      RAADU MICHAEL A DR
                      SAGDEEV ROALD Z DR
                      SONETT CHARLES P PROF
                      SUESS STEVEN T DR
                      VINOD S KRISHAN MRS DR

Members:

AHLUWALIA HARJIT SINGH DR	ANANTHAKRISHNAN S	ANDERSON KINSEY A PROF
BARNES AARON DR	BARROW COLIN H DR	BARTH CHARLES A PROF
BERTAUX J L DR	BLACKWELL DONALD E PROF	BLANDFORD ROGER DAVID DR
BLUM PETER PROF	BOCHSLER PETER	BONNET ROGER M DR
BRANDT JOHN C DR	BUECHNER JORG DR	CHAMBERLAIN JOSEPH W PROF
CHASSEFIERE ERIC	CHEN BIAO	CHITRE DATTAKUMAR M DR
CHITRE SHASHIKUMAR M DR	CHOU CHIH-KANG DR	COUTURIER PIERRE
CRAMER NEIL DR	CUPERMAN SAMI PROF	DE JAGER CORNELIS PROF
DELACHE PHILIPPE J DR	DOLGINOV ARKADY Z PROF DR	DRYER MURRAY DR
DULDIG MARCUS LESLIE DR	DURNEY BERNARD DR	DYSON JOHN E DR
ERGMA E V DR	ESHLEMAN VON R PROF	EVIATAR AHARON PROF
FAHR HANS JOERG PROF DR	FEYNMAN JOAN DR	FIELD GEORGE B PROF
GOLDMAN MARTIN V	GOSLING JOHN T DR	GRZEDZIELSKI STANISLAW PR
HABBAL SHADIA RIFAI	HARVEY CHRISTOPHER C DR	HERAS ANA M DR
HEYVAERTS JEAN DR	HOLLWEG JOSEPH V	HOLZER THOMAS EDWARD DR
HUMBLE JOHN EDMUND DR	INAGAKI SHOGO DR	JOKIPII J R PROF
JOSELYN JO ANN C DR	KAKINUMA TAKAKIYO T PROF	LAFON JEAN-PIERRE J DR
LAI SEBASTIANA	LEVY EUGENE H DR	LI XIAO-QING
LOTOVA N A DR	LUEST REIMAR PROF	LUNDSTEDT HENRIK DR
MACQUEEN ROBERT M DR	MANGENEY ANDRE DR	MASON GLENN M
MATSUURA OSCAR T DR	MAVROMICHALAKI HELEN DR	MELROSE DONALD B PROF
MENDIS DEVAMITTA ASOKA DR	MESTEL LEON PROF	MICHEL F CURTIS PROF
MOUSSAS XENOPHON PH D	PARESCE FRANCESCO DR	PARKER EUGENE N
PFLUG KLAUS DR	PNEUMAN GERALD W	READHEAD ANTHONY C S DR
REAY NEWRICK K DR	RICKETT BARNABY JAMES DR	RIDDLE ANTHONY C DR
ROACH FRANKLIN E	ROSNER ROBERT `	ROXBURGH IAN W PROF
RUSSELL CHRISTOPHER T	SARRIS EMMANUEL T PH D	SASTRI HANUMATH J DR
SAWYER CONSTANCE B DR	SCHATZMAN EVRY PROF	SCHERB FRANK PROF
SCHINDLER KARL PROF DR	SCHMIDT H U DR	SCHREIBER ROMAN
SCHWARTZ STEVEN JAY	SETTI GIANCARLO PROF	SHAWHAN STANLEY D DR
SHEA MARGARET A DR	SMITH DEAN F DR	STONE R G DR
STURROCK PETER A PROF	TRITAKIS BASIL P DR	VAINSTEIN L A DR

VAN ALLEN JAMES A PROF	VERHEEST FRANK PROF	VUCETICH HECTOR DR
WANG SHUNDE DR	WANG YI-MING DR	WATANABE TAKASHI DR
WELLER CHARLES S DR	WILD JOHN PAUL DR	WU SHI TSAN DR
YEH TYAN DR		

COMPOSITION OF COMMISSION 50 1991-1994

President   : MURDIN PAUL G DR

Vice-President(s) : BHATTACHARYYA J C PROF

Organizing Committee: ARDEBERG ARNE L PROF
                       BLANCO VICTOR M DR
                       COSTERO RAFAEL
                       CRAWFORD DAVID L DR
                       DAVIS JOHN PROF
                       GERGELY TOMAS ESTEBAN DR
                       ISOBE SYUZO DR
                       KOVALEVSKY JEAN DR
                       MCNALLY DEREK DR
                       SHCHEGLOV P V DR

Members:

ALY M KHAIRY PROF	ARIAS DE GREIFF J PROF	ARSENIJEVIC JELISAVETA
BARRETO LUIZ MUNIZ PROF	BENSAMMAR SLIMANE DR	BLANCO CARLO DR
BROWN ROBERT HAMILTON	BURSTEIN DAVID	CAYREL ROGER DR
COYNE GEORGE V DR	DAWE JOHN ALAN DR	DOMMANGET J DR
DUNKELMAN LAWRENCE	EDWARDS PAUL J DR	GALAN MAXIMINO J
GIBSON DAVID MICHAEL DR	GOEBEL ERNST DR	HELMER LEIF
HIDAJAT BAMBANG PROF DR	HOAG ARTHUR A DR	HUANG YIN-LIANG
JIANG SHI-YANG	KAHLMANN HANS CORNELIS DR	KOZAI YOSHIHIDE PROF
LEIBOWITZ ELIA M DR	MAHRA H S DR	MARKKANEN TAPIO DR
MARX SIEGFRIED DR	MATTIG W PROF DR	MCCARTHY MARTIN F DR
MENZIES JOHN W DR	NELSON BURT DR	OEZEL MEHMET EMIN DR
OSORIO JOSE J S P PROF	OWEN FRAZER NELSON DR	PANKONIN VERNON LEE DR
SANCHEZ FRANCISCO PROF	SCHILIZZI RICHARD T DR	SMITH F GRAHAM PROF
SPOELSTRA T A TH DR	TORRES CARLOS ALBERTO DR	TORRES CARLOS DR
TREMKO JOZEF DR	UPGREN ARTHUR R DR	VAN DEN BERGH SIDNEY PROF
WALKER MERLE F PROF	WAYMAN PATRICK A PROF	WHITEOAK J B DR
WOOLF NEVILLE J	WOSZCZYK ANDRZEJ PROF	WU MING-CHAN
ZHANG BAI-RONG		

COMPOSITION OF COMMISSION 51 1991-1994

President   : BROWN RONALD D PROF

Vice-President(s) : TARTER JILL C DR

Organizing Committee: CAMPBELL BRUCE DR
                       COLOMB FERNANDO R DR
                       DRAKE FRANK D PROF
                       HEIDMANN JEAN DR
                       HIRABAYASHI HISASHI DR
                       KLEIN MICHAEL J DR
                       MARX GYORGY PROF
                       PAPAGIANNIS MICHAEL D PRO
                       SLYSH VJACHOSLAV I DR

Members:

ALMAR IVAN PROF	AL-NAIMY HAMID M K DR	AL-SABTI ABDUL ADIM DR
AMBARTSUMIAN V A PROF DR	ANDO HIROYASU DR	BALAZS BELA A PROF DR
BALL JOHN A DR	BANIA THOMAS MICHAEL	BARBIERI CESARE PROF
BASU BAIDYANATH PROF	BASU DIPAK DR	BAUM WILLIAM A DR
BEAUDET GILLES DR	BECKMAN JOHN E PROF	BECKWITH STEVEN V W
BEEBE RETA FAYE DR	BENEST DANIEL DR	BERENDZEN RICHARD DR
BERNACCA P L PROF	BILLINGHAM JOHN	BIRAUD FRANCOIS DR
BLESS ROBERT C PROF	BOWYER C STUART PROF	BOYCE PETER B DR
BRACEWELL RONALD N PROF	BRODERICK JOHN DR	BURKE BERNARD F DR
CALVIN WILLIAM H DR	CAMPUSANO LUIS E	CARLSON JOHN B
CARR THOMAS D PROF	CHAISSON ERIC J PROF	CHOU KYONG CHOL PROF
CLARK THOMAS A DR	CONNES PIERRE DR	COUPER HEATHER MISS
CURRIE DOUGLAS G DR	DAIGNE GERARD	DARIUS JON DR
DAVIS MICHAEL M DR	DAWE JOHN ALAN DR	DE GRAAFF W DR
DE JAGER CORNELIS PROF	DE JONGE J K DR	DE LOORE CAMIEL PROF
DE VINCENZI DONALD DR	DELSEMME ARMAND H PROF DR	DICK STEVEN J
DIXON ROBERT S DR	DJORGOVSKI STANISLAV DR	DORSCHNER JOHANN DR
DOWNS GEORGE S DR	DYSON F J DR	ECCLES MICHAEL J DR
ELLIS GEORGE F R PROF	EPSTEIN EUGENE E DR	EVANS NEAL J, II ASS PROF
FAZIO GIOVANNI G DR	FEJES ISTVAN DR	FELDMAN PAUL A DR
FIELD GEORGE B PROF	FIRNEIS FRIEDRICH J DR	FIRNEIS MARIA G DR
FISHER PHILIP C	FREDRICK LAURENCE W PROF	FUJIMOTO MASA-KATSU DR
FUJIMOTO MITSUAKI DR	GATEWOOD GEORGE DIRECTOR	GEHRELS TOM PROF
GHIGO FRANCIS D DR	GINZBURG VITALY L PROF	GIOVANNELLI FRANCO DR
GODOLI GIOVANNI PROF	GOLDSMITH DONALD W DR	GOTT III J RICHARD
GOUDIS CHRISTOS D PROF	GREENBERG J MAYO DR	GREENSTEIN J L PROF
GREGORY PHILIP C DR	GULKIS SAMUEL DR	GUNN JAMES E PROF
GURM HARDEV S PROF	HADDOCK FRED T DR	HAISCH BERNHARD MICHAEL
HAJDUK ANTON DR	HARRINGTON ROBERT S DR	HARRISON EDWARD R PROF
HART MICHAEL H DR	HECK ANDRE DR	HEESCHEN DAVID S DR
HERCZEG TIBOR J PROF DR	HERSHEY JOHN L DR	HEUDIER JEAN-LOUIS DR
HOANG BINH DY DR	HOLLIS JAN MICHAEL DR	HOROWITZ PAUL PROF

COMPOSITION OF COMMISSION WGPSN 1991-1994

President   : AKSNES KAARE DR

Organizing Committee: BRAHIC ANDRE DR
                       FULCHIGNONI MARCELLO PROF
                       MAROV MIKHAIL YA PROF
                       MORRISON DAVID PROF
                       OWEN TOBIAS C PROF
                       SHEVCHENKO VLADISLAV V DR
                       SMITH BRADFORD A PROF
                       SODERBLOM DAVID R

3.   Geographical repartition of members   Members

within adhering countries 7260

Country : ALGERIA

SADAT RACHIDA DR

Country : ARGENTINA

AGUERO ESTELA L DR	ALTAVISTA CARLOS A DR	AQUILANO ROBERTO OSCAR DR
ARIAS ELISA FELICITAS	AZCARATE ISMAEL N DR	BAJAJA E DR
BASSINO LILIA P DR	BENVENUTO OMAR DR	BRANDI ELISANDE ESTELA DR
BRANHAM RICHARD L JR	CAPPA DE NICOLAU CRISTINA	CARESTIA REINALDO A DR
CARRANZA GUSTAVO J DR	CASTAGNINO MARIO DR	CERSOSIMO JUAN CARLOS DR
CLARIA JUAN DR	COLOMB FERNANDO R DR	DUBNER GLORIA DR
FEINSTEIN ALEJANDRO DR	FERNANDEZ SILVIA M DR	FERRER OSVALDO EDUARDO DR
FILLOY EMILIO MANUEL E E	FORTE JUAN CARLOS DR	GARCIA LAMBAS DIEGO DR
GIACANI ELSA BEATRIZ DR	HERNANDEZ CARLOS ALBERTO	IANNINI GUALBERTO DR
LANDI-DESSY J DR	LAPASSET EMILIO DR	LEVATO ORLANDO HUGO DR
LOPEZ CARLOS LIC	LOPEZ GARCIA ZULEMA L DR	LOPEZ JOSE A ING
LOPEZ-GARCIA FRANCISCO DR	LUNA HOMERO G DR	MACHADO MARCOS
MALARODA STELLA M DR	MANRIQUE WALTER T PROF	MARABINI RODOLFO JOSE
MARRACO HUGO G DR	MARTIN MARIA CRISTINA DR	MENDEZ MARIANO DR
MENDEZ ROBERTO H DR	MILONE LUIS A DR	MORRAS RICARDO DR
MORRELL NIDIA DR	MUZZIO JUAN C PROF	NIEMELA VIRPI S DR
OLANO CARLOS ALBERTO DR	ORELLANA ROSA BEATRIZ DR	ORSATTI ANA M DR
PERDOMO RAUL	PIACENTINI RUBEN DR	POEPPEL WOLFGANG G L DR
RABOLLI MONICA DR	RINGUELET ADELA E DR	ROVIRA MARTA GRACIELA
SAHADE JORGE PROF	SERSIC J L DR	SISTERO ROBERTO F DR
SOLIVELLA GLADYS R LIC	VAZQUEZ RUBEN ANGEL DR	VEGA E IRENE DR
VUCETICH HECTOR DR	ZADUNAISKY PEDRO E PROF	

Country : ARMENIA

AMBARTSUMIAN V A PROF DR	GURZADIAN G A PROF DR	HARUTYUNIAN HAIK A DR
KALLOGLIAN ARSEN T DR	KHACHIKIAN E YE PROF	MAGAKIAN TIGRAN Y DR
MELIK-ALAVERDIAN YU DR	MIRZOYAN L V DR PROF	MNATSAKANIAN MAMIKON A DR
NIKOGHOSSIAN ARTHUR G DR	PARSAMYAN ELMA S DR	PETROSIAN ARTASHES R DR
SANAMIAN V A DR	SHAHBAZIAN ROMELIA K DR	SHAKHBAZYAN YURIJ L DR
STEPANIAN JIVAN A DR	TOVMASSIAN H M DR	VARDANIAN R A DR

Country : AUSTRALIA

ABLES JOHN G DR   AITKEN DAVID K DR   ALLEN DAVID A DR
BAILEY JEREMY A   BALL LEWIS DR   BATTY MICHAEL DR
BESSELL MICHAEL S DR   BICKNELL GEOFFREY V DR   BIRCH PETER MR
BLAIR DAVID GERALD   BOLTON JOHN G   BOOTH ANDREW J
BOWEN EDWARD G DR   BOYLE BRIAN DR   BRAY ROBERT J DR
BROUW W N DR   BROWN RONALD D PROF   BURTON MICHAEL G DR
CALLY PAUL S DR   CANDY MICHAEL P MR   CANE HILARY VIVIEN
CANNON RUSSELL D DR   CARTER DAVID DR   CARVER JOHN H PROF
CASWELL JAMES L DR   CHAPMAN JESSICA DR   CHRISTIANSEN WILBUR PROF
COGAN BRUCE C DR   COLE TREVOR WILLIAM PROF   COUCH WARRICK DR
CRAM LAWRENCE EDWARD PROF   CRAMER NEIL DR   DA COSTA GARY STEWART DR
DAVIS JOHN PROF   DAWE JOHN ALAN DR   DOPITA MICHAEL ANDREW DR
DULDIG MARCUS LESLIE DR   DUNCAN ROBERT A PROF   DURRANT CHRISTOPHER J DR
EDWARDS PAUL J DR   EKERS RONALD D DR   ELFORD WILLIAM GRAHAM DR
ELLIS G R A PROF   ERICKSON WILLIAM C DR   EVANS ROBERT REV
FAULKNER DONALD J DR   FENTON K B DR   FRATER ROBERT H DR
FREEMAN KENNETH C PROF   GALLOWAY DAVID DR   GARDNER FRANCIS F DR
GASCOIGNE S C B DR   GILLINGHAM PETER MR   GINGOLD ROBERT ARTHUR DR
GODFREY PETER DOUGLAS DR   GOLLNOW H DR   GOTTLIEB KURT
GRAY PETER MURRAY   GREEN ELIZABETH M DR   GREENHILL JOHN DR
GREGG MICHAEL DAVID DR   HALL PETER J DR   HAMILTON P A DR
HARNETT JULIEINE DR   HARWOOD DENNIS MR   HATZIDIMITRIOU DESPINA DR
HAYNES RAYMOND F PROF   HORTON BRIAN H DR   HOSKING ROGER J PROF
HUMBLE JOHN EDMUND DR   HUNSTEAD RICHARD W DR   HYLAND A R HARRY DR
JAUNCEY DAVID L DR   KALNAJS AGRIS J DR   KEAY COLIN S L PROF
KESTEVEN MICHAEL J L DR   KILLEEN NEIL DR   KVIZ ZDENEK DR
LAMBECK KURT PROF   LARGE MICHAEL I DR   LOMB NICHOLAS RALPH DR
LOUGHHEAD RALPH E DR   LUCK J DR   LYNGA GOSTA DR
MALIN DAVID F MR   MANCHESTER RICHARD N DR   MATHEWSON DONALD S PROF
MCADAM W BRUCE DR   MCCONNELL DAVID DR   MCCRACKEN KENNETH G DR
MCCULLOCH PETER M DR   MCGEE RICHARD X DR   MCGREGOR PETER JOHN DR
MCLEAN DONALD J DR   MCNAUGHT ROBERT H   MEATHERINGHAM STEPHEN DR
MELROSE DONALD B PROF   MILLS BERNARD Y PROF   MILNE DOUGLAS K DR
MINNET HARRY C MR   MONAGHAN JOSEPH J DR   MORETON G E
MORGAN PETER DR   MULLALY RICHARD F DR   MURDOCH HUGH S DR
NELSON GRAHAM JOHN DR   NEWELL EDWARD B DR   NIKOLOFF IVAN DR
NORRIS JOHN DR   NORRIS RAYMOND PAUL   NULSEN PAUL DR
O'MARA BERNARD J PROF   O'SULLIVAN JOHN DAVID DR   PAGE ARTHUR MR
PETERSON BRUCE A DR   PIDDINGTON JACK H RES FEL   PRENTICE ANDREW J R DR
PROTHEROE RAYMOND J DR   QUINN PETER DR   REBER GROTE DR
REES DAVID ELWYN DR   REYNOLDS JOHN DR   ROBERTSON JAMES GORDON DR
ROBINSON BRIAN J DR   ROBINSON GARRY DR   ROBINSON JR RICHARD D DR
RODGERS ALEX W DR   ROSS JOHN E R DR   RUSSELL KENNETH S DR
SADLER ELAINE MARGARET   SAULT ROBERT DR   SAVAGE ANN DR
SHARMA DHARMA PAL DR   SHERIDAN K V DR   SHIMMINS ALBERT JOHN
SHOBBROOK ROBERT R DR   SIMS KENNETH P DR   SLEE O B DR
SMITH CRAIG H DR   SMITH LINDSEY F DR   SMITH ROBERT G DR
SPARROW JAMES G DR   STAVELEY-SMITH LISTER DR   STEEL DUNCAN I DR
STEWART RONALD T MR   STIBBS DOUGLAS W N PROF   STOREY JOHN W V DR
STOREY MICHELLE DR   TANGO WILLIAM J DR   TAYLOR KEITH DR

Country : AUSTRALIA (Follow on)

TAYLOR KENNETH N R PROF	THOMPSON KEITH DR	TUOHY IAN R DR
TURTLE A J DR	TZIOUMIS ANASTASIOS DR	VAN DER BORGHT RENE PROF
VAUGHAN ALAN DR	VISVANATHAN NATARAJAN DR	WATERWORTH MICHAEL DR
WATSON FREDERICK GARNETT	WATSON ROBERT DR	WELLINGTON KELVIN DR
WESTFOLD KEVIN C PROF	WHITE GRAEME LINDSAY DR	WHITEOAK J B DR
WICKRAMASINGHE D T DR	WILD JOHN PAUL DR	WILSON BRIAN G PROF
WILSON PETER R PROF	WOOD PETER R DR	WRIGHT ALAN E DR
ZAMBON GIULIO DR	ZEALEY WILLIAM J DR	

Country : AUSTRIA

AUNER GERHARD DR	BALAZS BELA A PROF DR	BREGER MICHEL PROF DR
DORFI ERNST ANTON DR	DVORAK RUDOLF DR	FERRARI D'OCCHIEPPO K DR
FIRNEIS FRIEDRICH J DR	FIRNEIS MARIA G DR	GOEBEL ERNST DR
HANSLMEIER ARNOLD	HARTL HERBERT DR	HAUPT HERMANN F PROF
HRON JOSEF DR	JACKSON PAUL DR	KOEBERL CHRISTIAN DR
LUSTIG GUENTER DR	MAITZEN HANS M DR	PFLEIDERER JORG PROF
POLNITZKY GERHARD DR	PROCHAZKA FRANZ V DR	RAKOS KARL D PROF
SCHNELL ANNELIESE DR	SCHOBER HANS J DR	SCHROLL ALFRED DR
SCHRUTKA-RECHTENSTAMM PR	STIFT MARTIN JOHANNES DR	STRASSMEIER KLAUS G DR
WEINBERGER RONALD DR	WEISS WERNER W DR	

Country : AZERBAIDZHAN

ABBASOV ALIK R DR	ASLANOV I A DR	EMINZADE T A DR
GUSEINOV O H PROF	GUSEJNOV RAGIM EH DR	KASUMOV FIKRET K O DR
SEIDOV ZAKIR F DR	SULTANOV G F ACAD	

Country : BELGIUM

ARNOULD MARCEL L DR | ARPIGNY CLAUDE PROF | BAECK NICOLE A L DR
BERTIAU FLOR C PROF | BIEMONT EMILE DR | BLOMME RONNY DR
BOSMAN-CRESPIN DENISE | BRIHAYE CHARLES C A DR | BURGER MARIJKE DR
CALLEBAUT DIRK K DR | COUTREZ RAYMOND A J PROF | CUGNON PIERRE DR
CUYPERS JAN DR | DE GREVE JEAN-PIERRE DR | DE LOORE CAMIEL PROF
DEBEHOGNE HENRI DR SC | DEHANT VERONIQUE DR | DEJAIFFE RENE J DR
DEJONGHE HERWIG BERT DR | DELBOUILLE LUC PROF | DELCROIX ANDRE J S DR
DEMARET JACQUES DR | DENIS CARLO DR | DENOYELLE JOZEF KIC
DINGENS P PROF DR | DOMMANGET J DR | DOSSIN F DR
ELST ERIC WALTER DR | GABRIEL MAURICE R DR | GERARD JEAN-CLAUDE M C DR
GODART ODON PROF | GONZE ROGER F J IR | GOOSSENS MARCEL DR
GOSSET ERIC DR | GREVESSE N DR | HENRARD JACQUES PROF
HENSBERGE HERMAN | HOUZIAUX L PROF | HUTSEMEKERS DAMIEN DR
JAMAR CLAUDE A J DR | KOECKELENBERGH ANDRE DR | LAMPENS PATRICIA DR
LAUSBERG ANDRE DR | LEMAITRE ANNE DR | MAGAIN PIERRE DR
MALAISE DANIEL J DR | MANFROID JEAN DR | MEIRE RAPHAEL
MELCHIOR PAUL J PROF DIR | MOERDIJK WILLY G DR | MONFILS ANDRE G PROF
MOONS MICHELE B M M | NICOLET MARCEL PROF | NOELS ARLETTE DR
OTTELET I J DR | OXENIUS JOACHIM DR | PAQUET PAUL EG DR
PAUWELS T DR | PERDANG JEAN M DR | REMY BATTIAU LILIANE G A
RENSON P F M DR | ROBE H A G DR | ROLAND GINETTE DR
SAUVAL A JACQUES DR | SCUFLAIRE RICHARD DR | SIMON PAUL C DR
SIMON RENE L E PROF | SINACHOPOULOS D DR | SMEYERS PAUL PROF
STERKEN CHRISTIAAN LEO DR | STEYAERT HERMAN PROF DR | SURDEJ JEAN M G
SVALGAARD LEIF DR | SWINGS JEAN-PIERRE DR | VAN DESSEL EDWIN LUDO DR
VAN RENSBERGEN WALTER DR | VERBEEK PAUL DR | VERHEEST FRANK PROF
VREUX JEAN MARIE DR | WAELKENS CHRISTOFFEL | ZANDER RODOLPHE DR

Country : BRAZIL

ABRAHAM ZULEMA DR | AGUIAR ODYLIO DENYS DR | ALDROVANDI RUBEN DR
ALDROVANDI S M VIEGAS DR | ARAUJO FRANCISCO X DE DR | BARBUY BEATRIZ DR
BARRETO LUIZ MUNIZ PROF | BARROSO JR JAIR | BATALHA CELSO CORREA DR
BENEVIDES SOARES P DR | BERMAN MARCELO S DR | BICA EDUARDO L D DR
BRAGA JOAO DR | BRUNINI ADRIAN DR | BUSKO IVO C DR
CAPELATO HUGO VICENTE DR | CHAN ROBERTO DR | CHIAN ABRAHAM CHIAN-LONG
CODINA LANDABERRY SAYD J | CORREIA EMILIA DR | COSTA JOAQUIM EDUARDO DR
DA COSTA JOSE MARQUES DR | DA COSTA NICOLAI L-A | DA ROCHA VIEIRA E DR
DA SILVA LICIO DR | DAMINELI NETO AUGUSTO DR | DE ALMEIDA AMAURY A DR
DE CARVALHO REINALDO DR | DE FREITAS PACHECO J A DR | DE LA REZA RAMIRO DR
DE SOUZA RONALDO DR | DOTTORI HORACIO A DR | DUCATI JORGE RICARDO DR
FERRAZ-MELLO S PROF DR | FREITAS MOURAO R R DR | GIACAGLIA GIORGIO E PROF
GOMES ALERCIO M DR | GOMIDE FERNANDO DE MELLO | GONZALES-A WALTER D DR
GRIJO DE OLIVEIRA A K DR | GRUENWALD RUTH DR | JABLONSKI FRANCISCO DR
JANOT-PACHECO EDUARDO DR | JAYANTHI UDAYA B DR | KAUFMANN PIERRE PROF
KEPLER S O | KOTANYI CHRISTOPHE DR | LAZZARO DANIELA DR
LEITE SCHEID PAULO DR | LEPINE JACQUES R D DR | LOPES DALTON DE FARIA DR

Country : BRAZIL (Follow on)

NICOLACI DA COSTA L-A	NOVELLO MARIO DR	OPHER REUVEN PROF
PALMEIRA RICARDO A R DR	PASTORIZA MIRIANI G DR	PELLEGRINI PAULO S S DR
PIAZZA LILIANA RIZZO	QUARTA MARIA LUCIA	QUAST GERMANO RODRIGO
RAO K RAMANUJA DR	REZENDE COSTA JOAQUIM DR	SANTOS NILTON OSCAR DR
SATO MASSAE DR	SAWANT HANUMANT S DR	SCALISE JR EUGENIO DR
SCHAAL RICARDO E DR	SCHUCH NELSON JORGE	SESSIN WAGNER DR
SINGH PATAN DEEN DR	STEINER JOAO E DR	STORCHI-BERGMAN THAISA DR
TAKAGI SHIGETSUGU DR	TORRES CARLOS ALBERTO DR	TSUCHIDA MASAYOSHI DR
VAZ LUIZ PAULO RIBEIRO	VIEIRA MARTINS ROBERTO DR	VILAS-BOAS JOSE W DR
VILHENA DE MORAES R DR	VILLELA THYRSO NETO DR	YOKOYAMA TADASHI DR

Country : BULGARIA

BACALOV MIHAIL	BONEV BONU K MR	BUYUKLIEV GEORGI DR
DERMENDJIEV VLADIMIR DR	DIKOVA SMILIANA D	DOBRITSCHEV V M MR
FILIPOV LATCHEZAR	GEORGIEV TSVETAN DR	GOLEV VALERY K DR
ILIEV ILIAN	IVANOV GEORGI R DR	IVANOVA VIOLETA DR
KALINKOV MARIN P DR	KJURKCHIEVA DIANA DR	KOLEV DIMITAR ZDRAVKOV
KOVACHEV B J DR	KRAICHEVA ZDRAVSKA DR	KUNCHEV PETER DR
MINEVA VENETA DR	MOMCHEV GOSPODIN	NICOLOV NIKOLAI S DR
NIKOLOV ANDREJ DR	PANOV KIRIL DR	PETROV GEORGY TRENDAFILOV
PETROV NIKOLAI	POPOV VASIL NIKOLOV	POPOVA MALINA D PROF DR
RADOSLAVOVA TSVETANKA	RAIKOVA DONKA DR .	RUSSEV RUSCHO DR
RUSSEVA TATJANA	SBIRKOVA-NATCHEVA T	SERAFIMOV KIRIL B ACAD
SHKODROV V G DR	SPASOVA NEDKA MARINOVA	STOEV ALEXEI
TOMOV ALEXANDER NIKOLOV	TOMOV TOMA V DR	TSVETKOV MILCHO K DR
TSVETKOV TSVETAN DR	TSVETKOVA KATIA	UMLENSKI VASIL
VELKOV KIRIL	VLADIMIROV SIMEON	YANKULOVA IVANKA DR
ZHEKOV SVETOZAR A DR	ZHELYAZKOV IVAN DR	ZLATEV SLAVEY

Country : CANADA

AIKMAN G CHRIS L	ANDREW BRYAN H DR	AUMAN JASON R PROF
AVERY LORNE W DR	BARKER PAUL K DR	BARNARD HANNES A J DR
BASTIEN PIERRE DR	BATTEN ALAN H DR	BEAUDET GILLES DR
BELL MORLEY B	BINETTE LUC	BISHOP ROY L DR
BLACKWELL ALAN TREVOR	BOCHONKO D RICHARD DR	BOLTON C THOMAS PROF
BOND JOHN RICHARD	BORRA ERMANNO F DR	BRANDIE GEORGE W DR
BROOKS RANDALL C DR	BROTEN NORMAN W	BURKE J ANTHONY DR
CALDWELL JOHN JAMES	CAMPBELL BRUCE DR	CANNON WAYNE H DR
CARIGNAN CLAUDE DR	CARLBERG RAYMOND GARY DR	CHAU WAI Y PROF
CLARK THOMAS ALAN DR	CLARKE THOMAS R DR	CLEMENT MAURICE J PROF
CLIMENHAGA JOHN L PROF	CLUTTON-BROCK MARTIN DR	COSTAIN CECIL C DR
COUTTS-CLEMENT CHRISTINE	COVINGTON ARTHUR E	CRABTREE DENNIS DR
CRAMPTON DAVID DR	DAVIDGE TIMOTHY J DR	DE ROBERTIS M M DR
DEMERS SERGE DR	DEWDNEY PETER E F DR	DOHERTY LORNE H DR
DOUGLAS R J MR	DULEY WALTER W PROF	DYER CHARLES CHESTER DR
EVANS NANCY REMAGE DR	FAHLMAN GREGORY G DR	FELDMAN PAUL A DR
FERNIE J DONALD PROF	FICH MICHEL DR	FITZGERALD M PIM PROF
FLETCHER J MURRAY	FONTAINE GILLES DR	FORBES DOUGLAS DR
FRIEL EILEEN D DR	GAETZ TERRANCE J DR	GAIZAUSKAS VICTOR DR
GALT JOHN A DR	GARRISON ROBERT F PROF	GOWER ANN C DR
GOWER J F R DR	GRAY DAVID F PROF	GREGORY PHILIP C DR
GRIFFITH JOHN S PROF	GRUNDMANN WALTER	GULLIVER AUSTIN FRASER DR
HALLIDAY IAN DR	HANES DAVID A DR	HARDY EDUARDO
HARRIS GRETCHEN L H DR	HARRIS WILLIAM E DR	HARROWER GEORGE A DR
HARTWICK F DAVID A DR	HARTZ THEODORE R DR	HAWKES ROBERT LEWIS DR
HENRIKSEN RICHARD N DR	HERZBERG GERHARD DR	HESSER JAMES E DR
HICKSON PAUL DR	HIGGS LLOYD A DR	HILL GRAHAM DR
HUBE DOUGLAS P DR	HUGHES VICTOR A PROF	HUTCHINGS JOHN B DR
INNANEN KIMMO A PROF	IRWIN ALAN W DR	IRWIN JUDITH DR
ISRAEL WERNER PROF	JEFFERS STANLEY DR	JONCAS GILLES DR
JONES JAMES DR	KAMPER KARL W DR	KENNEDY JOHN E PROF
KOEHLER JAMES A PROF	KRONBERG PHILIPP DR	KWOK SUN DR
LAKE KAYLL WILLIAM DR	LAMONTAGNE ROBERT DR	LANDECKER THOMAS L DR
LANDSTREET JOHN D PROF	LAPOINTE S M DR	LEAHY DENIS A DR
LEGG THOMAS H DR	LESTER JOHN B DR	LILLY SIMON J DR
LOCKE JACK L DR	LOWE ROBERT P DR	MACLEOD JOHN M DR
MACRAE DONALD A PROF	MADORE BARRY FRANCIS DR	MANN PATRICK J DR
MARLBOROUGH J M PROF	MARTIN PETER G PROF	MATTHEWS JAYMIE
MCCALL MARSHALL LESTER DR	MCCLURE ROBERT D PROF	MCCUTCHEON WILLIAM H PROF
MCDONALD J K PETRIE DR	MCINTOSH BRUCE A DR	MEIKSIN AVERY ABRAHAM DR
MENON T K PROF	MERRIAM JAMES B	MICHAUD GEORGES J DR
MILONE EUGENE F PROF	MITALAS ROMAS ASSOC PROF	MITCHELL GEORGE F DR
MOCHNACKI STEPHAN W DR	MOFFAT ANTHONY F J DR	MOFFAT JOHN W DR
MOORHEAD JAMES M DR	MORBEY CHRISTOPHER L	MORRIS STEPHEN C DR
MORTON DONALD C DR	NADEAU DANIEL DR	NAQVI S I H PROF
NEMEC JAMES	NICHOLLS RALPH W PROF	NOREAU LOUIS DR
ODGERS GRAHAM J DR	PACHNER JAROSLAV PROF	PATHRIA RAJ K PROF
PEDREROS MARIO DR	PERCY JOHN R PROF	PINEAULT SERGE DR
POECKERT ROLAND H DR	POPELAR JOSEF DR	PRITCHET CHRISTOPHER J DR
PRYCE MAURICE H L DR	PURTON CHRISTOPHER R DR	RACINE RENE DR

Country : CANADA (Follow on)

RAMSAY DONALD A DR | REED B CAMERON DR | RICE JOHN B DR
RICHARDSON E HARVEY DR | RICHER HARVEY B DR | ROBB RUSSEL M
ROCHESTER MICHAEL G PROF | ROGER ROBERT S DR | ROGERS CHRISTOPHER DR
ROTTENBERG J A DR | ROUTLEDGE DAVID DR | ROY JEAN-RENE
RUCINSKI SLAWOMIR M DR | SAWYER-HOGG HELEN B DR | SCARFE COLIN D DR
SCRIMGER J NORMAN DR | SEAQUIST ERNEST R PROF | SHUTER WILLIAM L H DR
SMYLIE DOUGLAS E DR | SREENIVASAN S RANGA PROF | STAGG CHRISTOPHER DR
STETSON PETER B. DR | SUKUMAR SUNDARAJAN DR | SUTHERLAND PETER G DR
TAPPING KENNETH F | TASSOUL JEAN-LOUIS PROF | TASSOUL MONIQUE DR
TATUM JEREMY B DR | TAYLOR A R DR | TREMAINE SCOTT DUNCAN
TURNER DAVID G DR | UNDERHILL ANNE B DR | VALLEE JACQUES P DR
VAN DEN BERGH SIDNEY PROF | VANDENBERG DON DR | VENKATESAN DORASWAMY DR
VINER MELVYN R DR | VOLK KEVIN DR | WAINWRIGHT JOHN DR
WALKER GORDON A H PROF | WEBSTER RACHEL | WEHLAU AMELIA DR
WEHLAU WILLIAM H PROF | WELCH DOUGLAS L DR | WELCH GARY A DR
WESEMAEL FRANCOIS DR | WESSON PAUL S DR | WILLIS ANTHONY GORDON DR
WOODSWORTH ANDREW W.DR | WOOLSEY E G | WRIGHT KENNETH O DR
YEE HOWARD K C DR | YEN JUI-LIN PROF | ZHANG CHENG-YUE
ZHANG SHENG-PAN | ZHUANG QIXIANG |

Country : CHILE

ALCAINO GONZALO DR | ALVAREZ HECTOR DR | ANGUITA CLAUDIO A DR
APARICI JUAN DR | BALDWIN JACK A DR | BLANCO VICTOR M DR
BOUCHET PATRICE DR | BRONFMAN LEONARDO DR | CAMPUSANO LUIS E
CARRASCO GUILLERMO DR | CELIS LEOPOLDO DR | COSTA EDGARDO DR
FAUNDEZ-ABANS M DR | FOUQUE PASCAL DR | GARAY GUIDO DR
GEISLER DOUGLAS P DR | GIEREN WOLFGANG P DR | GUTIERREZ-MORENO A DR MRS
INGERSON THOMAS DR | KRZEMINSKI WOJCIECH DR | KUNKEL WILLIAM E DR
LILLER WILLIAM DR | LOYOLA PATRICIO DR | MATHYS GAUTIER DR
MAY J | MAZA JOSE | MELNICK JORGE
MINTZ BLANCO BETTY MRS | MORENO HUGO PROF | NOEL FERNANDO
PHILLIPS MARK M DR | QUINTANA HERNAN DR | REIPURTH BO
ROTH MIGUEL R DR | RUBIO MONICA DR | RUIZ MARIA TERESA DR
SCHWARZ HUGO E | SUNTZEFF NICHOLAS B | TORRES CARLOS DR
VOGT NIKOLAUS DR | WALKER ALISTAIR ROBIN DR | WILLIAMS ROBERT E DR
WROBLEWSKI HERBERT DR

Country : CHINA PR

AI GUOXIANG	BAO KEREN	BIAN YU-LIN
BO SHU-REN	CAO CHANGXIN	CAO LIHONG
CAO SHENGLIN	CHEN BIAO	CHEN DAO-HAN
CHEN HONGSHENG	CHEN JIAN-SHENG	CHEN PEISHENG
CHEN XIAO-ZHONG	CHEN ZHEN	CHEN ZHENCHENG
CHEN ZUN-GUI	CHENG FU-HUA	CHENG FU-ZHEN
CHU HAN-SHU PROF	CHU YAOQUAN	CUI DOU-XING
CUI ZHEN-HUA	DAN XHI-XIANG	DENG ZUGAN DR
DI XIAO-HUA	DING YOU-JI	FAN YING
FANG CHENG	FENG HESHENG	FENG KE-JIA
FONG CHU-GANG	FU CHENG-QI	FU DELIAN
FU QI JUN	GAO BILIE	GAO BUXI
GONG SHOU-SHEN	GONG SHU-MO	GU XIAO-MA
GUO NEI-SHU DR	GUO QUAN SHI	HAN FU
HAN TIANQI	HAN WENJUN	HAN ZHENG-ZHONG
HANG HENG-RONG	HAO YUN-XIANG	HE MIAO-FU
HE XIANG-TAO	HSIANG YAN-YU	HU FU-XING
HU JING-YAO	HU NING-SHENG	HU WEN-RUI
HU ZHONG-WEI	HUA YING-MIN	HUANG BI-KUN
HUANG CHANG-CHUN	HUANG CHENG DR	HUANG JIE-HAO
HUANG KE-LIANG	HUANG KUN-YI	HUANG LIN
HUANG RUN-QIAN	HUANG TIANYI	HUANG TIE-QIN
HUANG YIN-LIANG	HUANG YONGWEI	HUANG YOU-RAN
JI HONG-QING	JI SHUCHEN DR	JIANG CHONG-GUO
JIANG DONG-RONG	JIANG SHI-YANG	JIANG SHUDING
JIANG XIAO-YUAN DR	JIANG YAO-TIAO	JIANG ZHAOJI
JIN BIAOREN DR	JIN SHEN-ZENG	JIN WEN-JING
KIMURA HIROSHI DR	LI CHUN-SHENG	LI DEPEI
LI DONG-MING	LI HONG-WEI	LI JING
LI NENG-YAO	LI QI-BIN	LI TING
LI TIPEI	LI WEI BAO	LI XIAO-QING
LI YUAN-JIE	LI ZHENG-XIN DR	LI ZHIAN DR
LI ZHIGANG	LI ZHI-FANG	LI ZHI-SEN
LI ZHONGYUAN	LI ZONG-WEI	LI ZONG-YUN
LIANG SHI-GUANG	LIANG ZHONG-HUAN	LIAO XINHAO DR
LIN YUANZHANG	LIU BAO-LIN	LIU CAIPIN
LIU CI-YUAN DR	LIU JINMING	LIU LIAO
LIU LIN	LIU LIN-ZHONG	LIU QINGYAO DR
LIU RU-LIANG	LIU XINPING PROF	LIU XUEFU
LIU YONG-ZHEN	LIU ZONGLI	LU BEN-KUI
LU CHUN-LIN	LU JU FU DR	LU RUWEI DR
LU TAN	LU YANG	LUO BAO-RONG
LUO DINGCHANG	LUO DING-JIANG	LUO SHI-FANG
LUO XIANHAN	MA ER	MA XING-YUAN
MA YU-QIAN	MAO WEI	MENG XINMIN
MIAO YONG-KUAN	MIAO YONG-RUI	MO JING-ER
NAN REN-DONG	PAN JUN-HUA	PAN LIANDE
PAN NING-BAO	PAN RONG-SHI	PENG QIU-HE
PENG YUN-LOU	QI GUAN RONG	QIAN BO-CHEN
QIAN SHAN-JIE	QIAN ZHI-HAN DR	QIAN ZHONG-YU

Country : CHINA PR (Follow on)

QIAO GUOJUN	QIN DAO	QIN SONG-NIAN
QIN ZHI-HAI	QIU PUZHANG ASS PROF	QIU YU-HAI
QU QIN-YUE	QUAN HEJUN	RONG JIAN-XIANG
SHEN CHANGJUN	SHEN KAIXIAN	SHEN LIANG-ZHAO
SHEN LONG-XIANG	SHEN PARN-AN	SHI GUANG-CHEN
SHI ZHONG-XIAN	SONG GUO-XUAN	SONG JIN-AN
SONG MU-TAO	SU BUMEI	SU DING-QIANG
SU HONG-JUN	SU QING-RUI DR	SU WAN-ZHEN
SUN JIN	SUN KAI	SUN YI-SUI
SUN YONGXIANG	TAN HUISONG	TANG YU-HUA
TONG FU	TONG YI	WAN LAI
WAN TONG-SHAN	WANG CHUAN-JIN	WANG DEYU
WANG DE-CHANG	WANG JIA-JI	WANG JIA-LONG
WANG JING-SHENG	WANG JING-XIU	WANG LAN-JUAN
WANG RENCHUAN	WANG SHOU-GUAN	WANG SHUI
WANG SHUNDE DR	WANG SI-CHAO	WANG YANAN
WANG YIMING	WANG ZHENG MING	WANG ZHEN-RU
WANG ZHEN-YI	WEI MINGZHI	WU HUAI-WEI
WU LIAN-DA	WU LIN-XIANG	WU MING-CHAN
WU NAILONG DR	WU SHENGYIN	WU SHOU-XIAN
WU XINJI	WU ZHIREN DR	XI ZE-ZONG
XIA JIONGYU	XIA XIAOYANG DR	XIA YI-FEI
XIAN DING-ZHANG	XIANG DELIN	XIANG SHOUPING
XIAO NAI-YUAN	XIE GUANG-ZHONG	XIE LIANGYUN
XIONG DA-RUN	XU AO-AO	XU BANG-XIN
XU JIA-YAN	XU JI-HONG DR	XU PEI-YUAN
XU PINXIN	XU TONG-QI	XU ZHENTAO
XU ZHI-CAI	YAN LIN-SHAN	YANG FUMIN
YANG JIAN	YANG LAN-TIAN	YANG SHI JIE
YANG TING-GAO	YAO BAO-AN	YAO JIN-XING
YE BINXUN	YE SHI-HUI	YE SHU-HUA
YE WENWEI	YI ZHAO-HUA	YIN JI-SHENG
YIN QI-FENG	YOU JIAN-QI	YOU JUNHAN
YUE ZENG-YUAN	ZENG QIN DR	ZHAI DI-SHENG
ZHAI ZAOCHENG	ZHANG BAI-RONG	ZHANG BIN
ZHANG FU JUN	ZHANG GUO-DONG	ZHANG HE-QI
ZHANG HUI	ZHANG JIA-LU	ZHANG JIA-XIANG
ZHANG JINTONG	ZHANG PEIYU	ZHANG XIU ZHONG
ZHANG YOUYI	ZHANG ZHEN-DA	ZHANG ZHEN-JIU
ZHAO GANG	ZHAO JUN-LIANG	ZHAO MING
ZHAO REN-YANG	ZHENG DA-WEI	ZHENG XUE-TANG
ZHENG YING	ZHENG YI-JIA	ZHOU BIFANG DR
ZHOU DAOQI	ZHOU HONG-NAN	ZHOU TI-JIAN
ZHOU YOU-YUAN	ZHOU ZHEN-PU	ZHU CI-SHENG
ZHU NENGHONG	ZHU SHI-CHANG	ZHU WEN-YAO
ZHU XINGFENG	ZHU YAOZHONG DR	ZHU YONG-HE
ZHUANG WEIFENG	ZOU HUI-CHENG	ZOU YI-XIN
ZOU ZHEN-LONG		

Country : CHINA R
_____

CHIUEH TZIHONG DR   CHOU CHIH-KANG DR   CHOU DEAN-YI DR
FU-SHONG KUO   HSIANG-KUANG TSENG   HUANG YINN-NIEN DR
HUANG YI-LONG DR   HWANG WOEI-YANN P PROF   LEE THYPHOON
LING CHIH-BING DR   NEE TSU-WEI DR   NG KIN-WANG
SHEN CHUN-SHAN   SUN WEI-HSIN DR   TING YEOU-TSWEN
TSAI CHANG-HSIEN DIRECTOR TSAO MO PROF   TSAY WEAN-SHUN DR
WU HSIN-HENG DR

Country : COLOMBIA
_____

ARIAS DE GREIFF J PROF   BRIEVA EDUARDO PROF

Country : CUBA
_____

ALVAREZ POMARES A O DR   BOYTEL JORGE DEL PINO DR   DOVAL JORGE PEREZ DR
GARCIA EDUARDO DEL POZO   RAMOS ISABEL FERRO DR   TABOADA RAMON RODRIGUEZ

Country : CZECHOSLOVAKIA

AMBROZ PAVEL DR	ANTALOVA ANNA	BICAK JIRI DR
BOUSKA JIRI DR	BUMBA VACLAV DR	BURSA MILAN DR
CEPLECHA ZDENEK DR	CHOCHOL DRAHOMIR	CHVOJKOVA WOYK E DR
FARNIK FRANTISEK	FISCHER STANISLAV DR	GRYGAR JIRI DR
HADRAVA PETR	HAJDUK ANTON DR	HAJDUKOVA MARIA
HANDLIROVA DAGMAR DR	HARMANEC PETR DR	HEFTY JAN DR
HEINZEL PETR DR	HEJNA LADISLAV DR	HEKELA JAN DR
HORAK TOMAS B DR	HORAK ZDENEK PROF DR	HORSKY JAN PROF
HRIC LADISLAV DR	HUDEC RENE DR	KAPISINSKY IGOR
KARLICKY MARIAN	KLECZEK JOSIP DR	KLOCOK LUBOMIR DR
KLOKOCNIK JAROSLAV DR	KLVANA MIROSLAV	KNOSKA STEFAN
KOPECKY MILOSLAV DR	KOTRC PAVEL	KOUBSKY PAVEL
KRESAK LUBOR DR	KRESAKOVA MARGITA DR	KRIVSKY LADISLAV DR
KRIZ SVATOPLUK DR	KULCAR LADISLAV DR	LETFUS VOJTECH DR
LOCHMAN JAN	MAYER PAVEL DR	MESZAROS ATTILA DR
MIKULASEK ZDENEK DR	MINAROVJECH MILAN	MRKOS ANTONIN DR
NEUZIL LUDEK DR	NOVOTNY JAN DR	NOVOTNY VACLAV
ONDERLICKA BEDRICH DR	OUHRABKA MIROSLAV DR	PADEVET VLADIMIR DR
PALOUS JAN DR	PALUS PAVEL DR	PAPOUSEK JIRI
PECINA PETR	PEREK LUBOS DR	PESEK RUDOLPH PROF
PITTICH EDUARD M DR	POKORNY ZDENEK DR	POLECHOVA PAVLA DR
PORUBCAN VLADIMIR DR	RAJCHL JAROSLAV DR	RUPRECHT JAROSLAV DR
RUSIN VOJTECH	RUZICKOVA-TOPOLOVA B DR	RYBANSKY MILAN
SEHNAL LADISLAV DR	SIDLICHOVSKY MILOS DR	SIMA ZDISLAV DR
SIMEK MILOS DR	SIROKY JAROMIR DR	SOBOTKA MICHAL DR
SOLC IVAN DR	SOLC MARTIN	STEFL VLADIMIR
STOHL JAN DR	SVESTKA JIRI DR	SVOREN JAN
SYKORA JULIUS DR	TLAMICHA ANTONIN DR	TREMKO JOZEF DR
VALNICEK BORIS DR	VANYSEK VLADIMIR PROF	VAVROVA ZDENKA DR
VETESNIK MIROSLAV DR	VONDRAK JAN DR	WEBROVA LUDMILA DR
ZACHAROV IGOR DR	ZIZNOVSKY JOZEF DR	ZVERKO JURAJ DR
ZVOLANKOVA JUDITA		

Country : DENMARK

ANDERSEN JOHANNES
CHRISTENSEN PER R DR
EINICKE OLE H DR
FRANDSEN SOEREN PROF
HANSEN LEIF LECTURER
HOEG ERIK DR
JONES JANET E DR
KJAERGAARD PER DR
KUSTAANHEIMO PAUL E PROF
MADSEN JES
NORDLUND AKE DR
OLSEN ERIK H
PEDERSEN HOLGER DR
PETHICK CHRISTOPHER J DR
SCHNOPPER HERBERT W DR
THEJLL PETER ANDREAS DR
WESTERGAARD NIELS J DR

BAERENTZEN JORN
CHRISTENSEN-DALSGAARD J
FABRICIUS CLAUS V
GAMMELGAARD PETER MAG SCI
HELMER LEIF
JOHANSEN KAREN T LEKTOR
JORGENSEN HENNING E PROF
KNUDE JENS KIRKESKOV DR
LILJE PER VIDAR BARTH DR
MOESGAARD KRISTIAN P
NORDSTROEM BIRGITTA DR
OLSEN FOGH H J
PEDERSEN OLAF PROF
REIZ ANDERS PROF
SOMMER-LARSEN JESPER DR
THOMSEN BJARNE B LECT
WIETH-KNUDSEN NIELS P DR

BRANDENBURG AXEL DR
CLAUSEN JENS VIGGO LEKTOR
FLORENTIN-NIELSEN RALPH
GYLDENKERNE KJELD DR
HELT BODIL E
JONES BERNARD J T DR
JORGENSEN UFFE GRAE DR
KRISTENSEN LEIF KAHL
LUND NIELS
NISSEN POUL E PROF
NORGAARD-NIELSEN HANS U
PAGEL BERNARD E J PROF
PETERSEN J O DR
RUDKJOBING MOGENS PROF
SORENSEN GUNNAR DR
ULFBECK OLE DR

Country : EGYPT

ABDEL HADY AHMED DR
ABULAZM MOHAMED SAMIR DR
AIAD A PROF
AWADALLA NABIL SHOUKRY DR
EL NAWAWAY M S DR
EL-SHAHAWY MOHAMAD PROF
GHOBROS ROSHDY AZER PROF
HASSAN S M PROF
KAMEL OSMAN M PROF
MIKHAIL FAHMY I PROF DR
OSMAN ANAS MOHAMED PROF
SOLIMAN MOHAMED AHMED DR
YOUSEF SHAHINAZ M DR

ABDELKAWI M ABUBAKR DR
AHMED IMAM IBRAHIM PROF
ALY M KHAIRY PROF
BAGHOS BALEGH B DR
EL-BASSUNY ALAWY A A DR
GALAL A A PROF
HAMDY M A M PROF
HELALI YHYA E DR
MAHMOUD FAROUK M A B DR
MIKHAL JOSEPH SIDKY PROF
SHALTOUT MESALAM A M PROF
TAWADROS MAHER JACOUB DR
YOUSSEF NAHED H PROF

ABOU-EL-ELLA MOHAMED S DR
AHMED MOSTAFA DR
AWAD MERVAT EL-SAID DR
BAKRY ABDEL AZIZ DR
EL-RAEY MOHAMED E PROF
GAMALELDIN ABDULLA I DR
HAMID S EL DIN PROF
ISSA ALI DR
MARIE M A DR
NAWAR SAMIR DR
SHARAF MOHAMED ADEL PROF
WANAS M I PROF

Country : ESTONIA

EELSALU HEINO DR
KIPPER TONU DR
SAAR ENN DR
VEISMANN UNO DR

EINASTO JAAN DR
NUGIS TIIT
SAPAR ARVED DR
VIIK TONU DR

ERGMA E V DR
PUSTYL'NIK IZOLD B DR
UUS UNDO DR

Country : FINLAND

DONNER KARL JOHAN	HAEMEEN ANTTILA KAARLE A	HAIKALA LAURI K
HUOVELIN JUHANI DR	JAAKKOLA TOIVO S	KARTTUNEN HANNU DR
KULTIMA JOHANNES	LUMME KARI A DR	MARKKANEN TAPIO DR
MATTILA KALEVI DR	MIKKOLA SEPPO DR	NIEMI AIMO
OJA HEIKKI DR	OTERMA LIISI PROF	PIIROLA VILPPU E DR
RAHUNEN TIMO	RAITALA JOUKO T	RIIHIMAA JORMA J DR
SILLANPAA AIMO KALEVI DR	TAKALO LEO O DR	TEERIKORPI VELI PEKKA DR
TIURI MARTTI PROF	TUOMINEN ILKKA V DR	URPO SEPPO I
VALTAOJA ESKO	VALTAOJA LEENA DR	VALTONEN MAURI J PROF
VILHU OSMI DR	ZHENG JIA-QING	

Country : FRANCE

ABOUDARHAM JEAN DR	ACKER AGNES PROF DR	AGRINIER BERNARD L MR
AIME C DR	ALECIAN GEORGES DR	ALLARD NICOLE DR
ALLEGRE CLAUDE PROF	ALLOIN DANIELLE DR	ALY JEAN JACQUES DR
AMARI TAHAR DR	ANDRILLAT HENRI L PROF	ANDRILLAT YVETTE DR
ARDUINI-MALINOVSKY M DR	ARIMOTO NOBUO DR	ARLOT JEAN-EUDES
ARNAUD JEAN PAUL	ARNAUD MONIQUE	ARTRU MARIE-CHRISTINE DR
ARTZNER GUY	ASSEO ESTELLE DR	ASSUS PIERRE DR
ATHANASSOULA EVANGELIE DR	AUBIER MONIQUE G DR	AUDOUZE JEAN PROF
AUGARDE RENEE DR	AURIERE MICHEL	AUVERGNE MICHEL
AVIGNON YVETTE DR	AZZOPARDI MARC DR	BACCHUS PIERRE PROF
BAGLIN ANNIE DR	BAIZE PAUL DR	BALKOWSKI-MAUGER CH DR
BALLEREAU DOMINIQUE DR	BALMINO GEORGES G DR	BALUTEAU JEAN-PAUL DR
BARANNE A DR	BARBIER-BROSSAT M DR	BARLIER FRANCOIS E DR
BARUCCI MARIA A DR	BAUDRY ALAIN DR	BEC-BORSENBERGER ANNICK
BEL NICOLE J DR	BELY-DUBAU FRANCOISE	BENAYDOUN JEAN-JACQUES DR
BENEST DANIEL DR	BENSAMMAR SLIMANE DR	BERGEAT JACQUES G DR
BERGER CHRISTIANE DR	BERGER JACQUES G DR	BERGERON JACQUELINE A DR
BERRUYER-DESIROTTE N DR	BERTAUX J L DR	BERTHOMIEU GABRIELLE DR
BERTOUT CLAUDE	BIENAYME OLIVIER DR	BIJAOUI ALBERT DR
BILLAUD GERARD J	BIRAUD FRANCOIS DR	BLAMONT JACQUES E PROF
BLAZIT ALAIN DR	BOCCHIA ROMEO DR	BOCKELEE-MORVAN DOMINIQUE
BOIGEY FRANCOISE	BOISCHOT ANDRE DR	BOISSE PATRICK DR
BOISSON CATHERINE DR	BOMMIER VERONIQUE DR	BONAZZOLA SILVANO DR
BONNEAU DANIEL	BONNET ROGER M DR	BONNET-BIDAUD J M DR
BORGNINO JULIEN DR	BOSMA ALBERT DR	BOTTINELLI LUCETTE DR
BOUCHER CLAUDE DR	BOUCHET FRANCOIS R DR	BOUGEARD MIREILLE L DR
BOUGERET J L DR	BOUIGUE R	BOULANGER FRANCOIS
BOULESTEIX JACQUES	BOULON JACQUES J DR	BOUVIER JEROME
BOYER RENE	BRAHIC ANDRE DR	BRETAGNON PIERRE DR
BRIOT DANIELLE DR	BRUNET JEAN-PIERRE DR	BRUSTON PAUL DR
BRYANT JOHN DR	BUECHER ALAIN DR	BURKHART CLAUDE DR
BURNAGE ROBERT	CALAME ODILE DR	CAMICHEL HENRI DR
CANAVAGGIA RENEE DR	CAPITAINE NICOLE	CAPLAN JAMES
CARQUILLAT JEAN-MICHEL	CASOLI FABIENNE DR	CASSE MICHEL DR
CASTETS ALAIN DR	CATALA CLAUDE DR	CAYREL DE STROBEL GIUSA
CAYREL ROGER DR	CAZENAVE ANNY DR	CELNIKIER LUDWIK DR
CESARSKY CATHERINE J DR	CESARSKY DIEGO A DR	CHALABAEV ALMAS DR
CHAMARAUX PIERRE DR	CHAMBE GILBERT	CHAPRONT JEAN DR
CHAPRONT-TOUZE MICHELLE	CHASSEFIERE ERIC	CHEVALIER CLAUDE DR
CHOLLET FERNAND DR	CHOPINET MARGUERITE DR	CLAIREMIDI JACQUES DR
COLIN JACQUES DR	COLLIN-SOUFFRIN SUZY DR	COMBES FRANCOISE DR
COMBES MICHEL	COMTE GEORGES DR	CONNES JANINE DR
CONNES PIERRE DR	CORNILLE MARGUERITE DR	COUPINOT GERARD DR
COURTES G PROF	COUTEAU PAUL PROF	COUTURIER PIERRE
COX PIERRE DR	CREZE MICHEL DR	CRIFO FRANCOISE DR
CROVISIER JACQUES	CRUVELLIER PAUL E DR	CUNY YVETTE J DR
DAIGNE GERARD	DAPPEN WERNER	DAVOUST EMMANUEL
DE BERGH CATHERINE DR	DE LA NOE JEROME DR	DE LAPPARENT-GURRIET V DR
DEBARBAT SUZANNE V DR	DEHARVENG JEAN-MICHEL DR	DEHARVENG LISE DR
DELABOUDINIERE J-P	DELACHE PHILIPPE J DR	DELANNOY JEAN DR

Country : FRANCE (Follow on)

DELHAYE JEAN PROF	DEMARCQ JEAN ING	DENISSE JEAN-FRANCOIS DR
DENNEFELD MICHEL	DESESQUELLES JEAN DR	DESPOIS DIDIER DR
DIVAN LUCIENNE DR	DOAZAN VERA DR	DOLEZ NOEL DR
DOLLFUS AUDOUIN PROF	DONAS JOSE DR	DOURNEAU GERARD DR
DOWNES DENNIS DR	DROSSART PIERRE DR	DUBAU JACQUES DR
DUBOIS MARC A	DUBOIS PASCAL DR	DUBOUT RENEE
DUCHESNE MAURICE DR	DUFAY MAURICE PROF	DUFLOT MARCELLE DR
DUMONT RENE DR	DUMONT SIMONE DR	DURIEZ LUC DR
DUROUCHOUX PHILIPPE	DURRET FLORENCE DR	DUVAL MARIE-FRANCE
DUVERT GILLES DR	EDELMAN COLETTE DR	EGRET DANIEL DR
EIDELSBERG MICHELE DR	ENCRENAZ PIERRE J DR	ENCRENAZ THERESE DR
FAUCHER PAUL DR	FAUROBERT-SCHOLL M DR	FEAUTRIER NICOLE DR
FEHRENBACH CHARLES PROF	FEISSEL MARTINE DR	FELENBOK PAUL DR
FERLET ROGER DR	FERNANDEZ JEAN-CLAUDE DR	FERRANDO PHILIPPE DR
FESTOU MICHEL C DR	FLOQUET MICHELE DR	FLORSCH ALPHONSE DR
FOING BERNARD H DR	FORT BERNARD P DR	FORVEILLE THIERRY DR
FOSSAT ERIC DR	FOY RENAUD DR	FRANCOIS PATRICK DR
FREIRE FERRERO RUBENS G	FRESNEAU ALAIN DR	FRIEDJUNG MICHAEL DR
FRINGANT ANNE-MARIE DR	FRISCH HELENE DR	FRISCH URIEL DR
FROESCHLE CHRISTIANE D DR	FROESCHLE CLAUDE DR	FROESCHLE MICHEL DR
GABRIEL ALAN H	GAIGNEBET JEAN DR	GALLOUET LOUIS DR
GAMBIS DANIEL DR	GARGAUD MURIEL DR	GARNIER ROBERT ING
GAUTIER DANIEL	GAY JEAN DR	GENOVA FRANCOISE DR
GEORGELIN YVON P DR	GEORGELIN YVONNE M DR	GERARD ERIC DR
GERBAL DANIEL DR	GERBALDI MICHELE DR	GERIN MARYVONNE DR
GILLET D DR	GIRAUD EDMOND	GOLDBACH CLAUDINE MME
GONCZI GEORGES	GORDON CHARLOTTE PROF	GORET PHILIPPE DR
GOUGUENHEIM LUCIENNE	GOUPIL MARIE JOSE	GOUTTEBROZE PIERRE DR
GRANVEAUD MICHEL	GREC GERARD	GRENIER ISABELLE DR
GRENIER SUZANNE	GREVE ALBERT DR	GRUDLER PIERRE
GRY CECILE DR	GUELIN MICHEL DR	GUERIN PIERRE DR
GUIBERT JEAN DR	GUINOT BERNARD R PROF	HALBWACHS JEAN LOUIS DR
HAMMER FRANCOIS DR	HARVEY CHRISTOPHER C DR	HAYLI AVRAM PROF
HECK ANDRE DR	HECQUET JOSETTE DR	HEIDMANN JEAN DR
HENDECOURT D' LOUIS DR	HENON MICHEL C DR	HENOUX JEAN-CLAUDE DR
HEUDIER JEAN-LOUIS DR	HEYVAERTS JEAN DR	HOANG BINH DY DR
HUA CHON TRUNG DR	HUBERT HENRI DR	HUBERT-DELPLACE A-M DR
IMBERT MAURICE DR	IRIGOYEN MAYLIS	ISRAEL GUY MARCEL DR
JACQ THIERRY	JACQUINOT PIERRE DR	JASCHEK CARLOS O R PROF
JASCHEK MERCEDES DR	JASNIEWICZ GERARD DR	JOLY FRANCOIS DR
JOLY MONIQUE	JOUBERT MARTINE	JOURNET ALAIN
JUNG JEAN DR	KAHANE CLAUDINE DR	KANDEL ROBERT S DR
KAZES ILYA DR	KLEIN KARL LUDWIG DR	KOCH-MIRAMOND LYDIE DR
KOURGANOFF VLADIMIR PROF	KOUTCHMY SERGE DR	KOVALEVSKY JEAN DR
KRIKORIAN RALPH DR	KUNTH DANIEL	LABEYRIE ANTOINE DR
LABEYRIE JACQUES DR	LACHIEZE-REY MARC	LACLARE F MR
LACROUTE PIERRE A PROF	LAFFINEUR MARIUS MR	LAFON JEAN-PIERRE J DR
LALLEMENT ROSINE DR	LAMY PHILIPPE DR	LANNES ANDRE DR
LANTOS PIERRE DR	LAQUES PIERRE DR	LASKAR JACQUES DR
LASOTA JEAN-PIERRE DR	LATOUR JEAN J	LAUNAY JEAN-MICHEL DR

Country : FRANCE (Follow on)
_____

LAURENT CLAUDINE DR	LAVAL ANNIE DR	LAZAREFF BERNARD DR
LE BORGNE JEAN FRANCOIS	LE BOURLOT JACQUES DR	LE CONTEL JEAN-MICHEL
LE DOURNEUF MARYVONNE	LE FEVRE OLIVIER DR	LE SQUEREN ANNE-MARIE DR
LEACH SYDNEY DR	LEBLANC YOLANDE DR	LEBRETON YVELINE DR
LEFEBVRE MICHEL DR	LEFEVRE JEAN DR	LEGER ALAIN DR
LELIEVRE GERARD DR	LEMAIRE JEAN-LOUIS DR	LEMAIRE PHILIPPE DR
LEMAITRE GERARD R DR	LENA PIERRE J PROF	LEORAT JACQUES DR
LEQUEUX JAMES DR	LEROY BERNARD DR	LEROY JEAN-LOUIS
LESAGE ALAIN DR	LESTRADE JEAN FRANCOIS DR	LEVASSEUR-REGOURD A C PR
LEVY JACQUES R DR	LORTET MARIE CLAIRE	LOSCO LUCETTE DR
LOUISE RAYMOND PROF	LOULERGUE MICHELLE DR	LUCAS ROBERT DR
LUMINET JEAN-PIERRE	LUNEL MADELEINE DR	MAGNAN CHRISTIAN DR
MAILLARD JEAN-PIERRE DR	MALHERBE JEAN MARIE DR	MANGENEY ANDRE DR
MARCELIN MICHEL	MARCHAL CHRISTIAN DR	MARIOTTI JEAN MARIE DR
MARTIN FRANCOIS DR	MARTIN NICOLE DR	MARTRES MARIE-JOSEPHE
MASNOU FRANCOISE DR	MASNOU J L DR	MATHEZ GUY
MAUCHERAT J DR	MAURICE ERIC N	MAURON NICOLAS DR
MAVRIDES STAMATIA DR	MAZURE ALAIN DR	MCCARROLL RONALD PROF
MEGESSIER CLAUDE DR	MEIN NICOLE DR	MEIN PIERRE
MEKARNIA DJAMEL DR	MELLIER YANNICK DR	MENEGUZZI MAURICE M DR
MENNESSIER MARIE-ODILE DR	MERAT PARVIZ	MERAT PARVIZ DR
MERCIER CLAUDE DR	MERLEAU-PONTY J PROF	MEYER CLAUDE DR
MEYER JEAN-PAUL DR	MIANES PIERRE DR	MICHARD RAYMOND DR
MIGNARD FRANCOIS DR	MILET BERNARD L DR	MILLET JEAN DR
MILLIARD BRUNO	MIRABEL IGOR FELIX DR	MOCHKOVITCH ROBERT DR
MONIN JEAN-LOUIS DR	MONNET GUY J DR	MONTES CARLOS DR
MONTMERLE THIERRY DR	MORANDO BRUNO L DR	MOREELS GUY DR
MOREL PIERRE JACQUES DR	MORRIS DAVID DR	MOUCHET MARTINE DR
MOURADIAN ZADIG M DR	MULHOLLAND J DERRAL DR	MULLER PAUL
MULLER RICHARD DR	NADAL ROBERT	NAHON FERNAND PROF
NGUYEN-QUANG RIEU DR	NITTMAN JOHANN	NOENS JACQUES-CLAIR DR
NOLLEZ GERARD DR	NOTTALE LAURENT	OBLAK EDOUARD
OCHSENBEIN FRANCOIS DR	OMNES ROLAND PROF	OMONT ALAIN PROF
PARCELIER PIERRE DR	PARISOT JEAN-PAUL	PATUREL GEORGES
PAUL JACQUES DR	PECKER JEAN-CLAUDE PROF	PEDERSEN BENT M DR
PEDOUSSAUT ANDRE	PELLAS PAUL DR	PELLET ANDRE
PELLETIER GUY DR PR	PEQUIGNOT DANIEL	PERAULT MICHEL
PERRIER CHRISTIAN DR	PERRIN JEAN MARIE DR	PERRIN MARIE-NOEL DR
PETIT GERARD DR	PETIT JEAN-MARC DR	PETON ALAIN DR
PETRINI DANIEL DR	PEYTURAUX ROGER H PROF	PHAM-VAN JACQUELINE MME
PICAT JEAN-PIERRE DR	PICK MONIQUE DR	PINEAU DES FORETS G DR
POQUERUSSE MICHEL	POULLE EMMANUEL PROF	POUQUET ANNICK DR
POYET JEAN-PIERRE DR	PRADERIE FRANCOISE DR	PRANTZOS NIKOS DR
PREVOT LOUIS DR	PREVOT-BURNICHON M L DR	PRIEUR JEAN-LOUIS DR
PROISY PAUL E DR	PROUST DOMINIQUE	PROVOST JANINE DR
PUGET JEAN-LOUP DR	QUERCI FRANCOIS R DR	QUERCI MONIQUE DR
RABBIA YVES DR	RAOULT ANTOINETTE DR	RAPAPORT MICHEL DR
RAYROLE JEAN R DR	REBEIROT EDITH DR	REEVES HUBERT PROF
REINISCH GILBERT DR	REQUIEME YVES DR	RIBES ELIZABETH DR
RIBES JEAN-CLAUDE DR	RICORT GILBERT DR	ROBILLOT JEAN-MAURICE DR

Country : FRANCE (Follow on)

ROBIN ANNIE C DR	ROBLEY R DR	ROCCA-VOLMERANGE BRIGITTE
ROQUES FRANCOISE DR	ROQUES SYLVIE DR	ROSCH JEAN PROF
ROSTAS FRANCOIS DR	ROTHENFLUG ROBERT DR	ROUAN DANIEL DR
ROUDIER THIERRY DR	ROUEFF EVELYNE M A DR	ROUSSEAU JEANINE DR
ROUSSEAU JEAN-MICHEL MR	ROZELOT JEAN P	RYTER CHARLES E DR
SAGNIER JEAN-LOUIS DR	SAHAL-BRECHOT SYLVIE DR	SAISSAC JOSEPH DR
SAMAIN DENYS DR	SAREYAN JEAN-PIERRE DR	SCHAEFFER RICHARD DR
SCHATZMAN EVRY PROF	SCHEIDECKER JEAN-PAUL DR	SCHMIEDER BRIGITTE DR
SCHNEIDER JEAN	SCHOLL HANS DR	SCHUMACHER GERARD DR
SEMEL MEIR DR	SERVAN BERNARD	SIBILLE FRANCOIS
SICARDY BRUNO DR	SIGNORE MONIQUE DR	SIMIEN FRANCOIS DR
SIMON GUY	SIMON JEAN-LOUIS MR	SIMON PAUL A DR
SIMONNEAU EDUARDO DR	SIVAN JEAN-PIERRE DR	SLEZAK ERIC DR
SOL HELENE DR	SORU-ESCAUT IRINA MRS	SOTIROVSKI PASCAL DR
SOUFFRIN PIERRE B DR	SOULIE GUY	SPIELFIEDEL ANNIE DR
SPITE FRANCOIS M DR	SPITE MONIQUE DR	STASINSKA GRAZYNA DR
STEENMAN-CLARK LOIS DR	STEFANOVITCH-GOMEZ A E DR	STEHLE CHANTAL DR
STEINBERG JEAN-LOUIS DR	STELLMACHER GOETZ	STELLMACHER IRENE DR
SYGNET JEAN FRANCOIS DR	TALON RAOUL DR	TARRAB IRENE
TATON RENE PROF	TCHANG-BRILLET LYDIA DR	TERRIEN JEAN
TERZAN AGOP DR	TEXEREAU JEAN M	THEVENIN FREDERIC DR
THIRY YVES R PROF	THOMAS CLAUDINE DR	TRAN MINH NGUYET DR
TRAN-MINH FRANCOISE DR	TRELLIS MICHEL DR	TROTTET GERARD DR
TRUONG-BACH	TULLY JOHN A DR	TURON C DR
VAKILI FARROKH DR	VALBOUSQUET ARMAND DR	VALIRON PIERRE DR
VALTIER JEAN-CLAUDE DR	VAN REGEMORTER HENRI DR	VAN'T VEER FRANS DR
VAN'T VEER-MENNERET CL DR	VAPILLON LOIC J DR	VAUCLAIR GERARD P DR
VAUCLAIR SYLVIE D DR	VAUGLIN ISABELLE DR	VEILLET CHRISTIAN
VELLI MARCO DR	VERDET JEAN-PIERRE DR	VERGEZ MADELEINE DR
VERON MARIE-PAULE DR	VERON PHILIPPE DR	VIAL JEAN-CLAUDE
VIALA YVES	VIALLEFOND FRANCOIS	VIDAL JEAN-LOUIS DR
VIDAL-MADJAR ALFRED DR	VIGIER JEAN-PIERRE DR	VILKKI ERKKI U
VILMER NICOLE DR	VITON MAURICE DR	VOLONTE SERGE DR
VU DUONG TUYEN DR	VUILLEMIN ANDRE DR	WALCH JEAN-JACQUES
WEIMER THEOPHILE P F DR	WENGER MARC	WENIGER SCHAME DR
WINK JOERN ERHARD DR	WLERICK GERARD DR	WOLTJER LODEWIJK PROF
ZAHN JEAN-PAUL DR	ZEIPPEN CLAUDE DR	ZOREC JEAN DR

Country : GEORGIA

BARTAYA R A DR	BORCHKHADZE TENGIZ M DR	CHKHIKVADZE IAKOB N
DOLIDZE MADONA V DR	DZHAPIASHVILI VICTOR P DR	DZIGVASHVILI R M DR
FISHKOVA LUISA M PROF	KALANDADZE N B DR	KHARADZE E K PROF
KHATISASHVILI ALFEZ SH DR	KHETSURIANI TSIALA S DR	KILADZE R I DR
KOGOSHVILI NATELA G	KUMSIASHVILY MZIA I DR	KUMSISHVILI J I DR
MAGNARADZE NINA G DR	SALUKVADZE G N DR	TOROSHLIDZE TEIMURAZ I DR

Country : GERMANY

ALBRECHT MIGUEL A DR	ALBRECHT RUDOLF DR	ALTENHOFF WILHELM J DR
ANZER ULRICH DR	APPENZELLER IMMO PROF	ARNAL MARCELO EDMUNDO DR
ARP HALTON DR	ASCHENBACH BERND PH D	AURASS HENRY DR
AXFORD W IAN PROF	BAADE DIETRICH DR	BAESSGEN MARTIN DR
BAHNER KLAUS DR	BALTHASAR HORST DR	BARROW COLIN H DR
BARWIG HEINZ	BASCHEK BODO PROF	BASTIAN ULRICH
BECK H G	BECK RAINER	BECKERS JACQUES M DR
BECKWITH STEVEN V W	BEHR ALFRED PROF EMERITUS	BENDER RALF DR
BENVENUTI PIERO DR	BERKHUIJSEN ELLY M DR	BEUERMANN KLAUS P PROF
BIEN REINHOLD DR	BIERMANN PETER L DR	BIRKLE KURT PH D
BLUM PETER PROF	BOEHME ANNELIES DR	BOEHME SIEGFRIED DR
BOEHNHARDT HERMANN DR	BOERNER GERHARD DR	BOERNGEN FREIMUT DR PH
BOHN HORST-ULRICH	BOHRMANN ALFRED PROF	BORGEEST ULF DR
BOSCHAN PETER DR	BRANDT PETER N	BRAUNINGER HEINRICH DR
BRAUNSFURTH EDWARD PH D	BREINHORST ROBERT A DR	BREITSCHWERDT DIETER DR
BREYSACHER JACQUES	BRINKMANN WOLFGANG	BROSCHE PETER PROF
BRUCH ALBERT	BRUZEK ANTON DR	BUECHNER JORG DR
BUES IRMELA D DR	BUTLER KEITH DR	CAMENZIND MAX DR
CARSENTY URI DR	CHE-BOHNENSTENGEL ANNE	CHINI ROLF
CRANE PHILIPPE	CULLUM MARTIN DR	DACHS JOACHIM PROF DR
DANZIGER I JOHN DR	DAUTCOURT G DR	DE BOER KLAAS SJOERDS DR
DE VEGT CH PROF DR	DEINZER W PROF DR	DETTMAR RALF-JURGEN DR
DEUBNER FRANZ-LUDWIG DR	DICK WOLFGANG DR	DIERCKSEN GEERD H F PH D
DOMKE HELMUT PH D	DORENWENDT KLAUS DR	DORSCHNER JOHANN DR
DRAPATZ SIEGFRIED W DR	DRECHSEL HORST DR	DUERBECK HILMAR W DR
DUSCHL WOLFGANG J DR	D'ODORICO SANDRO DR	EHLERS JURGEN PROF
EL EID MOUNIB DR	ELSAESSER HANS PROF	ELWERT GERHARD PROF
ENARD DANIEL DR	ENGELS DIETER DR	ENSLIN HEINZ DR
FAHR HANS JOERG PROF DR	FECHTIG HUGO DR	FEITZINGER JOHANNES PROF
FEIX GERHARD DR	FOSBURY ROBERT A E DR	FRICKE KLAUS DR
FRIED JOSEF WILHELM DR	FRIEDEMANN CHRISTIAN DR	FUCHS BURKHARD DR
FUERST ERNST DR	GAIL HANS-PETER DR	GEBLER KARL-HEINZ DR
GEFFERT MICHAEL DR	GEHREN THOMAS PH D	GENZEL REINHARD DR
GERHARD ORTWIN	GEYER EDWARD H PROF DR	GIESEKING FRANK DR
GLATZEL WOLFGANG DR	GLEISSBERG WOLFGANG PROF	GLIESE WIHELM PROF
GOETZ WOLDEMAR DR	GONDOLATSCH FRIEDRICH PRF	GRAHAM DAVID A
GRAHL BERND H DR	GREWING MICHAEL PROF	GROOTE DETLEF
GROSBOL PREBEN JOHNSON DR	GROSSMANN-DOERTH U DR	GROTEN ERWIN PROF
GROTH HANS G PROF DR	GRUEN EBERHARD DR	GUERTLER JOACHIN DR
GUESTEN ROLF	GUSSMANN E A DR	GUTCKE DIETRICH
HACHENBERG OTTO PROF DR	HAEFNER REINHOLD DR	HAERENDEL G DR
HAGEN HANS-JUERGEN DR	HAMMAN WOLF-RAINER	HAMMER REINER
HANUSCHIK REINHARD DR	HARRIS ALAN WILLIAM	HARTQUIST THOMAS WILBUR
HARVEY GALE A DR	HASER LEO N K DR	HASLAM C GLYN T DR
HAUBOLD HANS JOACHIM	HAUG EBERHARD DR	HAUG ULRICH PROF
HAUPT WOLFGANG DR	HAZLEHURST JOHN DR	HEBER ULRICH
HEFELE HERBERT PH D	HEINRICH INGE	HEMMLEB GERHARD DR
HEMPE KLAUS	HENKEL CHRISTIAN	HENNING THOMAS DR
HENSLER GERHARD PROF	HERMAN JACOBUS DR	HEROLD HEINZ
HERRMANN DIETER PROF DR	HILF EBERHARD R H PH D	HILLEBRANDT WOLFGANG PH D

Country : GERMANY (Follow on)

HIPPELEIN HANS H DR	HIRTH WOLFGANG ERNST PH D	HOEFLICH PETER DR
HOFFMANN MARTIN DR	HOFMANN WILFRIED DR	HOLWEGER HARTMUT PROF
HOPP ULRICH DR	HOREDT GEORG PAUL DR	HOUSE FRANKLIN C DR
HUCHTMEIER WALTER K DR	HUNGER KURT PROF	HUSFELD DIRK DR
IP WING-HUEN	ISSERSTEDT JOERG DR	JACKISCH GERHARD DR
JAEGER FRIEDRICH W PROF	JAHREISS HARTMUT DR	JENSCH A
JOCKERS KLAUS DR	JORDAN H L DR DIREKTOR	JUST ANDREAS DR
KAEHLER HELMUTH DR	KAFKA PETER	KALBERLA PETER
KANBACH GOTTFRIED DR	KASPER U DR	KAUFMANN JENS PETER DR
KAYSER RAINER DR	KEGEL WILHELM H PROF	KELLER HORST UWE DR
KELLER HUWE ULRICH DR	KENDZIORRA ECKHARD DR	KIPPENHAHN RUDOLF PROF
KIRK JOHN DR	KLARE GERHARD DR	KLEIN ULRICH
KNEER FRANZ DR	KNOLKER MICHAEL DR	KOEHLER H PROF DR
KOEHLER PETER	KOEPPEN JOACHIM DR	KOHOUTEK LUBOS DR
KOLLATSCHNY WOLFRAM DR	KOZASA TAKASHI	KRAEMER GERHARD DR
KRAUSE F DR	KRAUTTER JOACHIM DR	KREISEL E PROF
KREYSA ERNST	KRUEGEL ENDRIK DR	KRUEGER ALBRECHT DR
KUDRITZKI ROLF-PETER PH D	KUEHNE CHRISTOPH F	KUEHR HELMUT
KUENZEL HORST	KUNDT WOLFGANG PROF DR	KUNITZSCH PAUL PROF
KUNZE RUEDIGER DR	LABS DIETRICH PROF	LAMLA ERICH E DR
LANGER NORBERT DR	LEDERLE TRUDPERT DR	LEINERT CHRISTOPH DR
LEMKE DIETRICH DR	LENZEN RAINER DR	LESCH HAROLD DR
LIEBSCHER DIERCK-E DR	LOOSE HANS-HERMANN DR	LORENZ HILMAR
LUCY LEON B PROF	LUEST REIMAR PROF	LUEST RHEA DR
MANN GOTTFRIED DR	MARX SIEGFRIED DR	MATAS VLADIMIR R DR
MATERNE JUERGEN DR	MATTIG W PROF DR	MAUDER HORST PROF DR
MAUERSBERGER RAINER DR	MEBOLD ULRICH DR PROF	MEINIG MANFRED DR
MEISENHEIMER KLAUS DR	MERKLE FRITZ DR	METZ KLAUS DR
MEURS EVERT DR	MEYER FRIEDRICH DR	MEYER-HOFMEISTER E DR
MEZGER PETER G PROF	MOEHLMANN DIEDRICH	MOELLENHOFF CLAUS DR
MOORWOOD ALAN F M	MUELLER EWALD	MUENCH GUIDO PROF
MUNDT REINHARD DR	NECKEL HEINZ DR	NECKEL TH DR
NESIS ANASTASIOS DR	NEUKUM G DR	NOTNI P DR
OEGELMAN HAKKI B DR	OESTREICHER ROLAND	OETKEN L DR
OLEAK H DR	PAULDRACH ADALBERT W A DR	PAULINY TOTH IVAN K K DR
PETRI WINFRIED PROF DR	PFAU WERNER	PFENNIG HANS H DR
PFLUG KLAUS DR	PIERRE MARGUERITE DR	PILOWSKI K PROF DR
PINKAU K PROF	PITZ ECKHART DR	POHL ECKHARD DR
PORCAS RICHARD DR	PREUSS EUGEN DR	PRIESTER WOLFGANG PROF
PULS JOAHIM DR	RAEDLER K H DR	REFSDAL S PROF DR
REICH WOLFGANG	REIF KLAUS DR	REIMERS DIETER PROF
RICHTER G A DR	RICHTER JOHANNES PROF	RICHTLER TOM DR
RIEGER ERICH DR	RIPKEN HARTMUT W DR	RITTER HANS DR
ROEMER MAX PROF	ROESER HANS-PETER	ROESER HERMANN-JOSEF DR
ROESER SIEGFRIED DR	ROESSIGER SIEGFRIED DR	ROHLFS K PROF DR
ROSA DOROTHEA DR	ROSA MICHAEL RICHARD DR	ROTH-HOPPNER MARIA LUISE
RUBEN G PROF DR	RUDER HANNS	RUEDIGER GUNTHER DR
RUPPRECHT GERO DR	SCHAEFER GERHARD DR	SCHAIFERS KARL DR
SCHEFFLER HELMUT PROF	SCHILBACH ELENA DR	SCHILLER KARL PROF DR
SCHINDLER KARL PROF DR	SCHLICKEISER REINHARD DR	SCHLOSSER WOLFHARD PROF

Country : GERMANY (Follow on)

SCHLUETER A PROF DR	SCHLUETER DIETER PROF	SCHMADEL LUTZ D DR
SCHMAHL GUENTER PROF	SCHMEIDLER F PROF DR	SCHMIDT H U DR
SCHMIDT HANS PROF	SCHMIDT K H DR	SCHMIDT THOMAS DR
SCHMIDT WOLFGANG DR	SCHMIDT-KALER TH PROF	SCHMID-BURGK J DR PROF
SCHMITT DIETER DR	SCHNEIDER HARTMUT DR	SCHNEIDER PETER DR
SCHNUR GERHARD F O	SCHOEFFEL EBERHARD F DR	SCHOEMBS ROLF DR
SCHOENBERNER DETLEF PROF	SCHOENEICH W DR	SCHOENFELDER VOLKER DR
SCHOLZ GERHARD DR	SCHOLZ M PROF	SCHROEDER KLAUS PETER DR
SCHROEDER ROLF DR	SCHROETER EGON H PROF	SCHRUEFER EBERHARD DR
SCHUBART JOACHIM DR	SCHUESSLER MANFRED DR	SCHULTZ G V DR
SCHULZ HARTMUT DR	SCHULZ ROLF ANDREAS	SCHUMANN JOERG DIETER DR
SCHWAN HEINER DR	SCHWARTZ ROLF PH D	SEDLMAYER ERWIN DR
SEGGEWISS WILHELM PROF	SEITTER WALTRAUT C PROF	SENGBUSCH KURT V DR
SERAFIN RICHARD AUGUST	SETTI GIANCARLO PROF	SHARP CHRISTOPHER DR
SHAVER PETER A DR	SHERWOOD WILLIAM A DR	SIEBER WOLFGANG PH D
SIMON KLAUS PETER	SINGH H P	SNIJDERS MATTHEUS A J DR
SOFFEL MICHAEL DR	SOLF JOSEF DR	SOLLAZZO CLAUDIO
SPRUIT HENK C DR	SPURZEM RAINER DR	STAHL OTMAR RICHARD DR
STANGE LOTHAR	STAUBERT RUDIGER PROF DR	STAUDE HANS JAKOB PH D
STAUDE JUERGEN DR	STEFFEN MATTHIAS DR	STEFL STANISLAV DR
STEINERT KLAUS GUENTER DR	STEINLE HELMUT DR	STIX MICHAEL DR
STRASSL HANS L PROF	STROHMEIER WOLFGANG PROF	STUMPFF PETER PROF DR
STUTZI JUERGEN DR	SZOSTAK ROLAND DR	TACCONI LINDA J DR
TARENGHI MASSIMO DR	TENORIO-TAGLE G DR	THIELHEIM KLAUS O DR
THOMAS HANS-CHRISTOPH DR	TRAVING GERHARD PROF	TREDER H J PROF DR
TREFFTZ ELEONORE E DR	TREUMANN RUDOLF A DR	TRUEMPER JOACHIM PROF
TSCHARNUTER WERNER M DR	TUEG HELMUT DR	ULMSCHNEIDER PETER PROF
ULRICH BRUCE T PROF	ULRICH MARIE-HELENE D DR	UNSOELD ALBRECHT PROF
URBARZ H DR	VAN DER LAAN H PROF DR	VAN MOORSEL GUSTAAF DR
VOELK HEINRICH J PROF	VOIGT HANS H PROF	VOLLAND H DR
VON BORZESZKOWSKI H H DR	VON DER HEIDE JOHANN DR	VON HOERNER SEBASTIAN DR
VON WEIZSAECKER C F PROF	WALMSLEY C MALCOLM DR	WALTER HANS G DR
WALTER KURT PROF DR	WAMPLER E JOSEPH PROF	WARMELS REIN HERM DR
WATTENBERG D PROF	WEHRSE RAINER DR	WEIDEMANN VOLKER PROF
WEIGELT GERD DR	WEIGERT ALFRED PROF	WEISS ACHIM DR
WELLMANN PETER PROF DR	WENDKER HEINRICH J PROF	WENZEL W DR
WEST RICHARD M DR	WIEHR EBERHARD DR	WIELEBINSKI RICHARD PROF
WIELEN ROLAND PROF DR	WILSON P DR	WILSON RAYMOND N DR
WILSON THOMAS L DR	WINNEWISSER GISBERT DR	WITTMANN AXEL D DR
WITZEL ARNO DR	WOEHL HUBERTUS DR	WOLF BERNHARD PH D
WOLF RAINER E A DR	WUNNER GUENTER	YORKE HAROLD W DR
ZACHARIAS NORBERT DR	ZEILINGER WERNER W DR	ZEKL HANS WILHELM
ZERULL REINER H DR	ZIMMERMANN HELMUT DR	ZINNECKER HANS

Country : GREECE

ABBOTT WILLIAM N DR	ALISSANDRAKIS C PH D	ANTONACOPOULOS GREG PROF
ANTONOPOULOU E DR	ARABELOS DIMITRIOS DR	ARGYRAKOS JEAN PROF DR
ASTERIADIS GEORGIOS DR	AVGOLOUPIS STAVROS DR	BANOS COSMAS J DR
BANOS GEORGE J PROF	BARBANIS BASIL PROF	BELLAS-VELIDIS IOANNIS DR
BOZIS GEORGE PROF	CARANICOLAS NICHOLAS DR.	CAROUBALOS C A PROF
CONTADAKIS MICHAEL E DR	CONTOPOULOS GEORGE PROF	DANEZIS EMMANUEL DR
DAPERGOLAS A DR	DARA HELEN DR	DELIYANNIS JEAN DR
DIALETIS DIMITRIS DR	DIONYSIOU DEMETRIOS PROF	GEROYANNIS VASSILIS S DR
GOUDAS CONSTANTINE L PROF	GOUDIS CHRISTOS D PROF	HADJIDEMETRIOU JOHN D
HANTZIOS PANAYIOTIS DR	HIOTELIS NICOLAOS DR	KATSIS DEMETRIUS DR
KAZANTZIS PANAYOTIS DR	KOKKOTAS KONSTANTINOS DR	KONTIZAS EVANGELOS DR
KONTIZAS MARY DR	KYLAFIS NIKOLAOS D DR	LASKARIDES PAUL G ASSPROF
LIRITZIS IOANNIS DR	MACRIS CONSTANTIN J PROF	MARKELLOS VASSILIS V DR
MAVRAGANIS A G PROF	MAVRIDIS L N PROF	MAVROMICHALAKI HELEN DR
MERZANIDES CONSTANTINOS	MOUSSAS XENOPHON PH D	MOUTSOULAS MICHAEL PROF
NIARCHOS PANAYIOTIS PH D	NICOLAIDIS EFTHYMIOS DR	PAPAELIAS PHILIP DR
PAPATHANASOGLOU D DR	PAPAYANNOPOULOS TH DR	PERSIDES SOTIRIOS C
PETROPOULOS BASIL CH DR	PINOTSIS ANTONIS D DR	POLYMILIS CHRONIS DR
POULAKOS CONSTANTINE DR	PREKA-PAPADEMA P DR	PROKAKIS THEODORE J DR
ROVITHIS PETER DR	ROVITHIS-LIVANIOU HELEN	SARRIS ELEFTHERIOS PH D
SARRIS EMMANUEL T PH D	SEIMENIS JOHN DR	SEIRADAKIS JOHN HUGH DR
SPITHAS ELEFTERIOS N DR	SPYROU NICOLAOS PROF	STATHOPOULOU MARIA DR
SVOLOPOULOS SOTIRIOS PROF	TERZIDES CHARALAMBOS DR	THEODOSSIOU EFSTRATIOS DR
TRITAKIS BASIL P DR	TSAMPARLIS MICHAEL DR	TSIKOUDI VASSILIKI PH D
TSINGANOS KANARIS DR	TSIOUMIS ALEXANDROS DR	TSIROPOULA GEORGIA DR
VARDAVAS ILIAS MIHAIL	VARVOGLIS H DR	VEIS GEORGE PH D
VENTURA JOSEPH DR	VLACHOS DEMETRIUS G PROF	VLAHOS LOUKAS DR
VOGLIS NIKOS DR	XANTHAKIS JOHN N PROF	XANTHOPOULOS B C DR
XIRADAKI EVANGELIA DR	ZACHARIADIS THEODOSIOS DR	ZACHILAS LOUKAS DR
ZAFIROPOULOS BASIL DR	ZIKIDES MICHAEL C DR	

Country : HUNGARY

ALMAR IVAN PROF	BALAZS LAJOS G DR	BARCZA SZABOLCS DR
BARLAI KATALIN DR	CSADA IMRE K DR	DEZSO LORANT PROF
ERDI B DR	FEJES ISTVAN DR	GERLEI OTTO
GRANDPIERRE ATTILA DR	HORVATH ANDRAS DR	ILL MARTON J DR
ILLES ALMAR ERZSEBET DR	JANKOVICS ISTVAN DR	KALMAN BELA DR
KANYO SANDOR DR	KELEMEN JANOS	KOLLATH ZOLTAN DR
KOVACS AGNES DR	KOVACS GEZA DR	KUN MARIA DR
LOVAS MIKLOS	MARIK MIKLOS DR	MARX GYORGY PROF
OLAH KATALIN DR	PAPARO MARGIT DR	PATKOS LASZLO DR
SZABADOS LASZLO PH D	SZALAY ALEX DR	SZATMARY KAROLY DR
SZECSENYI-NAGY GABOR DR	SZEGO KAROLY DR	SZEIDL BELA DR
TOTH IMRE DR	VERES FERENC	ZSOLDOS ENDRE DR

Country : ICELAND

GUDMUNDSSON EINAR H	JULIUSSON EINAR DR	SAEMUNDSON THORSTEINN

Country : INDIA

ABHYANKAR KRISHNA D PROF	ACHARYA BANNANJE S DR	AGRAWAL P C DR
AHMAD FAROOQ DR	ALLADIN SALEH MOHAMED DR	ALURKAR S K DR
AMBASTHA A K DR	ANANTHAKRISHNAN S	ANSARI S M RAZAULLAH PROF
ANTIA H M DR	APPARAO K M V DR	ASHOK N M DR
AULUCK FAQIR CHAND PROF	BABU G S D	BAGARE S P DR
BALASUBRAMANIAN V DR	BALLABH G M DR	BANDYOPADHYAY A PROF
BANERJI SRIRANJAN DR	BANHATTI DILIP GOPAL DR	BASU BAIDYANATH PROF
BASU DIPAK DR	BHANDARI N DR	BHANDARI RAJENDRA DR
BHAT CHAMAN LAL DR	BHAT NARAYANA P DR	BHATIA PREM K DR
BHATIA R K DR	BHATIA V B DR	BHATNAGAR ARVIND DR
BHATNAGAR ASHOK KUMAR	BHATNAGAR K B DR	BHATT H C DR
BHATTACHARYA DIPANKAR	BHATTACHARYYA J C PROF	BHATTACHARYYA TARA DR
BHONSLE RAJARAM V PROF	BISWAS SUKUMAR DR	BODDAPATI G ANANDARAO DR
BONDAL KRISHNA RAJ DR	BUTI BIMLA PROF	CHAKRABARTI SANDIP K DR
CHAKRABORTY DEO K DR	CHANDRA SURESH DR	CHANDRASEKHAR T DR
CHAUBEY UMA SHANKAR DR	CHITRE SHASHIKUMAR M DR	COWSIK RAMANATH
DADHICH NARESH DR	DAMLE S V DR	DAS MRINAL KANTI
DAS P K DR	DATTA BHASKAR DR	DEGAONKAR S S DR
DESAI JYOTINDRA N	DESHPANDE AVINASH	DESHPANDE M R DR
DHAWAN VIVEK DR	DUORAH HIRA LAL DR	DURGAPRASAD N DR
DWARAKANATH K S	DWIVEDI BHOLA NATH DR	GAUR V P
GHOSH KAJAL KUMAR DR	GHOSH P DR	GHOSH S K DR
GIRIDHAR SUNETRA DR	GOKHALE MORESHWAR HARI PR	GOPALA RAO U V MR
GOSWAMI J N DR	GOYAL A N DR	GUPTA SUNIL K DR
GURM HARDEV S PROF	HASAN SAIYID STRAJUL	IYENGAR K V K PROF
IYER B R DR	JAIN RAJMAL DR	JAIN SURENDRA DR
JAYARAJAN A P MR	JOG CHANDA J DR	JOSHI G C DR
JOSHI MOHAN N PROF	JOSHI SURESH CHANDRA DR	JOSHI U C DR
KANDPAL CHANDRA D	KAPAHI VIJAY K DR	KAPOOR RAMESH CHANDER

Country : INDIA (Follow on)

KARANDIKAR R V PROF	KASTURIRANGAN K DR	KEMBHAVI AJIT K
KILAMBI G C DR	KOCHHAR R K DR	KOTHARI D S DR
KRISHNA GOPAL	KRISHNA SWAMY K S DR	KRISHNAMOHAN S DR
KRISHNAN THIRUVENKATA MR	KULKARNI PRABHAKAR V PROF	KULKARNI VASANT K DR
KUSHWAHA R S PROF	LAHIRI N C	LAL DEVENDRA
MAHRA H S DR	MALLIK D C V DR	MANCHANDA R K DR
MARAR T M K	MATHUR B S DR	MITRA A P DR
MOHAN CHANDER DR	MOHAN VIJAY DR	NARANAN S PROF
NARASIMHA DELAMPADY DR	NARAYANA J V	NARLIKAR JAYANT V PROF
NITYANANDA R DR	PADALIA T D DR	PADMANABHAN T DR
PANDE GIRISH CHANDRA PROF	PANDE MAHESH CHANDRA DR	PANDEY A K
PANDEY S K	PANDEY UMA SHANKAR DR	PARTHASARATHY M DR
PATI A K	PERAIAH ANNAMANENI DR	PRABHAKARAN NAYAR S R DR
PRABHU TUSHAR P	PRASANNA A R DR	PRATAP R DR
PUNETHA LALIT MOHAN DR	RADHAKRISHNAN V PROF	RAGHAVAN NIRUPAMA DR
RAJAMOHAN R DR	RAJU P K DR	RAKSHIT H PROF
RAM SAGAR DR	RAMADURAI SOURIRAJA DR	RAMAMURTHY SWAMINATHAN
RAMANA MURTHY P V DR	RANA NARAYAN CHANDRA DR	RAO A PRAMESH DR
RAO M N DR	RAO N KAMESWARA	RAO P VIVEKANANDA DR
RAO RAMACHANDRA V PROF	RAUTELA B S DR	RAY ALAK DR
RAY CHOUDHURI ARNAB DR	RAYCHAUDHURI AMALKUMAR DR	RAZDAN HIRALAL
RENGARAJAN T N DR	SAHA SWAPAN KUMAR DR	SAIKIA DHRUBA JYOTI DR
SANWAL BASANT BALLABH DR	SANWAL N B DR	SAPRE A K DR
SARMA M B K PROF	SARMA N V G PROF	SASTRI HANUMATH J DR
SASTRY CH V	SASTRY SHANKARA K	SAXENA A K DR
SAXENA P P DR	SCARIA K K DR	SEN S N DR
SHAH GHANSHYAM A DR	SHEVGAONKAR R K DR	SHUKLA K
SHUKRE C S DR	SINGH JAGDEV DR	SINGH KULINDER PAL DR
SINHA K DR	SINHA RAMESHWAR P	SINVHAL SHAMBHU DAYAL DR
SIVARAM C DR	SIVARAMAN K R DR	SREEKANTAN B V DR
SRINIVASAN G	SRIVASTAVA J B DR	SRIVASTAVA RAM KUMAR DR
STEPHENS S A DR	SUBRAHMANYA C R	SUBRAHMANYAM P V DR
SUBRAMANIAN KANDASWAMY DR	SWARUP GOVIND PROF	TALWAR SATYA P DR
TANDON JAGDISH NARAIN DR	TANDON S N PROF	TARAFDAR SHANKAR P DR
THAKUR RATNA KUMAR DR	TONWAR SURESH C PROF	TREHAN SURINDAR K PROF
TRIPATHI B M DR	UDAYA SHANKAR N DR	VAIDYA P C PROF
VARDYA M S DR	VARMA RAM KUMAR PROF	VASU-MALLIK SUSHMA DR
VATS HARI OM DR	VELUSAMY T DR	VENUGOPAL V R DR
VERMA R P DR	VERMA SATYA DEV DR	VERMA V K DR
VINOD S KRISHAN MRS DR	VISHVESHWARA C V PROF	VIVEKANAND M DR
VIVEKANANDA RAO		

Country : INDONESIA

DAWANAS DJONI N DR	HIDAJAT BAMBANG PROF DR	IBRAHIM JORGA
RADIMAN IRATIUS	RAHARTO MOEDJI	SIREGAR SURYADI DR
SUKARTADIREDJA DARSA	SUTANTYO WINARDI	WIRAMIHARDJA SUHARDJA DR

Country : IRAN

ADJABSHIRIZADEH ALI   ARDEBILI M REZA DR   GHEIDARI S NASSIRI DR
KALAFI MANOUCHER   KHALESSEH BAHRAM DR   KIASATPOOR AHMAD PROF
MALAKPUR IRADJ DR   SOBOUTI YOUSEF PROF   TEHERANY D

Country : IRAQ

ABDULLA SHAKER ABDUL AZIZ   AL-NAIMY HAMID M K DR   AL-SABTI ABDUL ADIM DR
JABBAR SABEH RHAMAN   JABIR NIAMA LAFTA   KADOURI TALIB HADI
SADIK AZIZ R DR   YOUNIS SAAD M

Country : IRELAND

CARROLL P KEVIN PROF   CAWLEY MICHAEL DR   DRURY LUKE O'CONNOR DR
ELLIOTT IAN DR   FAHY EDWARD F PROF   FEGAN DAVID J DR
FLORIDES PETROS S PROF   GRIMLEY PETER DR   HAYWOOD J
HOEY MICHAEL J DR   KENNEDY EUGENE T   KIANG TAO PROF
MCBREEN BRIAN PHILIP DR   MCCREA J DERMOTT   MCKEITH NIALL ENDA DR
MCKENNA LAWLOR SUSAN   O'CONNOR SEAMUS L DR   O'MONGAIN EON
O'SULLIVAN DENIS F   PORTER NEIL A PROF   RAY THOMAS P
REDFERN MICHAEL R DR   RUDDY VINCENT P DR   RUSSELL STEPHEN DR
WAYMAN PATRICK A PROF   WRIXON GERARD T DR

Country : ISRAEL

BARKAT ZALMAN PROF   BEKENSTEIN JACOB D DR   BRAUN ARIE
BROSCH NOAH DR   CUPERMAN SAMI PROF   DEKEL AVISHAI
ERSHKOVICH ALEXANDER PROF   EVIATAR AHARON PROF   FINZI ARRIGO DR
GLASNER SHIMON AMI   GOLDMAN ITZHAK DR   GOLDSMITH S DR
GRADSZTAJN ELI DR   HARPAZ AMOS DR   HORWITZ GERALD PROF
IBBETSON PETER AARON DR   JOSEPH J H DR   KATZ JOSEPH DR
KOVETZ ATTAY PROF   KOZLOVSKY B Z DR   LEIBOWITZ ELIA M DR
LIVIO MARIO PROF   MAZEH TSEVI DR   MEERSON BARUCH DR
MEIDAV MEIR DR   MEKLER YURI PROF   NEEMAN YUVAL PROF
NEMIROFF ROBERT DR   NETZER HAGAI PROF   OHRING GEORGE PROF
PEKERIS CHAIM LEIB PROF   PRIALNIK-KOVETZ DINA DR   RAKAVY GIDEON PROF
REGEV ODED DR   REPHAELI YOEL DR   SACK NOAM DR
SADEH D DR   SEGALUVITZ ALEXANDER DR   SHAVIV GIORA PROF
STEINITZ RAPHAEL PROF   TUCHMAN YTZHAK   VAGER ZEEV DR
VIDAL NISSIM V DR   YEIVIN Y PROF

Country : ITALY

ABRAMI ALBERTO PROF	ABRAMOWICZ MAREK DR	AIELLO SANTI DR
ALTAMORE ALDO	ANDREANI PAOLA MICHELA DR	ANGELETTI LUCIO DR
ANILE ANGELO M	ANTONELLO ELIO	ANTONUCCI ESTER DR
AURIEMMA GIULIO DR	BADIALI MASSIMO	BADOLATI ENNIO
BAFFA CARLO DR	BALDINELLI LUIGI DR	BALLARIO M C PROF
BANDIERA RINO DR	BANNI ALDO DR	BARATTA GIOVANNI BATTISTA
BARBARO G DR	BARBERIS BRUNO	BARBIERI CESARE PROF
BARBON ROBERTO PROF	BARLETTI RAFFAELE ENG	BARONE FABRIZIO DR
BARTOLINI CORRADO	BATTINELLI PAOLO DR	BATTISTINI PIERLUIGI DR
BEDOGNI ROBERTO	BELVEDERE GAETANO DR	BENACCHIO LEOPOLDO
BENDINELLI ORAZIO	BERNACCA P L PROF	BERTELLI GIANPAOLO DR
BERTIN GIUSEPPE PROF	BERTOLA FRANCESCO PROF	BETTONI DANIELA DR
BIANCHI LUCIANA	BIANCHINI ANTONIO DR	BIGNAMI GIOVANNI F
BLANCO CARLO DR	BODO GIANLUIGI DR	BONACCINI DOMENICO DR
BONANNO GIOVANNI DR	BONIFAZI ANGELO DR	BONO GIUSEPPE DR
BONOLI FABRIZIO	BONOMETTO SILVIO A DR	BRACCESI ALESSANDRO PROF
BRINI DOMENICO PROF	BROCATO ENZO DR	BROGLIA PIETRO DR
BUONANNO ROBERTO	BUSON LUCIO M DR	BUSSO MAURIZIO
CACCIANI ALESSANDRO PROF	CACCIARI CARLA DR	CACCIN BRUNO
CALOI VITTORIA DR	CALVANI MASSIMO DR	CANTU ALBERTO M DR
CAPACCIOLI MASSIMO DR	CAPACCIONI FABRIZIO DR	CAPPELLARO ENRICO DR
CAPRIA MARIA TERESA DR	CAPRIOLI GIUSEPPE PROF	CAPUTO FILIPPINA DR
CAPUZZO DOLCETTA ROBERTO	CARDINI DANIELA DR	CARPINO MARIO DR
CARUSI ANDREA	CASSATELLA ANGELO DR	CASTELLANI VITTORIO PROF
CASTELLI FIORELLA DR	CATALANO FRANCESCO A DR	CATALANO SANTO DR
CATARZI MARCO DR	CAVALIERE ALFONSO G PROF	CAVALLINI FABIO
CAZZOLA PAOLO DR	CELLINO ALBERTO DR	CEPPATELLI GUIDO DR
CERRONI PRISCILLA DR	CERRUTI-SOLA MONICA	CESTER BRUNO PROF
CEVOLANI GIORDANO	CHINCARINI GUIDO L DR	CHIOSI CESARE S DR
CHIUDERI CLAUDIO PROF	CHIUDERI-DRAGO FRANCA PR	CHIUMIENTO GIUSEPPE
CHLISTOVSKY FRANCA DR	CIATTI FRANCO DR	COLOMBO G PROF DR
COLUZZI REGINA DR	COMORETTO GIOVANNI	CONCONI PAOLO DR
CORADINI ANGIOLETTA	CORBELLI EDVIGE DR	COSMOVICI BATALLI C DR
COSTA ENRICO	CRISTALDI SALVATORE DR	CRISTIANI STEFANO DR
CRIVELLARI LUCIO	CUGUSI LEONINO DR	CURIR ANNA
CUTISPOTO GIUSEPPE DR	DALLAPORTA N PROF	DALL'OGLIO GIORGIO DR
DANESE LUIGI DR	DE BIASE GIUSEPPE A DR	DE FELICE FERNANDO DR
DE RUITER HANS RUDOLF	DE SABBATA V PROF DR	DE SANCTIS GIOVANNI
DE ZOTTI GIANFRANCO DR	DELLA VALLE MASSIMO DR	DELLI SANTI SAVERIO
DI COCCO GUIDO	DI FAZIO ALBERTO	DI MARTINO MARIO
DI SEREGO ALIGHIERI S DR	DI TULLIO GRAZIELLA DR	D'ANTONA FRANCESCA DR
EINAUDI GIORGIO	ELLIS GEORGE F R PROF	EMANUELE ALESSANDRO DR
FACONDI SILVIA ROSA DR	FALCHI AMBRETTA	FALCIANI ROBERTO DR
FALOMO RENATO DR	FANTI CARLA GIOVANNINI	FANTI ROBERTO
FARAGGIANA ROSANNA PROF	FARINELLA PAOLO DR	FEDERICI LUCIANA
FELLI MARCELLO DR	FERETTI LUIGINA	FERLUGA STENO DR
FERRARI ATTILIO DR	FERRARI TONIOLO MARCO	FERRERI WALTER
FERRINI FEDERICO	FICARRA ANTONINO DR	FOCARDI PAOLA DR
FODERA SERIIO GIORGIA DR	FOFI MASSIMO DR	FORTI GIUSEPPE DR
FORTINI TERESA DR	FRACASTORO MARIO G PROF	FRANCESCHINI ALBERTO

Country : ITALY (Follow on)

FRANCHINI MARIAGRAZIA DR	FULCHIGNONI MARCELLO PROF	FUSCO-FEMIANO ROBERTO
FUSI PECCI FLAVIO	GALEOTTI PIERO PROF	GALLETTA GIUSEPPE PROF
GALLETTO DIONIGI	GALLIANO PIER GIORGIO	GALLIANO PIER GIORGIO
GALLINO ROBERTO	GAUDENZI SILVIA DR	GAVAZZI GIUSEPPE DR
GEMMO ALESSANDRA DR	GIACHETTI RICCARDO PROF	GIALLONGO EMANUELE DR
GIANNONE PIETRO PROF	GIANNUZZI MARIA A DR	GIOVANARDI CARLO
GIOVANNELLI FRANCO DR	GIOVANNINI GABRIELE	GIURICIN GIULIANO
GODOLI GIOVANNI PROF	GOMEZ MARIA THERESA DR	GRATTON R G DR
GREGGIO LAURA DR	GREGORINI LORETTA	GRUBISSICH C PROF DR
GRUEFF GAVRIL DR	GUARNIERI ADRIANO DR	GUERRERO GIANANTONIO DR
HACK MARGHERITA PROF	HUANG SONG-NIAN DR	HUNT LESLIE DR
IIJIMA TAKASHI DR	KRANJC ALDO DR	LA PADULA CESARE
LAI SEBASTIANA	LANDI DEGL'INNOCENTI E PR	LANDI DEGL'INNOCENTI M
LANDINI MASSIMO PROF	LANDOLFI MARCO	LARI CARLO DR
LATTANZI MARIO G	LESCHIUTTA S PROF	LISEAU RENE DR
LISI FRANCO DR	LUCCHIN FRANCESCO	MACCAGNI DARIO
MACERONI CARLA	MAFFEI PAOLO PROF	MAGAZZU ANTONIO DR
MAGGIO ANTONIO DR	MAGNI GIANFRANCO	MALAGNINI MARIA LUCIA
MAMMANO AUGUSTO DR	MANARA ALESSANDRO A DR	MANCUSO SANTI PROF
MANDOLESI NAZZARENO	MANNINO GIUSEPPE PROF	MANTEGAZZA LUCIANO
MANTOVANI FRANCO	MARANO BRUNO	MARASCHI LAURA DR
MARDIROSSIAN FABIO	MARGONI RINO	MARILLI ETTORE DR
MARMOLINO CIRO	MARTINI ALDO DR	MASANI A PROF
MASSAGLIA SILVANO	MATTEUCCI FRANCESCA DR	MAZZITELLI ITALO DR
MAZZONI MASSIMO DR	MAZZUCCONI FABRIZIO DR	MERIGHI ROBERTO DR
MESSEROTTI MAURO DR	MESSINA ANTONIO	MEZZETTI MARINO
MICELA GIUSEPPINA DR	MILANI ANDREA	MILANO LEOPOLDO DR
MISSANA MARCO DR	MISSANA NATALE PROF	MOLARO PAOLO DR
MONSIGNORI FOSSI BRUNA DR	MORBIDELLI ROBERTO DR	MOROSSI CARLO
MOTTA SANTO DR	MUREDDU LEONARDO DR	NATALI GIULIANO DR
NATTA ANTONELLA DR	NESCI ROBERTO	NOBILI ANNA M
NOBILI L DR	NOCERA LUIGI DR	NOCI GIANCARLO PROF
OCCHIONERO FRANCO PROF	OLIVA ERNESTO DR	ORTOLANI SERGIO
PACINI FRANCO PROF	PADRIELLI LUCIA	PALAGI FRANCESCO
PALLA FRANCESCO	PALLAVICINI ROBERTO DR	PALUMBO GIORGIO G C DR
PANNUNZIO RENATO	PAOLICCHI PAOLO DR	PARMA PAOLA
PASIAN FABIO	PASINETTI LAURA E PROF	PASTORI LIVIO
PATERNO LUCIO PROF	PATRIARCHI PATRIZIO DR	PERES GIOVANNI DR
PERINOTTO MARIO PROF	PEROLA GIUSEPPE C DR	PERSI PAOLO
PETTINI MARCO	PICCA DOMENICO DR	PICCIONI ADALBERTO
PIGATTO LUISA DR	PINTO GIROLAMO PROF	PIOTTO GIAMPAOLLO
PIRO LUIGI DR	PIRRONELLO VALERIO	PIZZELLA G DR
PIZZICHINI GRAZIELLA	POLCARO V F	POLETTO GIANNINA PROF
POMA ANGELO DR	PORCEDDU IGNAZIO E P DR	PORETTI ENNIO
PREITE-MARTINEZ ANDREA DR	PROVERBIO EDOARDO PROF	PUCILLO MAURO DR
PUGLIANO ANTONIO PROF	QUESADA VINICIO DR	RAFANELLI PIERO DR
RAMELLA MASSIMO	RAMPAZZO ROBERTO DR	RANIERI MARCELLO
RENZINI ALVIO PROF	RIGHINI ALBERTO PROF	RIGUTTI MARIO PROF
ROBERTI GIUSEPPE DR	RODONO MARCELLO DR	ROMANO GIULIANO PROF
ROSINO LEONIDA PROF	ROSSI CORINNE DR	ROSSI LUCIO

Country : ITALY (Follow on)
_____

RUFFINI REMO	RUSCONI LUIGIA DR	RUSSO GUIDO DR
SABBADIN FRANCO DR	SAGGION ANTONIO PROF	SALINARI PIERO
SALVATI MARCO	SANTAMARIA RAFFAELE DR	SANTIN PAOLO DR
SARASSO MARIA DR	SCALTRITI FRANCO DR	SCARDIA MARCO
SCIAMA DENNIS W DR	SCIORTINO SALVATORE DR	SECCO LUIGI DR
SEDMAK GIORGIO PROF	SELVELLI PIERLUIGI DR	SEMENZATO ROBERTO
SERIO SALVATORE DR	SEVERINO GIUSEPPE	SILVESTRO GIOVANNI
SMALDONE LUIGI ANTONIO	SMRIGLIO FILIPPO PROF	SPADA GIANFRANCO DR
SPADARO DANIELE DR	STAGNI RUGGERO	STALIO ROBERTO DR
STANGA RUGGERO	STIAVELLI MASSIMO DR	STIRPE GIOVANNA M DR
STRAFELLA FRANCESCO	STRAZZULLA GIOVANNI	TAFFARA SALVATORE PROF
TAGLIAFERRI GIUSEPPE PROF	TANZELLA-NITTI GIUSEPPE	TANZI ENRICO G
TEMPESTI PIERO PROF	TERNULLO MAURIZIO DR	TOFANI GIANNI PROF
TOMASI PAOLO DR	TORELLI M DR	TORNAMBE AMEDEO
TORRICELLI GUIDETTA DR	TOSI MONICA	TOZZI GIAN PAOLO
TREVESE DARIO	TRINCHIERI GINEVRA	TRUSSONI EDOARDO
TURATTO MASSIMO DR	UBERTINI PIETRO	URAS SILVANO DR
VAGNETTI FAUSTO	VAIANA GIUSEPPE S DR	VALSECCHI GIOVANNI B DR
VERGNANO A PROF	VERNIANI FRANCO PROF	VETTOLANI GIAMPAOLO
VIETRI MARIO DR	VIGOTTI MARIO	VIOTTI ROBERTO DR
VIRGOPIA NICOLA PROF	VITTONE ALBERTO ANGELO	VITTORIO NICOLA
VLADILO GIOVANNI DR	WOLTER ANNA DR	ZAMORANI GIOVANNI
ZANINETTI LORENZO	ZAPPALA ROSARIO ALDO DR	ZAPPALA VINCENZO PROF
ZAVATTI FRANCO	ZITELLI VALENTINA DR	ZLOBEC PAOLO DR
ZUCCARELLO FRANCESCA		

Country : JAPAN

AIZU KO PROF	AKABANE KENJI A PROF	AKABANE TOKUHIDE DR
ANDO HIROYASU DR	AOKI SHINKO PROF	ARAFUNE JIRO DR
ARAI KENZO DR	AZUMA TAKAHIRO DR	BABA NAOSHI DR
CHIKADA YOSHIHIRO DR	DAISHIDO TSUNEAKI PROF	DEGUCHI SHUJI DR
DOTANI TADAYASU DR	EBISUZAKI TOSHIKAZU DR	ENOME SHINZO PROF
ERIGUCHI YOSHIHARU DR	FUJIMOTO MASAYUKI DR	FUJIMOTO MASA-KATSU DR
FUJIMOTO MITSUAKI DR	FUJISHITA MITSUMI DR	FUJITA YOSHIO PROF
FUJIWARA AKIRA DR	FUJIWARA TAKAO DR	FUKUDA ICHIRO
FUKUE JUN DR	FUKUGITA MASATAKA DR	FUKUI TAKAO DR
FUKUI YASUO DR	FUKUNAGA MASATAKA DR	FUKUSHIMA TOSHIO DR
FURUKAWA KIICHIRO DR	GOUDA NAOTERU DR	HABE ASAO
HACHISU IZUMI DR	HAGIO FUMIHIKO DR	HAMABE MASARU DR
HAMADA TETSUO PROF	HAMAJIMA KIYOTOSHI DR	HANAMI HITOSHI DR
HANAOKA YOICHIRO DR	HANAWA TOMOYUKI DR	HANDA TOSHIHIRO DR
HARA KEN NOSUKE DR	HARA TADAYOSHI DR	HARA TETSUYA DR
HASEGAWA ICHIRO DR	HASEGAWA TETSUO DR	HASHIMOTO MASA-AKI DR
HAYASHI CHUSHIRO PROF	HAYASHI MASAHIKO DR	HAYASHI SAEKO S DR
HIEI EIJIRO DR	HIRABAYASHI HISASHI DR	HIRAI MASANORI DR
HIRATA RYUKO	HIRAYAMA TADASHI PROF	HIROMOTO NORIHISA DR
HITOTSUYANAGI JUICHI PROF	HORI GENICHIRO PROF	HORIUCHI RITOKU DR
HOSHI REIUN DR	HOSOKAWA YOSHIMASA H PROF	HOSOYAMA KENNOSHUKE DR
ICHIKAWA SHIN-ICHI DR	ICHIKAWA TAKASHI	ICHIMARU SETSUO DR
IIJIMA SHIGETAKA PROF	IKEUCHI SATORU DR	INAGAKI SHOGO DR
INATANI JUNJI	INOUE HAJIME DR	INOUE MAKOTO DR
INOUE TAKESHI PROF	IRIYAMA JUN DR	ISHIDA GORO DR
ISHIDA KEIICHI PROF	ISHIGURO MASATO PROF	ISHIHARA HIDEKI DR
ISHIZAWA TOSHIAKI A PROF	ISHIZUKA TOSHIHISA DR	ISOBE SYUZO DR
ITO KENSAI A PROF	ITOH HIROSHI DR	ITOH MASAYUKI DR
ITOH NAOKI DR	IWASAKI KYOSUKE DR	IYE MASANORI DR
JUGAKU JUN DR	KABURAKI MASAKI PROF	KAI KEIZO DR
KAIFU NORIO DR	KAKINUMA TAKAKIYO T PROF	KAKUTA CHUICHI DR
KAMEYA OSAMU DR	KAMIJO FUMIO PROF DR	KAMINISHI KEISUKE PROF
KANEKO NOBORU DR	KAROJI HIROSHI DR	KASUGA TAKASHI
KATO MARIKO	KATO SHOJI PROF	KATO TAKAKO DR
KAWABATA KINAKI PROF	KAWABATA KIYOSHI	KAWABATA SHUSAKU PROF
KAWAGUCHI ICHIRO PROF	KAWARA KIMIAKI	KAWATA YOSHIYUKI DR
KIGUCHI MASAYOSHI DR	KII TSUNEO DR	KIKUCHI SADAEMON PROF
KINOSHITA HIROSHI DR	KITAMOTO SHUNJI DR	KITAMURA M PROF
KITAMURA SEIICHI DR	KOBAYASHI EISUKE DR	KOBAYASHI YUKISAYU
KODAIRA KEIICHI PROF	KODAMA HIDEO	KOGURE TOMOKAZU DR
KOJIMA MASAYOSHI DR	KONDO MASAAKI DR	KONDO MASAYUKI DR
KOSAI HIROKI	KOSHIBA MASA-TOSHI DR	KOSUGI TAKEO
KOYAMA KATSUJI	KOYAMA KO-U-ICHI DR	KOYAMA SHIN PROF DR
KOZAI YOSHIHIDE PROF	KUBO YOSHIO	KUBOTA JUN DR
KUNIEDA HIDEYO DR	KUROKAWA HIROKI DR	MAEDA KEI-ICHI DR
MAEDA KOITIRO	MAEHARA HIDEO DR	MAIHARA TOSHINORI DR
MAKINO FUMIYOSHI DR	MAKISHIMA KAZUO	MAKITA MITSUGU DR
MANABE SEIJI DR	MATSUDA TAKUYA PROF	MATSUI TAKAFUMI DR
MATSUMOTO MASAMICHI PROF	MATSUMOTO RYOJI DR	MATSUMOTO TOSHIO DR
MATSUMURA MASAFUMI DR	MATSUOKA MASARU DR	MIKAMI TAKAO DR

Country : JAPAN (Follow on)

MINESHIGE SHIN DR	MITSUDA KAZUHISA DR	MIYAJI SHIGEKI DR
MIYAMA SYOKEN	MIYAMOTO MASANORI DR	MIYAMOTO SIGENORI PROF
MIYAMOTO SYOTARO PROF DR	MIZUNO SHUN	MORIMOTO MASAKI DR
MORITA KAZUHIKO	MORIYAMA FUMIO PROF	MUKAI SONOYO DR
MUKAI TADASHI DR	MURAKAMI TOSHIO	NAGASAWA SHINGO PROF
NAGASE FUMIAKI DR	NAKADA YOSHIKAZU DR	NAKAGAWA NAOYA DR
NAKAGAWA TAKAO DR	NAKAGAWA YOSHINARI DR	NAKAGAWA YOSHITSUGU DR
NAKAI NAOMASA DR	NAKAI YOSHIHIRO	NAKAJIMA HIROSHI
NAKAJIMA KOICHI DR	NAKAMURA TAKASHI DR	NAKAMURA TSUKO DR
NAKAMURA YASUHISA	NAKANO SABURO DR	NAKANO SYUICHI
NAKANO TAKENORI DR	NAKAYAMA SHIGERU DR	NAKAZAWA KIYOSHI DR
NAMBU YASUSADA DR	NARIAI KYOJI PROF	NARITA SHINJI DR
NIIMI YUKIO	NISHI KEIZO DR	NISHIDA MINORU PROF
NISHIDA MITSUGU	NISHIMURA JUN DR	NISHIMURA MASAKI
NISHIMURA SHIRO DR	NISHIO MASANORI DR	NOGUCHI MASAFUMI DR
NOMOTO KEN'ICHI DR	OBI SHINYA PROF	ODA MINORU PROF
ODA NAOKI	OGAWARA YOSHIAKI	OGURA KATSUO DR
OHASHI TAKAYA DR	OHISHI MASATOSHI DR	OHKI KENICHIRO DR
OHTANI HIROSHI DR	OHTSUBO JUNJI DR	OHYAMA NOBORU PROF
OKAMOTO ISAO DR	OKAMURA SADANORI DR	OKAZAKI AKIRA DR
OKAZAKI ATSUO T DR	OKAZAKI SEICHI DR	OKUDA HARUYUKI DR PROF
OKUDA TORU	OKUMURA SACHIKO DR	ONAKA TAKASHI
ONO YORO PROF	OOE MASATSUGU DR	OSAKI TORU DR
OSAKI YOJI DR	OSAWA KIYOTERU DR	OWAKI NAOAKI DR
SABANO YUTAKA DR	SADAKANE KOZO DR	SAIJO KEIICHI
SAIO HIDEYUKI DR	SAITO KUNIJI PROF	SAITO MAMORU DR
SAITO SUMISABURO DR	SAITO TAKAO PROF	SAKAI JUNICHI
SAKASHITA SHIRO PROF	SAKURAI KUNITOMO PROF	SAKURAI TAKASHI DR
SAKURAI TAKEO T PROF	SASAKI MISAO	SASAKI TOSHIYUKI DR
SASAO TETSUO DR	SATO FUMIO DR	SATO HUMITAKA PROF
SATO KATSUHIKO PROF	SATO KOICHI DR	SATO NAONOBU PROF
SATO SHUJI DR	SATO YUZO DR	SEKI MUNEZO DR
SEKIGUCHI NAOSUKE PROF	SHIBAHASHI HIROMOTO DR	SHIBASAKI KIYOTO
SHIBATA KAZUNARI DR	SHIBATA SHINPEI DR	SHIBATA YUKIO DR
SHIMIZU MIKIO PROF	SHIMIZU TSUTOMU PROF EMER	SIMODA MAHIRO PROF
SINZI AKIRA M DR	SOFUE YOSHIAKI PROF	SOMA MITSURU DR
SUDA KAZUO PROF	SUEMATSU YOSHINORI DR	SUGAWA CHIKARA DR
SUGIMOTO DAIICHIRO PROF	SUGITANI KOJI DR	SUGIYAMA NAOSHI DR
SUTO YASUSHI DR	SUZUKI YOSHIMASA PROF	TABARA HIROTO DR
TAKABA HIROSHI DR	TAKADA-HIDAI MASAHIDE DR	TAKAGI KOJIRO PROF
TAKAHARA FUMIO DR	TAKAHARA MARIKO	TAKAKUBO KEIYA PROF
TAKAKURA TATSUO PROF EMER	TAKAMI HIDEKI DR	TAKANO TOSHIAKI DR
TAKARADA KATSUO DR	TAKASE BUNSHIRO PROF	TAKAYANAGI KAZUO PROF
TAKEDA HIDENORI DR	TAKEDA YOICHI DR	TAKENOUCHI TADAO DR
TAKEUTI MINE DR	TAMENAGA TATSUO DR	TAMURA SHINICHI DR
TANABE HIROYOSHI DR	TANABE KENJI DR	TANABE TOSHIHIKO DR
TANAKA MASUO DR	TANAKA RIICHIRO PROF	TANAKA WATARU DR
TANAKA YASUO DR	TANAKA YASUO PROF	TANAKA YUTAKA D DR
TANIGUCHI YOSHIAKI DR	TATEMATSU KENICHI DR	TAWARA YUZURU DR
TERASHITA YOICHI PROF	TOMIMATSU AKIRA DR	TOMISAKA KOHJI DR

Country : JAPAN (Follow on)
_____

TOMITA KENJI PROF	TOMITA KOICHIRO MR	TORAO MASAHISA
TOSA MAKOTO DR	TOSHIKI AIKAWA DR	TOTSUKA YOJI DR
TOYAMA KIYOTAKA	TSUBAKI TOKIO PROF	TSUBOI MASATO DR
TSUBOKAWA IETSUNE DR	TSUCHIYA ATSUSHI DR PROF	TSUJI TAKASHI
TSUNEMI HIROSHI DR	TSUNETA SAKU DR	UCHIDA JUICHI DR
UCHIDA YUTAKA PROF	UENO SUEO PROF	UESUGI AKIRA DR
UKITA NOBUHARU	UMEMURA MASAYUKI DR	UNNO WASABURO PROF
UTSUMI KAZUHIKO DR	WAKAMATSU KEN-ICHI DR	WAKO KOJIRO DR
WASHIMI HARUICHI DR	WATANABE JUN-ICHI DR	WATANABE TAKASHI DR
WATANABE TETSUYA	YABUSHITA SHIN A PROF	YABUUTI KIYOSHI PROF
YAMAGATA TOMOHIKO DR	YAMAGUCHI SHICHIRO	YAMAKOSHI KAZUO
YAMAMOTO TETSUO DR	YAMASAKI ATSUMA DR	YAMASHITA KOJUN DR
YAMASHITA YASUMASA PROF	YAMAZAKI AKIRA DR	YASUDA HARUO PROF DR
YOKOSAWA MASAYOSHI DR	YOKOYAMA KOICHI DR	YONEYAMA TADAOKI DR
YOSHIDA HARUO	YOSHIDA JUNZO PROF	YOSHII YUZURU DR
YOSHIMURA HIROKAZU DR	YOSHIMURA MOTOHIKO DR	YOSHIZAWA MASANORI DR
YUASA MANABU DR	YUMI SHIGERU PROF DR	

Country : KAZAKHSTAN
_____

DENISYUK EDVARD K DR	GENKIN IGOR L PROF DR	KARYGINA ZOYA V DR
KHARITONOV ANDREJ V DR	OBASHEV SAKEN O DR	OMAROV TUKEN B PROF
ROZHKOVSKIJ DIMITRIJ A	TEJFEL VIKTOR G DR	VILKOVISKIJ EMMANUIL Y DR

Country : KOREA DPR
_____

BAEK CHANG RYONG	BANG YONG GOL	CHA DU JIN
CHA GI UNG	CHIO CHOL ZONG	CHOI WON CHOL
DONG IL ZUN	HONG HYON IK	KANG GON IK
KANG JIN SOK	KIM JIK SU	KIM YONG HYOK DR
KIM YONG UK	KIM YUL	KIM ZONG DOK
LI GI MAN	LI GYONG WON	LI HYOK HO
LI J Y	LI SIN HYONG	LI SON JAE

Country : KOREA R

ANN HONG BAE DR	CHANG KYONGAE DR	CHO SE HYUNG DR
CHOE SEUNG URN DR	CHOI KYU-HONG	CHOU KYONG CHOL PROF
CHUN MUN-SUK DR	HONG SEUNG SOO DR	HYUN JONG-JUNE PROF
KANG YOUNG WOON DR	KIM CHULHEE DR	KIM KAP-SUNG DR
KIM KWANG-TAE DR	KIM TU HWAN	LEE HYUNG MOK DR
LEE SANG GAK	LEE SEE-WOO DR	LEE WOO-BAIK DR
LEE YONG-SAM DR	MINN YOUNG KEY DR	MOON SHIN HAENG DR
NHA IL-SEONG DR	OH KYU DONG DR	PARK HONG SUH DR
PARK MYEONG-GU DR	PARK SEOK JAE DR	SHIM WOON-TAIK PROF
SONG DOO JONG DR	SUH KYUNG-WON DR	WOO JONG OK
YU KYUNG-LOH PROF	YUN HONG-SIK PROF	

Country : LATVIA

ABELE MARIS K DR	ALKSNIS ANDREJS DR	BALKLAVS A E DR
DAUBE-KURZEMNIECE I A DR	DIRIKIS M A DR	FRANTSMAN YU L DR
LAPUSHKA K K DR	PLATAIS IMANT K DR	ZHAGAR YOURI H DR

Country : LITHUANIA

RUDZIKAS ZENONAS B	STRAIZYS V PROF DR	ZDANAVICIUS KAZIMERAS DR

Country : MALAYSIA

ILYAS MOHAMMAD DR	MAHAT ROSLI H DR	MAJID ABDUL BIN A H DR
MOHD ZAMBRI ZAINUDDIN DR	OTHMAN MAZLAN	

Country : MEXICO

ALLEN CHRISTINE	ARELLANO FERRO ARMANDO	BISIACCHI GIANFRANCO DR
CANTO JORGE DR	CARDONA OCTAVIO DR	CARRASCO LUIS DR
CHATTERJEE TAPAN K DR	CHAVARRIA-K CARLOS	CHAVIRA ENRIQUE SR
CHELLI ALAIN	CORNEJO ALEJANDRO A DR	COSTERO RAFAEL
CRUZ-GONZALEZ IRENE	DALTABUIT ENRIQUE DR	DE LA HERRAN V JOSE ENG
ECHEVARRIA JUAN DR	ECHEVERRIA ROMAN JUAN M	ESCALANTE VLADIMIR DR
FIERRO JULIETA	FIRMANI CLAUDIO A PROF	FRANCO JOSE DR
GALINDO TREJO JESUS DR	GARCIA-BARRETO JOSE A	GONZALEZ G
HACYAN SHAHEN DR	HERRERA MIGUEL ANGEL DR	KL'APP JAIME DR
KOENIGSBERGER GLORIA	LIZANO-SOBERON SUSANA DR	LOPEZ JOSE ALBERTO DR
MALACARA DANIEL	MARTINEZ MARIO DR	MENDEZ MANUEL DR
MENDOZA V EUGENIO E DR	MORENO CORRAL MARCO A DR	MORENO EDMUNDO DR
OBREGON DIAZ OCTAVIO J DR	PEIMBERT MANUEL DR	PENA JOSE
PENA MIRIAM DR	PENICHE ROSARIO DR	PEREZ-DE-TEJADA H A DR
PEREZ-PERAZA JORGE DR	PISMIS DE RECILLAS PARIS	POVEDA ARCADIO DR
RECILLAS-CRUZ ELSA DR	RODRIGUEZ LUIS F	ROSADO MARGARITA DR
RUELAS-MAYORGA R A DR	SARMIENTO-GALAN A F DR	SCHUSTER WILLIAM JOHN DR
SERRANO ALFONSO DR	TAPIA MAURICIO DR	TORRES-PEIMBERT SILVIA DR
WARMAN JOSEF DR		

Country : MOROCCO

EL BAKKALI LARBI DR

Country : NETHERLANDS

ACHTERBERG ABRAHAM DR	ANDRIESSE CORNELIS D DR	ATANASIJEVIC IVAN DR
BARTHEL PETER DR	BAUD BOUDEWIJN DR	BEGEMAN KOR G DR
BEINTEMA DOUWE A DR	BENNETT KEVIN DR	BLAAUW ADRIAAN PROF DR
BLEEKER JOHAN A M DR IR	BLOEMEN JOHANNES B G M DR	BOLAND WILFRIED
BONTEKOE ROMKE DR	BORGMAN JAN DR PROF	BOS ALBERT DR
BOSMA PIETER B DR	BRAES L L E DR	BRAUN ROBERT DR
BREGMAN JACOB D IR	BREUKERS R J L H DR	BRINKMAN BERT C DR
BURGER J J DR IR	BURTON W BUTLER DR	BUTCHER HARVEY R PROF DR
BYLEVELD WILLEM DR	CLAVEL JEAN	COLEMAN PAUL HENRY DR
DE BRUYN A GER'DR	DE GRAAF T DR	DE GRAAFF W DR
DE GRAAUW TH DR	DE GROOT T DR	DE JAGER CORNELIS PROF
DE JONG TEIJE DR	DE KORT JULES J DR	DE KORTE PIETER A J DR
DE VRIES CORNNELIS DR	DE ZEEUW PIETER T DR	DEERENBERG A J M DR
DEGEWIJ JOHAN DR	DEKKER E DR	DEUL ERIK DR
DOMINGO VICENTE DR	DOUGLAS NIGEL DR	FITTON BRIAN DR
FOLEY ANTHONY DR	FRIDLUND MALCOLM DR	FRITZOVA-SVESTKA L DR
GOEDBLOED JOHAN P PROF DR	GREENBERG J MAYO DR	HABING H J DR
HAMMERSCHLAG ROBERT H DR	HAMMERSCHLAG-HENSBERGE G	HEARN ANTHONY G DR
HEINTZE J R W DR	HEISE JOHN DR	HENRICHS HUBERTUS F DR
HERAS ANA M DR	HERMSEN WILLEM DR	HESKE ASTRID DR
HOEKSTRA ROEL DR	HOVENIER J W DR	HOYNG PETER DR
HUBENET HENRI DR	HUBER MARTIN C E DR	HULSBOSCH A N M DR
ICKE VINCENT DR	ISRAEL FRANK P DR	IVES JOHN CHRISTOPHER MR
JAKOBSEN PETER	JOURDAIN DE MUIZON M DR	KAASTRA JELLE S DR
KAHLMANN HANS CORNELIS DR	KATGERT PETER DR	KATGERT-MERKELIJN J K DR
KESSLER MARTIN F DR	KLINKHAMER FRANS DR	KRAAN-KORTEWEG RENEE C DR
KUIJPERS H JAN M E DR	KUPERUS MAX PROF DR	KWEE K K DR
LAMERS HENNY J G L M DR	LE POOLE RUDOLF S DR	LUB JAN DR
METCALFE LEO DR	MEWE R DR	MILEY G K DR
MULLER A B DR	MULLER C A PROF JR	MURPHY BRIAN WILLIAM DR
NAMBA OSAMU DR	NIEUWENHUIJZEN HANS DR	NORTH JOHN DAVID PROF
OLLONGREN A PROF DR	OLNON FRISO	OLTHOF HINDERICUS DR
OORT JAN H PROF	PATERSON-BEECKMANS F	PEACOCK ANTHONY DR
PEL JAN WILLEM DR	PERRYMAN MICHAEL A C DR	POTTASCH STUART R PROF
RAIMOND ERNST DR	ROELFSEMA PETER DR	ROOS NICOLAAS DR
ROSENBERG J DR	RUTTEN ROBERT J DR	SAHU KAILASH C DR
SANCISI RENZO DR	SANDERS ROBERT DR	SAVONIJE GERRIT JAN DR
SCHADEE AERT DR	SCHEEPMAKER ANTON DR	SCHILIZZI RICHARD T DR
SCHRIJVER C J DR	SCHRIJVER JOHANNES DR	SCHWARZ ULRICH J DR
SCHWEHM GERHARD DR	SHANE WILLIAM W DR	SMIT J A PROF
SPOELSTRA T A TH DR	STEVENS GERARD A DR	STROM RICHARD G DR
SVESTKA ZDENEK DR	SWANENBURG B N DR	TAKENS ROELF JAN DR
THE PIK-SIN PROF	TINBERGEN JAAP DR	TJIN-A-DJIE HERMAN R E DR
VAGHI SERGIO DR	VALENTIJN EDWIN A DR	VAN AGT S L TH J DR
VAN ALBADA TJEERD S DR	VAN BEEK FRANK PROF DR	VAN BUEREN HENDRIK G PROF

Country : NETHERLANDS (Follow on)

VAN DE HULST H C PROF DR	VAN DE KAMP PETER	VAN DE STADT HERMAN DR
VAN DEN HEUVEL EDWARD P J	VAN DEN OORD BERT H J DR	VAN DER HUCHT KAREL A DR
VAN DER HULST JAN M DR	VAN DER KLIS MICHIEL DR	VAN DER KRUIT PIETER C DR
VAN DIGGELEN J DR	VAN DISHOECK EWINE F DR	VAN DRIEL WILLEM DR
VAN DRIEL-GESZTELYI L DR	VAN DUINEN R J DR	VAN GENDEREN A M DR
VAN HERK G	VAN HOUTEN C J DR	VAN HOUTEN-GROENEVELD I
VAN NIEUWKOOP J DR IR	VAN PARADIJS JOHANNES DR	VAN WOERDEN HUGO PROF DR
VERBUNT FRANCISCUS DR	WATERS LAURENS B F M DR	WESSELIUS PAUL R DR
WIJNBERGEN JAN DR	WINKLER CHRISTOPH DR	ZWAAN CORNELIS PROF DR

Country : NEW ZEALAND

ALEXANDER MURRAY E DR	ALLEN WILLIAM	BAGGALEY WILLIAM J PROF
BATESON FRANK M OBE DR	BLOW GRAHAM L	BUDDING EDWIN DR
COTTRELL PETER LEDSAM	CRAIG IAN JONATHAN D DR	DODD RICHARD J DR
DOUGHTY NOEL A DR	GILMORE ALAN C MR	HEARNSHAW JOHN B DR
JONES ALBERT F MR	KERR ROY P PROF	MARINO BRIAN F ENG
ORCHISTON WAYNE DR	RUMSEY NORMAN J	SCHATTEN KENNETH H DR
SULLIVAN DENIS JOHN DR	TOBIN WILLIAM	TRODAHL HARRY JOSEPH DR
WALKER WILLIAM S G		

Country : NORWAY

AKSNES KAARE DR	ANDERSEN BO NYBORG DR	BRAHDE ROLF
CARLSSON MATS DR	ELGAROY OYSTEIN PROF	ENGVOLD ODDBJOERN DR
ERIKSEN GUNNAR PROF	ESSER RUTH DR	HAUGE OIVIND DR
HAVNES OVE DR	JENSEN EBERHART PROF	KJELDSETH-MOE OLAV DR
LEER EGIL PROF	MALTBY PER PROF	OESTGAARD ERLEND
PETTERSEN BJOERN RAGNVALD	RINGNES TRULS S DR	SOLHEIM JAN ERIK
STABELL ROLF DR	TRULSEN JAN K PROF	

Country : PERU

AGUILAR MARIA LUISA

Country : POLAND

BEM JERZY DR
CHOLONIEWSSKI JACEK DR
CZERNY BOZENA DR
DROZYNER ANDRZEJ
GASKA STANISLAW DR
GORGOLEWSKI STANISLAW PR
GRZEDZIELSKI STANISLAW PR
HELLER MICHAEL PROF
IWANOWSKA WILHELMINA PROF
JAROSZYNSKI MICHAL
JUSZKIEWICZ ROMAN
KOZIEL KAROL PROF DR
KREINER JERZY MAREK DR
KRUSZEWSKI ANDRZEJ PROF
KURPINSKA-WINIARSKA M DR
KUSUNOSE MASAAKI DR
MACIEJEWSKI ANDRZEJ J DR
MERGENTALER JAN PROF
MIETELSKI JAN S DR
MOSKALIK PAWEL DR
PROSZYNSKI MIECZYSLAW
RUDAK BRONISLAW
SCHREIBER ROMAN
SIENKIEWICZ RYSZARD DR
SITARSKI GRZEGORZ PROF
SOKOLOWSKI LECH
STEPIEN KAZIMIERZ DR
SYLWESTER JANUSZ
TYLENDA ROMUALD DR
USOWICS JERZY BOGDAN DR
WOSZCZYK ANDRZEJ PROF
ZIOLKOWSKI KRZYSZTOF DR

BORKOWSKI KAZIMIERZ M DR
CIURLA TADEUSZ
DOBRZYCKI JERZY PROF
DZIEMBOWSKI WOJCIECH PROF
GIL JANUSZ A PROF
GRABOWSKI BOLESLAW DR
HAENSEL PAWEL DR
HURNIK HIERONIM PROF
JAKIMIEC JERZY PROF
JARZEBOWSKI TADEUSZ DR
KALUZNY JANUSZ DR
KOZLOWSKI MACIEJ DR
KRELOWSKI JACEK DR
KRYGIER BERNARD DR
KURZYNSKA KRYSTYNA DR
LEHMANN MAREK DR
MADEJ JERZY
MICHALEC ADAM
MIKOLAJEWSKA JOANNA DR
OPOLSKI ANTONI PROF
ROMPOLT BOGDAN DR
RUDNICKI KONRAD PROF
SCHWARZENBERG-CZERNY A
SIKORA MAREK
SMAK JOSEPH I PROF
SOLTAN ANDRZEJ MARIA DR
STROBEL ANDRZEJ DR
SZAFRANIEC ROZALIA DR
UDALSKI ANDRZEJ DR
WINIARSKI MACIEJ
ZIEBA STANISLAW DR

BRZEZINSKI ALEKSANDER DR
CUGIER HENRYK DR
DOMINSKI IRENEUSZ DR
FLIN PIOTR
GLEBOCKI ROBERT PROF
GRUDZINSKA STEFANIA DR
HANASZ JAN DR
IWANISZEWSKA CECILIA DR
JAKS WALDEMAR DR
JERZYKIEWICZ MIKOLAJ DR
KOLACZEK BARBARA DR
KRASINSKI ANDRZEJ PROF
KREMPEC-KRYGIER JANINA DR
KUBIAK MARCIN A DR
KUS ANDRZEJ JAN DR
MACHALSKI JERZY DR
MASLOWSKI JOZEF DR
MICHALOWSKI TADEUSZ DR
MOCZKO JANUSZ DR
PACZYNSKI BOHDAN PROF
ROZYCZKA MICHAL
RYBKA PRZEMYSLAW DR
SEMENIUK IRENA DR
SIKORSKI JERZY DR
SMOLINSKI JAN DR
STAWIKOWSKI ANTONI DR
SYLWESTER BARBARA DR
TURLO ZYGMUNT DR
URBANIK MAREK DR
WNUK EDWIN
ZIOLKOWSKI JANUSZ DR

Country : PORTUGAL

CABRITA EZEQUIEL DR
DA COSTA ANTONIO A DR
LAGO MARIA TERESA V T PR
NUNES ROGERIO S DE SOUSA
SANTOS FILIPE D DR
VICENTE RAIMUNDO O PROF

CAMPOS L M BRAGA DA COSTA
DA SILVA A V C S
MAGALHAES ANTONIO A S ENG
OSORIO JOSE J S P PROF
TABORDA JOSE ROSA DR

COELHO BALSA MARIO C DR
DOS REIS M PROF
MARQUES MANUEL N DR
PASCOAL ANTONIO J B SCI
TAVARES J T L DR

Country : RUSSIA

ABALAKIN VICTOR K DR | AFANASJEVA PRASKOVYA M DR | AFANAS'EV VIKTOR L DR
AGEKJAN TATEOS A PROF | AKIM EFRAIM L DR | AKSENOV E P PROF DR
ANTIPOVA LYUDMILA DR | ANTONOV VADIM A DR | ARKHIPOVA V P DR
BABADZHANIANC MICHAIL DR | BABIN V G DR | BAGILDINSKIJ BRONISLAV K
BALEGA YURI YU DR | BATRAKOV YU V DR | BAZILEVSKY ALEXANDR T
BELINSKY VLADIMIR DR | BELKOVICH O I DR | BELYAEV NIKOLAJ A DR
BISNOVATYI-KOGAN G S DR | BLINOV N S DR | BOBROV M S DR
BOCHKAREV NIKOLAY G DR | BONDARENKO L N DR | BOYARCHUK A A DR
BOYARCHUK MARGARITA E DR | BREJDO IZABELLA I DR | BRONNIKOVA NINA M
BRUMBERG VICTOR A DR | BYSTROVA NATALIJA V DR | CHECHETKIN VALERIJ M DR
CHEREPASHCHUK A M PROF | CHERNOV GENNADIJ DR | CHERTOPRUD V E DR
CHISTYAKOV VLADIMIR E DR | CHUGAI NIKOLAI N DR | DADAEV ALEKSANDR N DR
DAGKESAMANSKY RUSTAM D DR | DANILOV VLADIMIR M DR | DEMIN V G PROF DR
DLUZHNEVSKAYA O B DR | DOKUCHAEV VYACHESLAV DR | DOKUCHAEVA OLGA D DR
DOLGINOV ARKADY Z PROF DR | DOROSHKEVICH ANDREI G DR | DOUBINSKIJ B A DR
DRAVSKIKH A F DR | EFREMOV YU I DR | EFREMOV YURY N DR
EMELIANOV NIKOLAJ V DR | ERPYLEV N P DR | ESIPOV VALENTIN F DR
EVDOKIMOV YU V DR | FADEYEV YURI A | FINKELSTEIN ANDREJ M DR
FOMENKO ALEXANDR F DR | FOMICHEV VALERI V DR | FOMIN VALERY A DR
FOMINOV ALEXANDR M DR | FRIDMAN ALEKSEY M DR | FROLOV M S DR
FURSENKO M A DR | GALIBINA I V DR | GALPERIN YU I PROF
GELFREIKH GEORGIJ B DR | GINZBURG VITALY L PROF | GLAGOLEVSKIJ JU V DR
GLEBOVA NINA I DR | GLUSHNEVA I N DR | GNEDIN YURIJ N DR
GNEVYSHEVA RAISA S DR | GORBATSKY VITALIJ G PROF | GOSACHINSKIJ I V DR
GREBENIKOV E A PROF DR | GRIGORJEV VICTOR M DR | GRISHCHUK L P DR
GROUSHINSKY N P PROF DR | GUBANOV VADIM S DR | GULYAEV A P DR
GULYAEV RUDOLF A DR | GURSHTEIN ALEXANDER A DR | HABIBULLIN SH T PROF DR
HAGEN-THORN VLADIMIR A DR | IDLIS G M DR | IKHSANOV ROBERT N DR
IKHSANOVA VERA N DR | IMSHENNIK V S DR | IOSHPA B A DR
IVANOV VSEVOLOD V DR PROF | IVANOV-KHOLODNY G S DR | IZVEKOV V A DR
KADLA ZDENKA I DR | KANAEV IVAN I DR | KARACHENTSEV I D DR
KARDASHEV N S DR | KARPINSKIJ VADIM N DR | KHOKHLOVA V L DR
KHOLSHEVNIKOV K V DR | KHOZOV GENNADIJ V | KHROMOV G S DR
KIM IRADIA S | KISELYOV ALEXEJ A DR | KISLYAKOV ALBERT G DR
KISSELEVA TAMARA P | KOCHAROV GRANT E PROF | KOKURIN YURIJ L DR
KOLESOV A K DR | KONONOVICH EDWARD V DR | KOPYLOV I M DR
KORCHAK A A DR | KOROVYAKOVSKIJ YURIJ P DR | KOSIN GENNADIJ S DR
KOSTINA LIDIJA D DR | KOSTYAKOVA ELENA B DR | KOSTYLEV K V DR
KOTELNIKOV V A ACAD | KRASINSKY GEORGE A DR | KRASSOVSKY V I DR
KSANFOMALITI L V DR | KUKLIN G V DR | KUMAJGORODSKAYA RAISA DR
KUMKOVA IRINA I DR | KURIL-CHIK V N DR | KURT V G DR
KUTUZOV S A DR | KUZMIN ARKADII D PROF DR | LAVROV M I PROF
LAVRUKHINA A K PROF DR | LEBEDINETS VLADIMIR N DR | LEIKIN G A DR
LIPOVETSKY V A | LIVSHITS M A DR | LOTOVA N A DR
LOZINSKAYA TAT'YANA A DR | LOZINSKIJ A M DR | MAKAROV VALENTINE I
MAKAROVA ELENA A DR | MANDZHOS ANDREJ V DR | MAROCHNIK L S PROF DR
MAROV MIKHAIL YA PROF | MASSEVICH ALLA G DR | MATVEYENKO L I DR
MERMAN G A DR | MERMAN NATALIA V DR | MIKHELSON NIKOLAJ N DR
MININ I N PROF | MITROFANOVA LYUDMILA A DR | MOGILEVSKIJ EH I DR
MOLCHANOV A P PROF | MOROZ V I PROF DR | MYACHIN VLADIMIR F DR

Country : RUSSIA (Follow on)

NADYOZHIN D K DR	NAGIRNER DMITRIJ I DR	NAGNIBEDA VALERY G DR
NAIDENOV VICTOR O	NAUMOV VITALIJ A DR	NEFEDEVA ANTONINA I PROF
NEMIRO ANDREJ A DR PROF	NIKITIN A A DR	NOSKOV BORIS N DR
NOVIKOV I D DR	NOVIKOV SERGEJ B DR	NOVOSELOV V S PROF DR
OBRIDKO VLADIMIR N DR	PAMYATNIKH A A DR	PARIJSKIJ N N PROF
PARIJSKIJ YU N DR	PAVLOVSKAYA E D DR	PETROV GENNADIJ M
PETROV GEORGIJ I PROF DR	PETROVSKAYA M S DR	PISKUNOV ANATOLY E
PISKUNOV NIKOLAI E DR	POLOZHENTSEV DIMITRIJ DR	POLYACHENKO VALERIJ L DR
POPOV VICTOR S DR	PORFIR'EV V V DR	POTTER HEINO I DR
PRODAN Y I DR	PROKOF'EVA IRINA A DR	PSKOVSKIJ JU P DR
PUSHKIN SERGEY B DR	RAZIN V A DR	RIZVANOV NAUFAL G DR
RUBASHEV BORIS M DR	RUSKOL EUGENIA L DR	RYABCHIKOVA TANYA DR
RYABOV YU A PROF DR	RYKHLOVA LIDIJA V DR	RYLOV VALERIJ S DR
RYUTOVA MARGARITA P DR	RYZHKOV NIKOLAI F DR	RZHIGA OLEG N DR
SAFRONOV VICTOR S DR	SAGDEEV ROALD Z DR	SAKHIBULLIN NAIL A DR
SAMUS NIKOLAI N DR	SAZHIN MICHAIL DR	SHAKURA NICHOLAJ I DR
SHANDARIN SERGEI F DR	SHARAF SH G DR	SHAROV A S DR
SHCHEGLOV P V DR	SHCHEGOLEV DIMITRIJ E DR	SHCHERBINA-SAMOJLOVA I DR
SHEFFER EUGENE K DR	SHEFOV NICOLAI N	SHEVCHENKO VLADISLAV V DR
SHIRYAEV ALEXANDER A DR	SHISHOV VLADIMIR I DR	SHOLOMITSKY G B DR
SHOR VIKTOR A DR	SHULOV OLEG S DR	SHUSTOV BORIS M DR
SIDORENKOV NIKOLAY S	SITNIK G F PROF	SKRIPNICHENKO VLADIMIR DR
SLYSH VJACHOSLAV I DR	SMOL'KOV GENNADIJ YA DR	SNEZHKO LEONID I
SOBOLEV V V DR	SOBOLEV VLADISLAV M DR	SOBOLEVA N S DR
SOCHILINA ALLA S DR	SOKOLSKY ANDREJ G DR	SOMOV BORIS V DR
SOROCHENKO R L DR	STANKEVICH KAZIMIR S DR	STREL'NITSKIJ VLADIMIR DR
STRUKOV IGOR A DR	SUNYAEV RASHID A DR	SVECHNIKOVA MARIA A DR
TATEVYAN S K DR	TEPLITSKAYA R B DR	TERENTJEVA ALEXANDRA K DR
TOKAREV YURIJ V DR	TOKOVININ ANDREJ A DR	TROITSKY V S PROF DR
TRUTSE YU L DR	TSEYTLIN NAUM M	TSYGAN ANATOLII I PROF
TUTUKOV A V DR	UDAL'TSOV V A DR	URASIN LIRIK A DR
VAINSTEIN L A DR	VARSHALOVICH DIMITRIJ PR	VASHKOV'YAK SOF'YA N DR
VASILEVA GALINA J DR	VEKSTEIN GREGORY DR	VITINSKIJ YURIJ I DR
VITYAZEV VENEAMIN V DR	VORONTSOV-VEL'YAMINOV B A	VOSHCHINNIKOV NICOLAI DR
VYALSHIN GENNADIJ F DR	YAROV-YAROVOJ M S DR	YAVNEL ALEXANDER A DR
YUNGELSON LEV R	ZAITSEV VALERII V DR	ZASOV ANATOLE V DR
ZEL'MANOV A L DR	ZHARKOV VLADIMIR N DR	ZHELEZNIAKOV VLADIMIR V
ZHEVAKIN S A PROF DR	ZHUGZHDA YUZEF D DR	ZLOTNIK ELENA YA DR

Country : SAUDI ARABIA

BOYDAG-YILDIZDOGDU F S DR   BROSTERHUS E B F DR   HAMZAOGLU ESAT E H DR
NIAZY ADNAN MOHAMMAD DR   TOPAKTAS LATIF A DR   TUFEKCIOGLU ZEKI DR

Country : SOUTH AFRICA

BAART EDWARD E PROF   BALONA LUIS ANTERO DR   BLOCK DAVID LAZAR PROF
CALDWELL JOHN A R   CHURMS JOSEPH   CILLIE G G PROF
COUSINS A W J DR   DE JAGER GERHARD PROF   DE JAGER OCKER C DR
ENGELBRECHT CHRISTIAN DR   EVANGELIDIS E DR   FAIRALL ANTHONY P PROF
FEAST MICHAEL W PROF   GAYLARD MICHAEL JOHN   GLASS IAN STEWART DR
HERS JAN MR   HIRST WILLIAM P   JARRETT ALAN H PROF
JONAS JUSTIN LEONARD   KILKENNY DAVID DR   KOEN MARTHINUS DR
KURTZ DONALD WAYNE DR   LANEY CLIFTON D DR   LLOYD EVANS THOMAS DR
MENZIES JOHN W DR   NICOLSON GEORGE D DR   OVERBEEK MICHIEL DANIEL
O'DONOGHUE DARRAGH DR   RAUBENHEIMER BAREND C PR   STOKER PIETER H
VAN DER WALT D J DR   WALRAVEN TH DR   WARGAU WALTER F DR
WARNER BRIAN PROF   WHITELOCK PATRICIA ANN DR

Country : SPAIN

ABAD ALBERTO J DR	ALCOLEA JAVIER DR	ALFARO EMILIO JAVIER
ALVAREZ PEDRO DR	ANDERNACH HEINZ DR	ANGLADA GUILLEM DR
APARICIO ANTONIO DR	ARRIBAS SANTIAGO DR	BACHILLER RAFAEL DR
BALLESTER JOSE LUIS DR	BARCIA ALBERTO DR	BARCONS XAVIER DR
BATTANER EDUARDO DR	BECKMAN JOHN E PROF	BELMONTE AVILES J A DR
BENAVENTE JOSE	BETANCORT-RIJO JUAN DR	BOLOIX RAFAEL DR
BONET JOSE A	BUITRAGO JESUS	BUJARRABAL VALENTIN
CALVO MANUEL	CAMARENA BADIA VICENTE PR	CANAL RAMON M DR
CARDUS ALMEDA J O MR	CASTANEDA HECTOR	CATALA POCH M A
CATALAN MANUEL DR	CEPA JORDI DR	CERNICHARO JOSE DR
CID PALACIOS RAFAEL PROF	CODINA VIDAL J M DR	COLLADOS MANUEL DR
COMA JUAN CARLOS	CORNIDE MANUEL	COSTA VICTOR DR
CUBARSI RAFAEL DR	DE CASTRO ANGEL DR	DE CASTRO ELISA
DE PASCUAL MARTINEZ M DR	DEL OLMO OROZCO A DR	DEL RIO GERARDO DR
DEL TORO INIESTA JOSE DR	DELGADO ANTONIO JESUS	DIAZ ANGELES ISABEL DR
DOCOBO DURANTEZ JOSE A	DULTZIN-HACYAN D DR	EIROA DE SAN FRANCISCO C
ELIPE SANCHEZ ANTONIO	ESTALELLA ROBERT	FABREGAT JUAN DR
FERNANDEZ-FIGUEROA M J DR	FERNANZ MARGARITA DR	FERRANDIZ JOSE MANUEL DR
FERRER MARTINEZ SEBASTIAN	FIGUERAS FRANCESCA DR	FUENSALIDA JIMENEZ J DR
GALAN MAXIMINO J	GARCIA DE LA ROSA JOSE I	GARCIA DE MARIA J M DR
GARCIA-BERRO ENRIQUE DR	GARCIA-PELAYO JOSE DR	GARRIDO RAFAEL
GARZON FRANCISCO DR	GILMOZZI ROBERTO	GIMENEZ ALVARO
GOMEZ GONZALEZ JESUS DR	GONZALEZ CAMACHO ANTONIO	GONZALEZ SERRANO J I DR
GONZALEZ-RIESTRA R DR	GORGAS GARCIA JAVIER DR	HERRERO DAVO ARTEMIO DR
HIDALGO MIGUEL A DR	ISERN JORGE DR	JORDI NEBOT CARME DR
LABAY JAVIER	LAHULLA J FORNIES DR	LAZARO CARLOS DR
LINDGREN HARRI	LING J DR	LOISEAU NORA DR
LOPEZ DE COCA M D P DR	LOPEZ ROSARIO DR	LOPEZ-ARROYO M
LOPEZ-GONZALEZ MARIA J DR	LOPEZ-MORENO JOSE JUAN	LOPEZ-PUERTAS MANUEL
LOPEZ-VALVERDE M A DR	MANCHADO ARTURO DR	MARCAIDE JUAN-MARIA DR
MARTINEZ ROGER CARLOS DR	MARTINEZ-GONZALEZ E DR	MARTIN-DIAZ CARLOS DR
MARTIN-LORON M DR	MARTIN-PINTADO JESUS	MASEGOSA GALLEGO J DR
MASSAGUER JOSEP PROF	MEDIAVILLA EVENCIO DR	MEDINA JOSE DR
MOLES MARIANO J DR	MOLINA ANTONIO	MORALES-DURAN CARMEN
MORENO FERNANDO DR	MORENO-INSERTIS FERNANDO	MUINOS JOSE L DR
MUNOZ-TUNON CASIANA	NUNEZ JORGE DR	ORTE ALBERTO
ORUS JUAN J PROF	PALLE PERE-LLUIS DR	PAREDES JOSE MARIA DR
PENSADO JOSE DR	PEREA-DUARTE JAIME D DR	PEREZ ENRIQUE DR
PEREZ FOURNON ISMAEL DR	PLANESAS PERE	PRIETO MERCEDES
QUIJANO LUIS	QUINTANA JOSE M DR	REBOLO RAFAEL DR
REGLERO-VELASCO VICTOR DR	REGO FERNANDEZ M DR	REGULO CLARA DR
ROCA CORTES TEODORO	RODRIGO RAFAEL	RODRIGUEZ ELOY DR
RODRIGUEZ-ESPINOSA JOSE	RODRIGUEZ-VILLAMIL R DR	ROLLAND ANGEL DR
ROMERO PEREZ M PILAR	ROSSELLO GASPAR	SALA FERRAN DR
SALAZAR ANTONIO DR	SALVADOR-SOLE EDUARDO	SANAHUJA BLAS
SANCHEZ ALMEIDA JORGE DR	SANCHEZ FRANCISCO PROF	SANCHEZ MANUEL
SANCHEZ-LAVEGA AGUSTIN DR	SANCHEZ-SAAVEDRA M LUISA	SANROMA MANUEL DR
SANSATURIO MARIA E DR	SANZ I SUBIRANA JAUME DR	SANZ JOSE L DR
SEIN-ECHALUCE M LUISA DR	SEQUEIROS JUAN DR	SEVILLA MIGUEL J DR
SIMO CHARLES DR	STEPPE HANS DR	TALAVERA A DR

Country : SPAIN (Follow on)
_____

THUM CLEMENS DR	TORRA JORDI DR	TORRELLES JOSE M DR
TORROJA J PROF	TRUJILLO BUENO JAVIER DR	VAZQUEZ MANUEL DR
VILCHEZ MEDINA JOSE M DR	VIVES TEODORO JOSE DR	WAMSTEKER WILLEM DR
ZAMBRANO ALEJANDRO DR	ZAMORANO JAIME DR	

Country : SWEDEN
_____

ADOLFSSON TORD DR	ALFVEN HANNES PROF	ARDEBERG ARNE L PROF
BAATH LARS B DR	BERGVALL NILS AKE SIGVARD	BJORNSSON CLAES-INGVAR
BOOTH ROY S PROF	CARLQVIST PER A DR	CATO B TORGNY DR
DRAVINS DAINIS PROF	EDLEN BENGT PROF	EDVARDSSON BENGT DR
ELLDER JOEL DR	ELVIUS AINA M PROF	ELVIUS TORD PROF EMERITUS
ERIKSSON KJELL DR	FAELTHAMMAR CARL GUNNE PR	FRANSSON CLAES
FREDGA KERSTIN PROF	FRISK URBAN DR	GAHM GOESTA F DR
GUSTAFSSON BENGT DR	HAHN GERHARD J DR	HANSSON NILS DR
HJALMARSON AKE G DR	HOEGBOM JAN A DR	HOEGLUND BERTIL PROF
HOLMBERG ERIK B PROF	JOERSAETER STEVEN DR	JOHANSSON LARS ERIK B DR
JOHANSSON LENNART DR	JOHANSSON SVENERIC DR	KOLLBERG ERIK L PROF
KRISTENSON HENRIK DR	KRISTIANSSON KRISTER PROF	LAGERKVIST CLAES-INGVAR
LAGERQVIST ALBIN PROF	LARSSON STEFAN DR	LARSSON-LEANDER G PROF
LAUBERTS ANDRIS DR	LAURENT BERTEL E PROF	LEHNERT B P PROF
LINDBLAD BERTIL A DR	LINDBLAD PER OLOF PROF	LINDE PETER DR
LINDEGREN LENNART DR	LODEN KERSTIN R DR	LODEN LARS OLOF PROF
LUNDSTEDT HENRIK DR	LUNDSTROM INGEMAR DR	LYTTKENS EJNAR DR
NILSON PETER DR	NORDH H LENNART DR	NYMAN LARS-AKE DR
OEHMAN YNGVE PROF	OJA TARMO PROF	OLBERG MICHAEL DR
OLOFSSON HANS	OLOFSSON S GOERAN DR	RAADU MICHAEL A DR
RICKMAN HANS DR	ROENNAENG BERNT O DR	ROSLUND CURT DR
ROSQUIST KJELL	RYDBECK GUSTAF H B DR	RYDBECK OLOF E H PROF
SAHAI RAGHVENDRA DR	SANDQVIST AAGE DR	SCHALEN CARL PROF
SCHARMER GOERAN BJARNE	SINNERSTAD ULF E PROF	SOEDERHJELM STAFFAN DR
STENHOLM BJOERN DR	STENHOLM LARS	SUNDELIUS BJOERN DR
SUNDMAN ANITA DR	SVENSSON ROLAND	SWENSSON JOHN W DR
VAN GRONINGEN ERNST DR	WALLENQUIST AAKE A E PROF	WESTERLUND BENGT E PROF
WIEDLING TOR DR	WINNBERG ANDERS DR	WRAMDEMARK STIG S O DR

Country : SWITZERLAND

BARTHOLDI PAUL DR
BEUTLER GERHARD PROF
BOCHSLER PETER
BURKI GILBERT PROF
CHMIELEWSKI YVES DR
DEBRUNNER HERMANN DR
FENKART ROLF P PROF DR
GOLAY MARCEL PROF
HAUCK BERNARD PROF
MAETZLER CHRISTIAN DR
MAYOR MICHEL PROF
MEYNET GEORGES DR
NORTH PIERRE
RUFENER FREDY G PROF
SCHMUTZ WERNER
SPAENHAUER ANDREAS MARTIN
TAMMANN G ANDREAS PROF DR
WALDMEIER MAX PROF DR
ZELENKA ANTOINE DR

BECKER WILHELM PROF
BINGGELI BRUNO
BONANOMI JACQUES DR
BUSER ROLAND DR
COURVOISIER THIERRY J-L
DRESSLER KURT PROF
FROEHLICH CLAUS
GOY GERALD PROF
LABHARDT LUKAS
MAGUN ANDREAS DR
MEGEVAND DENIS DR
MUELLER EDITH A PROF
NUSSBAUMER HARRY PROF
SCHANDA ERWIN PROF
SCHULER WALTER DR
STEINLIN ULI PROF
TREFZGER CHARLES F DR
WILD PAUL PROF

BENZ ARNOLD DR
BLECHA ANDRE BORIS G DR
BOUVIER PIERRE PROF
CAMERON LUZIUS MARTIN
DE MEDEIROS JOSE RENAN DR
DUERST JOHANNES DR
GEISS JOHANNES PROF
GRENON MICHEL DR
MAEDER ANDRE PROF
MARTINET LOUIS PROF
MERMILLIOD JEAN-CLAUDE DR
NICOLET BERNARD
PFENNIGER DANIEL DR
SCHMID HANS MARTIN DR
SOLANKI SAMI K DR
STENFLO JAN O DR
VOGEL MANFRED DR
XIA ZHIGUO DR

Country : TADZHIKISTAN

BABADZHANOV PULAT B DR
KISELEV NIKOLAI N DR

BIBARSOV RAVIL'SH DR

IBADINOV KHURSANDKUL DR

Country : TURKEY

AKCAYLI MELEK M A DR
ASLAN ZEKI DR
AYDIN CEMAL PROF DR
BOZKURT SUKRU DR
DIZER MUAMMER PROF
ENGINOL TURAN B DR
ESKIOGLU A NIHAT
GOELBASI ORHAN DR
GULMEN OMUR DR
IBANOGLU C DR
KIRAL ADNAN PROF
KIZILOGLU UEMIT DR
MENTESE HUSEYIN DR
OEZKAN MUSTAFA TUERKER DR
SEZER CENGIZ DR
YILMAZ FATMA DR

AKYOL MUSTAFA UNAL PROF
ATAC TAMER
BALLI EDIBE PROF
DEMIRCAN OSMAN DR
DOGAN NADIR PROF
ERCAN E NIHAL
EVREN SERDAR DR
GOKDOGAN NUZHET PROF
HAZER S DR
KANDEMIR GUELCIN
KIRBIYIK HALIL DR
KOCER DURCUN DR
OEKTEN ADNAN DR
OZGUC ATILA
TEKTUNALI H GOKMEN DR
YILMAZ NIHAL DR

ALPAR ALI DR
AVCIOGLU KAMURAN PROF DR
BOLCAL CETIN DR
DERMAN I ETHEM DR
ENGIN SEMANUR PROF
ERTAN A YENER DR
EZER-ERYURT DILHAN PROF
GUDUR N DR
HOTINLI METIN DR
KARAALI SALIH DR
KIZILOGLU NILGUEN DR
MARSOGLU A DR
OEZEL MEHMET EMIN DR
PEKUENLUE E RENNAN DR
TUNCA ZEYNEL DR

Country : UK

AARSETH SVERRE J DR	ADAM MADGE G DR	ADAMS DAVID J DR
ADAMSON ANDREW DR	ADE PETER A R DR	ALBINSON JAMES DR
ALEXANDER JOHN B	ALEXANDER PAUL DR	ALLAN PETER M
ALLEN ANTHONY JOHN DR	ANDERSON BRYAN DR	ANDREWS DAVID A DR
ANDREWS PETER J DR	ARDAVAN HOUSHANG DR	ARGUE A NOEL MR
ARGYLE ROBERT WILLIAM MR	ATHERTON PAUL DAVID	AXON DAVID
BAILEY MARK EDWARD	BALDWIN JOHN E DR	BARLOW MICHAEL J DR
BAROCAS VINICIO PROF	BARROW JOHN DAVID	BARROW RICHARD F DR
BARSTOW MARTIN ADRIAN DR	BARUCH JOHN DR	BASTIN JOHN A PROF
BATES BRIAN DR	BATES DAVID R PROF	BATH GEOFFREY T DR
BEALE JOHN S DR	BEGGS DENIS W MR	BELL BURNELL S JOCELYN DR
BELL KENNETH LLOYD DR	BELL KENNETH LLOYD DR	BELL STEVEN DR
BENN CHRIS R DR	BERGER MITCHELL DR	BERRINGTON KEITH ADRIAN
BINGHAM RICHARD G DR	BINNEY JAMES J DR	BLACKMAN CLINTON PAUL DR
BLACKWELL DONALD E PROF	BODE MICHAEL F	BOKSENBERG ALEC PROF
BONDI HERMANN PROF SIR	BONNOR W B PROF	BOYD ROBERT L F PROF SIR
BRAND PETER W J L DR	BRANDUARDI-RAYMONT G	BRANSON NICHOLAS J B A DR
BRIDGELAND MICHAEL DR	BROMAGE GORDON E DR	BROOKES CLIVE J DR
BROWN JOHN C PROF	BROWNE IAN W A DR	BROWNING PHILIPPA DR
BRUCK HERMANN A PROF	BRUCK MARY T DR	BUNCLARK PETER STEPHEN DR
BURGESS ALAN DR	BURGESS DAVID D PROF	BURTON WILLIAM M
BUTCHINS SYDNEY ADAIR	BUTLER C JOHN DR	BYRNE PATRICK B DR
CALLANAN PAUL DR	CAMERON ANDREW COLLIER DR	CAMPBELL JAMES W
CARR BERNARD JOHN	CARSON T R DR	CARSWELL ROBERT F DR
CATCHPOLE ROBIN M DR	CHARLES PHILIP ALLAN	CLARK DAVID H DR
CLARKE DAVID DR	CLEGG PETER E DR	CLEGG ROBIN E S DR
CLUBE S V M DR	COHEN RAYMOND J DR	COLES PETER DR
CONWAY ROBIN G DR	COOK ALAN H PROF	COOKE B A DR
COOKE JOHN ALAN	COUPER HEATHER MISS	CRAWFORD IAN ANDREW DR
CROOM DAVID L DR	CRUISE ADRIAN MICHAEL DR	CULHANE LEONARD PROF
CZERNY MICHAL DR	DAINTREE EDWARD J DR	DARIUS JON DR
DAVIDSON WILLIAM PROF	DAVIES PAUL CHARLES W	DAVIES RODNEY D PROF
DAVIS RICHARD J DR	DE GROOT MART DR	DENNISON P A DR
DEWHIRST DAVID W DR	DICKENS ROBERT J DR	DIEGO FRANCISCO DR
DISNEY MICHAEL J PROF	DONNISON JOHN RICHARD DR	DORMAND JOHN RICHARD DR
DOWNES ANN JULIET B	DOYLE JOHN GERARD	DREW JANET
DUFFETT-SMITH PETER JAMES	DUFTON PHILIP L DR	DUNLOP STORM
DWORETSKY MICHAEL M DR	DYSON JOHN E DR	ECCLES MICHAEL J DR
EDMUNDS MICHAEL GEOFFREY	EDWIN ROGER P	EFSTATHIOU GEORGE
EGGLETON PETER P DR	ELLIOTT KENNETH H DR	ELLIS RICHARD S
ELSMORE BRUCE DR	EMERSON DAVID	EMERSON JAMES P
EVANS ANEURIN	EVANS KENTON DOWER DR	EVANS ROGER G DR
FABIAN ANDREW C DR	FALLE SAMUEL A DR	FAWELL DEREK R DR
FIELD DAVID	FIELDER GILBERT DR	FLETT ALISTAIR M
FLOWER DAVID R DR	FONG RICHARD	FOX W E MR
FRENK CARLOS S	FURNISS IAN	GADSDEN MICHAEL DR
GARRINGTON SIMON DR	GARTON W R S PROF	GEAKE JOHN E DR
GENT HUBERT MR	GIETZEN JOSEPH W	GILMORE GERARD FRANCIS
GLENCROSS WILLIAM M DR	GODWIN JON GUNNAR DR	GOLDSWORTHY FREDERICK A
GONDHALEKAR PRABHAKAR DR	GOUGH DOUGLAS O DR	GRAINGER JOHN F DR

GRANT IAN P DR	GREEN DAVID DR	GREEN ROBIN M DR
GREEN SIMON F	GRIFFIN MATTHEW J DR	GRIFFIN RITA E M DR
GRIFFIN ROGER F DR	GRIFFITHS WILLIAM K	GUEST JOHN E DR
GULL STEPHEN F DR	GUTHRIE BRUCE N G DR	HADLEY BRIAN W
HALL ANDREW NORMAN	HANBURY BROWN ROBERT PROF	HARPER DAVID DR
HARRIS STELLA	HARRISON RICHARD A DR	HARTLEY KENNETH F DR
HASSALL BARBARA J M DR	HAWARDEN TIMOTHY G DR	HAWKING STEPHEN W PROF
HAWKINS MICHAEL R S	HAYWARD JOHN	HAZARD CYRIL DR
HEAVENS ALAN DR	HEDDLE DOUGLAS W O PROF	HEGGIE DOUGLAS C DR
HERMANS DIRK DR	HEWETT PAUL	HEWISH ANTONY PROF
HEY JAMES STANLEY DR	HIDE RAYMOND PROF	HILDITCH RONALD W DR
HILL PHILIP W DR	HILLS RICHARD E DR	HILTON JOHN DR
HOARE MELVIN DR	HOLDEN FRANK	HOLLOWAY NIGEL J DR
HOOD ALAN	HOOD ALAN DR	HOSKIN MICHAEL A DR
HOUGH JAMES DR	HOWARTH IAN DONALD	HOWSE H DEREK
HOYLE FRED SIR	HUGHES DAVID W DR	HUMMEL EDSHO
HUMPHRIES COLIN M DR	HUNT G E DR	HUTCHEON RICHARD J DR
HYSOM EDMUND J	IRELAND JOHN G DR	IRWIN MICHAEL JOHN DR
ISAAK GEORGE R PROF	JACKSON JOHN CHARLES DR	JAMES JOHN F MR
JAMES RICHARD A DR	JAMESON RICHARD F DR	JARDINE MOIRA MARY DR
JEFFERY CHRISTOPHER S DR	JENKINS CHARLES R	JENNINGS R E PROF
JENNISON ROGER C PROF	JONES BARRIE W DR	JONES DEREK H P DR
JORDAN CAROLE DR	JORDEN PAUL RICHARD	JOSEPH ROBERT D DR
JUPP ALAN H DR	KAHN FRANZ D PROF	KAISER THOMAS R PROF
KENDERDINE SIDNEY DR	KIBBLEWHITE EDWARD J DR	KING ANDREW R DR
KING HENRY C DR	KINGSTON ARTHUR E PROF	KING-HELE DESMOND G DR
KITCHIN CHRISTOPHER R DR	KOPAL ZDENEK PROF	KOSOVICHEV ALEXANDER
KROTO HAROLD PROF	LACEY CEDRIC DR	LAHAV OFER DR
LAING ROBERT	LANCASTER BROWN PETER	LANG JAMES DR
LASENBY ANTHONY	LAWRENCE ANDREW DR	LEAHY J PATRICK DR
LEE TERENCE J DR	LITTLE LESLIE T DR	LONGAIR M S PROF
LOVELL SIR BERNARD PROF	LUCEY JOHN DR	LYNAS-GRAY ANTHONY E
LYNDEN-BELL DONALD PROF	LYNE ANDREW G DR	LYTTLETON RAYMOND A PROF
MACCALLUM MALCOLM A H	MACDONALD GEOFFREY H DR	MACGILLIVRAY HARVEY T DR
MACKAY CRAIG D DR	MACKINNON ALEXANDER L	MADDISON RONALD CH DR
MADDOX STEPHEN DR	MAJOR JOHN DR	MALIN STUART
MALLIA EDWARD A DR	MAREK JOHN	MARSDEN PHILIP L PROF
MARSH JULIAN C D	MARSH THOMAS DR	MARSHALL KEVIN P
MARTIN ANTHONY R DR	MARTIN DEREK H PROF	MARTIN WILLIAM L DR
MASON HELEN E DR	MASON JOHN WILLIAM DR	MASON KEITH OWEN
MATHESON DAVID NICHOLAS	MCCREA WILLIAM SIR	MCDONNELL J A M PROF
MCHARDY IAN MICHAEL DR	MCKEITH CONAL D DR	MCMULLAN DENNIS DR
MCNALLY DEREK DR	MCWHIRTER R W PETER DR	MEABURN J DR
MEADOWS A JACK PROF	MEIKLE WILLIAM P S	MESSAGE PHILIP J DR
MESTEL LEON PROF	MILES HOWARD G MR	MILLAR THOMAS J DR
MILLER JOHN C DR	MILLS ALLAN A DR	MITTON JACQUELINE
MITTON SIMON DR	MOFFATT HENRY KEITH PROF	MONTEIRO TANIA S DR
MOORE DANIEL R DR	MOORE PATRICK DR	MORGAN BRIAN LEALAN
MORGAN DAVID H DR	MORISON IAN MR	MORRIS MICHAEL C
MORRISON LESLIE V DR	MOSS CHRISTOPHER DR	MOSS DAVID L DR

Country : UK (Follow on)

MURDIN PAUL G DR	MURRAY C ANDREW	MURRAY CARL D DR
MURRAY JOHN B DR	MUXLOW THOMAS	NANDY KASHINATH DR
NAPIER WILLIAM M DR	NAVARRO JULIO FERNANDO DR	NELSON ALISTAIR H DR
NEWTON GAVIN DR	NICHOLSON WILLIAM	OSBORNE JOHN L DR
O'BRIEN PAUL THOMAS DR	PADMAN RACHAEL	PAGE CLIVE G DR
PALMER PHILIP DR	PAPALOIZOU JOHN C B DR	PARKER EDWARD A DR
PARKER QUENTIN DR	PARKINSON JOHN H DR	PATNAIK ALOK DR
PAXTON HAROLD J B R	PEACH GILLIAN DR	PEACH JOHN V DR
PEACOCK JOHN ANDREW	PEARCE GILLIAN DR	PEDLAR ALAN DR
PENNY ALAN JOHN DR	PENSTON MARGARET	PERRY JUDITH J DR
PETFORD A DAVID DR	PETTINI MAX	PHILLIPS JOHN PETER
PHILLIPS KENNETH J H	PIKE CHRISTOPHER DAVID	PILKINGTON JOHN D H DR
PILLINGER COLIN DR	PONMAN TREVOR DR	PONSONBY JOHN E B DR
POOLEY GUY DR	POUNDS KENNETH A PROF	PRIEST ERIC R PROF
PRINGLE JAMES E DR	PRINJA RAMAN DR	PYE JOHN P DR
QUENBY JOHN J DR	RACKHAM THOMAS W DR	RAINE DEREK J DR
RAPLEY CHRISTOPHER G DR	RAWLINGS JONATHAN DR	RAWLINGS STEVEN DR
REAY NEWRICK K DR	REES MARTIN J PROF	RICHARDSON KEVIN J
RIJNBEEK RICHARD DR	RILEY JULIA M DR	RING JAMES PROF
ROBERTSON JOHN ALISTAIR	ROBERTSON NORNA DR	ROBINSON WILLIAM J DR
ROBSON IAN E DR	ROCHE PATRICK F DR	RONAN COLIN A
ROWAN-ROBINSON MICHAEL DR	ROWSON BARRIE DR	ROXBURGH IAN W PROF
ROY ARCHIE E PROF	RUNCORN S K PROF	SANFORD PETER WILLIAM MR
SAUNDERS RICHARD D E	SCARROTT STANLEY M DR	SCHEUER PETER A G DR
SCHILD HANSRUEDI	SCHUTZ BERNARD F PROF	SCHWARTZ STEVEN JAY
SCOTT PAUL F DR	SEATON MICHAEL J PROF	SEYMOUR P A H
SHAKESHAFT JOHN R DR	SHALLIS MICHAEL J DR	SHARPLES RAY DR
SIM MARY E MISS	SIMMONS JOHN FRANCIS L	SIMNETT GEORGE M
SIMONS STUART DR	SINCLAIR ANDREW T DR	SISSON GEORGE M MR
SKILLEN IAN DR	SKILLING JOHN DR	SKINNER GERALD DR
SMITH F GRAHAM PROF	SMITH GEOFFREY DR	SMITH HUMPHRY M
SMITH LINDA J	SMITH ROBERT CONNON DR	SMITH RODNEY M DR
SMYTH MICHAEL J DR	SOMERVILLE WILLIAM B DR	SORENSEN SOREN-AKSEL DR
SPARKS WILLIAM BRIAN	SPEER R J DR	SPENCER RALPH E DR
STANNARD DAVID DR	STEPHENSON F RICHARD DR	STEWART JOHN MALCOLM DR
STEWART PAUL DR	STICKLAND DAVID J DR	STOBIE ROBERT S DR
SUMMERS HUGH P DR	SUTHERLAND WILLIAM DR	SWEET PETER A PROF
SYKES-HART AVRIL B DR	TAVAKOL REZA	TAYLER ROGER J PROF
TAYLOR DONALD BOGGIA DR	TER HAAR DIRK	TERLEVICH ELENA DR
TERLEVICH ROBERTO JUAN	THOBURN CHRISTINE	THOMAS DAVID V DR
THOMAS PETER A DR	THOMASSON PETER DR	THOMPSON G I DR
THOMPSON MICHAEL J DR	TOUT CHRISTOPHER DR	TOZER DAVID C DR
TRITTON KEITH P DR	TRITTON SUSAN BARBARA	TURNER MARTIN J L DR
TWISS R Q DR	TWORKOWSKI ANDRZEJ S	UNGER STEPHEN DR
VAN BREDA IAN G DR	VAN DER RAAY HERMAN B	VAN LEEUWEN FLOOR DR
VECK NICHOLAS	WALKER DAVID DOUGLAS DR	WALKER EDWARD N MR
WALKER HELEN J	WALKER IAN WALTER	WALL JASPER V DR
WALLACE PATRICK T MR	WALLIS MAX K DR	WALSH DENNIS DR
WALTON NICHOLAS A DR	WARD HENRY DR	WARD MARTIN JOHN
WARNER PETER J DR	WARWICK ROBERT S DR	WATSON MICHEAL G DR

Country : UK (Follow on)
_____

WATT GRAEME DAVID	WEISS NIGEL O DR	WELLGATE G BERNARD MR
WHITE GLENN J	WHITROW GERALD JAMES PROF	WHITWORTH ANTHONY PETER
WICKRAMASINGHE N C PROF	WILCOCK WILLIAM L PROF	WILKINS GEORGE A DR
WILKINSON ALTHEA	WILKINSON PETER N DR	WILLIAMS DAVID A PROF
WILLIAMS IWAN P PROF	WILLIAMS PEREDUR M DR	WILLIS ALLAN J DR
WILLMORE A PETER PROF	WILLSTROP RODERICK V DR	WILSON LIONEL DR
WILSON MICHAEL JOHN DR	WILSON ROBERT PROF SIR	WOLFENDALE ARNOLD W PROF
WOLSTENCROFT RAMON D DR	WOOD ROGER DR	WOOLFSON MICHAEL M PROF
WORRALL GORDON DR	WORSWICK SUSAN	WYNNE CHARLES G PROF
YALLOP BERNARD D DR	YAU KEVIN K C DR	ZARNECKI JAN CHARLES DR
ZUIDERWIJK EDWARDUS J		

Country : UKRAINE
_____

ANDRIENKO DMITRY A DR	BABIN ARTHUR DR	BRATIJCHUK MATRONA V
BRAUDE SEMION YA PROF AG	CHEREDNICHENKO V I DR	CHERNEGA N A A DR
CHERNYKH N S DR	CHUGAJNOV P F DR	CHUVAEV K K DR
DIVARI N B DR	DOBRONRAVIN PETER DR	DUMA DMITRIJ P DR
FEDOROVA RIMMA T DR	GERSHBERG R E DR	GOPASYUK S I DR
GORDON ISAAC M DR	GREGUL A YA DR	GRININ VLADIMIR P DR
GURTOVENKO E A DR	IVANCHUK VICTOR I DR	IZOTOV YURI DR
KARETNIKOV VALENTIN G R	KASHSCHEEV B L PROF DR	KHARIN A S DR
KISLYUK VITALIJ S DR	KLIMISHIN I A PROF	KOLCHINSKIJ I G DR
KOLESNIK IGOR G DR	KOLESNIK L N DR	KOMAROV N S DR
KONIN V V DR	KONOPLEVA VARVARA P DR	KOSTIK ROMAN I
KOTOV VALERY DR	KOVAL I K DR	KRAMER KH N DR
KRUCHINENKO VITALIY G	KUROCHKA L N DR	LITVINENKO LEONID N DR
LUPISHKO DMITRIJ F	LYUBIMKOV LEONID S DR	LYUTY VICTOR M DR
MAKARENKO EKATERINA N DR	MEDVEDEV YURI A DR	MEN' A V DR
MIRONOV NIKOLAY T	MOISEEV I G DR	MOROZHENKO A V DR
MOROZHENKO N N DR	OGIR MAYA DR	ONEGINA A B DR
ORLOV MIKHAIL DR	PETROV G M DR	PETROV PETER P DR
PINIGIN GENNADIJ I DR	POLOSUKHINA-CHUVAEVA N DR	POLUPAN P N DR
PROKOF'EV VLADIMIR K PROF	PROKOF'EVA VALENTINA V DR	PRONIK I I DR
PRONIK V I DR	PUGACH ALEXANDER F DR	RACHKOVSKY D N DR
ROMANCHUK PAVEL R DR	ROMANOV YURI S DR	SANDAKOVA E V DR
SHAKHOVSKOJ NIKOLAY M DR	SHESTAKA IVAN S DR	SHUL'BERG A M DR
SHUL'MAN L M DR	STEPANIAN A A DR	STEPANIAN N N DR
STEPANOV ALEXANDER V DR	STESHENKO N V DR	STESHENKO N V DR
TARADY VLADIMIR K DR	TEREBIZH VALERY YU DR	TSAP T T DR
VOROSHILOV V I DR	YAKOVKIN N A DR	YANOVITSKIJ EDGARD G DR
YATSKIV YA S DR	ZHARKOVA VATENINA DR	ZOSIMOVICH IRINA D

Country : URUGUAY

FERNANDEZ JULIO A DR

Country : USA

AANNESTAD PER ARNE DR	ABBOTT DAVID C DR	ABLES HAROLD D DR
ABT HELMUT A DR	ACTON LOREN W DR	ADAMS A N MR
ADAMS FRED DR	ADAMS JAMES H JR DR	ADAMS THOMAS F DR
ADEL ARTHUR F PROF EMER	ADELMAN SAUL J DR	AGUILAR CHIU LUIS A DR
AHLUWALIA HARJIT SINGH DR	AHMAD IMAD ALDEAN DR	AIZENMAN MORRIS L DR
ALBERS HENRY PROF	ALEXANDER JOSEPH K	ALLAN DAVID W MR
ALLEN RONALD J DR	ALLER HUGH D DR	ALLER LAWRENCE HUGH
ALLER MARGO F DR	ALLEY CARROL O DR	ALTROCK RICHARD C DR
ALTSCHULER MARTIN D PROF	ALVAREZ MANUEL DR	AMBRUSTER CAROL DR
ANAND S P S DR	ANANTHARAMAIAH K R DR	ANDERSEN TORBEN BRENDER
ANDERSON CHRISTOPHER M DR	ANDERSON KINSEY A PROF	ANDERSON KURT S
ANDREW KENNETH L PROF	ANGEL J ROGER P PROF	ANGIONE RONALD J DR
ANTHONY-TWAROG BARBARA J	ANTIOCHOS SPIRO KOSTA	APPLEBY JOHN F
APPLETON PHILIP NOEL DR	ARMANDROFF TAFT E DR	ARMSTRONG JOHN THOMAS DR
ARNETT W DAVID PROF	ARNOLD JAMES R DR	ARNQUIST WARREN N DR
ARNY THOMAS T DR	ARONS JONATHAN	ARTHUR DAVID W G
ASCHWANDEN MARKUS DR	ASPIN COLIN DR	ASSOUSA GEORGE ELIAS DR
ATHAY R GRANT DR	ATREYA SUSHIL K	AUER LAWRENCE H DR
AUGASON GORDON C DR	AVRETT EUGENE H DR	AYRES THOMAS R
A'HEARN MICHAEL F DR	BAAN WILLEM A	BAARS JACOB W M DR
BABCOCK HORACE W DR	BACKER DONALD CH DR	BAGNUOLO WILLIAM G JR DR
BAGRI DURGADAS S	BAHCALL JOHN N PROF	BAHNG JOHN D R PROF
BAIRD SCOTT R	BAKER JAMES GILBERT DR	BAKER NORMAN H PROF
BALBUS STEVEN A DR	BALDWIN RALPH B	BALICK BRUCE PROF
BALIUNAS SALLIE L	BALL JOHN A DR	BALLY JOHN DR
BALONEK THOMAS J DR	BANDERMANN L W DR	BANIA THOMAS MICHAEL
BARDEEN JAMES M PROF	BARKER EDWIN S DR	BARKER TIMOTHY DR
BARNES AARON DR	BARNES III THOMAS G DR	BARNOTHY JENO DR PROF
BARRETT ALAN H PROF	BARRETT PAUL EVERETT DR	BARRY DON C DR
BARTEL NORBERT HARALD DR	BARTH CHARLES A PROF	BARVAINIS RICHARD DR
BASART JOHN P	BASH FRANK N PROF	BASRI GIBOR B
BATCHELOR DAVID ALLEN DR	BATSON RAYMOND MILNER DR	BAUER CARL A DR
BAUER WENDY HAGEN	BAUM WILLIAM A DR	BAUSTIAN W W MR
BAUTZ LAURA P DR	BAYM GORDON ALAN DR	BEARD DAVID B DR
BEAVERS WILLET I DR	BECHTOLD JILL DR	BECKER ROBERT A DR
BECKER ROBERT HOWARD	BECKER STEPHEN A	BECKLIN ERIC E DR
BEEBE HERBERT A	BEEBE RETA FAYE DR	BEER REINHARD DR
BEERS TIMOTHY C DR	BEGELMAN MITCHELL CRAIG	BELL BARBARA DR
BELL ROGER A DR	BELSERENE EMILIA P	BELTON MICHAEL J S DR
BENDER PETER L DR	BENEDICT GEORGE F DR	BENFORD GREGORY DR
BENNETT CHARLES L DR	BENSON PRISCILLA J DR	BENZ WILLY
BERENDZEN RICHARD DR	BERG RICHARD A DR	BERGE GLENN L DR
BERGSTRALH JAY T DR	BERMAN ROBERT HIRAM DR	BERNAT ANDREW PLOUS DR
BERTSCHINGER EDMUND DR	BETTIS DALE G PROF	BHAVSAR SUKETU P
BIDELMAN WILLIAM P PROF	BIEGING JOHN HAROLD DR	BIGNELL R CARL DR

Country : USA (Follow on)

BILLINGHAM JOHN	BILLINGS DONALD E PROF	BINZEL RICHARD P DR
BIRETTA JOHN ANTHONY DR	BIRKINSHAW MARK	BLACK JOHN HARRY DR
BLADES JOHN CHRIS DR	BLAHA MILAN DR	BLAIR GUY NORMAN DR
BLAIR WILLIAM P DR	BLANDFORD ROGER DAVID DR	BLASIUS KARL RICHARD DR
BLESS ROBERT C PROF	BLITZ LEO DR	BLITZSTEIN WILLIAM DR
BLOEMHOF ERIC E DR	BLONDIN JOHN M DR	BLUDMAN SIDNEY A PROF
BLUMENTHAL GEORGE R DR	BODENHEIMER PETER PROF	BOEHM KARL-HEINZ PROF
BOEHM-VITENSE ERIKA PROF.	BOESGAARD ANN M PROF	BOESHAAR GREGORY ORTH DR
BOGGESS ALBERT DR	BOGGESS NANCY W DR	BOHANNAN BRUCE EDWARD
BOHLIN J DAVID DR	BOHLIN RALPH C DR	BOLDT ELIHU DR
BOLEY FORREST I	BOND HOWARD E DR	BONSACK WALTER K PROF
BOOK DAVID L	BOOKBINDER JAY A DR	BOOKMYER BEVERLY B DR
BOPP BERNARD W DR	BORD DONALD JOHN	BORDERIES NICOLE
BORIAKOFF VALENTIN	BORNMANN PATRICIA L DR	BOSS ALAN P DR
BOWELL EDWARD L G DR	BOWEN GEORGE H DR	BOWERS PHILLIP F
BOWYER C STUART PROF	BOYCE PETER B DR	BOYNTON PAUL EDWARD DR
BRACEWELL RONALD N PROF	BRADSTREET DAVID H DR	BRANCH DAVID R DR
BRANDT JOHN C DR	BRANSCOMB L M DR	BRAULT JAMES W DR
BREAKIRON LEE ALLEN DR	BRECHER AVIVA DR PROF	BRECHER KENNETH PROF
BRECKINRIDGE JAMES B DR	BREGMAN JOEL N	BRIDLE ALAN H PROF
BRINKS ELIAS DR	BROADFOOT A LYLE DR	BRODERICK JOHN DR
BRODIE JEAN P	BROUCKE ROGER DR	BROWN ALEXANDER
BROWN DOUGLAS NASON	BROWN HARRISON DR	BROWN ROBERT HAMILTON
BROWN ROBERT L DR	BROWNLEE DONALD E PROF	BROWNLEE ROBERT R DR
BRUCATO ROBERT J	BRUECKNER GUENTER E DR	BRUGEL EDWARD W DR
BRUHWEILER FRED C JR	BRUNER MARILYN E DR	BRUNING DAVID H DR
BRUNK WILLIAM E DR	BUCCIARELLI BEATRICE DR	BUCHLER J ROBERT PROF
BUFF JAMES S DR	BUHL DAVID DR	BUNNER ALAN N DR
BURATTI BONNIE J DR	BURBIDGE E MARGARET PROF	BURBIDGE GEOFFREY R PROF
BURKE BERNARD F DR	BURKHEAD MARTIN S	BURLAGA LEONARD F DR
BURNS JACK O'NEAL JR	BURNS JOSEPH A PROF	BURROWS ADAM SETH
BURSTEIN DAVID	BUSCOMBE WILLIAM PROF	BUTA RONALD J DR
BUTLER DENNIS DR	BUTTERWORTH PAUL	BYARD PAUL L DR
BYRD GENE G DR	CAHN JULIUS H PROF	CAILLAULT JEAN PIERRE DR
CALVIN WILLIAM H DR	CAMERON ALASTAIR G W PROF	CAMERON WINIFRED S MRS
CAMPBELL ALISON DR	CAMPBELL BELVA G S DR	CAMPBELL DONALD B
CAMPBELL MURRAY F	CAMPINS HUMBERTO DR	CANFIELD RICHARD C DR
CANIZARES CLAUDE R PROF	CAPEN CHARLES F	CAPRIOTTI EUGENE R DR
CARBON DUANE F DR	CARDELLI JASON A DR	CARGILL PETER J DR
CARLETON NATHANIEL P DR	CARLSON JOHN B	CARNEY BRUCE WILLIAM
CAROFF LAWRENCE J	CARPENTER KENNETH G DR	CARPENTER LLOYD DR
CARR THOMAS D PROF	CARRUTHERS GEORGE R DR	CARTER WILLIAM EUGENE
CASERTANO STEFANO DR	CASH WEBSTER C JR	CASSINELLI JOSEPH P DR
CASTELAZ MICHEAL W DR	CASTELLI JOHN P	CASTOR JOHN I DR
CATON DANIEL B DR	CATURA RICHARD C DR	CAUGHLAN GEORGEANNE R
CEFOLA PAUL J DR	CENTRELLA JOAN M DR	CHAFFEE FREDERIC H DR
CHAISSON ERIC J PROF	CHAMBERLAIN JOSEPH M DR	CHAMBERLAIN JOSEPH W PROF
CHAMBLISS CARLSON R DR	CHAN KWING LAM	CHANDRA SUBHASH
CHANDRASEKHAR S PROF	CHANMUGAM GANESAR PROF	CHAPMAN CLARK R DR
CHAPMAN GARY A DR	CHAPMAN ROBERT D DR	CHEN KWAN-YU PROF

Country : USA (Follow on)

CHENG CHUNG-CHIEH DR	CHEVALIER ROGER A DR	CHITRE DATTAKUMAR M DR
CHIU HONG-YEE DR	CHIU LIANG-TAI GEORGE	CHRISTIAN CAROL ANN
CHRISTIANSEN WAYNE A	CHRISTODOULOU DMITRIS DR	CHRISTY JAMES WALTER DR
CHRISTY ROBERT F DR	CHU YOU-HUA	CHUBB TALBOT A DR
CHUPP EDWARD L DR	CHURCHWELL EDWARD B DR	CLARK ALFRED JR PROF
CLARK BARRY G DR	CLARK FRANK OLIVER DR	CLARK GEORGE W PROF
CLARK THOMAS A DR	CLARKE JOHN T	CLAUSSEN MARK J DR
CLAYTON DONALD D PROF	CLAYTON GEOFFREY C DR	CLAYTON ROBERT N DR
CLEMENS DAN P DR	CLIFTON KENNETH ST	CLINE THOMAS L DR
CLIVER EDWARD W	COCHRAN ANITA L DR	COCHRAN WILLIAM DAVID DR
COCKE WILLIAM JOHN PROF	CODE ARTHUR D	COFFEEN DAVID L DR
COFFEY HELEN E MS	COHEN JEFFREY M DR	COHEN JUDITH DR
COHEN LEON PROF	COHEN MARSHALL H PROF	COHEN MARTIN DR
COHEN RICHARD S	COHEN ROSS D DR	COHN HALDAN N
COLBURN DAVID S DR	COLGATE STIRLING A DR	COLLINS GEORGE W II PROF
COMBI MICHAEL R DR	COMINS NEIL FRANCIS	CONDON JAMES J DR
CONKLIN EDWARD K	CONNOLLY LEO PAUL	CONTI PETER S DR
COOK JOHN W	COOK KEM HOLLAND DR	CORBALLY CHRISTOPHER
CORBIN THOMAS ELBERT DR	CORDES JAMES M	CORDOVA FRANCE A D
CORLISS C H DR	CORWIN HAROLD G JR	COTTON WILLIAM D Jr
COULSON IAIN M DR	COUNSELMAN CHARLES C PROF	COWAN JOHN J DR
COWIE LENNOX LAUCHLAN DR	COWLEY ANNE P DR	COWLEY CHARLES R PROF
COX ARTHUR N DR	COX DONALD P PROF	CRAINE ERIC RICHARD DR
CRANE PATRICK C	CRANNELL CAROL JO DR	CRAWFORD DAVID L DR
CROCKER DEBORAH ANN DR	CRUIKSHANK DALE P DR	CRUTCHER RICHARD M DR
CUDABACK DAVID D DR	CUDWORTH KYLE MCCABE DR	CUFFEY J MR
CULVER ROGER BRUCE DR	CUNTZ MANFRED DR	CURRIE DOUGLAS G DR
CZYZAK STANLEY J DR	DAHN CONARD CURTIS DR	DALGARNO ALEXANDER PROF
DANBY J M ANTHONY DR	DANFORD STEPHEN C DR	DANKS ANTHONY C DR
DANLY LAURA DR	DATLOWE DAYTON DR	DAVID LAURENCE P DR
DAVIDSEN ARTHUR FALNES DR	DAVIDSON KRIS DR	DAVIES MERTON E MR
DAVIES ROGER L DR	DAVILA JOSEPH DR	DAVIS CECIL G JR
DAVIS LEVERETT JR PROF	DAVIS MARC DR	DAVIS MICHAEL M DR
DAVIS MORRIS S PROF	DAVIS ROBERT J DR	DAVIS SUMNER P DR
DE FREES DOUGLAS J DR	DE JONGE J K DR	DE PATER IMKE
DE VAUCOULEURS GERARD PR	DE VINCENZI DONALD DR	DE YOUNG DAVID S DR
DEARBORN DAVID PAUL S DR	DEEMING TERENCE J DR	DELSEMME ARMAND H PROF DR
DEMARQUE P PROF	DEMING LEO DRAKE DR	DENNIS BRIAN ROY DR
DENNISON EDWIN W DR	DENT WILLIAM A PROF	DEPRIT ANDRE PROF
DERE KENNETH PAUL	DERMOTT STANLEY F	DESPAIN KEITH HOWARD DR
DEUPREE ROBERT G DR	DEUTSCHMAN WILLIAM A DR	DEVINNEY EDWARD J DR
DEVORKIN DAVID H	DEWITT BRYCE S DR	DEWITT JOHN H JR
DEWITT-MORETTE CECILE PR	DIAMOND PHILIP JOHN DR	DICK STEVEN J
DICKE ROBERT H PROF	DICKEL HELENE R DR	DICKEL JOHN R
DICKEY JEAN O'BRIEN	DICKEY JOHN M	DICKINSON DALE F DR
DICKMAN ROBERT L DR	DICKMAN STEVEN R	DIETER NANNIELOU H DR
DINERSTEIN HARRIET L	DIXON ROBERT S DR	DJORGOVSKI STANISLAV DR
DOGGETT LEROY E DR	DOHERTY LOWELL R PROF	DOLAN JOSEPH F. DR
DONN BERTRAM D	DOSCHEK GEORGE A DR	DOUGLAS JAMES N PROF
DOUGLASS GEOFFREY G	DOWNES RONALD A DR	DOWNS GEORGE S DR

Country : USA (Follow on)

DOYLE LAURANCE R DR	DRAINE BRUCE T	DRAKE FRANK D PROF
DRAKE STEPHEN A	DREHER JOHN W	DRESSEL LINDA L
DRESSLER ALAN	DREVER RONALD W P DR	DRILLING JOHN S
DRYER MURRAY DR	DUBIN MAURICE DR	DUFOUR REGINALD JAMES
DULK GEORGE A PROF	DUNCAN DOUGLAS KEVIN DR	DUNCOMBE RAYNOR L DR
DUNHAM DAVID W	DUNKELMAN LAWRENCE	DUNN RICHARD B DR
DUPREE ANDREA K DR	DUPUY DAVID L DR	DURISEN RICHARD H DR
DURNEY BERNARD DR	DURRANCE SAMUEL T DR	DUTHIE JOSEPH G PROF
DUVALL THOMAS L JR	DWEK ELI	DYCK M DR
DYER EDWARD R DR	DYSON F J DR	EATON JOEL A DR
EDDY JOHN A DR	EDELSON RICK DR	EDMONDSON FRANK K PROF
EDWARDS ALAN CH DR	EDWARDS TERRY W	EICHHORN HEINRICH K DR
EICHLER DAVID DR	EILEK JEAN	ELITZUR MOSHE
ELLIOT JAMES L DR	ELMEGREEN BRUCE GORDON DR	ELMEGREEN DEBRA MELOY
ELSTE GUNTHER H DR	ELSTON WOLFGANG E PROF	ELVIS MARTIN S DR
EL-BAZ FAROUK DR	EMSLIE A GORDON	ENDAL ANDREW S DR
EPPS HARLAND WARREN PROF	EPSTEIN EUGENE E DR	EPSTEIN GABRIEL LEO DR
EPSTEIN ISADORE PROF	EPSTEIN RICHARD I DR	ESHLEMAN VON R PROF
ESKRIDGE PAUL B DR	ESPOSITO F PAUL PROF	ESPOSITO LARRY W
ETZEL PAUL B DR	EUBANKS THOMAS M DR	EVANS J V DR
EVANS JOHN W DR	EVANS NEAL J II ASS PROF	EVANS W DOYLE
EWEN HAROLD I DR	EWING MARTIN S	FABBIANO GIUSEPPINA
FABER SANDRA M PROF	FABRICANT DANIEL G	FALCO-ACOSTA EMILIO E DR
FALGARONE EDITH	FALK SYDNEY W JR DR	FALL S MICHAEL DR
FALLER JAMES E PROF	FALLON FREDERICK W DR	FANG LI-ZHI
FANSELOW JOHN LYMAN	FAULKNER JOHN PROF	FAY THEODORE D DR
FAZIO GIOVANNI G DR	FEDERMAN STEVEN ROBERT	FEIBELMAN WALTER A DR
FEIGELSON ERIC D DR	FEKEL FRANCIS C	FELDMAN PAUL DONALD DR
FELDMAN URI	FELDMAN ÜRI DR	FELTEN JAMES E DR
FERLAND GARY JOSEPH	FEYNMAN JOAN DR	FIALA ALAN D DR
FICHTEL CARL E DR	FIEDLER RALPH L DR	FIELD GEORGE B PROF
FIENBERG RICHARD T DR	FILIPPENKO ALEXEI V DR	FINDLAY JOHN W DR
FINK UWE DR	FINN G D DR	FIROR JOHN W DR
FISCHEL DAVID DR	FISCHER JACQUELINE	FISHER GEORGE HEWITT DR
FISHER J RICHARD	FISHER PHILIP C	FISHER RICHARD R DR
FISHMAN GERALD J	FITCH WALTER S DR	FITZPATRICK EDWARD L DR
FIX JOHN D DR	FLANNERY BRIAN PAUL DR	FLECK ROBERT CHARLES DR
FLEISCHER ROBERT DR	FLIEGEL HENRY F	FLORKOWSKI DAVID R DR
FOGARTY WILLIAM G DR	FOLTZ CRAIG B	FOMALONT EDWARD B DR
FONTENLA JUAN MANUEL DR	FORBES J E DR	FORBES TERRY G DR
FORD HOLLAND C RES PROF	FORD W KENT JR DR	FORMAN WILLIAM RICHARD DR
FORREST WILLIAM JOHN	FORSTER JAMES RICHARD DR	FORT DAVID NORMAN DR
FOUKAL PETER V DR	FOWLER WILLIAM A PROF	FOX KENNETH DR
FRANK JUHAN	FRANKLIN FRED A DR	FRANZ OTTO G DR
FRAZIER EDWARD N DR	FREDRICK LAURENCE W PROF	FREEDMAN WENDY L DR
FRENCH RICHARD G	FRIBERG PER	FRIEDLANDER MICHAEL PROF
FRIEDMAN HERBERT DR	FRIEDMAN SCOTT DAVID DR	FRIEND DAVID B DR
FRISCH PRISCILLA	FROGEL JAY ALBERT DR	FROST KENNETH J DR
FRYE GLENN M PROF	FTACLAS CHRIST	FUHR JEFFREY ROBERT DR
FURENLID INGEMAR K DR	GAISSER THOMAS K	GALLAGHER III JOHN S DR

Country : USA (Follow on)

GALLAGHER JEAN W DR	GALLET ROGER M	GAPOSCHKIN EDWARD M DR
GARCIA MICHAEL R DR	GARFINKEL BORIS DR	GARLICK GEORGE F DR
GARMANY CATHERINE D DR	GARMIRE GORDON P PROF	GARSTANG ROY H PROF
GARY DALE E	GARY GILMER ALLEN DR	GATEWOOD GEORGE DIRECTOR
GATLEY IAN	GAUME RALPH A DR	GAUSS F STEPHEN
GAUSTAD JOHN E PROF	GEBALLE THOMAS R DR	GEBBIE KATHARINE B DR
GEHRELS TOM PROF	GEHRZ ROBERT DOUGLAS DR	GELDZAHLER BERNARD J
GELLER MARGARET JOAN	GENET R M DR	GERGELY TOMAS ESTEBAN DR
GEROLA HUMBERTO DR	GEZARI DANIEL YSA DR	GHIGO FRANCIS D DR
GIACCONI RICCARDO PROF	GIAMPAPA MARK S	GIBSON DAVID MICHAEL DR
GIBSON JAMES	GICLAS HENRY L MR	GIERASCH PETER J DR
GIES DOUGLAS R DR	GIGAS DETLEF DR	GILLILAND RONALD LYNN
GILMAN PETER A DR	GILRA DAYA P DR	GINGERICH OWEN PROF
GIOIA ISABELLA M DR	GIOVANE FRANK	GIOVANELLI RICCARDO DR
GLASER HAROLD DR	GLASPEY JOHN W DR	GLASS BILLY PRICE DR
GLASSGOLD ALFRED E PROF	GLATZMAIER GARY A	GOEBEL JOHN H DR
GOLD THOMAS PROF	GOLDMAN MARTIN V	GOLDREICH P DR
GOLDSMITH DONALD W DR	GOLDSMITH PAUL F DR	GOLDSMITH PAUL F DR
GOLDSTEIN RICHARD M DR	GOLDSTEIN SAMUEL J PROF	GOLDWIRE HENRY C JR
GOLUB LEON DR	GOMES RODNEY D S DR	GOODE PHILIP R
GOODRICH ROBERT W DR	GOODY R M	GOPALSWAMY N DR
GORDON COURTNEY P PROF	GORDON KURTISS J PROF	GORDON MARK A DR
GORENSTEIN MARC V	GORENSTEIN PAUL DR	GOSLING JOHN T DR
GOSS W MILLER PROF	GOTT III J RICHARD	GOTTESMAN STEPHEN T DR
GOTTLIEB CARL A DR	GOULD ROBERT J PROF	GRABOSKE HAROLD C JR
GRADIE JONATHAN CAREY	GRADY CAROL ANNE DR	GRAHAM ERIC DR
GRAHAM JOHN A DR	GRANDI STEVEN ALDRIDGE DR	GRASDALEN GARY L DR
GRAUER ALBERT D	GRAYZECK EDWIN J DR	GREEN JACK PROF
GREEN LOUIS C PROF	GREEN RICHARD F DR	GREENBERG RICHARD DR
GREENSTEIN GEORGE PROF	GREENSTEIN J L PROF	GREGORY STEPHEN ALBERT DR
GREISEN KENNETH I PROF	GREYBER HOWARD D DR	GRIFFITHS RICHARD E DR
GRINDLAY JONATHAN E DR	GROSS PETER G PROF	GROSSMAN ALLEN S PROF
GROSSMAN LAWRENCE PROF	GROTH EDWARD J III	GUETTER HARRY HENDRIK
GUIDICE DONALD A DR	GUINAN EDWARD FRANCIS DR	GULKIS SAMUEL DR
GULL THEODORE R DR	GUNN JAMES E PROF	GURMAN JOSEPH B DR
GURSKY HERBERT DR	GUSTAFSON BO A S	GWINN CARL R DR
HABBAL SHADIA RIFAI	HACKWELL JOHN A DR	HADDOCK FRED T DR
HAGEN JOHN P	HAGFORS T DR	HAGYARD MONA JUNE
HAISCH BERNHARD MICHAEL	HAKKILA JON ERIC DR	HALL DONALD N DR
HALL DOUGLAS S DR	HALL R GLENN DR	HALLAM KENNETH L DR
HAMMOND GORDON L DR	HANISCH ROBERT J DR	HANKINS TIMOTHY HAMILTON
HANNER MARTHA S DR	HANSEN CARL J PROF	HANSEN RICHARD T MR
HANSON ROBERT B DR	HAPKE BRUCE W DR	HARDEBECK ELLEN G DR
HARDEE PHILIP	HARMER CHARLES F W MR	HARMER DIANNE L MRS
HARMS RICHARD JAMES DR	HARNDEN FRANK R Jr	HARRINGTON J PATRICK DR
HARRINGTON ROBERT S DR	HARRIS ALAN WILLIAM DR	HARRIS DANIEL E DR
HARRIS HUGH C	HARRISON EDWARD R PROF	HART MICHAEL H DR
HARTEN RONALD H DR	HARTKOPF WILLIAM I DR	HARTMANN LEE WILLIAM
HARTMANN WILLIAM K	HARTOOG MARK RICHARD DR	HARVEL CHRISTOPHER ALVIN
HARVEY JOHN W DR	HARVEY PAUL MICHAEL DR	HARWIT MARTIN PROF

Country : USA (Follow on)

HASAN HASHIMA DR	HASCHICK AUBREY	HASEGAWA TATSUHIKO DR
HATHAWAY DAVID H DR	HATZES ARTIE P DR	HAUSER MICHAEL G DR
HAVLEN ROBERT J DR	HAWKINS GERALD S DR	HAWKINS ISABEL DR
HAYES DONALD S DR	HAYMES ROBERT C PROF	HAYNES MARTHA P
HAZEN MARTHA L DR	HEAP SARA R DR	HEASLEY JAMES NORTON
HECHT JAMES H DR	HECKATHORN HARRY M	HECKMAN TIMOTHY M
HEDEMAN E RUTH MISS	HEESCHEN DAVID S DR	HEFFERLIN RAY A PROF
HEGYI DENNIS J ASSOC PROF	HEILES CARL PROF	HEINTZ WULFF D DR
HEISER ARNOLD M DR	HELFAND DAVID JOHN	HELFER H LAWRENCE PROF
HELIN ELEANOR FRANCIS	HELLWIG HELMUT WILHELM DR	HELMKEN HENRY F DR
HELOU GEORGE DR	HEMENWAY MARY KAY M DR	HEMENWAY PAUL D DR
HENIZE KARL G ASTRONAUT	HENRY RICHARD B C DR	HENRY RICHARD C PROF
HERBIG GEORGE H DR	HERBST ERIC DR	HERBST WILLIAM DR
HERCZEG TIBOR J PROF DR	HERNQUIST LARS ERIC DR	HERR RICHARD B DR
HERSHEY JOHN L DR	HERTZ PAUL L DR	HEWITT ADELAIDE
HEWITT ANTHONY V DR	HIBBS ALBERT R MGR PLANS	HILDEBRAND ROGER H
HILDNER ERNEST DR	HILL FRANK DR	HILL HENRY ALLEN DR
HILLIARD R DR	HILLS JACK G DR	HINKLE KENNETH H
HINTEREGGER HANS E DR	HINTZEN PAUL MICHAEL N DR	HJELLMING ROBERT M DR
HO PAUL T P	HOAG ARTHUR A DR	HOBBS LEWIS M DR
HOBBS ROBERT W DR	HODGE PAUL W PROF	HOESSEL JOHN GREG
HOFF DARREL BARTON	HOFFLEIT E DORRIT DR	HOFFMAN JEFFREY ALAN DR
HOGAN CRAIG J DR	HOGG DAVID E DR	HOLBERG JAY B
HOLLENBACH DAVID JOHN DR	HOLLIS JAN MICHAEL DR	HOLLOWELL DAVID EARL DR
HOLLWEG JOSEPH V	HOLMAN GORDON D	HOLT STEPHEN S
HOLZER THOMAS EDWARD DR	HONEYCUTT R KENT PROF	HOOGHOUDT B G IR
HOROWITZ PAUL PROF	HOUCK JAMES R	HOUK NANCY DR
HOUSE LEWIS L DR	HOWARD ROBERT F DR	HOWARD W MICHAEL DR
HOWARD WILLIAM E III DR	HOWELL STEVE BRUCE DR	HRIVNAK BRUCE J
HU ESTHER M DR	HUBBARD WILLIAM B PROF	HUBENY IVAN
HUCHRA JOHN PETER DR	HUDSON HUGH S DR	HUEBNER WALTER F DR
HUENEMOERDER DAVID P DR	HUGHES JOHN P DR	HUGHES PHILIP
HUGHES SHAUN	HUGUENIN G RICHARD	HUMMER DAVID G DR
HUMPHREYS ROBERTA M PROF	HUNDHAUSEN ARTHUR DR	HUNTEN DONALD M PROF
HUNTER CHRISTOPHER PROF	HUNTER DEIDRE ANN	HUNTER JAMES H PROF
HUNTRESS WESLEY T DR	HURFORD GORDON JAMES	HUT PIET
HYDER C L DR	IANNA PHILIP A	IBEN ICKO JR PROF
ILLING RAINER M E	ILLINGWORTH GARTH D DR	IMAMURA JAMES DR
IMHOFF CATHERINE L DR	IMPEY CHRISTOPHER D DR	IPSER JAMES R PROF
IRVINE WILLIAM M PROF	IRWIN JOHN B PROF	JACCHIA LUIGI G DR
JACKSON BERNARD V DR	JACKSON PETER DOUGLAS DR	JACKSON WILLIAM M DR
JACOBS KENNETH C DR	JACOBSEN THEODOR S PROF	JACOBY GEORGE H
JAFFE DANIEL T	JAFFE WALTER JOSEPH DR	JANES KENNETH A DR
JANICZEK PAUL M DR	JANSSEN MICHAEL ALLEN	JASTROW ROBERT
JEFFERIES JOHN T DR	JEFFERIES STUART DR	JEFFERYS WILLIAM H DR
JENKINS EDWARD B DR	JENKINS L F MS	JENKNER HELMUT DR
JENNER DAVID C DR	JEWELL PHILIP R DR	JOHNSON DONALD R DR
JOHNSON FRED M PROF DR	JOHNSON HOLLIS R PROF	JOHNSON HUGH M DR
JOHNSON TORRENCE V DR	JOHNSTON KENNETH J	JOKIPII J R PROF
JONES BARBARA	JONES BURTON DR	JONES DAYTON L

Country : USA (Follow on)
_____

JONES ERIC M	JONES FRANK CULVER DR	JONES HARRISON PRICE DR
JONES THOMAS WALTER DR	JORDAN STUART D DR	JOSELYN JO ANN C DR
JOSEPH CHARLES LYNN DR	JOSS PAUL CHRISTOPHER DR	JOY MARSHALL J DR
JUDGE PHILIP DR	JUNKKARINEN VESA T DR	JURA MICHAEL DR
JURGENS RAYMOND F	JURKEVICH IGOR DR	KAFATOS MINAS DR
KAFTAN MAY A DR	KAHLER STEPHEN W DR	KAITCHUCK RONALD H
KALER JAMES B PROF	KALKOFEN WOLFGANG DR	KAMMEYER PETER C DR
KAMP LUCAS WILLEM DR	KANDRUP HENRY EMIL DR	KANE SHARAD R DR
KAPLAN GEORGE H DR	KAPLAN J DR	KAPLAN LEWIS D DR
KAROVSKA MARGARITA DR	KARP ALAN HERSH DR	KARPEN JUDITH T
KATZ JONATHAN I	KAUFMAN MICHELE DR	KAULA WILLIAM M PROF
KAWALER STEVEN D DR	KEEL WILLIAM C	KEENAN PHILIP C PROF EMER
KEIL KLAUS DR	KEIL STEPHEN L	KELLER CHARLES F
KELLER GEOFFREY	KELLERMANN KENNETH I DR	KELLOGG EDWIN M DR
KENNICUTT ROBERT C JR	KENT STEPHEN M	KENYON SCOTT J DR
KERR FRANK J DR	KESSLER KARL G DR	KHARE BISHUN N DR
KIELKOPF JOHN F DR	KIMBLE RANDY A DR	KING DAVID S PROF
KING IVAN R PROF	KING R B DR	KING ROBERT WILSON JR DR
KINMAN THOMAS D DR	KINNEY ANNE L DR	KIPLINGER ALAN L DR
KIRBY KATE P DR	KIRKPATRICK RONALD C DR	KIRSHNER ROBERT PAUL DR
KISSELL KENNETH E DR	KLARMANN JOSEPH PROF	KLEIN MICHAEL J DR
KLEIN RICHARD I DR	KLEINMANN DOUGLAS E DR	KLEMOLA ARNOLD R DR
KLEMPERER W K DR	KLEPCZYNSKI WILLIAM J DR	KLIMCHUK JAMES A DR
KLINGLESMITH DANIEL A DR	KLIORE ARVYDAS JOSEPH DR	KLOCK B L DR
KNACKE ROGER F DR	KNAPP GILLIAN R DR	KNIFFEN DONALD A DR
KNOWLES STEPHEN H DR	KO HSIEN C PROF	KOCH DAVID G
KOCH ROBERT H DR	KOESTER DETLEV DR	KOHL JOHN L DR
KOJOIAN GABRIEL DR	KOLB EDWARD W DR	KONDO YOJI DR
KONIGL ARIEH DR	KOO DAVID C-Y DR	KOORNNEEF JAN DR
KOPP ROGER A DR	KORMENDY JOHN DR	KOUPELIS THEODOROS DR
KOUVELIOTOU CHRYSSA DR	KOVAR N S DR	KOVAR ROBERT P DR
KOWAL CHARLES THOMAS	KRAFT ROBERT P PROF	KRAUS JOHN D PROF
KRAUSHAAR WILLIAM L PROF	KREIDL TOBIAS J N	KRIEGER ALLEN S DR
KRISCIUNAS KEVIN DR	KRISS GERARD A DR	KRISTIAN JEROME DR
KROGDAHL W S DR	KROLIK JULIAN H	KRON GERALD E DR
KRON KATHERINE GORDON	KRON RICHARD G	KRUMM NATHAN ALLYN
KRUPP EDWIN C DR	KUHI LEONARD V PROF	KUHN JEFFERY RICHARD DR
KUIPER THOMAS B H DR	KULKARNI SHRINIVAS R DR	KULSRUD RUSSELL M DR
KUMAR C KRISHNA DR	KUMAR SHAILENDRA	KUMAR SHIV S PROF
KUNDU MUKUL R DR	KURFESS JAMES D	KURUCZ ROBERT L DR
KUTNER MARC LESLIE DR	KUTTER G SIEGFRIED DR	KWITTER KAREN BETH DR
LA BONTE BARRY JAMES	LA DOUS CONSTANZE A DR	LACY CLAUD H DR
LACY JOHN H DR	LADA CHARLES JOSEPH DR	LAIRD JOHN B DR
LALA PETR DR	LAMB DONALD QUINCY JR DR	LAMB FREDERICK K PROF
LAMB RICHARD C DR	LAMB SUSAN ANN DR	LAMBERT DAVID L PROF
LAMPTON MICHAEL	LANDE KENNETH PROF	LANDECKER PETER BRUCE DR
LANDMAN DONALD ALAN	LANDOLT ARLO U PROF	LANE ADAIR P
LANE ARTHUR LONNE DR	LANG KENNETH R ASST PROF	LANGER GEORGE EDWARD DR
LANGER WILLIAM DAVID DR	LANZ THIERRY DR	LARSON HAROLD P DR
LARSON RICHARD B PROF	LARSON STEPHEN M	LASALA GERALD J DR

Country : USA (Follow on)
―――――――――――

LASHER GORDON JEWETT DR	LASKER BARRY M DR	LATHAM DAVID W DR
LATTIMER JAMES M DR	LAUTMAN D A DR	LAWRENCE CHARLES R DR
LAWRENCE G M DR	LAWRIE DAVID G	LAYZER DAVID PROF
LEA SUSAN MAUREEN DR	LEACOCK ROBERT JAY	LEBOFSKY LARRY ALLEN
LEBOVITZ NORMAN R PROF	LECAR MYRON DR	LECKRONE DAVID S DR
LEE PAUL D DR	LEIBACHER JOHN DR	LEIGHTON R B PROF
LEISAWITZ DAVID DR	LEPP STEPHEN H DR	LESTER DANIEL F DR
LEUNG CHUN MING DR	LEUNG KAM CHING PROF	LEVINE RANDOLPH H DR
LEVISON HAROLD F DR	LEVREAULT RUSSELL M DR	LEVY EUGENE H DR
LEWIN WALTER H G PROF	LEWIS BRIAN MURRAY DR	LEWIS J S
LI NED C DR	LIANG EDISON P DR	LIBBRECHT K G DR
LIDDELL U MR	LIEBERT JAMES W DR	LIESKE JAY H DR
LILLEY EDWARD A PROF	LILLIE CHARLES F DR	LIN CHIA C PROF
LIN DOUGLAS N C DR	LINCOLN J VIRGINIA MISS	LINDSEY CHARLES ALLAN
LINGENFELTER RICHARD E	LINKE RICHARD ALAN DR	LINNELL ALBERT P PROF
LINSKY JEFFREY L DR	LINSLEY JOHN	LIPPINCOTT SARAH LEE DR
LIPSCHUTZ MICHAEL E DR	LISSAUER JACK J DR	LISZT HARVEY STEVEN
LITTLETON JOHN E	LITTLE-MARENIN IRENE R DR	LITVAK MARVIN M DR
LIU SOU-YANG DR	LIVINGSTON WILLIAM C	LO KWOK-YUNG DR
LOCANTHI DOROTHY DAVIS DR	LOCKMAN FELIX J	LOCKWOOD G WESLEY DR
LONG KNOX S DR	LONGMORE ANDREW J	LONSDALE CAROL J DR
LOPES-GAUTIER ROSALY DR	LORD STEVEN DONALD DR	LOREN ROBERT BRUCE DR
LOVAS FRANCIS JOHN DR	LOVELACE RICHARD V E DR	LOW BOON CHYE
LOW FRANK J DR	LU PHILLIP K DR	LUCK R EARLE DR
LUCKE PETER B DR	LUGGER PHYLLIS M	LUNDQUIST CHARLES A DR
LUTTERMOSER DONALD DR	LUTZ BARRY L DR	LUTZ JULIE H DR
LUTZ THOMAS E DR	LUYTEN WILLEM J PROF	LYNCH DAVID K
LYNDS BEVERLY T DR	LYNDS ROGER C DR	MACALPINE GORDON M
MACCACARO TOMMASO DR	MACCHETTO FERDINANDO DR	MACCONNELL DARRELL J DR
MACDONALD JAMES	MACK PETER DR	MACQUEEN ROBERT M DR
MACY WILLIAM WRAY DR	MAGALHAES ANTONIO MARIO	MAGNANI LORIS ALBERTO DR
MALINA ROGER FRANK DR	MALITSON HARRIET H MS	MALKAMAEKI LAURI J DR
MALKAN MATTHEW ARNOLD DR	MALVILLE J MCKIM PROF	MANSFIELD VICTOR N PROF
MARAN STEPHEN P DR	MARGON BRUCE H PROF	MARGRAVE THOMAS EWING JR
MARISKA JOHN THOMAS	MARK JAMES WAI-KEE DR	MARKERT THOMAS H DR
MARKOWITZ WILLIAM DR	MARSCHALL LAURENCE A	MARSCHER ALAN PATRICK
MARSDEN BRIAN G DR	MARSHALL HERMAN LEE DR	MARSTON ANTHONY PHILIP DR
MARTENS PETRUS C DR	MARTIN ROBERT N DR	MARTIN WILLIAM C DR
MARTINS DONALD HENRY DR	MARVIN URSULA B DR	MASON GLENN M
MASSEY PHILIP L	MASSON COLIN R	MATHER JOHN CROMWELL
MATHEWS WILLIAM G PROF	MATHIEU ROBERT D DR	MATHIS JOHN S PROF
MATSAKIS DEMETRIOS N	MATSON DENNIS L DR	MATSUSHIMA SATOSHI DR
MATTEI JANET AKYUZ DR	MATTHEWS CLIFFORD PROF	MATTHEWS HENRY E DR
MATTHEWS THOMAS A DR	MATZ STEVEN MICHEAL DR	MATZNER RICHARD A PROF
MAX CLAIRE E DR	MAXWELL ALAN DR	MAYALL MARGARET W
MAYALL NICHOLAS U ASTRON	MAYER CORNELL H	MAYFIELD EARLE B DR
MAZUREK THADDEUS JOHN DR	MCALISTER HAROLD A DR	MCCABE MARIE K MS
MCCAMMON DAN	MCCARTHY DENNIS D DR	MCCLAIN EDWARD F
MCCLINTOCK JEFFREY E DR	MCCLUSKEY GEORGE E JR DR	MCCORD THOMAS B DR
MCCRAY RICHARD DR	MCCROSKY RICHARD E DR	MCDONALD FRANK B DR

Country : USA (Follow on)

MCDONOUGH THOMAS R DR	MCELROY M B DR	MCFADDEN LUCY ANN DR
MCGIMSEY BEN Q JR DR	MCGRAW JOHN T DR	MCINTOSH PATRICK S
MCKEE CHRISTOPHER F PROF	MCLAREN ROBERT A DR	MCLEAN BRIAN JOHN
MCLEAN IAN S DR	MCMAHAN ROBERT KENNETH DR	MCMILLAN ROBERT S DR
MCNAMARA DELBERT H DR	MEAD JAYLEE MONTAGUE DR	MEECH KAREN DR
MEEKS M LITTLETON DR	MEIER DAVID L	MEIER ROBERT R
MEINEL ADEN B PROF	MEISEL DAVID D DR	MELBOURNE WILLIAM G DR
MELIA FULVIO DR	MELNICK GARY J	MELOTT ADRIAN L PROF
MENDIS DEVAMITTA ASOKA DR	MERTZ LAWRENCE N DR	MESZAROS PETER DR
MEYER DAVID M DR	MEYERS KARIE ANN	MEYLAN GEORGES DR
MICHALITSIANOS ANDREW	MICHEL F CURTIS PROF	MICKELSON MICHAEL E DR
MIDDLEHURST BARBARA M MS	MIHALAS BARBARA R WEIBEL	MIHALAS DIMITRI DR
MIKESELL ALFRED H MR	MILKEY ROBERT W DR	MILLER FREEMAN D PROF
MILLER HUGH R PROF	MILLER JOSEPH S PROF	MILLER RICHARD H DR
MILLIGAN J E	MILLIKAN ALLAN G MR	MILLIS ROBERT L DR
MISCONI NEBIL YOUSIF DR	MISNER CHARLES W PROF	MITCHELL KENNETH J DR
MITCHELL RICHARD MR	MITCHELL WALTER E JR	MODALI SARMA B DR
MODISETTE JERRY L PROF	MOFFETT THOMAS J PROF	MOLNAR MICHAEL R PROF
MONET ALICE K B DR	MONET DAVID G	MOOK DELO E PROF
MOORE ELLIOTT P PROF	MOORE RONALD L DR	MOOS HENRY WARREN DR
MORAN JAMES M DR	MORGAN JOHN ADRIAN	MORGAN THOMAS H DR
MORGAN WILLIAM W PROF	MORIARTY-SCHIEVEN G H DR	MORRIS CHARLES S
MORRIS MARK ROOT DR	MORRIS STEVEN DR	MORRISON DAVID PROF
MORRISON NANCY DUNLAP DR	MORRISON PHILIP PROF	MORTON G A DR
MOTZ LLOYD PROF	MOULD JEREMY R	MOUSCHOVIAS TELEMACHOS CH
MOZURKEWICH DAVID DR	MUELLER IVAN I PROF	MUFSON STUART LEE DR
MUINONEN KARRI DR	MUKHERJEE KRISHNA	MULDERS GERARD F W
MULLAN DERMOTT J DR	MULLER RICHARD A	MUMFORD GEORGE S PROF
MUMMA MICHAEL JON	MUNDY LEE G DR	MUNRO RICHARD H DR
MURDOCK THOMAS LEE	MURPHY ROBERT E DR	MURRAY STEPHEN S DR
MUSEN PETER DR	MUSIELAK ZDZISLAW E DR	MUSMAN STEVEN DR
MUTEL ROBERT LUCIEN	MUTSCHLECNER J PAUL DR	MYERS PHILIP C
NACOZY PAUL E DR	NARAYAN RAMESH DR	NATHER R EDWARD
NEFF JAMES EDWARD DR	NEFF JOHN S	NEFF SUSAN GALE DR
NEIDIG DONALD F DR	NELSON BURT DR	NELSON GEORGE DRIVER DR
NELSON JERRY E DR	NELSON ROBERT M	NESS NORMAN F DR
NEUGEBAUER GERRY DR	NEUPERT WERNER M DR	NEWBURN RAY L JR
NEWHALL X X DR	NEWMAN MICHAEL JOHN DR	NEWSOM GERALD H PROF
NEWTON ROBERT R DR	NEY EDWARD P PROF	NICHOLS-BOHLIN JOY DR
NICOLAS KENNETH ROBERT	NIEDNER MALCOLM B DR	NIEDNER MALCOLM DR
NIELL ARTHUR E DR	NILSSON CARL DR	NISHIMURA TETSUO DR
NOERDLINGER PETER D PROF	NOLL KEITH STEPHEN DR	NOONAN THOMAS W PROF
NORIEGA-CRESPO ALBERTO DR	NORMAN COLIN A PROF	NOVICK ROBERT
NOYES ROBERT W PROF	NUTH JOSEPH A III	ODELL ANDREW P
ODENWALD STEN F DR	OEGERLE WILLIAM R	OEMLER AUGUSTUS JR DR
OERTEL GOETZ K DR	OESTERWINTER CLAUS	OGELMAN HAKKI B DR
OKA TAKESHI DR	OKE J BEVERLEY PROF	OLIVER BERNARD M DR
OLIVER BERNARD M PROF	OLIVER JOHN PARKER DR	OLOWIN RONALD PAUL DR
OLSEN KENNETH H DR	OLSON EDWARD C PROF	ORLIN HYMAN DR
ORMES JONATHAN F DR	ORRALL FRANK Q PROF	ORTON GLENN S DR

Country : USA (Follow on)

OSBORN WAYNE DR	OSMER PATRICK S DR	OSTER LUDWIG F PROF DR
OSTERBROCK DONALD E PROF	OSTRIKER JEREMIAH P PROF	OSWALT TERRY D DR
OWEN FRAZER NELSON DR	OWEN TOBIAS C PROF	OWOCKI STANLEY PETER DR
OZERNOY LEONID M PROF	OZSVATH I PROF	O'CONNELL ROBERT F PROF
O'CONNELL ROBERT WEST DR	O'DEA CHRISTOPHER P DR	O'DELL CHARLES R DR
O'DELL STEPHEN L	O'HANDLEY DOUGLAS A DR	O'KEEFE JOHN A DR
O'LEARY BRIAN T	PACHOLCZYK ANDRZEJ G PROF	PACIESAS WILLIAM S DR
PAGE DON NELSON	PAGE THORNTON L DR	PALMER PATRICK E PROF
PAN XIAO-PEI	PANAGIA NINO DR	PANEK ROBERT J DR
PANG KEVIN	PANKONIN VERNON LEE DR	PAP JUDIT
PAPAGIANNIS MICHAEL D PRO	PAPALIOLIOS COSTAS DR	PARESCE FRANCESCO DR
PARISE RONALD A DR	PARKER EUGENE N	PARKER ROBERT A R
PARKINSON TRUMAN DR	PARKINSON WILLIAM H DR	PARRISH ALLAN DR
PARSONS SIDNEY B DR	PARTRIDGE ROBERT B PROF	PASACHOFF JAY M PROF
PASCU DAN DR	PAULS THOMAS ALBERT DR	PAYNE DAVID G
PEALE STANTON J PROF	PEARSON TIMOTHY J	PEEBLES P JAMES E
PEERY BENJAMIN F PROF	PELLERIN JR CHARLES J DR	PENZIAS ARNO A DR
PERKINS FRANCIS W DR	PERLEY RICHARD ALAN	PERRY CHARLES L DR
PERRY PETER M DR	PESCH PETER DR	PETERS GERALDINE JOAN DR
PETERS WILLIAM L III DR	PETERSON BRADLEY MICHAEL	PETERSON CHARLES JOHN DR
PETERSON DEANE M DR	PETERSON LAURENCE E PROF	PETERSON RUTH CAROL DR
PETRO LARRY DAVID	PETROSIAN VAHE PROF	PETTENGILL GORDON H PROF
PFEIFFER RAYMOND J	PHILIP A G DAVIS	PHILLIPS JOHN G PROF
PHILLIPS THOMAS GOULD DR	PICKLES ANDREW JOHN DR	PIER JEFFREY R DR
PIERCE A KEITH DR	PIERCE DAVID ALLEN DR	PILACHOWSKI CATHERINE DR
PILCHER CARL BERNARD DR	PINES DAVID PROF	PINGREE DAVID PROF
PINTO PHILIP ALFRED DR	PIPHER JUDITH L	PLAVEC MIREK J PROF
PLAVEC ZDENKA DR	PNEUMAN GERALD W	POGO ALEXANDER DR
POLAND ARTHUR I DR	POLIDAN RONALD S	POLLACK JAMES B DR
PONNAMPERUMA CYRIL PROF	POPPER DANIEL M PROF	PORTER JASON G DR
POTTER ANDREW E DR	PRADHAN ANIL DR	PRASAD SHEO S
PRAVDO STEVEN H	PRENDERGAST KEVIN H PROF	PRESS WILLIAM H DR
PRESTON GEORGE W DR	PRESTON ROBERT ARTHUR	PRICE MICHAEL J DR
PRICE R MARCUS DR	PRICE STEPHAN DONALD	PRINCE HELEN DODSON PROF
PROBSTEIN R F DR	PROFFITT CHARLES R DR	PROTHEROE WILLIAM M PROF
PRYOR CARLTON PHILIP DR	PUETTER RICHARD C DR	PURCELL EDWARD M PROF
PUSCHELL JEFFERY JOHN	PYPER SMITH DIANE M DR	QUIRK WILLIAM J DR
RABIN DOUGLAS MARK	RADICK RICHARD R DR	RADOSKI HENRY R DR
RAFERT JAMES BRUCE	RAHE JURGEN PROF	RAMATY REUVEN DR
RAMSEY LAWRENCE W DR	RANK DAVID M PROF	RANKIN JOANNA M DR
RAO K NARAHARI	RATNATUNGA KAVAN U	RAYMOND JOHN CHARLES
READHEAD ANTHONY C S DR	REASENBERG ROBERT D DR	REAVES GIBSON PROF
REEVES EDMOND M DR	REICHERT GAIL ANNE DR	REID MARK JONATHAN DR
REID NEILL	REITSEMA HAROLD J	RENSE WILLIAM A DR
REVELLE DOUGLAS ORSON DR	REYES FRANCISCO DR	REYNOLDS JOHN H PROF
REYNOLDS RONALD J DR	REYNOLDS STEPHEN P	RHODES EDWARD J JR
RICHARDS MERCEDES T DR	RICHARDSON R S	RICHSTONE DOUGLAS O DR
RICKARD JAMES JOSEPH DR	RICKARD LEE J DR	RICKER GEORGE R DR
RICKETT BARNABY JAMES DR	RIDDLE ANTHONY C DR	RIEGEL KURT W DR
RIEGLER GUENTER R DR	RIGHINI-COHEN GIOVANNA DR	RINDLER WOLFGANG PROF

Country : USA (Follow on)

RIVOLO ARTHUR REX	ROACH FRANKLIN E	ROARK TERRY P PROF
ROBBINS R ROBERT PROF	ROBERGE WAYNE G DR	ROBERTS DAVID HALL DR
ROBERTS MORTON S DR	ROBERTS WILLIAM W JR PROF	ROBERTSON DOUGLAS S
ROBINSON EDWARD LEWIS DR	ROBINSON I PROF	ROBINSON LEIF J
ROBINSON LLOYD B DR	RODDIER CLAUDE DR	RODDIER FRANCOIS PROF
RODMAN RICHARD B DR	ROEDER ROBERT C PROF	ROEMER ELIZABETH PROF
ROGERS ALAN E E DR	ROGERSON JOHN B PROF	ROGSTAD DAVID H DR
ROMAN NANCY G DR	ROMANISHIN WILLIAM DR	ROMNEY JONATHAN D DR
ROOD HERBERT J	ROOD ROBERT T DR	ROOSEN ROBERT G DR
ROOSEN ROBERT G DR	ROSE JAMES ANTHONY	ROSE WILLIAM K DR
ROSEN EDWARD DR	ROSENDHAL JEFFREY D DR	ROSNER ROBERT
ROSS DENNIS K PROF	ROSSI BRUNO B PROF	ROTS ARNOLD H DR
ROUNTREE JANET DR	ROUSE CARL A DR	ROUTLY PAUL M DR
RUBIN ROBERT HOWARD	RUBIN VERA C DR	RUDERMAN MALVIN A
RUDNICK LAWRENCE DR	RUGGE HUGO R DR	RULE BRUCE H
RUSSELL CHRISTOPHER T	RUSSELL JANE L DR	RUSSELL JOHN A PROF
RUST DAVID M DR	RYBICKI GEORGE B DR	RYDGREN ALFRED ERIC JR DR
SACKMANN I JULIANA DR	SADUN ALBERTO CARLO DR	SAFKO JOHN L
SAGAN CARL DR	SALISBURY J W DR	SALO HEIKKI
SALPETER EDWIN E PROF	SALZER JOHN JOSEPH DR	SAMEC RONALD G DR
SAMPSON DOUGLAS H PROF	SANDAGE ALLAN	SANDELL GORAN HANS L DR
SANDERS DAVID B DR	SANDERS W L PROF	SANDERS WILTON TURNER III
SANDFORD MAXWELL T II	SANDMANN WILLIAM HENRY	SANYAL ASHIT DR
SARAZIN CRAIG L DR	SARGENT ANNEILA I	SARGENT WALLACE L W DR
SARTORI LEO PROF	SASALOV DIMITAR D DR	SASLAW WILLIAM C PROF
SAVAGE BLAIR D DR	SAVEDOFF MALCOLM P PROF	SAWYER CONSTANCE B DR
SCALO JOHN MICHAEL	SCARGLE JEFFREY D DR	SCHAEFER BRADLEY E DR
SCHECHTER PAUL L DR	SCHERB FRANK PROF	SCHERRER PHILIP H DR
SCHILD RUDOLPH E DR	SCHILLER STEPHEN	SCHLEGEL ERIC MATTHEW DR
SCHLEICHER DAVID G DR	SCHLESINGER BARRY M DR	SCHLOERB F PETER
SCHMAHL EDWARD J DR	SCHMALBERGER DONALD C DR	SCHMELZ JOAN T DR
SCHMIDT EDWARD G	SCHMIDT MAARTEN PROF	SCHMIDTKE PAUL C DR
SCHNEIDER GLENN H DR	SCHNEPS MATTHEW H	SCHOOLMAN STEPHEN A DR
SCHRAMM DAVID N PROF	SCHREIER ETHAN J DR	SCHROEDER DANIEL J PROF
SCHUECKING E L DR	SCHULTE D H DR	SCHULTZ ALFRED BERNARD DR
SCHUTZ BOB EWALD	SCHWARTZ DANIEL A DR	SCHWARTZ PHILIP R DR
SCHWARTZ RICHARD D	SCHWARZSCHILD MARTIN PROF	SCHWEIZER FRANCOIS DR
SCONZO PASQUALE DR	SCOTT ELIZABETH L PROF	SCOTT EUGENE HOWARD
SCOTT JOHN S DR	SCOVILLE NICHOLAS Z	SEARLE LEONARD DR
SEARS RICHARD LANGLEY DR	SEEDS MICHAEL AUGUST DR	SEEGER CHARLES LOUIS III
SEEGER PHILIP A DR	SEGAL IRVING E DR	SEIDELMANN P KENNETH DR
SEIDEN PHILIP E	SEIELSTAD GEORGE A	SEKANINA ZDENEK DR
SEKIGUCHI KAZUHIRO DR	SELLWOOD JERRY A	SEWARD FREDERICK D
SHAFFER DAVID B DR	SHAFTER ALLEN W DR	SHAHAM JACOB PROF
SHAO CHENG-YUAN	SHAPERO DONALD C DR	SHAPIRO IRWIN I PROF
SHAPIRO MAURICE M PROF	SHAPIRO STUART L	SHAPLEY ALAN H
SHARA MICHAEL DR	SHARPLESS STEWART PROF	SHAW JAMES SCOTT DR
SHAW JOHN H PROF	SHAW R WILLIAM PROF	SHAWHAN STANLEY D DR
SHAWL STEPHEN J DR	SHAYA EDWARD J DR	SHEA MARGARET A DR
SHEELEY NEIL R DR	SHEFFIELD CHARLES DR	SHELUS PETER J DR

SHEN BENJAMIN S P PROF
SHER DAVID DR
SHIELDS GREGORY A DR

SHINE RICHARD A DR
SHIPMAN HENRY L DR
SHIVANANDAN KANDIAH DR

SHOEMAKER EUGENE M
SHORE BRUCE W
SHORE STEVEN N

SHOSTAK G SETH DR
SHU FRANK H PROF
SHULL JOHN MICHAEL

SHULL PETER OTTO DR
SILBERBERG REIN DR
SILK JOSEPH I PROF

SILVERBERG ERIC C DR
SIMKIN SUSAN M DR
SIMON GEORGE W DR

SIMON MICHAL PROF
SIMON NORMAN R PROF
SIMON THEODORE

SIMONSON S CHRISTIAN DR
SINTON WILLIAM M
SION EDWARD MICHAEL

SIRY JOSEPH W
SITKO MICHAEL L
SITTERLY CHARLOTTE M DR

SJOGREN WILLIAM L MR
SKALAFURIS ANGELO J
SKILLMAN EVAN D DR

SKUMANICH ANDRE PROF
SLADE MARTIN A III DR
SLETTEBAK ARNE PROF

SLOVAK MARK HAINES DR
SMITH ALEX G PROF
SMITH ANDREW M DR

SMITH BARHAM W DR
SMITH BRADFORD A PROF
SMITH BRUCE F DR

SMITH CHARLES DITTO
SMITH CLAYTON A JR DR
SMITH DEAN F DR

SMITH ELSKE V P DR
SMITH ERIC PHILIP DR
SMITH GRAEME H DR

SMITH HARDING E JR DR
SMITH HAYWOOD C DR
SMITH HORACE A

SMITH HOWARD ALAN
SMITH MALCOLM G DR
SMITH MYRON A ASST PROF

SMITH PETER L DR
SMITH VERNE V DR
SMITH WM HAYDEN PROF

SMOLUCHOWSKI ROMAN PROF
SMOOT III GEORGE F.
SNEDEN CHRISTOPHER A

SNELL RONALD L
SNOW THEODORE P PROF
SNYDER LEWIS E

SOBERMAN ROBERT K DR
SOBIESKI STANLEY DR
SODERBLOM DAVID R

SODERBLOM LARRY DR
SOFIA SABATINO PROF
SOIFER BARUCH T DR

SOLOMON PHILIP M DR
SONETT CHARLES P PROF
SONNEBORN GEORGE DR

SOWELL JAMES ROBERT DR
SPARKE LINDA
SPARKS WARREN M DR

SPENCER JOHN HOWARD
SPERGEL DAVID N DR
SPICER DANIEL SHIELDS DR

SPIEGEL E DR
SPINRAD HYRON PROF
SPITZER LYMAN JR DR

SRAMEK RICHARD A DR
STACEY GORDON J DR
STACHNIK ROBERT V

STAHLER SETVEN W DR
STAHR-CARPENTER M DR
STANDISH E MYLES DR

STANFORD SPENCER A
STANLEY G J
STARK ANTONY A

STARK GLEN DR
STARRFIELD SUMNER PROF
STEBBINS ROBIN

STECHER THEODORE P
STECKER FLOYD W DR
STEFANIK ROBERT DR

STEIGER W R PROF
STEIGMAN GARY PROF
STEIMAN-CAMERON THOMAS DR

STEIN JOHN WILLIAM
STEIN ROBERT F ASSOC PROF
STEIN WAYNE A PROF

STEINOLFSON RICHARD S DR
STELLINGWERF ROBERT F DR
STENCEL ROBERT EDWARD

STEPHENSON C BRUCE PROF
STEPINSKI TOMASZ DR
STERN ROBERT ALLAN

STIER MARK T
STINEBRING DANIEL R
STOCKMAN HERVEY S JR DR

STOCKTON ALAN N DR
STONE EDWARD C DR
STONE R G DR

STONE REMINGTON P S DR
STONE RONALD CECIL
STRAND KAJ AA DR

STRITTMATTER PETER A PROF
STROBEL DARRELL F
STROM KAREN M

STROM ROBERT G PROF
STROM STEPHEN E
STRONG IAN B DR

STRONG JOHN D PROF
STRONG KEITH T DR
STRUBLE MITCHELL F

STRUCK-MARCELL CURTIS J
STRYKER LINDA L
STURCH CONRAD R DR

STURROCK PETER A PROF
SUESS STEVEN T DR
SULENTIC JACK W DR

SULLIVAN WOODRUFF T III
SWANK JEAN HEBB
SWEIGART ALLEN V DR

SWEITZER JAMES STUART DR
SWENSON GEORGE W JR PROF
SWERDLOW NOEL PROF

SWIHART THOMAS L DR
SYKES MARK VINCENT DR
SYNNOTT STEPHEN P

SZEBEHELY VICTOR G PROF
SZKODY PAULA DR
TAAM RONALD EVERETT DR

TADEMARU EUGENE DR
TAFF LAURENCE G DR
TALBOT RAYMOND J JR DR

TANDBERG-HANSSEN EINAR A
TAPIA-PEREZ SANTIAGO
TAPLEY BYRON D DR

TARNSTROM GUY DR
TARTER C BRUCE DR
TARTER JILL C DR

Country : USA (Follow on)

TAYLOR DONALD J DR	TAYLOR JOSEPH H PROF	TEAYS TERRY J DR
TEDESCO EDWARD F	TELESCO CHARLES M DR	TERRELL NELSON JAMES JR
TERRILE RICHARD JOHN	TERZIAN YERVANT PROF	TESKE RICHARD G PROF
TEUBEN PETER J DR	THADDEUS PATRICK PROF	THIELEMANN FRIEDRICH-KARL
THOLEN DAVID J DR	THOMAS JOHN H PROF	THOMAS RICHARD N DR
THOMAS ROGER J DR	THOMPSON A RICHARD DR	THOMPSON LAIRD A DR
THOMPSON RODGER I PROF	THONNARD NORBERT DR	THORNE KIP S PROF
THORSTENSEN JOHN R	THRONSON HARLEY ANDREW JR	THUAN TRINH XUAN DR
TIFFT WILLIAM G PROF	TIMOTHY J GETHYN DR	TIPLER FRANK JENNINGS DR
TOHLINE JOEL EDWARD	TOKUNAGA ALAN TAKASHI DR	TOLBERT CHARLES R DR
TOLLER GARY N DR	TOMASKO MARTIN G DR	TOMBAUGH CLYDE W PROF
TONRY JOHN DR	TOOMRE ALAR DR	TOOMRE JURI
TOUSEY RICHARD DR	TOWNES CHARLES HARD DR	TRAFTON LAURENCE M DR
TRAUB WESLEY ARTHUR	TREFFERS RICHARD R	TREXLER JAMES H MR
TRIMBLE VIRGINIA L DR	TRIPICCO MICHAEL J DR	TROLAND THOMAS HUGH
TRURAN JAMES W JR	TSURUTA SACHIKO DR	TSVETANOV ZLATAN IVANOV
TUCKER WALLACE H DR	TULL ROBERT G	TULLY RICHARD BRENT DR
TURNER BARRY E DR	TURNER EDWIN L DR	TURNER KENNETH C DR
TURNER MICHAEL S	TWAROG BRUCE A	TYLER JR G LEONARD DR
TYLKA ALLAN J DR	TYSON JOHN A DR	TYTLER DAVID DR
ULICH BOBBY LEE	ULMER MELVILLE P PROF	ULRICH ROGER K PROF
UNDERWOOD JAMES H DR	UNWIN STEPHEN C	UOMOTO ALAN K DR
UPGREN ARTHUR R DR	UPSON WALTER L II DR	UPTON E K L DR
URRY CLAUDIA MEGAN DR	USHER PETER D DR	USON JUAN M DR
VAN ALLEN JAMES A PROF	VAN ALTENA WILLIAM F PROF	VAN BLERKOM DAVID J PROF
VAN BREUGEL WIL	VAN CITTERS GORDON W DR	VAN DORN BRADT HALE DR
VAN FLANDERN THOMAS DR	VAN GORKOM JACQUELINE H	VAN HAMME WALTER
VAN HORN HUGH M PROF	VAN HOVEN GERARD DR	VAN RIPER KENNETH A DR
VAN SPEYBROECK LEON P DR	VANDEN BOUT PAUL A	VANDERVOORT PETER O DR
VAUGHAN ARTHUR H DR	VEEDER GLENN J DR	VENKATAKRISHNAN P DR
VERSCHUUR GERRIT L PROF	VERTER FRANCES DR	VESECKY J F DR
VEVERKA JOSEPH DR	VILA SAMUEL C PROF	VILAS FAITH DR
VISHNIAC ETHAN T	VOGEL STUART NEWCOMBE DR	VOGT STEVEN SCOTT
VORPAHL JOAN A DR	VRBA FREDERICK J DR	VRTILEK JAN M DR
VRTILEK SAEQA DIL DR	WACKERNAGEL H BEAT DR	WADDINGTON C JAKE PROF
WADE RICHARD ALAN DR	WAGNER RAYMOND L DR	WAGNER ROBERT M DR
WAGNER WILLIAM J DR	WAGONER ROBERT V PROF	WALBORN NOLAN R DR
WALKER ALTA SHARON DR	WALKER ARTHUR B C JR PROF	WALKER MERLE F PROF
WALKER RICHARD L	WALKER ROBERT C DR	WALKER ROBERT M A PROF
WALLACE LLOYD V DR	WALLACE RICHARD K	WALLERSTEIN GEORGE PROF
WALTER FREDERICK M	WALTERBOS RENE A M DR	WANG HAIMIN DR
WANG YI-MING DR	WANNIER PETER GREGORY DR	WARD RICHARD A DR
WARD WILLIAM R DR	WARDLE JOHN F C PROF	WARES GORDON W DR
WARNER JOHN W DR	WARREN WAYNE H JR DR	WARWICK JAMES W DR
WASSERMAN LAWRENCE H DR	WASSON JOHN T	WATSON WILLIAM D PROF
WDOWIAK THOMAS J DR	WEAVER HAROLD F PROF	WEAVER THOMAS A DR
WEBB DAVID F	WEBBER JOHN C DR	WEBBINK RONALD F DR
WEBER STEPHEN VANCE	WEBSTER ADRIAN S DR	WEEDMAN DANIEL W PROF
WEEKES TREVOR C DR	WEGNER GARY ALAN	WEHINGER PETER A DR
WEHRLE ANN ELIZABETH DR	WEIDENSCHILLING S J DR	WEILER EDWARD J DR

Country : USA (Follow on)

WEILER KURT W DR	WEILL GILBERT M DR	WEINBERG J L DR
WEINBERG STEVEN DR	WEIS EDWARD W DR	WEISBERG JOEL MARK
WEISHEIT JON C DR	WEISSKOPF MARTIN CH DR	WEISSMAN PAUL ROBERT
WEISTROP DONNA DR	WELCH WILLIAM J PROF	WELLER CHARLES S DR
WELLS DONALD C III DR	WENTZEL DONAT G DR	WESSELINK ADRIAAN J DR
WEST DONALD K DR	WEST ROBERT ALAN	WESTERHOUT GART DR
WESTPHAL JAMES A PROF	WETHERILL GEORGE W	WEYMANN RAY J PROF
WHEELER J CRAIG PROF	WHEELER JOHN A DR	WHIPPLE ARTHUR L DR
WHIPPLE FRED L DR	WHITAKER EWEN A	WHITE NATHANIEL M DR
WHITE ORAN R DR	WHITE R STEPHEN PROF	WHITE RAYMOND E DR
WHITE RICHARD E	WHITE RICHARD L	WHITE SIMON DAVID MANION
WHITFORD ALBERT E PROF	WHITMORE BRADLEY C	WHITNEY BALFOUR S
WHITNEY CHARLES A PROF	WHITTET DOUGLAS C B DR	WHITTLE D MARK DR
WIDING KENNETH G DR	WIESE WOLFGANG L DR	WIITA PAUL JOSEPH
WILDEY ROBERT L PROF DR	WILKENING LAUREL L DR	WILKES BELINDA J
WILKINSON DAVID T	WILL CLIFFORD M DR	WILLIAMON RICHARD M
WILLIAMS BARBARA A	WILLIAMS CAROL A	WILLIAMS GLEN A DR
WILLIAMS JAMES G DR	WILLIAMS JOHN A DR	WILLIAMS THEODORE B DR
WILLNER STEVEN PAUL DR	WILLS BEVERLEY J DR	WILLS DEREK DR
WILLSON LEE ANNE DR	WILLSON ROBERT FREDERICK	WILSON ALBERT G DR
WILSON ANDREW S DR	WILSON CURTIS A	WILSON JAMES R DR
WILSON ROBERT E PROF	WILSON ROBERT W DR	WILSON WILLIAM J DR
WINCKLER JOHN R PROF	WINDHORST ROGIER A DR	WING ROBERT F PROF
WINGET DONALD E	WINKLER GERNOT M R DR	WINKLER KARL-HEINZ A DR
WINKLER PAUL FRANK DR	WISNIEWSKI WIESLAW Z	WITHBROE GEORGE L DR
WITT ADOLF N DR	WITTEN LOUIS PROF	WOLFE ARTHUR M PROF
WOLFF SIDNEY C DR	WOLFSON C JACOB	WOLFSON RICHARD DR
WOLSZCZAN ALEXANDER DR	WOOD DAVID B DR	WOOD F BRADSHAW PROF
WOOD III H J DR	WOOD JANET H DR	WOOD JOHN A DR
WOODWARD PAUL R DR	WOOLF NEVILLE J	WOOSLEY S E PROF
WOOTTEN HENRY ALWYN	WORDEN SIMON P DR	WORLEY CHARLES E DR
WORRALL DIANA MARY	WRAY JAMES D DR	WRIGHT EDWARD L DR
WRIGHT FRANCES W.DR	WRIGHT HELEN GREUTER	WRIGHT JAMES P DR
WRIGHT MELVYN C H DR	WROBEL JOAN MARIE DR	WU CHI CHAO DR
WU SHI TSAN DR	WYCKOFF SUSAN DR	WYLLER ARNE A PROF
WYNN-WILLIAMS C G DR	WYSE ROSEMARY F DR	YAHIL AMOS DR
YANG KE-JUN	YAPLEE B S	YEH TYAN DR
YEOMANS DONALD K DR	YODER CHARLES F	YORK DONALD G DR
YOSS KENNETH M DR	YOUNG ANDREW T DR	YOUNG ARTHUR DR
YOUNG JUDITH SHARN	YOUNG LOUISE GRAY DR	YU YAN DR
YUAN CHI PROF	ZABRISKIE F R PROF	ZARE KHALIL DR
ZARRO DOMINIC M DR	ZEILIK MICHAEL II DR	ZELLNER BENJAMIN H DR
ZENSUS J-ANTON DR	ZHANF SHOUZHONG DR	ZHANG ER-HO DR
ZINN ROBERT J DR	ZIRIN HAROLD DR	ZIRKER JACK B DR
ZOMBECK MARTIN V DR	ZUCKERMAN BEN M DR	

Country : UZBEKISTAN

BYKOV MIKLE F DR	KALMYKOV A M DR	LATYPOV A A DR
SLONIM E M DR	YULDASHBAEV TAIMAS S	

Country : VATICAN CITY STATE

BOYLE RICHARD P DR	CASANOVAS JUAN DR	COYNE GEORGE V DR
MCCARTHY MARTIN F DR	STOEGER WILLIAM R DR	

Country : VENEZUELA

BRUZUAL GUSTAVO	CALVET NURIA DR	FERRIN IGNACIO
FUENMAYOR FRANCISCO J DR	IBANEZ S MIGUEL H DR	MENDOZA CLAUDIO
STOCK JURGEN D		

Country : YUGOSLAVIA

ANGELOV TRAJKO	ARSENIJEVIC JELISAVETA	CADEZ ANDREJ DR
CADEZ VLADIMIR	DADIC ZARKO DR	DIMITRIJEVIC MILAN
DINTINJANA BOJAN DR	DJURASEVIC GOJKO	DJUROVIC DRAGUTIN M DR
DOMINKO FRAN PROF DR	JOVANOVIC BOZIDAR	KILAR BOGDAN DR
KNEZEVIC ZORAN	KUBICELA ALEKSANDAR DR	KUZMANOSKI MIKE
LAZOVIC JOVAN P PROF	LUKACEVIC ILIJA S DR	MILOGRADOV-TURIN JELENA
MILOVANOVIC VLADETA DR	MITIC LJUBISA A DR	NINKOVIC SLOBODAN
PAKVOR IVAN	PAVLOVSKI KRESIMIR	POPOVIC BOZIDAR PROF DR
POPOVIC GEORGIJE DR	PROTICH MILORAD B	RANDIC LEO PROF DR
RUZDJAK VLADIMIR DR	SADZAKOV SOFIJA DR	SALETIC DUSAN
SEGAN STEVO	SEVARLIC BRANISLAV M PROF	SIMOVLJEVITCH JOVAN L DR
SOLARIC NIKOLA	VINCE ISTVAN	VRSNAK BOJAN DR
VUJNOVIC VLADIS DR	VUKICEVIC K M PROF DR	ZWITTER TOMAZ

3.   Geographical repartition of members   Members

        within non adhering countries   41

Country : HONG KONG

YU XIN ALFRED DR

Country : LEBANON

PLASSARD J DR

Country : NIGERIA

AKUJOR CHIDI E	OKEKE PIUS N DR	OKOYE SAMUEL E PROF
ONUORA LESLEY IRENE DR	SCHMITTER EDWARD F DR	

Country : PAKISTAN

QUAMAR JAWAID

Country : PARAGUAY

TROCHE-BOGGINO A E DR

Country : PHILIPPINES

VINLUAN RENATO

Country : RUMANIA

BOTEZ ELVIRA DR	CARSMARU MARIA M DR	CRISTESCU CORNELIA G DR
DINESCU A DR	DRAMBA C PROF	DUMITRACHE CHRISTANA
LUNGU NICOLAIE DR	MARIS GEORGETA DR	MIHAILA IERONIM PROF
MIOC VASILE DR	NADOLSCHI V PROF DR	OPROIU TIBERIU DR
PAL ARPAD PROF DR	POP VASILE DR	PREDEANU IRINA DR
RUSU I DR	RUSU L DR	STANILA GEORGE DR
TIFREA EMILIA DR	TODORAN IOAN DR	TORO TIBOR PROF
URECHE VASILE DR		

Country : SINGAPORE
_____

WAN FOOK SUN   WILSON S J

Country : SRI LANKA
_____

DE SILVA L N K DR   MAHESWARAN MURUGESAPILLAI

Country : SYRIA
_____

YAO ZHENG-QIU

Country : THAILAND
_____

PORNCHAI P.-TANAKUN   SONGSATHAPORN RUANGSAK DR

Country : VIETNAM
_____

HUAN NGEYEN DUSH DR   TUNG NGUYEN MAU DR

4.   Alphabetical list of members

Note: In field "COM",

         "EC" means Member of the Executive Committee
              or Adviser to the Executive Committee
         "P" means President of Commission
         "V" means Vice-President of Commission
         "C" means Member of the Scientific Committee
              of the related Commission

AANNESTAD PER ARNE DR
DPT OF PHYSICS
ARIZONA STATE UNIVERSITY
TEMPE AZ 85287
USA
TEL 602 965 3644
TLF
TLX
EML
COM 34

AARSETH SVERRE J DR
INSTITUTE OF ASTRONOMY
THE OBSERVATORIES
MADINGLEY RD
CAMBRIDGE CB3 OHA
UK
TEL 223 622 04
TLF
TLX 817297 ASTRON G
EML
COM 33,37

ABAD ALBERTO J DR
DPT FISICA TEORICA
UNIVERSIDAD DE ZARAGOZA
E 50009 ZARAGOZA
SPAIN
TEL 76 35 7011
TLF
TLX 58198
EML
COM 07

ABALAKIN VICTOR K DR
PULKOVO OBSERVATORY
ACADEMY OF SCIENCES
10 KUTUZOV QUAY
196140 ST PETERSBURG
RUSSIA
TEL 298-2242
TLF
TLX 12261 FENIKS
EML
COM 04C,05,07,20

ABBASOV ALIK R DR
SCIENT & INDUSTRIAL ASS
OF COSMIC RESEARCH
LENIN PROSPEKT 159
370106 BAKU
AZERBAIDZHAN
TEL
TLF
TLX
EML
COM 10

ABBOTT DAVID C DR
JILA
UNIVERSITY OF COLORADO
BOX 440
BOULDER CO 80309 0440
USA
TEL
TLF
TLX
EML
COM 36

ABBOTT WILLIAM N DR
UNIVERSITY OF ATHENS
MICHALACOPOULOU 42
GR 115 28 ATHENS
GREECE
TEL 1 721 3352
TLF
TLX
EML
COM 22

ABDEL HADY AHMED DR

EGYPT
TEL
TLX
TLF
EML
COM

ABDELKAWI M ABUBAKR DR
ACADEMY OF SCIENTIFIC
RESEARCH AND TECHNOLOGY
101 KASR EL-EINI STREET
CAIRO
EGYPT
TEL
TLF
TLX
EML
COM

ABDULLA SHAKER ABDUL AZIZ
SARC
SCIENTIFIC RES COUNCIL
BOX 2441
JADIRIYAH BAGHDAD
IRAQ
TEL 1 776 5127
TLF
TLX 2187 BATHILMI IK
EML
COM 40

ABELE MARIS K DR
ASTRONOMICAL OBSERVATORY
LATVIAN STATE UNIVERSITY
RAINIS BUL 19
226098 RIGA
LATVIA
TEL
TLF
TLX
EML
COM 31

ABHYANKAR KRISHNA D PROF
DPT OF ASTRONOMY
UNIVERSITY OF OSMANIA
HYDERABAD 500 007
INDIA
TEL 85 1672
TLF
TLX
EML
COM 24,29,36,42

ABLES HAROLD D DR
US NAVAL OBSERVATORY
FLAGSTAFF STATION
BOX 1149
FLAGSTAFF AZ 86002
USA
TEL 602 779 5132
TLF
TLX 26230 ASTRO
EML
COM 09,25,28

ABLES JOHN G DR
CSIRO
DIVISION OF RADIOPHYSICS
BOX 76
EPPING NSW 2121
AUSTRALIA
TEL
TLF
TLX
EML
COM 40

ABOUDARHAM JEAN DR
OBSERVATOIRE DE PARIS
SECTION DE MEUDON
F 92195 MEUDON PPL CDX
FRANCE
TEL 1 45 07 7784
TLF 1 4507 7469
TLX 201571
EML MESIOA ABOU
COM 10,12

ABOU-EL-ELLA MOHAMED S DR
HELWAN OBSERVATORY
HELWAN
EGYPT
TEL 78 0645/2683
TLF
TLX 93070 HIAG UN
EML
COM 37

ABRAHAM ZULEMA DR
IAG/USP
CP 30627
01051 SAO PAULO SP
BRAZIL
TEL 11 577 8599
TLF 11 815 3848
TLX 1156735 IAGM BR
EML SPAN 47550 IAGUSP
COM

ABRAMI ALBERTO PROF
OAT
BOX SUCC TRIESTE 5
VIA TIEPOLO 11
I 34131 TRIESTE
ITALY
TEL
TLF
TLX
EML
COM 10,40

ABRAMOWICZ MAREK DR
SISSA
ST COSTIERA 11
MIRAMARE
I 34014 TRIESTE
ITALY
TEL 40 22 4281
TLF
TLX 460392 ICTP
EML
COM 48

ABT HELMUT A DR
KITT PEAK NTL OBS
BOX 26732
950 N CHERRY AVE
TUCSON AZ 85726 6732
USA
TEL 602 325 9215
TLF
TLX 0666-484 AURA NOAO
EML
COM 05,26P,29,30

ABULAZM MOHAMED SAMIR DR
HELWAN OBSERVATORY
HELWAN
EGYPT
TEL 78 0645/2683
TLF
EML
COM
TLX 93070 HIAG UN

ACHARYA BANNANJE S DR
TIFR/COSMIC RAY GROUP
HOMI BHABHA RD
COLABA
BOMBAY 400 005
INDIA
TEL 22 215 2971
TLF 22 215 2110
TLX 011 3009 TIFR IN
EML ACHARYA@TIFRVAX BITNET
COM 48

ACHTERBERG ABRAHAM DR
STERREKUNDIG INSTITUTE
BOX 80000
NL 3508 TA UTRECHT
NETHERLANDS
TEL 30 53 5213
TLF
TLX 40048 FYLUT NL
EML BITNET WNMACHT@HUTRUUO
COM

ACKER AGNES PROF DR
OBS DE STRASBOURG
11 RUE UNIVERSITE
F 67000 STRASBOURG
FRANCE
TEL 88 35 4300
TLF 88 25 0160
TLX 890506 STAROBS
EML
COM 34,46

ACTON LOREN W DR
LOCKHEED PALO ALTO RES LB
DPT 91 20 BLDG 255
3251 HANOVER ST
PALO ALTO CA 94304
USA
TEL 415 424 3267
TLF
TLX 346409 LMSC SUVL
EML
COM 12,44

ADAM MADGE G DR
DPT OF ASTROPHYSICS
UNIVERSITY OF OXFORD
SOUTH PARKS RD
OXFORD OX1 3RQ
UK
TEL
TLF
TLX
EML
COM 12

ADAMS A N MR
6549 N 35TH RD
ARLINGTON VA 22213
USA
TEL 703 532 8246
TLF
TLX
EML
COM

ADAMS DAVID J DR
DPT OF ASTRONOMY
UNIVERSITY OF LEICESTER
UNIVERSITY RD
LEICESTER LE1 7RH
UK
TEL 533 554 455
TLF
TLX 341198
EML
COM 48

ADAMS FRED DR
DPT OF PHYSICS
UNIVERSITY OF MICHIGAN
ANN ARBOR MI 48109 1090
USA
TEL 313 747 4320
TLF
TLX
EML 47509 FADAMS/FADAMS@UMIPHYS
COM 34

ADAMS JAMES H JR DR
NAVAL RESEARCH LABORATORY
CODE 4154 2
4555 OVERLOOK AVE SW
WASHINGTON DC 20375 5000
USA
TEL 202 767 2747
TLF
TLX
EML SPAN 11335 ADAMS
COM

ADAMS THOMAS F DR
LOS ALAMOS SCIENTIFIC LAB
MS 329 G 7
BOX 1663
LOS ALAMOS NM 87545
USA
TEL 505 667 6384
TLF
TLX
EML
COM

ADAMSON ANDREW DR
SCHOOL OF PHYSICS & ASTRO
LANCASHIRE POLYTECHNIC
CORPORATION ST
PRESTON PR1 2TQ
UK
TEL
TLF
TLX
EML
COM 33

ADE PETER A R DR
DPT OF PHYSICS
QUEEN MARY/WESTFIELD COLL
MILE END RD
LONDON E1 4NS
UK
TEL 1 980 4811
TLF
TLX 893750
EML
COM 40

ADEL ARTHUR F PROF EMER
BOX 942
FLAGSTAFF AZ 86002
USA
TEL 602 774 6597
TLF
EML
COM
TLX

ADELMAN SAUL J DR
DPT OF PHYSICS
THE CITADEL
CHARLESTON SC 29409
USA
TEL 803 792 6943
TLF
TLX
EML
COM 14C,25C,29

ADJABSHIRIZADEH ALI
CTR FOR ASTRON RESEARCH
UNIVERSITY OF TABRIZ
TABRIZ 51664
IRAN
TEL 41 32564
TLF
TLX
EML
COM

ADOLFSSON TORD DR
KRAGEHOLMSGATAN 12
S 216 19 MALMOE
SWEDEN
TEL 40 15 7586
TLF
EML
COM
TLX

AFANASJEVA PRASKOVYA M DR
PULKOVO OBSERVATORY
ACADEMY OF SCIENCES
10 KUTUZOV QUAY
196140 ST PETERSBURG
RUSSIA
TEL 298 22 42
TLF
TLX
EML
COM 31

AFANAS'EV VIKTOR L DR
SPECIAL ASTROPHYSICAL OBS
ACADEMY OF SCIENCES
NIZHNIJ ARKHYZ
357147 STAVROPOLSKIJ
RUSSIA
TEL
TLF
TLX
EML
COM 28,33

AGEKJAN TATEOS A PROF
ASTRONOMICAL OBSERVATORY
ST PETERSBURG UNIVERSITY
BIBLIOTECHNAJA PL 2
199178 ST PETERSBURG
RUSSIA
TEL
TLF
TLX
EML
COM 33,37

AGRAWAL P C DR
TIFR
HOMI BHABHA RD
COLABA
BOMBAY 400 005
INDIA
TEL 22 495 2311*393
TLF
TLX 0113009 TIFR IN
EML
COM 44,48

AGRINIER BERNARD L MR
CEA CEN
DAPNIA/SAP
BP 2
F 91191 GIF/YVETTE CDX
FRANCE
TEL
TLF
TLX
EML
COM

AGUERO ESTELA L DR
OBSERVATORIO ASTRONOMICO
DE CORDOBA
LAPRIDA 854
5000 CORDOBA
ARGENTINA
TEL 51 40613
TLF
TLX 51822 BUCOR
EML
COM 28

AGUIAR ODYLIO DENYS DR
INPE
CP 515
12201 S JOSE DOS CAMPOS
BRAZIL
TEL 123 41 8977*689
TLF 123 21 8743
TLX 123 3530 INPE BR
EML INPEDAS@BRFAPESP BITNE
COM 48

AGUILAR CHIU LUIS A DR
BOX 439027
SAN DIEGO CA 92143 9027
USA
TEL 706 674 4580
TLF
EML AGUILAR@ALFA UNAM UCAR EDU
COM 28,33
TLX

AGUILAR MARIA LUISA
UNSM
FAC CIENCIAS FISICAS
AV ARICA 830
LIMA 5
PERU
TEL 24 3961/52 1343
TLF
TLX
EML
COM

AHLUWALIA HARJIT SINGH DR
DPT PHYSICS & ASTRONOMY
UNIVERSITY OF NEW MEXICO
800 YALE BLVD NE
ALBUQUERQUE NM 87131
USA
TEL 505 277 2941
TLF
TLX 660461
EML
COM 10,48,49

AHMAD FAROOQ DR
DPT OF PHYSICS
UNIVERSITY OF KASHMIR
SRINAGAR 190 006
INDIA
TEL 71559
TLF
TLX
EML
COM 28

AHMAD IMAD ALDEAN DR
IMAD-AD DEAN INC
4323 ROSEDALE AVE
BETHESDA MD 20814
USA
TEL 301 565 4714
TLF
TLX
EML
COM 44

AHMED IMAM IBRAHIM PROF
DPT OF ASTRONOMY
FACULTY OF SCIENCES
CAIRO UNIVERSITY
GEZA
EGYPT
TEL
TLF
TLX
EML
COM

AHMED MOSTAFA DR
DEP OF ASTRONOMY
FACULTY OF SCIENCES
CAIRO UNIVERSITY
GEZA
EGYPT
TEL
TLF
TLX
EML
COM 07

AI GUOXIANG
BEIJING ASTRONOMICAL OBS
CAS
W SUBURB
BEIJING 100080
CHINA PR
TEL
TLF
TLX
EML
COM 09,10C,12

AIAD A PROF
DPT OF ASTRONOMY
FACULTY OF SCIENCES
CAIRO UNIVERSITY
GEZA ORMAN
EGYPT
TEL 86 9538
TLF
TLX
EML
COM 34,35,37,46

AIELLO SANTI DR
DPT DI FISICA
UNIVERSTIA DI FIRENZE
VIA L PANCALDO 3/45
I 50125 FIRENZE
ITALY
TEL 55 437 8540
TLF 55 43 5939
TLX
EML
COM

AIKMAN G CHRIS L
HERZBERG INST ASTROPHYS
DOMINION ASTROPHYS OBS
5071 W SAANICH RD
VICTORIA BC V8X 4M6
CANADA
TEL 604 388 3975
TLF 604 363 0045
TLX 049 7295
EML
COM 29

AIME C DR
DPT ASTROPHYSIQUE
UNIVERSITE DE NICE
PARC VALROSE
F 06034 NICE CDX
FRANCE
TEL 93 51 9100
TLF
TLX
EML
COM 09,12

AITKEN DAVID K DR
DPT OF PHYSICS/RAAF
UNIVERSITY OF MELBOURNE
PARKVILLE VIC 3052
AUSTRALIA
TEL 3 341 6818
TLF
TLX 35185 UNIMEL
EML
COM 34

AIZENMAN MORRIS L DR
NTL SCIENCE FOUNDATION
DIV ASTRONOMICAL SCIENCES
1800 G ST NW
WASHINGTON DC 20550
USA
TEL 202 357 7643
TLF
TLX
EML
COM 27,35

AIZU KO PROF
3-24-3 KATAHIRA
ASAO KU
KAWASAKI
JAPAN
TEL
TLF
TLX
EML
COM 33,40,47,48

AKABANE KENJI A PROF
TOKYO ASTRONOMICAL OBS
NAOJ
OSAWA MITAKA
TOKYO 181
JAPAN
TEL 0267-98-2831
TLF
TLX 3329005 TAO NRO J
EML
COM 34,40

AKABANE TOKUHIDE DR
HIDA OBSERVATORY
UNIVERSITY OF KYOTO
KAMITAKARA
GIFU 506 13
JAPAN
TEL 0578-6-2311
TLF
TLX
EML
COM 16

AKCAYLI MELEK M A DR
FACULTY OF SCIENCE
EGE UNIVERSITY
BOX 21
35100 BORNOVA IZMIR
TURKEY
TEL
TLF
TLX
EML
COM

AKIM EFRAIM L DR
INST OF APPLIED MATHS
ACADEMY OF SCIENCES
MIUSSKAJA SQ 4
125047 MOSCOW
RUSSIA
TEL 251 37 39
TLF
TLX
EML
COM 07

AKSENOV E P PROF DR
STERNBERG STATE ASTR INST

UNIVERSITETSKIJ PROSP 13
119899 MOSCOW
RUSSIA
TEL 139-28-58
TLF
TLX
EML
COM 07

AKSNES KAARE DR
INST THEORET ASTROPHYSICS
UNIVERSITY OF OSLO
BOX 1029
N 60315 OSLO 3
NORWAY
TEL 285 6515
TLF 285 6505
TLX 72705 ASTRO N
EML 21813 ROLF AKSNES
COM 06,07,20C,WGPSNP

AKUJOR CHIDI E
DPT OF PHYSICS & ASTRON
UNIVERSITY OF NIGERIA
NSUKKA ANAMBRA STATE
NIGERIA
TEL 42 77 1532
TLF
TLX 51496U LION NG
EML
COM 40

AKYOL MUSTAFA UNAL PROF
FACULTY OF EDUCATION
SELCUK UNIVERSITY
42090 KONYA
TURKEY
TEL
TLF
TLX
EML
COM

ALBERS HENRY PROF
62 PROSPECT ST

FALMOUTH MA 02540
USA
TEL 508 540 0978
TLF
TLX
EML
COM 45

ALBINSON JAMES DR
DPT OF PHYSICS
UNIVERSITY OF KEELE
STAFFS ST5 5BG
UK
TEL 782 621 111
TLF
TLX 36113 UNKLIB G
EML JANET JSA@UK AC KL PH STAR
COM 27

ALBRECHT MIGUEL A DR
ESO
KARL-SCHWARZSCHILDSTR 2
D 8064 GARCHING MUENCHEN
GERMANY
TEL 89 320 06346
TLF 89 320 06480
TLX 5282820 EO D
EML MALBRECHT@ESO ORG
COM 05

ALBRECHT RUDOLF DR
SPACE TELESCOPE EUROPEAN
COORDINATING FACILITY
KARL-SCHWARZSCHILD-STR 2
D 8046 GARCHING MUENCHEN
GERMANY
TEL 89 320 06287
TLF
TLX 528 282 22 EO D
EML
COM 09,25

ALCAINO GONZALO DR
INSTITUTO ISAAC NEWTON
CASILLA 8-9
SANTIAGO 9
CHILE
TEL 2 472 013
TLF
TLX c/o ESO 240853 ESOGO
EML
COM 28,37

ALCOLEA JAVIER DR
CTR ASTRON DE YEBES
OAN
APD 148
E 19090 GUADALAJARA
SPAIN
TEL 11 29 0311
TLF 11 29 0063
TLX
EML ALCOLEA@CAY ES
COM 34

ALDROVANDI RUBEN DR
INST DI FISICA TEORICA
RUA PAMPLONA 15
01405 SAO PAULO SP
BRAZIL
TEL 11 288 5643
TLF
TLX
EML
COM

ALDROVANDI S M VIEGAS DR
IAG
UNIVERSIDADE DE SAO PAULO
AV MIGUEL STEFANO 4200
04301 SAO PAULO SP
BRAZIL
TEL 11 577 8599
TLF
TLX 1136221 IAGM BR
EML IAGUSP%BRFAPESP ANSP
COM 34

ALECIAN GEORGES DR
OBSERVATOIRE DE PARIS
SECTION DE MEUDON
DAF
F 92195 MEUDON PPL CDX
FRANCE
TEL 1 45 34 7420
TLF
TLX 201571
EML
COM 29

ALEXANDER JOHN B
ROYAL GREENWICH OBS
HERSTMONCEUX CASTLE
HAILSHAM BN27 1RP
UK
TEL 323 833 171
TLX 87451
EML
TLF
COM

ALEXANDER JOSEPH K
NASA HEADQUARTERS
CODE P
OFFICE OF CHEIF SCIENTIST
WASHINGTON DC 20546
USA
TEL
TLF
TLX
EML
COM 40,44

ALEXANDER MURRAY E DR
DPT OF COMPUTER SCIENCE
UNIVERSITY OF OTAGO
BOX 56
DUNEDIN
NEW ZEALAND
TEL 64 3 479 8585
TLF 64 3 479 8577
TLX
EML MURRAY%OTAGO AC NZ@RELAY CS NE
COM 07

ALEXANDER PAUL DR
MULLARD RADIO ASTRON OBS
CAVENDISH LABORATORY
MADINGLEY RD
CAMBRIDGE CB3 OHE
UK
TEL 223 664 77
TLF
TLX 81292 CAVLAB G
EML PA25@UK AC CAM PHX
COM 40

ALFARO EMILIO JAVIER
INST ASTROFISICA
DE ANDALUCIA APD 3004
C/SANCHO PANZA S/N
E 18080 GRANADA
SPAIN
TEL 58 12 1311
TLF
TLX 78573 IAAG E
EML 16488 EMILIO/EMILIO@IAA ES
COM 27,37

ALFVEN HANNES PROF
DPT OF PLASMA PHYSICS
ROYAL INST OF TECHNOLOGY
S 100 44 STOCKHOLM 70
SWEDEN
TEL 87 87 7000
TLF
TLX 10389 KTHB
EML
COM 47,48

ALISSANDRAKIS C PH D
DPT OF ASTROPHYSICS
NTL UNIVERSITY OF ATHENS
PANEPISTIMIOPOLIS
GR 157 71 ATHENS
GREECE
TEL 1 723 5122
TLF
TLX
EML
COM 10,12

ALKSNIS ANDREJS DR
RADIOASTROPHYSICAL OBS
LATVIAN ACAD OF SCIENCES
TURGENEVA 19
226524 RIGA
LATVIA
TEL 226796 RIGQ
TLF
TLX
EML
COM 37

ALLADIN SALEH MOHAMED DR
DPT OF ASTRONOMY
UNIVERSITY OF OSMANIA
HYDERABAD 500 007
INDIA
TEL 71 116
TLF
TLX
EML
COM 28

ALLAN DAVID W MR
BUREAU OF STANDARDS
TIME & FREQUENCY DIV
CODE 524
BOULDER CO 80302
USA
TEL 303 497 5637
TLF
TLX 910 940 5906
EML
COM 31C

ALLAN PETER M
DPT OF ASTRONOMY
UNIVERSITY OF MANCHESTER
MANCHESTER M13 9PL
UK
TEL 612 737 121
TLF
TLX
EML
COM 47

ALLARD NICOLE DR
OBSERVATOIRE DE PARIS
SECTION DE MEUDON
F 92125 MEUDON PPL CDX
FRANCE
TEL 1 45 07 7449
TLF 1 45 07 7469
TLX 201571
EML 17670 ALLARD
COM 14

ALLEGRE CLAUDE PROF
INST PHYSIQUE DU GLOBE
4 PLACE JUSSIEU
F 75005 PARIS
FRANCE
TEL
TLF
TLX
EML
COM 15

ALLEN ANTHONY JOHN DR
ASTRONOMY UNIT
QUEEN MARY/WESTFIELD COLL
MILE END RD
LONDON E1 4NS
UK
TEL 1 980 4811
TLF
TLX 893750 QMCUOL G
EML allen@UK AC QMC MATHS
COM

ALLEN CHRISTINE
INSTITUTO DE ASTRONOMIA
UNAM
APDO POSTAL 70-264
04510 MEXICO DF
MEXICO
TEL
TLF
TLX
EML
COM 26,37

ALLEN DAVID A DR
AAO
OBSERVATORY
BOX 296
EPPING NSW 2121
AUSTRALIA
TEL 2 868 1666
TLF
TLX 23999 OSYD AA
EML
COM

ALLEN RONALD J DR
STSCI
HOMEWOOD CAMPUS
3700 SAN MARTIN DR
BALTIMORE MD 21218
USA
TEL 301 338 4574
TLF 301 338 5090
TLX
EML VJALLEN@STSCI EDU
COM 28,40

ALLEN WILLIAM
ADAMS LANE OBSERVATORY
46 ADAMS LANE
BLENHEIM
NEW ZEALAND
TEL 057 87258
TLF
TLX
EML
COM

ALLER HUGH D DR
DPT OF ASTRONOMY
UNIVERSITY OF MICHIGAN
DENNISON BLDG
ANN ARBOR MI 48109 1090
USA
TEL 313 764 3466
TLF
TLX
EML
COM 40

ALLER LAWRENCE HUGH
DPT OF ASTRONOMY
UNIVERSITY OF CALIFORNIA
MATH SCIENCES BLDG
LOS ANGELES CA 90024
USA
TEL 213 825 3515
TLF
TLX 910-342-7597
EML
COM 29,34,36

ALLER MARGO F DR
DPT OF ASTRONOMY
UNIVERSITY OF MICHIGAN
DENNISON BLDG
ANN ARBOR MI 48109 1090
USA
TEL 313 764 3465
TLF
TLX 810-223-6056
EML
COM 40

ALLEY CARROL O DR
ASTRONOMY PROGRAM
UNIVERSITY OF MARYLAND
COLLEGE PARK MD 20742
USA
TEL 301 454 3405
TLF
TLX 908787
EML
COM 31

ALLOIN DANIELLE DR
OBSERVATOIRE DE PARIS
SECTION DE MEUDON
DAEC
F 92195 MEUDON PPL CDX
FRANCE
TEL 1 45 07 7404
TLF 1 45 07 7469
TLX 201571
EML ALLOIN@FRMEU51
COM 28

ALMAR IVAN PROF
KONKOLY OBSERVATORY
THEGE U 13/17
BOX 67
H 1525 BUDAPEST
HUNGARY
TEL 1 75 5866/75 4122
TLF
TLX 227460
EML
COM 51

ALPAR ALI DR
DPT OF PHYSICS
MIDDLE EAST TECH UNIV
06531 ANKARA
TURKEY
TEL 41 22 337100/ 3259
TLF 41 22 36945
TLX 42761 ODTK TR
EML ALPAR@TRMETU
COM 27

ALTAMORE ALDO
ISTITUTO ASTRONOMICO
UNIVERSITA DI ROMA
VIA G M LANCISI 29
I 00161 ROMA
ITALY
TEL 6 84 42977
TLF
TLX 613255 INFRO
EML
COM

ALTAVISTA CARLOS A DR
OBSERVATORIO ASTRONOMICO
PASEO DEL BOSQUE
1900 LA PLATA (BS AS)
ARGENTINA
TEL 21 21 7308
TLF
TLX 31151 BULAP
EML
COM 07

ALTENHOFF WILHELM J DR
MPI FUER RADIOASTRONOMIE

AUF DEM HUEGEL 69
D 5300 BONN 1
GERMANY
TEL 228 52 5293
TLF
TLX 886440
EML
COM 33,34,40

ALTROCK RICHARD C DR
AIR FORCE GEOPHYSICS LAB
NTL SOLAR OBSERVATORY
BOX 62
SUNSPOT NM 88349
USA
TEL 505 434 7016
TLF 505 434 7029
TLX 156 1030
EML RALTROCK@SUNSPOT NOAO EDU
COM 10,12,36

ALTSCHULER MARTIN D PROF
DPT RAD THERAPY BOX 522
HOSP UNIV OF PENNSYLVANIA
3400 SPRUCE ST
PHILADELPHIA PA 19104
USA
TEL 215 662 6472
TLF
TLX
EML
COM 10,12

ALURKAR S K DR
PHYSICAL RESEARCH LAB
NAVRANGPURA
AHMEDABAD 380 009
INDIA
TEL 272 46 2129
TLF 272 44 5292
TLX 121-397 PRL IN
EML
COM

ALVAREZ HECTOR DR
DPT DE ASTRONOMIA
UNIVERSIDAD DE CHILE
CASILLA 36 D
SANTIAGO
CHILE
TEL 2 229 4101
TLF
TLX 440 001
EML
COM

ALVAREZ MANUEL DR
OBS ASTRONOMICO NACIONAL
BOX 439027
SAN YSIDRO CA 92027
USA
TEL 706 674 4580
TLX 56539 CICEME
EML ALVAREZ@ALFA ASTROSCU UNAM MX
TLF 706 667 4607
COM

ALVAREZ PEDRO DR
INST DE ASTROFISICA
DE CANARIAS
OBS DEL TEIDE
E 38071 LA LAGUNA
SPAIN
TEL
TLF
TLX 92640 IACE E
EML
COM 05,09

ALVAREZ POMARES A O DR
INST GEOPHYS & ASTRONOMY
CALLE 212 N 2906/29 Y 31
LISA
LA HABANA
CUBA
TEL 21 8416
TLF
TLX 0511240
EML
COM

ALY JEAN JACQUES DR
CEA CEN
DAPNIA/SAP
BP 2
F 91191 GIF/YVETTE CDX
FRANCE
TEL 1 69 08 4030
TLF 1 69 08 9266
TLX 604860
EML JJALY@SOLAR
COM 10

ALY M KHAIRY PROF
HELWAN OBSERVATORY
HELIOPOLIS CAIRO
EGYPT
TEL 44 8832
TLF
EML
COM 10,38,50
TLX

AL-NAIMY HAMID M K DR
SARC
SCIENTIFIC RES COUNCIL
BOX 2441
JADIRIYAH BAGHDAD
IRAQ
TEL 1 776 5127
TLF
TLX 212187
EML
COM 42,51

AL-SABTI ABDUL ADIM DR
DPT OF PHYSICS
UNIVERSITY OF BAGHDAD
SCIENCE COLLEGE
JADIRIYAH BAGHDAD
IRAQ
TEL 1 555 2340
TLF
TLX
EML
COM 38,51

AMARI TAHAR DR
OBSERVATOIRE DE PARIS
SECTION DE MEUDON
SECTION D'ASTROPHYSIQUE
F 92195 MEUDON PPL CDX
FRANCE
TEL 1 45 07 7760
TLF 1 45 07 7469
TLX 270912
EML AMARI@FRMEU51
COM 10

AMBARTSUMIAN V A PROF DR
BYURAKAN ASTROPHYSICAL
OBSERVATORY
378433 ARMENIA
ARMENIA
TEL 88 52 28 4580
TLF
TLX 412623
EML
COM 28,33,51

AMBASTHA A K DR
UDAIPUR SOLAR OBSERVATORY
11 VIDYA MARG
UDAIPUR 313 001
INDIA
TEL 25 626
TLF
TLX
EML
COM 10,33

AMBROZ PAVEL DR
ASTRONOMICAL INSTITUTE
CZECH ACADEMY OF SCIENCES
ONDREJOV OBSERVATORY
CS 251 65 ONDREJOV
CZECHOSLOVAKIA
TEL 204 85201
TLF 204 85314
TLX 121579 ASTR C
EML
COM 10

AMBRUSTER CAROL DR
DPT OF ASTRONOMY
VILLANOVA UNIVERSITY
MENDEL HALL
VILLANOVA PA 19085
USA
TEL
TLF
TLX
EML A?BRUSTER@VUVAXCOM
COM

ANAND S P S DR
APPLIED RESEARCH CORP
8201 CORPORATE DRIVE
SUITE 920
LANDOVER MD 20785
USA
TEL 301 459 8442
TLF
TLX
EML
COM 35

ANANTHAKRISHNAN S
TATA INST OF FUNDAMENTAL
RESEARCH
POONA UNIVERSITY CAMPUS
PUNE 411 007
INDIA
TEL 212 33 6105
TLF 212 33 5760
TLX 0145658 GMRT IN
EML uunet!shakti!gmrt!ananth
COM 49

ANANTHARAMAIAH K R DR
NRAO
VLA SITE
BOX 0
SOCORRO NM 87801 0387
USA
TEL 505 772 4306
TLF
TLX 9109881710
EML
COM 34,40

ANDERNACH HEINZ DR
INST DE ASTROFISICA
VIA LACTEA S/N
E 38200 LA LAGUNA
SPAIN
TEL 22 60 5200/37
TLF. 22 60 5210
TLX 92640 IAC E
EML HJA@IAC DNET NASA GOV
COM 05,40

ANDERSEN BO NYBORG DR
NORWEGIAN SPACE CENTRE
BOX 85
SMESTAD
N 0309 OSLO 3
NORWAY
TEL 2 52 38 00
TLF 47 2 522397
TLX 0056 78174 SPACEN
EML 21813 BANDERSEN
COM 10,12,44

ANDERSEN JOHANNES
COPENHAGEN UNIVERSITY OBS
BRORFELDEVEJ 23
DK 4340 TOLLOSE
DENMARK
TEL 53 48 8195
TLF 53 48 8755
TLX 44155 DANAST DK
EML
COM 30C,33,42

ANDERSEN TORBEN BRENDER
LOCKHEED PALO ALTO RES LB
OPTICAL DESIGN GR B254 E
3251 HANOVER ST
PALO ALTO CA 94304
USA
TEL
TLF
TLX
EML
COM

ANDERSON BRYAN DR
NRAL

JODRELL BANK
MACCLESFIELD SK11 9DL
UK
TEL
TLF
TLX
EML
COM

ANDERSON CHRISTOPHER M DR
WASHBURN OBSERVATORY
UNIVERSITY OF WISCONSIN
475 N CHARTER ST
MADISON WI 53706
USA
TEL 608 262 0492
TLF
TLX
EML
COM

ANDERSON KINSEY A PROF
SPACE SCIENCES LABORATORY
UNIVERSITY OF CALIFORNIA
BERKELEY CA 94720
USA
TEL 415 642 1313
TLF
TLX 910-3667945 UC SPACE
EML
COM 10,21,49

ANDERSON KURT S
DPT OF ASTRONOMY
NEW MEXICO STATE UNIV
LAS CRUCES NM 88003
USA
TEL 505 646 1032
TLF
TLX 210-983-0549 NMSUCI
EML
COM

ANDO HIROYASU DR
TOKYO ASTRONOMICAL OBS
NAOJ
OSAWA MITAKA
TOKYO 181
JAPAN
TEL 0422-32-5111
TLF
TLX 2822307 TAOMK J
EML
COM 12,27,51

ANDREANI PAOLA MICHELA DR
DPT DI ASTRONOMIA
UNIVERSITA DI PADOVA
VIC DELL OSSERVATORIO 5
I 35122 PADOVA
ITALY
TEL 49 829 3442
TLF 49 875 9840
TLX 432071 ASTROS I
EML 39003 ANDREANI SPAN
COM 47

ANDREW BRYAN H DR
CHIEF PROGRAM SERVICES
NTL RESEARCH COUNCIL
100 SUSSEX DR
OTTAWA ON K1A 0R6
CANADA
TEL 613 993 3731
TLF 613 952 6602
TLX 053 3145
EML
COM 34,40

ANDREW KENNETH L PROF
1637 MAY - 1002
WICHITA 67213
USA
TEL 316 243 8381
TLF
TLX
EML
COM 14

ANDREWS DAVID A DR
ARMAGH OBSERVATORY
COLLEGE HILL
ARMAGH BT61 9DG
UK
TEL
TLF
TLX
EML
COM

ANDREWS PETER J DR
ROYAL GREENWICH OBS
HERSTMONCEUX CASTLE
HAILSHAM BN27 1RP
UK
TEL 323 833 171
TLX 87451
EML
TLF
COM

ANDRIENKO DMITRY A DR
ASTRONOMICAL OBSERVATORY
KIEV STATE UNIVERSITY
OBSERVATORNAYA UL 3
252022 KIEV
UKRAINE
TEL 25 0775
TLF
TLX 132201
EML
COM 15

ANDRIESSE CORNELIS D DR
POGGENBECSKSTRAAT 31
NL 6813 KD ARNHEM
NETHERLANDS
TEL
TLF
TLX
EML
COM 34

ANDRILLAT HENRI L PROF
LAB ASTRONOMIE
USTL II
PLACE EUGENE BATAILLON
F 34095 MONTPELLIER CDX 5
FRANCE
TEL 67 14 3415
TLF 67 54 3079
TLX 490944 USTMONT F
EML
COM 29,34,46,47

ANDRILLAT YVETTE DR
LAB ASTRONOMIE
USTL II
PLACE EUGENE BATAILLON
F 34095 MONTPELLIER CDX 5
FRANCE
TEL 67 14 3412
TLF 67 54 3079
TLX 490944 USTMONT F
EML
COM 28,29,34

ANGEL J ROGER P PROF
STEWARD OBSERVATORY
UNIVERSITY OF ARIZONA
TUCSON AZ 85721
USA
TEL 602 621 6541
TLF
TLX 467175
EML
COM 25

ANGELETTI LUCIO DR
OAR
VIA DEL PARCO MELLINI 84
I 00136 ROMA
ITALY
TEL 6 34 7056
TLF
TLX
EML
COM

ANGELOV TRAJKO
INSTITUTE OF ASTRONOMY
UNIVERSITY OF BELGRADE
STUDENTSKI TRG 16
YU 11000 BEOGRAD
YUGOSLAVIA
TEL 11 638 715
TLF
TLX
EML
COM 35

ANGIONE RONALD J DR
DPT OF ASTRONOMY
SAN DIEGO STATE UNIV
SAN DIEGO CA 92182
USA
TEL 619 265 6183
TLF
TLX
EML
COM 21,25

ANGLADA GUILLEM DR
DPT ASTRON I METEOROLOG
UNIVERSIDAD DE BARCELONA
AVD DIAGONAL 647
E 08028 BARCELONA
SPAIN
TEL 3 402 1121
TLF 3 411 0873
TLX
EML 16488 50179 GUILLEM
COM 34

ANGUITA CLAUDIO A DR
DPT DE ASTRONOMIA
UNIVERSIDAD DE CHILE
CASILLA 36 D
SANTIAGO
CHILE
TEL 2 229 4101
TLF
TLX 440 001
EML
COM 08

ANILE ANGELO M
DPT MATEMATICA
CITTA UNIVERSITARIA
VIA A DÒRIA 6
I 95125 CATANIA
ITALY
TEL 95 33 0533
TLF
TLX
EML
COM

ANN HONG BAE DR
DPT EARTH SCIENCES
PUSAN NTL UNIVERSITY
KUM JONG KU
PUSAN 609 735
KOREA R
TEL 51 510 2705
TLF 51 513 7495
TLX
EML
COM 28

ANSARI S M RAZAULLAH PROF
DPT OF PHYSICS
ALIGARH MUSLIM UNIVERSITY
ALIGARH UP 202 002
INDIA
TEL 571 29 001
TLF
TLX 564230 AMU-IN
EML
COM 12,41VP,46

ANTALOVA ANNA
ASTRONOMICAL INSTITUTE
SLOVAK ACADEMY SCIENCES
CS 059 60 TATRANSKA LOMNI
CZECHOSLOVAKIA
TEL 969 96 7866/7/8
TLF 969 96 7656
TLX
EML
COM 10

ANTHONY-TWAROG BARBARA J
DPT PHYSICS & ASTRONOMY
UNIVERSITY OF KANSAS
LAWRENCE KS 66045
USA
TEL 913 864 4933
TLF
TLX
EML
COM 25

ANTIA H M DR
TIFR
HOMI BHABHA RD
COLABA
BOMBAY 400 005
INDIA
TEL 22 495 2311
TLF
TLX 011 3009 TIFR IN
EML
COM 12,35

ANTIOCHOS SPIRO KOSTA
NAVAL RESEARCH LABORATORY
CODE 4170 SA
4555 OVERLOOK AVE SW
WASHINGTON DC 20375 5000
USA
TEL 202 767 6199
TLF
TLX
EML
COM 10

ANTIPOVA LYUDMILA DR
INST OF ASTRONOMY
ACADEMY OF SCIENCES
PYATNITSKAYA UL 48
109017 MOSCOW
RUSSIA
TEL 231 06 80
TLF
TLX 411 576 ASCON SU
EML
COM 27,42

ANTONACOPOULOS GREG PROF
DPT OF ASTRONOMY
UNIVERSITY OF PATRAS
GR 261 10 RION
GREECE
TEL 61 99 1145
TLF
TLX
EML
COM 07

ANTONELLO ELIO
OSS ASTRONOMICO DI MILANO
VIA E BIANCHI 46
I 22055 MERATE
ITALY
TEL 59 2035
TLX
EML
TLF
COM 27

ANTONOPOULOU E DR
DPT OF ASTRONOMY
NTL UNIVERSITY OF ATHENS
PANEPISTIMIOPOLIS
GR 157 71 ZOGRAFOS
GREECE
TEL
TLF
TLX
EML
COM 42

ANTONOV VADIM A DR
ASTRONOMICAL OBSERVATORY
ST PETERSBURG UNIVERSITY
BIBLIOTECHNAJA PL 2
199178 ST PETERSBURG
RUSSIA
TEL
TLF
TLX
EML
COM 33

ANTONUCCI ESTER DR
IST DI FISICA
UNIVERSITA DI TORINO
CORSO D AZEGLIO 46
I 10125 TORINO
ITALY
TEL 11 65 7694
TLF
TLX 211041 INFNTO I
EML
COM 10C

ANZER ULRICH DR
MPI FUER PHYSIK UND
ASTROPHYSIK
KARL-SCHWARZSCHILD-STR 1
D 8046 GARCHING MUENCHEN
GERMANY
TEL 89 329 90
TLF
TLX 524629 ASTRO D
EML
COM 10

AOKI SHINKO PROF
TOKYO ASTRONOMICAL OBS
NAOJ
OSAWA MITAKA
TOKYO 181
JAPAN
TEL 422 32 5111
TLF
TLX 2822307 TAOMTK J
EML
COM 04,07,31,33

APARICI JUAN DR
DPT DE ASTRONOMIA
UNIVERSIDAD DE CHILE
CASILLA 36 D
SANTIAGO
CHILE
TEL 2 229 4101
TLF
TLX 440 005
EML
COM 09,40

APARICIO ANTONIO DR
INST DE ASTROFISICA
DE CANARIAS
OBS DEL TEIÖE
E 38200 LA LAGUNA
SPAIN
TEL 22 26 2211
TLF 22 26 3005
TLX 92640 IAC E
EML
COM 37

APPARAO K M V DR
TIFR
HOMI BHABHA RD
COLABA
BOMBAY 400 005
INDIA
TEL 22 219 111*341
TLF
TLX 011-3009 TIFR IN
EML
COM 48

APPENZELLER IMMO PROF
LANDESSTERNWARTE
KOENIGSTUHL
D 6900 HEIDELBERG 1
GERMANY
TEL 62 215 090
TLF 62 215 09202
TLX 461153 LSWHD D
EML CTO@DHDURZ1
COM 29,35

APPLEBY JOHN F
JPL/CALTECH
MS 183 301
4800 OAK GROVE DR
PASADENA CA 91109
USA
TEL 818 354 3943
TLF
TLX
EML
COM 16

APPLETON PHILIP NOEL DR
DPT OF PHYSICS
IOWA STATE UNIVERSITY
AMES IA 50011
USA
TEL 515 294 3667
TLF
TLX
EML BITNET S1 PNA@ISUMVS
COM

AQUILANO ROBERTO OSCAR DR
INSTITUTO DE FISICA
ROSARIO CONICET UNR
BV 27 DE FEBRERO 210 BIS
2000 ROSARIO
ARGENTINA
TEL 41 821769/72
TLF 41 25 7164
TLX 41817 CIROS AR
EML USUARIOS%IFIR EDU AR@UUNET UU
COM 42

ARABELOS DIMITRIOS DR
DPT GEODESY & SURVEYING
UNIVERSITY THESSALONIKI
UNIV BOX 503
GR 540 06 THESSALONIKI
GREECE
TEL 31 99 2693
TLF
TLX 412181 AUTH GR
EML
COM 19

ARAFUNE JIRO DR
INST COSMIC RAY RESEARCH
UNIVERSITY OF TOKYO
MIDORICHO TANASHI
TOKYO 188
JAPAN
TEL 0424 614131
TLF 0424 68 1438
TLX 2822371 ICRTUJ
EML
COM 48

ARAI KENZO DR
DPT OF PHYSICS
KUMAMOTO UNIVERSITY
2-39-1 KUROKAMI
KUMAMOTO 860
JAPAN
TEL 096-344-2111
TLF
TLX
EML
COM 35

ARAUJO FRANCISCO X DE DR
OBSERVATORIO NACIONAL
RUA GL BRUCE 586
SAO CRISTOVAO
20921 RIO DE JANEIRO RJ
BRAZIL
TEL 21 580 0235
TLF 21 580 0332
TLX 21288
EML USERFXA@LNCC BITNET
COM 29

ARDAVAN HOUSHANG DR
INSTITUTE OF ASTRONOMY
THE OBSERVATORIES
MADINGLEY RD
CAMBRIDGE CB3 0HA
UK
TEL 223 622 04
TLF
TLX 817297 ASTRON G
EML
COM

ARDEBERG ARNE L PROF
LUND OBSERVATORY
BOX 43
S 221 00 LUND
SWEDEN
TEL 46 10 7290
TLF
TLX 33199 OBSNOT S
EML
COM 28,33,45,50C

ARDEBILI M REZA DR
BOX 47415 341

BABOLSAR
IRAN
TEL
TLF
TLX
EML
COM

ARDUINI-MALINOVSKY M DR
CNES
2 PLACE MAURICE QUENTIN
F 75039 PARIS CDX 01
FRANCE
TEL
TLF
TLX
EML
COM 14

ARELLANO FERRO ARMANDO
INSTITUTO DE ASTRONOMIA
UNAM
APDO POSTAL 70-264
04510 MEXICO DF
MEXICO
TEL 905-548-5305
TLF
TLX 01760155 CICME
EML
COM 27,45

ARGUE A NOEL MR
INSTITUTE OF ASTRONOMY
THE OBSERVATORIES
MADINGLEY RD
CAMBRIDGE CB3 0HA
UK
TEL 223 622 04
TLF
TLX 817297 ASTRON G
EML
COM 24C,25,26

ARGYLE ROBERT WILLIAM MR
ROYAL GREENWICH OBS
MADINGLEY RD
CAMBRIDGE CB3 0EZ
UK
TEL 223 37 4783
TLF 223 37 4700
TLX 817235 RGOSTR G
EML RWA@UK AC RGO SRF
COM 08

ARGYRAKOS JEAN PROF DR
193 PATISSON ST
GR 112 53 ATHENS
GREECE
TEL 1 867 7000
TLF
EML
COM 08,41
TLX

ARIAS DE GREIFF J PROF
OBSERVATORIO NACIONAL
APDO 2584
BOGOTA 1 DE
COLOMBIA
TEL
TLF
TLX
EML
COM 04,50

ARIAS ELISA FELICITAS
UNIV NACIONAL DE LA PLATA
FCAG
1900 LA PLATA (BS AS)
ARGENTINA
TEL 21 21 7308
TLF
TLX
EML
COM 19

ARIMOTO NOBUO DR
OBSERVATOIRE DE PARIS
SECTION DE MEUDON
LAM
F 92195 MEUDON PPL CDX
FRANCE
TEL 1 45 34 7570
TLF
TLX 207571
EML
COM 35

ARKHIPOVA V P DR
STERNBERG STATE ASTR INST
UNIVERSITETSKIJ PROSP 13
119899 MOSCOW
RUSSIA
TEL 139-26-57
TLX
EML
TLF
COM 27,28,34

ARLOT JEAN-EUDES
BUREAU DES LONGITUDES
77 AVE DENFERT ROCHEREAU
F 75014 PARIS
FRANCE
TEL 1 40 51 2267
TLF
TLX
EML
COM 04,20C

ARMANDROFF TAFT E DR
KITT PEAK NTL OBS
BOX 26732
950 N CHERRY AVE
TUCSON AZ 85726 6732
USA
TEL 602 325 9382
TLF 602 325 9360
TLX 0666 484
EML ARMAND@NOAO EDU
COM 37

ARMSTRONG JOHN THOMAS DR
US NAVAL OBSERVATORY
ASTROMETRY DIV AD 5
34 & MASSACHUSETTS AVE NW
WASHINGTON DC 20392 5100
USA
TEL 202 653 1769
TLF
TLX
EML ATLAS¦ARMSTR@INTERF COM
COM 26

ARNAL MARCELO EDMUNDO DR
MPI FUER RADIOASTRONOMIE
AUF DEM HUGEL 69
D 5300 BONN 1
GERMANY
TEL
TLX
EML ¦
TLF
COM 40

ARNAUD JEAN PAUL
OBS MIDI PYRENEES
14 AVE E BELIN
F 31400 TOULOUSE CDX
FRANCE
TEL 61 33 2929
TLF 61 53 6722
TLX 530 776 F
EML
COM 09,12,25

ARNAUD MONIQUE
CEA CEN
DAPNIA/SAP
BP 2
F 91191 GIF/YVETTE CDX
FRANCE
TEL 1 69 08 7017
TLF
TLX 604860
EML SPAN 32779 ARNAUD
COM 48

ARNETT W DAVID PROF
ENRICO FERMI INSTITUTE
UNIVERSITY OF CHICAGO
933 E 56TH ST
CHICAGO IL 60637
USA
TEL 312 962 8208
TLF
TLX 9102215617
EML
COM 35

ARNOLD JAMES R DR
DPT OF CHEMSTRY
UCSD
B 017
LA JOLLA CA 92093 0216
USA
TEL 619 534 2908
TLF
TLX 9103371271
EML Bitnet jarnold@ucsd
COM 15

ARNOULD MARCEL L DR
IAAG
VRIJE UNIV BRUSSELS
CP 165
B 1050 BRUSSELS
BELGIUM
TEL 2 649 0030
TLF
TLX 23069 UNILIB
EML
COM 35,48

ARNQUIST WARREN N DR
8127 DELGANY AVE
PLAYA DEL REY CA 90291
USA
TEL 213 821 2724
TLF
TLX
EML
COM

ARNY THOMAS T DR
DPT PHYSICS & ASTRONOMY
UNIV OF MASSACHUSETTS
GRC B
AMHERST MA 01003
USA
TEL 413 545 2194
TLF
TLX
EML
COM 34

ARONS JONATHAN
ASTRONOMY DPT
UNIVERSITY OF CALIFORNIA
601 CAMPBELL HALL
BERKELEY CA 94720
USA
TEL 415 642 4730
TLF
TLX 820181 UCB AST RAL
EML
COM 48

ARP HALTON DR
MAX PLANCK INSTITUT FUER
PHYSIK UND ASTROPHYSIK
KARLSCHWARZSCHILDSTR 1
D 8046 GARCHING MUENCHEN
GERMANY
TEL
TLF
TLX
EML
COM 28

ARPIGNY CLAUDE PROF
INSTITUT D'ASTROPHYSIQUE
UNIVERSITE DE LIEGE
AVE COINTE 5
B 4000 COINTE-LIEGE
BELGIUM
TEL 41 52 9980*263
TLF 41 52 7474
TLX
EML
COM 15,36

ARRIBAS SANTIAGO DR
INST DE ASTROFISICA
DE CANARIAS
OBS DEL TEIDE
E 38200 LA LAGUNA
SPAIN
TEL 22 26 2211
TLF
TLX 92640
EML SPAN IAC SAM
COM

ARSENIJEVIC JELISAVETA
ASTRONOMICAL OBSERVATORY
VOLGINA 7
YU 11050 BEOGRAD
YUGOSLAVIA
TEL
TLF
TLX
EML
COM 25,27,50

ARTHUR DAVID W G
US GEOLOGICAL SURVEY
BRANCH OF ASTROGEOLOGY
2255 N GEMINI DR
FLAGSTAFF AZ 86001
USA
TEL
TLF
TLX
EML
COM 16

ARTRU MARIE-CHRISTINE DR
OBSERVATOIRE DE LYON
AVE CHARLES ANDRE
F 69561 S GENIS LAVAL CDX
FRANCE
TEL 78 56 0705
TLF 72 39 9791
TLX 310926
EML
COM 14,29

ARTZNER GUY
IAS
BP 10
F 91371 VERRIERES BUISSON
FRANCE
TEL 1 64 47 4309
TLF
TLX 600252
EML
COM

ASCHENBACH BERND PH D
MPI F PHYSIK & ASTROPHYS
INST F EXTRATERR PHYSIK
KARL-SCHWARZSCHILD-STR 1
D 8046 GARCHING MUENCHEN
GERMANY
TEL
TLF
TLX
EML
COM 44,48

ASCHWANDEN MARKUS DR
ASTRONOMY PROGRAM
UNIVERSITY OF MARYLAND
COLLEGE PARK MD 20742
USA
TEL 301 405 1525
TLF 301 314 9067
TLX
EML MARKUS@ASTRO UMD EDU
COM 10,40

ASHOK N M DR
PHYSICAL RESEARCH LAB
NAVRANGPURA
AHMEDABAD 380 009
INDIA
TEL 272 46 2129
TLF 272 44 5292
TLX 121397
EML
COM 09,25

ASLAN ZEKI DR
FACULTY OF SCIENCE
INONU UNIVERSITY
44069 MALATYA
TURKEY
TEL
TLF
TLX
EML
COM

ASLANOV I A DR
SHEMAKHA ASTROPHYSICAL
OBSERVATORY
AZER ACADEMY OF SCIENCES
373243 SHEMAKHA
AZERBAIDZHAN
TEL
TLF
TLX
EML
COM

ASPIN COLIN DR
JOINT ASTRONOMY CENTER
665 KOMOHANA ST
HILO HI 96720
USA
TEL 808 961 3756
TLX
EML CAA@JACH HAWAII EDU
TLF 808 961 6516
COM 25

ASSEO ESTELLE DR
CENTRE PHYSIQUE THEORIQUE
ECOLE POLYTECHNIQUE
F 91128 PALAISEAU CDX
FRANCE
TEL 1 69 41 8200
TLF
TLX 691596
EML
COM 48

ASSOUSA GEORGE ELIAS DR
545 BOYLSTON ST
SUITE 901
BOSTON MA 02116
USA
TEL
TLF
TLX
EML
COM 40

ASSUS PIERRE DR
OCA OBSERV DE NICE
BP 139
F 06003 NICE CDX
FRANCE
TEL 92 00 3086
TLF 92 00 3033
TLX
EML
COM 09

ASTERIADIS GEORGIOS DR
DPT GEODESY & SURVEYING
UNIVERSITY THESSALONIKI
UNIV BOX 503
GR 540 06 THESSALONIKI
GREECE
TEL 31 99 2693
TLF
TLX 412181 AUTH GR
EML
COM 27,33

ATAC TAMER
KANDILLI OBSERVATORY
BOGAZICI UNIVERSITY
CENGELKOY
81220 ISTANBUL
TURKEY
TEL 1 332 0240
TLF
TLX 26411 BOUN TR
EML
COM 10,29

ATANASIJEVIC IVAN DR
FACULTY OF SCIENCES
NL 6500 GL NIJMEGEN
NETHERLANDS
TEL
TLF
TLX
EML
COM

ATHANASSOULA EVANGELIE DR
OBSERVATOIRE DE MARSEILLE
2 PLACE LE VERRIER
F 13248 MARSEILLE CDX 04
FRANCE
TEL 91 95 9088
TLF
TLX 420241 F
EML
COM 28,33

ATHAY R GRANT DR
HIGH ALTITUDE OBSERVATORY
NCAR
BOX 3000
BOULDER CO 80307 3000
USA
TEL 303 497 1556
TLF
TLX
EML
COM 10,12,36

ATHERTON PAUL DAVID
ASTROPHYSICS GROUP
IMPERIAL COLLEGE
BLACKETT LABORATORY
LONDON SW7
UK
TEL
TLF
TLX
EML
COM 09

ATREYA SUSHIL K
DPT ATM & OCEANIC SCIENCE
UNIVERSITY OF MICHIGAN
SPACE RESEARCH BLDG
ANN ARBOR MI 48109 2143
USA
TEL 313 764 3335
TLF
TLX 8102236056
EML
COM 16

AUBIER MONIQUE G DR
OBSERVATOIRE DE PARIS
SECTION DE MEUDON
F 92195 MEUDON PPL CDX
FRANCE
TEL 1 45 34 7755
TLF
TLX 270912
EML
COM

AUDOUZE JEAN PROF
INSTITUT D'ASTROPHYSIQUE
98BIS BD ARAGO
F 75014 PARIS
FRANCE
TEL 1 43 20 1425
TLF
TLX
EML
COM 35,47,48

AUER LAWRENCE H DR
LOS ALAMOS NATIONAL LAB
MS F665
ESS 5
LOS ALAMOS NM 87545
USA
TEL 505 667 5824
TLF
TLX
EML
COM 36

AUGARDE RENEE DR
OBSERVATOIRE DE MARSEILLE
2 PLACE LE VERRIER
F 13248 MARSEILLE CDX 04
FRANCE
TEL 91 95 9088
TLF
TLX 420241
EML
COM

AUGASON GORDON C DR
NASA AMES RESEARCH CTR
MS 245 6
MOFFETT FIELD CA 94035
USA
TEL 415 694 4156
TLF
TLX
EML
COM

AULUCK FAQIR CHAND PROF
DPT PHYSICS & ASTROPHYS
UNIVERSITY OF DELHI
NEW DELHI 110 007
INDIA
TEL 11 291 8993
TLF
TLX
EML
COM 47

AUMAN JASON R PROF
DPT GEOPHYS & ASTRONOMY
UNIV OF BRITISH COLUMBIA
2075 WESBROOK PL
VANCOUVER BC V6T 1WS
CANADA
TEL 604 228 2892
TLF 604 228 6047
TLX
EML
COM 36

AUNER GERHARD DR
INSTITUT FUER ASTRONOMIE
TUERKENSCHANZSTR 17
A 1180 WIEN
AUSTRIA
TEL
TLF
TLX
EML
COM

AURASS HENRY DR
AKADEMIE DER WISSENSCHA
DER DDR ZENTRALINSTITUT
FUR ASTROPHYSIK
D 1501 TREMSDORF
GERMANY
TEL 2261
TLF
TLX 15420
EML
COM 10,40

AURIEMMA GIULIO DR
DPT DI FISICA
UNIVERSITA DI ROMA
PA MORO 2
I 00187 ROMA
ITALY
TEL 6 49 76336
TLF
TLX 613255~INFNRO
EML
COM

AURIERE MICHEL
OBS MIDI PYRENEES
14 AVE E BELIN
F 31400 TOULOUSE CDX
FRANCE
TEL 62 95 1969
TLF
TLX
EML
COM 37

AUVERGNE MICHEL
OCA OBSERV DE NICE
BP 139
F 06003 NICE CDX
FRANCE
TEL 93 89 0420
TLF
TLX 460004 OBSNICE F
EML
COM

AVCIOGLU KAMURAN PROF DR
UNIVERSITY OBSERVATORY
UNIVERSITY OF ISTANBUL
34452 ISTANBUL
TURKEY
TEL 1 522 3597
TLF
TLX
EML
COM

AVERY LORNE W DR
HERZBERG INST ASTROPHYS
NTL RESEARCH COUNCIL
100 SUSSEX DR
OTTAWA ON K1A 0R6
CANADA
TEL 613 993 6060
TLF 613 952 6602
TLX 053 3715
EML
COM 34,40

AVGOLOUPIS STAVROS DR
DPT OF ASTROPHYSICS
UNIVERSITY THESSALONIKI
GR 540 06 THESSALONIKI
GREECE
TEL 31 99 1357
TLF
TLX 0412181 AUTH GR
EML
COM 27

AVIGNON YVETTE DR
OBSERVATOIRE DE PARIS
SECTION DE MEUDON
F 92195 MEUDON PPL CDX
FRANCE
TEL 1 45 34 7771
TLF
TLX 200590
EML
COM 10,40

AVRETT EUGENE H DR
CENTER FOR ASTROPHYSICS
HCO/SAO
60 GARDEN ST
CAMBRIDGE MA 02138
USA
TEL 617 495 7423
TLF
TLX 921428 SATELLITE CAM
EML
COM 36

AWAD MERVAT EL-SAID DR
DPT OF ASTRONOMY
FACULTY OF SCIENCES
CAIRO UNIVERSITY
GEZA
EGYPT
TEL
TLF
TLX
EML
COM

AWADALLA NABIL SHOUKRY DR
HELWAN OBSERVATORY
HELWAN
EGYPT
TEL 78 0645/2683
TLF
TLX 93070 HIAG UN
EML
COM 42

AXFORD W IAN PROF
MPI FUER AERONOMIE
POSTFACH 20
D 3411 KATLENBURG LINDAU
GERMANY
TEL 555 64 1414
TLF
TLX 965527
EML
COM 15,34,48

AXUN DAVID
NRAL
JODRELL BANK
MACCLESFIELD SK11 9DL
UK
TEL 477 713 21
TLX 36149
EML
TLF
COM 25,40

AYDIN CEMAL PROF DR
DPT OF ASTRONOMY
UNIVERSITY OF ANKARA
FEN FAKULTESI
06100 BESEVLER
TURKEY
TEL 41 23 2105*94
TLF
TLX
EML
COM

AYRES THOMAS R
CASA
UNIVERSITY OF COLORADO
BOX 391
BOULDER CO 80309 0391
USA
TEL 303 492 5320
TLF
TLX 755842 JILA
EML
COM 12C,44

AZCARATE ISMAEL N DR
IAR

CC 5
1894 VILLA ELISA (BS AS)
ARGENTINA
TEL 21 4 3793
TLF
TLX 18052 CICYT AR
EML AZCARATE@UMA EDU AR
COM 34

AZUMA TAKAHIRO DR
DOKKYO UNIVERSITY
SOKA
SAITAMA 340
JAPAN
TEL 0489 42 1111
TLF
TLX 2972005
EML
COM 47

AZZOPARDI MARC DR
OBSERVATOIRE DE MARSEILLE
2 PLACE LE VERRIER
F 13248 MARSEILLE CDX 04
FRANCE
TEL 91 95 9088
TLF
TLX 420241 F
EML
COM 28,30

A'HEARN MICHAEL F DR
ASTRONOMY PROGRAM
UNIVERSITY OF MARYLAND
COLLEGE PARK MD 20742
USA
TEL 301 405 6076
TLF 301 314 9067
TLX 710 826 0352
EML MA@ASTRO UMD EDU
COM 05,15VP,20

BAADE DIETRICH DR
ST/ECF
ESO
KARL-SCHWARZSCHILD-STR 2
D 8046 GARCHING MUENCHEN
GERMANY
TEL 89 320 06388
TLF
TLX 05 282820 EO D
EML DBAADE@DGAESO51 BITNE
COM 27,29C,36

BAAN WILLEM A
ARECIBO OBSERVATORY
BOX 995
ARECIBO PR 00613
USA
TEL 809 878 2612
TLF
TLX 385638
EML
COM 48

BAARS JACOB W M DR
STEWARD OBSERVATORY
UNIVERSITY OF ARIZONA
TUCSON AZ 85721
USA
TEL 602 621 1515
TLF 602 621 1532
TLX 467175
EML BAARS@TUCSMT AS ARIZONA EDU
COM 34,40

BAART EDWARD E PROF
DPT OF PHYSICS
RHODES UNIVERSITY
BOX 94
GRAHAMSTOWN 6140
SOUTH AFRICA
TEL 0461-7128
TLF
TLX 244226
EML
COM 34,40C

BAATH LARS B DR
BOX 8481

S 439 00 ONSALA
SWEDEN
TEL
TLF
TLX 2400 ONSPACE S
EML
COM 40C

BABA NAOSHI DR
DPT OF PHYSICS
HOKKAIDO UNIVERSITY
KITA 13 NISHI 8
SAPPORO 060
JAPAN
TEL 011 716 2111
TLF 011 726 4336
TLX 932302 HOKUEN-J
EML A10156@JPNAC BITNET
COM 09

BABADZHANIANC MICHAIL DR
ASTRONOMICAL OBSERVATORY
ST PETERSBURG UNIVERSITY
194904 ST PETERSBURG
RUSSIA
TEL
TLF
TLX
EML
COM

BABADZHANOV PULAT B DR
ASTROPHYSICAL INSTITUTE
TADJIK ACAD OF SCIENCES
734670 DUSHANBE
TADZHIKISTAN
TEL
TLF
TLX
EML
COM 15,20,22C

BABCOCK HORACE W DR
MT WILSON & LAS CAMPANAS
OBSERVATORIES
813 SANTA BARBARA ST
PASADENA CA 91101
USA
TEL 818 577 1122
TLF
TLX
EML
COM 09

BABIN ARTHUR DR

UKRAINE
TEL
TLF
EML
TLX

COM 10

BABIN V G DR
SIBIZMIR
ACADEMY OF SCIENCES
664697 IRKUTSK 33
RUSSIA
TEL 6 02 65
TLF
TLX
EML
COM 10

BABU G S D
INDIAN INSTITUTE OF
ASTROPHYSICS
KORAMANGALA
BANGALORE 560 034
INDIA
TEL 812 56 9179
TLF
TLX 8452763 IIAB IN
EML
COM 45

BACALOV MIHAIL
PEOPLE'S ASTRONOMICAL OBS
& PLANETARIA
J BRUNO
BG DIMITROVGRAD
BULGARIA
TEL 39 13 797
TLF
TLX
EML
COM 46

BACCHUS PIERRE PROF
40 RUE HAUTE
F 77130 LA GDE PAROISSE
FRANCE
TEL 64 32 1315
TLF
EML
COM 08,26
TLX

BACHILLER RAFAEL DR
CTR ASTRON DE YEBES
OAN
APD 148
E 19080 GUADALAJARA
SPAIN
TEL 11 29 0311
TLF
TLX 23465 IGC E
EML
COM 34,40

BACKER DONALD CH DR
RADIO ASTRONOMY LAB
UNIVERSITY OF CALIFORNIA
601 CAMPBELL HALL
BERKELEY CA 94720
USA
TEL 415 642 5128
TLF
TLX 820181 UCB AST RAL
EML
COM 08,40C

BADIALI MASSIMO
IAS
CNR
VIA ENRICO FERMI 21
I 00044 FRASCATI
ITALY
TEL 6 942 5655
TLF 6 941 6847
TLX 610261 CNR FRA
EML MASSIMO@IRMIAS BITNE
COM 44

BADOLATI ENNIO
VIA GIUSEPPE COTRONEI 11
I 80129 NAPOLI
ITALY
TEL 81 24 3245
TLF
EML
COM 41
TLX

BAECK NICOLE A L DR
STERREKUNDIG OBSERV
RIJKSUNIVERSITEIT GENT
KRIJGSLAAN 281
B 9000 GENT
BELGIUM
TEL 91 64 4757
TLF 91 64 4995
TLX
EML BAECK@ASTRO RUG AC BE
COM

BAEK CHANG RYONG
DPT OF PHYSICS
KIM IL SUNG UNIVERSITY
TAESONG DISTRICT
PYONGYANG
KOREA DPR
TEL
TLF
TLX
EML
COM

BAERENTZEN JORN
ELMEHOJVEJ 66

DK 8270 HOJBJERG
DENMARK
TEL 86 27 2428
TLF
TLX
EML
COM

BAESSGEN MARTIN DR
ASTRONOMISCHES INSTITUT
TUEBINGEN
WALDHAEUSERSTR 64
D 7400 TUEBINGEN
GERMANY
TEL 707 129 5470
TLF 707 129 3458
TLX 7262714
EML AITMVX BAESSGEN
COM

BAFFA CARLO DR
OSS ASTROFISICO
DI ARCETRI
LARGO E FERMI 5
I 50125 FIRENZE
ITALY
TEL 55 437 8540
TLF 55 43 5939
TLX 572268 ARCETR I
EML baffa@astrfi cineca it
COM 09

BAGARE S P DR
INDIAN INSTITUTE OF
ASTROPHYSICS
KORAMANGALA
BANGALORE 560 034
INDIA
TEL 812 56 6585/497
TLF
TLX 845763 IIAB IN
EML
COM 10

BAGGALEY WILLIAM J PROF
DPT OF PHYSICS
UNIVERSITY OF CANTERBURY
PRIVATE BAG
CHRISTCHURCH 1
NEW ZEALAND
TEL 482 009*767
TLF
TLX 4144 UNICANT NZ
EML
COM 21,22

BAGHOS BALEGH B DR
HELWAN OBSERVATORY
HELWAN
EGYPT
TEL 78 0645/2683
TLF
EML
COM 07
TLX 93070 HIAG

BAGILDINSKIJ BRONISLAV K
PULKOVO OBSERVATORY
ACADEMY OF SCIENCES
10 KUTUZOV QUAY
196140 ST PETERSBURG
RUSSIA
TEL
TLF
TLX
EML
COM 08

BAGLIN ANNIE DR
OBSERVATOIRE DE PARIS
SECTION DE MEUDON
DASGAL
F 92195 MEUDON PPL CDX
FRANCE
TEL 1 47 05 7855
TLF
TLX
EML
COM 27,35,45

BAGNUOLO WILLIAM G JR DR
CHARA
GEORGIA STATE UNIVERSITY
ATLANTA GA 30303 3083
USA
TEL 404 651 2932
TLF
TLX
EML
COM 26

BAGRI DURGADAS S
NRAO
BOX 0
SOCORRO NM 87801 0387
USA
TEL 505 772 4011
TLX 910-988-1710
EML
TLF
COM 40

BAHCALL JOHN N PROF
INST FOR ADVANCED STUDY
SCHOOL OF NATURAL SCIENCE
OLDEN LN BLDG E
PRINCETON NJ 08540
USA
TEL 609 734 8054
TLF
TLX 837680
EML
COM 28,33

BAHNER KLAUS DR
MPI FUER ASTRONOMIE
ADOLF-KOLPING-STR 5
D 6903 NECKARGEMUEND
GERMANY
TEL 62 23 3735
TLF
TLX
EML
COM

BAHNG JOHN D R PROF
DPT PHYSICS & ASTRONOMY
NORTHWESTERN UNIVERSITY
DEARBORN OBSERVATORY
EVANSTON IL 60208
USA
TEL 312 491 8645
TLF
TLX
EML
COM 25,45

BAILEY JEREMY A
AAO
BOX 296
EPPING NSW 2121
AUSTRALIA
TEL 2 868 1666
TLX 23999 AAOSYD AA
EML
TLF
COM

BAILEY MARK EDWARD
DPT OF ASTRONOMY
UNIVERSITY OF MANCHESTER
MANCHESTER M13 9PL
UK
TEL 612 737 121
TLF 612 735 867
TLX 66517 UNIMAN
EML MEB@UK AC MAN AST STAR
COM 15,20,28

BAIRD SCOTT R
DPT PHYSICS & ASTRONOMY
BENEDICTINE COLLEGE
N 14
ATCHISON KS 66002 1499
USA
TEL 913 367 5340
TLF
TLX
EML
COM 36

BAIZE PAUL DR
6 RUE DAUBIGNY

F 75017 PARIS
FRANCE
TEL
TLF
TLX
EML
COM 26

BAJAJA E DR
IAR
CC 5
1864 VILLA ELISA (BS AS)
ARGENTINA
TEL 21 4 3793
TLX
EML
TLF
COM 28

BAKER JAMES GILBERT DR
14 FRENCH DR
BEDFORD NH 03102
USA
TEL 603 472 5860
TLF
EML
COM
TLX

BAKER NORMAN H PROF
DPT OF ASTRONOMY
COLUMBIA UNIVERSITY
PUPIN HALL 538 W 120TH ST
NEW YORK NY 10027
USA
TEL 212 280 3280
TLF
TLX 220094 COLU UR
EML
COM 05,27,35

BAKRY ABDEL AZIZ DR

EGYPT
TEL
TLX
TLF
EML
COM

BALASUBRAMANIAN V DR
TIFR/RADIO ASTRONOMY CTR
BOX 8
UDHAGAMANDALAM 643 001
INDIA
TEL 423 2032
TLF
TLX 8504 208 RAC IN
EML
COM 40

BALAZS BELA A PROF DR
DIREKTOR
COLLEGIUM HUNGARICUM
HOLLANDSTR 4
A 1020 WIEN 2
AUSTRIA
TEL 1 24 0581
TLF 1 62 96575
TLX
EML
COM 37,51

BALAZS LAJOS G DR
KONKOLY OBSERVATORY
THEGE U 13/17
BOX 67
H 1525 BUDAPEST
HUNGARY
TEL 1 75 5866/75 4122
TLF
TLX 227460 KONOB H
EML
COM 33

BALBUS STEVEN A DR
UNIVERSITY STATION
UNIVERSITY OF VIRGINIA
BOX 3818
CHARLOTTESVILLE VA 22903
USA
TEL 804 924 4897
TLF
TLX
EML BITNET sb@virginia
COM 33

BALDINELLI LUIGI DR
CP 1630
I 40100 BOLOGNA
ITALY
TEL 51 22 7002
TLF
EML
COM 25
TLX

BALDWIN JACK A DR
CERRO TOLOLO
INTERAMERICAN OBSERVATORY
CASILLA 603
LA SERENA
CHILE
TEL 51 21 3352
TLF 51 21 2466*342
TLX 359620301 AURA CT
EML
COM 28

BALDWIN JOHN E DR
MULLARD RADIO ASTRON OBS
CAVENDISH LABORATORY
MADINGLEY RD
CAMBRIDGE CB3 OHE
UK
TEL 223 664 77
TLF
TLX 81292 CAVLAB G
EML
COM 33,34,40,47

BALDWIN RALPH B
6190 GATEHOUSE DR SE

GRAND RAPIDS MI 49506
USA
TEL 619 949 6190
TLF
TLX
EML
COM

BALEGA YURI YU DR
SPECIAL ASTROPHYSICAL OBS
ACADEMY OF SCIENCES
NIZHNIJ ARKHYZ
357147 STAVROPOLSKIJ
RUSSIA
TEL 86 578 92501
TLF
TLX 123244 ZENIT SU
EML
COM 26C

BALICK BRUCE PROF
DPT OF ASTRONOMY
UNIVERSITY OF WASHINGTON
FM 20
SEATTLE WA 98195
USA
TEL 206 543 7683
TLF
TLX 4740096 UW UI
EML
COM

BALIUNAS SALLIE L
CENTER FOR ASTROPHYSICS
HCO/SAO
60 GARDEN ST
CAMBRIDGE MA 02138
USA
TEL 617 495 7415
TLF
TLX
EML
COM 12,29,36,44

BALKLAVS A E DR
RADIOASTROPHYSICAL OBS
LATVIAN ACAD OF SCIENCES
TURGENEVA 19
226524 RIGA
LATVIA
TEL
TLF
TLX
EML
COM 40

BALKOWSKI-MAUGER CH DR
OBSERVATOIRE DE PARIS
SECTION DE MEUDON
DAF
F 92195 MEUDON PPL CDX
FRANCE
TEL 1 45 34 7556
TLF
TLX 201571
EML
COM 28

BALL JOHN A DR
HAYSTACK OBSERVATORY
WESTFORD MA 01886
USA
TEL 617 692 4764
TLF
EML
COM 51
TLX

BALL LEWIS DR
THEORETICAL ASTROPHYSICS
UNIVERSITY OF SYDNEY
SYDNEY NSW 2006
AUSTRALIA
TEL 2 692 3241
TLF 2 660 2903
TLX
EML BALL@PHYSICS SU OZ AU
COM 40

BALLABH G M DR
DPT OF ASTRONOMY
UNIVERSITY OF OSMANIA
HYDERABAD 500 007
INDIA
TEL 71 951*247
TLF
TLX
EML
COM 24,28

BALLARIO M C PROF
OSS ASTRONOMICO
DI ARCETRI
VIA S LEONARDO
I 50100 FIRENZE
ITALY
TEL
TLF
TLX
EML
COM

BALLEREAU DOMINIQUE DR
OBSERVATOIRE DE PARIS
SECTION DE MEUDON
SECTION D'ASTROPHYSIQUE
F 92195 MEUDON PPL CDX
FRANCE
TEL 1 45 07 7854
TLF 1 45 07 7878
TLX 631987
EML MESIOA BALLEREAU
COM 29

BALLESTER JOSE LUIS DR
DPT FISICA
UNIVERSIDAD DE LAS ISLAS
BALEARES
E 07071 PALMA DE MALLORCA
SPAIN
TEL 71 17 3228
TLF 71 43 8028
TLX 69121 UNPM E
EML DFSULBO@PS UJB ES
COM 10

BALLI EDIBE PROF
UNIVERSITY RASATHANESI

ISTANBUL
TURKEY
TEL
TLF
TLX
EML
COM 10

BALLY JOHN DR
AT & BELL LABORATORIES
HOH L 245
HOLMDEL NJ 07733
USA
TEL 201 888 7124
TLF
TLX 219879 BTLH UR
EML jb@hohn-2 att com
COM 40

BALMINO GEORGES G DR
CNES/GRGS/BGI
18 AVE E BELIN
F 31055 TOULOUSE CDX
FRANCE
TEL 61 27 4427
TLF
TLX 531081 CNEST B F
EML
COM 07

BALONA LUIS ANTERO DR
SAAO
BOX 9
OBSERVATORY 7935
SOUTH AFRICA
TEL 47 0025
TLF
TLX 20309
EML
COM 25,27,30,36

BALONEK THOMAS J DR
DPT OF PHYSICS & ASTRON
COLGATE UNIVERSITY
HAMILTON NY 13346
USA
TEL 315 824 1000
TLF
TLX
EML BITNET TBALONEK@COLGATEI
COM 40

BALTHASAR HORST DR
UNIVERSTAETS STERNWARTE
GEISMARLANDSTRASSE 11
D 3400 GOETTINGEN
GERMANY
TEL 551 39 5048
TLF
TLX 96753
EML BITNET HBALTHA@DGOGWDG1
COM 12

BALUTEAU JEAN-PAUL DR
OBSERVATOIRE DE MARSEILLE
2 PLACE LE VERRIER
F 13248 MARSEILLE CDX 04
FRANCE
TEL 91 95 9088
TLF
TLX 420241 F
EML EARN OBSMRS@FROMRS51
COM 34

BANDERMANN L W DR
21131 GRENOLA DR
CUPERTINO CA 95014
USA
TEL
TLF
TLX
EML
COM

BANDIERA RINO DR
OSS ASTROFISICO
DI ARCETRI
LARGO E FERMI 5
I 50125 FIRENZE
ITALY
TEL 55 437 8540
TLF 55 43 5939
TLX 572268
EML BANDIERA@ASTRFI INFN IT
COM

BANDYOPADHYAY A PROF
RESEARCH DIVISION
BIRLA PLANETARIUM
96 JAWAHARLAL NEHRU RD
CALCUTTA 700 071
INDIA
TEL 33 281 515
TLF
TLX
EML
COM 04,41

BANERJI SRIRANJAN DR
DPT OF PHYSICS
UNIVERSITY OF BURDWAN
GOLOPBAG
BURDWAN 713 104
INDIA
TEL
TLF
TLX
EML
COM 47

BANG YONG GOL
PYONGYANG ASTRON OBS
ACADEMY OF SCIENCES DPRK
TAESONG DISTRICT
PYONGYANG
KOREA DPR
TEL
TLF
TLX
EML
COM 19

BANHATTI DILIP GOPAL DR
SCHOOL OF PHYSICS
MADURAI KAMARAJ UNIVERS
PALKALAINAGAR
MADURAI 625021
INDIA
TEL 85 252
TLF
TLX 445308 MKU IN
EML
COM 28,33,40,47

BANIA THOMAS MICHAEL
DPT OF ASTRONOMY
BOSTON UNIVERSITY
725 COMMONWEALTH AVE
BOSTON MA 02215
USA
TEL 617 353 3652
TLF
TLX 95129 BOS UNIV BSN
EML
COM 34,51

BANNI ALDO DR
STAZIONE ASTRONOMICA
VIA OSPEDALE 72
I 09124 CAGLIARI
ITALY
TEL 70 66 3544
TLX
EML VAXGA2 LASER
TLF 70 65 7657
COM 19

BANOS COSMAS J DR
ASTRONOMICAL INSTITUTE
NTL OBSERVATORY OF ATHENS
BOX 20048
GR 118 10 ATHENS
GREECE
TEL 1 346 1191
TLF
TLX 215530 OBSA GR
EML
COM 21

BANOS GEORGE J PROF
DPT OF PHYSICS/DAG
UNIVERSITY OF IOANNINA
GR 453 32 IOANNINA
GREECE
TEL 65 19 1697
TLF
TLX 322160
EML
COM

BAO KEREN
NANJING ASTRONOMICAL
INSTRUMENT FACTORY
NANJING
CHINA PR
TEL 25 46191
TLF
TLX 34136 GLYNJ c/o NAIF
EML
COM 09

BARANNE A DR
OBSERVATOIRE DE MARSEILLE
2 PLACE LE VERRIER
F 13248 MARSEILLE CDX 04
FRANCE
TEL 91 95 9088
TLF
TLX
EML
COM 09

BARATTA GIOVANNI BATTISTA
OAR
VIA DEL PARCO MELLINI 84
I 00136 ROMA
ITALY
TEL 6 34 7056
TLF
TLX
EML
COM 29

BARBANIS BASIL PROF
DPT OF ASTRONOMY
UNIVERSITY THESSALONIKI
GR 540 06 THESSALONIKI
GREECE
TEL 31 99 1357
TLF
TLX 412181
EML
COM 33

BARBARO G DR
OSS ASTRONOMICO DI PADOVA
VIC DELL OSSERVATORIO 5
I 35122 PADOVA
ITALY
TEL 49 66 1499
TLX 430176 UNPADU I
EML
TLF
COM

BARBERIS BRUNO
IST DI FISICA MATEMATICA
UNIVERSITA DI TORINO
VIA C ALBERTO 10
I 10123 TORINO
ITALY
TEL 11 53 9214
TLF
TLX
EML
COM 07,33,47

BARBIERI CESARE PROF
DPT DI ASTRONOMIA
UNIVERSITA DI PADOVA
VIC DELL OSSERVATORIO 5
I 35122 PADOVA
ITALY
TEL 49 66 1499
TLF
TLX 430176 UNPADU I
EML
COM 51

BARBIER-BROSSAT M DR
OBSERVATOIRE DE MARSEILLE
2 PLACE LE VERRIER
F 13248 MARSEILLE CDX 04
FRANCE
TEL 91 95 9088
TLF
TLX 420241 F
EML
COM 30,45

BARBON ROBERTO PROF
OSSERVATORIO ASTROFISICO
VIA DELL OSSERVATORIO 8
I 36012 ASIAGO
ITALY
TEL 42 46 2665
TLF
TLX 430110 SETURIST
EML
COM 28

BARBUY BEATRIZ DR
UNIVERSIDADE DE SAO PAULO
DEPT DE ASTRONOMIA
CP 30627
01051 SAO PAULO SP
BRAZIL
TEL 11 577 8599
TLF 11 276 3848
TLX 1156735 IAGM BR
EML
COM 29C

BARCIA ALBERTO DR
CTR ASTRON DE YEBES
OAN
APD 148
E 19080 GUADALAJARA
SPAIN
TEL 11 29 0311
TLF 11 29 0063
TLX
EML BARCIA CAY ES
COM 09

BARCONS XAVIER DR
DPT DE FISICA MODERNA
UNIVERSIDAD DE CANTABRIA
AVD LOS CASTROS S/N
E 39005 SANTANDER
SPAIN
TEL 42 20 1461
TLF 42 20 1402
TLX 35861 EDUCI E
EML BARCONS@CCUCVX UNICAN ES
COM 28

BARCZA SZABOLCS DR
KONKOLY OBSERVATORY
THEGE U 13/17
BOX 67
H 1525 BUDAPEST
HUNGARY
TEL 1 75 5866/75 4122
TLF
TLX 227460 KONOB H
EML
COM

BARDEEN JAMES M PROF
DPT OF PHYSICS
UNIVERSITY OF WASHINGTON
FM 15
SEATTLE WA 98195
USA
TEL 206 545 2394
TLF
TLX 4740096 UW UI
EML
COM 47

BARKAT ZALMAN PROF
RACAH INST OF PHYSICS
HEBREW UNIV OF JERUSALEM
JERUSALEM 91904
ISRAEL
TEL 2 584 490
TLF
TLX 25391
EML
COM

BARKER EDWIN S DR
ASTRONOMY DPT
UNIVERSITY OF TEXAS
RLM 15 308
AUSTIN TX 78712 1083
USA
TEL 512 471 4461
TLF
TLX 910-874-1351
EML
COM 15

BARKER PAUL K DR
DPT OF PHYSICS
YORK UNIVERSITY
4700 KEELE ST
NORTH YORK ON M3J 1P3
CANADA
TEL 416 736 2100
TLF 416 736 5386
TLX 065 24736
EML
COM

BARKER TIMOTHY DR
DPT PHYSICS & ASTRONOMY
WHEATON COLLEGE
NORTON MA 02766
USA
TEL 508 285 7722
TLF
TLX
EML
COM

BARLAI KATALIN DR
KONKOLY OBSERVATORY
THEGE U 13/17
BOX 67
H 1525 BUDAPEST
HUNGARY
TEL 1 75 5866/75 4122
TLF
TLX 227460 KONOB H
EML
COM

BARLETTI RAFFAELE ENG
OSS ASTROFISCIO
DI ARCETRI
LARGO E FERMI 5
I 50125 FIRENZE
ITALY
TEL 55 437 8540
TLF 55 43 5939
TLX 572268
EML
COM

BARLIER FRANCOIS E DR
OCA CERGA
AVE COPERNIC
F 06130 GRASSE
FRANCE
TEL 93 36 5849
TLF
TLX 470865
EML
COM 19

BARLOW MICHAEL J DR
DPT PHYSICS & ASTRONOMY
UNIVERSITY COLLEGE LONDON
GOWER ST
LONDON WC1E 6BT
UK
TEL 1 713 877 050
TLF
TLX 28722 UCPHYS G
EML
COM 34

BARNARD HANNES A J DR
DPT OF PHYSICS
UNIV OF BRITISH COLUMBIA
6224 AGRICULTURE RD
VANCOUVER BC V6T 2A6
CANADA
TEL 604 228 2894
TLF 604 228 5324
TLX 045 08576
EML
COM 14

BARNES AARON DR
NASA AMES RESEARCH CTR
MS 245 3
MOFFETT FIELD CA 94035
USA
TEL 415 694 5506
TLF
TLX
EML
COM 34,49

BARNES III THOMAS G DR
ASTRONOMY DPT
UNIVERSITY OF TEXAS
MCDONALD OBSERVATORY
AUSTIN TX 78712 1083
USA
TEL 512 471 4461
TLF
TLX 910-874-1351
EML
COM 25,27C

BARNOTHY JENO DR PROF
833 LINCOLN ST
EVANSTON IL 60201
USA
TEL 312 328 5729
TLF
EML
COM 47,48
TLX

BAROCAS VINICIO PROF
11 YEWLANDS AVE
FULWORD
PRESTON PR2 4QR
UK
TEL 772 719 249
TLF
TLX
EML
COM

BARONE FABRIZIO DR
DPT DI SCIENZE FISICHE
UNIVERSITA DI NAPOLI
MOSTRA D'OLTREMARE PAD 19
I 80125 NAPOLI
ITALY
TEL 81 725 3447
TLF 81 61 4508
TLX 720320
EML FBARONE@NAPOLI INFN IT
COM 42

BARRETO LUIZ MUNIZ PROF
OBSERVATORIO NACIONAL
RUA GL BRUCE 586
SAO CRISTOVAO
20921 RIO DE JANEIRO RJ
BRAZIL
TEL 21 580 7313
TLF 21 580 0332
TLX 02121288 OBSN
EML
COM 50

BARRETT ALAN H PROF
15 TREASURE RD
MARATHON FL 33050
USA
TEL
TLF
EML
COM 34,40
TLX

BARRETT PAUL EVERETT DR
NASA/GSFC
CODE 668 1
GREENBELT MD 20771
USA
TEL
TLF
TLX
EML
COM 25

BARROSO JR JAIR
OBSERVATORIO NACIONAL
RUA GL J CRISTINA 77
SAO CRISTOVAO
20921 RIO DE JANEIRO RJ
BRAZIL
TEL 21 580 7313*273
TLF 21 580 0332
TLX 021-21288
EML
COM 09

BARROW COLIN H DR
MAX PLANCK INSTITUT
MAX PLANCK STRASSE 2
POSTFACH 20
D 3411 KATLENBURG LINDAU
GERMANY
TEL
TLF
TLX
EML
COM 10,16,40,49

BARROW JOHN DAVID
ASTRONOMY CENTRE
UNIVERSITY OF SUSSEX
FALMER
BRIGHTON BN1 9QH
UK
TEL 273 60 6755
TLF
TLX 877159 UNISEX G
EML
COM 47

BARROW RICHARD F DR
PHYSICAL CHEMISTRY LAB
UNIVERSITY OF OXFORD
SOUTH PARKS RD
OXFORD OX1 3QZ
UK
TEL 865 533 22
TLF
TLX
EML
COM 14

BARRY DON C DR
DPT OF ASTRONOMY
UNIV SOUTHERN CALIFORNIA
LOS ANGELES CA 90089
USA
TEL 213 743 2764
TLF
TLX
EML
COM 29,45

BARSTOW MARTIN ADRIAN DR
DPT OF PHYSICS
UNIVERSITY OF LEICESTER
UNIVERSITY RD
LEICESTER LE1 7RH
UK
TEL 533 523 492
TLF 533 550 182
TLX
EML MAB@UK AC LE STAR/19838 MAB
COM 44

BARTAYA R A DR
ABASTUMANI ASTROPHYSICAL
OBSERVATORY
GEORGIAN ACAD OF SCIENCES
383762 ABASTUMANI
GEORGIA
TEL 237 ABASTUMANI
TLF
TLX 327409
EML
COM 45C

BARTEL NORBERT HARALD DR
CENTER FOR ASTROPHYSICS
HCO/SAO
60 GARDEN ST
CAMBRIDGE MA 02138
USA
TEL 617 495 9278
TLF
TLX 921428
EML
COM 40

BARTH CHARLES A PROF
LASP
UNIVERSITY OF COLORADO
BOX 392
BOULDER CO 80309 0392
USA
TEL 303 492 7502
TLF
TLX
EML
COM 49

BARTHEL PETER DR
KAPTEYN ASTRONOMICAL INST
UNIVERSITY OF GRONINGEN
BOX 800
NL 9700 AV GRONINGEN
NETHERLANDS
TEL 50 634073
TLF
TLX 53572 STARS NL
EML
COM 28,40,47

BARTHOLDI PAUL DR
OBSERVATOIRE DE GENEVE
CHEMIN DES MAILLETTES 51
CH 1290 SAUVERNY
SWITZERLAND
TEL 22 755 2611
TLF 22 755 3983
TLX 419209 OBS CH
EML bartho@obs unige ch
COM

BARTOLINI CORRADO
DPT DI ASTRONOMIA
UNIVERSITA DI BOLOGNA
VIA ZAMBONI 33
I 40126 BOLOGNA
ITALY
TEL 51 22 6677
TLF
TLX 211664
EML
COM 27,42

BARUCCI MARIA A DR
OBSERVATOIRE DE PARIS
SECTION DE MEUDON
F 92195 MEUDON PPL CDX
FRANCE
TEL 1 45 07 7639
TLF 1 45 07 7469
TLX 201571
EML MESIOA BARUCCI/17670 BARUCCI
COM 15C

BARUCH JOHN DR
DPT OF ELECTRICAL ENGIN
UNIVERSITY OF BRADFORD
BRADFORD BD7 1DP
UK
TEL 274 38 4024
TLF 274 39 1521
TLX 51309 UNIBFD G
EML UK AC BRADFORD ELECENG JEFB
COM 09

BARVAINIS RICHARD DR
HAYSTACK OBSERVATORY
WESTFORD MA 01886
USA
TEL 617 692 4764
TLF
EML
COM 40
TLX 948149

BARWIG HEINZ
INST F ASTRON & ASTROPHYS
UNIVERSITAET MUNCHEN
SCHEINERSTR 1
D 8000 MUENCHEN 80
GERMANY
TEL 89 989 021
TLF
TLX 529815 UNIVM D
EML
COM 09,27

BASART JOHN P
DPT OF PHYSICS
IOWA STATE UNIVERSITY
AMES IA 50011
USA
TEL 515 294 2663
TLF
TLX
EML
COM

BASCHEK BODO PROF
INSTITUT F THEORETISCHE
ASTROPHYSIK
IM NEUENHEIMER F 561
D 6900 HEIDELBERG 1
GERMANY
TEL 62 215 62837
TLF
TLX 461515 UNIHD D
EML
COM 36

BASH FRANK N PROF
ASTRONOMY DET
UNIVERSITY OF TEXAS
RLM
AUSTIN TX 78712 1083
USA
TEL 512 471 4461
TLF
TLX 910-874-1351
EML
COM 33,34,40

BASRI GIBOR B
ASTRONOMY DPT
UNIVERSITY OF CALIFORNIA
601 CAMPBELL HALL
BERKELEY CA 94720
USA
TEL 415 642 8198
TLF
TLX 820181 UCB ASTRAL UD
EML
COM 29,36

BASSINO LILIA P DR
UNIV NACIONAL DE LA PLATA
FCAG
1900 LA PLATA (BS AS)
ARGENTINA
TEL 21 21 7308
TLF 21 21 1761
TLX 31151 BULAP
EML LBASSINO@FCAGLP EDU AR
COM 28

BASTIAN ULRICH
ASTRONOMISCHES-RECHEN
INSTITUT
MOENCHHOFSTR 12-14
D 6900 HEIDELBERG 1
GERMANY
TEL 62 214 9026
TLF
TLX 461336 ARI HD D
EML
COM 24

BASTIEN PIERRE DR
DPT DE PHYSIQUE
UNIVERSITE DE MONTREAL
CP 6128 SUCC A
MONTREAL QC H3C 3J7
CANADA
TEL 514 343 7355
TLF 514 343 2071
TLX 055 62425 UDEMPHYSA
EML
COM 27

BASTIN JOHN A PROF
DPT OF PHYSICS
QUEEN MARY/WESTFIELD COLL
MILE END RD
LONDON E1 4NS
UK
TEL
TLF
TLX
EML
COM

BASU BAIDYANATH PROF
DPT APPLIED MATHEMATICS
UNIVERSITY OF CALCUTTA
92 A P C RD
CALCUTTA 700 009
INDIA
TEL
TLF
TLX
EML
COM 28,33,51

BASU DIPAK DR
DPT OF PHYSICS
UNIVERSITY OF WEST INDIES
ST AUGUSTINE
TRINIDAD WEST INDIES
INDIA
TEL
TLF
TLX
EML
COM 40,47,48,51

BATALHA CELSO CORREA DR
OBSERVATORIO NACIONAL
RUA GL BRUCE 586
SAO CRISTOVAO
20921 RIO DE JANEIRO RJ
BRAZIL
TEL 21 580 7181
TLF· 21 580 0332
TLX 021288
EML IAGUSP CELSO/CCB@LNCCUM
COM 29

BATCHELOR DAVID ALLEN DR
NASA/GSFC
CODE 633
GREENBELT MD 20771
USA
TEL 301 286 2988
TLF
TLX 892339
EML
COM 10

BATES BRIAN DR
DPT OF PURE & APPL PHYS
QUEEN'S UNIVERSITY
BELFAST BT7 1NN
UK
TEL 232 24 5133
TLF
TLX 74487
EML
COM

BATES DAVID R PROF
DPT OF APPLIED MATHS
& THEORETICAL PHYSICS
QUEEN'S UNIVERSITY
BELFAST BT7 1NN
UK
TEL
TLF
TLX
EML
COM 14,21

BATESON FRANK M OBE DR
ASTRONOMICAL RESEARCH LTD
BOX 3093
GREERTON TAURANGA
NEW ZEALAND
TEL 64-075-410-216
TLF
TLX 2880 CPO TG NZ
EML
COM 27,42

BATH GEOFFREY T DR
DPT OF ASTROPHYSICS
UNIVERSITY OF OXFORD
SOUTH PARKS RD
OXFORD OX1 3RQ
UK
TEL 865 511 336
TLF
TLX 83295
EML
COM 27,42

BATRAKOV YU V DR
INST OF THEORET ASTRONOMY
ACADEMY OF SCIENCES
N KUTUZOVA 10
191187 ST PETERSBURG
RUSSIA
TEL 272 40 23
TLF
TLX 121578 ITA SU
EML
COM 07,20

BATSON RAYMOND MILNER DR
US GEOLOGICAL SURVEY
BRANCH OF ASTROGEOLOGY
2255 N GEMINI DR
FLAGSTAFF AZ 86001
USA
TEL 602 556 7260
TLF 602 556 7090
TLX
EML NASAMAILRA%BATSON
COM 16

BATTANER EDUARDO DR
DPT FIS TEORICA & COSMOS
FAC DE CIENCIAS
AVD FUENTENUEVA
E 18002 GRANADA
SPAIN
TEL 20 22 12306
TLF
TLX
EML.
COM 16,28

BATTEN ALAN H DR
HERZBERG INST ASTROPHYS
DOMINION ASTROPHYS OBS
5071 W SAANICH RD
VICTORIA BC V8X 4M6
CANADA
TEL 604 388 0009
TLF 604 363 0045
TLX 049 7295
EML
COM 26,30,42,WGWWDA

BATTINELLI PAOLO DR
OAR
VIA DEL PARCO MELLINI 84
I 00136 ROMA
ITALY
TEL 6 34 7056
TLF 6 34 7802
TLX 626326 OAROMA I
EML 40061 BATTINELLI
COM 28

BATTISTINI PIERLUIGI DR
OSS ASTRONOMICO
UNIVERSITA DI BOLOGNA
VIA ZAMBONI 33
I 40126 BOLOGNA
ITALY
TEL 51 51 9593
TLF
TLX
EML
COM

BATTY MICHAEL DR
SCHOOL OF MATHS PHYSICS
COMPUTING & ELECTRONONICS
MACQUARIE UNIVERSITY
NSW 2109
AUSTRALIA
TEL
TLF
TLX
EML
COM 40

BAUD BOUDEWIJN DR
FOKKER BV
SPACE DIVISION
BOX 7600
NL 1117 ZJ SHIPHOL
NETHERLANDS
TEL 20 54 49111
TLF
TLX
EML
COM 33

BAUDRY ALAIN DR
OBSERVATOIRE DE BORDEAUX
BP 89
F 33270 FLOIRAC
FRANCE
TEL 56 86 4330
TLF 56 40 4251
TLX
EML
COM 34,40

BAUER CARL A DR
DPT OF ASTORNOMY
PENNSYLVANIA STATE UNIV
506 DAVEY LAB
UNIVERSITY PARK PA 16802
USA
TEL
TLF
TLX
EML
COM

BAUER WENDY HAGEN
WHITIN OBSERVATORY
WELLESLEY COLLEGE
WELLESLEY MA 02181
USA
TEL 617 235 0320
TLF
TLX
EML
COM 27,29

BAUM WILLIAM A DR
DPT OF ASTRONOMY
UNIVERSITY OF WASHINGTON
2124 NE PARK RAOD
SEATTLE WA 98195
USA
TEL
TLF
TLX
EML
COM 09,16,28,51

BAUSTIAN W W MR
KITT PEAK NTL OBS
BOX 26732
950 N CHERRY AVE
TUCSON AZ 85726 6732
USA
TEL
TLF
TLX
EML
COM

BAUTZ LAURA P DR
NTL SCIENCE FOUNDATION
DIV ASTRONOMICAL SCIENCES
1800 G ST NW
WASHINGTON DC 20550
USA
TEL 202 357 9488
TLF
TLX
EML
COM

BAYM GORDON ALAN DR
DPT OF PHYSICS
UNIVERSITY OF ILLINOIS
1110 W GREEN ST
URBANA IL 61801
USA
TEL 217 333 4363
TLF
TLX 910-830-6599 PHYSICS
EML
COM 35,48

BAZILEVSKY ALEXANDR T
VERNADSKY INST GEOCHEM &
ANALYTICAL CHEMISTRY
KOSYGIN STR 19
117334 MOSCOW
RUSSIA
TEL
TLF
TLX
EML
COM 16C

BEALE JOHN S DR
231 MARLBOROUGH RD

SWINDON SN3 1NN
UK
TEL 793 347 25
TLF
TLX
EML
COM

BEARD DAVID B DR
DPT PHYSICS & ASTRONOMY
UNIVERSITY OF KANSAS
LAWRENCE KS 66045
USA
TEL 913 864 3752
TLF
TLX
EML
COM 12,15,22

BEAUDET GILLES DR
DPT DE PHYSIQUE
UNIVERSITE DE MONTREAL
CP 6128 SUCC A
MONTREAL QC H3C 3J7
CANADA
TEL 514 343 6669
TLF 514 343 2071
TLX 055 62425
EML
COM 35,51

BEAVERS WILLET I DR
ERWIN W FICK OBSERVATORY
IOWA STATE UNIVERSITY
AMES IA 50011
USA
TEL 515 294 3667
TLF
TLX
EML
COM 26,30

BECHTOLD JILL DR
STEWARD OBSERVATORY
UNIVERSITY OF ARIZONA
TUCSON AZ 85721
USA
TEL 602 621 6533
TLF 602 428 2854
TLX
EML
COM 47

BECK H G
VEB CARL ZEISS
FORSCHUNGSZENTRUM
CARL-ZEISS STR 1
D 6900 JENA
GERMANY
TEL
TLF
TLX
EML
COM

BECK RAINER
MPI FUER RADIOASTRONOMIE
AUF DEM HUEGEL 69
D 5300 BONN 1
GERMANY
TEL 228 52 5320
TLF
TLX 886440
EML
COM 25,28,40

BECKER ROBERT A DR
BOX 4609
CARMEL CA 93921
USA
TEL
TLF
TLX
EML
COM

BECKER ROBERT HOWARD
DPT OF PHYSICS
UNIVERSITY OF CALIFORNIA
DAVIS CA 95616
USA
TEL 916 752 6921
TLF
TLX 910-531-0785 UC DAVS
EML
COM 48

BECKER STEPHEN A
LOS ALAMOS NATIONAL LAB
MS B220
BOX 1663
LOS ALAMOS NM 87545
USA
TEL 505 667 8931
TLF
TLX 660495
EML
COM 35

BECKER WILHELM PROF
ASTRONOMISCHES INSTITUT
UNIVERSITAET BASEL
VENUSSTRASSE 7
CH 4102 BINNINGEN
SWITZERLAND
TEL 61 22 7711
TLF
TLX
EML
COM 25,33,37

BECKERS JACQUES M DR
ESO
KARL SCHWARZSCHILDSTR 2
D 8046 GARCHING MUENCHEN
GERMANY
TEL 89 320 060
TLF 89 320 2362
TLX 05 28 282 22 EO D
EML
COM 10,12

BECKLIN ERIC E DR
DPT OF ASTRONOMY
UNIVERSITY OF CALIFORNIA
LOS ANGELES CA 90024
USA
TEL 213 206 0208
TLF
TLX
EML BECKLIN@UCLASTRO BITNET
COM 09,34

BECKMAN JOHN E PROF
INST DE ASTROFISICA
DE CANARIAS
OBS DEL TEIDE
E 38071 LA LAGUNA
SPAIN
TEL
TLF
TLX
EML
COM 12,29,34,47,51

BECKWITH STEVEN V W
MPI FUER ASTRONOMIE
KOENIGSTUHL 17
D 6900 HEIDELBERG 1
GERMANY
TEL 6221 528 210
TLF 6221 528 246
TLX
EML BECKWITH@MPIAHD MPI-HD MPG DE
COM 34,51

BEC-BORSENBERGER ANNICK
BUREAU DES LONGITUDES
77 AVE DENFERT ROCHEREAU
F 75014 PARIS
FRANCE
TEL 1 40 51 2273
TLF 1 46 33 2834
TLX
EML SPAN IAPC08 Borsenberger
COM 04,07,20

BEDOGNI ROBERTO
DPT DI ASTRONOMIA
UNIVERSITA DI BOLOGNA
CP 596
I 40100 BOLOGNA
ITALY
TEL 51 22 2956
TLF
TLX 211664 INFNBO I
EML
COM 27,34

BEEBE HERBERT A
DPT OF ASTRONOMY
NEW MEXICO STATE UNIV
LAS CRUCES NM 88003
USA
TEL 505 646 4438
TLF
TLX 910-983-0549 NMSUC
EML
COM 10,12

BEEBE RETA FAYE DR
DPT OF ASTRONOMY
NEW MEXICO STATE UNIV
BOX 4500
LAS CRUCES NM 88003
USA
TEL 505 646 1938
TLF
TLX
EML
COM 16,51

BEER REINHARD DR
JPL/CALTECH
MS 183 301
4800 OAK GROVE DR
PASADENA CA 91109
USA
TEL 818 354 4748
TLF
TLX
EML
COM 09,16

BEERS TIMOTHY C DR
PHYSICS & ASTRONOMY DPT
MICHIGAN STATE UNIVERSITY
EAST LANSING MI 48824
USA
TEL 517 353 4541
TLF
TLX
EML BEERS@MSUPA BITNET
COM 30

BEGELMAN MITCHELL CRAIG
JILA
UNIVERSITY OF COLORADO
BOX 440
BOULDER CO 80309 0440
USA
TEL 303 492 7856
TLF
TLX 755842 JILA
EML
COM 48

BEGEMAN KOR G DR
KAPTEYN ASTRONOMICAL INST
BOX 800
NL 9700 AV GRONINGEN
NETHERLANDS
TEL 50 63 4073
TLX 53572
EML KGB@RUGFX4 RUG NL
TLF 50 63 6100
COM 28

BEGGS DENIS W MR
INSTITUTE OF ASTRONOMY
THE OBSERVATORIES
MADINGLEY RD
CAMBRIDGE CB3 OHA
· UK
TEL 223 622 04
TLF
TLX 817297 ASTRON G
EML
COM

BEHR ALFRED PROF EMERITUS
ESCHENWEG 3
D 3406 BOVENDEN
GERMANY
TEL 551 8897
TLF
TLX
EML
COM 25

BEINTEMA DOUWE A DR
SPACE RESEARCH DPT
UNIVERSITY OF GRONINGEN
BOX 800
NL 9700 AV GRONINGEN
NETHERLANDS
TEL 50 11 6631
TLF
TLX 53572
EML
COM

BEKENSTEIN JACOB D DR
DPT OF PHYSICS
BEN GURION UNIVERSITY
BOX 653
BEERSHEVA 84105
ISRAEL
TEL 57 66 4271
TLF
TLX 5253 UNASI IL
EML
COM

BEL NICOLE J DR
OBSERVATOIRE DE PARIS
SECTION DE MEUDON
F 92195 MEUDON PPL CDX
FRANCE
TEL 1 45 34 7412
TLF
TLX 201571
EML
COM 12,34,47

BELINSKY VLADIMIR DR
LANDAU INST THEOR PHYSICS
ACADEMY OF SCIENCES
117940 MOSCOW
RUSSIA
TEL 137 32 44
TLF
TLX
EML
COM 47

BELKOVICH O I DR
ENGELHARDT ASTRONOMICAL
OBSERVATORY
OBSERVATORIA STATION
422526 KAZAN
RUSSIA
TEL 324827
TLF
TLX
EML
COM 21,22

BELL BARBARA DR
CENTER FOR ASTROPHYSICS
HCO/SAO
60 GARDEN ST
CAMBRIDGE MA 02138
USA
TEL 617 495 2688
TLF
TLX
EML
COM 10

BELL BURNELL S JOCELYN DR
DPT OF PHYSICS
THE OPEN UNIVERSITY
WALTON HALL
MILTON KEYNES MK7 6AA
UK
TEL 908 274 066
TLF 908 653 744
TLX
EML
COM

BELL KENNETH LLOYD DR
DPT OF APPLIED MATHS
& THEORETICAL PHYSICS
QUEEN'S UNIVERSITY
BELFAST BT7 1NN
UK
TEL 232 24 5133
TLF
TLX 74487 QUBADM
EML
COM ·

BELL KENNETH LLOYD DR
DPT OF APPLIED MATHS
& THEORETICAL PHYSICS
QUEEN'S UNIVERSITY
BELFAST BT7 1NN
UK
TEL 232 24 5133
TLF
TLX 74487 QUBADM
EML
COM

BELL MORLEY B
HERZBERG INST ASTROPHYS
NTL RESEARCH COUNCIL
100 SUSSEX DR
OTTAWA ON K1A 0R6
CANADA
TEL· 613 993 6060
TLF 613 952 6602
TLX 053 3715
EML
COM

BELL ROGER A DR
ASTRONOMY PROGRAM
UNIVERSITY OF MARYLAND
COLLEGE PARK MD 20742
USA
TEL 301 454 6282
TLF
TLX 887294
EML
COM 36,37,45

BELL STEVEN DR
DPT OF PHYSICS & ASTRON
UNIVERSITY OF ST ANDREWS
NORTH HAUGH
ST ANDREWS FIFE KY16 9SS
UK
TEL 334 761 61*8306
TLF 334 744 87
TLX 9312110846 SA G
EML ASRSB@STAR ST-AND AC UK
COM 42

BELLAS-VELIDIS IOANNIS DR
ASTRONOMICAL INSTITUTE
NTL OBSERVATORY OF ATHENS
BOX 20048
GR 118 10 ATHENS
GREECE
TEL 1 346 1191
TLF 1 346 3803
TLX
EML YBELLAS@GRATHUN1
COM 29

BELMONTE AVILES J A DR
INST DE ASTROFISICA
DE CANARIAS
OBS DEL TEIDE
E 38200 LA LAGUNA
SPAIN
TEL. 22 60 5200
TLF 22 26 3005
TLX 92640 IAC E
EML IAC JBA
COM 27,35

BELSERENE EMILIA P
MARIA MITCHELL OBS
3 VESTAL STREET
NANTUCKET MA 02554
USA
TEL 617 228 9273
TLF
TLX
EML
COM 27

BELTON MICHAEL J S DR
SOLAR SYSTEM PROGRAM
BOX 26732
950 N CHERRY AVE
TUCSON AZ 85726 6732
USA
TEL 602 327 5511
TLF
TLX 666-484 AURA KNPO TU
EML
COM 15C,16C

BELVEDERE GAETANO DR
IST DI ASTRONOMIA
CITTA UNIVERSITARIA
VIA A DORIA 6
I 95125 CATANIA
ITALY
TEL 95 33 0533
TLF
TLX 970359 ASTRCT I
EML
COM 10,27

BELYAEV NIKOLAJ A DR
INST OF THEORET ASTRONOMY
ACADEMY OF SCIENCES
N KUTUZOVA 10
191187 ST PETERSBURG
RUSSIA
TEL 279 06 67
TLF
TLX 121578 ITA SU
EML
COM

BELY-DUBAU FRANCOISE
OCA OBSERV DE NICE
BP 139
F 06003 NICE CDX
FRANCE
TEL 93 89 0420
TLF
TLX 460004 OBSNICE F
EML
COM 14

BEM JERZY DR
ASTRONOMICAL INSTITUTE
WROCLAW UNIVERSITY
UL KOPERNIKA 11
PL 51 622 WROCLAW
POLAND
TEL
TLF
TLX
EML
COM 08

BENACCHIO LEOPOLDO
OSS ASTRONOMICO DI PADOVA
VIC DELL OSSERVATORIO 5
I 35122 PADOVA
ITALY
TEL 49 66 1499
TLX 430176 UNPADU I
EML
TLF
COM 05

BENAVENTE JOSE
URBANIZACION LAS REDES
OCEANO ALANTICO 11
E 11500 PUERTO SANTA MARI
SPAIN
TEL
TLF
TLX
EML
COM 31

BENAYDOUN JEAN-JACQUES DR
OBSERVATOIRE DE GRENOBLE
CERMO/ASTROPHYSIQUE
BP 53X
F 38041 GRENOBLE CDX
FRANCE
TEL 76 51 4914
TLF 76 44 8821
TLX USMG 980134
EML BENAYDOUN@FRGAG51
COM 34

BENDER PETER L DR
JILA
UNIVERSITY OF COLORADO
BOX 440
BOULDER CO 80309 0440
USA
TEL 303 492 6793
TLF
TLX 755842 JILA
EML
COM 16,19,31

BENDER RALF DR
LANDESSTERNWARTE
KOENIGSTUHL 12
D 6900 HEIDELBERG 1
GERMANY
TEL 62 215 090
TLF 62 215 09202
TLX 491153 LSWHD
EML H36 DHDURZ1
COM 28

BENDINELLI ORAZIO
DPT DI ASTRONOMIA
UNIVERSITA DI BOLOGNA
VIA ZAMBONI 33
I 40126 BOLOGNA
ITALY
TEL 51 22 6677*956
TLF
TLX 211664 INFNBO I
EML
COM 28

BENEDICT GEORGE F DR
ASTRONOMY DPT
UNIVERSITY OF TEXAS
RLM 15 308
AUSTIN TX 78712 1083
USA
TEL 512 471 4461
TLF
TLX
EML
COM 24,28,44

BENEST DANIEL DR
OCA OBSERV DE NICE
BP 139
F 06003 NICE CDX
FRANCE
TEL 93 89 0420
TLF
TLX 460004 OBSNICE F
EML
COM 07,20,51

BENEVIDES SOARES P DR
INST ASTRON E GEOFISICO
CP 30627
01051 SAO PAULO SP
BRAZIL
TEL 11 275 3720
TLX 1136221 IAGM BR
EML
TLF
COM 08C

BENFORD GREGORY DR
DPT OF PHYSICS
UNIVERSITY OF CALIFORNIA
IRVINE CA 92717
USA
TEL 714 856 5147
TLF
TLX.
EML
COM 12,48

BENN CHRIS R DR
ROYAL GREENWICH OBS
HERSTMONCEUX CASTLE
HAILSHAM
EAST SUSSEX BN27 1RP
UK
TEL 323 833 171
TLF
TLX 87451 RGOBS GB
EML JANET CRB@UK AC RGO STAR
COM 05,40

BENNETT CHARLES L DR
NASA/GSFC
CODE 685
GREENBELT MD 20771
USA
TEL 301 286 3902
TLF
TLX
EML CHAMP BENNETT
COM 40,44,47

BENNETT KEVIN DR
ESA/ESTEC
SSD
BOX 299
NL 2200 AG NOORDWIJK
NETHERLANDS
TEL 17 19 83559
TLF
TLX 39098
EML KBENNETT@ESTEC
COM 44

BENSAMMAR SLIMANE DR
OBSERVATOIRE DE PARIS
SECTION DE MEUDON
F 92195 MEUDON PPL CDX
FRANCE
TEL 1 45 34 7835
TLF
TLX 270912
EML
COM 09,50

BENSON PRISCILLA J DR
WHITIN OBSERVATORY
WELLESLEY COLLEGE
WELLESLEY MA 02181
USA
TEL 617 235 0320
TLF
TLX
EML PBENSON@LUCY WELLESLEY EDU
COM 27,40,41,46

BENVENUTI PIERO DR
ST/ECF
C/O ESO
KARL-SCHWARZSCHILD-STR 2
D 8046 GARCHING MUENCHEN
GERMANY
TEL 89 320 06291
TLF
TLX 52828222 EO D
EML
COM

BENVENUTO OMAR DR
UNIV NACIONAL DE LA PLATA
FCAG
1900 LA PLATA (BS AS)
ARGENTINA
TEL 21 21 7308
TLF 21 25 5004
TLX 31151 BULAP AR
EML
COM 48

BENZ ARNOLD DR
INSTITUT FUER ASTRONOMIE
ETH ZENTRUM
CH 8092 ZUERICH
SWITZERLAND
TEL 1 256 4223
TLF 1 252 0192
TLX 53178 ETHBI CH
EML
COM 10C,40

BENZ WILLY
CENTER FOR ASTROPHYSICS
HCO/SAO
60 GARDEN ST
CAMBRIDGE MA 02138
USA
TEL 617 495 9889
TLF
TLX
EML
COM 35

BERENDZEN RICHARD DR
PRESIDENT'S OFFICE
THE AMERICAN UNIVERSITY
WASHINGTON DC 20016
USA
TEL 202 885 2121
TLF
TLX
EML
COM 41,51

BERG RICHARD A DR
NAVAL RESEARCH LABORATORY
BUILDING 56
HQ DEFENCE MAPPING AGENCY
WASHINGTON DC 20305
USA
TEL
TLF
TLX
EML
COM

BERGE GLENN L DR
CALTECH
MS 170 25
OWENS VALLEY RADIO OBS
PASADENA CA 91125
USA
TEL 818 356 6969
TLF
TLX 675425
EML
COM 16,40

BERGEAT JACQUES G DR
OBSERVATOIRE DE LYON
AVE CHARLES ANDRE
F 69561 S GENIS LAVAL CDX
FRANCE
TEL 78 56 0705
TLF 72 39 9791
TLX
EML
COM

BERGER CHRISTIANE DR
OCA CERGA
AVE COPERNIC
F 06130 GRASSE
FRANCE
TEL 93 36 5849
TLF
TLX 470865 F
EML
COM

BERGER JACQUES G DR
OBSERVATOIRE DE PARIS
61 AVE OBSERVATOIRE
F 75014 PARIS
FRANCE
TEL 1 40 51 2247
TLF
TLX 270776 OBS F
EML
COM 29

BERGER MITCHELL DR
DPT OF MATHEMATICAL SCI
UNIVERSITY OF ST ANDREWS
NORTH HAUGH
ST ANDREWS FIFE KY16 9SS
UK
TEL 334 761 61
TLF
TLX 76213 SAULIB GB
EML EARN%SOLAR MBERGER
COM 10

BERGERON JACQUELINE A DR
INSTITUT D'ASTROPHYSIQUE
98BIS BD ARAGO
F 75014 PARIS
FRANCE
TEL 1 43 20 1425
TLF 1 43 29 8673
TLX
EML IAPOBS BERGERON
COM 28,34,44,47,48,GS

BERGSTRALH JAY T DR
JPL/CLATECH
MS 183 301
4800 OAK GROVE DR
PASADENA CA 91109
USA
TEL 818 354 2296
TLF
TLX
EML
COM 16

BERGVALL NILS AKE SIGVARD
ASTRONOMICAL OBSERVATORY
BOX 515
S 751 20 UPPSALA
SWEDEN
TEL
TLX
EML
TLF
COM 28

BERKHUIJSEN ELLY M DR
MPI FUER RADIOASTRONOMIE
AUF DEM HUEGEL 69
D 5300 BONN 1
GERMANY
TEL
TLX 886440
EML
TLF
COM 28,33,34,40

BERMAN MARCELO S DR
RUA CANDIDO HARTMAN 575
AP 17 ED RENOIR
80430 CURITIBA PR
BRAZIL
TEL 41 224 6426
TLF 41 226 1679
TLX
EML
COM 47

BERMAN ROBERT HIRAM DR
DPT OF PHYSICS
MIT
BOX 165
CAMBRIDGE MA 02139
USA
TEL 617 253 1000
TLF
TLX
EML
COM

BERNACCA P L PROF
OSSERVATORIO ASTROFISICO
VIA DELL OSSERVATORIO 8
I 36012 ASIAGO
ITALY
TEL 42 46 2505
TLF
TLX 430110 SETOUR
EML
COM 26C,44,51

BERNAT ANDREW PLOUS DR
DPT OF COMPUTER SCIENCE
UNIVERSITY OF TEXAS
EL PASO TX 79968
USA
TEL 915 747 5494
TLF
TLX
EML
COM 34,36

BERRINGTON KEITH ADRIAN
DPT OF APPLIED MATHS
& THEORETICAL PHYSICS
QUEEN'S UNIVERSITY
BELFAST BT7 1NN
UK
TEL
TLF
TLX
EML
COM 14

BERRUYER-DESIROTTE N DR
OCA OBSERV DE NICE
BP 139
F 06003 NICE CDX
FRANCE
TEL 92 00 3011
TLF
TLX 460004 OBSNICE F
EML
COM

BERTAUX J L DR
SERVICE D'AERONOMIE
BP 3
F 91371 VERRIERES BUISSON
FRANCE
TEL 1 69 20 3116
TLF
TLX 602400
EML
COM 16,49

BERTELLI GIANPAOLO DR
DPT DI ASTRONOMIA
UNIVERSITA DI PADOVA
VIC DELL OSSERVATORIO 5
I 35122 PADOVA
ITALY
TEL 49 66 1499
TLF
TLX 432071 ASTROS I
EML
COM

BERTHOMIEU GABRIELLE DR
OCA OBSERV DE NICE
BP 139
F 06003 NICE COX
FRANCE
TEL 93 89 0420
TLF
TLX 460004 OBSNICE F
EML
COM 27,35

BERTIAU FLOR C PROF

WAVERSEBAAN 220
B 3030 HEVERLEE
BELGIUM
TLF
TEL
TLX
EML
COM 30

BERTIN GIUSEPPE PROF
SCUOLA NORMALE SUPERIORE
PIAZZA DEI CAVALIERI
I 56100 PISA
ITALY
TEL 50 59 7265
TLX 590548 SNSPI I
EML BERTIN@IPISNSVA
TLF
COM

BERTOLA FRANCESCO PROF
OSS ASTRONOMICO DI PADOVA
VIC DELL OSSERVATORIO 5
I 35122 PADOVA
ITALY
TEL 49 66 1499
TLX 430176 UNPADU I
EML
TLF
COM 28C,47

BERTOUT CLAUDE
INSTITUT D'ASTROPHYSIQUE
98BIS BD ARAGO
F 75014 PARIS
FRANCE
TEL 1 43 20 1425
TLF 1 43 29 8673
TLX
EML
COM 34,36

BERTSCHINGER EDMUND DR
DPT OF PHYSICS
MIT RM 6 207
BOX 165
CAMBRIDGE MA 02139
USA
TEL 617 253 5083
TLF 617 253 9798
TLX
EML EDBERT@ARCTURUS MIT EDI
COM 47

BESSELL MICHAEL S DR
MOUNT STROMLO & SIDING
SPRING OBSERVATORIES
PRIVATE BAG
WESTON CREEK PO ACT 2611
AUSTRALIA
TEL 62 49 0268
TLF 62 49 0233
TLX 62270 CANOPUS AA
EML BESSELL@MSO ANU EDU AU
COM 05,25,27,29VP

BETANCORT-RIJO JUAN DR
INST DE ASTROFISICA
DE CANARIAS
OBS DEL TEIDE
E 38200 LA LAGUNA
SPAIN
TEL 22 26 2211
TLF 22 26 3005
TLX 92640 IAC E
EML
COM 47

BETTIS DALE G PROF
TICOM
UNIVERSITY OF TEXAS
AUSTIN TX 78712 1083
USA
TEL
TLF
TLX
EML
COM 07

BETTONI DANIELA DR
OSS ASTRONOMICO DI PADOVA
VIC DELL OSSERVATORIO 5
I 35122 PADOVA
ITALY
TEL 49 66 1499
TLF
TLX 430176 UNPADU I
EML
COM 28

BEUERMANN KLAUS P PROF
INSTITUT FUR ASTRONOMIE &
ASTROPHYSIK ZU BERLIN
ERNST-REUTER-PL 7
D 1000 BERLIN 10
GERMANY
TEL
TLF
TLX
EML
COM

BEUTLER GERHARD PROF
ASTRONOMISCHES INSTITUT
UNIVERSITAET BERN
SIDLERSTRASSE 5
CH 3012 BERN
SWITZERLAND
TEL 31 65 8591
TLF 41 31 653869
TLX 912643 PIBE CH
EML BEUTLER@AIUB UNIBE CH
COM· 19

BHANDARI N DR
PHYSICAL RESEARCH LAB
NAVRANGPURA
AHMEDABAD 380 009
INDIA
TEL 272 46 2129
TLF 272 44 5292
TLX 0121397
EML
COM 22

BHANDARI RAJENDRA DR
RAMAN RESEARCH INSTITUTE
SADASHIVANAGAR
BANGALORE 560 080
INDIA
TEL 812 36 0122
TLF 812 34 0492
TLX 8452671 RRI IN
EML
COM 40

BHAT CHAMAN LAL DR
HEAD NRL HARL C 7
BHABHA ATOMIC RES CENTRE
SHASTRI NAGAR
JAMMU 180 004
INDIA
TEL
TLF
TLX
EML
COM

BHAT NARAYANA P DR
TIFR
HOMI BHABHA RD
COLABA
BOMBAY 400 005
INDIA
TEL 22 495 2311
TLF
TLX 0113009 TIFR IN
EML
COM

BHATIA PREM K DR
DPT OF MATHEMATICS
UNIVERSITY OF JODHPUR
JODHPUR 342 001
INDIA
TEL
TLF
TLX
EML
COM

BHATIA R K DR
DPT OF ASTRONOMY
UNIVERSITY OF OSMANIA
HYDERABAD 500 007
INDIA
TEL 71 951
TLF
TLX
EML
COM 16

BHATIA V B DR
DPT PHYSICS & ASTROPHYS
UNIVERSITY OF DELHI
NEW DELHI 110 007
INDIA
TEL 11 291 8993
TLF
TLX
EML
COM

BHATNAGAR ARVIND DR
UDAIPUR SOLAR OBSERVATORY
11 VIDYA MARG
UDAIPUR 313 001
INDIA
TEL. 25 626/23 861
TLF
TLX
EML
COM 10,12

BHATNAGAR ASHOK KUMAR
POSITIONAL ASTR CTR
P 546 BLOCK N 1ST FL
NEW ALIPORE
CALCUTTA 700 053
INDIA
TEL 33 450 321/493 541
TLF
TLX
EML
COM 04

BHATNAGAR K B DR
ZAKIR HUSSAIN COLLEGE
UNIVERSITY OF DELHI
AJMERI GATE
NEW DELHI 110 006
INDIA
TEL 11 52 2802
TLF 11 723 4544
TLX 3162442,3162431
EML
COM 07C

BHATT H C DR
INDIAN INSTITUTE OF
ASTROPHYSICS
KORAMANGALA
BANGALORE 560 034
INDIA
TEL 812 56 6585
TLF
TLX 845763 IIAB IN
EML
COM 34

BHATTACHARYA DIPANKAR
RAMAN RESEARCH INSTITUTE
SADASHIVANAGAR
BANGALORE 560 080
INDIA
TEL 812 34 0122
TLF 812 34 0492
TLX 845-2671
EML
COM

BHATTACHARYYA J C PROF
INDIAN INSTITUTE OF
ASTROPHYSICS
KORAMANGALA
BANGALORE 560 034
INDIA
TEL 812 56 6583/6585
TLF
TLX 845763 IIAB IN
EML
COM 09P,12,50VP

BHATTACHARYYA TARA DR
JOGAMAYA DEVI COLLEGE
92 SYAMAPRADAD MUKERJEE
CALCUTTA 700 026
INDIA
TEL
TLF
TLX
EML
COM 28

BHAVSAR SUKETU P
DPT PHYSICS & ASTRONOMY
UNIVERSITY OF KENTUCKY
LEXINGTON KY 40506 0055
USA
TEL 606 257 6722
TLF
TLX
EML
COM 47

BHONSLE RAJARAM V PROF
PHYSICAL RESEARCH LAB
NAVRANGPURA
AHMEDABAD 380 009
INDIA
TEL 272 46 2129
TLF 272 44 5292
TLX 121397 PRL IN
EML
COM 40

BIAN YU-LIN
BEIJING ASTRONOMICAL OBS
CAS
W SUBURB
BEIJING 100080
CHINA PR
TEL
TLF 1 256 1085
TLX 22040 BAOAS CN
EML BMABAO@ICA BEIJING CANET CN
COM

BIANCHI LUCIANA
OSS ASTRONOMICO DI TORINO
ST OSSERVATORIO 20
I 10025 PINO TORINESE
ITALY
TEL 11 84 2040
TLF
TLX 213236 TO ASTR I
EML
COM 34,44

BIANCHINI ANTONIO DR
OSSERVATORIO ASTROFISICO
VIA DELL OSSERVATORIO 8
I 36012 ASIAGO
ITALY
TEL 42 46 2665
TLF
TLX
EML
COM 27

BIBARSOV RAVIL'SH DR
ASTROPHYSICAL INSTITUTE
TADJIK ACAD OF SCIENCES
734670 DUSHANBE
TADZHIKISTAN
TEL
TLF
TLX
EML
COM 22

BICA EDUARDO L D DR
INSTITUTO DE FISICA
UFRGS
CP 15051
90000 PORTO ALEGRE RS
BRAZIL
TEL 512 36 4677
TLF
TLX 515730 CCUF BR
EML
COM 28

BICAK JIRI DR
DPT OF MATH PHYSICS
CHARLES UNIVERSITY
HOLESOVICKACH 2
CS 180 00 PRAHA 8
CZECHOSLOVAKIA
TEL 2 84 9951
TLF
TLX
EML
COM

BICKNELL GEOFFREY V DR
MOUNT STROMLO & SIDING
SPRING OBSERVATORIES
PRIVATE BAG
WODEN PO ACT 2606
AUSTRALIA
TEL 62 88 1111
TLF
TLX 62270 AA
EML NSSOCA PSI%MSSSO GEOFF
COM 47,48

BIDELMAN WILLIAM P PROF
WARNER & SWASEY OBS
CASE WESTERN RESERVE UNIV
CLEVELAND OH 44106
USA
TEL 216 368 6699
TLF
TLX
EML
COM 05,45

BIEGING JOHN HAROLD DR
STEWARD OBSERVATORY
UNIVERSITY OF ARIZONA
TUCSON AZ 85721
USA
TEL:
TLF
TLX
EML
COM 34,40

BIEMONT EMILE DR
INSTITUT D'ASTROPHYSIQUE
UNIVERSITE DE LIEGE
AVE COINTE 5
B 4000 COINTE-LIEGE
BELGIUM
TEL 41 52 9980
TLF 41 52 7474
TLX
EML
COM 14

BIEN REINHOLD DR
ASTRONOMISCHES RECHEN-
INSTITUT
MOENCHHOFSTR 12-14
D 6900 HEIDELBERG 1
GERMANY
TEL 62 214 9026
TLF
TLX 461336 ARIHD D
EML
COM 08,20

BIENAYME OLIVIER DR
OBSERVATOIRE DE BESANCON
BP 1615
F 25010 BESANCON CDX
FRANCE
TEL 81 66 6900
TLF 81 66 6944
TLX 361 144
EML BIENAYME@FROBES51
COM 33

BIERMANN PETER L DR
MPI FUER RADIOASTRONOMIE
AUF DEM HUEGEL 69
D 5300 BONN 1
GERMANY
TEL 228 52 5279
TLF
TLX 886440
EML
COM 28,40,48

BIGNAMI GIOVANNI F
IST DI FISICA COSMICA
CNR
15/A VIA BASSINI
I 20133 MILANO
ITALY
TEL 2 236 7587
TLF
TLX 313839 MUACNR I
EML
COM 47,48

BIGNELL R CARL DR
NRAO
VLA
BOX O
SOCORRO NM 87801 0387
USA
TEL 505 772 4242
TLF
TLX 910-988-1710
EML
COM 34,40

BIJAOUI ALBERT DR
OCA OBSERV DE NICE
BP 139
F 06003 NICE CDX
FRANCE
TEL 93 89 0420
TLF
TLX 460004 OBSNICE F
EML
COM 28,37

BILLAUD GERARD J
OCA CERGA
AVE COPERNIC
F 06130 GRASSE
FRANCE
TEL 93 36 5849
TLF
TLX 470865
EML
COM 08

BILLINGHAM JOHN
NASA AMES RESEARCH CTR
LIFE SCIENCE DIV
MOFFETT FIELD CA 94035
USA
TEL 415 694 5181
TLF
TLX 348408 NASA AMES MOF
EML
COM 51

BILLINGS DONALD E PROF
3E MOORE STREET
STATESBORO CA 30458
USA
TEL 912 764 7625
TLF
EML
COM 12
TLX

BINETTE LUC
CITA MCLENNAN LABS
UNIVERSITY OF TORONTO
60 ST GEORGE ST
TORONTO ON M5S 1A1
CANADA
TEL 416 978 8497
TLF 416 978 3921
TLX
EML
COM 28,34

BINGGELI BRUNO
ASTRONOMISCHES INSTITUT
UNIVERSITAET BASEL
VENUSSTRASSE 7
CH 4102 BINNINGEN
SWITZERLAND
TEL 61 22 7711
TLF
TLX
EML
COM 28

BINGHAM RICHARD G DR
ROYAL GREENWICH OBS
MADINGLEY RD
CAMBRIDGE CB3 0EZ
UK
TEL 223 83 3171
TLF
TLX 87451
EML
COM 09

BINNEY JAMES J DR
DPT THEORETICAL PHYSICS
UNIVERSITY OF OXFORD
1 KEBLE RD
OXFORD OX1 3NP
UK
TEL 865 273 979
TLF 865 273 974
TLX 83295 NUCLOX G
EML BINNEY%DIONYSOS THPHYS@PRG OXF
COM 28,33VP

BINZEL RICHARD P DR
DPT OF EARTH SCIENCE
MIT RM 54 426
BOX 165
CAMBRIDGE MA 02139
USA
TEL 617 253 6486
TLF
TLX
EML RPB@ASTRON MIT EDU
COM 15,16

BIRAUD FRANCOIS DR
OBSERVATOIRE DE PARIS
SECTION DE MEUDON
F 92195 MEUDON PPL CDX
FRANCE
TEL 1 45 07 7602
TLF
TLX 270912
EML
COM 06,40,51

BIRCH PETER MR
PERTH OBSERVATORY
BICKLEY WA 6076
AUSTRALIA
TEL 9 293 8255
TLF
TLX
EML
COM 15

BIRETTA JOHN ANTHONY DR
NRAO
BOX O
SOCORRO NM 87801 0387
USA
TEL 505 835 7302
TLX 910 988 1710
EML JBIRETTA@NRAO BITNET
TLF
COM 28,40

BIRKINSHAW MARK
DPT OF ASTRONOMY
HARVARD UNIVERSITY
60 GARDEN ST
CAMBRIDGE MA 02138
USA
TEL 617 495 9092
TLF.
TLX 921428
EML
COM 28,40,47

BIRKLE KURT PH D
MPI FUER ASTRONOMIE
KOENIGSTUHL
D 6900 HEIDELBERG 1
GERMANY
TEL
TLF
TLX
EML
COM 34

BISHOP ROY L DR
DPT OF PHYSICS
ACADIA UNIVERSITY
WOLFVILLE NS BOP 1X0
CANADA
TEL 902 542 2201
TLF
TLX
EML
COM 41

BISIACCHI GIANFRANCO DR
INSTITUTO DE ASTRONOMIA
UNAM
APDO POSTAL 70-264
04510 MEXICO DF
MEXICO
TEL 548 4537
TLF
TLX 1760155 CICME
EML
COM

BISNOVATYI-KOGAN G S DR
SPACE RESEARCH INSTITUTE
ACADEMY OF SCIENCES
PROFSOYUZNAYA UL 84/32
117810 MOSCOW
RUSSIA
TEL 333-31-22
TLF
TLX 411498 STARSU
EML
COM 35

BISWAS SUKUMAR DR
TIFR/COSMIC RAY GROUP
HOMI BHABHA RD
COLABA
BOMBAY 400 005
INDIA
TEL 22 219 111
TLF
TLX 113009 TIFR IN
EML
COM 48

BJORNSSON CLAES-INGVAR
STOCKHOLM OBSERVATORY

S 133 36 SALTSJOEBADEN
SWEDEN
TEL 87 17 0195
TLF 87 17 4719
TLX 12972 SOBSERV S
EML BJORNSSON@ASTRO SU SE
COM

BLAAUW ADRIAAN PROF DR
KAPTEYN ASTRONOMICAL INST
BOX 800
NL 9700 AV GRONINGEN
NETHERLANDS
TEL 50 63 4084
TLX 53572 STARS NL
EML
TLF
COM 24,33,37

BLACK JOHN HARRY DR
STEWARD OBSERVATORY
UNIVERSITY OF ARIZONA
TUCSON AZ 85721
USA
TEL 602 621 6531
TLF
TLX 467175
EML
COM 14,34

BLACKMAN CLINTON PAUL DR
CARLSTON LODGE
CAMPSIE RD
TORRANCE
GLASGOW G64 4HD
UK
TEL
TLF
TLX
EML
COM

BLACKWELL ALAN TREVOR
NTL RESEARCH COUNCIL
1771 BELVAL CRESCENT
ORLEANS ON K1C 6J6
CANADA
TEL 613 993 8521
TLF
TLX
EML
COM 22

BLACKWELL DONALD E PROF
DPT OF ASTROPHYSICS
UNIVERSITY OF OXFORD
SOUTH PARKS RD
OXFORD OX1 3RQ
UK
TEL 865 511 336
TLF
TLX
EML
COM 12,21,49

BLADES JOHN CHRIS DR
STSCI
HOMEWOOD CAMPUS
3700 SAN MARTIN DR
BALTIMORE MD 21218
USA
TEL 301 338 4805
TLF
TLX 6849101 STSCI UW
EML STSCIC BLADES
COM 34

BLAHA MILAN DR
NAVAL RESEARCH LABORATORY
CODE 4720
4555 OVERLOOK AVE SW
WASHINGTON DC 20375 5000
USA
TEL
TLF
TLX
EML
COM 14

BLAIR DAVID GERALD
DPT OF PHYSICS
UNIVERSITY W AUSTRALIA
NEDLANDS WA 6009
AUSTRALIA
TEL
TLF
TLX 92992 AA
EML
COM 40

BLAIR GUY NORMAN DR
9460 SW CHERAW CT
TUALATIN OR 97062
USA
TEL
TLF
EML
COM 34
TLX

BLAIR WILLIAM P DR
STSCI
HOMEWOOD CAMPUS
CHARLES & 34TH ST
BALTIMORE MD 21218
USA
TEL 301 338 8447
TLF
TLX 9102400225
EML SPAN SCIVAX WBLAIR
COM 42

BLAMONT JACQUES E PROF
CNES
2 PLACE MAURICE QUENTIN
F 75039 PARIS COX 01
FRANCE
TEL 1 45 08 7612
TLF
TLX 214674
EML
COM 12,15,16,21,44

BLANCO CARLO DR
IST DI ASTRONOMIA
CITTA UNIVERSITARIA
VIA A DORIA 6
I 95125 CATANIA
ITALY
TEL 95 33 0533
TLF
TLX 970359 ASTRCT I
EML
COM 20,36,50

BLANCO VICTOR M DR
CERRO TOLOLO
INTERAMERICAN OBSERVATORY
CASILLA 603
LA SERENA
CHILE
TEL 51 21 3352
TLF 51 21 2466*342
TLX 620 301 AURA CT
EML
COM 25,33,45,50C

BLANDFORD ROGER DAVID DR
CALTECH
MS 130 33
THEROETICAL ASTROPHYSICS
PASADENA CA 91125
USA
TEL· 213 356 4200
TLF
TLX 675429
EML
COM 40,48,49

BLASIUS KARL RICHARD DR
3839 MYRTLE
LONG BEACH CA 90807
USA
TEL
TLF
TLX
EML
COM

BLAZIT ALAIN DR
OCA OBSERV DU CALERN
CAUSSOLS
F 06460 S VALLIER THIEY
FRANCE
TEL 93 42 6270
TLF
TLX 461 402
EML
COM

BLECHA ANDRE BORIS G DR
16 RUE ET DUMONT
CH 1204 GENEVE
SWITZERLAND
TEL
TLF
EML blecha@obs.unige.ch
COM 25
TLX

BLEEKER JOHAN A M DR IR
SPACE RESEARCH LABORATORY
SRON
SORBONNELAAN 2
NL 3584 CA UTRECHT
NETHERLANDS
TEL 30 53 5600
TLF 30 54 0860
TLX 47224 ASTRO NL
EML
COM 44,48

BLESS ROBERT C PROF
DPT OF ASTRONOMY
UNIVERSITY OF WISCONSIN
475 N CHARTER ST
MADISON WI 53706
USA
TEL 608 262 1715
TLF
TLX
EML
COM 34,36,44,51

BLINOV N S DR
STERNBERG STATE ASTR INST
UNIVERSITETSKIJ PROSP 13
119899 MOSCOW
RUSSIA
TEL 139 10 49
TLF
TLX
EML
COM 19,31

BLITZ LEO DR
ASTRONOMY PROGRAM
UNIVERSITY OF MARYLAND
COLLEGE PARK MD 20742
USA
TEL 301 405 6650
TLF 301 314 9067
TLX 7108260352
EML blitz@astro.umd.edu
COM 28,33P,34

BLITZSTEIN WILLIAM DR
DPT ASTRON & ASTROPHYS
UNIV OF PENNSYLVANIA
DAVID RITTENHOUSE LAB E1
PHILADELPHIA PA 19104
USA
TEL 215 898 7899
TLF
TLX 834621
EML
COM 09,42

BLOCK DAVID LAZAR PROF
DPT COMPUT & APPL MATHS
WITWATERSRAND UNIVERSITY
PRIVATE BAG 3
WITS 2050
SOUTH AFRICA -
TEL 27 11 716 3761
TLF 27 11 716 3000
TLX 427125 SA
EML 076BLOK@WITSVMA WITS AC ZA
COM 28

BLOEMEN JOHANNES B G M DR
STERREWACHT
BOX 9513
NL 2300 RA LEIDEN
NETHERLANDS
TEL 71 27 5818
TLF
TLX 39058 ASTRO NL
EML
COM 33C

BLOEMHOF ERIC E DR
CENTER FOR ASTROPHYSICS
HCO/SAO
60 GARDEN ST
CAMBRIDGE MA 02138
USA
TEL 617 495 7314
TLF
TLX
EML BLOEMHOF@CFA
COM 40

BLOMME RONNY DR
OBSERVATOIRE ROYAL DE
BELGIQUE
AVE CIRCULAIRE 3
B 1180 BRUSSELS
BELGIUM
TEL 2 375 2484
TLF 2 374 9822
TLX 21565
EML RONNY@ASTRO OMA BE
COM 36

BLONDIN JOHN M DR
DPT PHYSICS & ASTRONOMY
UNIVERSITY NORTH CORALINA
PHILLIPS HALL CB 3255
CHAPEL HILL NC 27599
USA
TEL 919 962 3018
TLF 919 962 0480
TLX
EML BLONDIN@PHYSICS UNC EDU
COM 48

BLOW GRAHAM L
CARTER OBSERVATORY
BOX 2909
WELLINGTON
NEW ZEALAND
TEL 4 728 167
TLF
TLX NZ 30172 NATOBS
EML
COM 20

BLUDMAN SIDNEY A PROF
DPT OF PHYSICS
UNIV OF PENNSYLVANIA
PHILADELPHIA PA 19104
USA
TEL 215 898 8151
TLF
TLX 831908
EML
COM 35,47,48

BLUM PETER PROF
INSTITUT F ASTROPHYSIK
UNIVERSITAET BONN
AUF DEM HUEGEL 71
D 5300 BONN 1
GERMANY
TEL 228 73 3665
TLF
TLX 886440
EML
COM 49

BLUMENTHAL GEORGE R DR
LICK OBSERVATORY
UNIVERSITY OF CALIFORNIA
SANTA CRUZ CA 95064
USA
TEL 408 429 2005
TLF
TLX
EML BITNET george@portal
COM 28

BO SHU-REN
INST HISTORY NAT SCIENCE
1 GONG YUAN WEST RD
BEIJING
CHINA PR
TEL 1 55 7180
TLF
TLX
EML
COM 41

BOBROV M S DR
INST OF ASTRONOMY
ACADEMY OF SCIENCES
PYATNITSKAYA UL 48
109017 MOSCOW
RUSSIA
TEL 231-39-80
TLF
TLX 412623 SCSTP SU
EML
COM 16

BOCCHIA ROMEO DR
OBSERVATOIRE DE BORDEAUX
BP 89
F 33270 FLOIRAC
FRANCE
TEL 56 86 4330
TLF 56 40 4251
TLX
EML
COM 10,12,35

BOCHKAREV NIKOLAY G DR
STERNBERG STATE ASTR INST
UNIVERSITETSKIJ PROSP 13
119899 MOSCOW
RUSSIA
TEL
TLF
TLX
EML
COM 34

BOCHONKO D RICHARD DR
DPT OF MATHS & ASTRONOMY
UNIVERSITY OF MANITOBA
WINNIPEG MB R3T 2M8
CANADA
TEL 204 474 9501
TLF
TLX
EML
COM 27,46

BOCHSLER PETER
PHYSIKALISCHES INSTITUT
UNIVERSITAET BERN
SIDLERSTRASSE 5
CH 3012 BERN
SWITZERLAND
TEL 31 65 4419
TLF
TLX 32320 PHYBE CH
EML
COM 49

BOCKELEE-MORVAN DOMINIQUE
OBSERVATOIRE DE PARIS
SECTION DE MEUDON
F 92195 MEUDON PPL CDX
FRANCE
TEL 1 45 07 7605
TLF
TLX 270912
EML
COM 15,40

BODDAPATI G ANANDARAO DR
PHYSICAL RESEARCH LAB
NAVRANGPURA
ROOM 760
AHMEDABAD 380 009
INDIA
TEL 272 46 2129
TLF 272 44 5292
TLX 121397
EML
COM

BODE MICHAEL F
SCHOOL SCI & TECHNOLOGY
LIVERPOOL POLYTECHNIC
BYROM ST
LIVERPOOL L3 3AF
UK
TEL
TLF
TLX
EML
COM 34

BODENHEIMER PETER PROF
LICK OBSERVATORY
UNIVERSITY OF CALIFORNIA
SANTA CRUZ CA 95064
USA
TEL 408 429 2064
TLF
TLX
EML
COM 34,35

BODO GIANLUIGI DR
OSS ASTRONOMICO
DI TORINO
I 10025 PINO TORINESE
ITALY
TEL 11 84 1067
TLF
TLX 213236 TO ASTRI
EML
COM 36,39

BOEHM KARL-HEINZ PROF
DPT OF ASTRONOMY
UNIVERSITY OF WASHINGTON
FM 20
SEATTLE WA 98195
USA
TEL 206 543 2888
TLF
TLX 4740096
EML
COM 12,35,36

BOEHME ANNELIES DR
HEINRICH HERTZ INSTITUTE
SOLAR TERRESTR PHYSICS
TELEGRAFENBERG
D 1500 POTSDAM
GERMANY
TEL
TLF
TLX
EML
COM

BOEHME SIEGFRIED DR
FURTRANGLER STR 19
D 6900 HEIDELBERG 1
GERMANY
TEL 62 214 9026
TLF
EML
COM 36
TLX

BOEHM-VITENSE ERIKA PROF
DPT OF ASTRONOMY
UNIVERSITY OF WASHINGTON
FM 20
SEATTLE WA 98195
USA
TEL 206 543 4858
TLF
TLX
EML
COM 12,36

BOEHNHARDT HERMANN DR
INST FUER ASTRONOMIE
UND ASTROPHYIK
SCHEINERSTR 1
D 8000 MUENCHEN 80
GERMANY
TEL 89 922094 0
TLF
TLX
EML
COM 15

BOERNER GERHARD DR
MPI F PHYSIK & ASTROPHYS
FOEHRINGER RING 6
D 8000 MUENCHEN
GERMANY
TEL
TLF
TLX
EML
COM

BOERNGEN FREIMUT DR PH
ZTRL INST F ASTROPHYSIK
KARL-SCHWARZSCHILD-OBS
D 6901 TAUTENBURG
GERMANY
TEL 78 23530
TLF
TLX
EML
COM 20,28

BOESGAARD ANN M PROF
INSTITUTE FOR ASTRONOMY
UNIVERSITY OF HAWAII
2680 WOODLAWN DR
HONOLULU HI 96822
USA
TEL 808 948 8756
TLF
TLX 723-8459
EML
COM 29C,36

BOESHAAR GREGORY ORTH DR
STSCI
HOMEWOOD CAMPUS
3700 SAN MARTIN DR
BALTIMORE MD 21218
USA
TEL
TLF
TLX
EML
COM 28,34

BOGGESS ALBERT DR
NASA/GSFC
CODE 685
GREENBELT MD 20771
USA
TEL 301 286 5975
TLF
TLX
EML
COM 29,34,44

BOGGESS NANCY W DR
NASA HEADQUARTERS
CODE EZ
600 INDEPENDENCE AVE SW
WASHINGTON DC 20546
USA
TEL   202 453 1469
TLF
TLX
EML
COM   44

BOHANNAN BRUCE EDWARD
SOMMERS-BAUSCH OBS
UNIVERSITY OF COLORADO
BOX 391
BOULDER CO 80309 0391
USA
TEL   303 492 8782
TLF
TLX
EML:
COM:

BOHLIN J DAVID DR
NASA HEADQUARTERS
CODE EZ
600 INDEPENDENCE AVE SW
WASHINGTON DC 20546
USA
TEL   202 453 1466
TLF
TLX   89530
EML
COM

BOHLIN RALPH C DR
STSCI
HOMEWOOD CAMPUS
3700 SAN MARTIN DR
BALTIMORE MD 21218
USA
TEL   301 338 4804
TLF
TLX   6849101 STSCI UWI
EML
COM   34,44

BOHN HORST-ULRICH
FRAUNHOFER GES FHAK
WALDPARKSTR 41
D 8012 OTTOBRUNN
GERMANY
TEL   89 601 3086
TLX
EML
TLF
COM   10,12

BOHRMANN ALFRED PROF
SCHAERSTR 23
D 2050 HAMBURG 80
GERMANY
TEL   73 99800
TLF
EML
COM
TLX

BOIGEY FRANCOISE
IMTA LAB MECAN CELESTE
UNIVERSITE PARIS VI
4 PLACE JUSSIEU TOUR 66
F 75230 PARIS CDX 05
FRANCE
TEL
TLF
TLX
EML
COM   07

BOISCHOT ANDRE DR
OBSERVATOIRE DE PARIS
SECTION DE MEUDON
F 92195 MEUDON PPL CDX
FRANCE
TEL   1 45 07 7774
TLF
TLX   200590
EML
COM   40

BOISSE PATRICK DR
RADIOASTRONOMIE ENS
24 RUE LHOMOND
F 75231 PARIS CDX 05
FRANCE
TEL   1 43 29 1225
TLF   1 45 87 3489
TLX   202601
EML   29233 BOISSE
COM   34

BOISSON CATHERINE DR
OBSERVATOIRE DE PARIS
SECTION DE MEUDON
DAEC
F 92195 MEUDON PPL CDX
FRANCE
TEL   1 45 07 7436
TLF   1 45 07 7469
TLX   201571
EML   17733 BOISSON/BOISSON@FRMEU51
COM   28

BOKSENBERG ALEC PROF
ROYAL GREENWICH OBS
HERSTMONCEUX CASTLE
HAILSHAM BN27 1RP
UK
TEL   323 833 171
TLF
TLX   87451 RGOBSY G
EML
COM   28,44,47

BOLAND WILFRIED
NFRA
BOX 2
NL 7990 AA DWINGELOO
NETHERLANDS
TEL   52 19 7244
TLF
TLX   42043
EML
COM   34

BOLCAL CETIN DR
DPT OF PHYSICS
ISTANBUL UNIVERSITY
34459 VEZNECILER
TURKEY
TEL   1 332 0240
TLF
TLX   26401 BOUNTR
EML
COM

BOLDT ELIHU DR
NASA/GSFC
CODE 661
GREENBELT MD 20771
USA
TEL   301 286 5853
TLF
TLX   89675 NASCOM-GBLT
EML
COM

BOLEY FORREST I
DPT OF PHYSICS & ASTRON
DARTMOUTH COLLEGE
WILDER LABORATORY
HANOVER NH 03755
USA
TEL   603 646 2966
TLF
TLX
EML
COM

BOLOIX RAFAEL DR
REAL INST Y OBSERVATORIO
DE LA ARMADA
CECILIO PUJAZON S/N
E 11110 SAN FERNANDO
SPAIN
TEL
TLF
TLX
EML
COM   31

BOLTON C THOMAS PROF
DAVID DUNLAP OBSERVATORY
UNIVERSITY OF TORONTO
BOX 360
RICHMOND HILL ON L4C 4Y6
CANADA
TEL   416 884 9562
TLF   416 978 3921
TLX   069 86766 TOR
EML
COM   27,42

BOLTON JOHN G
39 PANORAMA CRESCENT
BUDERIM QLD 4556
AUSTRALIA
TEL   7 145 3374
TLF
EML
COM   40
TLX

BOMMIER VERONIQUE DR
OBSERVATOIRE DE PARIS
SECTION DE MEUDON
DAMAP
F 92195 MEUDON PPL CDX
FRANCE
TEL   1 45 07 7454
TLF
TLX   201571
EML   28726 BOMMIER
COM   10,12,14

BONACCINI DOMENICO DR
OSS ASTROFISICO
DI ARCETRI
LARGO E FERMI 5
I 50125 FIRENZE
ITALY
TEL   55 437 8540
TLF   55 43 5939
TLX
EML   38954 BONACCINI
COM

BONANNO GIOVANNI DR
OSS ASTROFISICO
CITTA UNIVERSITARIA
VIA A DORIA 6
I 95125 CATANIA
ITALY
TEL   95 33 0533
TLF   95 33 0592
TLX
EML   40297 GIOVANNI
COM   09

BONANOMI JACQUES DR
SOUS LES BUIS 28
CH 2068 HAUTERIVE
SWITZERLAND
TEL
TLF
EML
COM   31
TLX

BONAZZOLA SILVANO DR
OBSERVATOIRE DE PARIS
SECTION DE MEUDON
F 92195 MEUDON PPL CDX
FRANCE
TEL   1 45 07 7429
TLF
TLX   201571
EML
COM   42,48

BOND HOWARD E DR
STSCI
HOMEWOOD CAMPUS
3700 SAN MARTIN DR
BALTIMORE MD 21218
USA
TEL   301 338 4718
TLF
TLX   6849101
EML
COM   27,29

BOND JOHN RICHARD
CITA MCLENNAN LABS
UNIVERSITY OF TORONTO
60 ST GEORGE ST
TORONTO ON M5S 1A1
CANADA
TEL   416 978 6874
TLF   416 978 3921
TLX
EML   bond@utorphys bitnet
COM   47C

BONDAL KRISHNA RAJ DR
UTTAR PRADESH STATE
OBSERVATORY
PO MANORA PEAK 263 129
NAINITAL 263 129
INDIA
TEL   2136/2583
TLF
TLX
EML   ASTRONOMY NAINITAL
COM   10

BONDARENKO L N DR
STERNBERG STATE ASTR INST
UNIVERSITETSKIJ PROSP 13
119899 MOSCOW
RUSSIA
TEL   139-3721
TLF
TLX
EML
COM   16

BONDI HERMANN PROF SIR
CHURCHILL COLLEGE

CAMBRIDGE CB3 0DS
UK
TEL
TLF
TLX
EML
COM   35,47

BONET JOSE A
INST DE ASTROFISICA
DE CANARIAS
OBS DEL TEIDE
E 38071 LA LAGUNA
SPAIN
TEL
TLF
TLX
EML
COM

BONEV BONU K MR
PEOPLE'S ASTRONOMICAL OBS
& PLANETARIA IN BULGARIA
ST AVGUSTA TRAIANA 29/8
BG 6000 STARA ZAGORA
BULGARIA
TEL   42 43 183
TLF
TLX
EML
COM

BONIFAZI ANGELO DR
OSS ASTRONOMICO
UNIVERSITA DI BOLOGNA
CP 596
I 40100 BOLOGNA
ITALY
TEL
TLF
TLX
EML
COM

BONNEAU DANIEL
OCA OBSERV DU CALERN
CAUSSOLS
F 06460 S VALLIER THIEY
FRANCE
TEL   93 42 6270
TLF
TLX   461402
EML
COM   09,26

BONNET ROGER M DR
ESA
8-10 RUE MARIO NIKIS
F 75738 PARIS CDX 15
FRANCE
TEL   1 42 73 7107
TLF
TLX   202746
EML
COM   12,44,49

BONNET-BIDAUD J M DR
CEA CEN
DAPNIA/SAP
BP 2
F 91191 GIF/YVETTE CDX
FRANCE
TEL   1 69 08 9259
TLF   1 69 08 6577
TLX   604860
EML   32779 BOBI BOBI AT FRSAC11
COM   48

BONNOR W B PROF
1 SOUTH BANK TERRACE
SURBITON SURREY KT6 6DG
UK
TEL   1 399 1103
TLF
TLX
EML
COM   47

BONO GIUSEPPE DR
OAT
BOX SUCC TRIESTE 5
VIA TIEPOLO II
I 34131 TRIESTE
ITALY
TEL   40 319 9233
TLF   40 30 9418
TLX   461137 OAT I
EML   ASTRTS BONO/38439 BONO
COM   35

BONOLI FABRIZIO
OSS ASTRONOMICO
UNIVERSITA DI BOLOGNA
CP 596
I 40100 BOLOGNA
ITALY
TEL   51 22 2956
TLF
TLX   211664 INFNBO I
EML
COM

BONOMETTO SILVIO A DR
DPT DI FISICA G GALILEI
UNIVERSITA DI PADOVA
VIA MARZOLO 8
I 35131 PADOVA
ITALY
TEL   49 84 4111
TLF   49 84 4245
TLX
EML
COM   48

BONSACK WALTER K PROF
SUITE 298
5100 18 CLAYTON ROAD
CONCORD CA 94521
USA
TEL
TLF
TLX
EML
COM   29

BONTEKOE ROMKE DR
ESA/ESTEC
SSD
BOX 299
NL 2200 AG NOORDWIJK
NETHERLANDS
TEL   17 19 85160
TLF   31 1719 84690
TLX
EML   ROMKE@GUSPACE RUG NL
COM   28

BOOK DAVID L
NAVAL RESEARCH LABORATORY
CODE 4040
4555 OVERLOOK AVE SW
WASHINGTON DC 20375 5000
USA
TEL
TLF
TLX
EML
COM 12

BOOKBINDER JAY A DR
CENTER FOR ASTROPHYSICS
HCO/SAO MS 58
60 GARDEN ST
CAMBRIDGE MA 02138
USA
TEL 617 495 7058
TLF
TLX 921428 SATELLITE CAM
EML BOOKBIND@CFA227 HARVARD EDU
COM

BOOKMYER BEVERLY B DR
DPT PHYSICS & ASTRONOMY
CLEMSON UNIVERSITY
CLEMSON SC 29631
USA
TEL 803 656 3417
TLF
TLX
EML
COM 25,42

BOOTH ANDREW J
SCHOOL OF PHYSICS
UNIVERSITY OF SYDNEY
SYDNEY NSW 2006
AUSTRALIA
TEL
TLF
TLX
EML
COM

BOOTH ROY S PROF
ONSALA SPACE OBSERVATORY
GOETEBORG UNIVERSITY
S 439 00 ONSALA
SWEDEN
TEL 30 06 2590
TLF
TLX 2400 ONSPACE S
EML
COM 40

BOPP BERNARD W DR
DPT PHYSICS & ASTRONOMY
UNIVERSITY OF TOLEDO
2801 W BANCROFT ST
TOLEDO OH 43606
USA
TEL 419 537 2274
TLF
TLX
EML
COM 27,29,36,42

BORCHKHADZE TENGIZ M DR
ABASTUMANI ASTROPHYSICAL
OBSERVATORY
GEORGIAN ACAD OF SCIENCES
383762 ABASTUMANI
GEORGIA
TEL
TLF
TLX
EML
COM 28

BORD DONALD JOHN
DPT OF NATURAL SCIENCES
UNIVERSITY OF MICHIGAN
DEARBORN
DEARBORN MI 48128
USA
TEL 313 593 5483
TLF
TLX
EML
COM

BORDERIES NICOLE
JPL
MS 301 150
4800 OAK GROVE DR
PASADENA CA 91109
USA
TEL 818 354 8211
TLF
TLX 675429
EML
COM 07

BORGEEST ULF DR
HAMBURGER STERNWARTE
GOJENBERGSWEG 112
D 2050 HAMBURG 80
GERMANY
TEL 40 7252 4121
TLX 217884
EML ST40010@DHHUNI4 BITNET
TLF 40 7252 4198
COM 47

BORGMAN JAN DR PROF
KAPTEYN OBSERVATORY
WERKGROEP
MENSINGHEWEG 20
NL 9301 KA RODEN
NETHERLANDS
TEL
TLF
TLX
EML
COM 25,34

BORGNINO JULIEN DR
DPT ASTROPHYSIQUE
UNIVERSITE DE NICE
PARC VALROSE
F 06034 NICE CDX
FRANCE
TEL 93 51 9100
TLF
TLX
EML
COM 09

BORIAKOFF VALENTIN
NAIC
CORNELL UNIVERSITY
420 SPACE SCIENCES BLDG
ITHACA NY 14853 6801
USA
TEL 607 256 3734
TLF
TLX 932454
EML
COM 40

BORKOWSKI KAZIMIERZ M DR
INST OF RADIO ASTRONOMY
N COPERNICUS UNIVERSITY
UL CHOPINA 12/18
PL 87 100 TORUN
POLAND
TEL 48 56 783327
TLF 48 56 11651
TLX 552324 TRAO PL
EML KAZIK@PLTUMK11
COM

BORNMANN PATRICIA L DR
NOAA ERL R/E/SE
SPACE ENVIRONMENT LAB
325 BROADWAY
BOULDER CO 80303
USA
TEL 303 497 3532
TLF
TLX 45897 SOLTERWARN BDR
EML SPAN SELVAX pbornmann
COM 10,12

BORRA ERMANNO F DR
DPT DE PHYSIQUE
UNIVERSITE DE LAVAL
STE FOY QC G1K 7P4
CANADA
TEL 418 656 7405
TLF 418 656 2040
TLX 051 31621
EML
COM 25

BOS ALBERT DR
NFRA
BOX 2
NL 7990 AA DWINGELOO
NETHERLANDS
TEL· 52 19 7244
TLX 42043 SRZM NL
EML PSI002041521004 SYSTEM
TLF 31 5219 7332
COM 40

BOSCHAN PETER DR
ASTRONOMISCHES INSTITUT
UNIVERSITAT MUNSTER
WELHELM-KLEMM-STR 10
D 4400 MUENSTER
GERMANY
TEL 251 83 3561
TLF 251 83 3669
TLX 892529
EML PASO4 AT BMSWWU IA BITNET
COM

BOSMA ALBERT DR
OBSERVATOIRE DE MARSEILLE
2 PLACE LE VERRIER
F 13248 MARSEILLE CDX 04
FRANCE
TEL 91 95 9088
TLF
TLX 420241
EML
COM 28

BOSMA PIETER B DR
DPT PHYSICS & ASTRONOMY
FREE UNIVERSITY
DE BOELELAAN 1081
NL 1081 HV AMSTERDAM
NETHERLANDS
TEL 20 54 84139
TLF 20 46 1459
TLX
EML
COM 16

BOSMAN-CRESPIN DENISE
BVD D AVROY 68
BOX 093
B 4000 COINTE-LIEGE
BELGIUM
TEL 41 23 7486
TLF
TLX
EML
COM

BOSS ALAN P DR
DPT TERRESTR MAGNETISM
CARNEGIE INST WASHINGTON
5241 BROAD BRANCH RD NW
WASHINGTON DC 20015
USA
TEL 202 686 4402
TLF
TLX 440427
EML
COM. 16,35

BOTEZ ELVIRA DR
INSTITUT D'ENSEIGNEMENT
SUPERIEUR
13 RUE EM BODNARAS
R 5800 SUCEAVA
RUMANIA
TEL 98716147
TLF
TLX
EML
COM 46

BOTTINELLI LUCETTE DR
OBSERVATOIRE DE PARIS
SECTION DE MEUDON
RADIOASTRONOMIE
F 92195 MEUDON PPL CDX
FRANCE
TEL 1 45 07 7604
TLF
TLX 270912
EML
COM 28,40,46

BOUCHER CLAUDE DR
INSTITUT GEOGRAPHIQUE NTL
2 AVE PASTEUR
F 94160 SAINT MANDE
FRANCE
TEL 1 43 98 8000
TLF
TLX
EML
COM 19

BOUCHET FRANCOIS R DR
INSTITUT D'ASTROPHYSIQUE
98BIS BD ARAGO
F 75014 PARIS
FRANCE
TEL 1 43 20 1425
TLF 1 43 29 8673
TLX
EML BOUCHET@FRIAP51
COM

BOUCHET PATRICE DR
ESO
CASILLA 19001
SANTIAGO 19
CHILE
TEL 2 698 8757
TLF
TLX 240881
EML PBOUCHET@DGAESO51/ESOMCI PATO
COM 28

BOUGEARD MIREILLE L DR
OBSERVATOIRE DE PARIS
61 AVE OBSERVATOIRE
F 75014 PARIS
FRANCE
TEL 1 40 51 2226
TLF 1 40 51 2232
TLX 270775 OBS F
EML
COM 08

BOUGERET J L DR
OBSERVATOIRE DE PARIS
SECTION DE MEUDON
DESPA
F 92195 MEUDON PPL CDX
FRANCE
TEL 1 45 07 7704
TLF
TLX 204464
EML
COM 10,12,44

BOUIGUE R
11 RUE PELLETIER D'OISY
F 31400 TOULOUSE
FRANCE
TEL
TLF
EML
COM 24
TLX

BOULANGER FRANCOIS
RADIOASTRONOMIE ENS
24 RUE LHOMOND
F 75231 PARIS CDX 05
FRANCE
TEL
TLF
TLX
EML
COM 34

BOULESTEIX JACQUES
OBSERVATOIRE DE MARSEILLE
2 PLACE LE VERRIER
F 13248 MARSEILLE CDX 04
FRANCE
TEL 91 95 9088
TLF
TLX 420241F
EML
COM

BOULON JACQUES J DR
OBSERVATOIRE DE PARIS
61 AVE OBSERVATOIRE
F 75014 PARIS
FRANCE
TEL 1 40 51 2253
TLF
TLX 270776 OBS F
EML
COM 27,30,33

BOUSKA JIRI DR
DPT OF ASTRONOMY
CHARLES UNIVERSITY
SVEDSKA 8
CS 150 00 PRAHA 5
CZECHOSLOVAKIA
TEL 2 54 0395
TLF
TLX 121673 MFF
EML
COM 05,15

BOUVIER JEROME
CERMO
BP 68
F 38402 S MARTIN HERES CD
FRANCE
TEL 76 51 4790
TLF 76 44 8821
TLX
EML BOUVIER@FRGAG51 BITNET
COM 29,34

BOUVIER PIERRE PROF
OBSERVATOIRE DE GENEVE
CHEMIN DES MAILLETTES 51
CH 1290 SAUVERNY
SWITZERLAND
TEL 22 755 2611
TLF 22 755 3983
TLX 419209 OBS CH
EML
COM 29,37

BOWELL EDWARD L G DR
LOWELL OBSERVATORY
1400 W MARS HILL RD
BOX 1149
FLAGSTAFF AZ 86001
USA
TEL 602 774 3358
TLF
TLX
EML
COM 15,20

BOWEN EDWARD G DR
1/39 CLARKE STREET
NARRABEEN NSW 2101
AUSTRALIA
TEL 98 8565
TLX
EML
TLF
COM

BOWEN GEORGE H DR
DPT OF PHYSICS
IOWA STATE UNIVERSITY
AMES IA 50011
USA
TEL 515 294 7659
TLF
TLX
EML BITNET S1 GHB@ISUMVS
COM 27,36

BOWERS PHILLIP F
NAVAL RESEARCH LABORATORY
CODE 4134
4555 OVERLOOK AVE SW
WASHINGTON DC 20375 5000
USA
TEL 202 767 2495
TLF
TLX
EML
COM 40

BOWYER C STUART PROF
ASTRONOMY DPT
UNIVERSITY OF CALIFORNIA
601 CAMPBELL HALL
BERKELEY CA 94720
USA
TEL   415 642 1648
TLF
TLX   910-366 7945
EML
COM   21C,44,51

BOYARCHUK A A DR
INST OF ASTRONOMY
ACADEMY OF SCIENCES
PYATNITSKAYA UL 48
109017 MOSCOW
RUSSIA
TEL   70 95 231 2129
TLF   70 95 230 2081
TLX   411576 ASCON SU
EML   IAAS@NODE IAS MSK SU
COM   27,29,38,44,EC

BOYARCHUK MARGARITA E DR
INST OF ASTRONOMY
ACADEMY OF SCIENCES
PYATNITSKAYA UL 48
109017 MOSCOW
RUSSIA
TEL
TLF
TLX
EML
COM   27

BOYCE PETER B DR
AMERICAN ASTRON SOCIETY
2000 FLORIDA AVE NW
SUITE 300
WASHINGTON DC 20009
USA
TEL   202 328 2010
TLF
TLX   257588 AASW UR
EML
COM   09,16,51

BOYD ROBERT L F PROF SIR
41 CHURCH ST
LITTLEHAMPTON BN17 5PU
UK
TEL
TLF
EML
COM   44,48
TLX

BOYDAG-YILDIZDOGDU F S DR
KING SAUD UNIVERSITY
COLLEGE OF SCIENCE
BOX 2452
RIYADH 11495
SAUDI ARABIA
TEL
TLF
TLX
EML
COM

BOYER RENE
OBSERVATOIRE DE PARIS
SECTION DE MEUDON
DASOP
F 92195 MEUDON PPL CDX
FRANCE
TEL   1 45 07 7741
TLF
TLX   201571
EML
COM   10

BOYLE BRIAN DR
AAO
BOX 296
EPPING NSW 2121
AUSTRALIA
TEL   2 868 1666
TLF
TLX   23999 AAOSYD AA
EML
COM   47

BOYLE RICHARD P DR
VATICAN OBSERVATORY
I 00120 VATICAN CITY
VATICAN CITY STATE
TEL   6 698 5266
TLF
EML   BOYLE@ARIZRVAX BITNET
COM
TLX   504 2020 VATOBS VA

BOYNTON PAUL EDWARD DR
DPT OF ASTRONOMY
UNIVERSITY OF WASHINGTON
FM 20
SEATTLE WA 98195
USA
TEL
TLF
TLX
EML
COM

BOYTEL JORGE DEL PINO DR
INST GEOPHYS & ASTRONOMY
CALLE 212 N 2906/29 Y 31
LISA
LA HABANA
CUBA
TEL   21 8416/0644
TLF
TLX   511240 GEOAS CU
EML
COM   19

BOZIS GEORGE PROF
DPT THEORET MECHANICS
UNIVERSITY THESSALONIKI
GR 540 06 THESSALONIKI
GREECE
TEL   31 99 2845
TLF
TLX
EML
COM   07

BOZKURT SUKRU DR
OBSERVATORY
EGE UNIVERSITY
BOX 21
35100 BORNOVA IZMIR
TURKEY
TEL   51 18 0306
TLF
TLX
EML
COM

BRACCESI ALESSANDRO PROF
DPT DI ASTRONOMIA
UNIVERSITA DI BOLOGNA
VIA ZAMBONI 33
I 40126 BOLOGNA
ITALY
TEL   51 22 2956
TLF
TLX   211664 INFNBOI
EML
COM   28

BRACEWELL RONALD N PROF
DURAND 329 A
STANFORD UNIVERSITY
STANFORD CA 94305
USA
TEL   415 497 3545
TLF
TLX
EML
COM   40,51

BRADSTREET DAVID H DR
DPT OF PHYSICAL SCIENCE
EASTERN COLLEGE
ST DAVIDS PA 19087
USA
TEL   215 341 5945
TLF
TLX
EML
COM   42

BRAES L L E DR
STERREWACHT
BOX 9513
NL 2300 RA LEIDEN
NETHERLANDS
TEL   71 27 2727
TLX
EML
TLF
COM   46

BRAGA JOAO DR
INPE
CP 515
12201 S JOSE DOS CAMPOS
BRAZIL
TEL   123 41 8977*679
TLF   123 21 8743
TLX   123 353 INPE BR
EML   INPEDAS@BRFAPESR BITNET
COM   48

BRAHDE ROLF
INST THEORET ASTROPHYSICS
UNIVERSITY OF OSLO
BOX 1029
N 0315 BLINDERN OSLO 3
NORWAY
TEL   2 45 6508
TLF
TLX
EML
COM

BRAHIC ANDRE DR
OBSERVATOIRE DE PARIS
SECTION DE MEUDON
F 92195 MEUDON PPL CDX
FRANCE
TEL   1 45 07 7402
TLF
TLX   201571
EML
COM   16C,WGPSNC

BRANCH DAVID R DR
DPT PHYSICS & ASTRONOMY
UNIVERSITY OF OKLAHOMA
NORMAN OK 73019
USA
TEL   405 325 3961
TLF
TLX   9108306521
EML
COM

BRAND PETER W J L DR
ROYAL OBSERVATORY
BLACKFORD HILL
EDINBURGH EH9 3HJ
UK
TEL   316 673 321
TLX   72383 ROE EDIN G
EML
TLF
COM   34

BRANDENBURG AXEL DR
NORDITA
BLEGDAMSVEJ 17
DK 2100 COPENHAGEN O
DENMARK
TEL   31 42 1616
TLF   31 38 9157
TLX   15216 NBI DK
EML   BRANDENB@NORDITA DK
COM   10

BRANDI ELISANDE ESTELA DR
OBSERVATORIO ASTRONOMICO
PASEO DEL BOSQUE
1900 LA PLATA (BS AS)
ARGENTINA
TEL   21 21 7308
TLF
TLX
EML
COM   29,42

BRANDIE GEORGE W DR
ENVIRONMENTAL ENGINEERING
QUEEN'S UNIVERSITY
KINGSTON ON K7L 3N6
CANADA
TEL
TLF
TLX
EML
COM

BRANDT JOHN C DR
LASP
UNIVERSITY OF COLORADO
BOX 392
BOULDER CO 80309 0392
USA
TEL   303 492 3215
TLF
TLX   9109403441
EML
COM   15C,44,49

BRANDT PETER N
KIEPENHEUER INSTITUT FUER
SONNENPHYSIK
SCHONECKSTR 6
D 7800 FREIBURG BREISGAU
GERMANY
TEL   761 32864
TLF
TLX   7721552
EML
COM   10,12

BRANDUARDI-RAYMONT G
MULLARD SPACE SCIENCE LAB
UNIVERSITY COLLEGE LONDON
HOLMBURY ST MARY
DORKING SURREY RH5 6NT
UK
TEL   306 702 92
TLF
TLX   859185
EML
COM

BRANHAM RICHARD L JR
CENTRO REGIONAL DE INVEST
CIENTIFICAS Y TECNOL
CC 131
5500 MENDOZA
ARGENTINA
TEL   61 24 11794
TLF
TLX   55438 CYTME AR
EML
COM   07,20,24

BRANSCOMB L M DR
NTL BUREAU OF STANDARDS
WASHINGTON DC 20025
USA
TEL
TLF
EML
COM   14
TLX·

BRANSON NICHOLAS J B A DR
GENERAL BOARD OFFICE
THE OLD SCHOOLS
CAMBRIDGE CB2 1TT
UK
TEL   223 33 2250
TLF   223 33 2332
TLX   81240 CAMSPL G
EML
COM

BRATIJCHUK MATRONA V
SATELITE OBSERVATORY
UZHGOROD STATE UNIVERSITY
HORKIY 46
294000 UZHGOROD
UKRAINE
TEL   3 6065
TLF
TLX   274155 KNIGA
EML
COM

BRAUDE SEMION YA PROF AG
INSTITUTE OF RADIO ASTRON
UKRAINIAN ACAD OF SCIENCE
KRASNOZNAMENNAYA UL 4
310002 KHARKOV
UKRAINE
TEL   44 1092
TLF
TLX
EML
COM   40

BRAULT JAMES W DR
NTL SOLAR OBSERVATORY
BOX 26732
950 N CHERRY AVE
TUCSON AZ 85726 6732
USA
TEL   602 325 9363
TLF
TLX   666484 AURA NOAO TUC
EML
COM   09,12,14

BRAUN ARIE
RACAH INST OF PHYSICS
HEBREW UNIV OF JERUSALEM
JERUSALEM 91904
ISRAEL
TEL   2 584 521
TLF
TLX   25391 HU IL
EML
COM

BRAUN ROBERT DR
NFRA

BOX 2
NL 7990 AA DWINGELOO
NETHERLANDS
TEL   52 19 7244
TLF   52 19 7332
TLX   42043 SRZM NL
EML   RBRAUN@NFRA NL
COM   28

BRAUNINGER HEINRICH DR
MPI F PHYSIK & ASTROPHYS
INST F EXTRATERR PHYSIK
D 8046 GARCHING MUENCHEN
GERMANY
TEL   89 329 9566
TLF
TLX
EML
COM

BRAUNSFURTH EDWARD PH D
IM HAARMANNSBOCH 99 A
D 4630 BOCHUM 1
GERMANY
TEL
TLF
EML
COM   34
TLX

BRAY ROBERT J DR
31/126 CRIMEA ROAD
MARSFIELD NSW 2122
AUSTRALIA
TEL
TLF
TLX
EML
COM   10,12

BREAKIRON LEE ALLEN DR
US NAVAL OBSERVATORY
TIME SERVICE DPT
34 & MASSACHUSETTS AVE NW
WASHINGTON DC 20392 5100
USA
TEL   202 653 1888
TLF
TLX
EML   I navobsy@nardacva arpa
COM   24

BRECHER AVIVA DR PROF
35 MADISON ST
BELMONT MA 02178
USA
TEL 617 489 1386
TLF
EML
COM 15,16
TLX

BRECHER KENNETH PROF
DPT OF ASTRONOMY
BOSTON UNIVERSITY
725 COMMONWEALTH AVE
BOSTON MA 02215
USA
TEL 617 353 3423
TLF
TLX 95-1289 BIS UNIV BSN
EML
COM· 28,47,48

BRECKINRIDGE JAMES B DR
JPL/CALTECH
MS 183 301
4800 OAK GROVE DR
PASADENA CA 91109
USA
TEL 213 354 6785
TLF
TLX 675429
EML
COM 09,12

BREGER MICHEL PROF DR
INSTITUT FUER ASTRONOMIE
TUERKENSCHANZSTRASSE 17
A 1180 WIEN
AUSTRIA
TEL 1 345 3605
TLF 1 345 36015
TLX 133099 VIAST A
EML BREGER@AVIA UNA AC AT
COM 25C,27C,30

BREGMAN JACOB D IR
NFRA
BOX 2
NL 7990 AA DWINGELOO
NETHERLANDS
TEL 52 19 7244
TLX 42043
EML
TLF
COM 40

BREGMAN JOEL N
DPT OF ASTRONOMY
UNIVERSITY OF MICHIGAN
DENNISON BLDG
ANN ARBOR MI 48109 1090
USA
TEL 313 764 3440
TLF
TLX
EML JBREGMAN@ASTRO LSA UMICH EDU
COM

BREINHORST ROBERT A DR
ASTRONOMISCHES INSTITUT
STERNWARTE
AUF DEM HUEGEL 71
D 5300 BONN 1
GERMANY
TEL 228 73 3660
TLF
TLX
EML
COM 42

BREITSCHWERDT DIETER DR
MPI FUR KERNPHYSIK
POSTFACH 1039 80
D 6900 HEIDELBERG 1
GERMANY
TEL 62 215 16471
TLF 62 215 16324
TLX 461666
EML WINDS@DHOMPI5V (BITNET)
COM· 34

BREJDO IZABELLA I DR
PULKOVO OBSERVATORY
ACADEMY OF SCIENCES
10 KUTUZOV QUAY
196140 ST PETERSBURG
RUSSIA
TEL 297-94-59
TLF
TLX
EML
COM 09

BRETAGNON PIERRE DR
BUREAU DES LONGITUDES
77 AVE DENFERT ROCHEREAU
F 75014 PARIS
FRANCE
TEL 1 40 51 2269
TLF
TLX
EML
COM 04,07

BREUKERS R J L H DR
STERREWACHT
BOX 9513
NL 2300 RA LEIDEN
NETHERLANDS
TEL
TLF
TLX
EML
COM

BREYSACHER JACQUES
ESO
KARL-SCHWARZSCHILD STR 2
D 8046 GARCHING MUENCHEN
GERMANY
TEL 89 320 06224
TLF 89 320 2362
TLX 5282820
EML
COM 29

BRIDGELAND MICHAEL DR
INSTITUTE OF ASTRONOMY
THE OBSERVATORIES
MADINGLEY RD
CAMBRIDGE CB3 OHA
UK
TEL 223 33 7524
TLF
TLX 817297 ASTRON G
EML
COM 09

BRIDLE ALAN H PROF
NRAO
EDGEMONT RD
CHARLOTTESVILLE VA 22903
USA
TEL 804 296 0375
TLX 910-997-0174
EML
TLF
COM 40

BRIEVA EDUARDO PROF
OBSERVATORIO NACIONAL
APARTADO 2584
BOGOTA 1 DE
COLOMBIA
TEL 1 42 3786
TLF
TLX
EML
COM 07,46

BRIHAYE CHARLES C A DR
IAAG
VRIJE UNIV BRUSSELS
CP 165
B 1050 BRUSSELS
BELGIUM
TEL 2 687 6928
TLF
TLX
EML
COM

BRINI DOMENICO PROF
LABORATORIO TESRE
VIA CASTAGNOLI 1
I 40100 BOLOGNA
ITALY
TEL
TLF
TLX
EML·
COM

BRINKMAN BERT C DR
SPACE RESEARCH LABORATORY
SRON
SORBONNELAAN 2
NL 3584 CA UTRECHT
NETHERLANDS
TEL 30 53 5600
TLF 30 54 0860
TLX 47224 SRON NL
EML
COM 44

BRINKMANN WOLFGANG
MPI F PHYS & ASTROPHYSIK
INST F EXTRATERR PHYSIK
KARL-SCHWARZSCHILD-STR 1
D 8046 GARCHING MUENCHEN
GERMANY
TEL 89 329 9877
TLF
TLX 05215845 XTER D
EML
COM 28,34

BRINKS ELIAS DR
NRAO
BOX 0
SOCORRO NM 87801 0387
USA
TEL 505 835 7000
TLF 505 835 7027
TLX 910 9981710
EML EBRINKS@NRAO
COM 28,40

BRIOT DANIELLE DR
OBSERVATOIRE DE PARIS
61 AVE OBSERVATOIRE
F 75014 PARIS
FRANCE
TEL 1 40 51 2239
TLF
TLX 270776 OBS F
EML
COM

BROADFOOT A LYLE DR
LUNAR & PLANETARY LAB
UNIVERSITY OF ARIZONA
901 GOULD SIMPSON BLDG
TUCSON AZ 85721
USA
TEL· 602 621 4301
TLF
TLX. 910-952-1143
EML
COM 16,21

BROCATO ENZO DR
OSS DI TERAMO
COLLURANIA
I 64100 TERAMO
ITALY
TEL 861 21 0490
TLF 861 21 0493
TLX
EML
COM

BRODERICK JOHN DR
PHYSICS DPT
VIRGINIA TECH
BLACKSBURG VA 24061
USA
TEL 703 231 5321
TLF
TLX 9103331861 VPIBKS
EML
COM 40,51

BRODIE JEAN P
SPACE SCIENCES LABORATORY
UNIVERSITY OF CALIFORNIA
BERKELEY CA 94720
USA
TEL 415 642 1579
TLF
TLX 910-366-7945
EML
COM 28

BROGLIA PIETRO DR
OSS ASTRONOMICO DI MILANO
VIA E BIANCHI 46
I 22055 MERATE
ITALY
TEL 59 2035
TLX
EML
TLF
COM 42

BROMAGE GORDON E DR
RUTHERFORD APPLETON LAB
SPACE & ASTROPHYSICS DIV
BLDG R25/R68
CHILTON DIDCOT OX11 0QX
UK
TEL. 235 219 00
TLF
TLX 83159
EML
COM 14,34

BRONFMAN LEONARDO DR
DPT DE ASTRONOMIA
UNIVERSIDAD DE CHILE
CASILLA 36 D
SANTIAGO
CHILE
TEL 2 228 1941
TLF
TLX 440 001
EML
COM 33,40

BRONNIKOVA NINA M
PULKOVO OBSERVATORY
ACADEMY OF SCIENCES
10 KUTUZOV QUAY
196140 ST PETERSBURG
RUSSIA
TEL
TLF
TLX
EML·
COM 24

BROOKES CLIVE J DR
DPT OF MATHEMATICS
EARTH/SATELLITE RES UNIT
ASTON UNIVERSITY
BIRMINGHAM B4 7ET
UK
TEL 213 59 3611
TLF
TLX 335787
EML
COM 07

BROOKS RANDALL C DR
PHYSICS SCIENCES
NTL MUSEUM SCIENCE & TECH
BOX 9724
OTTAWA ONT K1G 5A3
CANADA
TEL 613 990 2804
TLF 613 991 3636
TLX
EML BROOKS@HUSKY1 STMARYS CA
COM 41

BROSCH NOAH DR
WISE OBSERVATORY
TEL AVIV UNIVERSITY
RAMAT AVIV
TEL AVIV 69978
ISRAEL
TEL 3 41 3788
TLF
TLX 342171 VERSY IL
EML BITNET H38@TAUNOS
COM 28,46

BROSCHE PETER PROF
OBSERVATORIUM HOHER LIST
UNIV STERNWARTE BONN
D 5568 DAUN
GERMANY
TEL 65 92 2150
TLF
TLX
EML
COM 19,24,26,28

BROSTERHUS E B F DR
LOCKHEED CITY
BOX 6308
JEDDAH
SAUDI ARABIA
TEL 2 656 2501*355
TLX
EML
TLF
COM

BROTEN NORMAN W
HERZBERG INST ASTROPHYS
NTL RESEARCH COUNCIL
100 SUSSEX DR
OTTAWA ON K1A 0R6
CANADA
TEL 613 993 6060
TLF 613 952 6602
TLX 053 3715
EML
COM 40

BROUCKE ROGER DR
7203 RUNNING ROPE CIRCLE

AUSTIN TX 78731
USA
TEL 512 345 6435
TLF
TLX
EML
COM 07

BROUW W N DR
CSIRO
ATNF
BOX 76
EPPING NSW 2121
AUSTRALIA
TEL
TLF
TLX
EML WBROUW@ATNF CSIRO AU
COM 08,40

BROWN ALEXANDER
JILA
UNIVERSITY OF COLORADO
BOX 440
BOULDER CO 80309 0440
USA
TEL 303 492 8962
TLF
TLX 755842 JILA
EML
COM 36,44

BROWN DOUGLAS NASON
DPT OF ASTRONOMY
UNIVERSITY OF WASHINGTON
FM 20
SEATTLE WA 98195
USA
TEL 206 543 4313*2888
TLF
TLX
EML
COM 25,27,29,36

BROWN HARRISON DR
3005 LA MANCHA DR

ALBUQUERQUE NM 87104
USA
TEL
TLF
TLX
EML
COM

BROWN JOHN C PROF
DPT OF PHYSCIS & ASTRON
UNIVERSITY OF GLASGOW
GLASGOW G12 8QQ
UK
TEL 413 305 182
TLF
TLX 777070 UNIGLA
EML
COM 10

BROWN ROBERT HAMILTON
JPL/CALTECH
MS 183 501
4800 OAK GROVE DR
PASADENA CA 91109
USA
TEL 818 354 2517
TLF
TLX
EML
COM 15,16,50

BROWN ROBERT L DR
NRAO
EDGEMONT RD
CHARLOTTESVILLE VA 22901
USA
TEL 804 296 0232
TLF
TLX 910-997-0174
EML
COM

BROWN RONALD D PROF
DPT OF CHEMISTRY
MONASH UNIVERSITY
WELLINGTON RD
CLAYTON VIC 3168
AUSTRALIA
TEL 3 565 4550
TLF 3 565 4597
TLX AA 32691
EML CHE265K@VAXC CC MONASH EDU AU
COM 34,51P

BROWNE IAN W A DR
NRAL
JODRELL BANK
MACCLESFIELD SK11 9DL
UK
TEL 477 713 21
TLX 36149
EML
TLF
COM 40

BROWNING PHILIPPA DR
DPT OF PURE & APPLIED PHY
UMIST
BOX 88
MANCHESTER M60 1QD
UK
TEL 612 363 311
TLF
TLX 666094
EML MCCPPB@UK AC UMRCC CMS
COM 10

BROWNLEE DONALD E PROF
DPT OF ASTRONOMY
UNIVERSITY OF WASHINGTON
FM 20
SEATTLE WA 98195
USA
TEL 206 543 2888
TLF
TLX
EML
COM 15,22

BROWNLEE ROBERT R DR
LOS ALAMOS NATIONAL LAB
MS F670
BOX 1663
LOS ALAMOS NM 87545
USA
TEL 505 662 6427
TLF
TLX
EML
COM 35,42

BRUCATO ROBERT J
CALTECH
MS 105 24
PALOMAR OBS
PASADENA CA 91125
USA
TEL 818 356 4035
TLF 818 568 1517
TLX 675425 OR 188192
EML
COM

BRUCH ALBERT
ASTRONOMISCHES INSTITUT
DER UNIVERSITAET MUENSTER
DOMAGKSTR 75
D 4400 MUENSTER
GERMANY
TEL
TLF
TLX
EML
COM

BRUCK HERMANN A PROF
CRAIGOWER
PENICUIK EH26 9LA
UK
TEL 968 759 18
TLF
TLX
EML
COM 25

BRUCK MARY T DR
ROYAL OBSERVATORY

BLACKFORD HILL
EDINBURGH EH9 3HJ
UK
TEL 316 673 321
TLF
TLX 72383
EML
COM 46

BRUECKNER GUENTER E DR
NAVAL RESEARCH LABORATORY
CODE 4160
4555 OVERLOOK AVE SW
WASHINGTON DC 20375 5000
USA
TEL 202 767 3287
TLF
TLX
EML
COM 10,12,44

BRUGEL EDWARD W DR
CASA
UNIVERSITY OF COLORADO
BOX 391
BOULDER CO 80309 0391
USA
TEL 303 492 4054
TLF
TLX
EML BRUGEL@CYGNUS COLORADO EDU
COM

BRUHWEILER FRED C JR
10102 GARDINER AVE
SILVER SPRING MD 20902
USA
TEL
TLF
TLX
EML
COM 29,34C,42,44

BRUMBERG VICTOR A DR
INST OF APPLIED ASTRONOMY
ACADEMY OF SCIENCES
ZDANOVSKAYA UL 8
197042 ST PETERSBURG
RUSSIA
TEL
TLF
TLX
EML
COM 04,07,31C

BRUNER MARILYN E DR
LOCKHEED PALO ALTO RES LB
DPT 91 20 BLDG 255
3251 HANOVER ST
PALO ALTO CA 94304
USA
TEL 415 858 4023
TLF
TLX 346409 LMSC
EML
COM 10,12,44

BRUNET JEAN-PIERRE DR
OBS MIDI PYRENEES
14 AVE E BELIN
F 31400 TOULOUSE CDX
FRANCE
TEL· 61 25 2101
TLF
TLX 503776
EML
COM 41

BRUNING DAVID H DR
STELLAR RESEARCH &
EDUCATION
BOX 1223
WAUKESHA WI 53187
USA
TEL
TLF
TLX
EML BITNEET dhbrun01@ulkyvx
COM 12,29

BRUNINI ADRIAN DR
IAG
UNIVERISDADE DE SAO PAULO
AV MIGUEL STEFANO 4200
04301 SAO PAULO SP
BRAZIL
TEL
TLF 11 276 3848
TLX 11 56735 IAGM BR
EML IAGUSP%FPSP HEPNET@LB BITNET
COM 07

BRUNK WILLIAM E DR
NASA HEADQUARTERS
CODE SL OFF SPACE SCIENCE
400 MARYLAND AVE S W
WASHINGTON DC 20546
USA
TEL   202 453 1596
TLF
TLX
EML
COM   15,16

BRUSTON PAUL DR
IAS
BP 10
F 91371 VERRIERES BUISSON
FRANCE
TEL   1 69 20 1060
TLF
TLX
EML
COM

BRUZEK ANTON DR
SCHWAIGHOFSTR 7
D 7800 FREIBURG BREISGAU
GERMANY
TEL   761 78522
TLF
TLX
EML
COM   10,12

BRUZUAL GUSTAVO
CIDA

BOX 264
MERIDA 5101 A
VENEZUELA
TEL   74 63 9930
TLF
TLX   74174 CIDA VC
EML   emsca!cida!bruzual@Sun COM
COM   28C

BRYANT JOHN DR
47 AVE FELIX FAURE
F 75015 PARIS
FRANCE
TEL   1 45 57 7647
TLF
EML
COM
TLX

BRZEZINSKI ALEKSANDER DR
SPACE RESEARCH CENTER
POLISH ACAD OF SCIENCES
UL ORDONA 21
PL 01 237 WARSAW
POLAND
TEL. 41 00 41
TLF   22 376 564
TLX   825670 CBKPL
EML   EARN "CBKPAN@PLEARN"
COM   19

BUCCIARELLI BEATRICE DR
STSCI
HOMEWOOD CAMPUS
3700 SAN MARTIN DR
BALTIMORE MD 21218
USA
TEL   301 338 4570
TLF   301 338 2617
TLX   684 9101 78724
EML   6559 BUCC/BUCC@STSCI EDU
COM   08

BUCHLER J ROBERT PROF
DPT OF PHYSICS
UNIVERSITY OF FLORIDA
GAINESVILLE FL 32611
USA
TEL   904 373 9942
TLF
TLX
EML
COM   35

BUDDING EDWIN DR
CARTER OBSERVATORY
BOX 2909
WELLINGTON 1
NEW ZEALAND
TEL   04-728-167
TLX   30172 NATOBS NZ
EML
TLF
COM   42,46

BUECHER ALAIN DR
OBS MIDI PYRENEES
9 R PONT DE LA MOUETTE
F 65200 BAGNERES BIGORRE
FRANCE
TEL   62 95 1969
TLF   62 95 1070
TLX   531625 OBSPIC F
EML
COM   10

BUECHNER JORG DR
ZNTRLINST F ASTROPHYSIK
ROSA LUXEMBURG STR 17A
D 1591 POTSDAM
GERMANY
TEL   23 75105
TLF
TLX   15471
EML
COM   10,49

BUES IRMELA D DR
REMEIS STERNWARTE

STERNWARTSTR 7
D 8600 BAMBERG
GERMANY
TEL   951 57708
TLF
TLX
EML   BUES@STERNWARTE UNI-ERLANGE
COM   29,36

BUFF JAMES S DR
DPT OF PHYSICS & ASTRON
DARTMOUTH COLLEGE
HANOVER NH 03755
USA
TEL
TLF
TLX
EML
COM

BUHL DAVID DR
NASA/GSFC
CODE 693
INFRARED & RADIO ASTRO BR
GREENBELT MD 20771
USA
TEL   301 286 8810
TLF
TLX
EML
COM

BUITRAGO JESUS
INST DE ASTROFISICA
DE CANARIAS
OBS DEL TEIDE
E 38071 LA LAGUNA
SPAIN
TEL   22 26 2211
TLF
TLX   92640
EML
COM

BUJARRABAL VALENTIN
CTR ASTRON DE YEBES
OAN
APD 148
E 19080 GUADALAJARA
SPAIN
TEL   11 22 3358
TLF
TLX
EML
COM   34,40

BUMBA VACLAV DR
ASTRONOMICAL INSTITUTE
CZECH ACADEMY OF SCIENCES
ONDREJOV OBSERVATORY
CS 251 65 ONDREJOV
CZECHOSLOVAKIA
TEL   204 85201
TLF   204 85314
TLX   121579 ASTR C
EML
COM   10,12,44

BUNCLARK PETER STEPHEN DR
INSTITUTE OF ASTRONOMY
THE OBSERVATORIES
MADINGLEY RD
CAMBRIDGE CB3 OHA
UK
TEL   223 33 7548
TLF
TLX   817297 ASTRON G
EML
COM   24

BUNNER ALAN N DR
PERKIN ELMER CORP
MS 897
100 WOOSTER HEIGHTS RD
DANBURY CT 06810 7859
USA
TEL   203 797 6339
TLF
TLX
EML
COM   42,44,48

BUONANNO ROBERTO
OAR
VIA DEL PARCO MELLINI 84
I 00136 ROMA
ITALY
TEL   6 34 52656
TLF
TLX
EML
COM   37

BURATTI BONNIE J DR
JPL
MS 183 501
4800 OAK GROVE DR
PASADENA CA 91109
USA
TEL 818 354 7427
TLF
TLX
EML
COM 15,16

BURBIDGE E MARGARET PROF
CASS
UCSD
C 011
LA JOLLA CA 92093 0216
USA
TEL 619 452 4477
TLF
TLX
EML
COM 28

BURBIDGE GEOFFREY R PROF
CASS
UCSD
C 011
LA JOLLA CA 92093 0216
USA
TEL 619 452 6626
TLF
TLX
EML
COM 28,35,40,47,48

BURGER J J DR IR
ESA/ESTEC
SSD
BOX 299
NL 2200 AG NOORDWIJK
NETHERLANDS
TEL 17 19 84404
TLF
TLX 39098
EML
COM

BURGER MARIJKE DR
OBSERVATOIRE ROYAL DE
BELGIQUE
AVE CIRCULAIRE 3
B 1180 BRUSSELS
BELGIUM
TEL 2 375 2484
TLF
TLX 21565 OBSBEL B
EML
COM 44

BURGESS ALAN DR
DPT APPLIED MATHS
& THEORETICAL PHYSICS
SILVER STREET
CAMBRIDGE CB3 9EW
UK
TEL
TLF
TLX
EML
COM 14,34

BURGESS DAVID D PROF
BLACKETT LABORATORY
IMPERIAL COLLEGE
PRINCE CONSORT RD
LONDON SW7 2BZ
UK
TEL 1 589 5111*6931
TLF
TLX
EML
COM

BURKE BERNARD F DR
DPT OF PHYSICS
MIT RM 26 335
BOX 165
CAMBRIDGE MA 02139
USA
TEL 617 253 2572
TLF
TLX 92-1473
EML
COM 33,34,40,44C,51

BURKE J ANTHONY DR
DPT OF PHYSICS
UNIVERSITY OF VICTORIA
BOX 1700
VICTORIA BC V8W 2Y2
CANADA
TEL 604 721 7743
TLF 604 721 7715
TLX
EML
COM

BURKHART CLAUDE DR
OBSERVATOIRE DE LYON
AVE CHARLES ANDRE
F 69561 S GENIS LAVAL CDX
FRANCE
TEL 78 56 0705
TLF 72 39 9791
TLX 310-926
EML
COM 29

BURKHEAD MARTIN S
ASTRONOMY DPT
INDIANA UNIVERSITY
SWAIN WEST 319
BLOOMINGTON IN 47405
USA
TEL 812 335 6917
TLF
TLX
EML
COM 37

BURKI GILBERT PROF
OBSERVATOIRE DE GENEVE
CHEMIN DES MAILLETTES 51
CH 1290 SAUVERNY
SWITZERLAND
TEL 22 755 2611
TLF 22 755 3983
TLX 419209 OBS CH
EML burki@obs.unige.ch
· COM 27,30P

BURLAGA LEONARD F DR
NASA/GSFC
CODE 692
GREENBELT MD 20771
USA
TEL
TLF 301 286 3271
TLX
EML
COM 15,49C

BURNAGE ROBERT
OHP
F 04870 S MICHEL OBS
FRANCE
TEL 92 76 6368
TLF
EML
COM 30
TLX 410690 OHP F

BURNS JACK O'NEAL JR
DPT PHYSICS & ASTRONOMY
UNIVERSITY OF NEW MEXICO
800 YALE BLVD NE
ALBUQUERQUE NM 87131
USA
TEL 505 277 2705
TLF
TLX
EML
COM 28

BURNS JOSEPH A PROF
CORNELL UNIVERSITY
THURSTON HALL
ITHACA NY 14850
USA
TEL 607 256 4875
TLF
TLX 937478
EML
COM 15,16,20

BURROWS ADAM SETH
DPT OF PHYSICS
STONY BROOK
STONY BROOK NY 11794
USA
TEL 516 246 6810
TLX
EML
TLF
COM 48

BURSA MILAN DR
ASTRONOMICAL INSTITUTE
CZECH ACADEMY SCIENCES
BUDECSKA 6
CS 120 23 PRAHA 2
CZECHOSLOVAKIA
TEL 2 25 0551
TLF
TLX 122486
EML
COM

BURSTEIN DAVID
DPT OF PHYSICS
ARIZONA STATE UNIVERSITY
TEMPE AZ 85287
USA
TEL
TLF
TLX
EML
COM 28,50

BURTON MICHAEL G DR
AAO
EPPING LABORATORY
BOX 296
EPPING NSW 2121
AUSTRALIA
TEL 2 868 1666
TLF 2 876 8536
TLX 23999 AAOSYD AA
EML PSI%NSSDCA· AAOEPP MGB
COM 34

BURTON W BUTLER DR
STERREWACHT
BOX 9513
NL 2300 RA LEIDEN
NETHERLANDS
TEL 71 27 2727
TLX 39058 ASTRONL
EML
TLF
COM 09,33,34

BURTON WILLIAM M
RUTHERFORD APPLETON LAB
SPACE & ASTROPHYSICS DIV
BLDG R25/R68
CHILTON DIDCOT OX11 OQX
UK
TEL· 235 219 00
TLF.
TLX 83159
EML
COM. 44

BUSCOMBE WILLIAM PROF
DPT PHYSICS & ASTRONOMY
NORTHWESTERN UNIVERSITY
DEARBORN OBSERVATORY
EVANSTON IL 60208
USA
TEL 312 491 7527
TLF
TLX
EML
COM 45,46

BUSER ROLAND DR
ASTRONOMISCHES INSTITUT
UNIVERSITAET BASEL
VENUSSTRASSE 7
CH 4102 BINNINGEN
SWITZERLAND
TEL 61 22 7711
TLF
TLX
EML
COM 25,45

BUSKO IVO C DR
INPE
CP 515
12200 S JOSE DOS CAMPOS
BRAZIL
TEL 123 22 9977*392
TLF 123 21 8743
TLX 011-33530 INPE BR
EML
COM 27

BUSON LUCIO M DR
OSS ASTRONOMICO DI PADOVA
VIC DELL OSSERVATORIO 5
I 35122 PADOVA
ITALY
TEL 49 66 1499
TLX 432071 ASTROS I
EML BUSON@ASTRPD INFNET
TLF
COM

BUSSO MAURIZIO
OSS ASTRONOMICO DI TORINO
ST OSSERVATORIO 20
I 10025 PINO TORINESE
ITALY
TEL 11 84 1067
TLF
TLX 213239 TO ASTR I
EML
COM 42

BUTA RONALD J DR
ASTRONOMY DPT
UNIVERSITY OF TEXAS
RLM 15 308
AUSTIN TX 78712 1083
USA
TEL 512 471 3466
TLF
TLX 910 874 1351
EML
COM 28

BUTCHER HARVEY R PROF DR
KAPTEYN ASTRONOMICAL INST
BOX 800
NL 9700 AV GRONINGEN
NETHERLANDS
TEL 59 08 19631
TLX 53767 KSWRO NL
EML
TLF
COM 28

BUTCHINS SYDNEY ADAIR
DPT OF CIVIL AVIATION STU
FAC OF MATHS
THE MINORIES
TOWER HILL EC3N 1JY
UK
TEL 71 722 7344
TLF
TLX
EML
COM

BUTI BIMLA PROF
PHYSICAL RESEARCH LAB
NAVRANGPURA
AHMEDABAD 380 009
INDIA
TEL 272 46 2129
TLF 272 44 5292
TLX 1216397 PRL IN
EML BUTI@PRL ERNET IN
COM 49P

BUTLER C JOHN DR
ARMAGH OBSERVATORY
COLLEGE HILL
ARMAGH BT61 9DG
UK
TEL 861 52 2928
TLF
TLX 747937 ARMOBS G
EML
COM 27,44

BUTLER DENNIS DR
YALE UNIVERSITY OBS
YALE STATION,
BOX 2023
NEW HAVEN CT 06520
USA
TEL
TLF
TLX
EML
COM 27,37

BUTLER KEITH DR
INSTITUT FUR ASTRONOMIE
UND ASTROPHYSIK
SCHEINERSTR 1
D 8000 MUENCHEN 80
GERMANY
TEL 89 989 021
TLF
TLX
EML
COM 29

BUTTERWORTH PAUL
NASA/GSFC
CODE 633
GREENBELT MD 20771
USA
TEL 301 286 3995
TLF
TLX
EML
COM 44

BUYUKLIEV GEORGI DR
NTL ASTRONOMICAL OBS
BULGARIAN ACAD SCIENCES
BOX 136
BG 4700 SMOLJAN
BULGARIA
TEL 73 41 599
TLF
TLX 48446 SST RZNBG
EML
COM 10

BYARD PAUL L DR
DPT OF ASTRONOMY
OHIO STATE UNIVERSITY
174 W 18TH AVE
COLUMBUS OH 43210 1106
USA
TEL 614 422 1773
TLF
TLX
EML
COM

BYKOV MIKLE F DR
ASTRONOMICAL INSTITUTE
UZBEK ACADEMY OF SCIENCES
700000 TASHKENT
UZBEKISTAN
TEL
TLF·
TLX
EML
COM 08

BYLEVELD WILLEM DR
OMNIVERSUM SPACE THEATRE
PRES KENNEDYLAAN 5
NL 2517 JK THE HAGUE
NETHERLANDS
TEL 70 54 7479
TLF 70 52 4280
TLX
EML
COM

BYRD GENE G DR
DPT OF PHYSICS & ASTRON
UNIVERSITY OF ALABAMA
BOX 1921
UNIVERSITY AL 35487 0324
USA
TEL 205 348 5050
TLF
TLX
EML
COM 28,37

BYRNE PATRICK B DR
ARMAGH OBSERVATORY
COLLEGE HILL
ARMAGH BT61 9DG
UK
TEL 861 52 2928
TLF
TLX 747937 ARMOBS G
EML PBB@STAR ARM GUB AC UK
COM 27

BYSTROVA NATALIJA V DR
PULKOVO OBSERVATORY
ACADEMY OF SCIENCES
10 KUTUZOV QUAY
196140 ST PETERSBURG
RUSSIA
TEL 297-9452
TLF
TLX
EML
COM 34

CABRITA EZEQUIEL DR
OBS ASTRONOMICO DE LISBOA
TAPADA DA LISBOA
P 1300 LISBOA
PORTUGAL
TEL 637351-634669
TLF
TLX
EML
COM 26

CACCIANI ALESSANDRO PROF
DPT DI FISICA
UNIVERSITA DI ROMA
PA MORO 2
I 00185 ROMA
ITALY
TEL 6 49 76265
TLF
TLX 613255 INFNRO
EML
COM

CACCIARI CARLA DR
OSS ASTRONOMICO
UNIVERSITA DI BOLOGNA
CP 596
I 40100 BOLOGNA
ITALY
TEL 51 25 9301*9401
TLF 51 25 9407
TLX 520634 INFNBO I
EML CACCUARU@ASTBO1 INFNET
COM

CACCIN BRUNO
DPT DI FISICA
VIA RAIMONDO SNC
UNIVERSITA TOR VERGATA
I 00173 ROMA
ITALY
TEL 6 79 792323
TLF
TLX 626382 FIUNTV
EML
COM 38

CADEZ ANDREJ DR
DPT OF PHYSICS
UNIVERSITY OF LJUBLJANA
JADRANSKA 19
YU 61000 LJUBLJANA
YUGOSLAVIA
TEL 61 265 061
TLF 61 217 281
TLX
EML ANDREJ CADEZ@UNI-LJ AC MAIL YU
COM

CADEZ VLADIMIR
INSTITUTE OF PHYSICS
BOX 57
YU 11001 BEOGRAD
YUGOSLAVIA
TEL 11 212 219
TLF
TLX 11002 INFIZ YU
EML
COM 10,12

CAHN JULIUS H PROF
DPT OF ASTRONOMY
UNIVERSITY OF ILLINOIS
1011 W SPRINGFIELD AVE
URBANA IL 61801
USA
TEL 217 333 3090
TLF
TLX
EML
COM

CAILLAULT JEAN PIERRE DR
DPT PHYSICS & ASTRONOMY
UNIVERSITY OF GEORGIA
ATHENS GA 30602
USA
TEL 404 542 2883
TLF
TLX 490 999 1619
EML
COM

CALAME ODILE DR
OCA CERGA
AVE COPERNIC
F 06130 GRASSE
FRANCE
TEL 93 36 5849
TLF
TLX 470865
EML
COM 07,16,20

CALDWELL JOHN A R
SAAO
BOX 9
OBSERVATORY 7935
SOUTH AFRICA
TEL 021 47 0025
TLF
TLX 57-20309
EML
COM 33

CALDWELL JOHN JAMES
DPT OF PHYSICS
YORK UNIVERSITY
4700 KEELE ST
NORTH YORK ON M3J 1P3
CANADA
TEL 416 736 2100
TLF 416 736 5386
TLX 065 24736
EML
COM 16

CALLANAN PAUL DR
DPT OF ASTROPHYSICS
UNIVERSITY OF OXFORD
KEBLE RD
OXFORD OX1 3RH
UK
TEL 865 273 293
TLF 865 273 418
TLX 83295 NUCLOX G
EML UK AC OX ASTRO (JANET)
COM 42

CALLEBAUT DIRK K DR
DPT OF PHYSICS
UNIVERSITY OD ANTWERPEN
UNIVERSITEITSPLEIN 1
B 2610 ANTWERPEN WILRIJK
BELGIUM
TEL 3 820 2457
TLF 3 820 2245
TLX 33646
EML
COM 35,37

CALLY PAUL S DR
DPT OF MATHEMATICS
MONASH UNIVERSITY
WELLINGTON RD
CLAYTON VIC 3168
AUSTRALIA
TEL 3 565 4471
TLF
TLX MONASH AA 32691
EML apml50f@vaxc cc monash edu au
COM 10

CALOI VITTORIA DR
IAS
CNR
CP 67
I 00044 FRASCATI
ITALY
TEL 6 942 5654
TLF 6 941 6847
TLX 610261 CNR FRA
EML
COM 35,37,

CALVANI MASSIMO DR
DPT DI ASTRONOMIA
UNIVERSITA DI PADOVA
VIC DELL OSSERVATORIO 5
I 35122 PADOVA
ITALY
TEL 49 66 1499
TLF
TLX 430176 UNPADU I
EML
COM 47

CALVET NURIA DR
CIDA
BOX 264
MERIDA 5101 A
VENEZUELA
TEL 74 63 9930
TLF
TLX 74174 CIDA VC
EML
COM 46

CALVIN WILLIAM H DR
DPT OF PHYSICS
UNIVERSITY OF WASHINGTON
FM 15
SEATTLE WA 98195
USA
TEL 206 328 1192
TLF 206 543 3041
TLX
EML WCALVIN@U WASHINGTON EDU
COM 51

CALVO MANUEL
DPT DE ASTRONOMIA
UNIVERSIDAD DE ZARAGOZA
E 50009 ZARAGOZA
SPAIN
TEL   35 7011
TLF
TLX
EML
COM

CAMARENA BADIA VICENTE PR
DPT MATEMATICA APLICADA
UNIVERSIDAD DE ZARAGOZA
E 50009 ZARAGOZA
SPAIN
TEL
TLF
TLX.
EML
COM

CAMENZIND MAX DR
LANDESSTERNWARTE
KOENIGSTUHL
D 6900 HEIDELBERG 1
GERMANY
TEL   62 215 09262
TLF   62 215 09202
TLX   461153
EML   AB4@DHDURZ1 BITNET
COM   48

CAMERON ALASTAIR G W PROF
CENTER FOR ASTROPHYSICS
HCO/SAO
60 GARDEN ST
CAMBRIDGE MA 02138
USA
TEL   617 495 5374
TLF
TLX
EML
COM   35,48

CAMERON ANDREW COLLIER DR
ASTRONOMY CENTRE
UNIVERSITY OF SUSSEX
FALMER
BRIGHTON BN1 9QH
UK
TEL   273 67 8117
TLF
TLX   877159 BHVTXS G
EML
COM   27

CAMERON LUZIUS MARTIN
ASTRONOMISCHES INSTITUT
UNIVERSITAET BASEL
VENUSSTRASSE 7
CH 4102 BINNINGEN
SWITZERLAND
TEL   61 22 7711
TLF
TLX
EML
COM   28

CAMERON WINIFRED S MRS
LA RANEHITA DE LA LUNA
BLDG 26
200 ROJO DR
SEDONA AZ 86336
USA
TEL
TLF
TLX
EML
COM   16

CAMICHEL HENRI DR
24 AVE C FLAMMARION

F 31500 TOULOUSE
FRANCE
TEL   61 48 9691
TLF
TLX
EML
COM   16

CAMPBELL ALISON DR
DPT PHYSICS & ASTRONOMY
JOHNS HOPKINS UNIVERSITY
CHARLES & 34TH ST
BALTIMORE MD 21218
USA
TEL   301 338 5186
TLF   301 338 8260
TLX
EML   AWC@STSCI BITNET SCIVAX   AWC
COM   26

CAMPBELL BELVA G S DR
DPT PHYSICS & ASTRONOMY
UNIVERSITY OF NEW MEXICO
800 YALE BLVD NE
ALBUQUERQUE NM 87131
USA
TEL   505 277 5148
TLF
TLX
EML   BITNET BEL@UNMB
COM

CAMPBELL BRUCE DR
4537 RITHEWOOD PLC
VICTORIA BC VX 4J9
CANADA
TEL
TLF
TLX
EML
COM   29,30,51C

CAMPBELL DONALD B
CORNELL UNIVERSITY

SPACE SCIENCES BLDG
ITHACA NY 14853
USA
TEL   607 255 5274
TLF
TLX   932454
EML
COM   16

CAMPBELL JAMES W
ROYAL OBSERVATORY
BLACKFORD HILL
EDINBURGH EH9 3HJ
UK
TEL
TLX
EML
TLF
COM

CAMPBELL MURRAY F
DPT PHYSICS & ASTRONOMY
COLBY COLLEGE
WATERVILLE ME 04901
USA
TEL   207 872 3251
TLF
TLX·
EML
COM   44

CAMPINS HUMBERTO DR
DPT OF ASTRONOMY
UNIVERSITY OF FLORIDA
211 SSRB
GAINESVILLE FL 32611
USA
TEL   904 392 3066
TLF
TLX
EML   campins@astro ufl.edu
COM   15

CAMPOS L M BRAGA DA COSTA
INST SUPERIOR TECNICO
AVE ROVISCO PAIS
P 1096 LISBOA CODEX
PORTUGAL
TEL   800525
TLF
TLX   63423 ISTUTL P
EML
COM

CAMPUSANO LUIS E
DPT DE ASTRONOMIA
UNIVERSIDAD DE CHILE
CASILLA 36 D
SANTIAGO
CHILE
TEL   2 229 4101
TLF
TLX   440 001
EML
COM   28,51

CANAL RAMON M DR
DPT FISICA DE ATMOSFERA
UNIVERSIDAD DE BARCELONA
AVD DIAGONAL 645
E 08028 BARCELONA
SPAIN
TEL
TLF
TLX
EML
COM   35C

CANAVAGGIA RENEE DR
OBSERVATOIRE DE PARIS
61 AVE OBSERVATOIRE
F 75014 PARIS
FRANCE
TEL   1 43 20 1210
TLF
TLX   270776 OBS F
EML
COM

CANDY MICHAEL P MR
PERTH OBSERVATORY

BICKLEY WA 6076
AUSTRALIA
TEL
TLF
TLX
EML
COM   06,07,15,20

CANE HILARY VIVIEN
DPT OF PHYSICS
UNIVERSITY OF TASMANIA
GPO BOX 252C
HOBART TAS 7001
AUSTRALIA
TEL 2 202 401
TLF 2 202 186
TLX AA58150
EML HILARY CANE@PHYS UTAS EDU AU
COM 10,33

CANFIELD RICHARD C DR
INSTITUTE FOR ASTRONOMY
UNIVERSITY OF HAWAII
2680 WOODLAWN DR
HONOLULU HI 96822
USA
TEL
TLF
TLX
EML
COM·

CANIZARES CLAUDE R PROF
CENTER FOR SPACE RESEARCH
MIT
RM 37-241
CAMBRIDGE MA 02139
USA
TEL 617 253 7480
TLF 617 253 0861
TLX 921473 MITCAM
EML brendap@space mit edu
COM

CANNON RUSSELL D DR
AAO
BOX 296
EPPING NSW 2121
AUSTRALIA
TEL 2 868 1666
TLF
TLX 23999 AAOSYD
EML
COM 28,37

CANNON WAYNE H DR
DPT PHYS/EARTH & ATM SCI
YORK UNIVERSITY
4700 KEELE ST
DOWNSVIEW ON M3J 1P3
CANADA
TEL 416 667 6410
TLF
TLX 065 24736
EML
COM

CANTO JORGE DR
INSTITUTO DE ASTRONOMIA
UNAM
APDO POSTAL 70-264
04510 MEXICO DF
MEXICO
TEL 5485305
TLF
TLX 1760155 CIC ME
EML
COM 34

CANTU ALBERTO M DR
IST CIBERNETICA/BIOFISICA
CNR
I 16032 CAMOGLI
ITALY
TEL 185 77 0646
TLF
TLX
EML
COM

CAO CHANGXIN
NANJING ASTRONOMICAL
INSTRUMENT FACTORY
NANJING
CHINA PR
TEL 25 46191
TLF
TLX 34136 GLYNJ CN
EML
COM 09

CAO LIHONG
PURPLE MOUNTAIN OBSERV
CAS
NANJING
CHINA PR
TEL 25 46700
TLF
TLX
EML
COM

CAO SHENGLIN
DPT OF ASTRONOMY
BEIJING NORMAL UNIVERSITY
BEIJING 100875
CHINA PR
TEL 1 201 2255
TLF
TLX 222701
EML
COM

CAPACCIOLI MASSIMO DR
DPT DI ASTRONOMIA
UNIVERSITA DI PADOVA
VIC DELL OSSERVATORIO 5
I 35122 PADOVA
ITALY
TEL 49 66 1499
TLF
TLX 430176 UNDAPU I
EML
COM 28

CAPACCIONI FABRIZIO DR
IAS
REPORTO DI PLANETOLOGIA
VIA DELL'UNIVERSITA 11
I 00185 ROMA
ITALY
TEL 6 44 56951
TLF 6 44 54969
TLX
EML
COM 15

CAPELATO HUGO VICENTE DR
INPE
CP 515
12201 S JOSE DOS CAMPOS
BRAZIL
TEL 123 22 9977
TLF
TLX 1233530 INPE BR
EML
COM

CAPEN CHARLES F
SOLIS LACUS OBSERVATORY
RT 2
BOX 262 E
CUBA MO 65453
USA
TEL
TLF
TLX
EML
COM

CAPITAINE NICOLE
OBSERVATOIRE DE PARIS
61 AVE OBSERVATOIRE
F 75014 PARIS
FRANCE
TEL 1 40 51 2231
TLF
TLX 270776 OBS F
EML
COM 04,19C

CAPLAN JAMES
OBSERVATOIRE DE MARSEILLE
2 PLACE LE VERRIER
F 13248 MARSEILLE CDX 04
FRANCE
TEL 91 95 9088
TLF
TLX
EML
COM 34

CAPPA DE NICOLAU CRISTINA
IAR
CC 5
1894 VILLA ELISA (BS AS)
ARGENTINA
TEL 21 4 3793
TLX
EML
TLF
COM 34

CAPPELLARO ENRICO DR
OSS ASTRONOMICO DI PADOVA
VIC DELL OSSERVATORIO 5
I 35122 PADOVA
ITALY
TEL 49 66 1499
TLX 432071 ASTROS I
EML SPAN ASTRPD CAPPELLARO
TLF
COM

CAPRIA MARIA TERESA DR
IAS
REPARTO DI PLANETOLOGIA
VIA DELL'UNIVERSITA 11
I 00185 ROMA
ITALY
TEL 6 44 56951
TLF 6 44 54969
TLX
EML TERESA@IRMIAS
COM 15,16

CAPRIOLI GIUSEPPE PROF
OAR
VIA TRIONFALE 204
I 00136 ROMA
ITALY
TEL 6 34 7050
TLF
TLX
EML
COM 31

CAPRIOTTI EUGENE R DR
DPT OF ASTRONOMY
OHIO STATE UNIVERSITY
505B ALPHEUS SMITH LAB
COLUMBUS OH 43210 1106
USA
TEL 614 422 1773
TLF
TLX
EML
COM 34

CAPUTO FILIPPINA DR
IAS
CNR
CP 67
I 00044 FRASCATI
ITALY
TEL 6 942 5651
TLF 6 941 6847
TLX 610261 CNR FRA
EML·
COM 35,37

CAPUZZO DOLCETTA ROBERTO
ISTITUTO ASTRONOMICO
UNIVERSITA DI ROMA
VIA G M LANCISI 29
I 00161 ROMA
ITALY
TEL 6 86 7525
TLF
TLX
EML
COM 34,37

CARANICOLAS NICHOLAS DR
DPT OF ASTRONOMY
UNIVERSITY THESSALONIKI
GR 540 06 THESSALONIKI
GREECE
TEL 31 99 1357/59
TLF
TLX
EML
COM 07

CARBON DUANE F DR
NASA AMES RESEARCH CTR
MS 258 5
MOFFETT FIELD CA 94035
USA
TEL 415 604 4413
TLF
TLX
EML DCARBON@NAS NASA GOV
COM 14,36

CARDELLI JASON A DR
DPT OF ASTRONOMY
UNIVERSITY OF WISCONSIN
475 N CHARTER ST
MADISON WI 53706
USA
TEL 608 262 7921
TLF
TLX
EML MADRAF CARDELLI
COM 34

CARDINI DANIELA DR
IAS
CNR
CP 67
I 00044 FRASCATI
ITALY
TEL 6 942 5655
TLF 6 941 6847
TLX 610261 CNR FRA
EML DANIELA@IRMIAS
COM 48

CARDONA OCTAVIO DR
INAOE
TONANTZINTLAZ
APDO POSTAL 216 y 51
72000 PUEBLA PUE
MEXICO
TEL 22 470 500
TLF
TLX
EML
COM

CARDUS ALMEDA J O MR
OBSERVATORIO DEL EBRO
E 43520 ROQUETES
SPAIN
TEL 77 50 0511
TLF
EML
COM
TLX

CARESTIA REINALDO A DR
OBSERVATORIO ASTRONOMICO
FELIX AGUILAR
AV BENAVIDEZ 8175 OESTE
5400 SAN JUAN
ARGENTINA
TEL 64 23 1615
TLF
TLX
EML
COM 08

CARGILL PETER J DR
NAVAL RESEARCH LABORATORY
CODE 4790
4555 OVERLOOK AVE SW
WASHINGTON DC 20375 5000
USA
TEL 202 767 4978
TLF 202 767 0631
TLX
EML CARGILL@PPDPI1 NRL NAVY MIL
COM 10

CARIGNAN CLAUDE DR
DPT DE PHYSIQUE
UNIVERSITE DE MONTREAL
CP 6128 SUCC A
MONTREAL QC H3C 3J7
CANADA
TEL 514 343 7355
TLF 514 343 2071
TLX 055 62425
EML CARIGNAN@CC UMONTREAL.CA
COM 28

CARLBERG RAYMOND GARY DR
DPT OF PHYSICS
YORK UNIVERSITY
4700 KEELE ST
TORONTO ON M3J 1P3
CANADA
TEL 416 667 3851
TLF
TLX 065 24736
EML
COM

CARLETON NATHANIEL P DR
CENTER FOR ASTROPHYSICS
HCO/SAO
60 GARDEN ST
CAMBRIDGE MA 02138
USA
TEL 617 495 7405
TLF
TLX 921428 SATELLITE CAM
EML
COM

CARLQVIST PER A DR
DPT OF PLASMA PHYSICS
ROYAL INST OF TECHNOLOGY
S 100 44 STOCKHOLM 70
SWEDEN
TEL 87 87 7697
TLF
TLX
EML
COM 10

CARLSON JOHN B
ARCHAEOASTRONOMY CENTER
BOX X
COLLEGE PARK MD 20740
USA
TEL 301 864 6637
TLF
TLX
EML
COM 41,51

CARLSSON MATS DR
INST THEORET ASTROPHYSICS
UNIVERSITY OF OSLO
BOX 1029
N 0315 BLINDERN OSLO 3
NORWAY
TEL
TLF
TLX
EML
COM 36

CARNEY BRUCE WILLIAM
DPT PHYSICS & ASTRONOMY
UNIVERSITY NORTH CAROLINA
204 PHILLIPS HALL 039A
CHAPEL HILL NC 27514
USA
TEL 919 962 3023
TLF
TLX
EML
COM 25,29,30,37

CAROFF LAWRENCE J
NASA AMES RESEARCH CTR
MS 245 6
SPACE SCIENCE DIV
MOFFETT FIELD CA 94035
USA
TEL 415 694 5523
TLF
TLX
EML
COM

CAROUBALOS C A PROF
LAB ELECTRONIC PHYSICS
NTL UNIVERSITY OF ATHENS
KTHRIA TYPA-ILISSIA
GR ATHENS 144
GREECE
TEL 1 72 44 096/11119
TLF
TLX 215530 OBSA GR
EML
COM 40

CARPENTER KENNETH G DR
NASA/GSFC
CODE 681
GREENBELT MD 20771
USA
TEL 301 286 3453
TLF
TLX 89675 K CARPENTER
EML SPAN 6172 HRSCARPENTER
COM 29,44

CARPENTER LLOYD DR
13902 RESIN CT
BOWIE MD 20720
USA
TEL
TLF
EML
COM
TLX

CARPINO MARIO DR
OSS ASTRONOMICO DI BRERA
VIA BRERA 28
I 20121 MILANO
ITALY
TEL 2 87 4444
TLF 2 72 00 1600
TLX
EML CARPINO@ASTMIB ASTRO IT
COM 07

CARQUILLAT JEAN-MICHEL
OBS MIDI PYRENEES
14 AVE E BELIN
F 31400 TOULOUSE COX
FRANCE
TEL 61 25 2101
TLF
TLX 530776 OBSTLSE
EML
COM 30

CARR BERNARD JOHN
SCHOOL OF MATHEMATICAL SC
QUEEN MARY/WESTFIELD COLL
MILE END RD
LONDON E1 4NS
UK
TEL 1 980 4811
TLF
TLX
EML
COM 47

CARR THOMAS D PROF
DPT OF ASTRONOMY
UNIVERSITY OF FLORIDA
211 SSRB
GAINESVILLE FL 32611
USA
TEL 904 392 2066
TLF
TLX
EML tacarr@uffsc bitnet
COM 40,51

CARRANZA GUSTAVO J DR
LAPRIDA 880
5000 CORDOBA
ARGENTINA
TEL 51 40613
TLF
TLX
EML
COM 28

CARRASCO GUILLERMO DR
DPT DE ASTRONOMIA
UNIVERSIDAD DE CHILE
CASILLA 36 D
SANTIAGO
CHILE
TEL 2 229 4002
TLF
TLX 440 001
EML
COM 08

CARRASCO LUIS DR
INSTITUTO DE ASTRONOMIA
UNAM
APDO POSTAL 70-264
04510 MEXICO DF
MEXICO
TEL 905-548-5305
TLF
TLX
EML
COM 33

CARROLL P KEVIN PROF
DPT OF PHYSICS
UNIVERSITY COLLEGE
BELFIELD
DUBLIN 4
IRELAND
TEL 1 693 244
TLF
TLX
EML
COM 14,44

CARRUTHERS GEORGE R DR
NAVAL RESEARCH LABORATORY
CODE 7123
SPACE SCIENCE DIVISION
WASHINGTON DC 20375 5000
USA
TEL 202 767 2764
TLF
TLX
EML
COM 15,34

CARSENTY URI DR
DLR NE-OE-PE
OBERPFAFFENHOFEN
D 8031 WESSLING
GERMANY
TEL 815 328 1328
TLF 815 32476
TLX
EML SPAN OOEPES Carsenty
COM 15

CARSMARU MARIA M DR
ASTRONOMICAL OBSERVATORY
CUTITUL DE ARGINT 5
BOX 28
R 75212 BUCHAREST
RUMANIA
TEL 236892
TLF
TLX 11882 ASTRO R
EML
COM 16

CARSON T R DR
DPT OF PHYSICS & ASTRON
UNIVERSITY OF ST ANDREWS
NORTH HAUGH
ST ANDREWS FIFE KY16 9SS
UK
TEL 334 761 61
TLF 334 744 87
TLX
EML
COM 35,36

CARSWELL ROBERT F DR
INSTITUTE OF ASTRONOMY
THE OBSERVATORIES
MADINGLEY RD
CAMBRIDGE CB3 OHA
UK
TEL 223 622 04
TLF
TLX 817297 ASTRON G
EML
COM 28

CARTER DAVID DR
MOUNT STROMLO & SIDING
SPRING OBSERVATORIES
PRIVATE BAG
WODEN PO ACT 2606
AUSTRALIA
TEL 62 88 1111
TLF
TLX 62270 AA
EML
COM 28

CARTER WILLIAM EUGENE
NGS
N/CG 114
ROCKVILLE MD 20852
USA
TEL 301 443 8423
TLF
TLX
EML
COM 19,31

CARUSI ANDREA
IAS
REPARTO DI PLANETOLOGIA
VIA DELL'UNIVERSITA 11
I 00185 ROMA
ITALY
TEL 6 44 56951
TLF 6 44 54969
TLX 610261 CNRFRA
EML CARUSI@IRMIAS
COM 15,20P,22

CARVER JOHN H PROF
AUSTRALIAN NTL UNIVERSITY
RES SCHOOL PHYS SCIENCE
BOX 4
CANBERRA ACT 2601
AUSTRALIA
TEL 62 492 476
TLF
TLX 62615 RPHYS
EML
COM 14,44

CASANOVAS JUAN DR
VATICAN OBSERVATORY

I 00120 VATICAN CITY
VATICAN CITY STATE
TEL 6 698 3411/5266
TLF
TLX 2020 VATOBS VA
EML
COM

CASERTANO STEFANO DR
DPT OF ASTRONOMY
UNIVERSITY OF ILLINOIS
1011 W SPRINGFIELD AVE
URBANA IL 61801
USA
TEL 217 333 9380
TLF
TLX
EML STEFANO@RIGEL ASTRO ULUC EDU
COM 28

CASH WEBSTER C JR
LASP
UNIVERSITY OF COLORADO
BOX 392
BOULDER CO 80309 0392
USA
TEL 303 492 8208
TLF
TLX
EML
COM· 48

CASOLI FABIENNE DR
RADIOASTRONOMIE ENS
24 RUE LHOMOND
F 75231 PARIS CDX 05
FRANCE
TEL 1 43 29 1225
TLF
TLX 202601
EML CASOLI@FRULM11
COM

CASSATELLA ANGELO DR
IAS
CNR
CP 67
I 00044 FRASCATI
ITALY
TEL 6 942 5655
TLF 6 941 6847
TLX
EML
COM 29C

CASSE MICHEL DR
CEA CEN
DAPNIA/SAP
BP 2
F 91191 GIF/YVETTE CDX
FRANCE
TEL
TLF
TLX
EML
COM 48

CASSINELLI JOSEPH P DR
DPT OF ASTRONOMY
UNIVERSITY OF WISCONSIN
475 N CHARTER ST
MADISON WI 53706
USA
TEL 608 262 1752
TLF
TLX 265452 UOFWISC MDS
EML
COM 36

CASTAGNINO MARIO DR
IAFE
CC 67 SUC 28
1428 BUENOS AIRES
ARGENTINA
TEL 1 781 6755
TLF 1 814 4299
TLX 17181 VERBA
EML CASTAGNINO@IAFE EDU AR
COM 47

CASTANEDA HECTOR
INST DE ASTROFISICA
DE CANARIAS
OBS DEL TEIDE
E 38200 LA LAGUNA
SPAIN
TEL 22 26 2211
TLF 22 26 3005
TLX 92640 IAC E
EML
COM 34

CASTELAZ MICHEAL W DR
DPT OF PHYSICS
E TENNESSEE STATE UNIV
BOX
JOHNSON CITY TN 37614
USA
TEL 615 461 7064
TLF
TLX
EML R29CASTZ@ETSU BITNET
COM 25

CASTELLANI VITTORIO PROF
ISTITUTO ASTRONOMICO
UNIVERSITA DI ROMA
VIA G M LANCISI 29
I 00161 ROMA
ITALY
TEL 6 86 7525
TLF
TLX
EML
COM 35,37

CASTELLI FIORELLA DR
OAT
BOX SUCC TRIESTE 5
VIA TIEPOLO 11
I 34131 TRIESTE
ITALY
TEL 40 79 3921
TLF
TLX 461137 OAT I
EML
COM 29

CASTELLI JOHN P
A F GEOPHYSICS LABORATORY
SPACE PHYSICS DIV PHP
HANSCOM AFB
BEDFORD MA 01731
USA
TEL
TLF
TLX
EML
COM

CASTETS ALAIN DR
OBSERVATOIRE DE GRENOBLE
CERMO/ASTROPHYSIQUE
BP 53X
F 38041 S MARTIN HERES CD
FRANCE
TEL 76 51 4786
TLF 76 44 8821
TLX
EML EARN CASTETS@FRAG51
COM 40

CASTOR JOHN I DR
LAWRENCE LIVERMORE LAB
L 23
BOX 808
LIVERMORE CA 94550
USA
TEL 415 422 4664
TLF
TLX 910-3868339 LLNL
EML
COM 35,36

CASWELL JAMES L DR
CSIRO
DIVISION OF RADIOPHYSICS
BOX 76
EPPING NSW 2121
AUSTRALIA
TEL 2 868 0222
TLF
TLX 26230 ASTRO AA
EML JCASWELL@ATNF CSIRO AU
COM 33,34,40

CATALA CLAUDE DR
OBSERVATOIRE DE PARIS
SECTION DE MEUDON
F 92195 MEUDON PPL CDX
FRANCE
TEL 1 45 07 7663
TLF
TLX 204464
EML
COM 29,36

CATALA POCH M A
DPT FISICA DE ATMOSFERA
UNIVERSIDAD DE BARCELONA
AVD DIAGONAL 645
E 08028 BARCELONA
SPAIN
TEL 3 330 7311*244
TLF
TLX
EML
COM 46

CATALAN MANUEL DR
REAL INST Y OBSERVATORIO
DE LA ARMADA
CECILIO PUJAZON S/N
E 11110 SAN FERNANDO
SPAIN
TEL 56 88 3548
TLF
TLX 76108
EML
COM 04,08,31

CATALANO FRANCESCO A DR
IST DI ASTRONOMIA
CITTA UNIVERSITARIA
VIA A DORIA 6
I 95125 CATANIA
ITALY
TEL 95 33 0533
TLF
TLX 970359 ASTRCT I
EML
COM 36

CATALANO SANTO DR
IST DI ASTRONOMIA
CITTA UNIVERSITARIA
VIA A DORIA 6
I 95125 CATANIA
ITALY
TEL 95 33 0533
TLF
TLX 970359 ASTRCT I
EML
COM 16,29,36,42

CATARZI MARCO DR
OSS ASTROPISICO
DI ARCETRI
LARGO E FERMI 5
I 50125 FIRENZE
ITALY
TEL 55 437 8540
TLF
TLX 572268 ARCETR I
EML 38954 CATARZI
COM 40

CATCHPOLE ROBIN M DR
ROYAL GREENWICH OBS
MADINGLEY RD
CAMBRIDGE CB3 0EZ
UK
TEL 223 37 4000
TLF 223 37 4700
TLX 26541/265871
EML CATHPOLE@UK AC CAM AST-STAR
COM 27,29

CATO B TORGNY DR
NORDISK TELESATELLITSTAT
BOX 107
S 457 00 TANUMSHEDE
SWEDEN
TEL 52 52 9155
TLF
TLX 20164 NORDSAT S
EML
COM

CATON DANIEL B DR
ASSISTANT PROFESSOR
PHYSICS AND ASTRONOMY
APPALACHIAN STATE UNIV
BOONE NC 28608
USA
TEL 704 262 2446
TLF
TLX 888370 OR 62671500
EML· BITNET CATONDB@APPSTATE
COM

CATURA RICHARD C DR
LOCKHEED PALO ALTO RES LB
DPT 91 20 BLDG 255
3251 HANOVER ST
PALO ALTO CA 94304
USA
TEL 415 858 4066
TLF
TLX 346409 LMSC SUVL
EML
COM 44,48

CAUGHLAN GEORGEANNE R
DPT OF PHYSICS
MONTANA STATE UNIVERSITY
BOZEMAN MT 59717
USA
TEL 406 994 6170
TLF
TLX
EML
COM 35,48

CAVALIERE ALFONSO G PROF
DPT DI ASTROFISICA
II UNIVERSITA DI ROMA
VIA ORAZIO RAIMONDO
I 00173 ROMA
ITALY
TEL
TLF
TLX
EML
COM 47,48

CAVALLINI FABIO
OSS ASTROFISICO
DI ARCETRI
LARGO E FERMI 5
I 50125 FIRENZE
ITALY
TEL 55 437 8540
TLF 55 43 5939
TLX 572268
EML
COM 12

CAWLEY MICHAEL DR
DPT OF PHYSICS
ST PATRICK'S COLLEGE
MAYNOOTH
CO KILDARE
IRELAND
TEL 1 285 222*X499
TLF
TLX
EML BITNET MCAWLEY@VAX1 MAY IE
COM

CAYREL DE STROBEL GIUSA
OBSERVATOIRE DE PARIS
SECTION DE MEUDON
F 92195 MEUDON PPL CDX
FRANCE
TEL 1 45 07 7863
TLF
TLX
EML
COM 29,36

CAYREL ROGER DR
OBSERVATOIRE DE PARIS
61 AVE OBSERVATOIRE
F 75014 PARIS
FRANCE
TEL 1 40 51 2251
TLF
TLX 270776 OBS F
EML
COM 29,36,50

CAZENAVE ANNY DR
CNES/GRGS
18 AVE E BELIN
F 31055 TOULOUSE CDX
FRANCE
TEL 61 27 4011
TLF
TLX 531081
EML
COM

CAZZOLA PAOLO DR
OSS ASTRONOMICO DI PADOVA
VIC DELL OSSERVATORIO 5
I 35122 PADOVA
ITALY
TEL 49 66 1499
TLX
EML
TLF
COM

CEFOLA PAUL J DR
MAIL STATION 64
C S DRAPER LAB
555 TECHNOLOGY SQ
CAMBRIDGE MA 02139
USA
TEL 617 258 1787
TLF
TLX
EML
COM 07

CELIS LEOPOLDO DR
DPT DE ASTRONOMIA
UNIVERSIDAD CATOLICA
CASILLA 6014
SANTIAGO
CHILE
TEL 2 552 2375
TLF 2 552 5692
TLX 440 001
EML
COM 25,45

CELLINO ALBERTO DR
OSS ASTRONOMICO
DI TORINO
I 10025 PINO TORINESE
ITALY
TEL 11 84 2040
TLF
TLX 213236 TOASTR
EML CELLINO@ASTTO2 INFN IT
COM 15

CELNIKIER LUDWIK DR
OBSERVATOIRE DE PARIS
SECTION DE MEUDON
F 92195 MEUDON PPL CDX
FRANCE
TEL 1 45 07 7410
TLF
TLX 201571
EML
COM

CENTRELLA JOAN M DR
DPT OF PHYSICS
DREXEL UNIVERSITY
PHILADELPHIA PA 19104
USA
TEL 215 895 2715
TLF
TLX
EML
COM·

CEPA JORDI DR
INST DE ASTROFISICA
DE CANARIAS
OBS DEL TEIDE
E 38200 LA LAGUNA
SPAIN
TEL 22 26 2211
TLF 22 26 3005
TLX 92640 IAC E
EML IAC JCN
COM 28

CEPLECHA ZDENEK DR
ASTRONOMICAL INSTITUTE
CZECH ACADEMY OF SCIENCES
ONDREJOV OBSERVATORY
CS 251 65 ONDREJOV
CZECHOSLOVAKIA
TEL 204 85201
TLF 204 85314
TLX 121579
EML
COM 15,22C

CEPPATELLI GUIDO DR
OSS ASTROFISICO
DI ARCETRI
LARGO E FERMI 5
I 50125 FIRENZE
ITALY
TEL 55 437 8540
TLF 55 43 5939
TLX 572268
EML
COM 12

CERNICHARO JOSE DR
CTR ASTRON DE YEBES
OAN
APO 148
E 19080 GUADALAJARA
SPAIN
TEL· 11 29 0311
TLF
TLX· 58279508 IRAM
EML
COM 34,40

CERRONI PRISCILLA ÓR
IAS
REPARTO DI PLANETOLOGIA
VIA DELL'UNIVERSITA 11
I 00185 ROMA
ITALY
TEL 6 49 56951
TLF 6 44 54969
TLX 610261 CNR FRA
EML BITNET CARUSI@IRMUNISA
COM 15

CERRUTI-SOLA MONICA
OSS ASTROFISICO
DI ARCETRI
LARGO E FERMI 5
I 50125 FIRENZE
ITALY
TEL 55 437 8540
TLF 55 43 5939
TLX 572268 ARCETR I
EML
COM 34

CERSOSIMO JUAN CARLOS DR
IAR
CC 5
1894 VILLA ELISA (BS AS)
ARGENTINA
TEL 21 4 3793
TLX
EML
TLF
COM 34

CESARSKY CATHERINE J DR
CEA CEN
DPHG/SAP
BP 2
F 91191 GIF/YVETTE CDX
FRANCE
TEL 1 69 08 3912
TLF
TLX 690860
EML
COM 34,33C,48C

CESARSKY DIEGO A DR
INSTITUT D'ASTROPHYSIQUE
98BIS BD ARAGO
F 75014 PARIS
FRANCE
TEL 1 43 20 1425
TLF 1 43 29 8673
TLX
EML
COM 33,34

CESTER BRUNO PROF
OAT
BOX SUCC TRIESTE 5
VIA TIEPOLO 11
I 34131 TRIESTE
ITALY
TEL 40 79 3921*221
TLF
TLX 461137 OAT I
EML
COM 26,42,45

CEVOLANI GIORDANO
FISBAT
CNR
VIA CASTAGNOLI 1
I 40126 BOLOGNA
ITALY
TEL 51 23 9593/94
TLF
TLX 511350
EML
COM 22

CHA DU JIN
PYONGYANG ASTRON OBS
ACADEMY OF SCIENCES DPRK
TAESONG DISTRICT
PYONGYANG
KOREA DPR
TEL
TLF
TLX
EML
COM 08

CHA GI UNG
PYONGYANG ASTRON OBS
ACADEMY OF SCIENCES DPRK
TAESONG DISTRICT
PYONGYANG
KOREA DPR
TEL
TLF
TLX
EML
COM

CHAFFEE FREDERIC H DR
MULT MIRROR TELESCOPE OBS
UNIVERSITY OF ARIZONA
TUCSON AZ 85721
USA
TEL
TLF
TLX
EML
COM

CHAISSON ERIC J PROF
STSCI
HOMEWOOD CAMPUS
3700 SAN MARTIN DR
BALTIMORE MD 21218
USA
TEL 301 338 4757
TLF
TLX 6849101
EML
COM 51

CHAKRABARTI SANDIP K DR
TIFR
HOMI BHABHA RD
COLABA
BOMBAY 400 005
INDIA
TEL 22 215 2971*305
TLF 22 215 2110
TLX 011 83009 TIFR IN
EML CHAKRABA@TIFRVAX BITNET
COM 48

CHAKRABORTY DEO K DR
DPT OF PHYSICS
UNIVERSITY OF RAVISHANKAR
RAIPUR 492 010
INDIA
TEL 27 064
TLF
TLX
EML
COM 48

CHALABAEV ALMAS DR
OHP
F 04870 S MICHEL OBS
FRANCE
TEL 92 76 6368
TLF
TLX 410690 OHP F
EML CHALABAEV@FRONI51
COM 28

CHAMARAUX PIERRE DR
OBSERVATOIRE DE PARIS
SECTION DE MEUDON
F 92195 MEUDON PPL CDX
FRANCE
TEL 1 45 07 7594
TLF
TLX 270912
EML
COM 28

CHAMBE GILBERT
OBSERVATOIRE DE PARIS
SECTION DE MEUDON
DASOP
F 92195 MEUDON PPL CDX
FRANCE
TEL 1 45 34 7793
TLF
TLX
EML
COM 10,12

CHAMBERLAIN JOSEPH M DR
ADLER PLANETARIUM
1300 S LAKE SHORE DR
CHICAGO IL 60605
USA
TEL 312 322 0325
TLF
TLX
EML
COM 08,31,46

CHAMBERLAIN JOSEPH W PROF
DPT SPACE PHYS & ASTRON
RICE UNIVERSITY
HOUSTON TX 77001
USA
TEL 713 527 8101
TLF
TLX 556457
EML
COM 16,21,49

CHAMBLISS CARLSON R DR
DPT PHYSICAL SCIENCES
KUTZTOWN UNIVERSITY
KUTZTOWN PA 19530
USA
TEL 215 683 4439
TLF
TLX
EML
COM 42

CHAN KWING LAM
APPLIED RESEARCH CORP
8201 CORPORATE DR
SUITE 920
LANDOVER MD 20785
USA
TEL 301 459 8442
TLF
TLX
EML
COM 12,35,36,40

CHAN ROBERTO DR
OBSERVATORIO NACIONAL
RUA GL J CRISTINO 77
SAO CRISTOVAO
20921 RIO DE JANEIRO RJ
BRAZIL
TEL 21 580 7313
TLF 21 580 0332
TLX 021 21 288
EML USERCHAN@LNCCVM
COM 35

CHANDRA SUBHASH
MIS PHILIPS LABS
345 SCARBOROUGH RD
BRIAR CLIFF NY 10510
USA
TEL
TLF
TLX
EML
COM

CHANDRA SURESH DR
DPT OF PHYSICS
UNIVERSITY OF GORAKHPUR
GORAKHPUR 273 009
INDIA
TEL
TLF
TLX
EML
COM 10,34

CHANDRASEKHAR S PROF
ASTROPHYS & SPACE RES LAB
UNIVERSITY OF CHICAGO
933 E 56TH ST
CHICAGO IL 60637
USA
TEL 312 962 7860
TLF
TLX
EML
COM 35,48

CHANDRASEKHAR T DR
PHYSICAL RESEARCH LAB
NAVRANGPURA
AHMEDABAD 380 009
INDIA
TEL 272 46 2129
TLF 272 44 5292
TLX 121397 PRL IN
EML
COM 15

CHANG KYONGAE DR
DPT PHYSICS & OPTICS
CHUNGJU UNIVERSITY
CHUNGJU
KOREA R
TEL
TLF
TLX
EML
COM 47

CHANMUGAM GANESAR PROF
DPT PHYSICS & ASTRONOMY
LOUISIANA STATE UNIV
BATON ROUGE LA 70803 4001
USA
TEL 504 388 6894
TLF
TLX
EML
COM 42

CHAPMAN CLARK R DR
PLANETARY SCIENCE INST
2030 E SPEEDWAY
SUITE 201
TUCSON AZ 85719
USA
TEL 602 881 0332
TLF
TLX
EML
COM 15,16

CHAPMAN GARY A DR
DPT PHYSICS & ASTRONOMY
SAN FERNANDO OBSERVATORY
CALIFORNIA STATE UNIV
NORTHRIDGE CA 91330
USA
TEL 818 885 2775
TLF
TLX
EML
COM 10,12

CHAPMAN JESSICA DR
CSIRO
AUSTRALIAN TELESCOPE
BOX 296
EPPING NSW 2121
AUSTRALIA
TEL 2 868 0222
TLF 2 868 0310
TLX 26230 ASTRO AA
EML JCHAPMAN@ATNF CSIRO AU
COM 33

CHAPMAN ROBERT D DR
LYNDON JOHNSON SPACE CTR
CODE PO311
HOUSTON TX 77058
USA
TEL
TLF
TLX
EML
COM 15,42,44

CHAPRONT JEAN DR
BUREAU DES LONGITUDES
77 AVE DENFERT ROCHEREAU
F 75014 PARIS
FRANCE
TEL 1 40 51 2271
TLF
TLX
EML
COM 04C,07C

CHAPRONT-TOUZE MICHELLE
BUREAU DES LONGITUDES
77 AVE DENFERT ROCHEREAU
F 75014 PARIS
FRANCE
TEL 1 40 51 2266
TLF
TLX
EML
COM 04,07,20

CHARLES PHILIP ALLAN
DPT OF ASTROPHYSICS
UNIVERSITY OF OXFORD
SOUTH PARKS RD
OXFORD OX1 3RQ
UK
TEL 865 511 336*506
TLF
TLX 83295 NUCLOX
EML
COM 44

CHASSEFIERE ERIC
SERVICE D'AERONOMIE
BP 3
F 91371 VERRIERES BUISSON
FRANCE
TEL 1 64 47 4211
TLF
TLX 602400
EML
COM 49

CHATTERJEE TAPAN K DR
FAC DE CIENCIAS
UNIVERISDAD A PUEBLA
APDO POSTAL 1152
72000 PUEBLA PUE
MEXICO
TEL 9122 33 0455
TLF 9122 44 8947
TLX
EML
COM 28

CHAU WAI Y PROF
DPT OF PHYSICS
QUEEN'S UNIVERSITY
KINGSTON ON K7L 3N6
CANADA
TEL 613 547 3526
TLF 613 545 6463
TLX
EML
COM

CHAUBEY UMA SHANKAR DR
UTTAR PRADESH STATE
OBSERVATORY
PO MANORA PEAK 263 129
NAINITAL 263 129
INDIA
TEL 2136/2583
TLF
TLX
EML
COM 42

CHAVARRIA-K CARLOS
INSTITUTO DE ASTRONOMIA
UNAM
APDO POSTAL 70-264
04510 MEXICO DF
MEXICO
TEL
TLF
TLX
EML
COM 37

CHAVIRA ENRIQUE SR
INAOE
TONANTZINTLAZ
APDO POSTAL 216 y 51
72000 PUEBLA PUE
MEXICO
TEL 47-05-00
TLF
TLX
EML
COM 27

CHECHETKIN VALERIJ M DR
INST OF APPLIED MATHS
ACADEMY OF SCIENCES
MIUSSKAJA SQ 4
125047 MOSCOW
RUSSIA
TEL 251 37 39
TLF
TLX
EML
COM 35,48

CHELLI ALAIN
INSTITUTO DE ASTRONOMIA
UNAM
APDO POSTAL 70-264
04510 MEXICO DF
MEXICO
TEL 548-3712/5306
TLF
TLX 1760155
EML
COM 09

CHEN BIAO
PURPLE MOUNTAIN OBSERV
CAS
NANJING
CHINA PR
TEL 25 46700
TLF
TLX 34144 PMONJ CN
EML
COM 10,12,49

CHEN DAO-HAN
PURPLE MOUNTAIN OBSERV
CAS
NANJING
CHINA PR
TEL 25 31096
TLF
TLX 34144 PMONJ CN
EML
COM 15,16

CHEN HONGSHENG
BEIJING ASTRONOMICAL OBS
CAS
W SUBURB
BEIJING 100080
CHINA PR
TEL
TLF
TLX 9053
EML
COM 40

CHEN JIAN-SHENG
BEIJING ASTRONOMICAL OBS
CAS
W SUBURB
BEIJING 100080
CHINA PR
TEL
TLF
TLX 22040 BAOAS CN
EML
COM 28C,47

CHEN KWAN-YU PROF
DPT OF ASTRONOMY
UNIVERSITY OF FLORIDA
211 SSRB
GAINESVILLE FL 32611
USA
TEL 904 392 2055
TLF
TLX
EML
COM 42

CHEN PEISHENG
YUNNAN OBSERVATORY
CAS
BOX 110
KUNMING 72946 YUNNAN
CHINA PR
TEL 871 2035
TLF
TLX
EML
COM 36

CHEN XIAO-ZHONG
BEIJING PLANETARIUM
138 XI WAI ST
BEIJING
CHINA PR
TEL 1 89 3003
TLF
TLX
EML
COM

CHEN ZHEN
PURPLE MOUNTAIN OBSERV
CAS
NANJING
CHINA PR
TEL 25 46700
TLF
TLX 34144 PMONJ CN
EML
COM 07,26,33

CHEN ZHENCHENG
BEIJING ASTRONOMICAL OBS
CAS
W SUBURB
BEIJING 100080
CHINA PR
TEL 1 28 1698
TLF
TLX 9053
EML
COM 10,28

CHEN ZUN-GUI
BEIJING PLANETARIUM
138 XI WAI ST
BEIJING
CHINA PR
TEL 1 89 3003
TLF
TLX
EML
COM 41

CHENG CHUNG-CHIEH DR
NAVAL RESEARCH LABORATORY
CODE 4175CC
4555 OVERLOOK AVE SW
WASHINGTON DC 20375 5000
USA
TEL 202 767 2350
TLF
TLX
EML
COM 10,12

CHENG FU-HUA
CENTER FOR ASTROPHYSICS
UNIV SCIENCE & TECHNOLOGY
HEFEI 230026 ANHUI
CHINA PR
TEL 551 33 1134*526
TLF
TLX 90028 USTC CN
EML
COM 47

CHENG FU-ZHEN
ASTROPHYSICS DIVISION
UNIV SCIENCE & TECHNOLOGY
HEFEI 230026 ANHUI
CHINA PR
TEL 551 33 1134*987
TLF
TLX 90028 USTC CN
EML
COM 47

CHEREDNICHENKO V I DR
KIEV POLYTECHNICAL INST
252056 KIEV
UKRAINE
TEL
TLF
EML
COM 15
TLX

CHEREPASHCHUK A M PROF
STERNBERG STATE ASTR INST
UNIVERSITETSKIJ PROSP 13
119899 MOSCOW
RUSSIA
TEL 139-38-38
TLF
TLX
EML
COM 27,42,45

CHERNEGA N A A DR
ASTRONOMICAL OBSERVATORY
KIEV SYAYE UNIVERSITY
OBSERVATORNAYA UL 3
252053 KIEV
UKRAINE
TEL 26 2391
TLF
TLX
EML
COM 08

CHERNOV GENNADIJ DR
IZMIRAN
ACADEMY OF SCIENCES
142092 TROITSK
RUSSIA
TEL 334 0902
TLF 334 0124
TLX 412623 SCSTP SU
EML
COM 10

CHERNYKH N S DR
CRIMEAN ASTROPHYS OBS
UKRAINIAN ACAD OF SCIENCE
NAUCHNY
334413 CRIMEA
UKRAINE
TEL 43 2945
TLF
TLX
EML
COM 15,20

CHERTOPRUD V E DR
HYDROMETEOROLOGICAL CTR
123376 MOSCOW
RUSSIA
TEL
TLF
TLX
EML
COM 10

CHEVALIER CLAUDE DR
OHP
F 04870 S MICHEL OBS
FRANCE ·
TEL 92 76 6368
TLF
TLX 410690 OHP F
EML
COM 35

CHEVALIER ROGER A DR
UNIVERSITY STATION
UNIVERSITY OF VIRGINIA
BOX 3818
CHARLOTTESVILLE VA 22903
USA
TEL 804 924 4889
TLF
TLX
EML
COM 34

CHE-BOHNENSTENGEL ANNE
SUELZBRACKRING 39A
D 2050 HAMBURG 80
GERMANY
TEL 40 723 8550
TLF
EML
COM
TLX

CHIAN ABRAHAM CHIAN-LONG
INPE
CP 515
12200 S JOSE DOS CAMPOS
BRAZIL
TEL 123 22 9977
TLF 123 21 8743
TLX 011-33530
EML
COM 48

CHIKADA YOSHIHIRO DR
NOBEYAMA RADIO OBS
NAOJ
MINAMIMAKI MURA
NAGANO 384 13
JAPAN
TEL 267-98-2831
TLF
TLX 3329005 TAONRO
EML
COM 40

CHINCARINI GUIDO L DR
OSS ASTRONOMICO DI MILANO
VIA E BIANCHI 46
I 22055 MERATE
ITALY
TEL 59 6412
TLX
EML
TLF
COM 28,47

CHINI ROLF
MPI FUER RADIOASTRONOMIE
AUF DEM HUEGEL 69
D 5300 BONN 1
GERMANY
TEL
TLX 886440
EML
TLF
COM 34,40

CHIO CHOL ZONG
PYONGYANG ASTRON OBS
ACADEMY OF SCIENCES DPRK
TAESONG DISTRICT
PYONGYANG
KOREA DPR
TEL
TLF
TLX
EML
COM 20

CHIOSI CESARE S DR
DPT DI ASTRONOMIA
UNIVERSITA DI PADOVA
VIC DELL OSSERVATORIO 5
I 35122 PADOVA
ITALY
TEL 49 66 1499
TLF
TLX 430176 UNDAPU I
EML
COM 35VP,37

CHISTYAKOV VLADIMIR E DR
USSURIISK SOLAR STATION
PRIMORSKY KRAY
692533 GORNOTAEZHNOE
RUSSIA
TEL 91121 USSURIISK
TLX 213954 SOLNZE
EML
TLF
COM 12

CHITRE DATTAKUMAR M DR
COMPUTER SCIENCES CORP
SYSTEM SCIENCES DIV
8728 COLESVILLE RD
SILVER SPRING MD 20910
USA
TEL
TLF
TLX
EML
COM 38C,47,49

CHITRE SHASHIKUMAR M DR
TIFR
HOMI BHABHA RD
COLABA
BOMBAY 400 005
INDIA
TEL 22 219 111
TLF
TLX 011-3009 TIFR IN
EML
COM 35,48,49

CHIU HONG-YEE DR
MUDD BLDG RM 828
COLUMBIA UNIVERSITY
NEW YORK NY 10027
USA
TEL
TLF
TLX
EML
COM 35

CHIU LIANG-TAI GEORGE
IBM
THOMAS J WATSON RES CTR
BOX 218
YORKTOWN HEIGHTS NY 10598
USA
TEL 914 945 2436
TLF
TLX
EML
COM 24

CHIUDERI CLAUDIO PROF
DPT DI ASTRONOMIA
UNIVERSITA DI FIRENZE
LARGO E FERMI 5
I 50125 FIRENZE
ITALY
TEL 55 27521
TLF 55 22 0039
TLX 572268 ARCETR I
EML
COM

CHIUDERI-DRAGO FRANCA PR
OSS ASTROFISICO
DI ARCETRI
LARGO E FERMI 5
I 50125 FIRENZE
ITALY
·TEL 55 437 8540
TLF 55 43 5939
TLX 572268 ARCETR I
EML
COM 10

CHIUEH TZIHONG DR
INST PHYSICS & ASTRONOMY
NTL CENTRAL UNIVERSITY
CHUNG LI
TAIWAN
CHINA R
TEL 3 422 7151*5341
TLF 3 425 1175
TLX
EML CHIUEH@PHYAST DNET NCU EDU TW
COM 10

CHIUMIENTO GIUSEPPE
OSS ASTRONOMICO DI TORINO
ST OSSERVATORIO 20
I 10025 PINO TORINESE
ITALY
TEL 11 84 1067
TLX 213236 TOASTR I
EML
TLF
COM 08,19

CHKHIKVADZE IAKOB N
ABASTUMANI ASTROPHYSICAL
OBSERVATORY
GEORGIAN ACAD OF SCIENCES
383762 ABASTUMANI
GEORGIA
TEL 2-78
TLF
TLX 327409
EML
COM 35

CHLISTOVSKY FRANCA DR
OSS ASTRONOMICO DI BRERA
VIA BRERA 28
I 20121 MILANO
ITALY
TEL 2 87 4444
TLF 2 72 00 1600
TLX
EML
COM 08

CHMIELEWSKI YVES DR
OBSERVATOIRE DE GENEVE
CHEMIN DES MAILLETTES 51
CH 1290 SAUVERNY
SWITZERLAND
TEL 22 755 2611
TLF 22 755 3983
TLX 419209 OBS CH
EML igor@obs unige ch
COM

CHO SE HYUNG DR
ISSA
YOOSUNG KOON
DAEJEON 305 348
KOREA R
TEL 042 851 1281
TLF 042 861 5610
TLX
EML
COM 40

CHOCHOL DRAHOMIR
ASTRONOMICAL INSTITUTE
SLOVAK ACADEMY SCIENCES
CS 059 60 TATRANSKA LOMNI
CZECHOSLOVAKIA
TEL 969 96 7866/7/8
TLF 969 96 7656
TLX 78277
EML
COM 42,44

CHOE SEUNG URN DR
DPT OF EARTH SCIENCE & ED
SEOUL NTL UNIVERSITY
SINLIM DONG GWANG GU
SEOUL 151 742
KOREA R
TEL
TLF
TLX
EML
COM

CHOI KYU-HONG
DPT ASTRON & METEOROLOGY
YONSEI UNIVERSITY
SUDAEMUN
SEOUL 120 749
KOREA R
TEL 02-392-0131
TLF
TLX
EML
COM 07,42

CHOI WON CHOL
PYONGYANG ASTRON OBS
ACADEMY OF SCIENCES DPRK
TEASONG DISTRICT
PYONGYANG
KOREA DPR
TEL 5-3134, 5-3239
TLF
TLX
EML
COM

CHOLLET FERNAND DR
OBSERVATOIRE DE PARIS
61 AVE OBSERVATOIRE
F 75014 PARIS
FRANCE
TEL 1 40 51 2205
TLF
TLX 270776 OBS F
EML
COM 04,08

CHOLONIEWSSKI JACEK DR
ASTRONOMICAL OBSERVATORY
WARSAW UNIVERSITY
AL UJAZDOWSKIE 4
PL 00 478 WARSAW
POLAND
TEL 294011
TLF
TLX 817063 OAUW PL
EML JCH@PLWAUW61 BITNET
COM

CHOPINET MARGUERITE DR
57 RUE THIERS

F 92100 BOULOGNE
FRANCE
TEL 1 47 61 1144
TLF
TLX
EML
COM 34

CHOU CHIH-KANG DR
ASTRONOMY,
NTL CENTRAL UNIVERSITY
CHUNG LI
TAIWAN ROC
CHINA R
TEL 3 425 1175
TLF
TLX
EML
COM 28,49

CHOU DEAN-YI DR
PHYSICS DPT
NTL TSING HUA UNIVERSITY
HSIN CHU 300043
CHINA R
TEL
TLF
TLX
EML BITNET DYCHOU@TWNCTU01
COM

CHOU KYONG CHOL PROF
DPT ASTRONOMY SPACE SCI
KYUNG HEE UNIVERSITY
YONG IN KUN
KYUNGGI DO 170 73
KOREA R
TEL 966-0061/5
TLF
TLX
EML
COM 51

CHRISTENSEN PER R DR
NIELS BOHR INSTITUTE
BLEGDAMSVEJ 17
DK 2100 COPENHAGEN O
DENMARK
TEL 31 42 1616
TLF 31 38 9157
TLX 15216 NBI DK
EML PERREX@NBIVAX NBI DK
COM

CHRISTENSEN-DALSGAARD J
INST OF PHYSICS & ASTRON
UNIVERSITY OF AARHUS
NY MUNKEGADE
DK 8000 AARHUS C
DENMARK
TEL 86 12 8899
TLF 86 20 2711
TLX 64767 AAUSCI DK
EML
COM 12,27C,35

CHRISTIAN CAROL ANN
CTR FOR EUV ASTROPHYSICS
UNIVERSITY OF CALIFORNIA
2150 KITTREDGE ST
BERKELEY CA 94720
USA
TEL
TLF
TLX
EML
COM 37

CHRISTIANSEN WAYNE A
DPT PHYSICS & ASTRONOMY
UNIVERSITY NORTH CAROLINA
204 PHILLIPS HALL 039A
CHAPEL HILL NC 27514
USA
TEL 919 962 3011
TLF
TLX
EML
COM 40

CHRISTIANSEN WILBUR PROF
RMB 436
MAC'S REEF RD
VIA BUNGENDORE 2621
AUSTRALIA
TLF
TEL 62 30 3287
TLX
EML
COM 40

CHRISTODOULOU DMITRIS DR
STEWARD OBSERVATORY
UNIVERSITY OF ARIZONA
TUCSON AZ 85721
USA
TEL 602 621 2288
TLF 602 621 1532
TLX 467175
EML BITNET PHCHRI@LSUMVS
COM 33

CHRISTY JAMES WALTER DR
HUGHES AIRCRAFT CO
1720 W NIONA PL
TUCSON AZ 85704
USA
TEL 602 297 1377
TLF
TLX
EML
COM 09,24,45

CHRISTY ROBERT F DR
CALTECH
MS 105 24
PALOMAR OBS
PASADENA CA 91125
USA
TEL 213 795 6811
TLF
TLX
EML
COM 27,35

CHU HAN-SHU PROF
PURPLE MOUNTAIN OBSERV
CAS
NANJING
CHINA PR
TEL 25 30 1096
TLF
TLX 34144 PMONJ CN
EML
COM 40

CHU YAOQUAN
ASTROPHYSICS DIVISION
UNIV SCIENCE & TECHNOLOGY
HEFEI 230026 ANHUI
CHINA PR
TEL 551 33 1134
TLF 551 33 1760
TLX 90028 USTC CN
EML
COM 05,28,47

CHU YOU-HUA
DPT OF ASTRONOMY
UNIVERSITY OF ILLINOIS
1011 W SPRINGFIELD AVE
URBANA IL 61801
USA
TEL 217 333 5535
TLF
TLX
EML
COM 34

CHUBB TALBOT A DR
5023 N 38TH ST
ARLINGTON VA 22207
USA
TEL
TLF
TLX
EML
COM 44,48

CHUGAI NIKOLAI N DR
INST OF ASTRONOMY
ACADEMY OF SCIENCES
PYATNITSKAYA UL 48
109017 MOSCOW
RUSSIA
TEL 231 21 29
TLF
TLX 411576 ASCON SU
EML
COM 28,36

CHUGAJNOV P F DR
CRIMEAN ASTROPHYS OBS
UKRAINIAN ACAD OF SCIENCE
NAUCHNY
334413 CRIMEA
UKRAINE
TEL 43 2945
TLF
TLX
EML
COM 25,27

CHUN MUN-SUK DR
DPT ASTRON & METEOROLOGY
YONSEI UNIVERSITY
SUDAEMUN
SEOUL 120 749
KOREA R
TEL 2-392-0131
TLF
TLX
EML
COM 37

CHUPP EDWARD L DR
DPT OF PHYSICS
UNIV OF NEW HAMPSHIRE
DEMERITT HALL
DURHAM NH 03824
USA
TEL 603 862 2750
TLF
TLX 950030
EML SOLMAX CHUPP
COM 10,48

CHURCHWELL EDWARD B DR
WASHBURN OBSERVATORY
UNIVERSITY OF WISCONSIN
475 N CHARTER ST
MADISON WI 53706
USA
TEL 608 262 7857
TLF
TLX 265452 UOFWISC MDS
EML
COM 33,34

CHURMS JOSEPH
SAAO
BOX 9
OBSERVATORY 7935
SOUTH AFRICA
TEL 21 47 0025
TLF
TLX
EML
COM 20,24

CHUVAEV K K DR
CRIMEAN ASTROPHYS OBS
UKRAINIAN ACAD OF SCIENCE
NAUCHNY
334413 CRIMEA
UKRAINE
TEL 43 2945
TLF
TLX
EML
COM 28

CHVOJKOVA WOYK E DR
ASTRONOMICAL INSTITUTE
CZECH ACADEMY SCIENCES
BUDECSKA 6
CS 120 23 PRAHA 2
CZECHOSLOVAKIA
TEL
TLF
TLX
EML
COM 12

CIATTI FRANCO DR
OSSERVATORIO ASTROFISICO
VIA DELL OSSERVATORIO 8
I 36012 ASIAGO
ITALY
TEL 42 46 2665
TLF
TLX 430110 SETURIST
EML
COM

CID PALACIOS RAFAEL PROF
DPT DE ASTRONOMIA
UNIVERSIDAD DE ZARAGOZA
E 50009 ZARAGOZA
SPAIN
TEL 35 7011
TLF
TLX
EML
COM 07

CILLIE G G PROF
4 MINSERIE ST
STELLENBOSCH 7600
SOUTH AFRICA
TEL 02231-3515
TLF
EML
COM 42
TLX

CIURLA TADEUSZ
ASTRONOMICAL INSTITUTE
WROCLAW UNIVERSITY
UL KOPERNIKA 11
PL 51 622 WROCLAW
POLAND
TEL
TLF
TLX
EML
COM 33

CLAIREMIDI JACQUES DR
OBSERVATOIRE DE BESANCON
41BIS AVE OBSERVATOIRE
F 25000 BESANCON CDX
FRANCE
TEL 81 66 6900
TLF
TLX 361144 F
EML CLAIREMI@FROBES51
COM 15,16,21

CLARIA JUAN DR
OBSERVATORIO ASTRONOMICO
DE CORDOBA
LAPRIDA 854
5000 CORDOBA
ARGENTINA
TEL 51 40613
TLF
TLX 51822 BUCOR
EML
COM 37C,42,45,47

CLARK ALFRED JR PROF
DPT MECH & AEROSPACE SCI
UNIVERSITY OF ROCHESTER
ROCHESTER NY 14627
USA
TEL
TLF
TLX
EML
COM

CLARK BARRY G DR
NRAO
VLA
BOX 0
SOCORRO NM 87801 0387
USA
TEL 505 772 4011
TLF
TLX 9109881710
EML
COM 40

CLARK DAVID H DR
SCIENCE DIV
SCI & ENGINEER RES COUNCI
NORTH STAR AVE
SWINDON SN2 1ET
UK
TEL 793 262 22
TLF
TLX 449466
EML
COM 40

CLARK FRANK OLIVER DR
DPT PHYSICS & ASTRONOMY
UNIVERSITY OF KENTUCKY
LEXINGTON KY 40506
USA
TEL 606 257 3376
TLF
TLX
EML
COM 34,40

CLARK GEORGE W PROF
CENTER FOR SPACE RESEARCH
MIT RM 37 611
BOX 165
CAMBRIDGE MA 02139
USA
TEL 617 253 5842
TLF
TLX
EML
COM 44,48C

CLARK THOMAS A DR
NASA/GSFC
CODE 974
GREENBELT MD 20771
USA
TEL 301 286 5957
TLF
TLX
EML
COM 51

CLARK THOMAS ALAN DR
DPT OF PHYSICS
UNIVERSITY OF CALGARY
2500 UNIVERSITY DR NW
CALGARY AB T2N 1N4
CANADA
TEL 403 284 5392
TLF 403 289 3331
TLX
EML
COM 12,44

CLARKE DAVID DR
DPT OF ASTRONOMY
UNIVERSITY OF GLASGOW
GLASGOW G12 8QQ
UK
TEL 413 398 855
TLF
TLX 778421
EML
COM 09

CLARKE JOHN T
NASA/GSFC
CODE 681
HUBBLE SPACE TELESCOPE
GREENBELT MD 20771
USA
TEL 301 286 5781
TLF
TLX 710-828-9716
EML
COM

CLARKE THOMAS R DR
MCLAUGHLIN PLANETARIUM
ROYAL ONTARIO MUSEUM
100 QUEENS PARK CRESCENT
TORONTO ON M5S 2C6
CANADA
TEL 416 998 8551
TLF
TLX
EML
COM

CLAUSEN JENS VIGGO LEKTOR
COPENHAGEN UNIVERSITY OBS
BRORFELDEVEJ 23
DK 4340 TOLLOSE
DENMARK
TEL 53 48 8195
TLF 58 48 8755
TLX 44155 DANAST
EML
COM 42

CLAUSSEN MARK J DR
NAVAL RESEARCH LABORATORY
CODE 4210
4555 OVERLOOK AVE SW
WASHINGTON DC 20375 5000
USA
TEL 202 767 0670
TLF
TLX
EML
COM

CLAVEL JEAN
ESA/ESTEC
SSD
BOX 299
NL 2200 AG NOORDWIJK
NETHERLANDS
TEL
TLF
TLX
EML
COM 28

CLAYTON DONALD D PROF
DPT PHYSICS & ASTRONOMY
CLEMSON UNIVERSITY
CLEMSON SC 29634 1911
USA
TEL 803 656 5299
TLF
TLX
EML BITNET Clayton@Clemson
COM

CLAYTON GEOFFREY C DR
NASA HEADQUARTERS
CODE EZ
600 INDEPENDENCE AVE SW
WASHINGTON DC 20546
USA
TEL 202 453 1469
TLF
TLX
EML SPAN POLLUX GCLAYTON
COM 15

CLAYTON ROBERT N DR
ENRICO FERMI INSTITUTE
UNIVERSITY OF CHICAGO
5640 S ELLIS AVE
CHICAGO IL 60637
USA
TEL 312 702 7777
TLF
TLX
EML
COM 15

CLEGG PETER E DR
ROYAL GREENWICH OBS
MADINGLEY RD
CAMBRIDGE CB3 0EZ
UK
TEL 223 33 7548
TLF
TLX 817297 ASTRON G
EML SPAN rlvad cavad oper
COM

CLEGG ROBIN E S DR
ROYAL GREENWICH OBS
MADINGLEY RD
CAMBRIDGE CB3 0EZ
UK
TEL
TLF
TLX
EML
COM 34

CLEMENS DAN P DR
DPT OF ASTRONOMY
BOSTON UNIVERSITY
725 COMMONWEALTH AVE
BOSTON MA 02215
USA
TEL 617 353 6140
TLF
TLX
EML
COM 33,40

CLEMENT MAURICE J PROF
DPT OF ASTRONOMY
UNIVERSITY OF TORONTO
60 ST GEORGE ST
TORONTO ON M5S 1A7
CANADA
TEL 416 978 4833
TLF 416 978 3921
TLX 069 86766
EML
COM

CLIFTON KENNETH ST
NASA/MSFC
CODE ES 63
HUNTSVILLE AL 35812
USA
TEL 205 453 2305
TLF
TLX 594416
EML
COM 22

CLIMENHAGA JOHN L PROF
DPT OF PHYSICS
UNIVERSITY OF VICTORIA
BOX 1700
VICTORIA BC V8W 2Y2
CANADA
TEL 604 721 7741
TLF 604 721 7715
TLX
EML
COM 29

CLINE THOMAS L DR
NASA/GSFC
CODE 661
LHEA
GREENBELT MD 20771
USA
TEL 301 286 8375
TLF
TLX 89675 NASCOM GBLT
EML SPAN 6197 CLINE
COM

CLIVER EDWARD W
A F GEOPHYSICS LABORATORY
SPACE PHYSICS DIV
HANSCOM AFB
BEDFORD MA 01731
USA
TEL 617 861 3975
TLF
TLX 928123 AFGL HANSCOM
EML
COM 10

CLUBE S V M DR
DPT OF ASTROPHYSICS
UNIVERSITY OF OXFORD
SOUTH PARKS RD
OXFORD OX1 3RQ
UK
TEL 865 511 336
TLF
TLX
EML
COM 15,22,24,33

CLUTTON-BROCK MARTIN DR
DPT OF MATHS & ASTRONOMY
UNIVERSITY OF MANITOBA
WINNIPEG MB R3T 2N2
CANADA
TEL 204 261 9255
TLF
TLX
EML
COM

COCHRAN ANITA L DR
ASTRONOMY DPT
UNIVERSITY OF TEXAS
RLM 15 308
AUSTIN TX 78712 1083
USA
TEL 512 471 1471
TLF
TLX 910 874 1351
EML BITNET as1j720@uta3081
COM 15,16

COCHRAN WILLIAM DAVID DR
ASTRONOMY DPT
UNIVERSITY OF TEXAS
RLM 15 308
AUSTIN TX 78712 1083
USA
TEL 512 471 4461
TLF
TLX
EML
COM 15,30

COCKE WILLIAM JOHN PROF
STEWARD OBSERVATORY
UNIVERSITY OF ARIZONA
TUCSON AZ 85721
USA
TEL 602 621 6540
TLF
TLX
EML
COM 47

CODE ARTHUR D
WASHBURN OBSERVATORY
UNIVERSITY OF WISCONSIN
475 N CHARTER ST
MADISON WI 53706
USA
TEL 608 262 9594
TLF
TLX
EML
COM 34,44

CODINA LANDABERRY SAYD J
OBSERVATORIO NACIONAL
RUA GL BRUCE 586
SAO CRISTOVAO
20921 RIO DE JANEIRO RJ
BRAZIL
TEL 21 580 7313*267
TLF 21 580 0332
TLX 21288
EML
COM 46

CODINA VIDAL J M DR
FABRA OBSERVATORY
GRAN VIA DE LOS CORTES
CATALANES 679
E 08013 BARCELONA
SPAIN
TEL 3 245 4766
TLF
TLX
EML
COM

COELHO BALSA MARIO C DR
RUA TRINDADE COELHO 21
20 DTO
P 3000 COIMBRA
PORTUGAL
TEL
TLF
TLX
EML
COM

COFFEEN DAVID L DR
BOX 151
HASTINGS HUDSON NY 10706
USA
TEL 914 478 2594
TLF
TLX
EML
COM

COFFEY HELEN E MS
NOAA
NGDC E/GC2
325 BROADWAY
BOULDER CO 80303
USA
TEL 303 497 6223
TLF
TLX 592811 NOAA MASC BDR
EML
COM 10

COGAN BRUCE C DR
MOUNT STROMLO & SIDING
SPRING OBSERVATORIES
PRIVATE BAG
WODEN PO ACT 2606
AUSTRALIA
TEL 62 88 1111
TLF
TLX 62270
EML
COM 05

COHEN JEFFREY M DR
DPT OF PHYSICS
UNIV OF PENNSYLVANIA
PHILADELPHIA PA 19104
USA
TEL
TLF
TLX
EML
COM 35,47,48

COHEN JUDITH DR
CALTECH
MS 105 24
PALOMAR OBS
PASADENA CA 91125
USA
TEL 818 356 4005
TLF
TLX
EML JLC@DEIMOS CALTECH EDU
COM

COHEN LEON PROF
DPT OF PHYSICS
HUNTERS COLLEGE
695 PARK AVE
NEW YORK NY 10021
USA
TEL 212 570 5696
TLF
TLX
EML
COM

COHEN MARSHALL H PROF
CALTECH
MS 105 24
PALOMAR OBS
PASADENA CA 91125
USA
TEL 213 356 4000
TLF
TLX 675425
EML
COM 34,40

COHEN MARTIN DR
RADIO ASTRONOMY LAB
UNIVERSITY OF CALIFORNIA
601 CAMPBELL HALL
BERKELEY CA 94720
USA
TEL 415 642 2833
TLF.
TLX· 820181 UCB AST RALUD
EML.
COM 27

COHEN RAYMOND J DR
NRAL
JODRELL BANK
MACCLESFIELD SK11 9DL
UK
TEL 477 713 21
TLF
TLX 36149
EML
COM 40

COHEN RICHARD S
INST FOR SPACE STUDIES

2880 BROADWAY
NEW YORK NY 10025
USA
TEL 212 678 5611
TLF
TLX
EML
COM 09,33,40

COHEN ROSS D DR
CASS
UCSD
C 011
LA JOLLA CA 92093 0216
USA
TEL 619 534 2664
TLF
TLX
EML
COM 28,47

COHN HALDAN N
ASTRONOMY DPT
INDIANA UNIVERSITY
SWAIN WEST 319
BLOOMINGTON IN 47405
USA
TEL 812 335 4174
TLF
TLX
EML
COM

COLBURN DAVID S DR
1944 WAVERLEY STREET
PALO ALTO CA 94301
USA
TEL
TLF
TLX
EML
COM

COLE TREVOR WILLIAM PROF
SCHOOL OF ELECTRICAL ENG
UNIVERSITY OF SYDNEY
SYDNEY NSW 2006
AUSTRALIA
TEL 2 692 2682
TLF
TLX
EML
COM 40

COLEMAN PAUL HENRY DR
KAPTEYN ASTRONOMICAL INST
BOX 800
NL 9700 AV GRONINGEN
NETHERLANDS
TEL 50 63 064
TLX 53572 STARS NL
EML BITNET GRUFF@HGRRUG5
TLF
COM 40

COLES PETER DR
ASTRONOMY UNIT
QUEEN MARY/WESTFIELD COLL
MILE END RD
LONDON E1 4AS
UK
TEL 1 719 755 481
TLF 1 819 819 587
TLX 893750
EML PCOLES@UK AC QMW STARLINK
COM 47

COLGATE STIRLING A DR
LOS ALAMOS SCIENTIFIC LAB
MS 8275
THEORETICAL DIV
LOS ALAMOS NM 87545
USA
TEL 505 667 2897
TLF
TLX
EML
COM

COLIN JACQUES DR
OBSERVATOIRE DE BORDEAUX
BP 89
F 33270 FLOIRAC
FRANCE
TEL 56 86 4330
TLF 56 40 4251
TLX
EML
COM 28,37,33

COLLADOS MANUEL DR
INST DE ASTROFISICA
DE CANARIAS
OBS DEL TEIDE
E 38200 LA LAGUNA
SPAIN
TEL 22 26 2211
TLF
TLX 92640
EML
COM 10,12

COLLINS GEORGE W II PROF
DPT OF ASTRONOMY
OHIO STATE UNIVERSITY
174 W 18TH AVE
COLUMBUS OH 43210 1106
USA
TEL 614 422 5467
TLF
TLX
EML
COM 42

COLLIN-SOUFFRIN SUZY DR
INSTITUT D'ASTROPHYSIQUE
98BIS BD ARAGO
F 75014 PARIS
FRANCE
TEL 1 43 20 1425
TLF 1 43 29 8673
TLX
EML
COM 34,48

COLOMB FERNANDO R DR
IAR.

CC 5
1894 VILLA ELISA (BS AS)
ARGENTINA
TEL 21 4 3793
TLF
TLX 18052 CICYT AR
EML
COM 34,40C,51C

COLOMBO G PROF DR
IST MECCANICA APPL
UNIVERSITA DI PADOVA
VIA F MARZOLO 9
I 35122 PADOVA
ITALY
TEL 49 66 1499
TLF
TLX
EML
COM 16

COLUZZI REGINA DR
OAR
VIA DEL PARCO MELLINI 84
I 00136 ROMA
ITALY
TEL 6 34 7056
TLF
TLX 626226 OA ROMA I
EML
COM 05,29,45

COMA JUAN CARLOS
REAL INST Y OBSERVATORIO
DE LA ARMADA
CECILIO PUJAZON S/N
E 11110 SAN FERNANDO
SPAIN
TEL
TLF
TLX
EML
COM 04

COMBES FRANCOISE DR
OBSERVATOIRE DE PARIS
SECTION DE MEUDON
DEMIRM
F 92195 MEUDON PPL CDX
FRANCE
TEL 1 45 07 7898
TLF
TLX 270912
EML
COM

COMBES MICHEL
OBSERVATOIRE DE PARIS
SECTION DE MEUDON
F 92195 MEUDON PPL CDX
FRANCE
TEL 1 45 07 7691
TLF
TLX 201571
EML
COM

COMBI MICHAEL R DR
SPACE PHYSICS RESEARCH LB
UNIVERSITY OF MICHIGAN
2455 HAYWARD ST
ANN ARBOR MI 48109 2143
USA
TEL 313 764 7226
TLF 313 747 3083
TLX
EML COMBI@SPRLC SPL UMICH EDU
COM 15,16

COMINS NEIL FRANCIS
DPT PHYSICS & ASTRONOMY
UNIVERSITY OF MAINE
BENNETT HALL
ORONO ME 04469
USA
TEL 207 581 1037
TLF
TLX
EML
COM 33

COMORETTO GIOVANNI
OSS ASTROFISICO
DI ARCETRI
LARGO E FERMI 5
I 50125 FIRENZE
ITALY
TEL 55 437 8540
TLF 55 43 5939
TLX 572268 ARCETR I
EML
COM

COMTE GEORGES DR
OBSERVATOIRE DE MARSEILLE
2 PLACE LE VERRIER
F 13248 MARSEILLE CDX 04
FRANCE
TEL 91 95 9088
TLF
TLX 420241 F
EML EARN "COMTE@FROMRS51"
COM 28

CONCONI PAOLO DR
OSS ASTRONOMICO DI MILANO
VIA E BIANCHI 46
I 22055 MERATE
ITALY
TEL 59 2035
TLX
EML
TLF
COM

CONDON JAMES J DR
NRAO
EDGEMONT RD
CHARLOTTESVILLE VA 22903
USA
TEL 804 296 0211
TLF
TLX 910-997-0174
EML
COM 40,47,48

CONKLIN EDWARD K
FORTH INC
111 N SEPULVEDA BLVD 300
MANHATTAN BEACH CA 90266
USA
TEL
TLF
TLX 275182 FORT UR
EML
COM 40

CONNES JANINE DR
CIRCE
BP 63
F 91400 ORSAY
FRANCE
TEL 1 69 28 7675
TLF
TLX 692166
EML
COM 16

CONNES PIERRE DR
SERVICE D'AERONOMIE
BP 3
F 91371 VERRIERES BUISSON
FRANCE
TEL 1 64 47 4277
TLF
TLX 602400
EML
COM 24,51

CONNOLLY LEO PAUL
DPT OF PHYSICS
CALIFORNIA STATE UNIV
5500 UNIVERSITY PARKWAY
SAN BERNARDINO CA 92407
USA
TEL 714 880 5400
TLF
TLX
EML
COM 25,27,35

CONTADAKIS MICHAEL E DR
DPT GEODESY & SURVEYING
UNIVERSITY THESSALONIKI
UNIV BOX 503
GR 540 06 THESSALONIKI
GREECE
TEL 31 99 2693
TLF
TLX 412181 AUTH GR
EML
COM 27

CONTI PETER S DR
JILA
UNIVERSITY OF COLORADO
BOX 440
BOULDER CO 80309 0440
USA
TEL 303 492 7789
TLF 303 492 5235
TLX 755842 JILA
EML 33833 PCONTI/PCONTI@JILA
COM 29C,36

CONTOPOULOS GEORGE PROF
DPT OF ASTRONOMY
NTL UNIVERSITY OF ATHENS
PANEPISTIMIOPOLIS
GR 157 71 ATHENS
GREECE
TEL 1 724 3211
TLF
TLX
EML
COM 07,28,33

CONWAY ROBIN G DR
NRAL
`JODRELL BANK
MACCLESFIELD SK11 9DL
UK
TEL 477 713 21
TLF
TLX 36149
EML
COM 40

COOK ALAN H PROF
DPT PHYS/UNIV CAMRIDGE
THE MASTER'S LODGE
SELWYN COLLEGE
CAMBRIDGE CB3 9DQ
UK
TEL 223 623 81*29
TLF
TLX 81292 CAVLAB
EML
COM 07,14

COOK JOHN W
8032 SLEEPY VIEW LN
SPRINGFIELD VA 22153
USA
TEL 202 767 2161
TLF
EML
COM 10,12
TLX

COOK KEM HOLLAND DR
LAWRENCE LIVERMORE LAB
L 401
BOX 808
LIVERMORE CA 94550
USA
TEL 510 423 4634
TLF 510 294 5512
TLX
EML KCOOK@IMAGER LLNL GOV
COM 28

COOKE B A DR
DPT OF PHYSICS
UNIVERSITY OF LEICESTER
X-RAY ASTRONOMY GROUP
LEICESTER LE1 7RH
UK
TEL 533 554 455*188
TLF
TLX 341664
EML
COM

COOKE JOHN ALAN
ROYAL OBSERVATORY

BLACKFORD HILL
EDINBURGH EH9 3HJ
UK
TEL 316 673 221
TLF
TLX 72383 ROEDIN G
EML
COM 09

CORADINI ANGIOLETTA
IAS
REPARTO DI PLANETOLOGIA
VIA DELL'UNIVERSITA 11
I 00185 ROMA
ITALY
TEL 6 49 56951
TLF
TLX 680489 CNR FRA
EML
COM

CORBALLY CHRISTOPHER
STEWARD OBSERVATORY
UNIVERSITY OF ARIZONA
VATICAN OBS RESEARCH GP
TUCSON AZ 85721
USA
TEL 602 621 3225
TLF
TLX 467175
EML
COM 29,45C

CORBELLI EDVIGE DR
OSS ASTROFISICO
DI ARCETRI
LARGO E FERMI 5
I 50125 FIRENZE
ITALY
TEL 55 437 8540
TLF 55 43 5939
TLX
EML EDVIGE●SISIFO ARCETRI ASTRO IT
COM 34

CORBIN THOMAS ELBERT DR
US NAVAL OBSERVATORY
ASTROMETRY DPT
34 & MASSACHUSETTS AVE NW
WASHINGTON DC 20392 5100
USA
TEL 202 653 1557
TLF
TLX 710-8221970
EML
COM 08C,24

CORDES JAMES M
CORNELL UNIVERSITY
SPACE SCIENCES BLDG
ITHACA NY 14853
USA
TEL 607 256 3734
TLX 932 458
EML
TLF
COM 40

CORDOVA FRANCE A D
DPT OF ASTRONOMY
PENNSYLVANIA STATE UNIV
525 DAVEY LAB
UNIVERSITY PARK PA 16802
USA
TEL
TLF
TLX
EML
COM 44

CORLISS C H DR
FOREST HILLS LABORATORY
2955 ALBEMARLE STREET NW
WASHINGTON DC 20008
USA
TEL 202 362 6085
TLF
TLX
EML
COM 14

CORNEJO ALEJANDRO A DR
INAOE
TONANTZINTLAZ
APDO POSTAL 216 y 51
72000 PUEBLA PUE
MEXICO
TEL 47-05-00
TLF
TLX
EML
COM 09,41

CORNIDE MANUEL
DPT DE ASTROFISICA
FAC DE FISICA
UNIVERSIDAD COMPLUTENSE
E 28040 MADRID
SPAIN
TEL 1 449 5316
TLF
TLX 47273 FFUC
EML
COM 29

CORNILLE MARGUERITE DR
OBSERVATOIRE DE PARIS
SECTION DE MEUDON
UA 812
F 92195 MEUDON PPL CDX
FRANCE
TEL 1 45 07 7455
TLF
TLX 201571
EML BITNET FRORS31
COM 14

CORREIA EMILIA DR
CRAAE/ESCOLA POLITECNICA
UNIVERSIDADE DE SAO PAULO
CP 8174
05508 SAO PAULO SP
BRAZIL
TEL 11 815 6289
TLF 11 815 4272
TLX 1180127 INPE BR
EML ECORREIA●BRUSP ANSP BR
COM 10

CORWIN HAROLD G JR
CALTECH
MS 100 22
IPAC
PASADENA CA 91125
USA
TEL 818 584 2902
TLF 818 584 2902
TLX 675429
EML info%ipac●hamlet
COM 28

COSMOVICI BATALLI C DR
IAS
CNR
I 00044 FRACASTI
ITALY
TEL 6 942 3801
TLF
TLX 610261 I
EML
COM 15

COSTA EDGARDO DR
DPT DE ASTRONOMIA
UNIVERSIDAD DE CHILE
CASILLA 36 D
SANTIAGO
CHILE
TEL 2 229 4101
TLF 2 271 2799
TLX 440 001
EML
COM 08,33

COSTA ENRICO
IAS
CNR
CP 67
I 00044 FRASCATI
ITALY
TEL 6 942 5655
TLF 6 941 6847
TLX 610261 I
EML
COM

COSTA JOAQUIM EDUARDO DR
CRAAE/PTE ESCOLA POLI USP
CP 8174 CEP
01051 SAO PAULO SP
BRAZIL
TEL 11 815 6289
TLF 11 815 4272
TLX 1180127 INPE BR
EML JERCOSTA●BRUSP ANSP BR
COM 10

COSTA VICTOR DR
INST ASTROFISICA
DE ANDALUCIA APD 2144
C/SANCHO PANZA S/N
E 18080 GRANADA
SPAIN
TEL 58 12 1311
TLF
TLX 78573 IAGG E
EML
COM

COSTAIN CECIL C DR
DIVISION OF PHYSICS
NTL RESEARCH COUNCIL
100 SUSSEX DR
OTTAWA ON K1A 0R6
CANADA
TEL 613 993 6060
TLF 613 952 6602
TLX
EML
COM

COSTERO RAFAEL
INSTITUTO DE ASTRONOMIA
UNAM
APDO POSTAL 70-264
04510 MEXICO DF
MEXICO
TEL 548-5305
TLF
TLX 1760155 CICME
EML
COM 34,50C

COTTON WILLIAM D Jr
NRAO

EDGEMONT RD
CHARLOTTESVILLE VA 22901
USA
TEL 804 296 0319
TLF
TLX 5105875482
EML
COM 40

COTTRELL PETER LEDSAM
DPT OF PHYSICS
UNIVERSITY OF CANTERBURY
PRIVATE BAG
CHRISTCHURCH 1
NEW ZEALAND
TEL   03 482 009
TLF
TLX   4144 NZ
EML
COM 29

COUCH WARRICK DR
AAO
BOX 296
EPPING NSW 2121
AUSTRALIA
TEL 2 868 1666
TLX 23999 AAOSYD AA
EML
TLF
COM 28

COULSON IAIN M DR
JOINT ASTRONOMY CENTER
UKIRT
665 KOMOHANA ST
HILO HI 96720
USA
TEL
TLF
TLX
EML
COM 27

COUNSELMAN CHARLES C PROF
DPT OF EARTH PLANET & SCI
MIT RM 54 620
BOX 165
CAMBRIDGE MA 02139
USA
TEL 617 253 7902
TLF
TLX 921473 MIT CAM
EML
COM 07,08,16

COUPER HEATHER MISS
COLLINS COTTAGE
LOWER RD
LOOSLEY ROW
BUCKS HP17 0PF
UK
TEL
TLF
TLX
EML
COM 46,51

COUPINOT GERARD DR
OBS MIDI PYRENEES
9 R PONT DE LA MOULETTE
F 65200 BAGNERES BIGORRE
FRANCE
TEL 62 95 1969
TLF 62 95 1969
TLX
EML
COM

COURTES G PROF
LAS
LES TROIS LUCS
TRAVERSE DU SIPHON
F 13012 MARSEILLE CDX4
FRANCE
TEL 91 05 5900
TLF 91 66 1855
TLX 410584
EML
COM 28,33,34,44

COURVOISIER THIERRY J-L
OBSERVATOIRE DE GENEVE
CHEMIN DES MAILLETTES 51
CH 1290 SAUVERNY
SWITZERLAND
TEL 22 755 2611
TLF 22 755 3983
TLX 419209 OBS CH
EML COURVOIS@OBS UNIGE CH
COM 28,44

COUSINS A W J DR
SAAO
BOX 9
OBSERVATORY 7935
SOUTH AFRICA
TEL 021-47-0025
TLF
TLX 5720309
EML
COM 25

COUTEAU PAUL PROF
OCA OBSERV DE NICE
BP 139
F 06003 NICE CDX
FRANCE
TEL 93 89 0420
TLF
TLX 460004 OBSNICE F
EML
COM 26

COUTREZ RAYMOND A J PROF
6 RUE EGIDE BOUVIER
B 1160 BRUSSELS
BELGIUM
TEL
TLF
TLX
EML
COM 40

COUTTS-CLEMENT CHRISTINE
DPT OF ASTRONOMY
UNIVERSITY OF TORONTO
60 ST GEORGE ST
TORONTO ON M5S 1A7
CANADA
TEL 416 978 5186
TLF 416 978 3921
TLX 069 86766
EML
COM 27

COUTURIER PIERRE
INSU
77 AVE DENFERT ROCHEREAU
F 75014 PARIS
FRANCE
TEL 1 40 51 2004
TLF 1 40 51 2101
TLX
EML IAPOBS COUTURIER
COM 49

COVINGTON ARTHUR E
131 COLLEGE STREET
KINGSTON ON K7L 4L7
CANADA
TEL
TLF
EML
COM 10,40
TLX

COWAN JOHN J DR
DPT PHYSICS & ASTRONOMY
UNIVERSITY OF OKLAHOMA
NORMAN OK 73019
USA
TEL 405 325 3961
TLF
TLX
EML
COM 35

COWIE LENNOX LAUCHLAN DR
INSTITUTE FOR ASTRONOMY
UNIVERSITY OF HAWAII
2680 WOODLAWN DR
HONOLULU HI 96822
USA
TEL
TLF
TLX
EML
COM 34,48

COWLEY ANNE P DR
DPT OF PHYSICS
ARIZONA STATE UNIVERSITY
TEMPE AZ 85287
USA
TEL 602 965 2919
TLF
TLX
EML
COM 29,42,45

COWLEY CHARLES R PROF
DPT OF ASTRONOMY
UNIVERSITY OF MICHIGAN
DENNISON BLDG
ANN ARBOR MI 48109 1090
USA
TEL 313 764 3437
TLF
TLX 810-2236056
EML
COM 29,36

COWSIK RAMANATH
TIFR
HOMI BHABHA RD
COLABA
BOMBAY 400 005
INDIA
TEL 22 219 111
TLF
TLX
EML
COM 28,48

COX ARTHUR N DR
LOS ALAMOS NATIONAL LAB
BOX 1663
LOS ALAMOS NM 87545
USA
TEL 505 667 7648
TLF
TLX 910-988-1773
EML
COM 12,27

COX DONALD P PROF
DPT OF ASTRONOMY
UNIVERSITY OF WISCONSIN
475 N CHARTER ST
MADISON WI 53706
USA
TEL 608 262 5916
TLF
TLX
EML
COM 34

COX PIERRE DR
OBSERVATOIRE DE MARSEILLE
2 PLACE LE VERRIER
F 13248 MARSEILLE CDX 04
FRANCE
TEL 91 95 9088
TLF 91 62 1190
TLX
EML COX@FROMRS51
COM 34

COYNE GEORGE V DR
DIRETTORE DELLA SPECOLA
VATICANA
I 00120 VATICAN CITY
VATICAN CITY STATE
TEL 6 698 3411
TLF
TLX 2020 VAT OBS VA
EML
COM 25,34,50

CRABTREE DENNIS DR
HERZBERG INST ASTROPHYS
DOMINION ASTROPHYS OBS
5071 W SAANICH RD
VICTORIA BC V8X 4M6
CANADA
TEL 604 388 0025
TLF 604 363 0045
TLX 049 7295
EML
COM

CRAIG IAN JONATHAN D DR
DPT APPLIED MATHEMATICS
UNIVERSITY OF WAIKATO
HAMILTON
NEW ZEALAND
TEL 62889
TLF
TLX
EML
COM 12

CRAINE ERIC RICHARD DR
WESTERN RESEARCH CO
5061 W CAMINO DE GIRASOL
TUCSON AZ 85745
USA
TEL 602 743 7377
TLX
EML
TLF
COM

CRAM LAWRENCE EDWARD PROF
SCHOOL OF PHYSICS
UNIVERSITY OF SYDNEY
SYDNEY NSW 2006
AUSTRALIA
TEL 2 692 2537
TLF 2 660 2903
TLX AA26169 UNISYD
EML LC@PHYSICS SU OZ AU
COM 12,36VP

CRAMER NEIL DR
SCHOOL OF PHYSICS
UNIVERSITY OF SYDNEY
SYDNEY NSW 2006
AUSTRALIA
TEL 2 692 3162
TLF 2 660 2903
TLX AA 26169
EML
COM 10,12,49

CRAMPTON DAVID DR
HERZBERG INST ASTROPHYS
DOMINION ASTROPHYS OBS
5071 W SAANICH RD
VICTORIA BC V8X 4M6
CANADA
TEL 604 388 3900
TLF 604 363 0045
TLX 049 7295
EML
COM 30,33,45

CRANE PATRICK C
NRAO
BOX 0
SOCORRO NM 87801 0387
USA
TEL 505 772 4011
TLX 910-988-1710
EML
TLF
COM 40,47

CRANE PHILIPPE
ESO
KARL-SCHWARZSCHILD-STR 2
D 8046 GARCHING MUENCHEN
GERMANY
TEL 89 792 098
TLF
TLX 528-28222 EO D
EML
COM 28,34,47

CRANNELL CAROL JO DR
NASA/GSFC
CODE 682
GREENBELT MD 2077I
USA
TEL 301 286 5007
TLF
TLX 89675
EML
COM 10,44

CRAWFORD DAVID L DR
KITT PEAK NTL OBS
BOX 26732
950 N CHERRY AVE
TUCSON AZ 85726 6732
USA
TEL 602 325 9346
TLF 602 325 9360
TLX 1561401 AURA UT
EML
COM 09,25,33,45,50C

CRAWFORD IAN ANDREW DR
DPT PHYSICS & ASTRONOMY
UNIVERSITY COLLEGE LONDON
GOWER ST
LONDON WC1E 6BT
UK
TEL 1 713 877 050*3498
TLF 1 713 807 145
TLX 28722
EML IAC@STAR UCL AC UK
COM 34

CREZE MICHEL DR
OBS DE STRASBOURG
11 RUE UNIVERSITE
F 67000 STRASBOURG
FRANCE
TEL 88 35 8216
TLF 88 25 0160
TLX OBSBES 361144 F
EML creze@simbad u-strasbg fr
COM 05C,24,33

CRIFO FRANCOISE DR
OBSERVATOIRE DE PARIS
SECTION DE MEUDON
DEPEG
F 92195 MEUDON PPL CDX
FRANCE
TEL 1 45 07 7834
TLF
TLX 201571
EML
COM 08,24

CRISTALDI SALVATORE DR
OSS ASTROFISICO
CITTA UNIVERSITARIA
VIA ARTALE ALAGONA 75
I 95126 CATANIA
ITALY
TEL 33 0734
TLF
TLX 970359 ASTRCT I
EML
COM 42

CRISTESCU CORNELIA G DR
ASTRONOMICAL OBSERVATORY
CUTITUL DE ARGINT 5
BOX 28
R 75212 BUCAREST
RUMANIA
TEL 23 68 92
TLF
TLX
EML
COM 15,20

CRISTIANI STEFANO DR
DPT DI ASTRONOMIA
UNIVERSITA DI PADOVA
VIC DELL OSSERVATORIO 5
I 35122 PADOVA
ITALY
TEL 49 66 1499
TLF
TLX 432071
EML SPAN 39003 CRISTIANI
COM 47

CRIVELLARI LUCIO
OAT
BOX SUCC TRIESTE 5
VIA TIEPOLO 11
I 34131 TRIESTE
ITALY
TEL 40 79 3221
TLF
TLX 461137 OAT I
EML
COM 36

CROCKER DEBORAH ANN DR
DPT OF PHYSICS & ASTRON
UNIVERSITY OF ALABAMA
BOX 870324
TUSCALOOSA AL 35387 0324
USA
TEL   205 348 3758
TLF
TLX
EML   CROCK@OKRA ASTR UA EDU
COM

CROOM DAVID L DR
RUTHERFORD APPLETON LAB
SPACE & ASTROPHYSICS DIV
BLDG R25/R68
CHILTON DIDCOT OX11 0QX
UK
TEL   235 219 00
TLF
TLX   83159
EML
COM   40

CROVISIER JACQUES
OBSERVATOIRE DE PARIS
SECTION DE MEUDON
F 92195 MEUDON PPL CDX
FRANCE
TEL   1 45 07 7599
TLF
TLX   270912
EML
COM   15,34,40

CRUIKSHANK DALE P DR
NASA AMES RESEARCH CTR
MS 245 6
MOFFETT FIELD CA 94035
USA
TEL
TLF
TLX
EML
COM   15,16C

CRUISE ADRIAN MICHAEL DR
RUTHERFORD APPLETON LAB
CHILBOLTON OBSERVATORY
DITTON PARK
SLOUGH SL3 9JX
UK
TEL
TLF
TLX
EML
COM   48

CRUTCHER RICHARD M DR
DPT OF ASTRONOMY
UNIVERSITY OF ILLINOIS
1011 W SPRINGFIELD AVE
URBANA IL 61801
USA
TEL   217 333 9581
TLF
TLX
EML
COM

CRUVELLIER PAUL E DR
LAS
TRAVERSE DU SIPHON
LES TROIS LUCS
F 13012 MARSEILLE
FRANCE
TEL   91 05 5900
TLF   91 66 1855
TLX   420584
EML
COM   34

CRUZ-GONZALEZ IRENE
INSTITUTO DE ASTRONOMIA
UNAM
APDO POSTAL 70-264
04510 MEXICO DF
MEXICO
TEL   905-548-5306
TLF
TLX
EML
COM

CSADA IMRE K DR
KONKOLY OBSERVATORY
THEGE U 13/17
BOX 67
H 1121 BUDAPEST
HUNGARY
TEL   1 66 426
TLF
TLX   227460
EML
COM   10

CUBARSI RAFAEL DR
DPT MATEMATICA APLICADA
UNIV POLITEC DE CATALUNYA
BOX 30002
E 08080 BARCELONA
SPAIN
TEL   3 401 6799
TLF   3 401 6801
TLX   52821 UPC-E
EML   MATRCM@MAT UPC ES
COM   33

CUDABACK DAVID D DR
RADIO ASTRONOMY LAB
UNIVERSITY OF CALIFORNIA
601 CAMPBELL HALL
BERKELEY CA 94720
USA
TEL   415 642 5724
TLF
TLX   820181 UCB AST
EML
COM   34,40

CUDWORTH KYLE MCCABE DR
YERKES OBSERVATORY
UNIVERSITY OF CHICAGO
BOX 258
WILLIAMS BAY WI 53191
USA
TEL   414 245 5555
TLF
TLX
EML
COM   24,33,37

CUFFEY J MR
DPT OF EARTH SCI & ASTRON
NEW MEXICO STATE UNIV
UNIVERSITY PARK NM 88001
USA
TEL
TLF
TLX
EML
COM   37

CUGIER HENRYK DR
ASTRONOMICAL INSTITUTE
WROCLAW UNIVERSITY
UL KOPERNIKA 11
PL 51 622 WROCLAW
POLAND
TEL   48 24 34
TLF
TLX   712791 UWR PL
EML
COM   36

CUGNON PIERRE DR
OBSERVATOIRE ROYAL DE
BELGIQUE
AVE CIRCULAIRE 3
B 1180 BRUSSELS
BELGIUM
TEL   2 375 2484
TLF
TLX   21565 OBSBEL
EML
COM   34

CUGUSI LEONINO DR
DPT DI SCIENZE FISICHE
UNIVERSITA DI CAGLIARI
VIA OSPEDALE 72
I 09100 CAGLIARI
ITALY
TEL   70 66 4770
TLF
TLX
EML
COM

CUI DOU-XING
CHANGCHUN ARTIFICIAL
SATELLITE OBSERVATORY
BOX 1067
CHANGCHUN
CHINA PR
TEL   42859
TLF
TLX   2421 CHANGCHUN
EML
COM   07

CUI ZHEN-HUA
BEIJING PLANETARIUM
138 XI WAI ST
BEIJING
CHINA PR
TEL· 1 89 3003
TLF
TLX
EML
COM   41,46

CULHANE LEONARD PROF
MULLARD SPACE SCIENCE LAB
UNIVERSITY COLLEGE LONDON
HOLMBURY ST MARY
DORKING SURREY RH5 6NT
UK
TEL   306 702 92
TLF
TLX   859185 UCMSSL G
EML
COM   10,44,48

CULLUM MARTIN DR
ESO
KARL SCHWARZSCHILD STR 2
D 8046 GARCHING MUENCHEN
GERMANY
TEL   89 320 060
TLF   89 320 2362
TLX   5282820 EO D
EML   EARN cullum@dgaeso51
COM   09C

CULVER ROGER BRUCE DR
DPT OF PHYISCS
COLORADO STATE UNIVERSITY
FORT COLLINS CO 80523
USA
TEL   303 491 6206
TLF
TLX   9109309000 ENGRCSUFT
EML
COM   26

CUNTZ MANFRED DR
JILA
UNIVERSITY OF COLORADO
BOX 440
BOULDER CO 80309 0440
USA
TEL
TLF
TLX
EML
COM   36

CUNY YVETTE J DR
OBSERVATOIRE DE PARIS
SECTION DE MEUDON
F 92195 MEUDON PPL CDX
FRANCE
TEL   1 45 07 7838
TLF
TLX
EML
COM   36C

CUPERMAN SAMI PROF
DPT OF PHYSICS & ASTRON
TEL AVIV UNIVERSITY
RAMAT AVIV
TEL AVIV 69978
ISRAEL
TEL   3 420 21/425 697
TLF
TLX   342171 VERSY IL
EML
COM   33,49

CURIR ANNA
OSS ASTRONOMICO DI TORINO
ST OSSERVATORIO 20
I 10025 PINO TORINESE
ITALY
TEL   11 84 1067
TLX   213236 TO ASTR I
EML
TLF   .
COM   48

CURRIE DOUGLAS G DR
ASTRONOMY PROGRAM
UNIVERSITY OF MARYLAND
COLLEGE PARK MD 20742
USA
TEL   301 454 3405
TLF
TLX
EML
COM   09,51

CUTISPOTO GIUSEPPE DR
OSS ASTROFISICO
CITTA UNIVERSITARIA
VIA A DORIA 6
I 95125 CATANIA
ITALY
TEL   95 33 7241
TLF   95 33 0592
TLX   970359 ASTRCT I
EML   40297 GIUSEPPE
COM   27

CUYPERS JAN DR
OBSERVATOIRE ROYAL DE
BELGIQUE
AVE CIRCULAIRE 3
B 1180 BRUSSELS
BELGIUM
TEL   2 375 2484
TLF
TLX   21565
EML
COM   15,25,27

CZERNY BOZENA DR
COPERNICUS ASTRON CENTER
POLISH ACAD OF SCIENCES
UL BARTYCKA 18
PL 00 716 WARSAW
POLAND
TEL   WARSAW 41 00 41
TLF
TLX
EML
COM

CZERNY MICHAL DR
DPT OF ASTRONOMY
UNIVERSITY OF LEICESTER
UNIVERSITY RD
LEICESTER LE1 7RH
UK
TEL   533 522 073
TLF
TLX   347250 LEICUN
EML
COM

CZYZAK STANLEY J DR
DPT OF PHYSICS
OHIO STATE UNIVERSITY
174 W 18TH AVE
COLUMBUS OH 43210
USA
TEL   614 422 6543
TLF
TLX
EML
COM   14,34

DA COSTA ANTONIO A DR
COMPLEXO INTERDISCIPLINAR
COMPLEXO INTERDISCIPLANAR
INSTITUTO SUPERIOR TECNIC
P 1096 LISBOA CODEX
PORTUGAL
TEL   351 1 3524 303
TLF   351 1 3524 372
TLX
EML
COM   48

DA COSTA GARY STEWART DR
AAO
BOX 296
EPPING NSW 2121
AUSTRALIA
TEL   2 868 1666
TLX   123999 AAOSYD
EML   DACOSTA%aaιcpp oz au@uunet uu
TLF
COM   37C

DA COSTA JOSE MARQUES DR
INPE
CP 515
12200 S JOSE DOS CAMPOS
BRAZIL
TEL   123 22 9977
TLF   123 21 8743
TLX   1133530 INPEBR
EML
COM   48

DA COSTA NICOLAI L-A
OBSERVATORIO NACIONAL
RUA GL BRUCE 586
SAO CRISTOVAO
20921 RIO DE JANEIRO RJ
BRAZIL
TEL   21 580 7181
TLF   21 580 0332
TLX
EML   LNDC@LNCCVM
COM   28,30C

DA ROCHA VIEIRA E DR
INSTITUTO DE FISICA
UFRGS
CP 15051
90000 PORTO ALEGRE RS
BRAZIL
TEL   512 21 7666
TLF
TLX   0511055 UFRSBR
EML
COM
     \

DA SILVA A V C S
OBSERVATORIO ASTRONOMICO
UNIVERSIDADE SANTA CLARA
P 3000 COIMBRA
PORTUGAL
TEL
TLF
TLX
EML
COM

DA SILVA LICIO DR
OBSERVATORIO NACIONAL
RUA GL BRUCE 586
SAO CRISTOVAO
20921 RIO DE JANEIRO RJ
BRAZIL
TEL   21 580 7313
TLF   21 580 0332
TLX   21288
EML
COM

DACHS JOACHIM PROF DR
ASTRONOMISCHES INSTITUT
RUHR-UNIVERSITAET
POSTFACH 102148
D 4630 BOCHUM 1
GERMANY
TEL   234 70 03454
TLF
TLX   825860
EML
COM   21,25

DADAEV ALEKSANDR N DR
PULKOVO OBSERVATORY
ACADEMY OF SCIENCES
10 KUTUZOV QUAY
196140 ST PETERSBURG
RUSSIA
TEL
TLF
TLX
EML
COM   26,42

DADHICH NARESH DR
IUCAA
POST BAG 4
GANESHKHIND PUNE 411 007
INDIA
TEL 21 233 6415
TLF 21 233 5760
TLX 0145658 GMRT IN
EML ROOT@IUCAA ERNET IN
COM 47,48

DADIC ZARKO DR
ZAVOD ZA POVIJEST
ZNANOSTI JAZU
ANTE KOVACICA 5
YU 41000 ZAGREB
YUGOSLAVIA
TEL 41 440 124
TLF
TLX
EML
COM 41

DAGKESAMANSKY RUSTAM D DR
LEBEDEV PHYSICAL INST
ACADEMY OF SCIENCES
LENINSKY PROSPEKT 53
117924 MOSCOW
RUSSIA
TEL 135 14 29
TLF
TLX 411479 NEOD SU
EML
COM 40

DAHN CONARD CURTIS DR
US NAVAL OBSERVATORY
FLAGSTAFF STATION
BOX 1149
FLAGSTAFF AZ 86002
USA
TEL 602 779 5132
TLF
TLX
EML
COM 24,25,34

DAIGNE GERARD
OBSERVATOIRE DE BORDEAUX
BP 89
F 33270 FLOIRAC
FRANCE
TEL 56 86 4330
TLF 56 40 4251
TLX
EML
COM 51

DAINTREE EDWARD J DR
NRAL
JODRELL BANK
MACCLESFIELD SK11 9DL
UK
TEL 477 713 21
TLX 36149
EML·
TLF
COM 40

DAISHIDO TSUNEAKI PROF
DPT OF SCIENCE
WASEDA UNIVERSITY
SHINJUKU KU TOKYO 160
JAPAN
TEL 2-203-4141
TLF
TLX 2323280 WASEDA J
EML
COM 40

DALGARNO ALEXANDER PROF
CENTER FOR ASTROPHYSICS
HCO/SAO
60 GARDEN ST
CAMBRIDGE MA 02138
USA
TEL 617 495 4403
TLF
TLX 921428
EML DALGARNO@CFA7
COM 14,34

DALLAPORTA N PROF
DPT DI ASTRONOMIA
UNIVERSITA DI PADOVA
VIC DELL OSSERVATORIO 5
I 35122 PADOVA
ITALY
TEL 49 66 1499
TLF
TLX
EML
COM

DALL'OGLIO GIORGIO DR
DPT DI FISICS
UNIVERSITA DI ROMA
PA MORO 2
I 00185 ROMA
ITALY
TEL 6 49 914271
TLF 6 495 7697
TLX 613255 INFN RO
EML VAXMA DALLAGLIO
COM 09

DALTABUIT ENRIQUE DR
INSTITUTO DE ASTRONOMIA
UNAM
APDO POSTAL 70-264
04510 MEXICO DF
MEXICO
TEL
TLF
TLX
EML
COM

DAMINELI NETO AUGUSTO DR
IAG
UNIVERSIDADE DE SAO PAULO
CP 30627
01051 SAO PAULO SP
BRAZIL
TEL
TLF
TLX
EML
COM

DAMLE S V DR
TIFR
HOMI BHABHA RD
COLABA
BOMBAY 400 005
INDIA
TEL 22 219 111
TLF
TLX 0113009 TIFR IN
EML
COM 48

DAN XHI-XIANG
SHANGHAI OBSERVATORY
CAS
80 NANDAN RD
SHANGHAI
CHINA PR
TEL 21 38 6191
TLF
TLX 33164 SHAO CN
EML
COM 09

DANBY J M ANTHONY DR
DPT OF MATHEMATICS
N CAROLINA STATE UNIV
RALEIGH NC 27695 8205
USA
TEL 919 737 3210
TLF
TLX
EML
COM 07

DANESE LUIGI DR
OSS ASTRONOMICO DI PADOVA

VIC DELL OSSERVATORIO 5
I 35122 PADOVA
ITALY
TEL 49 66 1499
TLF
TLX 430176 UNPADU-I
EML·
COM 47

DANEZIS EMMANUEL DR
DPT OF ASTROPHYSICS
NTL UNIVERSITY OF ATHENS
PANEPISTIMIOPOLIS
GR 157 83 ZOGRAFOS
GREECE
TEL 1 724 3414
TLF
TLX
EML
COM

DANFORD STEPHEN C DR
DPT OF PHYSICS & ASTRON
UNIVERSITY OF N CAROLINA
AT GREENSBORO
GREENSBORO NC 27412
USA
TEL 919 334 5669
TLF
TLX
EML DANFORD@UNCG BITNET
COM 25,27,37

DANILOV VLADIMIR M DR
DPT OF ASTRONOMY
URALSKIJ STATE UNIVERSITY
629983 SVERDLOVSK
RUSSIA
TEL 22 33 86
TLF
TLX
EML
COM

DANKS ANTHONY C DR
1315 PEACHTREE CT

BOWIE MD 20721
USA
TEL 301 249 8206
TLF
TLX
EML
COM 15,28,34

DANLY LAURA DR
STSCI
HOMEWOOD CAMPUS
3700 SAN MARTIN DR
BALTIMORE MD 21218
USA
TEL   301 338 4422
TLF   301 338 5090
TLX   6839101
EML   SCIVAX DANLY/DANLY@STSCI
COM   34

DANZIGER I JOHN DR
ESO
KARL-SCHWARZSCHILD-STR 2
D 8046 GARCHING MUENCHEN
GERMANY
TEL   89 320 060
TLX
EML
TLF   89 320 2362
COM

DAPERGOLAS A DR
ASTRONOMICAL INSTITUTE
NTL OBSERVATORY OF ATHENS
BOX 20048
GR 118 10 ATHENS
GREECE
TEL   1 346 1191
TLF
TLX
EML
COM   37

DAPPEN WERNER
OBSERVATOIRE DE PARIS
SECTION DE MEUDON
F 92195 MEUDON PPL CDX
FRANCE
TEL   1 45 34 7530
TLF
TLX
EML
COM

DARA HELEN DR
RES CENTER FOR ASTRONOMY
ACADEMY OF ATHENS
14 ANAGNOSTOPOULOU ST
GR 106 73 ATHENS
GREECE
TEL   1 361 3589
TLF
TLX
EML   EXAKAZO@GRATHUN1
COM   12

DARIUS JON DR
SCIENCE MUSEUM
LONDON SW7 2DD
UK
TEL   1 589 3456*643
TLF
EML
COM   41,46,51
TLX   21200 SCMLIB G

DAS MRINAL KANTI
DPT PHYSICS/DELHI UNIV
SRI VENKATESWARA COLLEGE
DHAULA KUAN
NEW DELHI 110 021
INDIA
TEL
TLF
TLX
EML
COM   35

DAS P K DR
INDIAN INSTITUTE OF
ASTROPHYSICS
KORAMANGALA
BANGALORE 560 034
INDIA
TEL   812 56 6585
TLF
TLX   845763 IIAB IN
EML
COM   47

DATLOWE DAYTON DR
LOCKHEED PALO ALTO RES LB
DPT 91 20 BLDG 255
3251 HANOVER ST
PALO ALTO CA 94304
USA
TEL   415 858 4074
TLF
TLX
EML
COM   10

DATTA BHASKAR DR
INDIAN INSTITUTE OF
ASTROPHYSICS
KORAMANGALA
BANGALORE 560 034
INDIA
TEL   812 56 6585/6497
TLF
TLX   845763 IIAB IN
EML
COM   47

DAUBE-KURZEMNIECE I A DR
RADIOASTROPHYSICAL OBS
LATVIAN ACAD OF SCIENCES
TURGENEVA 19
226524 RIGA
LATVIA
TEL   226796
TLF
TLX
EML
COM   37

DAUTCOURT G DR
ZNTRLINST F ASTROPHYSIK
STERNWARTE BABELSBERG
ROSA-LUXEMBURG-STR 17A
D 1502 POTSDAM
GERMANY
TEL
TLF
TLX
EML
COM   48

DAVID LAURENCE P DR
CENTER FOR ASTROPHYSICS
MS 4
60 GARDEN ST
CAMBRIDGE MA 02138
USA
TEL   617 495 7245
TLF   617 495 7356
TLX
EML   DAVID@CFA
COM

DAVIDGE TIMOTHY J DR
HERZBERG INST ASTROPHYS
DOMINION ASTROPHYS OBS
5071 W SAANICH RD
VICTORIA BC V8X 4M6
CANADA
TEL   604 363 0047
TLF
TLX
EML   DAVIDGE@DAO NRC CA
COM   28

DAVIDSEN ARTHUR FALNES DR
DPT PHYSICS & ASTRONOMY
JOHNS HOPKINS UNIVERSITY
CHARLES & 34TH ST
BALTIMORE MD 21218
USA
TEL   301 338 7370
TLF
TLX
EML
COM   28,44,48

DAVIDSON KRIS DR
SCHOOL OF PHYS & ASTRON
UNIVERSITY OF MINNESOTA
116 CHURCH ST SE
MINNEAPOLIS MN 55455
USA
TEL   612 373 7795
TLF
TLX
EML   KD@APS2 SPA UMN EDU
COM

DAVIDSON WILLIAM PROF
80 WEST CLOSE
FERNHURST
HASLEMERE
SURREY GU27 3JT
UK
TEL
TLF
TLX
EML
COM   47,48

DAVIES MERTON E MR
THE RAND CORPORATION
1700 MAIN ST
SANTA MONICA CA 90406
USA
TEL   213 393 0411
TLF
TLX
EML
COM   04,16C

DAVIES PAUL CHARLES W
SCHOOL OF PHYSICS
UNIVERSITY OF NEWCASTLE
NEWCASTLE/TYNE NE1 7RU
UK
TEL
TLF
TLX
EML
COM   47

DAVIES RODNEY D PROF
NRAL

JODRELL BANK
MACCLESFIELD SK11 9DL
UK
TEL   477 713 21
TLF   477 716 18
TLX   36149
EML
COM   28,33,34,40

DAVIES ROGER L DR
KITT PEAK NTL OBS
BOX 26732
950 N CHERRY AVE
TUCSON AZ 85726 6732
USA
TEL 602 325 9353
TLF
TLX 0666 484 AURA NOAO
EML
COM 47

DAVILA JOSEPH DR
NASA/GSFC
CODE 682 1
GREENBELT MD 20771
USA
TEL 301 286 8366
TLF
TLX
EML SOLAR IDAVILA
COM 10

DAVIS CECIL G JR
LOS ALAMOS NATIONAL LAB
MS D406
GROUP P-15
LOS ALAMOS NM 87545
USA
TEL 505 667 5908
TLF
TLX
EML
COM 35,36

DAVIS JOHN PROF
SYDNEY UNIV STELL INTER
THE PAUL WILD OBSERVATORY
BOX 94
NARRABRI NSW 2390
AUSTRALIA
TEL
TLF 67 959 270
TLX
EML
COM 09C,50C

DAVIS LEVERETT JR PROF
CALTECH
MS 405 47
PASADENA CA 91125
USA
TEL 818 356 4243
TLF
TLX
EML
COM 48

DAVIS MARC DR
ASTRONOMY DPT
UNIVERSITY OF CALIFORNIA
601 CAMPBELL HALL
BERKELEY CA 94720
USA
TEL 415 642 5156
TLF
TLX 820181 UCB AST
EML
COM 28,30,47

DAVIS MICHAEL M DR
ARECIBO OBSERVATORY
BOX 995
ARECIBO PR 00613
USA
TEL 809 878 2612
TLF
TLX 385638
EML
COM 40,47,48,51

DAVIS MORRIS S PROF
DPT PHYSICS & ASTRONOMY
UNIVERSITY NORTH CAROLINA
204 PHILLIPS HALL 039A
CHAPEL HILL NC 27514
USA
TEL 919 962 3011
TLF
TLX
EML
COM 05,07

DAVIS RICHARD J DR
NRAL
JODRELL BANK
MACCLESFIELD SK11 9DL
UK
TEL
TLX
EML
TLF
COM 40C

DAVIS ROBERT J DR
CENTER FOR ASTROPHYSICS
HCO/SAO MS 20
60 GARDEN ST
CAMBRIDGE MA 02138
USA
TEL 616 495 7335
TLF
TLX 921428 SATELLITE CAM
EML
COM 05,30,40,44

DAVIS SUMNER P DR
PHYSICS DPT
UNIVERSITY OF CALIFORNIA
BERKELEY CA 94720
USA
TEL 415 642 4857
TLF
TLX
EML
COM 14

DAVOUST EMMANUEL
OBS MIDI PYRENEES
14 AVE E BELIN
F 31400 TOULOUSE CDX
FRANCE
TEL 61 25 2101
TLF
TLX 530776
EML
COM

DAWANAS DJONI N DR
DPT OF ASTRONOMY
BANDUNG INSTITUTE OF TECH
JL GANESHA 10
BANDUNG 40132
INDONESIA
TEL 622 244 0252
TLF
TLX 28324 ITB BD
EML
COM 29

DAWE JOHN ALAN DR
ANU
SIDING SPRING OBSERVATORY
PRIVATE BAG
COONABARABRAN NSW 2857
AUSTRALIA
TEL 68 426 221
TLF
TLX 63945 AA CANOPUS
EML
COM 50,51

DE ALMEIDA AMAURY A DR
IAG
UNIVERSIDADE DE SAO PAULO
CP 30627
01065 SAO PAULO SP
BRAZIL
TEL 11 577 8599*38
TLF 11 276 3848
TLX 11 56735 IAGM BR
EML 47556 AMAURY
COM 15

DE BERGH CATHERINE DR
OBSERVATOIRE DE PARIS
SECTION DE MEUDON
DPT DE RECHERCHE SPATIALE
F 92195 MEUDON PPL CDX
FRANCE
TEL 1 45 07 7666
TLF
TLX 201571
EML MESIOA DEBERGH/DEBERGH@FRMEU5
COM 16VP

DE BIASE GIUSEPPE A DR
OAR
VIA DEL PARCO MELLINI 84
I 00136 ROMA
ITALY
TEL 6 34 7056
TLF
TLX
EML
COM

DE BOER KLAAS SJOERDS DR
ASTRONOMISCHES INSTITUT
UNIVERSITAET BONN
AUF DEM HUEGEL 71
D 5300 BONN 1
GERMANY
TEL 228 73 3656
TLF
TLX 886440
EML
COM 05,28,34

DE BRUYN A GER DR
NFRA
BOX 2
NL 7990 AA DWINGELOO
NETHERLANDS
TEL 52 19 7244
TLF
TLX 42043 SRZM NL
EML
COM 28

DE CARVALHO REINALDO DR
OBSERVATORIO NACIONAL
RUA GL BRUCE 586
SAO CRISTOVAO
20921 RIO DE JANEIRO RJ
BRAZIL
TEL 21 580 7181
TLF 21 580 0332
TLX 21288
EML 47556 REINALDO
COM 28

DE CASTRO ANGEL DR
OBS ASTRONOMICO NCL
APD 12354
ALFONSO XII-3
E 28014 MADRID
SPAIN
TEL 1 227 1935
TLF
TLX 234651 GCE
EML
COM 04

DE CASTRO ELISA
DPT DE ASTROFISICA
FAC DE FISICA
UNIVERSIDAD COMPLUTENSE
E 28040 MADRID
SPAIN
TEL 1 449 5316
TLF
TLX 47273 FFUC
EML
COM 29

DE FELICE FERNANDO DR
DPT DI FISICA G GALILEI
UNIVERSITA DI PADOVA
VIA MARZOLO 8
I 35131 PADOVA
ITALY
TEL 49 84 4278
TLF
TLX 430308 DFGGPDI
EML
COM 48

DE FREES DOUGLAS J DR
IBM ALMADEN RESEARCH CENT
DPT K 84/801
650 HARRY ROAD
SAN JOSE CA 95120-6099
USA
TEL 408 927 2854
TLF 408 927 2100
TLX
EML DEFREES@ALMVMA
COM 14

DE FREITAS PACHECO J A DR
IAG
UNIVERSIDADE DE SAO PAULO
CP 30627
01051 SAO PAULO SP
BRAZIL
TEL 21 717 3518
TLF
TLX
EML
COM

DE GRAAF T DR
INSTITUUT VOOR FONETISCHE
WETENSCHAPPEN
GROTE ROZENSTRAAT 31
NL 9712 TG GRONINGEN
NETHERLANDS
TEL
TLF
TLX
EML
COM 48

DE GRAAFF W DR
APPELGAARDE 117
NL 3992 JD HOUTEN
NETHERLANDS
TEL
TLF
TLX
EML
COM 51

DE GRAAUW TH DR
ESA/ESTEC
SSD
BOX 800
NL 2200 AG NOORDWIJK
NETHERLANDS
TEL
TLF
TLX
EML
COM

DE GREVE JEAN-PIERRE DR
IAAG
VRIJE UNIV BRUSSELS
CP 165
B 1050 BRUSSELS
BELGIUM
TEL 2 641 3498
TLF
TLX 61051 VUBCO
EML
COM 35,42

DE GROOT MART DR
ARMAGH OBSERVATORY
COLLEGE HILL
ARMAGH BT61 9DG
UK
TEL 861 52 2928
TLF
TLX 747937 ARMOBS G
EML
COM 27,29,42

DE GROOT T DR
STERREKUNDIG INSTITUTE
BOX 80000
NL 3508 TA UTRECHT
NETHERLANDS
TEL 30 53 5200
TLF
TLX 47224 ASTRO
EML
COM 40

DE JAGER CORNELIS PROF
SPACE RESEARCH LABORATORY
SRON
SORBONNELAAN 2
NL 3584 CA UTRECHT
NETHERLANDS
TEL 30 53 5600
TLF 30 54 0860
TLX 47224 SRON NL
EML XMKEESJ@HUTRUUO BITNE
COM 12,35,40,44,49,51

DE JAGER GERHARD PROF
DPT PHYSICS/ELECTRONICS
RHODES UNIVERSITY
BOX 94
GRAHAMSTOWN 6140
SOUTH AFRICA
TEL 0461-7128
TLF
TLX 244226
EML
COM

DE JAGER OCKER C DR
DPT OF PHYSICS
POTCHEFSTROOM UNIVERSITY
POTCHEFSTROOM 2520
SOUTH AFRICA
TEL 148 992418
TLF
TLX 34 6019
EML
COM

DE JONG TEIJE DR
ASTRONOMICAL INSTITUTE
UNIVERSITY OF AMSTERDAM
KRUISLAAN 403
NL 1098 SJ AMSTERDAM
NETHERLANDS
TEL 20 52 57491
TLF 20 52 57484
TLX 10262 HEF NL
EML
COM 33,34

DE JONGE J K DR
DPT OF ASTRONOMY
UNIVERSITY OF PITTSBURGH
RIVERVIEW PARK
PITTSBURGH PA 15214
USA
TEL
TLF
TLX
EML
COM 30,51

DE KORT JULES J DR
HOUTLAAN 4
NL 6500 GV NIJMEGEN
NETHERLANDS
TEL
TLF
EML
COM 42
TLX

DE KORTE PIETER A J DR
SPACE RESEARCH LABORATORY
SRON
SORBONNELAAN 2
NL 3584 CA UTRECHT
NETHERLANDS
TEL 30 53 5600
TLF 30 54 0860
TLX 47224
EML
COM

DE LA HERRAN V JOSE ENG
INSTITUTO DE ASTRONOMIA
APDO POSTAL 971
04510 MEXICO DF
MEXICO
TEL
TLF
TLX
EML
COM

DE LA NOE JEROME DR
OBSERVATOIRE DE BORDEAUX
BP 89
F 33270 FLOIRAC
FRANCE
TEL 56 86 4330
TLF 56 40 4251
TLX
EML
COM 28,34,40

DE LA REZA RAMIRO DR
OBSERVATORIO NACIONAL
RUA GL BRUCE 586
SAO CRISTOVAO
20921 RIO DE JANEIRO RJ
BRAZIL
TEL 21 580 7313
TLF 21 580 0332
TLX 21288
EML
COM

DE LAPPARENT-GURRIET V DR
INSTITUT D'ASTROPHYSIQUE
98BIS BD ARAGO
F 75014 PARIS
FRANCE
TEL 1 43 20 1425
TLF 1 43 29 8673
TLX
EML LAPPAREN@FRIAP51
COM

DE LOORE CAMIEL PROF
IAAG
VRIJE UNIV BRUSSELS
CP 165
B 1050 BRUSSELS
BELGIUM
TEL 2 641 3496
TLF
TLX 61051 VUBCO B
EML
COM 35,42,51

DE MEDEIROS JOSE RENAN DR
OBSERVATOIRE DE GENEVE
CHEMIN DES MAILLETTES 51
CH 1290 SAUVERNY
SWITZERLAND
TEL 22 755 2611
TLF 22 755 3983
TLX 419209 OBSQ CH
EML MEDEIROS@CGEUGE51
COM 35

DE PASCUAL MARTINEZ M DR
OBS ASTRONOMICO NCL
ALFONSO XII 3 & 5
E 28014 MADRID
SPAIN
TEL 1 227 0107
TLX 23475 IGC
EML
TLF
COM 20

DE PATER IMKE
ASTRONOMY DPT
UNIVERSITY OF CALIFORNIA
601 CAMPBELL HALL
BERKELEY CA 94720
USA
TEL 415 642 1947
TLF
TLX
EML
COM 15,16

DE ROBERTIS M M DR
DPT OF PHYSICS
YORK UNIVERSITY
4700 KEELE ST
NORTH YORK ON M3J 1P3
CANADA
TEL 416 736 2100*7761
TLF 416 736 5386
TLX 065 24736
EML FS300141@YUSOL BITNET
COM 28

DE RUITER HANS RUDOLF
IST DI RADIOASTRONOMIA
CNR
VIA IRNERIO 46
I 40126 BOLOGNA
ITALY
TEL 51 23 2856
TLF
TLX 211664 INFN BO I
EML
COM 40,47

DE SABBATA V PROF DR
IST DI FISICA
UNIVERSITA DI BOLOGNA
VIA IRNERIO 46
I 40100 BOLOGNA
ITALY
TEL 51 26 0991*051
TLF
TLX
EML
COM

DE SANCTIS GIOVANNI
OSS ASTRONOMICO DI TORINO
ST OSSERVATORIO 20
I 10025 PINO TORINESE
ITALY
TEL 11 84 1067
TLX 213236 TOASTR I
EML
TLF
COM 15,20

DE SILVA L N K DR
DPT OF MATHEMATICS
UNIVERSITY OF COLOMBO
COLOMBO 03
SRI LANKA
TEL
TLF
TLX
EML
COM 28

DE SOUZA RONALDO DR
IAG
UNIVERSIDADE DE SAO PAULO
CP 30627
01051 SAO PAULO SP
BRAZIL
TEL 11 577 8599
TLF
TLX 36221 IAGM BR
EML
COM

DE VAUCOULEURS GERARD PR
ASTRONOMY DPT
UNIVERSITY OF TEXAS
RLM 15 212
AUSTIN TX 78712 1083
USA
TEL 512 471 4461
TLF,
TLX
EML
COM 28,30,47

DE VEGT CH PROF DR
HAMBURGER STERNWARTE
GOJENSBERGSWEG 112
D 2050 HAMBURG 80
GERMANY
TEL 40 7252 4128/4112
TLX 21788
EML
TLF 40 7252 4198
COM 08,24P

DE VINCENZI DONALD DR
NASA AMES RESEARCH CTR
MS 245 1
MOFFETT FIELD CA 94035
USA
TEL
TLF
TLX
EML
COM 51

DE VRIES CORNNELIS DR
SRL
HUYGENS LAB
BOX 9504
NL 2300 RA LEIDEN
NETHERLANDS
TEL 71 27 5816
TLF
TLX
EML SRONCGJHLERUL2
COM

DE YOUNG DAVID S DR
KITT PEAK NTL OBS
BOX 26732
950 N CHERRY AVE
TUCSON AZ 85726 6732
USA
TEL 602 327 5511
TLF
TLX
EML
COM 40,48

DE ZEEUW PIETER T DR
SRL
HUYGENS LAB
BOX 9504
NL 2300 RA LEIDEN
NETHERLANDS
TEL 71 27 5879/5832
TLF 71 27 5819
TLX 38058 ASTRO NL
EML tim@rulhsy leidenuniv nl
COM 28

DE ZOTTI GIANFRANCO DR
DPT DI ASTRONOMIA
VIC DELL OSSERVATORIO 5
I 35122 PADOVA
ITALY
TEL 49 66 1499
TLF
TLX 430176 UNPADU I
EML
COM 47

DEARBORN DAVID PAUL S DR
LAWRENCE LIVERMORE LAB
L 23
BOX 808
LIVERMORE CA 94550
USA
TEL
TLF
TLX
EML
COM 35

DEBARBAT SUZANNE V DR
OBSERVATOIRE DE PARIS
61 AVE OBSERVATOIRE
F 75014 PARIS
FRANCE
TEL 1 40 51 2209
TLF
TLX 270776 OBS F
EML
COM 08,19,41P

DEBEHOGNE HENRI DR SC
OBSERVATOIRE ROYAL DE
BELGIQUE
AVE CIRCULAIRE 3
B 1180 BRUSSELS
BELGIUM
TEL 2 374 3801
TLF 2 374 9822
TLX 21565 B
EML
COM 15,20

DEBRUNNER HERMANN DR
PHYSIKALISCHES INSTITUT
UNIVERSITAET BERN
SIDLERSTRASSE 5
CH 3000 BERN
SWITZERLAND
TEL 31 65 4051
TLF
TLX 32320 CH
EML
COM 48

DEEMING TERENCE J DR
ICARUS RESEARCH

BOX 540205
HOUSTON TX 77254
USA
TEL 713 772 8414
TLF
TLX
EML
COM 41

DEERENBERG A J M DR
SRL
HUYGENS LAB
BOX 9504
NL 2300 RA LEIDEN
NETHERLANDS
TEL 71 27 2727
TLF
TLX
EML
COM

DEGAONKAR S S DR
PHYSICAL RESEARCH LAB
NAVRANGPURA
AHMEDABAD 380 009
INDIA
TEL 272 46 2129
TLF 272 44 5292
TLX 121397 PRL IN
EML
COM 40

DEGEWIJ JOHAN DR
MODDERMANSTRAAT 66
NL 2313 GS LEIDEN
NETHERLANDS
TEL
TLF
TLX
EML
COM 15,16

DEGUCHI SHUJI DR
NOBEYAMA RADIO OBS
NAOJ
MINAMIMAKI MURA
NAGANO 384 13
JAPAN
TEL 0267 98 2831 ˜
TLF
TLX 3329005 TAONRO J
EML
COM 34

DEHANT VERONIQUE DR
OBSERVATOIRE ROYAL DE
BELGIQUE
AVE CIRCULAIRE 3
B 1180 BRUSSELS
BELGIUM
TEL 2 373 0246
TLF 2 374 9822
TLX 21565 OBSBEL B
EML
COM 19

DEHARVENG JEAN-MICHEL DR
LAS
TRAVERSE DU SIPHON
LES TROIS LUCS
F 13012 MARSEILLE
FRANCE
TEL 91 05 5900
TLF 91 66 1855
TLX 420 584 F
EML
COM

DEHARVENG LISE DR
OBSERVATOIRE DE MARSEILLE
2 PLACE LE VERRIER
F 13248 MARSEILLE CDX 04
FRANCE
TEL 91 95 9088
TLF
TLX
EML
COM 34

DEINZER W PROF DR
UNIVERSITAETS-STERNWARTE

GEISMARLANDSTR 11
D 3400 GOETTINGEN
GERMANY
TEL 551 39 5044
TLF
TLX 96753
EML·
COM 35

DEJAIFFE RENE J DR
OBSERVATOIRE ROYAL DE
BELGIQUE
AVE CIRCULAIRE 3
B 1180 BRUSSELS
BELGIUM
TEL 2 375 2484
TLF 2 374 9822
TLX 21565 OBSBEL B
EML
COM 08,19

DEJONGHE HERWIG BERT DR
STERREKUNDIG OBSERV
RIJKSUNIVERSITEIT GENT
KRIJGSLJAN 281
B 9000 GENT
BELGIUM
TEL 91 22 5715
TLF 91 24 0634
TLX
EML
COM 28,33,37

DEKEL AVISHAI
DPT OF PHYSICS
HEBREW UNIV OF JERUSALEM
JERUSALEM 91904
ISRAEL
TEL 2 584 605
TLF
TLX
EML
COM 28,33,47

DEKKER E DR
MEIDOORNLAAN 13

NL 3461 ES LINSCHOTEN
NETHERLANDS
TEL 34 80 15406
TLF
TLX
EML
COM 41

DEL OLMO OROZCO A DR
INST ASTROFISICA
DE ANDALUCIA APD 2144
C/SANCHO PANZA S/N
E 18080 GRANADA
SPAIN
TEL 58 12 1311
TLF 58 81 4530
TLX 78573 IAAG E
EML CHONY@IAA ES/16488 CHONY
COM 28

DEL RIO GERARDO DR
OBS ASTRONOMICO NCL
ALFONSO XII-3
E 28014 MADRID
SPAIN
TEL 1 227 0107/1935
TLX 23465 IGCE
EML
TLF
COM

DEL TORO INIESTA JOSE DR
INST DE ASTROFISICA
DE CANARIAS
OBS DEL TEIDE
E 38200 LA LAGUNA
SPAIN
TEL 22 26 2211
TLF 922 26 3005
TLX 92640 IACE
EML SPAN IAC JTI
COM 10,12

DELABOUDINIERE J-P
IAS
BP 10
F 91371 VERRIERES BUISSON
FRANCE
TEL 1 64 47 4305
TLF
TLX 600252
EML
COM

DELACHE PHILIPPE J DR
OCA OBSERV DE NICE
BP 139
F 06003 NICE CDX
FRANCE
TEL 93 89 0420
TLF
TLX 460004 OBSNICE F
EML
COM 12,36,49

DELANNOY JEAN DR
IRAM
300 RUE DE LA PISCINE
F 38406 S MARTIN HERES CD
FRANCE
TEL 76 42 3383
TLF
TLX 980753
EML
COM 40

DELBOUILLE LUC PROF
INSTITUT D'ASTROPHYSIQUE
UNIVERSITE DE LIEGE
AVE COINTE 5
B 4000 COINTE-LIEGE
BELGIUM
TEL 41 52 9980
TLF 41 52 7474
TLX 41264 ASTRLG B
EML
COM 12

DELCROIX ANDRE J S DR
19A RUE E VANDERVELDE
19A RUE E VANDERVELDE
B 7230 FRAMERIES
BELGIUM
TEL
TLF
TLX
EML
COM

DELGADO ANTONIO JESUS
INST ASTROFISICA
DE ANDULUCIA APD 2144
PROFESOR ALBAREDA 1
E 18080 GRANADA
SPAIN
TEL 58 12 1300
TLF
TLX 78573 IAAG E
EML
COM 27,42

DELHAYE JEAN PROF
2 RUE DE LA PLEIADE
F 94240 L'HAY LES ROSES
FRANCE
TEL 1 46 64 5771
TLF
EML
COM 24,33
TLX

DELIYANNIS JEAN DR
DPT OF ASTROPHYSICS
NTL UNIVERSITY OF ATHENS
PANEPISTIMIOPOLIS
GR 157 83 ZOGRAFOS
GREECE
TEL 1 724 3414
TLF 1 722 8981
TLX
EML YANNIS@GRATHUN1
COM 12

DELLA VALLE MASSIMO DR
DPT DI ASTRONOMIA
UNIVERSITA DI PADOVA
VIC DELL OSSERVATORIO 5
I 35122 PADOVA
ITALY
TEL 49 66 1499
TLF
TLX 432071
EML INFNET 39003 DELLAVALLE
COM

DELLI SANTI SAVERIO
OSS ASTRONOMICO
UNIVERSITA DI BOLOGNA
CP 596
I 40100 BOLOGNA
ITALY
TEL 51 22 2956
TLF
TLX 211664 INFNBO I
EML
COM

DELSEMME ARMAND H PROF DR
DPT PHYSICS & ASTRONOMY
UNIVERSITY OF TOLEDO
2801 W BANCROFT ST
TOLEDO OH 43606
USA
TEL 419 537 2654
TLF
TLX
EML
COM 14,15,20,51

DEMARCQ JEAN ING

FRANCE
TEL
TLF
TLX
EML
COM

DEMARET JACQUES DR
158/033 AVE OBSERVATOIRE

B 4000 COINTE-LIEGE
BELGIUM
TEL 41 52 7261
TLF
TLX 41264 ASTROLIEGE
EML
COM 47

DEMARQUE P PROF
YALE UNIVERSITY OBS
BOX 6666
NEW HAVEN CT 06511
USA
TEL 203 432 3024
TLX 510 1012363
EML DEMARQUE@YALASTRO
TLF 203 432 5048
COM 12,35P,37

DEMERS SERGE DR
DPT DE PHYSIQUE
UNIVERSITE DE MONTREAL
CP 6128 SUCC A
MONTREAL QC H3C 3J7
CANADA
TEL 514 343 6718
TLF 514 343 2071
TLX 055 62425
EML
COM 27,28,37

DEMIN V G PROF DR
STERNBERG STATE ASTR INST
UNIVERSITETSKIJ PROSP 13
119899 MOSCOW
RUSSIA
TEL 139-36-81
TLF
TLX
EML
COM 07

DEMING LEO DRAKE DR
NASA/GSFC
CODE 693
GREENBELT MD 20771
USA
TEL 301 286 6519
TLF 301 286 3271
TLX 89675 NASCOM
EML LEPVAX YSLOD SPAN
COM 12

DEMIRCAN OSMAN DR
DPT OF ASTRONOMY
UNIVERSITY OF ANKARA
FEN FAKULTESI
06100 BESEVLER
TURKEY
TEL 41 212 6720
TLF
TLX 42761 OOTK TR
EML
COM 42

DENG ZUGAN DR
GRADUATE SCHOOL
DPT OF PHYSICS
BOX 3908
BEIJING 100039
CHINA PR
TEL 1 28 9461*89
TLF
TLX 22040 BAOAS CN
EML
COM

DENIS CARLO DR
INSTITUT D'ASTROPHYSIQUE
UNIVERSITE DE LIEGE
AVE COINTE 5
B 4000 COINTE-LIEGE
BELGIUM
TEL 41 52 9980
TLF 41 52 7474
TLX 41264
EML
COM

DENISSE JEAN-FRANCOIS DR
48 RUE MR LE PRINCE

F 75006 PARIS
FRANCE
TEL 1 43 29 4874
TLF
TLX
EML
COM 40

DENISYUK EDVARD K DR
ASTROPHYSICAL INSTITUTE
KAZAKH ACAD OF SCIENCES
480068 ALMA ATA
KAZAKHSTAN
TEL
TLF
TLX
EML
COM

DENNEFELD MICHEL
INSTITUT D'ASTROPHYSIQUE
98BIS BD ARAGO
F 75014 PARIS
FRANCE
TEL 1 43 20 1425
TLF. 1 43 29 8673
TLX
EML SPAN IAPOBS DENNEFELD
COM

DENNIS BRIAN ROY DR
NASA/GSFC
CODE 682
GREENBELT MD 20771
USA
TEL 301 286 6604
TLF
TLX 89675
EML
COM 10,44,48

DENNISON EDWIN W DR
985 CYNTHIA AVE
PASADENA CA 91107
USA
TEL 818 351 8751
TLF
TLX
EML
COM

DENNISON P A DR
TRINITY HALL
UNIVERSITY OF CAMBRIDGE
CAMBRIDGE CB3
UK
TEL
TLF
TLX
EML
COM

DENOYELLE JOZEF KIC
OBSERVATOIRE ROYAL DE
BELGIQUE
AVE CIRCULAIRE 3
B 1180 BRUSSELS
BELGIUM
TEL 2 375 2484
TLF 2 374 9822
TLX 21565 OBSBEL
EML
COM 25,33

DENT WILLIAM A PROF
DPT PHYSICS & ASTRONOMY
UNIV OF MASSACHUSETTS
GRC B
AMHERST MA 01003
USA
TEL 413 545 3665
TLF
TLX
EML
COM 40

DEPRIT ANDRE PROF
NTL BUREAU OF STANDARDS
APPLIED MATHEMATICS CTR
GAITHERSBURG MD 20899
USA
TEL 301 921 2631
TLF
TLX
EML
COM 04,07P

DERE KENNETH PAUL
NAVAL RESEARCH LABORATORY
CODE 4163
4555 OVERLOOK AVE SW
WASHINGTON DC 20375 5000
USA
TEL 202 767 2517
TLF
TLX
EML
COM 10

DERMAN I ETHEM DR
DPT OF ASTRONOMY
UNIVERSITY OF ANKARA
FEN FAKULTESI
06100 BESEVLER
TURKEY
TEL 41 23 6550/0109
TLF
TLX
EML
COM

DERMENDJIEV VLADIMIR DR
DPT OF ASTRONOMY
BULGARIAN ACAD SCIENCES
72 LENIN BLVD
BG 1784 SOFIA
BULGARIA
TEL 2 75 8927
TLF
TLX 23561 ECF BAN BG
EML
COM 10

DERMOTT STANLEY F
DPT OF ASTRONOMY
UNIVERSITY OF FLORIDA
224 SSRB
GAINESVILLE FL 32611
USA
TEL 904 392 2361
TLF 904 392 9605
TLX
EML DERMOTT@JUPITER ASTRO UF
COM 15,16,21

DESAI JYOTINDRA N
PHYSICAL RESEARCH LAB
NAVRANGPURA
ROOM 763
AHMEDABAD 380 009
INDIA
TEL 272 46 2129
TLF 272 44 5292
TLX 121397
EML
COM. 09

DESESQUELLES JEAN DR
UNIVERSITE LYON 1
CAMPUS LA DOUA
43 BD DU 11 NOVEMBRE
F 69621 VILLEURBANNE
FRANCE
TEL 78 89 8124
TLF
TLX 380273 IPN F
EML
COM 14

DESHPANDE AVINASH
RAMAN RESEARCH INSTITUTE
SADASHIVANAGAR
BANGALORE 560 080
INDIA
TEL 812 36 0122
TLF 812 34 0492
TLX 845 2671 RRI IN
EML
COM 40

DESHPANDE M R DR
PHYSICAL RESEARCH LAB
NAVRANGPURA
AHMEDABAD 380 009
INDIA
TEL 272 46 2129
TLF 272 44 5292
TLX 121397 PRL IN
EML
COM 25

DESPAIN KEITH HOWARD DR
LOS ALAMOS NATIONAL LAB
MS B220 X 2
BOX 1663
LOS ALAMOS NM 87545
USA
TEL 505 667 2388
TLF
TLX
EML
COM 35

DESPOIS DIDIER DR
OBSERVATOIRE DE BORDEAUX
BP 89
F 33270 FLOIRAC
FRANCE
TEL 56 86 4330
TLF 56 40 4251
TLX
EML EARN DESPOIS@FROBOR51
COM 40

DETTMAR RALF-JURGEN DR
RADIOASTRON INSTITUT DER
UNIVERSITAET BONN
AUF DEM HUEGEL 71
D 5300 BONN 1
GERMANY
TEL 228 73 5658
TLF 228 52 5229
TLX 886440
EML DETTMAR@BABSY MPIFR-BONN MPG D
COM 28

DEUBNER FRANZ-LUDWIG DR
INST F ASTRONOMIE &
ASTROPHYSIK
AM HUBLAND
D 8700 WUERZBURG
GERMANY
TEL 931 888 5030
TLF 931 706 297
TLX 68671 UNIWBG D
EML ASTI006@DWUUNI21 BITNET
COM 10,12VP

DEUL ERIK DR
STERREWACHT
BOX 9513
NL 2300 RA LEIDEN
NETHERLANDS
TEL 71 27 5880
TLX 39058 ASTRO NL
EML DEUL@HLERUL51/RULHL1 ·DEUL
TLF 71 27 5819
COM

DEUPREE ROBERT G DR
LOS ALAMOS NATIONAL LAB
MS F665 ESS 5
BOX 1663
LOS ALAMOS NM 87545
USA
TEL 505 667 8215
TLF
TLX
EML
COM 27,35

DEUTSCHMAN WILLIAM A DR
OREGON LASER CONSULTANTS
455 HILLSIDE AVE
KLAMATH FALLS OR 97601
USA
TEL 503 882 3295
TLF
TLX
EML
COM 15

DEVINNEY EDWARD J DR
100 UNION AVE

DELANCO NJ 08075
USA
TEL 609 764 1250
TLF
TLX
EML
COM

DEVORKIN DAVID H
NTL AIR & SPACE MUSEUM
SMITHSONIAN INSTITUTION
WASHINGTON DC 20560
USA
TEL 202 357 2828
TLF
TLX
EML
COM 41

DEWDNEY PETER E F DR
DOMINION RADIO ASTROPHYS
OBSERVATORY
BOX 248
PENTICTON BC V2A 6K3
CANADA
TEL 604 493 2277
TLF 604 493 7767
TLX 048 88127
EML
COM 34,40

DEWHIRST DAVID W DR
INSTITUTE OF ASTRONOMY
THE OBSERVATORIES
MADINGLEY RD
CAMBRIDGE CB3 0HA
UK
TEL 233 622 04
TLF
TLX 817297 ASTRON G
EML
COM 05,41

DEWITT BRYCE S DR
ASTRONOMY DPT
UNIVERSITY OF TEXAS
RLM 15 308
AUSTIN TX 78712 1083
USA
TEL 512 471 5055
TLF
TLX 9108741305
EML
COM 48

DEWITT JOHN H JR
3602 HOODS HILL RD
NASHVILLE TN 37215
USA
TEL 615 383 8272
TLF
EML
COM
TLX

DEWITT-MORETTE CECILE PR
ASTRONOMY DPT
UNIVERSITY OF TEXAS
RLM 15 308
AUSTIN TX 78712 1083
USA
TEL 512 471 1052
TLF
TLX 910-8741305
EML
COM

DEZSO LORANT PROF
HELIOPHYSICAL OBSERVATORY
BOC 30
H 4010 DEBRECEN
HUNGARY
TEL 52 11 015
TLF
TLX 72517 DEOBS L
EML
COM 10,12

DHAWAN VIVEK DR
RAMAN RESEARCH INSTITUTE
SADASHIVANAGAR
BANGALORE 560 080
INDIA
TEL 812 34 0122
TLF 812 34 0492
TLX 8452671 RRI IN
EML RRI@VIGYAN ERNET IN
COM 40

DI COCCO GUIDO
IST TE S R E
CNR
VIA DE CASTAGNOLI 1
I 40126 BOLOGNA
ITALY
TEL 51 51 9593
TLF
TLX 511350 CNR BO I
EML
COM 44

DI FAZIO ALBERTO
OAR
VIA DEL PARCO MELLINI 84
I 00136 ROMA
ITALY
TEL 6 34 7056
TLF
TLX 613103 PPRMT I
EML
COM 28,34,37

DI MARTINO MARIO
OSS ASTRONOMICO DI TORINO
ST OSSERVATORIO 20
I 10025 PINO TORINESE
ITALY
TEL 11 84 1067
TLF
TLX 213236 TO ASTR I
EML
COM 15

DI SEREGO ALIGHIERI S DR
OSS ASTROFISICO
DI ARCETRI
LARGO E FERMI 5
I 50125 FIRENZE
ITALY
TEL 55 437 8540
TLF 55 43 5939
TLX
EML ASTRFI SPERELLO
COM 28

DI TULLIO GRAZIELLA DR
OSS ASTRONOMICO DI PADOVA
VIC DELL OSSERVATORIO 5
I 35122 PADOVA
ITALY
TEL 49 66 1499
TLX
EML
TLF
COM

DI XIAO-HUA
PURPLE MOUNTAIN OBSERV
CAS
NANJING
CHINA PR
TEL 25 37609
TLF
TLX 34144 PMONJ CN
EML
COM 04

DIALETIS DIMITRIS DR
ASTRONOMICAL INSTITUTE
NTL OBSERVATORY OF ATHENS
BOX 20048
GR 118 10 ATHENS
GREECE
TEL 1 346 1191/1 804 0619
TLF
TLX 21 5530 OBSA GR
EML
COM 10

DIAMOND PHILIP JOHN DR
NRAO

BOX 0
SOCORRO NM 87801 0387
USA
TEL 505 835 7900
TLF
TLX 910988 1710
EML PDIAMOND@NRAO EDU
COM

DIAZ ANGELES ISABEL DR
DPT DE FISICA TEORICA
C-XI UNIVERSIDAD AUTONOMA
DE MADRID CANTOBLANCO
E 28049 MADRID
SPAIN
TEL 1 397 4223
TLF
TLX 27810
EML BITNET ADIAZ@EMDUAM11
COM 28

DICK STEVEN J
US NAVAL OBSERVATORY
34 & MASSACHUSETTS AVE NW
WASHINGTON DC 20392 5100
USA
TEL
TLX
EML
TLF
COM 08,41C,51

DICK WOLFGANG DR
OBSERVATORIUM HOHER LIST
UNIV STERNWARTE BONN
D 5568 DAUN
GERMANY
TEL 65 92 2150
TLF
TLX
EML
COM 24

DICKE ROBERT H PROF
DPT OF PHYSICS
PRINCETON UNIVERSITY
JOSEPH HENRY LABS
PRINCETON NJ 08544
USA
TEL 609 452 4317
TLF
TLX
EML
COM 47,48

DICKEL HELENE R DR
DPT OF ASTRONOMY
UNIVERSITY OF ILLINOIS
1011 W SPRINGFIELD AVE
URBANA IL 61801
USA
TEL 217 244 7044
TLF
TLX
EML LANIERIGEL ASTR VIUC EDU
COM 05,33,34,40

DICKEL JOHN R
UNIVERSITY OF ILLINOIS
349 ASTRONOMY BLDG
1002 W GREEN ST
URBANA IL 61801
USA
TEL 217 333 5532
TLF
TLX 910-245-2434 PURCH
EML
COM 16,33,34,40

DICKENS ROBERT J DR
RUTHERFORD APPLETON LAB
SPACE & ASTROPHYSICS DIV
BLDG R25/R68
CHILTON DIDCOT OX11 0QX
UK
TEL 235 219 00
TLF
TLX 83159
EML
COM 27,28,37

DICKEY JEAN O'BRIEN
JPL/CALTECH
MS 138 208
4800 OAK GROVE DR
PASADENA CA 91109
USA
TEL 818 354 3235
TLF
TLX 675429
EML
COM 04,16,19C,31

DICKEY JOHN M
DPT OF ASTRONOMY
UNIVERSITY OF MINNESOTA
116 CHURCH ST SE
MINNEAPOLIS MN 55455
USA
TEL 612 373 3308
TLF
TLX
EML
COM 28,34,40

DICKINSON DALE F DR
LOCKHEED PALO ALTO RES LB
DPT 92 20 BLDG 205
3251 HANOVER ST
PALO ALTO CA 94304
USA
TEL 415 424 2701
TLF
TLX
EML
COM

DICKMAN ROBERT L DR
DPT OF RADIO ASTRONOMY
UNIV OF MASSACHUSETTS
GRC
AMHERST MA 01003
USA
TEL 413 545 0925
TLF
TLX 95 5491
EML
COM 33,40

DICKMAN STEVEN R
DPT GEOLOGICAL SCIENCES
STATE UNIV OF NEW YORK
BINGHAMTON NY 13901
USA
TEL 607 777 4378
TLF
TLX
EML
COM 19C

DIEGO FRANCISCO DR
DPT PHYSICS & ASTRONOMY
UNIVERSITY COLLEGE LONDON
GOWER ST
LONDON WC1E 6BT
UK
TEL 1 713 877 050*3512
TLF 1 713 807 145
TLX 28722
EML FD@UK AC UCL STARLINK
COM 09

DIERCKSEN GEERD H F PH D
MPI F PHYSIK & ASTROPHYS
KARL-SCHWARZSCHILD-STR 1
D 8046 GARCHING MUENCHEN
GERMANY
TEL 89 329 90
TLX 524629 ASTRO D
EML
TLF
COM 14

DIETER NANNIELOU H DR
CLAY RD
N THERFORD VT 05054
USA
TEL 802 333 4079
TLF
TLX
EML
COM 33,40

DIKOVA SMILIANA D
DPT OF ASTRONOMY
BULGARIAN ACAD SCIENCES
72 LENIN BLVD
BG 1784 SOFIA
BULGARIA
TEL 2 75 8927
TLF
TLX 23561 ECF BAN
EML
COM 07

DIMITRIJEVIC MILAN
ASTRONOMICAL OBSERVATORY
VOLGINA 7
YU 11050 BEOGRAD
YUGOSLAVIA
TEL 11 419 357/421 875
TLF
TLX
EML EAOP021@YUBGSS21 BITNET
COM 14,36

DINERSTEIN HARRIET L
ASTRONOMY DPT
UNIVERSITY OF TEXAS
RLM 15 308
AUSTIN TX 78712 1083
USA
TEL 512 471 3449
TLF
TLX 910-874-1351
EML
COM 34

DINESCU A DR
INSTITUE OF GEODESY
PHOTOGRAMM CARTOGRAPHIE
1A BLVD DE L'EXPOSITION
R 78334 BUCAREST
RUMANIA
TEL
TLF
TLX
EML
COM

DING YOU-JI
YUNNAN OBSERVATORY
CAS
BOX 110
KUNMING 72946 YUNNAN
CHINA PR
TEL 871 2035
TLF
TLX 64040 YUOBS CN
EML
COM 10

DINGENS P PROF DR
KORTRIJKSE STEENWEG 763
B 9000 GENT
BELGIUM
TEL 91 22 1966
TLF
EML
COM 35
TLX

DINTINJANA BOJAN DR
ASTRONOMICAL OBSERVATORY
UNIVERSITY OF E KARDELJ
JADRANSKA 19
YU 61110 LJUBLJANA
YUGOSLAVIA
TEL 61 265 061
TLF 61 217 281
TLX
EML BOJAN DINTINJANA@UNI-LJ AC MAI
COM

DIONYSIOU DEMETRIOS PROF
HELLENIC AIR-FORCE ACAD
DEKELIA-ATTICA
18 AMASSIAS STR
GR 116 34 ATHENS
GREECE
TEL 1 723 8436/1 246 6366
TLF
TLX
EML
COM 47

DIRIKIS M A DR
ASTRONOMICAL OBSERVATORY
LATVIAN STATE UNIVERSITY
RAINIS BUL 19
226098 RIGA
LATVIA
TEL
TLF
TLX
EML
COM 20

DISNEY MICHAEL J PROF
PHYSICS DPT
UNIV WALES COLLEGE
BOX 913
CARDIFF CF1 3TH
UK   .
TEL 222 874 785
TLF 222 371 921
TLX 498635
EML
COM 34,48

DIVAN LUCIENNE DR
INSTITUT D'ASTROPHYSIQUE
98BIS BD ARAGO
F 75014 PARIS
FRANCE
TEL 1 43 20 1425
TLF 1 43 29 8673
TLX
EML
COM 29,45

DIVARI N B DR
ODESSA POLYTECHNICAL INST
270044 ODESSA
UKRAINE
TEL
TLF
TLX
EML
COM 21

DIXON ROBERT S DR
RADIO OBSERVATORY
OHIO STATE UNIVERSITY
2015 NEIL AVE
COLUMBUS OH 43210
USA
TEL 614 422 6789
TLF
TLX
EML
COM 05,40,51

DIZER MUAMMER PROF
KANDILLI OBSERVATORY
BOGAZICI UNIVERSITY
CENGELKOV
81220 ISTANBUL
TURKEY
TEL 1 332 0277
TLF
TLX
EML
COM 10

DJORGOVSKI STANISLAV DR
CALTECH
MS 105 24
PASADENA CA 91125
USA
TEL 818 356 4415
TLF
TLX 675425
EML SPAN GEORGE@6035
COM 22,27,51

DJURASEVIC GOJKO
ASTRONOMICAL OBSERVATORY
VOLGINA 7
YU 11050 BEOGRAD
YUGOSLAVIA
TEL 11 419 553
TLF
TLX
EML
COM

DJUROVIC DRAGUTIN M DR
DPT OF ASTRONOMY
FACULTY OF SCIENCES
STUDENTSKI TRG 16
YU 11000 BEOGRAD
YUGOSLAVIA
TEL 11 420 221
TLF
TLX
EML
COM 08,19

DLUZHNEVSKAYA O B DR
INST OF ASTRONOMY
ACADEMY OF SCIENCES
PYATNITSKAYA UL 48
109017 MOSCOW
RUSSIA
TEL 231 5461
TLF
TLX 412623 SCSTP SU
EML
COM 05VP,35,37

DOAZAN VERA DR
OBSERVATOIRE DE PARIS
61 AVE OBSERVATOIRE
F 75014 PARIS
FRANCE
TEL 1 40 51 2235
TLF
TLX 270776 OBS F
EML
COM 29,36

DOBRITSCHEV V M MR
DPT OF ASTRONOMY
BULGARIAN ACAD SCIENCES
7TH NOVEMBER ST 1
BG 1000 SOFIA
BULGARIA
TEL 2 7341
TLF 23561 ECFBAN BG
TLX 23561 ECFBAN BG
EML
COM

DOBRONRAVIN PETER DR
CRIMEAN ASTROPHYS OBS
UKRAINIAN ACAD OF SCIENCE
NAUCHNY
334413 CRIMEA
UKRAINE
TEL 43 2945
TLF
TLX
EML
COM 09,29

DOBRZYCKI JERZY PROF
HISTORY OF SCIENCE
POLISH ACAD OF SCIENCES
GWIAZDZISTA 27/169
PL 01 814 WARSAW
POLAND
TEL 33-22-03
TLF
TLX
EML
COM 41

DOCOBO DURANTEZ JOSE A
OBSERVATORIO ASTRONOMICO
RAMON MARIA ALLER
AVD DE LAS CIENCIAS S/N
E SANTIAGO DE COMPOSTELA
SPAIN
TEL
TLF
TLX
EML
COM 26

DODD RICHARD J DR
CARTER OBSERVATORY
BOX 2909
WELLINGTON 1
NEW ZEALAND
TEL 728-167
TLF
TLX 30172 NATOBS NZ
EML
COM

DOGAN NADIR PROF
DPT OF ASTRONOMY
UNIVERSITY OF ANKARA
FEN FAKULTESI
06100 BESEVLER
TURKEY
TEL
TLF
TLX
EML
COM 12

DOGGETT LEROY E DR
US NAVAL OBSERVATORY
NAUTICAL ALMANAC OFFICE
34 & MASSACHUSETTS AVE NW
WASHINGTON DC 20392 5100
USA
TEL   202 653 1572
TLF
TLX
EML
COM 04C,07

DOHERTY LORNE H DR
HERZBERG INST ASTROPHYS
NTL RESEARCH COUNCIL
100 SUSSEX DR
OTTAWA ON K1A 0R6
CANADA
TEL   613 993 6060
TLF   613 952 6602
TLX   053 3715
EML
COM

DOHERTY LOWELL R PROF
DPT OF ASTRONOMY
UNIVERSITY OF WISCONSIN
475 N CHARTER ST
MADISON WI 53706
USA
TEL   608 262 1249
TLF
TLX
EML
COM

DOKUCHAEV VYACHESLAV DR
INST NUCLEAR RESEARCH
ACADEMY OF SCIENCES
60TH ANIV OCT PROSPEKT 7A
117312 MOSCOW
RUSSIA
TEL   095 382 7678
TLF   095 292 6511
TLX   411051 INR SU
EML dokuchaev@inucres msk su
COM 34

DOKUCHAEVA OLGA D DR
STERNBERG STATE ASTR INST
UNIVERSITETSKIJ PROSP 13
119899 MOSCOW
RUSSIA
TEL
TLX
EML
TLF
COM 09,34

DOLAN JOSEPH F DR
NASA/GSFC
CODE 681
GREENBELT MD 20771
USA
TEL   301 286 5920
TLF
TLX 89675
EML
COM 25,44,48

DOLEZ NOEL DR
OBS MIDI PYRENEES
14 AVE E BELIN
F 31400 TOULOUSE CDX
FRANCE
TEL
TLF
TLX
EML
COM

DOLGINOV ARKADY Z PROF DR
IOFFE PHYSICAL TECH INST
ACADEMY OF SCIENCES
POLYTECHNICHESKAYA UL 26
194021 ST PETERSBURG
RUSSIA
TEL
TLF
TLX
EML
COM 49

DOLIDZE MADONA V DR
ABASTUMANI ASTROPHYSICAL
OBSERVATORY
GEORGIAN ACAD OF SCIENCES
383762 ABASTUMANI
GEORGIA
TEL
TLF
TLX
EML
COM 29

DOLLFUS AUDOUIN PROF
OBSERVATOIRE DE PARIS
SECTION DE MEUDON
F 92195 MEUDON PPL CDX
FRANCE
TEL   1 45 34 7530
TLF
TLX
EML
COM 10,16,20

DOMINGO VICENTE DR
ESA/ESTEC
SSD
BOX 299
NL 2200 AG NOORDWIJK
NETHERLANDS
TEL   17 19 83576
TLF   17 19 84698
TLX 39098
EML ESTC1 VDOMINGO/VDOMINGO@ESTEC
COM 44

DOMINKO FRAN PROF DR
SARANOVICEVA 11

YU 61000 LJUBLJANA
YUGOSLAVIA
TEL   61 322 210
TLF
TLX
EML
COM

DOMINSKI IRENEUSZ DR
ASTRONOMICAL LATITUDE OBS
BOROWIEC
BOX 62 035
PL 62 035 KORNIK
POLAND
TEL   POZNAN 170187
TLF
TLX 412623 AOS PL
EML
COM 31

DOMKE HELMUT PH D
ZNTRLINST F ASTROPHYSIK
STERNWARTE BABELSBERG
ROSA-LUXEMBURG-STR 17A
D 1502 POTSDAM
GERMANY
TEL
TLF
TLX
EML
COM 36

DOMMANGET J DR
OBSERVATOIRE ROYAL DE
BELGIQUE
AVE CIRCULAIRE 3
B 1180 BRUSSELS
BELGIUM
TEL   2 375 2484
TLF   2 374 9822
TLX 21565
EML
COM 24,26,50

DONAS JOSE DR
LAS
TRAVERSE DU SIPHON
LES TROIS LUCS
F 13012 MARSEILLE
FRANCE
TEL   91 05 5900
TLF   91 66 1855
TLX 420584 F
EML BITNEET DONAS@FRLASM51
COM 28

DONG IL ZUN
PYONGYANG ASTRON OBS
ACADEMY OF SCIENCES DPRK
TAESONG DISTRICT
PYONGYANG
KOREA DPR
TEL
TLF
TLX
EML
COM

DONN BERTRAM D
NASA/GSFC
CODE 691
GREENBELT MD 20771
USA
TEL   301 286 6859
TLF
TLX 89675
EML
COM 15,34

DONNER KARL JOHAN
OBS & ASTROPHYSICS LAB
UNIVERSITY OF HELSINKI
TAEHTITORNINMAKI
SF 00130 HELSINKI 13
FINLAND
TEL
TLF
TLX
EML
COM 28

DONNISON JOHN RICHARD DR
DPT OF MATH SCIENCES
GOLDSMITHS' COLLEGE
NEW CROSS
LONDON SE14 6NW
UK
TEL   1 716 927 171
TLF
TLX
EML
COM 20

DOPITA MICHAEL ANDREW DR
MOUNT STROMLO & SIDING
SPRING OBSERVATORIES
PRIVATE BAG
WODEN PO ACT 2606
AUSTRALIA
TEL   62 88 1111
TLF
TLX   62270 CANOPUS AA
EML
COM   34C

DORENWENDT KLAUS DR
PHYSIKALISCH-TECHNISCHES
BUNDESANSTALT
BUNDESALLEE 100
D 3300 BRAUNSCHWEIG
GERMANY
TEL   531 592 1210
TLF
TLX. 952822
EML
COM

DORFI ERNST ANTON DR
INSTITUT FUER ASTRONOMIE
UNIVERSITAET WIEN
TUERKENSCHANZSTR 17
A 1180 WIEN
AUSTRIA
TEL   1 345 3600
TLF
TLX 133099
EML
COM 42

DORMAND JOHN RICHARD DR
DPT OF MATHEMATICS
TEESSIDE POLYTECHNIC
MIDDLESBROUGH
CLEVELAND TS1 3BA
UK
TEL   642 218 121*4365
TLF
TLX
EML
COM   07

DOROSHKEVICH ANDREI G DR
INST OF APPLIED MATHS
ACADEMY OF SCIENCES
MIUSSKAJA SQ 4
125047 MOSCOW
RUSSIA
TEL
TLF
TLX
EML
COM 47

DORSCHNER JOHANN DR
UNIV STERNWARTE JENA
SCHILLERGAESSCHEN 2
D 6900 JENA
GERMANY
TEL 78 822 2637
TLX 5886134
EML
TLF·
COM 34,51

DOS REIS M PROF
OBSERVATORIO ASTRONOMICO
P 3000 COIMBRA
PORTUGAL
TEL
TLF
TLX
EML
COM

DOSCHEK GEORGE A DR
NAVAL RESEARCH LABORATORY
CODE 4170
4555 OVERLOOK AVE SW
WASHINGTON DC 20375 5000
USA
TEL 202 767 6473
TLF
TLX
EML
COM

DOSSIN F DR
INSTITUT D'ASTROPHYSIAUE
UNIVERSITE DE LIEGE
AVE COINTE 5
B 4000 COINTE-LIEGE
BELGIUM
TEL   41 52 9980
TLF   41 52 7474
TLX   41264 ASTRLG
EML
COM 15

DOTANI TADAYASU DR
ISAS
3-1-1 YOSHINODAI
SAGAMIHARA
KANAGAWA 229
JAPAN
TEL   427 51 3911
TLF   427 59 4253
TLX   J27758 ISAS ERO
EML DOTANI@ASTRO ISAS AC JP
COM 48

DOTTORI HORACIO A DR
INSTITUTO DE FISICA
IFRGS
CP 15051
90049 PORTO ALEGRE RS
BRAZIL
TEL   512 36 4677
TLF
TLX 511055 UFRS BR
EML
COM 28,34

DOUBINSKIJ B A DR
INST OF RADIO & ELECTRON
ACADEMY OF SCIENCES
103907 MOSCOW
RUSSIA
TEL
TLF
TLX
EML
COM 40

DOUGHTY NOEL A DR
DPT OF PHYSICS
UNIVERSITY OF CANTERBURY
PRIVATE BAG
CHRISTCHURCH 1
NEW ZEALAND
TEL
TLF
TLX
EML
COM 42

DOUGLAS JAMES N PROF
ASTRONOMY DPT
UNIVERSITY OF TEXAS
RLM 15 308
AUSTIN TX 78712 1083
USA
TEL 512 471 4461
TLF
TLX 910874-1351
EML
COM 40

DOUGLAS NIGEL DR
KAPTEYN STERRENWACHT
UNIVERSITY OF GRONINGEN
BOX 800
NL 9700 AV GRONINGEN
NETHERLANDS
TEL   59 08 28818
TLF   59 08 28800
TLX
EML NDOUGLAS@HRDKSW5
COM 09

DOUGLAS R J MR
PHYSICS DIVISION (M-36)
NTL RESEARCH COUNCIL
100 SUSSEX DR
OTTAWA ON K1A 0S1
CANADA
TEL   613 993 6060
TLF   613 952 6602
TLX
EML
COM 31

DOUGLASS GEOFFREY G
US NAVAL OBSERVATORY
34 & MASSACHUSETTS AVE NW
WASHINGTON DC 20392 5100
USA
TEL   202 653 1457
TLX   710 822 1970
EML
TLF   202 653 1497
COM 24,26

DOURNEAU GERARD DR
OBSERVATOIRE DE BORDEAUX
BP 89
F 33270 FLOIRAC
FRANCE
TEL   56 86 4330
TLF   56 40 4251
TLX
EML EARN DOURNEAU@FROBOR51
COM 07,20

DOVAL JORGE PEREZ DR
INST GEOPHYS & ASTRONOMY
CALLE 212 N 2906/29 Y 31
RPTO COROMELA/LA LISA
LA HABANA
CUBA
TEL   21 8435/0644
TLF
TLX 511240 GEOAS CU
EML
COM 20

DOWNES ANN JULIET B
MULLARD RADIO ASTRON OBS
CAVENDISH LABORATORY
MADINGLEY RD
CAMBRIDGE CB3 0HE
UK
TEL   233 664 77
TLF
TLX 81292
EML
COM

DOWNES DENNIS DR
IRAM
300 RUE DE LA PISCINE
F 38406 S MARTIN HERES CD
FRANCE
TEL 76 42 3383
TLF
TLX 980753
EML
COM 33,34,40

DOWNES RONALD A DR
STSCI
C/O USER SUPPORT BRANCH
3700 SAN MARTIN DR
BALTIMORE MD 21218
USA
TEL 301 338 4700
TLF·
TLX
EML SCIVAX DOWNES
COM 27

DOWNS GEORGE S DR
MIT LINCOLN LABORATORY
ROOM B285
BOX 73
LEXINGTON MA 02173
USA
TEL
TLF
TLX
EML
COM 40,51

DOYLE JOHN GERARD
ARMAGH OBSERVATORY
COLLEGE HILL
ARMAGH BT61 9OG
UK
TEL 861 52 2928
TLF
TLX 747937
EML
COM 36

DOYLE LAURANCE R DR
SETI INSTITUTE 245 7
NASA AMES RESEARCH CENTER
MAFFETT FIELD CA 94035
USA
TEL 415 604 3372
TLF
TLX
EML 24609 DOYLE SPAN
COM

DRAINE BRUCE T
PRINCETON UNIVERSITY OBS
PEYTON HALL
PRINCETON NJ 08544
USA
TEL 609 452 3574
TLF
TLX
EML
COM 34

DRAKE FRANK D PROF
DIV OF NATURAL SCIENCES
UNIVERSITY OF CALIFORNIA
SANTA CRUZ CA 95064
USA
TEL 408 429 2931
TLF
TLX 760 7936 UCSC UC
EML
COM 16,40,48,51C

DRAKE STEPHEN A
NASA/GSFC
CODE 668
GREENBELT MD 20771
USA
TEL 301 286 6962
TLF
TLX
EML
COM 36,40

DRAMBA C PROF
ASTRONOMICAL OBSERVATORY
CUTITUL DE ARGINT 5
BOX 28
R 75212 BUCAREST
RUMANIA
TEL 753998, 193407
TLF
TLX
EML
COM

DRAPATZ SIEGFRIED W DR
MPI F EXTRATERRESTRISCHE
PHYSIK
KARL-SCHWARZSCHILD-STR 1
D 8046 GARCHING MUENCHEN
GERMANY
TEL 89 329 9880
TLF
TLX 05215845
EML
COM 34

DRAVINS DAINIS PROF
LUND OBSERVATORY
BOX 43
S 221 00 LUND
SWEDEN
TEL 46 10 7297
TLF 46 10 4614
TLX 33199 OBSNOT S
EML dainis@astro lu se (internet)
COM 09,12,29,36C

DRAVSKIKH A F DR
PULKOVO OBSERVATORY
ACADEMY OF SCIENCES
10 KUTUZOV QUAY
196140 ST PETERSBURG
RUSSIA
TEL 297-94-52
TLF
TLX
EML
COM 08,40

DRECHSEL HORST DR
DR-REMEIS-STERNWARTE
ASTR INST UNIV ERLANGEN-N
STERNWARTSTR 7
D 8600 BAMBERG
GERMANY
TEL 951 57708
TLF
TLX 629830 UNIER D
EML DRECHSEL@STERNWARTE UNI-ERLANG
COM 42

DREHER JOHN W
DPT OF PHYSICS
MIT RM 26 315
BOX 165
CAMBRIDGE MA 02139
USA
TEL 617 253 8519
TLF
TLX 921473
EML
COM 09,34,40

DRESSEL LINDA L
DPT SPACE PHYS & ASTRON
RICE UNIVERSITY
BOX 1892
HOUSTON TX 77251 1892
USA
TEL 713 527 8101
TLF
TLX
EML
COM 28

DRESSLER ALAN
MT WILSON & LAS CAMPANAS
OBSERVATORIES
813 SANTA BARBARA ST
PASADENA CA 91101 1292
USA
TEL (818 304 0245
TLF
TLX 675425 CALTECH PSD
EML
COM 28,47C

DRESSLER KURT PROF
LAB PHYSIK CHEMIE
ETH ZENTRUM
CH 8092 ZUERICH
SWITZERLAND
TEL 1 256 4441
TLF
TLX 53178 ETHBI CH
EML
COM 14

DREVER RONALD W P DR
CALTECH
PASADENA CA 95064
USA
TEL
TLF 818 795 1547
EML
COM
TLX 675425 CALTECH PSD

DREW JANET
DPT OF ASTROPHYSICS
UNIVERSITY OF OXFORD
KEBLE RD
OXFORD OX1 3RH
UK
TEL
TLF
TLX
EML
COM

DRILLING JOHN S
DPT PHYSICS & ASTRONOMY
LOUISIANA STATE UNIV
BATON ROUGE LA 70803 4001
USA
TEL 504 388 6795
TLF
TLX
EML
COM 33

DROSSART PIERRE DR
OBSERVATOIRE DE PARIS
SECTION DE MEUDON
DESPA
F 92195 MEUDON PPL CDX
FRANCE
TEL 1 45 07 7664
TLF 1 45 07 2806
TLX 204464
EML FRMEU51@DROSSART
COM 16

DROZYNER ANDRZEJ
INSTITUTE OF ASTRONOMY
N COPERNICUS UNIVERSITY
CHOPINA 12/18
PL 87 100 TORUN
POLAND
TEL 260 17/ 260 37
TLF
TLX 0552234 ASTR PL
EML
COM 07

DRURY LUKE O'CONNOR DR
DIAS
SCHOOL OF COSMIC PHYSCIS
5 MERRION SQ\
DUBLIN 2
IRELAND
TEL 1 774 321
TLF
TLX 31687 DIAS
EML LD@DIASCP uucp
COM 48

DRYER MURRAY DR
NOAO ERL R/E/SE
SPACE ENVIRONMENT LAB
325 BROADWAY
BOULDER CO 80303
USA
TEL 303 497 3978
TLF
TLX 592811 NOAA MASC BDR
EML
COM 10,15,49

DUBAU JACQUES DR
OBSERVATOIRE DE PARIS
SECTION DE MEUDON
F 92195 MEUDON PPL CDX
FRANCE
TEL 1 45 07 7456
TLF
TLX 270912
EML
COM 10,14C

DUBIN MAURICE DR
NASA/GSFC
CODE 616
LAB FOR ATMOSPHERES
GREENBELT MD 20771
USA
TEL 301 286 5475
TLF 301 286 2630
TLX
EML
COM 21,22

DUBNER GLORIA DR
IAR
CC 5
1894 VILLA ELISA (BS AS)
ARGENTINA
TEL 21 4 3793
TLF
TLX 22414 CEDOC AR
EML
COM 34

DUBOIS MARC A
CEA CEN
DRFC
BP 6
F 92260 FONTENAY AUX ROSE
FRANCE
TEL 1 46 54 7881
TLF
TLX
EML
COM 10

DUBOIS PASCAL DR
OBS DE STRASBOURG
11 RUE UNIVERSITE
F 67000 STRASBOURG
FRANCE
TEL 88 35 4300
TLF 88 25 0160
TLX
EML
COM 05C,28

DUBOUT RENEE
OBSERVATOIRE DE LYON
AVE CHARLES ANDRE
F 69561 S GENIS LAVAL CDX
FRANCE
TEL 78 56 0705
TLF 72 39 9791
TLX 310926
EML
COM 25,34

DUCATI JORGE RICARDO DR
INSTITUTO DE FISICA
UFRGS
CP 15051
90000 PORTO ALEGRE RS
BRAZIL
TEL 512 36 4677
TLF
TLX 051-1055 UFRS BR
EML
COM 05,25,33,38C

DUCHESNE MAURICE DR
OBSERVATOIRE DE PARIS
61 AVE OBSERVATOIRE
F 75014 PARIS
FRANCE
TEL 1 43 20 1210
TLF
TLX
EML
COM 09

DUERBECK HILMAR W DR
ASTRONOMISCHES INSTITUT
UNIVERSITAT MUENSTER
DOMAGKSTR 75
D 4400 MUENSTER
GERMANY
TEL 251 83 3561
TLF
TLX
EML
COM 42

DUERST JOHANNES DR
LANGWIES
CH 8821 SCHONENBERG
SWITZERLAND
TEL 1 788 1785
TLF
EML
COM
TLX

DUFAY MAURICE PROF
UNIVERSITE LYON 1
CAMPUS LA DOUA
43 BD DU 11 NOVEMBRE
F 69621 VILLEURBANNE
FRANCE
TEL
TLF
TLX
EML
COM 14,21

DUFFETT-SMITH PETER JAMES
MULLARD RADIO ASTRON OBS
CAVENDISH LABORATORY
MADINGLEY RD
CAMBRIDGE CB3 OHE
UK
TEL 223 664 77
TLF
TLX 81292 CAVLAB G
EML
COM 40

DUFLOT MARCELLE DR
OBSERVATOIRE DE MARSEILLE
2 PLACE LE VERRIER
F 13248 MARSEILLE CDX 04
FRANCE
TEL 91 95 9088
TLF
TLX 420241
EML
COM 30,45

DUFOUR REGINALD JAMES
DPT SPACE PHYS & ASTRON
RICE UNIVERSITY
204K SPACE SCIENCE BLDG
HOUSTON TX 77001
USA
TEL 713 527 8101
TLF
TLX
EML
COM 28,34

DUFTON PHILIP L DR
DPT OF PURE & APPL PHYS
QUEEN'S UNIVERSITY
BELFAST BT7 1NN
UK
TEL 232 24 5133
TLF
TLX 74487
EML
COM 36

DULDIG MARCUS LESLIE DR
DPT OF PHYSICS
UNIVERSITY OF TASMANIA
GPO BOX 252C
HOBART TAS 7001
AUSTRALIA
TEL 2 202 022
TLF 2 202 410
TLX AA 58150 COSANT
EML DULGIG@PHYSVAX PHYS UTAS EDU A
COM 10C,49

DULEY WALTER W PROF
DPT OF PHYSICS
YORK UNIVERSITY
4700 KEELE ST
DOWNSVIEW ON M3J 1P3
CANADA
TEL 416 667 3040
TLF
TLX
EML
COM 34

DULK GEORGE A PROF
CASA
UNIVERSITY OF COLORADO
BOX 391
BOULDER CO 80309 0391
USA
TEL 303 492 8788
TLF
TLX
EML
COM 10,40

DULTZIN-HACYAN D DR
INST ASTROFISICA
DE ANDULUCIA APD 2144
PROFESOR ALBAREDA 1
E 18080 GRANADA
SPAIN
TEL 58 12 1300
TLF
TLX 78573 IAAGE
EML
COM · 28,47

DUMA DMITRIJ P DR
MAIN ASTRONOMICAL OBS
UKRAINIAN ACAD OF SCIENCE
GOLOSEEVO
252127 KIEV
UKRAINE
TEL 66 3110
TLF
TLX 131406 SKY SU
EML
COM 08

DUMITRACHE CHRISTANA
ASTRONOMICAL OBSERVATORY
CUTITUL DE ARGINT 5
BOX 28
R 75212 BUCHAREST 28
RUMANIA
TEL 90 236892
TLF
TLX 11882 ASTRO R
EML
COM 10

DUMONT RENE DR
OBSERVATOIRE DE BORDEAUX
BP 89
F 33270 FLOIRAC
FRANCE
TEL 56 86 4330
TLF 56 40 4251
TLX
EML
COM 21C

DUMONT SIMONE DR
INSTITUT D'ASTROPHYSIQUE
98BIS BD ARAGO
F 75014 PARIS
FRANCE
TEL 1 43 20 1425
TLF 1 43 29 8673
TLX
EML
COM 12,41

DUNCAN DOUGLAS KEVIN DR
STSCI
HOMEWOOD CAMPUS
3700 SAN MARTIN DR
BALTIMORE MD 21218
USA
TEL 301 338 4935
TLF
TLX
EML
COM 29

DUNCAN ROBERT A PROF
CSIRO
DIVISION OF RADIOPHYSICS
BOX 76
EPPING NSW 2121
AUSTRALIA
TEL
TLF
TLX
EML RDUNCAN@ATNF CSIRO AU
COM 10

DUNCOMBE RAYNOR L DR
AEROSPACE ENGINEERING DPT
UNIVERSITY OF TEXAS
WRW 414
AUSTIN TX 78712
USA
TEL 512 471 4250
TLF
TLX 704265 CSRUTX UD
EML
COM 04,05,07,08

DUNHAM DAVID W
COMPUTER SCIENCES CORP
SYSTEM SCIENCES DIV
8728 COLESVILLE RD
SILVER SPRING MD 20910
USA
TEL
TLF
TLX
EML
COM 04,20,26

DUNKELMAN LAWRENCE
LUNAR & PLANETARY LAB
UNIVERSITY OF ARIZONA
BOX 36241
TUCSON AZ 85721
USA
TEL 602 621 6963
TLF
TLX
EML
COM 09,12,21,44,50

DUNLOP STORM
140 STOCKS LANE
EAST WITTERING
CHICHESTER W SUS PO20 8NT
UK
TEL 243 670 354
TLF
TLX 9312-1107-40
EML TELECOM GOLD72 MAG100665
COM 27

DUNN RICHARD B DR
AIR FORCE GEOPHYSICS LAB
NTL SOLAR OBSERVATORY
SUNSPOT NM 88349
USA
TEL 505 434 1390
TLF
TLX
EML
COM 10,12

DUORAH HIRA LAL DR
DPT OF PHYSICS
UNIVERSITY OF GAUHATI
GUWAHATI 781014
INDIA
TEL 88 531
TLF
TLX
EML
COM 48

DUPREE ANDREA K DR
CENTER FOR ASTROPHYSICS
SOLAR & STELLAR DIVISION
60 GARDEN ST
CAMBRIDGE MA 02138
USA
TEL 617 495 7489
TLF
TLX 921428 SATELLITE CAM
EML
COM 34,36,44

DUPUY DAVID L DR
DPT OF PHYSICS
VIRGINIA MILITARY INST
LEXINGTON VA 24450
USA
TEL 703 464 7504
TLF
TLX
EML fpydupuy%faculty%vmi@1st vm1 e
COM 27,46

DURGAPRASAD N DR
TIFR
HOMI BHABHA RD
COLABA
BOMBAY 400 005
INDIA
TEL 22 219 111*342
TLF
TLX 0113009
EML
COM

DURIEZ LUC DR
LABORATOIRE D'ASTRONOMIE
1 IMPASSE OBSERVATOIRE
F 59000 LILLE
FRANCE
TEL 20 52 4424
TLF
TLX
EML EARN DURIEZ@FRCITL71
COM 07

DURISEN RICHARD H DR
ASTRONOMY DPT
INDIANA UNIVERSITY
SWAIN WEST 319
BLOOMINGTON IN 47405
USA
TEL 812 335 6921
TLF
TLX
EML
COM 35,42

DURNEY BERNARD DR
NTL SOLAR OBSERVATORY
BOX 26732
950 N CHERRY AVE
TUCSON AZ 85726 6732
USA
TEL 602 327 5511
TLF
TLX
EML
COM 49

DUROUCHOUX PHILIPPE
CEA CEN
DAPNIA/SAP
BP 2
F 91191 GIF/YVETTE CDX
FRANCE
TEL 1 69 08 3376
TLF
TLX 690860
EML
COM 48

DURRANCE SAMUEL T DR
DPT PHYSICS & ASTRONOMY
JOHNS HOPKINS UNIVERSITY
CHARLES & 34TH ST
BALTIMORE MD 21218
USA
TEL 301 338 8707
TLF
TLX 9102400225 JHU CAS
EML
COM 16

DURRANT CHRISTOPHER J DR
DPT APPLIED MATHEMATICS
UNIVERSITY OF SYDNEY
SYDNEY NSW 2006
AUSTRALIA
TEL 2 692 3373
TLF
TLX 20056 FISHLIB AA
EML
COM

DURRET FLORENCE DR
INSTITUT D'ASTROPHYSIQUE
98BIS BD ARAGO
F 75014 PARIS
FRANCE
TEL 1 43 20 1425
TLF 1 43 29 8673
TLX
EML IAPOBS FLORENCE
COM 28

DUSCHL WOLFGANG J DR
INSTITUT F THEORETISCHE
ASTROPHYSIK
IM NEUENHEIMER F 561
D 6900 HEIDELBERG 1
GERMANY
TEL 62 215 62967
TLF
TLX 46515 UNIHD D
EML BITNET CJO@DHDURZ1
COM 42

DUTHIE JOSEPH G PROF
DPT PHYSICS & ASTRONOMY
UNIVERSITY OF ROCHESTER
ROCHESTER NY 14627
USA
TEL
TLF
TLX
EML
COM 48

DUVAL MARIE-FRANCE
OBSERVATOIRE DE MARSEILLE
2 PLACE LE VERRIER
F 13248 MARSEILLE CDX 04
FRANCE
TEL 91 95 9088
TLF
TLX
EML
COM 28,46

DUVALL THOMAS L JR
NTL SOLAR OBSERVATORY
BOX 26732
950 N CHERRY AVE
TUCSON AZ 85726 6732
USA
TEL 602 325 9338
TLF
TLX 666-484 AURA-KPNO-
EML
COM 12

DUVERT GILLES DR
OBSERVATOIRE DE GRENOBLE
CERMO/ASTROPHYSIQUE
BP 53X
F 38041 S MARTIN HERES CD
FRANCE
TEL 76 51 4885
TLF 76 44 8821
TLX
EML BITNET DUVERT@FRGAG51
COM

DVORAK RUDOLF DR
INSTITUT FUER ASTRONOMIE
UNIVERSITAETSSTERNWARTE
TUERKENSCHANZSTR 17
A 1180 WIEN
AUSTRIA
TEL 1 345 3600
TLF
TLX
EML
COM 07,20

DWARAKANATH K S
RAMAN RESEARCH INSTITUTE
SADASHIVANAGAR
BANGALORE 560 080
INDIA
TEL 812 36 0122
TLF 812 34 0492
TLX 845 2671 RRI IN
EML
COM 40

DWEK ELI
NASA/GSFC
CODE 697
LEP
GREENBELT MD 20771
USA
TEL 301 286 6209
TLF
TLX
EML
COM 34

DWIVEDI BHOLA NATH DR
DPT OF APPLIED PHYSICS IT
UNIVERSITY BANARAS HINDU
VARANASI 221 005
INDIA
TEL
TLF
TLX 0545-208 TECH IN
EML
COM 10

DWORETSKY MICHAEL M DR
DPT PHYSICS & ASTRONOMY
UNIVERSITY COLLEGE LONDON
GOWER ST
LONDON WC1E 6BT
UK
TEL 1 713 877 050
TLF
TLX 28722
EML
COM 29

DYCK M DR
KITT PEAK NTL OBS
BOX 26732
950 N CHERRY AVE
TUCSON AZ 85726 6732
USA
TEL
TLF
TLX
EML
COM

DYER CHARLES CHESTER DR
SCARBOROUGH COLLEGE
UNIVERSITY OF TORONTO
PHYS SCS GR RM S 650
TORONTO ON M1C 1A4
CANADA
TEL 416 284 3318
TLF
TLX
EML
COM 47

DYER EDWARD R DR
3626 DAVIS ST NW
WASHINGTON DC 20007
USA
TEL
TLF
EML.
COM
TLX

DYSON F J DR
INST FOR ADVANCED STUDY
SCHOOL OF NATURAL SCIENCE
PRINCETON NJ 08540
USA
TEL 609 734 8055
TLF
TLX
EML
COM 40,51

DYSON JOHN E DR
DPT OF ASTRONOMY
UNIVERSITY OF MANCHESTER
MANCHESTER M13 9PL
UK
TEL 612 754 235
TLF
TLX 668932
EML
COM 34,49

DZHAPIASHVILI VICTOR P DR
ABASTUMANI ASTROPHYSICAL
OBSERVATORY,
GEORGIAN ACAD OF SCIENCES
383762 ABASTUMANI
GEORGIA
TEL
TLF
TLX
EML
COM 15,16

DZIEMBOWSKI WOJCIECH PROF
COPERNICUS ASTRON CENTER
POLISH ACAD OF SCIENCES
UL BARTYCKA 18
PL 00 716 WARSAW
POLAND
TEL
TLF
TLX
EML
COM 27,35

DZIGVASHVILI R M DR
ABATSUMANI ASTROPHYSICAL
OBSERVATORY
GEORGIAN ACAD OF SCIENCES
383762 ABASTUMANI
GEORGIA
TEL
TLF
TLX
EML
COM 33

D'ANTONA FRANCESCA DR
OAR

MONTE PORZIO
I 00040 MONTEPORZIO
ITALY
TEL 6 944 9019
TLF
TLX
EML
COM 35,37,42

D'ODORICO SANDRO DR
ESO
KARL-SCHWARZSCHILD-STR 2
D 8046 GARCHING MUENCHEN
GERMANY
TEL 89 320 060
TLX 528-28-222
EML
TLF 89 320 2362
COM 28,34

EATON JOEL A DR
ASTRONOMY DPT
INDIANA UNIVERSITY
SWAIN WEST 319
BLOOMINGTON IN 47405
USA
TEL 812 335 4176
TLF
TLX
EML
COM 42

EBISUZAKI TOSHIKAZU DR
KOMABA MEGURO-KU
TOKYO 153
JAPAN
TEL 81 3 3467 1171 (665)
TLF 81 3 3465 3925
TLX
EML
COM 28

ECCLES MICHAEL J DR
BALLENCRIEFF TOLL
SUNNYSIDE
BATHGATE EH48 4LD
UK
TEL 506 53 989
TLF
TLX 727484
EML
COM 51

ECHEVARRIA JUAN DR
OBS ASTRONOMICO NACIONAL
UNAM
APDO POSTAL 877
22800 ENSENADA B CALIF
MEXICO
TEL
TLF
TLX
EML
COM 42

ECHEVERRIA ROMAN JUAN M
OBS ASTRONOMICO NACIONAL
UNAM
APDO POSTAL 877
22860 ENSENADA B CALIF
MEXICO
TEL
TLF
TLX
EML
COM

EDDY JOHN A DR
3460 ASH AVE
BOULDER CO 80303
USA
TEL 303 497 1680
TLF
TLX 45694
EML
COM 10,41

EDELMAN COLETTE DR
BUREAY DES LONGITUDES
77 AVE DENFERT ROCHEREAU
F 75014 PARIS
FRANCE
TEL 1 40 51 2272
TLF
TLX
EML
COM 07,20

EDELSON RICK DR
CASA
UNIVERSITY OF COLORADO
BOX 391
BOULDER CO 80309 0391
USA
TEL 303 492 6784
TLF
TLX
EML ELROY RICK SPAN
COM 28,40,48

EDLEN BENGT PROF
DPT OF PHYSICS
UNIVERSITY OF LUND
SOELVEGATAN 14
S 223 62 LUND
SWEDEN
TEL 46 10 7730
TLF
TLX
EML
COM 14

EDMONDSON FRANK K PROF
ASTRONOMY DPT
INDIANA UNIVERSITY
SWAIN WEST 319
BLOOMINGTON IN 47405
USA
TEL 812 335 6918
TLF
TLX
EML
COM 20,30,33,41

EDMUNDS MICHAEL GEOFFREY
PHYSICS DPT
UNIV WALES COLLEGE
BOX 913
CARDIFF CF1 3TH
UK
TEL 222 874 785
TLF 222 371 921
TLX
EML
COM 28

EDVARDSSON BENGT DR
ASTRONOMICAL OBSERVATORY
BOX 515
S 751 20 UPPSALA
SWEDEN
TEL 18 51 2488
TLX 76024 UNIVUPS S
EML ASTBE@SEUDAC21 BITNET
TLF 18 52 7583
COM 36

EDWARDS ALAN CH DR
DPT PHYSICS & ASTRONOMY
UNIVERSITY OF GEORGIA
ATHENS GA 30602
USA
TEL
TLF
TLX
EML
COM 35

EDWARDS PAUL J DR
MOUNT STROMLO & SIDING
SPRING OBSERVATORIES
PRIVATE BAG
WODEN PO ACT 2606
AUSTRALIA
TEL 62 88 1111
TLF
TLX 68270 AA
EML
COM 25,27,48,50

EDWARDS TERRY W
DPT PHYSICS & ASTRONOMY
UNIVERSITY OF MISSOURI
COLUMBIA MO 65211
USA
TEL 314 882 3036
TLF
TLX
EML
COM 35

EDWIN ROGER P
DPT OF PHYSICS & ASTRON
UNIVERSITY OF ST ANDREWS
NORTH HAUGH
ST ANDREWS FIFE KY16 9SS
UK
TEL   334 761 61
TLF   334 744 87
TLX
EML
COM 09

EELSALU HEINO DR
TARTU ASTROPHYSICAL OBS
ESTONIAN ACAD OF SCIENCES
202400 TARTU 4
ESTONIA
TEL 14 34 28163
TLF 14 34 10205
TLX
EML
COM 25,30,41

EFREMOV YU I DR
INST OF APPLIED MATHS
ACADEMY OF SCIENCES
MIUSSKAJA SQ 4
125047 MOSCOW
RUSSIA
TEL
TLF
TLX
EML
COM

EFREMOV YURY N DR
STERNBERG STATE ASTR INST

UNIVERSITETSKIJ PROSP 13
119899 MOSCOW
RUSSIA
TEL 139 26 57
TLF
TLX
EML
COM 27,33,37

EFSTATHIOU GEORGE
INSTITUTE OF ASTRONOMY
THE OBSERVATORIES
MADINGLEY RD
CAMBRIDGE CB3 OHA
UK
TEL 223 622 04
TLF
TLX 817297 ASTRON G
EML
COM 28,47

EGGLETON PETER P DR
INSTITUTE OF ASTRONOMY
THE OBSERVATORIES
MADINGLEY RD
CAMBRIDGE CB3 OHA
UK
TEL 223 622 04
TLF
TLX 817297 ASTRON G
EML
COM 35,42C

EGRET DANIEL DR
OBS DE STRASBOURG
11 RUE UNIVERSITE
F 67000 STRASBOURG
FRANCE
TEL 88 35 4300
TLF 88 25 0160
TLX 890506 STAROBS
EML
COM 05,33,45C

EHLERS JURGEN PROF
MPI FUER PHYSIK UND
ASTROPHYSIK
KARL-SCHWARZSCHILD-STR 1
D 8046 GARCHING MUENCHEN
GERMANY
TEL 89 329 99444
TLF
TLX 524629 ASTRO D
EML
COM 47

EICHHORN HEINRICH K DR
DPT OF ASTRONOMY
UNIVERSITY OF FLORIDA
231 SSRB
GAINESVILLE FL 32611
USA
TEL 904 392 2052
TLF
TLX
EML
COM 07,08,24,26

EICHLER DAVID DR
ASTRONOMY PROGRAM
UNIVERSITY OF MARYLAND
COLLEGE PARK MD 20742
USA
TEL 301 454 6448
TLF
TLX 710-8260352
EML
COM 48

EIDELSBERG MICHELE DR
OBSERVATOIRE DE PARIS
SECTION DE MEUDON
F 92195 MEUDON PPL CDX
FRANCE
TEL 1 45 07 7562
TLF 1 45 07 7469
TLX
EML FRMEU051
COM 14

EILEK JEAN
DPT OF PHYSICS
NEW MEXICO TECH
CAMPUS STATION
SOCORRO NM 87801
USA
TEL 505 835 5433
TLF
TLX
EML
COM 48

EINASTO JAAN DR
TARTU ASTROPHYSICAL OBS
ESTONIAN ACAD OF SCIENCES
202444 TARTU
ESTONIA
TEL
TLF
TLX
EML
COM 28,33,37,47

EINAUDI GIORGIO
DPT DI ASTRONOMIA
UNIVERSITA DI FIRENZE
LARGO E FERMI 5
I 50125 FIRENZE
ITALY
TEL 55 27521
TLF 55 22 0039
TLX 572268 ARCETR I
EML
COM 12

EINICKE OLE H DR
COPENHAGEN UNIVERSITY OBS
BRORFELDEVEJ 23
DK 4340 TOLLOSE
DENMARK
TEL 53 48 8195
TLF 58 48 8755
TLX
EML
COM 08

EIROA DE SAN FRANCISCO C
OBS ASTRONOMICO NCL

ALFONSO XII-3
E 28014 MADRID
SPAIN
TEL 1 227 0107
TLF
TLX 49880 OANM
EML
COM

EKERS RONALD D DR
CSIRO
DIVISION OF RADIOPHYSICS
BOX 76
EPPING NSW 2121
AUSTRALIA
TEL 2 868 0222
TLF 2 868 0310
TLX 26230
EML REKERS@ATNF CSIRO AU
COM 28,40

EL BAKKALI LARBI DR
FAC DE SCIENCES
BOX 2121
TETOUAN
MOROCCO
TEL 974509
TLF 971763
TLX
EML
COM 07

EL EID MOUNIB DR
UNIVERSITATS-STERNWARTE
GEISMARLANDSTR 11
98 BIS, BOULEVARD ARAGO
D 3400 GOETTINGEN
GERMANY
TEL 551 39 5042
TLF
TLX 96753
EML
COM

EL NAWAWAY M S DR
ACADEMY OF SCIENTIFIC
RESEARCH AND TECHNOLOGY
101 KASR EL EINI STREET
CAIRO
EGYPT
TEL
TLF
TLX
EML
COM

ELFORD WILLIAM GRAHAM DR
DPT OF PHYSICS
UNIVERSITY OF ADELAIDE
BOX 498
ADELAIDE SA 5001
AUSTRALIA
TEL 2 228 5321
TLF
TLX 89141 UNIVQD AA
EML
COM 22

ELGAROY OYSTEIN PROF
INST THEORET ASTROPHYSICS
UNIVERSITY OF OSLO
BOX 1029
N 0315 BLINDERN OSLO 3
NORWAY
TEL 02-456-504
TLF·
TLX
EML
COM 40

ELIPE SANCHEZ ANTONIO
DPT FIS TIERRA & COSMOS
UNIVERSIDAD DE ZARAGOZA
E 50009 ZARAGOZA
SPAIN
TEL 76 35 7011
TLF
TLX 58198
EML
COM 07

ELITZUR MOSHE
DPT PHYSICS & ASTRONOMY
UNIVERSITY OF KENTUCKY
LEXINGTON KY 40506 0055
USA
TEL 606 257 4720
TLF
TLX
EML
COM 34

ELLDER JOEL DR
ONSALA SPACE OBSERVATORY
S 430 34 ONSALA
SWEDEN
TEL
TLF
EML
COM
TLX

ELLIOT JAMES L DR
DPT OF EARTH & PLANET SCI
MIT RM 54 422A
BOX 165
CAMBRIDGE MA 02139
USA
TEL 617 253 6308
TLF
TLX 921473 MIT CAM
EML
COM 16,20

ELLIOTT IAN DR
DUNSINK OBSERVATORY
DIAS
DUBLIN 15
IRELAND
TEL 1 387 959
TLF
TLX 31687 DIAS EI
EML
COM 12

ELLIOTT KENNETH H DR
DPT OF SPACE RESEARCH
UNIVERSITY OF BIRMINGHAM
BOX 363
BIRMINGHAM B15 2TT
UK
TEL 214 72 1301
TLF
TLX 338938
EML
COM 34

ELLIS G R A PROF
DPT OF PHYSICS
UNIVERSITY OF TASMANIA
GPO BOX 252C
HOBART TAS 7001
AUSTRALIA
TEL
TLF
TLX 58150
EML
COM 40

ELLIS GEORGE F R PROF
SISSA
ST COSTIERA 11
MIRAMARE
I 34014 TRIESTE
ITALY
TEL 40 22 4118
TLF
TLX 460 392 ICTP
EML
COM 47,51

ELLIS RICHARD S
DPT OF PHYSICS
UNIVERSITY OF DURHAM
SOUTH RD
DURHAM DH1 3LE
UK
TEL
TLF
TLX
EML
COM 28C,47

ELMEGREEN BRUCE GORDON DR
IBM
THOMAS J WATSON RES CTR
BOX 218
YORKTOWN HEIGHTS NY 10598
USA
TEL 914 945 2448
TLF 914 945 2141
TLX 137456
EML BGE@YKTVMT
COM 34,37

ELMEGREEN DEBRA MELOY
DPT PHYSICS & ASTRONOMY
VASSAR COLLEGE
POUGHKEEPSIE
NY 12601
USA
TEL 914 437 7356
TLF
TLX
EML ELMEGREEN@VASSAR EDU
COM 28,33,34

ELSAESSER HANS PROF
MPI FUR ASTRONOMIE
KOENIGSTUHL
D 6900 HEIDELBERG 1
GERMANY
TEL 62 215 28200
TLF
TLX
EML
COM 21,33

ELSMORE BRUCE DR
MULLARD RADIO ASTRON OBS
CAVENDISH LABORATORY
MADINGLEY RD
CAMBRIDGE CB3 0HE
UK
TEL 223 664 77
TLF
TLX 81292
EML
COM 19,24,40

ELST ERIC WALTER DR
OBSERVATOIRE ROYAL DE
BELGIQUE
AVE CIRCULAIRE 3
B 1180 BRUSSELS
BELGIUM
TEL
TLF 2 374 9822
TLX
EML
COM 20

ELSTE GUNTHER H DR
DPT OF ASTRONOMY
UNIVERSITY OF MICHIGAN
DENNISON BLDG
ANN ARBOR MI 48109 1090
USA
TEL 313 764 3444
TLF
TLX
EML
COM 10,12,36

ELSTON WOLFGANG E PROF
DPT OF GEOLOGY
UNIVERSITY OF NEW MEXICO
800 YALE BLVD NE
ALBUQUERQUE NM 87131
USA
TEL 505 277 5339
TLF
TLX 660461
EML
COM 16

ELVIS MARTIN S DR
CENTER FOR ASTROPHYSICS
HCO/SAO
60 GARDEN ST
CAMBRIDGE MA 02138
USA
TEL 617 495 7442
TLF
TLX 921428
EML
COM 28,44,47,48

ELVIUS AINA M PROF
STOCKHOLM OBSERVATORY

S 133 36 SALTSJOEBADEN
SWEDEN
TEL 87 17 0195
TLF 87 17 4719
TLX 12972 SOBBSERV S
EML ELVIUS@ASTRO SU SE
COM 28,34

ELVIUS TORD PROF EMERITUS
NORRLANDSGATAN 34F
S 752 29 UPPSALA
SWEDEN
TEL 18 10 0857
TLF
EML
COM 33,45
TLX

ELWERT GERHARD PROF
LEHRSTUHL F THEORETISCHE
ASTROPHYSIK
UNIVERSITAET TUEBINGEN
D 7400 TUEBINGEN
GERMANY
TEL 707 129 6483
TLF
TLX 7-262714 ALT D
EML
COM 10,40

EL-BASSUNY ALAWY A A DR
HELWAN OBSERVATORY
HELWAN
EGYPT
TEL 78 0645/2683
TLF
TLX 9703 HIAG UN
EML
COM 27,37

EL-BAZ FAROUK DR
ITEK OPTICAL SYSTEMS
10 MAGUIRE RD
LEXINGTON MA 02173
USA
TEL 617 276 2532
TLF
TLX 923456
EML
COM 16

EL-RAEY MOHAMED E PROF
DPT ENVIRONMENT STUDIES
INST GRADUATE STUD & RES
UNIVERSITY OF ALEXANDRIA
ALEXANDRIA
EGYPT
TEL
TLF
TLX
EML
COM 44

EL-SHAHAWY MOHAMAD PROF
DPT ASTRON & METEOROLOGY
FACULTY OF SCIENCEŞ
CAIRO UNIVERSITY
GEZA
EGYPT
TEL 27 2022
TLF
TLX 94372 UNCAI
EML
COM 19

EMANUELE ALESSANDRO DR
IAS
CNR
CP 67
I 00044 FRASCATI
ITALY
TEL 6 942 5655
TLF 6 941 6847
TLX 610261
EML POLIFEMO@IRMIAS
COM 48

EMELIANOV NIKOLAJ V DR
STERNBERG STATE ASTR INST
UNIVERSITETSKIJ PROSP 13
119899 MOSCOW
RUSSIA
TEL 139-37-64
TLF
TLX
EML
COM 07

EMERSON DAVID
ROYAL OBSERVATORY
BLACKFORD HILL
EDINBURGH EH9 3HJ
UK
TEL
TLX
EML
TLF
COM 28,40,46

EMERSON JAMES P
DPT OF PHYSICS
QUEEN MARY/WESTFIELD COLL
MILE END RD
LONDON E1 4NS
UK
TEL 1 980 4811
TLF
TLX 893750 QMCUOL G
EML
COM 34

EMINZADE T A DR
SHEMAKHA ASTROPHYSICAL
OBSERVATORY
AZER ACADEMY OF SCIENCES
373243 SHEMAKHA
AZERBAIDZHAN
TEL
TLF
TLX
EML
COM 35

EMSLIE A GORDON
DPT OF PHYSICS
UNIVERSITY OF ALABALA
HUNTSVILLE AL 35899
USA
TEL 205 895 6167
TLF
TLX
EML
COM 10

ENARD DANIEL DR
ESO
KARL-SCHARZSCHILDSTR 2
D 8046 GARCHING MUENCHEN
GERMANY
TEL 89 320 06251
TLF 89 320 2362
TLX
EML
COM

ENCRENAZ PIERRE J DR
RADIOASTRONOMIE ENS
24 RUE LHOMOND
F 75231 PARIS CDX 05
FRANCE
TEL 1 43 29 1235
TLF
TLX
EML
COM 34

ENCRENAZ THERESE DR
OBSERVATOIRE DE PARIS
SECTION DE MEUDON
GROUPE PLANETES
F 92195 MEUDON PPL CDX
FRANCE
TEL 1 45 07 7691
TLF
TLX 204464
EML
COM 15,16C

ENDAL ANDREW S DR
APPLIED RESEARCH CORP
8201 CORPORATE DR
SUITE 920
LANDOVER MD 20785
USA
TEL 301 459 8442
TLF
TLX
EML
COM 35

ENGELBRECHT CHRISTIAN DR
UNIVERSITY OF STELLENBOSC
PRETORIA
SOUTH AFRICA
TEL
TLF
EML
COM
TLX

ENGELS DIETER DR
STERNWARTE DER UNIVERS
HAMBURG
GOJENSBERGSWEG 112
D 2050 HAMBURG 80
GERMANY
TEL 40 725 24136
TLF 40 7252 4198
TLX 217884
EML ST20050 DHHUNI4 EARN
COM 09

ENGIN SEMANUR PROF
DPT OF ASTRONOMY
UNIVERISTY OF ANKARA
FEN FAKULTESI
06100 BESEVLER
TURKEY
TEL
TLF
TLX
EML
COM

ENGINOL TURAN B DR
INST FOR GRADUATE STUDIES
IN SCIENCE & ENGINEERING
BOGAZICI UNIVERSITY
TURKEY
TEL 1 163 1500
TLF 1 165 8480
TLX 26411 BOUN TR
EML ENGINOL F TR BOUN
COM 47

ENGVOLD ODDBJOERN DR
INST THEORET ASTROPHYSICS
UNIVERSITY OF OSLO
BOX 1029
N 0315 BLINDERN OSLO 3
NORWAY
TEL 47 2 856521
TLF 47 2 856505
TLX 2456515
EML 21813 55418 ODDBJORN
COM 09,10VP

ENOME SHINZO PROF
NOBEYAMA RADIO OBS
NAOJ
MINAMIMAKI MURA
NAGANO 384 13
JAPAN
TEL 81 267 98 2831
TLF 81 267 98 2884
TLX 3329005 NAONRO J
EML
COM 10,40

ENSLIN HEINZ DR
YIERLAENDER WEG 5
D 2057 REINBECK
GERMANY
TEL 40 722 3616*194
TLF
TLX
EML
COM 19,31

EPPS HARLAND WARREN PROF
DPT OF ASTRONOMY
UNIVERSITY OF CALIFORNIA
MATH SCI RM 8983
LOS ANGELES CA 90024
USA
TEL 213 825 3025
TLF
TLX 910-3427597
EML
COM

EPSTEIN EUGENE E DR
AEROSPACE CORPORATION
2118 PATRICIA AVE
LOS ANGELES CA 90009
USA
TEL 213 648 6798
TLX 664460
EML
TLF
COM 40,51

EPSTEIN GABRIEL LEO DR
NASA/GSFC
CODE 682
GREENBELT MD 20771
USA
TEL
TLF
TLX
EML
COM· 12,14

EPSTEIN ISADORE PROF
DPT OF ASTRONOMY
COLUMBIA UNIVERSITY
PUPIN HALL 538 W 120TH ST
NEW YORK NY 10027
USA
TEL 212 280 3280
TLF
TLX 125953 COLUMBIA
EML
COM 35

EPSTEIN RICHARD I DR
LOS ALAMOS NATIONAL LAB
MS 436
BOX 1663
LOS ALAMOS NM 87545
USA
TEL 505 667 9595
TLF
TLX
EML
COM

ERCAN E NIHAL
KANDILLI OBSERVATORY
BOGAZICI UNIVERSITY
CENGELKOY
81220 ISTANBUL
TURKEY
TEL 1 332 0240/41
TLF
TLX 26411 BOUN TR
EML
COM

ERDI B DR
ASTRONOMY DPT
EOTVOS UNIVERSITY
KUN BELA TER 2
H 1083 BUDAPEST
HUNGARY
TEL 1 41 019
TLF
TLX
EML
COM 07

ERGMA E V DR
DPT THEORET PHYS & ASTOPH
TARTU UNIVERSITY
ULIKOOLI 18
202400 TARTU
ESTONIA
TEL 73775
TLF
TLX 412623 SCSTP SU
EML
COM 35,49

ERICKSON WILLIAM C DR
DPT OF PHYSICS
UNIVERSITY OF TASMANIA
GPO BOX 252C
HOBART TAS 7001
AUSTRALIA
TEL 2 202 401
TLF 2 202 186
TLX AA58150
EML
COM 40

ERIGUCHI YOSHIHARU DR
DPT EARTH SCI & ASTRONOMY
UNIVERSITY OF TOKYO
MEGURO KU
TOKYO 153
JAPAN
TEL 03-467-1171x439
TLF
TLX 25510 UNITOKYO
EML
COM 35

ERIKSEN GUNNAR PROF
INST THEORET ASTROPHYSICS
UNIVERSITY OF OSLO
BOX 1029
N 0315 BLINDERN OSLO 3
NORWAY
TEL 02-45-65-15
TLF
TLX
EML
COM 40

ERIKSSON KJELL DR
ASTRONOMICAL OBSERVATORY
BOX 515
S 751 20 UPPSALA
SWEDEN
TEL 18 11 2488
TLF
TLX 76024 UNIVUPS S
EML
COM 36

ERPYLEV N P DR
INST OF ASTRONOMY
ACADEMY OF SCIENCES
PYATNITSKAYA UL 48
109017 MOSCOW
RUSSIA
TEL 231-54-61
TLF
TLX 412623 SCSTP SU
EML
COM 41

ERSHKOVICH ALEXANDER PROF
DPT GEOPHYS & PLANET SCI
TEL AVIV UNIVERSITY
TEL AVIV
ISRAEL
TEL 3 413 505
TLF
TLX 342171 VERSY IL
EML
COM 15

ERTAN A YENER DR
FACULTY OF SCIENCE
EGE UNIVERSITY
BOX 21
35100 BORNOVA IZMIR
TURKEY
TEL·
TLF
TLX
EML
COM

ESCALANTE VLADIMIR DR
INSTITUTO DE ASTRONOMIA
UNAM
APDO POSTAL 70-264
04510 MEXICO DF
MEXICO
TEL 525 5 485305
TLF 525 5 483712
TLX 017 60155 CICME
EML' VLADIMIR@ALFA ASTROS UNAM MX
COM 34

ESHLEMAN VON R PROF
DURAND 221
STANFORD UNIVERSITY
STANFORD CA 94305
USA
TEL 415 497 3531
TLF
TLX
EML
COM 16,40,49

ESIPOV VALENTIN F DR
STERNBERG STATE ASTR INST
117234 MOSCOW
RUSSIA
TEL
TLF
EML
COM 34
TLX

ESKIOGLU A NIHAT
DEVLET MUHENDISLIK
MIMARLIK AKADEMISI
ADAPAZARI SAKARYA
TURKEY
TEL
TLF
TLX
EML
COM 27

ESKRIDGE PAUL B DR
CENTER FOR ASTROPHYSICS
MS 81
60 GARDEN ST
CAMBRIDGE MA 02138
USA
TEL 617 496 7585
TLF
TLX
EML ESKRIDGE@CFA HARVARD EDU
COM

ESPOSITO F PAUL PROF
DPT OF PHYSICS
UNIVERSITY OF CINCINNATI
210 BRAUNSTEIN ML 11
CINCINNATI OH 45221 0111
USA
TEL 513 475 2233
TLF
TLX
EML
COM

ESPOSITO LARRY W
LASP
UNIVERSITY OF COLORADO
BOX 392
BOULDER CO 80309 0392
USA
TEL 303 492 7325
TLF
TLX
EML
COM 16

ESSER RUTH DR
INST MATHS & PHYSICAL SCI
UNIVERSITY OF TROMSO
BOX 953
N 9000 TROMSO
NORWAY
TEL 47 83 45164
TLF 47 83 89852
TLX 64124 SUROB N
EML RUTH-ESSER@ESUIT UIT NO
COM 12

ESTALELLA ROBERT
DPT FISICA DE ATMOSFERA
UNIVERSIDAD DE BARCELONA
AVD DIAGONAL 645
E 08028 BARCELONA
SPAIN
TEL 3 330 7311/298
TLF
TLX
EML
COM

ETZEL PAUL B DR
DPT OF ASTRONOMY
SAN DIEGO STATE UNIV
SAN DIEGO CA 92182
USA
TEL 619 495 6169
TLF 619 594 5485
TLX
EML ETZEL@MINTAKA SDSU EDU
COM 42

EUBANKS THOMAS M DR
7243 ARCHLAW DR
CLIFTON VA 22024 2126
USA
TEL
TLF
EML
COM
TLX

EVANGELIDIS E DR
PLASMA PHYSICS DIVISION
NUCOR / PELINDABA
PRIVATE BAG X256
PRETORIA 0001
SOUTH AFRICA
TEL 27-12-21-3311
TLF
TLX 30253 SA
EML
COM 33,36

EVANS ANEURIN
DPT OF PHYSICS
UNIVERSITY OF KEELE
KEELE ST5 5BG
UK
TEL 782 621 111
TLF
TLX 36113 UNKLIB G
EML
COM 27,34

EVANS J V DR
COMSAT LABORATORIES
22300 COMSAT DR
CLARKSBURG MD 20871
USA
TEL 301 428 4422
TLF
TLX 908753
EML
COM 12

EVANS JOHN W DR
1 BAYA RD
ELDORADO
SANTA FE NM 87503
USA
TEL
TLF
TLX
EML
COM

EVANS KENTON DOWER DR
DPT OF PHYSICS
UNIVERSITY OF LEICESTER
UNIVERSITY RD
LEICESTER LE1 7RN
UK
TEL 533 554 455
TLF
TLX 341664
EML
COM 40

EVANS NANCY REMAGE DR
SAL/ISTS
YORK UNIVERSITY
4700 KEELE ST
NORTH YORK ON L4K 3C8
CANADA
TEL
TLF 416 736 5386
TLX
EML
COM 27

EVANS NEAL J II ASS PROF
ASTRONOMY DPT
UNIVERSITY OF TEXAS
RLM 15 308
AUSTIN TX 78712 1083
USA
TEL 512 471 4461
TLF
TLX
EML
COM 34,51

EVANS ROBERT REV
57 TALBOT RD
HAZELBROOK NSW 2779
AUSTRALIA
TEL
TLF
EML
COM 28
TLX

EVANS ROGER G DR
RUTHERFORD APPLETON LAB
SPACE & ASTROPHYSICS DIV
BLDG R25/R68
CHILTON DIDCOT OX11 0QX
UK
TEL 235 219 00
TLF
TLX 83159 RUTHLB G
EML
COM 28

EVANS W DOYLE
390 EL CONEJO
LOS ALAMOS NM 87544
USA
TEL 505 667 3644
TLF
TLX
EML
COM 48

EVDOKIMOV YU V DR
ENGELHARDT ASTRONOMICAL
OBSERVATORY
OBSERVATORIA STATION
422526 KAZAN
RUSSIA
TEL
TLF
TLX
EML
COM 20

EVIATAR AHARON PROF
DPT GEOPHYS & PLANET SCI
TEL AVIV UNIVERSITY
TEL AVIV
ISRAEL
TEL 3 420 620
TLF
TLX 342171 VERSY IL
EML
COM 15,49

EVREN SERDAR DR
FACULTY OF SCIENCE
EGE UNIVERSITY
BOX 21
35100 BORNOVA IZMIR
TURKEY
TEL 51 18 0110*2322
TLF
TLX
EML EFEASTO1@TREARN BITNET
COM 27

EWEN HAROLD I DR
HILLCREST DRIVE 60
BEAVER
SOUTH DEERFIELD MA 01373
USA
TEL
TLF
TLX
EML
COM

EWING MARTIN S
YALE UNIVERSITY SECF
10 HILLHOUSE AVE 233DL
NEW HAVEN CT 06520
USA
TEL 203 432 4342
TLF 203 432 2797
TLX 9102508365 YALE
EML EWING@YALEVM YCC YALE EDU
COM 40

EZER-ERYURT DILHAN PROF
DPT OF PHYSICS
MIDDLE EAST TECH UNIV
06531 ANKARA
TURKEY
TEL 41 23 7100*3255
TLF
TLX 42761 ODTK TR
EML
COM 35

FABBIANO GIUSEPPINA
CENTER FOR ASTROPHYSICS
HCO/SAO
60 GARDEN ST
CAMBRIDGE MA 02138
USA
TEL 617 495 7204
TLF
TLX 921428 SATELLITE CAM
EML
COM 28

FABER SANDRA M PROF
LICK OBSERVATORY
UNIVERSITY OF CALIFORNIA
SANTA CRUZ CA 95064
USA
TEL 408 429 2944
TLF
TLX
EML
COM 28,33,47

FABIAN ANDREW C DR
INSTITUTE OF ASTRONOMY
THE OBSERVATORIES
MADINGLEY RD
CAMBRIDGE CB3 0HA
UK
TEL
TLF
TLX
EML
COM 48

FABREGAT JUAN DR
DPT MATEMATICA Y ASTRON
UNIVERSIDAD DE VALENCIA
BURJASOT
E 46100 VALENCIA
SPAIN
TEL 6 386 4573
TLF 6 386 4735
TLX
EML 16444 FABREGAT
COM 25

FABRICANT DANIEL G
CENTER FOR ASTROPHYSICS
HCO/SAO
60 GARDEN ST
CAMBRIDGE MA 02138
USA
TEL 617 495 7398
TLF
TLX 921428 SATELLITE CAM
EML
COM 09,28,44

FABRICIUS CLAUS V
COPENHAGEN UNIVERSITY OBS
BRORFELDEVEJ 23
DK 4340 TOLLOSE
DENMARK
TEL 53 48 8195
TLF 58 48 8755
TLX 44155 DANAST DK
EML
COM 08

FACONDI SILVIA ROSA DR
IST DI RADIOASTRONOMIA
CNR
VIA IRNERIO 46
I 40126 BOLOGNA
ITALY
TEL 51 28 7835
TLF 51 24 3130
TLX 520634 INFN BO
EML SFACONDI@ASTBO1 INFN IT
COM 40

FADEYEV YURI A
INST OF ASTRONOMY
ACADEMY OF SCIENCES
PYATNITSKAYA UL 48
109017 MOSCOW
RUSSIA
TEL 231-54-61
TLF
TLX 412623 SCSTP SU
EML
COM 27,35

FAELTHAMMAR CARL GUNNE PR
DPT OF PLASMA PHYSICS
ROYAL INST OF TECHNOLOGY
S 100 44 STOCKHOLM 70
SWEDEN
TEL 86 87 7685
TLF
TLX 10389 KTHB
EML
COM

FAHLMAN GREGORY G DR
DPT GEOPHYS & ASTRONOMY
UNIV OF BRITISH COLUMBIA
2075 WESBROOK PL
VANCOUVER BC V6T 1W5
CANADA
TEL 604 228 4891
TLF 604 228 6047
TLX
EML
COM

FAHR HANS JOERG PROF DR
INSTITUT FUR ASTROPHYSIK
DER UNIVERSITAET BONN
AUF DEM HUEGEL 71
D 5300 BONN 1
GERMANY
TEL 228 73 3677
TLF
TLX 886440
EML
COM 49

FAHY EDWARD F PROF
DPT OF PHYSICS
UNIVERSITY COLLEGE CORK
CORK
IRELAND
TEL 1 212 6871
TLF
TLX 26050
EML
COM

FAIRALL ANTHONY P PROF
DPT OF ASTRONOMY
UNIVERSITY OF CAPE TOWN
RONDEBOSCH 7700
SOUTH AFRICA
TEL 21 698531 x 629
TLF
TLX 5721439
EML
COM 28,30C,46

FALCHI AMBRETTA
OSS ASTROFISICO
DI ARCETRI
LARGO E FERMI 5
I 50125 FIRENZE
ITALY
TEL 55 275 2236
TLF
TLX 572268 ARCETR I
EML
COM 10

FALCIANI ROBERTO DR
DPT DI ASTRONOMIA
UNIVERSITA DI FIRENZE
LARGO E FERMI 5
I 50125 FIRENZE
ITALY
TEL 55 27521
TLF 55 22 0039
TLX 572268 ARCETR I
EML
COM 10,12

FALCO-ACOSTA EMILIO E DR
CENTER FOR ASTROPHYSICS
HCO/SAO MS 19
60 GARDEN ST
CAMBRIDGE MA 02138
USA
TEL 617 495 7131
TLF 617 495 7467
TLX 921428 SATELLITE CAM
EML FALCO@CFA HARVARD EDU
COM 28

FALGARONE EDITH
CALTECH
MS 405 47
DOWNES LAB OF PHYSICS
PASADENA CA 91125
USA
TEL 818 243 2438
TLF
TLX
EML
COM 34C

FALK SYDNEY W JR DR
ASTRONOMY DPT
UNIVERSITY OF TEXAS
RLM 15 308
AUSTIN TX 78712 1083
USA
TEL
TLF
TLX
EML
COM 34,47

FALL S MICHAEL DR
STSCI
HOMEWOOD CAMPUS
3700 SAN MARTIN DR
BALTIMORE MD 21218
USA
TEL
TLF
TLX
EML
COM 28,33,37,47

FALLE SAMUEL A DR
DPT OF APPL MATHEMATICS
UNIVERSITY OF LEEDS
LEEDS LS2 9JT
UK
TEL 532 431 751
TLF
TLX
EML
COM 34

FALLER JAMES E PROF
JILA/NBS
UNIVERSITY OF COLORADO
BOX 400
BOULDER CO 80309 0440
USA
TEL 303 492 8509
TLF
TLX 755842 JILA
EML
COM

FALLON FREDERICK W DR
NGS
N/CG 114 NOAA
6010 EXECUTIVE BLVD
ROCKVILLE MD 20852
USA
TEL 301 443 8424
TLF
TLX
EML
COM 24,31

FALOMO RENATO DR
OSS ASTRONOMICO DI PADOVA
VIC DELL OSSERVATORIO 5
I 35122 PADOVA
ITALY
TEL 49 66 1499
TLF
TLX 432071 ASTROS I
EML ASTRPD FALOMO
COM

FAN YING
DPT OF ASTRONOMY
BEIJING NORMAL UNIVERSITY
BEIJING 100875
CHINA PR
TEL 1 65 3531*6285
TLF
TLX
EML
COM 34

FANG CHENG
DPT OF ASTRONOMY
NANJING UNIVERSITY
NANJING
CHINA PR
TEL 25 34651*2882
TLF
TLX 34151 PRCNU CN
EML
COM 10,12

FANG LI-ZHI
DPT OF PHYSICS
UNIVERSITY OF ARIZONA
TUCSON AZ 85721
USA
TEL
TLF
TLX
EML
COM 47,48

FANSELOW JOHN LYMAN
JPL
MS 264 748
4800 OAK GROVE DR
PASADENA CA 91109
USA
TEL 213 354 6323
TLF
TLX 675429
EML
COM 24

FANTI CARLA GIOVANNINI
IST DI RADIOASTRONOMIA
CNR
VIA IRNERIO 46
I 40126 BOLOGNA
ITALY
TEL 51 23 2856/57
TLF
TLX 211664 INFN BO
EML
COM

FANTI ROBERTO
IST DI FISICA
UNIVERSITA DI BOLOGNA
VIA IRNERIO 46
I 40126 BOLOGNA
ITALY
TEL 51 23 2856/57
TLF
TLX 211664 INFN BO
EML
COM 40C

FARAGGIANA ROSANNA PROF
OAT
BOX SUCC TRIESTE 5
VIA TIEPOLO 11
I 34131 TRIESTE
ITALY
TEL 40 79 3921
TLF
TLX 461137 OAT I
EML
COM 29,36,44,45

FARINELLA PAOLO DR
IST DI MATEMATICA
UNIVERSITA DI PISA
VIA BUONARROTI 2
I 56127 PISA
ITALY
TEL 50 59 9524
TLF 50 59 9524
TLX 500371 CNUCE I
EML TWIN2@ICNUCEVM BITNET
COM 07,15,16

FARNIK FRANTISEK
ASTRONOMICAL INSTITUTE
CZECH ACADEMY OF SCIENCES
ONDREJOV OBSERVATORY
CS 251 65 ONDREJOV
CZECHOSLOVAKIA
TEL 204 85201
TLF 204 85314
TLX 121579
EML
COM 10

FAUCHER PAUL DR
OCA OBSERV DE NICE
BP 139
F 06003 NICE CDX
FRANCE
TEL 93 89 0420
TLF·
TLX 460004 OBSNICE F
EML
COM 14

FAULKNER DONALD J DR
MOUNT STROMLO & SIDING
SPRING OBSERVATORIES
PRIVATE BAG
WODEN PO ACT 2606
AUSTRALIA
TEL 62 88 1111
TLF
TLX 62270
EML
COM 34,35

FAULKNER JOHN PROF
LICK OBSERVATORY
UNIVERSITY OF CALIFORNIA
SANTA CRUZ CA 95064
USA
TEL 408 429 2815
TLF
TLX
EML
COM 35,42

FAUNDEZ-ABANS M DR	FAUROBERT-SCHOLL M DR	FAWELL DEREK R DR	FAY THEODORE D DR
DPT DE FISICA	OCA OBSERV DE NICE	UNIVERSITY OF LONDON OBS	TELEDYNE BROWN ENG
UNIVERSIDA DE SANTIAGO	BP 139	MILL HILL PARK	CUMMINGS RES PARK
CASILLA 307 2	F 06003 NICE CDX	LONDON NW7 2QS	MS 19
SANTIAGO	FRANCE	UK	HUNTSVILLE AL 35807
CHILE	TEL 92 00 3011	TEL	USA
TEL	TLF 92 00 3033	TLF	TEL
TLF	TLX 460004 OBSNICE F	TLX	TLF
TLX	EML FAUROB@FRONI51 BITNET	EML	TLX
EML	COM 36	COM	EML
COM			COM

FAZIO GIOVANNI G DR	FEAST MICHAEL W PROF	FEAUTRIER NICOLE DR	FECHTIG HUGO DR
CENTER FOR ASTROPHYSICS	SAAO	OBSERVATOIRE DE PARIS	SANSERWEG 3
HCO/SAO	BOX 9	SECTION DE MEUDON	
60 GARDEN ST	OBSERVATORY 7935	F 92195 MEUDON PPL CDX	D 6906 LEIMEN
CAMBRIDGE MA 02138	SOUTH AFRICA	FRANCE	GERMANY
USA	TEL 27 21 47 00 25	TEL 1 45 07 7552	TEL
TEL 617 495 7458	TLF 27 21 473639	TLF	TLF
TLF 617 495 7490	TLX 520309	TLX 201571	TLX
TLX 921428 SATELLITE CAM	EML	EML	EML
EML FAZIO@CFA BITNET	COM 27,28,29,33,37,45	COM 14	COM 15,21,22
COM 44VP,48,51			

FEDERICI LUCIANA	FEDERMAN STEVEN ROBERT	FEDOROVA RIMMA T DR	FEGAN DAVID J DR
DPT DI ASTRONOMIA	DPT PHYSICS & ASTRONOMY	NIKOLAEV OBSERVATORY	DPT OF PHYSICS
UNIVERSITA DI BOLOGNA	UNIVERSITY OF TOLEDO	UKRAINIAN ACAD OF SCIENCE	UNIVERSITY COLLEGE
CP 596	2801 W BANCROFT ST	327000 NIKOLAEV	BELFIELD
I 40100 BOLOGNA	TOLEDO OH 43606	UKRAINE	DUBLIN 4
ITALY	USA	TEL 37 5714	IRELAND
TEL 51 22 2956	TEL 419 537 2652	TLF	TEL 1 693 244
TLF	TLF	TLX	TLF
TLX 211664 INFN BO I	TLX	EML	TLX 32693 UCD EI
EML	EML	COM 08	EML
COM 14	COM 14,34		COM

FEHRENBACH CHARLES PROF	FEIBELMAN WALTER A DR	FEIGELSON ERIC D DR	FEINSTEIN ALEJANDRO DR
LES MAGNANARELLES	NASA/GSFC	DPT OF ASTRONOMY	OBSERVATORIO ASTRONOMICO
LOURMARIN	CODE 685	PENNSYLVANIA STATE UNIV	PASEO DEL BOSQUE
F 84160 CADENET	GREENBELT MD 20771	525 DAVEY LAB	1900 LA PLATA (BS AS)
FRANCE	USA	UNIVERSITY PARK PA 16802	ARGENTINA
TEL 90 68 0028	TEL 301 286 5272	USA	TEL 21 21 6357
TLF	TLF	TEL 814 865 0162	TLF 21 21 1761
TLX	TLX	TLF	TLX 31216 CESLA AR
EML	EML	TLX 842 510	EML AFEINSTEIN@FCAGLP EDU AR
COM 09,30,33,45	COM 10,27,34	EML INTERNET EDF@ASTRO PSU EDU	COM 25,37VP
		COM 40	

FEISSEL MARTINE DR	FEITZINGER JOHANNES PROF	FEIX GERHARD DR	FEJES ISTVAN DR
OBSERVATOIRE DE PARIS	ASTRONOMISCHES INSTITUT	RUHR UNIVERSITAET BOCHUM	FOMI SATELLITE
61 AVE OBSERVATOIRE	RUHR UNIVERSITAET BOCHUM	DEPT XII	GEODETIC OBSERVATORY
F 75014 PARIS	POSTFACH 102148	POSTFACH 102148	BOX 546
FRANCE	D 4630 BOCHUM 1	D 4630 BOCHUM 1	H 1373 BUDAPEST
TEL 1 40 51 2226	GERMANY	GERMANY	HUNGARY
TLF 1 40 51 2232	TEL 234 70 03450	TEL 234 70 02051	TEL
TLX 270776 OBS F	TLF	TLF	TLF
EML IAPOBS· FEISSEL/FEISSEL@FRIAP5	TLX 825860	TLX 0825860	TLX
COM 08,19C31	EML	EML	EML
	COM 28,33,34	COM 40	COM 51

FEKEL FRANCIS C
DYER OBSERVATORY
VANDERBILT UNIVERSITY
NASHVILLE TN 37235
USA
TEL 615 322 2804
TLF
TLX
EML
COM 26C,30,42

FELDMAN PAUL A DR
HERZBERG INST ASTROPHYS
NTL RESEARCH COUNCIL
100 SUSSEX DR
OTTAWA ON K1A 0R6
CANADA
TEL 613 993 6060
TLF 613 952 6602
TLX 053 3715
EML
COM 40,51

FELDMAN PAUL DONALD DR
DPT PHYSICS & ASTRONOMY
JOHNS HOPKINS UNIVERSITY
CHARLES & 34TH ST
BALTIMORE MD 21218
USA
TEL 301 338 7339
TLF
TLX 710-234-1090
EML
COM 15,21,44

FELDMAN URI
NAVAL RESEARCH LABORATORY
HOLBURT CTR FOR SPACES RE
4555 OVERLOOK AVE SW
WASHINGTON DC 20375 5000
USA
TEL 202 767 3286
TLF
TLX
EML
COM 12

FELDMAN URI DR
NAVAL RESEARCH LABORATORY
CODE 4174
4555 OVERLOOK AVE SW
WASHINGTON DC 20375 5000
USA
TEL
TLF
TLX
EML
COM

FELENBOK PAUL DR
OBSERVATOIRE DE PARIS
SECTION DE MEUDON
F 92195 MEUDON PPL CDX
FRANCE
TEL 1 45 07 7523
TLF
TLX 201571
EML
COM 09,29

FELLI MARCELLO DR
OSS ASTROFISICO
DI ARCETRI
LARGO E FERMI 5
I 50125 FIRENZE
ITALY
TEL 55 275 2240
TLF
TLX 572268
EML
COM 34,40

FELTEN JAMES E DR
NASA/GSFC
CODE 685
GREENBELT MD 20771
USA
TEL 301 552 1526
TLF
TLX
EML
COM 34,40,47,48

FENG HESHENG
YUNNAN OBSERVATORY
CAS
BOX 110
KUNMING 72946 YUNNAN
CHINA PR
TEL 871 2035
TLF
TLX
EML
COM

FENG KE-JIA
DPT OF ASTRONOMY
BEIJING NORMAL UNIVERSITY
BEIJING 100875
CHINA PR
TEL 1 65 3531*6967
TLF
TLX
EML
COM 10,46

FENKART ROLF P PROF DR
ASTRONOMISCHES INSTITUT
UNIVERSITAET BASEL
VENUSSTRASSE 7
CH 4102 BINNINGEN
SWITZERLAND
TEL 61 22 7711
TLF
TLX
EML
COM 33

FENTON K B DR
DPT OF PHYSICS
UNIVERSITY OF TASMANIA
GPO BOX 252C
HOBART TAS 7001
AUSTRALIA
TEL 2 202 411
TLF
TLX 58150 AA
EML
COM 48

FERETTI LUIGINA
IST DI RADIOASTRONOMIA
CNR
VIA IRNERIO 46
I 40126 BOLOGNA
ITALY
TEL 51 23 2856
TLF
TLX 520634 INFNBO I
EML
COM 40

FERLAND GARY JOSEPH
DPT OF ASTRONOMY
OHIO STATE UNIVERSITY
174 W 18TH AVE
COLUMBUS OH 43210 1106
USA
TEL 614 422 1773
TLF
TLX 8104821715
EML
COM 27,28

FERLET ROGER DR
INSTITUT D'ASTROPHYSIQUE
98BIS BD ARAGO
F 75014 PARIS
FRANCE
TEL 1 43 20 1425
TLF 1 43 29 8673
TLX
EML EARN FERLET@FRIAP51
COM 34

FERLUGA STENO DR
DPT DI ASTRONOMIA
UNIVERSITA DI TRIESTE
VIA TIEPOLO 11
I 34131 TRIESTE
ITALY
TEL 40 76 3912
TLF 40 30 9418
TLX 461137 OAT I
EML ASTRONET ASTRTS FERLUGA
COM 42

FERNANDEZ JEAN-CLAUDE DR
OCA OBSERV DE NICE
BP 139
F 06003 NICE CDX
FRANCE
TEL 93 89 0420
TLF
TLX 460004 OBSNICE F
EML
COM 15

FERNANDEZ JULIO A DR
DPT DE ASTRONOMIA
FACUT HUMANIDADES Y CIENC
TRISTAN NARVAJA 1674
MONTEVIDEO
URUGUAY
TEL 598 2 49 11 044
TLF
TLX UDELAR UY 26692
EML
COM 15,20,46

FERNANDEZ SILVIA M DR
OBSERVATORIO ASTRONOMICO
DE CORDOBA
LAPRIDA 854
5000 CORDOBA
ARGENTINA
TEL 51 40613
TLF
TLX 51822 BUCOR
EML
COM 07

FERNANDEZ-FIGUEROA M J DR
DPT DE ASTROFISICA
FAC DE FISICA
UNIVERSIDAD COMPLUTENSE
E 28040 MADRID
SPAIN
TEL 1 449 5316
TLF
TLX 47273 FF UC
EML
COM 29,46

FERNANZ MARGARITA DR
CSIC
CAMI DE STA BARABARA
E 17300 BLANES GIRONA
SPAIN
TEL 72 33 6101/02
TLF 972 33 7806
TLX 56372
EML EARN MARGA@CEAB ES
COM 35

FERNIE J DONALD PROF
DAVID OUNLAP OBSERVATORY
UNIVERSITY OF TORONTO
BOX 360
RICHMOND HILL ON L4C 4Y6
CANADA
TEL 416 884 9562
TLF 416 978 3921
TLX 069 86766 TELEXPERTS
EML fernie@centaur astro utoronto
COM 25,27,41

FERRANDIZ JOSE MANUEL DR
ETS
INGENIEROS INDUSTRIALES
PASEO DEL CAUCE S/N
E 47011 VALLADOLID
SPAIN
TEL 83 30 4700
TLF 83 39 2026
TLX
EML
COM

FERRANDO PHILIPPE DR
CEA CEN
DPHG/SAP
BP 2
F 91191 GIF/YVETTE CDX
FRANCE
TEL 1 69 08 2020
TLF
TLX 604860
EML BITNET FERRANDO@32779
COM

FERRARI ATTILIO DR
IST DI FISICA
UNIVERSITA DI TORINO
CORSO D AZEGLIO 46
I 10125 TORINO
ITALY
TEL 11 65 7694
TLF
TLX 211041 INFN TO I
EML
COM 40,48

FERRARI D'OCCHIEPPO K DR
OESTERREICHISCHE AKADEMIE
DER WISSENSCHAFTEN
DR IGNAZ-SEIPEL-PLATZ 2
A 1010 WIEN
AUSTRIA
TEL 5222 81991
TLF
TLX 01-12628
EML
COM 41,42

FERRARI TONIOLO MARCO
IAS
CNR
CP 67
I 00044 FRASCATI
ITALY
TEL 6 942 5651
TLF 6 941 6847
TLX 610261 CNR-FRA I
EML
COM 44

FERRAZ-MELLO S PROF DR
IAG
UNIVERSIDADE DE SAO PAULO
CP 30627
01065 SAO PAULO SP
BRAZIL
TEL 11 577 8599*27
TLF 11 276 3848
TLX 1156735 IAGM BR
EML SYLVIO@IAG USP ANSP BR
COM 07VP,20,46

FERRER MARTINEZ SEBASTIAN
DPT FIS TIERRA & COSMOS
UNIVERSIDAD DE ZARAGOZA
E 50009 ZARAGOZA
SPAIN
TEL 76 35 7011
TLF
TLX 58198
EML
COM 07

FERRER OSVALDO EDUARDO DR
UNIV NACIONAL DE LA PLATA
FCAG
1900 LA PLATA (BS AS)
ARGENTINA
TEL 21 21 7308
TLF
TLX
EML
COM 26,42

FERRERI WALTER
OSS ASTRONOMICO DI TORINO
ST OSSERVATORIO 20
I 10025 PINO TORINESE
ITALY
TEL 11 84 2040
TLF
TLX 213236 ASTR I
EML
COM 20

FERRIN IGNACIO
UNIVERSIDAD DE LOS ANDES
FACULTAD DE CIENCIAS
DPT DE FISICA
MERIDA 5101 A
VENEZUELA
TEL 74 63 9930
TLF
TLX
EML
COM 15

FERRINI FEDERICO
IST DI ASTRONOMIA
UNIVERSITA DI PISA
PIAZZA TORRICELLI 2
I 56100 PISA
ITALY
TEL 50 43343
TLF
TLX
EML
COM 28,34

FESTOU MICHEL C DR
OBS MIDI PYRENEES
14 AVE E BELIN
F 31400 TOULOUSE CDX
FRANCE
TEL 61 25 2101
TLF
TLX 530776
EML
COM 15

FEYNMAN JOAN DR
JPL
MS 144 218
4800 OAK GROVE DR
PASADENA CA 91109
USA
TEL 818 354 3454
TLF
TLX 675429
EML
COM 49

FIALA ALAN D DR
US NAVAL OBSERVATORY
NAUTICAL ALMANAC OFFICE
34 & MASSACHUSETTS AVE NW
WASHINGTON DC 20392 5100
USA
TEL 202 653 1274
TLF
TLX 710-822-1970
EML
COM 04,07,12

FICARRA ANTONINO DR
IST DI RADIOASTRONOMIA
CNR
VIA IRNERIO 46
I 40126 BOLOGNA
ITALY
TEL 51 23 2856
TLF
TLX 211664 INFN BO I
EML
COM

FICH MICHEL DR
DPT OF PHYSICS
UNIVERSITY OF WATERLOO
WATERLOO ON N2L 3G1
CANADA
TEL 519 885 1572
TLF 519 746 8115
TLX 069 55259
EML BITNET FICH@WATSC1
COM

FICHTEL CARL E DR
NASA/GSFC
CODE 660
GREENBELT MD 20771
USA
TEL 301 286 6281
TLF
TLX 89675
EML
COM 44,48

FIEDLER RALPH L DR
NAVAL RESEARCH LABORATORY
CODE 4210
4555 OVERLOOK AVE SW
WASHINGTON DC 20375 5000
USA
TEL 202 767 0644
TLF
TLX
EML FIEDLER@RIRA NRL NAVY MIL
COM

FIELD DAVID
SCHOOL OF CHEMISTRY
CANTOCKS CLOSE
BRISTOL BS8 1TS
UK
TEL 272 24 161*505
TLF
TLX 444174 BUPHYS
EML
COM 34

FIELD GEORGE B PROF
CENTER FOR ASTROPHYSICS
HCO/SAO
60 GARDEN ST
CAMBRIDGE MA 02138
USA
TEL 617 495 4721
TLF
TLX 921428 SATELLITE CAM
EML
COM 28,34,40,47,48,49,51

FIELDER GILBERT DR
WILLOW TREE
ELDROTH
AUSTWICK
LANCASTER LA2 8AH
UK
TEL
TLF
TLX
EML
COM 16

FIENBERG RICHARD T DR
SKY & TELESCOPE
49 BAY STATE ROAD
CAMBRIDGE MA 02138
USA
TEL 617 864 7360
TLF
TLX
EML SPAN CFA FIENBERG
COM 46

FIERRO JULIETA
INSTITUTO DE ASTRONOMIA
UNAM
APDO POSTAL 70-264
04510 MEXICO DF
MEXICO
TEL
TLF
TLX
EML
COM 34,46C

FIGUERAS FRANCESCA DR
DPT ASTRONOMIA
UNIVERSIDAD DE BARCELONA
AVD DIAGONAL 647
E 08028 BARCELONA
SPAIN
TEL 3 330 7311
TLF
TLX
EML D3FAFFSO●EBOUBO11
COM 33

FILIPOV LATCHEZAR
CTR LAB FOR SPACE RES
BULGARIAN ACAD SCIENCES
MOSKOVA ST 6
BG 1000 SOFIA
BULGARIA
TEL 87 0978
TLF
TLX 23351 CLSR BG
EML
COM

FILIPPENKO ALEXEI V DR
ASTRONOMY DPT
UNIVERSITY OF CALIFORNIA
601 CAMPBELL HALL
BERKELEY CA 94720
USA
TEL 415 642 1813
TLF
TLX 820181
EML BITNET ALEX●BKYAST
COM 06,28,47

FILLOY EMILIO MANUEL E E
IAR
CC 5
1894 VILLA ELISA (BS AS)
ARGENTINA
TEL 21 4 3793
TLX
EML
TLF
COM

FINDLAY JOHN W DR
NRAO
EDGEMONT RD
CHARLOTTESVILLE VA 22901
USA
TEL
TLX 910-997-0174
EML
TLF
COM 40

FINK UWE DR
LUNAR & PLANETARY LAB
UNIVERSITY OF ARIZONA
TUCSON AZ 85721
USA
TEL 602 621 2736
TLF
TLX 9109521143
EML
COM 14,16

FINKELSTEIN ANDREJ M DR
INST OF APPLIED ASTRONOMY
ACADEMY OF SCIENCES
ZDANOVSKAYA UL 8
197042 ST PETERSBURG
RUSSIA
TEL
TLF
TLX
EML
COM 40

FINN G D DR

USA
TEL
TLF
EML
TLX

COM 36

FINZI ARRIGO DR
DPT OF MATHEMATICS
IIT
TECHNION CITY
HAIFA 32000
ISRAEL
TEL
TLF
TLX 46406 TECON IT
EML
COM

FIRMANI CLAUDIO A PROF
INSTITUTO DE ASTRONOMIA
UNAM
APDO POSTAL 70-264
04510 MEXICO DF
MEXICO
TEL 905-548-3712
TLF
TLX 1760155 CICME
EML
COM 42

FIRNEIS FRIEDRICH J DR
INST INFO PROC/OEAW
SONNENFELSGASSE 19/2
A 1010 WIEN
AUSTRIA
TEL
TLF
TLX
EML
COM 24,51

FIRNEIS MARIA G DR
INSTITUT FUER ASTRONOMIE
TUERKENSCHANZSTR 17
A 1180 WIEN
AUSTRIA
TEL 1 345 3600
TLF
TLX
EML
COM 24,41C,51

FIROR JOHN W DR
HIGH ALTITUDE OBSERVATORY
NCAR
BOX 3000
BOULDER CO 80307 3000
USA
TEL 303 497 1600
TLF
TLX 45694
EML
COM

FISCHEL DAVID DR
EARTH OBSERVATION
SATELLITE COMPANY
4300 FORBES BLVD
LANHAM MD 20706
USA
TEL 301 552 0500
TLF
TLX 277685
EML
COM

FISCHER JACQUELINE
NAVAL RESEARCH LABORATORY
CODE 4138F
4555 OVERLOOK AVE SW
WASHINGTON DC 20375 5000
USA
TEL 202 767 3058
TLF
TLX
EML
COM 34

FISCHER STANISLAV DR
ASTRONOMICAL INSTITUTE
CZECH ACADEMY SCIENCES
BUDECSKA 6
CS 120 23 PRAHA 2
CZECHOSLOVAKIA
TEL   2 25 2438
TLF
TLX 122486 ASTRC
EML
COM

FISHER GEORGE HEWITT DR
INSTITUTE FOR ASTRONOMY
UNIVERSITY OF HAWAII
2680 WOODLAWN DR
HONOLULU HI 96822
USA
TEL 808 956 6665
TLF 808 988 2790
TLX
EML FISHER@UHIFA IFA HAWAII EDU
COM 10,12

FISHER J RICHARD
NOAO
BOX 2
GREEN BANK WV 24944
USA
TEL   304 456 2011
TLF
TLX 7109381530
EML
COM

FISHER PHILIP C
RUFFNER ASSOCIATES
BOX 7070
MENLO PARK CA 94026
USA
TEL
TLF
TLX
EML
COM 44,48,51

FISHER RICHARD R DR
HIGH ALTITUDE OBSERVATORY
NCAR
BOX 3000
BOULDER CO 80307 3000
USA
TEL 303 494 5151
TLF
TLX
EML
COM

FISHKOVA LUISA M PROF
ABASTUMANI ASTROPHYSICAL
OBSERVATORY
GEORGIAN ACAD OF SCIENCES
383762 ABASTUMANI
GEORGIA
TEL·
TLF
TLX
EML
COM 21

FISHMAN GERALD J
NASA/MSFC
CODE ES 62
ASTROPHYSICS BRANCH
HUNTSVILLE AL 35812
USA
TEL 205 453 0117
TLF
TLX
EML
COM 44

FITCH WALTER S DR
BOX 100

ORACLE AZ 85623
USA
TEL 602 896 2911
TLF
TLX 467175
EML
COM 27

FITTON BRIAN DR
ESA/ESTEC
ASTROPHYSICS DIV
BOX 299
NL 2200 AG NOORDWIJK
NETHERLANDS
TEL 25 24 4635
TLF
TLX
EML
COM 44

FITZGERALD M PIM PROF
DPT OF PHYSICS
UNIVERSITY OF WATERLOO
WATERLOO ON N2L 3G1
CANADA
TEL 519 885 1572
TLF 519 746 8115
TLX 069 55259
EML
COM 33,37

FITZPATRICK EDWARD L DR
PRINCETON UNIVERSITY OBS
PEYTON HALL
PRINCETON NJ 08544
USA
TEL 609 452 3702
TLF 609 243 7333
TLX
EML FITZ@ASTROVAX PRINCETON EDU
COM 29,36,45

FIX JOHN D DR
DPT PHYSICS & ASTRONOMY
UNIVERSITY OF IOWA
IOWA CITY IA 52240
USA
TEL 319 353 7064
TLF
TLX 910-525-1398
EML
COM

FLANNERY BRIAN PAUL DR
EXXON RES & ENGINEERING
ROUTE 22 EAST
ANNANDALE NJ 08801
USA
TEL 201 730 2540
TLF
TLX 136140 EXXONRES
EML
COM 34,35,42

FLECK ROBERT CHARLES DR
DPT OF MATH & PHYSICAL SC
EMBRY-RIDDLE AERON UNIVER
DAYTONA FL 32114
USA
TEL 904 226 6612
TLF 904 226 6713
TLX
EML
COM 34,46

FLEISCHER ROBERT DR
18712 KEEDYSVILLE RD
KEEDYSVILLE MD 21756 1112
USA
TEL
TLF
TLX
EML
COM 40

FLETCHER J MURRAY
HERZBERG INST ASTROPHYS
DOMINION ASTROPHYS OBS
5071 W SAANICH RD
VICTORIA BC V8X 4M6
CANADA
TEL 604 388 3905
TLF 604 363 0045
TLX 049 7295
EML
COM 09,26,30

FLETT ALISTAIR M
DPT OF PHYSICS
UNIVERSITY OF ABERDEEN
FRASER NOBLE BLDG
ABERDEEN AB9 2UE
UK
TEL 224 40 241
TLF
TLX 73458 UNIABN G
EML
COM 40

FLIEGEL HENRY F
3730 EL MORENO AVE
BOX 8682
LA CRESCENTA CA 91214
USA
TEL 213 648 7452
TLF 213 336 5076
TLX
EML
COM 19,31VP

FLIN PIOTR
ASTRONOMICAL OBSERVATORY
JAGIELLONIAN UNIVERSITY
UL ORLA 171
PL 30 244 KRAKOW
POLAND
TEL
TLF
TLX 0322297 UJ PL
EML
COM 28

FLOQUET MICHELE DR
OBSERVATOIRE DE PARIS
SECTION DE MEUDON
DEPEG
F 92195 MEUDON PPL CDX
FRANCE
TEL 1 45 07 7851
TLF
TLX
EML
COM 29

FLORENTIN-NIELSEN RALPH
COPENHAGEN UNIVERSITY OBS
BRORFELDEVEJ 23
DK 4340 TOLLOSE
DENMARK
TEL   53 48 8195
TLF   58 48 8755
TLX   44155 DANAST
EML
COM

FLORIDES PETROS S PROF
SCHOOL OF MATHEMATICS
TRINITY COLLEGE
DUBLIN 2
IRELAND
TEL   1 772 941
TLF
TLX   25442 TCD EI
EML
COM   41,47

FLORKOWSKI DAVID R DR
US NAVAL OBSERVATORY
34 & MASSACHUSETTS AVE NW
WASHINGTON DC 20392 5100
USA
TEL   202 653 1509
TLF   202 653 0587
TLX
EML   DRF@PHOBOS USNO NAVY MIL
COM   40

FLORSCH ALPHONSE DR
OBS DE STRASBOURG
11 RUE UNIVERSITE
F 67000 STRASBOURG
FRANCE
TEL   88 35 4300
TLF   88 25 0160
TLX   890506 STAROBS F
EML
COM   28,30,38

FLOWER DAVID R DR
DPT OF PHYSICS
UNIVERSITY OF DURHAM
SOUTH RD
DURHAM DH1 3LE
UK
TEL   913 742 145
TLF   913 743 749
TLX   537351
EML   DAVID FLOWER@DURHAM AC UK
COM   14,34VP

FOCARDI PAOLA DR
OSS ASTRONOMICO
UNIVERSITA DI BOLOGNA
CP 596
I 40100 BOLOGNA
ITALY
TEL   51 25 9301
TLF
TLX   520634 INFNBO I
EML   SPAN ASTBO3 PAOLA
COM   47

FODERA SERIIO GIORGIA DR
OSS ASTRONOMICO
UNIVERSITA DI PALERNO
PALAZZO DEI NORMANNI
I 90134 PALERMO
ITALY
TEL   91 65 70451
TLF   91 48 8900
TLX   910402 ASTROP I
EML   ASTROPA@IPACUC BITNET
COM   41

FOFI MASSIMO DR
OAR
VIA DEL PARCO MELLINI 84
I 00136 ROMA
ITALY
TEL   6 34 7056
TLF
TLX   626326 OAROMA I
EML   40061 FOFI
COM   12

FOGARTY WILLIAM G DR
IBM
NCMD
411 EAST WISCONSIN AVE
MILWAUKEE WI 53202
USA
TEL
TLF
TLX
EML
COM

FOING BERNARD H DR
IAS
BP 10
F 91371 VERRIERES BUISSON
FRANCE
TEL   1 64 47 4328
TLF
TLX   600252
EML   SPAN IAPOBS FOING
COM   10,29,44

FOLEY ANTHONY DR
NFRA
BOX 2
NL 7990 AA DWINGELOO
NETHERLANDS
TEL   52 19 7244
TLF   52 19 7332
TLX   42043 SRZM NL
EML   FOLRED@NFRA NL
COM   40

FOLTZ CRAIG B
MULT MIRROR TELESCOPE OBS
UNIVERSITY OF ARIZONA
TUCSON AZ 85721
USA
TEL   602 621 1269
TLF
-TLX   467175
EML
COM   28,30

FOMALONT EDWARD B DR
NRAO
BOX O
SOCORRO NM 87801 0387
USA
TEL
TLX   910-988-1710
EML
TLF
COM   40

FOMENKO ALEXANDR F DR
SPECIAL ASTROPHYSICAL OBS
ACADEMY OF SCIENCES
NIZHNIJ ARKHYZ
357147 STAVROPOLSKIJ
RUSSIA
TEL
TLF
TLX
EML
COM   09

FOMICHEV VALERI V DR
IZMIRAN
ACADEMY OF SCIENCES
142092 TROITSK
RUSSIA
TEL
TLF
TLX
EML
COM   12

FOMIN VALERY A DR
PULKOVO OBSERVATORY
ACADEMY OF SCIENCES
10 KUTUZOV QUAY
196140 ST PETERSBURG
RUSSIA
TEL
TLF
TLX
EML
COM   08

FOMINOV ALEXANDR M DR
INST OF THEORET ASTRONOMY
ACADEMY OF SCIENCES
N KUTUZOVA 10
191187 ST PETERSBURG
RUSSIA
TEL   278-88-90
TLF
TLX   121578 ITA SU
EML
COM   04

FONG CHU-GANG
SHANGHAI OBSERVATORY
CAS
80 NANDAN RD
SHANGHAI
CHINA PR
TEL   21 38 6191
TLF
TLX   33164 SHAO CN
EML
COM   07,19

FONG RICHARD
DPT OF PHYSICS
UNIVERSITY OF DURHAM
SOUTH RD
DURHAM DH1 3LE
UK
TEL   913 856 4971
TLF
TLX   537351
EML
COM   47

FONTAINE GILLES DR
DPT DE PHYSIQUE
UNIVERSITY OF MONTREAL
CP 6128 SUCC A
MONTREAL QC H3C 3J7
CANADA
TEL   514 343 6680
TLF   514 343 2071
TLX   055 62425
EML
COM   35,36

FONTENLA JUAN MANUEL DR
NASA/MSFC
CODE ES 52
HUNTSVILLE AL 35812
USA
TEL
TLF
TLX
EML
COM 12,36

FORBES DOUGLAS DR
SIR WILFRED GRENFELL COLL
MEMORIAL UNIVERSITY
NEWFOUNDLAND
CORNER BROOK NF A2H 6P9
CANADA
TEL 709 637 6295
TLF 709 637 6390
TLX 016 44190
EML DFORBES@KEAN UCS MUN CA
COM 37

FORBES J E DR
BOX 88120
INDIANAPOLIS IN 46208
USA
TEL
TLF
TLX
EML
COM 35

FORBES TERRY G DR
SPACE SCIENCE CENTER/EOS
UNIV OF NEW HAMPSHIRE
SC ENGIN RES BLDG
DURHAM NH 03857
USA
TEL 603 862 3872
TLF
TLX 950030 UNH PHYS
EML BITNET%"T_FORBES@UNHH"
COM 10C

FORD HOLLAND C RES PROF
STSCI
HOMEWOOD CAMPUS
3700 SAN MARTIN DR
BALTIMORE MD 21218
USA
TEL 301 338 4803
TLF
TLX
EML
COM 28,34,47

FORD W KENT JR DR
DPT TERRESTR MAGNETISM
CARNEGIE INST WASHINGTON
5241 BROAD BRANCH RD NW
WASHINGTON DC 20015
USA
TEL 202 966 0863
TLF
TLX 440427 MAGN UI
EML
COM 09,28

FORMAN WILLIAM RICHARD DR
SMITHSONIAN ASTROPHYS OBS
60 GARDEN ST
CAMBRIDGE MA 02138
USA
TEL 617 495 7210
TLF
TLX 921428
EML
COM 47,48

FORREST WILLIAM JOHN
DPT PHYSICS & ASTRONOMY
UNIVERSITY OF ROCHESTER
ROCHESTER NY 14627
USA
TEL 716 275 4343
TLF
TLX
EML FORREST@BORIS PAS ROCHEST
COM

FORSTER JAMES RICHARD DR
HAT CREEK RADIO OBSERV
RT 2
BOX 500
CASSEL CA 96016
USA
TEL 916 335 2364
TLF 916 335 3968
TLX
EML
COM 34

FORT BERNARD P DR
OBS MIDI PYRENEES
14 AVE E BELIN
F 31400 TOULOUSE CDX
FRANCE
TEL 61 25 2101
TLF
TLX 530776 F
EML
COM 09

FORT DAVID NORMAN DR
JPL
4800 OAK GROVE DR
PASADENA CA 91109
USA
TEL 818 354 9132
TLF
TLX
EML
COM 40

FORTE JUAN CARLOS DR
UNIV NACIONAL DE LA PLATA
FCAG
1900 LA PLATA (BS AS)
ARGENTINA
TEL 21 21 7308
TLF
TLX
EML
COM 25,28,37

FORTI GIUSEPPE DR
OSS ASTROFISICO
DI ARCETRI
LARGO E FERMI 5
I 50125 FIRENZE
ITALY
TEL 55 275 2236
TLF
TLX 572268 ARCETR
EML FORTI@ARCETRI ASTRO IT
COM 15,20,22

FORTINI TERESA DR
VIA F D GUERRAZZI 19
I 00152 ROMA
ITALY
TEL
TLF
EML
COM 10
TLX

FORVEILLE THIERRY DR
OBSERVATOIRE DE GRENOBLE
CERMO/ASTROPHYSIQUE
BP 53X
F 38041 S MARTIN HERES CD
FRANCE
TEL 76 51 4567
TLF 76 44 8821
TLX
EML FORVEILL@FRGAG51 BITNET
COM 40

FOSBURY ROBERT A E DR
ST/ECF
C/O ESO
KARL-SCHWARZSCHILD-STR 2
D 8046 GARCHING MUENCHEN
GERMANY
TEL 89 320 06235
TLF
TLX 52828222 EO D
EML
COM

FOSSAT ERIC DR
OCA OBSERV DE NICE
BP 139
F 06003 NICE CDX
FRANCE
TEL 93 89 0420
TLF
TLX 460004 OBSNICE F
EML
COM 10,12,35

FOUKAL PETER V DR
CAMBRIDGE RESEARCH &
INSTRUMENTATIONM INC
21 ERIE ST
CAMBRIDGE MA 02139
USA
TEL 617 491 2627
TLF
TLX
EML
COM 12

FOUQUE PASCAL DR
GRUPO ASTROFIS PONTIFICA
UNIVERSIDAD CATOLICA
CASILLA 104
SANTIAGO 22
CHILE
TEL
TLF
TLX
EML PFOUQUE@ASTROUC PUC CL
COM 28,40,47

FOWLER WILLIAM A PROF
CALTECH
MS 106 38
PASADENA CA 91125
USA
TEL 818 356 4272
TLF
TLX
EML
COM 35,48

FOX KENNETH DR
DPT PHYSICS & ASTRONOMY
UNIVERSITY OF TENNESSEE
503 PHYSICS
KNOXVILLE TN 37996 1200
USA
TEL   615 974 2288
TLF
TLX
EML
COM 16

FOX W E MR
BRITISH ASTRONOMICAL ASS
40 WINDSOR RD
NEWARK NOTTINGHAMSHIRE
UK
TEL 636 704 932
TLF
TLX
EML
COM 16

FOY RENAUD DR
OCA OBSERV DU CALERN
CAUSSOLS
F 06460 S VALLIER THIEY
FRANCE
TEL 93 42 6270
TLF
TLX 461402 CERGLOBS
EML
COM 09,29,36

FRACASTORO MARIO G PROF
VIA MONVISO 3

I 10025 PINO TORINESE
ITALY
TEL 11 84 0493
TLF
TLX
EML
COM 24,26,42

FRANCESCHINI ALBERTO
DPT DI FISICA G GALILAI
UNIVERSITA DI PADOVA
VIA MARZOLO 8
I 35131 PADOVA
ITALY
TEL 49 84 4111
TLF 49 84 4245
TLX
EML
COM 47,48

FRANCHINI MARIAGRAZIA DR
OAT
BOX SUCC TRIESTE 5
VIA TIEPOLO 11
I 34131 TRIESTE
ITALY
TEL 40 319 9111
TLF 40 30 9418
TLX 461137 OAT I
EML 38439 FRANCHINI
COM 29

FRANCO JOSE DR
INSTITUTO DE ASTRONOMIA
UNAM
APDO POSTAL 70-264
04510 MEXICO DF
MEXICO
TEL
TLF
TLX
EML
COM

FRANCOIS PATRICK DR
OBSERVATOIRE DE PARIS
SECTION DE MEUDON
DASGAL
F 92195 MEUDON PPL CDX
FRANCE
TEL 1 45 97 7867
TLF 1 45 07 7472
TLX 270912
EML
COM 29

FRANDSEN SOEREN PROF
INST OF PHYSICS & ASTRON
UNIVERSITY OF AARHUS
NY MUNKEGADE
DK 8000 AARHUS C
DENMARK
TEL 86 12 8899
TLF 86 20 2711
TLX 64767 AAUSCI DK
EML
COM 29,48

FRANK JUHAN
DPT PHYSICS & ASTRONOMY
LOUISIANA STATE UNIV
BATON ROUGE LA 70803 4001
USA
TEL 504 388 6845
TLF 504 388 5855
TLX
EML PHRANK@LSUVM
COM 42,48

FRANKLIN FRED A DR
CENTER FOR ASTROPHYSICS
PLANETARY SCIENCE DIV
60 GARDEN ST
CAMBRIDGE MA 02138
USA
TEL 617 495 7230
TLF
TLX
EML
COM 20

FRANSSON CLAES
STOCKHOLM OBSERVATORY

S 133 36 SALTSJOEBADEN
SWEDEN
TEL 87 17 0195
TLF 87 17 4719
TLX 12972 SOBSERV S
EML CLAES@ASTRO SU SE
COM 48

FRANTSMAN YU L DR
RADIOASTROPHYSICAL OBS
LATVIAN ACAD OF SCIENCES
TURGENEVA 19
226524 RIGA
LATVIA
TEL 226006
TLF
TLX
EML
COM 35,42

FRANZ OTTO G DR
LOWELL OBSERVATORY
1400 W MARS HILL RD
BOX 1149
FLAGSTAFF AZ 86001
USA
TEL 602 774 3358
TLF
TLX
EML
COM 24,26

FRATER ROBERT H DR
CSIRO
CHIEF DIV RADIOPHYSICS
BOX 76
EPPING NSW 2121
AUSTRALIA
TEL 2 868 0222
TLF
TLX 26230 ASTRO
EML
COM 40

FRAZIER EDWARD N DR
TRW

1 SPACE PARK
REDONDO BEACH CA 90278
USA
TEL 213 535 4723
TLF
TLX
EML
COM 12

FREDGA KERSTIN PROF
SWEDISH BOARD F SPACE ACT
BOX 4006
S 171 54 SOLNA
SWEDEN
TEL 87 33 6486
TLX 17128 SPACECO S
EML
TLF
COM 44

FREDRICK LAURENCE W PROF
UNIVERSITY STATION
UNIVERSITY OF VIRGINIA
BOX 3818
CHARLOTTESVILLE VA 22903
USA
TEL 804 924 4905
TLF
TLX 510-587-5453 (TWX)
EML
COM 24,26,42,51

FREEDMAN WENDY L DR
CARNEGIE OBSERVATORIES
INSTITUTION OF WASHINGTON
813 SANTA BARBARA ST
PASADENA CA 91101
USA
TEL 818 577 1122
TLF
TLX
EML wendy@mwlco caltech edu
COM 28

FREEMAN KENNETH C PROF
MOUNT STROMLO OBSERVATORY
PRIVATE BAG
WODEN PO
CANBERRA ACT 2606
AUSTRALIA
TEL 62 881 111
TLF
TLX 62270 CANOPUS AA
EML
COM 28C,30C,33,37

FREIRE FERRERO RUBENS G
IAS
BP 10
F 91371 VERRIERES BUISSON
FRANCE
TEL 1 64 47 4245
TLF
TLX 600252
EML
COM 29,36

FREITAS MOURAO R R DR
MUSEUM ASTR E CIENCIAS
RUA GL BRUCE 586
SAN CRISTOVAO
20921 RIO DE JANEIRO RJ
BRAZIL
TEL 21 580 7154/7204
TLF 21 580 0332
TLX 22653
EML
COM 20,26,41

FRENCH RICHARD G
DPT OF EARTH & PLANET SCI
MIT RM 54 422
BOX 165
CAMBRIDGE MA 02139
USA
TEL 617 253 3392
TLF
TLX
EML
COM

FRENK CARLOS S
DPT OF PHYSICS
UNIVERSITY OF DURHAM
SOUTH RD
DURHAM DH1 3LE
UK
TEL 913 856 4971
TLF
TLX 537351 DURLIB G
EML
COM 47

FRESNEAU ALAIN DR
OBS DE STRASBOURG
11 RUE UNIVERSITE
F 67000 STRASBOURG
FRANCE
TEL 88 35 8200
TLF 88 25 0160
TLX 890506 STAROBS F
EML
COM 24

FRIBERG PER
JOINT ASTRONOMY CENTER
UKIRT
665 KOMOHANA ST
HILO HI 96720
USA
TEL
TLF
TLX
EML FRIBERG@JACH HAWAII EDU
COM 40

FRICKE KLAUS DR
UNIVERSITAETSSTERNWARTE
UNIVERSITAET GOETTINGEN
GEISMARLANDSTR 11
D 3400 GOETTINGEN
GERMANY
TEL 551 39 5051
TLF
TLX 96753
EML
COM 28

FRIDLUND MALCOLM DR
ESA/ESTEC
ASTROPHYSICS DIV
BOX 299
NL 2200 AG NOORDWIJK
NETHERLANDS
TEL 17 19 84768
TLF 17 19 84690
TLX 39098
EML ESTCS51 MFRIDLUN
COM 34

FRIDMAN ALEKSEY M DR
INST OF ASTRONOMY
ACADEMY OF SCIENCES
PYATNITSKAYA UL 48
109017 MOSCOW
RUSSIA
TEL 231 5461
TLF
TLX 412623 SCSTP SU
EML
COM 33C

FRIED JOSEF WILHELM DR
MPI FUER ASTRONOMIE
KOENIGSTUHL
D 6900 HEIDELBERG 1
GERMANY
TEL 62 215 281
TLX 461789 MPIA D
EML
TLF
COM 28

FRIEDEMANN CHRISTIAN DR
UNIV STERNWARTE JENA
SCHILLERGAESSCHEN 2
D 6900 JENA
GERMANY
TEL 78 822 2637/27122
TLF
TLX 05886134
EML
COM 34

FRIEDJUNG MICHAEL DR
INSTITUT D'ASTROPHYSIQUE
98BIS BD ARAGO
F 75014 PARIS
FRANCE
TEL 1 43 20 1425
TLF 1 43 29 8673
TLX
EML FRIEDJUNG@FRIAP51
COM 27,29,42

FRIEDLANDER MICHAEL PROF
DPT OF PHYSICS
WASHINGTON UNIVERSITY
ST LOUIS MO 63130
USA
TEL 314 889 6279
TLF
TLX
EML
COM

FRIEDMAN HERBERT DR
NAVAL RESEARCH LABORATORY
CODE 7100
4555 OVERLOOK AVE SW
WASHINGTON DC 20375 5000
USA
TEL
TLF
TLX
EML
COM 10,12,40,44,48

FRIEDMAN SCOTT DAVID DR
DPT PHYSICS & ASTRONOMY
JOHNS HOPKINS UNIVERSITY
CHARLES & 34TH ST
BALTIMORE MD 21218
USA
TEL 301 338 5317
TLF 301 338 5494
TLX
EML SCIVAX CASA SCOTT
COM

FRIEL EILEEN D DR
HERZBERG INST ASTROPHYS
DOMINION ASTROPHYS OBS
5071 W SAANICH RD
VICTORIA BC V8X 4M6
CANADA
TEL 604 388 0062
TLF 604 363 0045
TLX 049 7295
EML
COM 29

FRIEND DAVID B DR
DPT OF PHYS & ASTRONOMY
WILLIAMS COLLEGE
WILLIAMSTOWN MA 01267
USA
TEL 413 597 2817
TLF
TLX
EML BITNET DBFRIEND@WILLIAMS
COM 36

FRINGANT ANNE-MARIE DR
OBSERVATOIRE DE PARIS
61 AVE OBSERVATOIRE
F 75014 PARIS
FRANCE
TEL 1 40 51 2248
TLF
TLX 270776 OBS F
EML
COM

FRISCH HELENE DR
OCA OBSERV DE NICE
BP 139
F 06003 NICE COX
FRANCE
TEL 93 89 0420
TLF
TLX 460004 OBSNICE F
EML
COM 36

FRISCH PRISCILLA
ASTRONOMY & ASTROPHYS CTR
UNIVERSITY OF CHICAGO
5640 S ELLIS AVE
CHICAGO IL 60637
USA
TEL 312 962 8211
TLF
TLX 910 221 5617
EML FRISCH@ODDJOB UCHICAGO EDU
COM 34

FRISCH URIEL DR
OCA OBSERV DE NICE
BP 139
F 06003 NICE CDX
FRANCE
TEL 93 89 0420
TLF
TLX 460004 OBSNICE F
EML
COM 36

FRISK URBAN DR
SPACE SCIENCE SECTION
SWEDISH SPACE CORPORATION
BOX 4207
S 171 04 SOLNA
SWEDEN
TEL 86 27 6200
TLF 89 87069
TLX
EML
COM 40,44

FRITZOVA-SVESTKA L DR
DOPPERSTRAAT 147
NL 3752 JC BUNSCHOTEN
NETHERLANDS
TEL 34 99 84403
TLF
TLX
EML
COM

FROEHLICH CLAUS
WORLD RADIATION CENTER
PHYSIKALISCH-METEOROL OBS
POSFACH 173
CH 7260 DAVOS-DORF
SWITZERLAND
TEL 83 52131
TLF 83 35058
TLX 853232 PMOD CH
EML
COM 12

FROESCHLE CHRISTIANE D DR
OCA OBSERV DE NICE
BP 139
F 06003 NICE CDX
FRANCE
TEL 93 89 0420
TLF
TLX 460004 OBSNICE F
EML
COM 15,36

FROESCHLE CLAUDE DR
OCA OBSERV DE NICE
BP 139
F 06003 NICE CDX
FRANCE
TEL 93 89 0420
TLF
TLX 460004 OBSNICE F
EML
COM 07C,20

FROESCHLE MICHEL DR
OCA CERGA
AVE COPERNIC
F 06130 GRASSE
FRANCE
TEL 93 35 5849
TLF
TLX
EML
COM 08

FROGEL JAY ALBERT DR
DPT OF ASTRONOMY
OHIO STATE UNIVERSITY
174 W 18TH AVE
COLUMBUS OH 43210 1106
USA
TEL 614 292 5651
TLF
TLX
EML FROGEL@OHSTPY BITNET
COM 28

FROLOV M S DR
INST OF ASTRONOMY
ACADEMY OF SCIENCES
PYATNITSKAYA UL 48
109017 MOSCOW
RUSSIA
TEL 231 54 61
TLF
TLX 412623 SCSTP SU
EML
COM 27

FROST KENNETH J DR
NASA/GSFC
CODE 600 2 BLDG 16
SPACE STATION OFFICE
GREENBELT MD 20771
USA
TEL 301 286 8824
TLF
TLX
EML
COM

FRYE GLENN M PROF
DPT OF PHYSICS
CASE WESTERN RESERVE UNIV
ROCK BDG
CLEVELAND OH 44106
USA
TEL 216 368 2997
TLF
TLX
EML
COM

FTACLAS CHRIST
PERKIN ELEMR CORP
MS 897
100 WOOSTER HEIGHTS RD
DANBURY CT 06810 7859
USA
TEL 203 797 6448
TLF
TLX
EML
COM 28

FU CHENG-QI
SHANGHAI OBSERVATORY
CAS
80 NANDAN RD
SHANGHAI
CHINA PR
TEL 21 38 6191
TLF
TLX 33164 SHAO CN
EML
COM 44

FU DELIAN
BEIJING ASTRONOMICAL OBS
CAS
W SUBURB
BEIJING 100080
CHINA PR
TEL 1 28 2070
TLF
TLX 22040 BAOAS CN
EML
COM 09

FU QI JUN
BEIJING ASTRONOMICAL OBS
CAS
W SUBURB
BEIJING 100080
CHINA PR
TEL 1 27 5580
TLF
TLX 22040 BAOAS CN
EML
COM 10

FUCHS BURKHARD DR
ASTRONOMISCHES-RECHEN
INSTITUT
MOENCHHOFSTR 12-14
D 6900 HEIDELBERG 1
GERMANY
TEL 62 214 9026
TLF
TLX 461336 ARIHD D
EML
COM 28,33

FUENMAYOR FRANCISCO J DR
UNIVERSIDAD DE LOS ANDES
FACULDAD DE CIENCIAS
DPT DE FISICA
MERIDA 5101 A
VENEZUELA
TEL 74 63 9930
TLF
TLX 74173 CDCH-ULA
EML
COM

FUENSALIDA JIMENEZ J DR
INST DE ASTROFISICA
DE CANARIAS
OBS DEL TEIDE
E 38200 LA LAGUNA
SPAIN
TEL 22 26 2211
TLF
TLX 92640
EML
COM

FUERST ERNST DR
MPI FUER RADIOASTRONOMIE
AUF DEM HUEGEL 69
D 5300 BONN 1
GERMANY
TEL
TLF
TLX 886440
EML
COM 40

FUHR JEFFREY ROBERT DR
NTL BUREAU OF STANDARDS
PHYSICS BLDG RM A267
GAITHERSBURG MD 20899
USA
TEL 301 975 3204
TLF 301 975 3038
TLX 197674 NIST UT
EML FUHR@NISTCS2
COM 14

FUJIMOTO MASAYUKI DR
FAC OF EDUCATION
NIIGATA UNIVERSITY
8050 IKARASHI 2
NIIGATA 950 21
JAPAN
TEL
TLF
TLX
EML
COM 28,35

FUJIMOTO MASA-KATSU DR
TOKYO ASTRONOMICAL OBS
NAOJ
OSAWA MITAKA
TOKYO 181
JAPAN
TEL 422 32 5111
TLF
TLX 2822307 TAOMTK J
EML
COM 31C,33,51

FUJIMOTO MITSUAKI DR
DPT OF PHYSICS
NAGOYA UNIVERSITY
FUROCHO CHIKUSA KU
NAGOYA 464
JAPAN
TEL
TLF
TLX
EML
COM 47,51

FUJISHITA MITSUMI DR
DIV OF EARTH ROTATION
NTL ASTRONO OBSERVATORY
2-12 HOSHIGAOKA
IWATE 023
JAPAN
TEL 197 24 7111
TLF
TLX 837628 ILSMIZ J
EML TELEMAIL ID MIZUSAWA
COM 08,19

FUJITA YOSHIO PROF
DPT OF ASTRONOMY
UNIVERSITY OF TOKYO
BUNKYO KU
TOKYO 113
JAPAN
TEL. 423-74-4186
TLF·
TLX
EML
COM 29

FUJIWARA AKIRA DR
DPT OF PHYSICS
KYOTO UNIVERSITY
SAKYO KU
KYOTO 606
JAPAN
TEL 075-751-2111
TLF
TLX 5422693 LIBKYU J
EML
COM 15,16,21

FUJIWARA TAKAO DR
KYOTO CITY UNIV OF ARTS
NISHIKYO KU
KYOTO 610 11
JAPAN
TEL 75 332 0701
TLF 75 332 0709
TLX
EML C53926@JPNKUDPC BITNET
COM 33

FUKUDA ICHIRO
KANAZAWA TECHNOLOGY INST
7-1 OGIGAOKA
NONOICHIMACHI
ISHIKAWA 921
JAPAN
TEL 0762-48-1100
TLF
TLX 5122456 KIT LC J
EML
COM 45

FUKUE JUN DR
ASTRONOMICAL INSTITUTE
OSAKA KYOIKU UNIVERSITY
4 88 MINAMIKAWAHORICHO
OSAKA 543
JAPAN
TEL
TLF
TLX
EML
COM

FUKUGITA MASATAKA DR
RES INST FUNDAMENTAL PHYS
KYOTO UNIVERSITY
KITASHIRAKAWA SAKYO KU
KYOTO 606
JAPAN
TEL 075 711 1381
TLF
TLX 5423179 RIFPK
EML
COM 28,47

FUKUI TAKAO DR
DPT OF LIBERAL ARTS
DOKKYO UNIVERSITY
SAKAE-MACHI 600
SOKA SAITAMA
JAPAN
TEL 0489-42-1111
TLF
TLX
EML
COM 47

FUKUI YASUO DR
DPT OF PHYSICS
NAGOYA UNIVERSITY
FUROCHO CHIKUSA KU
NAGOYA 464
JAPAN
TEL 052-781-5111
TLF
TLX 4477323 SCUNAG J
EML
COM 34,40

FUKUNAGA MASATAKA DR
ASTRONOMICAL INSTITUTE
TOHOKU UNIVERSITY
SENDAI AOBA
MIYAGI 980
JAPAN
TEL 22 222 1800
TLF
TLX
EML
COM 33

FUKUSHIMA TOSHIO DR
HYDROGRAPHIC DPT
GEODESY & GEOPHYSICS DIV
TSUKIJI 5 CHUO KU
TOKYO 104
JAPAN
TEL 3 541 3811
TLF
TLX 2522452 HDJODC J
EML GE MarkIII NODE-RC28 GGDHDJ
COM 04

FULCHIGNONI MARCELLO PROF
IAS
CNR
CP 67
I 00044 FRASCATI
ITALY
TEL 6 495 6951
TLF
TLX 680489 CNR FRA
EML
COM 15,WGPSNC

FURENLID INGEMAR K DR
DPT PHYSICS & ASTRONOMY
GEORGIA STATE UNIVERSITY
ATLANTA GA 30303 3083
USA
TEL 404 658 2932
TLF
TLX
EML
COM 26,29

FURNISS IAN
DPT PHYSICS & ASTRONOMY
UNIVERSITY COLLEGE LONDON
GOWER ST
LONDON WC1E 6BT
UK
TEL 1 713 877 050
TLF
TLX 28722 UCPHYS
EML
COM 34,44

FURSENKO M A DR
INST OF THEORET ASTRONOMY
ACADEMY OF SCIENCES
N KUTUZOVA 10
191187 ST PETERSBURG
RUSSIA
TEL 278-88-98
TLF
TLX 121578 ITA SU
EML
COM 04

FURUKAWA KIICHIRO DR
TOKYO ASTRONOMICAL OBS
NAOJ
OSAWA MITAKA
TOKYO 181
JAPAN
TEL
TLF
TLX
EML
COM 20

FUSCO-FEMIANO ROBERTO
IAS
CNR
CP 67
I 00044 FRASCATI
ITALY
TEL 6 942 5655
TLF
TLX 610261
EML
COM

FUSI PECCI FLAVIO
DPT DI ASTRONOMIA
UNIVERSITA DI BOLOGNA
CP 596
I 40100 BOLOGNA
ITALY
TEL 51 22 2956
TLF
TLX 211664 INFNBO-I
EML
COM 37

FU-SHONG KUO
DPT OF PHYSICS
NTL CENTRAL UNIVERSITY
CHUNG LI
CHINA R
TEL
TLF
TLX
EML:
COM

GABRIEL ALAN H
IAS
BP 10
F 91371 VERRIERES BUISSON
FRANCE
TEL 1 64 47 4315
TLF
TLX 600252
EML
COM 10,12,14C,44

GABRIEL MAURICE R DR
INSTITUT D'ASTROPHYSIQUE
UNIVERSITE DE LIEGE
AVE COINTE 5
B 4000 COINTE-LIEGE
BELGIUM
TEL 41 52 9980
TLF 41 52 7474
TLX 41264 ASTRLG
EML
COM 35

GADSDEN MICHAEL DR
DPT OF PHYSICS
UNIVERSITY OF ABERDEEN
FRASER NOBLE BLDG
ABERDEEN AB9 2UE
UK
TEL 224 57 4585
TLF 224 58 4776
TLX 73458 UNIABN G
EML
COM 21

GAETZ TERRANCE J DR
DPT OF ASTRONOMY
UNIV OF WESTERN ONTARIO
LONDON ON N6A 3K7
CANADA
TEL 519 661 3183
TLF 519 661 3486
TLX 064 7134
EML BITNET GAETZ@UWOVAX
COM

GAHM GOESTA F DR
STOCKHOLM OBSERVATORY
S 133 36 SALTSJOEBADEN
SWEDEN
TEL 87 17 0637
TLF 87 17 4719
TLX 12972
EML GAHM@ASTRO SU SE
COM 27

GAIGNEBET JEAN DR
OCA CERGA
AVE COPERNIC
F 06130 GRASSE
FRANCE
TEL 93 36 5849
TLF
TLX 470865 F
EML
COM 31

GAIL HANS-PETER DR
INST THEORET ASTROPHYSIK
DER UNIVERSITAET ·
IM NEUENHEIMER FELD 294
D 6900 HEIDELBERG 1
GERMANY
TEL
TLF
TLX
EML
COM 36

GAISSER THOMAS K
BARTOL RESEARCH INSTITUTE
UNIVERSITY OF DELAWARE
NEWARK DE 19716
USA
TEL 302 451 8111
TLF
TLX 510-666-0805 BARTOL
EML
COM 48

GAIZAUSKAS VICTOR DR
HERZBERG INST ASTROPHYS
NTL RESEARCH COUNCIL
100 SUSSEX DR
OTTAWA ON K1A 0R6
CANADA
TEL 613 993 7395
TLF 613 952 0974
TLX 053 3715 NRCOTT
EML 18584 VICGZ
COM 10P,12

GALAL A A PROF
HELWAN OBSERVATORY

HELWAN
EGYPT
TEL 78 0645/2683
TLF
TLX
EML
COM 10C

GALAN MAXIMINO J
M & G ENGS
S MARTIN DE PORRES 45
E 28035 MADRID
SPAIN
TEL 1 216 0995
TLX
EML
TLF
COM 09,50

GALEOTTI PIERO PROF
IST DI COSMO GEOFISICA
CNR
CORSO FIUME 4
I 10133 TORINO
ITALY
TEL 11 65 8979
TLF
TLX 224379 COSMOT I
EML
COM 48

GALIBINA I V DR
INST OF THEORET ASTRONOMY
ACADEMY OF SCIENCES
N KUTUZOVA 10
191187 ST PETERSBURG
RUSSIA
TEL 186 19 74
TLF
TLX 121578 ITA SU
EML
COM 07

GALINDO TREJO JESUS DR
INSTITUTO DE ASTRONOMIA
UNAM
APDO POSTAL 70-264
04510 MEXICO DF
MEXICO
TEL 905 5485305
TLF
TLX 1760155 CICME
EML BITNET JGAL@UNAMVM1
COM

GALLAGHER III JOHN S DR
LOWELL OBSERVATORY
1400 W MARS HILL RD
BOX 1149
FLAGSTAFF AZ 86001
USA
TEL· 602 734 3358
TLF
TLX
EML
COM 27,28C

GALLAGHER JEAN W DR
NIST
PHYSICS BUILDINO 221
ROOM A 323
GAITHERSBURG MD 20899
USA
TEL 301 975 2204
TLF. 301 926 0416
TLX 197 674 NISTUT
EML JWG@NISTCS2 BITNET
COM 14

GALLET ROGER M
964 7TH ST
BOULDER CO 80302
USA
TEL
TLF
TLX
EML
COM

GALLETTA GIUSEPPE PROF
DPT DI ASTRONOMIA
UNIVERSITA DI PADOVA
VIC DELL OSSERVATORIO 5
I 35122 PADOVA
ITALY
TEL 49 66 1499
TLF
TLX 430176 UNPADU I
EML
COM 28

GALLETTO DIONIGI
IST DI FISICA MATEMATICA
UNIVERSITA DI TORINO
VIA C ALBERTO 10
I 10123 TORINO
ITALY
TEL 11 53 9214
TLF
TLX
EML
COM 07,33,47

GALLIANO PIER GIORGIO
IST ELETTRONICO NAZIONALE
ST DELLA CACCE 91
I 10135 TORINO
ITALY
TEL 348 8933
TLF
TLX 211553 IENGF I
EML
COM 31

GALLIANO PIER GIORGIO
IST ELETTRONICO NAZIONALE
ST DELLA CACCE 91
I 10135 TORINO
ITALY
TEL 348 8933
TLF
TLX 211553 IENGF I
EML
COM 31

GALLINO ROBERTO
IST DI FISICA GENERALE

CORSO D AZEGLIO 46
I 10125 TORINO
ITALY
TEL 11 65 5103
TLF
TLX 21104 INFN TO
EML
COM 35,36,46

GALLOUET LOUIS DR
OBSERVATOIRE DE PARIS
61 AVE OBSERVATOIRE
F 75014 PARIS
FRANCE
TEL 1 40 51 2207
TLF
TLX 270776 OBS F
EML
COM 24,25

GALLOWAY DAVID DR
DPT OF APPLIED MATHS
UNIVERSITY OF SYDNEY
SYDNEY NSW 2006
AUSTRALIA
TEL 2 692 2222
TLF
TLX 26169 UNISYD AA
EML
COM 10

GALPERIN YU I PROF
SPACE RESEARCH INSTITUTE
ACADEMY OF SCIENCES
PROFSOJUSNAYA UL 84/32
117810 MOSCOW
RUSSIA
TEL 333-31-22
TLF
TLX 411498 STAR SU
EML
COM 21C

GALT JOHN A DR
DOMINION RADIO ASTROPHYS
OBSERVATORY
BOX 248
PENTICTON BC V2A 6K3
CANADA
TEL 604 493 2277
TLF 604 493 7767
TLX 048 88127
EML
COM 40

GAMALELDIN ABDULLA I DR
HELWAN OBSERVATORY
HELWAN
EGYPT
TEL 78 0645/2683
TLF
EML
COM 28
TLX

GAMBIS DANIEL DR
OBSERVATOIRE DE PARIS
61 AVE OBSERVATOIRE
F 75014 PARIS
FRANCE
TEL 1 40 51 2233
TLF
TLX 270776 OBS F
EML GAMBIS@FRIAP51
COM 19,31

GAMMELGAARD PETER MAG SCI
INST OF PHYSICS & ASTRON
UNIVERSITY OF AARHUS
NY MUNKEGADE
DK 8000 AARHUS C
DENMARK
TEL 86 12 8899
TLF 86 20 2711
TLX 64767 AAUSCI DK
EML
COM 15

GAO BILIE
NANJING ASTRONOMICAL
INSTRUMENT FACTORY
NANJING
CHINA PR
TEL 25 46191
TLF
TLX
EML
COM 09

GAO BUXI
INSTITUTE OF GEODESY &
GEOPHYSICS
XU DONG LU
WUCHANG HUBEI
CHINA PR
TEL 81 3805
TLF
TLX
EML
COM 08,19

GAPOSCHKIN EDWARD M DR
55 FARMCREST AVE
LEXINGTON MA 02173
USA
TEL 617 862 2538
TLF
EML
COM 07,19
TLX

GARAY GUIDO DR
DPT DE ASTRONOMIA
UNIVERSIDAD DE CHILE
CASILLA 36 D
SANTIAGO
CHILE
TEL
TLF
TLX
EML
COM 40

GARCIA DE LA ROSA JOSE I
INST DE ASTROFISICA
DE CANARIAS
OBS DEL TEIDE
E 38071 LA LAGUNA
SPAIN
TEL 22 26 2211
TLF
TLX 92640
EML
COM 10

GARCIA DE MARIA J M DR
DPT FISICA APLICADA
RONDA DE VALENCIA 3
E 28012 MADRID
SPAIN
TEL 1 336 7686
TLF 1 530 9244
TLX
EML
COM 42

GARCIA EDUARDO DEL POZO
INST GEOPHYS & ASTRONOMY
CALLE 212 N 2906 29 Y 31
LISA
LA HABANA
CUBA
TEL
TLF
TLX
EML
COM

GARCIA LAMBAS DIEGO DR
UNIV NACIONAL DE LA PLATA
FCAG
1900 LA PLATA (BS AS)
ARGENTINA
TEL 21 21 7308
TLF
TLX
EML
COM 28

GARCIA MICHAEL R DR
CENTER FOR ASTROPHYSICS
HCO/SAO
60 GARDEN ST
CAMBRIDGE MA 02138
USA
TEL 617 495 7169
TLF 617 495 7356
TLX
EML GARCIA@CFA200 HARVARD EDU
COM

GARCIA-BARRETO JOSE A
OBS ASTRONOMICO NACIONAL
UNAM
APDO POSTAL 877
22860 ENSENADA B CALIF
MEXICO
TEL 667-830-93
TLF
TLX
EML
COM

GARCIA-BERRO ENRIQUE DR
DPT FISICA APLICADA
UNIV POLITEC DE CATALUNYA
E 08032 BARCELONA
SPAIN
TEL 3 401 6898
TLF 3 401 6801
TLX 52821 UPC E
EML GARCIA@ETSECCPB UPC ES
COM 12

GARCIA-PELAYO JOSE DR
INST ASTROFISICA
DE ANDULUCIA APD 2144
PROFESOR ALBAREDA 1
E 18080 GRANADA
SPAIN
TEL 58 25 6103
TLF
TLX
EML
COM

GARDNER FRANCIS F DR
CSIRO
DIVISION OF RADIOPHYSICS
BOX 76
EPPING NSW 2121
AUSTRALIA
TEL 2 868 0222
TLF
TLX 26230 AA
EML
COM 34,40

GARFINKEL BORIS DR
YALE UNIVERSITY OBS
YALE STATION
BOX 2023
NEW HAVEN CT 06520
USA
TEL 203 436 3460
TLF
TLX
EML
COM 07,20

GARGAUD MURIEL DR
OBSERVATOIRE DE BORDEAUX
BP 89
F 33270 FLOIRAC
FRANCE
TEL 56 86 4330
TLF 56 40 4251
TLX
EML MURIEL@FROBOR51
COM 14

GARLICK GEORGE F DR
267 SOUTH BELOIT AVE
LOS ANGELES CA 90049
USA
TEL 213 472 3512
TLF
TLX
EML
COM

GARMANY CATHERINE D DR
JILA
UNIVERSITY OF COLORADO
BOX 440
BOULDER CO 80309 0440
USA
TEL 303 492 7836
TLF
TLX
EML
COM 29,42,45

GARMIRE GORDON P PROF
DPT OF ASTRONOMY
PENNSYLVANIA STATE UNVI
525 DAVEY LAB
UNIVERSITY PARK PA 16802
USA
TEL 814 865 0418
TLF
TLX 842510 PENNSTBSTR SC
EML
COM 48

GARNIER ROBERT ING
OBSERVATOIRE DE LYON
AVE CHARLES ANDRE
F 69561 S GENIS LAVAL CDX
FRANCE
TEL 78 56 0705
TLF 72 39 9791
TLX
EML
COM

GARRIDO RAFAEL
INST ASTROFISICA
DE ANDULUCIA APD 2144
PROFESOR ALBAREDA 1
E 18080 GRANADA
SPAIN
TEL 58 12 1311
TLF
TLX 78753
EML
COM 27

GARRINGTON SIMON DR
NRAL

JODRELL BANK
MACCLESFIELD SK11 9DL
UK
TEL 477 712 31
TLF 447 771 618
TLX 36149
EML JANET STG@UK AC MAN JB STAR
COM 40

GARRISON ROBERT F PROF
DAVID DUNLAP OBSERVATORY
UNIVERSITY OF TORONTO
BOX 360
RICHMOND HILL ON L4C 4Y6
CANADA
TEL 416 884 9562
TLF 416 978 3921
TLX 069 86766
EML
COM 29,45,47

GARSTANG ROY H PROF
JILA
UNIVERSITY OF COLORADO
BOX 440
BOULDER CO 80309 0440
USA
TEL 303 492 7795
TLF
TLX 755842 JILA
EML
COM 05,14

GARTON W R S PROF
BLACKETT LABORATORY
IMPERIAL COLLEGE
PRINCE CONSORT RD
LONDON SW7 2BZ
UK
TEL 1 233 21657
TLF
TLX 261503
EML
COM 14

GARY DALE E
CALTECH
MS 264 33
SA
PASADENA CA 91125
USA
TEL 818 356 3863
TLF
TLX 675425 CALTECH PSD
EML
COM

GARY GILMER ALLEN DR
NASA/MSFC
CODE ES 52
SPACE SCIENCE LAB
HUNTSVILLE AL 35812
USA
TEL 205 544 7609
TLF 205 544 5862
TLX 620 26079
EML SPAN SSL GARY
COM 10

GARZON FRANCISCO DR
INST DE ASTROFISICA
DE CANARIAS
OBS DEL TEIDE
E 38200 LA LAGUNA
SPAIN
TEL 22 26 2211
TLF
TLX 92640
EML SPAN IAC FGL
COM· 33

GASCOIGNE S C B DR
MOUNT STROMLO & SIDING
SPRING OBSERTORIES
PRIVATE BAG
WODEN PO ACT 2606
AUSTRALIA
TEL
TLF
TLX
EML
COM 27,28,37

GASKA STANISLAW DR
INSTITUTE OF ASTRONOMY
N COPERNICUS UNIVERSITY
UL CHOPINA 12/18
PL 87 100 TORUN
POLAND
TEL
TLF
TLX
EML
COM 07

GATEWOOD GEORGE DIRECTOR
ALLEGHENY OBSERVATORY
OBSERVATORY STATION
PITTSBURGH PA 15214
USA
TEL 412 321 2400
TLF
TLX
EML
COM 24,26,51

GATLEY IAN
NOAO
BOX 26732
950 N CHERRY AVE
TUCSON AZ 85726 6732
USA
TEL 602 327 5511
TLF
TLX 0666484 AUR
EML·
COM

GAUDENZI SILVIA DR
ISTITUTO ASTRONOMICO
UNIVERSITA DI ROMA
VIA G M LANCISI 29
I 00161 ROMA
ITALY
TEL 6 44 03734
TLF 6 44 03673
TLX 613255 INFNRO
EML 40058 GAUDENZI
COM 26

GAUME RALPH A DR
NAVAL RESEARCH LABORATORY
CODE 4210
4555 OVERLOOK AVE SW
WASHINGTON DC 20375 5000
USA
TEL 202 767 0670
TLF
TLX
EML GAUME@RIRA1 NRL NAVY MIL
COM

GAUR V P
UTTAR PRADESH STATE
OBSERVATORY
PO MANDRA PEAK 263 129
NAINITAL 263 129
INDIA
TEL 59 42 2136
TLF
TLX
EML
COM 12

GAUSS F STEPHEN
US NAVAL OBSERVATORY
34 & MASSACHUSETTS AVE NW
WASHINGTON DC 20392 5100
USA
TEL 202 653 1510
TLX
EML FSG@SICON USNO NAVY MIL
TLF 202 653 1497
COM 08,09

GAUSTAD JOHN E PROF
DPT OF ASTRONOMY
SWARTHMORE COLLEGE
SWARTHMORE PA 19081
USA
TEL 215 328 8271
TLF 215 328 8673
TLX
EML JGAUSTA1@CC SWARTHMORE EDU
COM 34

GAUTIER DANIEL
OBSERVATOIRE DE PARIS
SECTION DE MEUDON
F 92195 MEUDON PPL CDX
FRANCE
TEL 1 45 07 7707
TLF
TLX 201571
EML
COM 16C

GAVAZZI GIUSEPPE DR
OSS ASTRONOMICO DI BRERA
VIA BRERA 28
I 20121 MILANO
ITALY
TEL 2 87 4444
TLX
EML 39216 GAVAZZI
TLF 2 72 00 1600
COM 28

GAY JEAN DR
OCA CERGA
AVE COPERNIC
F 06130 GRASSE
FRANCE
TEL 93 36 5849
TLF
TLX 470865
EML
COM 09,34

GAYLARD MICHAEL JOHN
HARTEBEESTHOEK RADIOASTRO
OBSERVATORY
BOX 443
KRUGERSDORP 1740
SOUTH AFRICA
TEL 27 11 642 4692
TLF 27 11 642 2446
TLX 321006
EML
COM 40

GEAKE JOHN E DR
DPT OF PHYSICS
UMIST
BOX 88
MANCHESTER M60 1QD
UK
TEL 612 363 311
TLF
TLX 666094
EML
COM 16

GEBALLE THOMAS R DR
JOINT ASTRONOMY CENTER
UKIRT
665 KOMOHANA ST
HILO HI 96720
USA
TEL 808 961 3756
TLF
TLX 633135
EML NSSDCA PSI%UKTH TOM
COM 34

GEBBIE KATHARINE B DR
JILA
UNIVERSITY OF COLORADO
BOX 440
BOULDER CO 80309 0440
USA
TEL 303 492 7825
TLF
TLX 755842 JILA
EML
COM 36

GEBLER KARL-HEINZ DR
RADIOASTRONOMISCHES INST
DER UNIVERSITAET BONN
AUF DEM HUEGEL 71
D 5300 BONN 1
GERMANY
TEL 228 73 3662
TLF
TLX
EML
COM 40

GEFFERT MICHAEL DR
OBSERVATORIUM HOHER LIST
D 5568 DAUN
GERMANY
TEL 65 92 2150
TLF
TLX
EML
COM

GEHRELS TOM PROF
LUNAR & PLANETARY LAB
UNIVERSITY OF ARIZONA
TUCSON AZ 85721
USA
TEL 602 621 6970
TLF
TLX
EML
COM 15,16,20,25,34,51

GEHREN THOMAS PH D
INST ASTRON & ASTROPHYSIK
UNIVERSITAETS STERNWARTE
SCHEINERSTR 1
D 8000 MUENCHEN 80
GERMANY
TEL 89 989 021
TLF
TLX 529815 UNIVERS D
EML
COM 29

GEHRZ ROBERT DOUGLAS DR
DPT PHYSICS & ASTRONOMY
UNIVERSITY OF WYOMING
BOX 3905
LARAMIE WY 82071
USA
TEL 307 766 6176
TLF
TLX
EML
COM 25

GEISLER DOUGLAS P DR
CERRO TOLOLO
INTERAMERICAN OBSERVATORY
CASILLA 603
LA SERENA
CHILE
TEL 51 22 5415*208
TLF 51 22 5415*342
TLX 620 301 AURA CT
EML DGEISLER@NOAO EDU
COM 37

GEISS JOHANNES PROF
PHYSIKALISCHES INSTITUT
UNIVERSITAET BERN
SIDLERSTRASSE 5
CH 3012 BERN
SWITZERLAND
TEL 31 65 4402
TLF
TLX 32320
EML
COM 15,16

GELDZAHLER BERNARD J
NAVAL RESEARCH LABORATORY
CODE 4121 6
4555 OVERLOOK AVE SW
WASHINGTON DC 20375 5000
USA
TEL
TLF
TLX
EML
COM 40

GELFREIKH GEORGIJ B DR
PULKOVO OBSERVATORY
ACADEMY OF SCIENCES
10 KUTUZOV QUAY
196140 ST PETERSBURG
RUSSIA
TEL
TLF
TLX
EML
COM 10,40

GELLER MARGARET JOAN
CENTER FOR ASTROPHYSICS
HCO/SAO
60 GARDEN ST
CAMBRIDGE MA 02138
USA
TEL 617 495 7409
TLF
TLX 921428 SATELLITE CAM
EML
COM 28,47C

GEMMO ALESSANDRA DR
DPT DI ASTRONOMIA
UNIVERSITA DI PADOVA
VIC DELL OSSERVATORIO 5
I 35122 PADOVA
ITALY
TEL 49 66 1499
TLF 49 38 919
TLX 432071 ASTROS I
EML 39003 GEMMO
COM 33

GENET R M DR
FAIRBORN OBSERVATORY
3435 E EDGEWOOD ABE
MESA AZ 85204
USA
TEL 602 988 6561*223
TLF
TLX
EML
COM 25C,27

GENKIN IGOR L PROF DR
FAC OF PHYSICS
KAZAKH STATE UNIVERSITY
KOMSOMOLSKAYA 96
480012 ALMA ATA
KAZAKHSTAN
TEL 67-70-18
TLF
TLX
EML
COM 33

GENOVA FRANCOISE DR
CNES/SU/AS
2 PLACE MAURICE QUENTIN
F 75039 PARIS CDX 01
FRANCE
TEL 1 45 08 7741
TLF 1 45 08 7557
TLX 214674
EML
COM

GENT HUBERT MR
PROSPECT HOUSE
SHERWOOD LANE
WORCESTER WR2 4NX
UK
TEL 422 186
TLF
TLX
EML
COM 40

GENZEL REINHARD DR
MAX PLANCK INSTITUT FUER
PHYSIK UND ASTROPHYSIK
KARLSCHARZSCHILD STR 1
D 8046 GARCHING MUENCHEN
GERMANY
TEL 89 329 90
TLF
TLX 524629
EML
COM 34C,40

GEORGELIN YVON P DR
OBSERVATOIRE DE MARSEILLE
2 PLACE LE VERRIER
F 13248 MARSEILLE CDX 04
FRANCE
TEL 91 95 9088
TLF
TLX 420241 F
EML
COM 30,33,34

GEORGELIN YVONNE M DR
OBSERVATOIRE DE MARSEILLE
2 PLACE LE VERRIER
F 13248 MARSEILLE CDX 04
FRANCE
TEL 91 95 9088
TLF
TLX 420241 F
EML
COM 33

GEORGIEV TSVETAN DR
DPT OF ASTRONOMY
BULGARIAN ACAD SCIENCES
72 LENIN BLVD
BG 1784 SOFIA
BULGARIA
TEL 2 88 3503
TLF
TLX 23351 IKI BAN BG
EML
COM 28

GERARD ERIC DR
OBSERVATOIRE DE PARIS
SECTION DE MEUDON
F 92195 MEUDON PPL CDX
FRANCE
TEL 1 45 07 7607
TLF
TLX 270912
EML
COM 15,34,40C

GERARD JEAN-CLAUDE M C DR
INSTITUT D'ASTROPHYSIQUE
UNIVERSITE DE LIEGE
AVE COINTE 5
B 4000 COINTE-LIEGE
BELGIUM
TEL 41 52 9980
TLF 41 52 7474
TLX 41264
EML U2163JC AT BLIULG 11
COM 16

GERBAL DANIEL DR
OBSERVATOIRE DE PARIS
SECTION DE MEUDON
F 92195 MEUDON PPL CDX
FRANCE
TEL 1 45 07 7419
TLF
TLX 201571
EML
COM

GERBALDI MICHELE DR
INSTITUT D'ASTROPHYSIQUE
98BIS BD ARAGO
F 75014 PARIS
FRANCE
TEL 1 43 20 1425
TLF 1 43 29 8673
TLX
EML
COM 25,29C,45,46C

GERGELY TOMAS ESTEBAN DR
NTL SCIENCE FOUNDATION
DIV ASTRONOMICAL SCIENCES
1800 G ST NW
WASHINGTON DC 20550
USA
TEL 202 357 9696
TLF
TLX
EML
COM 10,40,50C

GERHARD ORTWIN
LANDESSTERNWARTE
KONIGSSTUHL
D 6900 HEIDELBERG 1
GERMANY
TEL 62 215 09222
TLF
TLX
EML
COM 28

GERIN MARYVONNE DR
RADIOASTRONOMIE ENS
24 RUE LHOMOND
F 75231 PARIS CDX 05
FRANCE
TEL 1 43 29 1225
TLF
TLX
EML GERIN@FRULM11
COM

GERLEI OTTO
HELIOPHYSICAL OBSERVATORY
BOC 30
H 4010 DEBRECEN
HUNGARY
TEL 52 11 015
TLX
EML
TLF
COM

GEROLA HUMBERTO DR
20390 KNOLLWOOD DRIVE
SARATOGA CA 95070
USA
TEL
TLF
EML
COM 34
TLX

GEROYANNIS VASSILIS S DR
DPT OF ASTRONOMY
UNIVERSITY OF PATRAS
GR 261 10 RION
GREECE
TEL
TLF
TLX
EML
COM 35

GERSHBERG R E DR
CRIMEAN ASTROPHYS OBS
UKRAINIAN ACAD OF SCIENCE
NAUCHNY
334413 CRIMEA
UKRAINE
TEL 43 2945
TLF
TLX
EML
COM 27C,29

GEYER EDWARD H PROF DR
OBSERVATORIUM HOHER LIST
UNIVERSITAET BONN
D 5568 DAUN/EIFEL
GERMANY
TEL 65 92 2150
TLF
TLX
EML
COM 26,27,42,45

GEZARI DANIEL YSA DR
NASA/GSFC
CODE 693
GREENBELT MD 20771
USA
TEL 301 286 3432
TLF
TLX
EML
COM 34,44

GHEIDARI S NASSIRI DR
DPT OF PHYSICS
UNIVERSITY OF SHIRAZ
BIRUNI OBSERVATORY
SHIRAZ 71459
IRAN
TEL
TLF
TLX
EML
COM 35

GHIGO FRANCIS D DR
NRAO

BOX 2
GREEN BANK WV 24944
USA
TEL
TLF
TLX
EML
COM 28,40,51

GHOBROS ROSHDY AZER PROF
HELWAN OBSERVATORY
HELWAN
EGYPT
TEL 78 0645
TLF
EML
COM
TLX

GHOSH KAJAL KUMAR DR
INDIAN INST OF ASTROPHYS
VAINU BAPPU OBSERVATORY
KAVALUR
ALANGAYAM NAA TN 635701
INDIA
TEL 530 676
TLF
TLX 8452763 IIAB IN
EML SHAKTI'VIGYAN'IIAP'GCA
COM 29

GHOSH P DR
TIFR
HOMI BHABHA RD
COLABA
BOMBAY 400 005
INDIA
TEL 22 219 111*260
TLF
TLX 011-3009
EML
COM 28

GHOSH S K DR
TIFR
HOMI BHABHA RD
COLABA
BOMBAY 400 005
INDIA
TEL 22 215 2311
TLF 22 215 2110
TLX 011 3009 TIFR IN
EML SWARNA@TIFRVAX BITNET
COM 25

GIACAGLIA GIORGIO E PROF
CRAAE/ESCOLA POLITECNICA
UNIVERSIDADE DE SAO PAULO
CP 8174
05508 SAO PAULO SP
BRAZIL
TEL 11 815 9322
TLF
TLX
EML
COM 07

GIACANI ELSA BEATRIZ DR
IAR
CC 5
1894 VILLA ELISA (BS AS)
ARGENTINA
TEL 21 4 3793
TLX 18052 CICYT AR
EML
TLF
COM 28,34

GIACCONI RICCARDO PROF
STSCI
HOMEWOOD CAMPUS
3700 SAN MARTIN DR
BALTIMORE MD 21218
USA
TEL 301 338 4711
TLF
TLX 6849101 ST SCI
EML
COM 44,48C

GIACHETTI RICCARDO PROF
DPT DI FISICA
UNIVERSITA DI FIRENZE
LARGO E FERMI 2
I 50100 FIRENZE
ITALY
TEL 55 229 8141
TLF
TLX 572570
EML
COM

GIALLONGO EMANUELE DR
OAR
I 00040 MONTEPORZIO
ITALY
TEL 6 944 9019
TLF
EML DECNET 17468 GIALLONGO
COM 47
TLX

GIAMPAPA MARK S
NTL SOLAR OBSERVATORY
BOX 26732
950 N CHERRY AVE
TUCSON AZ 85726 6732
USA
TEL 602 327 5511
TLF
TLX 0666484 AURA NOAOTUC
EML
COM 29,36

GIANNONE PIETRO PROF
OAR
VIA DEL PARCO MELLINI 84
I 00136 ROMA
ITALY
TEL 6 34 52794
TLF
TLX
EML
COM 35,42

GIANNUZZI MARIA A DR
DPT DI MATEMATICA
UNIV DI ROMA LA SAPIENZA
PIAZZA GRAMSCI 5
I 00041 ALBANO/LAZIALE
ITALY
TEL 6 932 1101
TLF
TLX
EML
COM

GIBSON DAVID MICHAEL DR
SITE MAN GEODSS FIELD ST
MIT LINCOLN LABORATORY
BOX 1707
SOCORRO NM 87801 1707
USA
TEL 505 679 4244
TLF
TLX
EML
COM 09,10,50

GIBSON JAMES
JPL/CALTECH ITT & FEC
MS 264 781
4800 OAK GROVE DR
PASADENA CA 91109
USA
TEL 818 354 2900
TLF
TLX 675429
EML
COM 15,20

GICLAS HENRY L MR
120 E ELM AVE
FLAGSTAFF AZ 86001
USA
TEL 602 774 4769
TLF
TLX
EML
COM 16,20,24

GIERASCH PETER J DR
DPT OF ASTRONOMY
CORNELL UNIVERSITY
SPACE SCIENCES BLDG
ITHACA NY 14853
USA
TEL 607 256 3507
TLF
TLX
EML
COM 16

GIEREN WOLFGANG P DR
GRUPO ASTROFIS PONTIFICA
UNIVERSIDAD CATOLICA
CASILLA 104
SANTIAGO 22
CHILE
TEL
TLF
TLX
EML WGIEREN@ASTROUC PUC CL
COM 27

GIES DOUGLAS R DR
DPT PHYSICS & ASTRONOMY
GEORGIA STATE UNIVERSITY
ATLANTA GA 30303 3083
USA
TEL 404 651 2932
TLF
TLX
EML BITNET PHYDDG@GSUVM1
COM 27

GIESEKING FRANK DR
OBSERVATORIUM HOHER LIST
UNIVERSITAET BONN
D 5568 DAUN/EIFEL
GERMANY
TEL
TLF
TLX
EML
COM 30

GIETZEN JOSEPH W
ROYAL GREENWICH OBS

HERSTMONCEUX CASTLE
HAILSHAM BN27 1RP
UK
TEL 323 813 171
TLF
TLX
EML
COM

GIGAS DETLEF DR
NASA/GSFC
CODE 910 2
GREENBELT MD 20771
USA
TEL
TLF
TLX
EML
COM 36

GIL JANUSZ A PROF
THE ASTRONOMICAL CENTER
LUBUSKA 2
PL 65 001 ZIELONA GORA
POLAND
TEL 48 68 663 85
TLF 48 68 663 85
TLX 0433467 WSPPL .
EML
COM 40

GILLET D DR
OHP
F 04870 S MICHEL OBS
FRANCE
TEL 92 76 6368
TLF
TLX 410690 OHP F
EML BITNET GILLET@FRONI51
COM

GILLILAND RONALD LYNN
STSCI
HOMEWOOD CAMPUS
3700 SAN MARTIN DR
BALTIMORE MD 21218
USA
TEL
TLF
TLX
EML
COM 10

GILLINGHAM PETER MR
AAO
PRIVATE BAG
COONABARABRAN NSW 2357
AUSTRALIA
TEL 68 421 122
TLX 63945AA
EML
TLF
COM 09

GILMAN PETER A DR
HIGH ALTITUDE OBSERVATORY
NCAR
BOX 3000
BOULDER CO 80307 3000
USA
TEL 303 497 1560
TLF
TLX
EML
COM 10

GILMORE ALAN C MR
MT JOHN OBSERVATORY
BOX 57
LAKE TEKAPO
NEW ZEALAND
TEL 64 5 056 813
TLF
TLX
EML
COM 06,20

GILMORE GERARD FRANCIS
INSTITUTE OF ASTRONOMY
THE OBSERVATORIES
MADINGLEY RD
CAMBRIDGE CB3 OHA
UK
TEL 223 622 04
TLF
TLX 81797 ASTRON G
EML
COM 30,33C

GILMOZZI ROBERTO
ESA IUE OBSERVATORY
APD 54065
E 28080 MADRID
SPAIN
TEL 1 401 9661
TLX 42555
EML
TLF
COM

GILRA DAYA P DR
SM SYSTEMS & RESEARCH CO
8401 CORPORATE DR
SUITE 450
LANDOVER MD 20785
USA
TEL 301 763 4483
TLF
TLX
EML
COM 29,34,44

GIMENEZ ALVARO
INST ASTROFISICA
DE ANDALUCIA APD 2144
PROFESOR ALBAREDA 1
E 18080 GRANADA
SPAIN
TEL 58 12 1311
TLF
TLX 78573 IAAG E
EML
COM 35,42C

GINGERICH OWEN PROF
CENTER FOR ASTROPHYSICS
HCO/SAO
60 GARDEN ST
CAMBRIDGE MA 02138
USA
TEL 617 495 7216
TLF 617 496 7564
TLX 921428 SATELLITE CAM
EML GINGER@CFA/CFA GINGER
COM 41,46

GINGOLD ROBERT ARTHUR DR
AUSTRALIAN NTL UNIVERSITY
ANU SUPERCPTR FACILITY
BOX 4
CANBERRA ACT 2600
AUSTRALIA
TEL 62 493 437
TLF
TLX
EML
COM 35

GINZBURG VITALY L PROF
LEBEDEV PHYSICAL INST
ACADEMY OF SCIENCES
LENINSKY PROSPEKT 53
117924 MOSCOW
RUSSIA
TEL
TLF
TLX
EML
COM 40,48,51

GIOIA ISABELLA M DR
INSTITUTE FOR ASTRONOMY
UNIVERSITY OF HAWAII
2680 WOODLAWN DR
HONOLULU HI 96822
USA
TEL
TLF
TLX 723 8459 UHAST HR
EML
COM 40

GIOVANARDI CARLO
OSS ASTROFISICO
DI ARCETRI
LARGO E FERMI 5
I 50125 FIRENZE
ITALY
TEL 55 275 2239
TLF
TLX 572268 ARCETR I
EML
COM 28

GIOVANE FRANK
NASA HEADQUARTERS
CODE SZ
600 INDEPENDENCE AVE SW
WASHINGTON DC 20546
USA
TEL 202 453 1472
TLF
TLX
EML 6646 FGIOVANE
COM 15,21

GIOVANELLI RICCARDO DR
DPT OF ASTRONOMY
CORNELL UNIVERSITY
SPACE SCIENCES BLDG
ITHACA NY 14853
USA
TEL
TLF
TLX
EML
COM 28,30,34

GIOVANNELLI FRANCO DR
IAS
CNR
CP 67
I 00044 FRASCATI
ITALY
TEL 6 942 5655
TLF
TLX 610261 CNRFRAI
EML
COM 29,42,51

GIOVANNINI GABRIELE
IST DI RADIOASTRONOMIA
CNR
VIA IRNERIO 46
I 40126 BOLOGNA
ITALY
TEL 51 23 2856
TLF
TLX 211664 INFN BO I
EML
COM 40

GIRAUD EDMOND
CENTRE PHYSIQUE THEORIQUE
CNRS
LUMINY CASE 07
F 13288 MARSEILLE CDX
FRANCE
TEL 91 26 9519
TLF
TLX 430838 CNRSLUM F
EML EARN GIRAUD@FRCPTM51
COM

GIRIDHAR SUNETRA DR
INDIAN INSTITUTE OF
ASTROPHYSICS
KORAMANGALA
BANGALORE 560 034
INDIA
TEL 812 56 6585
TLF
TLX 845763 IIAB IN
EML
COM 35

GIURICIN GIULIANO
DPT DI ASTRONOMIA
UNIVERSITA DI TRIESTE
VIA TIEPOLO 11
I 34131 TRIESTE
ITALY
TEL 40 76 8005
TLF
TLX 461137 OAT I
EML
COM 42,47

GLAGOLEVSKIJ JU V DR
SPECIAL ASTROPHYSICAL OBS
ACADEMY OF SCIENCES
NIZHNIJ ARKHYZ
357147 STAVROPOLSKIJ
RUSSIA
TEL 93-577
TLF
TLX
EML
COM 14,27,29,45

GLASER HAROLD DR
1346 BONITA ST
BERKELEY CA 94709
USA
TEL 415 527 1860
TLF
EML
COM 44
TLX

GLASNER SHIMON AMI
RACAH INST OF PHYSICS
HEBREW UNIV OF JERUSALEM
JERUSALEM 91904
ISRAEL
TEL 2 584 521
TLF
TLX 25391 HUIL
EML
COM

GLASPEY JOHN W DR
CANADA FRANCE HAWAII
TELESCOPE CORPORATION
BOX 1597
KAMUELA HI 96743
USA
TEL
TLF
TLX
EML GLASPEY@UHCFHT BITNET
COM

GLASS BILLY PRICE DR
DPT OF GEOLOGY
UNIVERSITY OF DELAWARE
NEWARK DE 19716
USA
TEL 302 451 8458
TLF
TLX
EML
COM 22

GLASS IAN STEWART DR
SAAO
BOX 9
OBSERVATORY 7935
SOUTH AFRICA
TEL 21 47 00 25
TLF
TLX 57 20309 SA
EML ISG@SAAO AC ZA
COM 09,25,28

GLASSGOLD ALFRED E PROF
DPT OF PHYSICS
NEW YORK UNIVERSITY
4 WASHINGTON PLACE
NEW YORK NY 10003
USA
TEL 212 598 2020
TLF
TLX 235128 NYU UR
EML
COM

GLATZEL WOLFGANG DR
MAX PLANCK INSTITUT FUER
PHYSIK UND ASTROPHYSIK
KARL-SCHWARZSCHILDSTR 1
D 8046 GARCHING MUENCHEN
GERMANY
TEL 89 329 90
TLF
TLX 524629
EML BITNET WOG @ DGAIPP1S
COM

GLATZMAIER GARY A
LOS ALAMOS NATIONAL LAB
MS F665
ESS 5
LOS ALAMOS NM 87545
USA
TEL 505 667 7647
TLF
TLX
EML
COM 10,12,35

GLEBOCKI ROBERT PROF
INST THEORETICAL PHYSICS
UNIVERSITY OF GDANSK
UL WITA STWOSZA 57
PL 80 952 GDANSK
POLAND
TEL 41-87-00
TLF
TLX 0512706 IFAS
EML
COM 36

GLEBOVA NINA I DR
INST OF THEORET ASTRONOMY
ACADEMY OF SCIENCES
N KUTUZOVA 10
191187 ST PETERSBURG
RUSSIA
TEL 278-88-98
TLF
TLX 121578 ITA SU
EML
COM 04

GLEISSBERG WOLFGANG PROF
BUCHENWEG 12
D 6374 OBERURSEL
GERMANY
TEL
TLF
TLX
EML
COM 10

GLENCROSS WILLIAM M DR
DPT PHYSICS & ASTRONOMY
UNIVERSITY COLLEGE LONDON
GOWER ST
LONDON WC1E 6BT
UK
TEL 1 713 877 050
TLF
TLX 28722 UCPHYS G
EML
COM

GLIESE WIHELM PROF
ASTRONMISCHES-RECHEN
INSTITUT HEIDELBERG
MOENCHHOFSTR 12-14
D 6900 HEIDELBERG 1
GERMANY
TEL 62 21 4050
TLF 62 21 40 52 97
TLX 461336 ARIHD D
EML
COM 24,26,33

GLUSHNEVA I N DR
STERNBERG STATE ASTR INST
UNIVERSITETSKIJ PROSP 13
119899 MOSCOW
RUSSIA
TEL 139-20-46
TLX
EML
TLF
COM 29

GNEDIN YURIJ N DR
IOFFE PHYSICAL TECH INST
ACADEMY OF SCIENCES
POLYTECHNICHESKAYA UL 26
194021 ST PETERSBURG
RUSSIA
TEL
TLF
TLX
EML
COM

GNEVYSHEVA RAISA S DR
PULKOVO OBSERVATORY
ACADEMY OF SCIENCES
10 KUTUZOV QUAY
196140 ST PETERSBURG
RUSSIA
TEL
TLF
TLX
EML
COM 10

GODART ODON PROF
9 AVE-DES CERISERS
B 1488 BOUSVAL
BELGIUM
TEL 10 61 3817
TLF
EML
COM 47
TLX

GODFREY PETER DOUGLAS DR
DPT OF CHEMISTRY
MONASH UNIVERSITY
WELLINGTON RD
CLAYTON VIC 3168
AUSTRALIA
TEL 3 541 0811
TLF
TLX 32691 AA
EML
COM 34

GODOLI GIOVANNI PROF
DPT DI ASTRONOMIA
UNIVERSITA DI FIRENZE
LARGO E FERMI 5
I 50125 FIRENZE
ITALY
TEL 55 27521
TLF 55 22 0039
TLX 572268 ARCETR I
EML
COM 10,12,27,51

GODWIN JON GUNNAR DR
UNIVERSITY OF OXFORD
85 CHERWELL DRIVE
OXFORD OX3 OND
UK
TEL 865 722 313
TLF 865 721 000
TLX
EML UK AC RL STAR JG
COM

GOEBEL ERNST DR
INSTITUT FUER ASTRONOMIE
UNIVERSITAET WIEN
TUERKENSCHANZSTR 17
A 1180 WIEN
AUSTRIA
TEL 1 345 360186
TLF
TLX
EML
COM 50

GOEBEL JOHN H DR
NASA AMES RESEARCH CTR
MS 244 7
SPACE SCIENCE DIV
MOFFETT FIELD CA 94035
USA
TEL 415 694 6525
TLF
TLX
EML
COM 29,34

GOEDBLOED JOHAN P PROF DR
INST VOOR PLASMAFYSICA
BOX 1207
NL 3430 BE NIEUWEGEIN
NETHERLANDS
TEL 34 02 31224
TLF 34 02 31204
TLX
EML GOEDBLOED@SARA NL
COM 10

GOELBASI ORHAN DR
FACUTLY OF SCIENCE
INONU UNIVERSITY
44069 MALATYA
TURKEY
TEL 821 21871
TLF 821 18133
TLX 66140
EML
COM

GOETZ WOLDEMAR DR
STERNWARTESTRASSE 23A
PSF 55 27/28
D 6400 SONNEBERG
GERMANY
TEL 96 74 3793
TLX 627180
EML
TLF
COM 27

GOKDOGAN NUZHET PROF
UNIVERSITY OBSERVATORY
UNIVERSITY OF ISTANBUL
34452 ISTANBUL
TURKEY
TEL 1 522 3597
TLF
TLX·
EML
COM 12,36

GOKHALE MORESHWAR HARI PR
INDIAN INSTITUTE OF
ASTROPHYSICS
KORAMANGALA
BANGALORE 560 034
INDIA
TEL 812 56 6585
TLF
TLX 845763 IIAB IN
EML
COM 10

GOLAY MARCEL PROF
OBSERVATOIRE DE GENEVE
CHEMIN DES MAILLETTES 51
CH 1290 SAUVERNY
SWITZERLAND
TEL 22 755 2611
TLF 22 755 3983
TLX 419209 OBS CH
EML golay@obs unige ch
COM 25,37,45C

GOLD THOMAS PROF
CRSR
CORNELL UNIVERSITY
SPACE SCIENCES BLDG
ITHACA NY 14853 6801
USA
TEL 607 256 5284
TLF
TLX 937478
EML
COM 16,40,44,47,48

GOLDBACH CLAUDINE MME
INSTITUT D'ASTROPHYSIQUE
98BIS BD ARAGO
F 75014 PARIS
FRANCE
TEL 1 43 20 1425
TLF 1 43 29 8673
TLX
EML
COM 14

GOLDMAN ITZHAK DR
DPT OF PHYSICS & ASTRON
TEL AVIV UNIVERSITY
RAMAT AVIV
TEL AVIV 69978
ISRAEL
TEL 3 545 0303
TLF
TLX 342171 VERSY IL
EML BITNET E17@TAUNOS
COM 42

GOLDMAN MARTIN V
CASA
UNIVERSITY OF COLORADO
BOX 391
BOULDER CO 80309 0391
USA
TEL 303 492 8896
TLF
TLX
EML
COM 12,49

GOLDREICH P DR
CALTECH
PASADENA CA 91125
USA
TEL 213 356 6193
TLF
EML PETER@CIIDEIMO
COM 07,16,33,34
TLX

GOLDSMITH DONALD W DR
INTERSTELLAR MEDIA
2153 RUSELL ST
BERKELEY CA 94705
USA
TEL 415 848 1989
TLF
TLX
EML
COM 34,47,48,51

GOLDSMITH PAUL F DR
DPT PHYSICS & ASTRONOMY
UNIV OF MASSACHUSETTS
GRC B 626
AMHERST MA 01003
USA
TEL
TLF
TLX
EML
COM 34,40

GOLDSMITH PAUL F DR
LEADMINE HILL ROAD 5

AMHERST MA 01002
USA
TEL 413 545 0925
TLF
TLX 95 5491
EML
COM

GOLDSMITH S DR
DPT OF PHYSICS & ASTRON
TEL AVIV UNIVERSITY
RAMAT AVIV
TEL AVIV 69978
ISRAEL
TEL 3 420 303
TLF
TLX 342171 VERSY
EML
COM

GOLDSTEIN RICHARD M DR
JPL/CALTECH
MS 183 701
4800 OAK GROVE DR
PASADENA CA 91109
USA
TEL 818 354 6999
TLF
TLX
EML
COM 16

GOLDSTEIN SAMUEL J PROF
UNIVERSITY STATION
UNIVERSITY OF VIRGINIA
BOX 3818
CHARLOTTESVILLE VA 22903
USA
TEL
TLF
TLX
EML
COM 34,40

GOLDSWORTHY FREDERICK A
DPT OF APPL MATHEMATICS
UNIVERSITY OF LEEDS
LEEDS LS2 9JT
UK
TEL 532 431 751
TLF
TLX
EML
COM 34

GOLDWIRE HENRY C JR
LAWRENCE LIVERMORE LAB
L 451
BOX 808
LIVERMORE CA 94550
USA
TEL 415 423 0160
TLF
TLX
EML
COM 40

GOLEV VALERY K DR
ASTRONOMICAL OBSERVATORY
BULGARIAN ACAD SCIENCES
BOX 36
BG 1504 SOFIA
BULGARIA
TEL 66 2324
TLF
TLX
EML
COM

GOLLNOW H DR
MOUNT STROMLO OBSERVATORY
PRIVATE BAG
WODEN PO
CANBERRA ACT 2606
AUSTRALIA
TEL
TLF
TLX
EML
COM

GOLUB LEON DR
CENTER FOR ASTROPHYSICS
HCO/SAO
60 GARDEN ST
CAMBRIDGE MA 02138
USA
TEL 617 495 7177
TLF
TLX
EML
COM

GOMES ALERCIO M DR
R GAVIAO PEIXOTO 13
AP 1401
ICARAI 24000
NITEROJ ERJ
BRAZIL
TEL
TLF
TLX
EML
COM

GOMES RODNEY D S DR
CORNELL UNIVERSITY
SPACE SCIENCES BLDG
ITHACA NY 14853
USA
TEL 607 255 4709
TLX
EML Gomes@astrosun TN CORNELL EDU
TLF
COM 07

GOMEZ GONZALEZ JESUS DR
PASEO IMPERIAL 29 6H
E MADRID 5
SPAIN
TEL
TLF
TLX
EML
COM 34,40C

GOMEZ MARIA THERESA DR
OSS ASTRONOMICO

I 80131 NAPOLI
ITALY
TEL
TLF
TLX
EML
COM 12

GOMIDE FERNANDO DE MELLO
INPE
CP 515
12225 S JOSE DOS CAMPOS
BRAZIL
TEL 123 22 9088
TLF
TLX 0113393 CTAE BR
EML
COM

GONCZI GEORGES
OCA OBSERV DE NICE
BP 139
F 06003 NICE CDX
FRANCE
TEL 93 89 0420
TLF
TLX 460004 OBSNICE F
EML
COM

GONDHALEKAR PRABHAKAR DR
RUTHERFORD APPLETON LAB
SPACE & ASTROPHYSICS DIV
BLDG R25/R68
CHILTON DIDCOT OX11 0QX
UK
TEL 235 219 00
TLF
TLX 83159
EML
COM 44

GONDOLATSCH FRIEDRICH PRF
ASTRONOMISCHES RECHEN
INSTITUT
MOENCHHOFSTR 12-14
D 6900 HEIDELBERG 1
GERMANY
TEL 62 21 49026
TLF
TLX
EML
COM 04

GONG SHOU-SHEN
SHANGHAI OBSERVATORY
CAS
80 NANDAN RD
SHANGHAI
CHINA PR
TEL 21 38 6191
TLF
TLX 33164 SHAO CN
EML
COM 09

GONG SHU-MO
PURPLE MOUNTAIN OBSERV
CAS
NANJING
CHINA PR
TEL 25 46700
TLF
TLX 34144 PMO CN
EML
COM 35,47

GONZALES-A WALTER D DR
INPE
CP 515
12200 S JOSE DOS CAMPOS
BRAZIL
TEL 123 22 9977
TLF 123 21 8743
TLX 011-33530 INPE BR
EML
COM 48

GONZALEZ CAMACHO ANTONIO
INST DE ASTRON & GEODESIA
FAC DE CIENCIAS MATEMAT
UNIVERSIDAD COMPLUTENSE
E 28040 MADRID
SPAIN
TEL 1 244 2501
TLF
TLX
EML
COM 07

GONZALEZ G
INAOE
TONANTZINTLAZ
APDO POSTAL 216 y 51
72000 PUEBLA PUE
MEXICO
TEL
TLF
TLX
EML
COM

GONZALEZ SERRANO J I DR
DPT DE FISICA MODERNA
UNIVERSIDAD DE CANTABRIA
AVD LOS CASTROS S/N
E 39005 SANTANDER
SPAIN
TEL 42 20 1450
TLF 42 20 1402
TLX 35861 EDUCI E
EML GSERRANO@CCUCVX UNICAN ES
COM 28

GONZALEZ-RIESTRA R DR
ESA IUE OBSEERVATORY
VILSPA
APD 54065
E 28080 MADRID
SPAIN
TEL 1 401 9661
TLF
TLX 42555 VILS E
EML EARN IUEHOT@DDAESA10
COM

GONZE ROGER F J IR
OBSERVATOIRE ROYAL DE
BELGIQUE
AVE CIRCULAIRE 3
B 1180 BRUSSELS
BELGIUM
TEL 2 375 2484
TLF 2 374 9822
TLX 21565
EML
COM 40

GOODE PHILIP R
DPT OF PHYSICS
NJ INST OF TECHNOLOGY
323 HIGH ST
NEWARK NJ 07102
USA
TEL 201 596 3562
TLF
TLX
EML
COM

GOODRICH ROBERT W DR
CALTECH
MS 105 24
PASADENA CA 91125
USA
TEL 818 356 3693
TLF
TLX
EML RWG@DEIMOS CALTECH EDU
COM 28

GOODY R M
CEPP
PIERCE HALL
29 OXFORD ST
CAMBRIDGE MA 02138
USA
TEL 617 495 4517
TLF
TLX
EML
COM 16

GOOSSENS MARCEL DR
ASTRONOMISCH INSTITUUT
KATHOLIEKE UNIV LEUVEN
CELESTIJNENLAAN 200 B
B 3030 HEVERLEE
BELGIUM
TEL
TLF
TLX
EML
COM 10

GOPALA RAO U V MR
SATELLITE METEOROLOGY
INDIAN METEOROLOGIC DPT
LODI ROAD / MAUSAM BHAVAN
NEW DELHI 110 003
INDIA
TEL
TLF
TLX
EML
COM

GOPALSWAMY N DR
ASTRONOMY PROGRAM
UNIVERSITY OF MARYLAND
COLLEGE PARK MD 20742
USA
TEL 301 454 6649
TLF
TLX 62891478
EML
COM 12,40

GOPASYUK S I DR
CRIMEAN ASTROPHYS OBS
UKRAINIAN ACAD OF SCIENCE
NAUCHNY
334413 CRIMEA
UKRAINE
TEL 43 2945
TLF
TLX
EML
COM 10,12

GORBATSKY VITALIJ G PROF
ASTRONOMICAL OBSERVATORY
ST PETERSBURG UNIVERSITY
BIBLIOTECHNAJA PL 2
198904 ST PETERSBURG
RUSSIA
TEL 257-94-91
TLF
TLX
EML
COM 27

GORDON CHARLOTTE PROF
11 RUE TOURNEFORT
F 75005 PARIS
FRANCE
TEL
TLF
EML
COM 12,36
TLX

GORDON COURTNEY P PROF
DPT OF ASTRONOMY
HAMPSHIRE COLLEGE
AMHERST MA 01002
USA
TEL
TLF
TLX
EML
COM 34

GORDON ISAAC M DR
INST RADIOPHYS & ELECTRON
310085 KHARKOV
UKRAINE
TEL
TLF
TLX
EML
COM

GORDON KURTISS J PROF
DPT OF ASTRONOMY
HAMPSHIRE COLLEGE
AMHERST MA 01002
USA
TEL 413 549 4600
TLF
TLX
EML
COM

GORDON MARK A DR
NRAO
CAMPUS BLDG 65
949 N CHERRY AVE
TUCSON AZ 85721 0655
USA
TEL 602 882 8250
TLF
TLX
EML
COM 33,34,40

GORENSTEIN MARC V
CENTER FOR ASTROPHYSICS
MS 42
60 GARDEN ST
CAMBRIDGE MA 02138
USA
TEL 617 495 9296
TLF
TLX
EML
COM

GORENSTEIN PAUL DR
CENTER FOR ASTROPHYSICS
HCO/SAO
60 GARDEN ST
CAMBRIDGE MA 02138
USA
TEL 617 495 7250
TLF
TLX 921428 SATELLITE CAM
EML
COM 16

GORET PHILIPPE DR
CEA CEN
DAPNIA/SAP
BP 2
F 91191 GIF/YVETTE CDX
FRANCE
TEL 1 69 08 4463
TLF
TLX 690860
EML
COM 47

GORGAS GARCIA JAVIER DR
DPT DE ASTROFISICA
FAC DE FISICA
UNIVERSIDAD COMPLUTENSE
E 28040 MADRID
SPAIN
TEL 1 549 5316
TLF
TLX 47273
EML EARN/BITNET W062@EEMDUCM11
COM 28

GORGOLEWSKI STANISLAW PR
INST OF RADIO ASTRONOMY
N COPERNICUS UNIVERSITY
UL CHOPINA 12/18
PL 87 100 TORUN
POLAND
TEL 20651 TORUN
TLF
TLX 0552324 TRAO PL
EML
COM 40

GOSACHINSKIJ I V DR
PULKOVO OBSERVATORY
ACADEMY OF SCIENCES
10 KUTUZOV QUAY
196140 ST PETERSBURG
RUSSIA
TEL 2979452
TLF
TLX 321262
EML
COM 34,40C

GOSLING JOHN T DR
LOS ALAMOS NATIONAL LAB
MS D438
ESS 8
LOS ALAMOS NM 87545
USA
TEL 505 667 5389
TLF
TLX 660495
EML
COM 49

GOSS W MILLER PROF
NRAO
VLA
BOX 0
SOCORRO NM 87801 0387
USA
TEL 505 772 4011
TLF
TLX 910-988-1710
EML
COM 28,34,40

GOSSET ERIC DR
INSTITUT D'ASTROPHYSIQUE
UNIVERSITE DE LIEGE
AVE COINTE 5
B 4000 COINTE-LIEGE
BELGIUM
TEL 41 52 9980
TLF 41 52 7474
TLX 41264 ASTRLG
EML U2141EG@BLIULG11
COM 47

GOSWAMI J N DR
PHYSICAL RESEARCH LAB ,
NAVRANGPURA
AHMEDABAD 380 009
INDIA
TEL 272 46 2129
TLF 272 44 5292
TLX 0121397 PRL IN
EML
COM 22

GOTT III J RICHARD
DPT ASTROPHYSICAL SCI
PRINCETON UNIVERSITY
PRINCETON NJ 08544 1001
USA
TEL 609 452 3813
TLF
TLX
EML
COM 51

GOTTESMAN STEPHEN T DR
DPT OF ASTRONOMY
UNIVERSITY OF FLORIDA
211 SSRB
GAINESVILLE FL 32611
USA
TEL 904 392 2050/2052
TLF
TLX 8108252308
EML
COM 28,33,40

GOTTLIEB CARL A DR
INST FOR SPACE STUDIES
2880 BROADWAY
NEW YORK NY 10025
USA
TEL 212 678 5566
TLX
EML
TLF
COM

GOTTLIEB KURT
46 JENNINGS STREET
CURTIN ACT 2605
AUSTRALIA
TEL 62 814 166
TLF
TLX
EML
COM

GOUDA NAOTERU DR
DPT OF PHYSICS
KYOTO UNIVERSITY
SAKYO KU
KYOTO 606
JAPAN
TEL 81 75 753 3844
TLF 81 75 753 3886
TLX 5422829 KUNSDP J
EML GOUDA@JPNRIFP BITNE
COM 47

GOUDAS CONSTANTINE L PROF
DPT OF MATHEMATICS
UNIVERSITY OF PATRAS
GR 261 10 RION
GREECE
TEL 61 99 1889
TLF
TLX 312239 EFAP GR
EML
COM 07,16

GOUDIS CHRISTOS D PROF
DPT OF PHYSICS
UNIVERSITY OF PATRAS
GR 261 10 PATRAS
GREECE
TEL
TLF
TLX
EML
COM 51

GOUGH DOUGLAS O DR
INSTITUTE OF ASTRONOMY
THE OBSERVATORIES
MADINGLEY RD
CAMBRIDGE CB3 0HA
UK
TEL 223 622 04
TLF
TLX 817297
EML
COM 27,35,36

GOUGUENHEIM LUCIENNE
OBSERVATOIRE DE PARIS
SECTION DE MEUDON
RADIOASTRONOMIE
F 92195 MEUDON PPL CDX
FRANCE
TEL 1 45 07 7604
TLF 1 45 07 7939
TLX
EML GOUGUENHEIM@FRMEU51
COM 28,30,46P

GOULD ROBERT J PROF
DPT OF PHYSICS
UCSD
B 01
LA JOLLA CA 92093 0216
USA
TEL 619 452 3649
TLF
TLX
EML
COM

GOUPIL MARIE JOSE
OBSERVATOIRE DE PARIS
SECTION DE MEUDON
DASGAL
F 92195 MEUDON PPL CDX
FRANCE
TEL 1 45 07 7880
TLF
TLX 270912
EML EARN GOUPIL@FRMEU51
COM 27,35

GOUTTEBROZE PIERRE DR
IAS
BP 10
F 91371 VERRIERES BUISSON
FRANCE
TEL 1 64 47 4204
TLF
TLX 600252
EML
COM

GOWER ANN C DR
DPT OF PHYSICS
UNIVERSITY OF VICTORIA
BOX 1700
VICTORIA BC V8W 2Y2
CANADA
TEL 604 721 7700
TLF 604 721 7715
TLX
EML
COM

GOWER J F R DR
1615 MCTAVISH ROAD RR 2
SIDNEY BC V8L 3S1
CANADA
TEL 604 656 5457
TLF
EML
COM 40
TLX

GOY GERALD PROF
OBSERVATOIRE DE GENEVE
CHEMIN DES MAILLETTES 51
CH 1290 SAUVERNY
SWITZERLAND
TEL 22 755 2611
TLF 22 755 3983
TLX 419209 OBS CH
EML goy@obs.unige.ch
COM 25

GOYAL A N DR
DPT OF MATHEMATICS
UNIVERSITY OF RAJASTHAN
JAIPUR 302 004
INDIA
TEL 74 060
TLF
TLX
EML
COM 24

GRABOSKE HAROLD C JR
LAWRENCE LIVERMORE LAB
L 23
BOX 808
LIVERMORE CA 94550
USA
TEL 415 422 7262
TLF
TLX
EML
COM

GRABOWSKI BOLESLAW DR
INSTITUTE OF PHYSICS
PEDAGOGICAL UNIVERSITY
UL OLESKA 48
PL 45 951 OPOLE
POLAND
TEL 358-41
TLF
TLX 0732230 WSP PL
EML
COM

GRADIE JONATHAN CAREY
INSTITUTE OF GEOPHYSICS
UNIVERSITY OF HAWAII
DIV PLANETARY GEOSCIENCES
HONOLULU HI 96822
USA
TEL 808 948 6488
TLF
TLX
EML
COM 15

GRADSZTAJN ELI DR
DPT OF PHYSICS & ASTRON
TEL AVIV UNIVERSITY
RAMAT AVIV
TEL AVIV 69978
ISRAEL
TEL
TLF
TLX
EML
COM

GRADY CAROL ANNE DR
DPT OF PHYSICS
THE CATHOLIC UNIVERSITY
OF AMERICA
WASHINGTON DC 20064
USA
TEL 202 319 5315
TLF
TLX
EML IUE GRADY ON SPAN
COM 29

GRAHAM DAVID A
MPI FUER RADIOASTRONOMIE
AUF DEM HUEGEL 69
D 5300 BONN 1
GERMANY
TEL 228 52 5282
TLX 886440
EML
TLF
COM 34,40

GRAHAM ERIC DR
BOC 9056
WOODLAND PARK
COLORADO 80866
USA
TEL
TLF
TLX
EML
COM 35

GRAHAM JOHN A DR
DPT TERRESTR MAGNETISM
CARNEGIE INST WASHINGTON
5241 BROAD BRANCH RD NW
WASHINGTON DC 20015
USA
TEL 202 966 0863
TLF
TLX 440427 MAGN UI
EML
COM 25,27,28

GRAHL BERND H DR
MPI FUER RADIOASTRONOMIE

AUF DEM HUEGEL 69
D 5300 BONN 1
GERMANY
TEL 225 73 112
TLF
TLX 886440
EML
COM

GRAINGER JOHN F DR
DPT OF PHYSICS
UMIST
BOX 88
MANCHESTER M60 1QD
UK
TEL   612 363 311
TLF
TLX   666094
EML
COM

GRANDI STEVEN ALDRIDGE DR
NOAO
BOX 26732
950 N CHERRY AVE
TUCSON AZ 85726 6732
USA
TEL   602 327 5511
TLF
TLX
EML
COM. 28

GRANDPIERRE ATTILA DR
KONKOLY OBSERVATORY
THEGE U 13/17
BOX 67
H 1525 BUDAPEST
HUNGARY
TEL   1 75 5866/75 4122
TLF   1 156 9640
TLX   277 460 KONOB H
EML   H697 KNO E ELLA UUCP
COM   10

GRANT IAN P DR
PEMBROKE COLLEGE

OXFORD OX1 1DW
UK
TEL   865 242 271
TLF
TLX
EML
COM   14,36

GRANVEAUD MICHEL
OBSERVATOIRE DE PARIS
61 AVE OBSERVATOIRE
LPTF
F 75014 PARIS
FRANCE
TEL   1 43 20 1210
TLF
TLX   270776 OBS F
EML
COM   31C

GRASDALEN GARY L DR
DPT PHYSICS & ASTRONOMY
UNIVERSITY OF WYOMING
BOX 3905 UN STA
LARAMIE WY 82071
USA
TEL   307 766 4385
TLF
TLX
EML
COM   27,28,34

GRATTON R G DR
OAR
VIA DELL'OSSERVATORIO
I 00040 MONTEPORZIO
ITALY
TEL   6 944 9019
TLF
TLX
EML   SPAN 17468 RAFFAELE
COM   29,37

GRAUER ALBERT D
DPT PHYSICS & ASTRONOMY
UALR
33RD & UNIVERSITY
LITTLE ROCK AR 72204
USA
TEL   501 569 3275
TLF
TLX
EML
COM   25

GRAY DAVID F PROF
DPT OF ASTRONOMY
UNIV OF WESTERN ONTARIO
LONDON ON N6A 3K7
CANADA
TEL   519 679 3184
TLF   519 661 3486
TLX   064 7134
EML   DFGRAY@UWO BITNET
COM   29,36C

GRAY PETER MURRAY
AAO
BOX 296
EPPING NSW 2121
AUSTRALIA
TEL   2 868 1666
TLX   23999 AAOSYD AA
EML
TLF
COM   09

GRAYZECK EDWIN J DR
DPT OF PHYSICS
UNIVERSITY OF NEVADA
4505 S MARYLAND PARKWAY
LAS VEGAS NV 89154
USA
TEL   702 739 3507
TLF
TLX
EML
COM   33

GREBENIKOV E A PROF DR
LOMONOSOV STATE UNIVER

117234 MOSCOW
RUSSIA
TEL
TLF
TLX
. EML
COM   07

GREC GERARD
DPT ASTROPHYSIQUE
UNIVERSITE DE NICE
PARC VALROSE
F 06034 NICE CDX
FRANCE
TEL   93 51 9100
TLF
TLX
EML
COM

GREEN DAVID DR
MULLARD RADIO ASTRON OBS
CAVENDISH LABORATORY
MADINGLEY RD
CAMBRIDGE CB3 OHE
UK
TEL   223 337 310
TLF   223 35 4599
TLX   81292 CAVLAB G
EML   DAG9@UK AC CAM PHX
COM   40

GREEN ELIZABETH M DR
MOUNT STROMLO & SIDING
SPRING OBSERVATORIES
PRIVATE BAG
WODEN PO ACT 2606
AUSTRALIA
TEL   62 88 1111
TLF
TLX   62270 AA
EML
COM   37

GREEN JACK PROF
DPT OF GEOLOGY
CALIF STATE UNIVERSITY
LONG BEACH CA 90840
USA
TEL   213 498 4809
TLF
TLX
EML
COM   16

GREEN LOUIS C PROF
HAVERFORD COLLEGE
7901 COLLEGE AVENUE
HAVERFORD PA 19041
USA
TEL   215 649 0265
TLX
EML
TLF
COM   14

GREEN RICHARD F DR
KITT PEAK NTL OBS
BOX 26732
950 N CHERRY AVE
TUCSON AZ 85726 6732
USA
TEL   602 325 9299
TLF
TLX   0666 484 AURA NOAO
EML
COM

GREEN ROBIN M DR
DPT OF ASTRONOMY
UNIVERSITY OF GLASGOW
GLASGOW G12 8QQ
UK
TEL   413 398 855
TLF
TLX   778421
EML
COM

GREEN SIMON F
PHYSICS LABORATORY
UNIVERSITY OF KENT
CANTERBURY CT2 7NR
UK
TEL   227 764 000*3780
TLF   227 762 616
TLX   965449
EML
COM   15

GREENBERG J MAYO DR
STERREWACHT
HUYGENS LAB
BOX 9513
NL 2300 RA LEIDEN
NETHERLANDS
TEL 71 27 5700
TLF
TLX 39058 ASTRO NL
EML
COM 15,21,34,51

GREENBERG RICHARD DR
LUNAR & PLANETARY LAB
UNIVERSITY OF ARIZONA
TUCSON AZ 85721
USA
TEL 602 621 6940
TLF
TLX
EML
COM 07,15,20

GREENHILL JOHN DR
DPT OF PHYSICS
UNIVERSITY OF TASMANIA
GPO BOX 252C
HOBART TAS 7001
AUSTRALIA
TEL 2 202 429
TLF 2 202 410
TLX AA 58150
EML GRNHIL@PHYSVAX PHYS UTAS EDU A
COM 48

GREENSTEIN GEORGE PROF
DPT OF ASTRONOMY
AMHERST COLLEGE
AMHERST MA 01002
USA
TEL 413 542 2075
TLF
TLX
EML
COM

GREENSTEIN J L PROF
CALTECH
MS 105 24
PALOMAR OBS
PASADENA CA 91125
USA
TEL 818 356 4006
TLF
TLX
EML
COM 29,36,51

GREGG MICHAEL DAVID DR
MOUNT STROMLO & SIDING
SPRING OBSERVATORIES
PRIVATE BAG
WESTON CREEK PO ACT 2611
AUSTRALIA
TEL. 62 49 0286
TLF 62 49 0233
TLX
EML GREGG@MSO ANU OZ AU
COM 28

GREGGIO LAURA DR
DPT DI ASTRONOMIA
UNIVERSITA DI BOLOGNO
CP 596
I 40125 BOLOGNA
ITALY
TEL 51 25 9413
TLF
TLX 520634 INFNBO 1
EML SPAN/DECNET 37928 LAURA
COM 35

GREGORINI LORETTA
IST DI RADIOASTRONOMIA
CNR
VIA IRNERIO 46
I 40126 BOLOGNA
ITALY
TEL 51 23 2856
TLF
TLX, 211664 INFN BOI
EML
COM 40

GREGORY PHILIP C DR
DPT OF PHYSICS
UNIV OF BRITISH COLUMBIA
224 AGRICULTURAL RD
VANCOUVER BC V6T 1W5
CANADA
TEL 604 228 6417
TLF 604 228 5324
TLX 045 08576
EML
COM 40,51

GREGORY STEPHEN ALBERT DR
DPT PHYSICS & ASTRONOMY
BOWLING GREEN STATE UNIV
BOWLING GREEN OH 43403
USA
TEL
TLF
TLX
EML
COM 47

GREGUL A YA DR
ASTRONOMICAL OBSERVATORY
KIEV STATE UNIVERSITY
OBSERVATORNAYA UL 3
252053 KIEV
UKRAINE
TEL 26 2391
TLF
TLX
EML
COM

GREISEN KENNETH I PROF
336 FOREST HOME DR
ITHACA NY 14850
USA
TEL 607 257 1650
TLF
TLX
EML
COM 48

GRENIER ISABELLE DR
CEA CEN
DAPNIA/SAP
BP 2
F 91191 GIF/YVETTE CDX
FRANCE
TEL 1 69 08 4400
TLF 1 69 08 6577
TLX 604860
EML GRENIER@FRSAC11/32779 GRENIER
COM 48

GRENIER SUZANNE
OBSERVATOIRE DE PARIS
SECTION DE MEUDON
F 92195 MEUDON PPL CDX
FRANCE
TEL 1 45 07 7841
TLF
TLX 201571
EML
COM

GRENON MICHEL DR
OBSERVATOIRE DE GENEVE
CHEMIN DES MAILLETTES 51
CH 1290 SAUVERNY
SWITZERLAND
TEL 22 755 2611
TLF 22 755 3983
TLX 419209 OBS CH
EML grenon@obs unige ch
COM 25,33,45C

GREVE ALBERT DR
IRAM
300 RUE DE LA PISCINE
F 38406 S MARTIN HERES CD
FRANCE
TEL 76 82 4931
TLF
TLX
EML
COM

GREVESSE N DR
INSTITUT D'ASTROPHYSIQUE
UNIVERSITE DE LIEGE
AVE COINTE 5
B 4000 COINTE-LIEGE
BELGIUM
TEL 41 52 9980
TLF 41 52 7474
TLX 41264
EML
COM 12,36

GREWING MICHAEL PROF
ASTRON INST UNIVERSITAET
WALDHAUSERSTR 64
D 7400 TUEBINGEN
GERMANY
TEL 707 129 2486
TLX 07262714 AIT D
EML
TLF
COM 25,34,40,44,48

GREYBER HOWARD D DR
10123 FALLS RD
POTOMAC MD 20854
USA
TEL
TLF
TLX
EML
COM 44,47,48

GRIFFIN MATTHEW J DR
DPT OF PHYSICS
QUEEN MARY/WESTFIELD COLL
MILE END RD
LONDON E1 4NS
UK
TEL 1 980 4811
TLF
TLX 893750
EML MJG@UK AC QMC STAR
COM

GRIFFIN RITA E M DR
INSTITUTE OF ASTRONOMY
THE OBSERVATORIES
MADINGLEY RD
CAMBRIDGE CB3 OHA
UK
TEL 223 622 04
TLF
TLX 817297 ASTRON G
EML
COM 29

GRIFFIN ROGER F DR
INSTITUTE OF ASTRONOMY
THE OBSERVATORIES
MADINGLEY RD
CAMBRIDGE CB3 OHA
UK
TEL 223 622 04
TLF
TLX 817297 ASTRON G
EML RFF1@UK AC CAM PHX
COM 05,29,30

GRIFFITH JOHN S PROF
DPT OF MATH SCIENCE
LAKEHEAD UNIVERSITY
THUNDER BAY ON P7B 5E1
CANADA
TEL 807 345 2121
TLF
TLX
EML
COM

GRIFFITHS RICHARD E DR
STSCI
HOMEWOOD CAMPUS
3700 SAN MARTIN DR
BALTIMORE MD 21218
USA
TEL 301 338 4824
TLF
TLX
EML scivax griffiths
COM 09,28,44,48

GRIFFITHS WILLIAM K
DPT OF PHYSICS
UNIVERSITY OF LEEDS
LEEDS LS2 9JT
UK
TEL
TLF
TLX
EML
COM 37

GRIGORJEV VICTOR M DR
SIBIZMIR
ACADEMY OF SCIENCES
664697 IRKUTSK 33
RUSSIA
TEL
TLF
TLX
EML
COM 09

GRIJO DE OLIVEIRA A K DR
OBSERVATORIO NACIONAL
RUA GL BRUCE 586
SAN CRISTOVAO
20921 RIO DE JANEIRO RJ
BRAZIL
TEL 21 580 7181
TLF 21 580 0332
TLX
EML
COM

GRIMLEY PETER DR
DPT OF PHYSICS
ST PATRICK'S COLLEGE
MAYNOOTH
CO KILDARE
IRELAND
TEL 1 285 222
TLF
TLX 31493
EML
COM

GRINDLAY JONATHAN E DR
CENTER FOR ASTROPHYSICS
HCO/SAO
60 GARDEN ST
CAMBRIDGE MA 02138
USA
TEL 617 495 7204
TLF 617 495 7356
TLX 921428 SATELLITE CAM
EML CFA "JOSH@CFA255"/6699 JOSH
COM 06P,37,48

GRININ VLADIMIR P DR
CRIMEAN ASTROPHYS OBS
UKRAINIAN ACAD OF SCIENCE
NAUCHNY
334413 CRIMEA
UKRAINE
TEL 43 2945
TLF
TLX
EML
COM 27,36

GRISHCHUK L P DR
STERNBERG STATE ASTR INST
UNIVERSITETSKIJ PROSP 13
119899 MOSCOW
RUSSIA
TEL 139-50-06
TLF
TLX
EML
COM 47

GROOTE DETLEF
HAMBURGER STERNWARTE
GOJENSBERGSWEG 112
D 2050 HAMBURG 80
GERMANY
TEL 40 725 24112
TLF
TLX
EML
COM

GROSBOL PREBEN JOHNSON DR
ESO
KARL SCHWARZSCHILD STR 2
D 8046 GARCHING MUENCHEN
GERMANY
TEL 89 320 06237
TLX 52828222 EOD
EML
TLF 89 320 2362
COM 05C,09

GROSS PETER G PROF
714 OXFORD RD
BALA CYNWYD PA 19004
USA
TEL
TLF
EML
COM
TLX

GROSSMAN ALLEN S PROF
ERWIN W FICK OBSERVATORY
IOWA STATE UNIVERSITY
AMES IA 50011
USA
TEL 515 294 3666
TLF
TLX
EML
COM

GROSSMAN LAWRENCE PROF
DPT GEOPHYSICAL SCIENCES
UNIVERSITY OF CHICAGO
5734 S ELLIS AVE
CHICAGO IL 60637
USA
TEL 312 962 8153
TLF
TLX
EML
COM 15,16

GROSSMANN-DOERTH U DR
KIEPENHEUER INSTITUT
FUER SONNENPHYSIK
SCHOENECKSTR 6
D 7800 FREIBURG BREISGAU
GERMANY
TEL 761 32864
TLF
TLX 7721552 KIS D
EML
COM

GROTEN ERWIN PROF
INST/PHYSIKALISCHE GEOD
PETERSENSTR 13
D 6100 DARMSTADT
GERMANY
TEL 615 116 3109
TLX 419579
EML
TLF
COM 19

GROTH EDWARD J III
DPT OF PHYSICS
PRINCETON UNIVERSITY
JADWIN HALL
PRINCETON NJ 08544
USA
TEL 609 452 4361
TLF
TLX
EML
COM

GROTH HANS G PROF DR
INST ASTRON & ASTROPHYS
UNIVERSITAT MUENCHEN
SCHEINERSTR 1
D 8000 MUENCHEN 80
GERMANY
TEL 89 989 021
TLF
TLX
EML
COM 29,36

GROUSHINSKY N P PROF DR
STERNBERG STATE ASTR INST
UNIVERSITETSKIJ PROSP 13
119899 MOSCOW
RUSSIA
TEL
TLX
EML
TLF
COM 07

GRUBISSICH C PROF DR
VI AOSTA 34/5
I 35142 PADOVA
ITALY
TEL 49 38 301
TLF
EML·
COM 37
TLX

GRUDLER PIERRE
OCA CERGA
AVE COPERNIC
F 06130 GRASSE
FRANCE
TEL 93 36 5849
TLF
TLX
EML
COM 08,31

GRUDZINSKA STEFANIA DR
INSTITUTE OF ASTRONOMY
N COPERNICUS UNIVERSITY
UL CHOPINA 12/18
PL 87 100 TORUN
POLAND
TEL 20655
TLF
TLX 0552234 ASTR PL
EML
COM 15

GRUEFF GAVRIL DR
IST DI RADIOASTRONOMIA
CNR
VIA IRNERIO 46
I 40126 BOLOGNA
ITALY
TEL 51 23 2856
TLF
TLX
EML
COM

GRUEN EBERHARD DR
MPI FUER KERNPHYSIK
POSTFACH 103 980
D 6900 HEIDELBERG 1
GERMANY
TEL 62 215 16478
TLX 461666 MPIHD D
EML
TLF
COM 15C,21,22

GRUENWALD RUTH DR
IAG
UNIVERSIDADE DE SAO PAULO
CP 30627
01051 SAO PAULO SP
BRAZIL
TEL 11 577 8599
TLF
TLX 36221 IAGM BR
EML
COM

GRUNDMANN WALTER
HERZBERG INST ASTROPHYS
DOMINION ASTROPHYS OBS
5071 W SAANICH RD
VICTORIA BC V8X 4M6
CANADA
TEL 604 388 3157
TLF 604 363 0045
TLX 049 7295
EML
COM 09

GRY CECILE DR
LAS
TRAVERSE DU SIPHON
LES TROIS LUCS
F 13012 MARSEILLE
FRANCE
TEL 91 05 5900
TLF 91 66 1855
TLX 420 584
EML EARN CECILE@FRLASM51
COM

GRYGAR JIRI DR
INSTITUTE OF PHYSICS
CZECH ACADEMY OF SCIENCES
NA SLOVANCE 2
CS 180 40 PRAHA 8
CZECHOSLOVAKIA
TEL
TLF
TLX
EML GRYGAR@CSPGAS11
COM 27,42

GRZEDZIELSKI STANISLAW PR
SPACE RESEARCH CENTER
POLISH ACAD OF SCIENCES
UL ORDONA 21
PL 01 237 WARSAW
POLAND
TEL
TLF
TLX
EML
COM 49

GU XIAO-MA
YUNNAN OBSERVATORY
CAS
BOX 110
KUNMING 72946 YUNNAN
CHINA PR
TEL 871 2035
TLF
TLX 64040 YUOBS CN
EML
COM 10,12

GUARNIERI ADRIANO DR
OSS ASTRONOMICO
VIA ZAMBONI 33
I 40126 BOLOGNA
ITALY
TEL 51 51 9593
TLX
EML
TLF
COM

GUBANOV VADIM S DR
PULKOVO OBSERVATORY
ACADEMY OF SCIENCES
10 KUTUZOV QUAY
196140 ST PETERSBURG
RUSSIA
TEL 297-94-81
TLF
TLX
EML
COM 08

GUDMUNDSSON EINAR H
RAUNVISINDASTOFNUN
HASKOLANS
DUNHAGA 3
IS 107 REYKJAVIK
ICELAND
TEL 1 21340
TLF
TLX 2307 ISINFO
EML
COM 47

GUDUR N DR
OBSERVATORY
EGE UNIVERSITY
BOX 21
35100 BORNOVA IZMIR
TURKEY
TEL 51 18 0110*2326
TLF
TLX
EML
COM

GUELIN MICHEL DR
IRAM
300 RUE DE LA PISCINE
F 38406 S MARTIN HERES CD
FRANCE
TEL 76 42 3383
TLF
TLX 980753 IRAM
EML
COM 34,40

GUERIN PIERRE DR
INSTITUT D'ASTROPHYSIQUE
98BIS BD ARAGO
F 75014 PARIS
FRANCE
TEL 1 43 20 1425
TLF 1 43 29 8673
TLX
EML
COM 16

GUERRERO GIANANTONIO DR
OSS ASTRONOMICO DI MILANO
VIA E BIANCHI 46
I 22055 MERATE
ITALY
TEL 59 2035
TLF
TLX
EML
COM 27

GUERTLER JOACHIN DR
UNIVERSITATS-STERNWARTE

SCHILLERGAESSCHEN 2
D 6900 JENA
GERMANY
TEL 78 822 2637
TLF 78 822 2345
TLX 588634 UNI DD
EML
COM 34

GUEST JOHN E DR
UNIVERSITY OF LONDON OBS
MILL HILL PARK
LONDON NW7 2QS
UK
TEL 1 959 7367
TLX 28722 UCPHYS
EML
TLF
COM 16

GUESTEN ROLF
MPI FUER RADIOASTRONOMIE
AUF DEM HUEGEL 69
D 5300 BONN 1
GERMANY
TEL· 228 52 5379
TLX
EML
TLF
COM· 34,40C

GUETTER HARRY HENDRIK
US NAVAL OBSERVATORY
FLAGSTAFF STATION
BOX 1149
FLAGSTAFF AZ 86002
USA
TEL 602 779 5132
TLF
TLX
EML
COM 25,37,45

GUIBERT JEAN DR
OBSERVATOIRE DE PARIS
61 AVE OBSERVATOIRE
F 75014 PARIS
FRANCE
TEL 1 40 51 2098
TLF
TLX 270776 OBS F
EML
COM 05,09,24

GUIDICE DONALD A DR
A F GEOPHYSICS LABORATORY
SPACE PHYSICS DIV
HANSCOM AFB
BEDFORD MA 01731
USA
TEL 617 861 3989
TLF
TLX
EML
COM 40

GUINAN EDWARD FRANCIS DR
DPT OF ASTRONOMY
VILLANOVA UNIVERSITY
VILLANOVA PA 19085
USA
TEL 215 527 2100
TLF
TLX
EML
COM 27,42C

GUINOT BERNARD R PROF
2 RUE DES SOUPIRS
F 77590 CHARTRETTES
FRANCE
TEL
TLF
TLX
EML
COM 19,31C

GULKIS SAMUEL DR
JPL
MS 169 506
4800 OAK GROVE DR
PASADENA CA 91109
USA
TEL 818 354 5708
TLF 818 354 2946
TLX
EML
COM 16,40,51

GULL STEPHEN F DR
MULLARD RADIO ASTRON OBS
CAVENDISH LABORATORY
MADINGLEY RD
CAMBRIDGE CB3 OHE
UK
TEL 223 664 77
TLF
TLX 81292
EML
COM 40

GULL THEODORE R DR
NASA/GSFC
CODE 680
LASP
GREENBELT MD 20771
USA
TEL 301 286 8060
TLF
TLX 710-8289716
EML:
COM 34,44

GULLIVER AUSTIN FRASER DR
DPT OF PHYSICS
UNIVERSITY OF BRANDON
2500 UNIVERSITY DR NW
BRANDON MB R7A 6A9
CANADA
TEL 204 728 9520
TLF
TLX 075 02721
EML
COM 42

GULMEN OMUR DR
OBSERVATORY
EGE UNIVERSITY
BOX 21
35100 BORNOVA IZMIR
TURKEY
TEL 51 18 0110
TLF
TLX
EML
COM

GULYAEV A P DR
STERNBERG STATE ASTR INST
UNIVERSITETSKIJ PROSP 13
119899 MOSCOW
RUSSIA
TEL 139-19-70
TLX
EML
TLF
COM 08

GULYAEV RUDOLF A DR
IZMIRAN
ACADEMY OF SCIENCES
142092 TROITSK
RUSSIA
TEL
TLF
TLX
EML
COM

GUNN JAMES E PROF
DPT ASTROPHYSICAL SCI
PRINCETON UNIVERSITY
PEYTON HALL
PRINCETON NJ 08544 1001
USA
TEL 609 452 3802
TLF
TLX
EML
COM 28,47,48,51

GUO NEI-SHU DR
NANJING ASTRONOMICAL
INSTRUMENT FACTORY
BOX 846
NANJING
CHINA PR
TEL 25 64 6191
TLF
TLX 34136
EML
COM

GUO QUAN SHI
PURPLE MOUNTAIN OBSERV
CAS
NANJING
CHINA PR
TEL 25 46700
TLF
TLX
EML
COM

GUPTA SUNIL K DR
TIFR
HOMI BHABHA RD
COLABA
BOMBAY 400 005
INDIA
TEL 22 215 2971*545
TLF 22 215 2110
TLX
EML GUPTASK@TIFRVAX BITNET
COM 33

GURM HARDEV S PROF
DPT ASTRONOMY & SPACE SCI
UNIVERSITY OF PANJABI
PATIALA 147 002
INDIA
TEL 73262*96
TLF
TLX
EML
COM. 27,35,46,51

GURMAN JOSEPH B DR
NASA/GSFC
CODE 602 6
GREENBELT MD 20771
USA
TEL 301 286 7599
TLF
TLX 89675
EML SPAN SOLMAX GURMAN
COM 10

GURSHTEIN ALEXANDER A DR
INST HIST OF SCI & TECHN
ACADEMY OF SCIENCES
STAROPANSKY 1/5
103012 MOSCOW
RUSSIA
TEL
TLF
TLX
EML
COM 16,41

GURSKY HERBERT DR
NAVAL RESEARCH LABORATORY
CODE 4100
4555 OVERLOOK AVE SW
WASHINGTON DC 20375 5000
USA
TEL 202 767 6343
TLF
TLX
EML.
COM 27,42,44,48

GURTOVENKO E A DR
MAIN ASTRONOMICAL OBS
UKRAINIAN ACAD OF SCIENCE
GOLOSEEVO
252127 KIEV
UKRAINE
TEL 66 1065
TLF
TLX 131406
EML
COM 10,12

GURZADIAN G A PROF DR
BYURAKAN ASTROPHYSICAL
OBSERVATORY
378433 ARMENIA
ARMENIA
TEL 88 52 28 3453
TLF
TLX
EML
COM 28,34,45

GUSEINOV O H PROF
INSTITUTE OF PHYSICS
NARIMANOV UL 33
370143 BAKU
AZERBAIDZHAN
TEL 39 3951
TLX
EML
TLF
COM 34,42,44,48

GUSEJNOV RAGIM EH DR
SHEMAKHA ASTROPHYSICAL
OBSERVATORY
AZER ACADEMY OF SCIENCES
373243 SHEMAKHA
AZERBAIDZHAN
TEL
TLF
TLX
EML
COM

GUSSMANN E A DR
ZNTRLINST F ASTROPHYSIK
STERNWARTE BABALSBERG
ROSA-LUXEMBURG-STR 17A
D 1502 POTSDAM
GERMANY
TEL
TLF
TLX
EML
COM 36

GUSTAFSON BO A S
DPT OF ASTRONOMY
UNIVERSITY OF FLORIDA
211 SSRB
GAINESVILLE FL 32611
USA
TEL 904 392 7677
TLF 904 392 5089
TLX
EML gustaf@astro ufl edu
COM 15,21,22

GUSTAFSSON BENGT DR
ASTRONOMICAL OBSERVATORY
BOX 515
S 751 20 UPPSALA
SWEDEN
TEL
TLX
EML
TLF
COM 29C,36

GUTCKE DIETRICH
CARL-ZEISS STR 1
D 6900 JENA
GERMANY
TEL
TLF
EML
COM 09
TLX

GUTHRIE BRUCE N G DR
ROYAL OBSERVATORY
BLACKFORD HILL
EDINBURGH EH9 3HJ
UK
TEL 316 673 321
TLF
TLX 72383 ROEDIN G
EML
COM 29

GUTIERREZ-MORENO A DR MRS
DPT DE ASTRONOMIA
UNIVERSIDAD DE CHILE
CASILLA 36 D
SANTIAGO
CHILE
TEL 2 229 4101/4002
TLF
TLX 440 001
EML
COM 25

GWINN CARL R DR
DPT OF PHYSICS
UNIVERSITY OF CALIFORNIA
SANTA BARBARA CA 93106
USA
TEL 805 961 2814
TLF 805 961 4170
TLX
EML cgwinn@voodoo ucsb edu
COM 40

GYLDENKERNE KJELD DR
COPENHAGEN UNIVERSITY OBS
BRORFELDEVEJ 23
DK 4340 TOLLOSE
DENMARK
TEL 53 48 8195
TLF 58 48 8755
TLX 44155
EML
COM 33,42

HABBAL SHADIA RIFAI
CENTER FOR ASTROPHYSICS
HCO/SAO
60 GARDEN ST
CAMBRIDGE MA 02138
USA
TEL 617 495 7348
TLF
TLX 921428
EML
COM 49

HABE ASAO
DPT OF PHYSICS
HOKKAIDO UNIVERSITY
KITA 10 NISHI 8
SAPPORO 060
JAPAN
TEL 11-711-2111
TLF
TLX
EML
COM 33

HABIBULLIN SH T PROF DR
DPT OF ASTRONOMY
KAZAN STATE UNIVERSITY
LENIN UL 18
420008 KAZAN
RUSSIA
TEL 323641
TLF
TLX
EML
COM 16

HABING H J DR
STERREWACHT
BOX 9513
NL 2300 RA LEIDEN
NETHERLANDS
TEL 71 27 2727
TLX 39058
EML
TLF
COM 33,34P

HACHENBERG OTTO PROF DR
RADIOASTRONOMISCHES INST
UNIVERSITAET BONN
AUF DEM HUEGEL 71
D 5300 BONN 1
GERMANY
TEL
TLF
TLX
EML
COM

HACHISU IZUMI DR
DPT AERONAUTICAL ENGINEER
KYOTO UNIVERSITY
KITASHIRAKAWA SAKYO KU
KYOTO 606
JAPAN
TEL 075-751-2111
TLF
TLX 05422693 LIBKYU J
EML
COM 35

HACK MARGHERITA PROF
OAT
BOX SUCC TRIESTE 5
VIA TIEPOLO 11
I 34131 TRIESTE
ITALY
TEL 40 79 3921
TLF
TLX 461137 OAT I
EML
COM 29,36,44,45

HACKWELL JOHN A DR
AEROSPACE CORPORATION
BOX 92957
LOS ANGELES CA 90009
USA
TEL 307 766 6296
TLX·
EML
TLF
COM 27,34

HACYAN SHAHEN DR
INSTITUTO DE ASTRONOMIA
UNAM
APDO POSTAL 70-264
04510 MEXICO DF
MEXICO
TEL 905-548-5305
TLF
TLX 1760155 CICME
EML.
COM 47

HADDOCK FRED T DR
DPT OF ASTRONOMY
UNIVERSITY OF MICHIGAN
DENNISON BLDG
ANN ARBOR MI 48109 1090
USA
TEL 313 764 3430
TLF
TLX
EML
COM 40,44,51

HADJIDEMETRIOU JOHN D
DPT THEORET MECHANICS
UNIVERSITY THESSALONIKI
GR 540 06 THESSALONIKI
GREECE
TEL 31 99 1410
TLF
TLX
EML
COM 07

HADLEY BRIAN W
ROYAL OBSERVATORY
BLACKFORD HILL
EDINBURGH EH9 3HJ
UK
TEL 316 688 296
TLX 72383 ROEDIN G
EML BWH@UK AC ROE STAR
TLF
COM 09

HADRAVA PETR
ASTRONOMICAL INSTITUTE
CZECH ACADEMY OF SCIENCES
ONDREJOV OBSERVATORY
CS 251 65 ONDREJOV
CZECHOSLOVAKIA
TEL 204 85201
TLF 204 85314
TLX
EML
COM 42

HAEFNER REINHOLD DR
UNIVERSITAETS STERNWARTE

SCHEINERSTR 1
D 8000 MUENCHEN 80
GERMANY
TEL 89 989 021
TLF
TLX
EML
COM 27

HAEMEEN ANTTILA KAARLE A
DPT OF ASTRONOMY
UNIVERSITY OF OULU
SF 90570 OULU 57
FINLAND
TEL
TLF 81 56 1278
TLX
EML
COM

HAENSEL PAWEL DR
COPERNICUS ASTRON CENTER
POLISH ACAD OF SCIENCES
UL BARTYCKA 18
PL 00 716 WARSAW
POLAND
TEL 410828
TLF
TLX 81 3978 ZAPLAN PL
EML
COM

HAERENDEL G DR
MPI F PHYSIK & ASTROPHYS
INST F EXTRATERR PHYSIK
D 8046 GARCHING MUENCHEN
GERMANY
TEL 89 329 93516
TLF 89 329 93569
TLX 05215845 XTER D
EML
'COM

HAGEN HANS-JUERGEN DR
HAMBURGER STERNWARTE

GOJENSBERGSWEG 112
D 2050 HAMBURG 80
GERMANY
TEL 40 725 24136
TLF 40 7252 4198
- TLX 217884
EML·
COM 47

HAGEN JOHN P
613 W PARK AVE
STATE COLLEGE PA 16803
USA
TEL 814 237 3031
TLF
EML
COM 10,40
TLX

HAGEN-THORN VLADIMIR A DR
ASTRONOMICAL OBSERVATORY
ST PETERSBURG UNIVERSITY
BIBLIOTECHNAJA PL 2
198904 ST PETERSBURG
RUSSIA
TEL 257-94-91
TLF
TLX
EML
COM 28

HAGFORS T DR
NAIC
CORNELL UNIVERSITY
SPACE SCIENCES BLDG
ITHACA NY 14853 6801
USA
TEL 607 256 3734
TLF
TLX 932454
EML
COM 16

HAGIO FUMIHIKO DR
KUMAMOTO INST TECHNOLOGY

IKEDA 4
KUMAMOTO
JAPAN
TEL 96 326 3111
TLF 96 326 3000
TLX
EML
COM 28

HAGYARD MONA JUNE
NASA/MSFC
CODE ES 52
HUNTSVILLE AL 35812
USA
TEL 205 453 5687
TLF
TLX 594416 NASA/MSFC HTV
EML
COM 10,12

HAHN GERHARD J DR
ASTRONOMICAL OBSERVATORY
BOX 515
S 751 20 UPPSALA
SWEDEN
TEL 18 52 5724
TLX 76024 UNIVUPS S
EML GERHARD@LABAN SUNET SE
TLF 18 52 7583
COM 20

HAIKALA LAURI K
OBS & ASTROPHYSICS LAB
UNIVERSITY OF HELSINKI
TAEHTITORNINMAKI
SF 00130 HELSINKI 13
FINLAND
TEL 191 2948
TLF
TLX 124690 UNIH SF
EML
COM

HAISCH BERNHARD MICHAEL
LOCKHEED PALO ALTO RES LB
DPT 91 20 BLDG 255
3251 HANOVER ST
PALO ALTO CA 94304
USA
TEL 415 858 4073
TLF
TLX 346409
EML
COM 27,36,51

HAJDUK ANTON DR
ASTRONOMICAL INSTITUTE
SLOVAK ACADEMY SCIENCES
CS CS 842 28 BRATISLAVA
CZECHOSLOVAKIA
TEL 7 37 5157
TLF
TLX 93373 SEIS
EML
COM 15,22,51

HAJDUKOVA MARIA
DPT OF ASTRONOMY
COMENIUS UNIVERSITY
MLYNSKA DOLINA
CS 842 15 BRATISLAVA
CZECHOSLOVAKIA
TEL 7 32 0003
TLF
TLX
EML
COM 22

HAKKILA JON ERIC DR
DPT MATHS ASTRONOMY & STA
MANKATO STATE UNIVERSITY
BOX 41
MANKATO MN 56002 8400
USA
TEL 507 389 1840
TLF
TLX
EML JHAKK@MSUS1
COM 33

HALBWACHS JEAN LOUIS DR
OBS DE STRASBOURG
11 RUE UNIVERSITE
F 67000 STRASBOURG
FRANCE
TEL 88 35 8200
TLF 88 25 0160
TLX 890506
EML EARN U01103@FRCCSC21
COM 26,30

HALL ANDREW NORMAN
DPT OF ASTROPHYSICS
UNIVERSITY OF OXFORD
SOUTH PARKS RD
OXFORD OX1 3RQ
UK
TEL
TLF
TLX
EML
COM 48

HALL DONALD N DR
INSTITUTE FOR ASTRONOMY
UNIVERSITY OF HAWAII
2680 WOODLAWN DR
HONOLULU HI 96822
USA
TEL 808 948 8312
TLF
TLX 723-8459
EML
COM

HALL DOUGLAS S DR
DYER OBSERVATORY
VANDERBILT UNIVERSITY
NASHVILLE TN 37235
USA
TEL 615 373 4897
TLF
TLX 554323
EML
COM 25,27,36,42

HALL PETER J DR
CSIRO AUSTRALIA TELESCOPE
NATIONAL FACULTY
BOX 94
NARRABRI NSW 2390
AUSTRALIA
TEL 67 959 205
TLF 67 959 255
TLX
EML PHALL@ATNF CSIRO AU
COM 40

HALL R GLENN DR
3612 SPRING ST
CHEVY CHASE MD 20815
USA
TEL 301 652 7221
TLF
EML
COM 19,31
TLX

HALLAM KENNETH L DR
701 E CAPITOL ST SE
WASHINGTON DC 20003
USA
TEL
TLF
EML
COM 09,44,45
TLX

HALLIDAY IAN DR
HERZBERG INST ASTROPHYS
NTL RESEARCH COUNCIL
100 SUSSEX DR
OTTAWA ON K1A 0R6
CANADA
TEL 613 990 0704
TLF 613 952 6602
TLX 053 3715
EML
COM 15,16,21,22

HAMABE MASARU DR
KISO OBSERVATORY
UNIVERSITY OF TOKYO
MITAKEMURA KISOGUN
NAGANO 397 01
JAPAN
TEL 26452-3360
TLF
TLX 3347577 KSOOBS J
EML
COM 28

HAMADA TETSUO PROF
DPT OF PHYSICS
IBARAKI UNIVERSITY
BUNKYO
MITO 310
JAPAN
TEL 0292-252-35-24
TLF
TLX
EML
COM

HAMAJIMA KIYOTOSHI DR
KISO OBSERVATORY
UNIVERSITY OF TOKYO
MITAKEMURA KISOGUN
NAGANO 397 01
JAPAN
TEL
TLF
TLX
EML
COM 33

HAMDY M A M PROF
HELWAN OBSERVATORY
HELWAN
EGYPT
TEL 78 0645/2683
TLF
TLX 93070 HIAG UN
EML
COM 27

HAMID S EL DIN PROF
DPT OF ASTRONOMY
FACULTY OF SCIENCES
CAIRO UNIVERSITY
GEZA
EGYPT
TEL
TLF
TLX
EML
COM 07

HAMILTON P A DR
DPT OF PHYSICS
UNIVERSITY OF TASMANIA
GPO BOX 252C
HOBART TAS 7001
AUSTRALIA
TEL 2 202 419
TLF
TLX 58150
EML
COM 40

HAMMAN WOLF-RAINER
INST THEOR PHYS & STERNW
UNIVERSITAET KIEL
OLSHAUSENSTR
D 2300 KIEL 1
GERMANY
TEL 431 880 4110
TLF
TLX 292706
EML
COM· 36

HAMMER FRANCOIS DR
OBSERVATOIRE DE PARIS
SECTION DE MEUDON
DAEC
F 92195 MEUDON PPL CDX
FRANCE
TEL 1 45 07 7408
TLF 1 45 07 2806
TLX
EML HAMMER@FRMEU51
COM 28

HAMMER REINER
KIEPENHEUER INSTITUT FUER
SONNENPHYSIK
SCHOENECKSTR 6
D 7800 FREIBURG BREISGAU
GERMANY
TEL 761 32864
TLF
TLX 7721552 KIS D
EML
COM 10,12

HAMMERSCHLAG ROBERT H DR
STERREKUNDIG INSTITUTE
BOX 80000
NL 3508 TA UTRECHT
NETHERLANDS
TEL 30 53 5218
TLX 40048 FYLUT NL
EML BITNET WNMMAIL@HUTRUUO
TLF
COM 09

HAMMERSCHLAG-HENSBERGE G
ASTRONOMICAL INSTITUTE
UNIVERSITY OF AMSTERDAM
KRUISLAAN 403
NL 1098 SJ AMSTERDAM
NETHERLANDS
TEL 20 52 57491/7492
TLF 20 52 57484
TLX 10262 HEF NL
EML A410INST@HASARA11 BITNET
COM· 42

HAMMOND GORDON L DR
DPT OF MATHEMATICS
UNIVERSITY OF S FLORIDA
ASTRONOMY PROGRAM
TAMPA FL 33620
USA
TEL 813 783 1226
TLF
TLX
EML DKGAWAA@CFRVM BITNET
COM 35

HAMZAOGLU ESAT E H DR
KING SAUD UNIVERSITY
COLLEGE OF SCIENCE
BOX 2455
RIYADH 11453
SAUDI ARABIA
TEL
TLF
TLX
EML
COM

HAN FU
PURPLE MOUNTAIN OBSERV
CAS
NANJING
CHINA PR
TEL 25 33738
TLF
TLX 34144 PMONJ CN
EML
COM 40

HAN TIANQI
INSTITUTE OF GEODESY &
GEOPHYSICS
XU DONG LU
WUCHANG HUBEI
CHINA PR
TEL 81 3712*570
TLF
TLX
EML
COM 19,31

HAN WENJUN
BEIJING ASTRONOMICAL OBS
CAS
W SUBURB
BEIJING 100080
CHINA PR
TEL 1 28 1698
TLF
TLX
EML
COM 40

HAN ZHENG-ZHONG
PURPLE MOUNTAIN OBSERV
CAS
NANJING
CHINA PR
TEL 25 33583
TLF
TLX 34144
EML
COM 44

HANAMI HITOSHI DR
COLL HUMANITIES SOCIAL SC
IWATE UNIVERSITY
UEDA 3
MORIOKA 020
JAPAN
TEL 0196 23 5171*2284
TLF 0196 54 2289
TLX
EML D12697@JPNKUDPC
COM 28,33,34

HANAOKA YOICHIRO DR
NOBEYAMA RADIO OBS
NAOJ
MINAMIMAKI MURA
NAGANO 384 13
JAPAN
TEL 81 267 63 4381
TLF 81 267 98 2506
TLX 3329005 NAONRO J
EML E52863@JPNKUDPC BITNET
COM 10

HANASZ JAN DR
INSTITUTE OF ASTRONOMY
N COPERNICUS UNIVERSITY
UL CHOPINA 12/18
PL 87 100 TORUN
POLAND
TEL 260-37
TLF
TLX 0552234 ASTR PL
EML
COM 10,40

HANAWA TOMOYUKI DR
DPT OF ASTROPHYSICS
NAGOYA UNIVERSITY
FUROCHO CHIKUSA KU
NAGOYA 464 01
JAPAN
TEL 52 781 6769
TLF
TLX 4477323
EML BITNET B42287@JPNKUDPC
COM 42

HANBURY BROWN ROBERT PROF
WHITE COTTAGE
PENTON MEWSEY
HANTS SP11 ORQ
UK
TEL 264 772 334
TLF
TLX
EML
COM 40

HANDA TOSHIHIRO DR
INST ASTRO FAC SCIENCE
UNIVERSITY OF TOKYO
OSAWA MITAKA
TOKYO 181
JAPAN
TEL 422 41 3735
TLF 422 41 3749
TLX 2822307 TAOMTK J
EML
COM 28,40

HANDLIROVA DAGMAR DR
N COPERNICUS OBSERVATORY
& PLANETARIUM
KRAVI HORA
CS 616 00 BRNO
CZECHOSLOVAKIA
TEL 5 74 4374
TLF
TLX
EML
COM

HANES DAVID A DR
DPT OF PHYSICS
QUEEN'S UNIVERSITY
KINGSTON ON K7L 3N6
CANADA
TEL 613 547 5750
TLF 613 545 6463
TLX
EML
COM 37

HANG HENG-RONG
PURPLE MOUNTAIN OBSERV
CAS
NANJING
CHINA PR
TEL 25 33583
TLF
TLX 34144 PMONJ CN
EML
COM 44,48

HANISCH ROBERT J DR
STSCI
HOMEWOOD CAMPUS
3700 SAN MARTIN DR
BALTIMORE MD 21218
USA   .
TEL 301 338 4910
TLF
TLX 6849101
EML SPAN SCIVAX HANISCH
COM 05,09,40

HANKINS TIMOTHY HAMILTON
DPT OF PHYSICS & ASTRON
NEW MEXICO TECH
CAMPUS STATION
SOCORRO NM 87801
USA
TEL 505 476 8011
TLF
TLX
EML
COM 40

HANNER MARTHA S DR
JPL
MS 183 601
4800 OAK GROVE DR
PASADENA CA 91109
USA
TEL 818 354 4100
TLF
TLX 675429
EML SPAN JPLSC8 MSH
COM 15,21P,22

HANSEN CARL J PROF
JILA
UNIVERSITY OF COLORADO
BOX 440
BOULDER CO 80309 0440
USA
TEL 303 492 7811
TLF
TLX 755842 JILA
EML
COM 27

HANSEN LEIF LECTURER
UNIVERSITY OBSERVATORY
OESTER VOLDGADE 3
DK 1350 COPENHAGEN K
DENMARK
TEL 31 14 1790
TLF 31 38 9157
TLX 44155 DANAST DK
EML
COM

HANSEN RICHARD T MR
ENGINEERING 138
VAMC
150 S HUNTINGTON AVE
BOSTON MA 02130
USA
TEL 617 734 2534
TLF
TLX
EML
COM 10

HANSLMEIER ARNOLD
INSTITUT FUER ASTRONOMIE
KARL-FRANZENS-UNIVERSITAT
UNIVERSITAETSPLATZ 5
A 8010 GRAZ
AUSTRIA
TEL 316 380 5275
TLF
TLX
EML.
COM 07,10

HANSON ROBERT B DR
LICK OBSERVATORY
UNIVERSITY OF CALIFORNIA
SANTA CRUZ CA 95064
USA
TEL 408 429 2755
TLF
TLX
EML
COM 24

HANSSON NILS DR
LUND OBSERVATORY
BOX 43
S 221 00 LUND
SWEDEN
TEL 46 10 7000
TLX 33533 LUNIVER S
EML
TLF
COM

HANTZIOS PANAYIOTIS DR
ASTRONOMICAL INSTITUTE
NTL OBSERVATORY OF ATHENS
BOX 20048
GR 118 10 ATHENS
GREECE
TEL 1 346 1191
TLF
TLX
EML
COM 42

HANUSCHIK REINHARD DR
ASTRONOMISCHES INSTITUT
RUHR UNIVERSITAT BOCHUM
POSTFACH 102148
D 4630 BOCHUM 1
GERMANY
TEL 234 70 03450
TLF
TLX 0825860
EML
COM 29

HAO YUN-XIANG
DPT OF ASTRONOMY
BEIJING NORMAL UNIVERSITY
BEIJING 100875
CHINA PR
TEL 1 65 6531*6285
TLF
TLX
EML
COM 09

HAPKE BRUCE W DR
DPT GEOL & PLANETARY SCI
UNIVERSITY OF PITTSBURGH
321 OLD ENGINEERING HALL
PITTSBURGH PA 15235
USA
TEL 412 624 4719
TLF
TLX
EML
COM 15

HARA KEN NOSUKE DR
SECOND TECHNICAL HIGH SCH
MAYAGINO KU
SENDAI 983
JAPAN
TEL
TLF
TLX
EML
COM 31,47

HARA TADAYOSHI DR
NTL ASTRONOMICAL OBS
DIV OF EARTH ROTATION
MIZUSAWA SHI
IWATE 023
JAPAN
TEL 197 24 7111
TLF
TLX 837628 ILSMIZ J
EML
COM

HARA TETSUYA DR
DPT OF PHYSICS
KYOTO SANGYO UNIVERSITY
KAMIGAMO
KYOTO 603
JAPAN
TEL 075-701-2151
TLF
TLX 5422661 KSU J
EML
COM 28

HARDEBECK ELLEN G DR
3106 TUMBLEWEED RD
BISHOP CA 93514
USA
TEL
TLF
EML
COM 34
TLX

HARDEE PHILIP
DPT OF PHYSICS & ASTRON
UNIVERSITY OF ALABAMA
BOX 1921
UNIVERSITY AL 35487 0324
USA
TEL 205 348 5050
TLF
TLX
EML
COM 40

HARDY EDUARDO
DPT DE PHYSIQUE
UNIVERSITE LAVAL
FAC SCIENCES & GENIE
LAVAL QC G1K 7P4
CANADA
TEL 418 656 2960
TLF 418 656 2040
TLX 051 31621
EML
COM 28,47

HARMANEC PETR DR
ASTRONOMICAL INSTITUTE
CZECH ACADEMY OF SCIENCES
ONDREJOV OBSERVATORY
CS 251 65 ONDREJOV
CZECHOSLOVAKIA
TEL 204 85201
TLF: 204 85314
TLX 121579
EML
COM· 27,42

HARMER CHARLES F W MR
NOAO
BOX 26732
950 N CHERRY AVE
TUCSON AZ 85726 6732
USA
TEL 602 327 5511
TLF
TLX 0666-484 AURA NOAO
EML.
COM: 09,29

HARMER DIANNE L MRS
NOAO
BOX 26732
950 N CHERRY AVE
TUCSON AZ 85726 6732
USA
TEL 602 325 9218
TLF
TLX
EML DIHARMER@NOAO EDU
COM. 09,29

HARMS RICHARD JAMES DR
APPLIED RESEARCH CORP
8201 CORPORATE DR
SUITE 920
LANDOVER MD 20785
USA
TEL 301 459 8442
TLF
TLX
EML
COM. 28,44,47

HARNDEN FRANK R Jr
CENTER FOR ASTROPHYSICS
HCO/SAO
60 GARDEN ST
CAMBRIDGE MA 02138
USA
TEL 617 495 7143
TLF
TLX 921428 SATELLITE CAM
EML
COM

HARNETT JULIEINE DR
SCHOOL OF PHYSICS
UNIVERSITY OF SYDNEY
SYDNEY NSW 2006
AUSTRALIA
TEL 2 692 2727
TLF 2 660 2903
TLX AA 26169
EML JHATNETT@ROBIN RP CSIRO AU
COM 28,40

HARPAZ AMOS DR
DPT OF PHYISICS/SPACE RES
IIT
TECHNION CITY
HAIFA 32000
ISRAEL
TEL 4 293 521
TLF
TLX
EML BITNET PHR89AH@TECHNION
COM

HARPER DAVID DR
ASTRONOMY UNIT
QUEEN MARY/WESTFIELD COLL
MILE END RD
LONDON E1 4NS
UK
TEL 1 719 755 492
TLF 1 819 819 587
TLX 893750
EML ADH@STAR QMW AC UK
COM 20

HARRINGTON J PATRICK DR
ASTRONOMY PROGRAM
UNIVERSITY OF MARYLAND
COLLEGE PARK MD 20742
USA
TEL 301 454 5944
TLF
TLX 7108260352
EML
COM 34

HARRINGTON ROBERT S DR
US NAVAL OBSERVATORY
34 & MASSACHUSETTS AVE NW
WASHINGTON DC 20392 5100
USA
TEL 202 653 1533
TLX
EML
TLF
COM 20,24,26,51

HARRIS ALAN WILLIAM
MPI FUER PHYS U ASTROPHY
INST FUER EXTRATER PHYSIK
D 8046 GARCHING MUENCHEN
GERMANY
TEL 89 329 90
TLF 89 329 93235
TLX 05 215845 XTERR D
EML SPAN 19457 AWH
COM 16,34

HARRIS ALAN WILLIAM DR
JPL
MS 183 501
4800 OAK GROVE DR
PASADENA CA 91109
USA
TEL 818 354 6741
TLF 818 354 0966
TLX 675429/9105883294/69
EML AWHARRIS@JPLGP JPL NASA
COM 15P,20

HARRIS DANIEL E DR
CENTER FOR ASTROPHYSICS
HCO/SAO
60 GARDEN ST
CAMBRIDGE MA 02138
USA
TEL 617 495 7148
TLF
TLX 921428
EML
COM 40

HARRIS GRETCHEN L H DR
DPT OF PHYSICS
UNIVERSITY OF WATERLOO
WATERLOO ON N2L 3G1
CANADA
TEL 519 885 1211
TLF 519 746 8115
TLX 069 55259
EML glharris@astro waterloo edu
COM 37C

HARRIS HUGH C
US NAVAL OBSERVATORY
FLAGSTAFF STATION
POX 1149
FLAGSTAFF AZ 86002
USA
TEL 602 779 5132
TLF
TLX
EML
COM 37

HARRIS STELLA
DPT OF PHYSICS
QUEEN MARY/WESTFIELD COLL
MILE END RD
LONDON E1 4NS
UK
TEL 1 980 4811*4050
TLF
TLX 893750
EML
COM 34

HARRIS WILLIAM E DR
DPT OF PHYSICS
MCMASTER UNIVERSITY
HAMILTON ON L8S 4M1
CANADA
TEL 416 525 9140
TLF 416 546 1252
TLX 061 8347
EML
COM 37

HARRISON EDWARD R PROF
DPT PHYSICS & ASTRONOMY
UNIV OF MASSACHUSETTS
GRC
AMHERST MA 01003
USA
TEL 413 545 2194
TLF
TLX
EML·
COM 47,51

HARRISON RICHARD A DR
RUTHERFORD APPLETON LAB
SPACE & ASTROPHYSICS DIV
BLDG R25/R68
CHILTON DIDCOT OX11 0QX
UK
TEL 235 446 497
TLF
TLX 83159 RUTHLB G
EML SPAN 19457 RAH
COM

HARROWER GEORGE A DR
204 1033 BELMONT AVENUE

VICTORIA BC V8S 3T4
CANADA
TEL
TLF
TLX
EML
COM

HART MICHAEL H DR
7301 MASONVILLE DR
ANNANDALE VA 22003
USA
TEL
TLF
EML
COM 51
TLX

HARTEN RONALD H DR
RCA ASTRO ELECTRONIC
TB-1
BOX 800
PRINCETON NJ 08540
USA
TEL 609 426 3551
TLF
TLX
EML
COM 34,40

HARTKOPF WILLIAM I DR
CHARA
GEORGIA STATE UNIVERSITY
ATLANTA GA 30303 3083   ·
USA
TEL 404 651 2932
TLF
TLX
EML
COM 24,26,33

HARTL HERBERT DR
INSTITUT FUER ASTRONOMIE

TECHNIKERSTR 15
A 6020 INNSBRUCK
AUSTRIA
TEL 5222 748 5263
TLF 5222 748 5252
TLX 533808 UNITE A
EML C70605@AINUNIO1
COM 34

HARTLEY KENNETH F DR
RUTHERFORD APPLETON LAB
SPACE & ASTROPHYSICS DIV
BLDG R25/R68
CHILTON DIDCOT OX11 OQX
UK
TEL 235 219 00
TLF
TLX 83159
EML
COM

HARTMANN LEE WILLIAM
CENTER FOR ASTROPHYSICS
HCO/SAO
60 GARDEN ST
CAMBRIDGE MA 02138
USA
TEL 617 495 7487
TLF
TLX
EML
COM 29,36

HARTMANN WILLIAM K
PLANETARY SCIENCE INST
2030 E SPEEDWAY
SUITE 201
TUCSON AZ 85719
USA
TEL 602 881 0332
TLF
TLX
EML
COM 15

HARTOOG MARK RICHARD DR
LICK OBSERVATORY
UNIVERSITY OF CALIFORNIA
SANTA CRUZ CA 95064
USA
TEL
TLF
TLX
EML
COM

HARTQUIST THOMAS WILBUR
MPI FUR PHYS & ASTROPHYS
KARL-SCHWARZSCHILD-STR 1
D 8046 GARCHING MUENCHEN
GERMANY
TEL 89 329 9838
TLX 05215845 XTERR D
EML
TLF
COM 34

HARTWICK F DAVID A DR
DPT OF PHYSICS
UNIVERSITY OF VICTORIA
BOX 1700
VICTORIA BC V8W 2Y2
CANADA
TEL 604 721 7742
TLF 604 721 7715
TLX
EML
COM

HARTZ THEODORE R DR
915 MOUNTAINVIEW AVENUE
OTTAWA ON K2B 5G3
CANADA
TEL 613 596 1211
TLF
TLX
EML
COM 40,44

HARUTYUNIAN HAIK A DR
BYURAKAN ASTROPHYSICAL
OBSERVATORY
378433 BYURAKAN
ARMENIA
TEL 88 52 28 3453/4142
TLF
TLX 411576 AS CON SU
EML
COM 36

HARVEL CHRISTOPHER ALVIN
6161 STEVEN'S FOREST RD
COLUMBIA MD 21045
USA
TEL 301 964 0211
TLF
EML
COM 05,37
TLX

HARVEY CHRISTOPHER C DR
OBSERVATOIRE DE PARIS
SECTION DE MEUDON
F 92195 MEUDON PPL CDX
FRANCE
TEL 1 45 07 7669
TLF
TLX 204464
EML MEUDON HARVEY
COM 44,49

HARVEY GALE A DR
INST F GESCHICHTE D
NATURWISSENSCHAFTEN
GOETHE UNIVERSITAET
D 6000 FRANKFURT
GERMANY
TEL
TLF
TLX
EML
COM 22

HARVEY JOHN W DR
NTL SOLAR OBSERVATORY
BOX 26732
950 N CHERRY AVE
TUCSON AZ 85726 6732
USA
TEL 602 327 5511
TLF 602 325 9278
TLX 1561401 AURA UT
EML JHARVEY@NOAO EDU
COM 10,12C

HARVEY PAUL MICHAEL DR
ASTRONOMY DPT
UNIVERSITY OF TEXAS
RLM 15 308
AUSTIN TX 78712 1083
USA
TEL 512 471 4461
TLF
TLX 910-874-1351
EML
COM 34,44

HARWIT MARTIN PROF
DPT OF ASTRONOMY
CORNELL UNIVERSITY
SPACE SCIENCES BLDG
ITHACA NY 14853
USA
TEL 607 256 4805
TLF
TLX
EML
COM 15,21,48

HARWOOD DENNIS MR
PERTH OBSERVATORY
BICKLEY WA 6076
AUSTRALIA
TEL 9 293 8255
TLF
TLX
EML
COM 24,25

HASAN HASHIMA DR
STSCI
HOMEWOOD CAMPUS
3700 SAN MARTIN DR
BALTIMORE MD 21218
USA
TEL 301 338 4519
TLF 301 338 4767
TLX 301 684 9101 STSCI
EML HASAN@STSCI
COM 28

HASAN SAIYID STRAJUL
INDIAN INSTITUTE OF
ASTROPHYSICS
KORAMANGALA
BANGALORE 560 034
INDIA
TEL 812 56 6585
TLF
TLX
EML
COM 10

HASCHICK AUBREY
HAYSTACK OBSERVATORY
WESTFORD MA 01886
USA
TEL 617 692 4764
TLF
EML
COM 40
TLX

HASEGAWA ICHIRO DR
4-18-5 FUJIWARADAI KITA
KIYA KU
KOBE 651-13
JAPAN
TEL 81 78 982 5255
TLF
TLX
EML
COM 15,20,22C

HASEGAWA TATSUHIKO DR
DPT OF PHYSICS
OHIO STATE UNIVERSITY
174 W 18TH AVE
COLUMBUS OH 43210
USA
TEL
TLF 614 292 7557
TLX
EML
COM

HASEGAWA TETSUO DR
INST OF ASTRONOMY
UNIVERSITY OF TOKYO
OSAWA MITAKA
TOKYO 181
JAPAN
TEL   422 41 3737
TLF
TLX 2822307
EML
COM 40

HASER LEO N K DR
MPI F EXTRATERR PHYSIK
D 8046 GARCHING MUENCHEN
GERMANY
TEL 89 329 9803
TLF
EML
COM 15
TLX 5215845 XTER D

HASHIMOTO MASA-AKI DR
DPT OF PHYSICS
COLLEGE OF GENERAL EDU
KYUSHU UNIV ROPPONMATSU
FUKUDOKA 810
JAPAN
TEL 092 771 4161 EXT 360
TLF 092 731 8745
TLX
EML E76051AJPNCCKU
COM 35

HASLAM C GLYN T DR
MPI FUER RADIOASTRONOMIE
AUF DEM HUEGEL 69
D 5300 BONN 1 .
GERMANY
TEL
TLF
TLX 886440
EML
COM 40

HASSALL BARBARA J M DR
ROYAL GREENWICH OBS
MADINGLEY RD
CAMBRIDGE CB3 OEZ
UK
TEL 223 33 7548
TLF
TLX 817297 ASTRON G
EML JANET BJMH@UK AC CAM AST-STAR
COM 42

HASSAN S M PROF
HELWAN OBSERVATORY
HELWAN
EGYPT
TEL 78 0645/2683
TLF
EML
COM 37
TLX· 93070

HATHAWAY DAVID H DR
NASA/MSFC
CODE ES 52
HUNTSVILLE AL 35812
USA
TEL 205 544 7610
TLF
TLX
EML
COM 10

HATZES ARTIE P DR
ASTRONOMY DPT
UNIVERSITY OF TEXAS
RLM 15 308
AUSTIN TX 78712 1083
USA
TEL 512 471 1473
TLF
TLX
EML ARTIE@ASTRO AS UTEXAS EDL
COM

HATZIDIMITRIOU DESPINA DR
AAO
EPPING LABORATORY
BOX 296
EPPING NSW 2121
AUSTRALIA
TEL 2 868 1666
TLF 2 876 8536/42
TLX 123 999 AAOSYD
EML DH@AAOEPP OZ AU
COM 37

HAUBOLD HANS JOACHIM
ZNTRLINST F ASTROPHYSIK
STERNWARTE BABELSBRG
ROSA LUXEMBURG STR 17A
D 1591 POSTDAM
GERMANY
TEL 7620
TLF
TLX 15305
EML
COM 48

HAUCK BERNARD PROF
INSTITUT D'ASTRONOMIE
UNIVERSITE DE LAUSANNE
CH 1290 CHAVANNES-D-BOIS
SWITZERLAND
TEL 22 755 2611
TLF 22 755 3983
TLX 27720 OBSG CH
EML hauck@obs unige ch
COM 05P,25,45

HAUG EBERHARD DR
MOZARTSTR 20
D 7430 METZINGEN
GERMANY
TEL 707 129 6483
TLF
TLX
EML
COM 10

HAUG ULRICH PROF
HAMBURGER STERNWARTE
GOJENSBERGSWEG 112
D 2050 HAMBURG 80
GERMANY
TEL 40 7252 4131
TLX 217884
EML
TLF
COM 21,33

HAUGE OIVIND DR
INST THEORET ASTROPHYSICS
UNIVERSITY OF OSLO
BOX 1029
N 0315 BLINDERN OSLO 3
NORWAY
TEL 245-65-06
TLF
TLX
EML
COM

HAUPT HERMANN F PROF
INSTITUT FUER ASTRONOMIE
DER UNIVERSITAET
UNIVERSITAETSPLATZ 5
A 8010 GRAZ
AUSTRIA
TEL 316 380 5271
TLF
TLX 31078A
EML
COM 15,20,38,46

HAUPT WOLFGANG DR
GIRONDELLE 105
D 4630 BOCHUM 1
GERMANY
TEL
TLF
TLX
EML
COM

HAUSER MICHAEL G DR
NASA/GSFC
CODE 680
LASP
GREENBELT MD 20771
USA
TEL 301 286 8701
TLF
TLX
EML STARS HAUSER
COM 21VP,44

HAVLEN ROBERT J DR
NRAO
BOX O
SOCORRO NM 87801 0387
USA
TEL 505 835 7330
TLX 910 9881710
EML BITNET RHAVLEN@NRAO
TLF 505 835 7027
COM

HAVNES OVE DR .
AURORAL OBSERVATORY
UNIVERSITY OF TROMSO
BOX 953
N 9001 TROMSO
NORWAY
TEL 83-86060
TLF
TLX 64124 AUROB N
EML
COM

HAWARDEN TIMOTHY G DR
ROYAL OBSERVATORY
BLACKFORD HILL
EDINBURGH EH9 3HJ
UK
TEL 316 673 321
TLF
TLX 72383 ROEDIN G
EML
COM 37

HAWKES ROBERT LEWIS DR
DPT OF PHYSICS
MOUNT ALLISON UNIVERSITY
SACKVILLE NB EOA 3CO
CANADA
TEL 506 364 2580
TLF 506 364 2216
TLX
EML BITNET A014@MTAM
COM 22

HAWKING STEPHEN W PROF
DPT APPLIED MATHS
& THEORETICAL PHYSICS
SILVER STREET
CAMBRIDGE CB3 9EW
UK
TEL. 223 351 645
TLF
TLX 81240 CAMSPL G
EML
COM 47,48

HAWKINS GERALD S DR
CONSUL 906
2400 VIRGINIA AVE NW
WASHINGTON DC 20037
USA
TEL 202 485 2050
TLF
TLX
EML
COM 22,41

HAWKINS ISABEL DR
CTR FOR EUV ASTROPHYSICS
UNIVERSITY OF CALIFORNIA
2150 KITTREDGE ST
BERKELEY CA 94720
USA
TEL 415 642 0816
TLF 415 643 7629
TLX
EML
COM 44

HAWKINS MICHAEL R S
ROYAL OBSERVATORY
BLACKFORD HILL
EDINBURGH EH9 3HJ
UK
TEL
TLX
EML
TLF
COM 33

HAYASHI CHUSHIRO PROF
MOMOYAMA YOGORO-CHO 1
FUSHIMI KU
KYOTO 612
JAPAN
TEL 075-611-1062
TLF
TLX
EML
COM 35,47

HAYASHI MASAHIKO DR
DPT OF ASTRONOMY
UNIVERSITY OF TOKYO
BUNKYO KU
TOKYO 113
JAPAN
TEL 3 812 2111
TLF
TLX 2722126 UTGAB J
EML
COM 40

HAYASHI SAEKO S DR
TOKYO ASTRONOMICAL OBS
NAOJ
OSAWA MITAKA
TOKYO 181
JAPAN
TEL 0422 41 3609
TLF 0422 41 3608
TLX
EML SAEKO@OPTIK MMTK NAO AC JP
COM 34

HAYES DONALD S DR
BOX 1907
SCOTTSDALE AZ 85252
USA
TEL 602 947 3572
TLF
EML
COM 25,45
TLX

HAYLI AVRAM PROF
OBSERVATOIRE DE LYON
AVE CHARLES ANDRE
F 69561 S GENIS LAVAL CDX
FRANCE
TEL 78 56 0705
TLF 72 39 9791
TLX 310916
EML
COM 33,41

HAYMES ROBERT C PROF
DPT SPACE PHYS & ASTRON
RICE UNIVERSITY
HOUSTON TX 77001
USA
TEL 713 527 4045
TLF
TLX 556457
EML
COM 48

HAYNES MARTHA P
DPT OF ASTRONOMY
CORNELL UNIVERSITY
SPACE SCIENCES BLDG
ITHACA NY 14853
USA
TEL 607 256 3734
TLF
TLX 932454
EML
COM 40

HAYNES RAYMOND F PROF
CSIRO
DIVISION OF RADIOPHYSICS
BOX 76
EPPING NSW 2121
AUSTRALIA
TEL 2 868 0276
TLF 2 868 0457
TLX ASTRO 26230
EML RHAYNES@RPEPPING OZ AU
COM 34,40

HAYWARD JOHN
DPT MATHS & COMPUTING
POLYTECHNIC OF WALES
PONTYPRIDD
MID GLAMORGAN CF38 2PJ
UK
TEL
TLF
TLX
EML
COM 10

HAYWOOD J
DPT OF TEACHER EDUCATION
TRINITY COLLEGE
DUBLIN 2
IRELAND
TEL 1 772 941
TLF
TLX
EML
COM 46

HAZARD CYRIL DR
INSTITUTE OF ASTRONOMY
THE OBSERVATORIES
MADINGLEY RD
CAMBRIDGE CB3 0HA
UK
TEL 223 622 04
TLF
TLX
EML
COM 40

HAZEN MARTHA L DR
CENTER FOR ASTROPHYSICS
HCO/SAO
60 GARDEN ST
CAMBRIDGE MA 02138
USA
TEL 617 495 3362
TLF
TLX
EML
COM 37

HAZER S DR
FACULTY OF SCIENCE
EGE UNIVERSITY
BOX 21
35100 BORNOVA IZMIR
TURKEY
TEL
TLF
TLX
EML
COM

HAZLEHURST JOHN DR
HAMBURGER STERNWARTE
GOJENSBERGSWEG 112
D 2050 HAMBURG 80
GERMANY
TEL
TLF
TLX
EML
COM 42

HE MIAO-FU
SHANGHAI OBSERVATORY
CAS
80 NANDAN RD
SHANGHAI
CHINA PR
TEL 21 38 6191
TLF
TLX 33164 SHAO CN
EML
COM 04,07C,20

HE XIANG-TAO
DPT OF ASTRONOMY
BEIJING NORMAL UNIVERSITY
BEIJING 100875
CHINA PR
TEL 1 65 6531*6285
TLF
TLX
EML
COM 28,47

HEAP SARA R DR
NASA/GSFC
CODE 672
GREENBELT MD 20771
USA
TEL
TLF
TLX
EML
COM

HEARN ANTHONY G DR
STERREKUNDIG INSTITUTE
BOX 80000
NL 3508 TA UTRECHT
NETHERLANDS
TEL 30 53 5200
TLF
TLX 40048 FYLUT NI
EML BITNET WNMTONY@HUTRUUD
COM 36,44

HEARNSHAW JOHN B DR
DPT OF PHYSICS
UNIVERSITY OF CANTERBURY
PRIVATE BAG
CHRISTCHURCH 1
NEW ZEALAND
TEL 348 2009*771
TLF
TLX 4144 UNICANT NZ
EML
COM 29,30C

HEASLEY JAMES NORTON
INSTITUTE FOR ASTRONOMY
UNIVERSITY OF HAWAII
2680 WOODLAWN DR
HONOLULU HI 96822
USA
TEL 808 948 6826
TLF
TLX 7238459 UHAST HR
EML
COM 36

HEAVENS ALAN DR
ROYAL OBSERVATORY
BLACKFORD HILL
EDINBURGH EH9 3HJ
UK
TEL 316 688 352
TLX 72383 ROEDIN G
EML AFH@UK AC ROE STAR
TLF
COM 47

HEBER ULRICH
INST THEOR PHYS & STERNW
UNIVERSITAET KIEL
LEIBNIZSTR
D 2300 KIEL 1
GERMANY
TEL 431 880 4103
TLF
TLX 292706
EML
COM 29,36

HECHT JAMES H DR
AEROSPACE CORPORATION
MS M2 255
BOX 92957
LOS ANGELES CA 90009
USA
TEL 213 336 7017
TLF
TLX
EML
COM 34

HECK ANDRE DR
OBS DE STRASBOURG
11 RUE UNIVERSITE
F 67000 STRASBOURG
FRANCE
TEL 88 35 8216
TLF 88 25 0160
TLX 890506 STAROBS F
EML heck@frccsc21 bitnet
COM 05C,25,51

HECKATHORN HARRY M
NAVAL RESEARCH LABORATORY
CODE 4143 2
4555 OVERLOOK AVE SW
WASHINGTON DC 20375 5000
USA
TEL 202 767 2764
TLF
TLX
EML
COM 09,44

HECKMAN TIMOTHY M
ASTRONOMY PROGRAM
UNIVERSITY OF MARYLAND
COLLEGE PARK MD 20742
USA
TEL 301 454 3001
TLF
TLX 7108260352 ASTR CORP
EML
COM 28

HECQUET JOSETTE DR
OBS MIDI PYRENEES
14 AVE E BELIN
F 31400 TOULOUSE CDX
FRANCE
TEL 61 25 2101
TLF
TLX 530776 F
EML
COM

HEDDLE DOUGLAS W O PROF
DPT OF PHYSICS
ROYAL HOLLOWAY COLLEGE
UNIVERSITY OF LONDON
EGHAM SURREY TW20 0EX
UK
TEL 784 353 51
TLF
TLX 935504
EML
COM 14

HEDEMAN E RUTH MISS
3440 ST JEFFERSON ST
FALLS CHURCH VA 22041
USA
TEL
TLF
EML
COM
TLX

HEESCHEN DAVID S DR
NRAO
EDGEMONT RD
CHARLOTTESVILLE VA 22901
USA
TEL
TLF
TLX 910-997-0174
EML
COM 28,40,51

HEFELE HERBERT PH D
MPI FUER ASTRONOMIE
KONIGSTUHL
D 6900 HEIDELBERG 1
GERMANY
TEL
TLF
TLX
EML
COM 05

HEFFERLIN RAY A PROF
DPT OF PHYSICS
SOUTHERN COLLEGE
DRAWER H
COLLEGEDALE TN 37315 0370
USA
TEL 615 238 2869
TLF
TLX
EML
COM 14

HEFTY JAN DR
OBSERVATORY OF THE SLOVAK
TECHNICAL UNIVERSITY
RADLINSKEHO 11
CS 813 68 BRATISLAVA
CZECHOSLOVAKIA
TEL 7 49 8047
TLF 7 52027
TLX 92720 ELFAKC
EML
COM 19

HEGGIE DOUGLAS C DR
DPT OF MATHEMATICS
UNIVERSITY OF EDINBURGH
KING'S BUILDINGS
EDINBURGH EH9 3JZ
UK
TEL 316 671 081
TLF
TLX 727442 UNIVEDG
EML D C HEGGIE@ED AC UK
COM 07,37

HEGYI DENNIS J ASSOC PROF
RANDALL LABORATORY
UNIVERSITY OF MICHIGAN
ANN ARBOR MI 48109 1090
USA
TEL 313 764 5448
TLF
TLX 810-2236056
EML
COM

HEIDMANN JEAN DR
OBSERVATOIRE DE PARIS
SECTION DE MEUDON
F 92195 MEUDON PPL CDX
FRANCE
TEL 1 45 07 7598
TLF
TLX 270912
EML
COM 28,40,47,51C

HEILES CARL PROF
ASTRONOMY DPT
UNIVERSITY OF CALIFORNIA
601 CAMPBELL HALL
BERKELEY CA 94720
USA
TEL 415 642 4510
TLF
TLX 820181 UCB AST RAL
EML
COM 33,34,40

HEINRICH INGE
ASTRONOMISCHES RECHEN-
INSTITUT
MOENCHHOFSTR 12-14
D 6900 HEIDELBERG 1
GERMANY
TEL 62 214 9026
TLF
TLX
EML
COM 05

HEINTZ WULFF D DR
DPT OF ASTRONOMY
SWARTHMORE COLLEGE
SWARTHMORE PA 19081
USA
TEL 215 447 7265
TLF
TLX
EML
COM 05,08,24,42

HEINTZE J R W DR
STERREKUNDIG INSTITUTE
BOX 80000
NL 3508 TA UTRECHT
NETHERLANDS
TEL 30 53 5235
TLX 40048 FYLUT NI
EML BITNET WNMMAIL@HUTRUUO
TLF
COM 29,30

HEINZEL PETR DR
ASTRONOMICAL INSTITUTE
CZECH ACADEMY OF SCIENCES
ONDREJOV OBSERVATORY
CS 251 65 ONDREJOV
CZECHOSLOVAKIA
TEL 204 85201
TLF 204 85314
TLX 121579 ASTR C
EML
COM 10,12

HEISE JOHN DR
SPACE RESEARCH LABORATORY
SRON
SORBONNELAAN 2
NL 3584 CA UTRECHT
NETHERLANDS
TEL 30 53 5600
TLF 30 54 0860
TLX 47224 ASTRO NL
EML
COM 44,48

HEISER ARNOLD M DR
DYER OBSERVATORY
VANDERBILT UNIVERSITY
BOX 1803 STA B
NASHVILLE TN 37235
USA
TEL 615 373 4897
TLF
TLX
EML
COM 27

HEJNA LADISLAV DR
FAC OF MATHS & PHYSIC
CHARLES UNIVERSITY
MALOSTRANSKE NAM 2/25
CS 110 00 PRAHA 1
CZECHOSLOVAKIA
TEL 2 53 2132
TLF 2 84 7688
TLX 121673 MFF C
EML LHEJNA@CSPGUK11
COM 12

HEKELA JAN DR
ASTRONOMICAL INSTITUTE
CZECH ACADEMY OF SCIENCES
ONDREJOV OBSERVATORY
CS 251 65 ONDREJOV
CZECHOSLOVAKIA
TEL 204 85201
TLF 204 85314
TLX
EML
COM 36

HELALI YHYA E DR
HELWAN OBSERVATORY
HELWAN
EGYPT
TEL 78 0645/2683
TLF
TLX
EML
COM 07

HELFAND DAVID JOHN
COLUMBIA ASTROPHYSICS LAB
COLUMBIA UNIVERSITY
538 WEST 120TH ST
NEW YORK NY 10027
USA
TEL 212 854 2150
TLF 212 316 9504
TLX 220094 COLU UR
EML BITNET DJHT@CUPHYD
COM 48

HELFER H LAWRENCE PROF
DPT PHYSICS & ASTRONOMY
UNIVERSITY OF ROCHESTER
ROCHESTER NY 14627
USA
TEL 716 275 4377
TLF
TLX
EML
COM 34

HELIN ELEANOR FRANCIS
JPL
MS 183 501
4800 OAK GROVE DR
PASADENA CA 91109
USA
TEL 818 354 4606
TLF
TLX 67 5429
EML
COM 15,20,22

HELLER MICHAEL PROF
POWSTANCOW WARSAWY 13/94
PL 33 110 TARNOW
POLAND
TEL
TLF
TLX
EML
COM 47

HELLWIG HELMUT WILHELM DR
FREQUENCY & TIME SYSTEMS
34 TOZER RD
BEVERLY MA 01915
USA
TEL 617 927 8220
TLF
TLX 940518
EML
COM 31

HELMER LEIF
COPENHAGEN UNIVERSITY OBS
BRORFELDEVEJ 23
DK 4340 TOLLOSE
DENMARK
TEL 53 48 8195
TLF 58 48 8758
TLX 44155
EML
COM 08C,50

HELMKEN HENRY F DR
CENTER FOR ASTROPHYSICS
HCO/SAO
60 GARDEN ST
CAMBRIDGE MA 02138
USA
TEL
TLF
TLX
EML
COM 44

HELOU GEORGE DR
CALTECH
MS 100 22
IPAC
PASADENA CA 91125
USA
TEL 818 584 2928
TLF
TLX 584 9945
EML BITNET HELOU%IPAC@HAMLET
COM 05,28,34,40,44

HELT BODIL E
UNIVERSITY OBSERVATORY
OESTER VOLDGADE 3
DK 1350 COPENHAGEN K
DENMARK
TEL 31 14 1790
TLF 31 38 9157
TLX 44155 DANAST DK
EML
COM 42

HEMENWAY MARY KAY M DR
ASTRONOMY DPT
UNIVERSITY OF TEXAS
RLM 15 308
AUSTIN TX 78712 1083
USA
TEL 512 471 1309
TLF 512 471 6016
TLX 910 874 1351
EML MARYKAY@ASTRO.AS UTEXAS EDU
COM .46

HEMENWAY PAUL D DR
ASTRONOMY DPT
UNIVERSITY OF TEXAS
RLM 15 308
AUSTIN TX 78712 1083
USA
TEL 512 471 4461
TLF
TLX:
EML
COM 08,20,24C

HEMMLEB GERHARD DR
ZENTRALINSTITUT FUR
PHYSIK DER ERDE
TELEGRAFENBERG A 17
D 1500 POTSDAM
GERMANY
TEL 4551
TLF
TLX 15305
EML
COM 19,31

HEMPE KLAUS
GRENZWEG 24 B

D 2057 REINBEK
GERMANY
TEL 40 710 5628
TLF
TLX
EML
COM

HENDECOURT D' LOUIS DR
GROUPE PHYSIQUE SOLIDES
UNIVERSITE PARIS VII
4 PLACE JUSSIEU TOUR 23
F 75251 PARIS CDX 05
FRANCE
TEL
TLF
TLX
EML
COM 14,21,34

HENIZE KARL G ASTRONAUT
NASA/JOHNSON SPACE CENTER
CODE CB
HOUSTON TX 77058
USA
TEL 713 483 2411
TLF
TLX
EML
COM 28,29,34,44,45

HENKEL CHRISTIAN
MPI FUER RADIOASTRONOMIE
AUF DEM HUEGEL 69
D 5300 BONN 1
GERMANY
TEL
TLF
TLX 886440
EML
COM 34,40

HENNING THOMAS DR
UNIVERSITY OBS JENA

SCHILLERGAESSCHEN 2
D 6900 JENA
GERMANY
TEL 78 27122
TLF 78 425 039
TLX 058 86134
EML PHYSFAK@PHYSIK UNI-JENA DB
COM 34

HENON MICHEL C DR
OCA OBSERV DE NICE
BP 139
F 06003 NICE CDX
FRANCE
TEL 93 89 0420
TLF
TLX 460004 OBSNICE F
EML
COM 07,33,37

HENOUX JEAN-CLAUDE DR
OBSERVATOIRE DE PARIS
SECTION DE MEUDON
F 92195 MEUDON PPL CDX
FRANCE
TEL 1 45 07 7803
TLF
TLX
EML
COM 10,44

HENRARD JACQUES PROF
DPT DE PHTSIQUE
FACULTES UNIVERSITAIRES
RUE DE BRUXELLES 61
B 5000 NAMUR
BELGIUM
TEL 81 22 9061
TLF 81 23 0391
TLX 59222
EML JHENRARD@BNANDP51
COM 04,07C,20

HENRICHS HUBERTUS F DR
ASTRONOMICAL INSTITUTE
UNIVERSITY OF AMSTERDAM
KRUISLANN 403
NL 1098 SJ AMSTERDAM
NETHERLANDS
TEL 20 52 57491
TLF 31 205 25 7484
TLX 10262 HEF NL
EML
COM 29

HENRIKSEN RICHARD N DR
DPT OF PHYSICS
QUEEN'S UNIVERSITY
KINGSTON ON K7L 3N6
CANADA
TEL 613 547 5536
TLF 613 545 6463
TLX
EML
COM 48

HENRY RICHARD B C DR
DPT PHYSICS & ASTRONOMY
UNIVERSITY OF OKLAHOMA
NORMAN OK 73019
USA
TEL 405 325 3961
TLF·
TLX
EML
COM 28,35

HENRY RICHARD C PROF
DPT PHYSICS & ASTRONOMY
JOHNS HOPKINS UNIVERSITY
CHARLES & 34TH ST
BALTIMORE MD 21218
USA
TEL 301 338 7350
TLF
TLX
EML
COM 21,48

HENSBERGE HERMAN
KONINKLIJKE STERRENWACHT

RINGLAAN 3
B 1180 BRUSSELS
BELGIUM
TEL 2 237 30284
TLF 2 237 49822
TLX 21565 OBSBEL
EML HERMAN@ASTRO OMO BE
COM 25,44   —

HENSLER GERHARD PROF
INST THEORICAL PHYS & OBS
UNIVERSITY OF KIEL
OLSHAUSENSTR 40
D 2300 KIEL 1
GERMANY
TEL 431 880 4125
TLF 431 880 4432
TLX 292706
EML PAS30@RZ UNI-KIEL DBP DE
COM 42

HERAS ANA M DR
ESA/ESTEC
SSD
BOX 299
NL 2200 AG NOORDWIJK
NETHERLANDS
TEL 17 19 85016
TLF 17 19 84698
TLX
EML ESTCS1 AHERAS/AHERAS@ESTEC
COM 49

HERBIG GEORGE H DR
INSTITUTE FOR ASTRONOMY
UNIVERSITY OF HAWAII
2680 WOODLAWN DR
HONOLULU HI 96822
USA
TEL 808 948 8312
TLF
TLX 723-8459
EML
COM 27,29

HERBST ERIC DR
DPT OF PHYSICS
OHIO STATE UNIVERSITY
174 W 18TH AVE
COLUMBUS OH 43210
USA
TEL 614 292 2653
TLF
TLX
EML
COM

HERBST WILLIAM DR
DPT OF ASTRONOMY
VAN VLECK OBSERVATORY
WESLEYAN UNIVERSITY
MIDDLETOWN CT 06457
USA
TEL   203 347 9411
TLF
TLX
EML
COM 33,37

HERCZEG TIBOR J PROF DR
DPT PHYSICS & ASTRONOMY
UNIVERSITY OF OKLAHOMA
NORMAN OK 73019
USA
TEL   405 325 3961
TLF
TLX
EML
COM 42,51

HERMAN JACOBUS DR
MAX PLANCK INSTITUT FUR
EXTRATERRESTRISCE PHYSIK
D 8046 GARCHING MUENCHEN
GERMANY
TEL
TLF
TLX
EML
COM 33

HERMANS DIRK DR
SCHOOL MATHS/STATISTICS
UNIVERSITY OF BIRMINGHAM
EDGBASTON
BIRMINGHAM B15 2TT
UK
TEL 214 14 3961
TLF 214 14 3907
TLX 337262 UOBHAM G
EML D F H HERMANS@UK AC BHAM
COM 10

HERMSEN WILLEM DR
SRL
HUYGENS LAB
BOX 9504
NL 2300 RA LEIDEN
NETHERLANDS
TEL 71 27 5810
TLF
TLX 39058 ASTRO NL
EML
COM

HERNANDEZ CARLOS ALBERTO
OBSERVATORIO ASTRONOMICO
PASEO DEL BOSQUE
1900 LA PLATA (BS AS)
ARGENTINA
TEL 21 21 7308
TLF
TLX
EML
COM

HERNQUIST LARS ERIC DR
LICK OBSERVATORY
UNIVERSITY OF CALIFORNIA
NTL SCIENCES II
SANTA CRUZ CA 95064
USA
TEL   408 425 4733
TLF   408 426 3115
TLX 760 7936
EML LARS@HELIOS UCSU EDU
COM 28

HEROLD HEINZ
THEORETISCHE ASTROPHYSIK
UNIVERSITY TUEBINGEN
AUF DER MORGENSTELLE 12,C
D 7400 TUEBINGEN
GERMANY
TEL 707 129 2043
TLF
TLX
EML
COM 14,36

HERR RICHARD B DR
DPT OF PHYSICS
UNIVERSITY OF DELAWARE
NEWARK DE 19716
USA
TEL 302 451 2673
TLF
TLX
EML
COM 27

HERRERA MIGUEL ANGEL DR
INSTITUTO ASTRONOMIA
UNAM
APDO POSTAL 70-264
04510 MEXICO DF
MEXICO
TEL 5 48 53 05
TLF
TLX 1760155 CICME
EML BITNET MANHER@UNAMVM1
COM

HERRERO DAVO ARTEMIO DR
INST DE ASTROFISICA
DE CANARIAS
OBS DEL TEIDE
E 38200 LA LAGUNA
SPAIN
TEL 22 26 2211
TLF
TLX 92640
EML
COM

HERRMANN DIETER PROF DR
ARCHENHOLD STERNWARTE
ALT TREPTOW 1
D 1193 BERLIN
GERMANY
TEL 272 8871*494
TLF
TLX
EML
COM 41

HERS JAN MR
BOX 48
SEDGEFIELD 6573
SOUTH AFRICA
TEL 4455 736
TLF
EML
COM 06,20,27,31
TLX

HERSHEY JOHN L DR
US NAVAL OBSERVATORY
34 & MASSACHUSETTS AVE NW
WASHINGTON DC 20392 5100
USA
TEL 202 653 1554
TLX 710-822-1970
EML
TLF
COM 24,26,51

HERTZ PAUL L DR
NAVAL RESEARCH LABORATORY
CODE 4121 5
4555 OVERLOOK AVE SW
WASHINGTON DC 20375 5000
USA
TEL 202 767 2438
TLF
TLX
EML
COM

HERZBERG GERHARD DR
HERZBERG INST ASTROPHYS
NTL RESEARCH COUNCIL
100 SUSSEX DR
OTTAWA ON K1A OR6
CANADA
TEL 613 990 0917
TLF 613 952 6602
TLX 053 3715
EML
COM 14,15,16,34

HESKE ASTRID DR
ESA/ESTEC
ASTROPHYSICS DIV
BOX 299
NL 2200 AG NOORDWIJK
NETHERLANDS
TEL 17 19 83173
TLF 17 19 84690
TLX 39098
EML ESTCS1 AHESKE / AHESKE@ESTEC
COM 29,40,44

HESSER JAMES E DR
HERZBERG INST ASTROPHYS
DOMINION ASTROPHYS OBS
5071 W SAANICH RD
VICTORIA BC V8X 4M6
CANADA
TEL 604 388 3974
TLF 604 363 0045
TLX 049 7295
EML
COM 14,27,37

HEUDIER JEAN-LOUIS DR
OCA OBSERV DE NICE
BP 139
F 06003 NICE CDX
FRANCE
TEL 92 00 3011
TLF 92 00 3033
TLX 460004 OBSNICE F
EML
COM 09,20,24,37,46,51

HEWETT PAUL
INSTITUTE OF ASTRONOMY
THE OBSERVATORIES
MADINGLEY RD
CAMBRIDGE CB3 OHA
UK
TEL 223 622 04
TLF
TLX 817297 ASTRON G
EML
COM 30,47

HEWISH ANTONY PROF
MULLARD RADIO ASTRON OBS
CAVENDISH LABORATORY
MADINGLEY RD
CAMBRIDGE CB3 OHE
UK
TEL   223 664 77
TLF
TLX 81292
EML
COM 40

HEWITT ADELAIDE
CASS
UCSD
C 011
LA JOLLA CA 92093.0216
USA
TEL 619 534 6627
TLF
TLX
EML INT hewitt%cass.span@ucsd edu
COM 28,47

HEWITT ANTHONY V DR
GE MEDICAL SYSTEMS
PET ENGINEERING
BOX 206
PEWAUKEE WI 53072
USA
TEL 414 320 5170
TLF
TLX
EML
COM 09,28

HEY JAMES STANLEY DR
4 SHORTLANDS CLOSE

EASTBOURNE BN22 OJE
UK
TEL
TLF
TLX
EML
COM 22,40

HEYVAERTS JEAN DR
OBSERVATOIRE DE PARIS
SECTION DE MEUDON
F 92195 MEUDON PPL CDX
FRANCE
TEL 1 45 07 7405
TLF
TLX 201571
EML
COM 49

HIBBS ALBERT R MGR PLANS
781 PROSPECT BL
PASADENA CA 91103
USA
TEL
TLF
EML
COM
TLX

HICKSON PAUL DR
DPT GEOPHYS & ASTRONOMY
UNIV OF BRITISH COLUMBIA
2219 MAIN MALL
VANCOUVER BC V6T 1W5
CANADA
TEL 604 228 2267
TLF 604 228 6047
TLX
EML
COM 28

HIDAJAT BAMBANG PROF DR
BOSSCHA OBSERVATORY

LEMBANG 40391
INDONESIA
TEL 229 6001
TLF
TLX 28234 BD ITB
EML
COM 26,34,46,50

HIDALGO MIGUEL A DR
FAC DE CIENCIAS FISICAS
UNIVERSIDAD DE ZARAGOZA
E 50009 ZARAGOZA
SPAIN
TEL
TLF
TLX
EML
COM

HIDE RAYMOND PROF
GEOPHYSICAL FLUID
DYNAMICS LABORATORY
METEOROLOGICAL OFFICE
BRACKNELL BERKS RG12 2SZ
UK
TEL 344 42242
TLF
TLX 849801
EML
COM 16,19

HIEI EIJIRO DR
TOKYO ASTRONOMICAL OBS
NAOJ
OSAWA MITAKA
TOKYO 181
JAPAN
TEL 0422-32-5111
TLF
TLX 2822307
EML
COM 10,12

HIGGS LLOYD A DR
DOMINION RADIO ASTROPHYS
OBSERVATORY
BOX 248
PENTICTON BC V2A 6K3
CANADA
TEL 604 493 2277
TLF 604 493 7767
TLX 048 88127
EML
COM 34,40

HILDEBRAND ROGER H
ENRICO FERMI INSTITUTE
UNIVERSITY OF CHICAGO
5640 S ELLIS AVE
CHICAGO IL 60637
USA
TEL 312 962 7581
TLF
TLX
EML
COM 34

HILDITCH RONALD W DR
DPT OF PHYSICS & ASTRON
UNIVERSITY OF ST ANDREWS
NORTH HAUGH
ST ANDREWS FIFE KY16 9SS
UK
TEL 334 761 61
TLF 334 744 87
TLX 9312110846 SA G
EML ASSRH@UK AC ST-AND STAR
COM 25,30,42C

HILDNER ERNEST DR
DIRECTOR R/S/SE
SPACE ENVIRONMENT LAB
325 BROADWAY
BOULDER CO 80303 3328
USA
TEL
TLF
TLX
EML
COM 10C,12

HILF EBERHARD R H PH D
PESTRUPSWEG 30

D 2900 OLDENBURG
GERMANY
TEL
TLF
TLX
EML
COM 35,40

HILL FRANK DR
NTL SOLAR OBSERVATORY
BOX 26732
950 N CHERRY AVE
TUCSON AZ 85726 6732
USA
TEL 602 323 4138
TLF 602 325 9278
TLX 1561401 AURA UT
EML FHILL@NOAO EDU
COM 12

HILL GRAHAM DR
HERZBERG INST ASTROPHYS
DOMINION ASTROPHYS OBS
5071 W SAANICH RD
VICTORIA BC V8X 4M6
CANADA
TEL. 602 388 3935
TLF 604 363 0045
TLX 049 7295
EML
COM 24,26,30,42C

HILL HENRY ALLEN DR
DPT OF PHYSICS
UNIVERSITY OF ARIZONA
BLDG 81
TUCSON AZ 85721
USA
TEL 602 621 6784
TLF
TLX 910-9521143
EML
COM 27

HILL PHILIP W DR
DPT OF PHYSICS & ASTRON
UNIVERSITY OF ST ANDREWS
NORTH HAUGH
ST ANDREWS FIFE KY16 9SS
UK
TEL 334 761 61
TLF 334 744 87
TLX 76213 SAULIB G
EML P W HILL@UK AC ST-ANDREWS
COM 25,27

HILLEBRANDT WOLFGANG PH D
MPI F PHYSIK & ASTROPHYS
KARL-SCHWARZSCHILD-STR 1
D 8046 GARCHING MUENCHEN
GERMANY
TEL 89 329 99409
TLX
EML
TLF
COM

HILLIARD R DR

USA
TEL
TLF
EML
COM 09
TLX

HILLS JACK G DR
LOS ALAMOS NATIONAL LAB
MS B228
THEORETICAL DIV T6
LOS ALAMOS NM 87545
USA
TEL 505 667 9152
TLF
TLX
EML
COM 37,42

HILLS RICHARD E DR
MULLARD RADIO ASTRON OBS
CAVENDISH LABORATORY
MADINGLEY RD
CAMBRIDGE CB3 OHE
UK
TEL 223 664 77
TLF
TLX 81282
EML
COM 40

HILTON JOHN DR
DPT OF MATH SCIENCES
GOLDSMITHS' COLLEGE
NEW CROSS
LONDON SE14 6NW
UK
TEL
TLF
TLX
EML
COM

HINKLE KENNETH H
KITT PEAK NTL OBS
BOX 26732
950 N CHERRY AVE
TUCSON AZ 85726 6732
USA
TEL 602 327 5511
TLF
TLX 0666-484 AURA NOAO T
EML
COM 29

HINTEREGGER HANS E DR
HAYSTACK OBSERVATORY
WESTFORD MA 01886
USA
TEL
TLF
TLX
EML
COM 44

HINTZEN PAUL MICHAEL N DR
NASA/GSFC
CODE 681
GREENBELT MD 20771
USA
TEL 301 286 5101
TLF
TLX
EML
COM 28

HIOTELIS NICOLAOS DR
DPT OF ASTRONOMY
NTL UNIVERSITY OF ATHENS
PANEPISTIMIOPOLIS
GR 157 83 ZOGRAFOS
GREECE
TEL 1 724 3414
TLF
TLX
EML NHIOT@GRATHUN1
COM 28

HIPPELEIN HANS H DR
MPI FUER ASTRONOMIE
KOENIGSTUHL
D 6900 HEIDELBERG 1
GERMANY
TEL
TLF
TLX
EML
COM 34

HIRABAYASHI HISASHI DR
NOBEYAMA RADIO OBS
NAOJ
MINAMIMAKI MURA
NAGANO 384 13
JAPAN
TEL 2679-8-2831
TLF
TLX 3329005
EML
COM 40,51C

HIRAI MASANORI DR
DPT EARTH SCI & ASTRONOMY
FUKUOKA UNIVERSITY OF EDU
729 MUNAKATA
FUKUOKA 811-41
JAPAN
TEL 094 032 2381
TLF
TLX
EML
COM 29

HIRATA RYUKO
DPT OF ASTRONOMY
KYOTO UNIVERSITY
KITASHIRAKAWA SAKYO KU
KYOTO 606
JAPAN
TEL
TLF
TLX
EML
COM 29

HIRAYAMA TADASHI PROF
TOKYO ASTRONOMICAL OBS
NAOJ
OSAWA MITAKA
TOKYO 181
JAPAN
TEL 0422-32-5111
TLF
TLX 2822307 TAOMTK-J
EML
COM 10,12

HIROMOTO NORIHISA DR
COMMUNICATIONS RES LAB
4-2-1 NUKUIKITAMACHI
KOGANEI
TOKYO 184
JAPAN
TEL 0423 27 7548
TLF 0423 27 6667
TLX 2832611 DEMPA J
EML HIROMOTO@BC CRL GO JP
COM 34

HIRST WILLIAM P
1 CLIFFORD CRESENT
BERGVLIET 7945
SOUTH AFRICA
TEL
TLF
TLX
EML
COM

HIRTH WOLFGANG ERNST PH D
THEODOR-HEUS STR 18
D 5354 WEILERSWIST
GERMANY
TEL
TLF
EML
COM
TLX

HITOTSUYANAGI JUICHI PROF
KATAHIRA 1-4-6-401
SENDAI 980
JAPAN
TEL 0222-27-9351
TLF
EML
COM. 35,36
TLX

HJALMARSON AKE G DR
ONSALA SPACE OBSERVATORY
GOETEBORG UNIVERSITY
S 439 00 ONSALA
SWEDEN
TEL 30 06 0653
TLF
TLX 2400 ONSPACE
EML
COM 28,34,40

HJELLMING ROBERT M DR
NRAO
BOX O
SOCORRO NM 87801 0387
USA
TEL 505 835 7273
TLF
TLX 910-988-1710
EML BITNET RHJELLMI@NRAO edu
COM 34,40,42

HO PAUL T P
SMITHSONIAN ASTROPHYS OBS
60 GARDEN ST
CAMBRIDGE MA 02138
USA
TEL 617 495 3627
TLX 921428
EML
TLF
COM 40

HOAG ARTHUR A DR
4410 E 14TH ST
TUCSON AZ 85711
USA
TEL. 602 795 8644
TLF
EML
COM 50
TLX

HOANG BINH DY DR
OBSERVATOIRE DE PARIS
SECTION DE MEUDON
LAM
F 92195 MEUDON PPL CDX
FRANCE
TEL 1 45 07 7445
TLF
TLX 201571
EML
COM 12,14,40,51

HOARE MELVIN DR
DPT OF ASTROPHYSICS
UNIVERSITY OF OXFORD
KEBLE RD
OXFORD OX1 3RH
UK
TEL 865 273 292
TLF 865 273 418
TLX 83295 NUCLOX G
EML MGH@UK AC OX ASTRO
COM 36

HOBBS LEWIS M DR
YERKES OBSERVATORY
UNIVERSITY OF CHICAGO
BOX 258
WILLIAMS BAY WI 53191
USA
TEL 414 245 5555
TLF
TLX
EML
COM 34

HOBBS ROBERT W DR
COMPUTER TECHN ASSOCIATES
6116 EXECUTIVE BOULEVARD
APT 800
ROCKVILLE MD 20852
USA
TEL
TLF
TLX
EML
COM 33,40

HODGE PAUL W PROF
DPT OF ASTRONOMY
UNIVERSITY OF WASHINGTON
FM 20
SEATTLE WA 98195
USA
TEL 206 543 2888
TLF
TLX 9104740096
EML
COM 22,28

HOEFLICH PETER DR
MAX PLANCK INSTITUT FUER
ASTROPHYSIK
KARL-SCHWARZSCHILDSTR 1
D 8046 GARCHING MUENCHEN
GERMANY
TEL 89 329 93249
TLF 89 329 93235
TLX 52 46 29 ASTRO D
EML PAH@DGAIPP1S
COM 29,36

HOEG ERIK DR
UNIVERSITY OBSERVATORY
OESTER VOLDGADE 3
DK 1350 COPENHAGEN K
DENMARK
TEL 31 14 1790 `
TLF 31 38 9157
TLX 44155 DANAST
EML
COM 08

HOEGBOM JAN A DR
STOCKHOLM OBSERVATORY
S 133 36 SALTSJOEBADEN
SWEDEN
TEL 87 17 0195
TLF 87 17 4719
EML HOGBOM@ASTRO SU SE
COM 40
TLX 12972 SOBSERV S

HOEGLUND BERTIL PROF
ONSALA SPACE OBSERVATORY
GOETEBORG UNIVERSITY
S 439 00 ONSALA
SWEDEN
TEL 30 06 0652
TLF
TLX 2400
EML
COM 34,40

HOEKSTRA ROEL DR
TPD/TNO/TH
BOX 155
NL 2600 AD DELFT
NETHERLANDS
TEL
TLF
TLX
EML
COM

HOESSEL JOHN GREG
WASHBURN OBSERVATORY
UNIVERSITY OF WISCONSIN
475 N CHARTER ST
MADISON WI 53706
USA
TEL 608 262 1752
TLF
TLX
EML
COM

HOEY MICHAEL J DR
DPT OF PHYSICS
UNIVERSITY COLLEGE
BELFIELD
DUBLIN 4
IRELAND
TEL 1 693 244
TLF
TLX
EML
COM

HOFF DARREL BARTON
CENTER FOR ASTROPHYSICS
PROJECT STAR
60 GARDEN ST
CAMBRIDGE MA 02138
USA
TEL 617 495 9798
TLF
TLX
EML
COM 46

HOFFLEIT E DORRIT DR
DPT OF ASTRONOMY
YALE UNIVERSITY
BOX 6666
NEW HAVEN CT 06520
USA
TEL
TLF
TLX
EML
COM 24,27

HOFFMAN JEFFREY ALAN DR
NASA/JOHNSON SPACE CENTER
CODE CB 4
HOUSTON TX 77058
USA
TEL 713 483 2411
TLF
TLX
EML
COM 44,48

HOFFMANN MARTIN DR
OBSERVATORIUM HOHER LIST
STERNWARTE DER
UNIVERSITAET BONN
D 5568 DAUN
GERMANY
TEL
TLF
TLX
EML
COM 42

HOFMANN WILFRIED DR
ASTRONOMISCHES RECHEN-
INSTITUT
MOENCHHOFSTR 12-14
D 6900 HEIDELBERG 1
GERMANY
TEL 62 214 9026
TLF
TLX 461336 ARIHD D
EML
COM 21

HOGAN CRAIG J DR
DPT OF ASTRONOMY
UNIVERSITY OF WASHINGTON
FM 20
SEATTLE WA 98195
USA
TEL
TLF
TLX
EML
COM

HOGG DAVID E DR
NRAO
EDGEMONT RD
CHARLOTTESVILLE VA 22901
USA
TEL 804 296 0220
TLX 910-997-0174
EML
TLF
COM 40

HOLBERG JAY B
LUNAR & PLANETARY LAB
UNIVERSITY OF ARIZONA
901 GOULD SIMPSON BLDG
TUCSON AZ 85721
USA
TEL 602 621 4301
TLF.
TLX 9109521143
EML
COM 16,44

HOLDEN FRANK
2 COLWICH CRESENT
KINGSTON HILL
STAFFORD ST16 3XP
UK
TEL 785 531 20
TLF
TLX
EML
COM 26

HOLLENBACH DAVID JOHN DR
NASA AMES RESEARCH CTR
MS 245 6
MOFFETT FIELD CA 94035
USA
TEL 415 997 6426
TLF
TLX
EML
COM 34

HOLLIS JAN MICHAEL DR
NASA/GSFC
CODE 930
GREENBELT MD 20771
USA
TEL
TLF
TLX
EML
COM 34,40,51

HOLLOWAY NIGEL J DR
SAFETY & RELIABILITY DIR
WIGSHAW LANE
CULCHETH
WARRINGTON WA3 4NE
UK
TEL 953 1244
TLF
TLX 629301
EML
COM 48

HOLLOWELL DAVID EARL DR
LOS ALAMOS NATIONAL LAB
MS B220 X2
LOS ALAMOS NM 87545
USA
TEL 505 667 4812
TLF
TLX
EML U095874@LANL GOV
COM 35

HOLLWEG JOSEPH V
DPT OF PHYSICS
UNIV OF NEW HAMPSHIRE
DEMERITT HALL
DURHAM NH 03824
USA
TEL 603 862 3869
TLF
TLX
EML
COM 10,49

HOLMAN GORDON D
NASA/GSFC
CODE 682
GREENBELT MD 20771
USA
TEL 301 286 7921
TLF
TLX
EML
COM 10

HOLMBERG ERIK B PROF
ENELIDEN 2
S 433 00 PARTILLE
SWEDEN
TEL 31 26 5842
TLF
EML
COM 25,28
TLX

HOLT STEPHEN S
NASA/GSFC
CODE 660
GREENBELT MD 20771
USA
TEL 301 286 8801
TLF
TLX
EML
COM 42,44

HOLWEGER HARTMUT PROF
INST THEOR PHYS & STERNW
UNIVERSITAET KIEL
OLSHAUSENSTR
D 2300 KIEL 1
GERMANY
TEL 431 88 04 107
TLF
TLX 292706
EML
COM 12,36

HOLZER THOMAS EDWARD DR
HIGH ALTITUDE OBSERVATORY
NCAR
BOX 3000
BOULDER CO 80307 3000
USA
TEL 303 497 1536
TLF
TLX 45694
EML
COM 10,36,49

HONEYCUTT R KENT PROF
ASTRONOMY DPT
INDIANA UNIVERSITY
SWAIN WEST 319
BLOOMINGTON IN 47405
USA
TEL 812 335 6916
TLF
TLX
EML
COM 09,42

HONG HYON IK
DPT OF PHYSICS
KIM IL SUNG UNIVERSITY
TAESONG DISTRICT
PYONGYANG
KOREA DPR
TEL
TLF
TLX
EML
COM 10

HONG SEUNG SOO DR
DPT OF ASTRONOMY
SEOUL NTL UNIVERSITY
KWANAK KU
SEOUL 151
KOREA R
TEL 877-2131
TLF
TLX 29664
EML
COM 21C,22,34

HOOD ALAN
DPT OD APPLIED MATHS
UNIVERSITY OF ST ANDREWS
NORTH HAUGH
ST ANDREWS FIFE KY16 9SS
UK
TEL 334 761 61
TLF
TLX 76213
EML
COM 10

HOOD ALAN DR
DPT MATHEMATICAL SCIENCE
UNIVERSITY OF ST ANDREWS
FIFE KY16 9SS
UK
TEL 334 762 13
TLF
TLX 932110846 SA G
EML AMSAH@SAVB ST-AND AC UK
COM

HOOGHOUDT B G IR
NAIC
CORNELL UNIVERSITY
SPACE SCIENCES BLDG
ITHACA NY 14853 6801
USA
TEL
TLF
TLX
EML
COM 09,40

HOPP ULRICH DR
MAX PLANCK INSTITUT FUR
ASTRONOMIE
KOENIGSTUHL 17
D 6900 HEIDELBERG 1
GERMANY
TEL 62 215 28224
TLF 62 215 28246
TLX 461789 MPIA D
EML HOPP@DHOMPI5V BITNET
COM 28

HORAK TOMAS B DR
VIRK
RIECNA 1
CS 815 58 BRATISLAVA
CZECHOSLOVAKIA
TEL 7 33 8451
TLF
TLX
EML
COM 42

HORAK ZDENEK PROF DR
VIETNAMSKA 2
CS 160 00 PRAHA 6
CZECHOSLOVAKIA
TEL
TLF
EML
COM
TLX

HOREDT GEORG PAUL DR
DFVLR
D 8031 WESSLING
GERMANY
TEL
TLF
TLX
EML
COM 16

HORI GENICHIRO PROF
DPT OF ASTRONOMY
UNIVERSITY OF TOKYO
BUNKYO KU
TOKYO 113
JAPAN
TEL 03-8122111x4251
TLF
TLX 33659 UTYOSCI J
EML
COM 07,33

HORIUCHI RITOKU DR
NTL INSTITUTE FOR
FUSION SCIENCE
NAGOYA 464 01
JAPAN
TEL 52 781 511
TLF 052 782 7106
TLX 0447-3691 NIFS
EML
COM 42

HOROWITZ PAUL PROF
DPT OF PHYSICS
HARVARD UNIVERSITY
60 GARDEN ST
CAMBRIDGE MA 02138
USA
TEL 617 495 3265
TLF
TLX 4992111
EML
COM 51

HORSKY JAN PROF
DPT OF THEORETICAL PHYS
PURKYNE UNIVERSITY
KOTLARSKA 2
CS 611 37 BRNO
CZECHOSLOVAKIA
TEL 5 1I12
TLF
TLX
EML
COM

HORTON BRIAN H DR
JACABRI ENT

BOX 309
GOOLWA SA 5214
AUSTRALIA
TEL 8 555 3376
TLF
TLX
EML
COM 12

HORVATH ANDRAS DR
TIT PLANETARIUM &
URANIA OBSERVATORY
BOX 46
H 1476 BUDAPEST
HUNGARY
TEL 1 33 425
TLF
TLX
EML
COM

HORWITZ GERALD PROF
RACAH INST OF PHYSICS
HEBREW UNIV OF.JERUSALEM
JERUSALEM 91904
ISRAEL
TEL 2 584 592
TLF
TLX
EML
COM

HOSHI REIUN DR
DPT OF PHYSICS
RIKKYO UNIVERSITY
NISHI-IKEBUKURO
TOSHIMA KU TOKYO 171
JAPAN
TEL 03-985-2414
TLF
TLX
EML
COM 35

HOSKIN MICHAEL A DR
CHURCHILL COLLEGE

CAMBRIDGE CB3 0DS
UK
TEL 223 35 8381
TLF
TLX
EML
COM 41

HOSKING ROGER J PROF
JAMES COOK UNIVERSITY
OF NORTH QUEENSLAND
TOWNSVILLE QLD 4811
AUSTRALIA
TEL 77 81 41113
TLF
TLX AA47009
EML
COM

HOSOKAWA YOSHIMASA H PROF
SAKIGAOKA 3-4-9
FUNABASHI CITY
CHIBA PREFECTURE 274
JAPAN
TEL 0474 48 6679
TLF
TLX
EML
COM

HOSOYAMA KENNOSHUKE DR
INTL LATITUDE OBSERVATORY
NAOJ
HOSHIGAOKA MIZUSAWA SHI
IWATE 023
JAPAN
TEL
TLF
TLX
EML
COM

HOTINLI METIN DR
UNIVERSITY RASATHANESI

ISTANBUL
TURKEY
TEL
TLF
TLX
EML
COM 12,36

HOUCK JAMES R
DPT OF ASTRONOMY
CORNELL UNIVERSITY
220 SPACE SCIENCE BLDG
ITHACA NY 14853
USA
TEL 607 256 4806
TLF 607 255 2365
TLX 937478 ITCA
EML
COM 21C

HOUGH JAMES DR
DPT OF PHYSICS & ASTRON
UNIVERSITY OF GLASGOW
GLASGOW G12 8QQ
UK
TEL 413 304 706
TLF 413 349 029
TLX 777070 UNIGLA
EML GWO4%GLA PH I1@UK AC RL BITNET
COM 09

HOUK NANCY DR
DPT OF ASTRONOMY
UNIVERSITY OF MICHIGAN
1045 PHYS-ASTRO BLDG
ANN ARBOR MI 48109 1090
USA
TEL 313 764 3436
TLF .
TLX
EML
COM 27,29,45

HOUSE FRANKLIN C DR
HEIDENREICHSTR 42

D 6100 DARMSTADT
GERMANY
TEL 615 142 2412
TLF
TLX
EML
COM

HOUSE LEWIS L DR
HIGH ALTITUDE OBSERVATORY
NCAR
BOX 3000
BOULDER CO 80307 3000
USA
TEL 303 494 5151
TLF
TLX
EML
COM 12,14,36

HOUZIAUX L PROF
INSTITUT DE MATHEMATIQUES
15 AV TILLEULS
B 4000 LIEGE
BELGIUM
TEL 41 66 9494
TLF 41 66 9493
TLX
EML U2141LH@BLIULG11
COM 29,34,44,46C

HOVENIER J W DR
DPT PHYSICS & ASTRONOMY
FREE UNIVERSITY
DE BOELELAAN 1081
NL 1081 HV AMSTERDAM
NETHERLANDS
TEL 20 54 02414
TLF
TLX
EML
COM 16

HOWARD ROBERT F DR
NTL SOLAR OBSERVATORY
BOX 26732
950 N CHERRY AVE
TUCSON AZ 85726 6732
USA
TEL 602-327-5511
TLF
TLX 0666484 AURA NOAOTUC
EML
COM 10,12

HOWARD W MICHAEL DR
LAWRENCE LIVERMORE LAB
L 297
BOX 808
LIVERMORE CA 94550
USA
TEL 415 422 4138
TLF
TLX
EML
COM

HOWARD WILLIAM E III DR
1653 QUAIL HOLLOW CT
MCLEAN
VA 22101 3234
USA
TEL 703 695 1447
TLF 703 697 6956
TLX
EML· howard%peo-mis-emhl army mil
COM 40

HOWARTH IAN DONALD
DPT PHYSICS & ASTRONOMY
UNIVERSITY COLLEGE LONDON
GOWER ST
LONDON WC1E 6BT
UK
TEL 1 713 877 050
TLF
TLX 28722
EML
COM 44

HOWELL STEVE BRUCE DR
PLANETARY SCIENCE INST
2030 E SPEEDWAY
SUITE 201
TUCSON AZ 85719
USA
TEL 602 881 0332
TLF 602 881 0335
TLX
EML 5470 HOWELL
COM 27

HOWSE H DEREK
12 BARNFIELD RD
RIVERHEAD
SEVENOAKS TN13 2AY
UK
TEL 732 454 366
TLF
TLX
EML
COM 41

HOYLE FRED SIR
102 ADMIRALS WALK
WEST CLIFF RD
BOUREMOUTH
DORSET BH2 5HF
UK
TEL
TLF
TLX
EML
COM 28,35,47,48

HOYNG PETER DR
SPACE RESEARCH LABORATORY
SRON
SORBONNELAAN 2
NL 3584 CA UTRECHT
NETHERLANDS
TEL 30 53 5600
TLF 30 54 0860
TLX 47224 SRON NI
EML
COM 10,12,44

HRIC LADISLAV DR
ASTRONOMICAL INSTITUTE
SLOVAK ACADEMY SCIENCES
CS 059 60 TATRANSKA LOMNI
CZECHOSLOVAKIA
TEL 969 96 7866/7/8
TLF 969 96 7656
TLX 78277
EML
COM 42

HRIVNAK BRUCE J
DPT OF PHYSICS
VALPARAISO UNIVERSITY
VALPARAISO IN 46383
USA
TEL 219 464 5379
TLF
TLX
EML
COM 30,42

HRON JOSEF DR
INSTITUT FUER ASTRONOMIE
UNIVERSITAET WIEN
TUERKENSCHANZSTR 17
A 1180 WIEN
AUSTRIA
TEL 1 345 3600
TLF
TLX 133099 vıast a
EML EARN AWIUNI11@A8201DAH
COM 29,33

HSIANG YAN-YU
BEIJING ASTRONOMICAL OBS
CAS
W SUBURB
BEIJING 100080
CHINA PR
TEL 1 28 1698
TLF
TLX 22040 BAOAS CN
EML
COM

HSIANG-KUANG TSENG
DPT OF PHYSICS
NTL CENTRAL UNIVERSITY
CHUNG LI
CHINA R
TEL
TLF
TLX
EML
COM

HU ESTHER M DR
INSTITUTE FOR ASTRONOMY
UNIVERSITY OF HAWAII
2680 WOODLAWN DR
HONOLULU HI 96822
USA
TEL 808 948 7190
TLF
TLX 723 8459 UHAST HR
EML INT hu@uhıfa ıfa hawaıı edu
COM 47

HU FU-XING
PURPLE MOUNTAIN OBSERV
CAS
NANJING
CHINA PR
TEL 25 46700
TLF
TLX
EML
COM 28

HU JING-YAO
BEIJING ASTRONOMICAL OBS
CAS
W SUBURB
BEIJING 100080
CHINA PR
TEL 1 28 1698
TLF
TLX 22040 BAOAS CN
EML
COM 09,25

HU NING-SHENG
NANJING ASTRONOMICAL
INSTRUMENT FACTORY
BOX 846
NANJING
CHINA PR
TEL 25 46191
TLF
TLX 34136 GLYNJ CN NAIF
EML
COM 08C,09

HU WEN-RUI
INSTITUTE OF MECHANICS
CAS
W SUBURB
BEIJING 100080
CHINA PR
TEL 1 28 4185
TLF
TLX 22474 ASCHI CN
EML
COM ·44

HU ZHONG-WEI
DEPT OF ASTRONOMY
NANJING UNIVERSITY
NANJING
CHINA PR
TEL 25 37651
TLF
TLX. 0909
EML
COM. 15,16

HUA CHON TRUNG DR
LAS
TRAVERSE DU SIPHON
LES TROIS LUCS
F 13012 MARSEILLE
FRANCE
TEL 91 05 5932
TLF 91 66 1855
TLX 420584 F
EML
COM 28,34

HUA YING-MIN
SHANGHAI OBSERVATORY
CAS
80 NANDAN RD
SHANGHAI 200030
CHINA PR
TEL 21 38 6191
TLF 21 38 4618
TLX 33164 SHAO CN
EML
COM 08,19

HUAN NGEYEN DUSH DR
UNIVERSITY OF VINH
PROVINCE NGH^E AN
VIETNAM
TEL
TLF
EML
COM
TLX

HUANG BI-KUN
PURPLE MOUNTAIN OBSERV
LIBRARY
CAS
NANJING
CHINA PR
TEL: 25 307521
TLF
TLX 34144 PMONJ CN
EML
COM 05

HUANG CHANG-CHUN
PURPLE MOUNTAIN OBSERV
CAS
NANJING
CHINA PR
TEL 25 46700/42817
TLF
TLX 34144 PMONJ CN
EML
COM 29,30

HUANG CHENG DR
SHANGHAI OBSERVATORY
CAS
80 NANDAN RD
SHANGHAI
CHINA PR
TEL 21 38 6191
TLF
TLX
EML
COM 07

HUANG JIE-HAO
ASTROPHYSICS INSTITUTE
NANJING UNIVERSITY
NANJING
CHINA PR
TEL
TLF
TLX 34151 PRCNU CN
EML
COM 28

HUANG KE-LIANG
DPT OF ASTRONOMY
NANJING UNIVERSITY
NANJING
CHINA PR
TEL 25 34651*2882
TLF
TLX 34151 PRCNU CN
EML.
COM· 28,48

HUANG KUN-YI
PURPLE MOUNTAIN OBSERV
CAS
NANJING
CHINA PR
TEL 25 32893
TLF
TLX 34144 PMONJ CN
EML
COM

HUANG LIN
BEIJING ASTRONOMICAL OBS
CAS
W SUBURB
BEIJING 100080
CHINA PR
TEL 1 28 1698
TLF
TLX 22040 BAOAS CN
EML
COM 25,45

HUANG RUN-QIAN
YUNNAN OBSERVATORY
CAS
BOX 110
KUNMING 72946 YUNNAN
CHINA PR
TEL 871 2035
TLF
TLX 64040 YUOBS CN
EML
COM 35,42

HUANG SONG-NIAN DR
ICTP CP 586
ST COSTIERA 11
MIRAMARE
I 34100 TRIESTE
ITALY
TEL
TLF
TLX
EML·
COM 28,33

HUANG TIANYI
DEPT OF ASTRONOMY
NANJING UNIVERSITY
NANJING
CHINA PR
TEL
TLF
TLX 34151 PRCNU CN
EML
COM 07

HUANG TIE-QIN
NANJING ASTRONOMICAL
INSTRUMENT FACTORY
NANJING
CHINA PR
TEL 25 46191
TLF·
TLX 34136 GLYNJ CN
EML
COM 09

HUANG YINN-NIEN DR
DPT OF POSTS & TELECOMMUN
MIN OF TRANSPORT & COMMUN
2 CHANGSHA ST SEC 1
TAIPEI 100
CHINA R
TEL
TLF
TLX
EML
COM:

HUANG YIN-LIANG
TIANJIN INST OF TECH
HONG QI NAN RD
TIANJIN 300191
CHINA PR
TEL 368787 256
TLF
TLX
EML
COM 50

HUANG YI-LONG DR
INST OF HISTORY
NTL TSING HUA UNIVERSITY
HSIN CHU 300043
CHINA R
TEL 35 71 6780
TLF
TLX
EML
COM

HUANG YONGWEI
BEIJING ASTRONOMICAL OBS
CAS
W SUBURB
BEIJING 100080
CHINA PR
TEL 1 28 1698
TLF
TLX 9053
EML
COM. 28

HUANG YOU-RAN
DPT OF ASTRONOMY
NANJING UNIVERSITY
NANJING
CHINA PR
TEL
TLF
TLX 34151
EML
COM 10

HUBBARD WILLIAM B PROF
LUNAR & PLANETARY LAB
UNIVERSITY OF ARIZONA
TUCSON AZ 85721
USA
TEL 602 621 6942
TLF
TLX 9109521143
EML
COM 16

HUBE DOUGLAS P DR
DPT OF PHYSICS
UNIVERSITY OF ALBERTA
EDMONTON AB T6G 2J1
CANADA
TEL 403 432 5410
TLF 403 432 4256
TLX 037 2979
EML
COM 30,42

HUBENET HENRI DR
STERREKUNDIG INSTITUTE

BOX 80000
NL 3508 TA UTRECHT
NETHERLANDS
TEL 30 53 5200
TLF
TLX 40048 FYLUT NL
EML
COM

HUBENY IVAN
JILA
UNIVERSITY OF COLORADO
BOX 440
BOULDER CO 80309 0440
USA
TEL 303 492 7838
TLF
TLX 755842 JILA
EML
COM 29,36

HUBER MARTIN C E DR
ESA/ESTEC
SSD
BOX 299
NL 2200 AG NOORDWIJK
NETHERLANDS
TEL 17 19 83552
TLF
TLX 39098
EML
COM 14,44

HUBERT HENRI DR
OBSERVATOIRE DE PARIS
SECTION DE MEUDON
DASGAL
F 92195 MEUDON PPL CDX
FRANCE
TEL 1 45 07 7850
TLF
TLX 270912
EML HUBERT@FRMEU51
COM 29

HUBERT-DELPLACE A-M DR
OBSERVATOIRE DE PARIS
SECTION DE MEUDON
F 92195 MEUDON PPL CDX
FRANCE
TEL 1 45 34 7856
TLF
TLX 270912
EML
COM 29

HUCHRA JOHN PETER DR
CENTER FOR ASTROPHYSICS
HCO/SAO
60 GARDEN ST
CAMBRIDGE MA 02138
USA
TEL 617 495 7375
TLF
TLX 921428 SATELLITE CAM
EML
COM 28,30,47

HUCHTMEIER WALTER K DR
MPI FUER RADIOASTRONOMIE
AUF DEM HUEGEL 69
D 5300 BONN 1
GERMANY
TEL 228 52 5215
TLX 886440
EML
TLF
COM 28,40

HUDEC RENE DR
ASTRONOMICAL INSTITUTE
CZECH ACADEMY OF SCIENCES
ONDREJOV OBSERVATORY
CS 251 65 ONDREJOV
CZECHOSLOVAKIA
TEL 204 85201
TLF 204 85314
TLX 121579
EML
COM

HUDSON HUGH S DR
DPT OF PHYSICS
UCSD
C 011
LA JOLLA CA 92093 0216
USA
TEL 619 452 4476
TLF
TLX
EML
COM 10

HUEBNER WALTER F DR
SOUTHWEST RESEARCH INST
6220 CULEBRA RD
BOX 28510
SAN ANTONIO TX 78284
USA
TEL 512 522 2730
TLF
TLX 244846
EML
COM 14,15C

HUENEMOERDER DAVID P DR
DPT OF ASTRONOMY
PENNSYLVANIA STATE UNIV
525 DAVEY LAB
UNIVERSITY PARK PA 16802
USA
TEL 814 865 6601
TLF
TLX
EML INTERNET dph@astro psu edu
COM 27,29

HUGHES DAVID W DR
DPT OF PHYSICS
THE UNIVERSITY
SHEFFIELD S3 7RH
UK
TEL 742 785 55
TLF
TLX 54348 ULSHEF 'G
EML
COM 15C,22

HUGHES JOHN P DR
CENTER FOR ASTROPHYSICS
HCO/SAO
60 GARDEN ST
CAMBRIDGE MA 02138
USA
TEL 617 495 7142
TLF
TLX 921428 SATELLITE CAM
EML BITNET HUGHES@CFA
COM

HUGHES PHILIP
DPT OF ASTRONOMY
UNIVERSITY OF MICHIGAN
DENNISON BLDG
ANN ARBOR MI 48109 1090
USA
TEL 313 764 3430
TLF
TLX
EML
COM 40

HUGHES SHAUN
CALTECH
DPT OF ASTRONOMY
PASADENA CA 91125
USA
TEL
TLF
TLX
EML
COM

HUGHES VICTOR A PROF
DPT OF PHYSICS
QUEEN'S UNIVERSITY
KINGSTON ON K7L 3N6
CANADA
TEL 613 547 6633
TLF 613 545 6463
TLX
EML HUGHESV@QUCDN BITNET
COM 33,34,40

HUGUENIN G RICHARD
MULTITECH CORPORATION
BOX 109
SOUTH DEERFIELD RES PARK
SOUTH DEERFIELD MA 01373
USA
TEL 413 665 8551
TLF
TLX 3719862 TRUB
EML
COM

HULSBOSCH A N M DR
STERRENKUNDIG INSTITUTE
KATHOLIEKE UNIVERSITEIT
TOERNDOIVELD
NL 6525 ED NIJMEGEN
NETHERLANDS
TEL 80 55 8833
TLF
TLX 48228
EML
COM 33,34,40

HUMBLE JOHN EDMUND DR
DPT OF PHYSICS
UNIVERSITY OF TASMANIA
GPO BOX 252C
HOBART TAS 7001
AUSTRALIA
TEL· 02 202 401
TLF, 02 202 410
TLX AA 58150
EML, JOHN HUMBLE@PHYS UTAS EDU AU
COM 49

HUMMEL EDSHO
NRAL
JODRELL BANK
MACCLESFIELD SK11 9DL
UK
TEL
TLF
TLX
EML
COM 28

HUMMER DAVID G DR
JILA
UNIVERSITY OF COLORADO
BOX 440
BOULDER CO 80309 0440
USA
TEL 303 492 7837
TLF
TLX 755842 JILA
EML
COM 34,36

HUMPHREYS ROBERTA M PROF
DPT OF ASTRONOMY
UNIVERSITY OF MINNESOTA
116 CHURCH ST SE
MINNEAPOLIS MN 55455
USA
TEL 612 373 9747
TLF
TLX
EML
COM 28,33,35,45

HUMPHRIES COLIN M DR
ROYAL OBSERVATORY
BLACKFORD HILL
EDINBURGH EH9 3HJ
UK
TEL 316 673 321
TLX 72383
EML
TLF
COM 09C

HUNDHAUSEN ARTHUR DR
HIGH ALTITUDE OBSERVATORY
NCAR
BOX 3000
BOULDER CO 80307 3000
USA
TEL
TLF
TLX
EML
COM

HUNGER KURT PROF
INST THEOR PHYS & STERNW
NEUE UNIV PHYSIK ZENTRUM
OLSHAUSENST 40 N61C
D 2300 KIEL 1
GERMANY
TEL 431 880 4110
TLF
TLX 292706
EML
COM 29,36

HUNSTEAD RICHARD W DR
SCHOOL OF PHYSICS
UNIVERSITY OF SYDNEY
SYDNEY NSW 2006
AUSTRALIA
TEL 2 692 3871
TLF
TLX 26169 UNISYD AA
EML
COM 28,40

HUNT G E DR
ELBURY
37 BLENHEIM ROAD
RAYNES PARK
LONDON SW20 9BA
UK
TEL 1 542 2374
TLF
TLX
EML
COM 16

HUNT LESLIE DR
OSS ASTROFISICO
DI ARCRETI CAISMI CNR
LARGO E FERMI 5
I 50125 FIRENZE
ITALY
TEL 55 275 2296
TLF 55 22 0039
TLX 572 268 ARCETR
EML HUNT@SISIFO ARCETRI ASTRO IT
COM 48

HUNTEN DONALD M PROF
LUNAR & PLANETARY LAB
UNIVERSITY OF ARIZONA
TUCSON AZ 85721
USA
TEL 602 621 4002
TLF
TLX
EML
COM 16,51

HUNTER CHRISTOPHER PROF
DPT OF MATHEMATICS
FLORIDA STATE UNIVERSITY
TALLAHASSEE FL 32306
USA
TEL 904 644 2488
TLF
TLX
EML
COM 28,33

HUNTER DEIDRE ANN
DPT TERRESTR MAGNETISM
CARNEGIE INST WASHINGTON
5241 BROAD BRANCH RD NW
WASHINGTON DC 20015
USA
TEL 202 966 0863
TLF
TLX 440427 MAGN UI
EML
COM

HUNTER JAMES H PROF
DPT OF ASTRONOMY
UNIVERSITY OF FLORIDA
211 SSRB
GAINESVILLE FL 32611
USA
TEL 904 392 1078
TLF
TLX
EML
COM 28,51

HUNTRESS WESLEY T DR
NASA HEADQUARTERS
CODE SL
600 INDEPENDENCE AVE SW
WASHINGTON DC 20546
USA
TEL 202 453 1588
TLF 202 426 1023
TLX
EML
COM 15

HUOVELIN JUHANI DR
OBS & ASTROPHYSICS LAB
UNIVERSITY OF HELSINKI
TAEHTITORNINMAKI
SF 00130 HELSINKI 13
FINLAND
TEL 191 2948
TLF 191 2952
TLX 124690 UNIH SF
EML HUOVELIN AT CC HELSINKI FI
COM 25

HURFORD GORDON JAMES
CALTECH
MS 264 33
PASADENA CA 91125
USA
TEL 818 356 3866
TLF
TLX 675425
EML
COM 10

HURNIK HIERONIM PROF
ASTRONOMICAL OBSERVATORY
A MICKIEWICZ UNIVERSITY
UL SLONECZNA 36
PL 60 286 POZNAN
POLAND
TEL 679 670
TLF
TLX
EML
COM 20

HUSFELD DIRK DR
INSTITUT FUR ASTRONOMIE
UND ASTROPHYSIK
SCHEINERSTR 1
D 8000 MUENCHEN 80
GERMANY
TEL 89 922 09440
TLF 89 922 09427
TLX
EML UH101BS@DMOLRZO1 EARN
COM 36

HUT PIET
INST FOR ADVANCED STUDY
SCHOOL OF NATURAL SCIENCE
PRINCETON NJ 08540
USA
TEL   609 734 8075
TLF
TLX   229734 IAS UR
EML
COM   37

HUTCHEON RICHARD J DR
DPT OF PHYSICS
UNIVERSITY OF LEICESTER
X-RAY ASTRONOMY GROUP
LEICESTER LE1 7RH
UK
TEL
TLF
TLX
EML
COM

HUTCHINGS JOHN B DR
HERZBERG INST ASTROPHYS
DOMINION ASTROPHYS OBS
5071 W SAANICH RD
VICTORIA BC V8X 4M6
CANADA
TEL   604 388 3909
TLF   604 363 0045
TLX   049 7295
EML
COM   27,34,36,42,44C

HUTSEMEKERS DAMIEN DR
INSTITUT D'ASTROPHYSIQUE
UNIVERSITE DE LIEGE
AVE COINTE 5
B 4000 COINTE-LIEGE
BELGIUM
TEL   41 52 9980
TLF   41 52 7474
TLX   41264 ASTRLG B
EML   U2141MG@BLIULG11
COM   34

HWANG WOEI-YANN P PROF
DPT OF PHYSICS
NTL TAIWAN UNIVERSITY
TAIPEI
TAIWAN 10764 ROC
CHINA R
TEL   2 363 0231*3159
TLF   2 363 7204/9984
TLX
EML
COM

HYDER C L DR
HIGH ALTITUDE OBSERVATORY
NCAR
BOX 3000
BOULDER CO 80307 3000
USA
TEL
TLF
TLX
EML
COM   10

HYLAND A R HARRY DR
DPT PHYSICS UNIV COLLEGE
UNIVER OF NEW SOUTH WALES
NORTHCOTT DR
CAMPBELL ACT 2600
AUSTRALIA
TEL   62 68 8787
TLF   6 268 8786
TLX   62030 ADFADM AA
EML
COM   25,29

HYSOM EDMUND J
8 EAST DRIVE
CALDECOTE
CAMBRIDGE CB3 7NZ
UK
TEL   223 954 211*137
TLF
TLX
EML
COM   09,41,51

HYUN JONG-JUNE PROF
DPT OF ASTRONOMY
SEOUL NTL UNIVERSITY
KWANAK KU
SEOUL 151
KOREA R
TEL   877-3010/2542
TLF
TLX
EML
COM

IANNA PHILIP A
UNIVERSITY STATION
UNIVERSITY OF VIRGINIA
BOX 3818
CHARLOTTESVILLE VA 22903
USA
TEL   804 924 4898
TLF
TLX
EML
COM   20,24,26

IANNINI GUALBERTO DR
OBSERVATORIO ASTRONOMICO
DE CORDOBA
LAPRIDA 854
5000 CORDOBA
ARGENTINA
TEL   51 40613
TLF
TLX
EML
COM

IBADINOV KHURSANDKUL DR
ASTROPHYSICAL INSTITUTE
TADJIK ACAD OF SCIENCES
734670 DUSHANBE
TADZHIKISTAN
TEL
TLF
TLX
EML
COM   15

IBANEZ S MIGUEL H DR
UNIVERSIDAD DE LOS ANDES
FACULDAD DE CIENCIAS
DPT DE FISICA
MERIDA 5101 A
VENEZUELA
TEL   74 63 9930/7477
TLF
TLX   74174 CIDA
EML
COM

IBANOGLU C DR
FACULTY OF SCIENCE
EGE UNIVERSITY
BOX 21
35100 BORNOVA IZMIR
TURKEY
TEL   51 18 0110*2332
TLF
TLX
EML
COM   42

IBBETSON PETER AARON DR
WISE OBSERVATORY
TEL AVIV UNIVERSITY
RAMAT AVIV
TEL AVIV 69978
ISRAEL
TEL   3 41 3788
TLF
TLX   342171 VERSY IL
EML
COM

IBEN ICKO JR PROF
DPT OF ASTRONOMY
UNIVERSITY OF ILLINOIS
1011 W SPRINGFIELD AVE
URBANA IL 61801
USA
TEL   217 333 3090
TLF
TLX   9102452434 AST
EML
COM   27,35C,37

IBRAHIM JORGA
DPT OF ASTRONOMY
BANDUNG INSTITUTE OF TECH
JL TAMANSARI 64
BANDUNG
INDONESIA
TEL
TLF
TLX
EML
COM

ICHIKAWA SHIN-ICHI DR
TOKYO ASTRONOMICAL OBS
NAOJ
OSAWA MITAKA
TOKYO 181
JAPAN
TEL   422 41 3604
TLF   422 41 3608
TLX   02822307 TAOMTK J
EML   ICHIKAWA@C1 MTK NAO AC JP
COM   28

ICHIKAWA TAKASHI
DPT OF ASTRONOMY
KYOTO UNIVERSITY
KITASHIRAKAWA SAKYO KU
KYOTO 606
JAPAN
TEL   75 751 3890
TLF
TLX
EML
COM   28

ICHIMARU SETSUO DR
DPT OF PHYSICS
UNIVERSITY OF TOKYO
BUNKYO KU
TOKYO 113
JAPAN
TEL   03-812-2111
TLF
TLX   UTPHYSIC J23472
EML
COM   48

ICKE VINCENT DR
STERREWACHT
BOX 9513
NL 2300 RA LEIDEN
NETHERLANDS
TEL 71 27 2727
TLX 39058 ASTRO NL
EML
TLF
COM 47

IDLIS G M DR
INST HIST OF SCI & TECH
ACACEMY OF SCIENCES
STAROPANSKY 1/5
103012 MOSCOW
RUSSIA
TEL 2281969
TLF
TLX·
EML
COM 41,51

IIJIMA SHIGETAKA PROF
MUSASHI INSTITUTE OF
TECHNOLOGY
TAMAZUTSUMI, SETAGAYA-KU
TOKYO 158
JAPAN
TEL 33 703 3111
TLF
TLX
EML
COM 19,31

IIJIMA TAKASHI DR
OSSERVATORIO ASTROFISICO
VIA DELL OSSERVATORIO 8
I 36012 ASIAGO
ITALY
TEL 42 46 2505
TLF 424 46 2884
TLX
EML IIJIMA@ASTRAS CINECA IT
COM 27

IKEUCHI SATORU DR
TOKYO ASTRONOMICAL OBS
NAOJ
OSAWA MITAKA
TOKYO 181
JAPAN
TEL 0422-32-5111
TLF
TLX 2822307
EML
COM 33,47

IKHSANOV ROBERT N DR
PULKOVO OBSERVATORY
ACADEMY OF SCIENCES
10 KUTUZOV QUAY
196140 ST PETERSBURG
RUSSIA
TEL
TLF
TLX
EML
COM 40

IKHSANOVA VERA N DR
PULKOVO OBSERVATORY
ACADEMY OF SCIENCES
10 KUTUZOV QUAY
196140 ST PETERSBURG
RUSSIA
TEL
TLF
TLX
EML
COM 40

ILIEV ILIAN
NTL ASTRONOMICAL OBS
BULGARIAN ACAD SCIENCES
BOX 136
BG 4700 SMOLJAN
BULGARIA
TEL 73 41 559
TLF
TLX 23561
EML
COM 14,35

ILL MARTON J DR
KONKOLY OBSERVATORY
TOTH KALMAN U 19
H 6501 BAJA
HUNGARY
TEL 79 12 110
TLX 281303
EML
TLF
COM

ILLES ALMAR ERZSEBET DR
KONKOLY OBSERVATORY
THEGE U 13/17
BOX 67
H 1525 BUDAPEST
HUNGARY
TEL 1 75 5866/75 4122
TLF
TLX 227460
EML
COM

ILLING RAINER M E
BALL AEROSPACE SYSTEMS
DIVISION
BOX 1062
BOULDER CO 80306
USA
TEL 303 939 5888
TLF
TLX
EML
COM 12

ILLINGWORTH GARTH D DR
LICK OBSERVATORY
UNIVERSITY OF CALIFORNIA
SANTA CRUZ CA 95064
USA
TEL
TLF
TLX
EML
COM 28,37

ILYAS MOHAMMAD DR
SCHOOL OF PHYSICS
UNIVERSITI SAINS MALAYSIA
11800 USM
PENANG
MALAYSIA
TEL 883822
TLF
TLX 40254 MA
EML
COM 04,09,46

IMAMURA JAMES DR
DPT OF PHYSICS
UNIVERSITY OF OREGON
EUGENE OR 97403
USA
TEL 503 686 5212
TLF
TLX
EML Imamura@astro UOREGON EDU
COM 42

IMBERT MAURICE DR
OBSERVATOIRE DE MARSEILLE
2 PLACE LE VERRIER
F 13248 MARSEILLE CDX 04
FRANCE
TEL 91 95 9088
TLF
TLX 420241
EML
COM 30,42

IMHOFF CATHERINE L DR
NASA/GSFC
CODE 684 9
IUE OBS
GREENBELT MD 20771
USA
TEL 301 286 5749
TLF
TLX 89675
EML
COM 44

IMPEY CHRISTOPHER D DR
STEWARD OBSERVATORY
UNIVERSITY OF ARIZONA
TUCSON AZ 85721
USA
TEL 602 621 6522
TLF
TLX 467175
EML impey@solpl as arizona edu
COM 28,46,47

IMSHENNIK V S DR
INST THEOR & EXPER PHYS
CHEREMUSHKINSKAYA UL 25
117259 MOSCOW
RUSSIA
TEL 123-02-92
TLX 411059 CERII SU
EML
TLF
COM 35

INAGAKI SHOGO DR
DPT OF ASTRONOMY
KYOTO UNIVERSITY
KITASHIRAKAWA SAKYO KU
KYOTO 606
JAPAN
TEL 075-751-2111
TLF
TLX 5422693 LIBKYU J
EML
COM 33,37,49

INATANI JUNJI
NOBEYAMA RADIO OBS
NAOJ
MINAMIMAKI MURA
NAGANO 384 13
JAPAN
TEL 267-98-2831
TLF
TLX 3329005 TAO NRO J
EML
COM 40

INGERSON THOMAS DR
CERRO TOLOLO
INTERAMERICAN OBSERVATORY
CASILLA 603
LA SERENA
CHILE
TEL 51 21 3352
TLF 51 21 2466*342
TLX 620 301 AURA CT
EML tingerson@noao.edu
COM

INNANEN KIMMO A PROF
DPT OF PHYSICS
YORK UNIVERSITY
4700 KEELE ST
NORTH YORK ON M3J 1P3
CANADA
TEL. 416 667 3837
TLF 416 736 5386
TLX 065 24736
EML
COM 33

INOUE HAJIME DR
INST SPACE & ASTRON SCI
UNIVERSITY OF TOKYO
MEGURO KU
TOKYO 153
JAPAN
TEL 03-467-1111
TLF
TLX 24550 J
EML
COM 44,48

INOUE MAKOTO DR
NOBEYAMA RADIO OBS
NAOJ
MINAMIMAKI MURA
NAGANO 384 13
JAPAN
TEL 0267-98-2831
TLF
TLX 3329005 TAONKKRO J
EML
COM 40

INOUE TAKESHI PROF
DPT OF PHYSICS
KYOTO SANGYO UNIVERSITY
KAMIGAMO
KYOTO 603
JAPAN
TEL 075-701-2151
TLF
TLX 5422661 KSU J
EML
COM

IOSHPA B A DR
IZMIRAN
ACADEMY OF SCIENCES
142092 TROITSK
RUSSIA
TEL 2321921
TLF
TLX 412623 SCP
EML
COM 10

IP WING-HUEN
MPI FUER ASTRONOMIE
D 3411 KATLENBURG LINDAU
GERMANY
TEL 0049-555-6416
TLF
TLX 0965527
EML
COM 15

IPSER JAMES R PROF
DPT OF PHYSICS
UNIVERSITY OF FLORIDA
WILLIAMSON HALL
GAINESVILLE FL 32611
USA
TEL 904 392 0521
TLF
TLX
EML
COM 48

IRELAND JOHN G DR
C/O 13 GORDEN ROAD
BELVEDERE KENT DA17 6EA
UK
TEL
TLF
EML
COM
TLX

IRIGOYEN MAYLIS
UNIVERSITE DE PARIS II
12 PLACE DU PANTHEON
F 75005 PARIS
FRANCE
TEL
TLF
TLX
EML
COM

IRIYAMA JUN DR
FAC OF ENGINEERING
CHUBU UNIVERSITY
1200 MATSUMOTO
KASUGAI-SHI,AICHI 487
JAPAN
TEL 568 51 1111
TLF
TLX
EML
COM 17

IRVINE WILLIAM M PROF
FIVE COLLEGE
RADIO ASTRONOMY OBSERV
B619 LEDERLE GRAD RES TWR
AMHERST MA 01003
USA
TEL 413 545 0733
TLF
TLX 955491 UNIV MASS AMS
EML
COM 15,16,34,51

IRWIN ALAN W DR
DPT OF PHYSICS
UNIVERSITY OF VICTORIA
BOX 1700
VICTORIA BC V8W 2Y2
CANADA
TEL 604 721 7700
TLF 604 721 7715
TLX
EML IRWIN@OTTO PHYS UNIC CA
COM 14,25,30

IRWIN JOHN B PROF
2744 N TYNDALL AVE
TUCSON AZ 85719
USA
TEL 602 623 7423
TLF
EML
COM 33,42
TLX

IRWIN JUDITH DR
DPT OF ASTRONOMY
UNIVERSITY OF TORONTO
60 ST GEORGE ST
TORONTO ON M5S 1A1
CANADA
TEL 416 978 5558
TLF 416 978 3921
TLX 062 18915
EML
COM 28

IRWIN MICHAEL JOHN DR
INSTITUTE OF ASTRONOMY
THE OBSERVATORIES
MADINGLEY RD
CAMBRIDGE CB3 0HA
UK
TEL 223 33 7548
TLF
TLX
EML MIKE@UK AC CAM AST-STAR
COM 24

ISAAK GEORGE R PROF
DPT OF PHYSICS
UNIVERSITY OF BIRMINGHAM
BOX 363
BIRMINGHAM B15 2TT
UK
TEL 214 72 1301
TLF
TLX 338938 SPAPHY G
EML
COM 35

ISERN JORGE DR
C/SEPULVEDA 83-6-3A
E 08015 BARCELONA
SPAIN
TEL
TLF
EML
COM 35
TLX

ISHIDA GORO DR
BROADCAST UNIVERSITY
23-11 AKABANE NISHI
1 CHOME KITA KU
TOKYO 115
JAPAN
TEL 03-909-3871
TLF
TLX
EML
COM 26

ISHIDA KEIICHI PROF
TOKYO ASTRONOMICAL OBS
NAOJ
OSAWA MITAKA
TOKYO 181
JAPAN
TEL 04-22-32-5211
TLF
TLX 2822307 TAOMTK J
EML
COM 37

ISHIGURO MASATO PROF
NOBEYAMA RADIO OBS
NAOJ
MINAMIMAKI MURA
NAGANO 384 13
JAPAN
TEL 0267 98 2831
TLF
TLX 3329005 NAONRO J
EML
COM 40

ISHIHARA HIDEKI DR
DPT OF PHYSICS
KYOTO UNIVERSITY
SAKYO KU
KYOTO 606
JAPAN
TEL· 075 753 3850
TLF:
TLX
EML ISHIHARA@NWS841 SCPHYS KYOTOUA
COM 47

ISHIZAWA TOSHIAKI A PROF
DPT OF ASTRONOMY
KYOTO UNIVRESITY
KITASHIRAKAWA SAKYO KU
KYOTO 606
JAPAN
TEL 075-751-2111
TLF
TLX 5422693 LIBKYU J
EML
COM

ISHIZUKA TOSHIHISA DR
DPT OF PHYSICS
IBARAKI UNIVERSITY
BUNKYO
MITO 310
JAPAN
TEL 0292-26-1621
TLF
TLX
EML
COM 35

ISOBE SYUZO DR
TOKYO ASTRONOMICAL OBS
NAOJ
OSAWA MITAKA
TOKYO 181
JAPAN
TEL 422 41 3645
TLF 422 41 3608
TLX 2822307 TAOMTK J
EML OISOBEX@CL MTK NAO AC JP
COM 06C,15,20,33,34,46C,

ISRAEL FRANK P DR
STERREWACHT
BOX 9513
NL 2300 RA LEIDEN
NETHERLANDS
TEL
TLX
EML
TLF
COM 28C,33,34,51

ISRAEL GUY MARCEL DR
SERVICE D'AERONOMIE
BP 3
F 91371 VERRIERES BUISSON
FRANCE
TEL 1 64 47 4289
TLF
TLX
EML
COM

ISRAEL WERNER PROF
DPT OF PHYSICS
UNIVERSITY OF ALBERTA
EDMONTON AB T6G 2J1
CANADA
TEL 403 432 3552
TLF 403 432 4256
TLX 037 2979
EML
COM 48

ISSA ALI DR
HELWAN OBSERVATORY
HELWAN
EGYPT
TEL 78 0645/2683
TLF
EML
COM 28,34
TLX 93070

ISSERSTEDT JOERG DR
INSTITUT FUER ASTRONOMIE
UND ASTROPHYSIK
AM HUBLAND
D 8700 WUERZBURG
GERMANY
TEL
TLF
TLX
EML
COM

ITO KENSAI A PROF
DPT OF PHYSICS
RIKKYO UNIVERSITY
NISHI-IKEBUKURO
TOKYO 171
JAPAN
TEL 03-985-2384
TLF
TLX
EML
COM 48

ITOH HIROSHI DR
DPT OF ASTRONOMY
KYOTO UNIVERSITY
KITASHIRAKAWA SAKYO KU
KYOTO 606
JAPAN
TEL 075-751-2111
TLF
TLX
EML
COM 34

ITOH MASAYUKI DR
ISAS
3-1-1 YOSHINODAI
SAGAMIHARA
KANAGAWA 229
JAPAN
TEL 427 51 3911
TLF 427 59 4253
TLX J27758 ISAS ERO
EML
COM 48

ITOH NAOKI DR
DPT OF PHYSICS
SOPHIA UNIVERSITY
7-1 KIOI-CHO CHIYODA KU
TOKYO 102
JAPAN
TEL 03-238-3431
TLF
TLX
EML
COM 35

IVANCHUK VICTOR I DR
ASTRONOMICAL OBSERVATORY
KIEV STATE UNIVERSITY
OBSERVATORNAYA UL 3
252053 KIEV
UKRAINE
TEL
TLF
TLX
EML
COM 10

IVANOV GEORGI R DR
DPT OF ASTRONOMY
UNIVERSITY OF SOFIA
ANTON IVANOV ST 5
BG 1126 SOFIA
BULGARIA
TEL 2 54 4852
TLF
TLX
EML
COM

IVANOV VSEVOLOD V DR PROF
ASTRONOMICAL OBSERVATORY
ST PETERSBURG UNIVERSITY
BIBLIOTECHNAJA PL 2
198904 ST PETERSBURG
RUSSIA
TEL 257-94-91
TLF
TLX
EML
COM 36

IVANOVA VIOLETA DR
ASTRONOMICAL OBSERVATORY
BULGARIAN ACAD SCIENCES
72 LENIN BLVD
BG 1784 SOFIA
BULGARIA
TEL 2 75 8927
TLF
TLX 23561 ECF BAN BG
EML
COM 07,15,20

IVANOV-KHOLODNY G S DR
IZMIRAN
ACADEMY OF SCIENCES
142092 MOSCOW
RUSSIA
TEL
TLF
TLX
EML
COM 21

IVES JOHN CHRISTOPHER MR
ESA/ESTEC
SSD
BOX 299
NL 2200 AG NOORDWIJK
NETHERLANDS
TEL 17 19 83629
TLF
TLX 39098
EML
COM

IWANISZEWSKA CECYLIA DR
INSTITUTE OF ASTRONOMY
N COPERNICUS UNIVERSITY
UL CHOPINA 12/18
PL 87 100 TORUN
POLAND
TEL 2-60-18
,TLF
TLX 0552234 ASTR PL
EML
COM 33,46C

IWANOWSKA WILHELMINA PROF
INSTITUTE OF ASTRONOMY
N COPERNICUS UNIVERSITY
UL CHOPINA 12/18
PL 87 100 TORUN
POLAND
TEL 260-18
TLF
TLX 86412 PL
EML
COM 33

IWASAKI KYOSUKE DR
KYOTO GAKUEN UNIVERSITY
NANJO
SOGABECHO KAMEOKA
KYOTO 621
JAPAN
TEL 7712 2 2001
TLF 7712 4 8150
TLX
EML
COM 16

IYE MASANORI DR
TOKYO ASTRONOMICAL OBS
NAOJ
OSAWA MITAKA
TOKYO 181
JAPAN
TEL 0422-32-511x313
TLF
TLX 2822307 TAOMTK J
EML
COM 33

IYENGAR K V K PROF
INDIAN INSTITUTE OF
ASTROPHYSICS
KORAMANGALA
BANGALORE 560 034
INDIA
TEL 812 53 0672/0676
TLF
TLX 845 2763 IIAB IN
EML
COM 25,34,44

IYER B R DR
RAMAN RESEARCH INSTITUTE
SADASHIVANAGAR
BANGALORE 560 080
INDIA
TEL 812 36 0122
TLF 812 34 0492
TLX 8452671 RRI IN
EML
COM 47

IZOTOV YURI DR
MAIN ASTRONOMICAL OBS
UKRAINIAN ACAD OF SCIENCE
GOLOSEEVO
252127 KIEV
UKRAINE
TEL 66 3110
TLF
TLX 131406 SKY SU
EML
COM 28

IZVEKOV V A DR
INST OF THEORET ASTRONOMY
ACADEMY OF SCIENCES
N KUTUZOVA 10
191187 ST PETERSBURG
RUSSIA
TEL 272 40 23
TLF
TLX 121578 ITA SU
EML
COM 07,20

JAAKKOLA TOIVO S
OBS & ASTROPHYSICS LAB
UNIVERSITY OF HELSINKI
TAEHTITORNINMAKI
SF 00130 HELSINKI 13
FINLAND
TEL 191 2907
TLF
TLX 124690 UNIH SF
EML
COM

JABBAR SABEH RHAMAN
SARC
SCIENTIFIC RES COUNCIL
BOX 2441
JADIRIYAH BAGHDAD
IRAQ
TEL 1 776 5127
TLF
TLX 213976 SRC IK
EML
COM 12,42

JABIR NIAMA LAFTA
SARC
SCIENTIFIC RES COUNCIL
BOX 2441
JADIRIYAH BAGHDAD
IRAQ
TEL 1 776 5127
TLF
TLX 213976 SRC IK
EML
COM 34

JABLONSKI FRANCISCO DR
INPE
CP 515
12201 S JOSE DOS CAMPOS
BRAZIL
TEL 123 41 8977
TLF 123 21 8743
TLX
EML
COM 27

JACCHIA LUIGI G DR
CENTER FOR ASTROPHYSICS
HCO/SAO
60 GARDEN ST
CAMBRIDGE MA 02138
USA
TEL 617 495 7213
TLF
TLX
EML
COM 22

JACKISCH GERHARD DR
ZNTRLINST F ASTROPHYSIK
STERNWARTE BABELSBERG
D 6400 SONNEBERG
GERMANY
TEL
TLF
TLX
EML
COM 41

JACKSON BERNARD V DR
CASS
UCSD
C 011
LA JOLLA CA 92093 0216
USA
TEL 619 534 3358
TLF
TLX
EML SOLAR BJACKSON/CASS01 BVJCME
COM 12,21

JACKSON JOHN CHARLES DR
16 THE PARK
NEWARK NG24 1SO
UK
TEL
TLF
TLX
EML
COM 48

JACKSON PAUL DR
INSTITUT FUER ASTRONOMIE
DER UNIVERSITAET WIEN
TUERKENSCHANZSTR 17
A 1180 WIEN
AUSTRIA
TEL
TLF
TLX
EML
COM 08

JACKSON PETER DOUGLAS DR
JG-CHESAPEAKE ASSOCIATES
BOX 1667
ANNAPOLIS MD 21404
USA
TEL 301 757 3613
TLF
TLX
EML
COM 33

JACKSON WILLIAM M DR
DPT OF CHEMISTRY
UNIVERSITY OF CALIFORNIA
ROOM 214
DAVIS CA 95616
USA
TEL 916 752 0503
TLF·
TLX
EML
COM 15

JACOBS KENNETH C DR
DPT OF PHYSICS
HOLLINS COLLEGE
BOX 9661
ROANOKE VA 24020
USA
TEL 703 362 6478
TLF
TLX
EML
COM

JACOBSEN THEODOR S PROF
6205 17TH PARK RD
SEATTLE WA 98115
USA
TEL 206 523 5245
TLF
EML
COM
TLX

JACOBY GEORGE H
KITT PEAK NTL OBS
BOX 26732
950 N CHERRY AVE
TUCSON AZ 85726 6732
USA
TEL· 602 325 9292
TLF
TLX
EML
COM 34

JACQ THIERRY
OBSERVATOIRE DE BORDEAUX
BP 89
F 33270 FLOIRAC
FRANCE
TEL 56 86 4330
TLF 56 40 4251
TLX
EML JACQ@FROBOR51
COM 34,40

JACQUINOT PIERRE DR
LABORATOIRE AIME COTTON
UNIVERSITE PARIS XI
BT 505
F 91405 ORSAY CDX
FRANCE
TEL
TLF
TLX
EML
COM 14

JAEGER FRIEDRICH W PROF
TELEGRAFENBERG A 33
D 1500 POTSDAM
GERMANY
TEL 4551
TLF
EML
COM
TLX

JAFFE DANIEL T
SPACE SCIENCES LABORATORY
UNIVERSITY OF CALIFORNIA
BERKELEY CA 94720
USA
TEL 415 642 1930
TLF
TLX
EML DTJ@ASTRO AS UTEXAS EDU
COM 34

JAFFE WALTER JOSEPH DR
STSCI
HOMEWOOD CAMPUS
3700 SAN MARTIN DR
BALTIMORE MD 21218
USA
TEL 301 338 4762
TLF
TLX 684 9101 STSCI
EML
COM 28,40,48

JAHREISS HARTMUT DR
ASTRONOMISCHES RECHEN-
INSTITUT
MOENCHHOFSTR 12-14
D 6900 HEIDELBERG 1
GERMANY
TEL 62 214 9026
TLF
TLX 461 336 ARIHD D
EML
COM 24C,33

JAIN RAJMAL DR
UDAIPUR SOLAR OBSERVATORY
11 VIDYA MARG
UDAIPUR 313 001
INDIA
TEL 25 626/27 457
TLF
TLX
EML
COM 10

JAIN SURENDRA DR
INDIAN INSTITUTE OF
ASTROPHYSICS
KORAMANGALA
BANGALORE 560 034
INDIA
TEL 812 56 9702
TLF
TLX 8452763 IIAB IN
EML
COM

JAKIMIEC JERZY PROF
ASTRONOMICAL INSTITUTE
WROCLAW UNIVERSITY
UL KOPERNIKA 11
PL 51 622 WROCLAW
POLAND
TEL 482434
TLF
TLX 0712791 UWRPL
EML
COM 10C

JAKOBSEN PETER
ESA/ESTEC
SSD
BOX 299
NL 2200 AG NOORDWIJK
NETHERLANDS
TEL 17 19 833614
TLF
TLX 39098
EML
COM

JAKS WALDEMAR DR
ASTRONOMICAL LATITUDE OBS
BOROWIEC
·BOX 62 035
PL 62 035 KORNIK
POLAND
TEL POZNAN 170187
TLF
TLX 0412623 AOS PL
EML
COM 19

JAMAR CLAUDE A J DR
IAL SPACE
UNIVERSITE DE LIEGE
AVE DU PRE AILY
B 4900 ANGLEUR-LIEGE
BELGIUM
TEL 41 67 6760
TLF
TLX 41320 IAL SP
EML
COM 44

JAMES JOHN F MR
SCHUSTER LABORATORY
UNIVERSITY OF MANCHESTER
MANCHESTER M13 9PL
UK
TEL 612 737 121
TLF
TLX
EML
COM 21

JAMES RICHARD A DR
DPT OF ASTRONOMY
UNIVERSITY OF MANCHESTER
MANCHESTER M13 9PL
UK
TEL
TLF
TLX
EML
COM 35

JAMESON RICHARD F DR
DPT OF ASTRONOMY
UNIVERSITY OF LEICESTER
UNIVERSITY RD
LEICESTER LE1 7RH
UK
TEL 533 554 455
TLF
TLX 341198
EML
COM

JANES KENNETH A DR
DPT OF ASTRONOMY
BOSTON UNIVERSITY
725 COMMONWEALTH AVE
BOSTON MA 02215
USA
TEL 617 353 2627
TLF
TLX 95-1289 BOS UNIV BSN
EML
COM 37C

JANICZEK PAUL M DR
US NAVAL OBSERVATORY
34 & MASSACHUSETTS AVE NW
WASHINGTON DC 20392 5100
USA
TEL 202 653 1569
TLF
TLX 710-822-1970
EML
COM 04,07

JANKOVICS ISTVAN DR
KONKOLY OBSERVATORY
THEGE U 13/17
BOX 67
H 1525 BUDAPEST
HUNGARY
TEL 1 75 5866/75 4122
TLF
TLX 227460
EML
COM

JANOT-PACHECO EDUARDO DR
IAG
UNIVERSIDADE DE SAO PAULO
CP 30627
01051 SAO PAULO SP
BRAZIL
TEL 11 577 8599
TLF
TLX 36221 IAGM BR
EML
COM

JANSSEN MICHAEL ALLEN
JPL
MS 183 301
4800 OAK GROVE DR
PASADENA CA 91109
USA
TEL 213 354 7247
TLF
TLX
EML·
COM· 40

JARDINE MOIRA MARY DR
ASTRONOMY CENTRE
UNIVERSITY OF SUSSEX
FALMER
BRIGHTON BN1 9QH
UK
TEL 273 60 6755*3085
TLF 273 67 8335
TLX 877159 BHV TXS G
EML MMJ@UK AC SUSSEX STARLINK
COM 10

JAROSZYNSKI MICHAL
ASTRONOMICAL OBSERVATORY
WARSAW UNIVERSITY
AL UJAZDOWSKIE 4
PL 00 478 WARSAW
POLAND
TEL 29 40 11
TLF
TLX 813978 ZAPAN PL
EML
COM 47

JARRETT ALAN H PROF
BOYDEN OBSERVATORY
BOX 334
BLOEMFONTEIN 9300
SOUTH AFRICA
TEL 051-37605
TLF
TLX 267666 SA
EML
COM 21,46

JARZEBOWSKI TADEUSZ DR
ASTRONOMICAL INSTITUTE
WROCLAW UNIVERSITY
UL KOPERNIKA 11
PL 51 622 WROCLAW
POLAND
TEL
TLF
TLX
EML
COM 27

JASCHEK CARLOS O G R PROF
OBS DE STRASBOURG
11 RUE UNIVERSITE
F 67000 STRASBOURG
FRANCE
TEL 88 35 8200
TLF· 88 25 0160
TLX 890506 STAROBS
EML
COM 05,26,33,42,45

JASCHEK MERCEDES DR
OBS DE STRASBOURG
11 RUE UNIVERSITE
F 67000 STRASBOURG
FRANCE
TEL 88 35 8200
TLF 88 25 0160
TLX 890506 STAROBS F
EML
COM 45

JASNIEWICZ GERARD DR
OBSERVATOIRE ASTRONOMIQUE
11 RUE UNIVERSITE
F 67000 STRASBOURG
FRANCE
TEL 88 35 8200
TLF 88 25 0160
TLX 890 506 STAROBS
EML EARN U01109@FRCCSC21
COM 33,42

JASTROW ROBERT
INST FOR SPACE STUDIES
2880 BROADWAY
NEW YORK NY 10025
USA
TEL
TLX
EML
TLF
COM 51

JAUNCEY DAVID L DR
CSIRO
DIVISION OF RADIOPHYSICS
BOX 76
EPPING NSW 2121
AUSTRALIA
TEL
TLF
TLX 26230 ASTRO
EML DJAUNCEY@ATNF CSIRO AU
COM 40,47C

JAYANTHI UDAYA B DR
INPE
CP 515
12201 S JOSE DOS CAMPOS
BRAZIL
TEL 123 41 8977*392
TLF 123 21 8743
TLX 123 3530 INPE BR
EML
COM

JAYARAJAN A P MR
INDIAN INSTITUTE OF
ASTROPHYSICS
KORAMANGALA
BANGALORE 560 034
INDIA
TEL 812 56 6585
TLF
TLX 845763 IIAB IN
EML
COM 09

JEFFERIES JOHN T DR
NOAO
BOX 26732
950 N CHERRY AVE
TUCSON AZ 85726 6732
USA
TEL 602 881 1960
TLF
TLX 0666484 AURANOAOTUC
EML
COM 12,36

JEFFERIES STUART DR
NTL SOLAR OBSERVATORY
BARTOL RESEARCH INST
950 N CHERRY AVE
TUCSON AZ 85726 6732
USA
TEL 602 323 4182
TLF
TLX 0666 484 AURANOAOTUC
EML STUARTJ@NOAO EDU/5355 STUARTJ
COM 12

JEFFERS STANLEY DR
CRESS PHYSICS DPT
YORK UNIVERSITY
4700 KEELE ST
DOWNSVIEW ON M3J 1P3
CANADA
TEL 416 667 3851
TLF
TLX
EML
COM 09,51

JEFFERY CHRISTOPHER S DR
UNIVERSITY OBSERVATORY
UNIVERSITY OF ST ANDREWS
BUCHANAN GARDEN
ST ANDREWS FIFE KY16 9LZ
UK
TEL
TLF
TLX
EML
COM 27

JEFFERYS WILLIAM H DR
ASTRONOMY DPT
UNIVERSITY OF TEXAS
RLM 15 308
AUSTIN TX 78712 1083
USA
TEL 512 471 4461
TLF
TLX
EML:
COM 07,24

JENKINS CHARLES R
ROYAL GREENWICH OBS
HERSTMONCEUX CASTLE
HAILSHAM BN27 1RP
UK
TEL 323 833 171
TLF
TLX 87451
EML
COM 40

JENKINS EDWARD B DR
PRINCETON UNIVERSITY OBS
PEYTON HALL
PRINCETON NJ 08544
USA
TEL 609 452 3826
TLF 609 258 1020
TLX 322409 ASTRO PRIN
EML ebj@astro princeton edu
COM 34,44C

JENKINS L F MS
YALE UNIVERSITY OBS
YALE STATION
BOX 2023
NEW HAVEN CT 06520
USA
TEL
TLF
TLX
EML
COM 34

JENKNER HELMUT DR
STSCI
HOMEWOOD CAMPUS
3700 SAN MARTIN DR
BALTIMORE MD 21218
USA
TEL 301 338 4842
TLF
TLX 6849101 STSCI
EML
COM 05,09

JENNER DAVID C DR
DPT OF ASTRONOMY
UNIVERSITY OF WASHINGTON
FM 20
SEATTLE WA 98195
USA
TEL 206 543 6182
TLF
TLX
EML
COM

JENNINGS R E PROF
DPT PHYSICS & ASTRONOMY
UNIVERSITY COLLEGE LONDON
GOWER ST
LONDON WC1E 6BT
UK
TEL 1 713 877 050
TLF
TLX 28722
EML
COM 34

JENNISON ROGER C PROF
ELECTRONICS LABORATORY
UNIVERSITY OF KENT
CANTERBURY CT2 7NT
UK
TEL
TLF
TLX 965449
EML
COM 22,40,51

JENSCH A
PESTALOZZISTR 9
D 6900 JENA
GERMANY
TEL
TLF.
EML
COM
TLX

JENSEN EBERHART PROF
INST THEORET ASTROPHYSICS
UNIVERSITY OF OSLO
BOX 1029
N 0315 BLINDERN OSLO 3
NORWAY
TEL 02-456502
TLF
TLX 72425N UNIOS
EML
COM 10

JERZYKIEWICZ MIKOLAJ DR
ASTRONOMICAL INSTITUTE
WROCLAW UNIVERSITY
UL KOPERNIKA 11
PL 51 622 WROCLAW
POLAND
TEL 48-24-34
TLF
TLX 0712791 UWR PL
EML
COM 25,27VP

JEWELL PHILIP R DR
NRAO
CAMPUS BLDG 65
949 N CHERRY AVE
TUCSON AZ 85721 0655
USA
TEL 602 882 8250
TLF
TLX 9102409524 NRAO TUC
EML BITNET PJEWELL@NRAO
COM 27,40

JI HONG-QING
INTL LATITUTDE STATION
TIANJIN
CHINA PR
TEL
TLF
EML
COM
TLX

JI SHUCHEN DR
BOX 110 KUNMING
YUNNAN PROVINCE
YUNNAN
CHINA PR
TEL 72 946
TLF
TLX 64040 YUOBS CN
EML
COM 40

JIANG CHONG-GUO
YUNNAN OBSERVATORY
CAS
BOX 110
KUNMING 72946 YUNNAN
CHINA PR
TEL 871 2035
TLF
TLX
EML
COM 08

JIANG DONG-RONG
SHANGHAI OBSERVATORY
CAS
80 NANDAN RD
SHANGHAI
CHINA PR
TEL 21 38 6191
TLF
TLX 33164 SHAO CN
EML
COM 33

JIANG SHI-YANG
BEIJING ASTRONOMICAL OBS
CAS
W SUBURB
BEIJING 100080
CHINA PR
TEL 1 28 1698
TLF
TLX 22040 BADAS CN
EML
COM 09,29,27,50

JIANG SHUDING
GRADUATE SCHOOL
UNIV SCIENCE & TECHNOLOGY
BOX 3908
BEIJING 100039
CHINA PR
TEL 1 81 7031*253
TLF
TLX
EML
COM 47

JIANG XIAO-YUAN DR
SHANGHAI OBSERVATORY
CAS
80 NANDAN RD
SHANGHAI 200030
CHINA PR
TEL 21 38 6191*49
TLF
TLX 33164 SHAO CN
EML
COM 41

JIANG YAO-TIAO
DPT OF ASTRONOMY
NANJING UNIVERSITY
NANJING
CHINA PR
TEL 25 34151
TLF
TLX
EML
COM 10

JIANG ZHAOJI
BEIJING ASTRONOMICAL OBS
CAS
W SUBURB
BEIJING 100080
CHINA PR
TEL
TLF
TLX 22040 BAOAS CN
EML
COM

JIN BIAOREN DR
DPT OF GEOPHYSICS
WUHAN TECHNICAL UNIVER
39 LO-YU RD
WUHAN 430070
CHINA PR
TEL 87 5571
TLF
TLX 40210 WTUSM CN
EML
COM

JIN SHEN-ZENG
BEIJING ASTRONOMICAL OBS
CAS
W SUBURB
BEIJING 100080
CHINA PR
TEL 1 28 1698
TLF
TLX 22040 BAOBS CN
EML
COM 40

JIN WEN-JING
SHANGHAI OBSERVATORY
CAS
80 NANDAN RD
SHANGHAI
CHINA PR
TEL. 21 38 6191
TLF
TLX 33164 SHAO CN
EML
COM 19C,31

JOCKERS KLAUS DR
MPI FUER AERONOMIE
POSTFACH 20
D 3411 KATLENBURG LINDAU
GERMANY
TEL 555 6411
TLX 965527
EML
TLF
COM 10,15

JOERSAETER STEVEN DR
STOCKHOLM OBSERVATORY
S 133 36 SALTSJOEBADEN
SWEDEN
TEL 81 64 463
TLF 87 17 4719
TLX
EML STEVEN@ASTRO SU SE
COM

JOG CHANDA J DR
DPT OF PHYSICS
INDIAN INSTITUTE SCIENCES
BANGALORE 560 012
INDIA
TEL 812 34 4411
TLF
TLX 845 8349 IISC IN
EML·
COM 28,33

JOHANSEN KAREN T LEKTOR
COPENHAGEN UNIVERSITY OBS
BRORFELDEVEJ 23
DK 4340 TOLLOSE
DENMARK
TEL 53 48 8195
TLF 58 48 8758
TLX
EML
COM

JOHANSSON LARS ERIK B DR
LADAEMNESGATAN 7
S 416 79 GOETEBORG
SWEDEN
TEL
TLF
EML
COM 40
TLX

JOHANSSON LENNART DR
ASTRONOMICAL OBSERVATORY
BOX 515
S 751 20 UPPSALA
SWEDEN
TEL 18 51 1274
TLF 18 52 7583
TLX
EML LABAN LJ
COM

JOHANSSON SVENERIC DR
DPT OF PHYSICS
UNIVERSITY OF LUND
SOELVEGATAN 14
S 223 62 LUND
SWEDEN
TEL 46 12 6097
TLF 46 10 4709
TLX 33533 LUNIVER S
EML ATOMSEJ@SELDC52
COM 14

JOHNSON DONALD R DR
NTL BUREAU OF STANDARDS
BLDG 221 RM A363
GAITHERSBURG MD 20899
USA
TEL 301 921 2828
TLF
TLX
EML
COM 14,40

JOHNSON FRED M PROF DR
PHYSICS & ASTRONOMY DPT
CALIFORNIA STATE UNIV
FULLERTON CA 92634
USA
TEL 714 773 3366
TLF
TLX
EML
COM 14,34

JOHNSON HOLLIS R PROF
ASTRONOMY DPT
INDIANA UNIVERSITY
SWAIN WEST 319
BLOOMINGTON IN 47405
USA
TEL 812 335 4172
TLF
TLX 272279
EML
COM 29,36

JOHNSON HUGH M DR
1017 NEWELL RD

PALO ALTO CA 94303
USA
TEL 415 326 7223
TLF
TLX
EML
COM 33,34

JOHNSON TORRENCE V DR
JPL
MS 183 301
4800 OAK GROVE DR
PASADENA CA 91109
USA
TEL 818 354 2761
TLF
TLX 67-5429
EML
COM 15,16

JOHNSTON KENNETH J
NAVAL RESEARCH LABORATORY
CODE 7134
4555 OVERLOOK AVE SW
WASHINGTON DC 20375 5000
USA
TEL 202 767 2351
TLF
TLX
EML
COM 04,08,24,34,40

JOKIPII J R PROF
LUNAR & PLANETARY LAB
UNIVERSITY OF ARIZONA
TUCSON AZ 85721
USA
TEL 602 621 4256
TLF
TLX
EML
COM 48,49

JOLY FRANCOIS DR
UNIVERSITE DE BORDEAUX

123 RUE LAMARTINE
F 33405 TALENCE
FRANCE
TEL 56 86 4330
TLF 56 40 4251
TLX
EML
COM 14,40

JOLY MONIQUE
OBSERVATOIRE DE PARIS
SECTION DE MEUDON
F 92195 MEUDON PPL CDX
FRANCE
TEL 1 45 34 7570
TLF
TLX 201571
EML
COM

JONAS JUSTIN LEONARD
DPT PHYSICS/ELECTRONICS
RHODES UNIVERSITY
BOX 94
GRAHAMSTOWN 6140
SOUTH AFRICA
TEL (461)22023
TLF
TLX 244226 RUANT SA
EML
COM 33,40

JONCAS GILLES DR
DPT DE PHYSIQUE
UNIVERSITE DE LAVAL
STE FOY QC G1K 7P4
CANADA
TEL 418 656 2652
TLF 418 656 2040
TLX 051 31621
EML BITNET 1150041@LAVALVX1
COM

JONES ALBERT F MR
31 RANUI RD
STOKE
NELSON
NEW ZEALAND
TEL 054-73-905
TLF
TLX
EML
COM 27

JONES BARBARA
CASS
UCSD
C 011
LA JOLLA CA 92093 0216
USA
TEL 714 452 4474
TLF
TLX
EML
COM 09

JONES BARRIE W DR
DPT OF PHYSICS
THE OPEN UNIVERSITY
WALTON HALL
MILTON KEYNES MK7 6AA
UK
TEL 908 653 378
TLF 908 653 744
TLX· 825061
EML BW_JONES@UK AC OPEN ACS VAX
COM 46

JONES BERNARD J T DR
NORDITA
BLEGDAMSVEJ 17
DK 2100 COPENHAGEN 0
DENMARK
TEL 31 42 1616
TLF 31 38 9157
TLX 15216 NBI DK
EML
COM 47

JONES BURTON DR
LICK OBSERVATORY
UNIVERSITY OF CALIFORNIA
SANTA CRUZ CA 95064
USA
TEL 408 429 2384
TLF
TLX
EML
COM 24

JONES DAYTON L
JPL
MS 238 700
4800 OAK GROVE DR
PASADENA CA 91109
USA
TEL 818 354 7774
TLF
TLX 675429
EML
COM 40

JONES DEREK H P DR
ROYAL GREENWICH OBS
MADINGLEY RD
CAMBRIDGE CB3 0EZ
UK
TEL 223 37 4000
TLF 223 37 4700
TLX 265451
EML
COM 24,33,37

JONES ERIC M
LOS ALAMOS NATIONAL LAB
MS F665
BOX 1663
LOS ALAMOS NM 87545
USA
TEL 505 667 6386
TLF
TLX
EML
COM 51

JONES FRANK CULVER DR
NASA/GSFC
CODE 665
GREENBELT MD 20771
USA
TEL 301 286 5506
TLF
TLX 710-828-9716
EML
COM 34,48

JONES HARRISON PRICE DR
KITT PEAK NTL OBS
SOLAR STATION
900 N CHERRY AVE
TUCSON AZ 85726 6732
USA
TEL 602 325 9354
TLF
TLX
EML
COM 10,12

JONES JAMES DR
DPT OF PHYSICS
UNIV OF WESTERN ONTARIO
LONDON ON N6A 5B9
CANADA
TEL 519 661 3183
TLF 519 661 3486
TLX 064 7134
EML
COM 22C

JONES JANET E DR
NORDITA
BLEGDAMSVEJ 17
DK 2100 COPENHAGEN 0
DENMARK
TEL 31 42 1616
TLF 31 38 9157
TLX 15216 NBI DK
EML
COM

JONES THOMAS WALTER DR
DPT OF ASTRONOMY
UNIVERSITY OF MINNESOTA
116 CHURCH ST SE
MINNEAPOLIS MN 55455
USA
TEL 612 373 3307
TLF
TLX
EML
COM· 28,48

JORDAN CAROLE DR
DPT THEORETICAL PHYSICS
UNIVERSITY OF OXFORD
1 KEBLE RD
OXFORD OX1 3NP
UK
TEL 865 532 81
TLF
TLX 83295 NUCLOX
EML
COM 12,14,29,44

JORDAN H L DR DIREKTOR
INSTITUT F PLASMAPHYSIK
KERNFORSCHUNGSANLAGE
JUELICH GMBH PF 365
D 5170 JUELICH 1
GERMANY
TEL
TLF
TLX
EML
COM 14

JORDAN STUART D DR
NASA/GSFC
CODE 682
LASP
GREENBELT MD 20771
USA
TEL 301 286 8811
TLF
TLX 89675
EML
COM 10,12,44C

JORDEN PAUL RICHARD
ROYAL GREENWICH OBS
MADINGLEY RD
CAMBRIDGE CB3 0EZ
UK
TEL 223 37 4000
TLF 223 37 4700
TLX 265451/265871MONREFG
EML AC RO-GREENWICH STARLINK
COM

JORDI NEBOT CARME DR
DPT ASTRONOMIA
UNIVERSIDAD DE BARCELONA
AVD DIAGONAL 647
E 08028 BARCELONA
SPAIN
TEL 3 330 7311
TLF
TLX
EML D3FACJNO@EBOUBO11
COM 25

JORGENSEN HENNING E PROF
UNIVERSITY OBSERVATORY
OESTER VOLDGADE 3
DK 1350 COPENHAGEN K
DENMARK
TEL 31 14 1790
TLF 31 38 9157
TLX 44155
EML HENNING@ASTRO KU DK
COM 38VP

JORGENSEN UFFE GRAE DR
NIELS BOHR INSTITUTE
BLEGDAMSVEJ 17
DK 2100 COPENHAGEN 0
DENMARK
TEL 31 42 1616
TLF 31 38 9157
TLX 15216
EML
COM

JOSELYN JO ANN C DR
NOAA ERL R/E/SE2
SPACE ENVIRONMENT LAB
325 BROADWAY
BOULDER CO 80303
USA
TEL 303 497 5147
TLF
TLX 888776 NOAA BLDR
EML SPAN SELVAX JJOSELYN
COM 10,49

JOSEPH CHARLES LYNN DR
DPT ASTROPHYSCIAL SCI
PRINCETON UNIVERSITY
PRINCETON NJ 08544 1001
USA
TEL   609 258 3808
TLF   609 258 1020
TLX
EML   CLJ@ASTRO PRINCETON EDU
COM   09

JOSEPH J H DR
DPT GEOPHYS & PLANET SCI
TEL AVIV UNIVERSITY
RAMAT AVIV
TEL AVIV
ISRAEL
TEL   3 420 633
TLF
TLX   342171 VERSY IL
EML
COM

JOSEPH ROBERT D DR
ASTROPHYSICS GROUP
IMPERIAL COLLEGE
BLACKETT LABORATORY
LONDON SW7 2BZ
UK
TEL   1 589 5111*6660
TLF
TLX   261503
EML
COM

JOSHI G C DR
UTTAR PRADESH STATE
OBSERVATORY
PO MANORA PEAK 263 129
NAINITAL 263 129
INDIA
TEL   59 421 2136
TLF
TLX
EML
COM   12

JOSHI MOHAN N PROF
TIFR/RADIO ASTRONOMY CTR
BOX 8
UDHAGAMANDALAM 643 001
INDIA
TEL   423 2032
TLF
TLX   8458488 TIFR IN
EML
COM   28,40,47

JOSHI SURESH CHANDRA DR
UTTAR PRADESH STATE
OBSERVATORY
PO MANORA PEAK 263 129
NAINITAL 263 129
INDIA
TEL   59 422136
TLF
TLX
EML
COM   25,29

JOSHI U C DR
PHYSICAL RESEARCH LAB
NAVRANGPURA
AHMEDABAD 380 009
INDIA
TEL   272 46 2129
TLF   272 44 5292
TLX   121397
EML
COM   25,28,37

JOSS PAUL CHRISTOPHER DR
DPT OF PHYSICS
MIT RM 6 203
BOX 165
CAMBRIDGE MA 02139
USA
TEL   617 243 4845
TLF
TLX
EML
COM   42,48

JOUBERT MARTINE
LAS
TRAVERSE DU SIPHON
LES TROIS LUCS
F 13012 MARSEILLE
FRANCE
TEL   91 05 5900
TLF   91 66 1855
TLX   420584
EML
COM   21

JOURDAIN DE MUIZON M DR
STERREWACHT
BOX 9513
NL 2300 RA LEIDEN
NETHERLANDS
TEL   71 27 5840
TLX   39058 ASTRO NL
EML   MUIZON@HLERUL51
TLF   71 27 5819
COM   34

JOURNET ALAIN
OCA CERGA
AVE COPERNIC
F 06130 GRASSE
FRANCE
TEL   93 36 5849
TLF
TLX   470865
EML
COM   07,08

JOVANOVIC BOZIDAR
FAC OF AGRICULTURE
INST WATERRANGING
VELJKA VLAHOVICA 2
YU 21000 NOVI SAD
YUGOSLAVIA
TEL   215 366
TLF
TLX
EML
COM   07,10

JOY MARSHALL J DR
NASA/MSFC
CODE ES 65
SPACE SCIENCE LAB
HUNTSVILLE AL 35812
USA
TEL   205 544 3423
TLF   205 544 7754
TLX
EML   SSL JOY/JOY@SSL MSFC NASA GOV
COM   28

JUDGE PHILIP DR
HIGH ALTITUDE OBSERVATORY
NCAR
BOX 3000
BOULDER CO 80307 3000
USA
TEL
TLF
TLX
EML
COM   36

JUGAKU JUN DR
TOKYO ASTRONOMICAL OBS
NAOJ
OSAWA MITAKA
TOKYO 181
JAPAN
TEL
TLF
TLX   2822307 TAOMTKJ
EML
COM   28,29,51

JULIUSSON EINAR DR
UNIVERSITY IF ICELAND
V/SUOURGOTU
IS 101 REYKJAVIK
ICELAND
TEL   1 79979
TLF
TLX   0501 2307 ISINFO
EML
COM   48

JUNG JEAN DR
THOMSON
173 BD HAUSSMANN
F 75379 PARIS CDX 08
FRANCE
TEL   1 45 61 9600
TLF
TLX   204780
EML
COM

JUNKKARINEN VESA T DR
CASS
UCSD
C 011
LA JOLLA CA 92093 0216
USA
TEL   619 534 0735
TLF
TLX
EML   SPAN 27783 VESA
COM   28,47

JUPP ALAN H DR
DPT APPL MATHS THEOR PHYS
UNIVERSITY OF LIVERPOOL
BOX 147
LIVERPOOL L69 3BX
UK
TEL   517 096 022
TLF
TLX   627095
EML
COM   07

JURA MICHAEL DR
DPT OF ASTRONOMY
UNIVERSITY OF CALIFORNIA
MATH SCIENCES BLDG
LOS ANGELES CA 90024
USA
TEL   213 825 4302
TLF
TLX
EML
COM   34

JURGENS RAYMOND F
JPL
MS 238 420
4800 OAK GROVE DR
PASADENA CA 91109
USA
TEL 818 354 4974
TLF
TLX 675429
EML
COM 16

JURKEVICH IGOR DR
3130 PORT WAY
ANNAPOLIS MD 21403
USA
TEL 202 767 2003
TLF
EML.
COM 42
TLX

JUST ANDREAS DR
INST FUER THEORETISCHE
PHYSIK DER UNIVERSITAET
ROBERT-MAYER-STR 10
D 6000 FRANKFURT MAIN 1
GERMANY
TEL 69 798 2636
TLF 69 798 8350
TLX
EML
COM 34

JUSZKIEWICZ ROMAN
COPERNICUS ASTRON CENTER
POLISH ACAD OF SCIENCES
UL BARTYCKA 18
PL 00 716 WARSAW
POLAND
TEL
TLF
TLX
EML
COM 47

KAASTRA JELLE S DR
SRL
HUYGENS LAB
BOX 9504
NL 2300 RA LEIDEN
NETHERLANDS
TEL 71 27 5818
TLF
TLX 39058 ASTRO NL
EML
COM

KABURAKI MASAKI PROF
4-24-9 KICHIJYOJI
MINAMI MUSASHINO
TOKYO 180
JAPAN
TEL
TLF
TLX
EML
COM 33

KADLA ZDENKA I DR
PULKOVO OBSERVATORY
ACADEMY OF SCIENCES
10 KUTUZOV QUAY
196140 ST PETERSBURG
RUSSIA
TEL
TLF
TLX
EML
COM 05,37

KADOURI TALIB HADI
SARC
SCIENTIFIC RES COUNCIL
BOX 2441
JADIRIYAH BAGHDAD
IRAQ
TEL 1 776 5127
TLF
TLX 213976 SRC IK
EML
COM 27,30,36,42

KAEHLER HELMUTH DR
HAMBURGER STERNWARTE
GOJENSBERGSWEG 112
D 2050 HAMBURG 80
GERMANY
TEL
TLX
EML
TLF
COM 35

KAFATOS MINAS DR
DPT OF PHYSICS
GEORGE MASON UNIVERSITY
FAIRFAX VA 22030
USA
TEL 703 323 2303
TLF
TLX
EML
COM 34,44,51

KAFKA PETER
MPI F PHYSIK & ASTROPHYS
INSTITUT FUR ASTROPHYSIK
KARL-SCHWARZSCHILD-STR 1
D 8046 GARCHING MUENCHEN
GERMANY
TEL 89 329 90
TLF
TLX 524629 ASTRO D
EML
COM 48,51

KAFTAN MAY A DR
NRAO

BOX 2
GREEN BANK WV 24944
USA
TEL
TLF
TLX 710-938-1530
EML
COM 34,40

KAHANE CLAUDINE DR
OBSERVATOIRE DE GRENOBLE
CERMO/ASTROPHYSIQUE
BP 53X
F 38041 S MARTIN HERES CD
FRANCE
TEL 76 51 4600
TLF 76 44 8821
TLX 980134 USMG
EML KAHANE@FRGAG51
COM

KAHLER STEPHEN W DR
A F GEOPHYSICS LABORATORY
SPACE PHYSICS DIV PHP
HANSCOM AFB
BEDFORD MA 01731
USA
TEL 617 861 3975
TLF
TLX
EML
COM· 10

KAHLMANN HANS CORNELIS DR
RADIO OBS WESTERBORK
SCHATTENBERG 1
NL 9433 TA ZWIGGELTE
NETHERLANDS
TEL 59 39 2421
TLF 59 39 2486
TLX
EML KAHLMANN_HANS@NFRA NL
COM 40,50

KAHN FRANZ D PROF
DPT OF ASTRONOMY
UNIVERSITY OF MANCHESTER
MANCHESTER M13 9PL
UK
TEL 612 737 121
TLF
TLX 668932 MCHRUL G
EML
COM 34,40,48

KAI KEIZO DR
TOKYO ASTRONOMICAL OBS
NAOJ
OSAWA MITAKA
TOKYO 181
JAPAN
TEL 0422-32-5111
TLF
TLX 3329005 TAONRO J
EML
COM 10,40

KAIFU NORIO DR
NOBEYAMA RADIO OBS
NAOJ
MINAMIMAKI MURA
NAGANO 384 13
JAPAN
TEL
TLF
TLX
EML
COM 34C,40

KAISER THOMAS R PROF
DPT OF PHYSICS
THE UNIVERSITY
SHEFFIELD S3 7RH
UK
TEL 742 785 55*4277
TLF
TLX 547216 UGSHEF G
EML
COM 22

KAITCHUCK RONALD H
DPT OF PHYSICS & ASTRON
BALL STATE UNIVERSITY
MUNCIE IN 47306
USA
TEL
TLF
TLX
EML
COM 42

KAKINUMA TAKAKIYO T PROF
INST FOR ATMOSPHERIC RES
NAGOYA UNIVERSITY
3-13 HONOHARA
TOYOKAWA AICHI 442
JAPAN
TEL 05338-6-3154
TLF
TLX
EML
COM 40,49

KAKUTA CHUICHI DR
ASTROGEODYNAMICS OBS
HOSHIGAOKA 2-12
MIZUSAWA SHI
IWATE 023
JAPAN
TEL 197 24 7111
TLF
TLX 837628 ILSMIZ J
EML
COM 19,31

KALAFI MANOUCHER
CTR FOR ASTRON RESEARCH
UNIVERISY OF TABRIZ
TABRIZ 51664
IRAN
TEL 41 32564
TLF
TLX
EML
COM 28

KALANDADZE N B DR
ABASTUMANI ASTROPHYSICAL
OBSERVATORY
GEORGIAN ACAD OF SCIENCES
383762 ABASTUMANI
GEORGIA
TEL 227
TLF
TLX 327409
EML
COM 33

KALBERLA PETER
RADIOASTRONOMISCHES INST
DER UNIVERSITAET BONN
AUF DEM HUEGEL 71
D 5300 BONN 1
GERMANY
TEL 228 73 3645
TLF
TLX 886440
EML
COM 05,40

KALER JAMES B PROF
UNIVERSITY OF ILLINOIS
103 ASTRONOMY BLDG
1002 W GREEN ST
URBANA IL 61801
USA
TEL 217 333 9382
TLF
TLX 910-2452434 AST
EML
COM 34

KALINKOV MARIN P DR
DPT OF ASTRONOMY
BULGARIAN ACAD SCIENCES
72 LENIN BLVD
BG 1784 SOFIA
BULGARIA
TEL 2 75 8927
TLF
TLX 22774 CLANP BG
EML
COM 28

KALKOFEN WOLFGANG DR
CENTER FOR ASTROPHYSICS
HCO/SAO
60 GARDEN ST
CAMBRIDGE MA 02138
USA
TEL 617 495 7285
TLF
TLX
EML
COM 12,36P

KALLOGLIAN ARSEN T DR
BYURAKAN ASTROPHYSICAL
OBSERVATORY
378433 BYURAKAN
ARMENIA
TEL 88 52 28 3453
TLF
TLX
EML
COM 28

KALMAN BELA DR
HELIOPHYSICAL OBSERVATORY
BOX 30
H 4010 DEBRECEN
HUNGARY
TEL 52 11 015
TLX 72517 DEOBS H
EML
TLF
COM 10,12

KALMYKOV A M DR
ASTRONOMICAL INSTITUTE
UZBEK ACADEMY OF SCIENCES
700000 TASHKENT
UZBEKISTAN
TEL
TLF
TLX
EML
COM

KALNAJS AGRIS J DR
MOUNT STROMLO & SIDING
SPRING OBSERVATORIES
PRIVATE BAG
WODEN PO ACT 2606
AUSTRALIA
TEL 62 88 1111*248
TLF
TLX 62270 AA
EML agris@mso anu oz
COM 33C

KALUZNY JANUSZ DR
ASTRONOMICAL OBSERVATORY
WARSAW UNIVERSITY
AL UJAZDOWSKIE 4
PL 00 478 WARSAW
POLAND
TEL 294011
TLF
TLX 817063
EML
COM 42

KAMEL OSMAN M PROF
DPT OF ASTRONOMY
FACULTY OF SCIENCES
CAIRO UNIVERSITY
GEZA
EGYPT
TEL
TLF
TLX
EML
COM

KAMEYA OSAMU DR
NOBEYAMA RADIO OBS
NAOJ
MINAMIMAKI MURA
NAGANO 384 13
JAPAN
TEL 267 98 2831
TLF
TLX 3329005 NAONRO J
EML
COM 19,40

KAMIJO FUMIO PROF DR
DPT OF ASTRONOMY
UNIVERSITY OF TOKYO
BUNKYO KU
TOKYO 113
JAPAN
TEL
TLF
TLX
EML
COM 34

KAMINISHI KEISUKE PROF
DPT OF PHYSICS
KUMAMOTO UNIVERSITY
2-39-1 KUROKAMI
KUMAMOTO 860
JAPAN
TEL 096-344-2111
TLF
TLX
EML
COM 35

KAMMEYER PETER C DR
US NAVAL OBSERVATORY
34 & MASSACHUSETTS AVE NW
WASHINGTON DC 20392 5100
USA
TEL 202 653 1563
TLX
EML
TLF
COM 07

KAMP LUCAS WILLEM DR
DPT OF ASTRONOMY
BOSTON UNIVERSITY
725 COMMONWEALTH AVE
BOSTON MA 02215
USA
TEL
TLF
TLX
EML
COM 36,37

KAMPER KARL W DR
DAVID DUNLAP OBSERVATORY
UNIVERSITY OF TORONTO
BOX 360
RICHMOND HILL ON L4C 4Y6
CANADA
TEL 416 884 9562
TLF 416 978 3921
TLX 069 86766
EML
COM

KANAEV IVAN I DR
PULKOVO OBSERVATORY
ACADEMY OF SCIENCES
10 KUTUZOV QUAY
196140 ST PETERSBURG
RUSSIA
TEL
TLF
TLX
EML
COM 24

KANBACH GOTTFRIED DR
MAX PLANCK INSTITUT FUR
EXTRATERR PHYSIK
D 8046 GARCHING MUENCHEN
GERMANY
TEL 89 329 9544
TLF
TLX 5215845 XTER D
EML BITNET GOK@DGAIPP1S
COM

KANDEL ROBERT S DR
LMD
ECOLE POLYTECHNIQUE
F 91128 PALAISEAU CDX
FRANCE
TEL 1 69 41 8200
TLF
TLX 691596
EML
COM 36

KANDEMIR GUELCIN
ISTANBUL TECHNICAL UNIV
FEN FAKULTESI, FIZIK B
MASLAK
34452 ISTANBUL
TURKEY
TEL 1 609 109
TLF
TLX
EML
COM

KANDPAL CHANDRA D
UTTAR PRADESH STATE
OBSERVATORY
PO MANORA PEAK 263 129
NAINITAL 263 129
INDIA
TEL 59 42 2136,2325
TLF
TLX
EML
COM 42

KANDRUP HENRY EMIL DR
DPT OF ASTRONOMY
UNIVERSITY OF FLORIDA
211 SSRB
GAINESVILLE FL 32611
USA
TEL 904 392 2681
TLF 904 392 9741
TLX
EML KANDRUP@ASTRO UFL EDU
COM 28,37,47

KANE SHARAD R DR
SPACE SCIENCES LABORATORY
UNIVERSITY OF CALIFORNIA
BERKELEY CA 94720
USA
TEL 415 642 1719
TLF
TLX 910-366-7945
EML
COM 10

KANEKO NOBORU DR
DPT OF PHYSICS
HOKKAIDO UNIVERSITY
KITA 10 NISHI 8
SAPPORO 060
JAPAN
TEL 11-716-2111
TLF
TLX 932510 HOKUSC J
EML
COM 28

KANG GON IK
PYONGYANG ASTRON OBS
ACADEMY OF SCIENCES DPRK
TAESONG DISTRICT
PYONGYANG
KOREA DPR
TEL
TLF
TLX
EML
COM 40

KANG JIN SOK
PYONGYANG ASTRON OBS
ACADEMY OF SCIENCES DPRK
TAESONG DISTRICT
PYONGYANG
KOREA DPR
TEL
TLF
TLX
EML
COM 10

KANG YOUNG WOON DR
DPT OF EARTH SCIENCE
KING SEJONG UNIVERSITY
89 KOONJA-DONG
SEOUL SUNGDONG 133 747
KOREA R
TEL 467 5121
TLF
TLX
EML
COM 42

KANYO SANDOR DR
KONKOLY OBSERVATORY
THEGE U 13/17
BOX 67
H 1525 BUDAPEST
HUNGARY
TEL 1 75 5866/75 4122
TLF
TLX
EML
COM 27

KAPAHI VIJAY K DR
TATA INST OF FUNDAMENTAL
RESEARCH
POONA UNIVERSITY CAMPUS
PUNE 411 007
INDIA
TEL 212 33 6105
TLF 212 33 5760
TLX 0145658 GMRT IN
EML uunet!shakti!gmrt!vijay
COM 28,40

KAPISINSKY IGOR
ASTRONOMICAL INSTITUTE
SLOVAK ACADEMY SCIENCES
DUBRAVSKA 9
CS 842 28 BRATISLAVA
CZECHOSLOVAKIA
TEL 7 37 5157
TLF
TLX. 093355
EML
COM 22

KAPLAN GEORGE H DR
US NAVAL OBSERVATORY
34 & MASSACHUSETTS AVE NW
WASHINGTON DC 20392 5100
USA
TEL 202 653 0722
TLF 202 653 0179
TLX
EML GHK@PHOBOS USNO NAVY MIL
COM 04,05,08

KAPLAN J DR
DPT OF PHYSICS
UNIVERSITY OF CALIFORNIA
LOS ANGELES CA 90024
USA
TEL
TLF
TLX
EML
COM 21

KAPLAN LEWIS D DR
ATMOSPH & ENVIRONMENTAL
RESEARCH INC
840 MEMORIAL DR
CAMBRIDGE MA 02139
USA
TEL 617 547 6207
TLF
TLX· 951417 AERC
EML
COM

KAPOOR RAMESH CHANDER
INDIAN INSTITUTE OF
ASTROPHYSICS
KORAMANGALA
BANGALORE 560 034
INDIA
TEL 812 56 6585
TLF
TLX 845763 IIAB IN
EML
COM 47,48

KARAALI SALIH DR
FACULTY OF SCIENCE
UNIVERSITY OF ISTANBUL
34452 UNIVERSITY
TURKEY
TEL 1 522 4200*610
TLF
TLX
EML
COM

KARACHENTSEV I D DR
SPECIAL ASTROPHYSICAL OBS
ACADEMY OF SCIENCES
ZELENCHUKSKAJA
357147 STAVROPOLSKIJ
RUSSIA
TEL
TLF
TLX
EML
COM 09,28,30,47

KARANDIKAR R V PROF
DPT OF ASTRONOMY
UNIVERSITY OF OSMANIA
HYDERABAD 500 007
INDIA
TEL 71 951
TLF
TLX
EML
COM 16,21

KARDASHEV N S DR
SPACE RESEARCH INSTITUTE
ACADEMY OF SCIENCES
PROFSOYUZNAYA UL 84/32
117810 MOSCOW
RUSSIA
TEL
TLF
TLX
EML
COM 40,51

KARETNIKOV VALENTIN G R
ASTRONOMICAL OBSERVATORY
ODESSA STATE UNIVERSITY
SHEVCHENKO PARK
270014 ODESSA
UKRAINE
TEL 25 0356
TLF
TLX
EML
COM 42

KARLICKY MARIAN
ASTRONOMICAL INSTITUTE
CZECH ACADEMY OF SCIENCES
ONDREJOV OBSERVATORY
CS 251 65 ONDREJOV
CZECHOSLOVAKIA
TEL 204 85201
TLF 204 85314
TLX 121579 ASTR C
EML
COM 10C,12

KAROJI HIROSHI DR
TOKYO ASTRONOMICAL OBS
NAOJ
OSAWA MITAKA
TOKYO 181
JAPAN
TEL 81 422 41 3643
TLF 81 422 41 3776
TLX 02822307 TAOMTK J
EML KAROJI@SXT1 MTK NAO AC JP
COM 28

KAROVSKA MARGARITA DR
CENTER FOR ASTROPHYSICS
HCO/SAO
60 GARDEN ST
CAMBRIDGE MA 02138
USA
TEL 617 495 7347
TLF
TLX
EML KAROVSKA@CFA
COM 27

KARP ALAN HERSH DR
SCIENTIFIC CENTER IBM
1530 PAGE MILL RD
PALO ALTO CA 94304
USA
TEL 415 855 3127
TLF
TLX
EML
COM 27,36

KARPEN JUDITH T
NAVAL RESEARCH LABORATORY
CODE 4175 K
4555 OVERLOOK AVE SW
WASHINGTON DC 20375 5000
USA
TEL 202 767 3441
TLF
TLX
EML
COM 10,12

KARPINSKIJ VADIM N DR
PULKOVO OBSERVATORY
ACADEMY OF SCIENCES
10 KUTUZOV QUAY
196140 ST PETERSBURG
RUSSIA
TEL
TLF
TLX
EML
COM 09,12C,44

KARTTUNEN HANNU DR
CENTRE FOR SCIENTIFIC
COMPUTING
BOX 40
SF 02101 ESPOO
FINLAND
TEL 457 2709
TLF 457 2302
TLX
EML HKARTTUNEN AT CSC FI
COM 15

KARYGINA ZOYA V DR
ASTROPHYSICAL INSTITUTE
KAZAKH ACAD OF SCIENCES
480068 ALMA ATA
KAZAKHSTAN
TEL
TLF
TLX
EML
COM 21

KASHSCHEEV B L PROF DR
KHARKOV INSTITUTE FOR
RADIOELECTRONICS
310059 KHARKOV
UKRAINE
TEL
TLF
TLX
EML
COM 22

KASPER U DR
ZNTRLINST F ASTROPHYSIK
STERNWARTE BABELSBERG
ROSA-LUXEMBURG-STR 17A
D 1502 POTSDAM
GERMANY
TEL
TLF
TLX
EML
COM 47

KASTURIRANGAN K DR
ISRO SATELLITE CTR
AIRPORT RD
VIMANAPURA POST
BANGALORE 560 017
INDIA
TEL 812 54 779
TLF
TLX 845325 & 769
EML
COM 44

KASUGA TAKASHI
NOBEYAMA RADIO OBS
NAOJ
MINAMIMAKI MURA
NAGANO 384 13
JAPAN
TEL 267-98-2831
TLF
TLX 3329005 TAO NRO J
EML
COM 40

KASUMOV FIKRET K O DR
PHYSICS INSTITUTE
NARIMANOVA PR 33
AKADEMGORODOC
370122 BAKU
AZERBAIDZHAN
TEL 39 6784
TLF
TLX
EML
COM 33

KATGERT PETER DR
STERREWACHT
BOX 9513
NL 2300 RA LEIDEN
NETHERLANDS
TEL 71 27 2727
TLX 39058
EML
TLF
COM 28

KATGERT-MERKELIJN J K DR
STERREWACHT
HUYGENS LAB
BOX 9513
NL 2300 RA LEIDEN
NETHERLANDS
TEL
TLF
TLX
EML
COM

KATO MARIKO
DPT OF ASTRONOMY
KEIO UNIVERSITY
HIYOSHI KOULOKU-KU
YOKOHAMA SHI 223
JAPAN
TEL 44-63-1111
TLF
TLX
EML
COM· 35

KATO SHOJI PROF
DPT OF ASTRONOMY
KYOTO UNIVERSITY
KITASHIRAKAWA SAKYO KU
KYOTO 606
JAPAN
TEL 075-751-2111
TLF
TLX 5422693 LIBKYU J
EML
COM 12,33,47

KATO TAKAKO DR
INST OF PLASMA PHYSICS
NAGOYA UNIVERSITY
FUROCHO CHIKUSA KU
NAGOYA 464
JAPAN
TEL 052-781-5111
TLF
TLX 0447-3691 IPPJNU J
EML
COM 14C

KATSIS DEMETRIUS DR
12 RUE VARNIS
GR 17124 NEA SMYRNE
GREECE
TEL 933 6014
TLF
EML
COM 07
TLX

KATZ JONATHAN I
DPT OF PHYSICS
WASHINGTON UNIVERSITY
ST LOUIS MO 63130
USA
TEL 314 889 6202
TLF
TLX
EML
COM 48

KATZ JOSEPH DR
RACAH INST OF PHYSICS
HEBREW UNIV OF JERUSALEM
JERUSALEM 91904
ISRAEL
TEL. 2 584 604
TLF
TLX· 25391 HUIL
EML
COM

KAUFMAN MICHELE DR
DPT OF PHYSICS
OHIO STATE UNIVERSITY
174 W 18TH AVE
COLUMBUS OH 43210
USA
TEL 614 422 5713
TLF
TLX
EML
COM 28

KAUFMANN JENS PETER DR
INSTITUT FUR ASTRONOMIE
TECHNISCHE UNIVERSITAT
HARDENBERGSTR 36
D 1000 BERLIN 12
GERMANY
TEL 30 314 5462
TLF
TLX 184262
EML·
COM

KAUFMANN PIERRE PROF
CRAAE/ESCOLA POLITECNICA
UNIVERSIDADE DE SAO PAULO
CP 8174
05508 SAO PAULO SP
BRAZIL
TEL 11 815 9322
TLF
TLX
EML
COM 10,12,40,51

KAULA WILLIAM M PROF
DPT OF EARTH & SPACE SCI
UNIVERSITY OF CALIFORNIA
LOS ANGELES CA 90024
USA
TEL
TLF
TLX
EML
COM 07,16

KAWABATA KINAKI PROF
DPT OF PHYSICS
NAGOYA UNIVERSITY
FUROCHO CHIKUSA KU
NAGOYA 464
JAPAN
TEL
TLF
TLX
EML
COM 40,47

KAWABATA KIYOSHI
DPT OF PHYSICS COLL SCI
SCIENCE UNIVERSITY TOKYO
1-3 KAGURAZAKA SHINJUKU
TOKYO
JAPAN
TEL 3-260-4271
TLF
TLX
EML
COM

KAWABATA SHUSAKU PROF
KYOTO GAKUEN UNIVERSITY
NANJO
SOGABECHO KAMEOKA
KYOTO 621
JAPAN
TEL 07712-2-2001
TLF
TLX
EML
COM 42

KAWAGUCHI ICHIRO PROF
DPT OF ASTRONOMY
KYOTO UNIVERSITY
KITASHIRAKAWA SAKYO KU
KYOTO 606
JAPAN
TEL
TLF
TLX
EML
COM 12

KAWALER STEVEN D DR
CTR SOLAR & SPACE RES
YALE UNIVERSITY
BOX 6666
NEW HAVEN CT 06511
USA
TEL 203 432 3012
TLF
TLX
EML BITNET kawaler@yalastro
COM

KAWARA KIMIAKI
TOKYO ASTRONOMICAL OBS
NAOJ
OSAWA MITAKA
TOKYO 181
JAPAN
TEL
TLF
TLX
EML
COM 25

KAWATA YOSHIYUKI DR
KANAZAWA TECHNOLOGY INST
7-1 OGIGAOKA
NONOICHOMACHI
ISHIKAWA 921
JAPAN
TEL
TLF
TLX
EML
COM

KAYSER RAINER DR
HAMBURGER STERNWARTE
GOJENSBERGSWEG 112
D 2050 HAMBURG 80
GERMANY
TEL 40 7252 4126
TLF 40 7252 4198
TLX 217884
EML ST40010@DHHUNI4 BITNET
COM 47

KAZANTZIS PANAYOTIS DR
DPT OF MATHEMATICS
UNIVERSITY OF PATRAS
GR 265 10 RION
GREECE
TEL 61 99 3456
TLF
TLX
EML
COM

KAZES ILYA DR
OBSERVATOIRE DE PARIS
SECTION DE MEUDON
SECTION ASTROPHYSIQUE
F 92195 MEUDON PPL CDX
FRANCE
TEL 1 45 07 7606
TLF
TLX
EML
COM 34,40

KEAY COLIN S L PROF
DPT OF PHYSICS
NEWCASTLE UNIVERSITY
NEWCASTLE NSW 2308
AUSTRALIA
TEL 49 21 5451/5440
TLF 49 21 6907
TLX 28194 NEWUN AA
EML PHCSLK@CC NEWCASTLE EDU AU
COM 15,22C

KEEL WILLIAM C
DPT OF PHYSICS & ASTRON
UNIVERSITY OF ALABAMA
BOX 1921
TUSCALOOSA AL 35487 0324
USA
TEL 205 348 5050
TLF
TLX
EML
COM 28

KEENAN PHILIP C PROF EMER
PERKINS OBSERVATORY
OHIO STATE UNIVERSITY
BOX 449
DELAWARE OH 43015
USA
TEL 614 363 1257
TLF
TLX 810-482-1715
EML
COM 29,45

KEGEL WILHELM H PROF
INST THEORETISCHE PHYSIK
UNIVERSITAT FRANKFURT
ROBERT-MAYER-STR 8-10
D 6000 FRANKFURT MAIN 1
GERMANY
TEL 69 798 2357
TLF
TLX 413932
EML
COM 34

KEIL KLAUS DR
DPT OF GEOLOGY
UNIVERSITY OF NEW MEXICO
800 YALE BLVD NE
ALBUQUERQUE NM 87131
USA
TEL 505 277 4204
TLF
TLX
EML
COM 15

KEIL STEPHEN L
AIR FORCE GEOPHYSICS LAB
NTL SOLAR OBSERVATORY
SACRAMENTO PEAK OBS
SUNSPOT NM 88349
USA
TEL 505 434 1390
TLF
TLX
EML
COM 12

KELEMEN JANOS
KONKOLY ·OBSERVATORY
THEGE U 13/17
BOX 67
H 1525 BUDAPEST
HUNGARY
TEL 1 75 5866/75 4122
TLF
TLX 227460 KONOB H
EML
COM

KELLER CHARLES F
LOS ALAMOS NATIONAL LAB
MS F665
BOX 1663
LOS ALAMOS NM 87545
USA
TEL 505 667 5648
TLF
TLX
EML
COM

KELLER GEOFFREY
DPT OF ASTRONOMY
OHIO STATE UNIVERSITY
174 W 18TH AVE
COLUMBUS OH 43210 1106
USA
TEL 614 422 6279
TLF
TLX
EML
COM

KELLER HORST UWE DR
MPI FUER AERONOMIE

POSTFACH 20
D 3411 KATLENBURG LINDAU
GERMANY
TEL 555 641 419
TLF
TLX 965527
EML
COM 15C,49C

KELLER HUWE ULRICH DR
OBSERVATORY & PLANETARIUM
NECKARSTRASSE 47
D 7000 STUTTGART 1
GERMANY
TEL 711 29 1004
TLX 721855 STBST D
EML
TLF
COM 46,51

KELLERMANN KENNETH I DR
NRAO
EDGEMONT RD
CHARLOTTESVILLE VA 22903
USA
TEL 804 296 0240
TLX 9109970174
EML NRAO KKELLERM
TLF 804 296 0278
COM 28,40,47,48,51

KELLOGG EDWIN M DR
CENTER FOR ASTROPHYSICS
HCO/SAO MS 3
60 GARDEN ST
CAMBRIDGE MA 02138
USA
TEL
TLF
TLX
EML
COM 48

KEMBHAVI AJIT K
TIFR
HOMI BHABHA RD
COLABA
BOMBAY 400 005
INDIA
TEL
TLF
TLX
EML
COM 47,48

KENDERDINE SIDNEY DR
MULLARD RADIO ASTRON OBS
CAVENDISH LABORATORY
MADINGLEY RD
CAMBRIDGE CB3 OHE
UK
TEL 223 664 77
TLF
TLX 81292
EML
COM 40

KENDZIORRA ECKHARD DR
ASTRONOMISCHES INSTITUT
DER UNIV TUEBINGEN
WALDHAEUSERSTR 64
D 7400 TUEBINGEN
GERMANY
TEL 707 129 6127
TLF 707 129 3458
TLX 7262714 AIT D
EML 29382 KENDZIORRA
COM

KENNEDY EUGENE T
SCHOOL OF PHYSICAL SCI
NTL INSTITUTE HIGHER EDU
GLASNEVIN
DUBLIN 9
IRELAND
TEL 1 370 071
TLF
TLX 30690 NIHFD
EML
COM 14

KENNEDY JOHN E PROF
323 LAKE CRESENT

SASKATOON SK S7H 3A1
CANADA
TEL 306 374 4614
TLF
TLX
EML
COM 41,46

KENNICUTT ROBERT C JR
DPT OF ASTRONOMY
UNIVERSITY OF MINNESOTA
116 CHURCH ST SE
MINNEAPOLIS MN 55455
USA
TEL 612 376 5224
TLF
TLX
EML
COM 28,34

KENT STEPHEN M
FERMILAB
MS 127
BOX 500
BATAVIA IL 60510
USA
TEL
TLF
TLX
EML
COM

KENYON SCOTT J DR
CENTER FOR ASTROPHYSICS
HCO/SAO
60 GARDEN ST
CAMBRIDGE MA 02138
USA
TEL 617 495 7235
TLF
TLX 92 1428
EML BITNET kenyon@cfa
COM 42

KEPLER S O
INSTITUTO DE FISICA
UFRGS
CP 15051
90049 PORTO ALEGRE RS
BRAZIL
TEL 512 36 4677
TLF
TLX 051-1055 UFRS BR
EML
COM 25,27

KERR FRANK J DR
ASTRONOMY PROGRAM
UNIVERSITY OF MARYLAND
COLLEGE PARK MD 20742
USA
TEL   301 454 6302
TLF
TLX 710-826-0352
EML
COM 28,33,34,40

KERR ROY P PROF
DPT OF PHYSICS
UNIVERSITY OF CANTERBURY
PRIVATE BAG
CHRISTCHURCH 1
NEW ZEALAND
TEL 348 2009
TLF
TLX NZ 4144
EML
COM

KESSLER KARL G DR
NTL BUREAU OF STANDARDS
A 505 ADMIN
GAITHERSBURG MD 20899
USA
TEL 301 921 3643
TLF
TLX 197674 TRT
EML
COM 14,31

KESSLER MARTIN F DR
ESA/ESTEC
ASTROPHYSICS DIV
BOX 299
NL 2200 AG NOORDWIJK
NETHERLANDS
TEL 17 19 83623
TLF 17 19 85434
TLX 39098
EML ESTCS1 MKESSLER
COM 44

KESTEVEN MICHAEL J L DR
CSIRO
DIVISION OF RADIOPHYSICS
BOX 76
EPPING NSW 2121
AUSTRALIA
TEL 2 868 0222
TLF
TLX 26230 ASTRO
EML MKEVSTEVE@ATNF CSIRO AU
COM 40

KHACHIKIAN E YE PROF
BYURAKAN ASTROPHYSICAL
OBSERVATORY
378433 BYURAKAN
ARMENIA
TEL 88 52 284142/283453
TLF 88 52 52 3640
TLX· 243344 ORION SU
EML
COM 28P

KHALESSEH BAHRAM DR
DPT OF PHYSICS 91735 654
UNIVERSITY OF FERDOWSI
SCHOOL OF SCIENCES
MASHHAD
IRAN
TEL 51 32021*64
TLF 51 87079
TLX 512271 FUON IR
EML
COM 42

KHARADZE E K PROF
ABASTUMANI ASTROPHYSICAL
OBSERVATORY
GEORGIAN ACAD OF SCIENCES
383762 ABASTUMANI
GEORGIA
TEL· 998891/225460
TLF
TLX 327409
EML
COM 33,34,45

KHARE BISHUN N DR
CRSR
CORNELL UNIVERSITY
306 SPACE SCIENCES BLDG
ITHACA NY 14853 6801
USA
TEL 607 256 3934
TLF
TLX
EML
COM

KHARIN A S DR
MAIN ASTRONOMICAL QBS
UKRAINIAN ACAD OF SCIENCE
GOLOSEEVO
252127 KIEV
UKRAINE
TEL 66 4765
TLF
TLX 132517 NEBO
EML
COM 05,08

KHARITONOV ANDREJ V DR
ASTROPHYSICAL INSTITUTE
KAZAKH ACAD OF SCIENCES
480068 ALMA ATA
KAZAKHSTAN
TEL
TLF
TLX
EML
COM

KHATISASHVILI ALFEZ SH DR
ABASTUMANI ASTROPHYSICAL
OBSERVATORY
GEORGIAN ACAD OF SCIENCES
383762 ABASTUMANI
GEORGIA
TEL
TLF
TLX
EML
COM 20

KHETSURIANI TSIALA S DR
ABASTUMANI ASTROPHYSICAL
OBSERVATORY
GEORGIAN ACAD OF SCIENCES
383762 ABASTUMANI
GEORGIA
TEL
TLF
TLX
EML
COM 12

KHOKHLOVA V L DR
INST OF ASTRONOMY
ACADEMY OF SCIENCES
PYATNITSKAYA UL 48
109017 MOSCOW
RUSSIA
TEL 231-54-61
TLF
TLX 412623 SCSTP SU
EML
COM 29,36

KHOLSHEVNIKOV K V DR
ASTRONOMICAL OBSERVATORY
ST PETERSBURG UNIVERSITY
BIBLIOTECHNAJA PL 2
198904 ST PETERSBURG
RUSSIA
TEL 257-94-88
TLF
TLX
EML
COM 07C

KHOZOV GENNADIJ V
ASTRONOMICAL OBSERVATORY
ST PETERSBURG UNIVERSITY
BIBLIOTECHNAJA PL 2
198904 ST PETERSBURG
RUSSIA
TEL 2-57-94-84
TLF
TLX 12168 PHOBOS
EML
COM 35

KHROMOV G S DR
INST OF ASTRONOMY
ACADEMY OF SCIENCES
PYATNITSKAYA UL 48
109017 MOSCOW
RUSSIA
TEL
TLF
TLX
EML
COM 34,41

KIANG TAO PROF
DUNSINK OBSERVATORY
DIAS
DUBLIN 15
IRELAND
TEL 1 387 911
TLF
TLX 31687 DIAS EI
EML
COM 20,41

KIASATPOOR AHMAD PROF
DPT OF PHYSICS
UNIVERSITY OF ESFAHAN
DANESHGAH E
ESFAHAN
IRAN
TEL 31 44321
TLF
TLX 31-2295 IRE U
EML
COM

KIBBLEWHITE EDWARD J DR
INSTITUTE OF ASTRONOMY
THE OBSERVATORIES
MADINGLEY RD
CAMBRIDGE CB3 0HA
UK
TEL
TLF
TLX
EML
COM

KIELKOPF JOHN F DR
DPT OF PHYSICS
UNIVERSITY OF LOUISVILLE
LOUISVILLE KY 40292
USA
TEL 502 588 6787
TLF
TLX
EML BITNET JFKIELO1@ULKYVX
COM 14

KIGUCHI MASAYOSHI DR
RES INST SCIENCE & TECH
KINKI UNIVERSITY
HIGASHI
OSAKA 577
JAPAN
TEL 06-721-2332
TLF
TLX
EML
COM 35

KII TSUNEO DR
ISAS
3-1-1 YOSHINODAI
SAGAMIHARA
KANAGAWA 229
JAPAN
TEL 427 51 3911 EX 2624
TLF 0427 59 4253
TLX J27758IASA ERO
EML
COM 48

KIKUCHI SADAEMON PROF
ASTRONOMICAL INSTITUTE
TOHOKU UNIVERSITY
SENDAI AOBA
MIYAGI 980
JAPAN
TEL
TLF
TLX
EML
COM

KILADZE R I DR
ABASTUMANI ASTROPHYSICAL
OBSERVATORY
GEORGIAN ACAD OF SCIENCES
383762 ABASTUMANI
GEORGIA
TEL
TLF
TLX
EML
COM 16

KILAMBI G C DR
DPT OF ASTRONOMY
UNIVERSITY OF OSMANIA
HYDERABAD 500 007
INDIA
TEL 71 951*247
TLF
TLX
EML
COM 37

KILAR BOGDAN DR
FAC OF GEODESY
UNIVERSITY OF LJUBLJANA
JAMOVA 2
YU 61000 LJUBLJANA
YUGOSLAVIA
TEL
TLF
TLX
EML
COM

KILKENNY DAVID DR
SAAO
BOX 9
OBSERVATORY 7935
SOUTH AFRICA
TEL 021-47-0025
TLF
TLX 57-20309 SA
EML
COM 25

KILLEEN NEIL DR
AAO
ATNF
BOX 76
EPPING NSW 2121
AUSTRALIA
TEL 2 868 0222
TLF 2 868 0400
TLX
EML NIKILLEEN@ATNF CSIRO AU
COM 40,48

KIM CHULHEE DR
DPT OF EARTH SCI EDU
CHONBUK NTL UNIVERSITY
CHONJU 560 756
KOREA R
TEL
TLF
TLX
EML
COM 27

KIM IRADIA S
STERNBERG STATE ASTR INST
117234 MOSCOW
RUSSIA
TEL
TLF
TLX
EML
COM 10C

KIM JIK SU
PYONGYANG ASTRON OBS
ACADEMY OF SCIENCES DPRK
TAESONG DISTRICT
PYONGYANG
KOREA DPR
TEL
TLF
TLX
EML
COM 47

KIM KAP-SUNG DR
DPT ASTRONOMY SPACE SCI
KYUNG HEE UNIVERSITY
YONG IN KUN
KYUNG KIE DO 449 900
KOREA R
TEL 331 280 2443
TLF 82 331 281 4964
TLX
EML
COM 10

KIM KWANG-TAE DR
DPT ASTRON & SPACE RES
CHUNGNAM NTL UNIVERSITY
DAEJOEN 304 764
KOREA R
TEL 042 821 5463
TLF
TLX
EML
COM 40

KIM TU HWAN
KOREA ASTR/SPACE SCI INST
36-1 WHAAM-DONG JUNG-GU
TAEJEON CHUNGCHUNGNAM-DO
TAEJEON 300 31
KOREA R
TEL 042-823-1497
TLF
TLX 45532 K
EML
COM 27

KIM YONG HYOK DR
PYONGYANG ASTRON OBS
ACADEMY OF SCIENCES DPRK
TAESONG DISTRICT
PYONGYANG
KOREA DPR
TEL 5-3134, 53239
TLF
TLX
EML
COM

KIM YONG UK
PYONGYANG ASTRON OBS
ACADEMY OF SCIENCES DPRK
TAESONG DISTRICT
PYONGYANG
KOREA DPR
TEL
TLF
TLX
EML
COM

KIM YUL
PYONGYANG ASTRON OBS
ACADEMY OF SCIENCES DPRK
TAESONG DISTRICT
PYONGYANG
KOREA DPR
TEL
TLF
TLX.
EML
COM

KIM ZONG DOK
PYONGYANG ASTRON OBS
ACADEMY OF SCIENCES DPRK
TAESONG DISTRICT
PYONGYANG
KOREA DPR
TEL
TLF.
TLX
EML
COM 14

KIMBLE RANDY A DR
DPT PHYSICS & ASTRONOMY
JOHNS HOPKINS UNIVERSITY
CHARLES & 34TH ST
BALTIMORE MD 21218
USA
TEL 301 338 8738
TLF
TLX 9102400225 JHU CASMD
EML SPAN SCIVAX CASA RAK
COM 44

KIMURA HIROSHI DR
PURPLE MOUNTAIN OBSEV
CAS
NANJING
CHINA PR
TEL 25 33921
TLF
TLX 34144 PMONJ CN
EML
COM 34

KING ANDREW R DR
DPT OF ASTRONOMY
UNIVERSITY OF LEICESTER
UNIVERSITY RD
LEICESTER LE1 7RH
UK
TEL 533 554 455
TLF
TLX 341198
EML
COM

KING DAVID S PROF
DPT PHYSICS & ASTRONOMY
UNIVERSITY OF NEW MEXICO
800 YALE BLVD NE
ALBUQUERQUE NM 87131
USA
TEL
TLF
TLX
EML
COM 35,41

KING HENRY C DR
TRILLIUM
206 WHITE LION ROAD
LITTLE CHALFONT
BUCKS HP7 9NU
UK
TEL
TLF
TLX
EML
COM 41

KING IVAN R PROF
ASTRONOMY DPT
UNIVERSITY OF CALIFORNIA
601 CAMPBELL HALL
BERKELEY CA 94720
USA
TEL 415 642 2206
TLF
TLX 820181 UCB ASTRAL
EML
COM 25,28,33,37

KING R B DR
BOX 725
MEDOCINO CA 95460
USA
TEL
TLF
EML
COM 14,29
TLX

KING ROBERT WILSON JR DR
DPT OF EARTH & PLANET SCI
MIT RM 54 620
BOX 165
CAMBRIDGE MA 02139
USA
TEL 617 253 7064
TLF
TLX 921473 MIT CAM
EML
COM 04

KINGSTON ARTHUR E PROF
DPT OF APPLIED MATHS
& THEORETICAL PHYSICS
QUEEN'S UNIVERSITY
BELFAST BT7 1NN
UK
TEL 232 24 5133
TLF
TLX 74487 QUB AMD G
EML
COM 14

KING-HELE DESMOND G DR
ROYAL AIRCRAFT ESTABL
FARNBOROUGH HANTS
UK
TEL 252 244 61
TLF
EML
COM 07
TLX

KINMAN THOMAS D DR
KITT PEAK NTL OBS
BOX 26732
950 N CHERRY AVE
TUCSON AZ 85726 6732
USA
TEL 602 327 5511
TLF
TLX 0666-484 AURA NOAO
EML·
COM· 28,33

KINNEY ANNE L DR
STSCI
HOMEWOOD CAMPUS
3700 SAN MARTIN DR
BALTIMORE MD 21218
USA
TEL 301 338 4831
TLF
TLX 684 9101 STSCI
EML 6549 KINNEY
COM

KINOSHITA HIROSHI DR
TOKYO ASTRONOMICAL OBS
NAOJ
OSAWA MITAKA
TOKYO 181
JAPAN
TEL 422 41 3615
TLF 422 41 3793
TLX 02822307 TAOMTK
EML KINOSHITA@C1 MTK NAO AC JP
COM 04VP,07C,20

KIPLINGER ALAN L DR
APAS DPR
UNIVERSITY OF COLORADO
BOX 391
BOULDER CO 80309 0391
USA
TEL 303 497 5892
TLF
TLX
EML SPAN 9555 AKIPLINGER
COM 10,27

KIPPENHAHN RUDOLF PROF
RAUTENBREITE 2
D 3400 GOETTINGEN
GERMANY
TEL 551 24714
TLF 551 22902
EML
COM 27,35
TLX

KIPPER TONU DR
TARTU ASTROPHYSICAL OBS
ESTONIAN ACAD OF SCIENCES
202444 TARTU
ESTONIA
TEL
TLF
TLX
EML
COM 09,14,29

KIRAL ADNAN PROF
UNIVERSITY OBSERVATORY
UNIVERSITY OF ISTANBUL
34452 ISTANBUL
TURKEY
TEL
TLF
TLX
EML
COM

KIRBIYIK HALIL DR
DPT OF PHYSICS
MIDDLE EAST TECH UNIV
06531 ANKARA
TURKEY
TEL 41 22 37100/3528
TLF
TLX 42761 ODTK TR
EML
COM

KIRBY KATE P DR
CENTER FOR ASTROPHYSICS
HCO/SAO
60 GARDEN ST
CAMBRIDGE MA 02138
USA
TEL 617 495 7237
TLF
TLX
EML·
COM 14

KIRK JOHN DR
MAX PLANCK INSTITUT
F PHYSIK & ASTROPHYSIK
KARLSCHWARZSCHILD STR 1
D 8046 GARCHING MUENCHEN
GERMANY
TEL 89 329 90
TLF
TLX
EML
COM 48

KIRKPATRICK RONALD C DR
LOS ALAMOS NATIONAL LAB
MS 220
BOX 1663
LOS ALAMOS NM 87545
USA
TEL 505 667 4812
TLF
TLX
EML
COM 34

KIRSHNER ROBERT PAUL DR
DPT OF ASTRONOMY
HARVARD UNIVERSITY
60 GARDEN ST
CAMBRIDGE MA 02138
USA
TEL 617 495 7390
TLF
TLX
EML
COM 28,34

KISELEV NIKOLAI N DR
ASTROPHYSICAL INSTITUTE
TADJIK ACAD OF SCIENCES
SVIRIDENKO UL 22
734042 DUSHANBE
TADZHIKISTAN
TEL
TLF
TLX
EML
COM 15C

KISELYOV ALEXEJ A DR
PULKOVO OBSERVATORY
ACADEMY OF SCIENCES
10 KUTUZOV QUAY
196140 ST PETERSBURG
RUSSIA
TEL
TLF
TLX
EML
COM 26

KISLYAKOV ALBERT G DR
INST OF APPLIED PHYSICS
ACADEMY OF SCIENCES
ULYANOV UL 46
603600 N NOVGOROD
RUSSIA
TEL
TLF
TLX
EML
COM 40

KISLYUK VITALIJ S DR
MAIN ASTRONOMICAL OBS
UKRAINIAN ACAD OF SCIENCE
GOLOSEEVO
252127 KIEV
UKRAINE
TEL
TLF
TLX 131406 SKY US
EML
COM 16,24

KISSELEVA TAMARA P
PULKOVO OBSERVATORY
ACADEMY OF SCIENCES
10 KUTUZOV QUAY
196140 ST PETERSBURG
RUSSIA
TEL
TLF
TLX
EML
COM 20

KISSELL KENNETH E DR
CHIEF SCIENTIST
RPS MAUI BOX 758
AMOS/MOTIF OBSERVATORIES
PAUNENE HI 96784
USA
TEL
TLF
TLX
EML
COM 09

KITAMOTO SHUNJI DR
FAC OF SCIENCES
OSAKA UNIVERSITY
MACHIKANEYAMA
TOYONAKA OSAKA 560
JAPAN
TEL 6 844 1151
TLF
TLX
EML BITNET KITAMOTO@JPNOSKFM
COM

KITAMURA M PROF
TOKYO ASTRONOMICAL OBS
NAOJ
OSAWA MITAKA
TOKYO 181
JAPAN
TEL 0422-32-5111
TLF
TLX 2822307 TAOMTK J
EML
COM 42

KITAMURA SEIICHI DR
DPT OF EARTH SCIENCE
SHIGA UNIVERSITY
2-5-1 HIRATSU
OTSU 520
JAPAN
TEL 775-37-0081
TLF
TLX
EML
COM

KITCHIN CHRISTOPHER R DR
HATFIELD POLYTECHNIC
OBSERVATORY
BAYFORDBURY
HERTFORD HERTS SG13 8LD
UK
TEL 992 558 451
TLF
TLX 262413
EML
COM 29,46

KIZILOGLU NILGUEN DR
DPT OF PHYSICS
MIDDLE EAST TECH UNIV
06531 ANKARA
TURKEY
TEL 41 223 7100*3268
TLF 41 286 8638
TLX 42761 ODTU TR
EML NLK@TRMETU BITNET
COM 35

KIZILOGLU UEMIT DR
DPT OF PHYSICS
MIDDLE EAST TECH UNIV
06531 ANKARA
TURKEY
TEL 41 22 37100*3275
TLF 41 28 68638
TLX 42761 ODTU TR
EML UMK@TRMETU BITNET
COM

KJAERGAARD PER DR
UNIVERSITY OBSERVATORY
OESTER VOLOGADE 3
DK 1350 COPENHAGEN K
DENMARK
TEL 31 14 1790
TLF 31 38 9157
TLX 44155 DANAST DK
EML
COM

KJELDSETH-MOE OLAV DR
INST THEORET ASTROPHYSICS
UNIVERSITY OF OSLO
BOX 1029
N 0315 BLINDERN OSLO 3
NORWAY
TEL 47-2-456510
TLF
TLX 72425 UNIOS N
EML
COM 10

KJURKCHIEVA DIANA DR
DPT OF PHYSICS
HIGHER PEDAGOGICAL INST
BG 9700 SHOUMEN
BULGARIA
TEL 63 15 1216
TLF
TLX
EML
COM 27,42

KLAPP JAIME DR
DPT DE FISICA
UNAM UNIDAD
APDO POSTAL 55-534
09340 IZTAPALAPAO DF
MEXICO
TEL 515 6442
TLF
TLX 1764186 KEBEME
EML
COM

KLARE GERHARD DR
LANDESSTERNWARTE
KOENIGSTUHL
D 6900 HEIDELBERG 1
GERMANY
TEL 62 211 0036
TLF
TLX
EML
COM 33

KLARMANN JOSEPH PROF
DPT OF PHYSICS
WASHINGTON UNIVERSITY
ST LOUIS MO 63130
USA
TEL 314 889 6299
TLF
TLX 650-2557719 MCI
EML
COM

KLECZEK JOSIP DR
ASTRONOMICAL INSTITUTE
CZECH ACADEMY OF SCIENCES
ONDREJOV OBSERVATORY
CS 251 65 ONDREJOV
CZECHOSLOVAKIA
TEL 204 85201
TLF 204 85314
TLX 121579 ASTR C
EML
COM 05,10

KLEIN KARL LUDWIG DR
OBSERVATOIRE DE PARIS
SECTION DE MEUDON
DASOP
F 92195 MEUDON PPL CDX
FRANCE
TEL 1 45 34 7761
TLF
TLX
EML SPAN MEUDON KLEIN
COM 10,12,40

KLEIN MICHAEL J DR
JPL/SSD
MS 303 401
4800 OAK GROVE DR
PASADENA CA 91109
USA
TEL 818 354 7132
TLF
TLX
EML
COM 51C

KLEIN RICHARD I DR
LAWRENCE LIVERMORE LAB
L 23
BOX 808
LIVERMORE CA 94550
USA
TEL 415 422 3548
TLF
TLX
EML
COM 36

KLEIN ULRICH
RADIOASTRONOMISCHES INST
DER UNIVERSITAET BONN
AUF DEM HUEGEL 71
D 5300 BONN 1
GERMANY
TEL 228 73 3644
TLF
TLX
EML
COM 28,40

KLEINMANN DOUGLAS E DR
HONEYWELL ELECTRO OPTICS
OPERATION
2 FORBES RD
LEXINGTON MA 02173
USA
TEL 617 863 3841
TLF
TLX 92-3477
EML
COM

KLEMOLA ARNOLD R DR
LICK OBSERVATORY
UNIVERSITY OF CALIFORNIA
UCSC
SANTA CRUZ CA 95064
USA
TEL 408 429 2907
TLF
TLX
EML
COM 20,24

KLEMPERER W K DR
NBS
ELECTROMAGNETIC FIELDS D
325 BROADWAY
BOULDER CO 80303
USA
TEL 303 497 3757
TLF
TLX 592811 NOAA MASC BDR
EML
COM

KLEPCZYNSKI WILLIAM J DR
US NAVAL OBSERVATORY

34 & MASSACHUSETTS AVE NW
WASHINGTON DC 20392 5100
USA
TEL 202 653 1521
TLF
TLX 710-822-1970
EML
COM 04,19,31C

KLIMCHUK JAMES A DR
CTR FOR SPACE SCIENCES &
ASTROPHYSICS
STANFORD UNIV ERL
STANFORD CA 94305 4055
USA
TEL 415 723 1765
TLF 415 725 2333
TLX
EML KLIMCHUK@FLARE STANFORD EDU
COM 10

KLIMISHIN I A PROF
PEDAGOGIC INSTITUTE
PUSHKIN UL 96 APT 66
284000 IVANOFRANKOVSK
UKRAINE
TEL
TLX·
EML
TLF
COM·

KLINGLESMITH DANIEL A DR
NASA/GSFC
CODE 684
GREENBELT MD 20771
USA
TEL 301 286 6541
TLF
TLX
EML
COM

KLINKHAMER FRANS DR
NIKHEF H CHEAF
BOX 41882
NL 1009 DB AMSTERDAM
NETHERLANDS
TEL 20 59 29444
TLF 20 59 25155
TLX
EML KLINKHAMER@NIKHEFH NIKHEF
COM 48

KLIORE ARVYDAS JOSEPH DR
JPL
4800 OAK GROVE DR
PASADENA CA 91109
USA
TEL 818 354 6164
TLX 675429
EML
TLF
COM

KLOCK B L DR
6601 S HOMESTAKE DR
BOWIE MD 20715
USA
TEL 301 262 1506
TLF
EML
COM 08,09,24
TLX

KLOCOK LUBOMIR DR
ASTRONOMICAL INSTITUTE
SLOVAK ACADEMY SCIENCES
CS 059 60 TATRANSKA LOMNI
CZECHOSLOVAKIA
TEL 969 96 7866/7/8
TLF 969 96 7656
TLX 78277
EML
COM 09

KLOKOCNIK JAROSLAV DR
ASTRONOMICAL INSTITUTE
CZECH ACADEMY OF SCIENCES
ONDREJOV OBSERVATORY
CS 251 65 ONDREJOV
CZECHOSLOVAKIA
TEL 204 85201
TLF 204 85314
TLX 121579 ASTR C
EML
COM 07

KLVANA MIROSLAV
ASTRONOMICAL INSTITUTE
CZECH ACADEMY OF SCIENCES
ONDREJOV OBSERVATORY
CS 251 65 ONDREJOV
CZECHOSLOVAKIA
TEL 204 85201
TLF 204 85314
TLX
EML
COM 10

KNACKE ROGER F DR
DPT OF EARTH & SPACE SCI
ASTRONOMY PROGRAM
SUNY AT STONY BROOK
STONY BROOK NY 11794 2100
USA
TEL 516 246 7673
TLF
TLX 5102287767
EML
COM 15,34

KNAPP GILLIAN R DR
DPT ASTROPHYSICAL SCI
PRINCETON UNIVERSITY
PRINCETON NJ 08544 1001
USA
TEL 609 452 3824
TLF
TLX
EML
COM 28,33,34

KNEER FRANZ DR
UNIVERSITATS STERNWARTE
GEISMARLANDSTRASSE 11
D 3400 GOETTINGEN
GERMANY
TEL 551 39 5042
TLF
TLX 96753
EML
COM 12

KNEZEVIC ZORAN
ASTRONOMICAL OBSERVATORY
VOLGINA 7
YU 11050 BEOGRAD
YUGOSLAVIA
TEL 11 419 357/421 875
TLF
TLX 72610 AOB YU
EML EAOPOO2@YUBGSS21 BITNET
COM 07,15C,20

KNIFFEN DONALD A DR
DPT OF PHYSICS
BOX 862
HAMPDEN SYDNEY COLLEGE
VA 23943
USA
TEL
TLF
TLX
EML
COM

KNOLKER MICHAEL DR
UNIV STERNWARTE
GOTTINGEN
GEISMARLANDSTRASSE 11
D 3400 GOETTINGEN
GERMANY
TEL 551 39 5046
TLF
TLX
EML
COM 12,35

KNOSKA STEFAN
ASTRONOMICAL INSTITUTE
SLOVAK ACADEMY SCIENCES
CS 059 60 TATRANSKA LOMNI
CZECHOSLOVAKIA
TEL 969 96 7866/7/8
TLF 969 96 7656
TLX
EML
COM 10

KNOWLES STEPHEN H DR
CODE 02
NAVSPASUR
DAHLGREN VA 22448 5180
USA
TEL
TLF
TLX
EML
COM 19,51

KNUDE JENS KIRKESKOV DR
UNIVERSITY OBSERVATORY
OESTER VOLDGADE 3
DK 1350 COPENHAGEN K
DENMARK
TEL 31 14 1790
TLF 31 38 9157
TLX 44155 DANAST DK
EML
COM 25C,34

KO HSIEN C PROF
DPT OF ELECT ENGINEERING
OHIO STATE UNIVERSITY
1958 NEIL AVE
COLUMBUS OH 43210
USA
TEL 614 422 2571
TLF
TLX 24-5334
EML
COM 40

KOBAYASHI EISUKE DR
SCIENCE INST OF OSAKA
459 SUGIMOTO CHO
SUMIYOSHI KU
OSAKA 558
JAPAN
TEL 06-692-1882
TLF
TLX
EML
COM

KOBAYASHI YUKISAYU
TOKYO ASTRONOMICAL OBS
NAOJ
OSAWA MITAKA
TOKYO 181
JAPAN
TEL 422 32 5111
TLF
TLX 2822307 TAOMTK
EML
COM 31

KOCER DURCUN DR
DPT OF ASTRONOMY
ISTANBUL UNIVERSITY
34452 ISTANBUL
TURKEY
TEL 1 522 3597
TLF 1 522 6123
TLX
EML
COM 51

KOCH DAVID G
NASA AMES RESEARCH CTR
MS 245 6
MOFFETT FIELD CA 94035
USA
TEL
TLF
TLX
EML
COM

KOCH ROBERT H DR
DPT ASTRON & ASTROPHYS
UNIV OF PENNSYLVANIA
DAVID RITTENHOUSE LAB
PHILADELPHIA PA 19104
USA
TEL 215 898 7882
TLF 215 898 9336
TLX 834621
EML RKOCH@PENNDRLS
COM 25,42C,51

KOCHAROV GRANT E PROF
IOFFE PHYSICAL TECH INST
ACADEMY OF SCIENCES
POLYTECHNICHESKAYA UL 26
194021 ST PETERSBURG
RUSSIA
TEL 247-91-67
TLF
TLX
EML
COM 48

KOCHHAR R K DR
INDIAN INSTITUTE OF
ASTROPHYSICS
KORAMANGALA
BANGALORE 560 034
INDIA
TEL 812 56 6585
TLF
TLX 845763 IIAB IN
EML
COM 28,35

KOCH-MIRAMOND LYDIE DR
CEA CEN
IRF/DPHG/SAP
BP 2
F 91191 GIF/YVETTE CDX
FRANCE
TEL 1 69 08 4329
TLF
TLX 690860
EML
COM 44,48

KODAIRA KEIICHI PROF
TOKYO ASTRONOMICAL OBS
NAOJ
OSAWA MITAKA
TOKYO 181
JAPAN
TEL 0422-32-5111
TLF
TLX 2822307 TAOMTK J
EML
COM 28,29,36

KODAMA HIDEO
COLL LIB ARTS KYOTO UNIV
YOSHIDA NIHONMATSU-CHO
SAKYO KU
KYOTO 606
JAPAN
TEL 075-7512111
TLF
TLX
EML
COM· 47

KOEBERL CHRISTIAN DR
INSTITUTE OF GEOCHEMISTRY
UNIVERSITY OF VIENNA
DR KARL-LUEGER-RING 1
A 1010 VIENNA
AUSTRIA
TEL 1 222 4300 2360
TLF.
TLX
EML
COM 15,22C,51

KOECKELENBERGH ANDRE DR
OBSERVATOIRE ROYAL DE
BELGIQUE
AVE CIRCULAIRE 3
B 1180 BRUSSELS
BELGIUM
TEL 2 373 0311
TLF 2 374 9822
TLX 21565 OBSBEL B
EML
COM 10

KOEHLER H PROF DR
SAUERBRUCHSTR 6
D 7920 HEIDENHEIM
GERMANY
TEL 732 144560
TLF
TLX
EML
COM 09

KOEHLER JAMES A PROF
DPT OF PHYSICS
UNIV OF SASKATCHEWAN
SASKATOON SK S7N 0W0
CANADA
TEL 306 966 6442
TLF
TLX
EML
COM

KOEHLER PETER
CARL ZEISS JENA GMBH
ASTRON INSTRUMENTS DIV
PO BOX 125
D 6900 JENA
GERMANY
TEL 78 588 2575
TLF 78 588 2083
TLX 587 452 CZG DD
EML
COM 09

KOEN MARTHINUS DR
SAAO
BOX 9
OBSERVATORY 7935
SOUTH AFRICA
TEL 2721 470025
TLF 2721 473639
TLX. 520309 SAAO SA
EML CK@SAAO AC ZA
COM 27

KOENIGSBERGER GLORIA
INSTITUTO DE ASTRONOMIA
UNAM
APDO POSTAL 70-264
04510 MEXICO DF
MEXICO
TEL 905-548-5305/06
TLF
TLX
EML
COM

KOEPPEN JOACHIM DR
INST F THEOR ASTROPHYSIK
DER UNIV HEIDELBERG
IM NEUENHEIMER FELD 561
D 6900 HEIDELBERG 1
GERMANY
TEL 62 215 62988
TLF
TLX 461515 UNIHD D
EML
COM 34

KOESTER DETLEV DR
DPT PHYSICS & ASTRONOMY
LOUISIANA STATE UNIV
BATON ROUGE LA 70803 4001
USA
TEL 504 388 2261
TLF
TLX 559184
EML
COM 35,36

KOGOSHVILI NATELA G
ASTROPHYSICAL OBSERVATORY
MOUNT KANOBILI
383762 ABASTUMANI
GEORGIA
TEL 283
TLF
TLX
EML
COM 28

KOGURE TOMOKAZU DR
TOGANO-O 1-10
HASHIMOTO
YAWATA
KYOTO 614
JAPAN
TEL 075-983-2984
TLF
TLX
EML
COM 29

KOHL JOHN L DR
CENTER FOR ASTROPHYSICS
HCO/SAO
60 GARDEN ST
CAMBRIDGE MA 02138
USA
TEL 617 495 7377
TLF
TLX 921428
EML
COM 14

KOHOUTEK LUBOS DR
HAMBURGER STERNWARTE
GOJENSBERGSWEG 112
D 2050 HAMBURG 80
GERMANY
TEL 40 7252 4112
TLX 217884
EML
TLF
COM 15,20,34

KOJIMA MASAYOSHI DR
SOLAR TERRESTRIAL LAB
NAGOYA UNIVERSITY
3-13 HONOHARA TOYOKAWA
AICHI 442
JAPAN
TEL 81 5338 6 3154
TLF 81 5338 6 0811
TLX 4322311 STELAB J
EML KOJIMA@STELAB NAGOYA-U AC JP
COM 40

KOJOIAN GABRIEL DR
DPT OF PHYSICS
UNIVERSITY OF WISCONSIN
EAU CLAIRE WI 54701
USA
TEL 715 836 3148
TLF
TLX
EML
COM 28,40

KOKKOTAS KONSTANTINOS DR
DPT ASTROPHY ASTRON MECH
UNIVERSITY THESSALONIKI
GR 540 06 THESSALONIKI
GREECE
TEL 31 99 1357
TLF 31 99 2777
TLX 412181 AUTH GR
EML CAAZ0104@GRTHEUN1 EARN
COM 47

KOKURIN YURIJ L DR
LEBEDEV PHYSICAL INST
ACADEMY OF SCIENCES
LENINSKY PROSPEKT 53
117924 MOSCOW
RUSSIA
TEL 135-03-60
TLF
TLX 411479 NEOD SU
EML
COM 08

KOLACZEK BARBARA DR
PLANETARY GEODESY DEPT
POLISH ACAD OF SCIENCES
UL BARTYCKA 18
PL 00 716 WARSAW
POLAND
TEL 41-00-41
TLF
TLX 815670 CBK PL
EML
COM 04,19P,31

KOLB EDWARD W DR
FERMILAB
MS 209
THEORETICAL ASTROPHYSICS
BATAVIA IL 60510
USA
TEL 312 840 4695
TLF
TLX 720481
EML
COM 47,48

KOLCHINSKIJ I G DR
MAIN ASTRONOMICAL OBS
UKRAINIAN ACAD OF SCIENCE
GOLOSEEVO
252127 KIEV
UKRAINE
TEL
TLF
TLX
EML
COM 24

KOLESNIK IGOR G DR
MAIN ASTRONOMICAL OBS
UKRAINIAN ACAD OF SCIENCE
GOLOSEEVO
252127 KIEV
UKRAINE
TEL 66 3110
TLF
TLX 131406 SKY SU
EML
COM 33,34

KOLESNIK L N DR
MAIN ASTRONOMICAL OBS
UKRAINIAN ACAD OF SCIENCE
GOLOSEEVO
252127 KIEV
UKRAINE
TEL 66 0869
TLF
TLX 131406
EML
COM 33

KOLESOV A K DR
ASTRONOMICAL OBSERVATORY
ST PETERSBURG UNIVERSITY
BIBLIOTECHNAJA PL 2
199178 ST PETERSBURG
RUSSIA
TEL
TLF
TLX
EML
COM 36

KOLEV DIMITAR ZDRAVKOV
NTL ASTRONOMICAL OBS
BULGARIAN ACAD SCIENCES
BOX 136
BG 4700 SMOLJAN
BULGARIA
TEL 73 41 559
TLF
TLX
EML
COM

KOLLATH ZOLTAN DR
KONKOLY OBSERVATORY
THEGE U 13/17
BOX 67
H 1525 BUPAPEST
HUNGARY
TEL 1 75 5866/75 4122
TLF
TLX 22 74 60 KONOB H
EML. H 697 KNO E ELLA UUCP
COM 27

KOLLATSCHNY WOLFRAM DR
UNIVERSITAETS STERNWARTE
GEISMARLANDSTRASSE 11
D 3400 GOETTINGEN
GERMANY
TEL 551 39 5067
TLF
TLX 96753
EML BITNET WKOLLAT@DGOGWDG1
COM 28

KOLLBERG ERIK L PROF
DPT OF PHYSICS
CHALMERS TECHNICAL UNIV
S 412 96 GOETEBORG
SWEDEN
TEL 31 81 0100
TLF
TLX 2400 ONSPACE
EML
COM

KOMAROV N S DR
ASTRONOMICAL OBSERVATORY
ODESSA STATE UNIVERSITY
SHEVCHENKO PARK
270014 ODESSA
UKRAINE
TEL 22 0396
TLF
TLX
EML
COM

KONDO MASAAKI DR
SENSHU UNIVERSITY
HIGASHI-MITA TAMA-KU
KAWASAKI SHI
KANAGAWA 214
JAPAN
TEL. 044 911 7131
TLF a86123@nacsis ac jp
TLX
EML
COM 48

KONDO MASAYUKI DR
TOKYO ASTRONOMICAL OBS
NAOJ
OSAWA MITAKA
TOKYO 181
JAPAN
TEL 0422-32-5111
TLF
TLX 2822307 TAOMTK J
EML
COM

KONDO YOJI DR
NASA/GSFC
CODE 684
GREENBELT MD 20771
USA
TEL 301 286 6247
TLF 30 286 7642
TLX
EML JUE KONDO
COM 34,36,42P,44

KONIGL ARIEH DR
ASTRONOMY & ASTROPHYS CTR
UNIVERSITY OF CHICAGO
5640 S ELLIS AVE
CHICAGO IL 60637
USA
TEL 312 702 7968
TLF
TLX 282131
EML SPAN lasr oddjob arieh
COM

KONIN V V DR
NIKOLAEV OBSERVATORY
UKRAINIAN ACAD OF SCIENCE
327000 NIKOLAEV
UKRAINE
TEL
TLF
TLX
EML
COM 08

KONONOVICH EDWARD V DR
STERNBERG STATE ASTR INST
UNIVERSITETSKIJ PROSP 13
119899 MOSCOW
RUSSIA
TEL
TLF
TLX
EML
COM 12,46

KONOPLEVA VARVARA P DR
MAIN ASTRONOMICAL OBS
UKRAINIAN ACAD OF SCIENCE
GOLOSEEVO
252127 KIEV
UKRAINE
TEL 66 3110
TLF
TLX 131406 SKY SU
EML
COM 15

KONTIZAS EVANGELOS DR
ASTRONOMICAL INSTITUTE
NTL OBSERVATORY OF ATHENS
BOX 20048
GR 118 10 ATHENS
GREECE
TEL 1 346 1191
TLF
TLX 215530 OBSA GR
EML
COM 28,36,37

KONTIZAS MARY DR
DPT OF ASTRONOMY
NTL UNIVERSITY OF ATHENS
PANEPISTIMIOPOLIS
GR 157 71 ZOGRAFOS
GREECE
TEL 1 723 5122
TLF
TLX
EML
COM 37

KOO DAVID C-Y DR
LICK OBSERVATORY
UNIVERSITY OF CALIFORNIA
NTL SCIENCES II
SANTA CRUZ CA 95064
USA
TEL 408 429 2130
TLF 408 426 3115
TLX 910 9971741 UNICAL
EML KOO@HELIOS UCSC EDU
COM 28,47

KOORNNEEF JAN DR
STSCI
HOMEWOOD CAMPUS
3700 SAN MARTIN DR
BALTIMORE MD 21218
USA
TEL 301 338 4802
TLF
TLX 6849101
EML
COM 34

KOPAL ZDENEK PROF
DPT OF ASTRONOMY
UNIVERSITY OF MANCHESTER
MANCHESTER M13 9PL
UK
TEL 612 737 121
TLF
TLX 668932 MCHRUL G
EML
COM 16,26,42

KOPECKY MILOSLAV DR
ASTRONOMICAL INSTITUTE
CZECH ACADEMY OF SCIENCES
ONDREJOV OBSERVATORY
CS 251 65 ONDREJOV
CZECHOSLOVAKIA
TEL 204 85201
TLF 204 85314
TLX 121579
EML
COM 10,12

KOPP ROGER A DR
LOS ALAMOS NATIONAL LAB
MS E531
BOX 1663
LOS ALAMOS NM 87545
USA
TEL 505 667 4398
TLF
TLX 660495 LOS ALAMOS
EML
COM

KOPYLOV I M DR
PULKOVO OBSERVATORY
ACADEMY OF SCIENCES
10 KUTUZOV QUAY
196140 ST PETERSBURG
RUSSIA
TEL 928 22 42
TLF
TLX
EML
COM 09,27

KORCHAK A A DR
IZMIRAN
ACADEMY OF SCIENCES
142092 TROITSK
RUSSIA
TEL
TLF
TLX
EML
COM

KORMENDY JOHN DR
INSTITUTE FOR ASTRONOMY
UNIVERSITY OF HAWAII
2680 WOODLAWN DR
HONOLULU HI 96822
USA
TEL 808 956 6680
TLF· 808 988 2790
TLX 723 8459
EML KORMENT+DY@UHIFA IFA HAWAII ED
COM 28,33,47

KOROVYAKOVSKIJ YURIJ P DR
SPECIAL ASTROPHYSICAL OBS
ACADEMY OF SCIENCES
NIZHNIJ ARKHYZ
357147 STAVROPOLSKIJ
RUSSIA
TEL
TLF
TLX
EML
COM 09

KOSAI HIROKI
TOKYO ASTRONOMICAL OBS
NAOJ
OSAWA MITAKA
TOKYO 181
JAPAN
TEL 0422 41 3638
TLF 0422 41 3698
TLX
EML
COM 20

KOSHIBA MASA-TOSHI DR
TOKAI UNIVERSITY
2-28 TOMIGAYA
SHIBUYA
TOKYO 151
JAPAN
TEL 03 3467 2211 EXT 483
TLF 03 485 4958
TLX
EML
COM 48

KOSIN GENNADIJ S DR
PULKOVO OBSERVATORY
ACADEMY OF SCIENCES
10 KUTUZOV QUAY
196140 ST PETERSBURG
RUSSIA
TEL·
TLF
TLX
EML
COM 08

KOSOVICHEV ALEXANDER
INSTITUTE OF ASTRONOMY
THE OBSERVATORIES
MADINGLEY RD
CAMBRIDGE CB3 OHA
UK
TEL 223 33 7548*7516
TLF 223 33 7523
TLX 817297 ASTRON G
EML agk@uk ac cam ast-star
COM 35

KOSTIK ROMAN I
MAIN ASTRONOMICAL OBS
UKRAINIAN ACAD OF SCIENCE
GOLOSEEVO
252127 KIEV
UKRAINE
TEL 66 4762
TLF
TLX 131406 SKY SU
EML
COM 10,12

KOSTINA LIDIJA D DR
PULKOVO OBSERVATORY
ACADEMY OF SCIENCES
10 KUTUZOV QUAY
196140 ST PETERSBURG
RUSSIA
TEL
TLF
TLX
EML
COM

KOSTYAKOVA ELENA B DR
STERNBERG STATE ASTR INST
117234 MOSCOW
RUSSIA
TEL
TLF
EML
COM 34
TLX

KOSTYLEV K V DR
ENGERHARDT ASTRONOMICAL
OBSERVATORY
OBSERVATORIA STATION
422526 KAZAN
RUSSIA
TEL
TLF
TLX
EML·
COM 22

KOSUGI TAKEO
TOKYO ASTRONOMICAL OBS
NAOJ
OSAWA MITAKA
TOKYO 181
JAPAN
TEL
TLF
TLX
EML
COM 10

KOTANYI CHRISTOPHE DR
NEPAE
UFSM
CIDADO UNIVERSITARIA
97100 SANTA MARIA RS
BRAZIL
TEL 55 226 1616
TLF
TLX 0552230 UFSM BR
EML
COM

KOTELNIKOV V A ACAD
INST OF RADIO & ELECTRON
ACADEMY OF SCIENCES
103907 MOSCOW
RUSSIA
TEL 203-60-78
TLF
TLX
EML
COM 40

KOTHARI D S DR
DPT OF PHYSICS
UNIVERSITY OF DELHI
NEW DELHI 110 007
INDIA
TEL 11 291 8993
TLF
TLX
EML
COM 35

KOTOV VALERY DR
CRIMEAN ASTROPHYS OBS
UKRAINIAN ACAD OF SCIENCE
NAUCHNY
334413 CRIMEA
UKRAINE
TEL 43 2945
TLF
TLX
EML
COM 12

KOTRC PAVEL
ASTRONOMICAL INSTITUTE
CZECH ACADEMY OF SCIENCES
ONDREJOV OBSERVATORY
CS 251 65 ONDREJOV
CZECHOSLOVAKIA
TEL 204 85201
TLF 204 85314
TLX 121579
EML
COM 10,12

KOUBSKY PAVEL
ASTRONOMICAL INSTITUTE
CZECH ACADEMY OF SCIENCES
ONDREJOV OBSERVATORY
CS 251 65 ONDREJOV
CZECHOSLOVAKIA
TEL 204 85201
TLF 204 85314
TLX
EML
COM 29,42

KOUPELIS THEODOROS DR
DPT PHYSICS & ASTRON
UNIVERSITY OF ROCHESTER
ROCHESTER NY 14627
USA
TEL 716 275 4389
TLF 716 275 8527
TLX
EML TKOU@UORDBV BITNET
COM 48

KOURGANOFF VLADIMIR PROF
20 AVE PAUL APELL

F 75014 PARIS
FRANCE
TEL 1 45 40 5053
TLF
TLX
EML
COM 46

KOUTCHMY SERGE DR
INSTITUT D'ASTROPHYSIQUE
98BIS BD ARAGO
F 75014 PARIS
FRANCE
TEL 1 43 20 1425
TLF 1 43 29 8673
TLX
EML
COM 10,12,21

KOUVELIOTOU CHRYSSA DR
NASA/MSFC
CODE ES 62
HUNTSVILLE AL 35812
USA
TEL 205 544 7711
TLF 205 544 5800
TLX
EML BATSE KOUVELIOTOU SSL KOUVEL
COM

KOVACHEV B J DR
DPT OF ASTRONOMY
BULGARIAN ACAD SCIENCES
72 LENIN BLVD
BG 1784 SOFIA
BULGARIA
TEL 2 75 8827
TLF
TLX 23561 ECF BAN BG
EML
COM 09,29

KOVACS AGNES DR
HELIOPHYSICAL OBSERVATORY
BOX 30
H 4010 DEBRECEN
HUNGARY
TEL 52 11 015
TLF
TLX 72517 DEOBS H
EML
COM 10

KOVACS GEZA DR
KONKOLY OBSERVATORY
THEGE U 13/17
BOX 67
H 1525 BUDAPEST
HUNGARY
TEL 1 75 5866/75 4122
TLF
TLX 227460
EML
COM

KOVAL I K DR
MAIN ASTRONOMICAL OBS
UKRAINIAN ACAD OF SCIENCE
GOLOSEEVO
252127 KIEV
UKRAINE
TEL 66 0869
TLF
TLX
EML
COM

KOVALEVSKY JEAN DR
OCA CERGA
AVE COPERNIC
F 06130 GRASSE
FRANCE
TEL 93 36 5849
TLF
TLX 470865 CERGA F
EML
COM 07,08C,24,31C,50C

KOVAR N S DR
DPT OF PHYSICS
UNIVERSITY OF HOUSTON
HOUSTON TX 77004
USA
TEL
TLF
TLX
EML
COM

KOVAR ROBERT P DR
9666 E ORCHARD DR
ENGLEWOOD CO 80111
USA
TEL 303 394 4494
TLF.
EML
COM
TLX

KOVETZ ATTAY PROF
DPT OF PHYSICS & ASTRON
TEL AVIV UNIVERSITY
RAMAT AVIV
TEL AVIV 69978
ISRAEL
TEL 3 420 234
TLF
TLX 342-171 VERSY IL
EML
COM 35,47

KOWAL CHARLES THOMAS
STSCI
HOMEWOOD CAMPUS
3700 SAN MARTIN DR
BALTIMORE MD 21218
USA
TEL
TLF
TLX
EML
COM 15,16,20

KOYAMA KATSUJI
INST SPACE & ASTRON SCI
UNIVERSITY OF TOKYO
MEGURO KU
TOKYO 153
JAPAN
TEL 3-467-1111
TLF
TLX 34757 ISASTRO J
EML
COM 48

KOYAMA KO-U-ICHI DR
RES INST SCIENCE & TECH
KINKI UNIVERSITY
HIGASHI
OSAKA 577
JAPAN
TEL 06 721 2332 EXT 4711
TLF 06 721 2353
TLX
EML
COM 35

KOYAMA SHIN PROF DR
FAC OF EDUCATION
KAGAWA UNIVERSITY
SAIWAI CHO TAKAMATSUSHI
KAGAWA 760
JAPAN
TEL 878-61-4141
TLF
TLX
EML
COM 12

KOZAI YOSHIHIDE PROF
TOKYO ASTRONOMICAL OBS
NAOJ
OSAWA MITAKA
TOKYO 181
JAPAN
TEL 422 41 3650
TLF 422 41 3690
TLX 2822307 TAOMTK
EML
COM 06,07,20,38,50,EC

KOZASA TAKASHI
MAX PLANCK INSTITUT FUR
KERNPHYSIK
PO BOX 103980
D 6900 HEIDELBERG 1
GERMANY
TEL 62 215 16271
TLF 62 215 16540
TLX
EML KOZASA@DHDMPI5
COM 34

KOZIEL KAROL PROF DR
ASTRON OBSERVATORY KRAKOW
UL 22 LIPCA 16
PL 43 460 WISLA
POLAND
TEL 32-42
TLX
EML
TLF
COM

KOZLOVSKY B Z DR
DPT OF PHYSICS & ASTRON
TEL AVIV UNIVERSITY
RAMAT AVIV
TEL AVIV 69978
ISRAEL
TEL
TLF
TLX
EML
COM 47

KOZLOWSKI MACIEJ DR
COPERNICUS ASTRON CENTER
POLISH ACAD OF SCIENCES
UL BARTYCKA 18
PL 00 716 WARSAW
POLAND
TEL 41 10 86
TLF
TLX 813978 ZAPAN PL
EML
COM 35,48

KRAAN-KORTEWEG RENEE C DR
KAPTEYN ASTRONOMICAL INST
BOX 800
NL 9700 AV GRONINGEN
NETHERLANDS
TEL 50 63 4045
TLF 50 63 6100
TLX
EML KRAAN@RUG NL
COM 28

KRAEMER GERHARD DR
ASTRONOMISCHES INSTITUT
DER UNIVERSITAET
PHILOSOPHENWEG 37
D 7400 TUEBINGEN
GERMANY
TEL
TLF
TLX
EML
COM 12,44

KRAFT ROBERT P PROF
LICK OBSERVATORY
UNIVERSITY OF CALIFORNIA
SANTA CRUZ CA 95064
USA
TEL 408 429 2991
TLF
TLX 910 9971741
EML·
COM 27,29,30,37,42

KRAICHEVA ZDRAVSKA DR
DPT OF ASTRONOMY
BULGARIAN ACAD SCIENCES
7TH NOVEMBER ST 1
BG 1000 SOFIA
BULGARIA
TEL 2 7341
TLF 23561 ECFBAN BG
TLX
EML
COM 42

KRAMER KH N DR
ASTRONOMICAL OBSERVATORY
ODESSA STATE UNIVERSITY
SHEVCHENKO PARK
270014 ODESSA
UKRAINE
TEL 22 0396
TLF
TLX
EML
COM 22

KRANJC ALDO DR
OSS ASTRONOMICO DI BRERA
VIA BRERA 28
I 20121 MILANO
ITALY
TEL 2 87 4444
TLX
EML
TLF 2 72 00 1600
COM

KRASINSKI ANDRZEJ PROF
COPERNICUS ASTRON CENTER
POLISH ACAD OF SCIENCES
UL BARTYCKA 18
PL 00 716 WARSAW
POLAND
TEL. 41 0828
TLF
TLX 813978
EML:
COM 47

KRASINSKY GEORGE A DR
INST OF APPLIED ASTRONOMY
ACADEMY OF SCIENCES
ZDANOVSKAYA UL 8
197042 ST PETERSBURG
RUSSIA
TEL
TLF
TLX
EML
COM 04,07

KRASSOVSKY V I DR
INST PHYSICS OF ATMOSPH
ACADEMY OF SCIENCES
PYSHEVSKY PER 3
109017 MOSCOW
RUSSIA
TEL 231-88-62
TLF
TLX
EML
COM

KRAUS JOHN D PROF
RADIO OBSERVATORY
OHIO STATE UNIVERSITY
2015 NEIL AVE
COLUMBUS OH 43210
USA
TEL 614 548 7895
TLF
TLX
EML
COM 40,51

KRAUSE F DR
ZNTRLINST F ASTROPHYSIK
ASTROPHYSIKALISCHES OBS
TELEGRAFENBERG
D 1500 POTSDAM
GERMANY
TEL
TLF
TLX
EML ELSTNER@VAX HMI DBP DE
COM 10

KRAUSHAAR WILLIAM L PROF
DPT OF PHYSICS
UNIVERSITY OF WISCONSIN
1150 UNIVERSITY AVE
MADISON WI 53706
USA
TEL 608 262 5916
TLF
TLX 265452
EML
COM 44

KRAUTTER JOACHIM DR
LANDESSTERNWARTE
KOENIGSTUHL
D 6900 HEIDELBERG 1
GERMANY
TEL 62 211 0036
TLF
TLX 461789 MPIA D
EML
COM 27,34,42

KREIDL TOBIAS J N
LOWELL OBSERVATORY
1400 W MARS HILL RD
BOX 1149
FLAGSTAFF AZ 86001
USA
TEL 602 774 3358
TLF
TLX
EML
COM 09

KREINER JERZY MAREK DR
INSTITUTE OF PHYSICS
PEDAGOGICAL UNIVERSITY
UL PODCHORAZYCH 2
PL 30 084 KRAKOW
POLAND
TEL 12 37 8286
TLF 12 372 2243
TLX 322444 WSP PL
EML
COM 27,42,46

KREISEL E PROF
EINSTEIN LABORATORIUM
ROSA-LUXEMBURG-STR 17A
D 1502 POTSDAM
GERMANY
TEL 762 225
TLF
TLX 15471
EML
COM 48

KRELOWSKI JACEK DR
INSTITUTE OF ASTRONOMY
N COPERNICUS UNIVERSITY
UL CHOPINA 12/18
PL 87 100 TORUN
POLAND
TEL 856-206-55
TLF
TLX 055-2234 ASTR PL
EML
COM

KREMPEC-KRYGIER JANINA DR
INSTITUTE OF ASTRONOMY
N COPERNICUS UNIVERSITY
UL CHOPINA 12/18
PL 87 100 TORUN
POLAND
TEL 260-18
TLF
TLX 0552234 ASTR PL
EML
COM 29

KRESAK LUBOR DR
ASTRONOMICAL INSTITUTE
SLOVAK ACADEMY SCIENCES
DUBRAVSKA 9
CS 842 28 BRATISLAVA
CZECHOSLOVAKIA
TEL. 7 37 5157
TLF
TLX 93373 SEIS
EML
COM· 15C,20C,22

KRESAKOVA MARGITA DR
ASTRONOMICAL INSTITUTE
SLOVAK ACADEMY SCIENCES
DUBRAVSKA 9
CS 842 28 BRATISLAVA
CZECHOSLOVAKIA
TEL 7 37 5157
TLF
TLX 93373 SEIS
EML
COM 15,22

KREYSA ERNST
MPI FUER RADIOASTRONOMIE

AUF DEM HUEGEL 69
D 5300 BONN 1
GERMANY
TEL 228 52 5269
TLF
TLX 886440
EML
COM 34,40

KRIEGER ALLEN S DR
RADIATION SCIENCE INC
BOX 293
BELMONT MA 02178
USA
TEL 617 494 0335
TLF
TLX
EML
COM

KRIKORIAN RALPH DR
INSTITUT D'ASTROPHYSIQUE
98BIS BD ARAGO
F 75014 PARIS
FRANCE
TEL 1 43 20 1425
TLF 1 43 29 8673
TLX
EML
COM 36

KRISCIUNAS KEVIN DR
JOINT ASTRONOMY CENTER
665 KOMOHANA ST
HILO HI 96720
USA
TEL 808 935 4332
TLF
TLX
EML KEVIN@JACH HAWAII EDU
COM 27

KRISHNA GOPAL
TATA INST OF FUNDAMENTAL
RESEARCH
POONA UNIVERSITY CAMPUS
PUNE 411 007
INDIA
TEL 212 33 7107
TLF 212 33 5760
TLX 0145658 GMRT IN
EML uunet!shakti!gmrt!gk
COM 28,38C,40

KRISHNA SWAMY K S DR
TIFR/ASTROPHYSICS GROUP
HOMI BHABHA RD
COLABA
BOMBAY 400 005
INDIA
TEL 22 219 111
TLF
TLX 113009 TIFR IN
EML
COM 15,34,36

KRISHNAMOHAN S DR
TIFR/RADIO ASTRONOMY CTR
BANGALORE 560 012
INDIA
TEL 812 36 4062
TLF
EML
COM 40
TLX 8458488 TIFR IN

KRISHNAN THIRUVENKATA MR
HELIOS ANTENNAS/ELECTRON
234 AVVAI SHANMUGHAM RD
GOPALAPURAM
MADRAS 600 086
INDIA
TEL 44 472 680
TLF
TLX
EML
COM 40

KRISS GERARD A DR
DPT PHYSICS & ASTRONOMY
JOHNS HOPKINS UNIVERSITY
CHARLES & 34TH ST
BALTIMORE MD 21218
USA
TEL 301 338 7679
TLF
TLX 9101300225 JHU CASMO
EML
COM 47

KRISTENSEN LEIF KAHL
INST OF PHYSICS & ASTRON
UNIVERSITY OF AARHUS
NY MUNKEGADE
DK 8000 AARHUS C
DENMARK
TEL 86 12 8899
TLF 86 20 2711
TLX
EML
COM 05,15,20

KRISTENSON HENRIK DR
DORTTININGGATAN 20
S 432 410 VARBERG
SWEDEN
TEL
TLF
EML
COM
TLX

KRISTIAN JEROME DR
MT WILSON & LAS CAMPANAS
OBSERVATORIES
813 SANTA BARBARA ST
PASADENA CA 91101
USA
TEL 818 577 1122
TLF
TLX
EML
COM

KRISTIANSSON KRISTER PROF
DPT OF PHYSICS
UNIVERSITY OF LUND
SOELVEGATAN 14
S 223 62 LUND
SWEDEN
TEL 46 10 7726
TLF
TLX
EML
COM 48

KRIVSKY LADISLAV DR
ASTRONOMICAL INSTITUTE
CZECH ACADEMY OF SCIENCES
ONDREJOV OBSERVATORY
CS 251 65 ONDREJOV
CZECHOSLOVAKIA
TEL 204 85201
TLF 204 85314
TLX 121579 ASTR C
EML
COM 10

KRIZ SVATOPLUK DR
ASTRONOMICAL INSTITUTE
CZECH ACADEMY OF SCIENCES
ONDREJOV OBSERVATORY
CS 251 65 ONDREJOV
CZECHOSLOVAKIA
TEL 204 85201
TLF 204 85314
TLX
EML
COM 42

KROGDAHL W S DR
DPT PHYSICS & ASTRONOMY
UNIVERSITY OF KENTUCKY
LEXINGTON KY 40506
USA
TEL 606 272 2659
TLF
TLX
EML
COM

KROLIK JULIAN H
DPT PHYSICS & ASTRONOMY
JOHNS HOPKINS UNIVERSITY
CHARLES & 34TH ST
BALTIMORE MD 21218
USA
TEL 301 338 7926
TLF
TLX
EML
COM

KRON GERALD E DR
PINECREST OBSERVATORY
AT QUEEN'S COURT
2929 PONI MOI RD
HONOLULU HI 96815
USA
TEL 808 922 1514
TLF
TLX
EML
COM 37,45

KRON KATHERINE GORDON
PINECREST OBSERVATORY
AT QUEEN'S COURT
2929 PONI MOI RD
HONOLULU HI 96815
USA
TEL 808 922 1514
TLF.
TLX
EML
COM 42

KRON RICHARD G
YERKES OBSERVATORY
UNIVERSITY OF CHICAGO
BOX 258
WILLIAMS BAY WI 53191
USA
TEL 312 236 5468
TLF
TLX.
EML
COM 28

KRONBERG PHILIPP DR
DPT OF ASTRONOMY
UNIVERSITY OF TORONTO
60 ST GEORGE ST
TORONTO ON M5S 1A7
CANADA
TEL 416 978 4971
TLF 416 978 3921
TLX 069 86766
EML
COM 40

KROTO HAROLD PROF
SCHOOL OF CHEMISTRY
UNIVERSITY OF SUSSEX
FALMER
BRIGHTON BNI 9QJ
UK
TEL 273 67 8329
TLF
TLX
EML BIT KAFE4@CLUSTER SUSSEX AC UK EML
COM 14

KRUCHINENKO VITALIY G
ASTRONOMICAL OBSERVATORY
KIEV STATE UNIVERSITY
OBSERVATORNAYA UL 3
252053 KIEV
UKRAINE
TEL
TLF
TLX
EML
COM 22,42

KRUEGEL ENDRIK DR
MPI FUER RADIOASTRONOMIE
AUF DEM HUEGEL 69
D 5300 BONN 1
GERMANY
TEL
TLF
TLX 886440
EML
COM 40

KRUEGER ALBRECHT DR
ZNTRLINST F ASTROPHYSIK

TELEGRAFENBERG
D 1500 POTSDAM
GERMANY
TEL 4551
TLF
TLX 15239
EML
COM 10,12,40

KRUMM NATHAN ALLYN
DPT OF PHYSICS
UNIVERSITY OF CINCINNATI
210 BRAUNSTEIN ML 11
CINCINNATI OH 45221 0111
USA
TEL 513 475 2232
TLF
TLX
EML
COM 28

KRUPP EDWIN C DR
GRIFFITH OBSERVATORY
2800 EAST OBS ROAD
LOS ANGELES CA 90027
USA
TEL 213 664 1181
TLF
TLX
EML
COM. 41,46

KRUSZEWSKI ANDRZEJ PROF
ASTRONOMICAL OBSERVATORY
WARSAW UNIVERSITY
AL UJAZDOWSKIE 4
PL 00 478 WARSAW
POLAND
TEL
TLF
TLX
EML
COM 42

KRYGIER BERNARD DR
INST OF RADIO ASTRONOMY
N COPERNICUS UNIVERSITY
UL GAGARINA 11
PL 87 100 TORUN
POLAND
TEL 873327
TLF 11651
TLX
EML
COM 40

KRZEMINSKI WOJCIECH DR
CARNEGIE INST WASHINGTON
LAS CAMPANAS OBSERVATORY
CASILLA 601
LA SERENA
CHILE
TEL 51 21 3032
TLF
TLX 645 227 AURA CT
EML
COM 27,42

KSANFOMALITI L V DR
SPACE RESEARCH INSTITUTE
ACADEMY OF SCIENCES
PROFSOYUZNAYA UL 84/32
117810 MOSCOW
RUSSIA
TEL 333-2322/3122
TLF
TLX 411498 STAR SU
EML
COM 16,51

KUBIAK MARCIN A DR
ASTRONOMICAL OBSERVATORY
WARSAW UNIVERSITY
AL UJAZDOWSKIE 4
PL 00 478 WARSAW
POLAND
TEL 295346/294011
TLF
TLX 813978 ZAPAN PL
EML
COM 27

KUBICELA ALEKSANDAR DR
ASTRONOMICAL OBSERVATORY
VOLGINA 7
YU 11050 BEOGRAD
YUGOSLAVIA
TEL 11 419 357
TLF
TLX
EML
COM 12

KUBO YOSHIO
HYDROGRAPHIC DPT
GEODESY & GEOPHYSICS DIV
TSUKIJI 5 CHUO KU
TOKYO 104
JAPAN
TEL 03-541-3811
TLF
TLX 02522222 JAHYD J
EML
COM 04C

KUBOTA JUN DR
KWASAN OBSERVATORY
KYOTO UNIVERSITY
YAMASHINA
KYOTO 607
JAPAN
TEL 75-581-1235
TLF
TLX 5422693 LIBKYU J
EML
COM 10

KUDRITZKI ROLF-PETER PH D
INST F ASTRON & ASTROPHYS
SCHEINERSTR 1
D 8000 MUENCHEN 80
GERMANY
TEL 89 989 021
TLF
TLX 529815 UNIVM D
EML
COM 36

KUEHNE CHRISTOPH F
KOEFLACHER STR 36

D 7928 GIENGEN/BRENZ
GERMANY
TEL 732 24448
TLF
TLX
EML
COM 09

KUEHR HELMUT
MPI FUER ASTRONOMIE
KOENIGSTUHL
D 6900 HEIDELBERG 1
GERMANY
TEL 62 215 281
TLF .
TLX 461789 MPIA D
EML
COM

KUENZEL HORST
DIESELSTR 13
D 1502 POTSDAM
GERMANY
TEL· 77318
TLF
EML
COM 10
TLX

KUHI LEONARD V PROF
DPT OF ASTRONOMY
UNIVERSITY OF MINNESOTA
116 CHURCH ST SE
MINNEAPOLIS MN 55455
USA
TEL
TLF
TLX
EML
COM 27,36

KUHN JEFFERY RICHARD DR
PHYSICS & ASTRONOMY DPT
MICHIGAN STATE UNIVERSITY
EAST LANSING MI 48824
USA
TEL 517 353 2986
TLF
TLX
EML KUHN@MSUPA BITNET
COM

KUIJPERS H JAN M E DR
STERREKUNDIG INSTITUTE
BOX 80000
NL 3508 TA UTRECHT
NETHERLANDS
TEL 30 53 5209
TLX 40048 FYLUT NL
EML KUIJPERS@HUTRUU51
TLF 30 53 1601
COM 40

KUIPER THOMAS B H DR
JPL
MS 169 5065
4800 OAK GROVE DR
PASADENA CA 91109
USA
TEL 818 354 5479
TLF
TLX. 675429
EML·
COM. 34,40,51

KUKLIN G V DR
SIBIZMIR
ACADEMY OF SCIENCES
664697 IRKUTSK 33
RUSSIA
TEL 6-02-65
TLF
TLX
EML
COM 10,12

KULCAR LADISLAV DR
PEDAGOGICKA FAKULTA
KATEDRA FYZIKY
TAJOVSKEHO 40
CS 975 49 BANSKA BYSTRICE
CZECHOSLOVAKIA
TEL 88 3 4553
TLF 88 3 3132
TLX
EML
COM 12

KULKARNI PRABHAKAR V PROF
PHYSICAL RESEARCH LAB
NAVRANGPURA
AHMEDABAD 380 009
INDIA
TEL 272 46 2129
TLF 272 44 5292
TLX 8458488 TIFR IN
EML
COM 09,21,25,40

KULKARNI SHRINIVAS R DR
CALTECH
MS 105 24
PASADENA CA 91125
USA
TEL 818 356 4010
TLF
TLX 188192 675425
EML
COM 40

KULKARNI VASANT K DR
TATA INST OF FUNDAMENTAL
RESEARCH
POONA UNIVERSITY CAMPUS
PUNE 411 007
INDIA
TEL 212 33 7107
TLF 212 33 5760
TLX 0145658 GMRT IN
EML uunet'shakti'gmrt'vasant
COM 40

KULSRUD RUSSELL M DR
DPT ASTROPHYSICAL SCI
PRINCETON UNIVERSITY
PRINCETON NJ 08544 1001
USA
TEL 609 683 2613
TLF
TLX
EML
COM 33,48

KULTIMA JOHANNES
GEOPHYSICAL OBSERVATORY
EISCAT
SF 99600 SOKANKYLAE
FINLAND
TEL
TLF
TLX
EML —
COM

KUMAJGORODSKAYA RAISA DR
SPECIAL ASTROPHYSICAL OBS
ACADEMY OF SCIENCES
NIZHNIJ ARKHYZ
357147 STAVROPOLSKIJ
RUSSIA
TEL 93-515
TLF
TLX
EML
COM

KUMAR C KRISHNA DR
DPT PHYSICS & ASTRONOMY
HOWARD UNIVERSITY
WASHINGTON DC 20059
USA
TEL 202 636 6245
TLF
TLX
EML
COM 34

KUMAR SHAILENDRA
4 CLOVERLEAF DRIVE
MARLBORO NJ 07746
USA
TEL
TLF
TLX
EML
COM

KUMAR SHIV S PROF
UNIVERSITY STATION
UNIVERSITY OF VIRGINIA
BOX 3818
CHARLOTTESVILLE VA 22903
USA
TEL 804 924 4896
TLF
TLX
EML
COM 16,35,36

KUMKOVA IRINA I DR
INST OF APPLIED ASTRONOMY
ACADEMY OF SCIENCES
ZDANOVSKAYA UL 8
197042 ST PETERSBURG
RUSSIA
TEL
TLF
TLX
EML
COM 24

KUMSIASHVILY MZIA I DR
ABASTUMANI ASTROPHYSICAL
OBSERVATORY
GEORGIAN ACAD OF SCIENCES
383762 ABASTUMANI
GEORGIA
TEL 252
TLF
TLX 327409
EML
COM 42

KUMSISHVILI J I DR
ABASTUMANI ASTROPHYSICAL
OBSERVATORY
GEORGIAN ACAD OF SCIENCES
383762 ABASTUMANI
GEORGIA
TEL 279
TLF
TLX 327409
EML
COM 26,27

KUN MARIA DR
KONKOLY OBSERVATORY
THEGE U 13/17
BOX 67
H 1525 BUDAPEST
HUNGARY
TEL 1 75 5866/75 4122
TLF
TLX 227460 KONOB H
EML.
COM 37

KUNCHEV PETER DR
DPT OF ASTRONOMY
UNIVERSITY OF SOFIA
ANTON IVANOV ST 5
BG 1126 SOFIA
BULGARIA
TEL 2 54 4852
TLF
TLX
EML
COM 28

KUNDT WOLFGANG PROF DR
INSTITUT F ASTROPHYSIK
& EXTRATERR FORSCHUNG
AUF DEM HUEGEL 71
D 5300 BONN 1
GERMANY
TEL 228 26 7400
TLF
TLX 886440
EML
COM 40,48

KUNDU MUKUL R DR
ASTRONOMY PROGRAM
UNIVERSITY OF MARYLAND
COLLEGE PARK MD 20742
USA
TEL 301 454 3005
TLF
TLX 710-826-0352
EML
COM 10,12,34,40

KUNIEDA HIDEYO DR
DPT OF ASTROPHYSICS
NAGOYA UNIVERSITY
FUROCHO CHIKUSA KU
NAGOYA 464
JAPAN
TEL 52 781 5111
TLF 52 781 3541
TLX 4477323 SCUNAG J
EML
COM

KUNITZSCH PAUL PROF
INST FOR SEMITIC STUDIES
UNIVERSITY OF MUNICH
VETERINAERSTRASSE 1
D 8000 MUENCHEN 22
GERMANY
TEL 89 2180*2352
TLF
TLX 916280
EML
COM 41

KUNKEL WILLIAM E DR
CARNEGIE INST WASHINGTON
LAS CAMPANAS OBSERVATORY
CASILLA 601
LA SERENA
CHILE
TEL 51 21 3032
TLF
TLX 645 227 AURA CT
EML
COM 25,27

KUNTH DANIEL
INSTITUT D'ASTROPHYSIQUE
98BIS BD ARAGO
F 75014 PARIS
FRANCE
TEL 1 43 20 1425
TLF 1 43 29 8673
TLX
EML
COM 28,34,47

KUNZE RUEDIGER DR
INST THEOR PHYSIK STERN
DER UNIVERSITAET KIEL
OLSHAUSENSTR 40
D 2300 KIEL 1
GERMANY
TEL 431 880 1575
TLF 431 880 4432
TLX 292706
EML PAS29@RZ UNI-KIEL DBP DE
COM 34

KUPERUS MAX PROF DR
STERREKUNDIG INSTITUTE
BOX 80000
NL 3508 TA UTRECHT
NETHERLANDS
TEL 30 53 5212
TLX 40048 FYLUT NL
EML BITNET WNMMAIL@HUTRUU0
TLF
COM 10,12

KURFESS JAMES D
NAVAL RESEARCH LABORATORY
CODE 4150
4555 OVERLOOK AVE SW
WASHINGTON DC 20375 5000
USA
TEL 202 767 3182
TLF
TLX
EML
COM

KURIL-CHIK V N DR
STERNBERG STATE ASTR INST

UNIVERSITETSKIJ PROSP 13
119899 MOSCOW
RUSSIA
TEL 139-10-30
TLF
TLX
EML
COM 40

KUROCHKA L N DR
ASTRONOMICAL OBSERVATORY
KIEV STATE UNIVERSITY
OBSERVATORNAYA UL 3
252053 KIEV
UKRAINE
TEL 26 2691
TLF
TLX
EML
COM 10,12

KUROKAWA HIROKI DR
HIDA OBSERVATORY
UNIVERSITY OF KYOTO
KAMITAKARA
GIFU 506 13
JAPAN
TEL 0578-6-2628
TLF
TLX
EML
COM 10

KURPINSKA-WINIARSKA M DR
ASTRONOMICAL OBSERVATORY
JAGIELLONIAN UNIVERSITY
UL ORLA 171
PL 30 244 KRAKOW
POLAND
TEL
TLF
TLX 0322297 UJ PL
EML
COM 42

KURT V G DR
SPACE RESEARCH INSTITUTE
ACADEMY OF SCIENCES
PROFSOYUZNAYA UL 84/32
117810 MOSCOW
RUSSIA
TEL 333-31-22
TLF
TLX 411498 STAR SU
EML
COM. 16,44,48

KURTZ DONALD WAYNE DR
DPT OF ASTRONOMY
UNIVERSITY OF CAPE TOWN
RONDEBOSCH 7700
SOUTH AFRICA
TEL 69 8531
TLF
TLX 521439
EML
COM 25,27C,45

KURUCZ ROBERT L DR
CENTER FOR ASTROPHYSICS
HCO/SAO
60 GARDEN ST
CAMBRIDGE MA 02138
USA
TEL 617 495 7429
TLF
TLX 921428
EML Kurucz@cfa harvard edu
COM 36

KURZYNSKA KRYSTYNA DR
ASTRONOMICAL OBSERVATORY
A MICKIEWICZ UNIVERSITY
UL SLONECZNA 36
PL 60 286 POZNAN
POLAND
TEL 61 679 670
TLF 61 535 535
TLX 413260 UAMPL
EML KURZASTR@PLPUAMM
COM 08

KUS ANDRZEJ JAN DR
INST OF RADIOASTRONOMY
N COPERNICUS UNIVERSITY
UL CHOPINA 12/18
PL 87 100 TORUN
POLAND
TEL 04856-20651
TLF
TLX 0552324 TRAO PL
EML
COM 40

KUSHWAHA R S PROF
DPT OF MATHEMATICS
UNIVERSITY OF JODHPUR
JODHPUR
INDIA
TEL
TLF
TLX
EML
COM 35,36

KUSTAANHEIMO PAUL E PROF
DANMARKS TEKN HOJSKOLE
LUNDTOFTEVEJ 7
DK 2800 LYNGBY
DENMARK
TEL 42 88 3022
TLF
TLX
EML
COM 07,28,47

KUSUNOSE MASAAKI DR
COPERNICUS ASTRON CENTER
POLISH ACAD OF SCIENCES
UL BARTYCKA 18
PL 00 716 WARSAW
POLAND
TEL 22 411 086
TLF 22 410 046
TLX 813978 ZAPLAN
EML KUSUNOSE@CAMK EDU PL
COM 48

KUTNER MARC LESLIE DR
DPT OF PHYSICS
RENSSELAER POLYTECHN INST
TROY NY 12180 3590
USA
TEL 518 266 6417
TLF
TLX
EML
COM 34,40

KUTTER G SIEGFRIED DR
NASA/GSFC
CODE 681
GREENBELT MD 20771
USA
TEL
TLF
TLX
EML
COM

KUTUZOV S A DR
ASTRONOMICAL OBSERVATORY
ST PETERSBURG UNIVERSITY
BIBLIOTECHNAJA PL 2
199164 ST PETERSBURG
RUSSIA
TEL.
TLF
TLX
EML
COM 33

KUZMANOSKI MIKE
INSTITUTE OF ASTRONOMY
UNIVERSITY OF BELGRADE
STUDENTSKI TRG 16
YU 11000 BEOGRAD
YUGOSLAVIA
TEL 11 638 715
TLF
TLX
EML
COM

KUZMIN ARKADII D PROF DR
LEBEDEV PHYSICAL INST
ACADEMY OF SCIENCES
LENINSKY PROSPEKT 53
117924 MOSCOW
RUSSIA
TEL
TLF
TLX 411479 NEOD SU
EML
COM 16,40,51

KVIZ ZDENEK'DR
SCHOOL OF PHYSICS
UNIVERSITY OF SOUTH WALES
BOX 1
KENSINGTON NSW 2033
AUSTRALIA
TEL 2 697 4578
TLF
TLX 26054 AA
EML
COM 22,25,42

KWEE K K DR
STERREWACHT
BOX 9513
NL 2300 RA LEIDEN
NETHERLANDS
TEL 71 27 2727
TLX: 39058 ASTRO NL
EML
TLF
COM 27,42

KWITTER KAREN BETH DR
DPT OF PHYS & ASTRONOMY
THOMPSON PHYSICS LAB
WILLIAMS COLLEGE
WILLIAMSTOWN MA 01267
USA
TEL 413 597 2272
TLF
TLX
EML
COM 34

KWOK SUN DR
DPT OF PHYSICS
UNIVERSITY OF CALGARY
2500 UNIVERSITY DR NW
CALGARY AB T2N 1N4
CANADA
TEL 403 284 5414
TLF 403 289 3331
TLX 038 21545
EML
COM 29,34,35,40

KYLAFIS NIKOLAOS D DR
DPT OF PHYSICS
UNIVERSITY OF CRETE
BOX 1527
GR 711 11 IRAKLION
GREECE
TEL 81 23 9757
TLF
TLX
EML
COM 34

LA BONTE BARRY JAMES
INSTITUTE FOR ASTRONOMY
UNIVERSITY OF HAWAII
2680 WOODLAWN DR
HONOLULU HI 96822
USA
TEL 808 948 6531
TLF
TLX 723-8459 UHAST HR
EML
COM 12

LA DOUS CONSTANZE A DR
NASA/GSFC
CODE 630 2
GREENBELT MD 20771
USA
TEL 301 286 9793
TLF
TLX 817297 ASTRON G
EML NSSDCA LADOUS
COM 42

LA PADULA CESARE
IAS
CNR
CP 67
I 00044 FRASCATI
ITALY
TEL 6 942 5655
TLF
TLX
EML
COM

LABAY JAVIER
DPT FISICA DE ATMOSFERA
UNIVERSIDAD DE BARCELONA
AVD DIAGONAL 645
E 08028 BARCELONA
SPAIN
TEL 3 330 7311
TLF
TLX
EML
COM 35

LABEYRIE ANTOINE DR
OCA CERGA
AVE COPERNIC
F 06130 GRASSE
FRANCE
TEL 93 36 5849
TLF
TLX 461402
EML
COM 09

LABEYRIE JACQUES DR
CEA CEN
CFR
BP 2
F 91191 GIF/YVETTE CDX
FRANCE
TEL 1 69 67 7828
TLF
TLX 691137
EML
COM

LABHARDT LUKAS
ASTRONOMISCHES INSTITUT
UNIVERSITAET BASEL
VENUSSTRASSE 7
CH 4102 BINNINGEN
SWITZERLAND
TEL 61 22 7711
TLF
TLX
EML
COM 25,45

LABS DIETRICH PROF
LANDESSTERNWARTE
KOENIGSTUHL
D 6900 HEIDELBERG 1
GERMANY
TEL 62 211 0036
TLF
TLX 461153 LSWHD D
EML
COM 12

LACEY CEDRIC DR
DPT OF ASTROPHYSICS
UNIVERSITY OF OXFORD
KEBLE RD
OXFORD OX1 3RH
UK
TEL 865 273 352
TLF 865 273 418
TLX 83295 NUCLOX G
EML· 19464 CGL (SPAN)
COM· 47

LACHIEZE-REY MARC
CEA CEN
DAPNIA/SAP
BP 2
F 91191 GIF/YVETTE CDX
FRANCE
TEL 1 69 08 6292
TLF
TLX 690860
EML
COM 47

LACLARE F MR
OCA CERGA
AVE COPERNIC
F 06130 GRASSE
FRANCE
TEL 93 36 5849
TLF
TLX
EML
COM 08

LACROUTE PIERRE A PROF
2 RUE D'ALISE
F 21000 DIJON
FRANCE
TEL 80 66 1154
TLF
EML
COM 08,24
TLX

LACY CLAUD H DR
DPT OF PHYSICS
UNIVERSITY OF ARKANSAS
104 PHYSICS BDG
FAYETTEVILLE AR 72701
USA
TEL 501 575 2506
TLF
TLX
EML
COM 42

LACY JOHN H DR
ASTRONOMY DPT
UNIVERSITY OF TEXAS
RLM 15 308
AUSTIN TX 78712 1083
USA
TEL 512 471 1469
TLF
TLX
EML
COM

LADA CHARLES JOSEPH DR
STEWARD OBSERVATORY
UNIVERSITY OF ARIZONA
TUCSON AZ 85721
USA
TEL 602 621 4878
TLF
TLX
EML
COM 34,37,40

LAFFINEUR MARIUS MR
21 BD BRUNE
F 74014 PARIS
FRANCE
TEL
TLF
EML
COM
TLX

LAFON JEAN-PIERRE J DR
OBSERVATOIRE DE PARIS
SECTION DE MEUDON
F 92195 MEUDON PPL CDX
FRANCE
TEL· 1 45 07 7858
TLF
TLX· 204464
EML
COM 28,33,34,49,51

LAGERKVIST CLAES-INGVAR
ASTRONOMICAL OBSERVATORY
BOX 515
S 751 20 UPPSALA
SWEDEN
TEL 18 11 3522
TLF
TLX 76024 UNIV UPS S
EML
COM 15,20

LAGERQVIST ALBIN PROF
INST OF THEORETICAL PHYS

VANADISVAEGEN 9
S 113 46 STOCKHOLM
SWEDEN
TEL 81 64 500
TLF
TLX 15433 FYSTO S
EML
COM 14

LAGO MARIA TERESA V T PR
GRUPO DE MATEM APLICADA
UNIVERSIDADE DO PORTO
RUA DAS TAIPAS 135
P 4000 PORTO
PORTUGAL
TEL 380313
TLF
TLX 28109
EML
COM 27,29,46

LAHAV OFER DR
INSTITUTE OF ASTRONOMY
THE OBSERVATORIES
MADINGLEY RD
CAMBRIDGE CB3 OHA
UK
TEL 223 33 7548
TLF 223 33 7523
TLX 817297 ASTRON G
EML OL1@AST-STAR CAM AC UK
COM 47

LAHIRI N C
INDIAN METEOROLOGIC DPT
P 546 BLOCK N
NEW ALIPORE
CALCUTTA 700 053
INDIA
TEL
TLF
TLX
EML
COM 04

LAHULLA J FORNIES DR
OBS ASTRONOMICO NCL

ALFONSO XII-3
E 28014 MADRID
SPAIN
TEL 1 227 0107
TLF
TLX 22465 IGC E
EML
COM

LAI SEBASTIANA
ISTITUTO DI ASTRONOMIA
VIA OSPEDALE 72
I 09100 CAGLIARI
ITALY
TEL 70 66 3544
TLX
EML
TLF
COM 46,49

LAING ROBERT
ROYAL GREENWICH OBS
HERSTMONCEUX CASTLE
HAILSHAM BN27 1RP
UK
TEL 323 833 171
TLX 87451 RGOBS G
EML
TLF
COM· 40

LAIRD JOHN B DR
DPT PHYSICS & ASTRONOMY
BOWLING GREEN STATE UNIV
BOWLING GREEN OH 43403
USA
TEL 419 372 7244
TLF
TLX
EML LAIRD@ANDY BGSU EDU
COM 29

LAKE KAYLL WILLIAM DR
DPT OF PHYSICS
QUEEN'S UNIVERSITY
KINGSTON ON K7L 3N6
CANADA
TEL 613 547 3020
TLF 613 545 6463
TLX
EML
COM 47

LAL DEVENDRA
PHYSICAL RESEARCH LAB
NAVRANGPURA
AHMEDABAD 380 009
INDIA
TEL 272 46 2129
TLF 272 44 5292
TLX
EML
COM

LALA PETR DR
OUTER SPACE AFFAIRS DIV
UNITED NATIONS SECRETARIA
NEW YORK NY 10017
USA
TEL
TLF
TLX
EML
COM 07

LALLEMENT ROSINE DR
SERVICE D'AERONOMIE
BP 3
F 91371 VERRIERES BUISSON
FRANCE
TEL 1 64 47 4235
TLF
TLX 602400
EML BITNET ROSINE@FRIAP51
COM

LAMB DONALD QUINCY JR DR
UNIV OF CHICAGO PRESS
UNIVERSITY OF CHICAGO
5801 S ELLIS AVE
CHICAGO IL 60637
USA
TEL 312 962 8203
TLF
TLX 269266
EML
COM 35,42,48

LAMB FREDERICK K PROF
DPT OF PHYSICS
UNIVERSITY OF ILLINOIS
1110 W GREEN ST
URBANA IL 61801
USA
TEL   217 333 6363
TLF
TLX   6502272050 MCI
EML
COM 48

LAMB RICHARD C DR
DPT OF PHYSICS
IOWA STATE UNIVERSITY
AMES IA 50011
USA
TEL   515 294 3873
TLF
TLX
EML. BITNET LAMB@ALISUVAX
COM

LAMB SUSAN ANN DR
DPT OF PHYSICS
UNIVERSITY OF MISSOURI
8001 NATURAL BRIDGE RD
ST LOUIS MO 63121
USA
TEL
TLF
TLX
EML
COM 35,48

LAMBECK KURT PROF
AUSTRALIAN NTL UNIVERSITY
RES SCHOOL EARTH SCIENCE
BOX 4
CANBERRA ACT 2600
AUSTRALIA
TEL   62 492 487
TLF
TLX 62693
EML
COM

LAMBERT DAVID L PROF
ASTRONOMY DPT
UNIVERSITY OF TEXAS
RLM 15 308
AUSTIN TX 78712 1083
USA
TEL 512 471 7438
TLF 512 471 6016
TLX 9108741351
EML DLL@ASTRO AS UTEXAS EDU
COM 14,29P,36

LAMERS HENNY J G L M DR
SPACE RESEARCH LABORATORY
SRON
SORBONNELAAN 2
NL 3584 CA UTRECHT
NETHERLANDS
TEL 30 53 5720
TLF 30 54 0860
TLX 47224 SRON NL
EML HENNYLE@ SRON RUU NL
COM 29,36,44

LAMLA ERICH E DR
BRUESSELERSTR 9
D 5300 BONN 1
GERMANY
TEL
TLF
TLX
EML
COM

LAMONTAGNE ROBERT DR
DPT DE PHYSIQUE
UNIVERSITY DE MONTREAL
CP 6128 SUCC A
MONTREAL QC H3C 3J7
CANADA
TEL 514 342 7273
TLF 514 343 2071
TLX 055 62425
EML 5007@CC UMONTREAL CA
COM 29

LAMPENS PATRICIA DR
OBSERVATOIRE ROYAL DE
BELGIQUE
AVE CIRCULAIRE 3
B 1180 BRUSSELS
BELGIUM
TEL   2 373 0263
TLF   2 374 9822
TLX 21565 OBSBEL
EML PATRICIA@ASTRO OMA BE
COM 26

LAMPTON MICHAEL
SPACE SCIENCES LABORATORY
UNIVERSITY OF CALIFORNIA
BERKELEY CA 94720
USA
TEL 415 642 3576
TLF
TLX 9103667945
EML
COM 48

LAMY PHILIPPE DR
LAS
TRAVERSE DU SIPHON
LES TROIS LUCS
F 13012 MARSEILLE
FRANCE
TEL 91 05 5932
TLF 91 66 1855
TLX 420584 F
EML
COM 15,21C,22

LANCASTER BROWN PETER
10A ST PETER'S ROAD

ALDEBURGH SUFFOLK
UK
TEL
TLF
TLX
EML
COM 15

LANDE KENNETH PROF
DPT OF PHYSICS
UNIV OF PENNSYLVANIA
PHILADELPHIA PA 19104
USA
TEL 215 898 8177
TLF
TLX
EML
COM

LANDECKER PETER BRUCE DR
HUGHES AIRCRAFT CO
SPACE & COMM GR BLDG S41
MS B322 BOX 92919
LOS ANGELES CA 90009
USA
TEL 213 648 0815
TLF
TLX 664480
EML
COM

LANDECKER THOMAS L DR
DOMINION RADIO ASTROPHYS
OBSERVATORY
BOX 248
PENTICTON BC V2A 6K3
CANADA
TEL 604 493 2277
TLF 604 493 7767
TLX 048 88127
EML
COM

LANDI DEGL'INNOCENTI E PR
DPT DI ASTRONOMIA
UNIVERSITA DI FIRENZE
LARGO E FERMI 5
I 50125 FIRENZE
ITALY
TEL 55 27521
TLF 55 22 0039
TLX 572268 ARCETR I
EML
COM 12C

LANDI DEGL'INNOCENTI M
OSS ASTROFISICO
DI ARCETRI
LARGO E FERMI 5
I 50125 FIRENZE
ITALY
TEL 55 275 2256
TLF
TLX 572268 ARCETR
EML
COM 12

LANDINI MASSIMO PROF
OSS ASTROFISICO
DI ARCETRI
LARGO E FERMI 5
I 50125 FIRENZE
ITALY
TEL 55 275 2247
TLF
TLX
EML
COM

LANDI-DESSY J DR
OBSERVATORIO ASTRONOMICO
DE CORDOBA
LAPRIDA 854
5000 CORDOBA
ARGENTINA
TEL 51 40613
TLF
TLX
EML
COM

LANDMAN DONALD ALAN
525 CAMINO LAGUNA VISTA

GOLETA CA 93117
USA
TEL
TLF
TLX
EML
COM 10,12,14

LANDOLFI MARCO
OSS ASTROFISICO
DI ARCETRI
LARGO E FERMI 5
I 50125 FIRENZE
ITALY
TEL 55 275 2256
TLF
TLX 572268 ARCETR
EML
COM 12

LANDOLT ARLO U PROF
DPT PHYSICS & ASTRONOMY
LOUISIANA STATE UNIV
BATON ROUGE LA 70803 4001
USA
TEL· 504 388 8276
TLF
TLX· 559184
EML
COM 25C,27,37,42

LANDSTREET JOHN D PROF
DPT OF ASTRONOMY
UNIV OF WESTERN ONTARIO
LONDON ON N6A 3K7
CANADA
TEL 519 679 3184
TLF 519 661 3486
TLX 064 7134
EML· QLANDSTR@UWOVAX BITNET
COM 25VP,29,36

LANE ADAIR P
DPT OF ASTRONOMY
BOSTON UNIVERSITY
725 COMMONWEALTH AVE
BOSTON MA 02215
USA
TEL 617 353 2633
TLF
TLX
EML
COM

LANE ARTHUR LONNE DR
JPL
4800 OAK GROVE DR
PASADENA CA 91109
USA
TEL 818 345 2725
TLX
EML
TLF
COM 15,16

LANEY CLIFTON D DR
SAAO
BOX 9
OBSERVATORY 7935
SOUTH AFRICA
TEL 470-025
TLF
TLX.
EML·
COM· 27

LANG JAMES DR
RUTHERFORD APPLETON LAB
SPACE & ASTROPHYSICS DIV
BLDG R25/R68
CHILTON DIDCOT OX11 0QX
UK
TEL 235 219 00
TLF
TLX 83159
EML
COM 14

LANG KENNETH R ASST PROF
DPT OF PHYSICS
TUFTS UNIVERSITY
ROBINSON HALL
MEDFORD MA 02155
USA
TEL 617 381 3390
TLF
TLX
EML
COM 10,40,41

LANGER GEORGE EDWARD DR
PHYSICS DPT
COLORADO COLLEGE
COLORADO SPRINGS CO 80903
USA
TEL 303 473 2233*578
TLF
TLX
EML
COM· 29

LANGER NORBERT DR
UNIVERSITAET STERNWARTE
GOETTINGEN
GEISMARLANDSTR 11
D 3400 GOETTINGEN
GERMANY
TEL. 551 39 5054
TLF. 551 39 5043
TLX 96753
EML· NLANGER@DGOGWDG1 BITNET
COM 35

LANGER WILLIAM DAVID DR
JPL
MS 169 506
4800 OAK GROVE DR
PASADENA CA 91109
USA
TEL 818 354 5823
TLF
TLX
EML
COM 34,40

LANNES ANDRE DR
OBS MIDI PYRENEES
14 AVE E BELIN
F 31400 TOULOUSE CDX
FRANCE
TEL 61 25 2101
TLF
TLX 530 776 F
EML
COM

LANTOS PIERRE DR
OBSERVATOIRE DE PARIS
SECTION DE MEUDON
DASOP
F 92195 MEUDON PPL CDX
FRANCE
TEL 1 45 07 7767
TLF
TLX
EML
COM 05,10,12,40

LANZ THIERRY DR
NASA/GSFC
CODE 681
LASP
GREENBELT MD 20771
USA
TEL
TLF
TLX
EML LANZ@STARS GSFC NASA GOV
COM 29

LAPASSET EMILIO DR
OBSERVATORIO ASTRONOMICO
DE CORDOBA
LAPRIDA 854
5000 CORDOBA
ARGENTINA
TEL 51 36876
TLF
TLX 51822 BUCOR
EML
COM 37,42

LAPOINTE S M DR
UNIVERSITE DU QUEBEC
2875 BLD LAURIER
STE FOY QC G1V 2M3
CANADA
TEL 418 657 3551
TLF
TLX 051 31623
EML
COM   ·

LAPUSHKA K K DR
ASTRONOMICAL OBSERVATORY
LATVIAN STATE UNIVERSITY
RAINIS BUL 19
226098 RIGA
LATVIA
TEL 223149/611984
TLF
TLX
EML
COM 24

LAQUES PIERRE DR
OBS MIDI PYRENEES
9 R PONT DE LA MOULETTE
F 65200 BAGNERES BIGORRE
FRANCE
TEL 62 95 1969
TLF
TLX 531625
EML
COM 09,51

LARGE MICHAEL I DR
SCHOOL OF PHYSICS
UNIVERSITY OF SYDNEY
SYDNEY NSW 2006
AUSTRALIA
TEL 2 692 2222
TLF
TLX 26169 UNISYD AA
EML
COM 40

LARI CARLO DR
IST DI RADIOASTRONOMIA
CNR
VIA IRNERIO 46
I 40126 BOLOGNA
ITALY
TEL 51 23 2856
TLF
TLX
EML
COM

LARSON HAROLD P DR
LUNAR & PLANETARY LAB
UNIVERSITY OF ARIZONA
TUCSON AZ 85721
USA
TEL 602 621 6943
TLF
TLX
EML
COM 15,16

LARSON RICHARD B PROF
DPT OF ASTRONOMY
YALE UNIVERSITY
BOX 6666
NEW HAVEN CT 06520
USA
TEL 203 436 3015
TLF
TLX
EML
COM· 28,33,35

LARSON STEPHEN M
LUNAR & PLANETARY LAB
UNIVERSITY OF ARIZONA
TUCSON AZ 85721
USA
TEL 602 621 4973
TLF
TLX 910-952-1143
EML
COM 15,16

LARSSON STEFAN DR
STOCKHOLM OBSERVATORY

S 133 36 SALTSJOEBADEN
SWEDEN
TEL 81 64 464
TLF 87 17 4719
TLX
EML LARSSON@ASTRO SU SE
COM 42

LARSSON-LEANDER G PROF
LUND OBSERVATORY
BOX 43
S 221 00 LUND
SWEDEN
TEL 46 10 7000
TLX 331990BSNOT S
EML
TLF
COM 37,42

LASALA GERALD J DR
DPT OF PHYSICS
UNIV OF SOUTHERN MAINE
96 FALMOUTH ST
PORTLAND ME 04103
USA
TEL 207 780 4557
TLF 207 780 4933
TLX
EML LASALA@PORTLAND
COM 45

LASENBY ANTHONY
MULLARD RADIO ASTRON OBS
CAVENDISH LABORATORY
MADINGLEY RD
CAMBRIDGE CB3 OHE
UK
TEL 223 664 77
TLF
TLX 81292 CAVLAB G
EML
COM 40

LASHER GORDON JEWETT DR
IBM
THOMAS J WATSON RES CTR
BOX 218
YORKTOWN HEIGHTS NY10598
USA
TEL
TLF
TLX
EML
COM 48

LASKAR JACQUES DR
BUREAU DES LONGITUDES
77 AVE DENFERT ROCHEREAU
F 75014 PARIS
FRANCE
TEL 1 40 51 2274
TLF
TLX
EML
COM 04,07

LASKARIDES PAUL G ASSPROF
DPT OF ASTRONOMY
NTL UNIVERSITY OF ATHENS
PANEPISTIMIOPOLIS
GR 157 71 ZOGRAFOS
GREECE
TEL 1 724 3211
TLF
TLX
EML
COM 25,27,35

LASKER BARRY M DR
STSCI
HOMEWOOD CAMPUS
3700 SAN MARTIN DR
BALTIMORE MD 21218
USA
TEL 301 338 4840
TLF
TLX 6849191 STSI
EML
COM 09,25,28,34

LASOTA JEAN-PIERRE DR
OBSERVATOIRE DE PARIS
SECTION DE MEUDON
F 92195 MEUDON PPL CDX
FRANCE
TEL 1 45 07 7416
TLF
TLX 201571
EML
COM 35,47

LATHAM DAVID W DR
CENTER FOR ASTROPHYSICS
HCO/SAO
60 GARDEN ST
CAMBRIDGE MA 02138
USA
TEL 617 495 7215
TLF
TLX 921428 SATELLITE CAM
EML LATHAM@CFA3
COM 26,30C,33,37

LATOUR JEAN J
OBS MIDI PYRENEES
14 AVE E BELIN
F 31400 TOULOUSE CDX
FRANCE
TEL 61 25 2101
TLF
TLX
EML
COM 35

LATTANZI MARIO G
OSS ASTRONOMICO DI TORINO
ST OSSERVATORIO 20
I 10025 PINO TORINESE
ITALY
TEL 11 84 1067
TLF
TLX 213236 TO ASTR 1
EML SPAN 39181 LATTANZI
COM 08,26

LATTIMER JAMES M DR
DPT OF EARTH & SPACE SCI
ASTRONOMY PROGRAM
SUNY AT STONY BROOK
STONY BROOK NY 11794 2100
USA
TEL 516 246 8223
TLF
TLX
EML
COM 48

LATYPOV A A DR
ASTRONOMICAL INSTITUTE
UZBEK ACADEMY OF SCIENCES
700052 TASHKENT
UZBEKISTAN
TEL 358102
TLF
TLX 116012 VREMJA
EML
COM 24

LAUBERTS ANDRIS DR
FOA 3
BOX 1165
S 581 11 LINKOPING
SWEDEN
TEL 13 11 8235
TLF. 13 13 1665
TLX 50073 foatre S
EML
COM 28

LAUNAY JEAN-MICHEL DR
OBSERVATOIRE DE PARIS
SECTION DE MEUDON
F 92195 MEUDON PPL CDX
FRANCE
TEL 1 45 07 7554
TLF
TLX 201571
EML
COM 14

LAURENT BERTEL E PROF
INST OF THEORETICAL PHYS

VANADISVAEGEN 9
S 113 46 STOCKHOLM
SWEDEN
TEL 81 64 500
TLF
TLX 15433 FYSTO S
EML
COM

LAURENT CLAUDINE DR
INSU
77 AVE DENFERT ROCHEREAU
F 75014 PARIS
FRANCE
TEL 1 40 51 2118
TLF
TLX
EML
COM 34

LAUSBERG ANDRE DR
INSTITUT D'ASTROPHYSIQUE
UNIVERSITE DE LIEGE
AVE COINTE 5
B 4000 COINTE-LIEGE
BELGIUM
TEL 41 52 9980
TLF 41 52 7474
TLX
EML
COM 28,47

LAUTMAN D A DR
CENTER FOR ASTROPHYSICS
HCO/SAO
60 GARDEN ST
CAMBRIDGE MA 02138
USA
TEL
TLF
TLX
EML
COM

LAVAL ANNIE DR
OBSERVATOIRE DE MARSEILLE
2 PLACE LE VERRIER
F 13248 MARSEILLE CDX 04
FRANCE
TEL 91 95 9088
TLF
TLX 420241 F
EML
COM 37

LAVROV M I PROF
ENGELHARDT ASTRONOMICAL
OBSERVATORY
OBSERVATORIA STATION
422526 KAZAN
RUSSIA
TEL
TLF
TLX
EML
COM 42

LAVRUKHINA A K PROF DR
INST OF GEOCHEMISTRY
ANALYTICAL CHEMISTRY
USSR ACADEMY OF SCIENCES
117334 MOSCOW
RUSSIA
TEL 137-75-38
TLF
TLX
EML
COM

LAWRENCE ANDREW DR
ASTRONOMY UNIT
QUEEN MARY/WESTFIELD COLL
MILE END RD
LONDON E1 4NS
UK
TEL 1 975 5481
TLF
TLX 893750
EML AL@UK AC QMC STAR
COM 28

LAWRENCE CHARLES R DR
CALTECH
MS 105 24
PASADENA CA 91125
USA
TEL 818 356 4976
TLF
TLX 675429
EML BITNET CRL@CITDEIMO
COM 40

LAWRENCE G M DR
LASP
UNIVERSITY OF COLORADO
BOX 392
BOULDER CO 80309 0392
USA
TEL
TLF
TLX
EML
COM 14

LAWRIE DAVID G
AEROSPACE CORPORATION
MS M4 041
BOX 92957
LOS ANGELES CA 90009
USA
TEL 213 648 6142
TLF
·TLX
EML
COM

LAYZER DAVID PROF
CENTER FOR ASTROPHYSICS
HCO/SAO MS 31
60 GARDEN ST
CAMBRIDGE MA 02138
USA
TEL
TLF
TLX
EML
COM 14,28,47

LAZAREFF BERNARD DR
OBSERVATOIRE DE GRENOBLE
CERMO/ASTROPHYSIQUE
BP 53X
F 38041 S MARTIN HERES CD
FRANCE
TEL 76 51 4600
TLF 76 44 8821
TLX 980134 F
EML
COM

LAZARO CARLOS DR
INST DE ASTROFISICA
DE CANARIAS
OBS DEL TEIDE
E 38200 LA LAGUNA
SPAIN
TEL 22 26 2211
TLF
TLX 92640
EML
COM 27

LAZOVIC JOVAN P PROF
DPT OF ASTRONOMY
FACULTY OF SCIENCES
STUDENTSKI TRG 16
YU 11000 BEOGRAD
YUGOSLAVIA
TEL 11 638 715
TLF
TLX:
EML
COM 07

LAZZARO DANIELA DR
OBSERVATORIO NACIONAL
RUA GL BRUCE 586
SAN CRISTOVAO
20921 RIO DE JANEIRO RJ
BRAZIL
TEL 21 580 7181
TLF 21 580 0332
TLX 2121288 OBSN BR
EML DAZA@LNCCVM
COM 20

LE BORGNE JEAN FRANCOIS
OBS MIDI PYRENEES
14 AVE E BELIN
F 31400 TOULOUSE CDX
FRANCE
TEL 61 25 2101
TLF
TLX 530 776 F
EML BITNET LEBORGNE@FROMP51
COM

LE BOURLOT JACQUES DR
OBSERVATOIRE DE PARIS
SECTION DE MEUDON
DAEC
F 92195 MEUDON PPL CDX
FRANCE
TEL 1 45 07 7566
TLF 1 45 07 7469
TLX
EML LEBOURLOT@FRMEU51
COM 14

LE CONTEL JEAN-MICHEL
OCA OBSERV DE NICE
BP 139
F 06003 NICE CDX
FRANCE
TEL 93 89 0420
TLF
TLX 460004 OBSNICE F
EML
COM 29

LE DOURNEUF MARYVONNE
OBSERVATOIRE DE PARIS
SECTION DE MEUDON
F 92195 MEUDON PPL CDX
FRANCE
TEL 1 45 07 7555
TLF
TLX 201571
EML
COM 14

LE FEVRE OLIVIER DR
OBSERVATOIRE DE PARIS
SECTION DE MEUDON
DAECC
F 92195 MEUDON PPL CDX
FRANCE
TEL.
TLF
TLX
EML
COM 28

LE POOLE RUDOLF S DR
STERREWACHT
BOX 9513
NL 2300 RA LEIDEN
NETHERLANDS
TEL 71 27 2727
TLX 39058 ASTRO NL
EML
TLF
COM 24

LE SQUEREN ANNE-MARIE DR
OBSERVATOIRE DE PARIS
SECTION DE MEUDON
F 92195 MEUDON PPL CDX
FRANCE
TEL 1 45 07 7595
TLF
TLX
EML
COM 34,40

LEA SUSAN MAUREEN DR
DPT PHYSICS & ASTRONOMY
SAN FRANSISCO STATE UNIV
1600 HOLLOWAY AVE
SAN FRANCISCO CA 94132
USA
TEL 405 469 1880
TLF
TLX
EML
COM 48

LEACH SYDNEY DR
OBSERVATOIRE DE PARIS
SECTION DE MEUDON
DAMAP
F 92195 MEUDON PPL CDX
FRANCE
TEL 1 45 07 7561
TLF 47 07 7469
TLX 201571
EML LEACH@FRMEU51
COM 14

LEACOCK ROBERT JAY
DPT OF ASTRONOMY
UNIVERSITY OF FLORIDA
211 SSRB
GAINESVILLE FL 32611
USA
TEL 904 392 2052
TLF
TLX
EML
COM 28

LEAHY J PATRICK DR
NRAL
JODRELL BANK
MACCLESFIELD SK11 9DL
UK
TEL 477 713 21
TLX 36149
EML JPL@UK AC MAN JB STAR
TLF 477 716 18
COM 40

LEAHY DENIS A DR
DPT OF PHYSICS
UNIVERSITY OF CALGARY
2500 UNIVERSITY DR NW
CALGARY AB T2N 1N4
CANADA
TEL 403 220 7192
TLF 403 289 3331
TLX
EML BITNET LEAHY@UNCAMULT
COM

LEBEDINETS VLADIMIR N DR
INST OF ASTRONOMY
ACADEMY OF SCIENCES
PYATNITSKAYA UL 48
109017 MOSCOW
RUSSIA
TEL
TLF
TLX
EML
COM 22

LEBLANC YOLANDE DR
OBSERVATOIRE DE PARIS
SECTION DE MEUDON
F 92195 MEUDON PPL CDX
FRANCE
TEL 1 45 07 7759
TLF
TLX
EML
COM 40

LEBOFSKY LARRY ALLEN
LUNAR & PLANETARY LAB
UNIVERSITY OF ARIZONA
TUCSON AZ 85721
USA
TEL 602 621 6947
TLF
TLX
EML
COM 15

LEBOVITZ NORMAN R PROF
DPT OF MATHEMATICS
UNIVERSITY OF CHICAGO
5734 S UNIVERSITY AVE
CHICAGO IL 60637
USA
TEL 312 753 8074
TLF
TLX
EML
COM 35

LEBRETON YVELINE DR
OBSERVATOIRE DE PARIS
SECTION DE MEUDON
F 92195 MEUDON PPL CDX
FRANCE
TEL 1 45 07 7859
TLF
TLX 201571
EML LEBRETON@FRMEU51
COM 35

LECAR MYRON DR
CENTER FOR ASTROPHYSICS
HCO/SAO
60 GARDEN ST
CAMBRIDGE MA 02138
USA
TEL 617 495 7251
TLF
TLX 921428 SATELLITE CAM
EML
COM 33

LECKRONE DAVID S DR
NASA/GSFC
CODE 681
GREENBELT MD 20771
USA
TEL 301 286 8904
TLF
TLX
EML
COM 29,44

LEDERLE TRUDPERT DR
ASTRONOMISCHES-RECHEN
INSTITUT
MOENCHHOFSTR 12-14
D 6900 HEIDELBERG 1
GERMANY
TEL 62 214 9026
TLF
TLX 461336 ARIHD D
EML
COM 04,05,08

LEE HYUNG MOK DR
DPT EARTH SCIENCES
PUSAN NTL UNIVERSITY
KUM JONG KU
PUSAN
KOREA R
TEL 51 510 2702
TLF 51 513 7495
TLX
EML
COM 33

LEE PAUL D DR
DPT PHYSICS & ASTRONOMY
LOUISIANA STATE UNIV
BATON ROUGE LA 70803 4001
USA
TEL
TLF
TLX
EML
COM

LEE SANG GAK
DPT OF ASTRONOMY
SEOUL NTL UNIVERSITY
KWANAK KU
SEOUL 151
KOREA R
TEL 877-2131/2139
TLF.
TLX
EML
COM 33,45,51

LEE SEE-WOO DR
DPT OF ASTRONOMY
SEOUL NTL UNIVERSITY
SEOUL CITY
KOREA R
TEL 877-2131-9x3308
TLF
TLX
EML
COM

LEE TERENCE J DR
ROYAL OBSERVATORY
HEAD OF TECHNOLOGY
BLACKFORD HILL
EDINBURGH EH9 3HJ
UK
TEL 316 673 321
TLF
TLX 72383 ROEDING UK
EML
COM 34

LEE THYPHOON
INST EARTH SCIENCES
ACADEMIA SINICA
BOX 23 59
TAIPEI 107
CHINA R
TEL 2 396 3211
TLF
TLX
EML
COM 15,35

LEE WOO-BAIK DR
ISSA
36-1 WHAAM DONG
YUSEONG GU
DAEJEON 305 348
KOREA R
TEL 042 861 5611
TLF 042 861 5610
TLX
EML
COM 42

LEE YONG-SAM DR
DPT ASTRONOMY & SPACE SCI
CHUNGBUK NTL UNIVERSITY
CHEONGJU 360 763
KOREA R
TEL 431 61 2314
TLF 431 67 4232
TLX
EML YSLEE@CBUCC CBNU AC KR
COM 42

LEER EGIL PROF
INST THEORET ASTROPHYSICS
UNIVERSITY OF OSLO
BOX 1029
N 0315 BLINDERN OSLO 3
NORWAY
TEL 2-456503
TLF 2-456505
TLX 64124 AUROB N
EML
COM

LEFEBVRE MICHEL DR
CNES/GRGS
18 AVE E BELIN
F 31055 TOULOUSE CDX
FRANCE
TEL
TLF
TLX
EML
COM

LEFEVRE JEAN DR
OCA OBSERV DE NICE
BP 139
F 06003 NICE CDX
FRANCE
TEL 93 89 0420
TLF
TLX 460004 OBSNICE F
EML
COM

LEGER ALAIN DR
GROUPE PHYSIQUE SOLIDES
UNIVERSITE PARIS VII
4 PLACE JUSSIEU TOUR 23
F 75251 PARIS CDX 05
FRANCE
TEL 1 43 36 2525*4673
TLF
TLX
EML
COM 14,21,34

LEGG THOMAS H DR
HERZBERG INST ASTROPHYS
NTL RESEARCH COUNCIL
100 SUSSEX DR
OTTAWA ON K1A 0R6
CANADA
TEL 613 993 6060
TLF 613 952 6602
TLX 053 3715
EML
COM 40

LEHMANN MAREK DR
ASTRONOMICAL LATITUDE OBS
BOROWIEC
BOX 62 035
PL 62 035 KORNIK
POLAND
TEL (61)170187
TLF
TLX 412623 AOS PL
EML
COM 04,08,19

LEHNERT B P PROF
DPT OF PLASMA PHYSICS
ROYAL INST OF TECHNOLOGY
S 100 44 STOCKHOLM 70
SWEDEN
TEL. 78 77 763
TLF
TLX 10389 KTHB S
EML
COM

LEIBACHER JOHN DR
NTL SOLAR OBSERVATORY
BOX 26732
950 N CHERRY AVE
TUCSON AZ 85726 6732
USA
TEL 602 325 9302
TLF 602 325 9305
TLX 066 484 AURA NOAO TU
EML JLEIBACHER@NOAO ARIZONA EDU
COM 10,12,36

LEIBOWITZ ELIA M DR
DPT OF PHYSICS & ASTRON
TEL AVIV UNIVERSITY
RAMAT AVIV
TEL AVIV 69978
ISRAEL
TEL 3 413 788
TLF
TLX 343171 VERSY IL
EML
COM 50

LEIGHTON R B PROF
CALTECH
MS 105 24
PALOMAR OBS
PASADENA CA 91125
USA
TEL 818 356 4286
TLF
TLX
EML
COM 12

LEIKIN G A DR
INST OF ASTRONOMY
ACADEMY OF SCIENCES
PYATNITSKAYA UL 48
109017 MOSCOW
RUSSIA
TEL 231-54-61
TLF
TLX 412623 SCSTP SU
EML .
COM 16

LEINERT CHRISTOPH DR
MPI FUER ASTRONOMIE
KOENIGSTUHL
D 6900 HEIDELBERG 1
GERMANY
TEL 62 215 28264
TLF
TLX 461789 MPIAD
EML
COM 21C,26

LEISAWITZ DAVID DR
NASA/GSFC
CODE 685
GREENBELT MD 20771
USA
TEL 301 286 2150
TLF 301 286 8709
TLX
EML 6168 LEISAWITZ
COM 33,34,37

LEITE SCHEID PAULO DR
OBSERVATORIO NACIONAL
RUA GL BRUCE 586
SAO CRISTOVAO
20921 RIO DE JANEIRO RJ
BRAZIL
TEL 21 580 7313
TLF 21 580 0332
TLX
EML
COM 27

LELIEVRE GERARD DR
OBSERVATOIRE DE PARIS
61 AVE OBSERVATOIRE
F 75014 PARIS
FRANCE
TEL 1 40 51 2255
TLF 1 40 51 2232
TLX 270776 OBS F
EML LELIEVRE@FRIAP51
COM 09VP,28

LEMAIRE JEAN-LOUIS DR
OBSERVATOIRE DE PARIS
SECTION DE MEUDON
F 92195 MEUDON PPL CDX
FRANCE
TEL 1 45 07 7563
TLF 1 45 07 7469
TLX
EML
COM 14

LEMAIRE PHILIPPE DR
IAS
BP 10
F 91371 VERRIERES BUISSON
FRANCE
TEL 1 64 47 4312
TLF
TLX 600252
EML
COM 44

LEMAITRE ANNE DR
DPT DE MATHEMATIQUES
FACULTES UNIVERSITAIRES
REMPART DE LA VIERGE 8
B 5000 NAMUR
BELGIUM
TEL 81 22 9061
TLF 81 23 0391
TLX 59222 FACNAM B
EML
COM 07

LEMAITRE GERARD R DR
OBSERVATOIRE DE MARSEILLE
2 PLACE LE VERRIER
F 13248 MARSEILLE CDX 04
FRANCE
TEL 91 95 9088
TLF
TLX
EML
COM. 09

LEMKE DIETRICH DR
MPI FUER ASTRONOMIE
KOENIGSTUHL
D 6900 HEIDELBERG 1
GERMANY
TEL 62 215 28259
TLF
TLX 461789 IMPIA-D
EML
COM

LENA PIERRE J PROF
OBSERVATOIRE DE PARIS
SECTION DE MEUDON
F 92195 MEUDON PPL CDX
FRANCE
TEL 1 45 07 7719
TLF
TLX 201571
EML
COM

LENZEN RAINER DR
MPI FUER ASTRONOMIE
KOENIGSTUHL
D 6900 HEIDELBERG 1
GERMANY
TEL
TLF
TLX
EML
COM 25

LEORAT JACQUES DR
OBSERVATOIRE DE PARIS
SECTION DE MEUDON
F 92195 MEUDON PPL CDX
FRANCE
TEL 1 45 07 7421
TLF
TLX 201571
EML
COM

LEPINE JACQUES R D DR
IAG
UNIVERSIDADE DE SAO PAULO
AV MIGUEL STEFANO 4200
04301 SAO PAULO SP
BRAZIL
TEL 11 275 3720
TLF
TLX 1136221
EML
COM 34,35,40

LEPP STEPHEN H DR
CENTER FOR ASTROPHYSICS
HCO/SAO
60 GARDEN ST
CAMBRIDGE MA 02138
USA
TEL 617 495 4086
TLF
TLX
EML
COM

LEQUEUX JAMES DR
RADIOASTRONOMIE ENS
24 RUE LHOMOND
F 75231 PARIS CDX 05
FRANCE
TEL 1 43 29 1215
TLF
TLX 202601
EML
COM 05,28,34,40,47

LEROY BERNARD DR
OBSERVATOIRE DE PARIS
SECTION DE MEUDON
DASOP
F 92195 MEUDON PPL CDX
FRANCE
TEL 1 45 07 7812
TLF
TLX
EML
COM 10

LEROY JEAN-LOUIS
OBS MIDI PYRENEES
14 AVE E BELIN
F 31400 TOULOUSE CDX
FRANCE
TEL 61 33 2929
TLF
TLX 530776 F
EML EARN LEROY@FRMEU51
COM 10,12

LESAGE ALAIN DR
OBSERVATOIRE DE PARIS
SECTION DE MEUDON
DASGAL
F 92195 MEUDON PPL CDX
FRANCE
TEL 1 45 07 7829
TLF 1 45 07 7878
TLX
EML LESAGE@FRMEU51/MESIOA LESAGE
COM 14

LESCH HAROLD DR
MPI FUR RADIOASTRONOMIE
AUF DEN HUGEL 69
D 5300 BONN 1
GERMANY
TEL 228 52 51
TLF
TLX
EML
COM 40

LESCHIUTTA S PROF
DPT ELECTRONICA
POLITECNICO
CORSO DUCA D ABRUZZI 24
I 10129 TURINO
ITALY
TEL 11 55 67235
TLF
TLX 220646 POLITO
EML BITNET LESCHIUTTA@ITOPOLI
COM 31

LESTER DANIEL F DR
ASTRONOMY DPT
UNIVERSITY OF TEXAS
RLM 15 308
AUSTIN TX 78712 1083
USA
TEL 512 471 3442
TLF
TLX 910 8741351
EML ARPA dfl@astro as utexas edu
COM

LESTER JOHN B DR
ERINDALE COLLEGE
UNIVERSITY OF TORONTO
DPT OF ASRONOMY
MISSISSAUGA L5L 1C6
CANADA
TEL 416 828 5356
TLF 416 828 5328
TLX
EML
COM 29

LESTRADE JEAN FRANCOIS DR
BUREAU DES LONGITUDES
77 AVE DENFERT ROCHEREAU
F 75014 PARIS
FRANCE
TEL 1 40 51 2265
TLF
TLX 270070
EML
COM 40

LETFUS VOJTECH DR
ASTRONOMICAL INSTITUTE
CZECH ACADEMY OF SCIENCES
ONDREJOV OBSERVATORY
CS 251 65 ONDREJOV
CZECHOSLOVAKIA
TEL 204 85201
TLF 204 85314
TLX
EML
COM

LEUNG CHUN MING DR
DPT OF PHYSICS
RENSSELAER POLYTECHN INST
TROY NY 12180 3590
USA
TEL 518 266 6318
TLF
TLX
EML
COM 34,40

LEUNG KAM CHING PROF
DPT OF PHYSICS & ASTRON
UNIVERSITY OF NEBRASKA
BEHLEN OBSERVATORY
LINCOLN NE 68588
USA
TEL 402 472 2770
TLF
TLX 484340 UNL
EML
COM 27,38C,42

LEVASSEUR-REGOURD A C PR
SERVICE D'AERONOMIE
BP 3
F 91371 VERRIERES BUISSON
FRANCE
TEL 1 64 47 4293
TLF 1 43 29 8673
TLX 602400
EML
COM 15,21C,22

LEVATO ORLANDO HUGO DR
COMPLEJO ASTRONOMICO
EL LEONCITO
CC 467
5400 SAN JUAN
ARGENTINA
TEL 64 22 5720
TLF 64 21 1475
TLX. 59134 ENTOP AR
EML MSSOCA ·PSI% IAFE LEVATO
COM 29,30,45VP

LEVINE RANDOLPH H DR
50 CARVER RD
NEWTON MA 02161
USA
TEL 617 965 5953
TLF
TLX
EML
COM

LEVISON HAROLD F DR
US NAVAL OBSERVATORY
FLAGSTAFF STATION
BOX 1149
FLAGSTAFF AZ 86002
USA
TEL 602 779 5132
TLF
TLX
EML PERCY LEVISON
COM

LEVREAULT RUSSELL M DR
DPT OF ASTRONOMY
VAN VLECK OBSERVATORY
WESLEYAN UNIVERSITY
MIDDLETOWN CT 06457
USA
TEL 203 347 9411
TLF
TLX
EML BITNET rlevreault@wesleyan
COM 40

LEVY EUGENE H DR
LUNAR & PLANETARY LAB
UNIVERSITY OF ARIZONA
TUCSON AZ 85721
USA
TEL 602 621 6962
TLF
TLX 9109521143
EML
COM 41,49

LEVY JACQUES R DR
OBSERVATOIRE DE PARIS
61 AVE OBSERVATOIRE
F 75014 PARIS
FRANCE
TEL 1 43 20 1210
TLF
TLX 270776 OBS F
EML
COM 41

LEWIN WALTER H G PROF
DPT OF PHYSICS
MIT RM 37 627
BOX 165
CAMBRIDGE MA 02139
USA
TEL 617 253 4282
TLF
TLX
EML
COM 44

LEWIS BRIAN MURRAY DR
ARECIBO OBSERVATORY
BOX 995
ARECIBO PR 00612
USA
TEL 809 878 2612
TLF
TLX
EML
COM 30

LEWIS J S
LUNAR & PLANETARY LAB
UNIVERSITY OF ARIZONA
TUCSON AZ 85721
USA
TEL 602 621 4972
TLF
TLX
EML
COM 16

LI CHUN-SHENG
DPT OF ASTRONOMY
NANJING UNIVERSITY
NANJING 210008
CHINA PR
TEL 25 34651*2882
TLF
TLX 34151 PRCNU CN
EML
COM 10,40

LI DEPEI
NANJING ASTRONOMICAL
INSTRUMENT FACTORY
NANJING
CHINA PR
TEL 25 46191
TLF
TLX 1131
EML
COM 09

LI DONG-MING
PURPLE MOUNTAIN OBSERV
CAS
NANJING
CHINA PR
TEL 25 46700
TLF
TLX
EML
COM 08

LI GI MAN
PYONGYANG ASTRON OBS
ACADEMY OF SCIENCES DPRK
TAESONG DISTRICT
PYONGYANG
KOREA DPR
TEL
TLF
TLX
EML
COM 04

LI GYONG WON
PYONGYANG ASTRON OBS
ACADEMY OF SCIENCES DPRK
TAESONG DISTRICT
PYONGYANG
KOREA DPR
TEL
TLF
TLX
EML
COM 40

LI HONG-WEI
DPT OF ASTRONOMY
NANJING UNIVERSITY
NANJING
CHINA PR
TEL 25 34651/34751
TLF
TLX 34151 PRCNU CN
EML
COM 40

LI HYOK HO
PYONGYANG ASTRON OBS
ACADEMY OF SCIENCES DPRK
TAESONG DISTRICT
PYONGYANG
KOREA DPR
TEL
TLF
TLX
EML
COM· 04

LI J Y
PYONGYANG ASTRON OBS
ACADEMY OF SCIENCES DPRK
TAESONG DISTRICT
PYONGYANG
KOREA DPR
TEL
TLF
TLX
EML
COM

LI JING
BEIJING ASTRONOMICAL OBS
CAS
W SUBURB
BEIJING 100080
CHINA PR
TEL 1 22 040
TLF
TLX
EML
COM 28,33

LI NED C DR
CALIFORNIA UNIVERSITY

6531 WITHWORTH RD
LOS ANGELES CA 90035
USA
TEL
TLF
TLX
EML
COM

LI NENG-YAO
PURPLE MOUNTAIN OBSERV
CAS
NANJING
CHINA PR
TEL 25 37609
TLF
TLX 34144 PMONJ CN
EML
COM 04,08

LI QI-BIN
BEIJING ASTRONOMICAL OBS
CAS
W SUBURB
BEIJING 100080
CHINA PR
TEL 1 28 1968
TLF
TLX 22040 BADAS CN
EML·
COM 05C,28,46,48

LI SIN HYONG
PYONGYANG ASTRON OBS
ACADEMY OF SCIENCES DPRK
TAESONG DISTRICT
PYONGYANG
KOREA DPR
TEL
TLF
TLX
EML
COM 25

LI SON JAE
PYONGYANG ASTRON OBS
ACADEMY OF SCIENCES DPRK
TAESONG DISTRICT
PYONGYANG
KOREA DPR
TEL
TLF
TLX
EML
COM 10

LI TING
NANJING ASTRONOMICAL
INSTRUMENT FACTORY
NANJING
CHINA PR
TEL 25 46191
TLF
TLX
EML
COM 09

LI TIPEI
INSTITUTE OF HIGH ENERGY
PHYSICS
BOX 918 3
BEIJING
CHINA PR
TEL 1 81 2971*464
TLF
TLX 22082 IHEP CN
EML
COM 44,48

LI WEI BAO
YUNNAN OBSERVATORY
CAS
BOX 110
KUNMING 72946 YUNNAN
CHINA PR
TEL 871 2035
TLF
TLX
EML
COM 10

LI XIAO-QING
PURPLE MOUNTAIN OBSERV
CAS
NANJING
CHINA PR
TEL 25 31096
TLF
TLX 34144 PMO NJ CN
EML
COM 28,49

LI YUAN-JIE
DPT OF PHYSICS
HUAZHONG UNIVERSITY OF
SCIENCE AND TECHNOLOGY
WUHAN
CHINA PR
TEL 87 0541
TLF
TLX 7122
EML
COM 48

LI ZHENG-XIN DR
SHANGHAI OBSERVATORY
CAS
80 NANDAN RD
SHANGHAI
CHINA PR
TEL 21 38 6191
TLF
TLX
EML
COM 19

LI ZHIAN DR
DPT OF ASTRONOMY
BEIJING NORMAL UNIVERSITY
BEIJING 100875
CHINA PR
TEL 1 201 2255*2618
TLF
TLX 222701 BNU CH
EML
COM 19

LI ZHIGANG
SHAANXI OBSERVATORY
CAS
LINTONG XIAN
SHAANXI
CHINA PR
TEL 33 2255
TLF
TLX 70121 CSAO CN
EML
COM 08,09

LI ZHI-FANG
SHANGHAI OBSERVATORY
CAS
80 NANDAN RD
SHANGHAI
CHINA PR
TEL 21 38 6191
TLF
TLX 33164 SHAO CN
EML
COM 08,47

LI ZHI-SEN
BEIJING ASTRONOMICAL OBS
CAS
W SUBURB
BEIJING 100080
CHINA PR
TEL 1 28 1698
TLF
TLX 9053
EML
COM 41

LI ZHONGYUAN
DPT EARTH & SPACE SCI
UNIV SCIENCE & TECHNOLOGY
HEFEI 230026 ANHUI
CHINA PR
TEL 551 33 1134
TLF
TLX 90028 USTC CN
EML
COM 42,44

LI ZONG-WEI
DPT OF ASTRONOMY
BEIJING NORMAL UNIVERSITY
BEIJING 100875
CHINA PR
TEL 1 65 6531*683
TLF
TLX
EML
COM 35,48

LI ZONG-YUN
DPT OF ASTRONOMY
NANJING UNIVERSITY
NANJING
CHINA PR
TEL
TLF
TLX· 34151 PRCNU CN
EML·
COM

LIANG EDISON P DR
DPT SPACE PHYS & ASTRON
RICE UNVIERSITY
BOX 1892
HOUSTON TX 77251 1892
USA
TEL 713 527 8101*3524
TLF 713 285 5143
TLX 556457
EML LIANG@VEGA RICE EDU
COM 48

LIANG SHI-GUANG
SHANGHAI OBSERVATORY
CAS
80 NANDAN RD
SHANGHAI
CHINA PR
TEL 21 38 6191
TLF
TLX 33164 SHAO CN
EML
COM 40

LIANG ZHONG-HUAN
BOX 18
LINTONG
XIAN
CHINA PR
TEL 33 2255
TLF
TLX 70121 CSAO CN
EML·
COM 31

LIAO XINHAO DR
DPT OF ASTRONOMY
NANJING UNIVERSITY
NANJING 210008
CHINA PR
TEL 25 63 7651*2884
TLF
TLX 34151 PRCNU CN
EML 210008
COM 07

LIBBRECHT K G DR
CALTECH
MS 264 33
BIG BEAR SOLAR OBS
PASADENA CA 91125
USA
TEL 818 356 3722
TLF
TLX 675425 CALTECH PSD
EML KGL@SUNDOG CALTECH EDU
COM

LIDDELL U MR
NASA HEADQUARTERS
SPACE SCI & APPLICATIONS
600 INDEPENDENCE AVE SW
WASHINGTON DC 20546
USA
TEL
TLF
TLX·
EML
COM

LIEBERT JAMES W DR
STEWARD OBSERVATORY
UNIVERSITY OF ARIZONA
TUCSON AZ 85721
USA
TEL 602 621 4513
TLF
TLX 621 41 410
EML LIEBERT@ARIZRVAX
COM 29,33,35,36

LIEBSCHER DIERCK-E DR
ZNTRLINST F ASTROPHYSIK
STERNWARTE BABELSBERG
ROSA-LUXEMBURG-STR 17A
D 1502 POTSDAM
GERMANY
TEL
TLF
TLX
EML
COM 47

LIESKE JAY H DR
JPL/CALTECH
MS 264 664
4800 OAK GROVE DR
PASADENA CA 91109
USA
TEL 818 354 3642
TLF 818 354 3437
TLX 675429
EML NAIF JHL
COM 04,07C,19,20

LILJE PER VIDAR BARTH DR
NORDITA
BLEGDAMSVEJ 17
DK 2100 COPENHAGEN O
DENMARK
TEL 31 42 1616
TLF 31 38 9157
TLX 15216 NBI DK
EML LILJE@NORDITA DK
COM 47

LILLER WILLIAM DR
INSTITUTO ISAAC NEWTON
CASILLA 8-9
VINA DEL MAR
CHILE
TEL 32 97 0864
TLF
TLX
EML
COM 15,28,34

LILLEY EDWARD A PROF
CENTER FOR ASTROPHYSICS
HCO/SAO
60 GARDEN ST
CAMBRIDGE MA 02138
USA
TEL 617 495 3971
TLF
TLX 921428 SATELLITE CAM
EML.
COM 40,51

LILLIE CHARLES F DR
TRW SPACE & TECHNOLOGY
1 SPACE PARK
REDONDO BEACH CA 90278
USA
TEL 213 812 2248
TLX 910-325-6611
EML
TLF
COM· 15,21

LILLY SIMON J DR
DPT OF ASTRONOMY
UNIVERSITY OF TORONTO
60 ST GEORGE ST
TORONTO ON M5S 1A1
CANADA
TEL 416 978 3150
TLF 416 978 3921
TLX 062 86766
EML·
COM 28,47

LIN CHIA C PROF
DPT OF MATHEMATICS
MIT
77 MASSACHUSETTS AVE
CAMBRIDGE MA 02139
USA
TEL 617 253 1796
TLF
TLX 921473 MIT CAM
EML
COM 28,33,34

LIN DOUGLAS N C DR
LICK OBSERVATORY
UNIVERSITY OF CALIFORNIA
SANTA CRUZ CA 95064
USA
TEL 408 429 2732
TLF
TLX
EML
COM

LIN YUANZHANG
BEIJING ASTRONOMICAL OBS
CAS
W SUBURB
BEIJING 100080
CHINA PR
TEL 1 28 1698
TLF
TLX 22040 BAOAS CN
EML
COM 10,12

LINCOLN J VIRGINIA MISS
2005 ALPINE DR
BOULDER CO 80304
USA
TEL 303 442 6757
TLF
TLX
EML
COM

LINDBLAD BERTIL A DR
LUND OBSERVATORY

BOX 43
S 221 00 LUND
SWEDEN
TEL 46 10 7000
TLF
TLX 33199 OBSNOT S
EML
COM 20,22,44

LINDBLAD PER OLOF PROF
STOCKHOLM OBSERVATORY
S 133 36 SALTSJOEBADEN
SWEDEN
TEL 87 17 0380
TLF 87 17 4719
EML LINDBLAD@ASTRO SU SE
COM 28,33,EC
TLX 12972 SOBSERV S

LINDE PETER DR
LUND OBSERVATORY
BOX 43
S 221 00 LUND
SWEDEN
TEL 46 10 4701
TLX 33199 OBSNOT
EML PETER@ASTRO LU SE
TLF
COM 05,25

LINDEGREN LENNART DR
LUND OBSERVATORY
BOX 43
S 221 00 LUND
SWEDEN
TEL 46 10 7000
TLF
TLX 33199 OBSNOT S
EML
COM 08C

LINDGREN HARRI
GRUPO NORDICO DEL OBS DEL
ROQUE DE LOS MUCHACHOS
APD 474
E 38700 SANTA CRUZ
SPAIN
TEL·
TLF
TLX
EML
COM 30

LINDSEY CHARLES ALLAN
INSTITUTE FOR ASTRONOMY
UNIVERSITY OF HAWAII
2680 WOODLAWN DR
HONOLULU HI 96822
USA
TEL 808 948 6526
TLF
TLX
EML
COM 15

LING CHIH-BING DR
INSTITUTE OF MATHEMATICS
ACADEMIA SINICA
BOX NO 143
TAIPEI
CHINA R
TEL
TLF
TLX.
EML
COM

LING J DR
OBSERVATORIA ASTRONOMICO
RAMON MARIA ALLER
AVD DE LAS CIENCIAS S/N
E SANTIAGO DE COMPOSTELA
SPAIN
TEL
TLF
TLX
EML
COM 26

LINGENFELTER RICHARD E
CASS
UCSD
C 011
LA JOLLA CA 92093 0216
USA
TEL 619 452 2464
TLF
TLX 9103371271 SIOCCAN
EML
COM

LINKE RICHARD ALAN DR
B ANDERSON LANE
PRINCETON NJ 08540
USA
TEL
TLF
EML
COM 34,40
TLX

LINNELL ALBERT P PROF
PHYSICS & ASTRONOMY DPT
MICHIGAN STATE UNIVERSITY
EAST LANSING MI 48824
USA
TEL 517 353 6670
TLF
TLX
EML
COM 35,36,42

LINSKY JEFFREY L DR
JILA
UNIVERSITY OF COLORADO
BOX 440
BOULDER CO 80309 0440
USA
TEL 303 492 7838
TLF
TLX 755842 JILA
EML
COM 12,36C,44

LINSLEY JOHN
DPT PHYSICS & ASTRONOMY
UNIVERSITY OF NEW MEXICO
800 YALE BLVD NE
ALBUQUERQUE NM 87131
USA
TEL 505 243 1924
TLF
TLX 910989
EML
COM 44,48

LIPOVETSKY V A
SPECIAL ASTROPHYSICAL OBS
ACADEMY OF SCIENCES
NIZHNIH ARKHYZ
357147 STAVROPOLSKIJ
RUSSIA
TEL 93-2-42
TLF
TLX
EML
COM 28

LIPPINCOTT SARAH LEE DR
DPT OF ASTRONOMY
SWARTHMORE COLLEGE
SPOUL OBSERVATORY
SWARTHMORE PA 19081
USA
TEL 215 543 9058
TLF
TLX
EML
COM 24,26,51

LIPSCHUTZ MICHAEL E DR
WETHERILL CHEMISTRY BLDG
PURDUE UNIVERSITY
W LAFAYETTE IN 47907
USA
TEL 317 494 5326
TLF
TLX 272 396
EML BITNET RNAAPUML@PURCCVM
COM 15

LIRITZIS IOANNIS DR
RES CENTER FOR ASTRONOMY
ACADEMY OF ATHENS
14 ANAGNOSTOPOULOU ST
GR 106 73 ATHENS
GREECE
TEL 1 361 3589
TLF 1 363 1606
TLX
EML
COM 10

LISEAU RENE DR
IAS
CNR
VIA G GALILEI CP 27
I 00044 FRASCATI
ITALY
TEL 6 942 3801
TLF
TLX 610261 CNR FRA I
EML
COM

LISI FRANCO DR
OSS ASTROFISICO
DI ARCETRI
LARGO E FERMI 5
I 50125 FIRENZE
ITALY
TEL 55 275 2289
TLF 55 22 0039
TLX 572268 I
EML SPAN 38954 LISI
COM

LISSAUER JACK J DR
DPT OF EARTH & SPACE SCI
ASTRONOMY PROGRAM
SUNY AT STONY BROOK
STONY BROOK NY 11794 2100
USA
TEL 516 632 8225
TLF
TLX
EML
COM 07,15,16

LISZT HARVEY STEVEN
NRAO
EDGEMOMT RD
CHARLOTTESVILLE VA 22901
USA
TEL 804 296 0344
TLF
TLX 910-997-0714
EML
COM 34

LITTLE LESLIE T DR
ELECTRONICS LABORATORY
UNIVERSITY OF KENT
CANTERBURY CT2 7NT
UK
TEL 227 668 22
TLF
TLX 965449 UKCLIB
EML
COM 40

LITTLETON JOHN E
DPT OF PHYSICS
WEST VIRGINIA UNIVERSITY
BOX 6023
MORGANTOWN WV 26506 6023
USA
TEL 304 293 3498
TLF
TLX 710-921-0309
EML
COM 35

LITTLE-MARENIN IRENE R DR
WHITIN OBSERVATORY
WELLESLEY COLLEGE
WELLESLEY MA 02181
USA
TEL 617 235 5303
TLF
TLX
EML llittle@lucy wellesley edu
COM 27,29,46

LITVAK MARVIN M DR
TRW INC 01/1260
1 SPACE PARK
REDONDO BEACH CA 90278
USA
TEL
TLF
TLX
EML
COM

LITVINENKO LEONID N DR
INSTITUTE OF RADIO ASTRON
UKRAINIAN ACAD OF SCIENCE
KRASNOZNAMENNAYA UL 4
310002 KHARKOV
UKRAINE
TEL 45 1009
TLF 32 0273
TLX 115144 METIL SU
EML
COM 40

LIU BAO-LIN
PURPLE MOUNTAIN OBSERV
CAS
NANJING
CHINA PR
TEL 25 42817/46700
TLF
TLX 34144 PMONJ CN
EML
COM 04

LIU CAIPIN
PURPLE MOUNTAIN OBSERV
CAS
NANJING
CHINA PR
TEL 25 42817
TLF
TLX 34144 PMONJ CN
EML
COM 36

LIU CI-YUAN DR
SHAANXI OBSERVATORY
CAS
LINTONG XIAN
SHAANXI
CHINA PR
TEL 33 2255
TLF 9237 3496
TLX 70121 CSAO
EML
COM 19,41

LIU JINMING
SHANGHAI OBSERVATORY
CAS
80 NANDAN RD
SHANGHAI
CHINA PR
TEL 21 38 6191
TLF
TLX 33164 SHAO CN
EML
COM 05,06,31

LIU LIAO
DPT OF PHYSICS
BEIJING NORMAL UNIVERSITY
BEIJING 100071
CHINA PR
TEL
TLF
TLX
EML
COM 47

LIU LIN
DPT OF ASTRONOMY
NANJING UNIVERSITY
NANJING
CHINA PR
TEL 25 34651*2882
TLF
TLX 34151 PRCNU CN
EML
COM

LIU LIN-ZHONG
PURPLE MOUNTAIN OBSERV
CAS
NANJING
CHINA PR
TEL 25 46700
TLF
TLX 34144 PMONT CN
EML
COM 15

LIU QINGYAO DR
YUNNAN OBSERVATORY
CAS
BOX 110
KUNMING 72946 YUNNAN
CHINA PR
TEL 871 2035
TLF
TLX 64040 YUOBS CN
EML
COM 42

LIU RU-LIANG
PURPLE MOUNTAIN OBSERV
CAS
NANJING
CHINA PR
TEL 25 42817/46700
TLF
TLX 34144 PMONJ CN
EML
COM 28,48

LIU SOU-YANG DR
COMPUTER SCIENCES CORP
SYSTEM SCIENCES DIV
8728 COLESVILLE RD
SILVER SPRING MD 20910
USA
TEL 301 589 1545
TLF
TLX
EML·
COM

LIU XINPING PROF
INSTITUTE OF MECHANICS
CAS
BEIJING 100080
CHINA PR
TEL 1 28 4185
TLF 1 25 61284
TLX 222554 MEHAS CN
EML
COM 10

LIU XUEFU
DPT OF ASTRONOMY
BEIJING NORMAL UNIVERSITY
BEIJING 100875
CHINA PR
TEL 1 65 6531*6285
TLF
TLX 8511
EML
COM 42

LIU YONG-ZHEN
GRADUATE SCHOOL
UNIV SCIENCE & TECHNOLOGY
BOX 3908
BEIJING 100039
CHINA PR
TEL 1 81 7031
TLF
TLX·
EML
COM 28,47

LIU ZONGLI
BEIJING ASTRONOMICAL OBS
CAS
W SUBURB
BEIJING 100080
CHINA PR
TEL
TLF
TLX 22040
EML
COM 15,27

LIVINGSTON WILLIAM C
NOAO/NSO
BOX 26732
950 N CHERRY AVE
TUCSON AZ 85726 6732
USA
TEL 602 327 5511
TLF
TLX 0666484 AURA NOAO TU
EML
COM 09,12

LIVIO MARIO PROF
DPT OF PHYSICS
IIT
TECHNION CITY
HAIFA 32000
ISRAEL
TEL 4 293 549
TLF
TLX 46650 TECLI IL
EML PHR81ML@TECHNION
COM 35

LIVSHITS M A DR
IZMIRAN
ACADEMY OF SCIENCES
142092 TROITSK
RUSSIA
TEL
TLF
TLX 412623 SCSTP SU
EML
COM 10

LIZANO-SOBERON SUSANA DR
INSTITUTO DE ASTRONOMIA
UNAM
APDO POSTAL 70-264
04510 MEXICO DF
MEXICO
TEL 011 525 548 53 05
TLF 011 525 548 3712
TLX 017 60155 CICME
EML LIZANO%ALFA ASTROSCU UNAM MX
COM 34

LLOYD EVANS THOMAS DR
SAAO
BOX 9
OBSERVATORY 7935
SOUTH AFRICA
TEL 021-47-0026
TLF
TLX 5720309 SA
EML
COM 37,45C

LO KWOK-YUNG DR
CALTECH
MS 105 24
OWENS VALLEY RADIO OBS
PASADENA CA 91125
USA
TEL 818 356 4415
TLF
TLX 675425 CALTECH PSD
EML
COM 28,34,40

LOCANTHI DOROTHY DAVIS DR
2180 PINECREST DR
ALTADENA CA 91001
USA
TEL 213 797 0629
TLF
EML·
COM 29
TLX

LOCHMAN JAN
ASTRONOMICAL INSTITUTE
CZECH ACADEMY OF SCIENCES
DVORAKOVA 298
CS 511 01 TURNOV
CZECHOSLOVAKIA
TEL 43 62 2622
TLF
TLX
EML
COM 09

LOCKE JACK L DR
250 BRAESIDE AVENUE
OTTAWA ON K1H 7J5
CANADA
TEL 613 523 0812
TLF
TLX
EML
COM 12,40

LOCKMAN FELIX J
NRAO
EDGEMONT RD
CHARLOTTESVILLE VA 22903
USA
TEL 804 296 0211
TLX 910-997-0174
EML
TLF
COM 33,34,40

LOCKWOOD G WESLEY DR
LOWELL OBSERVATORY
1400 W MARS HILL RD
BOX 1149
FLAGSTAFF AZ 86001
USA
TEL 607 774 3358
TLF
TLX
EML
COM 16,25,27

LODEN KERSTIN R DR
STOCKHOLM OBSERVATORY
S 133 36 SALTSJOEBADEN
SWEDEN
TEL 87 17 0195
TLF 87 17 4719
TLX 12972 SOBSERV S
EML LODEN_K@ASTRO SU SE
COM 26,33,45

LODEN LARS OLOF PROF
ASTRONOMICAL OBSERVATORY
BOX 515
S 751 20 UPPSALA
SWEDEN
TEL 18 11 4490
TLF
TLX 76024
EML
COM 26,33,37,51

LOISEAU NORA DR
INST DE ASTROFISICA
DE CANARIAS
OBS DEL TEIDE
E 38200 LA LAGUNA
SPAIN
TEL 22 605 237
TLF 22 605 210
TLX 92640 IACE E
EML NLL@IAC DNET NASA GOV
COM 40

LOMB NICHOLAS RALPH DR
SYDNEY OBSERVATORY
MUSEUM APPLIED ARTS & SCI
BOX K346
HAYMARKET NSW 2000
AUSTRALIA
TEL
TLF
TLX
EML
COM 20,46

LONG KNOX S DR
DPT PHYSICS & ASTRONOMY
JOHNS HOPKINS UNIVERSITY
CHARLES & 34TH ST
BALTIMORE MD 21218
USA
TEL 301 338 7391
TLF
TLX 9102400225 JHUCASMD
EML
COM 44

LONGAIR M S PROF
MULLARD RADIO ASTRON OBS
CAVENDISH LABORATORY
MADINGLEY RD
CAMBRIDGE CB3 OHE
UK
TEL 223 337 083
TLF 223 35 4599
TLX 81292 CAVLAB
EML msl@uk ac cam phy ras
COM 40,47,48

LONGMORE ANDREW J
JOINT ASTRONOMY CENTER
UKIRT
665 KOMOHANA ST
HILO HI 96720
USA
TEL 808 961 3756
TLF
TLX 633135
EML
COM 27

LONSDALE CAROL J DR
CALTECH
MS 100 22
IPAC
PASADENA CA 91125
USA
TEL 818 584 2929
TLF
TLX 67 5429
EML INTERNET CJL@IPAC CALTECH EDU
COM 05,28,47

LOOSE HANS-HERMANN DR
UNIVERSTAETS STERNWARTE
GEAISMARLANDSTRASSE 11
D 3400 GEOTTINGEN
GERMANY
TEL 551 395 056*953
TLF
TLX
EML
COM 28

LOPES DALTON DE FARIA DR
OBSERVATORIO NACIONAL
RUA GL JOSE CRISTINO 77
SAO CRISTOVAO
20921 RIO DE JANEIRO TJ
BRAZIL
TEL 21 580 7181
TLF 21 580 0332
TLX 021288
EML DFLO@LNCCVB BITNET
COM 29

LOPES-GAUTIER ROSALY DR
JPL
4800 OAK GROVE DR
PASADENA CA 91109
USA
TEL 818 393 4584
TLX 7401843 JPLR
EML
TLF 818 354 0966
COM 15,16,41

LOPEZ CARLOS LIC
OBSERVATORIO ASTRONOMICO
FELIX AGUILAR
AV BENAVIDEZ 8175 OESTE
5407 SAN JUAN
ARGENTINA
TEL 64 23 1494
TLF
TLX
EML
COM 24

LOPEZ DE COCA M D P DR
INST ASTROFISICA
DE ANDALUCIA APD 2144
PROFESOR ALBAREDA 1
E 18080 GRANADA
SPAIN
TEL 58 12 1311
TLF 58 11 4530
TLX 78573 IAAGE
EML PILAR@IAA ES
COM 27

LOPEZ GARCIA ZULEMA L DR
OBSERVATORIO ASTRONOMICO
FELIX AGUILAR
AV BENAVIDEZ 8175 OESTE
5407 MARQUESADO (S J)
ARGENTINA
TEL
TLF
TLX
EML
COM

LOPEZ JOSE A ING
OBSERVATORIO ASTRONOMICO
FELIX AGUILAR
AV BENAVIDEZ 8175 OESTE
5407 MARQUESADO (S J)
ARGENTINA
TEL 64 23 1494
TLF
TLX 59100 UNSJA AR
EML
COM 08C

LOPEZ JOSE ALBERTO DR
INSTITUTO DE ASTRONOMIA
UNAM
APDO POSTAL 877
22800 ENSENADA B CALIF
MEXICO
TEL· 667- 44580
TLF·
TLX 56539 CICEME
EML.
COM·

LOPEZ ROSARIO DR
DPT FISICA
UNIVERSADAD DE BARCELONA
AVD DIAGONAL 647
E 08028 BARCELONA
SPAIN
TEL 3 330 7311
TLF
TLX
EML BITNET D3FARLHO@EBOUBO11
COM 28

LOPEZ-ARROYO M
OBS ASTRONOMICO NCL

ALFONSO XII-5
E 28014 MADRID
SPAIN
TEL
TLF
TLX
EML
COM 12

LOPEZ-GARCIA FRANCISCO DR
OBSERVATORIO ASTRONOMICO
FELIX AGUILAR
AV BENAVIDEZ 8175 OESTE
5407 MARQUESADO (S J)
ARGENTINA
TEL 64 23 1494
TLF
TLX
EML
COM

LOPEZ-GONZALEZ MARIA J DR
INST ASTROFISICA
DE ANDALUCIA APD 3004
PROFESOR ALBAREDA 1
E 18080 GRANADA
SPAIN
TEL 58 12 1311
TLF 58 81 4530
TLX 78573 IAAG E
EML 16488   MARIAJOSE
COM 21

LOPEZ-MORENO JOSE JUAN
INST ASTROFISICA
DE ANDALUCIA APD 2144
PROFESOR ALBAREDA 1
E 18080 GRANADA
SPAIN
TEL 58 12 1300
TLF
TLX 78573 IAAG E
EML
COM 16,21

LOPEZ-PUERTAS MANUEL
INST ASTROFISICA
DE ANDALUCIA APD 2144
PROFESOR ALBAREDA 1
E 18080 GRANADA
SPAIN
TEL 58 12 1300
TLF
TLX. 78573 IAAG E
EML
COM 16,21

LOPEZ-VALVERDE M A DR
INST ASTROFISICA
DE ANDALUCIA APD 3004
PROFESOR ALBAREDA 1
E 18080 GRANADA
SPAIN
TEL 58 12 1311
TLF 58 81 4530
TLX 78573 IAAG E
EML 16488   VALVERDE
COM 16

LORD STEVEN DONALD DR
NASA AMES RESEARCH CTR
MS 245 6
MOFFETT FIELD CA 94035
USA
TEL 415 604 3571
TLF
TLX
EML LORD@GAL ARC NASA GOV
COM 28

LOREN ROBERT BRUCE DR
BOX 2915
SILVER CITY NM 88062
USA
TEL
TLF
TLX
EML
COM 34,40

LORENZ HILMAR
ZNTRLINST F ASTROPHYSIK
AKAD WISSENSCHAFTEN DDR
ROSA-LUXEMBURG-STR 17A
D 1502 POTSDAM
GERMANY
TEL
TLF
TLX
EML
COM 28,40

LORTET MARIE CLAIRE
OBSERVATOIRE DE PARIS
SECTION DE MEUDON
DAPHE
F 92195 MEUDON PPL CDX
FRANCE
TEL 1 45 07 7426
TLF
TLX 201571
EML
COM 05,28,34

LOSCO LUCETTE DR
FACULTE DES SCIENCES
F 25030 BESANCON CDX
FRANCE
TEL
TLF
EML
COM
TLX·

LOTOVA N A DR
IZMIRAN
ACADEMY OF SCIENCES
142092 TROITSK
RUSSIA
TEL
TLF
TLX
EML
COM 49

LOUGHHEAD RALPH E DR
CSIRO
DIV OF APPLIED PHYSICS
BOX 218
LINDFIELD NSW 2070
AUSTRALIA
TEL 2 467 6355
TLF
TLX 26296
EML
COM 10,12

LOUISE RAYMOND PROF
FACULTE DES SCIENCES
DEPT DE PHYSIQUE
33 RUE ST-LEU
F 80039 AMIENS
FRANCE
TEL
TLF
TLX
EML
COM· 34

LOULERGUE MICHELLE DR
OBSERVATOIRE DE PARIS
SECTION DE MEUDON
F 92195 MEUDON PPL CDX
FRANCE
TEL 1 45 07 7455
TLF
TLX 270912
EML
COM 14

LOVAS FRANCIS JOHN DR
NTL BUREAU OF STANDARDS
DIV 545
MOLECULAR SPECTROSCOPIC
WASHINGTON DC 20234
USA
TEL 301 921 2023
TLF
TLX 898993
EML
COM· 14,34

LOVAS MIKLOS
KONKOLY OBSERVATORY
THEGE U 13/17
BOX 67
H 1525 BUDAPEST
HUNGARY
TEL· 1 75 5866/75 4122
TLF
TLX. 227460 KONOB
EML
COM· 20

LOVELACE RICHARD V E DR
CORNELL UNIVERSITY
SPACE SCIENCES BLDG
ITHACA NY 14853
USA
TEL   607 256 3968
TLX
EML
TLF
COM   48

LOVELL SIR BERNARD PROF
NRAL
JODRELL BANK
MACCLESFIELD SK11 9DL
UK
TEL   477 713 21
TLX   36149
EML
TLF
COM   22,40,44,51

LOW BOON CHYE
HIGH ALTITUDE OBSERVATORY
NCAR
BOX 3000
BOULDER CO 80307 3000
USA
TEL   303 497 1553
TLF
TLX   45694
EML
COM   10

LOW FRANK J DR
4940 CALLE BARRIL

TUCSON AZ 85718
USA
TEL   602 621 2779
TLF
TLX
EML
COM   28,34,45

LOWE ROBERT P DR
DPT OF PHYSICS
UNIV OF WESTERN ONTARIO
LONDON ON N6A 3K7
CANADA
TEL   519 679 2917
TLF   519 661 3486
TLX   064 7134
EML
COM

LOYOLA PATRICIO DR
DPT DE ASTRONOMIA
UNIVERSIDAD DE CHILE
CASILLA 36 D
SANTIAGO
CHILE
TEL   2 229 4101
TLF
TLX   440 001
EML
COM   08

LOZINSKAYA TAT'YANA A DR
STERNBERG STATE ASTR INST
UNIVERSITETSKIJ PROSP 13
119899 MOSCOW
RUSSIA
TEL   139-10-30
TLF
TLX
EML
COM   34C,40

LOZINSKIJ A M DR
INST OF ASTRONOMY
ACADEMY OF SCIENCES
PYATNITSKAYA UL 48
109017 MOSCOW
RUSSIA
TEL   231-54-61
TLF
TLX   412623 SCSTP SU
EML
COM   24

LU BEN-KUI
PURPLE MOUNTAIN OBSERV
CAS
NANJING
CHINA PR
TEL   25 32893
TLF
TLX   34144 PMONJ CN
EML
COM   07,31

LU CHUN-LIN
PURPLE MOUNTAIN OBSERV
CAS
NANJING
CHINA PR
TEL   25 42700
TLF
TLX   34144 PMONJ CN
EML
COM   08

LU JU FU DR
ASTROPHYSICS DIVISION
UNIV SCIENCE & TECHNOLOGY
HEIFI 230026 ANHUI
CHINA PR
TEL   551 33 1134*527
TLF   551 33 1760
TLX   90028 USTC CN
EML
COM   48

LU PHILLIP K DR
DPT PHYSICS & ASTRONOMY
W CONNECTICUT STATE UNIV
181 WHITE ST
DANBURY CT 06810 7859
USA
TEL   203 797 4218
TLF
TLX
EML
COM   24,33,37

LU RUWEI DR
YUNNAN OBERVATORY
CAS
BOX 110
KUNMING 650011 YUNNAN
CHINA PR
TEL   871 2035
TLF   871 71845
TLX   64040 YUOBS
EML
COM   09

LU TAN
DPT OF ASTRONOMY
NANJING UNIVERSITY
NANJING
CHINA PR
TEL   25 34651*2882
TLF
TLX   34151 PRCNU CN
EML
COM   47,48

LU YANG
DPT OF ASTRONOMY
NANJING UNIVERSITY
NANJING
CHINA PR
TEL   25 34651*2882
TLF
TLX   34151 PRCNU CN
EML
COM   40

LUB JAN DR
STERREWACHT
HUYGENS LAB
BOX 9513
NL 2300 RA LEIDEN
NETHERLANDS
TEL   71 27 2727
TLF
TLX   39068 ASTRO NL
EML
COM   25C,27

LUCAS ROBERT DR
OBSERVATOIRE DE GRENOBLE
CERMO/ASTROPHYSIQUE
BP 68
F 38041 S MARTIN HERES CD
FRANCE
TEL   76 51 4600
TLF
TLX
EML
COM   34

LUCCHIN FRANCESCO
DPT DI FISICA G GALILEI
UNIVERSITA DI PADOVA
VIA MARZOLO 8
I 35131 PADOVA
ITALY
TEL   49 84 4333
TLF
TLX   430308 DF GGPDI
EML
COM   47

LUCEY JOHN DR
DPT OF PHYSICS
UNIVERSITY OF DURHAM
SOUTH RD
DURHAM DH1 3LE
UK
TEL
TLF
TLX
EML
COM   28

LUCK J DR
DIVISION OF NTL MAPPING
BOX 31
BELCOMMEN ACT 2626
AUSTRALIA
TEL   62 52 5172
TLF
TLX   62622 LUNAR AA
EML
COM   31C

LUCK R EARLE DR
WARNER & SWASEY OBS
CASE WESTERN RESERVE UNIV
CLEVELAND OH 44106
USA
TEL 216 368 6697
TLF
TLX
EML
COM 29,36

LUCKE PETER B DR
DPT PHYSICS & ASTRONOMY
MOUNT UNION COLLEGE
ALLIANCE OH 44601
USA
TEL 216 821 5320
TLF
TLX
EML
COM

LUCY LEON B PROF
ESO
KARL-SCHWARZSCHILD-STR 2
D 8046 GARCHING MUENCHEN
GERMANY
TEL 89 320 06249
TLF 89 320 2362
TLX 0528282-0
EML
COM 42

LUEST REIMAR PROF
MAX PLANCK INSTITUTE FOR
METEOROLOGY
BUNDESSTRASSE 55
D 2000 HAMBURG 13
GERMANY
TEL 40 41 1730
TLF 40 41 173298
TLX 211092
EML
COM 12,44,48,49

LUEST RHEA DR
MPI FUER PHYSIK UND
ASTROPHYSIK
K-SCHWARZSCHILDSTR 1
D 8046 GARCHING MUENCHEN
GERMANY
TEL 89 320 32990
TLF
TLX 524629 ASTROD
EML
COM 15

LUGGER PHYLLIS M
ASTRONOMY DPT
INDIANA UNIVERSITY
SWAIN WEST 319
BLOOMINGTON IN 47405
USA
TEL 812 335 6929
TLF
TLX
EML
COM 28

LUKACEVIC ILIJA S DR
DPT OF MECHANICS
FACULTY OF SCIENCES
STUDENTSKI TRG 16
YU 11000 BEOGRAD
YUGOSLAVIA
TEL 11 638 715
TLF
TLX
EML
COM

LUMINET JEAN-PIERRE
OBSERVATOIRE DE PARIS
SECTION DE MEUDON
F 92195 MEUDON PPL CDX
FRANCE
TEL 1 45 07 7423
TLF
TLX 201571
EML
COM 28,47,48

LUMME KARI A DR
OBS & ASTROPHYSICS LAB
UNIVERSITY OF HELSINKI
TAEHTITORNINMAKI
SF 00130 HELSINKI 13
FINLAND
TEL 191 2910
TLF
TLX
EML
COM 15,16,21

LUNA HOMERO G DR
IAR
CC 5
1894 VILLA ELISA (BS AS)
ARGENTINA
TEL 21 4 3793
TLX
EML
TLF
COM 25

LUND NIELS
DANISH SPACE RESEARCH INS
LUNDTOFTEVEJ 7
DK 2800 LYNGBY
DENMARK
TEL 42 88 2277
TLF
TLX 37198
EML
COM

LUNDQUIST CHARLES A DR
RESEARCH INSTITUTE
UNIVERSITY OF ALABAMA
BOX 209
HUNTSVILLE AL 35899
USA
TEL 205 895 6100
TLF
TLX
EML
COM 07

LUNDSTEDT HENRIK DR
LUND OBSERVATORY
BOX 43
S 221 00 LUND
SWEDEN
TEL 46 10 7294
TLX 33199 OBSNOT S
EML henrik@astro lu se
TLF 46 10 4614
COM 10,49

LUNDSTROM INGEMAR DR
LUND OBSERVATORY
BOX 43
S 221 00 LUND
SWEDEN
TEL 46 10 7300
TLX 33199
EML
TLF
COM 29

LUNEL MADELEINE DR
OBSERVATOIRE DE LYON
AVE CHARLES ANDRE
F 69561 S GENIS LAVAL CDX
FRANCE
TEL 78 56 0705
TLF 72 39 9791
TLX 310-926
EML
COM 33

LUNGU NICOLAIE DR
INSTITUTE POLITEHNIC
CATEDRA DE MATEMATICA
STR EMIL ISAC 15
R 3400 CLUJ NAPOCA
RUMANIA
TEL 951-17229
TLF
TLX
EML
COM

LUO BAO-RONG
YUNNAN OBSERVATORY
CAS
BOX 110
KUNMING 72946 YUNNAN
CHINA PR
TEL 871 2035
TLF
TLX
EML
COM 10

LUO DINGCHANG
BEIJING ASTRONOMICAL OBS
CAS
W SUBURB
BEIJING 100080
CHINA PR
TEL 1 27 5580
TLF
TLX 22040
EML
COM 31

LUO DING-JIANG
BEIJING ASTRONOMICAL OBS
CAS
W SUBURB
BEIJING 100080
CHINA PR
TEL 1 28 1698
TLF
TLX 22040 BAO ASCN
EML
COM 08,19

LUO SHI-FANG
SHANGHAI OBSERVATORY
CAS
80 NANDAN RD
SHANGHAI
CHINA PR
TEL 21 38 6191
TLF
TLX 33164 SHAO CN
EML
COM 19,31

LUO XIANHAN
DPT OF GEOPHYSICS
BEIJING UNIVERSITY
BEIJING 100071
CHINA PR
TEL 1 22 239
TLF
TLX
EML
COM 10,40

LUPISHKO DMITRIJ F
ASTRONOMICAL OBSERVATORY
KHARKOV UNIVERSITY
SUMSKAYA UL 35
310022 KHARKOV
UKRAINE
TEL 43 2428
TLF
TLX 125531 ICAR
EML
COM 15C

LUSTIG GUENTER DR
INSTITUT FUER ASTRONOMIE
KARL-FRANZENS-UNIVERSITAT
UNIVERSITATSPLATZ 5
A 8010 GRAZ
AUSTRIA
TEL 316 380 5272
TLF
TLX 0311662 UBGRZ
EML
COM 10,12

LUTTERMOSER DONALD DR
DPT OF PHYSICS
IOWA STATE UNIVERSITY
AMES IA 50010
USA
TEL
TLF
TLX
EML
COM 36

LUTZ BARRY L DR
LOWELL OBSERVATORY
1400 W MARS HILL RD
BOX 1149
FLAGSTAFF AZ 86001
USA
TEL 602 774 3358
TLF
TLX 6502358958 MCI
EML
COM 14,15,16

LUTZ JULIE H DR
PROGRAM IN ASTRONOMY
WASHINGTON STATE UNIV
PULLMAN WA 99164 2930
USA
TEL 509 335 3136
TLF
TLX 5107741091 WSUOIPPMA
EML
COM 45

LUTZ THOMAS E DR
PROGRAM IN ASTRONOMY
WASHINGTON STATE UNIV
PULLMAN WA 99164 2930
USA
TEL 509 335 3141
TLF
TLX 5107741091 WSUOIPPMA
EML
COM 24

LUYTEN WILLEM J PROF
SPACE SCIENCE CENTER
UNIVERSITY OF MINNESOTA
100 UNION ST SE
MINNEAPOLIS MN 55455
USA
TEL 612 373 3366
TLF
TLX
EML
COM 24,26,33

LYNAS-GRAY ANTHONY E
DPT PHYSICS & ASTRONOMY
UNIVERSITY COLLEGE LONDON
GOWER ST
LONDON WC1E 6BT
UK
TEL
TLF
TLX
EML
COM

LYNCH DAVID K
AEROSPACE CORPORATION
MS M2 226
BOX 92957
LOS ANGELES CA 90009
USA
TEL 213 648 6686
TLF
TLX 664460
EML
COM 09

LYNDEN-BELL DONALD PROF
INSTITUTE OF ASTRONOMY
THE OBSERVATORIES
MADINGLEY RD
CAMBRIDGE CB3 OHA
UK
TEL 223 622 04
TLF
TLX 817297 ASTRON G
EML
COM 28,33,37,47,48

LYNDS BEVERLY T DR
KITT PEAK NTL OBS
BOX 26732
950 N CHERRY AVE
TUCSON AZ 85726 6732
USA
TEL 602 325 9396
TLF
TLX 0666-484 AURA NOAO
EML
COM 28,34

LYNDS ROGER C DR
KITT PEAK NTL OBS
BOX 26732
950 N CHERRY AVE
TUCSON AZ 85726 6732
USA
TEL 602 327 5511
TLF
TLX
EML
COM 28

LYNE ANDREW G DR
NRAL
JODRELL BANK
MACCLESFIELD SK11 9DL
UK
TEL 477 713 21
TLX 36149
EML
TLF
COM 40

LYNGA GOSTA DR
MOUNT STROMLO & SIDING
SPRING OBSERVATORIES
PRIVATE BAG
WODEN PO ACT 2606
AUSTRALIA
TEL
TLF
TLX
EML GOSTA@MSO ANU EDU AU
COM 05,33,37,45

LYTTKENS EJNAR DR
SKOLGATAN 33B

S 752 21 UPPSALA
SWEDEN
TEL
TLF
TLX
EML
COM

LYTTLETON RAYMOND A PROF
INSTITUTE OF ASTRONOMY
THE OBSERVATORIES
MADINGLEY RD
CAMBRIDGE CB3 OHA
UK
TEL 223 622 04
TLF
TLX 817297 ASTRON G
EML
COM 15

LYUBIMKOV LEONID S DR
CRIMEAN ASTROPHYS OBS
UKRAINIAN ACAD OF SCIENCE
NAUCHNY
334413 CRIMEA
UKRAINE
TEL 43 2945
TLF
TLX
EML
COM 05,36

LYUTY VICTOR M DR
CRIMEAN STATION OF
STERNBERG INSTITUTE
NAUCHNY
334413 CRIMEA
UKRAINE
TEL 43 2945
TLF
TLX
EML
COM 42

MA ER
BEIJING ASTRONOMICAL OBS
CAS
W SUBURB
BEIJING 100080
CHINA PR
TEL 1 28 1698
TLF
TLX 22040 BAOAS CN
EML
COM 28

MA XING-YUAN
DPT OF GEOGRAPHY
BEIJING TEACHERS COLLEGE
BALIZHUANG
BEIJING
CHINA PR
TEL
TLF
TLX
EML
COM 46

MA YU-QIAN
INSTITUTE OF HIGH ENERGY
PHYSICS
BOX 918 3
BEIJING
CHINA PR
TEL 1 81 2971*464
TLF
TLX 22082 IHEP CN
EML·
COM 44,48

MACALPINE GORDON M
DPT OF ASTRONOMY
UNIVERSITY OF MICHIGAN
DENNISON BLDG
ANN ARBOR MI 48109 1090
USA
TEL 313 764 3433
TLF
TLX 810-223-6056
EML
COM 28

MACCACARO TOMMASO DR
CENTER FOR ASTROPHYSICS
HCO/SAO
60 GARDEN ST
CAMBRIDGE MA 02138
USA
TEL 617 495 7253
TLF
TLX 921428 SATELLITE CAM
EML
COM 48

MACCAGNI DARIO
IST DI FISICA COSMICA
CNR
VIA BASSINI 15
I 20133 MILANO
ITALY
TEL 2 298 237
TLF
TLX 313839 MUACNR I
EML
COM 48

MACCALLUM MALCOLM A H
SCHOOL OF MATHEMATICAL SC
QUEEN MARY/WESTFIELD COLL
MILE END RD
LONDON E1 4NS
UK
TEL· 1 980 4811
TLF
TLX· 893750 QMCUOL
EML·
COM 47

MACCHETTO FERDINANDO DR
STSCI
HOMEWOOD CAMPUS
3700 SAN MARTIN DR
BALTIMORE MD 21218
USA
TEL 301 338 4790
TLF
TLX 6849101
EML
COM 28,40,44,48

MACCONNELL DARRELL J DR
STSCI/CSC
HOMEWOOD CAMPUS
3700 SAN MARTIN DR
BALTIMORE MD 21218
USA
TEL 301 338 4800
TLF 301 338 4746
TLX
EML SCIVAX MACCONNELL
COM 24,33,45P

MACDONALD GEOFFREY H DR
ELECTRONICS LABORATORY
UNIVERSITY OF KENT
CANTERBURY CT2 7NT
UK
TEL 227 668 22*258
TLF
TLX 965449 UKCLIB
EML
COM 40

MACDONALD JAMES
DPT OF PHYSICS
UNIVERSITY OF DELAWARE
NEWARK DE 19716
USA
TEL· 302 451 2661
TLF·
TLX
EML
COM 40,42

MACERONI CARLA
OAR
VIA DEL PARCO MELLINI 84
I 00136 ROMA
ITALY
TEL
TLF
TLX
EML
COM 42

MACGILLIVRAY HARVEY T DR
ROYAL OBSERVATORY

BLACKFORD HILL
EDINBURGH EH9 3HJ
UK
TEL 316 673 321
TLF
TLX 72383 ROEDIN G
EML
COM

MACHADO LUIZ E DA SILVA
UNIV FED RIO DE JANEIRO
AV SERNAMBETIBA 3300 BL7
22 630 BARRA DA TIJUCA
20080 RIO DE JANEIRO RJ
BRAZIL
TEL 21 399 2589
TLF
TLX
EML
COM 20,24

MACHADO MARCOS
CNIE
AVENIDA MITRE 3100
1663 SAN MIGUEL (BS AS)
ARGENTINA
TEL 1 664 8371
TLF·
TLX· 17511 LANBA AR
EML
COM 10C

MACHALSKI JERZY DR
ASTRONOMICAL OBSERVATORY
JAGIELLONIAN UNIVERSITY
UL MAZOWIECKA 36/33
PL 30 019 KRAKOW
POLAND
TEL
TLF
TLX 0322297 UJ PL
EML
COM 40

MACIEJEWSKI ANDRZEJ J DR
INSTITUTE OF ASTRONOMY
N COPERNICUS UNIVERSITY
UL CHOPINA 12/18
PL 87 100 TORUN
POLAND
TEL 26018*53
TLF
TLX 552234 ASTR PL
EML MACIEJKA@PLTUMK11
COM 07

MACIEL WALTER J DR
IAG
UNIVERSIDADE DE SAO PAULO
CP 30627
01051 SAO PAULO SP
BRAZIL
TEL
TLF
TLX
EML
COM 34,46

MACK PETER DR
MDM OBSERVATORY
HC 04
BOX 7520
TUCSON AZ 85735
USA
TEL 602 620 5360
TLF
TLX 910 952 1116
EML
COM 09

MACKAY CRAIG D DR
INSTITUTE OF ASTRONOMY
THE OBSERVATORIES
MADINGLEY RD
CAMBRIDGE CB3 0HA
UK
TEL 223 622 04
TLF
TLX 817297 ASTRON G
EML
COM

MACKINNON ALEXANDER L
DPT OF ASTRONOMY
UNIVERSITY OF GLASGOW
GLASGOW G12 8QW
UK
TEL 413 398 855
TLF
TLX 777070 UNIGLA
EML
COM 10

MACLEOD JOHN M DR
HERZBERG INST ASTROPHYS
NTL RESEARCH COUNCIL
100 SUSSEX DR
OTTAWA ON K1A 0R6
CANADA
TEL   613 993 6060
TLF   613 952 6602
TLX   053 3715
EML
COM   34,40

MACQUEEN ROBERT M DR
DPT OF PHYSICS
RHODES COLLEGE
2000 N PARKWAY
MEMPHIS TENNESSEE 38112
USA
TEL   901 726 3000
TLF
TLX
EML
COM   10,49

MACRAE DONALD A PROF
DAVID DUNLAP OBSERVATORY
UNIVERSITY OF TORONTO
BOX 360
RICHMOND HILL ON L4C 4Y6
CANADA
TEL   416 884 9562
TLF   416 978 3921
TLX   069 86766
EML
COM   33,38,40

MACRIS CONSTANTIN J PROF
RES CENTER FOR ASTRONOMY
ACADEMY OF ATHENS
14 ANAGNOSTOPOULOU ST
GR 106 73 ATHENS
GREECE
TEL   1 361 3589
TLF
TLX
EML
COM   10

MACY WILLIAM WRAY DR
151 MELVILLE AVE
PALO ALTO CA 94304
USA
TEL
TLF
EML
COM
TLX

MADDISON RONALD CH DR
UNIVERSITY OF KEELE
2 CHURCH PLANTATION
KEELE PARK
KEELE STAFFS
UK
TEL   782 621 111
TLF
TLX
EML
COM   46

MADDOX STEPHEN DR
DPT OF ASTROPHYSICS
UNIVERSITY OF OXFORD
KEBLE RD
OXFORD OX1 3RH
UK
TEL   865 273 310
TLF   865 273 418
TLX   83295 NUCLOX G
EML   SJM@UK AC OX ASTRO/19464 SJM
COM   47

MADEJ JERZY
ASTRONOMICAL OBSERVATORY
WARSAW UNIVERSITY
AL UJAZDOWSKIE 4
PL 00 478 WARSAW
POLAND
TEL   4822-29-40-11
TLF
TLX
EML
COM   36

MADORE BARRY FRANCIS DR
DAVID DUNLAP OBSERVATORY
UNIVERSITY OF TORONTO
BOX 360
RICHMOND HILL ON L4C 4Y6
CANADA
TEL   416 884 9562
TLF   416 978 3921
TLX   069 86766
EML
COM   27,28

MADSEN JES
INST OF PHYSICS & ASTRON
UNIVERSITY OF AARHUS
NY MUNKEGADE
DK 8000 AARHUS C
DENMARK
TEL   86 12 8899
TLF   86 20 2711
TLX   64767 AAUSCI DK
EML
COM

MAEDA KEI-ICHI DR
DPT OF PHYSICS
WASEDA UNIVERSITY
OKUBO 3-4-1 SHINJUKU-KU
TOKYO 160
JAPAN
TEL   03 203 4141
TLF
TLX   2323280 WASEDA J
EML   BITNET maeda@jpnwas00
COM   47

MAEDA KOITIRO
DPT OF PHYSICS
HYOGO COLLEGE OF MEDICINE
NISHINOMIYA
HYOGO 663
JAPAN
TEL   798-45-6111
TLF
TLX
EML
COM

MAEDER ANDRE PROF
OBSERVATOIRE DE GENEVE
CHEMIN DES MAILLETTES 51
CH 1290 SAUVERNY
SWITZERLAND
TEL   22 755 2611
TLF   22 755 3983
TLX   419209 OBS CH
EML   MAEDER@CGEUGE11
COM   27,35C,37

MAEHARA HIDEO DR
OKAYAMA ASTROPHYSICAL OBS
NAOJ
KAMOGATA ASAKUCHI
OKAYAMA 719 02
JAPAN
TEL   86 544 2155
TLF   86 544 2360
TLX
EML   MAEHARA@KIBI DAO NAO AC JP
COM   45C

MAETZLER CHRISTIAN DR
PHYSIKALISCHES INSTITUT
UNIVERSITAET BERN
SIDLERSTRASSE 5
CH 3012 BERN
SWITZERLAND
TEL   31 65 8911
TLF
TLX   32320 PHYBE CH
EML
COM

MAFFEI PAOLO PROF
CATTEDRA DI ASTROFISCA
UNIVERSITA DI PERUGIA
VIA DELL'ELCE DI SOTTO
I 06100 PERUGIA
ITALY
TEL   75 45 647
TLF
TLX
EML
COM   27,51

MAGAIN PIERRE DR
INSTITUT D'ASTROPHYSIQUE
UNIVERSITE DE LIEGE
AVE COINTE 5
B 4000 COINTE-LIEGE
BELGIUM
TEL   41 52 9980
TLF   41 52 7474
TLX
EML
COM   29

MAGAKIAN TIGRAN Y DR
BYURAKAN ASTROPHYSICAL
OBSERVATORY
378433 BYURAKAN
ARMENIA
TEL   88 52 28 28 4142
TLF
TLX   411576 ASCON SU
EML
COM

MAGALHAES ANTONIO A S ENG
OBSERVATORIO ASTRONOMICO
UNIVERSIDADE DO PORTO
MONTE DA VIRGEM
P 4400 VILA NOVA GAIA
PORTUGAL
TEL   782-0404
TLF
TLX
EML
COM

MAGALHAES ANTONIO MARIO
DPT OF ASTRONOMY
UNIVERSITY OF WISCONSIN
475 N CHARTER ST
MADISON WI 53706
USA
TEL   608 263 4680
TLF
TLX   265 452 UW MAD
EML
COM

MAGAZZU ANTONIO DR
OSS ASTROFISICO
CITTA UNIVERSITARIA
VIA A DORIA 6
I 95125 CATANIA
ITALY
TEL 95 33 0533
TLF 95 33 0592
TLX 970359 ASTRCT I
EML ASTRCT ANTONIO
COM 29

MAGGIO ANTONIO DR
OSS ASTRONOMICO
UNIVERSITA DI PALERNO
PALAZZO DEI NORMANNI
I 90134 PALERMO
ITALY
TEL 91 657 0451
TLF. 91 48 8900
TLX 910402 ASTROP I
EML· ASTROPA@IPACUC BITNET
COM. 48

MAGNAN CHRISTIAN DR
GRAAL CC 72
USTL II
PLACE EUGENE BATAILLON
F 34095 MONTPELLIER CDX 5
FRANCE
TEL 67 14 3902
TLF
TLX
EML
COM

MAGNANI LORIS ALBERTO DR
DPT PHYSICS & ASTRONOMY
UNIVERSITY OF GEORGIA
ATHENS GA 30602
USA
TEL 404 542 2876
TLF 404 542 2492
TLX 414190
EML LORIS@JOVE PHYSAST UGA ED
COM

MAGNARADZE NINA G DR
DPT OF ASTRONOMY
TBILISI UNIVERSITY
380043 TBILISI
GEORGIA
TEL
TLF
TLX
EML
COM 07

MAGNI GIANFRANCO
IAS
VIA DELL'UNIVERSITA 11
I 00185 ROMA
ITALY
TEL
TLX
EML
TLF
COM·

MAGUN ANDREAS DR
PHYSIKALISCHES INSTITUT
UNIVERSITAET BERN
SIDLERSTRASSE 5
CH 3012 BERN
SWITZERLAND
TEL 31 65 8923
TLF
TLX
EML
COM

MAHAT ROSLI H DR
DPT OF ASTRONOMY
UNIVERSITY OF MALAYA
59100 KUALA LUMPUR
MALAYSIA
TEL
TLF
TLX
EML
COM

MAHESWARAN MURUGESAPILLAI
INST OF FUNDAMENT STUDIES
380/72 BAUDDHALOKA
MAWATHA
COLOMBO 7
SRI LANKA
TEL 1-597538
TLF
TLX 21700 IFS CE
EML·
COM 35

MAHMOUD FAROUK M A B DR
HELWAN OBSERVATORY
HELWAN
EGYPT
TEL 78 0645/2683
TLF·
EML
COM 27
TLX

MAHRA H S DR
UTTAR PRADESH STATE
OBSERVATORY
PO MANORA PEAK 263 129
NAINITAL 263 129
INDIA
TEL 59 42 2136/2583
TLF
TLX
EML
COM 09,16,20,27,50

MAIHARA TOSHINORI DR
DPT OF PHYSICS
KYOTO UNIVERSITY
SAKYO KU
KYOTO 606
JAPAN
TEL 075-751-2111
TLF
TLX 5422693 LIBKYU J
EML
COM 21,34

MAILLARD JEAN-PIERRE DR
INSTITUT D'ASTROPHYSIQUE
98BIS BD ARAGO
F 75014 PARIS
FRANCE
TEL 1 43 20 1425
TLF 1 43 29 8673
TLX
EML
COM 09,14,29

MAITZEN HANS M DR
INSTITUT FUER ASTRONOMIE
TUERKENSCHANZSTR 17
A 1180 WIEN
AUSTRIA
TEL 1 345 36094
TLX· 116222 PHYSI A
EML.
TLF
COM 25,29

MAJID ABDUL BIN A H DR ,
JABATAN UKUR DAN PEMETAAN
JALAN SEMARAK
50578 KUALA LUMPUR
MALAYSIA
TEL 03 2926327
TLF
TLX
EML
COM 04

MAJOR JOHN 'DR
DPT OF PHYSICS
UNIVERSITY OF DURHAM
SOUTH RD
DURHAM DH1 3LE
UK
TEL 913 742 111
TLF
TLX
EML
COM 09

MAKARENKO EKATERINA N DR
ASTRONOMICAL OBSERVATORY
ODESSA STATE UNIVERSITY
SHEVCHENKO PARK
270014 ODESSA
UKRAINE
TEL
TLF
TLX
EML
COM 27

MAKAROV VALENTINE I
KISLOVODSK STATION OF THE
PULKOVO OBSERVATORY
357741 KISLOVODSK
RUSSIA
TEL
TLF
TLX
EML
COM 10,12

MAKAROVA ELENA A DR
STERNBERG STATE ASTR INST
117234 MOSCOW
RUSSIA
TEL 139-1973
TLF
TLX 113037 JAPET
EML
COM 12

MAKINO FUMIYOSHI DR
ISAS
3-1-1 YOSHINODAI
SAGAMIHARA
KANAGAWA 229
JAPAN
TEL 81-427 51 3911
TLF 81-427 59 4253
TLX 34757 ISASTRO J
EML
COM·

MAKISHIMA KAZUO
INST SPACE & ASTRON SCI
UNIVERSITY OF TOKYO
MEGURO KU
TOKYO 153
JAPAN
TEL 03-467-1111x303
TLF
TLX 34757 ISASTRO J
EML
COM

MAKITA MITSUGU DR
KWASAN & HIDA OBS
KYOTO UNIVERSITY
YAMASHINA
KYOTO 607
JAPAN
TEL 075-581-1235
TLF
TLX
EML
COM 10,12

MALACARA DANIEL
CENTRO DE INVESTIGACIONES
EN OPTICA
APDO POSTAL 948
37000 LEON GTO
MEXICO
TEL 758-23
TLF
TLX
EML
COM

MALAGNINI MARIA LUCIA
OAT
BOX SUCC TRIESTE 5
VIA TIEPOLO 11
I 34131 TRIESTE
ITALY
TEL 40 79 3921
TLF
TLX 461137 OAT I
EML
COM

MALAISE DANIEL J DR
INSTITUT D'ASTROPHYSIQUE
UNIVERSITE DE LIEGE
AVE COINTE 5
B 4000 COINTE-LIEGE
BELGIUM
TEL 41 52 9980
TLF 41 52 7474
TLX
EML
COM 15,44

MALAKPUR IRADJ DR
INSTITUTE OF GEOPHYSICS
UNIVERSITY OF TEHRAN
KARGAR SHOMALI
TEHRAN 14394
IRAN
TEL 21 63 1081/3
TLF
TLX 215319 UTIG
EML
COM

MALARODA STELLA M DR
COMPLEJO ASTRONOMICO
EL LEONCITO
CC 467
5400 SAN JUAN
ARGENTINA
TEL 64 22 5718
TLF
TLX 59134 ENTOP AR
EML
COM 29,45

MALHERBE JEAN MARIE DR
OBSERVATOIRE DE PARIS
SECTION DE MEUDON
DASOP
F 92195 MEUDON PPL CDX
FRANCE
TEL 1 45 07 7796
TLF
TLX 201571
EML BITNET MALHERBE@FRMEU51
COM 10

MALIN DAVID F MR
AAO
BOX 296
EPPING NSW 2121
AUSTRALIA
TEL 2 868 1666
TLX 23999 AA
EML
TLF
COM 09

MALIN STUART
NATIONAL MARITIME MUSEUM
GREENWICH
LONDON SE10 9NF
UK
TEL 1 858 1167
TLX
EML
TLF
COM 41

MALINA ROGER FRANK DR
CTR FOR EUV ASTROPHYSICS
UNIVERSITY OF CALIFORNIA
2150 KITTREDGE ST
BERKELEY CA 94720
USA
TEL 510 643 5636
TLF 510 643 5660
TLX
EML RMALINA@SSL BERKELEY EDU
COM

MALITSON HARRIET H MS
13315 MAGELLAN AVE

ROCKVILLE MD 20853
USA
TEL 301 946 0496
TLF
TLX
EML
COM 10,44

MALKAMAEKI LAURI J DR
ELSON RESEARCH INC
BOX 167
SALEM NH 03079 9998
USA
TEL 603 893 0766
TLF 603 8932131
TLX
EML
COM 09

MALKAN MATTHEW ARNOLD DR
DPT OF ASTRONOMY
UNIVERSITY OF CALIFORNIA
LOS ANGELES CA 90024
USA
TEL 213 825 3404
TLF
TLX
EML MALKAN@ASTRO UCLA EDU
COM 44

MALLIA EDWARD A DR
DPT OF ASTROPHYSICS
UNIVERSITY OF OXFORD
SOUTH PARKS RD
OXFORD OX1 3RQ
UK
TEL
TLF
TLX
EML
COM

MALLIK D C V DR
INDIAN INSTITUTE OF
ASTROPHYSICS
KORAMANGALA
BANGALORE 560 034
INDIA
TEL 812 56 6585/6497
TLF
TLX 845763 IIAB IN
EML
COM 34,35

MALTBY PER PROF
INST THEORET ASTROPHYSICS
UNIVERSITY OF OSLO
BOX 1029
N 0315 BLINDERN OSLO 3
NORWAY
TEL 2-456509
TLF
TLX
EML
COM 10

MALVILLE J MCKIM PROF
CASA
UNIVERSITY OF COLORADO
BOX 391
BOULDER CO 80309 0391
USA
TEL 303 492 8788
TLF
TLX
EML.
COM 10

MAMMANO AUGUSTO DR
OSSERVATORIO ASTROFISICO
VIA DELL OSSERVATORIO 8
I 36012 ASIAGO
ITALY
TEL 42 46 2665
TLF
TLX
EML
COM 42

MANABE SEIJI DR
INTL LATITUDE OBSERVATORY
NAOJ
HOSHIGAOKA MIZUSAWA SHI
IWATE 023
JAPAN
TEL
TLF
TLX
EML
COM 19

MANARA ALESSANDRO A DR
OSS ASTRONOMICO DI BRERA
VIA BRERA 28
I 20121 MILANO
ITALY
TEL 2 87 4444
TLX
EML
TLF 2 72 00 1600
COM 20,44

MANCHADO ARTURO DR
INST DE ASTROFISICA
DE CANARIAS
OBS DEL TEIDE
E 38200 LA LAGUNA
SPAIN
TEL 22 26 2211
TLF 22 26 3005
TLX 92640 IAC E
EML
COM 34

MANCHANDA R K DR
TIFR
HOMI BHABHA RD
COLABA
BOMBAY 400 005
INDIA
TEL 22 219 111*336
TLF
TLX 113009 TIFR IN
EML
COM

MANCHESTER RICHARD N DR
CSIRO
DIVISION OF RADIOPHYSICS
BOX 76
EPPING NSW 2121
AUSTRALIA
TEL 2 868 0225
TLF
TLX 26320 ASTRO
EML RMANCHES@ATNF.CSIRO AU
COM 33,34,40

MANCUSO SANTI PROF
OSS ASTRONOMICO
DI CAPODIMONTE
VIA MOIARIELLO 16
I 80131 NAPOLI
ITALY
TEL 44 0101
TLF
TLX
EML
COM

MANDOLESI NAZZARENO
IST TE S R E
CNR
VIA DE CASTAGNOLI 1
I 40126 BOLOGNA
ITALY
TEL 51 23 8022
TLF
TLX 511350
EML·
COM 40,44,47

MANDZHOS ANDREJ V DR
SPECIAL ASTROPHYSICAL OBS
ACADEMY OF SCIENCES
NIZHNIJ ARKHYZ
357147 STAVROPOLSKIJ
RUSSIA
TEL 86 578 92501
TLF
TLX 123244 ZENIT
EML AMAND@SAO STAVROPOL SU
COM 47

MANFROID JEAN DR
INSTITUT D'ASTROPHYSIQUE
UNIVERSITE DE LIEGE
AVE COINTE 5
B 4000 COINTE-LIEGE
BELGIUM
TEL 41 52 9980
TLF 41 52 7474
TLX 41264 ASTRLG
EML
COM 25,34

MANGENEY ANDRE DR
OBSERVATOIRE DE PARIS
SECTION DE MEUDON
F 92195 MEUDON PPL CDX
FRANCE
TEL 1 45 07 7661
TLF
TLX
EML
COM 49

MANN GOTTFRIED DR
ZNTRLINST F ASTROPHYSIK
OBSERVATORIUM FUER SOLARE
RADIO ASTRONOMIE
D 1501 TREMSDORF
GERMANY
TEL
TLF
TLX 15420
EML
COM 10

MANN PATRICK J DR
DPT OF ASTRONOMY
UNIV OF WESTERN ONTARIO
LONDON ON N6A 3K7
CANADA
TEL 519 661 3183
TLF 519 661 3486
TLX 064 7134
EML BITNET 2014-562@UWOVAX
COM

MANNINO GIUSEPPE PROF
IST MATEMATICO

VIA CAMPI 181
I 41100 MODENA
ITALY
TEL
TLF
TLX
EML
COM 27

MANRIQUE WALTER T PROF
OBSERVATORIO ASTRONOMICO
FELIX AGUILAR
AV BENAVIDEZ 8175 OESTE
5407 MARQUESADO (S J)
ARGENTINA
TEL 64 23 1494
TLF
TLX
EML
COM 08

MANSFIELD VICTOR N PROF
DPT OF PHYSICS & ASTRON
COLGATE UNIVERSITY
HAMILTON NY 13346
USA
TEL 315 824 1000
TLF
TLX
EML
COM

MANTEGAZZA LUCIANO
OSS ASTRONOMICO DI MILANO
VIA E BIANCHI 46
I 22055 MERATE
ITALY
TEL 59 2035
TLF
TLX
EML
COM 27

MANTOVANI FRANCO
IST DI RADIOASTRONOMIA
CNR
VIA IRNERIO 46
I 40126 BOLOGNA
ITALY
TEL 51 23 2856
TLF
TLX 211664 INFNBO I
EML
COM

MAO WEI
YUNNAN OBSERVATORY
CAS
BOX 110   /
KUNMING 72946 YUNNAN
CHINA PR
TEL 871 2035
TLX
TLX
EML
COM 08

MARABINI RODOLFO JOSE
UNIV NACIONAL DE LA PLATA
FCAG
1900 LA PLATA (BS AS)
ARGENTINA
TEL 21 21 7308
TLF
TLX
EML
COM

MARAN STEPHEN P DR
NASA/GSFC
CODE 680
GREENBELT MD 20771
USA
TEL 301 286 8607
TLF
TLX 89675
EML
COM 15,40,44

MARANO BRUNO
DPT DI ASTRONOMIA
UNIVERSITA DI BOLOGNA
CP 596
I 40100 BOLOGNA
ITALY
TEL 51 22 2956
TLF
TLX 211664 INFNBO I
EML
COM 47

MARAR T M K
ISRO SATELLITE CTR
AIRPORT RD
VIMANAPURA POST
BANGALORE 560 017
INDIA
TEL   812 56 6251
TLF
TLX
EML
COM 44

MARASCHI LAURA DR
IST DI FISICA
UNIVERSITA DI MILANO
VIA CELORIA 16
I 20133 MILANO
ITALY
TEL   2 239 2275
TLF   2 236 6583
TLX
EML
COM

MARCAIDE JUAN-MARIA DR
INST ASTROFISICA
DE ANDALUCIA APD 2144
PROFESOR ALBAREDA 1
E 18080 GRANADA
SPAIN
TEL   58 12 1311
TLF
TLX 78573 IAAG E
EML
COM 40

MARCELIN MICHEL
OBSERVATOIRE DE MARSEILLE
2 PLACE LE VERRIER
F 13248 MARSEILLE CDX 04
FRANCE
TEL 91 95 9088
TLF
TLX 420241 F
EML
COM 28

MARCHAL CHRISTIAN DR
DPT ETUDES DE SYNTHESE
ONERA
F 92320 CHATILLON
FRANCE
TEL   1 46 57 1160
TLF
TLX 260907
EML
COM 07

MARDIROSSIAN FABIO
DPT DI ASTRONOMIA
UNIVERSITA DI TRIESTE
VIA TIEPOLO 11
I 34131 TRIESTE
ITALY
TEL 40 79 3921*221
TLF
TLX 461137 OAT I
EML
COM 42,47

MAREK JOHN
44 PERCY RD
WREXHAM CLWYD
UK
TEL
TLF
TLX
EML
COM 47

MARGON BRUCE H PROF
DPT OF ASTRONOMY
UNIVERSITY OF WASHINGTON
FM 20
SEATTLE WA 98195
USA
TEL 206 543 0089
TLF
TLX 4740096
EML MARGON@JANUS ASTRO WASHINGTON
COM

MARGONI RINO
OSSERVATORIO ASTROFISICO
VIA DELL OSSERVATORIO 8
I 36012 ASIAGO
ITALY
TEL 42 46 2665
TLF
TLX 430110 SETOUR
EML
COM

MARGRAVE THOMAS EWING JR
400 JOHNSON ST
VIENNA VA 22180
USA
TEL
TLF
EML
COM 27,51
TLX

MARIE M A DR
DPT OF ASTRONOMY
FACULTY OF SCIENCES
CAIRO UNIVERSITY
GEZA
EGYPT
TEL
TLF
TLX
EML
COM

MARIK MIKLOS DR
ASTRONOMY DPT
EOTVOS UNIVERSITY
KUN BELA TER 2
H 1083 BUDAPEST
HUNGARY
TEL 1 41 019
TLF
TLX
EML
COM 12,38

MARILLI ETTORE DR
OSS ASTROFISICO
CITTA UNIVERSITARIA
VIA A DORIA 6
I 95125 CATANIA
ITALY
TEL 95 33 0533
TLF
TLX 970359 ASTRCT I
EML
COM 12,29,42

MARINO BRIAN F ENG
AUCKLAND OBSERVATORY
BOX 72009
NORTHCOTE
AUCKLAND 9
NEW ZEALAND
TEL 649 486 951
TLF
TLX
EML
COM 42

MARIOTTI JEAN MARIE DR
OBSERVATOIRE DE PARIS
SECTION DE MEUDON
F 92195 MEUDON PPL CDX
FRANCE
TEL 1 45 07 7570
TLF
TLX 204464
EML BITNET MARIOTTI@FRMEU51
COM

MARIS GEORGETA DR
ASTRONOMICAL OBSERVATORY
CUTITUL DE ARGINT 5
BOX 28
R 75212 BUCURESTI 28
RUMANIA
TEL 90 236010
TLF
TLX 11882 ASTRO R
EML
COM 10

MARISKA JOHN THOMAS
NAVAL RESEARCH LABORATORY
CODE 4175M
4555 OVERLOOK AVE SW
WASHINGTON DC 20375 5000
USA
TEL 202 767 2605
TLF
TLX
EML
COM 10,12

MARK JAMES WAI-KEE DR
LAWRENCE LIVERMORE LAB
L 477
BOX 808
LIVERMORE CA 94550
USA
TEL 415 422 5931
TLF
TLX 910-386-8339 UCCLLL
EML
COM 33

MARKELLOS VASSILIS V DR
DPT ENFINEERING SCIENCE
UNIVERSITY OF PATRAS
GR 260 00 RION
GREECE
TEL 61 99 1465
TLF
TLX
EML
COM 07

MARKERT THOMAS H DR
CENTER FOR SPACE RESEARCH
MIT RM 37 621
BOX 165
CAMBRIDGE MA 02139
USA
TEL 617 253 5169
TLF
TLX
EML THM@SPACE MIT EDU
COM·

MARKKANEN TAPIO DR
OBS & ASTROPHYSICS LAB
UNIVERSITY OF HELSINKI
TAEHTITORNINMAKI
SF 00130 HELSINKI 13
FINLAND
TEL 090 8391
TLF
TLX
EML
COM 25,37,50

MARKOWITZ WILLIAM DR
APT 15 B
2800 E SUNRISE BLVD
FORT LAUDERDALE FL 33304
USA
TEL 305 563 2859
TLF
TLX
EML
COM 19,31

MARLBOROUGH J M PROF
DPT OF ASTRONOMY
UNIV OF WESTERN ONTARIO
LONDON ON N6A 3K7
CANADA
TEL 519 679 3184
TLF 519 661 3486
TLX 064 7134
EML
COM 36

MARMOLINO CIRO
DPT DI FISICA
UNIVERSTIA DI NAPOLI
MOSTRA D OLTREMARE PAD 19
I 80125 NAPOLI
ITALY
TEL 81 725 3428
TLF
TLX 720320 INFNNA I
EML
COM 12

MAROCHNIK L S PROF DR
SPACE RESEARCH INSTITUTE
ACADEMY OF SCIENCES
PROFSOJOSNAYA UL 84/32
117810 MOSCOW
RUSSIA
TEL 333-31-22
TLF
TLX 411498 STAR SU
EML
COM 33

MAROV MIKHAIL YA PROF
INST OF APPLIED MATHS
ACADEMY OF SCIENCES
MIUSSKAYA SQ 4
125047 MOSCOW
RUSSIA
TEL 095 250 0485
TLF 095 972 0737
TLX
EML
COM 16VP,44,51,WGPSNC

MARQUES DOS SANTOS P PROF
IAG
UNIVERSIDADE DE SAO PAULO
CP 30627
01051 SAO PAULO SP
BRAZIL
TEL 11 276 3941
TLF
TLX 36221 IAGM BR
EML
COM 28,40

MARQUES MANUEL N DR
OBSERVATORIO ASTRONOMICO

TAPADA DA AJUDA
P 1300 LISBOA 3
PORTUGAL
TEL
TLF
TLX
EML
COM

MARRACO HUGO G DR
UNIV NACIONAL DE LA PLATA
FACULTAD DE CIENCIAS
ASTRONOMICAS Y GEOFISICAS
1900 LA PLATA (BS AS)
ARGENTINA
TEL 21 21 7308
TLF
TLX 31151 BULAP AR
EML
COM 25,37

MARSCHALL LAURENCE A
DPT OF PHYSICS
GETTYSBURG COLLEGE
GETTYSBURG PA 17325
USA
TEL 717 337 1865
TLF
TLX
EML
COM 24,30

MARSCHER ALAN PATRICK
DPT OF ASTRONOMY
BOSTON UNIVERSITY
725 COMMONWEALTH AVE
BOSTON MA 02215
USA
TEL 617 353 5029
TLF
TLX 951289 BOS UNIV BSN
EML
COM 40

MARSDEN BRIAN G DR
CENTER FOR ASTROPHYSCIS
HCO/SAO
60 GARDEN ST
CAMBRIDGE MA 02138
USA
TEL 617 495 7244
TLF
TLX 7103206842 ASTROGRAM
EML CFAPS1 BRIAN
COM 06C,07,15,20C

MARSDEN PHILIP L PROF
DPT OF PHYSICS
UNIVERSITY OF LEEDS
LEEDS LS2 9JT
UK
TEL 532 431 751
TLF
TLX 556473 UNIDS
EML
COM

MARSH JULIAN C D
HATFIELD POLYTECHNIC
OBSERVATORY
BAYFORDBURY
HERTFORD HERTS SG13 8LD
UK
TEL 992 558 451
TLF
TLX 262413
EML
COM 46

MARSH THOMAS DR
DPT OF ASTROPHYSICS
UNIVERSITY OF OXFORD
KEBLE RD
OXFORD OX1 3RH
UK
TEL 865 273 303
TLF 865 273 418
TLX 83295 NUCLOX G
EML 19464 TRM
COM 42

MARSHALL HERMAN LEE DR
CTR FOR EUV ASTROPHYSICS
UNIVERSITY OF CALIFORNIA
2150 KITTREDGE ST
BERKELEY CA 94720
USA
TEL 415 643 5671
TLF 415 643 5660
TLX 910 414 061 UCBSAG
EML HERMANM@SSL BERELEY EDU
COM

MARSHALL KEVIN P
INSTITUTE OF ASTRONOMY
THE OBSERVATORIES
MADINGLEY RD
CAMBRIDGE CB3 OHA
UK
TEL 223 337 548
TLF
TLX 817297 ASTRON G
EML KPM@UK AC CAM AST-STAR
COM 37

MARSOGLU A DR
UNIVERSITY OBSERVATORY
UNIVERSITY OF ISTANBUL
34452 ISTANBUL
TURKEY
TEL 1 522 3597
TLF
TLX
EML
COM

MARSTON ANTHONY PHILIP DR
DPT PHYSICS & ASTRONOMY
DRAKE UNIVERSITY
DES MOINES IA 50311
USA
TEL 515 271 3034
TLF
TLX
EML TM9991R@DRAKE
COM 34

MARTENS PETRUS C DR
LOCKHEED PALO ALTO RES LB
DPT 91 30 BLDG 252
3251 HANOVER ST
PALO ALTO CA 94304
USA
TEL 415 354 5819
TLF 415 424 3994
TLX
EML 24707 MARTENS
COM 10

MARTIN ANTHONY R DR
UK CULHAM LABORATORY
RM F4/135
ABINGDON OX14 3DB
UK
TEL 235 21 840
TLF
TLX 83189
EML
COM 51

MARTIN DEREK H PROF
DPT OF PHYSICS
QUEEN MARY/WESTFIELD COLL
MILE END RD
LONDON E1 4NS
UK
TEL
TLF
TLX
EML
COM

MARTIN FRANCOIS DR
DPT ASTROPHYSIQUE
UNIVERSITE DE NICE
PARC VALROSE
F 06034 NICE CDX
FRANCE
TEL 93 51 9100
TLF
TLX 970281
EML
COM

MARTIN INACIO MALMONGE DR
INSTITUTO DE FISICA
UNIVERSIDADE EST CAMPINAS
CP 1170
13100 CAMPINAS SP
BRAZIL
TEL 192 39 1301
TLF
TLX 019-1150
EML
COM 48

MARTIN MARIA CRISTINA DR
IAR
CC 5
1894 VILLA ELISA (BS AS)
ARGENTINA
TEL 21 4 3793
TLX 31216 CESLA AR
EML
TLF
COM 28,51

MARTIN NICOLE DR
OBSERVATOIRE DE MARSEILLE
2 PLACE LE VERRIER
F 13248 MARSEILLE CDX 04
FRANCE
TEL 91 95 9088
TLF
TLX 420241
EML
COM 30

MARTIN PETER G PROF
CITA MCLENNAN LABS
UNIVERSITY OF TORONTO
60 ST GEORGE ST
TORONTO ON M5S 1A7
CANADA
TEL 416 978 6840
TLF 416 978 3921
TLX
EML
COM 34C

MARTIN ROBERT N DR
STEWARD OBSERVATORY
UNIVERSITY OF ARIZONA
TUCSON AZ 85721
USA
TEL 602 621 1539
TLF
TLX 467175
EML
COM 34,40

MARTIN WILLIAM C DR
NTL BUREAU OF STANDARDS
PHYSICS BLDG A167
GAITHERSBURG MD 20899
USA
TEL 301 921 2011
TLF
TLX
EML
COM 14

MARTIN WILLIAM L DR
ROYAL GREENWICH OBS
HERSTMONCEUX CASTLE
HAILSHAM BN27 1RP
UK
TEL 323 833 171
TLX 87451 RGOBSY G
EML
TLF
COM 27

MARTINET LOUIS PROF
OBSERVATOIRE DE GENEVE
CHEMIN DES MAILLETTES 51
CH 1290 SAUVERNY
SWITZERLAND
TEL 22 755 2611
TLF 22 755 3983
TLX 419209 OBS CH
EML martinet@obs unige ch
COM 28,33,46

MARTINEZ MARIO DR
DPT DE GEOFISICA
CIESE
APDO POSTAL 2732
22860 ENSENADA B CALIF
MEXICO
TEL
TLF
TLX
EML
COM

MARTINEZ ROGER CARLOS DR
INST DE ASTROFISICA
DE CANARIAS
OBS DEL TEIDE
E 38200 LA LAGUNA
SPAIN
TEL 22 26 2211
TLF
TLX 92640
EML
COM

MARTINEZ-GONZALEZ E DR
DPT DE FISICA MODERNA
UNIVRESIDAD DE CANTABRIA
AVD LOS CASTROS S/N
E 39005 SANTANDER
SPAIN
TEL 42 20 1468
TLF 42 20 1402
TLX 35861 EDUCI E
EML 16438 CCUCVX E_MARTINEZ
COM 47

MARTINI ALDO DR
IAS
CNR
CP 67
I 00044 FRASCATI
ITALY
TEL 6 942 5655
TLF
TLX
EML
COM

MARTINS DONALD HENRY DR
DPT PHYSICS & ASTRONOMY
UNIVERSITY OF ALASKA
3221 UAA DRIVE
ANCHORAGE AK 99508
USA
TEL 907 786 1238
TLF
TLX
EML
COM 09,37

MARTIN-DIAZ CARLOS DR
INST DE ASTROFISICA
DE CANARIAS
OBS DEL TEIDE
E 38200 LA LAGUNA
SPAIN
TEL 22 26 2211
TLF 22 26 3005
TLX 92640 IACE
EML SPAN IAC CMD
COM

MARTIN-LORON M DR
HERMANOS MIRALLES 14
E MADRID 1
SPAIN
TEL
TLF
EML
COM
TLX

MARTIN-PINTADO JESUS
CTR ASTRON DE YEBES
OAN
APD 148
E 19080 GUADALAJARA
SPAIN
TEL 11 22 3358
TLF
TLX
EML
COM 34,40

MARTRES MARIE-JOSEPHE
OBSERVATOIRE DE PARIS
SECTION DE MEUDON
F 92195 MEUDON PPL CDX
FRANCE
TEL 1 45 34 7530
TLF
TLX
EML
COM 10

MARVIN URSULA B DR
CENTER FOR ASTROPHYSICS
HCO/SAO
60 GARDEN ST
CAMBRIDGE MA 02138
USA
TEL 617 495 7270
TLF
TLX 921428 SATELLITE CAM
EML
COM 22

MARX GYORGY PROF
EOTVOS UNIVERSITY
PUSHKIN U 5-7
H 1088 BUDAPEST
HUNGARY
TEL 1 18 7902
TLF 118 0206
TLX 225459 ATOMF H
EML KUERTI@AWIRAF
COM· 35,51C

MARX SIEGFRIED DR
ZNTRLINST F ASTROPHYSIK
KARL-SCHWARZSCHILD OBS
D 6901 TAUTENBURG
GERMANY
TEL 78 23530
TLF
TLX 5886284 KSOT DD
EML
COM 50

MASANI A PROF
OSS ASTRONOMICO DI BRERA

VIA BRERA 28
I 20100 MILANO
ITALY
TEL
TLF
TLX
EML
COM 25,27,35

MASEGOSA GALLEGO J DR
INST ASTROFISICA
DE ANDALUCIA APD 2144
C/SANCHO PANZA S/N
E 18080 GRANADA
SPAIN
TEL 58 12 1311
TLF 58 81 4530
TLX 78573 IAAG E
EML
COM 28

MASLOWSKI JOZEF DR
ASTRONOMICAL OBSERVATORY
JAGIELLONIAN UNIVERSITY
UL ORLA 171
PL 30 244 KRAKOW
POLAND
TEL 34-10-41
TLF
TLX 0322297 UJ PL
EML
COM 40

MASNOU FRANCOISE DR
28 ALLEE GAMBAUBERIE
F 91190 GIF/YVETTE
FRANCE
TEL
TLF
TLX
EML
COM

MASNOU J L DR
OBSERVATOIRE DE PARIS
SECTION DE MEUDON
ER 176 DARC
F 92195 MEUDON PPL CDX
FRANCE
TEL 1 45 34 7570
TLF
TLX 201571
EML
COM

MASON GLENN M
ASTRONOMY PROGRAM
UNIVERSITY OF MARYLAND
COLLEGE PARK MD 20742
USA
TEL 301 454 2616
TLF
TLX 71-8261125
EML
COM 10,48,49

MASON HELEN E DR
DPT APPLIED MATHS
& THEORETICAL PHYSICS
SILVER STREET
CAMBRIDGE CB3 9EW
UK
TEL· 223 351 645
TLF
TLX 81240
EML
COM· 14

MASON JOHN WILLIAM DR
51 ORCHARD WAY
WEST BARNHAM
BOGNOR REGIS
WEST SUSSEX PO22 OHX
UK
TEL 243 553 244
TLF 243 554 272'
TLX
EML
COM 22

MASON KEITH OWEN
MULLARD SPACE SCIENCE LAB
UNIVERSITY COLLEGE LONDON
HOLMBURY ST MARY
DORKING SURREY RH5 6NT
UK
TEL 306 702 92
TLF
TLX 859185
EML
COM 48

MASSAGLIA SILVANO
IST DI FISICA GENERALE
CORSO D AZEGLIO 46
I 10125 TORINO
ITALY
TEL 11 65 7694
TLX 211041
EML
TLF
COM 36

MASSAGUER JOSEP PROF
DPT FISICA APLICADA
UNIV POLITEC DE CATALUNYA
E 08034 BARCELONA
SPAIN
TEL 3 401 6827
TLF 3 401 6090
TLX
EML MASSAGUER FA UPC ES
COM

MASSEVICH ALLA G DR
INST OF ASTRONOMY
ACADEMY OF SCIENCES
PYATNITSKAYA UL 48
109017 MOSCOW
RUSSIA
TEL 231-54-61
TLF
TLX 412623 SCSTP SU
EML
COM 35

MASSEY PHILIP L
KITT PEAK NTL OBS
BOX 26732
950 N CHERRY AVE
TUCSON AZ 85726 6732
USA
TEL 602 327 5511
TLF
TLX
EML.
COM 29

MASSON COLIN R
CENTER FOR ASTROPHYSICS
HCO/SAO
60 GARDEN ST
CAMBRIDGE MA 02138
USA
TEL 617 495 7000
TLF
TLX
EML
COM 34,40

MATAS VLADIMIR R DR
ASTRONOMISCHES-RECHEN
INSTITUT
MOENCHHOFSTR 12-14
D 6900 HEIDELBERG 1
GERMANY
TEL 62 214 9026
TLF
TLX
EML
COM 07

MATERNE JUERGEN DR
ARETINSTR 27
D 8000 MUENCHEN 90
GERMANY
TEL
TLF
TLX
EML
COM 47

MATHER JOHN CROMWELL
NASA/GSFC
CODE 685
LASP
GREENBELT MD 20771
USA
TEL 301 286 8720
TLF
TLX 89675
EML
COM 21,34,44,47

MATHESON DAVID NICHOLAS
RUTHERFORD APPLETON LAB
SPACE & ASTROPHYSICS DIV
BLDG R25/R68
CHILTON DIDCOT OX11 0QX
UK
TEL 235 219 00
TLF
TLX 83159
EML
COM 40

MATHEWS WILLIAM G PROF
LICK OBSERVATORY
UNIVERSITY OF CALIFORNIA
SANTA CRUZ CA 95064
USA
TEL. 408 429 2074
TLF
TLX
EML
COM 34

MATHEWSON DONALD S PROF
MOUNT STROMLO & SIDING
SPRING OBSERVATORIES
PRIVATE BAG
WODEN PO ACT 2606
AUSTRALIA
TEL 62 88 1111
TLF
TLX 62270 AA
EML
COM 28,33,34

MATHEZ GUY
OBS MIDI PYRENEES
14 AVE E BELIN
F 31400 TOULOUSE CDX
FRANCE
TEL 61 25 2101
TLF
TLX 530776 OBSTLSE F
EML
COM

MATHIEU ROBERT D DR
DPT OF ASTRONOMY
UNIVERSITY OF WISCONSIN
475 N CHARTER ST
MADISON WI 53706
USA
TEL 608 262 5679
TLF
TLX 265452
EML BITNET MATHIEU@WISCMAC3
COM 30,42

MATHIS JOHN S PROF
DPT OF ASTRONOMY
UNIVERSITY OF WISCONSIN
475 N CHARTER ST
MADISON WI 53706
USA
TEL 608 262 5994
TLF 608 263 0361
TLX 265452 UOFWISC MDS
EML 7309 MATHIS/MATHIS@WISCMACC
COM 34C

MATHUR B S DR
NTL PHYSICAL LABORATORY
TIME & FREQUENCY SECTION
HILLSIDE RD
NEW DELHI 110 012
INDIA
TEL 11 586 168
TLF
TLX 31-62454 RSD IN
EML
COM 31

MATHYS GAUTIER DR
ESO
CASILLA 19001
SANTIAGO 19
CHILE
TEL 2 698 87 57
TLF 2 695 42 63
TLX 240881 ESOGO CL
EML
COM 29,36

MATSAKIS DEMETRIOS N
US NAVAL OBSERVATORY
34 & MASSACHUSETTS AVE NW
WASHINGTON DC 20392 5100
USA
TEL 202 653 1823
TLX
EML
TLF
COM 19,31,40,51

MATSON DENNIS L DR
JPL
MS 183 501
4800 OAK GROVE DR
PASADENA CA 91109
USA
TEL 213 354 2984
TLF
TLX
EML
COM 15,16

MATSUDA TAKUYA PROF
DPT AERONAUTIC ENGINEER
KYOTO UNIVERSITY
KITASHIRAKAWA SAKYO KU
KYOTO 606
JAPAN
TEL 075-751-2111
TLF
TLX 05422693 LIBKYUJ
EML
COM 51

MATSUI TAKAFUMI DR
DPT EARTH & PLANETARY PHY
UNIVERSITY OF TOKYO
BUNKYO KU
TOKYO 113
JAPAN
TEL 3 3812 2111*4305
TLF 3 3818 3247
TLX 2722126 UTGAB
EML
COM 16

MATSUMOTO MASAMICHI PROF
FACULTY OF ENGINEERING
GIFU UNIVERSITY
GIFU 501 11
JAPAN
TEL·
TLF
TLX·
EML
COM 36

MATSUMOTO RYOJI DR
COLLEGE OF ARTS & SCIENCE
CHIBA UNIVERSITY
1-33 YAYOICHO
CHIBA 260
JAPAN
TEL 81 472 51 1111*2297
TLF 81 472 53 7409
TLX
EML· MATUMOTO@CUCAS C CHIBA-U AC JP
COM 48

MATSUMOTO TOSHIO DR
DPT OF PHYSICS
NAGOYA UNIVERSITY
FUROCHO CHIKUSA KU
NAGOYA 464
JAPAN
TEL 052-781-5111
TLF
TLX 4477323 SCUNAG J
EML
COM 21,47

MATSUMURA MASAFUMI DR
FAC OF EDUCATION
KAGAWA UNIVERSITY
SAIWAI CHO TAKAMATSUSHI
KAGAWA 760
JAPAN
TEL 878 61 4141*400
TLF 878 34 7144
TLX
EML F61210%OSAKA N1NET@KYU-CC CC K
COM 34

MATSUOKA MASARU DR
INST PHYS & CHEMICAL RES
KENKYUSHITSU
2-1 HIROSAWA
WAKO 351 01
JAPAN
TEL 0484-62-1111
TLF
TLX 02962818 RIKEN J
EML
COM 44,48

MATSUSHIMA SATOSHI DR
DPT OF ASTRONOMY
PENNSYLVANIA STATE UNIV
525 DAVEY LAB
UNIVERSITY PARK PA 16802
USA
TEL 814 865 0418
TLF
TLX·
EML
COM 12,36

MATSUURA OSCAR T DR
IAG
UNIVERSIDADE DE SAO PAULO
CP 30627
01051 SAO PAULO SP
BRAZIL
TEL: 11 275 3720
TLF
TLX 1136221 IAGM BR
EML
COM 10,15,49

MATTEI JANET AKYUZ DR
AAVSO
25 BIRCH ST
CAMBRIDGE MA 02138
USA
TEL 617 354 0484
TLF 617 354 0665
TLX
EML
COM 27C,42

MATTEUCCI FRANCESCA DR
IAS
CNR
CP 67
I 00044 FRASCATI
ITALY
TEL 6 942 5655
TLF
TLX
EML
COM 33C,35,37

MATTHEWS CLIFFORD PROF
DPT OF CHEMISTRY M/C111
UNIVERSITY OF ILLINOIS
BOX 4338
CHICAGO IL 60680
USA
TEL 312 996 3161
TLF. 312 996 0431
TLX
EML
COM 51

MATTHEWS HENRY E DR
JOINT ASTRONOMY CENTER
JCMT
665 KOMOHANA ST
HILO HI 96720
USA
TEL 808 961 3756
TLF
TLX 633135
EML
COM 40

MATTHEWS JAYMIE
DPT OF PHYSICS
UNIVERSITY OF MONTREAL
CP 6128 SUCC A
MONTREAL QC H3C 3J7
CANADA
TEL 514 343 6111*3214
TLF 514 343 2071
TLX 055 62425
EML MATTHEWS@ASTRO UMONTREAl
COM 27

MATTHEWS THOMAS A DR
ASTRONOMY PROGRAM
UNIVERSITY OF MARYLAND
COLLEGE PARK MD 20742
USA
TEL
TLF
TLX
EML
COM

MATTIG W PROF DR
KIEPENHEUER INSTITUT
FUER SONNENPHYSIK
SCHOENECKSTRASSE 6
D 7800 FREIBURG BREISGAU
GERMANY
TEL 761 32864
TLF
TLX 7721552 KIS D
EML·
COM 10,12,50

MATTILA KALEVI DR
OBS & ASTROPHYSICS LAB
UNIVERSITY OF HELSINKI
TAEHTITORNINMAKI
SF 00130 HELSINKI 13
FINLAND
TEL 191 2947
TLF
TLX 124690 UNIH SF
EML
COM 21,34,40

MATVEYENKO L I DR
SPACE RESEARCH INSTITUTE
ACADEMY OF SCIENCES
PROFSOJOSNAYA UL 84/32
117810 MOSCOW
RUSSIA
TEL 333-31-22
TLF
TLX 411498 STAR SU
EML
COM 40

MATZ STEVEN MICHEAL DR
DPT PHYSICS & ASTRONOMY
NORTHWESTERN UNIVERSITY
DEARBORN OBSERVATORY
EVANSTON IL 60208
USA
TEL 708 491 8643
TLF
TLX
EML MATZ@OSSENU ASTRO NWU EDU
COM 05,09,48

MATZNER RICHARD A PROF
ASTRONOMY DPT
UNIVERSITY OF TEXAS
RLM 15 308
AUSTIN TX 78712 1083
USA
TEL 512 471 5062
TLF
TLX
EML·
COM· 47

MAUCHERAT J DR
LAS
TRAVERSE DU SIPHON
LES TROIS 'LUCS
F 13012 MARSEILLE
FRANCE
TEL 91 05 5900
TLF 91 66 1855
TLX
EML
COM· 21

MAUDER HORST PROF DR
ASTRONOMISCHES INSTITUT

WALDHAUSER-STR 64
D 7400 TUEBINGEN
GERMANY
TEL
TLF
TLX
EML
COM 42

MAUERSBERGER RAINER DR
MPI FUR RADIOASTRONOMIE
AUF DEM HUGEL 69
D 5300 BONN 1
GERMANY
TEL 228 5251
TLX 886440
EML· P373MAU@MPIRBN MPIFR-BONN MPG
TLF· 49 228 525229
COM 40

MAURICE ERIC N
OBSERVATOIRE DE MARSEILLE
2 PLACE LE VERRIER
F 13248 MARSEILLE CDX 04
FRANCE
TEL 91 95 9088
TLF
TLX 420241 F
EML
COM 28,30

MAURON NICOLAS DR
OBS MIDI PYRENEES
14 AVE E BELIN
F 31400 TOULOUSE CDX
FRANCE
TEL 61 25 2101
TLF
TLX 530776 F
EML
COM

MAVRAGANIS A G PROF
DPT OF ENG SECT OF MECH
NTL TECHN UNIV/5 HEROES
POLYTECH AVE
GR 157 73 ZOGRAFOS
GREECE
TEL 1 643 3170
TLF
TLX
EML
COM 07

MAVRIDES STAMATIA DR
OBSERVATOIRE DE PARIS
SECTION DE MEUDON
DPT RADIOASTRONOMIE
F 92195 MEUDON PPL CDX
FRANCE
TEL· 1 45 07 7597
TLF
TLX
EML
COM· 28,47

MAVRIDIS L N PROF
DPT GEODETIC ASTRONOMY
UNIVERSITY THESSALONIKI
UNIV BOX 503
GR 540 06 THESSALONIKI
GREECE
TEL 31 99 2693
TLF
TLX
EML
COM 08,27,33,46,51

MAVROMICHALAKI HELEN DR
UNIVERSITY / PHYSICS DEPT
NUCLEAR PHYSICS SECTION
104 SOLONOS ST
GR 106 80 ATHENS
GREECE
TEL 1 363 9439
TLF
TLX
EML
COM 49

MAX CLAIRE E DR
LAWRENCE LIVERMORE LAB
L 413
BOX 808
LIVERMORE CA 94550
USA
TEL 415 422 5442
TLF
TLX 9103868339 UCLLLLVMR
EML
COM·

MAXWELL ALAN DR
CENTER FOR ASTROPHYSICS
HCO/SAO
60 GARDEN ST
CAMBRIDGE MA 02138
USA
TEL 617 495 9059
TLF
TLX
EML
COM 10,40

MAY J
OBS RADIOASTR DE MAIPU
UNIVERSIDAD DE CHILE
CASILLA 68
MAIPU
CHILE
TEL 2 229 4101
TLF
TLX 440 001
EML
COM 40

MAYALL MARGARET W
5 SPARKS ST
CAMBRIDGE MA 02138
USA
TEL 617 876 1563
TLF
TLX
EML
COM 27

MAYALL NICHOLAS U ASTRON
7206 E CAMINO VECINO

TUCSON AZ 85715
USA
TEL 602 886 2423
TLF
TLX
EML
COM

MAYER CORNELL H
1209 VILLAMAY BLVD
ALEXANDRA
VIRGINIA 22307
USA
TEL
TLF
TLX
EML
COM 16,40

MAYER PAVEL DR
DPT OF ASTRONOMY
CHARLES UNIVERSITY
SVEDSKA 8
CS 150 00 PRAHA 5
CZECHOSLOVAKIA
TEL 2 54 0395
TLF
TLX
EML
COM 25,42

MAYFIELD EARLE B DR
CALIFORNIAN POLYTECHNIC
STATE UNIVERSITY
1427 BAYVIEW HEIGHTS DR
LOS OSOS CA 93403
USA
TEL 805 528 5231
TLF
TLX
EML
COM

MAYOR MICHEL PROF
OBSERVATOIRE DE GENEVE
CHEMIN DES MAILLETTES 51
CH 1290 SAUVERNY
SWITZERLAND
TEL 22 755 2611
TLF 22 755 3983
TLX 419209 OBS CH
EML mayor@obs unige ch
COM 30,33C,37

MAZA JOSE
DPT DE ASTRONOMIA
UNIVERSIDAD DE CHILE
- CASILLA 36 D
SANTIAGO
CHILE
TEL 2 229 4101
TLF
TLX 440 001
EML MASA@UCHCECVM
COM 46

MAZEH TSEVI DR
WISE OBSERVATORY
TEL AVIV UNIVERSITY
RAMAT AVIV
TEL AVIV 69978
ISRAEL
TEL 3 54 50729
TLF
TLX 342171 VERSY IL
EML BITNET K23@TAUNOS
COM 30,42

MAZURE ALAIN DR
LAB ASTRONOMIE
USTL II
PLACE EUGENE BATAILLON
F 34095 MONTPELLIER CDX 5
FRANCE
TEL 67 52 3548
TLF
TLX 490944
EML
COM

MAZUREK THADDEUS JOHN DR
4580 NUECES DR

SANTA BARBARA CA 93110
USA
TEL
TLF
TLX
EML
COM 35,48

MAZZITELLI ITALO DR
IAS
CNR
CP 67
I 00044 FRASCATI
ITALY
TEL 6 942 1483
TLF
TLX 610261 CNRFRA
EML
COM 35

MAZZONI MASSIMO DR
DP DI ASTRONOMIA
UNIVERSTIA DI FIRENZE
LARGO E FERMI 5
I 50125 FIRENZE
ITALY
TEL 55 27521
TLF 55 22 0039
TLX 572268 ARCETR I
EML
COM

MAZZUCCONI FABRIZIO DR
OSS ASTROFISICO
DI ARCETRI
LARGO E FERMI 5
I 50125 FIRENZE
ITALY
TEL 55 275 2250
TLF
TLX
EML
COM

MCADAM W BRUCE DR
SCHOOL OF PHYSICS
UNIVERSITY OF SYDNEY
SYDNEY NSW 2006
AUSTRALIA
TEL 2 692 2222
TLF
TLX 26169 UNISYD
EML
COM 40

MCALISTER HAROLD A DR
DPT PHYSICS & ASTRONOMY
GEORGIA STATE UNIVERSITY
ATLANTA GA 30303 3083
USA
TEL 404 658 2932
TLF 404 651 2013
TLX
EML HAL@CHARA GSU EDU
COM 24,26C,51

MCBREEN BRIAN PHILIP DR
DPT OF PHYSICS
UNIVERSITY COLLEGE
BELFIELD
DUBLIN 4
IRELAND
TEL 1 693 244
TLF·
TLX 36293
EML
COM 28,48

MCCABE MARIE K MS
1617 S BERETANIA ST #801
HONOLULU HI 96826
USA
TEL 808 956 0923
TLF
TLX
EML
COM 10

MCCALL MARSHALL LESTER DR
DPT OF PHYSICS
YORK UNIVERSITY
4700 KEELE ST
NORTH YORK ON M3J 1P3
CANADA
TEL 416 736 2100
TLF 416 736 5386
TLX 065 24736
EML BITNET FS300050@YUSOL
COM· 34

MCCAMMON DAN
DPT OF PHYSICS
UNIVERSITY OF WISCONSIN
1150 UNIVERSITY AVE
MADISON WI 53706
USA
TEL 608 262 5916
TLF
TLX 265452 UOFWISC MDS
EML
COM

MCCARROLL RONALD PROF
UNIVERSITE DE BORDEAUX
LAB ASTROPHYSIQUE
123 RUE LAMARTINE
F 33405 TALENCE
FRANCE
TEL 56 86 4330
TLF 56 40 4251
TLX
EML
COM

MCCARTHY DENNIS D DR
US NAVAL OBSERVATORY
34 & MASSACHUSETTS AVE NW
WASHINGTON DC 20392 5100
USA
TEL 202 653 0066
TLF
TLX 710 822 1970
EML
COM 19C,31

MCCARTHY MARTIN F DR
SPECOLA VATICANA

I 00120 VATICAN CITY
VATICAN CITY STATE
TEL 6 698 3411
TLF
TLX 5042020 VAT OBS VA
EML
COM 25,33,45,46,50

MCCLAIN EDWARD F
4133 MAPLE RD
MORNINGSIDE MD 20746
USA
TEL 301 736 8933
TLF
EML
COM
TLX

MCCLINTOCK JEFFREY E DR
CENTER FOR ASTROPHYSICS
HCO/SAO
60 GARDEN ST
CAMBRIDGE MA 02138
USA
TEL 617 495 7136
TLF
TLX
EML
COM

MCCLURE ROBERT D PROF
HERZBERG INST ASTROPHYS
DOMINION ASTROPHYS OBS
5071 W SAANICH RD
VICTORIA BC V8X 4M6
CANADA
TEL 604 388 0230
TLF 604 363 0045
TLX 049 7295
EML
COM 30,45

MCCLUSKEY GEORGE E JR DR
ASTRONOMY DIV/MATHS DPT
LEIGH UNIVERSITY
BETHLEHEM PA 18015
USA
TEL 215 861 3721
TLF
TLX
EML
COM 42,44

MCCONNELL DAVID DR
AUSTRALIA TELESCOPE
NATIONAL FACILITY
BOX 94
NARRABRI NSW 2390
AUSTRALIA
TEL 67 959 205
TLF 67 959 255
TLX
EML DMCCONNE@RPEPPING OZ.AU
COM 40

MCCORD THOMAS B DR
PLANETARY GEOSCIENCES DIV
HAWAII INST OF GEOPHYSICS
2525 CORREA RD
HONOLULU HI 96822
USA
TEL 808 948 6488
TLF
TLX
EML
COM 15,16

MCCRACKEN KENNETH G DR
CSIRO COSSA
LIMESTONE AVE
BOX 225
DICKSON ACT 2602
AUSTRALIA
TEL 62 484 595
TLF
TLX 62003 AA
EML
COM 44

MCCRAY RICHARD DR
JILA
UNIVERSITY OF COLORADO
BOX 440
BOULDER CO 80309 0440
USA
TEL 303 492 7835
TLF
TLX
EML DICK@JILA
COM 34,48

MCCREA J DERMOTT
DPT OF MATHS/PHYSICS
UNIVERSITY COLLEGE
BELFIELD
DUBLIN 4
IRELAND
TEL 1 693 244
TLF
TLX
EML
COM 34,35,47

MCCREA WILLIAM SIR
ASTRONOMY CENTRE
UNIVERSITY OF SUSSEX
FALMER
BRIGHTON BN1 9QH
UK
TEL 273 60 6755
TLF
TLX. 877259 UNISEX G
EML
COM 28,47

MCCROSKY RICHARD E DR
CENTER FOR ASTROPHYSICS
HCO/SAO
60 GARDEN ST
CAMBRIDGE MA 02138
USA
TEL 617 495 7212
TLF
TLX
EML
COM 15,20,22

MCCULLOCH PETER M DR
DPT OF PHYSICS
UNIVERSITY OF TASMANIA
GPO BOX 252C
HOBART TAS 7001
AUSTRALIA
TEL 2 202 420
TLF
TLX 58150
EML
COM 40

MCCUTCHEON WILLIAM H PROF
DPT OF PHYSICS
UNIV OF BRITISH COLUMBIA
2075 WESBROOK MALL
VANCOUVER BC V6T 2A6
CANADA
TEL 604 228 3853
TLF 604 228 5324
TLX 045 08576 UBCPHYSICS
EML
COM

MCDONALD FRANK B DR
NASA/GSFC
CODE 660
GREENBELT MD 20771
USA
TEL
TLF
TLX
EML
COM

MCDONALD J K PETRIE DR
768 RICHMOND AVE
VICTORIA BC V8S 3Z1
CANADA
TEL
TLF
TLX
EML
COM

MCDONNELL J A M PROF
UNIT FOR SPACE SCIENCES
UNIVERSITY OF KENT
CANTERBURY CT2 7NR
UK
TEL 227 459 616
TLF
TLX 965449 UKCLIB G
EML
COM 15,21,22

MCDONOUGH THOMAS R DR
CALTECH
500 S OAK KNOLL NO 46
PASADENA CA 91101
USA
TEL 818 795 0147
TLX
EML
TLF
COM 51

MCELROY M B DR
DPT OF EARTH & PLANET SCI
HARVARD UNIVERSITY
60 GARDEN ST
CAMBRIDGE MA 02138
USA
TEL
TLF
TLX
EML
COM 16

MCFADDEN LUCY ANN DR
CALIFORNIA SPACE INST
UCSD
A 016
LA JOLLA CA 92093 0216
USA
TEL 619 534 3915
TLF
TLX
EML LMCFADDEN@UCSD EDU
COM 15

MCGEE RICHARD X DR
CSIRO
DIVISION OF RADIOPHYSICS
BOX 76
EPPING NSW 2121
AUSTRALIA
TEL 2 868 0222
TLF
TLX 26230 ASTRO
EML
COM 34

MCGIMSEY BEN Q JR DR
DPT PHYSICS & ASTRONOMY
GEORGIA STATE UNIVERSITY
ATLANTA GA 30303 3083
USA
TEL 404 658 2279
TLF
TLX
EML
COM

MCGRAW JOHN T DR
STEWARD OBSERVATORY
UNIVERSITY OF ARIZONA
TUCSON AZ 85721
USA
TEL 602 621 5381
TLF
TLX 467175
EML
COM 27

MCGREGOR PETER JOHN DR
MOUNT STROMLO & SIDING
SPRING OBSERVATORIES
PRIVATE BAG
WODEN PO ACT 2611
AUSTRALIA
TEL 62 88 1111
TLF
TLX 62270 CANOPUS AA
EML
COM 09,29,33,34

MCHARDY IAN MICHAEL DR
DPT OF PHYSICS
SOUTHAMPTON UNIVERSITY
ASTRO & SPACE PHYSICS GP
SOUTHAMPTON SO9 5NH
UK
TEL
TLF
TLX
EML
COM

MCINTOSH BRUCE A DR
HERZBERG INST ASTROPHYS
NTL RESEARCH COUNCIL
100 SUSSEX DR
OTTAWA ON K1A 0R6
CANADA
TEL 613 993 6060
TLF 613 952 6602
TLX 053 3715
EML
COM 22

MCINTOSH PATRICK S
NOAA ERL R/E/SE3
SPACE ENVIRONMENT LAB
325 BROADWAY
BOULDER CO 80303
USA
TEL 303 497 3795
TLF
TLX
EML SPAN 9555 PMCINTOSH
COM 10

MCKEE CHRISTOPHER F PROF
PHYSICS DPT
UNIVERSITY OF CALIFORNIA
BERKELEY CA 94720
USA
TEL 415 642 0805
TLF
TLX 820181 UCB AST RALUD
EML
COM 34

MCKEITH CONAL D DR
DPT OF PURE & APPL PHYS
QUEEN'S UNIVERSITY
BELFAST BT7 1NN
UK
TEL 232 24 5133
TLF
TLX 74487 QUB ADM
EML·
COM 34

MCKEITH NIALL ENDA DR
DPT OF PHYSICS
ST PATRICK'S COLLEGE
MAYNOOTH
CO KILDARE
IRELAND
TEL 1 285 222
TLF
TLX
EML
COM

MCKENNA LAWLOR SUSAN
DPT OF EXPERIMENTAL PHYS
ST PATRICK'S COLLEGE
MAYNOOTH
CO KILDARE
IRELAND
TEL 1 285 222
TLF
TLX 31493 SPCM EI
EML
COM 10,12,15,40,41

MCLAREN ROBERT A DR
INSTITUTE FOR ASTRONOMY
UNIVERSITY OF HAWAII
2680 WOODLAWN DR
HONOLULU HI 96822
USA
TEL 808 948 8312/8566
TLF
TLX 8459 UHAST HR
EML
COM

MCLEAN BRIAN JOHN
STSCI
HOMEWOOD CAMPUS
3700 SAN MARTIN DR
BALTIMORE MD 21218
USA
TEL 301 333 9101
TLF
TLX 6849101 STSCI
EML
COM 05,24

MCLEAN DONALD J DR
CSIRO
DIVISION OF RADIOPHYSICS
BOX 76
EPPING NSW 2121
AUSTRALIA
TEL 2 868 0222
TLF
TLX 26230 ASTRO AA
EML DMCLEAN@RP CSIRO AU
COM 10,40

MCLEAN IAN S DR
DPT OF ASTRONOMY
UNIVERSITY OF CALIFORNIA
405 HILGARD AVENUE
LOS ANGELES CA 90024
USA
TEL 213 825 1140
TLF 213 206 2096
TLX
EML
COM 09C,25C

MCMAHAN ROBERT KENNETH DR
DPT PHYSICS & ASTRONOMY
UNIVERSITY NORTH CAROLINA
PHILLIPS HALL CB 3255
CHAPEL HILL NC 27599 3255
USA
TEL 919 962 7168
TLF 919 962 0480
TLX
EML
COM·

MCMILLAN ROBERT S DR
LUNAR & PLANETARY LAB
UNIVERSITY OF ARIZONA
SPACE SCIENCES BLDG
TUCSON AZ 85721
USA
TEL 602 621 6968
TLF
TLX
EML NASAMAIL RSMCMILLAN
COM 30

MCMULLAN DENNIS DR
MULLARD RADIO ASTRON OBS
CAVENDISH LABORATORY
MADINGLEY RD
CAMBRIDGE CB3 OHE
UK
TEL
TLF
TLX
EML
COM 09

MCNALLY DEREK DR
UNIVERSITY OF LONDON OBS
MILL HILL PARK
LONDON NW7 2QS
UK
TEL 1 819 590 421
TLX 28722 UCPHYS G
EML RLVAD ZUVAD DMN
TLF 1 880 7145
COM 05,34,46,50C,EC

MCNAMARA DELBERT H DR
DPT PHYSICS & ASTRONOMY
BRIGHAM YOUNG UNIVERSITY
PROVO UT 84602
USA
TEL 801 378 2298
TLF
TLX
EML
COM 05,27,29,45

MCNAUGHT ROBERT H
SIDING SPRING OBSERVATORY
PRIVATE BAG
COONABARABRAN NSW 2357
AUSTRALIA
TLF 68 842 298
TEL 68 426 269
TLX CANOPUS AA 163945
EML NSSDCA PSI%AAOCBN UKES
COM 20

MCWHIRTER R W PETER DR
RUTHERFORD APPLETON LAB
SPACE & ASTROPHYSICS DIV
BLDG R25/R68
CHILTON DIDCOT OX11 OQX
UK
TEL 235 446 424
TLF
TLX 83159 RUTHLB G
EML
COM 14,44

MEABURN J DR
DPT OF ASTRONOMY
UNIVERSITY OF MANCHESTER
MANCHESTER M13 9PL
UK
TEL·
TLF
TLX
EML
COM 34

MEAD JAYLEE MONTAGUE DR
NASA/GSFC
CODE 680
GREENBELT MD 20771
USA
TEL 301 286 8543
TLF
TLX
EML
COM 05C,44,45

MEADOWS A JACK PROF
DPT ASTRONOMY & HISTORY
UNIVERSITY OF LEICESTER
UNIVERSITY RD
LEICESTER LE1 7RH
UK
TEL
TLF
TLX
EML
COM 05,16

MEATHERINGHAM STEPHEN DR
MOUNT STROMLO & SIDING
SPRING OBSERVATORIES
PRIVATE BAG
WODEN PO ACT 2606
AUSTRALIA
TEL 62 88 1111
TLF 62 49 0233
TLX 62270 CANOPOUS AA
EML SJM@MOGGIE ANU EDU AU
COM 33

MEBOLD ULRICH DR PROF
RADIOASTRONOMISCHES INST
DER UNIVERSITAT BONN
AUF DEM HUEGEL 71
D 5300 BONN 1
GERMANY
TEL
TLF
TLX
EML
COM 34,40

MEDIAVILLA EVENCIO DR
INST DE ASTROFISICA
DE CANARIAS
OBS DEL TEIDE
E 38200 LA LAGUNA
SPAIN
TEL 22 26 2211
TLF
TLX 92640
EML
COM 28

MEDINA JOSE DR
DPT FISICA
UNIVERSIDAD DE ALCALA
APD 20
E 28871 ALCALA DE HENARES
SPAIN
TEL 1 885 4940
TLF 1 885 4953
TLX
EML
COM 48

MEDVEDEV YURI A DR
ASTRONOMICAL OBSERVATORY
ODESSA STATE UNIVERSITY
SHEVCHENKO PARK
270014 ODESSA
UKRAINE
TEL 22 8442
TLF
TLX
EML
COM

MEECH KAREN DR
INSTITUTE FOR ASTRONOMY
UNIVERSITY OF HAWAII
2680 WOODLAWN DR
HONOLULU HI 96822
USA
TEL 808 956 6828
TLF 808 988 2790
TLX 8459 UHAST HR
EML MEECH@UHIFA IFA HAWAII EDU
COM 15

MEEKS M LITTLETON DR
MEEKS ASSOCIATES INC
BOX 643
LINCOLN MA 01773
USA
TEL 617 259 0093
TLF
TLX
EML
COM 40

MEERSON BARUCH DR
RACAH INST OF PHYSICS
HEBREW UNIVERSITY
BOX 4040
JERUSALEM 91904
ISRAEL
TEL 2 584 470
TLF 2 584 437
TLX 25391 HUIL
EML MEERSON@HUJIVMS
COM 10

MEGESSIER CLAUDE DR
OBSERVATOIRE DE PARIS
SECTION DE MEUDON
LAM
F 92195 MEUDON PPL CDX
FRANCE
TEL 1 45 07 7862
TLF
TLX 201571
EML
COM 29

MEGEVAND DENIS DR
OBSERVATOIRE DE GENEVE
CHEMIN DES MAILLETTES 51
CH 1290 SAUVERNY
SWITZERLAND
TEL 22 755 2611
TLF 22 755 3983
TLX 45 419209 OBSQ CH
EML MEGEVAND@SCSUN UNIGE CH
COM 09

MEIDAV MEIR DR
SCHOOL OF EDUCATION
TEL AVIV UNIVERSITY
TEL AVIV 69978
ISRAEL
TEL 3 54 50840
TLF 3 64 13944
TLX
EML
COM 46

MEIER DAVID L
JPL
MS 264 700
4800 OAK GROVE DR
PASADENA CA 91109
USA
TEL 213 354 5062
TLF
TLX 675429
EML DLM@CENA JPL NASA GOV/CEN
COM 28,48,40.

MEIER ROBERT R
NAVAL RESEARCH LABORATORY
CODE 4140
4555 OVERLOOK AVE SW
WASHINGTON DC 20375 5000
USA
TEL 202 767 2773
TLF
TLX
EML
COM 34

MEIKLE WILLIAM P S
ASTROPHYSICS GROUP
IMPERIAL COLLEGE
BLACKETT LABORATORY
LONDON SW7 2BZ
UK
TEL 1 589 5111
TLF
TLX 261503
EML
COM 28

MEIKSIN AVERY ABRAHAM DR
CITA MCLENNAN LABS
UNIVERSITY OF TORONTO
60 ST GEORGE ST
TORONTO ON M5S 1A7
CANADA
TEL 416 978 8494
TLF 416 978 3921
TLX
EML MEIKSIN@UTORPHYS
COM 48

MEIN NICOLE DR
OBSERVATOIRE DE PARIS
SECTION DE MEUDON
DASOP
F 92195 MEUDON PPL CDX
FRANCE
TEL 1 45 07 7801
TLF
TLX
EML
COM

MEIN PIERRE
OBSERVATOIRE DE PARIS
SECTION DE MEUDON
F 92195 MEUDON PPL CDX
FRANCE
TEL 1 45 07 7801
TLF
TLX 270912
EML
COM 05,10,12

MEINEL ADEN B PROF
JPL
MS 186 134
4800 OAK GROVE DR
PASADENA CA 91109
USA
TEL 818 354 6827
TLF
TLX
EML
COM 09,24

MEINIG MANFRED DR
ZNTRLINST F PHYSIK ERDE
TELEGRAFENBERG A17
D 1500 POTSDAM
GERMANY
TEL 4551
TLF
TLX 15305
EML
COM 19,31

MEIRE RAPHAEL
ASTRON STERRENWACHE
RIJKSUNIVERSITEIT GENT
WEIDESTR 11
B 9050 EVERGEM
BELGIUM
TEL 91 53 8755
TLF
TLX
EML
COM 07

MEISEL DAVID D DR
DPT PHYSICS & ASTRONOMY
STATE UNIVERSITY COLLEGE
SUNY
GENESEO NY 14454
USA
TEL 716 245 5284
TLF
TLX
EML
COM 15,22

MEISENHEIMER KLAUS DR
MPI FUER ASTRONOMIE
KONIGSTUHL 17
D 6900 HEIDELBERG 1
GERMANY
TEL 62 215 28206
TLF 62 215 28246
TLX 461 789 MPI D
EML MEISE@DHDMPI5V
COM 28

MEKARNIA DJAMEL DR
OCA OBSERV DE NICE
BP 139
F 06003 NICE CDX
FRANCE
TEL 92 00 3011
TLF 92 00 3033
TLX 460004
EML MEKARNIA@FRONI51
COM

MEKLER YURI PROF
DPT GEOPHYS & PLANET SCI
TEL AVIV UNIVERSITY
TEL AVIV
ISRAEL
TEL 3 413 505
TLF
TLX
EML
COM

MELBOURNE WILLIAM G DR
JPL
MS 238 540
4800 OAK GROVE DR
PASADENA CA 91109
USA
TEL 818 354 5071
TLF
TLX
EML
COM 07,19C,31

MELCHIOR PAUL J PROF DIR
OBSERVATOIRE ROYAL DE
BELGIQUE
AVE CIRCULAIRE 3
B 1180 BRUSSELS
BELGIUM
TEL 2 375 2484
TLF 2 374 8822
TLX 21565 OBSBEL B
EML
COM 08,19,31

MELIA FULVIO DR
DPT PHYSICS & ASTRONOMY
NORTHWESTERN UNIVERSITY
DEARBORN OBSERVATORY
EVANSTON IL 60208
USA
TEL 312 491 4568
TLF
TLX
EML SPAN nssdca 11340 melia
COM 42,44

MELIK-ALAVERDIAN YU DR
BYURAKAN ASTROPHYSICAL
OBSERVATORY
378433 ARMENIA
ARMENIA
TEL
TLF
TLX
EML
COM 35

MELLIER YANNICK DR
OBS MIDI PYRENEES
14 AVE E BELIN
F 31400 TOULOUSE CDX
FRANCE
TEL 61 25 2101
TLF
TLX 530776
EML
COM

MELNICK GARY J
CENTER FOR ASTROPHYSICS
HCO/SAO
60 GARDEN ST
CAMBRIDGE MA 02138
USA
TEL. 617 495 7388
TLF
TLX
EML
COM 30,34,44

MELNICK JORGE
DPT DE ASTRONOMIA
UNIVERSIDAD DE CHILE
CASILLA 36 D
SANTIAGO
CHILE
TEL 2 229 4101
TLF
TLX 440 001
EML
COM

MELOTT ADRIAN L PROF
DPT PHYSICS & ASTRONOMY
UNIVERSITY OF KANSAS
LAWRENCE KS 66045
USA
TEL 913 864 4626
TLF
TLX
EML BITNET melott@ukanvax
COM 47

MELROSE DONALD B PROF
DPT THEORETICAL PHYSICS
UNIVERSITY OF SYDNEY
SYDNEY NSW 2006
AUSTRALIA
TEL
TLF
TLX
EML
COM 10,12,48,49

MENDES DA COSTA ARACY DR
CRAAE/PTR ESCOLA POLI USP
CP 8174 CEP 05508
01051 SAO PAULO SP
BRAZIL
TEL 11 815 5936
TLX 1180127 INPE BR
EML AMDCOSTA@BRUSP BITNET
TLF 11 815 6289
COM 10

MENDEZ MANUEL DR
INSTITUTO DE ASTRONOMIA
UNAM
APDO POSTAL 70-264
04510 MEXICO DF
MEXICO
TEL
TLF
TLX
EML
COM

MENDEZ MARIANO DR
OBSERVATORIO ASTRONOMICO
PASEO DEL BOSQUE
1900 LA PLATA (BS AS.)
ARGENTINA
TEL 21 21 6357
TLF 21 21 1761
TLX 31151 BULAP AR
EML MMENDEZ@FCAGLP EDU AR
COM 37

MENDEZ ROBERTO H DR
IAFE
CC 67 SUC 28
1428 BUENOS AIRES
ARGENTINA
TEL 1 781 6755
TLX 22414 CEDOC AR
EML
TLF
COM 34

MENDIS DEVAMITTA ASOKA DR
EECS
UCSD
LA JOLLA CA 92093 0216
USA
TEL 619 452 2719
TLF
TLX
EML
COM 15,49

MENDOZA CLAUDIO
IBM VENEZUELA SCIENT CTR
BOX 388
CARACAS 1010 A
VENEZUELA
TEL 2 9088 697
TLF
TLX 23283 IBMVE VC
EML
COM

MENDOZA V EUGENIO E DR
INSTITUTO DE ASTRONOMIA
UNAM
APDO POSTAL 20-528
01000 MEXICO DF
MEXICO
TEL 683-3094
TLF
TLX
EML
COM 25,45,51

MENEGUZZI MAURICE M DR
CEA CEN
DAPNIA/SAP
BP 2
F 91191 GIF/YVETTE CDX
FRANCE
TEL 1 69 08 4438
TLF
TLX 690860
EML
COM

MENG XINMIN
YUNNAN OBSERVATORY
CAS
BOX 110
KUNMING 72946 YUNNAN
CHINA PR
TEL 871 2035
TLF
TLX 64040 YUOBS CN
EML
COM. 09

MENNESSIER MARIE-ODILE DR
LAB ASTRONOMIE
USTL II
PLACE EUGENE BATAILLON
F 34095 MONTPELLIER CDX 5
FRANCE
TEL 67 63 9144
TLF
TLX 490944 USTMONT F
EML
COM 24,27,33

MENON T K PROF
DPT GEOPHYS & ASTRONOMY
UNIV OF BRITISH COLUMBIA
2075 WESBROOK PL
VANCOUVER BC V6T 1W5
CANADA
TEL 604 228 2082
TLF 604 228 6047
TLX
EML
COM 28,34,37,40

MENTESE HUSEYIN DR
UNIVERSITY OBSERVATORY
UNIVERSITY OF ISTANBUL
34452 ISTANBUL
TURKEY
TEL 1 522 3597
TLF
TLX
EML
COM

MENZIES JOHN W DR
SAAO
BOX 9
OBSERVATORY 7935
SOUTH AFRICA
TEL 47-0025
TLF
TLX 5720309
EML
COM 25C,34,37,50

MEN' A V DR
INST RADIOPHYS & ELECTRON
UKRAINIAN ACAD OF SCIENCE
310085 KHARKOV
UKRAINE
TEL
TLF
TLX
EML
COM

MERAT PARVIZ
INSTITUT D'ASTROPHYSIQUE
98BIS BD ARAGO
F 75014 PARIS
FRANCE
TEL 1 43 20 1425
TLF 1 43 29 8673
TLX
EML
COM 47

MERAT PARVIZ DR
INSTITUT D'ASTROPHYSIQUE
98BIS BD ARAGO
F 75014 PARIS
FRANCE
TEL 1 43 20 1425
TLF 1 43 29 8673
TLX
EML
COM

MERCIER CLAUDE DR
OBSERVATOIRE DE PARIS
SECTION DE MEUDON
DASOP
F 92195 MEUDON PPL CDX
FRANCE
TEL· 1 45 07 7815
TLF
TLX 201571
EML
COM

MERGENTALER JAN PROF
ASTRONOMICAL INSTITUTE
UL KOPERNIKA 19
PL 51 617 WROCLAW
POLAND
TEL 48-23-29
TLF
TLX
EML
COM 10,12

MERIGHI ROBERTO DR
OSS ASTRONOMICO
UNIVERSITA DI BOLOGNA
CP 596
I 40100 BOLOGNA
ITALY
TEL 51 25 9401
TLF
TLX 520634 INFNBO I
EML SPAN 37929 MERIGHI
COM· 47

MERKLE FRITZ DR
ESO
KARLSCHWARZSCHILD-STR 2
D 8046 GARCHING MUENCHEN
GERMANY
TEL 89 320 060
TLF 89 320 2362
TLX 5828222
EML
COM 09C

MERLEAU-PONTY J PROF
5 RUE GL DE CASTELNAU
F 75015 PARIS
FRANCE
TEL
TLF
EML
COM 41
TLX

MERMAN G A DR
INST OF THEORET ASTRONOMY
ACADEMY OF SCIENCES
N KUTUZOVA 10
192187 ST PETERSBURG
RUSSIA
TEL
TLF
TLX
EML
COM 07

MERMAN NATALIA V DR
PULKOVO OBSERVATORY
ACADEMY OF SCIENCES
10 KUTUZOV QUAY
196140 ST PETERSBURG
RUSSIA
TEL
TLF
TLX
EML
COM

MERMILLIOD JEAN-CLAUDE DR
INSTITUT D'ASTRONOMIE
UNIVERSITE DE LAUSANNE
CH 1290 CHAVANNES-D-BOIS
SWITZERLAND
TEL 22 755 2611
TLF
TLX 27720 OBSG CH
EML mermio@obs.unige.ch
COM 05,30,37P

MERRIAM JAMES B
DPT GEOLOGICAL SCIENCES
UNIV OF SASKATCHEWAN
SASKATOON SK S7N 0W0
CANADA
TEL 306 966 5716
TLF
TLX
EML
COM 19

MERTZ LAWRENCE N DR
287 FAIRFIELD CT
PALO ALTO CA 94306
USA
TEL
TLF
TLX
EML
COM 09

MERZANIDES CONSTANTINOS
ZALOGOU 15
GR 654 03 KAVALA
GREECE
TEL 51 22 840
TLF
TLX
EML
COM

MESSAGE PHILIP J DR
DPT APPL MATHS THEOR PHYS
UNIVERSITY OF LIVERPOOL
BOX 147
LIVERPOOL L69 3BX
UK
TEL 517 096 022
TLF
TLX 627095
EML
COM 07,20

MESSEROTTI MAURO DR
SOLAR ASTROPHYSICS GROUP
ASTRONOMICAL OBSERVATORY
BASOVIZZA 302
I 34012 TRIESTE
ITALY
TEL 40 22 6176
TLF 40 22 6630
TLX 461137 OAT I
EML ASTRTS MESSEROTTI
COM 10

MESSINA ANTONIO
DPT DI ASTRONOMIA
UNIVERSTIA DI BOLOGNA
CP 596
I 40100 BOLOGNA
ITALY
TEL 51 22 2956
TLF
TLX
EML
COM

MESTEL LEON PROF
ASTRONOMY CENTRE
UNIVERSITY OF SUSSEX
FALMER
BRIGHTON BN1 9QH
UK
TEL 273 60 6755
TLF
TLX 877159 UNISEX G
EML
COM 35,48,49

MESZAROS ATTILA DR
DPT ASTRONOMY/ASTROPHYS
CHARLES UNIVERSITY
SVEDSKA 8
CS 150 00 PRAHA 5
CZECHOSLOVAKIA
TEL 2 54 0395
TLF
TLX
EML
COM 47

MESZAROS PETER DR
DPT OF ASTRONOMY
PENNSYLVANIA STATE UNIV
525 DAVEY LAB
UNIVERSITY PARK PA 16802
USA
TEL 814 865 0418
TLF
TLX 842510
EML
COM 34,47,48

METCALFE LEO DR
ESA/ESTEC
SSD
BOX 299
NL 2200 AG NOORDWIJK
NETHERLANDS
TEL 17 19 83616
TLF
TLX 39098
EML EARN LMETCALF@ESTEC
COM

METZ KLAUS DR
INSTITUT F ASTRONOMIE
& ASTROPHYSIK
SCHEINERSTR 1
D 8000 MUENCHEN 80
GERMANY
TEL 89 989 021
TLF
TLX
EML
COM 27

MEURS EVERT DR
MPE
GIESSENBACHSTRASSE
D 8046 GARCHING MUENCHEN
GERMANY
TEL
TLF
TLX
EML
COM

MEWE R DR
SPACE RESEARCH LABORATORY
SRON
SORBONNELAAN 2
NL 3584 CA UTRECHT
NETHERLANDS
TEL 30 53 5600
TLF 30 54 0860
TLX 47224
EML
COM 12,14,44

MEYER CLAUDE DR
OCA CERGA
AVE COPERNIC
F 06130 GRASSE
FRANCE
TEL 93 36 5849
TLF
TLX 470865
EML
COM· 26

MEYER DAVID M DR
DPT PHYSICS & ASTRONOMY
NORTHWESTERN UNIVERSITY
DEARBORN OBSERVATORY
EVANSTON IL 60208
USA
TEL 312 491 4516
TLF
TLX
EML
COM 47

MEYER FRIEDRICH DR
MPI FUER PHYSIK UND
ASTROPHYSIK
KARL-SCHWARZSCHILD-ST 1
D 8046 GARCHING MUENCHEN
GERMANY
TEL 89 329 90
TLF
TLX 524629 ASTRO D
EML
COM 12,48

MEYER JEAN-PAUL DR
CEA CEN
DAPNIA/SAP
BP 2
F 91191 GIF/YVETTE CDX
FRANCE
TEL 1 69 08 5025
TLF
TLX 690860
EML
COM 48

MEYERS KARIE ANN
NOAO
BOX 26732
950 N CHERRY AVE
TUCSON AZ 85726 6732
USA
TEL 602 325 9202
TLF 602 325 9360
TLX
EML KMEYERS@NOAO EDU
COM

MEYER-HOFMEISTER E DR
MPI F PHYSIK & ASTROPHYS

KARL-SCHWARZSCHILD-STR 1
D 8046 GARCHING MUENCHEN
GERMANY
TEL 89 329 90
TLF
TLX 524629 ASTRO D
EML
COM 35,42

MEYLAN GEORGES DR
STSCI
HOMEWOOD CAMPUS
3700 SAN MARTIN DR
BALTIMORE MD 21218
USA
TEL 301 338 4700
TLF 301 338 4788
TLX 05282 820 ES D
EML BITNET meylan@stc1
COM 30,37C,47

MEYNET GEORGES DR
OBSERVATOIRE DE GENEVE
CHEMIN DES MAILLETTES 51
CH 1290 SAUVERNY
SWITZERLAND
TEL 22 755 2611
TLF 22 755 3983
TLX 419209 OBSG CH
EML MEYNET@CGEUGE11
COM 35

MEZGER PETER G PROF
MPI FUER RADIOASTRONOMIE
AUF DEM HUEGEL 69
D 5300 BONN 1
GERMANY
TEL 228 52 5297
TLF 228 52 5229
TLX 886440
EML
COM 33,34,40C

MEZZETTI MARINO
OAT
BOX SUCC TRIESTE 5
VIA TIEPOLO 11
I 34131 TRIESTE
ITALY
TEL 40 79 3221
TLF
TLX 461137 OAT I
EML
COM 42,47

MIANES PIERRE DR
OBS MIDI PYRENEES
14 AVE E BELIN
F 31400 TOULOUSE CDX
FRANCE
TEL 61 25 2101
TLF
TLX
EML
COM 25

MIAO YONG-KUAN
DPT OF ASTRONOMY
NANJING UNIVERSITY
NANJING 210008
CHINA PR
TEL
TLF
TLX
EML
COM

MIAO YONG-RUI
SHAANXI,OBSERVATORY
CAS
LINTONG XIAN
SHAANXI
CHINA PR
TEL 33 2255
TLF
TLX 70121 CSAO CN
EML
COM 31

MICELA GIUSEPPINA DR
OSS ASTRONOMICO
UNIVERSITA DI PALERNO
PALAZZO DEI NORMANNI
I 90134 PALERMO
ITALY
TEL 91 657 0451
TLF 91 48 8900
TLX 910402 ASTROP I
EML ASTROPA@IPACUC
COM 48

MICHALEC ADAM
ASTRONOMICAL OBSERVATORY
JAGIELLONIAN UNIVERSITY
UL ORLA 171
PL 30 244 KRAKOW
POLAND
TEL 221817/3056
TLF
TLX 0322297 UJ PL
EML
COM 40

MICHALITSIANOS ANDREW
NASA/GSFC
CODE 684 1
GREENBELT MD 20771
USA
TEL 301 286 6177
TLF
TLX
EML
COM 10,34,44

MICHALOWSKI TADEUSZ DR
ASTRONOMICAL OBSERVATORY
A MICKIEWICZ UNIVERSITY
UL SLONECZNA 36
PL 60 286 POZNAN
POLAND
TEL 61 679 670
TLF 61 535 535
TLX 413260 UAM PL
EML MICHASTR@PLPUAM11
COM 15

MICHARD RAYMOND DR
OCA OBSERV DE NICE
BP 139
F 06003 NICE CDX
FRANCE
TEL 93 89 0420
TLF
TLX 460004 OBSNICE F
EML
COM 12

MICHAUD GEORGES J DR
250 DU FINISTERE
ST LAMBERT J4S 1P5
CANADA
TEL 514 343 6672
TLF
EML
COM 35C,36
TLX 055 62425 UDEMPHYSAS

MICHEL F CURTIS PROF
DPT PHYISCS & ASTRONOMY
RICE UNIVERSITY
BOX 1892
HOUSTON TX 77251 1892
USA
TEL 713 527 4925
TLF
TLX 556457
EML
COM 48,49

MICKELSON MICHAEL E DR
DPT OF PHYSICS & ASTRON
DENISON UNIVERSITY
GRANVILLE OH 43023
USA
TEL 614 587 6467
TLF
TLX
EML MICKELSON@DENISON
COM 14,16

MIDDLEHURST BARBARA M MS
16567 EL CAMINO REAL

HOUSTON TX 77062
USA
TEL
TLF
TLX
EML
COM 16

MIETELSKI JAN S DR
ASTRONOMICAL OBSERVATORY
JAGIELLONIAN UNIVERSITY
UL ORLA 171
PL 30 244 KRAKOW
POLAND
TEL   48-12-22-38-56
TLF
TLX   0322297 UJ PL
EML
COM

MIGNARD FRANCOIS DR
OCA CERGA
AVE COPERNIC
F 06130 GRASSE
FRANCE
TEL   93 36 5849
TLF
TLX   470865
EML
COM   07

MIHAILA IERONIM PROF
BUCHAREST UNIVERSITY
ACADEMIEI 14
R 70109 BUCAREST
RUMANIA
TEL   230819
TLF
TLX
EML
COM

MIHALAS BARBARA R WEIBEL
HIGH ALTITUDE OBSERVATORY
NCAR
BOX 3000
BOULDER CO 80307 3000
USA
TEL   304 494 5151
TLF
TLX
EML
COM

MIHALAS DIMITRI DR
DPT OF ASTRONOMY
UNIVERSITY OF ILLINOIS
1011 W SPRINGFIELD AVE
URBANA IL 61801
USA
TEL   217 333 3090
TLF
TLX
EML
COM   12,36

MIKAMI TAKAO DR
OSAKA GAKUIN UNIVERSITY
2-36-1 KISHIBE MINAMI
SUITA SHI
OSAKA 564
JAPAN
TEL   06-381-8434
TLF
TLX
EML
COM

MIKESELL ALFRED H MR
8316 WALDNUT RD NE
OLYMPIA WA 9850669550
USA
TEL   206 493 1457
TLF
TLX
EML
COM

MIKHAIL FAHMY I PROF DR
AIN SHAMS UNIVERSITY
FACULTY OF SCIENCES
CAIRO UNIVERSITY
CAIRO
EGYPT
TEL   57 5887
TLF
TLX   94070 USHMS UN
EML
COM

MIKHAL JOSEPH SIDKY PROF
HELWAN OBSERVATORY
HELWAN
EGYPT
TEL   78 0645/2683
TLF
EML
COM   16
TLX   93070 HIAG UN

MIKHELSON NIKOLAJ N DR
PULKOVO OBSERVATORY
ACADEMY OF SCIENCES
10 KUTUZOV QUAY
196140 ST PETERSBURG
RUSSIA
TEL   2-979-465
TLF
TLX
EML
COM   09

MIKKOLA SEPPO DR
TURKU UNIVERSITY OBS
TUORLA
SF 21500 PIIKKIO
FINLAND
TEL   21 43 5822
TLF   21 43 3767
TLX   62638 TYF SF
EML
COM   07,26,33,37

MIKOLAJEWSKA JOANNA DR
INSTITUTE OF ASTRONOMY
N COPERNICUS UNIVERSITY
UL CHOPINA 12/18
PL 87 100 TORUN
POLAND
TEL   56 26017 26018
TLF
TLX   552234 ASTR PL
EML
COM   42

MIKULASEK ZDENEK DR
N COPERNICUS OBSERVATORY
& PLANETARIUM
KRAVI HORA
CS 616 00 BRNO 16
CZECHOSLOVAKIA
TEL   5 74 4347
TLF
TLX
EML
COM   29,42

MILANI ANDREA
IST DI MATEMATICA
UNIVERSITA DI PISA
VIA BUONARROTI 2
I 56127 PISA
ITALY
TEL
TLF
TLX
EML
COM   07C,15,20

MILANO LEOPOLDO DR
DPT DI SCIENZE FISICHE
UNIVERSITA DI NAPOLI
MOSTRA D OLTREMARE PAD 19
I 80125 NAPOLI
ITALY
TEL   81 725 3447
TLF
TLX   720320
EML
COM   42

MILES HOWARD G MR
LANE PARK
PTIYNE
ST MINVER
WADEBRIDGE PL27 6PN
UK
TEL   208 863 153
TLF
TLX
EML
COM   22

MILET BERNARD L DR
OCA OBSERV DE NICE
BP 139
F 06003 NICE CDX
FRANCE
TEL   93 89 0420
TLF
TLX   460004 OBSNICE F
EML
COM   15,20,28,51

MILEY G K DR
STERREWACHT
NIELS BOHRWEG 2
BOX 9513
NL 2333 RA LEIDEN
NETHERLANDS
TEL· 71 27 5849
TLF
TLX   39058 ASTRO NL
EML
COM   28,40

MILKEY ROBERT W DR
STSCI
HOMEWOOD CAMPUS
3700 SAN MARTIN DR
BALTIMORE MD 21218
USA
TEL   301 338 4720
TLF
TLX   6849101
EML
COM. 12

MILLAR THOMAS J DR
DPT OF MATHEMATICS
UMIST
BOX 88
MANCHESTER M60 1QD
UK
TEL   612 363 311
TLF
TLX   666094
EML
COM   34

MILLER FREEMAN D PROF
DPT OF ASTRONOMY
UNIVERSITY OF MICHIGAN
DENNISON BLDG
ANN ARBOR MI 48109 1090
USA
TEL 313 764 3447
TLF
TLX
EML
COM 15

MILLER HUGH R PROF
DPT PHYSICS & ASTRONOMY
GEORGIA STATE UNIVERSITY
ATLANTA GA 30303 3083
USA
TEL 404 658 2279
TLF
TLX.
EML
COM 28

MILLER JOHN C DR
DPT OF ASTROPHYSICS
UNIVERSITY OF OXFORD
SOUTH PARKS RD
OXFORD OX1 3RQ
UK
TEL 865 511 336
TLF
TLX 83295 NUCLOX G
EML
COM 48

MILLER JOSEPH S PROF
LICK OBSERVATORY
UNIVERSITY OF CALIFORNIA
SANTA CRUZ CA 95064
USA
TEL 408 429 2135
TLF
TLX
EML
COM 25,28,34

MILLER RICHARD H DR
ASTRONOMY & ASTROPHYS CTR
UNIVERSITY OF CHICAGO
5640 S ELLIS AVE
CHICAGO IL 60637
USA
TEL 312 962 8201
TLF
TLX 6871133
EML
COM 28,33

MILLET JEAN DR
LAS
TRAVERSE DU SIPHON
LES TROIS LUCS
F 13012 MARSEILLE
FRANCE
TEL 91 05 5900
TLF 91 66 1855
TLX 420584 ASTROSP F
EML
COM

MILLIARD BRUNO
LAS
TRAVERSE DU SIPHON
LES TROIS LUCS
F 13012 MARSEILLE
FRANCE
TEL 91 05 5900
TLF 91 66 1855
TLX 420584 ASTROSP F
EML
COM

MILLIGAN J E
NASA/GSFC

INFRARED ASTROPHYSICS BR
GREENBELT MD 20771
USA
TEL
TLF
TLX
EML
COM

MILLIKAN ALLAN G MR
RESEARCH LAB B-59
EASTMAN KODAK CO
343 STATE ST
ROCHESTER NY 14650
USA
TEL
TLF
TLX
EML
COM 09

MILLIS ROBERT L DR
LOWELL OBSERVATORY
1400 W MARS HILL RD
BOX 1149
FLAGSTAFF AZ 86001
USA
TEL 602 774 3358
TLF
TLX
EML
COM 15,16,20

MILLS ALLAN A DR
DPT OF ASTRONOMY
UNIVERSITY OF LEICESTER
UNIVERSITY RD
LEICESTER LE1 7RH
UK
TEL 533 554 455
TLF
TLX
EML
COM

MILLS BERNARD Y PROF
SCHOOL OF PHYSICS
UNIVERSITY OF SYDNEY
SYDNEY NSW 2006
AUSTRALIA
TEL 2 692 2544
TLF
TLX 26169 UNISYD
EML
COM 28,40

MILNE DOUGLAS K DR
CSIRO
DIVISION OF RADIOPHYSICS
BOX 76
EPPING NSW 2121
AUSTRALIA
TEL 2 868 0222
TLF ·
TLX 26230 ASTRO
EML DMILNE@ATNF CSIRO AU
COM 34,40

MILOGRADOV-TURIN JELENA
INSTITUTE OF ASTRONOMY
UNIVERSITY OF BELGRADE
STUDENTSKI TRG 16
YU 11000 BEOGRAD
YUGOSLAVIA
TEL 11 638 715
TLF
TLX
EML
COM 40

MILONE EUGENE F PROF
DPT OF PHYSICS
UNIVERSITY OF CALGARY
2500 UNIVERSITY DR NW
CALGARY AB T2N 1N4
CANADA
TEL 403 220 5412
TLF 403 289 3331
TLX 038 21545
EML
COM 25,27,37,42

MILONE LUIS A DR
OBSERVATORIO ASTRONOMICO
DE CORDOBA
LAPRIDA 854
5000 CORDOBA
ARGENTINA
TEL 36876/40629
TLF
TLX 51822 BUCOR
EML
COM 27

MILOVANOVIC VLADETA DR
INSTITUT ZA GEODEZIJU
BULEVAR REVOLUCIJE 73
YU 11000 BEOGRAD
YUGOSLAVIA
TEL
TLX
EML
TLF
COM

MINAROVJECH MILAN
ASTRONOMICAL INSTITUTE
SLOVAK ACADEMY SCIENCES
CS 059 60 TATRANSKA LOMNI
CZECHOSLOVAKIA
TEL 969 96 7866/7/8
TLF 969 96 7656
TLX 78277
EML
COM 09

MINESHIGE SHIN DR
DPT OF PHYSICS
IBARAKI UNIVERSITY
BUNKYO
MITO 310
JAPAN
TEL
TLF
TLX·
EML
COM 42

MINEVA VENETA DR
DPT OF ASTRONOMY
BULGARIAN ACAD SCIENCES
72 LENIN BLVD
BG 1784 SOFIA
BULGARIA
TEL 2 75 8827
TLF
TLX 23561
EML
COM· 28

MININ I N PROF
ASTRONOMICAL OBSERVATORY
ST PETERSBURG UNIVERSITY
BIBLIOTECHNAJA PL 2
198904 ST PETERSBURG
RUSSIA
TEL 257-94-89
TLF
TLX
EML
COM 34

MINN YOUNG KEY DR
DPT ASTRONOMY SPACE SCI
KYUNG HEE UNIVERSITY
YONG IN KUN
KYUNGGI DO 170 73
KOREA R
TEL 02-764-6131
TLF
TLX
EML
COM 34,51

MINNET HARRY C MR
CSIRO
DIVISION OF RADIOPHYSICS
BOX 76
EPPING NSW 2121
AUSTRALIA
TEL
TLF
TLX
EML
COM

MINTZ BLANCO BETTY MRS
CERRO TOLOLO
INTERAMERICAN OBSERVATORY
CASILLA 603
LA SERENA
CHILE
TEL 51 21 3352
TLF 51 21 2466*342
TLX 620 301 AURA CT
EML
COM 20,25

MIOC VASILE DR
ASTRONOMICAL OBSERVATORY
STR CIRESILOR 19
R 3400 CLUJ NAPOCA
RUMANIA
TEL 951 11592
TLF
TLX
EML
COM 07

MIRABEL IGOR FELIX DR
CEA CEN
DAPNIA/SAP
BP 2
F 91191 GIF/YVETTE CDX
FRANCE
TEL 1 69 08 3492
TLF 1 69 08 6577
TLX 604860
EML 32773 MIRABEL
COM 28,33,40,51

MIRONOV NIKOLAY T
MAIN ASTRONOMICAL OBS
UKRAINIAN ACAD OF SCIENCE
GOLOSEEVO
252127 KIEV
UKRAINE
TEL 66 4759
TLF
TLX 131406 SKY
EML
COM 19C

MIRZOYAN L V DR PROF
BYURAKAN ASTROPHYSICAL
OBSERVATORY
378433 ARMENIA
ARMENIA
TEL
TLF
TLX
EML
COM 27,33

MISCONI NEBIL YOUSIF DR
FIT
328 W HIBISCUS
MELBOURNE 32901
USA
TEL 407 768 8000
TLF
TLX
EML
COM 21,22

MISNER CHARLES W PROF
ASTRONOMY PROGRAM
UNIVERSITY OF MARYLAND
COLLEGE PARK MD 20742
USA
TEL
TLF
TLX
EML
COM 47

MISSANA MARCO DR
OSS ASTRONOMICO DI BRERA
VIA CREMAGNANI 13/11
I 20059 VIMERCATE
ITALY
TEL
TLF
TLX
EML
COM

MISSANA NATALE PROF
VIA PUCCINI 2
I 10025 PINO TORINESE
ITALY
TEL
TLF
TLX
EML
COM

MITALAS ROMAS ASSOC PROF
DPT OF ASTRONOMY
UNIV OF WESTERN ONTARIO
LONDON ON N6A 5B9
CANADA
TEL 519 679 3184
TLF 519 661 3486
TLX 064 7134
EML
COM 35

MITCHELL GEORGE F DR
DPT OF ASTRONOMY
ST MARY'S UNIVERSITY
HALIFAX NS B3H 3C3
CANADA
TEL 902 429 9780
TLF 902 420 5561
TLX
EML
COM 34

MITCHELL KENNETH J DR
APPLIED RESEARCH CORP
8201 CORPORATE DR
SUITE 920
LANDOVER MD 20785
USA
TEL 301 459 8442
TLF
TLX
EML
COM 40

MITCHELL RICHARD MR
C/O DR SCHUSTER WILLIAMS
BOX 73027
SAN YSIDRO CA 92073-9027
USA
TEL
TLF
TLX
EML
COM

MITCHELL WALTER E JR
DPT OF ASTRONOMY
OHIO STATE UNIVERSITY
174 W 18TH AVE
COLUMBUS OH 43210 1106
USA
TEL 614 422 5554
TLF
TLX
EML
COM

MITIC LJUBISA A DR
ASTRONOMICAL OBSERVATORY
VOLGINA 7
YU 11050 BEOGRAD
YUGOSLAVIA
TEL 11 419 357/421 875
TLF·
TLX
EML
COM 08

MITRA A P DR
NTL PHYSICAL LABORATORY
HILLSIDE RD
NEW DELHI 110 012
INDIA
TEL 11 585 298/440
TLF
TLX 3162454 RSD IN
EML
COM

MITROFANOVA LYUDMILA A DR
PULKOVO OBSERVATORY
ACADEMY OF SCIENCES
10 KUTUZOV QUAY
196140 ST PETERSBURG
RUSSIA
TEL
TLF
TLX
EML
COM

MITSUDA KAZUHISA DR
ISAS
3-1-1 YOSHINODAI
SAGAMIHARA
KANAGAWA 229
JAPAN
TEL 427 51 3911
TLF 427 59 4253
TLX J34757 ISASTRO
EML
COM

MITTON JACQUELINE
8A CANTERBURY CLOSE
CAMBRIDGE CB4 3QQ
UK
TEL 223 355 924
TLF
EML
COM
TLX

MITTON SIMON DR
CAMBRIDGE UNIV PRESS
SHAFTSBURY RD
CAMBRIDGE CB2 2RU
UK
TEL 223 31 2393
TLF
TLX 817256 CUPCAM UK
EML
COM 05

MIYAJI SHIGEKI DR
COLLEGE OF ARTS & SCIENCE
CHIBA UNIVERSITY
1-33 YAYOICHO
CHIBA 260
JAPAN
TEL 472 51 1111
TLF
TLX
EML
COM 35,42,48

MIYAMA SYOKEN
DPT OF PHYSICS
KYOTO UNIVERSITY
SAKYO KU
KYOTO 606
JAPAN
TEL 075-751-2111
TLF
TLX
EML
COM 34

MIYAMOTO MASANORI DR
TOKYO ASTRONOMICAL OBS
NAOJ
OSAWA MITAKA
TOKYO 181
JAPAN
TEL 422 32 5111
TLF 422 32 1924
TLX 2822307 TAOMTK J
EML
COM 08C,24,33

MIYAMOTO SIGENORI PROF
DPT OF PHYSICS/FAC SCI
OSAKA UNIVERSITY
MACHIKANEYAMA
TOYONAKA OSAKA 560
JAPAN
TEL 06-844-1151
TLF
TLX
EML
COM 16,36,44,48

MIYAMOTO SYOTARO PROF DR
KASAN OBSERVATORY
KYOTO UNIVERSITY
YAMASHINA
KYOTO 607
JAPAN
TEL 075-581-1235
TLF
TLX
EML
COM

MIZUNO SHUN
KANAZAWA TECHNOLOGY INST
7-1 OGIGAOKA
NONOICHIMACHI
ISHIKAWA 921
JAPAN
TEL 0762-48-1100
TLF
TLX 5122456 KITLC J
EML
COM 34

MNATSAKANIAN MAMIKON A DR
BYURAKAN ASTROPHYSICAL
OBSERVATORY
378433 ARMENIA
ARMENIA
TEL 88 52 28 3453
TLF
TLX
EML
COM 36

MO JING-ER
PURPLE MOUNTAIN OBSERV
CAS
NANJING
CHINA PR
TEL 25 36967
TLF
TLX 34144 PMONJ CN
EML
COM 34

MOCHKOVITCH ROBERT DR
INSTITUT D'ASTROPHYSIQUE
98BIS BD ARAGO
F 75014 PARIS
FRANCE
TEL 1 43 20 1425
TLF 1 43 29 8673
TLX
EML BITNET MOCHKO@FRIAP 51
COM

MOCHNACKI STEPHAN W DR
DAVID DUNLAP OBSERVATORY
UNIVERSITY OF TORONTO
60 ST GEORGE ST
TORONTO ON M5S 1A7
CANADA
TEL 416 978 2016
TLF 416 978 3921
TLX 069 86766
EML
COM 42

MOCZKO JANUSZ DR
ASTRONOMICAL LATITUDE OBS
BOROWIEC
BOX 62 035
PL 63 120 KORNIK
POLAND
TEL
TLF·
TLX
EML
COM 19

MODALI SARMA B DR
SM SYSTEMS & RESEARCH CO
8401 CORPORATE DR
LANDOVER MD 20785
USA
TEL 301 459 3322
TLF
TLX
EML
COM

MODISETTE JERRY L PROF
18323 HEREFORD LN
HOUSTON TX 77058
USA
TEL
TLF
TLX
EML
COM 44

MOEHLMANN DIEDRICH
INST F KOSMOSFORSCHUNG
RUDOWER CHAUSSEE 5
D 1199 BERLIN
GERMANY
TEL 674 3485
TLF
TLX 113132
EML
COM 15,16

MOELLENHOFF CLAUS DR
LANDESSTERNWARTE
KOENIGSTUHL
D 6900 HEIDELBERG 1
GERMANY
TEL 62 211 0036
TLF
TLX 461789 MPIA D
EML
COM 35

MOERDIJK WILLY G DR
STERREKUNDIG OBSERV
ST PIETERSAALSTSTRAAT 171
B 9000 GENT
BELGIUM
TEL 91 22 1233
TLF
TLX
EML
COM

MOESGAARD KRISTIAN P
HISTORY OF SCIENCE INST
UNIVERSITY OF AARHUS
BYGADAN 1 / TORRILD
DK 8300 ODDER
DENMARK
TEL 86 53 1004
TLF
TLX
EML
COM 41C

MOFFAT ANTHONY F J DR
DPT DE PHYSIQUE
UNIVERSITE DE MONTREAL
CP 6128 SUCC A
MONTREAL QC H3C 3J7
CANADA
TEL   514 343 6682
TLF   514 343 2071
TLX   055 62425 UDEMPHYSAS
EML
COM 29,33,37

MOFFAT JOHN W DR
DPT OF PHYSICS
UNIVERSITY OF TORONTO
60 ST GEORGE ST
TORONTO ON M5S 1A7
CANADA
TEL   416 978 2949
TLF   416 978 3921
TLX   069 86766
EML·
COM

MOFFATT HENRY KEITH PROF
DPT APPLIED MATHS
& THEORETICAL PHYSICS
SILVER STREET
CAMBRIDGE CB3 9EW
UK
TEL   223 351 645
TLF
TLX   81249 CAMSPL G
EML
COM

MOFFETT THOMAS J PROF
DPT OF PHYSICS
PURDUE UNIVERSITY
WEST LAFAYETTE IN 47907
USA
TEL   317 494 5508
TLF
TLX
EML
COM 25C,27

MOGILEVSKIJ EH I DR
IZMIRAN
ACADEMY OF SCIENCES
142092 TROITSK
RUSSIA
TEL   232-19-31
TLF
TLX   412623 SCSTP SU
EML
COM 10

MOHAN CHANDER DR
DPT OF MATHEMATICS
UNIVERSITY OF ROORKEE
ROORKEE 247 667
INDIA
TEL   26
TLF
TLX
EML
COM 26,27,35

MOHAN VIJAY DR
UTTAR PRADESH STATE
OBSERVATORY
PO MANORA PEAK 263 129
NAINITAL 263 129
INDIA
TEL   59 42 2136
TLF
TLX
EML
COM 37

MOHD ZAMBRI ZAINUDDIN DR
CENTRE FOR FOUNDATION
UNIVERSITY OF MALAYA
59100 KUALA LUMPUR
MALAYSIA
TEL   03 7552744
TLF   603 7573661
TLX   MA39845
EML
COM

MOISEEV I G DR
CRIMEAN ASTROPHYS OBS
UKRAINIAN ACAD OF SCIENCE
NAUCHNY
334413 CRIMEA
UKRAINE
TEL   43 2945
TLF
TLX
EML
COM 10,40

MOLARO PAOLO DR
OAT
BOX SUCC TRIESTE 5
VIA TIFPOLO 11
I 34131 TRIESTE
ITALY
TEL   40 30 9342
TLF
TLX   461137 OAT I
EML
COM 29

MOLCHANOV A P PROF
ASTRONOMICAL OBSERVATORY
ST PETERSBURG UNIVERSITY
BIBLIOTECHNAJA PL 2
199178 ST PETERSBURG
RUSSIA
TEL
TLF
TLX
EML
COM 40

MOLES MARIANO J DR
INST ASTROFISICA
DE ANDALUCIA APD 2144
PROFESOR ALBAREDA 1
E 18080 GRANADA
SPAIN
TEL   58 12 1311
TLF
TLX   78573 IAAG E
EML
COM 28

MOLINA ANTONIO
INST ASTROFISICA
DE ANDALUCIA APD 2144
PROFESOR ALBAREDA 1
E 18080 GRANADA
SPAIN
TEL   58 12 1300
TLF
TLX   78573 IAAG E
EML
COM 16,21

MOLNAR MICHAEL R PROF
MOLNAR TECHNOLOGIES
3 STONINGHAM DR
WARREN NJ 07060
USA
TEL   201 580 1404
TLF
TLX
EML
COM

MOMCHEV GOSPODIN
ASTRONOMICAL OBSERVATORY
BOX 7
BG 8800 SLIVEN
BULGARIA
TEL   44 27 204
TLF
TLX
EML
COM 46

MONAGHAN JOSEPH J DR
DPT OF MATHEMATICS
MONASH UNIVERSITY
WELLINGTON RD
CLAYTON VIC 3168
AUSTRALIA
TEL   3 541 2563
TLF
TLX   MONASH AA 32691
EML
COM 35

MONET ALICE K B DR
US NAVAL OBSERVATORY
FLAGSTAFF STATION
BOX 1149
FLAGSTAFF AZ 86002
USA
TEL
TLF
TLX
EML
COM 19

MONET DAVID G
US NAVAL OBSERVATORY
FLAGSTAFF STATION
BOX 1149
FLAGSTAFF AZ 86002
USA
TEL   692 779 5132
TLF
TLX
EML percy dmonet
COM 24,33,44

MONFILS ANDRE G PROF
IAL SPACE
UNIVERSITE DE LIEGE
AVE DU PRE AILY
B 4900 ANGLEUR-LIEGE
BELGIUM
TEL   41 67 6668
TLF
TLX· 41320 IAL SP B
EML
COM 44

MONIN JEAN-LOUIS DR
OBSERVATOIRE DE GRENOBLE
CERMO/ASTROPHYSIQUE
BP 53X
F 38041 S MARTIN HERES CD
FRANCE
TEL   76 51 4786
TLF   76 44 8821
TLX   980134 F
EML MONIN@FRGAG51/GAGVX3 MONIN
COM 34

MONNET GUY J DR
OBSERVATOIRE DE LYON
AVE CHARLES ANDRE
F 69561 S GENIS LAVAL CDX
FRANCE
TEL 78 56 0705
TLF 72 39 9791
TLX 310926
EML
COM 33

MONSIGNORI FOSSI BRUNA DR
OSS ASTROFISICO
DI ARCETRI
LARGO E FERMI 5
I 50125 FIRENZE
ITALY
TEL 55 275 2239
TLF
TLX
EML
COM 10,12

MONTEIRO TANIA S DR
DPT OF MATHEMATICS
ROYAL HOLLOWAY COLLEGE
UNIVERSITY OF LONDON
EGHAM SURREY TW20 0EX
UK
TEL 784 344 55*3106
TLF
TLX
EML ASL0703@ULCC
COM

MONTES CARLOS DR
OCA OBSERV DE NICE
BP 139
F 06003 NICE CDX
FRANCE
TEL 93 89 0420
TLF
TLX 460004 OBSNICE F
EML
COM

MONTMERLE THIERRY DR
CEA CEN
DAPNIA/SAP
BP 2
F 91191 GIF/YVETTE CDX
FRANCE
TEL 1 69 08 5722
TLF
TLX 690860
EML 33590 MONTMERLE
COM

MOOK DELO E PROF
DPT OF PHYSICS & ASTRON
DARTMOUTH COLLEGE
HANOVER NH 03755
USA
TEL 603 646 2972
TLF
TLX
EML
COM

MOON SHIN HAENG DR
ISSA
36-1 WHAAM DONG
YUSEONG GU
DAEJEON 305 348
KOREA R
TEL 042 861 1497
TLF 042 881 5610
TLX
EML
COM 48

MOONS MICHELE B M M
DPT DE MATHEMATIQUE
FACULTES UNIVERSITAIRES
REMPART DE LA VIERGE 8
B 5000 NAMUR
BELGIUM
TEL 81 22 9061*2438
TLF 81 23 0391
TLX 59222 FACNAM B
EML
COM 07

MOORE DANIEL R DR
DPT OF MATHEMATICS
IMPERIAL COLLEGE
HUXLEY QUEEN'S GATE
LONDON SW7 2BZ
UK
TEL
TLF
TLX
EML
COM 35

MOORE ELLIOTT P PROF
JOINT OBSERVATORY
NEW MEXICO TECH
CAMPUS STATION
SOCORRO NM 87801
USA
TEL 505 835 5431
TLF
TLX
EML.
COM 15

MOORE PATRICK DR
FARTHINGS
39 WEST ST
SELSEY SUSSEX
UK
TEL 243 603 668
TLF
TLX
EML
COM 16

MOORE RONALD L DR
NASA/MSFC
CODE ES 52
SPACE SCIENCE LAB
HUNTSVILLE AL 35812
USA
TEL 205 453 0118
TLF
TLX· 59-4416 NASA MSFC HT
EML.
COM 12

MOORHEAD JAMES M DR
DPT OF ASTRONOMY
UNIV OF WESTERN ONTARIO
LONDON ON N6A 3K7
CANADA
TEL 519 679 3186
TLF 519 661 3486
TLX 064 7134
EML
COM

MOORWOOD ALAN F M
ESO
KARL-SCHWARZSCHILD-STR 2
D 8046 GARCHING MUENCHEN
GERMANY
TEL 89 320 06294
TLX 05 28 282 24 EO D
EML
TLF 89 320 2362
COM 28

MOOS HENRY WARREN DR
DPT PHYSICS & ASTRONOMY
JOHNS HOPKINS UNIVERSITY
CHARLES & 34TH ST
BALTIMORE MD 21218
USA
TEL 301 338 7337
TLF
TLX 7102341090
EML
COM 29,44

MORALES-DURAN CARMEN
INTA
DPT DE PROGR ESPACIALES
E 28850 TORREJON DE ARDOZ
SPAIN
TEL 1 675 0700
TLF
TLX 22026 INTA E
EML
COM

MORAN JAMES M DR
CENTER FOR ASTROPHYSICS
HCO/SAO
60 GARDEN ST
CAMBRIDGE MA 02138
USA
TEL 617 495 7477
TLF
TLX 921428 SATELLITE CAM
EML
COM 40C

MORANDO BRUNO L DR
BUREAU DES LONGITUDES
77 AVE DENFERT ROCHEREAU
F 75014 PARIS
FRANCE
TEL 1 40 51 2276
TLF 1 46 33 2834
TLX
EML DBLMCA MORANDO
COM 04C,07,20

MORBEY CHRISTOPHER L
HERZBERG INST ASTROPHYS
DOMINION ASTROPHYS OBS
5071 W SAANICH RD
VICTORIA BC V8X 4M6
CANADA
TEL 604 388 0220
TLF 604 363 0045
TLX 049 7295
EML
COM 26,30

MORBIDELLI ROBERTO DR
OSS ASTRONOMICO DI TORINO

ST OSSERVATORIO 20
I 10025 PINO TORINESE
ITALY
TEL 11 84 1067
TLF 11 84 1281
TLX 213236 TOASTR I
EML 39181 MORBIDELLI
COM 26

MOREELS GUY DR
OBSERVATOIRE DE BESANCON
41BIS AVE OBSERVATOIRE
F 25000 BESANCON CDX
FRANCE
TEL 81 50 2266
TLF
TLX 361144 F
EML
COM 46

MOREL PIERRE JACQUES DR
OCA OBSERV DE NICE
BP 139
F 06003 NICE CDX
FRANCE
TEL 93 89 0420
TLF
TLX 460004 OBSNICE F
EML
COM 26

MORENO CORRAL MARCO A DR
OBS ASTRONOMICO NACIONAL
UNAM
APDO POSTAL 877
22830 ENSENADA B CALIF
MEXICO
TEL 667 44580
TLF 667 44607
TLX 56539 CICEME
EML
COM 34

MORENO EDMUNDO DR
INSTITUTO DE ASTRONOMIA
UNAM
APDO POSTAL 70-264
04510 MEXICO DF
MEXICO
TEL 525 5 48 53 06
TLF 9 011 525 548 3712
TLX
EML
COM 28

MORENO FERNANDO DR
INST ASTROFISICA
DE ANDALUCIA APD 3004
PROFESOR ALBAREDA 1
E 18080 GRANADA
SPAIN
TEL 58 12 1311
TLF 58 81 4530
TLX 78573 IAAG E
EML 16488 FERNANDO
COM 16

MORENO HUGO PROF
DPT DE ASTRONOMIA
UNIVERSIDAD DE CHILE
CASILLA 36 D
SANTIAGO
CHILE
TEL 2 229 4101/4002
TLF
TLX 440 001
EML
COM 25

MORENO-INSERTIS FERNANDO
INST DE ASTROFISICA
DE CANARIAS
OBS DEL TEIDE
E 38071 LA LAGUNA
SPAIN
TEL 22 26 2211
TLF
TLX 92640
EML
COM 10,12

MORETON G E
15 5 THE ESPLANADE

BALMORAL BEACH NSW 2088
AUSTRALIA
TEL
TLF
TLX
EML
COM 10

MORGAN BRIAN LEALAN
BLACKETT LABORATORY
IMPERIAL COLLEGE
PRINCE CONSORT RD
LONDON SW7 2BZ
UK
TEL 1 589 5111
TLF
TLX 261503
EML
COM 09

MORGAN DAVID H DR
ROYAL OBSERVATORY
BLACKFORD HILL
EDINBURGH EH9 3HJ
UK
TEL 316 673 321
TLX 72383 ROEDIN G
EML
TLF
COM 21,34

MORGAN JOHN ADRIAN
AEROSPACE CORPORATION
MS M4 041
BOX 92957
LOS ANGELES CA 90009
USA
TEL
TLF
TLX
EML
COM 35

MORGAN PETER DR
CANBERRA COLL ADV EDUC
SCHOOL OF APPLIED SCIENCE
BOX 1
BELCONNEN ACT 2616
AUSTRALIA
TEL 62 52 2557
TLF
TLX 62267 CANCOL AA
EML
COM 19,31

MORGAN THOMAS H DR
NASA/JOHNSON SPACE CENTER
CODE SN3
HOUSTON TX 77058
USA
TEL 713 483 5039
TLF
TLX 762931
EML
COM 42,44

MORGAN WILLIAM W PROF
YERKES OBSERVATORY
UNIVERSITY OF CHICAGO
BOX 258
WILLIAMS BAY WI 53191
USA
TEL 414 245 5555
TLF
TLX
EML
COM 28,45

MORIARTY-SCHIEVEN G H DR
JPL
MS 169 506
4800 OAK GROVE DR
PASADENA CA 91109
USA
TEL 818 354 7894
TLF
TLX 675429
EML JPLRAG SCHIEVEN
COM

MORIMOTO MASAKI DR
NOBEYAMA RADIO OBS
NAOJ
MINAMIMAKI MURA
NAGANO 384 13
JAPAN
TEL 267 63 4372
TLF 267 98 2884
TLX 720 3329005 NAONRO J
EML MORIMOTO@NRO NAO AC JP
COM 34,38C,40P,51

MORISON IAN MR
NRAL
JODRELL BANK
MACCLESFIELD SK11 9DL
UK
TEL 477 713 21
TLX 36149
EML
TLF
COM 40

MORITA KAZUHIKO
DPT OF PHYSICS
HOKKAIDO UNIVERSITY
KITA 10 NISHI 8
SAPPORO 060
JAPAN
TEL 11-711-2111
TLF
TLX
EML
COM· 40

MORIYAMA FUMIO PROF
OSAKA GAKUIN UNIVERSITY
2-36-1 KISHIBE MINAMI
SUITA SHI
OSAKA 564
JAPAN
TEL 06-381-8434
TLF·
TLX
EML
COM 10,12,40

MOROSSI CARLO
OAT
BOX SUCC TRIESTE 5
VIA TIEPOLO 11
I 34131 TRIESTE
ITALY
TEL 40 76 8506
TLF
TLX 461137 OAT I
EML
COM 29,45

MOROZ V I PROF DR
SPACE RESEARCH INSTITUTE
ACADEMY OF SCIENCES
PROFSOJOSNAYA UL 84/32
117810 MOSCOW
RUSSIA
TEL 333 3122
TLF
TLX 411498 CTAP CY
EML
COM 15,16C,51

MOROZHENKO A V DR
MAIN ASTRONOMICAL OBS
UKRAINIAN ACAD OF SCIENCE
GOLOSEEVO
252127 KIEV
UKRAINE
TEL 66 3110
TLF
TLX 131406 SKY
EML
COM 16

MOROZHENKO N N DR
MAIN ASTRONOMICAL OBS
UKRAINIAN ACAD OF SCIENCE
GOLOSEEVO
252127 KIEV
UKRAINE
TEL 66 3110
TLF
TLX 131406 SKY
EML
COM 10

MORRAS RICARDO DR
IAR

CC 5
1894 VILLA ELISA (BS AS)
ARGENTINA
TEL 21 4 3793
TLF
TLX 18052 CICYR-AR
EML
COM 40

MORRELL NIDIA DR
UNIV NACIONAL DE LA PLATA
FCAG
1900 LA PLATA (BS AS)
ARGENTINA
TEL 21 21 7308
TLF 21 21 1761
TLX 31151 BULAP AR
EML NIDIA@FCAGLP EDU AR
COM 30

MORRIS CHARLES S
JPL
MS 300 319
4800 OAK GROVE DR
PASADENA CA 91109
USA
TEL 818 354 8074
TLF
TLX 675429 JPL USA
EML SPAN STANS CSM
COM

MORRIS DAVID DR
IRAM
300 RUE DE LA PISCINE
F 38406 S MARTIN HERES CD
FRANCE
TEL 76 42 3383
TLF
TLX 950753
EML
COM 40

MORRIS MARK ROOT DR
DPT OF ASTRONOMY
UNIVERSITY OF CALIFORNIA
MATH SCIENCES BLDG
LOS ANGELES CA 90024
USA
TEL 213 825 3320
TLF
TLX
EML
COM 33C,34,40,51

MORRIS MICHAEL C
ROYAL GREENWICH OBS
HERSTMONCEUX CASTLE
HAILSHAM
EAST SUSSEX BN27 1RP
UK
TEL 323 833 171
TLF
TLX 87451
EML JANET UK AC RGO BOWEN
COM 09

MORRIS STEPHEN C DR
HERZBERG INST ASTROPHYS
DOMINION ASTROPHYS OBS
5071 W SAANICH RD
VICTORIA BC V8X 4M6
CANADA
TEL 604 388 3976
TLF 604 363 0045
TLX 049 7295
EML
COM 25,35

MORRIS STEVEN DR
APT 2
2860 W 235TH ST
TORRANCE CA 90505
USA
TEL 213 530 8708
TLF
TLX
EML
COM

MORRISON DAVID PROF
NASA AMES RESEARCH CTR
MS 245 1
SPACE SCIENCE DIV
MOFFETT FIELD CA 94035
USA
TEL 415 604 5028
TLF 415 604 6779
TLX
EML DMORRISON@nasamail
COM 15,16P,WGPSNC

MORRISON LESLIE V DR
ROYAL GREENWICH OBS
MADINGLEY RD
CAMBRIDGE CB3 0EZ
UK
TEL 223 37 4000
TLF 223 37 4700
TLX 265451/265871
EML MERRLP@UK AC RO-GREENWICH SRF
COM 04,08P,19C,24

MORRISON NANCY DUNLAP DR
DPT PHYSICS & ASTRONOMY
UNIVERSITY OF TOLEDO
2801 W BANCROFT ST
TOLEDO OH 43606
USA
TEL 419 537 2659
TLF
TLX
EML
COM 27,29

MORRISON PHILIP PROF
DPT OF PHYSICS
MIT RM 6 205
BOX 165
CAMBRIDGE MA 02139
USA
TEL 617 253 5086
TLF
TLX
EML
COM 47,48,51

MORTON DONALD C DR
HERZBERG INST ASTROPHYS
NTL RESEARCH COUNCIL
100 SUSSEX DR
OTTAWA ON K1A 0R6
CANADA
TEL 613 993 6060
TLF 613 952 6602
TLX 053 3715
EML
COM 09,34,44

MORTON G A DR
1122 SKYCREST DR
APT 6
WALNUT CREEK CA 94595
USA
TEL 415 933 3802
TLF
TLX
EML
COM

MOSKALIK PAWEL DR
COPERNICUS ASTRON CENTER
POLISH ACAD OF SCIENCES
UL BARTYCKA 18
PL 00 716 WARSAW
POLAND
TEL
TLF
TLX
EML.
COM· 35

MOSS CHRISTOPHER DR
INSTITUTE OF ASTRONOMY
THE OBSERVATORIES
MADINGLEY RD
CAMBRIDGE CB3 0HA
UK
TEL 223 337 548
TLF
TLX 817297 ASTRON G
EML
COM 28

MOSS DAVID L DR
DPT OF MATHEMATICS
UNIVERSITY OF MANCHESTER
MANCHESTER M13 9PL
UK
TEL
TLF
TLX
EML
COM 35

MOTTA SANTO DR
DPT DI MATIMATICA
CITTA UNIVERSITARIA
VIA A DORIA 6
I 95125 CATANIA
ITALY
TEL 95 33 0533*668
TLF
TLX 970359 ASTRCT-I
EML
COM 10

MOTZ LLOYD PROF
DPT OF ASTRONOMY
COLUMBIA UNIVERSITY
PUPIN HALL 538 W 120TH ST
NEW YORK NY 10027
USA
TEL 212 280 3279
TLF
TLX
EML
COM

MOUCHET MARTINE DR
OBSERVATOIRE DE PARIS
SECTION DE MEUDON
DAEC
F 92195 MEUDON PPL CDX
FRANCE
TEL 1 45 07 7522
TLF 1 45 07 7469
TLX 201571
EML MELAMA MOUCHET/MOUCHET@FRMEU5
COM 42

MOULD JEREMY R
CALTECH
MS 105 24
PASADENA CA 91125
USA
TEL 818 356 4168
TLF
TLX
EML
COM 28,37

MOURADIAN ZADIG M DR
OBSERVATOIRE DE PARIS
SECTION DE MEUDON
DASOP
F 92195 MEUDON PPL CDX
FRANCE
TEL 1 45 07 7800
TLF
TLX
EML
COM 12

MOUSCHOVIAS TELEMACHOS CH
DPT OF ASTRONOMY
UNIVERSITY OF ILLINOIS
1011 W SPRINGFIELD AVE
URBANA IL 61801
USA
TEL 217 333 3090
TLF
TLX
EML
COM 34

MOUSSAS XENOPHON PH D
DPT OF ASTROPHYSICS
NTL UNIVERSITY OF ATHENS
PANEPISTIMIOPOLIS
GR 157 83 ATHENS
GREECE
TEL 1 723 5122/1 884 3877
TLF
TLX
EML MOUSSAS@GRATHON1 EARN ‘
COM 49

MOUTSOULAS MICHAEL PROF
DPT OF EARTH SCIENCES
NTL UNIVERSITY OF ATHENS
PANEPISTIMIOPOLIS
GR 157 84 ATHENS
GREECE
TEL 1 724 7569
TLF
TLX 215255 GR
EML
COM 16,51

MOZURKEWICH DAVID DR
US NAVAL OBSERVATORY
AD 5 BLDG 22 RM 220
34 & MASSACHUSETTS AVE NW
WASHINGTON DC 20392 5100
USA
TEL 202 653 0948
TLF
TLX
EML
COM

MRKOS ANTONIN DR
DPT OF ASTRONOMY
CHARLES UNIVERSITY
SVEDSKA 8
CS 150 00 PRAHA 5
CZECHOSLOVAKIA
TEL 2 54 0395
TLF
TLX 144307 KLET CZ
EML
COM 06,15,20

MUELLER EDITH A PROF
RENNWEG 15
CH 4052 BASEL
SWITZERLAND
TEL 61 42 3168
TLF
TLX
EML
COM 12,38,44

MUELLER EWALD
MPI F PHYS & ASTROPHYSIK
INSTITUT F ASTROPHYSIK
KARL-SCHWARZSCHILD-STR 1
D 8046 GARCHING MUENCHEN
GERMANY
TEL 89 329 90
TLF
TLX 524629 ASTRO D
EML
COM 35

MUELLER IVAN I PROF
GEODETIC SCI & SURVEYING
OHIO STATE UNIVERSITY
1958 NEIL AVE
COLUMBUS OH 43210 1247
USA
TEL 614 422 2269
TLF
TLX 245334
EML
COM 04,19,31C

MUENCH GUIDO PROF
MPI FUER ASTRONOMIE
KOENIGSTUHL
D 6900 HEIDELBERG 1
GERMANY
TEL 62 215 28210
TLF
TLX 461 789 MPIA D
EML
COM 33,34,36

MUFSON STUART LEE DR
ASTRONOMY DPT
INDIANA UNIVERSITY
SWAIN WEST 319
BLOOMINGTON IN 47405
USA
TEL 812 335 6927
TLF
TLX
EML
COM 34

MUINONEN KARRI DR
LOWELL OBSERVATORY
1400 W MARS HILL RD
BOX 1149
FLAGSTAFF AZ 86001
USA
TEL 602 774 3358
TLF 602 774 6296
TLX
EML S PERCY KOM
COM 21

MUINOS JOSE L DR
REAL INST Y OBSERVATORIO
DE LA ARMADA
CECILIO PUJAZON S/N
E 11110 SAN FERNANDO
SPAIN
TEL 56 88 3548
TLF 56 88 1732
TLX 76108 IOM E
EML
COM 08

MUKAI SONOYO DR
KANAZAWA TECHNOLOGY INST
7-1 OGIGAOKA
NONOICHIMACHI
ISHIKAWA 921
JAPAN
TEL 0762-48-1100
TLF
TLX 5122456 KITLCJ
EML
COM 21,36

MUKAI TADASHI DR
DPT OF EARTH SCIENCES
KOBE WOMEN'S UNIVERSITY
SUMA KU
KOBE 657
JAPAN
TEL 78 881 1212 EXT 4475
TLF 78 882 1549
TLX
EML E01142@JPNAC BITNET
COM 15,21C

MUKHERJEE KRISHNA
DPT PHYSICS & ASTRONOMY
UNIVERSITY OF KANSAS
LAWRENCE KS 66045
USA
TEL 913 864 4030
TLF
TLX
EML
COM

MULDERS GERARD F W
4519 EVERETT ST
KENSINGTON MD 20895
USA
TEL 301 564 0090
TLF
EML
COM
TLX

MULHOLLAND J DERRAL DR
OCA CERGA
AVE COPERNIC
F 06130 GRASSE
FRANCE
TEL 93 36 5849
TLF
TLX
EML
COM 07,16,20

MULLALY RICHARD F DR
SCHOOL OF ELECTRICAL ENG
UNIVERSITY OF SYDNEY
SYDNEY NSW 2006
AUSTRALIA
TEL
TLF
TLX
EML
COM

MULLAN DERMOTT J DR
BARTOL RESEARCH INSTITUTE
UNIVERSITY OF DELAWARE
NEWARK DE 19716
USA
TEL 301 398 3368
TLF
TLX 5106660805
EML
COM

MULLER A B DR
THOPMASLAAN 40
NL 5631 GM EINDHOVEN
NETHERLANDS
TEL 40 43 0322
TLF
EML
COM 25
TLX

MULLER C A PROF JR
ODINKSVELD 8
NL 7491 HD DELDEN
NETHERLANDS
TEL 54 07 2428
TLF
EML
COM
TLX

MULLER PAUL
3 RUE CHAUVAIN
F 06000 NICE
FRANCE
TEL
TLF
TLX
EML
COM 26

MULLER RICHARD A
LAWRENCE BERKELEY LAB
BLDG 50 RM 238
BERKELEY CA 94720
USA
TEL 415 486 5235
TLF
TLX
EML
COM 47,51

MULLER RICHARD DR
OBS MIDI PYRENEES
14 AVE E BELIN
F 31400 TOULOUSE CDX
FRANCE
TEL 62 95 0069
TLF
TLX
EML
COM 10,12

MUMFORD GEORGE S PROF
DPT OF EDUCATION
TUFTS UNIVERSITY
FILENE CENTER
MEDFORD MA 02155
USA
TEL 617 653 8923
TLF
TLX
EML
COM 25,27,42

MUMMA MICHAEL JON
NASA/GSFC
CODE 693
GREENBELT MD 20771
USA
TEL 301 286 6994
TLF
TLX
EML
COM 14,15,16

MUNDT REINHARD DR
MPI FUR ASTRONOMIE
KOENIGSTUHL
D 6900 HEIDELBERG 1
GERMANY
TEL 62 215 28227
TLF
TLX 461789
EML
COM

MUNDY LEE G DR
ASTRONOMY PROGRAM
UNIVERSITY OF MARYLAND
COLLEGE PARK MD 20742
USA
TEL
TLF
TLX
EML
COM 40

MUÑOZ-TUNON CASIANA
INST DE ASTROFISICA
DE CANARIAS
OBS DEL TEIDE
E 38200 LA LAGUNA
SPAIN
TEL 22 26 2211
TLF
TLX 92640
EML
COM. 28

MUNRO RICHARD H DR
2378 DENNISON LANE
BOULDER CO 80302
USA
TEL
TLF
TLX
EML
COM 12

MURAKAMI TOSHIO
INST SPACE & ASTRON SCI
UNIVERSITY OF TOKYO
MEGURU KU
TOKYO 153
JAPAN
TEL 03-467-1111x303
TLF
TLX 34757 ISASTRO J
EML
COM 48

MURDIN PAUL G DR
ROYAL OBSERVATORY
BLACKFORD HILL
EDINBURGH EH9 3HJ
UK
TEL 316 688 261
TLX
EML UK AC ROE STAR MURDIN
TLF 316 688 264
COM 27,50P

MURDOCH HUGH S DR
DPT OF ASTROPHYSICS
UNIVERSITY OF SYDNEY
SYDNEY NSW 2006
AUSTRALIA
TEL 2 692 2222
TLF
TLX 26169 UNISYD
EML
COM 40

MURDOCK THOMAS LEE
GENERAL RESEARCH CORP
DPT OF TECHNOLOGY
5 CHERRY HILL
DANVERS MA 01923
USA
TEL 617 777 6323
TLF
TLX
EML
COM 44

MUREDDU LEONARDO DR
STAZIONE ASTRONOMICA

VIA OSPEDALE 72
I 09100 CAGLIARI
ITALY
TEL 70 66 3544
TLF 70 65 7657
TLX
EML
COM

MURPHY BRIAN WILLIAM DR
STERREKUNDIG INSTITUTE
BOX 80000
NL 3508 TA ULTRECHT
NETHERLANDS
TEL 30 53 5203
TLX 40048 FYLUT
EML MURPHY@RUUNSC FYS RUU NL
TLF
COM

MURPHY ROBERT E DR
NASA HEADQUARTERS
CODE EEL
600 INDEPENDENCE AVE SW
WASHINGTON DC 20546
USA
TEL 202 453 1720
TLF
TLX· 89530 NASA WSH
EML
COM 16

MURRAY C ANDREW
DERWENT COTTAGE
12 DERWENT RD
MEADS
EASTBOURNE DN20 7PH
UK
TEL
TLF
TLX
EML
COM 08,22,24,33,37

MURRAY CARL D DR
SCHOOL OF MATHEMATICAL SC
QUEEN MARY/WESTFIELD COLL
MILE END RD
LONDON E1 4NS
UK
TEL 1 719 804 811
TLF
TLX 893750
EML
COM 20

MURRAY JOHN B DR
UNIVERSITY OF LONDON OBS
MILL HILL PARK
LONDON NW7 2QS
UK
TEL
TLX
EML.
TLF
COM

MURRAY STEPHEN S DR
CENTER FOR ASTROPHYSICS
HCO/SAO
60 GARDEN ST
CAMBRIDGE MA 02138
USA
TEL 617 495 7205
TLF
TLX 921428 SATELLITE CAM
EML
COM 09,28

MUSEN PETER DR
8804 ORBIT LANE
LANHAM MD 20801
USA
TEL 301 552 3848
TLF
TLX
EML
COM 07

MUSIELAK ZDZISLAW E DR
NASA/MSFC
CODE ES 52
SPACE SCIENCE LAB
HUNTSVILLE AL 35812
USA
TEL 205 544 7619
TLF
TLX 594416
EML SPAN SSL MUSIELAK
COM 10,36

MUSMAN STEVEN DR
NGS
N/CG 112
ROCKVILLE MD 20852
USA
TEL
TLF
TLX
EML
COM

MUTEL ROBERT LUCIEN
DPT PHYSICS & ASTRONOMY
UNIVERSITY OF IOWA
IOWA CITY IA 52242
USA
TEL 319 353 7205
TLF
TLX
EML
COM 40,42C

MUTSCHLECNER J PAUL DR
ASTRONOMY DPT
INDIANA UNIVERSITY
SWAIN WEST 319
BLOOMINGTON IN 47405
USA
TEL
TLF
TLX
EML
COM 36

MUXLOW THOMAS
NRAL

JODRELL BANK
MACCLESFIELD SK11 9DL
UK
TEL 477 713 21
TLF
TLX 36149
EML
COM 40

MUZZIO JUAN C PROF
OBSERVATORIO ASTRONOMICO
PASEO DEL BOSQUE
1900 LA PLATA (BS AS)
ARGENTINA
TEL 21 21 7308
TLF 21 1761/25 5004
TLX 31216 CESLA AR
EML jcmuzzio%fcaglp edu ar@uucpnet TLX
COM 28,33,37,46

MYACHIN VLADIMIR F DR
INST OF THEORET ASTRONOMY
ACADEMY OF SCIENCES
N KUTUZOVA 10
192187 ST PETERSBURG
RUSSIA
TEL
TLF
TLX
EML
COM 07

MYERS PHILIP C
CENTER FOR ASTROPHYSICS
HCO/SAO MS 42
60 GARDEN ST
CAMBRIDGE MA 02138
USA
TEL
TLF
TLX
EML
COM 34C,40

NACOZY PAUL E DR
FEDEREAL SPACE SYSTEMS
BOX 26712
AUSTIN TX 78755
USA
TEL 512 467 6659
TLF
TLX
EML
COM· 07,20

NADAL ROBERT
OBS MIDI PYRENEES
14 AVE E BELIN
F 31400 TOULOUSE CDX
FRANCE
TEL 61 25 2101
TLF
TLX 530776 F
EML
COM 41

NADEAU DANIEL DR
DPT DE PHYSIQUE
UNIVERSITE DE MONTREAL
CP 6128 SUCC A
MONTREAL QC H3C 3J7
CANADA
TEL 514 343 6676
TLF 514 343 2071
TLX· 055 62425
EML
COM 40

NADOLSCHI V PROF DR
COM ARDEOANI OF TESCANI
JUD
R BACAU
RUMANIA
TEL
TLF
TLX
EML
COM

NADYOZHIN D K DR
INST THEOR & EXPER PHYS

CHEREMUSHKINSKAJA UL 25
117259 MOSCOW
RUSSIA
TEL 123-02-92
TLF
TLX 411059 CERII SU
EML
COM 35

NAGASAWA SHINGO PROF
3-20-9 HIGASHI MACHI
KICHIJOJI
MUSASHINO SHI
TOKYO 180
JAPAN
TEL
TLF
TLX
EML
COM 10

NAGASE FUMIAKI DR
DPT OF ASTROPHYSICS
NAGOYA UNIVERSITY
FUROCHO CHIKUSA KU
NAGOYA 464
JAPAN
TEL 052-781-5111
TLF
TLX 4477323 SCUNAG J
EML
COM

NAGIRNER DMITRIJ I DR
ASTRONOMICAL OBSERVATORY
ST PETERSBURG UNIVERSITY
BIBLIOTECHNAJA PL 2
198904 ST PETERSBURG
RUSSIA
TEL 257-94-89
TLF
TLX
EML
COM 36

NAGNIBEDA VALERY G DR
ASTRONOMICAL OBSERVATORY
ST PETERSBURG UNIVERSITY
BIBLIOTECHNAJA PL 2
198904 ST PETERSBURG
RUSSIA
TEL 257-94-91
TLF
TLX
EML
COM 40

NAHON FERNAND PROF
25 AVE DE L'EUROPE
F 92310 SEVRES
FRANCE
TEL 1 45 34 1805
TLF
EML
COM 07,33
TLX

NAIDENOV VICTOR O
IOFFE PHYSICAL TECH INST
ACADEMY OF SCIENCES
POLYTECHNICHESKAYA UL 26
194021 ST PETERSBURG
RUSSIA
TEL
TLF·
TLX
EML
COM 48

NAKADA YOSHIKAZU DR
DPT OF ASTRONOMY
UNIVERSITY OF TOKYO
BUNKYO KU
TOKYO 113
JAPAN
TEL 03-812-2111
TLF
TLX
EML
COM 34

NAKAGAWA NAOYA DR
UNIV OF ELECTRO-COMMUNICA
CHOFU-SHI
TOKYO 182
JAPAN
TEL 0424-83-2161
TLF
TLX 2822446 UEC J
EML
COM

NAKAGAWA TAKAO DR
ISAS
3-1-1 YOSHINODAI
SAGAMIHARA
KANAGAWA 229
JAPAN
TEL 427 51 3911
TLF 427 59 4253
TLX J 34757 ISASTRO
EML NAKAGAWA@ASTRO ISAS AC JP
COM 34

NAKAGAWA YOSHINARI DR
CHIBA INST OF TECHNOLOGY
NARASHINO 275
JAPAN
TEL 0474 75 2111
TLF
EML
COM 10,49C,51
TLX

NAKAGAWA YOSHITSUGU DR
GEOPHYSICAL INSTITUTE
UNIVERSITY OF TOKYO
BUNKYO KU
TOKYO 113
JAPAN
TEL 3-812-2111x4304
TLF 3-818-3247
TLX 2722126 UTGAB J
EML a81938%tansei.cc u-tokyo ac jp
COM 16

NAKAI NAOMASA DR
NOBEYAMA RADIO OBS
NAOJ
MINAMIMAKI MURA
NAGANO 384 13
JAPAN
TEL 267 63 4367
TLF 267 63 4387
TLX 3329005 NAONRO J
EML NRONAKA@JPNNRO
COM 28

NAKAI YOSHIHIRO
KWASAN & HIDA OBS
KYOTO UNIVERSITY
YAMASHINA
KYOTO 607
JAPAN
TEL 075-581-1235
TLF
TLX
EML
COM 09

NAKAJIMA HIROSHI
NOBEYAMA RADIO OBS
NAOJ
MINAMIMAKI MURÀ
NAGANO 384 13
JAPAN
TEL 267-98-2034
TLF
TLX 3329005 TAONRO J
EML
COM. 10

NAKAJIMA KOICHI DR
FAC OF ECONOMICS
HITOTSUBASHI UNIVERSITY
NAKA 2-1 KUNITACHI
TOKYO 186
JAPAN
TEL 435 72 1101
TLF
TLX 2842107
EML
COM 08

NAKAMURA TAKASHI DR
DPT OF PHYSICS
KYOTO UNIVERSITY
SAKYO KU
KYOTO 606
JAPAN
TEL
TLF
TLX
EML
COM 35

NAKAMURA TSUKO DR
TOKYO ASTRONOMICAL OBS
NAOJ
OSAWA MITAKA
TOKYO 181
JAPAN
TEL 3 812 2111
TLF
TLX
EML
COM 15,20

NAKAMURA YASUHISA
DPT OF EARTH SCINECE
FUKUSHIMA UNIVERSITY
MATSUKAWA MACHI
FUKUSHIMA 960-12
JAPAN
TEL 0245485151/421
TLF
TLX
EML
COM 42

NAKANO SABURO DR
KOBINATO 1-21-7
BUNKYO KU
TOKYO 112
JAPAN
TEL
TLF
TLX
EML
COM

NAKANO SYUICHI
3-19 1 CHOMO
TAKENOKUCHI
SUMOTO
HYOGO KEN 656
JAPAN
TEL
TLF·
TLX
EML
COM 06,20

NAKANO TAKENORI DR
DPT OF PHYSICS
KYOTO UNIVERSITY
SAKYO KU
KYOTO 606
JAPAN
TEL 075-751-2111
TLF
TLX
EML
COM 35

NAKAYAMA SHIGERU DR
FACULTY OF GENERAL EDU
UNIVERSITY OF TOKYO
MEGURO KU
TOKYO 153
JAPAN
TEL· 03-467-1171
TLF
TLX
EML
COM 41

NAKAZAWA KIYOSHI DR
TOKYO INST OF TECH
OHOKAYAMA 2-12-1
MEGUROKU TOKYO 152
JAPAN
TEL
TLF
TLX
EML
COM 22,35

NAMBA OSAMU DR
MARCO POLOLAAN 319

NL 3526 GE UTRECHT
NETHERLANDS
TEL
TLF
TLX
EML
COM 10,12

NAMBU YASUSADA DR
DPT OF PHYSICS
KYOTO UNIVERSITY
SAKYO KU
KYOTO 606
JAPAN
TEL 81 75 753 3844
TLF 81 75 753 3866
TLX
EML NAMBU@JPNRIFP
COM 47

NAN REN-DONG
BEIJING ASTRONOMICAL OBS
CAS
W SUBURB
BEIJING 100080
CHINA PR
TEL 1 28 1698
TLF
TLX 22040 BAOBS CN
EML
COM 40

NANDY KASHINATH DR
ROYAL OBSERVATORY
BLACKFORD HILL
EDINBURGH EH9 3HJ
UK
TEL 316 673 321
TLF
TLX 72383 ROEDING
EML
COM 34,45

NAPIER WILLIAM M DR
ROYAL OBSERVATORY

BLACKFORD HILL
EDINBURGH EH9 3HJ
UK
TEL 316 673 321
TLF
TLX 72383 ROEDIN G
EML
COM 15,22

NAQVI S I H PROF
DPT PHYSICS & ASTRONOMY
UNIVERSITY OF REGINA
REGINA SK S4S 0A2
CANADA
TEL 306 584 4262
TLF 306 586 9862
TLX 071 2683 U R REG
EML
COM

NARANAN S PROF
TIFR
HOMI BHABHA RD
COLABA
BOMBAY 400 005
INDIA
TEL 22 219 111
TLF
TLX 0113009
EML
COM

NARASIMHA DELAMPADY DR
TIFR
HOMI BHABHA RD
COLABA
BOMBAY 400 005
INDIA
TEL 22 495 2311
TLF
TLX 3009 IN
EML
COM 35,36,47

NARAYAN RAMESH DR
STEWARD OBSERVATORY
UNIVERSITY OF ARIZONA
TUCSON AZ 85721
USA
TEL 602 621 2560
TLF
TLX 467175
EML
COM

NARAYANA J V
REGIONAL METEOROLOGICAL
OFFICE
4 COLLEGE RD
MADRAS 600 006
INDIA
TEL
TLF
TLX
EML
COM

NARIAI KYOJI PROF
TOKYO ASTRONOMICAL OBS
NAOJ
OSAWA MITAKA
TOKYO 181
JAPAN
TEL 0422-32-5111
TLF
TLX
EML
COM 36,42

NARITA SHINJI DR
DOSHISHA UNIVERSITY
KYOTO 602
JAPAN
TEL
TLF
TLX
EML
COM 35

NARLIKAR JAYANT V PROF
IUCAA
BOST BAG 4
GANESHKHIND
PUNE 411 007
INDIA
TEL 212 33 6415
TLF 212 33 5760
TLX 0145658 GMRT IN
EML jayant@iucaa ernet in
COM 28,46,47VP

NATALI GIULIANO DR
IAS
CNR
CP 67
I 00044 FRASCATI
ITALY
TEL 6 942 5655
TLF
TLX
EML:
COM

NATHER R EDWARD
ASTRONOMY DPT
UNIVERSITY OF TEXAS
RLM 15 308
AUSTIN TX 78712 1083
USA
TEL
TLF
TLX·
EML·
COM 27,42

NATTA ANTONELLA DR
CENTRO PER ASTRONOMIA IR
LARGO E FERMI 5
I 50125 FIRENZE
ITALY
TEL 55 275 2239
TLF
TLX 572268 ARCETRI
EML
COM

NAUMOV VITALIJ A DR
PULKOVO OBSERVATORY
ACADEMY OF SCIENCES
10 KUTUZOV QUAY
196140 ST PETERSBURG
RUSSIA
TEL 2982242
TLF
TLX
EML
COM 19,31

NAVARRO JULIO FERNANDO DR
DPT OF PHYSICS
UNIVERSITY OF DURHAM
SOUTH RD
DURHAM DH1 3LE
UK
TEL 913 742 194
TLF 913 743 749
TLX 537351 DURLIB G
EML JFN@STAR DUR AC UK
COM 28

NAWAR SAMIR DR
HELWAN OBSERVATORY
HELWAN
EGYPT
TEL 78 0645/2683
TLF
EML
COM 21
TLX

NECKEL HEINZ DR
HAMBURGER STERNWARTE
GOJENSBERGSWEG 112
D 2050 HAMBURG 80
GERMANY
TEL 40 7252 4130
TLF
TLX 217884
EML
COM 12,29

NECKEL TH DR
MPI FUER ASTRONOMIE

KOENIGSTUHL
D 6900 HEIDELBERG 1
GERMANY
TEL 62 215 28288
TLF
TLX 461789 MPIA D
EML
COM 33

NEE TSU-WEI DR
DPT OF PHYSICS
NTL CENTRAL UNIVERSITY
CHUNG LI
CHINA R
TEL
TLF
TLX
EML
COM

NEEMAN YUVAL PROF
DPT OF PHYSICS & ASTRON
TEL AVIV UNIVERSITY
RAMAT AVIV
TEL AVIV 69978
ISRAEL
TEL 3 425 411
TLF
TLX 342171 VERSY IL
EML
COM 47,48

NEFEDEVA ANTONINA I PROF
ENGELHARDT ASTRONOMICAL
OBSERVATORY
OBSERVATORIA STATION
422526 KAZAN
RUSSIA
TEL 324827
TLF
TLX
EML
COM 08

NEFF JAMES EDWARD DR
NASA/GSFC
CODE 681
GREENBELT MD 20771
USA
TEL 301 286 5781
TLF
TLX
EML JNEFF@STARS GSFC NASA GOV
COM

NEFF JOHN S
DPT PHYSICS & ASTRONOMY
UNIVERSITY OF IOWA
605 BROOKLAND PARK DR
IOWA CITY IA 52240
USA
TEL 319 353 4340
TLF
TLX
EML
COM 15,27,36,40

NEFF SUSAN GALE DR
NASA/GSFC
CODE 684 1
GREENBELT MD 20771
USA
TEL 301 286 5137
TLF
TLX 89675
EML NEFF@STARS GSFC NASA GOV
COM 44

NEIDIG DONALD F DR
AIR FORCE GEOPHYSICS LAB
NTL SOLAR OBSERVATORY
SUNSPOT NM 88349
USA
TEL 505 434 7000
TLF
TLX 0666 484 NOAO TUC
EML BITNET DNEIDIG@SUNSPOT NOAO EO
COM 10

NELSON ALISTAIR H DR
PHYSICS DPT
UNIV WALES COLLEGE
BOX 913
CARDIFF CF1 3TH
UK
TEL 222 874 785
TLF 222 371 921
TLX·
EML
COM 33

NELSON BURT DR
DPT OF ASTRONOMY
SAN DIEGO STATE UNIV
SAN DIEGO CA 92182
USA
TEL 619 265 6175
TLF
TLX
EML
COM 42,50

NELSON GEORGE DRIVER DR
DPT OF ASTRONOMY
UNIVERSITY OF WASHINGTON
FM 20
SEATTLE WA 98195
USA
TEL 206 543 6616
TLF 206 685 3218
TLX
EML PNELSON@COSMOS ASTRO WASHINTON
COM

NELSON GRAHAM JOHN DR
CSIRO
DIVISION OF RADIOPHYSICS
BOX 76
EPPING NSW 2121
AUSTRALIA
TEL 2 868 0222
TLF
TLX 26230
EML GNELSON@ATNF CSIRO AU
COM 10

NELSON JERRY E DR
WM KECK OBS/CARA
BOX 220
65-1120 MAMALAHOA HIGHWAY
KAMUELA HI 96743
USA
TEL. 808 885 7887
TLF 808 885 4464
TLX
EML
COM 09

NELSON ROBERT M
JPL
MS 183 501
4800 OAK GROVE DR
PASADENA CA 91109
USA
TEL 213 354 6893
TLF
TLX
EML
COM 51

NEMEC JAMES
DPT GEOPHYS & ASTRONOMY
UNIV OF BRITISH COLUMBIA
2075 WESBROOK PL
VANCOUVER BC V6T 1W5
CANADA
TEL 604 652 4517
TLF 604 228 6047
TLX
EML
COM 37

NEMIRO ANDREJ A DR PROF
PULKOVO OBSERVATORY
ACADEMY OF SCIENCES
10 KUTUZOV QUAY
196140 ST PETERSBURG
RUSSIA
TEL 2982242
TLF
TLX
EML
COM 08

NEMIROFF ROBERT DR
RACAH INST OF PHYSICS
HEBREW UNIV OF JERUSALEM
GIVAT RAM
JERUSALEM 91904
ISRAEL
TEL 2 584 928
TLF
TLX
EML
COM

NESCI ROBERTO
ISTITUTO ASTRONOMICO
UNIVERSITA DI ROMA
VIA LANCISI 29
I 00161 ROMA
ITALY
TEL 6 86 7525
TLF
TLX 613255 INFNRO
EML
COM 37

NESIS ANASTASIOS DR
KEIPENHEUER INSTITUT FUR
SONNENPHYSIK
SCHONECKSTRASSE 6
D 7800 FREIBURG BREISGAU
GERMANY
TEL 761 382067
TLF 761 32280
TLX 7721552 KISD
EML
COM 12

NESS NORMAN F DR
BARTOL RESEARCH INSTITUTE
UNIVERSITY OF DELAWARE
NEWARK DE 19716
USA
TEL 302 451 8116
TLF
TLX 510-6665
EML
COM 16,44

NETZER HAGAI PROF
SCHOOL OF PHY & ASTRON
TEL AVIV UNIVERSITY
RAMAT AVIV
TEL AVIV 69978
ISRAEL
TEL 3 54 50208
TLF
TLX 342171 VERSY IL
EML BITNET H31@TAUNOS
COM

NEUGEBAUER GERRY DR
CALTECH
MS 320 47
DOWNES LAB OF PHYSICS
PASADENA CA 91125
USA
TEL 818 356 4284
TLF
TLX 675425
EML
COM 34

NEUKUM G DR
D F V L R
NE-OE-PE
D 8031 WESSLING
GERMANY
TEL 815 328 731
TLF
TLX 0526419 DVLOP D
EML
COM 15,16

NEUPERT WERNER M DR
NASA/GSFC
CODE 680
GREENBELT MD 20771
USA
TEL 301 286 8169
TLF
TLX
EML
COM 10,44

NEUZIL LUDEK DR
ASTRONOMICAL INSTITUTE
CZECH ACADEMY OF SCIENCES
ONDREJOV OBSERVATORY
CS 251 65 ONDREJOV
CZECHOSLOVAKIA
TEL 204 85201
TLF 204 85314
TLX
EML
COM 21

NEWBURN RAY L JR
3226 EMERALD ILSE DR
GLENDALE CA 91206
USA
TEL
TLF
EML
COM 15,22
TLX

NEWELL EDWARD B DR
MOUNT STROMLO & SIDING
SPRING OBSERVATORIES
PRIVATE BAG
WODEN PO ACT 2606
AUSTRALIA
TEL 62 88 1111
TLF
TLX AA 62270 CANOPUS
EML
COM 37

NEWHALL X X DR
JPL
MS 238 332
4800 OAK GROVE DR
PASADENA CA 91109
USA
TEL 818 354 0000
TLF
TLX 192961003
EML SPAN LOGOS XXN
COM 04,19,31

NEWMAN MICHAEL JOHN DR
LOS ALAMOS NATIONAL LAB
MS B220 X 2
BOX 1663
LOS ALAMOS NM 87545
USA
TEL 505 667 7698
TLF
TLX
EML
COM 35

NEWSOM GERALD H PROF
DPT OF ASTRONOMY
OHIO STATE UNIVERSITY
174 W 18TH AVE
COLUMBUS OH 43210 1106
USA
TEL 614 422 7082
TLF
TLX
EML
COM 14,42

NEWTON GAVIN DR
DPT OF PHYSICS & ASTRON
UNIVERSITY OF GLASGOW
GLASGOW G12 8QQ
UK
TEL 413 398 855*4196
TLF 413 349 029
TLX 777070 UNIGLA
EML
COM 09

NEWTON ROBERT R DR
APPLIED PHYSICS LAB
JOHNS HOPKINS UNIVERSITY
JOHNS HOPKINS RD
LAUREL MD 20707
USA
TEL 301 953 7100
TLF
TLX
EML
COM

NEY EDWARD P PROF
DPT OF ASTRONOMY
UNIVERSITY OF MINNESOTA
116 CHURCH ST SE
MINNEAPOLIS MN 55455
USA
TEL 612 373 4687
TLF
TLX
EML
COM 21

NG KIN-WANG
INSTITUTE OF PHYSICS
ACADEMIA SINICA
TAIPEI
TAIWAN 11529
CHINA R
TEL 2 782 3075
TLF 2 783 4187
TLX
EML PHKWNG@TWNAS886
COM

NGUYEN-QUANG RIEU DR
OBSERVATOIRE DE PARIS
SECTION DE MEUDON
F 92195 MEUDON PPL CDX
FRANCE
TEL 1 45 34 7530
TLF
TLX 270912
EML
COM 34,40

NHA IL-SEONG DR
YONSEI UNIVERSITY OBS
134 SINCHON-DONG
SUDAEMUN
SEOUL 120
KOREA R
TEL 392-0131
TLF
TLX
EML
COM 38,42,46

NIARCHOS PANAYIOTIS PH D
DPT OF ASTRONOMY
NTL UNIVERSITY OF ATHENS
PANEPISTIMIOPOLIS
GR 157 71 ZOGRAFOS
GREECE
TEL 1 724 3414
TLF
TLX
EML
COM 27,42,51

NIAZY ADNAN MOHAMMAD DR
KACST
BOX 6086
RIYADH 11442
SAUDI ARABIA
TEL 1 488 3751
TLX
EML
TLF
COM

NICOLACI ·DA COSTA L-A
OBSERVATORIO NACIONAL
RUA GL BRUCE 586
SAO CRISTOVAO
20921 RIO DE JANEIRO RJ
BRAZIL
TEL 21 580 7313
TLF 21 580 0332
TLX 021-21288
EML
COM

NICOLET MARCEL PROF
IASB
AVE DEN DOORN 30
B 1180 BRUSSELS
BELGIUM   »
TEL 2 374 2948
TLX 21563 ESPACE B
EML
TLF
COM 12,21

NIEDNER MALCOLM DR
NASA/GSFC
GREENBELT MD 20771
USA
TEL 301 286 2000
TLF
EML
COM
TLX

NIEUWENHUIJZEN HANS DR
STERREKUNDIG INSTITUTE
BOX 80000
NL 3508 TA UTRECHT
NETHERLANDS
TEL 30 53 5237
TLX 40048 FYLUT NL
EML BITNET XMHANSEN@HUTRUUO
TLF
COM

NICHOLLS RALPH W PROF
DPT OF PHYSICS
YORK UNIVERSITY
4700 KEELE ST
NORTH YORK ON M3J 1P3
CANADA
TEL 416 736 5247
TLF 416 736 5386
TLX 065 24736
EML FS300003@YUSOL
COM 14C,29

NICOLAIDIS EFTHYMIOS DR
NTL RESEARCH FOUNDATION
48 VAS CONSTANTINOU AVE
GR 116 35 ATHENS
GREECE
TEL 1 721 0554
TLF 1 721 2729
TLX 224064
EML·
COM 41

NICOLOV NIKOLAI S DR
DPT OF ASTRONOMY
UNIVERSITY OF SOFIA
ANTON IVANOV ST 5
BG 1126 SOFIA
BULGARIA
TEL 2 54 4852
TLF
TLX
EML
COM 27,46

NIELL ARTHUR E DR
HAYSTACK OBSERVATORY
WESTFORD MA 01886
USA
TEL 617 692 4764
TLF
EML
COM
TLX 948149

NIIMI YUKIO
TOKYO ASTRONOMICAL OBS
NAOJ
OSAWA MITAKA
TOKYO 181
JAPAN
TEL 422 32 5111
TLF
TLX 2822307 TAOMTK J
EML
COM 31

NICHOLSON WILLIAM
ROYAL GREENWICH OBS
HERSTMONCEUX CASTLE
HAILSHAM BN27 1RP
UK
TEL
TLF
TLX
EML
COM 24

NICOLAS KENNETH ROBERT
NAVAL RESEARCH LABORATORY
CODE 4163
4555 OVERLOOK AVE SW
WASHINGTON DC 20375 5000
USA
TEL 202 767 2517
TLF
TLX
EML
COM 12

NICOLSON GEORGE D DR
HARTEBEESTHOEK RADIOASTRO
OBSERVATORY
BOX 443
KRUGERSDORP 1740
SOUTH AFRICA
TEL 27 11 642 4692
TLF 27 11 642 2446
TLX 321006 HART SA
EML
COM 40

NIEMELA VIRPI S DR
CALLE 51 ESQ 11
1894 VILLA ELISA (BS AS)
ARGENTINA
TEL 21 4 3793
TLF
TLX
EML
COM 29 42

NIKITIN A A DR
ASTRONOMICAL OBSERVATORY
ST PETERSBURG UNIVERSITY
BIBLIOTECHNAJA PL 2
198904 ST PETERSBURG
RUSSIA
TEL 293-22-62
TLF
TLX
EML
COM

NICHOLS-BOHLIN JOY DR
COMPUTOR SCIENCES CORP

10000 A AEROSPACE RD
LANHAM SEABROOK MD 20706
USA
TEL 301 794 1410
TLF.
TLX
EML 6890 NICHOLS
COM 44

NICOLET BERNARD
OBSERVATOIRE DE GENEVE
CHEMIN DES MAILLETTES 51
CH 1290 SAUVERNY
SWITZERLAND
TEL 22 755 2611
TLF· 22 755 3983
TLX 419209 OBS CH
EML nicolet@obs unige ch
COM 25,45

NIEDNER MALCOLM B DR
NASA/GSFC

LASP
GREENBELT MD 20771
USA
TEL
TLF
TLX
EML
COM 15

NIEMI AIMO
TURKU UNIVERSITY OBS
TUORLA
SF 21500 PIIKKIO
FINLAND
TEL 21 43 5822
TLF 21 43 3767
TLX 62638 TYF SF
EML
COM 09,19

NIKOGHOSSIAN ARTHUR G DR
BYURAKAN ASTROPHYSICAL
OBSERVATORY
378433 BYURAKAN
ARMENIA
TEL 88 52 28 3453
TLF
TLX 411576 ASCON SU
EML
COM 36

NIKOLOFF IVAN DR
PERTH OBSERVATORY
BICKLEY WA 6076
AUSTRALIA
TEL 9 293 1865
TLF
EML
COM 08
TLX

NIKOLOV ANDREJ DR
DPT OF ASTRONOMY
UNIVERSITY OF SOFIA
ANTON IVANOV ST 5
BG 1126 SOFIA
BULGARIA
TEL 2 54 4852
TLF
TLX 23296 SUKOR BG
EML
COM 27

NILSON PETER DR
ASTRONOMICAL OBSERVATORY
BOX 515
S 751 20 UPPSALA
SWEDEN
TEL
TLF
TLX
EML
COM

NILSSON CARL DR
CENTER FOR ASTROPHYSICS
HCO/SAO
60 GARDEN ST
CAMBRIDGE MA 02138
USA
TEL
TLF
TLX
EML
COM

NINKOVIC SLOBODAN
ASTRONOMICAL OBSERVATORY
VOLGINA 7
YU 11050 BEOGRAD
YUGOSLAVIA
TEL 11 419 357/421 875
TLF
TLX
EML
COM 33,38

NISHI KEIZO DR
TOKYO ASTRONOMICAL OBS
NAOJ
OSAWA MITAKA
TOKYO 181
JAPAN
TEL 0422-32-5111
TLF
TLX 02822307 TAOMTK J
EML
COM 10,12

NISHIDA MINORU PROF
DPT OF PHYSICS
KYOTO UNIVERSITY
SAKYO KU
KYOTO 606
JAPAN
TEL
TLF
TLX 5422693 LIBKYU J
EML
COM 33,35,47

NISHIDA MITSUGU
DPT OF LITERATURE
KOBE WOMEN'S UNIVERSITY
SUMA KU
KOBE 654
JAPAN
TEL 078-731-4416
TLF
TLX
EML
COM 33

NISHIMURA JUN DR
INST SPACE & AERON SCI
UNIVERSITY OF TOKYO
MEGURO KU
TOKYO 153
JAPAN
TEL 03-467-1111x388
TLF
TLX J 24550 SPACETKY
EML·
COM

NISHIMURA MASAKI
DPT OF PHYSICS
HOKKAIDO UNIVERSITY
KITA 10 NISHI 8
SAPPORO 060
JAPAN
TEL 11-71--2111
TLF
TLX
EML
COM

NISHIMURA SHIRO DR
TOKYO ASTRONOMICAL OBS
NAOJ
OSAWA MITAKA
TOKYO 181
JAPAN
TEL 422-32-5111
TLF
TLX 2822307 TAOMTK J
EML
COM 05,09,29

NISHIMURA TETSUO DR
STEWARD OBSERVATORY
UNIVERSITY OF ARIZONA
TUCSON AZ 85721
USA
TEL 602 621 2054
TLF
TLX 467175
EML
COM 21

NISHIO MASANORI DR
NOBEYAMA RADIO OBS
NAOJ
MINAMIMAKI MURA
NAGANO 384 13
JAPAN
TEL 81 267 63 4381
TLF 81 267 98 2506
TLX 3329005
EML
COM 40

NISSEN POUL E PROF
INST OF PHYSICS & ASTRON
UNIVERSITY OF AARHUS
NY MUNKEGADE
DK 8000 AARHUS C
DENMARK
TEL 86 12 8899
TLF 86 20 2711
TLX 64767 AAUSCI DK
EML
COM 29,37

NITTMAN JOHANN
ETUDES ET FABRICATION
DOWELL SCHLUMBERGER
BP 90
F 42003 ST ETIENNE
FRANCE
TEL 77 32 6423
TLF
TLX
EML
COM

NITYANANDA R DR
RAMAN RESEARCH INSTITUTE
SADASHIVANAGAR
BANGALORE 560 080
INDIA
TEL 812 36 0122
TLF 812 34 0492
TLX 8452671 RRI IN
EML
COM 28,48

NOBILI ANNA M
IST DI MATEMATICA
UNIVERSITA DI PISA
VIA BUONARROTI 2
I 56127 PISA
ITALY
TEL
TLF
TLX
EML
COM· 07,20

NOBILI L DR
DPT DI FISICA G GALILEI
UNIVERSITA DI PADOVA
VIA MARZOLO 8
I 35131 PADOVA
ITALY
TEL 49 84 4205*111
TLF.
TLX 430308 DFGGPDI
EML
COM·

NOCERA LUIGI DR
IST DI FISICA
ATOMICA E MOLECOLARE
VIA GIARDINO 7
I 56127 PISA
ITALY
TEL 50 50 1384
TLF 50 25175
TLX
EML BISTAB @ ICNUCEVM
COM 10

NOCI GIANCARLO PROF
DPT DI ASTRONOMIA
UNIVERSITA DI FIRENZE
LARGO E FERMI 5
I 50125 FIRENZE
ITALY
TEL 55 27521
TLF 55 22 0039
TLX 572268 ARCETR I
EML
COM·

NOEL FERNANDO
DPT DE ASTRONOMIA
UNIVERSIDAD DE CHILE
CASILLA 36 D
SANTIAGO
CHILE
TEL 2 229 4101
TLF
TLX 408 53
EML
COM 08C,31

NOELS ARLETTE DR
50 AVE DE LA PAIX
BOX 063
B 4030 GRIVEGNEE
BELGIUM
TEL 41 52 9980*7517
TLF
TLX
EML
COM 35,46

NOENS JACQUES-CLAIR DR
OBS MIDI PYRENEES
9 R PONT DE LA MOUETTE
F 65200 BAGNERES BIGORRE
FRANCE
TEL 62 95 1969
TLF 62 95 1070
TLX 531625
EML
COM 10

NOERDLINGER PETER D PROF
APPLIED RESEARCH CORP
8201 CORPORATE DR
SUITE 1120
LANDOVER MD 20785
USA
TEL 301 459 8442
TLF
TLX
EML
COM 47

NOGUCHI MASAFUMI DR
TOKYO ASTRONOMICAL OBS
NAOJ
OSAWA MITAKA
TOKYO 181
JAPAN
TEL 0422 32 5111
TLF
TLX 2822307 TAOMTK J
EML
COM 28

NOLL KEITH STEPHEN DR
STSCI
HOMEWOOD CAMPUS
3700 SAN MARTIN DR
BALTIMORE MD 21218
USA
TEL 410 338 5080
TLF
TLX
EML NOLL@STSCI EDU/STSCIC NOLL
COM 15,16

NOLLEZ GERARD DR
INSTITUT D'ASTROPHYSIQUE
98BIS BD ARAGO
F 75014 PARIS
FRANCE
TEL 1 48 20 1425
TLF 1 43 29 8673
TLX
EML
COM 14

NOMOTO KEN'ICHI DR
DPT OF ASTRONOMY
UNIVERSITY OF TOKYO
BUNKYO KU
TOKYO 113
JAPAN
TEL 33 812 2111*4255
TLF 33 813 9439
TLX 2722126 UTGABJ
EML nomoto@apsunl astron s.u
COM 35C,48

NOONAN THOMAS W PROF
SUNY AT BROCKPORT
DPT OF PHYSICS
BROCKPORT NY 14420
USA
TEL 716 395 5581
TLF
TLX
EML
COM. 28,47

NORDH H LENNART DR
STOCKHOLM OBSERVATORY
S 133 36 SALTSJOEBADEN
SWEDEN
TEL 87 17 0195
TLF 87 17 4719
EML NORDH@ASTRO SU SE
COM 34,44
TLX 12972

NORDLUND AKE DR
UNIVERSITY OBSERVATORY
OESTER VOLDGADE 3
DK 1350 COPENHAGEN K
DENMARK
TEL 31 14 1790
TLF 31 38 9157
TLX 44155 DANAST
EML
COM 12C,36

NORDSTROEM BIRGITTA DR
COPENHAGEN UNIVERSITY OBS
BRORFELDEVEJ 23
DK 4340 TIKKISE
DENMARK
TEL 53 48 8195
TLF 53 48 8755
TLX 44155 DANAST DK
EML BIRGITTA@ASTRO KU DK
COM· 30,33,42

NOREAU LOUIS DR
HERZBERG INST ASTROPHYS
NTL RESEARCH COUNCIL
100 SUSSEX DR
OTTAWA ON K1A OR6
CANADA
TEL 613 993 6060
TLF 613 952 6602
TLX 053 3715
EML
COM 28

NORGAARD-NIELSEN HANS U
UNIVERSITY OBSERVATORY
OESTER VOLDGADE 3
DK 1350 COPENHAGEN K
DENMARK
TEL 31 14 1790
TLF. 31 38 9157
TLX.
EML·
COM

NORIEGA-CRESPO ALBERTO DR
DPT OF ASTRONOMY
UNIVERSITY OF WASHINGTON
FM 20
SEATTLE WA 98195
USA
TEL 206 685 2155
TLF
TLX
EML AIPSAC@PHAST PHYS WASHINTON ED
COM

NORMAN COLIN A PROF
STSCI
HOMEWOOD CAMPUS
3700 SAN MARTIN DR
BALTIMORE MD 21218
USA
TEL 301 338 4895
TLF
TLX 6849101
EML STSCIC NORMAN
COM 28,33,34,44,47,48

NORRIS JOHN DR
MOUNT STROMLO & SIDING
SPRING OBSERVATORIES
PRIVATE BAG
WODEN PO ACT 2606
AUSTRALIA
TEL 62 88 1111
TLF
TLX AA 62270 CANOPUS
EML
COM 29

NORRIS RAYMOND PAUL
CSIRO
DIVISION OF RADIOPHYSICS
BOX 76
EPPING NSW 2121
AUSTRALIA
TEL 2 868 0222
TLF
TLX. 26230 ASTRO
EML RNORRIS@ATNF CSIRO AU
COM·

NORTH JOHN DAVID PROF
FILOSOFISCH INST
RIJKSUNIVERSITEIT
NL 9718 CW GRONINGEN
NETHERLANDS
TEL 59 07 1846
TLF 50 63 6160
TLX
EML
COM 41C

NORTH PIERRE
INSTITUT D'ASTRONOMIE
UNIVERSITE DE LAUSANNE
CH 1290 CHAVANNES-D-BOIS
SWITZERLAND
TEL 22 755 2611
TLF
TLX 27720 OBSG CH
EML north@obs unige ch
COM 45

NOSKOV BORIS N DR
STERNBERG STATE ASTR INST
UNIVERSITETSKIJ PROSP 13
119899 MOSCOW
RUSSIA
TEL
TLX
EML
TLF
COM 07

NOTNI P DR
ZNTRLINST F ASTROPHYSIK
STERNWARTE BABELSBERG
ROSA-LUXEMBURG-STR 17A
D 1502 POTSDAM
GERMANY
TEL
TLF
TLX
EML
COM 25,45

NOTTALE LAURENT
OBSERVATOIRE DE PARIS
SECTION DE MEUDON
DAF
F 92195 MEUDON PPL CDX
FRANCE
TEL 1 45 07 7403
TLF
TLX 201571
EML
COM 47

NOVELLO MARIO DR
CTR BRAS PESQUISAS FISIC
RUA DR XAVIER SIGAUD
150 URCA
22290 RIO DE JANEIRO RJ
BRAZIL
TEL 21 541 0337
TLF
TLX 21 22563
EML
COM 47

NOVICK ROBERT
DPT OF PHYSICS
COLUMBIA UNIVERSITY
538 W 120TH ST
NEW YORK NY 10027
USA
TEL 212 280 3293
TLF
TLX 22094 COLU UR
EML
COM 44,48

NOVIKOV I D DR
SPACE RESEARCH INSTITUTE
ACADEMY OF SCIENCES
PROFSOJOSNAYA UL 84/32
117810 MOSCOW
RUSSIA
TEL
TLF
TLX
EML
COM 47

NOVIKOV SERGEJ B DR
STERNBERG STATE ASTR INST
117234 MOSCOW
RUSSIA
TEL
TLF
TLX
EML
COM

NOVOSELOV V S PROF DR
ASTRONOMICAL OBSERVATORY
ST PETERSBURG UNIVERSITY
BIBLIOTECHNAJA PL 2
198904 ST PETERSBURG
RUSSIA
TEL 257-94-91
TLF
TLX
EML
COM 07

NOVOTNY JAN DR
FAC OF SCIENCE
MSARYK UNIVERSITY
KOTLARSKA 2
CS 611 37 BRNO
CZECHOSLOVAKIA
TEL
TLF
TLX
EML
COM 47

NOVOTNY VACLAV
ASTRONOMICAL INSTITUTE
CZECH ACADEMY OF SCIENCES
ONDREJOV OBSERVATORY
CS 251 65 ONDREJOV
CZECHOSLOVAKIA
TEL 204 85201
TLF 204 85314
TLX 121579
EML
COM 44

NOYES ROBERT W PROF
CENTER FOR ASTROPHYSICS
HCO/SAO
60 GARDEN ST
CAMBRIDGE MA 02138
USA
TEL 617 495 7424
TLF
TLX 921428 SATELLITE CAM
EML
COM 10,12,44

NUGIS TIIT
TARTU ASTROPHYSICAL OBS
ESTONIAN ACAD OF SCIENCES
202444 TARTU
ESTONIA
TEL
TLF
TLX
EML
COM 27,29C

NULSEN PAUL DR
MOUNT STROMLO & SIDING
SPRING OBSERVATORIES
PRIVATE BAG
WODEN PO ACT 2606
AUSTRALIA
TEL 62 88 1111
TLF
TLX 62270 AA
EML
COM 28,34,48

NUNES ROGERIO S DE SOUSA
GRUPO DE MATEM APLICADA
UNIVERSIDADE DO PORTO
RUA DAS TAIPAS 135
P 4000 PORTO
PORTUGAL
TEL 380313/769
TLF
TLX
EML
COM 09

NUNEZ JORGE DR
OBSERVATORIO FABRA
TIBIDADO
E 08022 BARCELONA
SPAIN
TEL 3 247 5736
TLF
TLX
EML
COM 24

NUSSBAUMER HARRY PROF
INSTITUT FUER ASTRONOMIE
ETH ZENTRUM
CH 8092 ZUERICH
SWITZERLAND
TEL 1 256 3631
TLF
TLX 53178 ETHBI CH
EML
COM 10,14C,34

NUTH JOSEPH A III
NASA/GSFC
CODE 691
LEP
GREENBELT MD 20771
USA
TEL 301 286 6364
TLF.
TLX
EML
COM 22,34

NYMAN LARS-AKE DR
ONSALA SPACE OBSERVATORY
GOETEBORG UNIVERSITY
S 439 00 ONSALA
SWEDEN
TEL. 30 06 0651
TLF
TLX 2400
EML
COM

OBASHEV SAKEN O DR
ASTROPHYSICAL INSTITUTE
KAZAKH ACAD OF SCIENCES
480068 ALMA ATA
KAZAKHSTAN
TEL
TLF
TLX 275
EML
COM

OBI SHINYA PROF
FACULTY OF GENERAL EDU
UNIVERSITY OF TOKYO
MEGURO KU
TOKYO 153
JAPAN
TEL:
TLF
TLX
EML
COM 14

OBLAK EDOUARD
OBSERVATOIRE DE BESANCON
41BIS AVE OBSERVATOIRE
F 25000 BESANCON CDX
FRANCE
TEL 81 50 3088
TLF
TLX 361144
EML
COM· 25,26,33

OBREGON DIAZ OCTAVIO J DR
DPT DE FISICA
UNAM UNIDAD
APDO POSTAL 55-534
09340 MEXICO DF
MEXICO
TEL· 686 0322
TLF
TLX 1764296 UAM ME
EML
COM

OBRIDKO VLADIMIR N DR
IZMIRAN
ACADEMY OF SCIENCES
142092 TROITSK
RUSSIA
TEL 232-1921
TLF
TLX 412623 SCSTP
EML
COM 10

OCCHIONERO FRANCO PROF
OAR
VIA DEL PARCO MELLINI 84
I 00136 ROMA
ITALY
TEL 6 34 52656
TLF
TLX·
EML
COM

OCHSENBEIN FRANCOIS DR
OBSERVATOIRE ASTRONOMIQUE
11 RUE UNIVERSITE
F 67000 STRASBOURG
FRANCE
TEL 88 35 8218
TLF 88 25 0160
TLX 890 506 STAROBSF
EML
COM 05

ODA MINORU PROF
WAKO SHI
SAITAMA 351 01
JAPAN
TEL 0484 62 1111
TLF 0484 67 5942
EML
COM 44,48,51
TLX 02962818 RIKEN J

ODA NAOKI
MICRO-ELECTRONICS RES LAB
NIPPON ELECTRIC COMPANY
4-1-1 MIYAZAKI MIYAMAE KU
KAWASAKI KANAGAWA 213
JAPAN
TEL
TLF
TLX
EML
COM 44

ODELL ANDREW P
ASTROPHYSICAL OBSERVATORY
N ARIZONA UNIVERSITY
FLAGSTAFF AZ 86011
USA
TEL
TLF
TLX
EML
COM 35

ODENWALD STEN F DR
NAVAL RESEARCH LABORATORY
CODE 4138 0
4555 OVERLOOK AVE SW
WASHINGTON DC 20375 5000
USA
TEL 202 767 3010
TLF
TLX
EML
COM

ODGERS GRAHAM J DR
HERZBERG INST ASTROPHYS
DOMINION ASTROPHYS OBS
5071 W SAANICH RD
VICTORIA BC V8X 4M6
CANADA
TEL 604 388 3977
TLF 604 363 0045
TLX 049 7295
EML
COM 09,27

OEGELMAN HAKKI B DR
MAX PLANCK INSTITUT FUR
EXTRATERESTRISHE PHYSICS
KARL-SCHWARZSCHILDSTR
D 8046 GARCHING MUENCHEN
GERMANY
TEL 89 329 93833
TLF 89 329 93569
TLX
EML HBO AT DGAIPP1S
COM

OEGERLE WILLIAM R
5924 BERWYN RD

COLLEGE PARK MD 20740
USA
TEL
TLF
TLX
EML
COM

OEHMAN YNGVE PROF
THULELEM 53
S 223 67 LUND
SWEDEN
TEL 46 14 3362
TLF
EML
COM·
TLX

OEKTEN ADNAN DR
UNIVERSITY OBSERVATORY
ISTANBUL UNIVERSITY
34452 ISTANBUL
TURKEY
TEL 1 522 3597
TLF 1 522 6123
TLX
EML JKO17@TRIUVM11
COM 10

OEMLER AUGUSTUS JR DR
YALE UNIVERSITY OBS
BOX 6666
NEW HAVEN CT 06511
USA
TEL 203 436 3460
TLF
TLX
EML
COM 28,47

OERTEL GOETZ K DR
AURA INC
1625 MASSACHUSETTS AVE NW
SUITE 701
WASHINGTON DC 20036
USA
TEL 202 483 2101
TLF
TLX
EML TELEMAIL G OERTEL
COM 44

OESTERWINTER CLAUS
BOX 1270
DAHLGREN VA 22448 5180
USA
TEL 703 663 2555
TLF
EML
COM 04,07
TLX

OESTGAARD ERLEND
DPT OF PHYSICS
UNIVERSITY OF TRONDHEIM
AVH
N 7055 DRAGVOLLM
NORWAY
TEL 07-920411 x 117
TLF
TLX
EML·
COM

OESTREICHER ROLAND
LANDESSTERNWARTE
KOENIGSTUHL
D 6900 HEIDELBERG 1
GERMANY
TEL 62 211 0036
TLF
TLX
EML
COM 25

OETKEN L DR
ZNTRLINST F ASTROPHYSIK
ASTROPHYSIK OBSERVATORIUM
TELEGRAFENBERG
D 1500 POTSDAM
GERMANY
TEL
TLF
TLX
EML
COM 14,30

OEZEL MEHMET EMIN DR
DPT OF PHYICS
CUKUROVA UNIVERSITY
01330 ADANA
TURKEY
TEL 71 13 33394*2480
TLF 71 12 1945
TLX 62934
EML
COM 40,48,50

OEZKAN MUSTAFA TUERKER DR
UNIVERSITY OBSERVATORY
ISTANBUL UNIVERSITY
34452 ISTANBUL
TURKEY
TEL 1 528 3847
TLF 1 522 6123
TLX
EML JK017@TRIUVM11
COM 42

OGAWARA YOSHIAKI
INST SPACE & ASTRON SCI
UNIVERSITY OF TOKYO
MEGURO KU
TOKYO 153
JAPAN
TEL 3-467-1111
TLF
TLX 34757 ISASTRO J
EML
COM 44,48

OGELMAN HAKKI B DR
DPT OF PHYSICS
UNIVERSITY OF WISCONSIN
1150 UNIVERSITY AVE
MADISON WI 53706
USA
TEL 608 265 2052
TLF. 608 263 0800
TLX 265452 UOFWISC MDS
EML 47452 OGELMAN
COM

OGIR MAYA DR
CRIMEAN OBSERVATORY
NAUCHNY
CRIMEA
UKRAINE
TEL 43 2945
TLF
TLX
EML
COM 10

OGURA KATSUO DR
COL OF LITERATURE
KOKUGAKUIN UNIVERSITY
HIGASHI 4-10-28
SHIBUYAKU TOKYO 150
JAPAN
TEL 298-42-6913
TLF
TLX 28899 SIBINBTH J
EML
COM 37

OH KYU DONG DR
DPT EARTH SCIENCE
CHONNAM NTL UNIVERSITY
KWANGJU CHONNAN
KOREA R
TEL
TLF
TLX
EML
COM 42

OHASHI TAKAYA DR
DPT OF PHYSICS
UNIVERSITY OF TOKYO
BUNKYO KU
TOKYO 113
JAPAN
TEL 3 812 2111
TLF 3 812 6938
TLX UTPHYSIC 23472 J
EML
COM

OHISHI MASATOSHI DR
NOBEYAMA RADIO OBS
NAOJ
MINAMIMAKI MURA
NAGANO 384 13
JAPAN
TEL 81 267 63 4373
TLF 81 267 98 2884
TLX 3329005 NAONRO J
EML NROOISI@JPNNRO BITNET
COM 40

OHKI KENICHIRO DR
TOKYO ASTRONOMICAL OBS
NAOJ
OSAWA MITAKA
TOKYO 181
JAPAN
TEL 0422-32-5111
TLF
TLX 2822307
EML
COM 10

OHRING GEORGE PROF
DPT GEOPHYS & PLANET SCI
TEL AVIV UNIVERSITY
TEL AVIV
ISRAEL
TEL
TLF
TLX
EML
COM

OHTANI HIROSHI DR
DPT OF ASTRONOMY
KYOTO UNIVERSITY
KITASHIRAKAWA SAKYO KU
KYOTO 606
JAPAN
TEL 075-751-2111
TLF
TLX 5422693 LIBKYU J
EML
COM 34

OHTSUBO JUNJI DR
FACULTY OF ENGINEERING
SHIZUOKA UNIVERSITY
3 CHOME JYOHOKU
HAMAMATSU 432
JAPAN
TEL 534 71 1171 EXT 585
TLF 534 75 1764
TLX
EML
COM 09

OHYAMA NOBORU PROF
FACULTY OF ENGINEERING
SHIZUOKA UNIVERSITY
3 CHOME JYOHOKU
HAMAMATSU 432
JAPAN
TEL
TLF
TLX
EML
COM 35

OJA HEIKKI DR
OBS & ASTROPHYSICS LAB
UNIVERSITY OF HELSINKI
TAEHTITORNINMAKI
SF 00130 HELSINKI 13
FINLAND
TEL 191 2942
TLF
TLX
EML
COM 46

OJA TARMO PROF
KVISTABERG OBSERVATORY

S 197 00 BRO
SWEDEN
TEL 75 84 0157
TLF
TLX
EML
COM 24,33,45

OKA TAKESHI DR
DPT OF CHEMISTRY
UNIVERSITY OF CHICAGO
5735 S ELLIS AVE
CHICAGO IL 60637
USA
TEL 312 962 7070
TLF
TLX
EML
COM 14

OKAMOTO ISAO DR
INTL LATITUDE OBSERVATORY
NAOJ
HOSHIGAOKA MIZUSAWA SHI
IWATE 023
JAPAN
TEL 0197-24-7111
TLF
TLX 8376-28 ILSMIZ J
EML
COM· 35

OKAMURA SADANORI DR
KISO OBSERVATORY
UNIVERSITY OF TOKYO
MITAKEMURA KISOGUN
NAGANO 397 01
JAPAN
TEL 0264-52-3360
TLF
TLX 3347577 KSOOBS J
EML
COM 28C

OKAZAKI AKIRA DR
DPT SCIENCE EDUCATION
GUNMA UNIVERSITY
MAEBASHI
GUNMA 371
JAPAN
TEL 0272 32 1611
TLF 0272 33 9231
TLX
EML
COM 42

OKAZAKI ATSUO T DR
COL OF GEN EDUCATION
HOKKAI GAKUEN UNIVERSITY
TOYOHIRA-KU
SAPPORO 062
JAPAN
TEL 011 841 1161*284.
TLF
TLX
EML A10935@JPNKUDPC BITNET
COM 29

OKAZAKI SEICHI DR
2-4-4 OSAWA
MITAKA
TOKYO 181
JAPAN
TEL 0422 31 6770
TLF
TLX
EML
COM

OKE J BEVERLEY PROF
CALTECH
MS 105 24
PASADENA CA 91125
USA
TEL 818 356 4007
TLF
TLX
EML
COM 28

OKEKE PIUS N DR
DPT OF PHYSICS & ASTRON
UNIVERSITY OF NIGERIA
NSUKKA ANAMBRA STATE
NIGERIA
TEL 42 77 1911
TLF
TLX
EML
COM 48

OKOYE SAMUEL E PROF
DPT OF PHYSICS & ASTRON
UNIVERSITY OF NIGERIA
NSUKKA ANAMBRA STATE
NIGERIA
TEL 042-770752
TLF
TLX
EML
COM 38,40,46,47,48

OKUDA HARUYUKI DR PROF
INST SPACE & ASTRON SCI
UNIVERSITY OF TOKYO
MEGURO KU
TOKYO 153
JAPAN
TEL 03-467-1111
TLF
TLX 24550 SPACETKY J
EML
COM 33,34

OKUDA TORU
INST OF EARTH SCIENCE
HOKKAIDO UNIV OF EDUCAT
1-2 HACHIMAN CHO
HAKODATE 040
JAPAN
TEL 0138-41-1121
TLF
TLX
EML
COM 44

OKUMURA SACHIKO DR
DPT EARTH SCI & ASTRON
UNIVERSITY OF TOKYO
MEGURO KU
TOKYO 153
JAPAN
TEL 81 3 3467 1171
TLF 81 3 34852904
TLX 2426728 TODAIK J
EML OKUMURA@KYOHOU C U-TOKYO
COM 40

OLAH KATALIN DR
KONKOLY OBSERVATORY
THEGE U 13/17
BOX 67
H 1525 BUDAPEST
HUNGARY
TEL 1 75 5866/75 4122
TLF
TLX 227460 KONOB H
EML
COM· 27,42

OLANO CARLOS ALBERTO DR
IAR
CC 5
1894 VILLA ELISA (BS AS)
ARGENTINA
TEL 21 4 3793
TLX
EML
TLF
COM 33

OLBERG MICHAEL DR
ONSALA SPACE OBSERVATORY
GOETEBORG UNIVERSITY
S 439 00 ONSALA
SWEDEN
TEL 30 06 0650
TLF 30 06 2621
TLX 2400 ONSPACE S
EML OLBERG@OSO CHALMERS SE
COM 40

OLEAK H DR
ZNTRLINST F ASTROPHYSIK
STERNWARTE BABELSBERG
ROSA-LUXEMBURG-STR 17A
D 1502 POTSDAM
GERMANY
TEL
TLF
TLX
EML
COM 28

OLIVA ERNESTO DR
OSS ASTROFISICO
DI ARCETRI
LARGO E FERMI 5
I 50125 FIRENZE
ITALY
TEL 55 275 2310
TLF (5 22 0039
TLX 572268
EML SPAN 38954 OLIVA
COM

OLIVER BERNARD M DR
NASA AMES RESEARCH CTR
MS 229 8
MOFFETT FIELD CA 94035
USA
TEL 415 694 5166
TLF
TLX 348 408 NASA AMES
EML
COM 51

OLIVER BERNARD M PROF
NASA AMES RESEARCH CTR
CHEIF SETI PROGRAM OFFICE
MOFFETT FIELD CA 94035
USA
TEL 415 694 5166
TLF
TLX
EML
COM 51

OLIVER JOHN PARKER DR
DPT OF ASTRONOMY
UNIVERSITY OF FLORIDA
211 SSRB
GAINESVILLE FL 32611
USA
TEL
TLF
TLX
EML
COM 42

OLLONGREN A PROF DR
STERREWACHT
DPT MATHS & COMPUTER SCI
BOX 9512
NL 2300 RA LEIDEN
NETHERLANDS
TEL 71 27 2727*5006
TLF
TLX 39058 ASTRO NL
EML
COM· 33,51

OLNON FRISO
NFRA
BOX 2
NL 7990 AA DWINGELOO
NETHERLANDS
TEL 52 19 7244
TLX 42043 SRZM NL
EML
TLF·
COM

OLOFSSON HANS
ONSALA SPACE OBSERVATORY
GOETEBORG UNIVERSITY
S 439 00 ONSALA
SWEDEN
TEL 30 06 0650
TLF
TLX 8542400 ONSPACE
EML
COM· 34

OLOFSSON S GOERAN DR
STOCKHOLM OBSERVATORY
S 133 36 SALTSJOEBADEN
SWEDEN
TEL 87 17 2639
TLF 87 17 4719
TLX 12972
EML OLOFSSON@ASTRO SU SE
COM

OLOWIN RONALD PAUL DR
DPT OF PHYS & ASTRONOMY
SAINT MARY'S COLLEGE
207 D GALILEO HALL
MORAGA CA 94575
USA
TEL 510 631 4428
TLF
TLX
EML· RPOLOWIN@GALILEO STMARYS-CA EO
COM 47

OLSEN ERIK H
COPENHAGEN UNIVERSITY OBS
BRORFELDEVEJ 23
DK 4340 TOLLOSE
DENMARK
TEL 53 48 8195
TLF· 58 48 8758
TLX 44155 DANAST DK
EML
COM 45

OLSEN FOGH H J
COPENHAGEN UNIVERSITY OBS
BRORFELDEVEJ 23
DK 4340 TOLLOSE
DENMARK
TEL 53 48 8195
TLF 58 48 8758
TLX 44155 DQNQST
EML
COM 08,46

OLSEN KENNETH H DR
GCS
BOX 1273
LYNNWOOD WA 98046 1273
USA
TEL
TLF
TLX
EML
COM

OLSON EDWARD C PROF
DPT OF ASTRONOMY
UNIVERSITY OF ILLINOIS
1011 W SPRINGFIELD AVE
URBANA IL 61801
USA
TEL 217 333 5531
TLF
TLX
EML
COM 42

OLTHOF HINDERICUS DR
ESA/ESTEC
SSD
BOX 299
NL 2200 AG NOORDWIJK
NETHERLANDS
TEL 17 19 86555
TLF
TLX 39098 ESTC NL
EML SPAN ESTEC 1 HOLTHOF
COM 44

OMAROV TUKEN B PROF
ASTROPHYSICAL INSTITUTE
KAZAKH ACAD OF SCIENCES
480068 ALMA ATA
KAZAKHSTAN
TEL 64-40-40
TLF
TLX
EML
COM 07

OMNES ROLAND PROF
LPTHE
UNIVERSITE PARIS XI
BT 211
F 91405 ORSAY CDX
FRANCE
TEL 1 69 41 7744
TLF
TLX 692166
EML
COM 47

OMONT ALAIN PROF
INSTITUT D'ASTROPHYSIQUE
98BIS BD ARAGO
F 75014 PARIS
FRANCE
TEL 1 43 20 1425
TLF 1 43 29 8673
TLX
EML
COM 14,34

ONAKA TAKASHI
DPT OF ASTRONOMY
UNIVERSITY OF TOKYO
BUNKYO KU
TOKYO 113
JAPAN
TEL 3-812-2111
TLF
TLX 33659 UTYOSCI J
EML
COM 34

ONDERLICKA BEDRICH DR
DPT OF ASTRONOMY
PURKYNE UNIVERSITY
KOTLARSKA 2
CS 611 37 BRNO
CZECHOSLOVAKIA
TEL 5 1112
TLF
TLX
EML
COM

ONEGINA A B DR
MAIN ASTRONOMICAL OBS
UKRAINIAN ACAD OF SCIENCE
GOLOSEEVO
252127 KIEV
UKRAINE
TEL 66 3744
TLF
TLX
EML
COM 24

ONO YORO PROF
DPT OF PHYSICS
HOKKAIDO UNIVERSITY
KITA 10 NISHI 8
SAPPORO 063
JAPAN
TEL
TLF
TLX
EML
COM

ONUORA LESLEY IRENE DR
DPT OF PHYSICS & ASTRON
UNIVERSITY OF NIGERIA
NSUKKA ANAMBRA STATE
NIGERIA
TEL
TLF
TLX 51496 ULIONS
EML
COM 40,46,47

OOE MASATSUGU DR
INTL LATITUDE OBSERVATORY
NAOJ
HOSHIGAOKA MIZUSAWA SHI
IWATE 023
JAPAN
TEL 0197-24-7111
TLF
TLX 837628 MIZ J
EML
COM 19

OORT JAN H PROF
PRES KENNEDYLAAN 169
NL 2343 GZ OEGSTGEEST
NETHERLANDS
TEL
TLF
TLX
EML
COM 28,33,40,47

OPHER REUVEN PROF
IAG
UNIVERSIDADE DE SAO PAULO
CP 30627
01051 SAO PAULO SP
BRAZIL
TEL 11 275 3720
TLF
TLX 1136221 IAGM BR
EML
COM

OPOLSKI ANTONI PROF
ASTRONOMICAL INSTITUTE
WROCLAW UNIVERSITY
UL KOPERNIKA 11
PL 51 622 WROCLAW
POLAND
TEL
TLF
TLX
EML
COM 27

OPROIU TIBERIU DR
ASTRONOMICAL OBSERVATORY
BLOC R-5 SC I & II AP 10
STR BUCIUM 25
R 3400 CLUJ NAPOCA
RUMANIA
TEL 951-62616
TLF
TLX
EML
COM

ORCHISTON WAYNE DR
27 MARIAN DRIVE

GISBORNE
NEW ZEALAND
TEL 83 832
TLF
TLX
EML
COM 41

ORELLANA ROSA BEATRIZ DR
OBSERVATORIO ASTRONOMICO
PASEO DEL BOSQUE
1900 LA PLATA (BS AS)
ARGENTINA
TEL 21 21 7308
TLF
TLX
EML· RORELLANA%FCAGLP EDU AR
COM 07

ORLIN HYMAN DR
NTL ACADEMY OF SCIENCES
NTL RESEARCH COUNCIL
2101 CONSTITUTION AVE NW
WASHINGTON DC 20418
USA
TEL.
TLF
TLX
EML
COM

ORLOV MIKHAIL DR
MAIN ASTRONOMICAL OBS
UKRAINIAN ACAD OF SCIENCE
GOLOSEEVO
252127 KIEV
UKRAINE
TEL 66 3110
TLF.
TLX 131406 SKY
EML
COM 29

ORMES JONATHAN F DR
NASA/GSFC
CODE 660
GREENBELT MD 20771
USA
TEL 301 286 8801
TLF
TLX 89675
EML SPAN LHEAVX ORMES
COM

ORRALL FRANK Q PROF
INSTITUTE FOR ASTRONOMY
UNIVERSITY OF HAWAII
2680 WOODLAWN DR
HONOLULU HI 96822
USA
TEL 808 948 8667
TLF
TLX 723-8459
EML
COM 10,12,36

ORSATTI ANA M DR
UNIV NACIONAL DE LA PLATA
FCAG
1900 LA PLATA (BS AS)
ARGENTINA
TEL 21 21 7308
TLF
TLX 31151 BULAP
EML AMO@FCAGLP EDU AR
COM 25

ORTE ALBERTO
CECILIO PUJAZON 22-3 A
E 11100 SAN FERNANDO
SPAIN
TEL 56 89 5441
TLF
TLX.
EML
COM

ORTOLANI SERGIO
OSS ASTRONOMICO DI PADOVA

VIC DELL OSSERVATORIO 5
I 35122 PADOVA
ITALY
TEL 49 66 1499
TLF
TLX 432071 ASTROS I
EML
COM 37

ORTON GLENN S DR
JPL
MS 183 301
4800 OAK GROVE DR
PASADENA CA 91109
USA
TEL 818 354 2460
TLF
TLX
EML
COM. 14

ORUS JUAN J PROF
DPT FISICA DE ATMOSFERA
UNIVERSIDAD DE BARCELONA
AVD DIAGONAL 645
E 08028 BARCELONA
SPAIN
TEL
TLF
TLX
EML
COM 07

OSAKI TORU DR
RYUKOKU UNIVERSITY
FUKAKUSA TSUKAMOTO
FUSHIMI KU
KYOTO 612
JAPAN
TEL 075-642-1111
TLF
TLX
EML
COM 34

OSAKI YOJI DR
DPT OF ASTRONOMY
UNIVERSITY OF TOKYO
BUNKYO KU
TOKYO 113
JAPAN
TEL 03-812-2111
TLF
TLX 33659 UTYOSCI J
EML
COM 35,42

OSAWA KIYOTERU DR
TOKYO ASTRONOMICAL OBS
NAOJ
OSAWA MITAKA
TOKYO 181
JAPAN
TEL 0422-32-5111
TLF
TLX
EML
COM

OSBORN WAYNE DR
DPT OF PHYSICS
CENTRAL MICHIGAN UNIV
MT PLEASANT MI 48859
USA
TEL 517 774 3321
TLF
TLX
EML 3Y2LW5G@CMUVM BITNET
COM 24,45,46

OSBORNE JOHN L DR
DPT OF PHYSICS
UNIVERSITY OF DURHAM
SOUTH RD
DURHAM DH1 3LE
UK
TEL 913 856 4971
TLF
TLX 537351 DURLIB G
EML
COM 34

OSMAN ANAS MOHAMED PROF
HELWAN OBSERVATORY

HELWAN
EGYPT
TEL 78 0645/2683
TLF
TLX
EML
COM· 28,37

OSMER PATRICK S DR
NOAO
BOX 26732
950 N CHERRY AVE
TUCSON AZ 85726 6732
USA
TEL 602 327 5511
TLF
TLX 666484
EML
COM

OSORIO JOSE J S P PROF
OBSERVATORIO ASTRONOMICO
UNIVERSIDADE DO PORTO
MONTE DA VIRGEM
P 4400 VILA NOVA DE GAIA
PORTUGAL
TEL 7820404
TLF
TLX 22367
EML
COM 07,08,46,50

OSTER LUDWIG F PROF DR
NTL SCIENCE FOUNDATION
DIV ASTRONOMICAL SCIENCES
1800 G ST NW
WASHINGTON DC 20550
USA
TEL 202 357 9857
TLF
TLX
EML
COM 12

OSTERBROCK DONALD E PROF
LICK OBSERVATORY
UNIVERSITY OF CALIFORNIA
SANTA CRUZ CA 95064
USA
TEL. 408 429 2605
TLF
TLX
EML
COM 28,34 40,41

OSTRIKER JEREMIAH P PROF
PRINCETON UNIVERSITY OBS
PEYTON HALL
PRINCETON NJ 08544
USA
TEL 609 258 3800
TLF
TLX 322409
EML jpo@astro.princeton.edu
COM 33,35,48P,51

OSWALT TERRY D DR
OPT OF PHYS & SPACE SCI
FLORIDA INST TECHNOLOGY
150 W UNIVERSITY BLVD
MELBOURNE FL 32901
USA
TEL 305 768 8098
TLF
TLX
EML
COM 26,27,35

OTERMA LIISI PROF
SIRKKALANKATU 31
SF 20700 TURKU
FINLAND
TEL 21 33 2081
TLF
TLX
EML
COM 20

OTHMAN MAZLAN
DPT OF PHYSICS
UNIVERSITI KEBANGSAAN
MALAYSIA
43600 BANGI SELANGOR
MALAYSIA
TEL 8250001
TLF
TLX 31496 UNIKEB MA
EML
COM 46

OTTELET I J DR
INSTITUT D'ASTROPHYSIQUE
UNIVERSITE DE LIEGE
AVE COINTE 5
B 4000 COINTE-LIEGE
BELGIUM
TEL 41 52 9980
TLF 41 52 7474
TLX 41264
EML
COM 16

OUHRABKA MIROSLAV DR
COLLEGE OF EDUCATION IN
HRADEC KRALOVE
NAM SVOBODY 301
CS 501 91 HRADEC KRALOVE
CZECHOSLOVAKIA
TEL 49 25226
TLF 49 25785
TLX
EML
COM

OVERBEEK MICHIEL DANIEL
BOX 212
EDENVALE 1610
SOUTH AFRICA
TEL 11 53 5447
TLF
TLX
EML
COM 20

OWAKI NAOAKI DR
DPT ASTRON & EARTH SCI
TOKYO GAKUGEI UNIVERSITY
KOGANEI
TOKYO 184
JAPAN
TEL
TLF
TLX
EML
COM 46

OWEN FRAZER NELSON DR
NRAO
VLA
BOX 0
SOCORRO NM 87801 0387
USA
TEL 505 772 4011
TLF
TLX 910-988-1710
EML
COM 28,40,50

OWEN TOBIAS C PROF
INSTITUTE FOR ASTRONOMY
UNIVERSITY OF HAWAII
2680 WOODLAWN DR
HONOLULU HI 96822
USA
TEL 808 956 8007
TLF 808 988 2790
TLX
EML
COM 16C,44,51,WGPSNC

OWOCKI STANLEY PETER DR
BARTOL RESEARCH INSTITUTE
UNIVERSITY OF DELAWARE
NEWARK DE 19716
USA
TEL 302 451 8357
TLF 302 451 1843
TLX 510 666 0805
EML BARTOL OWOCKI
COM 12,29,36

OXENIUS JOACHIM DR
IAAG
VRIJE UNIV BRUSSELS
CP 231
B 1050 BRUSSELS
BELGIUM
TEL 2 640 0015
TLF
TLX 23069 UNILIB B
EML
COM 36

OZERNOY LEONID M PROF
NASA/GSFC
CODE 665
LAB HIGH ENERGY ASTROPHYS
GREENBELT MD 20771
USA
TEL 301 286 8801
TLF
TLX
EML OZERNOY@LHEAVX GSFC NASA GOV
COM 34,47,48

OZGUC ATILA
KANDILLI OBSERVATORY
BUGAZICI UNIVERSITY
CENGELKOY
81220 ISTANBUL
TURKEY
TEL 1 332 0240/41
TLF
TLX
EML
COM 10

OZSVATH I PROF
UNIVERSITY OF TEXAS
PROGRAMS IN MATHEMAT SCI
BOX 830688
RICHARDSON TX 75083 0688
USA
TEL 214 690 2174
TLF
TLX
EML
COM 47

O'BRIEN PAUL THOMAS DR
DPT PHYSICS & ASTRONOMY
UNIVERSITY COLLEGE LONDON
GOWER ST
LONDON WC1E 6BT
UK
TEL 1 713 877 050
TLF 1 713 807 145
TLX 28722
EML PTO@UK AC UCL STAR
COM 44

O'CONNELL ROBERT F PROF
DPT PHYSICS & ASTRONOMY
LOUISIANA STATE UNIV
BATON ROUGE LA 70803 4001
USA
TEL 504 388 6848
TLF
TLX
EML
COM 48

O'CONNELL ROBERT WEST DR
UNIVERSITY STATION
UNIVERSITY OF VIRGINIA
BOX 3818
CHARLOTTESVILLE VA 22903
USA
TEL 804 924 7494
TLF
TLX 510-587-5453
EML
COM 28,47

O'CONNOR SEAMUS L DR
DPT OF PHYSICS
UNIVERSITY COLLEGE
BELFIELD
DUBLIN 4
IRELAND
TEL 1 693 244
TLF
TLX 32693 UCD EI
EML
COM

O'DEA CHRISTOPHER P DR
STSCI
HOMEWOOD CAMPUS
3700 SAN MARTIN DR
BALTIMORE MD 21218
USA
TEL 301 338 2590
TLF 301 338 5085
TLX 6849101
EML
COM 28,40

O'DELL CHARLES R DR
DPT SPACE PHYS & ASTRON
RICE UNIVERSITY
BOX 1892
HOUSTON TX 77251 1892
USA
TEL 713 527 8101
TLF
TLX 556457
EML
COM 09,15,34

O'DELL STEPHEN L
NASA/MSFC
CODE ES 65
SPACE SCIENCE LAB
HUNTSVILLE AL 35812
USA
TEL 205 544 7708
TLF
TLX
EML
COM 34

O'DONOGHUE DARRAGH DR
DPT OF ASTRONOMY
UNIVERSITY OF CAPE TOWN
RONDEBOSCH 7700
SOUTH AFRICA
TEL
TLF
TLX
EML
COM 27

O'HANDLEY DOUGLAS A DR
NASA HEADQUARTERS
CODE Z
600 INDEPENDENCE AVE SW
WASHINGTON DC 20546
USA
TEL 202 453 8932
TLF 202 426 0408
TLX
EML
COM 04,07

O'KEEFE JOHN A DR
NASA/GSFC
CODE 681
GREENBELT MD 20771
USA
TEL 301 286 8445
TLF
TLX 89675
EML.
COM 15,16,22

O'LEARY BRIAN T
FUTURE FOCAS
5136 E KAREN DR
SCOTTSDALE AZ 85254
USA
TEL
TLF
TLX
EML
COM

O'MARA BERNARD J PROF
DPT OF PHYSICS
UNIVERSITY OF QUEENSLAND
ST LUCIA
BRISBANE QLD 4067
AUSTRALIA
TEL
TLF
TLX
EML
COM 36

O'MONGAIN EON
DPT OF PHYSICS
UNIVERSITY COLLEGE
BELFIELD
DUBLIN 4
IRELAND
TEL 1 693 244
TLF
TLX 32693 UCD
EML
COM 44

O'SULLIVAN DENIS F
DIAS
SCHOOL OF COSMIC PHYSICS
5 MERRION SQ
DUBLIN 2
IRELAND
TEL 1 774 321
TLF
TLX 31687 DIAS EI
EML
COM 48

O'SULLIVAN JOHN DAVID DR
CSIRO
DIVISION OF RADIOPHYSICS
BOX 76
EPPING NSW 2121
AUSTRALIA
TEL
TLF
TLX. 26230 ASTRO
EML
COM· 40

PACHNER JAROSLAV PROF
606 55 WYNFORD HTS CR
TORONTO ON M3C 1L4
CANADA
TEL 416 447 1015
TLF
TLX
EML
COM 47

PACHOLCZYK ANDRZEJ G PROF
STEWARD OBSERVATORY
UNIVERSITY OF ARIZONA
TUCSON AZ 85721
USA
TEL 602 621 6928
TLF
TLX 467175
EML
COM 28,40,48

PACIESAS WILLIAM S DR
DPT OF PHYSICS
UNIVERSITY OF ALABAMA
HUNTSVILLE AL 35899
USA
TEL 205 544 7712
TLF
TLX 594416 ES62
EML
COM 44

PACINI FRANCO PROF
DPT DI ASTRONOMIA
UNIVERSITA DI FIRENZE
LARGO E FERMI 5
I 50125 FIRENZE
ITALY
TEL 55 27521
TLF 55 22 0039
TLX 572268 ARCETR-I
EML
COM 44,48C,51

PACZYNSKI BOHDAN PROF
COPERNICUS ASTRON CENTER
POLISH ACAD OF SCIENCES
UL BARTYCKA 18
PL 00 716 WARSAW
POLAND
TEL
TLF
TLX
EML
COM 35,42

PADALIA T D DR
UTTAR PRADESH STATE
OBSERVATORY
PO MANORA PEAK 263 129
NAINITAL 263 129
INDIA
TEL 59 42 2136
TLF
TLX
EML
COM 42

PADEVET VLADIMIR DR
ASTRONOMICAL INSTITUTE
CZECH ACADEMY OF SCIENCES
ONDREJOV OBSERVATORY
CS 251 65 ONDREJOV
CZECHOSLOVAKIA
TEL 204 85201
TLF 204 85314
TLX 121579
EML
COM· 22

PADMAN RACHAEL
MULLARD RADIO ASTRON OBS
CAVENDISH LABORATORY
MADINGLEY RD
CAMBRIDGE CB3 OHE
UK
TEL 223 664 77
TLF
TLX 81292
EML
COM 40

PADMANABHAN T DR
TIFR/PHYSICS GROUP
HOMI BHABHA RD
COLABA
BOMBAY 400 005
INDIA
TEL 22 495 2311
TLF
TLX 011-3009 TIFR IN
EML
COM 47

PADRIELLI LUCIA
IST DI RADIOASTRONOMIA
CNR
VIA IRNERIO 46
I 40126 BOLOGNA
ITALY
TEL 51 23 2856
TLF
TLX. 211664 INF BO I
EML
COM 40,47

PAGE ARTHUR MR
MT TAMBORINE OBSERVATORY
BOX 44
ASPLEY QLD 4034
AUSTRALIA
TEL 7 263 4813
TLF
TLX
EML
COM 25

PAGE CLIVE G DR
DPT OF PHYSICS
UNIVERSITY OF LEICESTER
UNIVERSITY RD
LEICESTER LE1 7RH
UK
TEL 533 554 455*23
TLF
TLX 341664 LUXRAY G
EML
COM 48

PAGE DON NELSON
DPT OF ASTRONOMY
PENNSYLVANIA STATE UNIV
104 DAVEY LAB
UNIVERSITY PARK PA 16802
USA
TEL 814 863 0163
TLF
TLX 842510
EML
COM 47

PAGE THORNTON L DR
NASA/JOHNSON SPACE CENTER

18639 POINT LOOKOUT DR
HOUSTON TX 77058
USA
TEL 713 483 3728
TLF
TLX
EML
COM 28,51

PAGEL BERNARD E J PROF
NORDITA
BLEGDAMSVEJ 17
DK 2100 COPENHAGEN O
DENMARK
TEL 31 42 1616
TLF 31 38 9157
TLX 15216 NBI DK
EML nordita@nbivax nbi dk
COM 29,34,36

PAKVOR IVAN
ASTRONOMICAL OBSERVATORY
VOLGINA 7
YU 11050 BEOGRAD
YUGOSLAVIA
TEL 11 419 357/421 875
TLF
TLX 72610 AOB YU
EML
COM 08

PAL ARPAD PROF DR
FAC OF MATHEMATICS
UNIVERSITY OF CLUJ-NAPOCA
STR RAKOCZI 72
R 3400 CLUJ NAPOCA
RUMANIA
TEL 951-16101/11592
TLF
TLX
EML
COM 07

PALAGI FRANCESCO
OSS ASTROFISICO
DI ARCETRI
LARGO E FERMI 5
I 50125 FIRENZE
ITALY
TEL 55 27521
TLF
TLX 572268 ARCETR I
EML
COM

PALLA FRANCESCO
OSS ASTROFISICO
DI ARCETRI
LARGO E FERMI 5
I 50125 FIRENZE
ITALY
TEL 55 275 2242
TLF
TLX 572268 ARCETR I
EML
COM 34

PALLAVICINI ROBERTO DR
OSS ASTROFISICO
DI ARCETRI
LARGO E FERMI 5
I 50125 FIRENZE
ITALY
TEL 55 275 2252
TLF
TLX 572268 ARCETR I
EML
COM 10,36C

PALLE PERE-LLUIS DR
INST DE ASTROFISICA
DE CANARIAS
OBS DEL TEIDE
E 38200 LA LAGUNA
SPAIN
TEL 22 26 2211
TLF 22 26 3005
TLX 92640
EML SPAN IAC PLP
COM 10,12

PALMEIRA RICARDO A R DR
INPE
CP 515
12200 S JOSE DOS CAMPOS
BRAZIL
TEL 123 22 9977
TLF 123 21 8743
TLX
EML
COM

PALMER PATRICK E PROF
ASTRONOMY & ASTROPHYS CTR
UNIVERSITY OF CHICAGO
5640 S ELLIS AVE
CHICAGO IL 60637
USA
TEL 312 962 7972
TLF
TLX 6871133
EML
COM 33,34,40

PALMER PHILIP DR
ASTRONOMY UNIT
QUEEN MARY/WESTFIELD COLL
MILE END RD
LONDON E1 4NS
UK
TEL 1 975 5462
TLF
TLX 893750
EML PHILIP @ QMC MATHS
COM 28

PALOUS JAN DR
ASTRONOMICAL INSTITUTE
CZECH ACADEMY SCIENCES
BUDECSKA 6
CS 120 23 PRAHA 2
CZECHOSLOVAKIA
TEL 2 25 8757
TLF
TLX 122486
EML
COM 33

PALUMBO GIORGIO G C DR
LABORATORIO TESRE
VIA CASTAGNOLI 1
I 40100 BOLOGNA
ITALY
TEL 51 51 9593
TLF
TLX 511350 CNR-80
EML
COM 28,48

PALUS PAVEL DR
FAC OF MATH & PHYSICS
DPT ASTRONOMY & ASTROPHYS
MLYNSKA DOLINA
CS 842 15 BRATISLAVA
CZECHOSLOVAKIA
TEL 7 72 3611
TLF 7 32 5882
TLX
EML
COM 10,12

PAMYATNIKH A A DR
INST OF ASTRONOMY
ACADEMY OF SCIENCES
PYATNITSKAYA UL 48
109017 MOSCOW
RUSSIA
TEL 231-54-61
TLF
TLX 412623 SCSTP SU
EML
COM 05,35

PAN JUN-HUA
NANJING ASTRONOMICAL
INSTRUMENTS FACTORY
BOX 846
NANJING
CHINA PR
TEL 25 46191
TLF
TLX 34136 GLYNJ CN
EML
COM

PAN LIANDE
SHAANXI OBSERVATORY
CAS
LINTONG XIAN
SHAANXI
CHINA PR
TEL 33 2255
TLF
TLX 70121 CSAO CN
EML
COM 10

PAN NING-BAO
BEIJING ASTRONOMICAL OBS
CAS
W SUBURB
BEIJING 100080
CHINA PR
TEL 1 28 1698
TLF
TLX
EML
COM

PAN RONG-SHI
SHANGHAI OBSERVATORY
CAS
80 NANDAN RD
SHANGHAI
CHINA PR
TEL 21 38 6191
TLF
TLX 33164 SHAO CN
EML
COM 24,28,47

PAN XIAO-PEI
CALTECH
MS 105 24
PASADENA CA 91125
USA
TEL 818 356 4015
TLF
TLX 67425 CALTECH PSD
EML XPP@DEIMOS CALTECH EDU
COM

PANAGIA NINO DR
STSCI
HOMEWOOD CAMPUS
3700 SAN MARTIN DR
BALTIMORE MD 21218
USA
TEL 301 338 4916
TLF
TLX 6849101 ST SCI
EML
COM 34

PANDE GIRISH CHANDRA PROF
126 ARYANAGAR
LUCKNOW 226 004
INDIA
TEL
TLF
EML
COM· 35
TLX

PANDE MAHESH CHANDRA DR
UTTAR PRADESH STATE
OBSERVATORY
PO MANORA PEAK 263 129
NAINITAL 263 129
INDIA
TEL 59 42 2136
TLF
TLX
EML
COM 12

PANDEY A K
UTTAR PRADESH STATE
OBSERVATORY
PO MANORA PEAK 263 129
NAINITAL 263 129
INDIA
TEL 59 42 2136
TLF
TLX
EML
COM 20,33,37

PANDEY S K
DPT OF PHYSICS
UNIVERSITY OF RAVISHANKAR
RAIPUR 492 010
INDIA
TEL 27 064
TLF
TLX
EML
COM

PANDEY UMA SHANKAR DR
DPT OF PHYSICS
UNIVERSITY OF GORAKHPUR
GORAKHPUR 273 009
INDIA
TEL 33 6601
TLF
TLX
EML
COM 42,46,48

PANEK ROBERT J DR
DPT OF ASTRONOMY
PENNSYLVANIA STATE UNIV
525 DAVEY LAB
UNIVERSITY PARK PA 16802
USA
TEL
TLF
TLX
EML
COM 36

PANG KEVIN
JPL
MS T11 823
4800 OAK GROVE DR
PASADENA CA 91109
USA
TEL 818 354 5392
TLF
TLX 675429
EML
COM 16

PANKONIN VERNON LEE DR
NTL SCIENCE FOUNDATION
DIV ASTRONOMICAL SCIENCES
1800 G ST NW
WASHINGTON DC 20550
USA
TEL 202 357 9696
TLF
TLX
EML
COM 34,40,50

PANNUNZIO RENATO
OSS ASTRONOMICO DI TORINO
ST OSSERVATORIO 20
I 10025 PINO TORINESE
ITALY
TEL 11 84 1067
TLX 213236 TOASTR I
EML
TLF 11 84 1281
COM 26

PANOV KIRIL DR
DPT OF ASTRONOMY
BULGARIAN ACAD SCIENCES
7TH NOVEMBER ST 1
BG 1000 SOFIA
BULGARIA
TEL 2 7341
TLF 23561 ECFBAN BG
TLX 23561 ECF BAN BG
EML
COM

PAOLICCHI PAOLO DR
IST DI ASTRONOMIA
UNIVERSITA DI PISA
PIAZZA TORRICELLI 2
I 56100 PISA
ITALY
TEL 50 43343
TLF
TLX
EML
COM 15,16

PAP JUDIT
JPL/CALTECH
MS 171 400
4800 OAK GROVE DR
PASADENA CA 91109
USA
TEL 818 354 2662
TLF 818 354 4707
TLX
EML jpap@solar stanford edu
COM 10

PAPAELIAS PHILIP DR
DPT OF PHYSICS
NTL UNIVERSITY OF ATHENS
PANEPISTIMIOPOLIC
GR 157 71 ZOGRAFOS
GREECE
TEL 1 723 5122
TLF
TLX 223815 UNIVGR
EML
COM·

PAPAGIANNIS MICHAEL D PRO
DPT OF ASTRONOMY
BOSTON UNIVERSITY
725 COMMONWEALTH AVE
BOSTON MA 02215
USA
TEL 617 353 2626
TLF
TLX
EML
COM 40,44,51C

PAPALIOLIOS COSTAS DR
CENTER FOR ASTROPHYSICS
HCO/SAO
60 GARDEN ST
CAMBRIDGE MA 02138
USA
TEL
TLF
TLX
EML
COM

PAPALOIZOU JOHN C B DR
SCHOOL OF MATHEMATICAL SC
QUEEN MARY/WESTFIELD COLL
MILE END RD
LONDON E1 4NS
UK
TEL
TLF
TLX
EML
COM 27,35

PAPARO MARGIT DR
KONKOLY OBSERVATORY
THEGE U 13/17
BOX 67
H 1525 BUDAPEST
HUNGARY
TEL 1 75 5866/75 4122
TLF
TLX 227460
EML
COM 27

PAPATHANASOGLOU D DR
DPT OF ASTRONOMY
NTL UNIVERSITY OF ATHENS
PANEPISTIMIOPOLIS
GR 157 71 ZOGRAFOS
GREECE
TEL 1 724 3414
TLF
TLX
EML
COM 12

PAPAYANNOPOULOS TH DR
DPT OF ASTRONOMY
NTL UNIVERSITY OF ATHENS
PANEPISTIMIOPOLIS
GR 157 83 ZOGRAFOS
GREECE
TEL 1 724 3414
TLF
TLX 223815 UNIV GR
EML SPM25@GRATHUN1
COM 28,33

PAPOUSEK JIRI
DPT OF ASTROPHYSICS
PURKYNE UNIVERSITY
KOTLARSKA 2
CS 611 37 BRNO
CZECHOSLOVAKIA
TEL 5 1112
TLF
TLX
EML
COM 27

PAQUET PAUL EG DR
OBSERVATOIRE ROYAL DE
BELGIQUE
AVE CIRCULAIRE 3
B 1180 BRUSSELS
BELGIUM
TEL 2 373 0211
TLF 2 374 9822
TLX 21565 OBSBEL
EML
COM 19,31C

PARCELIER PIERRE DR
OBSERVATOIRE DE PARIS
61 AVE OBSERVATOIRE
F 75014 PARIS
FRANCE
TEL 1 43 20 1210
TLF
TLX 270776 OBS F
EML
COM

PAREDES JOSE MARIA DR
DPT ASTRONOMIA
UNIVERSIDAD DE BARCELONA
AVD DIAGONAL 647
E 08028 BARCELONA
SPAIN
TEL 3 330 7311
TLF
TLX
EML D3FAJPPO @ EBOUBO11
COM 40

PARESCE FRANCESCO DR
STSCI
HOMEWOOD CAMPUS
3700 SAN MARTIN DR
BALTIMORE MD 21218
USA
TEL
TLF
TLX
EML
COM 21,49

PARIJSKIJ N N PROF
INST PHYSICS OF THE EARTH
ACADEMY OF SCIENCES
GRUZINSKAYA 10
123342 MOSCOW
RUSSIA
TEL 252-07-21
TLF
TLX 411196 IFZAN US
EML
COM

PARIJSKIJ YU N DR
PULKOVO OBSERVATORY
ACADEMY OF SCIENCES
10 KUTUZOV QUAY
196140 ST PETERSBURG
RUSSIA
TEL 2979452
TLF
TLX
EML
COM 40,51

PARISE RONALD A DR
NASA/GSFC
CODE 684 9
GREENBELT MD 20771
USA
TEL 301 286 3896
TLF
TLX
EML SPAN UIT PARISE
COM

PARISOT JEAN-PAUL
OBSERVATOIRE DE BORDEAUX
BP 89
F 33270 FLOIRAC
FRANCE
TEL 56 86 4330
TLF 56 40 4251
TLX
EML EARN <userid>@FROBOR51
COM 15,46

PARK HONG SUH DR
KOREAN NTL UNIVERSITY
OF EDUCATION
CHUNGWON-GUN
CHOONGBOOK 363 691
KOREA R
TEL 431 60 3903
TLF
TLX
EML
COM 42

PARK MYEONG-GU DR
DPT ASTRON & METEOROLOGY
COLLEGE OF NTL SCIENCES
KYUNGPOOK NTL UNIVERSITY
TAEGU 702 701
KOREA R
TEL 53 950 6364
TLF 53 957 0431
TLX
EML
COM 48

PARK SEOK JAE DR
1444-3 SHIN-LIM DONG
KWANAK KU
SEOUL 151 742
KOREA R
TEL
TLF
TLX
EML
COM

PARKER EDWARD A DR
ELECTRONICS LABORATORY
UNIVERSITY OF KENT
CANTERBURY CT2 7NT
UK
TEL 227 668 22
TLF
TLX 965449
EML
COM 40

PARKER EUGENE N
ASTROPHYS & SPACE RES LAB
UNIVERSITY OF CHICAGO
933 E 56TH ST
CHICAGO IL 60637
USA
TEL 312 962 7847
TLF
TLX 910-221-5617
EML
COM 34,48,49

PARKER QUENTIN DR
ROYAL OBSERVATORY
BLACKFORD HILL
EDINBURGH EH9 3HJ
UK
TEL 316 688 379
TLX 72383 ROEDIN G
EML
TLF
COM 28

PARKER ROBERT A R
NASA/JOHNSON SPACE CENTER
CODE CB
HOUSTON TX 77058
USA
TEL 713 483 2221
TLF
TLX
EML
COM

PARKINSON JOHN H DR
MULLARD SPACE SCIENCE LAB
UNIVERSITY COLLEGE LONDON
HOLMBURY ST MARY
DORKING SURREY RH5 6NT
UK
TEL 483 274111
TLF 483 287312
TLX 859185 UCMSSL G
EML
COM 10,44,48

PARKINSON TRUMAN DR
KITT PEAK NTL OBS
BOX 26732
950 N CHERRY AVE
TUCSON AZ 85726 6732
USA
TEL
TLF
TLX
EML
COM

PARKINSON WILLIAM H DR
CENTER FOR ASTROPHYSICS
HCO/SAO MS 50
60 GARDEN ST
CAMBRIDGE MA 02138
USA
TEL 617 495 4865
TLF 617 495 7052
TLX 921428
EML PARKINSON@CFA
COM 10,12,14VP,44

PARMA PAOLA
IST DI RADIOASTRONOMIA
CNR
VIA IRNERIO 46
I 40126 BOLOGNA
ITALY
TEL 51 23 2856
TLF
TLX
EML
COM 40

PARRISH ALLAN DR
STATE UNIV OF NEW YORK
UNIV OF MASSACHUSETTS
GRC 6191
AMHERST MA 01003
USA
TEL
TLF
TLX
EML
COM 40

PARSAMYAN ELMA S DR
BYURAKAN ASTROPHYSICAL
OBSERVATORY
378433 ARMENIA
ARMENIA
TEL 88 52 28 3453
TLF
TLX
EML
COM 27,37

PARSONS SIDNEY B DR
STSCI
HOMEWOOD CAMPUS
3700 SAN MARTIN DR
BALTIMORE MD 21218
USA
TEL 301 338 4807
TLF
TLX
EML
COM 29,45

PARTHASARATHY M DR
INDIAN INSTITUTE OF
ASTROPHYSICS
KORAMANGALA
BANGALORE 560 034
INDIA
TEL 812 56 6585/6497
TLF
TLX 845763 IIAB IN
EML
COM 27,29,42

PARTRIDGE ROBERT B PROF
HAVERFORD COLLEGE

HAVERFORD PA 19041
USA
TEL 215 896 1144
TLF.
TLX
EML
COM 47P

PASACHOFF JAY M PROF
HOPKINS OBSERVATORY
WILLIAMS COLLEGE
WILLIAMSTOWN MA 01267
USA
TEL 413 597 2105
TLF
TLX
EML BITNET PASACHOFF@WILLIAMS
COM 10,12C,40,46C

PASCOAL ANTONIO J B SCI
OBSERVATORIO ASTRONOMICO
UNIVERSIDADE DO PORTO
MONTE DA VIRGEM
P 4400 VILA NOVA DE GAIA
PORTUGAL
TEL 7820404
TLF
TLX
EML
COM

PASCU DAN DR
US NAVAL OBSERVATORY
34 & MASSACHUSETTS AVE NW
WASHINGTON DC 20392 5100
USA
TEL 202 653 1178
TLF
TLX
EML
COM 20,24

PASIAN FABIO
OAT
BOX SUCC TRIESTE 5
VIA TIEPOLO 11
I 34131 TRIESTE
ITALY
TEL 40 76 8005
TLF
TLX 461137 OAT I
EML
COM 09

PASINETTI LAURA E PROF
DPT DI FISICA
UNIVERSITA DI MILANO
VIA CELORIA 16
I 20133 MILANO
ITALY
TEL 2 239 2275*272
TLF 2 236 6583
TLX 334687 INFN MI
EML PASINETTI@MILANO INFN IT
COM 05,29,36,45,51

PASTORI LIVIO
OSS ASTRONOMICO DI MILANO
VIA E BIANCHI 46
I 22055 MERATE
ITALY
TEL 59 2035
TLX
EML
TLF
COM

PASTORIZA MIRIANI G DR
INSTITUTO DE FISICA
UFRGS
CP 15051
90000 PORTO ALEGRE RS
BRAZIL
TEL 512 36 4677
TLF·
TLX 0511055 UFRS BR
EML
COM 28

PATERNO LUCIO PROF
OSS ASTROFISICO
CITTA UNIVERSITARIA
VIA A DORIA 6
I 95125 CATANIA
ITALY
TEL 95 33 0533
TLF
TLX 970359 ASTRCT I
EML
COM 10,27

PATERSON-BEECKMANS F
VINCENT VAN GOGHLAAN 19
NL 2343 RH OEGSTGEEST
NETHERLANDS
TEL 71 17 0829
TLF
EML
COM 29
TLX

PATHRIA RAJ K PROF
DPT OF PHYSICS
UNIVERSITY OF WATERLOO
WATERLOO ON N2L 3G1
CANADA
TEL 519 885 1211
TLF. 519 746 8115
TLX 069 55259
EML
COM

PATI A K
INDIAN INSTITUTE OF
ASTROPHYSICS
KORAMANGALA
BANGALORE 560 034
INDIA
TEL 812 56 9702
TLF
TLX 0845 2763
EML
COM

PATKOS LASZLO DR
KONKOLY OBSERVATORY
THEGE U 13/17
BOX 67
H 1525 BUDAPEST
HUNGARY
TEL 1 75 5866/75 4122
TLF
TLX· 227460
EML
COM 42

PATNAIK ALOK DR
NRAL
JODRELL BANK
MACCLESFIELD SK11 9DL
UK
TEL 477 713 21
TLX
EML ALOK@UK AC MAN JB STAR
TLF 477 716 18
COM

PATRIARCHI PATRIZIO DR
OSS ASTROFISICO
DI ARCETRI
LARGO E FERMI 5
I 50125 FIRENZE
ITALY
TEL 55 275 2282
TLF
TLX 572268 ARCETR I
EML.
COM

PATUREL GEORGES
OBSERVATOIRE DE LYON
AVE CHARLES ANDRE
F 69561 S GENIS LAVAL CDX
FRANCE
TEL 78 56 0705
TLF 72 39 9791
TLX 310926
EML
COM 05,28

PAUL JACQUES DR
CEA CEN
DAPNIA/SAP
BP 2
F 91191 GIF/YVETTE CDX
FRANCE
TEL 1 69 08 4462
TLF
TLX
EML
COM

PAULDRACH ADALBERT W A DR
UNIVERSITAETS STERNWARTE
SCHEINERSTR 1
D 8000 MUENCHEN 80
GERMANY
TEL 89 922 09436
TLF
TLX
EML
COM

PAULINY TOTH IVAN K K DR
MPI FUER RADIOASTRONOMIE
AUF DEM HUEGEL 69
D 5300 BONN 1
GERMANY
TEL 228 52 5243
TLX 886440
EML
TLF
COM 40,48

PAULS THOMAS ALBERT DR
NAVAL RESEARCH LABORATORY
CODE 4130
4555 OVERLOOK AVE SW
WASHINGTON DC 20375 5000
USA
TEL
TLF
TLX
EML
COM 33,34,40

PAUWELS T DR
OBSERVATOIRE ROYAL DE
BELGIQUE
AVE CIRCULAIRE 3
B 1180 BRUSSELS
BELGIUM
TEL 2 375 2484
TLF 2 374 9822
TLX 21565 OBS BEL
EML THIERRY@ASTRO OMA BE
COM 07,20

PAVLOVSKAYA E D DR
STERNBERG STATE ASTR INST
117234 MOSCOW
RUSSIA
TEL
TLF
EML
COM 33
TLX

PAVLOVSKI KRESIMIR
HVAR OBSERVATORY
FACULTY OF GEODESY
KACICEVA 26
YU 41000 ZAGREB
YUGOSLAVIA
TEL 41 442 600
TLF
TLX
EML
COM 27,42

PAXTON HAROLD J B R
ROYAL GREENWICH OBS
HERSTMONCEUX CASTLE
HAILSHAM BN27 1RP
UK
TEL
TLF
TLX
EML
COM

PAYNE DAVID G
JPL
MS 264 748
4800 OAK GROVE DR
PASADENA CA 91109
USA
TEL
TLF
TLX
EML
COM 40

PEACH GILLIAN DR
DPT PHYSICS & ASTRONOMY
UNIVERSITY COLLEGE LONDON
GOWER ST
LONDON WC1E 6BT
UK
TEL 1 713 877 050
TLF
TLX 28722
EML
COM 14

PEACH JOHN V DR
DPT OF ASTROPHYSICS
UNIVERSITY OF OXFORD
SOUTH PARKS RD
OXFORD OX1 3RQ
UK
TEL 865 511 336
TLF
TLX
EML
COM

PEACOCK ANTHONY DR
ESA/ESTEC
SSD
BOX 299
NL 2200 AG NOORDWIJK
NETHERLANDS
TEL 17 19 83563
TLF 17 19 17400
TLX 39098
EML
COM 44

PEACOCK JOHN ANDREW
ROYAL OBSERVATORY
BLACKFORD HILL
EDINBURGH EH9 3HJ
UK
TEL 316 673 321
TLF
TLX 72383 ROEDIN G
EML
COM 47

PEALE STANTON J PROF
DPT OF PHYSICS
UNIVERSITY OF CALIFORNIA
SANTA BARBARA CA 93106
USA
TEL 805 961 2977
TLF
TLX.
EML
COM 07C

PEARCE GILLIAN DR
DPT OF ASTROPHYSICS
UNIVERSITY OF OXFORD
KEBLE RD
OXFORD OX1 3RH
UK
TEL 865 273 297
TLF
TLX 83295 NUCLOX G
EML
COM 35

PEARSON TIMOTHY J
CALTECH
MS 105 24
OWENS VALLEY RADIO OBS
PASADENA CA 91125
USA
TEL 818 356 4980
TLF
TLX 675425
EML
COM 40

PECINA PETR
ASTRONOMICAL INSTITUTE
CZECH ACADEMY OF SCIENCES
ONDREJOV OBSERVATORY
CS 251 65 ONDREJOV
CZECHOSLOVAKIA
TEL 204 85201
TLF 204 85314
TLX 122486
EML
COM 22

PECKER JEAN-CLAUDE PROF
COLLEGE DE FRANCE
3 RUE D'ULM
F 75331 PARIS CDX 05
FRANCE
TEL 1 44 27 1695
TLF
TLX
EML
COM 05,12,34,36,47

PEDERSEN BENT M DR
OBSERVATOIRE DE PARIS
SECTION DE MEUDON
F 92195 MEUDON PPL CDX
FRANCE
TEL 1 45 07 7809
TLF
TLX
EML
COM 10

PEDERSEN HOLGER DR
UNIVERSITY OBSERVATORY
OESTER VOLDGADE 3
DK 1350 COPENHAGEN K
DENMARK
TEL 31 14 1790
TLF 31 38 9157
TLX
EML
COM

PEDERSEN OLAF PROF
HISTORY OF SCIENCE INST
UNIVERSITY OF AARHUS
NY MUNKEGADE
DK 8000 AARHUS C
DENMARK
TEL 86 12 7188
TLF 86 20 2711
TLX
EML
COM 41

PEDLAR ALAN DR
NRAL
JODRELL BANK
MACCLESFIELD SK11 9DL
UK
TEL
TLX
EML
TLF
COM 40

PEDOUSSAUT ANDRE
OBS MIDI PYRENEES
14 AVE E BELIN
F 31400 TOULOUSE CDX
FRANCE
TEL 61 25 2101
TLF
TLX 530776 OBSTLSE
EML
COM· 29,30

PEDREROS MARIO DR
DPT OF PHYSICS
ST MARY'S UNIVERSITY
HALIFAX NS B3H 3C3
CANADA
TEL 902 420 5640
TLF 902 420 5561
TLX
EML
COM 25,37

PEEBLES P JAMES E
DPT OF PHYSICS
PRINCETON UNIVERSITY
JADWIN HALL
PRINCETON NJ 08544
USA
TEL 609 452 4386
TLF
TLX 499-3512
EML
COM 47

PEERY BENJAMIN F PROF
DPT PHYSICS & ASTRONOMY
HOWARD UNIVERSITY
WASHINGTON DC 20059
USA
TEL 202 636 6267
TLF
TLX
EML
COM 29

PEIMBERT MANUEL DR
INSTITUTO DE ASTRONOMIA
UNAM
APDO POSTAL 70-264
04510 MEXICO DF
MEXICO
TEL 905-548-5306
TLF
TLX 01760155 CIMCE
EML
COM 28,33,34

PEKERIS CHAIM LEIB PROF
DPT OF APPLIED MATHS
WEIZMANN INSTITUTE OF SCI
BOX 26
REHOVOT 76100
ISRAEL
TEL 8 483 292
TLF
TLX 361900
EML
COM

PEKUENLUE E RENNAN DR
FACULTY OF SCIENCE
EGE UNIVERSITY
BOX 21
35100 BORNOVA IZMIR
TURKEY
TEL 222295
TLF
TLX
EML
COM

PEL JAN WILLEM DR
KAPTEYN STERREWACHT
WERKGROEP
MENSINGHEWEG 20
NL 9301 KA RODEN
NETHERLANDS
TEL·
TLF·
TLX 53767 KSW RO NL
EML
COM 25

PELLAS PAUL DR
LABORATOIRE MINERALOGIE
61 RUE BUFFON
F 75005 PARIS
FRANCE
TEL 1 47 07 2824
TLF
TLX
EML
COM 15

PELLEGRINI PAULO S S DR
OBSERVATORIO NACIONAL
RUA GL JOSE CRISTINO 77
SAO CRISTOVAO
20921 RIO DE JANEIRO RJ
BRAZIL
TEL 21 580 3683
TLF 21 580 0332
TLX 21288
EML
COM 30

PELLERIN JR CHARLES J DR
NASA HEADQUARTERS
CODE EZ
600 INDEPENDENCE AVE SW
WASHINGTON DC 20546
USA
TEL 202 453 1437
TLF
TLX·
EML
COM

PELLET ANDRE
OBSERVATOIRE DE MARSEILLE
2 PLACE LE VERRIER
F 13248 MARSEILLE CDX 04
FRANCE
TEL 91 95 9088
TLF·
TLX 420241 F
EML
COM

PELLETIER GUY DR PR
OBSERVATOIRE DE GRENOBLE
CERMO/ASTROPHYSIQUE
BP 53X
F 38041 S MARTIN HERES CD
FRANCE
TEL 76 51 4570
TLF. 76 44 8821
TLX
EML
COM

PENA JOSE
INSTITUTO DE ASTRONOMIA
UNAM
APDO POSTAL 70-264
04510 MEXICO DF
MEXICO
TEL 905-5484537
TLF (905)5483712
TLX
EML BITNET PENAS@UNAMVM1
COM

PENA MIRIAM DR
INSTITUTO DE ASTRONOMIA
UNAM
APDO POSTAL 70-264
04510 MEXICO DF
MEXICO
TEL· 525 548 5305
TLF 525 548 3712
TLX
EML MIRIAM@ALFA ASTROSCU UNAH
COM 34

PENG QIU-HE
DPT OF ASTRONOMY
NANJING UNIVERSITY
NANJING
CHINA PR
TEL 25 34651*2882
TLF
TLX 34151 PRCNU CN
EML
COM 28,48

PENG YUN-LOU
DPT OF ASTRONOMY
NANJING UNIVERSITY
NANJING
CHINA PR
TEL 25 37551*2882
TLF
TLX 34151 PRCNU CN
EML
COM 40

PENICHE ROSARIO DR
INSTITUTO DE ASTRONOMIA
UNAM
APDO POSTAL 70-264
04510 MEXICO DF
MEXICO
TEL 905-5484537
TLF (905)5483712
TLX
EML BITNET PENAS @ UNAMVM1
COM

PENNY ALAN JOHN DR
RUTHERFORD APPLETON LAB
SPACE & ASTROPHYSICS DIV
BLDG R25/R68
CHILTON DIDCOT OX11 0QX
UK
TEL 235 219 00
TLF
TLX 83159 RUTHBL G
EML
COM 09,25,37

PENSADO JOSE DR
OBS ASTRONOMICO NCL
ALFONSO X11-5
E 28014 MADRID
SPAIN
TEL 1 227 0107
TLX
EML
TLF.
COM

PENSTON MARGARET
ROYAL GREENWICH OBS
MADINGLEY RD
CAMBRIDGE CB3 0EZ
UK
TEL 223 37 4000
TLF 223 37 4700
TLX
EML
COM

PENZIAS ARNO A DR
AT&T BELL LABORATORIES
RM 6A-409
600 MOUNTAIN AVENUE
MURRAY HILL NJ 07974
USA
TEL 201 582 3361
TLF
TLX 13-8650 OR 219348
EML
COM 34,40,47

PEQUIGNOT DANIEL
OBSERVATOIRE DE PARIS
SECTION DE MEUDON
DAF
F 92195 MEUDON PPL CDX
FRANCE
TEL 1 45 07 7438
TLF
TLX 201571
EML
COM 34

PERAIAH ANNAMANENI DR
INDIAN INSTITUTE OF
ASTROPHYSICS
KORAMANGALA
BANGALORE 560 034
INDIA
TEL 812 56 6585/6497
TLF
TLX 845763 IIAB IN
EML
COM 36C

PERAULT MICHEL
RADIOASTRONOMIE ENS
24 RUE LHOMOND
F 75231 PARIS CDX 05
FRANCE
TEL 1 43 29 1225
TLF 1 45 87 3489
TLX 202601
EML DECNET IAPOBS PERAULT
COM 34

PERCY JOHN R PROF
ERINDALE COLLEGE
UNIVERSITY OF TORONTO
DIVISION OF SCIENCE
MISSISSAUGA ON L5L 1C6
CANADA
TEL 416 828 5343
TLF 416 828 5328
TLX
EML PERCY@UTORPHYS
COM 27P,46VP

PERDANG JEAN M DR
INSTITUT D'ASTROPHYSIQUE
UNIVERSITE DE LIEGE
AVE COINTE 5
B 4000 COINTE-LIEGE
BELGIUM
TEL 41 52 9980
TLF 41 52 7474
TLX 41264 ASTRLG B
EML
COM

PERDOMO RAUL
UNIV NACIONAL DE LA PLATA
FCAG
1900 LA PLATA (BS AS)
ARGENTINA
TEL 21 38 810
TLF
TLX 31151 BULAP
EML
COM

PEREA-DUARTE JAIME D DR
INST ASTROFISICA
DE ANDALUCIA APD 2144
PROFESOR ALBAREDA 1
E 18080 GRANADA
SPAIN
TEL 58 12 1311
TLF
TLX 78573 IAAG E
EML JAIME@IAA ES
COM 28

PEREK LUBOS DR
ASTRONOMICAL INSTITUTE
CZECH ACADEMY SCIENCES
BUDECSKA 6
CS 120 23 PRAHA 2
CZECHOSLOVAKIA
TEL 2 25 4234
TLF
TLX 122486
EML
COM 33,51

PERES GIOVANNI DR
OSS ASTROFISICO
CITTA UNIVERSITARIA
VIA A DORIA 6
I 95125 CATANIA
ITALY
TEL 91 33 0533
TLF 91 33 0592
TLX 970359 ASTRCT I
EML GIANNIP@ASTRCT
COM 10

PEREZ ENRIQUE DR
INST DE ASTROFISICA
DE CANARIAS
OBS DEL TEIDE
E 38200 LA LAGUNA
SPAIN
TEL 22 26 2211
TLF 22 26 3005
TLX 92640 IAC E
EML IAC EPS/EPJ@IAC ES
COM 28

PEREZ FOURNON ISMAEL DR
INST DE ASTROFISICA
DE CANARIAS
OBS DEL TEIDE
E 38200 LA LAGUNA
SPAIN
TEL 22 26 2211
TLF· 22 26 3005
TLX 92640 IAC E
EML
COM· 28,40

PEREZ-DE-TEJADA H A DR
INSTITUTO DE GEOFISICA
UNAM
APDO POSTAL 877
22860 ENSENADA B CALIF
MEXICO
TEL 706-674-0601
TLF
TLX
EML
COM· 15

PEREZ-PERAZA JORGE DR
INAOE
TONANTZINTLAZ
APDO POSTAL 216 y 51
72000 PUEBLA PUE
MEXICO
TEL 47-04-19
TLF
TLX
EML·
COM·

PERINOTTO MARIO PROF
DPT DI ASTRONOMIA
UNIVERSITA DI FIRENZE
LARGO E FERMI 5
I 50125 FIRENZE
ITALY
TEL 55 27521
TLF 55 22 0039
TLX 572268 ARCETR
EML
COM 34

PERKINS FRANCIS W DR
PLASMA PHYSICS LAB
PRINCETON UNIVERSITY
BOX 451
PRINCETON NJ 08544
USA
TEL 609 683 2603
TLF
TLX 5106852399
EML
COM 49C

PERLEY RICHARD ALAN
NRAO
BOX 0
SOCORRO NM 87801 0387
USA
TEL 505 772 4011
TLF
TLX 910-988-1710
EML
COM 40

PEROLA GIUSEPPE C DR
ISTITUTO ASTRONOMICO
UNIVERSITA DI ROMA
VIA G M LANCISI 29
I 00161 ROMA
ITALY
TEL 6 86 7525
TLF
TLX 613255 INFNRO
EML
COM 48

PERRIER CHRISTIAN DR
OBSERVATOIRE DE LYON
AVE CHARLES ANDRE
F 69561 S GENIS LAVAL CDX
FRANCE
TEL 78 56 0705
TLF 72 39 9791
TLX 310926 OBSLYON F
EML EARN perrier@frgag51
COM

PERRIN JEAN MARIE DR
LAS
TRAVERSE DU SIPHON
LES TROIS LUCS
F 13012 MARSEILLE
FRANCE
TEL 91 05 5900
TLF 91 66 1855
TLX 420 584 E ASTROSP
EML BITNET PERRIN@FRLASM51
COM 21

PERRIN MARIE-NOEL DR
OBSERVATOIRE DE PARIS
61 AVE OBSERVATOIRE
F 75014 PARIS
FRANCE
TEL 1 40 51 2245
TLF
TLX 270776 OBS F
EML
COM 29

PERRY CHARLES L DR
DPT PHYSICS & ASTRONOMY
LOUISIANA STATE UNIV
BATON ROUGE LA 70803 4001
USA
TEL 504 388 8287
TLF
TLX 559184
EML·
COM 25,30,33,45

PERRY JUDITH J DR
INSTITUTE OF ASTRONOMY
THE OBSERVATORIES
MADINGLEY RD
CAMBRIDGE CB3 OHA
UK
TEL 223 622 04
TLF.
TLX 817297
EML
COM

PERRY PETER M DR
COMPUTER SCIENCES CORP
SYSTEM SCIENCES DIV
8728 COLESVILLE RD
SILVER SPRING MD 20910
USA
TEL 301 589 1545
TLF
TLX
EML
COM 44

PERRYMAN MICHAEL A C DR
ESA/ESTEC
SSD
BOX 299
NL 2200 AG NOORDWIJK
NETHERLANDS
TEL 17 19 83615
TLF 17 19 84690
TLX 39098
EML ESTCS1 MPERRRMA
COM 08,09,24,47

PERSI PAOLO
IAS
CNR
CP 67
I 00044 FRASCATI
ITALY
TEL 6 942 5655
TLF
TLX 610261 CNR FRA
EML
COM 34

PERSIDES SOTIRIOS C
DPT OF ASTRONOMY
UNIVERSITY THESSALONIKI
GR 540 06 THESSALONIKI
GREECE
TEL 31 99 1357
TLF
TLX
EML
COM 47

PESCH PETER DR
NTL SCIENCE FOUNDATION
DIV ASTRONOMICAL SCIENCES
1800 G ST NW
WASHINGTON DC 20550
USA
TEL 202 357 7622
TLF
TLX
EML
COM 33

PESEK RUDOLPH PROF
ASTRONOMICAL INSTITUTE
CZECH ACADEMY SCIENCES
BUDECSKA 6
CS 120 23 PRAHA 2
CZECHOSLOVAKIA
TEL
TLF
TLX
EML
COM 51

PETERS GERALDINE JOAN DR
SPACE SCIENCES CENTER
UNIV SOUTHERN CALIFORNIA
UNIVERSITY PARK
LOS ANGELES CA 90089 1341
USA
TEL 213 743 6962
TLF
TLX 4720490 USC LSA
EML
COM 29,36,42,44

PETERS WILLIAM L III DR
ASTRONOMY DPT
UNIVERSITY OF TEXAS
RLM 15 308
AUSTIN TX 78712 1083
USA
TEL
TLF
TLX
EML·
COM 28,34,40

PETERSEN J O DR
UNIVERSITY OBSERVATORY
OESTER VOLDGADE 3
DK 1350 COPENHAGEN K
DENMARK
TEL 31 14 1790
TLF 31 38 9157
TLX
EML
COM 27

PETERSON BRADLEY MICHAEL
DPT OF ASTRONOMY
OHIO STATE UNIVERSITY
174 W 18TH AVE
COLUMBUS OH 43210 1106
USA
TEL 614 422 7886
TLF
TLX
EML
COM 28

PETERSON BRUCE A DR
MOUNT STROMLO & SIDING
SPRING OBSERVATORIES
PRIVATE BAG
WODEN PO ACT 2606
AUSTRALIA
TEL 62 88 1111
TLF
TLX 62270
EML
COM 47,48

PETERSON CHARLES JOHN DR
DPT PHYSICS & ASTRONOMY
UNIVERSITY OF MISSOURI
223 PHYSICS BLDG
COLUMBIA MO 65211
USA
TEL 314 882 3217
TLF
TLX
EML
COM 28,37,41

PETERSON DEANE M DR
DPT OF EARTH & SPACE SCI
ASTRONOMY PROGRAM
SUNY AT STONY BROOK
STONY BROOK NY 11794 2100
USA
TEL 516 632 8223
TLF 516 632 6240
TLX
EML DPETERSON@ASTRO SUNYSB EDU
COM 26

PETERSON LAURENCE E PROF
CASS
UCSD
C 011
LA JOLLA CA 92093 0216
USA
TEL 619 452 3461
TLF
TLX 910-337-1271 SIOCEAN
EML
COM 44,48

PETERSON RUTH CAROL DR
607 MARION PL
PALO ALTO CA 94301
USA
TEL 415 321 1281
TLF
TLX
EML
COM 29,30

PETFORD A DAVID DR
DPT OF ASTROPHYSICS
UNIVERSITY OF OXFORD
SOUTH PARKS RD
OXFORD OX1 3RQ
UK
TEL 865 511 336
TLF
TLX
EML
COM 09

PETHICK CHRISTOPHER J DR
NORDITA
BLEGDAMSVEJ 17
DK 2100 COPENHAGEN O
DENMARK
TEL 31 42 1616
TLF 31 38 9157
TLX 15216 NBIOK
EML PETHICK@NBIVAX NBI DK
COM 48

PETIT GERARD DR
BIPM
PAVILLON DE BRETEUIL
F 92312 SEVRES CDX
FRANCE
TEL 1 45 07 7067
TLF 1 45 34 2021
TLX 631351
EML BIPM@FRMEU51
COM 31

PETIT JEAN-MARC DR
OCA OBSERV DE NICE
BP 139
F 06003 NICE CDX
FRANCE
TEL 92 00 3089
TLF 92 00 3033
TLX
EML PETIT@FRONI51
COM 07,16

PETON ALAIN DR
OBSERVATOIRE DE MARSEILLE
2 PLACE LE VERRIER
F 13248 MARSEILLE CDX 04
FRANCE
TEL 91 95 9088
TLF
TLX
EML
COM

PETRI WINFRIED PROF DR
UNTERLEITEN 2
POSTFACH 106
D 8162 SCHLIERSEE
GERMANY
TEL 80 266 428
TLF
TLX
EML
COM 41

PETRINI DANIEL DR
OCA OBSERV DE NICE
BP 139
F 06003 NICE CDX
FRANCE
TEL 93 89 0420
TLF
TLX 460004 OBSNICE F
EML
COM 14

PETRO LARRY DAVID
STSCI
HOMEWOOD CAMPUS
3700 SAN MARTIN DR
BALTIMORE MD 21218
USA
TEL 301 338 4501
TLF
TLX 6849101
EML
COM

PETROPOULOS BASIL CH DR
RES CENTER FOR ASTRONOMY
ACADEMY OF ATHENS
14 ANAGNOSTOPOULOU ST
GR 106 73 ATHENS
GREECE
TEL 1 361 3589
TLF
TLX
EML
COM 14,16

PETROSIAN ARTASHES R DR
BYURAKAN ASTROPHYSICAL
OBSERVATORY
378433 BYURAKAN
ARMENIA
TEL 88 52 28 3453
TLF 88 52 28 4142
TLX 243344 ORION SU
EML
COM 28

PETROSIAN VAHE PROF
CTR FOR SPACE SCIENCES &
ASTROPHYSICS
STANFORD UNIV ERL 304
STANFORD CA 94305 4055
USA
TEL 415 497 1435
TLF
TLX
EML
COM 10,34,47,48

PETROV G M DR
NIKOLAEV OBSERVATORY
UKRAINIAN ACAD OF SCIENCE
327000 NIKOLAEV
UKRAINE
TEL 36 1824
TLF
TLX
EML
COM 08

PETROV GENNADIJ M
INST OF RADIO & ELECTRON
ACADEMY OF SCIENCES
MARKS AVENJU 18
103907 MOSCOW
RUSSIA
TEL
TLF
TLX
EML
COM

PETROV GEORGIJ I PROF DR
SPACE RESEARCH INSTITUTE
ACADEMY OF SCIENCES
PROFSOJOSNAYA UL 84/32
117810 MOSCOW
RUSSIA
TEL
TLF
TLX
EML
COM

PETROV GEORGY TRENDAFILOV
DPT OF ASTRONOMY
BULGARIAN ACAD SCIENCES
72 LENIN BLVD
BG 1784 SOFIA
BULGARIA
TEL 2 75 8927
TLF
TLX 23561 ECF BAN BG
EML
COM 28

PETROV NIKOLAI
ASTRONOMICAL OBSERVATORY
BOX 120
BG 9000 VARNA
BULGARIA
TEL. 52 22 2890
TLF
TLX
EML
COM

PETROV PETER P DR
CRIMEAN ASTROPHYS OBS
UKRAINIAN ACAD OF SCIENCE
NAUCHNY
334413 CRIMEA
UKRAINE
TEL 43 2945
TLF
TLX
EML
COM 09,27

PETROVSKAYA M S DR
INST OF THEORET ASTRONOMY
ACADEMY OF SCIENCES
N KUTUZOVA 10
191187 ST PETERSBURG
RUSSIA
TEL 121578 ITA SU
TLF
TLX
EML
COM 07,37

PETTENGILL GORDON H PROF
CENTER FOR SPACE RESEARCH
MIT 37 241
BOX 165
CAMBRIDGE MA 02139
USA
TEL 617 253 7501
TLF
TLX 92-1473
EML
COM 16,40

PETTERSEN BJOERN RAGNVALD
INST THEORET ASTROPHYSICS
UNIVERSITY OF OSLO
BOX 1029
N 0315 BLINDERN OSLO 3
NORWAY
TEL 02-45-65-01
TLF
TLX 72705 ASTRO N
EML
COM 27

PETTINI MARCO
OSS ASTROFISICO
DI ARCETRI
LARGO E FERMI 5
I 50125 FIRENZE
ITALY
TEL· 55 275 2282
TLF
TLX 572268 ARCETR
EML
COM 14

PETTINI MAX
ROYAL GREENWICH OBS
HERSTMONCEUX CASTLE
HAILSHAM BN27 1RP
UK
TEL 323 833 171
TLX 87451 RGOBSY G
EML
TLF
COM

PEYTURAUX ROGER H PROF
INSTITUT D'ASTROPHYSIQUE
98BIS BD ARAGO
F 75014 PARIS
FRANCE
TEL 1 43 20 1425
TLF 1 43 29 8673
TLX
EML
COM 12

PFAU WERNER
UNIVERSITY OBSERVATORY
SCHILLERGAESSCHEN 2
D 6900 JENA
GERMANY
TEL 78 588 61347
TLF
TLX
EML
COM 25

PFEIFFER RAYMOND J
8 BARBARA LANE
TITUSVILLE NJ 08560
USA
TEL 609 883 4612
TLF
EML
COM 25
TLX

PFENNIG HANS H DR
MPI F PHYSIK & ASTROPHYS
KARL-SCHWARZSCHILD-STR 1
D 8046 GARCHING MUENCHEN
GERMANY
TEL 89 329 99435
TLX 524629 ASTRO D
EML
TLF
COM 14

PFENNIGER DANIEL DR
OBSERVATOIRE DE GENEVE
CHEMIN DES MAILLETTES 51
CH 1290 SAUVERNY
SWITZERLAND
TEL 22 755 2611
TLF 22 755 3983
TLX 419209 OBS CH
EML BITNET PFENNIGE@CGEUGE54
COM 28

PFLEIDERER JORG PROF
INSTITUT FUER ASTRONOMIE

TECHNIKERSTR 15
A 6020 INNSBRUCK
AUSTRIA
TEL 5222 748 5251
TLF
TLX
EML
COM 21

PFLUG KLAUS DR
ZNTRLINST F ASTROPHYSIK
SONNENOBSERVATORIUM
EINSTEINTURM
D 1500 POTSDAM
GERMANY
TEL
TLF
TLX
EML
COM 10,12,49

PHAM-VAN JACQUELINE MME
OCA CERGA
AVE COPERNIC
F 06130 GRASSE
FRANCE
TEL 93 36 5849
TLF
TLX 470865
EML
COM 08

PHILIP A G DAVIS
1125 OXFORD PL
SCHENECTADY NY 12308
USA
TEL 518 374 5636
TLF
TLX
EML AGDP@UNION
COM 05,25,30,33,37,45

PHILLIPS JOHN G PROF
ASTRONOMY DPT
UNIVERSITY OF CALIFORNIA
601 CAMPBELL HALL
BERKELEY CA 94720
USA
TEL 415 642 5275
TLF
TLX
EML
COM 14,36

PHILLIPS JOHN PETER
DPT OF PHYSICS
QUEEN MARY/WESTFIELD COLL
MILE END RD
LONDON E1 4NS
UK
TEL 1 980 4811
TLF
TLX 893750 QMEUOL G
EML.
COM 34

PHILLIPS KENNETH J H
RUTHERFORD APPLETON LAB
SPACE & ASTROPHYSICS DIV
BLDG R25/R68
CHILTON DIDCOT OX11 0QX
UK
TEL 235 219 00
TLF
TLX
EML
COM 10,12,44

PHILLIPS MARK M DR
CERRO TOLOLO
INTERAMERICAN OBSERVATORY
CASILLA 603
LA SERENA
CHILE
TEL 51 21 3352
TLF 51 21 2466*342
TLX 620 301 AURA CT
EML
COM 28,35

PHILLIPS THOMAS GOULD DR
CALTECH
MS 320 47
PASADENA CA 91125
USA
TEL 818 356 4278
TLF
TLX
EML
COM 34,40

PIACENTINI RUBEN DR
OBSERVATORIO ASTRONOMICO
DE ROSARIO
CC 606
2000 ROSARIO
ARGENTINA
TEL 41 63084
TLF
TLX 41817 CIROS AR
EML RUBEN%IFIR EDU.AR@UUNET UU NET EML.
COM 14

PIAZZA LILIANA RIZZO
CRAAE/ESCOLA POLITECNICA
UNIVERSIDADE DE SAO PAULO
CP 8174
05508 SAO PAULO SP
BRAZIL
TEL 11 813 3222
TLF 11 815 4272
TLX 1180127 INPE BR
EML.
COM

PICAT JEAN-PIERRE DR
OBS MIDI PYRENEES
14 AVE E BELIN
F 31400 TOULOUSE CDX
FRANCE
TEL 61 25 2101
TLF
TLX
EML
COM 09

PICCA DOMENICO DR
DPT DI FISICA
UNIVERSITA DI BARI
VIA G AMENDOLA 173
I 70123 BARI
ITALY
TEL 80 24 3215
TLF 80 24 2434
TLX
EML
COM 19

PICCIONI ADALBERTO
OSS ASTRONOMICO
UNIVERSITA DI BOLOGNA
CP 596
I 40100 BOLOGNA
ITALY
TEL 51 22 2956
TLF
TLX 211664 INFNBO I
EML
COM 42

PICK MONIQUE DR
OBSERVATOIRE DE PARIS
SECTION DE MEUDON
DASOP
F 92195 MEUDON PPL CDX
FRANCE
TEL 1 45 07 7811
TLF
TLX 200590
EML
COM 10,40

PICKLES ANDREW JOHN DR
INSTITUTE FOR ASTRONOMY
UNIVERSITY OF HAWAII
2680 WOODLAWN DR
HONOLULU HI 96822
USA
TEL 808 948 6756
TLF 808 988 2790
TLX 7238459 UHAST HR
EML pickles@uhifa ifa hawaii edu
COM 28

PIDDINGTON JACK H RES FEL
CSIRO NML

LINDFIELD
SYDNEY NSW 2070
AUSTRALIA
TEL 2 467 6211
TLF
TLX 26296 AA
EML
COM 10,48

PIER JEFFREY R DR
US NAVAL OBSERVATORY
FLAGSTAFF STATION
BOX 1149
FLAGSTAFF AZ 86002
USA
TEL 602 779 5132
TLF
TLX
EML
COM 33

PIERCE A KEITH DR
NTL SOLAR OBSERVATORY
BOX 26732
950 N CHERRY AVE
TUCSON AZ 85726 6732
USA
TEL 602 327 5511
TLF
TLX
EML
COM 07,12

PIERCE DAVID ALLEN DR
7706 WASTLAWN AVE
LOS ANGELES CA 90045
USA
TEL
TLF
TLX
EML
COM 20

PIERRE MARGUERITE DR
ESO
KARL-SCHWARZSCHILD STR 2
D 8046 GARCHING MUENCHEN
GERMANY
TEL 89 320 06293
TLF 89 320 2362
TLX 5282820 EO D
EML BITNET PIERRE@ESOMC1
COM

PIGATTO LUISA DR
OSS ASTRONOMICO DI PADOVA
VIC DELL OSSERVATORIO 5
I 35122 PADOVA
ITALY
TEL 49 66 1499
TLX 432071
EML
TLF
COM 41

PIIROLA VILPPU E DR
OBS & ASTROPHYSICS LAB
UNIVERSITY OF HELSINKI
TAEHTITORNINMAKI
SF 00130 HELSINKI 13
FINLAND
TEL 191 2801
TLF
TLX 124690 UNIH SF
EML
COM 25,27,42

PIKE CHRISTOPHER DAVID
RUTHERFORD APPLETON LAB
SPACE & ASTROPHYSICS DIV
BLDG R25/R68
CHILTON DIDCOT OX11 0QX
UK
TEL 235 219 00
TLF
TLX
EML
COM

PILACHOWSKI CATHERINE DR
KITT PEAK NTL OBS
BOX 26732
950 N CHERRY AVE
TUCSON AZ 85726 6732
USA
TEL 602 327 5511
TLF
TLX 0666484 AURA NOAOTUC
EML EARN "CATYP@NOAO EDU"
COM 29C,37

PILCHER CARL BERNARD DR
4316 ELLICOTT STREET NW
WASHINGTON DC 20016
USA
TEL
TLF
EML
COM 15
TLX

PILKINGTON JOHN D H DR
ROYAL GREENWICH OBS
HERSTMONCEUX CASTLE
HAILSHAM BN27 1RP
UK
TEL 323 841 139
TLX· 87451 RGOBSY G
EML
TLF
COM 19,31

PILLINGER COLIN DR
DPT OF EARTH SCIENCES
THE OPEN UNIVERSITY
WALTON HALL
MILTON KEYNES NK7 6AA
UK
TEL 908 65 2119
TLF 908 65 5910
TLX
EML
COM 15,16,22

PILOWSKI K PROF DR
GEODAETISCHES INSTITUT
TECHNISCHE UNIVERSITAET
NIENBURGER STR 1
D 3000 HANNOVER
GERMANY
TEL
TLF
TLX
EML
COM 08,33

PINEAU DES FORETS G DR
OBSERVATOIRE DE PARIS
SECTION DE MEUDON
DAEC
F 92195 MEUDON PPL CDX
FRANCE
TEL 1 45 07 7454
TLF 1 45 07 7469
TLX 631987
EML MESIOA FORETS/FORETS@FRMEU51
COM 34

PINEAULT SERGE DR
DPT DE PHYSIQUE
UNIVERSITE DE LAVAL
SAINTE-FOY QC G1K 7P4
CANADA
TEL 416 656 3901
TLF 418 656 2040
TLX
EML
COM

PINES DAVID PROF
DPT OF PHYSICS
UNIVERSITY OF ILLINOIS
1110 W GREEN ST
URBANA IL 61801
USA
TEL 217 333 0115
TLF
TLX 9103806599 PHYSICS S
EML
COM 35

PINGREE DAVID PROF
BROWN UNIVERSITY

PO BOX 1900
PROVIDENCE RI 02912
USA
TEL 401 863 2101
TLF
TLX
EML
COM 41

PINIGIN GENNADIJ I DR
NIKOLAEV OBSERVATORY
UKRAINIAN ACAD OF SCIENCE
327001 NIKOLAEV
UKRAINE
TEL
TLF
TLX
EML
COM 08

PINKAU K PROF
MPI FUER PLASMAPHYSIK
D 8046 GARCHING MUENCHEN
GERMANY
TEL 89 329 9342
TLF
EML
COM 44,48
TLX 05-215-808

PINOTSIS ANTONIS D DR
DPT OF ASTRONOMY
NTL UNIVERSITY OF ATHENS
PANEPISTIMIOPOLIS
GR 157 71 ZOGRAFOS
GREECE
TEL 1 724 3414
TLF
TLX·
EML
COM 35

PINTO GIROLAMO PROF
OSS ASTRONOMICO DI PADOVA

VIC DELL OSSERVATORIO 5
I 35122 PADOVA
ITALY
TEL 49 66 1499
TLF
TLX
EML
COM

PINTO PHILIP ALFRED DR
CENTER FOR ASTROPHYSICS
HCO/SAO
60 GARDEN ST
CAMBRIDGE MA 02138
USA
TEL 617 495 7174
TLF
TLX
EML PINTO@CFA HARVARD EDU
COM 36,48

PIOTTO GIAMPAOLLO
DPT DI ASTRONOMIA
UNIVERSITA DI PADOVA
VIC DELL OSSERVATORIO 5
I 35122 PADOVA
ITALY
TEL 49 66 1499
TLF 49 38 919
TLX 432071 ASTROS I
EML
COM

PIPHER JUDITH L
DPT PHYSICS & ASTRONOMY
UNIVERSITY OF ROCHESTER
ROCHESTER NY 14627
USA
TEL 716 275 4402
TLF
TLX
EML
COM 44

PIRO LUIGI DR
IAS
CNR
CP 67
I 00044 FRASCATI
ITALY
TEL 6 942 4589
TLF 6 941 6847
TLX 610261 CNR FRA
EML 40607 PIRO
COM 48

PIRRONELLO VALERIO
OSS ASTROFISICO
CITTA UNIVERSITARIA
VIA A DORIA 6
I 95125 CATANIA
ITALY
TEL 95 33 0533
TLF
TLX 970359 ASTRCT I
EML
COM

PISKUNOV ANATOLY E
INST OF ASTRONOMY
ACADEMY OF SCIENCES
PYATNITSKAYA UL 48
109017 MOSCOW
RUSSIA
TEL 231-54-61
TLF
TLX 412623 SCSTP SU
EML
COM 37

PISKUNOV NIKOLAI E DR
INST OF ASTRONOMY
ACADEMY OF SCIENCES
PYATNITSKAYA UL 48
109017 MOSCOW
RUSSIA
TEL
TLF
TLX
EML
COM 36

PISMIS DE RECILLAS PARIS
INSTITUTO DE ASTRONOMIA
UNAM
APDO POSTAL 70-264
04510 MEXICO DF
MEXICO
TEL 905-548-5306
TLF
TLX 1760155 CICME
EML
COM 28,33,34

PITTICH EDUARD M DR
ASTRONOMICAL INSTITUTE
SLOVAK ACADEMY SCIENCES
DUBRAVSKA 9
CS 842 28 BRATISLAVA
CZECHOSLOVAKIA
TEL. 7 37 5157
TLF
TLX 93373 SEIS
EML
COM 15,20

PITZ ECKHART DR
MPI FUER ASTRONOMIE
KOENIGSTUHL
D 6900 HEIDELBERG 1
GERMANY
TEL 62 215 281
TLF
TLX 461789 MPIA D
EML
COM 21

PIZZELLA G DR
DPT DI FISICA
UNIVERSITA DI ROMA
PA MORO 2
I 00185 ROMA
ITALY
TEL 6 49 40156
TLF
TLX 613255 INFNRO
EML
COM

PIZZICHINI GRAZIELLA
IST TESRE
CNR
VIA DE CASTAGNOLI 1
I 40126 BOLOGNA
ITALY
TEL 51 51 9593
TLF.
TLX 511350 CNR BO
EML
COM. 05,24,45

PLANESAS PERE
CTR ASTRON DE YEBES
OAN
APD 148
E 19080 GUADALAJARA
SPAIN
TEL   11 22 3358
TLF
TLX
EML
COM   40

PLASSARD J DR
KSARA OBSERVATORY
KSARA
LEBANON
TEL
TLF
EML
COM
TLX

PLATAIS IMANT K DR
RADIOASTROPHYSICAL OBS
LATVIAN ACAD OF SCIENCES
TURGENEVA 19
226524 RIGA
LATVIA
TEL   932088
TLF
TLX
EML
COM   24,37

PLAVEC MIREK J PROF
DPT OF ASTRONOMY
UNIVERSITY OF CALIFORNIA
MS 8979
LOS ANGELES CA 90024
USA
TEL   213 825 1672
TLF
TLX
EML
COM   29,35,42

PLAVEC ZDENKA DR
DPT OF ASTRONOMY
UNIVERSITY OF CALIFORNIA
405 HILGARD AVE
LOS ANGELES CA 90024
USA
TEL   213 206 8596
TLF
TLX
EML
COM   22

PNEUMAN GERALD W
HIGH ALTITUDE OBSERVATORY
NCAR
BOX 3000
BOULDER CO 80307 3000
USA
TEL   303 497 1000
TLF
TLX   45694
EML
COM   10,49

POECKERT ROLAND H DR
DEFENCE RESEARCH
ESTABLISHMENT PACIFIC
FMO CFB ESQUIMALT
VICTORIA BC VOS 1BO
CANADA
TEL
TLF
TLX
EML
COM

POEPPEL WOLFGANG G L DR
IAR
CC 5
1894 VILLA ELISA (BS AS)
ARGENTINA
TEL   21 4 3793
TLF
TLX   18052 CICYT-AR
EML
COM· 34

POGO ALEXANDER DR
MT WILSON & LAS CAMPANAS
OBSERVATORIES
813 SANTA BARBARA ST
PASADENA CA 91101
USA
TEL   213 577 1122
TLF
TLX
EML
COM   41

POHL ECKHARD DR
STERNWARTE NUERNBERG
REGIOMONTANUSWEG 1
D 8500 NUERNBERG 20
GERMANY
TEL   911 593 540
TLX
EML
TLF
COM

POKORNY ZDENEK DR
N COPERNICUS OBSERVATORY
& PLANETARIUM
KRAVI HORA
CS 616 00 BRNO 16
CZECHOSLOVAKIA
TEL   5 74 4374
TLF
TLX
EML
COM   16,46

POLAND ARTHUR I DR
NASA/GSFC
CODE 682
GREENBELT MD 20771
USA
TEL   301 286 7334
TLF
TLX   89675
EML
COM   10

POLCARO V F
IAS
CNR
CP 67
I 00044 FRASCATI
ITALY
TEL   6 942 5651
TLF
TLX   610261
EML
COM   29

POLECHOVA PAVLA DR
OBS AND PLANETARIUM
PETRIN 205
CS 118 46 PRAHA 1
CZECHOSLOVAKIA
TEL   2 53 53513
TLF
TLX
EML
COM   05

POLETTO GIANNINA PROF
OSS ASTROFISICO
DI ARCETRI
LARGO E FERMI 5
I 50125 FIRENZE
ITALY
TEL   55 275 2252
TLF
TLX   572268 ARCETR I
EML
COM   10

POLIDAN RONALD S
NASA/GSFC
CODE 681 O
LASP
GREENBELT MD 20771
USA
TEL   301 286 5039
TLF   301 286 8709
TLX   89675
EML   (SPAN)STARS POLIDAN
COM   29,42C,44

POLLACK JAMES B DR
NASA MAES RESEARCH CTR
MS 245 3
SPACE SCIENCE DIV
MOFFETT FIELD CA 94035
USA
TEL   415 694 5530
TLF
TLX
EML
COM   16,51

POLNITZKY GERHARD DR
INSTITUT FUER ASTRONOMIE
UNIVERSITAET WIEN
TUERKENSCHANZSTR 17
A 1180 WIEN
AUSTRIA
TEL   1 345 36090
TLF
TLX   116222 PHYSI A
EML
COM   08,22

POLOSUKHINA-CHUVAEVA N DR
CRIMEAN ASTROPHYS OBS
UKRAINIAN ACAD OF SCIENCE
NAUCHNY
334413 CRIMEA
UKRAINE
TEL   43 2945
TLF
TLX
EML
COM

POLOZHENTSEV DIMITRIJ DR
PULKOVO OBSERVATORY
ACADEMY OF SCIENCES
10 KUTUZOV QUAY
196140 ST PETERSBURG
RUSSIA
TEL   298-22-42
TLF
TLX
EML
COM   08C,24C

POLUPAN P N DR
ASTRONOMICAL OBSERVATORY
KIEV STATE UNIVERSITY
OBSERVATORNAYA UL 3
252053 KIEV
UKRAINE
TEL 26 0908
TLF
TLX 132201
EML
COM 10

POLYACHENKO VALERIJ L DR
INST OF ASTRONOMY
ACADEMY OF SCIENCES
PYATNITSKAYA UL 48
109017 MOSCOW
RUSSIA
TEL 95 231 3980
TLF 95 230 2081
TLX. 411576 ASCON SU
EML IAAS●NODE IAS MSK SU
COM 33

POLYMILIS CHRONIS DR
DPT OF ASTRONOMY
NTL UNIVERSITY OF ATHENS
PANEPISTIMIOPOLIS
GR 157 83 ZOGRAFIG
GREECE
TEL 1 724 3414
TLF
TLX
EML· SPMSO●GRATHUN1
COM. 33

POMA ANGELO DR
ISTITUTO DI ASTRONOMIA

VIA OSPEDALE 72
I 09100 CAGLIARI
ITALY
TEL 70 66 3544
TLF
TLX 790326 OSSAST
EML
COM. 08,19

PONMAN TREVOR DR
DPT OF SPACE RESEARCH
UNIVERSITY OF BIRMINGHAM
BOX 363
BIRMINGHAM B15 2TT
UK
TEL
TLF
TLX
EML
COM

PONNAMPERUMA CYRIL PROF
DPT OF CHEMISTRY
UNIVERSITY OF MARYLAND
COLLEGE PARK MD 20472
USA
TEL
TLF.
TLX
EML
COM 51

PONSONBY JOHN E B DR
NRAL
JODRELL BANK
MACCLESFIELD SK11 9DL
UK
TEL 477 713 21
TLF
TLX 36149
EML
COM 40,51

POOLEY GUY DR
MULLARD RADIO ASTRON OBS
CAVENDISH LABORATORY
MADINGLEY RD
CAMBRIDGE CB3 0HE
UK
TEL 223 664 77
TLF
TLX 81292
EML
COM 40

POP VASILE DR
FAC OF MATHEMATICS
UNIVERSITY OF CLUJ-NAPOCA
STR KOGALNICEANU 1
R 3400 CLUJ-NAPOCA
RUMANIA
TEL 40 95 141726
TLF 40 95 111905
TLX
EML
COM 27

POPELAR JOSEF DR
GEODETIC SURVEY DIVISION
CANADA CENTRE FOR SURVEY
615 BOOTH ST
OTTAWA ON K1A 0E9
CANADA
TEL
TLF
TLX.
EML·
COM. 19,31

POPOV VASIL NIKOLOV
DPT OF ASTRONOMY
BULGARIAN ACAD SCIENCES
72 LENIN BLVD
BG 1784 SOFIA
BULGARIA
TEL 2 75 8927
TLF
TLX
EML
COM 28

POPOV VICTOR S DR
PULKOVO OBSERVATORY
ACADEMY OF SCIENCES
10 KUTUZOV QUAY
196140 ST PETERSBURG
RUSSIA
TEL
TLF
TLX
EML
COM 30

POPOVA MALINA D PROF DR
DPT OF ASTRONOMY
BULGARIAN ACAD SCIENCES
72 LENIN BLVD
BG 1784 SOFIA
BULGARIA
TEL 2 75 8927
TLF
TLX
EML
COM 27,37

POPOVIC BOZIDAR PROF DR
BULV JNA 152/32
YU 11000 BEOGRAD
YUGOSLAVIA
TEL 11 692 352
TLF
EML
COM 07,20
TLX

POPOVIC GEORGIJE DR
ASTRONOMICAL OBSERVATORY
VOLGINA 7
YU 11050 BEOGRAD
YUGOSLAVIA
TEL 11 419 357
TLF
TLX
EML
COM. 26

POPPER DANIEL M PROF
DPT OF ASTRONOMY
UNIVERSITY OF CALIFORNIA
LOS ANGELES CA 90024
USA
TEL 213 825 3622
TLF
TLX 9103427597
EML
COM 42

POQUERUSSE MICHEL
OBSERVATOIRE DE PARIS
SECTION DE MEUDON
DESPA
F 92195 MEUDON PPL CDX
FRANCE
TEL 1 45 07 7530
TLF
TLX 204464
EML
COM 10,12

PORCAS RICHARD DR
MPI FUER RADIOASTRONOMIE
AUF DEM HUEGEL 69
D 5300 BONN 1
GERMANY
TEL· 228 52 5282
TLX 886440
EML
TLF
COM 40

PORCEDDU IGNAZIO E P DR
ISTITUTO DI ASTRONOMIA
VIA OSPEDALE 72
I 09124 CAGLIARI
ITALY
TEL 70 66 3544
TLF 70 65 7657
TLX
EML 40588 PORCEDDU
COM 34

PORETTI ENNIO
OSS ASTRONOMICO DI MILANO

VIA E BIANCHI 46
I 22055 MERATE
ITALY
TEL 59 6412
TLF
TLX
EML PORETTI●ASTMIB INFN IT
COM

PORFIR'EV V V DR
PEDAGOGOC INSTITUTE
MINISTRY OF EDUCATION
107846 MOSCOW
RUSSIA
TEL
TLF
TLX
EML
COM 35

PORNCHAI P -TANAKUN
DPT OF PHYSICS
CHULALONGKORN UNIVERSITY
10330 BANGKOK
THAILAND
TEL 2 252 7985
TLF
TLX 20217 UNICHUL TH
EML
COM 36

PORTER JASON G DR
NASA/MSFC
CODE E5 52
SPACE SCIENCE LAB
HUNTSVILLE AL 35812
USA
TEL 205 544 7607
TLF 205 544 5862
TLX 62026079 ESL
EML SSL PORTERJ
COM· 10

PORTER NEIL A PROF
DPT OF PHYSICS
UNIVERSITY COLLEGE
BELFIELD
DUBLIN 4
IRELAND
TEL 1 693 244*211
TLF
TLX 32693 UCDEI
EML
COM 41,48

PORUBCAN VLADIMIR DR
ASTRONOMICAL INSTITUTE
SLOVAK ACADEMY SCIENCES
DUBRAVSKA 9
CS 842 28 BRATISLAVA
CZECHOSLOVAKIA
TEL 7 37 5157
TLF
TLX 93373 SEIS
EML
COM 22

POTTASCH STUART R PROF
KAPTEYN ASTRONOMICAL INST
BOX 800
NL 9700 AV GRONINGEN
NETHERLANDS
TEL 50 11 6641
TLX 53572 STARS NL
EML
TLF
COM 34C,36

POTTER ANDREW E DR
NASA/JOHNSON SPACE CENTER
CODE SN3
HOUSTON TX 77058
USA
TEL 713 483 5276
TLF 713 483 5347
TLX
EML SN POTTER
COM 16

POTTER HEINO I DR
PULKOVO OBSERVATORY
ACADEMY OF SCIENCES
10 KUTUZOV QUAY
196140 ST PETERSBURG
RUSSIA
TEL 298-22-42
TLF
TLX
EML
COM 24

POULAKOS CONSTANTINE DR
RES CENTER FOR ASTRONOMY
ACADEMY OF ATHENS
14 ANAGNOSTOPOULOU ST
GR 106 73 ATHENS
GREECE
TEL
TLF
TLX
EML
COM

POULLE EMMANUEL PROF
ECOLE NATLE DES CHARTES
19 RUE DE LA SORBONNE
F 75005 PARIS
FRANCE
TEL 1 45 89 4857
TLF
TLX
EML
COM 41

POUNDS KENNETH A PROF
DPT OF PHYSICS
UNIVERSITY OF LEICESTER
UNIVERSITY RD
LEICESTER LE1 7RH
UK
TEL 533 554 455*151
TLF
TLX 341664 LUXRAYG
EML
COM 06,44,48

POUQUET ANNICK DR
OCA OBSERV DE NICE
BP 139
F 06003 NICE CDX
FRANCE
TEL 93 89 0420
TLF
TLX 460004 OBSNICE F
EML
COM

POVEDA ARCADIO DR
INSTITUTO DE ASTRONOMIA
UNAM
APDO POSTAL 70-264
04510 MEXICO DF
MEXICO
TEL 550-5805
TLF
TLX 1760155 CICME
EML
COM 26,28,35,37

POYET JEAN-PIERRE DR
OBS MIDI PYRENEES
14 AVE E BELIN
F 31400 TOULOUSE CDX
FRANCE
TEL 61 25 2101
TLF
TLX
EML
COM

PRABHAKARAN NAYAR S R DR
DPT OF PHYSICS
UNIVERSITY OF KERALA
KARIYAVATTOM
TRIVANDRUM 695 581
INDIA
TEL 471 41 8920
TLF
TLX
EML
COM 46

PRABHU TUSHAR P
INDIAN INSTITUTE OF
ASTROPHYSICS
KORAMANGALA
BANGALORE 560 034
INDIA
TEL 812 56 6585
TLF
TLX 845763 IIAB IN
EML
COM 28

PRADERIE FRANCOISE DR
OBSERVATOIRE DE PARIS
SECTION DE MEUDON
DPT RECHERCHE SPATIALE
F 92195 MEUDON PPL CDX
FRANCE
TEL 1 45 07 7651
TLF
TLX 204464
EML
COM 29,36

PRADHAN ANIL DR
DPT OF ASTRONOMY
OHIO STATE UNIVERSITY
174 W 18TH AVE
COLUMBUS OH 43210 1106
USA
TEL 614 292 5850
TLF 614 292 2928
TLX 755842
EML PRADHAN@OHSTPY BITNET
COM

PRANTZOS NIKOS DR
INSTITUT D'ASTROPHYSIQUE
98BIS BD ARAGO
F 75014 PARIS
FRANCE
TEL 1 43 20 1425
TLF 1 43 29 8673
TLX
EML NIKOS@FRIAP51
COM

PRASAD SHEO S
LOCKHEED PALO ALTO RES LB
DPT 91 20 BLDG 255
3251 HANOVER ST
PALO ALTO CA 94304
USA
TEL 415 424 2659
TLF 415 424 3333
TLX
EML. PRASAD%LPARL1
COM 34

PRASANNA A R DR
PHYSICAL RESEARCH LAB
NAVRANGPURA
AHMEDABAD 380 009
INDIA
TEL 272 46 2129
TLF 272 44 5292
TLX 021-397 PRL IN
EML
COM 48

PRATAP R DR
INSTITUTE OF APPLIED
SCIENCES
COCHIN 682 317
INDIA
TEL
TLF
TLX
EML
COM

PRAVDO STEVEN H
JPL
MS 168 222
4800 OAK GROVE DR
PASADENA CA 91109
USA
TEL 818 354 4134
TLF
TLX 910-588-3294
EML
COM

PREDEANU IRINA DR
ASTRONOMICAL OBSERVATORY
CUTITUL DE ARGINT 5
BOX 28
R 75212 BUCHAREST
RUMANIA
TEL 23 68 92
TLF
TLX 11882 ASTRO R
EML
COM 16

PREITE-MARTINEZ ANDREA DR
IAS
CNR
CP 67
I 00044 FRASCATI
ITALY
TEL 6 942 5655
TLF
TLX 610261
EML
COM 34

PREKA-PAPADEMA P DR
LAB OF ASTROPHYSICS
NTL UNIVERSITY OF ATHENS
PANEPISTIMIOPOLIS
GR 157 83 ATHENS
GREECE
TEL 1 723 5122
TLF
TLX
EML SPM75@GRATHUN1
COM 10

PRENDERGAST KEVIN H PROF
DPT OF ASTRONOMY
COLUMBIA UNIVERSITY
PUPIN HALL 538 W 120TH ST
NEW YORK NY 10027
USA
TEL 212 280 3280
TLF
TLX
EML
COM 28

PRENTICE ANDREW J R DR
DPT OF MATHEMATICS
MONASH UNIVERSITY
WELLINGTON RD
CLAYTON VIC 3168
AUSTRALIA
TEL
TLF
TLX
EML
COM 35

PRESS WILLIAM H DR
CENTER FOR ASTROPHYSICS
HCO/SAO
60 GARDEN ST
CAMBRIDGE MA 02138
USA
TEL 617 495 4908
TLF
TLX 921428 SATELLITE CAM
EML
COM 28,47

PRESTON GEORGE W DR
MT WILSON & LAS CAMPANAS
OBSERVATORIES
813 SANTA BARBARA ST
PASADENA CA 91101
USA
TEL 818 577 1122
TLF
TLX
EML
COM 30,45

PRESTON ROBERT ARTHUR
JPL
MS 138 307
4800 OAK GROVE DR
PASADENA CA 91109
USA
TEL 213 354 6895
TLF
TLX 675429
EML
COM 40

PREUSS EUGEN DR
MPI FUER RADIOASTRONOMIE
AUF DEM HUEGEL 69
D 5300 BONN 1
GERMANY
TEL· 228 52 51
TLF
TLX 886440
EML·
COM 40,48

PREVOT LOUIS DR
OBSERVATOIRE DE MARSEILLE
2 PLACE LE VERRIER
F 13248 MARSEILLE CDX 04
FRANCE
TEL 91 95 9088
TLF
TLX 420241
EML
COM 30

PREVOT-BURNICHON M L DR
OBSERVATOIRE DE MARSEILLE
2 PLACE LE VERRIER
F 13248 MARSEILLE CDX 04
FRANCE
TEL 91 95 9088
TLF
TLX 420241
EML
COM 28

PRIALNIK-KOVETZ DINA DR
DPT OF GEOPHYSICS
TEL AVIV UNIVERSITY
TEL AVIV
ISRAEL
TEL 3 545 0633
TLF
TLX 342171 VERSY IL
EML BITNET B13@TAUNOS
COM 15,35

PRICE MICHAEL J DR
SCIENCE APPLICATIONS
5151 E BROADWAY
SUITE 1100
TUCSON AZ 85711
USA
TEL 602 748 7400
TLF
TLX
EML
COM

PRICE R MARCUS DR
DPT PHYSICS & ASTRONOMY
UNIVERSITY OF NEW MEXICO
800 YALE BLVD NE
ALBUQUERQUE NM 87131
USA
TEL 505 277 2616
TLF
TLX
EML
COM 33,34,40

PRICE STEPHAN DONALD
2 POLLEY RD
WESTFORD MA 01886
USA
TEL 617 861 4552
TLF
EML
COM 44
TLX

PRIEST ERIC R PROF
DPT OF APPLIED MATHS
UNIVERSITY OF ST ANDREWS
NORTH HAUGH
ST ANDREWS FIFE KY16 9SS
UK
TEL 334 761 61*8156
TLF 334 744 87
TLX 76213 SAULIB
EML ERIC@CS ST-ANDREWS AC UK
COM 10C,12

PRIESTER WOLFGANG PROF
INSTITUT F ASTROPHYSIK
AUF DEM HUEGEL 71
D 5300 BONN 1
GERMANY
TEL 228 73 3671
TLF
TLX 886440
EML
COM 33,40

PRIETO MERCEDES
INST DE ASTROFISICA
DE CANARIAS
OBS DEL TEIDE
E 38071 LA LAGUNA
SPAIN
TEL 22 26 2211
TLF
TLX 92640
EML
COM

PRIEUR JEAN-LOUIS DR
OBS MIDI PYRENEES
14 AVE E BELIN
F 31400 TOULOUSE CDX
FRANCE
TEL 61 33 2929
TLF 61 53 6722
TLX 530776
EML PRIEUR AT FROMP51 BITNET
COM 28

PRINCE HELEN DODSON PROF
4800 FILLMORE AVE
ALEXANDRIA VA 22311
USA
TEL 703 578 1000
TLF
TLX
EML
COM

PRINGLE JAMES E DR
INSTITUTE OF ASTRONOMY
THE OBSERVATORIES
MADINGLEY RD
CAMBRIDGE CB3 OHA
UK
TEL 223 622 04
TLF
TLX, 817297 ASTRON G
EML
COM 27,42

PRINJA RAMAN DR
DPT PHYSICS & ASTRONOMY
UNIVERSITY COLLEGE LONDON
GOWER ST
LONDON WC1E 6BT
UK
TEL 1 713 877 050
TLF
TLX 28722
EML BITNET RKP@UK AC UCL STARLINK
COM 29

PRITCHET CHRISTOPHER J DR
DPT OF PHYSICS
UNIVERSITY OF VICTORIA
BOX 1700
VICTORIA BC V8W 2Y2
CANADA
TEL 604 721 7704
TLF 604 721 7715
TLX
EML
COM 09,28,37

PROBSTEIN R F DR
DPT MECHANICAL ENGINEERG
MIT
BOX 165
CAMBRIDGE MA 02139
USA
TEL 617 253 2240
TLF
TLX 921473 MIT CAM
EML
COM

PROCHAZKA FRANZ V DR
INSTITUT FUER FERSTUDIEN
UNIVERSITAET
KLAGENFURT
A 9020 KLAGENFURT
AUSTRIA
TEL 422 5317
TLF
TLX
EML
COM 24,51

PRODAN Y I DR
STERNBERG STATE ASTR INST
UNIVERSITETSKIJ PROSP 13
119899 MOSCOW
RUSSIA
TEL 139-55-43
TLX
EML
TLF
COM

PROFFITT CHARLES R DR
STSCI
HOMEWOOD CAMPUS
3700 SAN MARTIN DR
BALTIMORE MD 21218
USA
TEL 301 338 4572
TLF 301 338 5090
TLX
EML PROFFITT@STSCI EDU
COM 35

PROISY PAUL E DR
OBSERVATOIRE DE LYON
AVE CHARLES ANDRE
F 69561 S GENIS LAVAL CDX
FRANCE
TEL 78 56 0705
TLF 72 39 9791
TLX
EML
COM 15

PROKAKIS THEODORE J DR
ASTRONOMICAL INSTITUTE
NTL OBSERVATORY OF ATHENS
BOX 20048
GR 118 10 ATHENS
GREECE
TEL 1 346 1191/1 804 0619
TLF
TLX 21 5530
EML
COM 10,12,41

PROKOF'EV VLADIMIR K PROF
CRIMEAN ASTROPHYS OBS
UKRAINIAN ACAD OF SCIENCE
NAUCHNY
334413 CRIMEA
UKRAINE
TEL 43 2945
TLF
TLX
EML
COM 14,44

PROKOF'EVA IRINA A DR
PULKOVO OBSERVATORY
ACADEMY OF SCIENCES
10 KUTUZOV QUAY
196140 ST PETERSBURG
RUSSIA
TEL
TLF
TLX
EML
COM

PROKOF'EVA VALENTINA V DR
CRIMEAN ASTROPHYS OBS
UKRAINIAN ACAD OF SCIENCE
NAUCHNY
334413 CRIMEA
UKRAINE
TEL 43 2945
TLF
TLX
EML
COM 09

PRONIK I I DR
CRIMEAN ASTROPHYS OBS
UKRAINIAN ACAD OF SCIENCE
NAUCHNY
334413 CRIMEA
UKRAINE
TEL 43 2945
TLF
TLX
EML
COM 28,34

PRONIK V I DR
CRIMEAN ASTROPHYS OBS
UKRAINIAN ACAD OF SCIENCE
NAUCHNY
334413 CRIMEA
UKRAINE
TEL 43 2945
TLF
TLX
EML
COM 28

PROSZYNSKI MIECZYSLAW
COPERNICUS ASTRON CENTER
POLISH ACAD OF SCIENCES
UL BARTYCKA 18
PL 00 716 WARSAW
POLAND
TEL
TLF
TLX
EML
COM 44

PROTHEROE RAYMOND J DR
DPT OF PHYSICS
UNIVERSITY OF ADELAIDE
BOX 498
ADELAIDE SA 5001
AUSTRALIA
TEL 8 228 5996
TLF
TLX 89141 UNIVAD AA
EML
COM 48

PROTHEROE WILLIAM M PROF
DPT OF ASTRONOMY
OHIO STATE UNIVERSITY
174 W 18TH AVE
COLUMBUS OH 43210 1106
USA
TEL 614 422 7891
TLF
TLX
EML
COM·

PROTICH MILORAD B
ASTRONOMICAL OBSERVATORY
VOLGINA 7
YU 11050 BEOGRAD
YUGOSLAVIA
TEL 11 402 365
TLF
TLX
EML
COM 20

PROUST DOMINIQUE
OBSERVATOIRE DE PARIS
SECTION DE MEUDON
DAPHE
F 92195 MEUDON PPL CDX
FRANCE
TEL 1 45 07 7411
TLF
TLX 201571
EML
COM 28

PROVERBIO EDOARDO PROF
ISTITUTO DI ASTRONOMIA
VIA OSPEDALE 72
I 09100 CAGLIARI
ITALY
TEL 70 66 3544
TLF 70 65 7657
TLX
EML PROVERBIO@ASTRCA ASTRO IT
COM 08,19,31P,46,41

PROVOST JANINE DR
OCA OBSERV DE NICE
BP 139
F 06003 NICE CDX
FRANCE
TEL 93 89 0420
TLF
TLX 460004 OBSNICE F
EML
COM 27,35

PRYCE MAURICE H L DR
DPT OF PHYSICS
UNIV OF BRITISH COLUMBIA
2075 WESBROOK PL
VANCOUVER BC V6T 1W5
CANADA
TEL 604 228 6417
TLF 604 228 5324
TLX 045 08576
EML
COM

PRYOR CARLTON PHILIP DR
DPT PHYSICS & ASTRONOMY
RUTGERS UNIVERSITY
BOX 849
PISCATAWAY NJ 08854 0849
USA
TEL 908 932 5462
TLF 908 932 4343
TLX
EML PRYOR@PRYOR RUTGERS EDU
COM

PSKOVSKIJ JU P DR
STERNBERG STATE ASTR INST
UNIVERSITETSKIJ PROSP 13
119899 MOSCOW
RUSSIA
TEL 139-37-21
TLF
TLX
EML
COM 27,34

PUCILLO MAURO DR
OAT
BOX SUCC TRIESTE 5
VIA TIEPOLO 11
I 34131 TRIESTE
ITALY
TEL 40 79 3921
TLF
TLX 461137 OAT I
EML
COM 05,09

PUETTER RICHARD C DR
CASS
UCSD
C 011
LA JOLLA CA 92093 0216
USA
TEL 619 534 4995
TLF
TLX
EML
COM

PUGACH ALEXANDER F DR
MAIN ASTRONOMICAL OBS
UKRAINIAN ACAD OF SCIENCE
GOLOSEEVO
252127 KIEV
UKRAINE
TEL 66 4771
TLF
TLX 131406 SKY SU
EML
COM 27

PUGET JEAN-LOUP DR
RADIOASTRONOMIE ENS
24 RUE LHOMOND
F 75231 PARIS CDX 05
FRANCE
TEL 1 43 29 1225
TLF
TLX 270912
EML
COM 34,47

PUGLIANO ANTONIO PROF
IST UNIVERSITARIO NAVALE
VIA ACTON 38
I 80133 NAPOLI
ITALY
TEL 81 551 2330
TLF 81 552 1485
TLX 710417
EML
COM 08

PULS JOAHIM DR
INSITUT FUR ASTRONOMIE
UND ASTROPHYSIK
SCHEINERSTR 1
D 8000 MUENCHEN 80
GERMANY
TEL 89 922 09436
TLF 89 922 09427
TLX
EML UH101AW@DMOLRZO1/EARN
COM 36

PUNETHA LALIT MOHAN DR
UTTAR PRADESH STATE
OBSERVATORY
PO MANORA PEAK 263 129
NAINITAL 263 129
INDIA
TEL 59 42 2136
TLF
TLX
EML
COM

PURCELL EDWARD M PROF
DPT OF PHYSICS
HARVARD UNIVERSITY
60 GARDEN ST
CAMBRIDGE MA 02138
USA
TEL 617 495 2860
TLF
TLX.
EML
COM 51

PURTON CHRISTOPHER R DR
DOMINION RADIO ASTROPHYS
OBSERVATORY
BOX 248
PENTICTON BC V2A 6K3
CANADA
TEL 604 493 2277
TLF 604 493 7767
TLX 048 88127
EML
COM

PUSCHELL JEFFERY JOHN
MARTIN MARIETTA
103 CHESAPEAKE PARK PLAZA
E460
BALTIMORE MD 21220
USA
TEL 301 682 0885
TLF
TLX 908225
EML
COM 40

PUSHKIN SERGEY B DR
TIME & FREQUENCY SERVICE
GOSSTANDARD USSR
117049 MOSCOW
RUSSIA
TEL
TLF
TLX
EML
COM 31

PUSTYL'NIK IZOLD B DR
TARTU ASTROPHYSICAL OBS
ESTONIAN ACAD OF SCIENCES
202444 TARTU
ESTONIA
TEL 33439
TLF
TLX
EML
COM 42

PYE JOHN P DR
DPT OF PHYSICS
UNIVERSITY OF LEICESTER
UNIVERSITY RD
LEICESTER LE1 7RH
UK
TEL 533 554 455*23
TLF
TLX 341664 LUXRAY G
EML·
COM

PYPER SMITH DIANE M DR
DPT OF PHYSICS
UNIVERSITY OF NEVADA
4505 S MARYLAND PARKWAY
LAS VEGAS NV 89154
USA
TEL
TLF
TLX
EML
COM

QI GUAN RONG
SHANNKI OBSERVATORY
CAS
LINTONG XIAN
SHAANXI
CHINA PR
TEL 33 2255
TLF
TLX 70121 CSAO CN
EML
COM 08,31

QIAN BO-CHEN
SHANGHAI OBSERVATORY
CAS
80 NANDAN RD
SHANGHAI
CHINA PR
TEL 21 38 6191
TLF
TLX 33164 SHAO CN
EML
COM 37

QIAN SHAN-JIE
BEIJING ASTRONOMICAL OBS
CAS
W SUBURB
BEIJING 100080
CHINA PR
TEL 1 28 2194
TLF
TLX 22040 BAOAS CN
EML
COM 40

QIAN ZHI-HAN DR
SHANGHAI OBSERVATORY
CAS
80 NANDAN RD
SHANGHAI
CHINA PR
TEL 21 38 6191
TLF
TLX 33164 SHAO CN
EML
COM 08

QIAN ZHONG-YU
BEIJING ASTRONOMICAL OBS
CAS
W SUBURB
BEIJING 100080
CHINA PR
TEL
TLF
TLX 22040 BAOAS CN
EML
COM 33

QIAO GUOJUN
DPT OF GEOPHYSICS
BEIJING UNIVERSITY
BEIJING 100871
CHINA PR
TEL
TLF 1 25 64095
TLX 22239 PKUNI CN
EML
COM 42

QIN DAO
PURPLE MOUNTAIN OBSERV
CAS
NANJING
CHINA PR
TEL 25 46700
TLF
TLX 34144 PMONJ CN
EML
COM 24

QIN SONG-NIAN
YUNNAN OBSERVATORY
CAS
BOX 110
KUNMING 72946 YUNNAN
CHINA PR
TEL 871 2035
TLF
TLX
EML
COM

QIN ZHI-HAI
DPT OF ASTRONOMY
NANJING UNIVERSITY
NANJING
CHINA PR
TEL 25 34651*2882
TLF
TLX 0909
EML
COM 34

QIU PUZHANG ASS PROF
YUNNAN OBSERVATORY
CAS
BOX 110
KUNMING 72946 YUNNAN
CHINA PR
TEL 871 2035
TLF
TLX 64040 YUOBS CN
EML
COM 09,51

QIU YU-HAI
BEIJING ASTRONOMICAL OBS
CAS
W SUBURB
BEIJING 100080
CHINA PR
TEL
TLF
TLX 22040 BAOAS CN
EML
COM 40

QU QIN-YUE
DPT OF ASTRONOMY
NANJING UNIVERSITY
NANJING
CHINA PR
TEL 25 37551*2741
TLF
TLX 34151 PRCNU CN
EML
COM 35,47,48C

QUAMAR JAWAID
D 19 STAFF TOWN
UNIVERSITY OF KARACHI
KARACHI 3201
PAKISTAN
TEL 46 54 91
TLF
TLX
EML
COM

QUAN HEJUN
SHANGHAI OBSERVATORY
CAS
80 NANDAN RD
SHANGHAI
CHINA PR
TEL 21 38 6191
TLF
TLX 33164 SHAO CN
EML
COM 41

QUARTA MARIA LUCIA
IAG
UNIVERSIDADE DE SAO PAULO
CP 30627
01051 SAO PAULO SP
BRAZIL
TEL 11 577 8599
TLF 11 815 3848
TLX 11 36221 IAGM BR
EML
COM

QUAST GERMANO RODRIGO
OBSERVATORIO NACIONAL
RUA COLONEL RENNO 07
CP 21
37500 ITAJUBA MG
BRAZIL
TEL 35 622 0788
TLF
TLX· 031 2603
EML.
COM

QUENBY JOHN J DR
BLACKETT LABORATORY
IMPERIAL COLLEGE
PRINCE CONSORT RD
LONDON SW7 2BZ
UK
TEL 1 589 5111*6661
TLF
TLX 261503
EML
COM

QUERCI FRANCOIS R DR
OBS MIDI PYRENEES
14 AVE E BELIN
F 31400 TOULOUSE CDX
FRANCE
TEL 61 33 2929
TLF 61 53 6722
TLX. 530776 F OBSTLSE
EML FROMP51
COM 14,29,36

QUERCI MONIQUE DR
OBS MIDI PYRENEES
14 AVE E BELIN
F 31400 TOULOUSE CDX
FRANCE
TEL 61 33 2929
TLF 61 53 6722
TLX. 530776 F OBSTLSE
EML FROMP51
COM 29,36

QUESADA VINICIO DR
STAZIONE ASTRONOMICIA
VIA OSPEDALE 72
I 09124 CAGLIARI
ITALY
TEL 70 66 3544
TLX
EML VAXCA2 QUESADA
TLF 70 65 7657
COM 22

QUIJANO LUIS
REAL INST Y OBSERVATORIO
DE LA ARMADA
CECILIO PUJAZON S/N
E 11110 SAN FERNANDO
SPAIN
TEL 56 88 3548
TLF
TLX 76108 IOM E
EML
COM 08,20,24

QUINN PETER DR
MOUNT STROMLO & SIDING
SPRING OBSERVATORIES
PRIVATE BAG
WESTON CREEK PO ACT 2611
AUSTRALIA
TEL 62 49 0272
TLF 62 49 0233
TLX
EML PJQ@MINUET ANU OZ AU
COM 28

QUINTANA HERNAN DR
GRUPO ASTROFIS PONTIFICA
UNIVERSIDAD CATOLICA
CASILLA 104
SANTIAGO 22
CHILE
TEL 2 775 474
TLF 2 552 5692
TLX 240 395
EML HQUINTANA@ASTROUC PUC CL
COM 05,28,30,48,51

QUINTANA JOSE M DR
INST ASTROFISICA
DE ANDALUCIA APD 2144
PROFESOR ALBAREDA 1
E 18080 GRANADA
SPAIN
TEL 58 12 1300
TLF
TLX 78573 IAAG E
EML
COM 51

QUIRK WILLIAM J DR
LAWRENCE LIVERMORE LAB
L 35
BOX 808
LIVERMORE CA 94550
USA
TEL 415 422 1852
TLF
TLX
EML
COM

RAADU MICHAEL A DR
DPT OF PLASMA PHYSICS
ROYAL INST OF TECHNOLOGY
S 100 44 STOCKHOLM 70
SWEDEN
TEL 87 87 7000
TLF
TLX 10389 KTHB STOCKHOLM
EML
COM 10,49C

RABBIA YVES DR
OCA CERGA
AVE COPERNIC
F 06130 GRASSE
FRANCE
TEL 93 36 5849
TLF
TLX 470865 CERGA
EML EARN YVRABB@FRONI51
COM

RABIN DOUGLAS MARK
NTL SOLAR OBSERVATORY
BOX 26732
950 N CHERRY AVE
TUCSON AZ 85726 6732
USA
TEL 602 325 9331
TLF 602 325 9360
TLX 0666484 AURA NOAOTUC
EML RABIN@NOAO EDU
COM 10,12

RABOLLI MONICA DR
UNIV NACIONAL DE LA PLATA
FCAG
1900 LA PLATA (BS AS)
ARGENTINA
TEL 21 21 7308
TLF 1 786 8114
TLX
EML MRABOLLI%PSI@IAFE%SSL SPAN
COM 33

RACHKOVSKY D N DR
CRIMEAN ASTROPHYS OBS
UKRAINIAN ACAD OF SCIENCE
NAUCHNY
334413 CRIMEA
UKRAINE
TEL 43 2945
TLF
TLX 192
EML
COM 36

RACINE RENE DR
DPT DE PHYSIQUE
UNIVERSITE DE MONTREAL
CP 6128 SUCC A
MONTREAL QC H3C 3J7
CANADA
TEL 514 343 6718
TLF 514 343 2071
TLX 055 61359 RZLPNUM ML
EML
COM 09

RACKHAM THOMAS W DR
39 MEADOW AVE
GOOSTREY
CREWE CW4 8LS
UK
TEL 477 330 04
TLF
TLX
EML
COM

RADHAKRISHNAN V PROF
RAMAN RESEARCH INSTITUTE
SADASHIVANAGAR
BANGALORE 560 080
INDIA
TEL 812 34 0522
TLF 812 34 0492
TLX 8452671
EML
COM 34,40,48,EC

RADICK RICHARD R DR
AIR FORCE GEOPHYSICS LAB
NTL SOLAR OBSERVATORY
SUNSPOT NM 88349
USA
TEL 505 434 1390
TLF
TLX
EML
COM 12

RADIMAN IRATIUS
BOSSCHA OBSERVATORY
LEMBANG 40391
INDONESIA
TEL 229 6001
TLF
TLX
EML
COM

RADOSKI HENRY R DR
AFOSR/NP
BUILDING 410
BOLLING AIR FORCE BASE
WASHINGTON DC 20332
USA
TEL 202 767 4906
TLF
TLX
EML
COM 21

RADOSLAVOVA TSVETANKA
DPT OF ASTRONOMY
BULGARIAN ACAD SCIENCES
72 LENIN BLVD
BG 1784 SOFIA
BULGARIA
TEL 2 75 8927
TLF
TLX
EML
COM

RAEDLER K H DR
ZNTRLINST F ASTROPHYSIK
ROSA-LUXEMBURG-STR 17A
D 1502 POTSDAM
GERMANY
TEL
TLF
TLX
EML
COM 35

RAFANELLI PIERO DR
OSS ASTRONOMICO DI PADOVA
VIC DELL OSSERVATORIO 5
I 35122 PADOVA
ITALY
TEL 49 66 1499
TLF
TLX
EML
COM 28

RAFERT JAMES BRUCE
DPT PHYSICS & SPACE SCI
FLORIDA INST TECHNOLOGY
150 W UNIVERSITY BLVD
MELBOURNE FL 32901
USA
TEL
TLF
TLX
EML
COM 42

RAGHAVAN NIRUPAMA DR
2133 INDIAN INSTITUTE
OF TECHNOLOGY
CAMPUS
NEW DELHI 110 029
INDIA
TEL
TLF
TLX
EML
COM

RAHARTO MOEDJI
DPT OF ASTRONOMY
BANDUNG INSTITUTE OF TECH
JL GANESHA 10
BANDUNG 40132
INDONESIA
TEL 622 244 0252
TLF 622 243 8388
TLX 28324 BD
EML
COM 33

RAHE JURGEN PROF
NASA HEADQUARTERS
CODE EL
600 INDEPENDENCE AVE SW
WASHINGTON DC 20546
USA
TEL 202 453 1590
TLF 202 426 1023
TLX 4974843
EML
COM 15C,16C,42,44

RAHUNEN TIMO
TAMPERE SAERKAENNIEMI OY
SAERKAENNIEMI
SF 33410 TAMPERE
FINLAND
TEL 31 31 333
TLF
TLX
EML
COM 42

RAIKOVA DONKA DR
DPT OF ASTRONOMY
BULGARIAN ACAD SCIENCES
7TH NOVEMBER ST 1
BG 1000 SOFIA
BULGARIA
TEL 2 7341
TLF 23561 ECFBAN BG
TLX 23561 ECF BAN BG
EML
COM

RAIMOND ERNST DR
NFRA
BOX 2
NL 7990 AA DWINGELOO
NETHERLANDS
TEL 52 19 7244
TLF 52 19 7332
TLX 42043 SRZM NL
EML EXR@FRA NL
COM 05,08,34,40

RAINE DEREK J DR
DPT OF ASTRONOMY
UNIVERSITY OF LEICESTER
UNIVERSITY RD
LEICESTER LE1 7RH
UK
TEL 533 554 455
TLF
TLX 341198 LEICUL
EML
COM

RAITALA JOUKO T
DPT OF ASTRONOMY
UNIVERSITY OF OULU
SF 90570 OULU 57
FINLAND
TEL 81 35 21 06
TLF 81 56 1278
TLX 32375
EML
COM

RAJAMOHAN R DR
INDIAN INSTITUTE OF
ASTROPHYSICS
KORAMANGALA
BANGALORE 560 034
INDIA
TEL 812 56 6497/6585
TLF
TLX 845763 IIAB IN
EML
COM 20,51

RAJCHL JAROSLAV DR
ASTRONOMICAL INSTITUTE
CZECH ACADEMY OF SCIENCES
ONDREJOV OBSERVATORY
CS 251 65 ONDREJOV
CZECHOSLOVAKIA
TEL 204 85201
TLF 204 85314
TLX 121579
EML
COM 22

RAJU P K DR
INDIAN INSTITUTE OF
ASTROPHYSICS
KORAMANGALA
BANGALORE 560 034
INDIA
TEL 812 56 6585
TLF
TLX 845763 IIAB IN
EML
COM

RAKAVY GIDEON PROF
EINSTEIN INST OF PHYSICS
HEBREW UNIV OF JERUSALEM
JERUSALEM 91904
ISRAEL
TEL
TLF
TLX
EML
COM

RAKOS KARL D PROF
INSTITUT FUER ASTRONOMIE
UNIVERSITAET WIEN
TUERKENSCHANZSTR 17
A 1180 WIEN
AUSTRIA
TEL 1 345 36095
TLF
TLX 133099 VIAST A
EML
COM 09,26,27,42

RAKSHIT H PROF
BENGAL ENGINEERG COLLEGE
SIBPORE
HEWRAH
INDIA
TEL
TLF
TLX
EML
COM

RAM SAGAR DR
INDIAN INSTITUTE OF
ASTROPHYSICS
KORAMANGALA
BANGALORE 560 034
INDIA
TEL 812 56 6585/6497
TLF
TLX 845763 IIAB IN
EML
COM 37

RAMADURAI SOURIRAJA DR
TIFR
HOMI BHABHA RD
COLABA
BOMBAY 400 005
INDIA
TEL 22 495 2311
TLF
TLX 011 3009 TIFR IN
EML shakti¦t¦fr¦tap15@uunet uu net
COM· 35,46,48

RAMAMURTHY SWAMINATHAN
CASA
UNIVERSITY OF OSMANIA
HYDERABAD 500 007
INDIA
TEL 85 1672
TLF
TLX
EML
COM

RAMANA MURTHY P V DR
TIFR
HOMI BHABHA RD
COLABA
BOMBAY 400 005
INDIA
TEL 22 495 2979
TLF
TLX 011 3009 TIFR IN
EML
COM·

RAMATY REUVEN DR
NASA/GSFC
CODE 665
LAB HIGH ENERGY ASTROPHYS
GREENBELT MD 20771
USA
TEL 301 286 8715
TLF
TLX
EML
COM 40

RAMELLA MASSIMO
OAT
BOX SUCC TRIESTE 5
VIA TIEPOLO 11
I 34131 TRIESTE
ITALY
TEL   40 76 8506
TLF
TLX   461137 OAT I
EML
COM   47

RAMOS ISABEL FERRO DR
INST GEOPHYS & ASTRONOMY
CALLE 212 N 2906/29 Y 31
LISA
LA HABANA
CUBA
TEL   21 8416/0644
TLF
TLX   511240 GEOAS CU
EML
COM

RAMPAZZO ROBERTO DR
OSS ASTRONOMICO DI BRERA
VIA BRERA 28
I 20121 MILANO
ITALY
TEL   2 87 4444
TLF   39 59 8492
TLX
EML   RAMPAZZO@ASTMIB ASTRO IT
COM   28

RAMSAY DONALD A DR
HERZBERG INST ASTROPHYS
NTL RESEARCH COUNCIL
100 SUSSEX DR
OTTAWA ON K1A OR6
CANADA
TEL   613 990 0919
TLF· 613 952 0974
TLX· 053 3715
EML   BITNET·DAR@NRCVMO1
COM   20

RAMSEY LAWRENCE W DR
DPT OF ASTRONOMY
PENNSYLVANIA STATE UNIV
525 DAVEY LAB
UNIVERSITY PARK PA 16802
USA
TEL   814 865 0418
TLF
TLX
EML
COM   09,36

RANA NARAYAN CHANDRA DR
TIFR/THEOR ASTRO GROUP
HOMI BHABHA RD
COLABA
BOMBAY 400 005
INDIA
TEL   22 495 2972
TLF
TLX   011 3009
EML
COM   28

RANDIC LEO PROF DR
GEODETICAL FACULTY
GUNDULICEVA 54
YU 41000 ZAGREB
YUGOSLAVIA
TEL   41 446 675
TLF
TLX
EML
COM   19,31

RANIERI MARCELLO
IAS
CNR
CP 67
I 00044 FRASCATI
ITALY
TEL   6 942 5655
TLF
TLX
EML
COM

RANK DAVID M PROF
LICK OBSERVATORY
UNIVERSITY OF CALIFORNIA
SANTA CRUZ CA 95064
USA
TEL   408 429 2277
TLF
TLX
EML
COM

RANKIN JOANNA M DR
PHYSICS DPT
UNIVERSITY OF VERMONT
A405 COOK BUILDING
BURLINGTON VT 05405
USA
TEL   802 656 2644
TLF
TLX   510 299 0021
EML   RANKIN%MERLIN UVM-GEN UVM EDU
COM

RAO A PRAMESH DR
TATA INST OF FUNDAMENTAL
RESEARCH
POONA UNIVERSITY CAMPUS
PUNE 411 007
INDIA
TEL   212 33 7107
TLF: 212 33 5760
TLX. 0145658 GMRT IN
EML   uunet!shakti!gmrt!pramesh
COM   10,40

RAO K NARAHARI
DPT OF PHYSICS
OHIO STATE UNIVERSITY
174 W 18TH AVE
COLUMBUS OH 43210
USA
TEL   614 422 6505
TLF
TLX
EML
COM   14

RAO K RAMANUJA DR
C/O DR K SURENDRA
RUA CEL JOAO CURSINO 210
APT 92 VILA ADYANA
12200 S JOSE DOS CAMPOS
BRAZIL
TEL
TLF.
TLX
EML
COM

RAO M N DR
PHYSICAL RESEARCH LAB
NAVRANGPURA
AHMEDABAD 380 009
INDIA
TEL   272 46 2129
TLF   272 44 5292
TLX   121397
EML
COM· 16

RAO N KAMESWARA
INDIAN INSTITUTE OF
ASTROPHYSICS
KORAMANGALA
BANGALORE 560 034
INDIA
TEL   812 56 6585
TLF
TLX
EML
COM   27,29

RAO P VIVEKANANDA DR
DPT OF ASTRONOMY
UNIVERSITY OF OSMANIA
HYDERABAD 500 007
INDIA
TEL   71 951
TLF
TLX
EML
COM   25

RAO RAMACHANDRA V PROF
ISRO SATELLITE CENTER
PEENYA
BANGALORE 560 058
INDIA
TEL
TLF
TLX
EML
COM   44

RAOULT ANTOINETTE DR
OBSERVATOIRE DE PARIS
SECTION DE MEUDON
DASOP
F 92195 MEUDON PPL CDX
FRANCE
TEL   1 45 07 7766
TLF
TLX   200590
EML
COM· 10,12,40

RAPAPORT MICHEL DR
OBSERVATOIRE OE BORDEAUX
BP 89
F 33270 FLOIRAC
FRANCE
TEL   56 86 4330
TLF   56 40 4251
TLX
EML·
COM   20,21

RAPLEY CHRISTOPHER G DR
MULLARD SPACE SCIENCE LAB
UNIVERSITY COLLEGE LONDON
HOLMBURY ST MARY
DORKING SURREY RH5 6NT
UK
TEL. 306 702 92
TLF
TLX   859185
EML
COM

RATNATUNGA KAVAN U
NASA/GSFC
CODE 630
SDCD
GREENBELT MD 20771
USA
TEL 301 286 6276
TLF 301 286 3221
TLX 089675 NASCOM GBLT
EML KAVAN@OFTBIT, SDCCL KAVAN
COM 05,30,33

RAUBENHEIMER BAREND C PR
COSMIC RAY RESEARCH UNIT
POTCHEFSTROOM UNIVERSITY
POTCHEFSTROOM 2520
SOUTH AFRICA
TEL 01481-27511
TLF
TLX 421363
EML
COM 48

RAUTELA B S DR
UTTAR PRADESH STATE
OBSERVATORY
PO MANORA PEAK 263 129
NAINITAL 263 129
INDIA
TEL 59 42 2136
TLF
TLX
EML
COM 29,45

RAWLINGS JONATHAN DR
DPT OF ASTROPHYSICS
UNIVERSITY OF OXFORD
KEBLE RD
OXFORD OX1 3RH
UK
TEL 338 652 73292
TLF 865 273 418
TLX 83295 NUCLOX G
EML JR@UK AC OX ASTRO
COM 34

RAWLINGS STEVEN DR
DPT OF ASTROPHYSICS
UNIVERSITY OF OXFORD
KEBLE RD
OXFORD OX1 3RH
UK
TEL 865 273 303
TLF 865 273 418
TLX 83295 NUCLOX G
EML SR@UK AC OX ASTRO
COM 47

RAY ALAK DR
TIFR
HOMI BHABHA RD
COLABA
BOMBAY 400 005
INDIA
TEL 22 215 2971
TLF 22 215 2110
TLX 1183009 TIFR IN
EML AKR@TIFRVAX
COM 35

RAY CHOUDHURI ARNAB DR
DPT OF PHYSICS
INDIAN INSTITUTE OF SCI
BANGALORE 560 012
INDIA
TEL 812 34 4411
TLF
TLX
EML
COM 10

RAY THOMAS P
DIAS
SCHOOL OF COSMIC PHYSICS
5 MERRION SQ
DUBLIN 2
IRELAND
TEL 1 774 321
TLF
TLX 31687 DIAS EI
EML
COM 40

RAYCHAUDHURI AMALKUMAR DR
PRESIDENCY COLLEGE
COLLEGE ST
CALCUTTA 73
INDIA
TEL
TLF
TLX
EML
COM 47

RAYMOND JOHN CHARLES
CENTER FOR ASTROPHYSICS
HCO/SAO
60 GARDEN ST
CAMBRIDGE MA 02138
USA
TEL
TLF
TLX
EML
COM 34

RAYROLE JEAN R DR
OBSERVATOIRE DE PARIS
SECTION DE MEUDON
F 92195 MEUDON PPL CDX
FRANCE
TEL 1 45 07 7789
TLF
TLX
EML
COM 10

RAZDAN HIRALAL
BHABHA ATOMIC RES CTR
ZAKURA SRINIGAR
KASHMIR 190 006
INDIA
TEL
TLF
TLX
EML
COM 48

RAZIN V A DR
RADIOPHYSICAL RESEARCH
INSTITUTE
LYADOV UL 25/14
603600 N NOVGOROD
RUSSIA
TEL 36-72-94
TLF
TLX
EML
COM 40

READHEAD ANTHONY C S DR
CALTECH
ROBINSON BLDG
PASADENA CA 91125
USA
TEL 213 356 4972
TLF
TLX 675425 CALTECH PSD
EML
COM 40,49

REASENBERG ROBERT D DR
CENTER FOR ASTROPHYSICS
HCO/SAO RM B 217
60 GARDEN ST
CAMBRIDGE MA 02138
USA
TEL 617 495 7108
TLF
TLX 921428 SATELITE CAM
EML
COM 04

REAVES GIBSON PROF
DEPT OF ASTRONOMY
UNIV SOUTHERN CALIFORNIA
LOS ANGELES CA 90089 1342
USA
TEL 213 743 2039
TLF
TLX
EML
COM 28

REAY NEWRICK K DR
ASTROPHYSICS GROUP
IMPERIAL COLLEGE
BLACKETT LABORATROY
LONDON SW7 2BZ
UK
TEL 1 589 5111*6669
TLF
TLX 261503 IMPCOL
EML
COM 09,49,51

REBEIROT EDITH DR
OBSERVATOIRE DE MARSEILLE
2 PLACE LE VERRIER
F 13248 MARSEILLE CDX 04
FRANCE
TEL 91 95 9088
TLF
TLX 420241
EML
COM 28,30,33

REBER GROTE DR
C/O POST OFFICE
BOTHWELL TAS 7030
AUSTRALIA
TEL 2 23 7371
TLF
TLX
EML
COM 40

REBOLO RAFAEL DR
INST DE ASTROFISICA
DE CANARIAS
OBS DEL TEIDE
E 38200 LA LAGUNA
SPAIN
TEL 22 26 2211
TLF
TLX 92640
EML SPAN IAC RRL
COM 29

RECILLAS-CRUZ ELSA DR
INSTITUTO DE ASTRONOMIA
UNAM
APDO POSTAL 70-264
04510 MEXICO DF
MEXICO
TEL
TLF
TLX
EML
COM

REDFERN MICHAEL R DR
DPT OF PHYSICS
UNIVERSITY COLLEGE
GALWAY
IRELAND
TEL 41 24411
TLF
TLX 50023
EML
COM 09

REED B CAMERON DR
DPT OF PHYSICS
ST MARY'S UNIVERSITY
HALIFAX NS B3H 3C3
CANADA
TEL 902 420 5830
TLF 902 420 5561
TLX
EML BITNET REED@STMARYS
COM

REES DAVID ELWYN DR
CSIRO
DIVISION OF RADIOPHYSICS
BOX 76
EPPING NSW 2121
AUSTRALIA
TEL 2 868 0493
TLF 2 868 0411
TLX
EML drees@rp csiro au
COM 10,12

REES MARTIN J PROF
INSTITUTE OF ASTRONOMY
THE OBSERVATORIES
MADINGLEY RD
CAMBRIDGE CB3 0HA
UK
TEL 223 622 04
TLF
TLX 817297 ASTRON G
EML
COM 44,47C,48,51

REEVES EDMOND M DR
NASA HEADQUARTERS
CODE EM
600 INDEPENDENCE AVE SW
WASHINGTON DC 20546
USA
TEL 202 453 1571
TLF
TLX 89530
EML
COM 10,12,44

REEVES HUBERT PROF
CEA CEN
SEP-SES BAT 28
BP 2
F 91191 GIF/YVETTE CDX
FRANCE
TEL 1 69 08 5159
TLF
TLX
EML
COM 10,35,47C,48

REFSDAL S PROF DR
HAMBURGER STERNWARTE
GOJENSBERGSWEG 112
D 2050 HAMBURG 80
GERMANY
TEL 40 7252 4124
TLF
TLX 217884
EML
COM 42,47

REGEV ODED DR
DPT OF PHYSICS
IIT
TECHNION CITY
HAIFA 32000
ISRAEL
TEL 4 293 992
TLF 4 221 514
TLX
EML PHR910R@TECHNION
COM

REGLERO-VELASCO VICTOR DR
DPT MATEMATICA Y ASTRON
UNIVERSIDAD DE VALENCIA
BURJASOT
E 46100 VALENCIA
SPAIN
TEL 6 386 4326
TLF 6 386 4302
TLX 61071
EML
COM 25,42

REGO FERNANDEZ M DR
DPT DE ASTROFISICA
FAC DE FISICA
UNIVERSIDAD COMPLUTENSE
E 28040 MADRID
SPAIN
TEL 1 449 5316
TLF
TLX 47273 FF UC
EML
COM 29

REGULO CLARA DR
INST DE ASTROFISICA
DE CANARIAS
OBS DEL TEIDE
E 38200 LA LAGUNA
SPAIN
TEL 22 26 2211
TLF
TLX 92640
EML SPAN IAC ·CRR
COM 10,12

REICH WOLFGANG
MPI FUER RADIOASTRONOMIE
AUF DEM HUEGEL 69
D 5300 BONN 1
GERMANY
TEL
TLX 886440
EML
TLF
COM 40

REICHERT GAIL ANNE DR
NASA/GSFC
CODE 666
GREENBELT MD 20771
USA
TEL 301 286 5307
TLF 301 286 3391
TLX
EML IUE REICHERT/ZBGAR@SCFVM
COM 28,48

REID MARK JONATHAN DR
CENTER FOR ASTROPHYSICS
HCO/SAO
60 GARDEN ST
CAMBRIDGE MA 02138
USA
TEL 617 495 7470
TLF
TLX 921428 SATELLITE CAM
EML
COM 40

REID NEILL
CALTECH
PALOMAR OBS MS 105 24
PASADENA CA 91101
USA
TEL 818 356 6586
TLF
TLX
EML
COM 33

REIF KLAUS DR
RADIOASTRONOMISCHES INST
DER UNIVERSITAET BONN
AUF DEM HUEGEL 71
D 5300 BONN 1
GERMANY
TEL 228 73 3657
TLF
TLX
EML u 145ref%mpiehu%unido uucp
COM 33,40

REIMERS DIETER PROF
HAMBURGER STERNWARTE
UNIVERSITAET HAMBURG
GOJENSBERGSWEG 112
D 2050 HAMBURG 80
GERMANY
TEL 40 7252 4112
TLF
TLX 217884
EML
COM 29,36

REINISCH GILBERT DR
OCA OBSERV DE NICE
BP 139
F 06003 NICE CDX
FRANCE
TEL 93 89 0420
TLF
TLX
EML
COM

REIPURTH BO
ESO
CASILLA 19001
SANTIAGO 19
CHILE
TEL 2 698 8757
TLF
TLX 240881
EML
COM 34

REITSEMA HAROLD J
BALL AEROSPACE SYSTEMS
DIVISION
BOX 1062
BOULDER CO 80306
USA
TEL 303 441 5026
TLF
TLX
EML
COM 15,20

REIZ ANDERS PROF
LOVSPRINGSVEJ 3 B
DK 2920 CHARLOTTENLUND
DENMARK
TEL 31 63 2536
TLF
EML
COM 08,35
TLX

REMY BATTIAU LILIANE G A
CONSEIL DE LA RECHERCHE
UNIVERSITE DE LIEGE
7 PLACE DU XX AOUT
B 4000 COINTE-LIEGE
BELGIUM
TEL 41 42 0080
TLF
TLX 41397 UNIV ULG
EML
COM 05,15

RENGARAJAN T N DR
TIFR/IR ASTRONOMY
HOMI BHABHA RD
COLABA
BOMBAY 400 005
INDIA
TEL 22 219 111
TLF
TLX 011-3009 TIFR IN
EML
COM 34,48

RENSE WILLIAM A DR
CASA
UNIVERSITY OF COLORADO
BOX 391
BOULDER CO 80309 0391
USA
TEL 303 492 0111
TLF
TLX
EML
COM 44

RENSON P F M DR
INSTITUT D'ASTROPHYSIQUE
UNIVERSITE DE LIEGE
AVE COINTE 5
B 4000 COINTE-LIEGE
BELGIUM
TEL 41 52 9980
TLF 41 52 7474
TLX
EML U2148AP@BLIULG11
COM 05,27

RENZINI ALVIO PROF
DPT DI ASTRONOMIA
VIA ZAMBONI 33
I 40126 BOLOGNA
ITALY
TEL 51 22 2956
TLF
TLX 211664 INFNBO
EML ASTB03 ALVIO
COM 35C,37

REPHAELI YOEL DR
SCHOOL OF PHYSICS
TEL AVIV UNIVERSITY
RAMAT AVIV
TEL AVIV 69978
ISRAEL
TEL
TLF
TLX 342171 VERSY IL
EML
COM 28

REQUIEME YVES DR
OBSERVATOIRE DE BORDEAUX
BP 89
F 33270 FLOIRAC
FRANCE
TEL 56 86 4330
TLF 56 40 4251
TLX
EML
COM 08C,24

REVELLE DOUGLAS ORSON DR
METEOROLOGY PROGRAM
GEOGRAPHY DPT
815 DAVIS HALL
DEKALB IL 60115
USA
TEL 815 753 0631
TLF
TLX
EML
COM 15,22

REYES FRANCISCO DR
DPT OF ASTRONOMY
UNIVERSITY OF FLORIDA
211 SSRB
GAINESVILLE FL 32611
USA
TEL 904 392 2049
TLF
TLX
EML BITNET FREYES@UFPINE
COM 40

REYNOLDS JOHN DR
AAO
ATNF
BOX 76
EPPING NSW 2121
AUSTRALIA
TEL 2 868 0222
TLF 2 868 0400
TLX 26230 ASTRO AA
EML JREYNOLD@ATNF CSIRO AU
COM 08,40

REYNOLDS JOHN H PROF
PHYSICS DPT
UNIVERSITY OF CALIFORNIA
BERKELEY CA 94720
USA
TEL 415 642 4863
TLF
TLX 9103667114
EML
COM

REYNOLDS RONALD J DR
DPT OF PHYSICS
UNIVERSITY OF WISCONSIN
1150 UNIVERSITY AVE
MADISON WI 53706
USA
TEL 608 262 5916
TLF
TLX
EML
COM 34

REYNOLDS STEPHEN P
DPT OF PHYSICS
N CAROLINA STATE UNIV
BOX 8202
RALEIGH NC 27695 8202
USA
TEL 919 737 7751
TLF
TLX
EML
COM

REZENDE COSTA JOAQUIM DR
CRAAE/ESCOLA POLITECNICA
UNIVERSIDADE DE SAO PAULO
CP 8174
05508 SAO PAULO SP
BRAZIL
TEL 11 815 6289
TLF 11 815 4272
TLX 1180127 INPE BR
EML
COM 10

RHODES EDWARD J JR
11801 KILLIMORE AVE
NORTHRIDGE CA 91326
USA
TEL
TLF
EML
COM·
TLX

RIBES ELIZABETH DR
OBSERVATOIRE DE PARIS
SECTION DE MEUDON
F 92195 MEUDON PPL CDX
FRANCE
TEL 1 45 07 7786
TLF
TLX
EML
COM

RIBES JEAN-CLAUDE DR
INSU
77 AVE DENFERT ROCHEREAU
F 75014 PARIS
FRANCE
TEL 1 43 20 1330
TLF
TLX
EML.
COM 40

RICE JOHN B DR
DPT PHYSICS & ASTRONOMY
UNIVERSITY OF BRANDON
2500 UNIVERSITY DR NW
BRANDON MB R7A 6A9
CANADA
TEL 204 727 9693
TLF
TLX 075 02721
EML
COM

RICHARDS MERCEDES T DR
UNIVERSITY STATION
UNIVERSITY OF VIRGINIA
BOX 3818
CHARLOTTESVILLE VA 22903
USA
TEL. 804 924 4895
TLF 804 924 3104
TLX
EML. MTR8R@VIRGINIA EDU
COM 42

RICHARDSON E HARVEY DR
HERZBERG INST ASTROPHYS
DOMINION ASTROPHYS OBS
5071 W SAANICH RD
VICTORIA BC V8X 4M6
CANADA
TEL 604 388 0001
TLF 604 363 0045
TLX 049 7295
EML
COM 09

RICHARDSON KEVIN J
DPT OF PHYSICS
QUEEN MARY/WESTFIELD COLL
MILE END RD
LONDON E1 4NS
UK
TEL
TLF
TLX
EML
COM

RICHARDSON R S
GRIFFITH OBSERVATORY
BOX 27787
LOS FELIX STATION
LOS ANGELES CA 90027
USA
TEL
TLF
TLX
EML
COM

RICHER HARVEY B DR
DPT GEOPHYS & ASTRONOMY
UNIV OF BRITISH COLUMBIA
2075 WESBROOK PL
VANCOUVER BC V6T 1W5
CANADA
TEL 604 228 4134
TLF 604 228 6047
TLX
EML
COM 28,37

RICHSTONE DOUGLAS O DR
DPT OF ASTRONOMY
UNIVERSITY OF MICHIGAN
DENNISON BLDG
ANN ARBOR MI 48109 1090
USA
TEL 313 764 3441
TLF
TLX
EML D_Richstone@ub cc umich edu
COM 28

RICHTER G A DR
ZNTRLINST F ASTROPHYSIK
STERNWARTE BABELSBERG
ROSA-LUXEMBURG-STR 17A
D 6400 SONNEBERG
GERMANY
TEL 96 74 2287
TLF
TLX 6288180
EML
COM 27

RICHTER JOHANNES PROF
INST F EXPERIMENT PHYSIK
PHYSIKZENTRUM
OLSHAUSENSTRASSE
D 2300 KIEL 1
GERMANY
TEL 431 880 3835
TLF
TLX 292706
EML
COM 14

RICHTLER TOM DR
STERNWARTE DER UNIV BONN
AUF DEM HUGEL 71
D 5300 BONN 1
GERMANY
TEL 228 73 3669
TLX 886440
EML 45228090071 RICHTLER
TLF
COM 37

RICKARD JAMES JOSEPH DR
BOX 777
BORREGO SPRINGS CO 92004
USA
TEL 714 767 5462
TLF
EML
COM
TLX

RICKARD LEE J DR
NAVAL RESEARCH LABORATORY
CODE 4138 RRD
4555 OVERLOOK AVE SW
WASHINGTON DC 20375 5000
USA
TEL 202 767 2495
TLF
TLX
EML
COM 34,40

RICKER GEORGE R DR
CENTER FOR SPACE RESEARCH
MIT RM 37 527
BOX 165
CAMBRIDGE MA 02139
USA
TEL 617 253 7532
TLF
TLX 92-14-73
EML
COM

RICKETT BARNABY JAMES DR
DPT OF ELECTRICAL ENGIN
& COMPUTER SCIENCE
UCSD
LA JOLLA CA 92093 0216
USA
TEL 619 452 2731
TLF
TLX
EML
COM 40,49

RICKMAN HANS DR
ASTRONOMICAL OBSERVATORY
BOX 515
S 751 20 UPPSALA
SWEDEN
TEL 18 11 3522
TLX 76024 UNIVUPS
EML
TLF
COM 15C,20C

RICORT GILBERT DR
DPT ASTROPHYSIQUE
UNIVERSITE DE NICE
PARC VALROSE
F 06034 NICE CDX
FRANCE
TEL 93 51 9100
TLF
TLX
EML·
COM

RIDDLE ANTHONY C DR
700 GRANT PL

BOULDER CO 80302
USA
TEL 303 447 8127
TLF
TLX
EML
COM 49

RIEGEL KURT W DR
NTL SCIENCE FOUNDATION
3019 N OAKLAND ST
ARLINGTON VA 22207
USA
TEL 202 357 9450
TLF
TLX
EML kriegel@note nsf gov
COM 33

RIEGER ERICH DR
MAX PLANCK INSTITUT FUER
EXTRATERRESTRISCHE PHYSIK
KARL SCHWARWSCHILDSTR 1
D 8046 GARCHING MUENCHEN
GERMANY
TEL 89 329 93511
TLF 89 329 93539
TLX
EML EIR@DGAIPP1S
COM 10

RIEGLER GUENTER R DR
NASA HEADQUARTERS
CODE SZE
600 INDEPENDENCE AVE SW
WASHINGTON DC 20546
USA
TEL 202 453 1435
TLF
TLX 89530
EML AMES "GRIEGLER@NASAMAIL"
COM 05,44

RIGHINI ALBERTO PROF
DPT DI ASTRONOMIA
UNIVERSITA DI FIRENZE
LARGO E FERMI 5
I 50125 FIRENZE
ITALY
TEL 55 27521
TLF 55 22 0039
TLX 572268 ARCETRI I
EML
COM

RIGHINI-COHEN GIOVANNA DR
DPT OF EARTH & SPACE SCI
ASTRONOMY PROGRAM
SUNY AT STONY BROOK
STONY BROOK NY 11794 2100
USA
TEL
TLF
TLX
EML
COM 12,34,44

RIGUTTI MARIO PROF
OSS ASTRONOMICO
DI CAPODIMONTE
VIA MOIARIELLO 16
I 80131 NAPOLI
ITALY
TEL 440101
TLF
TLX
EML
COM 12

RIIHIMAA JORMA J DR
DPT OF ASTRONOMY
UNIVERSITY OF OULU
SF 90570 OULU 57
FINLAND
TEL
TLF 81 56 1278
TLX
EML
COM 40,51

RIJNBEEK RICHARD DR
DPT MATHEMATICAL SCIENCE
UNIVERSITY OF ST ANDREWS
FIFE KY16 9SS
UK
TEL 334 761 61*8186
TLF 334 744 87
TLX
EML RIJNBEEK@CS ST_ANDREWS AC UK
COM 10

RILEY JULIA M DR
MULLARD RADIO ASTRON OBS
CAVENDISH LABORATORY
MADINGLEY RD
CAMBRIDGE CB3 OHE
UK
TEL 223 664 77
TLF
TLX 81292
EML
COM 40

RINDLER WOLFGANG PROF
UNIVERSITY OF TEXAS
UTD
BOX 830688
RICHARDSON TX 75083 0688
USA
TEL 214 690 2885
TLF
TLX 791-880
EML
COM 47

RING JAMES PROF
BLACKETT LABORATORY
IMPERIAL COLLEGE
PRINCE CONSORT RD
LONDON SW7 2BZ
UK
TEL 1 589 5111
TLF
TLX 261503
EML
COM 09

RINGNES TRULS S DR
INST THEORET ASTROPHYSICS
UNIVERSITY OF OSLO
BOX 1029
N 0315 BLINDERN OSLO 3
NORWAY
TEL 472-456-503
TLF
TLX 72425 UNIOS N
EML
COM

RINGUELET ADELA E DR
49 342
1900 LA PLATA (BS AS)
ARGENTINA
TEL 21 3 1063
TLF
EML
COM 29
TLX 31151 BULAP AR

RIPKEN HARTMUT W DR
DARA GERMAN SPACE AGENCY
KOENIGWINTERERSTRASSE
D 5300 BONN 3
GERMANY
TEL 228 44 7313
TLX
EML
TLF
COM 21,22,49VP

RITTER HANS DR
MAX PLANCK INSTITUT FUER
PHYSIK UND ASTROPHYSIK
KARL SCHARZSCHILD STR 1
D 8046 GARCHING MUENCHEN
GERMANY
TEL
TLF
TLX
EML
COM 35,42

RIVOLO ARTHUR REX
DPT ASTRON & ASTROPHYS
UNIV OF PENNSYLVANIA
PHILADELPHIA PA 19104
USA
TEL 215 898 6250
TLF
TLX 358300
EML
COM 47

RIZVANOV NAUFAL G DR
ENGELHARDT ASTRONOMICAL
OBSERVATORY
OBSERVATORIA STATION
422526 KAZAN
RUSSIA
TEL 324827
TLF
TLX
EML
COM 24

ROACH FRANKLIN E
1241 SADDLEBACK
COTTONWOOD AZ 86326
USA
TEL
TLF
EML
COM 21,49
TLX

ROARK TERRY P PROF
UNIVERSITY OF WYOMING
BOX 3434E
LARAMIE
WYOMING 82071
USA
TEL 307 766 4121
TLF
TLX
EML
COM

ROBB RUSSEL M
DPT OF PHYSICS & ASTRON
UNIVERSITY OF VICTORIA
BOX 1700
VICTORIA BC V8W 2Y2
CANADA
TEL 604 721 7750
TLF 604 721 7715
TLX
EML BITNET ROBB@UVPHYS
COM 42

ROBBINS R ROBERT PROF
ASTRONOMY DPT
UNIVERSITY OF TEXAS
RLM 15 308
AUSTIN TX 78712 1083
USA
TEL 512 471 7312
TLF
TLX
EML
COM 34,46C

ROBE H A G DR
INSTITUT D'ASTROPHYSIQUE
UNIVERSITE DE LIEGE
AVE COINTE 5
B 4000 COINTE-LIEGE
BELGIUM
TEL 41 52 9980
TLF 41 52 7474
TLX 41264 ASTRLG
EML
COM·

ROBERGE WAYNE G DR
DPT OF PHYSICS
RENSSELAER POLYTECHN INST
TROY NY 12180 3590
USA
TEL 518 276 6454
TLF
TLX
EML ROBERGE@ORION PHYS RPI EDU
COM 34

ROBERTI GIUSEPPE DR
IST DI FISICA
UNIVERSITA DI NAPOLI
MOSTRA D OLTREMARE PAD 19
I 80125 NAPOLI
ITALY
TEL
TLF
TLX
EML
COM 12

ROBERTS DAVID HALL DR
DPT OF PHYSICS
BRANDEIS UNIVERSITY
WALTHAM MA 02254
USA
TEL 617 647 2846
TLF
TLX 703013
EML
COM 40,47

ROBERTS MORTON S DR
NRAO
EDGEMONT RD
CHARLOTTESVILLE VA 22903
USA
TEL
TLX
EML. NRAO MROBERTS
TLF
COM 28,33,38C,40,EC

ROBERTS WILLIAM W JR PROF
DPT OF APPLIED MATHS
UNIVERSITY OF VIRGINIA
THORNTON HALL
CHARLOTTESVILLE VA 22901
USA
TEL 804 924 1038
TLF
TLX
EML
COM 28,33,34

ROBERTSON DOUGLAS S
NGS
N/CG 114 NOAA
11400 ROCKVILLE PIKE
ROCKVILLE MD 20852
USA
TEL 301 443 8423
TLF
TLX
EML
COM 19C,31,40

ROBERTSON JAMES GORDON DR
DPT OF ASTROPHYSICS
SCHOOL OF PHYSICS
UNIVERSITY OF SYDNEY
SYDNEY NSW 2006
AUSTRALIA
TEL
TLF
TLX
EML
COM 28,40

ROBERTSON JOHN ALISTAIR
DPT OF APPLIED MATHS
UNIVERSITY OF ST ANDREWS
NORTH HAUGH
ST ANDREWS FIFE KY16 9SS
UK
TEL 334 761 61
TLF
TLX 76213
EML
COM 42

ROBERTSON NORNA DR
DPT OF PHYSICS & ASTRON
UNIVERSITY OF GLASGOW
GLASGOW G12 8QQ
UK
TEL 413 398 855
TLF 413 349 029
TLX 777070 UNIGLA
EML GWO5@UK AC GLA PH I1
COM 09

ROBILLOT JEAN-MAURICE DR
OBSERVATOIRE DE BORDEAUX
BP 89
F 33270 FLOIRAC
FRANCE
TEL 56 86 4330
TLF 56 40 4251
TLX
EML
COM

ROBIN ANNIE C DR
OBSERVATOIRE DE BESANCON
BP 1615
F 25010 BESANCON CDX
FRANCE
TEL 81 66 6941
TLF 81 66 6944
TLX· 361 144
EML ROBIN@FROBES51/OBSBEA ANNIE
COM 33

ROBINSON BRIAN J DR
CSIRO
DIVISION OF RADIOPHYSICS
BOX 76
EPPING NSW 2121
AUSTRALIA
TEL 2 868 0222
TLF
TLX 26230 ASTRO AA
EML
COM. 33,34,40

ROBINSON EDWARD LEWIS DR
ASTRONOMY DPT
UNIVERSITY OF TEXAS
RLM 15 308
AUSTIN TX 78712 1083
USA
TEL 512 471 3401
TLF
TLX
EML
COM 25,27,42

ROBINSON GARRY DR
DPT PHYSICS UNIV COLLEGE
UNIVER OF NEW SOUTH WALES
NORTHCOTT DR
CAMPBELL ACT 2600
AUSTRALIA
TEL 62 68 8800
TLF 6 268 8786
TLX
EML GARRY@PHADFA PH ADFA OZ A
COM 34

ROBINSON I PROF
UNIVERSITY OF TEXAS
MS BE 32
BOX 688
RICHARDSON TX 75080
USA
TEL 214 690 2176
TLF
TLX
EML.
COM 47

ROBINSON JR RICHARD D DR
AAO
BOX 296
EPPING NSW 2121
AUSTRALIA
TEL 2 868 1666
TLX 23999
EML
TLF·
COM· 10,40

ROBINSON LEIF J
SKY & TELESCOPE
49 BAY STATE RD
CAMBRIDGE MA 02238
USA
TEL 617 864 7360
TLF
TLX
EML
COM 46,51

ROBINSON LLOYD B DR
LICK OBSERVATORY
UNIVERSITY OF CALIFORNIA
SANTA CRUZ CA 95064
USA
TEL 408 429 2437
TLF
TLX
EML
COM 09

ROBINSON WILLIAM J DR
DPT OF MATHEMATICS
UNIVERSITY OF BRADFORD
BRADFORD BD7 1DP
UK
TEL. 274 73 3466
TLF
TLX
EML
COM 07

ROBLEY R DR
9 ALLEE F VERDIER
F 31000 TOULOUSE
FRANCE
TEL 61 52 2273
TLF
EML
COM 21
TLX

ROBSON IAN E DR
SCHOOL OF PHYSICS & ASTRO
LANCASHIRE POLYTECHNIC
CORPORATION ST
PRESTON PR1 2TQ
UK
TEL 772 221 41*2188
TLF
TLX 677409 LANPOL
EML
COM

ROCA CORTES TEODORO
INST DE ASTROFISICA
DE CANARIAS
OBS DEL TEIDE
E 38071 LA LAGUNA
SPAIN
TEL 22 26 2211
TLF
TLX 92640
EML
COM 10,12

ROCCA-VOLMERANGE BRIGITTE
INSTITUT D'ASTROPHYSIQUE
98BIS BD ARAGO
F 75014 PARIS
FRANCE
TEL 1 43 20 1425
TLF 1 43 29 8673
TLX
EML
COM

ROCHE PATRICK F DR
DPT OF ASTROPHYSICS
UNIVERSITY OF OXFORD
KEBLE RD
OXFORD OX1 3RH
UK
TEL 865 273 338
TLF 865 273 418
TLX 83295 NUCLOX G
EML PFROUK AC OXFORD ASTRPHYSICS
COM 34

ROCHESTER MICHAEL G PROF
DPT OF EARTH SCIENCES
MEMORIAL UNIVERSITY OF
NEWFOUNDLAND
ST JOHNS NF A1B 3X7
CANADA
TEL 709 737 7565
TLF
TLX 016 4101
EML
COM 19

RODDIER CLAUDE DR
NOAO/ADP
BOX 26732
950 N CHERRY AVE
TUCSON AZ 85726 6732
USA
TEL
TLF
TLX 1561401
EML
COM 09

RODDIER FRANCOIS PROF
NOAO/ADP
BOX 26732
950 N CHERRY AVE
TUCSON AZ 85726 6732
USA
TEL 602 325 9220
TLF
TLX 0666484 AURA NOAO TU
EML
COM 09,12

RODGERS ALEX W DR
MOUNT STROMLO & SIDING
SPRING OBSERVATORIES
PRIVATE BAG
WODEN PO ACT 2606
AUSTRALIA
TEL 62 88 1111
TLF
TLX 62270 CANOPUS AA
EML
COM 27,29,46

RODMAN RICHARD B DR
65 LOCUST AVE
LEXINGTON MA 02173
USA
TEL 617 861 8149
TLF
TLX
EML
COM

RODONO MARCELLO DR
IST DI ASTRONOMIA
CITTA UNIVERSITARIA
VIA A DORIA 6
I 95125 CATANIA
ITALY
TEL 95 33 7308
TLF 95 33 0533
TLX 970359 ASTRCT I
EML ASTRCT MARCELLO
COM 27C,36,42VP

RODRIGO RAFAEL
INST ASTROFISICA
DE ANDALUCIA APD 2144
PROFESOR ALBAREDA 1
E 18080 GRANADA
SPAIN
TEL 58 12 1300
TLF
TLX 78573 IAAG E
EML
COM 16,21

RODRIGUEZ ELOY DR
INST ASTROFISICA
DE ANDALUCIA APD 3004
PROFESOR ALBAREDA 1
E 18080 GRANADA
SPAIN
TEL 58 12 1311
TLF 58 81 4530
TLX 78573 IAAG E
EML 16488 ELOY
COM 27

RODRIGUEZ LUIS F
INSTITUTO DE ASTRONOMIA
UNAM
APDO POSTAL 70-264
04510 MEXICO DF
MEXICO
TEL 905-548-5306
TLF
TLX 1760155 CICME
EML
COM 34C,40

RODRIGUEZ-ESPINOSA JOSE
INST DE ASTROFISICA
DE CANARIAS
OBS DEL TEIDE
E 38200 LA LAGUNA
SPAIN
TEL 22 26 2211*449
TLF 22 26 3005
TLX 92640 IACE
EML SPAN 28844 IRE
COM 28

RODRIGUEZ-VILLAMIL R DR
REAL INST Y OBSERVATORIO
DE LA ARMADA
CECILIO PUJAZON S/N
E 11110 SAN FERNANDO
SPAIN
TEL 56 88 3548
TLF 56 89 9302
TLX 76108 IOM E
EML
COM 07

ROEDER ROBERT C PROF
DPT OF PHYSICS
SOUTHWESTERN UNIVERSITY
UNIVERSITY AVENUE
GEORGETOWN TX 78626
USA
TEL 512 863 1633
TLF
TLX 910-350-1677
EML
COM 40,47

ROELFSEMA PETER DR
KAPTEYN ASTRONOMICAL INST
BOX 800
NL 9700 AV GRONINGEN
NETHERLANDS
TEL 50 63 4043
TLF 50 63 6100
TLX
EML PJUTROHGRRUG
COM 34,40

ROEMER ELIZABETH PROF
LUNAR & PLANETARY LAB
UNIVERSITY OF ARIZONA
TUCSON AZ 85721
USA
TEL 602 621 2897
TLF
TLX 467175
EML
COM 06C,15,20,24

ROEMER MAX PROF
INST F ASTROPHYSIK &
EXTRATERR FORSCHUNG
AUF DEM HUEGEL 71
D 5300 BONN 1
GERMANY
TEL 228 73 3670
TLF
TLX
EML
COM 10

ROENNAENG BERNT O DR
ONSALA SPACE OBSERVATORY
GOETEBORG UNIVERSITY
S 439 00 ONSALA
SWEDEN
TEL 30 06 2637
TLF
TLX 2400
EML
COM 40

ROESER HANS-PETER
MPI FUER RADIOASTRONOMIE
AUF DEM HUEGEL 69
D 5300 BONN 1
GERMANY
TEL 228 52 5265
TLF
TLX 886440
EML
COM 34,40

ROESER HERMANN-JOSEF DR
MPI FUER ASTRONOMIE
KOENIGSTUHL
D 6900 HEIDELBERG 1
GERMANY
TEL 62 215 28206
TLF
TLX 461789 MPIA D
EML
COM 28

ROESER SIEGFRIED DR
ASTRONOMISCHES RECHEN-
INSTITUT
MOENCHHOFSTR 12-14
D 6900 HEIDELBERG 1
GERMANY
TEL 62 214 9026
TLF
TLX 461336 ARIHD D
EML
COM 08,20,24

ROESSIGER SIEGFRIED DR
ZENTRALINST FUER ASTROPHY
STERNWARTE SONNEBERG
D 6400 SONNEBERG
GERMANY
TEL 96 74 2287
TLF: 96 74 2836
TLX 627180
EML.
COM· 05

ROGER ROBERT S DR
DOMINION RADIO ASTROPHYS
OBSERVATORY
BOX 248
PENTICTON BC V2A 6K3
CANADA
TEL 604 493 2277
TLF 604 493 7767
TLX. 048 88127
EML
COM 34,40

ROGERS ALAN E E DR
HAYSTACK OBSERVATORY

WESTFORD MA 01886
USA
TEL 617 692 4764
TLF
TLX 948149 HAYSTACK WFRD
EML
COM 34,40

ROGERS CHRISTOPHER DR
DOMINION RADIO ASTROPHYS
OBSERVATORY
BOX 248
PENTICTON BC V2A 6K3
CANADA
TEL 604 493 2277
TLF 604 493 7767
TLX 048 88127
EML CROGERS@DRAO NRC CA
COM

ROGERSON JOHN B PROF
DPT ASTROPHYSICAL SCI
PRINCETON UNIVERSITY
PEYTON HALL
PRINCETON NJ 08544 1001
USA
TEL 609 452 3806
TLF
TLX· 322409
EML·
COM

ROGSTAD DAVID H DR
JPL
MS 264 748
4800 OAK GROVE DR
PASADENA CA 91109
USA
TEL
TLF
TLX
EML
COM 40

ROHLFS K PROF DR
RUHR UNIVERSITAET BOCHUM
INSTITUT FUR ASTROPHYSIK
POSTFACH 102 148
D 4630 BOCHUM 1
GERMANY
TEL 234 70 05802
TLF
TLX 0825860
EML
COM 33,34,40

ROLAND GINETTE DR
INSTITUT D'ASTROPHYSIQUE
UNIVERSITE DE LIEGE
AVE COINTE 5
B 4000 COINTE-LIEGE
BELGIUM
TEL 41 52 9980
TLF 41 52 7474
TLX 41254 ASTRLG B
EML
COM 12

ROLLAND ANGEL DR
INST ASTROFISICA
DE ANDALUCIA APD 2144
PREFESOR ALBAREDA 1
E 18080 GRANADA
SPAIN
TEL· 58 12 1300
TLF
TLX 78573
EML
COM

ROMAN NANCY G DR
4260 N PARK AVE
APT 306W
CHEVY CHASE MD 20815
USA
TEL 301 656 6092
TLF.
TLX
EML
COM 05,44,45

ROMANCHUK PAVEL R DR
ASTRONOMICAL OBSERVATORY
KIEV STATE UNIVERSITY
OBSERVATORNAYA UL 3
252053 KIEV
UKRAINE
TEL
TLF
TLX
EML
COM 10

ROMANISHIN WILLIAM DR
DPT PHYSICS & ASTRONOMY
UNIVERSITY OF OKLAHOMA
NORMAN OK 73019
USA
TEL 405 325 3961
TLF
TLX
EML
COM

ROMANO GIULIANO PROF
V S ANTONIO DA PADOVA 7
I 31100 TREVISO
ITALY
TEL
TLF
EML
COM 27
TLX

ROMANOV YURI S DR
ASTRONOMICAL OBSERVATORY
ODESSA STATE UNIVERSITY
SHEVCHENKO PARK
270014 ODESSA
UKRAINE
TEL· 22 0396
TLF
TLX
EML
COM 27,30

ROMERO PEREZ M PILAR
INST DE ASTRON & GEODESIA
FAC DE CIENCIAS MATEMAT
UNIVERSIDAD COMPLUTENSE
E 28040 MADRID
SPAIN
TEL 1 244 2501
TLF
TLX
EML
COM 04

ROMNEY JONATHAN D DR
NRAO
EDGEMONT RD
CHARLOTTESVILLE VA 22903
USA
TEL 804 296 0242
TLX 910-997-0174
EML
TLF
COM 40

ROMPOLT BOGDAN DR
ASTRONOMICAL INSTITUTE
WROCLAW UNIVERSITY
UL KOPERNIKA 11
PL 51 622 WROCLAW
POLAND
TEL 071-48-24-34
TLF·
TLX 0712791 UWR PL
EML
COM 10

RONAN COLIN A
FLAT 6, BOURNE CT
THE BOURNE
HASTINGS
EAST SUSSEX TN34 3UZ
UK
TEL 424 446 362
TLF
TLX
EML
COM 41

RONG JIAN-XIANG
DPT OF ASTRONOMY
NANJING UNIVERSITY
NANJING
CHINA PR
TEL 25 34651*2882
TLF
TLX 34151 PRCNU CN
EML
COM 33

ROOD HERBERT J
INST FOR ADVANCED STUDY
SCHOOL OF NATURAL SCIENCE
PRINCETON NJ 08540
USA
TEL
TLF
TLX
EML
COM 28

ROOD ROBERT T DR
UNIVERSITY STATION
UNIVERSITY OF VIRGINIA
BOX 3818
CHARLOTTESVILLE VA 22903
USA
TEL 804 924 4904
TLF
TLX
EML
COM 35,51

ROOS NICOLAAS DR
STERREWACHT
BOX 9513
NL 2300 RA LEIDEN
NETHERLANDS
TEL 71 275 864
TLF
TLX
EML ROOS@HLERUL151 BITNET
COM 28

ROOSEN ROBERT G DR
RAINBOW OBSERVATORY
RR1
BOX 5068
PAHOA HI 96778
USA
TEL
TLF
TLX
EML
COM 21,22

ROOSEN ROBERT G DR
3760 TEXAS ST 24
SAN DIEGO CA 92014
USA
TEL 619 295 8724
TLF
EML
COM 21
TLX

ROQUES FRANCOISE DR
OBSERVATOIRE DE PARIS
SECTION DE MEUDON
LAM
F 92195 MEUDON PPL CDX
FRANCE
TEL 1 45 07 7409
TLF 1 45 07 7469
TLX 201571
EML MESIOA ROQUES/ROQUES@FRMEU51
COM 16

ROQUES SYLVIE DR
OBS MIDI PYRENEES
14 AVE E BELIN
F 31400 TOULOUSE CDX
FRANCE
TEL 61 25 2101
TLF
TLX 530776 OBSTLSE
EML
COM

ROSA DOROTHEA DR
EMIL-KURZ STR 4
D 8045 ISMANING
GERMANY
TEL 89 964 299
TLF
TLX
EML
COM

ROSA MICHAEL RICHARD DR
ESO
ST/ECF
KARL-SCHWARZSCHILD-STR 2
D 8046 GARCHING MUENCHEN
GERMANY
TEL 89 320 060
TLF 89 320 2362
TLX 528-282-22-EO D
EML
COM 28,34C

ROSADO MARGARITA DR
INSTITUTO DE ASTRONOMIA
UNAM
APDO POSTAL 70-264
04510 MEXICO DF
MEXICO
TEL 905-548-5306
TLF
TLX
EML
COM 28,34

ROSCH JEAN PROF
OBS MIDI PYRENEES
14 AVE E BELIN
F 31400 TOULOUSE CDX
FRANCE
TEL 62 95 1969
TLF
TLX 531625 F
EML
COM 09,10,16

ROSE JAMES ANTHONY
DPT PHYSICS & ASTRONOMY
UNIVERSITY NORTH CAROLINA
204 PHILLIPS HALL 039A
CHAPEL HILL NC 27514
USA
TEL 919 962 7170
TLF
TLX
EML
COM 28,29

ROSE WILLIAM K DR
ASTRONOMY PROGRAM
UNIVERSITY OF MARYLAND
COLLEGE PARK MD 20742
USA
TEL 301 299 2777
TLF
TLX
EML
COM 34

ROSEN EDWARD DR
DPT OF HISTORY
CITY COLLEGE OF NY
NEW YORK NY 10031
USA
TEL
TLF
TLX
EML
COM

ROSENBERG J DR
STATE UNIV OF UTRECHT
HEIDELBERLAAN 8
NL 3584 CA UTRECHT
NETHERLANDS
TEL 30 53 5124
TLF
TLX
EML
COM

ROSENDHAL JEFFREY D DR
NASA HEADQUARTERS
CODE Z
ASS DIRECTOR SPACE EXPLO
WASHINGTON DC 20546
USA
TEL 202 453 9181
TLF 202 426 0408
TLX
EML
COM 44

ROSINO LEONIDA PROF
OSS ASTRONOMICO DI PADOVA
VIC DELL OSSERVATORIO 5
I 35122 PADOVA
ITALY
TEL 49 66 1499
TLX 430176 UNPADU
EML
TLF
COM 06,27,34,37

ROSLUND CURT DR
DPT OF ASTRONOMY
CHALMERS TECHNICAL UNIV
S 412 96 GOETEBORG
SWEDEN
TEL 31 81 0100
TLF
TLX
EML
COM 25,46

ROSNER ROBERT
ASTRONOMY & ASTROPHYS CTR
UNIVERSITY OF CHICAGO
5640 S ELLIS AVE
CHICAGO IL 60637
USA
TEL
TLF
TLX
EML
COM· 48,49

ROSQUIST KJELL
INST OF THEORETICAL PHYS
VANADISVAEGEN 9
S 113 46 STOCKHOLM
SWEDEN
TEL 82 28 160*225
TLF
TLX 15433 FYSTO S
EML
COM 47

ROSS DENNIS K PROF
DPT OF PHYSICS
IOWA STATE UNIVERSITY
AMES IA 50011
USA
TEL 515 294 6010
TLF
TLX
EML
COM

ROSS JOHN E R DR
DPT OF PHYSICS
UNIVERSITY OF QUEENSLAND
ST LUCIA
BRISBANE QLD 4067
AUSTRALIA
TEL 7 377 3429
TLF
TLX 40315 UNIVQLD AA
EML
COM 14,36

ROSSELLO GASPAR
DPT FISICA DE ATMOSFERA
UNIVERSIDAD DE BARCELONA
AVD DIAGONAL 645
E 08028 BARCELONA
SPAIN
TEL
TLF
TLX
EML
COM 04

ROSSI BRUNO B PROF
CENTER FOR SPACE RESEARCH
MIT RM 37 667
BOX 165
CAMBRIDGE MA 02139
USA
TEL 617 253 4283
TLF
TLX 92-1473
EML
COM 48

ROSSI CORINNE DR
ISTITUTO ASTRONOMICO
UNIVERSITA DI ROMA
VIA G M LANCISI 29
I 00161 ROMA
ITALY
TEL 6 44 03734
TLF 6 44 03673
TLX 613255 INFNRO
EML. 40058 ROSSI
COM 29

ROSSI LUCIO
IAS
CNR
CP 67
I 00044 FRASCATI
ITALY
TEL 6 942 5651/2/3
TLF
TLX 610261 CNR FRA
EML
COM 29

ROSTAS FRANCOIS DR
OBSERVATOIRE DE PARIS
SECTION DE MEUDON
F 92195 MEUDON PPL CDX
FRANCE
TEL 1 45 07 7565
TLF
TLX 201571
EML
COM 14

ROTH MIGUEL R DR
CARNEGIE INST WASHINGTON
LAS CAMPANAS OBSERVATORY
CASILLA 601
LA SERENA
CHILE
TEL 51 21 3032
TLF
TLX 645 227 AURA CT
EML MROTH@UCHCEVM
COM

ROTHENFLUG ROBERT DR
CEA CEN
DAPNIA/SAP
BP 2
F 91191 GIF/YVETTE CDX
FRANCE
TEL 1 69 08 4327
TLF
TLX 604806
EML ROTHENFLUG@32779 DECNET CERN
COM

ROTH-HOPPNER MARIA LUISE
HAMBURGER STERNWARTE
GOJENSBERGSWEG 112
D 2050 HAMBURG 80
GERMANY
TEL 40 7252 4112
TLX 217884
EML
TLF
COM 37

ROTS ARNOLD H DR
NRAO
BOX 0
SOCORRO NM 87801 0387
USA
TEL
TLF
TLX
EML
COM 28

ROTTENBERG J A DR
291 BAYVIEW AVE
SUITE 110C
WILLOWDALE ON M2K 1E8
CANADA
TEL
TLF
TLX
EML
COM

ROUAN DANIEL DR
OBSERVATOIRE DE PARIS
SECTION DE MEUDON
F 92195 MEUDON PPL CDX
FRANCE
TEL 1 45 07 7715
TLF 1 45 07 2806
TLX
EML MEGASY ROUAN ROUAN@FRMEU51
COM 34

ROUDIER THIERRY DR
OBS MIDI PYRENEES
14 AVE E BELIN
F 31400 TOULOUSE CDX
FRANCE
TEL 61 25 2101
TLF
TLX
EML
COM 10,12

ROUEFF EVELYNE M A DR
OBSERVATOIRE DE PARIS
SECTION DE MEUDON
DAF
F 92195 MEUDON PPL CDX
FRANCE
TEL 1 45 07 7435
TLF
TLX 201571
EML
COM 14

ROUNTREE JANET DR
6001 WYNNWOOD RD
BETHESDA MD 20816
USA
TEL
TLF
TLX
EML ROUNTREE@NSSDCA GSFC NASA
COM 09,27,37,45

ROUSE CARL A DR
627 15 TH STR
DEL MAR CA 92014
USA
TEL 619 455 4015
TLF
EML.
COM 35
TLX 695065

ROUSSEAU JEANINE DR
OBSERVATOIRE DE LYON
AVE CHARLES ANDRE
F 69561 S GENIS LAVAL CDX
FRANCE
TEL 78 56 0705
TLF 72 39 9791
TLX 310926
EML
COM

ROUSSEAU JEAN-MICHEL MR
OBSERVATOIRE DE BORDEAUX
BP 89
F 33270 FLOIRAC
FRANCE
TEL 56 86 4330
TLF 56 40 4251
TLX
EML
COM 08

ROUTLEDGE DAVID DR
DT OF ELECTRICAL ENGEENRG
UNIVERSITY OF ALBERTA
EDMONTON AB T6G 2J7
CANADA
TEL 403 432 5668
TLF
TLX 037 2979
EML
COM

ROUTLY PAUL M DR
US NAVAL OBSERVATORY
34 & MASSACHUSETTS AVE NW
WASHINGTON DC 20392 5100
USA
TEL   202 653 1532
TLX
EML
TLF
COM 38

ROVIRA MARTA GRACIELA
IAFE
CC 67 SUC 28
1428 BUENOS AIRES
ARGENTINA
TEL   1 781 6755
TLX
EML
TLF
COM 12,36

ROVITHIS PETER DR
ASTRONOMICAL INSTITUTE
NTL OBSERVATORY OF ATHENS
BOX 20048
GR 118 10 ATHENS 306
GREECE
TEL 1 346 3803
TLF
TLX 215530 OBSA GR
EML
COM 42

ROVITHIS-LIVANIOU HELEN
DPT OF ASTROPHYSICS
NTL UNIVERSITY OF ATHENS
PANEPISTIMIOPOLIS
GR 157 71 ZOGRAFOS
GREECE
TEL 1 724 3414
TLF
TLX
EML
COM 42

ROWAN-ROBINSON MICHAEL DR
SCHOOL OF MATHEMATICAL SC
QUEEN MARY/WESTFIELD COLL
MILE END RD
LONDON E1 4NS
UK
TEL
TLF
TLX
EML
COM 47,51

ROWSON BARRIE DR
NRAL
JODRELL BANK
MACCLESFIELD SK11 9DL
UK
TEL 477 713 21
TLX 36149
EML
TLF
COM 40

ROXBURGH IAN W PROF
SCHOOL OF MATHEMATICAL SC
QUEEN MARY/WESTFIELD COLL
MILE END RD
LONDON E1 4NS
UK
TEL 1 980 4811
TLF
TLX
EML
COM 10,34,35,42,47,49

ROY ARCHIE E PROF
DPT OF ASTRONOMY
UNIVERSITY OF GLASGOW
GLASGOW G12 8QQ
UK
TEL 413 398 855*502
TLF
TLX 778421 GLASUL
EML
COM 07C,46

ROY JEAN-RENE
DPT DE PHYSIQUE
UNIVERSITE LAVAL
FAC SCIENCES 2 GENIE
LAVAL QC G1K 7P4
CANADA
TEL 418 656 5816
TLF 418 656 2040
TLX 051 31621 UNILAVAL
EML
COM

ROZELOT JEAN P
OCA CERGA
AVE COPERNIC
F 06130 GRASSE
FRANCE
TEL 93 36 5849
TLF
TLX 470865
EML
COM 10

ROZHKOVSKIJ DIMITRIJ A
ASTROPHYSICAL INSTITUTE
KAZAKH ACAD OF SCIENCES
480068 ALMA ATA
KAZAKHSTAN
TEL 62-40-40
TLF
TLX
EML
COM 21,34

ROZYCZKA MICHAL
ASTRONOMICAL OBSERVATORY
WARSAW UNIVERSITY
AL UJAZDOWSKIE 4
PL 00 478 WARSAW
POLAND
TEL
TLF
TLX 813978 ZAPAN PL
EML
COM 34

RUBASHEV BORIS M DR
PULKOVO OBSERVATORY
ACADEMY OF SCIENCES
10 KUTUZOV QUAY
196140 ST PETERSBURG
RUSSIA
TEL
TLF
TLX
EML
COM 10

RUBEN G PROF DR
ZNTRLINST F ASTROPHYSIK
ROSA-LUXEMBURG-STR 17A
D 1502 POTSDAM
GERMANY
TEL
TLX
EML
TLF
COM 35,44

RUBIN ROBERT HOWARD
NASA AMES RESEARCH CTR
MS 245 6
MOFFETT FIELD CA 94035
USA
TEL 415 965 5528
TLF
TLX 348408
EML
COM 40,51

RUBIN VERA C DR
DPT TERRESTR MAGNETISM
CARNEGIE INST WASHINGTON
5241 BROAD BRANCH RD NW
WASHINGTON DC 20015
USA
TEL 202 966 0863
TLF
TLX 440427 MAGN UI
EML
COM 28,30,33,47

RUBIO MONICA DR
DPT DE ASTRONOMIA
UNIVERSIDAD DE CHILE
CASILLA 36 D
SANTIAGO
CHILE
TEL 2 229 4101
TLF. 2 271 2799
TLX 440 001
EML
COM 40

RUCINSKI SLAWOMIR M DR
SPACE ASTROPHYSICAL LAB
YORK UNIVERSITY
4700 KEELE ST
TORONTO ON M3J 1P3
CANADA
TEL. 416 665 3311
TLF
TLX 065 24736
EML BITNET FS300516@YUSOL
COM 36,42

RUDAK BRONISLAW
INSTITUTE OF ASTRONOMY
N COPERNICUS UNIVERSITY
UL CHOPINA 12/18
PL 87 100 TORUN
POLAND
TEL 26037*10
TLF
TLX 813978 ZAPAN PL
EML
COM

RUDDY VINCENT P DR
REGIONAL TECHNICAL COLL
ROSSA AVE
CORK
IRELAND
TEL
TLF
TLX
EML
COM 47

RUDER HANNS
LEHRSTUHL F THEORET ASTRO
PHYSIK DER UNIV TUEBINGEN
AUF DER MORGENSTELLE 12,C
D 7400 TUEBINGEN
GERMANY
TEL   707 129 2487
TLF
TLX
EML
COM 09,14,19,24,44

RUDERMAN MALVIN A
DPT OF PHYSICS
COLUMBIA UNIVERSITY
NEW YORK NY 10027
USA
TEL 212 280 3317
TLF
TLX
EML
COM

RUDKJOBING MOGENS PROF
INST OF PHYSICS & ASTRON
UNIVERSITY OF AARHUS
NY MUNKEGADE
DK 8000 AARHUS C
DENMARK
TEL 86 12 8899
TLF 86 20 2711
TLX· 64767 AAUSCI DK
EML
COM 45

RUDNICK LAWRENCE DR
DPT OF ASTRONOMY
UNIVERSITY OF MINNESOTA
116 CHURCH ST SE
MINNEAPOLIS MN 55455
USA
TEL 612 373 5457
TLF
TLX·
EML
COM 40,47

RUDNICKI KONRAD PROF
ASTRONOMICAL OBSERVATORY
JAGIELLONIAN UNIVERSITY
UL ORLA 171
PL 30 244 KRAKOW
POLAND
TEL 22 69 33
TLF
TLX
EML
COM 28

RUDZIKAS ZENONAS B
INST OF THEORET PHYS/ASTR
LITHUENIAN ACADEMY OF SCI
POZELOS 54
232600 VILNIUS
LITHUANIA
TEL· 122 62 0939
TLF
TLX 261141 LMA SU
EML
COM 14C

RUEDIGER GUNTHER DR
ZNTRLINST F ASTROPHYSIK
ROSA LUXEMBURG STR 17A
D 1591 POTSDAM
GERMANY
TEL 37 33 77138
TLF 069 15471
TLX
EML
COM 10

RUELAS-MAYORGA R A DR
INSTITUTO DE ASTRONOMIA
UNAM
APDO POSTAL 70-264
04510 MEXICO DF
MEXICO
TEL 525 548 5306
TLF 525 548 3712
TLX
EML RARM@ALFA ASTROSCU UNAM M>
COM 33

RUFENER FREDY G PROF
OBSERVATOIRE DE GENEVE
CHEMIN DES MAILLETTES 51
CH 1290 SAUVERNY
SWITZERLAND
TEL 22 755 2611
TLF 22 755 3983
TLX 419209 OBS CH
EML
COM 25

RUFFINI REMO
DPT DI FISICA
UNIVERSITA DI ROMA
PA MORO 2
I 00185 ROMA
ITALY
TEL 6 49 76304
TLF
TLX· 613255 INFNRO I
EML
COM 47,48

RUGGE HUGO R DR
AEROSPACE CORPORATION
MS M2 226
BOX 92957
LOS ANGELES CA 90009
USA
TEL 213 648 7086
TLF
TLX
EML
COM

RUIZ MARIA TERESA DR
DPT DE ASTRONOMIA
UNIVERSIDAD DE CHILE
CASILLA 36 D
SANTIAGO
CHILE
TEL 2 229 4101
TLF
TLX 440 001
EML
COM 33

RULE BRUCE H
HALE OBSERVATORIES
2205 MONTE VISTA ST
PASADENA CA 91107
USA
TEL 818 794 6593
TLF
TLX
EML
COM

RUMSEY NORMAN J
21 MALONE RD
LOWER HUTT
NEW ZEALAND
TEL 4 69 6787
TLF
EML
COM
TLX

RUNCORN S K PROF
SCHOOL OF PHYSICS
UNIVERSITY OF NEWCASTLE
NEWCASTLE/TYNE NE1 7RU
UK
TEL 632 325 11
TLF
TLX 53654 UNINEW G
EML
COM 16,19

RUPPRECHT GERO DR
ESO
KARL SCHWARZSCHILDSTR 2
D 8046 GARCHING MUENCHEN
GERMANY
TEL 89 320 06355
TLF 89 320 2362
TLX 52 82 820 EO D
EML GRUPPREC@ESO ORG
COM 09

RUPRECHT JAROSLAV DR
ASTRONOMICAL INSTITUTE
CZECH ACADEMY SCIENCES
BUDECSKA 6
CS 120 23 PRAHA 2
CZECHOSLOVAKIA
TEL 2 25 8757
TLF 2 25 5010
TLX. 122486
EML
COM 37

RUSCONI LUIGIA DR
DPT DI ASTRONOMIA
UNIVERSITA DI TRIESTE
VIA TIEPOLO 11
I 34131 TRIESTE
ITALY
TEL 40 79 4863
TLF
TLX 461137 OAOTI
EML
COM 09

RUSIN VOJTECH
ASTRONOMICAL INSTITUTE
SLOVAK ACADEMY SCIENCES
CS 059 60 TATRANSKA LOMNI
CZECHOSLOVAKIA
TEL 969 96 7866/7/8
TLF 969 96 7656
TLX 8078277 AUSSAV C
EML
COM 10,12

RUSKOL EUGENIA L DR
INST PHYSICS OF THE EARTH
ACADEMY OF SCIENCES
GRUZINSKAYA 10
123342 MOSCOW
RUSSIA
TEL 252 0726
TLF
TLX 411196 IFZAN SU
EML
COM 16

RUSSELL CHRISTOPHER T
INST OF GEOPHYS & PLANET
UNIVERSITY OF CALIFORNIA
LOS ANGELES CA 90024
USA
TEL 213 825 3188
TLF
TLX 910-342-6981
EML
COM 49

RUSSELL JANE L DR
NAVAL RESEARCH LABORATORY
CODE 4130R
4555 OVERLOOK AVE SW
WASHINGTON DC 20375 5000
USA
TEL. 202 767 0171
TLF
TLX
EML
COM 08,24,26,40,51

RUSSELL JOHN A PROF
DPT OF ASTRONOMY
UNIV SOUTHERN CALIFORNIA
UNIVERSITY PARK
LOS ANGELES CA 90089
USA
TEL 213 743 0231
TLF
TLX
EML
COM 22

RUSSELL KENNETH S DR
UK SCHMIDT TELESCOPE
AAO
PRIVATE BAG
COONABARABRAN NSW 2357
AUSTRALIA
TEL 68 426 311
TLF 68 846 298
TLX CANOPUS AA 163945
EML NSSDCA PSI%AAOCBN KSR
COM 20

RUSSELL STEPHEN DR
DIAS
SCHOOL OF COSMIC PHYSICS
5 MERRION SQ
DUBLIN 2
IRELAND
TEL 1 774 321
TLF 1 682 003
TLX 31687 DIAS EI
EML 29536 SR/SR@DIAS IE
COM 34

RUSSEV RUSCHO DR
DPT OF ASTRONOMY
UNIVERSITY OF SOFIA
ANTON IVANOV ST 5
BG 1126 SOFIA
BULGARIA
TEL. 2 54 4852
TLF
TLX
EML
COM 27

RUSSEVA TATJANA
DPT OF ASTRONOMY
BULGARIAN ACAD SCIENCES
72^{4}LENIN BLVD
BG 1784 SOFIA
BULGARIA
TEL 2 75 8927
TLF
TLX 23561 ECF BAN BG
EML
COM 37

RUSSO GUIDO DR
DPT DI SCIENZE FISICHE
UNIVERSITA DI NAPOLI
MOSTRA D OLTREMARE PAD 19
I 80125 NAPOLI
ITALY
TEL 81 725 3447
TLF
TLX 720320
EML
COM 05,42

RUST DAVID M DR
APPLIED PHYSICS LAB
JOHNS HOPKINS UNIVERSITY
JOHNS HOPKINS RD
LAUREL MD 20707
USA
TEL 301 953 5414
TLF
TLX 89-548 APL JHU LAUR
EML
COM 10

RUSU I DR
ASTRONOMICAL OBSERVATORY
CUTITUL DE ARGINT 5
BOX 28
R 75212 BUCAREST 28
RUMANIA
TEL 23 63 01
TLF
TLX
EML
COM 08,19

RUSU L DR
STR MITROPOLITUL IOSIF 47
R 75217 BUCAREST 28
RUMANIA
TEL
TLF
TLX
EML
COM

RUTTEN ROBERT J DR
STERREKUNDIG INSTITUTE

BOX 80000
NL 3508 TA UTRECHT
NETHERLANDS
TEL 30 53 5200
TLF
TLX 40048 FYLUT NL
EML BITNET WNMMAIL@HUTRUUO
COM 12,29,36

RUZDJAK VLADIMIR DR
INSTITUTE OF PHYSICS
UNIVERSITY OF ZAGREB
BOX 304
YU 41001 ZAGREB
YUGOSLAVIA
TEL
TLF
TLX
EML
COM 10

RUZICKOVA-TOPOLOVA B DR
ASTRONOMICAL INSTITUTE
CZECH ACADEMY OF SCIENCES
ONDREJOV OBSERVATORY
CS 251 65 ONDREJOV
CZECHOSLOVAKIA
TEL 204 85201
TLF 204 85314
TLX 121579
EML
COM 10

RYABCHIKOVA TANYA DR
INST OF ASTRONOMY
ACADEMY OF SCIENCES
PYATUITSKAYA UL 48
109017 MOSCOW
RUSSIA
TEL
TLF
TLX
EML
COM 36

RYABOV YU A PROF DR
DPT OF MATHEMATICS
MADI
LENINGRADSKY PROSP 64
125319 MOSCOW
RUSSIA
TEL 1550326
TLF
TLX
EML
COM 07

RYBANSKY MILAN
ASTRONOMICAL INSTITUTE
SLOVAK ACADEMY SCIENCES
CS 059 60 TATRANSKA LOMNI
CZECHOSLOVAKIA
TEL 969 96 7866/7/8
TLF 969 96 7656
TLX 80-78277 AUSAV C
EML
COM· 10,12

RYBICKI GEORGE B DR
CENTER FOR ASTROPHYSICS
HCO/SAO
60 GARDEN ST
CAMBRIDGE MA 02138
USA
TEL 617 495 7452
TLF
TLX. 92-1428
EML
COM 33,36

RYBKA PRZEMYSLAW DR
INST HISTORY OF SCIENCES
POLISH ACAD OF SCIENCES
NOWY SWIAT 72
PL 00 330 WARSAW
POLAND
TEL
TLF
TLX
EML
COM 41

RYDBECK GUSTAF H B DR
ONSALA SPACE OBSERVATORY
GOETEBORG UNIVERSITY
S 439 00 ONSALA
SWEDEN
TEL 30 06 2081
TLF
TLX
EML
COM 28,40

RYDBECK OLOF E H PROF
ONSALA SPACE OBSERVATORY
GOETEBORG UNIVERSITY
S 439 00 ONSALA
SWEDEN
TEL· 30 06 2081
TLF
TLX 8542400 ONSPACE
EML·
COM 40

RYDGREN ALFRED ERIC JR DR
BOEING AEROSPACE CO
MS 87 08
BOX 3999
SEATTLE WA 98124 2499
USA
TEL 206 773 2155
TLF
TLX
EML
COM 25

RYKHLOVA LIDIJA V DR
INST OF ASTRONOMY
ACADEMY OF SCIENCES
PYATNITSKAYA UL 48
109017 MOSCOW
RUSSIA
TEL 231-54-61
TLF
TLX 412623 SCSTP SU
EML
COM 19

RYLOV VALERIJ S DR
SPECIAL ASTROPHYSICAL OBS
ACADEMY OF SCIENCES
NIZHNIJ ARKHYZ
357147 STAVROPOLSKIJ
RUSSIA
TEL
TLF
TLX
EML
COM 09

RYTER CHARLES E DR
CEA CEN
DAPNIA/SAP
BP 2
F 91191 GIF/YVETTE CDX
FRANCE
TEL 1 69 08 3912
TLF
TLX
EML
COM

RYUTOVA MARGARITA P DR
INSTITUTE OF NUCLEAR PHYS
ACADEMY OF SCIENCES
SIBERIAN DIV
630090 NOVOSIBIRSK
RUSSIA
TEL 383 2 359943
TLF 383 2 352163
TLX 133116 ATOM SU
EML RYUTOVA@VXINPB INP NSK SU
COM 10

RYZHKOV NIKOLAI F DR
PULKOVO OBSERVATORY
ACADEMY OF SCIENCES
10 KUTUZOV QUAY
196140 ST PETERSBURG
RUSSIA
TEL
TLF
TLX
EML
COM 40

RZHIGA OLEG N DR
INST OF RADIO & ELECTRON
ACADEMY OF SCIENCES
103907 MOSCOW
RUSSIA
TEL
TLF
TLX
EML·
COM

SAAR ENN DR
TARTU ASTROPHYSICAL OBS
ESTONIAN ACAD OF SCIENCES
202444 TARTU
ESTONIA
TEL
TLF
TLX
EML saar@aai tartu ew su
COM. 33,47

SABANO YUTAKA DR
ASTRONOMICAL INSTITUTE
TOHOKU UNIVERSITY
SENDAI AOBA
MIYAGI 980
JAPAN
TEL 222 22 1800
TLF
TLX
EML
COM 34

SABBADIN FRANCO DR
OSSERVATORIO ASTROFISICO
VIA DELL OSSERVATORIO 8
I 36012 ASIAGO
ITALY
TEL 42 46 2665
TLF
TLX SETUR 430110
EML
COM 34

SACK NOAM DR
DPT OF THEORETICAL PHYS
HEBREW UNIV OF JERUSALEM
JERUSALEM 91904
ISRAEL
TEL
TLF
TLX
EML
COM

SACKMANN I JULIANA DR
CALTECH
KELLOGG RADIATION LAB
PASADENA CA 91125
USA
TEL 818 356 4256
TLF
TLX
EML
COM 35

SADAKANE KOZO DR
ASTRONOMICAL INSTITUTE
OSAKA KYOIKU UNIVERSITY
4 88 MINAMIKAWAHORICHO
OSAKA 543
JAPAN
TEL
TLF
TLX
EML
COM 29C

SADAT RACHIDA DR
CRAAG
BP 63
BOUZAREAH
ALGER
ALGERIA
TEL 279 1443
TLF 213 2 799862
TLX 61447 DZ
EML
COM 47

SADEH D DR
DPT OF PHYSICS & ASTRON
TEL AVIV UNIVERSITY
RAMAT AVIV
TEL AVIV 69978
ISRAEL
TEL 3 42 0553
TLF
TLX 34271 VERSY
EML
COM

SADIK AZIZ R DR
SARC
SCIENTIFIC RES COUNCIL
BOX 2441
JADIRIYAH BAGHDAD
IRAQ
TEL 1 776 5127
TLF
TLX 213976 SRC
EML
COM· 27,42

SADLER ELAINE MARGARET
AAO
BOX 296
EPPING NSW 2121
AUSTRALIA
TEL 2 868 1666
TLX 123999 AAOSYD AA
EML·
TLF
COM 28

SADUN ALBERTO CARLO DR
BRADLEY OBSERVATORY
AGNES SCOTT COLLEGE
DECATUR GA 30030
USA
TEL 404 371 6265
TLF
TLX
EML PHYAAS@GSUVM1 BITNET
COM 28

SADZAKOV SOFIJA DR
ASTRONOMICAL OBSERVATORY
VOLGINA 7
YU 11050 BEOGRAD
YUGOSLAVIA
TEL 11 419 357/421 875
TLF
TLX
EML
COM 08,19

SAEMUNDSON THORSTEINN
RAUNVISINDASTOFNUN
HASKOLANS
DUNHAGA 3
IS 107 REYKJAVIK
ICELAND
TEL 1 21340
TLF
TLX 2307 ISINFO
EML
COM 10

SAFKO JOHN L
DPT PHYSICS & ASTRONOMY
UNIVERSITY S CAROLINA
COLUMBIA SC 29208
USA
TEL 803 777 6466
TLF
TLX UNIVSCAROL CLB
EML
COM 46

SAFRONOV VICTOR S DR
INST PHYSICS OF THE EARTH
ACADEMY OF SCIENCES
GRUZINSKAYA 10
123242 MOSCOW
RUSSIA
TEL 252-07-26
TLF
TLX 411196 IFZAN SU
EML
COM 16

SAGAN CARL DR
CORNELL UNIVERSITY

302 SPACE SCIENCE BLDG
ITHACA NY 14853
USA
TEL 607 256 4971
TLF
TLX 937478
EML
COM 16,51

SAGDEEV ROALD Z DR
SPACE RESEARCH INSTITUTE
ACADEMY OF SCIENCES
PROFSOJOSNAYA UL 84/32
117810 MOSCOW
RUSSIA
TEL 333 14 66
TLF
TLX 411498 STAR SU
EML
COM 15,44,49C

SAGGION ANTONIO PROF
DPT DI FISICA G GALILEI
UNIVERSITA DI PADOVA
VIA MARZOLO 8
I 35131 PADOVA
ITALY
TEL· 49 84 4254
TLF
TLX 430308 DFGGPDI
EML
COM

SAGNIER JEAN-LOUIS DR
BUREAU DES LONGITUDES
77 AVE DENFERT ROCHEREAU
F 75014 PARIS
FRANCE
TEL 1 40 51 2261
TLF
TLX
EML
COM 07,20

SAHA SWAPAN KUMAR DR
INDIAN INSTITUTE OF
ASTROPHYSICS
KORAMANGALA
BANGALORE 560 034
INDIA
TEL 812 56 9902
TLF
TLX 845-2763 IIAB IN
EML
COM

SAHADE JORGE PROF
OBSERVATORIO ASTRONOMICO
CC 677
1900 LA PLATA (BS AS)
ARGENTINA
TEL 1 774 9310
TLX 31151 BULAP AR
EML
TLF 1 774 5703
COM 29,38P,42,44

SAHAI RAGHVENDRA DR
DPT OF PHYSICS
CHALMERS TECHNICAL UNIV
UNIV OF GOTEBORG
S 412 96 GOETEBORG
SWEDEN
TEL 31 72 3139
TLF
TLX 2369 CHALBIB S
EML BITNET TFARS@SECTHF51
COM

SAHAL-BRECHOT SYLVIE DR
OBSERVATOIRE DE PARIS
SECTION DE MEUDON
DAMAP
F 92195 MEUDON PPL CDX
FRANCE
TEL 1 45 07 7442
TLF 1 45 07 7469
TLX 201571
EML MESIOA SAHAL/SAHAL@FRMEU51
COM 10,14C

SAHU KAILASH C DR
KAPTEYN ASTRONOMICAL INST

BOX 800
NL 9700 AV GRONINGEN
NETHERLANDS
TEL 50 63 4073
TLF 50 63 6100
TLX 53572 STARS NL
EML KAILASH@HGRRUG5
COM 34

SAIJO KEIICHI
DPT OF PHYSICAL SCIENCES
NATIONAL SCIENCE MUSEUM
7-20 UENO PARK TAITO KU
TOKYO 110
JAPAN
TEL 3-822-0111
TLF
TLX
EML
COM 42

SAIKIA DHRUBA JYOTI DR
TATA INST OF FUNDAMENTAL
RESEARCH
POONA UNIVERSITY CAMPUS
PUNE 411 007
INDIA
TEL 212 33 7107
TLF 212 33 5760
TLX 0145658 GMRT IN
EML uunet!shakti!gmrt!djs
COM 40

SAIO HIDEYUKI DR
ASTRONOMICAL INSTITUTE
TOHOKU UNIVERSITY
SENDAI AOBA
MIYAGI 980
JAPAN
TEL
TLF
TLX
EML
COM 35

SAISSAC JOSEPH DR
OBS MIDI PYRENEES
14 AVE E BELIN
F 65200 BAGNERES BIGORRE
FRANCE
TEL 62 95 1969
TLF
TLX 531625 S
EML
COM 16

SAITO KUNIJI PROF
TOKYO ASTRONOMICAL OBS
NAOJ
OSAWA MITAKA
TOKYO 181
JAPAN
TEL
TLF
TLX
EML
COM 10,36

SAITO MAMORU DR
DPT OF ASTRONOMY
KYOTO UNIVERSITY
KITASHIRAKAWA SAKYO KU
KYOTO 606
JAPAN
TEL 0757512111x3904
TLF
TLX 5422693 LIBKYU J
EML
COM·

SAITO SUMISABURO DR
KWASAN OBSERVATORY
KYOTO UNIVERSITY
YAMASHINA
KYOTO 607
JAPAN
TEL 075-581-1235
TLF
TLX 5422693 LIBKYU J
EML
COM

SAITO TAKAO PROF
GEOPHYS INST SCIENCE FAC
TOHOKU UNIVERSITY
AOBA AOBA-KU
SENDAI 980
JAPAN
TEL 22 268 4508
TLF 22 268 4508
TLX MTA1S5C@JPNTOHOK
EML
COM· 15

SAKAI JUNICHI
FAC OF ENGINEERING
TOYAMA UNIVERSITY
TOYAMA 930
JAPAN
TEL· 0764-41-1271
TLF
TLX·
EML
COM 12

SAKASHITA SHIRO PROF
DPT OF PHYSICS
HOKKAIDO UNIVERSITY
KITA 10 NISHI 8
SAPPORO 060
JAPAN
TEL. 011-716-2111
TLF
TLX
EML
COM 35

SAKHIBULLIN NAIL A DR
DPT OF ASTRONOMY
KAZAN STATE UNIVERSITY
LENIN UL 18
420008 KAZAN
RUSSIA
TEL 32-36-41
TLF
TLX
EML
COM 36

SAKURAI KUNITOMO PROF
DPT OF PHYSICS
KANAGAWA UNIVERSITY
KANAGAWAKU
YOKOHAMA 221
JAPAN
TEL 45 481 5661
TLF
TLX
EML
COM 10,51

SAKURAI TAKASHI DR
TOKYO ASTRONOMICAL OBS
NAOJ
OSAWA MITAKA
TOKYO 181
JAPAN
TEL 422 32 5111
TLF.
TLX 2822307 TAOMTK J
EML
COM 10C,12

SAKURAI TAKEO T PROF
DPT AERONAITIC ENGINEER
KYOTO UNIVERSITY
KITASHIRAKAWA SAKYO KU
KYOTO 606
JAPAN
TEL. 75-7512111x5792
TLF
TLX 05422693 LIBKYUJ
EML
COM

SALA FERRAN DR
DPT FISICA
UNIVERSIDAD DE BARCELONA
AVD DIAGONAL 647
E 08028 BARCELONA
SPAIN
TEL· 3 330 7311
TLF
TLX
EML
COM 28,33

SALAZAR ANTONIO DR
REAL INST Y OBSERVATORIO
DE LA ARMADA
CECILIO PUJAZON S/N
E 11110 SAN FERNANDO
SPAIN
TEL 56 88 3548
TLF 56 89 9302
TLX 76108 IOM E
EML
COM 04

SALETIC DUSAN
ASTRONOMICAL OBSERVATORY
VOLGINA 7
YU 11050 BEOGRAD
YUGOSLAVIA
TEL 11 157 022
TLF
TLX
EML·
COM 08

SALINARI PIERO
OSS ASTROFISICO
DI ARCETRI
LARGO E FERMI 5
I 50125 FIRENZE
ITALY
TEL 55 275 2231
TLF
TLX 572268
EML
COM 34

SALISBURY J W DR
US GEOLOGICAL SURVEY
927 NATIONAL CENTER
RESTON VA 22092
USA
TEL 703 860 6668
TLF
TLX 92178
EML
COM

SALO HEIKKI
JPL
MS 183 501
4800 OAK GROVE DR
PASADENA CA 91109
USA
TEL 818 354 3833
TLF 358 81 561278
TLX
EML SALO@FINFUN BITNET
COM

SALPETER EDWIN E PROF
CORNELL UNIVERSITY
NEWMAN LAB OF NUCLEAR STU
ITHACA NY 14853
USA
TEL 607 256 3302
TLF
TLX 937478
EML
COM 34,35,40,48C

SALUKVADZE G N DR
ABASTUMANY ASTROPHYSICAL
OBSERVATORY
GEORGIAN ACAD OF SCIENCES
383762 ABASTUMANI
GEORGIA
TEL
TLF
TLX
EML
COM. 26,37

SALVADOR-SOLE EDUARDO
DPT FISICA DE ATMOSFERA
UNIVERSIDAD DE BARCELONA
AVD DIAGONAL 645
E 08028 BARCELONA
SPAIN
TEL 3 330 7311
TLF
TLX
EML
COM 28,47

SALVATI MARCO
OSS ASTROFISICO
DI ARCETRI
LARGO E FERMI 5
I 50125 FIRENZE
ITALY
TEL 55 275 2268
TLF
TLX 572230 ARCETR I
EML
COM 48

SALZER JOHN JOSEPH DR
DPT OF ASTRONOMY
VAN VLECK OBSERVATORY
WESLEYAN UNIVERSITY
MIDDLETOWN CT 06457
USA
TEL 203 347 9411*2827
TLF· 203 344 7981
TLX
EML SLAZ@PARCHA ASTRO WESLEYAN EDU
COM 47

SAMAIN DENYS DR
IAS
BP 10
F 91371 VERRIERES BUISSON
FRANCE
TEL 1 64 47 4304
TLF
TLX 600252
EML·
COM 12

SAMEC RONALD G DR
JI HOLCOMB OBSERVATORY
BUTLER UNIVERSITY
4600 SUNSET AVE
INDIANAPOLIS IN 46208
USA
TEL. 317 283 9282
TLF 317 283 9519
TLX
EML SAMEC@BUTLERU BITNET
COM 42

SAMPSON DOUGLAS H PROF
DPT OF ASTRONOMY
PENNSYLVANIA STATE UNIV
525 DAVEY LAB
UNIVERSITY PARK PA 16802
USA
TEL 814 865 0261
TLF
TLX 842510
EML.
COM.

SAMUS NIKOLAI N DR
INST OF ASTRONOMY
ACADEMY OF SCIENCES
PYATNITSKAYA UL 48
109017 MOSCOW
RUSSIA
TEL 231-54-61
TLF·
TLX 412623 SCSTP SU
EML
COM 27C,30,37

SANAHUJA BLAS
DPT FISICA DE ATMOSFERA
UNIVERSIDAD DE BARCELONA
AVD DIAGONAL 645
E 08028 BARCELONA
SPAIN
TEL 3 330 7311*298
TLF
TLX
EML
COM 28,46

SANAMIAN V A DR
BYURAKAN ASTROPHYSICAL
OBSERVATORY
378433 ARMENIA
ARMENIA
TEL 88 52 28 3453
TLF
TLX
EML
COM 40

SANCHEZ ALMEIDA JORGE DR
INST DE ASTROFISICA
DE CANARIAS
OBS DEL TEIDE
E 38200 LA LAGUNA
SPAIN
TEL 22 26 2211
TLF 22 26 3005
TLX 92640 IAC E
EML IAC JOS/JOS●IAS ES
COM 10,12,29

SANCHEZ FRANCISCO PROF
INST DE ASTROFISICA
DE CANARIAS
OBS DEL TEIDE
E 38071 LA LAGUNA
SPAIN
TEL 22 26 2211
TLF
TLX 92640 IAC E
EML
COM 21,50

SANCHEZ MANUEL
REAL INST Y OBSERVATORIO
DE LA ARMADA
CECILIO PUJAZON S/N
E 11110 SAN FERNANDO
SPAIN
TEL 56 88 3548
TLF
TLX 76108
EML
COM 08,19

SANCHEZ-LAVEGA AGUSTIN DR
DPT FISICA APLICADA
ETS INGENIEROS IND Y TEL
AVD URQUIJO S/N
E 18013 BILBAO
SPAIN
TEL 4 441 6400*353
TLF 4 441 4041
TLX
EML
COM 16

SANCHEZ-SAAVEDRA M LUISA
FAC DE CIENCIAS
UNIVERSIDAD DE GRANADA
E 18080 GRANADA
SPAIN
TEL 58 20 2212
TLF
TLX
EML
COM 21,33,34

SANCISI RENZO DR
KAPTEYN ASTRONOMICAL INST
BOX 800
NL 9700 AV GRONINGEN
NETHERLANDS
TEL 50 11 6695
TLX 53572 STARS NL
EML
TLF
COM 28,34,51

SANDAGE ALLAN
MT WILSON & LAS CAMPANAS
OBSERVATORIES
813 SANTA BARBARA ST
PASADENA CA 91101
USA
TEL 818 577 1122
TLF
TLX
EML
COM

SANDAKOVA E V DR
KIEV STATE UNIVERSITY
ASTRONOMICAL OBSERVATORY
252053 KIEV
UKRAINE
TEL
TLF
TLX
EML
COM

SANDELL GORAN HANS L DR
655 KOMOHAWA ST

HILO HI 96720
USA
TEL
TLF
TLX
EML
COM 34,40

SANDERS DAVID B DR
INSTITUTE FOR ASTRONOMY
UNIVERSITY OF HAWAII
2680 WOODLAWN DR
HONOLULU HI 96822
USA
TEL 808 948 7399
TLF 808 988 2790
TLX 7238459 UHAST HR
EML SANDERS●UHIFA IFA HAWAII EDU
COM 28,40

SANDERS ROBERT DR
KAPTEYN ASTRONOMICAL INST
BOX 800
NL 9700 AV GRONINGEN
NETHERLANDS
TEL 50 11 6695
TLX 53572 STARS NL
EML
TLF
COM 28

SANDERS W L PROF
DPT OF ASTRONOMY
NEW MEXICO STATE UNIV
BOX 4500
LAS CRUCES NM 88003
USA
TEL 505 646 4914
TLF
TLX
EML
COM 24,37

SANDERS WILTON TURNER III
DPT OF PHYSICS
UNIVERSITY OF WISCONSIN
1150 UNIVERSITY AVE
MADISON WI 53706
USA
TEL 608 262 5916
TLF
TLX
EML
COM 44,48

SANDFORD MAXWELL T II
LOS ALAMOS SCIENTIFIC LAB
LOS ALAMOS NM 87545
USA
TEL 505 667 6384
TLF
EML
COM
TLX

SANDMANN WILLIAM HENRY
DPT OF PHYSICS
HARVEY MUDD COLLEGE
CLAREMONT CA 91711
USA
TEL 714 621 8024
TLF
TLX·
EML
COM 27

SANDQVIST AAGE DR
STOCKHOLM OBSERVATORY
S 133 36 SALTSJOEBADEN
SWEDEN
TEL 87 17 0380
TLF 87 17 4719
TLX 12972 SOBSERV S
EML SANDQVIS●ASTRO SU SE
COM 33,34,46C

SANFORD PETER WILLIAM MR
DPT PHYSICS & ASTRONOMY
UNIVERSITY COLLEGE LONDON
GOWER ST
LONDON WC1E 6BT
UK
TEL
TLF
TLX
EML
COM

SANROMA MANUEL DR
DPT FISICA
UNIVERSIDAD DE BARCELONA
AVD DIAGONAL 647
E 08028 BARCELONA
SPAIN
TEL 3 330 7311
TLF
TLX
EML
COM 28

SANSATURIO MARIA E DR
ETS
INGENIEROS INDUSTRIALES
PASEO DEL CAUCE S/N
E 47011 VALLADOLID
SPAIN
TEL 83 30 4899
TLF· 83 39 2026
TLX
EML
COM 07

SANTAMARIA RAFFAELE DR
IST UNIVERSITARIO NAVALE
VIA ACTON 38
I 80133 NAPOLI
ITALY
TEL 81 547 5135
TLF 81 552 1485
TLX 710417
EML
COM

SANTIN PAOLO DR
OAT
BOX SUCC TRIESTE 5
VIA TIEPOLO 11
I 34131 TRIESTE
ITALY
TEL 40 79 3921
TLF
TLX 461137 OAT I
EML
COM

SANTOS FILIPE D DR
CENTRO DE FISICA NUCLEAR
UNIVERSIDADE DE LISBOA
AV PROF GAMMA PINTO Nº2
P 1699 LISBOA CODEX
PORTUGAL
TEL 351 1 7950 790
TLF 351 1 7656 22
TLX 62593 P
EML SANTOS@PTIFM
COM. 35

SANTOS NILTON OSCAR DR
OBSERVATORIO NACIONAL
RUA GL BRUCE 586
SAO CRISTOVAO
20921 RIO DE JANEIRO RJ
BRAZIL
TEL 21 580 0235
TLF 21 580 0332
TLX
EML.
COM. 48

SANWAL BASANT BALLABH DR
UTTAR PRADESH STATE
OBSERVATORY
PO MANORA PEAK 263 129
NAINITAL 263 129
INDIA
TEL 59 42 2136/2583
TLF
TLX
EML
COM 29

· SANWAL N B DR
DPT OF ASTRONOMY
UNIVERSITY OF OSMANIA
HYDERABAD 500 007
INDIA
TEL 71 951*247
TLF
TLX
EML
COM 30,42,45

SANYAL ASHIT DR
7505 RIDGEWELL CT
BELTSVILLE MD 20705
USA
TEL
TLF
EML
COM 27,42
TLX

SANZ I SUBIRANA JAUME DR
DPT MATEMATICA APLICADA
UNIV POLITEC DE CATALUNYA
BOX 30002
E 08080 BARCELONA
SPAIN
TEL 3 401 6799
TLF 3 401 6801
TLX 281 UPC E
EML MATJSS@MAT UPC ES
COM 33

SANZ JOSE L DR
DPT DE FISICA MODERNA
UNIVERSIDAD DE CANTABRIA
AVD LOS CASTROS S/N
E 39005 SANTANDER
SPAIN
TEL 42 20 1452
TLF
TLX· 35681 EDUCI E
EML
COM 47

SAPAR ARVED DR
TARTU ASTROPHYSICAL OBS
ESTONIAN ACAD OF SCIENCES
202444 TARTU
ESTONIA
TEL
TLF
TLX
EML
COM 36C,47

SAPRE A K DR
DPT OF PHYSICS
UNIVERISTY OF RAVISHANKAR
RAIPUR 429 010
INDIA
TEL 27 064
TLF
TLX
EML
COM 28

SARASSO MARIA DR
OSS ASTRONOMICO DI TORINO
ST OSSERVATORIO 20
I 10025 PINO TORINESE
ITALY
TEL 11 84 1067
TLX 213236 TOASTR I
EML 39181 SARASSO
TLF 11 84 1281
COM 05,08

SARAZIN CRAIG L DR
UNIVERSITY STATION
UNIVERSITY OF VIRGINIA
BOX 3818
CHARLOTTESVILLE VA 22903
USA
TEL 804 924 4903
TLF 804 924 3104
TLX
EML CLS7I@VIRGINIA EDU
COM 28,34

SAREYAN JEAN-PIERRE DR
OCA OBSERV DE NICE
BP 139
F 06003 NICE CDX
FRANCE
TEL 93 89 0420
TLF
TLX 460004 OBSNICE F
EML
COM 27,29

SARGENT ANNEILA I
CALTECH
MS 320 47
DOWNES LAB OF PHYSICS
PASADENA CA 91125
USA
TEL 818 356 6622
TLF
TLX 675425
EML
COM 33,34,40

SARGENT WALLACE L W DR
CALTECH
MS 105 24
PASADENA CA 91125
USA
TEL 818 356 4055
TLF
TLX 675425 CALTECH PSD
EML
COM 28,47

SARMA M B K PROF
DPT OF ASTRONOMY
UNIVERSITY OF OSMANIA
HYDERABAD 500 007
INDIA
TEL 65 228
TLF
TLX
EML
COM 25,27

SARMA N V G PROF
RAMAN RESEARCH INSTITUTE
SADASHIVANAGAR
BANGALORE 560 080
INDIA
TEL 812 36 0122
TLF 812 34 0492
TLX 8452671 RRI IN
EML
COM 34,40

SARMIENTO-GALAN A F DR
INST DE ASTRONOMIA UNAM
UNAM
APDO POSTAL 70-264
04510 MEXICO DF
MEXICO
TEL 548 5305
TLF
TLX 1760155 CICME
EML ANSAR@UNAMVM1 bitnet
COM

SARRIS ELEFTHERIOS PH D
ASTRONOMICAL INSTITUTE
NTL OBSERVATORY OF ATHENS
BOX 20048
GR 118 10 ATHENS
GREECE
TEL 1 346 1191
TLF
TLX 215530 OBSA GR
EML
COM

SARRIS EMMANUEL T PH D
DPT OF ELECT ENGINEERING
DEMOCRITOS UNIV OF THRACE
GR 671 00 XANTHI
GREECE
TEL 54 12 6948
TLF
TLX 452312 POLX GR
EML
COM 49

SARTORI LEO PROF
DPT OF PHYSICS & ASTRON
UNIVERSITY OF NEBRASKA
BEHLEN OBSERVATORY
LINCOLN NE 68588
USA
TEL
TLF
TLX
EML
COM 48

SASAKI MISAO
RES INST FOR THEORET PHYS
HIROSHIMA UNIVERSITY
RIRONKEN
TAKEHARA 725
JAPAN
TEL 08462-2-2362
TLF
TLX
EML
COM 47

SASAKI TOSHIYUKI DR
OKAYAMA ASTROPHYSICAL OBS
NAOJ
KAMOGATA ASAKUCHI
OKAYAMA 719 02
JAPAN
TEL 86544 2155
TLF 86544 2360
TLX
EML
COM 28

SASALOV DIMITAR D DR
CENTER FOR ASTROPHYSICS
HCO/SAO
60 GARDEN ST
CAMBRIDGE MA 02138
USA
TEL 617 495 7451
TLF 617 495 7049
TLX
EML SASSELOV@CFA HARVARD EDU
COM

SASAO TETSUO DR
INTL LATITUDE OBSERVATORY
NAOJ
HOSHIGAOKA MIZUSAWA SHI
IWATE 023
JAPAN
TEL 197-24-7111
TLF
TLX 837628 ILSMIZJ
EML
COM 19C

SASLAW WILLIAM C PROF
UNIVERSITY STATION
UNIVERSITY OF VIRGINIA
BOX 3818
CHARLOTTESVILLE VA 22903
USA
TEL 804 924 4892
TLF
TLX
EML
COM 28,48

SASTRI HANUMATH J DR
INDIAN INSTITUTE OF
ASTROPHYSICS
KORAMANGALA
BANGALORE 560 034
INDIA
TEL 812 56 6585
TLF
TLX 845763 IIAB IN
EML
COM 49

SASTRY CH V
INDIAN INSTITUTE OF
ASTROPHYSICS
KORAMANGALA
BANGALORE 560 034
INDIA
TEL 812 56 6585
TLF
TLX
EML
COM 40

SASTRY SHANKARA K
DPT OF ASTRONOMY
UNIVERSITY OF OSMANIA
HYDERABAD 500 007
INDIA
TEL 71 951
TLF
TLX
EML
COM 28

SATO FUMIO DR
DPT ASTRON & EARTH SCI
TOKYO GAKUGEI UNIVERSITY
KOGANEI
TOKYO 184
JAPAN
TEL 0423-25-2111
TLF
TLX
EML
COM 34,40

SATO HUMITAKA PROF
DPT OF PHYSICS
UNIVERSITY OF KYOTO
SAKYO KU
KYOTO 606
JAPAN
TEL
TLF
TLX
EML
COM 47

SATO KATSUHIKO PROF
DPT OF PHYSICS
UNIVERSITY OF TOKYO
BUNKYO KU
TOKYO 113
JAPAN
TEL 3 812 2111*4207
TLF 3 56 89 0465
TLX 23472 UTPHYSIC
EML sato%tkyvax hepnet@lbl bitnet
COM 35,44,47C

SATO KOICHI DR
INTL LATITUDE OBSERVATORY
NAOJ
HOSHIGAOKA MIZUSAWA SHI
IWATE 023
JAPAN
TEL 0197-24-7111
TLF
TLX 837628 ILSMIZJ
EML
COM 08,19

SATO MASSAE DR
IAG
UNIVERSIDADE DE SAO PAULO
AV MIGUEL STEFANO 4200
04301 SAO PAULO SP
BRAZIL
TEL 11 577 8599
TLF 11 276 3848
TLX 36224 IAGM BR
EML 47556 MASSAE
COM· 20

SATO NAONOBU PROF
AKITA UNIVERSITY
1-1 TEGATA GAKUENCHO
AKITA 010
JAPAN
TEL 0188-33-5261
TLX
EML
TLF
COM 27,41

SATO SHUJI DR
TOKYO ASTRONOMICAL OBS
NAOJ
OSAWA MITAKA
TOKYO 181
JAPAN
TEL 0422 41 3643
TLF 0422 41 3776
TLX
EML
COM 34

SATO YUZO DR
4-8-19 OSAWA
MITAKA
TOKYO 181
JAPAN
TEL
TLF
TLX
EML
COM

SAULT ROBERT DR
AAO
ATNF
BOX 76
EPPING NSW 2121
AUSTRALIA
TEL 2 868 0222
TLF 2 868 0310
TLX 26230
EML RSAULT@ATNF CSIRO AUU
COM 09

SAUNDERS RICHARD D E
MULLARD RADIO ASTRON OBS
CAVENDISH LABORATORY
MADINGLEY RD
CAMBRIDGE CB3 OHE
UK   .
TEL 223 664 77
TLF
TLX 81292
EML
COM 40

SAUVAL A JACQUES DR
OBSERVATOIRE ROYAL DE
BELGIQUE
AVE CIRCULAIRE 3
B 1180 BRUSSELS
BELGIUM
TEL 2 375 2484
TLF 2 374 9822
TLX 21565 OBSBEL B
EML
COM 12

SAVAGE ANN DR
UK SCHMIDT TELESCOPE
AAO
PRIVATE BAG
COONABARABRAN NSW 2357
AUSTRALIA
TEL
TLF
TLX
EML
COM 28,40,47

SAVAGE BLAIR D DR
DPT OF ASTRONOMY
UNIVERSITY OF WISCONSIN
475 N CHARTER ST
MADISON WI 53706
USA
TEL 608 262 3072
TLF
TLX 265452 UOFWISC-MDS
EML
COM 34,44C

SAVEDOFF MALCOLM P PROF
DPT PHYSICS & ASTRONOMY
UNIVERSITY OF ROCHESTER
BAUSCH AND LOMB BLDG
ROCHESTER NY 14627
USA
TEL 716 275 4357
TLF
TLX 978374 UNIBOOK ROC
EML
COM 34,35,48

SAVONIJE GERRIT JAN DR
ASTRONOMICAL INSTITUTE
UNIVERSITY OF AMSTERDAM
KRUISLAAN 403
NL 1098 SJ AMSTERDAM
NETHERLANDS
TEL 20 52 57491
TLF 20 52 57484
TLX 10262 HEF NL
EML
COM 35,42

SAWANT HANUMANT S DR
INPE
CP 515
12200 S JOSE DOS CAMPOS
BRAZIL
TEL 123 22 9977
TLF 123 21 8743
TLX 1233530
EML
COM 40

SAWYER CONSTANCE B DR
850 20TH ST # 705
325 BROADWAY
BOULDER CO 80302
USA
TEL
TLF
TLX
EML
COM 10,49

SAWYER-HOGG HELEN B DR
DAVID DUNLAP OBSERVATORY
UNIVERSITY OF TORONTO
BOX 360
RICHMOND HILL ON L4C 4Y6
CANADA
TEL 416 884 9562
TLF 416 978 3921
TLX 069 86766
EML
COM 27,37

SAXENA A K DR
INDIAN INSTITUTE OF
ASTROPHYSICS
KORAMANGALA
BANGALORE 560 034
INDIA
TEL 812 56 6585/6497
TLF
TLX 845763 IIAB IN
EML
COM 09

SAXENA P P DR
DPT OF MATHS & ASTRONOMY
UNIVERSITY OF LUCKNOW
LUCKNOW
INDIA
TEL
TLF
TLX
EML
COM 21,46

SAZHIN MICHAIL DR
STERNBERG STATE ASTR INST
UNIVERSITETSKIJ PROSP 13
119899 MOSCOW
RUSSIA
TEL 95 939 5006
TLX
EML SNN@SAI MSK SU
TLF
COM 47

SBIRKOVA-NATCHEVA T
PLANETARIUM AND PUBLIC
ASTRONOMICAL OBSERVATORY
BOX 136
BG 4700 SMOLJAN
BULGARIA
TEL 30 22 953
TLF
TLX
EML
COM 41,46

SCALISE JR EUGENIO DR
INPE
CP 515
12200 S JOSE DOS CAMPOS
BRAZIL
TEL 123 22 9977
TLF 123 21 8743
TLX 34061 INPE BR
EML
COM 40

SCALO JOHN MICHAEL
ASTRONOMY DPT
UNIVERSITY OF TEXAS
RLM 15 308
AUSTIN TX 78712 1083
USA
TEL 512 471 4461
TLF
TLX
EML
COM 34,35

SCALTRITI FRANCO DR
OSS ASTRONOMICO DI TORINO
ST OSSERVATORIO 20
I 10025 PINO TORINESE
ITALY
TEL 11 84 1067
TLX· 213236 TOASTR I
EML
TLF 11 84 1281
COM 15,42

SCARDIA MARCO
OSS ASTRONOMICO DI MILANO
VIA E BIANCHI 46
I 22055 MERATE
ITALY
TEL· 59 2035
TLX
EML
TLF
COM 26

SCARFE COLIN D DR
DPT OF PHYSICS
UNIVERSITY OF VICTORIA
BOX 3055
VICTORIA BC V8W 3P6
CANADA
TEL 604 721 7740
TLF 604 721 7715
TLX
EML SCARFE@UVPHYS UVIC CA
COM 26C,30VP,42

SCARGLE JEFFREY D DR
NASA AMES RESEARCH CTR
MS 245 3
MOFFETT FIELD CA 94035
USA
TEL 415 694 6330
TLF
TLX
EML
COM 48,51

SCARIA K K DR
INDIAN INSTITUTE OF
ASTROPHYSICS
KORAMANGALA
BANGALORE 560 034
INDIA
TEL 812 56 6585
TLF
TLX 845763 IIAB IN
EML
COM 37

SCARROTT STANLEY M DR
DPT OF PHYSICS
UNIVERSITY OF DURHAM
SOUTH RD
DURHAM DH1 3LE
UK
TEL
TLF.
TLX
EML
COM 34

SCHAAL RICARDO E DR
CRAAE/ESCOLA POLITECNICA
UNIVERSIDADE DE SAO PAULO
CP 8174
05508 SAO PAULO SP
BRAZIL
TEL 11 815 6289
TLF 11 815 4272
TLX 1180127 INPE BR
EML ECORREIA@BRUSP BITNET
COM 40

SCHADEE AERT DR
STERREKUNDIG INSTITUTE
BOX 80000
NL 3508 TA UTRECHT
NETHERLANDS
TEL 30 53 5200
TLF
TLX· 40048 FYLUT NL
EML BITNET/WNMAAERS@HUTRUUO
COM 14

SCHAEFER BRADLEY E DR
NASA/GSFC
CODE 661
GREENBELT MD 20771
USA
TEL 301 286 6955
TLF
TLX
EML
COM 27,41

SCHAEFER GERHARD DR
MAX PLANCK INSTITUT FUER
ASTROPHYSIK
KARL-SCHWARZSCHILDSTR 1
D 8046 GARCHING MUENCHEN
GERMANY
TEL· 89 329 93247
TLF 89 329 93235
TLX 524629 ASTRO D
EML GOS@DGAIPP1S BITNET
COM 48

SCHAEFFER RICHARD DR
CEA CEN
PHYSIQUE THEORIQUE
BP 2
F 91191 GIF/YVETTE CDX
FRANCE
TEL 1 69 08 7376
TLF
TLX 690641
EML
COM

SCHAIFERS KARL DR
STEINBACHWEG 37
D 6900 HEIDELBERG 1
GERMANY
TEL 62 218 01511
TLF
TLX
EML
COM

SCHALEN CARL PROF
LUND OBSERVATORY
BOX 43
S 221 00 LUND
SWEDEN
TEL 46 10 7000
TLX 33199 OBSNOT S
EML
TLF
COM 34

SCHANDA ERWIN PROF
PHYSIKALISCHES INSTITUT
INIVERSITAET BERN
SIDLERSTRASSE 5
CH 3012 BERN
SWITZERLAND
TEL 31 65 8910
TLF
TLX 32320
EML
COM

SCHARMER GOERAN BJARNE
STOCKHOLM OBSERVATORY
S 133 36 SALTSJOEBADEN
SWEDEN
TEL 87 17 0195
TLF 87 17 4719
TLX 12972
EML SCHARMER@ASTRO SU SE
COM 36

SCHATTEN KENNETH H DR
DPT OF PHYSICS
VICTORIA UNIVERSITY
PRIVATE BAG
WELLINGTON
NEW ZEALAND
TEL
TLF
TLX
EML
COM 10,35,48

SCHATZMAN EVRY PROF
OBSERVATOIRE DE PARIS
SECTION DE MEUDON
DASGAL
F 92195 MEUDON PPL CDX
FRANCE
TEL 1 45 07 7873
TLF
TLX 201571
EML
COM 34,35,47,48,49,51

SCHECHTER PAUL L DR
DPT OF PHYSICS
MIT RM 6 206
BOX 165
CAMBRIDGE MA 02139
USA
TEL 617 253 0690
TLF
TLX
EML SCHECH@ACHERNAR MIT EDU
COM 28,33,47

SCHEEPMAKER ANTON DR
COSMIC RAY WORKING GROUP
HUYGENS LAB
WASSENAARSEWEG 78
NL 2300 RA LEIDEN
NETHERLANDS
TEL
TLF
TLX
EML
COM

SCHEFFLER HELMUT PROF
CARL-ORFF-WEG 16
D 6906 LEIMEN 3
GERMANY
TEL
TLF
TLX
EML
COM

SCHEIDECKER JEAN-PAUL DR
OCA OBSERV DE NICE
BP 139
F 06003 NICE CDX
FRANCE
TEL 93 89 0420
TLF
TLX 460004 OBSNICE F
EML
COM

SCHERB FRANK PROF
DPT OF PHYSICS
UNIVERSITY OF WISCONSIN
1150 UNIVERSITY AVE
MADISON WI 53706
USA
TEL 608 262 6879
TLF
TLX
EML
COM 34,49

SCHERRER PHILIP H DR
CTR FOR SPACE SCIENCES &
ASTROPHYSICS
STANFORD UNIV ERL
STANFORD CA 94305 4055
USA
TEL 415 497 1505
TLF
TLX 348402 STANFRD STNU
EML
COM

SCHEUER PETER A G DR
MULLARD RADIO ASTRON OBS
CAVENDISH LABORATORY
MADINGLEY RD
CAMBRIDGE CB3 OHE
UK
TEL 223 66477*344
TLF.
TLX 81292
EML
COM 34,40,47,48C

SCHILBACH ELENA DR
ZENTRALINST FUR ASTROPHYS
ROSA LUXEMBURG STR 17A
D 1591 POTSDAM
GERMANY
TEL
TLF
TLX
EML
COM 05,24

SCHILD HANSRUEDI
DPT PHYSICS & ASTRONOMY
UNIVERSITY COLLEGE LONDON
GOWER ST
LONDON WC1E 6BT
UK
TEL 1 713 877 050*3475
TLF. 1 713 807 145
TLX 28722 UCPHYS G
EML
COM 35,37

SCHILD RUDOLPH E DR
CENTER FOR ASTROPHYSICS
HCO/SAO
60 GARDEN ST
CAMBRIDGE MA 02138
USA
TEL 617 495 7426
TLF
TLX 921428 SATELLITE CAM
EML
COM 29,45,51

SCHILIZZI RICHARD T DR
NFRA
BOX 2
NL 7990 AA DWINGELOO
NETHERLANDS
TEL 52 19 7244
TLF
TLX 42043 SRZM NL
EML
COM 40,48,50

SCHILLER KARL PROF DR
PIRSCHWEG 6
D 6072 DREIEICH BUCHSCHLA
GERMANY
TEL
TLF
EML
COM
TLX

SCHILLER STEPHEN
PHYSICS DPT
SOUTH DAKOTA STATE UNIV
BOX 2219 ROOM 310B
BROOKINGS SD 57007
USA
TEL 605 688 4293
TLF
TLX
EML
COM 42

SCHINDLER KARL PROF DR
INST FUER THEORET PHYSIK
RUHR-UNIVERSITAET BOCHUM
D 4630 BOCHUM 1
GERMANY
TEL
TLF
TLX
EML
COM 10,49

SCHLEGEL ERIC MATTHEW DR
NASA/GSFC
CODE 668
LAB HIGH ENERGY ASTROPHYS
GREENBELT MD 20771
USA
TEL 301 286 6636
TLF 301 286 3391
TLX
EML ERIC@HEASFS GSFC NASA GOV
COM 27

SCHLEICHER DAVID G DR
LOWELL OBSERVATORY
1400 W MARS HILL RD
BOX 1149
FLAGSTAFF AZ 86001
USA
TEL 602 774 3358
TLF
TLX
EML
COM 15,16,46

SCHLESINGER BARRY M DR
ST SYSTEMS CORP (STX)
4400 FORBES BLVD
LANHAM MD 20706
USA
TEL
TLX
EML
TLF
COM

SCHLICKEISER REINHARD DR
MPI FUER RADIOASTRONOMIE
AUF DEM HUEGEL 69
D 5300 BONN 1
GERMANY
TEL 228 5251
TLF
TLX 886440
EML
COM 40

SCHLOERB F PETER
DPT PHYSICS & ASTRONOMY
UNIV OF MASSACHUSETTS
GRC
AMHERST MA 01003
USA
TEL 413 545 4303
TLF
TLX 955491
EML
COM 15,16

SCHLOSSER WOLFHARD PROF
ASTRONOMISCHES INSTITUT
POSTFACH 102148
D 4630 BOCHUM 1
GERMANY
TEL 234 70 03454
TLX 0825860
EML
TLF
COM 46

SCHLUETER A PROF DR
MPI FUER PLASMAPHYSIK
D 8046 GARCHING MUENCHEN
GERMANY
TEL 89 329 9347
TLF
EML
COM 05,10
TLX 5215808 IPP D

SCHLUETER DIETER PROF
INST THEOR PHYS & STERNW
NEUE UNIV PHYSIK ZENTRUM
OLSHAUSENST GEB N 61C
D 2300 KIEL 1
GERMANY
TEL 431 880 4109
TLF
TLX
EML
COM

SCHMADEL LUTZ D DR
ASTRONOMISCHES RECHEN
INSTITUT
MOENCHHOFSTR 12-14
D 6900 HEIDELBERG 1
GERMANY
TEL 62 214 9026
TLF
TLX 461336 ARIHD D
EML
COM 05C,20

SCHMAHL EDWARD J DR
ASTRONOMY PROGRAM
UNIVERSITY OF MARYLAND
COLLEGE PARK MD 20742
USA
TEL 301 454 6074
TLF
TLX
EML
COM 10,12

SCHMAHL GUENTER PROF
UNIVERSITAETSSTERNWARTE
GEISMARLANDSTR 11
D 3400 GOETTINGEN
GERMANY
TEL 551 39 5061
TLX 96753
EML
TLF
COM

SCHMALBERGER DONALD C DR
THE ALBANY ACADEMY
ACADEMY RD
ALBANY NY 12208
USA
TEL 518 465 1461
TLF
TLX
EML
COM 36

SCHMEIDLER F PROF DR
MAUERKIRCHERSTR 17
D 8000 MUENCHEN 80
GERMANY
TEL·
TLF
TLX
EML
COM 08

SCHMELZ JOAN T DR
NASA/GSFC
CODE 602 6
SMM XRP
GREENBELT MD 20771
USA
TEL   301 220 4164
TLF   301 220 4171
TLX   248496
EML   SOLMAX JTS
COM   10

SCHMID HANS MARTIN DR
INSTITUT FUER ASTRONOMIE
ETH ZENTRUM
CH 8092 ZUERICH
SWITZERLAND
TEL   1 256 3633
TLF   41 1 262 0003
TLX   81 73 79 EHHG CH
EML   SCHMID@IFA ETHZ CH
COM   42

SCHMIDT EDWARD G
DPT OF PHYSICS & ASTRON
UNIVERSITY OF NEBRASKA
LINCOLN NE 68588 0111
USA
TEL   402 472 2788
TLF
TLX
EML
COM   25,27

SCHMIDT H U DR
MPI FUER PHYSIK UND
ASTROPHYSIK
KARL-SCHWARZSCHILD-STR 1
D 8046 GARCHING MUENCHEN
GERMANY
TEL   89 329 99413/4
TLF
TLX   524629 ASTRO D
EML
COM   10,15,49

SCHMIDT HANS PROF
UNIVERSITAETSSTERNWARTE
AUF DEM HUEGEL 71
D 5300 BONN 1
GERMANY
TEL
TLX
EML
TLF
COM   33,42

SCHMIDT K H DR
ZNTRLINST F ASTROPHYSIK
STERNWARTE BABELSBERG
ROSA-LUXEMBURG-STR 17A
D 1502 POTSDAM
GERMANY
TEL
TLF
TLX
EML
COM   05,28,33,44

SCHMIDT MAARTEN PROF
CALTECH
MS 105 24
ASTRONOMY DPT
PASADENA CA 91125
USA
TEL   818 356 4204
TLF
TLX   675425
EML
COM   15,28,33,40,47

SCHMIDT THOMAS DR
RUDOLF-STEINER-SCHULE

AN DER STIFTSKIRCHE 13
D 4800 BIELEFELD 1
GERMANY
TEL   521 88 0407
TLF
TLX
EML
COM   34,46

SCHMIDT WOLFGANG DR
KIEPENHEUER INSTITUT
FUR SONNENPHYSIK
SCHONECKSTRASSE 6
D 7800 FREIBURG BREISGAU
GERMANY
TEL   761 382067
TLF   761 32 280
TLX   7721552 KIS D
EML
COM   12

SCHMIDTKE PAUL C DR
DPT OF PHYSICS
ARIZONA STATE UNIVERSITY
TEMPE AZ 85287
USA
TEL   602 965 2918
TLF
TLX
EML   BITNET SCHMIDTKE@ASUCPS
COM   26,42

SCHMIDT-KALER TH PROF
ASTRONISCHES INSTITUT
RUHR-UNIVERSITAET BOCHUM
STEINHUEGEL 105
D 5810 WITTEN
GERMANY
TEL   23 470 03454
TLF
TLX   0825860
EML
COM   33,34,45

SCHMID-BURGK J DR PROF
MPI FUER RADIOASTRONOMIE

AUF DEM HUEGEL 69
D 5300 BONN 1
GERMANY
TEL   228 52 5271
TLF
TLX   886440
EML
COM   34,36

SCHMIEDER BRIGITTE DR
OBSERVATOIRE DE PARIS
SECTION DE MEUDON
F 92195 MEUDON PPL CDX
FRANCE
TEL   1 45 07 7817
TLF
TLX
EML
COM   10C

SCHMITT DIETER DR
UNIV STERNWARTE
GEISMARLANDSTRASSE 11
D 3400 GOETTINGEN
GERMANY
TEL   551 39 5046
TLF
TLX   96753
EML   BITNET dschmit@dgogwdg1
COM   12

SCHMITTER EDWARD F DR
DPT OF PHYSICS
UNIVERSITY OF LAGOS
AKOKA
LAGOS
NIGERIA
TEL   1 83 7864
TLF
TLX
EML
COM   46

SCHMUTZ WERNER
INSTITUT FUER ASTRONOMIE
ETH ZENTRUM
CH 8092 ZUERICH
SWITZERLAND
TEL
TLF
TLX
EML
COM   36

SCHNEIDER GLENN H DR
STCTI
HOMEWOOD CAMPUS
3700 SAN MARTIN DR
BALTIMORE MD 21218
USA
TEL   301 338 4717
TLF
TLX   684 9101 STSCI
EML   GSCHNEIDER@SCIVAX
COM

SCHNEIDER HARTMUT DR
UNIVERSITY STERNWARTE
GEISMARLANDSTR 11
D 3400 GOETTINGEN
GERMANY
TEL   551 39 5042
TLX   96753
EML   BITNET HSCHNEI@DGOGWDG1
TLF   551 39 5043
COM   27

SCHNEIDER JEAN
OBSERVATOIRE DE PARIS
SECTION DE MEUDON
F 92195 MEUDON PPL CDX
FRANCE
TEL   1 45 07 7430
TLF
TLX   201517
EML   EAR "SCHNEIDER@FRMEUSI"
COM   47,51

SCHNEIDER PETER DR
MAX PLANCK INSTITUT
FUR PHYSIK/ASTROPHYSIK
KARL-SCHARZSCHILD-STR 1
D 8046 GARCHING MUENCHEN
GERMANY
TEL   89 329 90
TLF
TLX   524629
EML
COM   47

SCHNELL ANNELIESE DR
INSTITUT FUER ASTRONOMIE
UNIVERSITAT WIEN
TUERKENSCHANZSTR 17
A 1180 WIEN
AUSTRIA
TEL 1 345 36093
TLF
TLX
EML
COM

SCHNEPS MATTHEW H
CENTER FOR ASTROPHYSICS
HCO/SAO
60 GARDEN ST
CAMBRIDGE MA 02138
USA
TEL 617 495 7472
TLF
TLX 921428 SATELLITE CAM
EML
COM

SCHNOPPER HERBERT W DR
DANISH SPACE RESEARCH INS
LUNDTOFTEVEJ 7
DK 2800 LYNGBY
DENMARK
TEL 42 88 2277
TLF
TLX 37198 DANRU
EML
COM 48

SCHNUR GERHARD F O
ASTRONOMISCHES INSTITUT
RUHR UNIVERSITAET
POSTFACH 102148
D 4630 BOCHUM 1
GERMANY
TEL
TLF
TLX
EML
COM

SCHOBER HANS J DR
INSTITUT FUER ASTRONOMIE
UNIVERSITAETSPLATZ 5
A 8010 GRAZ
AUSTRIA
TEL 316 380 5273
TLX 31078 OBSLGZ
EML
TLF
COM 10,12,15,20,42,51

SCHOEFFEL EBERHARD F DR
MERIANERSTR 42
D 8600 BAMBERG
GERMANY
TEL
TLF
EML
COM. 42
TLX

SCHOEMBS ROLF DR
INSTITUT FUER ASTRONOMIE
UND ASTROPHYSIK
SCHEINERSTR 1
D 8000 MUENCHEN 80
GERMANY
TEL 89 989 021
TLF
TLX
EML
COM 27

SCHOENBERNER DETLEF PROF
INST THEOR PHYS & STERNW

OLSHAUSENSTRASSE
D 2300 KIEL 1
GERMANY
TEL 431 880 4100
TLF
TLX 292706
EML
COM 35,36

SCHOENEICH W DR
ZNTRLINST F ASTROPHYSIK
ROSA-LUXEMBURG-STR 17A
D 1502 POTSDAM
GERMANY
TEL
TLX
EML
TLF.
COM 25,44

SCHOENFELDER VOLKER DR
MPI F EXTRATERRESTRISCHE
PHYSIK
D 8046 GARCHING MUENCHEN
GERMANY
TEL 89 329 9578
TLF
TLX 5215845 XTER D
EML
COM

SCHOLL HANS DR
OCA OBSERV DE NICE
BP 139
F 06003 NICE CDX
FRANCE
TEL 93 89 0420
TLF
TLX
EML
COM 07,15,20

SCHOLZ GERHARD DR
ZNTRLINST F ASTROPHYSIK
AKAD WISSENSCHAFTEN DDR
ROSA-LUXEMBURG-STR 17A
D 1502 POTSDAM
GERMANY
TEL
TLF
TLX
EML
COM 29

SCHOLZ M PROF
INST F THEORETISCHE
ASTROPHYS DER UNIVERSITAT
IM NEUENHEIMER FELD 561
D 6900 HEIDELBERG 1
GERMANY
TEL
TLF
TLX
EML ' B15@DHDURZ2 BITNET
COM 36

SCHOOLMAN STEPHEN A DR
LOCKHEED PALO ALTO RES LB
3251 HANOVER ST
PALO ALTO CA 94304
USA
TEL
TLX
EML
TLF
COM

SCHRAMM DAVID N PROF
ASTRONOMY & ASTROPHYS CTR
UNIVERSITY OF CHICAGO
5640 S ELLIS AVE
CHICAGO IL 60637
USA
TEL 312 962 8202
TLF
TLX 6871133 UNCGO UW
EML
COM 35,47,48C

SCHREIBER ROMAN
INSTITUTE OF ASTRONOMY
N COPERNICUS UNIVERSITY
UL CHOPINA 12/18
PL 87 100 TORUN
POLAND
TEL 48 5626017
TLF
TLX 0552234 ASTR PL
EML SCHREIBE@PLTUMK11
COM 49

SCHREIER ETHAN J DR
STSCI
HOMEWOOD CAMPUS
3700 SAN MARTIN DR
BALTIMORE MD 21218
USA
TEL 301 338 4740
TLF
TLX
EML
COM· 48

SCHRIJVER C J DR
STERREKUNDIG INSTITUTE
BOX 80000
NL 3508 TA UTRECHT
NETHERLANDS
TEL 30 53 5224
TLX
EML KSCHRIJVER@SOLAR
TLF 30 53 1601
COM 10,36

SCHRIJVER JOHANNES DR
SPACE RESEARCH LABORATORY
SRON
SORBONNELAAN 2
NL 3584 CA UTRECHT
NETHERLANDS
TEL 30 53 5600
TLF 30 54 0860
TLX 47224 ASTRO NL
EML
COM 14

SCHROEDER DANIEL J PROF
DPT PHYSICS & ASTRONOMY
BELOIT COLLEGE
BELOIT WI 53511
USA
TEL 608 365 3391
TLF
TLX
EML
COM 09,46

SCHROEDER KLAUS PETER DR
HAMBURGER STERNWARTE
GOJENSBERGSWEG 112
D 2050 HAMBURG 80
GERMANY
TEL 40 7252 4141
TLX. 217884
EML
TLF 40 7252 4198
COM 29

SCHROEDER ROLF DR
MOEDERKENWEG 37
D 2050 HAMBURG 80
GERMANY
TEL
TLF
EML
COM
TLX

SCHROETER EGON H PROF
KIEPENHEUER INSTITUT
FUER SONNENPHYSIK
SCHOENECKSTRASSE 6
D 7800 FREIBURG BREISGAU
GERMANY
TEL 761 32864
TLF.
TLX 7721552 KIS D
EML
COM 10

SCHROLL ALFRED DR
SONNENOBSERVATORIUM
KANZELHOEHE
A 9521 TREFFEN
AUSTRIA
TEL 424 82717
TLF
TLX 45699 SOLOBS A
EML
COM

SCHRUEFER EBERHARD DR
INSTITUT FUER ASTROPHYSIK
UNIVERSITAET BONN
AUF DEM HUEGEL 71
D 5300 BONN 1
GERMANY
TEL 228 73 3390
TLF
TLX 886440
EML
COM

SCHRUTKA-RECHTENSTAMM PR
WILLERGASSE 27/4/7
A 1238 WIEN
AUSTRIA
TEL 1 884 8132
TLF
TLX
EML
COM 20

SCHUBART JOACHIM DR
ASTRONOMISCHES RECHEN-
INSTITUT
MOENCHHOFSTR 12-14
D 6900 HEIDELBERG 1
GERMANY
TEL 62 214 9026
TLF
TLX 461336 ARIHD D
EML
COM 07,15,20

SCHUCH NELSON JORGE
OBSERVATORIO NACIONAL
UFSM/CTRO TECNOLOGIA
CIDADE UNIVERSITARIA
97100 SANTA MARIA RS
BRAZIL
TEL 55 226 1616
TLF
TLX 0552230 UFSM
EML
COM 40,47,51

SCHUECKING E L DR
DPT OF PHYSICS
NEW YORK UNIVERSITY
NEW YORK NY 10003
USA
TEL
TLF
TLX
EML
COM 28,47

SCHUESSLER MANFRED DR
KIEPENHEUER-INSTITUT
FUER SONNENPHYSIK
SCHOENECKSTR 6
D 7800 FREIBURG BREISGAU
GERMANY
TEL 761 32864
TLF
TLX 7721552 KIS D
EML
COM 10C,12

SCHULER WALTER DR
SONNENRAIN 15
CH 4533 RIEDHOLZ
SWITZERLAND
TEL 65 23 20 55
TLF
TLX
EML
COM 31

SCHULTE D H DR
ITEK CORPORATION

10 MAGUIRE RD
LEXINGTON MA 02173
USA
TEL
TLF
TLX
EML
COM

SCHULTZ ALFRED BERNARD DR
STSCI/CSC
HOMEWOOD CAMPUS
3700 SAN MARTIN DR
BALTIMORE MD 21218
USA
TEL 301 338 5044
TLF
TLX 684 9101 STSCI
EML SCIVAX ·SCHULTZ
COM 09

SCHULTZ G V DR
MPI FUER RADIOASTRONOMIE
AUF DEM HUEGEL 69
D 5300 BONN 1
GERMANY
TEL 228 52 5291
TLX 886440
EML
TLF
COM 09,28,34,40,44,47

SCHULZ HARTMUT DR
ASTRONOMISCHES INSTITUT
UNIVERSITAT BOCHUM
POSTFACH 10 21 48
D 4630 BOCHUM 1
GERMANY
TEL 234 70 03454
TLF
TLX 0825860
EML
COM 28

SCHULZ ROLF ANDREAS
MPI FUER RADIOASTRONOMIE

AUF DEM HUEGEL 69
D 5300 BONN 1
GERMANY
TEL 228 52 5232
TLF
TLX 886440
EML
COM 34,40

SCHUMACHER GERARD DR
OCA OBSERV DU CALERN
CAUSSOLS
F 06460 S VALLIER THIEY
FRANCE
TEL 93 42 6270
TLF
TLX· 460004
EML
COM

SCHUMANN JOERG DIETER DR
OBSERVATORIUM HOHER LIST
UNIV STERNWARTE BONN
D 5568 DAUN
GERMANY
TEL 65 92 2937
TLF
TLX
EML
COM 09

SCHUSTER WILLIAM JOHN DR
INSTITUTO DE ASTRONOMIA
UNAM
APDO POSTAL 877
22860 ENSENADA B CALIF
MEXICO
TEL 706 67 83093
TLF
TLX 56539 CICE ME
EML
COM· 20,25

SCHUTZ BERNARD F PROF
PHYSICS DPT
UNIV WALES COLLEGE
BOX 913
CARDIFF CF1 3TH
UK
TEL 222 874 785
TLF 222 371 921
TLX 498635
EML
COM 35

SCHUTZ BOB EWALD
CENTER FOR SPACE RESEARCH
UNIVERSITY OF TEXAS
AUSTIN TX 78712 1083
USA
TEL 512 471 1356
TLF
TLX 704265 CSRUTX UD
EML
COM 19

SCHWAN HEINER DR
ASTRONOMISCHES RECHEN-
INSTITUT
MOENCHHOFSTR 12-14
D 6900 HEIDELBERG 1
GERMANY
TEL 62 214 9026
TLF
TLX 461336 ARIHD D
EML·
COM 04C,08C

SCHWARTZ DANIEL A DR
CENTER FOR ASTROPHYSICS
HCO/SAO
60 GARDEN ST
CAMBRIDGE MA 02138
USA
TEL 617 495 7232
TLF
TLX
EML
COM 44,48

SCHWARTZ PHILIP R DR
NAVAL RESEARCH LABORATORY
CODE 4138
4555 OVERLOOK AVE SW
WASHINGTON DC 20375 5000
USA
TEL 202 767 3391
TLF
TLX
EML
COM 27,34,40

SCHWARTZ RICHARD D
DPT OF PHYSICS
UNIVERSITY OF MISSOURI
8001 NATURAL BRIDGE RD
ST LOUIS MO 63121
USA
TEL 314 553 5025
TLF
TLX 447658 UMSL BOOKSTOR
EML
COM 34

SCHWARTZ ROLF PH D
MPI FUER RADIOASTRONOMIE
AUF DEM HUEGEL 69
D 5300 BONN 1
GERMANY
TEL 228 52 5303
TLX
EML
TLF
COM

SCHWARTZ STEVEN JAY
ASTRONOMY UNIT
QUEEN MARY/WESTFIELD COLL
MILE END RD
LONDON E1 4NS
UK
TEL 1 980 4811*3849
TLF
TLX
EML
COM 12,44,49

SCHWARZ HUGO E
ESO
CASILLA 19001
SANTIAGO 19
CHILE
TEL 2 121 3249
TLF
TLX 240853
EML
COM

SCHWARZ ULRICH J DR
KAPTEYN ASTRONOMICAL INST
BOX 800
NL 9700 AV GRONINGEN
NETHERLANDS
TEL 50 11 6695
TLX 53572 STARS NL
EML
TLF
COM 28,34,40

SCHWARZENBERG-CZERNY A
ASTRONOMICAL OBSERVATORY
WARSAW UNIVERSITY
AL UJAZDOWSKIE 4
PL 00 478 WARSAW
POLAND
TEL 29 40 11
TLF
TLX 813978 ZAPAN PL
EML
COM 27

SCHWARZSCHILD MARTIN PROF
PRINCETON UNIVERSITY OBS
PEYTON HALL
PRINCETON NJ 08544
USA
TEL 609 452 3812
TLF
TLX
EML
COM 35

SCHWEHM GERHARD DR
ESA/ESTEC
SSD
BOX 299
NL 2200 AG NOORDWIJK
NETHERLANDS
TEL 17 19 865555
TLF 17 19 17400
TLX 39098
EML
COM 21,44

SCHWEIZER FRANCOIS DR
DPT TERRESTR MAGNETISM
CARNEGIE INST WASHINGTON
5241 BROAD BRANCH RD NW
WASHINGTON DC 20015
USA
TEL 202 966 0863
TLF
TLX 440427 MAGN UI
EML
COM 28

SCIAMA DENNIS W DR
SISSA
ST COSTIERA 11
MIRAMARE
I 34014 TRIESTE
ITALY
TEL 40 22 4118
TLF
TLX 460392
EML
COM 28,47,48

SCIORTINO SALVATORE DR
OSS ASTRONOMICO
UNIVERSITA DI PALERNO
PALAZZO DEI NORMANNI
I 90134 PALERMO
ITALY
TEL 91 657 0451
TLF 91 48 8900
TLX 910402 ASTROP I
EML ASTROPA@IPACUC
COM 48

SCONZO PASQUALE DR
29 OLD MYSTIC ST

ARLINGTON MA 02174
USA
TEL 617 646 9315
TLF
TLX
EML
COM 07

SCOTT ELIZABETH L PROF
STATISTICS DPT
UNIVERSITY OF CALIFORNIA
367 EVANS HALL
BERKELEY CA 94720
USA
TEL 415 642 2777
TLF
TLX 910-366-7114 UC BERK
EML
COM 47

SCOTT EUGENE HOWARD
NASA/GSFC
CODE 684 9
GREENBELT MD 20771
USA
TEL 301 286 8746
TLF
TLX
EML
COM 34

SCOTT JOHN S DR
STEWARD OBSERVATORY
UNIVERSITY OF ARIZONA
TUCSON AZ 85721
USA
TEL
TLF
TLX
EML
COM 40,48

SCOTT PAUL F DR
MULLARD RADIO ASTRON OBS
CAVENDISH LABORATORY
MADINGLEY RD
CAMBRIDGE CB3 OHE
UK
TEL 223 664 77
TLF
TLX 81292
EML
COM 40

SCOVILLE NICHOLAS Z
DPT PHYSICS & ASTRONOMY
UNIV OF MASSACHUSETTS
GRC
AMHERST MA 01003
USA
TEL   413 545 0789
TLF
TLX
EML
COM 28,34

SCRIMGER J NORMAN DR
DPT OF ASTRONOMY
ST MARY'S UNIVERSITY
HALIFAX NS B3H 3C3
CANADA
TEL 902 420 5633
TLF 902 420 5561
TLX
EML
COM

SCUFLAIRE RICHARD DR
INSTITUT D'ASTROPHYSIQUE
UNIVERSITE DE LIEGE
AVE COINTE 5
B 4000 COINTE-LIEGE
BELGIUM
TEL 41 52 9980
TLF 41 52 7474
TLX 41264 ASTRLG B
EML
COM 27,35

SEAQUIST ERNEST R PROF
DPT OF ASTRONOMY
UNIVERSITY OF TORONTO
60 ST GEORGE ST
TORONTO ON M5S 1A7
CANADA
TEL 416 978 3146
TLF 416 978 3921
TLX 069 86766
EML
COM 40

SEARLE LEONARD DR
HALE OBSERVATORIES
813 SANTA BARBARA ST
PASADENA CA 91101
USA
TEL 818 304 u220
TLX
EML
TLF
COM 28

SEARS RICHARD LANGLEY DR
DPT OF ASTRONOMY
UNIVERSITY OF MICHIGAN
DENNISON BLDG
ANN ARBOR MI 48109 1090
USA
TEL 313 763 3295
TLF
TLX
EML
COM 35

SEATON MICHAEL J PROF
DPT PHYSICS & ASTRONOMY
UNIVERSITY COLLEGE LONDON
GOWER ST
LONDON WC1E 6BT
UK
TEL 1 713 877 050
TLF
TLX 28722
EML
COM 12,14,34,36C

SECCO LUIGI DR
DPT DI ASTRONOMIA
UNIVERSITA DI PADOVA
VIC DELL OSSERVATORIO 5
I 35122 PADOVA
ITALY
TEL 49 66 1499
TLF.
TLX 430176 UNPADU I
EML
COM

SEDLMAYER ERWIN DR
INST FUER ASTRONOMIE &
ASTROPHYSIK DER TECHN UNI
ERNST-REUTER-PLATZ 7
D 1000 BERLIN 10
GERMANY
TEL
TLF
TLX
EML
COM 36

SEDMAK GIORGIO PROF
DPT DI ASTRONOMIA
UNIVERSITA DI TRIESTE
VIA TIEPOLO 11
I 34131 TRIESTE
ITALY
TEL 40 79 4863
TLF
TLX 461137
EML
COM 05,09

SEEDS MICHAEL AUGUST DR
ASTRONOMY PROGRAM
FRANKLIN/MARSHALL COLLEGE
LANCASER PA 17604 3003
USA
TEL 717 291 3800
TLF 717 291 4143
TLX
EML BITNET%"M_SEEDS@FANDM"
COM 27,46

SEEGER CHARLES LOUIS III
SAN FRANCISCO STATE UNIV
473 JAMES RD
PALO ALTO CA 94306
USA
TEL 415 493 6005
TLF
TLX
EML
COM 51

SEEGER PHILIP A DR
LOS ALAMOS NATIONAL LAB
MS H805
BOX 1663
LOS ALAMOS NM 87545
USA
TEL 505 667 8843
TLF
TLX
EML
COM

SEGAL IRVING E DR
DPT OF MATHEMATICS
MIT RM 2 224
BOX 165
CAMBRIDGE MA 02139
USA
TEL 617 253 4985
TLF
TLX
EML
COM 47

SEGALUVITZ ALEXANDER DR
BOX 659
KEFAR SAVA
ISRAEL
TEL
TLF
TLX
EML
COM

SEGAN STEVO
INSTITUTE OF ASTRONOMY
UNIVERSITY OF BELGRADE
STUDENTSKI TRG 16
YU 11000 BEOGRAD
YUGOSLAVIA
TEL
TLF
TLX
EML
COM 07

SEGGEWISS WILHELM PROF
OBSERVATORIUM HOHER LIST
UNIVERSITAETS-STERNWARTE
D 5568 DAUN EIFEL
GERMANY
TEL 65 92 2150
TLF
TLX
EML
COM 29,33,42

SEHNAL LADISLAV DR
ASTRONOMICAL INSTITUTE
CZECH ACADEMY OF SCIENCES
ONDREJOV OBSERVATORY
CS 251 65 ONDREJOV
CZECHOSLOVAKIA
TEL 204 85201
TLF 204 85314
TLX 121579
EML
COM 07

SEIDELMANN P KENNETH DR
US NAVAL OBSERVATORY
34 & MASSACHUSETTS AVE NW
WASHINGTON DC 20392 5100
USA
TEL 202 653 1545
TLF 202 653 1744
TLX 7108221970
EML USNAO@PHOBOS USNO NAVY MIL
COM 04C,07,20

SEIDEN PHILIP E
IBM
THOMAS J WATSON RES CTR
BOX 218
YORKTOWN HEIGHTS NY 10598
USA
TEL 914 945 1424
TLF
TLX 137456
EML
COM 28,47

SEIDOV ZAKIR F DR
SHEMAKHA ASTROPHYSICAL
OBSERVATORY
AZER ACADEMY OF SCIENCES
373243 SHEMAKHA
AZERBAIDZHAN
TEL
TLF
TLX
EML
COM 35

SEIELSTAD GEORGE A
NRAO
BOX 2
GREEN BANK WV 24944
USA
TEL 304 456 2301
TLX 710-938-1530
EML
TLF
COM 40,47,48,51

SEIMENIS JOHN DR
DPT OF MATHEMATICS
UNIVERSITY OF THE AEGEAN
GR 832 00 SAMOS
GREECE
TEL 27 33 3896
TLF 27 33 3896
TLX 294268 VASM GR
EML JSEIM@GRATHUN1
COM 33

SEIN-ECHALUCE M LUISA DR
DPT DE MATEMATICA APLIC
UNIVERSIDAD DE ZARAGOZA
AVD MARIA ZAMBRANO 50
E 50009 ZARAGOZA
SPAIN
TEL 76 51 8143
TLF 76 56 5852
TLX 58198
EML
COM 07

SEIRADAKIS JOHN HUGH DR
DPT OF ASTRONOMY
UNIVERSITY THESSALONIKI
GR 540 06 THESSALONIKI
GREECE
TEL 31 99 1357
TLF
TLX
EML
COM 40,51

SEITTER WALTRAUT C PROF
ASTRONOMISCHES INSTITUT
DOMAGKSTR 75
D 4400 MUENSTER
GERMANY
TEL 251 83 3561
TLX 892529
EML
TLF
COM 45

SEKANINA ZDENEK DR
JPL
EARTH & SPACE SCI DIV
4800 OAK GROVE DR
PASADENA CA 91109
USA
TEL 818 354 7589
TLF
TLX
EML
COM 15,20,22

SEKI MUNEZO DR
DPT OF EARTH SCIENCES
TOHOKU UNIVERSITY
KAWAUCHI
SENDAI 980
JAPAN
TEL 222 12 1800
TLF
TLX
EML
COM 34

SEKIGUCHI KAZUHIRO DR
DPT OF ASTRONOMY
NEW MEXICO STATE UNIV
BOX 4500
LAS CRUCES NM 88003
USA
TEL 505 646 2613
TLF
TLX
EML
COM

SEKIGUCHI NAOSUKE PROF
MUSASHIDAI 3-16-8
FUCHU
TOKYO 183
JAPAN
TEL
TLX
EML
TLF
COM 19

SELLWOOD JERRY A
DPT PHYSICS & ASTRONOMY
RUTGERS UNVIERSITY
BOX 849
PISCATAWAY NJ 08854 0849
USA
TEL
TLF 908 932 4343
TLX 703528
EML
COM 28,33

SELVELLI PIERLUIGI DR
OAT
BOX SUCC TRIESTE 5
VIA TIEPOLO 11
I 34131 TRIESTE
ITALY
TEL 40 79 3221
TLF
TLX 461137 OQT I
EML
COM 44

SEMEL MEIR DR
OBSERVATQIRE DE PARIS
SECTION DE MEUDON
F 92195 MEUDON PPL CDX
FRANCE
TEL 1 45 07 7790
TLF
TLX
EML·
COM 10,12

SEMENIUK IRENA DR
ASTRONOMICAL OBSERVATORY
WARSAW UNIVERSITY
AL UJAZDOWSKIE 4
PL 00 478 WARSAW
POLAND
TEL 29-40-11/12
TLF
TLX. 815548 OAUW
EML
COM 42

SEMENZATO ROBERTO
DPT DI FISICA G GALILEI
UNIVERSITA DI PADOVA
VIA MARZOLO 8
I 35131 PADOVA
ITALY
TEL 49 84 4247
TLF
TLX 430308 DF GGPD I
EML
COM

SEN S N DR
INDIAN ASSOCIATION FOR
THE CULTIVATION OF SCI
JADAVPUR
INDIA
TEL
TLF
TLX
EML
COM

SENGBUSCH KURT V DR
MPI FUER PHYSIK UND
ASTROPHYSIK
KARL-SCHWARZSCHILD-STR 1
D 8046 GARCHING MUENCHEN
GERMANY
TEL
TLF
TLX
EML
COM 35

SEQUEIROS JUAN DR
DPT FISICA
UNIVERSIDAD DE ALCALA DE
HENARES APD 20
E 28801 ALCALA DE HENARES
SPAIN
TEL 1 889 4940
TLF 18 89 4953
TLX
EML
COM 44

SERAFIMOV KIRIL B ACAD
DPT OF ASTRONOMY
BULGARIAN ACAD SCIENCES
72 LENIN BLVD
BG 1784 SOFIA
BULGARIA
TEL 2 75 8927
TLF
TLX 23561 ECF BAN
EML
COM

SERAFIN RICHARD AUGUST
THERESENSTRASSE 39a

D 4200 OBERHAUSEN 1
GERMANY
TEL
TLF
TLX
EML
COM

SERIO SALVATORE DR
OSS ASTRONOMICO
UNIVERSITA DI PALERNO
PALAZZO DEI NORMANNI
I 90134 PALERMO
ITALY
TEL 91 59 2451
TLF
TLX 910402 ASTROP I
EML BITNET ASTROPA@IPACUC
COM

SERRANO ALFONSO DR
INSTITUTO DE ASTRONOMIA
UNAM
APDO POSTAL 70-264
04510 MEXICO DF
MEXICO
TEL
TLF
TLX
EML
COM 05

SERSIC J L DR
OBSERVATORIO ASTRONOMICO
DE CORDOBA
LAPRIDA 854
5000 CORDOBA
ARGENTINA
TEL 51 25072
TLF
TLX 51-822 BUCOR OBSASTR
EML
COM ~28,47

SERVAN BERNARD
OBSERVATOIRE DE PARIS
61 AVE OBSERVATOIRE
F 75014 PARIS
FRANCE
TEL 1 40 51 2236
TLF
TLX 270776 OBS F
EML
COM 09

SESSIN WAGNER DR
INPE
CP 515
DPTO DE ASTRONOMIA
12200 S JOSE DOS CAMPOS
BRAZIL
TEL 123 22 9088
TLF
TLX 011 73437 ZWO-24-73
EML
COM 07

SETTI GIANCARLO PROF
ESO
KARL-SCHWARZSCHILD-STR 2
D 8046 GARCHING MUENCHEN
GERMANY
TEL 89 320 060
TLF 89 320 2362
TLX 52828222 EO D
EML
COM 28,40,47,48,49

SEVARLIC BRANISLAV M PROF
ASTRONOMICAL OBSERVATORY
VOLGINA 7
YU 11050 BEOGRAD
YUGOSLAVIA
TEL 11 419 357
TLF
TLX
EML
COM 08,19

SEVERINO GIUSEPPE
OSS ASTRONOMICO
DI CAPODIMONTE
VIA MOIARIELLO 16
I 80131 NAPOLI
ITALY
TEL 81 44 0101
TLF
TLX
EML
COM 12

SEVILLA MIGUEL J DR
INST DE ASTRON Y GEODESIA
FAC DE CIENCIAS MATEMAT
UNIVERSIDAD COMPLUTENSE
E 28040 MADRID
SPAIN
TEL 1 244 2501
TLF
TLX
EML
COM 19

SEWARD FREDERICK D
CENTER FOR ASTROPHYSICS
HCO/SAO
60 GARDEN ST
CAMBRIDGE MA 02138
USA
TEL 617 495 7282
TLF
TLX
EML
COM 48

SEYMOUR P A H
57 HERMITAGE RD
PLYMOUTH DEVON
UK
TEL
TLF
TLX
EML
COM

SEZER CENGIZ DR
FACULTY OF SCIENCE
EGE UNIVERSITY
BOX 21
35100 BORNOVA IZMIR
TURKEY
TEL 51 18 0110
TLF
TLX
EML
COM

SHAFFER DAVID B DR
NASA/GSFC
CODE 621 9
GREENBELT MD 20771
USA
TEL 301 286 6434
TLF
TLX
EML
COM 40

SHAFTER ALLEN W DR
DPT OF ASTRONOMY
SAN DIEGO STATE UNIV
SAN DIEGO CA 92182
USA
TEL 617 594 6170
TLF
TLX
EML SHAFTER@PROTEUS SDSU EDU
COM 42

SHAH GHANSHYAM A DR
INDIAN INSTITUTE OF
ASTROPHYSICS
KORAMANGALA
BANGALORE 560 034
INDIA
TEL 812 56 6585/6497
TLF
TLX 845-763 IIAB IN
EML
COM 34

SHAHAM JACOB PROF
DPT OF PHYSICS
COLUMBIA UNIVERSITY
NEW YORK NY 10027
USA
TEL 212 280 3349
TLF
TLX 220094 COLU UR
EML
COM 48

SHAHBAZIAN ROMELIA K DR
BYURAKAN ASTROPHYSICAL
OBSERVATORY
378433 BYURAKAN
ARMENIA
TEL 88 52 28 3453
TLF
TLX
EML
COM 28

SHAKESHAFT JOHN R DR
MULLARD RADIO ASTRON OBS
CAVENDISH LABORATORY
MADINGLEY RD
CAMBRIDGE CB3 OHE
UK
TEL 223 664 77
TLF
TLX 81292 CAVLAB G
EML
COM 05,28,40

SHAKHBAZYAN YURIJ L DR
BYURAKAN ASTROPHYSICAL
OBSERVATORY
378433 BYURAKAN
ARMENIA
TEL 88 52 28 3435
TLF
TLX
EML
COM 09,

SHAKHOVSKOJ NIKOLAY M DR
CRIMEAN ASTROPHYS OBS
UKRAINIAN ACAD OF SCIENCE
NAUCHNY
334413 CRIMEA
UKRAINE
TEL 43 2945
TLF
TLX
EML
COM 25

A288

SHAKURA NICHOLAJ I DR
STERNBERG STATE ASTR INST
117234 MOSCOW
RUSSIA
TEL
TLF
EML
COM 42,48
TLX

SHALLIS MICHAEL J DR
DPT OF ASTROPHYSICS
UNIVERSITY OF OXFORD
SOUTH PARKS RD
OXFORD OX1 3RQ
UK
TEL
TLF
TLX
EML
COM 12

SHALTOUT MESALAM A M PROF
HELWAN OBSERVATORY
HELWAN
EGYPT
TEL 78 0645/2683
TLF
TLX
EML
COM

SHANDARIN SERGEI F DR
INST FOR PHYSICS PROBLEMS

KOSYGIN 2
117334 MOSCOW
RUSSIA
TEL 137 32 48
TLF
TLX 113451 MAGNIT
EML
COM 47C

SHANE WILLIAM W DR
BOX 43
NL 6580 AA MALDEN
NETHERLANDS
TEL 80 58 2483
TLF
EML
COM 33,34
TLX

SHAO CHENG-YUAN
CENTER FOR ASTROPHYSICS
HCO/SAO
60 GARDEN ST
CAMBRIDGE MA 02138
USA
TEL 617 495 7212
TLF
TLX
EML
COM 22,34

SHAPERO DONALD C DR
NTL ACADEMY OF SCIENCES
NTL RESEARCH COUNCIL
2101 CONSTITUTION AVE NW
WASHINGTON DC 20418
USA
TEL 202 334 3520
TLF
TLX 248664
EML BITNET DSHAPERO@NAS
COM

SHAPIRO IRWIN I PROF
CENTER FOR ASTROPHYSICS
HCO/SAO RM P 209
60 GARDEN ST
CAMBRIDGE MA 02138
USA
TEL 617 495 7100
TLF
TLX 921428 SATELLITE CAM
EML
COM 04,07,16,19

SHAPIRO MAURICE M PROF
205 YOAKUM
PKW 2 1720
ALEXANDRIA VA 22304
USA
TEL 703 370 1985
TLF
TLX
EML
COM 48,51

SHAPIRO STUART L
CRSR
CORNELL UNIVERSITY
SPACE SCIENCES BLDG
ITHACA NY 14853 6801
USA
TEL 607 256 4936
TLF
TLX
EML
COM 34

SHAPLEY ALAN H
NOAA
325 BROADWAY
BOULDER CO 80303
USA
TEL
TLF
TLX
EML
COM 10

SHARA MICHAEL DR
STSCI
HOMEWOOD CAMPUS
3700 SAN MARTIN DR
BALTIMORE MD 21218
USA
TEL 301 338 4743
TLF
TLX 6849101 STSCI UW
EML
COM 27

SHARAF MOHAMED ADEL PROF
DPT OF ASTRONOMY
FACULTY OF SCIENCES
CAIRO UNIVERSITY
GEZA
EGYPT
TEL
TLF
TLX
EML
COM·

SHARAF SH G DR
INST OF THEORET ASTRONOMY
ACADEMY OF SCIENCES
N KUTUZOVA 10
192187 ST PETERSBURG
RUSSIA
TEL
TLF
TLX
EML
COM 07

SHARMA DHARMA PAL DR
DPT OF PHYSICS
UNIVERSITY OF TASMANIA
GPO BOX 252C
HOBART TAS 7001
AUSTRALIA
TEL 2 202 428
TLF 2 202 410
TLX AA 58150
EML SHARMA@PHYSVAX PHYS UTAS EDU A
COM 27

SHAROV A S DR
STERNBERG STATE ASTR INST

UNIVERSITETSKIJ PROSP 13
119899 MOSCOW
RUSSIA
TEL 139-26-57
TLF
TLX
EML
COM 06,21,33,37

SHARP CHRISTOPHER DR
MAX PLANCK INSTITUT
FUR PHYSIK/ASTROPHYSIK
KARL-SCHWARSCHILD-STR 1
D 8046 GARCHING MUENCHEN
GERMANY
TEL 89 329 90
TLF
TLX 524629
EML
COM 14,15

SHARPLES RAY DR
DPT OF PHYSICS
UNIVERSITY OF DURHAM
SOUTH RD
DURHAM DH1 3LE
UK
TEL
TLF
TLX
EML
COM· 28

SHARPLESS STEWART PROF
DPT PHYSICS & ASTRONOMY
UNIVERSITY OF ROCHESTER
ROCHESTER NY 14627
USA
TEL 716 275 4389
TLF
TLX
EML
COM 34,45

SHAVER PETER A DR
ESO

KARL-SCHWARTZSCHILD-STR 2
D 8046 GARCHING MUENCHEN
GERMANY
TEL· 89 320 060
TLF 89 320 2362
TLX 52828222 EOD
EML
COM 28,34,40,47C,48

SHAVIV GIORA PROF
DPT OF PHYSICS
IIT
TECHNION CITY
HAIFA 32000
ISRAEL
TEL
TLF
TLX
EML
COM 35,42,47,48

SHAW JAMES SCOTT DR
DPT PHYSICS & ASTRONOMY
UNIVERSITY OF GEORGIA
ATHENS GA 30602
USA
TEL   404 542 2485
TLF
TLX
EML
COM

SHAW JOHN H PROF
DPT OF ASTRONOMY
OHIO STATE UNIVERSITY
174 W 18TH AVE
COLUMBUS OH 43210 1106
USA
TEL 614 422 7968
TLF
TLX
EML
COM

SHAW R WILLIAM PROF
105 HALCYON HILL

ITHACA NY 14850
USA
TEL   607 257 1948
TLF
TLX
EML
COM

SHAWHAN STANLEY D DR
DPT PHYSICS & ASTRONOMY
UNIVERSITY OF IOWA
IOWA CITY IA 52242
USA
TEL 319 353 3294
TLF
TLX
EML
COM 49

SHAWL STEPHEN J DR
DPT PHYSICS & ASTRONOMY
UNIVERSITY OF KANSAS
LAWRENCE KS 66045
USA
TEL
TLF
TLX
EML
COM 25,34,37

SHAYA EDWARD J DR
ASTRONOMY PROGRAM
UNIVERSITY OF MARYLAND
COLLEGE PARK MD 20742
USA
TEL
TLF
TLX
EML
COM 28,47

SHCHEGLOV P V DR
STERNBERG STATE ASTR INST

UNIVERSITETSKIJ PROSP 13
119899 MOSCOW
RUSSIA
TEL 139-19-73
TLF
TLX
EML
COM 09,34,50C

SHCHEGOLEV DIMITRIJ E DR
PULKOVO OBSERVATORY
ACADEMY OF SCIENCES
10 KUTUZOV QUAY
196140 ST PETERSBURG
RUSSIA
TEL
TLF
TLX
EML
COM

SHCHERBINA-SAMOJLOVA I DR
INST OF SCIENCE & TECH
125219 MOSCOW
RUSSIA
TEL 1554237
TLF
EML
COM 05
TLX

SHEA MARGARET A DR
A F GEOPHYSICS LABORATORY
SPACE PHYSICS DIV PHC
HANSCOM AFB
BEDFORD MA 01732
USA
TEL
TLF
TLX
EML
COM 10,49

SHEELEY NEIL R DR
NAVAL RESEARCH LABORATORY
CODE 4172
4555 OVERLOOK AVE SW
WASHINGTON DC 20375 5000
USA
TEL 202 767 2777
TLF
TLX
EML
COM 10,12

SHEFFER EUGENE K DR
STERNBERG STATE ASTR INST
UNIVERSITETSKIJ PROSP 13
119899 MOSCOW
RUSSIA
TEL 1392046
TLX
EML
TLF
COM

SHEFFIELD CHARLES DR
EARTH SATELLITE CORP
7222 47TH STREET
(CHEVY CHASE)
WASHINGTON DC 20815
USA
TEL   301 951 0104
TLF
TLX 248618 ESCO UR
EML
COM 44

SHEFOV NICOLAI N
INST PHYSICS OF ATMOSPH
ACADEMY OF SCIENCES
PYZHEVSKY 3
109017 MOSCOW
RUSSIA
TEL
TLF
TLX
EML
COM 21

SHELUS PETER J DR
ASTRONOMY DPT
UNIVERSITY OF TEXAS
RLM 15 316
AUSTIN TX 78712 1083
USA
TEL 512 471 3339
TLF
TLX 910-874-1351
EML
COM 20,24

SHEN BENJAMIN S P PROF
DPT OF ASTRONOMY E1
UNIV OF PENNSYLVANIA
PHILADELPHIA PA 19104
USA
TEL 215 898 8176
TLF
TLX
EML
COM

SHEN CHANGJUN
PURPLE MOUNTAIN OBSERV
CAS
NANJING
CHINA PR
TEL 25 46700
TLF
TLX 34144 PMONJ CN
EML
COM 09

SHEN CHUN-SHAN
ASTRONOMICAL STY OF CHINA
NTL TSING HUA UNIVERSITY
HSIN CHU 300043
CHINA R
TEL 35 71 9039
TLF
TLX
EML
COM 46,51

SHEN KAIXIAN
SHAANXI OBSERVATORY
CAS
LINTONG XIAN
SHAANXI
CHINA PR
TEL 33 2255
TLF
TLX 70121 CSAO CN
EML
COM· 08

SHEN LIANG-ZHAO
BEIJING ASTRONOMICAL OBS
CAS
W SUBURB
BEIJING 100080
CHINA PR
TEL
TLF
TLX 22040 BAOAS CN
EML
COM 42

SHEN LONG-XIANG
BEIJING ASTRONOMICAL OBS
CAS
W SUBURB
BEIJING 100080
CHINA PR
TEL 1 28 1698
TLF
TLX 22040 BAOAS CN
EML
COM 12

SHEN PARN-AN
NANJING ASTRONOMICAL
INSTRUMENT FACTORY
BOX 846
NANJING
CHINA PR
TEL 25 46191
TLF
TLX 34136 GLYNJ c/o NAIF
EML
COM 09

SHER DAVID DR
BOX 9624

CINCINNATI OH 452098
USA
TEL 513 871 8850
TLF
TLX
EML
COM 33,37

SHERIDAN K V DR
17B/23 THORNTON STREET
DARLING POINT NSW 2027
AUSTRALIA
TEL
TLF
EML
COM 40
TLX

SHERWOOD WILLIAM A DR
MPI FUER RADIOASTRONOMIE
AUF DEM HUEGEL 69
D 5300 BONN 1
GERMANY
TEL 228 52 5362
TLX 886440
EML
TLF
COM 27,28,34

SHESTAKA IVAN S DR
ASTRONOMICAL OBSERVATORY
ODESSA STATE UNIVERSITY
SHEVCHENKO PARK
270014 ODESSA
UKRAINE
TEL
TLF
TLX
EML
COM 22

SHEVCHENKO VLADISLAV V DR
STERNBERG STATE ASTR INST

UNIVERSITETSKIJ PROSP 13
119899 MOSCOW
RUSSIA
TEL
TLF
TLX
EML
COM 16,WGPSNC

SHEVGAONKAR R K DR
DPT OF ELECTRICAL ENGG
IIT
POWAI
BOMBAY
INDIA
TEL 22 578 2545*2440
TLF 22 578 3480
TLX 011 72313
EML
COM 40

SHI GUANG-CHEN
PURPLE MOUNTAIN OBSERV
CAS
NANJING
CHINA PR
TEL 25 33921
TLF
TLX 34144 PMOAS CN
EML
COM 08,24

SHI ZHONG-XIAN
BEIJING ASTRONOMICAL OBS
CAS
W SUBURB
BEIJING 100080
CHINA PR
TEL 1 28 1698
TLF
TLX 9053
EML
COM 10

SHIBAHASHI HIROMOTO DR
DPT OF ASTRONOMY
UNIVERSITY OF TOKYO
BUNKYO KU
TOKYO 113
JAPAN
TEL 03-812-2111
TLF
TLX 33659 UTYOSCI J
EML
COM 35

SHIBASAKI KIYOTO
NOBEYAMA RADIO OBS
NAOJ
MINAMIMAKI MURA
NAGANO 384 13
JAPAN
TEL
TLF
TLX
EML
COM 10

SHIBATA KAZUNARI DR
DPT OF EARTH SCIENCES
AICHI UNIV OF EDUCATION
1 HIROSAWA
KARIYA 448
JAPAN
TEL 566363111, 596
TLF
TLX
EML
COM

SHIBATA SHINPEI DR
DPT OF PHYSICS
YAMAGATA UNIVERSITY
KOJIRAKAWA
YAMAGATA 990
JAPAN
TEL 236 31 1421
TLF
TLX
EML BITNET B26416@JPNKUDPC
COM

SHIBATA YUKIO DR
RES INST SCIENTIFIC MEAST
TOHOKU UNIVERSITY
ARAMAKI
SENDAI 980
JAPAN
TEL
TLF
TLX
EML
COM 35

SHIELDS GREGORY A DR
ASTRONOMY DPT
UNIVERSITY OF TEXAS
RLM 15 212
AUSTIN TX 78712 1083
USA
TEL 512 471 4461
TLF
TLX 910-874-1351
EML
COM 28,34,48

SHIM WOON-TAIK PROF
236-53 SINDAN-DONG
JOONG-KU
SEOUL 100
KOREA R
TEL
TLF
TLX
EML
COM

SHIMIZU MIKIO PROF
ISAS
3-1-1 YOSHINODAI
SAGAMIHARA
KANAGAWA 229
JAPAN
TEL 0427 51 3911
TLF
TLX
EML
COM 15,16,51

SHIMIZU TSUTOMU PROF EMER
TERADA OOTANTI 26-16
JOYO SHI
KYOTO FU 610 01
JAPAN
TEL
TLF
TLX
EML
COM 16,33

SHIMMINS ALBERT JOHN
18 PAGE ST
ALBERT PARK VIC 3206
AUSTRALIA
TEL 3 690 3803
TLF
EML
COM 40
TLX

SHINE RICHARD A DR
LOCKHEED PALO ALTO RES LB
DPT 91 30 BLDG 256
3170 PORTER DR
PALO ALTO CA 94304 1211
USA
TEL 415 858 4135
TLF
TLX
EML
COM 10,12,36

SHIPMAN HENRY L DR
DPT OF PHYSICS
UNIVERSITY OF DELAWARE
NEWARK DE 19716
USA
TEL 302 451 2986
TLF
TLX
EML
COM 36,46

SHIRYAEV ALEXANDER A DR
INST OF THEORET ASTRONOMY
ACADEMY OF SCIENCES
N KUTUZOVA 10
191187 ST PETERSBURG
RUSSIA
TEL 7-812-272 40 23
TLF
TLX 12578 ITA SU
EML
COM 04

SHISHOV VLADIMIR I DR
LEBEDEV PHYSICAL INST
ACADEMY OF SCIENCES
LENINSKY PROSPEKT 53
117924 MOSCOW
RUSSIA
TEL
TLF
TLX
EML
COM

SHIVANANDAN KANDIAH DR
NAVAL RESEARCH LABORATORY
CODE 4138 S
4555 OVERLOOK AVE SW
WASHINGTON DC 20375 5000
USA
TEL 202 767 2749
TLF
TLX 202-767-6473
EML
COM 09,44,47

SHKODROV V G DR
DPT OF ASTRONOMY
BULGARIAN ACAD SCIENCES
72 LENIN BLVD
BG 1784 SOFIA
BULGARIA
TEL 2 75 8927
TLF
TLX 23761 ECF BAN BG
EML
COM 15,20

SHOBBROOK ROBERT R DR
DPT OF ASTRONOMY
UNIVERSITY OF SYDNEY
SYDNEY NSW 2006
AUSTRALIA
TEL 2 692 3604
TLF
TLX 26169 UNISYD AA
EML
COM 27,37

SHOEMAKER EUGENE M
US GEOLOGICAL SURVEY
BRANCH OF ASTROGEOLOGY
2255 N GEMINI DR
FLAGSTAFF AZ 86001
USA
TEL 602 527 7181
TLF
TLX
EML
COM 15,16,20

SHOLOMITSKY G B DR
SPACE RESEARCH INSTITUTE
ACADEMY OF SCIENCES
PROFSOJOSNAYA UL 84/32
117810 MOSCOW
RUSSIA
TEL 333-31-22
TLF
TLX 411498 STAR SU
EML
COM 40,44

SHOR VIKTOR A DR
INST OF THEORET ASTRONOMY
ACADEMY OF SCIENCES
N KUTUZOVA 10
191187 ST PETERSBURG
RUSSIA
TEL
TLF
TLX 121578
EML
COM 15,20C

SHORE BRUCE W
LAWRENCE LIVERMORE LAB
BOX 808
LIVERMORE CA 94550
USA
TEL 415 447 1100
TLF
TLX
EML
COM 14

SHORE STEVEN N
ASTROPHYSICS RESEARCH CTR
NEW MEXICO TECH
CAMPUS STATION
SOCORRO NM 87801
USA
TEL 505 835 5792
TLF
TLX
EML
COM 28,29,45

SHOSTAK G SETH DR
1372 CUERNAVACA CIRC
MOUNTAIN VIEW CA 94040
USA
TEL 415 967 8193
TLF
EML
COM 28,51
TLX

SHU FRANK H PROF
ASTRONOMY DPT
UNIVERSITY OF CALIFORNIA
601 CAMPBELL HALL
BERKELEY CA 94720
USA
TEL 415 642 2529
TLF
TLX
EML
COM 33,34,42

SHUKLA K
DPT MATHS & ASTRONOMY
UNIVERSITY OF LUCKNOW
LUCKNOW UP
INDIA
TEL
TLF
TLX
EML
COM 41

SHUKRE C S DR
RAMAN RESEARCH INSTITUTE
SADASHIVANAGAR
BANGALORE 560 080
INDIA
TEL 812 36 0122
TLF 812 34 0492
TLX 8425671 RRI IN
EML
COM 48

SHULL JOHN MICHAEL
JILA
UNIVERSITY OF COLORADO
BOX 440
BOULDER CO 80309 0440
USA
TEL 303 492 7827
TLF
TLX
EML
COM 34

SHULL PETER OTTO DR
DPT OF PHYSICS
OKLAHOMA STATE UNIVERSITY
STILLWATER OK 74078 0444
USA
TEL 405 744 5785
TLF
TLX
EML PHYSPOS@OSUCC BITNET
COM 34

SHULOV OLEG S DR
ASTRONOMICAL OBSERVATORY
ST PETERSBURG UNIVERSITY
BIBLIOTECHNAJA PL 2
199178 ST PETERSBURG
RUSSIA
TEL
TLF
TLX
EML
COM

SHUL'BERG A M DR
ASTRONOMICAL OBSERVATORY
ODESSA STATE UNIVERSITY
SHEVCHENKO PARK
270014 ODESSA
UKRAINE
TEL 25 0356
TLF·
TLX
EML
COM 26,42

SHUL'MAN L M DR
MAIN ASTRONOMICAL OBS
UKRAINIAN ACAD OF SCIENCE
GOLOSEEVO
252127 KIEV
UKRAINE
TEL
TLF
TLX 131406 SKY SU
EML
COM 15

SHUSTOV BORIS M DR
INST OF ASTRONOMY
ACADEMY OF SCIENCES
PYATNITSKAYA UL 48
109017 MOSCOW
RUSSIA
TEL 231-54-61
TLF
TLX 412623 SCSTP SU
EML
COM 34,35

SHUTER WILLIAM L H DR
DPT OF PHYSICS
UNIV OF BRITISH COLUMBIA
6224 AGRICULTURE RD
VANCOUVER BC V6T 2A6
CANADA
TEL 604 228 4269
TLF 604 228 5324
TLX 045 08576
EML
COM 33,34,40,51

SIBILLE FRANCOIS
OBSERVATOIRE DE LYON
AVE CHARLES ANDRE
F 69561 S GENIS LAVAL CDX
FRANCE
TEL 78 56 0705
TLF 72 39 9791
TLX 310926
EML
COM

SICARDY BRUNO DR
OBSERVATOIRE DE PARIS
SECTION DE MEUDON
EUROPA
F 92195 MEUDON PPL CDX
FRANCE
TEL 1 45 07 7409
TLF 1 45 07 7469
TLX 201571
EML SICARDY@FRMEU51/MESIOA SICARD
COM 16

SIDLICHOVSKY MILOS DR
ASTRONOMICAL INSTITUTE
CZECH ACADEMY SCIENCES
BUDECSKA 6
CS 120 23 PRAHA 2
CZECHOSLOVAKIA
TEL 2 25 8757
TLF 2 25 5010
TLX 122486
EML
COM 07

SIDORENKOV NIKOLAY S
HYDROMETEOROLOGICAL CTR

123376 MOSCOW
RUSSIA
TEL
TLF
TLX
EML
COM 19

SIEBER WOLFGANG PH D
FACHHOCHSCHULE NIEDERRHEI
FACHBEREICH ELECTR
REINARZSTR 49 69
D 4150 KREFELD 1
GERMANY
TEL 215 18 220
TLF
TLX
EML
COM 40

SIENKIEWICZ RYSZARD DR
COPERNICUS ASTRON CENTER
POLISH ACAD OF SCIENCES
UL BARTYCKA 18
PL 00 716 WARSAW
POLAND
TEL 411086
TLF
TLX 813878 ZAPAN PL
EML
COM 35

SIGNORE MONIQUE DR
RADIOASTRONOMIE ENS
24 RUE LHOMOND
F 75231 PARIS CDX 05
FRANCE
TEL 1 45 29 1225
TLF
TLX
EML
COM 35,41,47,48

SIKORA MAREK
COPERNICUS ASTRON CENTER
POLISH ACAD OF SCIENCES
UL BARTYCKA 18
PL 00 716 WARSAW
POLAND
TEL
TLF
TLX
EML
COM 48

SIKORSKI JERZY DR
INST THEORETICAL PHYSICS
UNIVERSITY OF GDANSK
UL WITA STWOSZA 57
PL 80 952 GDANSK
POLAND
TEL
TLF
TLX 0512706 IFAS PL
EML
COM

SILBERBERG REIN DR
NAVAL RESEARCH LABORATORY
CODE 4154
4555 OVERLOOK AVE SW
WASHINGTON DC 20375 5000
USA
TEL 202 767 2803
TLF
TLX
EML
COM 10,34,48

SILK JOSEPH I PROF
ASTRONOMY DPT
UNIVERSITY OF CALIFORNIA
601 CAMPBELL HALL
BERKELEY CA 94720
USA
TEL 415 642 2113
TLF
TLX 820181 UCB AST
EML
COM 34,47

SILLANPAA AIMO KALEVI DR
TURKU UNIVERSITY OBS
TUORLA
SF 21500 PIIKKIO
FINLAND
TEL 21 43 5822
TLF 21 43 3767
TLX
EML AIMOSILL@KONTU UTU FI
COM 28

SILVERBERG ERIC C DR
MCDONALD OBSERVATORY
UNIVERSITY OF TEXAS
BOX 1337
FORT DAVIS TX 79734
USA
TEL
TLF
TLX
EML
COM·

SILVESTRO GIOVANNI
IST DI FISICA
UNIVERSITA DI TORINO
CORSO D AZEGLIO 46
I 10125 TORINO
ITALY
TEL 11 65 8623
TLF
TLX 211041 INFNTO
EML
COM 34,35,44

SIM MARY E MISS
ROYAL OBSERVATORY
BLACKFORD HILL
EDINBURGH EH9 3HJ
UK
TEL 316 673 321
TLF
TLX 72383 ROEDIN G
EML
COM 09

SIMA ZDISLAV DR
ASTRONOMICAL INSTITUTE
CZECH ACADEMY SCIENCES
BUDECSKA 6
CS 120 23 PRAHA 2
CZECHOSLOVAKIA
TEL 2 25 8757
TLF 2 25 5010
TLX 66-122486
EML
COM 07,42

SIMEK MILOS DR
ASTRONOMICAL INSTITUTE
CZECH ACADEMY OF SCIENCES
ONDREJOV OBSERVATORY
CS 251 65 ONDREJOV
CZECHOSLOVAKIA
TEL 204 85201
TLF 204 85314
TLX 121579
EML·
COM 22

SIMIEN FRANCOIS DR
OBSERVATOIRE DE LYON
AVE CHARLES ANDRE
F 69561 S GENIS LAVAL CDX
FRANCE
TEL 78 56 0705
TLF 72 39 9791
TLX 310926
EML
COM 28

SIMKIN SUSAN M DR
PHYSICS & ASTRONOMY DPT
MICHIGAN STATE UNIVERSITY
EAST LANSING MI 48824
USA
TEL 517 353 4540
TLF
TLX
EML
COM 28

SIMMONS JOHN FRANCIS L
31 HAVELOCK STREET

GLASGOW G11 5HA
UK
TEL
TLF
TLX
EML
COM 42

SIMNETT GEORGE M
DPT OF SPACE RESEARCH
UNIVERSITY OF BIRMINGHAM
BOC 363
BIRMINGHAM B15 2TT
UK
TEL
TLF
TLX
EML
COM 10

SIMO CHARLES DR
FAC DE MATEMATICAS
UNIVERSIDAD DE BARCELONA
AV JOSE ANTONIO 585
E 08028 BARCELONA
SPAIN
TEL
TLF
TLX
EML
COM

SIMODA MAHIRO PROF
DEPT ASTRON/EARTH SCI
TOKYO GAKUGEI UNIVERSITY
KOGANEI
TOKYO 184
JAPAN
TEL 0423-25-2111
TLF
TLX
EML
COM 37

SIMON GEORGE W DR
AIR FORCE GEOPHYSICS LAB
NTL SOLAR OBSERVATORY
SUNSPOT NM 88349
USA
TEL 505 434 1390
TLF
TLX
EML
COM 12

SIMON GUY
OBSERVATOIRE DE PARIS
SECTION DE MEUDON
F 92195 MEUDON PPL CDX
FRANCE
TEL 1 45 07 7787
TLF
TLX
EML
COM 10,12

SIMON JEAN-LOUIS MR
BUREAU DES LONGITUDES
77 AVE DENFERT ROCHEREAU
F 75014 PARIS
FRANCE
TEL 1 43 20 1210
TLF
TLX
EML
COM 04,07

SIMON KLAUS PETER
INST F ASTRON & ASTROPHYS
DER UNIVERSITAET MUENCHEN
SCHEINERSTR 1
D 8000 MUENCHEN 80
GERMANY
TEL 89 989 021
TLF
TLX 529815 UNIVM D
EML
COM 36

SIMON MICHAL PROF
DPT OF EARTH & SPACE SCI
ASTRONOMY PROGRAM
SUNY AT STONY BROOK
STONY BROOK NY 11794 2100
USA
TEL 516 246 7672
TLF
TLX 510-228-7767
EML
COM

SIMON NORMAN R PROF
DPT OF PHYSICS & ASTRON
UNIVERSITY OF NEBRASKA
BEHLEN OBSERVATORY
LINCOLN NE 68588 0111
USA
TEL 402 472 2788
TLF
TLX
EML
COM

SIMON PAUL A DR
1 RUE MORTE BOUTEILLE
F 78140 VELIZY
FRANCE
TEL
TLF
EML
COM 40,44
TLX

SIMON PAUL C DR
IASB
AVE CIRCULAIRE 3
B 1180 BRUSSELS
BELGIUM
TEL 2 375 1579
TLF
TLX 21563
EML
COM 44

SIMON RENE L E PROF
INSTITUT D'ASTROPHYSIQUE
UNIVERSITE DE LIEGE
AVE COINTE 5
B 4000 COINTE-LIEGE
BELGIUM
TEL 41 52 9980
TLF 41 52 7474
TLX 41264 ASTRLG B
EML
COM 47

SIMON THEODORE
INSTITUTE FOR ASTRONOMY
UNIVERSITY OF HAWAII
2680 WOODLAWN DR
HONOLULU HI 96822
USA
TEL 808 948 8968
TLF
TLX 723-8459 UHAST HR
EML
COM 29,36

SIMONNEAU EDUARDO DR
INSTITUT D'ASTROPHYSIQUE
98BIS BD ARAGO
F 75014 PARIS
FRANCE
TEL 1 43 20 1425
TLF 1 43 29 8673
TLX
EML
COM 36

SIMONS STUART DR
SCHOOL OF MATHEMATICAL SC
QUEEN MARY/WESTFIELD COLL
MILE END RD
LONDON E1 4NS
UK
TEL 1 719 804 811
TLF
TLX
EML
COM 34

SIMONSON S CHRISTIAN DR
1061 RUSSELL AVE

LOS ALTOS CA 94022
USA
TEL 415 968 0473
TLF
TLX
EML
COM 33

SIMOVLJEVITCH JOVAN L DR
DPT OF ASTRONOMY
FACULTY OF SCIENCES
STUDENTSKI TRG 16
YU 11000 BEOGRAD
YUGOSLAVIA
TEL 11 638 715
TLF
TLX
EML
COM

SIMS KENNETH P DR
SYDNEY OBSERVATORY
OBSERVATORY PARK
SYDNEY NSW 2000
AUSTRALIA
TEL
TLF
TLX
EML
COM 08,24

SINACHOPOULOS D DR
OBSERVATOIRE ROYAL DE
BELGIQUE
AVE CIRCULAIRE 3
B 1180 BRUSSELS
BELGIUM
TEL 2 373 0291
TLF 2 374 9822
TLX 21565 OBSBEL
EML DIMITRIS@ASTRO OMA BE
COM 26

SINCLAIR ANDREW T DR
ROYAL GREENWICH OBS
HERSTMONCEUX CASTLE
HAILSHAM BN27 1RP
UK
TEL 323 833 171
TLF
TLX 87451 RGOBSY G
EML
COM 07,20

SINGH H P
INST F REINE ANGEWANDTE
KERNPHYSIK UNVIERSTAET
OLSHAUSENSTRASSE 40
D 2300 KIEL 1
GERMANY
TEL
TLF
TLX
EML PKE06@RZ UNI-KIEL DBP DE
COM 51

SINGH JAGDEV DR
INDIAN INSTITUTE OF
ASTROPHYSICS
KORAMANGALA
BANGALORE 560 034
INDIA
TEL 812 56 6585/6497
TLF
TLX 845763 IIAB IN
EML
COM 12

SINGH KULINDER PAL DR
TIFR
HOMI BHABHA RD
COLABA
BOMBAY 400 005
INDIA
TEL 22 495 2971
TLF
TLX 011 3009
EML BITNET uunet!shakti!tifr!root
COM

SINGH PATAN DEEN DR
IAG
UNIVERSIDADE DE SAO PAULO
CP 30627
01051 SAO PAULO SP
BRAZIL
TEL 11 275 3720
TLF
TLX 011 36221 IAGM BR
EML IAGUSP@BRFAPESP
COM 34

SINHA K DR
UTTAR PRADESH STATE
OBSERVATORY
PO MANORA PEAK 263 129
NAINITAL 263 129
INDIA
TEL 59 42 2136
TLF
TLX CABLE   ASTRONOMY
EML
COM 10,12,14

SINHA RAMESHWAR P
TATA INST OF FUNDAMENTAL
RESEARCH
POONA UNIVERSITY CAMPUS
PUNE 411 007
INDIA
TEL 212 33 7107
TLF 212 33 5760
TLX 0145 658 GMRT IN
EML SINHA@GMRT ERNET IN
COM 40

SINNERSTAD ULF E PROF
STOCKHOLM OBSERVATORY
S 133 36 SALTSJOEBADEN
SWEDEN
TEL 87 17 0195
TLF 87 17 4719
TLX
EML
COM 29,45

SINTON WILLIAM M
INSTITUTE FOR ASTRONOMY
UNIVERSITY OF HAWAII
2680 WOODLAWN DR
HONOLULU HI 96822
USA
TEL 808 948 8007
TLF
TLX
EML
COM 16

SINVHAL SHAMBHU DAYAL DR
4/3 SNEHALATAGANG
INDORE 452 003
INDIA
TEL
TLF
EML
COM 27,42
TLX

SINZI AKIRA M DR
HYDROGRAPHIC DPT
GEODESY & GEOPHYSICS DIV
TSUKIJI 5 CHUO KU
TOKYO 104
JAPAN
TEL
TLF
TLX
EML
COM 04

SION EDWARD MICHAEL
DPT OF ASTRONOMY
VILLANOVA UNIVERSITY
VILLANOVA PA 19085
USA
TEL 215 645 4822
TLF
TLX
EML: Scion@scivax stsci edu
COM 35,42,45

SIREGAR SURYADI DR
DPT OF ASTRONOMY
BANDUNG INSTITUTE OF TECH
JL GANESHA 10
BANDUNG 40132
INDONESIA
TEL 84 254*476
TLF
TLX 28324 ITB BANDUNG
EML
COM

SIROKY JAROMIR DR
PALACKY UNIVERSITY
DPT PHYSICS & ASTRONOMY
LENIN ST 26
CS 771 46 OLOMOUC
CZECHOSLOVAKIA
TEL 68 22451
TLF
TLX
EML
COM 46

SIRY JOSEPH W
4438 42ND ST NW
WASHINGTON DC 20016
USA
TEL
TLF
EML
COM 07
TLX

SISSON GEORGE M MR
PLANETREES
WALL
HEXHAM NE46 4EQ
UK
TEL 434 814 34
TLF
TLX·
EML
COM

SISTERO ROBERTO F DR
OBSERVATORIO ASTRONOMICO
DE CORDOBA
LAPRIDA 854
5000 CORDOBA
ARGENTINA
TEL 40613/36876
TLF
TLX 51822 BUCOR
EML
COM 42,47

SITARSKI GRZEGORZ PROF
SPACE RESEARCH CENTER
POLISH ACAD OF SCIENCES
UL ORDONA 21
PL 01 237 WARSAW
POLAND
TEL 410041
TLF
TLX 815670 CBK PL
EML
COM 20

SITKO MICHAEL L
DPT OF PHYSICS
UNIVERSITY OF CINCINNATI
210 BRAUNSTEIN ML 11
CINCINNATI OH 45221 0111
USA
TEL
TLF
TLX
EML
COM 28,34

SITNIK G F PROF
STERNBERG STATE ASTR INST
UNIVERSITETSKIJ PROSP 13
119899 MOSCOW
RUSSIA
TEL 139-19-73
TLF.
TLX
EML
COM 10,12,36

SITTERLY CHARLOTTE M DR
3711 BRANDYWINE ST NW

WASHINGTON DC 20016
USA
TEL 202 966 9044
TLF
TLX
EML
COM 12,14

SIVAN JEAN-PIERRE DR
LAS
TRAVERSE DU SIPHON
LES TROIS LUCS
F 13012 MARSEILLE
FRANCE
TEL 91 05 5900
TLF 91 66 1855
TLX 420584 ASTROSP
EML
COM 34

SIVARAM C DR
INDIAN INSTITUTE OF
ASTROPHYSICS
KORAMANGALA
BANGALORE 560 034
INDIA
TEL 812 56 6585/6497
TLF
TLX 845763 IIAB IN
EML
COM 51

SIVARAMAN K R DR
INDIAN INSTITUTE OF
ASTROPHYSICS
KORAMANGALA
BANGALORE 560 034
INDIA
TEL 812 56 6585
TLF
TLX 845763 IIAB IN
EML
COM 12C,15

SJOGREN WILLIAM L MR
JPL/CALTECH
MS 264 664
4800 OAK GROVE DR
PASADENA CA 91109
USA
TEL 818 354 4868
TLF
TLX 675421
EML
COM 16

SKALAFURIS ANGELO J
NAVAL RESEARCH LABORATORY
CODE 5307
4555 OVERLOOK AVE SW
WASHINGTON DC 20375 5000
USA
TEL 302 767 3227
TLF
TLX
EML
COM

SKILLEN IAN DR
DPT OF ASTRONOMY
UNIVERSITY OF LEICESTER
UNIVERSITY RD
LEICESTER LE1 7RH
UK
TEL
TLF
TLX
EML
COM

SKILLING JOHN DR
DPT APPLIED MATHS
& THEORETICAL PHYSICS
SILVER STREET
CAMBRIDGE CB3 9EW
UK
TEL
TLF
TLX
EML
COM 34,48

SKILLMAN EVAN D DR
DPT OF ASTRONOMY
UNIVERSITY OF MINNESOTA
116 CHURCH ST SE
MINNEAPOLIS MN 55455
USA
TEL 612 624 4523
TLF 612 626 2029
TLX
EML SKILLMAN@AST1 SPA UMN EDU
COM 28,40

SKINNER GERALD DR
SCHOOL PHYSICS/RESEARCH
UNIVERSITY OF BIRMINGHAM
BOX 363
BIRMINGHAM B15 2TT
UK
TEL 214 14 6450
TLF
TLX 338938
EML SPAN 19457 BHVAD GKS
COM

SKRIPNICHENKO VLADIMIR DR
INST OF APPLIED ASTRONOMY
ACADEMY OF SCIENCES
ZDANOVSKAYA UL 8
197042 ST PETERSBURG
RUSSIA
TEL
TLF
TLX
EML
COM 07

SKUMANICH ANDRE PROF
HIGH ALTITUDE OBSERVATORY
NCAR
BOX 3000
BOULDER CO 80307 3000
USA
TEL 303 497 1528
TLF
TLX 45694
EML
COM 12,36

SLADE MARTIN A III DR
JPL
MS 264 737
4800 OAK GROVE DR
PASADENA CA 91109
USA
TEL 818 354 6538
TLF
TLX
EML
COM 40

SLEE O B DR
CSIRO
DIVISION OF RADIOPHYSICS
BOX 76
EPPING NSW 2121
AUSTRALIA
TEL 2 868 0222
TLF
TLX 26230 ASTRO
EML
COM 40

SLETTEBAK ARNE PROF
PERKINS OBSERVATORY
OHIO STATE UNIVERSITY
BOX 449
DELAWARE OH 43015
USA
TEL 614 363 1257
TLF
TLX
EML
COM 29,33,45

SLEZAK ERIC DR
OCA OBSERV DE NICE
BP 139
F 06003 NICE CDX
FRANCE
TEL 92 00 3124
TLF 92 00 3033
TLX 470865
EML SLEZAK@FRONI51/17499 SLEZAK
COM 28

SLONIM E M DR
ASTRONOMICAL INSTITUTE
UZBEK ACADEMY OF SCIENCES
700000 TASHKENT
UZBEKISTAN
TEL
TLF
TLX
EML
COM 10

SLOVAK MARK HAINES DR
DPT OF ASTRONOMY
UNIVERSITY OF WISCONSIN
475 N CHARTER ST
MADISON WI 53706
USA
TEL 608 262 7542
TLF
TLX 265452 UOFWISC MDS
EML MADRAF SLOVAK
COM 09,42

SLYSH VJACHOSLAV I DR
SPACE RESEARCH INSTITUTE
ACADEMY OF SCIENCES
PROFSOJOSNAYA UL 84/32
117810 MOSCOW
RUSSIA
TEL
TLF
TLX
EML
COM 40,51C

SMAK JOSEPH I PROF
COPERNICUS ASTRON CENTER
POLISH ACAD OF SCIENCES
UL BARTYCKA 18
PL 00 716 WARSAW
POLAND
TEL 41 00 41
TLF 22 41 08 28
TLX 813978 ZAPAN PL
EML camk@lwauw61 bitnet
COM 26,27,42

SMALDONE LUIGI ANTONIO
DPT DI FISICA
UNIVERSITA DI NAPOLI
MOSTRA D OLTREMARE PAD 19
I 80125 NAPOLI
ITALY
TEL 81 725 3428
TLF
TLX 720320 INFNNA I
EML
COM 10

SMEYERS PAUL PROF
ASTRONOMISCH INSTITUUT
KATHOLIEKE UNIV LEUVEN
CELESTIJNENLAAN 200B
B 3300 HEVERLEE
BELGIUM
TEL 16 20 0656
TLF
TLX 25715 KULBI B
EML
COM 27,35

SMIT J A PROF
STERREKUNDIG INSTITUTE
BOX 80000
NL 3508 TA UTRECHT
NETHERLANDS
TEL 30 53 5200
TLX
EML
TLF
COM

SMITH ALEX G PROF
DPT OF ASTRONOMY
UNIVERSITY OF FLORIDA
211 SSRB
GAINESVILLE FL 32611
USA
TEL 904 392 6135
TLF
TLX
EML
COM 40

SMITH ANDREW M DR
NASA/GSFC
CODE 681
GREENBELT MD 20771
USA
TEL 301 286 8648
TLF
TLX
EML
COM

SMITH BARHAM W DR
LOS ALAMOS NATIONAL LAB
MS D436
BOX 1663
LOS ALAMOS NM 87545
USA
TEL 505 667 1585
TLF
TLX
EML
COM 34,48

SMITH BRADFORD A PROF
LUNAR & PLANETARY LAB
UNIVERSITY OF ARIZONA
TUCSON AZ 85721
USA
TEL 602 621 6930
TLF
TLX 910-952-1143
EML
COM 15,16C,44,WGPSNC

SMITH BRUCE F DR
NASA AMES RESREACH CTR
MS 245 3
THEORETICAL STUDIES BR
MOFFETT FIELD CA 94035
USA
TEL 415 694 5515
TLF
TLX
EML
COM 28

SMITH CHARLES DITTO
NATIONAL SOARING MUSEUM
R D # 3 HARRIS HILL
ELMIRA NY 14903
USA
TEL 607 734 3128
TLF
TLX
EML
COM 09,25

SMITH CLAYTON A JR DR
US NAVAL OBSERVATORY
34 & MASSACHUSETTS AVE NW
WASHINGTON DC 20392 5100
USA
TEL 202 653 1511
TLX 7108221970
EML
TLF
COM 08VP,24

SMITH CRAIG H DR
DPT PHYSICS UNIV COLLEGE
UNIVER OF NEW SOUTH WALES
NORTHCOTT DR
CAMPBELL ACT 2600
AUSTRALIA
TEL 62 68 8790
TLF 6 268 8786
TLX· 62030 ADFADM AA
EML CRAIG@PHADFA PH ADFA OZ AU
COM 34

SMITH DEAN F DR
BERKELEY RESEARCH ASS
290 GREEN ROCK DRIVE
BOULDER CO 80302
USA
TEL 303 444 1922
TLF
TLX
EML
COM 10,40,49

SMITH ELSKE V P DR
COLL HUMANITIES/SCIENCES
VIRGINIA COMMON UNIV
900 PARK AVENUE
RICHMOND VA 23284
USA
TEL 804 257 1674
TLF
TLX
EML
COM

SMITH ERIC PHILIP DR
NASA/GSFC
CODE 681
LASP
GREENBELT MD 20771
USA
TEL 301 286 8549
TLF 301 286 8709
TLX
EML ESMITH@HUBBLE GSFC NASA GOV
COM 28

SMITH F GRAHAM PROF
NRAL
JODRELL BANK
MACCLESFIELD SK11 9DL
UK
TEL 477 713 21
TLX 36149
EML
TLF 477 716 18
COM 38C,40,50

SMITH GEOFFREY DR
DPT OF ASTROPHYSICS
UNIVERSITY OF OXFORD
SOUTH PARKS RD
OXFORD OX1 3RQ
UK
TEL 865 511 336
TLF
TLX
EML
COM 14

SMITH GRAEME H DR
LICK OBSERVATORY
UNIVERSITY OF CALIFORNIA
SANTA CRUZ CA 95064
USA
TEL
TLF
TLX
EML
COM 29,37

SMITH HARDING E JR DR
CASS
UCSD
C 011
LA JOLLA CA 92093 0216
USA
TEL   419 534 4558
TLF   419 534 2294
TLX
EML   27778 HARDING (SPAN)
COM   28,47

SMITH HAYWOOD C DR
DPT OF ASTRONOMY
UNIVERSITY OF FLORIDA
211 SSRB
GAINESVILLE FL 32611
USA
TEL   904 392 1079
TLF
TLX
EML
COM   28

SMITH HORACE A
PHYSICS & ASTRONOMY DPT
MICHIGAN STATE UNIVERSITY
EAST LANSING MI 48824
USA
TEL   517 353 6784
TLF
TLX
EML
COM

SMITH HOWARD ALAN
NTL AIR & SPACE MUSEUM
SMITHSONIAN INSTITUTION
WASHINGTON DC 20560
USA
TEL
TLF   202 786 2262
TLX   264 729
EML
COM   34,44,51

SMITH HUMPHRY M
23 NORMANDALE
BEXHILL ON SEA TN39 3LU
UK
TEL   424 21 4288
TLF
EML
COM   19,31
TLX

SMITH LINDA J
DPT PHYSICS & ASTRONOMY
UNIVERSITY COLLEGE LONDON
GOWER ST
LONDON WC1E 6BT
UK
TEL   1 713 877 050*788
TLF
TLX   28722 UCPHYS G
EML
COM   44

SMITH LINDSEY F DR
1 KENNEDY RD
AUSTINMER NSW 2515
AUSTRALIA
TEL   4 267 5366
TLF
TLX
EML
COM

SMITH MALCOLM G DR
JOINT ASTRONOMY CENTER
UKIRT
665 KOMOHANA ST
HILO HI 96720
USA
TEL   808 961 3756
TLF
TLX   708-633-135
EML
COM   28

SMITH MYRON A ASST PROF
NTL SCIENCE FOUNDATION
DIV ASTRONOMICAL SCIENCES
1800 G ST NW
WASHINGTON DC 20550
USA
TEL
TLF
TLX
EML
COM   27C,29,30

SMITH PETER L DR
CENTER FOR ASTROPHYSICS
HCO/SAO MS 50
60 GARDEN ST
CAMBRIDGE MA 02138
USA
TEL   617 495 4984
TLF
TLX   921428 SATELLITE CAM
EML
COM   12,14C,34,44

SMITH ROBERT CONNON DR
ASTRONOMY CENTRE
UNIVERSITY OF SUSSEX
FALMER
BRIGHTON BN1 9QH
UK
TEL   273 60 6755*3101
TLF
TLX   877159 BHVTXS G
EML
COM   35,42

SMITH ROBERT G DR
DPT PHYSICS UNIV COLLEGE
UNIVER OF NEW SOUTH WALES
NORTHCOTT DR
CAMPBELL ACT 2600
AUSTRALIA
TEL   62 68 8746
TLF   6 268 8786
TLX   ADFADM AA 62030
EML   RGS@PHADFA PH ADFA OZ AU
COM   34

SMITH RODNEY M DR
DPT OF PHYSICS
UNIVERSITY OF DURHAM
SOUTH RD
DURHAM DH1 3LE
UK
TEL
TLF
TLX
EML
COM   47

SMITH VERNE V DR
ASTRONOMY DPT
UNIVERSITY OF TEXAS
RLM 15 308
AUSTIN TX 78712 1083
USA
TEL   512 471 3351
TLF
TLX
EML
COM   29

SMITH WM HAYDEN PROF
DPT OF PHYSICS
WASHINGTON UNIVERSITY
MCDONNEL CTR SPACE SCI
ST LOUIS MO 63130
USA
TEL   314 889 6574
TLF
TLX
EML
COM   14

SMOLINSKI JAN DR
INSTITUTE OF ASTRONOMY
N COPERNICUS UNIVERSITY
UL CHOPINA 12/18
PL 87 100 TORUN
POLAND
TEL
TLF
TLX
EML
COM   29

SMOLUCHOWSKI ROMAN PROF
ASTRONOMY DPT
UNIVERSITY OF TEXAS
RLM 15 314
AUSTIN TX 78712 1083
USA
TEL   512 471 1305
TLF
TLX   910-874-1351
EML
COM   15,16

SMOL'KOV GENNADIJ YA DR
SIBIZMIR
ACADEMY OF SCIENCES
664697 IRKUTSK 33
RUSSIA
TEL
TLF
TLX
EML
COM   10,40

SMOOT III GEORGE F
LAWRENCE BERKELEY LAB
1 CYCLOTRON RD
BLDG 50 230
BERKELEY CA 94720
USA
TEL   415 486 5237
TLF
TLX
EML
COM   47

SMRIGLIO FILIPPO PROF
ISTITUTO ASTRONOMICO
UNIVERSITA DI ROMA
VIA G M LANCISI 29
I 00161 ROMA
ITALY
TEL   6 84 42977
TLF
TLX
EML
COM

SMYLIE DOUGLAS E DR
DPT EARTH & ATMOSPH SCI
YORK UNIVERSITY
4700 KEELE ST
DOWNSVIEW ON M3J 1P3
CANADA
TEL 416 736 5245
TLF
TLX
EML.
COM 31

SMYTH MICHAEL J DR
ROYAL OBSERVATORY
BLACKFORD HILL
EDINBURGH EH9 3HJ
UK
TEL 316 673 321
TLX 72383
EML
TLF
COM 09,25

SNEDEN CHRISTOPHER A
ASTRONOMY DPT
UNIVERSITY OF TEXAS
RLM 15 308
AUSTIN TX 78712 1083
USA
TEL 512 471 4461
TLF
TLX
EML
COM 29

SNELL RONALD L
FIVE COLLEGE
RADIO ASTRONOMY OBSERV
B619 LEDERLE DRAD RES TWR
AMHERST MA 01003
USA
TEL 413 545 1949
TLF
TLX 955491
EML
COM 34

SNEZHKO LEONID I
SPECIAL ASTROPHYSICAL OBS
ACADEMY OF SCIENCES
NIZHNIJ ARKHYZ
357147 STAVROPOLSKIJ
RUSSIA
TEL 93513
TLF
TLX 297140 ZENIT
EML
COM 09,36

SNIJDERS MATTHEUS A J DR
ASTRONOMISCHES INSTITUT
WALDHAEUSER STRASSE 64
D 7400 TUEBINGEN 1
GERMANY
TEL 707 129 2486
TLF
TLX 07262 714 AIT D
EML
COM 36

SNOW THEODORE P PROF
CASA CB 391
UNIVERSITY OF COLORADO
BOX 391
BOULDER CO 80309 0391
USA
TEL 303 492 6857
TLF
TLX
EML
COM 29,34,44

SNYDER LEWIS E
DPT OF ASTRONOMY
UNIVERSITY OF ILLINOIS
1011 W SPRINGFIELS AVE
URBANA IL 61801
USA
TEL 217 333 5530
TLF
TLX 910-245-2434
EML
COM 15,51

SOBERMAN ROBERT K DR
DPT ASTRON & ASTROPHYS
UNIV OF PENNSYLVANIA
DAVID RITTENHOUSE LAB
PHILADELPHIA PA 19104
USA
TEL 215 898 8176
TLF
TLX
EML
COM 21,22C

SOBIESKI STANLEY DR
NASA/GSFC
CODE 673
GREENBELT MD 20771
USA
TEL
TLF
TLX
EML
COM 42

SOBOLEV V V DR
ASTRONOMICAL OBSERVATORY
ST PETERSBURG UNIVERSITY
BIBLIOTECHNAJA PL 2
199178 ST PETERSBURG
RUSSIA
TEL
TLF
TLX
EML
COM 34,36

SOBOLEV VLADISLAV M DR
PULKOVO OBSERVATORY
ACADEMY OF SCIENCES
10 KUTUZOV QUAY
196140 ST PETERSBURG
RUSSIA
TEL 298-22-42
TLF
TLX
EML
COM 12

SOBOLEVA N S DR
PULKOVO OBSERVATORY
ACADEMY OF SCIENCES
10 KUTUZOV QUAY
196140 ST PETERSBURG
RUSSIA
TEL
TLF
TLX
EML
COM 40

SOBOTKA MICHAL DR
ASTRONOMICAL INSTITUTE
CZECH ACADEMY OF SCIENCES
ONDREJOV OBSERVATORY
CS 251 65 ONDREJOV
CZECHOSLOVAKIA
TEL 204 85201
TLF 204 85314
TLX
EML
COM 10

SOBOUTI YOUSEF PROF
DPT OF PHYSICS
UNIVERSITY OF SHIRAZ
BIRUNI OBSERVATORY
SHIRAZ 71459
IRAN
TEL 71 57339
TLF
TLX
EML
COM 28,33,35

SOCHILINA ALLA S DR
INST TO THEORET ASTRONOMY
ACADEMY OF SCIENCES
N KUTUZOVA 10
191187 ST PETERSBURH
RUSSIA
TEL 278-88-98
TLF
TLX 121578 ITA SU
EML
COM 04

SODERBLOM DAVID R
STSCI
HOMEWOOD CAMPUS
3700 SAN MARTIN DR
BALTIMORE MD 21218
USA
TEL 301 338 4830
TLF
TLX 6849101 STSCI
EML
COM· 29,WGPSNC

SODERBLOM LARRY DR
US GEOLOGICAL SURVEY
BRANHC OF ASTROGEOLOGY
2255 N GEMINI DR
FLAGSTAFF AZ 86001
USA
TEL
TLF
TLX
EML
COM 16

SOEDERHJELM STAFFAN DR
LUND OBSERVATORY
BOX 43
S 221 00 LUND
SWEDEN
TEL 46 10 7303
TLF
TLX 33199 OBSNOT S
EML
COM 08,42

SOFFEL MICHAEL DR
LEHR & FORSCHUNSBEREICH
THEOR ASTROPHYSIK UNIV
AUF DER MORGNSTELLE 10C
D 7400 TUEBINGEN
GERMANY
TEL 707 412 92043
TLF 707 129 5400
TLX 726 2867 UTNA D
EML SOFFEL●TAT PHYSIK UNI-TUEBI
COM 07,19

SOFIA SABATINO PROF
YALE UNIVERSITY OBS
BOX 6666
NEW HAVEN CT 06511
USA
TEL 203 436 3460
TLX 710-465-3041
EML
TLF
COM 34,35,44,48

SOFUE YOSHIAKI PROF
INST OF ASTRONONOMY
UNIVERSITY OF TOKYO
OSAWA MITAKA
TOKYO 181
JAPAN
TEL 81 422 41 3734
TLF 81 422 41 3749
TLX 2822307
EML. y sifyi%tansei cc u-tokyo ac
COM 34,40,51

SOIFER BARUCH T DR
CALTECH
MS 320 47
DOWNES LAB OF PHYSICS
PASADENA CA 91125
USA
TEL 818 356 6626
TLF
TLX 675425
EML
COM

SOKOLOWSKI LECH
ASTRONOMICAL OBSERVATORY
JAGIELLONIAN UNIVERSITY
UL ORLA 171
PL 30 244 KRAKOW
POLAND
TEL 012 22 3856
TLF
TLX 0322723 UJ PL
EML
COM 47

SOKOLSKY ANDREJ G DR
INST OF THEORET ASTRONOMY
ACADEMY OF SCIENCES
N KUTUZOVA 10
191187 ST PETERSBURG
RUSSIA
TEL 812 279 0667
TLF 812 272 7968
TLX 121578 ITA SU
EML sokolsky@1111 spb su
COM 05,07C,20

SOL HELENE DR
OBSERVATOIRE DE PARIS
SECTION DE MEUDON
DAEC
F 92195 MEUDON PPL CDX
FRANCE
TEL 1 45 07 7428
TLF
TLX 20157
EML BITNET SOL@FRMEU51
COM

SOLANKI SAMI K DR
INSTITUT FUER ASTRONOMIE
ETH ZENTRUM
CH 8092 ZUERICH
SWITZERLAND
TEL 1 256 3810
TLF 41 1 251 21 72
TLX
EML SOLANKI@IFA ETHZ CH
COM 10,12

SOLARIC NIKOLA
HVAR OBSERVATORY
FACULTY OF GEODESY
KACICEVA 26
YU 41000 ZAGREB
YUGOSLAVIA
TEL 41 521 548
TLF
TLX
EML
COM 08

SOLC IVAN DR
ASTRONOMICAL INSTITUTE
GROUP OF OPTICS
DVORAKOWA 298
CS 511 01 TURNOV
CZECHOSLOVAKIA
TEL 43 62 2622
TLF
TLX 121673 MFF
EML
COM

SOLC MARTIN
DPT ASTRONOMY/ASTROPHYS
CHARLES UNIVERSITY
SVEDSKA 8
CS 150 00 PRAHA 5
CZECHOSLOVAKIA
TEL 2 54 0395
TLF
TLX
EML
COM 15,34,41

SOLF JOSEF DR
MPI FUER ASTRONOMIE
KOENIGSTUHL
D 6900 HEIDELBERG 1
GERMANY
TEL 62 215 28226
TLF
TLX 461789 MPIA D
EML
COM

SOLHEIM JAN ERIK
INST MATHS & PHYSICAL SCI
UNIVERSITY OF TROMSO
BOX 953
N 9001 TROMSO
NORWAY
TEL 083-86060
TLF
TLX 64124
EML
COM 42,46NR

SOLIMAN MOHAMED AHMED DR
HELWAN OBSERVATORY
HELWAN
EGYPT
TEL 78 0645/2683
TLF
EML
COM 10,27
TLX 93070 HIAG UN

SOLIVELLA GLADYS R LIC
UNIV NACIONAL DE LA PLATA
FCAG
1900 LA PLATA (BS AS)
ARGENTINA
TEL 21 21 7308
TLF 21 38 810
TLX 31151 BULAP AR
EML GLADYS%FCAGLP EDU AR@UUNET UU
COM 30

SOLLAZZO CLAUDIO
EUROPEAN SPACE OPERATIONS
CENTER
ROBERT-BOSCH-STR 5
D 6100 DARMSTADT
GERMANY
TEL 615 18861
TLF
TLX 419453
EML
COM

SOLOMON PHILIP M DR
DPT OF EARTH & SPACE SCI
ASTRONOMY PROGRAM
SUNY AT STONY BROOK
STONY BROOK NY 11794 2100
USA
TEL 516 246 8383
TLF
TLX 510-228-7767
EML
COM 33,34

SOLTAN ANDRZEJ MARIA DR
COPERNICUS ASTRON CENTER
POLISH ACAD OF SCIENCES
UL BARTYCKA 18
PL 00 716 WARSAW
POLAND
TEL
TLF
TLX
EML
COM 28

SOMA MITSURU DR
TOKYO ASTRONOMICAL OBS
NAOJ
OSAWA MITAKA
TOKYO 181
JAPAN
TEL 0422 41 3788
TLF
TLX 02822307
EML
COM 04,08,20

SOMERVILLE WILLIAM B DR
DPT PHYSICS & ASTRONOMY
UNIVERSITY COLLEGE LONDON
GOWER ST
LONDON WC1E 6BT
UK
TEL 1 713 827 050
TLF
TLX 28722
EML
COM 14,34

SOMMER-LARSEN JESPER DR
NIELS BOHR INSTITUTE
BLEGDAMSVEJ 17
DK 2100 COPENHAGEN O
DENMARK
TEL 31 42 1616
TLF 31 38 9157
TLX
EML
COM

SOMOV BORIS V DR
STERNBERG STATE ASTR INST
UNIVERSITETSKIJ PROSP 13
119899 MOSCOW
RUSSIA
TEL 7 95 939 16 44
TLX 411483 MGU SU
EML SNN@SAI MSK SU
TLF 7 95 939 01 26
COM 10

SONETT CHARLES P PROF
LUNAR & PLANETARY LAB
UNIVERSITY OF ARIZONA
TUCSON AZ 85721
USA
TEL 602 621 6935
TLF
TLX 9109521143
EML
COM 16,49C

SONG DOO JONG DR
ISSA
36-1 WHAAM DONG
YUSEONG GU
DAEJEON 305 348
KOREA R
TEL 42 861 1502
TLF 42 861 5610
TLX 45532 TOTOROK K
EML DJSONG%APISS ISSA RE KR@GARAM
COM 47

SONG GUO-XUAN
SHANGHAI OBSERVATORY
CAS
80 NANDAN RD
SHANGHAI
CHINA PR
TEL 21 38 6191
TLF
TLX 33164 SHAO CN
EML
COM 28,33

SONG JIN-AN
SHAANXI OBSERVATORY
CAS
LINTONG XIAN
SHAANXI
CHINA PR
TEL 33 2255
TLF
TLX 70121 CSAO CN
EML
COM 31

SONG MU-TAO
PURPLE MOUNTAIN OBSERV
CAS
NANJING
CHINA PR
TEL 25 46700
TLF
TLX
EML
COM 12

SONGSATHAPORN RUANGSAK DR
DPT OF PHYSICS
CHIANG MAI UNIVERSITY
CHIANG MAI 50002
THAILAND
TEL 221934*135
TLF
TLX 43553 UNICHIM TH
EML
COM

SONNEBORN GEORGE DR
NASA/GSFC
CODE 681
LASP
GREENBELT MD 20771
USA
TEL 301 286 3665
TLF
TLX 89675
EML SPAN 6471 SONNEBORN
COM 29,44

SORENSEN GUNNAR DR
INST OF PHYSICS & ASTRON
UNIVERSITY OF AARHUS
NY MUNKEGADE
DK 8000 AARHUS C
DENMARK
TEL 86 12 8899
TLF 86 20 2711
TLX
EML
COM 14

SORENSEN SOREN-AKSEL DR
DPT COMPUTER SCIENCE
UNIVERSITY COLLEGE LONDON
LONDON WC1E 6BT
UK
TEL
TLF
TLX
EML
COM

SOROCHENKO R L DR
LEBEDEV PHYSICAL INST
ACADEMY OF SCIENCES
LENINSKY PROSPEKT 53
117924 MOSCOW
RUSSIA
TEL 135-01-71
TLF
TLX 411479 NEOD SU
EML
COM 40

SORU-ESCAUT IRINA MRS
OBSERVATOIRE DE PARIS
SECTION DE MEUDON
F 92195 MEUDON PPL CDX
FRANCE
TEL 1 45 34 7530
TLF
TLX
EML
COM

SOTIROVSKI PASCAL DR
OBSERVATOIRE DE PARIS
SECTION DE MEUDON
F 92195 MEUDON PPL CDX
FRANCE
TEL 1 45 07 7802
TLF
TLX 270912
EML
COM 10,12

SOUFFRIN PIERRE B DR
OCA OBSERV DE NICE
BP 139
F 06003 NICE CDX
FRANCE
TEL 93 89 0420
TLF
TLX 460004 OBSNICE F
EML
COM 12,35,36

SOULIE GUY
OBSERVATOIRE DE BORDEAUX
BP 89
F 33270 FLOIRAC
FRANCE
TEL 56 86 4330
TLF 56 40 4251
TLX
EML
COM

SOWELL JAMES ROBERT DR
GEORGIA INSTITUTE
OF TECHNOLOGY
SCHOOL OF PHYSICS
ATLANTA GA 30332
USA
TEL 404 894 3628
TLF
TLX
EML JS58@HYDRA GATECH EDU
COM 26

SPADA GIANFRANCO DR
TESRE
CNR
VIA DE CASTAGNOLI 1
I 40126 BOLOGNA
ITALY
TEL 51 51 9593
TLF
TLX 511350 CNR BO
EML
COM 44,48

SPADARO DANIELE DR
OSS ASTROFISICO
CITTA UNIVERSITARIA
VIA A DORIA 6
I 95125 CATANIA
ITALY
TEL 95 33 0533
TLF 95 33 0592
TLX 970359 ASTRCT I
EML 40297 DANIELE
COM 10

SPAENHAUER ANDREAS MARTIN
ASTRONOMISCHES INSTITUT
UNIVERSITAET BASEL
VENUSSTRASSE 7
CH 4102 BINNINGEN
SWITZERLAND
TEL 61 22 7711
TLF
TLX
EML
COM

SPARKE LINDA
WASHBURN OBSERVATORY
UNIVERSITY OF WISCONSIN
475 N CHARTER ST
MADISON WI 53706
USA
TEL 608 262 3071
TLF
TLX
EML SPARKE@WISCMAC3 BITNET
COM 33

SPARKS WARREN M DR
LOS ALAMOS NATIONAL LAB
MS F669
BOX 1663
LOS ALAMOS NM 87545
USA
TEL 505 667 4922
TLF
TLX
EML
COM 35,42

SPARKS WILLIAM BRIAN
ROYAL GREENWICH OBS
HERSTMONCEUX CASTLE
HAILSHAM BN27 1RP
UK
TEL 323 833 171
TLX 87451
EML
TLF
COM 28

SPARROW JAMES G DR
AERONAUTICAL RESEARCH
LABORATORIES
4331
MELBOURNE VIC 3001
AUSTRALIA
TEL 3 647 7623
TLF
TLX 39391 ARL AA
EML
COM 21

SPASOVA NEDKA MARINOVA
DPT OF ASTRONOMY
BULGARIAN ACAD SCIENCES
72 LENIN BLVD
BG 1784 SOFIA
BULGARIA
TEL 2 75 8927
TLF
TLX
EML
COM

SPEER R J DR
DPT OF PHYSICS
IMPERIAL COLLEGE
PRINCE CONSORT RD
LONDON SW7 2BZ
UK
TEL 1 589 5111
TLF
TLX 261503 IMPCOL
EML
COM 44

SPENCER JOHN HOWARD
NAVAL RESEARCH LABORATORY
CODE 4134
4555 OVERLOOK AVE SW
WASHINGTON DC 20375 5000
USA
TEL 202 767 3050
TLF
TLX
EML
COM 40

SPENCER RALPH E DR
NRAL
JODRELL BANK
MACCLESFIELD SK11 9DL
UK
TEL 477 713 21
TLF
TLX 36149 JODREL G
EML
COM 40

SPERGEL DAVID N DR
PRINCETON UNIVERSITY OBS
PEYTON HALL
PRINCETON NJ 08544
USA
TEL 609 258 3589
TLF
TLX
EML DNS@ASTRO PRINCETON EDU
COM 33

SPICER DANIEL SHIELDS DR
NASA/GSFC
CODE 682
GREENBELT MD 20771
USA
TEL 301 286 7334
TLF
TLX
EML
COM 10,12

SPIEGEL E DR
DPT OF ASTRONOMY
COLUMBIA UNIVERSITY
PUPIN HALL 538 W 120TH ST
NEW YORK NY 10027
USA
TEL
TLF
TLX
EML
COM 33,35,36

SPIELFIEDEL ANNIE DR
OBSERVATOIRE PARIS
SECTION DE MEUDON
DAMAP
F 92195 MEUDON PPL CDX
FRANCE
TEL 1 45 07 7453
TLF 1 45 07 7469
TLX 201571
EML SPILFILD@FRMEU51
COM 14

SPINRAD HYRON PROF
ASTRONOMY DPT
UNIVERSITY OF CALIFORNIA
601 CAMPBELL HALL
BERKELEY CA 94720
USA
TEL 415 642 2078
TLF
TLX
EML
COM 15,28

SPITE FRANCOIS M DR
OBSERVATOIRE DE PARIS
SECTION DE MEUDON
F 92195 MEUDON PPL CDX
FRANCE
TEL 1 45 07 7840
TLF
TLX 270912
EML
COM 05,29,36

SPITE MONIQUE DR
OBSERVATOIRE DE PARIS
SECTION DE MEUDON
F 92195 MEUDON PPL CDX
FRANCE
TEL 1 45 07 7839
TLF
TLX 270912
EML'
COM 29,36

SPITHAS ELEFTERIOS N DR
DPT OF ASTRONOMY
NTL UNIVERSITY OF ATHENS
PANEPISTIMIOPOLIS
GR 157 71 ZOGRAFOS
GREECE
TEL 1 724 3414
TLF
TLX
EML
COM

SPITZER LYMAN JR DR
PRINCETON UNIVERSITY OBS
PEYTON HALL
PRINCETON NJ 08544
USA
TEL 609 452 3809
TLF
TLX 322409
EML
COM 34,44

SPOELSTRA T A TH DR
NFRA
BOX 2
NL 7991 PD DWINGELOO
NETHERLANDS
TEL 52 19 7244
TLX 42043 SRZM NL
EML
TLF
COM 08,40,50

SPRUIT HENK C DR
MAX PLANCK INSTITUT
KARLSCHWARSCHILD STR 1
D 8046 GARCHING MUENCHEN
GERMANY
TEL 89 329 90
TLF.
TLX 524629 ASTRO D
EML
COM 10,36

SPURZEM RAINER DR
INSTITUT FUR THEORETISCHE
PHYSIK UND STERNWARTE
OLSHAUSENSTR 40
D 2300 KIEL 1
GERMANY
TEL 431 880 1574
TLF 431 880 4432
TLX 292706
EML PAS28@RZ UNI-KIEL DBP DE
COM 37

SPYROU NICOLAOS PROF
DPT OF ASTRONOMY
UNIVERSITY THESSALONIKI
GR 540 06 THESSALONIKI
GREECE
TEL 31 99 2658
TLF
TLX 412181
EML
COM 47

SRAMEK RICHARD A DR
NRAO
BOX 0
SOCORRO NM 87801 0387
USA
TEL 505 772 4011
TLX 9109881710
EML
TLF
COM 40

SREEKANTAN B V DR
TIFR
HOMI BHABHA RD
COLABA
BOMBAY 400 005
INDIA
TEL 22 219 111
TLF
TLX 011-3009
EML
COM

SREENIVASAN S RANGA PROF
DPT OF PHYSICS
UNIVERSITY OF CALGARY
2500 UNIVERSITY DR NW
CALGARY AB T2N 1N4
CANADA
TEL 403 284 5385
TLF 403 289 3331
TLX
EML
COM 35

SRINIVASAN G
RAMAN RESEARCH INSTITUTE
SADASHIVANAGAR
BANGALORE 560 080
INDIA
TEL 812 36 0122
TLF 812 34 0492
TLX· 8452671 RRI IN
EML
COM 48VP

SRIVASTAVA J B DR
UTTAR PRADESH STATE
OBSERVATORY
PO MANORA PEAK 263 129
NAINITAL 263 129
INDIA
TEL 59 42 2136
TLF
TLX
EML
COM 42

SRIVASTAVA RAM KUMAR DR
UTTAR PRADESH STATE
OBSERVATORY
PO MANORA PEAK 263 129
NAINITAL 263 129
INDIA
TEL 59 42 2136
TLF
TLX
EML
COM 27,42

STABELL ROLF DR
INST THEORET ASTROPHYSICS
UNIVERSITY OF OSLO
BOX 1029
N 0315 BLINDERN OSLO 3
NORWAY
TEL 2-456-530
TLF
TLX 72705 ASTRO N
EML
COM

STACEY GORDON J DR
DPT OF ASTRONOMY
CORNELL UNIVERSITY
SPACE SCIENCES BLDG
ITHACA NY 14853
USA
TEL 607 255 5900
TLF 607 255 5875
TLX
EML STACEY@ASTROSUN TN EDU
COM

STACHNIK ROBERT V
NASA HEADQUARTERS
ASTROPHYSICS DIV
CODE SZ
WASHINGTON DC 20546
USA
TEL 202 453 1469
TLF
TLX
EML CHAMP STACHNIK
COM 44

STAGG CHRISTOPHER DR
DPT OF PHYSICS & ASTRON
UNIVERSITY OF CALGARY
2500 UNIVERSITY DR NW
CALGARY AB T2N 1N4
CANADA
TEL 403 220 7423
TLF 403 289 3331
TLX 038 21545
EML CRSTAGG@UNCAMULT
COM 42

STAGNI RUGGERO
OSSERVATORIO ASTROFISICO
VIA DELL OSSERVATORIO 8
I 36012 ASIAGO
ITALY
TEL 42 46 2665
TLF
TLX 430110 SETOUR I
EML
COM

STAHL OTMAR RICHARD DR
LANDESSTERNWARTE
KOENIGSTUHL
D 6900 HEIDELBERG 1
GERMANY
TEL 62 215 09232
TLF ·
TLX 461153
EML BITNET BP2@DHDURZ1
COM

STAHLER SETVEN W DR
DPT OF PHYSICS
MIT
BOX 165
CAMBRIDGE MA 02139
USA
TEL 617 253 0905
TLF 617 253 9798
TLX
EML
COM 34

STAHR-CARPENTER M DR
1101 HILL TOP RD
CHARLOTTESVILLE VA 22903
USA
TEL 804 293 7063
TLF
EML
COM 40
TLX

STALIO ROBERTO DR
DPT DI ASTRONOMIA
UNIVERSITA DI TRIESTE
VIA TIEPOLO 11
I 34131 TRIESTE
ITALY
TEL 40 79 3921*221
TLF
TLX 461137 OAT I
EML
COM 29,36,51

STANDISH E MYLES DR
JPL
MS 264 664
4800 OAK GROVE DR
PASADENA CA 91109
USA
TEL 818 354 3959
TLF
TLX
EML
COM 04C,07,20

STANFORD SPENCER A
DPT OF ASTRONOMY
UNIVERSITY OF WISCONSIN
475 N CHARTER ST
MADISON WI 53706
USA
TEL 608 262 1298
TLF
TLX·
EML
COM

STANGA RUGGERO
DPT DI ASTRONOMIA
UNIVERSITA DI FIRENZE
LARGO E FERMI 5
I 50125 FIRENZE
ITALY
TEL 55 27521
TLF 55 22 0039
TLX 572269 ARCETR I
EML
COM 34

STANGE LOTHAR
TECHNICAL UNIVERSITY
DRESDEN
MOMMSENSTR 13
D 8027 DRESDEN
GERMANY
TEL 51 463 4652
TLF
TLX 02278
EML
COM 08

STANILA GEORGE DR
MITROPOLIT GRIGORE 28
BOX 28
R 75218 BUCAREST 28
RUMANIA
TLF
TEL 23 68 92
TLX
EML
COM 19,31

STANKEVICH KAZIMIR S DR
RADIOPHYSICAL RESEARCH
INSTITUTE
LYADOV UL 25/14
603600 N NOVGOROD
RUSSIA
TEL· 38-90-91
TLF
TLX
EML·
COM

STANLEY G J
BOX 1348
CARMEL VALLEY CA 93924
USA
TEL· 408 659 2940
TLF
EML
COM 40
TLX

STANNARD DAVID DR
NRAL
JODRELL BANK
MACCLESFIELD SK11 9DL
UK
TEL 477 713 21
TLF
TLX 36149 JODREL G
EML
COM 40

STARK ANTONY A
AT & BELL LABORATORIES
HOH L 231
HOLMDEL NJ 07733
USA
TEL 201 949 4842
TLF
TLX
EML
COM

STARK GLEN DR
WHITIN OBSERVATORY
WELLESLEY COLLEGE
DPT OF PHYSICS
WELLESLEY MA 02181
USA
TEL 617 235 0320
TLF
TLX
EML gstark@lucy wellesley edu
COM 14

STARRFIELD SUMNER PROF
DPT OF PHYSICS
ARIZONA STATE UNIVERSITY
TEMPE AZ 85287
USA
TEL 602 965 3561
TLF
TLX 667391 ARIZ ST U TMP
EML
COM· 27,35,42

STASINSKA GRAZYNA DR
OBSERVATOIRE DE PARIS
SECTION DE MEUDON
F 92195 MEUDON PPL CDX
FRANCE
TEL 1 45 07 7422
TLF
TLX 201571
EML
COM

STATHOPOULOU MARIA DR
DPT OF ASTROPHYSICS
NTL UNIVERSITY OF ATHENS
PANEPISTIMIOPOLIS
GR 157 83 ZOGRAFOS
GREECE
TEL 1 724 3414
TLF 1 722 8981
TLX
EML MARSTATH@GRATHUN1
COM 12

STAUBERT RUDIGER PROF DR
ASTRONOMISCHES INSTITUT
UNIVERSITAET TUEBINGEN
WALDHAUSERSTR 64
D 7400 TUEBINGEN
GERMANY
TEL 707 129 4980
TLF
TLX 7262714 AIT D
EML
COM 44,48

STAUDE HANS JAKOB PH D
MPI FUER ASTRONOMIE
KOENIGSTUHL
D 6900 HEIDELBERG 1
GERMANY
TEL 62 215 28229
TLF
TLX 461789 MPIA D
EML
COM 21

STAUDE JUERGEN DR
ZNTRLINST F ASTROPHYSIK
SONNENOBSERVATORIUM
EINSTEINTURM
D 1500 POTSDAM
GERMANY
TEL
TLF
TLX
EML
COM 12C

STAVELEY-SMITH LISTER DR
CSIRO
PO BOX 76
EPPING NSW 2121
AUSTRALIA
TEL 2 868 0222
TLF. 2 868 0310
TLX 26230 ASTRO AA
EML LSTAVELE@ATNF CSIRO AU
COM 28

STAWIKOWSKI ANTONI DR
INSTITUTE OF ASTRONOMY
N COPERNICUS UNIVERSITY
UL CHOPINA 12/18
PL 87 100 TORUN
POLAND
TEL
TLF
TLX
EML
COM 29

STEBBINS ROBIN
JILA
UNIVERSITY OF COLORADO
BOX 440
BOULDER CO 80309 0440
USA
TEL 303 492 6073
TLF
TLX 755842
EML
COM 12

STECHER THEODORE P
NASA/GSFC
CODE 680
GREENBELT MD 20771
USA
TEL 301 286 8718
TLF
TLX
EML
COM 29,44,34

STECKER FLOYD W DR
NASA/GSFC
CODE 660
LAB HIGH ENERGY ASTROPHYS
GREENBELT MD 20771
USA
TEL 301 286 6057
TLF
TLX
EML
COM 33,47,48

STEEL DUNCAN I DR
UK SCHMIDT TELESCOPE
AAO
PRIVATE BAG
COONABARABRAN NSW 2357
AUSTRALIA
TEL 68 842 298
TLF. 68 426 282
TLX
EML DIS@AAOCBN OZ AU
COM 15C,20,22C,42

STEENMAN-CLARK LOIS DR
OCA OBSERV DE NICE
BP 139
F 06003 NICE CDX
FRANCE
TEL 93 89 0420
TLF
TLX 460004 OBSNICE F
EML
COM 14

STEFANIK ROBERT DR
OAK RIDGE OBSERVATORY
HARVARD SMITHSONIAN CTR
PINNACLE ROAD
HARVARD MA 01451
USA
TEL 617 495 7070
TLF
TLX
EML SPAN cfaz stefanik
COM 30

STEFANOVITCH-GOMEZ A E DR
OBSERVATOIRE DE PARIS
SECTION DE MEUDON
F 92195 MEUDON PPL CDX
FRANCE
TEL 1 45 07 7843
TLF
TLX 201571
EML
COM 33

STEFFEN MATTHIAS DR
INST FUR THEOR PHYSIK
OLSHAUSENSTRASSE 40
D 2300 KIEL 1
GERMANY
TEL 431 880 4101
TLF
TLX 292706
EML
COM 12,29,36,40

STEFL STANISLAV DR
ESO
KARL-SCHWARZSCHILD STR 2
D 8046 GARCHING MUENCHEN
GERMANY
TEL 89 320 06243
TLF 89 320 2362
TLX 05 282820 EO D
EML ESO SSTEFL/SSTEFL@ESO ORG
COM 29

STEFL VLADIMIR
DPT THEORET PHYS & ASTRO
PURKYNE UNIVERSITY
KOTLARSKA 2
CS 611 37 BRNO
CZECHOSLOVAKIA
TEL 5 1112
TLF
TLX
EML
COM 46

STEHLE CHANTAL DR
OBSERVATOIRE DE PARIS
SECTION DE MEUDON
DPT ATOMES & MOLECULES
F 92195 MEUDON PPL CDX
FRANCE
TEL 1 45 07 7453
TLF
TLX 201571
EML
COM 14

STEIGER W R PROF
CALTECH SUBMILL OBS
BOX 4339
HILO HI 96720
USA
TEL
TLF
TLX
EML
COM

STEIGMAN GARY PROF
DPT OF PHYSICS
OHIO STATE UNIVERSITY
174 W 18TH AVE
COLUMBUS OH 43210
USA
TEL. 614 292 1999
TLF
TLX 8104821715
EML
COM 47,48

STEIMAN-CAMERON THOMAS DR
NASA AMES RESEARCH CTR
MS 245 3
THEORETICAL STUDIES BR
MOFFETT FIELD CA 94035
USA
TEL 415 694 3120
TLF
TLX
EML SPAN GAL TOMSC
COM 42

STEIN JOHN WILLIAM
555 HILL ST

SEWICKLEY PA 15143
USA
TEL 412 741 4182
TLF
TLX
EML
COM 24,26,51

STEIN ROBERT F ASSOC PROF
PHYSICS & ASTRONOMY DPT
MICHIGAN STATE UNIVERSITY
EAST LANSING MI 48824
USA
TEL 517 353 8661
TLF
TLX
EML
COM 36

STEIN WAYNE A PROF
SCHOOL OF PHYS & ASTRON
UNIVERSITY OF MINNESOTA
116 CHURCH ST SE
MINNEAPOLIS MN 55455
USA
TEL 612 373 9963
TLF
TLX
EML
COM

STEINBERG JEAN-LOUIS DR
OBSERVATOIRE DE PARIS
SECTION DE MEUDON
F 92195 MEUDON PPL CDX
FRANCE
TEL 1 45 07 7696
TLF
TLX 204464
EML
COM 40,44

STEINER JOAO E DR
INPE
CP 515
AV DOS ASTRONAUTAS 1758
12200 S JOSE DOS CAMPOS
BRAZIL
TEL 123 22 9977
TLF
TLX
EML
COM 42,44,48

STEINERT KLAUS GUENTER DR
LORHMANN OBSERVATORY
DRESDEN UNIVERSITY OF TEC
MOMMSENSTR 13
D 8027 DRESDEN
GERMANY
TEL 51 463 4097
TLF 51 463 7106
TLX 2278 TEUNI DD
EML
COM 24

STEINITZ RAPHAEL PROF
DPT OF PHYSICS
BEN GURION UNIVERSITY
BOX 653
BEERSHEVA 84105
ISRAEL
TEL 57 70 985
TLF
TLX
EML
COM

STEINLE HELMUT DR
MAX PLANCK INSTITUT
EXTRATERRESTRISCHE PHYSIK
D 8046 GARCHING MUENCHEN
GERMANY
TEL 89 329 99470
TLF 89 329 9569
TLX 215845 EXTER D
EML BITNET HCS@DGAIPP1S
COM

STEINLIN ULI PROF
ASTRONOMISCHES INSTITUT
UNIVERSITAET BASEL
VENUSSTRASSE 7
CH 4102 BINNINGEN
SWITZERLAND
TEL 61 22 7711
TLF
TLX
EML
COM 25,33,45

STEINOLFSON RICHARD S DR
SOUTHWEST RESEARCH INST
6220 CULEBRA RD
SAN ANTONIO TX 78228 2510
USA
TEL 512 522 2822
TLF 512 647 4325
TLX 244046
EML SWRI RICH/RICH@129 162 150 30
COM

STELLINGWERF ROBERT F DR
LOS ALAMOS NATIONAL LAB
MS F645
X DIV
LOS ALAMOS NM 87545
USA
TEL 505 667 4370
TLF
TLX
EML
COM 27,35

STELLMACHER GOETZ
INSTITUT D'ASTROPHYSIQUE
98BIS BD ARAGO
F 75014 PARIS
FRANCE
TEL 1 43 20 1425
TLF 1 43 29 8673
TLX
EML
COM 10

STELLMACHER IRENE DR
BUREAU DES LONGITUDES
77 AVE DENFERT ROCHEREAU
F 75014 PARIS
FRANCE
TEL 1 43 20 1210
TLF
TLX
EML
COM 07,20

STENCEL ROBERT EDWARD
CASA
UNIVERSITY OF COLORADO
BOX 391
BOULDER CO 80309 0391
USA
TEL 303 492 7178
TLF
TLX
EML
COM 29,42,44

STENFLO JAN O DR
INSTITUT FUER ASTRONOMIE
ETH ZENTRUM
CH 8092 ZUERICH
SWITZERLAND
TEL 1 256 3804
TLF 41 1 262 0003
TLX 817379 EHHG CH
EML STENFLO●IFA ETHZ CH
COM 10,12P

STENHOLM BJOERN DR
LUND OBSERVATORY
BOX 43
S 221 00 LUND
SWEDEN
TEL 46 10 7306
TLF 46 10 4614
TLX 33199
EML bjorn●astro lu se
COM 34,46

STENHOLM LARS
ASTRONOMICAL OBSERVATORY
BOX 515
S 751 20 UPPSALA
SWEDEN
TEL 18 11 2488
TLF 18 11 1853
TLX 76024 UNIVUPS S
EML stenholm●laban uv se
COM

STEPANIAN A A DR
CRIMEAN ASTROPHYS OBS
UKRAINIAN ACAD OF SCIENCE
NAUCHNY
334413 CRIMEA
UKRAINE
TEL 43 2945
TLF
TLX
EML
COM 28,48

STEPANIAN JIVAN A DR
BYURAKAN ASTROPHYSICAL
OBSERVATORY
378433 BYURAKAN
ARMENIA
TEL 88 52 28 3435
TLF
TLX
EML
COM 28

STEPANIAN N N DR
CRIMEAN ASTROPHYS OBS
UKRAINIAN ACAD OF SCIENCE
NAUCHNY
334413 CRIMEA
UKRAINE
TEL 43 2945
TLF
TLX
EML
COM 10

STEPANOV ALEXANDER V DR
CRIMEAN ASTROPHYS OBS
UKRAINIAN ACAD OF SCIENCE
NAUCHNY
334247 KATZIVELY RT 22
UKRAINE
TEL 72 7906
TLF
TLX 222192 VOSHOD
EML
COM 10C

STEPHENS S A DR
TIFR
HOMI BHABHA RD
COLABA
BOMBAY 400 005
INDIA
TEL 22 219 111
TLF
TLX 011-3009 TIFR IN
EML
COM 48

STEPHENSON C BRUCE PROF
WARNER & SWASEY OBS
CASE WESTERN RESERVE UNIV
CLEVELAND OH 44106
USA
TEL 216 368 3728
TLF
TLX
EML
COM 33,45

STEPHENSON F RICHARD DR
DPT OF PHYSICS -
UNIVERSITY OF DURHAM
SOUTH RD
DURHAM DH1 3LE
UK
TEL 913 856 4971*208
TLF
TLX 537351 DURLIB G
EML
COM 19,41

STEPIEN KAZIMIERZ DR
ASTRONOMICAL OBSERVATORY
WARSAW UNIVERSITY
AL UJAZDOWSKIE 4
PL 00 478 WARSAW
POLAND
TEL 29-40-11
TLF
TLX
EML
COM 27,36

STEPINSKI TOMASZ DR
LUNAR & PLANETARY INST
3303 NASA RD
HOUSTON TX 77058
USA
TEL 713 486 2170
TLX
EML LPI STEPINSKI
TLF
COM

STEPPE HANS DR
IRAM
AVD DIVINA PASTORA 7
BLOQUE 6/2B
E 18012 GRANADA
SPAIN
TEL
TLF
TLX
EML
COM

STERKEN CHRISTIAAN LEO DR
IAAG
VRIJE UNIV BRUSSELS
CP 165
B 1050 BRUSSELS
BELGIUM
TEL 2 641 3469
TLF
TLX 61051 VUBCO
EML
COM 25C,27

STERN ROBERT ALLAN
LOCKHEED PALO ALTO RES LB
DPT 91 20 BLDG 255
3251 HANOVER ST
PALO ALTO CA 94304
USA
TEL 415 858 4072
TLF
TLX
EML
COM 44

STESHENKO N V DR
CRIMEAN ASTROPHYS OBS
UKRAINIAN ACAD OF SCIENCE
NAUCHNY
334413 CRIMEA
UKRAINE
TEL 43 2945
TLF
TLX
EML
COM 09,10

STESHENKO N V DR
CRIMEAN ASTROPHYS OBS
UKRAINIAN ACAD OF SCIENCE
NAUCHNY
334413 CRIMEA
UKRAINE
TEL 43 2945
TLF
TLX
EML
COM 09,10

STETSON PETER B DR
HERZBERG INST ASTROPHYS
DOMINION ASTROPHYS OBS
5071 W SAANICH RD
VICTORIA BC V8X 4M6
CANADA
TEL 604 339 0001
TLF 604 363 0045
TLX 049 7295
EML
COM 37

STEVENS GERARD A DR
SPACE RESEARCH LABORATORY
SRON
SORBONNELAAN 2
NL 3584 CA UTRECHT
NETHERLANDS
TEL 30 53 5600
TLF 30 54 0860
TLX
EML.
COM

STEWART JOHN MALCOLM DR
DPT APPLIED MATHS
& THEORETICAL PHYSICS
SILVER STREET
CAMBRIDGE CB3 9EW
UK
TEL 223 351 645
TLF
TLX
EML
COM 47

STEWART PAUL DR
DPT OF MATHEMATICS
UNIVERSITY OF MANCHESTER
MANCHESTER M13 9PL
UK
TEL
TLF
TLX
EML
COM 40

STEWART RONALD T MR
CSIRO
DIVISION OF RADIOPHYSICS
BOX 76
EPPING NSW 2121
AUSTRALIA
TEL 2 868 0222
TLF
TLX 26230
EML RSTEWART@ATNF CSIRO AU
COM 10,40

STEYAERT HERMAN PROF DR
STERREKUNDIG OBSERV
RIJKSUNIVERSITEIT GENT
KRIJGSLAAN 281
B 9000 GENT
BELGIUM
TEL 91 22 5715*2572
TLF 91 24 0634
TLX
EML
COM

STIAVELLI MASSIMO DR
VIA F CORRIDONI 25
I 56100 PISA
ITALY
TEL· 50 48806
TLF
EML STIAVELL@OGAES051/39196 STIAV
COM 28
TLX

STIBBS DOUGLAS W N PROF
MOUNT STROMLO & SIDING
SPRING OBSERVATORIES
PRIVATE BAG
WODEN PO ACT 2606
AUSTRALIA
TEL. 62 88 1111
TLF 62 49 0233
TLX 62270 CANOPUS AA
EML DWNS@MSO ANU EDU AU
COM 33,35,36

STICKLAND DAVID J DR
RUTHERFORD APPLETON LAB
SPACE & ASTROPHYSICS DIV
BLDG R25/R68
CHILTON DIDCOT OX11 0QX
UK
TEL 235 219 00
TLF
TLX 83159
EML
COM

STIER MARK T
PERKIN ELMER CORP
MS 897
100 WOOSTER HEIGHTS RD
DANBURY CT 06810 7859
USA
TEL 203 797 5708
TLF
TLX 965954
EML
COM 44

STIFT MARTIN JOHANNES DR
INSTITUT FUER ASTRONOMIE
TUERKENSCHANZSTR 17
A 1180 WIEN
AUSTRIA
TEL 1 345 36096
TLF
TLX
EML
COM

STINEBRING DANIEL R
DPT OF PHYSICS
PRINCETON UNIVERSITY
JADWIN HALL
PRINCETON NJ 08544
USA
TEL 609 452 5578
TLF
TLX
EML
COM

STIRPE GIOVANNA M DR
OSS ASTRONOMICO
UNIVERSITA DI BOLOGNA
VIA ZAMBONI 33
I 40126 BOLOGNA
ITALY
TEL 51 25 9301
TLF 51 25 9407
TLX
EML STIRPE@ASTBO3 CINECA IT
COM 28

STIX MICHAEL DR
KIEPENHEUER-INSTITUT
FUER SONNENPHYSIK
SCHOENECKSTR 6
D 7800 FREIBURG BREISGAU
GERMANY
TEL
TLF
TLX 7721552 KIS D
EML
COM 10,12

STOBIE ROBERT S DR
ROYAL OBSERVATORY
BLACKFORD HILL
EDINBURGH EH9 3HJ
UK
TEL 316 673 321
TLX 72383 ROEDIN G
EML
TLF
COM 27

STOCK JURGEN D
CIDA
BOX 264
MERIDA 5101 A
VENEZUELA
TEL 74 63 9930
TLX 74174
EML
TLF
COM 24C,25,30,45

STOCKMAN HERVEY S JR DR
STSCI
HOMEWOOD CAMPUS
3700 SAN MARTIN DR
BALTIMORE MD 21218
USA
TEL 301 338 4820
TLF
TLX 6849101 STSCI UW
EML
COM 25,44,48

STOCKTON ALAN N DR
INSTITUTE FOR ASTRONOMY
UNIVERSITY OF HAWAII
2680 WOODLAWN DR
HONOLULU HI 96822
USA
TEL
TLF
TLX
EML
COM

STOEGER WILLIAM R DR
SPECOLA VATICANA
I 00120 VATICAN CITY
VATICAN CITY STATE
TEL 6 698 3411
TLF
EML
COM 47
TLX. 504-2020 VATOBS VA

STOEV ALEXEI
PEOPLE'S ASTRONOMICAL OBS
& PLANETARIA IN BULGARIA
YURI GAGARIN
BG 6000 STARA ZAGORA
BULGARIA
TEL 42 43 183
TLF
TLX
EML
COM 16,41,46

STOHL JAN DR
ASTRONOMICAL INSTITUTE
SLOVAK ACADEMY SCIENCES
DUBRAVSKA 9
CS 842 28 BRATISLAVA
CZECHOSLOVAKIA
TEL 7 37 5157
TLF
TLX 093355
EML
COM 15,22P

STOKER PIETER H
COSMIC RAY RESEARCH UNIT
POTCHEFSTROOM UNIVERSITY
POTCHEFSTROOM 2520
SOUTH AFRICA
TEL 27-1481-25360
TLF
TLX 421363
EML
COM 10

STONE EDWARD C DR
JPL
MS 180 904
4800 OAK GROVE DR
PASADENA CA 91109
USA
TEL 818 354 3405
TLF 818 393 4218
TLX
EML
COM 16

STONE R G DR
NASA/GSFC
CODE 690
LEP
GREENBELT MD 20771
USA
TEL 301 286 8631
TLF
TLX 710-82089716
EML
COM 40,44,49

STONE REMINGTON P S DR
LICK OBSERVATORY
MOUNT HAMILTON CA 95140
USA
TEL 408 274 1809
TLF
TLX
EML
COM 25,28

STONE RONALD CECIL
US NAVAL OBSERVATORY
FLAGSTAFF STATION
BOX 1149
FLAGSTAFF AZ 86002
USA
TEL 602 779 5132
TLF
TLX
EML
COM 08,24

STORCHI-BERGMAN THAISA DR
INSTITUTO DE FISICA
UFRGS
CP 15051
91500 PORTO ALEGRE RS
BRAZIL
TEL 512 36 4677
TLF
TLX 515730 CCUF BR
EML
COM 28

STOREY JOHN W V DR
DPT OF PHYSICS
UNIV NEW SOUTH WALES
BOX 1
KENSINGTON NSW 2033
AUSTRALIA
TEL 2 697 4591
TLF
TLX 26054 AA
EML
COM 09

STOREY MICHELLE DR
DEPT THEORETICAL PHYSICS
UNIVERSITY OF SYDNEY
SYDNEY NSW 2006
AUSTRALIA
TEL 2 692 2538
TLF 2 660 2903
TLX
EML
COM 40

STRAFELLA FRANCESCO
DPT DI FISICA
UNIVERSITA DI LECCE
CP 193
I 73100 LECCE
ITALY
TEL 83 2627/247
TLF
TLX 860830 UNSTLE I
EML
COM

STRAIZYS V PROF DR
INST OF THEORET PHYS/ASTR
LITHUENIAN ACADEMY OF SCI
GOSTAUTO 12
2600 VILNIUS
LITHUANIA
TEL 122 61 613440
TLF 122 61 818464
TLX 261141 LMA SU
EML
COM 25C,45,51

STRAND KAJ AA DR
3200 ROWLAND PL NW
WASHINGTON DC 20008
USA
TEL 202 966 0495
TLF
EML
COM 24,26
TLX

STRASSL HANS L PROF
ASTRON INST UNIV MUENSTER
DOMAGKSTR 75
D 4400 MUENSTER
GERMANY
TEL 251 86 2463
TLF
TLX
EML .
COM

STRASSMEIER KLAUS G DR
INSTITUT FUER ASTRONOMIE
UNIVERSITY OF VIENNA
TUERKENSCHANZSTR 17
A 1180 WIEN
AUSTRIA
TEL 1 340 31695
TLF 1 340 31615
TLX 133099
EML STRASSMEIER@AVIA UNA AC AT
COM 27

STRAZZULLA GIOVANNI
OSS ASTROFISICO
CITTA UNIVERSITARIA
VIA A DORIA 6
I 95125 CATANIA
ITALY
TEL 95 33 0533
TLF
TLX 970359 ASTRCT I
EML
COM

STREL'NITSKIJ VLADIMIR DR
INST OF ASTRONOMY
ACADEMY OF SCIENCES
PYATNITSKAYA UL 48
109017 MOSCOW
RUSSIA
TEL 231-54-61
TLF
TLX 412623 SCSTP SU
EML
COM 14

STRITTMATTER PETER A PROF
STEWARD OBSERVATORY
UNIVERSITY OF ARIZONA
TUCSON AZ 85721
USA
TEL 602 621 6532
TLF
TLX
EML
COM 35

STROBEL ANDRZEJ DR
INSTITUTE OF ASTRONOMY
N COPERNICUS UNIVERSITY
UL CHOPINA 12/18
PL 87 100 TORUN
POLAND
TEL 260-18
TLF
TLX 0552234 ASTR PL
EML
COM 33,45

STROBEL DARRELL F
DPT EARTH & PLANETARY SCI
JOHNS HOPKINS UNIVERSITY
CHARLES & 34TH ST
BALTIMORE MD 21218
USA
TEL
TLF
TLX
EML
COM· 16

STROHMEIER WOLFGANG PROF
VOLKFELDSTR 5
D 8600 BAMBERG
GERMANY
TEL 951 55394
TLF
EML
COM 25,27,42
TLX

STROM KAREN M
DPT OF ASTRONOMY
UNIV OF MASSACHUSETTS
GRC 518 B 6732
AMHERST MA 01003
USA
TEL 413 545 2290
TLF
TLX
EML
COM 27

STROM RICHARD G DR
NFRA
BOX 2
NL 7990 AA DWINGELOO
NETHERLANDS
TEL 52 19 7244
TLF
TLX 42043 SRZM NL
EML
COM· 28,34,40

STROM ROBERT G PROF
LUNAR & PLANETARY LAB
UNIVERSITY OF ARIZONA
TUCSON AZ 85721
USA
TEL 602 621 2720
TLF
TLX 9109521143
EML
COM 16,28

STROM STEPHEN E
DPT PHYSICS & ASTRONOMY
UNIV OF MASSACHUSETTS
GRC 518 B
AMHERST MA 01003
USA
TEL· 418 545 2290
TLF
TLX
EML
COM 27,36

STRONG IAN B DR
LOS ALAMOS NATIONAL LAB
MS 436
BOX 1663
LOS ALAMOS NM 87545
USA
TEL 505 667 4823
TLF
TLX
EML
COM 48

STRONG JOHN D PROF
ASTRON RESEARCH FACILITY
UNIV OF MASSACHUSETTS
AMHERST MA 01003
USA
TEL
TLF
TLX·
EML
COM 16

STRONG KEITH T DR
LOCKHEED PALO ALTO RES LB
DPT 91 30 BLDG 255
3251 HANOVER ST
PALO ALTO CA 94304
USA
TEL 415 354 5136
TLF 415 424 3333
TLX 346409 LMSC
EML
COM 10

STRUBLE MITCHELL F
DPT ASTRON & ASTROPHYS
UNIV OF PENNSYLVANIA
DAVID RITTENHOUSE LAB
PHILADELPHIA PA 19104
USA
TEL 215 243 8176
TLF
TLX
EML
COM 47

STRUCK-MARCELL CURTIS J
DPT OF PHYSICS
IOWA STATE UNIVERSITY
AMES IA 50011
USA
TEL 515 294 5440
TLF
TLX
EML
COM

STRUKOV IGOR A DR
SPACE RESEARCH INSTITUTE
ACADEMY OF SCIENCES
PROFSOJOSNAYA UL 84/32
117810 MOSCOW
RUSSIA
TEL 333 14 66
TLF
TLX. 411498 STAR SU
EML
COM 40,47

STRYKER LINDA L
ASTRONOMY DPT
ARIZONA STATE UNIVERSITY
BOX 37100
PHOENIX AZ 85069
USA
TEL 602 543 6000
TLF
TLX
EML STRYKER@ASUCPS BITNET
COM

STUMPFF PETER PROF DR
MPI FUER RADIOASTRONOMIE
AUF DEM HUEGEL 69
D 5300 BONN 1
GERMANY
TEL 228 52 5360
TLX 886440
EML
TLF
COM

STURCH CONRAD R DR
STSCI/CSC
HOMEWOOD CAMPUS
3700 SAN MARTIN DR
BALTIMORE MD 21218
USA
TEL 301 338 4856
TLF
TLX
EML
-COM 33

STURROCK PETER A PROF
CTR FOR SPACE SCIENCES &
ASTROPHYSICS
STANFORD UNIV ERL
STANFORD CA 94305 4055
USA
TEL 415 723 1438
TLF
TLX 3484 STANFRD STNU
EML
COM 10,48,49,51

STUTZI JUERGEN DR
I PHYSIKALISCHE INSTITUT
UNIVERSTAET KOELN
ZUELPICHERSTR 77
D 5000 KOELN 41
GERMANY
TEL 221 47 03494/3567
TLF 221 47 05162
TLX
EML STUTZKI@PH1 UNI-KOELN DE
COM

SU BUMEI
YUNNAN OBSERVATORY
CAS
BOX 110
KUNMING 72946 YUNNAN
CHINA PR
TEL 871 2035
TLF
TLX 64040 YUOBS CN
EML
COM 34

SU DING-QIANG
NANJING ASTRONOMICAL
INSTRUMENT FACTORY
JIANGSU PROVINCE
CHINA PR
TEL 41191
TLF
TLX
EML
COM 09

SU HONG-JUN
PURPLE MOUNTAIN OBSERV
CAS
NANJING
CHINA PR
TEL 25 36967
TLF
TLX 34144 PMONT CN
EML
COM 28

SU QING-RUI DR
PURPLE MOUNTAIN OBSERV
CAS
NANJING
CHINA PR
TEL 25 64 2817
TLF 25 30 1459
TLX
EML
COM 10

SU WAN-ZHEN
PURPLE MOUNTAIN OBSERV
CAS
NANJING
CHINA PR
TEL 25 33583
TLF
TLX 34144 PMONJ CN
EML
COM 44

SUBRAHMANYA C R
TATA INST OF FUNDAMENTAL
RESEARCH
POONA UNIVERSITY CAMPUS
PUNE 411 007
INDIA
TEL 212 33 7107
TLF 212 33 5760
TLX 0145658 GMRT IN
EML uunet!shakti!gmrt!crs
COM· 47

SUBRAHMANYAM P V DR
DPT OF ASTRONOMY
UNIVERSITY OF OSMANIA
HYDERABAD 500 007
INDIA
TEL 71 951*247
TLF
TLX
EML
COM 28

SUBRAMANIAN KANDASWAMY DR
TATA INST OF FUNDAMENTAL
RESEARCH
POONA UNIVERSITY CAMPUS
PUNE 411 007
INDIA
TEL 212 33 7107
TLF 212 33 5760
TLX 0145658 GMRT IN
EML uunet!shakti!gmrt!kandu
COM

SUDA KAZUO PROF
ASTRONOMICAL INSTITUTE
TOHOKU UNIVERSITY
SENDAI AOBA
MIYAGI 980
JAPAN
TEL 0222-22-1800
TLF
TLX
EML
COM 35

SUEMATSU YOSHINORI DR
TOKYO ASTRONOMICAL OBS
NAOJ
OSAWA MITAKA
TOKYO 181
JAPAN
TEL 81 422 41 3705
TLF 81 422 41 3700
TLX 028 22307 TAOMTK J
EML SUEMATSU@SPOT MTK NAO AC JP
COM 12

SUESS STEVEN T DR
NASA/MSFC
CODE ES 52
SPACE SCIENCE LAB
HUNTSVILLE AL 35812
USA
TEL 205 453 2824
TLF
TLX
EML
COM 49C

SUGAWA CHIKARA DR
HANANOI 1586-25
KASHIWA SHI
CHIBA KEN 277
JAPAN
TEL 471 33 3825
TLF
TLX
EML
COM 19

SUGIMOTO DAIICHIRO PROF
DPT EARTH SCI & ASTRON
UNIVERSITY OF TOKYO
KOMABA
MEGUROKU TOKYO 153
JAPAN
TEL 03 467 1171
TLF 3 465 3925
TLX 2426728 TODAIK J
EML B89761%TANSEI CCU-TOKYO JUNET@
COM 35,37,42

SUGITANI KOJI DR
COLL OF GENERAL EDUCAT
NAGOYA CITY UNIVERSITY
MIZUHO KU
NAGOYA 467
JAPAN
TEL 81 52 872 5846
TLF 81 52 882 3075
TLX
EML D43000@JPNKUDPC BITNET
COM 40

SUGIYAMA NAOSHI DR
DPT OF PHYSICS
UNIVERSITY OF TOKYO
HONGO 7-3-1 BUNKYO-KU
KYOTO 113
JAPAN
TEL 81 3 3812 2111*4191
TLF 81 3 5689 0465
TLX 23472 UTPHYSIC J
EML 41610 NSUGIYAMA
COM 47

SUH KYUNG-WON DR
DPT ASTRONOMY & SPACE SCI
CHUNGBUK NTL UNIVERSITY
CHEONGJU 360 763
KOREA R
TEL 431 61 2315
TLF 431 67 4232
TLX
EML KWSUH@CBUCC CBNU AC KR
COM 34

SUKARTADIREDJA DARSA
PLANETARIUM DKI
JL CIKINI RAYA 73
JAKARTA 10330
INDONESIA
TEL 21 377 530
TLF
TLX
EML
COM 10,46

SUKUMAR SUNDARAJAN DR
DOMINION RADIO ASTROPHYS
OBSERVATORY
BOX 248
PENTICTON BC V2A 6K3
CANADA
TEL 604 493 2277
TLF 604 493 7767
TLX 048 88127
EML
COM 40

SULENTIC JACK W DR
DPT OF PHYSICS & ASTRON
UNIVERSITY OF ALABAMA
BOX 1921
TUSCALOOSA AL 35487 0324
USA
TEL 205 348 5050
TLF
TLX 810729 5845
EML JSULENTIC@UA1VM
COM 28

SULLIVAN DENIS JOHN DR
DPT OF PHYSICS
VICTORIA UNIVERSITY
PRIVATE BAG
WELLINGTON
NEW ZEALAND
TEL 721000
TLF
TLX
EML
COM 25

SULLIVAN WOODRUFF T III
DPT OF ASTRONOMY
UNIVERSITY OF WASHINGTON
FM 20
SEATTLE WA 98195
USA
TEL
TLF
TLX
EML
COM 28,40,41,51

SULTANOV G F ACAD
SHEMAKHA ASTROPHYSICAL
OBSERVATORY
AZER ACADEMY OF SCIENCES
373243 SHEMAKHA
AZERBAIDZHAN
TEL
TLF
TLX
EML
COM 07,20

SUMMERS HUGH P DR
JET JOINT UNDERTAKING
CULHAM LABORATORY
ABINGDON OX14 3EA
UK
TEL 235 28 822
TLF
TLX 837505 JETEUR G
EML
COM 14

SUN JIN
DPT OF ASTRONOMY
BEIJING NORMAL UNIVERSITY
BEIJING 100875
CHINA PR
TEL 1 65 6531*6285
TLF
TLX
EML
COM 34

SUN KAI
DPT OF GEOPHYSICS
BEIJING UNIVERSITY
BEIJING 100071
CHINA PR
TEL
TLF
TLX 22239 PKUNI CN
EML
COM 10

SUN WEI-HSIN DR
INST PHYSICS & ASTRONOMY
NTL CENTRAL UNIVERSITY
CHUNG LI
TAIWAN
CHINA R
TEL 3 422 7151*5302
TLF 3 425 1175
TLX
EML SUN@PHYAST DNET NCU EDU TW
COM 44

SUN YI-SUI
DPT OF ASTRONOMY
NANJING UNIVERSITY
NANJING
CHINA PR
TEL 25 37551
TLF
TLX 34151 PRCNU CN
EML
COM 07

SUN YONGXIANG
INSTITUTE OF GEODESY &
GEOPHYSICS
XU DONG LU
WUHAN
CHINA PR
TEL
TLF
TLX
EML
COM 19

SUNDELIUS BJOERN DR
DPT OF ASTRONOMY
CHALMERS TECHNICAL UNIV
S 412 96 GOETEBORG
SWEDEN
TEL 31 72 3141
TLF
TLX
EML BJORN@OSO CHALMERS SE
COM 28

SUNDMAN ANITA DR
STOCKHOLM OBSERVATORY
S 133 36 SALTSJOEBADEN
SWEDEN
TEL 87 17 0634
TLF 87 17 4719
TLX 12972 SWEDEN
EML
COM 41,42

SUNTZEFF NICHOLAS B
CERRO TOLOLO
INTERAMERICAN OBSERVATORY
CASILLA 603
LA SERENA
CHILE
TEL 51 21 3352
TLF 51 21 2466*342
TLX 620 301 AURA CT
EML
COM 29

SUNYAEV RASHID A DR
SPACE RESEARCH INSTITUTE
ACADEMY OF SCIENCES
PROFSOJUSNAYA UL 84/32
117810 MOSCOW
RUSSIA
TEL 7 095 230 2585
TLF 7 095 310 7023
TLX 411498 STAR SU
EML
COM 44,47,48C

SURDEJ JEAN M G
INSTITUT D'ASTROPHYSIQUE
UNIVERSITE DE LIEGE
AVE COINTE 5
B 4000 COINTE-LIEGE
BELGIUM
TEL 41 52 9980
TLF 41 52 7474
TLX 41264 ASTRLG B
EML
COM 15,47

SUTANTYO WINARDI
BOSSCHA OBSERVATORY
LEMBANG 40391
INDONESIA
TEL 229 6001
TLF
TLX
EML
COM

SUTHERLAND PETER G DR
DPT OF PHYSICS
MCMASTER UNIVERSITY
HAMILTON ON L8S 4M1
CANADA
TEL 416 525 9140
TLF 416 546 1252
TLX 061 8347
EML
COM

SUTHERLAND WILLIAM DR
DPT OF ASTROPHYSICS
UNIVERSITY OF OXFORD
KEBLE RD
OXFORD OX1 3RH
UK
TEL 865 273 310
TLF 865 273 418
TLX 83295 NUCLOX G
EML WJS@UK AC OX ASTRO/19464 WJS
COM 47

SUTO YASUSHI DR
UJI RESEARCH CENTER
YUKAWA INST THEORICAL PHY
KYOTO UNIVERSITY
UJI 611
JAPAN
TEL 81 774 20 7432
TLF 81 774 33 6226
TLX
EML SUTO@JPNKEKVX
COM 47

SUZUKI YOSHIMASA PROF
23-1 NAKAJIMA
HIRONOMACHI
UJI 611
JAPAN
TEL
TLF
TLX
EML
COM 34

SVALGAARD LEIF DR
HERTOGENLAAN 31

B 3202 LUBBEEK
BELGIUM
TEL
TLF
TLX
EML
COM

SVECHNIKOVA MARIA A DR
DPT OF ASTRONOMY
URALSKIJ STATE UNIVERSITY
620083 SVERDLOVSK
RUSSIA
TEL
TLF
TLX
EML
COM 42

SVENSSON ROLAND
STOCKHOLM OBSERVATORY
S 133 36 SALTSJOEBADEN
SWEDEN
TEL 81 64 472
TLF 87 17 4719
EML SVENSSON@ASTRO SU SE
COM 48
TLX

SVESTKA JIRI DR
OBS AND PLANETARIUM
PETRIN 205
CS 118 46 PRAHA 1
CZECHOSLOVAKIA
TEL 2 53 53513
TLF
TLX
EML
COM 22,46

SVESTKA ZDENEK DR
SPACE RESEARCH LABORATORY
SRON
SORBONNELAAN 2
NL 3584 CA UTRECHT
NETHERLANDS
TEL 30 53 5600
TLF 30 54 0860
TLX 47224 ASTRO NL
EML
COM 10,12

SVOLOPOULOS SOTIRIOS PROF
DPT OF ASTROPHYSICS
NTL UNIVERSITY OF ATHENS
PANEPISTIMIOPOLIS
GR 157 71 ZOGRAFOS
GREECE
TEL 1 724 3414
TLF
TLX
EML
COM 29,33,41

SVOREN JAN
ASTRONOMICAL INSTITUTE
SLOVAK ACADEMY SCIENCES
CS 059 60 TATRANSKA LOMNI
CZECHOSLOVAKIA
TEL 969 96 7866/7/8
TLF 969 96 7656
TLX 78277 AU SAV CS
EML
COM 15,20

SWANENBURG B N DR
SRL
HUYGENS LAB
BOX 9504
NL 2300 RA LEIDEN
NETHERLANDS
TEL 71 27 2727
TLF
TLX 39058 ASTRO NL
EML
COM

SWANK JEAN HEBB
NASA/GSFC
CODE 661
GREENBELT MD 20771
USA
TEL 301 286 6188
TLF
TLX 89675 NASCOM GBLT
EML
COM 48

SWARUP GOVIND PROF
TATA INST OF FUNDAMENTAL
RESEARCH
POONA UNIVERSITY CAMPUS
PUNE 411 007
INDIA
TEL 212 33 6111
TLF 212 33 5760
TLX 0145658 GMRT IN
EML uunet!shakt↑!gmrt!gswarup
COM 38,40

SWEET PETER A PROF
DPT OF ASTRONOMY
UNIVERSITY OF GLASGOW
GLASGOW G12 8QW
UK
TEL 413 398 855
TLF
TLX
EML
COM 35

SWEIGART ALLEN V DR
NASA/GSFC
CODE 681
GREENBELT MD 20771
USA
TEL 301 286 6274
TLF
TLX 89675
EML
COM 35

SWEITZER JAMES STUART DR
ASTRONOMY & ASTROPHYS CTR
UNIVERSITY OF CHICAGO
5640 S ELLIS AVE
CHICAGO IL 60637
USA
TEL 312 702 7783
TLF 312 702 8212
TLX
EML JSS@ODDJOB UCHICAGO EDU
COM

SWENSON GEORGE W JR PROF
DPT ELECT & COMPUTER ENG
UNIVERSITY OF ILLINOIS
1406 W GREEN ST
URBANA IL 61801
USA
TEL 217 333 4498
TLF
TLX
EML
COM 40

SWENSSON JOHN W DR
DPT OF THEORETICAL PHYS
UNIVERSITY OF LUND
SOELVEGATAN 14 A
S 223 62 LUND
SWEDEN
TEL 10 96 9686
TLF
TLX
EML
COM 12,29

SWERDLOW NOEL PROF
ASTRONOMY & ASTROPHYS CTR
UNIVERSITY OF CHICAGO
5640 S ELLIS AVE
CHICAGO IL 60637
USA
TEL 312 962 7969
TLF
TLX
EML
COM 41

SWIHART THOMAS L DR
STEWARD OBSERVATORY
UNIVERSITY OF ARIZONA
TUCSON AZ 85721
USA
TEL 602 621 6525
TLF
TLX
EML
COM 36

SWINGS JEAN-PIERRE DR
INSTITUT D'ASTROPHYSIQUE
UNIVERSITE DE LIEGE
AVE COINTE 5
B 4000 COINTE-LIEGE
BELGIUM
TEL 41 52 9980
TLF 41 52 7474
TLX 41264 ASTRLG B
EML
COM 09,14,29,EC

SYGNET JEAN FRANCOIS DR
INSTITUT D'ASTROPHYSIQUE
98BIS BD ARAGO
F 75014 PARIS
FRANCE
TEL 1 43 20 1425
TLF 1 43 29 8673
TLX
EML SYGNET@FRIAP51
COM 33

SYKES MARK VINCENT DR
STEWARD OBSERVATORY
UNIVERSITY OF ARIZONA
TUCSON AZ 85721
USA
TEL 602 621 2054
TLF
TLX 467175
EML MSYKES@AS ARIZONA EDU
COM 15

SYKES-HART AVRIL B DR
DPT OF ASTROPHYSICS
UNIVERSITY OF OXFORD
SOUTH PARKS RD
OXFORD OX1 3RQ
UK
TEL
TLF
TLX
EML
COM

SYKORA JULIUS DR
ASTRONOMICAL INSTITUTE
SLOVAK ACADEMY SCIENCES
CS 059 60 TATRANSKA LOMNI
CZECHOSLOVAKIA
TEL 969 96 7866/7/8
TLF 969 96 7656
TLX 78277 AUSAV CZ
EML
COM 10

SYLWESTER BARBARA DR
ASTRONOMICAL INSTITUTE
WROCLAW UNIVERSITY
UL KOPERNIKA 11
PL 51 622 WROCLAW
POLAND
TEL 483238
TLF
TLX 0712791 UWRPL
EML
COM 10

SYLWESTER JANUSZ
ASTRONOMICAL INSTITUTE
WROCLAW UNIVERSITY
UL KOPERNIKA 11
PL 51 622 WROCLAW
POLAND
TEL 48 18 01
TLF
TLX
EML
COM 10

SYNNOTT STEPHEN P
JPL
MS 264 686
4800 OAK GROVE DR
PASADENA CA 91109
USA
TEL 818 354 6933
TLF
TLX
EML
COM 16,20

SZABADOS LASZLO PH D
KONKOLY OBSERVATORY
THEGE U 13/17
BOX 67
H 1525 BUDAPEST
HUNGARY
TEL 1 75 5866/75 4122
TLF
TLX 227460 KONOBH
EML
COM 26,27,42

SZAFRANIEC ROZALIA DR
UL KOPERNIKA 27
PL 31 501 KRAKOW
POLAND
TEL
TLF
EML
COM 42
TLX

SZALAY ALEX DR
DPT ATOMIC PHYSICS
EOTVOS UNIVERSITY
PUSHKIN U 5-7
H 1088 BUDAPEST
HUNGARY
TEL
TLF
TLX
EML
COM 47C

SZATMARY KAROLY DR
JATE UNIVERSITY
DPT OF PHYSICS
DOM TER 9
H 6720 SZEGED
HUNGARY
TEL 36 62 11622
TLF 36 62 12921
TLX
EML H2674SZA@ELLA HU
COM 27

SZEBEHELY VICTOR G PROF
AEROSPACE ENGINEERING DPT
UNIVERSITY OF TEXAS
WRW 414
AUSTIN TX 78712
USA
TEL 512 471 4239
TLF
TLX 9108741305
EML
COM 07,33

SZECSENYI-NAGY GABOR DR
ASTRONOMY DPT
EOTVOS UNIVERSITY
KUN BELA TER 2
H 1083 BUDAPEST
HUNGARY
TEL 1 41 1019
TLF
TLX
EML·
COM 27,36,37,46

SZEGO KAROLÝ DR
CENTRAL RESEARCH INST
FOR PHYSICS
BOX 49
H 1525 BUDAPEST
HUNGARY
TEL 1 551 682
TLF 1 696 567
TLX 224722
EML
COM 15

SZEIDL BELA DR
KONKOLY OBSERVATORY
THEGE U 13/17
BOX 67
H 1525 BUDAPEST
HUNGARY
TEL 1 75 5866/75 4122
TLF
TLX 227460 KONOB
EML
COM 27C

SZKODY PAULA DR
DPT OF ASTRONOMY
UNIVERSITY OF WASHINGTON
FM 20
SEATTLE WA 98195
USA
TEL 206 543 1988
TLF
TLX
EML
COM 25,27,42C

SZOSTAK ROLAND DR
INST FUR DIDAKTIK PHYSIK
WESTFALISCHE WILHELMS UNI
WILHELM-KLEMM-STR 10
D 4400 MUENSTER
GERMANY
TEL 251 83 9386/7
TLF 251 83 2090
TLX 892529
EML
COM 46

TAAM RONALD EVERETT DR
DPT PHYSICS & ASTRONOMY
NORTHWESTERN UNIVERSITY
DEARBORN OBSERVATORY
EVANSTON IL 60208
USA
TEL· 312 491 7528
TLF
TLX
EML
COM 35,42

TABARA HIROTO DR
FAC OF EDUCATION
UTSUNOMIYA UNIVERSITY
MINEMACHI
UTSUNOMIYA 321
JAPAN
TEL 0286-36-1515
TLF
TLX
EML
COM 40

TABOADA RAMON RODRIGUEZ
INST GEOPHYS & ASTRONOMY
CALLE 212 N 2906 29 Y 31
LISA
LA HABANA
CUBA
TEL
TLF
TLX
EML
COM

TABORDA JOSE ROSA DR
FACULTY OF SCIENCES
ASTRONOMICAL OBSERVATORY
R ESCOLA POLITECNICA 58
P 1200 LISBOA
PORTUGAL ,
TEL
TLF
TLX
EML
COM 07,46

TACCONI LINDA J DR
MPI FUER EXTRATERRESTRISC
PHYSIK
KARLSCHWARZSCHILDSTR 2
D 8046 GARCHING MUENCHEN
GERMANY
TEL 89 329 93289
TLF 89 329 93569
TLX 5215845 XTER D
EML LINDA@DGAMPE BITNAT
COM 28

TADEMARU EUGENE DR
DPT OF ASTRONOMY
UNIV OF MASSACHUSETTS
AMHERST MA 01002
USA
TEL
TLF
TLX
EML
COM

TAFF LAURENCE G DR
STSCI
HOMEWOOD CAMPUS
3700 SAN MARTIN DR
BALTIMORE MD 21218
USA
TEL 301 338 4799
TLF 301 338 4767
TLX 6849101 STSCI
EML LGTAFF@SCIVAX STSCI EDU
COM 08

TAFFARA SALVATORE PROF
VIA CALZA 5BIS
I 35128 PADOVA
ITALY
TEL 49 807 1624
TLF
EML
COM
TLX

TAGLIAFERRI GIUSEPPE PROF
OSS ASTROFISICO
DI ARCETRI
LARGO E FERMI 5
I 50125 FIRENZE
ITALY
TEL
TLF
TLX
EML
COM

TAKABA HIROSHI DR
KASHIMA SPACE RES CENTER
CRL
893-1 HIRAI
IBARAKI 314
JAPAN
TEL 81 299 82 1211
TLF 81 299 83 5728
TLX 7604038 CRL UC
EML KASHIMA@ALTAIR GSFC NASA C
COM 40

TAKADA-HIDAI MASAHIDE DR
RES INST OF CIVILIZATION
TOKAI UNIVERSITY
1117 KITAKANAME
KANAGAWA 259-12
JAPAN
TEL 0463-58-1211
TLF
TLX· 2423402 UNITOK J
EML
COM· 29,51

TAKAGI KOJIRO PROF
DPT OF PHYSICS
TOYAMA UNIVERSITY
3190 GOFUKU
TOYAMA 930
JAPAN
TEL 0764-234716
TLF
TLX
EML
COM· 40

TAKAGI SHIGETSUGU DR
DPT DE FISICA-CCE-UFRN
CAMPUS UNIVERSITARIO
59000 NATAL RN
BRAZIL
TEL
TLF
TLX
EML
COM

TAKAHARA FUMIO DR
DPT OF PHYSICS
TOKYO METROPOLITAN UNIVER
MINAMIOHSAWA 1-1
HACHIOJI TOKYO 192-03
JAPAN
TEL 426 77 1111
TLF 426 77 2483
TLX
EML.
COM· 47,48

TAKAHARA MARIKO
DPT OF ASTRONOMY
UNIVERSITY OF TOKYO
BUNKYO KU
TOKYO 113
JAPAN
TEL 03-812-2111
TLF
TLX 33659 UTYOSCI
EML
COM 35

TAKAKUBO KEIYA PROF
ASTRONOMICAL INSTITUTE
TOHOKU UNIVERSITY
SENDAI AOBA
MIYAGI 980
JAPAN
TEL 222-22-1800
TLF
TLX 852246 THUCOM J
EML
COM 34,40

TAKAKURA TATSUO PROF EMER
DPT OF ASTRONOMY
UNIVERSITY OF TOKYO
BUNKYO KU
TOKYO 113
JAPAN
TEL 03-812-2111
TLF
TLX 33659 UTYOSCI J
EML
COM 10,40,44

TAKALO LEO O DR
TURKU UNIVERSITY OBS
TUORLA
SF 21500 PIIKKIO
FINLAND
TEL 21 43 5822
TLF 21 43 3767
TLX
EML TAKALO@KONTU UTU FI
COM 28

TAKAMI HIDEKI DR
COMMUNICATIONS RES LAB
4-2-1 NUKUIKITAMACHI
KOGANEI
TOKYO 184
JAPAN
TEL 423 27 6875
TLF 423 27 6696
TLX
EML TAKAMI@ORION BC CRL GO JP
COM 28

TAKANO TOSHIAKI DR
NOBEYAMA RADIO OBS
NAOJ
MINAMIMAKI MURA
NAGANO 384 13
JAPAN
TEL 267 63 4487
TLF 267 63 4444
TLX
EML TAKANO@NRO NAO AC JP
COM 10,34,40

TAKARADA KATSUO DR
KYOTO INST OF TECHNOLOGY
MATSUGASAKI
SAKYO KU
KYOTO 606
JAPAN
TEL 757 91 3211
TLF
TLX
EML
COM

TAKASE BUNSHIRO PROF
TOKYO ASTRONOMICAL OBS
NAOJ
OSAWA MITAKA
TOKYO 181
JAPAN
TEL 422 32 5111
TLF
TLX 2822307 TAOMTK
EML
COM 28

TAKAYANAGI KAZUO PROF
ISAS
3-1-1 YOSHINODAI
SAGAMIHARA
KANAGAWA 229
JAPAN
TEL 427 51 3911
TLF
TLX 24550 SPACETKY J
EML
COM 14

TAKEDA HIDENORI DR
DPT AERONAUTIC ENGINEER
KYOTO UNIVERSITY
SAKYO KU
KYOTO 606
JAPAN
TEL
TLF
TLX
EML
COM 15

TAKEDA YOICHI DR
INST OF ASTRONOMY
UNIVERSITY OF TOKYO
OSAWA MITAKA
TOKYO 181
JAPAN
TEL 81 422 41 3739
TLF 81 422 41 3749
TLX 2822307 TAOMTK J
EML
COM 36

TAKENOUCHI TADAO DR
1-28-30 KICHIJYOJI
KITA MACHI
MUSASHINO SHI
TOKYO 180
JAPAN
TEL
TLF
TLX
EML
COM

TAKENS ROELF JAN DR
ASTRONOMICAL INSTITUTE
UNIVERSITY OF AMSTERDAM
KRUISLAAN 403
NL 1098 SJ AMSTERDAM
NETHERLANDS
TEL 20 52 57491
TLF 20 52 57484
TLX 10262 HEF NL
EML
COM

TAKEUTI MINE DR
ASTRONOMICAL INSTITUTE
TOHOKU UNIVERSITY
SENDAI AOBA
MIYAGI 980
JAPAN
TEL 22 222 1800
TLF 22 261 2806
TLX 852246 THUCOM J
EML I4AOS4D@JPNTOHOK
COM 27C

TALAVERA A DR
ESA IUE OBSERVATORY
VILSPA
APD 50727
E 28080 MADRID
SPAIN
TEL 1 813 1100
TLF 1 813 1139
TLX 42555 VILSE
EML
COM 29

TALBOT RAYMOND J JR DR
THE AEROSPACE CORPORATION

1927 CURTIS AVENUE
REDONDO BEACH CA 90278
USA
TEL 213 379 9927
TLF
TLX
EML
COM 28

TALON RAOUL DR
CESR
BP 4346
F 31029 TOULOUSE CDX
FRANCE
TEL 61 55 6666
TLF 61 55 6701
TLX 531329 UNSPAT F
EML
COM 10

TALWAR SATYA P DR
DPT OF PHYSICS
UNIVERSITY OF DELHI
NEW DELHI 110 007
INDIA
TEL 11 291 8993
TLF
TLX
EML
COM

TAMENAGA TATSUO DR
FAC OF EDUCATION
MIE UNIVERSITY
TSU-SHI
MIE 514
JAPAN
TEL
TLF
TLX
EML
COM 10

TAMMANN G ANDREAS PROF DR
ASTRONOMISCHES INSTITUT
UNIVERSITAET BASEL
VENUSSTRASSE 7
CH 4102 BINNINGEN
SWITZERLAND
TEL 61 22 7711
TLF 61 271 7810
TLX
EML
COM 27,28C,33,47

TAMURA SHINICHI DR
DPT OF ASTRONOMY
TOHOKU UNIVERSITY
ARAMAKI
SENDAI 980
JAPAN
TEL 222-22-1800
TLF
TLX 852246 THUCOM J
EML
COM 34

TAN HUISONG
YUNNAN OBSERVATORY
CAS
BOX 110
KUNMING 72946 YUNNAN
CHINA PR
TEL 871 2035
TLF
TLX 64040 YUOBS CN
EML
COM 42

TANABE HIROYOSHI DR
TOKYO ASTRONOMICAL OBS
NAOJ
OSAWA MITAKA
TOKYO 181
JAPAN
TEL 0422-32-5111
TLF
TLX 02822307 TAOMTK J
EML
COM 15,21

TANABE KENJI DR
OKAYAMA UNIV OF SCIENCE
1-1 RIDAI-CHO
OKAYAMA 700
JAPAN
TEL 0862 52 3161
TLF 0862 55 3847
TLX
EML
COM 47

TANABE TOSHIHIKO DR
TOKYO ASTRONOMICAL OBS
NAOJ
OSAWA MITAKA
TOKYO 181
JAPAN
TEL 422 32 5111
TLF
TLX
EML
COM

TANAKA MASUO DR
INST OF ASTRONOMY
UNIVERSITY OF TOYKO
OSAWA MITAKA
TOKYO 181
JAPAN
TEL 81 422 41 3743
TLF 81 422 41 3749
TLX
EML
COM 34

TANAKA RIICHIRO PROF
FAC OF EDUCATION
NIIGATA UNIVERSITY
8050 IKARASHI 2
NIIGATA 950 21
JAPAN
TEL
TLF
TLX
EML
COM 40

TANAKA WATARU DR
DPT OF ASTRONOMY
UNIVERSITY OF TOKYO
BUNKYO KU
TOKYO 113
JAPAN
TEL 03-812-2111
TLF
TLX
EML
COM

TANAKA YASUO DR
DPT OF PHYSICS
IBARAKI UNIVERSITY
BUNKYO
MITO 310
JAPAN
TEL 292-26-1621x372
TLF
TLX
EML
COM 48

TANAKA YASUO PROF
INST SPACE & ASTRON SCI
UNIVERSITY OF TOKYO
MEGURO KU
TOKYO 153
JAPAN
TEL 03-467-1111
TLF
TLX J24550 SPACE TKY
EML
COM 44

TANAKA YUTAKA D DR
KOBE YAMATE WOMEN'S
JUNIOR COLLEGE
SUWAYAMA CHUO-KU
KOBE 650
JAPAN
TEL 78 341 6060
TLF
TLX
EML
COM 28

TANDBERG-HANSSEN EINAR A
NASA/MSFC
CODE ES 01
HUNTSVILLE AL 35812
USA
TEL 205 544 7578
TLF
TLX
EML
COM 10C,12

TANDON JAGDISH NARAIN DR
DPT PHYSICS & ASTROPHYS
UNIVERSITY OF DELHI
NEW DELHI 110 007
INDIA
TEL 11 252 1521
TLF
TLX
EML
COM 10

TANDON S N PROF
IUCAA
PO BOX 4
GANESHKHIND
PUNE 411 007
INDIA
TEL 212 33 6105
TLF 212 33 5760
TLX
EML
COM 25

TANG YU-HUA
DPT OF ASTRONOMY
NANJING UNIVERSITY
NANJING
CHINA PR
TEL 25 37651
TLF
TLX 0909
EML
COM 10

TANGO WILLIAM J DR
SCHOOL OF PHYSICS
UNIVERSITY OF SYDNEY
SYDNEY NSW 2006
AUSTRALIA
TEL 2 692 3953
TLF
TLX 26169 UNISYD AA
EML
COM 09C

TANIGUCHI YOSHIAKI DR
KISO OBSERVATORY
UNIVERSITY OF TOKYO
MITAKEMURA KISOGUN
NAGANO 397 01
JAPAN
TEL 264 52 3360
TLF
TLX 334 7577 KSOOBS J
EML
COM 28

TANZELLA-NITTI GIUSEPPE
ROMAN ATHENAEUM OF THE
HOLY CROSS
S GIROLAMO DELA CARITA 64
I 00186 ROME
ITALY
TEL
TLF
TLX
EML
COM

TANZI ENRICO G
IST DI FISICA COSMICA
CNR
VIA BASSANI 15
I 20133 MILANO
ITALY
TEL 2 298 237
TLF
TLX
EML
COM

TAPIA MAURICIO DR
INSTITUTO DE ASTRONOMIA
UNAM
APDO POSTAL 877
22860 ENSENADA B CALIF
MEXICO
TEL 4-08-80/8-30-93
TLF
TLX 56739 CICEME
EML
COM·

TAPIA-PEREZ SANTIAGO
MIT LINCOLN LABORATORY
4 MAGUIRE ROAD
LEXINGTON MA 02173
USA
TEL 617 981 0832
TLF 617 981 0991
TLX
EML LOONEY TAPIA (SPAN)
COM 25

TAPLEY BYRON D DR
AEROSPACE ENGINEERING DPT
UNIVERSITY OF TEXAS
WRW 402
AUSTIN TX 78712
USA
TEL 512 471 1356
TLF
TLX
EML
COM 19

TAPPING KENNETH F
HERZBERG INST ASTROPHYS
NTL RESEARCH COUNCIL
100 SUSSEX DR
OTTAWA ON K1A 0R6
CANADA
TEL 613 991 5842
TLF 613 952 6602
TLX 053 3715
EML
COM 10

TARADY VLADIMIR K DR
MAIN ASTRONOMICAL OBS
UKRAINIAN ACAD OF SCIENCE
GOLOSEEVO
252127 KIEV
UKRAINE
TEL 66 2286
TLF
TLX 131406 SKY SU
EML
COM 19

TARAFDAR SHANKAR P DR
TIFR
HOMI BHABHA RD
COLABA
BOMBAY 400 005
INDIA
TEL 22 219 111
TLF
TLX 0113009 TIFR IN
EML TARAFDAR@TIFRVAX
COM 34,36

TARENGHI MASSIMO DR
ESO
KARL-SCHWARZSCHILD-STR 2
D 8046 GARCHING MUENCHEN
GERMANY
TEL 89 320 06236
TLX 52828223 EO D
EML
TLF 89 320 2362
COM

TARNSTROM GUY DR
MIT LINCOLN LABORATORY
BOX 73
LEXINGTON MA 02173
USA
TEL 617 863 5500
TLF
TLX 923355
EML
COM

TARRAB IRENE
OBSERVATOIRE DE PARIS
61 AVE OBSERVATOIRE
F 75014 PARIS
FRANCE
TEL 1 40 51 2237
TLF
TLX 270776 OBS P
EML
COM

TARTER C BRUCE DR
LAWRENCE LIVERMORE LAB
L 295
BOX 808
LIVERMORE CA 94550
USA
TEL 415 422 4169
TLF
TLX
EML
COM

TARTER JILL C DR
NASA AMES RESEARCH CTR
SETI OFFICE
MOFFETT FIELD CA 94035
USA
TEL 415 604 5727
TLF 415 968 5830
TLX 348408 NASA AMES MOF
EML TARTER@BKYAST BERKELEY EDU
COM 40,47,51VP

TASSOUL JEAN-LOUIS PROF
DPT DE PHYSIQUE
UNIVERSITE DE MONTREAL
CP 6128 SUCC A
MONTREAL QC H3C 3J7
CANADA
TEL 514 343 7274
TLF 514 343 2071
TLX 055 62425
EML
COM

TASSOUL MONIQUE DR
DPT DE PHYSIQUE
UNIVERSITE DE MONTREAL
CP 6128 SUCC A
MONTREAL QC H3C 3J7
CANADA
TEL 514 343 7274
TLF 514 343 2071
TLX 055 62425
EML
COM 35

TATEMATSU KENICHI DR
NOBEYAMA RADIO OBS
NAOJ
MINAMIMAKI MURA
NAGANO 384 13
JAPAN
TEL 267 63 4378
TLF 267 98 2884
TLX 3329005 NAONRO J
EML kt@tansei cc u-tokyo ac jp
COM 40

TATEVYAN S K DR
INST OF ASTRONOMY
ACADEMY OF SCIENCES
PYATNITSKAYA UL 48
109017 MOSCOW
RUSSIA
TEL 231-54-61
TLF
TLX 412623 SCSTP SU
EML
COM 07

TATON RENE PROF
CENTRE ALEXANDRE KOYRE
12 RUE COLBERT
F 75002 PARIS
FRANCE
TEL 1 42 97 5245
TLF
TLX
EML
COM 41

TATUM JEREMY B DR
CLIMENHOGA OBSERVATORY
UNIVERSITY OF VICTORIA
BOX 1700
VICTORIA BC V8W 2Y2
CANADA
TEL 604 721 7750
TLF 604 721 7715
TLX
EML
COM 14,15,20

TAVAKOL REZA
SCHOOL OF MATHEMATICAL SC
QUEEN MARY/WESTFIELD COLL
MILE END RD
LONDON E1 4NS
UK
TEL
TLF
TLX
EML
COM 51

TAVARES J T L DR
AVE DIAS OA SILVA
173 R/C ESQ
P 3000 COIMBRA
PORTUGAL
TEL
TLF
TLX
EML
COM

TAWADROS MAHER JACOUB DR
HELWAN OBSERVATORY
HELWAN
EGYPT
TEL 78 0645/2683
TLF
TLX 93070 HIAG
EML
COM 07

TAWARA YUZURU DR
DPT OF ASTROPHYSICS
NAGOYA UNIVERSITY
FUROCHO CHIKUSA KU
NAGOYA 464
JAPAN
TEL 52 781 5111
TLF 52 781 3541
TLX 4477323 SCUNGY J
EML
COM

TAYLER ROGER J PROF
ASTRONOMY CENTRE
UNIVERSITY OF SUSSEX
FALMER
BRIGHTON BN1 9QH
UK
TEL   273 60 6755
TLF
TLX   877159 UNISEX G
EML
COM   35,47,48

TAYLOR A R DR
DPT OF PHYSICS
UNIVERSITY OF CALGARY
2500 UNIVERSITY DR NW
CALGARY AB T2N 1N4
CANADA
TEL   403 220 5385
TLF   403 289 3331
TLX   082 1545
EML   BITNET ARTAYLOR@UNCAMULT
COM   40

TAYLOR DONALD BOGGIA DR
ROYAL GREENWICH OBS
HERSTMONCEUX CASTLE
HAILSHAM BN27 1RP
UK
TEL   323 833 272
TLF
TLX. 87451 RGOBSY G
EML
COM   07,20

TAYLOR DONALD J DR
DPT OF PHYSICS & ASTRON
UNIVERSITY OF NEBRASKA
LINCOLN NE 68588
USA
TEL   402 472 3686
TLF
TLX
EML
COM

TAYLOR JOSEPH H PROF
DPT OF PHYSICS
PRINCETON UNIVERSITY
JADWIN HALL
PRINCETON NJ 08544
USA
TEL   609 452 4368
TLF
TLX   4993512
EML
COM

TAYLOR KEITH DR
AAO
PO BOX 296
EPPING NSW 2121
AUSTRALIA
TEL
TLF
TLX
EML
COM

TAYLOR KENNETH N R PROF
105A COPELAND RD
BEECROFT NSW 2119
AUSTRALIA
TEL
TLF
TLX
EML
COM   34

TCHANG-BRILLET LYDIA DR
OBSERVATOIRE DE PARIS
SECTION DE MEUDON
DAMAP
F 92195 MEUDON PPL CDX
FRANCE   .
TEL   1 45 07 7576
TLF   1 45 07 7469
TLX   631987
EML   BRILLET@FRMEU51
COM   14

TEAYS TERRY J DR
NASA/GSFC
CODE 684 9
IUE OBS
GREENBELT MD 20771
USA
TEL   301 286 5749
TLF   301 286 7642
TLX
EML   IUESOC TEAYS
COM   42

TEDESCO EDWARD F
JPL
MS 183 501
4800 OAK GROVE DR
PASADENA CA 91109
USA
TEL   818 354 4739
TLF
TLX
EML
COM   15C,22C,51

TEERIKORPI VELI PEKKA DR
TURKU UNIVERSITY OBS
TUORLA
SF 21500 PIIKKIO
FINLAND
TEL   21 43 5822
TLF   21 43 3767
TLX   62638 TYF SF
EML
COM

TEHERANY D
83 AVENUE REY

TEHERAN
IRAN
TEL
TLF
TLX
EML
COM

TEJFEL VIKTOR G DR
ASTROPHYSICAL INSTITUTE
KAZAKH ACAD OF SCIENCES
480068 ALMA ATA
KAZAKHSTAN
TEL   68-30-53
TLF
TLX
EML
COM   16,51

TEKTUNALI H GOKMEN DR
UNIVERSITY OBSERVATORY
UNIVERSITY OF ISTANBUL
34452 ISTANBUL
TURKEY
TEL   1 522 3597
TLF
TLX
EML
COM

TELESCO CHARLES M DR
NASA/MSFC
CODE ES 63
SPACE SCIENCE LAB
HUNTSVILLE AL 35812
USA
TEL   205 544 7723
TLF
TLX   594416
EML
COM   28

TEMPESTI PIERO PROF
ISTITUTO ASTRONOMICO
UNIVERSITA DI ROMA
VIA G M LANCISI 29
I 00161 ROMA
ITALY
TEL   6 84 42977
TLF
TLX
EML
COM   27

TENORIO-TAGLE G DR
MPI F PHYSIK & ASTROPHYS
KARL-SCHWARZSCHILD-STR 1
D 8046 GARCHING MUENCHEN
GERMANY
TEL   89 320 90
TLX   524629 ASTRO D
EML
TLF
COM   34

TEPLITSKAYA R B DR
SIBIZMIR
ACADEMY OF SCIENCES
664697 IRKUTSK 33
RUSSIA
TEL   6-23-65
TLF
TLX
EML
COM   12

TER HAAR DIRK
BOX 10
349 MIDDLE ST
PETWORTH GU28 ORY
UK
TEL
TLF
TLX
EML
COM

TERASHITA YOICHI PROF
KANAZAWA TECHNOLOGY INST
7-1 OGIGAOKA
NONOICHIMACHI
ISHIKAWA 921
JAPAN
TEL
TLF
TLX
EML
COM   05

TEREBIZH VALERY YU DR
CRIMEAN STATION OF
STERNBERG INSTITUTE
NAUCHNY
334413 CRIMEA
UKRAINE
TEL 43 2945
TLF
TLX 222192 VOSHOD
EML
COM

TERENTJEVA ALEXANDRA K DR
INST OF ASTRONOMY
ACADEMY OF SCIENCES
PYATNITSKAYA UL 48
109017 MOSCOW
RUSSIA
TEL 231-54-61
TLF
TLX 412623 SCSTP SU
EML
COM 15,22

TERLEVICH ELENA DR
ROYAL GREENWICH OBS
HERSTMONCEUX CASTLE
HAILSHAM BN27 1RP
UK
TEL 323 833 171
TLF
TLX
EML ET@STARLINK RO GREENWICH AC UK
COM

TERLEVICH ROBERTO JUAN
ROYAL GREENWICH OBS

HERSTMONCEUX CASTLE
HAILSHAM BN27 1RP
UK
TEL
TLF
TLX
EML
COM 28

TERNULLO MAURIZIO DR
OSS ASTROFISICO
CITTA UNIVERSITARIA
VIA A DORIA 6
I 95125 CATANIA
ITALY
TEL 95 33 0533
TLF 95 33 0592
TLX 970359 ASTRCT I
EML ASTRCT MAURIZIO
COM 10

TERRELL NELSON JAMES JR
LOS ALAMOS NATIONAL LAB
MS D436 ESS 9
BOX 1663
LOS ALAMOS NM 87545
USA
TEL 505 667 2044
TLF
TLX
EML
COM 48

TERRIEN JEAN
103 RUE DE VERSAILLES
F 92410 VILLE D'AVRAY
FRANCE
TEL 1 47 09 1034
TLF
TLX
EML
COM

TERRILE RICHARD JOHN
JPL
MS 183 501
4800 OAK GROVE DR
PASADENA CA 91109
USA
TEL
TLF
TLX
EML
COM 16

TERZAN AGOP DR
OBSERVATOIRE DE LYON
AVE CHARLES ANDRE
F 69561 S GENIS LAVAL CDX
FRANCE
TEL 78 56 0705
TLF 72 39 9791
TLX 310926
EML
COM 27,37

TERZIAN YERVANT PROF
CORNELL UNIVERSITY
SPACE SCIENCES BLDG
ITHACA NY 14853
USA
TEL 607 256 4935
TLX 932454
EML
TLF
COM 28,34,40,51

TERZIDES CHARALAMBOS DR
DPT OF ASTRONOMY
UNIVERSITY THESSALONIKI
GR 540 06 THESSALONIKI
GREECE
TEL 31 99 1357
TLF
TLX
EML
COM 33

TESKE RICHARD G PROF
DPT OF ASTRONOMY
UNIVERSITY OF MICHIGAN
DENNISON BLDG
ANN ARBOR MI 48109 1090
USA
TEL 313 764 3398
TLF
TLX
EML
COM 10

TEUBEN PETER J DR
ASTRONOMY PROGRAM
UNIVERSITY OF MARYLAND
COLLEGE PARK MD 20742
USA
TEL 301 405 3001
TLF 301 314 9067
TLX 710 826 0352
EML
COM

TEXEREAU JEAN M
OCA CERGA
AVE COPERNIC
F 06130 GRASSE
FRANCE
TEL 93 36 5849
TLF
TLX
EML
COM

THADDEUS PATRICK PROF
CENTER FOR ASTROPHYSICS
HCO/SAO
60 GARDEN ST
CAMBRIDGE MA 02138
USA
TEL 617 495 7340
TLF
TLX 921428 SATELLITE CAM
EML
COM 34,51

THAKUR RATNA KUMAR DR
DPT OF PHYSICS
UNIVERSITY OF RAVISHANKAR
RAIPUR 492 010
INDIA
TEL 27 064
TLF
TLX
EML
COM 28

THE PIK-SIN PROF
ASTRONOMICAL INSTITUTE
UNIVERSITY OF AMSTERDAM
KRUISLAAN 403
NL 1098 SJ AMSTERDAM
NETHERLANDS
TEL 20 25 27491
TLF 20 52 57484
TLX 10262 HEF NL
EML
COM 33,34,37

THEJLL PETER ANDREAS DR
NORDITA
BLEGDAMSVEJ 17
DK 2100 COPENHAGEN O
DENMARK
TEL 31 42 1616*339
TLF 31 38 9157
TLX
EML THEJLL@NORDITA DK
COM 36

THEODOSSIOU EFSTRATIOS DR
DPT OF PHYSICS
NTL UNIVERSITY OF ATHENS
PANEPISTIMIOPOLIS
GR 157 83 ZOGRAFOS
GREECE
TEL 1 724 3414
TLF
TLX
EML
COM

THEVENIN FREDERIC DR
OCA OBSERV DE NICE
BP 139
F 06003 NICE CDX
FRANCE
TEL 92 00 3011
TLF
TLX
EML THEVENIN@FRONI51
COM 29

THIELEMANN FRIEDRICH-KARL
CENTER FOR ASTROPHYSICS
HCO/SAO MS 10
60 GARDEN ST
CAMBRIDGE MA 02138
USA
TEL 617 495 4979
TLF 617 495 5374
TLX
EML fkt@cfa 3 bitnet
COM 35

THIELHEIM KLAUS O DR
ABTEILUNG MATHEM PHYSIK
UNIVERSITAET KIEL
OLSHAUSENSTR 40/60
D 2300 KIEL 1
GERMANY
TEL 431 880 3216
TLF
TLX 292979
EML
COM 33

THIRY YVES R PROF
UNIVERSITE PARIS VI
4 PLACE JUSSIEU TOUR 66
F 75230 PARIS CDX 05
FRANCE
TEL 1 43 36 2525
TLF
TLX
EML
COM 07

THOBURN CHRISTINE
ROYAL GREENWICH OBS
HERSTMONCEUX CASTLE
HAILSHAM BN27 1RP
UK
TEL 323 833 171
TLF
TLX 87451
EML
COM 08

THOLEN DAVID J DR
INSTITUTE FOR ASTRONOMY
UNIVERSITY OF HAWAII
2680 WOODLAWN DR
HONOLULU HI 96822
USA
TEL 808 956 6930
TLF 808 988 2790
TLX 8459 UHAST HR
EML tholen@uhifa ifa hawaii edu
COM 06,15,16C,20

THOMAS CLAUDINE DR
BIPM
PAVILLON DE BRETEUIL
F 92312 SEVRES CDX
FRANCE
TEL 1 45 07 7073
TLF 1 45 34 2021
TLX 631351
EML BIPM@FRMEU51
COM 31

THOMAS DAVID V DR
DPT OF PHYSICS
UNIVERSITY OF EDINBURGH
MAYFIELD RD
EDINBURGH EH9 3JZ
UK
TEL 316 505 307
TLF 316 624 712
TLX 727442 UNIVED G
EML
COM 08,24

THOMAS HANS-CHRISTOPH DR
MPI FUER PHYSIK UND
ASTROPHYSIK
KARL-SCHWARZSCHILD-STR 1
D 8046 GARCHING MUENCHEN
GERMANY
TEL
TLF
TLX
EML
COM 35

THOMAS JOHN H PROF
DPT MECH & AEROSPACE SCI
UNIVERSITY OF ROCHESTER
ROCHESTER NY 14627
USA
TEL 716 275 4083
TLF
TLX
EML
COM 10,12

THOMAS PETER A DR
ASTRONOMY CENTRE
UNIVERSITY OF SUSSEX
FALMER
BRIGHTON BN1 9QH
UK
TEL 273 60 6755*3099
TLF 273 67 8097
TLX 877159 BHVTXS G
EML PETERT@SYMA SUSSEX AC UK
COM 28

THOMAS RICHARD N DR
1155 TIMBERLANE
PINEBROOK HILL
BOULDER CO 80302
USA
TEL 303 443 9290
TLF
TLX
EML
COM 12,36

THOMAS ROGER J DR
NASA/GSFC
CODE 682
GREENBELT MD 20771
USA
TEL 301 286 7921
TLF
TLX
EML
COM 10,44

THOMASSON PETER DR
NRAL
JODRELL BANK
MACCLESFIELD SK11 9DL
UK
TEL 477 713 21
TLX 36149
EML
TLF
COM 40

THOMPSON A RICHARD DR
NRAO
VLBA PROJECT
2015 IVY RD
CHARLOTTESVILLE VA 22903
USA
TEL 804 296 0211
TLF
TLX
EML
COM 34,40

THOMPSON G I DR
7 MACDOWALL RD
EDINBURGH EH9 3HJ
UK
TEL
TLF
TLX
EML
COM

THOMPSON KEITH DR
DPT OF PHYSICS
MONASH UNIVERSITY
WELLINGTON RD
CLAYTON VIC 3168
AUSTRALIA
TEL 3 565 4000
TLF
TLX 32691 AA
EML
COM 27,42

THOMPSON LAIRD A DR
DPT OF ASTRONOMY
UNIVERSITY OF ILLINOIS
1011 W SPRINGFIELD AVE
URBANA IL 61801
USA
TEL 217 333 3090
TLF
TLX
EML
COM 28,47

THOMPSON MICHAEL J DR
ASTRONOMY UNIT
QUEEN MARY/WESTFIELD COLL
MILE END RD
LONDON E1 4NS
UK
TEL 1 719 755 462
TLF 1 819 819 587
TLX 893750
EML MTHOMPSO@SOLAR BITNET
COM 10,35

THOMPSON RODGER I PROF
STEWARD OBSERVATORY
UNIVERSITY OF ARIZONA
TUCSON AZ 85721
USA
TEL 602 621 6527
TLF
TLX 467175
EML
COM

THOMSEN BJARNE B LECT
INST OF PHYSICS & ASTRON
UNIVERSITY OF AARHUS
NY MUNKEGADE
DK 8000 AARHUS C
DENMARK
TEL 86 12 8899
TLF 86 20 2711
TLX
EML
COM

THONNARD NORBERT DR
ATOM SCIENCES
114 RIDGEWAY CENTER
OAK RIDGE TN 37830
USA
TEL 615 483 1113
TLF
TLX
EML
COM 28,34

THORNE KIP S PROF
CALTECH
MS 130 33
PASADENA CA 91125
USA
TEL 213 356 4598
TLF
TLX 675425
EML
COM 48

THORSTENSEN JOHN R
DPT OF PHYSICS & ASTRON
DARTMOUTH COLLEGE
HANOVER NH 03755
USA
TEL 603 646 2869
TLF
TLX
EML
COM

THRONSON HARLEY ANDREW JR
DPT PHYSICS & ASTRONOMY
UNIVERSITY OF WYOMING
LARAMIE WY 82071
USA
TEL 307 766 6150
TLF
TLX
EML
COM 34

THUAN TRINH XUAN DR
UNIVERSITY STATION
UNIVERSITY OF VIRGINIA
BOX 3818
CHARLOTTESVILLE VA 22903
USA
TEL 804 924 4894
TLF
TLX
EML
COM 28,47

THUM CLEMENS DR
IRAM
AVD DIVINA PASTORA 7
BLOQUE 6/2B
E 18012 GRANADA
SPAIN
TEL 58 48 0413
TLF
TLX 78521 IRAM E
EML
COM 40

TIFFT WILLIAM G PROF
STEWARD OBSERVATORY
UNIVERSITY OF ARIZONA
TUCSON AZ 85721
USA
TEL 602 621 6532
TLF
TLX 467175
EML
COM 28,47

TIFREA EMILIA DR
ASTRONOMICAL OBSERVATORY
CUTITUL DE ARGENT 5
BOX 28
R 75212 BUCAREST 28
RUMANIA
TEL 23-60-10
TLF
TLX 09-26-29
EML
COM 10

TIMOTHY J GETHYN DR
CTR FOR SPACE SCIENCES &
ASTROPHYSICS
STANFORD UNIV ERL 314
STANFORD CA 94305 4055
USA
TEL 415 497 0059
TLF
TLX 348402 STANFRD STNU
EML
COM

TINBERGEN JAAP DR
KAPTEYN STERREWACHT
WERKGROEP
MENSINGHEWEG 20
NL 9301 KA RODEN
NETHERLANDS
TEL 59 08 19631
TLF
TLX 53767 KSWRO NL
EML
COM 25

TING YEOU-TSWEN
ASTRONOMY SECTION
CENTRAL WEATHER BUREAU
64 KUNG YUEN RD
TAIPEI 100
CHINA R
TEL 2 371 3181*281
TLF
TLX
EML
COM 04

TIPLER FRANK JENNINGS DR
DPT OF PHYSICS
TULANE UNIVERSITY
NEW ORLEANS LA 70118
USA
TEL
TLF
TLX
EML
COM 47

TIURI MARTTI PROF
HELSINKI UNIV TECHNOLOGY
RADIO LABORATORY
OTAKAARI 5 A
SF 02150 ESPOO 15
FINLAND
TEL 451 2545
TLF
TLX 122771 RORTA SF
EML
COM

TJIN-A-DJIE HERMAN R E DR
KOEKOELAAN 106
NL 1403 EJ BUSSUM
NETHERLANDS
TEL 21 59 17076
TLF
EML
COM 27,35
TLX 16460 FACWN NL

TLAMICHA ANTONIN DR
ASTRONOMICAL INSTITUTE
CZECH ACADEMY OF SCIENCES
ONDREJOV OBSERVATORY
CS 251 65 ONDREJOV
CZECHOSLOVAKIA
TEL 204 85201
TLF 204 85314
TLX 121579 ASTR CZ
EML
COM 10,40

TOBIN WILLIAM
DPT OF PHYSICS
UNIVERSITY OF CANTERBURY
PRIVATE BAG
CHRISTCHURCH 1
NEW ZEALAND
TEL 64 3 642 531
TLF 64 3 642 999
TLX
EML w tobin@canterbury ac nz
COM 33

TODORAN IOAN DR
ASTRONOMICAL OBSERVATORY
STR CIRESILOR 19
R 3400 CLUJ NAPOCA
RUMANIA
TEL
TLF
TLX
EML
COM 25,42

TOFANI GIANNI PROF
OSS ASTROFISICO
DI ARCETRI
LARGO E FERMI 5
I 50125 FIRENZE
ITALY
TEL 55 275 2217
TLF
TLX 572268 ARCETRI
EML
COM 40

TOHLINE JOEL EDWARD
DPT PHYSICS & ASTRONOMY
LOUISIANA STATE UNIV
BATON ROUGE LA 70803 4001
USA
TEL 504 388 6851
TLF
TLX 559184
EML
COM 35

TOKAREV YURIJ V DR
RADIOPHYSICAL RESEARCH
INSTITUTE
LYADOV UL 25
603600 N NOVGOROD
RUSSIA
TEL 97 8312 360188
TLF 97 8312 369902
TLX
EML
COM 34

TOKOVININ ANDREJ A DR
STERNBERG STATE ASTR INST
UNIVERSITETSKIJ PROSP 13
119899 MOSCOW
RUSSIA
TEL 939 33 18
TLX
EML
TLF
COM 26,30

TOKUNAGA ALAN TAKASHI DR
INSTITUTE FOR ASTRONOMY
UNIVERSITY OF HAWAII
2680 WOODLAWN DR
HONOLULU HI 96822
USA
TEL 808 956 6691
TLF
TLX 723 8459 UHAST HR
EML TOKUNAGA@GALILEO IFA HAWAII ED
COM 25

TOLBERT CHARLES R DR
UNIVERSITY STATION
UNIVERSITY OF VIRGINIA
BOX 3818
CHARLOTTESVILLE VA 22903
USA
TEL 804 924 7494
TLF
TLX
EML
COM 25,38,40,51

TOLLER GARY N DR
NASA/GSFC
CODE 685 3
GREENBELT MD 20771
USA
TEL
TLF
TLX
EML
COM 21

TOMASI PAOLO DR
IST DI RADIOASTRONOMIA
CNR
VIA IRNERIO 46
I 40126 BOLOGNA
ITALY
TEL 51 23 2856
TLF
TLX
EML
COM 40

TOMASKO MARTIN G DR
LUNAR & PLANETARY LAB
UNIVERSITY OF ARIZONA
SPACE SCIENCES BLDG
TUCSON AZ 85721
USA
TEL 602 621 6969
TLF
TLX
EML
COM

TOMBAUGH CLYDE W PROF
DPT OF ASTRONOMY
NEW MEXICO STATE UNIV
BOX 4500
LAS CRUCES NM 88003
USA
TEL 505 646 2107
TLF
TLX
EML
COM 16

TOMIMATSU AKIRA DR
DPT OF PHYSICS
NAGOYA UNIVERSITY
FUROCHO CHIKUSA KU
NAGOYA 464
JAPAN
TEL
TLF
TLX
EML
COM 47,48

TOMISAKA KOHJI DR
FAC OF EDUCATION
NIIGATA UNIVERSITY
8050 IKARASHI 2
NIIGATA 950 21
JAPAN
TEL 25 262 7269
TLF
TLX
EML BITNET c30841@jpnkudpc
COM 33

TOMITA KENJI PROF
UJI RESEARCH CENTER
YUKAWA INST THEORICAL PHY
KYOTO UNIVERSITY
UJI 611
JAPAN
TEL
TLF
TLX
EML tomita@yisun1 kyoto-u ac jp
COM 47

TOMITA KOICHIRO MR
11-12 20 YOGA
SETAGAYAKU
TOKYO 158
JAPAN
TEL 037000066
TLF
TLX
EML
COM 15,20,22

TOMOV ALEXANDER NIKOLOV
DPT OF ASTRONOMY
BULGARIAN ACAD SCIENCES
72 LENIN BLVD
BG 1784 SOFIA
BULGARIA
TEL 2 75 8927
TLF
TLX
EML
COM

TOMOV TOMA V DR
NAT ASTRONOMICAL OBS
BULGARIAN ACAD SCIENCES
BOX 136
BG 4700 SMOLJAN
BULGARIA
TEL 73 41 599
TLF
TLX 48446
EML
COM 29

TONG FU
PURPLE MOUNTAIN OBSERV
CAS
NANJING
CHINA PR
TEL 25 33921
TLF
TLX 34144 PMOAS CN
EML
COM 04C,07

TONG YI
DPT OF ASTRONOMY
BEIJING NORMAL UNIVERSITY
19 XINJISKOW OUT ST
BEIJING 100875
CHINA PR
TEL 1 65 6531*6285
TLF
TLX
EML
COM 28,33

TONRY JOHN DR
DPT OF MATHEMATICS
MIT RM 6 204A
77 MASSACHUSETTS AVE
CAMBRIDGE MA 02139
USA
TEL· 617 253 7528
TLF
TLX
EML JT@ALCOR MIT EDU
COM 28,30,47

TONWAR SURESH C PROF
TIFR
HOMI BHABHA RD
COLABA
BOMBAY 400 005
INDIA
TEL 22 495 2311
TLF
TLX 011 3009 TIFR IN
EML
COM

TOOMRE ALAR DR
DPT OF MATHEMATICS
MIT RM 2 372
77 MASSACHUSETTS AVE
CAMBRIDGE MA 02139
USA
TEL 617 253 4326
TLF
TLX
EML
COM 28,33

TOOMRE JURI
JILA/DAG
UNIVERSITY OF COLORADO
BOX 440
BOULDER CO 80309 0440
USA
TEL 303 492 7854
TLF
TLX
EML
COM 33,35,36

TOPAKTAS LATIF A DR
KING SAUD UNIVERSITY
COLLEGE OF SCIENCE
BOX 2455
RIYADH 11453
SAUDI ARABIA
TEL
TLF
TLX
EML
COM

TORAO MASAHISA
410-11
SENPUKU GA OKA 2-11-9
SUSONO CITY
JAPAN
TEL
TLF
TLX
EML
COM 19

TORELLI M DR
OAR
VIA DEL PARCO MELLINI 84
I 00136 ROMA
ITALY
TEL 6 34 7056
TLF
TLX
EML
COM 12

TORNAMBE AMEDEO
IAS
CNR
CP 67
I 00044 FRASCATI
ITALY
TEL 6 942 5655
TLF
TLX
EML
COM 37

TORO TIBOR PROF
INSTITUTE OF THEORETICAL
PHYSICS
UNIVERSITY TIMISUARA
TIMISUARA
RUMANIA
TEL 40 61 30823
TLF
TLX
EML
COM 51

TOROSHLIDZE TEIMURAZ I DR
ABASTUMANI ASTROPHYSICAL
OBSERVATORY
GEORGIAN ACAD OF SCIENCES
383762 ABASTUMANI
GEORGIA
TEL
TLF
TLX
EML
COM 21

TORRA JORDI DR
DPT FISICA
UNIVERSIDAD DE BARCELONA
AVD DIAGONAL 647
E 08028 BARCELONA
SPAIN
TEL 3 330 7311
TLF
TLX
EML
COM 33

TORRELLES JOSE M DR
INST ASTROFISICA
DE ANDALUCIA APD 3004
PREFESOR ALBAREDA 1
E 18080 GRANADA
SPAIN
TEL 58 12 1311
TLF 58 81 4530
TLX
EML 16488 TORRELLES/TORRELLES@IAA EML
COM 34

TORRES CARLOS ALBERTO DR
OBSERVATORIO NACIONA
RUA CORONEL RENNO 07
CP 21
37500 ITAJUBA MG
BRAZIL
TEL 35 622 0788
TLF
TLX 031-2603
EML
COM 27,50

TORRES CARLOS DR
DPT DE ASTRONOMIA
UNIVERSIDAD DE CHILE
CASILLA 36 D
SANTIAGO
CHILE
TEL 2 229 4101
TLF
TLX 440 001
EML
COM 20,50

TORRES-PEIMBERT SILVIA DR
INSTITUTO DE ASTRONOMIA
UNAM
APDO POSTAL 70-264
04510 MEXICO DF
MEXICO
TEL 905-548-5306
TLF
TLX 1760155 CIC ME
EML
COM 34,46

TORRICELLI GUIDETTA DR
OSS ASTROFISICO
DI ARCETRI
LARGO E FERMI 5
I 50125 FIRENZE
ITALY
TEL 55 275 2260
TLF
TLX 572268
EML
COM

TORROJA J PROF
CATEDRA DE ASTRONOMIA
FAC DE CIENCIAS
UNIVERSIDAD COMPLUTENSE
E 28040 MADRID
SPAIN
TEL
TLF
TLX
EML
COM

TOSA MAKOTO DR
ASTRONOMICAL INSTITUTE
TOHOKU UNIVERSITY
SENDAI AOBA
MIYAGI 980
JAPAN
TEL 0222-22-1800
TLF
TLX 852246 THUCOM J
EML
COM 33

TOSHIKI AIKAWA DR
ASTRONOMICAL INSTITUTE
TOHOKU GAKUIN UNIVERSITY
ICHINAZAKA IZUMI-KU
SENDAI 981 31
JAPAN
TEL 22 375 1111*318
TLF 22 375 4040
TLX
EML
COM

TOSI MONICA
OSS ASTRONOMICO
UNIVERSITA DI BOLOGNA
CP 596
I 40100 BOLOGNA
ITALY
TEL 51 22 2956
TLF
TLX 211664 INFN BO I
EML
COM 34

TOTH IMRE DR
KONKOLY OBSERVATORY
THEGE U 13/17
BOX 67
H 1525 BUDAPEST
HUNGARY
TEL 1 75 5866/75 4122
TLF 1 156 9640
TLX 22 74 60 KONOB H
EML H697KNO@ELLA UUCP
COM 15

TOTSUKA YOJI DR
INST COSMIC RAY RESEARCH
UNIVERSITY OF TOKYO
MIDORICHO TANASHI
TOKYO 188
JAPAN
TEL 81 42 46 14131
TLF 81 42 46 81438
TLX 02822371 ICRTU J
EML TOTSUKA@JPNUTINS BITNET
COM 48

TOUSEY RICHARD DR
NAVAL RESEARCH LABORATORY
CODE 7140
4555 OVERLOOK AVE SW
WASHINGTON DC 20375 5000
USA
TEL 202 767 3441
TLF
TLX
EML
COM 12,14

TOUT CHRISTOPHER DR
INSTITUTE OF ASTRONOMY
THE OBSERVATORIES
MADINGLEY RD
CAMBRIDGE CB3 0HA
UK
TEL· 223 337 548
TLF 223 337 523
TLX
EML CT23@PHOENIX CAMBRIDGE AC UK
COM 42

TOVMASSIAN H M DR
BYURAKAN ASTROPHYSICAL
OBSERVATORY
375433 ARMENIA
ARMENIA
TEL 88 52 28 3453
TLF
TLX·
EML
COM 28,40,44,51

TOWNES CHARLES HARD DR
PHYSICS DPT
UNIVERSITY OF CALIFORNIA
RM 557 BIRGE HALL
BERKELEY CA 94720
USA
TEL 415 642 1128
TLF
TLX
EML
COM 34,40,51

TOYAMA KIYOTAKA
HOKKAIDO INFORMATION UNIV
NISHINOPPORO 59-2
EBETSU 069
JAPAN
TEL 81 11 385 4411
TLF 81 11 384 0134
TLX
EML
COM 28

TOZER DAVID C DR
SCHOOL OF PHYSICS
UNIVERSITY OF NEWCASTLE
NEWCASTLE/TYNE NE1 7RU
UK
TEL
TLF
TLX
EML
COM

TOZZI GIAN PAOLO
OSS ASTROFISICO
DI ARCETRI
LARGO E FERMI 5
I 50125 FIRENZE
ITALY
TEL 55 275 2250
TLF
TLX 572268 ARCETR I
EML
COM 14

TRAFTON LAURENCE M DR
ASTRONOMY DPT
UNIVERSITY OF TEXAS
RLM 15 308
AUSTIN TX 78712 1083
USA
TEL 512 471 1476
TLF
TLX
EML
COM 16

TRAN MINH NGUYET DR
OBSERVATOIRE DE PARIS
SECTION DE MEUDON
F 92195 MEUDON PPL CDX
FRANCE
TEL 1 45 07 7447
TLF
TLX 270912
EML
COM 14

TRAN-MINH FRANCOISE DR
OBSERVATOIRE DE PARIS
SECTION DE MEUDON
DASGAL
F 92195 MEUDON PPL CDX
FRANCE
TEL 1 45 07 7553
TLF 1 45 07 7469
TLX 201571
EML
COM 16

TRAUB WESLEY ARTHUR
CENTER FOR ASTROPHYSICS
HCO/SAO
60 GARDEN ST
CAMBRIDGE MA 02138
USA
TEL 617 495 7406
TLF
TLX 921428 SATELLITE CAM
EML
COM 09,44

TRAVING GERHARD PROF
INSTITUT FUER
THEORETISCHE ASTROPHYSIK
NEUENHEIMER FELD 561
D 6900 HEIDELBERG 1
GERMANY
TEL 62 215 62815
TLF
TLX 461515
EML
COM 36

TREDER H J PROF DR
ZNTRLINST F ASTROPHYSIK
STERNWARTE BABELSBERG
ROSA-LUXEMBURG-STR 17A
D 1502 POTSDAM
GERMANY
TEL 762225
TLF
TLX 15471
EML
COM 28,47

TREFFERS RICHARD R
ASTRONOMY DPT
UNIVERSITY OF CALIFORNIA
601 CAMPBELL HALL
BERKELEY CA 94720
USA
TEL 415 642 4223
TLF
TLX
EML
COM 34

TREFFTZ ELEONORE E DR
MPI F PHYSIK UND
ASTROPHYSIK
KARL-SCHWARZSCHILD-STR 1
D 8046 GARCHING MUENCHEN
GERMANY
TEL 89 329 90
TLF
TLX 524629 ASTRO D
EML
COM 14

TREFZGER CHARLES F DR
ASTRONOMISCHES INSTITUT
UNIVERSITAET BASEL
VENUSSTRASSE 7
CH 4102 BINNINGEN
SWITZERLAND
TEL 61 22 7711
TLF
TLX
EML
COM 33

TREHAN SURINDAR K PROF
DPT OF MATHEMATICS
UNIVERSITY OF PANJAB
CHANDIGARH 160 014
INDIA
TEL 29938
TLF
TLX
EML
COM

TRELLIS MICHEL DR
OCA OBSERV DE NICE
BP 139
F 06003 NICE CDX
FRANCE
TEL 93 89 0420
TLF
TLX 460004 OBSNICE F
EML
COM 10

TREMAINE SCOTT DUNCAN
CITA MCLENNAN LABS
UNIVERSITY OF TORONTO
60 ST GEORGE ST
TORONTO ON M5S 1A1
CANADA
TEL 416 978 6879
TLF. 416 978 3921
TLX
EML
COM 28,47

TREMKO JOZEF DR
ASTRONOMICAL INSTITUTE
SLOVAK ACADEMY SCIENCES
CS 059 60 TATRANSKA LOMNI
CZECHOSLOVAKIA
TEL 969 96 7866/7/8
TLF 969 96 7656
TLX 78277 AUSAV CZ
EML
COM 27,42,50

TREUMANN RUDOLF A DR
MPI F PHYS & ASTROPHYSIK
INST F EXTRATERR PHYSIK
D 8046 GARCHING MUENCHEN
GERMANY
TEL 89 329 9831
TLF
TLX 5215845 XTER D
EML
COM 10

TREVESE DARIO
ISTITUTO ASTRONOMICO
UNIVERSITA DI ROMA
VIA G M LANCISI 29
I 00161 ROMA
ITALY
TEL 6 44 03734
TLF 6 440 3673
TLX 613255 INFNRO
EML
COM 47

TREXLER JAMES H MR
1921 SO ABREGO DR

GREENVALLEY AZ 85614 1403
USA
TEL
TLF
TLX
EML
COM

TRIMBLE VIRGINIA L DR
DPT OF PHYSICS
UNIVERSITY OF CALIFORNIA
IRVINE CA 92717
USA
TEL 714 856 6948
TLF 714 725 2174
TLX
EML VTRIMBLE@UCI EDU
COM 26,28VP,35,42,47,48C

TRINCHIERI GINEVRA
OSS ASTROFISICO
DI ARCETRI
LARGO E FERMI 5
I 50125 FIRENZE
ITALY
TEL 55 275 2230
TLF
TLX 572268 ARCETR I
EML
COM 28

TRIPATHI B M DR
UTTAR PRADESH STATE
OBSERVATORY
PO MANORA PEAK 263 129
NAINITAL 263 129
INDIA
TEL 59 42 2136
TLF
TLX CABLE   ASTRONOMY
EML
COM 12

TRIPICCO MICHAEL J DR
ASTRONOMY PROGRAM
UNIVERSITY OF MARYLAND
COLLEGE PARK MD 20742
USA
TEL 301 405 1548
TLF
TLX
EML MIKET@ASTRO UMD EDU
COM 37

TRITAKIS BASIL P DR
DPT ASTRONOMY & APPL MATH
ACADEMY OF ATHENS
14 ANAGNOSTOPOULOU ST
GR 106 73 ATHENS
GREECE
TEL 1 361 3589
TLF
TLX
EML
COM 10,49

TRITTON KEITH P DR
ROYAL GREENWICH OBS
MADINGLEY RD
CAMBRIDGE CB3 0EZ
UK
TEL 223 37 4000
TLF 223 37 4700
TLX 817235 RGOSTR G
EML
COM 40

TRITTON SUSAN BARBARA
ROYAL OBSERVATORY
BLACKFORD HILL
EDINBURGH EH9 3HJ
UK
TEL 316 673 321
TLF
TLX 72383 ROEDIN G
EML
COM 05

TROCHE-BOGGINO A E DR
INST DE CIENCIAS BASICAS
UNIV NACIONAL DE ASUNCION
CC 1039-1804
ASUNCION
PARAGUAY
TEL
TLF
TLX
EML
COM 46

TRODAHL HARRY JOSEPH DR
DPT OF PHYSICS
VICTORIA UNIVERSITY
PRIVATE BAG
WELLINGTON
NEW ZEALAND
TEL 721-000
TLF
TLX
EML
COM 25

TROITSKY V S PROF DR
RADIOPHYSICAL RESEARCH
INSTITUTE
LYADOV UL 25/14
603600 N NOVGOROD
RUSSIA
TEL 36-04-40
TLF
TLX
EML .
COM 16,40,51

TROLAND THOMAS HUGH
DPT PHYSICS & ASTRONOMY
UNIVERSITY OF KENTUCKY
LEXINGTON KY 40506
USA
TEL 606 257 8620
TLF
TLX
EML
COM 40

TROTTET GERARD DR
OBSERVATOIRE DE PARIS
SECTION DE MEUDON
DASOP
F 92195 MEUDON PPL CDX
FRANCE
TEL 1 45 07 7808
TLF
TLX
EML
COM 10

TRUEMPER JOACHIM PROF
MPI F EXTRATERRESTRISCHE
PHYSIK
D 8046 GARCHING MUENCHEN
GERMANY
TEL 89 329 93559
TLF 89 329 93569
TLX 5215845 XTER D
EML
COM 44P,48

TRUJILLO BUENO JAVIER DR
INST DE ASTROFISICA
DE CANARIAS
OBS DEL TEIDE
E 38200 LA LAGUNA
SPAIN
TEL 22 60 5266
TLF 22 60 5210
TLX 92640 IAC E
EML IAC JTB
COM 12

TRULSEN JAN K PROF
INST MATHS & PHYSICAL SCI
UNIVERSITY OF TROMSO
BOX 953
N 9001 TROMSO
NORWAY
TEL
TLF
TLX
EML
COM

TRUONG-BACH
OBSERVATOIRE DE PARIS
SECTION DE MEUDON
F 92195 MEUDON PPL CDX
FRANCE
TEL 1 45 07 7897
TLF 1 45 07 7893
TLX 270912
EML TBACH@FRMEU51
COM 40

TRURAN JAMES W JR
DPT OF ASTRONOMY
UNIVERSITY OF ILLINOIS
1011 W SPRINGFIELD AVE
URBANA IL 61801
USA
TEL 217 333 3090
TLF
TLX
EML
COM· 35,48

TRUSSONI EDOARDO
IST DI COSMO GEOFISICA
CNR
CORSO FIUME 4
I 10133 TORINO
ITALY
TEL 11 65 7694/8979
TLF
TLX 211041 INFNTO
EML
COM· 48

TRUTSE YU L DR
INST PHYSICS OF ATMOSPH
ACADEMY OF SCIENCES
PYZHEVSKY 3
109017 MOSCOW
RUSSIA
TEL
TLF
TLX
EML
COM 21

TSAI CHANG-HSIEN DIRECTOR
ASTRON SOCIETY OF CHINA

TAIPEI OBSERVATORY
TAIPEI 104
CHINA R
TEL
TLF
TLX
EML
COM

TSAMPARLIS MICHAEL DR
DPT OF ASTROPHYSICS
NTL UNIVERSITY OF ATHENS
PANEPISTIMIOPOLIS
GR 157 73 ZOGRAFOS
GREECE
TEL 1 724 3414*211
TLF 1 962 4430
TLX 223815 UNIV GR
EML RICH@GRATHUN1
COM 47

TSAO MO PROF
NO 47 SEC 3
HSIN-I RD
TAIPEI 106
CHINA R
TEL 2 704 7795
TLF
TLX
EML
COM 19

TSAP T T DR
CRIMEAN ASTROPHYS OBS
UKRAINIAN ACAD OF SCIENCE
NAUCHNY
334413 CRIMEA
UKRAINE
TEL 43 2945
TLF
TLX
EML
COM 12

TSAY WEAN-SHUN DR
INST PHYSICS & ASTRONOMY
NTL CENTRAL UNIVERSITY
CHUNG LI
TAIWAN 32054
CHINA R
TEL 3 422 7151*5335/5300
TLF 3 425 1175
TLX
EML TSAY@PHYAST DNET NCU.EDU T\
COM 26

TSCHARNUTER WERNER M DR
INST THEOR ASTROPHYSICS
DER UNIVERSITAET
IM NEUENHEIMER FELD 561
D 6900 HEIDELBERG 1
GERMANY
TEL 62 215 62815
TLF
TLX
EML
COM 35

TSEYTLIN NAUM M
RADIOPHYSICAL RESEARCH
INSTITUTE
LYADOV UL 25/14
603600 N NOVGOROD
RUSSIA
TEL 36-01-29
TLF
TLX 1113 LUNA
EML
COM

TSIKOUDI VASSILIKI PH D
DPT OF PHYSICS
UNIVERSITY OF IOANNINA
GR 453 32 IOANNINA
GREECE
TEL 65 19 1084
TLF
TLX
EML
COM

TSINGANOS KANARIS DR
DPT OF PHYSICS
UNIVERSITY OF CRETE
BOX 1527
GR 711 10 HERAKLION
GREECE
TEL 81 23 9757*154
TLF 81 23 9735
TLX 262860
EML TSINGAN@IESL FORTH GR
COM 10

TSIOUMIS ALEXANDROS DR
DPT GEODETIC ASTRONOMY
UNIVERSITY THESSALONIKI
UNIV BOX 503
GR 540 06 THESSALONIKI
GREECE
TEL 31 99 2693
TLF
TLX
EML
COM 27,33

TSIROPOULA GEORGIA DR
ASTRONOMICAL INSTITUTE
NTL OBSERVATORY OF ATHENS
BOX 20048
GR 118 10 ATHENS
GREECE
TEL 1 346 1191
TLF
TLX 215530 OBS GR
EML
COM 12

TSUBAKI TOKIO PROF
DPT OF EARTH SCIENCE
SHIGA UNIVERSITY
2-5-1 HIRATSU
OHTSU 520
JAPAN
TEL 0775-37-0081
TLF
TLX
EML
COM 10,12

TSUBOI MASATO DR
NOBEYAMA RADIO OBS
NAOJ
MINAMIMAKI MURA
NAGANO 384 13
JAPAN
TEL 81 267 63 4314
TLF 81 267 98 2927
TLX 3329005 NAONRO J
EML NRROTSUB@JPNNRO BITNET
COM 40

TSUBOKAWA IETSUNE DR
INTL LATITUDE OBSERVATORY
NAOJ
HOSHIGAOKA MIZUSAWA SHI
IWATE 023
JAPAN
TEL 0197247111
TLF
TLX 837628
EML
COM 19

TSUCHIDA MASAYOSHI DR
IAG
UNIVERSIDADE DE SAO PAULO
AV MIGUEL STEFANO 4200
04301 SAO PAULO SP
BRAZIL
TEL 11 577 8599
TLF 11 276 3848
TLX 36221 IAGM BR
EML MASA%IAGUSP@BRFAPESP BITNET
COM 20

TSUCHIYA ATSUSHI DR PROF
OMACHI 4-2-18
KAMAKURA 248
JAPAN
TEL
TLF
TLX
EML
COM 31

TSUJI TAKASHI
INST OF ASTRONOMY
UNIVERSITY OF TOKYO
OSAWA MITAKA
TOKYO 181
JAPAN
TEL 0422-32-5111
TLF
TLX 02822307 TAOMTK J
EML
COM 36C

TSUNEMI HIROSHI DR
FAC OF SCIENCES
OSAKA UNIVERSITY
MACHIKANEYAMA
TOYONAKA OSAKA 560
JAPAN
TEL 6 844 1151
TLF
TLX
EML BITNET TSUNEMI@JPNOSKFM
COM 44

TSUNETA SAKU DR
INST OF ASTRONOMY
UNIVERSITY OF TOKYO
OSAWA MITAKA
TOKYO 181
JAPAN
TEL 422 32 4710
TLF
TLX 2822307 TAOMIK J
EML STSUNETA NASAMAIL
COM 10

TSURUTA SACHIKO DR
DPT OF PHYSICS
MONTANA STATE UNIVERSITY
BOZEMAN MT 59715
USA
TEL 406 994 3614
TLF
TLX
EML
COM 48

TSVETANOV ZLATAN IVANOV
ASTRONOMY PROGRAM
UNIVERSITY OF MARYLAND
COLLEGE PARK MD 20742
USA
TEL 301 454 0878
TLF 301 454 2298
TLX 710 826 0352
EML ztsvetanov@astro umd edu
COM

TSVETKOV MILCHO K DR
DPT OF ASTRONOMY
BULGARIAN ACAD SCIENCES
72 LENIN BLVD
BG 1784 SOFIA
BULGARIA
TEL 2 75 8927
TLF
TLX 23561 ECF BAN BG
EML
COM 05,27,37

TSVETKOV TSVETAN DR
DPT OF ASTRONOMY
UNIVERSITY OF SOFIA
ANTON IVANOV ST 5
BG 1126 SOFIA
BULGARIA
TEL 2 54 4852
TLF
TLX
EML
COM

TSVETKOVA KATIA
ASTRONOMICAL OBSERVATORY
BULGARIAN ACAD SCIENCES
72 LENIN BLVD
BG 1784 SOFIA
BULGARIA
TEL 2 75 8927
TLF
TLX 23561 ECF BAN BG
EML
COM 37

TSYGAN ANATOLII I PROF
IOFFE PHYSICAL TECH INST
ACADEMY OF SCIENCES
POLYTECHNICHESKAYA UL 26
194021 ST PETERSBURG
RUSSIA
TEL 812 247 9326
TLF 812 247 1963
TLX 121453 FTIAN SU
EML VARSH@EO PTI SPB SU
COM 48

TUCHMAN YTZHAK
RACAH INST OF PHYSICS
HEBREW UNIV OF JERUSALEM
JERUSALEM 91904
ISRAEL
TEL 2 584 417
TLF
TLX 25391 HUIL
EML
COM

TUCKER WALLACE H DR
BOX 266
BONSALL CA 92003
USA
TEL 619 728 7103
TLF
EML
COM
TLX

TUEG HELMUT DR
ALFRED-WEGENER INSTITUT
FUR POLARFORSCHUNG
COLUMBUS CENTER
D 2850 BREMERHAVEN
GERMANY
TEL
TLF
TLX
EML
COM 09

TUFEKCIOGLU ZEKI DR
DPT OF ASTRONOMY
KING ABDULAZIZ UNIV
BOX 9028
JEDDAH 21413
SAUDI ARABIA
TEL
TLF
TLX
EML
COM

TULL ROBERT G
ASTRONOMY DPT
UNIVERSITY OF TEXAS
RLM 15 308
AUSTIN TX 78712 1083
USA
TEL 512 471 3337
TLF
TLX 910-874-1351
EML
COM 09

TULLY JOHN A DR
OCA OBSERV DE NICE
BP 139
F 06003 NICE CDX
FRANCE
TEL 93 89 0420
TLF
TLX 460004 OBSNICE F
EML
COM

TULLY RICHARD BRENT DR
INSTITUTE FOR ASTRONOMY
UNIVERSITY OF HAWAII
2680 WOODLAWN DR
HONOLULU HI 96822
USA
TEL 808 948 8606
TLF
TLX 723-8459 UHAST HR
EML
COM 28,47

TUNCA ZEYNEL DR
FACULTY OF SCIENCE
EGE UNIVERSITY
BOX 21
35100 BORNOVA IZMIR
TURKEY
TEL 51 18 0110*2332
TLF
TLX
EML
COM

TUNG NGUYEN MAU DR
COMMITTEE FOR SPACE RES
AND APPLICATION
NGHIA DO-TU LIEM
HANOI
VIETNAM
TEL
TLF
TLX
EML
COM 04

TUOHY IAN R DR
BRITISH AEROSP AUSTRALIA
14 PARK WAY
TECHNOLOGY PARK
SALISBURY 5095
AUSTRALIA
TEL 8 343 22111
TLF 8 349 6629
TLX 88342
EML
COM

TUOMINEN ILKKA V DR
OBS & ASTROPHYSICS LAB
UNIVERSITY OF HELSINKI
TAEHTITORNINMAKI
SF 00130 HELSINKI 13
FINLAND
TEL 191 2946
TLF
TLX
EML TUOMINEN@FINUH
COM 10,29,35,36

TURATTO MASSIMO DR
OSS ASTRONOMICO DI PADOVA
VIC DELL OSSERVATORIO 5
I 35122 PADOVA
ITALY
TEL 49 66 1499
TLF
TLX 432071
EML SPAN 39003 TURATTO
COM

TURLO ZYGMUNT DR
INSTITUTE OF ASTRONOMY
N COPERNICUS UNIVERSITY
UL CHOPINA 12/18
PL 87 100 TORUN
POLAND
TEL
TLF
TLX
EML
COM 40

TURNER BARRY E DR
NRAO
EDGEMONT RD
CHARLOTTESVILLE VA 22901
USA
TEL 804 296 0337
TLX 910-997-0174
EML
TLF
COM 34,40

TURNER DAVID G DR
DPT OF ASTRONOMY
ST MARY'S UNIVERSITY
HALIFAX NS B3H 3C3
CANADA
TEL 902 429 9780*2254
TLF 902 420 5561
TLX
EML
COM 27,37

TURNER EDWIN L DR
PRINCETON UNIVERSITY OBS
PEYTON HALL
PRINCETON NJ 08544
USA
TEL 609 258 3577
TLF 608 258 1020
TLX 322409 ASTRO PRIN
EML z1t@ASTROUVAV PRINCETON EDU
COM 28,47,51

TURNER KENNETH C DR
NTL SCIENCE FOUNDATION
DIV ASTRONOMICAL SCIENCES
1800 G ST NW
WASHINGTON DC 20550
USA
TEL
TLF
TLX
EML KTURNER@NOTE NSF GOV
COM 05C,34,40,51

TURNER MARTIN J L DR
DPT OF PHYSICS
UNIVERSITY OF LEICESTER
X-RAY ASTRONOMY GROUP
LEICESTER LE1 7RH
UK
TEL 533 554 455
TLF
TLX 341664 LUXRAY G
EML
COM

TURNER MICHAEL S
ASTRONOMY & ASTROPHYS CTR
UNIVERSITY OF CHICAGO
5640 S ELLIS AVE
CHICAGO IL 60637
USA
TEL 312 962 7974
TLF
TLX 6871133 UNCGO VW
EML
COM 47

TURON C DR
OBSERVATOIRE DE PARIS
SECTION DE MEUDON
DASGAL
F 92195 MEUDON PPL CDX
FRANCE
TEL 1 45 07 7837
TLF 1 45 07 7878
TLX.
EML TURON@FRMEU51/MESIOA TURON
COM 08,24VP,33

TURTLE A J DR
DPT OF PHYSICS
UNIVERSITY OF SYDNEY
SYDNEY NSW 2006
AUSTRALIA
TEL 2 692 2222
TLF
TLX 26169 UNISYD
EML
COM 40

TUTUKOV A V DR
INST OF ASTRONOMY
ACADEMY OF SCIENCES
PYATNITSKAYA UL 48
109017 MOSCOW
RUSSIA
TEL 231-54-61
TLF
TLX 412623 SCSTP SU
EML
COM 27,35C,42C

TWAROG BRUCE A
DPT PHYSICS & ASTRONOMY
UNIVERSITY OF KANSAS
LAWRENCE KS 66045
USA
TEL 913 864 5163
TLF
TLX
EML
COM 37

TWISS R Q DR
C/O A R BOSCHI
96A HOLLAND RDOSCHI
LONDON W14 8BD
UK
TEL
TLF
TLX
EML
COM

TWORKOWSKI ANDRZEJ S
SCHOOL OF MATHEMATICAL SC
QUEEN MARY/WESTFIELD COLL
MILE END RD
LONDON E1 4NS
UK
TEL 1 980 4822
TLF
TLX 893750
EML
COM

TYLENDA ROMUALD DR
INSTITUTE OF ASTRONOMY
N COPERNICUS UNIVERSITY
UL CHOPINA 12/18
PL 87 100 TORUN
POLAND
TEL 26018 x10
TLF
TLX 552234 ASTR PL
EML
COM 27

TYLER JR G LEONARD DR
RADAR ASTRONOMY INST
STANFORD UNIVERSITY
STANFORD CA 94305 4035
USA
TEL 415 497 3535
TLF
TLX
EML
COM 16

TYLKA ALLAN J DR
NAVAL RESEARCH LABORATORY
CODE 4154 GAMMA & COSMIC
RAY ASTROPHYSICS BRANCH
WASHINGTON DC 20375 5000
USA
TEL 202 767 2200
TLF 202 767 6473
TLX
EML 11335 TYLKA
COM 48

TYSON JOHN A DR
BELL LABS
RM 1D-316
600 MOUNTAIN AVE
MURRAY HILL NJ 07974
USA
TEL 201 582 6028
TLF
TLX 138650 BELL LABS MUH
EML
COM 21,28,47

TYTLER DAVID DR
CASS
UCSD
C 011
LA JOLLA CA 92093 0216
USA
TEL 619 534 3460
TLF
TLX 5106010681 CASS UQ
EML
COM 47

TZIOUMIS ANASTASIOS DR
AAO
ATNF
BOX 76
EPPING NSW 2121
AUSTRALIA
TEL 2 868 0222
TLF 2 868 0310
TLX 26230 ASTRO
EML ATZIOUMI@ATNF CSIRO AU
COM 40

UBERTINI PIETRO
IAS
CNR
CP 67
I 00044 FRASCATI
ITALY
TEL 6 942 5132
TLF
TLX 610261 CNRFRA
EML
COM

UCHIDA JUICHI DR
TOHOKU GAKUEN UNIVERSITY
TAGAYO UNIVERSITY
MIYAGI 985
JAPAN
TEL
TLF
TLX
EML
COM 35

UCHIDA YUTAKA PROF
DPT OF ASTRONOMY
UNIVERSITY OF TOKYO
BUNKYO KU
TOKYO 113
JAPAN
TEL
TLF
TLX
EML
COM 12

UDALSKI ANDRZEJ DR
ASTRONOMICAL OBSERVATORY
WARSAW UNIVERSITY
AL UJAZDOWSKIE 4
PL 00 478 WARSAW
POLAND
TEL 294011
TLF
TLX 817063 OAUW PL
EML VDALSKI@PLWAVW61 BITNET
COM

UDAL'TSOV V A DR
LEBEDEV PHYSICAL INST
ACADEMY OF SCIENCES
LENINSKY PROSPEKT 53
117924 MOSCOW
RUSSIA
TEL 135 85 60
TLF
TLX 411479 NEOD SU
EML
COM 40

UDAYA SHANKAR N DR
RAMAN RESEARCH INSTITUTE
SADASHIVANAGAR
BANGALORE 560 080
INDIA
TEL 812 36 0122
TLF 812 34 0492
TLX 0845 2671 RRI IN
EML RRI@VIGYAN ERNET IN
COM 40

UENO SUEO PROF
KANAZAWA TECHNOLOGY INST
7-1 OGIGAOKA
NONOICHIMACHI
ISHIKAWA 921
JAPAN
TEL 0762-48-1100
TLF
TLX 5122456 KIY LCJ
EML
COM 36

UESUGI AKIRA DR
DPT OF ASTRONOMY
KYOTO UNIVERSITY
KITASHIRAKAWA SAKYO KU
KYOTO 606
JAPAN
TEL
TLF
TLX
EML
COM 05,36

UKITA NOBUHARU
NOBEYAMA RADIO OBS
NAOJ
MINAMIMAKI MURA
NAGANO 384 13
JAPAN
TEL 267-98-2831
TLF
TLX 3329005
EML
COM 40

ULFBECK OLE DR
NIELS BOHR INSTITUTE
BLEGDAMSVEJ 17
DK 2100 COPENHAGEN O
DENMARK
TEL 31 42 1616
TLF 31 38 9157
TLX
EML
COM

ULICH BOBBY LEE
KAMAN AEROSPACE CORP
5055 E BROADWAY BLVD
TUCSON AZ 85711
USA
TEL 602 748 2038
TLF
TLX
EML
COM 09

ULMER MELVILLE P PROF
DPT PHYSICS & ASTRONOMY
NORTHWESTERN UNIVERSITY
DEARBORN OBSERVATORY
EVANSTON IL 60208
USA
TEL 312 491 5633
TLF
TLX 9102310040
EML
COM

ULMSCHNEIDER PETER PROF
INST FUER THEORETISCHE
ASTROPHYSIK
IM NEUENHEIMER FELD 561
D 6900 HEIDELBERG 1
GERMANY
TEL 62 215 62837
TLF
TLX 461515 UNIHD D
EML I98@DHDURZ1
COM 36

ULRICH BRUCE T PROF
HILTENSPERGERSTR 93
D 8000 MUENCHEN 40
GERMANY
TEL
TLF
EML
COM 25,40
TLX

ULRICH MARIE-HELENE D DR
ESO
KARL-SCHWARSZCHILD-STR 2
D 8046 GARCHING MUENCHEN
GERMANY
TEL 89 320 06229
TLX 52828222 EOD
EML
TLF 89 320 2362
COM 28C,34,40

ULRICH ROGER K PROF
DPT OF ASTRONOMY
UNIVERSITY OF CALIFORNIA
8931 MSB
LOS ANGELES CA 90024
USA
TEL 213 825 4270
TLF
TLX
EML
COM 35

UMEMURA MASAYUKI DR
TOKYO ASTRONOMICAL OBS
NAOJ
OSAWA MITAKA
TOKYO 181
JAPAN
TEL 81 422 41 3731
TLF 81 422 41 3746
TLX 028 22307 TAOMTK J
EML UMEMURA@UME MTK NAO AC JP
COM 47

UMLENSKI VASIL
DPT OF ASTRONOMY
BULGARIAN ACAD SCIENCES
72 LENIN BLVD
BG 1784 SOFIA
BULGARIA
TEL 2 75 8927
TLF
TLX 23561 ECF BAN BG
EML
COM

UNDERHILL ANNE B DR
4696 WEST 10TH AVE #301
VANCOUVER BC V6R 2J5
CANADA
TEL 604 224 3552
TLF
EML
COM 29,36,44
TLX

UNDERWOOD JAMES H DR
LAWRENCE BERKELEY LAB
UNIVERSITY OF CALIFORNIA
X RAY OPTICS LAB 80 101
BERKELEY CA 94720
USA
TEL 415 486 4958
TLF
TLX 9103662037
EML
COM 10,44

UNGER STEPHEN DR
ROYAL GREENWICH OBS
HERSTMONCEUX CASTLE
HAILSHAM
EAST SUSSEX BN27 1RP
UK
TEL 323 833 171
TLF
TLX 87451
EML UK AC RGO STAR
COM 28,40

UNNO WASABURO PROF
RES INST SCIENCE & TECH
KINKI UNIVERSITY
HIGASHI
OSAKA 577
JAPAN
TEL 06-721-2332
TLF
TLX
EML
COM 12,35,36

UNSOELD ALBRECHT PROF
INST THEOR PHYS & STERNW
NEUE UNIV PHYSIK ZENTRUM
OLSHAUSENST GEB N 61C
D 2300 KIEL 1
GERMANY
TEL 431 880 4205
TLF
TLX
EML
COM

UNWIN STEPHEN C
CALTECH
MS 105 24
PASADENA CA 91125
USA
TEL 213 356 4973
TLF
TLX 675425
EML
COM 40

UOMOTO ALAN K DR
DPT PHYSICS & ASTRONOMY
JOHNS HOPKINS UNIVERSITY
CHARLES & 34TH ST
BALTIMORE MD 21218
USA
TEL 301 338 8594
TLF
TLX
EML AU@STSCI BITNET
COM

UPGREN ARTHUR R DR
DPT OF ASTRONOMY
VAN VLECK OBSERVATORY
WESLEYAN UNIVERSITY
MIDDLETOWN CT 06457
USA
TEL 203 347 9411
TLF
TLX
EML
COM 24,26,33,37,45,50

UPSON WALTER L II DR
PRINCETON UNIVERSITY OBS
PEYTON HALL
PRINCETON NJ 08544
USA
TEL
TLF
TLX
EML
COM 44

UPTON E K L DR
DPT OF ASTRONOMY
UNIVERSITY OF CALIFORNIA
LOS ANGELES CA 90024
USA
TEL
TLF
TLX
EML
COM

URAS SILVANO DR
ISTITUTO DI ASTRONOMIA

VIA OSPEDALE 72
I 09100 CAGLIARI
ITALY
TEL 70 71 1246
TLF
TLX 790326 OSSAST I
EML
COM

URASIN LIRIK A DR
ENGELHARDT ASTRONOMICAL
OBSERVATORY
OBSERVATORIA STATION
422526 KAZAN
RUSSIA
TEL 32-48-27
TLF
TLX
EML
COM 34

URBANIK MAREK DR
ASTRONOMICAL OBSERVATORY
JAGIELLONIAN UNIVERSITY
UL ORLA 171
PL 30 244 KRAKOW
POLAND
TEL 4812-221877
TLF
TLX 0322297 UJ PL
EML
COM 28

URBARZ H DR
ASTRONOMISCHES INSTITUT
DER UNIVERSITAET
AUSSENSTELLE WEISSENAU
D 7980 RAVENSBURG/RASTHAL
GERMANY
TEL 751 61621
TLF
TLX
EML
COM

URECHE VASILE DR
FAC OF MATHEMATICS
UNIVERSITY OF CLUJ-NAPOCA
STR M KOGALNICEANU 1
R 3400 CLUJ NAPOCA
RUMANIA
TEL 951-16101/11592
TLF
TLX
EML
COM 25,42

URPO SEPPO I
HELSINKI UNIV TECHNOLOGY
RADIO LABORATORY
OTAKAARI 5 A
SF 02150 ESPOO 15
FINLAND
TEL. 451 2548
TLF
TLX 122771 RORTA SF
EML
COM 10,40

URRY CLAUDIA MEGAN DR
STSCI
HOMEWOOD CAMPUS
3700 SAN MARTIN DR
BALTIMORE MD 21218
USA
TEL 301 338 4593
TLF
TLX· 6849101 STSCI
EML SCIVAX CMU/CMU@STSCI BITNET
COM

USHER' PETER D DR
DPT OF ASTRONOMY
PENNSYLVANIA STATE UNIV
507 DAVEY LAB
UNIVERSITY PARK PA 16802
USA
TEL 814 865 3509
TLF
TLX 842510 PENNSTBSTRCG
EML
COM 27

USON JUAN M DR
NRAO
VLA
BOX O
SOCORRO NM 87801 0387
USA
TEL 505 835 7237
TLF
TLX 910 988 1710
EML
COM 40,47

USOWICS JERZY BOGDAN DR
INST OF RADIO ASTRONOMY
N COPERNICUS UNIVERSITY
UL CHOPINA 12/18
PL 87 100 TORUN
POLAND
TEL
TLF
TLX
EML·
COM

UTSUMI KAZUHIKO DR
DPT OF ASTRONOMY
HIROSHIMA UNIVERSITY
HIGASHI SENDA MACHI
HIROSHIMA 730
JAPAN
TEL 082-241-1221
TLF
TLX
EML
COM 29

UUS UNDO DR
TARTU ASTROPHYSICAL OBS
ESTONIAN ACAD OF SCIENCES
202444 TARTU
ESTONIA
TEL
TLF
TLX
EML
COM 12,35

VAGER ZEEV DR
DPT OF PHYSICS
WEIZMANN INSTITUTE OF SCI
BOX 26
REHOVOT 76100
ISRAEL
TEL
TLF
TLX
EML
COM

VAGHI SERGIO DR
ESA/ESTEC
PHA
BOX 299
NL 2200 AG NOORDWIJK
NETHERLANDS
TEL 17 19 83453
TLF
TLX 39098
EML
COM 20

VAGNETTI FAUSTO
DPT DI FISICA
II UNIVERSITA DI ROMA
VIA ORAZIO RAIMONDO
I 00173 ROMA
ITALY
TEL 6 24 990431
TLF 6 249 90300
TLX 626382 FIUNTV I
EML VAGNETTI@VAXTOV INFN IT
COM 47

VAIANA GIUSEPPE S DR
OSS ASTRONOMICO
UNIVERSITA DI PALERNO
PALAZZO DEI NORMANNI
I 90134 PALERMO
ITALY
TEL 91 42 2588
TLF
TLX
EML
COM

VAIDYA P C PROF
34\SHARDA NAGA
PALDI
AHMEDABAD 380 007
INDIA
TEL 272 41 3322
TLF
TLX
EML
COM 47

VAINSTEIN L A DR
LEBEDEV PHYSICAL INST
ACADEMY OF SCIENCES
LENINSKY PROSPEKT 53
117924 MOSCOW
RUSSIA
TEL 135-22-50
TLF
TLX
EML
COM 49

VAKILI FARROKH DR
OCA OBSERV DU CALERN
CAUSSOLS
F 06460 S VALLIER THIEY
FRANCE
TEL 93 42 6270
TLF 93 09 2613
TLX 461402 F
EML VAKILI@FRONI51
COM 09,36

VALBOUSQUET ARMAND DR
OBS DE STRASBOURG
11 RUE UNIVERSITE
F 67000 STRASBOURG
FRANCE
TEL 88 35 4300
TLF 88 25 0160
TLX
EML
COM 24,26,51

VALENTIJN EDWIN A DR
KAPTEYN ASTRONOMICAL INST

BOX 800
NL 9700 AV GRONINGEN
NETHERLANDS
TEL 50 11 6695
TLF
TLX
EML
COM 28

VALIRON PIERRE DR
OBSERVATOIRE DE GRENOBLE
CERMO/ASTROPHYSIQUE
BP 53X
F 38041 S MARTIN HERES CD
FRANCE
TEL 76 51 4787
TLF 76 44 8821
TLX 980753 IRAM F
EML BITNET VALIRON@FRGAG51
COM

VALLEE JACQUES P DR
HERZBERG INST ASTROPHYS
NTL RESEARCH COUNCIL
100 SUSSEX DR
OTTAWA ON K1A OR6
CANADA
TEL 613 993 6060
TLF 613 952 6602
TLX 053 3715
EML
COM 40,51

VALNICEK BORIS DR
ASTRONOMICAL INSTITUTE
CZECH ACADEMY OF SCIENCES
ONDREJOV OBSERVATORY
CS 251 65 ONDREJOV
CZECHOSLOVAKIA
TEL 204 85201
TLF 204 85314
TLX 121579
EML
COM 09,10,44

VALSECCHI GIOVANNI B DR
IAS
REPARTO DI PLANETOLOGIA
VIA DELL'UNIVERSITA 11
I 00185 ROMA
ITALY
TEL 6 44 56951
TLF
TLX CNR FRA 610261
EML BITNET GIOVANNI@IRMIAS
COM 07,15,20

VALTAOJA ESKO
TURKU UNIVERSITY OBS
TUORLA
SF 21500 PIIKKIO
FINLAND
TEL 21 43 5822
TLF 21 43 3767
TLX 62638 TYF
EML
COM 40

VALTAOJA LEENA DR
TURKU UNIVERSITY OBS
TUORLA
SF 21500 PIIKKIO
FINLAND
TEL 21 43 5822
TLF 21 43 3767
TLX
EML LVALTAOJA@KONTU UTU FI
COM

VALTIER JEAN-CLAUDE DR
OCA OBSERV DE NICE
BP 139
F 06003 NICE CDX
FRANCE
TEL 93 89 0420
TLF
TLX 460004 OBSNICE F
EML
COM 27,29

VALTONEN MAURI J PROF
TURKU UNIVERSITY OBS
TUORLA
SF 21500 PIIKKIO
FINLAND
TEL 21 43 5822
TLF 21 43 3767
TLX 62638 TYF SF
EML
COM 07,09,26,28,33,40,44

VAN AGT S L TH J DR
STERRENKUNDIG INSTITUTE
KATHOLIEKE UNIVERSITEIT
TOERNOOIVELD
NL 6525 ED NIJMEGEN
NETHERLANDS
TEL 80 55 8833
TLF
TLX 48228 WINAT NL
EML
COM 27

VAN ALBADA TJEERD S DR
KAPTEYN ASTRONOMICAL INST
BOX 800
NL 9700 AV GRONINGEN
NETHERLANDS
TEL 50 11 6695
TLX 53572 STARS NL
EML
TLF
COM 28

VAN ALLEN JAMES A PROF
DPT PHYSICS & ASTRONOMY
UNIVERSITY OF IOWA
IOWA CITY IA 52242
USA
TEL 319 353 4531
TLF
TLX
EML
COM 10,16,21,49

VAN ALTENA WILLIAM F PROF
YALE UNIVERSITY OBS

BOX 6666
NEW HAVEN CT 06511
USA
TEL 203 436 8318
TLF 203 432 5048
TLX 910 250 8365
EML VANALTEN@YALASTRO
COM 08,24C,26,37

VAN BEEK FRANK PROF DR
DPT OF MECHAN ENGINEERING
TECHNICAL UNIV OF DELFT
MEKELWEG 2
NL 2628 CD DELFT
NETHERLANDS
TEL 15 78 5396
TLF
TLX
EML
COM 44

VAN BLERKOM DAVID J PROF
DPT OF ASTRONOMY
UNIV OF MASSACHUSETTS
AMHERST MA 01002
USA
TEL
TLF
TLX
EML
COM

VAN BREDA IAN G DR
ROYAL GREENWICH OBS
HERSTMONCEUX CASTLE
HAILSHAM BN27 1RP
UK
TEL 323 833 171
TLF
TLX 87451
EML
COM

VAN BREUGEL WIL
RADIO ASTRONOMY LAB
UNIVERSITY OF CALIFORNIA
601 CAMPBELL HALL
BERKELEY CA 94720
USA
TEL 415 642 5275
TLF
TLX
EML
COM

VAN BUEREN HENDRIK G PROF
MEIDOORNLAAN 13
NL 3461 ES LINSCHOTEN
NETHERLANDS
TEL 34 80 15406
TLF
EML
COM
TLX

VAN CITTERS GORDON W DR
NTL SCIENCE FOUNDATION
DIV ASTRONOMICAL SCIENCES
1800 G ST NW
WASHINGTON DC 20550
USA
TEL
TLF
TLX
EML
COM 09

VAN DE HULST H C PROF DR
STERREWACHT
BOX 9513
NL 2300 RA LEIDEN
NETHERLANDS
TEL 71 14 8333
TLF
TLX 39058
EML
COM 21,34,40,44

VAN DE KAMP PETER
AMSTEL 244

NL 1017 AK AMSTERDAM
NETHERLANDS
TEL 20 22 3377
TLF
TLX
EML
COM 24,26,51

VAN DE STADT HERMAN DR
SPACE RESEARCH DPT
BOX 800
NL 9700 AV GRONINGEN
NETHERLANDS
TEL 50 11 6695
TLX 53572
EML
TLF
COM

VAN DEN BERGH SIDNEY PROF
HERZBERG INST ASTROPHYS
DOMINION ASTROPHYS OBS
5071 W SAANICH RD
VICTORIA BC V8X 4M6
CANADA
TEL 604 388 3924
TLF 604 363 0045
TLX 049 7295
EML
COM 28,37,50

VAN DEN HEUVEL EDWARD P J
ASTRONOMICAL INSTITUTE
UNIVERSITY OF AMSTERDAM
KRUISLAAN 403
NL 1098 SJ AMSTERDAM
NETHERLANDS
TEL 20 52 57491
TLF 20 52 57484
TLX 10262 HEF NL
EML
COM 35,38,42,48

VAN DEN OORD BERT H J DR
STERREKUNDIG INSTITUTE

BOX 80000
NL 3500 TA UTRECHT
NETHERLANDS
TEL 30 53 5200
TLF 30 53 1601
TLX 40048 FYLUT
EML OORD@FYS RUU NL
COM 10

VAN DER BORGHT RENE PROF
31 THE PROMENADE
ISLE OF CAPRI
SURFERS PARADISE 4217
AUSTRALIA
TEL 385712
TLX
EML
TLF
COM 35

VAN DER HUCHT KAREL A DR
SPACE RESEARCH LABORATORY
SRON
SORBONNELAAN 2
NL 3584 CA UTRECHT
NETHERLANDS
TEL 30 53 5600
TLF 30 54 0860
TLX 47224
EML
COM 26,29,44

VAN DER HULST JAN M DR
KAPTEYN ASTRONOMICAL INST
BOX 800
NL 9700 AV GRONINGEN
NETHERLANDS
TEL 50 63 4054
TLF 50 63 4033
TLX 53572 STARZ NL
EML VDHULST@HGRRUG5
COM 28,40C

VAN DER KLIS MICHIEL DR
ASTRONOMICAL INSTITUTE
UNIVERSITY OF AMSTERDAM
KRUISLAAN 403
NL 1098 SJ AMSTERDAM
NETHERLANDS
TEL 20 52 57498/7491/7492
TLF 20 52 57484
TLX 10262 HEF NL
EML EARN/BITNET A41OMVDK@HASAI
COM

VAN DER KRUIT PIETER C DR
KAPTEYN ASTRONOMICAL INST
BOX 800
NL 9700 AV GRONINGEN
NETHERLANDS
TEL 50 63 4073
TLX 53572 STARS NL
EML
TLF
COM 28,33,40

VAN DER LAAN H PROF DR
ESO
KARL SCHWARZSCHILDSTR 2
D 8046 GARCHING MUENCHEN
GERMANY
TEL 89 320 06227
TLF 89 320 2362
TLX 5 282 8220 EO D
EML HVDLAAN@ESOMCO HQ ESO ORG
COM 28,34,40,47

VAN DER RAAY HERMAN B
DPT OF PHYSICS
UNIVERSITY OF BIRMINGHAM
BOX 363
BIRMINGHAM B15 2TT
UK
TEL 214 72 1301
TLF
TLX 228938 SPAPHY G
EML
COM 35

VAN DER WALT D J DR
DPT OF PHYSICS
POTCHEFSTROOM UNIVERSITY
POTCHEFSTROOM
SOUTH AFRICA
TEL 27 148 99 2408
TLF 27 148 99 2421
TLX 3 46019 SA
EML FISJUDW@PUKVM1 PUK AC.ZA
COM 48

VAN DESSEL EDWIN LUDO DR
OBSERVATOIRE ROYAL DE
BELGIQUE
AVE CIRCULAIRE 3
B 1180 BRUSSELS
BELGIUM
TEL 2 673 5366
TLF 2 374 9822
TLX 21565 OBSBEL
EML
COM· 26C,30

VAN DIGGELEN J DR
OBSERVATORY UTRECHT
AETSVELDSELAAN 12
NL 1381 EA WEESP
NETHERLANDS
TEL
TLF
TLX
EML
COM

VAN DISHOECK EWINE F DR
STERREWACHT
BOX 9513
NL 2300 RA LEIDEN
NETHERLANDS
TEL 71 27 5874/5835
TLF 71 27 5819
TLX 39058 ASTRO NL
EML
COM 34

VAN DORN BRADT HALE DR
CENTER FOR SPACE RESEARCH
MIT RM 37 581
BOX 165
CAMBRIDGE MA 02139
USA
TEL 617 253 7550
TLF
TLX 921473 MITCAM
EML
COM

VAN DRIEL WILLEM DR
ASTRONOMICAL INSTITUTE
UNIVERSITY OF AMSTERDAM
KRUISLAAN 403
NL 1098 SJ AMSTERDAM
NETHERLANDS
TEL 20 52 57488
TLF 20 52 57404
TLX 10262 HEFNL
EML VANDRIEL@HUTRUU51 BITNET
COM 28

VAN DRIEL-GESZTELYI L DR
STERREKUNDIG INSTITUTE
BOX 80000
NL 3508 TA UTRECHT
NETHERLANDS
TEL
TLX
EML
TLF
COM 10

VAN DUINEN R J DR
FOKKER BV
SPACE DIVISION
BOX 7600
NL 1117 ZJ SCHIPHOL
NETHERLANDS
TEL 20 54 42030
TLF
TLX 12227
EML
COM 44

VAN FLANDERN THOMAS DR
VF ASSOCIATES

6327 WESTERN AVE NW
WASHINGTON DC 20015
USA
TEL 202 363 3860
TLF
TLX
EML
COM 04,15,16,20,51

VAN GENDEREN A M DR
STERREWACHT
POSTBUS 9513
NL 2300 RA LEIDEN
NETHERLANDS
TEL 71 27 2727
TLX 31476 ASTRO NL
EML
TLF
COM 27,28

VAN GORKOM JACQUELINE H
NRAO
BOX 0
SOCORRO NM 87801 0387
USA
TEL 505 772 4302
TLX 910-997-0174
EML
TLF
COM 28,34,40

VAN GRONINGEN ERNST DR
ASTRONOMICAL OBSERVATORY
BOX 515
S 751 20 UPPSALA
SWEDEN
TEL
TLF
TLX
EML
COM

VAN HAMME WALTER
DPT PHYSICS & ASTRONOMY
FLORIDA ITL UNIVERSITY
UNIVERSITY PARK
MIAMI FL 33199
USA
TEL
TLF
TLX
EML
COM 42

VAN HERK G
STERREWACHT
POSTBUS 9513
NL 2300 RA LEIDEN
NETHERLANDS
TEL 71 27 2727
TLX
EML
TLF
COM

VAN HORN HUGH M PROF
DPT PHYSICS & ASTRONOMY
UNIVERSITY OF ROCHESTER
ROCHESTER NY 14627
USA
TEL 716 275 4344
TLF
TLX
EML
COM 35

VAN HOUTEN C J DR
STERREWACHT
POSTBUS 9513
NL 2300 RA LEIDEN
NETHERLANDS
TEL 71 27 2727
TLF
TLX 39058 ASTRO NL
EML
COM 20

VAN HOUTEN-GROENEVELD I
STERREWACHT

POSTBUS 9513
NL 2300 RA LEIDEN
NETHERLANDS
TEL 71 27 2727
TLF
TLX 39058 ASTRO NL
EML
COM 20

VAN HOVEN GERARD DR
DPT OF PHYSICS
UNIVERSITY OF CALIFORNIA
IRVINE CA 92717
USA
TEL 714 856 5145
TLF
TLX 683322 IRIN
EML
COM 10,12

VAN LEEUWEN FLOOR DR
ROYAL GREENWICH OBS
HERSTMONCEUX CASTLE
HAILSHAM
EAST SUSSEX BN27 1RP
UK
TEL
TLF
TLX
EML
COM 08

VAN MOORSEL GUSTAAF DR
ESO
KARLSCHWARZSCHILD STR 2
D 8046 GARCHING MUENCHEN
GERMANY
TEL 89 320 06362
TLF 89 320 2362
TLX 52828222 EO D
EML
COM 28

VAN NIEUWKOOP J DR IR
PRINSESSELAAN 12

NL 7316 CN APELDOORN
NETHERLANDS
TEL
TLF
TLX
EML
COM 40

VAN PARADIJS JOHANNES DR
ASTRONOMICAL INSTITUTE
UNIVERSITY OF AMSTERDAM
KRUISLAAN 403
NL 1098 SJ AMSTERDAM
NETHERLANDS
TEL 20 52 57491
TLF 20 52 57484
TLX 10262 HEF NL
EML
COM 42

VAN REGEMORTER HENRI DR
OBSERVATOIRE DE PARIS
SECTION DE MEUDON
F 92195 MEUDON PPL CDX
FRANCE
TEL 1 45 07 7444
TLF
TLX 201571
EML
COM 14,36

VAN RENSBERGEN WALTER DR
IAAG
VRIJE UNIV BRUSSELS
CP 165
B 1050 BRUSSELS
BELGIUM
TEL 2 641 3497
TLF
TLX   ,
EML
COM 14

VAN RIPER KENNETH A DR
LOS ALAMOS NATIONAL LAB
MS B 226 X 6
BOX 1663
LOS ALAMOS NM 87545
USA
TEL 505 667 8104
TLF
TLX
EML
COM 35,48

VAN SPEYBROECK LEON P DR
CENTER FOR ASTROPHYSICS
HCO/SAO
60 GARDEN ST
CAMBRIDGE MA 02138
USA
TEL 617 495 7233
TLF
TLX
EML
COM 44

VAN WOERDEN HUGO PROF DR
KAPTEYN ASTRONOMICAL INST
BOX 800
NL 9700 AV GRONINGEN
NETHERLANDS
TEL 50 11 6695
TLX 53572 STARS NL
EML
TLF
COM 28,33,34,40

VANDEN BOUT PAUL A
NRAO
EDGEMONT RD
CHARLOTTESVILLE VA 22903
USA
TEL 804 296 0241
TLF
TLX 910-997-0174
EML
COM 34,40

VANDENBERG DON DR
DPT OF PHYSICS
UNIVERSITY OF VICTORIA
BOX 1700
VICTORIA BC V8W 2Y2
CANADA
TEL· 604 721 7739
TLF 604 721 7715
TLX
EML
COM 35C,37C

VANDERVOORT PETER O DR
ASTRONOMY & ASTROPHYS CTR
UNIVERSITY OF CHICAGO
5640 S ELLIS AVE
CHICAGO IL 60637
USA
TEL 312 962 8209
TLF
TLX
EML
COM 33

VANYSEK VLADIMIR PROF
DPT OF ASTRONOMY
CHARLES UNIVERSITY
SVEDSKA 8
CS 150 00 PRAHA 5
CZECHOSLOVAKIA
TEL 2 54 0395
TLF
TLX 121673 MFF
EML
COM 15,34,47

VAN'T VEER FRANS DR
OBSERVATOIRE DE PARIS
61 AVE OBSERVATOIRE
F 75014 PARIS
FRANCE
TEL 1 40 51 2221
TLF
TLX 270776 OBS P
EML
COM 10,36,42

VAN'T VEER-MENNERET CL DR
OBSERVATOIRE DE PARIS
61 AVE OBSERVATOIRE
F 75014 PARIS
FRANCE
TEL 1 40 51 2249
TLF
TLX
EML
COM 29,36

VAPILLON LOIC J DR
OBSERVATOIRE DE PARIS
SECTION DE MEUDON
F 92195 MEUDON PPL CDX
FRANCE
TEL 1 45 07 7623
TLF
TLX 201571
EML
COM

VARDANIAN R A DR
BYURAKAN ASTROPHYSICAL
OBSERVATORY
378433 ARMENIA
ARMENIA
TEL 88 52 28 4142
TLF
TLX
EML
COM 25

VARDAVAS ILIAS MIHAIL
DPT OF PHYSICS
UNIVERSITY OF CRETE
BOX 1527
GR 711 11 IRAKLION
GREECE
TEL 81 23 6589
TLF 02 416 7902
TLX 262728
EML
COM 36

VARDYA M S DR
TIFR
HOMI BHABHA RD
COLABA
BOMBAY 400 005
INDIA
TEL 22 219 111*221
TLF
TLX 011-3009 TIFR IN
EML
COM 35,36

VARMA RAM KUMAR PROF
PHYSICAL RESEARCH LAB
NAVRANGPURA
AHMEDABAD 380 009
INDIA
TEL 272 46 2129
TLF 272 44 5292
TLX 0121397 PRL IN
EML
COM 28

VARSHALOVICH DIMITRIJ PR
IOFFE PHYSICAL TECH INST
ACADEMY OF SCIENCES
POLYTECHNICHESKAYA UL 26
194021 ST PETERSBURG
RUSSIA
TEL 247-22-55
TLF
TLX
EML
COM 14,34,51

VARVOGLIS H DR
DPT OF ASTRONOMY
UNIVERSITY THESSALONIKI
GR 540 06 THESSALONIKI
GREECE
TEL 31 99 1357
TLF
TLX 412181
EML
COM 07

VASHKOV'YAK SOF'YA N DR
STERNBERG STATE ASTR INST

UNIVERSITETSKIJ PROSP 13
119899 MOSCOW
RUSSIA
TEL 139-37-64
TLF
TLX 113037 JAPET
EML
COM 07

VASILEVA GALINA J DR
PULKOVO OBSERVATORY
ACADEMY OF SCIENCES
10 KUTUZOV QUAY
196140 ST PETERSBURG
RUSSIA
TEL
TLF
TLX
EML
COM 12

VASU-MALLIK SUSHMA DR
INDIAN INSTITUTE OF
ASTROPHYSICS
KORAMANGALA
BANGALORE 560 034
INDIA
TEL 812 56 9179/9180
TLF
TLX 845763 IIAB IN
EML
COM 29,36

VATS HARI OM DR
PHYSICAL RESEARCH LAB
NAVRANGUVA
AHMEDABAD 9
INDIA
TEL 272 46 2129
TLF 272 44 5292
TLX, 01216397
EML IPS@PRL ERNET IN
COM 40

VAUCLAIR GERARD P DR
OBS MIDI PYRENEES
14 AVE E BELIN
F 31400 TOULOUSE CDX
FRANCE
TEL 61 25 2101
TLF
TLX 530776
EML
COM 35C

VAUCLAIR SYLVIE D DR
OBS MIDI PYRENEES
14 AVE E BELIN
F 31400 TOULOUSE CDX
FRANCE
TEL 61 25 2101
TLF
TLX
EML
COM 46

VAUGHAN ALAN DR
SCHOOL OF MATHS/PHYSICS
COMPUTING AND ELECTRONICS
MACQUARIE UNIVERSITY
MACQUARIE 2109
AUSTRALIA
TEL 2 805 8904
TLF 2 805 8983
TLX MACUNI AA 122377
EML ALANV@MACASTRO MPCS MQ OZ AU
COM 40

VAUGHAN ARTHUR H DR
PERKIN-ELMER CORP
7421 ORANGEWOOD AVE
GARDEN GROVE CA 92641
USA
TEL 714 895 1667
TLF
TLX
EML
COM 10,12,25,36

VAUGLIN ISABELLE DR
OBSERVATOIRE DE LYON
AVE CHARLES ANDRE
F 69561 S GENIS LAVAL CDX
FRANCE
TEL 72 39 9098
TLF 72 39 9791
TLX 310926
EML VAUGLIN@CASTOR UNIV-LYON1 FR
COM 28

VAVROVA ZDENKA DR
KLET OBSERVATORY
CESKE BUDEJOVICE
CS BEZRUCOVA 4
CZECHOSLOVAKIA
TEL 33 73274
TLF
TLX
EML
COM 20

VAZ LUIZ PAULO RIBEIRO
OBSERVATORIO ASTRONOMICO
DEPTO DE FISICA-ICEX-UFMG
CP 702
30161 BELO HORIZONTE MG
BRAZIL
TEL 31 441 2541
TLF
TLX 312308 UFMG BR
EML
COM 42

VAZQUEZ MANUEL DR
INST DE ASTROFISICA
DE CANARIAS
OBS DEL TEIDE
E 38071 LA LAGUNA
SPAIN
TEL
TLF
TLX 92640 IACE E
EML
COM 51

VAZQUEZ RUBEN ANGEL DR
OBSERVATORIO ASTRONOMICO
LA PLATA
1900 LA PLATA (BS AS)
ARGENTINA
TEL 21 21 7308
TLF
TLX 31216 CESLA AR
EML RVAZQUEZ@FCAGLP EDU AR
COM 37

VECK NICHOLAS
MARCONI RESEARCH CENTRE
WEST HANNINGFIELD RD
GT BADDOW
CHELMSFORD ESSEX CM2 8HN
UK
TEL 245 733 31
TLF 245 752 44
TLX 995016 GECRES G
EML YE08%a gec-nrc co uk@uel-cs
COM 10

VEEDER GLENN J DR
JPL
MS 183 501
4800 OAK GROVE DR
PASADENA CA 91109
USA
TEL 213 354 7388
TLF
TLX
EML
COM 15

VEGA E IRENE DR
OBSERVATORIO ASTRONOMICO
PASEO DEL BOSQUE
1900 LA PLATA (BS AS)
ARGENTINA
TEL 21 21 7308
TLF
TLX 31151 BULAP AR
EML
COM 33

VEILLET CHRISTIAN
OCA CERGA
AVE COPERNIC
F 06130 GRASSE
FRANCE
TEL 93 36 5849
TLF
TLX 470865
EML
COM 07,19,20,31

VEIS GEORGE PH D
GEODESY LABORATORY
NATL TECHNICAL UNIVERSITY
PANEPISTIMIOPOLIS
GR 157 73 ZOGRAFOS
GREECE
TEL 1 724 3414
TLF
TLX
EML
COM

VEISMANN UNO DR
TARTU ASTROPHYSICAL OBS
ESTONIAN ACAD OF SCIENCES
202444 TARTU
ESTONIA
TEL
TLF
TLX
EML
COM

VEKSTEIN GREGORY DR
INSTITUTE OF NUCLEAR PHYS
ACADEMY OF SCIENCES
SIBERIAN DIV
630090 NOVOSIBIRSK
RUSSIA
TEL 383 2 359407
TLF 383 2 352163
TLX 133116 ATOM SU
EML VEKSTEIN@VXINPA INP NSK SU
COM 10

VELKOV KIRIL
ASTRONOMICAL OBSERVATORY
BULGARIAN ACAD SCIENCES
72 LENIN BLVD
BG 1784 SOFIA
BULGARIA
TEL 2 75 8927
TLF
TLX 23561 ECF BQN BG
EML
COM 09,10

VELLI MARCO DR
OBSERVATOIRE DE PARIS
SECTION DE MEUDON
DESPA
F 92195 MEUDON PPL CDX
FRANCE
TEL 1 45 07 7659
TLF 1 45 07 2806
TLX
EML VELLI@FRMEU51/17710 VELLI
COM 10

VELUSAMY T DR
TIFR/RADIO ASTRONOMY CTR
BOX 8
UDHAGAMANDALAM 643 001
INDIA
TEL 423 2651/2032
TLF
TLX 8458488 TIFR IN
EML
COM 40C

VENKATAKRISHNAN P DR
NASA/MSFC
CODE ES 52
HUNTSVILLE AL 35812
USA
TEL 205 544 9404
TLF
TLX
EML
COM 12

VENKATESAN DORASWAMY DR
DPT OF PHYSICS
UNIVERSITY OF CALGARY
2500 UNIVERSITY DR NW
CALGARY AB T2N 1N4
CANADA
TEL 403 931 2366
TLF 403 289 3331
TLX
EML
COM 10

VENTURA JOSEPH DR
DPT OF PHYSICS
UNIVERSITY OF CRETE
BOX 1527
GR 711 11 IRAKLION
GREECE
TEL 81 23 9757
TLF.
TLX 262728
EML VENTURA@GREARN
COM

VENUGOPAL V R DR
TIFR/RADIO ASTRONOMY CTR
BOX 8
UDHAGAMANDALAM 643 001
INDIA
TEL 423 2651/2032
TLF
TLX 0853-241 RAC IN
EML
COM 33,40,51

VERBEEK PAUL DR
GEORGE MINNELAAN 50
B 9830 S MARTENS-LATEM
BELGIUM
TEL 9 82 6119
TLF
TLX
EML
COM

VERBUNT FRANCISCUS DR
STERREKUNDIG INSTITUTE
BOX 80000
NL 3508 TA UTRECHT
NETHERLANDS
TEL 89 32 99833
TLF
TLX 05 21845 XTER D
EML BITNET FWV@DGAIPP1S
COM

VERDET JEAN-PIERRE DR
OBSERVATOIRE DE PARIS
61 AVE OBSERVATOIRE
F 75014 PARIS
FRANCE
TEL 1 40 51 2206
TLF
TLX
EML
COM 41

VERES FERENC
KONKOLY OBSERVATORY
THEGE U 13/17
BOX 67
H 1525 BUDAPEST
HUNGARY
TEL 1 755 866
TLF 1 569 640
TLX 227460
EML
COM

VERGEZ MADELEINE DR
OBS MIDI PYRENEES
9 R PONT DE LA MOUETTE
F 65200 BAGNERES BIGORRE
FRANCE
TEL 62 95 1969
TLF
TLX 531625 OBSPIC F
EML
COM 10

VERGNANO A PROF
OSS ASTRONOMICO DI TORINO
ST OSSERVATORIO 20
I 10025 PINO TORINESE
ITALY
TEL 11 81 1061
TLF 11 84 1281
TLX
EML
COM

VERHEEST FRANK PROF
INST THEORET MECHANIKA
RIJKSUNIVERSITEIT GENT
KRIJGSLAAN 281
B 9000 GENT
BELGIUM
TEL 91 22 5715
TLF 91 24 0634
TLX 12 754 RUGENT B
EML VERHEEST@TEOMECH RUG AC BE
COM 10,27,49

VERMA R P DR
TIFR
HOMI BHABHA RD
COLABA
BOMBAY 400 005
INDIA
TEL 22 219 111
TLF
TLX 0113009 TIFR IN
EML
COM 25

VERMA SATYA DEV DR
DPT PHYSICS & SPACE SCI
UNIVERSITY SCHOOL OF SCI
GUJARAT UNIVERSITY
AHMEDABAD 380 009
INDIA
TEL 272 44 0920
TLF
TLX
EML
COM

VERMA V K DR
UTTAR PRADESH STATE
OBSERVATORY
PO MANORA PEAK 263 129
NAINITAL 263 129
INDIA
TEL 59 42 2136
TLF
TLX. 5942 2401
EML
COM 10

VERNIANI FRANCO PROF
DPT DI FISICA
CNR
VIA IRNERIO 46
I 40126 BOLOGNA
ITALY
TEL 51 26 0991
TLF
TLX 211664
EML
COM 22

VERON MARIE-PAULE DR
OHP
F 04870 S MICHEL OBS
FRANCE
TEL 92 76 6368
TLF
EML
COM 28
TLX 410690 OHP F

VERON PHILIPPE DR
OHP
F 04870 S MICHEL OBS
FRANCE
TEL 92 76 6368
TLF
TLX 410690 OHP F
EML
COM 28,40

VERSCHUUR GERRIT L PROF
4802 BROOKSTONE TERRACE
BOWIE MD 20715
USA
TEL
TLF
TLX
EML
COM 33,34,40,51

VERTER FRANCES DR
NASA/GSFC
CODE 685
GREENBELT MD 20771
USA
TEL 301 286 7860
TLF
TLX
EML
COM 40

VESECKY J F DR
RADAR ASTRONOMY INST
STANFORD UNIVERSITY
233 DURAND
STANFORD CA 94305 4035
USA
TEL
TLF
TLX
EML
COM

VETESNIK MIROSLAV DR
DPT OF ASTROPHYSICS
PURKYNE UNIVERSITY
KOTLARSKA 2
CS 611 37 BRNO
CZECHOSLOVAKIA
TEL 5 74 0500
TLF 5 74 0108
TLX
EML ASTRID@CSPUNI12 EARN
COM 33,42

VETTOLANI GIAMPAOLO
IST DI RADIOASTRONOMIA
CNR
VIA IRNERIO 46
I 40126 BOLOGNA
ITALY
TEL· 51 23 2856
TLF
TLX 211664 INFN BO I
EML
COM 47

VEVERKA JOSEPH DR
CORNELL UNIVERSITY
312 SPACE SCI BLDG
ITHACA NY 14853
USA
TEL 607 256 3507
TLX 937478
EML.
TLF
COM 15,16

VIAL JEAN-CLAUDE
IAS
BP 10
F 91371 VERRIERES BUISSON
FRANCE
TEL 1 64 47 4217
TLF
TLX 600252
EML
COM 10,12,44

VIALA YVES
OBSERVATOIRE DE PARIS
SECTION DE MEUDON
DENIRM
F 92195 MEUDON PPL CDX
FRANCE
TEL 1 45 07 7912
TLF
TLX 270912
EML
COM 34

VIALLEFOND FRANCOIS
OBSERVATOIRE DE PARIS
SECTION DE MEUDON
DENIRM
F 92195 MEUDON PPL CDX
FRANCE
TEL 1 45 07 7905
TLF 1 45 07 7893
TLX 270912
EML BITNET FVIALLEF@FRMEU51
COM 34

VICENTE RAIMUNDO O PROF
FACULDADE CIENCIAS LISBOA
RUA MESTRE AVIZ 30 R/C .
P 1495 LISBOA
PORTUGAL
TEL 2112666
TLX
EML
TLF
COM 19,31

VIDAL JEAN-LOUIS DR
UNIVERSITE MONTPELLIER II
USTL II
PLACE EUGENE BATAILLON
F 34095 MONTPELLIER CDX 5
FRANCE
TEL 67 14 3901
TLF
TLX 490944 USTMONT
EML
COM 34

VIDAL NISSIM V DR
INST FOR SCIENCES & TECHN
5 ARLOZOROV ST
GIVAT SHMUEL 51905
ISRAEL
TEL 3 532 2490
TLF
TLX
EML
COM 48

VIDAL-MADJAR ALFRED DR
INSTITUT D'ASTROPHYSIQUE
98BIS BD ARAGO
F 75014 PARIS
FRANCE
TEL 1 43 20 1425
TLF 1 43 29 8673
TLX
EML
COM 34,44

VIEIRA MARTINS ROBERTO DR
OBSERVATORIO NACIONAL
RUA GL BRUCE 586
SAO CRISTOVAO
20921 RIO DE JANEIRO RJ
BRAZIL
TEL 21 580 7313
TLF 21 580 0332
TLX 021-21288
EML
COM 07,20

VIETRI MARIO DR
OSS ASTRONOMICO
I 00040 MONTE CATONE
ITALY
TEL 6 944 9019
TLF
EML VIETRI@ASTRMP ASTRO IT
COM
TLX

VIGIER JEAN-PIERRE DR
INSTITUT H POINCARE
11 RUE P & M CURIE
F 75005 PARIS
FRANCE
TEL
TLF
TLX
EML
COM

VIGOTTI MARIO
IST DI RADIOASTRONOMIA
CNR
VIA IRNERIO 46
I 40126 BOLOGNA
ITALY
TEL 51 23 2856
TLF
TLX
EML
COM

VIIK TONU DR
TARTU ASTROPHYSICAL OBS
ESTONIAN ACAD OF SCIENCES
202444 TARTU
ESTONIA
TEL 4 1181
TLF
TLX
EML
COM 36

VILA SAMUEL C PROF
DPT OF ASTRONOMY
UNIV OF PENNSYLVANIA
33RD & WALNUT STREETS
PHILADELPHIA PA 19104
USA
TEL 215 898 5994
TLF
TLX
EML
COM 35

VILAS FAITH DR
NASA/JOHNSON SPACE CENTER
CODE SN3
HOUSTON TX 77058
USA
TEL 713 483 5056
TLF
TLX
EML SN VILAS
COM 15,40

VILAS-BOAS JOSE W DR
CRAAE/PTR ESCOLA POLI USP
CP 8174 CEP 05508
01051 SAO PAULO SP
BRAZIL
TEL 11 815 5936
TLF 11 815 6289
TLX 1180127 INPE BR
EML JWDSVBOA@BRUSP BITNET
COM 40

VILCHEZ MEDINA JOSE M DR
INST DE ASTROFISICA
DE CANARIAS
OBS DEL TEIDE
E 38200 LA LAGUNA
SPAIN
TEL 22 26 2211
TLF
TLX 92640
EML
COM.

VILHENA DE MORAES R DR
INPE
CP 515
12200 S JOSE DOS CAMPOS
BRAZIL
TEL 123 22 9088
TLF
TLX 01173437 ZWO-24-73
EML
COM 07

VILHU OSMI DR
OBS & ASTROPHYSICS LAB
UNIVERSITY OF HELSINKI
TAEHTITORNINMAKI
SF 00130 HELSINKI 13
FINLAND
TEL
TLF
TLX
EML
COM 29,35,36,44

VILKKI ERKKI U
4 PARC DE LA LONDE
F 76130 MONT SAINT AIGNAN
FRANCE
TEL
TLF
TLX
EML
COM 24

VILKOVISKIJ EMMANUIL Y DR
ASTROPHYSICAL INSTITUTE
KAZAKH ACAD OF SCIENCES
480068 ALMA ATA
KAZAKHSTAN
TEL
TLF
TLX
EML
COM 35

VILLELA THYRSO NETO DR
INPE
CP 515
12201 S JOSE DOS CAMPOS
BRAZIL
TEL 123 41 8977 *278
TLF
TLX
EML
COM

VILMER NICOLE DR
OBSERVATOIRE DE PARIS
SECTION DE MEUDON
DASOP
F 92195 MEUDON PPL CDX
FRANCE
TEL 1 45 07 7806
TLF
TLX 200590
EML SPAN MEUDON VILMER
COM 10,12

VINCE ISTVAN
ASTRONOMICAL OBSERVATORY
VOLGINA 7
YU 11050 BEOGRAD
YUGOSLAVIA
TEL 11 419 357/421 875
TLF
TLX 72610 AOB YU
EML EAOPOO1@YUBGSS21 BITNET
COM

VINER MELVYN R DR
DPT OF METALL ENGIN
QUEEN'S UNIVERSITY
KINGSTON ON K7L 3N6
CANADA
TEL
TLF 613 545 6463
TLX
EML
COM 34,40

VINLUAN RENATO
UNIVERSITY OF SOUTHERN
PHILIPPINES
OBRERO DAVAO CITY 9501
PHILIPPINES
TEL
TLF
TLX
EML
COM 10

VINOD S KRISHAN MRS DR
INDIAN INSTITUTE OF
ASTROPHYSICS
KORAMANGALA
BANGALORE 560 034
INDIA
TEL 812 56 6585/6497
TLF
TLX 845763 IIAB IN
EML
COM 10,49C

VIOTTI ROBERTO DR
IAS
CNR
CP 67
I 00044 FRASCATI
ITALY
TEL 6 942 5655
TLF 6 941 6847
TLX 610261 CNRFRA
EML VIOTTI@IRMIAS BITNET
COM 27,29,44

VIRGOPIA NICOLA PROF
DPT DI MATEMATICA
UNIV DI·ROMA LA SAPIENZA
CITTA UNIVERSITARIA
I 00185 ROMA
ITALY
TEL
TLF
TLX
EML
COM

VISHNIAC ETHAN T
ASTRONOMY DPT
UNIVERSITY OF TEXAS
RLM 15 308
AUSTIN TX 78712 1083
USA
TEL 512 471 1429
TLF
TLX
EML
COM 47

VISHVESHWARA C V PROF
RAMAN RESEARCH INSTITUTE
SADASHIVANAGAR
BANGALORE 560 080
INDIA
TEL 812 36 0122
TLF 812 34 0492
TLX 8452671 RRI IN
EML
COM 47

VISVANATHAN NATARAJAN DR
MOUNT STROMLO & SIDING
SPRING OBSERVATORIES
PRIVATE BAG
WODEN PO ACT 2606
AUSTRALIA
TEL 62 88 1111
TLF
TLX 62270 TLG CANOPUS AA
EML
COM 25,28,34

VITINSKIJ YURIJ I DR
PULKOVO OBSERVATORY
ACADEMY OF SCIENCES
10 KUTUZOV QUAY
196140 ST PETERSBURG
RUSSIA
TEL 298-22-42
TLF
TLX
EML
COM 10,12

VITON MAURICE DR
LAS
TRAVERSE DU SIPHON
LES TROIS LUCS
F 13012 MARSEILLE
FRANCE
TEL 91 05 5900
TLF 91 66 1855
TLX
EML
COM

VITTONE ALBERTO ANGELO
OSS ASTRONOMICO
DI CAPODIMONTE
VIA MOIARIELLO 16
I 80131 NAPOLI
ITALY
TEL 81 44 0101
TLF
TLX·
EML
COM

VITTORIO NICOLA
ISTITUTO ASTRONOMICO
UNIVERSITA DI ROMA
VIA G M LANCISI 29
I 00161 ROMA
ITALY
TEL
TLF
TLX
EML
COM

VITYAZEV VENEAMIN V DR
ASTRONOMICAL OBSERVATORY
ST PETERSBURG UNIVERSITY
BIBLIOTECHNAJA PL 2
198904 ST PETERSBURG
RUSSIA
TEL
TLF
TLX
EML
COM

VIVEKANAND M DR
RAMAN RESEARCH INSTITUTE
SADASHIVANAGAR
BANGALORE 560 080
INDIA
TEL 812 36 0122
TLF 812 34 0492
TLX 8452671 RRI IN
EML
COM 40

VIVEKANANDA RAO
CASA
UNIVERSITY OF OSMANIA
HYDERABAD 500 007
INDIA
TEL 85 1672
TLF
TLX
EML
COM

VIVES TEODORO JOSE DR
CTR ASTRON HISPANO ALEMAN
REINA 66 9°B
CORREOS 511
E 04002 ALMERIA
SPAIN
TEL 23 0988
TLF
TLX 78812 DSAZ E
EML
COM

VLACHOS DEMETRIUS G PROF
DPT GEODESY & SURVEYING
UNIVERSITY THESSALONIKI
FACULTY OF ENGINEERING
GR 540 06 THESSALONIKI
GREECE
TEL 31 99 1520
TLF
TLX 412181 AUTH GR
EML
COM

VLADILO GIOVANNI DR
OAT
BOX SUCC TRIESTE 5
VIA TIEPOLO 11
I 34131 TRIESTE
ITALY
TEL 40 30 9342
TLF
TLX 461137 OAT I
EML
COM 29

VLADIMIROV SIMEON
ASTRONOMICAL OBSERVATORY
BULGARIAN ACAD SCIENCES
BOX 15
BG 1309 SOFIA
BULGARIA
TEL 2 75 8927
TLF
TLX 23561 ECF BAN BG
EML
COM 09,46

VLAHOS LOUKAS DR
DPT OF ASTROPHYSICS
UNIVERSITY THESSALONIKI
GR 540 06 THESSALONIKI
GREECE
TEL· 31 99 1357
TLF
TLX 0412181 AVTH GR
EML
COM

VOELK HEINRICH J PROF
MPI FUER KERNPHYSIK
POSTFACH 103 980
D 6900 HEIDELBERG 1
GERMANY
TEL 62 215 16295
TLX 461666
EML
TLF
COM 14,48

VOGEL MANFRED DR
INSTITUT FUER ASTRONOMIE
ETH ZENTRUM
CH 8092 ZUERICH
SWITZERLAND
TEL 1 256 3806
TLF 1 262 0003
TLX 817379 EHHG CH
EML VOGEL@CZHETHSA
COM 34

VOGEL STUART NEWCOMBE DR
ASTRONOMY PROGRAM
UNIVERSITY OF MARYLAND
COLLEGE PARK
MARYLAND 20742
USA
TEL
TLF
TLX
EML
COM 40

VOGLIS NIKOS DR
DPT OF ASTROPHYSICS
NTL UNIVERSITY OF ATHENS
PANEPISTIMIOPOLIS
GR 157 83 ZOGRAFOS
GREECE
TEL 1 724 3414
TLF
TLX 223815 UNIV GR
EML node GRATHUN1,userid SPM70
COM 28,47

VOGT NIKOLAUS DR
GRUPO ASTROFIS PONTIFICA
UNIVERSIDAD CATOLICA
CASILLA 104
SANTIAGO 22
CHILE
TEL
TLF
TLX
EML NVOGT@ASTROUC PUC CL
COM 27,29,51

VOGT STEVEN SCOTT
LICK OBSERVATORY
UNIVERSITY OF CALIFORNIA
SANTA CRUZ CA 95064
USA
TEL 408 429 2844
TLF
TLX 910-598-4408
EML
COM 29

VOIGT HANS H PROF
NIKOLAUSBERGER WEG 74
GEISMARLANDSTR 11
D 3400 GOETTINGEN
GERMANY
TEL 551 55879
TLF
TLX
EML
COM

VOLK KEVIN DR
DPT OF PHYSICS & ASTRON
UNIVERSITY OF CALGARY
2500 UNIVERSITY DR NW
CALGARY AB T2N 1N4
CANADA
TEL 403 931 2366
TLF 403 289 3331
TLX
EML
COM 34

VOLLAND H DR
ASTRONOMISCHES INSTITUT
DER UNIVERSITAET
AUF DEM HUEGEL 71
D 5300 BONN 1
GERMANY
TEL 228 73 3674
TLF
TLX 0886440
EML
COM

VOLONTE SERGE DR
ESA
8-10 RUE MARIO NIKIS
F 75738 PARIS CDX 15
FRANCE
TEL
TLF
TLX
EML
COM 12,14

VON BORZESZKOWSKI H H DR
EINSTEIN-LABORATORIUM
AKAD WISSENSCHAFTEN DDR
ROSA-LUXEMBURG-STR 17A
D 1502 POTSDAM
GERMANY
TEL 762225
TLF
TLX
EML
COM 47

VON DER HEIDE JOHANN DR
ALARDUSSTR 12

D 2000 HAMBURG 20
GERMANY
TEL 40 491 4016
TLF
TLX
EML
COM 08

VON HOERNER SEBASTIAN DR
KRUMMENACKERSTR 186
D 7300 ESSLINGEN
GERMANY
TEL
TLF
EML
COM 51
TLX

VON WEIZSAECKER C F PROF
MAXIMILLIANSTR 15
D 8130 STARNBERG
GERMANY
TEL
TLF
EML
COM
TLX

VONDRAK JAN DR
ASTRONOMICAL INSTITUTE
CZECH ACADEMY SCIENCES
BUDECSKA 6
CS 120 23 PRAHA 2
CZECHOSLOVAKIA
TEL 2 25 8757
TLF 2 25 5010
TLX 66 122486
EML ASTOSS@CSEARN
COM 19VP

VORONTSOV-VEL'YAMINOV B A
STERNBERG STATE ASTR INST

117234 MOSCOW
RUSSIA
TEL
TLF
TLX
EML
COM 28,34

VOROSHILOV V I DR
MAIN ASTRONOMICAL OBS
UKRAINIAN ACAD OF SCIENCE
GOLOSEEVO
252127 KIEV
UKRAINE
TEL 66 3110
TLF
TLX 131406 SKY SU
EML '
COM 33

VORPAHL JOAN A DR
748 23RD ST
SANTA MONICA CA 90402
USA
TEL
TLF
EML
COM
TLX

VOSHCHINNIKOV NICOLAI DR
ASTRONOMICAL OBSERVATORY
ST PETERSBURG UNIVERSITY
BIBLIOTECHNAJA PL 2
198904 ST PETERSBURG
RUSSIA
TEL 428 4162
TLF 121481 LSU SU
TLX
EML
COM 34

VRBA FREDERICK J DR
US NAVAL OBSERVATORY
FLAGSTAFF STATION
BOX 1149
FLAGSTAFF AZ 86002
USA
TEL 602 779 5132
TLF
TLX
EML
COM 09,25C,34

VREUX JEAN MARIE DR
INSTITUT D'ASTROPHYSIQUE
UNIVERSITE DE LIEGE
AVE COINTE 5
B 4000 COINTE-LIEGEE
BELGIUM
TEL 41 52 9980
TLF 41 52 7474
TLX 41264
EML
COM 29

VRSNAK BOJAN DR
HVAR OBSERVATORY
FACULTY OF GEODESY
KACICEVA 26
YU 41000 ZAGREB
YUGOSLAVIA
TEL 41 442 600*335
TLF 41 445 410
TLX
EML BOJAN VRSNAK@UNI-FG AC MAIL YU
COM 10

VRTILEK JAN M DR
APPLIED SCIENCES DIV
HARVARD UNIVERSITY
29 OXFORD ST
CAMBRIDGE MA 02138
USA
TEL 617 495 0589
TLF
TLX
EML SPAN CFARGZ VRTILEK
COM 21,28

VRTILEK SAEQA DIL DR
CENTER FOR ASTROPHYSICS
HCO/SAO MS 4
60 GARDEN ST
CAMBRIDGE MA 02138
USA
TEL 617 495 7400
TLF 617 495 7356
TLX 921428 SATELLITE CAM
EML SVRTILEK@CFA
COM 44

VU DUONG TUYEN DR
BUREAU DES LONGITUDES
77 AVE DENFERT ROCHEREAU
F 75014 PARIS
FRANCE
TEL 1 45 07 2262
TLF
TLX
EML
COM 20

VUCETICH HECTOR DR
UNIV NACIONAL DE LA PLATA
DPT FISICA
CCN 67
1900 LA PLATA (BS AS)
ARGENTINA
TEL 21 3 9061
TLF 21 25 2006
TLX 31151 BULAP AR
EML ATINA!FISILP!VUCETICH
COM 49

VUILLEMIN ANDRE DR
LAS
TRAVERSE DU SIPHON
LES TROIS LUCS
F 13012 MARSEILLE
FRANCE
TEL 91 05 5900
TLF 91 66 1855
TLX
EML
COM

VUJNOVIC VLADIS DR
INSTITUTE OF PHYSICS
UNIVERSITY OF ZAGREB
BOX 304
YU 41001 ZAGREB
YUGOSLAVIA
TEL 41 271 211
TLF
TLX 22203 IFS YU
EML
COM 14,46

VUKICEVIC K M PROF DR
DPT OF ASTRONOMY
FACULTY OF SCIENCES
STUDENTSKI TRG 16
YU 11000 BEOGRAD
YUGOSLAVIA
TEL 11 638 715
TLF
TLX
EML
COM 12

VYALSHIN GENNADIJ F DR
PULKOVO OBSERVATORY
ACADEMY OF SCIENCES
10 KUTUZOV QUAY
196140 ST PETERSBURG
RUSSIA
TEL
TLF
TLX
EML
COM 10

WACKERNAGEL H BEAT DR
51 BROADMOOR HILLS DR
COLORADO SPRINGS CO 80906
USA
TEL 303 554 3801
TLF
TLX
EML
COM 04,31

WADDINGTON C JAKE PROF
DPT OF PHYSICS
UNIVERSITY OF MINNESOTA
116 CHURCH ST SE
MINNEAPOLIS MN 55455
USA
TEL 612 624 2566
TLF
TLX 910-576-2955
EML
COM

WADE RICHARD ALAN DR
DPT OF ASTRONOMY/ASTROPHY
PENNSYLVANIA STATE UNIV
525 DABEY LAB
UNIVERSITY PARK PA 16802
USA
TEL
TLF
TLX
EML wade@astro psu edu
COM· 42

WAELKENS CHRISTOFFEL
ASTRONOMISCH INSTITUUT
KATHOLIEKE UNIV LEUVEN
CELESTIJNENLAAN 200B
B 3030 HEVERLEE
BELGIUM
TEL 16 20 0656
TLF
TLX 25715
EML
COM 27

WAGNER RAYMOND L DR
ROCWELL INTERNATIONAL
BOX 19000
SAN BERNARDINO
CA 92423 9000
USA
TEL
TLF
TLX
EML
COM

WAGNER ROBERT M DR
LOWELL OBSERVATORY
1400 W MARS HILL RD
BOX 1149
FLAGSTAFF AZ 86001
USA
TEL 602 779 0106
TLF
TLX
EML
COM

WAGNER WILLIAM J DR
NASA HEADQUARTERS
CODE SS
SPACE PHYSICS DIV
WASHINGTON DC 20546
USA
TEL
TLF
TLX
EML
COM

WAGONER ROBERT V PROF
VARIAN PHYSICS BLDG
STANFORD UNIVERISTY
STANFORD CA 94305
USA
TEL   415 723 4561
TLF
TLX   348402
EML
COM   47

WAINWRIGHT JOHN DR
DPT OF APPLIED MATH
UNIVERSITY OF WATERLOO
WATERLOO ON N2L 3G1
CANADA
TEL   519 885 1211
TLF   519 746 8115
TLX   069 55259
EML
COM   47

WAKAMATSU KEN-ICHI DR
FACULTY OF ENGINEERING
GIFU UNIVERSITY
GIFU 501 11
JAPAN
TEL   582-30-1111
TLF
TLX
EML
COM   28

WAKO KOJIRO DR
INTL LATITUDE OBSERVATORY
NAOJ
HOSHIGAOKA MIZUSAWA SHI
IWATE 023
JAPAN
TEL
TLF
TLX
EML
COM

WALBORN NOLAN R DR
STSCI
HOMEWOOD CAMPUS
3700 SAN MARTIN DR
BALTIMORE MD 21218
USA
TEL   301 338 4915
TLF
TLX   6849101 STSCI UW
EML
COM   45C

WALCH JEAN-JACQUES
OCA CERGA
AVE COPERNIC
F 06130 GRASSE
FRANCE
TEL   93 36 5849
TLF
TLX   470865 CERGA F
EML
COM   07

WALDMEIER MAX PROF DR
SWISS FEDERAL OBSERVATORY

WIRZENWEID 15
CH 8053 ZUERICH
SWITZERLAND
TEL
TLF
TLX
EML
COM   10,12

WALKER ALISTAIR ROBIN DR
CERRO TOLOLO
INTERAMERICAN OBSERVATORY
CASILLA 603
LA SERENA
CHILE
TEL   51 21 3352
TLF   51 21 2466*342
TLX   621 301 AURA CT
EML
COM   09,15,25

WALKER ALTA SHARON DR
US GEOLOGICAL SURVEY
927 NATIONAL CENTER
RESTON VA 22092
USA
TEL   703 648 6387
TLF   703 648 6684
TLX
EML
COM   16

WALKER ARTHUR B C JR PROF
CTR FOR SPACE SCIENCES &
ASTROPHYSICS
STANFORD UNIV ERL 310
STANFORD CA 94305 4055
USA
TEL   415 497 1486
TLF
TLX
EML
COM

WALKER DAVID DOUGLAS DR
DPT PHYSICS & ASTRONOMY
UNIVERSITY COLLEGE LONDON
GOWER ST
LONDON WC1E 6BT
UK
TEL   1 713 877 050*3510
TLF   1 713 807 145
TLX   28722
EML   DDW@UK AC ULC STARLINK
COM   09

WALKER EDWARD N MR
ROYAL GREENWICH OBS
HERSTMONCEUX CASTLE
HAILSHAM BN27 1RP
UK
TEL   323 833 171
TLX   87451
EML
TLF
COM   27

WALKER GORDON A H PROF
DPT GEOPHYS & ASTRONOMY
UNIV OF BRITISH COLUMBIA
2075 WESBROOK PL
VANCOUVER BC V6T 1W5
CANADA
TEL   604 228 4133
TLF   604 228 6047
TLX
EML
COM   09,30,34,37,45

WALKER HELEN J
RUTHERFORD APPLETON LAB
SPACE & ASTROPHYSICS DIV
BLDG R25/R68
CHILTON DIDCOT OX11 OQX
UK
TEL   235 821 900
TLF   235 445 848
TLX   83159 RUTHLB G
EML
COM

WALKER IAN WALTER
DPT OF ASTRONOMY
UNIVERSITY OF GLASGOW
GLASGOW G12 8QQ
UK
TEL.
TLF
TLX
EML
COM   07

WALKER MERLE F PROF
LICK OBSERVATORY
UNIVERSITY OF CALIFORNIA
SANTA CRUZ CA 95064
USA
TEL   408 429 2526
TLF
TLX
EML
COM   09,27,37,50

WALKER RICHARD L
US NAVAL OBSERVATORY
FLAGSTAFF STATION
BOX 1149
FLAGSTAFF AZ 86002
USA
TEL   602 774 6623
TLF
TLX
EML
COM   26,42

WALKER ROBERT C DR
NRAO
BOX O
SOCORRO NM 87801 0387
USA
TEL   505 835 7247
TLF
TLX   910 988 1710
EML   BITNET CWALKER@NRAO
COM   40

WALKER ROBERT M A PROF
DPT OF PHYSICS
WASHINGTON UNIVERSITY
BOX 1105
ST LOUIS MO 63130
USA
TEL   314 889 6225
TLF
TLX
EML
COM   16

WALKER WILLIAM S G
14 APPLEYARD CRES
AUCKLAND 5
NEW ZEALAND
TEL 09-548-736
TLF
EML
COM 27,42
TLX

WALL JASPER V DR
ROYAL GREENWICH OBS
MADINGLEY RD
CAMBRIDGE CB3 OEZ
UK
TEL 223 37 4000
TLF 223 37 4700
TLX 265451/265871
EML 19463 CAVAD JVW
COM 40

WALLACE LLOYD V DR
KITT PEAK NTL OBS
BOX 26732
950 N CHERRY AVE
TUCSON AZ 85726 6732
USA
TEL 602 327 5511
TLF
TLX
EML
COM 16,21

WALLACE PATRICK T MR
RUTHERFORD APPLETON LAB
SPACE & ASTROPHYSICS DIV
BLDG R25/R68
CHILTON DIDCOT OX11 OQX
UK
TEL 235 445 472
TLF
TLX 83159
EML
COM 05,08,09

WALLACE RICHARD K
LOS ALAMOS NATIONAL LAB
MS B257
X 7
LOS ALAMOS NM 87545
USA
TEL 505 667 5000
TLF
TLX
EML
COM

WALLENQUIST AAKE A E PROF
ASTRONOMICAL OBSERVATORY
NORRLANDSGATAN 34 D
S 752 20 UPPSALA
SWEDEN
TEL 18 13 5685
TLX
EML
TLF
COM 25,37

WALLERSTEIN GEORGE PROF
DPT OF ASTRONOMY
UNIVERSITY OF WASHINGTON
FM 20
SEATTLE WA 98195
USA
TEL 206 543 2888
TLF
TLX
EML
COM 27,29

WALLIS MAX K DR
DPT APPLIED MATHS/ASTRON
UNIVERSITY COLLEGE
BOX 78
CARDIFF CF1 1XL
UK
TEL 222 442 11
TLF
TLX 488635
EML
COM 15,21,51

WALMSLEY C MALCOLM DR
MPI FUER RADIOASTRONOMIE
AUF DEM HUEGEL 69 ´
D 5300 BONN 1
GERMANY
TEL 228 52 5305
TLX 886440
EML
TLF
COM 34,40

WALRAVEN TH DR
BOX 98
ORANGE FREESTATE 9850
SOUTH AFRICA
TEL
TLF
EML
COM 25,27
TLX

WALSH DENNIS DR
NRAL
JODRELL BANK
MACCLESFIELD SK11 9DL
UK
TEL 477 71321
TLF
TLX 36149
EML
COM 40,44

WALTER FREDERICK M
DPT OF EARTH & SPACE SCI
ASTRONOMY PROGRAM
SUNY AT STONY BROOK
STONY BROOK NY 11794 2100
USA
TEL 516 632 8221
TLF
TLX
EML
COM 36

WALTER HANS G DR
ASTRONOMISCHES RECHEN
INSTITUT
MOENCHHOFSTR 12-14
D 6900 HEIDELBERG 1
GERMANY
TEL 62 214 9026
TLF
TLX 461336 ARIHD D
EML
COM 08,24

WALTER KURT PROF DR
ASTRONOMISCHES INSTITUT
DER UNIVERSITAET
WALDHAUSERSTR 64
D 7400 TUEBINGEN
GERMANY
TEL 707 129 6126
TLF
TLX 7262714 AIT D
EML
COM 42

WALTERBOS RENE A M DR
DPT OF ASTRONOMY
NEW MEXICO STATE UNIV
BOX 4500
LAS CRUCES NM 88003
USA
TEL 505 646 6522
TLF
TLX
EML
COM 28

WALTON NICHOLAS A DR
DPT PHYSICS & ASTRONOMY
UNIVERSITY COLLEGE LONDON
GOWER ST
LONDON WC1E 6BT
UK
TEL 1 713 877 050*476
TLF 1 713 807 145
TLX 28722 UCPHYS G
EML NAW UK AC UCL STAR
COM 34

WAMPLER E JOSEPH PROF
ESO
KARL-SCHWARZSCHILD-STR 2
D 8046 GARCHING MUENCHEN
GERMANY
TEL 89 320 06297
TLX 52828222 EO D
EML
TLF 89 320 2362
COM 09

WAMSTEKER WILLEM DR
ESA IUE GROUND STATION
VILSPA
APO 54065
E 28080 MADRID
SPAIN
TEL 1 401 9661
TLF
TLX 42555
EML
COM 44C

WAN FOOK SUN
DPT OF MATHEMATICS
NTL UNIVERSITY SINGAPORE
KENT RIDGE
SINGAPORE 0511
SINGAPORE
TEL 772-2742
TLF
TLX
EML
COM

WAN LAI
SHANGHAI OBSERVATORY
CAS
80 NANDAN RD
SHANGHAI
CHINA PR
TEL 21 38 0696
TLF
TLX 33164 SHAO CN
EML
COM 24,37

WAN TONG-SHAN
SHANGHAI OBSERVATORY
CAS
80 NANDAN RD
SHANGHAI
CHINA PR
TEL 21 38 6191
TLF
TLX 33164 SHAO CN
EML
COM 40

WANAS M I PROF
DPT OF ASTRONOMY
FACULTY OF SCIENCES
CAIRO UNIVERSITY
GEZA ORMAN
EGYPT
TEL
TLF
TLX
EML
COM 48

WANG CHUAN-JIN
PURPLE MOUNTAIN OBSERV
CAS
NANJING
CHINA PR
TEL 25 46700
TLF
TLX 34144 PMONJ CN
EML
COM 25

WANG DEYU
PURPLE MOUNTAIN OBSERV
CAS
NANJING
CHINA PR
TEL 25 42817/46700
TLF
TLX 34144 PMONJ CN
EML
COM 48

WANG DE-CHANG
PURPLE MOUNTAIN OBSERV
CAS
NANJING
CHINA PR
TEL 25 64 6700/4205
TLF
TLX 34144 PMONJ CN
EML
COM 22,41

WANG HAIMIN DR
CALTECH
MS 264 33
PASADENA CA 91125
USA
TEL 818 356 3858
TLF
TLX
EML HAIMIN@SUNSOG CALTECH EDU
COM 10

WANG JIA-JI
SHANGHAI OBSERVATORY
CAS
80 NANDAN RD
SHANGHAI
CHINA PR
TEL 21 38 6191
TLF
TLX 33164 SHAO CN
EML
COM 24C

WANG JIA-LONG
BEIJING ASTRONOMICAL OBS
CAS
W SUBURB
BEIJING 100080
CHINA PR
TEL
TLF
TLX 22040 BAOBS CN
EML
COM 10

WANG JING-SHENG
YUNNAN OBSERVATORY
CAS
BOX 110
KUNMING 72946 YUNNAN
CHINA PR
TEL 871 2035
TLF
TLX 64040 YUOBS CN
EML
COM 40

WANG JING-XIU
BEIJING ASTRONOMICAL OBS
CAS
W SUBURB
BEIJING 100080
CHINA PR
TEL 1 28 1698
TLF
TLX 22040 BAOBS CN
EML
COM 10,12C

WANG LAN-JUAN
SHANGHAI OBSERVATORY
CAS
80 NANDAN RD
SHANGHAI
CHINA PR
TEL 21 38 6191
TLF
TLX 33164 SHAO CN
EML
COM 09

WANG RENCHUAN
CENTER FOR ASTROPHYSICS
UNIV SCIENCE & TECHNOLOGY
HEFEI 230026 ANHUI
CHINA PR
TEL 551 33 1134
TLF
TLX 90028 USTC CN
EML
COM 47

WANG SHOU-GUAN
BEIJING ASTRONOMICAL OBS
CAS
W SUBURB
BEIJING 100080
CHINA PR
TEL 1 28 1261
TLF
TLX 22040 BADAS CN
EML
COM 38,40,48

WANG SHUI
DEPT EARTH & SPACE SCI
UNIV SCIENCE & TECHNOLOGY
HEFEI 230026 ANHUI
CHINA PR
TEL 551 33 1134*209
TLF
TLX 4430
EML
COM 44

WANG SHUNDE DR
BEIJING ASTRONOMICAL OBS
CAS
W SUBURB
BEIJING 100080
CHINA PR
TEL 1 256 1264
TLF
TLX 22040 BAOAS CN
EML
COM 46,49

WANG SI-CHAO
PURPLE MOUNTAIN OBSERV
CAS
NANJING
CHINA PR
TEL 25 44205
TLF
TLX 34144 PMONJ CN
EML
COM 15

WANG YANAN
NANJING ASTRONOMICAL
INSTRUMENT FACTORY
BOX 846
NANJING
CHINA PR
TEL 25 46191
TLF
TLX 34136,GLYNJ c/o NAIF
EML
COM 09

WANG YIMING
YUNNAN OBSERVATORY
CAS
BOX 110
KUNMING 72946 YUNNAN
CHINA PR
TEL 871 2035
TLF
TLX
EML
COM 09

WANG YI-MING DR
NAVAL RESEARCH LABORATORY
CODE 4172 W
4555 OVERLOOK AVE SW
WASHINGTON DC 20375 5000
USA
TEL 202 767 6202
TLF
TLX
EML
COM 10,48,49

WANG ZHENG MING
SHAANXI OBSERVATORY
CAS
LINTONG XIAN
SHAANXI
CHINA PR
TEL 33 2255
TLF
TLX 70121 CSAO CN
EML
COM 09,19

WANG ZHEN-RU
DEPT OF ASTRONOMY
NANJING UNIVERSITY
NANJING
CHINA PR
TEL 25 37551*2685
TLF
TLX 34151 PRCNU CN
EML
COM 48

WANG ZHEN-YI
PURPLE MOUNTAIN OBSERV
CASERVATORY
NANJING
CHINA PR
TEL 25 46700
TLF
TLX 34144 PMONJ CN
EML
COM 12

WANNIER PETER GREGORY DR
JPL
MS 169 506
4800 OAK GROVE DR
PASADENA CA 91109
USA
TEL 818 354 3347
TLF
TLX 67-5429
EML
COM 34,40

WARD HENRY DR
DPT OF PHYSICS & ASTRON
UNIVERSITY OF GLASGOW
GLASGOW G12 8QQ
UK
TEL 413 398 855*4705
TLF 413 349 029
TLX 777070 UNIGLA
EML GW10@UK AC GLA PH I1
COM 09

WARD MARTIN JOHN
INSTITUTE OF ASTRONOMY
THE OBSERVATORIES
MADINGLEY RD
CAMBRIDGE CB3 OHA
UK
TEL
TLF
TLX
EML
COM 28,42

WARD RICHARD A DR
LAWRENCE LIVERMORE LAB
L 23 A DIV
BOX 808
LIVERMORE CA 94550
USA
TEL 415 423 2679
TLF
TLX 910-386-8339 UCLLL
EML
COM 35

WARD WILLIAM R DR
JPL
MS 183 501
4800 OAK GROVE DR
PASADENA CA 91109
USA
TEL
TLF
TLX
EML
COM

WARDLE JOHN F C PROF
DPT OF PHYSICS
BRANDEIS UNIVERSITY
WALTHAM MA 02154
USA
TEL 617 647 2889
TLF
TLX
EML
COM 40

WARES GORDON W DR
73 PERKINS ST
WEST NEWTON MA 02165
USA
TEL
TLF
EML
COM 14
TLX

WARGAU WALTER F DR
DPT OF MATHS &ASTRONOMY
UNIVERSITY OF S AFRICA
BOX 392
PRETORIA 0001
SOUTH AFRICA
TEL 27-12-440-2133
TLF
TLX 350068 TA UNISA TTX
EML
COM 42

WARMAN JOSEF DR
INSTITUTO DE ASTRONOMIA
UNAM
APDO POSTAL 70-264
04510 MEXICO DF
MEXICO
TEL
TLF
TLX
EML
COM

WARMELS REIN HERM DR
ESO
KARL SCHWARZSCHILDSTR 2
D 8046 GARCHING MUENCHEN
GERMANY
TEL 89 3200 6292
TLF 89 320 2362
TLX
EML RWARMELS@ESO ORG
COM 40

WARNER BRIAN PROF
DPT OF ASTRONOMY
INST OF THEOR PHYSICS
RONDEBOSCH 7700
SOUTH AFRICA
TEL 6502391
TLF
TLX 521439
EML
COM 27,42

WARNER JOHN W DR
PERKIN ELMER CORP
MS 892
100 WOOSTER HEIGHTS RD
DANBURY CT 06810 7859
USA
TEL 203 796 7919
TLF
TLX
EML
COM 28,44

WARNER PETER J DR
MULLARD RADIO ASTRON OBS
CAVENDISH LABORATORY
MADINGLEY RD
CAMBRIDGE CB3 OHE
UK
TEL 223 664 77
TLF
TLX 81292 CAVLAB G
EML
COM 40

WARREN WAYNE H JR DR
NASA/GSFC
CODE 681
LAB ASTRON & SOLAR PHYS
GREENBELT MD 20771
USA
TEL
TLF
TLX
EML W3WHW@SCFMVS GSFC NASA GOV
COM 05C,25,37,45

WARWICK JAMES W DR
RADIOPHYSICS CORP
5475 WESTERN AVE
BOULDER CO 80301
USA
TEL 303 447 9524
TLF
TLX
EML
COM 12,40

WARWICK ROBERT S DR
DPT OF PHYSICS
UNIVERSITY OF LEICESTER
UNIVERSITY RD
LEICESTER LE1 7RH
UK
TEL 533 554 455
TLF
TLX 341664 LUXRAY G
EML
COM

WASHIMI HARUICHI DR
INST FOR ATMOSPHERIC RES
NAGOYA UNIVERSITY
3-13 HONOHARA
TOYOKAWA AICHI 442
JAPAN
TEL 05338-6-3154
TLF
TLX 4322311
EML
COM

WASSERMAN LAWRENCE H DR
LOWELL OBSERVATORY
1400 W MARS HILL RD
BOX 1149
FLAGSTAFF AZ 86001
USA
TEL 602 774 3358
TLF
TLX
EML
COM 16,20C,24

WASSON JOHN T
INST OF GEOPHYS & PLANET
UNIVERSITY OF CALIFORNIA
LOS ANGELES CA 90024
USA
TEL 213 825 1986
TLF
TLX
EML
COM 15,16

WATANABE JUN-ICHI DR
TOKYO ASTRONOMICAL OBS
NAOJ
OSAWA MITAKA
TOKYO 181
JAPAN
TEL 81 422 41 3614
TLF 81 422 41 3608
TLX 028 22307 TAOMTK J
EML OWATANA@CL MTK NAO AC JP
COM 15

WATANABE TAKASHI DR
INST FOR ATMOSPHERIC RES
NAGOYA UNIVERSITY
3-13 HONOHARA
TOYOKAWA 442
JAPAN
TEL 05338 6 3154
TLF 05338 6 0811
TLX 4322311 RIANAG J
EML
COM 49

WATANABE TETSUYA
TOKYO ASTRONONICAL OBS
NAOJ
OSAWA MITAKA
TOKYO 181
JAPAN
TEL 422-32-5111
TLF
TLX
EML
COM 36

WATERS LAURENS B F M DR
SRON
BOX 800
NL 9700 AV GRONINGEN
NETHERLANDS
TEL 50 63 4090
TLX
EML RENSW@RUG NL
TLF 50 63 4033
COM 36

WATERWORTH MICHAEL DR
UNIVERSITY OF TASMANIA
GPO BOS 252C
HOBART TAS 7001
AUSTRALIA
TEL 2 202 418
TLF
TLX 58150 AA
EML
COM 29

WATSON FREDERICK GARNETT
UK SCHMIDT TELESCOPE
AAO
PRIVATE BAG
COONABARABRAN NSW 2357
AUSTRALIA
TEL 68 421 622
TLF
TLX 63945 CANOPUS AA
EML
COM 09,51

WATSON MICHEAL G DR
DPT OF PHYSICS
UNIVERSITY OF LEICESTER
UNIVERSITY RD
LEICESTER LEI 7RH
UK
TEL 533 523 553
TLF 533 550 182
TLX
EML NGW@UK AC LE STAR
COM

WATSON ROBERT DR
DPT OF PHYSICS
UNIVERSITY OF TASMANIA
GPO BOX 252C
HOBART TAS 7001
AUSTRALIA
TEL 2 202 415
TLF 2 202 410
TLX AA 58150
EML
COM 27

WATSON WILLIAM D PROF
DPT OF PHYSICS
UNIVERSITY OF ILLINOIS
1110 W GREEN ST
URBANA IL 61801
USA
TEL 217 333 7240
TLF
TLX
EML
COM

WATT GRAEME DAVID
ROYAL OBSERVATORY
BLACKFORD HILL
EDINBURGH EH9 3HJ
UK
TEL
TLF
TLX
EML
COM 34

WATTENBERG D PROF
LINDERHOFSTR 57

D 1147 BERLIN
GERMANY
TEL 527 7772
TLF
TLX
EML
COM 41

WAYMAN PATRICK A PROF
DUNSINK OBSERVATORY
DIAS
DUBLIN 15
IRELAND
TEL 1 387 911
TLF
TLX 31687 DIAS
EML
COM 05,33,50

WDOWIAK THOMAS J DR
PHYSICS DPT
UNIVERSITY OF ALABAMA
BIRMINGHAM AL 35294
USA
TEL 205 934 4736
TLF
TLX 888826 UAB BHM
EML
COM 15

WEAVER HAROLD F PROF
ASTRONOMY DPT
UNIVERSITY OF CALIFORNIA
601 CAMPBELL HALL
BERKELEY CA 94720
USA
TEL
TLF
TLX 820181 UCS AST
EML
COM 15,33,34,37

WEAVER THOMAS A DR
LAWRENCE LIVERMORE LAB
L 17
BOX 808
LIVERMORE CA 94550
USA
TEL 415 423 1850
TLF
TLX
EML
COM 35,48

WEBB DAVID F
PHILLIPS LAB/PHG
BOSTON COLLEGE
HANSCOM AFB MA 01731
USA
TEL 617 377 3970
TLF 617 377 3160
TLX
EML AFGLSC WEBB
COM 10

WEBBER JOHN C DR
INTERFEROMETRICS INC
150 LEESBURG PIKE
VIENNA VA 22180
USA
TEL 703 790 8500
TLF
TLX
EML
COM

WEBBINK RONALD F DR
DPT OF ASTRONOMY
UNIVERSITY OF ILLINOIS
1011 W SPRINGFIELD AVE
URBANA IL 61801
USA
TEL 217-333-9582
TLF
TLX 910-245-2434 AST
EML
COM 27,35,42C

WEBER STEPHEN VANCE
LAWRENCE LIVERMORE LAB
L 477
BOX 808
LIVERMORE CA 94550
USA
TEL 415 422 5433
TLF
TLX 910-386-8339
EML
COM 36

WEBROVA LUDMILA DR
ASTRONOMICAL INSTITUTE
CZECH ACADEMY SCIENCES
BUDECSKA 6
CS 120 23 PRAHA 2
CZECHOSLOVAKIA
TEL 2 25 5287
TLF
TLX 122486
EML
COM 31

WEBSTER ADRIAN S DR
JOINT ASTRONOMY CENTER
UKIRT
665 KOMOHANA ST
HILO HI 96720
USA
TEL 808 961 3756
TLF
TLX 633135
EML
COM 47,48

WEBSTER RACHEL
CITA MCLENNAN LABS
UNIVERSITY OF TORONTO
60 ST GEORGE ST
TORONTO ON M5S 1A1
CANADA
TEL 416 978 8496
TLF 416 978 3921
TLX
EML BITNET WEBSTER@UTORPHYS
COM 47

WEEDMAN DANIEL W PROF
DPT OF ASTRONOMY
PENNSYLVANIA STATE UNIV
525 DAVEY LAB
UNIVERSITY PARK PA 16802
USA
TEL 814 865 0418
TLF
TLX 842510
EML
COM 28

WEEKES TREVOR C DR
FRED LAWRENCE WHIPPLE OBS
HARVARD-SMITHSONIAN CTR
AMADO AZ 85645 0097
USA
TEL 602 629 6741
TLF
TLX
EML
COM

WEGNER GARY ALAN
DPT OF PHYSICS & ASTRON
DARTMOUTH COLLEGE
WILDER LABORATORY
HANOVER NH 03755
USA
TEL 603 646 2359
TLF
TLX
EML
COM 29,30

WEHINGER PETER A DR
DPT OF PHYSICS
ARIZONA STATE UNIVERSITY
ASTRONOMY PROGRAM
TEMPE AZ 85287
USA
TEL 602 965 4063
TLF
TLX 140289 HALLEY ASU UT
EML BITNET WEHINGER@ASUCPS
COM 15,28,29

WEHLAU AMELIA DR
DPT OF ASTRONOMY
UNIV OF WESTERN ONTARIO
LONDON ON N6A 3K7
CANADA
TEL 519 679 3186
TLF 519 661 3486
TLX 064 7134
EML
COM 27,37

WEHLAU WILLIAM H PROF
DPT OF ASTRONOMY
UNIV OF WESTERN ONTARIO
LONDON ON N6A 3K7
CANADA
TEL 519 679 3184
TLF 519 661 3486
TLX 064 7134
EML
COM 27,29,42

WEHRLE ANN ELIZABETH DR
JPL
MS 168 427
4800 OAK GROVE DR
PASADENA CA 91109
USA
TEL 818 354 1672
TLF
TLX
EML DEIMOS AEW/AEW@CITDEIMOI CALT
COM 40,44,48

WEHRSE RAINER DR
INST F THEOR ASTROPHYSIK
IM NEUENHEIMER FELD 561
D 6900 HEIDELBERG 1
GERMANY
TEL 62 215 62837
TLF
TLX 461515 UNIHD D
EML
COM 36C

WEI MINGZHI
BEIJING ASTRONOMICAL OBS
CAS
W SUBURB
BEIJING 100080
CHINA PR
TEL 1 28 1698
TLF
TLX 22040 BAOBS CN
EML
COM 40

WEIDEMANN VOLKER PROF
INST THEOR PHYS & STERNW
NEUE UNIV PHYSIK ZENTRUM
OLSHAUSENST GEB N 61C
D 2300 KIEL 1
GERMANY
TEL 431 880 4110
TLF.
TLX 292706
EML
COM 05,36

WEIDENSCHILLING S J DR
PLANETARY SCIENCE INST
2030 E SPEEDWAY
SUITE 201
TUCSON AZ 85719
USA
TEL 602 881 0332
TLF
TLX
EML
COM 15,16

WEIGELT GERD DR
MPI FUER RADIOASTRONOMIE
AUF DEM HUEGEL 69
D 5300 BONN 1
GERMANY
TEL 228 52 5243
TLF 228 52 5229
TLX 886440
EML P561GWE@MPIFR-BONN MPG DE
COM 40

WEIGERT ALFRED PROF
HAMBURGER STERNWARTE

GOJENSBERGSWEG 112
D 2050 HAMBURG 80
GERMANY
TEL 40 7224 4112
TLF
TLX 217884
EML
COM 35,42

WEILER EDWARD J DR
PRINCETON UNIVERSITY OBS
PEYTON HALL
PRINCETON NJ 08544
USA
TEL
TLF.
TLX
EML
COM 40,42,44

WEILER KURT W DR
NAVAL RESEARCH LABORATORY
CODE 4131
4555 OVERLOOK AVE SW
WASHINGTON DC 20375 5000
USA
TEL 202 767 0292
TLF
TLX
EML SPAN 11334 WEILER
COM 28,34,40,44

WEILL GILBERT M DR
SPOT IMAGE CORPORATION
1897 PRESTON WHITE DRIVE
RESTON VA 22091 4326
USA
TEL 703 620 2200
TLF
TLX 4993073
EML
COM 21

WEIMER THEOPHILE P F DR
OBSERVATOIRE DE PARIS
61 AVE OBSERVATOIRE
F 75014 PARIS
FRANCE
TEL 1 43 20 1210
TLF
TLX
EML
COM 16

WEINBERG J L DR
SPACE ASTRONOMY LAB
UNIVERSITY OF FLORIDA
1810 NW 6TH ST
GAINESVILLE FL 32609
USA
TEL   904 392 5450
TLF
TLX   810 825 2308 SPACELA
EML
COM   21,22,44

WEINBERG STEVEN DR
ASTRONOMY DPT
UNIVERSITY OF TEXAS
RLM 15 308
AUSTIN TX 78712 1083
USA
TEL   512 471 4394
TLF
TLX   910-874-1305
EML
COM   47

WEINBERGER RONALD DR
INSTITUT FUER ASTRONOMIE
TECHNIKERSTRASSE 25
A 6020 INNSBRUCK
AUSTRIA
TEL   5222 7485 251
TLF
TLX
EML
COM

WEIS EDWARD W DR
DPT OF ASTRONOMY
VAN VLECK OBSERVATORY
WESLEYAN UNIVERSITY
MIDDLETOWN CT 06457
USA
TEL   203 347 9411
TLF
TLX
EML
COM   26

WEISBERG JOEL MARK
DPT PHYSICS & ASTRONOMY
CARLETON COLLEGE
NORTHFIELD MN 55057
USA
TEL   507 663 4367
TLF
TLX
EML
COM

WEISHEIT JON C DR
DPT SPACE PHYS & ASTRON
RICE UNIVERSITY
BOX 1892
HOUSTON TX 77251 1892
USA
TEL
TLF
TLX
EML
COM   34,48

WEISS ACHIM DR
MAX PLANCK INSTITUT FUR
ASTROPHYSIK
KARL-SCHWARZSCHILDSTR 1
D 8046 GARCHING MUENCHEN
GERMANY
TEL   89 329 900
TLF   89 329 93235
TLX   524629 ASTRO D
EML   ACW@DGAIPP1S
COM   35

WEISS NIGEL O DR
DPT APPLIED MATHS
& THEORETICAL PHYSICS
SILVER STREET
CAMBRIDGE CB3 9EW
UK
TEL   223 351 645
TLF
TLX   81240
EML
COM   12,35

WEISS WERNER W DR
INSTITUT FUER ASTRONOMIE
DER UNIVERSITAET WIEN
TUERKENSCHANZSTR 17
A 1180 WIEN
AUSTRIA
TEL   1 345 360
TLF
TLX   116222 PHYSI A
EML   EARN "WEISS@AVIA UNA AC AT"
COM   09,25,27,29

WEISSKOPF MARTIN CH DR
NASA/MSFC
CODE ES 65
HUNTSVILLE AL 35812
USA
TEL   205 453 3238
TLF
TLX
EML
COM   48

WEISSMAN PAUL ROBERT
JPL
MS 183 601
4800 OAK GROVE DR
PASADENA CA 91109
USA
TEL   818 354 2636
TLF
TLX   675429
EML
COM   15,20

WEISTROP DONNA DR
DPT OF PHYSICS
UNIVERSITY OF NEVADA
4505 S MARYLAND PARKWAY
LAS VEGAS NV 89154
USA
TEL
TLF
TLX
EML
COM   25,33

WELCH DOUGLAS L DR
DPT OF PHYSICS & ASTRON
MCMASTER UNIVERSITY
HAMILTON ON L8S 4M1
CANADA
TEL   416 525 9140*3186
TLF   416 546 1252
TLX   061 8347
EML   WELCH@PHYSUN PHYSICS MCMASTER
COM   27

WELCH GARY A DR
DPT OF ASTRONOMY
ST MARY'S UNIVERSITY
HALIFAX NS B3H 3C3
CANADA
TEL   902 429 9780
TLF   902 420 5561
TLX
EML
COM   28

WELCH WILLIAM J PROF
RADIO ASTRONOMY LAB
UNIVERSITY OF CALIFORNIA
601 CAMPBELL HALL
BERKELEY CA 94720
USA
TEL   415 642 6679
TLF
TLX   820181 UCB AST RAL
EML
COM   40,51

WELLER CHARLES S DR
DAVID TAYLOR RES CENTER
CODE 1402
BETHESDA MD 20084 5000
USA
TEL   301 227 1274
TLF
TLX
EML
COM   49

WELLGATE G BERNARD MR
CANEHEATH HOUSE
ARLINGTON
POLEGATE BN26 6SJ
UK
TEL
TLF
TLX
EML
COM

WELLINGTON KELVIN DR
CSIRO
DIVISION OF RADIOPHYSICS
BOX 76
EPPING NSW 2121
AUSTRALIA
TEL   2 868 0222
TLF
TLX   26230 AA
EML   KWELLING@RP CSIRO AU
COM   40

WELLMANN PETER PROF DR
INST FUER ASTRONOMIE &
ASTROPHYSIK
SCHEINERSTR 1
D 8000 MUENCHEN 80
GERMANY
TEL
TLF
TLX
EML
COM   36,42

WELLS DONALD C III DR
NRAO

EDGEMONT RD
CHARLOTTESVILLE VA 22901
USA
TEL   804 296 0211
TLF
TLX
EML
COM   05

WENDKER HEINRICH J PROF
HAMBURGER STERNWARTE
GOJENSBERGSWEG 112
D 2050 HAMBURG 80
GERMANY
TEL   40 7252 4112
TLX   217884
EML
TLF
COM   34,40

WENGER MARC
OBS DE STRASBOURG
11 RUE UNIVERSITE
F 67000 STRASBOURG
FRANCE
TEL   88 35 8219
TLF   88 25 0160
TLX
EML   SIMBAD   WENGER/WENGER@SIMBAD U
COM   05

WENIGER SCHAME DR
OBSERVATOIRE DE PARIS
SECTION DE MEUDON
F 92195 MEUDON PPL CDX
FRANCE
TEL   1 45 34 7530
TLF
TLX
EML
COM   14,21,29

WENTZEL DONAT G DR
ASTRONOMY PROGRAM
UNIVERSITY OF MARYLAND
COLLEGE PARK MD 20742
USA
TEL   301 405 1518
TLF
TLX
EML   WENTZEL@ASTRO UMD EDU
COM   10,12,46C,48

WENZEL W DR
ZNTRLINST F ASTROPHYSIK
STERNWARTE SONNEBERG
D 6400 SONNEBERG
GERMANY
TEL
TLF
TLX
EML
COM   27

WESEMAEL FRANCOIS DR
DPT DE PHYSIQUE
UNIVERSITE DE MONTREAL
CP 6128 SUCC A
MONTREAL QC H3C 3J7
CANADA
TEL   514 343 7355
TLF   514 343 2071
TLX   055 62425 UDEMPHYSAS
EML
COM

WESSELINK ADRIAAN J DR
143 FALLS RD
BETHANY CT 06525
USA
TEL   203 393 3297
TLF
TLX
EML
COM   24,25,27,42

WESSELIUS PAUL R DR
SRON

BOX 800
NL 9700 AV GRONINGEN
NETHERLANDS
TEL   50 634 074
TLF   50 634 033
TLX
EML   PAUL@GUSPACE RUG NL
COM   25,34,44,45

WESSON PAUL S DR
DPT OF PHYSICS
UNIVERSITY OF WATERLOO
WATERLOO ON N2L 3G1
CANADA
TEL   519 885 1211
TLF   519 746 8115
TLX   069 55259
EML
COM   21,47,51

WEST DONALD K DR
NASA/GSFC
CODE 684 1
GREENBELT MD 20771
USA
TEL
TLF
TLX
EML
COM   42

WEST RICHARD M DR
ESO
KARL-SCHWARZSCHILD-STR 2
D 8046 GARCHING MUENCHEN
GERMANY
TEL   89 320 06276
TLF   89 320 2362
TLX   52828220 ESO D
EML   ESOMC1   RICHARD
COM   06VP,09C,20C,46

WEST ROBERT ALAN
JPL
MS 183 301
4800 OAK GROVE DR
PASADENA CA 91109
USA
TEL   818 354 0479
TLF
TLX   675429
EML
COM   16

WESTERGAARD NIELS J DR
DANISH SPACE RESEARCH INS
LUNDTOFTEVEJ 7
DK 2800 LYNGBY
DENMARK
TEL   22 88 2277
TLF
TLX   37198 DANRU DK
EML
COM

WESTERHOUT GART DR
US NAVAL OBSERVATORY
SCIENTIFIC DIRECTOR
34 & MASSACHUSETTS AVE NW
WASHINGTON DC 20392 5100
USA
TEL   202 653 1513
TLF
TLX   7108221970
EML
COM   05C,08,24,33,40

WESTERLUND BENGT E PROF
ASTRONOMICAL OBSERVATORY
BOX 515
S 751 20 UPPSALA
SWEDEN
TEL   18 13 5157
TLF
TLX   76024 UNIV UPPSS
EML
COM   28,33,45

WESTFOLD KEVIN C PROF
MONASH UNIVERSITY
WELLINGTON RD
CLAYTON VIC 3168
AUSTRALIA
TLF
TEL   3 541 3080
TLX   32691 AA
EML
COM   40,48

WESTPHAL JAMES A PROF
CALTECH
MS 170 25
PASADENA CA 91125
USA
TEL   213 356 4900
TLF
TLX
EML
COM   09,44

WETHERILL GEORGE W
DPT TERRESTR MAGNETISM
CARNEGIE INST WASHINGTON
5241 BROAD BRANCH RD NW
WASHINGTON DC 20015
USA
TEL   202 966 0863
TLF
TLX   440427 MAGN UI
EML
COM   15,16,22,51

WEYMANN RAY J PROF
STEWARD OBSERVATORY
UNIVERSITY OF ARIZONA
TUCSON AZ 85721
USA
TEL   602 621 2375
TLF
TLX   467175
EML
COM   34

WHEELER J CRAIG PROF
ASTRONOMY DPT
UNIVERSITY OF TEXAS
RLM 15 308
AUSTIN TX 78712 1083
USA
TEL   512 471 4461
TLF
TLX   910-874-1351
EML
COM   35,42,48

WHEELER JOHN A DR
DPT OF PHYSICS
PRINCETON UNIVERSITY
JADWIN HALL
PRINCETON NJ 08544
USA
TEL 609 258 4400
TLF 609 258 1124
TLX 4993512
EML DWNS@MSO ANU.OZ AU
COM 47,48

WHIPPLE ARTHUR L DR
MCDONALD OBSERVATORY
UNIVERSITY OF TEXAS
AUSTIN TX 78712 1351
USA
TEL 512 471 6332
TLF
TLX
EML BITNET asag105@uta3081
COM 07,20

WHIPPLE FRED L DR
CENTER FOR ASTROPHYSICS
HCO/SAO
60 GARDEN ST
CAMBRIDGE MA 02138
USA
TEL 617 495 7200
TLF
TLX
EML
COM 15,20,22

WHITAKER EWEN A
LUNAR & PLANETARY LAB
UNIVERSITY OF ARIZONA
TUCSON AZ 85721
USA
TEL 602 621 2888
TLF
TLX 910-952-1143
EML
COM 16,41

WHITE GLENN J
DPT OF PHYSICS
QUEEN MARY/WESTFIELD COLL
MILE END RD
LONDON E1 4NS
UK
TEL 1 980 4811*4045
TLF
TLX 893750
EML
COM 34

WHITE GRAEME LINDSAY DR
CSIRO
DIVISION OF RADIOPHYSICS
BOX 76
EPPING NSW 2121
AUSTRALIA
TEL 2 868 0222*420
TLF
TLX 26230 ASTRO AA
EML
COM 24,41

WHITE NATHANIEL M DR
LOWELL OBSERVATORY
1400 W MARS HILL RD
BOX 1149
FLAGSTAFF AZ 86001
USA
TEL 602 774 3358
TLF
TLX
EML
COM 25

WHITE ORAN R DR
7590 ROAD 39

MANCOS CO 80307
USA
TEL 303 533 7318
TLF
TLX
EML
COM

WHITE R STEPHEN PROF
IGPP
UNIVERSITY OF CALIFORNIA
RIVERSIDE CA 92521
USA
TEL 714 787 4503
TLF
TLX
EML
COM

WHITE RAYMOND E DR
STEWARD OBSERVATORY
UNIVERSITY OF ARIZONA
TUCSON AZ 85721
USA
TEL 602 621 6528
TLF
TLX 467175
EML
COM 33,37

WHITE RICHARD E
DPT OF ASTRONOMY
SMITH COLLEGE
CLARK SCIENCE CENTER
NORTHAMPTON MA 01063
USA
TEL 413 584 2700
TLF
TLX
EML
COM

WHITE RICHARD L
STSCI
HOMEWOOD CAMPUS
3700 SAN MARTIN DR
BALTIMORE MD 21218
USA
TEL 301 338 4797
TLF
TLX
EML
COM 34,36

WHITE SIMON DAVID MANION
STEWARD OBSERVATORY
UNIVERSITY OF ARIZONA
TUCSON AZ 85721
USA
TEL 602 621 6530
TLF
TLX
EML
COM 28,47

WHITELOCK PATRICIA ANN DR
SAAO
BOX 9
OBSERVATORY 7935
SOUTH AFRICA
TEL 470025
TLF
TLX 57-20309
EML
COM 27,33,34

WHITEOAK J B DR
CSIRO
AUSTR TELESCOPE NTL FAC
BOX 76
EPPING NSW 2121
AUSTRALIA
TEL 2 868 0226
TLF 2 868 0310
TLX 26230
EML JWHITEOA@ATNF CSIRO AU
COM 33,34,40VP,50

WHITFORD ALBERT E PROF
LICK OBSERVATORY
UNIVERSITY OF CALIFORNIA
SANTA CRUZ CA 95064
USA
TEL 408 429 2149
TLF
TLX
EML
COM 28

WHITMORE BRADLEY C
STSCI
HOMEWOOD CAMPUS
3700 SAN MARTIN DR
BALTIMORE MD 21218
USA
TEL 301 338 4713
TLF
TLX
EML
COM 28

WHITNEY BALFOUR S
1102 E MISSOURI
NORMAN OK 73071
USA
TEL 405 321 3547
TLF
EML
COM
TLX

WHITNEY CHARLES A PROF
CENTER FOR ASTROPHYSICS
HCO/SAO
60 GARDEN ST
CAMBRIDGE MA 02138
USA
TEL 617 495 7451
TLF
TLX
EML
COM

WHITROW GERALD JAMES PROF
41 HOME PARK RD
WIMBLEDON
LONDON SW19 7HS
UK
TEL 1 947 343 467
TLF
TLX
EML
COM 41,47

WHITTET DOUGLAS C B DR
DPT OF PHYSICS
RENSSELAER POLYTECHN INST
TROY NY 12180 3590
USA
TEL   518 276 6310
TLF   518 276 6680
TLX
EML
COM   33,34

WHITTLE D MARK DR
UNIVERSITY STATION
UNIVERSITY OF VIRGINIA
BOX 3818
CHARLOTTESVILLE VA 22903
USA
TEL   864 924 4900
TLF
TLX   910 997 0174 NRAO
EML
COM

WHITWORTH ANTHONY PETER
DPT APPLIED MATHS/ASTRON
UNIVERSITY COLLEGE
BOX 78
CARDIFF CF1 1XL
UK
TEL   222 442 11
TLF
TLX   498635 ULIBCFG
EML
COM   34

WICKRAMASINGHE D T DR
AUSTRALIAN NTL UNIVERSITY
DPT OF APPLIED MATHS
BOX 4
CANBERRA ACT 2600
AUSTRALIA
TEL
TLF
TLX
EML
COM

WICKRAMASINGHE N C PROF
DPT APPLIED MATHS/ASTRON
UNIVERSITY COLLEGE
BOX 78
CARDIFF CF1 1XL
UK
TEL   222 442 11
TLF
TLX   498635 ULIBCFG
EML
COM   34,36,40

WIDING KENNETH G DR
NAVAL RESEARCH LABORATORY
CODE 7144
4555 OVERLOOK AVE SW
WASHINGTON DC 20375 5000
USA
TEL   202 767 2605
TLF
TLX
EML
COM

WIEDLING TOR DR
OSTRA VILLAVAGEN 15
S 611 36 NYKOPING
SWEDEN
TEL
TLF
TLX
EML
COM

WIEHR EBERHARD DR
UNIVERSITAETS STERNWARTE
GEISMARLANDSTR 11
D 3400 GOETTINGEN
GERMANY
TEL   551 39 5053
TLF
TLX   96753
EML
COM   10

WIELEBINSKI RICHARD PROF
MPI FUER RADIOASTRONOMIE
AUF DEM HUEGEL 69
D 5300 BONN 1
GERMANY
TEL   228 52 5300
TLX   886440
EML
TLF
COM   25,28,33,40,51

WIELEN ROLAND PROF DR
ASTRONOMISCHES-RECHEN
MOENCHHOFSTR 12-14
D 6900 HEIDELBERG 1
GERMANY
TEL   62 214 9026
TLX   461 336 ARIHD D
EML
TLF
COM   04,05,08,28,33,37

WIESE WOLFGANG L DR
NTL BUREAU OF SATNDARDS
DIV 842 RM A267 BLDG 221
GAITHERSBURG MD 20899
USA
TEL   301 975 3201
TLF   301 975 3038
TLX   898493
EML
COM   14P

WIETH-KNUDSEN NIELS P DR
SVEND TROSTSVEJ 12

DK 1912 FREDERIKSBERG C
DENMARK
TEL   31 24 9131
TLF
TLX
EML
COM   26,31

WIITA PAUL JOSEPH
DPT PHYSICS & ASTRONOMY
GEORGIA STATE UNIVERSITY
ATLANTA GA 30303 3083
USA
TEL   404 658 2932
TLF
TLX
EML
COM   28

WIJNBERGEN JAN DR
LAB VOOR RUIMTEONDERZOEK
HOOGBOUW WSN
BOX 800
NL 9700 AV GRONINGEN
NETHERLANDS
TEL   50 11 6660
TLF
TLX   53572 STARS NL
EML
COM

WILCOCK WILLIAM L PROF
SCHOOL OF PHYSICAL &
MOLECULAR SCIENCES
UNIV COLLEGE OF N WALES
BANGOR GWYNEDD LL57 2UW
UK
TEL   248 35 1151
TLF
TLX   61100
EML
COM   09

WILD JOHN PAUL DR
CSIRO
LIMESTONE AVE
BOX 225
DICKSON ACT 2602
AUSTRALIA
TEL
TLF
TLX
EML
COM   10,40,49

WILD PAUL PROF
ASTRONOMISCHES INSTITUT
UNIVERSITAET BERN
SIDLERSTRASSE 5
CH 3012 BERN
SWITZERLAND
TEL   31 65 8596
TLF
TLX   32320 PHYBE CH
EML
COM   20,28

WILDEY ROBERT L PROF DR
ASTROPHYSICAL OBSERVATORY
N ARIZONA UNIVERSITY
FLAGSTAFF AZ 86011
USA
TEL   602 523 2661
TLF
TLX
EML
COM   16

WILKENING LAUREL L DR
ADMIN BLDG 601
UNIVERSITY OF ARIZONA
TUCSON AZ 85721
USA
TEL   602 626 3513
TLF
TLX
EML
COM   15

WILKES BELINDA J
CENTER FOR ASTROPHYSICS
HCO/SAO
60 GARDEN ST
CAMBRIDGE MA 02138
USA
TEL   617 495 7268
TLF
TLX   921428
EML
COM

WILKINS GEORGE A DR
WINDWARD
HIGHER BROOK MEADOW
SIDFORD
SIDMOUTH DEVON EX10 9SS
UK
TEL 395 579 641
TLF
TLX
EML G A WILKINS@UK AC EXETER
COM 04,05C,19,31

WILKINSON ALTHEA
DPT OF ASTRONOMY
UNIVERSITY OF MANCHESTER
MANCHESTER M13 9PL
UK
TEL 612 737 121
TLF
TLX
EML
COM 28

WILKINSON DAVID T
PRINCETON UNIVERSITY
JADWIN HALL
BOX 708
PRINCETON NJ 08544
USA
TEL 609 452 4406
TLF
TLX
EML
COM 47

WILKINSON PETER N DR
NRAL

JODRELL BANK
MACCLESFIELD SK11 9DL
UK
TEL 477 71321
TLF
TLX 36149
EML
COM 40

WILL CLIFFORD M DR
DPT OF PHYSICS
WASHINGTON UNIVERSITY
ST LOUIS MO 63130
USA
TEL 314 889 6244
TLF
TLX
EML
COM 47,48

WILLIAMON RICHARD M
FERNBANK SCIENCE CENTER
156 HEATON PARK DRIVE
156 HEATON PARK DRIVE
ATLANTA GA 30307
USA
TEL 404 378 4313
TLF
TLX
EML
COM 27,42,46

WILLIAMS BARBARA A
DPT OF PHYSICS
UNIVERSITY OF DELAWARE
NEWARK DE 19716
USA
TEL 302 451 6526
TLF
TLX
EML
COM 28

WILLIAMS CAROL A
DPT OF MATHEMATICS
UNIVERSITY OF S FLORIDA
TAMPA FL 33620
USA
TEL 813 974 2643
TLF
TLX
EML
COM 04,07,24

WILLIAMS DAVID A PROF
DPT OF MATHEMATICS
UMIST
BOX 88
MANCHESTER M60 1QD
UK
TEL 612 363 311
TLF
TLX 666094
EML
COM 34

WILLIAMS GLEN A DR
DPT OF PHYSICS
CENTRAL MICHIGAN UNIV
MT PLEASANT MI 48858
USA
TEL 517 774 3365
TLF
TLX
EML 32NSQSV@CMUVM
COM 42

WILLIAMS IWAN P PROF
ASTRONOMY UNIT
QUEEN MARY/WESTFIELD COLL
MILE END RD
LONDON E1 4NS
UK
TEL 1 719 755 452
TLF
TLX 893750
EML IPW@UK AC QMW MATHS
COM 15,16,20,22VP,51

WILLIAMS JAMES G DR
JPL
MS 264 700
4800 OAK GROVE DR
PASADENA CA 91109
USA
TEL 818 354 6466
TLF
TLX 910 588 3269 JPL
EML
COM 04,16,19,20

WILLIAMS JOHN A DR
DPT OF PHYSICS
ALBION COLLEGE
ALBION MI 49224
USA
TEL 517 629 5511
TLF
TLX
EML
COM 45

WILLIAMS PEREDUR M DR
ROYAL OBSERVATORY
BLACKFORD HILL
EDINBURGH EH9 3HJ
UK
TEL 316 673 321
TLX 72383 ROEDIN G
EML
TLF
COM 29

WILLIAMS ROBERT E DR
CERRO TOLOLO
INTERAMERICAN OBSERVATORY
CASILLA 603
LA SERENA
CHILE
TEL 51 21 3352
TLF 51 21 2466*342
TLX 645 227 AURA CT
EML
COM 28,34,42

WILLIAMS THEODORE B DR
DPT PHYSICS & ASTRONOMY
RUTGERS UNIVERSITY
BOX 849
PISCATAWAY NJ 08854 0849
USA
TEL 201 932 2516
TLF
TLX
EML
COM 28

WILLIS ALLAN J DR
DPT PHYSICS & ASTRONOMY
UNIVERSITY COLLEGE LONDON
GOWER ST
LONDON WC1E 6BT
UK
TEL 1 713 877 050
TLF
TLX 28722
EML
COM 34,44

WILLIS ANTHONY GORDON DR
DOMINION RADIO ASTROPHYS
OBSERVATORY
BOX 248
PENTICTON BC V2A 6K3
CANADA
TEL 604 493 2277
TLF 604 493 7767
TLX 048 88127
EML
COM 40

WILLMORE A PETER PROF
SPACE RESEARCH DEPT
UNIVERSITY OF BIRMINGHAM
BOX 363
BIRMINGHAM B15 2TT
UK
TEL 214 72 1301
TLF
TLX 338938 SPAPHY G
EML
COM

WILLNER STEVEN PAUL DR
CENTER FOR ASTROPHYSICS
HCO/SAO
60 GARDEN ST
CAMBRIDGE MA 02138
USA
TEL 617 495 7123
TLF
TLX 921428 SATELLITE CAM
EML
COM 34,44

WILLS BEVERLEY J DR
ASTRONOMY DPT
UNIVERSITY OF TEXAS
RLM 15 308
AUSTIN TX 78712 1083
USA
TEL   512 471 3424
TLF
TLX   910-874-1351
EML
COM   28,40

WILLS DEREK DR
ASTRONOMY DPT
UNIVERSITY OF TEXAS
RLM 15 308
AUSTIN TX 78712 1083
USA
TEL   512 471 4461
TLF
TLX   910-874-1351
EML
COM   28,40

WILLSON LEE ANNE DR
DPT OF PHYSICS
IOWA STATE UNIVERSITY
AMES IA 50011
USA
TEL   515 294 6765
TLF
TLX   910-520-1157
EML
COM   27,35,36C

WILLSON ROBERT FREDERICK
DPT OF PHYSICS
TUFTS UNIVERSITY
MEDFORD MO 02155
USA
TEL   617 628 5000
TLF·
TLX
EML
COM   40,51

WILLSTROP RODERICK V DR
INSTITUTE OF ASTRONOMY
THE OBSERVATORIES
MADINGLEY RD
CAMBRIDGE CB3 OHA
UK
TEL   223 622 04
TLF
TLX   817297 ASTRON G
EML
COM   25,30

WILSON ALBERT G DR
BOX 1871
SEBASTOPOL CA 95473
USA
TEL
TLF
EML
COM   28,47
TLX

WILSON ANDREW S DR
ASTRONOMY PROGRAM
UNIVERSITY OF MARYLAND
COLLEGE PARK MD 20742
USA
TEL   301 454 6061
TLF
TLX   7108260352
EML
COM   40

WILSON BRIAN G PROF
UNIVERSITY OF QUEENSLAND
55 WALCOTT STREET
ST LUCIA QLD 4067
AUSTRALIA
TEL   7 377 2200
TLF
TLX   40315 UNIQLD AA
EML
COM

WILSON CURTIS A
ST JOHN'S COLLEGE
BOX 1671
ANNAPOLIS MD 21404
USA
TEL   301 263 2371
TLF
TLX
EML
COM   41

WILSON JAMES R DR
737 SOUTH M
LIVERMORE CA 94550
USA
TEL   415 422 1659
TLF
EML
COM   48
TLX

WILSON LIONEL DR
ENV SCIENCE DPT
LANCASTER UNIVERSITY
LANCASTER LA1 4YQ
UK
TEL   524 652 01*4075
TLF
TLX   65111 LQNCULG
EML
COM   27

WILSON MICHAEL JOHN DR
DPT OF APPL MATHEMATICS
UNIVERSITY OF LEEDS
LEEDS L52 9JTT
UK
TEL
TLF
TLX
EML
COM

WILSON P DR
INST ANGEWANDTE GEQOAESIE
RICHARD-STRAUSS-ALLEE 11
D 6000 FRANKFURT MAIN 70
GERMANY
TEL   69 633 3260
TLX   413592
EML
TLF
COM   19C,21

WILSON PETER R PROF
DPT OF APPLIED MATHS
UNIVERSITY OF SYDNEY
SYDNEY NSW 2006
AUSTRALIA
TEL
TLF
TLX
EML
COM   10,12C,36

WILSON RAYMOND N DR
ESO
KARL-SCHWARZSCHILD-STR 2
D 8046 GARCHING MUENCHEN
GERMANY
TEL   89 320 06274
TLF   89 320 2362
TLX   528282-0 EO D
EML
COM

WILSON ROBERT E PROF
DPT OF ASTRONOMY
UNIVERSITY OF FLORIDA
211 SSRB
GAINESVILLE FL 32611
USA
TEL   904 392 1182
TLF
TLX
EML
COM   35,42

WILSON ROBERT PROF SIR
DPT PHYSICS & ASTRONOMY
UNIVERSITY COLLEGE LONDON
GOWER ST
LONDON WC1E 6BT
UK
TEL   1 713 807 154
TLF
TLX   28722
EML
COM   14,44

WILSON ROBERT W DR
AT & BELL LABORATORIES
HOH L 239
BOX 400
HOLMDEL NJ 07733
USA
TEL   201 949 3803
TLF
TLX
EML
COM   34,40

WILSON S J
DPT OF MATHEMATICS
NTL UNIVERSITY SINGAPORE
KENT RIDGE
SINGAPORE 0511
SINGAPORE
TEL
TLF
TLX
EML
COM   36

WILSON THOMAS L DR
MPI FUER RADIOASTRONOMIE
AUF DEM HUEGEL 69
D 5300 BONN 1
GERMANY
TEL   228 52 5378
TLF   228 52 5229
TLX   886440
EML
COM   34,40,51

WILSON WILLIAM J DR
JPL
MS 168 327
4800 OAK GROVE DR
PASADENA CA 91109
USA
TEL 818 354 5699
TLF
TLX
EML
COM 40

WINCKLER JOHN R PROF
SCHOOL OF PHYS & ASTRON
UNIVERSITY OF MINNESOTA
116 CHURCH ST SE
MINNEAPOLIS MN 55455
USA
TEL 612 373 4688
TLF
TLX
EML
COM

WINDHORST ROGIER A DR
DPT OF PHYSICS
ARIZONA STATE UNIVERSITY
TEMPE AZ 85287 1504
USA
TEL 602 965 7143
TLF
TLX 156 1058
EML
COM 09,28,40,47

WING ROBERT F PROF
DPT OF ASTRONOMY
OHIO STATE UNIVERSITY
174 W 18TH AVE
COLUMBUS OH 43210 1106
USA
TEL 614 422 7876
TLF
TLX
EML
COM 27,29,45C

WINGET DONALD E
ASTRONOMY DPT
UNIVERSITY OF TEXAS
RLM 15 308
AUSTIN TX 78712 1083
USA
TEL 512 471 4461
TLF
TLX
EML
COM

WINIARSKI MACIEJ
ASTRONOMICAL OBSERVATORY
JAGIELLONIAN UNIVERSITY
UL ORLA 171
PL 30 244 KRAKOW
POLAND
TEL
TLF
TLX
EML
COM 25

WINK JOERN ERHARD DR
IRAM
300 RUE DE LA PISCINE
F 38406 S MARTIN HERES CD
FRANCE
TEL 76 42 3383
TLF
TLX
EML
COM 40

WINKLER CHRISTOPH DR
ESA/ESTEC
SSD
BOX 299
NL 2200 AG NOORDWIJK
NETHERLANDS
TEL 17 19 83591
TLF 17 19 84690
TLX 39098
EML CWINKLER@ESTEC
COM 44

WINKLER GERNOT M R DR
US NAVAL OBSERVATORY
TIME SERVICE DPT
34 & MASSACHUSETTS AVE NW
WASHINGTON DC 20392 5100
USA
TEL 202 653 1520
TLF
TLX 710 822 1970 NAVOBSY
EML
COM 04,19,31

WINKLER KARL-HEINZ A DR
LOS ALAMOS NATIONAL LAB
MS B218 X DOT
BOX 1663
LOS ALAMOS NM 87545
USA
TEL
TLF
TLX
EML
COM 35

WINKLER PAUL FRANK DR
DPT OF PHYSICS
MIDDLEBURY COLLEGE
MIDDLEBURY VT 05753
USA
TEL 802 388 3711
TLF
TLX 353249
EML
COM

WINNBERG ANDERS DR
ONSALA SPACE OBSERVATORY
GOETEBORG UNIVERSITY
S 439 00 ONSALA
SWEDEN
TEL 30 06 0653
TLF
TLX 2400 ONSPACE S
EML
COM 34,40

WINNEWISSER GISBERT DR
UNIVERSITAT ZU KOLN
I PHYSIKALISCHES INST
UNIVERSITATSSTRASSE 14
D 5000 KOELN 41
GERMANY
TEL 211 47 03567
TLF
TLX
EML
COM 14,34,40

WIRAMIHARDJA SUHARDJA DR
BOSSCHA OBSERVATORY
LEMBANG 40391
INDONESIA
TEL 229 6001
TLF 224 40699
EML
COM
TLX 28324 BD

WISNIEWSKI WIESLAW Z
LUNAR & PLANETARY LAB
UNIVERSITY OF ARIZONA
TUCSON AZ 85721
USA
TEL 602 621 6956
TLF
TLX 910-952-1143
EML
COM 15,25,27

WITHBROE GEORGE L DR
CENTER FOR ASTROPHYSICS
HCO/SAO
60 GARDEN ST
CAMBRIDGE MA 02138
USA
TEL 617 495 7438
TLF
TLX
EML
COM

WITT ADOLF N DR
DPT PHYSICS & ASTRONOMY
UNIVERSITY OF TOLEDO
2801 W BANCROFT ST
TOLEDO OH 43606
USA
TEL 419 537 2709
TLF
TLX 810 442 1633
EML
COM 21,34

WITTEN LOUIS PROF
DPT OF PHYSICS
UNIVERSITY OF CINCINNATI
210 BRAUNSTEIN ML 11
CINCINNATI OH 45221 0111
USA
TEL 513 475 6492
TLF
TLX
EML
COM

WITTMANN AXEL D DR
UNIVERSITAETS-STERNWARTE
GEISMARLANDSR 11
D 3400 GOETTINGEN
GERMANY
TEL 551 39 5042
TLF
TLX 96753
EML
COM 10,12

WITZEL ARNO DR
MPI FUER RADIOASTRONOMIE
AUF DEM HUEGEL 69
D 5300 BONN 1
GERMANY
TEL 228 52 5211
TLF
TLX 886440
EML
COM 40

WLERICK GERARD DR
OBSERVATOIRE DE PARIS
SECTION DE MEUDON
F 92195 MEUDON PPL COX
FRANCE
TEL 1 45 07 2240
TLF
TLX
EML
COM 09,28

WNUK EDWIN
ASTRONOMICAL OBSERVATORY
A MICKIEWICZ UNIVERSITY
UL SLONECZNA 36
PL 60 286 POZNAN
POLAND
TEL
TLF
TLX
EML
COM 07

WOEHL HUBERTUS DR
KIEPENHEUER INSTITUT
FUER SONNENPHYSIK
SCHOENECKSTRASSE 6
D 7800 FREIBURG BREISGAU
GERMANY
TEL 761 32864
TLF
TLX 7721552 KIS D
EML
COM 09,10,12,36

WOLF BERNHARD PH D
LANDESSTERNWARTE
KOENIGSTUHL
D 6900 HEIDELBERG 1
GERMANY
TEL 62 211 0036
TLF
TLX
EML
COM 29

WOLF RAINER E A DR
MPI FUER ASTRONOMIE
KOENIGSTUHL
D 6900 HEIDELBERG 1
GERMANY
TEL 62 215 281
TLF
TLX 461789 MPIA D
EML
COM

WOLFE ARTHUR M PROF
CASS
UCSD
C 011
LA JOLLA CA 92093 0216
USA
TEL 619 534 7435
TLF 619 534 6316
TLX
EML AWOLFE@UCSD
COM

WOLFENDALE ARNOLD W PROF
DPT OF PHYSICS
UNIVERSITY OF DURHAM
SOUTH RD
DURHAM DH1 3LE
UK
TEL 913 742 160
TLF 913 743 749
TLX 537351 DURLIB G
EML
COM 48C

WOLFF SIDNEY C DR
KITT PEAK NTL OBS
BOX 26732
950 N CHERRY AVE
TUCSON AZ 85726 6732
USA
TEL 602 327 5511
TLF
TLX 666-484 AURA NOAO
EML
COM· 29,36

WOLFSON C JACOB
LOCKHEED PALO ALTO RES LB
DPT 91 30 BLDG 256
3251 HANOVER ST
PALO ALTO CA 94304 1191
USA
TEL 415 424 2855
TLF 415 424 3994
TLX
EML
COM

WOLFSON RICHARD DR
DPT OF PHYSICS
MIDDLEBURY COLLEGE
MIDDLEBURY VT 05753
USA
TEL 802 388 3711
TLF
TLX
EML WOLFSON@MIDD BITNET
COM 10

WOLSTENCROFT RAMON D DR
ROYAL OBSERVATORY
BLACKFORD HILL
EDINBURGH EH9 3HJ
UK
TEL 316 673 321
TLF
TLX 72383
EML
COM 21,34,48,51

WOLSZCZAN ALEXANDER DR
ARECIBO OBSERVATORY
BOX 995
ARECIBO PR 00613
USA
TEL 809 878 2612
TLF
TLX 385638
EML
COM 34,40

WOLTER ANNA DR
OSS ASTRONOMICO DI BRERA
VIA BRERA 28
I 20121 MILANO
ITALY
TEL 2 87 4444
TLX
EML ASTMIB WOLTER
TLF 39 59 8492
COM 48

WOLTJER LODEWIJK PROF
OHP
F 04870 S MICHEL OBS
FRANCE
TEL 92 76 6368
TLF 92 76 6295
EML
COM 10,33,34,40,47,48C
TLX 410690 OHP F

WOO JONG OK
KOREAN NTL UNIVERSITY
OF EDUCATION
CHUNGWON-GUN
CHUNGBUK 320 23
KOREA R
TEL 431 60 3712
TLF
TLX
EML
COM 25,46

WOOD DAVID B DR
6 TURNING MILL RD
LEXINGTON MA 02173
USA
TEL
TLF
TLX
EML
COM 42

WOOD F BRADSHAW PROF
DPT OF ASTRONOMY
UNIVERSITY OF FLORIDA
211 SSRB
GAINESVILLE FL 32611
USA
TEL 904 392 2059
TLF.
TLX
EML
COM 38,42

WOOD III H J DR
ASTRONOMY DPT
INDIANA UNIVERSITY
SWAIN WEST 319
BLOOMINGTON IN 47405
USA
TEL
TLF
TLX
EML
COM 29

WOOD JANET H DR
ASTRONOMY DPT
UNIVERSITY OF TEXAS
RLM 15 308
AUSTIN TX 78712 1083
USA
TEL 512 471 3432
TLF 512 471 6016
TLX 910 874 1351
EML EDU UTEXAS AS ASTRO JHW
COM 42

WOOD JOHN A DR
CENTER FOR ASTROPHYSICS
HCO/SAO
60 GARDEN ST
CAMBRIDGE MA 02138
USA
TEL 617 495 7278
TLF
TLX 921428 SATELLITE CAM
EML
COM 15,16,22

WOOD PETER R DR
MOUNT STROMLO & SIDING
SPRING OBSERVATORIES
PRIVATE BAG
WODEN PO ACT 2606
AUSTRALIA
TEL   62 88 1111
TLF
TLX   62270 CANOPUS AA
EML
COM   27,35

WOOD ROGER DR
ROYAL GREENWICH OBS
HERSTMONCEUX CASTLE
HAILSHAM BN27 1RP
UK
TEL   323 833 171*3391
TLX   87451 RGOBSY G
EML
TLF
COM

WOODSWORTH ANDREW W DR
HERZBERG INST ASTROPHYS
DOMINION ASTROPHYS OBS
5071 W SAANICH RD
VICTORIA BC V8X 4M6
CANADA
TEL   604 388 0024
TLF   604 363 0045
TLX   049 7295
EML
COM   40

WOODWARD PAUL R DR
DPT OF ASTRONOMY
UNIVERSITY OF MINNESOTA
116 CHURCH ST SE
MINNEAPOLIS MN 55455
USA
TEL
TLF
TLX
EML
COM   33,34

WOOLF NEVILLE J
STEWARD OBSERVATORY
UNIVERSITY OF ARIZONA
TUCSON AZ 85721
USA
TEL
TLF
TLX
EML
COM   34,50

WOOLFSON MICHAEL M PROF
DPT OF PHYSICS
UNIVERSITY OF YORK
HESLINGTON YORK YO1 5DD
UK
TEL   904 598 61
TLF
TLX   57933 YORKULG
EML
COM   15,16,21,22

WOOLSEY E G
1909 LAUDER DRIVE
OTTAWA ON K2A 1A9
CANADA
TEL
TLF
TLX
EML
COM

.

WOOSLEY S E PROF
LICK OBSERVATORY
UNIVERSITY OF CALIFORNIA
SANTA CRUZ CA 95064
USA
TEL   408 429 2976
TLF
TLX
EML
COM   28,35

WOOTTEN HENRY ALWYN
NRAO
EDGEMONT RD
CHARLOTTESVILLE VA 22901
USA
TEL   804 296 0211
TLX   510-587-5482
EML
TLF
COM   34,40

WORDEN SIMON P DR
6757 N 27TH STREET
ARLINGTON VA 22213
USA
TEL
TLF
EML
COM   09,12
TLX

WORLEY CHARLES E DR
US NAVAL OBSERVATORY
ASTRON & ASTROPHYSICS DIV
34 & MASSACHUSETTS AVE NW
WASHINGTON DC 20392 5100
USA
TEL   202 653 1588
TLF
TLX
EML   EQB@PHOBOS USNO NAVY MIL
COM   05,24,26VP

WORRALL DIANA MARY
CENTER FOR ASTROPHYSICS
HCO/SAO
60 GARDEN ST
CAMBRIDGE MA 02138
USA
TEL   617 495 7139
TLF
TLX   921428 SATELLITE CAM
EML
COM   28,48

WORRALL GORDON DR
BIRDSWOOD
EARDISLEY HEREFDS
UK
TEL
TLF
EML
COM
TLX

WORSWICK SUSAN
ROYAL GREENWICH OBS
MADINGLEY RD
CAMBRIDGE CB3 0EZ
UK
TEL   223 37 4000
TLF   223 37 4700
TLX   265451/265871MONREFG
EML
COM   09

WOSZCZYK ANDRZEJ PROF
INSTITUTE OF ASTRONOMY
N COPERNICUS UNIVERSITY
UL CHOPINA 12/18
PL 87 100 TORUN
POLAND
TEL   2-60-18
TLF
TLX   00552234 ASTR PL
EML
COM   15,16,50

WRAMDEMARK STIG S O DR
LUND OBSERVATORY

BOX 43
S 221 00 LUND
SWEDEN
TEL   46 10 7303
TLF
TLX   33199 OBSNOT S
EML
COM   25,33,37

WRAY JAMES D DR
ASTRONOMY DPT
UNIVERSITY OF TEXAS
RLM 15 212
AUSTIN TX 78712 1083
USA
TEL
TLF
TLX
EML
COM   44

WRIGHT ALAN E DR
AAO
PRIVATE BAG
BOX 276
PARKES NSW 2870
AUSTRALIA
TEL   68 623 677
TLF
TLX   63999 QASAR
EML   AWRIGHT@ATNF CSIRO AU
COM   05,40

WRIGHT EDWARD L DR
DPT OF ASTRONOMY
UNIVERSITY OF CALIFORNIA
LOS ANGELES CA 90024
USA
TEL   213 825 5755
TLF
TLX
EML   BONNIE WRIGHT
COM   34,47

WRIGHT FRANCES W DR
DPT OF ASTRONOMY
HARVARD UNIVERSITY
60 GARDEN ST
CAMBRIDGE MA 02138
USA
TEL   617 495 2647
TLF
TLX
EML
COM   27

WRIGHT HELEN GREUTER
THOMAS HOUSE APT 517
1330 MASSACHUSETTS AVE
WASHINGTON DC 20005
USA
TEL
TLF
TLX
EML
COM 41

WRIGHT JAMES P DR
NTL SCIENCE FOUNDATION
DIV ASTRONOMICAL SCIENCES
1800 G ST NW
WASHINGTON DC 20550
USA
TEL 202 357 7639
TLF
TLX
EML
COM

WRIGHT KENNETH O DR
HERZBERG INST ASTROPHYS
DOMINION ASTROPHYS OBS
5071 W SAANICH RD
VICTORIA BC V8X 4M6
CANADA
TEL 604 388 3157
TLF 604 363 0045
TLX 049 7295
EML
COM 36,42

WRIGHT MELVYN C H DR
RADIO ASTRONOMY LAB
UNIVERSITY OF CALIFORNIA
601 CAMPBELL HALL
BERKELEY CA 94720
USA
TEL 415 642 0420
TLF
TLX
EML
COM

WRIXON GERARD T DR
NTL MICROELECTRON RES CTR
UNIVERSITY COLLEGE CORK
CORK
IRELAND
TEL 1 215 08375
TLF
TLX 26050
EML
COM

WROBEL JOAN MARIE DR
NRAO
BOX 0
SOCORRO NM 87801 0387
USA
TEL 505 835 7000
TLX 910 988 1710
EML JWROBEL@NRAO EDU
TLF 505 835 7027
COM 28,40

WROBLEWSKI HERBERT DR
DPT DE ASTRONOMIA
UNIVERSIDAD DE CHILE
CASILLA 36 D
SANTIAGO
CHILE
TEL 2 229 4101
TLF
TLX 440 001
EML
COM 20,24

WU CHI CHAO DR
STSCI/CSC
HOMEWOOD CAMPUS
3700 SAN MARTIN DR
BALTIMORE MD 21218
USA
TEL 301 338 4770
TLF
TLX U S A
EML
COM 34,44

WU HSIN-HENG DR
DPT OF PHYSICS
NTL CENTRAL UNIVERSITY
CHUNG LI
CHINA R
TEL
TLF
TLX
EML
COM 12,37,45

WU HUAI-WEI
SHANGHAI OBSERVATORY
CAS
80 NANDAN RD
SHANGHAI
CHINA PR
TEL 21 38 6191
TLF
TLX 33164 SHAO CN
EML
COM 40

WU LIAN-DA
PURPLE MOUNTAIN OBSERV
CAS
NANJING
CHINA PR
TEL 25 32893
TLF
TLX 34144 PMONJ CN
EML
COM 07

WU LIN-XIANG
DPT OF GEOPHYSICS
BEIJING UNIVERSITY
BEIJING 100071
CHINA PR
TEL
TLF
TLX
EML
COM 09,10,12

WU MING-CHAN
YUNNAN OBSERVATORY
CAS
BOX 110
KUNMING 72946 YUNNAN
CHINA PR
TEL 871 2035
TLF
TLX 64040 YUOBS CN
EML
COM 10,50

WU NAILONG DR
BEIJING ASTRONOMICAL OBS
CAS
W SUBURB
BEIJING 100080
CHINA PR
TEL 1 256 1265
TLF 1 256 1085
TLX 22040
EML BMAMAO@ICA BEIJING CANET CN
COM 40

WU SHENGYIN
BEIJING ASTRONOMICAL OBS
CAS
W SUBURB
BEIJING 100080
CHINA PR
TEL
TLF
TLX 22040 BADAS CN
EML
COM 40

WU SHI TSAN DR
SCHOOL OF ENGINEERING
UNIVERSITY OF ALABAMA
HUNTSVILLE AL 35899
USA
TEL 205 895 6413
TLF
TLX
EML
COM 10,49

WU SHOU-XIAN
SHAANXI OBSERVATORY
CAS
LINTONG XIAN
SHAANXI
CHINA PR
TEL 33 55951
TLF
TLX 70121 CSAO CN
EML
COM 19,31

WU XINJI
DPT OF GEOPHYSICS
BEIJING UNIVERSITY
BEIJING 100071
CHINA PR
TEL 1 28 2471*3929
TLF
TLX 22239 PKUNI CN
EML
COM 40

WU ZHIREN DR
SHANGHAI SCIENTIFICAL &
TECH EDU PUBLISHING HOUSE
393 GUAN SHEN YAUN RD
SHANGHAI 200233
CHINA PR
TEL 21 36 5791
TLF
TLX
EML
COM 05

WUNNER GUENTER
LEHRSTUHL F THEORET ASTRO
PHYSIK DER UNIV TUEBINGEN
AUF DER MORGENSTELLE 12,C
D 7400 TUEBINGEN
GERMANY
TEL 707 129 2487
TLF
TLX 7262714 AIT D
EML
COM 14,44

WYCKOFF SUSAN DR
DPT OF PHYSICS
ARIZONA STATE UNIVERSITY
ASTRONOMY PROGRAM
TEMPE AZ 85287
USA
TEL 602 965 3561
TLF
TLX 140289 HALLEY ASU UT
EML WYCKOFF@ASYCOS BITNET
COM 15,29,45

WYLLER ARNE A PROF
RR9 LOT 14
ARROYO HONDO WEST
SANTA FE NM 87505
USA
TEL
TLF 505 473 5849
TLX
EML
COM 09,12,36

WYNNE CHARLES G PROF
INSTITUTE OF ASTRONOMY
THE OBSERVATORIES
MADINGLEY RD
CAMBRIDGE CB3 OHA
UK
TEL 223 3375
TLF
TLX 817297 ASTRON G
EML
COM 09

WYNN-WILLIAMS C G DR
INSTITUTE FOR ASTRONOMY
UNIVERSITY OF HAWAII
2680 WOODLAWN DR
HONOLULU HI 96822
USA
TEL 808 948 8807
TLF
TLX
EML
COM 28,34

WYSE ROSEMARY F DR
DPT PHYSICS & ASTRONOMY
JOHNS HOPKINS UNIVERSITY
CHARLES & 34TH ST
BALTIMORE MD 21218
USA
TEL 410 516 5392
TLF 410 516 8260
TLX
EML JHMAIL WYSER WYSE
COM 33

XANTHAKIS JOHN N PROF
RES CENTER FOR ASTRONOMY
ACADEMY OF ATHENS
14 ANAGNOSTOPOULOU ST
GR 106 73 ATHENS
GREECE
TEL 1 361 3589
TLF
TLX
EML
COM 10

XANTHOPOULOS B C DR
DPT OF PHYSICS
UNIVERSITY OF CRETE
BOX 1527
GR 711 11 IRAKLION
GREECE
TEL 81 23 5576
TLF
TLX 262728
EML
COM 47

XI ZE-ZONG
INST HISTORY NAT SCIENCE
1 GONG YUAN WEST RD
BEIJING
CHINA PR
TEL 1 55 7180
TLF
TLX
EML
COM 41

XIA JIONGYU
INSTITUTE OF GEODESY &
GEOPHYSICS
XU DONG LU
WUCHANG HUBEI
CHINA PR
TEL
TLF
TLX
EML
COM 19

XIA XIAOYANG DR
DPT OF PHYSICS
TIANJIN NORMAL UNIVERSITY
TIANJIN 300074
CHINA PR
TEL 22 71 6989
TLF
TLX
EML
COM 28

XIA YI-FEI
DPT OF ASTRONOMY
NANJING UNIVERSITY
NANJING
CHINA PR
TEL 25 34651*2882
TLF
TLX 34151 PRCNU CN
EML
COM 08

XIA ZHIGUO DR
INSTITUT FUER ASTRONOMIE
ETH ZENTRUM
CH 8092 ZUERICH
SWITZERLAND
TEL 1 256 3813
TLF
TLX 817379 EHHG CH
EML
COM 40

XIAN DING-ZHANG
PURPLE MOUNTAIN OBSERV
CAS
NANJING
CHINA PR
TEL 25 37609
TLF
TLX 34144 PMONJ CN
EML
COM 04

XIANG DELIN
PURPLE MOUNTAIN OBSERV
CAS
NANJING
CHINA PR
TEL 25 33738
TLF
TLX 34144 PMONJ CN
EML
COM 33,34

XIANG SHOUPING
ASTROPHYSICS DIVISION
UNIV SCIENCE & TECHNOLOGY
HEFEI 230026 ANHUI
CHINA PR
TEL 551 33 1134
TLF
TLX
EML
COM 47

XIAO NAI-YUAN
DPT OF ASTRONOMY
NANJING UNIVERSITY
NANJING
CHINA PR
TEL 25 34651*2882
TLF
TLX 34151 PRCNU CN
EML
COM 19

XIE GUANG-ZHONG
YUNNAN OBSERVATORY
CAS
BOX 110
KUNMING 72946 YUNNAN
CHINA PR
TEL 871 2035
TLF
TLX 64040 YUOBS CN
EML
COM

XIE LIANGYUN
INSTITUTE OF GEODEDY &
GEOPHYSICS
XU DONG LU
WUCHANG HUBEI
CHINA PR
TEL
TLF
TLX
EML
COM 08

XIONG DA-RUN
PURPLE MOUNTAIN OBSERV
CAS
NANJING
CHINA PR
TEL 25 42817
TLF
TLX 34144 PMONJ CN
EML
COM 27,35

XIRADAKI EVANGELIA DR
DPT OF ASTROPHYSICS
UNIVERSITY OF ATHENS
PANEPISTIMIOPOLIS
GR ATHENS
GREECE
TEL 1 72 35 122/9628 306
TLF
TLX
EML
COM 37

XU AO-AO
DPT OF ASTRONOMY
NANJING UNIVERSITY
NANJING
CHINA PR
TEL
TLF
TLX   34151 PRC NU CN
EML
COM 10

XU BANG-XIN
DPT OF ASTRONOMY
NANJING UNIVERSITY
NANJING
CHINA PR
TEL
TLF
TLX
EML
COM 08,31

XU JIA-YAN
SHAANXI OBSERVATORY
CAS
LINTONG XIAN
SHAANXI
CHINA PR
TEL 33 2255
TLF
TLX
EML
COM 19

XU JI-HONG DR
URUMQI ASTRONOMICAL STAT
CAS
XINJIANG 830011
CHINA PR
TEL 33 5757
TLF
TLX 79152 KWTWZ CN
EML 830011
COM 07

XU PEI-YUAN
INST ELECTRONIC PHYSICS
SUST
JIA DING
SHANGHAI
CHINA PR
TEL 21 95 1602
TLF
TLX
EML
COM 40

XU PINXIN
PURPLE MOUNTAIN OBSERV
CAS
NANJING
CHINA PR
TEL 25 32893
TLF
TLX 34144 PMONJ CN
EML
COM 07,22

XU TONG-QI
SHANGHAI OBSERVATORY
CAS
80 NANDAN RD
SHANGHAI 200030
CHINA PR
TEL 21 38 6191
TLF
TLX 33164 SHAO CN
EML
COM 08,19

XU ZHENTAO
PURPLE MOUNTAIN OBSERV
CAS
NANJING
CHINA PR
TEL 25 31096
TLF
TLX 34144 PMONJ CN
EML
COM 10,41

XU ZHI-CAI
PURPLE MOUNTAIN OBSERV
CAS
NANJING
CHINA PR
TEL 25 46700
TLF
TLX
EML
COM 40

YABUSHITA SHIN A PROF
DPT APPLIED MATHS & PHYS
KYOTO UNIVERSITY
SAKYO KU
KYOTO 606
JAPAN
TEL 75 751 2111
TLF
TLX
EML
COM 15,20,34

YABUUTI KIYOSHI PROF
20 TANAKA HIGASKI
HINOKUCH MACHI
KYOTO 606
JAPAN
TEL
TLF
TLX
EML
COM 41

YAHIL AMOS DR
DPT OF EARTH & SPACE SCI
ASTRONOMY PROGRAM
SUNY AT STONY BROOK
STONY BROOK NY 11794 2100
USA
TEL 516 246 6545
TLF
TLX 510-228-7767
EML
COM

YAKOVKIN N A DR
ASTRONOMICAL OBSERVATORY
KIEV STATE UNIVERSITY
OBSERVATORNAYA UL 3
252053 KIEV
UKRAINE
TEL
TLF
TLX
EML
COM 10

YALLOP BERNARD D DR
ROYAL GREENWICH OBS
MADINGLEY RD
CAMBRIDGE CB3 0EZ
UK
TEL 223 37 4735
TLF 223 37 4700
TLX 817235 RGOSTR G
EML BDY@UK AC RGO SRF
COM 04P

YAMAGATA TOMOHIKO DR
TOKYO ASTRONOMICAL OBS
NAOJ
OSAWA MITAKA
TOKYO 181
JAPAN
TEL 0422 32 5111
TLF
TLX 02822307 TAOMTK J
EML
COM 28

YAMAGUCHI SHICHIRO
FACULTY OF ENGINEERING
GIFU UNIVERSITY
YANAGIDO
GIFU 501 11
JAPAN
TEL 582-30-1111
TLF
TLX
EML
COM

YAMAKOSHI KAZUO
INST COSMIC RAY RESEARCH
UNIVERSITY OF TOKYO
MIDORICHO TANASHI
TOKYO 188
JAPAN
TEL 0424-61-4131
TLF
TLX 02822371 ICRTU J
EML
COM 21,22

YAMAMOTO TETSUO DR
ISAS
3-1-1 YOSHINODAI
SAGAMIHARA
KANAGAWA 229
JAPAN
TEL 0427 51 3911
TLF
TLX 24550 SPACETKY J
EML
COM 21

YAMASAKI ATSUMA DR
DPT OF GEOSCIENCE
NATIONAL DEFENSE ACADEMY
HASHIRIMIZU
YOKOSUKA 239
JAPAN
TEL 81 468 41 3810
TLF 81 468 44 5902
TLX
EML
COM 42C

YAMASHITA KOJUN DR
ISAS
3-1-1 YOSHINODAI
SAGAMIHARA
KANAGAWA 229
JAPAN
TEL 427 51 3911
TLF
TLX
EML
COM 21,44

YAMASHITA YASUMASA PROF
TOKYO ASTRONOMICAL OBS
NAOJ
OSAWA MITAKA
TOKYO 181
JAPAN
TEL
TLF
TLX   2822307 TAOMTK J
EML
COM 25,29,45

YAMAZAKI AKIRA DR
HYDROGRAPHIC DPT
GEODESY & GEOPHYSICS DIV
TSUKIJI 5 CHUO KU
TOKYO 104
JAPAN
TEL   03-541-3811
TLF
TLX   0 252 2222 JAHYD J
EML
COM 04,08

YAN LIN-SHAN
SHANGHAI OBSERVATORY
CAS
80 NANDAN RD
SHANGHAI
CHINA PR
TEL   21 38 6191
TLF
TLX   33164 SHAO CN
EML
COM 26

YANG FUMIN
SHANGHAI OBSERVATORY
CAS
80 NANDAN RD
SHANGHAI
CHINA PR
TEL   21 38 6191
TLF
TLX   33164 SHAO CN
EML
COM 19

YANG JIAN
PURPLE MOUNTAIN OBSERV
CAS
NANJING
CHINA PR
TEL   25 46700
TLF
TLX   34144 PMONJ CN
EML
COM 40

YANG KE-JUN
GEOPHYSICAL INSTITUTE
UNIVERSITY OF ALASKA
FAIRBANKS AK 99775 0800
USA
TEL
TLF
TLX
EML
COM 31

YANG LAN-TIAN
DPT OF PHYSICS
HUAZHONG NORMAL UNIV
WUHAN
CHINA PR
TEL   75 601*300/401
TLF
TLX
EML
COM 47,48

YANG SHI JIE
PURPLE MOUNTAIN OBSERV
CAS
NANJING
CHINA PR
TEL   25 46700
TLF
TLX   34144 PMO NJ CN
EML
COM 09

YANG TING-GAO
SHAANXI ASTRONOMICAL POBS
BOX 18
LINTONG
WIAN
CHINA PR
TEL
TLF
TLX
EML
COM 24

YANKULOVA IVANKA DR
DPT OF ASTRONOMY
UNIVERSITY OF SOFIA
ANTON IVANOV ST 5
BG 1126 SOFIA
BULGARIA
TEL   2 54 4852
TLF
TLX
EML
COM

YANOVITSKIJ EDGARD G DR
MAIN ASTRONOMICAL OBS
UKRAINIAN ACAD OF SCIENCE
GOLOSEEVO
252127 KIEV
UKRAINE
TEL   66 3110
TLF
TLX   131406 SKY SU
EML
COM 36

YAO BAO-AN
SHANGHAI OBSERVATORY
CAS
80 NANDAN RD
SHANGHAI
CHINA PR
TEL   21 38 6191
TLF
TLX   33164 SHAO CN
EML
COM 27

YAO JIN-XING
PURPLE MOUNTAIN OBSERV
CAS
NANJING
CHINA PR
TEL   25 46700
TLF
TLX
EML
COM 10

YAO ZHENG-QIU
DPT OF SCIENTIFIC COOP
SSRG
BOX 4470
DAMASCUS
SYRIA
TEL
TLF
TLX
EML
COM 09

YAPLEE B S
8 CREST VIEW CT
ROCKVILLE MD 20854
USA
TEL   301 762 0935
TLF
TLX
EML
COM

YAROV-YAROVOJ M S DR
MATHEMATICS DPT
MVTU
VTORAYA BAUMANSKAYA 5
107005 MOSCOW
RUSSIA
TEL   267-03-92
TLF
TLX   111572
EML
COM 07

YASUDA HARUO PROF DR
TOKYO ASTRONOMICAL OBS
NAOJ
OSAWA MITAKA
TOKYO 181
JAPAN
TEL
TLF
TLX
EML
COM 08

YATSKIV YA S DR
MAIN ASTRONOMICAL OBS
UKRAINIAN ACAD OF SCIENCE
GOLOSEEVO
252127 KIEV
UKRAINE
TEL   66 3110
TLF
TLX   131406 SKY SU
EML
COM 08,19

YAU KEVIN K C DR
DPT OF PHYSICS
UNIVERSITY OF DURHAM
SOUTH RD
DURHAM DH1 3LE
UK
TEL   913 742 153
TLF   913 743 749
TLX   537351 DURLIB G
EML   YAU@UK AC DUR STAR
COM 41

YAVNEL ALEXANDER A DR
METEORITE COMMITTEE
ACADEMY OF SCIENCES
ULIANOVOJ M UL 3 K 1
117313 MOSCOW
RUSSIA
TEL   1377538
TLF
TLX
EML
COM 15,22

YE BINXUN
YUNNAN OBSERVATORY
CAS
BOX 110
KUNMING 72946 YUNNAN
CHINA PR
TEL 871 2035
TLF
TLX 64040 YUOBS CN
EML
COM 09

YE SHI-HUI
PURPLE MOUNTAIN OBSERV
CAS
NANJING
CHINA PR
TEL 25 46700
TLF
TLX 34144 PMONTJ CN
EML
COM 10

YE SHU-HUA
SHANGHAI OBSERVATORY
CAS
80 NANDAN RD
SHANGHAI 200030
CHINA PR
TEL 21 38 6191
TLF 21 38 4618
TLX 33164 SHAO CN
EML
COM 08,19,31C,38,40C,EC

YE WENWEI
INSTITUTE OF SEISMOLOGY
STATE SEISMO BUREAU
XIAO HONG SHAN WUHAN
HUBEI 230026 ANHUI
CHINA PR
TEL 81 3401
TLF
TLX
EML
COM

YEE HOWARD K C DR
DPT OF ASTRONOMY
UNIVERSITY OF TORONTO
60 ST GEORGE ST
TORONTO ON M5S 1A1
CANADA
TEL 416 978 4833
TLF 416 978 3921
TLX 069 86766
EML HYEE@UTORPHYS BITNET
COM

YEH TYAN DR
HIGH ALTITUDE OBSERVATORY
NCAR
BOX 3000
BOULDER CO 80307 3000
USA
TEL 303 497 5401
TLF
TLX
EML
COM 10,49

YEIVIN Y PROF
SCHOOL OF PHYS & ASTRON
TEL AVIV UNIVERSITY
RAMAT AVIV
TEL AVIV 69978
ISRAEL
TEL
TLF
TLX
EML
COM

YEN JUI-LIN PROF
DPT OF ELECTRICAL ENGIN
UNIVERSITY OF TORONTO
60 ST GEORGE ST
TORONTO ON M5S 1A4
CANADA
TEL 416 978 8756
TLF 416 978 3921
TLX 069 86766
EML
COM

YEOMANS DONALD K DR
JPL
MS 301 150 G
4800 OAK GROVE DR
PASADENA CA 91109
USA
TEL 818 354 2127
TLF 818 393 1159
TLX 675429 JPL COMM PSD
EML 5122 DKY
COM 15,20VP,22,41

YI ZHAO-HUA
DPT OF ASTRONOMY
NANJING UNIVERSITY
NANJING
CHINA PR
TEL
TLF
TLX
EML
COM. 07

YILMAZ FATMA DR
UNIVERSITY OBSERVATORY
UNIVERISTY OF ISTANBUL
34452 ISTANBUL
TURKEY
TEL
TLF
TLX
EML
COM

YILMAZ NIHAL DR
DPT OF ASTRONOMY
UNIVERSITY OF ANKARA
FEN FAKULTESI
06100 BESEVLER
TURKEY
TEL 41 23 6550
TLF
TLX
EML
COM

YIN JI-SHENG
BEIJING ASTRONOMICAL OBS
CAS
W SUBURB
BEIJING 100080
CHINA PR
TEL 1 28 1203
TLF
TLX 22040 BAOAS CN
EML
COM 25

YIN QI-FENG
DPT OF GEOPHYSICS
BEIJING UNIVERSITY
BEIJING 100071
CHINA PR
TEL 1 28 2471*3888
TLF
TLX 22239 PKUNI CN
EML
COM 40

YODER CHARLES F
JPL
MS 183 150
4800 OAK GROVE DR
PASADENA CA 91109
USA
TEL 818 354 2444
TLF
TLX 617-5429
EML
COM 16

YOKOSAWA MASAYOSHI DR
DPT OF PHYSICS
IBARAKI UNIVERSITY
2-1-1 BUNKYO MITO
IBARAKI 310
JAPAN
TEL 292 26 1621
TLF
TLX
EML
COM

YOKOYAMA KOICHI DR
INTL LATITUDE OBSERVATORY
NAOJ
HOSHIGAOKA MIZUSAWA SHI
IWATE 023
JAPAN
TEL 0197-24-7111
TLF
TLX 837628
EML
COM 19

YOKOYAMA TADASHI DR
UNIVERSIDADE ESTADUAL
PAULISTA
CP 178
13500 RIO CLARO
BRAZIL
TEL 195 34 0122
TLF
TLX 011-31870
EML
COM 07

YONEYAMA TADAOKI DR
2-1-16 HIBARIGAOKA-KITA
HOYA-SHI
TOKYO 202
JAPAN
TEL
TLF
TLX
EML
COM

YORK DONALD G DR
ASTRONOMY & ASTROPHYS CTR
UNIVERSITY OF CHICAGO
5640 S ELLIS AVE
CHICAGO IL 60637
USA
TEL 312 962 8930
TLF
TLX 910-221-5617
EML
COM 34

YORKE HAROLD W DR
INST ASTRONOIME/ASTROPHY
UNIVERSITAET WUERZBURG
AM HUBLAND
D 8700 WUERZBURG
GERMANY
TEL
TLF
TLX
EML
COM 34,35,36

YOSHIDA HARUO
TOKYO ASTRONOMICAL OBS
NAOJ
OSAWA MITAKA
TOKYO 181
JAPAN
TEL 422 41 3614
TLF 442 41 3793
TLX 02822307 TAOMTK
EML YOSHIDA@C1 MTK NAO AC JP
COM 07

YOSHIDA JUNZO PROF
DPT OF PHYSICS
KYOTO SANGYO UNIVERSITY
KAMIGAMO
KYOTO 603
JAPAN
TEL 75 7012151
TLF
TLX 5422661 KSUJ
EML
COM 07

YOSHII YUZURU DR
TOKYO ASTRONOMICAL OBS
NAOJ
OSAWA MITAKA
TOKYO 181
JAPAN
TEL 422 32 5111
TLF
TLX 2822307 TAOMTK J
EML
COM 33C,47

YOSHIMURA HIROKAZU DR
DPT OF ASTRONOMY
UNIVERSITY OF TOKYO
BUNKYO KU
TOKYO 113
JAPAN
TEL 03-812-2111
TLF.
TLX 33659 UTYOSCI
EML
COM 10,12C

YOSHIMURA MOTOHIKO DR
DPT OF PHYSICS
TOHOKU UNIVERSITY
ARAMAKI
SENDAI 980
JAPAN
TEL 81 22 222 1800
TLF 81 22 225 1891
TLX
EML
COM 47

YOSHIZAWA MASANORI DR
TOKYO ASTRONOMICAL OBS
NAOJ
OSAWA MITAKA
TOKYO 181
JAPAN
TEL 422 32 5111
TLF 422 32 1924
TLX 2822307 TAOMTK
EML
COM 08C

YOSS KENNETH M DR
DPT OF ASTRONOMY
UNIVERSITY OF ILLINOIS
1011 W SPRINGFIELD AVE
URBANA IL 61801
USA
TEL 217 333 3295
TLF
TLX
EML
COM 30,45

YOU JIAN-QI
PURPLE MOUNTAIN OBSERV
CAS
NANJING
CHINA PR
TEL 25 46700
TLF
TLX 34144 PMONJ CN
EML
COM 10,12

YOU JUNHAN
ASTROPHYSICS DIVISION
UNIV SCIENCE & TECHNOLOGY
HEIFI 230026 ANHUI
CHINA PR
TEL 551 33 1134
TLF
TLX 90028 USTC CN
EML
COM 48

YOUNG ANDREW T DR
DPT OF ASTRONOMY
SAN DIEGO STATE UNIV
SAN DIEGO CA 92182-0334
USA
TEL 619 265 5817
TLF
TLX
EML ATY@MINTAKA SDSU EDU
COM 16,25P

YOUNG ARTHUR DR
DPT OF ASTRONOMY
SAN DIEGO STATE UNIV
SAN DIEGO CA 92182
USA
TEL 619 265 6167
TLF
TLX
EML
COM

YOUNG JUDITH SHARN
FIVE COLLEGE
RADIO ASTRONOMY OBSERV
B619 LEDERLE GRAD RES TWR
AMHERST MA 01003
USA
TEL 413 545 0789
TLF
TLX· 95-5491
EML
COM 28

YOUNG LOUISE GRAY DR
DPT OF ASTRONOMY
SAN DIEGO STATE UNIV
SAN DIEGO CA 92182
USA
TEL 619 287 8890
TLF
TLX
EML
COM 14,16

YOUNIS SAAD M
SARC
SCIENTIFIC RES COUNCIL
BOX 2441
JADIRIYAH BAGHDAD
IRAQ
TEL 1 776 5127
TLF
TLX 2187 BATHILMI IK
EML
COM 24,33,34,40

YOUSEF SHAHINAZ M DR
DPT OF ASTRONOMY
FACULTY OF SCIENCES
CAIRO UNIVERSITY
GEZA
EGYPT
TEL
TLF
TLX
EML
COM

YOUSSEF NAHED H PROF
DPT OF ASTRONOMY
FACULTY OF SCIENCES
CAIRO UNIVERSITY
GEZA
EGYPT
TEL 58 6041
TLF
TLX
EML
COM 12

YU KYUNG-LOH PROF
DPT OF ASTRONOMY
SEOUL NTL UNIVERSITY
KWANAK KU
SEOUL 151
KOREA R
TEL
TLF
TLX
EML
COM 08

YU XIN ALFRED DR
DPT OF APLLIED MATHS
HONG KONG POLYTECHNIC
HUNG HOM
KOWLOON
HONG KONG
TEL 3 766 6951
TLF 3 362 9045
TLX
EML
COM

YU YAN DR
DPT OF ASTRONOMY
UNIVERSITY OF ILLINOIS
1002 W GREEN ST
URBANA IL 61801
USA
TEL 217 244 1187
TLF
TLX
EML YU@DENEB.ASTRO UIUC EDU
COM 14

YUAN CHI PROF
DPT OF PHYSICS
CITY COLLEGE OF NY
138 ST CONVENE AVE
NEW YORK NY 10031
USA
TEL 212 690 6823
TLF
TLX
EML
COM 33

YUASA MANABU DR
DPT OF MATH & PHYSICS
KINKI UNIVERSITY
HIGASHI
OSAKA 577
JAPAN
TEL·
TLF
TLX
EML
COM 07,20

YUE ZENG-YUAN
DPT OF GEOPHYSICS
BEIJING UNIVERSITY
BEIJING 100071
CHINA PR
TEL
TLF
TLX
EML
COM

YULDASHBAEV TAIMAS S
ASTRONOMICAL INSTITUTE
UZBEK ACADEMY OF SCIENCES
700052 TASHKENT
UZBEKISTAN
TEL
TLF
TLX
EML
COM

YUMI SHIGERU PROF DR
KEYAKIDAI 1-12-2
KIYAMACHO
MIYAKIGUN
SAGA 841 02
JAPAN
TEL
TLF
TLX
EML
COM 19

YUN HONG-SIK PROF
DPT OF ASTRONOMY
SEOUL NTL UNIVERSITY
KWANAK KU
SEOUL 151
KOREA R
TEL 877-2130 x2542
TLF
TLX
EML
COM 10,12

YUNGELSON LEV R
INST OF ASTRONOMY
ACADEMY OF SCIENCES
PYATNITSKAYA UL 48
109017 MOSCOW
RUSSIA
TEL 231-54-61
TLF
TLX 412623 SCSTP SU
EML
COM 35

ZABRISKIE F R PROF
RD 1

ALEXANDRIA PA 16611
USA
TEL 814 669 4483
TLF
TLX
EML
COM

ZACHARIADIS THEODOSIOS DR
RES CENTER FOR ASTRONOMY
ACADEMY OF ATHENS
14 ANAGNOSTOPOULOU ST
GR 106 73 ATHENS
GREECE
TEL
TLF
TLX
EML EXAKA20@GRATHUN1
COM 10

ZACHARIAS NORBERT DR
HAMBURGER STERNWARTE
GOJENSBERGSWEG 112
D 2050 HAMBURG 80
GERMANY
TEL 40 7252 4112
TLX 217884
EML ZACHARIAS@DKRZ-HAMBURG DBP DE
TLF 40 7252 4198
COM 24

ZACHAROV IGOR DR
ASTRONOMICAL INSTITUTE
CZECH ACADEMY OF SCIENCES
ONDREJOV OBSERVATORY
CS 251 65 ONDREJOV
CZECHOSLOVAKIA
TEL 204 85201
TLF 204 85314
TLX 121579
EML
COM 09

ZACHILAS LOUKAS DR
DPT OF CHEMISTRY
UNIVERSITY OF CRETE
BOX 1527
GR 714 09 IRAKLION CRETE
GREECE
TEL 81 21 2453
TLF 81 23 8468
TLX
EML ZACHILAS@TALOS CC UCH GR
COM 33

ZADUNAISKY PEDRO E PROF
UNIVERSIDAD BUENOS AIRES
FAC CIENCIAS EXACTAS MATH
CIUDAD UNIVERSITARIA PAB1
1428 BUENOS AIRES
ARGENTINA
TEL
TLF
TLX
EML
COM 20

ZAFIROPOULOS BASIL DR
DPT OF PHYSICS
UNIVERSITY OF PATRAS
GR 261 10 RION
GREECE
TEL 61 99 1973
TLF 61 99 1909
TLX 312447 UNPA GR
EML
COM 07

ZAHN JEAN-PAUL DR
OBS MIDI PYRENEES
14 AVE E BELIN
F 31400 TOULOUSE CDX
FRANCE
TEL 61 25 2101
TLF
TLX 530776 F
EML
COM 35,36

ZAITSEV VALERII V DR
INST OF APPLIED PHYSICS
ACADEMY OF SCIENCES
ULYANOV UL 46
603600 N NOVGOROD
RUSSIA
TEL
TLF
TLX
EML
COM 40

ZAMBON GIULIO DR
25 URAWA ROAD
DUNCRAIG 6023
AUSTRALIA
TEL 9 447 8849
TLF
EML
COM
TLX

ZAMBRANO ALEJANDRO DR
REAL INST Y OBSERVATORIO
DE LA ARMADA
CECILIO PUJAZON S/N
E 11110 SAN FERNANDO
SPAIN
TEL 56 88 3548
TLF 56 89 9302
TLX 76108 IOM E
EML
COM 04

ZAMORANI GIOVANNI
IST DI RADIOASTRONOMIA
CNR
VIA IRNERIO 46
I 40126 BOLOGNA
ITALY
TEL 51 23 2856
TLF
TLX 211664 INFN BO I
EML
COM 47,48

ZAMORANO JAIME DR
DPT DE ASTROFISICA
FAC C FISICAS
UNIVERSIDAD COMPLUTENSE
E 28040 MADRID
SPAIN
TEL 1 449 5316
TLF
TLX 47272
EML
COM 28

ZANDER RODOLPHE DR
INSTITUT D'ASTROPHYSIQUE
UNIVERSITE DE LIEGE
AVE COINTE 5
B 4000 COINTE-LIEGEE
BELGIUM
TEL 41 52 9980
TLF 41 52 7474
TLX 41264 ASTRLG B
EML
COM

ZANINETTI LORENZO
IST DI FISICA GENERALE
CORSO D AZEGLIO 46
I 10125 TORINO
ITALY
TEL 11 65 7694
TLX 211041 INFNTO I
EML
TLF
COM

ZAPPALA ROSARIO ALDO DR
IST DI ASTRONOMIA
CITTA UNIVERSITARIA
VIA A DORIA 6
I 95125 CATANIA
ITALY
TEL 33 0533*493
TLF
TLX 970359 ASTRCT I
EML
COM 10

ZAPPALA VINCENZO PROF
OSS ASTRONOMICO DI TORINO

ST OSSERVATORIO 20
I 10025 PINO TORINESE
ITALY
TEL 11 84 1067
TLF 11 84 1281
TLX 213236 TO ASTR I
EML
COM 15C,20

ZARE KHALIL DR
1180 AWALT DR
MOUNTAIN VIEW CA 94040
USA
TEL 415 940 1881
TLF
EML
COM 07
TLX

ZARNECKI JAN CHARLES DR
UNIT FOR SPACE SCIENCES
UNIVERSITY OF KENT
CANTERBURY CT2 7NR
UK
TEL 227 764 000
TLF 227 762 616
TLX 965449 UKLIB
EML jcz@uk ac ukc
COM 15,44

ZARRO DOMINIC M DR
NASA/GSFC
CODE 602 6
BLDG 7
GREENBELT MD 20771
USA
TEL 301 286 2039
TLF
TLX 89675
EML SOLAR DZARRO
COM 12

ZASOV ANATOLE V DR
STERNBERG STATE ASTR INST

UNIVERSITETSKIJ PROSP 13
119899 MOSCOW
RUSSIA
TEL
TLF
TLX
EML
COM 28

ZAVATTI FRANCO
DPT DI ASTRONOMIA
UNIVERSITA DI BOLOGNA
VIA ZAMBONI 33
I 40126 BOLOGNA
ITALY
TEL 51 22 2956
TLF
TLX 211664 INFNBO I
EML
COM 28

ZDANAVICIUS KAZIMERAS DR
INST OF THEORET PHYS/ASTR
LITHUENIAN ACADEMY OF SCI
POZELOS 54
232600 VILNIUS
LITHUANIA
TEL 122 61 3440
TLF
TLX 261141 LM1 SU
EML
COM 45C

ZEALEY WILLIAM J DR
UNIVERSITY OF WOLLONGONG
PHYSICS DEPT
BOX 1144
WOLLONGONG NSW 2500
AUSTRALIA
TEL 42 27 0555
TLF
TLX 29022 AA
EML
COM 09,34,46

ZEILIK MICHAEL II DR
DPT PHYSICS & ASTRONOMY
UNIVERSITY OF NEW MEXICO
800 YALE BLVD NE
ALBUQUERQUE NM 87131
USA
TEL 505 277 4442
TLF
TLX
EML
COM 34,42,46

ZEILINGER WERNER W DR
ESO
KARL SCHWARZSCHILDSTR 2
D 8046 GARCHING MUENCHEN
GERMANY
TEL 89 320 06247
TLF 89 320 06480
TLX 5282820 EO D
EML WERNER@DGAES051 BITNET
COM 28

ZEIPPEN CLAUDE DR
OBSERVATOIRE DE PARIS
SECTION DE MEUDON
F 92195 MEUDON PPL CDX
FRANCE
TEL 1 45 07 7443
TLF
TLX 201571
EML
COM 14

ZEKL HANS WILHELM
TON BELLER GMBH
BURGSTRASSE 22
D 6140 BENSHEIM 3
GERMANY
TEL 625 17 3001
TLF
TLX 468352
EML
COM

ZELENKA ANTOINE DR
DACHSLENBERGSTR 56

CH 8180 BUELACH
SWITZERLAND
TEL
TLF
TLX
EML
COM 10,12

ZELLNER BENJAMIN H DR
STSCI
HOMEWOOD CAMPUS
3700 SAN MARTIN DR
BALTIMORE MD 21218
USA
TEL
TLF
TLX
EML
COM 15

ZEL'MANOV A L DR
STERNBERG STATE ASTR INST
UNIVERSITETSKIJ PROSP 13
119899 MOSCOW
RUSSIA
TEL
TLX
EML
TLF
COM 47

ZENG QIN DR
PURPLE MOUNTAIN OBSERV
CAS
NANJING
CHINA PR
TEL 25 30 8516
TLF
TLX 34144 PMONJ CN
EML
COM 14,34

ZENSUS J-ANTON DR
NRAO

BOX 0
SOCORRO NM 87801 0387
USA
TEL 505 835 7348
TLF 505 835 7027
TLX 910-988-1710
EML BITNET AZENSUS@NRAO EDU
COM 40

ZERULL REINER H DR
RUHR-UNIVERSITAET BOCHUM
BEREICH EXTRATERR PHYSIK
D 4630 BOCHUM 1
GERMANY
TEL 234 70 04576
TLF
TLX 0825860
EML
COM 21

ZHAGAR YOURI H DR
ASTRONOMICAL OBSERVATORY
LATVIAN STATE UNIVERSITY
RAINIS BUL 19
226098 RIGA
LATVIA
TEL 13 2 223149
TLF
TLX 161171 TEMA SU
EML
COM

ZHAI DI-SHENG
BEIJING ASTRONOMICAL OBS
CAS
W SUBURB
BEIJING 100080
CHINA PR
TEL 1 28 1698
TLF
TLX 22040 BAOAS CN
EML
COM. 42C

ZHAI ZAOCHENG
SHANGHAI OBSERVATORY
CAS
80 NANDAN RD
SHANGHAI
CHINA PR
TEL 21 38 6191
TLF
TLX 33164 SHAO CN
EML
COM 31

ZHANF SHOUZHONG DR
414 WEST 120 ST
APT 401
NEW YORK NY 10027
USA
TEL 212 666 4689
TLF
TLX
EML
COM 41

ZHANG BAI-RONG
YUNNAN OBSERVATORY
CAS
BOX 110
KUNMING 72946 YUNNAN
CHINA PR
TEL 871 2035
TLF
TLX 64040 YUOBS CN
EML
COM 10,50

ZHANG BIN
DPT OF GEOPHYSICS
BEIJING UNIVERSITY
BEIJING 100071
CHINA PR
TEL
TLF
TLX
EML
COM 33

ZHANG CHENG-YUE
DPT OF PHYSICS
UNIVERSITY OF CALGARY
2500 UNIVERSITY DR NW
CALGARY AB T2N 1N4
CANADA
TEL 403 220 5385
TLF 403 220 3643
TLX
EML
COM 34

ZHANG ER-HO DR
ASTRONOMY DPT
UNIVERSITY OF TEXAS
RLM 15 220
AUSTIN TX 78712 1083
USA
TEL 512 471 4462
TLF
TLX
EML
COM· 42

ZHANG FU JUN
SHANGHAI OBSERVATORY
CAS
80 NANDAN RD
SHANGHAI
CHINA PR
TEL 21 38 6191
TLF
TLX
EML
COM 40

ZHANG GUO-DONG
BEIJING ASTRONOMICAL OBS
CAS
W SUBURB
BEIJING 100080
CHINA PR
TEL
TLF
TLX
EML
COM 19

ZHANG HE-QI
PURPLE MOUNTAIN OBSERV
CAS
NANJING
CHINA PR
TEL 25 46700
TLF
TLX 34144 PMONJ CN
EML
COM 10,48

ZHANG HUI
SHAANXI OBSERVATORY
CAS
LINTONG XIAN
SHAANXI
CHINA PR
TEL 33 2255
TLF
TLX 70121 CSAO CN
EML
COM 08

ZHANG JIA-LU
ASTROPHYSICS DIVISION
UNIV SCIENCE & TECHNOLOGY
HEFEI 230026 ANHUI
CHINA PR
TEL 551 33 1134
TLF
TLX 90028 USTC CN
EML
COM 47,48

ZHANG JIA-XIANG
PURPLE MOUNTAIN OBSERV
CAS
NANJING
CHINA PR
TEL 25 46700
TLF
TLX 34144 PMONJ CN
EML
COM 20C

ZHANG JINTONG
INSTITUTE OF GEODESY &
GEOPHYSICS
XU DONG LU
WUHAN HUBEI
CHINA PR
TEL
TLF
TLX
EML
COM 31,42

ZHANG PEIYU
PURPLE MOUNTAIN OBSERV
CAS
NANJING
CHINA PR
TEL 25 37521
TLF
TLX 34144 PMONJ CN
EML
COM 41

ZHANG SHENG-PAN
2 ASSINIBOINE ROAD
SUITE 720
DOWNSVIEW ON M3J 1L1
CANADA
TEL
TLF
TLX
EML
COM 07

ZHANG XIU ZHONG
SHANGHAI OBSERVATORY
CAS
80 NANDAN RD
SHANGHAI 200030
CHINA PR
TEL 21 38 6191
TLF 21 38 4618
TLX 33164 SHAO CN
EML
COM 09

ZHANG YOUYI
PURPLE MOUNTAIN OBSERV
CAS
NANJING
CHINA PR
TEL 25 46700
TLF
TLX 34144 PMONJ CN
EML
COM 09

ZHANG ZHEN-DA
DPT OF ASTRONOMY
NANJING UNIVERSITY
NANJING
CHINA PR
TEL 25 34651*2882
TLF
TLX 34151 PRCNU CN
EML
COM 10

ZHANG ZHEN-JIU
DPT OF PHYSICS
HUAZHONG NORMAL UNIV
WUHAN
CHINA PR
TEL 75 601
TLF
TLX 6908
EML
COM 47,48

ZHAO GANG
SHANGHAI OBSERVATORY
CAS
80 NANDAN RD
SHANGHAI
CHINA PR
TEL 21 38 6191
TLF
TLX 33164 SHAO CN
EML
COM 31

ZHAO JUN-LIANG
SHANGHAI OBSERVATORY
CAS
80 NANDAN RD
SHANGHAI
CHINA PR
TEL 21 38 6191
TLF·
TLX 33164 SHAO CN
EML
COM 33,37C

ZHAO MING
SHANGHAI OBSERVATORY
CAS
80 NANDAN RD
SHANGHAI
CHINA PR
TEL 21 38 6191
TLF
TLX 33164 SHAO CN
EML
COM 19

ZHAO REN-YANG
BEIJING ASTRONOMICAL OBS
CAS
W SUBURB
BEIJING 100080
CHINA PR
TEL
TLF
TLX
EML
COM 10

ZHARKOV VLADIMIR N DR
INST PHYSICS OF THE EARTH
ACADEMY OF SCIENCES
GRUZINSKAYA 10
123342 MOSCOW
RUSSIA
TEL 2545251
TLF
TLX 411196 IFZAN
EML
COM 16

ZHARKOVA VATENINA DR
DPT OF PHYSICS
KIEV STATE UNIVERSITY
GLUSHKOV PR 6
252022 KIEV
UKRAINE
TEL 26 1212
TLF 227 4482
TLX 131406 SKY SU
EML
COM 10

ZHEKOV SVETOZAR A DR
CTR LAB FOR SPACE RES
BULGARIAN ACAD SCIENCES
MOSKAVA ST 6
BG 1000 SOFIA
BULGARIA
TEL 2 88 3503
TLF
TLX 23351 IKIBAN BG
EML
COM

ZHELEZNIAKOV VLADIMIR V
INST OF APPLIED PHYSICS
ACADEMY OF SCIENCES
ULYANOV UL 46
603600 N NOVGOROD
RUSSIA
TEL
TLF
TLX
EML
COM 40

ZHELYAZKOV IVAN DR
FAC OF PHYSICS
UNIVERSITY OF SOFIA
ANTON IVANOV ST 5
BG 1126 SOFIA
BULGARIA
TEL 2 54 4852
TLF
TLX 23296 SUKO R BG
EML
COM 10

ZHENG DA-WEI
SHANGHAI OBSERVATORY
CAS
80 NANDAN RD
SHANGHAI
CHINA PR
TEL 21 38 6191
TLF
TLX 33164 SHAO CN
EML
COM 19

ZHENG JIA-QING
TUORLA OBSERVATORY
UNIVERSITY OF TURKU
SF 20520 TURKU
FINLAND
TEL
TLF
TLX
EML
COM 07

ZHENG XUE-TANG
DPT OF APPLIED PHYSICS
EAST CHINA INSTITUTE OF
TECHNOLOGY
NANJING 210014
CHINA PR
TEL
TLF
TLX
EML
COM 07

ZHENG YING
PURPLE MOUNTAIN OBSERV
CAS
NANJING
CHINA PR
TEL 25 46700
TLF
TLX 34144 PMONJ CN
EML
COM 31

ZHENG YI-JIA
BEIJING ASTRONOMICAL OBS
CAS
W SUBURB
BEIJING 100080
CHINA PR
TEL
TLF
TLX 22040 BAOBS CN
EML
COM 40

ZHEVAKIN S A PROF DR
RADIOPHYSICAL RESEARCH
INSTITUTE
LYADOV UL 25/14
603600 N NOVGOROD
RUSSIA
TEL 36-67-51
TLF
TLX
EML
COM 35

ZHOU BIFANG DR
CENTER FOR ASTRONOMICAL
INSTRUMENTS RESEARCH
182 BANCANG
NANJING 210042
CHINA PR
TEL 25 64 6191
TLF
TLX 34025 ISSAS CN
EML
COM 09

ZHOU DAOQI
DPT OF GEOPHYSICS
BEIJING UNIVERSITY
BEIJING 100071
CHINA PR
TEL 1 28 2471*3888
TLF
TLX 22239 PKUNI
EML
COM 10,12,42

ZHOU HONG-NAN
DPT OF ASTRONOMY
NANJING UNIVERSITY
NANJING
CHINA PR
TEL 25 34651*2882
TLF
TLX 34151 PRCNU CN
EML
COM 07,42

ZHOU TI-JIAN
DPT OF GEOPHYSICS
BEIJING UNIVERSITY
BEIJING 100071
CHINA PR
TEL 1 28 2471*3888
TLF
TLX. 22239 PKUNI
EML
COM 40

ZHOU YOU-YUAN
ASTROPHYSICS DIVISION
UNIV SCIENCE & TECHNOLOGY
HEFEI 230026 ANHUI
CHINA PR
TEL 551 33 1134
TLF
TLX 90028 USTC CN
EML
COM 28,47

ZHOU ZHEN-PU
PURPLE MOUNTAIN OBSERV
CAS
NANJING
CHINA PR
TEL 25 33738
TLF
TLX 34114 PMONTJ CN
EML
COM 34

ZHU CI-SHENG
DPT OF ASTRONOMY
NANJING UNIVERSITY
NANJING
CHINA PR
TEL 25 37551*2882
TLF
TLX 34151 PRCNU CN
EML
COM 42

ZHU NENGHONG
SHANGHAI OBSERVATORY
CAS
80 NANDAN RD
SHANGHAI
CHINA PR
TEL 21 38 6191
TLF
TLX 33164 SHAO CN
EML
COM 09

ZHU SHI-CHANG
DPT OF PHYSICS
SHANGHAI TEACHERS UNIV
10 GILLIN RD
SHANGHAI
CHINA PR
TEL 21 384 301
TLF
TLX 9016
EML
COM 47

ZHU WEN-YAO
SHANGHAI OBSERVATORY
CAS
80 NANDAN RD
SHANGHAI
CHINA PR
TEL 21 38 6191
TLF
TLX 33164 SHAO CN
EML
COM 07

ZHU XINGFENG
ASTROPHYSICS DIVISION
UNIV SCIENCE & TECHNOLOGY
HEFEI 230026 ANHUI
CHINA PR
TEL 551 33 1134
TLF
TLX 90028 USTC CN
EML
COM 47

ZHU YAOZHONG DR
INST OF GEODESY & GEOPHYS
54 XU DONG RD
WUCHANG
HUBEI
CHINA PR
TEL 81 3401
TLF
TLX
EML
COM 19

ZHU YONG-HE
BEIJING ASTRONOMICAL OBS
CAS
W SUBURB
BEIJING 100080
CHINA PR
TEL· 1 28 1698
TLF
TLX 22040 BAOBS CN
EML
COM 19

ZHUANG QIXIANG
LAB FOR BASIC STANDARDS
NTL RESEARCH COUNCIL
OTTAWA ONT K1A OR6
CANADA
TEL
TLF
TLX
EML
COM 31

ZHUANG WEIFENG
BEIJING ASTRONOMICAL OBS
CAS
W SUBURB
BEIJING 100080
CHINA PR
TEL
TLF
TLX
EML
COM 41

ZHUGZHDA YUZEF D DR
IZMIRAN
ACADEMY OF SCIENCES
142092 TROITSK
RUSSIA
TEL
TLF
TLX
EML
COM 10,12

ZIEBA STANISLAW DR
ASTRONOMICAL OBSERVATORY
JAGIELLONIAN UNIVERSITY
UL ORLA 171
PL 30 244 KRAKOW
POLAND
TEL 223856, 221877
TLF
TLX 0322297 UJ PL
EML
COM 40,47

ZIKIDES MICHAEL C DR
DPT OF·ASTRONOMY
NTL UNIVERSITY OF ATHENS
PANEPISTIMIOPOLIS
GR 157 71 ZOGRAFOS
GREECE
TEL 1 724 3414
TLF
TLX
EML
COM

ZIMMERMANN HELMUT DR
UNIVERSITAETS-STERNWARTE

SCHILLERGAESSCHEN 2
D 6900 JENA
GERMANY
TEL 78 27122
TLF
TLX
EML
COM 34,46

ZINN ROBERT J DR
DPT OF ASTRONOMY
YALE UNIVERSITY
BOX 6666
NEW HAVEN CT 06520
USA
TEL 203 436 3460
TLF
TLX
EML
COM 28,37

ZINNECKER HANS
INSTITUT FUER ASTRONOMIE
& ASTROPHYSIK
AM HUBLAND
D 8700 WUERZBURG
GERMANY
TEL 931 888 5031
TLF 931 706 297
TLX
EML
COM 26C

ZIOLKOWSKI JANUSZ DR
COPERNICUS ASTRON CENTER
POLISH ACAD OF SCIENCES
UL BARTYCKA 18
PL 00 716 WARSAW
POLAND
TEL
TLF
TLX
EML
COM 35,42

ZIOLKOWSKI KRZYSZTOF DR
SPACE RESEARCH CENTER
POLISH ACAD OF SCIENCES
UL ORDONA 21
PL 01 237 WARSAW
POLAND
TEL 22410041
TLF
TLX
EML
COM 20

ZIRIN HAROLD DR
CALTECH
MS 264 33
PASADENA CA 91125
USA
TEL 818 356 3857
TLF
TLX
EML
COM 10,12,14

ZIRKER JACK B DR
AIR FORCE GEOPHYSICS LAB
NTL SOLAR OBSERVATORY
SUNSPOT NM 88349
USA
TEL 505 434 1390
TLF
TLX
EML
COM 12

ZITELLI VALENTINA DR
DPT DI ASTRONOMIA
UNIVERSITA DI BOLOGNA
CP 596
I 40100 BOLOGNA
ITALY
TEL 51 25 9301
TLF
TLX 520634 INFN I
EML SPAN 37929
COM

ZIZNOVSKY JOZEF DR
ASTRONOMICAL INSTITUTE
SLOVAK ACADEMY SCIENCES
CS 059 60 TATRANSKA LOMNI
CZECHOSLOVAKIA
TEL 969 96 7866/7/8
TLF 969 96 7656
TLX 78277
EML
COM 25

ZLATEV SLAVEY
ASTRONOMICAL OBSERVATORY
OF KARDGALI
BG 6600 KARDGALI
BULGARIA
TEL 36 12 595
TLF
TLX 47421
EML
COM

ZLOBEC PAOLO DR
OAT
BOX SUCC TRIESTE 5
VIA TIEPOLO 11
I 34131 TRIESTE
ITALY
TEL 40 79 3921
TLF
TLX 461137 OAT I
EML
COM 10,40

ZLOTNIK ELENA YA DR
INST OF APPLIED PHYSICS
ACADEMY OF SCIENCES
ULYANOV UL 46
603600 N NOVGOROD
RUSSIA
TEL 7 8 8312 363519
TLF 7 8 8312 362081
TLX 412580 FIZIK SU
EML
COM 40

ZOMBECK MARTIN V DR
CENTER FOR ASTROPHYSICS
HCO/SAO
60 GARDEN ST
CAMBRIDGE MA 02138
USA
TEL 617 495 7227
TLF
TLX 921428 SATELLITE CAM
EML
COM 44,48

ZOREC JEAN DR
INSTITUT D'ASTROPHYSIQUE
98BIS BD ARAGO
F 75014 PARIS
FRANCE
TEL 1 43 20 1425
TLF 1 43 29 8673
TLX
EML
COM 29

ZOSIMOVICH IRINA D
INSTITUTE OF HISTORY
UKRAINIAN ACAD OF SCIENCE
KIROV UL 4
252001 KIEV
UKRAINE
TEL 29 0272
TLF
TLX
EML
COM 41

ZOU HUI-CHENG
SHANGHAI OBSERVATORY
CAS
80 NANDAN RD
SHANGHAI
CHINA PR
TEL 21 38 6191
TLF
TLX 33164 SHAO CNCN
EML
COM 44

ZOU YI-XIN
BEIJING ASTRONOMICAL OBS
CAS
W SUBURB
BEIJING 100080
CHINA PR
TEL 1 28 1261
TLF
TLX 22040
EML
COM 10

ZOU ZHEN-LONG
BEIJING ASTRONOMICAL OBS
CAS
W SUBURB
BEIJING 100080
CHINA PR
TEL
TLF
TLX 22040 BAOAS CN
EML
COM 47

ZSOLDOS ENDRE DR
KONKOLY OBSERVATORY
THEGE U 13/17
BOX 67
H 1525 BUDAPEST
HUNGARY
TEL 1 75 5866/75 4122
TLF 1 156 9640
TLX 22 74 60 KONOB H
EML H697KON@ELLA UUCP
COM 27

ZUCCARELLO FRANCESCA
IST DI ASTRONOMIA
CITTA UNIVERSITARIA
VIA A DORIA 6
I 95125 CATANIA
ITALY
TEL 330 533
TLF
TLX 970359 ASTRCT I
EML
COM

ZUCKERMAN BEN M DR
DPT OF ASTRONOMY
UNIVERSITY OF CALIFORNIA
LOS ANGELES CA 90024
USA
TEL 213 825 9338
TLF
TLX 910 342 7597
EML
COM 27,34,40,51

ZUIDERWIJK EDWARDUS J
ROYAL GREENWICH OBS
MADINGLEY RD
CAMBRIDGE CB3 OEZ
UK
TEL 223 37 4868
TLF
TLX
EML
COM 42,47

ZVERKO JURAJ DR
ASTRONOMICAL INSTITUTE
SLOVAK ACADEMY SCIENCES
CS 059 60 TATRANSKA LOMNI
CZECHOSLOVAKIA
TEL 969 96 7866/7/8
TLF 969 96 7656
TLX 78277
EML
COM 29

ZVOLANKOVA JUDITA
ASTRONOMICAL INSTITUTE
SLOVAK ACADEMY SCIENCES
DUBRAVSKA 9
CS 842 28 BRATISLAVA
CZECHOSLOVAKIA
TEL 7 37 5157
TLF
TLX 93373 SEIS
EML
COM 22

ZWAAN CORNELIS PROF DR
STERREKUNDIG INSTITUTE

BOX 80000
NL 3508 TA UTRECHT
NETHERLANDS
TEL 30 53 5223
TLF 30 53 5201
TLX 40048 FYLUT NL
EML ZWAAN@FYS RUU NL
COM 10,12,36C

ZWITTER TOMAZ
ASTRONOMICAL OBSERVATORY
UNIVERSITY OF E KARDELJ
JADRANSKA 19
YU 61110 LJUBLJANA
YUGOSLAVIA
TEL 61 265 061
TLF 61 217 281
TLX
EML ZWITTER@ITSSISSA BITNET
COM 42